AUTOMOTIVE SERVICE

Inspection, Maintenance, and Repair

AUTOMOTIVE SERVICE

Inspection, Maintenance, and Repair

TIM GILLES

Professor

Santa Barbara City College

ASE Master Automotive Technician, ASE Master Automotive Machinist

DELMAR

™

THOMSON LEARNING

Africa • Australia • Canada • Denmark • Japan • Mexico • New Zealand • Philippines
Puerto Rico • Singapore • Spain • United Kingdom • United States

NOTICE TO THE READER

Cover photo courtesy of: David Getty
Cover Design: Brucie Rosch

Delmar Staff

Publisher: Alar Elken	Production Coordinator: Karen Smith
Acquisitions Editor: Vernon Anthony	Production Manager: Mary Ellen Black
Developmental Editor: Denise Denisoff	Art and Design Coordinator: Cheri Plasse
Project Editor: Cori Filson	Editorial Assistant: Betsy Hough

Printed in the United States of America
6 7 8 9 10 XXX 03 02 01

For more information, contact Delmar, 3 Columbia Circle, PO Box 15015, Albany, NY 12212-0515; or find us on the World Wide Web at http://www.delmar.com

International Division List

Asia
Thomson Learning
60 Albert Street, #15-01
Albert Complex
Singapore 189969
Tel: 65 336 6411
Fax: 65 336 7411

Japan:
Thomson Learning
Palaceside Building 5F
1-1-1 Hitotsubashi, Chiyoda-ku
Tokyo 100 0003 Japan
Tel: 813 5218 6544
Fax: 813 5218 6551

Australia/New Zealand:
Nelson/Thomson Learning
102 Dodds Street
South Melbourne, Victoria 3205
Australia
Tel: 61 39 685 4111
Fax: 61 39 685 4199

UK/Europe/Middle East
Thomson Learning
Berkshire House
168-173 High Holborn
London
WC1V 7AA United Kingdom
Tel: 44 171 497 1422
Fax: 44 171 497 1426

Latin America:
Thomson Learning
Seneca, 53
Colonia Polanco
11560 Mexico D.F. Mexico
Tel: 525-281-2906
Fax: 525-281-2656

Canada:
Nelson/Thomson Learning
1120 Birchmount Road
Scarborough, Ontario
Canada M1K 5G4
Tel: 416-752-9100
Fax: 416-752-8102

For permission to use material from this text or product contact us by
Tel (800) 730-2214; Fax (800) 730-2215; www.thomsonrights.com

Library of Congress Cataloging-in-Publication Data
Gilles, Tim, 1951 -
    Automotive service inspection, maintenance, and repair / Tim
Gilles.
       p.   cm.
    Includes index.
    ISBN 0-8273-7354-6 (alk. paper)
    1. Automobiles—Maintenance and repair. I. Title
TL152.G516   1999
629.28'72—dc21   97-46812
                              CIP

TABLE OF CONTENTS

SECTION 1
■ THE AUTOMOBILE INDUSTRY

SECTION 2
■ SHOP PROCEDURES, SAFETY, TOOLS, AND EQUIPMENT

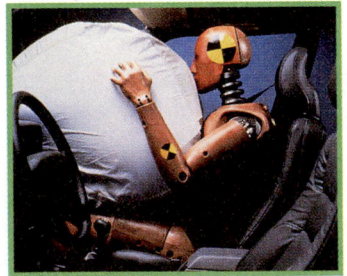

SECTION 3
■ VEHICLE INSPECTION (LUBRICATION/SAFETY CHECK)

SECTION 4
■ ENGINE OPERATION AND SERVICE

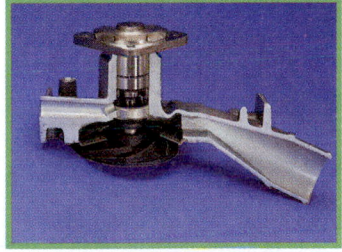

SECTION 5
■ COOLING AND FUEL SYSTEM THEORY AND SERVICE

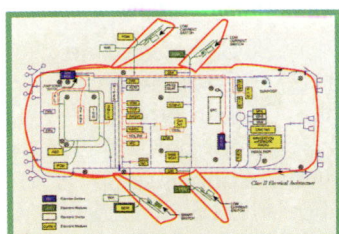

SECTION 6
ELECTRICAL SYSTEM THEORY AND SERVICE

SECTION 7
■ ENGINE PERFORMANCE DIAGNOSIS: THEORY AND SERVICE

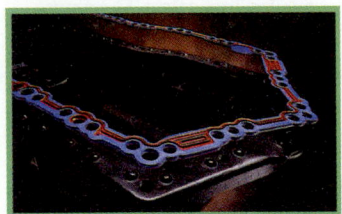

SECTION 8

■ AUTOMOTIVE ENGINE SERVICE AND REPAIR

SECTION 9
■ MISCELLANEOUS CHASSIS THEORY AND SERVICE

CHAPTER 51
Brakes Fundamentals 683

CHAPTER 52
Brake Service 706

CHAPTER 53
Bearings, Seals, and Greases 733

CHAPTER 54
Tire and Wheel Theory 753

SECTION 10
SUSPENSION, STEERING, ALIGNMENT

SECTION 11
▬ DRIVE TRAIN

SECTION 12
◼ COMFORT SYSTEMS AND VEHICLE ELECTRONICS

PREFACE

This book evolved as a result of the author's participation in a successful articulation program between local high schools and the community college where he has been a teacher for over 20 years. Input from 31 reviewers, many shop owners, and over 100 teachers helped to shape the book from its inception to its final form.

The transportation industry is vast, with one in every six people in the country making something of a contribution to it. These include people of many levels and abilities. With that in mind, the introductory fundamentals chapters are written at a low level for all of the students while the service chapters are for those who have mastered the introductory material. The author's philosophy as a teacher is to challenge the best students and the rest will be brought to a higher level.

The text can be used in a variety of educational situations. It can be used as a basic text in any automotive repair class, to train entry-level or apprentice technicians or to prepare more experienced technicians for ASE certification. All automotive terms, abbreviations, and acronyms used in this text comply with the SAE Technical Standards Board Publication *SAE J1930*.

The text is divided into 76 chapters covering the eight fundamental automotive ASE certification areas. In addition to coverage of the usual repairs performed in almost any automotive repair facility, the reader is introduced to the most frequently performed inspection and service procedures, from safety inspections to tire and wheel service.

Included in this text are chapters on safety, hand tools, and vehicle maintenance and lubrication that are more comprehensive than any found in most comparable texts. The accompanying lab manual emphasizes those jobs done in service stations, fast lube outlets, or mass merchandisers (such as Sears, Goodyear, Firestone, K-Mart, or Montgomery Ward).

Automobiles have become so complex in the last twenty years that to remain competent, many of today's technicians specialize on one or more systems of the car. Basic theory of all automotive and light truck systems is covered so that service personnel will understand the function of the parts being serviced. This book is comprehensive in that it deals with the entire car and aims to teach theory of vehicle systems at an introductory student level, followed by service, diagnosis, and light repairs at a more advanced student level.

Most of the systems used in automobiles today are strikingly similar. Repair techniques, universal to all automobiles, are discussed. Procedures or conditions unique to specific automobiles are purposely avoided. The reader is encouraged to refer to the proper repair manual for the specific vehicle in question.

The primary objective of *Automotive Service: Inspection, Maintenance, and Repair* is to help the reader develop confidence in both thinking skills and problem-solving ability. One unique aspect of automotive education is that many automotive graduates matriculate to other professions and trades, such as engineering or construction. They will find much of the material learned in automotive classes to be very valuable and useful in their chosen fields. This aspect of the student's education is especially valuable when one considers how high school industrial arts programs have been scaled back in recent years. Dealing with such things as tools, soldering, basic electrical repairs, and repairing broken fasteners helps to provide some measure of practical education.

The tremendous decline in the number of corner service stations has resulted in a loss of those jobs formerly available in abundance to students. The accompanying lab workbook contains service jobs students should be able to perform before enrolling in an advanced automotive specialty area class. Successful service personnel who possess necessary basic automotive skills must continually learn new things so that they may progress into other (higher paying) specialty areas.

It is not unusual for "advanced" students to enter community college programs who, although they have met the course prerequisites by taking high school shop classes, did not learn how to perform fundamental service procedures properly. Advanced students who slip through classes without learning the basics can blemish the records of even the finest school programs. One complaint from business owners is that new employees are often not prepared well enough in the basics. Better students will be at a high entry level following the completion of this text. They will also be considered to be at a suitable skill level for entry as an apprentice in the specialty area of automotive repair in an independent repair shop or dealership.

ACKNOWLEDGMENTS

The author extends special thanks to the following individuals and companies:

- Denise Denisoff, Delmar Publishers Developmental Editor, for her positive attitude, encouragement, professional assistance, and tireless dedication to this project.
- Lori Hilfinger, for her encouragement and extra effort in organizing the art package.
- Two friends and colleagues, Chuck Rockwood of Ventura College and Wayne Olson of Golden West College for their continuous input and support.
- Members of the North American Council of Automotive Teachers (NACAT) and California Automotive Teachers (CAT) who provided a vast amount of input.
- Bob Freudenberger for his efforts in creating the Section Openers.
- Stu Bradholdt for his help with the Science Notes.

The contributions of the following reviewers of the text are gratefully acknowledged:

Chuck Rockwood, Ventura College, Ventura, CA; Robert Abbey, Santa Barbara City College, Santa Barbara, CA; Bob Stockero, Santa Barbara City College, Santa Barbara, CA; Stu Bradholdt, Santa Barbara City College, Santa Barbara, CA; Joe Bradbury, Arroyo Grande High School, Arroyo Grande, CA; Ken Welch, Saddleback College, Mission Viejo, CA; Frederick G. Heath, Heath and Associates, Louisville, KY; David Lannom, Lincoln Technical Institute, Grand Prairie, TX; Stephan Baldwin, Sullivan High School, Sullivan, IN; Richard Caperton, James L. Walker Vo-Tech, Naples, FL; James Pinto, Porter and Chester Institute, Wethersfield, CT; Quentin Swan, California Department of Education, Los Angeles, CA; Charles Statz, Temple Jr. College, Temple, TX; Ronnie Bush, Tennessee Tech Center at Jackson, Jackson, TN; Patrick Devlin, Montgomery College, Rockville, MD; Brad Walsh, Mark Keppel High School, Alhambra, CA; Steve Ford, Santa Barbara City College, Santa Barbara, CA; Kim Edward Ray, Northwest Career Center, Dublin OH; Robert Wenzlaff, Fullerton College, Fullerton, CA; Walter Bertotti, Porter and Chester Institute, Watertown, CT; Earl Friedell, DeKalb Technical Institute, Clarkston, GA; Paul Pate, Community College of Southern Nevada, Boulder City, NV; Johnny Beason, Murray County High School, Chatsworth, GA; Jim Zimmerman, Union Grove High School/Gateway Technical College, Union Grove, WI; Michael T. Maher, Fox Valley Career Center, DeKalb, IL; Tommy Wilson, Columbia Technical Institute, Fort Mitchell, AL; Daniel Wickware, Northeast Texas Community College, Mt. Pleasant, TX; Photographers: Kurt Rhody; Tim Gilles; Models: Bridget Berg; Gary Semerdjian; Janet Hill; Julio A. Limon; Hsi Lin; Julianne Casey; Mark Kono; Jeff Felske; Kurt Reed; and William Tomlin.

The author expresses sincere appreciation to the following companies and individuals who supplied illustrations, technical information, or other assistance during the preparation of *Automotive Service: Inspection, Maintenance, and Repair.*

AERA; Tim Waters, All Pro/Bumper to Bumper; Guy Avellon, ASTM, Committee on Fasteners; Automotive Lift Institute; Bob Barkhouse, Yuba College; Tom Birch, Yuba College; Dave Brainerd, Santa Barbara City College; Caesar's Auto Supply, Santa Barbara, CA; Charlie Camp; Antonio Casillas, Santa Barbara High School; Joe Schuit, Channel City Engineering, Santa Barbara, CA; Clay Furtaw, DeWalt Machinery Co.; Boyce Dwiggins, Delmar Publishers; Dema Elgin, Elgin Racing Cams, Redwood City, CA; EngineTech, Inc., Carrollton, TX; Jack Erjavec, Delmar Publishers; Rick Escalambre, Skyline College; Ray Fausel, Professor Emeritus, Cal-State Los Angeles; Jim Halderman, Sinclair Community College; Frederick G. Heath, Heath and Associates, Louisville, KY; Ed Hernandez, Santa Barbara City College; Barry Hollembeak, Delmar Publishers; Jim Hughes, Rio Hondo College; Don Knowles, Delmar Publishers; Mark Kreger, Ertel Products Co.; Dr. Norman Laws, Professor Emeritus, Chicago State University; Randy Ingle, General Motors Technical Instructor; Al Santini, College of DuPage; Toby's Engine Parts, Santa Barbara, CA; Phil Unander, Larry's Auto Parts, Santa Barbara, CA; Dr. Jay Webster, Professor Emeritus, Long Beach State

DEDICATION

The completion of this book was made possible with help from a great many individuals. *Automotive Service: Inspection, Maintenance, and Repair* is dedicated to them and to my family, especially my daughter and editorial assistant, Terri Gilles. Her organizational skills and able assistance in preparing and organizing the art package made the completion of this book possible.

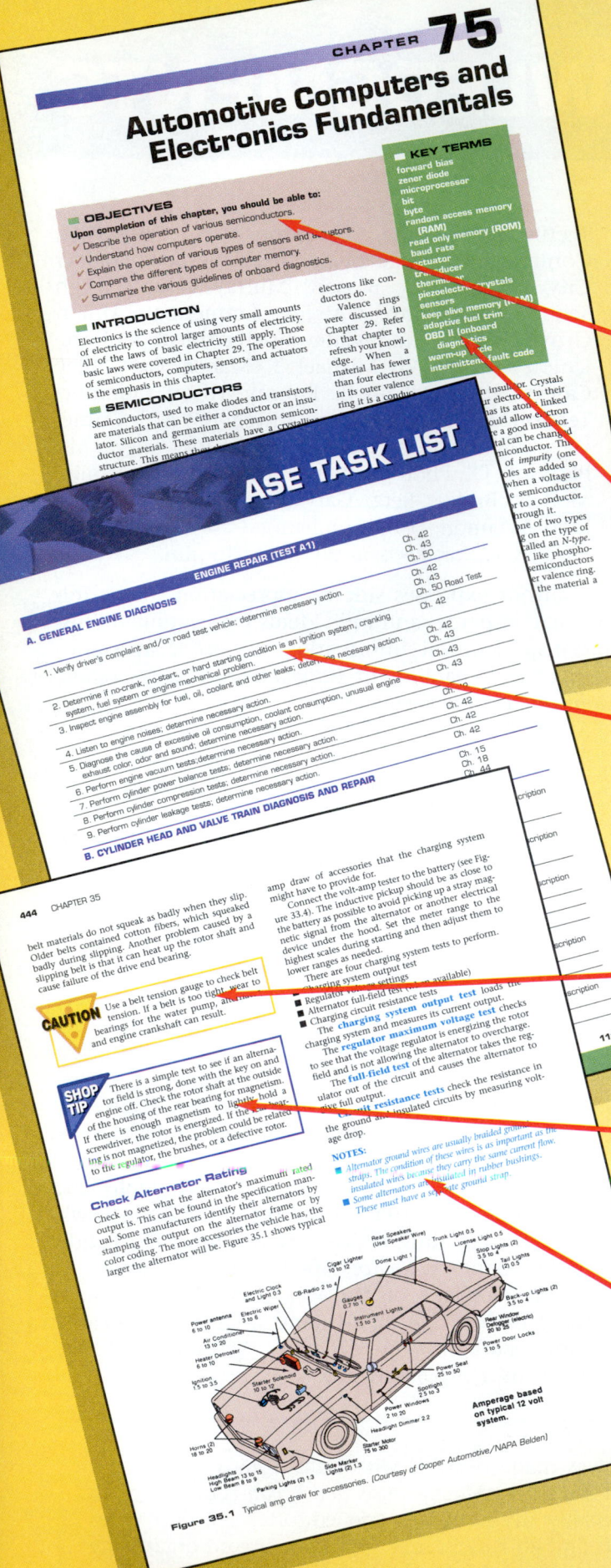

FEATURES OF THE TEXT

OBJECTIVES

Each chapter begins with a list of the most important points discussed in the chapter. This list of objectives is intended to provide the student with a general idea of what he/she will be studying.

KEY TERMS

Each chapter contains a list of new terms to know. These terms are highlighted in bold in the text. Descriptions of these terms can be found in the glossary.

ASE TASK LISTS

ASE Task Lists are provided in the appendix and in the Instructor's Guide for every chapter in which ASE tasks are studied. The purpose of these lists is to provide the student and instructor with a direct correlation between text content and information found on each of the eight ASE Automotive Tests.

CAUTIONS

Cautions are urgent warnings that personal injury or property damage could occur if careful preventative steps are not taken.

SHOP TIPS

Appropriate shop tips are described throughout the text. These tips provide shortcuts and emphasize fine-tuning procedures to shop practices commonly performed by experienced technicians.

NOTES

Notes are used throughout the text to highlight especially important topics.

SAFETY NOTES

Safety is the number one priority in an automotive shop. There are numerous safety notes contained throughout the text. Most of the safety notes have been taken from real world experiences and many are accompanied by actual case histories.

CASE HISTORIES

Case histories are presented throughout the text. These true stories recount actual automotive situations encountered by the author in over 25 years in automotive service. Case histories present the reader with examples of the pattern of critical thinking skills required to diagnose automotive problems.

SCIENCE, HISTORY, COMPUTER SYSTEM, AND MATH NOTES

These notes are included when interesting topics relating to them are covered in the text. The objective of these features is to pique the student's interest and show a correlation between their automotive studies and these areas of learning.

LAB WORKSHEET REFERENCES

References are made to worksheets in the lab manual that are applicable to the topic discussed. These references are intended to guide the study to an activity that demonstrates how the theory learned relates to various automotive practices and procedures.

REVIEW QUESTIONS

These questions guide the student to the most important points in the chapter and act as a check for understanding of the material. Each chapter's review questions are presented in the same order in which they appear in the chapter. This provides an instructor with the flexibility to assign portions of the chapter to read and then follow up with a few of the study questions.

ASE STYLE REVIEW QUESTIONS

These questions are designed to provide preparation for the Certification Examinations administered by the *National Institute for Automotive Service Excellence (ASE)*. The ASE test is a task-oriented test (the test taker is supposed to be able to relate to shop-oriented questions), so the practice tests should help the student in becoming familiar with accepted trade procedures.

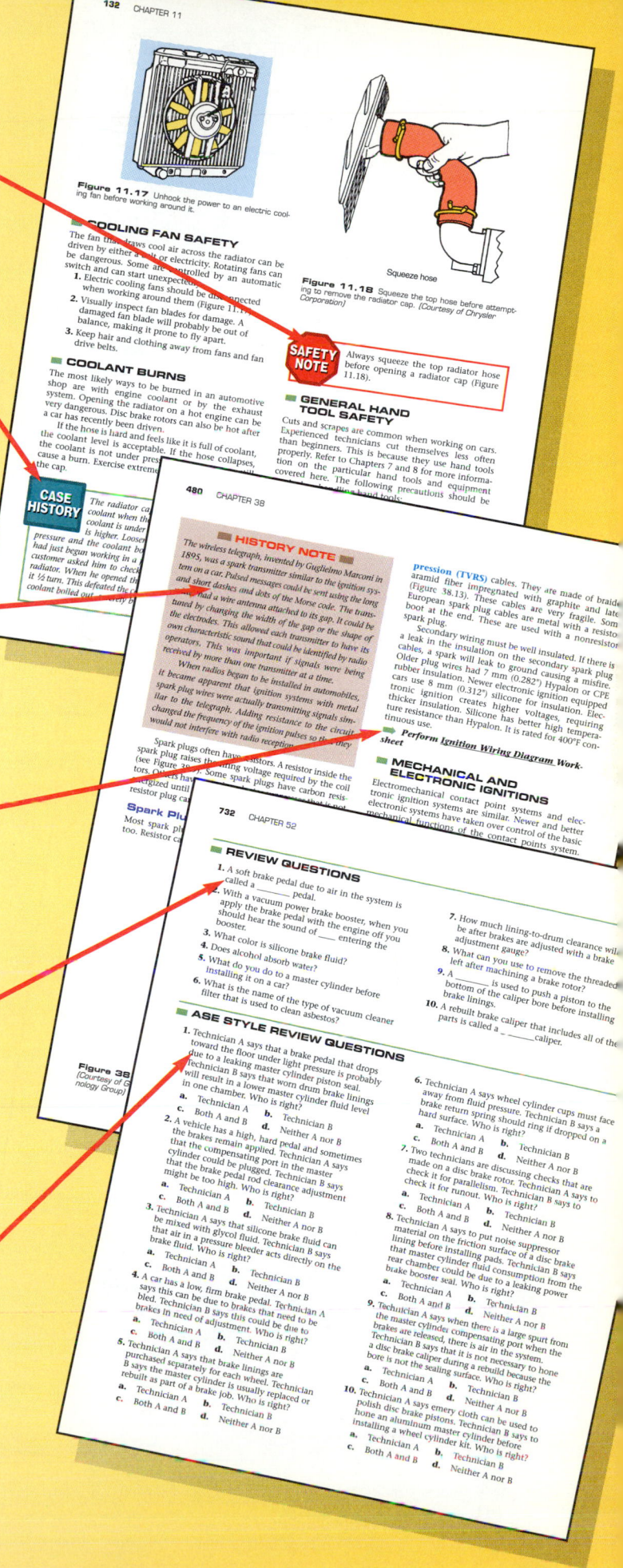

Instructor OK _____ Score _____

ASE LAB PREPARATION WORKHEET # 8-11
Checking Engine Oil Pressure

Name _____ Class _____

OBJECTIVE:
Upon completion of this assignment, you will be able to connect an oil pressure gauge to measure an engine oil pressure. This task will help prepare you to pass the ASE certification examination in engine repair.

DIRECTIONS:
Before beginning this lab assignment, review the worksheet completely. Fill in the information in the spaces provided as you complete each task.

TOOLS AND EQUIPMENT REQUIRED:
Safety glasses, fender covers, hand tools, flare nut wrenches, oil pressure gauge, shop towel

PROCEDURE:
Vehicle year _____ Make _____ Model _____
Repair Order # _____ Engine _____ # of Cylinders _____

1. Locate the oil pressure specification in the service manual and record the information below.
   Which manual was used? _____
   The specification was located on page number _____
   Minimum _____ psi at _____ rpm
   Maximum _____ psi at _____ rpm
2. Open the hood and place fender covers on the fenders and front body parts.
   Yes _____ No _____
3. Engine temperature:
   Cold _____
   Warm _____
4. The vehicle is equipped with an:
   Oil pressure gauge _____
   Indicator light _____
5. Check the operation of the oil pressure light. Turn the ignition key on and the engine off. Does the indicator light glow?
   Yes _____ No _____

■ 443

▲ **LAB # 1**
RACK/JACKS/SAFETY/SPEC MANUALS

Teaching aids/Handouts:
Safety tests (due today)
safety transparencies (in file with safety tests)
Wheel lug torque handout
Hydraulics Transparency
Gousha Lift Transparencies - cover lift points only

Lesson:
Students will meet in the classroom at the beginning of each lab day. We will see who brought in cars and reassign jobs to any groups where the assigned person did not bring a car. The LTA or teacher will inform the security guard at the kiosk that no permits are to be honored unless they have a handwritten note from the instructor giving permission to be late. The reason for this is that the people who are tardy will already have had their project reassigned to another person in the group.

❑ Check add cards to see that they have been processed by admissions
❑ Spend about 45 minutes covering lift points, the Car Care Guide (lift points only). Demo the lifts and jacks.
❑ Show the lab manual, go over lab assignment sheet #1, and the lab schedule. There is a lab assignment sheet for each week's lab. All labs must be completed before other lab projects can be undertaken.
❑ Describe the difference between 10 ton, 2 ton, and 1 1/2 ton Jacks, use transparency.
❑ Tell personal stories related to safety as you read each safety test question and put the answer on the board for them to copy onto class afterward for the safety presentation

Go to lab for the demo and return to class afterward for the safety presentation

❑ Safety test from book. Pass out the safety test and have them fill them in. Put all answers to safety test on board. Most of them were covered last time during the service station safety section. Substituting "acid" for "water" works well. See how many of them write the wrong word down because they were not paying attention. Try it again and you'll probably still catch some of them. Have them correct each others' test and then fix any wrong answers.
❑ Describe the Starburst symbol on a can of oil.
❑ Tell people bringing oil changes to look up their oil capacity in the Car Care Guide first, or bring 5 quarts to be safe. Next class is lecture/demo only. Oil changes will be half of the class after it.

Find one student who wants to bring in a car with oil and a filter to have demonstrated next class (lecture day).

❑ After the safety test have a shop tour.
❑ If time remains after the tour, they can complete their jacking and lifting assignments and get their shop workbooks signed off

Demo:
Show them how to locate lift points for their vehicle in the Car Care Guide. Look up the lift points for the demo shop cars, Xerox them and have them inside the windshield so you can show the beginners where the lift points are and what the sheet that they should be xeroxing looks like.

Part III: Lesson Plans - Fundamentals Of Automotive Servicing ■ 000

SUPPLEMENTS

■ **Instructor's Resource Kit**—contains the Instructor's Guide, full-color transparencies, and a computerized test bank.

■ **Instructor's Guide**—the Instructor's Guide contains answers to review questions, lesson plans (lecture outlines), lab exercises, ASE correlations, and supplementary handout materials.

■ **Lab Manual**—the Lab Manual available for use with this book can be used as either a text or for self-training. Worksheets define each lab procedure, presented in increasing levels of difficulty. A variety of illustrations is used to help the reader understand the jobs. Each project or lab assignment is built upon the next in a logical sequence in much the same manner as science instructional programs are constructed. The reader completes one task before progressing to the next one.

TBA (tires, batteries, and accessories service) is the primary emphasis of the Lab Manual. The most performed service jobs involving basic tools and shop procedures related to vehicle and fluid inspection are covered in the first section. Minor repairs are covered in the second section. Lab Manual tasks are designed to be completed in 1½ hours so they can fit into a school schedule.

■ **Computerized Test Bank**—contains over 1,500 questions, including sample ASE tests. This test bank is designed for easy tailoring and contains graphics.

The Automobile Industry

THEN AND NOW: THE AUTOMOBILE INDUSTRY

On January 29, 1886 Karl Benz of Mannheim, Germany patented the world's first automobile, the three-wheeled *Benz Motorwagen*. Later that same year, Gottlieb Daimler of Cannstadt, Germany built a four-wheeled car. Its 1.5 hp engine had 50% more power than the Benz; the horsepower race had begun. In 1900, Benz's company became the biggest auto maker in the world, building 603 cars.

Long before Benz's patent, there were ingenious automotive inventors and tinkerers in the United States. But the 1896 *Duryea* of Massachusetts was the first car to be produced for sale in the United States, followed shortly by the *Haynes and Winton*. In 1900, Ransom E. Olds of Detroit became the first to mass-produce automobiles in America, the curved-dash "merry Oldsmobile" of the song "Come Away With Me, Lucille."

Henry Ford was the first to produce the automobile in mass quantities. His grand idea was to build a car that everyone could afford. In 1903, the current Ford Motor Company was founded. The first Model T was sold in 1908. In its 19-year run, 15 million copies of that rugged, simple automobile were produced. This record was not surpassed until the Volkswagen Beetle did so in the early 1970s.

Another early automotive giant was General Motors, which in 1908 bought Buick, Oldsmobile, Cadillac, and Oakland (which would become Pontiac). Within two years, thirty firms had been brought under the GM umbrella, including eleven auto makers.

Walter P. Chrysler merged Willys and Maxwell-Chalmers in 1920. The first car to bear the Chrysler name went on sale in 1924 and was a huge success.

A current trend among foreign car makers is to build assembly plants in the United States. This idea is far from new. In 1888, Steinway & Sons, the New York piano maker, obtained the rights to all Daimler patents in the United States. It produced engines and cars in this country between 1905 and 1907.

Today, between one in nine working Americans build, sell, or fix and maintain motor vehicles. Automobile dealerships account for 29% of the retail business conducted in the United States. Automotive-related business represents about a third of a trillion dollars worth of our nation's economy in an average year. It is estimated that $150 billion of that total is spent on parts, repairs, and maintenance.

The 1886 Benz Patent Motorwagen. *(Courtesy of Mercedes-Benz of N.A., Inc.)*

A delivery to a modern dealership. *(Courtesy of Mercedes-Benz of N.A., Inc.)*

CHAPTER 1

Introduction to the Automobile

■ OBJECTIVES

Upon completion of this chapter, you should be able to:

✓ Describe the differences between the unibody design and frame and body design.

✓ Tell how the four-stroke cycle engine operates.

✓ Understand the purposes of the major engine support systems.

✓ Describe the parts of front and rear wheel drive powertrains.

✓ Explain major events in the history of the automobile.

■ KEY TERMS

horseless carriages
chassis
unibody
reciprocating engine
piston rings
four-stroke cycle
stroke
atomized
butterfly valve
ignition timing
sensors
actuators
FWD/RWD
torque
clutch
transaxle
Otto-cycle
Selden patent
Corporate Average Fuel Economy (CAFE)

■ INTRODUCTION

Automobiles have been on the scene for over 100 years. Early vehicles were built on the principle of the horse and wagon and were called **horseless carriages** (Figure 1.1). Continuing developments in vehicle design since those early years have resulted in today's high tech vehicle (Figure 1.2). Some of those developments are outlined later in this chapter.

Today, there are over 130 million passenger cars in the United States alone (1.5 per household). This is just over one-third of the cars in the world. The automobile is a direct or indirect source of employment for one of every nine workers in the country. Americans drive an average of 7767 miles per year.

The automobile includes several systems. These systems include the body and suspension, the engine, the electrical system, the powertrain, emission controls, and accessory systems.

■ BODY AND CHASSIS

The **chassis** (pronounced "chassy") includes the suspension. It supports the engine and the car body. The chassis also includes the *frame,* the *brakes,* and *steering* components. Older cars had heavy frames, but most cars are built today with what is known as a **unibody** design (Figure 1.3). This design has a floorpan and a small subframe section in the front and rear. The *body* includes the interior of the car, windows, door latches, the body electrical system, and the body accessories.

Figure 1.1 A horseless carriage. *(Courtesy of Ford Motor Company)*

Figure 1.2 Today's automobile. *(Courtesy of Chrysler Corporation)*

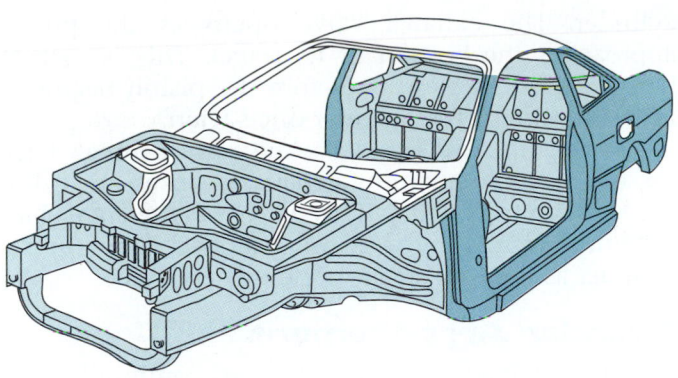

Figure 1.3 Unibody construction.

■ ENGINE PARTS AND OPERATION

Most of today's automobiles use spark-ignited *four-stroke reciprocating* gasoline engines. A limited number of cars use diesel engines. Figure 1.4 shows a cutaway of a modern gasoline engine. In-depth coverage of engine operation is found in Chapter 15. A quick overview is given here.

A reciprocating gasoline engine has a round piston in a cylinder, a connecting rod, and a crankshaft. The principle of its operation is simple. The piston moves up in the cylinder, compressing a mixture of air and fuel in front of it. Compressing the air and fuel makes it very flammable. When the piston reaches the top of its travel, the *air/fuel mixture* is ignited. As the piston is pushed down in the cylinder by the expanding gases, it pushes on the rod, forcing the crankshaft to rotate.

Power is taken from the rotation of the crankshaft to propel the car. As the crankshaft turns, the piston is returned to the top of the cylinder to repeat the cycle again. The continuing up-and-down motion of the piston is why the engine is called a **reciprocating engine**.

The burning mixture is sealed into the cylinder on the top end by a *cylinder head* and a *head gasket* (Figure 1.5). The cylinder head has intake and exhaust *ports*. The intake port allows the flow of the air/fuel mixture into the cylinder. The exhaust port allows the escape of the exhaust gases after the mixture has been burned. Each port is sealed by a valve that is opened by a lobe on the camshaft and closed by a spring (Figure 1.6). The piston is sealed to the cylinder with **piston rings** that slide against the cylinder wall as the piston moves up and down.

Four-Stroke Cycle

The **four-stroke cycle** is described here using a single cylinder engine (Figure 1.7). Automobile engines actually have multiple cylinders. The movement of the piston from the top of its travel to the bottom of its travel is called a **stroke**. Each cycle required to burn the air/fuel mixture has four strokes. Hence the name, four-stroke cycle.

During the *intake stroke*, the piston is pulled down by the turning crankshaft, creating a vacuum above it. Because the intake valve is open while the piston is moving down, the air/fuel mixture is drawn into the cylinder through the intake valve port. The mixture is supplied to the cylinder by the fuel system. Gasoline is especially combustible when one part of it is **atomized** with about 15 parts of air. Atomization makes the mixture like fog.

Figure 1.4 A modern gasoline-powered engine. [Courtesy of Chrysler Corporation]

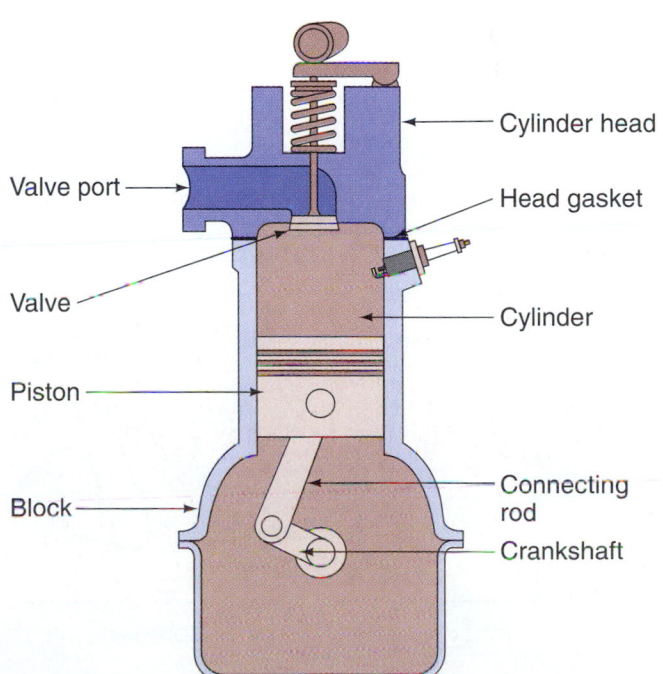

Figure 1.5 Engine parts. [Courtesy of Ford Motor Company]

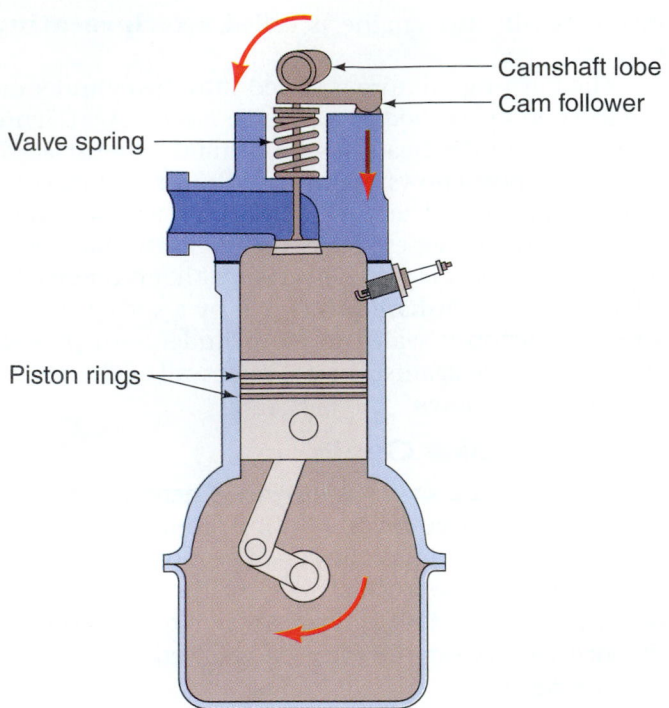

Camshaft lobe

Cam follower

Valve spring

Piston rings

Figure 1.6 The valve is opened by a lobe on the camshaft and closed by a spring. *(Courtesy of Ford Motor Company)*

The piston moves back up in the cylinder on the *compression stroke*, compressing the air/fuel mixture, making it far more combustible. As the piston approaches the top of its travel, a spark plug ignites the mixture.

During the *power stroke* the burning fuel expands rapidly, forcing the piston to move back down in the cylinder. The exhaust valve opens as the piston approaches the bottom of its travel. This is so that burning gases can escape before the piston begins to move upward in the cylinder once again.

During the *exhaust stroke* the piston moves back up, forcing any remaining exhaust gas from the cylinder through the open exhaust valve. As the crankshaft continues to rotate, the piston goes back down in the cylinder as the four-stroke cycle repeats itself.

Cylinder Arrangement

Car engines have multiple cylinders, commonly 4, 6, or 8 (Figure 1.8). Cylinder blocks have rows of cylinders that are arranged in either an *inline* or a "V" arrangement or are *opposed* to each other.

■ ENGINE SUPPORT SYSTEMS

Several subsystems support engine operation. They are the *cooling system*, the *fuel system*, the *lubrication system*, the *electrical system*, including the *charging*, *starting*, *ignition*, and *computer systems*, and the *exhaust system*. These systems are covered in detail in later chapters. To provide improvements in fuel economy and exhaust emissions, new cars also have sophisticated computerized systems that operate fuel and emission control systems.

The Cooling System

As an engine operates, it creates a great deal of heat that is wasted and must be carried away by the cooling system (Figure 1.9) so the engine does not get too hot. *Coolant*, also known as *anti-freeze*, is circulated by the *water pump* through *water jackets* in the engine's

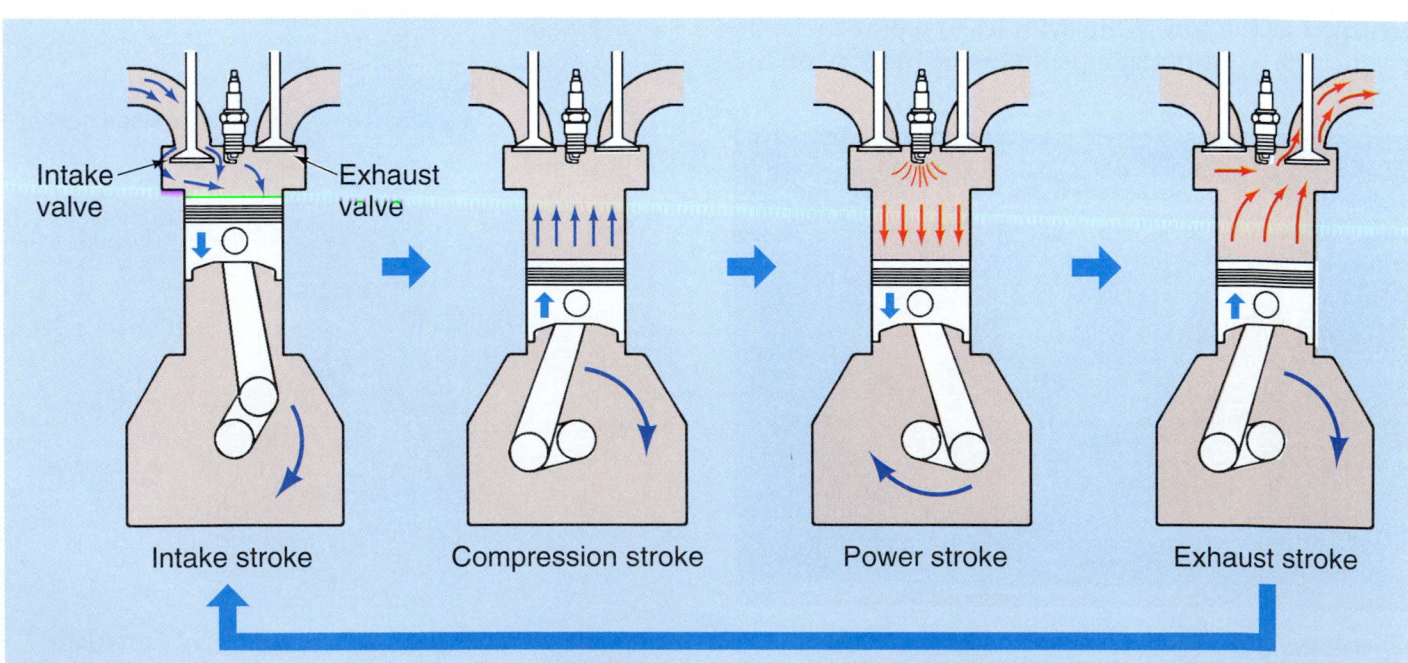

Intake valve

Exhaust valve

Intake stroke   Compression stroke   Power stroke   Exhaust stroke

Figure 1.7 The four-stroke cycle.

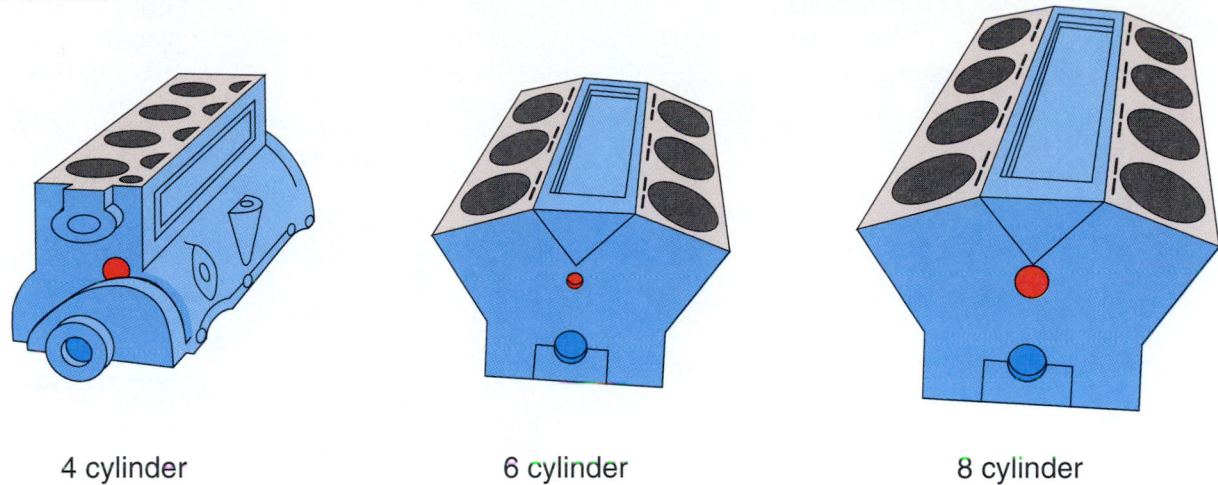

4 cylinder   6 cylinder   8 cylinder

Figure 1.8 Common cylinder block arrangements. *(Courtesy of Ford Motor Company)*

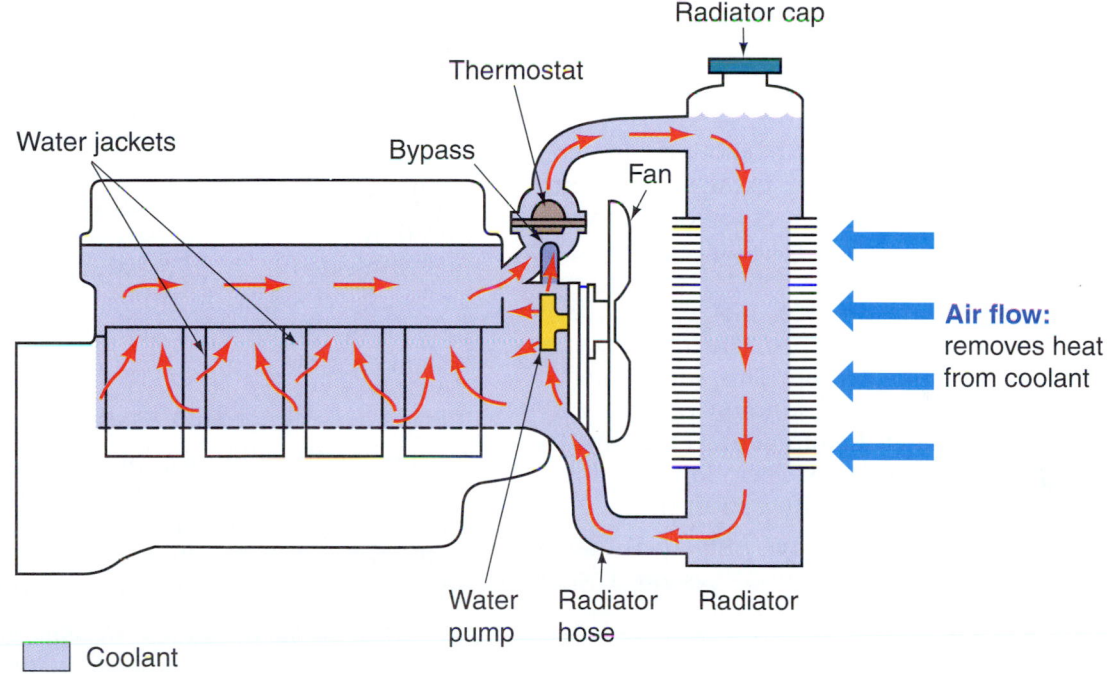

Figure 1.9 Parts of the cooling system. *(Reproduced by permission of Deere & Company, ©1991. Deere & Company. All rights reserved)*

cylinder block. It carries heat to the radiator where it can be carried away by the outside air.

The cooling system's thermostat maintains the coolant at a constant temperature. It also speeds the warm-up of the engine so emission controls and the heater can operate. The engine experiences less wear once it is at operating temperature.

Fuel System

The fuel system is responsible for supplying the correct air/fuel mixture to the cylinder. Liquid gasoline does not burn. It must first be mixed with air in the correct proportion to form a vapor. The *air/fuel ratio* on gaso-line engines ranges from about 12:1 (12 parts of air to 1 part of fuel) to 15:1.

There are three types of fuel delivery systems used on automobile four-stroke cycle engines. They are the *carburetor, gasoline fuel injection,* and *diesel fuel injection.* Carburetors are used on many cars built through about the 1985 model year. Because of the need for more exact control of the fuel system for control of exhaust emissions, fuel injection systems have replaced carbu-retors.

The engine draws in a great deal of air as it runs. This is called engine vacuum. Air rushing through the carburetor draws fuel into it on its way to the cylinders.

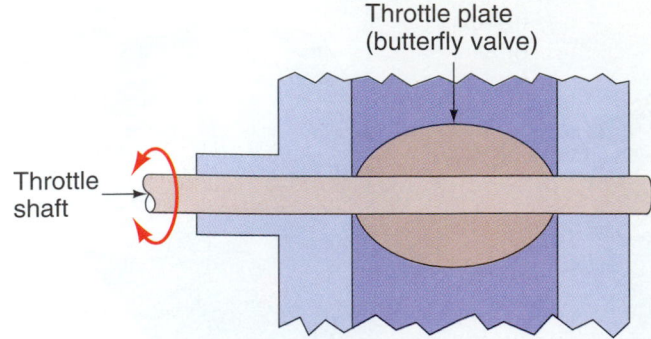

Figure 1.10 A butterfly valve controls air flow into the engine.

When the driver steps on the gas pedal (*throttle*), a **butterfly valve** opens to let in more air, and thus, more fuel (Figure 1.10). Opening the throttle valve allows in more air and raises engine speed.

The Carburetor. A carburetor mixes fuel and air in response to how much air is flowing through it (Figure 1.11). A mechanical fuel pump supplies the carburetor reservoir with fuel.

Gasoline Injection. Fuel injection systems use fuel injectors to spray fuel into the airstream flowing into the engine (Figure 1.12). Newer automobiles (since about 1985) use fuel injection fuel delivery systems operated by a computer.

An electric fuel pump provides fuel at a constant pressure to electronic fuel injectors. The injectors stay open a specified amount of time. This allows fuel to spray out of them in the exact amount called for by the operating condition at the moment. Open throttle allows the injectors to remain open longer; closed throttle leaves them open for only a short time.

Diesel Injection. In a diesel engine, only air is drawn into the cylinder through the air intake system. The air

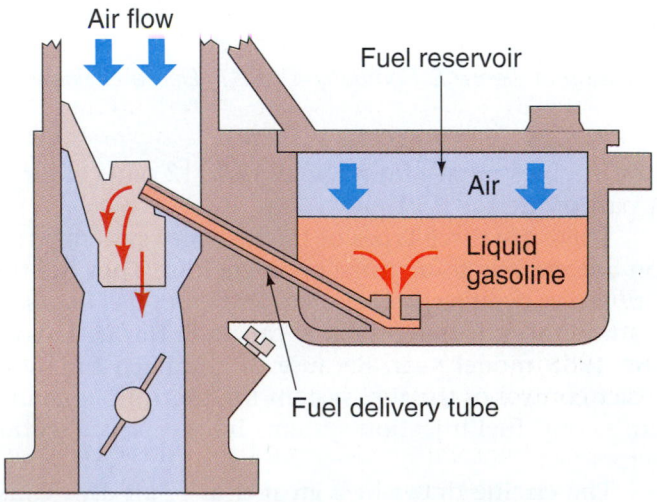

Figure 1.11 A carburetor mixes air and fuel. (*Courtesy of Ford Motor Company*)

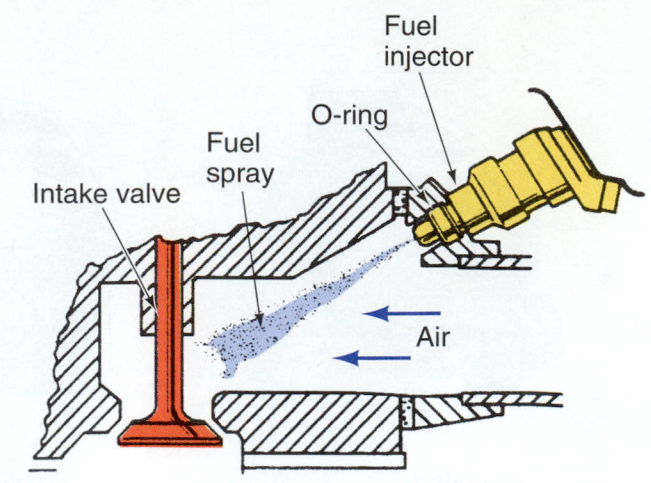

Figure 1.12 Fuel injection. (*Courtesy of General Motors Corporation, Service Technology Group*)

is compressed to about one-half of the size that it would be compressed in a gasoline engine. This results in a great deal of heat. The heat would cause fuel to ignite during compression if it were already in the cylinder. Diesel engines do not have spark plugs or an ignition system, like gasoline engines. Instead, they rely on the injection system for **ignition timing**. When diesel fuel is injected by high pressure injectors into the hot air in the cylinder, it ignites instantly (Figure 1.13). Because of the high pressure in the cylinder, the fuel system on a diesel must be under very high pressure.

The Lubrication System

The engine has a lubrication system that moves pressurized oil to all areas of the engine (Figure 1.14). A pump pulls oil out of the oil pan and forces it through a filter before it is distributed to the engine's parts. The oil prevents moving parts from touching each other.

NOTE: In theory, during a 1000-mile trip a properly operating lubrication system will allow about as much wear between parts as occurs during the first 15 seconds of engine operation in the morning before oil has reached all of the engine's parts.

The Electrical System

The engine electrical system includes the *ignition system, starting system, charging system,* and *computer system.* The body electrical system includes lighting and wiring systems.

The Ignition System. The ignition system has the job of creating and distributing a timed spark to the engine's cylinders (Figure 1.15). Through a process called electromagnetic induction (see Chapter 29), a voltage of 5000 to about 100,000 volts (on some of the newer systems) is created. The voltage causes a spark to jump a gap at the spark plugs to ignite the

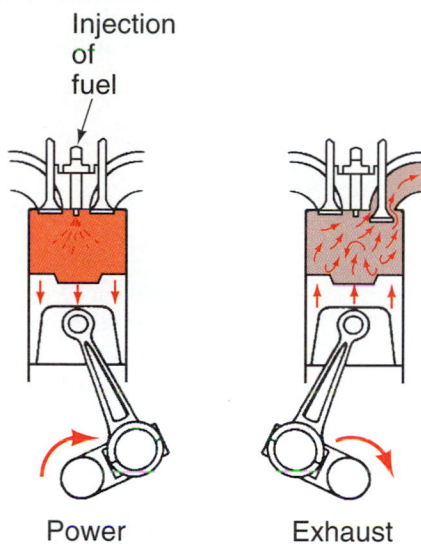
Injection of fuel

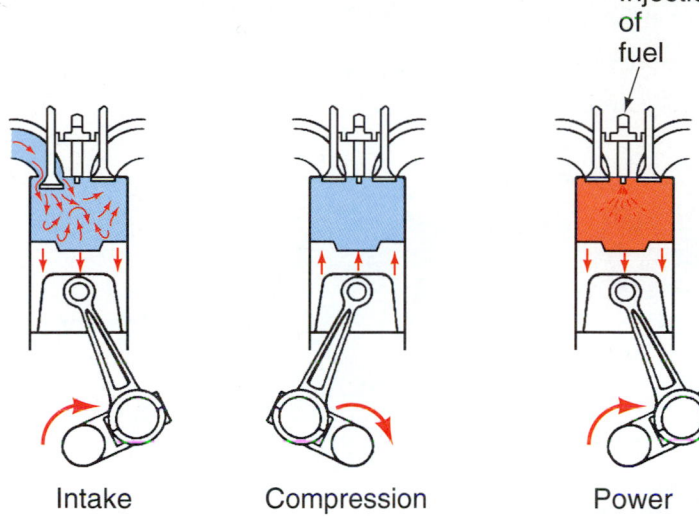

| Intake | Compression | Power | Exhaust |

Figure 1.13 Diesel four-stroke cycle.

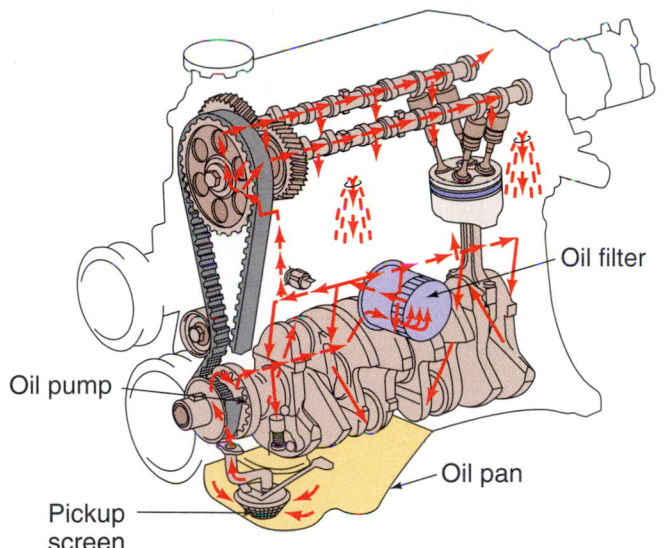

Oil filter

Oil pump

Oil pan

Pickup screen

Figure 1.14 Engine lubrication system.

air/fuel mixture. The spark is timed to occur just before the top of the compression stroke. This is called *ignition timing*.

The Starting System. The starting system has an electric motor mounted low on the rear of the engine. It has a small *pinion gear* on it that meshes with a large *ring gear* on the engine's flywheel (Figure 1.16). The flywheel is bolted to the rear of the engine's crankshaft. The motor draws electrical current through a large cable from the car's battery. When the starter operates, the pinion gear turns the flywheel. This causes the crankshaft to rotate, drawing in air and fuel to start the engine.

The Charging System. As the vehicle is operated, electricity is drawn from the charging system to operate the ignition system, body electrical accessories, or lighting. The charging system (Figure 1.17) includes an

alternator, driven by a belt on the engine's crankshaft pulley. The alternator produces electrical current and forces it into the battery to recharge it. A voltage regulator (sometimes inside the alternator) senses battery voltage. It switches the alternator on or off depending on charging requirements.

The Computer System. Modern automobiles have a substantial amount of on-board electronics. Vehicle electronics has become a specialty repair area with high earning potential. Qualified automotive electronics technicians command excellent pay. The computer system manages the operation of fuel injection, ignition, emission system components, automatic transmission shifting, anti-lock brakes, and body electrical accessories. Many of today's cars have several computers that manage these systems.

A main *computer*, called a *powertrain control module (PCM)* on late model cars, controls the operation of all of the system components. **Sensors** react to temperature, air flow, engine load, road speed, and oxygen content in the exhaust stream. The various sensors send voltage signals to the computer. The computer analyzes the data and makes compensating adjustments using **actuators**. Operation of the computer system is covered in Chapter 69.

The Exhaust System

The exhaust system carries exhaust from the engine to the rear of the car. It also quiets sound. The exhaust manifold, pipes, a muffler, a catalytic converter, and sometimes a resonator make up the exhaust system components (Figure 1.18).

The Emission Control System

Since the mid 1960s various emission devices have been installed on cars. Pollutants enter the air when there is incomplete combustion of fuel. The computer

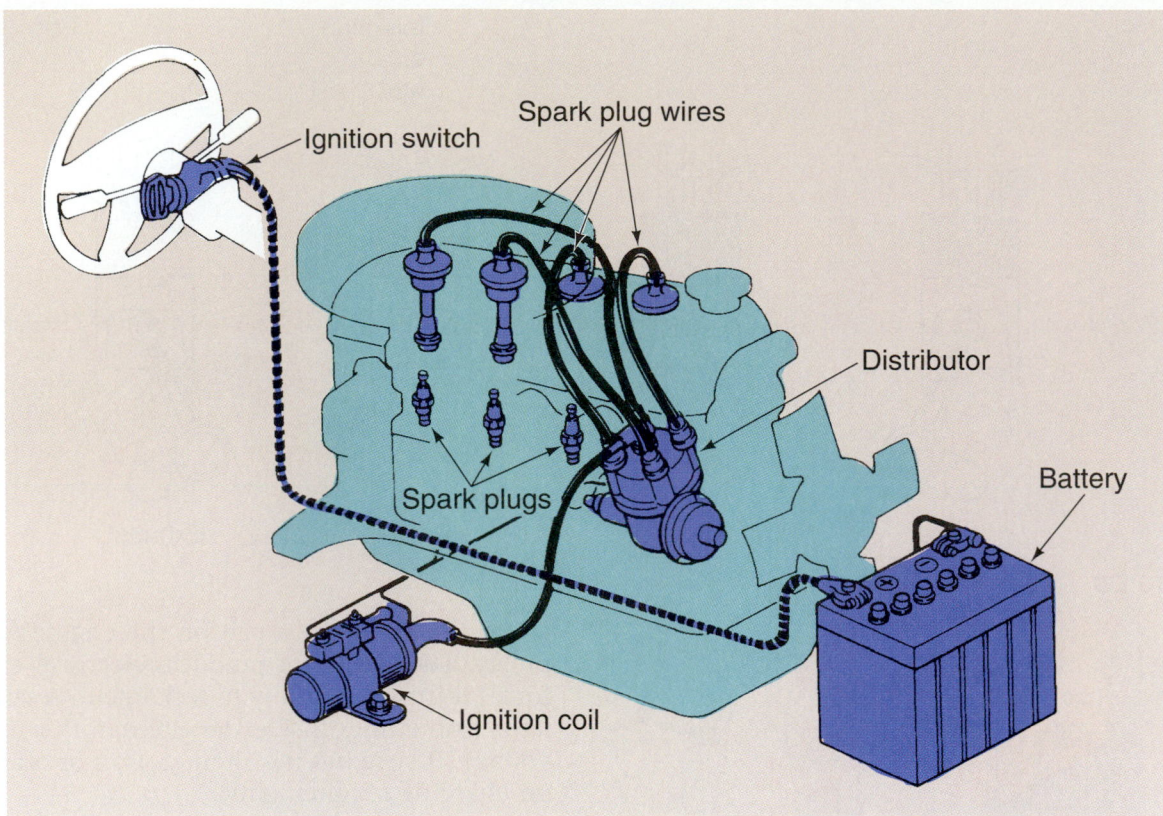

Figure 1.15 The ignition system provides a timed spark to the cylinder.

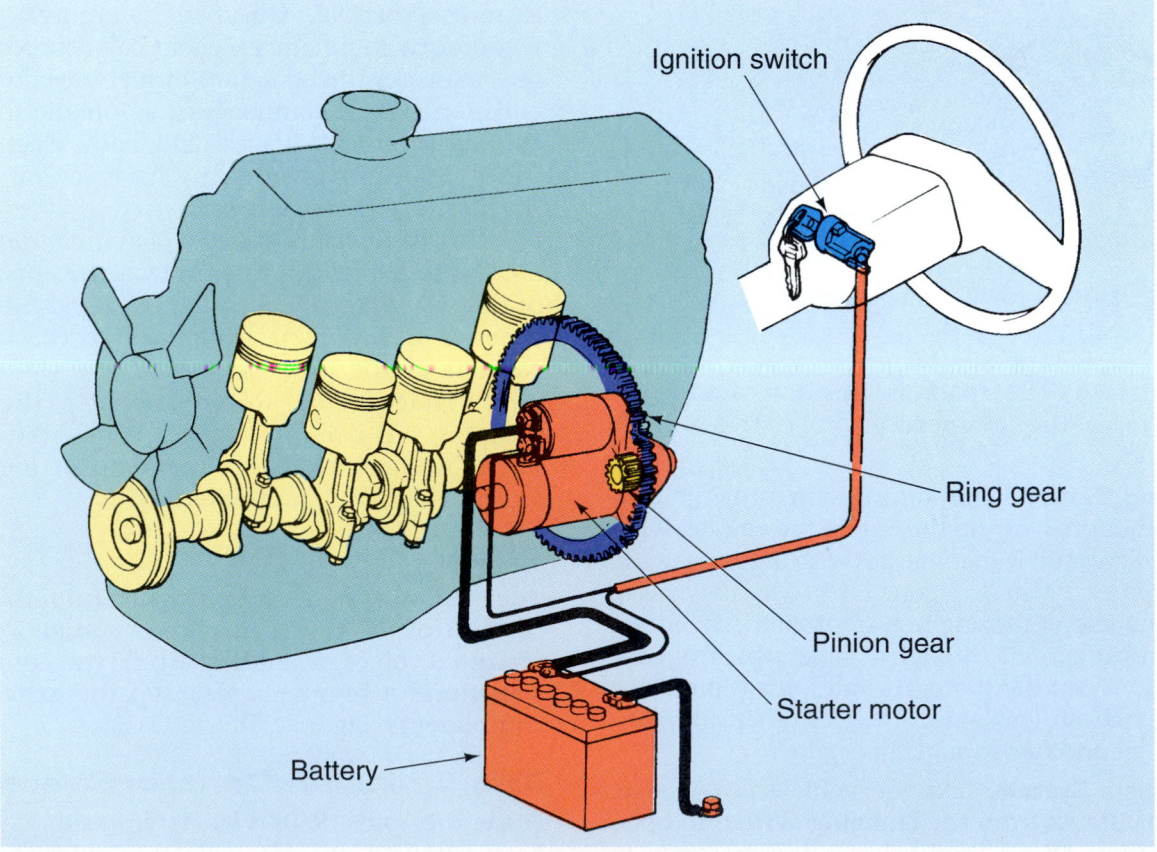

Figure 1.16 The starter motor turns the engine's crankshaft.

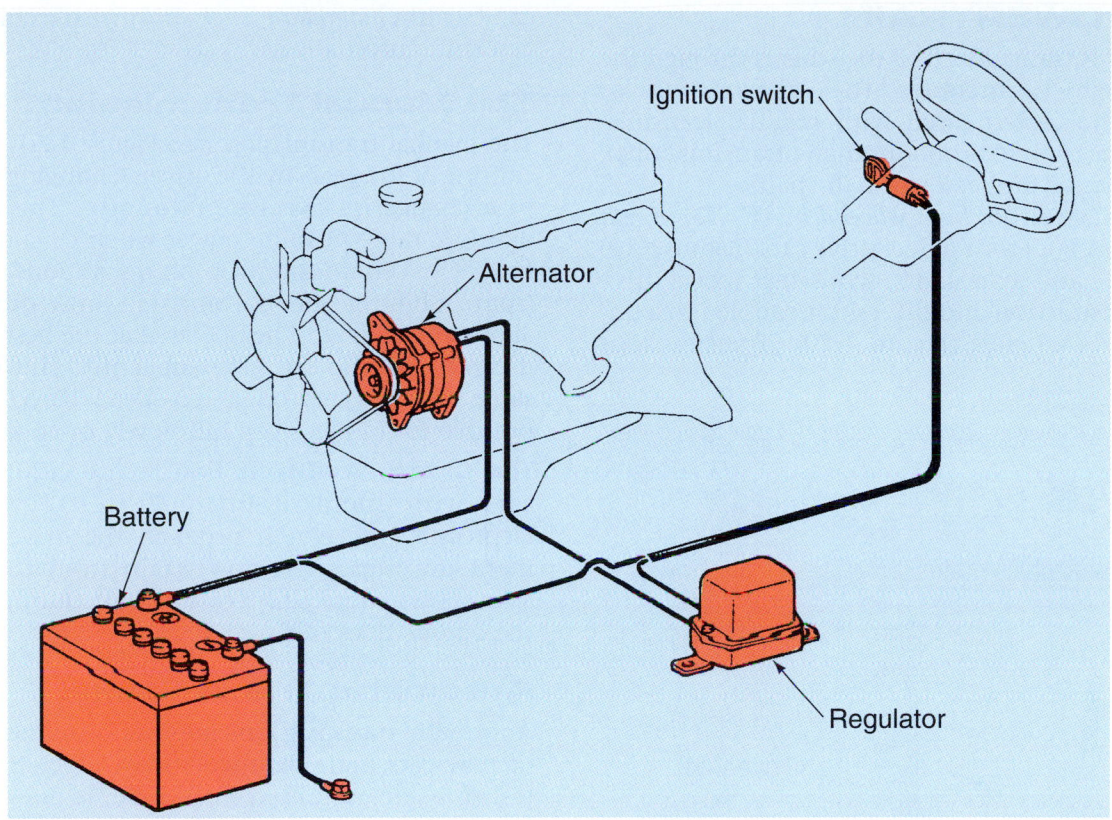

Figure 1.17 The charging system recharges the battery during engine operation.

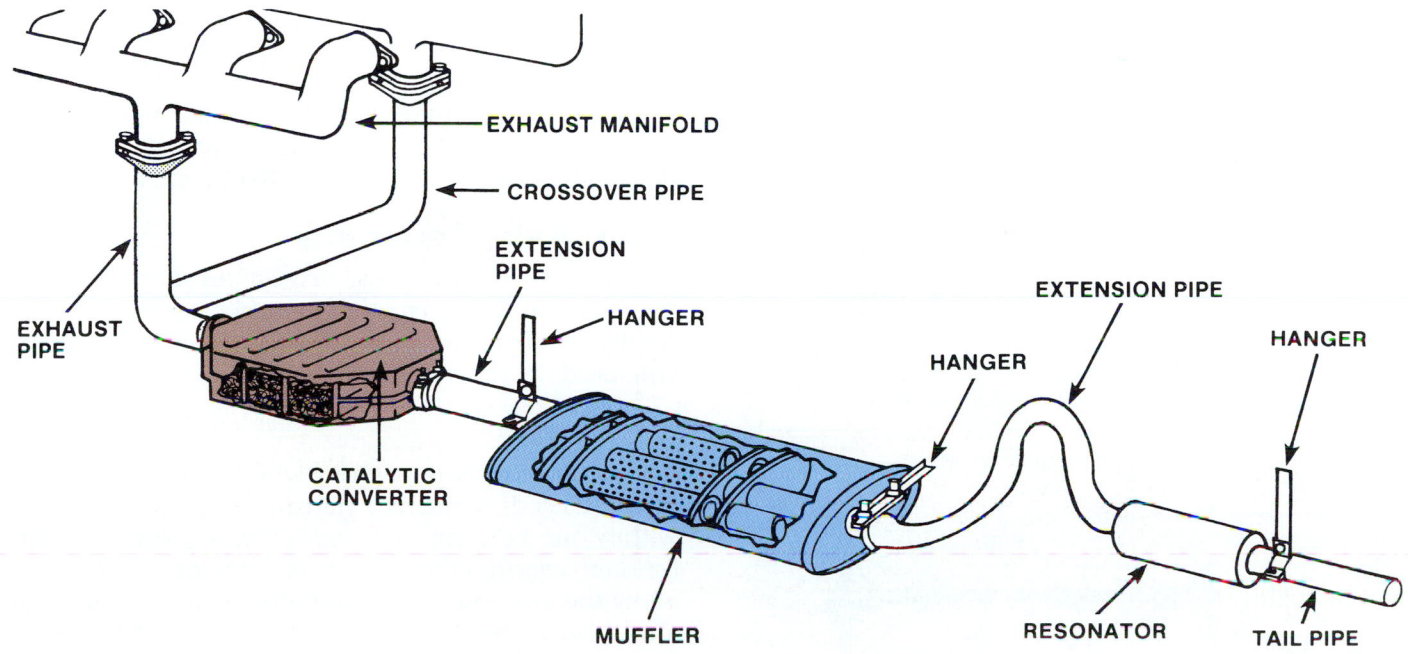

Figure 1.18 Parts of the exhaust system.

system provides the most complete combustion during all driving conditions. The emission system's purpose is to reduce or eliminate any remaining pollutants in the engine's exhaust. In addition to the control of exhaust pollutants, fuel vapors are controlled. Operation of these systems is covered in Chapter 37.

■ THE POWERTRAIN

Engine power is transmitted to the wheels through the *powertrain*, which includes the *transmission* or **transaxle**, the clutch (used with manual transmissions) or *torque converter* (on automatic transmissions), and the *differential* and axles or half-shafts.

Vehicles have either *front wheel drive* (**FWD**) or *rear wheel drive* (**RWD**). Front wheel drive cars (Figure 1.19) use a transaxle and axle shafts, while rear wheel drive cars use a transmission and driveshaft coupled to a differential and rear axles (Figure 1.20). Transmissions

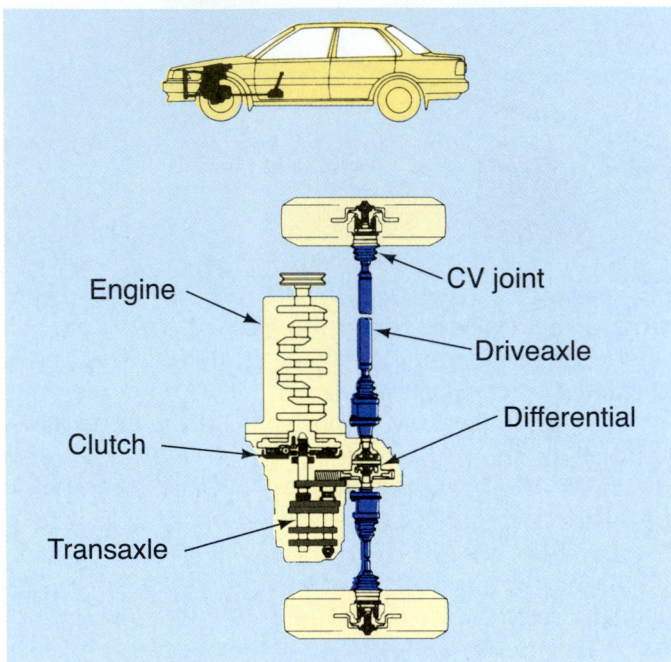

Figure 1.19 Front wheel drive.

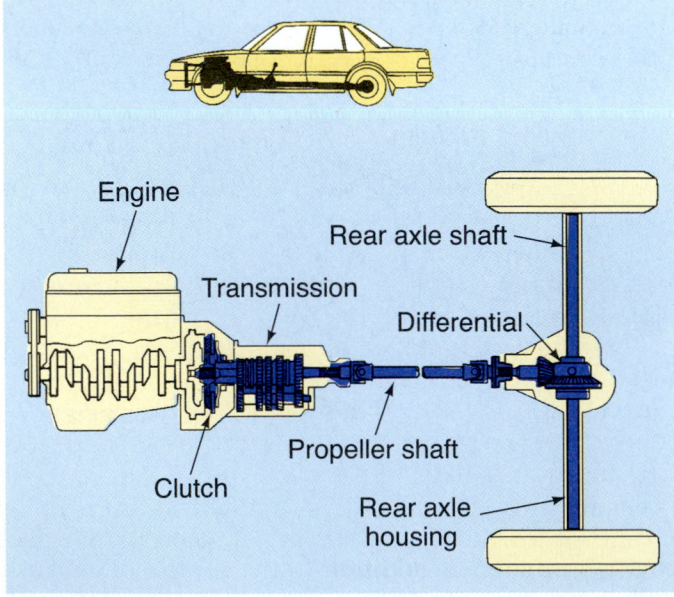

Figure 1.20 Rear wheel drive.

can be either *standard* (*stick shift*) with a clutch, or they can shift automatically.

The Manual Transmission

The manual transmission (see Figure 1.20) provides for shifting of the gears by the driver. Changing gears results in a change in leverage or **torque**. The car's engine develops more pushing power when its crankshaft pins run at a certain number of revolutions per minute (rpm). Shifting the transmission into different gears allows the engine to be operated at the best rpm so that it can propel the car more efficiently. This is much the same as the gears on a ten-speed bicycle making it possible to climb a steep hill slowly or go fast downhill.

The Clutch. A **clutch** (see Figure 1.20) is used on cars with standard (manual shift) transmissions. Depressing the clutch pedal in the driver's compartment uncouples the powertrain from the engine. It also allows some slippage to occur during take-off so the engine does not stall.

Automatic Transmission

Automatic transmissions are found on more than half of new cars manufactured today. Once an automatic transmission is shifted into drive, it does not require further shifting unless a direction change is desired. The torque converter (a fluid clutch) allows stopping and starting automatically. Internal hydraulic pressure is used to shift gears automatically in response to changes in vehicle speed and engine load.

Driveshaft

A *driveshaft*, or propeller shaft, is used on rear wheel drive cars to transfer power to the rear axle. It is a hollow metal tube that has a universal (swivel) joint at each end.

Rear Axle Assembly

The rear axle assembly has drive axles that go to each rear wheel and a *differential* assembly (see Figure 1.20). A differential allows the rear wheels to rotate at different speeds as the vehicle goes around corners.

Transaxle

A **transaxle** is used on front wheel drive vehicles (see Figure 1.19). It contains a transmission and differential within one housing. *Axle half-shafts* (driveshafts) with *constant velocity joints* (*CV joints*) are on each end to allow the axle to move up and down and operate at an angle. Their operation is covered in detail in Chapter 65.

■ ACCESSORY SYSTEMS

Accessory systems are also known as *comfort systems*. Included are *air conditioning* and *heating*, *power seats* and *power windows*, and *cruise control*. As automobiles have been improved over time, more and better features have been added to them.

HISTORY AND DEVELOPMENT OF THE AUTOMOBILE

Steam-Powered Vehicles

The first automobiles were powered by steam engines. The first steam engine was developed in 1698. In 1801 a steam-powered stage coach carried six people a distance of 6 miles. By 1825 a steam-powered bus was driven between Paris and Vienna at speeds of up to 22 miles per hour. Early vehicles had to operate at low speeds because of the poor quality of the roads. The early vehicles were also of large size and weight. Railroads were the solution to this problem.

The steam engine design is called an *external combustion engine* (Figure 1.21). This means that the fuel is burned outside of the engine. Locomotives used steam engines. Coal, wood, or oil is burned in a firebox under a boiler filled with water. The water is turned into steam, which pushes against the piston. The piston drives the wheels by way of a connecting rod (Figure 1.22).

In the early days of the automobile, over 60,000 Stanley Steamers were sold. Steam engines do not require a transmission because they can be started, stopped, or reversed easily. Steam power did have its problems, however. Inventors of the time realized that an engine that did not use coal for fuel must be developed. The steam-powered automobile was gradually replaced with the gasoline-powered forerunner of today's power plant.

Early Gasoline Engines

In 1876, Dr. Nicolas Otto developed the first slow-speed, four-stroke, internal combustion engine. Today, the gasoline type four-stroke piston engine is still called the **Otto-cycle** engine. Compared to previous internal combustion engine designs using the same amount of fuel, Otto's four-stroke cycle engine weighed less, ran much faster, and required less cylinder displacement to produce the same horsepower. A few years later, an Otto-cycle engine would power a motorcycle and then, a horseless carriage.

In 1885, Gottlieb Daimler patented a high-speed engine that ran on petroleum. Daimler is credited as the "Father of the Automobile." Today's engines are patterned after Daimler's invention. During the same

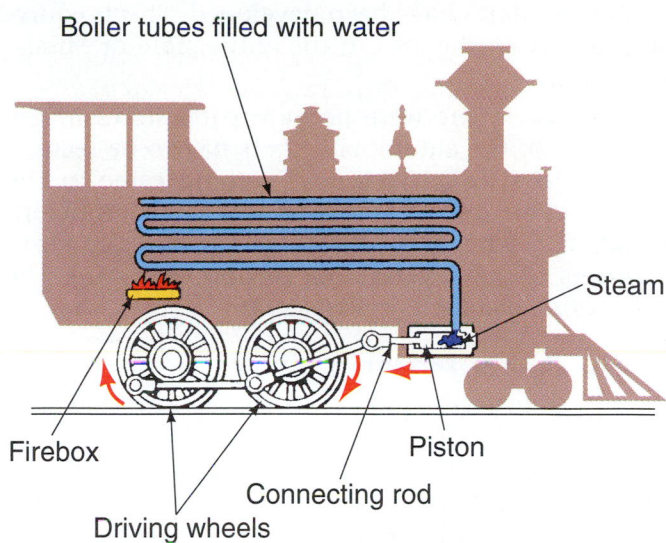

Figure 1.22 A steam-powered locomotive.

year, Carl Benz of Mannheim, Germany, patented a gasoline engine tricycle. The following year several engineers developed motor carriages. Benz developed a car with a one-cylinder, water-cooled engine capable of speeds of 10 miles an hour. Its use on streets was restricted to certain hours only.

The first car to be imported to the United States was a Benz, shown at the 1893 World's Fair in Chicago. You may recognize the names of Daimler and Benz from the corporate name of the company that manufactures Mercedes-Benz automobiles, Daimler-Benz.

Early cars were horseless carriages with single cylinder engines and chain drive (see Figure 1.1). They had tall wire wheels and solid tires. They were steered by a handle, rather than a steering wheel. Cars were so rare in 1900 that Barnum and Bailey Circus gave one top billing in its show. By 1919, 90% of cars still looked like carriages. Running boards were used until the late 1930s. Some cars, like the Volkswagen Bug still used them into modern times.

Early gasoline engines were underpowered, having few cylinders and low compression (about half of today's compression). Later engines of four or six cylinders replaced the early single cylinder engines. The cylinders were most often cast in pairs. The engines had very heavy connecting rods and pistons and ran at slow speeds.

As the automobile was developed, there came a need for better materials. Thus, the automobile contributed to scientific developments and provided the financial means to accomplish them. Present day manufacturing methods including mass production of interchangeable parts and the assembly line are a result of early developments required to produce engines and automobiles.

Cars have resulted in great changes to our world. In less than 100 years, 3.8 million miles of roads in

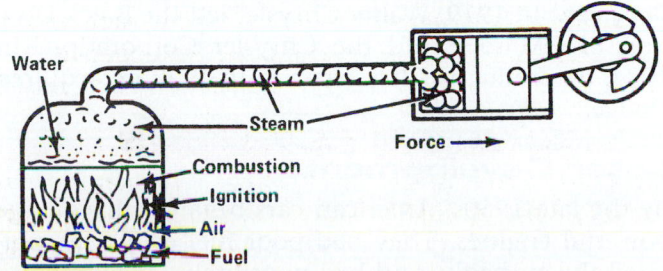

Figure 1.21 An external combustion steam engine.

the United States have been developed. This is equivalent in size to the area of the entire state of Mississippi.

Auto Parts. There were no auto parts stores in the early days of the automobile. Parts had to be secured from the individual factories. Nothing came in the same size on any of the cars. Bolts had different threads and wheels and tires were of different sizes. Engineers of the best factories met and organized the *Society of Automotive Engineers (SAE)*.

Early Auto Racing

The first auto race was held in Chicago in 1895 as a promotion for the automobile. Cars improved at a rapid pace after about 1910. Speed and reliability contests were held, the most famous being the Indianapolis 500, which was started in 1913.

Early Transmissions

Early cars had the transmission on the rear axle. Later on, the transmission was attached to the rear of the engine.

Carburetors

Early carburetors had a wick saturated with gasoline that air was drawn through. Later carburetors had a bowl full of gasoline. The carburetor was usually in a difficult location for repair. This was because it had to be lower than the gasoline tank, located under the driver's seat. One advancement was to put the tank in the rear of the car and use a fuel pump to move it to the carburetor. These cars could have a shorter intake manifold, which meant that atomized fuel could be more easily delivered to the cylinder.

Fuel Pumps

In 1915, Stewart Warner developed the vacuum tank, which was used on almost all cars to feed fuel from the tank. In 1928, the electric and mechanical fuel pumps, which were attached to the engine, improved the way fuel was moved.

Lubrication System

Early engines used a drip oiler. Later cars had mechanical oiling with individual oil tubes to the bearings and cylinders. Later engines used a combination of splash oiling and pressure lubrication.

Tires

In 1900, a Frenchman named Michelin made the first pneumatic (air filled) tires. Michelin's early tires had a single tube. In 1903, a casing and tube were developed. Tires ranged in diameter from 28 to 42 inches. Changing a tire was a big and dirty job. The tires had to be pumped full of air with a hand pump. By 1919, all cars were equipped with cord tires. In 1924, the balloon (low-pressure) tire was introduced.

Electrical Systems

Some of the early cars had 8-, 12-, or 24-volt electrical systems. Almost all of the early electrical systems combined the starter and generator in one unit. In 1915, separate units started to show up on cars and the 6-volt battery became standard. Battery ignition systems, which were combined with the generator became popular. Today's cars use 12-volt batteries, which became standard in the middle 1950s.

Starter System

Early engines had to be hand cranked to get them started. In 1912, Charles F. Kettering developed the electric starter motor.

Early American Automobiles

Selden Patent. In 1879, a United States patent, known as the **Selden patent**, covering all gasoline powered, self-propelled road vehicles was issued to George B. Selden. Selden did not actually invent the automobile, but all American companies had to pay licensing fees to him anyway. Henry Ford was a defendant in a case that was carried through the Michigan courts and on to the United States Supreme Court. In 1911, the patent was ruled to be void.

In 1892, Charles Duryea and his brother, Frank, built the first American car that would actually run. Alexander Winton sold the first automobile in the United States in 1898. Ransom Olds produced his Curved Dash Olds. A fire consumed all but the prototype model for the 425 that were produced that year.

Model T Ford. Henry Ford built his first car in 1893. Henry Ford produced the Model T Ford between 1908 and 1926. It was one single universal car, all of the same color, black. One thousand of them were produced on the assembly line per day. Over 15 million Model Ts were produced in all. By 1926 over two-thirds of the cars on the road were Model Ts.

General Motors. General Motors' Billy Durant had the idea of producing a variety of cars, "one for every purse and purpose." Durant was a good promoter, but a poor businessman. He was removed by GM when Dupont took over the company. Alfred P. Sloan became the new president of General Motors.

Chrysler. In 1919, Walter Chrysler left the Buick Division of GM to found the Chrysler Corporation. In 1924, he produced Chryslers and in 1928 he acquired Dodge.

Later Developments

By the late 1950s, American cars began to have large powerful engines. They had poor fuel economy, polluted the air badly, and had poor brakes compared to later cars. In the mid sixties, large engines like the 427

(7 liter) produced by Ford and Chevrolet, had over 400 horsepower. This was called the "muscle car" era.

Emission controls were added to cars starting in the early 1960s. These were brought about by clean air legislation. Horsepower output of engines was reduced as a result of the changes brought to these engines by emission controls.

Fuel Economy Standards

During the 1960s and early 1970s, a gallon of gasoline could be purchased in the United States for about 25 to 30 cents. An oil embargo by Middle Eastern countries in 1973 caused the price of gasoline to quadruple. In 1975, the U.S. Congress passed fuel economy standards. **Corporate Average Fuel Economy (CAFE)** is the name of these standards. Fuel economy for different models sold by each manufacturer are averaged to come up with the CAFE number. The CAFE standard

at this writing is 27.5 miles per gallon. When a manufacturer makes cars that get poor fuel economy, they pay a heavy penalty to the government. When a consumer buys a "gas guzzler," a tax is included in the purchase price.

Modern Developments

Today's automobile has benefited greatly from technological developments made in the military and space programs. The automobile of 1965, while still having the same basic parts, is a dinosaur compared to today's high tech vehicle. Today's car has a very good power-to-weight ratio, which means it gets good gas mileage and goes faster compared to comparably sized cars of the sixties. Computerized controls manage engine and body systems for comfort and efficiency. Yearly advancements are made that make vehicles safer and more reliable.

■ REVIEW QUESTIONS

1. What is the name of the automotive chassis design that has a floorpan and a small subframe section in the front and rear?

2. When air and fuel is compressed it becomes very_____.

3. What does reciprocating mean?

4. The valve is opened by a _____ on the camshaft.

5. What are the names of the parts that seal the combustion gases between the piston and the cylinder?

6. The movement of the piston from the top of its travel to the bottom of its travel is called a _____.

7. List the four strokes of the four-stroke cycle in order.

8. The air/fuel mixture is _____ as the piston moves up on the compression stroke.

9. What is another name for anti-freeze?

10. The air/fuel ratio on gasoline engines ranges from about _____:1 to _____:1.

11. What is the name of the butterfly valve that controls the amount of air entering the engine?

12. The starter motor pinion gear drives the _____ gear on the crankshaft's flywheel.

13. What is the name of the part that switches the alternator on and off?

14. The parts of a computer control system that react to temperature, airflow, engine load, road speed, and oxygen content in the exhaust stream are called _____ .

15. What was the name of the patent for the automobile?

■ ASE STYLE REVIEW QUESTIONS

1. Technician A says that the intake valve is closed during the power stroke. Technician B says that the exhaust valve is closed during the compression stroke. Who is right?

   a. Technician A   b. Technician B
   c. Both A and B   d. Neither A nor B

2. Technician A says that carburetors have replaced fuel injection systems. Technician B says that fuel injection systems are better for controlling exhaust emissions. Who is right?

   a. Technician A   b. Technician B
   c. Both A and B   d. Neither A nor B

3. Technician A says that diesel engines do not have spark plugs. Technician B says that diesel timing is controlled by when the fuel is injected. Who is right?

   a. Technician A   b. Technician B
   c. Both A and B   d. Neither A nor B

4. Technician A says that an ignition spark is 12 volts. Technician B says that the spark happens just after the power stroke begins. Who is right?

   a. Technician A   b. Technician B
   c. Both A and B   d. Neither A nor B

5. Technician A says that most rear wheel drive cars have a transaxle. Technician B says that front wheel drive cars have two driveshafts. Who is right?

   a. Technician A   **b.** Technician B

   c. Both A and B   **d.** Neither A nor B

6. Technician A says that gears are changed to increase engine torque. Technician B says that the majority of new cars have automatic transmissions. Who is right?

   a. Technician A   **b.** Technician B

   c. Both A and B   **d.** Neither A nor B

7. Technician A says that a steam engine is an external combustion engine. Technician B says that a four-stroke piston engine is an internal combustion engine. Who is right?

   a. Technician A   **b.** Technician B

   c. Both A and B   **d.** Neither A nor B

8. Technician A says that an Otto-cycle engine is a compression ignition engine. Technician B says that the diesel cycle engine has its ignition timing controlled by the spark plug. Who is right?

   a. Technician A   **b.** Technician B

   c. Both A and B   **d.** Neither A nor B

9. Technician A says that Model Ts were all black. Technician B says that the 12-volt electrical system has been standard on American cars since 1940. Who is right?

   a. Technician A   **b.** Technician B

   c. Both A and B   **d.** Neither A nor B

10. Technician A says that CAFE standards govern vehicle emissions. Technician B says that CAFE standards govern fuel economy. Who is right?

    a. Technician A   **b.** Technician B

    c. Both A and B   **d.** Neither A nor B

CHAPTER 2

Automotive Careers and Technician Certification

■ INTRODUCTION

Demand for Technicians

There are over 200 million motor vehicles in the United States alone. Due partly to the increased complexity of vehicles, the demand for skilled technicians is high and will continue to climb.

Today's automotive technician must be able to perform a multitude of tasks requiring several different types of skills. It is said that a mechanic must be a "jack of all trades." The ability to repair automobiles includes skills in plumbing, metal working and welding, electrical, electronics/computer/radio, and air conditioning.

The repair and service of automobiles is a challenging and rewarding field. Most important of all is to enjoy the work. If you like the work and have the talent, you will never be bored and will continually learn new things. Following are some of the jobs that automotive training can prepare you for.

■ AUTOMOTIVE CAREER OPPORTUNITIES

Service Stations

In the past, almost all auto mechanics began their careers in service stations (also called gas stations or filling stations) (Figure 2.1). Twenty-five years ago there were several times more service stations than there are today. Virtually all of them did service and repair work. Today, most service stations have mini-marts instead of service bays. Even so, there are still many service stations today that have a healthy "back room business."

Another change has been that gas stations used to be full service. People would have underhood checks done on their cars at each fill-up. Today's gas stations are mostly self-service and often the consumer neglects his or her car as a result. Lube outlets and mass merchandisers are filling the void in the periodic service business left by the closing of many former service stations.

Figure 2.1 Service stations often perform automotive repairs.

Lubrication/Safety Shops

Many repair facilities have a "quick-lube" section to accommodate maintenance services. There are also specialty shops that only perform lubrication services (Figure 2.2). A *lubrication specialist* performs maintenance and safety services on the automobile. Changing fluids, belts and hoses, and inspecting and replacing safety items are included. A thorough description of lubrication and safety service will be found in the early chapters in this book.

New Car Dealerships

New car dealers have a service department where warranty and customer pay repairs can be made (Figure 2.3). Dealerships also perform **pre-delivery inspections (PDI)** and *warranty repairs,* when the car is still new and a problem related to its manufacture develops. Warranty coverage is usually only given before a car has reached a certain age and a certain amount of mileage, 5 years and 50,000 miles, for instance. Dealerships often have body shops and used car departments also.

Figure 2.2 Some shops specialize in quick lubrication and safety service.

Automotive manufacturers all have training schools for updating the skills of their dealership technicians. A dealer technician has the advantage of being able to attend these schools. Some classes last a day, others can last a week or longer. Examples of classes are:
- a one-week class on specialized vehicle electronics
- a two-day class on battery, starting, and charging system.

The schools are usually located in large metropolitan areas. When a technician from another town attends a class in the metropolitan area, the dealer often provides him or her with a car to drive. He or she is also reimbursed for food and lodging.

Independent Repair Shops

Independent repair shops can be one person specialty shops or very large full coverage repair shops (Figure 2.4). They often specialize in one or more makes of vehicles. It is not uncommon for a dealer technician to start a competing business on one make of vehicle, after leaving the employment of a dealership. One of the things that customers like about the smaller establishments is that they get to actually talk to the person who works on their car. In a dealership or large independent shop, the technicians do not normally have the opportunity to talk with their customers.

Operation of Large Shops

The **service manager** is responsible for the operation of the service department (Figure 2.5). This sometimes includes the management of the parts department, although dealerships often have parts managers, too. The service manager supervises the service writers, dispatcher, and technicians. He or she is also responsible for the management of warranty communications between the manufacturer, dealer, and consumer. Customer complaints and communication are important aspects of the job.

Figure 2.3 New car dealers have service and repair departments specializing in one or more makes of vehicles.

Figure 2.4 Independent repair shops perform repairs on one or more makes of vehicles.

Figure 2.5 In a large repair facility, the service manager is responsible for the operation of the service department.

Figure 2.7 Fleet operations usually have their own repair facilities.

The link between the technician and the customer is the **service writer/service advisor**. He or she greets the customer, listens to the complaint, interprets it, and then writes the repair order for the job (Figure 2.6). The repair order includes the estimated cost of the repairs. When the cost of the repair goes over the estimate that the customer was informed of, the service writer will call the customer to explain the need for more repairs and get authorization to proceed. He or she also works with the technician to verify that the customer complaint is repaired.

In many states, consumer protection laws state that the customer must give consent for additional repairs before the additional work can be completed. The technician depends on the service writer to provide an accurate description of the problem so that the correct repair can be made at the estimated cost.

The **service dispatcher** is the person who organizes the repair orders and dispatches them to the technicians in the service bays. He or she keeps track of how work is progressing. This position requires a good deal

of diplomacy. Because the technicians are paid on a commission basis, the jobs that they are assigned can make a big difference in earnings, even between technicians working in the same establishment.

Some large shops have a **shop foreman** who keeps repair work on track. He or she must be able to diagnose or troubleshoot difficult problems and must keep up on the latest repair procedures. Some large repair shops have a *quality control technician* who checks the work of the technicians and handles customer complaints. This is part of the shop foreman's job description in other shops.

Fleet Shop

A fleet is a group of several vehicles owned by a company, utility, or municipality (government) (Figure 2.7). A taxi cab company, rental car company, or the telephone company are examples of fleets. Preventive maintenance is scheduled ahead of time. Public safety is the most important consideration, so safety inspections are done on a regular basis. Some **fleet shops** sub-contract major repairs to another specialty shop. Others perform all of their own repairs. Fleet technicians often repair diesel and alternative fuel vehicles, too.

■ TECHNICIAN CERTIFICATION AND LICENSING

There is a tremendous and growing shortage of trained automotive technicians. Vehicles have become very complex, and the education and knowledge required of today's technician is much more than that required to repair and service vehicles produced 25 years ago.

ASE Certification

Due to the complexity of today's vehicles, technicians often choose one or more areas to specialize in. There are eight specialty areas under the automotive repair umbrella. Many technicians are certified through the voluntary certification program administered by the

Figure 2.6 The service writer works with the customer.

Test Title		Test Content
Engine Repair (80 Questions)	Test A1	Valve train, cylinder head, and block assemblies; lubricating, cooling, ignition, fuel and carburetion, exhaust, and battery and starting systems
Automatic Transmissions/Transaxle (50 Questions)	Test A2	Automatic Transmissions and Transaxles; Controls and linkages; hydraulic and mechanical systems
Manual Drive Train and Axles (40 Questions)	Test A3	Manual Transmissions, clutches, front and rear drive systems
Suspension and Steering (40 Questions)	Test A4	Manual and power steering, suspension systems, alignment, and wheel and tires
Brakes (55 Questions)	Test A5	Drum, disc, combination, and parking brake systems; power assist and hydraulic systems; ABS
Electrical/Electronic Systems (50 Questions)	Test A6	Batteries; starting, charging, lighting, and signaling systems; electrical instruments and accessories
Heating and Air conditioning (50 Questions)	Test A7	Refrigeration, heating and ventilating, A/C controls, refrigerant recovery
Engine Performance (70 Questions)	Test A8	Oscilloscopes and exhaust analyzers; emission control and charging systems; cooling, ignition, fuel and carburetion, exhaust, and battery and starting systems

Figure 2.8 ASE tests cover the eight automotive specialty areas. Test content is listed here. *(Courtesy of ASE)*

National Institute for Automotive Service Excellence (ASE). The eight automotive repair areas of specialization are:

- Engine Repair
- Engine Performance
- Heating and Air Conditioning
- Electrical Systems
- Automatic Transmissions and Transaxles
- Standard Transmissions and Axles
- Brakes
- Suspensions

Tests are given twice each year, usually at a school in the community. To become certified in one of the specialty areas, you must pass between 60% to 70% of the questions (depending on the difficulty of the particular test). Figure 2.8 shows the number of test questions and the content of the specialty areas.

ASE sends test results that tell what percentage of test questions you passed in each area of the test. When you pass the test, a wall certificate and pocket protector with your certification dates are sent to you.

Certified technicians can be identified by a shoulder patch that says "Automotive Technician."

A **master automobile technician** is a journey level professional who is certified in all eight of the ASE areas of specialization. A master technician has a different shoulder patch that says "Master Automobile Technician" (Figure 2.9).

Work Experience Requirement

You must have two years of automotive work experience (school training can count as one of the years). If you lack the experience requirement, you can still take the test. You will receive your test results and ASE will certify you as soon as you meet the experience requirement.

Figure 2.9 ASE certification patch.

Why Certify

Employers often complain that the technicians and machinists they hire are unqualified. Many employers now ask for ASE certification when they advertise a job opening. Although some employees continue to resist certification, it does provide a technician with a means of showing a prospective employer that a technician has work experience and has developed sufficient knowledge about automobiles to pass a test in an automotive specialty area. Certification also demonstrates that a technician can read—something all the best technicians of the future must be able to do.

Advanced Engine Performance Specialist

One of ASE's other certifications is called the *advanced engine performance specialist* (L1). To qualify to take this test, a technician must first be certified in the regular automotive categories of engine performance and electrical systems. This test covers several areas: general powertrain diagnosis, computerized engine controls diagnosis, ignition system diagnosis, fuel and air induction system diagnosis, emission control system diagnosis, and emission I/M (inspection and maintenance) failure diagnosis.

Engine Machinist

Automotive machine shops also have an ASE certification program. Originally sponsored by the **Automotive Engine Rebuilders' Association (AERA)**, there are three areas of specialization:

- Assembly Specialist (M1)
- Cylinder Block Specialist (M2)
- Cylinder Head Specialist (M3)

Machinists who qualify in all three categories are known as *Master Engine Machinists*. Engine Machinist Tests are administered as part of the regular ASE technician test series.

Other ASE Test Areas

There are also certification categories in Heavy-Duty Truck, Auto Body, and Paint.

To receive a bulletin of information about ASE certification write to:

> ASE
> 13505 Dulles Technology Drive
> Herndon, VA 22071-3415

Certification or License

Certification is a *voluntary process*. When the government requires a certain testing or skill level before allowing specialized work to be performed, this is called **licensing**. Emission testing and repair licensing is one area in which state governments have become heavily involved. Some state governments use the ASE advanced engine performance specialist certification as a prerequisite to licensing their emission technicians. Other states have their own testing programs.

■ TECHNICIAN SKILL LEVELS AND PAY

Flat Rate/Commission

Master technicians are often paid on a **flat rate/commission** basis. Each job has a listing in the *Parts and Time Guide* or *Flat Rate Manual* (see Figure 4.16) that gives the amount of time it is supposed to take to complete. If the technician takes an hour and a half to complete a job that is listed as a two-hour job, he or she is still paid for the two hours.

The opposite is also true. If it takes three hours to complete the job, it still only pays two hours. Technicians call this "flagging time." On a really good day, a technician might earn twelve hours of pay for an eight-hour day. On a poor day, four hours' pay might be all that is earned.

When a flat-rate work fails, the job is done again by the technician at no charge to the customer. A problem deep inside a recently rebuilt engine could require removal of the engine, a disaster for the technician who first performed the job.

For income protection, the best technicians sometimes have a "guarantee" in which the repair shop agrees to pay a minimum amount per pay period. In reality, the guarantee rarely comes into play.

Automobiles have become so complex in recent years that technicians often earn more money than others who are in positions of management within the dealership. If a successful technician decides that a move into management is what he or she would like, management is often reluctant to have them make the move because they are too valuable in the shop.

An experienced master technician earning $50,000 a year is probably responsible for $110,000 to $120,000 per year of labor earnings for the business. This does not include the markup on the sale of required parts.

Apprentice Technician

Some areas have unions that have formal apprenticeship programs. Apprentices work under journey level, or master technicians for a specified number of years before they work on their own. In areas where there is no formal apprenticeship, many repair shops and dealers use an apprenticeship system within their shops.

One scenario for a one-year apprenticeship in a dealership works like this: a new technician is hired who has had some previous automotive work experience, but is not a master technician. The apprentice will work for two months with each of the master technicians in the dealership. The apprentice is paid an hourly wage by the dealership during this first year.

As their incentive for mentoring the apprentice in the above system, the master technicians are paid for all of the hours that they earn in combination with the apprentice. In other words, they have free help for the two-month period in exchange for giving instructions and sharing their expertise.

Another system used by some shops is the concept of a "team." Team members have several levels of expertise. They share pay for the jobs completed, with each team member being paid at a different level for the hours flagged. In this case, the apprentice at the bottom of the scale earns considerably less than the lead technician.

■ OTHER AREAS OF SPECIALIZATION

Tire and Wheel Shops

Some repair shops specialize in tires and wheels (Figure 2.10). An ASE technician certified in brakes and

Figure 2.10 A tire and wheel specialty shop. *(Courtesy of Bridgestone/Firestone, Inc.)*

suspension systems is usually employed by these shops and performs wheel alignment and suspension repairs related to tire wear. Sometimes brakes and alignment or other repair services are performed in the larger shops. A tire shop will have one or more *tire and wheel specialists* who mount and balance tires.

Muffler Shops

Muffler shops specialize in repairs and replacement of the vehicle's exhaust system (Figure 2.11). An *exhaust system specialist* fabricates new pipes and installs them. Muffler shops have large tubing benders for making new exhaust pipes from straight lengths of pipe. The new pipes and mufflers are welded together in place on the car. Muffler hangers are installed to keep the parts from vibrating and becoming damaged. Some muffler shops also do other things like quick lubes, tires, and brake and suspension work.

■ OTHER AUTOMOTIVE CAREERS

Auto Parts

An *auto parts specialist* needs to have a basic knowledge of the operation of the vehicle. Some of the best parts people have repaired cars and have very good diagnostic abilities. In addition to automobile knowledge,

Figure 2.11 Muffler shops specialize in repairs and replacement of the vehicle's exhaust system.

a parts person must have good customer relations and telephone skills. When things get hectic (telephones ringing and a line of customers waiting), a parts person must be able to work without losing patience. Good reading skills and the ability to effectively use computer and parts books are all part of the skills required of the successful parts persons (Figure 2.12).

Auto Body Technician

When auto body shops are associated with auto repair shops the work area is located in a separate area. This is because of noise, dust, and vapors. An auto body technician fixes accident damage and paints vehicles. Specialized skills and training are required of a true professional in this area. Auto body shops are now highly regulated due to clean air laws.

Small Business Opportunities

Many independent repair shops are owned and operated by families. It is very common to find a husband

(a)

(b)

Figure 2.12 An auto parts specialist identifies and orders replacement parts using (a) parts books, or (b) a computer.

Figure 2.13 Owners of automotive businesses often belong to professional associations. *(Courtesy of Automotive Service Association)*

and wife working together managing and operating a business. Several kinds of abilities are required, including leadership and management talents. Public relations skills are very important to the success of a small business. The most talented auto technician will quickly go out of business if he or she lacks adequate people skills.

Hiring and firing employees is another aspect of running a business. The owner must assess the character of a potential employee and if he or she does not work out, the owner must deal with terminating that person.

Other types of abilities required in managing a small business include payroll and bookkeeping. Some shops hire accounting firms for these duties.

Industry Professional Associations

Successful shop owners have leadership ability and are usually involved in automotive small business associations such as the **Automotive Service Association (ASA)** or **Automotive Service Councils (ASC)** (Figure 2.13). Dealers often belong to the **National Automobile Dealers Association (NADA)**. These associations provide a way to meet other professionals with similar concerns. Local chapter monthly meeting agendas include legal, business, and training items. The association at the state and national level works to influence legislation that affects automotive businesses. Grouping together also provides increased benefits to the group members for items such as insurance for medical and worker's compensation.

■ ON THE JOB AS AN AUTOMOTIVE TECHNICIAN

Successful automotive technicians have several attributes besides the ability to repair automobiles. The success of a business depends on the performance of

quality repairs. But a customer must also feel that he or she is being fairly treated.

Customer Relations

Technicians need to communicate effectively. In a large repair operation, this can be through a service writer. It is important that information be clearly written on a repair order. A customer's impression of the repair job and the business in general is affected by how professional the repair order appears. The repair order delivered to the customer must be neat and clean, without spelling or grammatical errors.

In a smaller business, technicians often communicate directly with customers (Figure 2.14). The success of the business can hinge on how considerate the technician is. Car maintenance is usually a planned and expected part of the customer's life. But sometimes a customer is unhappy because of a car repair that is unplanned and expensive. People skills are something that a good technician continually works to improve.

Appearance is another attribute that is important in customer relations. Many customers are bothered by dirty clothes and a lack of grooming. If you appear to lack a pride in your appearance, the customer feels that you do not treat their vehicle with respect either.

Every time a customer leaves a car at a repair shop, he or she is inconvenienced by having to arrange alternate transportation. Reliability is another attribute of a successful technician. A business depends on regular attendance by the technician so that promised repairs can be completed. Coming to work on time is also required. Repair shops sell the technician's time.

Customers often bring in their cars before going to work. In most shops, technicians begin working at 8 A.M. Under the flat rate system, it is beneficial to show up early and be ready to start work. If a technician is late, the service dispatcher might give him or her a less desirable job. The other technicians will be

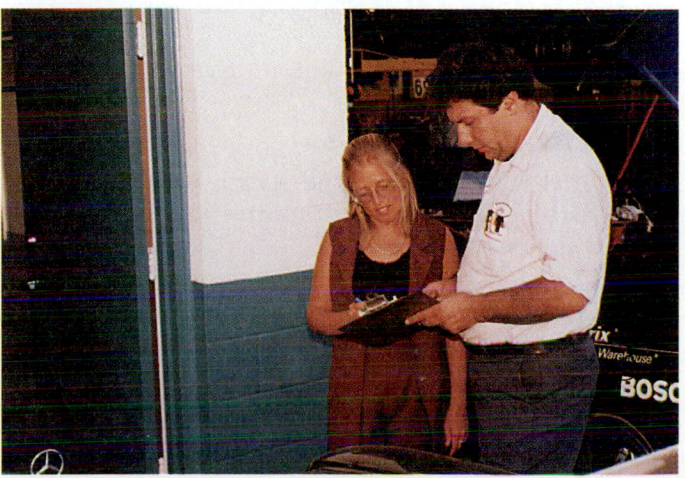

Figure 2.14 In a small business, the technician often communicates directly with the customer.

able to begin their second jobs sooner and will probably complete more work for the day. If all of the shop's business is completed by 3 P.M., the technician who was late will have flagged fewer hours than the others.

Getting along with fellow employees and supervisors is also very important. One employee who always has something negative to say can poison the environment of an otherwise happy shop. A supervisor will usually favor the person with the good attitude.

NATEF School Certification

ASE also certifies schools with automotive programs. Schools certified in all eight ASE areas are *Master Certified* (Figure 2.15). For this certification, called **NATEF** (National Automotive Technicians Education Foundation), a team consisting of a certified team leader and technicians from local repair shops tours the campus. The team evaluates the facilities, tools and equipment, curriculum content and hours, and administrative support. School certifications such as the one provided by NATEF provide one means for a prospective student to evaluate a school he or she might like to attend for further automotive study.

Cooperative Work Experience Training

Schools that are NATEF certified have cooperative work experience programs. Students attending school can earn school credit while working part-time. A contract is signed with the employer stating that the student will perform work in certain specialty areas. For instance, a student who wants to learn more about brakes will get to perform brake repairs while at work.

■ SUMMARY

Today's automotive technician must have a multitude of skills. There are many types of specialties under the

Figure 2.15 The National Automotive Technicians Education Foundation (NATEF) evaluates school automotive programs for certification by ASE.

automotive umbrella. Many skilled technicians are ASE certified. Technicians are often paid under the "flat-rate" commission system. There are large and small automotive businesses. Some are corporate and some are independently owned.

■ REVIEW QUESTIONS

1. List five skills that an automotive technician must have.

2. What does PDI stand for?

3. What is the name of the person who greets the customer in a large automotive repair shop or dealership?

4. What is the name of the person who assigns the jobs to the technicians in a large shop?

5. What is the name of the kind of shop run by taxi cab companies or utilities?

6. List the eight ASE automobile specialty service areas that make up a master certification.

7. What is a technician called who is certified in all eight ASE areas?

8. What are the three ASE Engine Machinist areas of specialization?

9. Which is a voluntary process, certification or licensing?

10. What is the automotive system of pay that is based on the amount of work completed called?

Shop Procedures, Safety, Tools, and Equipment

THEN AND NOW: TOOLS AND DIAGNOSTIC EQUIPMENT

The tools used to fix and maintain motor vehicles have changed almost as much the cars themselves. In the early years, the only tool available to remove a nut or bolt was a crude "spanner" (an open-end wrench). Ownership of a set of spanners and an assortment of adjustable "monkey" wrenches, hammers, screwdrivers, and pliers were enough to qualify a person as a mechanic.

Diagnostic equipment was limited to intelligence and ingenuity, both of which are still essential today. Various devices for determining useful troubleshooting information started to show up in the early decades of the 20th century. The vacuum gauge was considered essential, as was the compression tester. In the 1950s, the dwell meter (which told how long the ignition points were closed), the tachometer, the timing light, and the volt/ohmmeter became standard pieces of equipment in every service technician's toolbox. Possession of a secondary ignition oscilloscope and a distributor machine made a shop a high-tech diagnostic center.

Today's wrenches, whether the combination or socket type, are precisely made and finely finished in a fashion that reminds one of jewelry. Pneumatic tools, from impact guns to air ratchets and chisels, are of high quality and have become affordable. Hands and muscles are saved as these tools assist with the hard work.

Modern automobiles rely on complex electronic engine management systems and diagnostic equipment has evolved into a computer-age level of sophistication. Large console-type engine analyzers check everything from compression to ignition performance. Extremely useful and convenient hand-held scan tools plug into the vehicle's diagnostic connector (now standardized under OBD II regulations). These instruments instantly provide trouble codes and sensor voltages and values. Miniaturized lab scopes (also called "digital storage oscilloscopes" and sometimes referred to as a "visual voltmeter") catch electronic glitches that would be invisible to any other piece of equipment.

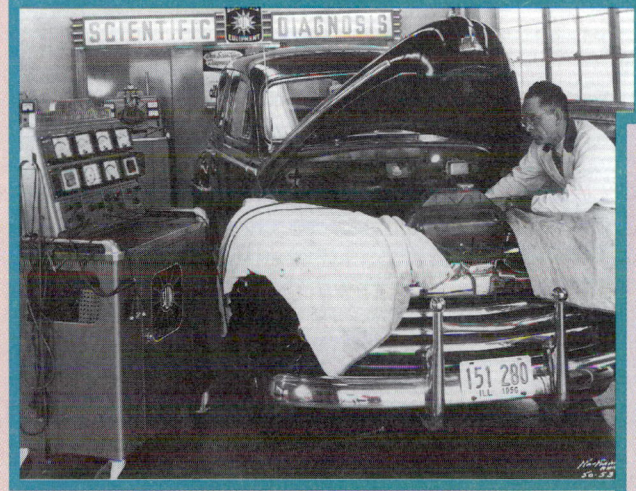

Automotive diagnostics, circa 1950. *(Courtesy of Olgivy, Adams, and Rinehart for Snap-on Tools Company)*

A modern computer system scan tool. *(Courtesy of Olgivy, Adams, and Rinehart for Snap-on Tools Company)*

Shop Management, Service Records, and Parts

■ INTRODUCTION

This chapter is about business practices. It describes such things as customer relations, filling out repair orders, and interviewing customers regarding problems with their cars. Also included is an explanation of how businesses buy parts and the discounts involved.

■ CUSTOMER RELATIONS

Automotive repair shops are in business to make a profit. If a shop does not make a profit, it goes out of business. Good relations with customers is the key to a successful business. It is important to make a good first impression with the customer so that he or she is confident in the shop (Figure 3.1). In other than large dealerships, most

Figure 3.1 The customer's first impression is important.

technicians will sometimes need to communicate directly with the customer.

Dealerships have professional salespersons who greet the customer and write up the repair order. These people are called service writers or service advisors. When a customer does not have a complaint about a car's performance but is simply maintaining it properly, an untrained service writer will not be a problem. But one of the complaints often heard about large dealerships is that the customer had to take the car back again in order to have the original problem corrected. In fact, in several surveys, the percentage of complaints not correctly diagnosed and repaired on the initial visit is nearly half. This problem can often be traced to faulty diagnosis by an undertrained service writer.

Service writing in smaller shops is usually done by the owner (who is also a technician). In this case, the customer gets to speak directly with the technician about a car's problem, which often results in a satisfied customer after only one trip to the shop. Part of the solution to the problem for larger shops is to hire service writers who understand the operation of the various systems of the car.

The other side of the coin is that product service training for new model vehicles is more available and is used more often by dealership personnel. This, and the ease of warranty service, often means that the dealership is the best choice for new vehicle repairs.

Telephone Service

Although the owner or manager of a business often answers the telephone, technicians in small shops usually share telephone responsibilities also. The phone should be answered promptly and courteously, stating the name of the business, and the person who is speaking.

Most often there is a phone extension in the repair area. Although it is sometimes an annoyance to have to stop a job to answer a telephone call, the phone is an important source of business. The telephone often presents the customer with his or her first impression of the business. If the customer's first impression is that the business is professionally managed, things will go easier from that point on.

SERVICE RECORDS

A service record is written for every car entering a repair shop. A multiple copy, numbered **service record, repair order (R.O.)**, or **work order (W.O.)** is used for legal, tax, and general recordkeeping purposes. One of the R.O. copies is given to the technician. It gives a listing of the repairs needed and is used for making notations of repairs completed and items needing attention.

One of the copies of the repair order includes the cost estimate and is for the customer. The remaining copy is for the shop's files.

The repair order (Figure 3.2) is important for several reasons:

- It fully identifies the customer and the vehicle.
- It gives the technician an idea of the reason the car is in the shop for repairs.
- It tells the shop's hourly labor rate.
- It gives the customer an estimate of the cost of the repair.
- It gives the time the vehicle will be ready for the customer.
- The signature of the customer gives approval for the repair and agrees to pay for the shop's services when the job is completed.
- It is a legal document.

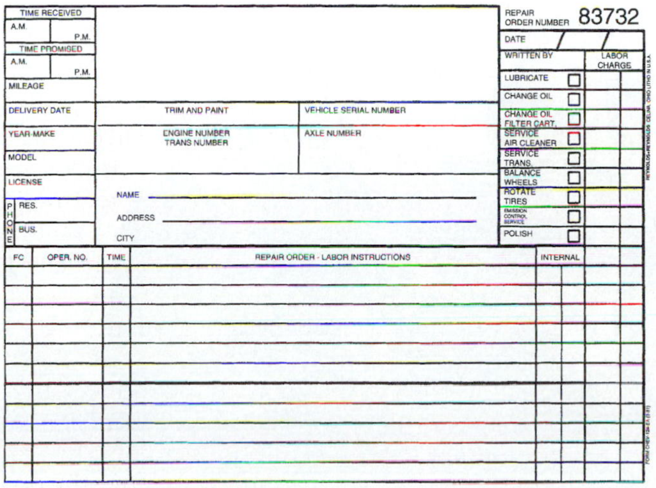
TIME RECEIVED A.M. P.M.
TIME PROMISED A.M. P.M.
MILEAGE
DELIVERY DATE
YEAR-MAKE
MODEL
LICENSE
PHONE RES. BUS.
TRIM AND PAINT
ENGINE NUMBER TRANS NUMBER
VEHICLE SERIAL NUMBER
AXLE NUMBER
NAME
ADDRESS
CITY
REPAIR ORDER NUMBER 83732
DATE
WRITTEN BY
LABOR CHARGE
LUBRICATE
CHANGE OIL
CHANGE OIL FILTER CART.
SERVICE AIR CLEANER
SERVICE TRANS.
BALANCE WHEELS
ROTATE TIRES
EMISSION CONTROL SERVICE
POLISH
FC | OPER. NO. | TIME | REPAIR ORDER - LABOR INSTRUCTIONS | INTERNAL

Figure 3.2 A repair order.

Repair orders are numbered for future reference. Also entered on the repair order are:

- The odometer reading giving the car's mileage
- The make of the vehicle (Ford)
- The model of the vehicle (Mustang)
- The model year of the vehicle (1996)
- The vehicle license number (1VAL239)

Information that will identify the customer includes his or her:

- Name
- Address
- Residence telephone number
- Business telephone number
- Signature

Below the area that contains customer information is room for labor instructions. This area will be filled in with the customer's complaint, possible cause(s), and repairs to be performed. The customer will need to be questioned carefully about the symptoms. He or she often has a preconceived notion of what is wrong that is incorrect.

An example of questions that might be asked if the complaint is that the car runs hot would be:

- Does it become hot right away after it is started?
- Does it become hot only at freeway speeds?
- Does the radiator consume coolant?
- Does it become hot enough to boil over?
- Have you heard any unusual noises from the engine?
- Has the cooling system had recent service?
- Has the ignition system been serviced recently?

There is a structured method of questioning that can be followed when asking questions of the customer.

- First ask for a general description of the problem.
- Ask whether it happens in the front, back, or under the hood.
- Identify symptoms: Do you hear, feel, or smell something?
- With the information you have learned, use your technical skills to identify the problem.

Computer Records

Most successful repair shops use personal computers for keeping records, maintaining a running inventory and ordering parts, and tracking employee productivity (Figure 3.3). Notes in the computer's memory can include personal notes regarding a particular customer and what occurred during their most recent visit to the shop.

> **COMPUTER NOTE**
>
> *A computer locates a customer's records using either the vehicle's license number, the owner's name, or the owner's address. The license number is the most popular means of access.*

Figure 3.3 Most shops use a personal computer for shop management activities.

Even though a shop may keep records on a computer, a *hard copy* record is still used by most shops for the technician. Although some shops have a computer terminal in the service bay, this is uncommon.

It is important that a shop keep written records of all repairs and recommendations. Litigation (court cases) often results in a loss in court due to a lack of adequate records. A completed R.O. includes a written estimate of the cost of the repair and a record of phone conversations with an owner when an earlier approved estimate requires updating.

CARING FOR THE CAR

Keep the Car Clean

Be sure that your hands, shoes, and clothing are clean before getting into a customer's car. A dirty steering wheel or carpet will guarantee an unhappy customer, no matter how good of a repair job was done. Work shoes often have grease on their soles. Many shops use carpet mats made of paper to help keep a customer's carpet clean (Figure 3.4).

Fender covers. Fender covers are used whenever underhood work is done (Figure 3.5). Greasy hands or

Figure 3.4 A carpet mat protects the car's carpet from grease.

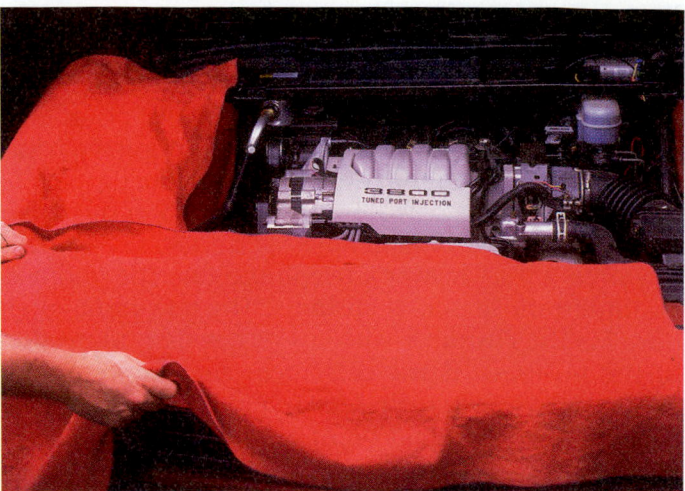

Figure 3.5 Fender covers are used when working under the hood.

brake fluid can ruin the finish on a vehicle. Fender covers also protect against scratches from items such as belt buckles. Cars cost many thousands of dollars. When working on cars, be sure to respect them as much as the owner does.

LINEN SERVICE

Most shops have weekly linen service for shop towels and uniforms (Figure 3.6). The shop is billed for the cost of any shop towels that have signs of acid on them. Shop towels are usually dyed red. When they are exposed to battery acid or other acids, they turn blue. This alerts the linen service to the probability that the rags will disintegrate in the laundry (Figure 3.7).

Uniforms are sometimes owned by the shop. Other times they are rented from the linen service. Some shops get shirt and pants service and others get shop coats or coveralls (Figure 3.8). Shop clothing is often made of materials that are resistant to battery acid.

Figure 3.6 Shops have linen service for uniforms and shop towels. *(Courtesy of Mission Industries, Santa Barbara, CA)*

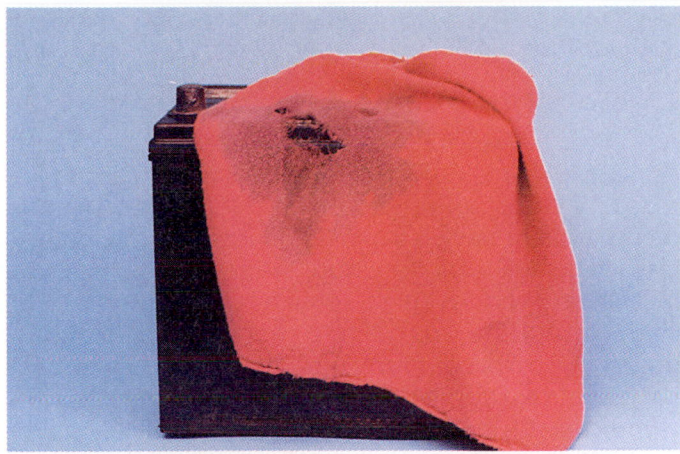

Figure 3.7 This shop towel was damaged by battery acid.

Figure 3.8 Coveralls and a shop coat.

■ WHOLESALE AND RETAIL DISTRIBUTION OF AUTO PARTS

The purpose of this section is to familiarize the reader with some of the terms used in the automotive service business. Approximately one in every six jobs is related to the automobile. The automotive **aftermarket** is supplied from the manufacturer to the customer through a large parts and service *distribution* system (Figure 3.9). Both the **original equipment manufacturers (OEMs)** and a large number of independent parts manufacturers sell parts to over 1000 **warehouse distributors (WDs)** throughout the continent. Warehouse distributors are large distribution centers who sell to auto parts **wholesalers**, known in the industry as **jobbers**.

Jobbers

Jobbers (Figure 3.10) sell parts, accessories, and tools to several different markets. More than one-half of a typical jobber's customers are the independent repair shops and service establishments. Fleet companies, farmers, trade and industrial accounts, and occasionally, new car dealerships are some of the other customers a jobber might have. The **do-it-yourself (DIY)** business of a typical parts business accounts for over one-fourth of a jobber's customers. Some parts, such as filters and ignition parts, sell at a higher volume. In the do-it-yourself market, filters (oil, air, and fuel) account for over half of all parts sales. Parts manufacturers use codes to tell which parts average more units of sale than others (Figure 3.11). For instance, code 1 parts average 4 or more units of sale per year, while code 8 parts average less than .1 unit of sale per year. An average parts store will carry parts listed as code 1 through code 5. This will provide a parts inventory

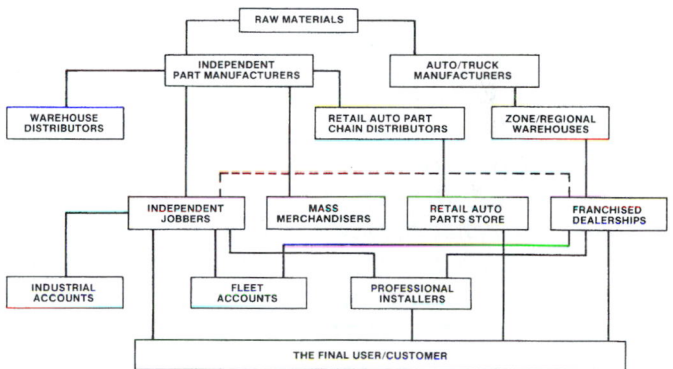

Figure 3.9 The auto parts supply network.

Figure 3.10 A wholesale and retail parts store.

Class Code	Number of Parts	Probability of Sale	#of Parts likely to sell	#of Parts likely not to sell
1	1,800	99.6%	1,793	7
2	700	97.0%	679	21
3	1,300	91.9%	1,195	105
4	3,400	77.8%	2,645	755
5	6,000	52.8%	3,170	2,830
6	7,000	31.3%	2,190	4,810
7	8,000	16.1%	1,290	6,870
8	50,000	4.9%	2,450	47,550

Figure 3.11 Probability of sale by part class. *(Courtesy of All Pro and Cooperative Computing)*

of about 75% of full coverage. Parts that are not often sold increase the cost of doing business. These parts will have to be marked up by a higher percentage in order to compensate.

Services provided to customers are often a prime element in gaining repeat business. Prompt delivery of phoned-in parts orders is provided to regular wholesale customers. Often a parts jobber will have a machine shop, too (Figure 3.12). These machine shops perform such jobs as engine machining and rebuilding, flywheel surfacing, and hydraulic press work.

Retail Chain Stores

Many parts retailers are managed by national or regional retail chains. *Chain stores* usually deal in faster moving items such as alternators, starters, tools, and car accessories. They often do not have as large an inventory of diverse parts as a jobber store. Some retail stores are owned by one or several owners, but they buy everything through the distribution system of the main company. Prices, especially on sale items, are sometimes lower than retail.

Mass merchandisers, such as large department store chains, sell fast-moving, popular items such as oil and filters, windshield wipers, and tune-up items to the

DIY market. They often have a large service facility for doing minor repairs on automobiles (Figure 3.13).

Dealership Parts Departments

Auto dealers have their own parts departments. A good dealership parts department will have a large organized inventory of original parts. Dealer parts departments usually have a separate counter for the technicians who work in the shop. Some dealerships have phone connections or an intercom in the service bay so the technician can call in a parts order before walking to the parts counter. Some dealerships even have personnel that bring the cars to the service bay for the technician. This is one way that the technician can be more productive in a busy shop.

The operation of a dealership's parts business is often determined by its owner. Some dealers are very aggressive in pricing and services offered to the independent market and others are not. Discounts are offered in varying degrees to wholesale accounts. It is not uncommon for dealerships to deliver parts to independent repair shops within a 50-mile radius in some markets. Although they are competitors, independent parts establishments often have a give-and-take arrangement with the dealers, supplying each other with parts when they are needed in a hurry.

Service Stations

A service business is very concerned with the cost of its high volume parts. A shop that sells five oil filters a day might be able to purchase them in quantity at half the cost and keep them in *stock* (on the premises). The same holds true with **tires, batteries, and accessories (TBA)**. Service stations often buy TBA through the oil company whose products they sell, at very good discounts. A higher volume shop will often have computerized inventory, where parts are automatically ordered by telephone modem as the supply falls below a certain level.

Figure 3.12 A parts store with a machine shop.

Figure 3.13 Mass merchandisers often have large service facilities.

Parts Pricing

Prices are usually determined by the price the jobber pays. The ability to find good prices on parts is an important part of a jobber's business. Jobbers are sometimes affiliated with cooperative groups. This gives them the ability to band together so that they can secure better prices by buying in huge quantities. They still continue to be independent owners, however, and may buy from whomever they choose.

Parts are priced to the customer according to either list, net, or variable pricing (Figure 3.14). The **list** price, also called the **retail** price, is the suggested price of an item and is what the customer paying for car repair pays. In reality, very few stores charge list price to the DIY customer who walks into a parts store. Most parts are discounted, although in varying amounts. The **net** price is the price that the wholesale auto repair customer pays.

Part Number	List	User Net	Dealer	Part Number	List	User Net	Dealer
5188	3.08	2.62	2.05	5282	20.94	17.80	13.94
5189	6.55	5.57	4.36	5283	22.40	19.04	14.92
5193B	13.48	11.46	8.98	5284	13.69	11.63	9.12
5195C	14.81	12.59	9.87	5287A	14.02	11.92	9.34
5197D	13.60	11.56	9.06	5288C	17.66	15.01	11.78
5197D	17.99	15.29	11.98	5289B	32.96	28.01	21.95
5198D	16.32	13.87	10.87	5290A	13.80	11.73	9.19
5200C	12.67	10.77	8.45	5291A	13.28	11.29	8.85
5201B	16.57	14.08	11.03	5292B	28.81	24.50	19.19
5202B	16.55	14.06	11.02	52393B	16.41	13.95	10.93
5204	18.73	15.92	12.47	5294	12.36	10.51	8.23
5208A	13.66	11.62	9.10	5296A	15.93	13.54	10.62
5210C	15.11	12.84	10.06	5297	10.94	9.30	7.29
5211E	20.19	17.16	13.45	5298	2.70	2.30	1.80

Figure 3.14 Example of list, net, and dealer prices.

Some items like oil, oil filters, coolant, and spark plugs are priced by most establishments with little or no mark-up, because most customers are aware of their prices. These parts, called **loss leaders** or **cost leaders**, are considered to be part of the cost of promoting a business (Figure 3.15). *Variably priced* items are those that are slow movers and not readily available to the customer. These can be marked up a larger amount to help compensate for cost leaders.

When a service or repair business (installer) buys auto parts from a jobber, the price the installer is charged is sometimes dependent on the volume of business that the installer does with that parts establishment. More volume on a monthly basis can mean a higher percentage of discount to the shop. The part is marked up to the retail price for the customer who brings in a car for repair. This system can make a difference in the profitability of a service business.

Replacement Parts

Factory replacement parts are categorized as **OE (original equipment)**. **Stock** means the part is the same as intended by the manufacturer. *Aftermarket* is a broad term that refers to parts that are sold by the non-OE market. Many OE parts are manufactured by the same manufacturer as aftermarket parts.

Figure 3.15 Oil, oil filters, coolant, and spark plugs are loss leaders or cost leaders.

■ REVIEW QUESTIONS

1. What are three names for the written record that is kept of a customer's service visit?
2. What is the most popular way of identifying a customer when entering information into the computer?
3. What are two things that fender covers protect the paint from?
4. What are wholesale parts sellers called?
5. What does DIY stand for?
6. What does TBA stand for?
7. What is another name for list price?
8. The price that a wholesale business pays for a part is called _____ pricing.
9. What does OE mean?
10. _____ is a broad term that refers to parts that are sold by the non-OE market.

ASE STYLE REVIEW QUESTIONS

1. Technician A says that using a computer to generate a repair order saves a good deal of paper. Technician B says that a hard copy of a repair order is needed for the customer and for the technician. Who is right?

   a. Technician A   **b.** Technician B
   c. Both A and B   **d.** Neither A nor B

2. Technician A says that a repair order contains a written estimate of the cost of a repair. Technician B says that a repair order lists a record of telephone conversations with the customer. Who is right?

   a. Technician A   **b.** Technician B
   c. Both A and B   **d.** Neither A nor B

3. Technician A says that red shop towels turn blue if they come into contact with battery acid. Technician B says that shop clothing is often made of materials that are resistant to battery acid. Who is right?

   a. Technician A   **b.** Technician B
   c. Both A and B   **d.** Neither A nor B

CHAPTER 4

Locating Service Information and Specifications

OBJECTIVES

Upon completion of this chapter, you should be able to:

✔ Describe the types of service information available.

✔ Locate repair information in shop manuals.

✔ Locate repair information on CD-ROM.

✔ Understand how microfiche works.

✔ Select the correct manual for estimating the time it will take to perform a task.

KEY TERMS

vehicle identification number (VIN)
underhood emission control label
service manual
owner's manual
microfiche
CD-ROM
R&R (remove and replace)
Flat-rate Manual
technical service bulletin (TSB)

INTRODUCTION

In 1965, the amount of material a mechanic needed to be able to read to be very knowledgeable about automobiles was about 5000 pages. According to CARQUEST Corporation, in 1974 the amount of published information for automobile repair was less than a quarter of a million pages. In 1993, that amount was 1,251,016, almost five times as many pages. Information is doubling two times a year. Today, a technician needs to have access to much, much more service information. Today, the amount of information a technician needs to have access to is approximately 1,601,300,400 (1.6 billion) pages.

Repair manuals in 1965 were relatively small. Service information for only one of today's make of vehicle usually consists of more than one manual, or information stored on compact disks.

There are many procedures to be learned to successfully service and repair automobiles. This chapter deals with how to locate service information. Several types of service information are available, including traditional service manuals, computer libraries, and microfiche.

SERVICE INFORMATION

Shop Manuals

Shop manuals, also called repair manuals, are books that give written instructions for repairs. Illustrations provide helpful hints. This section will describe the various types of shop manuals.

Before working on any make of vehicle for the first time, it is a good idea to consult a repair manual to find any helpful hints or precautions that can make your task more successful. Repair manuals are written for experienced technicians and make the assumption that certain procedures are understood and will be followed.

Keep in mind that it is important to keep parts organized when you are disassembling a component or system. Be sure to label parts for reassembly. As you progress in your automobile repair skills, you will see why it is important to be methodical. When in doubt, always check the repair manual. Even with the help of a good manual, it is very difficult to put back together something that you did not take apart yourself. It is especially helpful to note the location of accessory brackets, such as those for the alternator, smog pump, power steering pump, or air conditioning compressor. After removing brackets from an engine, reassemble them to the unit they hold, so you do not forget where they belong.

SHOP TIP It can be helpful to take a Polaroid photograph of the engine compartment or of an assembled component before disassembly of something unfamiliar.

Identifying the Vehicle

Even though you may know the year and model of the vehicle, sometimes you may need to know more about the specific vehicle. Each vehicle comes with a **vehicle identification number (VIN)**. Passenger vehicles manufactured for sale in the United States (domestic cars) have the VIN located on a plate on top of the dashboard, just under the windshield, on the driver's side of the vehicle (Figure 4.1). Since 1981,

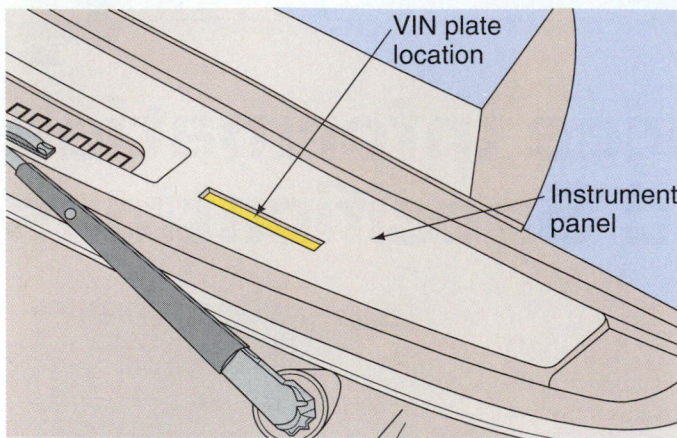

Figure 4.1 The vehicle identification number (VIN). *[Courtesy of Chrysler Corporation]*

each VIN has been made up of seventeen numbers and letters. Each letter or digit stands for something. For instance, on domestic cars the engine code is the eighth character and the year of the vehicle is the tenth character (Figure 4.2). Check the manufacturer's service manual for the meaning of each character for that make of car.

Underhood Label. Vehicles produced since 1972 are equipped with an **underhood emission control label** (Figure 4.3). This label gives useful information to the technician, as well as to an emission control specialist. It contains such information as ignition timing specifications, emission control devices, engine size, vacuum hose routing, and valve adjustment specifications.

➡ *Perform **Underhood Label Specifications** Worksheet*

```
1FABP43F2FZ100001
```
VEHICLE IDENTIFICATION NUMBER

① ⎫
Ⓕ ⎬ — World Manufacturing Iden-
Ⓐ ⎭   tifier
Ⓑ —— Restraint System Type
Ⓟ —— Constant "P"
④ ⎫
③ ⎬ — Line, Series, Body Type
Ⓕ —— Engine Type
② —— Check Digit
Ⓕ —— Model Year
Ⓩ —— Assembly Plant
① ⎫
⓪ ⎪
⓪ ⎬ — Production Sequence Num-
⓪ ⎪   ber
① ⎭

Figure 4.2 Each digit in the VIN stands for something. *[Courtesy of Ford Motor Company]*

VEHICLE EMISSION CONTROL INFORMATION – ACURA NSX
ENGINE FAMILY MHN3.0V5F3AX   TWC H025
EVAPORATIVE FAMILY 91FK   EGR SMPI
DISPLACEMENT 2977 cm³
         182CID
                    CATALYST
TUNE UP SPECIFICATIONS
  TUNE UP CONDITIONS:
     ENGINE AT NORMAL OPERATING TEMPERATURE
     ALL ACCESSORIES TURNED OFF, COOLING FAN OFF
     TRANSMISSION IN NEUTRAL
  ADJUSTMENTS TO BE MADE IN ACCORDANCE WITH
  INDICATIONS GIVEN IN SHOP MANUAL

IDLE SPEED	5 SPEED TRANSMISSION	800±50 rpm
	AUTOMATIC	800±50 rpm
IGNITION TIMING AT IDLE		15±2° BTDC
VALVE LASH	SETTING POINTS BETWEEN	IN. 0.17±0.02mm cold
	CAMSHAFT AND ROCKER ARM	EX. 0.19±0.02mm cold

NO OTHER ADJUSTMENTS NEEDED.

THIS VEHICLE CONFORMS TO U.S EPA AND STATE OF CALIFORNINIA
REGULATIONS APPLICABLE TO 1991 MODEL YEAR NEW MOTOR VEHICLE

Figure 4.3 An underhood emission label. *[Courtesy of American Honda Motor Co., Inc.]*

■ MANUFACTURERS' SERVICE MANUALS

Manufacturers publish their own service manuals (Figure 4.4). These are designed to be used by technicians in the dealership. They cover one year and model of vehicle. Dealer technicians have access to a full set of service manuals, listing every service operation in detail. Manufacturer's manuals are the most detailed service information for one particular year and make of car. Because of this limitation, they are costly. Independent repair shops do not usually own manufacturer's service manuals, unless one make of vehicle is repaired often. Figure 4.5 shows a table of contents from a manufacturers' service manual. One section of the car is covered in each section. For instance, brakes are covered in Section 5 and engines are covered in Section 6.

Three types of information are contained in shop manuals.

■ Diagnostic or troubleshooting information (Figure 4.6)
■ Step-by-step repair information
■ Specification charts (Figure 4.7)

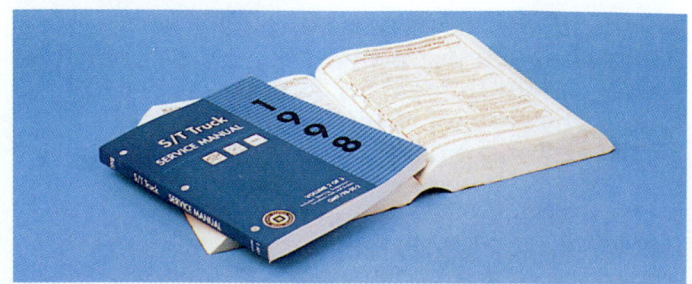

Figure 4.4 A manufacturer's service manual.

TABLE OF CONTENTS	SECTION NUMBER
GENERAL INFOR. AND LUBE	
General Information	0A
Maintenance and Lubrication	0B
HEATING AND AIR CONDITIONING	
Heating and Ventilation (Non-A/C)	1A
Air Conditioning System	1B
V-5 A/C Compressor Overhaul	1D3
BUMPERS AND FRONT BODY PANELS	
Bumbers (See 10-4)	
Front End Body Panels (See 10-5)	
STEERING, SUSPENSION, TIRES, AND WHEELS	
Diagnosis	3
Wheel Alignment	3A
Power Steering Gear & Pump	3B1
Front Suspension	3C
Rear Suspension	3D
Tires and Wheels	3E
Steering Column, On-Vehicle Service	3F
Steering Column - Std., Unit Repair	3F1
Steering Column - Tilt, Unit Repair	3F2
DRIVE AXLES	
Drive Axles	4D
BRAKES	
General Information - Diagnosis and On-car Service	5
Compact Master Cylinder	5A1
Disc Brake Caliper	5B2
Drum Brake - Anchor Plate	5C2
Power Brake Booster Assembly	5D2
ENGINES	
General Information	6
2.0 Liter L-4 Engine	6A1
3.1 Liter V6 Engine	6A3
Cooling System	6B
Fuel Systems	6C
Engine Electrical - General	6D
Battery	6D1
Cranking System	6D2
Charging System	6D3
Ignition System	6D4
Engine Wiring	6D5
Driveability and Emissions - General	6E
Driveability and Emissions - TBI	6E2
Driveability and Emissions - PFI	6E3
Exhaust System	6F

TABLE OF CONTENTS	SECTION NUMBER
TRANSAXLE	
Auto. Transaxle On-Car Service	7A
Auto. Trans. - Hydraulic Diagnosis	3T40-HD
Auto. Trans. - Unit Repair	3T40
Man. Trans. On-Car Service	7B
5-Sp. 5TM40 Man. Trans. Unit Repair	7B1
5-Sp. Isuzu Man. Trans. Unit Repair	7B2
Clutch	7C
CHASSIS ELECTRICAL, INSTRUMENT PANEL & WIPER/WASHER	
Electrical Diagnosis	8A
Lighting & Horns	8B
Instrument Panel & Console	8C
Windshield Wiper/Washer	8E5
ACCESSORIES	
Audio System	9A
Cruise Control	9B
Engine Block Heater	9E
BODY SERVICE	
General Body Service	10-1
Stationary Glass	10-2
Underbody	10-3
Bumpers	10-4
Body Front End	10-5
Doors	10-6
Rear Quarters	10-7
Body Rear End	10-8
Roof & Convertible Top	10-9
Seats	10-10
Safety Belts	10-11
Body Wiring	10-12
Unibody Collision Repair	11-1
Welded Panel Replacement	11-2
INDEX	
Alphabetical Index	

Figure 4.5 The table of contents from a typical manufacturer's service manual. *(Courtesy of General Motors Corporation, Service Technology Group)*

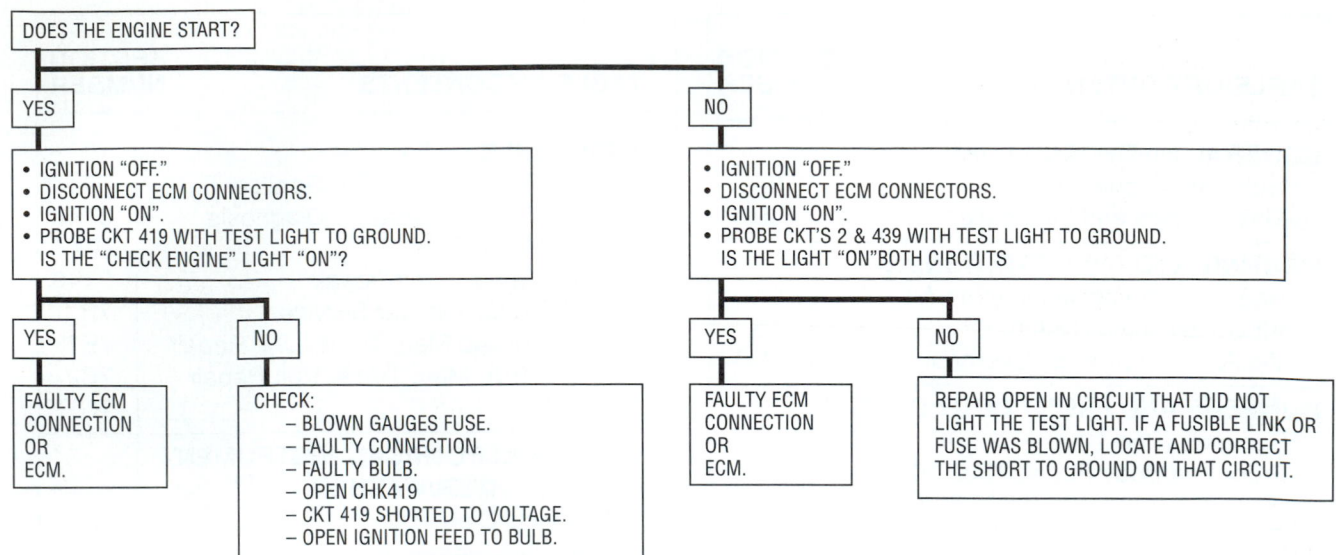

Figure 4.6 Typical step-by-step troubleshooting chart from a service manual. *(Courtesy of General Motors Corporation, Service Technology Group)*

TORQUE SPECIFICATIONS

Fan Assembly Motor-to-Fan	3.3 N • m (29 Lbs. In.)
Fan-to-Radiator Support Bolt	9 N • m (80 Lbs. In.)
Hose Clamps	
Heater Hose	1.7 N • m (15 Lbs. In.)
Radiator Hose	3.4 N • m (30 Lbs. In.)
Lower Air Deflector to Impact Bar	25 N • m (18 Lbs. Ft.)
Throttle Body Inlet Pipe Bolt 3.1L (VIN T)	25 N • m (18 Lbs. Ft.)
Trans. Oil Cooler Fittings at Radiator	27 N • m (20 Lbs. Ft.)
Trans. Oil Cooler Pipe Connections (Alum. Radiator)	20 N • m (15 Lbs. Ft.)
Radiator Outlet Pipe To Block 2.0l (VIN K)	27 N • m (20 Lbs. Ft.)
Radiator-to-Radiator Support Bolts	10 N • m (90 Lbs. In.)
Thermostat Housing Bolts	
2.0L (VIN K & 3.1L (VIN T)	27 N • m (20 Lbs. Ft.)
Coolant Pump-to-Block Bolts	
2.0L (VIN K)	25 N • m (19 Lbs. Ft.)
3.1L (VIN T)	24 N • m (18Lbs. Ft.)
Coolant Pump-to-Front Cover Bolts	
3.1L (VIN T)	10 N • m (90 Lbs. In.)
Coolant Pump Pulley-to-Pump Bolts	
2.0L (VIN K)	24 N • m (17 Lbs. Ft.)
3.1L (VIN T)	21 N • m (15 Lbs. Ft.)
Surge Tank Bolts	4 N • m (35 Lbs. In.)
Surge Tank Pipe To Block Bolt 3.1L (VIN T)	8 N • m (70 Lbs. In.)
Temperature Sending or Gage Unit	27 N • m (20 Lbs. Ft.)

Figure 4.7 A specification chart from a service manual. *(Courtesy of General Motors Corporation, Service Technology Group)*

Figure 4.8 Generic service manuals. *(Courtesy of Mitchell International)*

■ GENERIC SERVICE MANUALS

Generic manuals (Figure 4.8) (dealing with all makes of either foreign or domestic cars) are produced by several companies. There are usually two types of generic manuals: those dealing with mechanical repairs, such as engines and transmissions, and those devoted to the areas of fuel/emission/ignition/air conditioning. Separate manuals are available for light trucks.

CONTENTS:

ENGINES
Section 5

CLUTCHES
Section 6

DRIVE AXLES
Section 7

TRANSMISSION SERVICING
Section 12

| GENERAL INFORMATION |
| ACURA |
| AUDI |
| BMW |
| CHRYSLER/MITSUBISHI |
| DAIHATSU |
| FORD MOTOR CO. |
| GENERAL MOTORS |
| GEO |
| HONDA |
| HYUNDAI |
| INFINITI |
| ISUZU |
| JAGUAR |
| LEXUS |
| MAZDA |
| MERCEDES-BENZ |
| NISSAN |
| SAAB |
| SUBARU |
| SUZUKI |
| TOYOTA |
| VOLKSWAGEN |
| VOLVO |

Figure 4.9 In the front of a repair manual is an index of cars by maker. *(Courtesy of Mitchell International)*

General Motors Engines
6-171

3.0L, 3.8L & 3.8L "3800" V6 (Cont.)

ENGINE SPECIFICATIONS

VALVE SPRINGS

Engine	Free Length In. (mm)	PRESSURE Lbs. @ In. (Kg @ mm)	
		Valve Closed	Valve Open
3.8L (VIN C)	2.03 51.6	100-110@1.73 (45-49@44)	214-136@1.30 (97-61@33)
3.0L & 3.8L (VIN 3)	2.03 51.6	85-95@1.73 (39-42@44)	175-195@1.34 (79.1-88.2@34.04)
3.8L (VIN 7)	2.03 (51.6)	74-82@173 (33-37@44)	175-195@1.34 (79.1-88.2@34.04)

CAMSHAFT

Engine	Journal Diam. In. (mm)	Clearance In. (mm)	Lobe Lift In. (mm)
3.0L	1.785-1.786 (45.34-45.36)	.0005-.0025 (.013-.064)	Int .210 (5.334) Exh .240 (6.096)
3.8L (VIN C)	1.785-1.786 (45.34-45.36)	.0005-.0025 (.013-.064)	[1].272 (6.909)
3.8L (VIN 3)	1.785-1.786 (45.34-45.36)	.0005-.0025 (.013-.064)	[1].245 (6.223)
3.8L (VIN 7)	1.785-1.786 (45.34-45.36)	.0005-.0025 (.013-.064)	.....

[1] - Specification applies to both intake and exhaust.

CAUTION: Following specifications apply only to 3.0L (VIN L), 3.8L (VIN 3), 3.8L (VIN 7) and 3.8 "3800" (VIN C) engines.

TIGHTENING SPECIFICATIONS

Application	Ft. Lbs. (N.m)
Camshaft Sprocket Bolts	20 (27)
Balance Shaft Retainer Bolts	27 (37)
Balance Shaft Gear Bolt	45 (61)
Connecting Rod Bolts	45 (61)
Cylinder Head Bolts	[1] 60 (81)
Exhaust Manifold Bolts	37 (50)
Flywheel-to-Crankshaft Bolts	60 (81)
Front Engine Cover Bolts	22 (30)
Harmonic Balancer Bolt	219 (298)
Intake Manifold Bolts	[2]
Main Bearing Cap Bolts	100 (136)
Oil Pan Bolts	14 (19)
Outlet Exhaust Elbow-to-Turbo Housing	13 (17)
Pulley-to-Harmonic Balancer Bolts	20 (27)
Outlet Exhaust Right Side Exhaust Manifold-to-Turbo Housing	20 (27)
Rocker Arm Pedestal Bolts	37 (51)
Timing Chain Damper Bolt	14 (19)
Water Pump Bolts	13 (18)

[1] - Maximum torque is given. Follow specified procedure and sequence.
[2] - Tighten bolts to 80 INCH lbs. (9 N.m).

Figure 4.10 Some of the pages are devoted to specifications for different systems. *(Courtesy of Mitchell International)*

Using the Manuals

Comprehensive generic repair manuals include sections for each make of car arranged in alphabetical order by maker. An index of cars is located at the front of the manual (Figure 4.9). On the first page of each section is an index of service operations that can be used to locate a particular service procedure. Many manuals have separate sections for component repair such as carburetors, starters, alternators, transmissions, or differentials. A few pages of each section are devoted to specifications including tuneup, electrical, wheel alignment, engine tightening, and valves (Figure 4.10).

➡ *Perform Service Manual Worksheet*

■ LUBRICATION SERVICE MANUAL

The manual that is used most during lubrication/safety checks is a lubrication **service manual** (Figure 4.11). This manual has all of the most used specifications. It is arranged in three basic sections:

- ■ Lubrication and Maintenance Information
- ■ Capacities
- ■ Underhood Service Information

The lubrication and maintenance section includes lube and maintenance instructions, lubrication diagrams and specifications, vehicle lift points, and preventative maintenance intervals.

The capacities section includes cooling and air conditioning system capacities, cooling system air bleed locations, wheel and tire specifications, and wheel lug torque specifications.

The underhood service information section includes underhood service instructions, specifications for engine tune-up, mechanical, electrical, and fuel systems, diagrams, and belt tensions.

➡ *Perform Maintenance Specifications Worksheet*

■ OWNER'S MANUAL

An **owner's manual** is a booklet that comes with a new car. In addition to the information it contains about the operation of the vehicle's accessories, it also contains some information that can be useful to a technician. Features and safety are usually covered first. A maintenance schedule tells the owner about recommended service intervals. Sometimes an owner's manual contains instructions for maintenance for owners who want to do some maintenance themselves.

■ OWNERS' WORKSHOP MANUALS

Locating repair information for some cars can be more difficult. One company, Haynes, publishes owners' workshop manuals (Figure 4.12). These manuals, often sold by parts stores, cover several years of one model of car. Chilton's manuals also provide owner workshop manuals that go back as far as 1940.

■ MICROFICHE

Some service manuals have been published on **microfiche**, a small plastic film card that is magnified by a microfiche reader (Figure 4.13). The advantage is that microfiche takes up very little space. Many microfiche readers have a copying feature so that the service information can be taken to the repair area. Some parts departments also use microfiche. In recent years, most companies that previously used microfiche have changed to using computers.

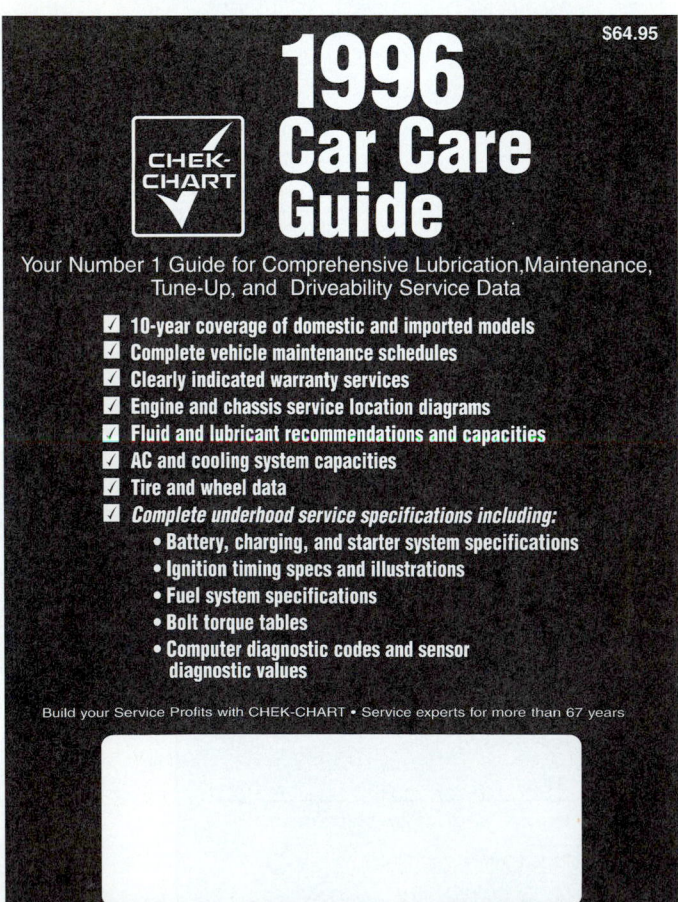

Figure 4.11 A lubrication service manual. *(Courtesy of Chek-Chart Publications)*

Figure 4.12 Owner's workshop manuals cover several years of one car model.

Figure 4.13 Microfiche stores information in a small space that is magnified by a microfiche reader. *(Courtesy of Mitchell International)*

➡ *Perform __Microfiche__ Worksheet*

■ CD-ROM COMPACT DISK SERVICE INFORMATION

The very popular new source of service information is the **CD-ROM** (compact-disk-read-only-memory) system (Figure 4.14). Due to the tremendous amount of information required to service new vehicles, this electronic media is rapidly gaining a foothold in the ser-

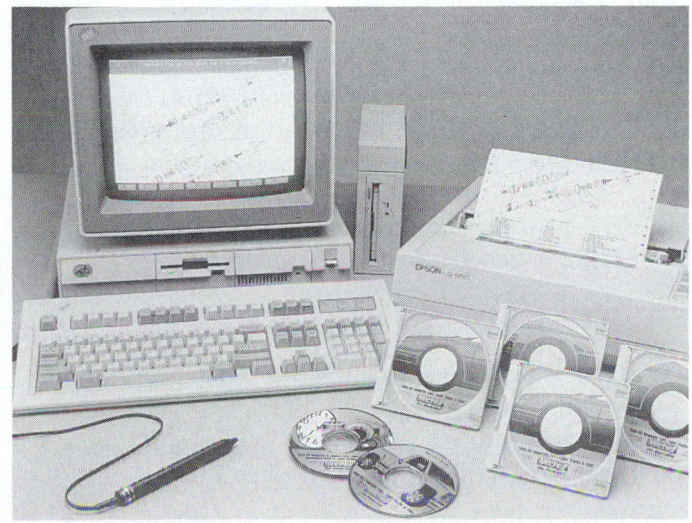

Figure 4.14 A computerized information library uses compact disks capable of storing a vast quantity of information. *(Courtesy of Mitchell International)*

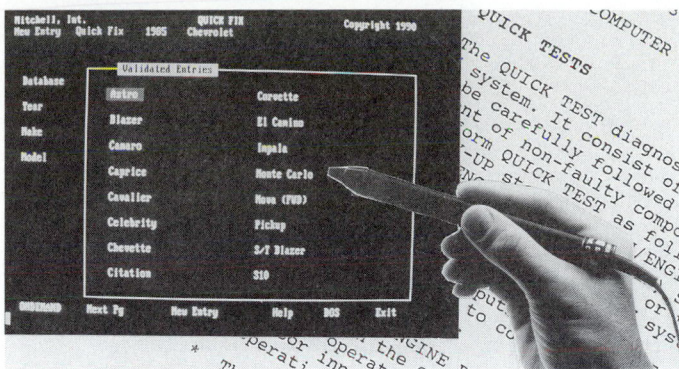

Figure 4.15 Selections are made from a menu on the screen. *(Courtesy of Mitchell International)*

vice industry. Several companies provide information on compact disks (CDs). Compact disk systems also provide labor time estimates.

> ### ■ COMPUTER NOTE ■
>
> *CDs store a great deal of information that can be accessed by a personal computer with a CD player. The total library is stored on several CDs. New ones are sent periodically throughout the year. A subscription to the service includes new updates.*

The systems are easy to use and information is easy to access. The technician uses the computer's keyboard, a mouse, or a light pen to make selections from a menu on the screen (Figure 4.15).

When a technician locates the information in the computer, it can be read off of the screen or printed out by the computer printer and can be taken to the service bay or workbench.

➡ *Perform __CD-ROM__ Worksheet*

■ PARTS AND LABOR ESTIMATING (FLAT-RATE) MANUALS

When parts fail they are replaced with either new or rebuilt parts. When the technician replaces a part, this is known as an **R&R (remove and replace)** job. The estimated time it should take a professional technician to perform an R&R job is listed in the **Flat-rate Manual**, also called a *Parts and Time Guide* (Figure 4.16). The flat-rate manual also lists the estimated cost of parts. A service writer or a business owner will use it to give a fairly accurate estimate to a customer.

➡ *Perform __Flat-Rate Manual__ Worksheet*

■ TECHNICAL SERVICE BULLETINS

Manufacturers and industry professional associations such as the Automotive Engine Rebuilders Association

Figure 4.16 A flat-rate manual. *(Courtesy of Mitchell International)*

(AERA) or Automatic Transmission Rebuilders Association (ATRA) publish monthly **technical service bulletins (TSBs)** to subscribers and members. In some of these bulletins, members report of experiences discovered through trial and error. Other times, information is contributed by manufacturers. The AERA technical committee writes technical bulletins from material contributed by the membership. An example of one of their technical service bulletins is shown in Figure 4.17. Manufacturers make extensive use of TSBs.

■ HOT LINE SERVICES

Hot line services are those that provide answers to service information by telephone. Manufacturers provide help by telephone for the technicians in their dealerships. There are subscription services for independents to be able to get repair information by phone.

Some manufacturers also have phone modem systems that can transmit computer information from the car to another location. The car's diagnostic link is connected to the modem. The technician in the service bay runs a test sequence on the automobile. The system downloads the latest updated repair information on that particular model of car. If that does not repair the problem, a technical specialist at the manufacturer's location will review the data and propose a repair.

> ### ■ COMPUTER NOTE ■
> *Technicians make extensive use of the internet. Free internet groups made up of several thousand technicians share diagnostic tips for solving tough problems.*

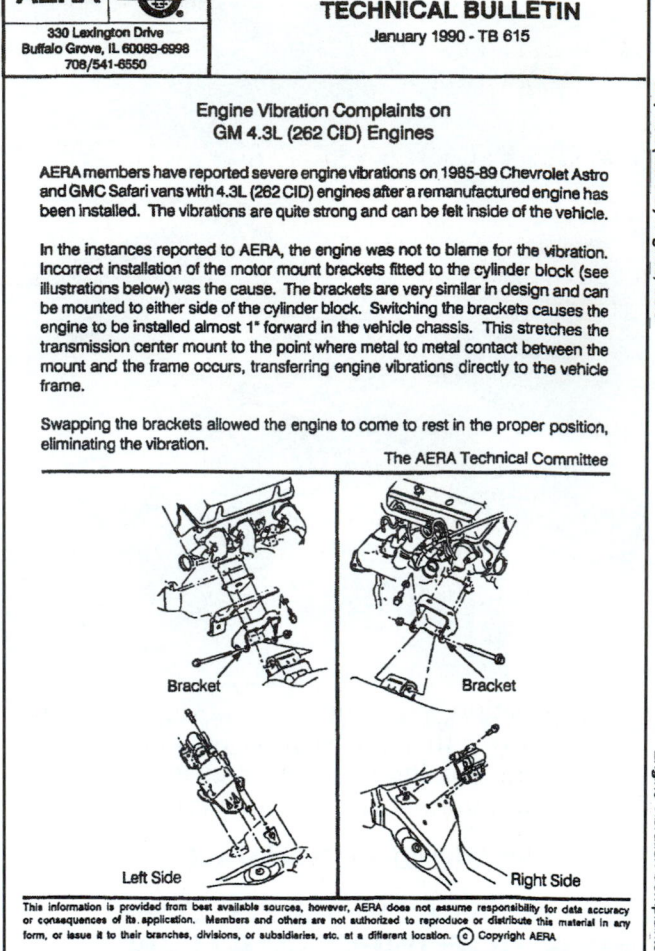

Figure 4.17 A technical service bulletin. *(Courtesy of AERA)*

■ TRADE MAGAZINES

There are many professional magazines that technicians subscribe to. Examples are *Motor, Automotive Rebuilder, Motor Service, Motor Age, Import Service, Transmission Digest,* and *Precision Machine Shop*. These magazines usually include information from the latest technical service bulletins. They also include a mail-in reader service postcard. The reader can circle numbers on this card that relate to items about which he or she would like to receive more information.

■ REVIEW QUESTIONS

1. What number is used by manufacturers to identify a particular vehicle?

2. Which character of the VIN number on a domestic car tells the engine code?

3. What kind of information is included on an underhood label?

4. Which type of manual is the most comprehensive one for a particular vehicle?

5. What are the three types of information found in lubrication service manuals?

6. What is the name of the type of service information that requires a reader or magnifier to enlarge the information and put it on a screen?

7. What is the name of the kind of service information that is stored on compact disks?

8. What is the name of the manual that is used to determine the length of time to complete a job?

9. What are bulletins called that are produced by manufacturers and associations?

10. List six names of professional trade magazines.

■ ASE STYLE REVIEW QUESTIONS

1. Technician A says during disassembly, reassemble brackets to a component for easier reassembly. Technician B says to take a polaroid photo of a complicated component before disassembling it. Who is right?

   a. Technician A   b. Technician B
   c. Both A and B   d. Neither A nor B

2. Technician A says that an underhood label can include specifications for valve adjustment. Technician B says that vacuum hose routing is included on an underhood label. Who is right?

   a. Technician A   b. Technician B
   c. Both A and B   d. Neither A nor B

3. Technician A says that step-by-step repair information is included in the owner's manual provided with a new vehicle. Technician B says there are usually two types of generic service manuals: mechanical and tune-up. Who is right?

   a. Technician A   b. Technician B
   c. Both A and B   d. Neither A nor B

4. Technician A says that information can be printed from a CD-ROM system. Technician B says that information can be printed from a microfiche system. Who is right?

   a. Technician A   b. Technician B
   c. Both A and B   d. Neither A nor B

5. Technician A says that a flat-rate manual gives the amount of time it should take to complete a repair job. Technician B says that a flat-rate technician will be paid for the amount of time it actually takes to complete the job. Who is right?

   a. Technician A   b. Technician B
   c. Both A and B   d. Neither A nor B

CHAPTER 5

Measuring Tools and Systems

■ OBJECTIVES

Upon completion of this chapter, you should be able to:

✔ Perform measurements using any of the tools in this chapter.

✔ Have a working understanding of the metric and English systems of measurement.

■ KEY TERMS

British Imperial (U.S.)
  system
metric system
pounds per square inch
  (psi)
foot-pounds/Newtons
plastigauge
outside diameter (O.D.)
inside diameter (I.D.)
liquid crystal display (LCD)

■ INTRODUCTION

This chapter deals with the common measuring instruments and systems of measurement used in repairing automobiles. The two systems of measurement that an automotive technician must understand are the **British Imperial (U.S.) system** and the **metric system**.

The British system, using fractions and decimals, has been the basic measuring system in England, the United States, and Canada until recently. This system is based on inches, feet, and yards. In the twelfth century, one inch was decreed to be a length equal to three barley corns end to end. One yard was the distance from King Henry's nose to the end of his thumb. The basis for this system is somewhat ridiculous, but we have used it for so many years that it is hard to dismiss.

American manufacturers are slowly changing their tooling to the metric system to be competitive in the rest of the world, which uses the metric system. Technicians working on domestic automobiles are seeing more and more metric parts, and will need to own tools of both metric and English sizes.

■ METRIC SYSTEM

The international system (S.I.) of measurement is known as the metric system. The basis of the metric system is the meter, which is 39.37" long, slightly longer than a yard.

Metric Engine Size Measurement

The metric measurement of volume is the liter. A liter is the quantity of liquid that will fill a cube $\frac{1}{10}$ of a meter long on each edge. A liter is slightly larger than a quart.

NOTE: *There are 61.02 cubic inches in one liter.*

Because one *liter* is equal to 1000 *cubic centimeters (cc),* a 2000 cc engine is also called a 2 liter engine.

Using the approximate measurement of 60 cu. in. to the liter, two liters are equal to about 120 cubic inches.

Weight Measurement

The metric unit of weight is the *gram,* which is the weight of the amount of water it takes to fill a cube that is $\frac{1}{100}$ of a meter long on each side. A thumbtack weighs about 1 gram. The *kilogram* is the most common use of the gram. A kilogram equals about 2.2 pounds. All metric measurement units are related. For instance, 1000 cc of water weighs 1 kilogram.

Pressure Measurement

The metric system equivalent to **pounds per square inch (psi)** is expressed as *kilograms per square centimeter.* Both of these measurements are used in measuring atmospheric pressure at sea level.

■ Atmospheric pressure is one *BAR* in the metric system (1 kilogram per square centimeter).

■ Atmospheric pressure in the English system is called one atmosphere (14.7 pounds per square inch).

Temperature Measurement

On the Fahrenheit scale, water freezes at 32° and boils at 212°. In the metric system, temperature is measured in degrees *centigrade,* more often expressed as degrees *Celsius.* The temperature at which water freezes is zero degrees Celsius (0°C). Water boils at 100°C.

Torque Measurement

In a later chapter, torque wrenches are discussed. Torque readings are expressed in **foot-pounds** in the English system. The metric equivalent is expressed in **Newtons**. Newtons are approximately $\frac{1}{10}$ of a kilogram.

A conversion chart from the English system to the metric system is included in the appendix, at the back of the book.

Metric Measuring

The metric system is based on the number 10. It is a very easy system to use, because multiplying and dividing simply involves adding or subtracting zeros. For example:

- $\frac{1}{100}$ (0.01) of a meter is a centimeter (cm).
- 1000 meters is a kilometer (km).
- $\frac{1}{1000}$ (0.001) of a meter is a millimeter (mm).

Because we know that 0.25 mm equals 0.010", we can make close approximations between the metric and U.S. systems when dealing with part sizes. For instance:

- a 0.75 mm oversize piston is 0.030" oversize.
- 0.50 mm undersize crankshaft bearings are 0.020" undersize in the inch system.

The appendix at the end of most repair manuals contains a metric conversion chart.

NOTE: *To convert to millimeters when the inch value is known, multiply by 25.4. To convert to inches when the metric value is known, divide by 25.4.*

■ MEASURING TOOLS

The common *steel rule* is used to make approximate measurements. Common rulers used by technicians measure both in the metric system and in fractions of one inch (Figure 5.1). There are several common types of rulers. A *tape measure* is handy to have in a toolbox. The pocket-style 6" steel rule is also a popular tool. Rulers may be used with dividers or calipers to transfer measurements. Uses of the ruler will be shown in subsequent chapters.

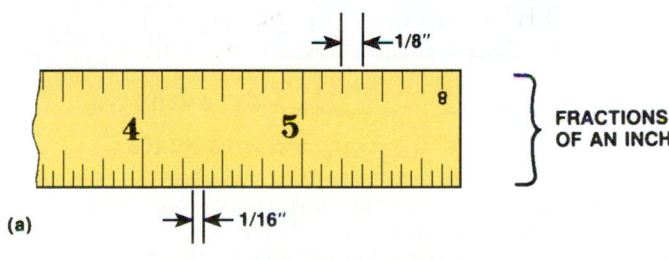

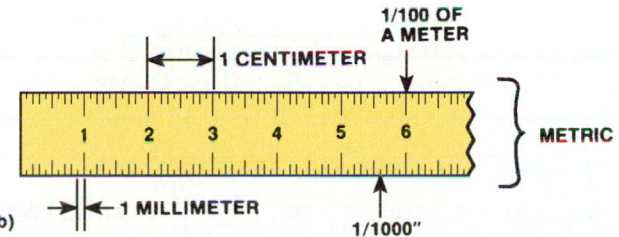

Figure 5.1 (a) Common rule measurements. (b) A metric ruler.

Figure 5.2 Flat and wire types of feeler gauges. *(Courtesy of Snap-on Tools Company, Copyright owner)*

A *metric ruler* has lines that indicate hundredths and thousandths of a meter (see Figure 5.1b).

- The hundredths are called *centimeters*.
- The thousandths are called *millimeters*.
- The numbered marks are centimeters.
- Each of the ten small lines between the numbers is one millimeter.

Thickness gauges are commonly called *feeler gauges*. They can be either the flat or wire type (Figure 5.2). Feeler gauges are commonly used to measure valve clearance, piston ring side clearance, crankshaft end play, and the gaps of spark plugs, piston rings, and ignition points.

When measuring clearance with a feeler gauge, there should be a slight drag when the feeler strip is moved through the area of clearance. If a feeler gauge that is 0.001" thicker will not fit into the gap, the correct measurement has been made. In industry, this is commonly called a "go/no-go" measurement.

Plastigauge is a product used to measure oil clearance in bearings and oil pumps. It is a strip of plastic that deforms when crushed. The deformed plastic gives an idea of the amount of clearance present. Its use is covered in detail in Chapter 23.

■ PRECISION MEASURING TOOLS

The following tools are available in both English and metric types. Measurements in industry are commonly made in thousandths of an inch, or 0.025 mm. This is one of the advantages of the inch system. The metric system measurement of 0.1 mm is four-thousandths of an inch (0.004"), too large for precision measurement. The next closest metric measurement (0.1 mm) would be $\frac{1}{10}$ as large, or 0.0004", too precise for most of the industry.

Vernier Caliper

The vernier caliper was developed in the 17th century. A vernier caliper is one that has a movable scale that runs parallel to a fixed scale. The tool shown in Figure 5.3 will measure **outside diameter (O.D.)**, and **inside diameter (I.D.)** from 0 to 7 inches in thousandths of an inch, or from 1 to 180 millimeters. The inside measurement is somewhat limited by the length of the jaws, so it cannot be used deep in a hole. The vernier caliper shown here can also measure depth. When the cost of a vernier caliper is compared to the

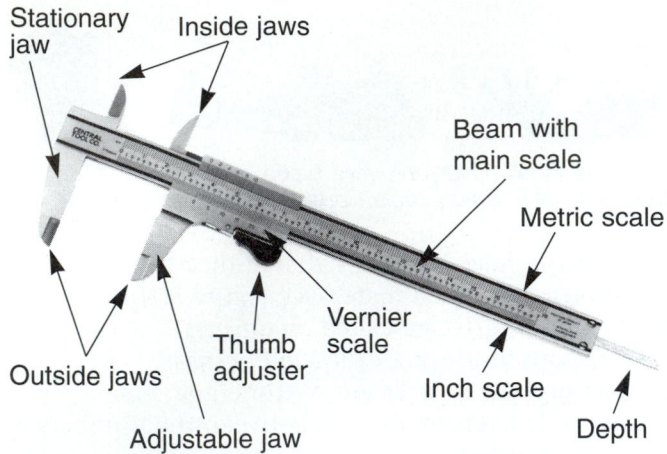

Figure 5.3 A vernier caliper. *[Courtesy of Central Tools Inc.]*

cost of an entire set of micrometers (which are covered later in this chapter), the reason for its popularity becomes apparent. A vernier caliper is the most versatile measuring tool. Every technician should own one.

How to Use a Vernier Caliper

The following explanation of the main and vernier scales deals with the inch-standard vernier caliper. The same principles apply to metric verniers, but measurements are made in decimal parts of a millimeter. The main parts of the scale are shown in Figure 5.3. A fine-adjustment nut found on some vernier calipers is helpful in getting a proper feel.

Main Scale. The main scale of the caliper is divided into inches. Each inch is divided into 10 parts of 1/10" (0.1") each (Figure 5.4). It is customary to read precision measurements in thousandths of an inch (0.001"). Therefore, each of these 1/10" measurements is expressed as 0.100" (one hundred thousandths).

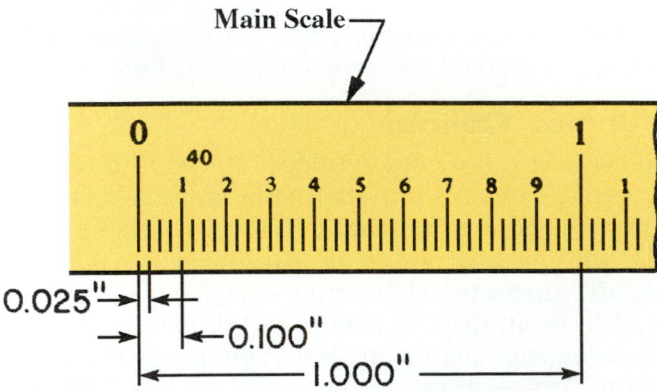

Figure 5.4 Measurements on the main scale were 1/40 of an inch, or 0.02 5" (twenty-five thousandths). *[Courtesy of C. Thomas Olivo Associates, Fundamentals of Machine Technology]*

The area between the 0.100" increments is further divided into fourths (1/4 of 0.100"). Each of these lines is equal to 0.025".

Vernier Scale. The vernier scale divides each of the 0.025" sections on the *main scale* into 25 parts, each one being 0.001" (Figure 5.5). Measurements are determined by combining readings on the sliding vernier scale with those on the main scale. There are actually only 24 divisions on the main scale to the 25 on the sliding vernier, and only one line will line up exactly with another at one time. A magnifying glass is helpful for reading the vernier lines, but in the examples here they are highlighted.

Determining the Measurement. Locate the line on the main scale that the zero on the sliding vernier scale is lined up with, or is just beyond. In Figure 5.6, the reading on the main scale is 0.025". If the zero lined up *exactly* on the third line, then 0.075" would be the measurement. If the vernier scale zero did not line up exactly with a line on the main scale, then find the number on the vernier scale that lines up perfectly with *any* line on the main scale. Only one line on the vernier scale will line up perfectly. In Figure 5.7a, it is the line representing 0.005". The main scale reading is 0.100", so the reading is 0.105". In Figure 5.7b, the

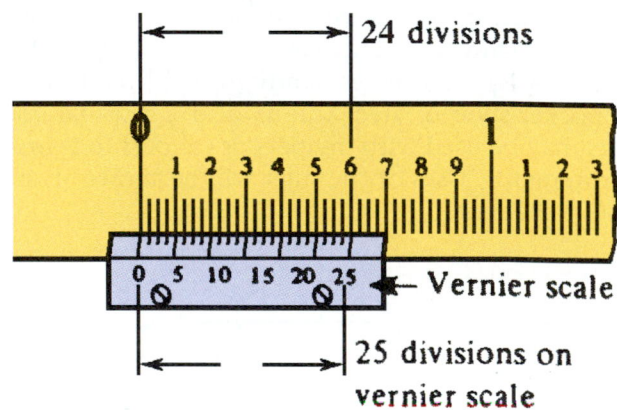

Figure 5.5 The vernier scale divides each of the main scale measurements into 25 parts. *[Courtesy of C. Thomas Olivo Associates, Fundamentals of Machine Technology]*

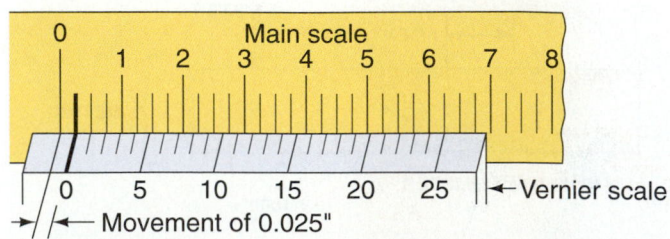

Figure 5.6 Measuring 0.025" on a vernier scale. *[Courtesy of C. Thomas Olivo Associates, Fundamentals of Machine Technology]*

main scale reading is 0.275" plus the vernier scale reading of 0.012" for a total of 0.287".

NOTE: *The main scale also shows 4", so the total reading is 4.287".*

An alternative to a standard vernier caliper is the *dial vernier* shown in Figure 5.8. It registers the 0.001" part of the measurement on its dial. This type of caliper is very popular, but more expensive.

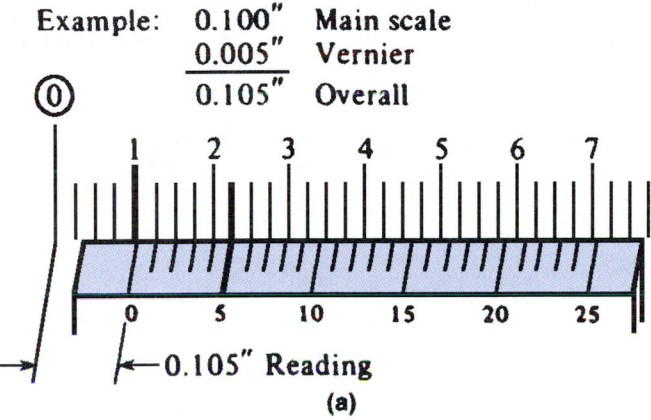

Example: 0.100″ Main scale
          0.005″ Vernier
          ────────
          0.105″ Overall

←0.105″ Reading
(a)

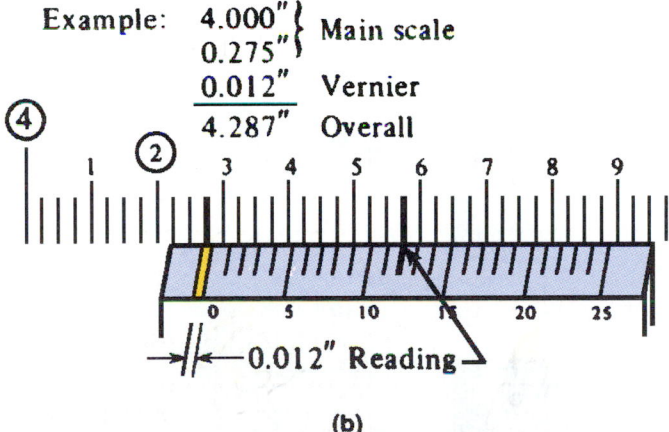

Example: 4.000″⎱ Main scale
          0.275″⎰
          0.012″ Vernier
          ────────
          4.287″ Overall

←0.012″ Reading
(b)

Figure 5.7 Select the vernier scale line that lines up exactly with any line on the main scale. *(Courtesy of C. Thomas Olivo Associates*, Fundamentals of Machine Technology*)*

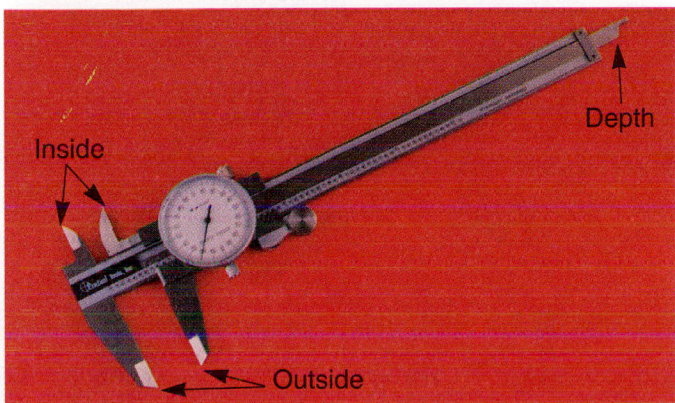

Inside / Outside / Depth

Figure 5.8 A dial vernier caliper. *(Courtesy of Central Tools, Inc.)*

Sometimes a vernier caliper main scale is divided into 50 sections instead of 25. There is a scale for reading inside measurements and a separate scale for reading outside measurements. There is only one division between the 0.100" line and the 0.200" line. Each of these divisions represents 0.050" instead of 0.025". Reading this vernier caliper is otherwise the same as reading the 25-section caliper. Be aware of which type you are reading.

A *computerized caliper* (Figure 5.9) works the same way as an ordinary caliper but its measurement is shown on a **liquid crystal display (LCD)** to an accuracy of 0.0005. Measurements can be in either inch or metric.

Micrometer

Micrometers are often called "mikes" for short. Figure 5.10 shows a set of *outside micrometers*. Each one has a range of only one inch. Because of the expense of a complete set of mikes, they are found mostly in machine shops.

Micrometers have several advantages over other types of measuring instruments. They are clear and easy to read. They measure consistently and accurately, and they have a built-in adjustment to compensate for

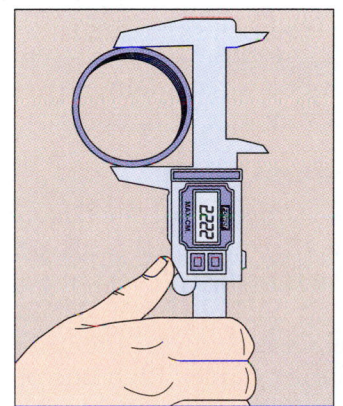

Fig. 1: Outside measurement.

Fig. 2: Inside measurement.

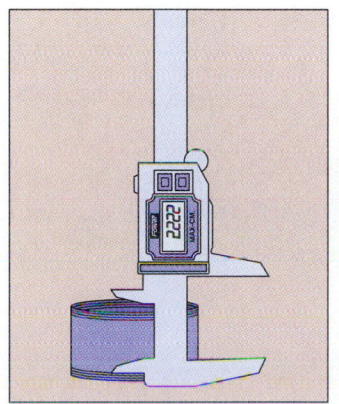

Fig. 3: "Step" measurement.

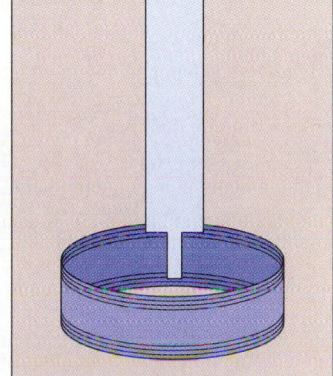

Fig. 4: Depth measurement.

Figure 5.9 A computerized caliper. *(Courtesy of Fred V. Fowler Co., Inc.)*

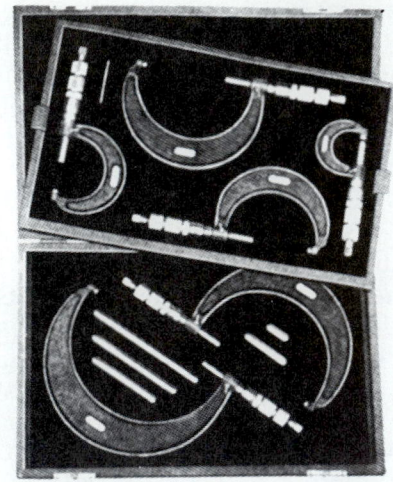

Figure 5.10 A set of outside micrometers with gauge blocks. *[Courtesy of K-D Tools]*

wear. A cutaway view of an outside micrometer is illustrated in Figure 5.11.

Some micrometers have extra features:

■ A ratchet, which sees that spindle pressure is always consistent.
■ The lock nut, which can hold a reading so it cannot be changed accidentally.
■ The decimal equivalent chart stamped on the frame.
■ Some micrometers have a digital readout. This feature may also give metric readings.

If a micrometer spindle must be moved a long way to adjust it, roll the thimble on your arm or hand.

Reading a Micrometer

The micrometer operates on a simple principle (Figure 5.12). The spindle (Figure 5.12a) has 40 threads per inch, so one revolution of the thimble will advance or retract the spindle ¼₀ of an inch (0.025") (Figure 5.12b). The sleeve, or barrel, is laid out in the same fashion as the frame on a vernier caliper. Each line on the barrel represents 0.025", and a new line is uncovered each time the thimble is turned one revolution. Measurement amounts are shown in Figure 5.13.

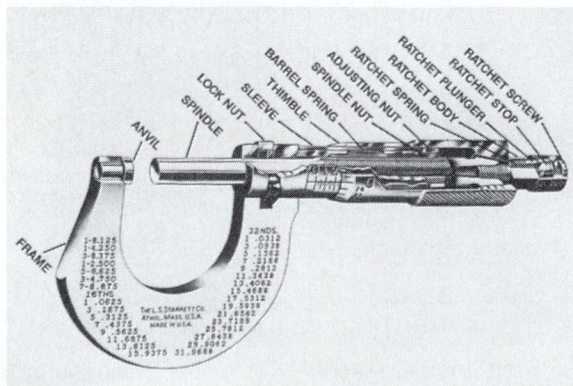

Figure 5.11 A cutaway view of an outside micrometer. *[Courtesy of The L. S. Starrett Company]*

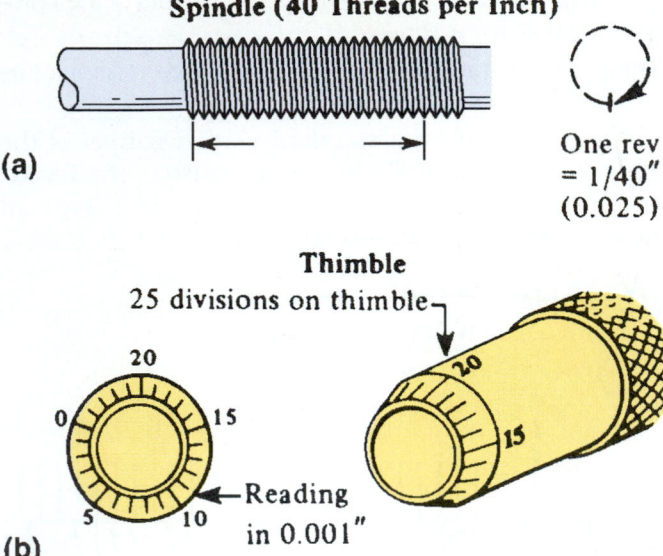

Figure 5.12 Relationships between (a) micrometer spindle pitch (40 threads to the inch); spindle/thimble movement (¼₀", or 0.025", per revolution); and (b) graduations on the thimble of an inch-standard micrometer. *[Courtesy of C. Thomas Olivo Associates, Fundamentals of Machine Technology]*

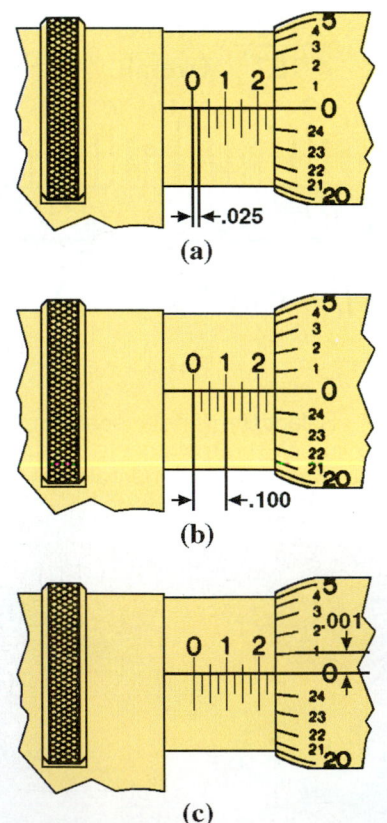

Figure 5.13 Micrometer measurements: (a) Each line on the barrel is 0.025" (one revolution of the thimble). (b) Each number on the barrel is 0.100" (four revolutions of the thimble). (c) Readings on the thimble are 0.001" (¼₂₅ of a revolution of the thimble). *[Courtesy of AE Clevite Engine Parts]*

As each line on the thimble passes the zero line on the hub, the mike opens another 0.001". To read the exact number of thousandths, note the number on the thimble that lines up with the zero line. In Figure 5.14, this amount is 0.003". If you add 0.350 to 0.003, you get 0.353", which is the actual measurement in this reading.

NOTE: *The measurement reflects the actual size of the part.*

In Figure 5.15, the reading is 0.283". The reading is obtained by adding the barrel measurement to the thimble measurement. The reading is rounded off to the nearest 1/1000 of an inch (0.001").

The micrometer is held as shown in Figure 5.16. An experienced technician will develop a feel for the micrometer. Compare your readings to those of an experienced person until you are consistent. The readings might be off by as much as 0.001". When you have developed a consistent feel, you can calibrate your micrometers. Calibration is the process of setting a precision measuring device to read at zero. You can set your micrometers to read at zero by using a special *gauge block* that comes with the micrometer.

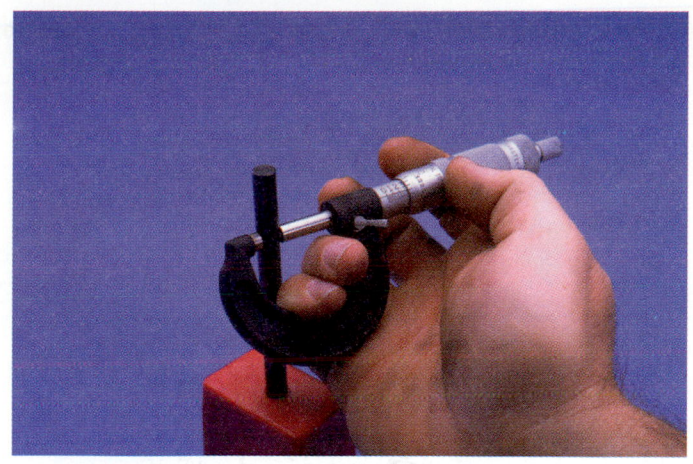

Figure 5.16 The proper way to hold a small micrometer.

Normally, a reading will be rounded off to the nearest 1/1000 of an inch (0.001"). But sometimes an estimate can be made to the nearest ten thousandths (0.0001") (Figure 5.17). This estimate is made by gauging the distance between the index line and the last number passed on the thimble.

Micrometer Vernier Scale

Some micrometers also have a *vernier scale* for making readings to 0.0001" (called *tenths*). Most measurements need not be more accurate than 0.001", but in case it is needed, the vernier scale provides the capability for more accuracy. The vernier scale divides a thousandth of an inch by 1/10, into tenths of thousandths of an inch (Figure 5.18). There are 10 graduations on the top of the micrometer barrel that occupy the same space as 9 graduations on the thimble. The difference between the width of one of the 9 spaces on the thimble and one of the 10 spaces on the barrel is 1/10 of one space.

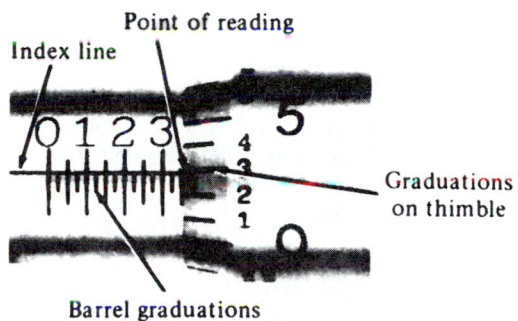

Figure 5.14 Reading a micrometer. *(Courtesy of C. Thomas Olivo Associates*, Fundamentals of Machine Technology*)*

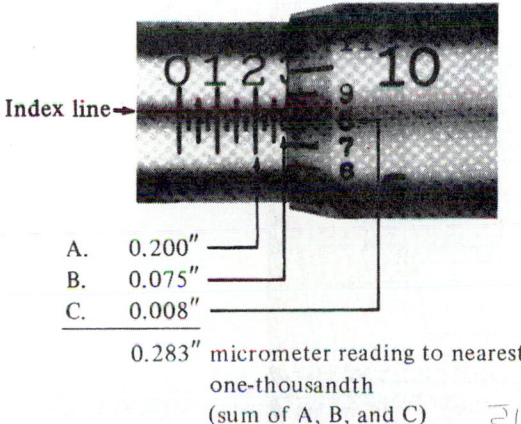

A. 0.200"
B. 0.075"
C. 0.008"

0.283" micrometer reading to nearest one-thousandth (sum of A, B, and C)

Figure 5.15 Reading a micrometer to the nearest 0.001". *(Courtesy of C. Thomas Olivo Associates*, Fundamentals of Machine Technology*)*

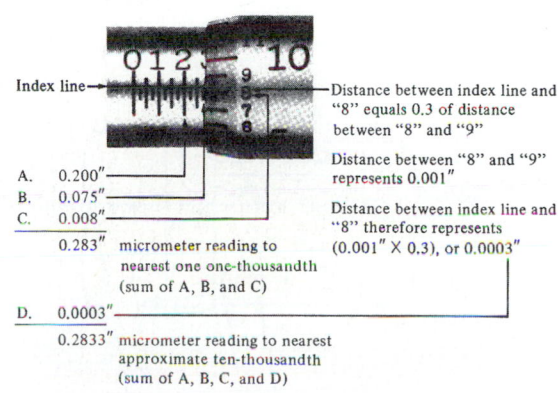

Index line

A. 0.200"
B. 0.075"
C. 0.008"

0.283" micrometer reading to nearest one one-thousandth (sum of A, B, and C)

D. 0.0003"

0.2833" micrometer reading to nearest approximate ten-thousandth (sum of A, B, C, and D)

Distance between index line and "8" equals 0.3 of distance between "8" and "9"

Distance between "8" and "9" represents 0.001"

Distance between index line and "8" therefore represents (0.001" × 0.3), or 0.0003"

Figure 5.17 Estimating an approximate reading to four places (0.0002"). *(Courtesy of C. Thomas Olivo Associates*, Fundamentals of Machine Technology*)*

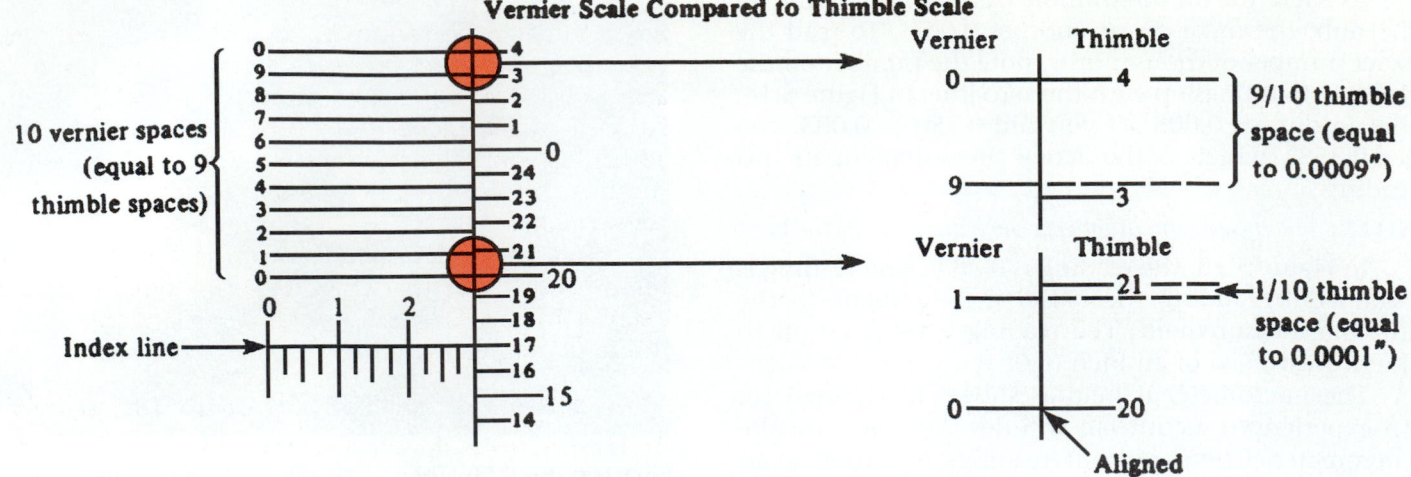

Figure 5.18 A micrometer vernier scale. *[Courtesy of C. Thomas Olivo Associates*, Fundamentals of Machine Technology*]*

Because each graduation on the thimble represents 0.001", the difference between graduations on the barrel and thimble is 0.0001". If the zero lines up with a line, the reading is accurate to three decimal places (Figure 5.19).

If no line on the thimble lines up exactly with the zero line, the vernier scale is used. To read to tenths, simply find the line on the vernier scale that lines up with any number on the thimble, and then add this number to your reading. The example in Figure 5.20 reads 0.3813".

Metric Micrometer

Except for the graduations on the hub and thimble, metric micrometers look just the same as English mikes. One turn of the thimble turns the spindle 0.5 mm, as opposed to 0.025". The hub is graduated in millimeters and half millimeters (Figure 5.21). The lines below the index line are the half millimeters

(0.5 mm). Two revolutions of the thimble equal 1.0 mm on the hub. The thimble usually has 50 divisions. Every fifth line has a number from 0 to 45. Each graduation on the thimble is equal to 0.01 mm ($\frac{1}{100}$ millimeter).

The metric micrometer is read in the same manner as its English counterpart. Just add the millimeter and half millimeter readings on the barrel to $\frac{1}{100}$ (0.01) millimeters on the thimble.

The vernier scale feature gives the ability to read to two-thousandths of a millimeter (0.002 mm). There are only 5 divisions on this scale, compared to 10 on the English vernier scale. When one of the lines on the vernier scale lines up with *any* of the lines on the thimble, add that amount to the reading.

Combination Digital Mikes

There are combination micrometers that give both metric and English readings. There are two versions

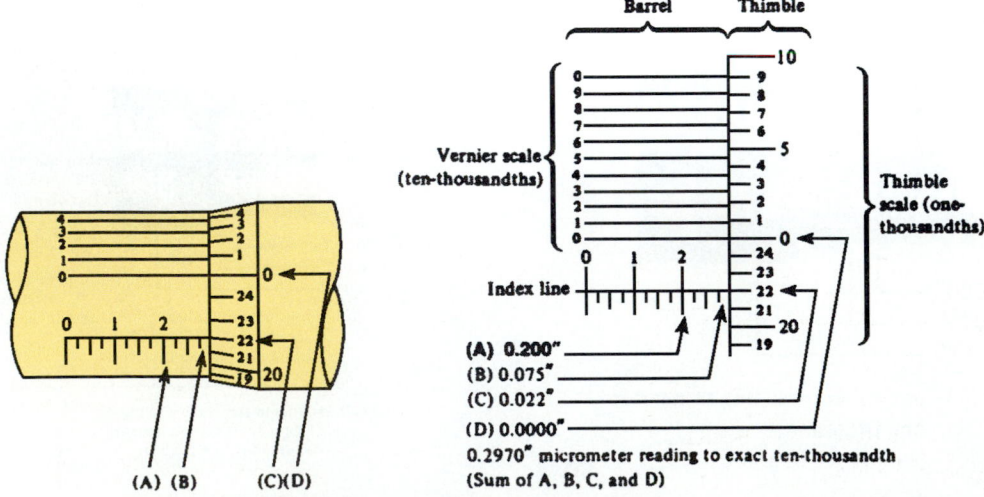

Figure 5.19 A reading of exactly 0.2970". *[Courtesy of C. Thomas Olivo Associates*, Fundamentals of Machine Technology*]*

available. One of them gives inch readings in the conventional manner, while the other reads the metric system. LCD readings taken from the *digital readout window* (Figure 5.22) are accurate to 0.0001 and are convertible between inch and metric.

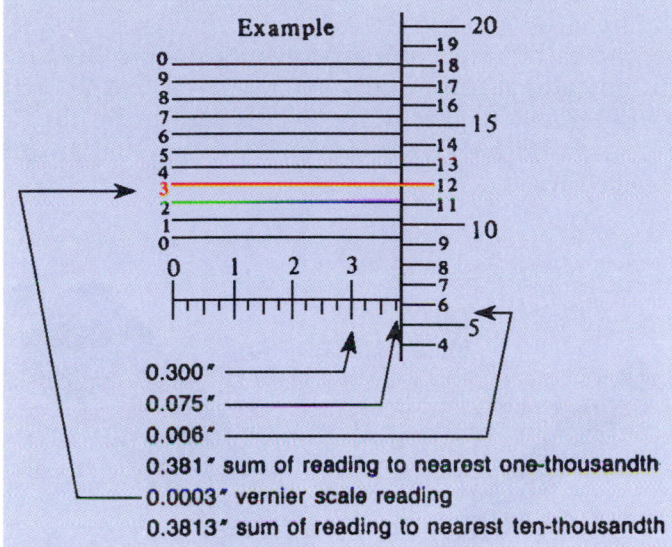

0.300″
0.075″
0.006″
0.381″ sum of reading to nearest one-thousandth
0.0003″ vernier scale reading
0.3813″ sum of reading to nearest ten-thousandth

Figure 5.20 The vernier scale indicates 0.0003". *(Courtesy of C. Thomas Olivo Associates,* Fundamentals of Machine Technology*)*

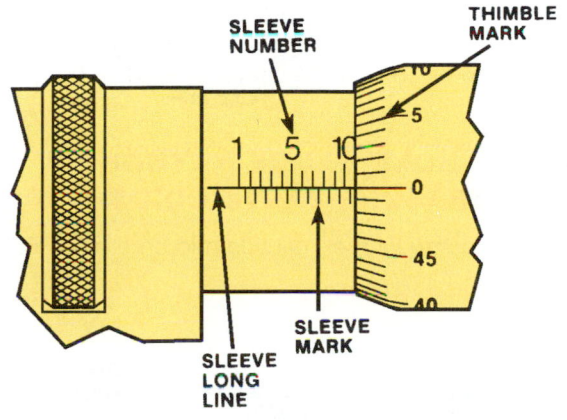

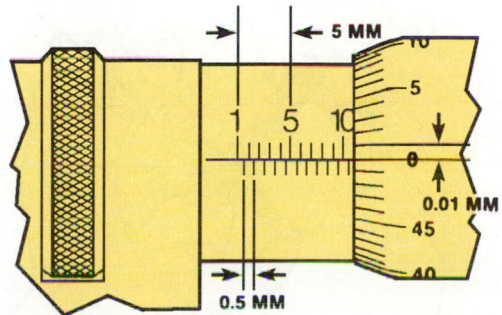

Figure 5.21 Metric micrometer markings and graduations. *(Courtesy of Ford Motor Company)*

Inside Micrometers

An inside micrometer can be used to measure cylinder bores and main and rod bearing bores. The thimble of an inside micrometer does not move as freely on the barrel as the thimble of an outside micrometer. The added friction helps keep the reading from changing. Reading an inside micrometer accurately requires some practice.

Inside micrometers have extension rods to make them the proper size. They have handles for use in deep cylinders. Figure 5.23 shows an example of an inside micrometer reading.

Telescoping and Split-Ball Gauges

An inside micrometer cannot work in cylinders smaller than about 2", so telescoping or small-hole gauges are used. These are known as *transfer gauges*, because the measurements they make are read with an outside micrometer.

The split-ball gauge is used on small holes such as valve guides. Be sure to take your reading at 90° to the split in the ball (Figure 5.24).

Figure 5.22 A digital micrometer. *(Courtesy of Fred V. Fowler Co., Inc.)*

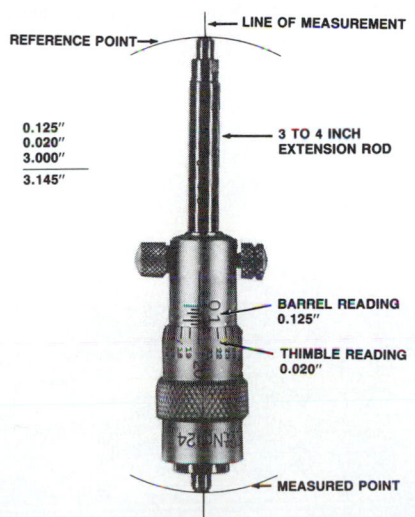

Figure 5.23 Taking an inside micrometer measurement. *(Courtesy of C. Thomas Olivo Associates,* Fundamentals of Machine Technology*)*

Telescoping gauges can be used with a micrometer to measure cylinders (Figure 5.25), bearing bores, and so on.

Dial Indicators

A dial indicator can measure movements such as end-play of crankshafts and valve guide wear. It can be used to measure valve-in-head depth and cylinder and main bearing bores. These processes are covered in subsequent chapters.

An indicator (Figure 5.26) is a very sensitive instrument consisting of small gears activated by spindle movement. It can measure in inches or millimeters. Movement is transmitted through a small gear to an indicating hand on a dial. The dial can be either balanced or continuous (Figure 5.27). Long-range indicators are equipped with revolution counters.

Indicators are comparison instruments because an indicator reading must be compared to a known measurement (Figure 5.28). When measuring thrust (forward and backward movement), no comparison is necessary.

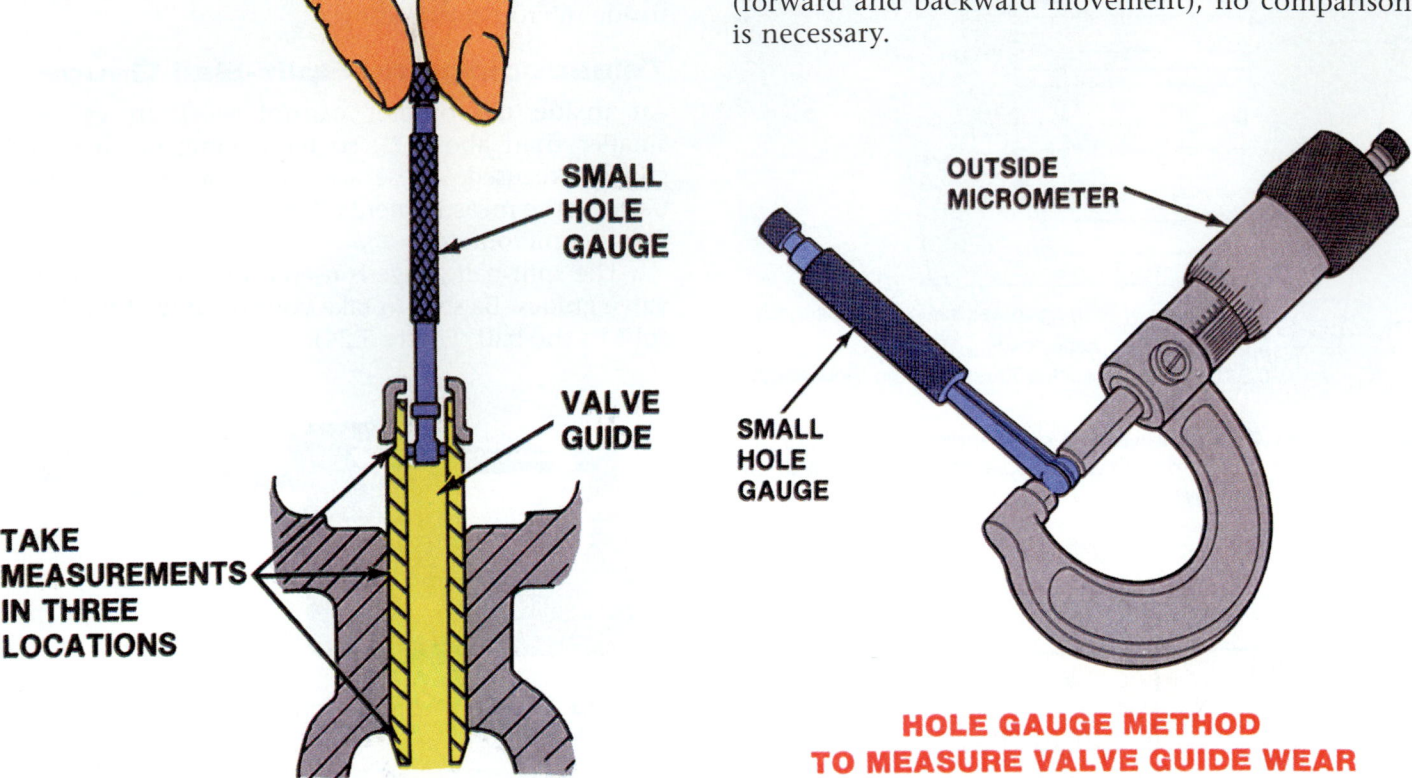

Figure 5.24 Small-hole gauges. After setting the gauge to size, read the size with an outside micrometer at 90° to the split in the ball. *(Courtesy of Ford Motor Company)*

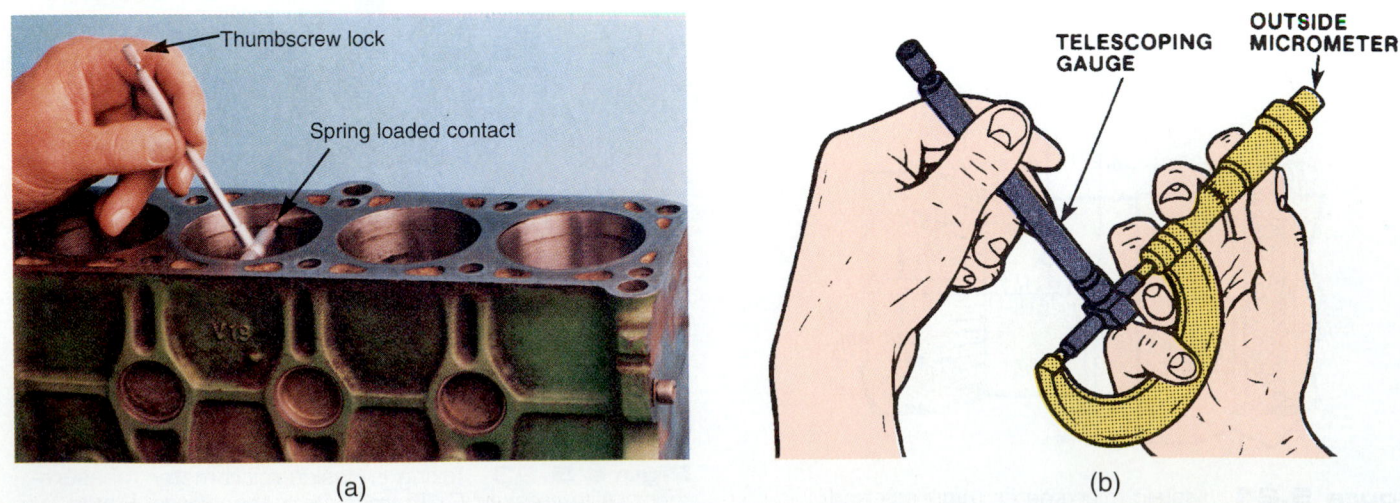

(a)   (b)

Figure 5.25 (a) A telescoping gauge in a cylinder bore. (b) The gauge is read with an outside micrometer.

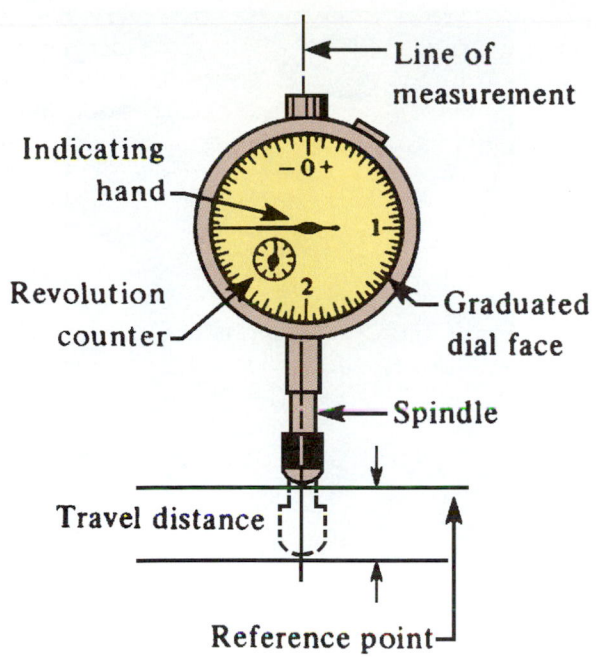

Figure 5.26 Dial indicator nomenclature. *[Courtesy of C. Thomas Olivo Associates*, Fundamentals of Machine Technology*]*

A special indicator and micrometer setting fixture used for measuring cylinder bores is shown in Figure 5.29a. The tool is a *comparator gauge*. It has a micrometer setting fixture (Figure 5.29b). If the setting fixture is set to read a cylinder bore of 4.375, then when the gauge on the indicator reads "0" it is actually reading 4.375".

One type of indicator has a magnetic base. Another type has various attachments for clamping it to the workpiece (Figure 5.30). The most popular and versatile dial indicator among technicians is the vise-grip indicator shown in Figure 5.31. It has many uses in the specialty areas of auto repair.

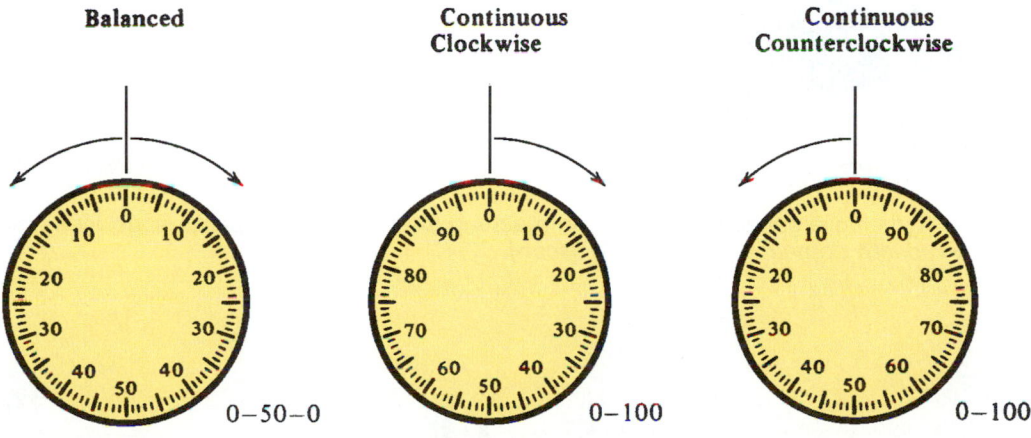

Figure 5.27 Dial indicator faces. *[Courtesy of C. Thomas Olivo Associates*, Fundamentals of Machine Technology*]*

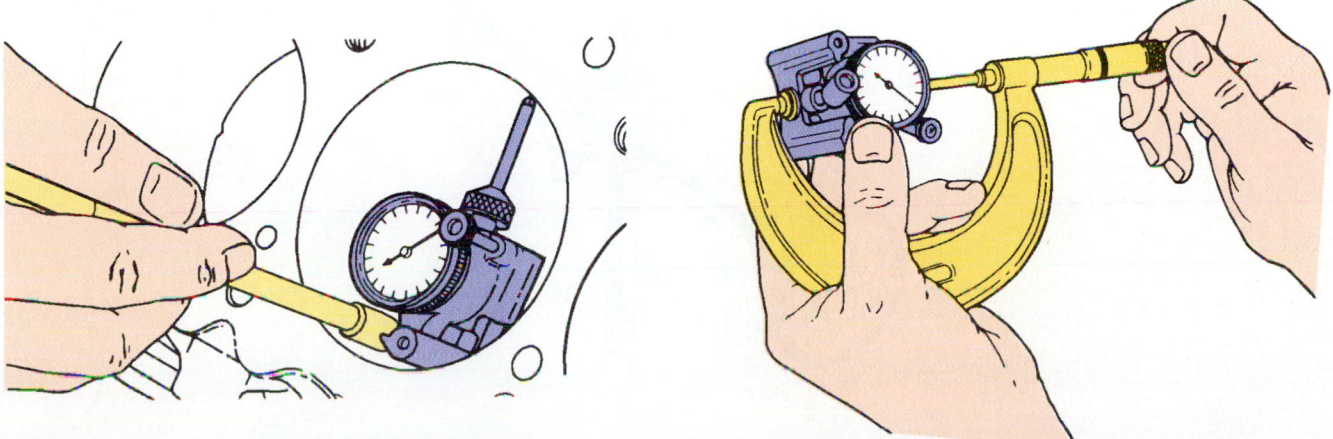

Figure 5.28 A dial indicator is measured for comparison. *[Courtesy of General Motors Corporation, Service Technology Group]*

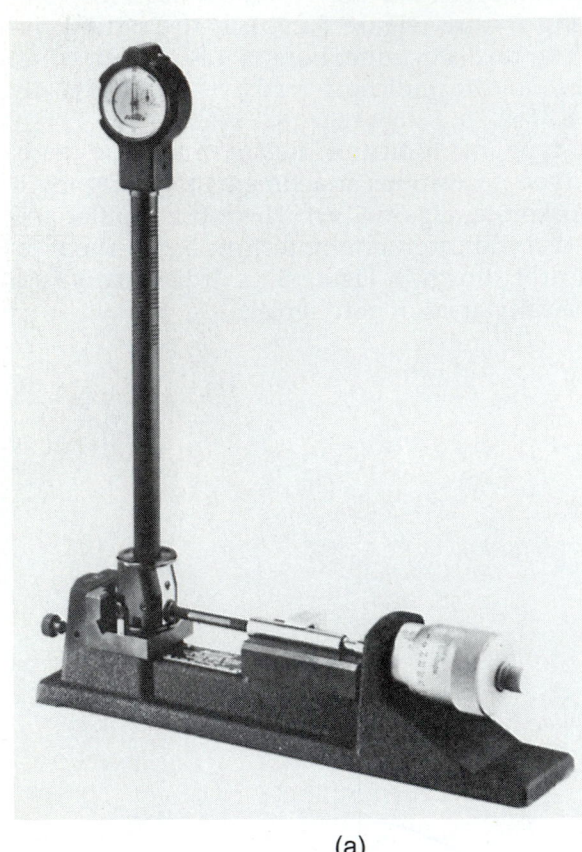

(a)

(b)

Figure 5.29 (a) A dial bore gauge is calibrated in a setting fixture. (b) A dial bore gauge measuring a cylinder bore. *(Courtesy of Sunnen Products Company, St. Louis, Missouri)*

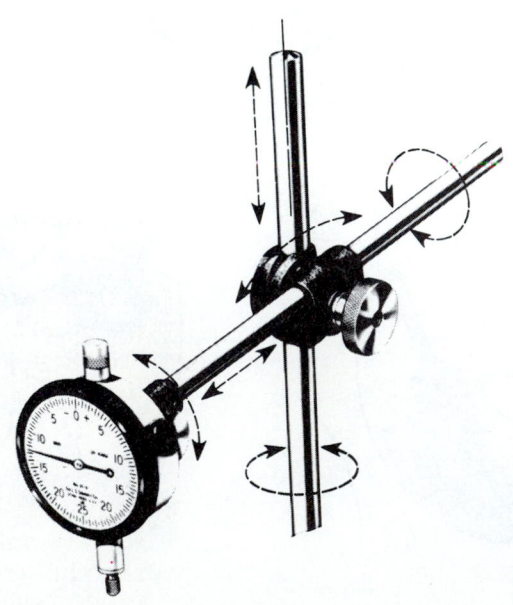

Figure 5.30 Various positions of a universal dial indicator set. *(Courtesy of C. Thomas Olivo Associates,* Fundamentals of Machine Technology*)*

Figure 5.31 A vise-grip indicator in use.

■ REVIEW QUESTIONS

1. On precision measuring instruments, how is $\frac{1}{10}$" expressed with decimals?

2. Name the tool used to calibrate a micrometer.

3. Which scale on the micrometer is used to measure to 0.0001"?

4. When reading a small-hole gauge, how is the micrometer reading taken in relation to the split in the ball, on the split, or 90° to it?

5. Name an instrument that would be used to measure endplay of a shaft?

6. What is the extra gauge that long-range dial indicators have?

7. Convert 5 inches to millimeters.

8. Convert 127 millimeters to inches.

9. Approximately how many cubic inches are there in one liter?

10. Approximately how many cubic inches equals 2.7 liters?

■ ASE STYLE REVIEW QUESTIONS

1. Technician A says that one liter equals 1000 cubic centimeters. Technician B says the most versatile precision measuring instrument for an automotive technician is the vernier caliper. Who is right?

   a. Technician A   **b.** Technician B
   c. Both A and B   **d.** Neither A nor B

2. Technician A says that one revolution of the micrometer thimble equals $\frac{1}{40}$ inch. Technician B says that one revolution of the micrometer thimble equals 0.025 inch. Who is right?

   a. Technician A   **b.** Technician B
   c. Both A and B   **d.** Neither A nor B

3. Technician A says that the English system of measurement is based on the number 10. Technician B says that calipers, dividers, and telescopic gauges are precision measuring instruments. Who is right?

   a. Technician A   **b.** Technician B
   c. Both A and B   **d.** Neither A nor B

4. Technician A says that a micrometer can be used to measure I.D., O.D., and depth from 0" to 7". Technician B says that a feeler gauge would be used to measure valve lash (clearance). Who is right?

   a. Technician A   **b.** Technician B
   c. Both A and B   **d.** Neither A nor B

5. Technician A says a gauge block is used to measure valve lash (clearance). Technician B says that the vernier scale on a micrometer measures to 0.001". Who is right?

   a. Technician A   **b.** Technician B
   c. Both A and B   **d.** Neither A nor B

VERNIER CALIPER PRACTICE

Enter the vernier caliper readings below.

1.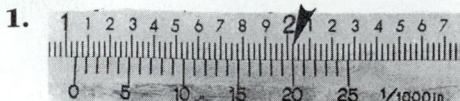

2.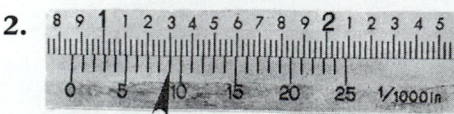

3.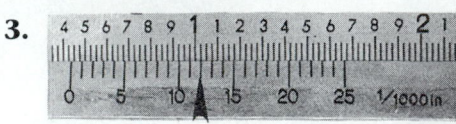

4.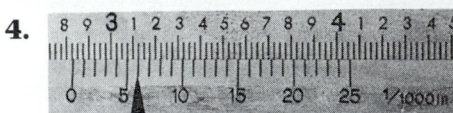

5.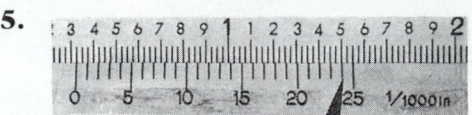

6.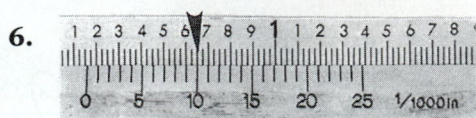

7.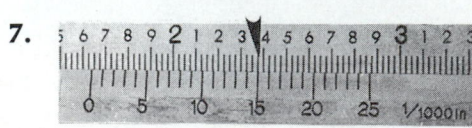

8.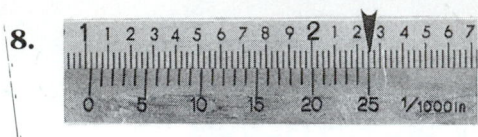

MICROMETER PRACTICE

Enter the micrometer readings below.

1.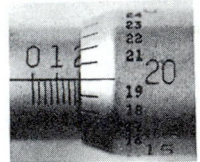

2.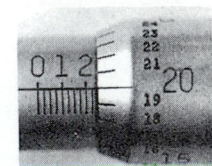

3.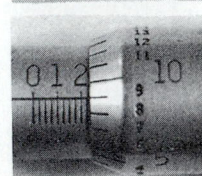

4.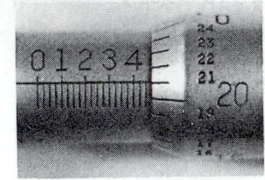

5.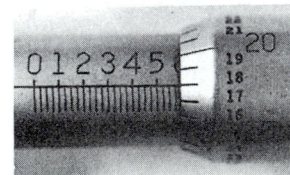

6.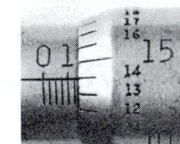

7.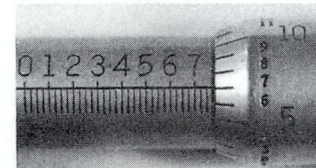

8.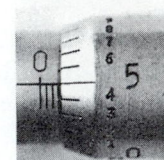

CHAPTER 6

Hardware, Fasteners, Drills, and Thread Repair

■ **KEY TERMS**

capscrew
temper
anneal
clamping force
elastic limit
bolt grade
American National
  Standards Institute
  (ANSI)
International Standards
  Organization (ISO)
hydrostatic lock
through-hardened
  washers
case-hardened washers
tap drill
electrical discharge
  machining (EDM)

■ OBJECTIVES

Upon completion of this chapter, you should be able to:

✔ List fasteners and the methods in which they are used for repair.

✔ Describe the various methods of thread repair.

✔ Describe characteristics of hand drills.

■ INTRODUCTION

A good automotive technician must know many things about fasteners (nuts and bolts) and different methods of repairing them when they break, and how not to break them. This chapter will deal with these parts and procedures.

■ CHARACTERISTICS OF FASTENERS

A *bolt* is an externally threaded fastener that is used with a nut. When it is used without a nut in a threaded or blind hole application to hold components together, it is known as a **capscrew**. During manufacturing, screw threads are either rolled or cut. *Rolled screw threads* are larger than the shank of the bolt (the area between the thread and the head of the bolt). To determine the size of a fastener, measure the screw thread.

Standard threads come in both coarse and fine threads. The fine thread is called *national fine*; the coarse thread is *national coarse*. These are classified, respectively, as UNF and UNC. The "U" stands for *unified*; it means that the thread interchanges with British threads. A UNF thread call-out for a ⅜" diameter thread would be ⅜ × 24. It would have 24 threads per inch. A UNC would be ⅜ × 16. It would have 16 threads per inch.

Metric screw threads are also coarse or fine. The last number in the call-out represents the thread pitch in millimeters. It is different than the threads per inch in the unified system. With metric screw threads, the thread pitch is the *distance* between threads in millimeters. This varies between 1.0 and 2.0 and depends on the diameter of the fastener. If the thread pitch is 1.0 this means there is 1 mm between each screw thread. An 8 mm fine thread could be either 8 × 1.0 or 8 × 1.25; a coarse thread would be 8 × 1.5. Notice that the lower pitch number denotes a finer thread.

Several different kinds of fasteners are used in automobiles. They are selected according to purpose and price. Fasteners are made of different types and grades of metals, and some are hardened.

■ To toughen, or **temper**, a metal, it is heated to a specific temperature and then quenched.

■ To soften, or **anneal**, a metal, it is heated and allowed to cool slowly.

Bolts can be broken or damaged by overtightening, bottoming out, or being forced into a thread that does not match.

Bolt Stretch

When a bolt is properly tensioned it will be "spring loaded" against the part it is fastening. Bolts are purposely stretched like this to provide **clamping force** or "spring tension" on the parts being held together. Usually a bolt is stretched to 70% of its **elastic limit** when properly tightened. The elastic limit of a fastener is reached when it will no longer return to its original shape when loosened. A torque wrench is used to tighten fasteners to a specified tension. Correct tightening methods and torque wrenches are covered in Chapter 45, Engine Sealing and Gaskets.

When a bolt is overtightened, it becomes overstretched. A stretched bolt can be identified when a

OVERSTRETCHED

Figure 6.1 When a bolt is overstretched, the screw thread distorts. A nut will not turn easily on the bolt. *(Reproduced with permission of Premier Farnell Corp.)*

nut can turn easily down the bolt threads until it encounters the stretched area, where it will become hard to turn (Figure 6.1).

Fastener Grades

Fasteners are of different quality grades, depending on their intended use. One fastener's use might call for flexibility, while another's might call for high strength. Some new vocabulary terms are necessary for understanding fastener quality:

■ **Tensile strength** is the maximum stress a material can withstand without breaking.
■ *Yield point* is the maximum stress a material can withstand and still be able to return to its original form without damage.
■ *Ultimate strength* is about 10% higher than the yield point (for high-quality steel). That is the point at which the fastener breaks.

The heads of capscrews or bolts have markings to identify their grades (Figure 6.2). On customary (inch) bolts, there are radial lines that indicate strength. Count the lines and add two to determine the strength of the bolt. These are known as SAE **bolt grades** and are evaluated by the **American National Standards Institute (ANSI)** standard, which is used to measure a fastener's tensile strength. Metric bolts are numbered. The higher the number, the greater the strength of the bolt.

In addition to the ANSI standard, there is another standard for fastener quality. The **International Standards Organization (ISO)** defines fastener quality in terms of tensile strength and yield strength. This standard is used for metric fasteners and is anticipated to eventually replace all other grading standards. When interpreting an ISO fastener grade number:

■ The first number is the tensile strength.
■ The second number is the yield strength.

For example, a metric bolt marked 8.8 has a tensile strength of 800 MPa (115,000 psi) and a yield strength of 80% of 800 MPa. A 10.9 bolt has a tensile strength of 1000 MPa (145,000 psi) and a yield strength of 90% of 1000 MPa. Markings are required only on bolts classed 8.8 or higher.

Metric fastener strength is called **property class** rather than *grade* (as with SAE fasteners). SAE grades and property classes are comparable:

SAE Grade	Property Class
Grade 2	5.8
Grade 5	8.8
Grade 7	9.8
Grade 8	10.9

In appendix A at the end of the book, there are charts describing the tensile strengths and the torque recommendations for various sizes and grades of fasteners. These torque recommendations are *not* for gasketed joints or for soft materials.

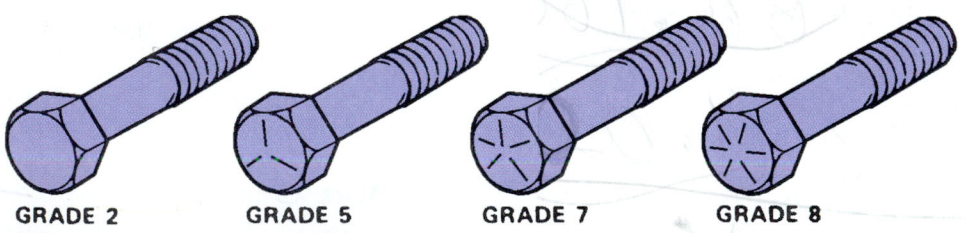

GRADE 2   GRADE 5   GRADE 7   GRADE 8

Customary (inch) bolts – Identification marks correspond to bolt strength – Increasing numbers represent increasing strength.

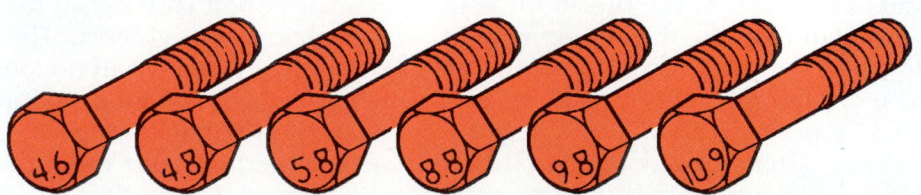

Metric Bolts – Identification class numbers correspond to bolt strength – Increasing numbers represent increasing strength.

Figure 6.2 Bolt grade markings. *(Courtesy of General Motors Corporation, Service Technology Group)*

NOTE: *The strongest bolts are not always the best choice. A high-grade fastener might be best for a front suspension bolt but may not be suitable for use in an engine. Using a higher grade of fastener allows a higher torque to be used. But if torque is not increased, there is no increase in the clamp load, whether a higher bolt strength is used or not. Also, higher strength bolts have less resistance to fatigue than softer bolts if they are undertightened.*

Automotive bolts are usually grade 5 or 6 (or 9.8 or 10.9 for metric use). Grade 6 bolts are usually used on engine main bearing caps. Grade 8 screws are used for flywheels and flexplates.

Thread Lubricants

An antiseize compound is used where a bolt might become difficult to remove after a period of time—for example, in an aluminum block, or on exhaust manifold bolts (Figure 6.3). A chart in appendix B (at the back of the book) shows how much to reduce torque when thread lubricants are used.

Lubricants also introduce the possibility of a **hydrostatic lock**, where oil is trapped in a blind hole. When the bolt contacts the oil, it cannot compress it; therefore, the bolt cannot be properly tightened. A cracked part can result.

Nuts

The grade of nut used must match the grade of bolt that it is used with. Manufacturers use several different nut markings to denote grade identification. Some are marked on the top and others are marked on the sides (Figure 6.4). Grade 8 nuts are always marked, but grade 5 nuts sometimes are not.

Reuse of Nuts

A nut must be slightly softer than the bolt so that its threads can flow to fit the threads of the bolt (Figure 6.5). This thread distortion is permanent. Bolts are normally reused in the engine rebuilding process. Nuts, however, lose some effectiveness with repeated use. Reusing the nut does not allow as good a match as the first use. So there is more friction between the threads and an accurate torque reading cannot be achieved.

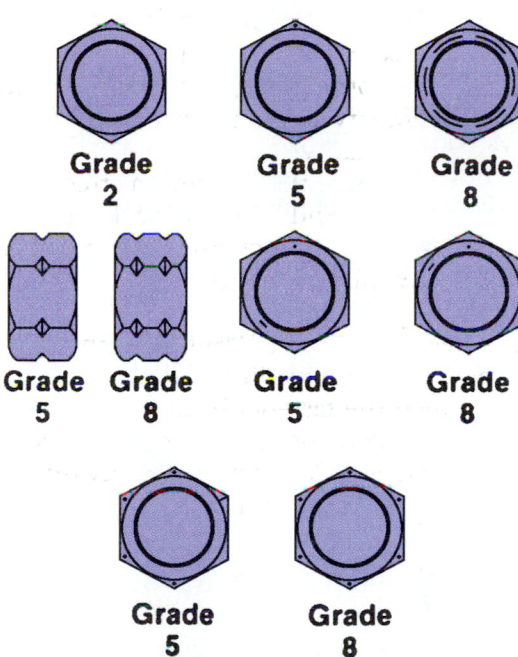

Figure 6.4 Nut grade markings. *(Courtesy of Bowman Distribution, Barnes Group Inc.)*

Figure 6.3 An antiseize compound is used where a bolt might be difficult to remove. *(Courtesy of Loctite Corporation)*

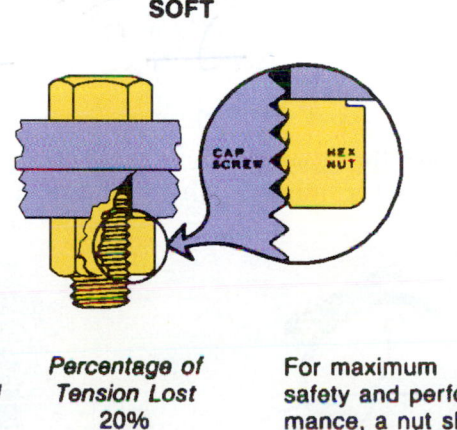

Times Nut Was Reused	Percentage of Tension Lost	For maximum safety and performance, a nut should be discarded once it is removed
1st	20%	
2nd	33%	
5th	47%	
9th	60%	

Figure 6.5 A nut conforms to the thread of a bolt, so it loses more holding ability each time it is reused. *(Reproduced with permission of Premier Farnell Corp.)*

According to Bowman Distribution, a nut acts much like the coils of a spring when an assembly is tightened. Each thread in the nut will carry progressively less load:

- The first thread of a coarse fastener (the one against the work being clamped) will support about 38% of the total load on the bolt.
- The second supports about 25%.
- The third supports about 18%.

The first thread will be into the plastic range (yield) and the outer threads may not even touch the bolt at all.

Lock nuts, like the type used on some rocker arm studs, are deformed at the top (Figure 6.6). These nuts lose some of their locking ability with repeated reuse. Do not remove and replace them any more than is necessary.

Another type of lock nut is called a *castle nut*. It has slots at the top that align with a hole in the male screw thread. A cotter pin is inserted as a safety assurance that the nut cannot come loose. This kind of nut is commonly used to retain front wheel bearings and steering linkage parts.

A *wing nut* is a nut that is tightened by hand only. Only minor tightening torque is necessary with a wing nut. Examples of its use include holding an air cleaner housing together or fastening a battery hold-down bracket.

Washers

Use of the proper washer is also necessary to achieve the correct load on a bolt. Use a **through-hardened** heat-treated flat **washer**. These washers are hardened throughout. **Case-hardened washers** are hard on the surface only; the core of the washer is soft and will compress, allowing the connection to lose clamping force. Washers are also used to provide a hardened surface for a bolt to act against. This is especially important when the material to be clamped is not steel.

NOTE: *Every 0.001" of bolt relaxation will cause a loss of about 30,000 psi of clamping force.*

Removing a Stud

Studs have threads on both ends. They are often found on carburetor mounts and exhaust manifolds. To remove a stud, a special stud puller may be used (see Figure 7.43). When a puller is not available, use two nuts with a lock washer between them. Turn the nut that is farthest from the end of the stud to remove the stud.

■ FASTENER FAILURES

Fatigue breaks account for about 75% of fastener problems. A bolt becomes fatigued from working back and forth when it is too loose. This problem occurs when the assembly is not properly torqued to stretch the bolt.

Shear or *torsion* breaks result from the use of a poor grade of fastener, too much friction, or an improper thread fit. Technicians sometimes fail to clean bolt threads in the block thoroughly.

SHOP TIP Threaded holes in the block must be chased (cleaned) with a tap (Figure 6.7) prior to reassembly, especially when the block has been hot tanked. The holes and the capscrew should receive a light coating of oil.

Bolts can be broken when they are *bottomed out*. Normally, bolts should turn into the thread for four to six turns, or about 1½ times the diameter of the screw thread. If a bolt is bottomed out, the sudden stop of the bolt is felt, instead of the "elastic" feeling of normal bolt stretch. Just before a bolt breaks off, it will become slightly easier to turn. This is a clue to stop turning, loosen the bolt, and replace it with a new one.

Crossed threads are usually the result of improper assembly procedures.

SHOP TIP When two parts are fastened together, do not completely tighten *any* of the fasteners until *all* of them have been started into their threads. With all the bolts loose, a hole that does not line up can be aligned using a rolling head pry bar (Figure 6.8). This tool is also handy in aligning motor mounts and transmission cross-members.

■ DRILL BITS

Drill bits are used to make or enlarge holes in metal and to help remove broken fasteners. A drill bit has two *flutes*, which provide a channel for the cutting chips to escape into during drilling (Figure 6.9). They also allow cutting fluid to reach the cutting edge of the drill.

The web that separates the flutes of the drill increases in size from the point to the shank. The web can be so large at its point that it will be necessary to drill a *pilot hole* with a smaller drill bit (Figure 6.10). A center punch is used to locate the web (or point) of the drill. Only a small area of the *land* (the spiral area on the drill body) is equal to the diameter of the drill. The rest of the land is *relieved* (made smaller) to reduce friction.

Figure 6.6 The locking nut used on rocker arm studs is distorted at the top. *(Reproduced with permission of Premier Farnell Corp.)*

Figure 6.7 Chase bolt holes with a tap to ensure correct clamping force after reassembly.

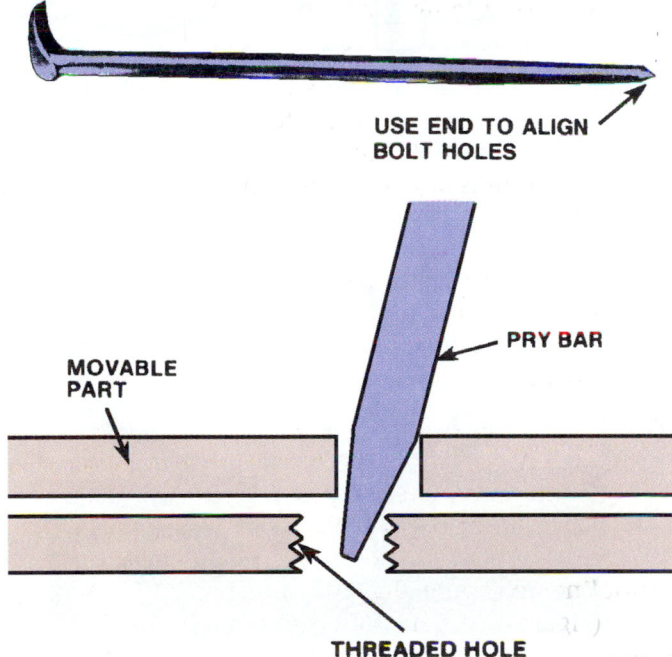

USE END TO ALIGN
BOLT HOLES

PRY BAR

MOVABLE
PART

THREADED HOLE

Figure 6.8 A rolling head prybar can be used to align parts. *(Courtesy of OTC-SPX Corp. Aftermarket Tool & Equipment Group)*

The tip of a drill bit must be sharp. Its angle will be flatter for hard metals such as steel, and steeper for softer metals. A drill bit that is sharpened "off center" (Figure 6.11) will drill a hole that is too large. Be careful to make the angles equal. For iron and steel, the bit is cut to a 59° angle (Figure 6.12a). The backside of the cutting lip must slope off, as shown in Figure 6.12b, to produce the chisel edge on the drill. Too little clearance causes the drill to rub, instead of cut. Too much clearance, and the bit cuts too rapidly, and may break or chip.

Hand Sharpening a Drill Bit

Hand sharpening the drill bit requires some practice. Hold the drill against the wheel at a 59° angle, and rotate it in an arc movement (Figure 6.13). The angle and length of the lip can be checked with a gauge (Figure 6.14). A properly ground drill will give off even amounts of chips from both sides of the metal as it cuts (Figure 6.15).

SHOP TIP Be sure to quench the bit often to keep it cool. If it gets too hot, the temper can be drawn from it.

If much precision drilling is to be done, several manufacturers have drilling handbooks available.

Drill Speed and Lubricants

To prevent drill bit wear, keep as close as possible to the recommended drilling speed. Bits are usually made of *high-speed steel*; they will be marked "HS" or "HSS" on their shanks.

SHOP TIP During grinding, different metals give off different color sparks:
- High-speed steel drill bits give off a dull red spark.
- Less expensive carbon steel drill bits give off white sparks.

Use carbon steel drill bits at half the cutting speed of high-speed steel because their resistance to heat is lower.

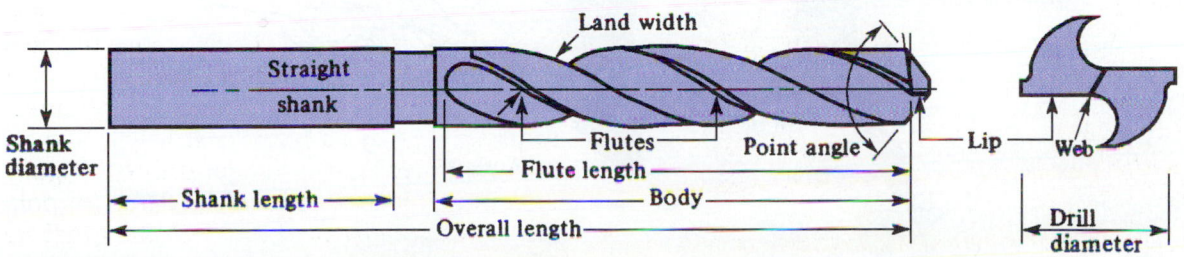

Land width

Straight
shank

Shank
diameter

Shank length

Flutes

Flute length

Body

Overall length

Point angle

Lip

Web

Drill
diameter

Figure 6.9 Parts of a drill bit. *(Courtesy of C. Thomas Olivo Associates, Fundamentals of Machine Technology)*

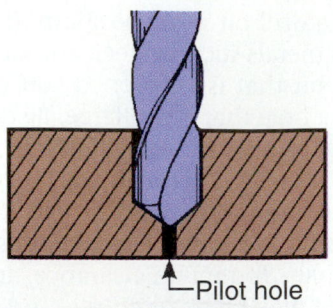

Figure 6.10 Before drilling with a large drill bit, drill a pilot hole. *(Courtesy of C. Thomas Olivo Associates, Fundamentals of Machine Technology)*

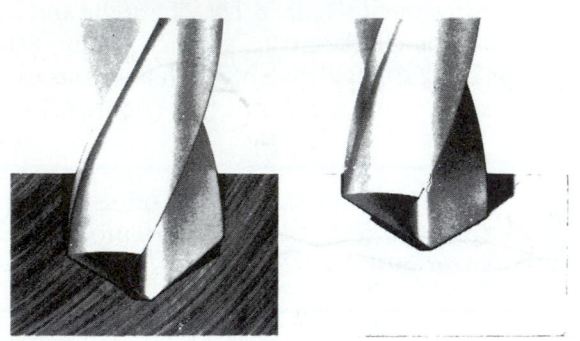

Figure 6.11 Results of incorrect drill sharpening. *(Courtesy of C. Thomas Olivo Associates, Fundamentals of Machine Technology)*

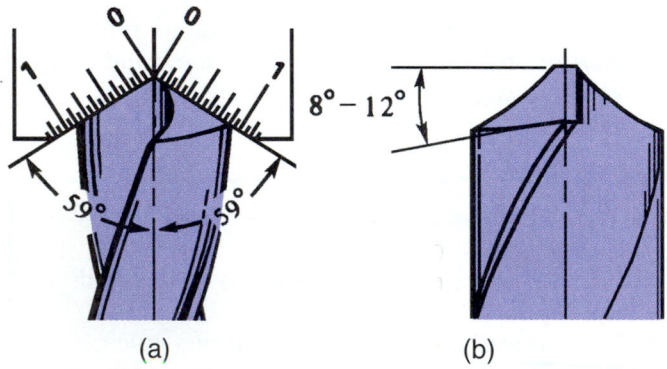

(a)   (b)

Figure 6.12 (a) Point angle for iron or steel. (b) The cutting lip must slope off as shown here. *(Courtesy of C. Thomas Olivo Associates, Fundamentals of Machine Technology)*

Approximate HS drill speeds when drilling mild steel are:

- 1500 rpm for ¼"
- 1000 rpm for ⅜"
- 750 rpm for ½"

Drill aluminum with the same size drill bit at an rpm twice as fast. Soft cast iron is generally drilled at the speed required for mild steel.

- When drilling steel, use a cutting oil.
- Cast iron and aluminum can be cut dry.

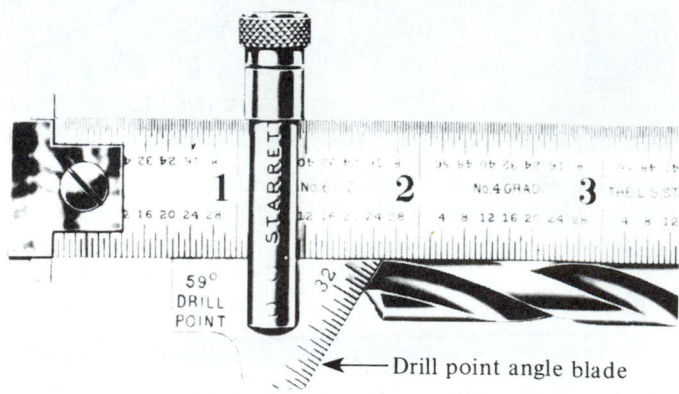

Figure 6.14 Check the angle on a drill gauge. *(Courtesy of C. Thomas Olivo Associates, Fundamentals of Machine Technology)*

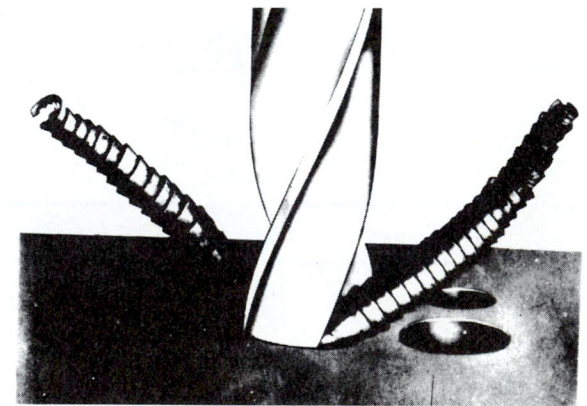

Figure 6.15 A properly ground drill bit will chip equally from both flutes. *(Courtesy of C. Thomas Olivo Associates, Fundamentals of Machine Technology)*

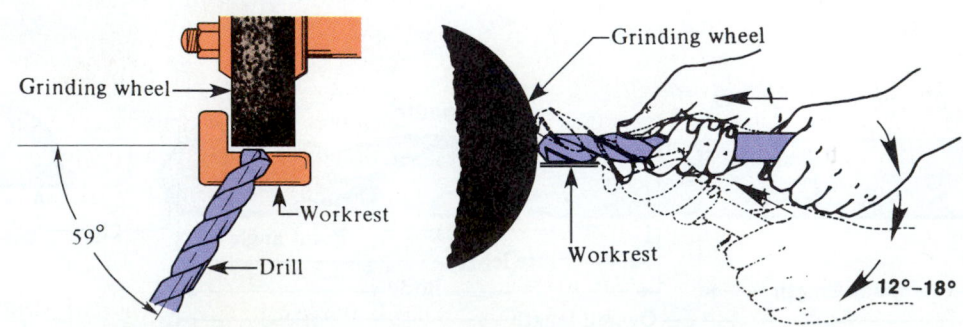

Figure 6.13 Hold the drill on the workrest at about 59° to the wheel and rotate the front of the drill upward. *(Courtesy of C. Thomas Olivo Associates, Fundamentals of Machine Technology)*

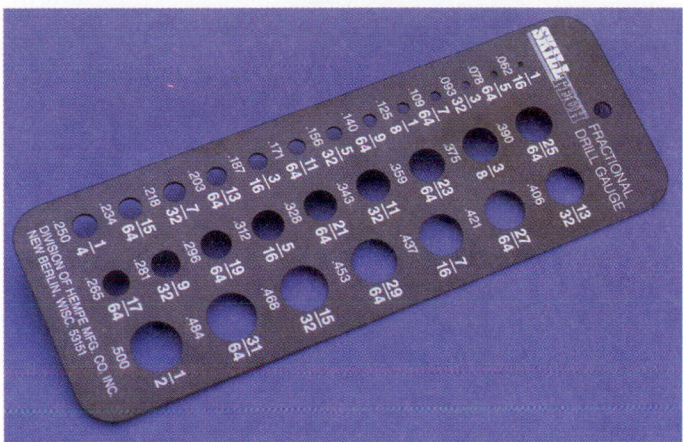

Figure 6.16 A drill gauge can be used to determine the size of a drill bit.

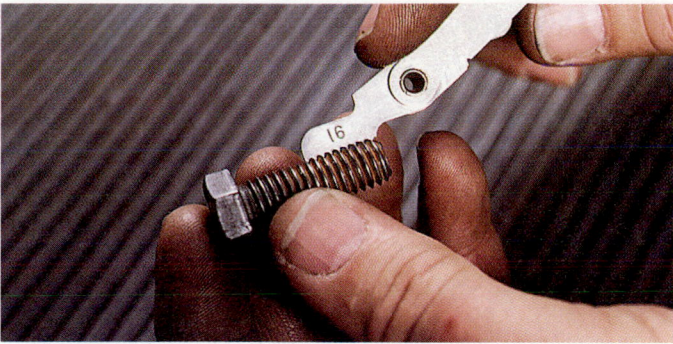

Figure 6.17 This screw has 16 threads per inch.

Drill Size

Drills are sized in a number of different ways. Most technicians own a *fractional set* of drills called a *drill index*. Sizes in the drill index increase by increments of ¹⁄₆₄". Less expensive drill indices are also available with drill sizes increasing in ¹⁄₃₂" or ¹⁄₁₆" increments. Decimal equivalents are also printed on the index. The size of a drill is printed on the shank of the drill bit. It is not unusual for a drill to slip in the chuck, erasing the size marking. The holes in the drill index correspond to the correct size of the drill. A micrometer or a special drill gauge with holes of the specified size can also be used to determine the size of a drill bit (Figure 6.16).

Drills also come in *letter* and *number* sizes. Letter-size drills increase in size from A (0.234") to Z (0.413"). They are larger than the largest number-sized drill.

Drills sized by number commonly range from 80 (0.0135") to 1, which is 0.228" (almost ¼"). The most widely used number-sized drills usually are numbers 1 to 60 (0.040"). Commonly used metric drills range from 0.20 mm to 16.00 mm in size.

■ TAPS AND THREADS

A tap is used to cut internal threads in a previously drilled hole or to *chase* (clean) existing threads. Standard and metric both have coarse and fine threads for each diameter of fastener. To determine the screw thread of a hole, screw in the proper tap and then read the size on the tap. A special *thread-pitch gauge* can be used, but this method is not always dependable; some screws have different diameters, but the same number of threads per inch (Figure 6.17).

Types of Taps

There are three types of taps available for any given thread size (Figure 6.18).
■ The *tapered* tap helps pilot the tap into the hole.
■ The *plug* tap is a standard tap.

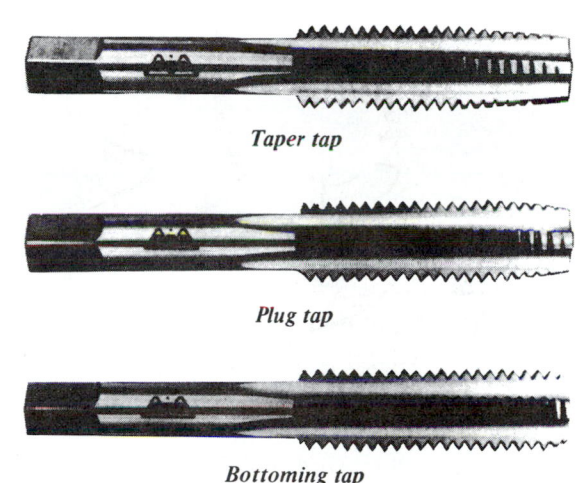

Taper tap

Plug tap

Bottoming tap

Figure 6.18 Three types of taps are available for each screw thread. The plug tap is found in most mechanics' tap sets. *[Courtesy of C. Thomas Olivo Associates, Fundamentals of Machine Technology]*

■ A *bottom* tap makes threads all the way to the bottom of a blind hole, and it chases threads to clean them.
A blind hole is one that does not go all the way through a part (Figure 6.19).

>
> **CASE HISTORY** *An apprentice was chasing the threads in a cylinder block after removing it from the cleaning tank. He tapped out all of the holes, including those for the oil gallery plugs. Oil gallery plugs have tapered pipe threads. The apprentice turned the tapered pipe tap through the entire length of the thread until it turned easily. When the new oil gallery plugs were installed, they could not become tight because the threads in the block were now too big. The block had to be repaired.*

Pipe Threads

Pipe threads are used for heater outlets in the block and intake manifold, as well as for oil and coolant temperature sending units (Figure 6.20). Pipe taps have tapered

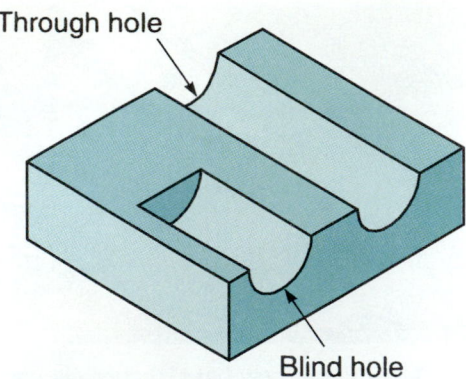

Figure 6.19 A blind hole does not go all of the way through the part.

Figure 6.20 Pipe threads are found on threaded plugs, heater outlets, and oil and coolant temperature sending units.

threads (Figure 6.21), designed to wedge against each other. Therefore, when tapping pipe thread, do not continue turning the tap until it turns easily. Because the size of a pipe thread is determined by the inside diameter of the piece of pipe, a ½" *NPT* (*National Pipe Taper*) tap will be quite a bit larger than ½" (Figure 6.22).

Determining Tap Drill Size

Before tapping a thread, a hole of the correct size must be drilled. A **tap drill** usually provides about 75% of a full thread. This allows some margin for error and keeps the tap from binding during the tapping operation. The tap drill chart (Figure 6.23) shows which drill to use. A ⅜"-by-24 (24 threads per inch) tap would require a Q (0.332") drill. From a fractional-size drill index, select the next largest size available. In this case, the right choice is an ¹¹⁄₃₂" (0.343"), which is only 0.011" larger than a letter Q drill (see appendix C for size conversion). If a tap drill chart is not available, select the largest drill bit that will fit into a nut that fits over the bolt that the hole is being drilled or tapped for.

NOTE: *Do not use a smaller drill; the tap could bind and break. A female screw thread that is only 50% of full depth has been shown to have so much strength that the bolt would strip before the thread would strip.*

Tapping a Hole

When tapping a hole, advance the tap in a clockwise direction. Then, back off about ¼" turn to break off any

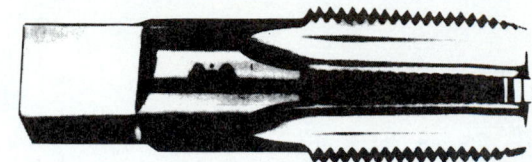

Figure 6.21 A pipe tap for NPT threads. *(Courtesy of C. Thomas Olivo Associates, Fundamentals of Machine Technology)*

ACTUAL PIPE THREAD SIZES

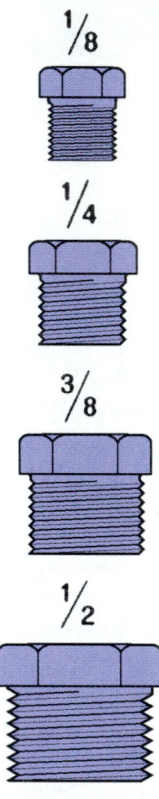

Figure 6.22 Pipe threads are sized according to the I.D. (inside diameter) of the pipe. *(Courtesy of Weatherhead Division, Dana Corporation)*

metal chips that might accumulate. These chips could ruin the new thread. The chips can be felt breaking as the tap is backed off. The broken chips gather in the flutes of the tap (the lower area next to the cutting teeth) (Figure 6.24). Thread lubricants are required when tapping steel and most nonferrous metals; cast iron can be tapped dry.

Removing a Broken Tap

When a tap is broken, try to drive it counterclockwise with a centerpunch or a special tap extractor to remove it (Figure 6.25). If the broken tap will not come out of the hole, try to break it up with a centerpunch. If this does not work, a process called **electrical discharge machining (EDM)** can be performed by a machine shop that has the necessary machinery. EDM erodes the fastener, leaving the thread intact.

Size	Threads per inch			Outside Diameter Inches	Tap Drill Approx. 75% Full Thread	Decimal Equivalent of Tap Drill
	NC	NF	NS			
0	...	80	...	.0600	3/64	.0469
1	...	...	56	.0730	54	.0550
1	64	...	...	.0730	53	.0595
1	...	72	...	.0730	53	.0595
2	56	...	...	.0860	50	.0700
2	...	64	...	.0860	50	.0700
3	48	...	...	.0990	47	.0785
3	...	56	...	.0990	45	.0820
4	...	...	32	.1120	45	.0820
4	...	...	36	.1120	44	.0860
4	40	...	...	.1120	43	.0890
4	...	48	...	.1120	42	.0935
5	...	...	36	.1250	40	.0980
5	40	...	...	.1250	38	.1015
5	...	44	...	.1250	37	.1040
6	32	...	...	.1380	36	.1065
6	...	...	36	.1380	34	.1110
6	...	40	...	.1380	33	.1130
8	...	...	30	.1640	30	.1285
8	32	...	...	.1640	29	.1360
8	...	36	...	.1640	29	.1360
8	...	...	40	.1640	28	.1405
10	24	...	...	.1900	25	.1495
10	...	...	28	.1900	23	.1540
10	...	...	30	.1900	22	.1570
10	...	32	...	.1900	21	.1590
12	24	...	...	.2160	16	.1770
12	...	28	...	.2160	14	.1820
12	...	...	32	.2160	13	.1850
1/4	20	...	...	.2500	7	.2010
1/4	...	28	...	.2500	3	.2130
5/16	18	...	...	.3125	F	.2570
5/16	...	24	...	.3125	I	.2720
3/8	16	...	...	.3750	5/16	.3125
3/8	...	24	...	.3750	Q	.3320
7/16	14	...	...	.4375	U	.3680
7/16	...	20	...	.4375	25/64	.3906
1/2	13	...	...	.5000	27/64	.4219
1/2	...	20	...	.5000	29/64	.4531
9/16	12	...	...	.5625	31/64	.4844
9/16	...	18	...	.5625	33/64	.5156
5/8	11	...	...	.6250	17/32	.5312
5/8	...	18	...	.6250	37/64	.5781
3/4	10	...	...	.7500	21/32	.6562
3/4	...	16	...	.7500	11/16	.6875
7/8	9	...	...	.8750	49/64	.7656

Figure 6.23 A tap drill chart for inch-sized drills. A size conversion chart is located in the appendix. *(Courtesy of The L.S. Starrett Company)*

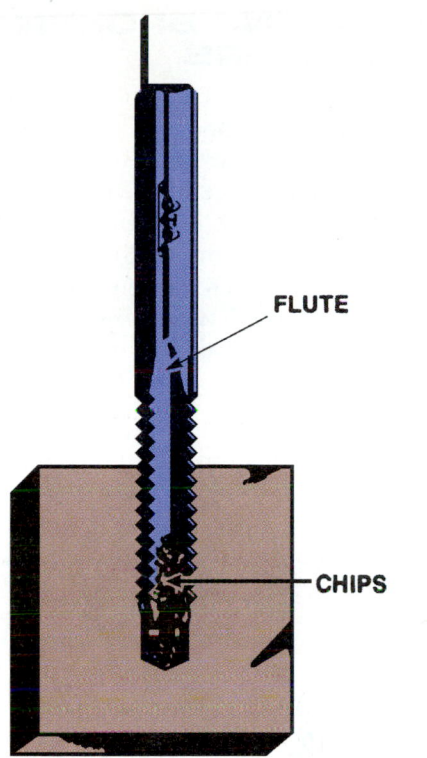

Figure 6.24 Chips are gathered in the flutes of the tap. *(Courtesy of C. Thomas Olivo Associates, Fundamentals of Machine Technology)*

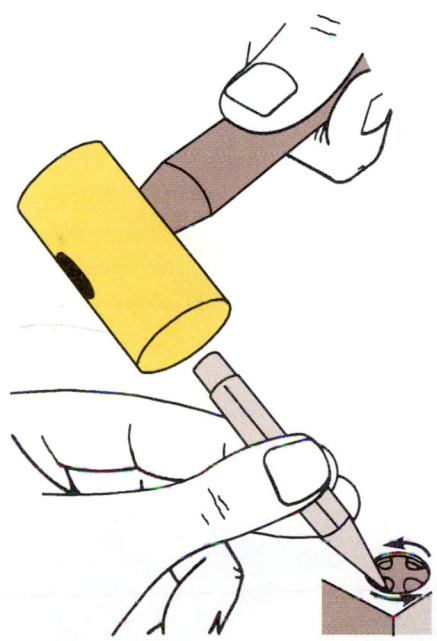

Figure 6.25 Try to remove a broken tap with a centerpunch or a chisel.

■ DIES

Dies are used to make external threads on a round rod. When making a thread, the die is advanced and then turned backward like when using a tap. The die can be broken if it is forced. A die is especially useful for chasing *burred* threads. When shortening a bolt, first screw a die onto the bolt threads. After shortening and chamfering the end of the thread, removing the die will clean up the end of the thread.

■ REPAIRING BROKEN FASTENERS

Every technician has to remove a broken bolt occasionally. If a broken bolt has not bottomed out in the hole, it can sometimes be removed by driving it in a counterclockwise direction with a chisel, just as in removing a broken tap. A bolt that has broken off above the surface (Figure 6.26) may be removable with pliers or a stud extractor (see Figure 7.43). A left-hand drill bit can be used with a reversible drill motor to remove a broken bolt that is not bottomed out.

SHOP TIP Heat, applied with a torch, will often make removal of a broken stud or bolt easier. Applying wax (paraffin) to quench after heating lubricates the screw thread, making removal easier.

Screw Extractors

There are two common types of screw extractors (Figure 6.27). Figure 6.27a shows an extractor that is commonly known as an *easy out*. The other type of an extractor has flutes (Figure 6.27b). Both are used in the same manner.

The side of an easy out is stamped with the correct size of drill bit to use with it. A hole is drilled and the easy out is pounded into it (Figure 6.28). Turning the easy out counterclockwise will remove the broken part (if the remaining screw thread is not bound up).

Extractors are hardened, which makes them difficult to remove when they break. A broken extractor

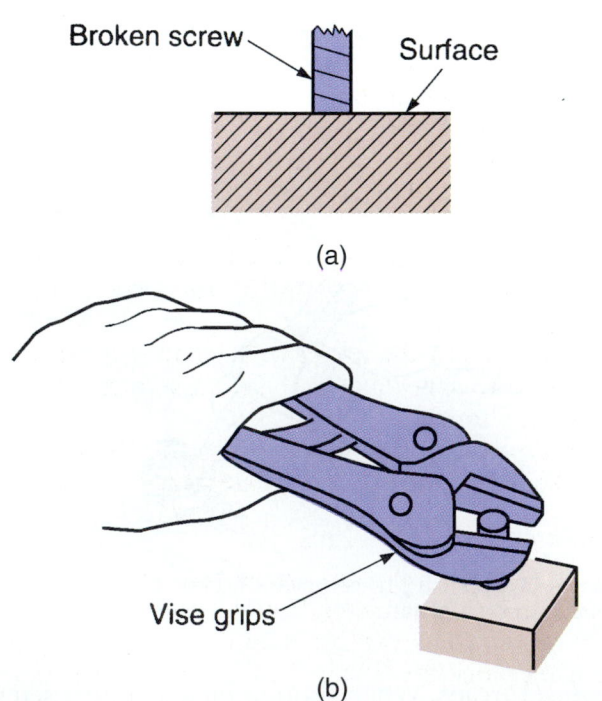

Figure 6.26 (a) A screw broken off above the work surface. (b) Removing a broken screw with vise grips. *(a, Courtesy of Lisle Tools Corp)*

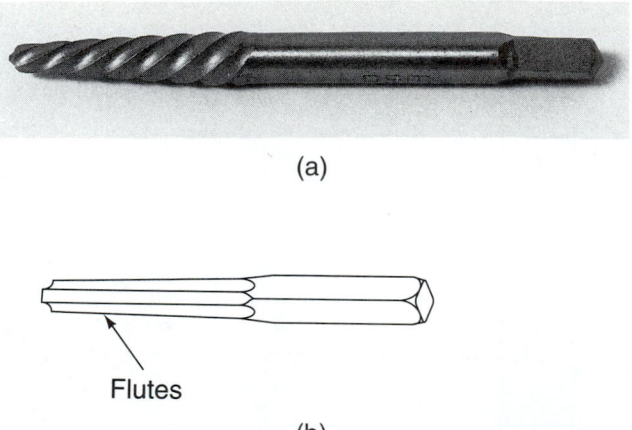

Figure 6.27 Two types of screw extractors: (a) An easy out, (b) A fluted extractor. *(Courtesy of U.S. Navy)*

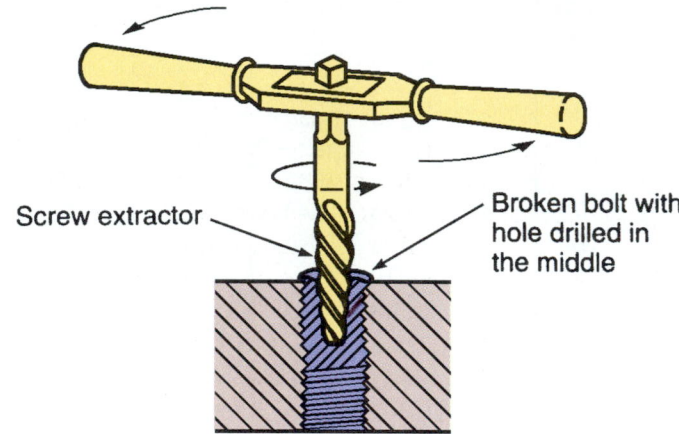

Figure 6.28 Using a screw extractor to remove a broken screw.

can sometimes be removed by heating it with a torch and letting it cool slowly. This softens the metal for drilling. If an extractor cannot be removed by this method, a machine shop can erode the screw with the EDM process discussed earlier.

A fluted extractor set includes several sizes of fluted extractors, the correct sized drill bits, splined hex nuts, and drill guides (Figure 6.29). To remove a broken fastener with a fluted extractor set:

■ First, make a centerpunch mark *precisely* in the center of the fastener (Figure 6.30a).

■ Then, drill a hole of a precise size into the broken fastener (Figure 6.30b). The drilled hole should go all the way through the bottom of the fastener. The first hole should be drilled with a small drill bit. Then, drill with the largest drill bit in the set that is smaller than the fastener thread being removed.

NOTE: *If the fastener is broken off below the surface of the workpiece, a drill guide in the extractor set can be used to align the hole with the center of the broken screw.*

■ Drive the extractor into the hole with a brass hammer (Figure 6.30c).

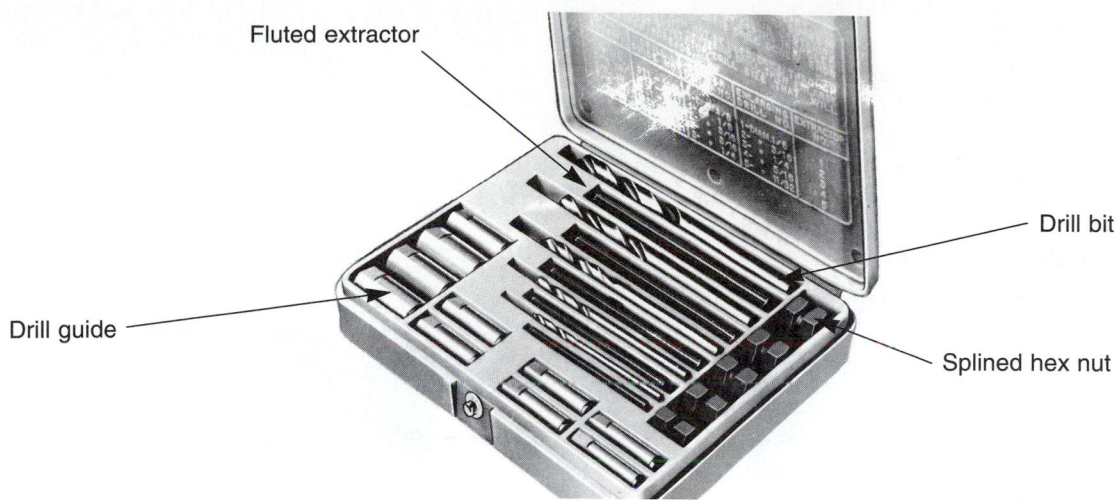

Fluted extractor

Drill bit

Drill guide

Splined hex nut

Figure 6.29 A fluted screw extractor set.

■ Then, unscrew the broken fastener (Figure 6.30d). Use as large an extractor as possible, so that the walls of the drilled stud will become as thin and flexible as possible. The extractor set includes a nut that fits over the extractor. Slide the nut down the flutes. Then grip the nut with a pair of wrenches (to exert equal force) and turn the extractor counterclockwise.

If the extractor begins to shear off, it should be removed from the drilled hole before it breaks off. The hole can then be heated with a torch (Figure 6.31), and the broken fastener cooled by dripping water onto it or applying wax (paraffin) to it. Then, try the extractor again.

Some machine shops will remove a broken fastener by placing a nut on top of it. Then, the inside of the nut is welded to the fastener with nickel welding rod (ni-rod). A wrench is then used on the nut to unscrew the fastener.

Drilling and Retapping

Sometimes a broken fastener can be repaired by drilling it out. The hole is then carefully tapped. If this fails, the next step would be to drill and then tap to the next larger thread size. When using an oversized screw thread, it is often necessary to drill an oversized hole in the part that the bolt fastens to (an exhaust manifold, for example).

Thread Inserts

When a screw thread is stripped, a better repair is to use a *thread insert* to preserve the original thread size (Figure 6.32). There are several varieties of thread inserts. They all require drilling and tapping a larger hole. The insert is then threaded into place.

The *heli-coil* method is shown in Figure 6.33. A heli-coil is an oversized spring coil of stainless steel (Figure 6.34). To use a heli-coil:
■ First drill an oversized hole.
■ Tap the hole.

■ Apply loctite or a similar adhesive to the insert.
■ Install it in the hole with a special tool. There is a drive tang at the bottom of the heli-coil that locks to the tool during installation. It is notched for easy removal.
■ Break the driving tang off the bottom of the heli-coil with a hammer and punch.

Figure 6.35b shows the appearance of a thread before it is repaired; Figure 6.35c shows the appearance of a repaired thread.

Locking Inserts

Heli-coil also makes a screw lock insert for use with permanent studs. The center coil has a series of flats that hold the insert in place. One type of insert has a locking feature that is activated by pounding it into place. This method—the *Slimsert* method—requires a special, stepped drill.
■ Tap the hole.
■ Screw the insert into place; the top of the slimsert should be approximately 0.015" below the work surface.
■ Strike the top of the driver with a hammer until the nylon washer touches the work surface. This locks the slimsert in place.

Another type of locking insert, called the *keensert*, has locking keys that are pounded into place.

Spark Plug Inserts

There are inserts available for repairing spark plug threaded holes in cylinder heads. Installation is the same as other inserts, but one caution must be observed. Some spark plugs have tapered seats while some are flat and require a gasket. Be sure to install the correct insert that corresponds to the rest of the spark plugs in the engine. Installing the insert from the combustion chamber side often works well for gasketed plugs

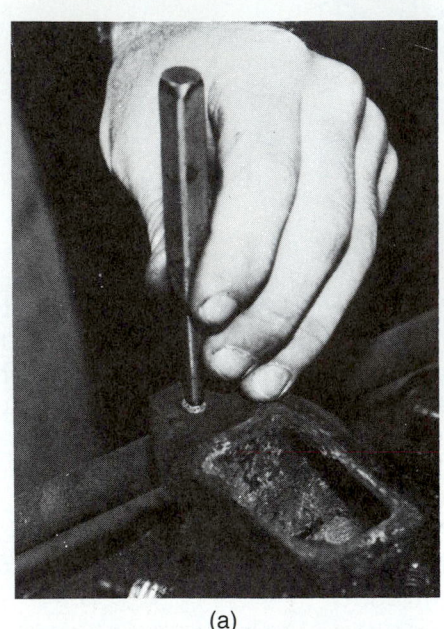

(a)

(b)

(c)

(d)

Figure 6.30 Removing a broken stud or bolt:
(a) The stud is punched exactly on center.
(b) A hole is drilled precisely on center.
(c) The extractor is pounded into the hole.
(d) The nut is slipped onto the extractor, and two wrenches are used to turn the nut counterclockwise.

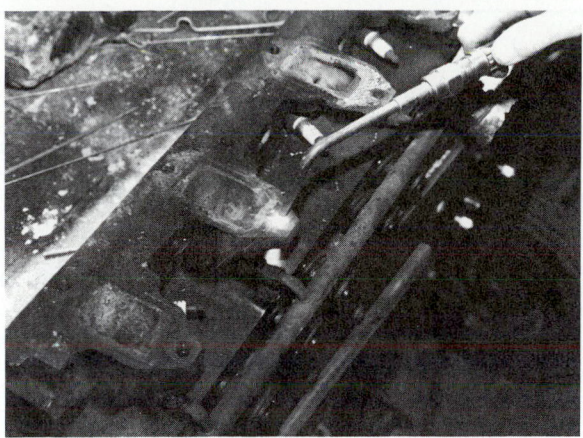

Figure 6.31 If the fastener will not come out, heat the drilled hole with a torch.

Figure 6.32 This internal thread is "stripped."

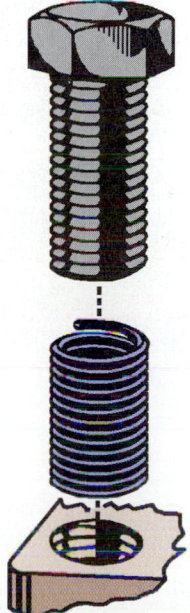

Figure 6.33 A heli-coil thread insert. *(Courtesy of Heli-Coil)*

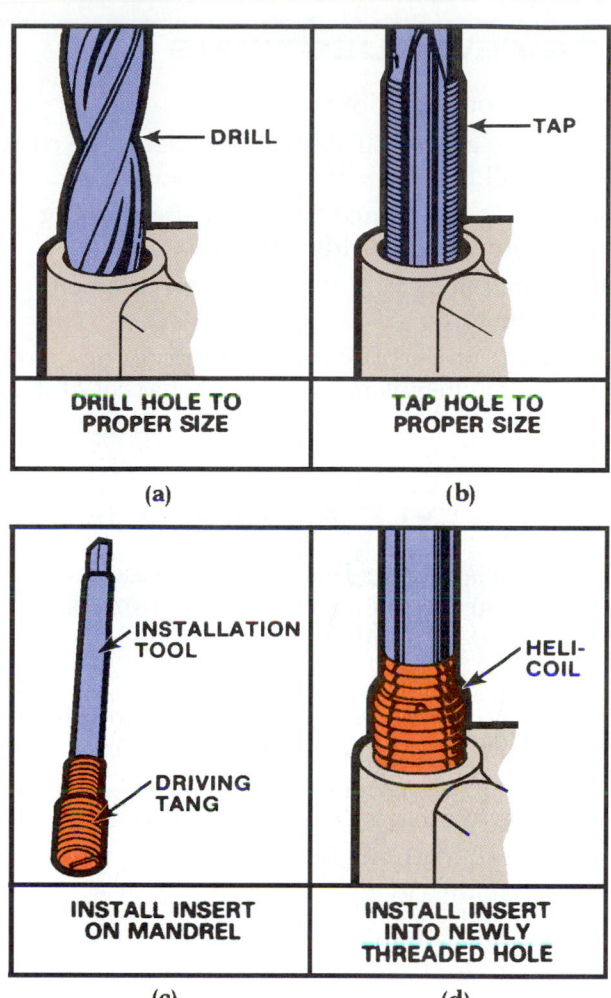

(a) DRILL — DRILL HOLE TO PROPER SIZE

(b) TAP — TAP HOLE TO PROPER SIZE

(c) INSTALLATION TOOL — DRIVING TANG — INSTALL INSERT ON MANDREL

(d) HELI-COIL — INSTALL INSERT INTO NEWLY THREADED HOLE

Figure 6.34 The heli-coil repair method. *(Courtesy of General Motors Corporation, Service Technology Group)*

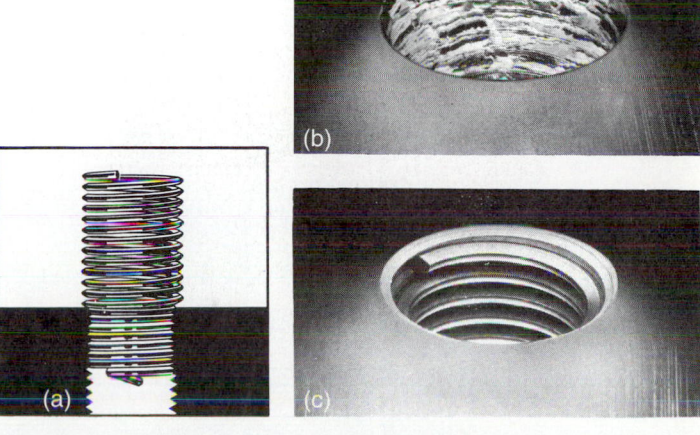

Figure 6.35 A heli-coil: (a) The coil is larger than the threaded hole that has been tapped for it. (b) The appearance of a thread before being repaired with a thread insert. (c) A repaired thread. *(Courtesy of Heli-Coil)*

■ REVIEW QUESTIONS

1. Name four types of taps.
2. Refer to a tap drill size chart. What is the correct tap drill for a ¼ × 20 screw thread?
3. Why is a tap turned backward after turning forward when cutting a thread?
4. To what percent of its elastic limit is a bolt usually torqued?
5. If a nut turns easily for a few threads on a bolt and then begins to turn hard, what could the problem be?
6. What cutting lubricant, if any, is used when cutting cast iron?
7. What is one of the trade names for a replacement thread insert?
8. Name two more drill classifications.
9. Name one place where pipe threads are found in an engine.
10. What is the name of the machine shop process that erodes a broken fastener without damaging the screw thread?

■ ASE STYLE REVIEW QUESTIONS

1. Technician A says that drill sizes are classified by letters. Technician B says that drill sizes are classified by numbers. Who is right?
   a. Technician A   b. Technician B
   c. Both A and B   d. Neither A nor B
2. Technician A says the "20" in ¼ × 20 signifies the total number of threads the fastener has. Technician B says that the ¼ signifies the socket size of the fastener head. Who is right?
   a. Technician A   b. Technician B
   c. Both A and B   d. Neither A nor B
3. Technician A says that a hole can be drilled in cast iron using no lubricant. Technician B says a hole can be drilled in aluminum using no lubricant. Who is right?
   a. Technician A   b. Technician B
   c. Both A and B   d. Neither A nor B
4. Technician A says when using a thread lubricant, the level of torque must be reduced. Technician B says that a grade 8 bolt has 8 radial lines on its head. Who is right?
   a. Technician A   b. Technician B
   c. Both A and B   d. Neither A nor B
5. Technician A says that a non-hardened nut loses clamping ability each time it is retorqued. Technician B says that a capscrew loses clamping ability each time it is retorqued. Who is right?
   a. Technician A   b. Technician B
   c. Both A and B   d. Neither A nor B

CHAPTER 7

Shop Tools

■ **KEY TERMS**

hand tools
flare nuts
R&R
filings
collets

■ OBJECTIVES

Upon completion of this chapter, you should be able to:

✔ Identify hand tools, air tools, and pullers used in an automotive repair shop.

✔ Understand the proper uses of hand tools, air tools, and pullers.

✔ Use shop tools safely.

■ INTRODUCTION

An automotive service technician uses many tools and pieces of equipment. This chapter will deal with the various types of **hand tools**, and their proper uses and advantages. A technician who works on commission (flat rate) will purchase many of the smaller, personal tools to make a job go easier or faster.

■ TOOLS OF THE TRADE

Tools of the trade are classified either as **hand tools** *portable power tools,* or *equipment.* Major pieces of equipment that are shared among all of the business's employees are typically owned by the employer. Hand tools and portable power tools that would be likely owned by an employee are covered in this section.

■ HAND TOOLS

The term *hand tools* refers to such tools as sockets, wrenches, and screwdrivers. Most automotive technicians own their own set of hand tools. A toolbox is a source of pride. A true professional keeps tools clean and well organized. It is not unusual to have several thousand dollars invested in tools (Figure 7.1). Loaning of tools between employees in a shop is generally not encouraged. If a technician finds the need to borrow a tool more than once, serious consideration should be given to the purchase of that tool. The master technician will often allow an apprentice to use his or her tools, but expects that the apprentice will start assembling his or her own set of tools.

One of the questions a prospective employer often asks during a job interview is, "Do you own a set of tools?" In fact, this is sometimes a condition of employment. The employer might want to inspect the tools to see whether the applicant has taken professional care of the tool set. A well-maintained toolbox indicates "pride in workmanship."

Neat and clean work habits show a professional attitude. Hand tools should be wiped clean of oil and grease following use. Experienced technicians usually stay considerably cleaner than beginners. Much of this cleanliness can be attributed to the use of shop towels in keeping tools clean during a job. Greasy tools tend to move grease around the shop (and into a customer's car).

Wrenches

Most screw or bolt heads are hexagonal (having six sides). The wrench size used on *hex heads* will almost

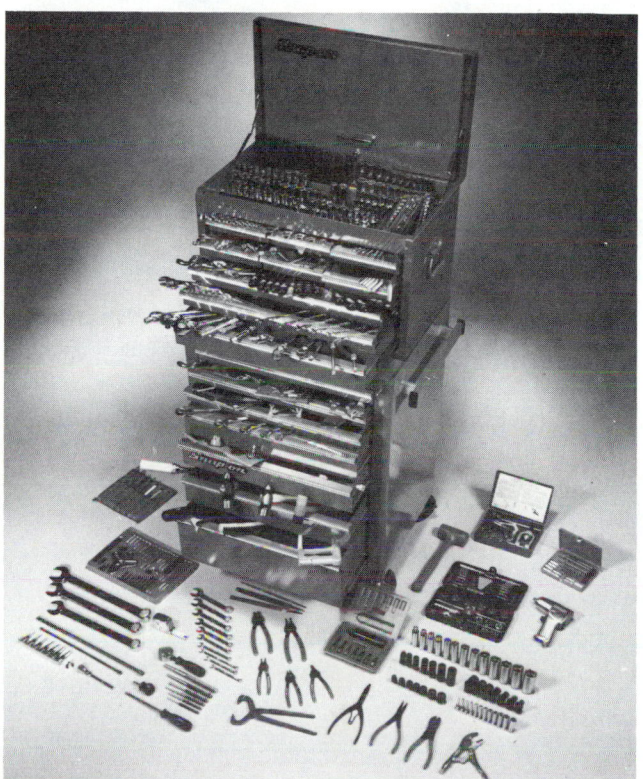

Figure 7.1 A professional toolbox. *(Courtesy of Snap-on Tools Company, Copyright Owner)*

always be either standard (British Imperial/U.S. Customary System) or metric. These measuring systems are covered in Chapter 5.

Wrenches are typically organized in racks so they can be easily located (Figure 7.2). There are several types of wrenches (Figure 7.3). The favorite among professional technicians is the *combination wrench*. Both ends of a combination wrench are the same size. One end is a *box wrench* and the other end is an *open end wrench*.

The box end of the wrench is used for loosening or tightening a fastener, while the open end is used for turning it. The box end of a combination wrench is angled to provide clearance for hands (Figure 7.4). The open end of the wrench is angled 15° so that it can be flipped over in tight quarters to provide access to the flats on the head of a nut or bolt (Figure 7.5).

Box wrenches are either *6-point* or *12-point* (Figure 7.6). The best choice for a box wrench is a 12-point because it is more versatile, fitting on the head of a bolt in more positions. Moving the wrench to a different position requires only 30° of movement of the wrench head. A 6-point box-end is more difficult to use because it must be moved 60°. A drawback to twelve-point wrenches and sockets is that they round off bolt heads more easily.

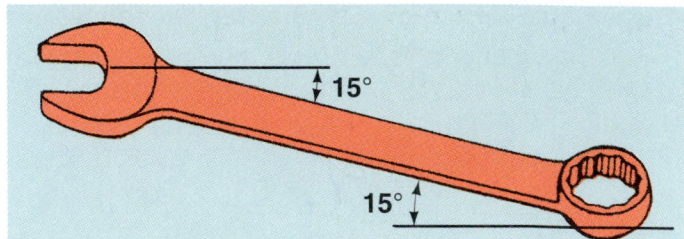

Figure 7.4 On a combination wrench, the box end is angled up and the open end is angled to the side. *(Courtesy of Snap-on Tools Company, Copyright Owner)*

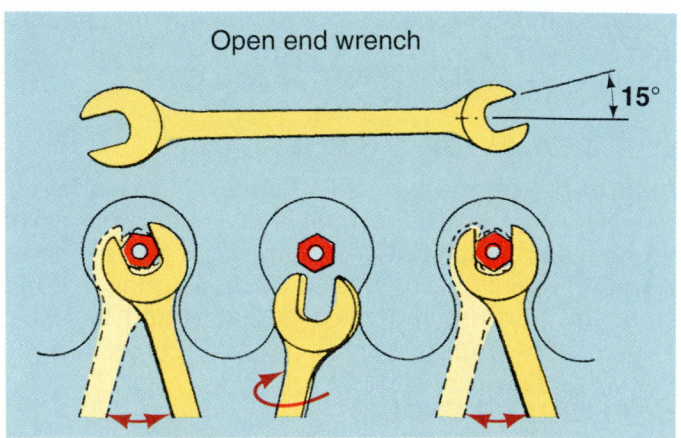

Figure 7.5 How to use an open end wrench. *(Reproduced by permission of Deere & Company, ©1996. Deere & Company. All rights reserved)*

Figure 7.2 Wrenches are organized so they can be easily located. *(Photo provided by Mac Tools and is a representative of products manufactured by Mac Tools or products obtained from another source)*

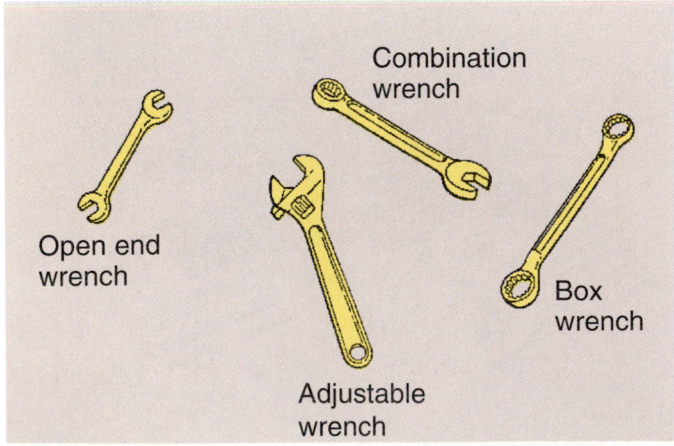

Figure 7.3 Common wrenches for bolts, screws, and nuts. *(Reproduced by permission of Deere & Company, ©1996. Deere & Company. All rights reserved)*

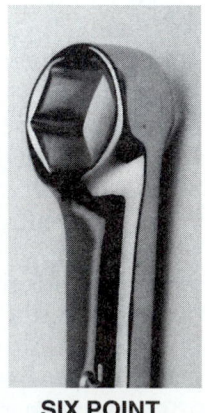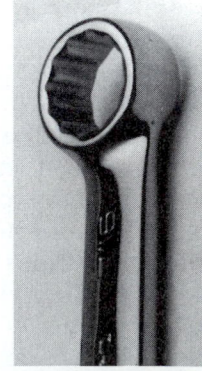

Figure 7.6 Box wrenches are either 6-point or 12-point. *(Courtesy of Snap-on Tools Company, Copyright Owner)*

A *flare-nut wrench,* or *line wrench,* is a special box wrench used for tightening or loosening the fittings on "flared" fuel or brake lines. These hollow fittings are called **flare nuts** (Figure 7.7a). The flare-nut wrench has an opening on one of its sides that allows it to slip over a fuel line (Figure 7.7b). Flare nuts often become rounded off whenever a common open end wrench is used instead of a flare-nut wrench. Two wrenches must be used when tightening or loosening a flare fitting so that the fuel line or a part is not damaged (Figure 7.7c).

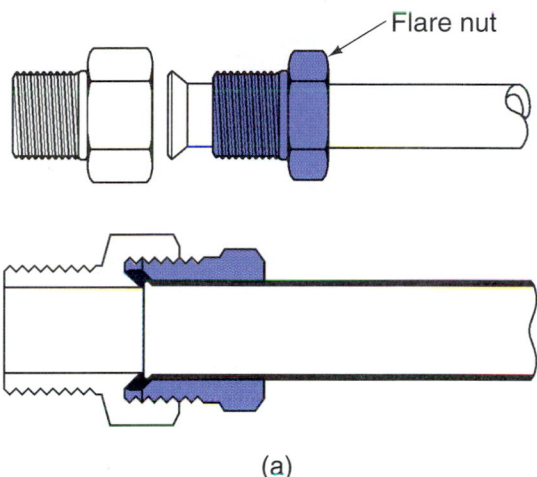

(a)

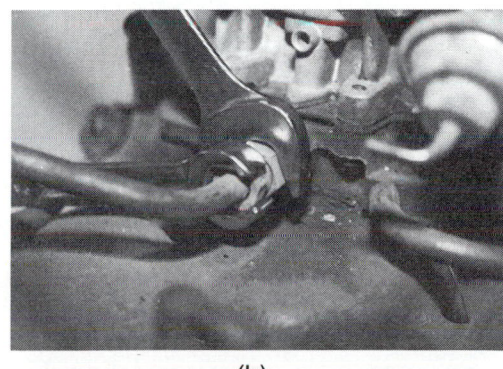

(b)

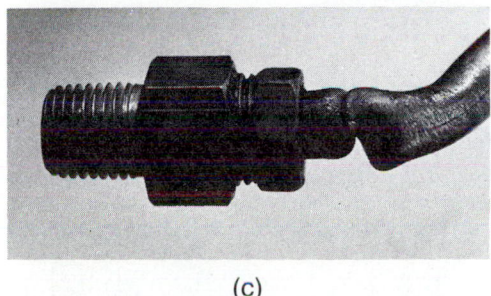

(c)

Figure 7.7 (a) Tubing fittings are called flare nuts. (b) A flare nut wrench used with an open end wrench to loosen a fuel line. (c) This fuel line was damaged when the apprentice tried to loosen the flare fitting without using two wrenches. *(a, Courtesy of Weatherhead Division of Dana Corporation)*

A *ratcheting box wrench* (Figure 7.8) is sometimes used for loosening or tightening special applications such as muffler clamps or tie-rod-end clamps on the steering linkage. This tool is especially handy when a nut is used on a very long bolt, where a deep socket would not be deep enough. The highest quality ratcheting box wrenches are very compact and can fit into tight spots. To change the direction of rotation, turn the wrench over and use it the opposite way.

An *adjustable wrench* (Figure 7.9), commonly called a *crescent wrench* after the original inventor of the tool, has one movable jaw. Different sizes are available. The medium-sized wrench is often used to adjust wheel bearings and perform other jobs on parts where substantial torque is not used. An adjustable wrench is a handy tool to keep in a vehicle for emergencies, but is hardly ever the tool of choice when working on automobiles. The tool has a tendency to slip and round off the head of a bolt and it is too bulky to fit easily into tight spaces. Figure 7.9 shows the right and wrong ways to use an adjustable end wrench.

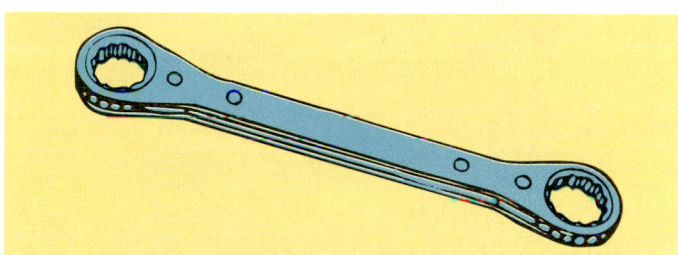

Figure 7.8 A ratcheting box wrench. *(Courtesy of Snap-on Tools Company, Copyright Owner)*

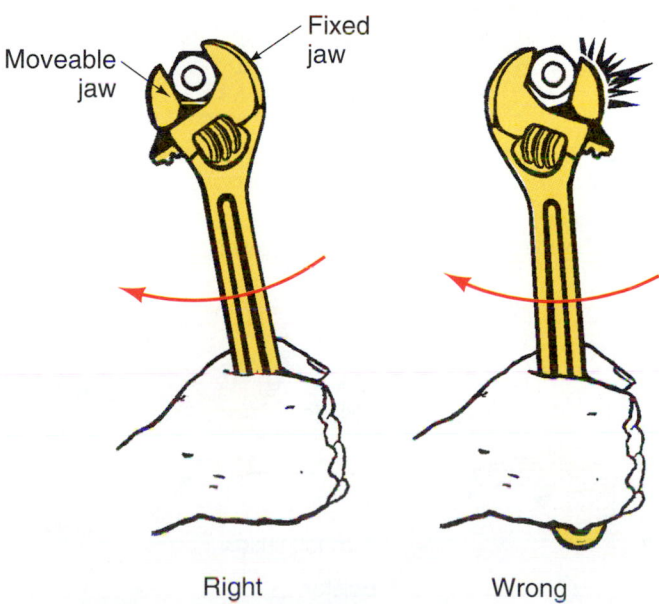

Right   Wrong

Figure 7.9 The right and wrong ways to use an adjustable end wrench. *(Courtesy of General Motors Corporation, Service Technology Group)*

An *Allen wrench* is a hex tool used to turn set screws and capscrews. An Allen wrench can be L-shaped, or it can be mounted on a socket driver. Figure 7.10 shows several types of socket drivers.

Screwdrivers

Screwdrivers are available with several different styles of heads (Figure 7.11). The two most common screwdrivers are the standard *flat* tip blade (Figure 7.12) and the *Phillips* head (Figure 7.13).

Be sure to select a screwdriver that is the correct thickness and width for the slot in the screw head. Using a screwdriver with the wrong size blade can ruin the head of a screw. The blade on a standard screwdriver should fill up the slot in the screw head. A worn screwdriver blade can be reground on the grinder. Grind in the proper direction to maintain the squareness of the blade (Figure 7.14). Be sure to quench often.

NOTE: *It is important that the blade be quenched often during grinding so the hardness of the metal is not changed.*

Some tool manufacturers warranty their screwdriver blades, so a worn or damaged one can be exchanged for a new one.

A Phillips head screwdriver pushes on four areas, rather than two like the standard flat tip. With a Phillips head, its number denotes its size. A #1 and a #2 are the most common sizes used in automobile repair. A #1 Phillips head is smaller than a #2.

SHOP TIP If a Phillips head is slipping in the screw slot, try putting some valve lapping compound on it.

A screwdriver that looks very similar to the Phillips is the *Reed and Prince*. Do not use a Reed and Prince on automotive screws. It has a deeper screw slot and the walls that separate the slots are tapered. It was developed for the furniture industry because the standard flat tip screwdriver would tend to slide off and cut

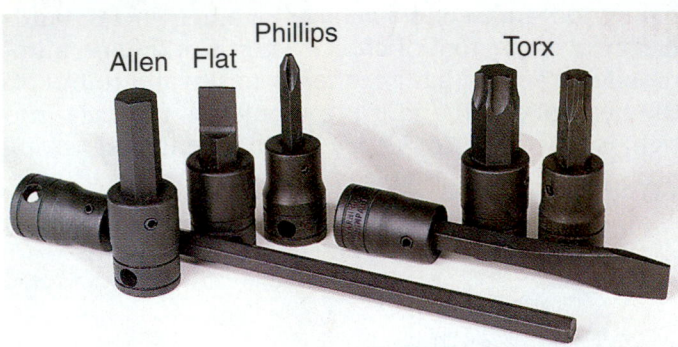

Figure 7.10　A variety of socket drivers. *(Courtesy of Snap-on Tools Company, Copyright Owner)*

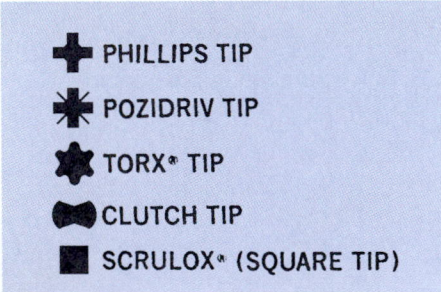

PHILLIPS TIP
POZIDRIV TIP
TORX® TIP
CLUTCH TIP
SCRULOX® (SQUARE TIP)

Figure 7.11　Miscellaneous screwdriver tips. *(Courtesy of Snap-on Tools Company, Copyright Owner)*

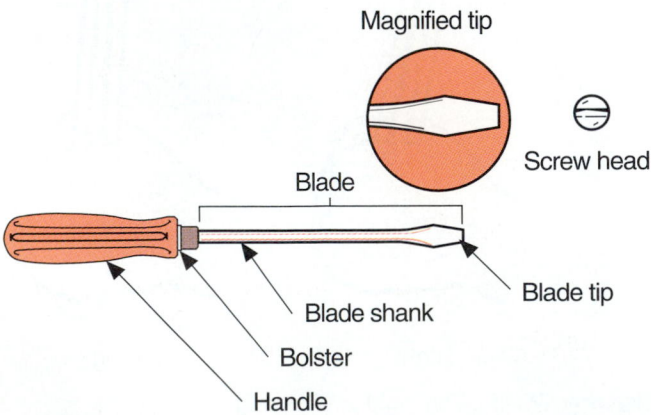

Magnified tip
Screw head
Blade
Blade shank
Blade tip
Bolster
Handle

Figure 7.12　A standard flat tip screwdriver.

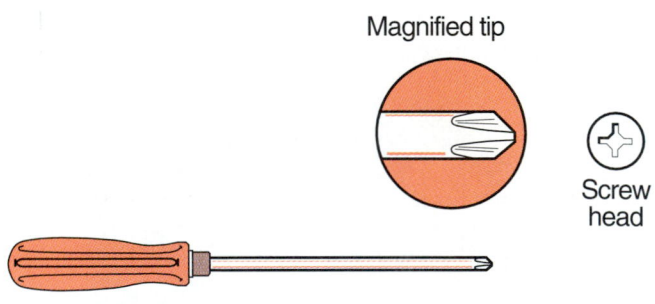

Magnified tip
Screw head

Figure 7.13　A Phillips head screwdriver.

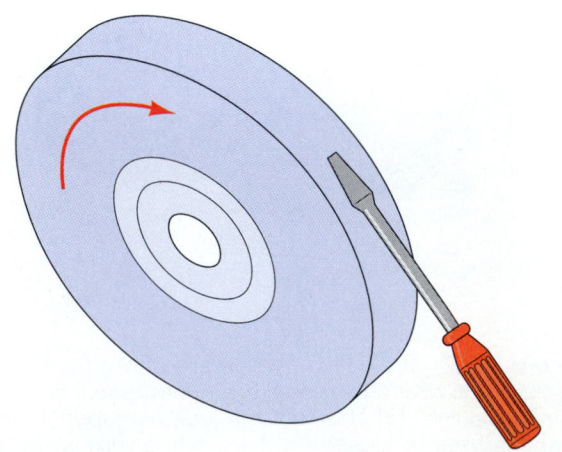

Figure 7.14　The proper way to grind a screwdriver.

upholstery. The Reed and Prince is also used in IBM equipment.

A *Pozidriv* is also similar to a Phillips head, but it has little wings on the inside that provide a slightly better grip and hold onto fasteners for easier assembly work. Because Phillips and Pozidriv screws are different, usually a Pozidriv's handle is a different color so it is not confused with a Phillips. If two screwdrivers from the same manufacturer have different colored handles, this could be a clue that something is different about them.

In 1980, there were two billion Phillips screws used in automobile production. Because the Phillips only pushes on four areas, tool bit breakage was common. In the early eighties, the industry changed to *Torx* head fasteners (Figure 7.15). The Torx design provides greater contact with the head of the screwdriver and less chance of stripping the screw head. It was first used on *door strike plates* in the early 1960s. Torx can be either internal drive (like a screwdriver) or external drive (like a wrench).

A *clutch head* or *butterfly* screwdriver is found on some older vehicle body parts, such as door latches.

A *Scrulox* screwdriver has a square tip to fit screws found on campers, boats, and truck bodies.

Screwdrivers come in many lengths, from *stubby* (very short) to extra long (sometimes more than 2 feet long). A long screwdriver is usually easier to use, given sufficient access space. Offset screwdrivers are used when space does not permit the use of a stubby screwdriver.

NOTE: *Do not use a screwdriver as a prybar, unless it was designed for such use.*

The handle on a good screwdriver is usually large and comfortable. Handles on premium screwdrivers are made of hard plastic that will survive a lifetime of use. Some tool manufacturers provide replacement blades while reusing the old handle.

An *impact screwdriver* is very effective in removing very tight screws (Figure 7.16). The tool is held against the screw slot while twisting it in the direction desired. When the tool is struck with a hammer, the spring-loaded tool breaks the screw loose. The $^3/_8$" socket drive end makes it possible to use the set's screwdriver bits with a $^3/_8$" drive tool.

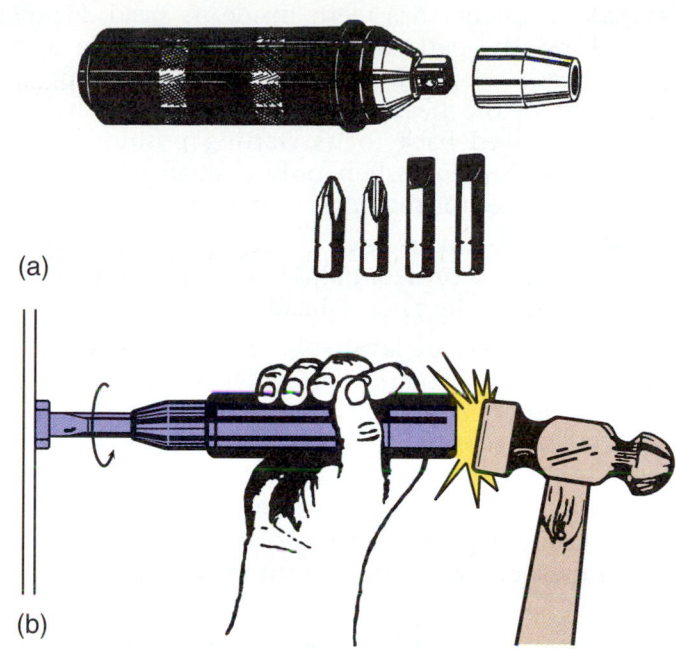

(a)

(b)

Figure 7.16 (a) An impact screwdriver set. (b) An impact screwdriver turns automatically when struck with a hammer. *(a, Courtesy of K-D Tools; b, Courtesy of Lisle Corporation)*

Sockets, Ratchets, and Breaker Bars

Whenever possible, a socket is used instead of a wrench because it is faster. A socket is used with either an *air impact wrench* (see Figure 7.44) a *ratchet* (Figure 7.17), or a *flex handle* (breaker bar). All of these tools allow for a change in direction so that a bolt can be either loosened or tightened.

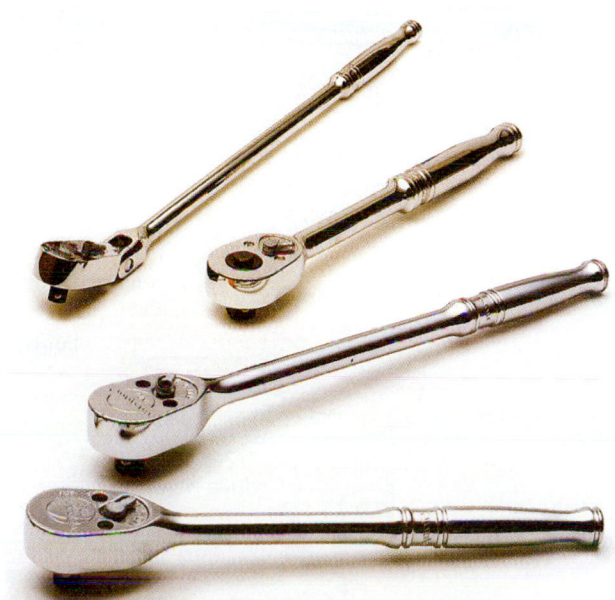

Figure 7.17 *An assortment of ratchets. (Courtesy of Snap-on Tools Company, Copyright Owner)*

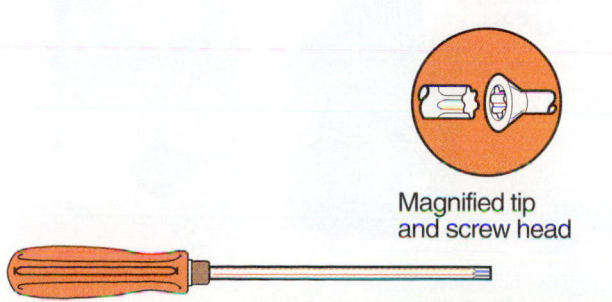

Magnified tip and screw head

Figure 7.15 A Torx screwdriver. *(Courtesy of Stanley Works)*

Ratchet. A ratchet has teeth inside its head. Figure 7.18a shows the parts that make up the inside of a ratchet head. The teeth allow for a small amount of movement of the head of a bolt before the ratchet handle is moved back to its starting position. This allows for tightening when only a small amount of room is available for the ratchet handle to move.

Flex Handle (breaker bar). When a nut or bolt is extremely tight, a breaker bar (Figure 7.18b) is used so that the teeth in the ratchet head are not damaged.

Sockets. Sockets are available in many sizes and styles. There are two size designations: one refers to the square ratchet *drive end* and the other refers to the *hex socket end* that fits the bolt head (Figure 7.19). A typical toolbox will contain sets of both *standard* (fractional) and *metric* bolt head sockets. The drive end is always an inch size, whether a socket hex end is standard or metric. That way different ratchets and exten-

sions are not required when switching from a standard to a metric size.

Socket Drive Sizes. The most common drive sizes used on automobiles are the ¼", the ⅜", and the ½". The ⅜" drive is the most versatile for automobile work. A set of ⅜" drive sockets usually includes hex (bolt head) sizes ranging from ⅜" to ¾".

Socket Adapters. Occasionally, a ratchet of a different drive size is needed but is not available, so a *socket adapter* is used (Figure 7.20). For instance, a ⅜" to ¼" drive adapter might be used to drive a ⁵⁄₁₆" socket head because a ⁵⁄₁₆" socket would not normally be found in a ⅜" drive socket assortment (which usually includes ⅜" to ¾" only).

> **CAUTION** Adapters are often used with driving tools that are too large for the socket being driven (a ½" ratchet driving a ⅜" socket, for instance). Be certain to use the correct torque to avoid breaking a tool.

Plug Drivers. Do not confuse a socket adapter with a plug driver. The plug driver is solid. An adapter, which has a spring loaded ball to hold the socket, will probably break if it is used to loosen a rusted gallery plug (which has a tapered locking thread). There are many fittings that use threaded tapered pipe plugs, especially in engines. Plug drivers (see Figure 7.20) are used here.

Extension Bar. When driving a socket with a ratchet or an air wrench, an *extension* is usually used (Figure 7.21). An extension allows the technician to have his or her hands clear of parts, preventing injury. Common sizes of extensions are 1", 3", 6", and 10". Very long extensions (two or three feet) are used with a swivel socket in transmission **R&R** (remove and replace) jobs and other specialized applications. One of the extensions shown in Figure 7.21 shows a *wobble drive*

(a)

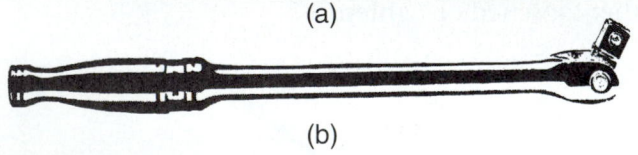

(b)

Figure 7.18 (a) Internal parts of a ratchet head make up a ratchet repair kit. (b) A breaker bar prevents the need for a ratchet repair kit. *(Courtesy of Snap-on Tools Company, Copyright Owner)*

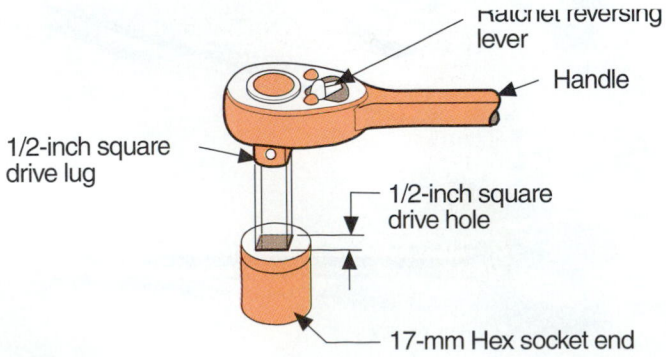

Ratchet reversing lever

Handle

1/2-inch square drive lug

1/2-inch square drive hole

17-mm Hex socket end

Figure 7.19 A ½" drive ratchet used with a 17-millimeter hex socket.

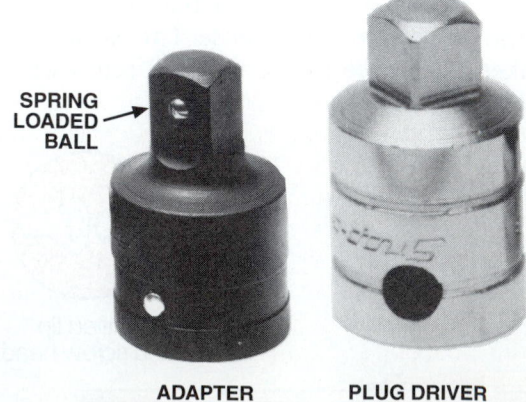

SPRING LOADED BALL

ADAPTER   PLUG DRIVER

Figure 7.20 A ½" to ³⁄₈" socket adapter and a male pipe plug driver. *(Courtesy of Snap-on Tools Company, Copyright Owner)*

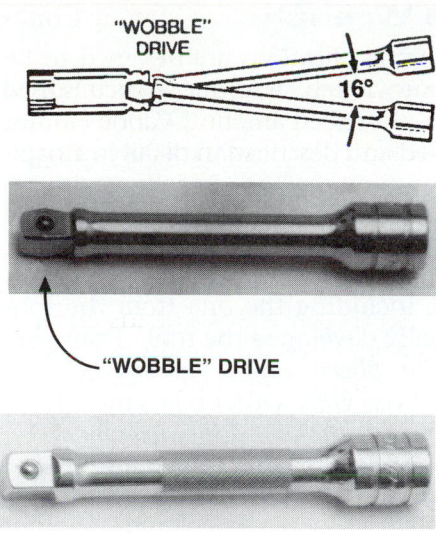

Figure 7.21 Standard and wobble extensions. The wobble allows the socket to pivot 16°. *(Courtesy of Snap-on Tools Company, Copyright Owner)*

that allows the socket to pivot 16° to allow easier access to bolts.

Socket Differences. When the proper tools are used, a professional will rarely ruin a fastener. Like wrenches, the bolt end of most sockets is *6- or 12-point*. A 6-point socket is the tool of choice because it is less likely to "round off" a fastener head. *Eight-point* sockets are available for driving square heads. Sockets are kept in order according to size in the same drawer or in the top of the toolbox so that they can be easily located when doing a job (Figure 7.22). Socket clip rails and trays are used to keep things in order.

Deep Sockets. A set of *deep (bolt-clearance) sockets* is usually part of a complete tool set, too. Figure 7.23 shows how a deep socket provides clearance when removing a nut from a long stud.

Special Use Sockets

A popular tool is a *swivel socket* that is an integral part of an individually sized socket. This tool is commonly known as a *wobble socket*. It is a much better choice than a universal swivel, which is used between a regular socket and a ratchet or extension so that it can pivot. Wobbles that are designed to be used with air-powered impact wrenches are very popular (see Figure 7.44)

Special, thicker, *impact sockets* are available for use with air tools (see Figure 7.45). Also, while standard sockets are chrome plated, impact and industrial sockets are not. A regular socket used with an air tool can result in a broken socket.

A *crowfoot socket* is a wrench head that has been adapted so that it can be used with a ratchet and extension. Some crowfoot wrenches are open-end and some

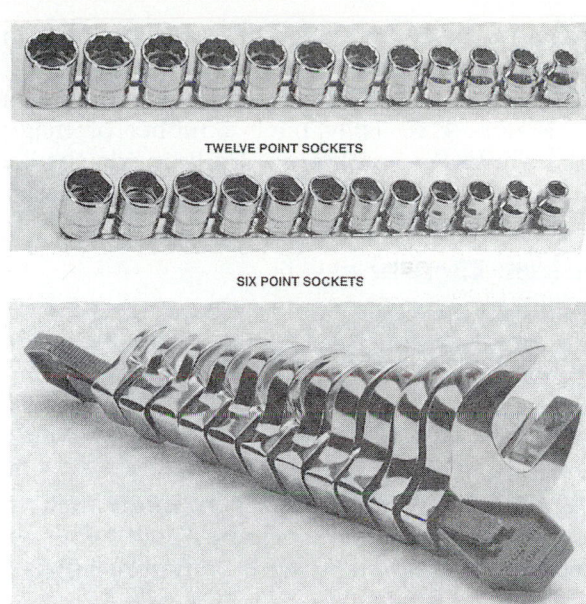

CROWFOOT SOCKETS ON A WRENCH HOLDER

IMPACT SOCKETS

EIGHT POINT SOCKETS

Figure 7.22 Sockets are kept in order so they can be easily located when working. *(Courtesy of Snap-on Tools Company, Copyright Owner)*

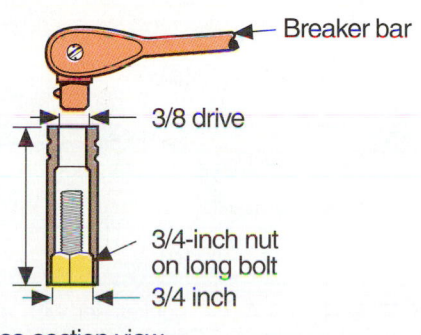

Figure 7.23 A deep socket used with a breaker bar to remove a nut from a long stud.

are flare-nut box wrenches. They can be stored in an organized fashion using the holder shown in Figure 7.22.

Spark plug sockets for automobiles come in two sizes, ⅝" or ¹³⁄₁₆". They have a rubber or magnetic insert to hold the spark plug firmly inside the socket because spark plugs are ceramic and can easily be cracked if dropped.

Socket Drivers

Socket drivers are very popular and are used for a variety of applications (see Figure 7.10). These tools, which can be driven either by a ratchet or air impact wrench, are available with various heads such as *hex (Allen head), Phillips, Pozidrive, clutch head (butterfly), torx, and the standard flat tip.*

A popular driver for small work is one that resembles a screwdriver handle (Figure 7.24a). The driver, sometimes called a *nutdriver*, is commonly used with small ¼" drive sockets.

Another popular driver is a *speed handle* (Figure 7.24b). It is an excellent tool for bench work on engines and transmissions. While disassembling these parts with an air tool is popular, re-assembly with an air tool is risky. A speed handle is almost as fast as using an air tool, but it minimizes the possibility of cross-threading or overtightening a fastener.

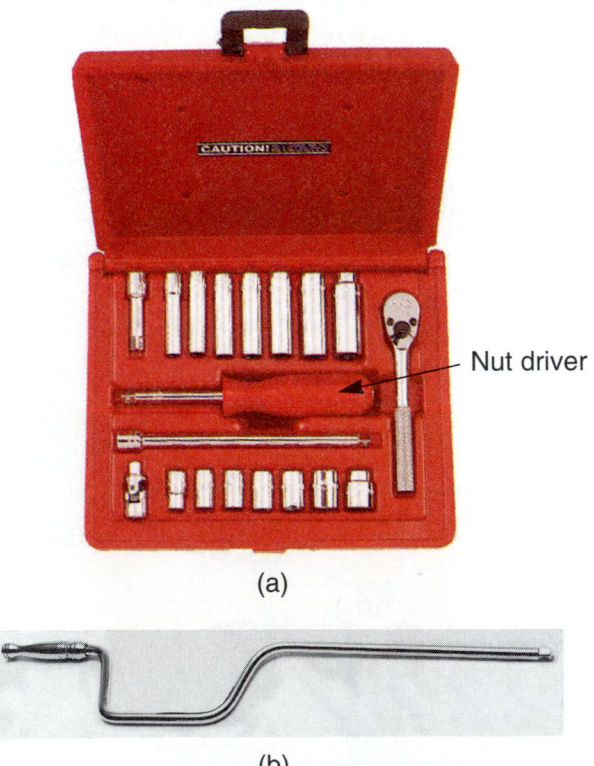

(a)

(b)

Figure 7.24 (a) A ¼" socket set with a nutdriver handle. (b) A speed handle. *(a, Photo provided by Mac Tools and is representative of products manufactured by Mac Tools or products obtained from another source; b, Courtesy of Snap-on Tools Company, Copyright Owner)*

Torque Wrench

Most fastener applications are designed to be tightened with a torque wrench. A torque wrench is used to tighten a fastener a prescribed amount. Various torque wrenches are illustrated and described in detail in Chapter 41.

Pliers

Pliers come in many types and sizes (Figure 7.25). It is not uncommon for the same tool to have more than one name, including the one from the tool company that originally developed the tool. Examples of this are the *rib joint pliers*, called *channel-locks*, and *locking pliers*, called *vise grips*. Other pliers include *cutting pliers*, *needle nose*, and *slip-joint utility (combination)* pliers.

Pliers are often misused for tightening and loosening fasteners. The result is most often a rounded-off fastener head. The proper use of a plier is dictated by its design. Most often a plier is for cutting, bending, or positioning a part (such as a cotter pin) or for stripping an electrical wire. The use of too small a plier for the size of the job can result in damage to the plier's jaws.

Channel-Locks. Channel-lock pliers, also called rib joint pliers, have tongue and groove rows that provide a means of jaw size adjustment. When the adjustment of the jaws is changed, they remain parallel for use.

Cutters. Cutters, such as diagonals (side cutters) are also popular in automotive work. They are used for cutting wire and removing and installing cotter pins (see Chapter 41). *Compound leverage cutters*, also called *bolt cutters*, have an extra fulcrum that increases the leverage on the jaws. Jaws on compound cutters move a smaller distance with handle movement than ordinary pliers.

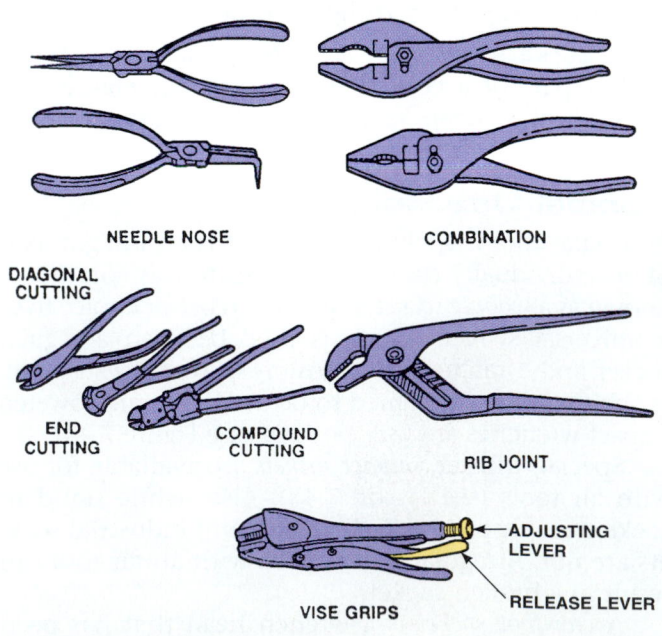

Figure 7.25 Different types of pliers. *(Courtesy of Ford Motor Company)*

Locking Pliers. Commonly called *vise grips*, locking pliers have locking jaws that are adjusted by turning a screw at the end of the plier handle (see Figure 7.25). The better quality vise grips also have a release lever. Tightly closing the jaws locks them, while pressing on the release lever releases them.

Needle Nose Pliers. *Needle nose* pliers are also called *long nose pliers* (see Figure 7.25). They are handy for positioning clips and very small parts.

NOTE: *Needle nose pliers are not designed to be used for bending. They can be damaged when overstressed.*

Special Use Pliers

Special use pliers include those used to remove and install spring-loaded hose clamps or pliers used to remove corroded battery bolts.

Lineman's Pliers. Lineman's pliers are used by electricians but are often seen in a technician's or machinist's toolbox. They have a cutting edge and gripping teeth. An area on the inside of the jaws is used to remove insulation from household electrical wiring (Figure 7.26).

High Leverage Combination Pliers. High leverage combination pliers (Figure 7.27) are also popular for cutting or gripping. They require only ¼ the normal application of effort on the handle. Much movement of the

handle results in little movement at the jaws. These are excellent pliers when a very strong grip is needed.

Snap Ring Pliers. A snap ring is used to secure a bearing to a shaft or hold it inside of a housing. Snap rings are *external* when they retain a bearing on a shaft. They are *internal* when they hold a bearing in a housing. Snap ring pliers come in several types to match the different types of snap rings available. Figure 7.28 shows three types of snap rings and some of their uses.

Hammers

Hammers come in various sizes, weights, and materials (Figure 7.29a). A common hammer, the *ball-peen* hammer, comes in several different weights. The handle should be gripped near the end and the face of the hammer should squarely contact the object being struck.

Soft-faced brass and *soft plastic* hammers are also popular. One type of soft plastic hammer has a cavity in its head that is filled with metal shot (Figure 7.29b). This hammer, called a *dead-blow* hammer, will not bounce back upon impact.

Chisels and Punches

Chisels and punches, used with a hammer, have different applications. A chisel is used for cutting metal

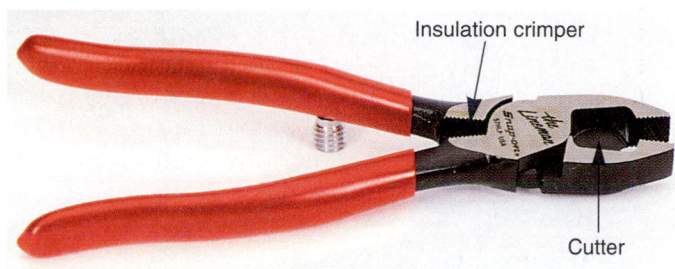

Figure 7.26 Lineman's pliers. *(Courtesy of Snap-on Tools Company, Copyright Owner)*

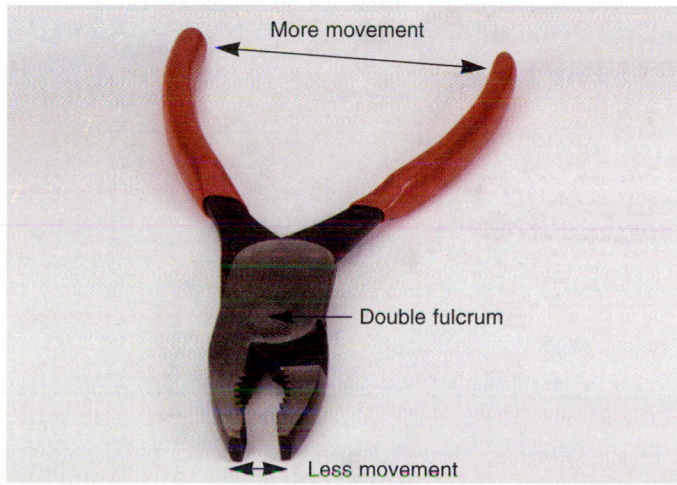

Figure 7.27 High leverage combination pliers. *(Courtesy of Snap-on Tools Company, Copyright Owner)*

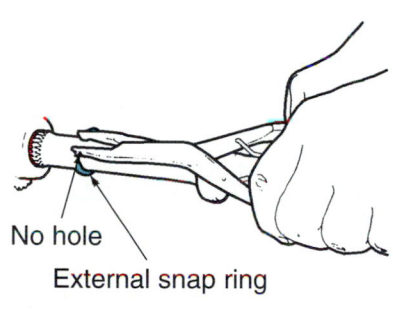

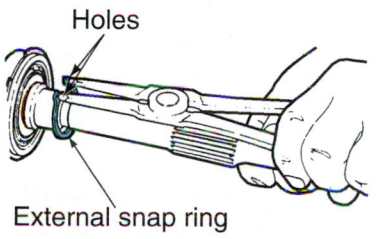

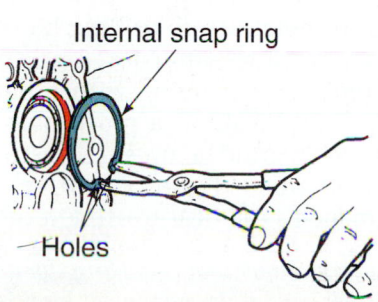

Figure 7.28 Types of snap ring pliers. *(Courtesy of Ford Motor Company)*

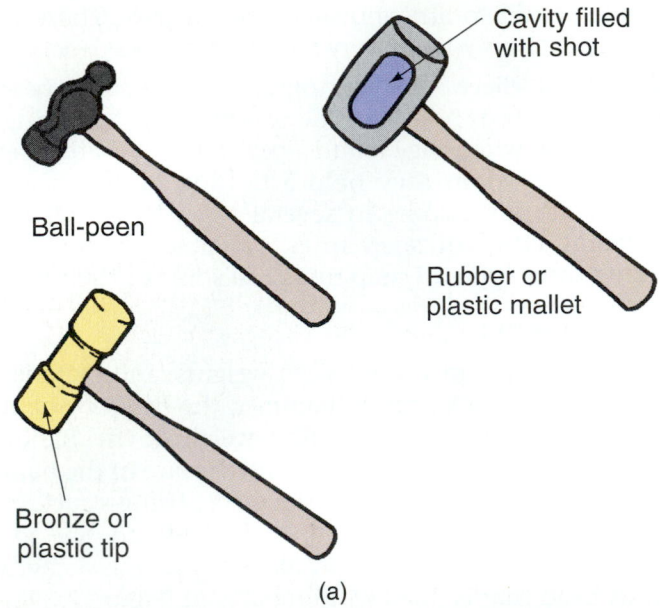

Cavity filled with shot

Ball-peen

Rubber or plastic mallet

Bronze or plastic tip

(a)

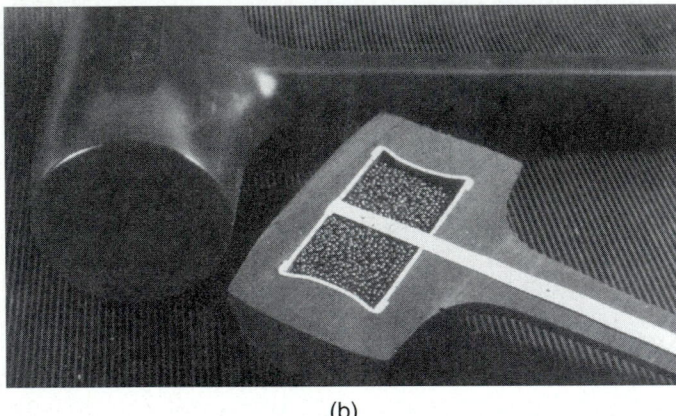

(b)

Figure 7.29 (a) Several types of hammers. (b) A dead blow hammer has a cavity full of metal shot. *(a, Courtesy of Ford Motor Company; b, Courtesy of Snap-on Tools Company, Copyright Owner)*

while a punch is used to drive a pin or key, or to make an indentation in metal. Several shapes of chisels are shown in Figure 7.30.

Various types of punches are shown in Figure 7.31. A starting punch is used to start driving a pin or rivet from a hole. The pin punch is then used to finish driving it out. A center punch is used to make an indentation in metal before drilling a hole.

Punches and chisels often develop a "mushroomed" head from repeated pounding with a hammer (Figure 7.32a). Chips can break off a mushroomed tool head, causing injury. Regrind a mushroomed tool head to maintain the chamfer (Figure 7.32b), quenching often to avoid damage to the heat treatment of the tool.

NOTE: *Using a hammer that is too large for a punch or chisel can result in chipping of the punch or chisel head. According to Snap-on Tools, a hammer head should be no more than ⅜" larger in diameter than the head of the punch or chisel.*

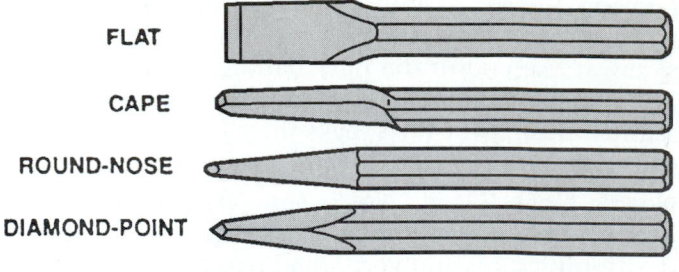

FLAT

CAPE

ROUND-NOSE

DIAMOND-POINT

Figure 7.30 Types of chisels. *(Courtesy of Ford Motor Company)*

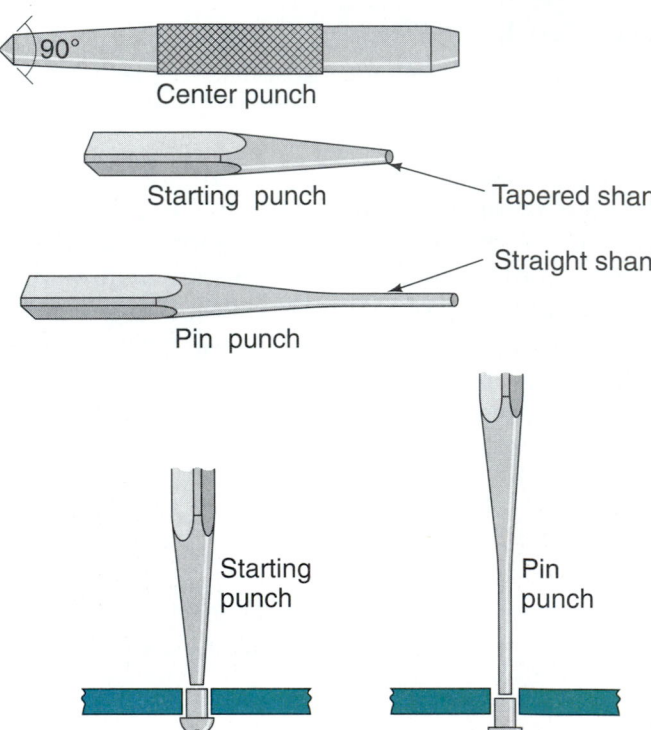

90°

Center punch

Starting punch

Tapered shank

Straight shank

Pin punch

Starting punch

Pin punch

Figure 7.31 Types of punches. *(Courtesy of Ford Motor Company)*

The cutting edge of the chisel must be maintained in sharp condition (Figure 7.32c). The edge is ground to a slight radius (curve).

The chisel can be held with a holder when cutting metal (Figure 7.32d).

 **SAFETY NOTE** Be sure to always use eye protection when using hammers and chisels.

Hacksaws

A hacksaw is used to cut metal. The following are some tips on correct hacksaw usage.

- The blade cuts on the forward stroke only. Applying pressure during the return stroke will dull the blade. Be sure the blade is properly installed with

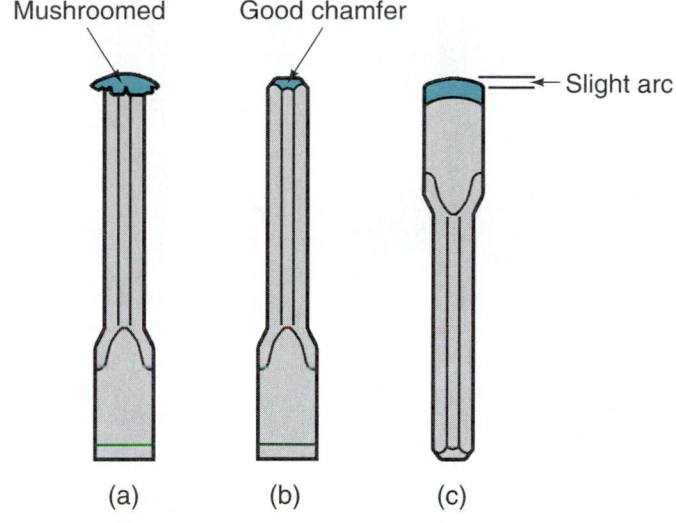

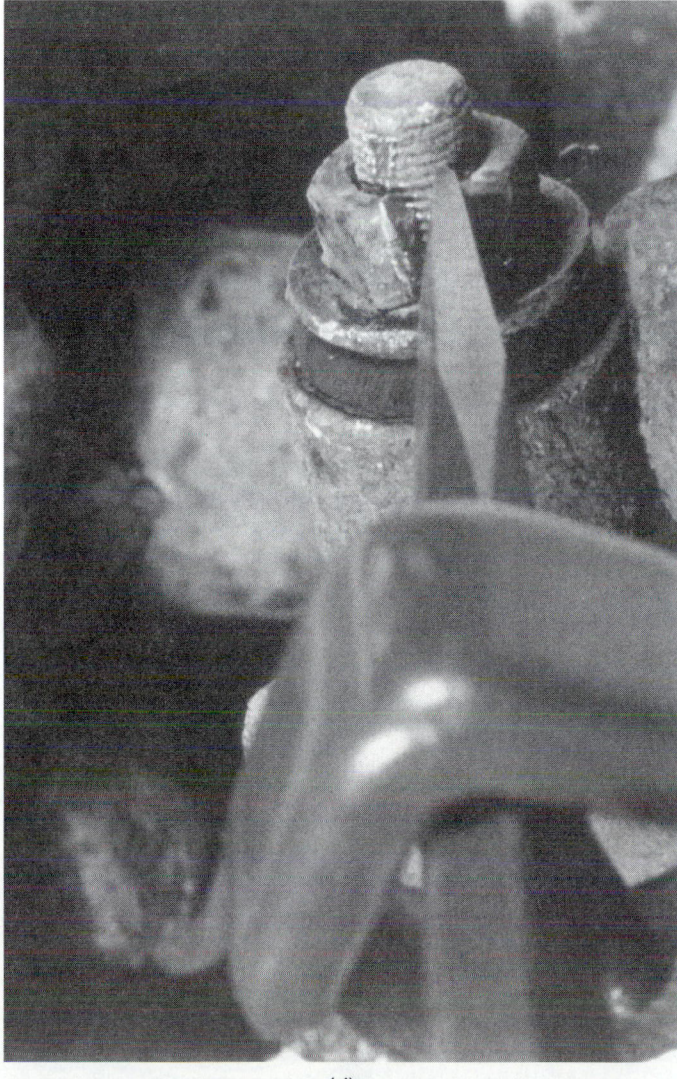

Figure 7.32 Chisel maintenance; (a) A mushroomed chisel head. (b) The head should have a chamfer. (c) The cutting edge of the chisel has a slight radius. (d) A chisel can be held with a chisel holder when cutting metal. *(Courtesy of Snap-on Tools Company, Copyright Owner)*

the teeth facing away from the handle (Figure 7.33a) so that they cut during the forward stroke.

- Use the entire blade when cutting so that the entire blade gets a chance to wear evenly.
- When cutting steel, cutting oil will make cutting easier and extend the life of the blade.
- Some hacksaw blades have fewer teeth for faster cutting of thick metals.
- For sheet metal, the hacksaw blade must have more teeth so that the sheet metal does not drop between the teeth, damaging them (Figure 7.33b). The blade should have at least two teeth in contact with the metal at all times (Figure 7.33c).

Files

Files are used to shape, roughen, or smooth metal. They vary in size and roughness and can be of either a round, half-round, square, triangular, or flat shape (Figure 7.34a).

- To avoid puncture wounds, a file should always be used with a handle on the tang.
- A file cuts on the forward stroke. Push down gently on it when filing. Then, release pressure on the return stroke.
- Filing softer metal, like aluminum or brass, calls for a coarser file.
- For filing steel, use a fine file.
- From roughest to finest, the names of file cuts are *bastard cut, second cut,* and *smooth.*

The diagonal rows of teeth on a file are either *single cut* or *double cut.* A double cut file has two rows of teeth that cross each other on an angle.

Filings are the pieces of metal that come off of the workpiece during filing. If they are allowed to remain

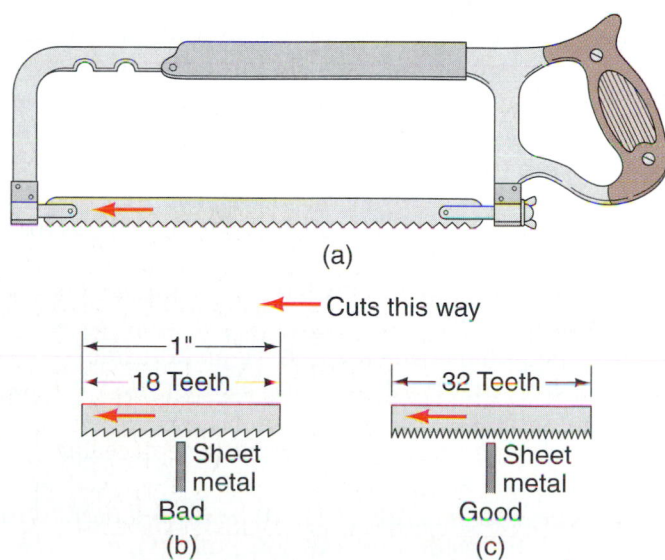

Figure 7.33 Hacksaws: (a) The blade is installed with the teeth facing forward. (b) A hacksaw blade that is too coarse will not work well with sheet metal. (c) A fine toothed blade is used with thin metal.

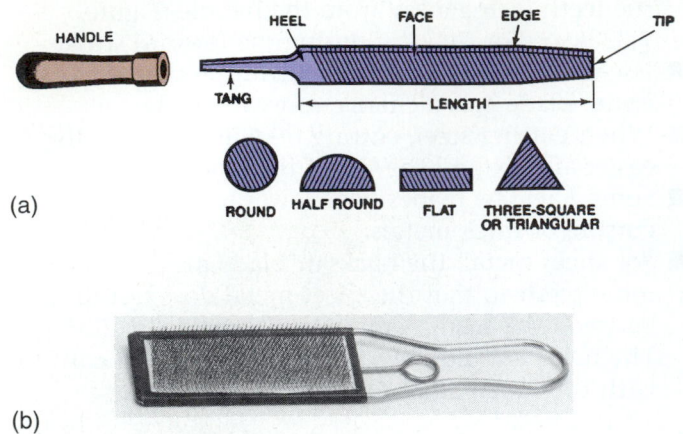

(a)

(b)

Figure 7.34 (a) Files come in several shapes. (b) A file card. *(a, Courtesy of Ford Motor Company; b, Courtesy of Snap-on Tools Company, Copyright Owner)*

trapped on the surface of the file, they can damage the filed surface and limit the file's effectiveness. Proper care of a file will prolong its life.

- A *file card* is used to clean the teeth of a file (Figure 7.34b).
- Putting chalk on a file's teeth will help avoid loading up the teeth when filing soft materials such as aluminum or brass.
- Store a file carefully so that it will not rust or be nicked by other tools.
- Keep files clean and free of oil or grease.

■ PULLERS

There are many types of *pullers* used in automotive work, ranging from small to large. Pullers are used to remove or install pressed-fit gears, bushings, bearings, or other parts from shafts. Specific uses of many of the pullers discussed here are covered in later chapters.

Pullers are either *hydraulic* or *manual*. Manual pullers have a bridge yoke (Figure 7.35a) or a bar-type yoke (Figure 7.35b). A *slide hammer* or a *pressure screw* is threaded through the *yoke*.

- A *bridge-type yoke* pushes against the outside of the part being pulled.
- Some *bar-type yoke* pullers have jaws and others use bolts or **collets**.
- Jaw-type pullers sometimes have an adjustable clamp bolt that holds the jaws against the work for a more reliable pull.
- The end of the pressure screw has a replaceable hardened point.

On many pullers, the jaws can be turned around to accommodate inside and outside pulls (Figure 7.36). Jaws of different lengths and sizes can be installed to make the puller more versatile. The size of the jaw depends on the reach and spread required for the pulling application. Pressure screws of different

BRIDGE YOKE

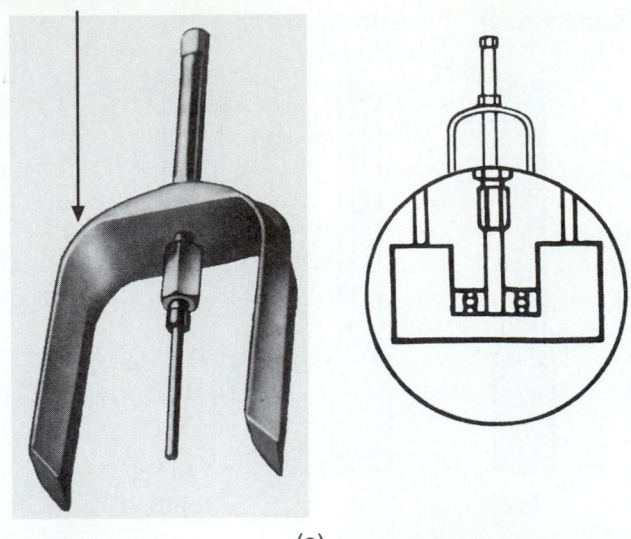

(a)

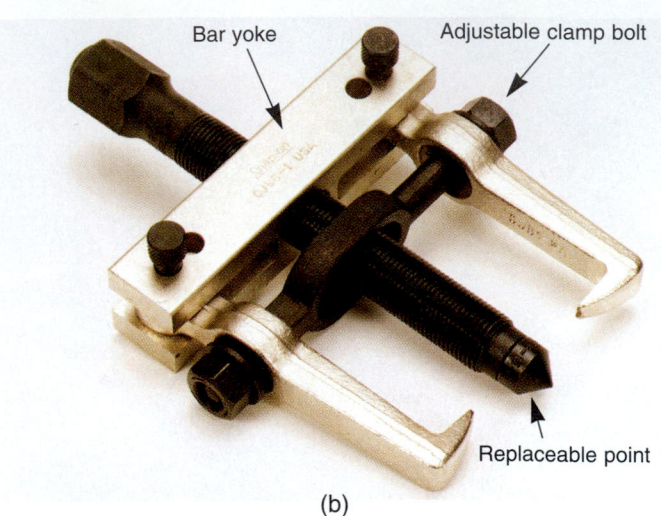

(b)

Figure 7.35 Pullers. (a) A bridge-yoke puller. (b) A bar-yoke puller. *(Courtesy of Snap-on Tools Company, Copyright Owner)*

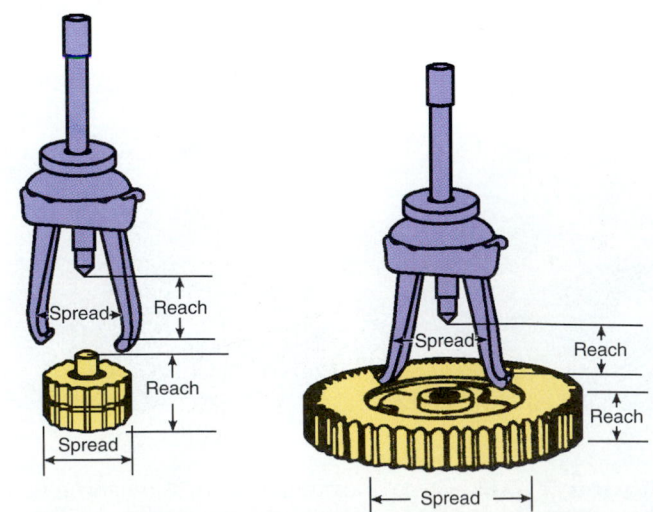

Figure 7.36 Puller jaws installed for inside and outside pulls. *(Courtesy of Snap-on Tools Company, Copyright Owner)*

lengths are installed to make it possible for the puller to use different length jaws.

Some pullers use bolts that are screwed into threaded holes in a gear or pulley (Figure 7.37). Others use bolts screwed into parts such as a vibration damper on the front of the crankshaft (Figure 7.38). The vibration damper puller is also called a *flange-type puller*. It can also be used for pulling steering wheels. It works on parts that have either two or three threaded holes. Slotted holes in the puller body allow it to accommodate many different size parts.

A *slide hammer puller* is another common tool. The ones shown in Figure 7.39 can be used to pull a bear-

ing from a bored hole that it is pressed into. They are adjustable with either a collet or puller jaws. To remove the bearing, slide the weight back hard until it hits the flat area on the rear of the puller.

A larger puller is the hydraulic-type puller, often called a *porta-power* (Figure 7.40).

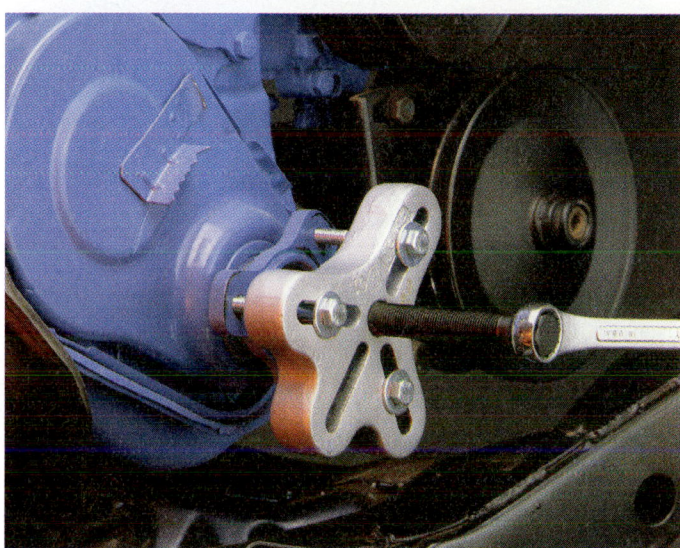

Figure 7.38 A flange-type puller can be used to pull a vibration damper or a steering wheel. *(Courtesy of OTC-SPX Corp. Aftermarket Tool & Equipment Group)*

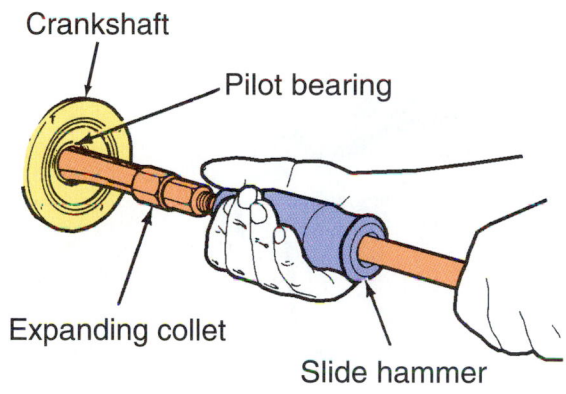

Crankshaft
Pilot bearing
Expanding collet
Slide hammer

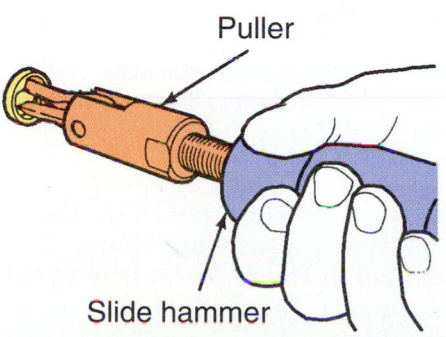

Puller
Slide hammer

Figure 7.39 Two types of slide hammer pullers. *(Courtesy of Ford Motor Company)*

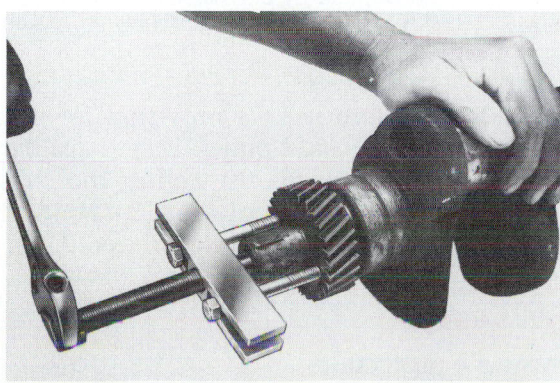

Figure 7.37 A gear and pulley puller uses bolts that thread into tapped holes in the part. *(Courtesy of OTC-SPX Corp. Aftermarket Tool & Equipment Group)*

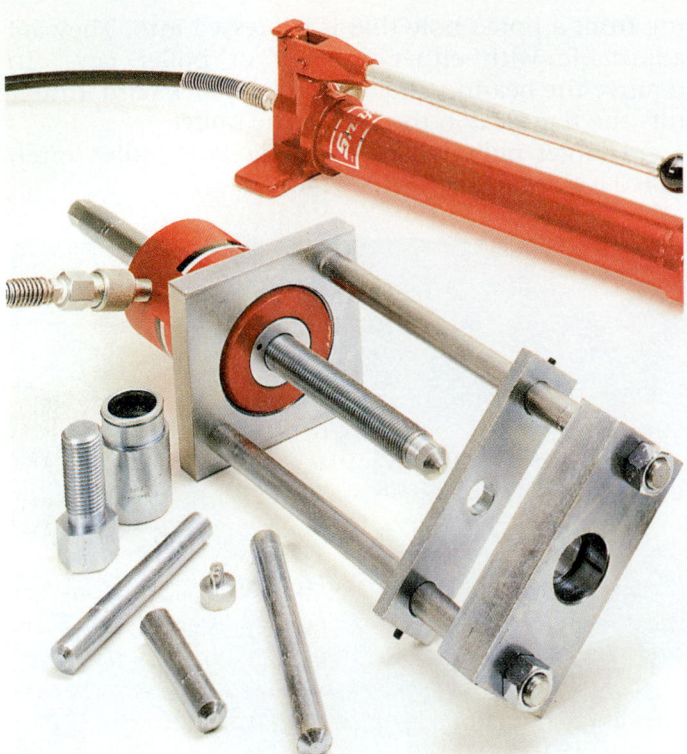

Figure 7.40 A hydraulic puller. *(Courtesy of Snap-on Tools Company, Copyright Owner)*

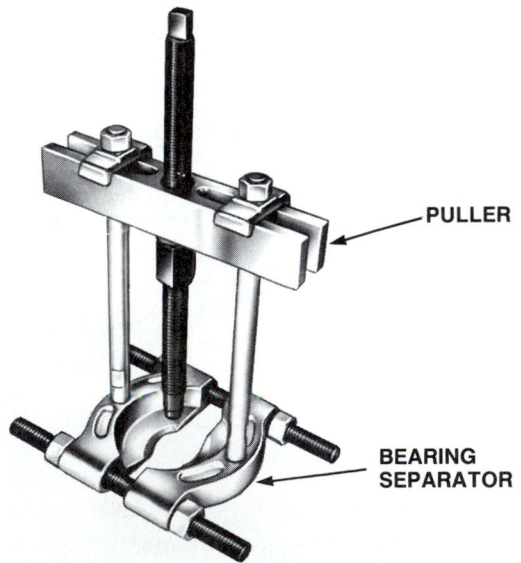

Figure 7.41 A bearing separator plate and bar yoke puller. *(Courtesy of OTC-SPX Corp. Aftermarket Tool & Equipment Group)*

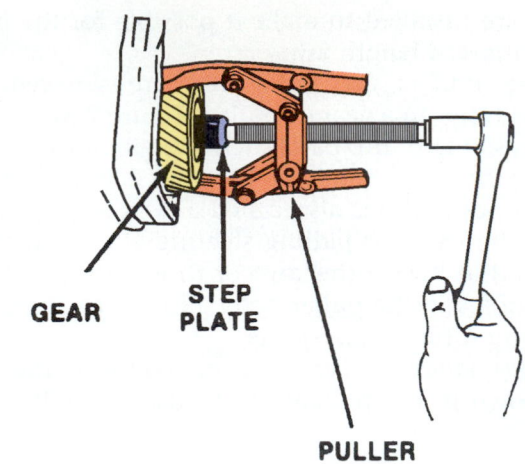

Figure 7.42 The step plate protects the end of the shaft from damage. *(Courtesy of Ford Motor Company)*

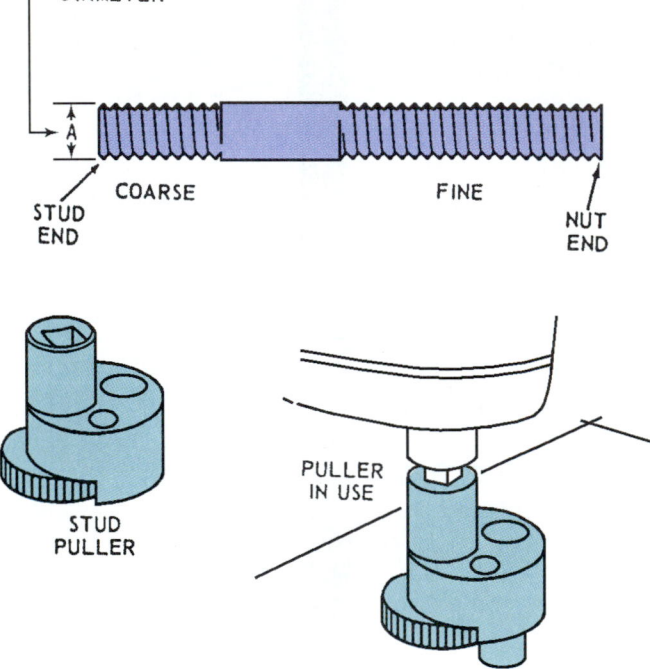

Figure 7.43 A stud puller is used to loosen and tighten studs. *(Reproduced by permission of Deere & Company, ©1996. Deere & Company. All rights reserved)*

One of the most versatile pullers is the *bearing separator* and bar-type puller setup shown in Figure 7.41. The puller used in Figure 7.42 is being used with a *step plate* to protect the end of the shaft from damage during pulling.

Stud pullers (Figure 7.43) are commonly used in engine repair. A stud is a fastener that is threaded on both ends. It therefore lacks a provision for using a wrench to turn it. The stud puller wedges against the stud (usually in an area in the center that has no threads) and loosens it. The tool works best if it is held all the way down against the work surface.

Puller Safety

1. Wear eye protection.
2. When heating a part to help free it, do not heat the jaws of the puller. This could change the temper of the metal.

3. Be sure the pressure screw is clean and lubricated before using an impact wrench.

4. Be sure the removable point is installed on the puller shaft tip.

5. Be sure the puller is aligned so it is perpendicular to the part being pulled.

6. Do not use a puller with damaged or worn parts.

7. Use the correct size puller so overloading is avoided.

8. Use a 3-jaw puller instead of a 2-jaw puller when possible.

■ AIR TOOLS

Compressed air is used in practically all automotive shops. Air tools are great timesavers for technicians. There are many air-operated tools including air drills, air hydraulic jacks, and other specialized pieces of equipment that are explained in other chapters. Smaller, air-operated hand tools are explained here.

The *air compressor* provides air at a regulated pressure of 90 to 150 psi (pounds per square inch). Air tool manufacturers recommend regulating air pressure to 90 psi to get the best performance and reliability from air tools. Chapter 11 deals with air compressors.

SAFETY NOTE Blowguns for blowing off parts are regulated so that they do not produce more than 35 psi. Rubber-tipped blowguns, used to blow into brake or automatic transmission fluid passageways or engine oil galleries, do not have this safety feature. A worker should not use these tools until proper instructions on their safe use is understood.

There are many air tools. Only the impact wrench, air ratchet, air chisel, die grinder, and air drill are discussed here.

Impact Wrenches

The air impact wrench is one of the technician's most popular tools. Impact wrenches are available in various sizes with different size socket drives. A ³⁄₈" drive impact wrench is very popular. It is generally used with a *wobble impact socket* (Figure 7.44).

CAUTION When using a wobble socket, do not turn on the impact wrench unless it is installed on a nut or bolt. The socket can fly off the impact wrench, possibly causing an injury.

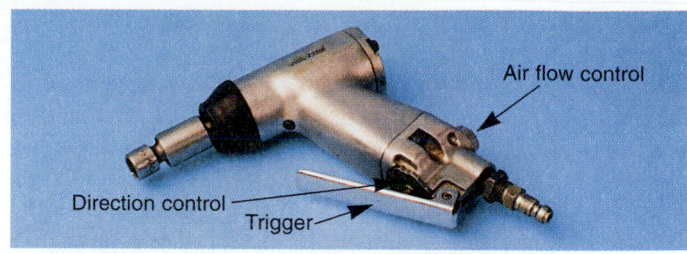

Figure 7.44 A ³⁄₈" impact wrench with a universal impact socket.

A ½" drive impact wrench is used to loosen large, very tight bolts. Special, extra-thick impact sockets must be used with the impact wrench (Figure 7.45). Regular sockets can crack or explode (see Figure 11.22).

Impact wrenches are especially useful around exhaust bolts. The "impact" caused by the tool tends to shake the nuts and bolts loose instead of twisting them off. Impact wrenches are also handy on vibration damper and flywheel bolts because the crankshaft does not need to be restrained from turning. Wheel lug nuts can also be loosened on a raised car without holding the wheel from turning.

To allow the maximum amount of air assist, the air flow control knob is turned counterclockwise to open a passageway. When an especially tight bolt is encountered, it might be helpful to put an extra amount of air-motor oil into the wrench through the air inlet to help seal the vanes in the tool.

Some air systems have automatic oilers to provide lubrication for air tools. Electric impact wrenches are also available.

Always use approved impact sockets, not chrome sockets. Be sure that the socket is secured to the air tool. A clip at the end of the tool's square drive can become worn so that it no longer holds the tool. When the impact wrench fails to loosen a fastener, use a large breaker bar.

NOTE: *For larger, extremely tight fasteners, use a 1" impact wrench or a planetary gear head multiplier to gain additional leverage.*

Gear multipliers with a ½" input (female) end and a ¾" output (male) end can multiply input torque by from

REGULAR   IMPACT

Figure 7.45 Impact and regular sockets. *(Courtesy of Snap-on Tools Corporation)*

3.3:1 up to 6:1 (Figure 7.46a). *Planetary gears* are used extensively in automatic transmissions. A simple planetary gear set includes four components: the *sun gear*, the *planetary carrier* or *cage*, *three pinion gears*, and the *ring gear* (Figure 7.46b). In order to operate, one component of the planetary gear set must be held while another part is turned. The output is the remaining component. The handle of the gear multiplier is the ring gear that is held when turning the center of the tool.

Some gear multipliers can be stacked to increase torque even more (Figure 7.46c). The need for this kind of torque would only be necessary for very heavy-duty applications, such as truck and heavy equipment. Torque is discussed in detail in Chapter 41.

Air Ratchet

Another popular tool is the air ratchet shown in Figure 7.47. It is an excellent tool for use in confined places;

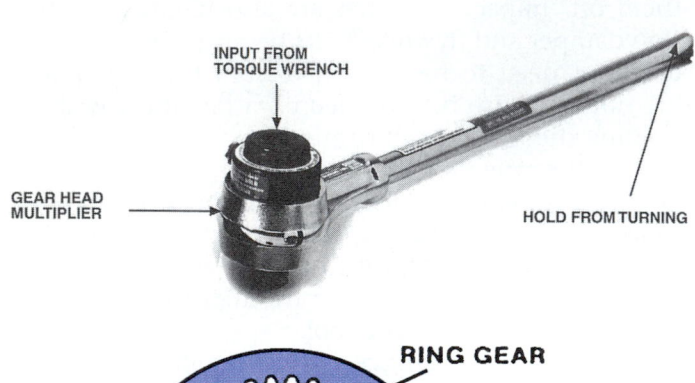

INPUT FROM TORQUE WRENCH

GEAR HEAD MULTIPLIER

HOLD FROM TURNING

(b)

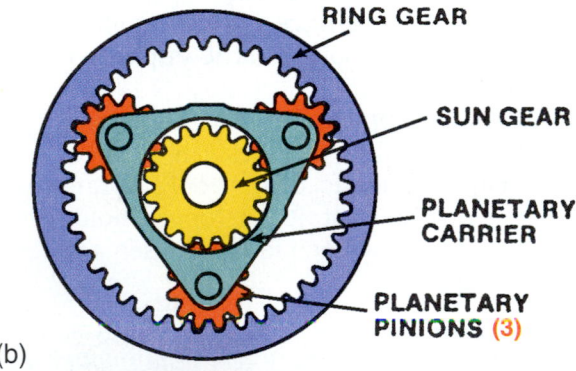

RING GEAR

SUN GEAR

PLANETARY CARRIER

PLANETARY PINIONS (3)

(c)

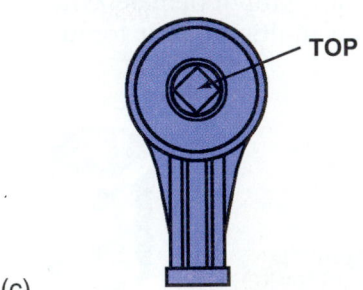

TOP

Figure 7.46 (a) A gear head multiplier, (b) Parts of a planetary gear set, (c) A drive hole in the top of the tool makes it possible to use more than one multiplier. *(a and c, Courtesy of Snap-on Tools Company, Copyright Owner)*

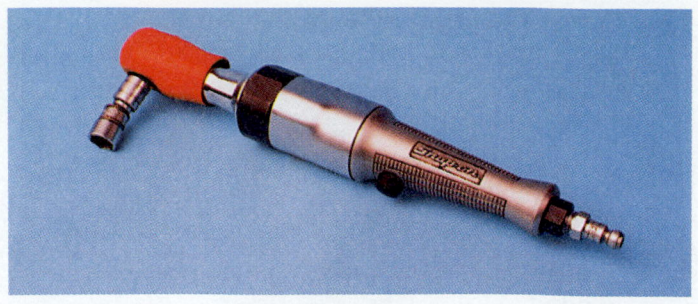

Figure 7.47 An air ratchet.

⅜" and ¼" drive sizes are common. A disadvantage about this tool is that it is noisy.

Air Chisel

The air chisel, or *air hammer*, is a miniature jackhammer (Figure 7.48). It is often used to drive valve guides out of cylinder heads, and it is useful in front-end repair and muffler work. There are many attachments available for a variety of uses.

 **SAFETY NOTE** Before pulling the trigger, be sure to have the tool bit against the workpiece. Otherwise, the tool might fly out of the gun. Be sure to wear eye protection.

Die Grinder

Die grinders (Figure 7.49) turn at very high speeds (about 20,000 rpm). They are used to remove burrs using either small cutters or grinding wheels. Carbide-tipped cutters are also available. Use a cutter rather than a grinding wheel if the workpiece is a soft metal. Be sure

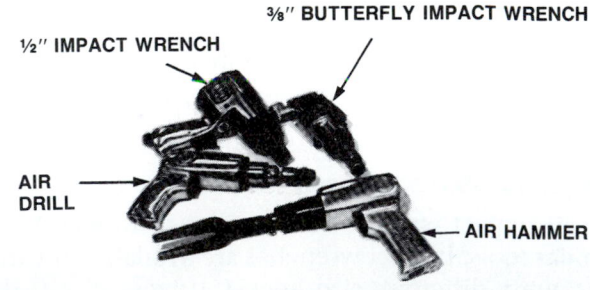

½" IMPACT WRENCH

⅜" BUTTERFLY IMPACT WRENCH

AIR DRILL

AIR HAMMER

Figure 7.48 Miscellaneous air tools. *(Courtesy of Chicago Pneumatic)*

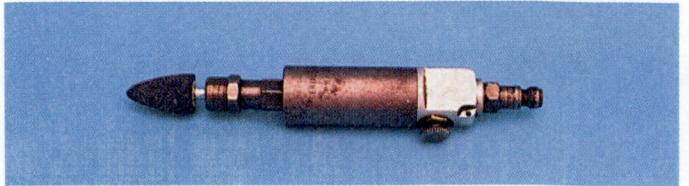

Figure 7.49 A die grinder.

that the grinding wheel or wire wheel is rated for a high enough rpm to match the top speed of the die grinder.

Air Drill

An air drill fitted with a small wire brush is very effective for removing gaskets or carbon (Figure 7.50). A special wire wheel is available for high rpm use. Called an "encapsulated wire wheel," it has molded plastic in between the wires of the brush. The plastic material wears away to expose only the tips of the wire.

NOTE: *Some air drills turn at too slow an rpm for effective wire brush cleaning.*

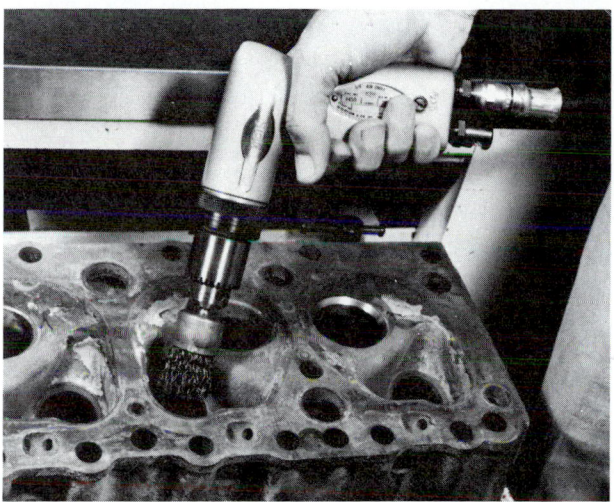

Figure 7.50 Cleaning the combustion chamber with an air drill. *(Courtesy of Sioux Tools, Inc., Sioux City, IA)*

SAFETY NOTE
■ Be sure to use face protection. The tool is very loud, so ear protection is also a good idea.
■ Some head gaskets contain asbestos. Do not remove head gaskets with the wire brush.

For maintenance purposes, air tools that are not lubricated through an automatic lubricating system should have a few drops of air-motor oil put into the air inlet each time they are used.

■ SPECIAL SERVICE TOOLS

There are many special tools required for some specialty area service and repair operations. Manufacturers identify special tools for the repair of specific components in their service manuals. Special tool usage is discussed in this text in those chapters dealing with the repair of specific components.

The hardware chapter, Chapter 6, contains information on specialty tools for extracting and repairing broken fasteners. Information in that chapter includes drill bits and taps and dies. The fuel system service chapter, Chapter 27, contains information on the flaring tool used to repair fuel lines. Check the index for the tool desired.

■ REVIEW QUESTIONS

1. The head of an open-end wrench is angled at _____°.

2. The name of the tool used to loosen and tighten fuel line fittings is a _____ nut wrench.

3. What is the name of a very short screwdriver?

4. An _____ screwdriver is pounded on with a hammer to loosen a screw.

5. For quickness and avoiding cross threading, a _____ handle is used when assembling parts.

6. An extra fulcrum provides some types of pliers or cutters with extra _____ .

7. To give a dead blow (one that does not bounce) a soft-faced hammer will have metal shot in its _____ .

8. What is the name of the type of puller that uses a heavy weight that is slid against its handle?

9. A _____ separator is a tool that is used with a puller to pull a bearing from a shaft.

10. A sun gear, _____ gear, and planetary pinions are the different gears in a gear head multiplier.

■ ASE STYLE REVIEW QUESTIONS

1. Technician A says that a 6-point wrench is easier to use in tight places than a 12-point. Technician B says that a ratchet is used to loosen fasteners that are very tight. Who is right?

  a. Technician A   **b.** Technician B

  c. Both A and B   **d.** Neither A nor B

2. Technician A says a special wire brush is needed if it will be rotated at high rpm. Technician B says that some wire brushes have plastic molded into them. Who is right?

  a. Technician A   **b.** Technician B

  c. Both A and B   **d.** Neither A nor B

3. Technician A says that a hacksaw cuts only on the forward stroke. Technician B says a fine tooth hacksaw blade (one with many teeth) would be the best choice for cutting thick steel. Who is right?

   a. Technician A   **b.** Technician B

   c. Both A and B   **d.** Neither A nor B

4. Technician A says that an 8-point socket can be used on square drive heads. Technician B says that a 12-point socket can be used on a 6-point head. Who is right?

   a. Technician A   **b.** Technician B

   c. Both A and B   **d.** Neither A nor B

5. Technician A says that when using an impact wrench to remove a bolt from the front of an engine's crankshaft, the crankshaft must be held to keep it from turning. Technician B says when using an impact wrench to remove lug nuts on a raised vehicle, the brakes must be applied. Who is right?

   a. Technician A   **b.** Technician B

   c. Both A and B   **d.** Neither A nor B

General Shop Equipment

■ KEY TERMS

American Society of
  Mechanical Engineers
  (ASME)
cherry picker
engine sling
chainfall
runout
arcing
brushes
arc welding
MIG welding
inert gas shield
oxyacetylene welding

■ SHOP EQUIPMENT

Tools of the trade are classified either as *hand tools, portable power tools,* or *equipment.* Many types of equipment are owned by a typical automotive repair shop. Some are small and do not require power. Others are powered by electricity, hydraulic pressure, or compressed air.

Major pieces of equipment that are shared among all of the company's employees are typically owned by the employer. A well-equipped general shop would have most of the equipment required to assist the technician.

■ HYDRAULIC EQUIPMENT

Example of hydraulic equipment include jacks, shop cranes (engine hoists), presses, and lifts (covered in Chapter 10). It is possible to increase pressure by hydraulic means when a small piston is used to move a larger one. Figure 8.1 shows hydraulic fluid pressure of 5 psi acting on a 2 square inch area to produce a force of 10 pounds. When that 5 psi is applied to a 4 square inch piston, the force that results is doubled to 20 pounds. The amount of movement required of the jack or press handle (input piston) will be more, but less effort will be needed.

There are many hydraulic lifting devices and supporting equipment in use in repair shops. Their operating principles are very similar. Only a few of the most popular ones have been selected for coverage in this chapter. Most of this equipment is described in an ANSI/**ASME** (American National Standards Institute)/(**American Society of Mechanical Engineers**) standard. This national standard lists the design, construction, and maintenance requirements for these devices.

Hydraulic Jacks

The hydraulic floor jack, called a *service jack* (Figure 8.2), is an important piece of equipment owned by most shops. It is used to raise and lower the vehicle and to help position heavier components, such as engines and transmissions.

Position the jack under an area of the vehicle frame, or at one of the correct lift points shown in a service manual (see Figure 11.24). Many vehicles do not have frames and may be damaged if lifted improperly. Do *not* jack under the vehicle floorpan, or under front-end linkages that can be bent.

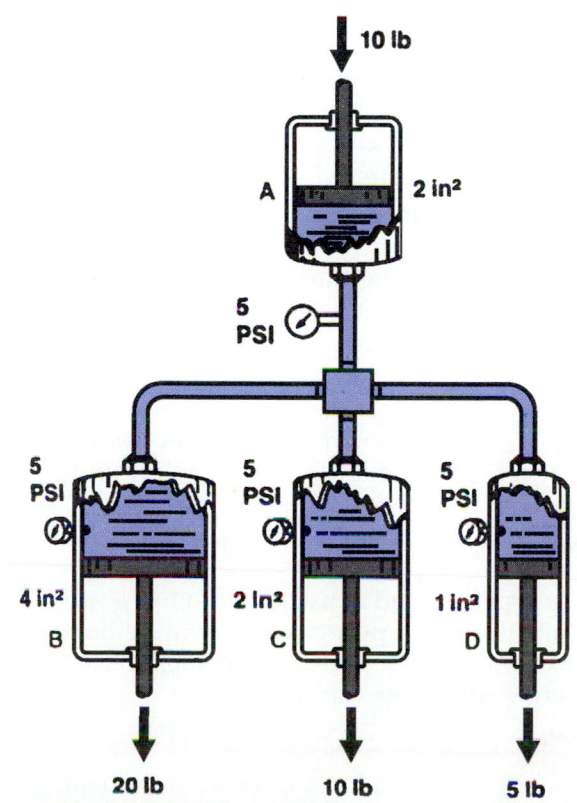

Figure 8.1 Using hydraulic pressure. *(Courtesy of Ford Motor Company)*

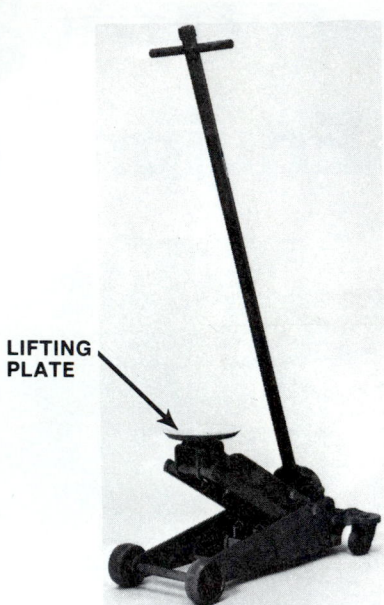

Figure 8.2 A hydraulic floor service jack. *(Courtesy of Hein-Werner)*

 Position the jack so that its wheels can roll as the vehicle is lifted. Otherwise, the lifting plate may slip on the frame, or the jack may tip over.

When lifting a vehicle from the rear, it can usually be lifted from the center (Figure 8.3). Service jack safety is covered in detail in Chapter 11.

Support Stands (Jackstands)

There are several different types of *vehicle support stands*, most often called jackstands. One type is shown in Figure 8.4.

The load should be transferred from the floor jack to the jackstand. The floor jack should be left in position as a secondary load support.

 A car should always be positioned on jackstands when it is jacked up for service.

Another type of stand is used to stabilize a lifted vehicle when heavy components are being removed. This tall jackstand, called a *high reach supplemental stand*, has a screw adjustable top (Figure 8.5).

 Be careful not to raise the vehicle off the lift when raising the adjustment on a high reach stand.

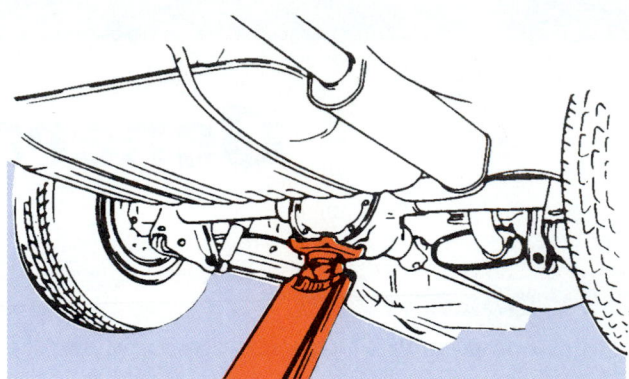

Figure 8.3 The rear of the vehicle can usually be lifted at the center. *(Courtesy of General Motors Corporation, Service Technology Group)*

Figure 8.4 Support stands.

Figure 8.5 A high reach stand is used when the car is raised on a lift. *(Courtesy of Branick Industries, Inc.)*

Creepers. A creeper has small wheels and saves wear and tear on the body when working in uncomfortable positions or working under a car that is resting on

Figure 8.6 A floor creeper. *(Courtesy of Lisle Corporation)*

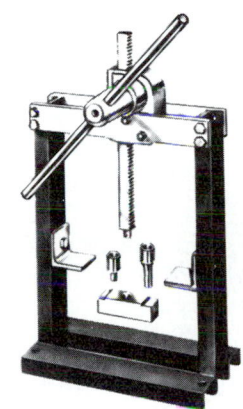

Figure 8.7 An arbor press. *(Courtesy of Snap-on Tools Company, Copyright Owner)*

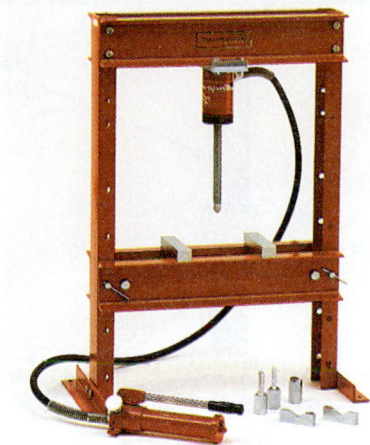

Figure 8.8 A large 50-ton press. *(Courtesy of Snap-on Tools Company, Copyright Owner)*

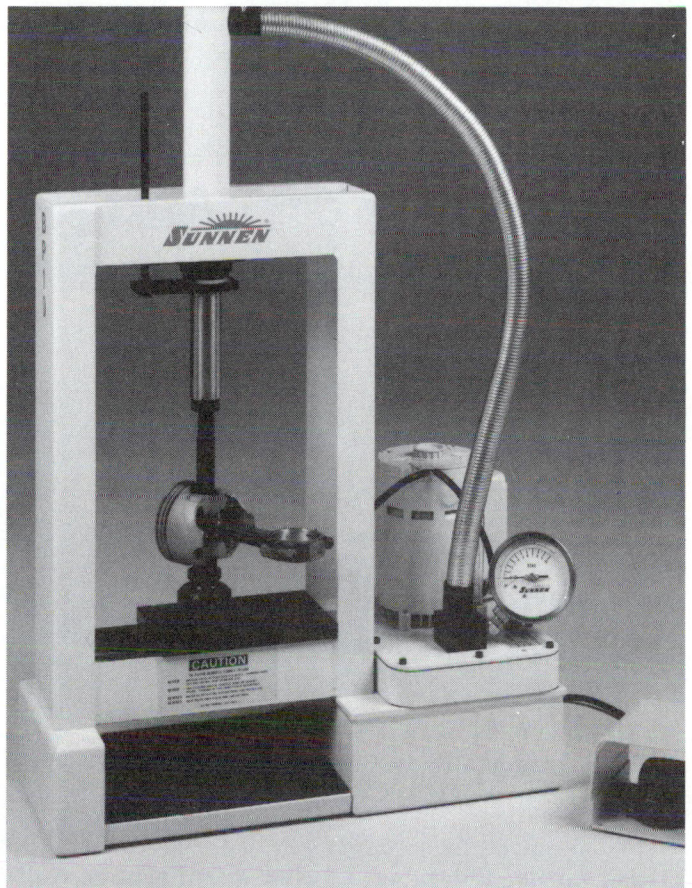

Figure 8.9 An electric, hydraulic press. *(Courtesy of Sunnen Products Company, St. Louis, Missouri)*

vehicle support stands. There are several special kinds of creepers. A floor creeper is shown in Figure 8.6. To avoid having someone accidentally step on it:

- Be sure to stand a floor creeper on end when not in use.
- When taking a break in the middle of a repair job push the creeper under the raised car.

Presses

There are many sizes of presses. Some are mechanical, such as the arbor press (Figure 8.7), and some are hydraulic. The largest press is the most versatile, having both a fast and slow pumping speed, and the ability to separate axle bearings, which sometimes require 25 tons of pressure (Figure 8.8).

A 10-ton *electric/hydraulic* press is found in some shops. It is used for such things as pressing wrist pins into and out of connecting rods (Figure 8.9). This press does not require pumping of the handle. An electric motor provides pressure that is controlled by a control handle.

There are many special-use press fixtures available. For press work a *bearing separator plate* is often used (see Figure 7.41). Be sure to support it where the bolt holds the two halves of the tool together (Figure 8.10). If the separator is installed in the press 90° to the correct position, the bolts will be bent and the tool can be damaged.

Shop Cranes (Hydraulic Engine Hoists)

A shop crane is commonly referred to as a **cherry picker** (Figure 8.11). It is used to remove the engine from the car.

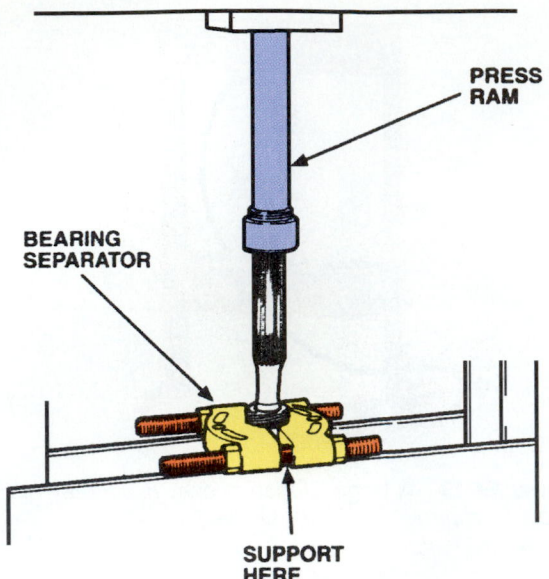

Figure 8.10 Support the bearing separator under the bolts so the tool is not damaged. *(Courtesy of Ford Motor Company)*

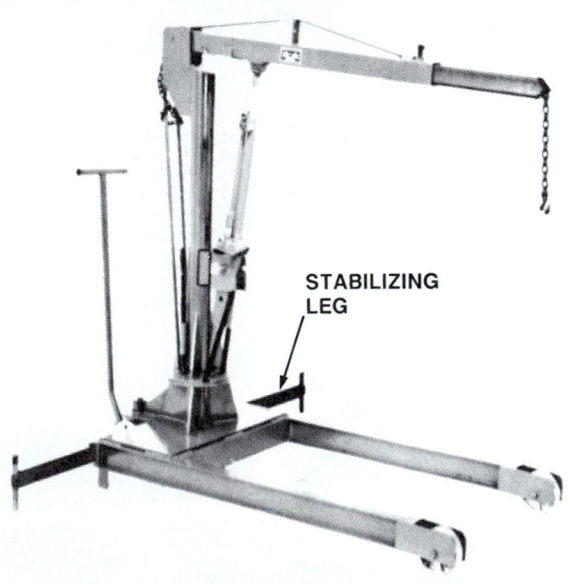

Figure 8.11 A hydraulic shop crane. *(Courtesy of Branick Industries, Inc.)*

Engine Lifting Sling. An **engine sling** (Figure 8.12) is used to hook the shop crane to the engine block.

Lifting Sling Safety

1. The sling must be tightened against the block, because the holding bolts can be overly stressed and break (see Figure 11.26).
2. The bolts should go into the block to a depth at least 1½ times the diameter of the fastener.

Figure 8.12 An engine sling. *(Courtesy of Snap-on Tools Company, Copyright Owner)*

Be careful that the sling does not contact the distributor or the carburetor, which can be damaged during engine removal.

Chainfalls

A **chainfall**, or chain hoist, is also a common tool for removing the engine from the car. It is usually hung from an overhead structural beam. The beam can be part of the building structure or a portable I-beam structure with wheels. When the chainfall is used for engine removal, the car's hood must first be removed. It is more difficult to maneuver an engine back into position during reinstallation using a chainfall.

Engine Stands. The engine stand provides a convenient means of turning the engine over for disassembly and reassembly (Figure 8.13). When an engine is removed from a car, it is a good idea to mount the engine on an engine stand immediately.

 **SAFETY NOTE** Inexpensive engine stands are widely available. Be sure to purchase one that is designed to carry the maximum load that you might encounter. Do not use a light-duty stand for a heavy engine.

The stand's universal mounting adapter (Figure 8.14) is designed to fit almost any engine. It is bolted to the rear of the engine in place of the bell housing or automatic transmission converter housing.

First, mount the universal mounting adapter to the engine. Then, install the engine and mounting adapter on the engine stand. For more information on the engine stand, see Chapter 40.

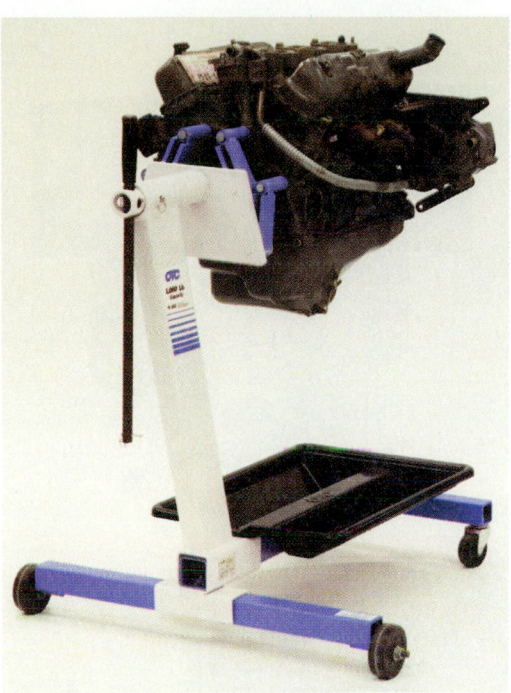

Figure 8.13 An engine stand. *(Courtesy of OTC-SPX Corp. Aftermarket Tool & Equipment Group)*

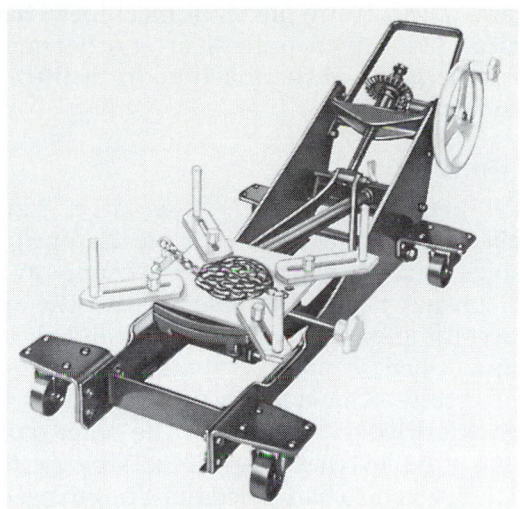

Figure 8.15 A low transmission jack.

AUTOMATIC
TRANSMISSION
FLEX PLATE

ENGINE STAND
MOUNTING PLATE

Figure 8.14 An engine stand mounting plate. Note the washers installed under the bolt heads.

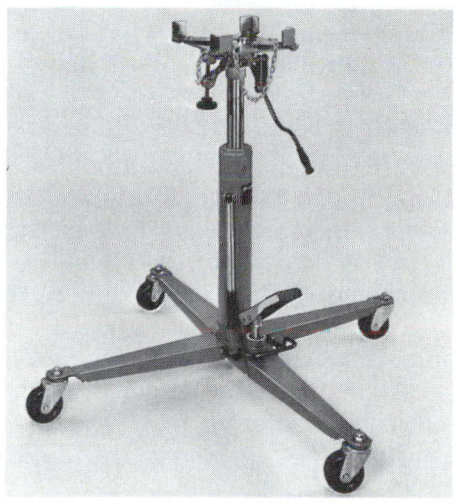

Figure 8.16 A tall hydraulic transmission jack. *(Courtesy of Hein-Werner)*

and the other is used with the car raised on a lift (Figure 8.16).

 **SAFETY NOTE** It is not safe to work on an engine that is hanging from a shop crane. An engine stand provides a margin for safety, as well as convenience.

SAFETY NOTE When using a transmission jack with a car raised on a lift, be sure to use high reach stands to stabilize the vehicle. Also be careful not to raise the car off the lift pads as you raise the transmission. For safety reasons, it is preferable to use a wheel-contact lift when removing the engine or transmission.

Transmission Jacks

The transmission jack is used to remove automatic and manual transmissions from vehicles. There are two major types of transmission jacks. One kind is used when the car is on support stands (Figure 8.15)

■ SHOP ELECTRIC MACHINERY

Much of the machinery in a shop is powered by electricity. Electric machinery includes anything with a motor. Motors that power large machines usually run at a constant speed. Many pieces of machinery have different sized belt-driven pulleys or another means of adjusting the speed of the machine to fit the application (see Figure 8.18).

Drill Press

The drill press (Figure 8.17) is a large, stationary machine used to drill parts that are clamped to the press table. There are various types of clamps available.

Drill presses have adjustable speeds. The speed is varied according to the type of metal and the size of hole being drilled. Figure 8.18 shows two types of variable-speed presses. One type has optional pulley grooves for the v-belt drive (Figure 8.18a). The other type (varispeed) has a pulley that changes sizes by means of a crank. The speed on a vari-speed must be changed while the drill press motor is running (Figure 8.18b).

A cutting lubricant is used to drill all metals except cast iron or aluminum, which can be drilled dry, or cooled with kerosene. For information on drilling speeds, drill bits, and lubricants, see Chapter 6.

Drill Motors/Portable Power Tools

There are a variety of sizes of hand-held drill motors (Figure 8.19). The part that grips the drill bit is called the *spindle chuck*, or chuck (Figure 8.20). A drill bit is tightened between the jaws of the chuck with a *chuck key*. A drill motor is classified according to the largest drill bit that its chuck can accommodate. The following are approximate speed ranges for different sized drill motors. Notice that the larger drill bits turn at slower speeds (see Chapter 6).

- ¼" drill motor—1500–2000 rpm (revolutions per minute)
- ⅜" drill motor—1000 rpm
- ½" drill motor—500 rpm

Some drills are variable speed; some are also reversible. Drills usually have a button located next to the trigger that can be used to hold the drill in the "on" position. To shut the drill off, just press the trigger again.

Heavy-Duty Power Tool Considerations

Power tools that are designed to be used by professionals are different than those that are used by the homeowner or hobbyist. That is why they cost more. A tool that will work well in a home environment will probably not survive the challenges of the workplace. Figure 8.21 shows the parts of a drill motor.

Bearings. Bearings in a drill motor are located at the front and rear of the motor armature, and at the spindle chuck. Three types of bearings are used in power tools: ball, roller, and powdered metal sleeves. Motor armatures on heavy-duty tools use ball bearings, which have almost unmeasurable free play.

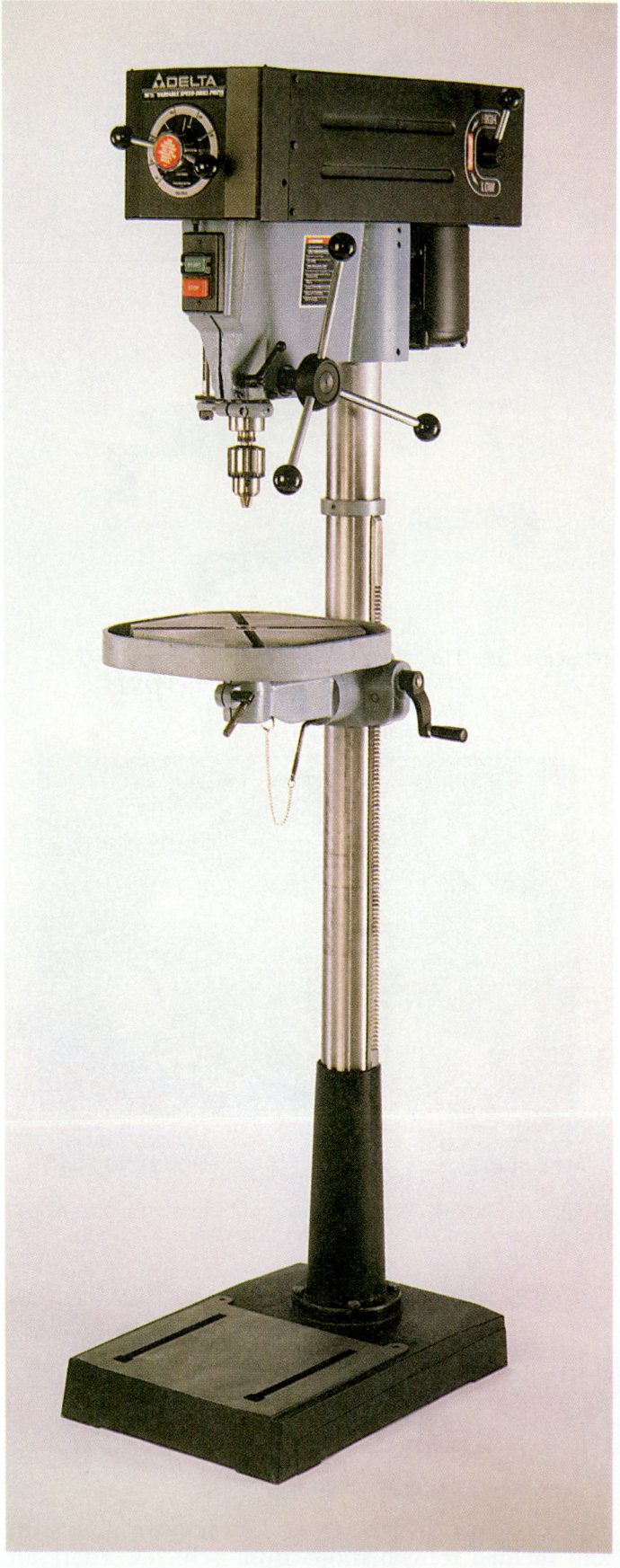

Figure 8.17 A drill press. *(Courtesy of Delta Int'l. Machinery Corp.)*

(a)

(b)

Figure 8.18 Two types of variable-speed drill presses: (a) Pulley type. (b) Vari-speed type. *(Courtesy of Clausing Industrial, Inc.)*

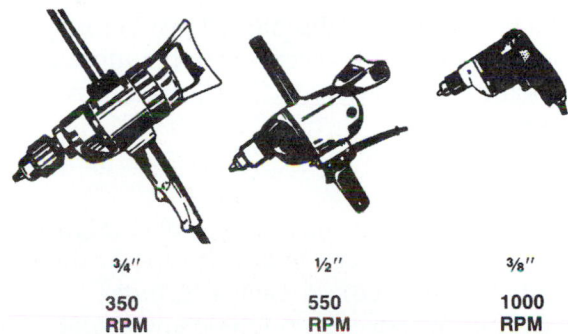

¾″	½″	⅜″
350 RPM	550 RPM	1000 RPM

Figure 8.19 Various handheld drill motors. *(Courtesy of Snap-on Tools Company, Copyright Owner)*

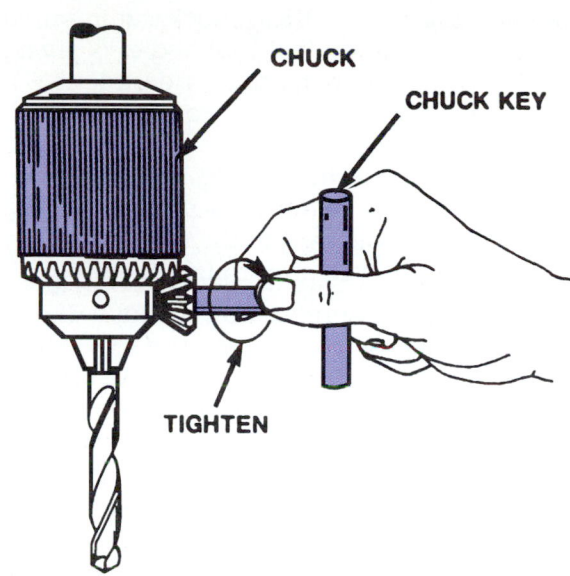

Figure 8.20 The drill bit is tightened into the chuck with a chuck key.

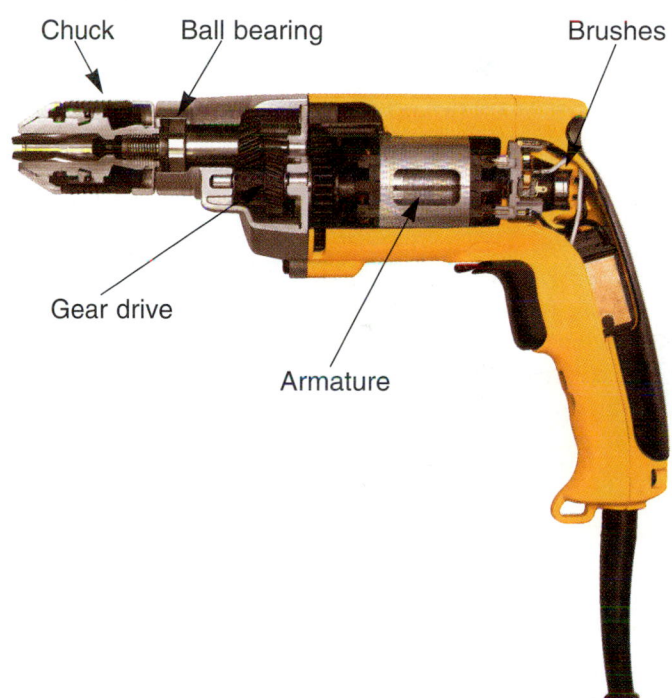

Figure 8.21 A cutaway of a drill motor. *(Courtesy of Black and Decker)*

The spindle chuck of a drill is often under high side-to-side and end loads. A quality commercial drill uses a *ball bearing chuck*, while consumer drills use sleeve bearings.

Chuck. The chuck on a professional drill is one of the most expensive parts of the drill. For durability, the jaws of the chuck, the parts that contact the drill or tool bit, are made of expensive case hardened steel. More precise machining is done in a commercial drill chuck. Chuck **runout**, measured at a point 1″ from the jaws, is the amount that the chuck wobbles. It is allowed to be up to 0.010″ in a consumer drill, but only half that amount in a commercial drill.

Housing. Weight and durability are factors in the design of portable power tools. Plastic is usually used for the housings of modern tools. Plastic is an insulator, so the danger of electrical shock is minimized. The plastic used for professional tool bodies is often Super Tough™ nylon. It is practically indestructible and is resistant to oil, solvents, and corrosives, as well as heat.

Pilot pins align the handle, gear housing, and motor housing of a commercial power tool. Separate sets of screws hold one part to the next to ensure alignment of the parts, even after rough handling.

Power cord. A consumer's power tool will probably have a short power cord. When the tool is used, an extension cord will be needed. A professional portable power tool usually has an 8–10' power cord, which almost always eliminates the need for an extension cord. The insulation on a commercial quality cord is made of rubber or a synthetic material that stays flexible when cold. Household power tools have a molded plastic cord protector that is part of the cord. The cord protector is located where the cord meets the tool. Professional cords have a separate rubber cord protector. This makes it easy to replace a damaged cord.

Switch. The switch in a heavy-duty tool will also be more expensive, designed to protect against a dusty or dirty environment. It is also designed to withstand repeated cycles of the switch.

Motor. The motor is designed to be small and light in weight, yet with more power and able to withstand overloading. **Brushes** carry electricity to the *motor armature* through *commutator segments*, or bars. **Arcing** within the motor, when sparks jump between poor connections of the brush and commutator, must be minimized in a good power tool. The motor's armature usually has more commutator bars to resist arcing. The small amount of radial play on a ball bearing armature also minimizes arcing of the brushes. Brushes that contact the commutator bars are more precisely positioned than in a consumer tool. The brush holder is usually brass. A provision is often made to replace the brushes without having to take the tool apart.

During manufacture of the armature, resin is applied to the armature windings to protect against heat. It soaks into every crevice, making the armature into a solid unit that can withstand high rpm without the wires rubbing against each other, which would result in a short circuit.

Gears. Gears in a commercial tool can be made of machined heat-treated wrought steel or powdered metal. Powdered metal is less expensive because gears made by this process do not require machining.

Battery-Operated Tools

Battery-operated tools, especially drill motors (Figure 8.22), are popular in shops. Some of the higher quality

Figure 8.22 A cutaway of a battery-operated drill. *(Courtesy of Black and Decker)*

ones have a *clutch* that releases during overtightening or when a drill bit catches on the work being drilled. The clutch feature can prevent breakage of fasteners and drill bits. The versatility provided by not having to run an electrical cord is also a desirable feature.

A typical professional setup includes a separate battery charger and two batteries.

> **SHOP TIP** Keep one battery in the charger while other is in use.

If a battery is fully discharged just before putting it in the charger, it will accept a charge more fully. It is not good for the battery to be allowed to remain in a state of discharge.

Grinder

Almost every shop has a grinder (Figure 8.23). Grinders can be either the bench type or the pedestal type (floor mounted). A typical grinder is equipped with a grinding wheel on one side and a wire brush for cleaning parts on the other side. Different size motors are available. The motor should be powerful enough to maintain speed under pressure. The grinding wheel is kept square by dressing it with the tool shown in Figure 8.24.

The grinder can be used to sharpen tools. A *water pot* is attached to the front of the grinder. The metal being ground must be constantly quenched. If the metal is allowed to become too hot, either of two things can occur:

1. Metal that is heated and then quenched will be made more brittle.
2. Metal that is heated and allowed to cool slowly will be softened (see Chapter 6).

Grinder safety precautions are outlined in Chapter 11.

■ OTHER ELECTRIC EQUIPMENT

Battery Charger

The *battery charger* (Figure 8.25) can be used to charge batteries as well as to help the car's battery start the engine. Some battery chargers have instructions that recommend against using the unit for starting the car. Be sure to check the instructions for the unit you are using and be sure the cables are clamped to the proper battery terminal. Chapter 11 covers battery charger safety.

Soldering Tools

A soldering iron or soldering gun (Figure 8.26) is used to melt *solder* when joining electrical wires. Soldering is covered in Chapter 34.

■ OTHER SHOP EQUIPMENT

The valve grinder, boring bar, cylinder hone, brake lathe, and wheel balancer are other pieces of equipment covered in greater detail in other chapters. The

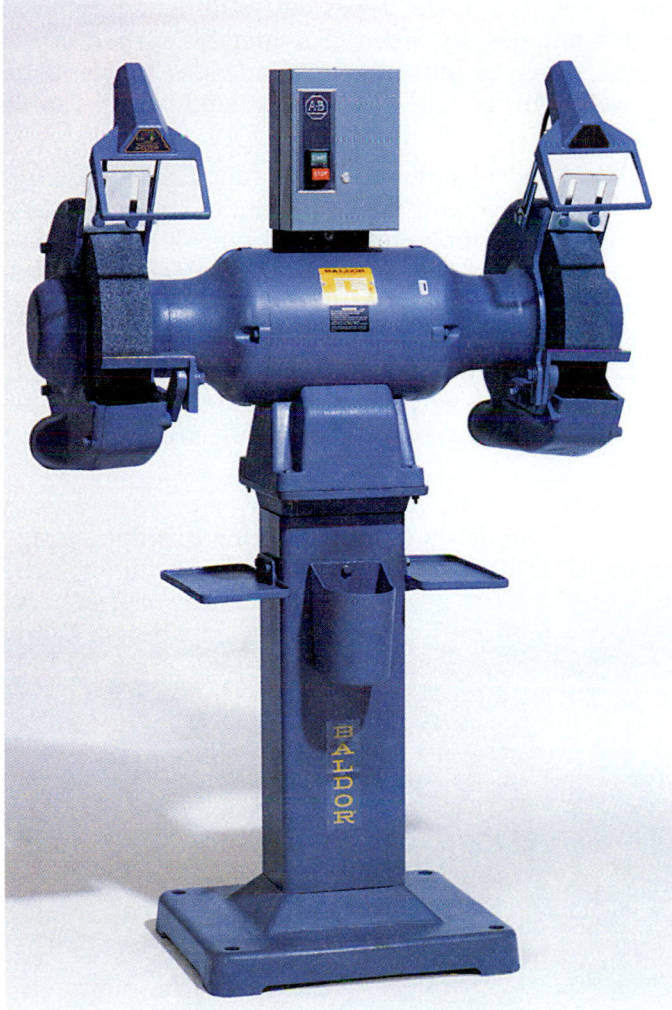

Figure 8.23 A large grinder. *(Courtesy of Baldor Electric Company)*

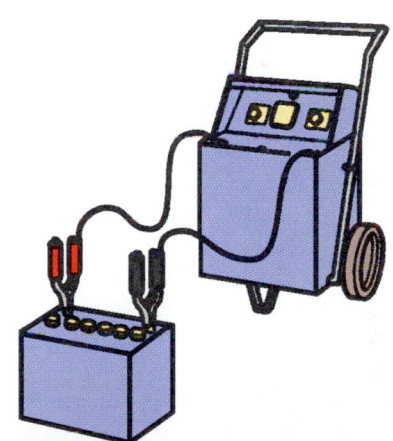

Figure 8.25 A battery charger. *(Courtesy of Chrysler Corporation)*

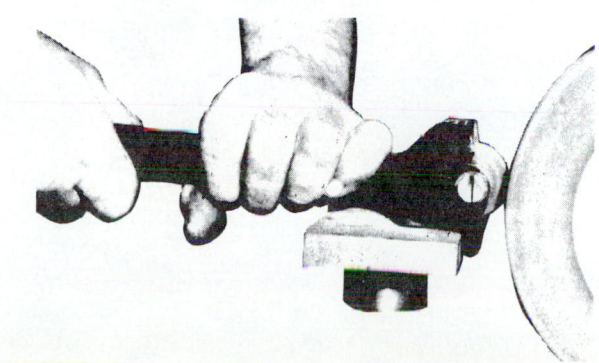

Figure 8.24 Dressing a grinding wheel. *(Courtesy of C. Thomas Olivo Associates, Fundamentals of Machine Technology)*

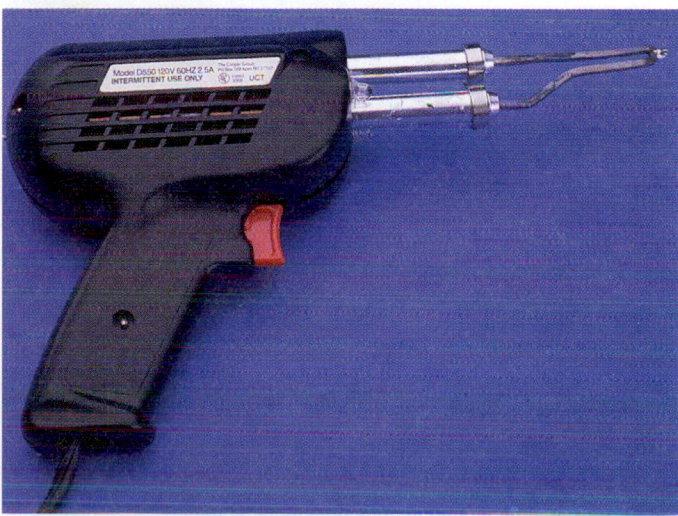

Figure 8.26 A soldering gun.

rest of this chapter deals with miscellaneous pieces of general shop equipment.

Vises

Every shop has a vise (Figure 8.27). Some vises have pivoting bases that allow them to be turned on the bench. The jaws of automotive vises generally have teeth, so a set of brass jaw caps is a good investment. They will prevent the vise from leaving a mark on the workpiece.

Vise Safety

1. Be careful not to tighten the vise too much at its wide-open position or the vise might break.
2. Never use a piece of pipe on the handle of a vise; it can break the vise.
3. Do not hammer on a vise, except on the anvil portion that some vises have.

Engine Analyzer

An engine analyzer or scope, is useful in diagnosing engine problems as well as in tuning a car after an overhaul. The scope can help to diagnose compression problems, uneven balance between cylinders, worn timing chains, and air leaks. Be sure to keep the scope leads (wires) out of the fan.

Tire Changer

Shops that sell or repair tires have a *tire changer* (Figure 8.28). The wheel rim is mounted securely to the machine either by an air-operated or mechanical clamp. Some tire changers are totally mechanical, but most are air assisted. Air, under a high pressure is used to unseat the bead of the tire from the wheel rim. A bar is inserted under the wheel rim. The bar is rotated by air pressure to remove the tire from the wheel.

Welding Equipment

There are two main types of welding equipment: electric and gas. In both types, metal is melted into a *puddle*, which is moved by the operator to complete the weld. This is a mechanical skill and requires some practice.

Arc Welding. Electric welding is called **arc welding**. It is inexpensive and fast, but usually requires a 220-volt electrical hook-up (which some shops do not have). A ground (negative) clamp is hooked to the piece to be welded. A welding rod attached to the positive clamp is "struck" against the workpiece in a motion similar to striking a match. The electric arc melts the metal when electrons flow across the arc between the positive and negative electrodes.

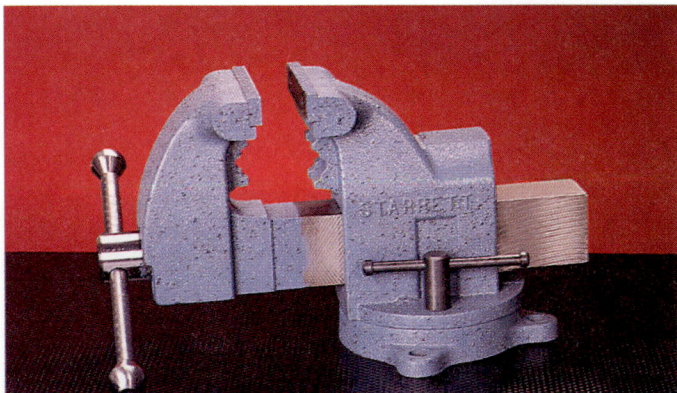

Figure 8.27 A bench vise with pivoting base and soft jaw caps. *(Courtesy of The L.S. Starrett Company)*

Figure 8.28 A tire changer. *(Courtesy of Hunter Engineering Company)*

MIG Welding. A very popular type of arc welding in automotive shops is **MIG welding** (Figure 8.29). MIG stands for *metal inert gas*. An **inert gas shield** (argon, helium, or CO_2) is applied around the arc area during welding to prevent oxidation of the metal. A MIG welder has a spool of wire that is automatically fed through the positive end of the torch. When aluminum wire is used, this kind of welding can be used on aluminum. Aluminum cannot be welded using ordinary arc welding. MIG welding is especially popular for welding sheet metal body panels.

Oxyacetylene Welding. Gas welding is called **oxyacetylene welding**. Two compressed gases, oxygen and acetylene, are stored in separate steel cylinders (Figure 8.30). Hoses connect the cylinders to the torch

Welding Safety

1. Arc welding produces *ultra violet rays*. A helmet with the appropriate shade of lens must be used to prevent damage to the eye.

2. Welding can result in splatter of molten metal. Be sure to wear appropriate protective gear.

3. Bystanders must have appropriate protective gear, too.

4. Cutting with a torch can result in a great deal of flying molten metal. Be sure that flammable materials are not in the area.

Figure 8.29 A MIG welder. *(Photo provided by Mac Tools and is representative of products manufactured by Mac Tools or products obtained from another source)*

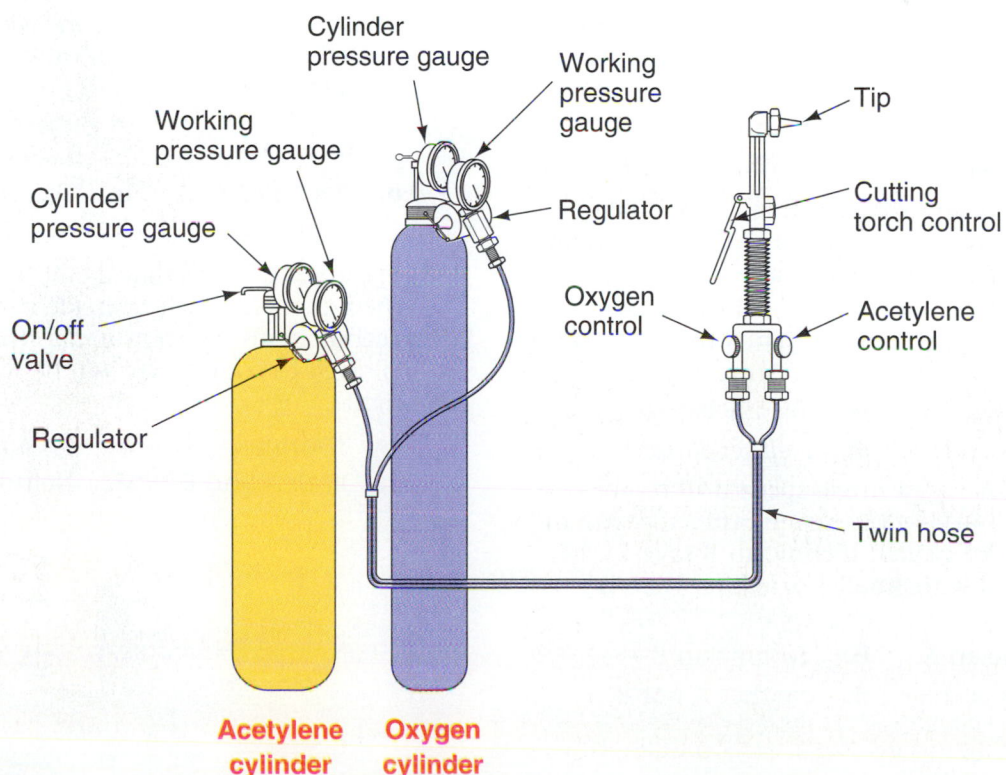

Figure 8.30 Oxyacetylene gas welding and cutting equipment.

where the gases are mixed. When the gases are mixed in the correct proportion and ignited, intense heat is created at the tip of the torch. The heat is intense enough to melt steel for welding or cutting.

Each cylinder is equipped with a regulator that controls the amount of pressure of the gas coming out of the cylinder. The regulator has two gauges: one shows the pressure inside of the cylinder and the other shows the pressure going to the torch.

Oxyacetylene welding has an advantage over arc welding in that it is more versatile. Gas welding equip-

ment can be used to cut and heat metals. Technicians often use the torch to heat rusty or seized parts, such as exhaust components, for easier removal. Pressed fit parts can be heated for easier installation. With a *rose-bud tip* (which puts out a great deal of heat) cylinder heads can even be straightened. An obvious disadvantage is that there is an open flame, which can result in burns and start fires.

■ REVIEW QUESTIONS

1. What is a cherry picker?

2. A _____ is a chain or cable attached to the block to help hoist an engine out of a car.

3. The bolts holding an engine sling should go into the block to a depth at least _____ times the diameter of the fastener.

4. When mounting an engine on an engine stand, the mounting head adapter should be attached to the engine before placing it on the _____.

5. Name two metals that do not require a lubricant when they are drilled.

6. When metal being ground is not quenched and is allowed to cool slowly, what happens to it?

7. A battery _____ can be used to assist the battery in starting the car.

8. Jaw caps or soft jaws are used when clamping a part in a _____ to avoid marring the work surface.

9. What type of welding would you use for cutting or heating, oxyacetylene or arc?

10. What is the name of a kind of welding that uses an inert gas shield?

■ ASE STYLE REVIEW QUESTIONS

1. Technician A says an arbor press is a type of hydraulic press. Technician B says to use a pipe on the handle of a vise to increase clamping force. Who is right?
   a. Technician A   b. Technician B
   c. Both A and B   d. Neither A nor B

2. Technician A says when a shop crane is being moved while it is carrying an engine, the engine should be positioned as low as possible. Technician B says a chainfall is easier to use when installing an engine than a hydraulic shop crane. Who is right?
   a. Technician A   b. Technician B
   c. Both A and B   d. Neither A nor B

3. Technician A says a single-speed drill motor with a ½" chuck will turn at a faster speed than one with a ⅜" chuck. Technician B says a faster speed is used with smaller drill bits. Who is right?
   a. Technician A   b. Technician B
   c. Both A and B   d. Neither A nor B

4. Technician A says it is possible to increase pressure by hydraulic means when a large piston is used to move a smaller one. Technician B says if a force of 10 pounds acts on a 2 square inch area, hydraulic fluid pressure of 5 psi is produced. Who is right?
   a. Technician A   b. Technician B
   c. Both A and B   d. Neither A nor B

5. Technician A says that metal that is heated and then quenched will be made more brittle. Technician B says that metal that is heated and allowed to cool slowly will be softened. Who is right?
   a. Technician A   b. Technician B
   c. Both A and B   d. Neither A nor B

Cleaning Equipment and Methods

KEY TERMS

blow-by gases
sludge
alkaline
acid
pH scale
base
labor intensive
caustic soaps
superheated
ferrous
scale

◼ OBJECTIVES

Upon completion of this chapter, you should be able to:

✔ Use cleaning tools and equipment safely and properly.

✔ Describe the best cleaning method for a particular application.

◼ INTRODUCTION

This chapter will deal with methods for cleaning various parts of vehicles. There are different types of materials that require cleaning. Some are hard deposits and others are dirt, oils, greases, and contaminants.

The insides of engines and gear cases such as transmissions and differentials contain oils and contaminants that cannot be washed down drains. Cleaning methods differ, depending on whether materials to be cleaned are considered to be toxic or not.

The internal combustion engine in the modern automobile produces many byproducts during operation. **Blow-by gases** (gases that leak past the piston rings into the crankcase) contain acids, as well as burned and unburned fuel. Moisture also contributes to the accumulation of **sludge** inside the engine (Figure 9.1). In addition, carbon builds up in the combustion chambers because of oil that leaks past the rings from the crankcase (Figure 9.2). Oil that leaks out of the engine combines with dirt and road grime to make a greasy film on the outside of the engine. Also, mineral deposits can build up in the cooling system.

It is estimated that approximately ⅓ of a shop's expenses can be attributed to cleaning the engine. This chapter deals with the cleaning equipment used when preparing an engine for rebuilding.

◼ GENERAL SHOP HOUSEKEEPING PRACTICES

Good housekeeping practices are essential when cleaning automotive parts. A clean, orderly shop is vital for impressing upon the public that a shop is thorough and competent and will do a professional job on its

Figure 9.1 Water and byproducts of combustion combine with engine oil to form sludge. *(Courtesy of Texaco's magazine, LUBRICATION)*

Figure 9.2 Carbon deposits form in the combustion chamber. *(Courtesy of Dana Corp., Perfect Circle Division)*

work. Of even more importance, however, is the health and safety of anyone in the shop area.

Slippery floors are dangerous. To avoid the possibility of a dangerous slip and fall, immediately clean up slippery spills. Examples of slippery spills include coolant, lubricants, solvents, glass beads, or steel shot. Preventing spills from occurring in the first place is best. When spills do occur, they must be dealt with immediately. This will also keep the mess from spreading around the shop. Parts that are wet with solvent should be blown off into the solvent tank (Figure 9.3) or allowed to air dry before being moved.

> **SHOP TIP** Wet parts can be carried from the solvent tank in a drain pan to prevent solvent from dripping onto the floor (Figure 9.4).

Figure 9.3 When drying parts with compressed air, blow solvent back into the solvent tank.

Figure 9.4 To avoid dripping solvent on the floor, carry wet parts in a drain pan.

When an engine is to be removed from a vehicle, oil and coolant should be drained from it first. The oil filter holds oil, so it should be removed, too.

A spill often occurs when an engine block mounted on a stand is turned upside-down during disassembly. There is usually some coolant still left in the block. Positioning a drain pan under the engine before turning it upside-down can control the spill.

CLEANING METHODS

There are several categories of automotive materials that require cleaning. Water-soluble deposits (dirt), organic soils, scale, or rust all require different methods of cleaning. Cleaning methods include:
■ Wet cleaning that includes water and/or chemical solutions
■ Abrasive cleaning
■ Thermal cleaning

The most common of the three methods is wet cleaning. Methods and materials covered in this chapter will be those that would be the most popular ones used in repair and machine shops. Specific information about cleaning materials is best obtained from the supplier of the equipment.

Chemical Cleaning

Chemical cleaning in engine repair is of three main types:
■ **Alkaline**
■ **Acid**
■ Solvents

> ### ■ SCIENCE NOTE ■
> The strength (or alkalinity) of a solution is measured on the **pH scale**, which ranges from 1 through 14.
> ■ *Pure water is rated at 7.*
> ■ *Solutions that are below 7 on the scale are acidic.*
> ■ *A pH rating of 1 is a strong acid.*
> ■ *A pH rating of above 7 is an alkaline cleaner, or **base**.*
> ■ *A strong alkaline cleaner has a pH rating above 10.*

Soap and Chemicals. To clean *soils*, a chemical must wet the material. *Soap* is a wetting agent. After wetting, the soap suspends dirt so it can be washed away. Dirt is the only water-soluble material found on automotive parts.

Organic soils include petroleum byproducts, gasket sealers and paints, carbon and other byproducts of combustion. These materials *cannot* be effectively cleaned with water. Chemicals are used to make these soluble so that water will be able to wash them off.

Cleaning with Bases

Alkaline materials cut grease very well and work best when heated. Most automotive soaps are alkaline. Using soap and hot enough water temperatures are both very important when cleaning grease. Soap lifts grease and makes cleaning easier and much more effective. Temperature is also very important. Remember how much faster hands clean up when using hot water instead of cold?

Cleaning with Acids

Acids are useful only in removing rust and scale. Acid will not cut grease. Before rust can be removed, any oil or grease must first be removed with an alkaline material. When an acid is used to remove rust or scale, it also removes a small amount of the base metal. The chemical used to remove scale deposits from the vehicle's cooling system is an example of an acid used in automotive repair.

Cleaning with Solvents

Solvents are of three types:
- Water-based
- Mineral spirits (Stoddard solvent)
- Chlorinated hydrocarbons (carburetor cleaner)

■ CLEANING THE OUTSIDE OF THE ENGINE

Before a component such as an engine or transmission is removed from a car, it can be cleaned by one of several methods. Pressure washing and steam cleaning are methods of grease removal whose popularity has declined recently due to environmental considerations. The sewer drain must be able to capture all of the hazardous contaminants that would otherwise go into the sewer. Also, these cleaning methods are **labor intensive**, which means that they require an operator. They can also cause damage to air conditioning and electrical components and connections.

Hot Pressure Washer

Although it is possible to remove engine grease by pressure washing with cold water, hot water washing is much more effective. Hot high-pressure washing (Figure 9.5) is done with water heated to 190°–210°F, but not boiling. This type of cleaner, commonly found in coin-operated car washes, is very effective in removing external grease from the engine. Water is heated by a kerosene or natural gas flame as it passes through a coil (Figure 9.6). If the temperature of the water is lowered to save energy costs, either the pressure or the amount of soap must be increased to compensate.

Figure 9.5 A hot high-pressure washer. *[Courtesy of Landa Water Cleaning Systems]*

Figure 9.6 Water passes through a coil and is heated by a flame. *[Courtesy of Landa Water Cleaning Systems]*

Some of the advantages to the hot pressure washer over other methods of cleaning are:
- It quickly heats water to 200°F.
- It does not damage paint, provided the temperature is properly set. (It may lift paint that is in poor condition.)
- Because there is no cloud of steam, visibility is relatively good.
- Very high pressures of from 200 to 1000 psi are possible.
- Small objects are more easily cleaned with a pressure washer than with a steam cleaner.

Cautions when using a hot pressure washer:
- Grab the pressure gun at the two grips designed for your hands to avoid being burned.
- Keep both hands on the pressure gun so that the front of the gun does not fly up in the air, possibly burning a bystander.
- Wear a face shield, rubber gloves, and a rubber bib apron when cleaning parts. The hot mixture is often splashed back at you by accident.

Steam Cleaner

A steam cleaner (Figure 9.7) is a super water heater. Water is heated in a coil by either oil, kerosene, or natural gas in the same way it is in a hot pressure washer. The water, under low pressure, is provided by a water pump. A sketch of a simple steam cleaner is shown in Figure 9.8. Pressure increases to between 80 and 120 psi as the water is heated.

■ SCIENCE NOTE ■

Every pound of pressure on water at 212°F increases the boiling point by about 3°F. In the same manner as a pressure cooker used for cooking, the steam cleaner can heat water to well above its normal boiling point.

The temperature of the steam mixture should be between 290° and 320°F. When this **superheated** water finally leaves the end of the steam gun the pressure on it suddenly drops, causing it to boil instantly, or *flash*. The "exploding" water that results is very effective in removing grease.

Ideal cleaning is achieved when the mixture leaving the gun is about 20% steam and 80% water. Steam without water is a poor cleaner. Too much steam can cause the coil to plug up sooner and is also hard on the steam cleaner hose.

Figure 9.7 A steam cleaner. *(Courtesy of Landa Water Cleaning Systems)*

Operating a Steam Cleaner

There are several important points to remember when using a steam cleaner. The most important points are those dealing with safety. Grabbing the steam gun at any point other than the two grips designed for your hands can result in a severe burn. Keep both hands on the steam gun so that the front of the gun does not fly up in the air, possibly burning a bystander. Be sure to wear a face shield, rubber gloves, and a rubber bib apron when steaming parts. The hot mixture is often splashed back at you by accident. Sometimes the mixture contains **caustic soaps**, which can be hazardous to you.

Any time the flame is turned on, there should be water flowing through the machine. Failure to observe this precaution can cause scaling of the coil, which can ultimately ruin the steam cleaner. When the coil becomes scaled, it cannot conduct heat properly. Scale formed as minerals in the local water supply settle out of the water during heating and become deposited in the coil of the steam cleaner. Scale buildup can be minimized by using softened water.

■ SCIENCE NOTE ■

Scale is formed in one of two ways. In one process, scale forms due to the reaction of calcium, magnesium, or ferrous ions with bicarbonate ion HCO_3 in water. Calcium and bicarbonate ions react to deposit solid calcium carbonate (a poor heat conductor). The other process is due to the reaction of the same calcium, magnesium, or ferrous ions with soap in water. Soaps are sodium "salts" of hydrocarbon acids in which the acidic hydrogen has been replaced by sodium. For example, sodium acetate (CH_3COONA) is the sodium salt of acetic acid (CH_3COOH). When sodium salts (soaps) are added to water containing calcium, magnesium, or ferrous ions insoluble calcium, magnesium, or ferrous salts (soaps) are deposited by displacing the sodium ion from the original soap. If water is not flowing through the machine when the flame is turned on, these salts or the previously mentioned carbonates become baked onto the coil.

To ensure safety remember the following:
- Be sure to lay out the hose so that it has no kinks in it. Pulling on a kinked hose can cause it to break, and then someone can be burned.
- Be sure that water is coming out the end of the steam gun before turning on the heat.
- Be sure to turn off the flame and let the water run until cool before turning off the machine completely. The soap control valve should be shut off at the same time the flame is. That way any soap that remains will be flushed through the coil as the machine cools down. It takes approximately

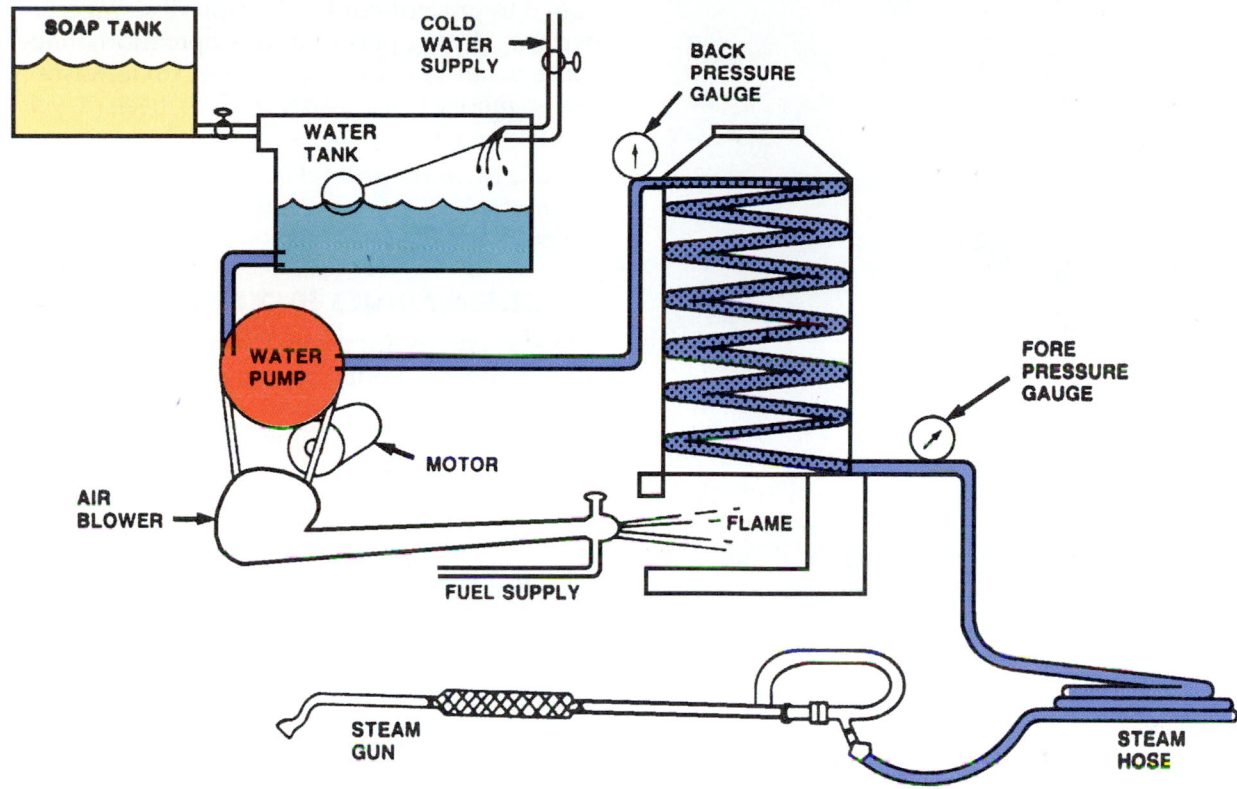

Figure 9.8 Steam cleaner components.

2–5 minutes for the soap to finish passing through the steam coil.

Most good steam cleaners have a *blow-down* procedure for scale removal. Blow-down is part of a regular maintenance schedule necessary for proper operation of the machine. Instructions for this procedure are in the operating manual.

Application of Soaps before Disassembly. Cleaning materials can be applied through the cleaning machine. They can also be brushed onto the parts or they can be sprayed on with a solvent siphon gun (Figure 9.9).

SHOP TIP A plastic outboard motor gasoline tank makes a handy soap storage container. Soap does not corrode the plastic as it would a metal tank. A siphon spray gun can be attached to the fuel outlet of the tank. Soap is siphoned out of the tank when compressed air is directed through the blowgun.

Cleaning with Solvent Chemicals before Disassembly

Ordinary stoddard solvent or cleaning chemicals such as "gunk" are sometimes used to clean the outside of engines and transmissions. Aerosol spray-on foam-type cleaners are also available.

Cleaning chemicals work better when applied to warm parts. Scrape off any large accumulation of dirt and apply the cleaner. Allow it to penetrate the grease deposits for five minutes. Hose the loosened grease off with water. The chemical will only penetrate to a certain depth, so reapplication of the cleaner might be necessary.

When using cleaning solvents use eye protection. Do not breathe vapors and clean only in areas with adequate ventilation. Wear a respirator if using a blow-gun to apply a chemical. (Nontoxic chemicals are

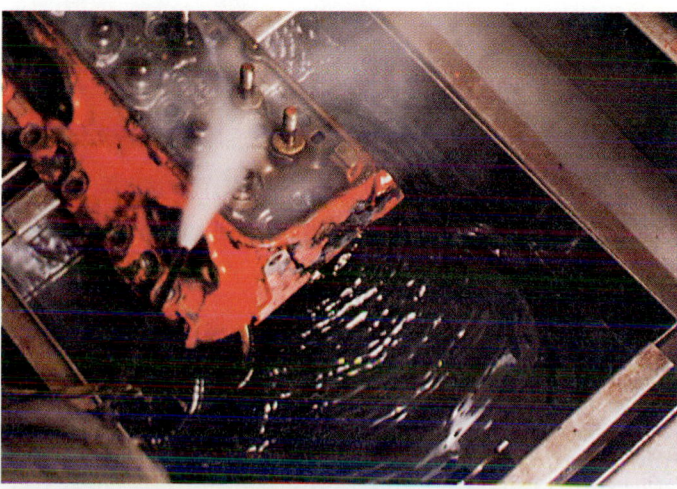

Figure 9.9 Spraying with a siphon blowgun.

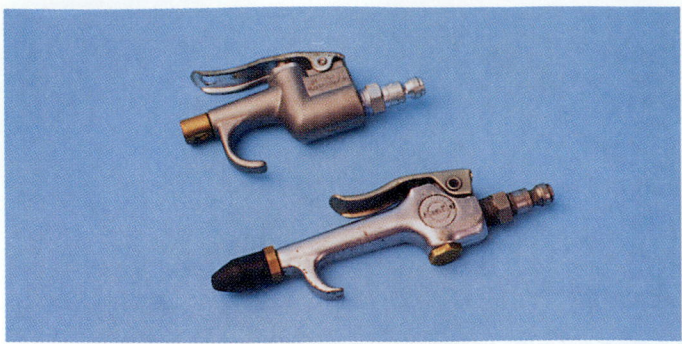

Figure 9.10 A safety blowgun (top) and a rubber-tipped high pressure blow gun (below).

available.) Be sure to use a nonflammable solvent because some solvents can catch fire on a hot engine. Do not spray flammable solvents if near any pilot lights. Chapter 11 describes some cautions about skin care and solvents.

Air Blowguns

Air blowguns can be used to blow off parts. There are two main types of blowguns. The *safety blowgun* (Figure 9.10) restricts pressure to around 35 psi for air-drying parts. When more pressure is required, a *rubber-tipped blowgun*, which allows full shop air pressure, is used.

SAFETY NOTE
Be especially careful when using the high-pressure, rubber-tipped blowgun. Compressed air at full shop pressure can be dangerous and can actually penetrate your skin. Do not blow air against your skin. Wear eye protection and use care when blowing into holes. The liquid will come back out at you. Remember to blow down and away from yourself.

■ ENVIRONMENTAL CONCERNS WITH ENGINE CLEANING

Cleaning engines when they are installed in a vehicle that runs involves cleaning dirt from the outside of the engine. Cleaning oily engine parts that have been disassembled presents another problem. The inside of a dirty engine can harbor hazardous waste. When cleaning the inside of the engine, the steam cleaner and pressure washer are to be avoided. Engine bearings are made of lead and other metals. As bearings wear, the metal is deposited in the engine oil. Used engine oil also contains other contaminants that cannot be allowed to enter the sewer.

Some localities have environmental laws that require the liquid from cleaning operations to be con-

tained to prevent contamination of sewage. Shops that perform these open cleaning operations have special waste water systems. Disposal of toxic waste is costly. Many methods of disposal have been developed by shop operators.

NOTE: *Even materials that are called biodegradable will become hazardous waste when they pick up hazardous materials during use.*

■ CLEANING INTERNAL PARTS

There are many methods for cleaning parts that have been removed from the engine. There are also additives for removing sludge from the inside of the crankcase of a running engine. The vehicle should *not* be driven while these additives are flushing the engine, because they do not have proper lubricating characteristics. Engine oil is changed immediately following the flushing procedure.

Methods of cleaning the inside of the engine include *chemical cleaning, abrasive cleaning,* and *thermal cleaning.* These cleaning methods must all keep contaminants contained for proper toxic waste handling.

■ CHEMICAL CLEANING

Chemical cleaning includes solvent tanks, small parts cleaners (carburetor cleaner), hot and cold tanks, spray washers, jet washers, ultrasonic cleaners, and salt baths.

Solvent Cleaning

Solvents are either mineral spirits, Stoddard solvent, carburetor cleaner, or water-based solvent. Once a solvent has been used, they all come under hazardous waste laws. Of these, water-based solvents are easiest to dispose of because they can be evaporated to reduce their volume prior to disposal. Stoddard solvent is commonly recycled and, although the use of gloves is recommended, it is not a major irritant to most people's skin. Carburetor cleaners are the most difficult of these three solvents to dispose of and fumes can be hazardous.

Solvent Tanks

The *solvent tank* or *cold tank* (Figure 9.11) is used for cleaning grease off smaller parts. Cold solvent does a good job of cutting grease. Although solvent would clean better if it was hot, evaporation and fuming would result.

SHOP TIP Scrape heavy deposits of grease from parts before cleaning them in the solvent tank.

Solvent is applied with a parts brush, which has stiff bristles that will not be softened by the solvent.

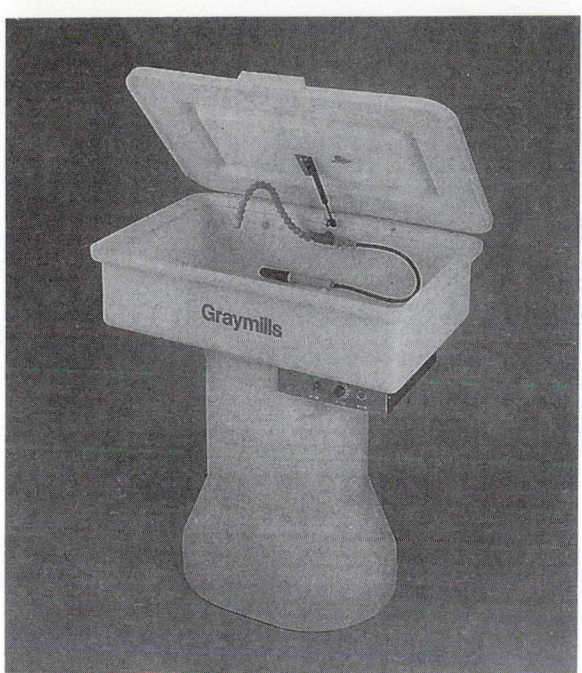

Figure 9.11 A solvent cleaning tank. *(Courtesy of Graymills)*

Figure 9.12 Carburetor cleaner.

Some solvent tanks have agitators to keep the solvent moving. Agitation reduces cleaning time. Dirt settles in sediment trays in the bottom of the tank where it is periodically cleaned out.

When solvent becomes too oily, it is usually recycled. Companies that specialize in waste handling pump the old solvent from the tank and replace it with clean recycled solvent. Solvents are often flammable. Chapter 11 includes a safety section about dealing with cleaning solvents.

Carburetor Cleaner

Carburetor cleaner *(chlorinated hydrocarbon)* is another type of chemical parts cleaner that is especially effective in cleaning aluminum, zinc, and cast metals (pot metal).

SCIENCE NOTE

Chlorinated hydrocarbons like chloroethane or dicloroethane are effective parts cleaners because of their ability to dissolve oils and greases. The organic or hydrocarbon portion of the oil or grease dissolves in the hydrocarbon portion of the chlorinated hydrocarbon, while the inorganic or metallic part of the oil or grease dissolves in the chlorinated part. Chlorinated hydrocarbons also do not react with the metal parts being cleaned.

Carburetor cleaner typically comes in a 5-gallon container. A parts basket like the one in Figure 9.12 is used to hold small parts. Most of these baskets have a small hook on the side that allows the basket to be hung on the top of the tank, out of the solution while parts drip

dry. Agitation, which reduces cleaning time, is accomplished by lifting the basket and turning it a few times while the parts soak. Some more expensive tanks for carburetor cleaner have lids that automatically agitate the solution.

SHOP TIP Put a 1"–2" thick layer of water on top of the carburetor cleaner. It floats on top of the cleaner and keeps it from evaporating.

SAFETY NOTE Carburetor cleaner is highly concentrated and dangerous to skin, eyes, and clothing. It is dangerous to breathe and can be absorbed through the skin, so rubber gloves are worn when handling it. Rinse parts in the basket thoroughly with water before handling them. Wear eye protection.

Carburetor cleaner is washed off with water. Be careful not to use water pressure that is too high; it might redirect carburetor cleaner to your skin or eyes. Keep carburetor cleaner off the floor. It eats through many kinds of shoe soles.

Water-Based Chemical Cleaning. Water-based chemicals are usually available in either liquid or dry forms. Powder weighs less so it costs less to ship. If a spill occurs, powder can be swept up and used. Spilled liquid would be wasted.

Caustic Cleaning

One cleaning solution for **ferrous** (iron and steel) materials is a mixture of water and caustic soda (lye) heated to about 190°. This strong alkaline mixture is higher than 10 on the pH scale. Brass, copper, or bronze can be cleaned in full strength solutions because they do not react with the caustic.

NOTE: *If aluminum is cleaned in the solution designed for ferrous metals, it reacts and slowly dissolves.*

Special solutions that are not as strong are available for cleaning aluminum. Some of the aluminum solutions include brighteners. Shops that do a high volume of parts cleaning will have separate cleaning tanks for aluminum and ferrous materials.

 **CASE HISTORY** *An apprentice mistakenly put an aluminum cylinder block into a full strength caustic solution. When another employee removed it from the soak tank one hour later, it was a very dark grey.*

SHOP TIP If you cannot tell by a visual inspection whether a painted engine part is aluminum or ferrous, use a magnet. Ferrous materials are magnetic, aluminum is not.

Cleaning with a caustic solution leaves metals completely clean. Parts should be rinsed immediately to avoid continuing chemical reaction on the metal. Rusting begins almost immediately, so lubricate all machined areas as soon as the part is dry.

CAUTION Caustic soda will eat your skin and can cause blindness. Chapter 11 contains a safety section about caustics.

Hot Tanks

The *hot tank*, or *soak tank*, is one of the oldest methods of cleaning engine parts. Materials are cleaned by soaking them for a period of time of from 1–8 hours, depending on the strength of the cleaning solution. Although hot tank cleaning takes a long time, it is popular in small shops because the equipment is inexpensive.

One of the disadvantages of a simple hot tank is that the user has to devise a method of finding items that are soaking in it. An *engine sling* or chain is bolted to the block. When a piece of wire fastened to the side of the hot tank somewhere above the water level is attached to the sling, it can be easily located to retrieve the part from the bottom of the tank. Some of the more expensive hot tanks have lifting tables or baskets built into them (Figure 9.13).

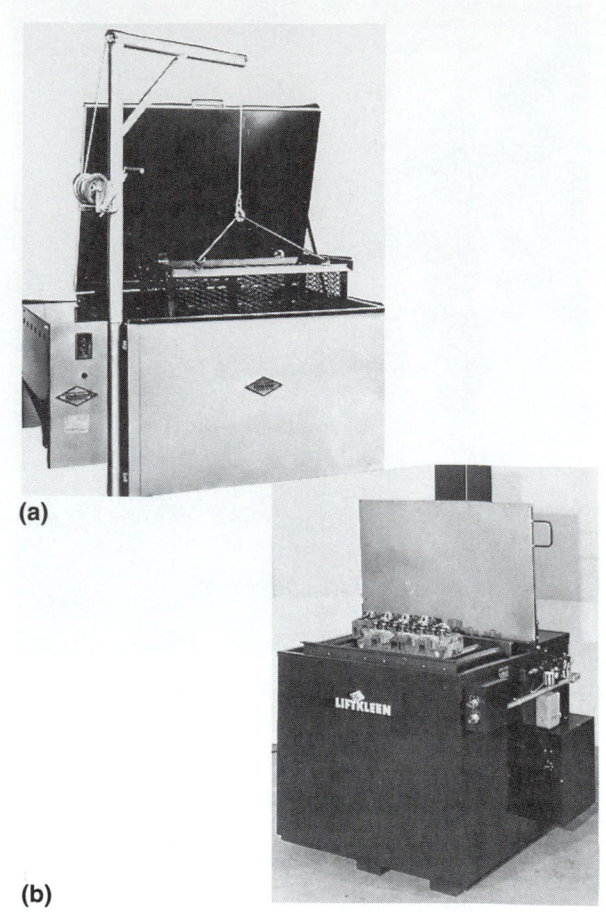

(a)

(b)

Figure 9.13 Two types of hot dip tanks. (a) A hot tank with a basket. (b) A hot tank with a lifting table. *(a, Courtesy of Kwik-Way; b, Courtesy of Graymills)*

Spray Washers

A more expensive type of hot tank that is often found in automotive machine shops is the *hot spray washer*, often called a *jet washer* (Figure 9.14). A jet washer is like an automatic dishwasher. It sprays the cleaning solution, heated to 180°F or hotter, from spray heads mounted in a long pipe. The engine block is mounted on a rotating cleaning platform. The spray heads are positioned to spray the solution at the block from the sides, above, and below. Some two-stage spray washers include a soak tank, too.

One problem the spray washer has is that of foaming, which lengthens cleaning time. Foaming occurs if a cleaning cycle is attempted before the cleaning solution is sufficiently heated. A chemical is added to caustic solutions to prevent foaming. An *agitator* diverts some of the pump water to the bottom of the tank where it stirs up any caustic that may have settled out of the solution. A control lever is positioned in the agitator position for the first five minutes of operation each day.

Advantages of a hot spray tank are:

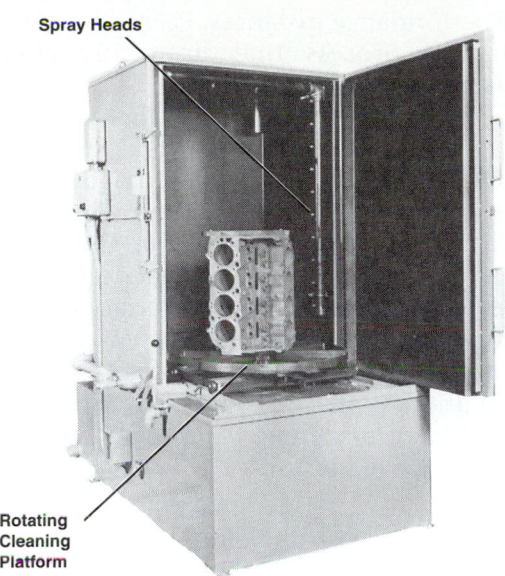

Spray Heads

Rotating
Cleaning
Platform

Figure 9.14 A jet washer. *(Courtesy of Storm Vulcan Mattoni)*

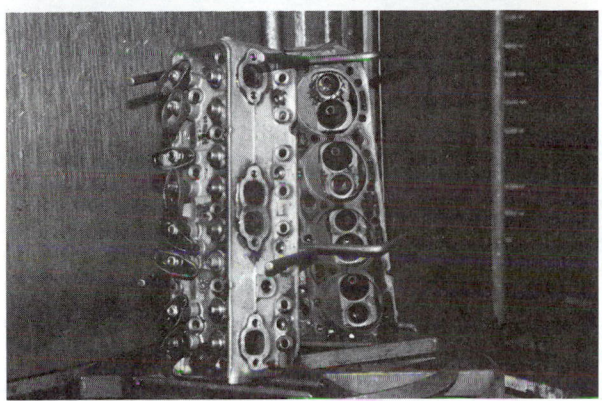

Figure 9.15 These cylinder heads were in the hot tank for 5–10 minutes. All paint and most of the dirt were removed.

1. Speed—cleaning is completed in 10–20 minutes.
2. Safety—chances of splashing are reduced. The machine will not operate unless the door is closed.
3. Drying—the temperature of the solution results in very quick drying of the parts.
4 Newer machines are more energy efficient.

A disadvantage is that spray washers do not clean oil galleries as well as the hot soak tank does.

Another type of washer has machinery that throws a large volume of cleaning solution at parts, rather than spraying it through nozzles. The advantage here is that the solution hits the dirty parts with a greater impact than the steady stream of a spray washer.

Cylinder heads can usually be hot-tanked before disassembly. This confines all oil and grease to the tank, helping keep the shop clean.

Be sure to rinse and lubricate all parts immediately after cleaning. If valve springs are allowed to rust, they will be ruined. New valve springs have a protective coating that is removed by the hot tank. Ideally, heads would be disassembled before hot-tanking, but this is not common trade practice. Figure 9.15 shows before-and-after pictures of the hot-tanked cylinder heads after 5–10 minutes in a hot spray tank.

NOTE: *Following cleaning in a caustic tank, oil galleries must be cleaned with a brush after the plugs sealing the ends of the galleries are removed. Over many miles of driving under various maintenance conditions, grime builds up in the oil galleries. This material is loosened during the cleaning process. If it is not physically removed, it will end up ruining new engine bearings and possibly more.*

Caustic Tank Hazardous Waste

Engine bearings must be removed before the block is put into caustic. Bearing lead that settles with other heavy materials in the sludge in the bottom of the tank causes it to become toxic waste.

Rebuilders have several methods of controlling or minimizing toxic waste. Skimming or overflowing the tank is done to remove oily contaminants from the surface of the solution. Some tanks have filtration systems. Others use evaporators to reduce the volume of the sludge. Sludge can be neutralized and solidified with chemicals so that they can be disposed of as non-hazardous waste.

NOTE: *Different communities have different laws governing the handling and disposal of waste materials. Be sure to check with the local agency in charge.*

Removing Scale

A chemical cooling system cleanser can be used to help remove **scale** from the radiator and water jackets *before* removing the engine.

Manual Cleaning Methods

A wire wheel (Figure 9.16) mounted on a bench grinder, on a pedestal grinder, or on a drill motor is the most common tool used for carbon removal.

Figure 9.16 A wire wheel mounted on a grinder.

Handheld wire brushes are available for cleaning valve guides and oil galleries in the block, head, and crankshaft. Sandpaper is often used to clean machined surfaces of the head and block. *Scotchbrite™* pads are also used, either by hand or machine.

Abrasive Cleaning

Materials to be cleaned by abrasive cleaning methods must be free of oil and grease, which can interfere with the proper operation of an abrasive cleaning machine. Following pre-cleaning, two types of abrasive blasting are used for various cleaning applications. *Shot* is round and *grit* is sharp and angular. Several blast materials are used by engine rebuilders for cleaning parts. Steel shot and glass beads are used for automotive part cleaning when removal of the surface of the material being cleaned is not desired.

Beads and shot come in various sizes, depending on the application. Smaller beads are used where there are tight corners and crevices (like screw threads or gear teeth) to be cleaned. Large beads are better for cleaning flat surfaces or loosening heavy deposits. Other shot materials, like stainless steel, ceramic, plastic, walnut shells, or aluminum may be used for special applications. Plastic chips, for instance, are used on plastics and soft metals, or when there is a chance of trapped shot dislodging and destroying an engine or a transmission.

Beads or shot may be used to improve the strength of parts. While ordinary shot blast cleaning does impart some strength to a part, it is not *peening*. Peening specifically for strength is done only in heavy-duty or high-performance instances. Peening for strength is a controlled process that uniformly compresses stressed areas of a part.

Steel, which is heavier than glass, is used more often for peening. It peens more intensely and lasts several times as long as glass beads before it is worn out.

Grit, used for heavy-duty cleaning, often uses the same blasting material, called *media*. The shape of the media is angular, rather than spherical so it etches (removes) material from the part surface during cleaning. It provides excellent surface preparation prior to painting. Steel grit and aluminum oxide are the most common grit materials. Grit blasting also causes stress in the surface of machined parts. It is not widely used in the automotive industry.

Glass Bead Blasters

Glass bead blasting is a very effective means of removing carbon. *Bead blasters* (Figure 9.17) are found in most automotive machine shops. The finish left by the beads improves the surface of the material by removing flaws and stress spots. It is also said to provide an ideal bearing surface. Bead-blasted parts are shown in Figure 9.18.

A costly drawback to glass bead blasting is that the machine is *labor intensive* (requires an operator). Other popular cleaning methods, like shot blasting, do not require a machine operator. To operate the bead blaster, compressed air draws in glass beads and directs them at the parts to be cleaned. Most units have a foot pedal to control the flow of air and beads. The operator watches through a window and aims the nozzle at the parts.

Figure 9.17 A glassbead blaster with a reclaimer. *(Courtesy of Winona Van Norman)*

Figure 9.18 One half of this cylinder head was bead blasted. *(Courtesy of Dee Blast Corporation, Stevensville, MI)*

Glass bead cleaning is often done improperly. Bead blasting a part that has oil galleries is not advisable. Oil galleries often have blind spots where beads can become lodged during cleaning (Figure 9.19). Trapped beads can later escape into engine oil. Glass beads that get into engine oil can embed in soft engine bearings. The bearings can swell, eliminating oil clearance or causing abrasive wear.

CASE HISTORY *A machinist cleaned an aluminum cylinder head using a glass bead blaster. The engine was reassembled and installed in the car. The customer drove the car home (30 miles). The next morning the engine would not crank over. Upon disassembly, it was discovered that the engine bearings were embedded with glass beads. The beads swelled the bearings, taking up the normal bearing clearance and preventing the crankshaft from turning.*

Glass beads can cause abrasive wear, especially to aluminum. *Excessive* blasting can round out ring grooves, which must be square for proper ring sealing. Pistons are commonly cleaned with a bead blaster but

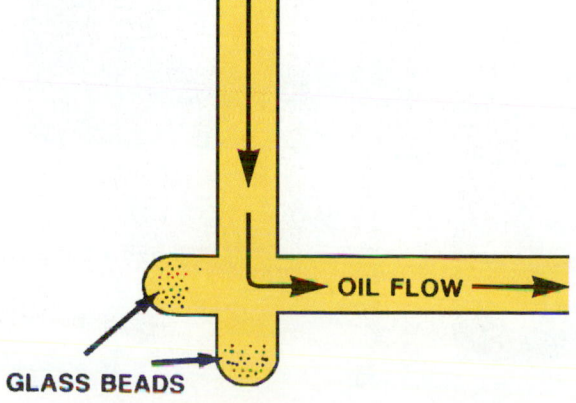

GLASS BEADS OIL FLOW

Figure 9.19 Glass beads can become trapped in blind oil galleries during cleaning.

this should never be done unless the connecting rod has first been removed. Glass beads that get in between the piston pin and bore can cause wear to the aluminum piston pin bores. Beads left in piston ring grooves can interfere with proper ring function.

Following cleaning, glass beads must be thoroughly removed from all parts. It is important that the air supply to the bead blaster be dry or beads will stick together, plugging up the machine. To clean parts after blasting, first use compressed air from the blowgun inside the blaster cabinet. Do not use solvent, which tends to stick the beads to each other. Parts can be dried in an oven to make removal of glass beads easier. A *tumbler*, used most often for removing steel shot following shot peening (covered later in this chapter), can also be used to remove glass beads (Figure 9.20). Machined areas on ferrous metals should be lubricated immediately to prevent rusting. The following precautions should be taken when using a bead blaster:
- Do not accidentally blast the window in the blaster cabinet. The result is a "frosted" glass that must be replaced.
- Do not blast parts unless the reclaim motor is on. The entire shop will be filled with glass dust.
- Some bead blasters have separate reclaim cabinets (see Figure 9.17). When glass beads become too small to be useful any longer, they are separated out by the reclaimer. A catch tray in the bottom of

Figure 9.20 A tumbler. Engine parts are placed in the center of the tires. As the tires roll, glass beads are shaken out of them. *(Courtesy of Winona Van Norman)*

the reclaimer fills with dust and must be emptied periodically (Figure 9.21).

■ Two rubber gloves extend into the cabinet for holding the blaster nozzle (Figure 9.22) and maneuvering parts. Try not to hold parts with the gloves. The fingers of the gloves get holes worn into them (Figure 9.23). A small spring clamp is effective for holding small parts. Do not use the bead blaster if there are holes in the gloves.

Figure 9.21 A reclaim tray full of worn-out glass beads.

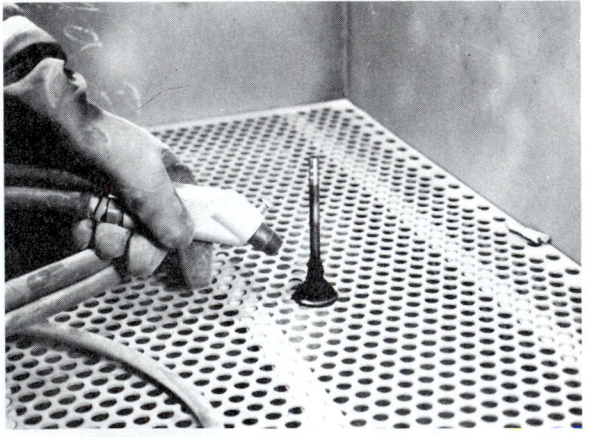

Figure 9.22 The blast nozzle held with a rubber glove.

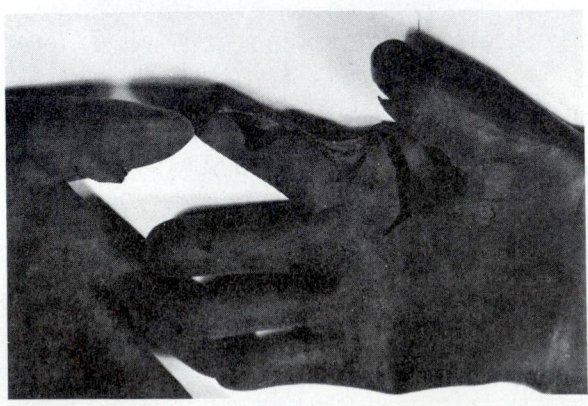

Figure 9.23 Worn bead-blaster gloves.

Airless Blasters

An airless centrifugal blasting machine, also called a *shot blaster* (Figure 9.24), uses an impeller in a sealed cabinet to scatter steel shot at a part from above and below. Shot blasting has recently become more popular due to environmental concerns with other cleaning methods. While this cleaning process is not labor intensive, two extra operations are required: pre-cleaning of oil and grease and removal of trapped shot from engine parts before beginning reassembly. A tumbler is used to remove any steel shot that remains in the part.

Shot blasting is widely used to clean ferrous parts such as the block, heads, and sometimes crankshafts and connecting rods. It is done before the parts are re-machined because the process distorts machined surfaces. Shot blasting can damage valve springs; it changes their spring rate. Most shops use 100% steel shot, although some like to use a mixture of steel and aluminum shot to give a brighter looking surface.

Shot comes in different sizes and hardnesses. Popular sizes range from 110 (smaller) to 230 (larger). The most popular size is 170.

NOTE: *Changing the size of the shot affects the amount of impact on the part; larger shot has more impact.*

Using a 110 shot results in an impact that is more than four times less than using a 230 shot. However, a small shot works better for lighter cleaning and smaller surfaces. As shot wears, it becomes smaller. Old, spent shot is automatically separated from the remaining shot. New shot is added and the result is a mixture of larger and smaller sizes.

Figure 9.24 A shot blaster. *(Courtesy of Winona Van Norman)*

The most widely used steel shot has a Rockwell C hardness of 40–50 Rc (hardness is measured using a scale called the Rockwell scale with values listed as Rc). Harder shot is used for heat treated parts and for peening.

THERMAL CLEANING

Many rebuilders use *thermal cleaning*; a cleaning procedure in which a *pyrolytic* (high-temperature) oven (Figure 9.25) cooks oil and grease, turning it to ash. The hard, dry deposit that is left on the part is removed by shot blasting or jet washing.

There are two types of thermal ovens: convection and open flame. A *convection oven* is a flameless, insulated oven that is heated by burners from the bottom. Parts are not exposed to flame and heat up gradually as the oven heats up. This is said to be an advantage in that the gradual heating of parts gives less chance of warpage. Depending on the size of the oven and the quantity of parts being cleaned, the cleaning cycle lasts from 1–4 hours.

Temperatures for cleaning ferrous metals are about 700°F. Aluminum, which softens at about 650°F, is cleaned at about 450°F.

An *open flame oven* is like a rotisserie. Parts are mounted in a cage that avoids hot spots by slowly rotating the parts directly over a flame. The average temperature of the flame is about 1100°F, but the temperature of the air inside the oven is only about 500°, allowing aluminum and ferrous metals to be cleaned together. After about 10 minutes of exposure to the flame, it goes out and a 20-minute baking cycle begins. The total cleaning cycle lasts about ½ hour.

With oven cleaning, three processes are actually used: pre-cleaning, baking, and post-cleaning. Shot blasting is the choice for post-cleaning recommended by most oven companies because parts can be blasted without waiting for them to cool down as would be required with jet washing. Ash that remains is soft and the time in the shot blaster is only a few minutes. Fifteen minutes in the tumbler finishes the process.

Advantages of thermal cleaning are:

1. Lower cost: not labor intensive, no operator is necessary except for transferring parts between the oven, blaster, and tumbler.
2. While lead or heavy metals still remain in the ash residue, there is a lower volume of hazardous waste to dispose of than with some of the other cleaning methods.
3. The inside of oil galleries in the block are thoroughly cleaned.
4. Heat turns rust and scale in the water jackets to powder. These materials would normally need to be removed with acid.
5. Carbon deposits in manifolds and combustion chambers are loosened. These deposits resist other cleaning methods.
6. Shot blasting removes stress raisers in the surface of parts. This can strengthen the part, lessening the tendency to crack.
7. Aluminum welding is easier after open flame cleaning because contaminants are so thoroughly cleaned.
8. Warped cylinder heads can be straightened in the oven (see Chapter 42).

Gallery plugs should be removed before cleaning ash with a shot blaster to prevent shot from being trapped in oil galleries. Studs in aluminum cylinder heads should be removed before oven cleaning. They will be difficult to remove afterwards.

Air pollution from vaporization of contaminants is dealt with by oxidizing the pollutants in the smokestack. Minimum temperature in the smokestack is from 1400°–2200°F so there is no visible soot or unburned hydrocarbons.

VIBRATORY PARTS CLEANERS

A *vibratory parts cleaner* (Figure 9.26) is a vibrating tub that uses large beads of ceramic, aluminum, or plastic in a cleaning solvent (usually Stoddard solvent). It is very effective on valves and valve springs and does not require a machine operator. It should be installed in a soundproof room. When the tub shakes, it is noisy.

OTHER CLEANING METHODS

Other cleaning methods are used in larger and non-automotive applications. *Salt bath* cleaning systems are used mostly by large production engine rebuilders. Parts are soaked in a bath of 650°F molten oxidizing salt for 5–10 minutes and then rinsed.

Ultrasonic cleaning is used commonly by jewelers and dentists. It cleans by sound waves cycling through water and detergent that is heated to between 120°

Figure 9.25 A cleaning furnace. *(Courtesy of Pollution Control Products Co.)*

and 140°F. As the sound waves cycle at about 40,000 times per second, bubbles open and collapse, tearing contaminants away from the metal. Some rebuilders use this method for small parts cleaning.

Figure 9.26   A vibratory parts cleaner. *(Courtesy of C&M Cleaning Systems)*

Marking Clean Parts

Clean parts can be marked with number or letter stamps or with a colored paint marker (Figure 9.27). Colored paint markers have white or yellow paint stored in a tube with a rolling ball point. A rubber bulb on the end is used to prime the ball point. The paint is resistant to oil and solvents. Parts like blocks, heads, or crankshafts can be readily labeled.

Figure 9.27   Labeling parts with a paint marker. *(Courtesy of K-line Industries, Inc.)*

◼ REVIEW QUESTIONS

1. What happens when superheated water comes out of the steam cleaning gun into the air?

2. Why is the hose laid out before using the steam cleaner?

3. If the nozzle of a bead blaster is aimed at the window, it frosts the _____.

4. Why should the nozzle not be aimed at the glove that is holding the part being blasted?

5. Be sure to rinse and _____ to protect machined areas of ferrous metals after hot-tanking.

6. How can a metal be tested to see if it is ferrous or nonferrous?

7. An OHC head has oil galleries. This makes it impractical to clean these heads with a glass bead _____.

8. What is another name for the airless blaster?

9. _____ and open flame are two types of thermal cleaning ovens.

10. Rust and _____ are two types of materials that thermal cleaning removes that other cleaning methods do not.

◼ ASE STYLE REVIEW QUESTIONS

1. Technician A says that hot pressure washing uses water that is heated above its normal boiling point while under pressure. Technician B says to shut off the water on the steam cleaner before turning off the flame. Who is right?

    a.   Technician A   **b.**   Technician B
    c.   Both A and B   **d.**   Neither A nor B

2. Technician A says that if the temperature of the hot tank liquid is too low, cleaning time will be longer. Technician B says that if the temperature of the hot tank liquid is too low a spray-type hot tank will foam over. Who is right?

    a.   Technician A   **b.**   Technician B
    c.   Both A and B   **d.**   Neither A nor B

3. Technician A says that larger shot hits a part with more impact. Technician B says that parts can be made stronger by shot peening. Who is right?

    a.   Technician A   **b.**   Technician B
    c.   Both A and B   **d.**   Neither A nor B

4. The temperature of the caustic cleaner in the hot tank should be about:

    a.   100°   **b.**   150°
    c.   190°   **d.**   250°

5. Technician A says that steel shot blasting is commonly used to clean aluminum. Technician B says that on a steam cleaner, the water is shut off immediately after shutting off the gas. Who is right?

    a.   Technician A   **b.**   Technician B
    c.   Both A and B   **d.**   Neither A nor B

Lifting Equipment and Air Compressors

■ **INTRODUCTION**

Hydraulic equipment is used for lifting vehicles and heavy parts. Portable equipment, like a hydraulic floor jack, is used in all automotive shops. That equipment is covered in Chapter 8. Most shops also have hydraulic or electric lifts for lifting vehicles high in the air for convenience when they are worked on. This chapter deals with lifts and the air compressors that operate some of them.

The Automotive Lift Institute (ALI) is a trade association that promotes the safe design, construction, installation, operation, maintenance, and repair of automotive lifts. The ALI has several safety materials available.

Automotive *lifts* are often called *hoists* or *racks*. The two main categories are the *in-ground* lift and the *above ground* (surface mount) lift. When the lifting mechanism of a lift is located below the floor, the lift is an in-ground lift.

■ **LIFT TYPES**

Within the two categories of lifts (surface mount and in-ground) there are several styles, but two main types:

■ The *frame-contact* lift (Figure 10.1a–c).
■ The *wheel-contact* (drive-on) lift (Figure 10.2).

There are also in-ground lifts called *axle engaging* lifts, and other lifts called *pad lifts* (covered later). An advantage to frame-contact, axle engaging, or pad lifts is that the wheels hang free, which makes it easier to perform tire, brake, and suspension work. These lifts usually provide better access to the underside of the vehicle.

■ **FRAME-CONTACT LIFTS**

Frame-contact lifts (see Figure 10.1a–c) have adapters at the end of adjustable *lift arms*. The lift adapters

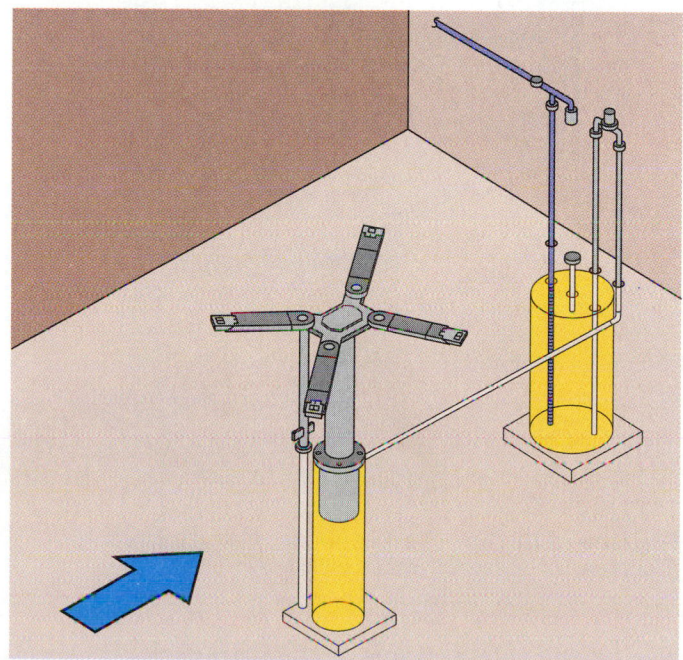

Figure 10.1a Single-post frame-contact lift. *(Courtesy of Automotive Lift Institute)*

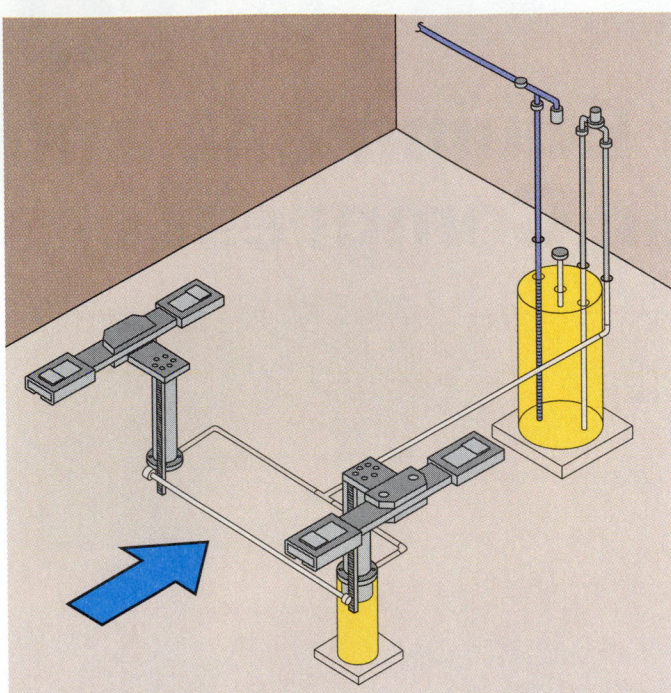

Figure 10.1b Two-post frame-contact lift. *(Courtesy of Automotive Lift Institute)*

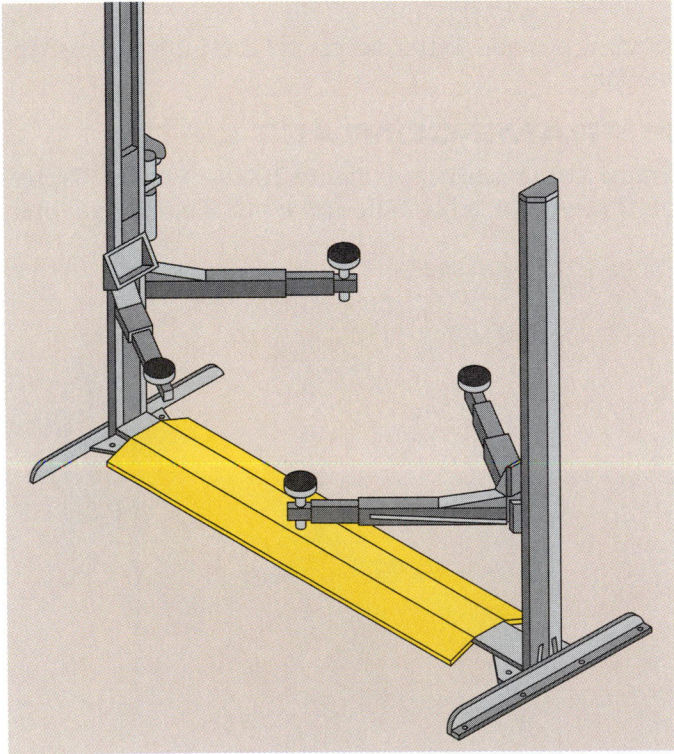

Figure 10.1c Surface mount frame-contact lift. *(Courtesy of Automotive Lift Institute)*

contact the vehicle frame at the manufacturer's specified lift points (at the rocker panels or on a section of the frame) (Figure 10.3). Be sure to check the service manual before trying to lift a car. Sometimes when a

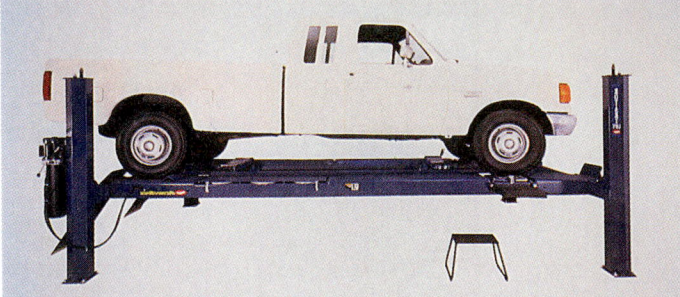

Figure 10.2 Wheel-contact lifts. *(Courtesy of Dover)*

car is lifted improperly, body, suspension, or steering components can be bent. The windshield might even pop out or the vehicle might fall.

In 1992 the Society of Automotive Engineers adopted a standard (SAE J2184) for vehicle lift points. Some manufacturers label their vehicles with decals depicting the recommended lift points. These are located inside the passenger side front door. Permanent markings on the underside of the chassis also mark the lift points. These markings can include a hole, a boss, or a ¾" depression of a triangle.

Some adapters (sometimes called *foot pads*) flip up. These adapters are adjustable to several positions to accommodate different heights between the lift points in the front and rear of the vehicle and to provide clearance between the rocker panel and the lift arm (Figure 10.4). Some foot pads have screw threads that allow them to be positioned higher or lower (Figure 10.5).

> **SAFETY NOTE** Adapters must be placed at the manufacturer's recommended lifting points and they must be set to raise the vehicle in a level plane or the vehicle will be unstable.

Some adapters have rubber pads. Be sure that they are in good condition and that they are not covered with oil or grease that could make them slippery.

Passenger cars with unibody construction are lifted at points on the rocker panel (just under the bottom outside edge of the car body). There are special rocker panel lifts, called pad lifts (Figure 10.6), that can be used for these cars and cars with perimeter frames. Do not use pad lifts for trucks.

Special Auxiliary Adapters/Extenders

Sometimes extenders are used on the adapters (Figure 10.7). If the lift points on the vehicle are undercoated, a special adapter might be needed when using a lift

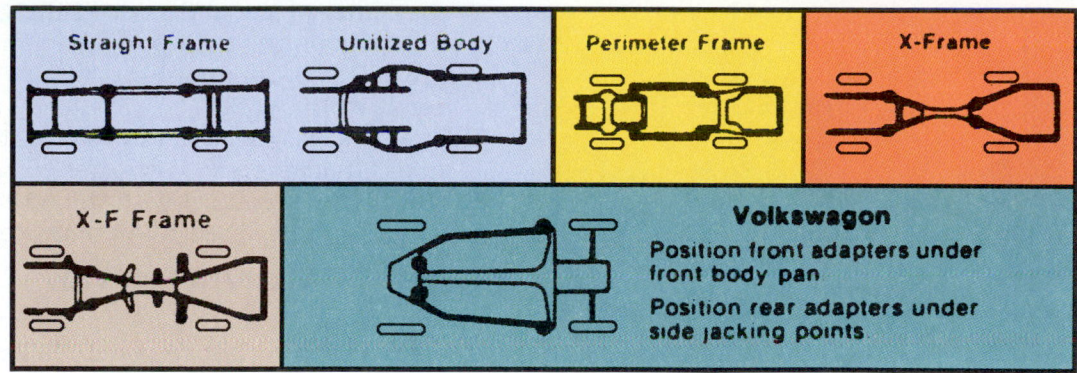

Figure 10.3 Check the service manual for the correct lift points for a particular vehicle. *(Courtesy of Automotive Lift Institute)*

Figure 10.4 Adapters flip up to accommodate different frame heights. *(Courtesy of Automotive Lift Institute)*

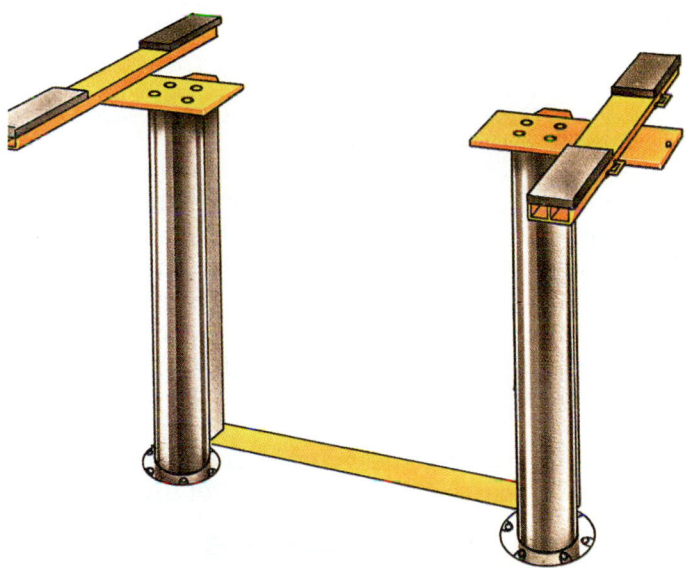

Figure 10.6 A rocker panel (pad) lift. *(Courtesy of Automotive Lift Institute)*

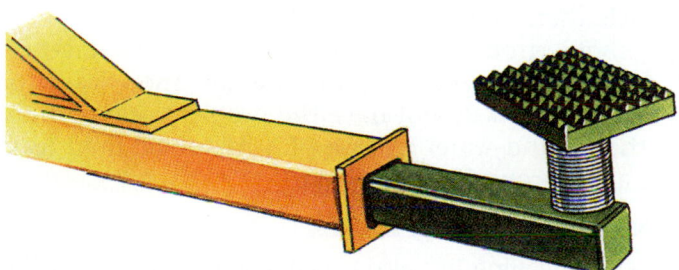

Figure 10.5 Some adapters have a screw adjustment. *(Courtesy of Automotive Lift Institute)*

that has steel adapters. Damaging the undercoating can void the vehicle's rust protection warranty. Some sport utility vehicles, light trucks, and vans require special adapters to provide clearance between the lift arm and the rocker panel. Most lift manufacturers make these available. Do not use makeshift extenders.

■ WHEEL-CONTACT LIFTS

A wheel-contact lift is one in which all four of the vehicle's wheels are supported (see Figure 10.2). Some of these "drive-on" lifts are of the in-ground type. Others are the four-post, surface mount type. The four-

Figure 10.7 Extenders can be used with pickup trucks and vans. *(Courtesy of Automotive Lift Institute)*

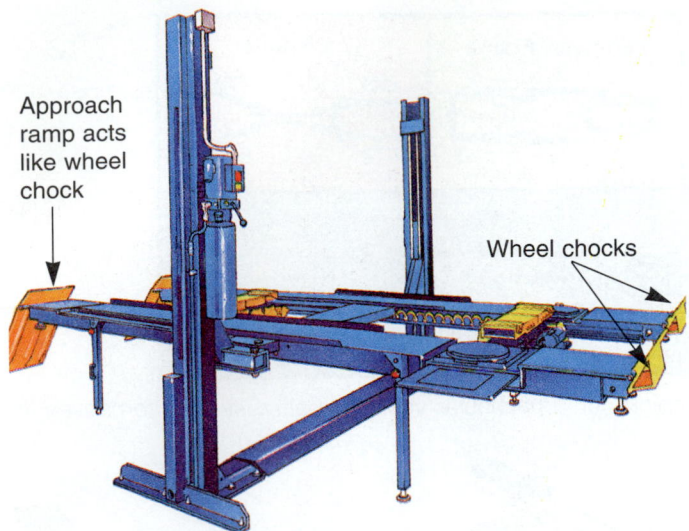

Figure 10.8 This lift has safety stops at each end. *(Courtesy of Automotive Lift Institute)*

post type has the advantage of easier installation and removal. Four-post wheel engaging lifts are especially popular in service shops, engine and transmission repair, wheel alignment, and muffler shops.

When driving a car onto one of these lifts be sure that the tires are an equal distance from the edges of the ramps. Wheel-contact lifts have secondary stops for roll-off protection at the front and rear (Figure 10.8). After spotting the vehicle, always use manual *wheel chocks* to prevent the car from rolling.

WHEEL-FREE JACKS

On a wheel-contact lift, the wheels are supported unless a *wheel-free* jack is used (Figure 10.9). These jacks, which are air powered or hydraulic, are used to raise either end of the car for wheels-free work. After the vehicle is raised, the jack is lowered onto a mechanical safety latch. When using one of these jacks, keep hands clear and be sure to extend each of the lift arms an equal amount to avoid uneven loading. Be sure that a wheels-free jack is lowered all of the way before driving into or out of the lift.

Some advantages of wheel contact lifts are:
- The vehicle may be raised safely, even without the engine in it. Trying to lift such a vehicle can be dangerous on a frame-contact lift, because the unbalanced vehicle might tip off the lift.

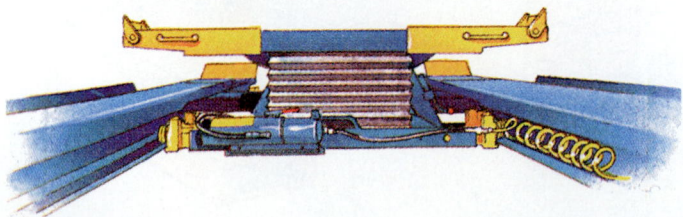

Figure 10.9 A wheels-free air jack. *(Courtesy of Automotive Lift Institute)*

- The center of the underside of the vehicle is accessible on the twin-post hydraulic and four-post types. This is important for exhaust system and transmission work.
- For quick, easy service the vehicle requires no set-up of lift adapters.
- Wheels are in contact with the wheel ramps. This is necessary for wheel alignment work.

IN-GROUND LIFTS

Most in-ground lifts used for raising automobiles have either one or two *pistons*, often called *posts*. In two-post lifts, the pistons are located in one of two ways:
- One in front of the other
- Side by side

Newer in-ground systems must be enclosed with double walls to prevent oil from leaking from them into the ground.

Frame-Engaging Lifts

In-ground, frame-contact lifts are either *single-post* or *two-post* (see Figure 10.1a & b). The two-post style is either *drive-through* (Figure 10.10) or *drive-over* (Figure 10.11). Drive-through lifts are more open in the center to allow easier access to the underside of the vehicle. The vehicle is driven between the lift arms.

The lift arms on a drive-over lift are located closer together so the vehicle can be driven over them without bumping into them. As vehicles have been downsized, drive-through lifts have become more popular. Using a drive-over lift to work on the underside of a small car is very difficult because access is extremely limited. Two-post lifts provide better access to the underside of the vehicle than single-post hoists do.

Above ground lifts also have the advantage of being relatively portable, in case the shop loses its lease. They also do not have the problem of pollution of the ground water in case of a hydraulic fluid leak like an in-ground hoist has.

Axle Engaging Lifts

An axle engaging lift, also called a *suspension contact* lift has at least two posts (pistons) positioned front to rear (Figure 10.12). The front post is movable so that it can accommodate vehicles with different length wheelbases.

The front post of the lift has adjustable arms that are positioned just under the lower control arm, as far out toward the wheel as possible. The rear post engages the

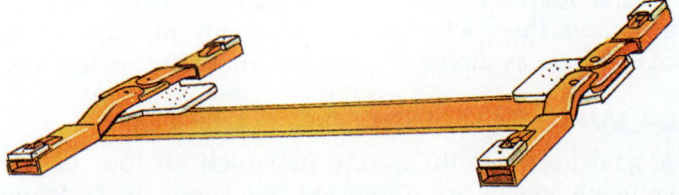

Figure 10.10 A drive-through lift. *(Courtesy of Automotive Lift Institute)*

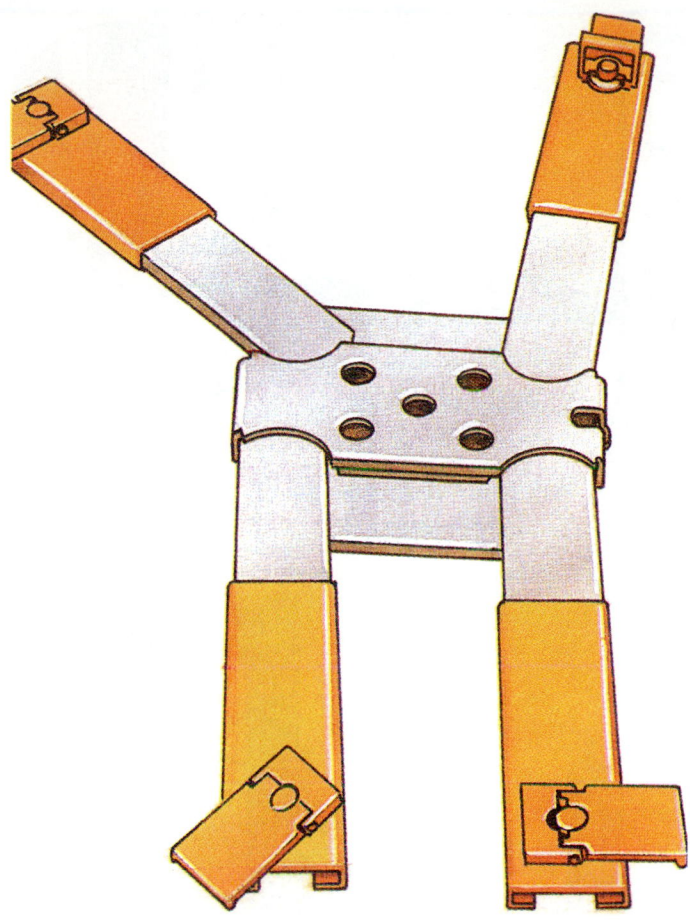

Figure 10.11 A drive-over lift. *(Courtesy of Automotive Lift Institute)*

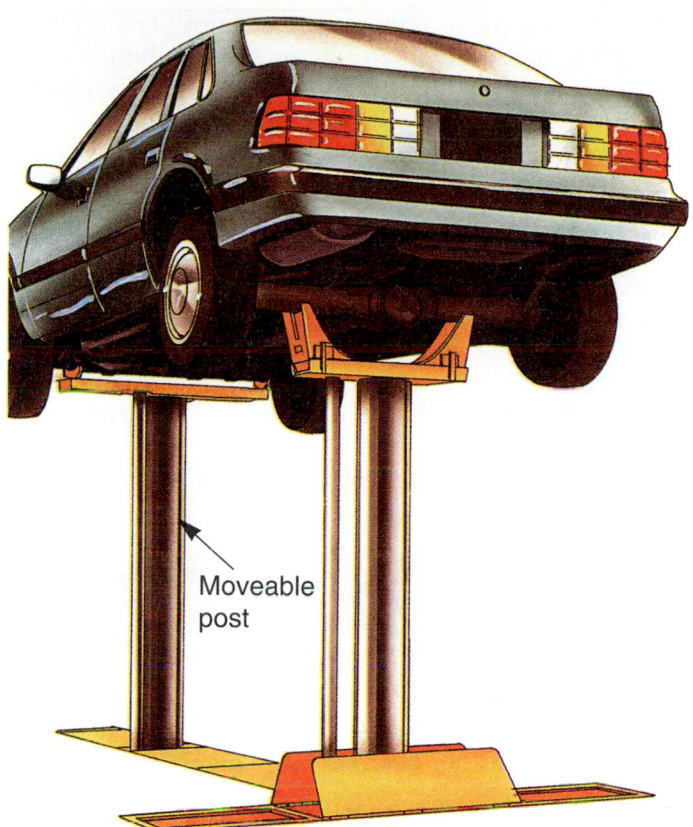

Moveable post

Figure 10.12 An axle engaging or suspension contact lift. *(Courtesy of Automotive Lift Institute)*

rear axle housing on rear wheel drive vehicles. Some front wheel drive cars have a *support rail* between the rear axles that allow the use of this type of lift, but this type of lift is not recommended for a majority of front wheel drive cars. Other lifts of this style provide the option of lifting the rear of the vehicle by the wheels (this option works for front wheel drive vehicles).

Semi-Hydraulic and Fully Hydraulic Lifts

In-ground lifts that are powered by compressed air are either semi-hydraulic or fully hydraulic. Other in-ground lifts use an electric motor driven pump to pressurize the hydraulic oil. Air pressure is not necessary in this system. Semi-hydraulic lifts have a self-contained air/oil reservoir. Fully hydraulic lifts have a separate air/oil reservoir. When compressed air is put into the tank containing the hydraulic oil, it pressurizes the oil to lift the vehicle. There are separate oil and air controls on fully hydraulic lifts (Figure 10.13). Semi-hydraulic lifts use only one control:

■ The air control is opened first.

■ When the air has had a chance to pressurize the air/oil tank, the second control handle is moved to allow the pressurized oil to lift the car.

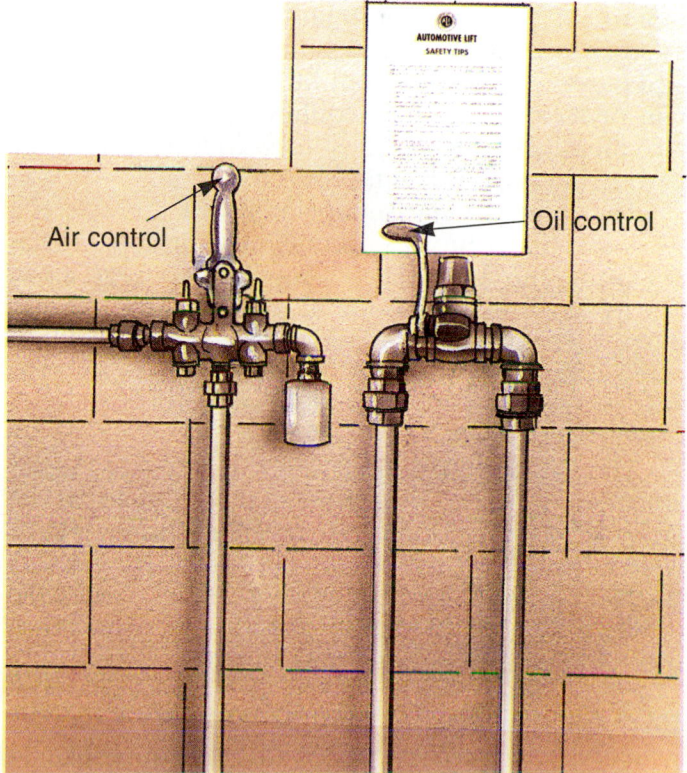

Air control   Oil control

Figure 10.13 Fully hydraulic hoists have two controls. *(Courtesy of Automotive Lift Institute)*

When the shop's air compressor has not been on long enough to build adequate air pressure, when a compressor is too small, or when the shop has been using too much air, the lift will work very slowly.

In-Ground Lift Maintenance

If an in-ground lift vibrates while lifting the vehicle or will not raise to its full height, this could be due to an oil leak. If you hear air escaping or see or suspect an oil leak, stop using the lift immediately and release the air pressure.

Lifts are usually repaired only by qualified service personnel. Sometimes an experienced technician or business owner will perform maintenance to lifts, however. Always follow the lift manufacturer's maintenance instructions.

NOTE: *Be sure that all of the air pressure is exhausted from the lift before attempting to check or add oil.*

Remove the fill plug carefully by hand. Do not use an impact wrench. There could still be air pressure in the air/oil tank. Use the specified oil only.

■ SURFACE MOUNT LIFTS

Many newer lifts (since 1970), called surface mount lifts, are mounted completely above the floor (see Figures 10.1C and 10.2). An electric motor operates either a *screw drive* (Figure 10.14) or a *hydraulic pump and cylinders*.

Ease of installation is one advantage of a surface mount lift. They can be removed and reinstalled in another location if a business owner is renting a shop and moves. A surface mount lift is also easier to maintain or repair because it requires no excavation. Surface mount lifts are also less likely to leak oil into the ground (an environmental concern).

The most popular surface mounted lift style is a two-post, drive-through, frame engaging type. It has two lifting carriages that each support two swing arms. The carriages are synchronized so that they go up and down together. Either a steel chain, cables, synchronized motors, or hydraulic circuits are used to keep the two carriages moving together. Figure 10.15 shows a chain synchronizer.

The synchronizing parts and the driving system are attached to the posts either overhead or across the floor. When they are across the floor, sometimes a wide groove is cut in the concrete. Parts that connect the two posts are submerged in the groove so that the floor remains unobstructed. This is especially convenient when a transmission jack is used.

Surface Mount Lift Maintenance

To bolt a surface mount lift, the floor must have at least 5" of solid concrete. The bolts that hold the lift to the concrete floor should be inspected periodically for

Figure 10.14 A screw drive hoist. *(Courtesy of Automotive Lift Institute)*

Figure 10.15 This lift has a chain type synchronizer. *(Courtesy of Automotive Lift Institute)*

tightness (Figure 10.16). They can and do vibrate loose. If cracks develop in the concrete, use of the lift should be discontinued until it can be checked by a *professional*. Always follow the lift maintenance instructions.

■ LIFT SAFETY

Lifts have an excellent safety record but unfortunately vehicles occasionally fall off of them. When a vehicle comes down by accident, this is usually due to carelessness, misuse, or neglected maintenance. Training in the use of the lift is mandatory before attempting to lift a car.

ANSI (American National Standards Institute) and ALI (Automotive Lift Institute) have set the American National Standard (ALOIM-1994) for automotive lifts. This standard lists safety requirements for operation, inspection, and maintenance of lifts. It requires

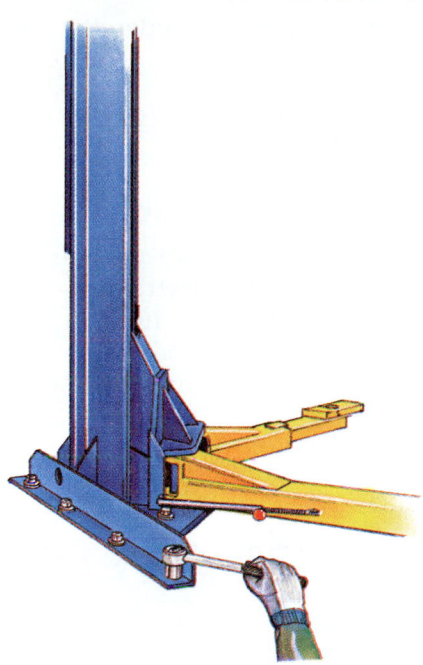

Figure 10.16 Periodically check the bolts of a surface mounted lift for tightness. *(Courtesy of Automotive Lift Institute)*

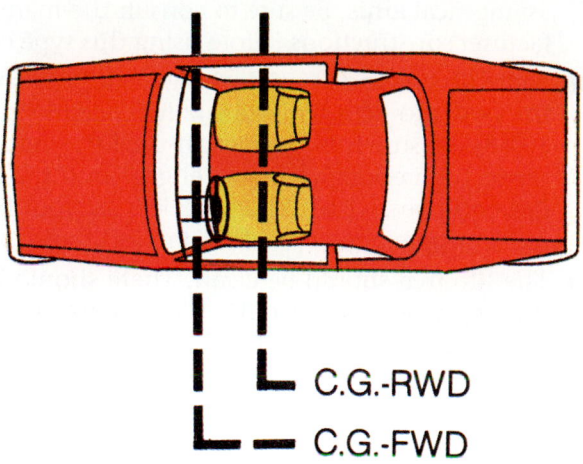

Figure 10.17 Center of gravity (C.G.) positions for front wheel drive (FWD) and rear wheel drive (RWD). *(Courtesy of Automotive Lift Institute)*

annual inspection of automotive lifts by a qualified lift inspector. This annual inspection is designed to keep automotive lifts in good operating condition.

Center of Gravity

Find the **center of gravity** of the vehicle and position it over the posts of the lift. The center of gravity is the point between the front and the rear wheels where the weight will be distributed evenly. Different positions are used for front wheel drive and rear wheel drive cars (Figure 10.17).

According to the *Automotive Lift Institute*:
- On rear wheel drive (RWD) cars, the center of gravity is usually below the driver's seat.
- On front wheel drive (FWD) cars, the center of gravity is usually slightly in front of the driver's seat, beneath the steering wheel.

 **SAFETY NOTE** You are lifting a vehicle that weighs several thousand pounds. The car must be correctly spotted on the lift so that the *center of gravity* is correct.

NOTE: *On a two-post above ground hoist, do not position the vehicle to the front or rear of the posts just so that the door can be opened easier. Positioning the **center of gravity** is very important.*

On one- or two-post lifts, position the vehicle's center of gravity over the posts. On four-post models, position the car equally between the front and rear.

Lift Operation Safety

- If there are any problems with the lift, do not use it. See your supervisor immediately. Do not take chances.
- When lifting, first raise the vehicle until its wheels are about 6" off the ground. Then, jounce the vehicle and double check the contact between the adapters and the frame to be sure the vehicle is safely engaged.
- Be certain that all four lift pads are contacting their lift points and bearing a load. It is not unusual for three lift arms to be touching the car with the fourth one free to move. If a lift arm can be moved after the car is in the air, the car is unevenly loaded. Lower the car and reposition the arm.
- If a lift arm is positioned improperly, lower the vehicle slowly to the ground and reposition the arm.

NOTE: *Some lifts have swing arm restraints that hold the unloaded arm in position to prevent accidental movement. These restraints are not designed to prevent the car from falling if it is not properly positioned.*

- When performing vehicle repairs on a vehicle on a frame contact lift, do not use a large prybar or do anything else that might knock the vehicle off of the adapters. When tight bolts are encountered, it is best to use an air impact wrench on them.
- Be sure that the lift contact points on the vehicle are in good condition and that there is no oil or grease on them.
- Some lifts have different length arms in the front than they do in the rear. These are called

asymetrical arms. Be sure to consult the manufacturer's instructions before using this type of lift.

■ Some lifts have a safety locking device that holds the post should a hydraulic failure occur (Figure 10.18). Be sure it is engaged. If the lift is not raised high enough for it to engage or if the lift is not equipped with a safety device, use four high reach supplementary stands (tall jackstands).

■ The lift area should be clean. There should be no grease or oil on the floor. Hoses, extension cords, and tools should be in the places where they belong.

■ Insurance companies usually prohibit customers from being in the lift area. Do not allow the customer to drive his/her own car onto the lift.

■ Be sure that the lift has adequate capacity to lift the weight of the vehicle. If the vehicle contains any loads inside, in the trunk, or in the bed of a pickup, the center of gravity will be affected and the vehicle will be unsafe to lift.

■ Be sure that the lift is all of the way down before attempting to drive a car into or out of it.

■ Before lowering a vehicle, be sure to alert anyone nearby. Be certain that no tools or equipment are below the car. All of the car's doors should be closed.

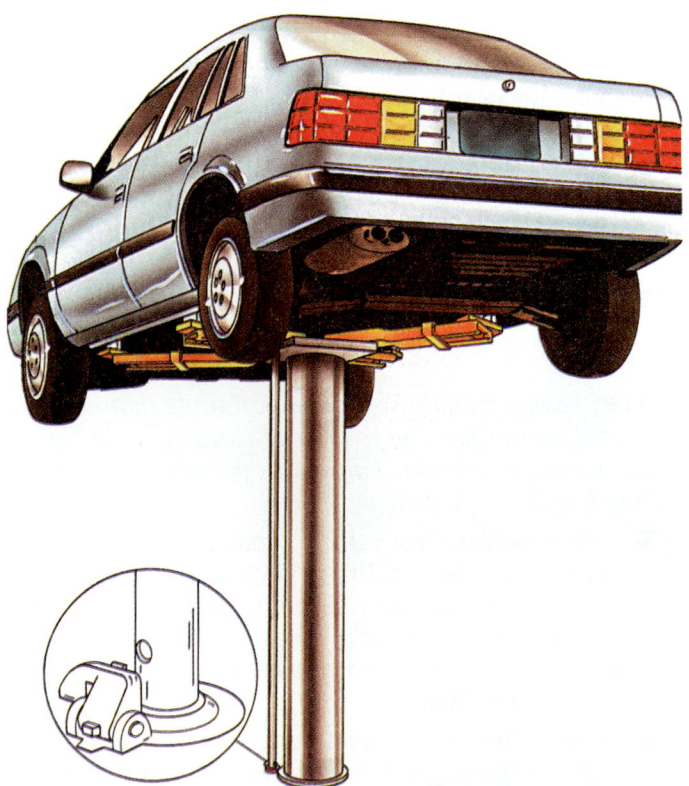

Figure 10.18 A locking device is engaged when the lift is at its highest travel. (Courtesy of Automotive Lift Institute)

Before lifting a vehicle consider the items listed in Figure 10.19.

➡ *Perform __Raise a Vehicle with a Frame-Contact Lift__ Worksheet*

■ AIR COMPRESSORS

Compressed air is used to power tools in the shop in much the same way that electricity is used. One of its advantages is that there is no danger of electrical shock, especially in auto body work where water is sometimes used when sanding. Compressed air can be

AUTOMOTIVE LIFT
SAFETY TIPS

Post these safety tips where they will be a constant reminder to your lift operator. For information specific to the lift, always refer to the lift manufacturer's manual.

1. Inspect your lift daily. Never operate if it malfunctions or if it has broken or damaged parts. Repairs should be made with original equipment parts.

2. Operating controls are designed to close when released. Do not block open or override them.

3. Never overload your lift. Manufacturer's rated capacity is shown on nameplate affixed to the lift.

4. Positioning of vehicle and operation of the lift should be done only by trained and authorized personnel.

5. Never raise vehicle with anyone inside it. Customers or by-standers should not be in the lift area during operation.

6. Always keep lift area free of obstructions, grease, oil, trash and other debris.

7. Before driving vehicle over lift, position arms and supports to provide unobstructed clearance. Do not hit or run over lift arms, adapters, or axle supports. This could damage lift or vehicle.

8. Load vehicle on lift carefully. Position lift supports to contact at the vehicle manufacturer's recommended lifting points. Raise lift until supports contact vehicle. Check supports for secure contact with vehicle. Raise lift to desired working height. CAUTION: If you are working under vehicle, lift should be raised high enough for locking device to be engaged.

9. Note that with some vehicles, the removal (or installation) of components may cause a critical shift in the center of gravity, and result in raised vehicle instability. Refer to the vehicle manufacturer's service manual for recommended procedures when vehicle components are removed.

10. Before lowering lift, be sure tool trays, stands, etc. are removed from under vehicle. Release locking devices before attempting to lower lift.

11. Before removing vehicle from lift area, position lift arms and supports to provide an unobstructed exit (See Item #7).

These "Safety Tips," along with "Lifting it Right," a general lift safety manual, are presented as an industry service by the Automotive Lift Institute. For more information on this material, write to: ALI, P.O. Box 33116, Indialantic, FL 32903-3116.

Look For This Label on all Automotive Service Lifts.

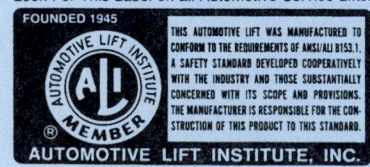

Figure 10.19 Automotive lift safety tips. (Courtesy of Automotive Lift Institute)

used to blow off parts, to apply paint with a spray gun, power hand tools, or in an air jack for lifting vehicles.

Air compressors resemble small engines (Figure 10.20). They have one or more pistons and one-way check valves. When a piston moves down, a check valve allows air to be drawn into the compressor's cylinder. When the piston moves back up, another check valve directs the air into a holding tank, called a *receiver tank* (Figure 10.21). Larger compressors are often *two-stage*, which means that air is compressed in two stages. Two-stage compressors are used when pressures over 100 psi are required.

Air Compressor Size

When shopping for an air compressor, size is an important consideration. Each air tool consumes a certain amount of air during use. Air compressor *capacity* is dependent on the number of tools that will be in use in the shop at one time and their air requirements.

Air compressors are equipped with *pressure switches* to shut the compressor off when the receiver tank reaches a set pressure. As air in the tank is consumed, the pressure drops and the switch turns the compressor back on again. An air compressor is operating at its maximum practical capacity when it runs for 7 out of every 10 minutes. This is referred to as a 70% **duty cycle**.

There is more than one rating for air compressor capacity. The best measurement of useful capacity is **free air delivery**, measured in **standard cubic feet per minute (SCFM)**. Most small air tools are designed to run at 90 psi, so it is necessary to know the air pressure at which the free air delivery was measured. According to the *Compressed Air and Gas Institute*, a good compressor choice for an automotive shop would be one with 15–22 SCFM of free air delivery at 150 psi. This amount of output is commonly achieved by a 5–10 hp two-stage, two-cylinder compressor. The air requirement depends on the number of service bays in the shop.

Air Compressor Tank

The correct receiver tank for an air compressor is an important consideration. A large tank takes longer to fill initially, but the compressor will not need to run as often to refill it. Large capacity air tools such as auto body tools or a glass bead blaster require large receiver tanks.

Where to put the compressor is also a consideration because compressors take up a good deal of floor space and they are noisy. The compressor is often

Figure 10.20 An air compressor resembles a small gasoline engine. *(Courtesy of The Campbell Group)*

Figure 10.21 The receiver tank acts as a reservoir for the air. *(Courtesy of The Campbell Group)*

installed outside of a building or in a storeroom where noise will not be a factor. Most receivers for compressors larger than 5 horsepower are *horizontal*. Tanks for smaller compressors are sometimes *vertical*, with the compressor mounted on top. This can present balance problems though, especially with larger compressors, and the tank must be bolted to the floor. Horizontal and vertical compressor tanks work equally well.

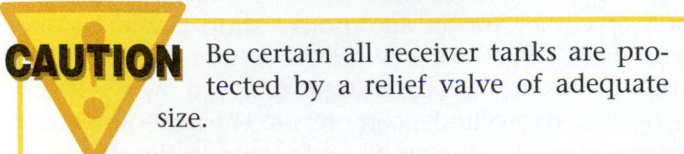 **CAUTION** Be certain all receiver tanks are protected by a relief valve of adequate size.

Compressor Motors

Permanently installed air compressors used in most shops are driven by 5 to 10 horsepower (hp) motors. Larger compressors usually require a 220-volt electrical supply. Portable compressors under 5 hp that run on 110V power are also available. Some compressors are driven by gasoline engines. These are usually used on portable compressors or permanent installations on trucks for use out in the field. Electric motors are less expensive, less noisy, cost less to operate, and they do not pollute the air in the shop.

Air Lines

It is important that air delivery lines be large enough so that enough air gets from the compressor to the tool. When lines are too small, the pressure must be raised to compensate. Raising the pressure heats the air, which is hard on tools and causes the compressor to work harder. When a line is longer than 150 feet in length, it is a good idea to use a minimum of 1" pipe size.

Each output line from the air feed system is called a *drop*. When several drops are to be fed from one manifold air line, it is a good practice for the manifold line to form a complete loop around the shop.

Be sure all air lines are at least as big as the outlet on the compressor and that there are no small diameter sections of line between the compressor and the tool. A small line acts as an orifice, dropping pressure at the tool. This happens when someone installs a *hose coupling* of the wrong size. A $5/16$" hose with a $1/4$" coupling will flow less air after the air passes the restriction caused by the smaller coupling. This also happens when someone mistakenly installs an air regulator that is too small; for instance, a $3/8$" regulator in a $1/2$" air line. Figure 10.22 shows a typical air-tool setup.

Air Compressor Maintenance

- Water is produced as outside air is condensed by the compressor into the receiver tank. The tank requires periodic *bleeding*, or *blow down*, to drain the water that accumulates. Moisture in the air supply line is reduced if the receiver tank is drained daily. A valve or faucet is usually located at the bottom of the tank. Some compressed-air systems have water filters (traps) at the outlet of the compressor to remove the moisture in the air. Some shops have oilers in the line to provide lubrication for air tools. Other air systems have special dryers to provide high quality air for spray painting.

- The oil level in the compressor should be checked regularly, especially if it appears that there is oil leakage. The oil should be changed every three months.

- It is important that the compressor breathes clean air, much the same as with an internal combustion engine. The air filter must be cleaned regularly. If it becomes restricted, the compressor will overheat.

- Check the drive belts regularly for signs of wear or looseness. A loose belt will slip and wear the pulley.

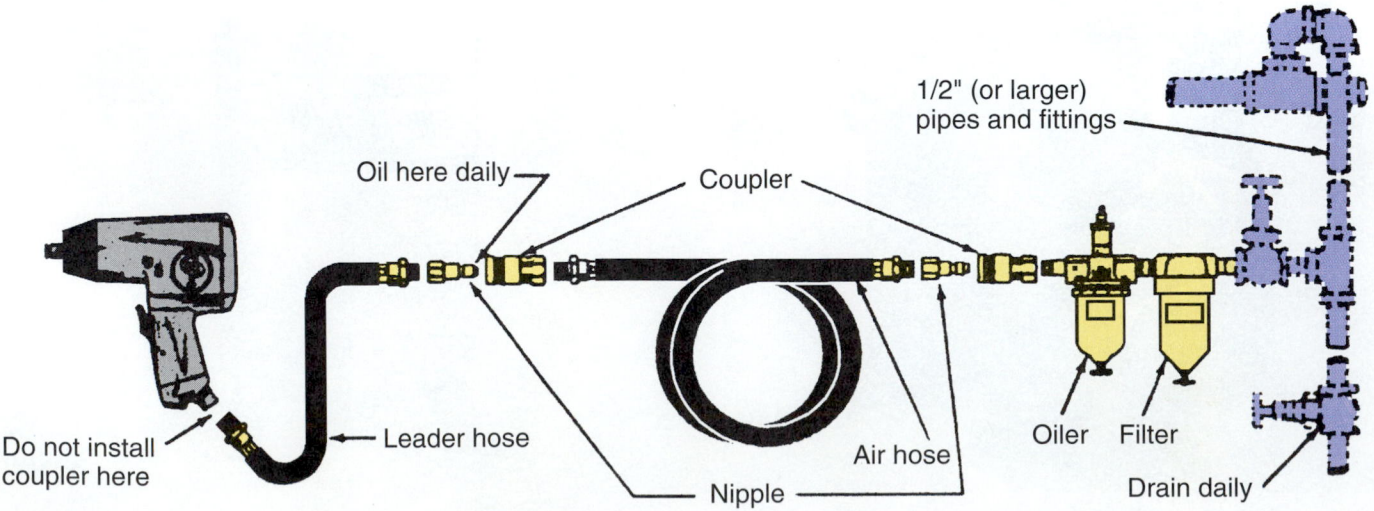

Figure 10.22 A typical setup for use of an air tool. *(Courtesy of Chicago Pneumatic)*

REVIEW QUESTIONS

1. The two main categories of lifts are the in-ground lift and the _____ lift.
2. What kind of jack is used to raise the wheels off of a wheel contact lift?
3. A single-post lift is handy for transmission removal. True or false?
4. What kind of lift has two controls, one for air and one for hydraulic pressure?
5. What kind of lift would you want to install if you were renting a building and had a short term lease, in-ground or surface mount?
6. Two ways that a surface mounted lift is driven are with a hydraulic pump and cylinders and _____ drive.
7. If one of the lift arms can be moved after the vehicle is in the air, what would you do?
8. When would a two-stage air compressor be used?
9. If an air compressor runs for 6 out of every 10 minutes, what is its duty cycle?
10. The useful capacity of an air compressor is measured in standard _____ _____ per minute.

ASE STYLE REVIEW QUESTIONS

1. Technician A says that if a car is lifted improperly, parts of the suspension or steering could be bent. Technician B says that lifting some cars incorrectly might cause the windshield to pop out. Who is right?
   - **a.** Technician A
   - **b.** Technician B
   - **c.** Both A and B
   - **d.** Neither A nor B
2. Technician A says that a frame-contact lift is the best choice for raising a car that has the engine out of it. Technician B says that unibody cars can be lifted from the edge of the car body at the rocker panel. Who is right?
   - **a.** Technician A
   - **b.** Technician B
   - **c.** Both A and B
   - **d.** Neither A nor B
3. Technician A says that on rear wheel drive cars, the center of gravity is usually just below the driver's seat. Technician B says that on front wheel drive cars, the center of gravity is usually just below the driver's seat. Who is right?
   - **a.** Technician A
   - **b.** Technician B
   - **c.** Both A and B
   - **d.** Neither A nor B
4. Technician A says to position a car on a two-post surface mount lift so that the driver's side door can be opened all of the way. Technician B says to raise the car until its wheels are about 6' off the ground and shake it to see if it is safely mounted. Who is right?
   - **a.** Technician A
   - **b.** Technician B
   - **c.** Both A and B
   - **d.** Neither A nor B
5. Technician A says that using a ⅜" regulator in the middle of a ½" air line will result in less airflow through the lines. Technician B says that water must be periodically bled from an air compressor. Who is right?
   - **a.** Technician A
   - **b.** Technician B
   - **c.** Both A and B
   - **d.** Neither A nor B

Shop Safety

■ **KEY TERMS**

Class A fire
Class B fire
Class C fire
Class D fire
flash point
backfire
popback
greasesweep
**Hazard Communication
rules**
manifest
**Material Safety Data
Sheet (MSDS)**
dermatitis
barrier creams
HEPA vacuum
electrolyte

■ OBJECTIVES

Upon completion of this chapter, you should be able to:

✔ Use shop tools and equipment safely.

✔ Understand safety rules.

■ INTRODUCTION

The number one priority of any business is the health and safety of its employees. Shop safety issues will be covered here.

■ A safety test is included at the end of the chapter.

■ As you read this chapter, realize that the situations described can and do occur, sometimes often.

■ Case histories presented throughout the book are true. Pay extra attention to the safety precautions described.

■ Safety precautions are presented for each piece of equipment described in this book.

The employer is responsible for safety training for each employee and for providing a safe working environment. According to federal law, each shop will have a safety training program. A shop owner in Los Angeles was prosecuted in the death of an employee for not providing proper safety training.

The employee is the one who is actually responsible for his or her own safety and the safety of others. Accidents that occur in an automotive shop often happen because safety considerations are not as obvious when repairing automobiles as they are in such trades as roofing or carpentry. Accidents are often caused by carelessness resulting from a lack of experience or knowledge or from being in a hurry and taking short cuts.

Injury accidents are often caused by someone other than the person injured. The person who caused the accident often suffers from guilt from the harm they have caused another. In the event of an accident, be sure to inform your instructor or supervisor, who knows what procedures to follow.

Injured persons often suffer from *shock* and should not be left unattended. When an injury does not appear to be serious enough to call an ambulance, a companion should go with the injured party to seek professional help.

The American Red Cross offers thorough first-aid training. Every shop should have someone trained to handle emergencies.

■ GENERAL PERSONAL SAFETY

A first-aid kit (Figure 11.1) contains items for treating some of the small cuts and abrasions that often happen. Fires and accidents involving lifts and battery chargers happen occasionally in automotive shops. But the most common injuries are mostly preventable injuries involving the *back or eyes*.

Wearing eye protection will prevent most eye injuries, so when using machinery eye protection is mandatory. Several types of eye protection are shown in Figure 11.2.

Eye Protection Safety

Parts can explode when they are pressed or pounded on. Rotating tools can throw pieces of metal or grit, causing eye injuries. Safety goggles or a face shield should be worn when using shop tools and equipment. Because eye injuries are so common in an automotive shop, continual use of safety glasses or goggles is recommended. Prescription safety glasses are an advantage because the user always wears them. Eye protection is emphasized for your protection!

Face shields (see Figure 11.2c) are convenient because they can stay with the piece of equipment. They are also easily adjusted to your head.

The following list gives some hints for preventing eye and back injuries. Eye protection must be worn:

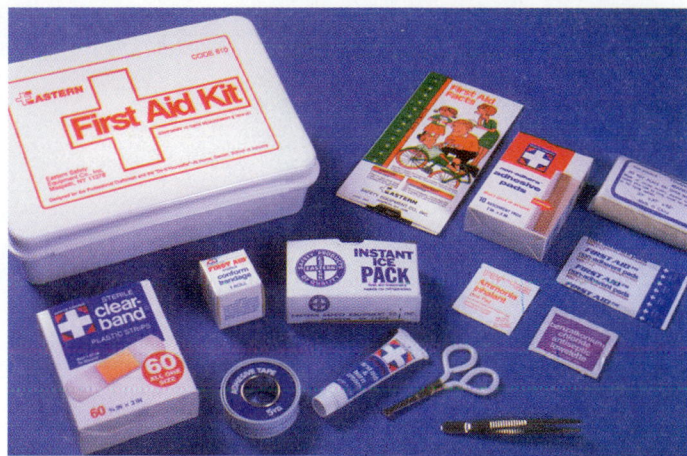

Figure 11.1 A first-aid kit.

- Whenever working around moving parts and machinery (grinding, cutting, drilling, washing, etc.)
- When blowing off parts with compressed air
- When working under the vehicle
- When working on air conditioning
- When flame-cutting or welding

If your eye is accidentally contaminated with a dangerous liquid flush thoroughly in an eyewash fountain (Figure 11.3) or other water source

Back Protection Safety

Following safe lifting procedures will prevent most back injuries (Figure 11.4). The following are safety precautions to be used when lifting:

- Be sure to get help when moving heavy items. The normal tendency is to say that items are not that heavy and you hate to ask somebody for help. If something is in an awkward position for lifting, leverage and the position your back is in can make it easier for an injury to occur.

- If an item is too heavy to lift, use the appropriate equipment.
- Before moving a heavy item, plan the route that the item will be carried and how it is to be set back down when you get there.
- Lift slowly.
- Do not jerk or twist your back. Shift your feet instead.
- Bend your knees and *lift with your legs, not your back!* Also, keep your lower back straight when lifting. Think about thrusting your stomach out.

CASE HISTORY *A technician lifted a spare tire out of the trunk of a car. As he reached forward and tugged the tire up and out of the trunk, he felt a small pop in his lower back. The result was a herniated disk in his lower back, an injury that will affect him for the rest of his life.*

Ear Protection

Damage to ears happens due to exposure to loud noises over a period of time. When loud air tools are used, ear protection should be worn.

NOTE: *Once it has been damaged, your hearing will not recover.*

Clothing and Hair

Clothing (shirt tails) or hair that hangs out can get caught in moving machinery or under a creeper. Keep long hair tied or under a cap. Shirt tails should be tucked in or shop clothing can be worn over the shirt.

Shoes or Boots

Leather shoes or boots offer much better protection than tennis shoes or sandals. Soles are available that resist slipping and are resistant to damage from petroleum

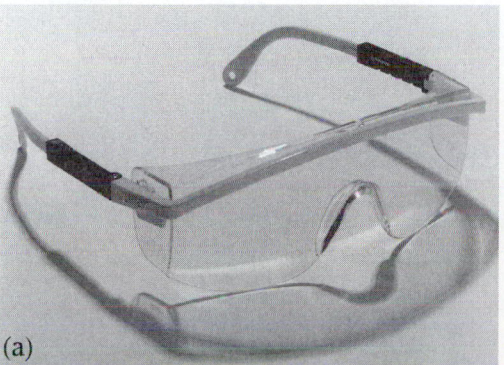
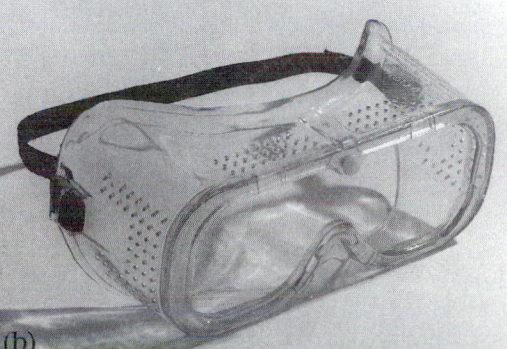
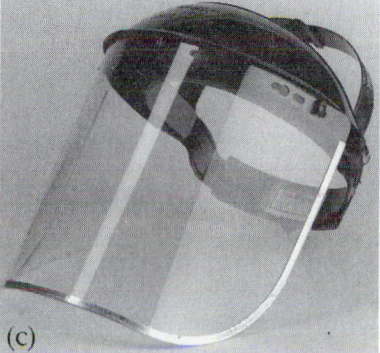

(a)   (b)   (c)

Figure 11.2 Eye Protection. (a) Safety glasses; (b) Goggles; (c) Face shield. *(Courtesy of Goodson Shop Supplies)*

Figure 11.3 An eyewash fountain. *(Courtesy of Western Emergency Equipment)*

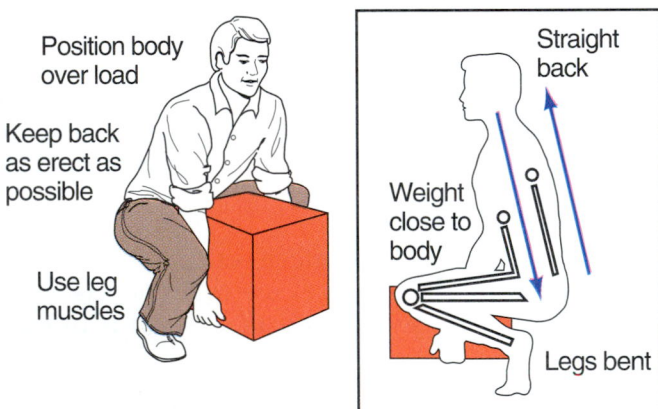

Position body over load

Keep back as erect as possible

Use leg muscles

Straight back

Weight close to body

Legs bent

Figure 11.4 Lifting precautions.

products. Boots and shoes that have the toe reinforced with steel inserts are widely available.

Hand Protection

Cuts and scrapes are common when repairing automobiles. Be sure to keep your fingers away from moving machinery and hot surfaces. Solvents used to clean parts can damage skin. Some types of solvents can actually penetrate your skin. Use the correct type of glove when handling chemicals (Figure 11.5). Some inexpensive gloves will actually allow some chemicals to pass through them. The use of gloves is becoming more common in the automotive repair industry.

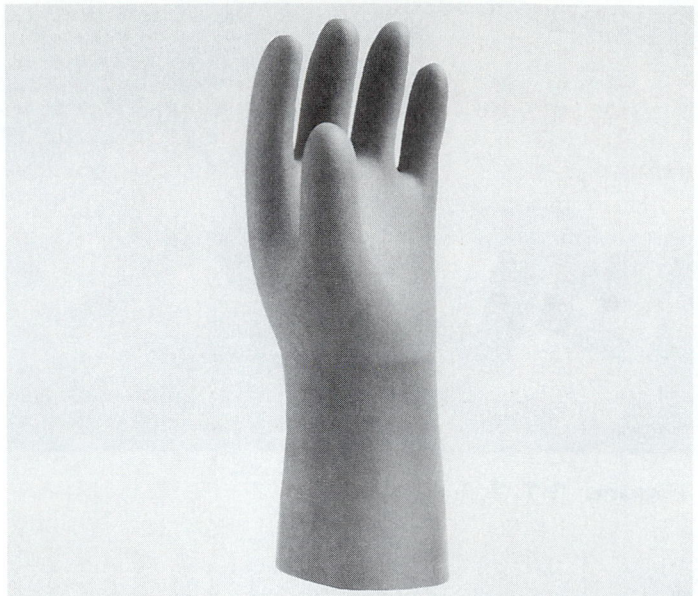

Figure 11.5 Use the correct type of glove when handling chemicals. *(Courtesy of Siebe North, Inc.)*

FIRE SAFETY

Two major items should be considered when dealing with fires. A little common sense is important. If the fire is burning so dangerously that your personal safety is jeopardized, leave the area immediately and call for help. If you can safely remove the source of fuel or heat to a fire, do so. This might include shutting off fuel to a fuel fire or disconnecting the electrical source from an electrical fire.

FIRE EXTINGUISHERS

A fire extinguisher is a portable tank that contains water or foam, a chemical, or a gas (Figure 11.6). There

Figure 11.6 Fire extinguishers.

are four kinds of fires, each calling for a different type of fire extinguisher (Figure 11.7).

- A **Class A fire** is one that can be *put out* with water. Such items as paper and wood make up these kinds of fires.
- A **Class B fire** is one in which there are *flammable liquids* such as grease, oil, gasoline, or paint.
- A **Class C fire** is *electrical*.
- A **Class D fire** is a *flammable metal* such as magnesium or potassium.
- Either *carbon dioxide* (CO_2) or a dry chemical fire extinguisher can be used on *Class B and C fires.*

A gauge on the top of the fire extinguisher tells whether it is fully charged or if the charge pressure has leaked off. Fire extinguishers in business establishments are periodically inspected by the local fire department or a fire extinguisher service company, but be sure to check the gauge on the extinguisher on a regular basis.

SAFETY NOTE Be sure to have a large enough fire extinguisher on hand. An engine compartment fire often requires a larger extinguisher, depending on how far the fire has progressed.

For auto repair shops, the size should be at least a 20B (this number designation is explained later). A popular fire extinguisher that people keep in their vehicles or boats is a 2A-10BC.

- The 2A means that the extinguisher can put out a 2 cubic foot fire of a material like paper.
- The 10B means that it can put out 10 square feet of burning liquid.
- The C means that electrical fires can be extinguished.

	Class of Fire	Typical Fuel Involved	Type of Extinguisher
Class **A** Fires (green)	**For Ordinary Combustibles** Put out a Class A fire by lowering its temperature or by coating the burning combustibles.	Wood Paper Cloth Rubber Plastics Rubbish Upholstery	Water[1] Foam* Multipurpose dry chemical[4]
Class **B** Fires (red)	**For Flammable Liquids** Put out a Class B fire by smothering it. Use an extinguisher that gives a blanketing flame-interrupting effect; cover whole flaming liquid surface.	Gasoline Oil Grease Paint Lighter fluid	Foam* Carbon dioxide[5] Halogenated agent[6] Standard dry chemical[2] Purple K dry chemical[3] Multipurpose dry chemical[4]
Class **C** Fires (blue)	**For Electrical Equipment** Put out a Class C fire by shutting off power as quickly as possible and by always using a nonconducting extinguishing agent to prevent electric shock.	Motors Appliances Wiring Fuse boxes Switchboards	Carbon dioxide[5] Halogenated agent[6] Standard dry chemical[2] Purple K dry chemical[3] Multipurpose dry chemical[4]
Class **D** Fires (yellow)	**For Combustible Metals** Put out a Class D fire of metal chips, turnings, or shaving by smothering or coating with a specially designed extinguishing agent.	Aluminum Magnesium Potassium Sodium Titanium Zirconium	Dry powder extinguishers and agents only

*Cartridge-operated water, foam, and soda-acid types of extinguishers are no longer manufactured. These extinguishers should be removed from service when they become due for their next hydrostatic pressure test.

Notes:
(1) Freeze in low temperatures unless treated with antifreeze solution, usually weighs over 20 pounds, and is heavier than any other extinguisher mentioned.
(2) Also called ordinary or regular dry chemical. (solution bicarbonate)
(3) Has the greatest initial fire-stopping power of the extinguishers mentioned for class B fires. Be sure to clean residue immediately after using the extinguisher so sprayed surfaces will not be damaged. (potassium bicarbonate)
(4) The only extinguishers that fight A, B, and C class fires. However, they should not be used on fires in liquified fat or oil of appreciable depth. Be sure to clean residue immediately after using the extinguisher so sprayed surfaces will not be damaged. (ammonium phosphates)
(5) Use with caution in unventilated, confined spaces.
(6) May cause injury to the operator if the extinguishing agent (a gas) or the gases produced when the agent is applied to a fire is inhaled.

Figure 11.7 Guide to Fire Extinguisher Selection.

Figure 11.8 Keep combustibles in safety containers.

Locate and check the type of fire extinguisher(s) in your shop. They should be located in a place *other* than where the most likely start of a fire would be. For instance, do not locate a fire extinguisher right over the welding bench or next to the solvent tank. If a fire began there, you would not be able to get to the extinguisher. Locating fire extinguishers on both sides of a common work area is a good idea.

■ FLAMMABLE MATERIALS

Greasesweep and rags soaked in oil or gasoline should be stored in covered metal containers (Figure 11.8). Keeping oily materials separated from air prevents them from self-igniting, a process called *spontaneous combustion*. Flammable materials should be stored in a flammable storage cabinet (Figure 11.9).

Used greasesweep is kept in a flammable storage container because it may be reused until it becomes

Figure 11.9 An approved flammable storage cabinet. *(Courtesy of Justrite Manufacturing Co.)*

saturated (wet). Saturated greasesweep must be disposed of using a licensed waste hauler.

■ FUEL FIRES

Gasoline is a major cause of automotive fires. Liquid gasoline is not what catches on fire. Rather, it is the *vapors* that are so dangerous. Gasoline vapors are heavier than air, so they can collect in low places in the shop. They can be ignited by a spark from a light switch, a motor, electrical wires that have been accidentally crossed, or a dropped shop light.

There are two kinds of shop lights that are acceptable. One of them has a fluorescent bulb that is enclosed in a plastic tube (Figure 11.10a). The other uses a special incandescent bulb that is rubber coated and shatter proof (Figure 11.10b).

Gasoline Safety

1. Gasoline should be stored only in an approved safety container (see Figure 11.8) and never in a glass jar.

2. Never use gasoline to clean floors or parts. Parts cleaning solvent (Stoddard solvent is the industry term) has a higher **flash point** than gasoline. A flammable liquid's flash point is the temperature at which the vapors will ignite when brought into contact with an open flame.

3. Careless cigarette smoking or failing to immediately and thoroughly clean gasoline spills can contribute to a dangerous situation. People get used to working around gasoline and then begin to ignore how dangerous it can be.

4. Do not attempt to siphon gasoline with your mouth. Accidental breathing of gasoline into the lungs can be fatal.

To be extinguished, a fuel fire must first have its source shut off. The fire will go out if the source of oxygen is removed or if the temperature is lowered below the fuel's flash point. Firefighters can use a fog

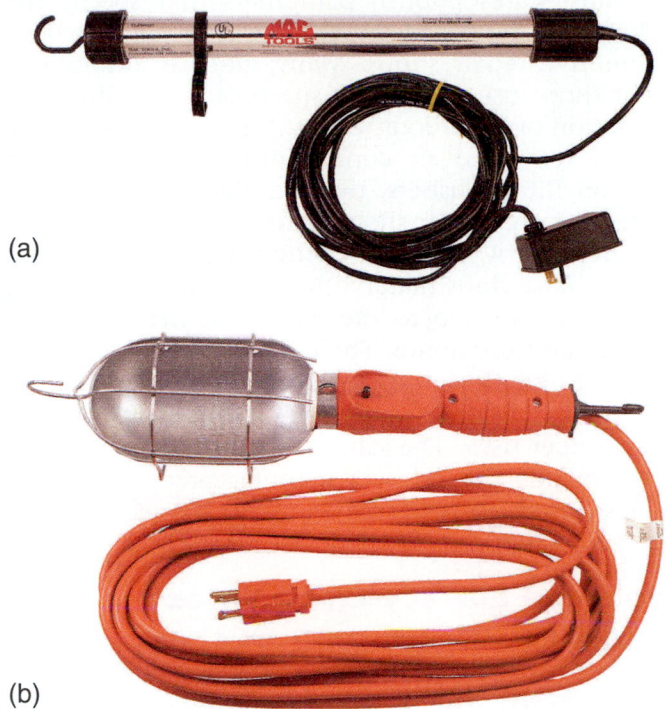

(a)

(b)

Figure 11.10 Two kinds of drop lights: (a) A fluorescent tube safety light. (b) The light shown here should be used with a shatterproof bulb. *(Photos provided by Mac Tools and are representative of products manufactured by Mac Tools or products obtained from another source)*

of water to cool a fuel fire and cut off its oxygen, but shop personnel should be knowledgeable about fire extinguishers and how to use them.

NOTE: *Do **not** attempt to douse a fuel fire with water. Water will cause the fire to spread because fuel is lighter than water and floats on top of it.*

CASE HISTORY
A technician completed the installation of a rebuilt engine. When he started the engine, the float in the carburetor stuck. The results were gasoline pouring out of the carburetor, and an air/fuel mixture that was too rich. When the engine "popped back" through the carburetor, a fire began. The technician grabbed a nearby hose and attempted to douse the fire. The water only served to spread the flames. The result was a total loss of the vehicle.

NOTES:

■ ***Backfire*** *is a term for combustion that occurs in the vehicle's exhaust (after it leaves the cylinder).*

■ ***Popback*** *is a term for combustion occurring in the fuel induction system (before it enters the cylinder).*

ELECTRICAL FIRES

Electrical fires are prevented by disconnecting the battery before working on the electrical system or around electrical components, such as the starter or alternator (Figure 11.11). Unbolt the ground cable first. This prevents the possibility of a spark occurring if a wrench accidentally completes a circuit between a "hot" cable and ground.

NOTE: *The ground cable is the one that is bolted to the engine block. Do not assume that ground is the negative cable. On some older vehicles, the positive cable is ground.*

If there is an electrical fire, the *battery must be disconnected* as fast as possible so the fire can be put out. Another advantage to removing the ground cable is that an electric cooling fan cannot accidentally come on while working near it.

CASE HISTORY
An apprentice was removing a clutch from a car. After raising the vehicle on the lift, he began to remove the starter motor while the battery was still hooked to the circuit. He did not disconnect the wiring, which was in an inaccessible location on the top of the starter. When the unbolted starter came loose, bare electrical terminal ends became wedged against the frame, which resulted in a major electrical fire. The technician working next to him rushed over with a bolt cutter and cut the battery ground cable near where it was bolted to the block. Damage to the vehicle was limited to a wiring loom and the alternator. The apprentice went to the emergency room to be treated for burned hands and respiratory irritation from inhaling noxious smoke when the insulation on the wires caught fire.

Figure 11.11 Remove the battery ground cable first.

■ SHOP HABITS

The shop can be kept cleaner if a technician gets into the habit of using a shop rag when working. Greasy, oily tools and hands should be wiped clean, preventing the mess from being spread around the rest of the shop.

Common sense around the shop dictates that spills be cleaned up as soon as they happen. Oil and solvent spills can be cleaned up with an absorbent material, such as rice hull ash or kitty litter **greasesweep**. Greasesweep is swept up and reused until it becomes too wet. It is better to use it when it is slightly wet, because the dust that results when using new greasesweep is avoided.

NOTE: *Coolant spills are best cleaned with a squeegee. Mixing water and coolant with oily greasesweep makes a muddy mess.*

One of the results of environmental regulation is that greasesweep is classified as hazardous if it is used to soak up used motor oil. Waste disposal companies provide a service where superabsorbent cloths are used to soak up spills. The disposal companies then collect the cloths for proper treatment.

NOTE: *Because saturated greasesweep could be flammable, it must be stored in a metal can with a lid.*

■ HAZARDOUS MATERIALS

Some materials routinely used in the shop may be dangerous to your health. In addition, many chemicals irritate skin. Cautions about skin and eye protection are covered throughout this chapter. A material is considered *hazardous* if it causes illness, injury, or death or pollutes water, air, or land.

Hazardous materials that automotive technicians commonly come into contact with are:

■ Cleaning chemicals (carburetor cleaner, solvents, caustics, acids)
■ Battery acid
■ Fuels
■ Paints and thinners
■ Used oil and fluids
■ Heavy metals
■ Anti-freeze/coolant
■ Asbestos (brakes and clutches)
■ Refrigerants

In the United States, the Occupational Safety and Health Administration (OSHA) regulates the use of many of these materials. The Environmental Protection Agency (EPA) regulates their disposal. OSHA's **Hazard Communication rules** summarize these regulations.

All businesses that generate hazardous waste must develop a *hazardous waste policy*. Each hazardous waste generator must have an *EPA identification number*. When waste is transported for disposal, a licensed waste hauler must be used to transport and dispose of the waste. A copy of a written **manifest** (an EPA form) must be kept by the shop.

Many states publish pamphlets that specifically address the proper disposal of wastes generated in automotive repair shops. Many modern automotive repair shops use equipment specifically to reduce the generation of hazardous wastes. Some of these include coolant recyclers, air conditioning refrigerant recyclers, oil filter crushers, pyrolytic baking ovens, and waste oil furnaces for shop heating.

The OSHA Hazard Communication Program requires hazardous material manufacturers to publish information related to the hazardous materials they produce and distribute. The employer provides information to employees by labeling hazardous materials and their containers and by training employees in their proper uses. The label on a hazardous material must include the product manufacturer's name and address, its chemical name and trade name, as well as safety information about the chemical.

Material Safety Data Sheet

For any hazardous materials used, the employer must also make available to employees a **Material Safety Data Sheet (MSDS)** (Figure 11.12). An

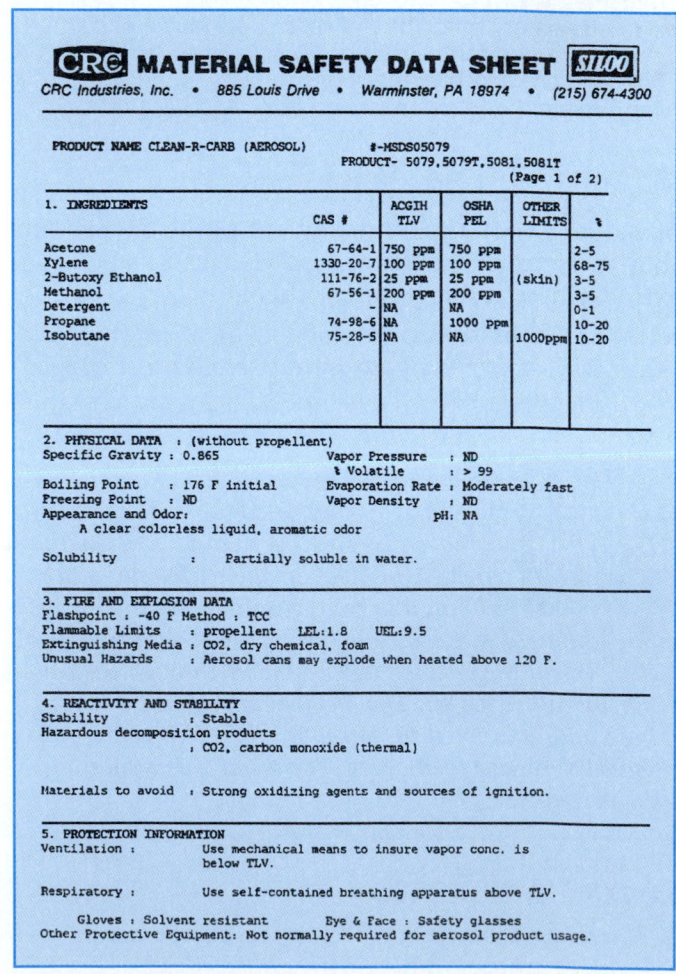

Figure 11.12 A Material Safety Data Sheet (MSDS). *[Courtesy of CRC Industries]*

MSDS tells many things about a hazardous material including:

- Chemical identification (its trade name and technical name).
- Hazardous ingredients in it and its permissible exposure limit (PEL) or threshhold limit value (TLV).
- Physical data telling the material's appearance, what it smells like, whether it evaporates, and if its fumes are heavier or lighter than air.
- Ignition or explosion temperature, the type of fire extinguisher to use, and personal protective gear to wear.
- Health hazards such as dizziness, headache, skin rash, nausea, and the best choice of first-aid options.
- Reactivity data tells whether the material reacts with other materials to cause a fire or explosion. This also tells under what conditions the material might become unstable.
- Procedures for cleaning up spills.

A supplier of a hazardous material is required by law to provide an MSDS (upon request).

 **SAFETY NOTE** Be aware of the materials you are dealing with. Remember, no one will be more concerned for your safety and health than **you!**

■ HOT TANK SAFETY PRECAUTIONS

Caustic or Acid Cautions

In addition to burns that can occur due to the high temperature of the solution, caustic solution from a hot tank can also eat your skin and cause blindness.

Hot Tank Safety

1. Always wear gloves and face protection when working around caustics.
2. If caustic contacts your skin, rinse immediately and seek medical attention.
3. If caustic gets into an eye, the caustic must be flushed from the eye immediately or blindness may result. Be sure to lift the eyelid and flush under it with water for 15 minutes. Continue repeated 15-minute washings while medical attention is sought.
4. Articles should be lifted from the tank with a lifting device. People have fallen into large hot tanks while trying to retrieve objects by hand.
5. Be sure that lifting slings are securely fastened to items in the tank. Failure to do so can cause a dangerous splash.

■ CLEANING SOLVENT SAFETY PRECAUTIONS

Cleaning Solvent Cautions

Cleaning solvents are used extensively in auto repair. Their use is generally safe provided the user has knowledge of hazards associated with them. A flammable liquid's flash point is the temperature at which it will catch fire. Stoddard solvent has a relatively high flash point, but fires can still result.

Cleaning Solvent Safety

1. Use eye protection.
2. Do not breathe vapors.
3. Clean only in areas with adequate ventilation.
4. Wear a respirator if using a blowgun to apply a chemical. (Nontoxic chemicals are available.)
5. Do not clean with gasoline. It is very explosive when it vaporizes.
6. Some solvents can catch fire on a hot engine. Be sure to use a nonflammable solvent.
7. Do not spray solvents if near any pilot lights.
8. Use a brass or nylon brush instead of a steel wire brush to avoid the possibility of an accidental spark.
9. The lid on the solvent tank should always be kept closed when not in use to prevent a fire hazard.
10. Do not smoke or weld anywhere near solvent.

■ SKIN CARE SAFETY PRECAUTIONS

Many solvents are damaging to skin. **Dermatitis** (irritated skin) can result from exposure. If hands are burned or swollen from solvent contact, seek medical attention.

- Do not put hand creams on the skin as treatment. Hand cream can seal in the solvent irritant, causing even more damage.
- Be sure the gloves you use do not allow penetration by the solvent being used. Neoprene, an oil resistant artificial rubber commonly used for hoses and gloves, protects against some types of solvents, but not others. Latex, a material commonly used in making the surgical gloves popular with many mechanics, will also allow penetration by some chemicals.
- Do not mix other chemicals with cleaning solvents.

To aid in cleaning hands, **barrier creams** are often rubbed on the skin prior to working in a greasy environment. Although these creams are quite effective in making hand cleaning easier, they do not protect the skin against penetration by solvents and are not a substitute for gloves.

◼ CARBURETOR CLEANER SAFETY

Carburetor Cleaner Safety Precautions

Carburetor cleaner is highly concentrated and dangerous to skin, eyes, and clothing. It is dangerous to breathe and can be absorbed through the skin, so rubber gloves are worn when handling it. Rinse thoroughly with water before handling parts in the parts basket.

Carburetor Cleaner Safety

1. Wear eye protection.
2. Wear gloves. Carburetor cleaner is a strong irritant to skin.
3. Carburetor cleaner is washed off with water. Be careful not to use water pressure that is too high; it might redirect carburetor cleaner to your skin or eyes.
4. Keep carburetor cleaner off the floor. It eats through many kinds of shoe soles.

◼ BREATHING SAFETY

There are many sources of breathing hazards. These can include paints, cleaning chemicals, burned wire insulation, grinding dust, asbestos, vehicle exhaust, and battery gas. Some of these hazards are only breathed by accident. Others are in the air and must be guarded against. Some are only dangerous after long-term exposure.

Sometimes you will need to work around a running engine. Be sure the parking brake is firmly applied and try to stand to the side of a running vehicle. Exhaust gas can contain large amounts of carbon monoxide. Protect yourself and others from exhaust gas by using an exhaust system (Figure 11.13).

Figure 11.13 Use an exhaust ventilation system when working around a running engine.

Asbestos

Asbestos, found in some brakes, clutches, and cylinder head gaskets is dangerous to breathe. In fact, the respirator shown in Figure 11.14 will probably not completely protect your lungs from asbestos fibers. When working on brakes or clutches or when removing head gaskets from cylinder head surfaces, one of two approved methods can be used.

◼ One asbestos cleaning method is to use a special apparatus (called a **HEPA vacuum**). It vacuums asbestos and captures the particles in a special filter. A regular vacuum will not do. Asbestos will go right through it.

◼ Another approved asbestos cleaning method is to use water with an organic solvent or wetting agent.

Dry brushing is expressly prohibited during brake cleaning. Using a drill-operated wire brush to clean fiber head gaskets can also release asbestos into the air.

Most brake shops own a HEPA vacuum (Figure 11.15). HEPA stands for *high efficiency particulate arresting*. A HEPA vacuum completely encloses the brake drum or disk assembly. Gloves and a window allow the technician to clean the brake or clutch assembly. After using the vacuum to suck up loose material, other particles are dislodged using compressed air and are sucked into the vacuum.

 Be sure to inspect the enclosure to see that it is tight against the brake backing plate before beginning work.

Asbestos Disposal

A vacuum filter that is full must be wetted with a mist of water before removing it from the vacuum. Filters

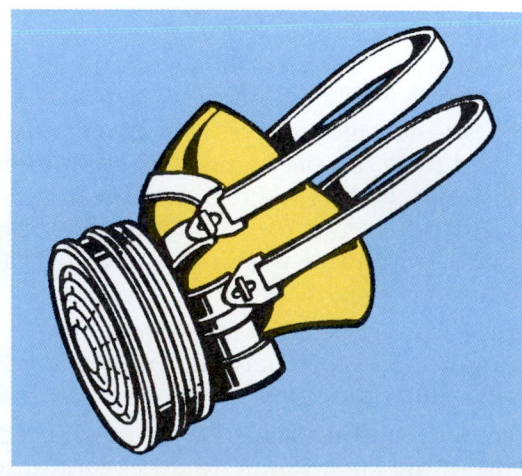

Figure 11.14 This respirator will not provide complete protection from asbestos.

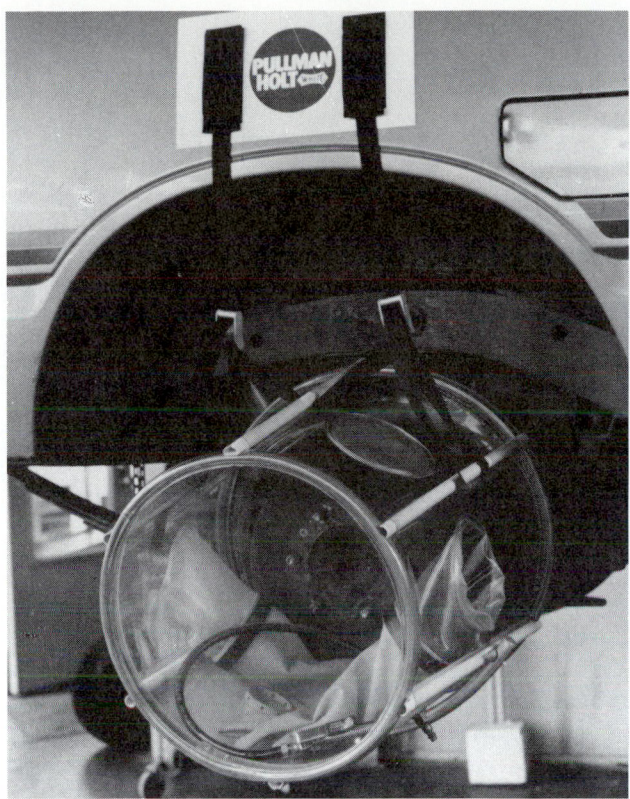

Figure 11.15 A HEPA vacuum can vacuum asbestos without letting it escape into the shop environment. *(Courtesy of Pullman, Holt, White)*

must be placed in impermeable containers, labelled, and disposed of in the manner prescribed by law.

When using the liquid method of asbestos protection, the wetting agent must flow through the brake drum *before* the drum is removed. When a drum is removed first, a quantity of asbestos can escape into the surrounding air. A catch basin is positioned under the brake assembly. After the drum is removed, the brake parts are thoroughly wetted before starting to clean. If the system has a filter it must be disposed of in the same way as a HEPA filter.

An alternative method for shops that do fewer than five pairs of brakes or clutches in one week allows a low pressure water system or fine mist spray bottle to be used to wet the parts before beginning work. All parts are then wiped clean with a cloth, which must be disposed of in an approved, properly labeled container or laundered in a way that prevents the release of asbestos fibers. Dry brushing is prohibited in this method also.

◼ ELECTRICAL SAFETY PRECAUTIONS

Electric Shock

Twelve-volt DC (direct current) electrical systems, like the ones used in automobiles, do not cause serious

electrical shock (unless the engine has a distributorless ignition). Shop equipment, however, is powered by either 110V or 220V alternating current, which can be very dangerous.

Electrical Safety

1. Do not stand in water when using electric tools.
2. Be sure that a tool is turned off before plugging it in so that a spark does not jump from the outlet to the plug.
3. Electrical wall outlets should be properly grounded.
4. Three-wire electrical tools are the best choice for commercial work. The extra terminal is for ground (Figure 11.16). If you use a homeowner-type tool with a two-wire plug it should be *double-insulated*.
5. Be sure to observe electrical wire color coding when repairing or replacing power cords on electric tools.

CASE HISTORY *An electrical plug on a drill with a metal housing was damaged and needed to be replaced. A technician bought a new 3-wire plug, cut and stripped the wires, and installed it on the cord. He had been working on cars for many years and was used to the color code used around automotive batteries. Traditionally in automobile wiring, black is ground and red is positive but in commercial wiring the **green wire is ground**. When he connected the black wire to the ground terminal, he was actually hooking one of the hot leads to ground (also the housing of the metal drill). When he plugged the drill into the wall, he was holding it in his hand and received a dangerous electrical shock. Luckily, he was not seriously injured.*

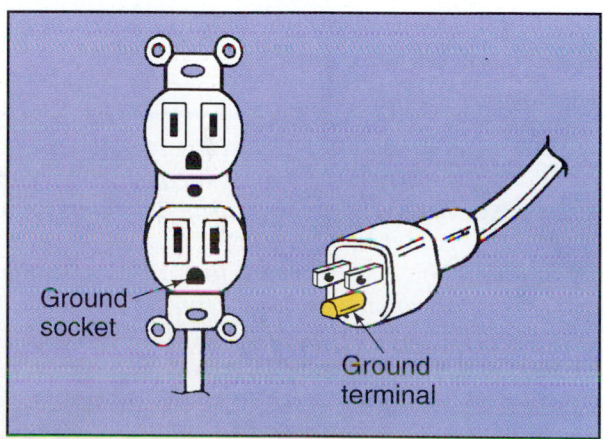

Figure 11.16 The third wire terminal is for ground.

Figure 11.17 Unhook the power to an electric cooling fan before working around it.

■ COOLING FAN SAFETY

The fan that draws cool air across the radiator can be driven by either a belt or electricity. Rotating fans can be dangerous. Some are controlled by an automatic switch and can start unexpectedly.

1. Electric cooling fans should be disconnected when working around them (Figure 11.17).

2. Visually inspect fan blades for damage. A damaged fan blade will probably be out of balance, making it prone to fly apart.

3. Keep hair and clothing away from fans and fan drive belts.

■ COOLANT BURNS

The most likely ways to be burned in an automotive shop are with engine coolant or by the exhaust system. Opening the radiator on a hot engine can be very dangerous. Disc brake rotors can also be hot after a car has recently been driven.

If the hose is hard and feels like it is full of coolant, the coolant level is acceptable. If the hose collapses, the coolant is not under pressure but steam can still cause a burn. Exercise extreme caution when opening the cap.

 **CASE HISTORY** *The radiator cap keeps pressure on the coolant when the engine heats up. When coolant is under pressure its boiling point is higher. Loosening the cap removes the pressure and the coolant boils instantly. A student had just begun working in a full-service gas station. A customer asked him to check the coolant level in the radiator. When he opened the radiator cap, he turned it ½ turn. This defeated the cap's safety feature and the coolant boiled out, severely burning him.*

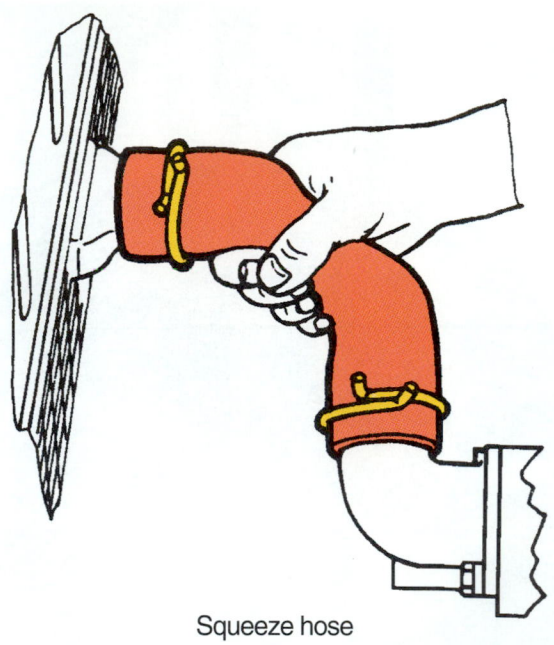

Squeeze hose

Figure 11.18 Squeeze the top hose before attempting to remove the radiator cap. *[Courtesy of Chrysler Corporation]*

 **SAFETY NOTE** Always squeeze the top radiator hose before opening a radiator cap (Figure 11.18).

■ GENERAL HAND TOOL SAFETY

Cuts and scrapes are common when working on cars. Experienced technicians cut themselves less often than beginners. This is because they use hand tools properly. Refer to Chapters 7 and 8 for more information on the particular hand tools and equipment covered here. The following precautions should be used when handling hand tools:

1. Before using a tool, inspect it to see that it is not damaged or cracked.

2. Maintain tools in safe working condition.

3. Do not use worn or broken tools.

4. When using screwdrivers, do not hold small components in your hand in case the blade slips.

5. Do not put sharp tools in your pocket.

6. Be sure to always use eye protection when using hammers and chisels.

7. Mushroomed chisels and punches should be reground before use (Figure 11.19).

8. Be sure that a file has a handle installed on its tang before using it (Figure 11.20).

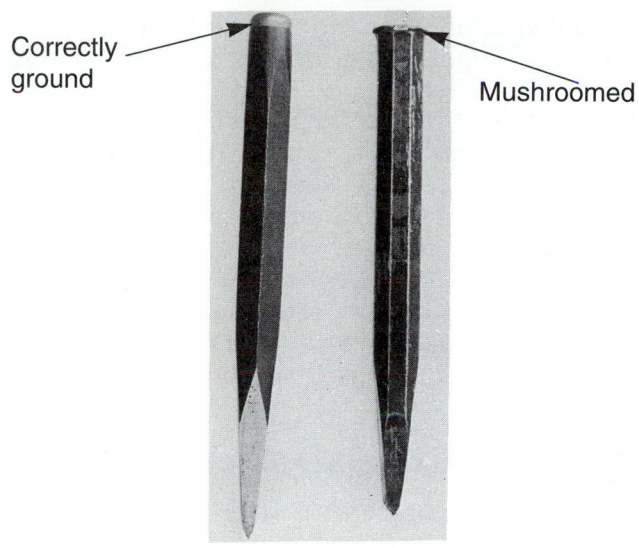

Figure 11.19 Regrind a chisel with a mushroomed head.

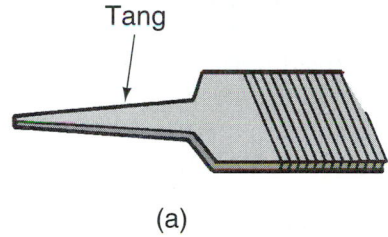

(a)

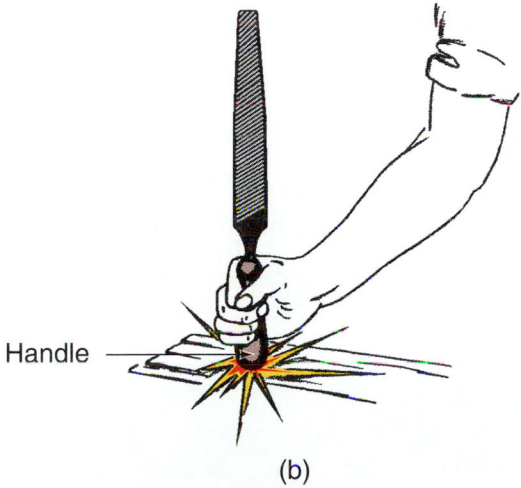

(b)

Figure 11.20 Install a handle on the tang of a file. *(Courtesy of General Motors Corporation, Service Technology Group)*

◾ VISE SAFETY

Be careful not to tighten the vise too much at its wide-open position or the vise might break. Never use a piece of pipe on the handle of a vise; it can break the vise.

◾ PULLER SAFETY

Pullers can be dangerous, especially if used improperly. Parts that are being separated can be under a good deal of pressure. They can explode, especially when using an impact wrench or a hydraulic puller. The following are some general tips on safe use with pullers:

- ◾ Wear eye protection.
- ◾ Be sure the puller is aligned so it is perpendicular to the part being pulled.
- ◾ Do not use a puller with damaged or worn parts.
- ◾ Use the correct size puller so overloading is avoided.
- ◾ Use a 3-jaw puller instead of a 2-jaw puller when possible.

◾ MACHINERY SAFETY

There are many types or machinery used in automotive shops. Common sense will prevent most injuries and accidents.

- ◾ *Do not talk* to someone who is operating a machine.
- ◾ Do not talk to someone when you are operating a machine.
- ◾ Be cautious when working around a running engine or rotating machinery. Fingers can be cut off by a moving belt and pulley.

NOTE: *Sometimes, machinery rotating at high speed does not appear to be moving. One example of this is household lighting. House and shop light appear to be constant. However, they are powered by alternating current so they actually flicker at 60 times per second. This can produce a strobe effect on moving machinery.*

◾ ELECTRIC DRILL SAFETY

Drilling Safety Precautions

Drills are used often in repairing automobiles. As with other types of machinery, knowledge and common sense will prevent accidents from occurring. The following are some tips on correct use of electric drills:

- ◾ *Always* wear eye protection.
- ◾ *Release pressure occasionally* to allow chips to break off before they can become too long and dangerous.
- ◾ A drill bit may catch when it starts to break through the work being drilled. *Be sure that sheet metal is clamped to the work table.* Let up the pressure on the bit as it starts to break through the bottom of the hole.
- ◾ If the drill grabs the work, *shut off the drill.* Never grab the moving work.
- ◾ *Be sure to remove the chuck key* from the chuck before drilling.
- ◾ *Never* stand in water when drilling. Standing in water increases the danger of electrical shock.

◾ GRINDER SAFETY

Grinder Safety Precautions

The grinder is a part of the machine inventory of all automotive shops. Serious accidents can occur if the grinder is not used properly. If you understand the

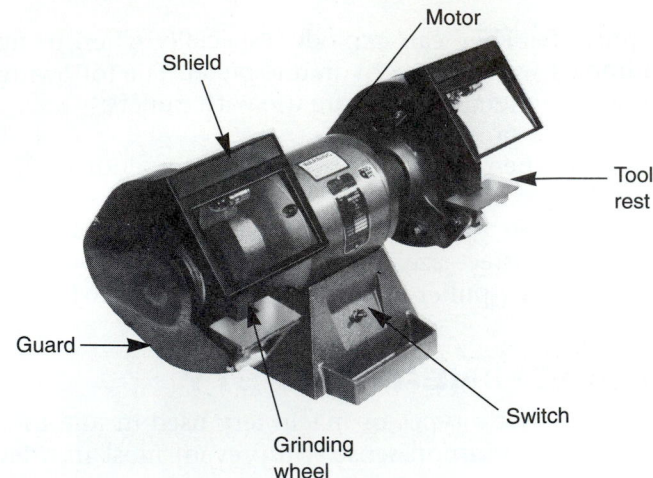

Figure 11.21 Position the tool rest as close to the grinding wheel as possible. *(Courtesy of Snap-on Tools Corporation, Copyright Owner)*

following safety precautions about the grinder, it can be used safely and effectively.

■ *Stand to the side when starting the motor.* The grinding wheel is more likely to explode during start-up because of the inertia of the wheel.
■ *Wear face protection.*
■ *Position the tool rest as close to the wheel as possible,* so that nothing can get trapped between the wheel and the tool rest (Figure 11.21).
■ Use the shield installed on the grinder.
■ *Do not grind on the side of the grinding wheel.*

WIRE WHEEL SAFETY

Wire Wheel Safety Precautions

Most grinders are set up with a wire wheel on one side and a grinding wheel on the other. The wire wheel is used to remove carbon, paint, or rust from parts and to deburr freshly machined parts. To avoid injury, use common sense when dealing with a wire wheel and follow the following precautions:

■ Use face protection.
■ Wear leather gloves if they are available.
■ Do not push too hard against the wheel. Severe finger injuries can result if your hand slips.
■ Wire wheels can damage soft aluminum surfaces.

COMPRESSED AIR SAFETY

Compressed Air Tool Safety Precautions

Compressed air is very useful to a technician, but can be dangerous when used improperly. Horseplay has no place in a shop. A blast of air can result in a broken eardrum. Blowing compressed air into a body orifice can result in death. The following are compressed air tool safety precautions:

■ Always wear eye protection when blowing off parts. Pieces of debris can be blown into eyes.

■ Do not blow air against your skin; the high-pressure compressed air used in auto repair shops can penetrate skin.
■ Compressed air is used to power chassis grease guns. Pressurized grease can penetrate skin.
■ Hold onto the air hose when uncoupling an air line so it does not fly through the air.
■ When possible, bleed off the air from an air line before uncoupling an air hose.
■ Rubber-tipped blowguns are not safety regulated like other normal blowguns (see Figure 9.10). When held against a part, full shop air pressure is available at the tip.

IMPACT WRENCH SAFETY

Air impact wrenches are used extensively in auto repair. They save time and make work easier when used with knowledge and common sense. The following are some safety tips when using impact wrenches:

■ Be careful of loose clothing or hair that might become tangled in the tool.
■ Use approved *impact* sockets, not ordinary chrome sockets, which can break (Figure 11.22).
■ Be sure that the socket is secured to the air tool. The clip at the end of the tool's square drive can become worn so that it no longer holds the tool.
■ Do not turn on the impact wrench unless the socket is installed on a nut or bolt, especially when using a wobble socket. The socket can fly off the impact wrench, possibly causing an injury (see Figure 7.44).
■ When the impact wrench fails to loosen a fastener, use a large breaker bar.

AIR CHISEL SAFETY

Examples of uses for air chisels include cutting metal, driving on valve guides, riveting, and replacing front suspension parts. An injury can occur if an air chisel is not used properly.

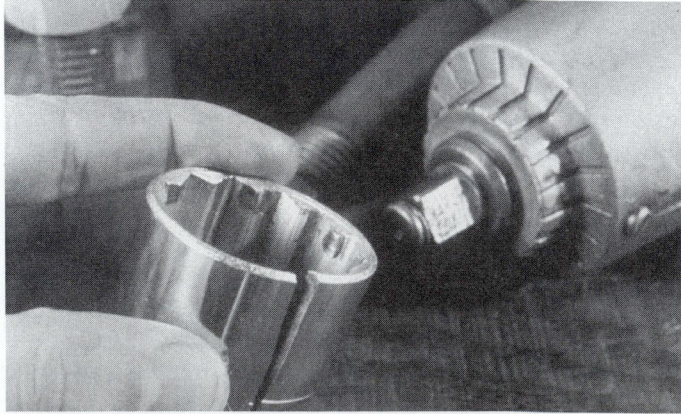

Figure 11.22 Using a regular chrome socket on an impact wrench can result in a broken socket. *(Courtesy of Snap-on Tools Corporation, Copyright Owner)*

Figure 11.23 Hold the tool bit against the workpiece before pulling the trigger on the air chisel.

- Be sure to wear eye protection.
- Before pulling the trigger, be sure to have the tool bit against the workpiece (Figure 11.23). Otherwise, the tool might fly out of the air chisel.

DIE GRINDER/AIR DRILL SAFETY

Air drills and die grinders are used with small grinding wheels, burrs, and wire brushes. A die grinder is similar to an air drill but turns at a much higher speed. There are several safety precautions to use when using these tools:

- The tool is very loud. Use ear protection.
- Die grinders turn at speeds of up to 20,000 rpm. When using a grinding wheel or wire wheel at high speeds, be sure that the wheel is rated for such use.
- Some head gaskets contain asbestos. Do not remove head gaskets using a wire brush and die grinder or air drill.

PRESS SAFETY

A hydraulic press can be very dangerous, because parts being separated are under such pressure that they can explode. Common automotive presses develop hydraulic force in the range of 20 to 40 tons (40,000–80,000 pounds). Bearings have been known to explode like a bomb, injuring the user with shrapnel. Be sure to use common sense.

- Use applicable safety guards.
- Wear face protection.
- When a press has fast and slow speeds, use the fast speed to press parts whenever possible. This position does not provide full hydraulic pressure.
- Use extreme caution as the pressure applied to the part becomes higher.

GENERAL LIFTING SAFETY

When lifting a vehicle, be sure that the lift or jack contacts the frame at the recommended lift point (Figure 11.24). To be sure that the vehicle is firmly supported, push on its fenders before going under it.

When lifting a vehicle on a frame-contact hoist, place the lift adapters at the specified locations and raise the vehicle about 6". Then, shake it to see that it is firmly placed.

HYDRAULIC FLOOR JACK (SERVICE JACK) SAFETY

Hydraulic floor jacks (service jacks) are used extensively in automotive repair. Accidents do not occur often, but when they do they can be life threatening. Cars are often serviced while raised on a portable service jack, but this is a dangerous practice. Cars have fallen and crushed people who used jacks that failed. Follow these safety precautions when using a hydraulic floor jack:

- A hydraulic service jack should be used to raise and lower a vehicle only.
- Always use vehicle support stands (jackstands) (Figure 11.25).
- Use support stands in pairs.
- Be sure that the support stands are positioned in the recommended location on the frame (see Figure 11.24).
- Be sure to use support stands only on a level concrete surface. On a hot day, the legs can dig into asphalt and cause an accident.

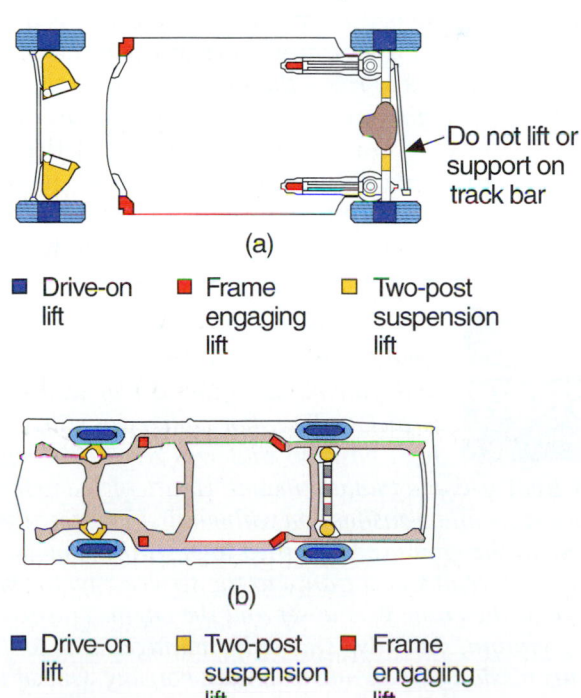

Figure 11.24 Lift a vehicle from one of the lift points shown in the shop service manual.

Figure 11.25 Always use vehicle support stands when a vehicle is jacked up.

 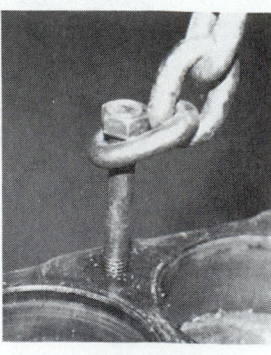

RIGHT   WRONG

Figure 11.26 The lifting fixture must be firmly tightened against the part to be lifted. The bolt on the right will be under extreme stress, which can cause it to break.

> **SHOP TIP** Place a piece of thick plywood or a steel plate under support stands that are used on asphalt.

It is foolish to crawl under a vehicle that is not resting *solidly* on vehicle support stands.

■ SHOP CRANE (ENGINE HOIST) SAFETY

When using a shop crane (cherry picker), several safety precautions are necessary. An engine is very heavy. It is important that the center of gravity be observed. When the engine is removed from the car, it is usually raised 3–4 feet off the ground to clear the radiator grill. *It is very dangerous to move the shop crane with the center of gravity this high because the crane can tip over easily.* Lower the engine close to the ground, and then roll the crane in as straight a line as possible to avoid tipping it over. This precaution is especially critical when the engine and transmission have been removed at the same time.

> **CASE HISTORY** *An apprentice had just removed an engine and transmission from a rear wheel drive pickup. The shop crane was still in the high position that was required to clear the front grill and radiator mount. He attempted to move the engine and transmission without first lowering them close to the ground. As he tried to push the shop crane over a ledge at the entrance to the service bay, the balance of the crane was upset and the engine crashed to the ground. Luckily, the only damage was to the transmission oil pan and extension housing. One of the legs of the shop crane was bent so badly that it was taken out of service.*

- It is very important that the bolts that hold the engine sling to the engine block are not too short. The amount of thread on the bolt that enters the block should be of an amount that is at least 1½ times the diameter of the screw thread (usually about 6 turns).
- Bolts used to attach a lifting fixture to a heavy part must be firmly tightened against the device. Use spacers if necessary (Figure 11.26).
- Be sure to lower the engine as soon as possible before attempting to roll the crane to a different location.
- When an engine has been removed from a vehicle, be careful when raising the vehicle in the air while on a frame-contact lift. The vehicle will be rear-end heavy when the weight of the engine is removed from it. The result can be that the vehicle falls off the lift.

■ TRANSMISSION JACK SAFETY

A transmission jack is used when the vehicle is raised in the air on a lift. The safest use of this jack is when the vehicle is on a wheel-contact lift. When a vehicle is on a frame-contact lift, there is a danger that raising the transmission can result in the vehicle being lifted off the adapters on the lift. If its balance is upset, the vehicle could be knocked off of the lift.

■ BATTERY SAFETY

As a battery charges it gives off explosive hydrogen gas. Sparks must be avoided around batteries, which occasionally explode (Figure 11.27). The most common cause of battery explosions is when someone unhooks a battery charger without turning it off. A battery explosion can cause:

Figure 11.27 Careless use of the battery charger caused this battery to explode.

- Damage to skin (from acid thrown by the explosion)
- Permanent or temporary hearing loss
- Blindness

Battery Safety

1. Always wear eye protection when working around batteries.
2. Unplug the charger before connecting or disconnecting any of the cables.
3. Be sure to shut off the battery charger before unplugging its 110V power cord, or a spark could happen when the electricity tries to jump from the outlet to the plug.

Jump-Starting a Car

When jump-starting a car, follow the proper procedures and wear eye protection (Figure 11.28). A battery that has been charging normally (the host vehicle) could have trapped hydrogen gas under its cell caps or the battery terminal clamp. Be sure that no sparks occur near it. Always connect the dead battery last and hook up the ground cable to the engine block away from the battery.

Battery Acid Safety

Battery acid (**electrolyte**) is a chemical combination of *sulfuric acid* and *water*. Battery acid can harm the skin, eyes, and clothing.

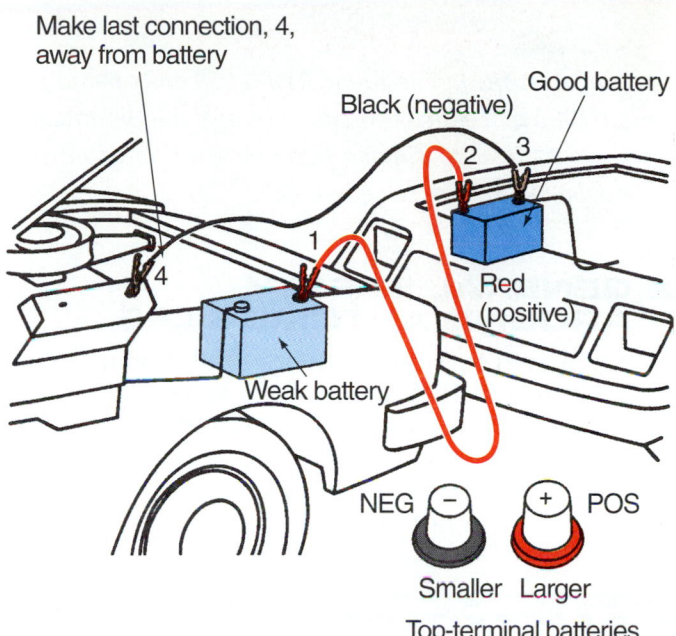

Figure 11.28 Correct hookup for jump-starting a car.

 **SAFETY NOTE** If battery acid gets on your skin, immediately wash the area with water for at least 15 minutes.

Baking soda can be used to neutralize battery acid. After carrying or servicing a battery, keep your hands out of your pockets. The outside of a battery is usually covered with acid resulting from the mist from the vent caps during charging. The acid from your hands will cause holes to develop in your pockets after your clothes are washed.

■ REFRIGERANT SAFETY

Refrigerant is used in automotive air conditioning systems. It is a vapor at room temperature and becomes liquid only when kept under pressure. When pressure is removed from it (such as when you disconnect one of the air conditioning lines), it boils vigorously as it turns into a vapor.

 **SAFETY NOTE** Liquid refrigerant contained in the air conditioning system will instantly **freeze** anything that it comes into contact with.

When working around refrigerants:
- Wear gloves to protect your skin.
- Wear goggles to protect your eyes. Refrigerant can cause instant blindness.

■ GENERAL SAFETY AROUND AUTOMOBILES

Do not leave things on the floor of the shop that people can trip on. Hydraulic service jacks (floor jacks) should be left with the handle up, rather than hanging down low where someone could trip on it. Floor creepers should be stood vertically against a wall or workbench when not in use.

Before a Test Drive

Before driving a customer's car, remember to check the operation of the brakes and look at the condition of the tires. Do not test drive a car with obvious safety hazards until they are corrected. It makes no sense to test drive a car with dangerous brakes.

Working Around Belts

One of the most common farm injuries is lost fingers. Farm machinery has many belts. Fingers are often cut off when they are caught between a belt and pulley.

Be sure that a helper understands what you are asking him or her to do. Assuming that they understand can result in an accident.

Before attempting a fan belt adjustment, be sure that the keys are out of the ignition. If someone cranks the engine over, fingers can be cut off.

■ SAFETY TEST

Choose your answers to the Safety Test from this list:

1½	charger	eye or face	hydraulic	manual	talk
6	data sheet	fire	hydraulic jack	picker	vapor
air	disconnected	flash	impact	pipe	water
air	drill	gasoline	instructor	reground	water
asbestos	drilling	green	instructor	rest	water
B	electrical	grinders	jack	side	wrench
back, eyes	end	ground	jack	spontaneous	
blowguns	explosions	handle	legs	squeezing	

1. If you should become injured while working in the shop, inform your _____ immediately.
2. The most common injuries in an automotive shop involve the _____ and _____ .
3. Lift with your _____, not your back.
4. If a _____ is burning out of control, leave the area immediately and call for help.
5. A flammable liquid fire is called a Class ____ fire.
6. Keep oily materials separated from air to prevent _____ combustion.
7. What form of gasoline is the most dangerous? liquid, vapor (circle one).
8. Never use _____ to clean floors or parts.
9. The temperature at which a flammable liquid's vapors will ignite when brought into contact with an open flame is called the _____ point.
10. Disconnect the battery _____ cable before working on a vehicle electric system.
11. Before an _____ fire can be extinguished, the battery must be disconnected.
12. Do not attempt to put out a gasoline fire with _____ .
13. The information sheet about a hazardous material is called a Material Safety _____ _____ (MSDS).
14. If caustic liquid (from the hot tank) gets into your eyes flush with _____ , especially under the eyelids.
15. A HEPA vacuum is a special vacuum for _____ .
16. The color of the wiring for ground on commercial and household electrical wiring is _____ .
17. Electric cooling fans should be _____ before working around them.
18. Before opening a radiator, test for pressure in the system by _____ the top radiator hose.
19. Mushroomed punches or chisels should be _____ before use.
20. Always wear _____ protection when using hammers, chisels, pullers, batteries, air conditioning machinery, compressed air, and other hazardous situations.
21. Always use a _____ on the tang of a file.

22. Compressed _____ can penetrate skin and must be used cautiously.

23. Do not _____ to someone who is operating a machine.

24. A _____ bit has a tendency to grab the work when it just begins to break through the metal.

25. When _____ sheet metal or large pieces of metal on the drill press, make sure they are clamped to the work table.

26. Stand to the _____ when starting the grinder.

27. Position the tool _____ as close to the grinding wheel as possible.

28. Rubber-tipped _____ work at full shop air pressure.

29. When using an air impact wrench, be sure to use an extra thick socket, called an _____ socket.

30. When an air impact _____ is used with a wobble socket, the socket can fly off if the tool is started when the socket is not on the bolt head.

31. When using an _____ chisel, be sure to have the tool bit against the workpiece before you squeeze the trigger.

32. Die _____ turn at speeds of up to 20,000 rpm. Be sure to use wheels that are rated for that high a speed.

33. Because they can develop force of from 20–40 tons, _____ presses can cause parts to explode.

34. Locate the correct lift points in the service _____ before using a hydraulic service jack or lift.

35. When lifting a vehicle on a frame-contact lift, place the lift adapters at the specified locations and raise the vehicle about _____". Then, shake it to see that it is firmly placed.

36. Use a _____ _____ to raise and lower a vehicle only.

37. A jacked-up vehicle should be placed firmly on _____ stands.

38. When moving an engine that is mounted on a shop crane, be sure the engine is lowered as far as possible to keep the center of _____ low.

39. Bolts that attach an engine sling or engine stand adapter to the block should enter the block at least _____ turns.

40. When using a transmission _____ on a car that is on a frame-contact lift, be careful not to raise the car off the lift adapters as you raise the transmission.

41. The most common cause of battery _____ is when someone unhooks the battery charger without turning it off.

42 Shut off and unplug the battery _____ before unhooking either of the cables.

43. If acid gets on the skin or eyes, immediately flush with _____ for at least 15 minutes.

44. If refrigerant from an air conditioning system contacts your skin or eyes it will _____ them.

45. The keys should be out of the ignition any time a _____ is being inspected or adjusted.

46. When a floor creeper is not in use, it should be stood on _____ so it is not accidentally stepped on.

Vehicle Inspection (Lubrication/Safety Check)

THEN AND NOW: SAFETY

In 1769, a French army engineer named Nicolas Cugnot mounted a steam engine on a wooden wagon in an effort to create a better way to transport a cannon. In the process, he became the first person in history to produce a motor vehicle. During a test, his invention went out of control and ran into a wall, making him the first person in the world to have an automobile accident.

The survival of the driver and passengers has been looked upon as a serious matter since the beginning of the automotive era. In the 1920s, steering wheels replaced tillers. This safety development was followed by *safety glass*, a glass that crumbles instead of breaking into the knife-like fragments of ordinary broken glass. Four-wheel brakes were also developed. These helped stop a vehicle much sooner. All of these milestones saved many thousands of lives and prevented countless injuries. But a car was still an essentially hostile environment for a human being. There were 4000 Americans killed in automobile accidents in 1913. By 1930, the death toll was 32,900. In some years since, over 50,000 have died—that is almost one fatality every 10 minutes!

In 1956, Ford Motor Company offered two safety options, seat belts and padded dashboards. Very few customers opted for them. In 1958, Volvos were sold with seat belts as standard equipment, followed the next year with shoulder harnesses. By 1966, the U.S. government introduced uniform vehicle safety regulations, one of which made dual circuit braking systems mandatory. This feature prevented total brake failures. Another safety regulation required seat belts on every car sold in America. Unfortunately, most people considered them a nuisance and never put them on. To remedy this, the seat belt interlock was mandated in 1974. It prevented the car from starting unless your seat belt was fastened. The public outcry against this inconvenience was so strong that the government rescinded the law.

Cars are much safer now than they were in the early days of motoring. The traffic death rate per 10,000 registered vehicles went down 94% between 1912 and 1995. This means where we once had thirty-three fatalities, we now have two.

Anti-lock brakes and *air bags* are the biggest news on the safety front today. By pulsing the brakes, anti-lock brakes help avoid a skid that can result in an accident. If you are unfortunate enough to be involved in an accident, an air bag inflates instantly to cushion the impact and improve your chances of survival.

Volvo introduced the shoulder harness in 1959. *(Courtesy of Volvo Cars of North America)*

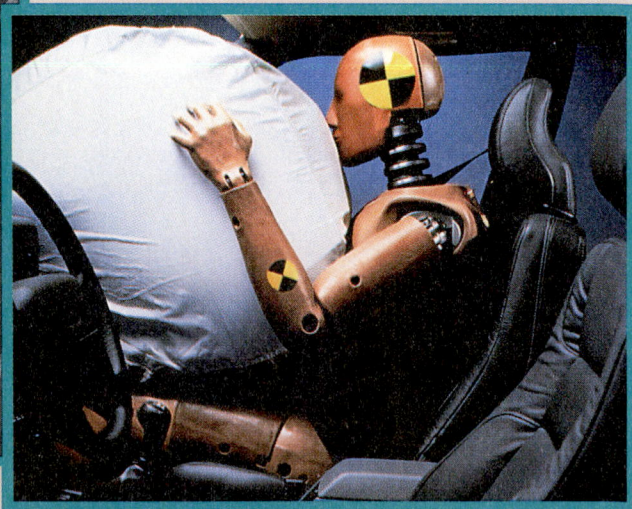

A passenger air bag. *(Courtesy of Volvo Cars of North America)*

CHAPTER 12

Engine Lubrication

■ OBJECTIVES

Upon completion of this chapter, you should be able to:

✔ Describe engine lubrication under different service conditions.

✔ Select the correct engine oil to use.

✔ Describe the operation of different types of oil filters.

■ INTRODUCTION

In theory, all moving parts are separated by a thin layer of oil (Figure 12.1). Oil is supplied to engine parts by an oil pump (Figure 12.2). If oil is properly maintained with no dirt allowed to accumulate in it, then very little wear should occur. Under normal conditions, the only time a breakdown in lubrication occurs is when the engine is first started in the morning. The crankshaft sits on the bearing until a wedge of oil is re-established after pressurized oil reaches the bearing (Figure 12.3). A few seconds can pass before the engine's oil pump can distribute oil to the entire engine (see Chapter 19). During this condition, called **dry start**, parts can rub and wear results.

NOTE: *The amount of wear that can happen during this short amount of time is equivalent to that caused by hundreds of miles of freeway driving.*

■ ENGINE OIL

Engine oil is more than just basic crude. It contains a complicated additive package. Crude oil contains wax, most of which is removed during the refining process. The first oil additives were developed in the 1930s to deal with residual wax in the oil. More additives were developed during World War II. Over the years, this additive package has been improved. The result is that engines last longer and longer.

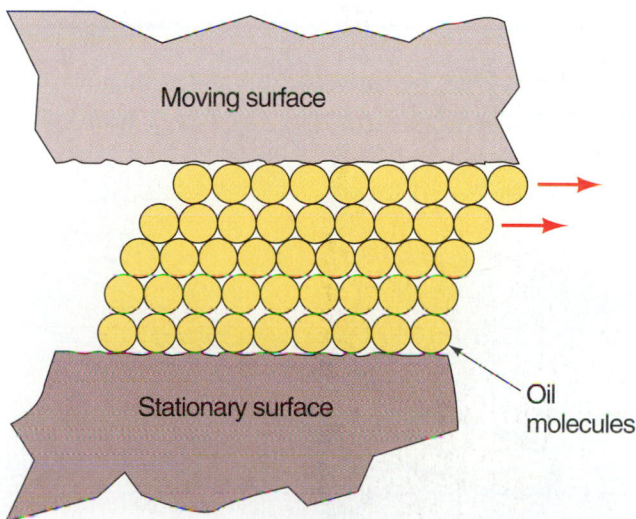

Figure 12.1 Moving parts are separated by a thin film of oil. *(Courtesy of Federal-Mogul Corporation)*

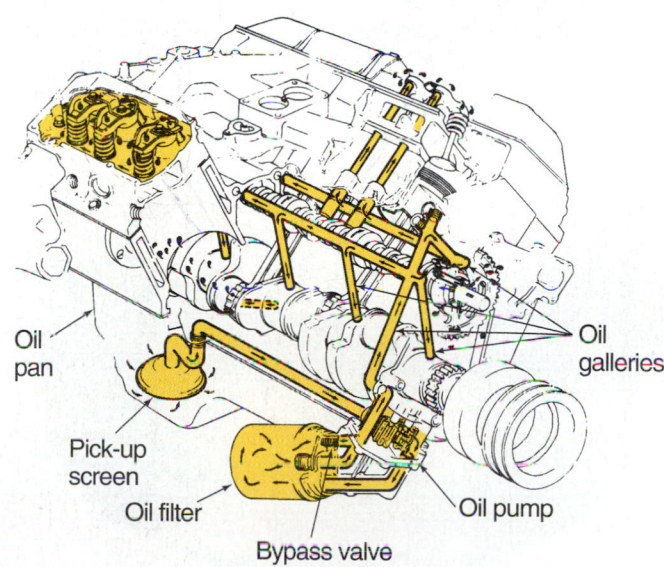

Figure 12.2 Oil is circulated through the system by an oil pump. *(Courtesy of Chrysler Corporation)*

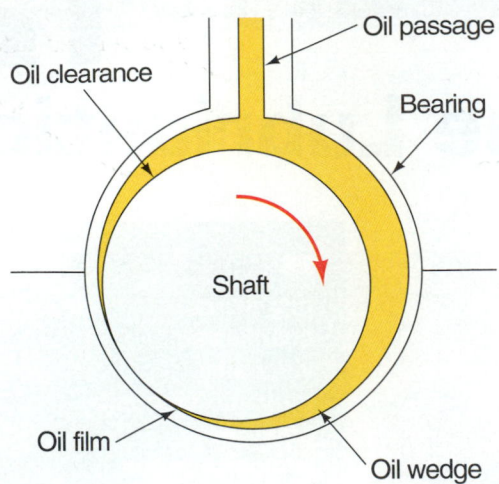

Figure 12.3 A wedge of oil lifts the shaft off of the bearing.

Besides lubricating, engine oil cools, cleans, and prevents rust from forming inside the engine. Oil also fills hydraulic lifters and helps to seal the piston rings against the walls of the cylinder.

Oil Level

The correct oil level is designed to keep the oil pick-up screen (see Figure 12.2) below the level of the oil under all operating conditions. If the oil level is allowed to drop too low, serious engine damage can result. The crankshaft bearings can be damaged (Figure 12.4) or the piston can become *scuffed*.

The **crankcase** is the area at the bottom of the engine enclosed by the oil pan. If there is *too much* oil in the crankcase, the crankshaft can dip into the oil as it spins. The oil is thrown onto the cylinder walls by

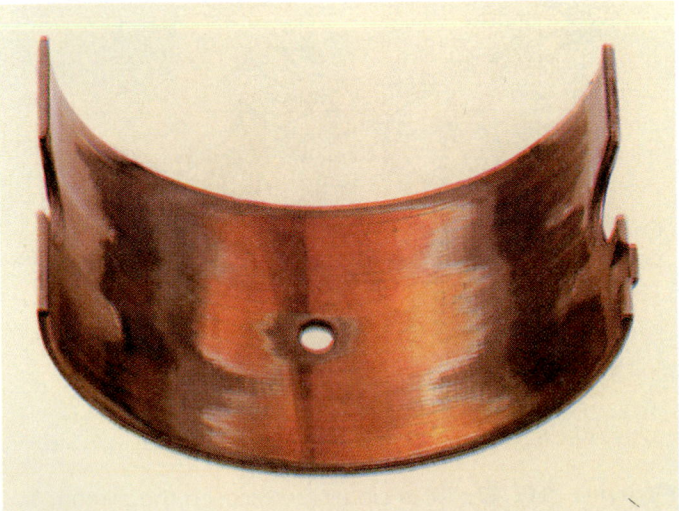

Figure 12.4 A worn crankshaft bearing. *(Courtesy of AE Clevite Engine Parts)*

the crankshaft in such an amount that the oil rings are overwhelmed. The result can be excessive smoke. Chapter 40 has more information on oil consumption. When the crankshaft hits the oil it churns it up, aerating it (mixing it with air). Aerated oil does not provide for sufficient oil pressure and can result in collapsed hydraulic lifters, or even a broken crankshaft. The oil *dipstick* is located somewhere on the engine block. It has markings that indicate "full" and "add" (Figure 12.5). When the oil level is at the add line, the crankcase is one quart low.

NOTE: *Be sure that the engine dipstick is not being confused with the dipstick for an automatic transmission. The engine dipstick will be located somewhere on the engine block. The transmission dipstick often has instructions printed on it regarding fluid type and checking procedure. Also, the transmission fluid "add line" indicates one* **pint,** *instead of one* **quart.**

When checking the engine oil level:
- The vehicle should be on a level surface.
- The engine should be warm.
- The engine should have been off for at least five minutes, giving the oil a chance to drain back to the oil pan.
- Be sure the dipstick is pushed all of the way down against its seat.
- If the oil level reading on the dipstick is unclear, try looking at the backside of the dipstick, or dip it into the oil and attempt to read it again.
- If the oil level is low, check to see if the vehicle is

SHOP TIP Polish a dipstick with emery cloth or the wire wheel on a grinder motor to make it easier to read.

due for service before adding a quart of new oil.

➡ *Perform **Check Engine Oil Level** Worksheet*

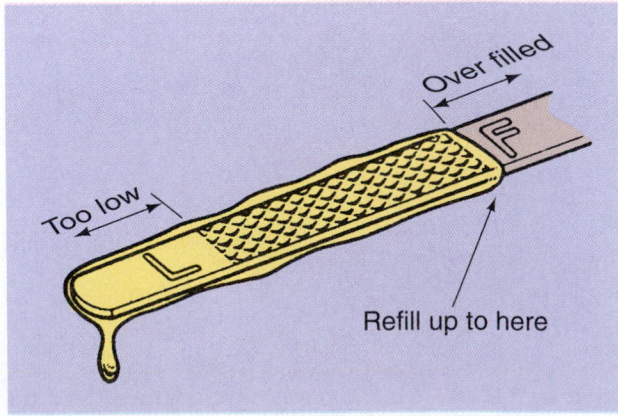

Figure 12.5 The dipstick indicates the oil level in the crankcase.

Oil Viscosity

Viscosity is the thickness or body of an oil. The Society of Automotive Engineers (SAE) designates viscosity ratings. The SAE classification was adopted by the petroleum industry in 1911. An oil with a viscosity rating of 30 will have "SAE 30" displayed on the oil container. Viscosity, which is measured at 212°F (100°C), is higher if the oil is thicker. For instance, an SAE 10 oil is thinner than an SAE 30 oil. The higher the oil's viscosity is, the stickier it is and the more resistant to flow it is.

When a "W" accompanies the rating, it means that the oil's viscosity has been tested at 0°F also; this oil is said to be a *winter-grade* oil.

Multiple Viscosity Oils

Some oil has only one viscosity rating but most new engine oils are **multiple viscosity** (*multi-vis*) (Figure 12.6). The following is an interpretation of a typical multi-vis oil (SAE 10W-30):

SAE = Society of Automotive Engineers

10W = The viscosity of the oil when measured at 0° (the "W" means winter grade).

30 = The viscosity of the oil when measured at 212°.

SAE 10W-30 has a base rating of 10 when cold. It will flow freely at temperatures as low as −20°F. Oil nor-

mally thins out when heated. Because it becomes thinner, its viscosity number becomes lower. An additive package containing **polymers** is blended into the oil. Polymers expand when heated, so the oil actually maintains its viscosity to the point where it is equal to what a hot SAE 30 oil would be (Figure 12.7a). This ability to resist change in viscosity as an oil heats up is called its **viscosity index**. Figure 12.7b shows recommended SAE viscosity grades for different climates.

A multi-vis oil helps combat engine wear because it flows more quickly to the bearings when the engine is first started—especially important during cold

Figure 12.6 A multi-viscosity oil. *(Courtesy of Quaker State Corporation)*

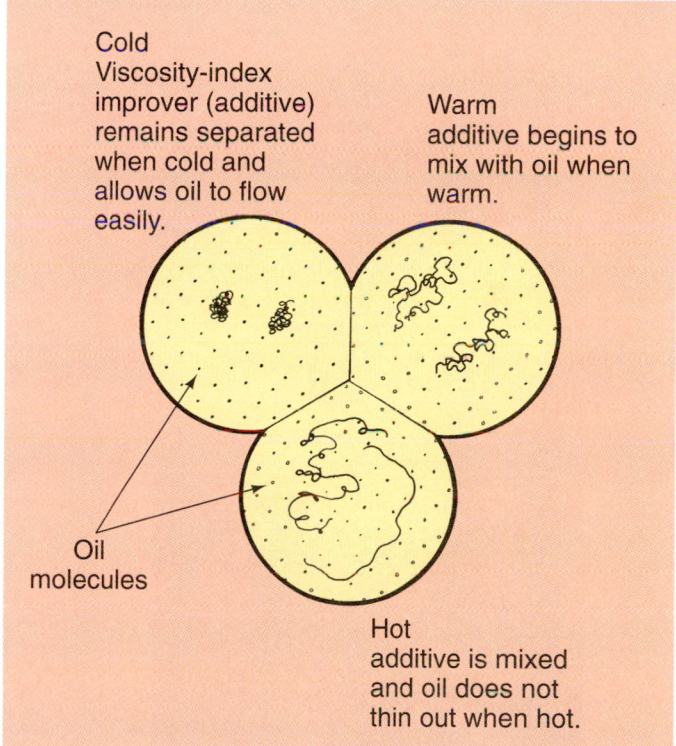

(a)

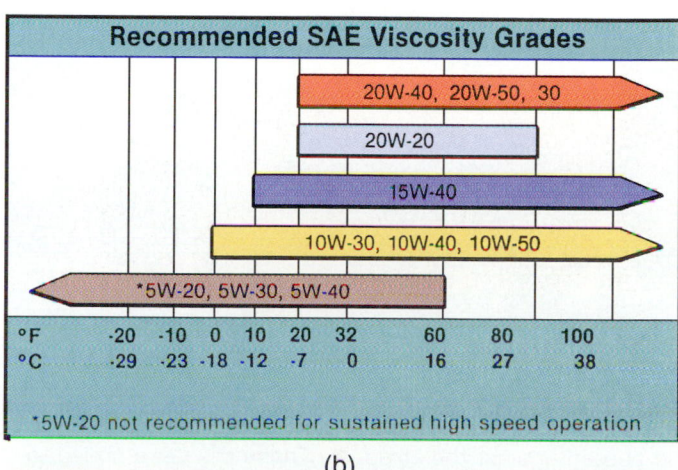

(b)

Figure 12.7 (a) How multiple viscosity oil performs. (b) SAE viscosity recommendations for various climates. *(Courtesy of Allied Signal Automotive Aftermarket)*

weather. The oil pick-up has a screen with a relief opening that opens, allowing oil that is too thick or dirty to pass through to engine parts (Figure 12.8a and b). The theory is that thick or dirty oil is better than no oil at all. Thick oil can also bypass the oil filter (see "Oil Filter" later in this chapter). In the past, some manufacturer's recommendations included SAE 30 in temperate climates (above 40°) because it flows acceptably at that temperature. Figure 12.8c shows how oils of different viscosity ratings flow.

Oil Pressure

Pressure cannot develop unless there is a resistance to flow. As an engine wears, clearance between its crank-

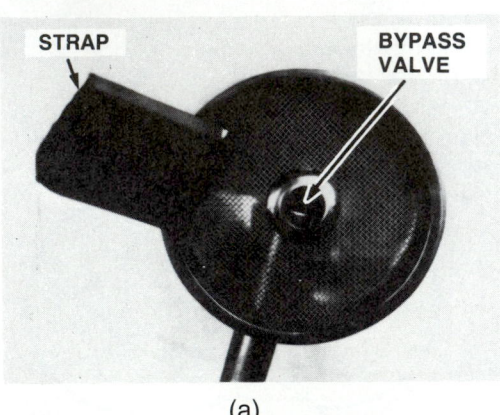

(a)

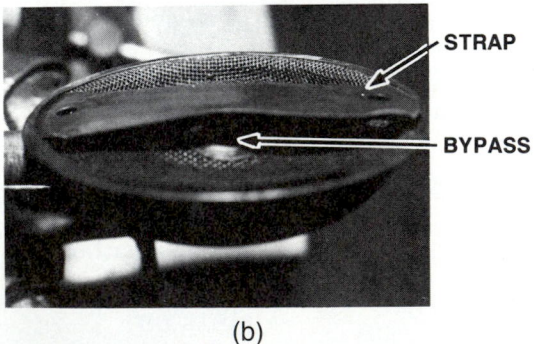

(b)

(c)

Figure 12.8 (a) This pump screen has a bypass valve. (b) This pump screen is defective; the bypass should be seated against the strap. (c) These oils were chilled to –35°C for 16 hours. The photo was taken 30 seconds after the caps were removed from the containers. (a, Courtesy of Federal-Mogul Corporation; c, Courtesy of Imperial Oil Limited)

shaft and bearings increases. The oil pump can no longer pump sufficient oil to fill this extra clearance. The result is that oil pressure is low whenever the engine rpm drops to idle speed. As the engine is accelerated, the oil pump (which is driven by the camshaft or the crankshaft) turns faster, so oil pressure will rise back to normal. Because a multi-vis SAE 20W-50 oil is thicker both when hot and cold, it can provide higher oil pressure to an older, idling engine. But this same oil will flow more slowly (than an SAE 10W-40) to the bearings of a new engine when the engine is first started in the morning.

■ OIL SERVICE RATINGS

The label on the oil container also includes its **American Petroleum Institute (API)** service rating (Figure 12.9a). The API sets oil service specifications. Ratings progress from *SA* to *SJ*. Straight mineral oil, or recycled oil without additives, is classified "SA." This grade of oil is good only for very light-duty applications. The rating system progresses from SB through SJ oils for the most recent manufacturer's requirements (there is no SI classification). SJ oil has many high-quality additives and will work well in any engine. The "S" means that the oil is for engines with *spark ignition*. Diesel engines use oils rated from CA to CF. The "C" means that the oil is for engines with *compression ignition*.

A new testing procedure has recently been implemented. Oils are randomly tested by the *Chemical Manufacturers Association CMA* for five criteria. Oils that pass these tests are labeled with a *starburst symbol* (Figure 12.9b).

Energy-efficient Oils

The *ASTM American Society for Testing Materials* certifies an oil as *energy conserving (EC)* if it passes certain tests. These oils can provide a 1% to 4% improvement in miles per gallon over regular oil. While improvements of this smaller amount would be practically impossible for a car owner to observe, these oils help the manufacturers to meet government fuel economy standards. An energy conserving oil that states "ECII" on its label will have better than a 2.7% fuel economy increase over regular oil. Be sure to use a high quality oil. The use of an SA oil can cause serious damage to a modern engine (Figure 12.9c).

■ OIL ADDITIVES

Oils (other than SA) contain an additive package that can make up as much as ⅓ of the volume of the oil. These additives have very important functions. Some of the additives found in engine oil are:
- *Pour-point depressants* that allow the oil to flow in very cold weather
- *Corrosion and rust inhibitors* that help the oil to stick to metal surfaces

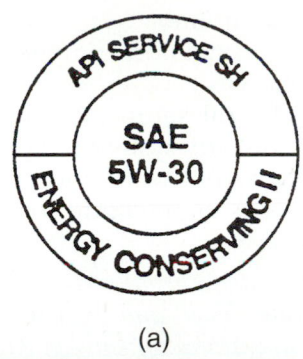

(a)

(b)

(c)

- *Anti-foam additives*
- *Friction modifiers*
- *Oxidation inhibitors*
- *Anti-wear additives*

The chemical, *zinc dithiophosphate*, that is most often used as an oxidation inhibitor also serves as an anti-wear additive. When the oil film becomes too thin or starts to break down under load, this is known as **boundary lubrication**. The primary job of anti-wear additives is to protect the cam and the lifters, which are under a severe load in today's small high compression engines. During boundary lubrication, the zinc additive combines chemically with engine metals to provide a lubricating layer.

Detergents and *dispersants* are added to oil to keep small particle contaminants suspended in the oil (Figure 12.9d). If they are large enough, they will be trapped by the oil filter. But if they are too small, they will not be removed from the engine until the oil is replaced.

As oil decomposes at high engine temperatures, a chemical reaction between oil and oxygen causes a gummy mixture to form that can plug oil passages. Detergents make these deposits oil soluble so that they can be suspended in the oil. Detergents in oils are similar to those found in household cleaning products. The difference is that they are oil soluble rather than water soluble.

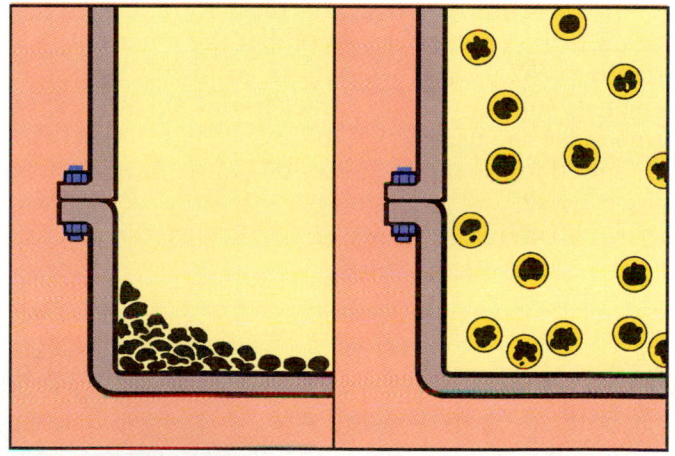

Non detergent oil | Detergent oil

Sludge particles remain separated in oil, tend to collect in low spots forming sludge deposits | Particles suspended, become trapped in filter

(d)

Figure 12.9 (a) The API rating is printed on the container. (b) The starburst symbol. (c) The lifter on the right is severely worn because it was used with an SA oil. (d) Detergent suspends particles so they can be trapped by the filter. *(a and b Courtesy of The Valvoline Co.; c, Courtesy of Texaco's Magazine LUBRICATION; d, Courtesy of Allied Signal Automotive Aftermarket)*

Sludge is a mixture of moisture, oil, and contaminants from combustion (Figure 12.10). It can clog the oil screen and oil lines if allowed to accumulate. Moisture accumulates from condensation and is also a product of combustion. When an engine gets warm enough, the positive crankcase ventilation (PCV) system removes the water vapors. Short trips allow the accumulation of sludge. Accumulated fuel and moisture do not get a chance to evaporate in an engine that never fully warms up.

NOTE: *It takes 10 to 15 minutes for the cylinder walls to warm to the point where evaporation begins to take place.*

The oil level on the dipstick can be artificially high due to a high water and gas content.

Dispersants cause the oil to become discolored because they keep sludge-forming particles in suspension in the oil. Excessively rich air/fuel mixtures can cause soot contamination, which turns oil black.

Oil without additives (SA oil) is referred to as *nondetergent* oil. Some older technicians suggest using nondetergent oil until the first oil change. The use of lower quality oil was recommended in the late 1950s to help the older-style, plated piston rings to seat. Today, the recommendation is to use the warranty-approved product for the entire life of the engine. Most service technicians use SH oil exclusively.

Oil oxidizes at temperatures in excess of 250°, becoming thicker and forming varnish deposits. The oxidation rate of oil actually doubles with every 20° rise in temperature above 140° until about 800° when carbon forms. *Anti-oxidants* combat the effects of heat on the oil. If the oxidation inhibitor becomes depleted, the oil becomes thicker and thicker. Detergents make the varnish oil-soluble so that they remain suspended in the oil. Excessive oil temperatures can be caused by:

- Lean air/fuel mixtures
- Retarded ignition timing
- Using the air conditioner
- An automatic transmission-equipped vehicle pulling a trailer

When possible, oil temperature should be kept below 220°F. The cooling system temperature required for water to vaporize out of the oil is 185–195°F. The ideal cooling system temperature for oil is therefore about 195°F.

NOTE: *Oil temperature is usually 10°F to 25°F hotter than coolant temperature. Fuel and moisture are effectively removed at oil temperatures of between 215°F and 220°F.*

■ SYNTHETIC OILS

Synthetic oils exhibit outstanding low-temperature characteristics, increased fuel economy and power, lower oil consumption, and a better viscosity index. They have no waxes, most of which must be removed from conventional oils during the refining process. The primary drawback with synthetics is their price. *Semi-synthetic oils* are a mixture of conventional oil and synthetic. They are less expensive than synthetics and share some of their improved properties.

Synthetic oils contain about the same amount of additives as conventional oils. These additives become depleted with use. Therefore, synthetics require the same oil change interval as conventional oils. The manufacturer's warranty requirements can be met by synthetics as long as the correct change interval, viscosity, and quality rating recommendations are followed.

Synthetics can also harden seals in some engines made prior to 1980. If your older engine has been rebuilt since then, it can safely use a synthetic.

■ CHANGING ENGINE OIL

A factor in the life of an engine is whether the engine oil is drained and refilled at regular intervals. Oil does not wear out, but it becomes diluted and contaminated. In a cold engine, contaminants condense on

Figure 12.10 The sludge in this engine shows that it suffered from a lack of maintenance. Many byproducts result from the burning of fuel.

the cylinder walls and migrate to the crankcase where they are mixed with the engine oil. Oil also becomes contaminated when an engine is run with a missing or dirty air filter. If an air cleaner is carelessly changed or serviced, dirt can actually enter the intake manifold. Leaking vacuum accessories and lines can also result in dirty air being drawn into the engine's cylinders. Following are some of the benefits that are obtained by changing the oil:

- An oil's additives are depleted over time and changing the oil replenishes them.
- Because the oil filter takes out only particles larger than a certain size, periodic oil changes help to clean the smaller contaminants from the oil.
- Sludge is removed.
- Unburned fuel and acids are removed with the oil.

Oil Change Intervals

Internal combustion engines produce many byproducts during operation. **Blow-by gases** (gases that leak past the piston rings into the crankcase) contain acids as well as unburned and burned fuel. When moisture combines with these materials, sludge forms inside the engine. If the engine is never fully warmed up or if oil is not changed often enough, more sludge develops. A vehicle driven primarily on the freeway—where it runs more efficiently—can have less frequent oil change intervals (7500 miles or more) than one that is driven primarily on short trips.

NOTE: For city driving, it is a good practice to change the oil every 90 days or 3000 miles. Oil is inexpensive insurance against engine damage.

Changing Brands of Oil

Oils with the same API service rating are usually compatible. But today's oils can be made up of ⅓ additives, so a chemical reaction is possible. Although the chemical reactions have *not* been found to be serious:

- It is a good idea to avoid mixing brands of oils *between* oil changes.
- Changing brands of oil is best done when the oil is being changed.
- When there is not a choice of brands available, it is better to add any brand of high grade oil than to operate the engine with a low crankcase oil level.

Changing Oil

It is best to change the oil while it is still hot because the detergents in the oil will suspend contaminants, allowing them to be removed with the oil. While the oil is draining, the filter can be changed. Check the lubrication manual for the capacity of oil that the crankcase holds and the API and viscosity classifications of the oil.

Sometimes a service technician can be distracted by a phone call or a customer with a problem and acci-

dentally forget to refill the crankcase with oil. This can result in serious engine damage.

> **SHOP TIP** If an engine's oil is to be changed, remove the oil filler cap and place it somewhere obvious. Make a habit of not replacing it until all of the new oil has been added after the oil change

When changing the oil:
- Check the condition of the drain plug gasket (Figure 12.11). The gasket is usually made of either nylon or a soft metal such as aluminum or copper. Drain plug gaskets are usually reusable, but at the first sign of wear or damage they should be replaced.
- Be careful not to strip the threads in the oil pan.

> **SHOP TIP** When starting a thread into a hole, first turn counterclockwise (direction of loosening) until you feel the thread drop into place. Then, tighten the plug.

- Always turn the drain plug all of the way into the thread *by hand*. Only use a wrench to torque the plug when its gasket has begun to be compressed against the oil pan. This will avoid the possibility of accidentally stripping a thread.
- If a stripped thread is encountered, *self-tapping drain plugs* are available (Figure 12.12).

Gasket

Figure 12.11 Check the condition of the gasket on the drain plug.

Figure 12.12 A self tapping drain plug and gasket.

■ Be sure to run the engine and check for leaks following completion of the oil change.
■ Close the hood and check to see that the latch is secure.
■ Wipe off any grease or handprints that might have accidentally gotten on the hood or steering wheel.

Mileage Service Record. When oil is changed, it is customary to put a reminder of the mileage somewhere where the customer will be able to refer to it. This is usually a door record sticker, placed on the driver's side door jamb or a transparent sticker placed on the upper inside of the windshield.

➡ *Perform Oil Change Worksheet*

■ OIL FILTER

An engine's oil filter prevents harmful abrasive particles in the oil from damaging internal parts of the engine. Most oil filters are the *spin-on* type. The sheet metal shell of the filter contains a filtering element, usually made of pleated paper. Oil flows through the paper from the outside to the inside, taking advantage of the entire surface area of the paper for maximum flow (Figure 12.13). A metal tube in the center of the

filter keeps the paper element from collapsing inward when the oil is thick or contaminated (Figure 12.14).

Full-Flow Oil Filter

Today's passenger cars use the **full-flow oil filter** system. In full-flow systems, all of the oil supplied by the pump is *supposed* to flow through the oil filter on its way to lubricate the engine bearings (Figure 12.15). The filter used is made of tough, treated paper that has a minimum resistance to flow. It is called a *surface-type* filter.

Because all of the oil supplied by the pump must first flow through the filter before it can reach the bearings, a full-flow filter must have a **bypass valve** (Figure 12.16). The bypass valve opens to let the unfiltered oil flow to the engine when:

■ The oil is cold and thick.
■ The filtering material is plugged due to a lack of proper maintenance.

Some engines have a built-in bypass valve. Otherwise, it is incorporated into the filter.

■ The bypass valve on most passenger car oil filters will open when resistance to flow reaches about 8 psi. At normal oil operating temperature, oil flowing at four gallons per minute through a paper filter will have a pressure drop that is less than 1 psi.
■ Full-flow filters filter out only relatively large particles. According to General Motors, their AC full-flow paper oil filter elements will trap and hold any particle that is larger than 40 microns (100 microns is equal to 0.004"). A human hair is usually about 60 to 80 microns in diameter so the size of filtered particles would be visible to the eye. Particles smaller than the filter can trap cause only minor wear when proper service intervals are maintained. They remain in the oil, held in suspension by the detergents and dispersants. If the filtering material was dense enough to filter out all of the smaller particles too, it would not be able to allow enough oilflow to be able to function as a full-flow filter.

Oil flows into the filter through the small holes that are in a circle in the filter base. It flows out through

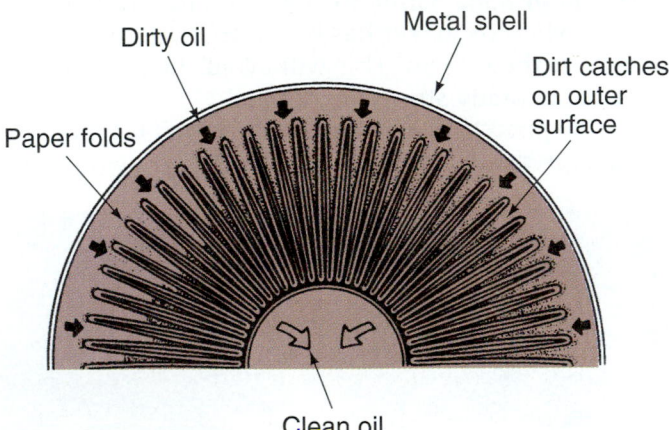

Figure 12.13 Oil flows through the paper in the filter. *(Courtesy of Dana Corporation)*

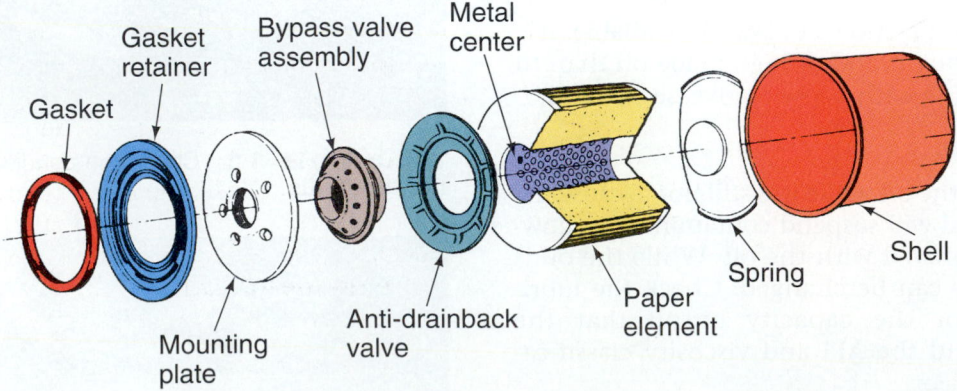

Figure 12.14 A metal tube in the center of the paper element keeps it from collapsing when pressure is high. *(Courtesy of General Motors Corporation, Service Technology Group)*

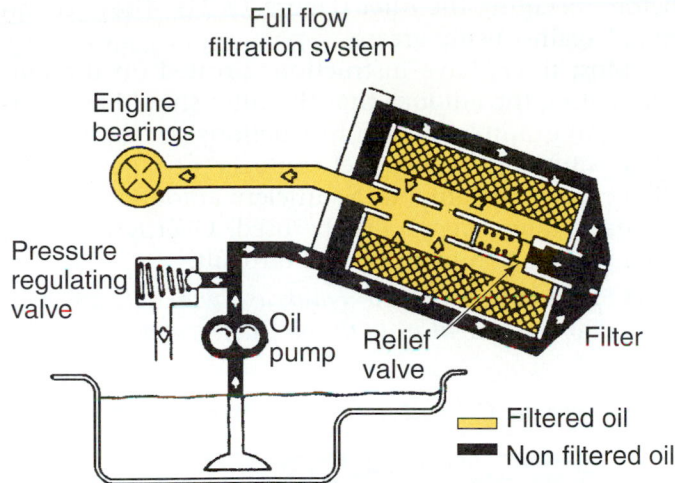

Full flow
filtration system

Engine bearings

Pressure regulating valve

Oil pump

Relief valve

Filter

Filtered oil
Non filtered oil

Figure 12.15 In a full-flow system, all oil is pumped to the filter. *(Courtesy of Federal-Mogul Corporation)*

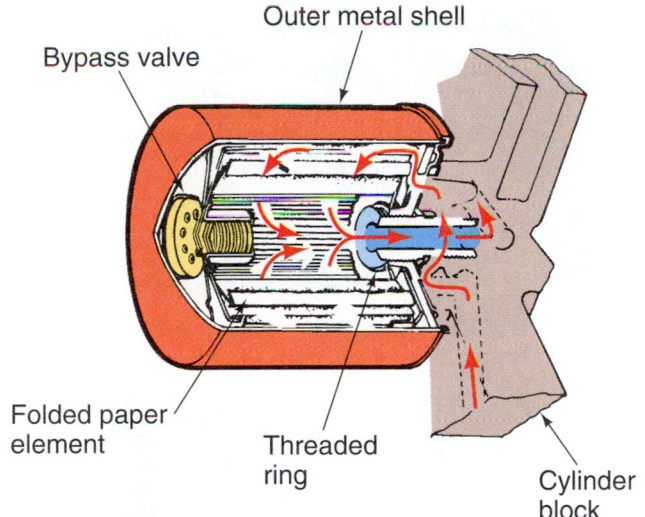

Outer metal shell

Bypass valve

Folded paper element

Threaded ring

Cylinder block

Figure 12.16 An oil filter has a bypass valve. *(Courtesy of Chrysler Corporation)*

Figure 12.17 Oil flows into the filter through the small holes and out through the large center one. *(Courtesy of Allied Signal Automotive Aftermarket)*

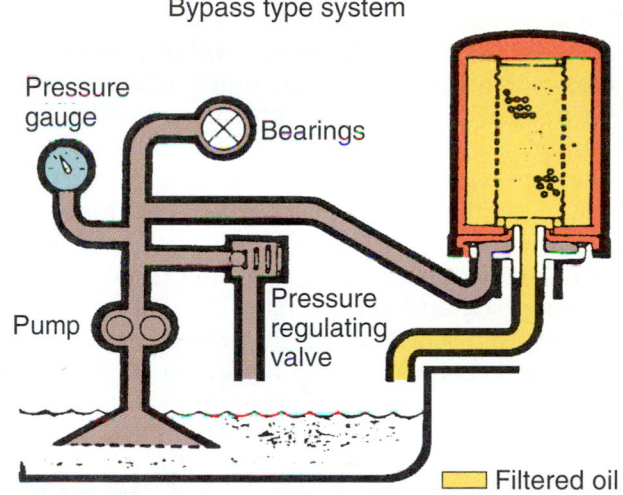

Bypass type system

Pressure gauge

Bearings

Pump

Pressure regulating valve

Filtered oil

Figure 12.18 Heavy-duty filtering systems use a bypass filter that filters only a small amount of the oil at any one time. *(Courtesy of Allied Signal Automotive Aftermarket)*

the large center hole (Figure 12.17). Many oil filters are mounted horizontally on the engine block. When the engine is shut off, about half of the oil in these filters can empty out through the filter inlet holes.

Horizontally mounted filters must have an **anti-drainback valve** to prevent this from happening (see Figure 12.14). If the drainback valve fails or if the wrong filter is used, the lubrication system will be starved of oil until the filter is filled back up. On engines with hydraulic lifters or overhead cam engines that have hydraulic timing chain tensioners this will result in a noisy condition in the morning.

Bypass Filters

Supplemental add-on filters and oil filters used on heavy trucks are often of the **bypass** or **partial-flow oil filter** design (Figure 12.18). A bypass filter traps very fine materials and allows very little oil to flow

through it. It only filters about 10% of the oil at one time. Oil is picked up at any pressurized point and allowed to "leak" through the parallel path provided by the filter on its way back to the oil pan. Little, if any, pressure drop occurs from this leak because the amount of flow is so little.

Bypass filter elements are made of such materials as cotton fiber, shredded paper, stacked high-density filter papers, and wood chips, which can remove particles down to one micron in size (one thirty-nine millionth of an inch). They are called *depth-type* filters. These filters are capable of removing sludge and were used in older vehicles that used straight mineral (SA) oil (which produced sludge rapidly). Today's oils do not allow the formation of sludge unless detergent/dispersants have been depleted.

Variations in Filters

Oil filters are identified by a number that is printed on the metal shell of the oil filter and on the filter box.

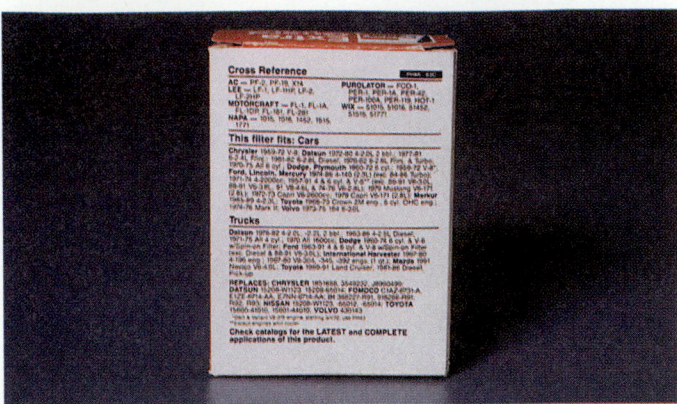

Figure 12.19 An application chart and cross reference. *(Courtesy of Allied Signal Automotive Aftermarket)*

Usually, printed on the box will be an application chart that tells which cars a filter will fit (Figure 12.19). A cross reference is often included that tells which other manufacturers' filters interchange with a particular filter.

Be sure to follow the filter manufacturer's catalog recommendations when selecting an oil filter. There are metric and U.S. threads found in a mix in both import and domestic filters, so a filter might fit on the mounting base and yet not have the correct thread. A loose thread fit could result in serious engine damage if the filter becomes loose. A tight thread fit can result in a stripped thread. Some filters have anti-drainback valves and bypass valves and others do not.

■ CHANGING THE OIL FILTER

The sheet metal shell on an oil filter is very thin. It can easily become collapsed with improper handling, making it difficult to remove. Use a professional quality *oil filter wrench* with a pivoting handle (Figure 12.20). This tool allows for easy positioning at the *base* of the filter. The sheet metal filter case is supported at the base by the heavy mounting plate (see Figure 12.14), so it will not collapse.

There is a rubber O-ring on the bottom of the filter that seals it against the engine block. Inspect it to see that it is in place on the filter and is undamaged. The Automotive Filter Manufacturer's Council recommends that the filter's O-ring be lubricated with oil

before installing the filter (Figure 12.21). They recommend against using grease.

Most filters have instructions printed on the outside stating the amount that the filter should be tightened. An example of such instructions is: "tighten the filter ¾ turn after the gasket contacts the block". If the filter cannot be tightened a sufficient amount with bare hands, a filter wrench can be used. Overtightening a filter will make future removal difficult.

NOTE: *When refilling the crankcase with oil, an extra amount of oil equivalent to the oil filter's capacity must be added.*

➡ *Perform Oil Filter Change Worksheet*

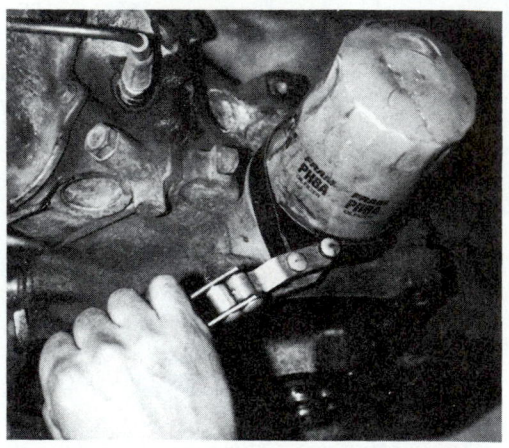

Figure 12.20 Position the filter wrench at the base of the filter.

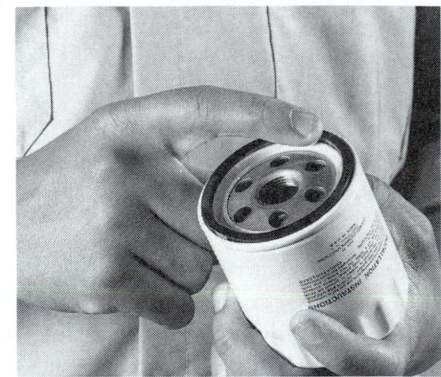

Figure 12.21 Lubricate the filter O-ring before installing the filter.

■ REVIEW QUESTIONS

1. Under normal conditions, when is the only time when an oil film does not separate engine parts?

2. Give two possible problems that can occur if the oil level in the crankcase is too high.

3. When the oil level is below the "add" line on the dipstick, how much oil should be added?

4. How will an engine dipstick be different than an automatic transmission dipstick?

5. What does SAE stand for?

6. What is the name of the part of the oil additive package that swells, causing the oil to maintain its viscosity as it heats up?

7. What causes oil oxidation?

8. What is the name of the oil filter part that allows the oil to get to the engine bearings even if the filter is plugged?

9. If sludge forms in a modern engine, what additive is probably depleted?

10. When figuring the amount of oil to add to the crankcase, add the amount specified in the specification manual plus some extra to compensate for what?

ASE STYLE REVIEW QUESTIONS

1. Technician A says that a thick oil has a higher viscosity than a thin oil. Technician B says that a multi-viscosity oil can reach bearings faster when the engine is first started than a single viscosity oil. Who is right?

   a. Technician A   **b.** Technician B

   c. Both A and B   **d.** Neither A nor B

2. Technician A says that detergents keep small particles suspended in the oil so that the filter can filter them out. Technician B says that detergents clean the surfaces within the engine. Who is right?

   a. Technician A   **b.** Technician B

   c. Both A and B   **d.** Neither A nor B

3. Which of the following are not reasons for changing engine oil?

   a. To remove small particles from the oil

   b. To remove sludge from the oil

   c. To replenish additives that have been depleted

   d. To replace worn out oil with new oil

   e. To remove unburned fuel and acids from the oil

4. Technician A says that when the oil is cold and thick, it can bypass the oil filter. Technician B says that when the oil filter is dirty and restricted, oil can bypass it? Who is right?

   a. Technician A   **b.** Technician B

   c. Both A and B   **d.** Neither A nor B

5. Technician A says that the type of oil filtering system used on modern cars is the full-flow system. Technician B says the type of filter that filters out sludge and very small particles in engine oil is called a full-flow filter. Who is right?

   a. Technician A   **b.** Technician B

   c. Both A and B   **d.** Neither A nor B

6. Technician A says an anti-drainback valve keeps a vertically mounted full-flow filter from emptying when the engine is shut off. Technician B says to position an oil filter wrench at the outer end of the filter. Who is right?

   a. Technician A   **b.** Technician B

   c. Both A and B   **d.** Neither A nor B

7. Technician A says to loosen the oil pan drain plug, turn it clockwise. Technician B says the best API service classification for automotive engine oil is "SA." Who is right?

   a. Technician A   **b.** Technician B

   c. Both A and B   **d.** Neither A nor B

8. Which of the following is *not* true when installing an oil filter?

   a. Tighten the filter an additional ⅔ to ¾ turn after the filter O-ring touches the mount surface.

   b. Compare the old filter mount surface to the new filter.

   c. Be sure the rubber O-ring is dry.

   d. Start the engine after adding oil and check for leaks.

9. Technician A says that the "W" in SAE 10W-40 stands for "weight." Technician B says an SAE 40 oil has a lower viscosity than an SAE 20W-50 at engine operating temperature. Who is right?

   a. Technician A   **b.** Technician B

   c. Both A and B   **d.** Neither A nor B

10. Technician A says that when oil oxidizes, it becomes heavier and forms varnish deposits. Technician B says that in-town driving is easier on the engine than driving on the freeway. Who is right?

   a. Technician A   **b.** Technician B

   c. Both A and B   **d.** Neither A nor B

CHAPTER 13

Underhood and Body Inspection (Car on Ground)

■ OBJECTIVES

Upon completion of this chapter, you should be able to:

✔ Pinpoint safety items in need of attention.

✔ Perform on-ground brake system checks.

✔ Perform on-ground steering and suspension checks.

✔ Perform an exhaust system inspection.

✔ Be knowledgeable about car body service.

✔ Perform fuel systems checks.

✔ Perform cooling system checks.

✔ Perform electrical systems checks.

✔ Safely and competently perform a complete underhood service.

■ KEY TERMS

pre-delivery inspection (PDI)
Department of Transportation (DOT)
dry park check
automatic transmission fluid (ATF)
PCV system
hydrometer
motor mounts

■ INTRODUCTION

During a vehicle inspection, items in need of repairs are located and documented. A lubrication/safety service includes underhood and underbody inspections. This chapter will deal with the *underhood and body inspection,* which is done while the car is on the ground. The next chapter describes an underbody inspection.

New car dealers often have a technician in charge of **pre-delivery inspections (PDI)**, lube/safety service, and tire and wheel service. Non-dealer shops perform the same jobs, with the exception of PDIs. Some shops have senior personnel or the owner of a small business working the "lube rack" because they are experienced in spotting needed repairs. Other businesses use less skilled people to perform this function, which sometimes proves to be shortsighted when repair opportunities are lost.

In a full-service facility (where all systems of the car are repaired), lube personnel sometimes earn a commission. Job responsibilities include inspection, diagnosis, and documenting problems with a vehicle. It is usually an inconvenience to the customer to arrange to leave a car for repairs or service. When contacted, the customer often decides to have needed repairs performed while the car is still in the shop. Thus, the lube person who highlights needed repairs can be a source of additional business. The customer is better served because needed repairs may be spotted before a road failure occurs or a potential safety hazard becomes dangerous.

■ BRAKE SYSTEM INSPECTION

Brakes are probably the most obvious safety item on a customer's mind. The customer expects the service specialist to give a complete appraisal regarding the condition of his or her brakes. Only obvious conditions are inspected. A complete brake inspection is usually *not* included in a lube service unless an inspection warrants further investigation. If the wheels need to be removed there is usually an extra charge, which is sometimes applied to the cost of a future brake job done by the same shop. Also, when a wheel is removed, the shop assumes a liability.

As part of the lube job, the following items in the brake system are inspected. There is also a good deal of related information that a service specialist should know in order to give the vehicle more than just a cursory brake inspection. That information is covered in Chapter 52, which deals with brakes.

Brake Pedal Travel and Feel

Brake pedal travel should be checked to see that it is not too low (Figure 13.1), and that the pedal has the proper feel when applied. Proper operation and condition of the master cylinder can be checked by holding light pressure on the pedal. It should not slowly

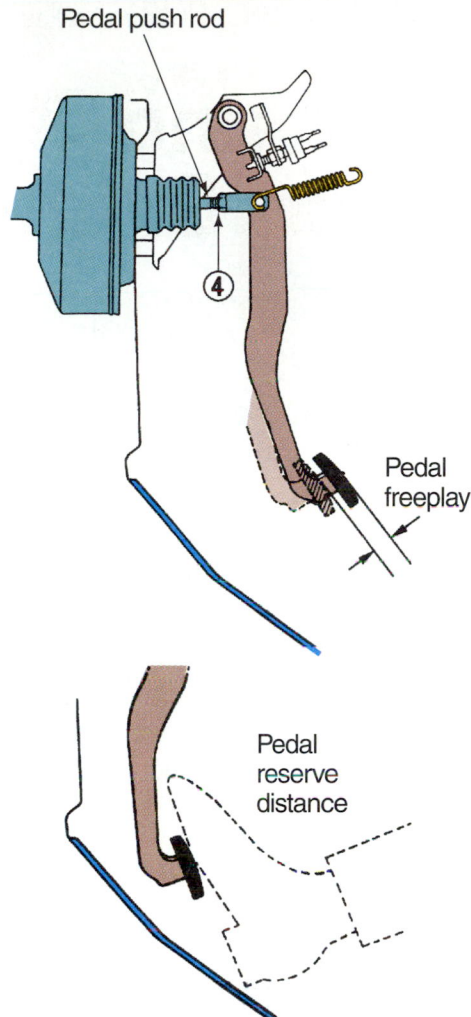

Figure 13.1 Check brake pedal travel.

slip toward the floor. If it does, an internal seal leak is indicated.

NOTE: *If a vehicle is to be taken on a test drive to check out a complaint involving the brakes, be sure to check the master cylinder fluid level and the brake pedal height before driving the car. A vehicle with faulty brakes should never be driven.*

Brake adjustment can be checked by pumping the pedal rapidly. If the height of the pedal rises on the second application, the brakes need an adjustment. If the pedal feels soft and "spongy" there is probably air in the system, which can be removed by bleeding the brakes (see Chapter 52).

Emergency Brake Travel and Adjustment

The brake system is required by law to include a parking brake (also called *emergency brake*) that operates independently of the hydraulic *service brakes*. Apply the parking brake to see that it is fully applied at about ½ of the full travel of the handle or pedal. The parking brake pedal or hand brake should be firm. When fully applied, it should work well enough to hold the car when stopped on a hill. If the service brakes require adjustment, the parking brake travel will also be excessive. After a brake adjustment, the emergency brake should have proper travel.

NOTE: *The service brakes should always be adjusted before attempting to adjust the emergency brake.*

Inspecting Brake Fluid

There are two styles of master cylinders. One type has a reservoir that is made of plastic and has a snap on or screw on cap(s) (Figure 13.2). The other type is made of cast iron or aluminum and has a metal *bail* that snaps in place to hold the cover. Sometimes the bail only moves in one direction (Figure 13.3). Clean around the top of the cover before removing it so the fluid does not become contaminated.

There is a diaphragm that is attached to the inside of the cover that keeps air, dirt, and moisture from

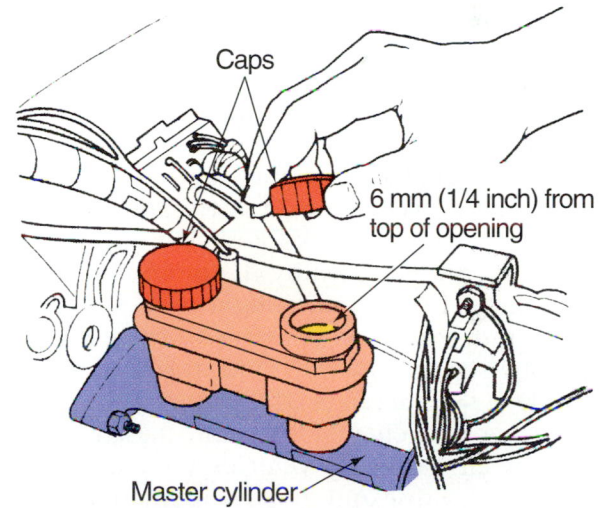

Figure 13.2 Master cylinder reservoir caps. *(Courtesy of Chrysler Corporation)*

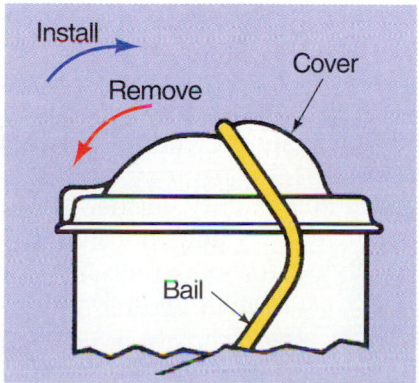

Figure 13.3 This bail can only be removed in one direction.

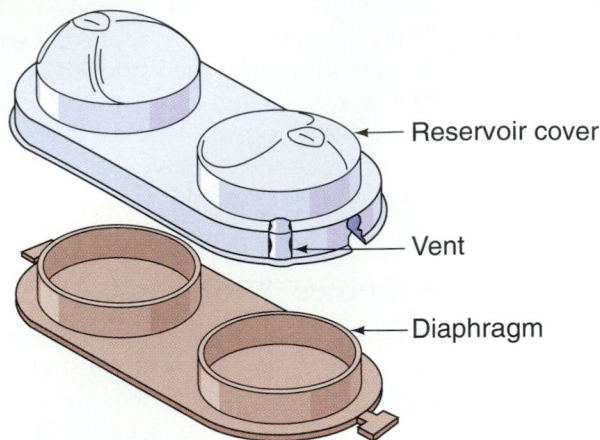

Figure 13.4 A diaphragm keeps moisture out of the brake fluid. Check the reservoir vent to see that it is not plugged. *(Courtesy of Allied Signal Automotive Aftermarket/Bendix)*

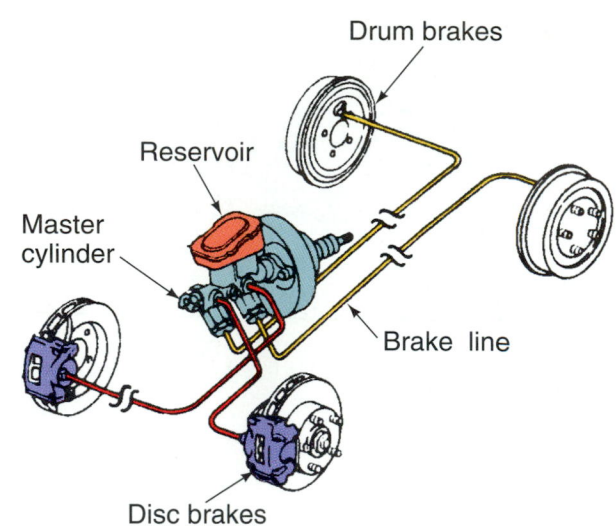

Figure 13.5 A disc and drum combination brake system. *(Courtesy of General Motors Corporation, Service Technology Group)*

contacting the brake fluid (Figure 13.4). Be sure that the diaphragm is not damaged. During a fluid level inspection, the reservoir cover vent can be easily checked to be sure that it is not obstructed. The cover should not be left off of the master cylinder for any period of time because brake fluid absorbs moisture rapidly. Exposure to the air can result in a lowering of its boiling point, which can result in brake failure.

When fluid level is low, a more complete brake inspection is required. Check for obvious signs of leakage, which might help identify the problem. See if the master cylinder is wet, indicating a leak. The backing plates on drum brakes can be wet with brake fluid. The tires will often be streaked with fluid and dirt. Sometimes the outside of the tire can even show signs of leakage.

There are two chambers in the master cylinder reservoir of all passenger vehicles built since 1967. Most cars have a disc and drum combination system, with disc brakes on the front and drum brakes on the rear (Figure 13.5). Disc brake cars usually have two chambers that are different sizes. The disc brake chamber is the larger of the two reservoirs. When inspecting the level of the brake fluid, a service specialist can often tell if the brake system needs a more thorough inspection. On cars that have disc brakes, when the disc brake pads wear, the fluid level in the master cylinder drops.

When brake fluid is dirty, further inspection of the system is warranted. Dirty fluid indicates that the system has not been serviced for some time. See Chapter 51 for a more complete explanation of brakes and fluids.

NOTE: *Be careful when checking or adding brake fluid. Brake fluid can damage a car's paint. Always use a fender cover when working around the master cylinder. Be sure to wash off any accidental brake fluid spills immediately with water or denatured alcohol. Some technicians keep a spray bottle of water handy for such spills.*

Be sure to use only an approved brake fluid with a high boiling temperature. The fluid should be approved by the **Department of Transportation (DOT)** and say "DOT" on the label. When adding brake fluid, the new fluid must be of DOT 3 (normal drum/disc service) or DOT 4 (heavy-duty drum/disc service). DOT 5 fluid has a silicone base and should not be mixed with other types of brake fluid.

NOTE: *Accidental contamination of the brake hydraulic system with oil or transmission fluid can cause serious damage to the rubber parts in the brake system. Also, wipe off the top of the master cylinder before removing it to prevent contamination of the fluid.*

➡ ***Perform Master Cylinder Brake Fluid Level Worksheet***

Some cars with manual transmissions have a clutch master cylinder located next to the brake master cylinder. It uses brake fluid, also. Check the fluid level and inspect the clutch master cylinder the same way as for a brake master cylinder.

Disc Lining Inspection

If the master cylinder fluid level in the large reservoir of a tandem master cylinder (one with two reservoirs) is low, inspect the front brake linings for wear. The lining should be as thick as the metal back of the lining (also called the "shoe"). There are several systems to warn the driver that brake pads are worn. One system uses an electronic sensor that causes a dash light to illuminate. Another warning system uses an electronic sensor that causes the pedal to pulsate when the pads become too worn. The oldest warning system is called the audible sensor system. A metal tab on the disc brake lining causes noise to warn the driver of lining wear (Figure 13.6).

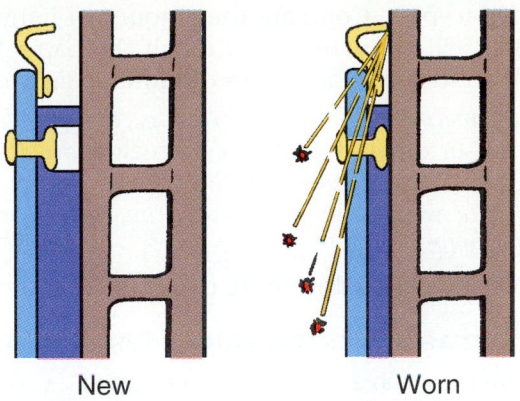

New   Worn

Figure 13.6 When the lining is worn too much, the metal warning sensor contacts the brake rotor.

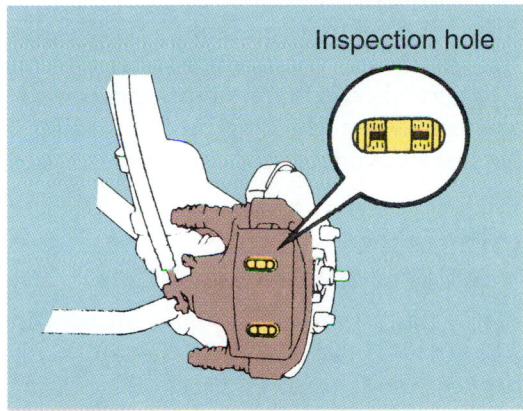

Inspection hole

Figure 13.7 Some calipers have an inspection hole to see if the lining is worn excessively.

The linings can usually be seen through an opening in the outside of the brake caliper (Figure 13.7). It is usually necessary to pull the wheel to perform the inspection. If the front brakes are worn, be sure to inspect the rear brakes for wear also.

■ ON-GROUND STEERING AND SUSPENSION CHECKS

Inspection of the steering and suspension systems is another part of a lubrication/safety service. Look for obvious defects, such as loose steering parts and worn tires. More detailed inspections are covered in Chapter 14, Undercar-Service and also in Section 10, Steering and Suspension Systems.

Steering Looseness

Freeplay of the steering wheel is the first item to check. With the wheels straight ahead, there should only be a minimal amount of freeplay (less than 30 mm). Excessive looseness can be checked easily by performing a **dry park check**. With the tires on the ground, have a helper repeatedly turn the steering wheel back and forth (about 3" or so). Look under the car to see if there are any signs of looseness between steering link-

age parts. The linkage parts should allow pivoting only. Slack between parts will show up during this test because of the resistance to the tires as they try to turn against the ground.

Steering Fluid Level

A typical power steering fluid reservoir is a part of the pump, which is mounted on the front of the engine and is driven by a belt (Figure 13.8). The level of the fluid is checked with the engine off. The filler cap often has a dipstick, which includes "hot" and "cold" fill levels (Figure 13.9). Fluid level is normally checked when hot with the engine off. The level will be lower when the fluid is cold.

Inspect the power steering hoses and connections to see that there are no leaks. When adding fluid to the reservoir, be sure to use the one specified by the manufacturer. Some power steering systems use **automatic transmission fluid (ATF)**. Others use a specially formulated fluid. Check the power steering system for fluid leaks at the hoses, the pump, or the steering gear.

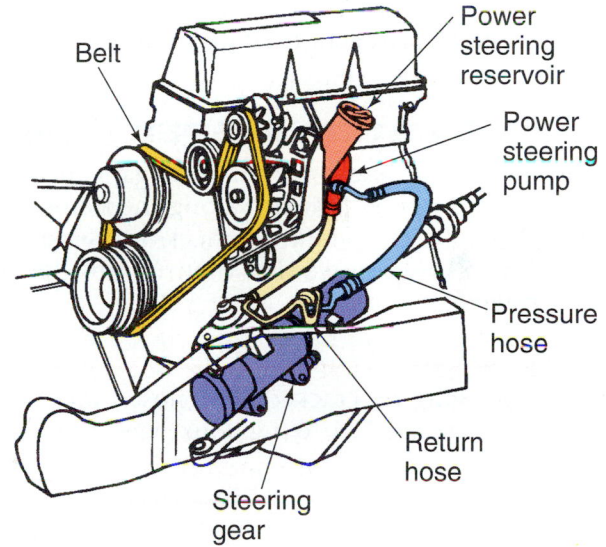

Belt   Power steering reservoir

Power steering pump

Pressure hose

Return hose

Steering gear

Figure 13.8 Parts of a power steering system include the pump and fluid reservoir.

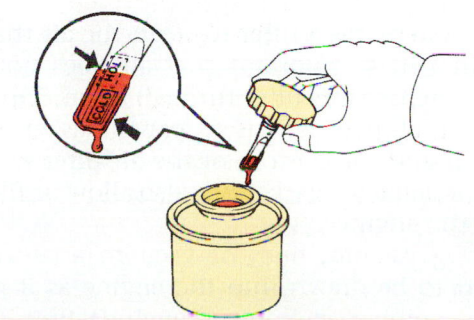

Figure 13.9 The dipstick for the power steering has hot and cold levels.

Manual steering gears use different fluids. Consult the service manual for the lubrication requirements for the gearbox and the location of the inspection and fill plug.

➡ *Perform **Check Power Steering Fluid Level Worksheet***

Check Shock Absorbers (Car on Ground)

Shock absorbers, commonly called *shocks*, are designed to control the initial movement and subsequent oscillations of the spring during jounce (downward movement of the car) and rebound (movement of the car back up). Shock absorbers are checked twice: once while the car is on the ground and again when it is in the air.

Push down on each of the vehicle's bumpers to check resistance of the shocks. Do not press on the fenders. They can become greasy and might get dented when you press on them. Normal shocks will allow the spring to rebound only once. This is only a cursory check of shocks. A more complete check can be done to shocks while the car is in the air. That information is covered in the section of the text on Undercar Service. See Section 10 for more information on shocks.

■ FUEL SYSTEM INSPECTION

Flexible hoses are used to connect gaps between sections of steel tubing that run along the frame from the fuel tank to the engine. Check for leaks and inspect all fuel hoses to see that they are not cracked, hard, or leaking.

Hard or cracked fuel hoses must be replaced! A dangerous fuel fire can result.

Check to see that the gas cap fits properly and that its gasket (if it has one) is in good condition.

Air Filter

The fuel system has a filter to clean the air that enters the engine. It is important that air does not have a chance to bypass this filter through a leak. On engines with carburetors, there is a paper gasket that fits between it and the bottom of the air filter housing. A missing or damaged gasket can also allow unfiltered air to enter the engine.

Leaking vacuum lines or vacuum accessories can allow dirt to be drawn into the engine as it operates. Dirt is the number one cause of engine failure. A light is shined through a paper air filter to check it. Look to see that there are no small holes in the paper that would allow dirt to pass. Compare the amount of light that a new filter will allow to pass through the paper to give yourself an idea of whether an old filter is dirty or not.

NOTE: *Sometimes, when an air filter is changed, dirt from the outside of the dirty filter is carelessly allowed to enter the engine. The equivalent of two aspirin tablets of dirt causes more wear than that caused by 75,000 miles of normal driving.*

Air filters are covered in more detail in Chapter 28.

Crankcase Ventilation System

Most engines have a positive crankcase ventilation system, called a **PCV system**. The PCV valve is usually serviced and replaced as part of a tune-up. Its service is covered in detail in Chapter 41. The system must have a means of drawing in filtered air. This sometimes includes an auxiliary air filter that can require service also.

If the crankcase ventilation system is operating properly, there should be a slight vacuum at the oil filler opening when the engine is running at idle. There are testers available for checking this (see Chapter 41).

■ COOLING SYSTEM INSPECTION

The radiator has a pressure cap and an overflow tank (Figure 13.10). The radiator pressure cap increases the pressure on the coolant so that its boiling point will be higher than normal (see Chapter 20). The pressure rating of the cap is stamped on its top. When opening the

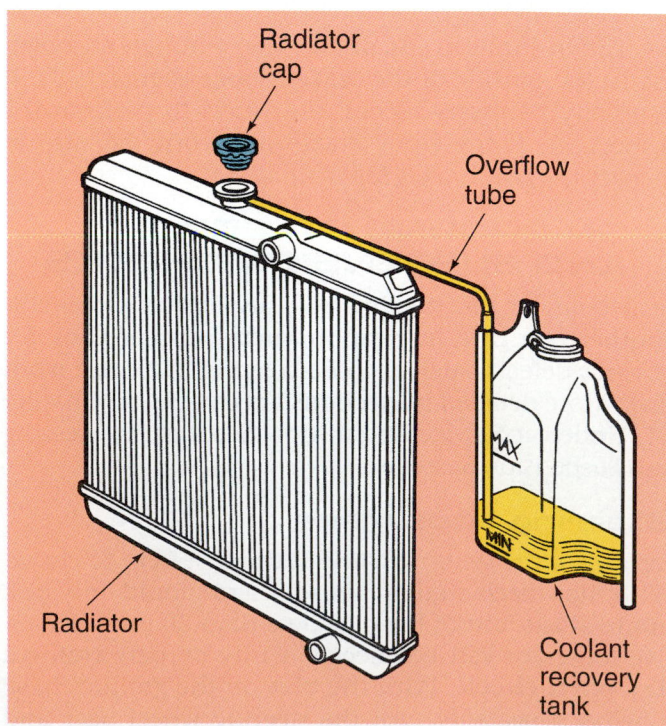

Figure 13.10 A radiator and recovery tank. *(Courtesy of Chrysler Corporation)*

radiator cap, there is a safety catch that stops it from turning after one quarter turn. To remove it the rest of the way, it must be pushed down and turned.

> **SAFETY NOTE** Before opening the radiator cap on a hot engine, squeeze the top radiator hose to be sure that the coolant is not under pressure (see Figure 11.18). Some aftermarket flex hoses cannot be squeezed. It is not a good idea to check coolant level on a hot engine. The coolant is best checked when cold. A gallon of 50/50 coolant and water mixture will expand by about ⅓ pint when heated to engine operating temperature (about 195°F).

Remove the radiator cap and visually inspect it. Check its pressure relief valve, sealing gaskets, and vacuum valve (Figure 13.11) (The operation of these parts is discussed in Chapter 20.) The lower sealing gasket is made of rubber and can become brittle or indented over a period of use. Dirty coolant can interfere with the operation of the vacuum valve and rubber seal. Clean these areas of the cap before reinstalling it if it is in good condition (Figure 13.12).

Coolant Inspection

Visually inspect the coolant. Coolant can be several different colors. See that the level is correct and it appears clean. The strength of the coolant can be checked with a coolant **hydrometer** (Figure 13.13). A coolant hydrometer is similar to a battery acid

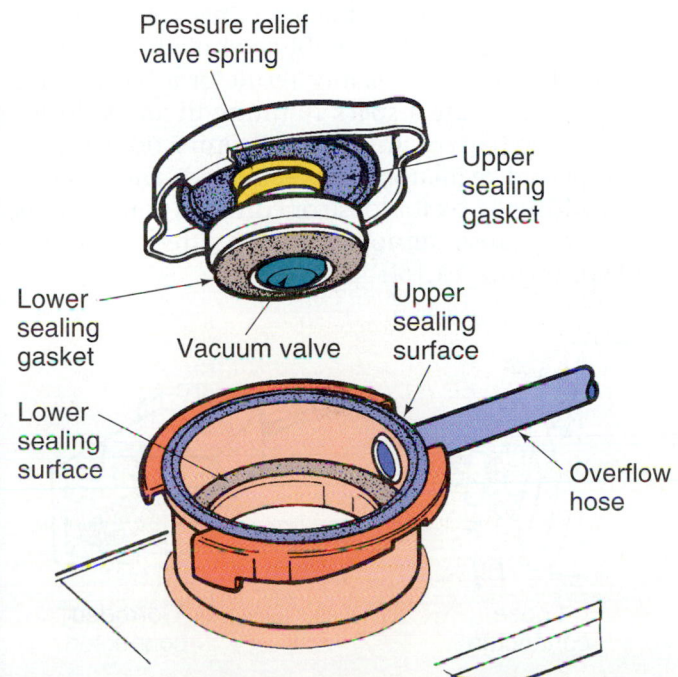

Figure 13.11 Remove and inspect the parts of the radiator cap. *(Courtesy of Chrysler Corporation)*

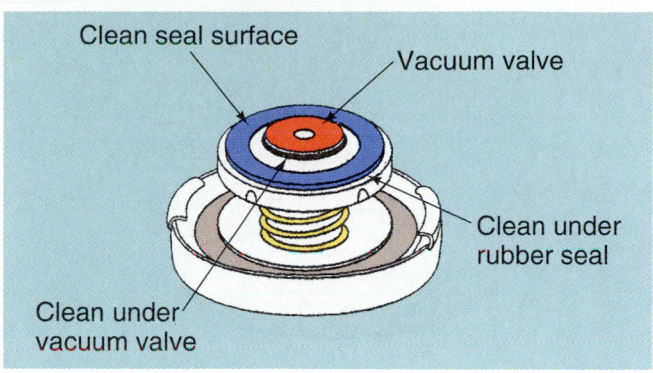

Figure 13.12 If the cap is in good condition, clean the vacuum valve and rubber seal. *(Courtesy of Chrysler Corporation)*

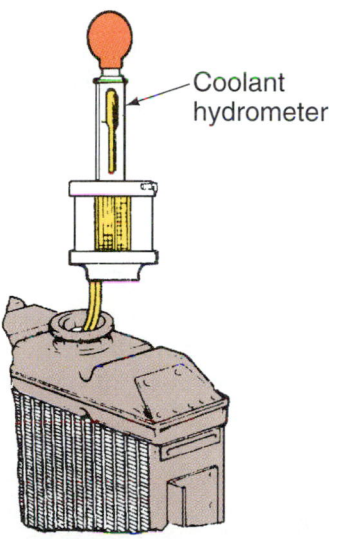

Figure 13.13 A coolant hydrometer is used to check the strength of the coolant. *(Courtesy of Chrysler Corporation)*

hydrometer (used to measure a battery's state of charge). Coolant is drawn out of the radiator and sampled to compare its weight to that of pure water (see Chapter 21). One kind of coolant hydrometer has several different weight balls in a small chamber. The strength of the coolant is determined by how many of the balls in the chamber float in the coolant. The more that float, the stronger the coolant concentration. A mixture of water and coolant provides better protection than pure coolant.

➡ *Perform **Check and Correct Coolant Level Worksheet***

Cooling System Leaks

Look for leaks in the cooling system. When the coolant level is low, a pressure test can be done to determine whether the leak is internal (inside the engine) or external (see Chapter 21).

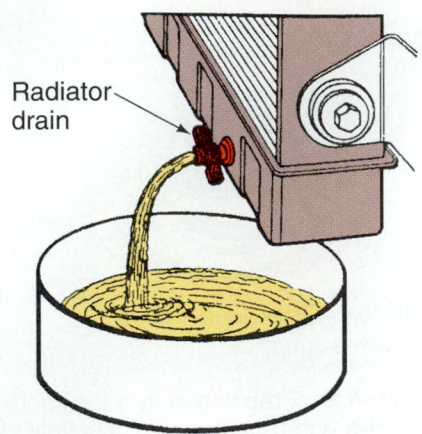

Figure 13.14 Drain some coolant from the radiator. *(Courtesy of Chrysler Corporation)*

NOTE: *When looking for leaks, be aware that the vehicle's air conditioner collects moisture and allows it to drain from the passenger compartment to the ground through a hose. It is not unusual for a customer to complain of a coolant leak when the only problem is that the air conditioning compressor is turned on.*

Inspect Cooling System Hoses

Check the upper and lower radiator hoses to see that they are not swollen or cracked.

- Oil damages rubber cooling system hoses. Oil leaks usually do the most damage to lower hoses.
- The upper radiator hose gets the hottest water, so it is the most prone to failure from heat.
- If a hose is collapsed, look for a defective radiator cap. Check both heater hoses, too. When a heater hose is hard only at the end, it can sometimes be fixed by cutting an inch or so off of the hose and clamping it back on.

NOTE: *If a hose is to be removed, coolant will probably escape. Drain some coolant from the radiator into a clean container first (Figure 13.14).*

- A customer would probably be irritated if coolant is allowed to escape and he has to pay for new coolant; especially if it has recently been changed.
- A pinch plier can be clamped on the hose to keep coolant from escaping during this procedure.
- Coolant that has leaked will create a mess that will require unnecessary cleanup.
- In some communities coolant is classified as toxic. It must be captured in a container.

■ BELT INSPECTION

Check the condition of the belts by looking at the underside to see that they are not cracked or glazed. The tension of a belt should be checked with a *belt tension gauge,* but a quick inspection during a safety check can be performed with the engine off (Figure 13.15). A quick method of checking belt tension is to try to turn

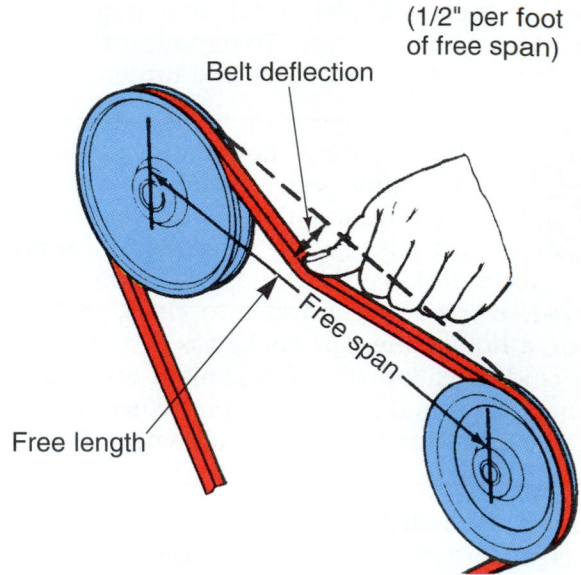

Figure 13.15 Inspect belt tension. *(Courtesy of Chrysler Corporation)*

a pulley by hand to see if the belt holds or slips. If the alternator belt tension is too loose, the alternator will not be able to keep the battery charged. Belts can be either V-belts or V-ribbed belts. More complete information on belts is included in Chapter 22.

➡ *Perform Identify and Inspect Accesory Drive Belts Worksheet*

■ ELECTRICAL SYSTEM INSPECTION

Give the entire electrical system a visual inspection, looking for obvious problems. Check for bare or burned wires that could possibly complete a circuit to ground. This would probably result in a blown fuse, but could also cause a spark resulting in an explosion of fuel or the battery. Inspect the connections on the alternator and regulator (if so equipped) and ignition system wiring. Look for loose or corroded connections, which can cause annoying intermittent electrical problems (Figure 13.16).

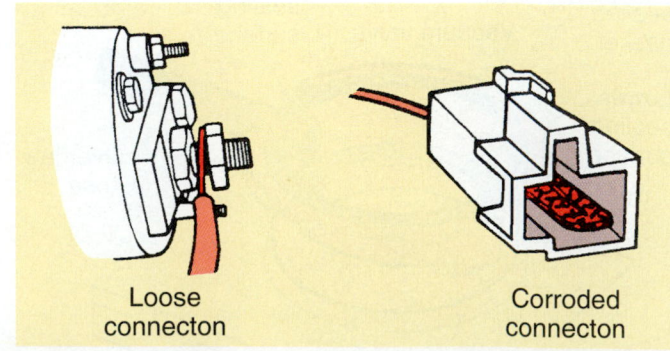

Loose connecton   Corroded connecton

Figure 13.16 Look for loose or corroded connections. *(Courtesy of General Motors Corporation, Service Technology Group)*

Battery Check

Visually inspect the battery (Figure 13.17). Look for damage or corrosion to the cable clamps. Cleaning the cable clamps is not usually done during a lubrication/safety service, but is part of a tune-up or maintenance service. Check the condition of the battery hold down and hold-down clamps.

NOTE: *If the terminal clamps are to be removed for any reason, be sure to remove the ground cable first. This will eliminate the chance of a spark that could cause a battery to explode. Computer memories can be preserved using a small battery connected through the cigarette lighter (see Figure 31.8).*

Battery Electrolyte Level

Battery *electrolyte* is composed of sulfuric acid and water. When a battery is recharged, some water may evaporate from the electrolyte mixture. On many batteries, clean water can be added to replenish the electrolyte (see Chapter 31). Twelve-volt batteries are used on modern cars. A 12-volt battery has six cells. Some batteries have fill holes (see Figure 13.17). If the battery has caps that are easily removable, inspect the electrolyte level. The electrolyte level should be up to the split ring that can be seen inside the fill hole (Figure 13.18). If it is low on charge, do not fill the battery to the *full indicator*. The electrolyte will expand and spill out of the battery during charging.

■ Battery acid can be neutralized with baking soda. A mixture of baking soda and water may be used to clean the top of a battery.

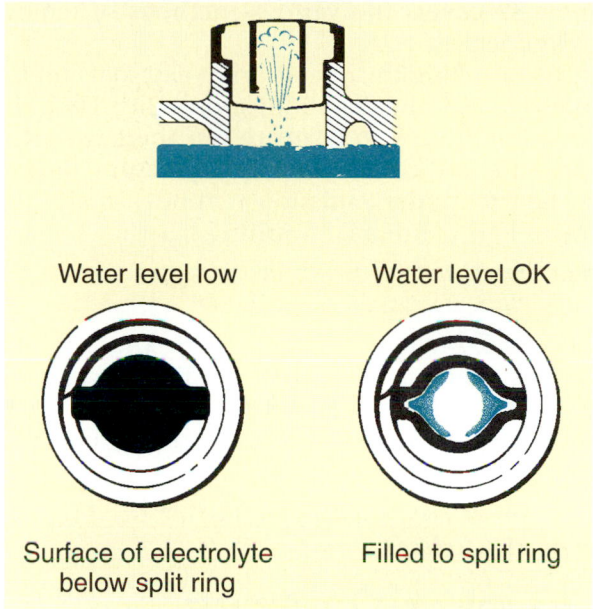

Water level low   Water level OK

Surface of electrolyte   Filled to split ring
below split ring

Figure 13.18 The battery electrolyte level is correct when the water level reaches the split ring. *(Courtesy of General Motors Corporation, Service Technology Group)*

■ In addition to being dangerous to skin and eyes, battery acid can damage a car's paint. Be sure to install fender covers on the vehicle when working on the battery. Wash off any spilled electrolyte immediately.

➡ *Perform **Battery Visual Inspection** Worksheet*

■ BODY CHECKS

Check Operation of Lights

Inspect the operation of all of the lights on the vehicle. On some cars, the ignition switch might have to be turned on for some of the lamps to operate. Check all lenses for cracks. A crack in a lens will allow moisture to get in, causing corrosion and failure of the circuit. A bulb is usually removed by pushing it in and turning it counterclockwise in its socket (Figure 13.19).

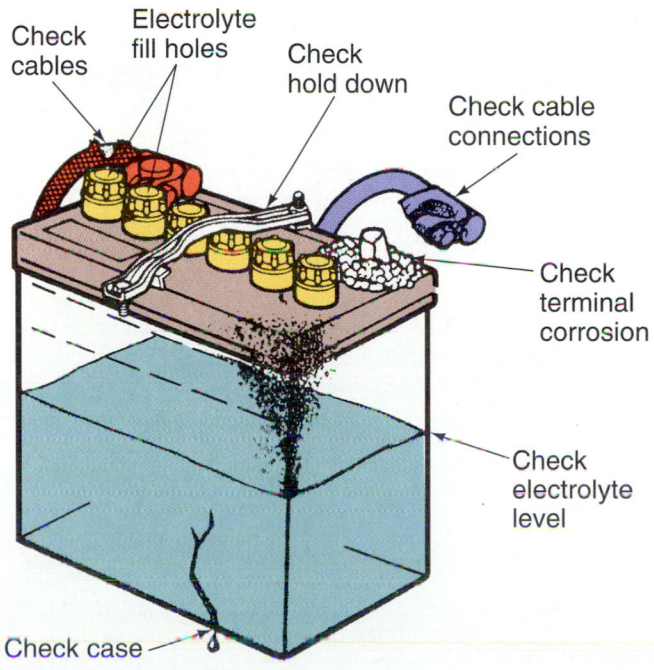

Check cables
Electrolyte fill holes
Check hold down
Check cable connections
Check terminal corrosion
Check electrolyte level
Check case

Figure 13.17 Inspect the battery, cable connections, and hold down.

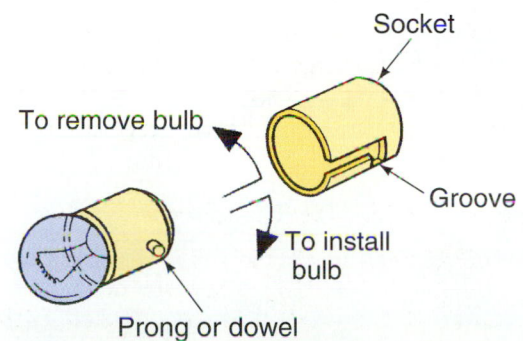

Socket
To remove bulb
Groove
To install bulb
Prong or dowel

Figure 13.19 A bulb is removed by pushing and turning counterclockwise. *(Reprinted by permission of Nissan Motor Corporation in USA)*

Chapter 37 covers the various methods of removing bulbs and lenses.

After checking the operation of external and internal bulbs, check the dash indicator lights. Dash lights that require the key to be on to test them include the oil pressure indicator, water temperature indicator, safety belt reminder, emission reminder, malfunction indicator light, and brake warning light.

NOTE: *Burned-out bulbs can be caused by excessive voltage, so a charging system check may be called for.*

To check obvious problems with the aim of the headlights, shine them against a wall. Check low and high beam operation and that they appear to be level. If an adjustment is called for, headlight aiming instructions are provided in Chapter 32.

➡ *Perform __Inspect Operation of the Lighting System__ Worksheet*

■ VISIBILITY CHECKS

Visibility items are often overlooked during a lubrication safety/service job. These include the windshield and windows, mirrors, wipers, and defroster. Windows are checked for damage, condition of weatherstripping rubber, and operation. Power windows are operated to see if they operate too slowly, indicating a problem needing attention.

Mirrors are inspected to see that they are installed and in good condition. They should not be cracked or fogged and should be fastened securely to the vehicle.

The operation of the defroster is a visibility item. The blower motor is tested to see that it works. Cable operation is tested by operating the heater/blower selector levers. Duct hoses should also be inspected when possible to see that they are in good condition.

Windshield Wipers

Parts of the windshield wiper system are shown in Figure 13.20. The wipers are checked for the condition of

the blades. Wiper arms should be installed so that they do not hit the trim around the window. They should have sufficient tension to contact the entire glass surface.

Check the level in the *windshield washer reservoir* (Figure 13.21). Washer fluid that cleans and prevents freezing is used to fill the reservoir. The windshield wiper reservoir on most new cars has a small rotary pump. When the switch is activated by the driver, a small impeller pushes washer fluid through small hoses to the nozzles that are below the windshield. These nozzles sometimes become plugged. They can be cleaned with a small piece of wire or very small drill bit. The windshield washer should work on the driver's and passenger's side of the windshield and should be aimed correctly. There should be a sufficient volume of washer fluid to clean the window.

Wiper Blades

Windshield wipers have a blade, an arm, and a spring (Figure 13.22). Blades can become dull from sliding on dirt or ice, or they can become torn or hard and inflexible. They are made of rubber, which deteriorates as it ages or is exposed to the sun, heat radiated from hot windshield glass, wind, cold weather, and road oils.

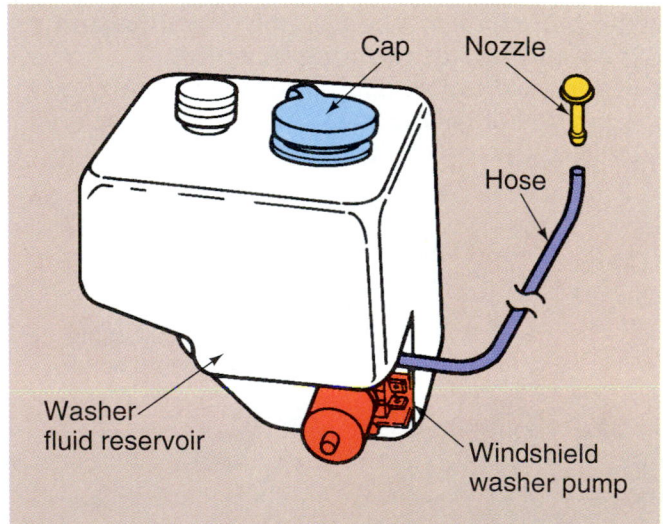

Figure 13.21 A windshield cleaning fluid reservoir. *(Courtesy of Volkswagen of America, Inc.)*

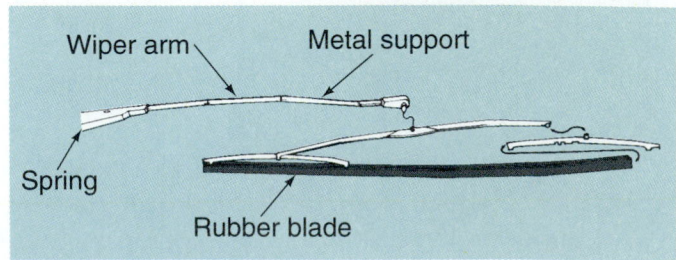

Figure 13.22 The wiper assembly has an arm, a blade, and a spring. *(Courtesy of Chrysler Corporation)*

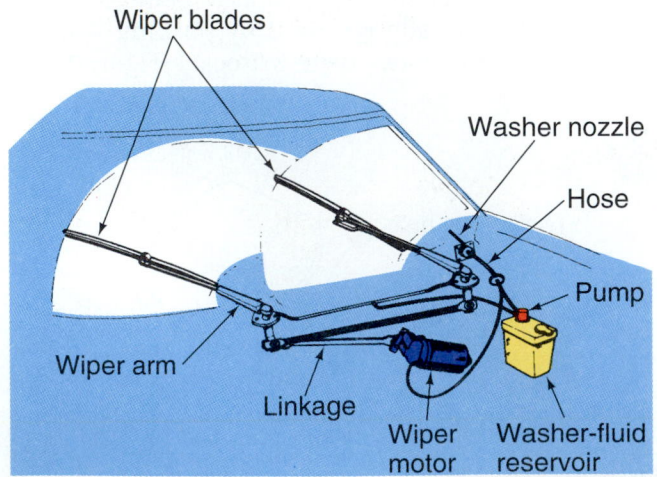

Figure 13.20 Parts of the windshield wiper system. *(Courtesy of Trico Products Corporation)*

Sometimes, wiper blade wear is evident to the eye and other times the wiper just stops doing a good job of cleaning the window. When the blade no longer does a good job of cleaning the window, try cleaning it with soap and water. Wet the window and try it again. If it still leaves a smear, it should be replaced.

SHOP TIP Some technicians clean wiper blades with very fine wet-or-dry sandpaper during a service job.

Wiper blades must contact the window evenly across the entire blade. Occasionally the spring on the wiper arm breaks, requiring replacement of the entire arm. If the metal on the blade assembly is bent or a pivot point is corroded, the blade will not lay evenly on the window. In this case, the entire blade assembly is replaced.

Removing and Replacing Wiper Blades

Wiper blades, sometimes called refills, squeegees, or cartridges are usually replaced once or twice a year, depending on conditions. Most wiper blade assemblies have replaceable inserts (Figure 13.23), usually purchased in pairs. Sometimes a wiper arm is not serviceable and must be replaced as an assembly.

There are three designs of wiper blade refill assemblies (Figure 13.24). One uses a metal clip on the end and another has a plastic release button. Foreign cars often use a system with a special clip, sometimes only available at the dealer. The foreign types can be replaced with one of the more common types by simply buying the entire blade assembly the first time wipers are replaced. Subsequent refills will consist of the refill only.

NOTE: *It is easy to accidentally scratch a window when replacing a wiper blade. To avoid this, remove the entire blade assembly before attempting to replace a refill. When the refilled blade assembly is reinstalled again, pull on it to see that it is firmly in place.*

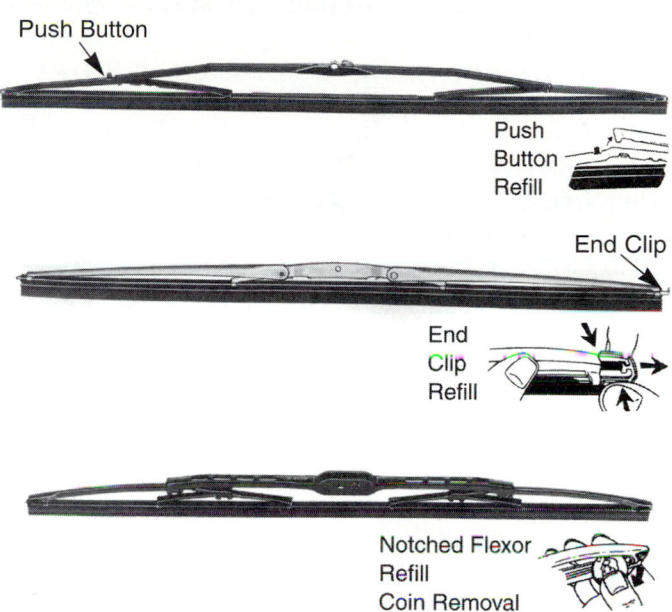

Figure 13.24 Three designs of wiper blade refills. *[Courtesy of Cooper Automotive, Inc.]*

There are several types of connector designs used to fasten the blade assembly to the wiper arm (Figure 13.25). One common type is the *bayonet connector*, which has a small round dimple on its top that fits into a hole to help hold the blade assembly on the arm. There is either a spring catch or a lever on the underside of the arm that is released to disengage the dimple from the hole. Another style uses a *pin connector*. Other styles include the *inner lock connector*, the *screw-type*

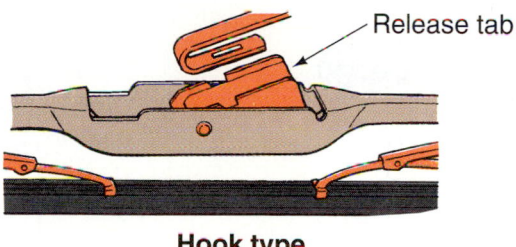

Hook type

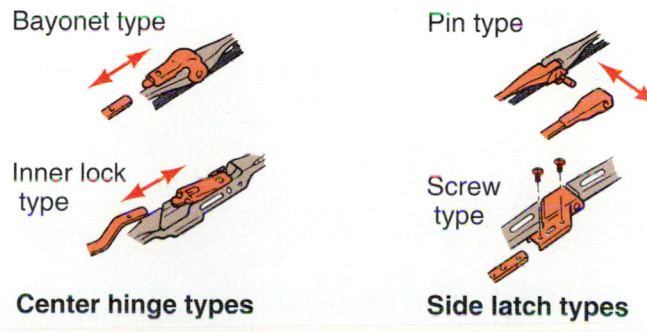

Figure 13.25 Several connector designs fasten the blade assembly to the wiper arm. *[Courtesy of Chrysler Corporation]*

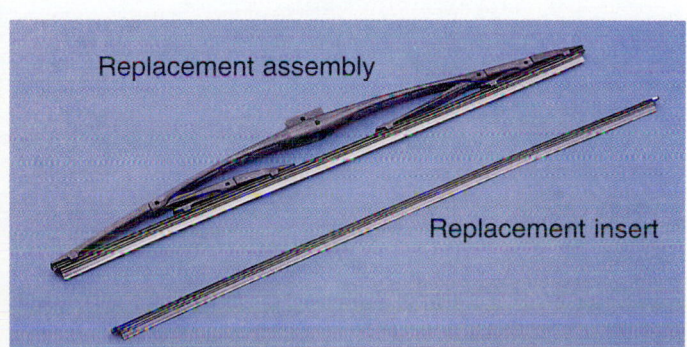

Figure 13.23 Replaceable wiper blades.

connector, the *shallow hook connector*, and the *shepherd's crook connector*.

⮕ *Perform **Refillable Wiper Blade Service** Worksheet*

Removing and Replacing Wiper Arms

Wiper *arms* are mounted on a splined post or a metric screw shaft (Figure 13.26). The splined posts either have a release lever called a clip tab or a slide latch (Figure 13.27). To remove the wiper arm, raise it as high as it goes and release the clip tab or tap on the slide latch with a screwdriver. The arm can be carefully pried off using two screwdrivers, prying a little on one side and then on the other. There is also a special tool available for prying off splined wiper arms. The metric screw shafts use a lock nut.

SHOP TIP The paint on the car can easily be chipped during this operation. Put some cardboard down under the wiper post when prying in case something accidentally slips

Most wiper arms can be installed at any point on the spline. When reinstalling a wiper arm, it should be positioned so that it cannot hit the frame around the windshield (Figure 13.28). When the wipers are turned off, most have electrical systems with a provision that allows the arm to continue to move until it reaches full travel at the bottom. A relay then shuts off the motor. Position the arm so that the blade is about an

Figure 13.26 Wiper arms are mounted on a splined post or a metric screw shaft. *(Courtesy of Cooper Automotive, Inc.)*

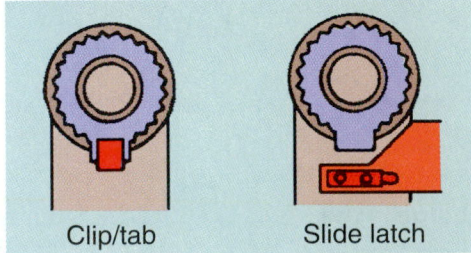

Clip/tab   Slide latch

Figure 13.27 Splined posts use a clip tab or a slide latch. *(Courtesy of Cooper Automotive, Inc.)*

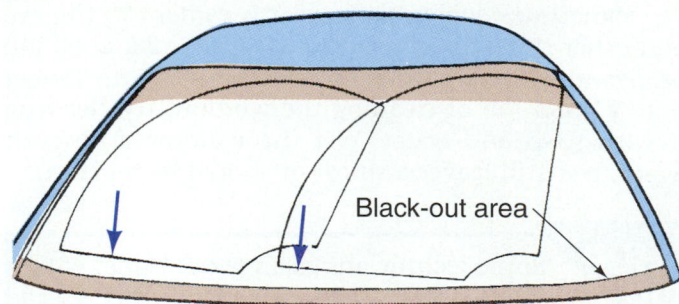

Black-out area

Figure 13.28 Position the wiper arm so that it cannot hit the window frame. *(Courtesy of Chrysler Corporation)*

inch from the windshield frame. Then, cycle the wipers to see that the blade assembly does not hit the side frame at the other end of its travel.

⮕ *Perform **Visibility Checklist** Worksheet*

■ OTHER SAFETY CHECKS AND SERVICE

Check the operation of the horn. Then, check the operation of the hood latches. There is a *primary hood latch* and also a *safety catch* in case the hood is not all the way latched. Check the operation of both. The safety catch should hold when tested. If it lets loose when pulled against, it is not working properly and will require attention. Be sure to document this on the repair order form. Its pivot points can be lubricated.

The door striker plate is lubricated with wax lubricant (called *Door Ease*) that comes in stick form. Engine oil is applied to the door latch rotor.

Door locks are lubricated with dry graphite. Because it is not sticky, it does not attract dust like oily products do. The dust would interfere with operation of the tumblers in the lock.

Check the seat belts to see that they are in place and operating correctly.

In addition to maintenance for the sake of vehicle upkeep, the owner will be especially interested in items related to the safety of his or her family. The shop owner or service manager is often in a position of trust. The recommendations of the shop will be what the owner must rely on for his or her decision making. Never fail to document (write down) safety related items on the repair order. Safety is the highest priority in vehicle repair.

General Safety Check

The safety checklist in Figure 13.29 is a simple version of that done by a professional facility. Its purpose is to provide an apprentice with a basic familiarity with those parts that will be inspected during a professional lubrication and safety service. Locating and repairing these items can please customers and increase their loyalty.

SAFETY CHECKLIST (Car on Ground)		
Safety Check:	OK	Needs Attention
Lights	_____	_____
Brake pedal travel	_____	_____
Wiper blade condition	_____	_____
V-belt tension and condition	_____	_____
Tire pressure check (cold)	_____	_____
Tire Condition (excessive wear)	_____	_____
Hood latch	_____	_____
Fuel hose condition	_____	_____
Exhaust leaks (listen)	_____	_____
Horn	_____	_____
Brake fluid level	_____	_____
Emergency brake adjustment	_____	_____
Battery water check and fill	_____	_____
Lug nut torque check (ft.-lb.)	_____	_____
Door latches (lube)	_____	_____
Electrical wiring	_____	_____
Hose condition (radiator, heater)	_____	_____
Power steering hoses	_____	_____
Defroster	_____	_____
Mirrors (condition and tightness)	_____	_____
Interior and dash lights	_____	_____
Seat belts (not loose or damaged)	_____	_____

Figure 13.29 Safety checklist.

Automatic Transmission Fluid Check

Transmissions are *not* supposed to consume fluid, but leaks sometimes develop. Follow these precautions when checking ATF:

■ When the car first arrives at the shop, lift the hood and check the ATF level while the fluid is still hot. Fluid expands as the transmission warms up (Figure 13.30).

■ Be sure the parking brake is firmly set and the vehicle is on level ground. Automatic transmission fluid level is checked with the engine running. The transmission must be in a specified gear position.

NOTE: *Most transmission fluid levels are checked with the transmission in park. Some transmissions are checked in neutral and will give a different dipstick reading when checked in park. If fluid is overfilled it can become aerated (foamed). Foamed fluid can self-siphon out the dipstick tube or overflow vent, resulting in a low fluid level and possible transmission failure.*

The dipstick is located behind the engine. It fits into a large round tube. When reading the ATF level on the dipstick, be aware that the add mark represents one *pint*, and not one quart as the engine oil dipstick does.

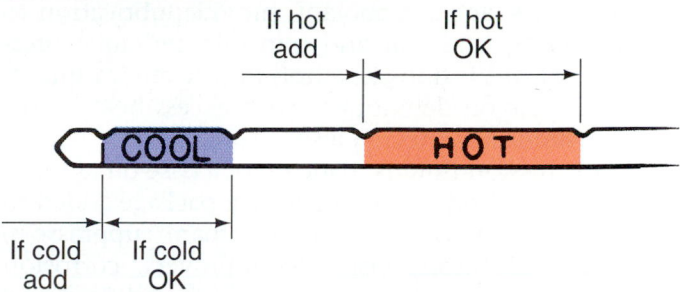

Figure 13.30 Automatic transmission fluid expands as it warms up.

It is important that the proper procedure be followed when checking fluid level. Checking the fluid level using the incorrect procedure could result in an overfilled or underfilled transmission. If a transmission is overfilled, the fluid will be whipped and filled with air bubbles. Just as is the case with brake fluid, air is compressible, while fluid is not. The aerated fluid will not transmit pressure as well as normal fluid.

Fluid Condition

Under normal use, fluids can last for a very long time. Under extreme conditions (severe service) such as hot weather driving, trailer towing, or stop-and-go city driving, the heat that develops can damage the fluid. When checking the fluid level also inspect its condition. A burnt smell could indicate possible transmission damage. Some fluids (synthetics or partial synthetics) have an unusual smell to begin with, so check with a transmission specialist if in doubt.

If the fluid has a milky appearance, water or coolant may have gotten into it. Check to see that the dipstick has been properly seated at the top of the dipstick tube. Wipe the dipstick to see if the fluid on it feels sticky. This could indicate excessive oxidation of the fluid due to heat. Changing the fluid might solve the problem, but a complete overhaul will probably be needed.

Always check the manufacturer's minimum service interval requirement and use this as a rule of thumb for normal transmission fluid service. Most of today's automatic transmissions do not require periodic adjustments, so "changing the fluid and adjusting the bands" has become a thing of the past.

➡ *Perform Check Automatic Transmission Fluid (ATF) Level Worksheet*

Automatic Transmission Fluid

The oil that is used in automatic transmissions is called *fluid*. Commonly called "ATF" (automatic transmission fluid), it is colored with a red dye to prevent confusion with engine oils. The red color also helps in determining the source of leaks. Transmission fluid serves several purposes. It must transmit engine torque to the wheels, act as a coolant, provide lubrication to moving parts, and seal and transmit hydraulic pressure. All of this is done in a hot, hostile environment. Just as engine oil deteriorates with excess heat, transmission fluids are affected also.

Transmission fluid is made up of a base oil (a high-quality mineral oil) with an additive package added to it. The additive package contains foam suppressors, oxidation inhibitors, viscosity improvers, corrosion and rust inhibitors, detergents, friction modifiers, and seal swelling agents. Also included are additives that make the fluid compatible with the various parts in the transmission.

Transmission parts are made of such materials as steel, aluminum, bronze, and also some non-metal friction materials. Different manufacturers use different friction materials for their transmission clutches, so there are different types of fluids available that have different friction characteristics. Most transmission fluids include friction modifiers. These include types *A, CJ, H, Dexron-III,* and *Mercon.* The use of *Type F* or *Type G* fluid results in a firm shift because it does not have friction modifiers. This will not damage the transmission, but may cause a driver complaint from the harsher shift feel. Type F fluid is recommended for 1976 and earlier Ford vehicles.

NOTE: *Be sure to check the service manual for the proper type of ATF to use. Using a friction-modified fluid in a transmission designed for nonmodified fluids can cause slippage and transmission damage.*

Dexron-III fluid is a fluid that can be used in most of today's transmissions. Some manufacturers recommend a different fluid but list Dexron-II or Dexron-III as a secondary choice. Mercon is the fluid used in late model Fords and some Chrysler products.

Tires

Tires can be checked for obvious wear when the car is on the ground. It is easier to check for nails and obvious tire damage when the car is in the air. Lug nut torque is checked while the car is on the ground. Check for the correct specifications in the repair manual.

Check Tire Pressure

Tires develop heat as the car is driven. Driving on tires that are low on air pressure is very damaging internally to the tire. A tire typically loses about 2 pounds of air pressure a month, so periodic refilling is necessary. By the time a tire appears to be underinflated, it will be way below specifications and tire damage may already have occurred. Tire pressures should be checked monthly (or preferably, weekly), when the tire is cold. Because the tire heats up during operation, the air in the tire gets hot. This causes more pressure in the tire. Every 10°F change in temperature results in a change in tire pressure of about 1 psi.

The correct pressures are listed in the owner's manual or on a sticker on the driver's side door jamb (Figure 13.31). Also, check the tire sidewall for the maximum safe operating pressure. Gauges that are part of the air hose filler at a service station are frequently inaccurate due to abuse. Use a dial or pocket type pencil gauge. Press the gauge firmly onto the valve stem. The scale will pop out the end of the gauge, showing the amount of pressure in the tire (Figure 13.32). When checking the air pressure in the tires, inspect each valve stem carefully to see that it is not cracked or leaking. It is customary to replace these valve stems whenever a tire is replaced.

➡ *Perform **Check and Adjust Tire Pressures** Worksheet*

Motor Mounts

The engine and the transmission are mounted to the frame with vulcanized rubber and steel mounts (Figure 13.33). Engine mounts, commonly called **motor mounts**, can fail due to unusual stress and strain or because they have been contaminated by leaking oil. On some classic cars, when mounts fail the throttle linkage can bind up causing an unsafe condition. Later model motor mounts have a built-in limiter made of

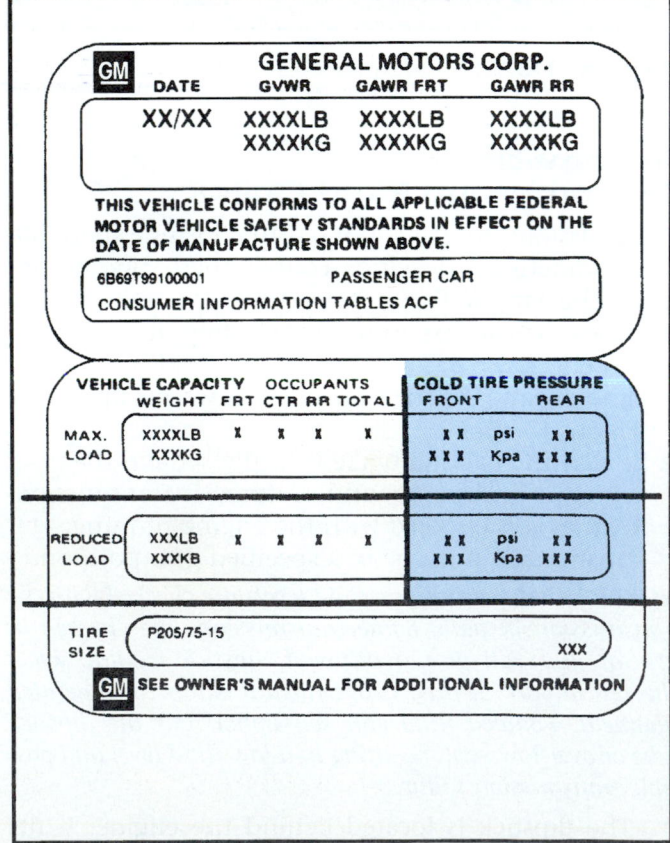

Figure 13.31 The correct tire pressure is listed on the door placard. *(Courtesy of General Motors Corporation, Service Technology Group)*

Scale extension

Read pressure here

Air check

Figure 13.32 The scale pops out of the gauge to show the tire pressure. *(Courtesy of the U.S. Government)*

metal that prevents excessive movement of the engine when a mount fails. Newer cars also use throttle cables instead of linkage.

A loose or broken motor mount can cause vibration during acceleration. This is especially true during engagement of the clutch on a vehicle with a manually shifted transmission.

To test for broken motor mounts, two people are needed.

■ First, set the emergency brake firmly and double-check to see that it holds the vehicle against the torque of the engine.

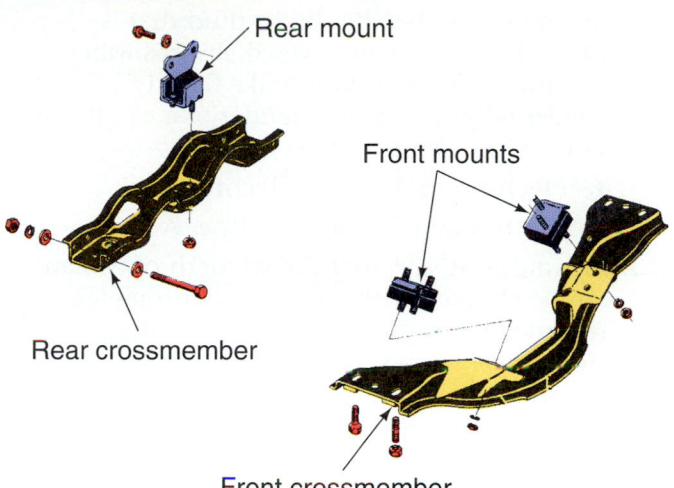

Rear mount

Front mounts

Rear crossmember

Front crossmember

Figure 13.33 Typical motor mounts for a rear wheel drive engine. *(Courtesy of Chrysler Corporation)*

■ While a helper puts the car in gear with the engine running, observe the mount on one side of the engine to see if the engine lifts off it.

 **SAFETY NOTE** Be sure that you are at one side of the vehicle, and not in front of it.

■ Then, have the helper shift the transmission into reverse and check the mount on the other side of the engine.

To replace motor mounts, they are unbolted and the engine is raised off them. The mount is replaced with a new one (see Chapter 50).

■ REVIEW QUESTIONS

1. What is the minimum standard brake fluid that can be used in automobiles?

2. What is the job of a shock absorber?

3. What checks are done to a gas cap?

4. What can be done to the top radiator hose to see if the radiator is under pressure before removing its cap?

5. When a stream of water leaks out from under the passenger compartment of the car, what could be the cause?

6. Which radiator hose is most likely to suffer heat damage, the top or bottom?

7. When a technician attempts to turn a fan pulley, what is being tested?

8. When installing a new wiper arm on the splined post, if it is installed in the wrong position, what can happen?

9. How many hood latches should a car have?

10. List two of the different ATF types available.

■ ASE STYLE REVIEW QUESTIONS

1. Technician A says that if the brake pedal height rises on the second application, the brakes probably need an adjustment. Technician B says that if the large chamber on a tandem master cylinder is low on fluid, the car probably has worn front disk brake linings. Who is right?

   a. Technician A   **b.** Technician B
   c. Both A and B   **d.** Neither A nor B

2. Technician A says that brake fluid that is allowed to remain uncovered absorbs water. Technician B says that if brake fluid is accidentally spilled on a fender of a car, it can damage the paint. Who is right?

   a. Technician A   **b.** Technician B
   c. Both A and B   **d.** Neither A nor B

3. Technician A says that the strength of coolant can be checked with a coolant hydrometer. Technician B says that before removing a radiator cap, you should always squeeze the top radiator hose to see if it is hard due to hot coolant under pressure. Who is right?

   a. Technician A   **b.** Technician B
   c. Both A and B   **d.** Neither A nor B

4. Technician A says that if the belt tension is too loose, a dead battery can result. Technician B says that when a terminal clamp is being removed from a battery, the positive cable should be removed first. Who is right?

   a. Technician A   **b.** Technician B
   c. Both A and B   **d.** Neither A nor B

5. Technician A says that battery electrolyte level is full when the liquid level is just above the top of the plates. Technician B says that the lower radiator hose is more likely to be damaged by oil than by heat. Who is right?

   a. Technician A   **b.** Technician B
   c. Both A and B   **d.** Neither A nor B

6. Technician A says that tire pressure should be checked when the tire is hot. Technician B says a typical tire loses two pounds of air pressure in a month. Who is right?

   a. Technician A   **b.** Technician B
   c. Both A and B   **d.** Neither A nor B

7. Technician A says that ATF is normally dyed red in color. Technician B says it is normal for an automatic transmission to consume a small amount of oil under normal use. Who is right?

   a. Technician A   **b.** Technician B
   c. Both A and B   **d.** Neither A nor B

8. The brake pedal is applied during a safety service. Technician A says that brake adjustment is being checked. Technician B says that master cylinder function is being checked. Who is right?

   a. Technician A   **b.** Technician B
   c. Both A and B   **d.** Neither A nor B

9. Which of the following is *not* true when checking fluid level in all automatic transmissions?

   a. The vehicle is parked on level ground.
   b. The fluid is hot.
   c. The transmission is in neutral.
   d. The engine is idling.

10. Which of the following statements about power steering fluid is *not* true?

   a. Fluid level is checked with the engine off.
   b. Automatic transmission fluid can be used in all power steering systems.
   c. Some power steering pumps have a dipstick.
   d. Fluid level is higher when hot.

Undercar Inspection And Service

■ INTRODUCTION

After completing inspection and maintenance under the hood, the vehicle is raised in the air to perform an undercar inspection. The oil and filter are usually changed while the vehicle is in the air (see Chapter 12). After reading instructions in Chapter 10 for lifting the vehicle, raise it in the air.

When performing undercar service, practice developing an efficient routine. After raising the car on the hoist, start at the front on the passenger side and work around the car, finishing up at the front again. Undercar service, including draining the oil, is done prior to underhood service. Then, the technician can refill the crankcase from the top while completing underhood services.

■ TIRE VISUAL INSPECTION

Visually inspect the tires for problems. Inspect the tire tread for signs of wear. A properly inflated tire will contact the road evenly (Figure 14.1). If the center of the tread is worn more than the outside edges, the tire has probably been run with too much air pressure (Figure 14.2). A more typical situation is to find wear on both outside edges of a tire's tread (Figure 14.3). This indicates an underinflated tire. When a car has been driven through corners at excessive speeds, wear to the outside edges of the tread can also result.

With the car in the air, rotate each tire slowly around while inspecting the tread area for nails, glass, screws, tears, and bulges. **Tread wear indicators** (*wear bars*) are molded into the tire's tread (Figure

14.4). Tires should have at least ¹⁄₁₆" of tread remaining. The wear bars are at this depth. When the wear bar shows across the entire tread, the tire is worn beyond legal limits and should be replaced.

Look for *scalloped* wear or bald spots that would indicate the need for a wheel balance or a suspension system repair (Figure 14.5). Uneven wear on the tire tread or a *feathered edge* across the entire tread indicates the need for a wheel alignment (see Chapter 60). When a car is aligned, front and rear suspension and steering parts are adjusted to minimize tire wear by allowing the

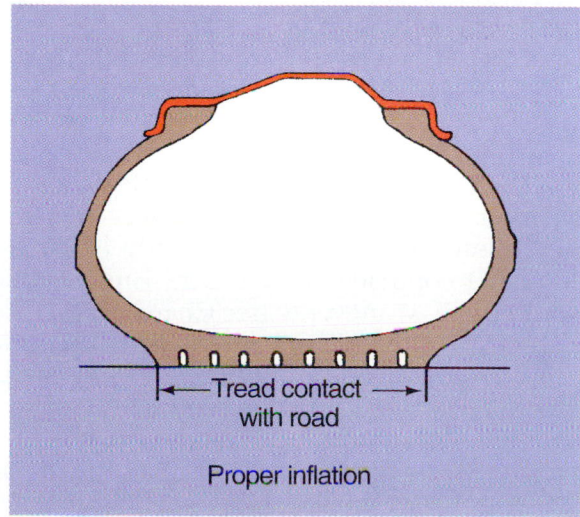

Figure 14.1 A tire with the correct air pressure. *(Courtesy of General Motors Corporation, Service Technology Group)*

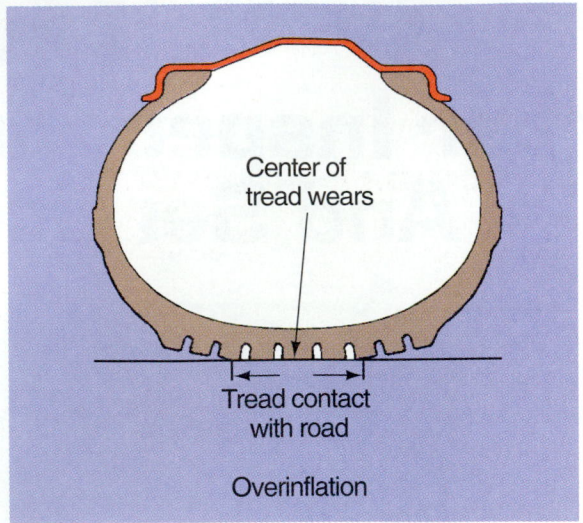

Figure 14.2 A tire with excessive air pressure. *(Courtesy of General Motors Corporation, Service Technology Group)*

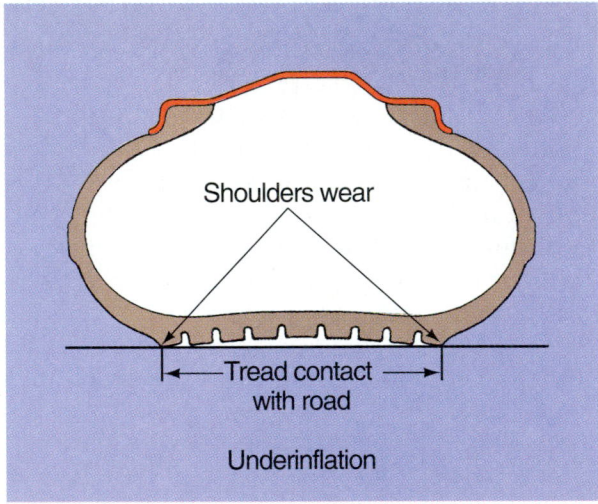

Figure 14.3 A tire with low air pressure. *(Courtesy of General Motors Corporation, Service Technology Group)*

tread of the tires to sit as flat as possible on the road during most driving conditions. Correct adjustment will also allow the car to *track* properly (go straight, with no pull to one side or the other). Tires are rotated so that they will wear evenly (see Chapter 55).

Inspect the tire sidewall for cracks and signs of tread separation (bubbles). With the car in the air, spin the tires to see if they appear to be "true". **Run-out** is the term that defines the up-and-down or side-to-side movement of a rotating tire. Check all of the lug bolts to see that none are damaged, missing, or broken and that they are torqued (tight). See that the wheels all spin with the same amount of effort. If one is tight, further investigation for a problem will be necessary.

➡ *Perform **Tire Inspection** Worksheet*

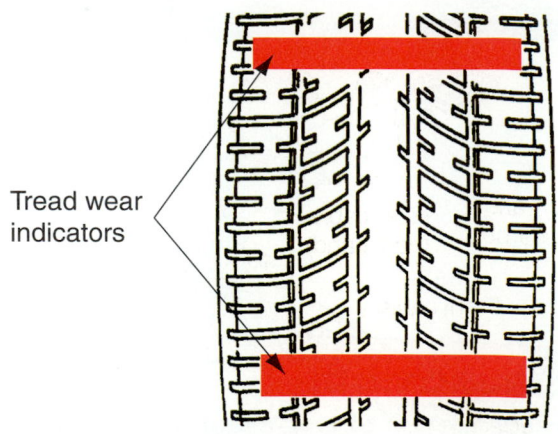

Figure 14.4 Tread wear indicators. *(Courtesy of Oldsmobile Division, General Motors Corporation)*

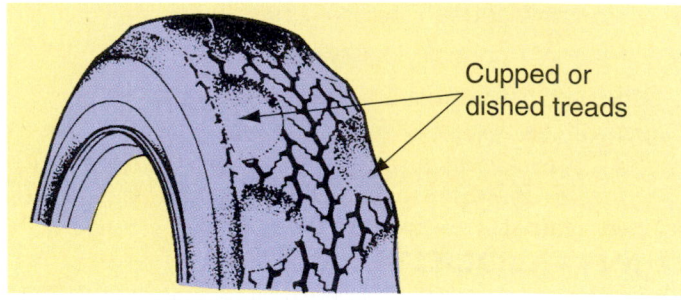

Figure 14.5 Bald spots and scalloped wear indicate loose parts or a need for a wheel balance. *(Courtesy of Ford Motor Company)*

▮ UNDERCAR BRAKE CHECKS

Because the axles must be able to move up and down with the springs, flexible rubber brake lines are used. Most vehicles have three rubber brake hoses: two in the front and one in the rear (Figure 14.6). When cars have a solid rear axle, only one rubber brake hose is needed in the rear. Cars with independent rear suspensions have two hoses on the rear.

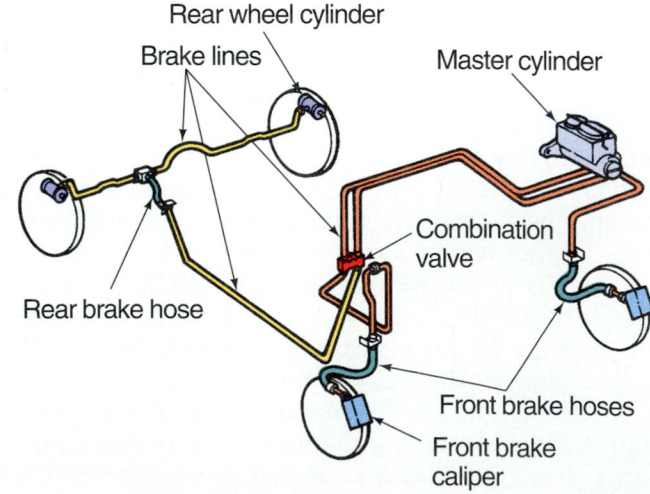

Figure 14.6 Hydraulic lines and hoses connect the master cylinder to the wheel brake units.

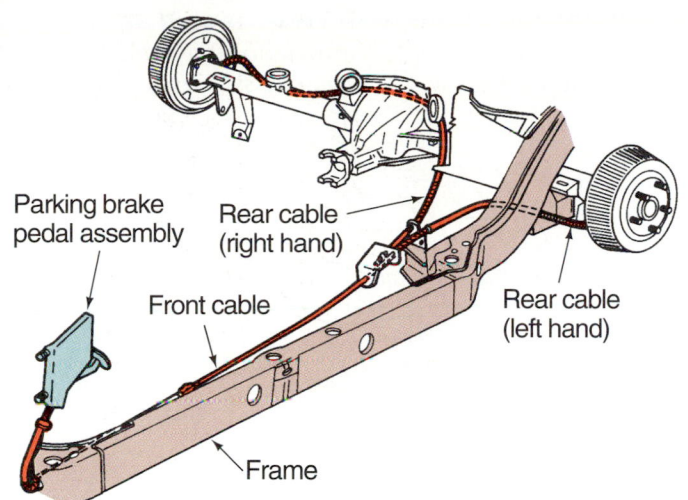

Figure 14.7 An emergency brake cable.

Inspect brake hoses for cracks or signs of leakage. Check to see that the front hoses have not been rubbing on tires. Check also to see that the rear hose is not too close to an exhaust pipe where it can be burned (or rubbed against).

Metal brake lines are checked for leaks at all threaded connections. Look for any signs of leakage along the fluid lines or on brake backing plates (see Chapter 52). Metal lines are usually fastened to the frame or axle. Check to be sure that they are properly secured.

Look at the emergency brake cable (Figure 14.7) to see if there is any sign of rust. If the cable requires lubrication, see Chapter 52 for an explanation.

EXHAUST SYSTEM INSPECTION

Part of a safety inspection is to check the condition of the exhaust system. Exhaust leaks can result in carbon monoxide poisoning of the passengers riding in the vehicle. With the vehicle running, listen for leaks at the exhaust manifold, muffler, and pipes (Figure 14.8). With the car in the air, check for broken muffler hangers that could allow a muffler to break loose. Exhaust components falling from a vehicle could cause a seri-

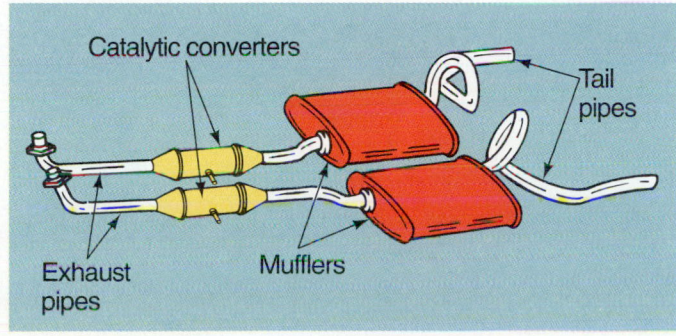

Figure 14.8 Listen for leaks in the exhaust system.

ous accident. Look to see that a muffler that was replaced has not been positioned near a fuel or brake line, where it might rub or cause excessive heat.

➡️ Perform *Inspect Exhaust System* Worksheet

CHASSIS LUBRICATION

When two parts move against each other, they pivot on a bushing or bearing. Pivot points such as *tie rod ends, ball joints,* and *constant velocity (CV) joints,* are sealed with rubber boots (Figure 14.9). Joints with boots are usually permanently sealed, but there are many older cars and newer light trucks that still have fittings provided for lubrication. To be certain, always check the lubrication chart showing lube points for the vehicle (Figure 14.10).

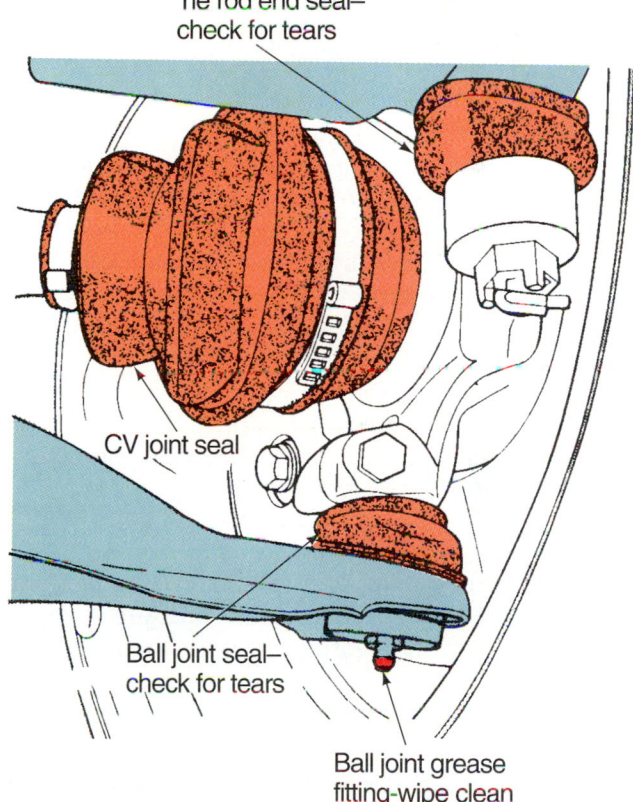

Figure 14.9 Pivot points sealed with rubber boots. (Courtesy of Chrysler Corporation)

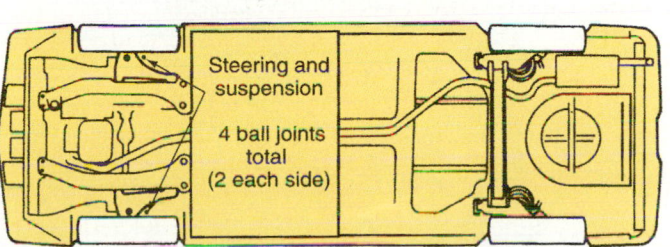

Figure 14.10 The location of lube fittings is found in shop service manuals. (Courtesy of General Motors Corporation, Service Technology Group)

Grease fittings are called **zerk fittings** or *standard nipple* type fittings. They are threaded to fit into holes in the part to be lubricated (Figure 14.11). The thread on a fitting is usually tapered pipe thread, which forms a wedged tight seal as the fitting is tightened into its hole (see Chapter 6). Zerk fittings have a one-way spring-loaded check valve. The valve allows grease to enter the joint, but not to leak back out. Be sure to wipe off the end of the zerk fitting before applying grease to it. Dirt can plug up the check valve, allowing grease to leak out and water and dirt to leak in.

Some vehicles have threaded plugs instead of zerk fittings (Figure 14.12). To apply grease, an adapter can be threaded into the hole. The plug is replaced after grease is applied. Metal plugs can be reused, but rubber or plastic plugs should be replaced. Plugs can also be replaced with new zerk fittings. Rubber-tipped adapters are also available for lubricating through threaded holes.

Grease Guns

Handheld grease guns (Figure 14.13) are available that can be used safely by anyone. Grease cartridges are available to provide a convenient grease gun refill. Cartridges are an advantage when special lubricants or multiple greases are required for the same car. Professional lubrication equipment uses a high pressure grease gun, powered by compressed air (Figure 14.14).

 **SAFETY NOTE** Be careful not to apply the grease gun against skin or an infection can result.

Grease gun nozzles (Figure 14.15) have teeth that grip the fitting tightly when grease pressure is applied. A grease fitting is often easier to get at when the wheels are turned to the side. If the ignition switch is left unlocked, it is easy to turn the wheels when the car is in the air with the weight of the vehicle off of them. A pressurized grease gun may be necessary unless the joint being greased is unloaded on a wheel-contact hoist.

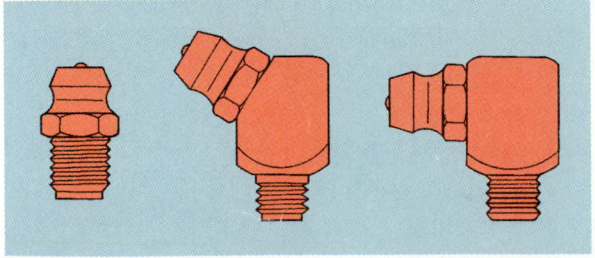

Figure 14.11 Zerk fittings. *[Reproduced by permission of Deere & Company, 1992. Deere & Company. All rights reserved.]*

Fill the joint until fresh grease is seen at the vent hole in its rubber boot. Rubber boots usually have a small hole that allows air or excessive grease to bleed

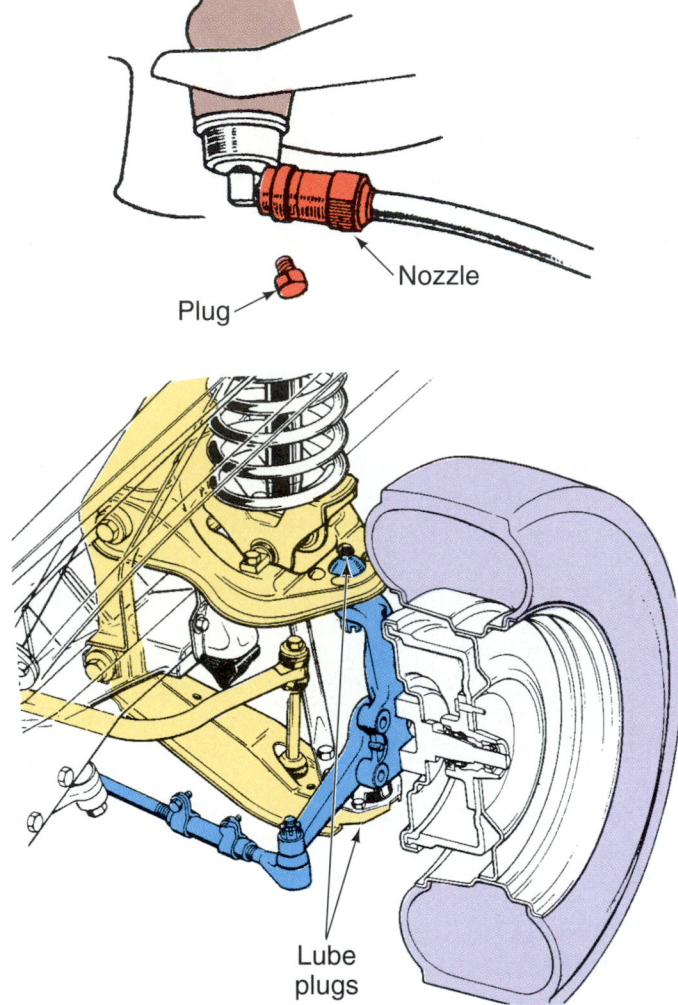

Plug   Nozzle

Lube plugs

Figure 14.12 Threaded plugs are used instead of zerk fittings in some vehicles. *[Lower courtesy of Chrysler Corporation]*

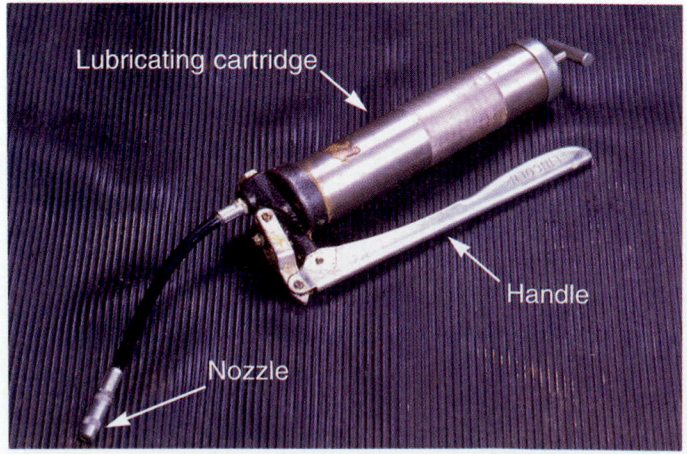

Lubricating cartridge

Handle

Nozzle

Figure 14.13 A handheld grease gun.

off. The boot can still be damaged, however, if too much pressure is allowed to enter the joint at once.

> **SHOP TIP** If lubricant will not go into a fitting, the joint may be frozen up. Try turning the wheels from side to side while applying pressurized grease. If this fails, put the car back on the ground and push on the bumper before trying to add grease again.

Sometimes the zerk fitting is frozen or plugged and must be replaced.

Grease

The grease used for automobile chassis lubrication is liquid petroleum oil that has been made semi-fluid by adding a metallic soap **thickening agent**, such as aluminum or lithium. Grease does not leak or flow out of a bearing like an oil would. This is what makes it desirable as a lubricant.

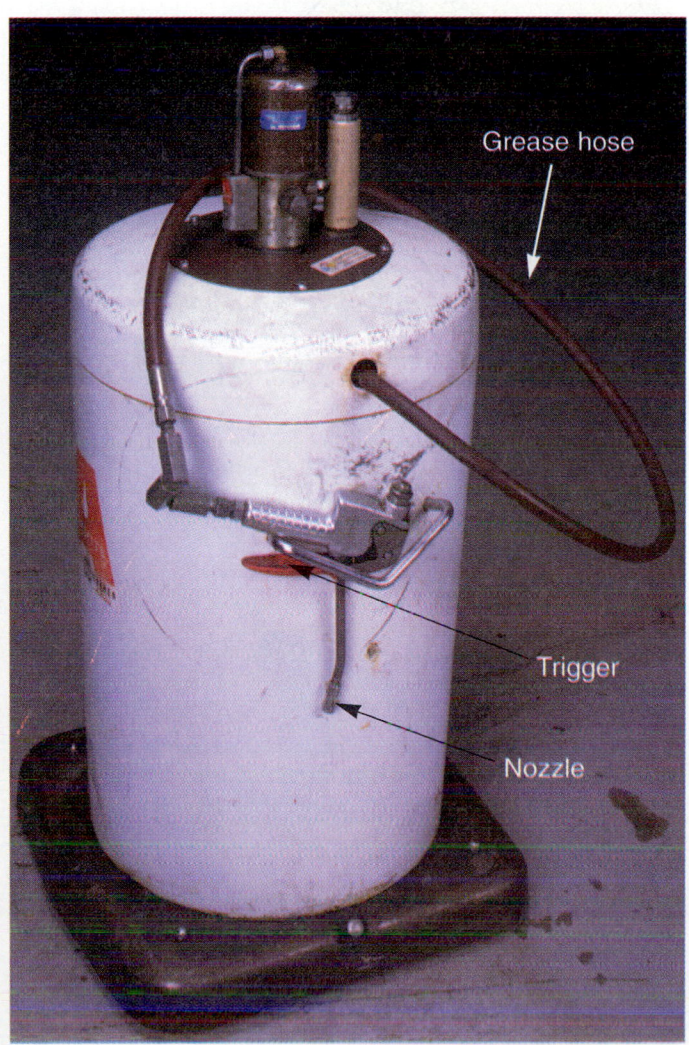

Figure 14.14 An air-operated grease gun.

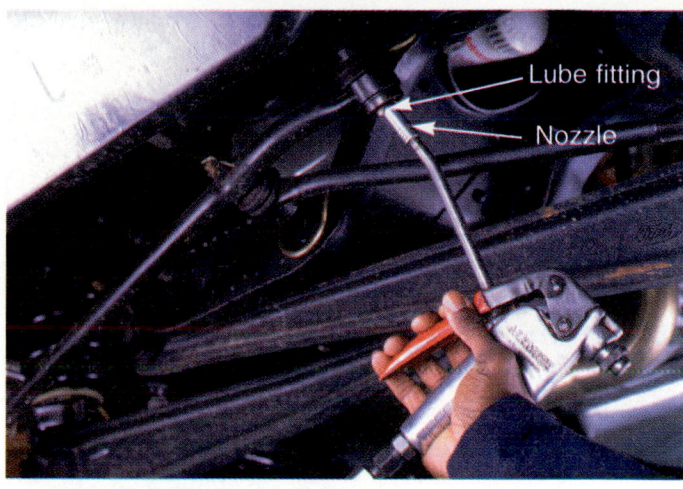

Figure 14.15 The lube gun nozzle grips the fitting.

Chassis lubricant is grease with a consistency that allows it to be applied through a zerk fitting with a grease gun. It is highly resistant to being diluted or washed away with water.

A *multipurpose grease*, which is the most common type used in service shops, satisfies the requirements of chassis, wheel bearing, and universal joint lubricants. Multipurpose does not mean all-purpose; it only meets certain requirements. There are several types of greases, which are discussed in greater detail in Chapter 53. The lubricant most often used in chassis lubrication is a *multipurpose lithium-based grease*.

Some greases contain solid lubricant materials such as *molybdenum (moly)* or *graphite*. These are often used to lubricate speedometer cables, emergency brake cables, splines, and leaf springs.

➡ *Perform **Lubricate Suspension Fittings Worksheet***

■ SUSPENSION AND STEERING CHECKS

Inspect Ride Height

Before raising the vehicle, look to see if it appears to be riding low or leaning to one side or the other. If anything appears out of the ordinary, check for tire wear and then perform a *ride height check* (see Figure 61.5). Wheel alignment angles change when the ride height of the vehicle changes. Lower ride height due to worn springs often results in front tire wear on the inside of the tread.

Rubber Bushings

Rubber bushings separate many pivoting suspension parts. The bushings stretch but do not move, so the friction that there would be between moving parts is not there. A bushing can be visually checked for signs of cracking. Use a prybar to pry on parts that have rubber bushings to see if there is any sign of looseness.

NOTE: *Rubber parts should not be lubricated by oil, which will rapidly deteriorate a rubber bushing.*

SAFETY NOTE Bushing replacement is done frequently by tire and wheel shops. Proper training is required and this job should not be attempted without trained assistance.

Steering Linkage Checks

Check for looseness in front end parts with the car in the air. Twist the tie rods and other steering linkages (Figure 14.16), trying to move them up and down. Parts should be free to move when twisted by hand, but should not move up and down (looseness). Check the condition of lubrication seal rubber boots. Permanently sealed joints with torn boots should be replaced, especially if dirty rain water has gotten into the joint.

Ball Joints

Ball joints (see Figure 14.12) are checked for excessive wear and for torn grease seals. Look for unusual tire wear. Worn ball joints or other front suspension parts that cause looseness can result in cupped wear of the tire tread. If the tires are cupped, check the ball joints as described in Chapter 46. A torn or missing ball joint seal will require replacement of the joint by a front suspension specialist. Bad shock absorbers or a tire out of balance can also result in cupped wear.

Shock Inspection—Undercar

Shock absorbers are located at each corner of the vehicle, at the spring (Figure 14.17). Their purpose is to dampen oscillations of the springs. A worn shock absorber will allow the wheel to *tramp*, or hop up and down on the road (Figure 14.18).

There are rubber **rebound bumpers** bolted to the lower control arm or chassis. These rubber bumpers come into use only when the suspension has

reached the full limit of its travel. They do not normally function unless the car has been driven in extremely rough conditions or if the shock absorbers are worn out.

SHOP TIP Worn or damaged rebound bumpers often indicate a need for shock absorber replacement. Check to see if they have come into contact with the frame of the vehicle.

Inspect the shock absorbers for leaks or signs of physical damage. Inspect the rubber bushings and mounting bolts for signs of wear or looseness. The appearance of a slight amount of moisture at the shock absorber seal is normal. But the outside of a shock should not show any drips or be entirely wet. Physical damage to a shock is cause for replacement of *both* shocks on that end of the vehicle. Shocks are always

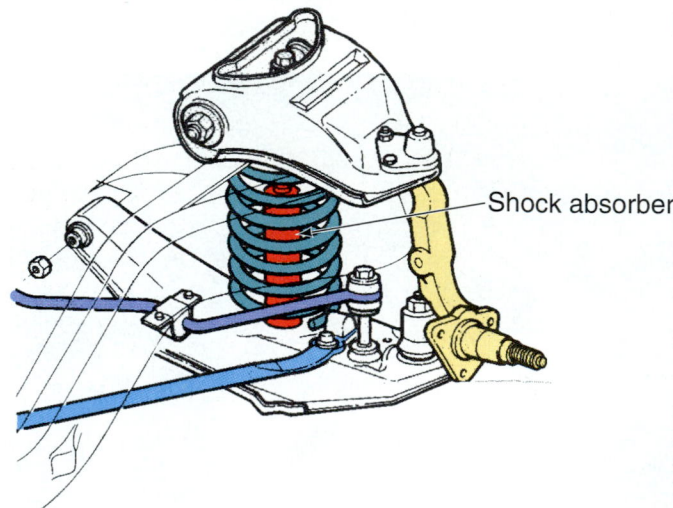

Shock absorber

Figure 14.17 A shock absorber. *(Courtesy of Ford Motor Company)*

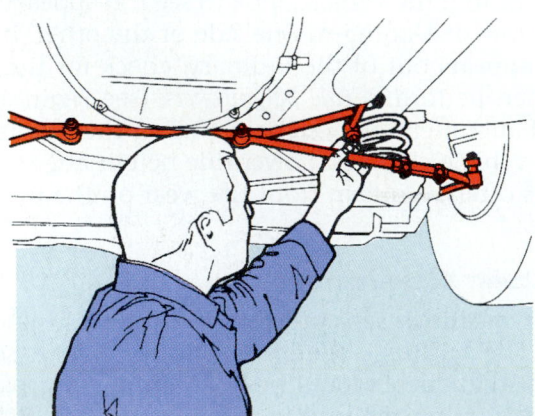

Figure 14.16 Twist and shake steering linkages to test for looseness. *(Courtesy of Ford Motor Company)*

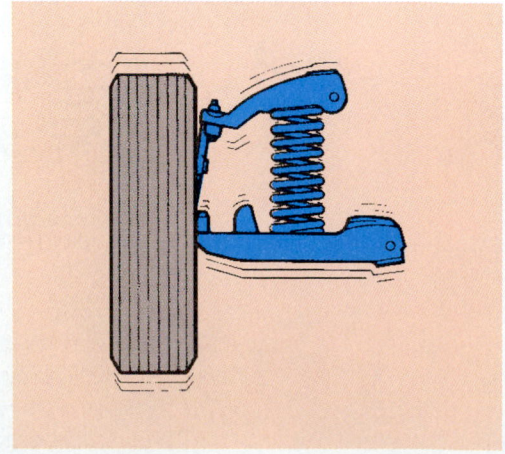

Figure 14.18 A worn shock absorber will allow wheel tramp. *(Courtesy of Ford Motor Company)*

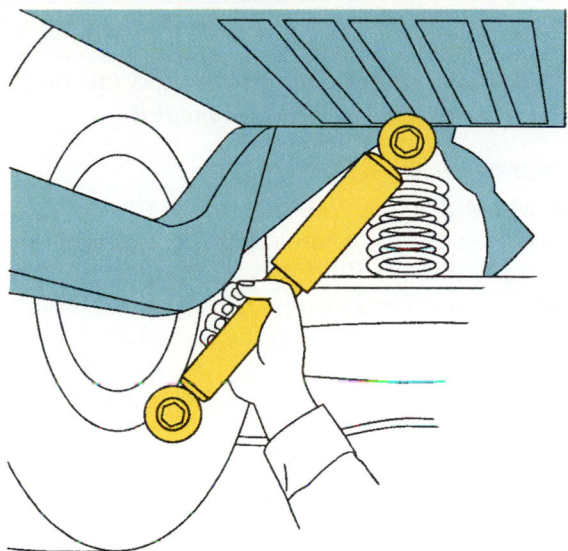

Figure 14.19 Check a shock by loosening one end and pulling it both ways through its full travel. *(Courtesy of Federal-Mogul Corporation)*

replaced in pairs. If shocks need to be replaced, follow the procedure in Chapter 57.

If there is a reason to question the condition of a shock, unbolt it from the bottom. Then, push it and pull it through its regular travel (Figure 14.19). Resistance should be felt in both directions, through the entire travel of the shock.

NOTE: *The wheels must be supported when removing shocks from rear wheel drive cars with coil springs.*

■ DRIVE LINE CHECKS

Inspect Clutch and Transmission Linkage

Vehicles use various methods to shift gears and disengage the clutch. Some use cables; others use linkage or a hydraulic master and slave cylinder. Even cars with automatic transmissions must have some form of linkage or cables for shifting. Visually inspect the ends of linkages to see that rubber or brass bushings are not worn or broken. Grasp linkages and try to move them to see if they are loose or worn.

Look for signs of leakage on a clutch master and slave cylinder.

Inspect for Transmission and Axle Leaks

Inspect the transmission or transaxle for leaks, especially if a unit is low on oil or fluid. Figure 14.20 shows the location of the filler plug and drain plugs on a manual transmission. Check the service manual for the correct type of lubricant to use. Transmissions and transaxles can use ATF (automatic transmission fluid), engine oil, or gear oil.

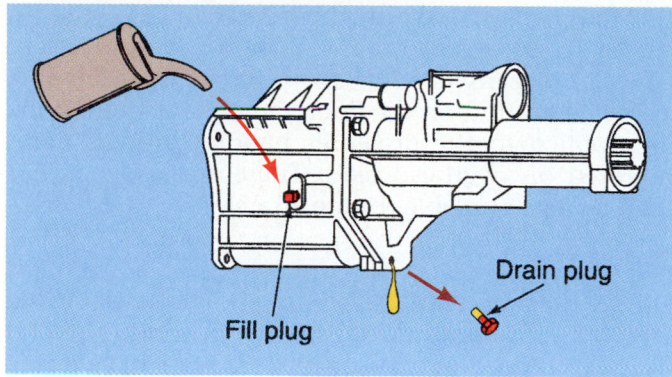

Figure 14.20 Location of manual transmission fill and drain plugs.

Rear wheel drive cars have a *differential* between the rear wheels of the car that splits the engine's torque equally to the drive wheels (see Figure 68.18). There is a vent on the top of the axle housing. A hose is usually installed from the vent into the trunk to protect against water entry. Check to see that the hose is not damaged.

Check the differential and its rear axles for leakage of oil. Figure 14.21 shows the location of typical differential fill plugs. The oil level should come to the bottom of the threaded fill plug hole (Figure 14.22). If there is a leak, check to see that the vent is not plugged.

> **SHOP TIP** To tell the difference between gear oil and brake fluid, apply water to the wet area. Brake fluid is water soluble, gear oil is not.

When gear oil leaks onto a rear brake assembly, the brakes will need to be replaced after repairing the axle leak. Chapter 69 includes more information on axles.

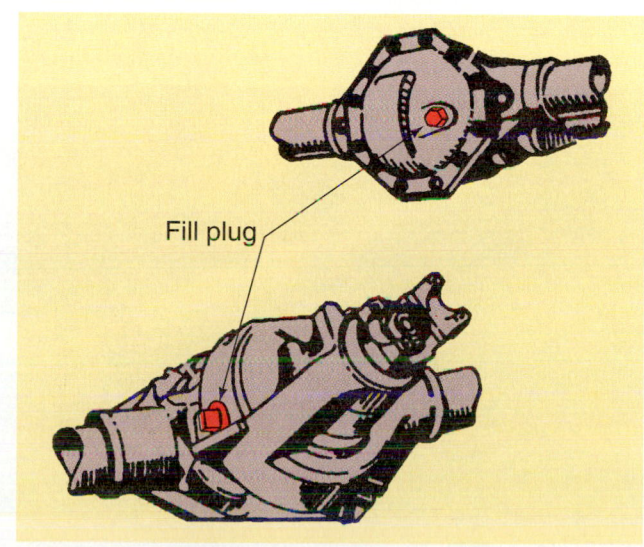

Figure 14.21 Typical locations of differential fill plugs.

Many shops have an air-operated gear oil dispenser. It has a gauge that tells how much lubricant has been dispensed. Some shops have dispensers that are rolled around on the floor. Other shops have bulk oil supplied to the dispenser handle through hoses mounted on overhead reels. There are also portable hand pumps available (Figure 14.23).

Gear Oil

Differentials in rear wheel drive cars have curved **hypoid gears** and use SAE 90 E.P. gear oil. The **E.P.** means **extreme pressure**. Hypoid gears have a ring gear and a pinion gear that intersect below the centerline of the ring gear (see Chapter 68). This is so that the hump in the floor of the vehicle can be lower. The teeth on hypoid gears are spiraled and curved. They "wipe" across each other on a very small area of the gear tooth. The tremendous load that results calls for the use of E.P. gear oil.

NOTE: *E.P. gear oils contain zinc. They are dark in color and have a strong odor.*

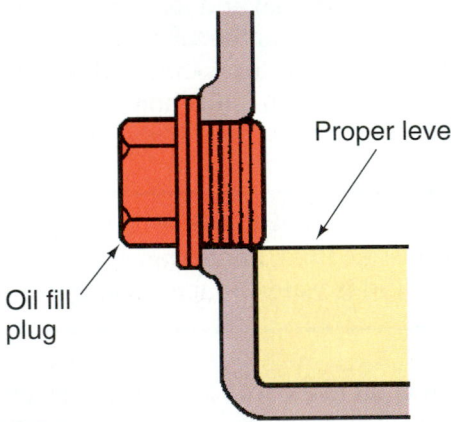

Figure 14.22 The fluid level should be at the bottom of the fill plug hole. *(Courtesy of American Honda Motor Co., Inc.)*

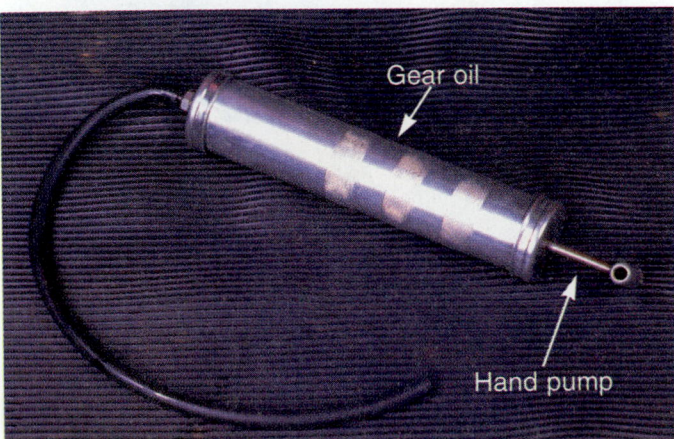

Figure 14.23 Small pump for pumping transmission lubricant.

Some differentials are of the *limited-slip* or *positraction* type. These axles, which usually have an I.D. tag that identifies them, require a special *limited-slip lubricant* that contains a *friction modifier*.

Transaxles

Front wheel drive (FWD) cars have a transaxle at the front of the vehicle. A low viscosity engine oil or automatic transmission fluid can be used to lubricate the differential section because they do not use hypoid gears. Figure 14.24 shows the location of a typical transaxle's drain and fill plugs. Be sure to check the service manual for the proper lubricant to use.

CV Joints

Front wheel drive cars have CV joints at each end of their two drive axles (see Figure 70.20). These flexible joints require special grease that is sealed permanently into the joint with a flexible air-tight rubber seal, commonly called a boot. Sometimes CV joint boots tear or rub on something and become damaged. Inspect the boots to see if they show signs of leakage. If so, the boot must be replaced *immediately*. If the leak has been allowed to continue for a period of time, the axle has probably been subjected to water (especially when driving in the rain). Dirt is carried by the water into the joint, where damage quickly results. Replacement of CV joint seals usually requires removal of the axle and disassembly of the entire joint (see Chapter 71).

U-Joints

On rear wheel drive cars there is a *driveshaft* that carries power between the transmission and the differential (Figure 14.25). The driveshaft has a *universal joint*, also called a *U-joint*, at each end. Two-piece driveshafts often have a U-joint in the center too. To test a U-joint for signs of damage, grasp the driveshaft near one end and work it up and down while watching each U-joint for signs of looseness (Figure 14.26). Inspect the seals at each end of the U-joint for signs of rust from the bearing cups.

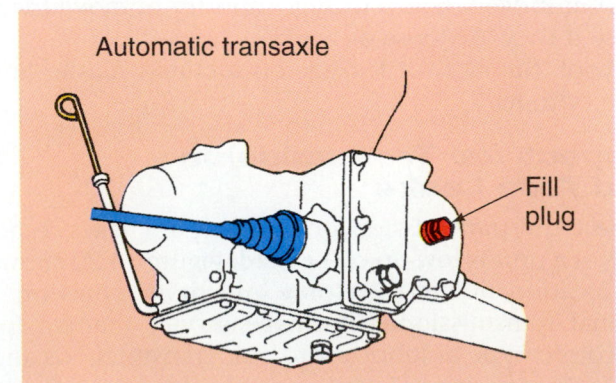

Figure 14.24 Typical location of transaxle fill plug.

Replacement aftermarket universal joints usually have a zerk fitting for providing lubrication after installation and periodic lubrication thereafter. Universal

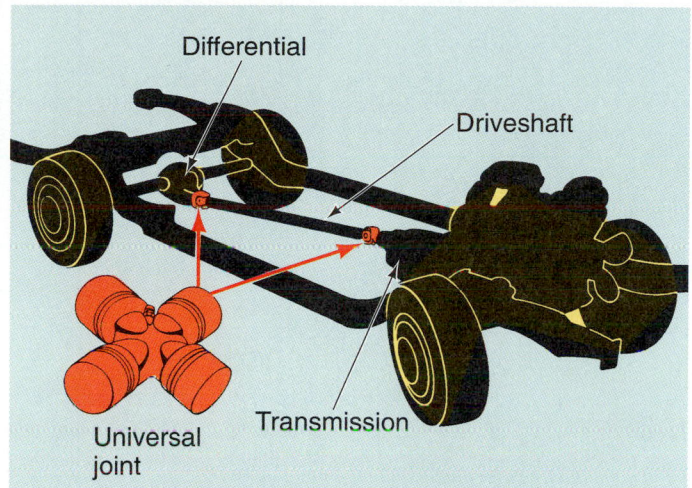

Figure 14.25 Rear wheel drive cars have a driveshaft carrying power between the transmission and the differential. *(Courtesy of Precision Universal Joint Corp., Chicago, IL)*

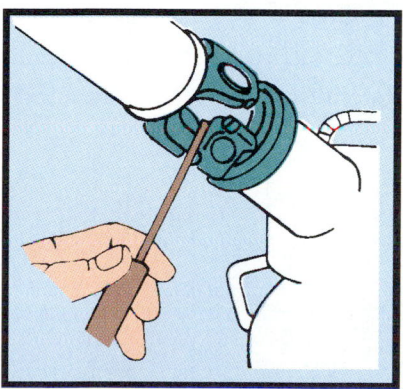

Check for rust

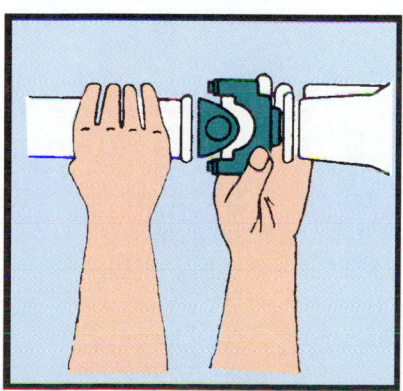

Use hand force to check for looseness

Figure 14.26 Testing U-joints for signs of obvious damage. *(Courtesy of Precision Universal Joint Corp., Chicago, IL)*

joint grease or multipurpose grease is added to a U-joint with a lube gun adapter that has a special pressure control attachment to avoid damage to the seal. Squeeze the handle using short, quick, grips on the handle to supply grease in small spurts until evidence of the grease appears at the seals. Further service of the U-joint is covered in Chapter 69.

Sometimes, long driveshafts are made up of two pieces and have a center support bearing that is surrounded by rubber (Figure 14.27). If the bearing fails and seizes up, the rubber will be torn away from the bearing. The rubber can also deteriorate due to age or contamination with oil. The driveshaft will operate at the wrong angle and vibration or noise will result. The vibration occurs most often during hard acceleration.

Exhaust Inspection

Exhaust systems rust because of the acids and moisture in the engine's exhaust. An engine that is never fully warmed up will rust from moisture that pools in the lowest spots in the exhaust system. Holes usually develop in the lowest spots in exhaust systems.

Listen for leaks while the engine is running. Check for leaks at all pipe connections and at low spots in the system. A stethoscope can be used. Remove the metal end from the stethoscope and use just the hose to listen. Additional information on exhaust system service is found in Chapter 28.

When the Lubrication Safety Service is completed, fill out a door record sticker and put it on the door jamb. A door sticker has a place for listing the mileage when this service was performed. Also fill in the kind of oil that was used.

➡ *Perform **Complete Lubrication/Safety Service** Worksheet*

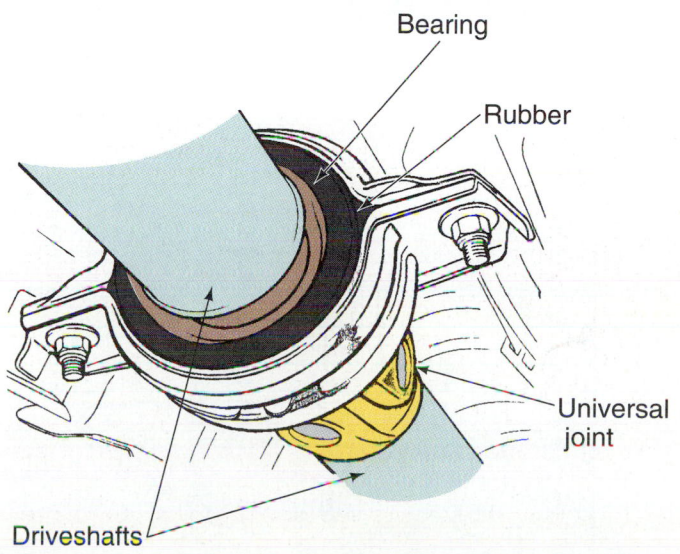

Figure 14.27 A center support bearing. *(Courtesy of Chrysler Corporation)*

CERTIFIED CAR CARE SERVICE

Customer Name _____

Address _____ City _____ Zip Code _____ Phone _____

Date _____ Year _____ Time _____ Model _____ License Number _____ Odometer Reading _____

Vehicle _____

ELECTRICAL SYSTEM CHECKS
- ___ Wiring Visual Inspection
- ___ Battery
- ___ Top Off Water Level
- ___ Posts and Cables
    - ___ Clean ___ Corroded
- ___ Damaged
- Battery Condition
    - ___ Good
    - ___ Recharge ___ Replace

LIGHTS
- ___ Park ___ Brake
- ___ Signal ___ Emergency
- ___ Dash Lights Back-up
- Headlight Operation
- ___ High Left ___ High Right
- ___ Low Left ___ Low Right
- ___ Horn Operation

FUEL SYSTEM CHECKS
- ___ Condition of Hoses
- ___ Gas Cap Condition
- ___ Air Cleaner
- ___ Crankcase Vent Filter
- ___ Fuel Filter (miles until change suggested) ___

COOLING SYSTEM CHECKS
- ___ Level
- ___ Strength of Coolant
- ___ (Protection to ___ °)
- ___ No Leaks
- Condition of Hoses
- ___ Radiator
- ___ Heater
- ___ Thermostat Bypass
- ___ Hose (if so equipped)

BRAKE INSPECTION
- ___ Pedal Travel
- ___ Emergency Brake
- ___ Brake Hoses and Lines
- ___ Master Brake Cylinder-
- ___ Fluid Level and Condition

ON-GROUND STEERING, SUSPENSION, DRIVE LINE CHECKS
- ___ Steering Wheel Freeplay
- ___ Power Steering Fluid Level
- ___ Shock Absorber Bounce Test
    - ___ Good ___ Unsafe
- ___ Front
- ___ Rear
- ___ No Squeaks
- ___ Ride Height Check
- ___ Check ATF Level

VISIBILITY
- ___ Mirrors
- ___ Wiper Blades
- ___ Wiper Operation ___ fast ___ slow
- ___ Washer Fluid and Pump
- ___ Clean and Inspect all Glass

Pressure Test
- ___ Radiator
- ___ Cap
- ___ Condition of Coolant
- ___ Pump Belt

UNDERCAR SERVICE

- ___ Drain Crankcase (if ordered)
- ___ Remove and replace oil filter
- Inspect Under Car for
    - ___ Oil, Gasoline, and Coolant Leaks
    - ___ Check Crankcase Oil Level
    - ___ Check Oil Filter for Leaks

INFLATE AND CHECK TIRES
- Inflate to ___ lb.
- Tire Condition:
    - ___ Good ___ Fair ___ Unsafe
- ___ RF
- ___ LF
- ___ RR
- ___ LR
- ___ Inflate and Check Spare
    - ___ Good ___ Fair ___ Unsafe

DRIVE LINE CHECKS
- ___ Check Universal or CV Joints
- Inspect Gear Cases
    - ___ Differential
    - ___ Transmission
- ___ Replace Drain Plugs
- ___ Inspect Motor Mounts
- ___ Lubricate Door and Hood Hinge and Latches

UNDERCAR FUEL SYSTEM CHECKS
- ___ Condition of Fuel Hoses
- ___ Condition of Fuel Tank

SUSPENSION AND STEERING
- ___ Inspect Steering Linkage
- ___ Inspect Shock Absorbers
- ___ Inspect Suspension
- ___ Bushings
- ___ Clean Lubrication Fittings
- ___ Lubricate Fittings
- ___ Ball Joints
- ___ Inspect Ball Joint Seals
- ___ Ball Joint Wear
- ___ Inspection
- ___ Inspect Ride Height

EXHAUST SYSTEM CHECKS
- ___ Mufflers and Pipes
- ___ Pipe Hangers
- ___ Exhaust Leaks
- ___ Heat Riser

FINAL VEHICLE PREPARATION
- ___ Replace Crankcase Oil
- ___ Clean Windows, Vacuum Interior
- ___ Fill Out and Affix Door Jamb
- ___ Record to Door Post
- ___ Complete a Repair Order

Figure 14.28 A complete lubrication, maintenance, and inspection worksheet.

After completing all of the worksheets covered in the book up to this point, the items in a complete lubrication safety service can be satisfactorily accomplished in an acceptable amount of time. Items on the worksheet in Figure 14.28 are representative of those that would be found on a typical commercial lubrication service record. There is a space on the worksheet called "items needing attention." This space is used to note any items requiring further service or repair.

■ REVIEW QUESTIONS

1. Which tire tread wear is more typical: outside edges or center?

2. When the wear bars on a tire's tread are even with the remaining tread, how much tread depth remains?

3. How many flexible brake hoses do most cars have?

4. Lubrication fittings are commonly called _____ fittings?

5. Before squirting grease into a fitting,_____ off the fitting.

6. Three uses that a multipurpose grease is designed to be able to serve are chassis, wheel bearing, and _____ joint.

7. Loose parts, such as worn ball joints, can cause _____ tire wear.

8. If a rubber bumper has been damaged because the suspension reached its limit of travel, what could be the cause?

9. Unbolt the lower end of a _____ _____ and move it through its travel to see if it is still operating properly.

10. What does the "E.P." in a gear oil label stand for?

■ ASE STYLE REVIEW QUESTIONS

1. Technician A says that most cars today have several lubrication fittings. Technician B says that if a ball joint will not accept grease, try turning the wheel from side to side while applying grease. Who is right?
   a. Technician A   b. Technician B
   c. Both A and B   d. Neither A nor B

2. Technician A says that an advantage to a hand-held grease gun is that cartridges of special use greases can be used. Technician B says that grease is made of soap and oil. Who is right?
   a. Technician A   b. Technician B
   c. Both A and B   d. Neither A nor B

3. Technician A says that when a tire has been run for a long time with too much air pressure, wear often occurs on its outside edges. Technician B says that if a rubber boot on a steering or suspension joint becomes torn, the entire joint will have to be replaced. Who is right?
   a. Technician A   b. Technician B
   c. Both A and B   d. Neither A nor B

4. Technician A says that you can tell the difference between a rear axle leak and a rear brake cylinder leak because a brake fluid leak will wash off with water. Technician B says a car with independent rear suspension will have four flexible brake hoses. Who is right?
   a. Technician A   b. Technician B
   c. Both A and B   d. Neither A nor B

5. Technician A says that wax stick lube is used on door latches and strike plates. Technician B says that liquid lubricants should not be used in door locks. Who is right?
   a. Technician A   b. Technician B
   c. Both A and B   d. Neither A nor B

6. All of the following are true about hypoid differentials *except*:
   a. They use 90W gear lube.
   b. Some of them use engine oil.
   c. They use oil with EP additives.
   d. Some of them require special limited-slip gear oil.

7. Technician A says that rubber bushings should be lubricated with oil. Technician B says that steering linkage parts should be able to move up and down. Who is right?
   a. Technician A   b. Technician B
   c. Both A and B   d. Neither A nor B

8. Technician A says that the name of the part found at each end of a rear wheel drive driveshaft is called a CV joint. Technician B says that the joints at the ends of front wheel drive axles are called universal joints. Who is right?

a. Technician A   **b.** Technician B
c. Both A and B   **d.** Neither A nor B

9. Technician A says that a multipurpose lubricant is used in a CV joint. Technician B says that a small hole in a CV joint boot is normal. Who is right?

a. Technician A   **b.** Technician B
c. Both A and B   **d.** Neither A nor B

10. Technician A says that bad shock absorbers are replaced in pairs. Technician B says that a moly or graphite lubricant would be used to lubricate an emergency brake cable. Who is right?

a. Technician A   **b.** Technician B
c. Both A and B   **d.** Neither A nor B

Engine Operation and Service

THEN AND NOW: ENGINE EVOLUTION

A Frenchman named Etienne Lenoir built the first internal combustion engine in 1860. It ran on *town gas* that was used for lighting in those days. In general, this engine was very different from the engines of today. It was not until 1876 that German inventor Nikolas Otto produced the original four-stroke cycle engine of the type used today.

Early engines did not put out much power for their size. The original single-cylinder Benz of 1886, for instance, displaced 58 cubic inches (954 cc—almost a liter). Yet it produced only three-quarters of a horsepower and ran at a mere 400 rpm.

There are numerous ways of increasing the power and fuel efficiency of an engine. Raising the compression ratio can result in a substantial increase. But when this was tried with ordinary gasoline, severe detonation occurred. In 1922, Thomas Midgley, Jr. discovered that adding *tetraethyl lead* to gasoline quieted the explosions in the cylinders. This was the birth of high octane gasoline.

The design of the combustion chamber and valve train are both extremely important in determining the power an engine produces. Flathead engines, with their valves alongside the cylinders, were the design of choice for many decades. They were simple and dependable. The later development of *overhead valves (OHV)* increased power immensely. Overhead valve engines even appeared on some early cars, such as the 1910 Buick.

Another means of increasing power is for the engine to spin faster without damage. Early design engines had heavy cast iron pistons. The change to much lighter aluminum helped allow higher engine rpm. Higher rpm is also made possible by the *overhead camshaft (OHC)*, which eliminates trouble-prone pushrods and, in many cases, rocker arms. The OHC advance is credited to French car maker Adolph Clement in 1902.

Force-feeding pressurized air into the engine's cylinders results in performance equal or greater to that of an engine that is much larger. A supercharger was used with some success on sports cars of the 1920s and turbochargers began to be used on cars beginning in the 1960s. Detonation and added complications are drawbacks to these systems.

Today, multi-valve, computer-managed, fuel-injected, high-rpm engines produce amazing amounts of power for their displacement. For example, the Acura Type R with its V-Tech valve system pumps 195 hp at 8000 rpm out of a mere 110 cubic inches (1.8 liters).

The 1910 Buick had overhead valves. *(Courtesy of Bob Freudenberger)*

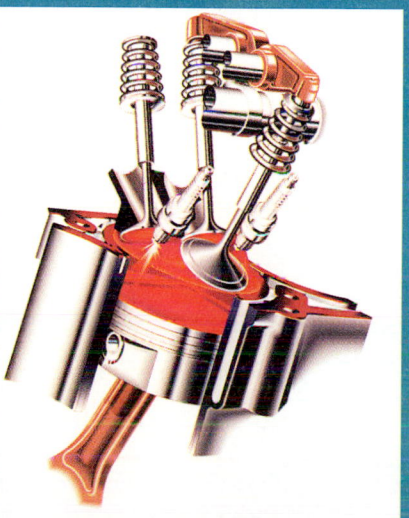

A Mercedes-Benz V6 combustion chamber. *(Courtesy of Mercedes-Benz of N.A., Inc.)*

Introduction to the Engine

■ KEY TERMS

poppet style valve
stoichiometric
blow-by
upper end
valve train
pushrod engine
overhead cam engine
runners
crankcase
long block/short block
end play
bearing clearance
piston slap

■ OBJECTIVES

Upon completion of this chapter, you should be able to:

✔ Explain the principles of internal combustion engine operation.

✔ Identify internal combustion engine parts by name.

✔ Describe the function of engine parts.

■ INTRODUCTION

The first chapter in this book gave a basic description of engine operation. More in-depth information is given in this chapter. Such things as firing orders, valve adjustment, oil pressure testing, and engine diagnosis are covered in later chapters.

■ BASIC ENGINE OPERATION

Most modern automobiles use a *four-stroke reciprocating* gasoline engine. The *four-stroke cycle* is described here using a single cylinder engine. Automobile engines actually have multiple cylinders.

A simple reciprocating engine has a cylinder, a piston, a connecting rod, and a crankshaft. If the cylinder is compared to a cannon, the piston is a round plug that can be compared to a cannonball. The cylinder is the cannon. In a spark-ignited internal combustion engine, a mixture of air and fuel is compressed in the cylinder.

Fuel that is burned in a "gasoline engine" must be a liquid that vaporizes easily (such as gasoline, methanol, or ethanol), or a flammable gas (such as propane or natural gas). When the air/fuel mixture is compressed and then burned, it pushes a piston down in a cylinder. This action turns a crankshaft that powers the car (see Figure 1.5).

A cylinder head closes off the end of the cylinder. A piston is connected to the crankshaft by a connecting rod and a piston pin (also called a *wrist* pin). This arrangement allows the piston to return to the top of the cylinder, making continuous rotary motion of the crankshaft possible. The piston is sealed to the cylinder wall by piston rings.

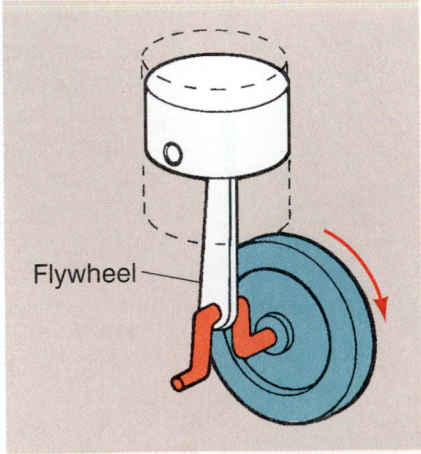

Figure 15.1 A flywheel is installed at the rear of the crankshaft. *(Courtesy of Ford Motor Company)*

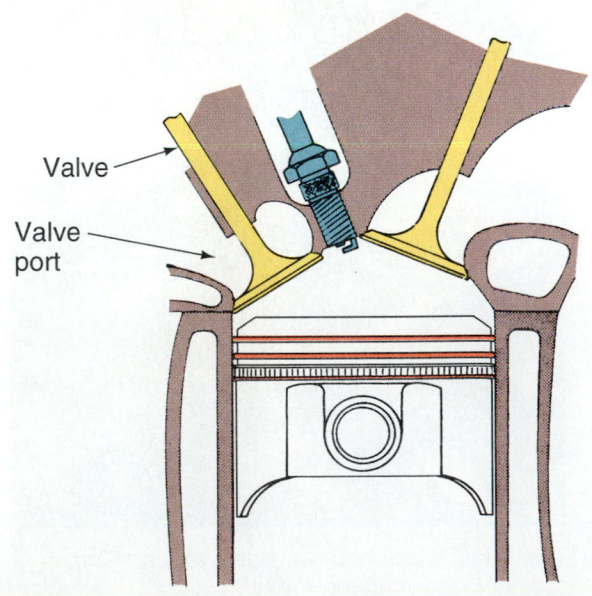

Figure 15.2 Valves seal off the valve ports. *(Courtesy of Federal-Mogul Corporation)*

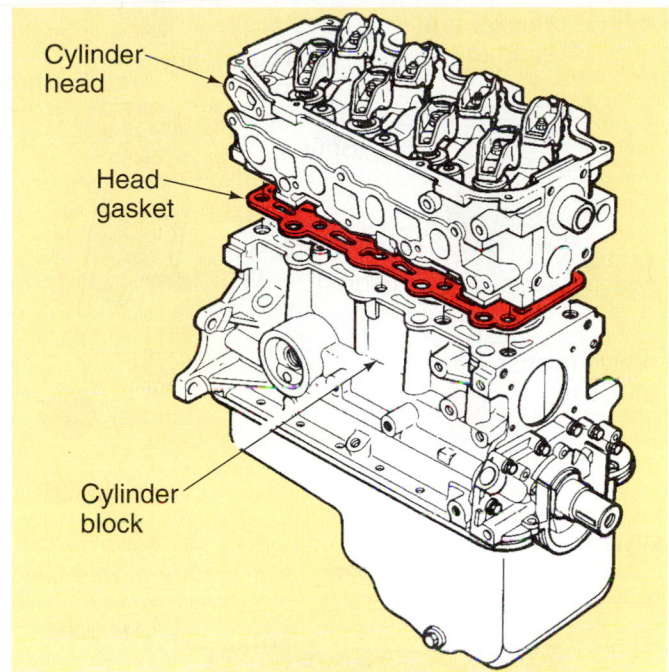

Cylinder head

Head gasket

Cylinder block

Figure 15.3 A head gasket seals the head to the block. *(Courtesy of Federal-Mogul Corporation)*

Because of the powerful pulses on the piston as the fuel is burned in the cylinder, a heavy *flywheel* is bolted to the rear of the crankshaft (Figure 15.1). The weight of the flywheel blends together the power pulses into one continuous motion of the crankshaft.

The cylinder head has an intake valve port for each cylinder that allows an air and fuel mixture to flow into the cylinder. An exhaust valve port in the head allows the burned gases to flow out. Each port is sealed off by a **poppet style valve** (Figure 15.2). The head is sealed to the cylinder block with a *head gasket* (Figure 15.3). The opening of the valves is controlled by the *camshaft* (see Figure 1.6). The camshaft is turned by the crankshaft with gears or sprockets driven by a belt or chain.

◼ FOUR-STROKE ENGINE OPERATION

Piston Travel

The piston travel in the cylinder is limited in both directions. The upper limit of piston travel is called *TDC (top dead center)* and the lower limit is called *BDC (bottom dead center)*.

Piston Stroke

A stroke is the movement of the piston from TDC to BDC, or from BDC to TDC. There are four strokes in one *four-stroke cycle* of the engine. They are called the intake stroke, compression stroke, power stroke, and exhaust stroke.

Intake Stroke.
◼ Gasoline will not burn unless it is mixed with the correct amount of air. It is very explosive when one part is mixed with about 15 parts of air.
◼ As the crankshaft turns, it pulls the rod and piston down in the cylinder. This action creates a suction known as *engine vacuum*, which draws in a mixture of air and fuel through the open intake valve (see Figure 1.7).
◼ About 10,000 gallons of air is drawn in for every 1 gallon of fuel. The air/fuel mixture is supplied by the *carburetor* or *fuel injection* system. The ideal mixture for the combined purposes of engine performance, emission control, and fuel economy is about 14.7 to 1. This is called a **stoichiometric** mixture.
◼ Most vehicles since the early 1980s have fuel injection systems with computer controls. The computer monitors the oxygen content in the vehicle's exhaust and then adjusts the fuel supply to provide the correct amount of fuel and air for each intake stroke.

Compression Stroke.
◼ The compression stroke begins at BDC after the intake stroke is completed. The intake valve closes as the piston moves up in the cylinder, compressing the air/fuel mixture (see Figure 1.7).
◼ Compressing the mixture of air and fuel into a smaller area heats it and makes it easier to burn.

Power Stroke.
◼ As the piston approaches TDC on its compression stroke, the compressed air/fuel mixture becomes very explosive (see Figure 1.7). When the ignition system produces a spark at the spark plug, the air/fuel mixture ignites.
◼ As the air/fuel mixture burns, it expands, forcing the piston to move down in the cylinder until it reaches BDC. The action of the piston turns the crankshaft to power the car.
◼ Some leakage of gases past the rings occurs during the power stroke. This leakage called **blow-by**, causes pressure in the area around the crankshaft (Figure 15.4).

Exhaust Stroke.
◼ As the piston nears BDC on the power stroke the exhaust valve opens, allowing the burned gases to escape. Because the burning gases are still expanding, they are forced out through the open valve.
◼ As the crankshaft continues to turn past BDC, the piston moves up in the cylinder, helping to force the remaining exhaust gases out through the open exhaust valve (see Figure 1.7).
◼ A few degrees after the piston passes TDC, the exhaust valve closes.

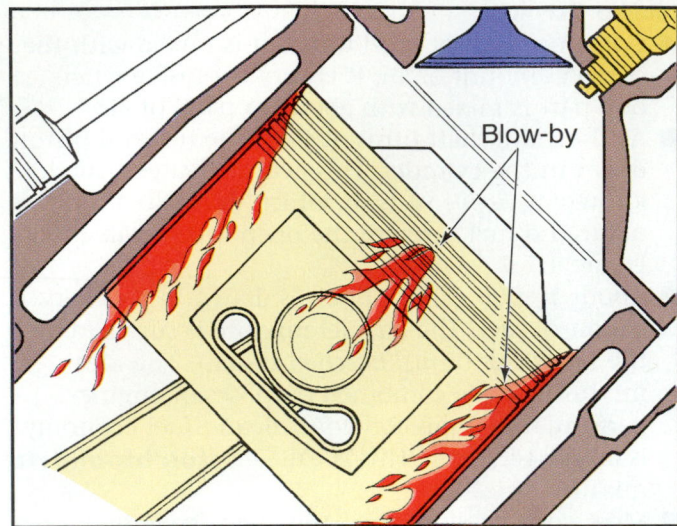

Figure 15.4 Leakage of gases past the ring is known as blow-by. *(Courtesy of General Motors Corporation, Service Technology Group)*

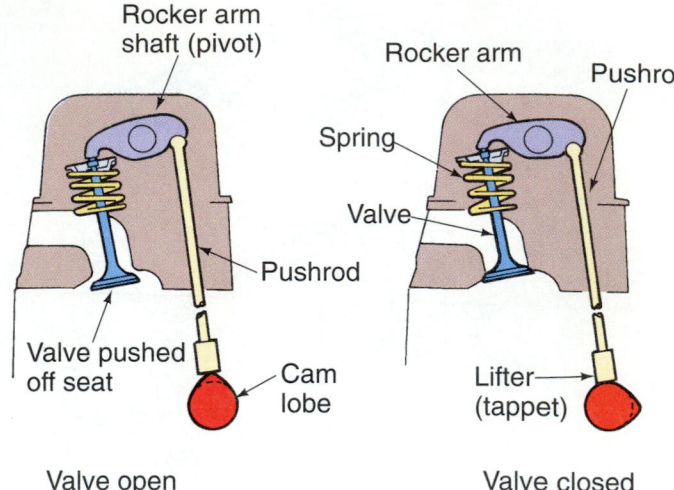

Figure 15.5 The cam lobe opens the valve. *(Courtesy of Ford Motor Company)*

The entire four-stroke cycle repeats itself, starting again as the piston moves down on the intake stroke.
■ One four-stroke cycle requires two 360° revolutions (720°) of the crankshaft.
■ While the crankshaft turns twice, the intake and exhaust valves each open once.
■ Ignition occurs once.

■ ENGINE UPPER END

Parts of the **upper end** of the engine include the cylinder head(s) and valve train.

■ VALVE TRAIN

The **valve train** includes the parts that open and close the valves. These parts include the camshaft, lifters, pushrods, rocker arms, valves, and springs. Valves are opened by the camshaft, commonly called the *"cam"*, which is considered to be the "heart" of the engine. The cam has eccentric (off-center) *lobes* that push against valve train parts, causing valves to open at precise times (Figure 15.5). When the lobe turns away from the lifter, the valve spring closes the valve. The camshaft has one cam lobe for each intake and exhaust valve, so a six-cylinder engine would have twelve cam lobes.

The cam can be located either in the block or in the cylinder head. The cam in block style is called a **pushrod engine** (see Figure 15.5). When the cam is located on the top of the head, this is called an **overhead cam engine** (Figure 15.6). Some engines have *dual overhead cams*. These have a cam for the intake valves and another for the exhaust valves.

Depending on the design of head used, some of the valve train parts are different.

On many engines the camshaft may also include another gear that drives the ignition distributor and

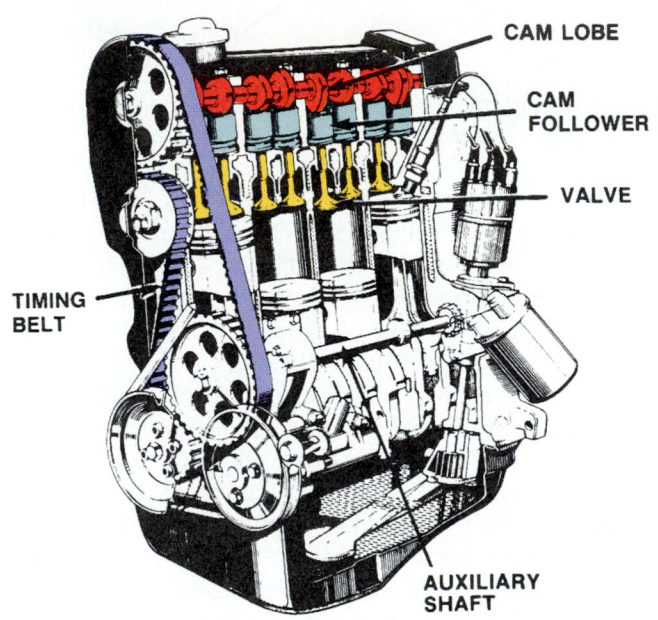

Figure 15.6 Cutaway of a four-cylinder OHC engine with the cam positioned over the valve. *(Courtesy of Chrysler Corporation)*

oil pump. The distributor is usually meshed to the oil pump so the one gear operates them both.

When an engine uses a carburetor, the camshaft also includes another eccentric lobe. This lobe drives its mechanical fuel pump. The fuel pump eccentric has a rounder shape than an ordinary cam lobe. The fuel pump rocker arm is pushed up against spring pressure by the eccentric.

On a pushrod engine, *valve lifters*, sometimes called *tappets*, fit into bores above the cam lobes. The lifters act on *pushrods* and *rocker arms* to open the valves.

Rocker arms are mounted on top of the cylinder head. They take the up and down motion of the pushrod and transfer it to the valve. As the rocker arm rocks up it pushes the valve down in the opposite direction.

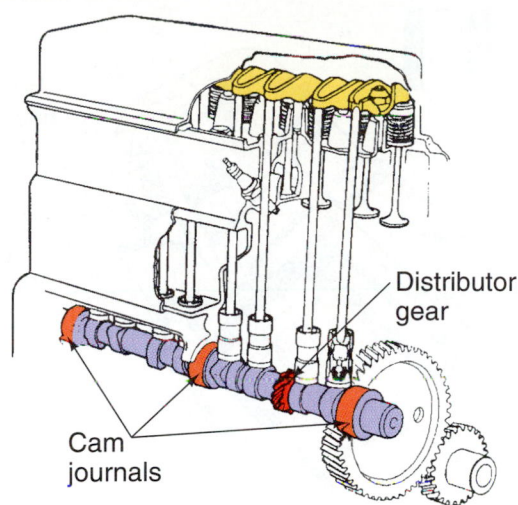

Figure 15.7 Cam journals ride inside of cam bearings to support the camshaft.

Pushrods are usually hollow tubes. One end of the pushrod fits into a socket in the lifter and the other fits into one end of the rocker arm. Overhead cam engines do not use pushrods.

Camshaft journals are machined surfaces that ride inside of cam bearings that are pressed into bores in the block on a pushrod engine (Figure 15.7). On an overhead cam engine, the cam journals ride in bores in the cylinder head.

◼ CYLINDER HEAD

The cylinder head(s) bolts to the top of the engine block, sealing off the cylinders (see Figure 15.3). The term *combustion chamber* refers to the small chambers in the cylinder heads that are located just above the pistons (Figure 15.8). The spark plugs are threaded into holes in the combustion chambers.

Valve Parts

Refer to Figure 15.9 for the locations of valve parts that are explained here. There are usually two valves per cylinder located in the combustion chambers. Some engines have three or four valves per cylinder (see Figure 16.9). The *intake valve* is the larger of the two valves because the incoming air/fuel mixture takes up a lot of space and is not under higher pressure like the exhaust (see Figure 15.8).

NOTE: *Intake and exhaust valves are made of different materials. Intake valves are usually magnetic but the exhaust valves, which are made of stainless steel, are not. Check an intake and exhaust valve with a magnet to verify this. You might also find that some exhaust valves are magnetic on the stem, but not on the valve head.*

- ◼ *Valve seats* are the machined areas under the valves. The valve and its seat fit against each other to provide a tight seal.
- ◼ The *valve port* is the area under the seat.
- ◼ The *valve face* is the machined area that the valve seat rides against.
- ◼ The *valve head* is the part of the valve that is exposed to combustion.
- ◼ The *valve margin* is located between the valve head and the valve face. If the margin of the valve is too thin (after excessive grinding), the valve can become too hot when exposed to combustion and it will burn.
- ◼ The *valve stem* is smooth and polished and fits into the valve guide.

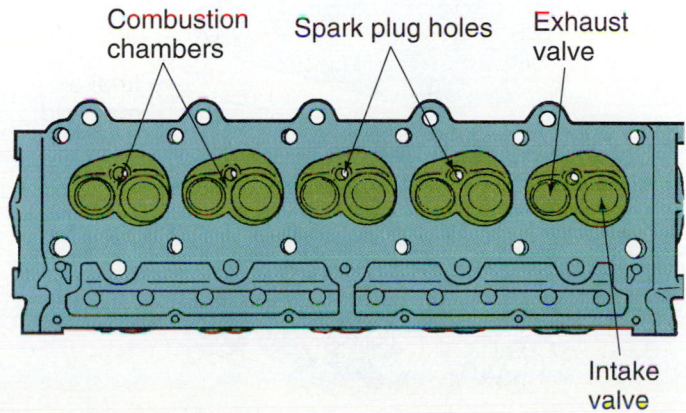

Figure 15.8 Combustion chambers on the bottom side of a cylinder head. *(Courtesy of Chrysler Corporation)*

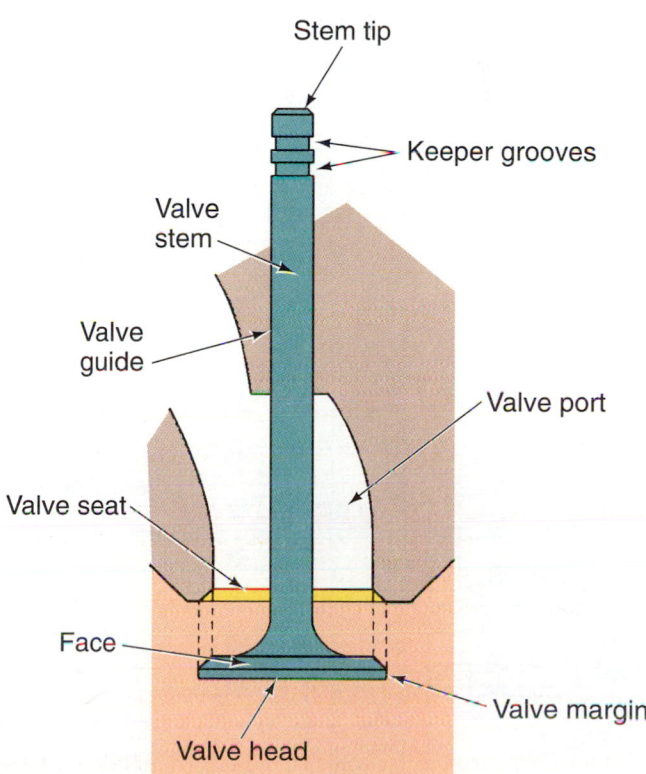

Figure 15.9 The valve face seals against the valve seat. *(Courtesy of Federal-Mogul Corporation)*

- *Keeper grooves* are at the top of the stem. *Keepers* are the small, half-moon pieces that keep the valve spring attached to the top of the valve.
- The *valve stem tip* is usually hardened on exhaust valves to protect against wear.

NOTE: *The exhaust valve has to open against the pressure of combustion. Because of this, exhaust lobes and related parts usually fail before intake parts.*

Other Valve Parts

- A *valve spring* closes the valve after the camshaft opens it.
- A *valve spring retainer* fits on top of the spring and is held to the stem tip by the keepers.
- *Valve guides* are bores in the bottom of the valve port that the valves fit into. The valves slide up and down in the valve guides (Figure 15.10).
- A *valve guide seal* is positioned on each of the valves. It keeps excessive oil from entering the valve guide. Defective valve guide seals can result in a serious oil consumption problem (see Figure 15.10).
- The *valve cover* encloses the valve spring area of the cylinder head (Figure 15.11). It is made of plastic or sheet metal and has a *valve cover gasket* sealing it to the head. Some valve covers are sealed with silicone sealant instead of a gasket.

Manifolds

The *intake manifold* is bolted to the side of a head or in between cylinder heads (Figure 15.12). The manifold provides a channel to the intake valve ports in the

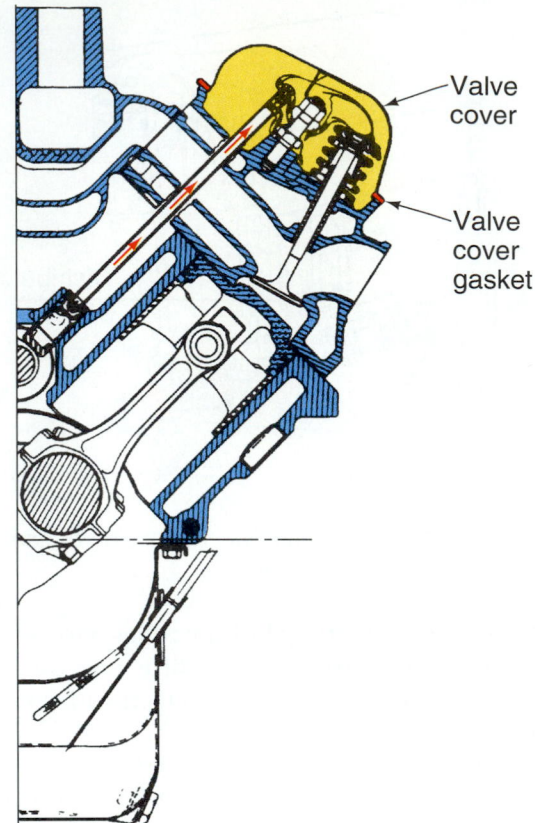

Figure 15.11 The valve cover encloses the top of the head. *[Courtesy of General Motors Corporation, Service Technology Group]*

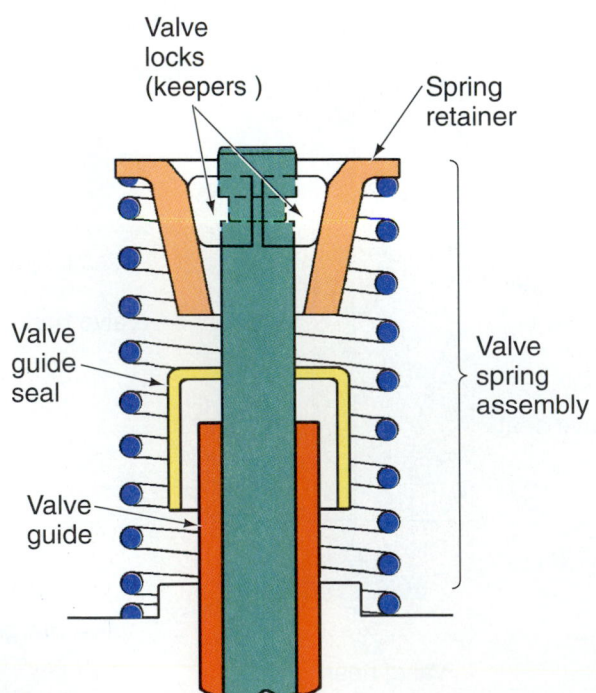

Figure 15.10 A valve guide and seal. *[Courtesy of Federal-Mogul Corporation]*

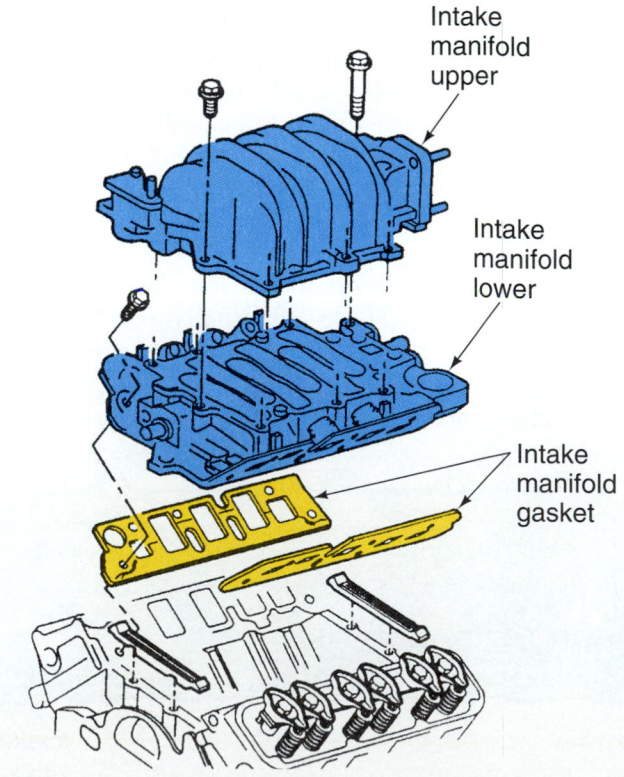

Figure 15.12 Intake manifold. *[Courtesy of General Motors Corporation, Service Technology Group]*

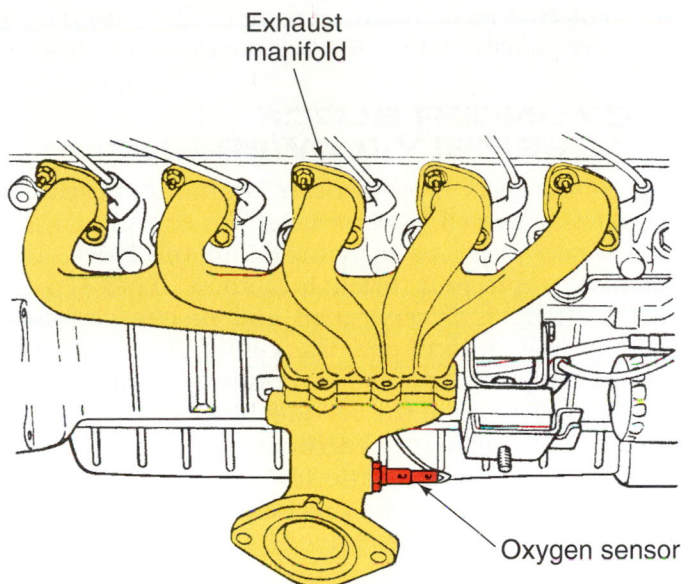

Figure 15.13 Exhaust manifold. *(Courtesy of Chrysler Corporation)*

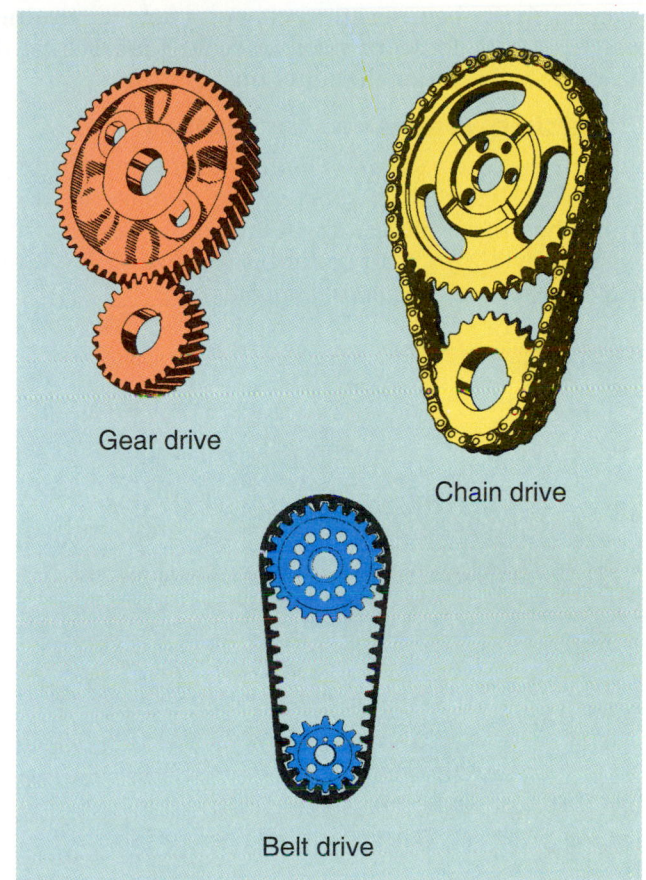

Figure 15.14 There are half as many teeth on the crank drive as on the cam drive. *(Courtesy of Federal-Mogul Corporation)*

head. A carburetor or fuel injection parts supply fuel to mix with the air that enters the **runners** of the manifold.

Exhaust manifolds are bolted to the cylinder head (Figure 15.13). They provide a channel to carry exhaust gases away from the engine. Exhaust manifolds must carry hot exhaust gases so they are usually made of cast iron. Some manifolds are constructed of steel tubing.

■ ENGINE FRONT

The front of the engine has a camshaft drive, a timing cover or front cover, and the crankshaft vibration damper or pulley.

Cam Drive

The camshaft is driven by the crankshaft by two *timing gears*, or by *sprockets* used with a *timing chain* or *timing belt* (Figure 15.14).

NOTES:

■ *The crankshaft must turn twice for every one turn of the cam. For that reason, there are half as many teeth on the crank drive as on the cam drive.*

■ *An engine running at 3000 rpm has to open and close a valve 1500 times per minute (25 times per second)! Each spark plug must also fire this frequently.*

Timing Cover

When an engine has a chain or gears, they are lubricated with engine oil. The timing cover on these engines must seal against oil leakage. In this case, it will have a full-round seal that rides on the sealing surface of the vibration damper or pulley (Figure 15.15). When the engine has a timing belt, the seal will be on

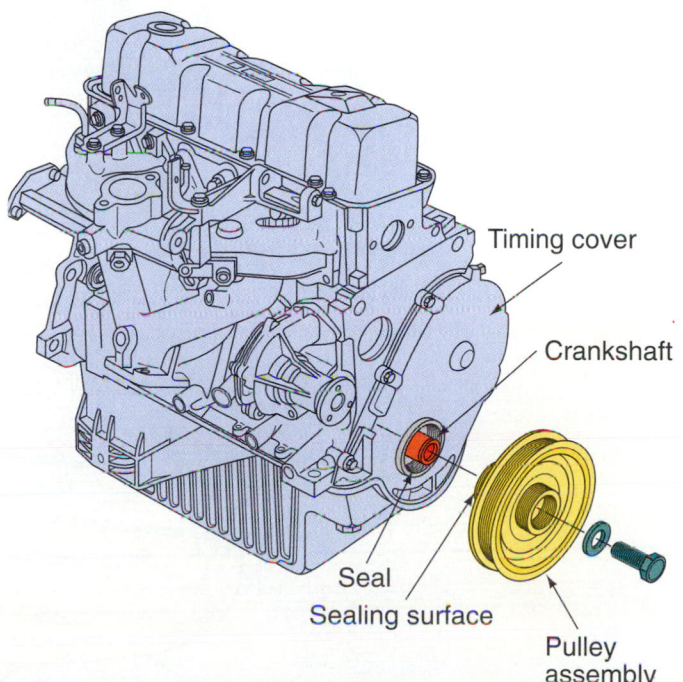

Figure 15.15 Timing cover and pulley. *(Courtesy of Ford Motor Company)*

the crankshaft, behind the belt sprockets. A timing cover for a belt does not need to seal in lubricants; it just needs to keep the elements out.

Vibration Damper or Pulley

Some engines use a *vibration damper* to minimize vibrations in the crankshaft, preventing damage to it. A damper has three pieces; an outer and an inner ring, separated by a thin strip of rubber (Figure 15.16). The damper is also called a *harmonic balancer*.

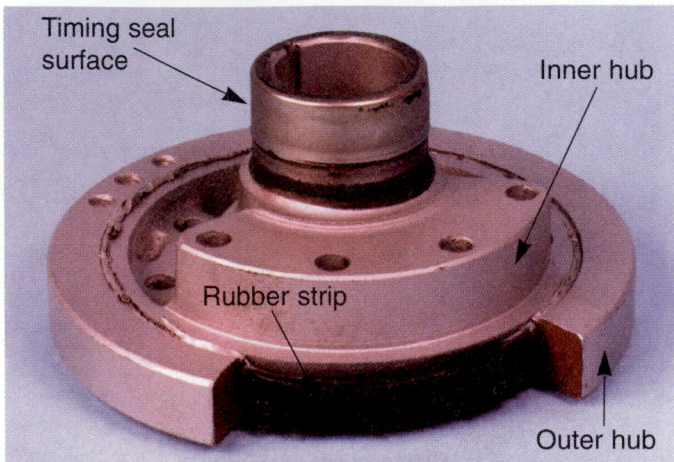

Figure 15.16 This cutaway of a damper shows the three pieces.

Some engines do not use a damper. They just have a one-piece *pulley* to which the belt pulleys can be bolted.

■ CYLINDER BLOCK ASSEMBLY (LOWER END)

The *cylinder block* is an intricate casting that includes oil galleries, as well as *water jackets* for coolant.

Cylinder bores are machined into the block. The pistons are precisely fitted into these bores (Figure 15.17). There is a very small and precise clearance between the pistons and the cylinders so that the pistons can move up and down easily without rocking.

The lower area of the cylinder block surrounded by the oil pan is called the **crankcase** because the crankshaft is located there (Figure 15.18).

Main bearing bores are precisely align-bored in the lower end of the block to provide a mounting place for the crankshaft.

Main bearing caps bolt to the bottom of the block and make up the lower half of the main bearing bores. The main bearing caps are removable, but they must be replaced in exactly the same position.

Bearing bores have *bearing inserts* installed in them to support the crankshaft (Figure 15.19). The bearing inserts are replaceable. They are in halves. One is the upper, fitting into the engine block. The other is the lower insert, fitting into the bearing cap.

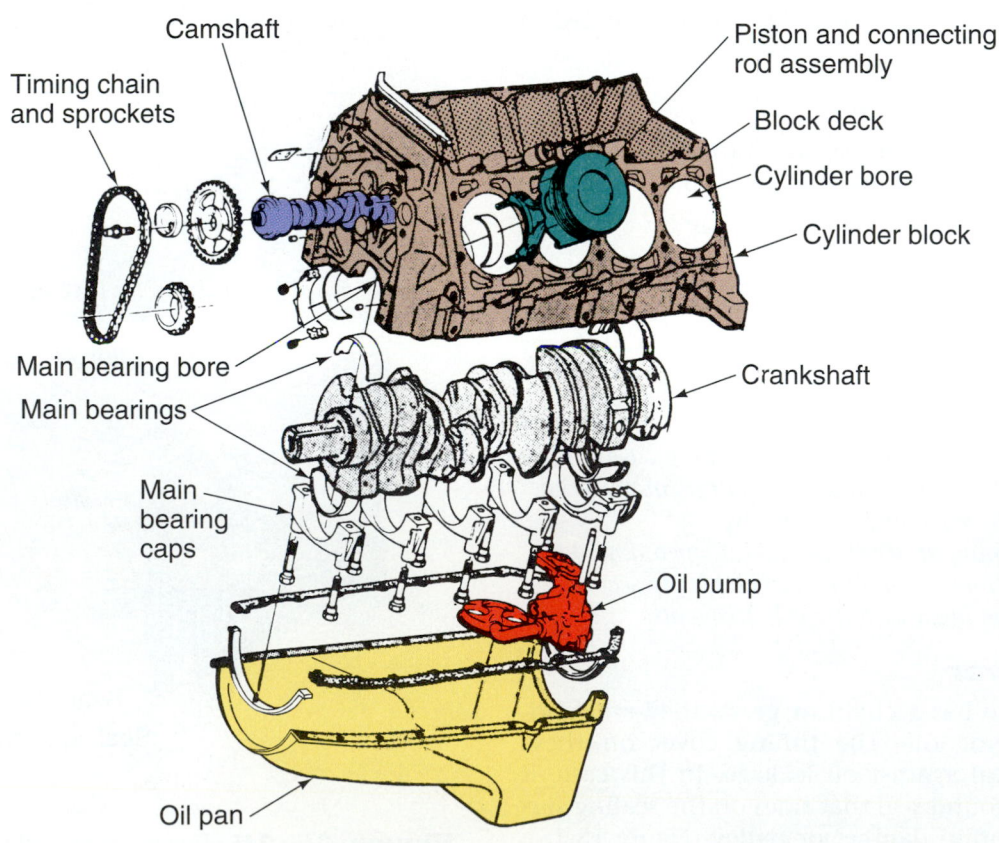

Figure 15.17 Lower end parts. *(Courtesy of General Motors Corporation, Service Technology Group)*

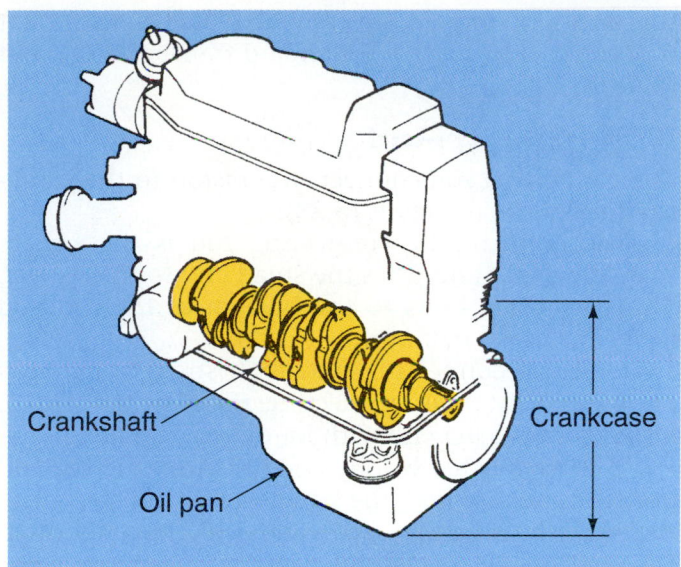

Figure 15.18 The oil pan encloses the crankcase. *(Courtesy of Federal-Mogul Corporation)*

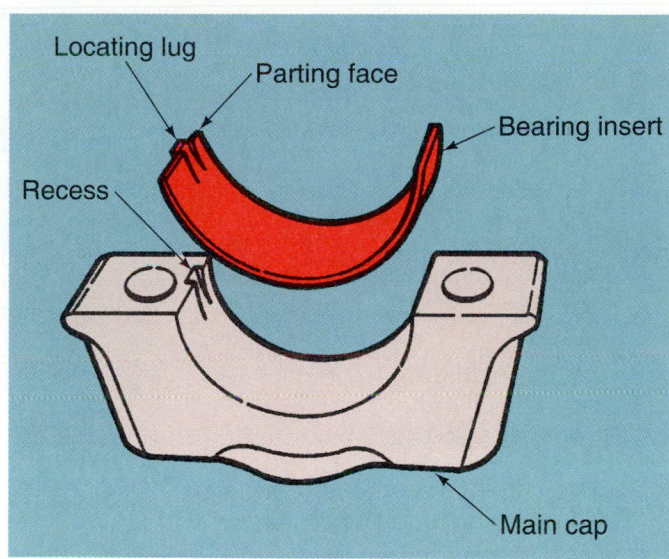

Figure 15.19 Bearing insert. *(Courtesy of Federal-Mogul Corporation)*

The block *deck* is the top surface of the block that supports the cylinder head. A *cylinder head gasket* fits between the head and the deck.

At the factory, the camshaft holes, cylinder bores, cylinder-head mounting surface, all threaded holes, and all gasket surfaces are machined automatically and in perfect alignment to each other.

■ SHORT BLOCK AND LONG BLOCK

A complete block assembly with the entire *valve train* (cylinder heads and related parts) included is called a **long block**.

The cylinder block assembly (without the heads installed) is called a **short block** (see Figure 15.17). The short block for a pushrod engine includes the

crankshaft, piston and rod assembly, camshaft and cam drive components, and all bearings. Short blocks are not usually available for OHC engines.

Crankshaft

The *crankshaft* changes the *reciprocating* (up and down) motion of the pistons to rotation. Its polished, machined bearing surfaces for the main and rod bearings are called *journals* (Figure 15.20).

The *main bearing journals* are in a row down the center of the crankshaft. The front and rear crankshaft journals and all journals in between are main journals.

One of the main bearings has *flanges* (sides) that ride on corresponding surfaces on the crankshaft to limit the fore and aft movement (**end play**) of the crankshaft (Figure 15.21). This bearing is called the *thrust main*.

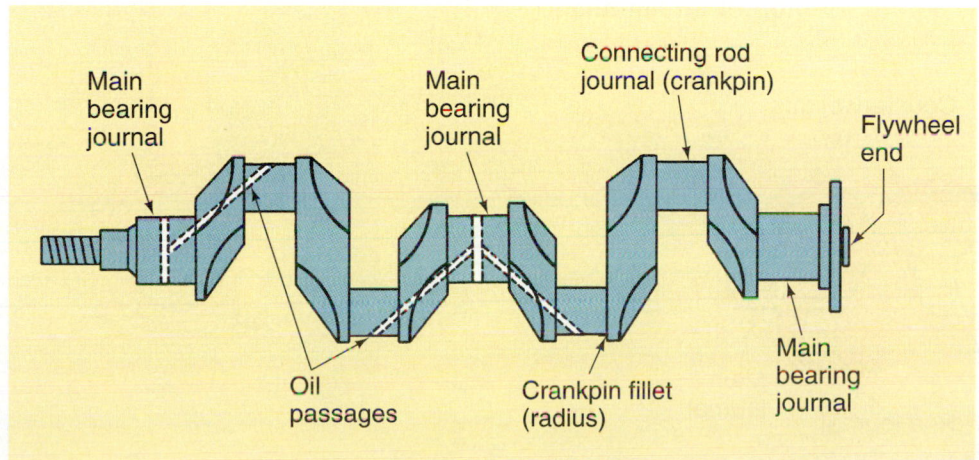

Figure 15.20 Crankshaft parts. *(Courtesy of Ford Motor Company)*

Figure 15.21 The thrust surface controls fore and aft movement of the crankshaft. *(Courtesy of Ford Motor Company)*

Rod journals are offset from the main journals. Connecting rods are bolted to the rod journals.

Oil comes to the main bearings through *oil galleries* in the block. There are oil holes in the upper bearing inserts to allow the oil to enter the bearing so that metal-to-metal contact between the bearing and the crankshaft is avoided.

Lubrication oil holes are drilled from the main bearing journals to the rod bearing journals to provide pressurized lubrication.

Bearing clearance is the name of the space between the bearing insert and the crankshaft.

Counterweights are part of the crankshaft. When the crankshaft rotates, their weight offsets the weight of the connecting rods on the opposite side (Figure 15.22).

The *snout* is the front of the crankshaft. Mounted on the snout are the crankshaft gear or sprocket and the damper or pulley.

The back of the crankshaft has a *flange* to which the flywheel is bolted. On the rear of the crankshaft, the center of the flange has a precision hole. This hole is used to align the front of the transmission with the engine. It will either locate the hub of an automatic transmission torque converter or it will hold a pilot bushing or bearing that a standard transmission input shaft fits into.

Connecting Rod

The connecting rod connects the piston to the crankshaft rod journal (Figure 15.23).

The *beam* of the connecting rod is the I-beam shaped part that attaches the small end to the *big end*.

The *rod cap* bolts to the bottom of the rod and forms the lower half of the big end.

On most connecting rods high strength *rod bolts* are pressed into the connecting rod. The rod cap is fastened to the big end with hardened *rod nuts*.

A *connecting rod bearing insert* fits into the big end of the connecting rod, the same way an insert fits into the main bearing cap (see Figure 15.19). It is fitted with bearing clearance, similar to the clearance in the main bearings.

Piston

The pressure created by the expanding gases during combustion is transferred to the *piston*. It in turn pushes on the connecting rod to turn the crankshaft. The piston also has a piston pin and piston rings (Figure 15.24).

The top of the piston is called the *head* or *crown*. The force of combustion acts on this part. Pistons on modern engines are made of aluminum, which has a melting point of about 1225°F. The temperature during combustion can approach 4500°F. The head of the piston must be thick so its surface can withstand this high temperature without melting.

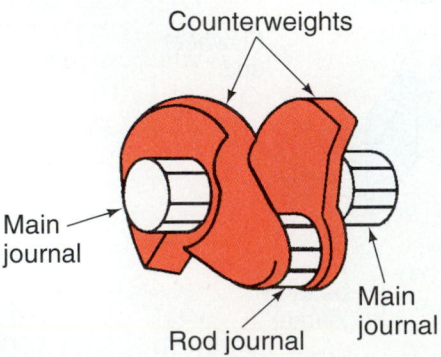

Figure 15.22 Crankshaft counterweights. *(Courtesy of Federal-Mogul Corporation)*

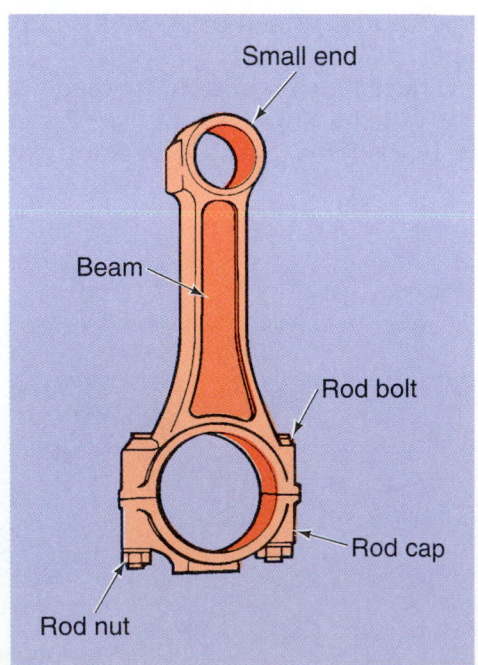

Figure 15.23 Connecting rod. *(Courtesy of Ford Motor Company)*

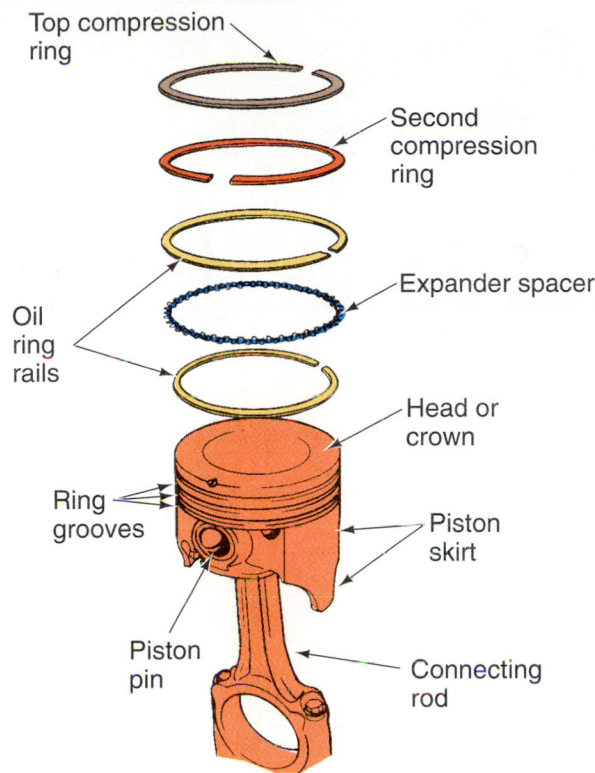

Figure 15.24 Piston and rings. *(Courtesy of General Motors Corporation, Service Technology Group)*

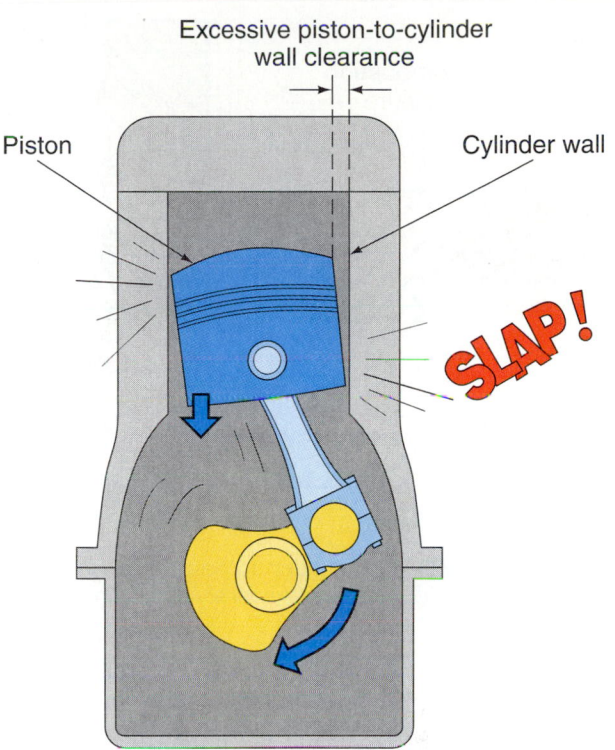

Figure 15.25 Piston slap. *(Courtesy of Ford Motor Company)*

The head of the piston is made in various shapes to help combustion and control the compression ratio of the engine (covered later).

There are three *piston ring grooves* machined into the side of the upper part of the piston (see Figure 15.24). The top two grooves are for the compression rings. The bottom groove is for the oil control ring. It has a slot, or holes, in the back of it that allows oil to return to the crankcase.

Between the piston ring grooves are *ring lands* (see Figure 19.20). This is the area of aluminum that separates the rings from each other.

Below the bottom ring on the side of the piston is the piston skirt (see Figure 15.24). The skirt glides on the cylinder wall on a thin layer of oil. There is only a small amount of *piston clearance* between the skirt and cylinder wall so that the piston cannot rock or "**slap**" (Figure 15.25). The piston is designed to allow for control of expansion after the engine warms up.

A *pin hole* is machined in each side of the piston to support the piston pin.

The *pin boss* is the reinforced area that surrounds the piston pin hole (see Figure 19.20).

Piston Pin

The *small end* of the connecting rod is attached to the piston pin or *wrist pin*, which fits into holes in the piston. Some connecting rods have a bushing for the wrist pin, but many wrist pins are pressed fit into the rod and pivot in the piston.

Piston Rings

There are three piston rings on a piston (see Figure 15.24). The top two rings are *compression rings*. They seal between the piston ring grooves and the cylinder wall, keeping combustion pressure from entering the crankcase (blow-by). The rings are spring loaded to hold them in position against the cylinder wall. During combustion, the pressure of the expanding gases forces the piston ring against the cylinder wall (Figure 15.26). This pressure, combined with the sealing action of the small amount of oil that is in the pores of the cylinder wall, ensures a very good seal.

The *ring gap* is the space where the ends of the piston ring come together after installation on the piston. The ring is constructed in an arc that is larger than the cylinder bore. When the ring is compressed, it will form a spring-loaded circle that matches the cylinder bore.

The lower ring is the *oil control ring* (see Figure 15.24). It scrapes oil that lubricates between the cylinder wall and piston skirt. Oil must be kept from entering the combustion chamber where it would be burned. There are slots in the oil ring that allow oil to reach the rear of the ring groove where it can return to the crankcase through the slot or holes (Figure 15.27). Excessive

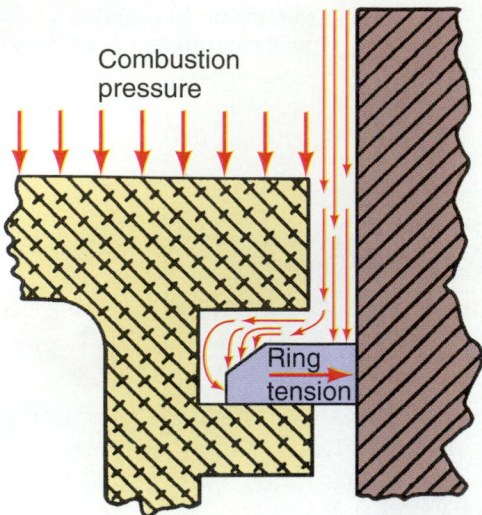

Figure 15.26 Combustion forces the ring against the cylinder wall. *[Courtesy of Federal-Mogul Corporation]*

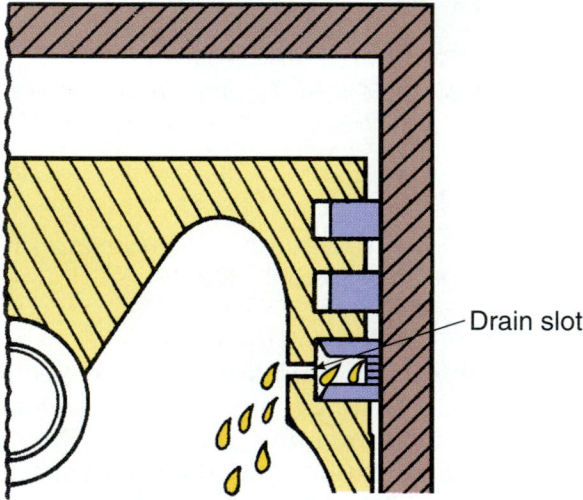

Figure 15.27 Oil is scraped off of the cylinder wall by the oil ring and returned to the crankcase through holes in the piston.

leakage past the oil ring results in oil consumption and blue smoke from the exhaust.

Oil Seals

Oil seals are installed on the front and rear of the crankshaft to keep oil in the crankcase. Seals are either one piece or two halves. The lip of the seal rides on a smooth surface on the crankshaft. The *rear main seal* is the seal at the rear of the crankshaft. The *front seal* is located in the timing cover or the front of the engine block.

Oil Pan

The oil pan is a stamped sheet metal or plastic part that encloses the crankcase (see Figure 15.18). It pro-

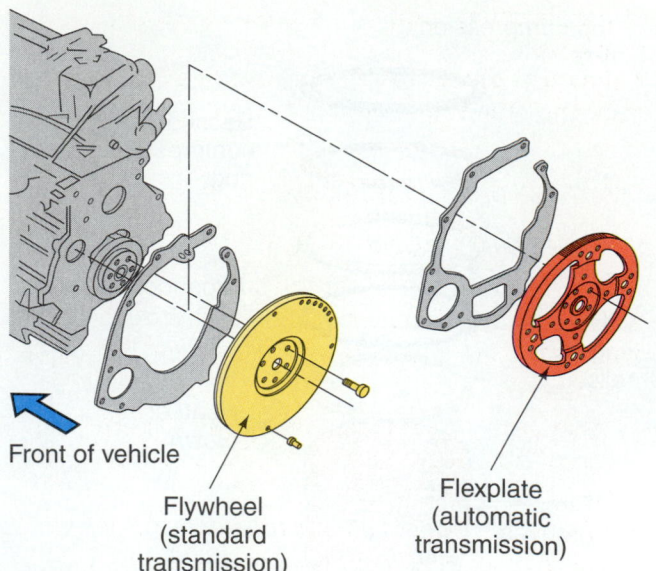

Figure 15.28 A flywheel and flexplate. *[Courtesy of Ford Motor Company]*

vides a reservoir where the engine oil is cooled as air passes across its surface.

Flywheel

A *flywheel* is used with a standard transmission (Figure 15.28). It is mounted on the rear of the engine's crankshaft. The flywheel does three things:

- The weight of the flywheel helps in carrying the crankshaft beyond BDC after the power stroke and smoothes out the power impulses of multiple cylinders.
- A ring gear mounted on the circumference of the flywheel provides a gear drive for the starter motor. Ring gears are sometimes damaged by faulty starter motors; they can be replaced easily while the engine is out of the car.
- A flywheel also provides a surface for the clutch to work upon.

When an automatic transmission is used, a torque converter and *flexplate* take the place of the flywheel and clutch (see Figure 15.28). The torque converter provides the necessary weight. The flexplate is simply a flat piece of steel, usually with a starter ring gear on its outside diameter.

■ SUMMARY

- Most of today's cars and light trucks are powered by Otto-cycle engines. During one four-stroke cycle, the intake, compression, power, and exhaust strokes are completed. This action takes 720° or two crankshaft revolutions.
- The camshaft turns once and the crankshaft turns twice during one four-stroke cycle.
- The up and down motion of the piston is changed

to usable rotary motion by the connecting rod and crankshaft.

■ A flywheel gives momentum to the crankshaft and smoothes the impulses between power strokes.

■ The camshaft and valve train control the engine's intake and exhaust flow in and out of the cylinders.

■ REVIEW QUESTIONS

1. How many times does each valve open during 720° of crankshaft rotation?

2. What does OHC mean?

3. Which valve is usually not magnetic, intake or exhaust?

4. When an engine is running at 6000 rpm, how many times does each valve have to open and close every second?

5. What is another name for the harmonic balancer?

6. What is the term for an engine block assembly that is sold without cylinder heads?

7. Beside spring pressure, what forces the piston ring against the cylinder wall?

8. If there is too much clearance between the piston and the cylinder wall, what happens?

9. What do the top two piston rings do?

10. Draw a sketch showing which direction the piston moves during each of the strokes of the four-stroke cycle. Show intake and exhaust valve positions as open or closed for each stroke (only show those strokes where a valve is open for the entire stroke).

■ ASE STYLE REVIEW QUESTIONS

1. Technician A says that the crankshaft turns 720° during one four-stroke cycle. Technician B says that the camshaft turns 360° during one four-stroke cycle. Who is right?

   a. Technician A　**b.** Technician B
   c. Both A and B　**d.** Neither A nor B

2. Technician A says that the camshaft on a typical six-cylinder engine would have six lobes. Technician B says that the cam gear has half as many teeth as the crank gear. Who is right?

   a. Technician A　**b.** Technician B
   c. Both A and B　**d.** Neither A nor B

3. Technician A says that the oil pump is sometimes driven by the camshaft. Technician B says the distributor is sometimes driven by the camshaft. Who is right?

   a. Technician A　**b.** Technician B
   c. Both A and B　**d.** Neither A nor B

4. Technician A says that the exhaust valve is larger than the intake valve. Technician B says that the intake valve must open against more pressure than the exhaust valve. Who is right?

   a. Technician A　**b.** Technician B
   c. Both A and B　**d.** Neither A nor B

5. Technician A says that the intake valve is open during part of the compression stroke. Technician B says that the exhaust valve is open during part of the compression stroke. Who is right?

   a. Technician A　**b.** Technician B
   c. Both A and B　**d.** Neither A nor B

CHAPTER 16

Engine Classifications

OBJECTIVES

Upon completion of this chapter, you should be able to:

✔ Explain various engine classifications and systems.

✔ Know the various differences in cylinder heads.

✔ Describe differences in operation between gasoline and diesel four-stroke piston engines.

✔ Explain the operation of two-stroke and Wankel rotary engines.

KEY TERMS

cylinder banks
firing order
companion cylinders
NO_x (oxides of nitrogen)
bimetal engine
electrolysis
L-head
I-Head
overhead cam engine (OHC)
single overhead cam (SOHC)/dual overhead cam (DOHC)
crossflow head
stratified charge
diesel-cycle
compression ratio
Wankel rotary engine
two-stroke engine
zero emission vehicle (ZEV)
hybrid vehicle

INTRODUCTION

A service technician needs an understanding of the basic designs and configurations of automobile engines to be able to intelligently use service manuals or communicate with customers or peers. Examples of engine-related terms found in a service manual might include reference to a 1.6L OHC V-6 engine, or the term "crankcase capacity." This chapter deals with the terms related to engines. After reading it, you should be able to look under the hood of an automobile and describe what type of engine it has.

ENGINE CLASSIFICATIONS

Piston engines all have the same basic parts, as we learned in the last chapter. But differences in design can affect how you will go about repairing them. Understand the following engine classifications as you read this chapter.

- Cylinder arrangement
- Cooling system: air or liquid
- Valve location: head or block
- Combustion chamber design
- Cam location: head or block
- Fuel used: gasoline or alternative fuel
- Ignition system: spark or compression
- Number of strokes per cycle: two or four

CYLINDER ARRANGEMENT

An automobile engine has three, four, five, six, or more cylinders. Cylinders are arranged in several ways; *inline*, in a "V" arrangement, or *opposed* to each other (Figure 16.1). Modern engines come with three, four, five, six, eight, ten, or twelve cylinders.

An inline engine has all of its cylinders lined up in one row. Some inline engines have the cylinders vertically arranged and some of them are slanted at various angles. This is so that they can fit into an engine compartment with a lower hood line for less wind resistance.

A V-type engine looks like the letter "V" when looked at from the end. V-type cylinders are cast in two rows, called the left and right **cylinder banks**. Left and right banks are identified when viewed from the flywheel end of the engine. V-8 blocks are cast with the cylinder banks separated by a 90° angle. V-6 blocks have either 60° or 90° between banks. The V-6 shown in Figure 16.2 is a 60° V.

HISTORY NOTE

Older cars had inline eight cylinder engines called "straight eights."

There are *big block* and *small block* V-8 engine designs. Smaller, lighter blocks are more popular in passenger cars because of their fuel efficiency. An *intake manifold* covers the area between the heads known as the *valley*.

The V cylinder arrangement has an advantage over the in-line arrangement when an engine has more than four cylinders. The V-type engine is considerably shorter in length, and a completely assembled V-type engine can typically be lighter in weight than an inline engine. Connecting rods from two cylinders on opposite sides of the engine will share one crankpin (Figure 16.3). Thus, the engine block requires fewer main bearing supports.

Opposed engines, sometimes called "pancake" engines, have their cylinders arranged so they face each other from opposite sides of the crankshaft. Opposed engines are especially suited for smaller areas. They have been found on such cars as Porsche and Volkswagen, Suburu, and G.M.'s Corvair of the early 1960s.

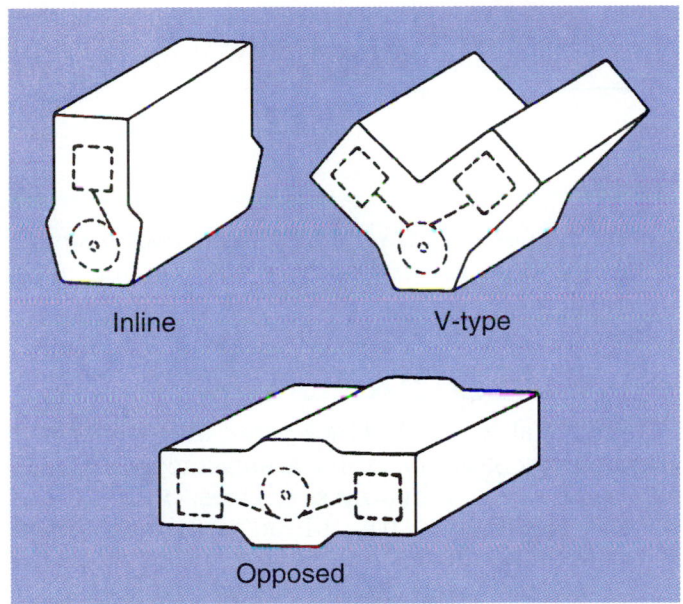

Figure 16.1 Cylinder arrangements. *(Reproduced by permission of Deere & Company, 1996. Deere & Company. All rights reserved)*

Figure 16.2 A 60° V-6 engine. *(Courtesy of Chrysler Corporation)*

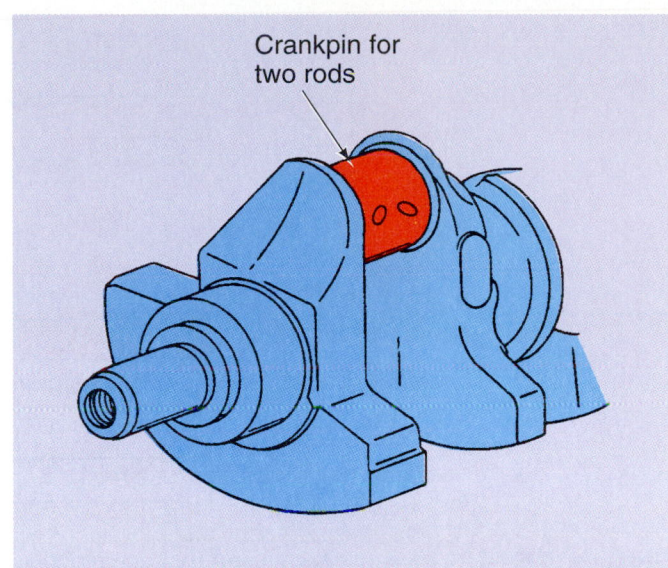

Figure 16.3 A crankpin for a V-8 engine is double wide so it can fit two connecting rods. *(Courtesy of General Motors Corporation, Service Technology Group)*

■ FIRING ORDER

To make a smooth-running engine, multiple-cylinder engines have their power strokes spaced at specified intervals. In a four-cylinder engine, one cylinder starts a power stroke at every 180° of crankshaft rotation (Figure 16.4). This interval between power strokes is known as the *ignition interval*.

Within two turns of the crankshaft, all of the engine's cylinders will have fired once. The order in which the cylinders fire is known as the **firing order**. The firing order does not usually follow the order of cylinder numbering. It is not unusual for two V-6 or V-8 engines to have different firing orders. Several different engine firing orders are shown in Figure 38.8. Sometimes the firing order is found cast into the surface of the intake manifold for easy reference.

Companion Cylinders

Any engine with an even number of cylinders will have pairs of cylinders called **companions cylinders**, or *running mates*. The pistons go up and down in pairs. When one piston is starting its power stroke, its companion piston will be at the start of its intake stroke. To find out which cylinders are companions, take the first half of the engine's firing order and place it above the second half. For a V-8 with a 1-8-4-3-6-5-7-2 firing order, put numbers 1, 8, 4, and 3 above numbers 6, 5, 7, and 2. The resulting pairs are companions.

first revolution,	1	8	4	3
second revolution,	6	5	7	2

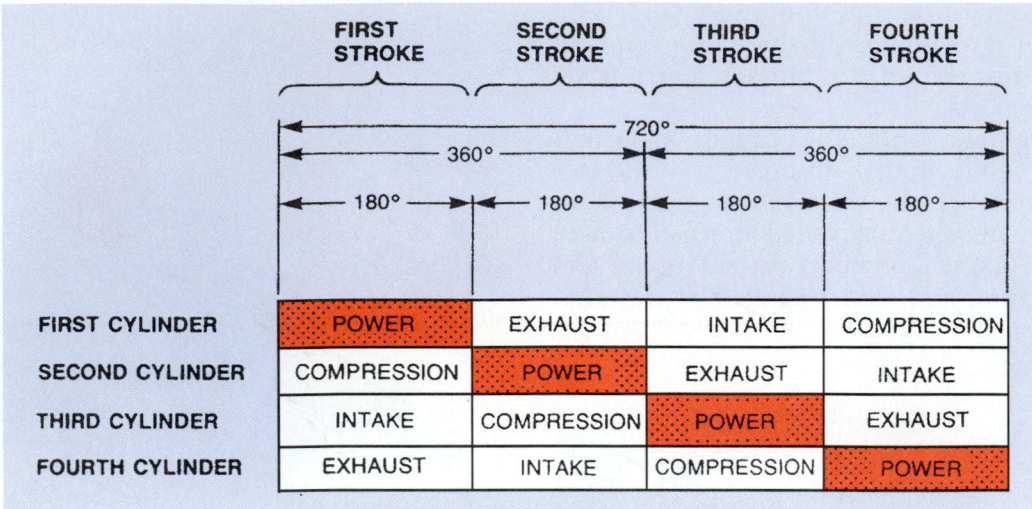

	FIRST STROKE	SECOND STROKE	THIRD STROKE	FOURTH STROKE
FIRST CYLINDER	POWER	EXHAUST	INTAKE	COMPRESSION
SECOND CYLINDER	COMPRESSION	POWER	EXHAUST	INTAKE
THIRD CYLINDER	INTAKE	COMPRESSION	POWER	EXHAUST
FOURTH CYLINDER	EXHAUST	INTAKE	COMPRESSION	POWER

Figure 16.4 A four-cylinder engine has one cylinder on a power stroke every 180° of crankshaft rotation.

Remember, the crankshaft turns two revolutions (720°) to complete one four-stroke cycle. The first half of the firing order represents one crankshaft revolution (360°). The second half of the firing order represents the second revolution of the crankshaft (360°).

In the above example, when cylinder number 7 is beginning its intake stroke, cylinder number 4 is beginning its power stroke. This 8-cylinder engine would have one power stroke every 90° of its 720° four-stroke cycle.

$$\frac{720}{8 \text{ (cyl.)}} = 90°$$

ENGINE COOLING

Engines use either of two types of cooling systems: *air cooling* or *liquid cooling*. Air-cooled engines are found on lawnmowers, motorcycles, and some automobiles. Cooling happens when air is circulated over cooling fins that are cast on the outside of engine parts. Because of higher running temperatures that cause increased **NO_X (oxides of nitrogen)** emissions (a major component in photochemical smog), production of these engines has been curtailed in recent years.

Liquid-cooled engines have *water jackets* to cool the areas around all cylinders and valve-seat areas. Coolant is pumped through the system by a coolant pump, commonly called a *water pump*. A *thermostat* controls the flow of coolant between the engine and radiator to maintain temperature at a specified level. Figure 16.5 shows parts of a cooling system.

A *coolant/anti-freeze* mixture with a concentration of about 50% water and 50% coolant provides freezing and boiling protection. The coolant is designed to prevent rust and electrolysis, which causes corrosion. In the **bimetal engine** found in most of today's cars, the combination of iron cylinder blocks and alu-minum cylinder heads is found. These two dissimilar metals promote **electrolysis**, or the creation of an electrical current. Electrolysis causes much faster deterioration of the metals.

VALVE LOCATION

Engines are also classified by where valves are located. There are two common valve arrangements for internal combustion four-stroke engines: **L-head** and **I-head** (Figure 16.6). The I-head is used in today's automobiles.

L-Head

Many automobiles until the early 1950s had L-head engines. The valve configuration resembles the letter L upside down (see Figure 16.6a). These engines are also called *flatheads* or *sidevalves* and are still used in lawnmowers, generators, and other industrial engines. The advantage of the L-head is that it is less expensive to manufacture. Its main disadvantage is that it produces more smog than other engine designs. The flathead is also limited in its compression ratio and valve lift (the height of the valve opening). Increased valve lift would require more clearance in the combustion chamber, which would lower compression.

I-Head

The overhead valve (OHV) engine found in today's cars is known as an I-head or *valve-in-head* engine (see Figure 16.6b). It breathes better than the flathead because it has a more direct path of air/fuel flow. I-heads also create less smog because they have less surface area in the combustion chamber; more surface area causes more quenched, unburned fuel. Unlike the L-head, an I-head valve job can be performed with the head on the workbench. The head may be cleaned

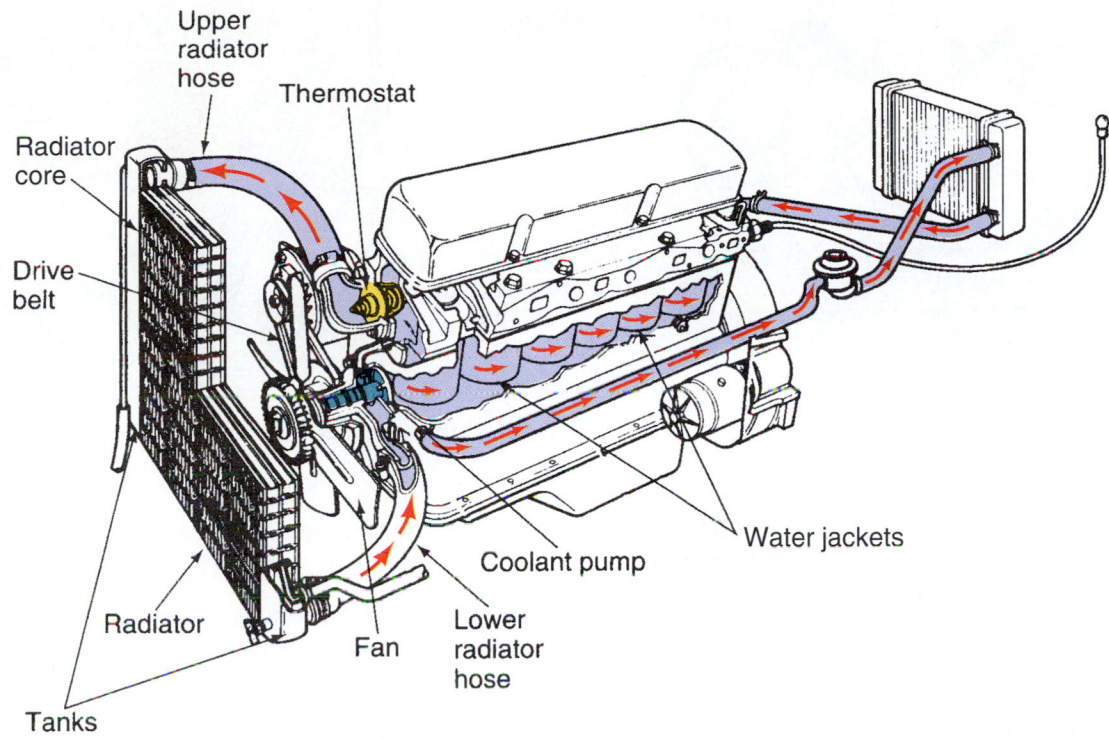

Figure 16.5 Cooling system parts and coolant flow.

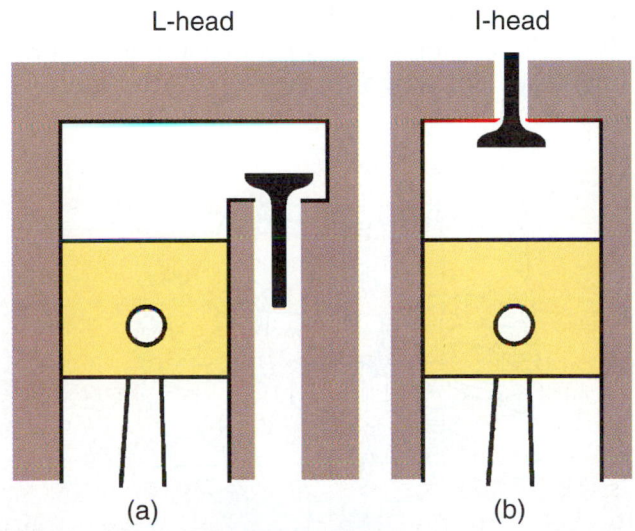

(a)   (b)

Figure 16.6 (a) An L-head, or flathead, engine has the valves in the block. (b) When the valves are in the cylinder head, the engine is known as an I-head engine. *(Reproduced by permission of Deere & Company, 1996. Deere & Company. All rights reserved)*

before it is reinstalled so that grinding grit will not enter the engine.

Higher compression is possible with the I-head because, unlike the flathead, the combustion chamber does not need extra volume to accommodate the valves. Increased valve lift is possible because, as the valve opens, the piston is moving down in the cylinder (on the intake stroke) or is already at BDC (on the

exhaust stroke). Increasing the valve opening to a certain point is necessary to allow enough air/fuel mixture into the cylinder to develop maximum power. The more air/fuel mixture that is packed into the cylinder, the more power will be developed. This is called volumetric efficiency, and it is the reason that *turbocharging* is so effective in producing extra power from relatively small engines. In turbocharged engines, a sophisticated air pump powered by exhaust gas actually forces more air/fuel mixture into the cylinder at higher engine speeds (see Chapter 30).

■ CAMSHAFT LOCATION

The camshaft on I-head engines is located in either the cylinder head or in the cylinder block. The cam-in-block engine is called a *pushrod engine* and the cam-in-head design is called an **overhead cam engine (OHC)**.

In a pushrod engine, the camshaft acts on pushrods that operate rocker arms to open the valves (see Figure 15.13). In late model vehicles, pushrods are found most often on V-type engines.

A more popular type of valve operating arrangement for in-line engines is the *overhead cam* design, or OHC (see Figure 15.15). This type of engine has the camshaft mounted on top of the cylinder head, just above the valve. The OHC is popular for high speed operation. It has the advantage of having fewer parts and less weight.

Some OHC engines have a **single overhead cam (SOHC)**. Each cylinder is provided with two separate lobes to operate the intake and exhaust valves.

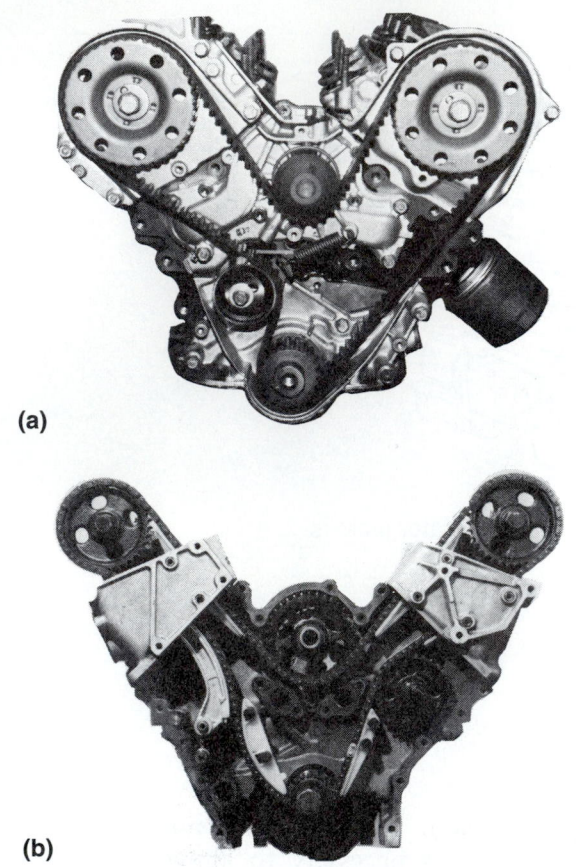

(a)

(b)

Figure 16.7 V-type overhead cam engines. (a) Belt-driven overhead cam V-6. (b) Chain-driven overhead cam V-8.

High-performance OHC engines often have two cams per head. This engine design is called **dual overhead cam (DOHC)**. One cam operates the intake valves; the other operates the exhaust valves.

To drive the cam, the OHC engine uses a long chain or belt from the crankshaft sprocket to the camshaft sprocket (Figure 16.7). Some OHC engines use an *auxiliary shaft* to drive the ignition distributor; others use the crankshaft to drive it. OHC was limited in the past to smaller, inline engines, except for its use in luxury or racing automobiles. In recent years, belt-driven OHC engines have become commonplace.

■ OTHER CYLINDER HEAD VARIATIONS

When intake and exhaust manifolds are on opposite sides on an inline engine, the head is called a **crossflow head** (Figure 16.8). This design improves engine breathing (how efficiently intake and exhaust mixtures can flow through it). Crossflow heads have a coolant passage that provides the intake manifold with heat to help vaporize the fuel.

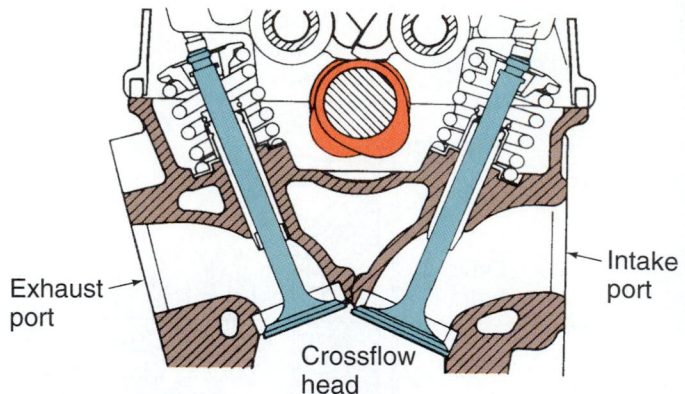

Exhaust port

Intake port

Crossflow head

Figure 16.8 Crossflow head. (*Courtesy of Chrysler Corporation*)

High-Performance Breathing Arrangements

Multiple Valve Heads. High-performance late model engines commonly use three or four valves per cylinder (Figure 16.9). Some exotic engines even have five or six valves per cylinder. The use of multiple valves has become popular due to its improved higher rpm breathing and reduced valve weight.

■ A greater amount of flow area for a given amount of valve lift is possible compared to two valve heads.

■ When an engine will be operated at high engine rpm, the light weight of valves becomes more important.

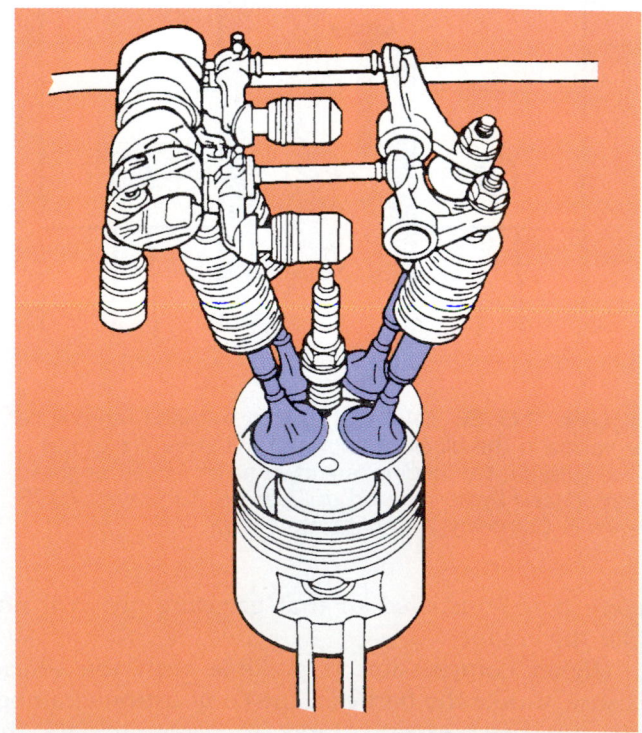

Figure 16.9 Four-valve combustion chamber. (*Courtesy of Ford Motor Company*)

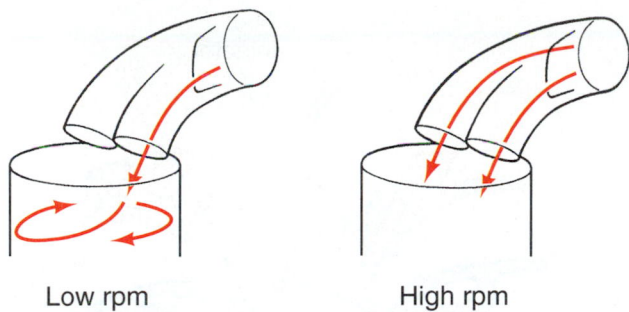

Figure 16.10 At low rpm, velocity and swirl are maintained. At high rpm, there is high flow. *[Courtesy of American Honda Motor Co., Inc.]*

■ Combining smaller combustion chambers (because of multivalves) with a more central spark plug location has decreased the chances for an engine to knock. This allows the use of higher compression ratios, and delivers more power.

To burn very lean air/fuel mixtures, the fuel must be mixed well. At high speeds, there is plenty of turbulence so this is not a problem. But multivalve heads tend to allow fuel to fall out of the mixture at low speeds. Some multivalve heads use control valves that cause only one intake valve to open at low rpm and another to open at higher rpm. This helps maintain velocity and swirl at low speed and high flow at high speed (Figure 16.10).

Other multivalve heads use two intake manifold runners per cylinder. These manifolds are variably tuned using a butterfly control valve to control air flow.

■ COMBUSTION CHAMBER DESIGNS

The principal combustion chamber designs are the *hemi*, the *wedge* (Figure 16.11), the *pent-roof*, and the **stratified charge** (Honda's CVCC) (Figure 16.12) combustion chambers. There are also newer combustion chamber designs with "D" or heart shapes.

The wedge is the most common combustion chamber design. It has a squish/quench area that causes

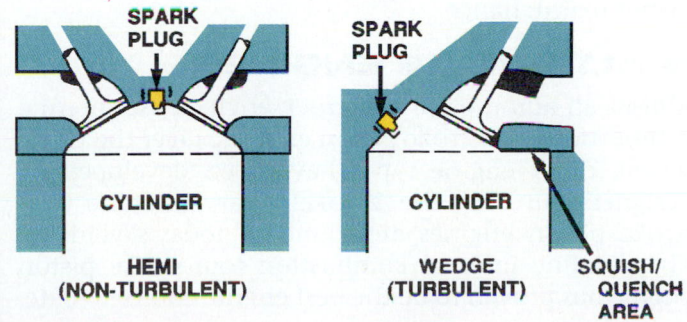

Figure 16.11 Hemi and wedge combustion chambers.

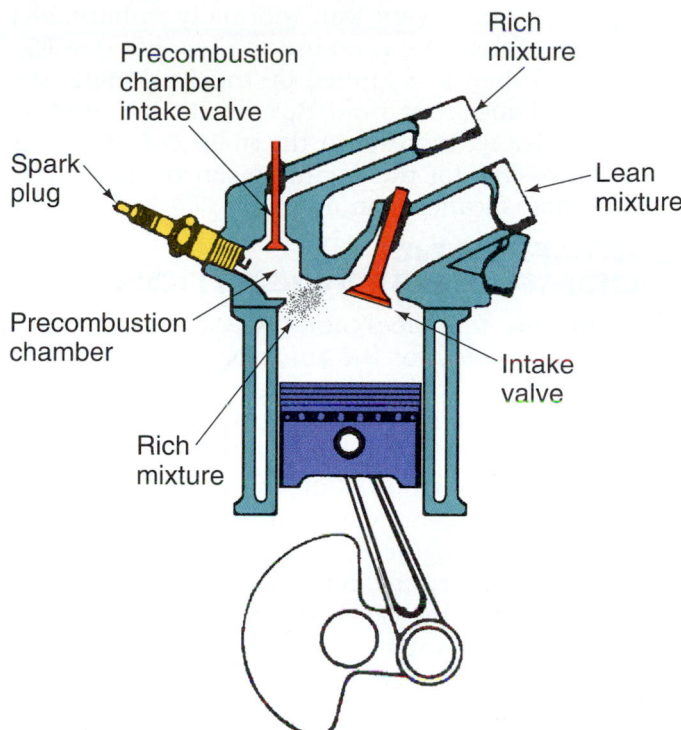

Figure 16.12 Stratified charge engine. *[Courtesy of American Honda Motor Co., Inc.]*

movement (*turbulence*) of the air/fuel mixture and cooling of the gases to prevent abnormal combustion. This movement causes more complete burning at lower speeds with less chance of detonation.

There are *turbulent* and *nonturbulent* combustion chambers. Turbulent combustion chambers, like the wedge, can cause air and fuel to separate from each other at high speeds. A nonturbulent combustion chamber, the hemispherical (hemi) design, is more efficient for high-speed use. Because the mixture is centered near the spark plug, the flame spreads evenly. A hemi chamber also allows the use of bigger valves. Sometimes hemis have a tendency to "spark knock" when using lower octane fuels (see Chapter 25).

Diesel engines have no chamber in the cylinder head itself. The combustion chamber side of the head is virtually flat. Turbulence and squish in the cylinder are controlled by the shape of the piston head.

A *pent-roof* combustion chamber is shaped like a "V." This design is popular for use with *four-valve per cylinder* designs. The pent-roof and other newer designs are designed for more efficient combustion and better emission control. In a *high swirl chamber*, like in the wedge chamber, areas on the head surface are raised to cause a planned turbulence of the air/fuel mixture.

The stratified charge design is the type pioneered by Honda. The name comes from the stratification, or layering of different densities of air/fuel mixtures. Honda and Mitsubishi use a very small amount of rich mixture to ignite a very lean (normally unburnable) mixture in a small pre-combustion chamber (see Figure 16.12). When it is ignited by the spark plug, the advancing flame front from this small, rich mixture ignites the leaner mixture in the main cylinder. This makes it possible for the engine to run on an air/fuel mixture that is leaner than normal.

■ SPARK AND COMPRESSION IGNITION

Although this text does not deal specifically with diesel engines, most of the automobile engine information included here applies to light-duty diesel engines found in some passenger cars and light trucks. **Diesel-cycle** and four-cycle gasoline engines share the same basic principles of operation. The difference is in the way the fuels are ignited.

The gasoline engine is called a *spark ignition (S.I.)* engine. A spark is created in the ignition system. A distributor, geared to the camshaft, times and distributes the spark to the spark plug at exactly the correct instant. Late model engines have computer-controlled spark ignition. Many new engines have *distributorless ignition systems (D.I.S.)* with ignition coils that are triggered by the computer in response to a signal from a camshaft or crankshaft sensor.

Diesel Engine

The diesel engine was invented by Rudolph Diesel in 1892 in Germany. Diesel engines, which can be either two- or four-stroke cycle, are used extensively in heavy equipment and were not used in automobiles until the 1930s. The operation and appearance of the diesel engine is very similar to the gasoline engine.

A diesel is a *compression ignition (C.I.)* engine. It does *not* use a spark to ignite the fuel. Basically, when air is compressed and fuel is injected into it, the fuel ignites.

Compression ratio is the comparison between the volume of the cylinder and combustion chamber when the piston is at TDC and BDC (see Figure 17.4). Diesel compression ratios can be in the neighborhood of 20:1. Gasoline engine compression ratios are usually from 8:1 to 10:1.

When air is compressed, it heats up. Because an air/fuel mixture explodes if it is compressed too much, the diesel engine compresses only air. Diesel fuel does not burn at room temperature. But when it is injected into the cylinder at the exact moment ignition is desired, it burns easily in the hot environment of the compressed air (approximately 1000°F).

In place of the distributor that is used with a carburetor or electronic fuel injection system, a diesel uses either mechanical injectors operated by a camshaft or a precision fuel distributor and individual injectors (Figure 16.13). With either type of injector the pressure of the fuel must be very high in order to

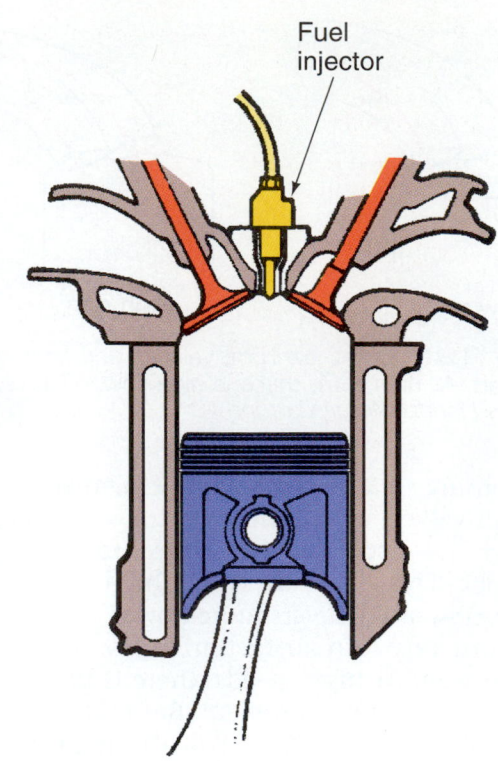

Fuel injector

Figure 16.13 A diesel engine has a timed high-pressure fuel injector to control the point of ignition. *(Courtesy of Federal-Mogul Corporation)*

overcome the pressures in the cylinders that are reached during the compression stroke.

The diesel can run at very lean mixtures at idle, and is generally about one-third more efficient on fuel, although it produces less power than a gas engine. In gasoline engines, the amount of air is changed to control speed and power. In the diesel, the amount of air remains the same while the fuel mixture is changed to control speed and power. The mixture can be as rich as 20:1 under load and as lean as about 80:1 at idle.

Problems with the diesel are its high particulate emissions (soot), and the high temperature of combustion, which produce high levels of NO_x emissions. Diesels also have starting problems in cold weather, and require more frequent oil changes and other owner maintenance.

■ ALTERNATE ENGINES

Almost all automotive and truck engines use internal combustion four-stroke piston engines. Over the years, several other engine types have been developed by designers but only the Wankel rotary and the two-stroke piston engines are found in today's vehicles. The gasoline internal combustion four-stroke piston engine has proven to be the best engine choice to date.

The Wankel Rotary Engine

The **Wankel** engine is also known as the **rotary engine**. It operates on the four-stroke cycle, although there are actually no strokes. Automotive rotary engines have two rotors that rotate inside of a chamber that looks like a modified figure eight (Figure 16.14). The rotor has three sides that act as pistons. While one of the chambers is experiencing intake, the others will be doing other parts of the cycle. Thus, one revolution of the crankshaft produces the equivalent of three power strokes.

As the rotor turns, the end of one of the lobes moves past the intake port drawing in fuel and air. Turning further, the mixture is compressed as it nears the spark plug. The spark plug ignites the air/fuel mixture and the rotor continues revolving until the exhaust port is uncovered. When the exhaust has escaped, the rotor is in position above the intake port to begin the cycle again.

Rotary engines do not have pistons that have to start and stop moving hundreds of times per second at high rpm like reciprocating engines do. There are no poppet valves to open and close either. This means that these engines run very smoothly at higher rpm.

Rotary engines require complicated emission control systems. The result is that they are not as fuel efficient as they could be. The engine has been in limited use in Mazdas. If readily available alternative clean-burning fuels become a reality, the rotary engine could become a popular choice.

Two-Stroke Cycle

Some new automobile engines will use the two-stroke cycle. A **two-stroke engine** can be made smaller and lighter than a four-stroke engine of comparable size. Two-strokes, used for years in outboards, chainsaws, and motorcycles, use a mixture of oil and gasoline for lubrication of the crank, rod, and piston. Some of the new designs for automobiles have crankcases lubricated with pumped oil.

A two-stroke engine has a power stroke every crankshaft revolution. The two-stroke cycle begins with the piston at TDC on the power stroke. The cylinder has intake and exhaust ports, which are openings in the side of the cylinder (Figure 16.15). As the piston reaches the bottom of the power stroke, the exhaust port is opened to release exhaust gases. Shortly after the exhaust port opens, the intake port opens and the air/fuel mixture is pushed into the cylinder. This action also helps to push the exhaust out. As the piston moves up on its compression stroke, both the intake and exhaust ports are covered.

Older two-stroke engines used a mixture of oil and fuel. This mixture lubricated the *lower end* (crankshaft and bearings) as it flowed through the crankcase on its way to the cylinder. New direct injection two strokes use fuel injectors to put fuel into the combustion chamber. Air is pushed into the cylinder using a supercharger (see Chapter 28). The crankcase is pressure lubricated in these engines just like in four-stroke engines. Some diesel truck engine designs also use a two-stroke cycle.

Electric Motors

An electric motor, or multiple motors can be used to power a vehicle. The U.S. Environmental Protection Agency is very interested in clean air and electric vehicles are called **zero emission vehicles (ZEVs)**. Manufacturers are working on solving some of the problems related to electric vehicles.

Technology has not yet reached the point where several concerns would be met. The distance that the vehicle can travel before a recharge makes electric vehicles viable for short distance commutes. It takes several hours to recharge batteries after they are run

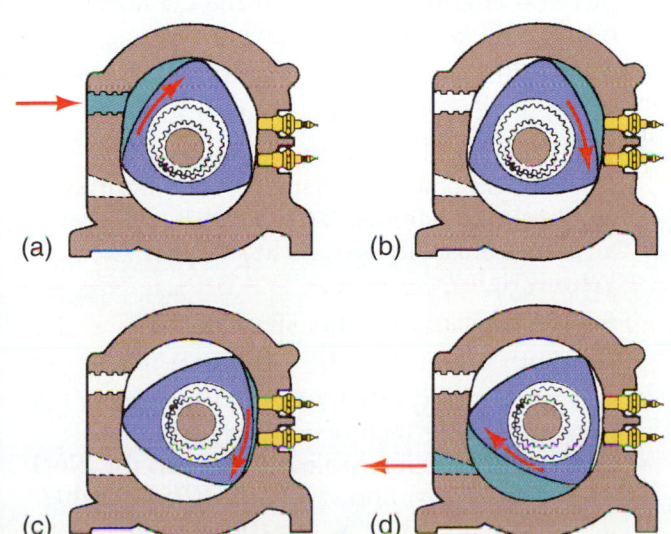

Figure 16.14 Rotary engine cycle.

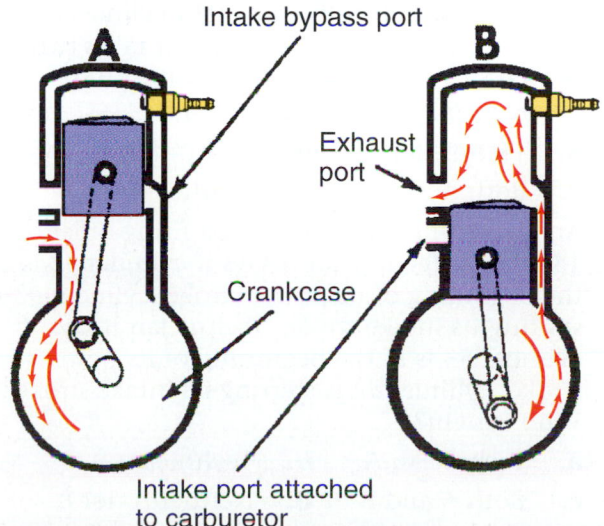

Figure 16.15 A two-stroke cycle engine. *(Courtesy of Sun Electric Corporation)*

down. The vehicle must carry many heavy batteries. If there is an accident, the acid in the batteries becomes a safety factor. There is no free heat for a heater like in a liquid-cooled engine. Also, the initial cost of electric vehicles will be high until the problems are solved and they are mass produced.

Hybrid vehicles are those that use more than one power source. An electric vehicle might have a small gasoline engine that is used with it. There are many hybrid vehicle ideas, including flywheel and hydraulic pressure energy storage.

Many types of vehicles have been invented. But whether they come to production depends on many things. The availability and distribution of fuel, cost of the vehicle, customer satisfaction, driveability and performance, and fuel economy are all considerations.

Alternative fuel vehicles are used extensively in fleet operations such as utility companies. Specialized training in these technologies is available.

■ SUMMARY

- Piston engines share common parts but there are many different design variations.
- Cylinders are arranged inline, in a V, or opposed to each other. The most popular automotive engines have either four, six, or eight cylinders.
- Cylinders rows, called banks, are determined from the flywheel end of the engine.
- The crankshaft turns two revolutions (720°) to complete one four-stroke cycle. The first half of the firing order represents one crankshaft revolution (360°). The second half of the firing order represents the second revolution of the crankshaft (360°).
- Engines use either liquid or air cooling.
- The two camshaft designs are pushrod and overhead.
- An engine operating at 3000 rpm on the freeway has to open and close each valve 25 times per second.
- Alternatives to the four-stroke piston engine are not yet viable.

■ REVIEW QUESTIONS

1. What are three ways that cylinders are arranged in a block?
2. What are the rows of cylinders in a V-type block called?
3. What is the name of the area between the heads on a V-type block?
4. List two firing orders for a V-8 and for a V-6 (see Figure 38.8).
5. If a four-cylinder engine has a firing order that is 1342, which cylinder is on its exhaust stroke when cylinder #2 is on its intake stroke?
6. What is the name of the design that uses a small amount of rich air/fuel mixture to ignite a leaner mixture?
7. What kind of engine has a combustion chamber that is flat?
8. What is the name of the engine design that is called compression ignition?
9. What kind of engine is a Wankel, reciprocating or rotary?
10. What kind of engine is often found in chainsaws and outboard motors?

■ ASE STYLE REVIEW QUESTIONS

1. Technician A says that it takes two revolutions of the crankshaft to fire all eight cylinders on a V-8. Technician B says that it takes two revolutions of the crankshaft to fire all six cylinders on a V-6. Who is right?
   a. Technician A b. Technician B
   c. Both A and B d. Neither A nor B
2. An eight-cylinder engine has a firing order of 18436572. Technician A says if cylinder #2 is at the beginning of its power stroke, cylinder #3 is starting its intake stroke. Technician B says if cylinder #3 is at the beginning of its power stroke, cylinder #2 is starting its intake stroke. Who is right?
   a. Technician A b. Technician B
   c. Both A and B d. Neither A nor B
3. Technician A says that the valve arrangement found in modern passenger cars is the I-head.

Technician B says that the camshaft in a pushrod engine is located on the cylinder head. Who is right?
   a. Technician A b. Technician B
   c. Both A and B d. Neither A nor B
4. Technician A says that the hemispherical combustion chamber shape is called a turbulent combustion chamber. Technician B says that an inline-6 usually weighs less than a V-6. Who is right?
   a. Technician A b. Technician B
   c. Both A and B d. Neither A nor B
5. Technician A says that a V-8 DOHC engine has four camshafts. Technician B says that two pistons that are next to each other in the block are called companion cylinders. Who is right?
   a. Technician A b. Technician B
   c. Both A and B d. Neither A nor B

CHAPTER 17

Engine Size and Measurements

■ OBJECTIVES

Upon completion of this chapter, you should be able to:

✔ Describe various ways of measuring engine size.

✔ Understand the effects of engine compression ratio.

✔ Explain the principles of engine power and efficiency.

✔ Relate torque to horsepower.

✔ Understand the operation of various types of dynamometers.

■ KEY TERMS

bore
piston stroke
crank throw
oversquare
undersquare
displacement
engine displacement
swept volume
compression ratio
compression pressure
force
work
foot-pounds
watts
joules
energy
inertia
momentum
power
British thermal units (Btu)
horsepower
dynamometer
prony brake
road horsepower
mechanical efficiency
thermal efficiency
brake mean effective
  pressure (BMEP)

■ INTRODUCTION

This chapter provides an understanding of various engine size and performance measurements. Methods of understanding and measuring engine power output are also discussed.

■ ENGINE SIZE MEASUREMENTS

Engines have different numbers of cylinders and different sizes of pistons and cylinders. An engine's size is determined by the volume of air that its pistons displace in the cylinders. You will need to know an engine's size to be able to order parts for it. You will also need to know an engine's size to be able to look up wear specifications for it. A sample general engine specifications chart is shown in Figure 17.1.

Cylinder Bore and Stroke

Cylinder **bores** in automobile engines usually range from about 3.5" to 4". The bore size is the diameter of the cylinder and is measured across the cylinder (Figure 17.2).

The **piston stroke** is the distance that the piston moves from TDC to BDC (see Figure 17.2). The stroke is controlled by the length of the throw of the rod journal (Figure 17.3). The length of the **crank throw** is one-half of the total stroke.

Specifications
GENERAL ENGINE SPECIFICATIONS

Year	Engine, Liter/ CID①	VIN Code ②	Fuel System	Bore & Stroke, Inch (Millimeters)	Compression Ratio	Horse- power @ RPM	Torque Ft. Lbs. @ RPM	Normal Oil Pressure. psi③
DODGE & PLYMOUTH								
1996	2.4L/4-148	B	SFI	3.45 x 3.98 (87.5 x 101)	9.4	150 @ 5200	167 @ 4000	25-80
	3.0L/V6-181	3	SFI	3.59 x 2.99 (91.1 x 76)	8.85	150 @ 5200	176 @ 4000	35-75
CHRYSLER, DODGE & PLYMOUTH								
1996	3.3L/V6-201	R	SFI	3.66 x 3.19 (93 x 81)	8.9	162 @ 4800	194 @ 3600	30-80
	3.8L/V6-231	L	SFI	3.78 x 3.43 (96 x 87)	8.9	162 @ 4400	213 @ 3600	30-80

① — CID-Cubic Inch Displacement
② — The eighth digit of the VIN denotes engine code.
③ — At 3000 RPM.

Figure 17.1 A service manual specifications chart. *(Courtesy of Hearst Motors Publications)*

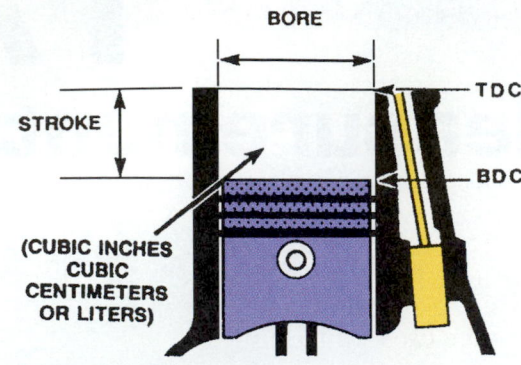

Bore² x stroke x 0.7854 x number of cylinders

Figure 17.2 Cylinder terms.

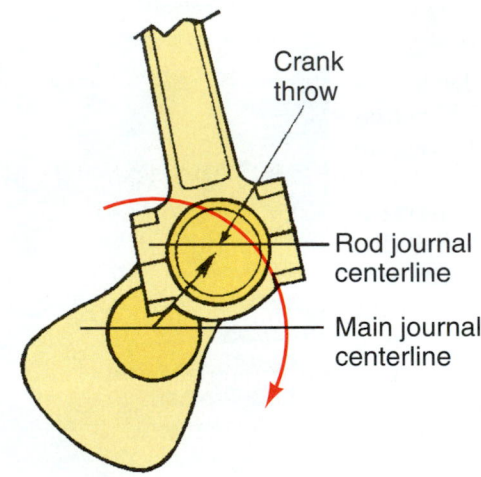

Figure 17.3 Crank throw.

The average engine has a stroke of from 3.5" to 4". In the service manual, bore and stroke are listed together. The bore is the first measurement and the stroke comes last. For instance, 4.000" × 3.500" would mean a 4" bore and 3½" stroke (see Figure 17.1).

When an engine has a cylinder bore that is larger than its stroke, this is called an **oversquare** engine. When the bore is less than the stroke, it is an **undersquare** engine.

Characteristics of oversquare engines (short stroke):
- Faster revving
- Lack low speed torque
  Characteristics of undersquare engines (long stroke):
- Slower revving
- Good low speed torque (make good engines for work trucks).

Displacement

Piston or cylinder **displacement** refers to the volume that the piston displaces in the cylinder as it moves from TDC to BDC (see Figure 17.2). It is measured in cubic inches, cubic centimeters, or liters.

The easiest formula to use to determine a cylinder's displacement is:

$$\text{Bore}^2 \times \text{Stroke} \times 0.7854$$

Engine displacement is determined by multiplying the cylinder displacement by the number of cylinders.

The engine's total displacement is determined by multiplying:

$$\text{Bore}^2 \times \text{Stroke} \times 0.7854 \times \text{number of cylinders}$$

The larger the displacement, the larger the engine.

SHOP TIP Appendix D is a chart showing the displacements of various engines. As an experiment, add 0.062" to the bore size of your engine. This is approximately the amount that the displacement of your engine would increase by if the cylinders were rebored to maximum oversize for new pistons.

In late model vehicles, volume is described in liters or cubic centimeters (converting between metric system measurements and English system measurements is covered in Chapter 5).

Another means of determining displacement is to multiply the **swept volume** of the cylinder by the number of cylinders. Swept volume is determined by multiplying:

$$\frac{\text{Bore}}{2} \times \frac{\text{Bore}}{2} \times \text{Stroke} \times \pi\,(3.14) = \text{Swept Volume}$$

or

$$\frac{(\pi \times B\,(\text{squared}) \times \text{stroke}) \times \#cyl}{4}$$

■ COMPRESSION RATIO

The compression ratio determines how much the air and fuel are compressed on the compression stroke. As the piston moves from BDC to TDC in a gasoline engine, the mixture is compressed to about ⅛ of the volume it occupied when the piston was at BDC. In this case, the **compression ratio** is said to be 8:1 (Figure 17.4). This means the maximum volume of the cylinder is eight times the cylinder's minimum volume. If the mixture was compressed to 1/12 its original volume, the compression ratio would then be 12:1. In a diesel engine, the compression ratio is much higher (from about 17:1 to over 20:1).

Gasoline engines of the "muscle car era" (approximately 1958–1970) had higher compression ratios, sometimes as high as 12:1. A higher compression ratio can increase an engine's power and fuel economy. Older cars with high compression engines used leaded gasoline that had a higher octane rating. Higher

8 to 1

Figure 17.4 Compression ratio. The mixture is compressed to about ⅛ of its original volume. *(Courtesy of Ford Motor Company)*

compression engines also produce more exhaust emissions, so today's engines have lower compression ratios (about 8:1 to 9:1).

Compression Ratio and Engine Power

During combustion the potential energy of the air/fuel mixture is turned into thermal (heat) energy and kinetic energy. Compression ratio has an effect on the amount of power an engine can produce by increasing the thermal efficiency of the engine. Squeezing the mixture into a smaller space results in higher combustion pressure and more expansion of the mixture throughout the power stroke. More of the heat energy of the fuel is converted to work, as less heat is allowed to escape from the engine.

In theory, each point of change is worth about 4% change in horsepower, proportional throughout the engine operating range.

■ **MATH NOTE** ■

Calculating an Engine's Compression Ratio

To calculate an engine's compression ratio, add together:
■ *The volume of the cylinder*
■ *Piston deck height volume*
■ *Piston volume*
■ *Compressed head gasket volume*
■ *Combustion chamber volume*
　　Total #1＿＿＿＿＿
The sum of these is divided by the sum of:
■ *The piston deck height volume*
■ *Cylinder volume*
■ *Compressed head gasket volume*
■ *Combustion chamber volume*
　　Total #2＿＿＿＿＿
Total #1 divided by Total #2 = Compression Ratio

Compression Pressure

Compression pressure is the amount of pressure made by the piston moving up in the cylinder in pounds per square inch (psi) or kilopascals (kPa). Gasoline engines typically produce compression pressure of 125–175 psi. The pressure is measured by installing a compression tester into the spark plug hole and cranking the engine through several revolutions (see Chapter 42). If compression pressure is low, a burned valve, a broken piston ring, or a blown head gasket could be causing a leak.

■ PHYSICAL PRINCIPLES OF WORK

Force

Force is measured in pounds, kilograms, or newtons. When something is pushed, pulled, or lifted against, force is produced. It is defined as any action that changes, or tends to change, the position of something.

Work

Work is when an object is moved against a resistance or opposing force. The movement can be either lifting or sliding (Figure 17.5). Work is measured in **foot-pounds** or **watts**. In the metric system, work is measured in *meter-kilograms*, or **joules**. The formula for work is:

$$\text{Force} \times \text{Distance} = \text{Work}$$

One *foot-pound* is when 1 pound is moved for a distance of 1 foot. For example:
■ Moving a 20-pound weight 50 feet results in 1000 foot-pounds of work (Figure 17.6).
■ Lifting a 10 kilogram weight a distance of 1 meter results in 10 meter-kilograms of work.

The four-stroke cycle can be used to illustrate work:
■ Intake stroke—the air/fuel mixture has work done on it to get it into the cylinder.
■ Compression stroke—work is done as the mixture is compressed.

Work = Force x Distance

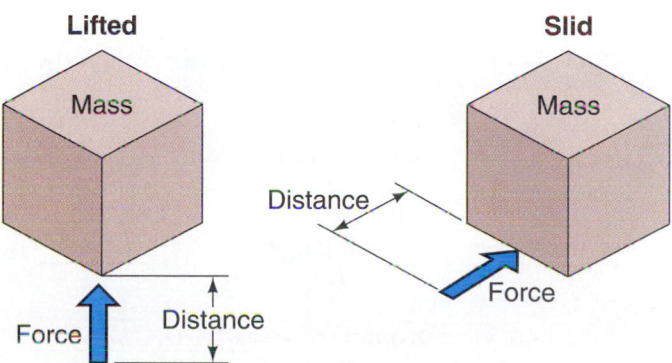

Figure 17.5 When work is performed, a mass is slid or lifted a certain distance.

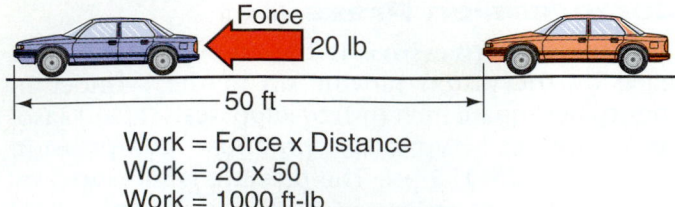

Work = Force x Distance
Work = 20 x 50
Work = 1000 ft-lb

Figure 17.6 Work measured in foot-pounds.

- Power stroke—the expanding air/fuel mixture works on the piston and crankshaft.
- Exhaust stroke—work is performed as the exhaust gas is expelled from the engine.

Energy

Energy is the ability to do work, or the ability to produce a motion against a resistance. When a weight is lifted, energy is stored in it. Dropping it performs work.

Inertia

Inertia is the tendency of a body to keep its state of rest or motion. The larger the mass, the more it is affected by inertia. Inertia and energy are stored in a flywheel.

Momentum

When a body is in motion it has **momentum**. Momentum is a product of a body's mass and speed. A body going in a straight line will keep going the same direction at the same speed if no other forces act on it.

Power

Power is how fast work is done or how fast motion is produced against a resistance.

■ TORQUE

Torque, the ability to make power, is defined as the tendency of force to rotate a body on which it acts. Tightening a bolt is a use of torque (Figure 17.7).

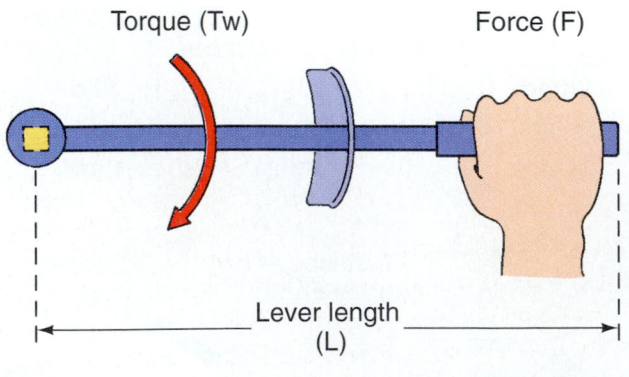

Torque (Tw)   Force (F)

Lever length (L)

Torque (Tw) = L x F

Figure 17.7 Tightening a bolt is a use of torque. *(Reproduced by permission of Deere & Company, 1991. Deere & Company. All rights reserved)*

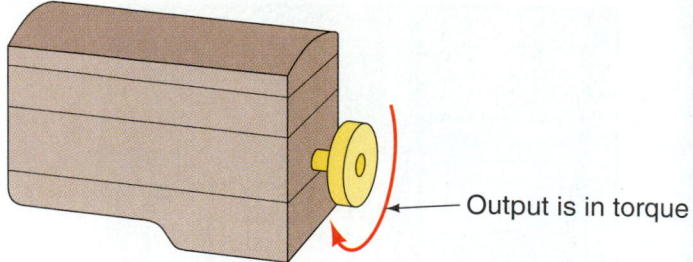

— Output is in torque

Figure 17.8 Torque is the amount of turning force exerted by the crankshaft.

Torque in an engine is the amount of turning force exerted by the crankshaft (Figure 17.8).

Torque is measured in foot-pounds in the English system and Newton-meters in the metric system.

NOTE: *Although the measurement of torque is commonly referred to in foot-pounds it should actually be expressed as pounds-feet to distinguish it from work.*

Engine torque varies with rpm. A force of 1 pound exerted at a distance of 1 foot from the center of a crankshaft results in 1 foot-pound of torque. The pulling ability of a car from a standing start depends on its engine's torque. This means that torque should be high at lower speeds.

To convert a torque reading to newton-meters: multiply ft.-lb. × 1.356.

To convert a torque reading to ft.-lb.: multiply newton-meters × 0.737.

Heat

Heat is another form of energy. It is measured in **British thermal units (Btu)**. One Btu is the amount of heat required to heat one pound of water by 1°F.

One *joule* is an equivalent value that compares heat energy (Btu) to mechanical energy (ft.-lb.).

$$1 \text{ Btu} = 778 \text{ ft.-lb.}$$

■ HORSEPOWER

Horsepower is the measurement of an engine's ability to perform work. One horsepower equals 33,000 foot-pounds of work per minute. When measuring one horsepower of work, it is the amount of work required to lift 550 pounds one foot in one second (Figure 17.9). In Figure 17.1 notice that the torque and horsepower readings are given at a specific rpm.

In the metric system, horsepower is measured in *watts*. One watt is the power to move one newton-meter per second. Because this is so small a measurement, *kilowatts* (kw) are used. One horsepower equals 0.746 kw.

Horsepower is a measure of work performed in a straight line. Torque measures force in a rotating direction.

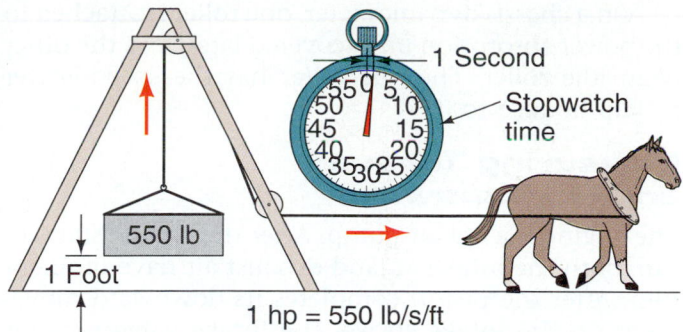

Figure 17.9 One horsepower is the amount of work required to lift 550 pounds one foot in one second.

Power produced at the crankshaft is called *gross horsepower*. Accessories that rob power include the alternator (charging system), air conditioning, coolant pump, cooling fan, power steering, and smog pump. These absorb about 25% of the power available at the crankshaft. The power that remains for use is called *net horsepower*. Power is also lost through friction in the driveline (transmission and differential), and due to wind resistance, vehicle weight, tires, and weather.

There are several measurements of engine horsepower:

■ *Brake horsepower (bhp)* is the usable horsepower at the crankshaft.
■ *Indicated horsepower (ihp)* is the amount of pressure made in the combustion chambers. It is measured with special instruments and varies throughout the four-stroke cycle. This is a theoretical measurement, which does not consider friction losses.
■ *Frictional horsepower (fhp)* is the power lost due to friction. It is the difference between brake horsepower and indicated horsepower (Figure 17.10). It considers the power needed to compress the air/fuel mixture and friction between engine parts such as the piston rings and cylinder walls.
■ *Net horsepower* is the maximum power available from the engine when all the accessories are turned on.
■ *Gross horsepower (ghp)* is the power available with only the water pump and alternator using power.

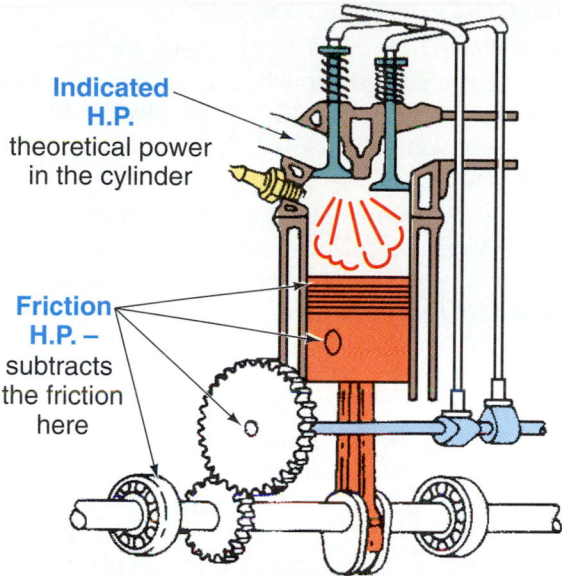

Figure 17.10 Frictional horsepower is the difference between brake horsepower and indicated horsepower. *(Reproduced by permission of Deere & Company, 1991. Deere & Company. All rights reserved)*

■ DYNAMOMETER

An engine's output can be measured using a **dynamometer**, commonly called a *dyno*. The engine must be loaded (braked) to measure the torque it can produce. Depending on the type of dyno, braking can be done by electricity, hydraulics, or friction.

A simple dynamometer that uses friction is called a **prony brake** (Figure 17.11). An arm pushes on a scale to provide a reading in pounds. When the length of the arm is known, the measurement can be converted to foot-pounds or newton-meters.

Engine Dynamometer

An engine dyno measures horsepower coming out of the engine. The horsepower measured is called brake horsepower because the dyno acts as a brake on the engine's crankshaft.

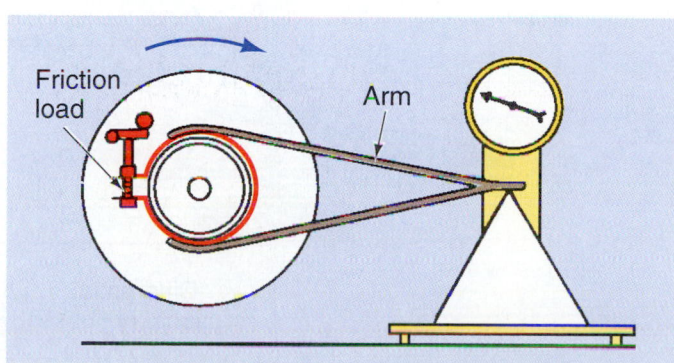

Figure 17.11 A prony brake. *(Reproduced by permission of Deere & Company, 1991. Deere & Company. All rights reserved)*

Chassis Dynamometer

A chassis dyno (Figure 17.12) measures horsepower available at the car's drive wheels. This is called **road horsepower** (Figure 17.13). It is always less than brake horsepower because of friction losses through the drive line.

NOTE: *Chassis dynamometers are used in emission control testing programs because testing of oxides of nitrogen can only be done when the engine is under load. This is called* **loaded mode testing**. *Emission testing is covered in Chapter 41.*

On a chassis dynamometer, one roller is attached to the power absorption unit (covered later) and the other is an idle roller. The idle roller has the speedometer pickup hooked to it.

Measuring Torque and Horsepower

The engine is a big air pump. After the engine starts to run, both the intake air and exhaust air have momentum. After the piston completes its downward movement on the intake stroke, the intake valve remains

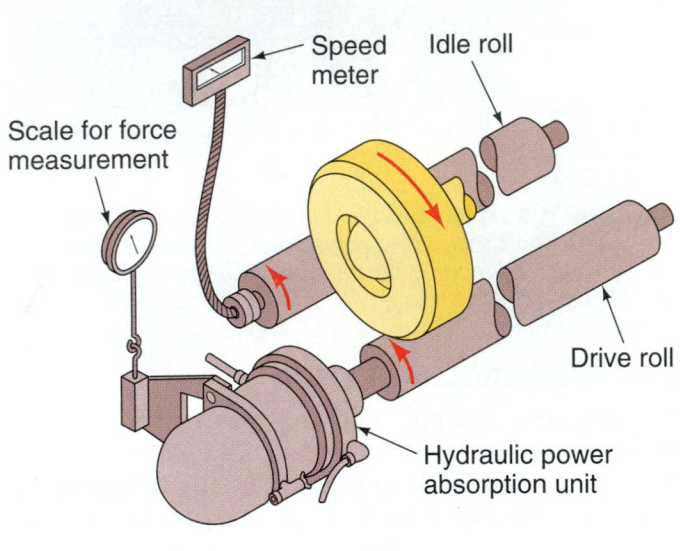

(a)

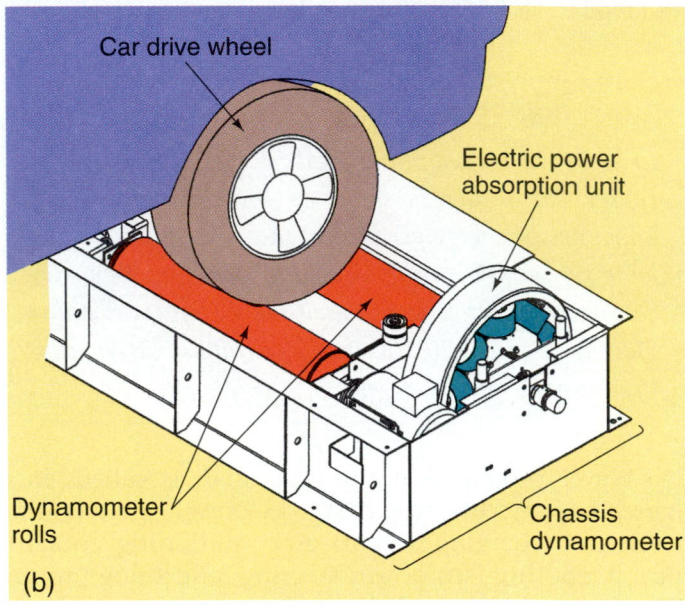

(b)

Figure 17.12 The rollers on a chassis dynamometer run off of the car's drive wheels. *[Courtesy of Clayton Industries]*

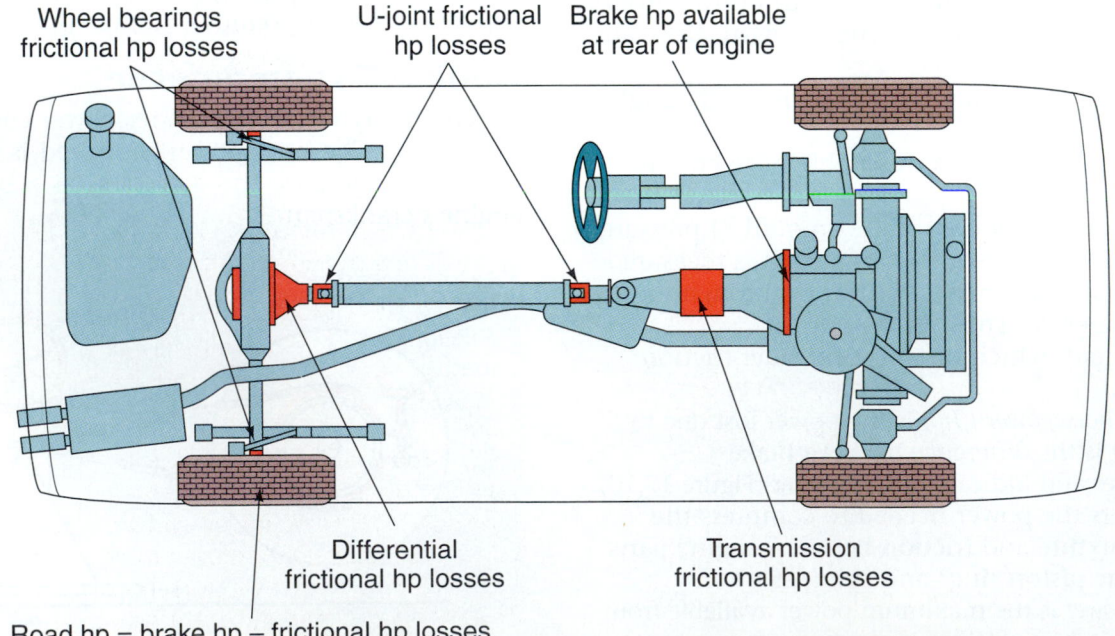

Figure 17.13 Road horsepower is always less than brake horsepower because of friction losses through the drive line. *[Courtesy of Mazda Motor Corporation]*

open for a period of time into the compression stroke (see Figure 18.31). Momentum of the incoming air/fuel mixture continues to fill the cylinder, even after the piston starts moving upward on the compression stroke.

After the piston finishes its upward movement on the exhaust stroke and begins moving down on the intake stroke, the exhaust valve remains open for a period of time. This period, when both the intake and exhaust valves are open, is called valve overlap (see Figure 18.32).

At higher speeds, the momentum of the exhaust gas moving through the exhaust system continues to pull exhaust from the cylinder. The vacuum created behind the moving exhaust stream also helps draw in fresh air and fuel through the open intake valve. This is called the *scavenging effect*.

An engine that does not breathe well will not produce as much power. Engines produce torque and horsepower in different amounts as rpm changes. These amounts are listed on a performance chart (Figure 17.14). Fuel consumption can also be predicted.

In designing an engine, selection of the camshaft, intake manifold, exhaust system, and valves is made to optimize horsepower and torque curves. A passenger car engine has more torque available at lower rpm. Racing engines make high horsepower at high speeds but operate poorly at low speeds. The idea is to have maximum torque and horsepower available across the widest rpm band of use.

NOTE: The 5.7 liter ZR-1 Corvette engine produces 385 lb.-ft. of torque at 5200 rpm and 405 hp at 5800 rpm. It is a dual overhead cam V-8 engine with four valves per cylinder (Figure 17.15). It has primary and secondary induction systems so that engine breathing will be acceptable at low and high rpm.

Figure 17.15 The 5.7 liter ZR-1 Corvette engine produces 385 lb.-ft. of torque at 5200 rpm and 405 hp at 5800 rpm. *(Courtesy of General Motors Corporation, Service Technology Group)*

Dynamometer Operation

Some dynos have computers that plot horsepower and torque curves. To make an engine perform work during a dyno test, the engine is put under load using a power absorption device. Most automotive dynamometer absorption units are fluid types controlled by the amount of water that enters the device.

The power absorption unit is a fluid coupling consisting of two members: a turbine and a stator (see Figure 66.5). The turbine tries to move the water, but the stator prevents it from moving. The load unit is like a torque wrench that measures the load applied. The amount of load put on the engine and the amount of torque it produces are used to calibrate horsepower.

The formula for horsepower is:

$$\frac{\text{Torque} \times \text{rpm}}{5250} = \text{hp}$$

An engine tested at 2625 rpm that develops 500 foot-pounds of torque produces 250 horsepower.

$$\text{Torque (500)} \times \text{rpm (2625)} = 1{,}312{,}500$$

$$\frac{1{,}312{,}500}{5250} = \text{hp (250)}$$

Torque readings are made at every 500 rpm. An engine that is warmed to operating temperature produces its best horsepower and has its lowest loss due to friction. Air intake temperature also makes a difference in the amount of horsepower produced; intake air at a lower temperature produces a higher oxygen content, which makes better combustion. More air and fuel can enter the engine in a colder (smaller) environment.

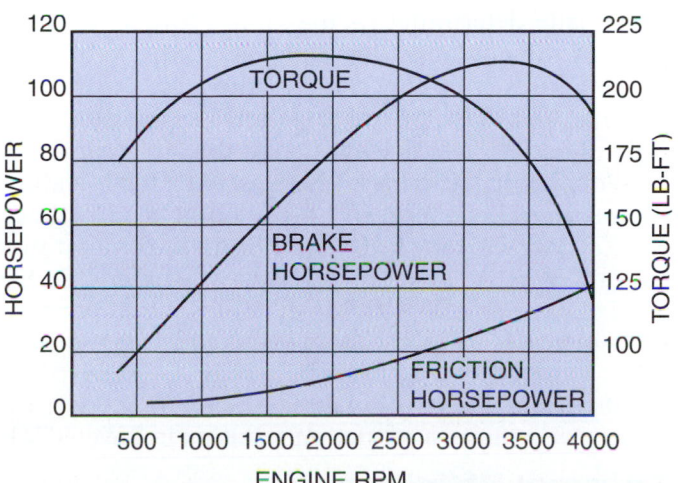

Figure 17.14 A performance chart lists torque and horsepower for an engine at a given rpm.

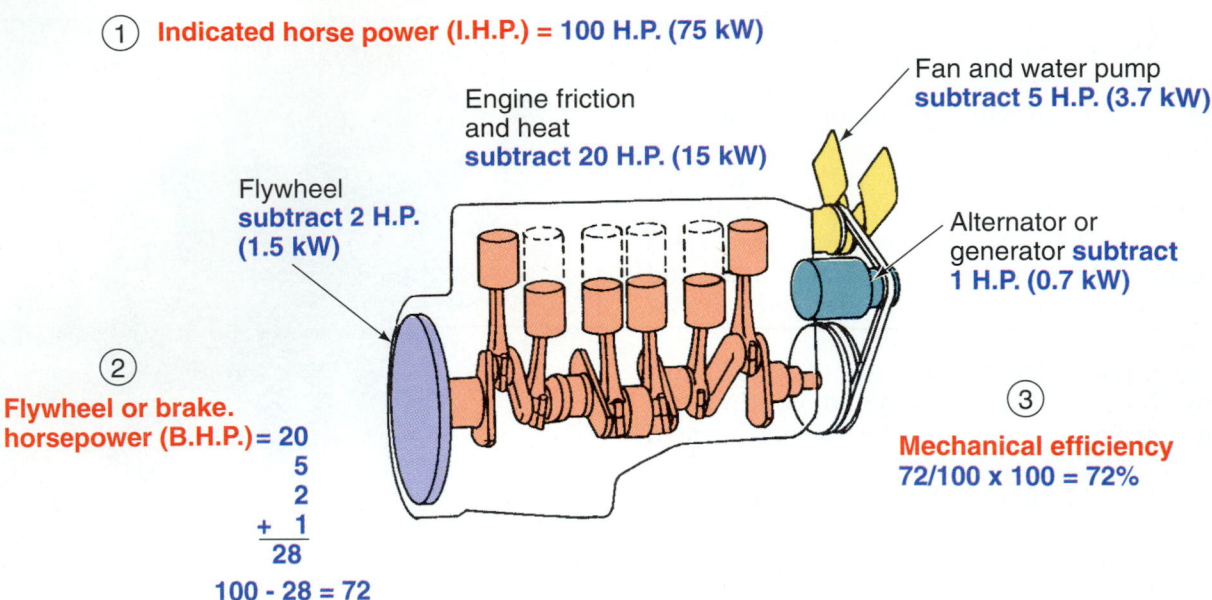

① **Indicated horse power (I.H.P.) = 100 H.P. (75 kW)**

Engine friction and heat **subtract 20 H.P. (15 kW)**

Fan and water pump **subtract 5 H.P. (3.7 kW)**

Flywheel **subtract 2 H.P. (1.5 kW)**

Alternator or generator **subtract 1 H.P. (0.7 kW)**

②
Flywheel or brake. horsepower (B.H.P.)= 20
```
        5
        2
      + 1
       28
```
100 - 28 = 72

③
**Mechanical efficiency
72/100 x 100 = 72%**

Figure 17.16 Mechanical efficiency. *(Reproduced by permission of Deere & Company, 1991. Deere & Company.*

ENGINE EFFICIENCY

To rate an engine in terms of efficiency, both the output and the input must be expressed in a common value. There are three types of engine efficiency measurements: **mechanical efficiency**, *volumetric efficiency*, and **thermal efficiency**. An efficiency measurement is a value less than 100%. The difference between the efficiency measurement and 100% is the amount of loss.

Mechanical efficiency describes all of the ways friction is lost in an engine (Figure 17.16). Horsepower is a value that can be used to compare the mechanical efficiency of two engines. Brake horsepower (bhp) divided by the indicated horsepower gives the mechanical efficiency of the engine.

The formula is:

$$\text{Mechanical efficiency} = \frac{\text{bhp (engine output)}}{\text{ihp (engine input)}}$$

If an engine had 100 indicated horsepower and 80 brake horsepower it would have a mechanical efficiency of 80%.

$$\text{Mechanical efficiency} = \frac{80 \text{ bhp}}{100 \text{ ihp}} = 80\%$$

Volumetric Efficiency

The measurement comparing the volume of airflow actually entering the engine with the maximum that theoretically could enter it (this is the same as the displacement) is called volumetric efficiency (V.E.) (Figure 17.17). Volumetric efficiency determines the engine's maximum torque output.

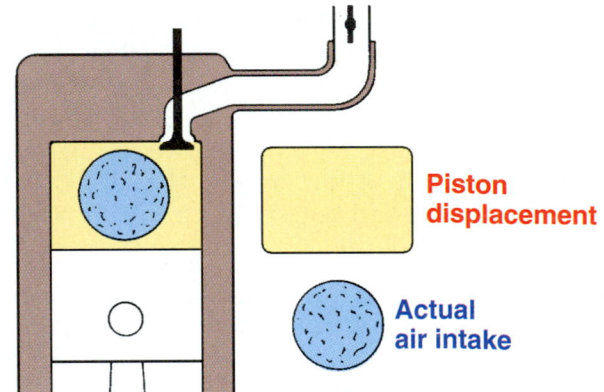

Piston displacement

Actual air intake

Figure 17.17 Volumetric efficiency. *(Reproduced by permission of Deere & Company, 1991. Deere & Company. All rights reserved)*

The rpm at which the engine does its best breathing usually determines its maximum torque. Remember that the engine is like a big air pump. Actually, a pump uses energy to compress air; an internal combustion engine gets its power from expanding air. Heat is what makes the air expand.

Volumetric efficiency changes with temperature, engine speed, load, and throttle opening. For instance, at 2000 rpm V.E. may be 85%, while at 4000 rpm it may be only 60%. At lower engine speeds, the engine has enough time to fill with air at atmospheric pressure. With increases in speed, there is less time for the air to move through the intake and exhaust systems. This results in decreased volumetric efficiency. Closing the throttle also causes a restriction resulting in lowered V.E.

Thermal Efficiency

Thermal efficiency is the ratio of how effectively an engine converts a fuel's heat energy into usable work.

1. ENGINE DEVELOPS 100 FLYWHEEL OR BRAKE H.P. (75 kW) PER HOUR
2. 2545 × 100 B.H.P. = 254,500 B.T.U. (75 kW)
3. FUEL BURNED PER HOUR = 800,000 B.T.U. (234 kW)
4. BRAKE THERMAL EFFICIENCY =
$$\frac{254,500}{800,000} \times 100 = 31.8\%$$

Figure 17.18 Brake thermal efficiency for a typical engine. *(Reproduced by permission of Deere & Company, 1991. Deere & Company. All rights reserved)*

EFFICIENCIES OF DIFFERENT ENGINES	
Gasoline Engine	25–28%
Diesel Engine	35–38%
Aircraft Gas Turbine	33–35%
Liquid Fuel Rocket	46–47%
Rotary Engine	20–22%
Steam Locomotive	10–12%

Figure 17.20 Overall efficiencies of different types of engines.

Each fuel has a certain amount of heat or thermal energy. Gasoline's thermal energy varies between fuels, but its average is about 19,000 Btu per pound. The energy of the fuel has the potential to produce a certain amount of work when burned in the engine. This thermal efficiency is a theoretical value.

A more useful measurement of thermal efficiency is called *brake thermal efficiency*. This is the brake horsepower converted to Btu, divided by the fuel's heat input in Btu with the result multiplied by 100. Figure 17.18 shows how the formula works in a typical engine.

In a spark ignition engine only about 1/4 of the energy from the burning of the fuel is converted to work at the crankshaft. The remainder of the energy is wasted as heat; part of it goes out the exhaust or is lost to the air. The other part is carried off by the cooling system (Figure 17.19). If the thermal efficiency of a gasoline engine was doubled, its fuel economy would also double.

Diesel fuel has more heat energy than gasoline (19,000–20,000 Btu/lb.). Diesel engines also have higher compression ratios. Both of these are reasons why diesels get better fuel economy than gasoline engines. Passenger car diesel engines have lower power and weigh more than gasoline engines. They do not perform as well. The overall efficiencies of different types of engines are shown in Figure 17.20.

■ MEAN EFFECTIVE PRESSURE

The pressure within the cylinder increases during the compression stroke and becomes highest after ignition. The peak pressure in the cylinder should occur at about 10° to 20° after top dead center (ATDC). This is similar to where a bicycle rider would want to apply pressure to the bicycle pedal. As the piston moves down on the power stroke, the pressure drops once again (Figure 17.21).

There are two kinds of mean effective pressure. The average pressure within the cylinder is called *indicated mean effective pressure (IMEP)*. This calculation requires equipment usually found only in laboratories.

Brake mean effective pressure (BMEP) is a term commonly heard in automobile racing. It is calculated from the horsepower reading on a dynamometer.

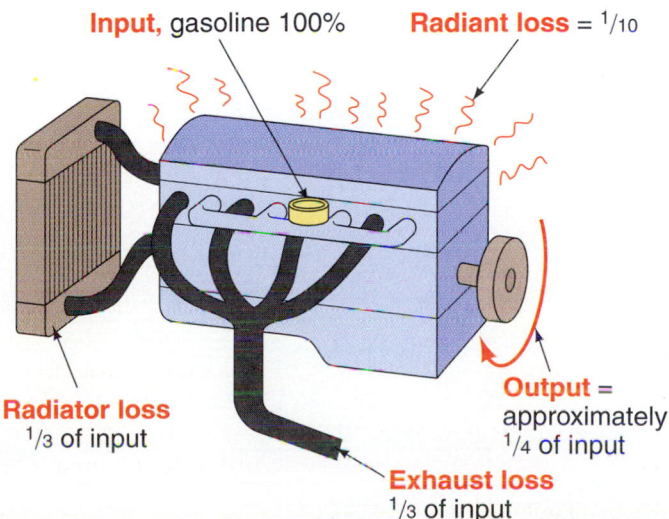

Figure 17.19 A gasoline engine loses a majority of its heat energy. *(Courtesy of DCA Educational Products)*

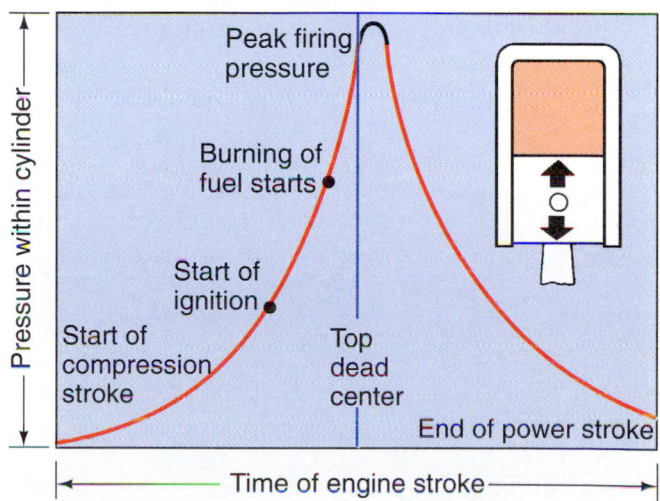

Figure 17.21 Pressure in the cylinder is called mean effective pressure. *(Reproduced by permission of Deere & Company, 1991. Deere & Company. All rights reserved)*

◼ REVIEW QUESTIONS

1. When an engine has a cylinder bore that is larger than its stroke, what is this called?

2. What is the name for the volume that the piston displaces in the cylinder as it moves from TDC to BDC?

3. If the mixture is compressed to about $\frac{1}{10}$ of the volume it occupied when the piston was at BDC, what is the compression ratio?

4. About how much change in horsepower does each point of change in compression ratio result in?

5. What is typical gasoline engine compression pressure?

6. Moving a 10-pound weight 50 feet results in _____ foot-pounds of work.

7. What is the name for the amount of turning force exerted by the crankshaft?

8. What is the term for work, pounds-feet or foot-pounds?

9. What is the name of the piece of equipment used to measure an engine's output?

10. What is the name of the measurement comparing the volume of air flow actually entering the engine with the maximum that theoretically could enter it?

◼ ASE STYLE REVIEW QUESTIONS

1. Technician A says that long stroke engines have more low end torque than shorter stroke engines. Technician B says that reboring a 302 cubic inch (5 liter) engine 0.060" oversize would result in an increase in engine size to 327 cubic inches. Who is right?
   a. Technician A   b. Technician B
   c. Both A and B   d. Neither A nor B

2. Technician A says engine rpm makes a difference in the amount of torque and horsepower output. Technician B says the length of the crank throw is twice the total stroke. Who is right?
   a. Technician A   b. Technician B
   c. Both A and B   d. Neither A nor B

3. Technician A says a dynamometer absorption unit is like a torque wrench for measuring engine output. Technician B says more air and fuel can enter the engine in warmer weather. Who is right?
   a. Technician A   b. Technician B
   c. Both A and B   d. Neither A nor B

4. Technician A says a spark ignition engine converts about 75% of the energy from the burning of the fuel to work at the crankshaft. Technician B says diesel fuel has more heat energy than gasoline. Who is right?
   a. Technician A   b. Technician B
   c. Both A and B   d. Neither A nor B

5. Technician A says passenger car diesel engines do not perform as well as gasoline engines of the same weight. Technician B says the amount of usable horsepower at the crankshaft is called brake horsepower. Who is right?
   a. Technician A   b. Technician B
   c. Both A and B   d. Neither A nor B

CHAPTER 18

Engine Upper End

■ OBJECTIVES

Upon completion of this chapter, you should be able to:

✔ Identify all of the parts of the engine's upper end.

✔ Understand the difference between cylinder head designs.

✔ Understand the variations in camshaft design.

✔ Describe the different camshaft grinds and the uses they are best suited for.

✔ Identify the different cam drive arrangements.

✔ Describe the difference between freewheeling and non-freewheeling engines.

■ KEY TERMS

integral seats (and guides)
induction hardened valve seats
poppet valves
positive stop
end thrust
production engine
stock
base circle
lift
duration
valve overlap
zero-lash
core
backlash
freewheeling engine
interference engine (non-freewheeling engine)

■ INTRODUCTION

In earlier chapters you became familiar with basic parts of the engine and the different types of engines. This chapter goes into more detail about the *upper end* of the engine (Figure 18.1). The upper end includes:

■ Cylinder head or heads

■ Valve train (camshaft and cam drive)

Intake and exhaust manifolds, which are sometimes considered to be a part of the upper end, are covered in another chapter. In later chapters you will learn to diagnose and repair engines. This chapter helps prepare you for these areas.

■ CYLINDER HEAD CONSTRUCTION

Cylinder heads are made of either cast iron or aluminum. A *bare head* is a head without any of its loose parts installed. The rest of the loose parts include the intake and exhaust valves, the keepers and retainers, the valve guide seals, the valve springs, and rocker arms. Parts of the cylinder head are shown in Figure 18.2. Sometimes a head is cracked or unrepairable and the rest of the parts are good. A bare head would be purchased in this instance.

■ VALVE GUIDES

The valve guide is a hole in the cylinder head that the valve slides up and down in. There are two types of guides: *integral* and *replaceable insert* (Figure 18.3). An integral valve guide is a hole bored into an iron cylinder head. Integral guides lower production costs but cannot be used in aluminum heads.

Some iron heads and all aluminum heads have pressed-in guides, called replaceable inserts. Insert guides are made of iron or bronze alloy.

Valve guides become worn, which usually results in oil consumption when combined with a bad valve guide seal (Figure 18.4).

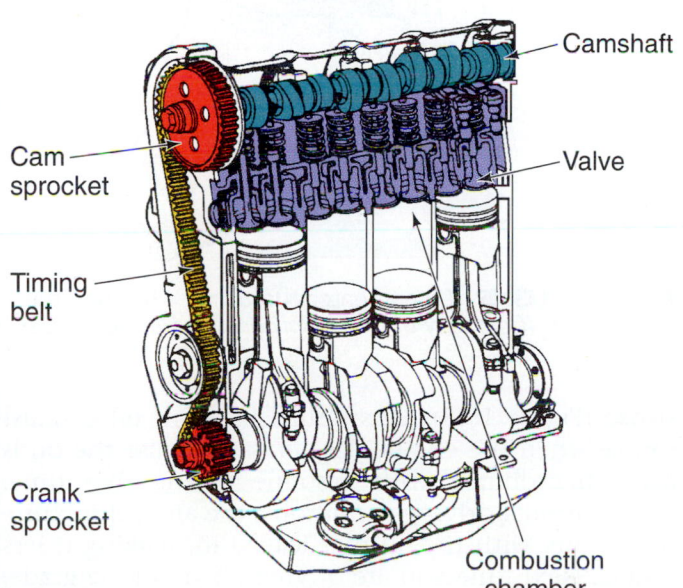

Figure 18.1 The upper end. *(Courtesy of Chrysler Corporation)*

Camshaft

Cam sprocket

Valve

Timing belt

Crank sprocket

Combustion chamber

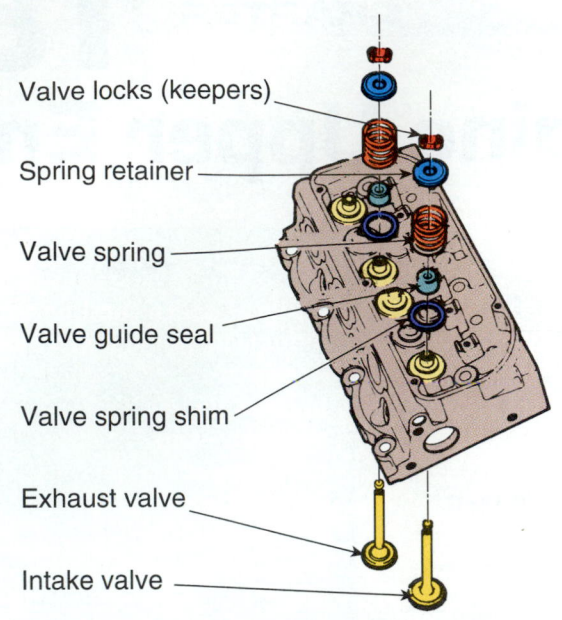

Valve locks (keepers)

Spring retainer

Valve spring

Valve guide seal

Valve spring shim

Exhaust valve

Intake valve

Figure 18.2 Parts of a cylinder head. *(Courtesy of Ford Motor Company)*

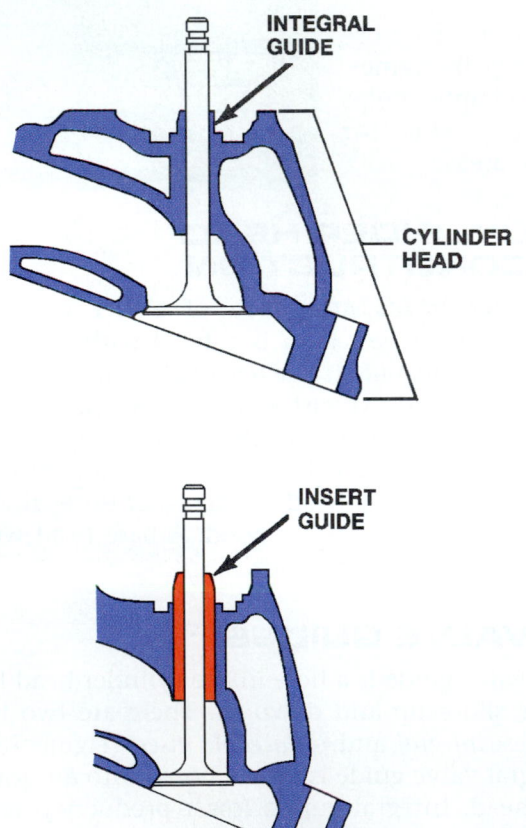

INTEGRAL GUIDE

CYLINDER HEAD

INSERT GUIDE

Figure 18.3 Valve guides are integral or replaceable. *(Courtesy of Federal-Mogul Corporation)*

■ **VALVE GUIDE SEALS**

Leaking valve guide seals account for nearly half of oil consumption complaints. A faulty intake seal allows oil to be drawn into the cylinder during the intake

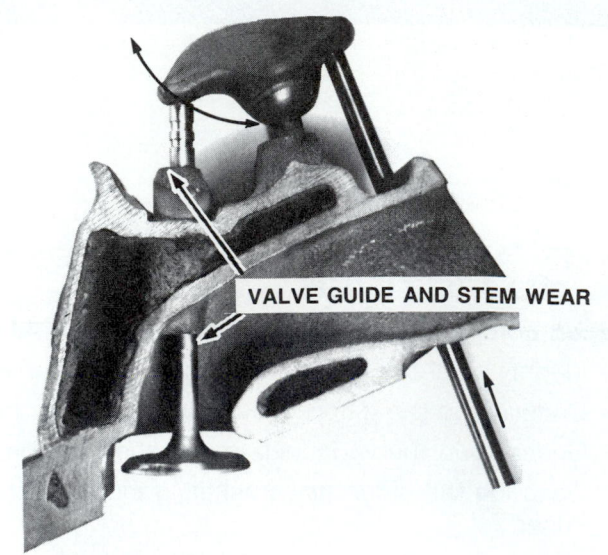

VALVE GUIDE AND STEM WEAR

Figure 18.4 The valve guide and valve stem wear as shown.

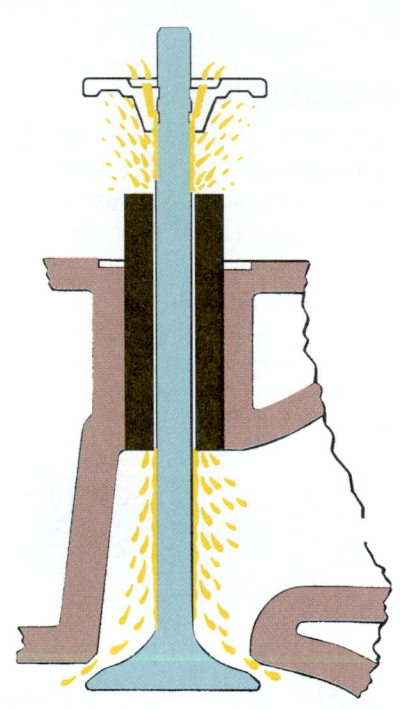

Figure 18.5 Oil can enter the cylinder through the valve guide. *(Courtesy of Dana Corporation-Perfect Circle Division)*

stroke (Figure 18.5). This results in burnt oil exhaust smoke when the engine decelerates because the oil is *not* being diluted with gasoline during that time. Under normal driving conditions, the air/fuel mixture would mix with the oil that leaked in, making it less visible. Next time you are driving down a long grade, notice how many cars in front of you have exhaust smoke. These are the ones that are using oil through the valve guides.

Besides causing oil consumption, a leaking intake seal can result in carbon buildup on the stem of the valve (which interferes with engine breathing) (Figure 18.6). The carbon buildup interrupts the flow of incoming air and fuel as it passes the irregular carbon surface. This causes turbulence in the air flow, which reduces volumetric efficiency (breathing ability) and decreases power.

Even though the exhaust guide is not exposed to engine vacuum on the intake stroke, it can also cause oil consumption. Figure 18.7 shows how the exhaust gas moving past the exhaust valve causes a vacuum that pulls oil down the guide and into the hot exhaust. This is called the *atomizer effect*.

Figure 18.6 Carbon buildup on the neck of the valve interferes with breathing.

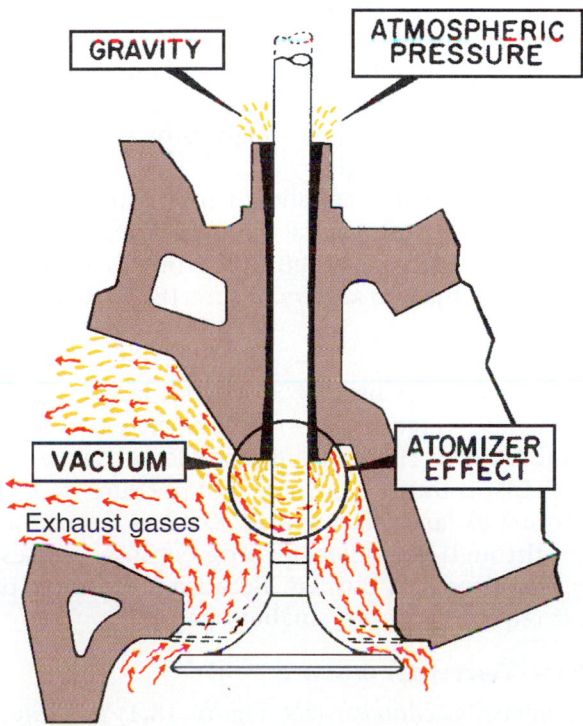

COMBUSTION CHAMBER

Figure 18.7 The atomizer effect results in oil being drawn down the exhaust guide as exhaust gas moving past the bottom of the valve guide creates a vacuum. *(Courtesy of Dana Corporation-Perfect Circle Division)*

Valve seals come in three types: *umbrella, positive,* and *O-ring* (Figure 18.8). An *umbrella seal* fits inside the valve spring (see Figure 15.10). It fits the valve snugly and rides up and down with it, keeping oil from entering the valve guide. If an umbrella seal becomes brittle with age, it can become loose on the valve stem, which allows oil to leak down the valve guide.

Sometimes umbrella seals break up, floating in the oil until the pieces are picked up by the oil pump. This can result in engine damage when the oil pump locks up.

A *positive valve seal* fits on the top of the valve guide. The valve moves up and down inside of a positive valve guide seal (Figure 18.9). Positive seals are used on overhead cam engines because the camshaft is heavily lubricated right above the valves. Occasionally, positive seals are also used on pushrod engines.

An *O-ring seal* is the type used by General Motors. It fits under the keepers, inside of the spring retainer (Figure 18.10). Oil is pumped through the pushrod to a hole in the rocker arm and spills onto the retainer. Without the O-ring seal, oil would be able to leak down the stem and into the guide.

Valve Guide Seal Materials

Valve guide seals are made of several materials having different resistances to high temperatures. Although

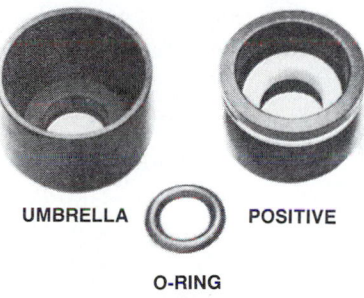

UMBRELLA   POSITIVE

O-RING

Figure 18.8 Three types of valve guide seals. *(Courtesy of Federal-Mogul Corporation)*

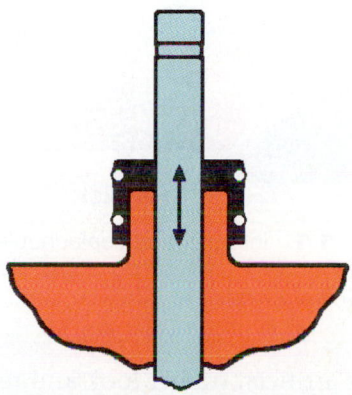

Figure 18.9 A positive valve seal fits snugly on the valve guide. *(Courtesy of Mercedes-Benz of North America, Inc.)*

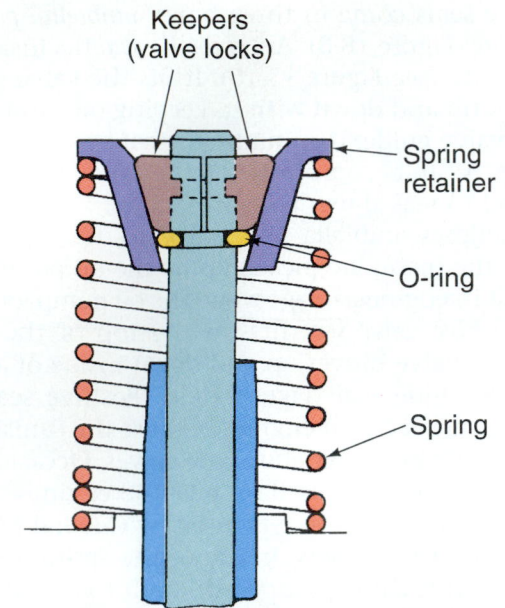

Figure 18.10 An O-ring seal fits beneath the keepers. *(Courtesy of Federal-Mogul Corporation)*

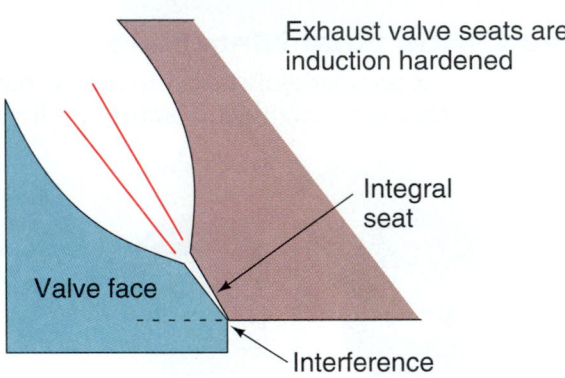

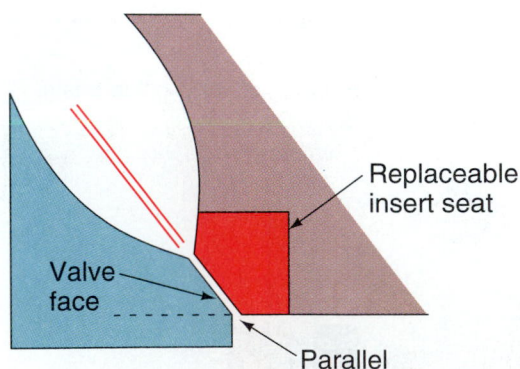

Figure 18.11 Integral and replaceable valve seats. *(Courtesy of General Motors Corporation, Service Technology Group)*

most kinds of artificial rubber look and feel alike, they are not. The premium seals resist heat up until about 440°F. Lower quality seals resist heat only to 250°F. The hottest area, at the combustion chamber end of the

valve stem, runs at a temperature in excess of 1300°F, so you can see that resistance to temperature is important.

Premium Viton and Teflon seals are used for the highest temperature applications. These materials are found only in positive valve guide seals. Umbrella seals are made of nitrile (used only on the intake side) or silicone; lower temperature materials.

Valve Seats

Valve seats are usually ground to a 45° angle, although some older heads have 30° seats. Like valve guides, seats can either be part of the head (integral) or replaceable inserts (Figure 18.11). **Integral seats** are only found in cast iron heads. *Replaceable seats* are found in original equipment aluminum heads and in iron heads that have been repaired by a machine shop. The inserts are made from iron, *stellite* (hard seats), or bronze.

Induction Hardening of Seats

Integral seats on heads manufactured since about 1970 are **induction hardened** to reduce valve seat wear. Induction hardening hardens only the seat area of the head. One advantage to integral seats is that the temperature they operate at is approximately 150°F cooler than replaceable seat inserts.

■ VALVES

Automotive valves are **poppet valves**, which were first used in steam engines. They were designed to require little or no lubrication and they operate well in high temperatures. To match the valve seat, the valve face angle is usually 45°, although some valves (mostly intakes) have a 30° angle.

Parts of the valve are shown in Figure 18.12. The valve is machined on the face surface, the tip, and the stem. There is a groove, or multiple grooves, in the valve stem near the tip that *keepers* (Figure 18.13) fit into.

Valve Size

The intake valve is usually 35%–40% larger than the exhaust valve (Figure 18.14). Intake air and fuel take up quite a bit of space, requiring a larger valve opening. Exhaust is under a great deal of pressure, so it does not require as large an opening. Exhaust gas can actually go through a smaller opening because it takes up less space after it is burned. Because it is under pressure, it requires an even smaller opening.

Valve Temperature

The *combustion chamber* (see Figure 18.1) is subjected to severe conditions during engine operation. Valves run at high temperatures, opening and closing at half engine speed. Exhaust seats run at about 800°F. The neck area of the exhaust valve sometimes runs at temperatures in excess of 1300°F. This is above the

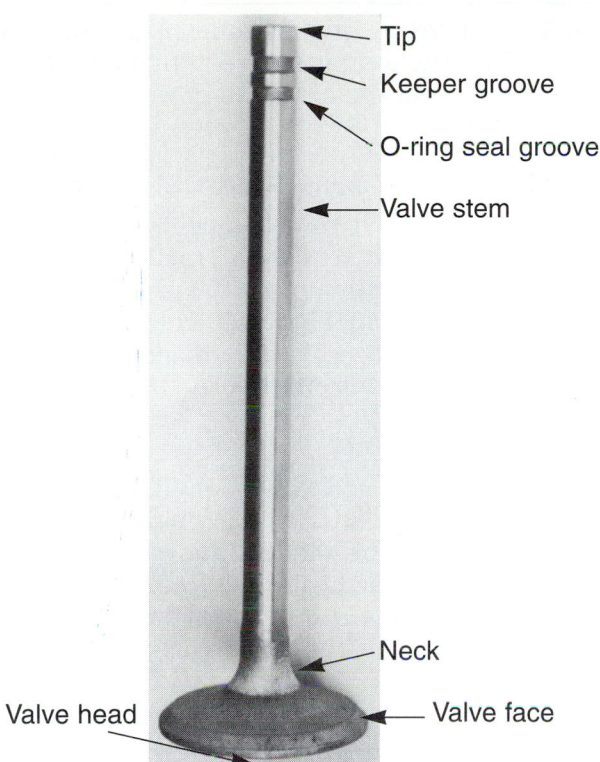

Figure 18.12 labels:
- Tip
- Keeper groove
- O-ring seal groove
- Valve stem
- Neck
- Valve head
- Valve face

Figure 18.12 Parts of a poppet valve.

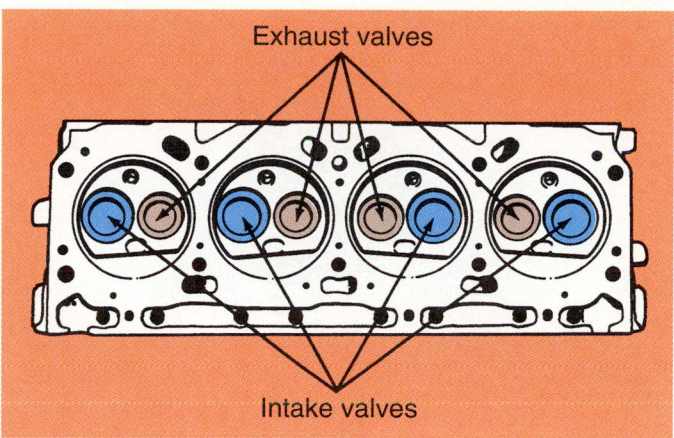

Exhaust valves

Intake valves

Figure 18.14 Intake valves are larger than exhaust valves. *(Courtesy of General Motors Corporation, Service Technology Group)*

melting temperature of aluminum, and it is red hot for these steel valves. Figure 18.15 shows typical temperatures of a passenger car exhaust valve.

The intake valve runs at a temperature that is usually below 1000°F. It is cooler than the exhaust valve because it is bathed in cooler intake air and fuel during every intake stroke. Water jackets are in the head behind the valve seat and around the valve guide (Figure 18.16). The heat in the valves is transmitted from the valve stem to the valve guide and then into the engine coolant. Valves depend on good contact between the guide and seat for proper cooling.

Valve Materials

Intake and exhaust valves are usually made of different materials. Exhaust valves are constructed of higher

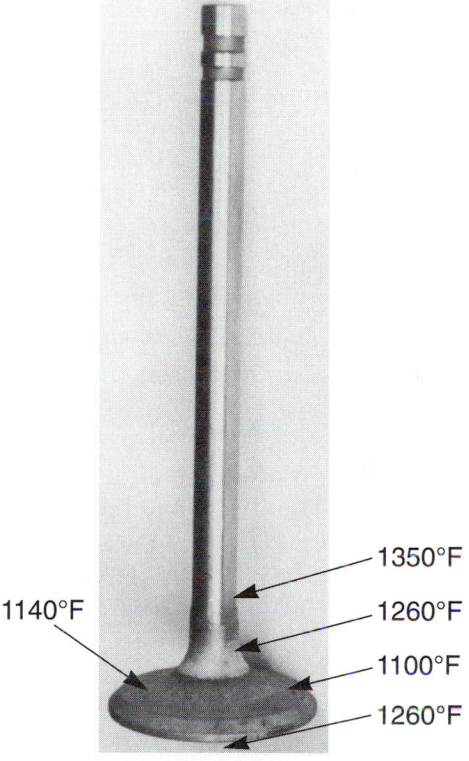

- 1140°F
- 1350°F
- 1260°F
- 1100°F
- 1260°F

Figure 18.15 Typical temperatures of an automotive exhaust valve.

Figure 18.13 Keepers. *(Courtesy of Ed Iskenderian Racing Cams)*

Figure 18.16 Water jackets are behind the valve seat and around the valve stem.

quality steel. Even though they are smaller than intake valves, they cost more money.

■ The *austenitic stainless steel* that exhaust valves are constructed of is usually nonmagnetic.
■ The carbon or alloy steel that intake valves are made of is magnetic.

■ SCIENCE NOTE ■

Iron is magnetic because it has unpaired bonding electrons. The interatomic distance between iron atoms is just right for these unpaired electrons to line up and reinforce each other. Stainless steel, an iron/chromium/nickel alloy is not magnetic even though both chromium and iron have unpaired electrons. Stainless steel is not magnetic because the iron/chromium interatomic distances are not right for the unpaired electrons to line up and reinforce each other.

The valve stem is often of a different material than the valve head. The two materials are *spin-welded* together. Two-piece welded valves typically have a *forged stainless steel* head welded to a *high carbon steel* stem about ⅔ down from the stem tip.

■ SCIENCE NOTE ■

Steel is either carbon steel or alloy steel. Ninety percent of all steel produced is carbon steel, which contains no other metals. Alloy steels contain carbon but also different metals used to give the steel different properties. Stainless steels contain chromium and nickel. Chromium improves hardness and resistance to corrosion, while nickel adds toughness. Other mineral additives such as vanadium and manganese provide the properties of springiness and wear resistance. Different rates of quenching also change the properties of steel. For flexibility, steel is cooled slowly. This results in large smooth steel crystals. If a good cutting edge was the desired property, the steel would be hardened by quenching rapidly, which causes small, jagged steel crystals.

Valve stems are often chrome plated, so the weld is not visible. The weld is so strong that failure of valves usually occurs at the neck area of the valve, not at the weld.

Valve Stem and Face Coatings

Valve stems have two common types of protective coating to prevent stem wear. *Chrome plating*, which reduces stem wear by up to 80% over 50,000 miles, is commonly done in North America. In Europe, the popular stem coating is *bath nitriding*, which leaves a black finish that is as effective as chrome.

Valve Burning

The *margin* of the valve gives an idea of how much of the valve remains after valve grinding. If the margin is too thin, it will heat up and the valve will burn (Figure 18.17). A burned valve will cause a misfire at idle speed. Because it runs so much hotter, the exhaust valve is usually the one that burns (Figure 18.18).

SHOP TIP To test for a burned exhaust valve, hold a piece of paper over the exhaust pipe as the engine idles. You will see it suck into the pipe every time there is an intake stroke. This is because the air and fuel filling the cylinder draws on the exhaust stream through the opening (burned area) in the exhaust valve.

Sodium-Filled Valves

Some heavy-duty engines use *sodium-filled valves* (Figure 18.19). They have hollow stems filled with sodium, with the weld near the stem tip. The sodium moves within the valve stem and carries heat from the neck of the valve to the cooler valve stem area that is surrounded by the water jacket. Maximum valve temperatures are actually cut by about 350°F.

■ SCIENCE NOTE ■

Metallic sodium is a solid powder in the valve stem until it reaches 208°F. It is a liquid at engine operating temperature. Materials that are good conductors of electricity are also excellent heat conductors. Thermal energy is transported rapidly from one part of the metal to another by the conduction (valence) electrons, which roam freely throughout the material when solid or liquid.

The hollow cavity in the valve is actually only about half full of sodium. The movement of the sodium in the valve stem helps to carry off heat from

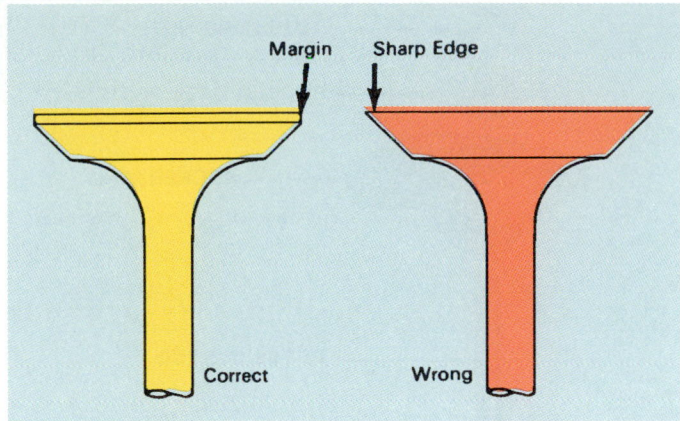

Figure 18.17 If the margin is too thin, the valve can burn. *(Courtesy of Federal-Mogul Corporation)*

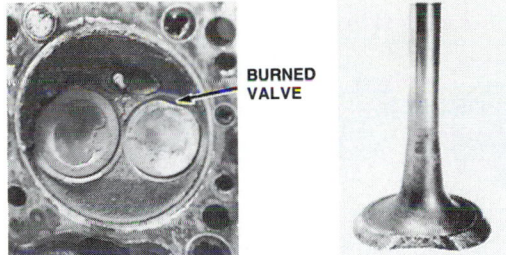

Figure 18.18 A burned valve.

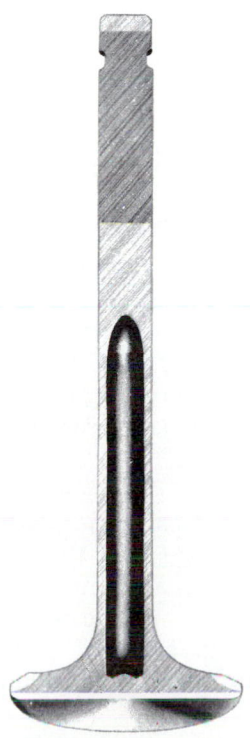

Figure 18.19 A sodium-filled valve. *(Courtesy of Federal-Mogul Corporation)*

Figure 18.20 An exhaust valve stem with a welded-on tip.

the neck of the valve to the cooler valve stem. Sodium-filled valves can fail if valve stem-to-guide clearance becomes excessive because the heat cannot dissipate from the valve stem.

> **CAUTION** Metallic sodium will burst into flame if it comes into contact with moisture, so it is dangerous to grind or break these valves. Sodium is safe so long as it is contained within the valve.

Valve Tip

The tip of the valve is hardened. The exhaust valve must open against very high pressure in the cylinder during the power stroke, so many exhaust valve stems have a welded-on hardened tip (Figure 18.20). An engine that has mechanical valve adjustment would tend to *mushroom* the end of the stem if it is not hardened or if the valve clearance becomes excessive.

■ RETAINERS AND KEEPERS

The retainer and valve keepers, or *locks*, hold the spring and valve together. Keepers fit into the groove in the valve stem. The retainer, which resembles a large washer, holds the spring against the spring seat on the top of the cylinder head. Some retainers have a feature that causes the valve to rotate. These retainers are called *valve rotators* (Figure 18.21).

■ VALVE SPRINGS

After the valve is opened by the valve train, a valve spring is needed to close it. Some engines have heavier valve springs than others. On camshafts with higher

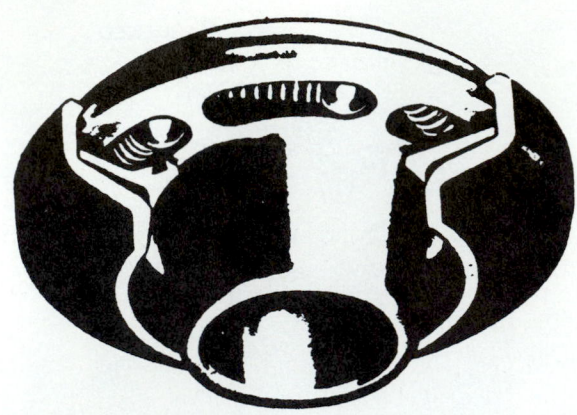

Figure 18.21 A valve rotator. *(Courtesy of Federal-Mogul Corporation)*

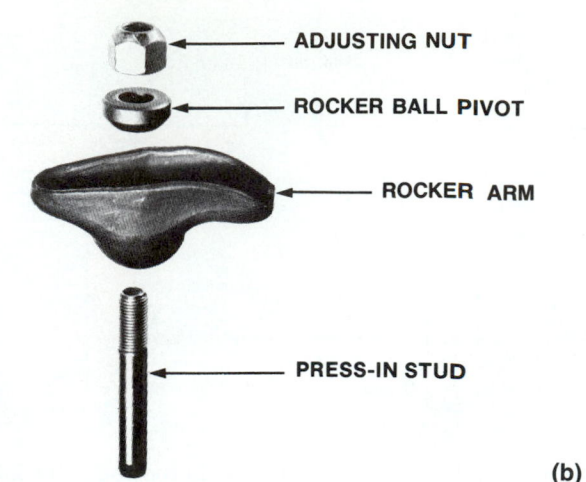

ADJUSTING NUT

ROCKER BALL PIVOT

ROCKER ARM

PRESS-IN STUD

(b)

Figure 18.22 A stud-mounted rocker arm. *(Courtesy of Federal-Mogul Corporation)*

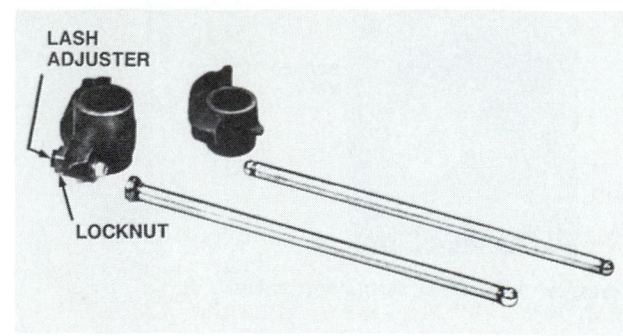

LASH ADJUSTER

LOCKNUT

Figure 18.23 Adjustable and nonadjustable shaft-mounted rocker arms. *(Courtesy of Dana Corporation-Perfect Circle Division)*

valve lift (covered later in this chapter), more spring pressure is often required. The thickness of the spring is limited by how close together its coils are. When the valve opens, the spring could stack against itself causing damage.

If a stronger spring is required, two springs are used (one inside the other). The spring on the outside is wound in a direction opposite the inside spring. When a spring is compressed, it tends to bend inward. Think about how a "slinky" bends to go down stairs. Having the two springs wound in different directions tends to minimize this tendency.

Some springs have a flat wound *dampener* on the inside to dampen spring vibrations. Springs vibrate at a certain frequency that can result in engine damage if allowed to continue. Think about how an opera singer can cause a glass to break when a note of the right pitch is sung. The dampening coil is wound in a direction opposite to the valve spring also.

■ ROCKER ARMS

Rocker arms either pivot on a rocker shaft or on rocker studs. *Shaft-mounted rocker arms* are usually made of iron. Older engines used more expensive cast iron rocker arms. Most modern engines use the less expensive stamped steel rocker arms.

Stud-mounted, or *pedestal-mounted*, rocker arms are made of steel stamped in a "canoe" shape. They are less expensive than the shaft-mounted type, and are commonly replaced if they are worn. The rocker arm shown in Figure 18.22 has a ball pivot and self-locking adjusting nut mounted above it.

Stud-mounted rocker arms are mounted on a stud that is pressed or threaded into the head. One kind of stud has a shoulder that the adjusting pivot stops against when the rocker stud is torqued. This is called a **positive stop** arrangement and requires no valve adjustment. Figure 18.23 compares adjustable and nonadjustable rocker arms. Nonadjustable rockers can only be used in positive stop engines.

Rocker Arm Ratio

Rocker arms sometimes have different *rocker arm ratios* (Figure 18.24). Rocker ratio is usually 1.4:1 to 1.75:1, depending on the manufacturer. Changing the ratio can cause movement of the lifter at the cam lobe to be *increased* at the valve, making it possible for an engine to be designed with smaller cam lobes.

Pushrods

Cam-in-block engines use pushrods to transmit motion from the lifters to the rocker arms (Figure 18.25). Most pushrods are hollow. Some have oil holes to provide lubrication to the rocker arms and valves (Figure 18.26). The ends of pushrods have many shapes, including concave (indented) and convex (protruding).

Some engines use a *guide plate* to limit side movement of the pushrods. Other engines have small holes in the cylinder head or intake manifold that act as guides.

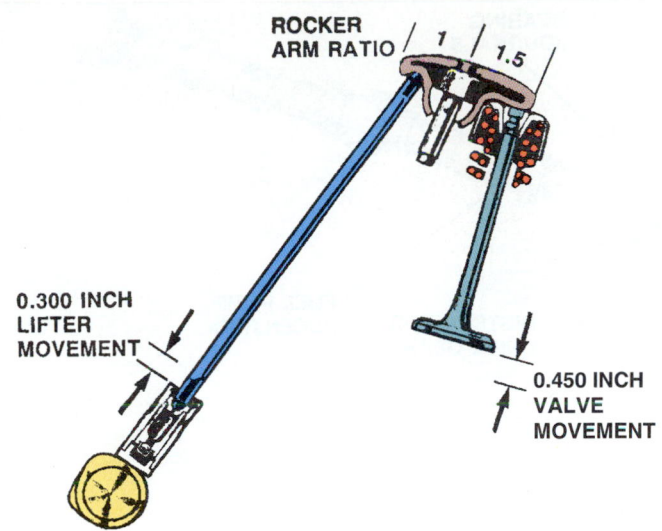

Figure 18.24 Rocker arm ratio. *(Courtesy of General Motors Corporation, Service Technology Group)*

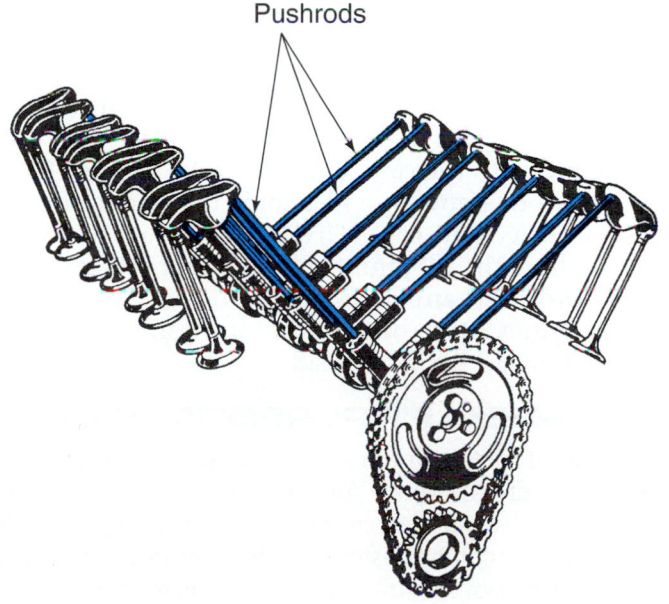

Figure 18.25 Pushrods transmit motion from the lifters to the rocker arms. *(Courtesy of Wolverine/Blue Racer, Osseo, Michigan, USA. A Crane Technologies Group Company)*

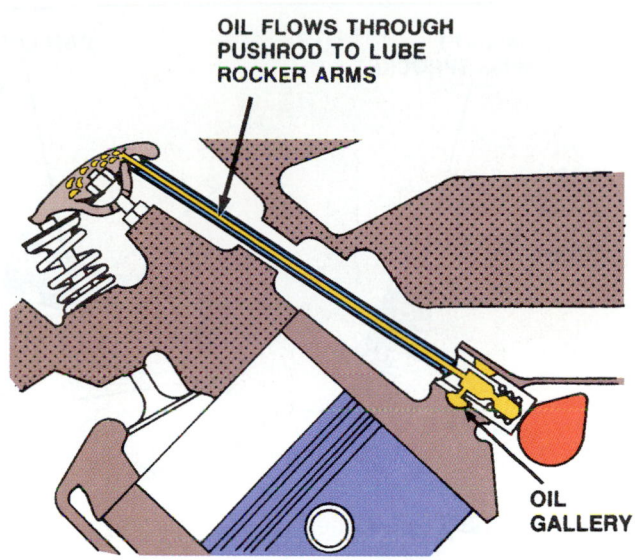

Figure 18.26 The stud-mounted rocker arm is lubricated through hollow pushrods.

Overhead Cam Valve Operation

Some overhead cam (OHC) engines use rocker arms. One end of the rocker arm rests against the valve stem, and the other end sits on a ball pivot. The cam lobe pushes on a pad in the middle of the rocker arm. The pivot is usually equipped with threads for adjusting valve clearance, or *lash* (Figure 18.27). Other OHC engines have cam lobes that act directly on the valves. These cam lobes are larger than lobes on engines with rocker arms. This is because there is no rocker arm ratio to multiply movement caused by the lobe. Lash

is usually adjusted on this type of engine by adding or subtracting shims (called *lash pad adjusters*).

CAMSHAFT

Camshaft Construction

The camshaft, commonly called "the cam," controls the opening and closing of the valves. It is usually made of hardened cast iron, although cams used with roller lifters are made of steel. *Cam journals* support the camshaft at several places. The number of *lobes* a camshaft has depends on how many valves the engine has.

NUMBER OF CAMS AND LOBES

Most engines have two cam lobes for each cylinder. One lobe operates the intake valve, and the other operates the exhaust valve.

Many engines use only one camshaft, although some use more than one. Some engines use three or four valves per cylinder. To increase the engine's breathing ability (volumetric efficiency), many racing engines and many of the latest small high-performance stock engines have four valves and four cam lobes on two camshafts for each cylinder. Opposed engines have only half as many cam lobes because the lobes of one cylinder are shared with its companion cylinder on the opposite side of the engine.

Intake and exhaust lobes are sometimes different from each other. When there is more than one camshaft on a cylinder head, one camshaft will have only intake lobes and the other will have only exhaust lobes.

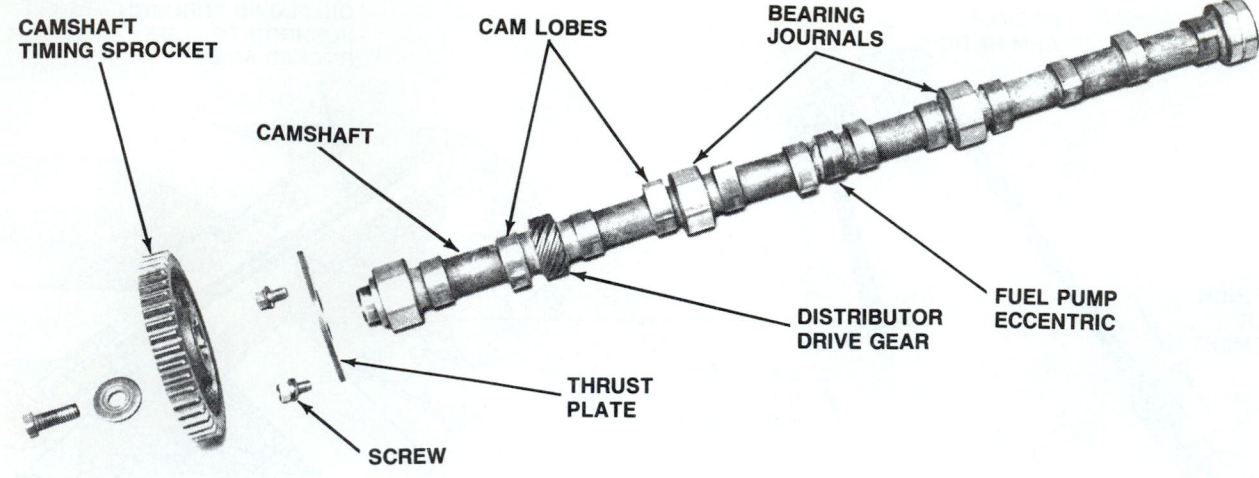

Figure 18.27 Camshaft parts. *(Courtesy of Ford Motor Company)*

■ FUEL PUMP ECCENTRIC

When an engine has a carburetor, there is usually an eccentric (off-center lobe) for the mechanical fuel pump. Fuel injected engines use electric fuel pumps and do not need this eccentric. Some engines have an eccentric that is not part of the cam, but bolts to the front of the cam sprocket (Figure 18.28).

■ DISTRIBUTOR AND OIL PUMP DRIVES

Because it turns at the same speed, the distributor is often driven off of the camshaft. The *distributor gear* is sometimes made from a softer material such as aluminum, plastic, or brass. A failure of one of these soft gears would not be as likely to damage the cam.

The oil pump is usually driven by a shaft that connects to the bottom of the distributor shaft. As the oil pump is turned, the resistance created by the pump-

Figure 18.28 The fuel pump eccentric in its lowest position. *(Courtesy of General Motors Corporation, Service Technology Group)*

ing of the oil causes the cam to thrust toward the rear of the engine.

NOTE: *Excessive **end thrust** can also cause the distributor timing to advance or retard as the cam moves fore and aft against the distributor gear.*

Some camshafts have a *thrust plate* to control end thrust (Figure 18.29). Excessive end thrust can also cause damage to the cam and lifters. The timing chain on a pushrod engine provides some thrust control.

Some overhead cam (OHC) engines use an extra shaft to drive the oil pump and distributor. It is called an *auxiliary shaft* and is driven by the crank. In other OHC engines, the distributor and oil pump are mounted on the cylinder head.

■ CAMSHAFT PERFORMANCE

The camshaft determines how well an engine breathes at a particular engine rpm. The shape of the cam lobe, or *cam grind*, is what determines when the engine will breathe best. It can be designed to have its best operation at maximum power and high speed, or for better fuel economy and low speed operation. Production engines deliver a balance between these two objectives. A **production engine** is an engine produced at the factory. Original equipment cams are referred to as **stock** cams.

Engine Breathing

Basically, the more air and fuel drawn into the cylinder in the proper proportion (not too rich or lean), the more power an engine can produce. This is called volumetric efficiency. Many of today's four- and six-cylinder engines are as powerful as eight-cylinder engines of the past. Some of these smaller engines use multiple-valve combustion chambers, along with other modifications to increase breathing ability. These engines are referred to as *naturally aspirated* engines. Other engines equipped with turbochargers or superchargers have much improved breathing abil-

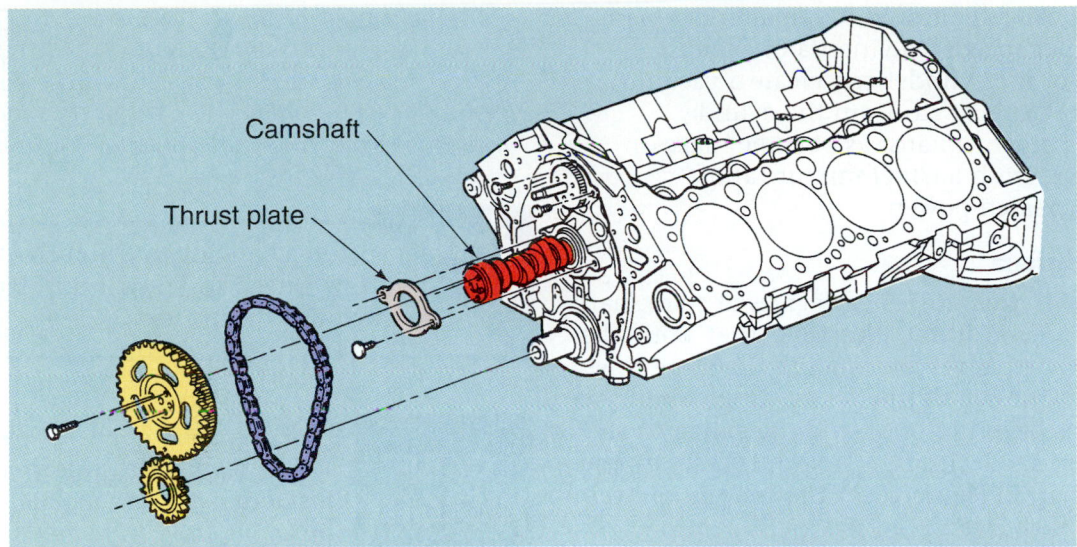

Figure 18.29 Some camshafts have a thrust plate to control end thrust. *(Courtesy of General Motors Corporation, Service Technology Group)*

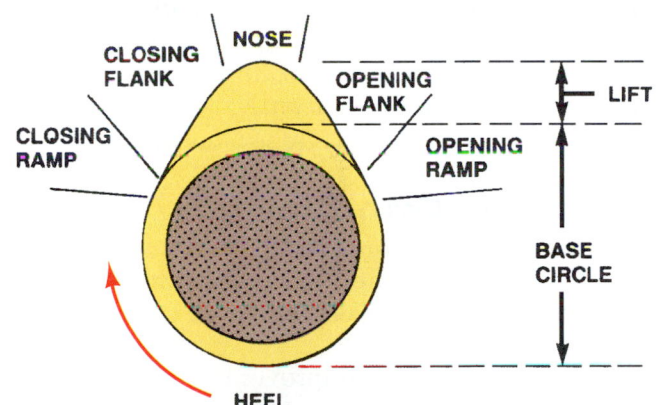

Figure 18.30 Parts of a cam lobe.

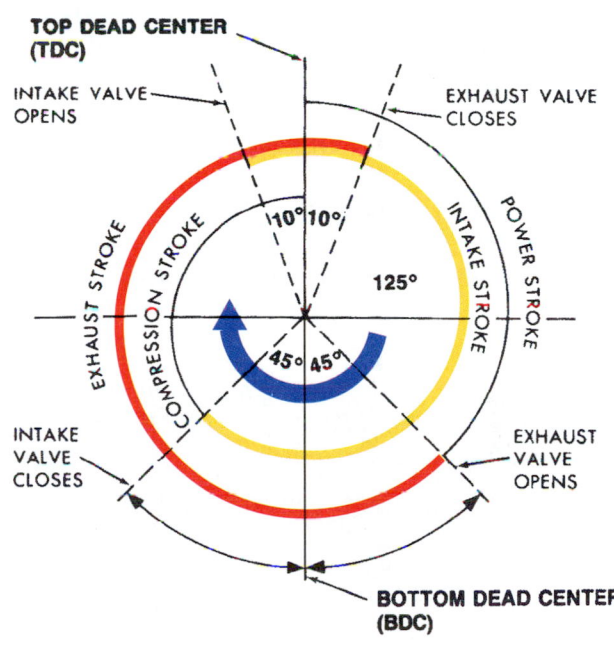

Figure 18.31 A valve timing chart. *(Courtesy of General Motors Corporation, Service Technology Group)*

ity and, therefore, produce more power. These are called artificially aspirated engines.

Cam Lobe Design

Cam lobe design differs from engine to engine, depending on the performance desired. The cam lobe has several parts (Figure 18.30). The **base circle** is what the cam would be if there were no lobe. The lobe has an opening and closing *flank* on each side of its nose.

Each cam lobe is ground to a precise contour. Lobe shape is always a compromise. The two factors of cam lobe profile are called **lift** and **duration**.
- Lift is the height to which the lobe raises the lifter.
- Duration is the number of degrees of *crankshaft* travel while the valve is off its seat.

▬ VALVE TIMING

Figure 18.31 shows an example of a valve timing chart. Several points should be noted:
- The intake valve must open *before* TDC, when the piston starts down on the intake stroke.

- The intake valve stays open long after BDC (into the compression stroke) to take advantage of the velocity of the incoming gases, which helps to "ram" additional air/fuel mixture into the cylinder.
- The exhaust valve must open long before the end of the power stroke in order to bleed off the pressure of expanding gases in the cylinder before the piston tries to move up against that pressure on its exhaust stroke. It does no harm to open the exhaust valve at this point because most of the power of the exploding gases has been delivered to the piston by about halfway through the power stroke.

■ The exhaust valve is allowed to remain open past TDC. At higher speeds, the inertia of exhaust moving out of the cylinder can create a vacuum, sucking more exhaust gas from the cylinder. Removing as much exhaust as possible makes room for more fresh air/fuel mixture and, therefore, increased power.

Valve Overlap

Valve overlap is the number of degrees of crankshaft rotation that occurs during the time that both the intake and exhaust valves are open at the same time. This happens at the top of the intake stroke when the intake valve has begun to open and the exhaust valve has not yet finished closing (Figure 18.32). In the valve timing chart in Figure 18.31 the amount of overlap is 20°.

■ Additional overlap provides more efficient cylinder filling at high rpm, but causes lower engine vacuum, poorer performance, and less fuel economy at idle and low speed. A high overlap cam may have high valve lift. It is designed for high performance, high rpm use.

■ Less overlap results in more pressure in the cylinder at lower rpm. This results in greater torque at low rpm. The "RV" or "high gas mileage" cams fit this profile. It is designed for low rpm use.

NOTE: *RV means uses such as off-road and does not apply to motorhomes that would be travelling mostly at highway speeds.*

NOTE: *Overlap is what gives high-performance engines their distinctive "lope" at idle. A cam grind that will benefit performance at high rpm will hurt performance at low rpm. The reason for this is the inertia (momentum) of the gases. The low vacuum associated with lope at idle can lead to problems such as inconsistent vacuum power brake applications. Overlap is also a factor that can lead to clearance problems between the valves and pistons.*

SHOP TIP Before changing to a non-stock camshaft, you will need to know information that is not included in this text.

CASE HISTORY *While rebuilding an engine, a student installed a performance camshaft. After installing the engine and starting it, he discovered that it wouldn't idle. Engine vacuum was low at lower engine speeds. Power didn't come on until an unacceptably high rpm. Fuel economy was terrible. He had to buy an auxiliary vacuum reservoir so his power brakes would work properly. When he finally decided to replace the camshaft, he was much happier with his car's performance.*

Changes in Cam Timing

Most cams are ground with 4°–6° advance for better low-end torque and to compensate for timing chain stretch. The results of changes in cam timing are as follows:

■ Advanced cam timing improves low and mid-range torque.

■ Retarded cam timing improves high rpm power.

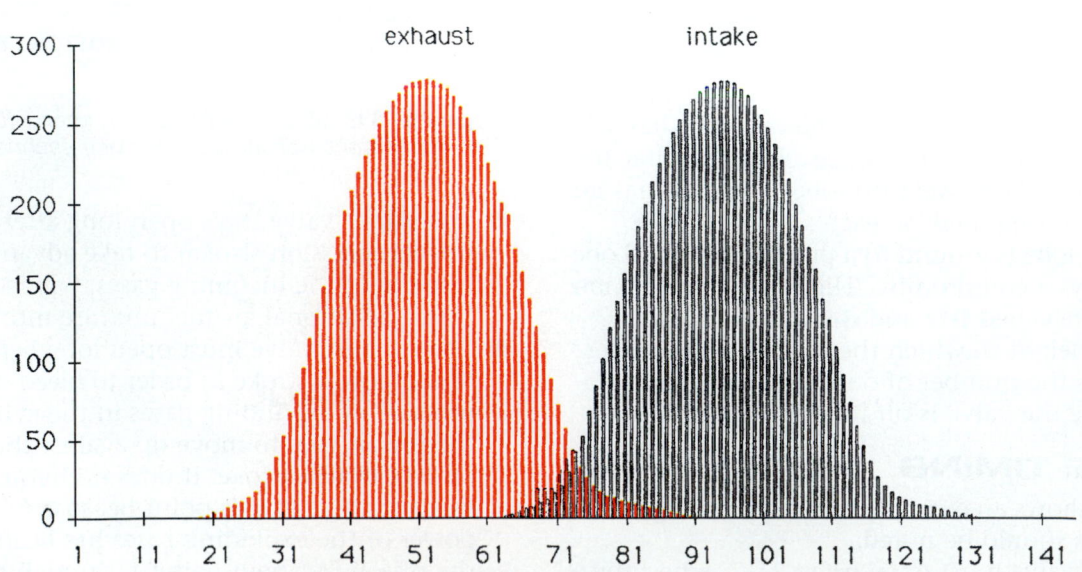

Figure 18.32 Valve overlap. *[Courtesy of Elgin Racing Cams]*

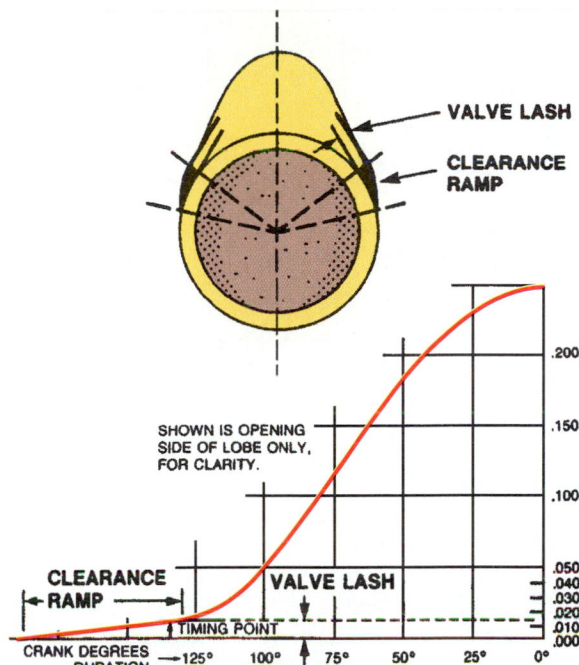

Figure 18.33 A clearance ramp cushions the opening and closing of the valves. *(Courtesy of Ed Isk-enderian Racing Cams)*

Mechanical Valve Lash

With solid lifter cams, valve adjustment is very important. Solid cams have a clearance ramp at the beginning and end of each lobe. This clearance ramp can be compared to a wedge. Its purpose is to cushion the opening and closing of the valve (Figure 18.33). A change of 0.001" in valve lash will result in approximately 3° of change in duration. Too much clearance will cause the lifter to miss the clearance ramps altogether. This in turn, can cause extra shock loads that can seriously damage valve train parts.

■ VALVE LIFTERS

The lifter is made of the same material as the camshaft. Lifters are also known as cam followers, or flat *tappets*. They can be either mechanical or hydraulic. *Mechanical lifters* are also called *solids* or *tappets*.

Hydraulic lifters are found on almost all later-model engines with pushrods, and also on some OHC engines.

NOTE: *A new lifter will often have a small, shallow pinhole in its base. This hole is normal and is a mark left during hardness testing.*

Lifter and Cam Lobe Relationship

The lifter and cam lobe have a unique relationship. On most pushrod engines, the lifter face has a convex shape (protruded face), and it is offset from the center of the cam lobe (Figure 18.34). In addition, the cam lobe is tapered from approximately 0.0007" to 0.002" across its face. These factors are all introduced to cause

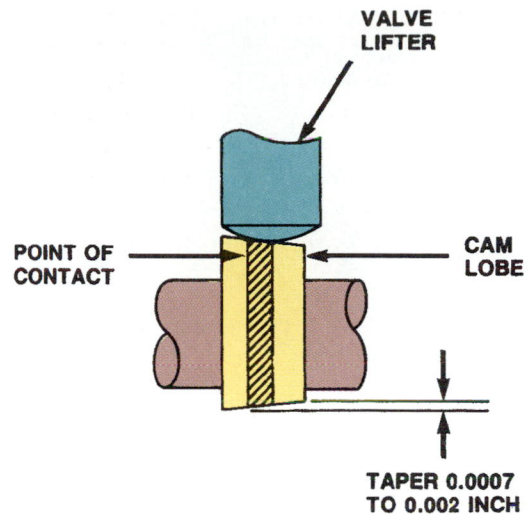

Figure 18.34 The lifter face is convex and is offset from the center of the cam lobe.

Figure 18.35 Two examples of wear that results when a lifter fails to spin. *(Courtesy of TRW, Inc.)*

the lifter to spin. Spinning is desirable because it helps to dissipate the tremendous load—up to 100,000 psi—that the lifter is subjected to. Figure 18.35 shows a lifter that failed to spin.

The lifter's convex shape also helps to prevent *edge loading*, which can destroy a cam lobe. Because only the center of the lifter contacts the face of the cam lobe, normal wear occurs near the center of the cam lobe (Figure 18.36). If the lifter face contacts the outside edge of the cam lobe, severe loading causes rapid wear of the cam lobe making the lifter face become concave. A concave lifter will wear the edges of a cam lobe (Figure 18.37).

■ HYDRAULIC LIFTERS

Hydraulic lifters automatically maintain a condition called **zero-lash**. This is a point where there is no clearance and no interference between parts of the valve opening mechanism. Hydraulic lifters eliminate the need for periodic valve adjustments, are quiet, and reduce unnecessary valve train wear.

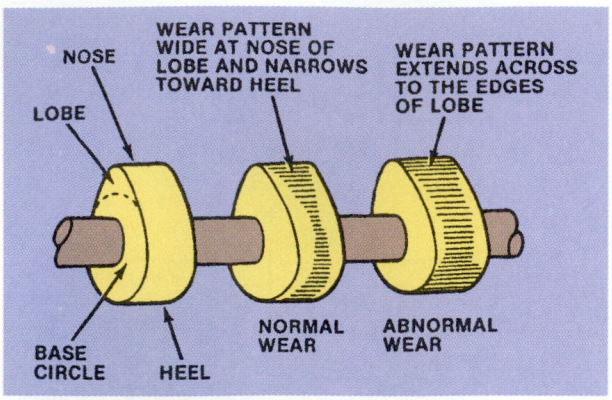

Figure 18.36 Normal and abnormal camshaft wear. *(Courtesy of Ford Motor Company)*

Hydraulic Lifter Construction

The lifter consists of a plunger, a plunger return spring, and a check valve assembly (Figure 18.38). The plunger is made of steel that is chrome plated to resist wear and corrosion. Lifters from different manufacturers sometimes have different outward appearances, even though they are for the same engine.

Hydraulic Lifter Operation

Whenever clearance occurs in the valve train, a spring between the plunger and lifter body causes the lifter to

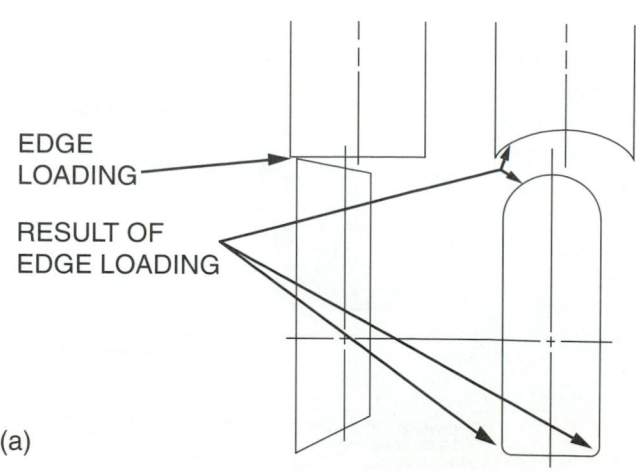

(a)

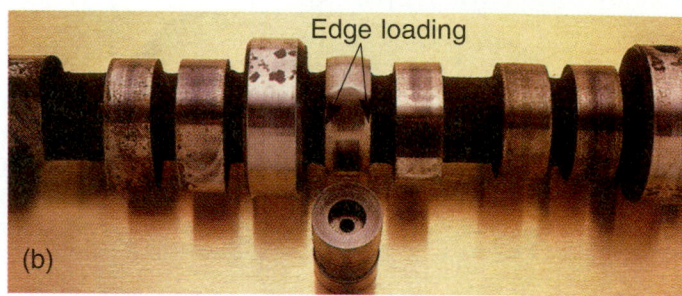

(b)

Figure 18.37 (a) A convex lifter prevents edge loading. (b) A severely worn camshaft. *(Courtesy of TRW, Inc.)*

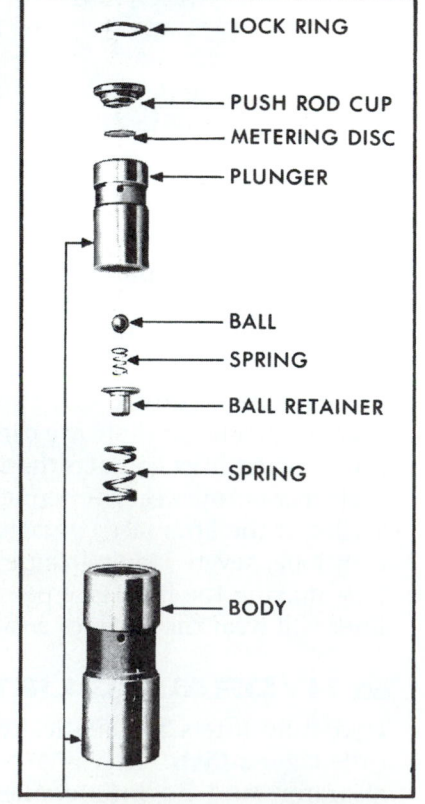

PLUNGER AND BODY ARE FITTED PAIRS AND MUST NOT BE MISMATED.

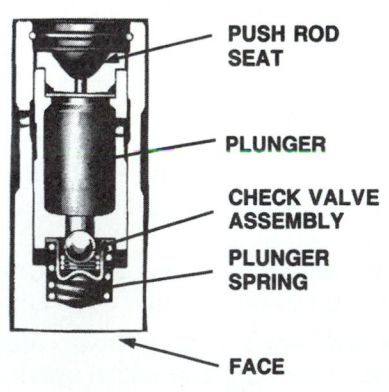

Figure 18.38 Parts of a hydraulic valve lifter. *(Courtesy of General Motors Corporation, Service Technology Group)*

expand (Figure 18.39). Oil under pressure fills the cavity created under the plunger. A very small amount of leakage of oil between the plunger and the lifter body allows a lifter with too much oil in it to leak down. The lifter "*leaks down*" as the other valve train parts expand when the engine warms up.

Lifter leak-down is very small, because there is only about 0.0002" (count the zeros) clearance between the plunger and the lifter body.

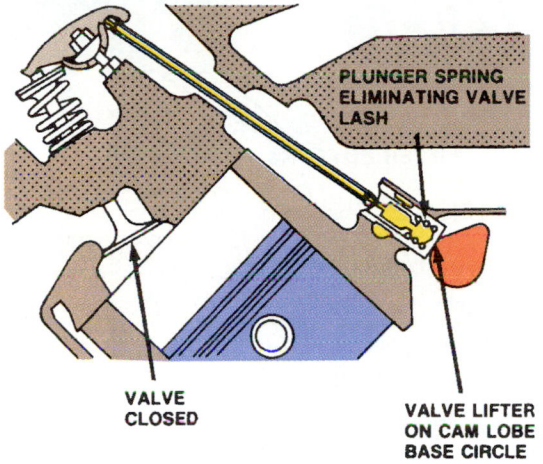

VALVE
CLOSED

PLUNGER SPRING
ELIMINATING VALVE
LASH

VALVE LIFTER
ON CAM LOBE
BASE CIRCLE

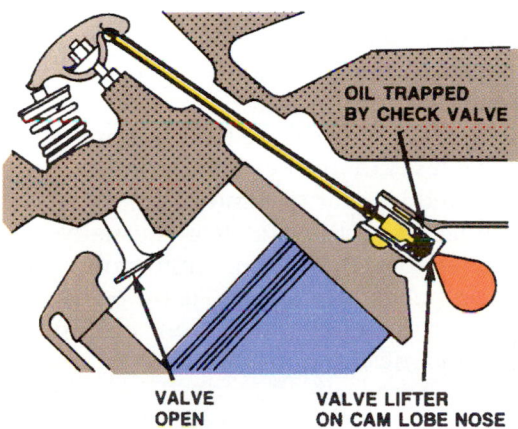

VALVE
OPEN

OIL TRAPPED
BY CHECK VALVE

VALVE LIFTER
ON CAM LOBE NOSE

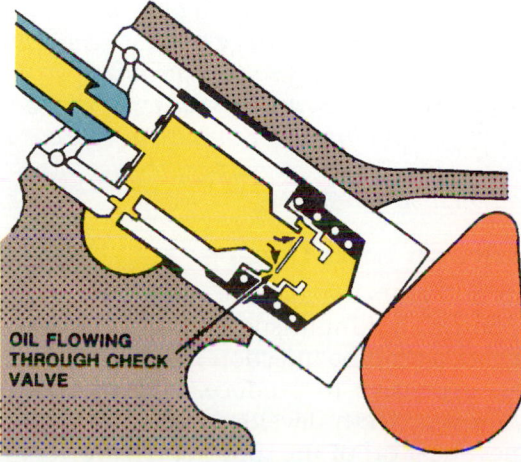

OIL FLOWING
THROUGH CHECK
VALVE

Figure 18.39 Hydraulic lifter operation.

MATH NOTE	
0.200	*two hundred thousandths*
0.020	*twenty thousandths*
0.002	*two thousandths*
0.0002	*two tenths of a thousandth*

This small clearance is beneficial because dirt particles are usually larger than 0.0002" so they cannot easily lodge between the lifter body and plunger.

Overhead Cam Hydraulic Lash Adjusters. An overhead cam (OHC) with hydraulic lash adjusters is shown in Figure 18.40. It works in the same manner that an ordinary hydraulic lifter works.

Overhead Camshaft Lubrication

An OHC camshaft does not have oil thrown off from the valve lifters like a cam-in-block engine does. Some OHC engines have hollow camshafts with oil holes on each cam lobe to provide pressure lubrication. These camshafts have long lives as long as the oil holes are kept clean. Periodic oil changes are especially important in these engines.

ROLLER CAM AND LIFTERS

In the mid-1930s, the *roller lifter* was developed (Figure 18.41). In earlier years, roller lifters have been used in diesels and race cars. They are now being used in production gasoline engines. A roller lifter can accept a much higher rate of movement than a flat tappet without wear to the lifter or cam lobe. The roller lifter

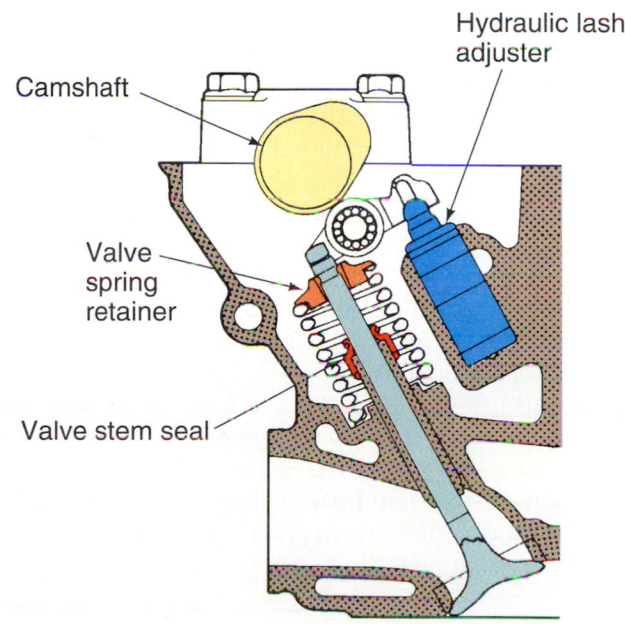

Hydraulic lash
adjuster

Camshaft

Valve
spring
retainer

Valve stem seal

Figure 18.40 An overhead cam head with hydraulic lash adjusters. *(Courtesy of Hyundai Motor America)*

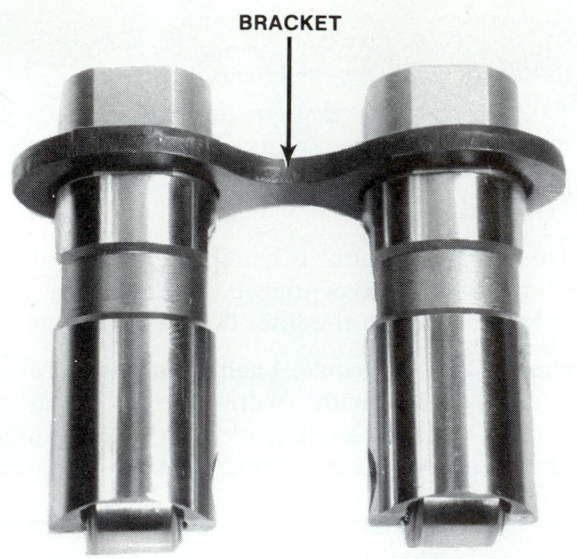

Figure 18.41 Roller lifters are kept from turning in their bores.

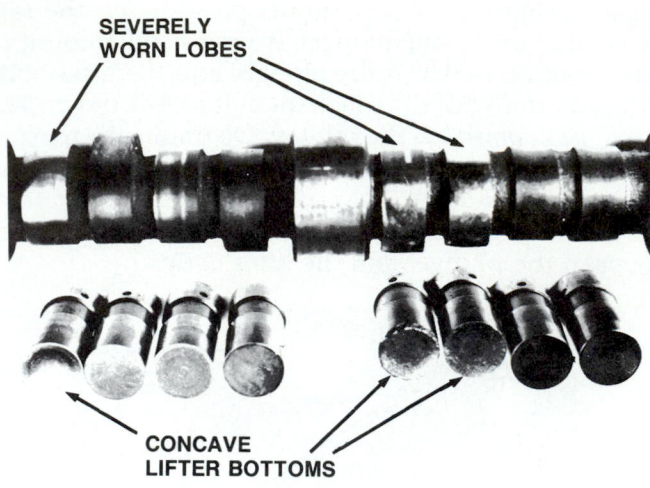

Figure 18.42 This cam has several worn lobes. Notice that the corresponding lifters are concave. [Courtesy of Federal-Mogul Corporation]

cuts valve train friction almost in half, resulting in an increase in horsepower and fuel economy.

Roller lifters need a means to keep them from turning. If the lifter could turn sideways, the roller would not work and damage would result. Either a pin, or a bracket, holds two lifters in the proper plane.

A *roller camshaft* is made of steel. The lobes have a different shape than standard cam lobes, which allows faster acceleration rates (the valve can be opened faster).

REGRINDING CAMS

Cam *lobes* can be reground. But cam *journals* rarely wear, so they are not reground. Although reground lobes are smaller, they can be reground to the same specifications as the original cam. By removing more metal from the bottom of the lobe, it can be shaped to the same contour as the original cam. A reground cam usually costs much less than a new cam. These cams have characteristics similar to a new cam, and are acceptable if they are properly remanufactured. A reground cam should retain substantial hardness.

New and reground cams are made of cast iron coated with a dark material that provides resistance to scuffing, especially during the break-in period between the cam and lifter.

Sometimes cam lobes wear severely (Figure 18.42). This usually happens on the exhaust lobe because it has to open against combustion pressure. An old cam is usually not acceptable as a **core** if more than two of its lobes are worn "flat."

NOTE: *A core is a part that is returned by a customer when a rebuilt or remachined part is provided.*

Flat cam lobes may be welded to build up enough metal for regrinding. It is usually impractical to repair a cam that requires welding on more than two lobes unless the cam is from a heavy-duty vehicle.

CAM DRIVES

The cam is driven by the crankshaft by one of three methods (see Figure 16.7):
- Gear drive
- Sprockets and timing chain
- Sprockets and timing belt

Gear drives are more positive than a chain (less chance of excessive **backlash** or clearance between the parts) but are limited by gear size and cost.

To reduce noise the cam gear is usually made of a soft material, either fiber or aluminum. Gear-drive engines use helical gears, (see Figure 64.9), the crank drives the cam in the direction opposite to the direction of crank rotation.

NOTES:
- *A conversion kit that changes a chain drive to a gear drive is available from racing parts distributors. It uses idler gears to make the cam and crank rotate in the same direction. A disadvantage of these gear drives is that they have spur gears so they are noisy.*
- *In some V-type engines used for racing or in expensive automobiles, a multiple gear arrangement drives dual overhead cams. This is impractical for use in conventional passenger cars.*

Sprockets and Gears

Sprockets and gears are made of steel, iron, aluminum, or fiber. Some aluminum sprockets have nylon teeth. These sprockets can be injection molded, which makes them less expensive to produce. When a chain used with nylon gear teeth develops slack, the nylon teeth sometimes shear off of the sprocket (Figure 18.43). The broken teeth can stick in the oil-pump relief valve.

Figure 18.43 Some of the nylon teeth are missing from the camshaft sprocket of this timing set.

Nylon gears should not be used in heavy-duty vehicles, such as trucks or high performance cars with standard transmissions. A worn sprocket (Figure 18.44) should not be used with a new chain; it must be replaced.

Timing Chains

The two types of chains are the *roller chain* and the *silent chain* (Figure 18.45). *Double roller chains* are usually used on overhead cam engines.

Pushrod engines usually use a silent chain. New silent chains are of the *large pin* design. These "floppy" chains are twice as strong as older silent chains. They are less susceptible to misalignment and overload problems than roller chains, due to greater flexibility. The large pin silent chain eliminates chordal action; a problem that roller chains have. Chordal action means the chain operates on a constantly changing diameter, which causes varying camshaft speeds.

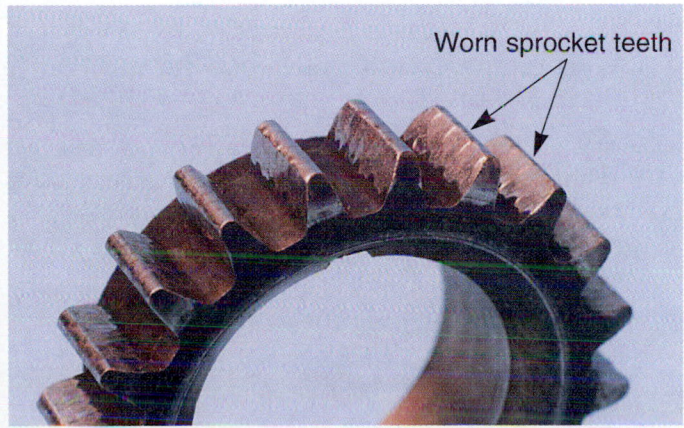

Figure 18.44 The teeth on this timing sprocket show wear. It should not be reused.

(a) SILENT

(b) ROLLER

Figure 18.45 Timing chain types.

Freewheeling and Non-freewheeling Engines

As an engine ages, its timing chain stretches. In fact, this is really the main obstacle to long engine life. If the valve timing is wrong or the cam sprocket keyway becomes stripped in a **freewheeling engine**, the pistons and valves cannot come into contact with each other (as piston-to-valve clearance is always maintained).

Some engines will experience piston-to-valve interference if the timing chain or belt skips or breaks. These engines are called **non-freewheeling** or **interference engines**. When this happens on a pushrod engine, pushrods will usually bend (Figure 18.46). The valves on a pushrod engine might survive,

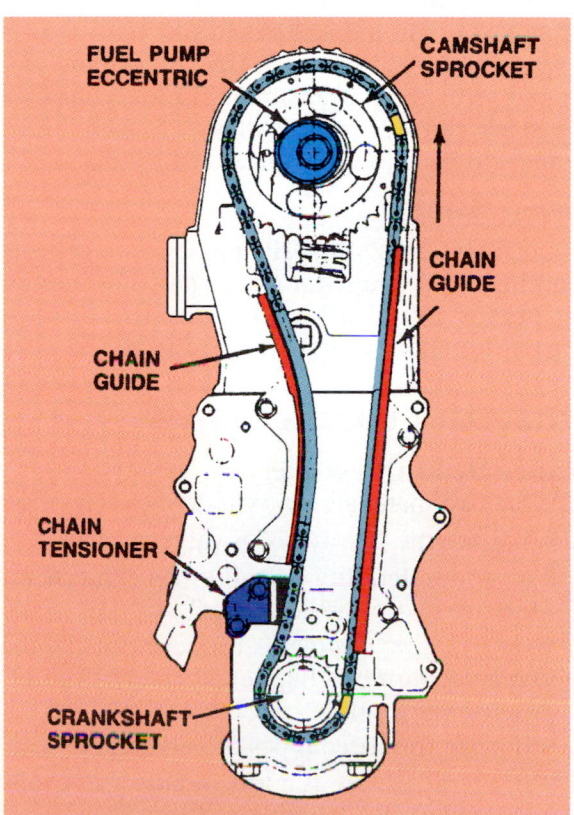

Figure 18.46 Parts of an OHC chain drive. [*Courtesy of Nissan Motors*]

but on OHC engines, the valves must bend (since there are no pushrods to bend under the stress).

Timing chains have a drive, or tension, side and an opposite side where slack accumulates. Long chains, such as those used on OHC engines, use chain tensioners to take up slack as the chain stretches (Figure 18.47). Some chain tensioners are spring loaded and hydraulic. Timing chains in these engines can make noise until the engine runs for a few seconds and the tensioner gets oil pressure (making it move against the chain free play).

Figure 18.47 This pushrod was bent when a valve collided with a piston. *(Courtesy of Federal-Mogul Corporation)*

CASE HISTORY

A high mileage non-freewheeling OHC engine with a hydraulic chain tensioner suddenly failed on start-up. The owner had parked the car and left the transmission in gear. The emergency brake was broken. Someone nudged the bumper while parallel parking in front of the car. This moved the car backward, turning its engine over backward, too. Because the engine had a hydraulic chain tensioner, there was too much slack in the chain without the engine running. Its chain skipped cam timing. When the engine was started, all exhaust valves became bent at the same angle (Figure 18.48).

In the preceding case history story, the valves were bent because the cam timing became too late, and the exhaust valves were still open too far when the pistons reached TDC on their exhaust strokes. The same thing can happen to an engine that has a chain that is not equipped with a tensioner. In an engine that is equipped with a non-tensioner timing chain, excess chain slack can occur and allow the cam timing to change (skip teeth on the sprockets).

Timing Belts

The camshafts on some OHC engines are driven by a cogged belt drive, called a *timing belt*. Oil pumps and water pumps can also be driven by the timing belt. Compared to chain drives, belt drives are:

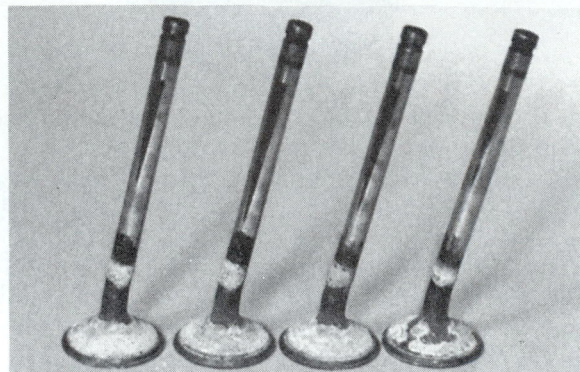

Figure 18.48 These exhaust valves from a 4-cylinder OHC engine were all bent when the timing chain skipped.

- Quieter
- Less costly
- Lighter in weight
- Easier to replace
- Lubrication-free
- Stretch-free

NOTE: *A non-freewheeling engine often has a decal on the valve cover stating that the belt must be changed at least every 60,000 miles. The mileage amount varies. Many late model belts are designed to go even longer. More information of timing belts is covered in Chapter 22.*

■ REVIEW QUESTIONS

Cylinder Head Questions

1. What two materials can cylinder heads be made of?
2. What are the two types of valve guides?
3. If a car decelerating on a hill has exhaust smoke, what are two possible causes?
4. The _____ effect is the term for the reason that a vacuum can happen at the bottom of the exhaust valve guide.
5. What are the three types of valve guide seals?
6. What temperature is a low-quality valve guide seal good until? A high-quality seal?
7. What is the name of the process that hardens integral valve seats?
8. What can happen to a valve that has a margin that is too thin?
9. Why would a spring have two coils?
10. What is the name of the type of rocker arm stud that does not require a valve adjustment?

Camshaft/Cam Drive Questions

1. What are three things that the camshaft can drive besides opening the valves?
2. What causes the camshaft to thrust toward the rear of the engine?
3. An engine without a turbocharger or supercharger is called a naturally _____ engine.

4. What is the term for the time when both the intake and the exhaust valves are open at the same time?

5. If the intake valve opens 15° before TDC and the exhaust valve closes 10° after TDC, how much valve overlap does the camshaft have?

6. What is the name for when there is no clearance between parts of the valve opening mechanism?

7. What is the name for the kind of lifter that must be held from turning?

8. List three ways that a camshaft can be driven.

9. The name for a part that is returned to replace a rebuilt or remachined part is a _____ .

10. When an engine breaks a timing belt but cannot possibly have interference between the valves and pistons, this type of engine is called a _____ engine.

◼ ASE STYLE REVIEW QUESTIONS

Cylinder Head Questions

1. Technician A says that integral valve guides are found only in aluminum heads. Technician B says leaking valve guide seals account for nearly one-half of all oil consumption complaints. Who is right?
   - **a.** Technician A
   - **b.** Technician B
   - **c.** Both A and B
   - **d.** Neither A nor B

2. Technician A says that a leaking valve guide seal can result in oil consumption. Technician B says that a leaking valve guide seal can result in engine breathing difficulties. Who is right?
   - **a.** Technician A
   - **b.** Technician B
   - **c.** Both A and B
   - **d.** Neither A nor B

3. Technician A says that an umbrella seal moves up and down with the valve stem. Technician B says that a positive valve guide seal is attached to the top of the valve guide. Who is right?
   - **a.** Technician A
   - **b.** Technician B
   - **c.** Both A and B
   - **d.** Neither A nor B

4. Technician A says that exhaust valves are larger than intake valves. Technician B says that intake valves run at cooler temperatures than exhaust valves. Who is right?
   - **a.** Technician A
   - **b.** Technician B
   - **c.** Both A and B
   - **d.** Neither A nor B

5. Technician A says that a valve spring with an inner and outer spring will have both coils wound in the same direction. Technician B says that exhaust valves are usually magnetic. Who is right?
   - **a.** Technician A
   - **b.** Technician B
   - **c.** Both A and B
   - **d.** Neither A nor B

Camshaft/Cam Drive Questions

1. Technician A says that the oil pump usually drives the distributor. Technician B says that duration is the number of degrees that the camshaft turns while the valve is closed. Who is right?
   - **a.** Technician A
   - **b.** Technician B
   - **c.** Both A and B
   - **d.** Neither A nor B

2. Technician A says that an engine with a high overlap cam will idle smoothly and not make as much power at higher speeds. Technician B says that an RV cam is designed to be used at highway speeds. Who is right?
   - **a.** Technician A
   - **b.** Technician B
   - **c.** Both A and B
   - **d.** Neither A nor B

3. Technician A says that the lifter on a pushrod engine is supposed to be concave on its bottom. Technician B says that the cam lobe on a pushrod engine is tapered. Who is right?
   - **a.** Technician A
   - **b.** Technician B
   - **c.** Both A and B
   - **d.** Neither A nor B

4. Technician A says that another name for a lifter is a tappet. Technician B says that the lifter should contact the entire width of the cam lobe. Who is right?
   - **a.** Technician A
   - **b.** Technician B
   - **c.** Both A and B
   - **d.** Neither A nor B

5. Technician A says that a flat tappet lifter is supposed to spin in its bore. Technician B says that a roller lifter is supposed to spin in its bore. Who is right?
   - **a.** Technician A
   - **b.** Technician B
   - **c.** Both A and B
   - **d.** Neither A nor B

CHAPTER 19

Engine Lower End and Lubrication System Theory

■ OBJECTIVES

Upon completion of this chapter, you should be able to:

✔ Describe the related theory of all of the parts that make up the lower end.

✔ Tell how a cylinder block is made and understand the functions of its parts.

✔ Understand how pistons are constructed and the reasons for their various design considerations.

✔ Discuss the various types of piston rings and be able to make the correct choice when selecting rings for an engine.

✔ Know about the various types of engine bearings.

✔ Describe the parts of the crankshaft and their functions.

■ KEY TERMS

lower end
cylinder bore taper wear
torsional vibration
bearing spread
bearing crush
cam ground piston
bypass valve

■ INTRODUCTION

The crankshaft assembly, the piston, and the rod are referred to as the **lower end**. This chapter deals with all lower end parts, various service procedures performed on them, and diagnosis of part failures. Also included are coverage of the engine block and lubrication system.

■ CYLINDER BLOCK CONSTRUCTION

The cylinder block, which is made of cast iron or aluminum, is cast around a sand mold called a *core* (Figure 19.1). The core is suspended in a container with a liner that will provide the shape for the outside surface of the engine block. The core is supported at several points around the outside of the core box, which leaves core holes in the finished block. When molten iron is poured into the core box, the heat of the casting process cooks the sand. When the casting cools, the sand breaks up. The casting is shaken out. Any remaining sand is washed away through the core holes leaving the finished casting, complete with water jackets. The core holes are closed off with core plugs (Figure 19.2).

■ CORE PLUGS

Core plugs are usually made of steel or brass, although rubber and copper expandable plugs are available, too. Brass core plugs are superior because they do not rust.

Brass plugs are not used in new cars because of their extra cost, and because new engines are filled with coolant, which prevents rust. Core plugs are also known as expansion plugs, welsh plugs, freeze plugs, or soft plugs.

■ CYLINDER BORE

Cylinders are bored in the block. They wear in a *taper* and *out-of-round* fashion. The top of the cylinder receives less lubrication, and it is subjected to the high pressure of the piston rings against the cylinder wall as the air/fuel mixture is ignited. This causes **cylinder bore taper wear** (Figure 19.3), which forms the ring ridge at the top of the ring travel.

■ CYLINDER SLEEVES

Some engines use replaceable cylinder bores, called wet or dry *sleeves*. Aluminum blocks usually have permanently installed iron cylinder sleeves. Sometimes a cylinder wall is cracked or rusted or the block has a serious defect (see Figure 47.3). If all the other cylinders are in good shape, the damaged cylinder can be bored oversize to accept a *dry* sleeve installed with an interference fit. Sleeves usually come in 1/8" and 3/32" wall thicknesses. Some heavy-duty engines use removable *wet* sleeves. These differ from conventional sleeves in that wet sleeves only contact the block at the upper and lower ends. The middle portion of the sleeves are

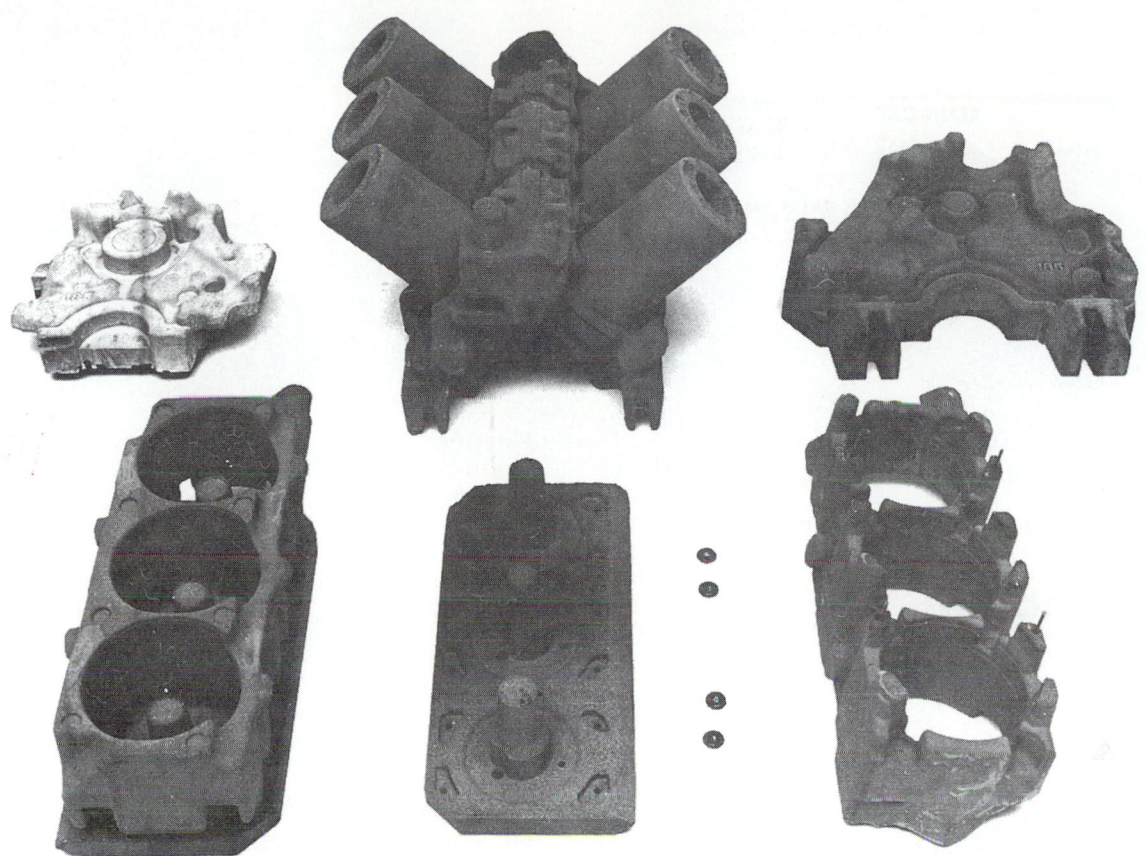

Figure 19.1 The casting process. Casting cores. *(Courtesy of General Motors Corporation, Service Technology Group)*

Core plugs

Figure 19.2 Core plugs.

exposed to coolant in the water jackets so the sleeves must be sealed at each end.

◼ MAIN BEARING CAPS

Main bearing bores are bored at the factory with the bearing caps in place. Main caps are not interchangeable and must be returned to the same bearing bores

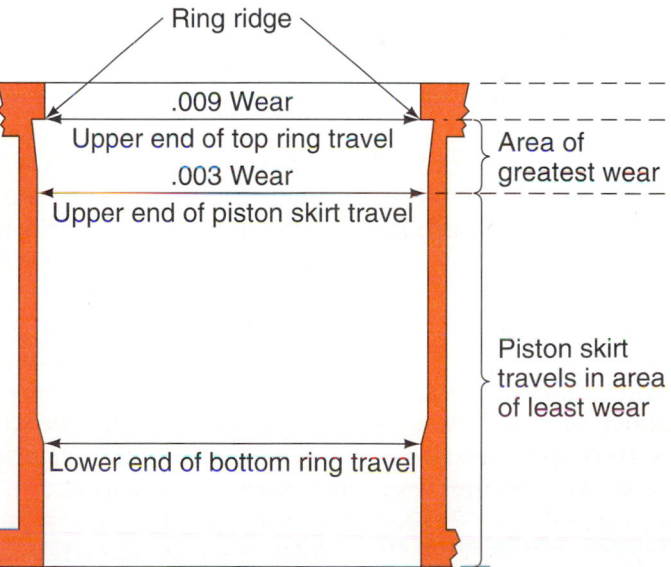

Ring ridge

.009 Wear

Upper end of top ring travel

.003 Wear

Upper end of piston skirt travel

Area of greatest wear

Piston skirt travels in area of least wear

Lower end of bottom ring travel

Figure 19.3 A ring ridge. *(Courtesy of Dana Corporation-Perfect Circle Division)*

from which they were removed. Some heavy-duty engines use four-bolt main caps instead of two-bolt main caps for extra strength (Figure 19.4).

Figure 19.4 Four-bolt main bearing caps.

LIFTER BORES

Lifter holes are bored in the block on engines with camshafts in the block (pushrod engines). They have very little clearance to the lifters and the lifters spin in the lifter bores.

CRANKSHAFT DESIGN

The crankshaft (see Figure 15.20) converts the up-and-down (reciprocating) motion of the pistons to rotary motion. Its polished bearing surfaces are called *journals*. The journals that support the crankshaft as it turns in the block are called *main bearing journals*. Those bearing journals that are in line on the same axis as the front and rear journals are all mains.

Journals that are offset from the main bearing journal centerline are called *rod journals* or *crankpins*. Connecting rod journals transfer up-and-down motion between the crankshaft and connecting rod. Rod journals are also known as crankpins. As described in Chapter 15, connecting rod journals are offset at 90° angles for eight cylinders, 120° angles for even-fire six cylinders, and 180° angles for four cylinders.

The crankshaft has oil passages drilled from the main journals to the rod journals. This allows oil under pressure to reach the connecting rod journal. With proper maintenance, crankshafts often last until a complete engine overhaul with very little wear.

Counterweights

Opposite each rod journal on the crankshaft is a counterweight that precisely balances the combined rotating mass of the offset rod journals and the rod (see Figure 15.22). The counterweight is much heavier than the rod journal to compensate for the opposite reciprocating weight of the connecting rods, bearings, and piston assembly.

An 8-cylinder, V-type crank sometimes has fewer main bearing journals than an inline 6-cylinder. Figure

Figure 19.5 Crankshaft design differences. From top to bottom: A V-8 crank, an inline 6 crank, and an inline 4 crank.

19.5 compares the crankshafts of an inline 4, an inline 6, and a V-8. V-8 engine rod journals are wide enough for two connecting rods (see Figure 16.3).

A V-6 that has 60° between the banks, like the smaller V-6, is *even-fired* (a cylinder fires every 120° of crank rotation). Some 90° V-6 engines are also even-fired, and some are not. To achieve 120° even-firing with a 90° V-6 block requires the crankpins to be offset. The amount of offset varies between manufacturers. Figure 19.6 shows the individual offset rod journals found on some General Motors V-6 engines.

Cast or Forged Cranks

Crankshafts are either cast or forged. Forged cranks are stronger, but more costly. Cast cranks are of high enough quality to do an adequate job.

NOTE: *The cast crank has a very narrow parting line where the casting mold halves were joined. The forged crank has a wider, ground-off area. The difference between the two types can be seen in Figure 19.7.*

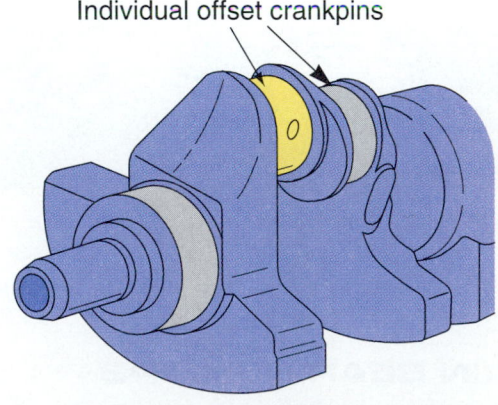

Individual offset crankpins

Splayed crankpin

Figure 19.6 A V-6 crankshaft with splayed crankpins. *(Courtesy of General Motors Corporation, Service Technology Group)*

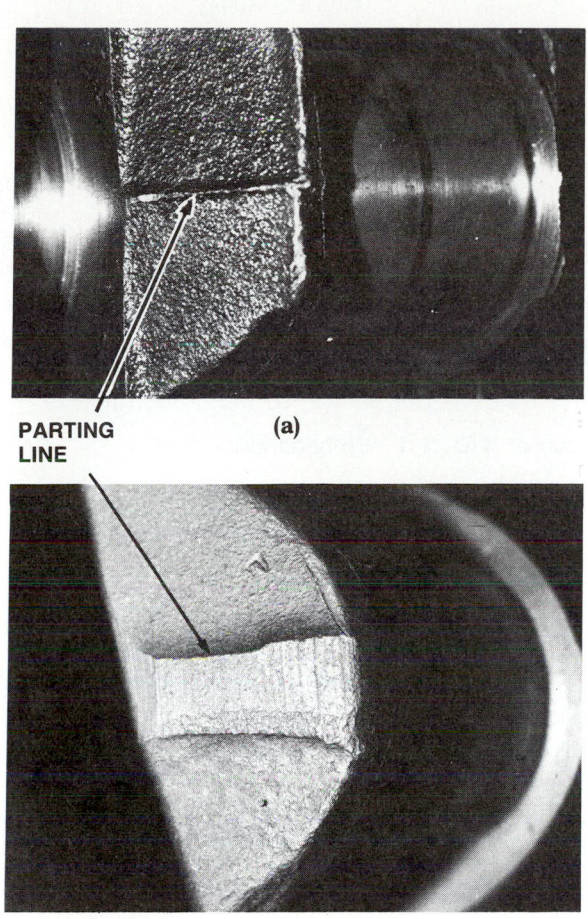

Figure 19.7 (a) Cast crank. (b) Forged crank.

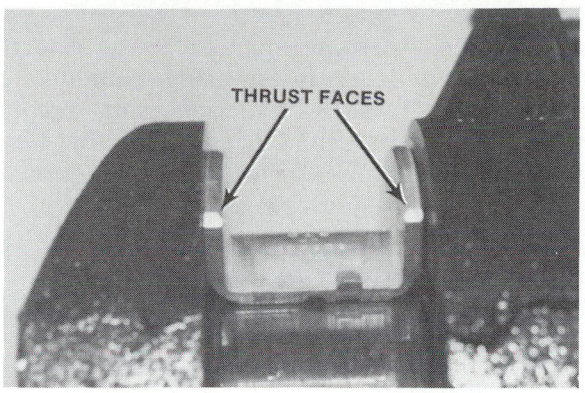

Figure 19.8 Main bearing thrust insert. *(Courtesy of Ford Motor Company)*

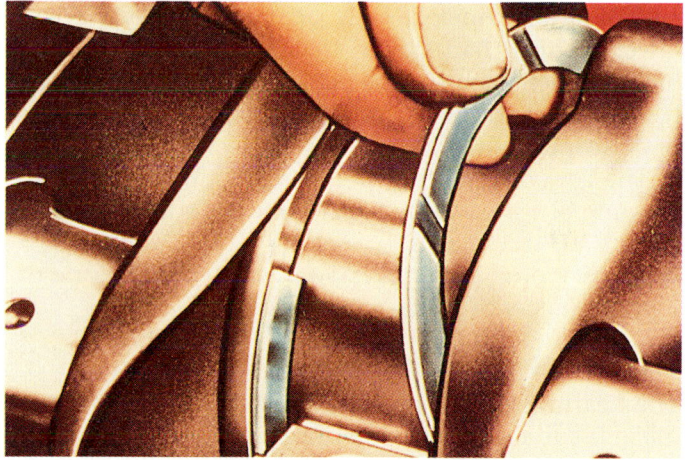

Figure 19.9 Thrust inserts. *(Courtesy of Ford Motor Company)*

Cast cranks must have larger counterweights because cast metal is not as dense as forged metal, so therefore it is lighter.

CRANKSHAFT END THRUST

The crankshaft is pushed forward by pressure in the torque converter or the release spring pressure of the clutch. This is called *end thrust*. One of the crankshaft main bearings has a precision bearing surface ground on its sides, called thrust surfaces (see Figure 15.11). A *flanged thrust bearing* fits between the crankshaft thrust surfaces (Figure 19.8) controlling back-and-forth movement (end thrust). Sometimes thrust insert half-washers are used instead of a flanged bearing (Figure 19.9).

DIRECTION OF CRANKSHAFT ROTATION

The Society of Automotive Engineers (SAE) has a standard for engine rotation. Automotive engines rotate counterclockwise when viewed from the flywheel end of the engine. *Transverse* mounted engines also follow this standard. A *longitudinal* mounted engine (called

front-to-rear or north-to-south) would turn clockwise when viewed from the front.

VIBRATION DAMPER

During combustion, the force on the piston twists the crankshaft and the crankshaft tries to straighten itself out again. It overcorrects, twisting back in the other direction. Like a tuning fork, these *oscillations* occur for several cycles before fading out. When oscillations from several cylinders are occurring at the same time, a **torsional vibration** can actually cause the crankshaft to break. Timing chain and sprocket wear can also result. Most of the vibration occurs at the front of the crankshaft because the flywheel is resistant to oscillations.

The vibration damper or *harmonic balancer* at the front of the crankshaft dampens torsional vibration. It has a heavy outer *inertia ring* and an *inner hub* separated by a synthetic rubber strip (see Figure 15.16). The two parts stretching against the rubber strip absorb vibrations. Four-cylinder engines usually have only a pulley and do not use a damper. Six- and eight-cylinder engines usually have dampers.

CRANKSHAFT HARDNESS

Crankshafts used by some manufacturers (mostly imports and heavy-duty) are specially hardened. If these cranks are reground, the manufacturers recommend that they be rehardened. Crankshafts that have *not* been previously hardened can suffer misalignment if they are subjected to hardening.

NOTE: *Most crankshafts are not specially hardened, but they tend to work-harden with use. Nickel in the crankshaft comes to the surface. A used, polished crankshaft will have a yellow tint because of this nickel.*

BEARINGS

Bearing Inserts

Bearing inserts are made from many different materials. Dissimilar metals slide against each other with lower friction than similar metals, so bearings usually have a steel back with a soft alloy surface. Bearings have been made with surfaces of babbit, copper-lead, or aluminum alloy.

- The copper-lined bearings are premium bearings.
- Some bearings are made of solid aluminum alloy with no steel backs.
- Multilayered bearings have steel backs with one or more layers of lining material (Figure 19.10).

Bearing Properties

The bearing surface has three primary properties that make it suitable for use in an engine:

- *Embeddability* (Figure 19.11), helps prevent crankshaft wear. When foreign particles are forced between the bearing and journal, they are absorbed by the softer bearing material. When the particle is too large to be completely absorbed, it creates a high spot on the bearing.
- *Conformability* is when the crankshaft journal surface is damaged and the bearing metal "flows" to conform to it (Figure 19.12).
- *Fatigue strength* is the bearing's ability to withstand intermittent loads without deteriorating.

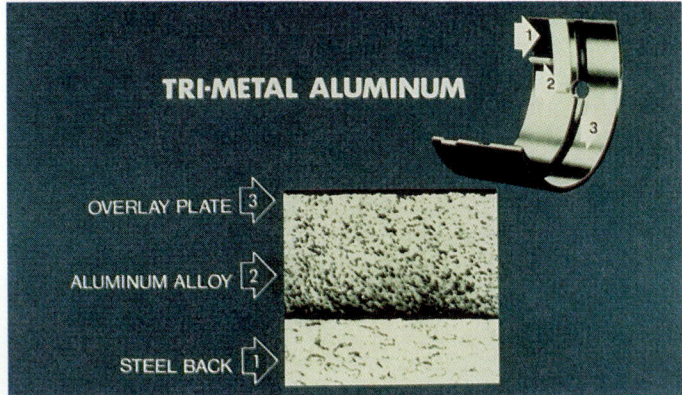

Figure 19.10 Multilayered steel-backed bearings. [Courtesy of AE Clevite Engine Parts]

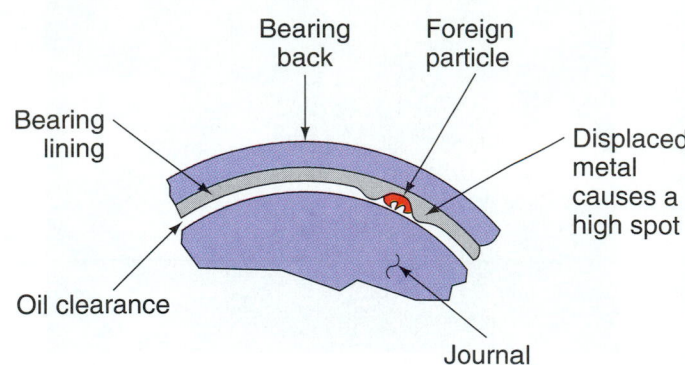

Figure 19.11 Embeddability.

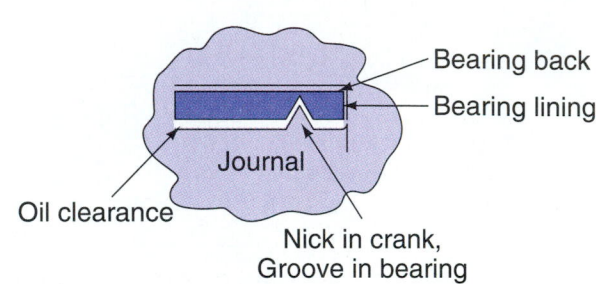

Figure 19.12 Conformability.

Different bearing materials are best for specific uses. Babbitt bearings, which have the best embeddability, were widely used in the past, but the loads of modern engines call for more load-carrying capacity.

Bearings are subjected to normal loads from three sources:

- Pressure from the flame front against the piston
- Centrifugal force from the rotating weight of the rod and crankpin
- Inertia from the up-and-down motion of the piston and rod assembly

Most bearing manufacturers prefer tri-metal bearings, which can carry at least three times the load of babbitt bearings and are acceptable in the other categories. Aluminum bearings have the best load-carrying capacity but are poor in embeddability.

Locating Lug and Oil Groove

Bearing inserts are positioned in the bearing bore by a locating lug (tang) or by a *dowel* (Figure 19.13).

NOTE: *Locating lugs are used to locate or position the bearing laterally; not to keep it from spinning.*

The lug is located on the parting face. It fits into a recess in the bearing bore. When a dowel is used, it fits into a hole in the bearing back.

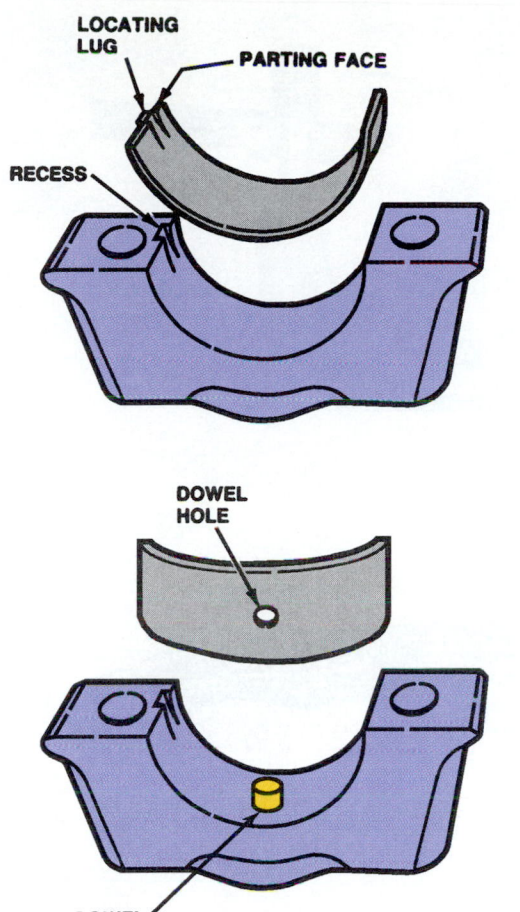

Figure 19.13 Bearings are located in the bearing bore by a locating lug or a dowel. *(Courtesy of Federal-Mogul Corporation)*

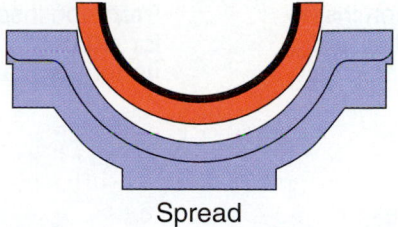

Figure 19.14 Bearing spread permits the bearing to be snapped into place. *(Courtesy of Sunnen Products Company, St. Louis, Missouri)*

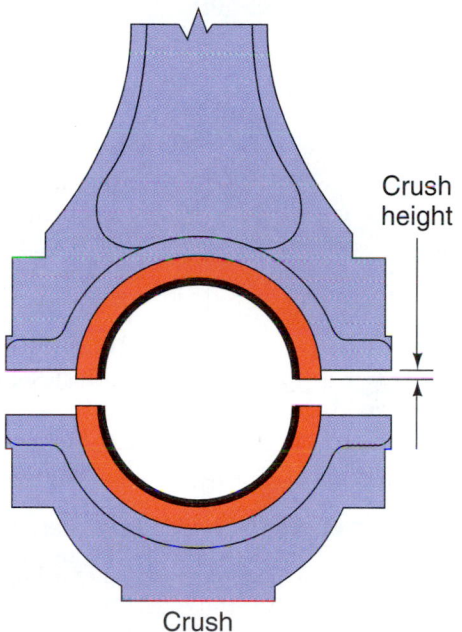

Figure 19.15 Bearing crush prevents bearing movement in the bore. *(Courtesy of Sunnen Products Company, St. Louis, Missouri)*

Bearing Spread and Crush

Bearing spread (Figure 19.14) is where the measurement across the parting face is slightly larger than the diameter of the bearing bore. Spread allows the bearing to snap into place and remain in place during assembly of the engine.

Bearing crush is when the bearing extends slightly above the parting line of the bearing bore half by about 0.0005–0.0015" (Figure 19.15). Crush promotes better heat transfer and prevents the bearing from spinning in the bore. Excessive crush would distort the bearing at its parting lines. Insufficient crush would allow the bearing to move back and forth in the housing because there will not be enough radial pressure on the bearing.

Bearing Undersizes

Bearings come in standard sizes, and in 0.010", 0.020", and 0.030" undersizes. The size is usually marked on the back of the bearing. Undersized bearings are used when a crankshaft is reground. They are also used on some new engines to compensate for errors in production.

Undersized bearings have a thicker bearing back to compensate for the smaller, reground crank (Figure 19.16a).

NOTE: *A crankshaft that is ground to a 0.030" undersize has had 0.015" of metal removed from its surface so both of its new bearing inserts will be 0.015" thicker (Figure 19.16b).*

Bearings are also available in 0.001" and 0.002" undersizes, but not 0.011" and 0.012".

Cam Bearings

Most cam bearings are made from seamless steel tubing with lining material bonded to the inside. Heavy-duty cam bearings are made in flat strips. The layer, or layers, of lining material are bonded to the strip, and the strip is rolled to form the bearing. The seam will be either interlocking or straight (Figure 19.17).

■ CONNECTING RODS

Connecting rods are usually made of forged or cast steel. Some racing rods are made of forged aluminum. Forged rods are stronger than cast rods, but casting

Precision insert for a standard size crankshaft

Precision insert for a 0.030" undersize crankshaft

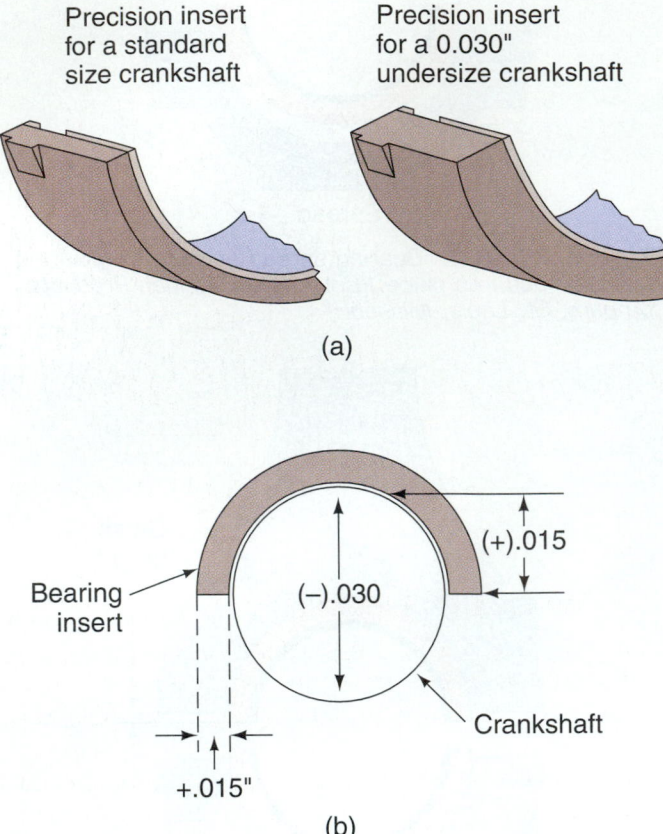

(a)

Bearing insert

(+).015

(−).030

+.015"

Crankshaft

(b)

Figure 19.16 (a) An undersized bearing insert is thicker. (b) A 0.030" undersized crankshaft will require bearing inserts that are 0.015" thicker (on the bearing back).

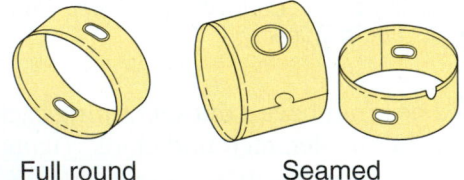

Full round   Seamed

Figure 19.17 Cam bearings are either seamless or they have an interlocking or straight seam. *(Courtesy of Federal-Mogul Corporation)*

techniques have been improved to the point that some late model passenger car engines use cast rods because they cost less. Rods are formed in an I-beam shape for strength.

The big end of each rod is precisely ground to achieve perfect oil clearance when the rod is installed with its bearing inserts on the crank journal with its bearing inserts. Rod caps are not interchangeable. If they are interchanged, the oil clearances of the bearings can vary greatly and the crank might not even be able to turn.

■ Rod bolts usually have slightly enlarged shanks to hold them tight in the rod (Figure 19.18). The cap has precise holes that line up with this enlarged

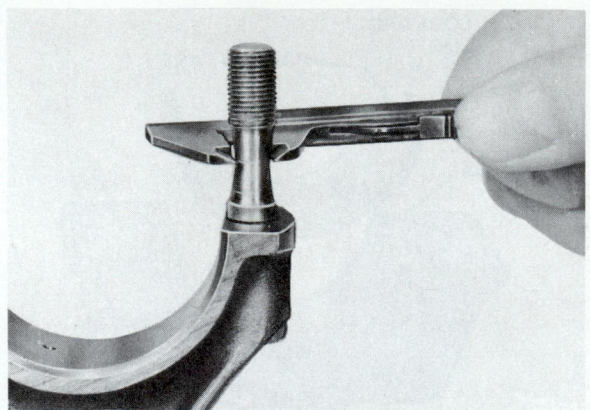

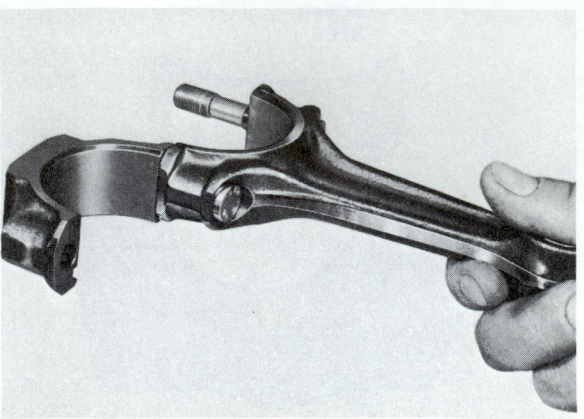

Figure 19.18 The shanks of rod bolts fit tightly in the rod cap for precise alignment of the machined halves. *(Courtesy of Mercedes Benz of North America, Inc.)*

area of the bolt to prevent misalignment of the rod and cap when they are assembled.

■ The notches cut in the rod for the bearing-locating lugs face each other when correctly installed. Incorrect installation of the rod cap can cause uneven bearing wear.

Rod Oil Holes

Some rods have a "spit" hole to provide cylinder wall lubrication at low engine speeds. Inline engines often have a hole drilled in the rod above the rod bolt so that each rod can lubricate its own cylinder. Some V-type engines have a groove between the rod and cap to lubricate the cylinders on the opposite bank (Figure 19.19).

NOTE: Most late model engines do not have oil spit holes but depend on increased rod side clearance to throw oil onto the cylinder walls at low speed.

■ PISTONS

Piston Parts

Older pistons were made of cast iron, but today's pistons are cast or forged aluminum. Aluminum is an

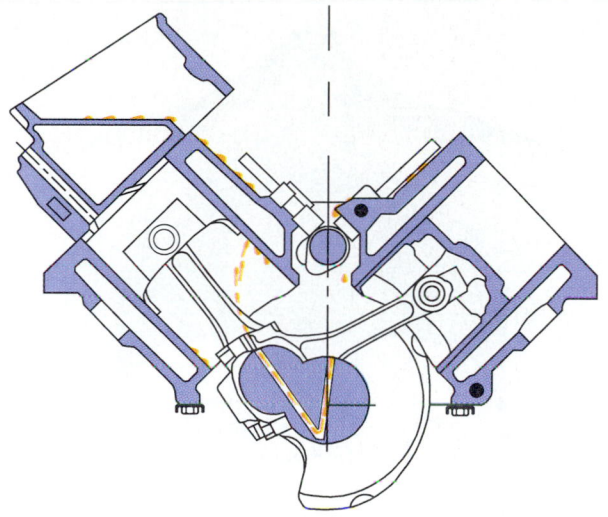

Figure 19.19 On V-type engines, the rod oil hole lubricates the cylinder on the opposite bank. *(Courtesy of General Motors Corporation, Service Technology Group)*

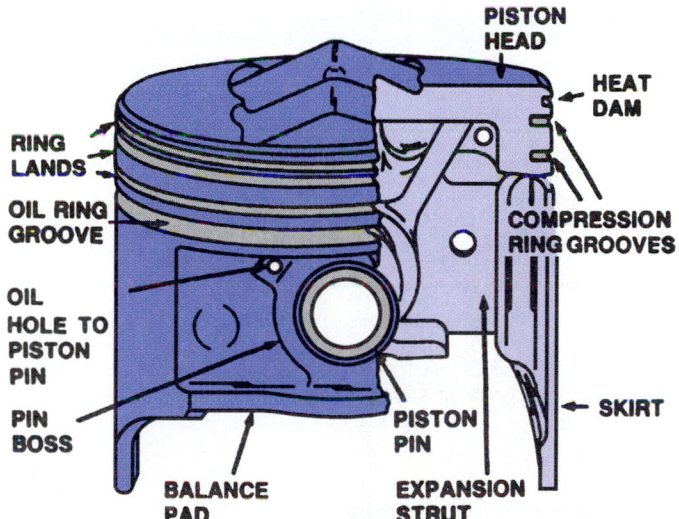

Figure 19.20 Parts of a piston. *(Courtesy of Dana Corporation-Perfect Circle Division)*

excellent material for pistons because of its strength, light weight, and ability to dissipate heat. The piston undergoes remarkable stresses. It starts and stops twice in each revolution of the crank. At a highway speed of 3000 rpm, the piston must change direction (start and stop) 6000 times per minute or 100 times per second. Obviously it must be as light as possible. The top of the piston is exposed to the heat of combustion and must be able to get rid of heat in a hurry or it will melt. Figure 19.20 shows the parts of a piston.

PISTON HEAD

The piston head (or *crown*) is round; the skirt is usually oval. The diameter of the piston head is smaller than the diameter of the skirt by about 0.019" to 0.048" (most are about 0.022" less) (Figure 19.21).

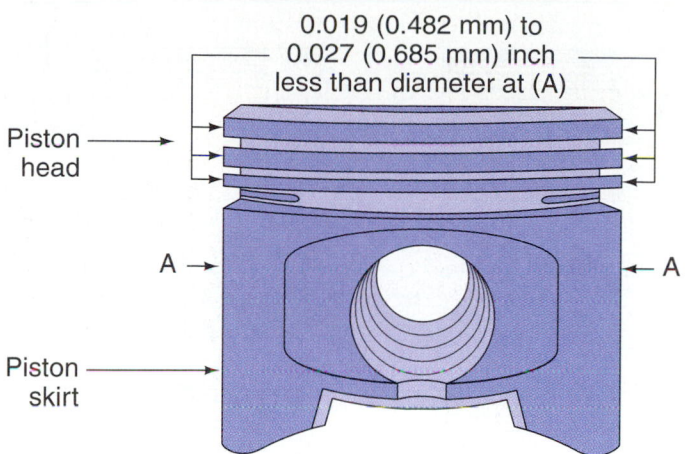

Figure 19.21 The piston head is smaller than the skirt. *(Courtesy of Chrysler Corporation)*

PISTON RING GROOVES

The top piston ring is positioned as high as possible on the piston to lessen piston slap, help ring durability, and lower emissions. The top ring groove suffers the most abuse.

The holes in the *oil ring groove* allow excess cylinder wall oil to return to the crankcase by way of the inside of the piston (see Figure 15.27).

Heat Transfer

Crown heat is transferred through the piston rings to the water jackets behind the cylinder walls. Some pistons have a slot below the oil groove, and some have a heat dam above the top groove, to keep crown heat from traveling down to the top ring groove. Sometimes oil return holes are long slots that also help to prevent the transfer of heat from the piston crown to the skirt.

Piston Head Shapes

Some manufacturers use many different piston head shapes available for the same engine to allow variations in compression ratios (Figure 19.22). High-compression pistons can only be installed in one direction in the cylinder (Figure 19.23). If they are installed backward, the piston crown will interfere with the combustion chamber or valves.

CAST AND FORGED PISTONS

Cast pistons are the most common, although forged pistons are available for heavy-duty or high performance use. Forged pistons have a *dense* grain structure (Figure 19.24) and are said to be 70% stronger than cast pistons. Forged pistons also dissipate heat better (Figure 19.25). Cast pistons are not strong enough for repeated use above 5000 rpm. Cast pistons have a *porous* grain structure (see Figure 19.28).

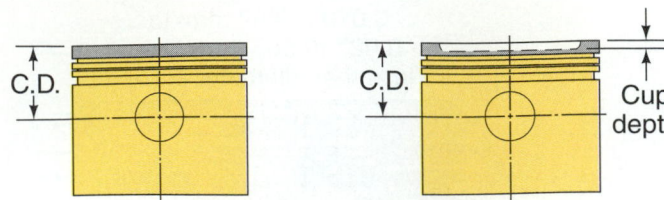

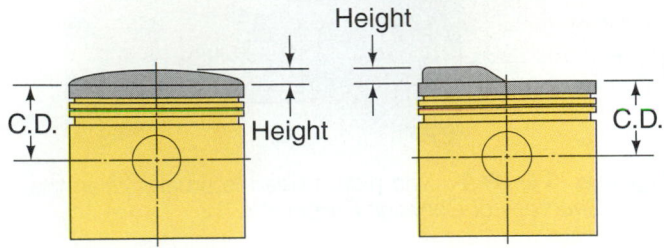

C.D. = Compression distance

Figure 19.22 Common piston head shapes. *(Courtesy of Federal-Mogul Corporation)*

Figure 19.23 A cutaway of a combustion chamber and cylinder showing the relationship between a high-compression piston and the valve and cylinder head. *(Courtesy of Ed Iskenderian Racing Cams)*

Pistons in newer vehicles are often *hypereutectic*, which refers to the silicon content of the aluminum. This allows them to be made lighter. A hypereutectic piston cannot withstand the tensile loads that a forged piston can, but has better wear characteristics, such as in the ring groove and pin bore areas. They also have lower rates of thermal expansion and can run with lower clearances.

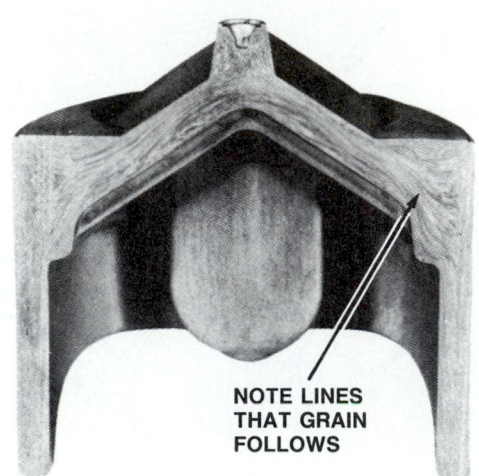

Figure 19.24 A forged piston has a dense grain structure. Note that it has a grain structure similar to wood.

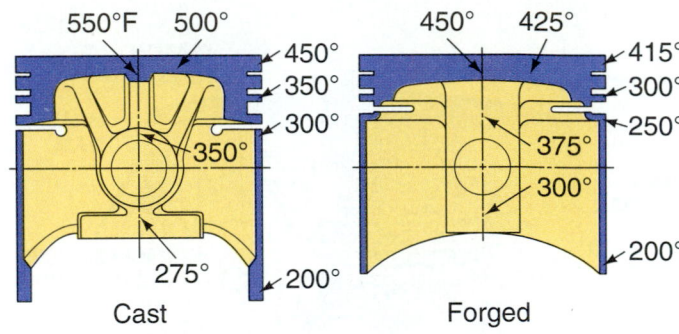

Temperature distribution

Figure 19.25 A forged piston dissipates heat better than a cast piston. *(Courtesy of Federal-Mogul Corporation)*

■ PISTON SKIRT

The surface of the piston skirt is purposely left slightly rough, with small machine marks less than 0.001" deep. This somewhat rough surface is superior for retaining lubrication.

An aluminum piston expands at about twice the rate of the cast-iron block. Three things can be done to control this expansion:

■ Most piston skirts are tapered (Figure 19.26). They are larger in diameter at the bottom because the bottom of the piston runs cooler than the top. At operating temperature, the piston will fit the cylinder wall properly. Many newer pistons are *barrel shaped*; smaller at both the top and bottom.

■ The piston skirt is **cam ground** (Figure 19.27). This allows it to fit the cylinder with only 0.0005" to 0.0025" of clearance. The smaller the clearance, the less chance of piston "slap" and the oval cylinder wear that it causes. Expansion takes place perpendicular (at a 90° angle) to the piston skirt.

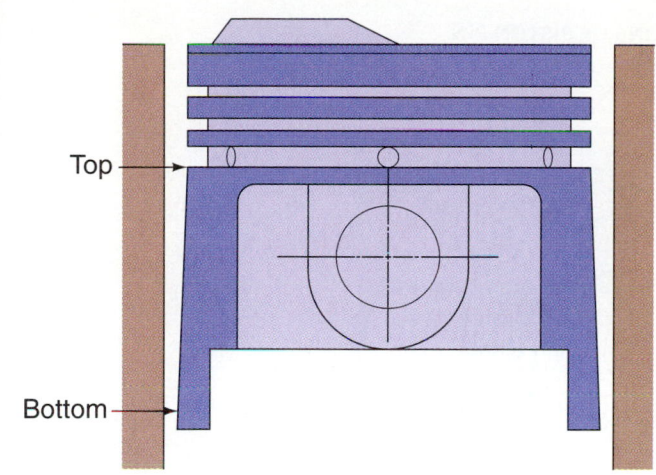

Figure 19.26 Piston skirt taper. *(Courtesy of Federal-Mogul Corporation)*

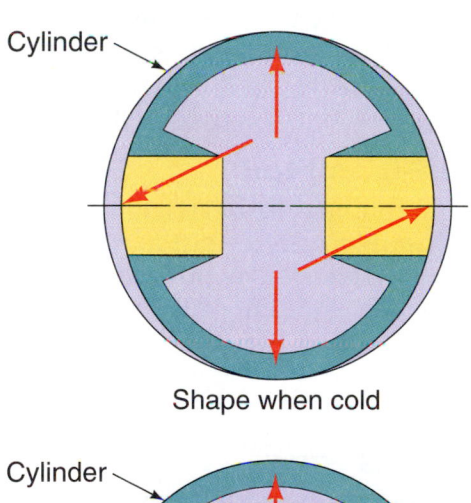

Shape when cold

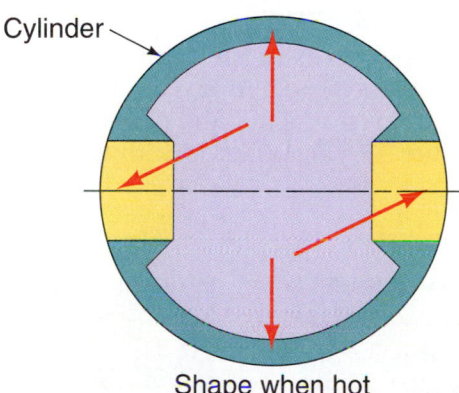

Shape when hot

Figure 19.27 Cam ground piston skirt. *(Courtesy of Sunnen Products Company, St. Louis, Missouri)*

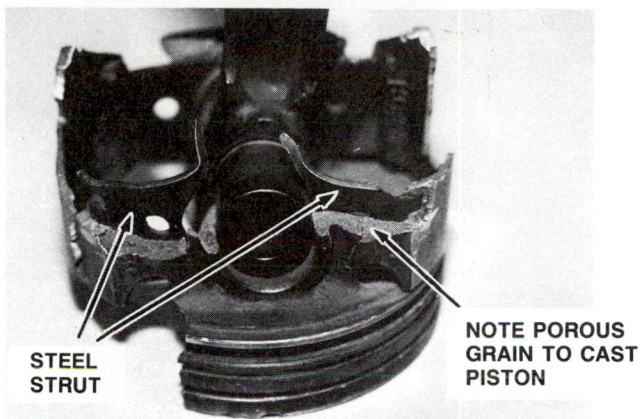

STEEL STRUT

NOTE POROUS GRAIN TO CAST PISTON

Figure 19.28 This piston was destroyed when an overtightened connecting rod bolt gave way. Notice the steel expansion strut around the wrist pin.

TRUNK PISTON   SLIPPER SKIRT

Figure 19.29 Trunk- and slipper-skirt pistons.

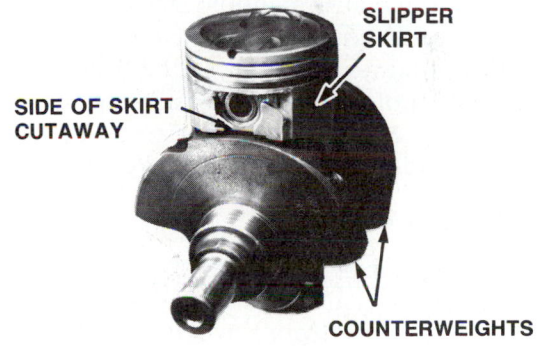

SLIPPER SKIRT

SIDE OF SKIRT CUTAWAY

COUNTERWEIGHTS

Figure 19.30 The slipper-skirt piston is cut away to clear the crankshaft counterweights.

■ Cast pistons often have struts of spring-loaded steel cast into them (Figure 19.28). Struts help the piston resist expansion as it warms up. Newer cast pistons do not have struts and require about 0.005" extra clearance. The broken piston in the photo is the result of overtightening of a rod bolt during engine assembly.

Skirt Types

Figure 19.29 shows slipper-skirt and trunk-skirt pistons. A full-skirt piston used on longer stroke engines is known as a *trunk piston*. Most of today's automotive pistons are slipper-skirt pistons. A *slipper-skirt* is designed to clear the crankshaft counterweights on engines with short strokes (Figure 19.30). The surfaces of the piston skirts that are 90° to the wrist pin are called *thrust surfaces*.

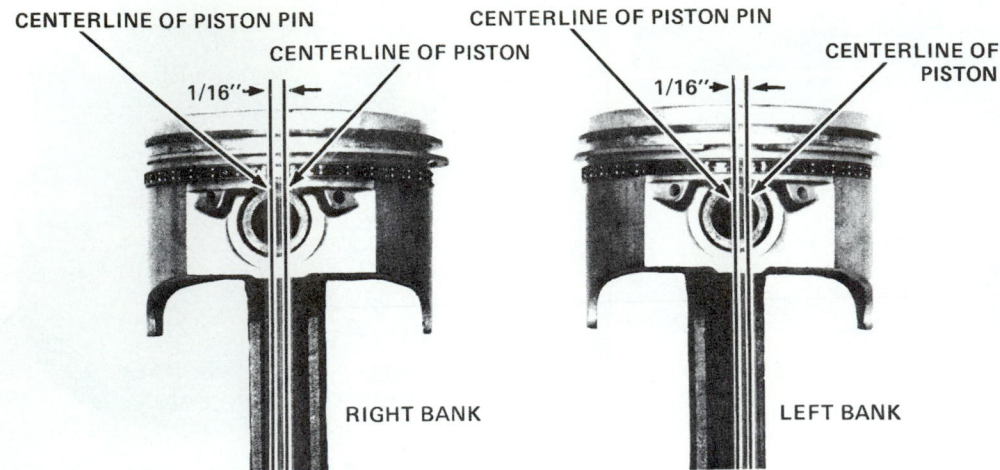

Figure 19.31 The piston pin is often offset. *(Courtesy of General Motors Corporation, Service Technology Group)*

◼ PISTON PIN OFFSET

Piston pins are often offset approximately ¹⁄₁₆" (0.062) from the piston centerline (Figure 19.31). The connecting rod changes to the other side between the compression and power strokes. When the piston rocks from one skirt to the other at TDC, the side of the piston that is "thrust" against the cylinder wall changes.

A piston usually has a direction-identifying notch that faces the front of the engine. On V-type engines, the pins on one bank of pistons will be offset to the intake manifold side of the cylinder, and the pins in the opposite bank will be offset toward the lower side of the cylinder (exhaust manifold side of the cylinder).

Some *crowned head (pop-up)* pistons used in high compression engines (see Figure 19.23) are not offset at all, so that all of the pistons for a V-type engine can be identical. The engine will not turn over if these pistons are accidentally installed backward.

Piston Pin Height

The piston pin can be placed at different heights also (see Figure 19.22). This dimension, called *compression height*, is the distance from the center of the pin bore to the flat area of the top of the piston.

NOTE: *When pistons from engines of the same bore have different displacements (because of different crankshaft strokes), do not interchange them without checking this critical dimension.*

To calculate compression height, measure the distance to the top of the piston from the top of the pin hole. Add one-half of the pin diameter to this amount.

◼ WRIST PINS

The piston is attached to the connecting rod with a wrist pin (also called a *piston pin*). The wrist pin is lubricated from either a hole in the top of the pin boss (see Figure 19.20) or through an angle-drilled hole that runs

from the oil ring groove to the pin boss. The pin boss is the metal area of the piston that surrounds the wrist pin. Modern automobile piston pins are either pressed-fit in the connecting rod or full-floating (Figure 19.32)

Pressed-Fit in Rod

The most common method of attaching the wrist pin is to have it *float* (pivot) in the piston with a very close clearance (about 0.0005"). A pin with this fit feels almost tight when cold but will move freely when hot. The pin is pressed into the rod with a force of about two tons (4000 lb). The *interference* fit is about 0.001".

Full-Floating Pins

A full-floating pin floats in the "eyes" of both the rod and piston. Retainers (lock rings) are required to prevent the pin from sliding out and damaging the cylinder wall (Figure 19.33). When lock rings come out of

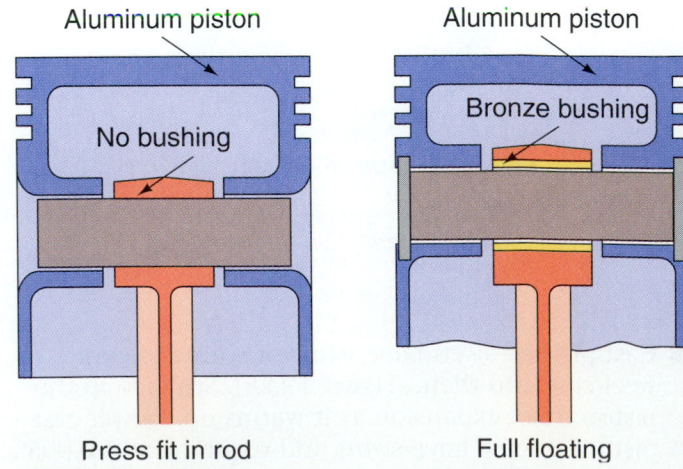

Figure 19.32 Piston pins are either pressed into the rod or they pivot in both the piston and rod. *(Courtesy of Sunnen Products Company, St. Louis, Missouri)*

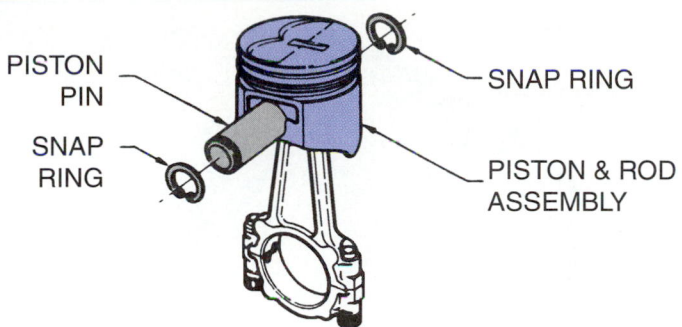

PISTON PIN

SNAP RING

SNAP RING

PISTON & ROD ASSEMBLY

Figure 19.33 Lock rings are installed on full-floating pins with the open end facing down. *[Courtesy of General Motors Corporation, Service Technology Group]*

their grooves during service, the cause is usually excessive crankshaft end thrust, a tapered rod journal, or a misaligned connecting rod.

■ PISTON RINGS

Piston rings do a truly remarkable job in the engine. Most engines today use two compression rings and one oil ring (see Figure 15.24). The top ring is exposed to the flame of combustion during every power stroke. In addition to sealing the tremendous combustion pressures, piston rings also help cool the piston and control oil consumption.

■ COMPRESSION RINGS

Compression rings are forced against the cylinder wall by combustion pressure at the top and back of the ring (see Figure 15.26). The *top ring* controls most of the sealing of combustion. It operates under a load of 110,000 psi and a temperature of 600°F. As long as the top ring seals, the piston stays in good condition. When excessive blow-by and particles of combustion begin to leak by along with the flame, the oil rings will become stuck and plugged up and the piston and rings will rapidly deteriorate.

NOTE: A square ring groove is a must. A scratch in the bottom of a ring groove can leak as much as a more obvious scratch in a cylinder wall.

The *second ring* runs at 35,000 psi and a temperature of 300°F. It performs two functions, serving as a backup for the compression ring above it and the oil ring below it. It captures what little pressure escapes past the top ring. It also scrapes oil back down the cylinder wall where it can be returned to the crankcase.

Compression rings are cast in groups. They are installed on a mandrel and machined out-of-round on a lathe. A slot is milled, which becomes the ring gap. When the ring is compressed closing the large gap, the ring forms a spring-loaded circle. Rings are made as narrow as possible in order to keep them light in weight so they will not flutter. Flutter occurs when

inertia from high speed causes the ring to stay against the top of the groove, causing leakage, blow-by, and ultimately, a broken ring. Increased crankcase pressure can also help to unseat the ring from the bottom of the groove causing flutter.

Low-Tension Rings

Most of the friction within an engine is between the valve lifter and cam lobe or between the piston rings and cylinder walls. To improve fuel economy, *low-tension piston rings* were introduced in 1985. Also called *low-friction* or *shallow groove rings*, these have reduced cylinder wear rates considerably.

Compression rings account for about 20% of the total drag on a cylinder wall and the oil ring accounts for 60%. Low friction ring sets have reduced ring pressures by 60% for the oil ring and 10% for the compression rings.

■ COMPRESSION RING DESIGN

Several ring designs are shown in Figure 19.34. The simplest cast-iron rings are square. Other rings are *taper faced* or have *chamfers* or *counterbores* that cause the ring to twist when compressed. The intent of these features is to cause the ring to contact the cylinder wall with only a narrow part of its face (the part of the ring that slides against the cylinder wall). This creates higher sealing pressure on the power stroke while maintaining lower friction on the other three strokes. These rings will seat more quickly and their downward scraping action will help to control oil. There are several designs of piston ring. Their theory of operation is beyond the scope of this text.

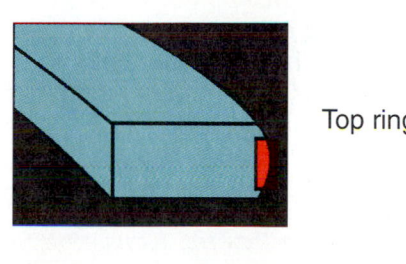

Top ring

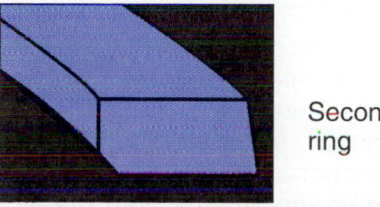

Second ring

Figure 19.34 Typical compression ring designs. *[Courtesy of Federal-Mogul Corporation]*

■ COMPRESSION RING MATERIALS AND COATINGS

Piston rings are either plain cast iron, moly, or chrome.

Cast-Iron Rings

Rings are often made of plain cast iron. Iron rings are used in re-ring jobs when the cylinder walls have not been rebored. Iron rings are very forgiving of cylinder wall imperfections and they are very popular with technicians because they seat easily. When a plain cast iron ring is machined on the surface that rides on the cylinder wall, some manufacturers leave a threaded finish on the ring. The peaks of thread help the ring to seat rapidly to the bore. The threads also hold oil for break-in.

NOTE: *The threaded finish can often be seen on the upper part of the face of a taper-faced second ring when an engine is disassembled for a rebuild.*

Chrome-plated and moly rings do not have the threaded finish because they are lapped at the factory.

NOTE: *Where premium rings are used in the top groove, iron rings are often used in the second groove.*

Iron rings can become loose and their spring tension is reduced at higher temperatures.

Moly Rings

Some premium cast-iron rings, called *moly rings*, have a groove machined on their faces. A plasma spray gun is used to fill the groove with molten molybdenum (Figure 19.35), which solidifies as soon as it contacts the ring. After machining, the moly is about 0.004–0.006" thick. The groove is cut into the ring because the moly does not stick to the ring's base material as well as chrome does. The groove keeps the moly from coming off the ring.

Moly top rings have become popular on a majority of today's new cars. Their high melting temperature makes them extremely resistant to scuffing.

> #### ■ SCIENCE NOTE ■
>
> *Molybdenum has a melting temperature of 4750°F; the melting temperature of iron is 2250°F. When molten molybdenum solidifies on the ring surface, it becomes molybdenum oxide. This oxide is a ceramic and is very hard, unlike the metal molybdenum.*

Moly rings are called self-lubricating; that is, they have a porous surface that helps them retain lubricant for less wear. They are also said to seat instantly because they are preground to make them perfectly round. Moly rings are ground rather than lapped; this prevents the porous surface from filling with lapping abrasive. As development of these rings has progressed, the porosity (5%) of the surface has become

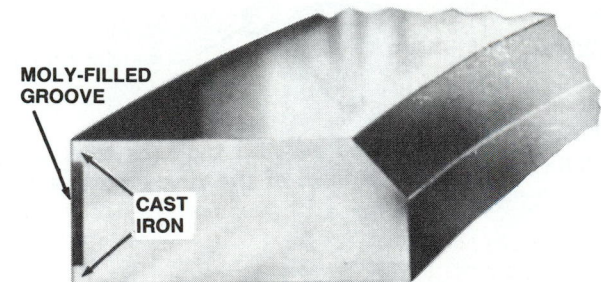

Figure 19.35 A moly-faced ring. *(Courtesy of Dana Corporation-Perfect Circle Division)*

less and less. Older moly rings, with more porosity (20%), trapped dirt to become abrasive.

Chrome Rings

Some rings are chrome plated (Figure 19.36). These rings have about twice the resistance to abrasive wear as moly rings but are difficult, or impossible, to seat in worn (oval) cylinders. They are lapped at the factory when new so they can seat in true cylinder bores. Chrome-plated rings are especially recommended for engines used where excessive dirt will be encountered. They should *not* be used with either propane or natural gas fuels. They also have a problem that moly does not have—moly will not weld, but chrome can overheat and weld to the cylinder bore (scuffing). Chrome rings typically last for the life of the engine with virtually no wear.

Premium Ring Combination

Moly has become more popular than chrome for passenger car use because it wears very well under normal conditions. A common piston ring combination found on passenger cars uses a moly barrel-faced top ring, a reverse torsion second ring, and a three-piece chrome oil ring (Figure 19.37).

High-Strength Rings

Ductile Iron Rings. Special ductile iron rings are often used in the top groove to withstand the higher temperatures in some of today's high heat engines. Ductile

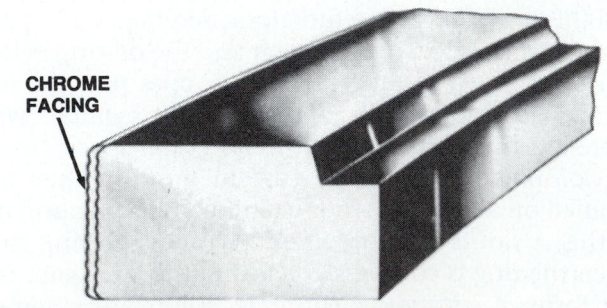

Figure 19.36 A chrome-faced ring. *(Courtesy of Dana Corporation-Perfect Circle Division)*

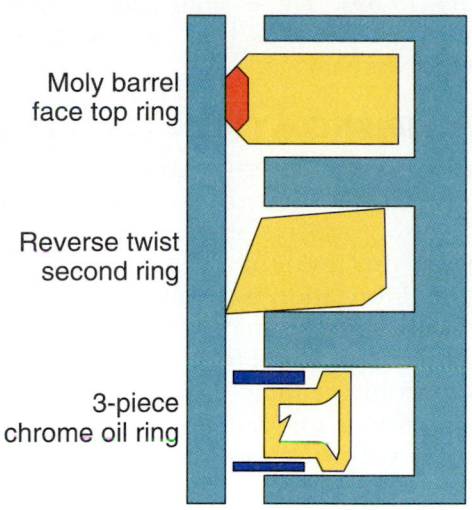

Figure 19.37 Typical piston ring set.

- Moly barrel face top ring
- Reverse twist second ring
- 3-piece chrome oil ring

iron, also called *nodular iron*, is about twice as strong as plain cast iron. It is always coated with moly or chrome.

Steel Rings. Steel rings are made from steel wire. They are coated with chrome or plasma moly because they would be prone to scuffing on cast iron cylinder walls. The width of these rings is so narrow (most are 1.2 mm, or 0.047") that a groove for moly is not easily machined; so these rings are usually chrome plated. These rings cost less to manufacture than ductile iron rings and provide the same advantage in ductility.

> **SHOP TIP** To test a moly or chrome ring to see if it is ductile iron, simply twist it. If it is regular iron it will break in two.

Plasma Ring Coatings

Plasma Ceramic. A newer ring face coating is plasma ceramic.

> ### SCIENCE NOTE
>
> *A plasma ceramic ring facing is a composition of titanium oxide and aluminum oxide. It is applied to nodular iron rings with a plasma torch at 30,000°F to 40,000°F. With this process, materials that are not normally sprayable can be sprayed. This provides a combination of metal and ceramic that has even higher resistance to wear than chrome.*

According to TRW, plasma ceramic rings are five times as strong as a stock ring. The rings are said to resist detonation damage, cause less cylinder wear, and have excellent break-in characteristics. Cylinder preparation is the same as for moly rings.

OIL CONTROL RINGS

Oil control rings do a remarkable job. The oil ring runs at a temperature of 250°F. Older engines often used three or four compression rings and oil rings at the top and bottom of the piston skirt.

- According to Sealed Power Corporation, if $\frac{1}{10}$ of a drop of oil were consumed during each power stroke in a vehicle driven at 60 mph, an 8-cylinder engine could consume about 90 quarts of oil in one 600-mile trip. Actual oil consumption per power stroke in an average engine is from $\frac{1}{1000}$ to $\frac{2}{1000}$ of a drop.
- An engine that uses a single drop of oil with every power stroke would use 1 quart of oil every 2 miles.
- Oil consumption increases with engine speed. A typical engine uses about seven times as much oil at 70 mph as it does at 40 mph, depending on the rear axle ratio.
- Vacuum during deceleration increases with compression ratio. Earlier, lower compression smog control engines developed only about 20 inches of vacuum, but today's higher compression engines can develop up to 25 inches. This makes the oil ring's job that much more difficult.

Oil rings fail if they become plugged because of improper maintenance, temperatures too high or too low, fuel with too much lead or impurities, or improper air/fuel mixtures.

Most engines have only one oil ring located below the two compression rings. There are several oil ring designs. Some are single-piece cast types, but most oil rings in passenger cars are the three-piece type. These consist of a stainless steel expander and two chrome rails (Figure 19.38). Oil rings rarely have problems because they operate in a relatively cool, well-lubricated environment.

The three-piece oil ring prevents oil from being drawn around the sides or back of the ring. Most of the oil is actually scraped off by the lower rail, so barely any oil must actually pass through the expander to the inside of the piston.

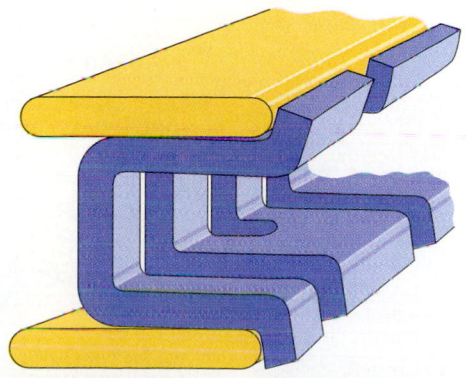

Figure 19.38 A three-piece oil control ring. *(Courtesy of Dana Corporation-Perfect Circle Division)*

The oil ring contributes the most friction to the engine. The chief advantage of the three-piece oil ring is that it can operate with less tension and, therefore, less friction. Low tension rings have rails that are not as deep and a weaker expander is used.

■ ENGINE BALANCING

When moving parts are not in balance, the result can be compared to a washing machine with all the clothes on the wrong side of the tub. The result is vibration and worn parts.

Every time engine speed doubles, the force that results from the imbalance is multiplied by four. An imbalance of ¼ ounce (7 grams or about one sheet of paper) at a 4" radius on a rotating part creates forces of:

■ 7 lb. at 2000 rpm
■ 28 lb. at 4000 rpm
■ 63 lb. at 6000 rpm
■ 112 lb. at 8000 rpm

Counterweights are added to the crankshaft to counteract the force of up or down vibration. They are positioned so they arrive at exactly TDC when the piston is arriving at BDC and vice versa.

An engine can be balanced to prevent vibration. Pistons and rods are weighed on an accurate scale (see Figure 48.30). Material is removed from the heavier parts until they weigh the same as the lightest parts. Balancing theory is covered in detail in Chapter 48.

Balance Shafts

Some of today's engines have used two counterweighted balance shafts that are driven by the crankshaft in opposite directions at twice the crankshaft speed (Figure 19.39). These shafts, also called *silent shafts*, have counterweights that are timed to cancel out the engine's imbalance.

■ THE LUBRICATION SYSTEM

The lubrication system is perhaps the least understood area of engine repair. Too often, when engines are repaired, important aspects of this system are neglected. This neglect can result in a very expensive failure.

In a typical lubrication system (Figure 19.40), oil is drawn in through the pump screen, and then forced by the pump through an oil filter to the engine's *main oil gallery*. Oil from the main gallery provides lubrication to the camshaft and cam bearings. In a typical pushrod engine with stud-mounted rocker arms, oil is also pumped to the hydraulic lifters. From the lifters, the oil is pumped through hollow pushrods to the rocker arms and valve guides.

■ The crankshaft receives oil at its main bearings.
■ Oil is pumped through drilled oil galleries from the main journals to the rod journals.
■ The cylinder wall and piston are lubricated by oil thrown or sprayed from the rod journals.
■ The timing chain, the distributor, and the oil pump drive gear are all splash lubricated.

■ OIL PUMPS

The oil pump usually has a drive shaft that is driven by a gear on the camshaft, although some oil pumps are driven by the crankshaft. Pumps that are driven by the cam gear are usually connected to the bottom of the distributor by a driveshaft with a tang or hex on its end (Figure 19.41). On the bottom of most distribu-

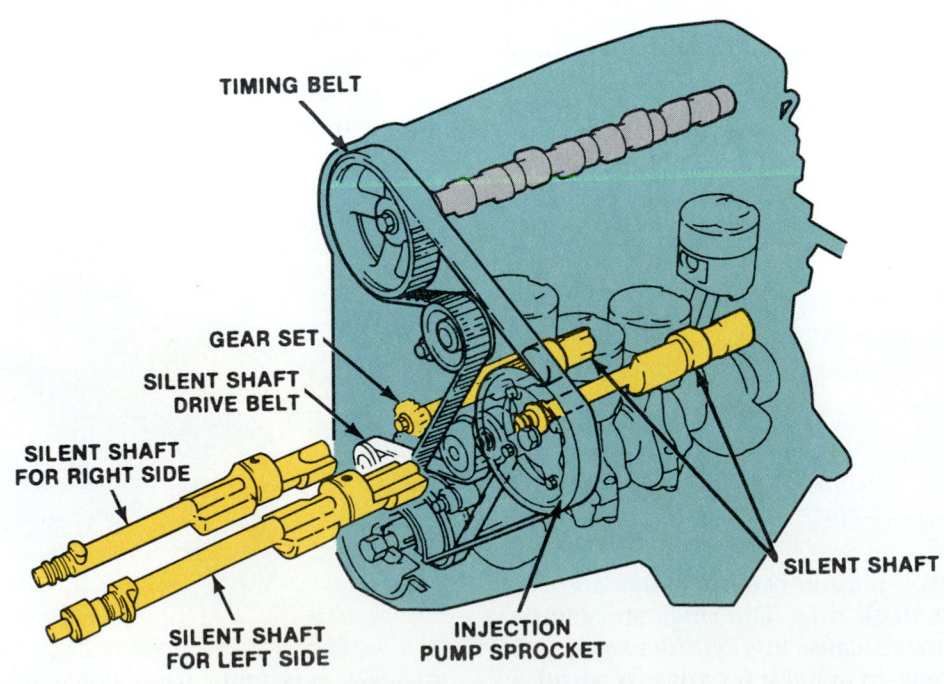

TIMING BELT
GEAR SET
SILENT SHAFT DRIVE BELT
SILENT SHAFT FOR RIGHT SIDE
SILENT SHAFT FOR LEFT SIDE
INJECTION PUMP SPROCKET
SILENT SHAFT

Figure 19.39 Balance shafts, also called silent shafts. *(Courtesy of Ford Motor Company)*

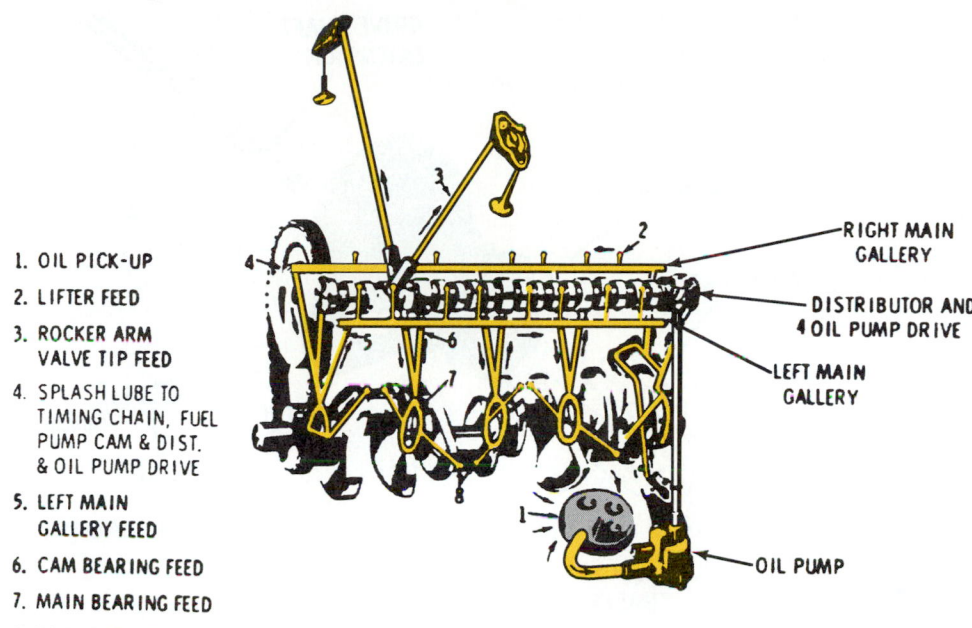

1. OIL PICK-UP
2. LIFTER FEED
3. ROCKER ARM VALVE TIP FEED
4. SPLASH LUBE TO TIMING CHAIN, FUEL PUMP CAM & DIST. & OIL PUMP DRIVE
5. LEFT MAIN GALLERY FEED
6. CAM BEARING FEED
7. MAIN BEARING FEED
8. ROD BEARING FEED

RIGHT MAIN GALLERY
DISTRIBUTOR AND OIL PUMP DRIVE
LEFT MAIN GALLERY
OIL PUMP

Figure 19.40   A typical lubrication system. *(Courtesy of General Motors Corporation, Service Technology Group)*

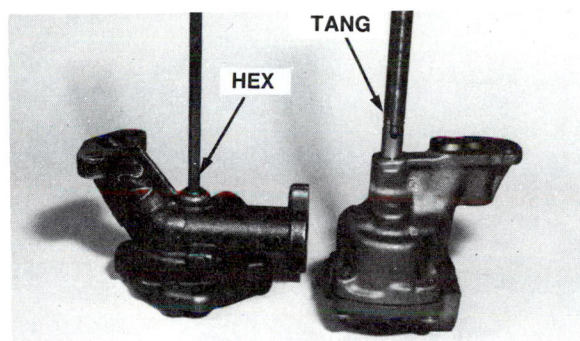

TANG
HEX

Figure 19.41   Two common distributor oil pump drives.

tors, there is a gear that meshes with the gear on the camshaft.

There are three types of oil pumps (Figure 19.42). They are:

- The *external gear* (Figure 19.42a)
- The *rotor* (Figure 19.42b)
- The *internal/external gear or gerotor* (which is similar to a rotor pump) (Figure 19.42c)

Internal/external pumps save space. They are driven by the crankshaft. External gear pumps and rotor pumps are the most common types. Cast-iron pumps are considered to be better than aluminum pumps because they wear less, expand less, and their relief valves are less likely to stick.

PRESSURE RELIEF VALVE

The oil pump must be capable of supplying oil under sufficient pressure to all areas of the engine *at idle speed*. The faster the pump turns, the more oil it pumps. For this reason, it must have a relief valve to

relieve excessive pressure when the pump turns at higher speeds (Figure 19.43).

- Most oil pump relief valves limit maximum oil pressure by diverting excess oil back to the inlet side of the pump.
- The maximum oil pressure in an engine is controlled by the tension of the relief valve spring.

OIL PUMP SCREEN BYPASS VALVE

Most screens have a **bypass valve** that can open when the screen becomes plugged (Figure 19.44), or when the oil is cold and too thick to flow freely. The bypass valve is normally seated against the *cross strap* on the screen. If a screen bypass valve has opened and remains open, the screen will have to be replaced. When the screen is restricted and the bypass opens, a large (about ½") opening results. This opening allows large particles to enter the pump because oil goes to the pump before it goes to the filter (Figure 19.45).

OIL PRESSURE

Proper lubrication in an engine is achieved only by the distribution of clean oil under pressure. Oil *pressure* can only be created when there is *resistance to the flow* of oil from the oil pump (see Pascal's law, Chapter 51).

The correct amount of bearing clearance is very important. If there is too much clearance, the engine will have low oil pressure at idle and will probably *knock*. If there is too little oil clearance, the crankshaft could seize or a bearing could be burned.

A *positive displacement pump*, like those used in automobiles, has a fixed output per revolution. The

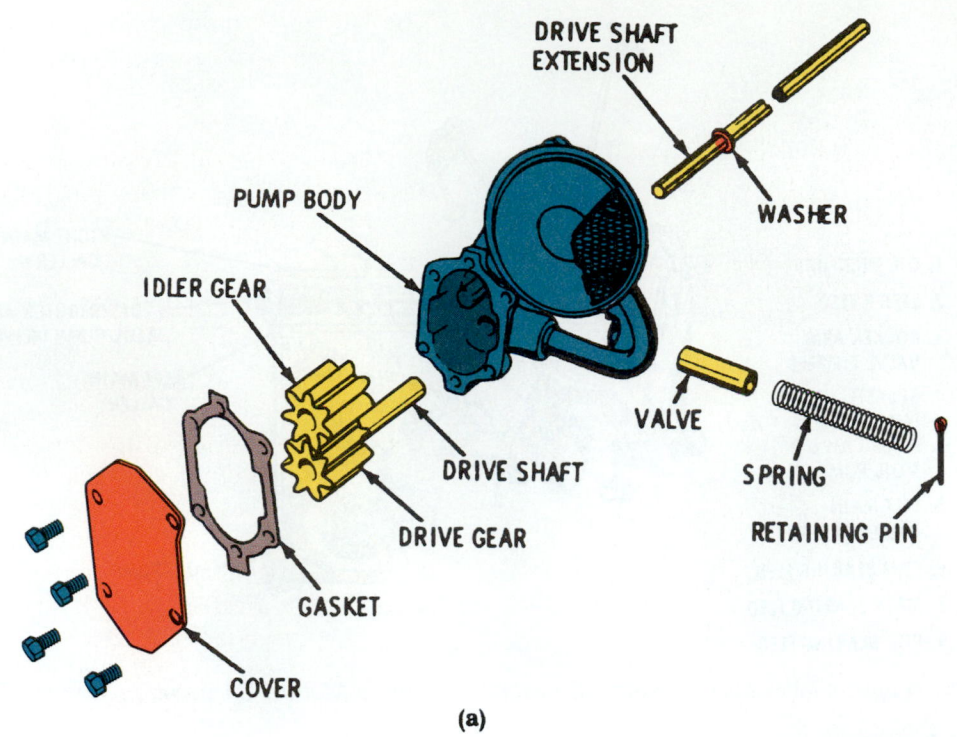

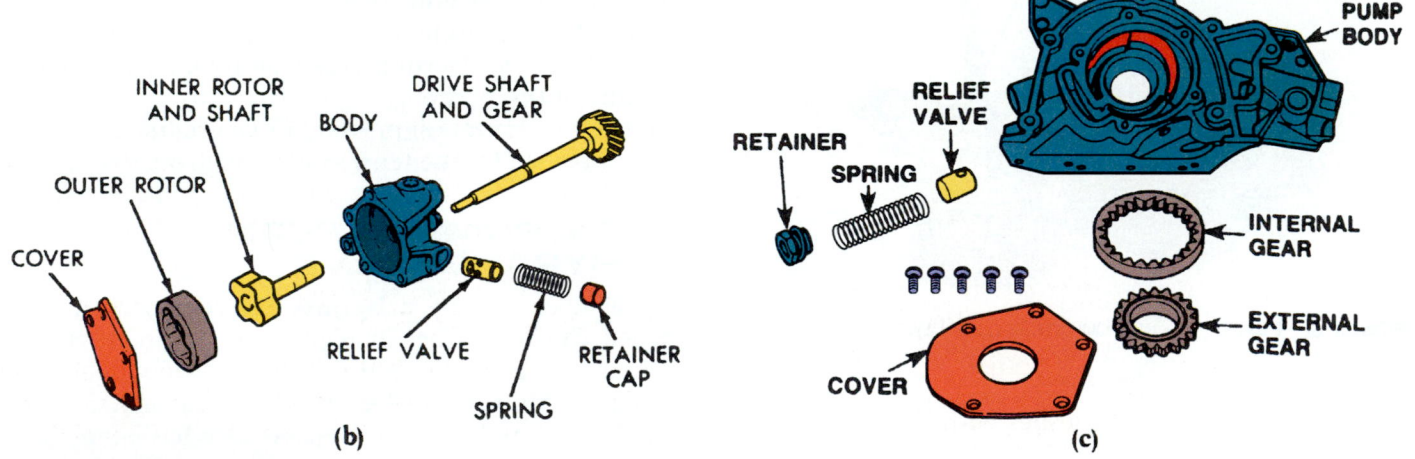

Figure 19.42 Types of oil pumps. (a) External gear pump. (b) Rotor pump. (c) Internal/external gear pump. *(a and c, Courtesy of General Motors Corporation, Service Technology Group; b, Courtesy of Chrysler Corporation)*

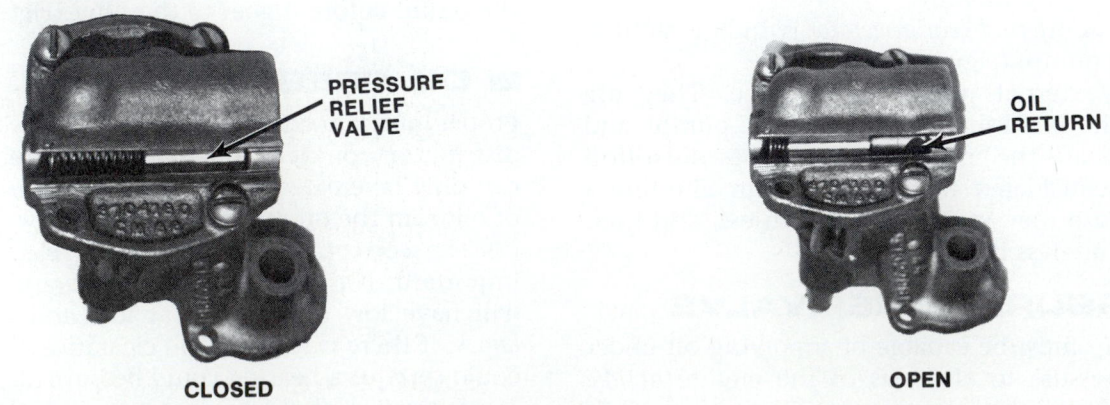

Figure 19.43 Oil pump pressure relief valve operation.

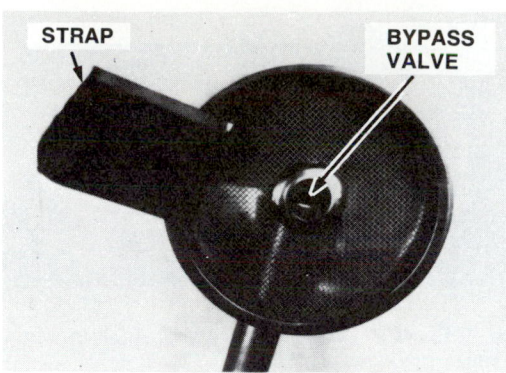

Figure 19.44 This pump screen has a bypass valve. *(Courtesy of Federal-Mogul Corporation)*

(a)

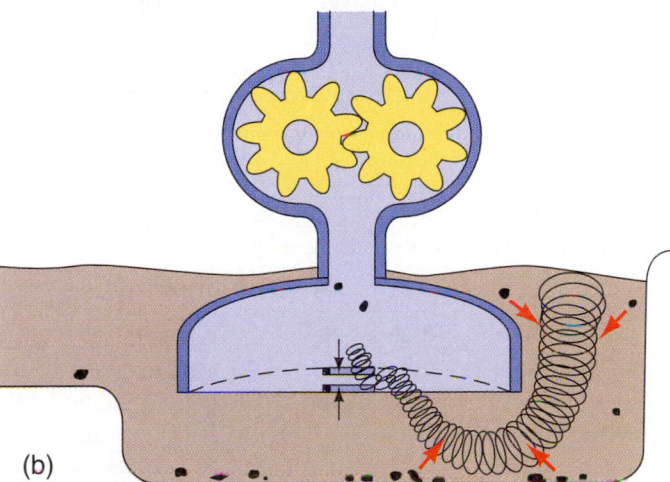

(b)

Figure 19.45 (a) This pump screen is full of large pieces of foreign material. (b) When the bypass valve opens on a cold morning, the foreign material will be sucked into the oil pump. *(Courtesy of Federal-Mogul Corporation)*

faster it turns, the more oil it pumps. If correct bearing clearances are not maintained, oil will not reach all areas of the engine when the engine is idling. At idle speed, the pump is turning slowly and cannot fill the system if clearances are excessive.

Excessive oil *clearance* near the oil pump—at a main bearing, for instance—will result in insufficient oil pressure at the rocker arms or valve lifters, which are farther away from the pump. Noisy hydraulic lifters are often an indication of an oil pressure problem.

NOTE: *Most manufacturers recommend a minimum of 10 psi of oil pressure for every 1000 rpm. Some engines have different requirements so check manufacturer's specifications.*

Satisfactory oil pressure at idle is normally in the neighborhood of 25 psi, but pressure indicator lights do not normally come on until pressure drops below 10 psi.

◼ HIGH-VOLUME OIL PUMPS

The output per revolution of an oil pump depends on the diameter and thickness of the rotors or gears. Oil pumps with about 20%–25% more volume are available. High-volume pumps deliver more oil per revolution due to the increased size of the pump cavity (Figure 19.46). These pumps are able to provide more oil to a worn engine at idle. Because they pump a larger volume of oil per revolution, this results in an increase in idling oil pressure in such an engine. However, they may not necessarily provide any other added advantages to a passenger car engine. Consider the application carefully before replacing a standard pump with a high-volume one.

NOTE: *Dry Sump System. Some racing engines use a dry sump system in which the oil is held in a large holding tank, which uses a high volume external oil pump, usually mounted on the front of the crankshaft. An oil cooler might also be used in this system. Oil is pumped from the pan to the holding tank. A system of this type holds more oil. The problem of foaming and aeration that would occur if the oil was stored in the crankcase is eliminated.*

◼ WINDAGE TRAY AND BAFFLES

At high speeds, the revolving crankshaft creates wind. This can cause air pockets around the oil pump screen, which can cause the pump to lose its prime. The objec-

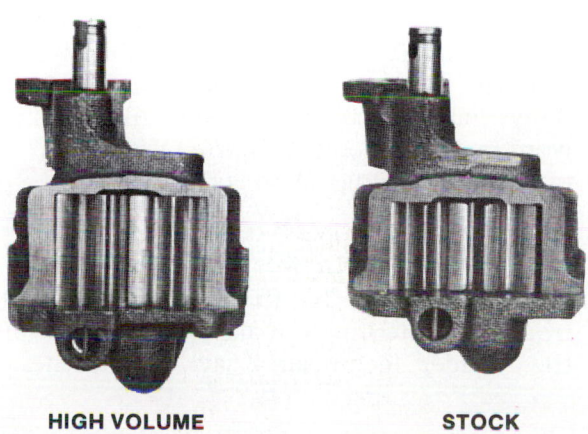

HIGH VOLUME **STOCK**

Figure 19.46 A high-volume oil pump is larger than a stock pump. *(Courtesy of Federal-Mogul Corporation)*

tive is to keep the pump screen continuously immersed in oil. A windage tray helps to prevent these air pockets (Figure 19.47). Some windage trays bolt onto the main bearing caps; others are built into the oil pan.

Some oil pans also have built-in baffles that help to keep the oil from sloshing back and forth when driving over hills or bumps or during hard cornering or braking. Be sure to check for pieces of foreign material trapped under a windage tray or baffle.

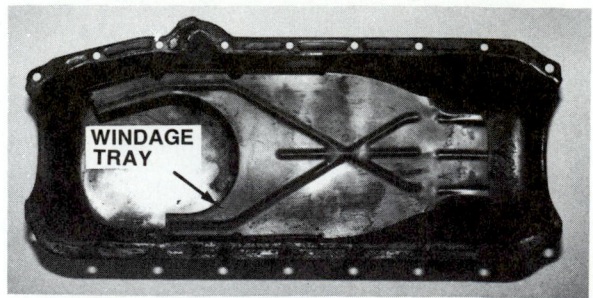

Figure 19.47 An oil pan with a built-in windage tray.

■ REVIEW QUESTIONS

1. What are two ways that cylinder bores wear?
2. When a cylinder wall is damaged, what is the name of the part that is installed to correct it?
3. What is the force that causes the crankshaft to move to the front or rear called?
4. The SAE standard for crankshaft rotation says that crankshafts will rotate (circle one) clockwise, counterclockwise when viewed from the flywheel end.
5. What are two names for the part that is on the front of the crankshaft to control vibrations?
6. What is the name for the bearing design feature that allows it to snap in place and remain there during engine assembly.
7. What keeps the bearing from spinning in its bore?
8. At 6000 rpm, how many times does the piston have to start and stop in one second?
9. Name three types of ring facings.
10. What part usually drives the oil pump drive shaft?

■ ASE STYLE REVIEW QUESTIONS

1. Technician A says that when a cylinder wears in a tapered fashion, a ridge is left at the top of the cylinder. Technician B says that a ridge forms at the top of the cylinder because the rings slowly push the metal up from the cylinder wall below it. Who is right?
   a. Technician A
   b. Technician B
   c. Both A and B
   d. Neither A nor B

2. Technician A says that a replacement bearing insert for a 0.020" undersized crankshaft would be 0.010" thicker than a standard one. Technician B says that a 0.010" undersized crankshaft that is worn by 0.001" could be refitted with 0.011" undersized bearings. Who is right?
   a. Technician A
   b. Technician B
   c. Both A and B
   d. Neither A nor B

3. Technician A says that the piston pin is offset from the centerline to make it easier to install in the cylinder. Technician B says that notches cut into piston rings are to cause them to twist. Who is right?
   a. Technician A
   b. Technician B
   c. Both A and B
   d. Neither A nor B

4. Technician A says that the oil pump screen has a relief valve in it in case the screen is plugged or the oil is too thick. Technician B says that excessive main bearing clearance can result in low oil pressure that happens only at high speeds. Who is right?
   a. Technician A
   b. Technician B
   c. Both A and B
   d. Neither A nor B

5. Technician A says that a longitudinally mounted engine would turn counterclockwise when viewed from the front. Technician B says that transverse engines turn clockwise when viewed from the flywheel end. Who is right?
   a. Technician A
   b. Technician B
   c. Both A and B
   d. Neither A nor B

Cooling and Fuel System Theory and Service

THEN AND NOW: WATER PUMPS

Henry Ford's Model T moved North America into the age of the automobile, selling 15 million between 1909 and 1927. But in spite of the clever engineering that made it so practical, the "Tin Lizzie," as the Model T was nicknamed, did not have a water pump. The liquid coolant in those days was plain water, sometimes combined with a little alcohol to prevent ice from forming. It was circulated by natural *convection*—just like hot air, hot liquids rise. The water absorbed heat from the block, causing it to rise and enter the top of the radiator. As it cooled, it sank to the bottom entering the lower part of the engine. While this thermosiphon principle worked well enough for normal service, heavier hauling jobs required spending a few dollars for an aftermarket add-on water pump.

Water pumps became standard on almost all other cars after the Model T. They move a huge volume of coolant—from a couple of hundred gallons per hour at idle speed, to several thousand in a big V8 cruising on the freeway.

A number of engines today have a water pump powered by an overhead cam's timing belt. An advantage is that the engine cannot run without the pump spinning, preventing one possibility of overheating. Also, the engine can be made a little shorter because the pump pulley is eliminated—an important consideration with today's crowded engine compartments. Another recent advance in driving the pump is the serpentine accessory belt, most examples of which are self-tensioning. Serpentine belts take up less space, last longer, run more quietly and are less likely to slip than the traditional "V" belt.

The biggest change in water pump design occurred when the ceramic seal was adopted many years ago. This seal increased the life expectancy of the pump, adding far more resistance to wear and abrasion than the rubber or leather types used previously.

Premature pump failure can also be prevented using quality anti-freeze, which contains additives that lubricate and protect the water pump shaft seal. Changing the coolant at reasonable intervals will also add to the life of the pump. Water pump replacement is a fact of life, however. Sooner or later almost every car will start to leak at its water pump seal.

An aftermarket water pump on a Ford Model T. *(Courtesy of Bob Freudenberger)*

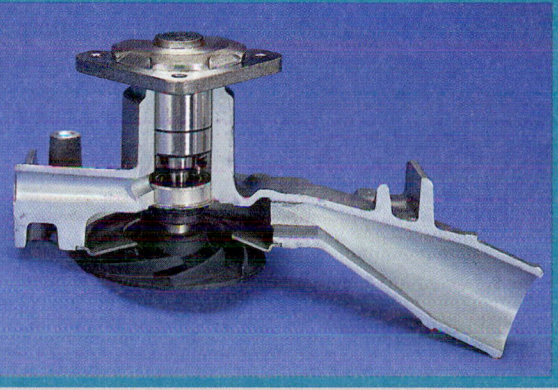

A modern Ford water pump. *(Courtesy of Keller Crescent Corporation, Airtex Division)*

Cooling System Theory

■ INTRODUCTION

Combustion temperatures sometimes reach as high as 4500°F; average temperatures of combustion are close to 2000°F. Aluminum pistons melt at about 1225°F, so this heat must be carried off rapidly to prevent engine damage (Figure 20.1). Most of this heat is not used to propel the piston and must be carried off by the cooling system.

Theory of the cooling system is covered here because it is an essential but often neglected area of vehicle maintenance. A thorough knowledge of the operation of the system provides the service technician with the ability to serve the customer's needs (and sell profitable service).

■ LIQUID AND AIR COOLING

Some smaller engines are only cooled by air, but most automotive cooling systems use liquid coolant to carry off engine heat. The engine block and cylinder head have cast-in water jackets surrounding its cylinders and combustion chambers (see Figure 16.5).

■ LIQUID COOLING SYSTEM PARTS

A liquid cooling system is made up of the following parts (Figure 20.2):

■ The water pump, or coolant pump circulates water throughout the system.

■ The radiator conducts heat away from the coolant.

■ Radiator hoses connect the parts of the cooling system.

■ The fan pulls air through the radiator when the car is not moving fast enough to move the air.

■ The thermostat controls the temperature of the coolant.

■ Coolant is circulated to the heater in the passenger compartment.

■ COOLING SYSTEM CIRCULATION

Coolant exits from the bottom of the radiator after it is cooled. It is drawn into the coolant pump (commonly called a *water pump*) (Figure 20.3), which pumps it into the block and head where it picks up heat. The coolant is then returned to the top of the radiator, where it is cooled again.

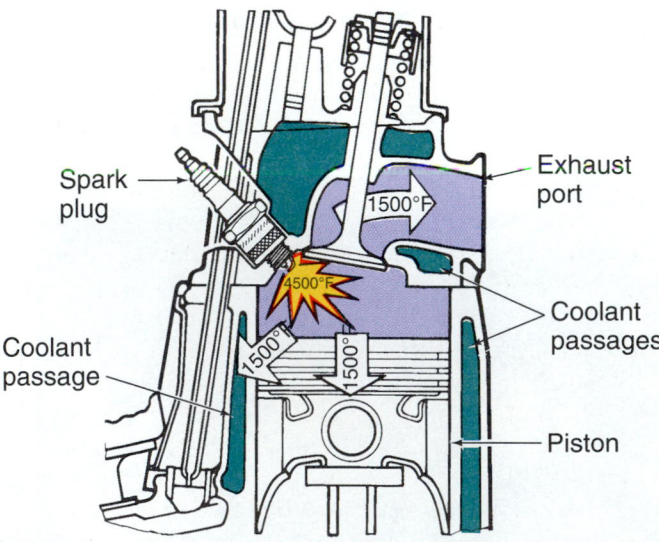

Figure 20.1 Heat must be removed from the engine by the cooling system. *(Courtesy of Chrysler Corporation)*

Spark plug
Exhaust port
1500°F
4500°F
1500°
1500°
1500°
Coolant passages
Coolant passage
Piston

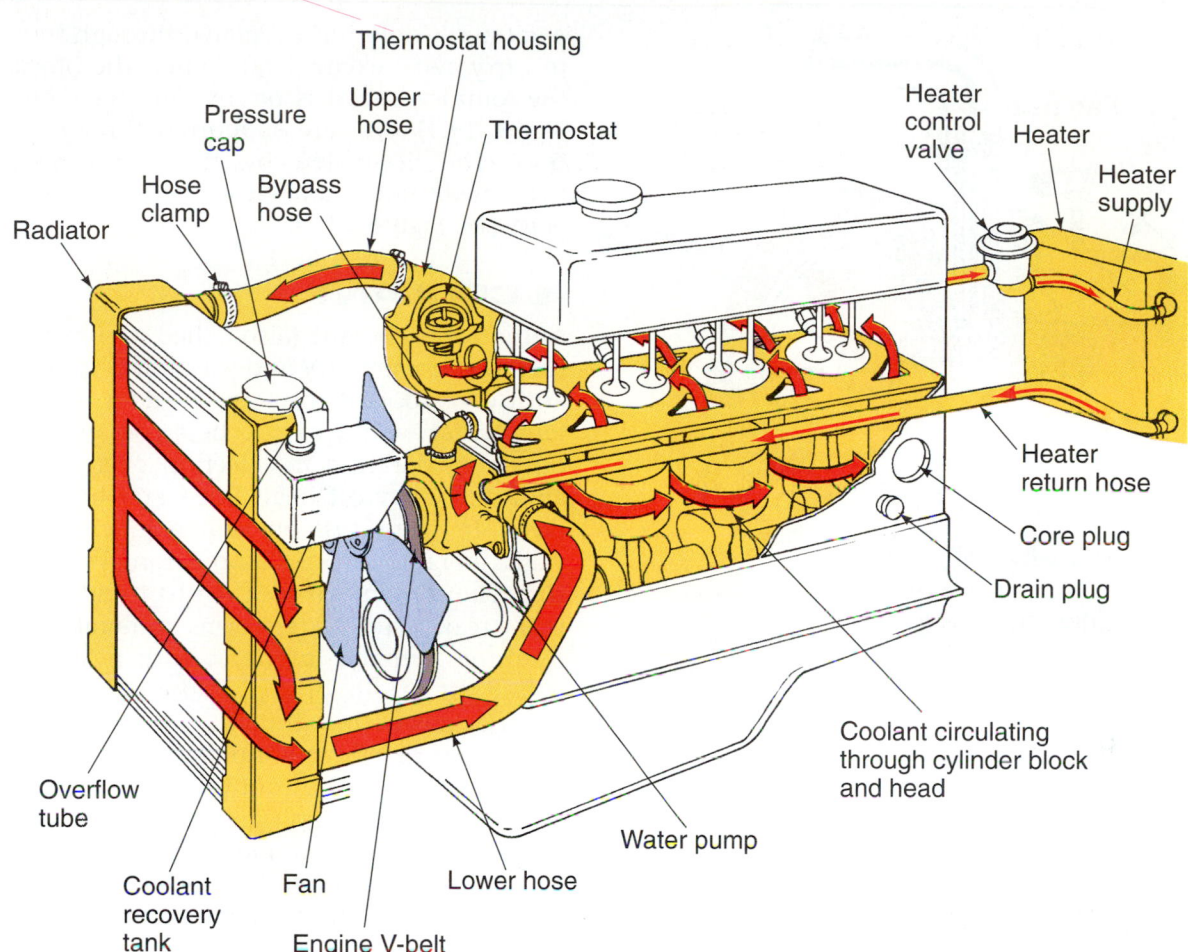

Figure 20.2 Parts of a liquid cooling system. *(Courtesy of The Gates Rubber Company)*

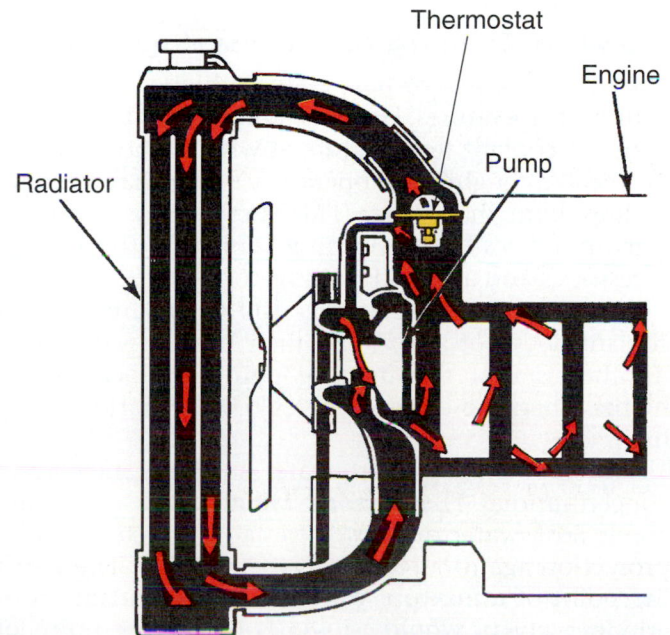

Figure 20.3 Coolant exits the bottom of the radiator and is drawn into the water pump. *(Courtesy of General Motors Corporation, Service Technology Group)*

Water Pump

The coolant pump, usually called a water pump, is a centrifugal impeller-type pump. It is mounted on the front of the engine and is belt-driven by the crankshaft. Usually, the alternator uses the same belt and has a provision for adjusting the belt tension. If the belt slips, the pump will not pump as much coolant.

An *impeller* fits in the circular pump housing. As it turns, centrifugal force draws coolant into the pump inlet near the center of the pump (Figure 20.4). It exits from the outlet at the outside of the pump. Centrifugal pumps are *variable displacement*, which means that their output changes with speed. The pump is sometimes called upon to circulate as much as 7500 gallons of coolant in one hour.

The water pump has a spring-loaded, permanent seal to keep water from leaking out. The water pump cutaway in Figure 20.5 shows the relationship between the bearing and its spring-loaded *mechanical seal*.

The water pump housing is either aluminum or cast iron. Cast iron is more popular in rebuilt pumps because aluminum is subject to erosion. As the aluminum erodes, the pump cavity becomes larger. The reduced

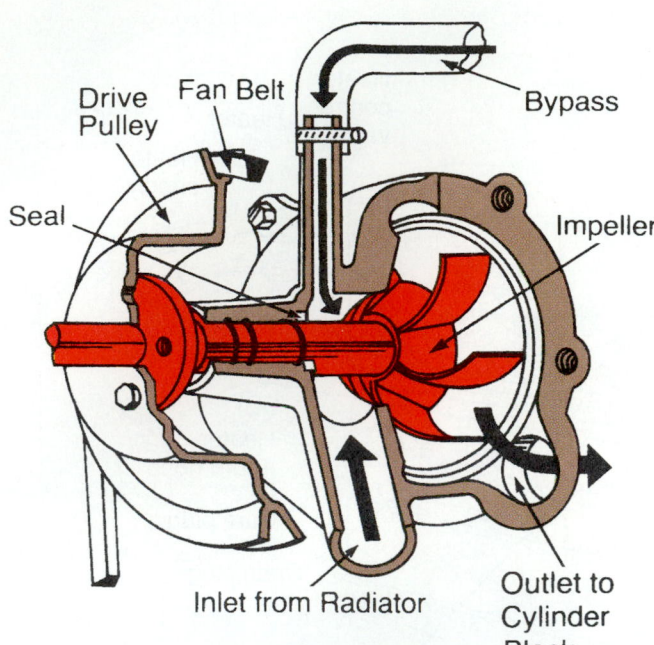

Figure 20.4 Parts of a water pump. *(Courtesy of The Gates Rubber Company)*

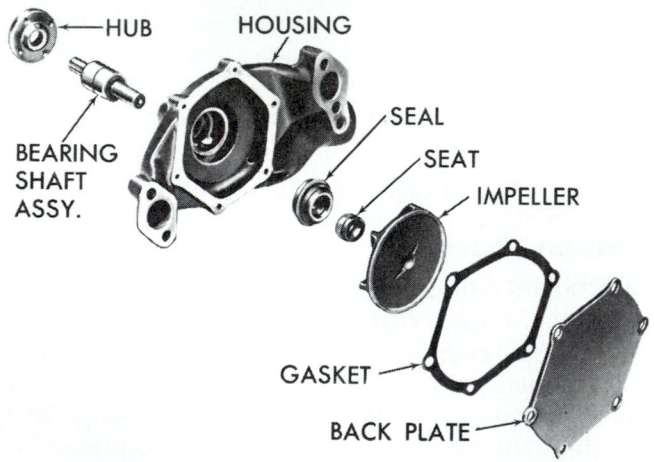

Figure 20.5 Parts of a water pump. *(Courtesy of General Motors Corporation, Service Technology Group)*

pump efficiency can cause air pockets. Severe erosion will also result in serious internal or external leaks.

Thermostat

The temperature of the coolant must be regulated to be neither too hot nor too cold. A thermostat (see Figure 20.3) regulates the temperature of coolant leaving the block. When the coolant temperature is too low, it is trapped in the block until it heats up sufficiently to cause the thermostat to open.

Water Jackets

Hollow water jackets are cast into all of the areas of the block and heads that are subjected to high tem-

peratures. Coolant is circulated through these passages to carry away excess heat. When the block is cast at the foundry, metal is poured around sand cores (see Figure 19.1) that are supported through core openings. The holes left by these core supports, are machined smooth and closed with interference fit *core plugs* (see Figure 19.2).

■ COOLANT

Automotive coolant (also called *anti-freeze*) is a mixture of *ethylene glycol* and water. It is called *permanent* antifreeze/coolant because it does not evaporate. Pure water would be the best coolant to use if the only consideration in selection of a coolant was its ability to carry off heat. But water has other limitations. It forms *rust* on iron engine parts. Cylinder walls vibrate during combustion, breaking off the rust. The coolant carries the rust to other cooling system areas. Hard water forms mineral deposits. The resulting *corrosion* interferes with heat transfer even before the results of it plug the radiator and fill the waterjackets with sediment.

The cooling system is often neglected, especially in frost-free areas of the country. People in more temperate areas often use tap water to fill their radiators. This can lead to plugged radiators, rusted radiator cap springs, rusted core plugs, plugged heater cores, rusted radiator hose reinforcement springs, and a host of other problems. Most of these problems could be avoided by simply using a coolant/water mixture instead of pure water.

Coolant Boiling and Freezing Points

The freezing point of pure water is higher than some winter temperatures, and its boiling point is lower than the coolant temperature at which a hot (but not overheated) engine can operate. Coolant has a higher boiling point than water. This is especially an advantage on today's hotter running vehicles with multiple accessories and air conditioning.

One drawback to using 100% coolant is that coolant does not absorb as much heat as water does. Another is that the freezing point will actually be higher when the ethylene glycol concentration is too high.

Coolant is mixed with water until it is the correct concentration. The factory radiator fill is usually about 50% water and 50% coolant, which provides protection against freezing down to –34°F. The freezing point of a mixture with a 70% concentration of ethylene glycol would be –85°F. Its boiling point of 235°F is 11°F higher than would be provided by a 50-50 mixture. A 70% concentration is the maximum that should be used. The chart in Figure 20.6 shows the effects of various coolant concentrations on boiling and freezing temperatures.

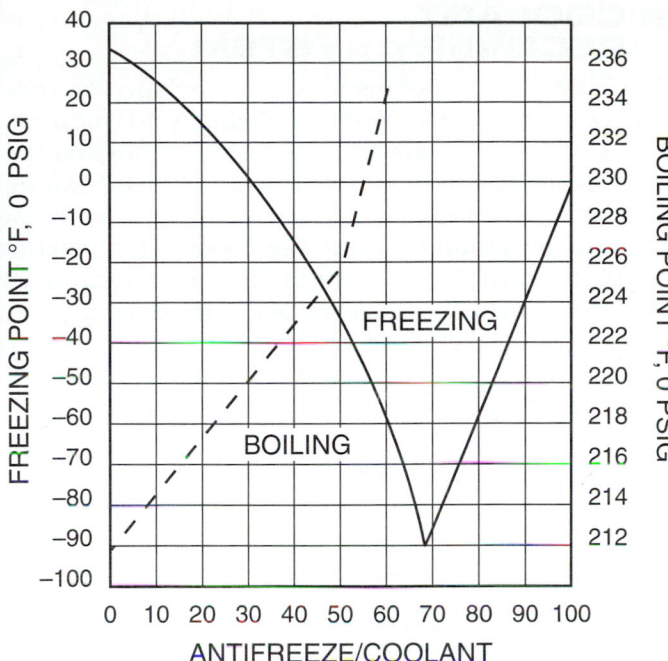

Figure 20.6 The percentage of coolant changes its freezing and boiling points.

NOTE: *Some automobiles do not have a temperature gauge, relying instead on a high temperature warning lamp. A switch in a sending unit closes when the engine temperature reaches a certain temperature. This temperature is usually well above water's boiling point of 212°F at sea level. If pure water was used on these systems, the warning light might never get a chance to work. The water would have already boiled out before the switch reached a temperature high enough to close it.*

Coolant Additive Package

There are a variety of additives that are included in automotive coolant. Because of the tendency for water in the coolant to corrode or rust metal, the addition of inhibitors is required. Besides inhibitors, coolant contains absorbers and buffers, cavitation retarders, corrosion inhibitors for aluminum, electrical galvanic activity preventers, antifoam additives, and dyes.

The additive package is made up of:

- *Borax* (sodium tetra borate), which absorbs acid.
- *Phosphate* (dipotassium phosphate), which is a buffer of acid and retards cavitation. It is good for cast iron.
- *Sodium molybdate*, which is an expensive corrosion inhibitor for heat rejecting aluminum (when

aluminum, such as in the water pump, is heating the coolant).

- *Sodium silicate* for aluminum protection.
- *Sodium nitrate*, which resists solder corrosion.
- *Trizoles*, which are chelants that grab and form a film over metal ions like brass and copper to prevent a galvanic cell.
- *Sodium hydroxide*, which is used to keep the pH between 10 and 11.
- *Phosphorescent dye* that will show up under ultraviolet light. The colors vary and are for consumer identification only.
- *Water* is used to make the additives soluble.

Bimetal Engines

Cylinder heads on automobile engines are often made of aluminum. Cylinder blocks are made of cast iron. *Electrolysis* happens when there are two dissimilar metals in a liquid that can carry an electrical current. Just like in a battery, one of the metals acts like a positive plate and the other acts like a negative plate. During electrolysis, metal leaves one of the "plates" (aluminum) and is deposited on the other (iron). To protect aluminum from corrosion, use a coolant that specifically states that it can be used with aluminum. A *corrosion inhibitor* additive is required to protect the aluminum.

▇ RADIATOR CAP

A radiator pressure cap is used to seal the opening on top of the radiator (see Figure 13.11). Its functions are to pressurize the coolant and keep it from surging out as it is circulated through the radiator.

Cooling systems are pressurized for two reasons.

- The efficiency of the water pump is increased.
- Putting pressure on the coolant raises the boiling point of the liquid in the cooling system.

When the system is not under pressure, the water pump does not always remain full of coolant. *Aeration* is the result, which causes a loss of pump efficiency of about 15%.

Normal engine operating temperatures are between 180° and 212°F. The typical coolant/water mixture boils at somewhere between 220° and 230°F without pressure. When the cooling system is overworked, temperatures can rise as high as 270°F without the coolant boiling. Pure water boils at 212°F at sea level, but its boiling point is less than that at higher altitudes because atmospheric pressure on the coolant is less.

NOTE: *Each pound of pressure on the coolant raises its boiling point by about 3°F.*

A 15-pound cap will raise the boiling point of the water by about 45°F to 257°F.

212°F (boiling point of water at sea level)
+ 45°F (3° × 15 lb)
―――――
257°F

Also, if coolant is mixed with water, the boiling point will be higher.

Radiator Cap Valves

The two valves in the radiator cap are the *pressure* valve and the *vacuum* valve (Figure 20.7)

Pressure Valve. Most pressure caps pressurize the system from 15-17 pounds. The large spring in the cap is what maintains this pressure by forcing the rubber *pressure seal* against the seat in the *radiator filler neck* (see Figure 13.11). The pressure valve maintains pressure on the coolant even if the engine is shut off.

Vacuum Valve. The vacuum valve protects the radiator as the temperature of the coolant drops. Hot liquid shrinks to occupy less space. This could create a vacuum on the inside of the radiator if the spring-loaded vacuum valve did not open to prevent it. Excess coolant (that went into the recovery tank during warm-up) is drawn back into the radiator.

Some radiator cap vacuum valves are spring-loaded and some are loose in the cap.

- A *constant pressure* radiator cap has a spring-loaded vacuum valve. It allows the system to begin building pressure as soon as the coolant starts to expand.
- An *atmospheric pressure* radiator cap has no spring on the vacuum valve. It does not build up pressure until the system gets hot.

■ COOLANT RECOVERY SYSTEM

Coolant recovery systems have been included on most new cars since 1969. Figure 20.8 shows a typical system. Coolant expands by over 1/10 of its original volume as it reaches engine operating temperature. Expanding coolant goes into the recovery tank instead of onto the ground. When the coolant temperature drops, the vacuum valve opens, drawing coolant back into the system from the recovery tank. Because air never enters the cooling system, a properly operating recovery tank system increases overall cooling efficiency by 10%.

The top radiator seal (see Figure 13.11) prevents steam from escaping when the cap is turned 1/2 turn to the pressure release position. It also seals when the coolant temperature drops in the radiator, causing coolant to be drawn in from the coolant recovery tank.

■ RADIATOR

Radiators are designed with either of two flow patterns; **downflow** or **crossflow** (Figure 20.9). Both designs work equally well but most automobiles use the crossflow design. It can be made wider with a lower profile, allowing for a lower hoodline (less wind resistance).

A radiator has top and bottom, or side *tanks* that are attached to a *core* that contains *tubes* with air *fins* attached to them (Figure 20.10). Coolant circulates through the radiator tubes. When air travels across the fins, it carries off the heat that is transferred from the coolant to the tubes.

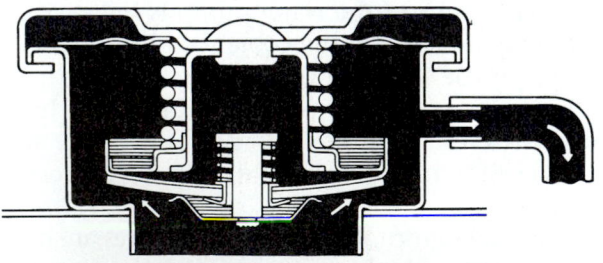

Pressure valve

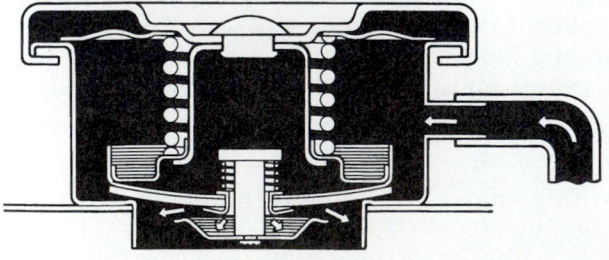

Vacuum valve

Figure 20.7 Pressure valve and vacuum valve operation.

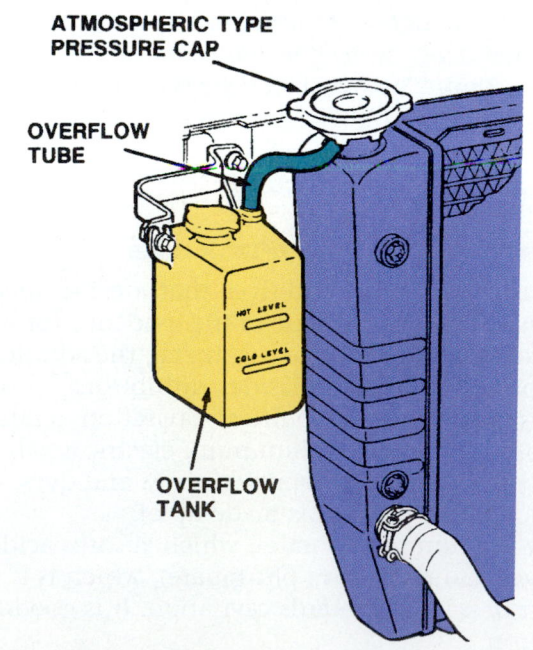

Figure 20.8 A coolant recovery system. *(Courtesy of Ford Motor Company)*

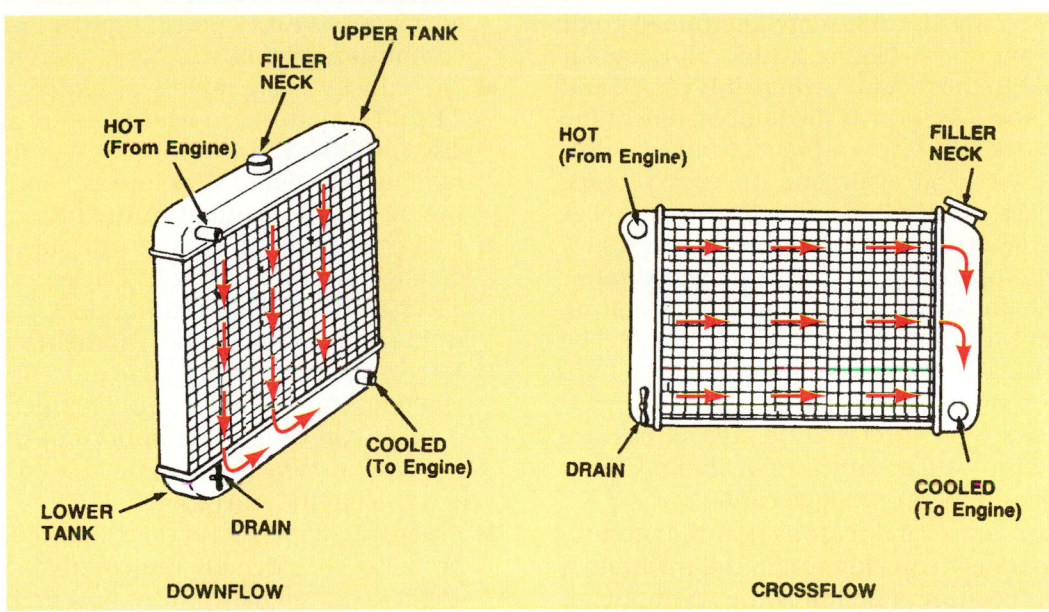

Figure 20.9 Radiator coolant flow.

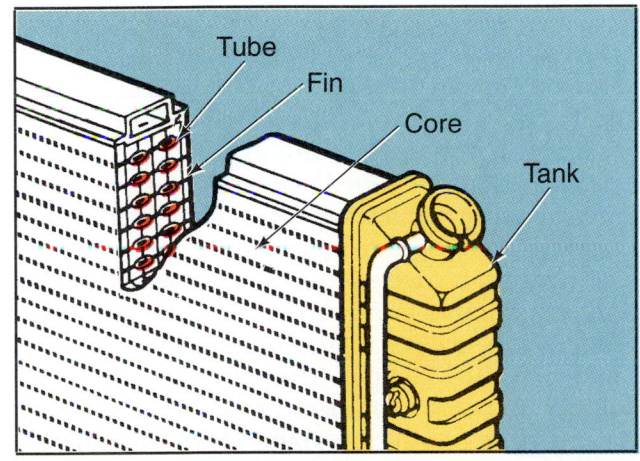

Figure 20.10 Tanks are attached to the radiator core. [Courtesy of General Motors Corporation, Service Technology Group]

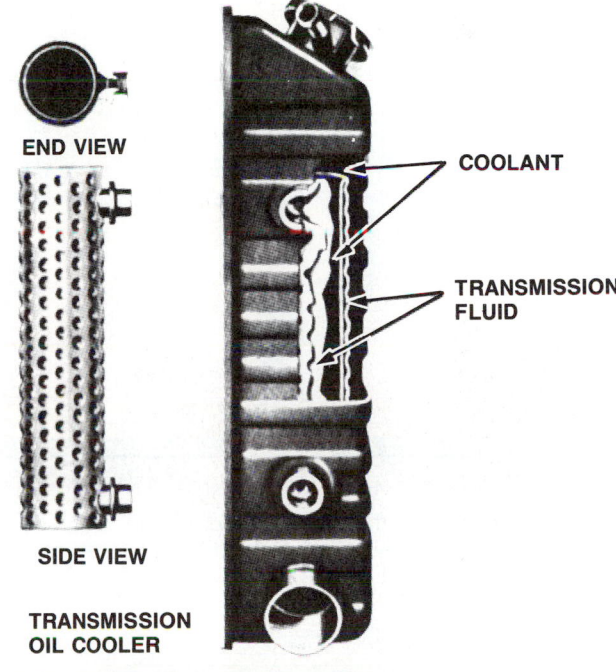

Figure 20.11 An automatic transmission cooler in the radiator. [Courtesy of General Motors Corporation, Service Technology Group]

Radiator cores are made of copper or aluminum. Many radiator *tanks* are brass and are soldered to the copper radiator core. The radiators on many newer cars have plastic tanks and vacuum brazed aluminum cores.

The vacuum brazed aluminum core/plastic tank radiator combination used in many new vehicles has a good life expectancy and is proving to be durable. But these radiators are not as tolerant of poor quality coolant, neglected maintenance, and low coolant level as are copper/brass radiators.

Transmission Heat Exchanger

A car with an automatic transmission usually has a **heat exchanger** in the radiator (Figure 20.11). This part is also called a **transmission oil cooler**.

■ REGULATING ENGINE TEMPERATURE

Thermostat

The thermostat is usually located where the coolant leaves the top of the block, just below the upper radiator hose (see Figure 20.2). When the engine is cold, the thermostat remains closed, trapping the coolant in the block. The coolant is circulated in the block by the water pump until it warms up to a predetermined point and the thermostat opens (Figure 20.12). Hot coolant can then circulate to the radiator, where it is cooled and returned to the block. As soon as the temperature of the coolant in the block falls below a predetermined point, the thermostat closes once again and the cycle repeats.

The *wax pellet* thermostats used in automobiles have an expansive wax compound that expands at a predictable rate to open the valve in the thermostat. A thermostat is rated according to the engine operating temperature that it is supposed to start to open at. The temperature of the thermostat is stamped on the bulb on its bottom in either Celsius or Fahrenheit (Figure 20.13). Thermostats of several different ratings are available. The temperature rating of a thermostat is somewhere above 180°F on modern cars.

It is important for several reasons that the thermostat allow the coolant to quickly reach and maintain a high enough temperature. The following are results of an engine that runs too hot:

■ Its oil becomes thinner. Thinner oil does not provide the same level of lubrication and can

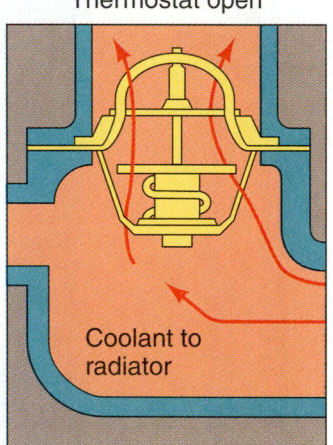

Thermostat closed   Thermostat open

Bypass

Coolant to water pump   Coolant to radiator

Figure 20.12 When the engine is cold, the thermostat is closed. When it warms to the temperature rating of the thermostat, it opens. *(Courtesy of The Gates Rubber Company)*

180°F   87°F

Figure 20.13 The rating of the thermostat is stamped on the bottom.

escape more easily past the piston rings, into the combustion chamber.

■ Oil oxidizes more rapidly at temperatures above 240°F. Oxidation causes the oil to thicken and can also result in varnish buildup. Varnish can cause hydraulic lifters to stick and fail and can also plug the oil return channels in the oil rings.

■ Excessive engine temperatures can result in engine damage. A hot engine can cause pistons to expand excessively. When a piston scuffs, it can actually weld to the cylinder wall momentarily.

■ **Detonation** or **preignition** occurs when fuel explodes due to high engine temperatures. Engine damage can result. It is important that the fuel in the combustion chamber be able to burn evenly.

When the engine runs too cold:

■ An engine experiences poor fuel economy and produces considerably more emissions when cold. Metal parts, such as pistons and rings, conform to the cylinder walls when warm. When they are cold, they do not fit as well. Blow-by, which is exhaust gas that escapes past the rings during combustion (see Figure 15.4), increases when the engine is cold.

■ The engine will experience far less wear if it is brought quickly to operating temperature.

■ If a car has a *computer feedback* fuel system, that system will not operate until its sensors have all reached a predetermined operating temperature. This will cause the fuel management (computer) system to continue to operate in the **open loop** mode (this information is covered in a later chapter). High emissions and poor fuel economy result when an engine does not reach operating temperature.

■ Sludge in the oil will increase if the engine is running at too cold a temperature for too long. Sludge is a mixture of moisture, byproducts of combustion, and water (see Figure 9.1). Considering that about 1 gallon of water is produced for each gallon of fuel burned, it is especially important that the engine be run at a temperature that results in the evaporation of condensation.

An engine will run without a thermostat in an emergency. But in addition to the problems already mentioned, leaving it out can cause an engine to run hot. Although the coolant temperature is relatively low, the coolant moves too quickly through the water jackets to absorb heat from the engine's parts.

When an engine is warm, the thermostat can be open either partially or fully, fluctuating to maintain a constant temperature. Should the cooling requirements increase in hot weather or under a load, the thermostat will open all the way.

No matter how hot the coolant is, a properly operating thermostat remains closed until its predetermined opening point. A common misconception is that installing a lower temperature rating thermostat will result in a cooler running engine. This practice will simply result in the engine operating temperature being lower during cooler weather. Another practice based on faulty logic is to install a thermostat with a higher opening temperature in an attempt to cause the engine to warm up faster. The engine will warm up at the same rate. Its operating temperature will simply be higher.

Types of Thermostats

There are two types of *wax pellet* thermostats commonly used in pressurized cooling systems: the *diaphragm* and the *positive piston* actuator. Figure 20.12 shows how a positive piston actuator thermostat operates. There is sometimes a small hole drilled in the thermostat that allows air to escape during filling of the cooling system. It also allows a small amount of coolant circulation around the wax sensor for more accurate temperature control.

Thermostat Bypass

When the thermostat is closed, the coolant is circulated in the block by the coolant pump. There is a passage (either a hose or a designed-in passage) that allows the coolant to circulate when the coolant is cold and the thermostat is closed. This is called a **thermostat bypass** (see Figure 20.5). Bypass hoses are usually molded hose.

■ TEMPERATURE WARNING LIGHT OR GAUGE

A *warning light* or *temperature gauge* is located on the instrument panel. When the temperature level of the coolant reaches unsafe levels, the light or gauge signals the driver of the problem. A **sending unit** for the gauge, also called an *engine coolant temperature sensor*, is screwed into one of the coolant passageways in the head or manifold. It is bathed in coolant and sends the temperature signal to the gauge.

A sending unit can signal a slow rise in temperature or a fast rise due to a stuck-closed thermostat, but it only works effectively when submerged in coolant. Sometimes the coolant boils out or is lost due to a fast leak. With no coolant in the cooling system, the sensor might not get hot enough to register the excessive temperature.

■ SCIENCE NOTE ■

Cavitation *happens when air bubbles form in the coolant. It can occur when coolant boils, when air gets into the coolant, or when suction is restricted and over-pumping occurs. If the water pump tries to move more coolant than is possible, a pressure drop at the suction side of the pump results in cavitation on the pump impeller. Bubbles can also form from vibration of the cylinder walls*

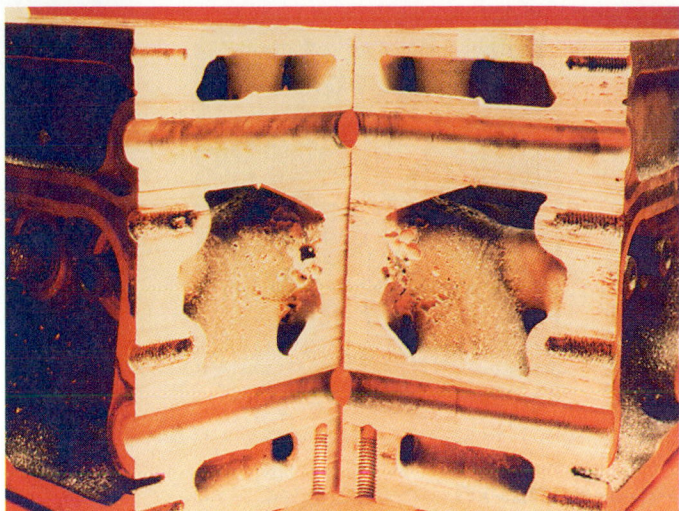

Figure 20.14 Cavitation erosion in an aluminum cylinder head. *[Courtesy of Prestone Products Corporation]*

during combustion, especially during detonation. Bursting bubbles can create pressures of up to 60,000 psi as they burst. As boiling coolant travels through the engine, the collapsing bubbles chip off metal.

Figure 20.14 shows an example of cavitation erosion in an aluminum cylinder head. The flaking aluminum can plug a radiator. Phosphate additives in the coolant help to control cavitation. Good cooling system maintenance, including good anti-freeze, a good pressure cap, and tight connections can help control cavitation.

■ FANS

Belt-driven fans are usually mounted on the front of the coolant pump, which is driven by a belt that rides in a pulley groove on the crankshaft. They are found on older vehicles. Fans powered by electric motors and controlled by engine temperature are found on most late model vehicles (Figure 20.15).

The fan's purpose is to draw air through the radiator when the vehicle is not moving fast enough to provide enough air circulation. It is really only necessary at idle and low speeds.

Electric Fans

With an electric fan, the fan motor is switched on and off as the engine temperature rises and falls. The fan might also be switched on to cool the air conditioning condenser. The fan can also operate after the engine is shut off. A timer controls this function so that the battery will not run dead. Aftermarket add-on fan kits are also available.

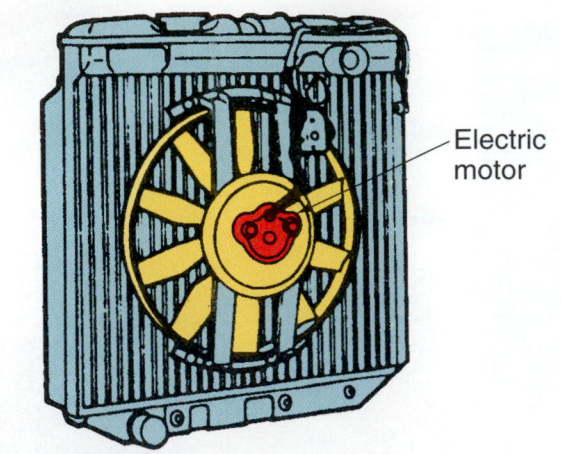

Figure 20.15 An electric fan.

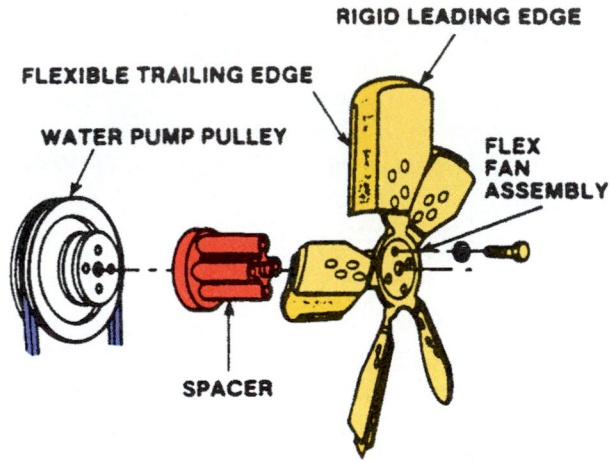

Figure 20.16 A flex fan.

The disadvantage of a belt-driven fan is that at higher speeds, when the fan is no longer required, it can rob the engine of horsepower. Manufacturers have developed several ways to minimize this. Fan blades are made of aluminum, flexible plastic, or steel.

Flex Fan

A flex fan has blades with a flexible trailing edge and a rigid leading edge (Figure 20.16). Flex fan blades have a high angle at low speeds. At high speeds the blades flatten out, reducing the horsepower required to turn them.

CAUTION A cracked or bent flex fan should be replaced. Flex fans can break up. There have been several recalls on them.

Besides being a safety hazard, a fan blade can cut a power steering or radiator hose, a brake line, or other part. Any time a fan blade loses a piece, the fan will become unbalanced. This can cause water pump failure.

■ FAN CLUTCH

A **fan clutch** reduces horsepower requirements. They are found especially on vehicles with air conditioning. There are two kinds of fan clutches: the temperature-regulated and the speed-sensitive (viscous).

The *temperature-controlled (thermal) fan clutch*, generally used with a heavy-duty fan blade, is the type most used in original equipment. It is usually controlled by a **bimetal coil spring** (Figure 20.17), a

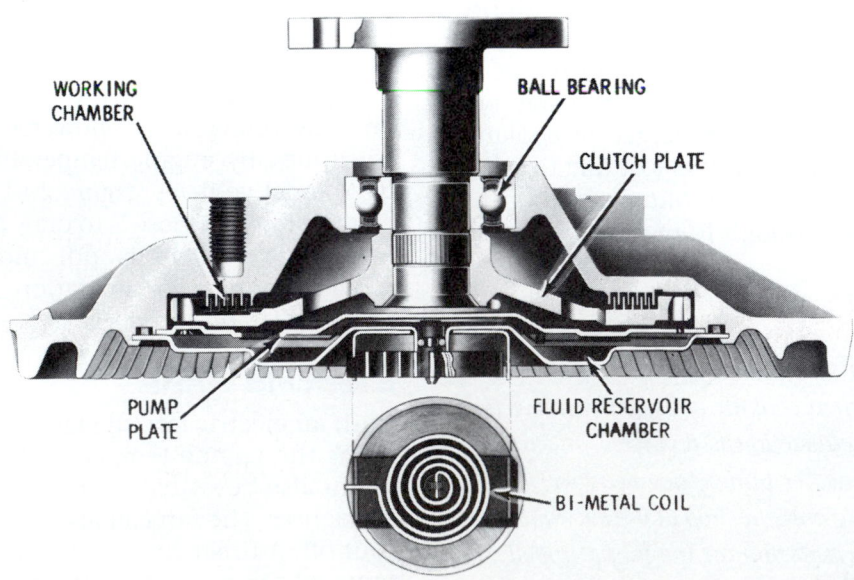

Figure 20.17 A cutaway of a fan clutch. *[Courtesy of General Motors Corporation, Service Technology Group]*

thermostatic coil consisting of two types of metal wound together. When the coil is heated it expands. When it cools it shrinks. The fan only works when the engine is hot. When the engine is cold, the fan free-wheels.

When the air coming through the radiator is hot, the bimetal spring causes an internal valve to open; silicone fluid from a reservoir expands to fill the working chamber and engages the clutch. When the air coming through the radiator is cool enough, the bimetal spring cools, the valve opens, and the silicone fluid moves back into the reservoir to disengage the clutch.

Thermal fan clutches also respond to engine rpm through slippage in the fluid chamber. Slippage occurs when ram air through the radiator from vehicle movement is sufficient for cooling.

A *speed-sensitive (viscous) fan clutch* slips when the resistance of air coming through the fan becomes higher. This fan is similar in operation to the flex fan in that it uses some horsepower at all times. A viscous fan clutch is not as efficient as a thermal fan clutch. A thermal fan clutch reacts not only to engine temperature, but also to rpm. So it would be the clutch of choice for high performance or heavy-duty cooling applications.

■ RADIATOR SHROUD

The radiator shroud makes the fan far more effective at pulling air through the radiator. If the fan shroud is damaged or missing, the vehicle could overheat in traffic.

■ HEATER CORE

The **heater core** is a small heat exchanger that engine coolant is circulated through (Figure 20.18). It

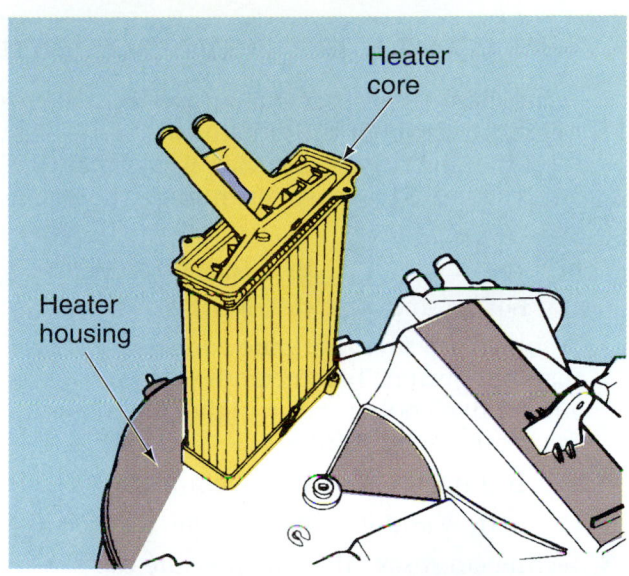

Figure 20.18 A heater core. *(Courtesy of Chrysler Corporation)*

is usually located inside the driver's compartment (see Figure 20.2). A blower motor passes air across the fins of the heater core, transferring heat from the engine to the passenger compartment.

The heater core is supplied with engine coolant through two *heater hoses*. One hose carries coolant leaving the engine. A valve installed in that hose controls the flow of coolant to the heater core when the heater dash controls are set for "heat." A return hose carries the coolant back to the engine for reheating.

■ REVIEW QUESTIONS

1. Automotive coolant is made of _____ and water.

2. Cylinder blocks are made of either iron or _____ .

3. When an electrical current develops between two dissimilar types of metal in the cooling system, this is called _____ .

4. _____ is the name of a coolant additive that protects aluminum.

5. What is the normal operating temperature of an engine?

6. For each pound of pressure on the coolant, approximately how much will its boiling point increase?

7. What is the name of the small valve in the center of a radiator pressure cap?

8. Two names for the part of the radiator that automatic transmission fluid flows through are the transmission oil cooler and the heat _____ .

9. _____ in the thermostat expands to cause it to open.

10. What passage allows coolant to circulate within the block when the thermostat is closed?

■ ASE STYLE REVIEW QUESTIONS

1. Technician A says that the purpose of the fan is to pull air through the radiator when the engine is on the highway. Technician B says that the fan is supposed to push air through the radiator. Who is right?

   a. Technician A   **b.** Technician B
   c. Both A and B   **d.** Neither A nor B

2. Technician B says that pure coolant cools better than pure water. Technician B says that automotive coolant has a higher boiling point than water. Who is right?

   a. Technician A   **b.** Technician B
   c. Both A and B   **d.** Neither A nor B

3. Technician A says that the pressure cap can raise the boiling temperature of the coolant. Technician B says that the small valve in the center of the radiator cap is the vacuum valve. Who is right?

   a. Technician A   **b.** Technician B
   c. Both A and B   **d.** Neither A nor B

4. Technician A says that installing a thermostat with a lower temperature rating will help an engine to run cooler in hot weather. Technician B says that a stuck-closed thermostat can cause a computer-controlled fuel system to run rich. Who is right?

   a. Technician A   **b.** Technician B
   c. Both A and B   **d.** Neither A nor B

5. Technician A says that radiator cores are made of copper and brass. Technician B says that radiator cores are made of aluminum and plastic. Who is right?

   a. Technician A   **b.** Technician B
   c. Both A and B   **d.** Neither A nor B

CHAPTER 21

Cooling System Service

OBJECTIVES

Upon completion of this chapter, you should be able to:

✔ Diagnose cooling system problems.

✔ Service all parts of the cooling system.

KEY TERMS

rod out a radiator
coolant hydrometer
hydrolocked engine
thermoplastic seizure
block-check tester

INTRODUCTION

This chapter deals with cooling system problem diagnosis, maintenance, and repairs. Cooling systems are generally quite dependable. However, they do require periodic maintenance. Cooling system service is one of the best values for the customer in terms of preventative maintenance. Working on the cooling system is usually not very difficult and is sometimes very profitable. This chapter describes those repairs and services.

DIAGNOSING COOLING SYSTEM PROBLEMS

Maintaining Correct Engine Temperature

The engine is designed to run at a predetermined temperature. There are several possible causes of incorrect engine temperature including leaks, overheating, and overcooling. If an engine does not get warm enough, emissions rise, fuel economy suffers, the heater does not work, and the engine can suffer excessive wear. Overheating will cause serious damage to an engine. The following are some of the cooling system problems that can affect engine temperature (Figure 21.1).

Coolant Level. This is the first thing to check.

Restricted Radiator. Rust and scale form in the cooling system when it has been neglected.

NOTE: *A plugged radiator will usually cause overheating on the highway. The restricted circulation can cause the coolant to be pumped out the radiator overflow.*

Stuck Thermostat. A stuck thermostat will cause an engine to overheat or to not heat enough. If it is stuck partially open, the engine can overheat on highway trips and not warm up as soon as it should in town. A thermostat that sticks closed will cause the engine to overheat very quickly (in the time that it normally takes for your heater to start to work in the morning).

SAFETY NOTE

■ When checking coolant, the engine must be off. The coolant must be cold enough so that it is not under pressure. Squeeze the top radiator hose before opening the system (see Figure 11.18). If it is hard, do not open the cap. See your instructor.

■ If you let the pressure off of the coolant, it will boil. This is a dangerous situation! Yet, circumstances sometimes call for a hot system to be opened. When opening the radiator on a system that is hot, fold a shop towel and place it over the radiator cap (Figure 21.2). Holding pressure down on the shop towel, use it to turn the radiator cap ¼ turn until its first stop. Letting pressure off of the shop towel will allow any remaining pressure on the coolant to escape. Remember, the maximum pressure on any radiator is only about 15 to 17 pounds. This is not so much that you cannot hold pressure against it.

■ Check to see that the cap is loose before turning it further and removing it.

Late Ignition Timing. Ignition timing that is retarded more than 2° or 3° from specifications can cause engine overheating. Retarded timing can also be caused by a vacuum advance that is adjusted improperly or a mechanical advance that is sticking (see Chapter 36).

Loose Belt Tension. If the belt that drives the water pump is too loose, overheating can result. Two things will be apparent if this is the case:

■ The belt will probably squeal as it slips.

■ The charging system light will probably come on or the battery will go dead. The same belt usually drives the water pump and the alternator. Adjust belt tension as shown in Chapter 22.

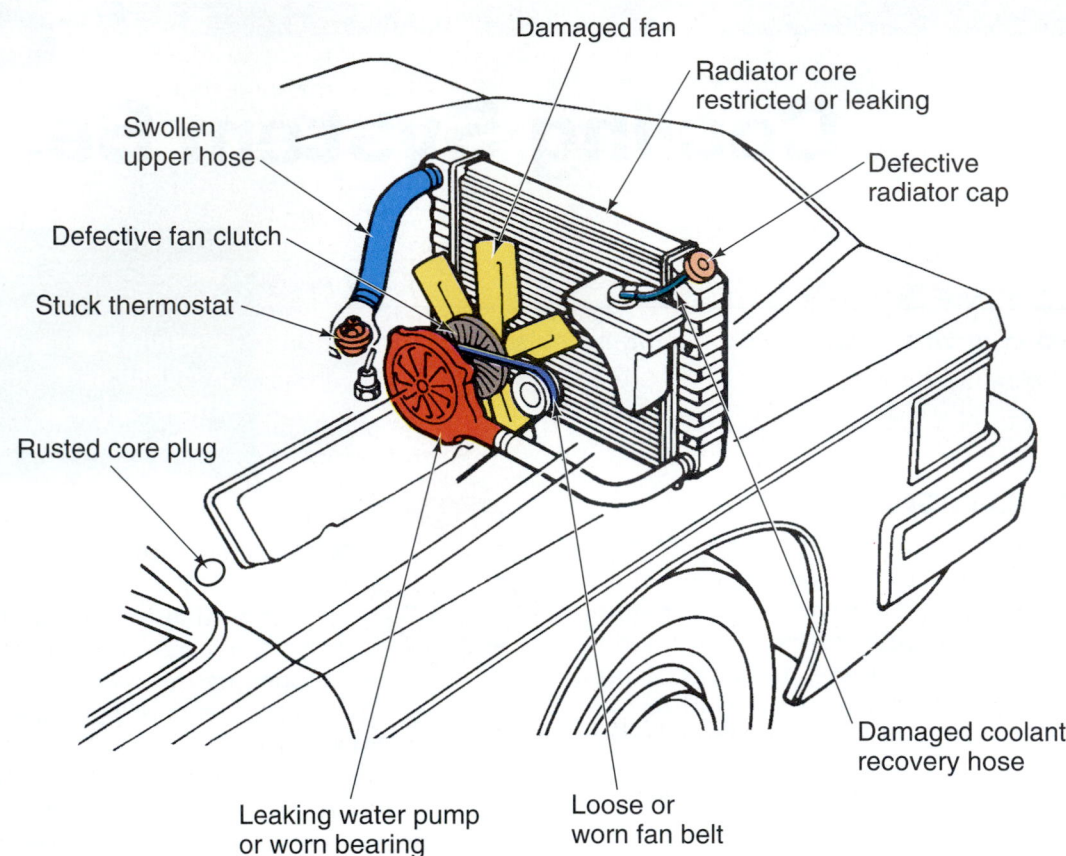

Damaged fan

Radiator core
restricted or leaking

Swollen
upper hose

Defective
radiator cap

Defective fan clutch

Stuck thermostat

Rusted core plug

Damaged coolant
recovery hose

Leaking water pump
or worn bearing

Loose or
worn fan belt

Figure 21.1 Common problems with a cooling system. *(Courtesy of General Motors Corporation, Service Technology Group)*

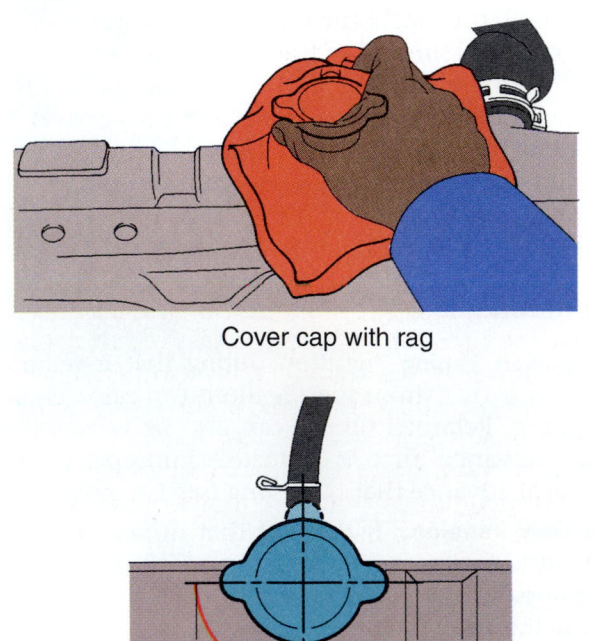

Cover cap with rag

Turn slowly counterclockwise

Figure 21.2 Cover the radiator cap with a rag and turn it slowly counterclockwise while holding down against spring pressure.

Bad Water Pump. Water pumps can fail because of a failure in the bearing, seal, or impeller. The pump has a *static seal* and *bearing* that can fail. The bearing is permanently sealed and can be damaged by excessive belt tension. When a bearing fails, the static seal is also damaged, which can result in a coolant leak. The gasket to the back plate can also leak.

Fan Shroud. When a fan shroud is loose, broken, or missing, the engine can overheat. The fan cannot be efficient without it on most of today's high temperature engines.

Frozen Coolant. When coolant has not been properly maintained, it can freeze in cold temperatures. When an attempt is made to start the engine in the morning, the engine could be seized. This is because the coolant in the water pump has frozen solid. Removing the pump drive belt and starting the engine will confirm your diagnosis.

As coolant begins to freeze, it gets slushy as ice crystals start to form. Slushy coolant can plug the radiator in a cold engine. As the coolant that is trapped in the block begins to heat up, boilover can occur.

Cooling Fan. When a cooling fan does not work properly, the engine can overheat. This problem will be more acute when idling or driving in town.

Exhaust Blockage. A partially blocked catalytic converter or exhaust system can contribute to overheating and loss of power.

Inoperative EGR Valve. Detonation can result from an inoperative EGR valve (see Chapter 38). This can contribute to engine overheating.

▉ RADIATOR CAP

Radiator Cap Inspection

One inexpensive item that a vehicle often needs is a new radiator cap. The rubber seal on the pressure cap can become worn or damaged with age, the pressure spring can rust, or the radiator filler neck seat can be damaged. A damaged cap can allow pressure to escape from the radiator. If pressure is not maintained on the coolant, its boiling point can drop from 40° to 50°F. This causes the radiator to boil over in hot weather or when driving in high altitude.

Several things are checked on a radiator cap (see Figure 13.11):

- Check to see that the rubber seal is not torn or imprinted so that it no longer provides an effective seal.
- Inspect the pressure valve spring for rust damage and freedom of movement.
- Inspect the vacuum valve to see that it is not stuck, broken, or plugged, and that the seal is not damaged. Vacuum valves can be of two types: spring loaded and ones with no spring.
- The cap should have the correct pressure specified by the manufacturer.

Radiator Pressure Tester. A pressure tester (Figure 21.3) can be used to test the cap's pressure valve.

- First, visually inspect the sealing surfaces and the vacuum valve.
- Moisten the sealing surface with water to help it seal during testing.
- Install the radiator cap on the adapter. Attach the pressure tester to the other end of the adapter.

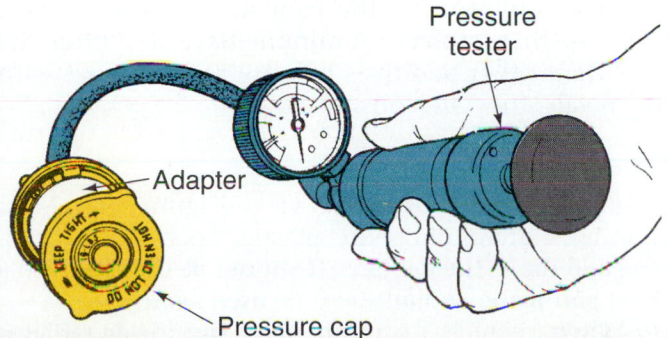

Figure 21.3 A pressure tester is used to test the pressure cap. *(Courtesy of Chrysler Corporation)*

- Pump the handle until the gauge reaches its highest point. The needle should reach a high point and remain constant. The pressure rating of the cap is stamped or printed on the top of the radiator cap. This is the point where the pressure cap should relieve pressure.
- Before removing the cap, release pressure from the system by pushing the tester hose sideways at the cap.

➡ *Perform Pressure Test a Radiator Cap Worksheet*

Radiator Inspection

Inspect the radiator for leaks, flaking, crushed or bent fins or damage to the filler neck seat. Look for obstructions to air flow such as bugs and so forth, which can be washed out with water from the engine side of the radiator. Inspect the overflow hose and passage, if possible.

> **CASE HISTORY**
>
> *A technician was performing an underhood inspection on a car. He noticed that the upper radiator hose was collapsed. When he removed the radiator cap, the hose swelled up to its correct size. An inspection of the radiator cap's vacuum valve showed that it was in good condition. He connected a hand vacuum pump to the hose going to the overflow tank and discovered that there was an obstruction. Further investigation of the hose showed that it had been pinched under the battery box when it was removed during recent body work.*

Radiator Flow

A radiator can be tested to see that it will allow a sufficient amount of water to flow through it. Some specialty radiator shops have equipment for measuring the amount of flow through the core, but this equipment is not usually available in general repair shops. A service technician will usually be able to judge a radiator's condition.

With the radiator full, watch the flow of coolant when the lower hose is removed from the radiator. If flow is sufficient, coolant should fill the entire radiator opening as it flows out.

Another test for radiator flow involves feeling the outside temperature of a radiator in various areas from top to bottom. If the radiator is allowing a sufficient volume of coolant to flow, the fins nearest the radiator inlet (the hose that goes to the thermostat) will feel hot to the touch. The areas near the radiator outlet will feel cooler. Cold spots indicate a restriction in the cold area of the radiator.

Cleaning a Radiator Core

Radiator cores are made of copper or aluminum. If the fins of the core are in good condition, a restricted radiator can be disassembled and **rodded out** by a radiator shop. Rods are forced through the tubes to clean them out.

Replacing a Radiator Core

Cooling efficiency suffers if the tubes have become plugged or if the fins are corroded or broken loose from the tubes. Check the condition of the fins by rubbing something gently against them (be careful not to bend them). If they are in good condition, they will be rigid and will not flake away.

When the cooling fin metal rots, the radiator will require a new core. Many radiator *tanks* are brass and are soldered to the copper radiator core. On copper/brass radiators, the top and bottom tanks are reused by soldering them to the new core. The filler neck is also soldered to the brass tank. This is done in a radiator shop.

The radiators on many newer cars have plastic tanks and vacuum brazed aluminum cores. It is especially important that these radiators be periodically maintained as they are more prone to corrosion/ electrolysis than copper/brass radiators.

Radiator Damage

A radiator can be damaged if a fan blade collides with it. This can happen when a fan clutch or coolant pump bearing fails. If the engine has a broken motor mount, the engine can move excessively, which can tear radiator hoses or allow contact between the fan and the radiator. A radiator that is not too badly damaged can be soldered by a radiator shop.

Most automatic transmissions have heat exchangers (oil coolers) built into the radiator. Look for damage to the transmission oil cooler lines and fittings, where applicable.

Transmission Heat Exchanger Leaks

If a leak develops in a heat exchanger, transmission fluid can be pumped into the radiator when the engine is running. The engine's crankshaft drives the transmission pump. With the engine running, pressure inside of the radiator heat exchanger is about 35 psi. Radiator cap pressure is only about 15 psi so transmission fluid migrates to the radiator.

After the engine is shut off, the pressure in the radiator is still 15 psi. Because the transmission has no pressure when the engine is not running, coolant can enter the transmission. The coolant gums up the transmission, meaning it will probably have to be rebuilt.

■ COOLANT SERVICE

After it has been in use for a period of time, coolant loses some of its protective ability and can become corrosive from contaminants it has picked up.

Coolant Inspection

Perform a visual inspection of the coolant while it is cold. Open the radiator cap and check inside the filler neck with your finger. Look for deposits of grease, dirt, or rust. Look for *corrosion bloom*, a white deposit that attaches to the tops of the tubes. Deposits like these indicate a need for a coolant change because the coolant additives probably have been depleted. If coolant appears to be rusty or contaminated, a radiator flush will be required.

Checking Coolant Conductivity

The coolant's conductivity can be checked with a voltmeter.
■ Ground the positive probe by attaching it to the radiator.
■ Insert the negative probe in the coolant.
■ A reading of 0.2 volt or less is good.

If the voltmeter reads 0.5 volt or more, the system should be flushed and refilled with new coolant to prevent corrosion of the metal parts in the system. Besides the internal problems this condition can cause, an inaccurate coolant temperature sensor reading can result when the cooling system charge is above 0.4 volt. The coolant solution should be close to 50% strength to provide adequate protection against electrolysis.

Coolant Change Interval

Most car manufacturers recommend that coolant be changed *at least* every three years or 30,000 miles or more. Byproducts of combustion contaminants can get into the coolant by leaking past the head gasket. Additives in the coolant package also become depleted and can be replenished by changing the coolant. There are also coolant recycling machines that clean and treat old coolant before returning it to use.

When mixing coolant and water, very hard water should not be used, especially with aluminum heads. Phosphate corrosion inhibitors can drop out of coolant with very hard water. Use distilled water instead.

Older vehicles require more frequent coolant changes. Inhibitors in the coolant wear out. Silicate additives that protect aluminum have a shorter life span than other additives because as they react with the metals, they are consumed.

Draining Coolant

A radiator usually has a drain valve (Figure 21.4). Most of today's drain plugs are plastic. Loosen the drain plug and drain the coolant. It should be drained into a clean pan if the coolant is to be used again.

When there is no drain plug, the lower radiator hose is removed to drain the radiator. Twist the hose before trying to pull it off of the connection. Be careful

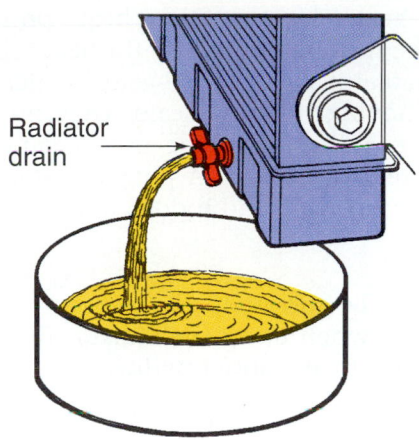

Figure 21.4 A radiator drain valve. *(Courtesy of Chrysler Corporation)*

not to damage the radiator. On copper/brass radiators, the hose connection is soldered to the lower tank. This seal can become broken or the connection (which is soft brass) can be easily deformed.

Sometimes a screwdriver can be used to separate a stuck hose from the connection. If it does not separate easily it will be necessary to cut the hose (Figure 21.5). A special hose cutting knife is available from tool manufacturers. Be especially careful not to cut through the thin brass on the inlet or outlet of a heater core.

There are also drain plugs on the side of the block (Figure 21.6). They are tapered pipe plugs and often have become rusted to the block and are difficult to remove. If they are accessible and come out easily, remove them to drain the block. Use sealer on them on reinstallation.

Cooling System Flush

When tap water is used instead of coolant, mineral deposits and dirt can build up in the water jackets (Figure 21.7). When the cooling system has been neglected, or the owner wants to have periodic maintenance performed on the cooling system, a back flush can be performed without removing the radiator from the car. A back flush is when water is run through the system backwards from its normal direction of flow.

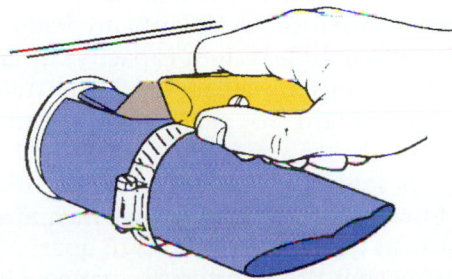

Figure 21.5 If a radiator hose will not come off without forcing it, cutting the hose might be necessary.

When flushing a radiator, it is handy to install a "flushing-tee" to the heater hose that comes out of the heater (Figure 21.8). One of the heater hoses runs from the firewall (bulkhead) to the water pump. The one that runs from the firewall to the top of the engine is the one to install the T into. Install it in a relatively straight section of hose. A splash tube is installed on the top of the radiator. The coolant should be captured and disposed of in the proper manner. The T is left in the hose and a cap is threaded on it to seal it off.

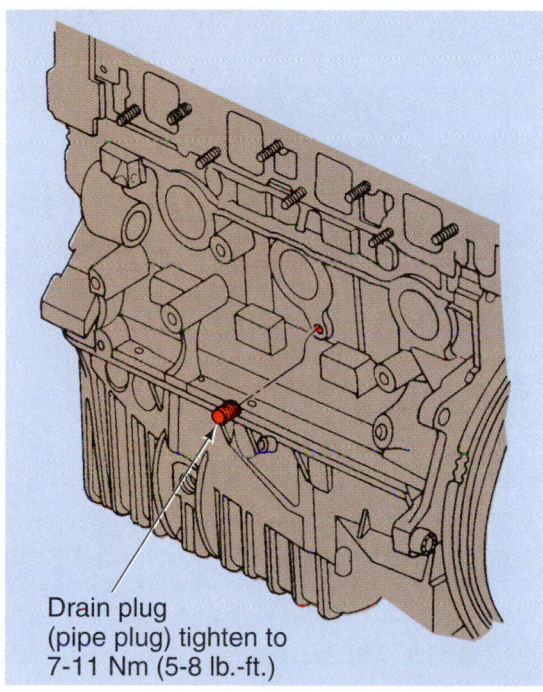

Figure 21.6 A drain plug in the block has tapered pipe threads. *(Courtesy of Ford Motor Company)*

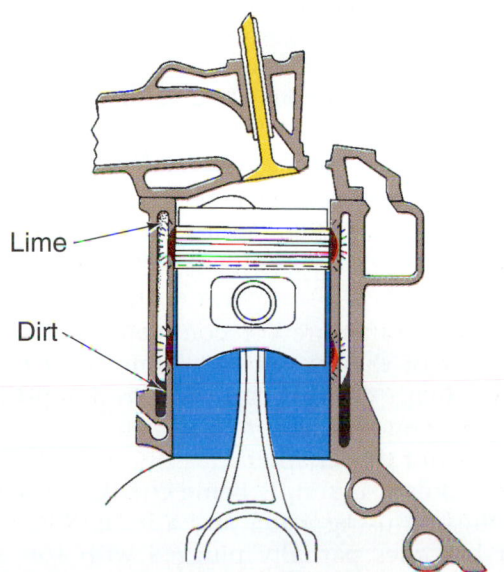

Figure 21.7 Using tap water results in dirt and mineral buildup in the water jackets. *(Reproduced by permission of Deere & Company. ©1992, Deere & Company. All rights reserved)*

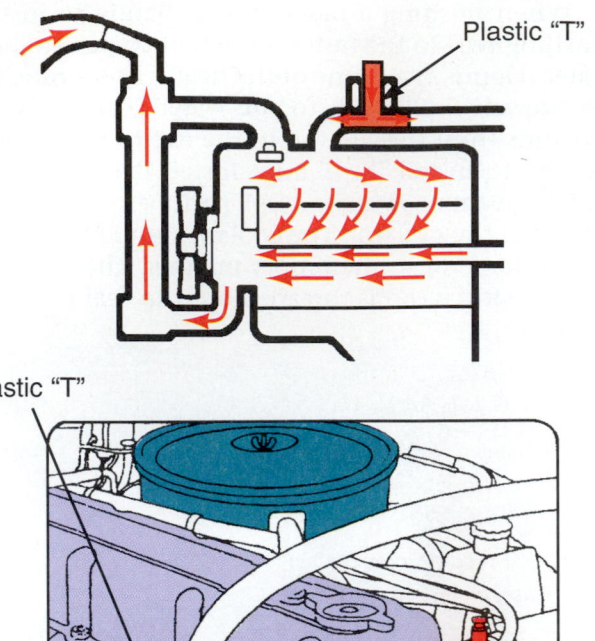

Plastic "T"

Plastic "T"

Figure 21.8 Install a plastic T in the upper heater hose. *(Courtesy of Prestone Products Corporation)*

NOTE: *Many communities have regulations governing how coolant is to be disposed of. Be sure to follow regulations in your area. Ethylene glycol is biodegradable when new. But when it is used, corrosion has taken place in the engine and heavy metals may be present in it. Besides being poisonous, that is why it is considered to be hazardous.*

Many manufacturers recommend the periodic replacement of the thermostat. Removal of the thermostat is usually done as part of a flushing procedure. Thermostat removal and replacement is covered in another part of this chapter.

If the coolant is simply being changed, the system can be flushed using water and a hose. Sometimes a radiator becomes partially plugged with soft sludge. Using a commercial chemical cleaner, the radiator can be flushed without removing it from the car.

With a hose running into the radiator, the engine runs at idle while the system is flushed. Move the dashboard heater levers to the "heat" position to be sure that the heater core will be flushed, too.

Some radiator cleaners are acids, which are effective for removing rust and scale. They must be neutralized with a base following use or damage to the cooling system can result. A plugged radiator can be removed from the car and rodded out.

CAUTION Be sure that you wear safety goggles when using the radiator flush chemical. It can cause blindness.

Run the car for the required period of time. If the cleaning chemical used requires a neutralizer, add it to the water remaining in the radiator after flushing is complete. The liquid remaining in the block is 100% water.

Special flushing guns and pressure flushing machines are available. They use air pressure to force water backward through the system to break loose layers of dirt and deposits.

Check the Coolant Condition and Strength

The strength of coolant can be checked using a **coolant hydrometer** (see Figure 13.13), which compares the weight of coolant to the weight of pure water. Draw some coolant into the hydrometer and read the gauge. Check instructions on the tester.

Coolant that is too concentrated can be diluted with water. If coolant strength is very weak, a coolant flush and change is recommended. A slightly weak concentration can be strengthened by draining off a quart of coolant and adding straight coolant. After running the engine, the strength is checked again.

➡ *Perform __Check Coolant Strength__ Worksheet*

Coolant Concentration

Coolant is mixed with water until it is the correct concentration. The maximum concentration that should be used is 70% coolant and 30% water. Look up the cooling system capacity in the manual so that you can add the correct mixture of coolant and water (approximately 50% of cooling system capacity). If an engine has a cooling system capacity of 16 quarts, use 2 gallons of coolant to get an approximate 50% concentration.

Coolant is sold in gallon containers. It is available in quarts, but is more expensive that way. If a cooling system holds 13 quarts, 1½ gallons (6 quarts) will provide a mixture that fits into the ⁴⁰⁄₀₀ range. Unless the winter weather in the area is especially harsh, this will provide a good mixture.

Which Coolant to Use

Many of today's engines are bimetal; that is, they have an aluminum head or heads and an iron cylinder block. These engines can suffer severe corrosion if the correct additives are not used. The coolant additive that protects aluminum is *sodium silicate*. These coolants have labeling that states "for use with aluminum."

If an engine's cooling system does *not* contain any aluminum parts, a good rule of thumb is to use a coolant without silicate additives. Silicate additives can cause problems when used in too high a concentration, such as when a silicate additive is added to a coolant that already has a high silicate concentration. Also, as the ratio of coolant to water is increased, the solubility of the silicates decreases.

- Silicates in a heavy concentration can gel on heat transfer surfaces. Engine overheating and poor heater operation can result. Sand granules can also form that are abrasive, resulting in coolant pump leaks.
- Coolant should be used before the shelf life date printed on the container expires. If coolant with silicate additives is stored for too long it will get gummy.

■ SCIENCE NOTE ■

Silicon is the most abundant element in the universe. It is the most important industrial semi-metal, its major use being in electronic components such as transistors. Glass is another material made from silicon. When silicon is dispersed in automotive coolant it protects aluminum, too.

SAFETY NOTE

- Ethylene glycol coolant is poisonous. Ingestion of 4 ounces is sufficient to kill a human being. It has a sweet taste so it is especially attractive to animals.
- Be careful when handling extremely cold ethylene glycol. It can freeze-burn skin when it has been stored in the trunk of a car during very cold weather.
- Ethylene glycol can be ignited by a flame at 474°F. Research by General Motors has shown that an explosion can actually occur if a mist of pressurized ethylene glycol coolant mixture is sprayed on an open flame.

■ SCIENCE NOTE ■

A hydrocarbon in which one of the hydrogen atoms has been replaced with a hydroxyl group (OH) is an alcohol. The hydroxyl group bestows waterlike properties to the hydrocarbon (gasoline) so alcohols are soluble in water. Glycols are dihydroxy alcohols (they contain 2 OH groups instead of one). Coolant (ethylene glycol) is made from ethane ($CH_3\, CH_3$) by replacing a hydrogen on each carbon with a hydroxyl group. When ethylene glycol coolant freezes it forms a slushy mass, rather than a solid block like water. In the absence of anti-freeze, the 9 volume expansion that takes place when water freezes would generate a force of 30,000 psi at –22°C. This is sufficient force to damage an engine block.

Brake fluid is a a chemical relative to coolant. It is a polyglycol, a hydrocarbon containing two or more hydroxyl groups making it also soluble in water.

➡ *Perform Flush Cooling System, Add Anti-Freeze Worksheet*

■ THERMOSTAT

Testing the Thermostat

Testing the thermostat in the car can be done using a thermometer attachment with a handheld multimeter or by putting a thermometer into the coolant while the engine runs and observing it while the engine warms up. With the engine running, the coolant will begin to show signs of movement when the thermostat opens. Note the temperature at that point.

If the coolant begins to circulate at a temperature that is different than the rating of the thermostat, the thermostat is defective and must be replaced. If the temperature does not rise to the proper point, the thermostat could be stuck open or missing completely.

Thermostat Check after Removal

After a thermostat is removed, it can be tested. To check its operation, suspend it in a pan of hot water. Use a thermometer while heating the water (Figure 21.9). If the thermostat is fully closed when cold and opens all of the way within a few degrees of its rating, it is good. Most thermostats start to open at between 188°F and 195°F. They are fully open at 212°F (water's boiling point).

The thermostat should open at least ¼" or more when immersed in boiling water. The thermostat should be fully closed at room temperature. Hold it up to the light to see if it is. If light comes through the thermostat at the sealing surface, replace it.

Remove the Thermostat

Drain some coolant into a clean container until the coolant level is below the thermostat housing. Remove the upper radiator hose connection from the thermostat housing. Loosen the housing bolts and remove the

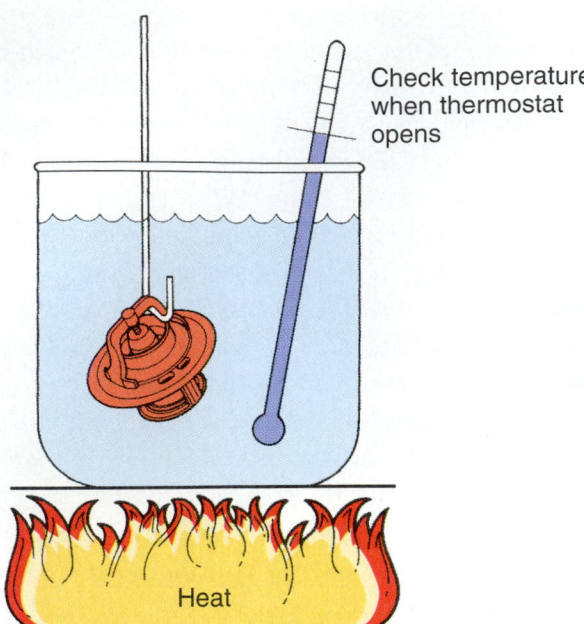

Figure 21.9 Checking thermostat operation. *(Courtesy of Ford Motor Company)*

Recessed area

Figure 21.10 Be sure that the thermostat fits into the recess in the housing or manifold.

housing. Remove the gasket and scrape it carefully from the surface of the housing and the mounting surface on the engine. If any gasket remains on either of the surfaces, there will probably be a coolant leak after reassembly. Some engines use a rubber O-ring to seal a thermostat housing.

Replacing the Thermostat

Compare the new thermostat to the old one. They are of different sizes, types, and temperature ratings. The temperature rating is stamped on the sensing bulb on the bottom of the thermostat (see Figure 20.12). The temperature bulb faces into the block. When replacing a thermostat, be sure that the thermostat fits into the groove in the block or outlet housing (Figure 21.10). Install the gasket.

SHOP TIP When a paper gasket is used and the recess is in the thermostat housing, it is a good practice to position the thermostat into the recess and glue the gasket to hold it in place. If it falls out of its groove during installation, the outlet housing can be cracked or a coolant leak will result. Before tightening the water outlet housing, try to rock it back and forth to be sure it is flush. Housings are often cracked during this step.

Reinstall the thermostat housing. Refill the system and run the engine or pressure test to check for leaks. When the engine has reached operating temperature make sure the thermostat opens. You should be able to see coolant circulating within the radiator.

CAUTION If the radiator is filled to the top with coolant and the engine is run without the radiator cap in place, the coolant will expand and spill over as the engine warms up.

Another way of checking thermostat operation is to feel the top radiator hose or use a thermometer or multimeter with a temperature probe to confirm that the coolant is warming up. If the engine is overheating but the top hose is still cool to the touch, the thermostat is stuck closed and must be replaced.

➡ *Perform **R & R a Thermostat** Worksheet*

Bleeding Air from the System

There is sometimes a small hole drilled in the thermostat that allows air to escape during filling of the cooling system. Some cooling systems are difficult to fill without trapping air. Overheating can result unless the air is bled off. Because of aerodynamics, many of the new cars have radiators that are lower than some of the other cooling system parts. These other parts trap air that cannot be bled out from the radiator cap as it normally would.

SHOP TIP To bleed a cooling system, remove a heater hose at the highest place in the system while filling the radiator (Figure 21.11). Removing a heater hose bleeds air from the heater core, too. If a flushing T has been installed in the top hose, unscrew its cap and raise the hose until it is higher than the heater core. Then, top off the coolant until it comes out the hole.

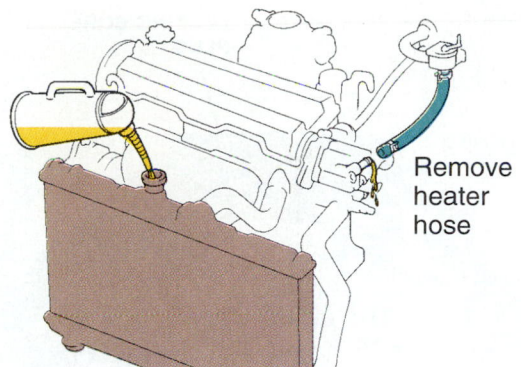

Figure 21.11 When refilling the system, bleed air off by removing a hose from the highest place in the system.

Check the service manual for instructions on bleeding the system. Some procedures call for removing a temperature sending unit or a thermal vacuum switch.

Always recheck to see that the system is full and all air has been purged from the system before releasing the vehicle to a customer.

■ LOCATING LEAKS

Whenever possible, locate a leak before starting a repair procedure. Leaks can be external or internal. External leaks are usually obvious and easily observable. Internal leaks can be through a leaking gasket or a crack. Tests for leaks are covered in this section.

Pressure Tester

When an internal or external leak is suspected, a pressure tester is helpful in locating it (Figure 21.12). Sometimes when trying to determine whether there is an internal leak or not, the engine must be tested at different temperatures. Cracks often leak when the engine is cold but not after it warms up. Also, although it is easier to pressure test a cold engine, leaks sometimes will not show up when the engine is cold. Test stubborn leaks with the engine both hot and cold.

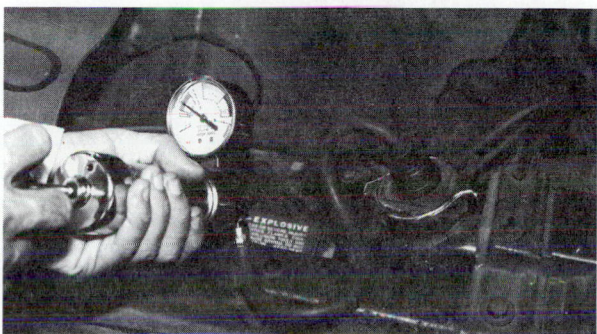

Figure 21.12 A pressure tester installed on a radiator filler neck.

NOTE: *Perform the pressure test with the engine off. The pressure tester does not blow off like a radiator cap does. The cooling system can be damaged if pressure is allowed to rise above normal system pressure.*

Pump on the handle of the pressure tester to pressurize the cooling system to the pressure marked on cap. After five minutes, the pressure on the gauge should remain steady, which indicates no leakage. If the gauge pressure drops and an external leak is not apparent, an internal leak is indicated.

NOTE: *A very small leak may not be evident in this short a time. Also, some pressure drop can occur as the coolant shrinks if the cooling system temperature drops.*

■ EXTERNAL LEAKS

If the pressure reading drops during a pressure test, first look for signs of external leakage.

SHOP TIP When a leak is in a position on the engine where it is not easy to see, an inspection mirror can usually be used with a flashlight to help to locate it. When the flashlight is shined on the mirror to bounce the light onto the area of the suspected leak, the leak will be visible in the inspection mirror

Check for external leaks at the heater core, heater hoses, radiator hoses, thermostat housing, core plugs, radiator, hole on the bottom of the water pump (see Figure 21.23), and carburetor. It is not uncommon for a leak that appears to be coming from the back of the engine (between the engine and transmission) to be running down the side of the block and along the side of the oil pan. There are core plugs behind the flywheel on some (but not all) engines. A black light tester (see Chapter 43) can be helpful in determining the location of difficult leaks.

CASE HISTORY *A student wanted to replace leaking core plugs on the back of the block between the engine and the transmission. He removed the automatic transmission from the car and removed the flex-plate from the back of the crankshaft. Unfortunately, the block he was working on did not have core plugs on the back. When he pressurized the cooling system, he found that a core plug behind one of the side motor mounts was leaking. Coolant was running down the edge of the oil pan to the rear of the block, where it appeared to be coming from behind the flywheel. Using a pressure tester and flashlight to locate the leak before attempting to repair it would have saved a good deal of needless work.*

Coolant Outlet Housing Inspection

Inspect the coolant outlet (thermostat) housing for leaks or damage. Aluminum housings often suffer electrolysis damage. When possible, replace the housing with one made of the same material as the rest of the engine block or the cylinder head that it bolts to. For instance, install an iron housing on an iron head and an aluminum housing on an aluminum head. Be sure to position the hose clamp so as not to allow corrosion between the hose and the housing (Figure 21.13).

Core Plug Inspection

When the cooling system has not been serviced regularly or when someone fills the system with water after a leak has been repaired, corrosion inside the system can occur. Core plugs are made of steel (unless they have been replaced with brass or stainless steel). They commonly rust out and begin to leak (Figure 21.14).

Core plugs are usually located on the sides of the block. Sometimes they are found on the front and rear of the block, too. Use a mirror and a flashlight to inspect core plugs for signs of rust or leakage.

➡ *Perform **Pressure Test Cooling System** Worksheet*

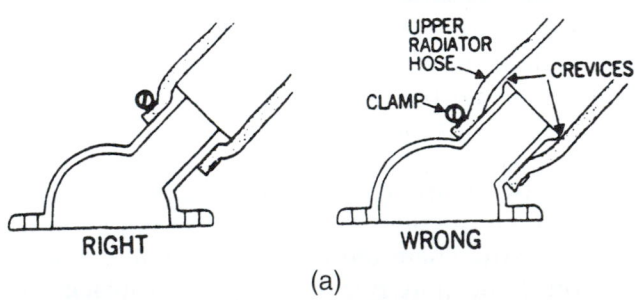

(a)

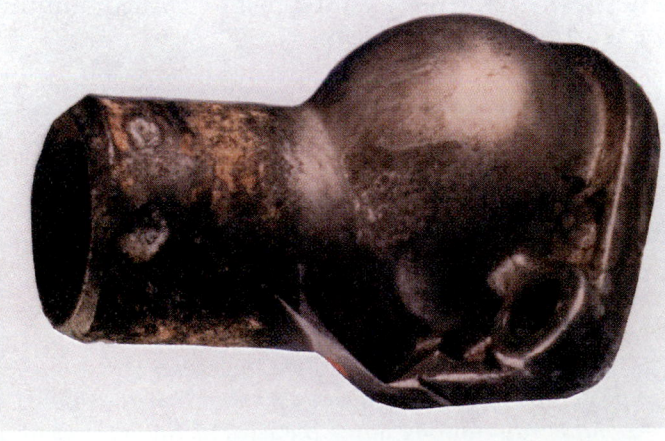

(b)

Figure 21.13 (a) Position the hose clamp near the ridge on the outlet housing. (b) This aluminum water outlet became corroded when the clamp was positioned improperly. *(Courtesy of Prestone Products Corporation)*

Figure 21.14 This core plug behind the flywheel was leaking.

◼ INTERNAL LEAKS

A leaking head gasket or a crack in a cylinder head or bore can result in an internal leak (Figure 21.15). When there is an internal leak, coolant will flow into the cylinder during the intake stroke and when the engine is off. During combustion, exhaust gas is forced into the cooling system and can appear as bubbles in the radiator (Figure 21.16). There are several tests that can be done to confirm an internal leak.

Bubble Test

Look for bubbles in the radiator when the engine is warm and under a load. Rapidly accelerating the engine is usually enough of a load to produce the bubbles.

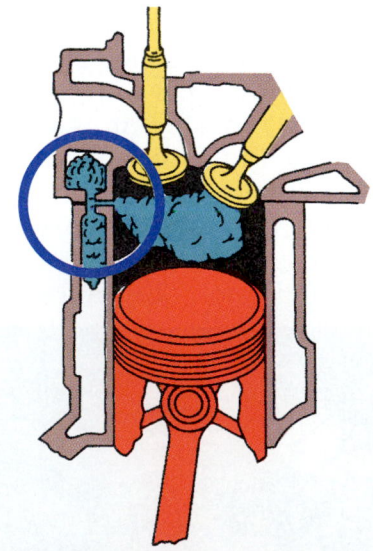

Figure 21.15 When there is an internal leak, coolant will flow into the cylinder when the engine is off and during the intake stroke. During combustion, exhaust gas migrates into the cooling system. *(Reproduced by permission of Deere & Company. ©1992, Deere & Company. All rights reserved)*

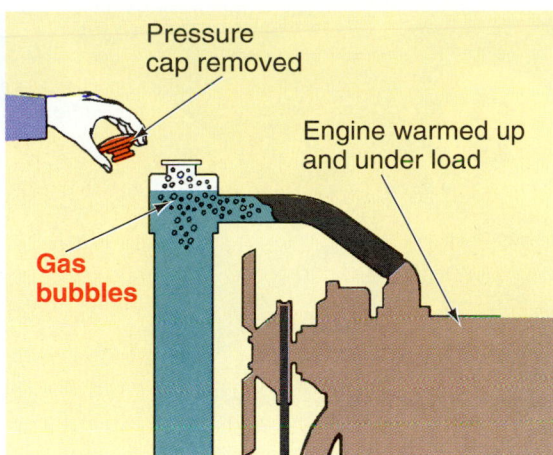

Figure 21.16 Exhaust gas leaking into the cooling system can result in bubbles in the radiator. *(Reproduced by permission of Deere & Company. ©1992, Deere & Company. All rights reserved)*

NOTE: *Cracks tend to leak more when the engine is cold. After warm-up, the crack closes.*

The radiator cap will blow off when the pressure from the combustion leak exceeds radiator cap pressure. To see if combustion pressure is indicated, put the radiator overflow hose into a container of water while the engine runs (Figure 21.17). If bubbles are evident, combustion pressure is getting in.

Bubbles could also be present because the cooling system is drawing in air. To eliminate this possibility shut off the engine, loosen the drive belt to the coolant pump, and repeat the test. If the bubbles disappear, air was getting into the pump.

When there is air in the system, corrosion occurs at about three times the normal rate. Air can leak into the cooling system through a leak in the lower radiator hose. The lower hose is the suction hose, where coolant is drawn into the pump. Air can leak in even though

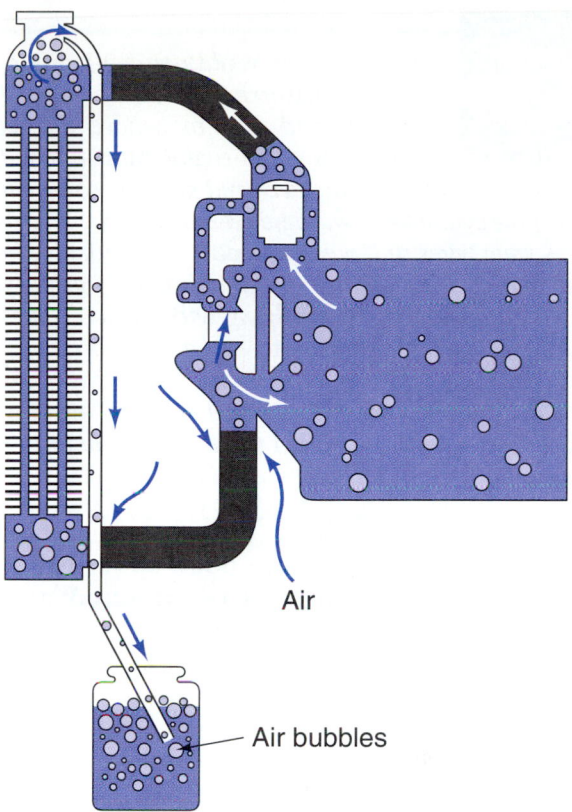

Figure 21.17 Air leaking into the system results in air bubbles coming out of the overflow tube. *(Reproduced by permission of Deere & Company. ©1992, Deere & Company. All rights reserved)*

water may not leak out. To test a cooling system for air leakage, tape the filler neck of the radiator closed. Put a hose from the radiator overflow pipe into a jar of water. With the engine running, look for bubbles in the jar.

Hydrostatic Lock

Sometimes an internal leak can result in one or more cylinders filling up with coolant after the engine is shut off. This happens because the radiator cap continues to exert pressure on the coolant, even though the engine is off. If the engine stops with a piston

---

SHOP TIP Here is a very effective test that can be done for final confirmation of an internal combustion leak before removing the cylinder heads:

- Loosen or remove the water pump belt and remove the thermostat. Reinstall the thermostat housing.
- Unhook the top radiator hose from the radiator and fill the hose with water and put a thermometer in it.
- Run the engine and look for bubbles in the coolant before the water's boiling point is reached. On a V-type engine, remove the radiator hose and look into the thermostat housing to see which side of the engine the bubbles are coming from to pinpoint the bank that has the leak.

SHOP TIP Remove spark plugs on a hydrolocked engine. Crank the engine to let the coolant out so that the car can be driven to the repair shop.

CAUTION If you smell gasoline after the spark plugs are removed, do not crank the engine. A malfunction in a fuel system could have resulted in gasoline locking up the engine.

A student's car would not crank over after he had been working on it in the shop. The instructor told him to remove the spark plugs to see if the engine turned over. The student turned the engine over and gasoline came out of one of the cylinders. It was ignited by one of the unhooked spark plug wires that was lying near the spark plug hole. The burning gasoline was sprayed onto the back of one of the students who was helping with the job. The student panicked and started to run across the shop. He was tackled by the instructor and the fire was extinguished from the back of his jacket and his hair. He was o.k. but he was lucky. Do not run if you are on fire. Drop and roll on the ground to extinguish it.

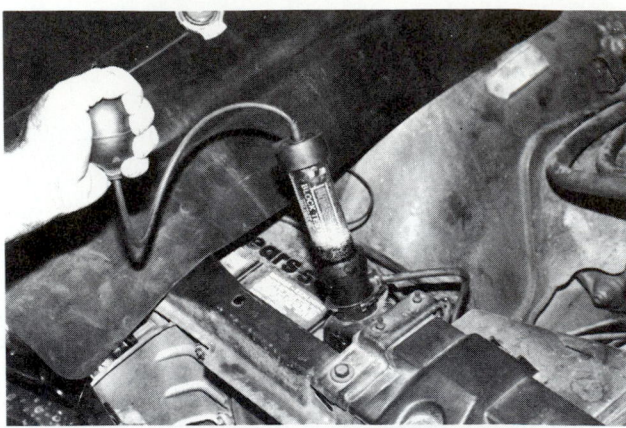

Figure 21.19 A block check test samples air above the coolant.

down in the cylinder while both of its valves are closed, the engine will be **hydrolocked** and the crankshaft will not be able to turn.

Sometimes a leaking head gasket or a cracked cylinder allows a concentrated mixture of ethylene glycol coolant to leak into the crankcase (Figure 21.18). The result is varnish-like oil that can plug oil rings and ruin valve guide seals. This sticky substance can actually seize the crankshaft (called **thermoplastic seizure**). The problem will happen again if the source of the leak is not found and repaired. Then, the engine and cooling system must be flushed.

Block Check Test

A leaking head gasket will not always show up on a pressure test. Another means of testing for leakage of

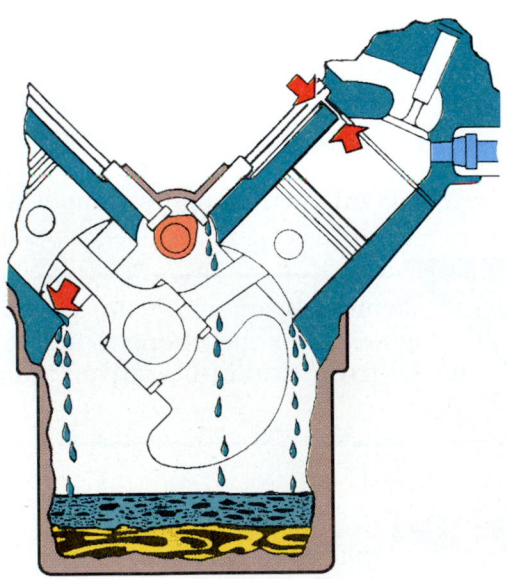

Figure 21.18 An internal leak can result in coolant entering the crankcase. *(Reproduced by permission of Deere & Company. ©1992, Deere & Company. All rights reserved)*

exhaust gas into the cooling system is the **block-check tester** (Figure 21.19). The tester samples air in the filler neck of the radiator. Unlike the pressure tester, the block check is used with the engine running. If there is carbon monoxide (CO) exhaust gas in the radiator, the color of the tester fluid will change. Carbon monoxide is a byproduct of combustion, so the tester will not work if the leaking cylinder's spark plug is not firing. Also, if compression is too low or if coolant entering the cylinder causes the plug not to fire, the tester will not give a reading.

A small amount of a special blue fluid is used to fill the tester to its fill line. If there is exhaust gas in the coolant, it will react with the tester fluid. First, the fluid turns from blue to green. Then it turns yellow.

This test is done on a warm engine. Be sure that the thermostat is open. First, the level in the radiator must be lowered if necessary until it is 2" below the top of the filler neck. This is done with the engine off. With the engine idling, place the tester on the radiator filler neck and pump the bulb several times to suck *air* from above the coolant.

NOTE: *As the coolant gets hotter, its level will rise. Letting coolant into the test fluid will ruin the fluid and void the test.*

If the results show no exhaust gas in the coolant after performing the following test, try the tester at the opening of the car's exhaust to see how the fluid reacts. When the fluid is left exposed to air, it will return to its original blue color.

➡ *Perform __Block-Check Test__ Worksheet*

Infrared Analyzer

An infrared exhaust analyzer can also be used to check to see if there is exhaust gas in the coolant (Figure 21.20). The analyzer probe is inserted into the neck of the radiator, in the air space above the coolant. The analyzer samples air, drawing it in through the probe.

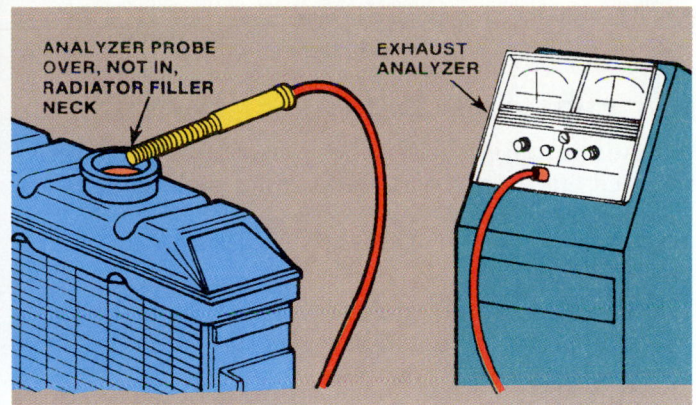

Figure 21.20 An infrared analyzer can be used to test for a combustion leak into the cooling system. *(Courtesy of Chrysler Corporation)*

Be careful that the coolant is not accidentally sucked into the probe.
- This method works whether the mixture has been burned or not.
- Load the engine by accelerating it in gear with the brakes on for three seconds or less.
- Hold the probe over the radiator filler neck to check for hydrocarbons in the coolant. Be sure not to suck coolant into the tester probe.
- If there is CO in the system, gas must be getting in during combustion.

RECOVERY TANK SERVICE

An overflow (recovery) tank (see Figure 20.4) is a good feature to add to an older car if it does not already have one. A replacement tank can also be installed on newer cars with a damaged or leaking tank. Some of the less expensive tanks have a single molded plastic mounting hole in the tank. Original systems that have these holes have several of them for support. Be sure to buy an original equipment one or one with a high quality bracket.

NOTE: *One gallon of water weighs about 8 pounds. When the tank vibrates on rough roads, it can be torn from its mount.*

If the radiator cap needs to be replaced on a recovery tank system, be sure the new cap includes a seal that works against the top of the filler neck, too. The hose from the top of the radiator allows coolant to escape when pressure exceeds the rating of the cap. With a recovery tank it is recycled instead of just going to the ground.

Recovery tanks help decrease corrosion. When the coolant level in the radiator is allowed to drop below the level of the tops of the cooling tubes, *solder bloom* corrosion can occur. This corrosion is lead oxide that forms when oxygen in the air reacts with the solder at the top of the tubes.

COOLING SYSTEM REPAIRS
Replacing Core Plugs

Cup-type core plugs are the most common type. To replace a core plug, pound it sideways with a *blunt* drift punch (Figure 21.21a). A sharp punch will penetrate the core plug. The objective is to turn the plug sideways inside of its bore so that it can be easily removed. When it is sideways, it can be pulled with pliers (Figure 21.21b). Try not to let it go into the water jacket.

NOTE: *Do not leave an old core plug inside of the block. This is an unprofessional practice that can result in further cooling system damage.*

Sometimes a slide hammer with a hook can also be used to remove a core plug. Sometimes the plug is so rusty that removal is difficult.

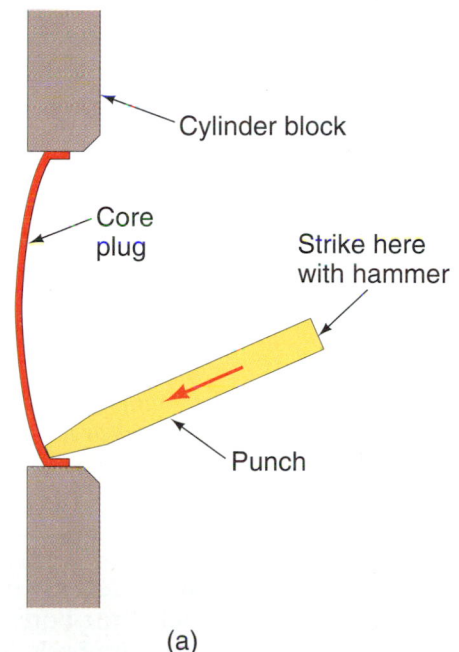

(a)

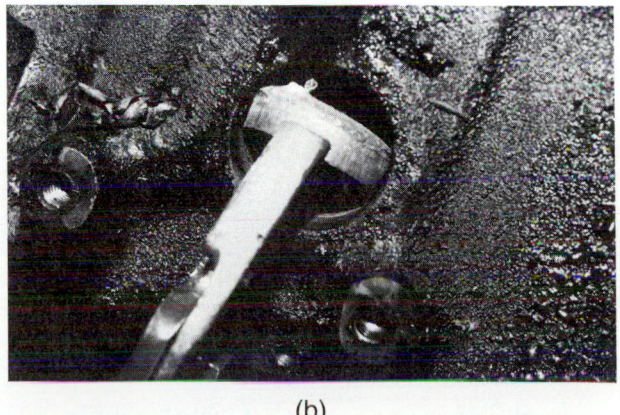

(b)

Figure 21.21 Removing a core plug. (a) Use a blunt punch to knock the core plug sideways. (b) Remove it with pliers.

> **CAUTION** Sometimes a core plug is positioned on the block with very little clearance to the back of wall of a cylinder. Be careful not to pound the core plug against the cylinder wall.

Core Plug Installation

Before installing a core plug, clean the opening in the block with emery cloth. Put some sealer on the sides of a new plug and also on the side that will be exposed to the coolant. Pound the core plug in with a driver or a socket that fits *loosely* into the inside diameter of the plug. Check to see that the driver contacts the core plug on its inside surface (not on its outer sealing edge). Be sure the core plug goes straight into the hole.

> **CASE HISTORY** *An apprentice technician was overhauling an engine and removed the core plugs. The wall of one of the cylinders was positioned directly behind a core plug. The core plug became wedged between the block and the cylinder wall so he forced it out. A cylinder wall is not very rigid and can easily become distorted when something such as a core plug is forced against it.*
>
> *During reassembly, the apprentice attempted to reinstall the piston and rings but the piston would not go into the cylinder. When the shop master technician measured the cylinder, he found that it was out-of-round by 0.005". The piston clearance specification was 0.002". The block had to be sent out to a machine shop for boring and honing.*

A core plug seals on its outer lip. It is correctly installed when the lip is against the bore. Figure 21.22 shows one manufacturer's recommendation. Pound it in until the outside sealing edge is just below the chamfer that is on the outside of the core plug bore.

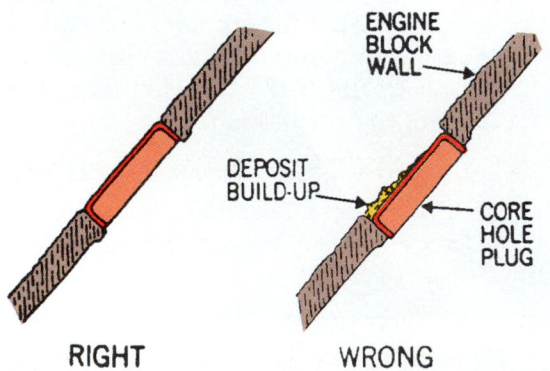

Figure 21.22 Installing a core plug. Install the core plug until it is all of the way into the hole. *(Courtesy of Prestone Products Corporation)*

NOTE: Some technicians prefer to replace steel core plugs with brass ones that will not corrode.

■ COOLANT PUMP SERVICE

Coolant pumps, usually called water pumps, are often replaced after many years and miles of service, when they begin to leak or make noise.

Leaking Pump

Leakage from a water pump will usually be visible from the weep hole in the bottom of the pump (Figure 21.23). A coolant pump leak can also appear to be from the lower radiator hose. Use a mirror to look at the weep hole. Pressurize the system if necessary with a pressure tester.

- A pump seal can fail because of bearing failure, corrosion of the shaft, or dirt.
- The seal can become red hot if it is run without coolant when the radiator boils over. Adding cold water to an overheated system causes the hot seal to crack.
- When a vehicle has been allowed to sit with a dry or dirty cooling system, the water pump seal will sometimes stick to the shaft. When the engine is started, it breaks loose, resulting in a leak.

Worn Bearing

Sometimes a worn bearing will result from the failure of the seal. With the engine running, a stethoscope can be used to listen to a bad bearing. Before replacing a pump, loosen the drive belt and feel for roughness in the bearing. It should turn freely and smoothly, without end play.

Worn or Broken Impeller

Sometimes a pump impeller can be loose or broken, but this is rare. Look for water pump action in the radiator with the engine warm and running.

Coolant Pump Replacement

There are various types of pump installations on new vehicles. When in doubt about the procedure to follow for removal of the pump, consult the service manual

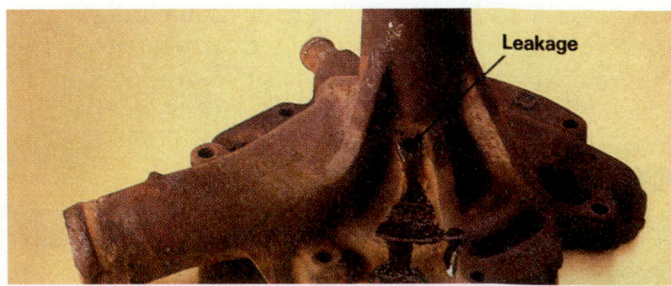

Figure 21.23 The pump seal has failed when leakage is evident at the vent hole. *(Courtesy of Federal-Mogul Corporation)*

for the vehicle. Sometimes other accessory belts must be removed so that the belt that drives the pump can be removed or installed. Other times they will not have to be removed. Simply loosen the bolts *slightly* so that the belts can be loosened. Unbolt the brackets to the air conditioning compressor or power steering pump if necessary.

> **SHOP TIP** Drain the coolant into a clean drain pan (especially if it is to be reused). Sometimes a customer has recently had the coolant changed. They will not be happy to pay for new coolant again. Coolant is also costly to dispose of.

Remove all of the bolts and the pump. Clean all remnants of the old gasket off of the engine. Be careful not to gouge aluminum parts.

Inspect the Old Pump

Inspect the old pump. Pump impellers are made either of steel or plastic. Carefully inspect the impeller for erosion, looseness, or breakage. If pieces break off the impeller they will go into the block and, possibly, the radiator. It may be necessary to disassemble the radiator to be sure that all pieces are removed.

Selecting a Replacement Pump

Replacement water pumps are usually new, although sometimes rebuilt pumps are available at reduced cost. Pumps are rebuilt by rebuilding companies. They are not rebuilt by service technicians. The best rebuilds will have all new internal parts and simply reuse the old housing. When rebuilders reuse bearings, they have an expensive tester that listens to them for noise.

Pump housings are made of either aluminum or cast iron. Aluminum is more susceptible to corrosion so cast iron is more popular for rebuilt pumps. Some pumps come with a lifetime guarantee.

When buying a rebuilt pump, the old one must be turned in as a core. A core is an old rebuildable part with a value that is added to the price of the rebuilt part. If a core is damaged, the core charge paid when the part was purchased will not be refunded.

Compare the new pump with the old one (before leaving the parts store, if possible).

> **SHOP TIP** In some applications that use serpentine belts, the water pump is driven from the bottom side, using the back of the belt. The water pump **turns the opposite direction** of a V-belt driven pump. Be sure the correct pump is installed or the **impeller** will be rotating backward, resulting in engine overheating.

Installing the New Pump

A water pump often has a steel plate bolted to its back. The plate gasket sometimes dries out during shipping. If this happens the bolts will be loose, causing a leak. Before installing a new pump, it is a good idea to remove the bolts and cement both sides of the plate gasket.

When replacing a pump, be sure that all gasket material is thoroughly removed and that any o-rings, hoses, or gaskets are not damaged or forced during assembly. Be sure that the screws that hold the cover on the back of the pump are tight. Use sealer to glue the gasket to the water pump. Sometimes a chemical gasket is used. Be sure that the surfaces of the pump and block have been cleaned of all oil and coolant so that the chemical can stick.

When tightening the pump, torque the screws in a criss-cross pattern. Be sure the pump is perfectly flat against the block before tightening any fasteners.

Refilling the System

It is a good idea to fill the cooling system with water and pressurize it with a pressure tester before completing the reassembly of the belts and radiator. If there is a leak, it can be easily fixed at that point without wasting newly added coolant.

After the job is completed, the engine needs to be run. When the thermostat has opened the water level will probably drop as the coolant leaves the radiator to fill up the empty cylinder block.

Refill the radiator. Inspect the hose from the radiator filler neck to the recovery tank. Be sure the hose is in good condition and tight on its fittings. Fill the recovery tank to about ½ full.

■ FANS

There are different types of fans. Procedures for checking them vary.

■ FAN INSPECTION

- An out-of-balance fan assembly can lead to water pump shaft and bearing failure.
- A leaking fan clutch, bent or broken fan, or a cocked or cracked aluminum fan spacer are all possible causes of pump failure. Be sure to clean all mating surfaces and tighten the fan bolts *evenly* to avoid causing a cocked assembly.

■ FAN CLUTCH INSPECTION

Inspect the fluid clutch for leaks and to see if it is loose or frozen. There are several ways to test the temperature-controlled clutch:

Physical Tests

- First, with the engine off, turn the fan by hand. There should be a slight resistance, but the fan should turn without the roughness that would indicate a bad bearing.

Figure 21.24 A leaking fan clutch. *(Courtesy of Federal-Mogul Corporation)*

- Rock the fan up and down to see if it is too loose.
- If there is a buildup of greasy dirt in the bearing areas of the clutch, the silicone fluid has probably leaked out (Figure 21.24). If the bearing in the clutch fails, replace the clutch; otherwise, the resulting imbalance will ruin the coolant pump.

Engine Running Tests

Block off the radiator and run the engine with the air conditioner operating to help warm the coolant.

- When the engine is cool, the fan will not pull much air.
- As the engine warms, there should be a noticeable increase in the noise level from the fan.
- If the fan clutch does not engage before the temperature gauge shows hot, it must be replaced.
- When a warm engine is shut off, the fan can turn a small amount, but it should not continue to freewheel. If it turns more than four or five turns, it is probably defective.
- When the engine cools down after the radiator is unblocked, the fan should disengage.
- The operation of the clutch can be checked with a timing light (Figure 21.25). Manufacturers' service

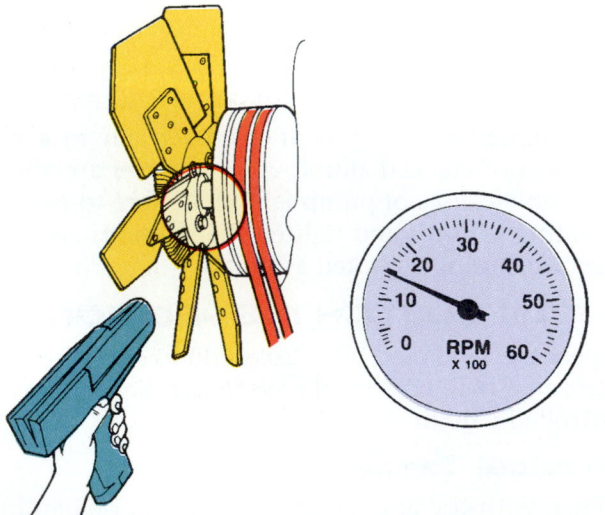

Figure 21.25 The operation of a fan clutch can be checked with a timing light. *(Courtesy of Ford Motor Company)*

manuals give the procedure and engine speed specifications.

SHOP TIP A handy way to check fan clutch engagement is to write a number on the engine side of each fan blade with a marking crayon, and then point a timing light at the fan. If the drive pulleys are of equal size, the numbers will be stationary if the fan is locked up. If the clutch is slipping, the numbers will run backward.

■ ELECTRIC COOLING FAN SERVICE

Electric fans are turned on and off in response to a signal from a coolant temperature switch screwed into the radiator, water outlet housing, or a water jacket in the engine block or head. When the engine temperature goes above a predetermined temperature, the fan comes on to provide extra cooling.

If the fan is not working, look for an obvious cause such as a disconnected wire. Then, check the fuse panel to see if the fuse is burned out. Next, check the switch to see if it is operating properly using the following procedure:

- When the engine is cold, disconnect the electrical connector to the coolant temperature switch. Use an ohmmeter to read across the two terminals of the switch. It should show an open switch (infinite resistance).
- With the wires to the switch connected, run the engine until it is warm. The fan should come on indicating a switch that is good.
- If the fan does not come on, disconnect the wires to the switch. With the ohmmeter, the switch should now show continuity (low resistance) indicating a closed switch. If not, replace the switch.

This will involve checking the coolant temperature sensor and any relays that apply.

SAFETY NOTE Be sure to disconnect an electric fan motor whenever working it.

■ HEATER CORE SERVICE

A heater core can leak or become plugged. The core must be removed to repair it or replace it in the event of a leak. This is sometimes a big job, requiring the removal of many parts under the dash.

NOTE: If the windshield becomes more fogged when the defroster is turned on, this can be a symptom of a leaking heater core. But momentary fogging of the windshield is common in combination heater/air conditioning systems when the heater is first turned on (even when the heater core is good). Operate the air conditioning with the heater control in the defrost position to dry the inside of the windshield.

The heater core is supplied with engine coolant through two *heater hoses*. One hose carries coolant leaving the engine (Figure 21.26). A return hose carries the coolant back to the engine for reheating. If both heater hoses are removed during service or repair, mark one of them so that they can be returned to their proper positions.

On older cars, a valve installed in that hose controls the flow of coolant to the heater core when the heater dash controls are set for "heat." Newer cars have coolant flowing through the heater core at all times. Heat demands are controlled by controlling the doors to the ducts around the heater core.

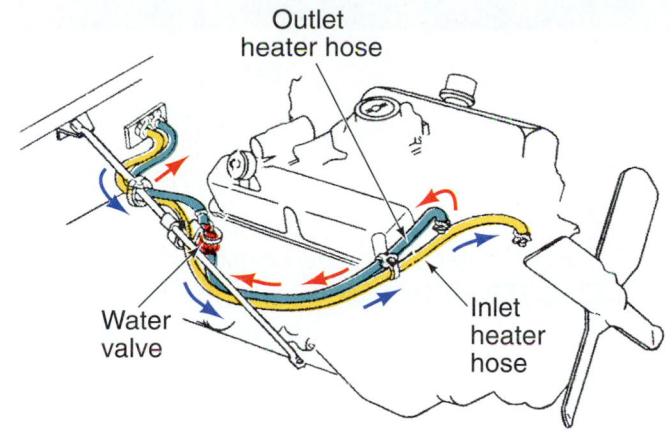

Figure 21.26 Direction of coolant flow to the heater core. *(Courtesy of Chrysler Corporation)*

REVIEW QUESTIONS

1. When the bottom of a radiator core feels considerably colder to the touch than its top, it is probably plugged. (True or False)

2. When a radiator is disassembled and cleaned out, this is called _____ out.

3. When measuring the voltage of coolant, what is the limit before the system should be flushed out?

4. What is the name of the tool that compares the density of water to that of coolant?

5. What type of cleaner removes rust and scale?

6. Ethylene glycol coolant is dangerous to animals because it is _____ .

7. Why is there a small hole in some thermostats?

8. What is the name of an old, rebuildable part that is turned in to the parts house when a rebuilt part is bought?

9. What is name of the test that uses colored liquid to find exhaust gas in the cooling system?

10. When an engine will not turn over because a cylinder has become full of coolant, this is called _____ lock.

ASE STYLE REVIEW QUESTIONS

1. Technician A says a car that overheats only on the freeway probably has a stuck-closed thermostat. Technician B says that the thermostat can be left out of the engine with no ill effects. Who is right?
   - **a.** Technician A
   - **b.** Technician B
   - **c.** Both A and B
   - **d.** Neither A nor B

2. Technician A says that coolant leaves the engine through the top radiator hose. Technician B says that if cold water is added to a hot cooling system, the water pump could start to leak. Who is right?
   - **a.** Technician A
   - **b.** Technician B
   - **c.** Both A and B
   - **d.** Neither A nor B

3. A transmission heat exchanger has failed. Technician A says if the engine is running, oil will flow from the transmission to the radiator. Technician B says that when the engine is off, coolant will flow into the transmission.

Who is right?
   - **a.** Technician A
   - **b.** Technician B
   - **c.** Both A and B
   - **d.** Neither A nor B

4. Two technicians are attempting to locate a leak by using a pressure tester on the radiator. Technician A says to use a mirror and flashlight to locate a leak if the location cannot be seen. Technician B says to run the engine when performing the test. Who is right?
   - **a.** Technician A
   - **b.** Technician B
   - **c.** Both A and B
   - **d.** Neither A nor B

5. Technician A says when replacing a thermostat the temperature sensing bulb faces out of the block. Technician B says engines with aluminum heads should use anti-freeze with special (silicate) additives. Who is right?
   - **a.** Technician A
   - **b.** Technician B
   - **c.** Both A and B
   - **d.** Neither A nor B

Automotive Belts

■ INTRODUCTION

The topic of this chapter is the theory and service of all types of belts. Accessories are usually driven with a belt from the crankshaft (Figure 22.1). Pumps and air conditioning compressors are driven either by a V-belt or a V-ribbed serpentine belt. On some engines, the camshaft is also driven by a belt, called a timing belt.

■ BELT MATERIAL

Belts are very strong and flexible with **tensile cords** to provide strength (Figure 22.2). The overcord material on the top of the belt is made of **neoprene** or another kind of oil-resistant artificial rubber. The undercord of the belt is the area beneath the tensile cords. It supports the cord and transfers loads to the pulleys. Sometimes, the undercords have a cord support platform with textile cords running perpendicular to the tensile cords (see Figure 22.2). Tensile cords are used to prevent the belt from sagging in the middle, which results in uneven load distribution and early belt failure.

■ V-BELT

Early cars used a flat drive belt. The V-belt was invented in 1917. It has more surface area in contact with the pulley groove than a flat belt of the same width would have.

V-belts must be the correct size:

■ A belt should extend slightly out of the pulley groove (Figure 22.3). The belt has a cord member to provide strength. If the belt rides too high in the groove, the belt will wear below the cord member.

■ When a belt rides below the edge of the groove this indicates either a worn belt, a worn pulley groove, or a belt that is too small.

■ A belt that is too small for the pulley will bottom out in the pulley groove, and the sides of the belt will not grab.

■ Some small diameter pulleys would cause severe bending stress on a belt. In this case, a notched belt is used.

Belt Cords

Most V-belts have polyester tensile cords. The strength of a belt is determined by the placement of the belt's tensile cords.

■ **High cordline belts** (Figure 22.4) are stronger but require more material to manufacture.

■ *Center cord* belts are cheaper, but do not last as long.

High cord belts have about 40% more cords because the cord is at a wider part of the belt. In SAE tests, high cord belts lasted about four times longer than center cord belts.

In past years, the edges of premium belts were covered with a fabric cover to protect them from the elements. Today's belts have no cover and do not show wear as easily. Sometimes they can appear good, even though they are ready to fail.

Some engines use dual belts to drive accessories; these must be replaced in pairs when they are worn.

■ V-RIBBED BELTS

V-ribbed belts are ribbed on one side (Figure 22.5) and flat on the other. Figure 22.6 compares the pulleys for V-belts and V-ribbed belts. The thinness of the belt makes it more flexible so it can bend around smaller pulleys and also be bent backward so both sides can be used to transmit power. Usually the ribbed side matches the pulley grooves of accessories and the flat side goes against a spring-loaded tensioning roller. But the flat side is also capable of transmitting power.

■ SERPENTINE BELT DRIVE

V-ribbed belts are used in newer cars in a conventional manner or in a **serpentine** drive. Serpentine V-ribbed

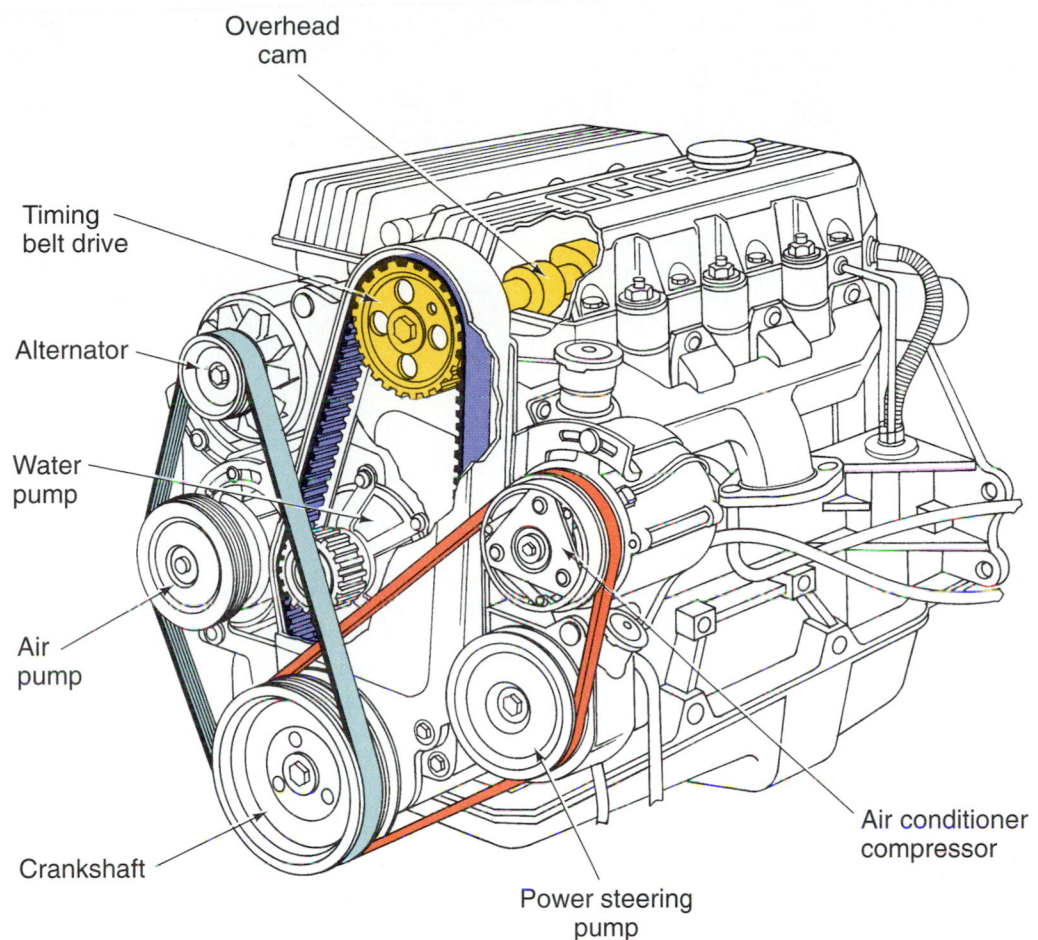

Figure 22.1 Belt-driven accessories and camshaft. *(Courtesy of The Gates Rubber Company)*

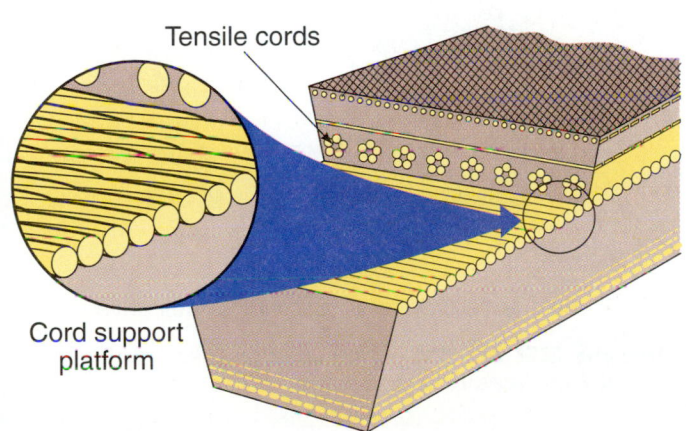

Figure 22.2 Tensile cords and the cord support platform. *(Courtesy of The Gates Rubber Company)*

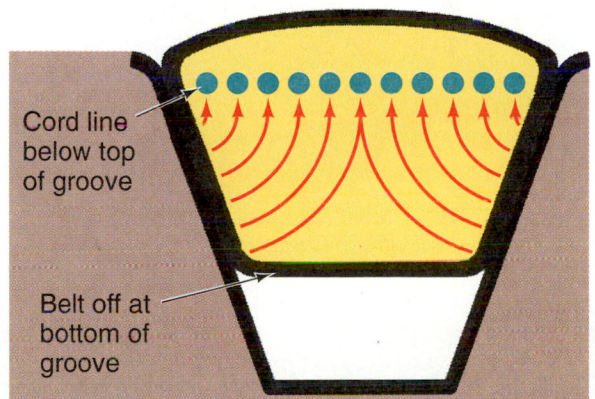

Figure 22.3 This V-belt fits well in the pulley groove. *(Courtesy of The Gates Rubber Company)*

belts, which first appeared in the late 1970s, are used on many new engines today. One belt is used to operate all accessories (Figure 22.7). The belts are called serpentine because they follow a snake-like path, weaving around the various pulleys. Compared to V-belts, serpentine belts are easier to install, take up less space

(Figure 22.8), transmit power more efficiently, and last longer.

▧ TIMING BELTS

On some overhead cam engines, a timing belt drives the camshaft. Timing belts were introduced in the

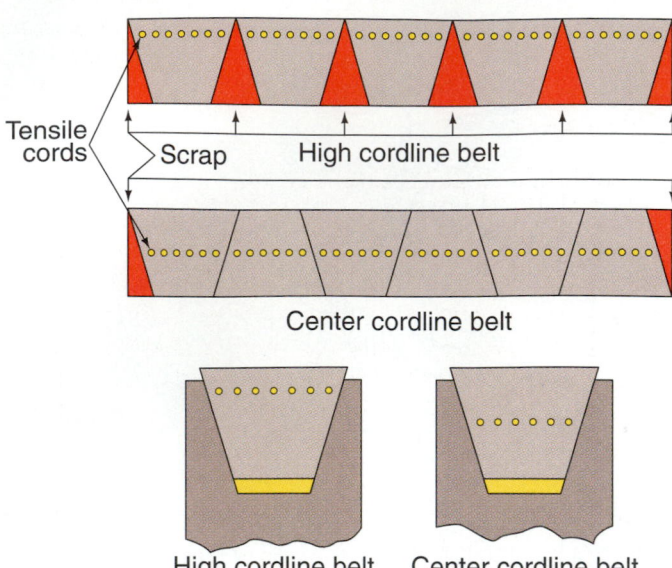

Figure 22.4 High cordline belts are stronger but require more material to manufacture. The higher cord position is evident when viewing the belt from the side. *(Courtesy of The Gates Rubber Company)*

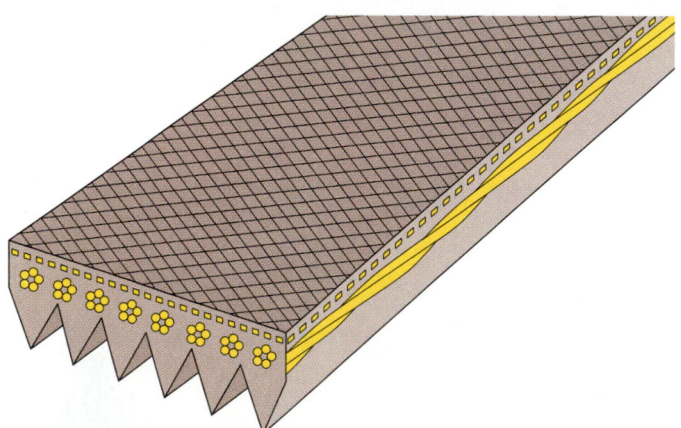

Figure 22.5 Construction of a V-ribbed belt. *(Courtesy of The Gates Rubber Company)*

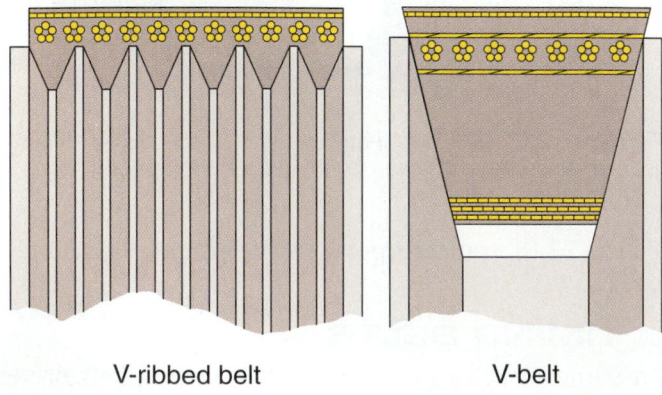

Figure 22.6 Comparison of a V-ribbed belt and a V-belt in their pulleys. *(Courtesy of The Gates Rubber Company)*

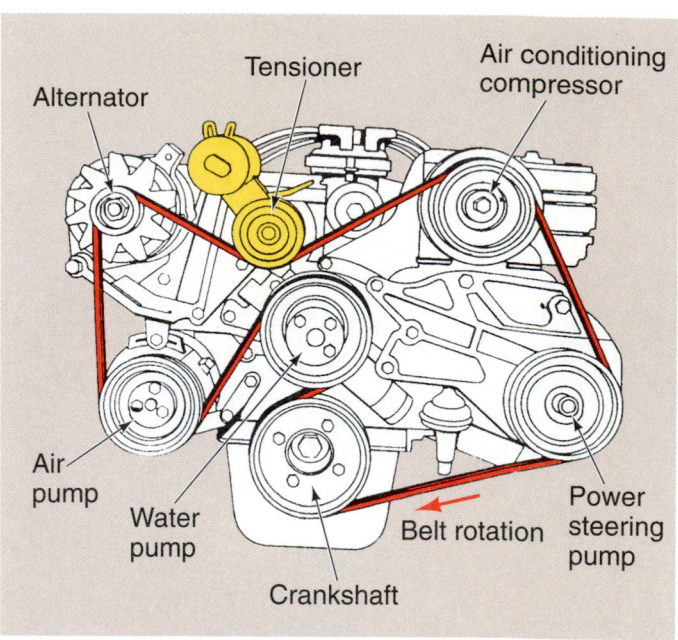

Figure 22.7 A serpentine V-ribbed belt drive. *(Courtesy of The Gates Rubber Company)*

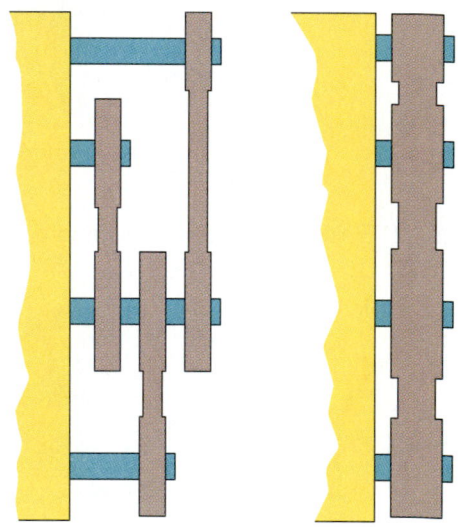

Figure 22.8 Serpentine belts save space. *(Courtesy of The Gates Rubber Company)*

1960s. Compared to a timing chain, they are quieter, do not require lubrication, are more efficient, and resist stretching. Both the inner cogged surface and the outer flat surface can be used to drive accessories. The timing belt sometimes drives a coolant pump (Figure 22.9), an oil pump, and/or a balance shaft.

Timing belts have a very strong fiberglass cordline and rubber impregnated molded teeth (Figure 22.10). The fiberglass cords are stronger than the polyester cords found in conventional belts, but they are more fragile. Bending them in any direction can cause them to break.

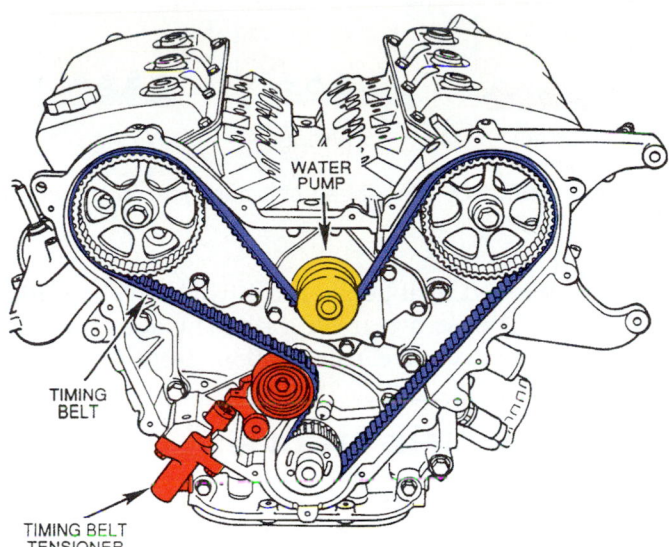

Figure 22.9 Both sides of a timing belt can drive accessories. *(Courtesy of Chrysler Corporation)*

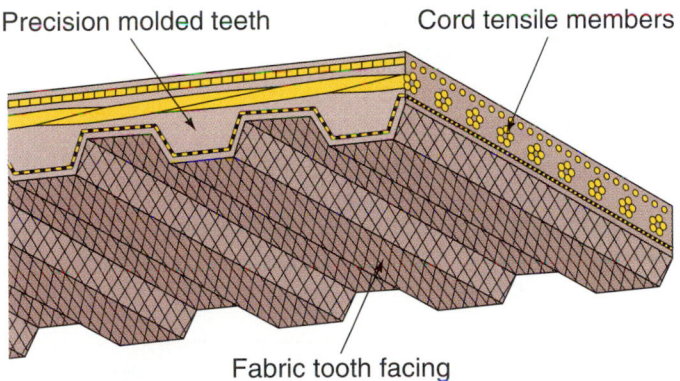

Figure 22.10 Parts of a timing belt. *(Courtesy of The Gates Rubber Company)*

The rubber material in timing belts can be either high temperature neoprene or HSN (highly saturated nitrile). HSN is a superior material for high performance applications. A comparison of belt materials by the Gates Rubber Company gives the following estimation of belt life at sustained high temperatures:

neoprene: 200 hours

high-temperature neoprene: 500 hours

HSN: 1500 hours

More information on timing belts can be found in Chapter 18.

◼ DRIVE BELT SERVICE

Accessory drive belts and timing belts are very strong and dependable, when they are replaced at reasonable intervals. A failed water pump drive belt can cause engine failure if the owner ignores the problem and continues to drive the car until the engine overheats

severely. Because belts are so important, it is advisable to change them periodically *before* they fail, regardless of appearance. Belt failures rise significantly after four years of use, so that is the recommended interval for replacement.

◼ BELT INSPECTION AND ADJUSTMENT

Belts often appear to the eye to be good, even though they are about to fail. Belt failure can result in a loss of power to turn the accessories and the coolant pump.

Check all drive belts for wear, cracks, or damage (Figure 22.11). A belt that has any of these conditions should be replaced. Be certain that there is sufficient tension on the belts.

> **SHOP TIP** With the engine off, check water pump/alternator belt tension by trying to turn the pulley on the alternator by hand. If it slips, the belt is too loose or glazed and must be replaced.

When a belt slips on the pulley, it becomes glazed. A glazed belt must be replaced because it will not grip the pulley tightly, even if tension is at specification.

Inspecting Belts

Normally, a V-belt contacts only the sides of the pulley groove. When a V-belt is the wrong size or has become worn from slipping, the bottom of the belt can contact the pulley groove bottom (see Figure 22.3). Cracks on

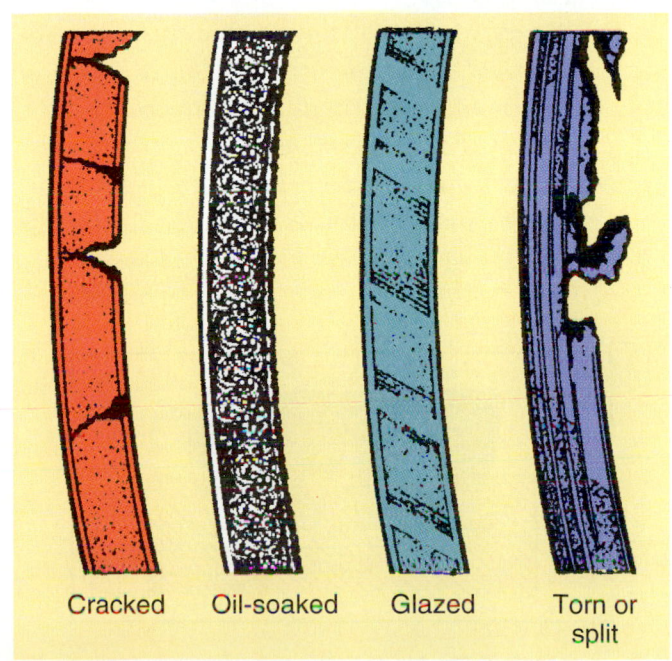

Cracked   Oil-soaked   Glazed   Torn or split

Figure 22.11 A worn belt can result in serious engine damage. *(Courtesy of Chrysler Corporation)*

the face of an old V-ribbed belt are normal, but replacement is suggested.

Inspect the pulley grooves for oil, rust, or wear. Smaller pulleys usually show wear first.

NOTE: *Surface finish of the pulley is more important with a V-ribbed pulley than with a V-type pulley.*

Some shops use belt dressings to help a belt to grip. The problem with these substances is that they are sticky and attract dirt, which wears the pulley groove. Some belt dressings actually attack and damage the belt material.

Belt Alignment

Inspect belt alignment before disassembly. V-belt pulleys must be in alignment within 1/16" for each foot of the distance between the pulleys. Misalignment can be due to the pulley shafts being unparallel or due to an accessory being located improperly. Misalignment can cause rapid belt and pulley wear and thrown belts. When pulleys are out of alignment, the belt will chirp. Noise is a result of vibration. In this case, the vibration results because the belt must continually slide into the groove as it moves against the pulley (Figure 22.12).

■ SCIENCE NOTE ■

When an object vibrates it causes the molecules of air surrounding it to compress and then expand. This cycle moves through the air much like a ripple through water. If the compression/expansion wave vibration is at or near the natural frequency of the part, damage can occur.

The movement of the air can be compared to a wave with the crests corresponding to forward movement (expansion) and the troughs corresponding to backward movement (contraction). The distance between the crests is called a wavelength. The number of crests that bypass a given point in one second is called the frequency of the vibration. The shorter the wavelength, the smaller the frequency and vice-versa.

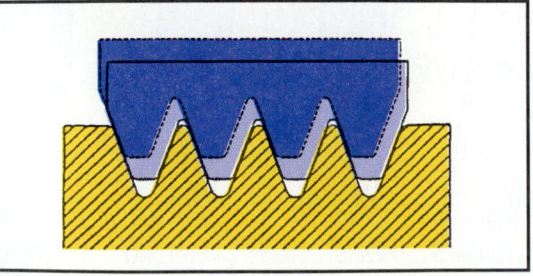

Figure 22.12 Noise happens when a misaligned belt vibrates in the pulley groove. *(Courtesy of The Gates Rubber Company)*

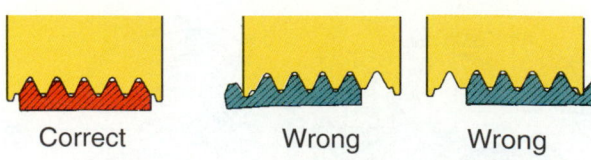

Correct   Wrong   Wrong

Figure 22.13 A V-ribbed belt can walk off the pulley if it is misaligned.

A misaligned V-ribbed belt can walk off the pulley (Figure 22.13) or can tear off a rib. The accessory can sometimes be realigned using 0.030" front end alignment shims.

■ REPLACING BELTS

SAFETY NOTE Before replacing a belt, disconnect the battery to prevent an electric cooling fan from accidentally coming on. Also, the output terminal on the alternator will be hot when the battery is connected. Prying against the outside of the alternator can result in an accidental short circuit.

Loosen bolts on any accessories that act as drive belt adjusters and remove the belt. Compare the new belt to the old belt. If no belt is available, use a piece of string in the pulley groove to estimate the replacement belt size. Belts usually change in size in 1/2" increments, which is reflected in the part number of the belt.

SAFETY NOTE Be sure the new belt is the right size. A belt that is too long can rub on a radiator hose or fuel line after tension is adjusted.

Dual drive V-belts are used on some high load requirement pulleys, such as high output alternators or air conditioning compressors. These belts must be replaced with matched set belts. Matched set belts are molded together and will provide equal tension because they are the same length.

SHOP TIP If a belt seems too small to install, try installing it in a different order. For instance, if the belt will not go over the water pump pulley last, try installing it over the alternator pulley last.

V-Ribbed Belt Replacement

Most V-ribbed belts have constantly spring-loaded tensioning idler pulleys that contact the smooth backside of the belt (Figure 22.14). A few others have "locked

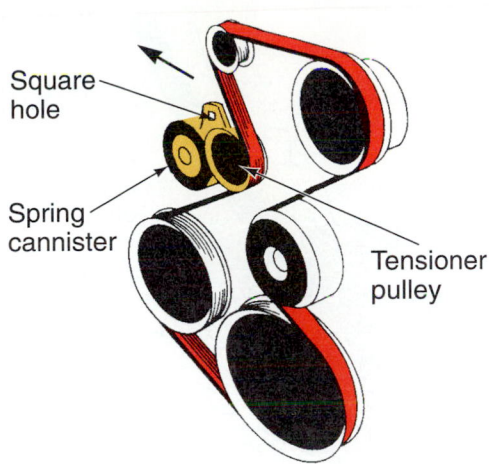

Figure 22.14 To install a V-ribbed belt with a spring loaded belt tensioner, pull the pulley in the direction of the arrow to release belt tension. *(Courtesy of The Gates Rubber Company)*

center" drives. In these, tension is controlled in one of the following manners:

- A tensioner pulley with an off-center bolt (Figure 22.15)
- An adjustable **jackscrew** (Figure 22.16)
- An accessory such as the alternator (Figure 22.17)

Before removing a serpentine belt make a sketch of how it is installed. It can be complicated figuring out the path of one of these belts. Most vehicles have an underhood diagram label. Belt manufacturers produce catalogs with belt routing diagrams for the various makes of vehicles. These are available from parts stores and are usually free of charge to customers. Belt routing information is sometimes available in the owner's manual, too.

When the belt information is not available, the following tips might be useful:

- Remember that both sides of the belt can be used to drive accessories. The V-grooved side of the belt will mesh with pulley grooves. The flat side of the belt will go against a flat pulley.
- The belt is routed around the outside of the pulleys and is drawn in and looped around the smooth pulleys (Figure 22.18).
- Only one pulley can be threaded incorrectly. It is always near the center and is a smooth pulley. If the belt does not seem to fit properly one way, try another way (Figure 22.19).

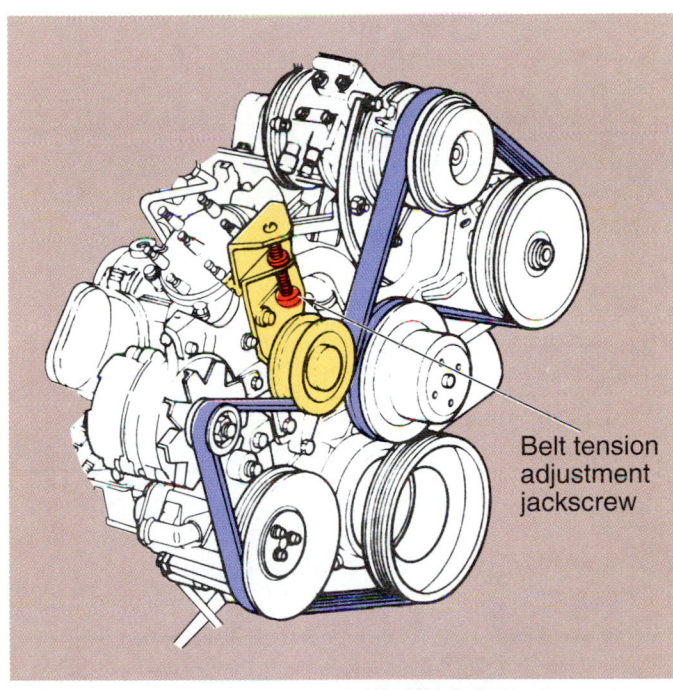

Figure 22.16 This belt is tensioned by an adjustable screw. *(Courtesy of Ford Motor Company)*

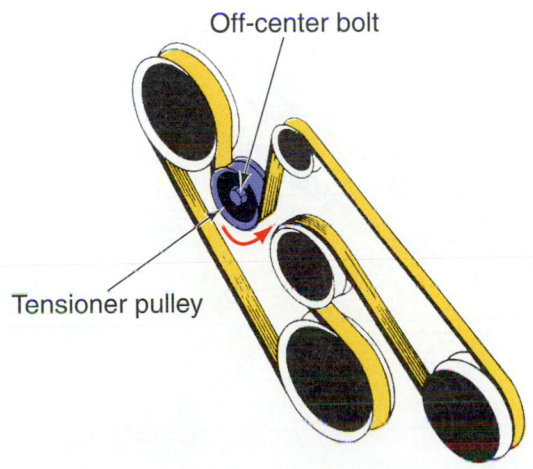

Figure 22.15 To loosen tension on this belt, turn the off-center bolt in the direction of the arrow. *(Courtesy of The Gates Rubber Company)*

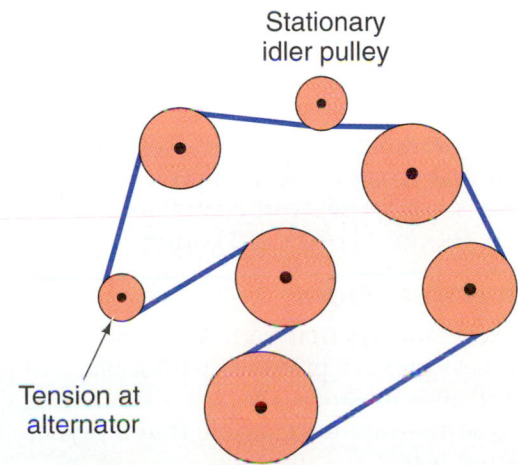

Figure 22.17 This belt is tensioned by the alternator. *(Courtesy of The Gates Rubber Company)*

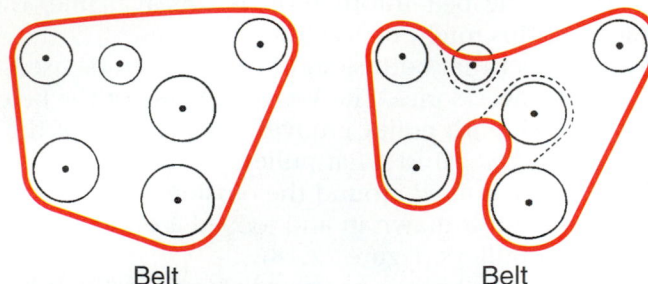

Figure 22.18 The belt is drawn in from the outside and looped around smooth pulleys. *(Courtesy of The Gates Rubber Company)*

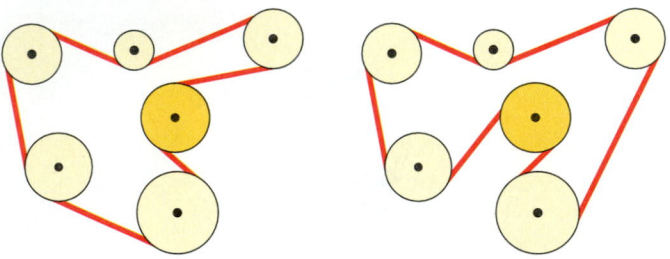

If this routing doesn't work ...   try this

Figure 22.19 Different ways to position a serpentine belt. *(Courtesy of The Gates Rubber Company)*

SHOP TIP   When a belt is replaced, the old belt can be stored in the trunk for emergencies. If a belt fails, power to all accessories and the coolant pump will be lost.

■ BELT TENSION

Belt tension is important for long belt life. Belts stretch slightly in the first few minutes of operation, and then remain constant in length. If they are overtightened, parts can be overloaded.

- Too much belt tension can cause failure of the coolant pump bearing, the alternator, or the front main bearing.
- Improperly tightened belts can also cause water pump and alternator bearing noise.
- Loose belts can cause overheating and abnormal combustion, noise, or a dead battery.

Tightening a Belt

There are several ways that belts are tensioned:
- Some accessories are provided with a place to pry against (Figure 22.20).
- Brackets sometimes have a hole that a prybar can be inserted into.
- Manufacturers often provide some provision for tightening, such as a ½" square hole in the

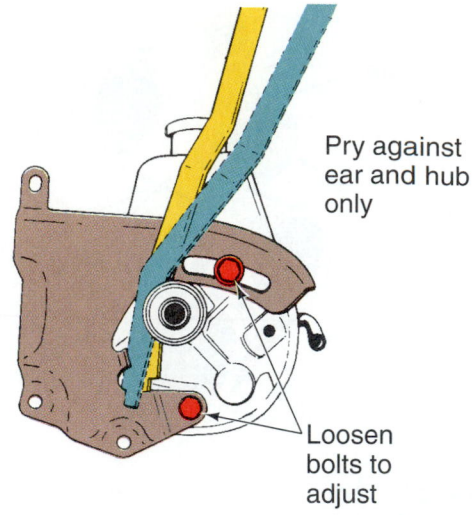

Pry against ear and hub only

Loosen bolts to adjust

Pulley removed to show bolts

Figure 22.20 When tightening the power steering belt, pry only in the designated areas. *(Courtesy of General Motors Corporation, Service Technology Group)*

mounting bracket for a ½" breaker bar to be inserted into (Figure 22.21).
- Sometimes a jackscrew adjustment is provided (see Figure 22.16).

When there is no provision for tightening the belt, pry against a strong area of the accessory. Be sure that the outside of an alternator or power steering pump is not accidentally damaged. Power steering reservoirs are made of sheet metal. Prying against unreinforced sheet metal will cause a dent, which can damage internal components of the pump. Do not pry on power steering reservoir or any other delicate part.

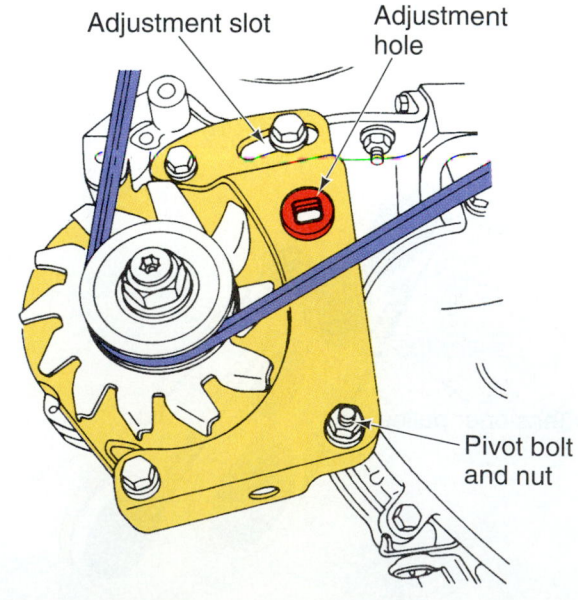

Adjustment slot   Adjustment hole

Pivot bolt and nut

Figure 22.21 Use a breaker bar or ratchet in the square hole to tighten the belt. *(Courtesy of Chrysler Corporation)*

V-Belt Tension

New belts stretch slightly during the first few minutes of operation. After that, their length will remain constant. When adjusting a new V-belt, set the tension about 15 pounds higher than the recommended specification. After running the engine for about fifteen or twenty minutes, recheck and adjust the belt tension using a belt tension gauge (Figure 22.22).

V-Ribbed Belt Tension

V-ribbed belts usually use more tension than V-belts. Maximum tension should be limited to 30 pounds per rib, checked at a splice free area.

■ Use a *click-type tension gauge* for V-ribbed and timing belts (Figure 22.23).

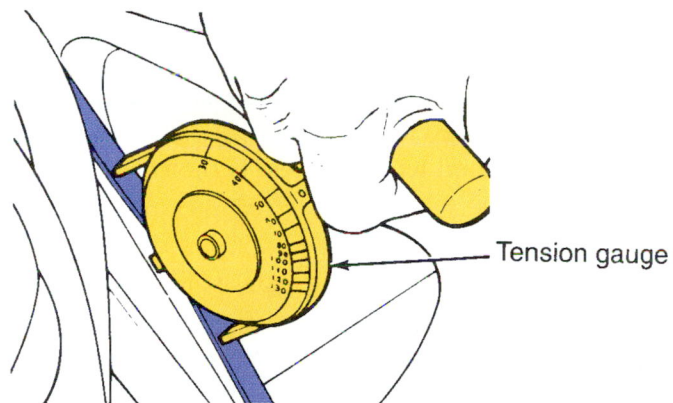

Figure 22.22 Use a belt tension gauge when adjusting the belt. *(Courtesy of Chrysler Corporation)*

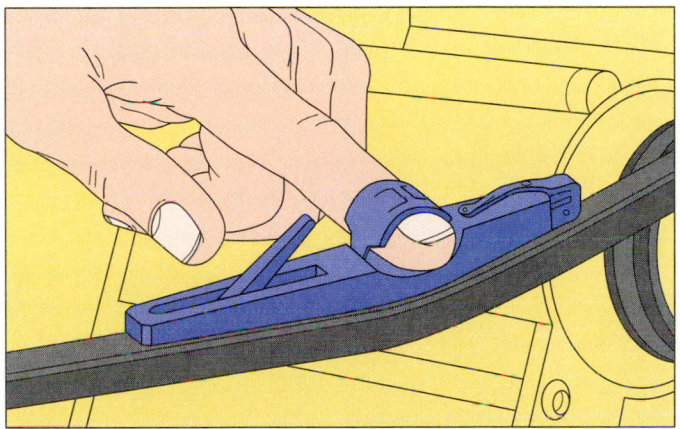

Figure 22.23 A click-type belt tension gauge is used for measuring tension on V-ribbed and timing belts. *(Courtesy of The Gates Rubber Company)*

■ After a new belt is installed, run the engine. Then, loosen the tensioner bolt and re-tighten it.

After some initial tension is lost, V-ribbed belts will maintain 20 pounds of tension per rib for a long time. Used belts should have 15 to 20 pounds of tension per rib.

➡ *Perform __Belt Service__ Worksheets*

■ TIMING BELT SERVICE

Poor alignment, improper tension, and worn sprockets will all result in increased timing belt fabric wear. After the fabric begins to wear, cracks will start to appear at the base of the teeth on the belt. Next, the tooth will separate from the belt (Figure 22.24). This is the most common type of belt failure.

NOTE: *The timing cover serves an important purpose, especially during bad weather. Snow can blow in on the warm engine of a parked vehicle where it melts and freezes up again on the timing sprocket.*

CASE HISTORY *After a trip, a man parked his vehicle outside for the night. During the night there was a snow storm with heavy winds. When he attempted to start his car in the morning, the engine would not turn over. His son had replaced the timing belt and the timing cover had not been properly reinstalled. Snow blew into the space between the sprocket and belt melted on the hot engine and then froze again during the night.*

To inspect the condition of a timing belt, twist it gently. Do not turn it more than 90°. The tensile cords in timing belts can be damaged. Replacement of timing belts is covered in Chapter 46. Be sure that all of the small rubber gaskets that separate the parts of a multiple piece timing cover are reinstalled properly.

Figure 22.24 The teeth have come off of this belt. *(Courtesy of Federal-Mogul Corporation)*

■ REVIEW QUESTIONS

1. What are the two types of accessory drive belts?

2. Which belt has more surface area in contact with the pulley groove, a V-belt or a flat belt of the same width?

3. Which cord design is used to make a better belt, high cordline or center cordline?

4. What is the name of the belt drive that is snake-like?

5. If there is a ½" square hole in the alternator bracket, what tool would you use to tighten the drive belt?

6. After a new V-belt is installed and the engine has been run for 15 minutes, what should be done?

7. Before removing a serpentine belt, what should be done?

8. What is the maximum tension per rib of a new V-ribbed belt?

9. What kind of belt tension gauge is used for V-ribbed belts and timing belts?

10. A timing belt should not be twisted more than _____° because the tensile cord could be damaged.

■ ASE REVIEW QUESTIONS

1. Technician A says that a V-belt can be used to drive from both sides. Technician B says a timing belt can be used to drive from both sides. Who is right?
   - **a.** Technician A
   - **b.** Technician B
   - **c.** Both A and B
   - **d.** Neither A nor B

2. Technician A says that a glazed belt should be retightened. Technician B says the surface finish of a pulley is more important with a V-belt than a V-ribbed belt. Who is right?
   - **a.** Technician A
   - **b.** Technician B
   - **c.** Both A and B
   - **d.** Neither A nor B

3. Technician A says that adjusting a belt to too high a tension can result in a worn upper front main engine bearing. Technician B says that too much belt tension can ruin a coolant pump bearing. Who is right?
   - **a.** Technician A
   - **b.** Technician B
   - **c.** Both A and B
   - **d.** Neither A nor B

4. Technician A says that belt failures rise significantly after two years of age. Technician B says to pry on a power steering reservoir when adjusting belt tension. Who is right?
   - **a.** Technician A
   - **b.** Technician B
   - **c.** Both A and B
   - **d.** Neither A nor B

5. Technician A says that a loose belt can cause engine overheating. Technician B says a loose belt can cause a dead battery. Who is right?
   - **a.** Technician A
   - **b.** Technician B
   - **c.** Both A and B
   - **d.** Neither A nor B

CHAPTER 23

Automotive Hoses

■ OBJECTIVES

Upon completion of this chapter, you should be able to:

✔ Tell what size a hose is.

✔ Explain the different types of hose materials, clamps, and their applications.

✔ Inspect all types of hoses for damage.

✔ Correctly remove and replace all types of hoses.

■ KEY TERMS

vulcanizing
abrasion
banjo fittings
worm gear clamp
rolled edge clamps

■ INTRODUCTION

There are more than ten different types of hoses found on modern automobiles (Figure 23.1). Hoses are found on the radiator, heater, automatic transmission, fuel and emission systems, brake system, lubrication system, and air conditioning system. This chapter deals with the theory and service of all automotive hoses.

■ HOSE THEORY

Hoses consist of an inner rubber tube, reinforcement, and the outer rubber cover, which are bonded together with adhesives. They have different kinds of tubes and covers and are reinforced differently to withstand different amounts of pressure.

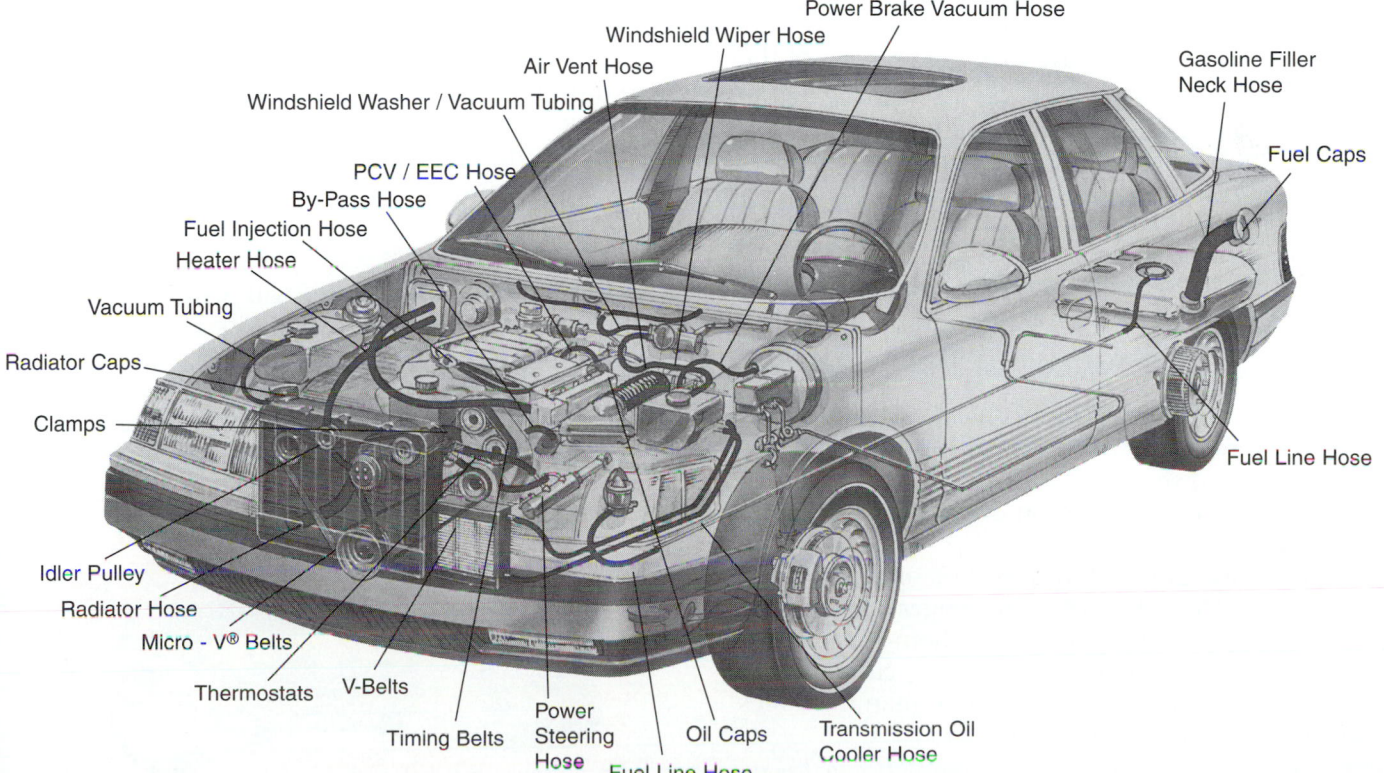

Windshield Wiper Hose
Power Brake Vacuum Hose
Air Vent Hose
Gasoline Filler Neck Hose
Windshield Washer / Vacuum Tubing
PCV / EEC Hose
By-Pass Hose
Fuel Caps
Fuel Injection Hose
Heater Hose
Vacuum Tubing
Radiator Caps
Clamps
Fuel Line Hose
Idler Pulley
Radiator Hose
Micro - V® Belts
Thermostats
V-Belts
Timing Belts
Power Steering Hose
Fuel Line Hose
Oil Caps
Transmission Oil Cooler Hose

Figure 23.1 Many types of hose and tubing are found on today's vehicles. (*Courtesy of The Gates Rubber Company*)

■ HOSE SIZE

Hoses are sized according to their inside diameter (called I.D.), the same way that pipe is (see Figure 6.21).
■ Common heater hose sizes are ⅝" and ¾".
■ Common fuel hose and line sizes are 5⁄16" and ⅜" for pressure hoses and ¼" for return lines.
None of these sizes correspond to the outside diameter (O.D.) of the hose. A chart in Appendix E at the end of the book shows actual sizes of flare fittings and pipe thread.

■ UNREINFORCED HOSE

Hoses used for vacuum, windshield washer, and drains are made of unreinforced rubber. These hoses are not under much strain but can present safety problems if they fail (when an engine stalls in an intersection, for instance). They harden with age and are routinely replaced.

■ RADIATOR HOSE

Radiator hoses are designed to have a burst strength of between five and six times the working pressure of the cooling system.

■ HOSE TYPES

There are three kinds of hoses: the *straight*, the *curved*, and the *universal* (Figure 23.2). Examples of uses of the straight hose include fuel or vacuum lines. It kinks when it is bent too much, so it is not used when sharp bends are encountered.

Curved Hose

Curved hose, also called *molded* hose, is preformed with the required bends. The hose is heated on a mandrel of the desired shape to vulcanize it. **Vulcanizing** is a process that cures rubber and causes it to set in a predetermined shape.

Molded hoses are used in over 90% of replacement radiator hoses. They often come in longer lengths than the desired replacement. Some of them are printed with cutoff line marks so that they can be cut for use in several different applications. The hose manufacturer's catalog tells which line to cut on.

NOTE: *Always exercise care when trimming a hose to length.*

This type of inventory control is done whenever possible so that a business can keep fewer part numbers in stock. Otherwise, prices would have to be higher to compensate for the different cataloging and packaging.

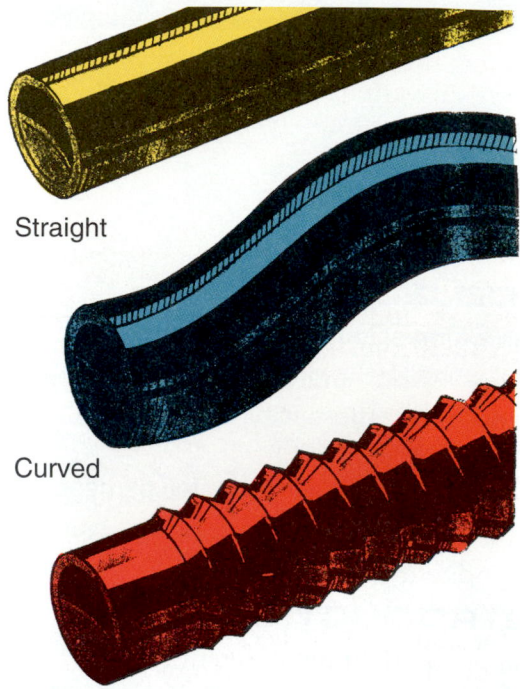

Straight

Curved

Flex or Universal

Figure 23.2 Three kinds of hose. *(Courtesy of The Gates Rubber Company)*

Universal Hose

Universal hose can be clamped on one end and then bent until it assumes the desired shape. It is reinforced with wire so that it will not collapse. Its disadvantage is that the ridges in it resist the flow of water, so pump efficiency is hampered.

Formable Hose

Bending conventional hose will cause it to kink, restricting water flow. Formable hose has a metal wire inside of it (Figure 23.3). This allows formable hose to be bent without kinking it. This type of hose is popular when replacing formed hoses, such as those found in the cooling and emission control systems.

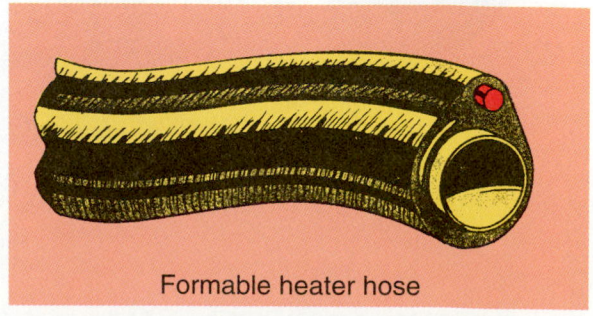

Formable heater hose

Figure 23.3 This hose can be bent without distorting. *(Courtesy of The Gates Rubber Company)*

Bypass Hoses

Hoses for the thermostat bypass (Figure 23.4) are produced in various lengths. They are molded to the proper curve but must be cut to length for the application.

Fuel Hose

The fuel delivery system from the fuel tank to the engine uses metal lines (tubing). These are connected to the tank and engine by rubber hoses. One line delivers fuel from the tank and the other is a return line. Fuel delivery lines range in size from 5/16" to 3/8". Return or vapor hoses are usually 1/4".

For safety reasons, fuel hose must also have toughness to resist **abrasion** (rubbing). The inside of fuel hoses and vapor hoses must be resistant to attack from ethanol and methanol fuels. The outside cover must be oil and temperature resistant.

The outside of a fuel hose is usually temperature rated for about 300°F. Because fuel is relatively cool, the inside of a fuel hose does not require a high temperature rating. The inside of the hose should not be subjected to temperatures in excess of 257°F.

NOTE: *Do not use fuel hose for an application in which hot oil will be used, such as for a transmission cooler line.*

Carburetor Fuel Hose

Carburetor fuel systems are under low pressure. The hose consists of a braided reinforcement with a synthetic rubber cover and inner tube. Fuel hose is only rated at 50–75 psi. It is made of a special compound that does not react to gasoline or alcohols.

Fuel Injection Hose

Fuel injection systems usually run under much higher pressure than carburetor systems. Pressure in some of these systems sometimes reaches 175 psi. Special hose and special hose clamps (covered later) must be used with these systems. Fuel injection hose has a burst strength in excess of 900 psi. On many systems, the hoses are crimped to the tubing, like high pressure power steering hoses.

Transmission Oil Cooler Hose

Hot oil is circulated from the automatic transmission to its cooler (located inside or outside of the radiator). When hoses are used to connect metal tubing in these systems, they are subjected to two of rubber's biggest enemies: oil and heat. Any rubber product is broken down by oil, shortening its life. Rubber products suffer a shortened life of about 50% for every 18°F increase in temperature. Transmission oil cooler (TOC) hose can withstand a constant temperature of up to 300°F. Its pressure rating is 450 psi.

Compare fuel hose's ratings for temperature and pressure and you can see why fuel hose should not be used in place of transmission oil cooler hose. Both of these types of hose look identical from the outside. Figure 23.5 shows the differences between these hoses internally.

CASE HISTORY *A technician installed an auxiliary transmission oil cooler on a motor home. Because of the heavy loads and high temperatures encountered by an automatic transmission in a motor home, this is a common installation. He installed the cooler in series with the existing cooler in the radiator. The connections were made with fuel hose. After using the motor home for one season, a leak developed in one of the hoses. The transmission fluid leaked onto a hot exhaust manifold, and in the resulting fire the motor home suffered serious damage. Luckily, no one was hurt.*

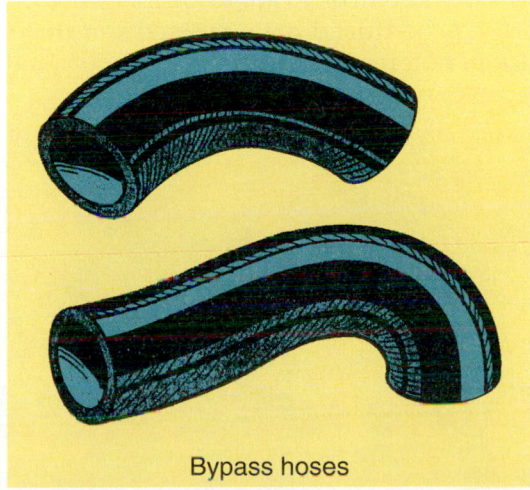

Bypass hoses

Figure 23.4 Thermostat bypass hoses. *(Courtesy of The Gates Rubber Company)*

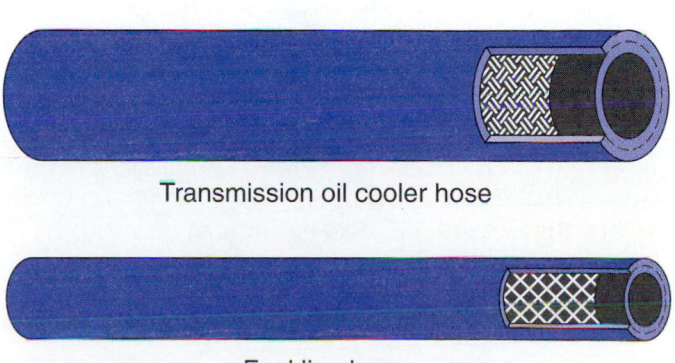

Transmission oil cooler hose

Fuel line hose

Figure 23.5 Transmission oil cooler hose and fuel hose look identical from the outside. *(Courtesy of The Gates Rubber Company)*

Power Steering Hose

There are usually two power steering hoses. One is a pressure line and the other is a return line (Figure 23.6). The return hose is not under the high pressure that the pressure side is. The hose on the pressure side is made of reinforced, oil-resistant synthetic rubber that is crimped to metal tubing. It must be able to withstand temperatures of 300° F and handle pressures up to 1500 psi. Some pressure hoses also have a flow reducer, an orifice to help control flow. The return line is usually connected to the power steering pump with a hose clamp. It must be able to withstand high temperatures, but operate under low pressure. It must also be resistant to damage by power steering fluid.

Brake Hose

Brake hoses provide a flexible connection to the wheels, which must be able to turn back and forth during steering. They are called upon to carry very high pressures. Because brake hoses are a safety item, federal motor vehicle standards require that they have a burst strength of at least 5000 psi.

Brake hoses come in different lengths and have fittings swaged or crimped on the ends (Figure 23.7). The hose is made of rubber reinforced with woven fabric, usually rayon. The fabric's job is to prevent the hose from expanding under pressure. The federal safety standard requires that the outside of a brake hose be labeled with either *HR*, which means that the hose expands normally or *HL*, which means that the hose has a lower amount of expansion.

The hose has an inner layer of rubber that comes into contact with the brake fluid. Most brake hoses have two *plies* that make up the middle fabric layers of the hose.

The *jacket* of the hose is the outside layer that protects the plies. The outside of a hose is subjected to damage from ozone in the atmosphere, which can cause the rubber to crack.

Federal safety standards also call for the hose to have either raised ribs or two 1/16" stripes painted on it. The stripes are provided so that during installation, the technician can tell whether the hose is twisted excessively. Excessive twist in a hose can cause it to be damaged internally.

Internal damage is difficult to diagnose. Depending on which way the internal rubber flap is torn (see Figure 52.11), the hose could cause delayed engagement of the brake or act as a one-way valve preventing fluid pressure from releasing from the brake after application. Internal damage also often results when a brake caliper is allowed to hang on a hose during a brake job. These hoses often become damaged when a technician clamps one off using vise grips.

Brake *tubing* always has male fittings on both ends. Brake *hose* fittings have either female threads, male threads, or banjo fittings. **Banjo fittings**, named because they resemble a banjo, allow brake fluid to make a 90° turn in a tight space (Figure 23.8). Brake fluid goes into the banjo fitting and exits through a hollow bolt. A copper gasket is used to seal both sides of the fitting.

NOTE: *The copper gasket, which is often reusable, is easily lost during a brake service. The best practice is to keep a supply of these gaskets on hand and replace them any time a brake hose is removed.*

Tapered pipe threads are not usually found in brake systems. This is because the female part of the fitting can split when overtightened due to the taper of the thread. Male compression fittings are used instead, usually with copper washers that are compressed when the fittings are torqued. Other times the end of the male thread has a tapered *seat* that seals at the bottom of a female fitting.

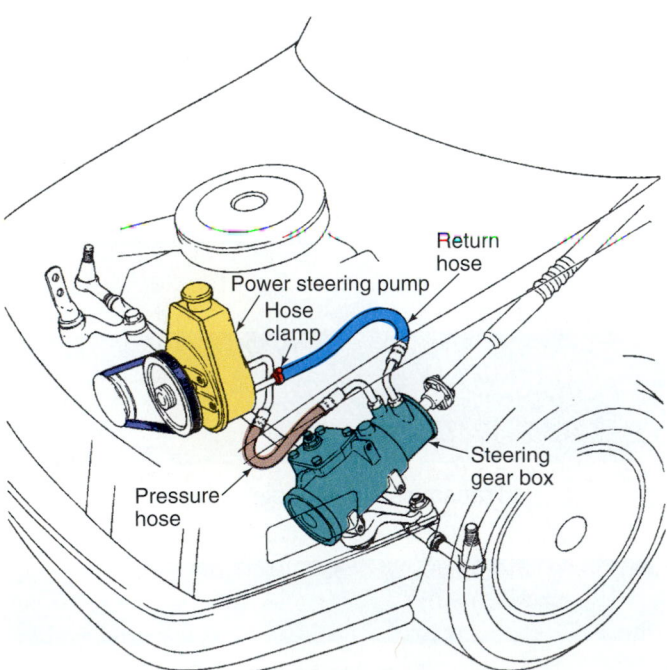

Figure 23.6 Power steering hoses. Pressure hoses are crimped. Return hoses, under less pressure, usually have hose clamps. *(Courtesy of Moog Automotive, Inc.)*

Return hose
Power steering pump
Hose clamp
Pressure hose
Steering gear box

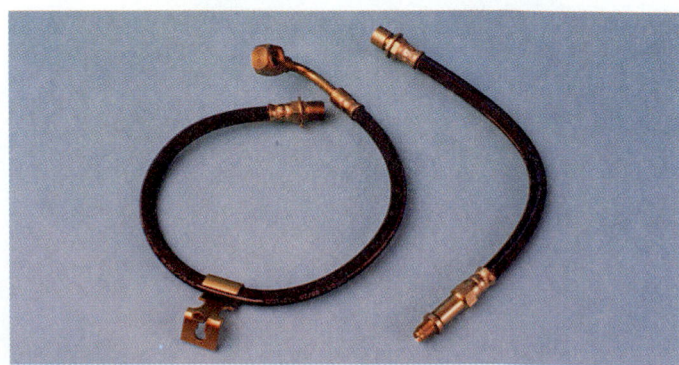

Figure 23.7 Brake hoses have fittings crimped on their ends.

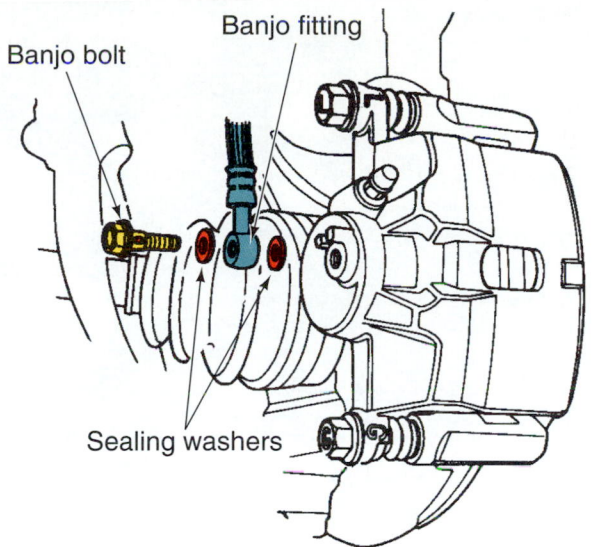

Figure 23.8 A banjo fitting makes tight turns possible. *(Courtesy of American Honda Motor Co., Inc.)*

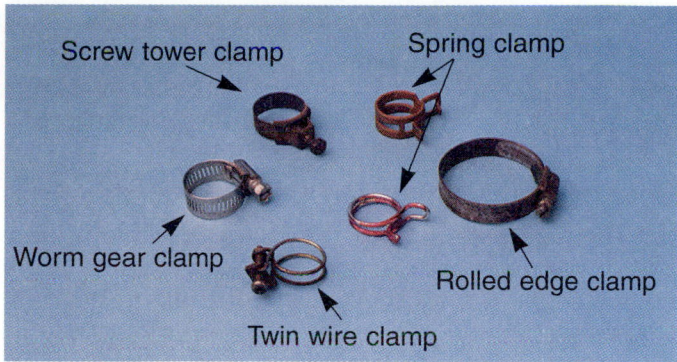

Figure 23.9 *Several types of hose clamps.*

SHOP TIP Once a compression fitting with a copper washer is tightened, the hose is no longer free to rotate. Therefore, the compression fitting must be installed before the other end of the hose is installed.

Air Conditioning Hose

Air conditioning systems carry liquid and vapor under high pressure. As with power steering systems, these high pressure hoses are reinforced and have crimped connectors. High quality coatings are also used on the hoses. Oil circulates along with the refrigerant in the system and temperatures are high. Newer air conditioning systems use R134A refrigerant. R134A consists of a considerably smaller molecule than the older R-12 (Freon) refrigerant does. Consequently, R134A refrigerant can leak through the older air conditioning hoses. As a result, today's new car refrigerant hoses incorporate a barrier of nylon or similar material to prevent leakage.

■ HOSE CLAMPS

Clamps and hoses are used to connect two metal lines together. There are several styles of clamps (Figure 23.9). One of the most popular hose clamps for all applications (except fuel injection) is the **worm gear clamp**. It is usually reusable and is easy to install. A worm gear clamp is shown in the lower right-hand corner of the photo.

Twin wire clamps are strong hose clamps, but can cut into the hose if tightened too far. **Rolled edge clamps** (Figure 23.10) are designed so that they will not cut into a hose. They are a good hose clamp for

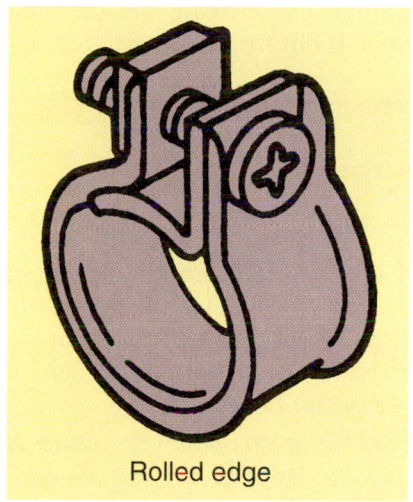

Figure 23.10 A rolled edge hose clamp.

fuel injection, where fuel pressures are higher than those found in the cooling system.

Screw tower clamps are strong clamps too, but are difficult to remove. Technicians usually cut these off with cutters. *Spring clamps*, used as original equipment by some manufacturers for cooling system hoses, are not used for most fuel system applications (but they come with replacement in-line fuel filters on engines with carburetors). They do not provide a strong enough clamping force for fuel injection applications and can cut into a hose, causing a fuel leak.

■ HOSE INSPECTION

Check the condition of all radiator, heater, and fuel hoses. Hoses are normally firm, yet flexible. Look for leaks, stiffness, sponginess, hidden rot, rubbed or burned areas, and oil soaking. Hoses that appear to be cracked, swollen, hard, soft, or broken should be replaced.

Hard Hoses

When checking hoses for hardness try to bend them, especially at the ends.

SAFETY NOTE Be careful when bending a hose at the end. A bad hose might break, allowing dangerous fuel or hot, pressurized coolant to escape.

The heater core, a small radiator, is located inside the driver's compartment. It has two hoses to carry engine coolant: one coming from the engine and one for return (Figure 23.11). Heater hoses often become brittle due to heat and aging.

SHOP TIP Sometimes cutting off the first inch of hose will eliminate a hardened section.

Oil Damage

Rubber hoses that are not used in fuel- or oil-related environments can deteriorate rapidly when they come into contact with oil. Oil, an enemy of rubber hoses, can cause them to swell. Be sure to fix any oil leaks. Also, water pump lubricants found in some cooling system additives contain petroleum, which can damage the inside of the hose.

NOTE: *Defective hoses do not always outwardly appear so. Deterioration on the inside of the hose can cause small particles to flake off and fall into the coolant (Figure 23.12).*

Check the rubber on the inside of the hose to see that it is not deteriorating. Occasionally, a relatively low-mileage vehicle will experience the failure of a hose that appears to be good on the outside.

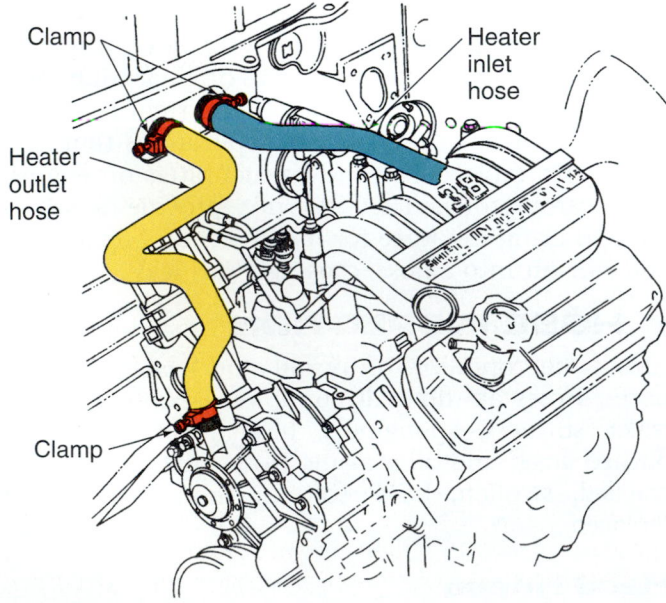

Figure 23.11 The heater has an inlet and an outlet hose. *(Courtesy of Ford Motor Company)*

Figure 23.12 A hose that is defective on the inside. *(Courtesy of The Gates Rubber Company)*

Electrochemical Damage and Radiator Hoses

A newly discovered condition discovered independently by both Gates and Goodyear has been found to lead to hose failure. It is called *electrochemical degradation* or *ECD*. ECD occurs when the hose, engine coolant, radiator, engine, and fittings form a galvanic cell, or battery. This causes small cracks in the inside of the hose, allowing coolant to penetrate into the hose reinforcement. New coolant hose that is electrochemically resistant is made of *ethylene propylene rubber (EPDM)*.

To check for ECD damage, squeeze the hose in several places to see if the rubber has the same consistent feel throughout. Be sure to thoroughly check both hoses. The upper radiator hose (engine outlet end) suffers the most abuse because it is exposed to fully heated coolant before it reaches the radiator. Lower hoses, on the other hand, are more difficult to inspect and are often overlooked.

Some lower radiator hoses have a coil of wire in them. The coil is installed at the factory so that the hose does not collapse during the factory coolant fill procedure. A rusty cooling system will often result in pieces breaking off and circulating in the system. If the wire is missing, coolant circulation, especially at highway speeds, can be hampered. Whenever one of these hoses is removed during service, check to see that the wire is not rusted.

SAFETY NOTE Any questionable hose should be replaced. If a hose fails, an engine can be ruined.

REPLACING HOSES

When a coolant hose is to be replaced, first drain off coolant to a level below the height of the hose. A hose clamp is used to fasten a hose to its connection. Hoses are usually easily removed by loosening the clamp and twisting gently. If the clamp is rusted in place, cut it off with side cutters or tinsnips.

The worm gear clamp is the most popular clamp used in the aftermarket. Most worm gear clamps have screw heads with both a screwdriver slot and a hex head (Figure 23.13). They can be removed with either a screwdriver or a socket.

Some car makers use a spring-type clamp for radiator connections. Special pliers with grooves in the jaws accommodates the clamp for removal (Figure 23.14).

When a hose does not come loose from its fitting easily, do not force it. This can damage the fitting, radiator, or heater core outlet. Cut it off carefully with a knife (see Figure 21.5). Most EPDM hoses tend to bond to metals, and during removal they *must* be cut off.

Clean the fitting of corrosion or pieces of old hose (Figure 23.15). Hose connections on the radiator and engine have a small ridge that helps seal the hose. Install the hose and position the clamp as close to the

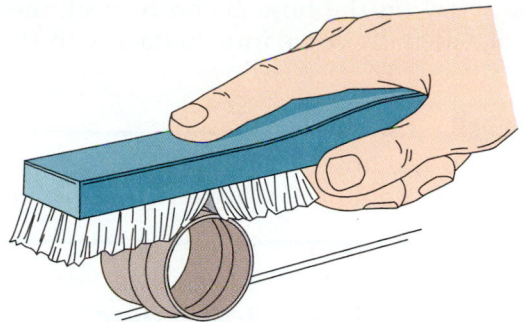

Figure 23.15 Clean the hose fitting before installing the new hose. *[Courtesy of The Gates Rubber Company]*

ridge on the connection as possible to prevent the possibility of corrosion (see Figure 21.12). Positioning the clamp on top of the ridge can cut the hose.

Be sure that all hose clamps are in good condition and are tight.

NOTE: *When new hoses are installed, they take a set following the engine's first heating and cooling cycle. This sometimes leaves the clamps loose and they must be retightened.*

Because a molecule of air is smaller than a molecule of coolant, air can be drawn in through a loose connection (even if the opening is not large enough to let coolant seep out). This is especially true with the lower radiator hose and heater return hose because they are *suction hoses*.

SAFETY NOTE Check to be sure that the hose does not contact fan belts, fuel lines, or the fan.

➡ *Perform **R & R a Radiator Hose** Worksheet*

Replacing a Heater Hose

Sometimes the heater hoses on a car are two different diameters. The size is determined by the inside diameter (I.D.) of the hose. Hose comes in long rolls. Be sure that you select the correct size hose before cutting it off of the main hose roll.

Twist the old heater hose to loosen it, and cut it carefully with a sharp knife if it does not remove easily. When installing the new section of hose, check the following:

- Be certain that the hose does not interfere with manifolds, belts, or spark plug wiring.
- Check to see that the hose will not be damaged by movement of the engine or accessories.

Position the screw side of the clamp for easy access. On lower radiator hoses, an extra long screwdriver can be used from the top of the engine compartment. Be

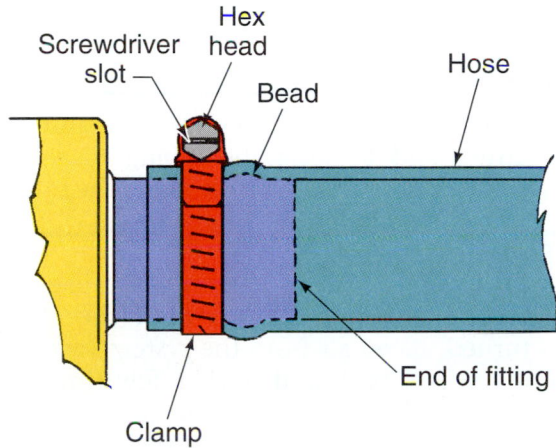

Figure 23.13 Worm gear clamps usually have a hex head and a screwdriver slot. *[Courtesy of The Gates Rubber Company]*

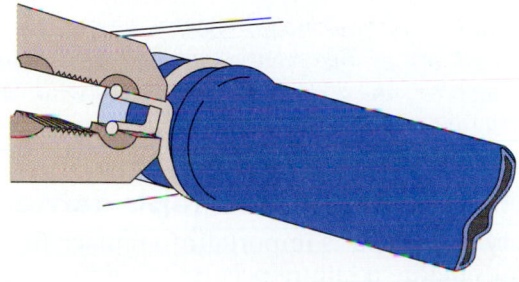

Figure 23.14 Spring clamps are removed using pliers with a groove in their jaws. *[Courtesy of The Gates Rubber Company]*

sure the screw on the hose clamp is positioned so it cannot accidentally come into contact with the cooling fan.

SHOP TIP If the hose is difficult to install, apply a small amount of soap to the connection.

➡️ **Perform _Replace a Heater Hose Worksheet_**

Bypass Hose Replacement

Obtain the correct replacement hose and compare it to the existing hose (Figure 23.16).

CAUTION Do not try to substitute an unmolded hose for a molded bypass hose. It can fold when bent, restricting coolant flow.

Power Steering Hose Service

When checking a power steering hose, look for signs of leakage or dampness at the connections. Also, look for signs of deterioration, such as cracks, signs of rubbing, or swelling. A failed pressure hose can leak, blowing oil onto an exhaust manifold where it can cause a fire.

Get the correct replacement hose and replace the old one. Parts stores have catalogs that have extensive illustrated listings for power steering hoses. It is also a good idea to compare the new hose to the old hose to be sure that the length and fittings are identical.

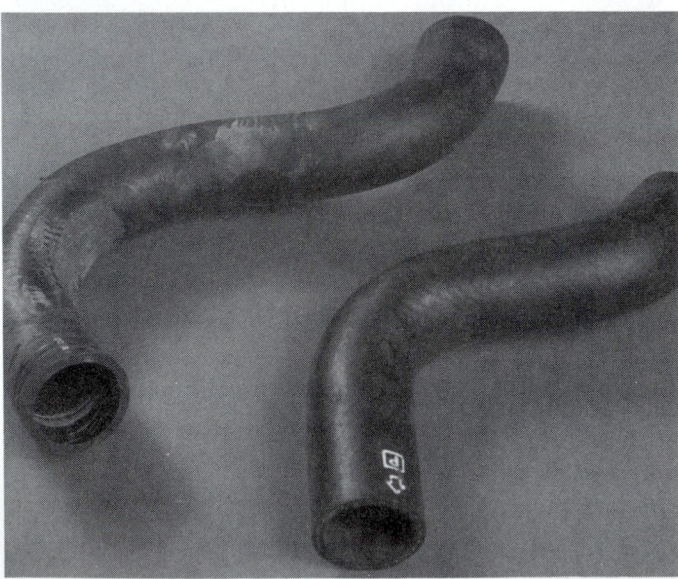

Figure 23.16 Compare the old hose to the new hose before installing it.

CAUTION Power steering fluid, which is not flammable at engine operating temperature, is flammable at temperatures a little above 300°F (lower than the temperature of an exhaust manifold).

SHOP TIP Sometimes a replacement hose is not available. Some businesses have special equipment for making hydraulic hoses. Check with a business that deals with heavy truck, industrial, or farm implement accounts.

Refilling the Power Steering System

Be sure to use the correct fluid when refilling the system. Earlier systems traditionally used automatic transmission fluid, which is red in color. Many late model vehicles use power steering fluid, which has a lower viscosity. It is colored either yellow or yellow/green.

After refilling the system, run the engine and check for leaks. On hoses that use O-rings, new O-rings should be used.

SHOP TIP If a flared fitting leaks, try loosening the fitting and twisting the tubing to seat it against the flare seat. Then retighten the fitting.

Air in the power steering system will cause erratic operation or a groaning sound when the steering wheel is turned. Bleed air from the system by turning the steering wheel back and forth a few times. This should be done with the tires in the air. Otherwise a flat spot can be worn on the tires. If the fluid appears to be foamy after this, let it sit for several minutes and allow the foam to disappear. In extreme cases, a vacuum pump can be applied to the top of the power steering reservoir filler neck to remove stubborn air.

NOTE: _Some cars have hydro boost power brakes, which are operated off of the power steering pump. On these systems, depress the brakes several times while turning the steering wheel to help rid the system of air._

Fuel Injection Hose Replacement

For safety reasons, it is important to replace fuel injection hoses before a failure occurs.

An original equipment manufacturer (OEM) requirement is that the outside of fuel injection hose be stamped with "_Fluoroelastomer_" or _EFI_ in yellow

SAFETY NOTE Be sure to use the correct fuel hose designed to withstand the pressure of the fuel injection system.

print. Always use the specified hose and do not connect high pressure lines with hoses and clamps. Instead, use hoses with crimped ends.

Fuel injection lines maintain pressure even when the engine is off. This is so the engine can easily restart without the system having to build pressure first.

SAFETY NOTE
- Be sure to carefully relieve the system of fuel pressure before replacing a fuel line or fuel filter (see applicable service manual or a manufacturer's hose and tubing catalog).
- Fuel hose must always be routed so that it is at least 4" (100 mm) away from any part of the exhaust system. Catalytic converters (see Chapter 40) run hotter than the rest of the exhaust system. Fuel hoses should not be any closer than 10" (284 mm) from a catalytic converter.

SAFETY NOTE When hose clamps and hose are used, be sure to position the hose clamp behind the ridge on the fitting, not on it (Figure 23.17). Failure to do this can result in a cut hose and dangerous fuel leak.

Repairing damaged tubing with a section of hose is covered in Chapter 24. Nylon tubing used in fuel injection systems is covered in that chapter also.

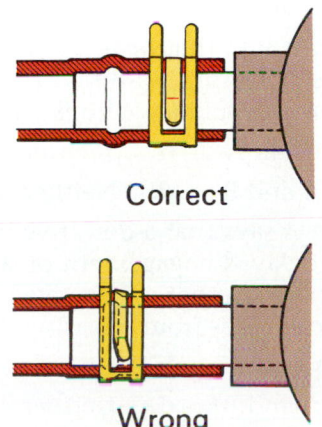

Figure 23.17 Installing a fuel hose clamp on the ridge can damage the hose.

Air Conditioning Hose Service

Factory-installed air conditioning systems come with a very wide variety of hose styles and configurations. Air conditioning shops commonly make up their own lines from bulk hose and a selection of fittings kept in stock. Parts stores keep a supply of the more common fittings on hand. Usually, the old fitting can be reused. The crimp connection must be cut off carefully to avoid damage to the fitting (Figure 23.18). The hoses used are identified using a chart like the one in Figure 23.19, which compares the actual sizes of the hoses.

In the past, worm gear hose clamps were used with *3-barbed fittings* when making repairs (Figure 23.20).

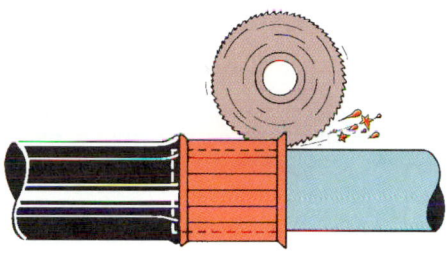

Figure 23.18 The crimp connection is carefully cut off of the old hose. *(Courtesy of The Gates Rubber Company)*

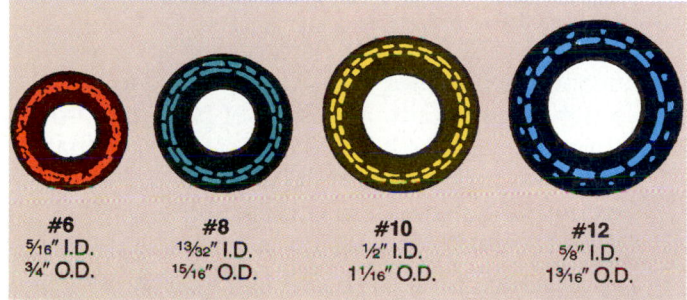

Figure 23.19 A refrigerant hose size chart. *(Courtesy of CARQUEST Refrigerant Hose Make-up Guide)*

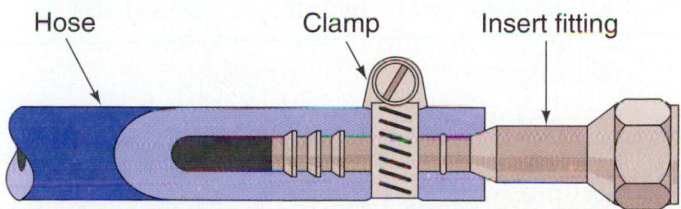

Figure 23.20 A worm gear hose clamp used with a 3-barbed fitting.

This method of repair is no longer recommended and must be absolutely avoided with R134A beadlock fittings.

There are two styles of crimp used with air conditioning hoses (Figure 23.21). When crimping a fitting onto an R134A line a *bubble crimp* is used. A *face crimp* is recommended to be used with 3-barbed fittings. Hydraulic press crimping or less expensive hand crank crimping tools are available.

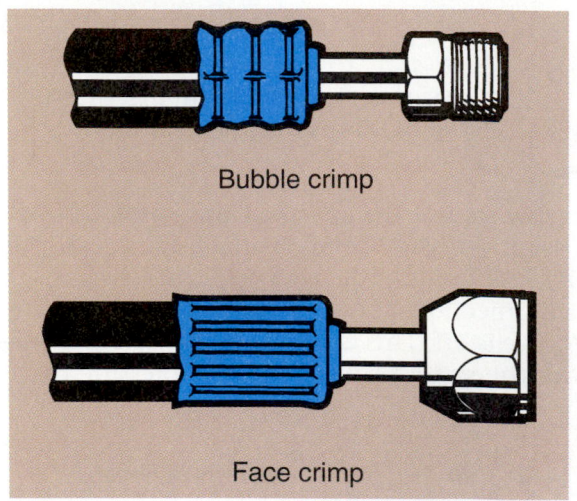

Figure 23.21 Two styles of crimps. *[Courtesy of CARQUEST Refrigerant Hose Make-up Guide]*

■ REVIEW QUESTIONS

1. Hoses are sized by their inside/outside (circle one) diameter?

2. What special feature allows formable hose to bend without kinking?

3. The inside of transmission oil cooler hose can withstand constant temperatures of up to ____ °.

4. When there are two hoses in a power steering system, what are they for?

5. What is the name of the type of hose clamp to use for fuel injection system hoses?

6. What is a drawback to screw tower hose clamps?

7. Which radiator hose suffers the most abuse, the upper or the lower? Why?

8. If a hose does not come off of a radiator fitting easily, what should be done?

9. What can happen if a regular hose is substituted for a molded bypass hose?

10. Can power steering fluid catch fire?

■ ASE STYLE REVIEW QUESTIONS

1. Technician A says that clamps used on neophrene radiator hoses do not have to be retightened because this kind of hose does not shrink like EPDM hose. Technician B says that fuel hose can be used to replace a damaged section of transmission cooler tubing. Who is right?

   a. Technician A   **b.** Technician B

   c. Both A and B   **d.** Neither A nor B

2. Technician A says that brake hose has either raised ribs or stripes so it will not be twisted during installation. Technician B says that a worm gear clamp is the best style to use in a fuel injection system. Who is right?

   a. Technician A   **b.** Technician B

   c. Both A and B   **d.** Neither A nor B

3. Technician A says that for a hose clamp to work properly, it should be positioned on top of the sealing ridge on the connection. Technician B says that fuel injection hoses have pressure in them even when the engine is off. Who is right?

   a. Technician A   **b.** Technician B

   c. Both A and B   **d.** Neither A nor B

4. Technician A says that heat is a major enemy of rubber hoses. Technician B says that oil is a major enemy of rubber. Who is right?

   a. Technician A   **b.** Technician B

   c. Both A and B   **d.** Neither A nor B

5. Technician A says that a defective brake hose can cause delayed engagement of the brake. Technician B says that a defective brake hose can prevent brakes from releasing. Who is right?

   a. Technician A   **b.** Technician B

   c. Both A and B   **d.** Neither A nor B

CHAPTER 24

Automotive Plumbing: Tubing and Pipe

■ OBJECTIVES

Upon completion of this chapter, you should be able to:

✔ Describe the different types of tubing used on automobiles.

✔ Understand the different types of tubing connections.

✔ Repair damaged tubing.

■ KEY TERMS

work hardens
double walled tubing
furnace brazing
double flare
ISO (International Standards Organization) flare
inverted flare nut
compression fittings
ferrule
union
street elbow
close nipple
long nipple
collet
pipe dies

■ INTRODUCTION

Tubing and pipe are found on automobiles and on shop equipment. This chapter covers how to service various types of tubing and pipe and what types of plumbing parts are available. Also covered are the different types of connectors used with tubing and pipe. Understanding these topics is just one of the considerations that separates good technicians from average ones.

■ TUBING

Tubing found on automobiles, often called *line*, does not have threads at its ends like pipe. Tubing can be made of either copper, steel, or plastic. Different tubing materials are selected depending on the intended use. Manufacturers choose the least expensive alternative that will do the job safely and dependably. The engineering rule of thumb for tubing selection is that the maximum pressure that the material will be exposed to should be less than ⅕ the rated pressure of the tubing.

NOTES:

■ *Tubing's size is determined by its outside diameter (O.D.).*

■ *Pipe and hose are sized by their inside diameter (I.D.).*

Compare three pieces of ½" diameter pipe: one plastic, another steel, and the other copper. All of them have the same inside diameter, but are very different in appearance (Figure 24.1).

Copper Tubing

Copper tubing can be either soft or rigid. It is softer and easier to bend than steel. Soft copper with a 0.020" wall thickness can withstand 1000 psi of pressure. This is sufficient strength for *low-pressure* applications, like carburetor fuel systems, vacuum lines, oil pressure gauge lines, and other lines that carry lubricants. Copper

work hardens, which means that it gets brittle with repeated flexing. Therefore, it should not be used where flexing will be encountered.

Copper tubing should not be used for brake lines. It is *single walled* and does not have sufficient burst strength. It also work hardens from vibration, which can cause it to crack. Copper also corrodes easily and holes develop from electrolysis.

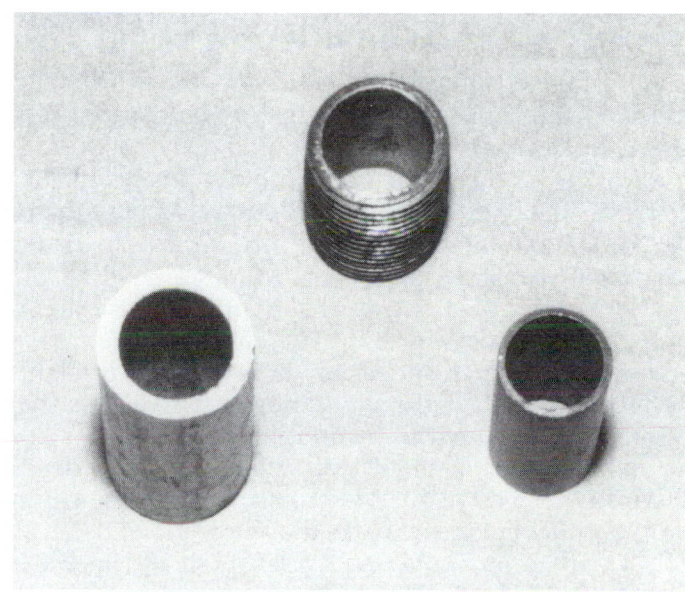

Figure 24.1 Comparison of inside and outside diameters of ¹/₂" plastic, steel, and copper pipe.

SAFETY NOTE Brakes and power steering systems develop very high pressures, sometimes in excess of 1000 psi. Do not use copper for these applications.

Plastic Tubing

Plastic tubing is used for vacuum lines or oil pressure lines. Polyethylene tubing has a rated pressure of only 200 psi. Nylon tubing is used on some fuel injected vehicles and for air brake lines on trucks (not hydraulic lines). Nylon has a rating of 300 psi. Plastic tubing is not used where it will be exposed to heat. Plastic can burn through if it comes into contact with an exhaust manifold, for instance. Fittings used with plastic tubing are covered later in this chapter.

Steel Tubing

Steel tubing, galvanized to prevent rust, can be used almost anywhere on an automobile. A flaring tool (covered later) is used to flare the ends of the tubing. Tubing diameter ranges from ⅛" to ⅜".

Steel tubing used for brake lines is **double walled tubing** so that it can bend easily. To make the tubing, copper-plated sheet steel is used. One type of tubing, called *seamless tubing*, is made by rolling the steel sheet until it has two layers and one seam, providing a double wall. Then, the copper is fused to this seamed tube using a **furnace brazing** process. When the copper melts fusing to the tubing, the seam disappears.

Another type of brake tubing is made in two different layers with the seams separated by at least 120°. Furnace brazing is used to bind the copper into the seams of this tubing also.

Most brake tubing used on automobiles has an outside diameter of 3/16". This is the size used for disc brakes. Drum brake systems sometimes use ¼" O.D. tubing.

An SAE standard for brake tubing dictates that it must be able to be bent in a full circle around a mandrel five times its diameter and not kink or be damaged in any way.

Sometimes there is a steel coil around the outside of a brake tube (Figure 24.2). The coil, called "armor," is there to protect the line from damage where it might be prone to rub against something. It also allows the tubing to be bent without kinking.

When tubing is subjected to vibration or flexing, it is usually coiled (Figure 24.3). This is especially common on lines below the brake master cylinder. Because it is mounted on the car body and the brake lines are secured to the frame, flexing must occur there.

Tubing Fittings

There are many types of fittings that are used to join tubing to components. Be sure when using fittings

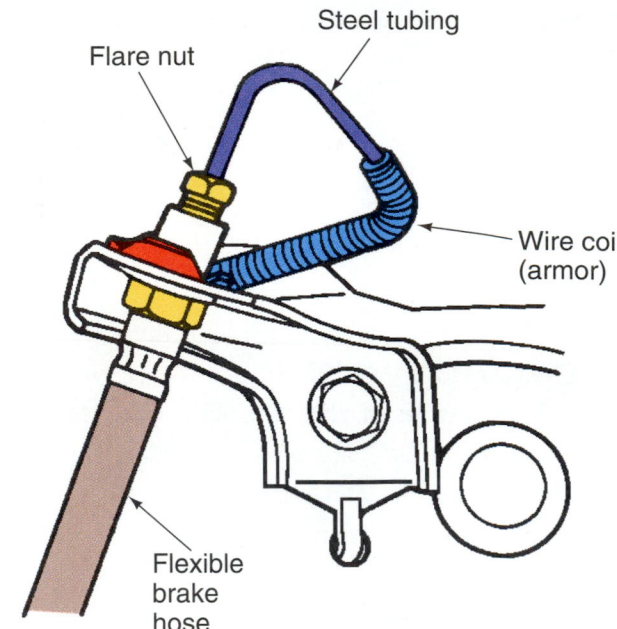

Figure 24.2 Armor protects the brake tubing from abrasion. *(Courtesy of Brake Parts Inc.)*

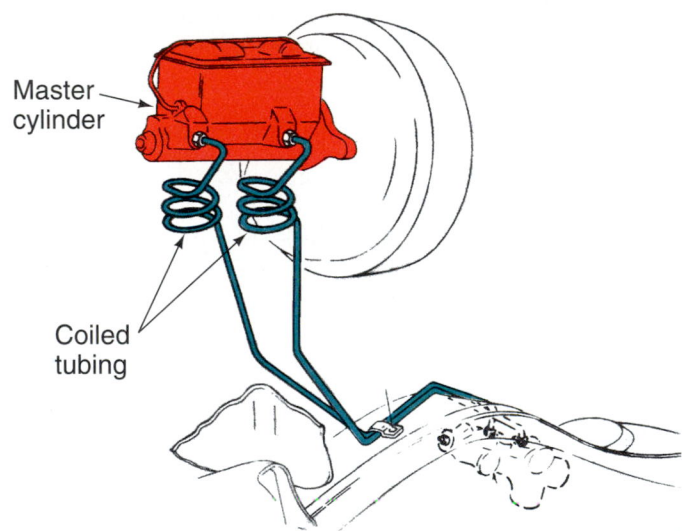

Figure 24.3 Tubing is coiled to prevent damage from vibration. *(Courtesy of General Motors Corporation, Service Technology Group)*

that they are the correct ones for the application. Flare fittings, compression fittings, and pipe fittings all have different types of threads and seats within them.

NOTE: *Threads can be either male or female. Female threads are those that are internal. Male threads are external.*

Sometimes manufacturers use connectors with oversized threads to prevent assembly errors. When replacing or adapting a part, this sometimes causes problems. *Step-up or step-down adapters* (Figure 24.4) are available for making these connections.

Connectors (Figure 24.5) are used between tubing and a part like the oil pump, carburetor, fuel pump, or brake parts. They can also serve as adapters between different types of fittings. An example of this would be if a pipe thread and flare fitting were to be joined.

> ⚠ **CAUTION** When two parts with different types of threads are accidentally joined, damaged threads and a leak can result.

When nylon lines are connected to steel lines, one of two types of *push connectors* is used. These are covered later in this chapter.

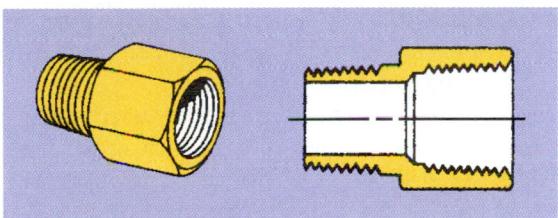

Figure 24.4 Adapter fittings allow different sizes and types of fittings to be connected together. *(Courtesy of Weatherhead Div., Dana Corp.)*

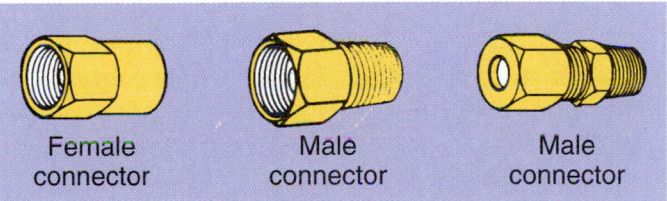

Female connector   Male connector   Male connector

Figure 24.5 Connectors are used for joining tubing to parts. *(Courtesy of Plews/Edelmann Division, Stant Corporation)*

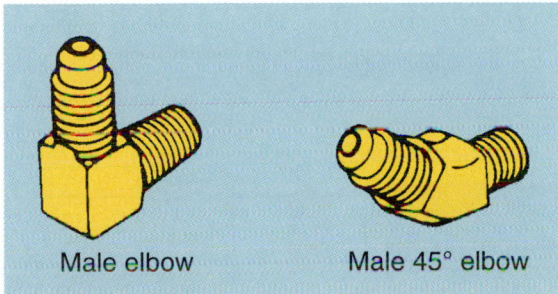

Male elbow   Male 45° elbow

Figure 24.6 Elbows are used for making sharp bends. *(Courtesy of Plews/Edelmann Division, Stant Corporation)*

An *elbow* is used when a sharp turn is made. Bending the tubing takes too much space to do. Elbows can be either 30°, 45°, or 90° (Figure 24.6).

■ FLARED CONNECTIONS

Flare fittings are sometimes used when two steel fuel lines are connected. A flare is suited for high pressure applications and *must* be used for brakes or power steering. Any type of tubing that can be formed with a flaring tool (copper, aluminum, or steel) can have a flare installed on it.

Figure 24.7 shows the relationship between the parts of a flared connection. The end of the line is tapered outward. This is called a *flare*. The flare on the end of the line is wedged between the flare fitting and the flare nut on the line. Long and short flare nut designs are available. The long nut is used when the part is subject to vibrations because it supports the tubing further away from the connection.

There are two kinds of flares used in automobiles, the *SAE type 45°* **double flare** or the **ISO (International Standards Organization) flare**, also called a *bubble flare* (Figure 24.8). SAE automotive fittings are flared at

Flare nut

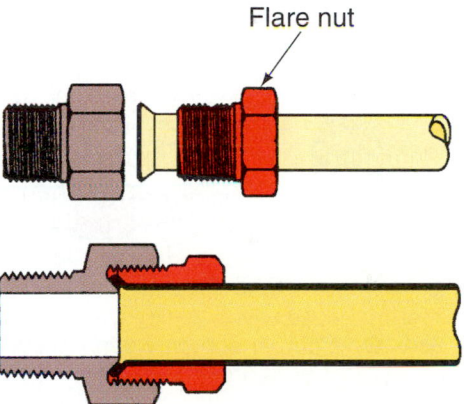

Figure 24.7 Relationship between parts of a flared coupling. *(Courtesy of Weatherhead Div., Dana Corp.)*

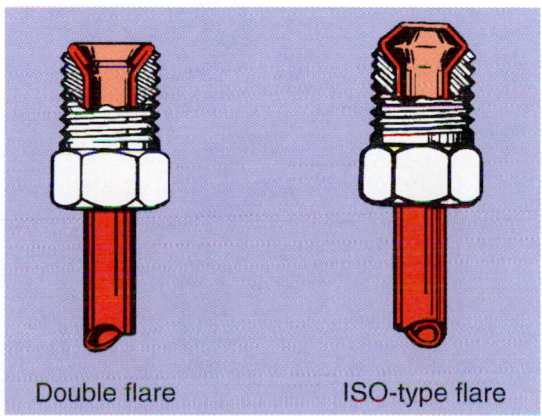

Double flare   ISO-type flare

Figure 24.8 Two types of flares used in automobiles. *(Courtesy of General Motors Corporation, Service Technology Group)*

45°. There are also 37° SAE flares. Be sure you are using SAE 45° fittings.

A 45° SAE type double flare is usually used with an **inverted flare nut** (Figure 24.9). The inverted-type flare nut is more common on automobiles. The standard flare is found in such applications as household natural gas lines.

The flare can be either a single or double flare. A single flare is not used on small automotive tubing because it will split the tubing (Figure 24.10). A double flare is a two-step process (described later).

ISO flares have been found on automobiles since the early 1980s. A bubble or ridge is formed in the line a short way back from its end. A single flaring operation can be done without danger of splitting the line (like with a single 45° flare).

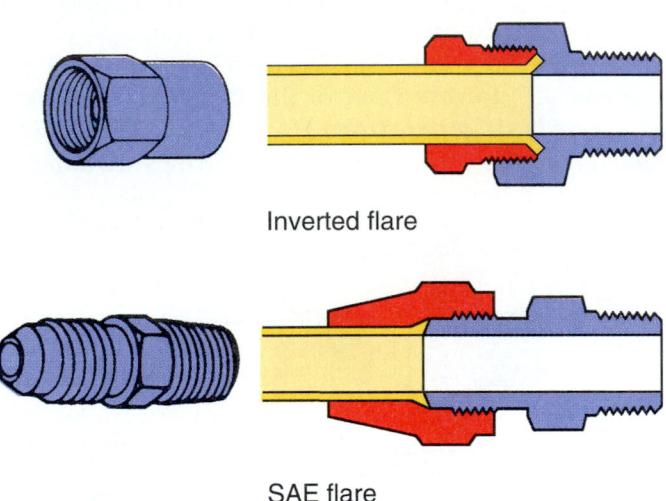

Inverted flare

SAE flare

Figure 24.9 Comparison between inverted and SAE flares. *(Courtesy of Plews/Edelmann Division, Stant Corporation)*

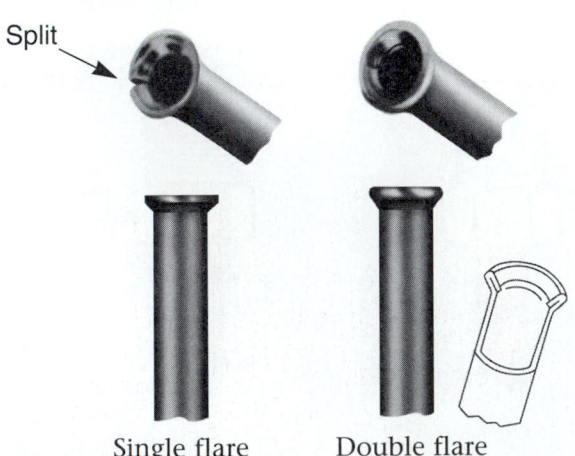

Split

Single flare   Double flare

Figure 24.10 A single flare will result in a split in the tubing. *(Courtesy of General Motors Corporation, Service Technology Group)*

■ COMPRESSION FITTINGS

Flareless fittings, called **compression fittings** are also found on automobiles (Figure 24.11). One kind uses a brass *sleeve*, called a **ferrule**. The sleeve is installed loosely on the tubing. When the compression nut is drawn against the fitting with a wrench, the sleeve is pinched tightly into the wall of the tubing to provide the seal.

To install a compression fitting:

- First, slide the nut onto the tubing.
- Next, slide a ferrule over the line to be compressed.
- Then, slide the sleeve onto the tubing.
- Finally, insert the tubing as far into the fitting as possible. Hold it in that position as you use one wrench to tighten the nut and another wrench to hold the fitting.
- After both halves of the fittings contact, they are tightened further 1¼ turns. This compresses the sleeve into the tubing to form the seal.

Compression fittings should not be used on high-pressure applications such as brakes or power steering systems. There are specialized compression fittings available for high-pressure applications.

Compression fittings can also be used with rigid plastic tubing. With softer plastic tubing an insert is put inside the end of the tubing so it does not get crushed when the sleeve is compressed (Figure 24.12).

Another kind of compression fitting, called a *double compression fitting*, does not have a separate sleeve but compresses the front part of the nut against the tubing (Figure 24.13). After both halves of the fittings contact, they are tightened further 1½ turns.

O-ring connections are sometimes used to seal fittings. An O-ring is an artificial rubber donut that is used with straight threads to seal when compressed (Figure 24.14).

Unions

A **union** joins two pieces of tubing together (Figure 24.15). This type of fitting is often used on vacuum or

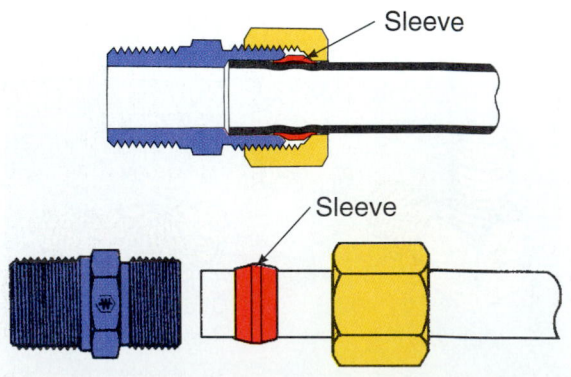

Sleeve

Sleeve

Figure 24.11 When a compression fitting is tightened, the sleeve compresses into the tubing to make the seal. *(Courtesy of Weatherhead Div., Dana Corp.)*

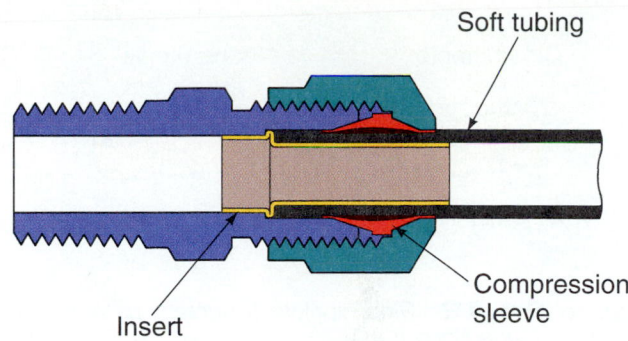

Figure 24.12 An insert is used when a compression fitting is used with plastic tubing. *(Courtesy of Weatherhead Div., Dana Corp.)*

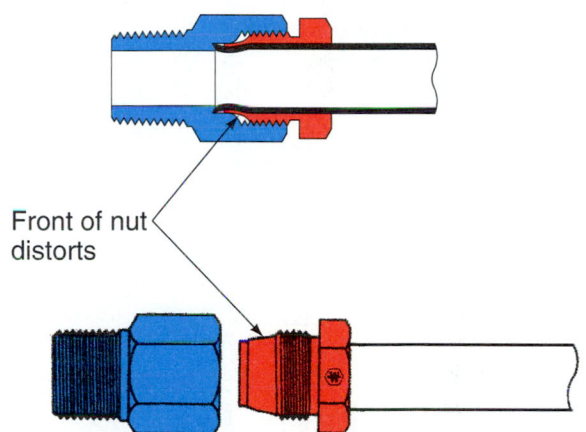

Figure 24.13 When a double compression fitting is tightened, the front part of the nut compresses into the tubing. *(Courtesy of Weatherhead Div., Dana Corp.)*

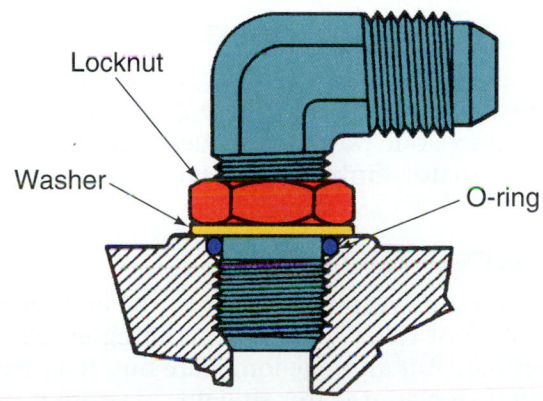

Figure 24.14 An O-ring seals between parts. *(Courtesy of Weatherhead Div., Dana Corp.)*

air pressure lines. It can be disassembled without having to turn the tubing. It is a good repair for the higher pressure that runs in automatic transmission cooler lines.

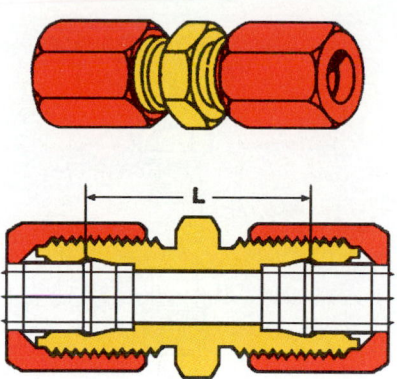

Figure 24.15 A union. The length that is removed from the tubing should equal L. *(Courtesy of Weatherhead Div., Dana Corp.)*

■ PIPE FITTINGS

On automobiles, pipe threads are used for heater outlets in the block and intake manifold, oil gallery and coolant drain plugs, as well as for oil and coolant temperature sending units (Figure 24.16). Pipe fittings are also used for compressed air lines in the shop.

Pipe fittings on copper, brass, or iron pipe use tapered threads that wedge together as they are tightened. An air tight seal is formed between the two tapered threads. The pressure capability of the joint is rated at 1000 psi.

The size of a pipe thread is determined by the inside diameter (I.D.) of the piece of pipe. The outside diameter (O.D.) of a ½" *NPT (National Pipe Taper)* thread will be quite a bit larger than ½" (Figure 24.17).

Two pipes can be joined together with a pipe coupling (Figure 24.18a). When a pipe is joined to two other sections of pipe, the fitting is called a "T" (Figure 24.18b). A female pipe coupling that makes a turn is called an *elbow* (Figure 24.18c). When there is a male thread on one end, the coupling is called a **street elbow** (Figure 24.18d).

A small section of pipe is called a nipple. A **close nipple** has tapered threads on each end that join in

Figure 24.16 Pipe threads are found on threaded plugs, heater outlets, and oil and coolant temperature sending units.

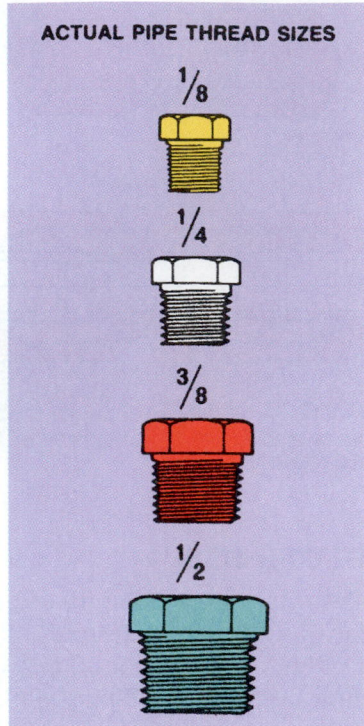

ACTUAL PIPE THREAD SIZES

Figure 24.17 Pipe threads are sized according to the I.D. (inside diameter) of the pipe so the thread diameter is larger than the callout size. *(Courtesy of Weatherhead Div., Dana Corp.)*

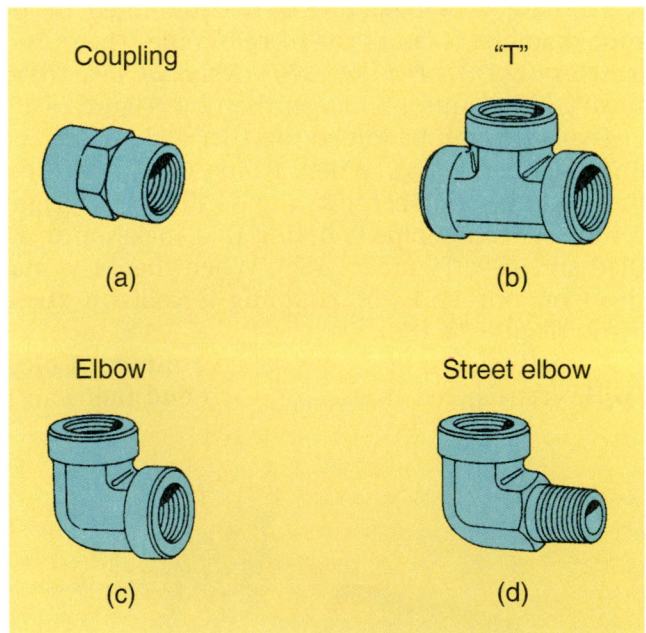

Coupling   "T"

(a)   (b)

Elbow   Street elbow

(c)   (d)

Figure 24.18 Different types of pipe connections. *(Courtesy of Weatherhead Div., Dana Corp.)*

the middle (Figure 24.19a). A **long nipple** has a section of plain pipe separating the threads (Figure 24.19b).

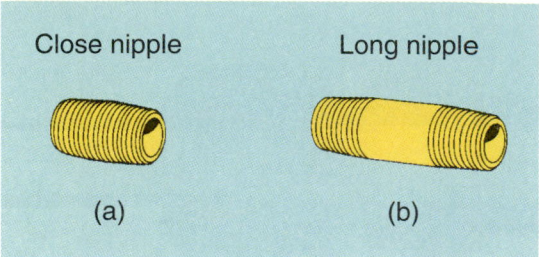

Close nipple   Long nipple

(a)   (b)

Figure 24.19 Pipe nipples. *(Courtesy of Weatherhead Div., Dana Corp.)*

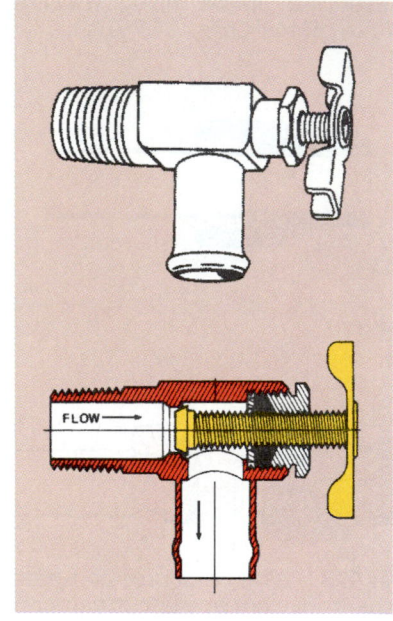

FLOW →

Figure 24.20 A shutoff valve. *(Courtesy of Weatherhead Div., Dana Corp.)*

■ SHUTOFF VALVES

Shutoff valves, also called *drain cocks*, are often found on the bottom of a radiator (Figure 24.20). Turning the end of the valve one way or the other opens it or closes it. Plastic radiator tanks usually have a plastic drain valve.

■ TUBING SERVICE

Tubing comes in precut lengths with the ends already on or in 25 foot rolls. Precut brake tubing usually has one short flare nut and one long flare nut. If a piece of tubing to be replaced is not straight, measure it using a piece of string. This will help to determine the correct length, including bends.

When unrolling bulk tubing, be careful not to kink it. Unroll it in the same direction that it was rolled. When fittings are to be installed on the line, add an additional ⅛" at each end for the flare to be formed (¼" total).

When loosening a flare fitting, a tubing wrench or flare nut wrench is used. Flared lines are held against a seat in the fitting. *Always* use two wrenches (see Figure 7.7b). The second wrench holds the female part of the fitting. If the female fitting is allowed to turn when the male flare fitting is turned with a wrench, the line will become kinked (see Figure 7.7c).

Damaged steel fuel lines can be cut and repaired or new lines can be fabricated using a flaring tool. Use only seamless steel tubing. Do not replace steel tubing with copper.

Cutting Tubing

When cutting tubing to length, be sure to cut it so it is square on the end. It is best to use a tubing cutter (Figure 24.21). A hacksaw will leave a rough edge that might not be square. When the end is dressed off with a file, metal chips can get into the end of the tube. If these are not thoroughly cleaned out, serious damage to a component can result.

The tubing cutter is first tightened against, then rolled around the tubing. The handle is tightened to advance the cutter as the tubing is cut. Do not over-tighten it or the cutter will cut through too soon and the tubing could be damaged.

A burr usually remains on the end of the tubing after cutting it. This should be removed with the reamer blade that most tubing cutters are equipped with. Be sure to remove any chips from the end of the tubing after completing the cut.

Some tubing cutters are very small. In tight quarters these can be used for repairing damaged tubing on the vehicle. The tubing cutter shown in Figure 24.22 is especially handy for this.

Bending Tubing

When bending tubing, remember that too sharp of a bend will result in a kink or restriction. Tubing can be bent with a tubing bender (Figure 24.23). It can also be bent by holding it over a large piece of pipe and slowly forming it using the tubing that is being replaced as a guide when possible. Be careful not to bend the tubing too sharply.

It is better to install fittings and flare both ends before bending tubing. Otherwise, if the bend is too close to the flare, the flaring tool will not be able to clamp to the line. It is best if the bend is not too close to the flare. Leave at least a couple of inches when possible.

A bending spring (Figure 24.24) can be installed over the tubing. It will help keep the tubing from becoming kinked during the bend. When using a bending spring, one end of the line must not have its fitting installed yet. This is so that the spring can be installed and removed from the line.

Steel lines should not have long, straight runs. They are difficult to remove and replace and they can fatigue at the connections. When a short section of

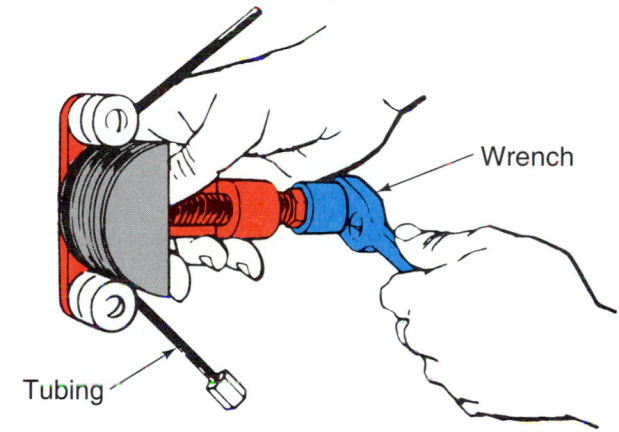

Figure 24.23 A tubing bender.

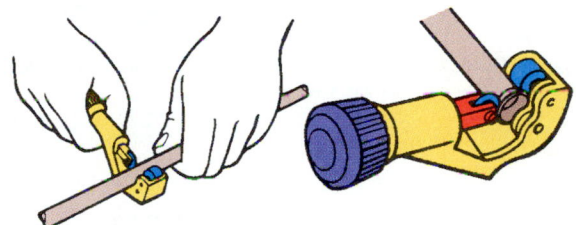

Figure 24.21 Use a tubing cutter to cut tubing. *(Courtesy of Chrysler Corporation)*

Figure 24.22 A small tubing cutter is handy for working in tight spaces on the vehicle.

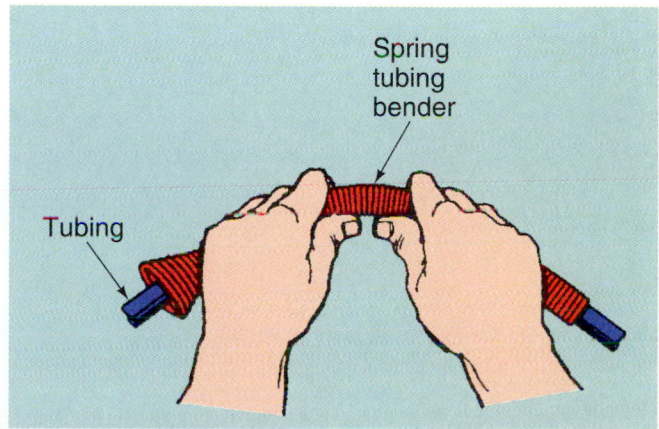

Figure 24.24 A bending spring helps prevent the tubing from becoming kinked.

tubing is replaced, at least one end should have a bend so that the connection can flex (Figure 24.25).

Long runs should be supported with clamps. Heavy connections such as distribution blocks must be securely mounted. The ends of the tubing should align with the fitting. Be sure that the threads can be turned all the way into the fitting easily.

Flaring the Ends of Tubing

Tubing is flared with either a double flare or an ISO (metric double) flare. Different tools are required to perform these jobs. Be sure to use the same kind of flare as was used on the original line. The different types of flares are not interchangeable.

Double flaring is a two-step procedure. The fuel line is clamped in a special tool while its end is formed. The tube is then folded over itself to complete the double flare.

- First, slip the fitting onto the line.
- Select the correct size hole in the flaring tool bar.
- Clamp the line in the flaring tool bar. It should extend out of the bar by the width of the flaring tool adapter (Figure 24.26).
- A threaded flaring cone and clamp are used to form the end of the tubing (Figure 24.27).
- Insert the adapter in the end of the tubing and tighten down on it with the threaded flaring tool until it bottoms out (Figure 24.28a).
- Remove the adapter and tighten the flaring tool against the line again to complete the flare (Figure 24.28b).

If the flare is not formed properly, you will have to cut the end off of the line and form another flare.

Remember:

- Always put the fitting on the line before flaring it.
- Leave enough space between a bend and the flared fitting so that the fitting can slide.

➡ **Perform *Metal Tubing Service* Worksheet**

Thread sealers or Teflon tape are not necessary when using flare fittings. A flare fitting seals internally and the threads should not be exposed to liquids. If the threads get wet, the flare is not a quality joint.

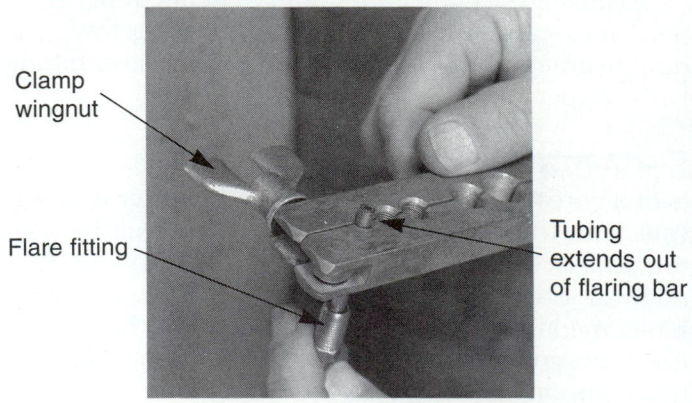

Figure 24.26 Clamp the tubing in the flaring bar with it protruding about the thickness of the adapter.

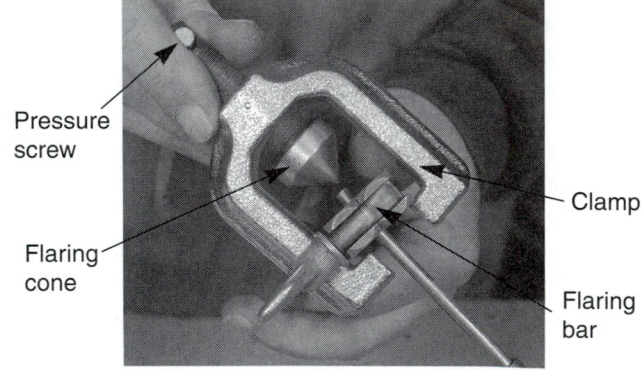

Figure 24.27 A threaded flaring cone and clamp are used to form the end of the tubing.

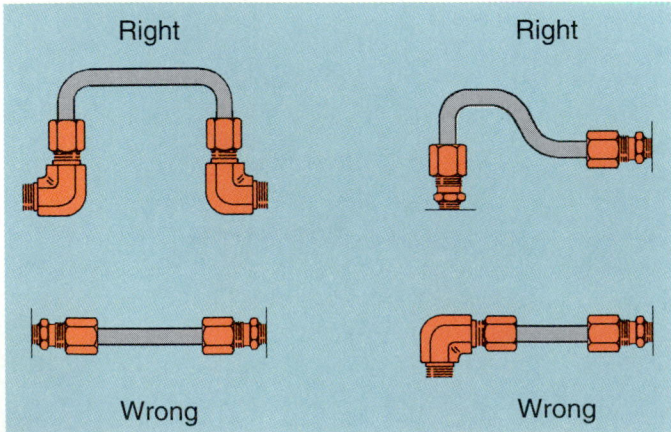

Figure 24.25 A bend in a rigid length of tubing can prevent vibration damage. *[Courtesy of Weatherhead Div., Dana Corp.]*

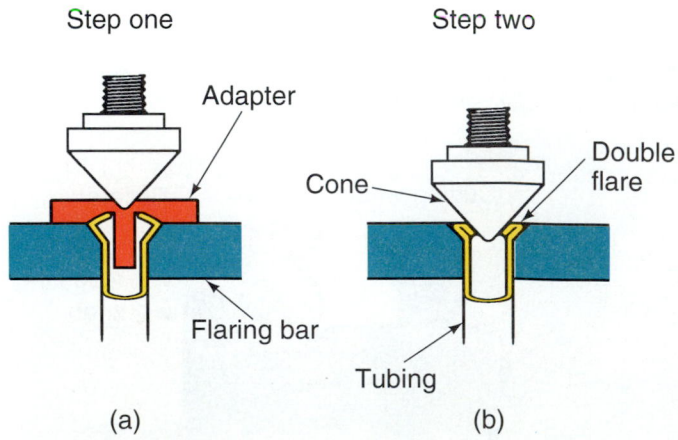

Figure 24.28 (a) The adapter is forced into the end of the tube. (b) The flaring cone completes the double flare. *[Courtesy of Weatherhead Div., Dana Corp.]*

■ INSTALLING TUBING

When you install a length of tubing, leave the first fitting loose after threading it into the fitting. After starting the fitting on the other end of the line, tighten them both. Hold the fitting with a wrench while tightening the flare nut. Do not overtighten the fittings. After the flare is brought into contact with the fitting, turn it an additional ⅙ turn only.

ISO Flaring

To perform a bubble flare (ISO flare), a different flaring tool is used (Figure 24.29). The tubing is inserted through the end of the clamping nut. When the clamping nut is tightened into threads in the flaring tool body, a **collet** is compressed to grip the outside of the tubing. Tightening the pressure screw forms the bubble on the tubing.

Union Repairs

Sometimes a union is used to repair a damaged line (see Figure 24.15). A section of the line slightly less than the length of the fitting is cut from the old line. Some unions require the ends of the line to be flared first. Others use compression fittings. With compression fittings, flaring is not required.

A union is a better repair than hose for higher pressure (30–35 psi) lines. When there is enough straight line remaining behind a flare, cut the kink from the line and install a union to couple the two pieces of line together.

Using Hoses to Repair Tubing

When a section of steel fuel line is damaged, it is best to replace it with a new section of line with flared ends and fittings. Besides being stronger and more reliable, steel lines allow better cooling of the fuel.

Sometimes a small section of damaged fuel tubing is replaced with hose. When using fuel hose to repair a damaged section of metal fuel line, use a section of hose that is 4" longer than the section of tubing removed. A raised bead is formed on each end of the

tubing with a flaring tool. The hose is installed with 2 inches of hose overlapping each end of the tubing. The clamp is positioned ⅛" from each end of the tubing.

If a section of tubing is removed that is over 6", the recommended safe practice is to use two sections of hose separated by a piece of steel tubing. Be sure to secure the tubing so that it cannot rub against body parts. Following completion of the repair, start the engine to check for leaks. After one week, the clamp should be retightened because hoses take a set and the clamp can become loose.

■ TRANSMISSION OIL COOLER LINE REPAIRS

The best repair for a transmission cooler line is a union. But these lines are often repaired with hose. For repairs to transmission cooler lines, be sure to use hose that is rated for high temperature, pressure, and oil. Do not use ordinary fuel hose, which deteriorates in the hostile environment of the hot oil. When a transmission cooler line is repaired using a hose, power steering hose can be used if transmission cooler hose is not available.

If a rubber hose is used to repair transmission lines, be sure to flare both ends of the metal tubing that is being joined. Otherwise, the high pressure will blow the hose off the lines. Over time, the edges of a double flare can cut the rubber hose as it works against the flare. Performing the first part of a double flare on the ends of the lines (see Figure 24.28a) will present a smoother edge for the hose to squeeze against.

CASE HISTORY *A technician repaired a damaged transmission cooler line by cutting out the bad section, flaring both ends of it, and installing a piece of hose with worm gear hose clamps. After a few thousand miles, the sharp edge of the single flared line cut through the rubber hose, resulting in a serious fluid leak. Luckily, the owner spotted smoke and stopped the car before a fire or damaged transmission could occur.*

■ NYLON FUEL INJECTION TUBING

Some vehicles use nylon fuel line, which must first be soaked in boiling water to allow it to stretch during installation. When nylon line is connected to steel line, a push connector is used.

- To remove a "hairpin" push connector (Figure 24.30), separate the clip legs by about ⅛" and pull the triangular tab to remove the clip. Then, firmly and gently pull the nylon line from the steel line.
- To remove a "duckbill" connector (Figure 24.31), narrow-jawed pliers are used to compress both retaining clips on the side of the push connector at

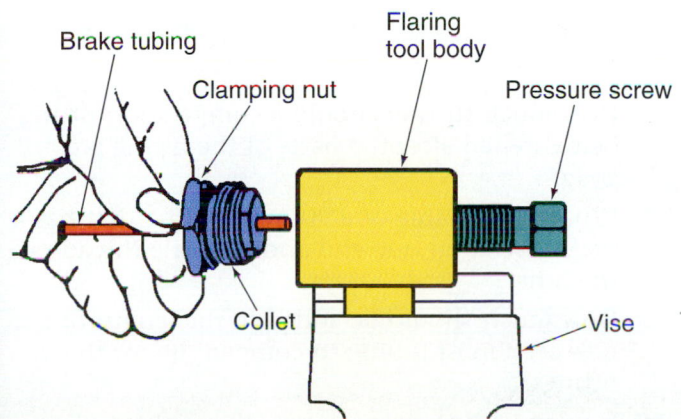

Figure 24.29 An ISO flaring tool.

Brake tubing

Flaring tool body

Clamping nut

Pressure screw

Collet

Vise

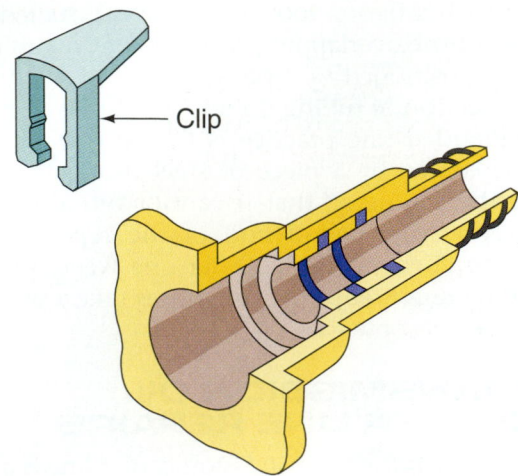

Figure 24.30 A hairpin push connector. Separate the clip legs by about 1/8" and pull the triangular tab to remove the clip.

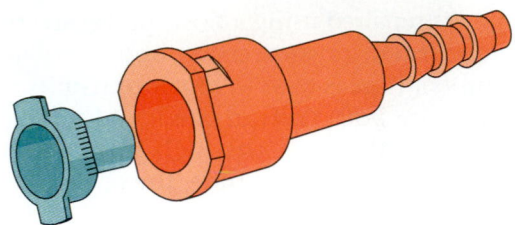

Figure 24.31 A duckbill connector. Use narrow-jawed pliers to compress both retaining clips on the side of the push connector at once.

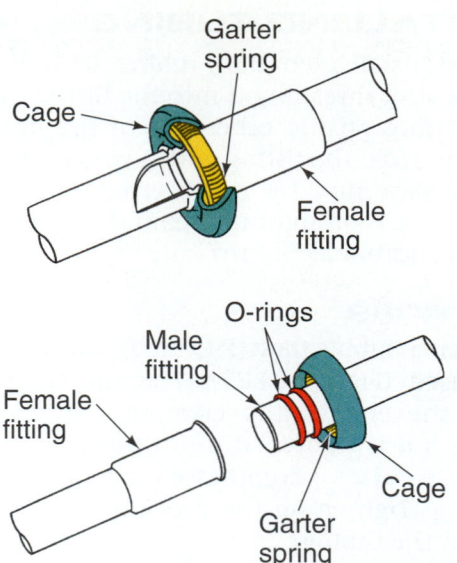

Details of spring lock (garter) connector

Figure 24.32 Some nylon lines are coupled together with a "spring lock" (garter) connector.

once. The retaining clip remains on the metal tube when the connector is removed.

Some nylon lines are coupled together with a "spring lock" connector (Figure 24.32). The flared end on the female fitting of this coupling lodges behind the garter spring inside the cage on the male fitting. This holds the two couplings together.

There are two O-rings on the male coupling. They are made of fluoroelastomer. When replaced, it is important that the correct O-rings be used. A special tool is used to release the garter spring so that the two halves of the coupling can be released. Sliding the tool toward the garter spring causes a ledge on the tool to go under the spring, raising it so the flare can clear it.

If the garter spring is damaged or missing, it can be replaced in the coupling. Lubricate the new O-rings with engine oil before installation. Be sure that the flare is all of the way under the garter spring.

▪ PIPE SERVICE

Pipe is cut with a pipe cutter, which resembles a large tubing cutter. **Pipe dies** are used to form the threads on the outside of the pipe. A thread sealer is used between the threads. Teflon tape is often used (see Figure 46.27). After hand-tightening the pieces of pipe, tighten further a minimum of 2½ turns.

▪ REVIEW QUESTIONS

1. List three materials that tubing can be made of.
2. What happens to copper when it is repeatedly flexed?
3. Why is tubing sometimes coiled when installed?
4. What are two types of flares used on automobiles?
5. What kind of flare is most often found on cars, inverted or standard?
6. What is the name of the fitting that uses a sleeve?

7. How much further should a compression fitting be tightened after the parts of the fitting are snug?
8. What is the name of a 90° pipe fitting that has a male thread on one end and a female thread on the other?
9. How much should be added to the length of a new section of tubing to compensate for the fittings?
10. How much is a flared connection tightened after the nut becomes snug?

■ ASE STYLE REVIEW QUESTIONS

1. Technician A says that copper or steel can be used for brake lines. Technician B says a single flare is not used because it will split the tubing. Who is right?

   a. Technician A   **b.** Technician B

   c. Both A and B   **d.** Neither A nor B

2. Technician A says to use Teflon tape or thread sealer with flared connections. Technician B says to use fuel hose for repairing transmission cooler lines. Who is right?

   a. Technician A   **b.** Technician B

   c. Both A and B   **d.** Neither A nor B

3. Technician A says that tubing's size is determined by its outside diameter. Technician B says that pipe and hose are sized by their outside diameter. Who is right?

   a. Technician A   **b.** Technician B

   c. Both A and B   **d.** Neither A nor B

4. Technician A says that steel tubing is galvanized to provide strength. Technician B says that pipe fittings have tapered threads. Who is right?

   a. Technician A   **b.** Technician B

   c. Both A and B   **d.** Neither A nor B

5. Technician A says that tubing is threaded at its ends. Technician B says it is best to install flare fittings on tubing after it has been bent. Who is right?

   a. Technician A   **b.** Technician B

   c. Both A and B   **d.** Neither A nor B

Fuels, Octane, Abnormal Combustion

■ OBJECTIVES

Upon completion of this chapter, you should be able to:

✔ Understand how petroleum is refined.

✔ Describe the different characteristics of various blends of gasolines.

✔ Know the effects of the different types of abnormal combustion.

✔ Decide on the best choice of gasoline or diesel fuel for a vehicle.

✔ Diagnose rich and lean fuel mixture problems.

✔ Describe the advantages and disadvantages of various types of alternative fuels.

■ INTRODUCTION

Motorists often have questions about the fuel used in their cars. There are several kinds of fuels used in motor vehicles. A service technician should have a basic understanding of these fuels and their characteristics. This chapter deals with gasoline, diesel, and alternative fuels. Also included are discussions of rich and lean air/fuel mixtures and abnormal combustion. These conditions can result in engine damage, poor fuel economy, and poor performance (driveability problems).

■ CRUDE OIL

Raw petroleum, also called crude oil, is used to make such products as gasoline, diesel fuel, motor oil, solvents, and LPG (liquified petroleum gas), among others. Figure 25.1 lists many of the products made from oil.

Crude oil is pumped from wells in the ground (Figure 25.2). It is a variable combination of large and small hydrocarbons together with petroleum gas, hydrogen sulfide, and water. The surplus petroleum gas can be used directly for fuel for the refinery.

The liquid *hydrocarbons*, consisting of approximately 12% hydrogen and 82% carbon, are of many kinds. They are like a keg of nails of many different sizes all mixed together. They must be sorted before they can be used. Hydrogen is a light gas vapor. Carbon is a heavy black solid.

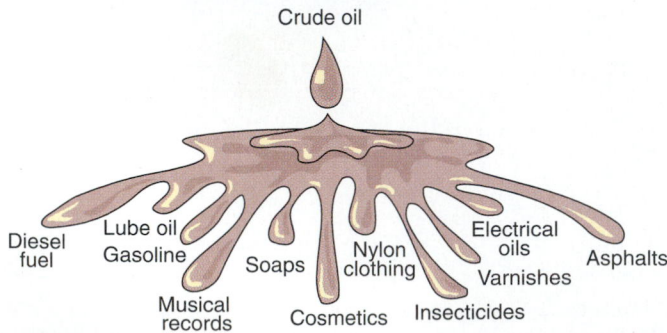

Figure 25.1 Some of the products made from oil.

Crude oil

Diesel fuel — Lube oil — Gasoline — Soaps — Musical records — Nylon clothing — Cosmetics — Insecticides — Electrical oils — Varnishes — Asphalts

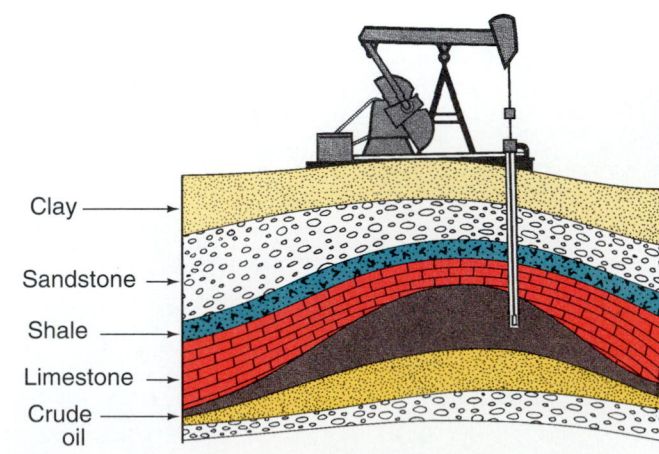

Figure 25.2 Oil is pumped from deep in the ground. *(Courtesy of Quaker State Corporation)*

Clay
Sandstone
Shale
Limestone
Crude oil

These materials as well as hydrogen sulfide and water are removed during refining.

Oil is separated into many useful products at the oil refinery. During refining, the crude is first heated by pumping it through pipes in hot furnaces into a fractionating column. Light hydrocarbon molecules are separated from heavier ones in these tall narrow columns sometimes called distillation towers. They are what make up the distinct skyline of oil refineries (Figure 25.3).

The refining process breaks the crude down into different parts, called *fractions*. Crude is not a single substance, like water. Water has one single boiling point (212°F). Crude oil fractions have different boiling points ranging from about 100° to 700°F. (Figure 25.4).

The fractioning tower has draw pipes at different heights for pulling the desired petroleum materials out of the tower (Figure 25.5). The lightest products are a gas at room temperature. They are taken from the top of the fractioning tower as the crude is boiled in the refining process (Figure 25.6). After that comes gasoline. The remaining fractions have higher and higher boiling points. The heaviest products boil last and are taken from the bottom where the temperature is the highest. They are tarlike and are at "the bottom of the barrel."

Some of the fractions are used as raw materials to be blended into gasoline to correct octane, emissions, volatility, and storage life. The next fractions can sometimes be used directly, such as kerosene or diesel. The last fraction, with a very high boiling point can be used as fuel oil or as a basic stock for lubricants. They can also be reformulated by heat or catalysts to produce more of the lighter stocks. The distillation

Figure 25.3 Fractionating or distillation towers are the tall columns that highlight the skyline of oil refineries. *(Courtesy of Chevron)*

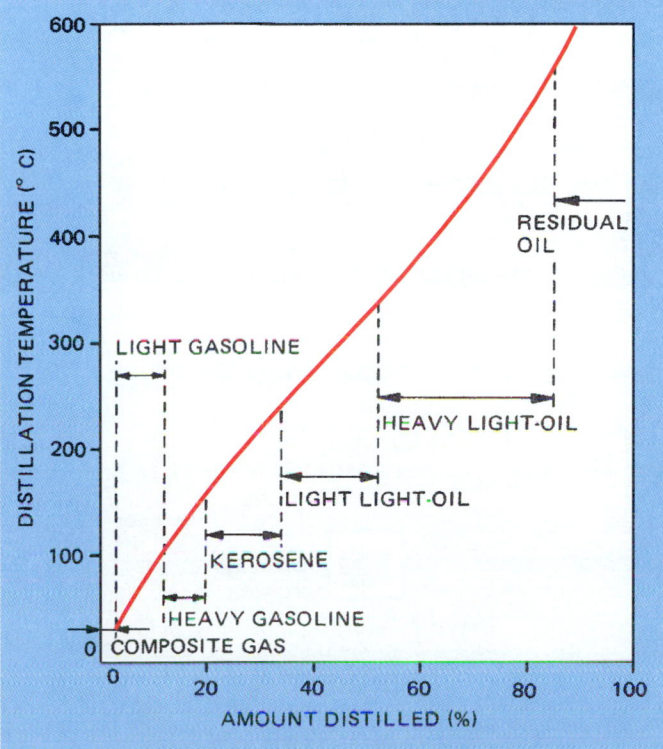

Figure 25.4 Crude oil fractions have different boiling points ranging from about 100° to 700°F.

Figure 25.5 Draw pipes at different heights pull lighter and heavier materials from the tower. *(Courtesy of Chevron)*

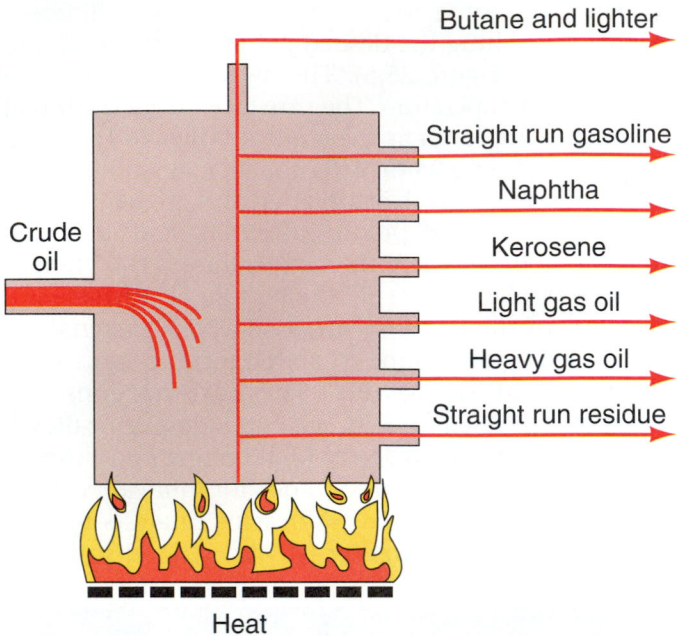

Butane and lighter

Straight run gasoline

Naphtha

Kerosene

Light gas oil

Heavy gas oil

Straight run residue

Crude oil

Heat

Figure 25.6 Lighter products come off of the tower closer to the top. *(Courtesy of Quaker State Corporation)*

process is repeated in other plants as the oil is further refined for its most efficient use (Figure 25.7).

Fractions from different geographical locations vary widely. Some crudes are light in color and flow easily. Other heavier crudes must be heated to make them flow. Some fractions are only useful for a particular product, such as gasoline or diesel. Pennsylvania grade crude, found only in Pennsylvania, New York, West Virginia, and Ohio, yields more high quality lubricating oil. Crude oil is measured and sold by the barrel, which is 42 gallons. Figure 25.8 shows the percentage of various products coming from an average barrel of petroleum.

■ GASOLINE

Gasoline makes up a substantial percentage of the crude oil. The amount depends on the location of the well and oil field. Gasoline is a very flammable hydrocarbon (HC). During combustion, the hydrogen and carbon combine with oxygen. Combining hydrogen and oxygen produces water (H_2O) and combining carbon and oxygen produces carbon dioxide (CO_2). These harmless gases are what would result if all of the

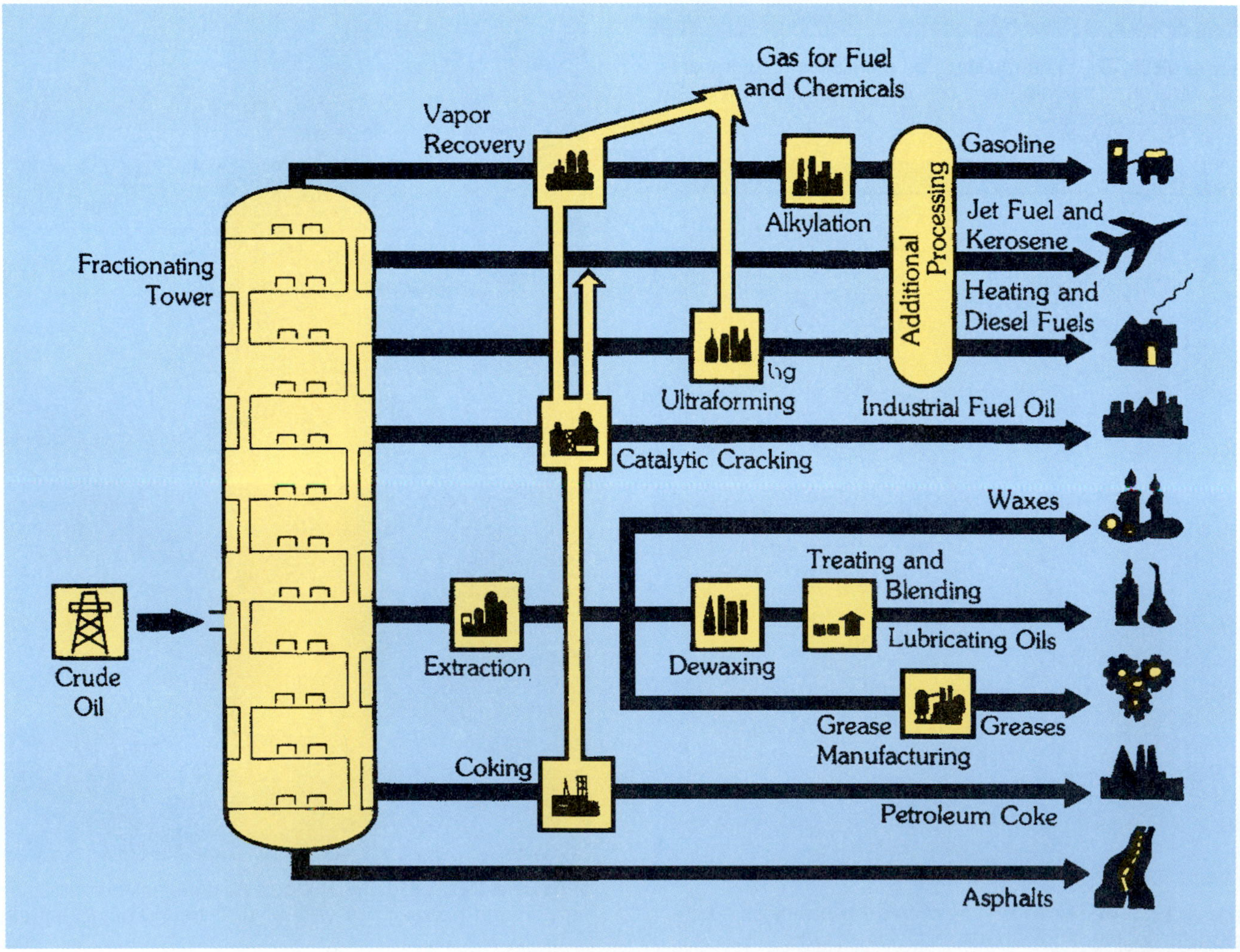

Gas for Fuel and Chemicals

Vapor Recovery

Alkylation

Gasoline

Jet Fuel and Kerosene

Additional Processing

Ultraforming

Heating and Diesel Fuels

Fractionating Tower

Catalytic Cracking

Industrial Fuel Oil

Waxes

Treating and Blending

Extraction

Dewaxing

Lubricating Oils

Crude Oil

Grease Manufacturing

Greases

Coking

Petroleum Coke

Asphalts

Figure 25.7 Oil refining process. *(Courtesy of American Petroleum Institute)*

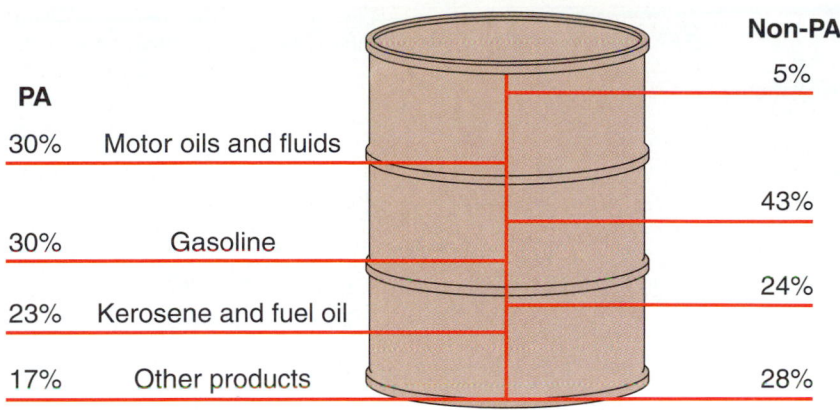

Figure 25.8 Average yields from a U.S. barrel of crude oil. The left side (PA) is Pennsylvania grade crude. The right side is an average of all others. *(Courtesy of Quaker State Corporation)*

hydrogen and carbon in the fuel combined with oxygen during combustion. But incomplete combustion results in some HC and some CO (instead of CO_2) (see Chapter 41). Emission controls are installed on the car to deal with these by-products of combustion.

> **CAUTION** Gasoline is very dangerous because it is so flammable. Gasoline *vapor* mixed with air is actually what is so flammable. Be careful around electrical equipment, hot metal, and open flames such as are found with natural gas and propane appliances and equipment. Gasoline should never be used as a cleaning solvent.

Gasoline Characteristics

Two characteristics of gasoline that are important to engine operation are volatility and resistance to spark knock.

Volatility is a measure of how easy a fuel evaporates. For fuel to burn properly in the cylinder, its vaporization point should be near the temperature in the intake manifold. If liquid fuel that is not atomized enters the cylinders, it will not burn. This wastes gasoline, which produces extra hydrocarbon and carbon monoxide emissions in the exhaust. Unburned gasoline also washes lubricating oil off of the cylinder walls, increasing wear.

When fuel does not vaporize easily enough, this is called *low volatility*. Low volatility results in:
- Difficulty in cold starting
- Poor driveability in cool weather
- Unequal fuel distribution
- Increased spark plug and combustion chamber deposits

If a fuel vaporizes too easily this is called *high volatility*. The resulting **vapor lock** causes the engine to stall because liquid fuel will not reach the carburetor. Other problems that can result are:

- High evaporative emissions
- Poor driveability when hot
- Lower fuel economy

Port fuel injection reduces volatility problems because the fuel is injected on or near the intake valves. The valves are hot and aid vaporization.

Refiners blend gasolines for the season and for the geographic area in which they will be used. These gasolines have different *vapor pressures*. Gasoline blended for summer use is less volatile (does not burn as easily). In higher altitude areas, fuels must have higher volatility because they can boil at lower temperatures. During unseasonable weather, cars can experience problems related to their fuel. Fuel that was refined for use in the summer that is used during the winter can cause hard starting. If winter fuel is used in the summer, vapor lock and carburetor icing can result.

Vapor lock happens to a carbureted car when fuel boils in the fuel line. Unlike liquid fuel, vapor is compressible. This means that the fuel cannot be pumped to the carburetor, so the engine stalls. After the fuel line cools sufficiently, the engine will run again. Fuel injected engines use electric fuel pumps that keep the fuel under higher pressure. Higher pressure raises the fuel's boiling point so vapor lock does not occur.

Measuring Volatility

The American Society for Testing and Materials (ASTM) has six volatility classes for gasoline, AA, B, C, D, and E. AA is the least volatile (Figure 25.9). In Figure 25.9 the 10% standard for AA gasoline means that 10% of the fuel would be evaporated before it reaches a temperature of 158°F, and so on. For all volatility classes, gasoline will have evaporated by 437°F.

Volatility is measured in one of three ways. The best known standard is called **Reid vapor pressure (RVP)**. During the RVP test, a sample of gasoline sealed in a metal chamber with a pressure measuring device is submerged in 100°F water. More volatile fuels

Vapor Pressure/ Distillation Class	Distillation Temperatures			End Point Maximum °F	Vapor Pressure psi/Max.	Vapor Lock Protection Class	Vapor-Liquid Ratio of 20 °F Min.
	10% Evap. Maximum °F	50% Evap. °F	90% Evap. Maximum °F				
AA	158	170–250	374	437	7.8	1	140
A	158	170–250	374	437	9.0	2	133
B	149	170–245	374	437	10.0	3	124
C	140	170–240	365	437	11.5	4	116
D	131	150–235	365	437	13.5	5	105
E	122	150–230	365	437	15.0	6	95

ASTM D 4814 Gasoline Volatility Requirements

Figure 25.9 Gasoline volatility requirements. *(Courtesy of Changes in Gasoline III, Downstream Alternatives, Inc. [1996])*

vaporize more easily, creating more pressure. The vapor pressure is measured in pounds per square inch. Because other ways of measuring volatility are becoming more popular, RVP is becoming known as VP.

In the United States, volatility standards for fuel are required by the EPA (Environmental Protection Agency) from June 1 to September 15 at retail gasoline stations. Gasoline vapor pressure must be below 9.0 psi. In southern high exhaust emission areas, called *non-attainment areas*, VP must be lower than 7.8 psi. VP for gasohol can be up to 1 psi higher.

■ AIR/FUEL MIXTURE

For an engine to start and run with good driveability and no internal damage to engine parts, it must have the correct air/fuel mixture. The air/fuel ratio of the engine is measured by weight in pounds. The desirable air/fuel mixture is about 15 pounds of air to one pound of fuel. A 15:1 air/fuel ratio is 9000 gallons of air to 1 gallon of fuel.

A normal air/fuel mixture for high power is about 12 parts of air to 1 part of fuel (Figure 25.10). This is a 12:1 mixture and is called *rich*. Maximum power occurs from 12:1–12.5:1 air/fuel ratio. The ratio for maximum economy is 15:1–16:1, called *lean*.

A computer feedback fuel system tailors the air/fuel mixture to adapt to various engine conditions. Stoichiometric (14.7:1) is the best air/fuel ratio for the most complete combustion for emission purposes (see Chapter 41).

Figure 25.11 shows air/fuel ratios that are called for under different operating conditions.

Rich Air/Fuel Mixture

A rich mixture results when there is too much fuel for the amount of air. 9:1 would be a rich mixture. A slight-

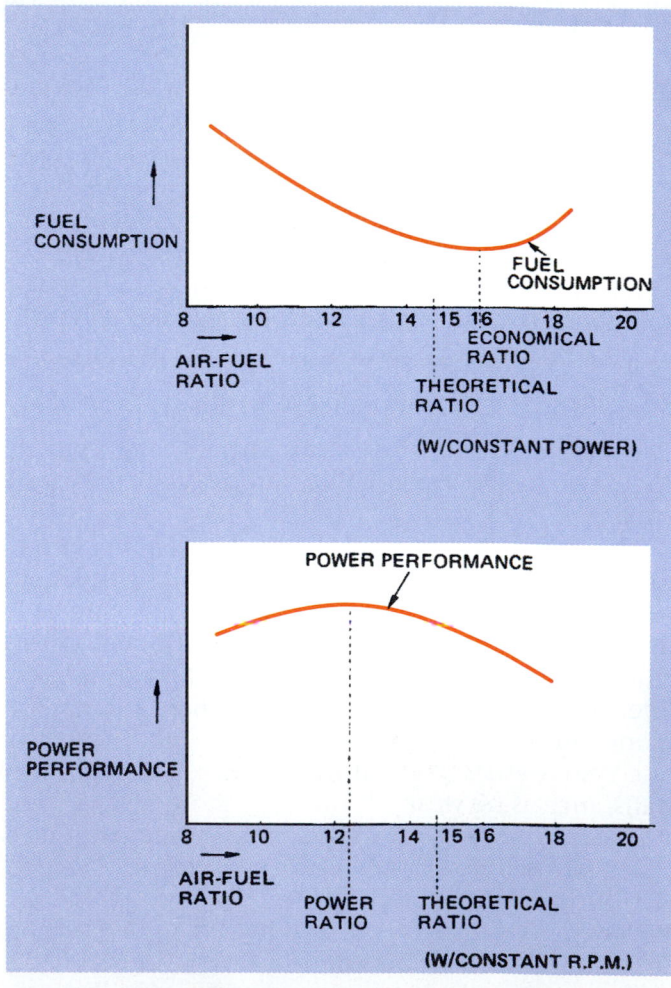

Figure 25.10 Engine operation with various air/fuel mixtures.

ly rich mixture (12:1 for instance) will improve power but will result in poor fuel economy and increased exhaust emissions.

ENGINE OPERATING CONDITION	AIR/FUEL RATIO (AIR:FUEL)
Starting (Air temperature approx. 0°C)	Approx. 1 : 1
Starting (Air temperature approx. 20°C)	Approx. 5 : 1
Idling	Approx. 11 : 1
Running slow	12 – 13 : 1
Accelerating	Approx. 8 : 1
Max. output (full load)	12 – 13 : 1
Running at medium (economical) speed	16 – 18 : 1

Figure 25.11 Air/fuel ratios for different operating conditions.

Lean Air/Fuel Mixture

An excessively lean mixture has a larger amount of air, 20:1 for example. With mixtures leaner than about 16:1 driveability suffers. A car's **surge** on the highway is a symptom of a lean condition. Surge is when the car does not keep constant power. Although it may not feel as though the engine has a misfire, the car will feel as though it is a boat riding on swells in the ocean. As the mixture becomes leaner, the engine will not run.

Figure 25.12 This spark plug was overheated by operating with a lean air/fuel mixture. (Courtesy of Cooper Automotive/Champion Spark Plug)

SHOP TIP Lean mixtures can result in burned parts, such as head gaskets. Pointed spark plug electrodes are often the first indication of the problem (Figure 25.12).

NOTE: *A leaner mixture results in a higher idle speed. Listen to a chain saw or a lawn mower as it runs out of gas. The speed of the engine will increase momentarily before the engine quits. This is because the mixture is becoming leaner as the engine runs out of gas.*

GASOLINE ENGINE RUN-ON

When an engine continues to run, even after the ignition key is turned off, this is called **dieseling** or *run-on*. It is called dieseling because a diesel engine's fuel ignites without a spark plug simply because of heat caused by pressure in the cylinder. In a gasoline engine with a carburetor, if the engine runs on, the cause is usually too high an idle speed or too lean an air/fuel mixture. Run-on can damage an engine. It can result in a broken crankshaft if allowed to continue. Putting the transmission in gear with the brakes applied when shutting off the engine can stop run-on. Fuel injected cars shut off the fuel to the injectors when the key is off so those engines do not run on.

SPARK KNOCKS, CARBON NOISE, ABNORMAL COMBUSTION

When the air/fuel mixture is ignited during normal combustion, a **flame front** travels across the combustion chamber and pushes the piston down in the cylinder (Figure 25.13). The flame front travels at from 50 to 250 meters per second (depending on rpm and load) to *push* the piston down in the cylinder. This is very fast but it is *not an explosion*.

- During cranking, pressures in the cylinder are usually around 140–170 psi.
- Following ignition, cylinder pressures reach about 400 psi at TDC and peak at around 600 psi around 15° after TDC.
- Normal burning of the fuel takes about ⅓₀₀th of a second.
- By the time the piston has traveled halfway down its bore, the flame has consumed most of the fuel and oxygen.

During normal combustion, the air/fuel mixture burns in a controlled manner to steadily force the piston down in the cylinder. Abnormal combustion can cause noise, shock damage, and burning of parts. It can be caused by:

- Cylinder temperatures that are too high
- Too lean an air/fuel mixture
- Engine overheating
- A driver that lugs an engine

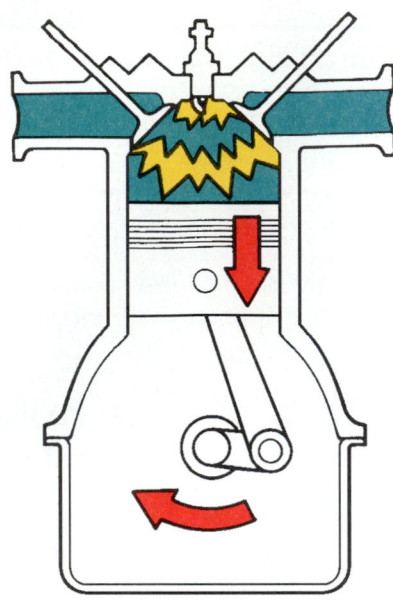

Figure 25.13 A flame front pushes the piston down in the cylinder. *(Courtesy of Ford Motor Company)*

Sounds caused by spark knocks can be mistaken for worn parts. Damage can result if these sounds are ignored.

■ ABNORMAL COMBUSTION

There are two common **abnormal combustion** conditions that can cause *spark knock* and engine damage: **preignition** and **detonation**. Sometimes engine knock is inaudible due to the absorption of the sound by the piston and cylinder head. Other times the rattling that results from detonation can be very loud.

NOTE: *The noise from spark knock is caused by the vibration of the combustion chamber walls.*

Occasional minor spark knock does not usually result in engine damage. During spark knock, there is a measurable loss of power on a dynamometer.

Variations in the flame front during normal combustion, preignition, and detonation are shown in Figure 25.14. Preignition and detonation are slightly different.

COMPRESSION	SPARK IGNITION	COMBUSTION	COMBUSTION CONTINUED	DETONATION

DETONATION

COMPRESSION	PREIGNITION	COMBUSTION & SPARK IGNITION	TWO FLAMES	DETONATION

PREIGNITION

COMPRESSION	SPARK IGNITION	COMBUSTION	COMBUSTION CONTINUED	COMBUSTION COMPLETED

NORMAL COMBUSTION

Figure 25.14 Normal and abnormal combustion. *(Courtesy of Cooper Automotive)*

Preignition

Preignition allows the high heat of combustion to remain for too long on engine parts. The rise in combustion chamber temperature can cause parts such as valves, pistons, and head gaskets to be burned. The burning effects of preignition are shown in Figure 23.15.

Preignition is commonly called *ping*. It occurs when the air/fuel mixture ignites before the regular spark occurs. Causes of preignition include:

■ Spark plugs of too high a heat range
■ Hot spots in the combustion chamber (from sharp edges on valves, head gaskets, or hot carbon particles)

Burn Time

The burn time of the air/fuel mixture is critical in an engine. Excessively lean air/fuel mixtures can result in burned parts similar to the damage caused by preignition. The longer a part is exposed to the flame, the more heat it absorbs. Lean mixtures burn more slowly than rich mixtures, so the metal parts are exposed to the flame for a longer period of time.

NOTES:

■ *One lawnmower manufacturer recommends that regular fuel only be used in its mowers. Premium fuel has a longer burn time so engine valves could be burned.*
■ *Excessively advanced ignition timing can cause detonation and a burned piston, but will not cause an increase in engine temperature.*
■ *Excessively retarded ignition timing will cause an engine to overheat, but will not burn a piston.*

When burned parts are discovered, look for a plugged fuel filter or an electric fuel pump that is failing. The lean mixture that results can raise temperatures excessively.

Detonation

Detonation (self-ignition due to pressure increase) occurs when the unburned portion of the air/fuel mixture explodes violently *after* spark ignition has occurred. The explosion causes parts to break (Figure 25.16).

Detonation is a race between the flame front and heat build-up during combustion. If burning of the fuel is completed before the temperature reaches the point of detonation, abnormal combustion will not occur. Late model engines often have centrally located or multiple spark plugs to help speed up normal combustion.

One common cause of detonation is over-advanced ignition timing. The explosion is instantaneous. This results in a loss of power and can cause engine damage. Ignition timing is best if it is as advanced as possible without causing detonation.

Many computer controlled engines now include a **detonation sensor** (see Figure 75.31), also called a **knock sensor**. When it senses the vibration caused by detonation, the computer retards the spark timing until the vibration goes away. Late model computers actually learn what the best spark timing is by trying to advance the timing until detonation occurs, then backing timing off slightly.

Another common cause of detonation that occurs at low or moderate speeds is the use of a fuel with too low an octane rating in a high compression engine. Octane is a measure of how quickly a gasoline burns in an engine. The faster the burn, the higher the temperature and the greater the chance for detonation. The low octane fuel burns too quickly.

Abnormal combustion can also be caused by cylinder temperatures that are too high, too lean an air/fuel mixture, lugging, and overheating.

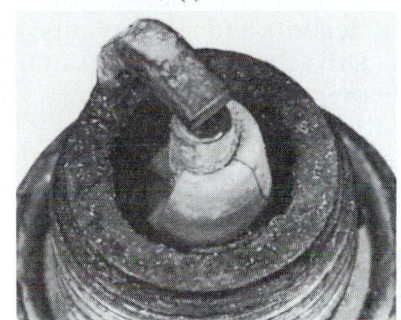

Figure 25.16 The explosive results of detonation. *(Courtesy of Cooper Automotive/Champion Spark Plug)*

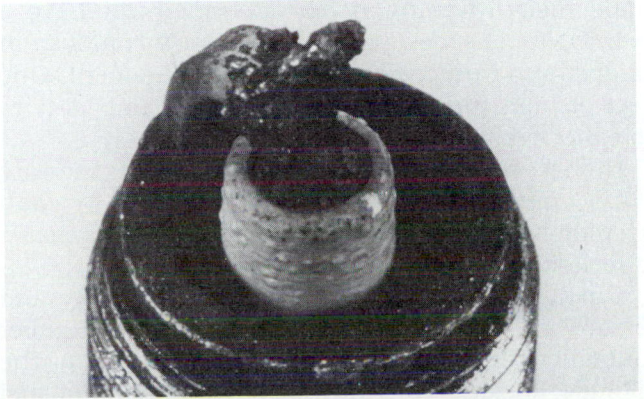

Figure 25.15 The burning effects of preignition. *(Courtesy of Denso Spark Plugs)*

A leading source of detonation is an inoperative EGR valve (exhaust gas recirculation) (see Chapter 41). This can be confirmed while running the engine at high idle (1800–2000 rpm) while the knock occurs. Manually open the valve (if possible). If the noise disappears, the valve is at fault.

 Be careful of hot exhaust components when working around the EGR valve.

A temperature-controlled air cleaner stuck in the heat-on position or a malfunctioning early fuel evaporation system (see Chapter 41) can also cause detonation.

■ EXCESSIVE CARBON BUILDUP

In a car with relatively low mileage, carbon buildup in the combustion chambers can cause an increase in the compression ratio (see Figure 43.17). This problem is not as common since the introduction of unleaded gas, which leaves fewer deposits, but some late model engines develop carbon problems in less than 10,000 miles.

There are two kinds of deposits: oil-based and carbonaceous:

■ **Oil-based deposits** are the traditional gummy, black ones like those sometimes found on intake valves. They are caused when oil and heat come together.

■ **Carbonaceous deposits** are from fuel. They are called *cauliflower deposits* because of their resemblance to the vegetable. These deposits are not as thick as oil deposits and are hard, dry, and tougher to remove. Driveability problems can result from them. Oil companies are developing fuels to minimize this problem.

Service procedures for removing carbon deposits are covered in Chapter 27.

■ REGULAR VERSUS PREMIUM FUELS

Increasing pressure in the combustion chamber can increase engine power. But increasing the pressure in the engine raises the likelihood of engine knocking. Gasoline can be refined so it will not knock as easily. This is called high **octane.**

Most gasoline stations offer three octane grades of unleaded gasoline: *regular* (87 octane), *midgrade* (89 octane), and *premium* (93 octane). Higher compression engines require a higher octane fuel. The amount of octane needed is called the engine *octane number requirement* (ONR) (Figure 25.17).

Octane, a measurement of a fuel's ability to resist explosion during combustion, compares the *anti-knock*

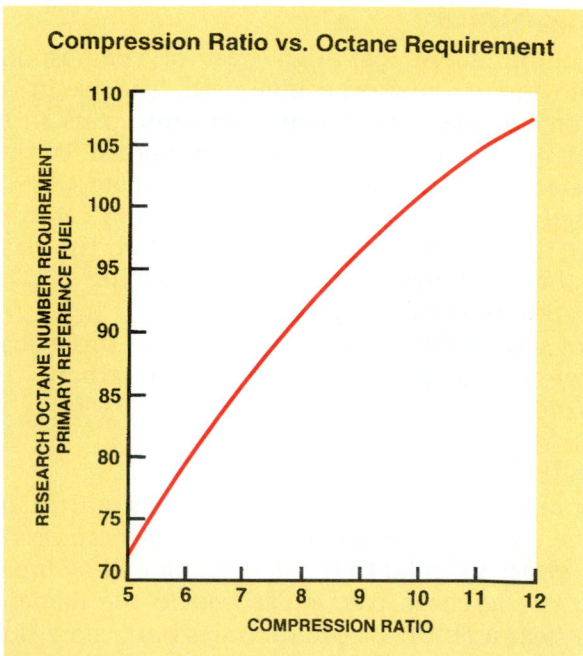

Figure 25.17 Octane requirement goes up with increased compression ratio. *[Courtesy of Changes in Gasoline III, Downstream Alternatives, Inc. [1996]]*

qualities of different fuels. Most cars manufactured since the early 1970s can use regular gas. Check the manufacturer's recommendation.

NOTE: *Using premium gasoline in a car designed to run on regular will not harm it, but the premium costs considerably more and no advantage is gained.*

Several things determine higher or lower octane requirements (Figure 25.18). When ignition timing is advanced, a higher octane number is required. In fact, when an engine pings slightly, retarding the ignition timing two degrees will sometimes take care of the problem. There is a limit to this practice, however. Retarding the timing too far will result in engine overheating and will cause worse fuel economy.

Air/fuel mixtures of 14.7:1 require the highest octane fuel. Anything richer or leaner will have a lower ONR. The design of the engine's combustion chamber also influences the octane requirement. One effect of high turbulence combustion chambers is to decrease ONR.

Higher combustion temperatures will increase ONR. Combustion temperatures are affected by cooling system temperature, inlet air temperature, exhaust gas recirculation, and intake manifold heating.

Changes in outside air temperature, humidity, and pressure all have an effect on the octane number requirement. If you drive in the high altitude of the Colorado Rockies, you will notice that premium fuel has a lower octane number. This is because barometric pressure and the oxygen content of the air is lower.

Lower Octane Requirements

- High altitude (lower barometric pressure)
- High humidity (wet air)
- Low compression ratio
- Part throttle EGR system operation
- Ignition timing retard controlled
- Low inlet air temperature
- Engine is under light load
- Air/fuel mixture richer or leaner than 14.7 : 1

Higher Octane Requirements

- Advanced ignition timing
- High compression ratio
- High air density (due to barometric pressure or turbocharging)
- Low humidity (dry air)
- Engine under higher load or lugging
- High inlet air temperature
- High cooling system temperature
- Lean air/fuel mixture
- 14.7:1 air/fuel mixture

Figure 25.18 Factors influencing octane requirement.

■ SCIENCE NOTE ■

Octane requirements are a measure of how easily gasoline can be ignited. The higher the octane, the harder the fuel is to ignite. Generally, the higher the octane, the greater the number of hydrocarbons containing larger numbers of carbon atoms. More carbon atoms per hydrocarbon means that more oxygen and more heat are needed to burn the fuel. As air temperatures fall, lower octane fuels can be used. This is also true at higher altitudes where there is less oxygen. With increases in pressure come increases in temperature, requiring higher octane fuels.

Many of the variables that affect the octane number requirement are taken into consideration in the design of the engine's fuel and ignition system. In addition to having a knock sensor, a barometric pressure sensor sends signals to a computer. Changes are then made in the ignition timing and air/fuel mixture to compensate.

■ OCTANE STANDARDS

The *American Society for Testing and Materials* (*ASTM*) sets gasoline quality standards. ASTM standards are voluntary, but because of their effects on air quality, EPA and some state standards now require all or part of these standards. The measurement of gasoline octane

quality most often used is the *Antiknock Index* (*AKI*). It is an average of the two ways that octane can be measured; the *research octane number* (*RON*) and the *motor octane number* (*MON*). Both the RON and MON measure the same things but do it at different speeds, temperatures, and spark advances.

The research method (RON) gives a higher reading for the same fuel as the motor method (typically eight to ten numbers higher). It affects low speed knock and engine run-on (dieseling). The motor method (MON) gives a measurement of how much engine knock will be present under heavy loads, such as when passing or climbing hills.

The antiknock index is stated as *(R+M)/2*. This is the number required by law to be listed on the octane decal on the gasoline pump. A third method of rating octane is the *road method*, which is not often used.

■ GASOLINE ADDITIVES

Gasoline additives are expensive and are added in minute quantities to fuel. A small amount is all that is necessary. One figure given in the publication *Changes in Gasoline III* (Downstream Alternatives, Inc., ©1996) is that 100 pounds of additive could be used to treat

Gasoline Additives

Additive	Purpose
Detergents/deposit control additives*	Eliminate or remove fuel system deposits
Anti-icers	Prevent fuel-line freeze up
Fluidizer oils	Used with deposit control additives to control intake valve deposits
Corrosion inhibitors	To minimize fuel system corrosion
Anti-oxidants	To minimize gum formation of stored gasoline
Metal deactivators	To minimize the effect of metal-based components that may occur in gasoline
Lead replacement additives	To minimize exhaust valve seat recession

* Deposit control additives can also control/reduce intake valve deposits.

Figure 25.19 Common gasoline additives and their purposes. *[Courtesy of Changes in Gasoline III, Downstream Alternatives, Inc. [1996]]*

20,000 gallons of gasoline. A large effect on the quality of gasoline has come from the use of detergents and deposit control additives. These are used to keep port fuel injectors from becoming fouled. According to the American Petroleum Institute (API), deposit control additives have been required by law since 1995 in all fifty states. Commonly used gasoline additives are listed in Figure 25.19.

Antiknock compounds such as lead were used in the past. Lead is a poison and also causes problems with catalytic converters used in the car's emission system. Lead was eliminated from the fuel of cars with catalytic converters in the middle 1970s. Government regulations resulted in the elimination of all lead from automobile fuel in 1996. Unleaded fuel can be refined in a more costly manner that results in a higher octane.

■ OXYGENATED FUELS

Oxygenated fuels are gasolines blended with ethers or alcohols. Ethyl alcohol (ethanol) at a 10% concentration and MTBE (methyl-tertiary butyl ether) at a 15% concentration are the most common oxygenates. These fuels are blended to enhance octane. They also provide wintertime control of carbon monoxide because of more complete combustion.

The most used alcohol/gasoline mixture usually contains about 10% ethyl alcohol (ethanol), which can be made from grain. This used to be called **gasohol** in the late 1970s. It was first used to extend gasoline supplies during shortages. Ethanol is about 35% oxygen, so a 10% concentration adds about 3.5% oxygen to the mixture. There is a slight difference in odor noticed during a fill-up. A 10% ethanol mixture will raise an 87 octane fuel by at least 2.5 octane numbers. Figure 25.20 shows the octane values of gasoline and different oxygenates.

Gasohol with less than 10% alcohol does not require any changes to the fuel system, although a fuel filter or two might need to be changed due to the cleaning effect the alcohol has on the fuel tank. When alcohol was first used in automobiles, there was a problem with some of those vehicles produced before 1980. Rubber parts in the fuel system were not designed to be used with alcohol and swelled up. Parts included hoses and the seat of the needle valve in the carburetor float bowl inlet. New materials were used in both manufacturing and the aftermarket to combat these effects.

Oxygenates suspend water in the fuel and tend to keep it from accumulating in the gas tank. Gasoline cannot hold much water, so it separates and accumulates at the bottom of the tank. Alcohol and ether attract and hold water. In fact, the water removers that are on the market that are added to fuel tanks contain alcohol. The following are the amounts of

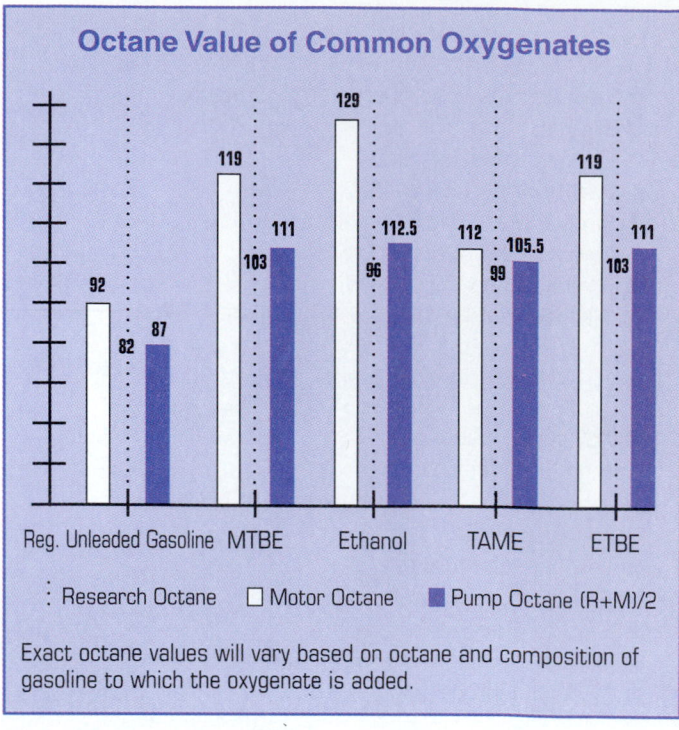

Figure 25.20 Octane values of gasoline and different oxygenates. *(Courtesy of Changes in Gasoline III, Downstream Alternatives, Inc. [1996])*

water that one gallon of each fuel can hold in suspension:
■ Gasoline = 0.5 teaspoon
■ 15% MTBE/gasoline = 1.5 teaspoons
■ 10% ethanol = 12 teaspoons

■ REFORMULATED GASOLINES

The way gasolines are refined can have a large effect on air pollution. Reformulated gasolines clean the air by providing more complete combustion. This kind of gasoline is required by the EPA in U.S. cities with the worst air pollution. The use of reformulated gasoline in older cars can result in damage to older rubber fuel lines. Reformulated gasoline has less energy content so fuel economy will be less, also.

Gasoline Fuel Economy

Several variables can change a vehicle's fuel economy (Figure 25.21). Driving conditions such as air temperature, hill climbing, head winds, and poor driving habits can contribute to a drop in economy. Engine condition and tire air pressure can also be factors. During the winter, fuel economy can be 10 to 20% lower. Fuel economy can be influenced by the energy content of the gasoline. During the summer, gasoline energy content is higher (Figure 25.22). A 10% ethanol mixture contains 3.4% less energy than pure gasoline.

Factors That Influence Fuel Economy of Individual Vehicles

Factor	Fuel Economy Impact	
	Average	Maximum
Ambient temperature drop from 77°F to 20°F	−5.3%	−13.0%
20 mph head wind	−2.3%	−6.0%
7% road grade	−1.9%	−25.0%
27 mph vs 20 mph stop and go driving pattern	−10.6%	−15.0%
Aggressive versus easy acceleration	−11.8%	−20.0%
Tire pressure of 15 psi versus 26 psi	−3.3%	−6.0%

Figure 25.21 Fuel economy variables. *[Courtesy of Changes in Gasoline III, Downstream Alternatives, Inc. [1996]]*

	Summer-Grade Btu	Winter-Grade Btu
Maximum	117,000	114,000
Minimum	113,000	108,500
%	3.4	5.0

Difference between summer maximum and winter minimum: 7.26%

Figure 25.22 Gasoline energy content. *[Courtesy of Changes in Gasoline III, Downstream Alternatives, Inc. [1996]]*

■ SCIENCE NOTE ■

During the combustion of gasoline, the hydrocarbon fuel is converted by oxygen to CO_2 and H_2O. Chemically, this process is called oxidation (the addition of oxygen to a chemical compound). When gasoline is burned, oxygen is added to the hydrocarbon molecule (a molecule contains only hydrogen and carbon) to form CO_2 and H_2O. The heat produced from the combustion process comes from the addition of oxygen. This is the reason using gasohol results in fewer miles to the gallon. Gasohol contains 10% ethyl alcohol (CH_3CH_2OH). Since ethyl alcohol already contains oxygen, it has been partially oxidized already. Therefore, it produces less heat during combustion. In fact, running your car on pure alcohol would require 75% more fuel than running it on gasoline.

Methanol

Methanol is methyl alcohol that can be burned in an internal combustion engine. But it produces only about ½ the energy that gasoline does. So the fuel system needs to be adjusted to provide an air/fuel ratio of about 6.5:1. Methanol has been used for years in race cars and will probably become more widely used in passenger cars. It can be made from coal, natural gas, oil shale, wood, or garbage.

Characteristics of Alcohol Fuels

■ A major disadvantage to alcohol fuels when compared to gasoline is that they are invisible when burning. That is why race car drivers wear fire suits and emergency crews douse them quickly with fire extinguishers after an accident, even though the racing fans cannot actually see the fire.

■ Methanol is very corrosive and is poisonous. It attacks aluminum, some plastics, and other materials. Gas tanks must be lined with resistant materials and fuel systems must be totally made of noncorrosive materials.

■ DIESEL ENGINES AND DIESEL FUEL

Diesel engines burn diesel oil as fuel. A diesel engine is called a compression ignition engine. A four-stroke diesel engine operates in a similar manner to a gasoline engine. The difference is that ignition in a diesel is controlled by the injection of fuel into the cylinder.

A diesel has a very high compression ratio (about 16:1 to 20:1) (see Figure 17.4). If fuel were drawn in and compressed as it is in a gasoline engine, it would self-ignite before the piston could reach TDC. In a diesel, the mixture is injected by a high pressure injector at the instant that ignition is desired (see Figure 16.13). The compressed, hot air in the cylinder causes the fuel to vaporize and burn.

Diesel fuel is light oil that is refined as part of the same process that makes gasoline. It has several properties that make it useful as a fuel.

Diesel Volatility

While gasoline has high volatility, diesel fuel has low volatility. It is safe at room temperature and only gives off vapors if heated. Besides their reliability and fuel economy, one major reason that diesel engines are so heavily used in boats is that they are much safer than gasoline in case of a fuel leak.

Diesel Grades

There are two grades for automotive diesel fuel.

■ Grade number 1-D, known as *number 1 diesel* is more volatile, is thinner, and is used in very low temperatures.

■ Grade number 2-D, known as *number 2 diesel* is of a lower volatility and is used for most automotive driving conditions.

Diesel fuel must be able to flow easily through the fuel system and be sprayed by the injectors, so it must be of low viscosity (Figure 25.23). Viscosity must be just high enough to provide lubrication to the fuel injectors and fuel pump. Number one diesel is only used in very cold weather because number two tends to become too thick as it gets colder. But number one might not provide sufficient lubrication.

Cloud Point

Diesel fuels contain some paraffin (wax). In very cold weather, the paraffin can separate from the fuel. This is called the **cloud point** because the fuel will appear cloudy when the wax separates out. The paraffin can clog fuel filters, causing the engine not to run.

Cetane Rating

The temperature at which a diesel fuel will ignite is called its ignition point. Diesel has good ignition quality, which means that it will burn without detonation soon after it is injected into the cylinder.

The **cetane number** of a diesel fuel describes how easily the fuel will ignite. The test number is derived by comparing how well a fuel burns with a sample of cetane. Cetane is a colorless, liquid hydrocarbon with excellent ignition quality. Its cetane rating is 100.

With gasoline octane ratings, the higher the octane number, the more resistant the fuel is to knocking. Diesel cetane ratings work in an opposite fashion (Figure 25.24). The higher the cetane rating, the easier it ignites (Figure 25.25). Fuels with a high cetane number burn as soon as they are injected, so no spark knock occurs. Number 45 is an average cetane value

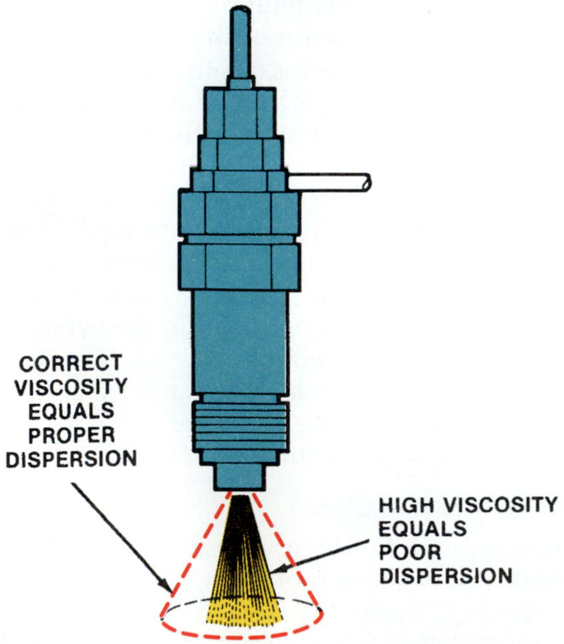

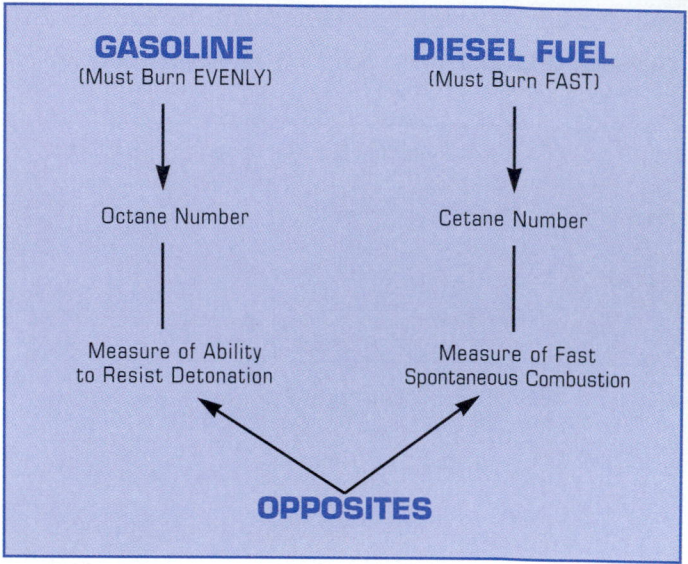

Figure 25.24 Octane and cetane numbers are opposites. *[Reproduced by permission of Deere & Company. ©1992, Deere & Company. All rights reserved]*

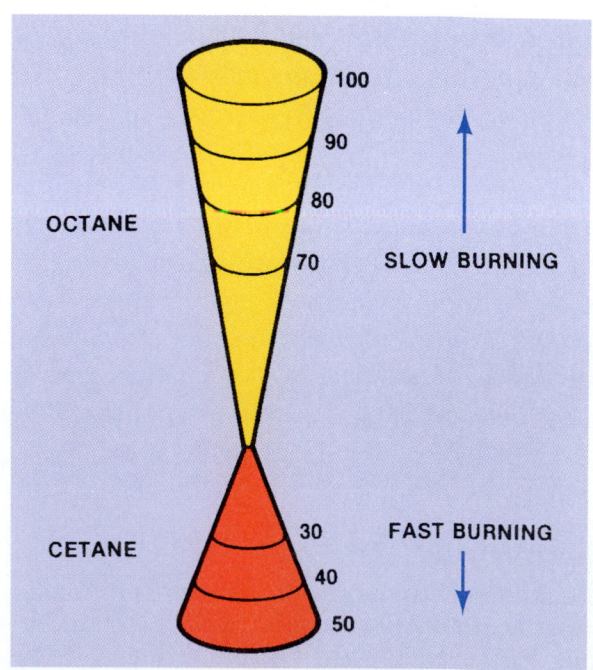

Figure 25.23 Diesel fuel must be of the correct viscosity. *[Courtesy of Ford Motor Company]*

Figure 25.25 Gasoline with a higher octane rating burns slower. Diesel with higher cetane rating burns faster. *[Courtesy of Ford Motor Company]*

for #2 diesel and is what most auto manufacturers recommend.

The lower the cetane number, the higher the temperature required to ignite the fuel. If the cetane number is too low for the conditions, the fuel will take too long to light and will build up in the cylinder. Engine knocking can occur under this condition when the fuel is burned in too much quantity and out of control. Figure 25.26 lists some of the problems caused by a fuel with too low a cetane rating.

Diesel Maintenance

Water is an enemy to diesels and must not be allowed to accumulate in the fuel system. It can rust the internal parts of the fuel injection system and can plug filters. Some diesel cars have a water detector in the fuel tank that completes a circuit to a dash light when the water level in the tank reaches a certain point. Other cars have fuel and water separators, which must be periodically drained.

More frequent oil changes are required with diesels. They generally have a larger capacity oil pan too. It is not uncommon to find oil change intervals of 3000 miles. If a car is driven in severe service conditions, the oil will need to be changed more often. Under a heavy load, a diesel makes soot, which is abrasive and is hard on engine parts. While heavy trucks have transmissions with many gears and are designed to pull heavy loads, passenger cars have only the normal number of gear ranges.

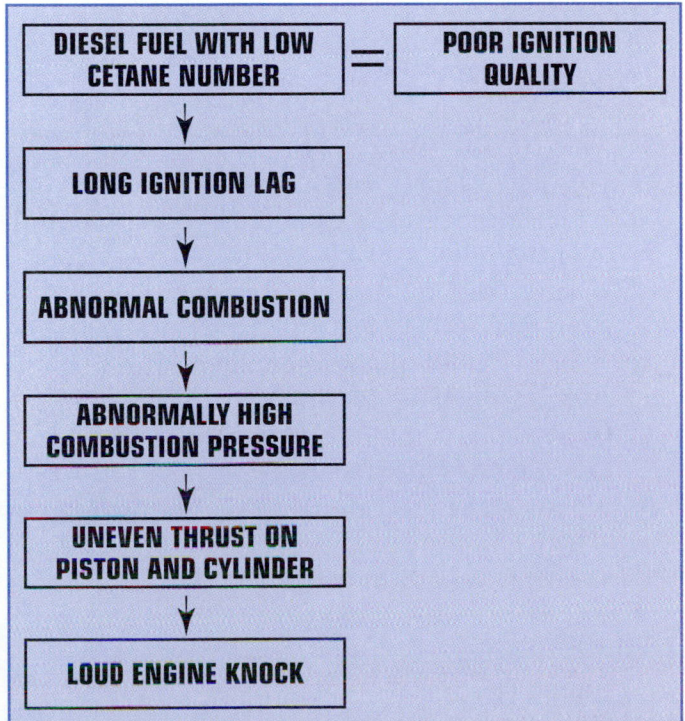

Figure 25.26 Problems caused by a fuel with too low a cetane rating. *(Courtesy of Ford Motor Company)*

■ ALTERNATIVE FUELS

Alternative fuels are those other than gasoline and diesel fuel. LP gas and alcohol are currently in use in automobiles and trucks. Hydrogen may be used in the future.

LP Gas

Liquified petroleum gas (LPG) is similar to gasoline chemically and is a product of gasoline refining. It can also be obtained from natural gas. LPG is mostly propane but contains a small percentage of butane (up to 8%). Because of this, many people simply refer to LPG as propane. LPG is a vapor above −40°F, rather than a liquid. It is called "liquified" because it is stored as a liquid in a bottle under pressure. The pressure increases its boiling point and turns it into a liquid (Figure 25.27). It requires a different fuel system to meter the gas vapor into the engine. Because the fuel is already under pressure, no fuel pump is required. When a carburetor is used, it is similar to a gasoline carburetor. The main difference is that there is no float bowl (fuel reservoir) (Figure 25.28).

LPG burns cleanly and produces fewer emissions than gasoline. It is often used in fleets, such as utility companies, taxis, buses, and for indoor machinery such as forklifts. Its heat energy per volume is less than that of gasoline. But it has higher octane and therefore, can be used in engines with higher compression.

Compressed Natural Gas

Compressed natural gas (CNG) is another fuel that is popular with fleets. It is stored under pressure (3000 psi) in a large cylinder. The cylinders are very strong and proponents of natural gas fuels have done many tests to prove their safety. The size of the cylinder is one reason why it is not in common use in passenger cars. Another reason is that it takes a long time to fill a compressed gas tank compared to filling an ordinary

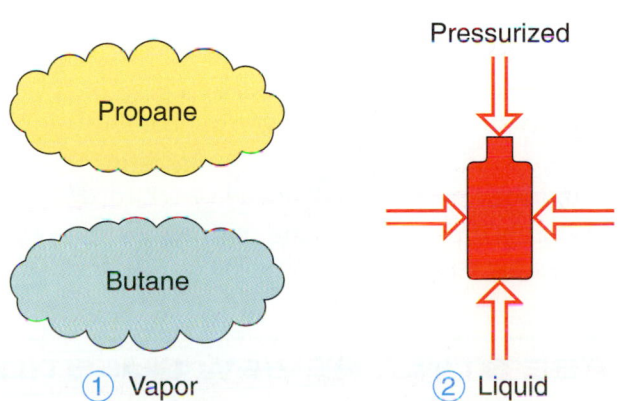

Figure 25.27 Pressure increases the boiling point of a gas and turns it into a liquid. *(Reproduced by permission of Deere & Company. ©1991, Deere & Company. All rights reserved.)*

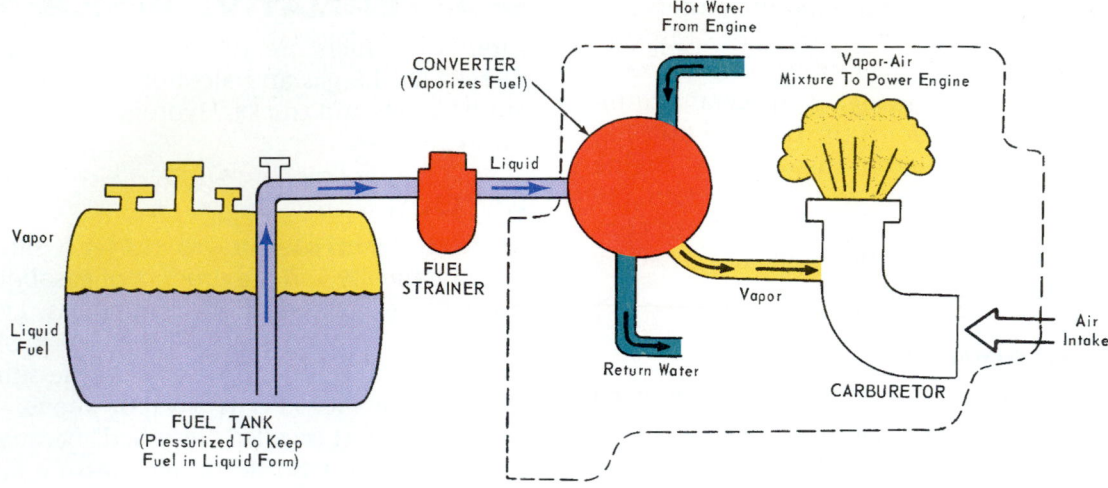

Figure 25.28 A liquified petroleum gas (LPG) carburetor has no float bowl. *(Reproduced by permission of Deere & Company, ©1992. Deere & Company. All rights reserved.)*

gasoline tank. Natural gas is the cleanest burning of the alternative fuels, with 99% less carbon monoxide and 85% less hydrocarbon emissions than gasoline. Because it burns cleanly, the oil change interval can be increased and spark plugs last longer. Natural gas is available in vast quantities in North America.

Hydrogen

Hydrogen is one of the most abundant elements in the universe. It is highly flammable and is a promising fuel for the future. It is an ideal fuel, producing no emissions other than water and carbon dioxide, which is a nonpolluting and nonpoisonous gas. The sun is an example of burning hydrogen.

■ SCIENCE NOTE ■

Hydrogen is difficult to store due to the small size of its molecule. It can be produced from water (H_2O) by electrolysis. Electrolysis is a process where electrical current is applied to water. Producing hydrogen in this manner is expensive. Hydrogen is not very reactive unless it is subjected to high temperatures or pressures. Hydrogen and oxygen mixtures are very explosive when sparked or heated to high temperatures.

■ REVIEW QUESTIONS

1. What is the term for how easily a fuel evaporates?

2. What are the two types of abnormal combustion?

3. What is the term that describes a gasoline's ability to resist explosion during combustion?

4. What kind of ignition timing can cause an engine to overheat, advanced or retarded?

5. What is it called when an engine continues to run after its ignition is shut off?

6. Which kind of fuel system prevents engine run-on, carburetor or fuel injection?

7. What is the term that describes how easily a diesel fuel will ignite?

8. Which engine requires more frequent oil changes, diesel or gasoline?

9. What happens to propane or natural gas when they are pressurized?

10. What kind of fuel can natural gas be used to make?

■ ASE STYLE REVIEW QUESTIONS

1. Technician A says that cars with fuel injection are not as likely to vapor lock. Technician B says that when timing is correct, the fuel explodes and pushes the piston down in the cylinder.

Who is right?

a. Technician A   b. Technician B
c. Both A and B   d. Neither A nor B

2. Technician A says that detonation can result in burned parts. Technician B says that preignition can result in broken parts. Who is right?
   - **a.** Technician A
   - **b.** Technician B
   - **c.** Both A and B
   - **d.** Neither A nor B

3. Technician A says that a car that has a rich air/fuel mixture will surge on the highway. Technician B says a lean mixture can result in burned parts. Who is right?
   - **a.** Technician A
   - **b.** Technician B
   - **c.** Both A and B
   - **d.** Neither A nor B

4. Technician A says that alcohol fuels can carry more water than straight gasoline. Technician B says that alcohol fuels have a higher octane number. Who is right?
   - **a.** Technician A
   - **b.** Technician B
   - **c.** Both A and B
   - **d.** Neither A nor B

5. Technician A says that a 9:1 air/fuel mixture is lean. Technician B says that retarded ignition timing can cause an engine to overheat. Who is right?
   - **a.** Technician A
   - **b.** Technician B
   - **c.** Both A and B
   - **d.** Neither A nor B

CHAPTER 26

Fuel System Fundamentals

■ OBJECTIVES

Upon completion of this chapter, you should be able to:

✔ Explain the operation of the various carburetor systems.

✔ Compare fuel injection to carburetion.

✔ Identify the different types of fuel injection.

✔ Describe the design and function of electronic fuel injection components.

✔ Understand how a computer feedback system works.

■ KEY TERMS

filter strainer
sock
intake manifold vacuum
atomization
vaporization
venturi
feedback carburetors
barrel
pulse width
throttle-body injection (TBI) system
port injection
speed density systems
air density system
sensors
actuators
open loop
closed loop fuel control

■ INTRODUCTION

The fuel system must be able to deliver a proper mixture of air and fuel to the engine so that it can be burned efficiently. It must also be able to store enough fuel so that the car can complete a trip of a few hundred miles. This chapter provides an overview of the operation, uses, and advantages of the different types of fuel delivery systems.

■ FUEL SYSTEM

The entire fuel system from the tank to the intake valve is called the fuel delivery system. It includes a storage tank, a pump, one or more filters, fuel lines, and hoses, used to deliver fuel to the engine's carburetor or fuel injection system (Figure 26.1). The job of the fuel induction system, whether it be fuel injection or carburetion, is to provide the engine with the correct mixture of burnable air/fuel mixture in all driving ranges from idle to wide open throttle (WOT). The combustion of gasoline and other fuels is covered in Chapter 25. The operation of carburetors and the

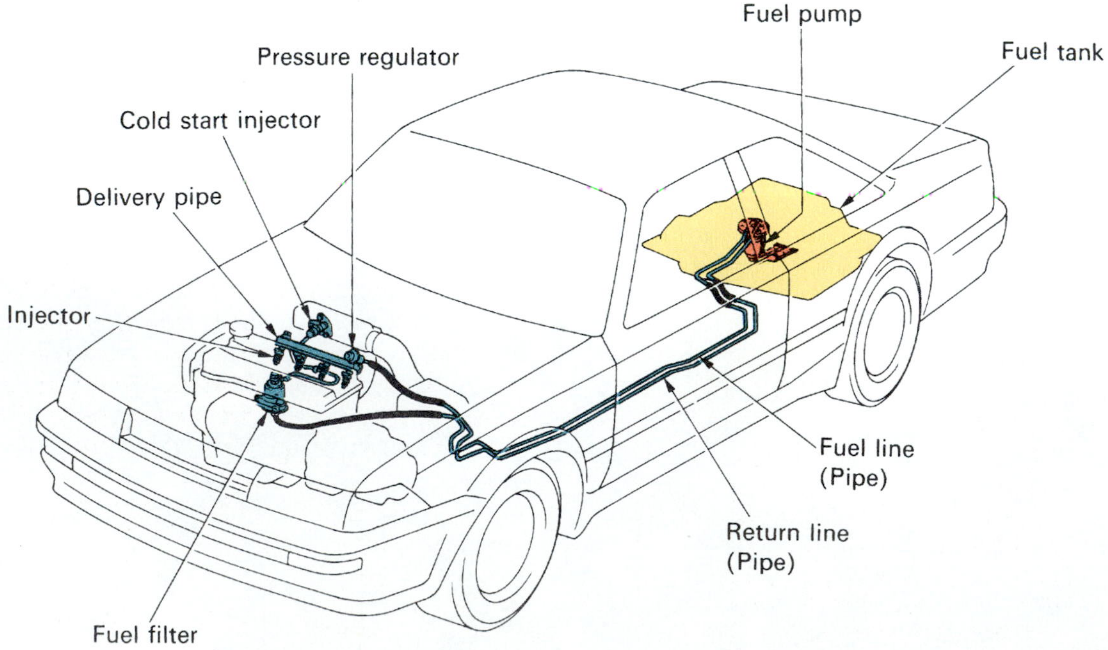

Figure 26.1 Parts of a typical fuel system.

major types of fuel injection systems are covered in this chapter, following coverage of their fuel supply systems.

■ FUEL TANKS

Fuel tanks on most cars hold between 12 and 20 gallons of fuel. Tanks are made of corrosion resistant *galvanized* steel or plastic. Galvanized metal has zinc applied to it by electrolysis in much the same manner that chrome is applied to bumpers. This makes it resistant to rust.

The tank is made in two sections. Before the two halves are welded together, a vertical sheet metal *baffle* is installed in the tank (Figure 26.2). Fuel has a tendency to slosh back and forth in the tank when the car is driven. The baffle prevents this.

The fuel tank is usually hung in the back of the vehicle. On older vehicles it was located under the trunk. In an effort to make vehicles safer in the event of a rear end collision, gas tanks are now usually in front of the rear axle.

Hanger straps hold the tank in place. The fuel *pickup tube* is installed through a hole in the top of the tank. It is usually positioned about ½" from the bottom of the tank so that dirt and water will not be drawn into the fuel system. Water and gasoline do not mix. Because water is heavier than gasoline, it accumulates at the bottom of the tank where it can cause problems in below freezing temperatures.

A fuel tank *cluster* assembly includes a fuel pickup tube, the fuel gauge sending unit and sometimes, a fuel pump. Most cars have an *in-tank filter* installed at the end of the pickup tube (Figure 26.3). It is usually made of woven plastic, sometimes called a **filter strainer.** Sometimes, over a period of years these filters plug up, requiring replacement. When an older car has a cloth-type filter, this is called a **sock.** The term is often used to describe a strainer.

The fuel gauge has a float attached to a sliding contact. As the float moves up or down the contact position changes, varying the signal that the fuel gauge receives. An electric fuel pump that runs while immersed in gasoline is included in most late model fuel tank clusters.

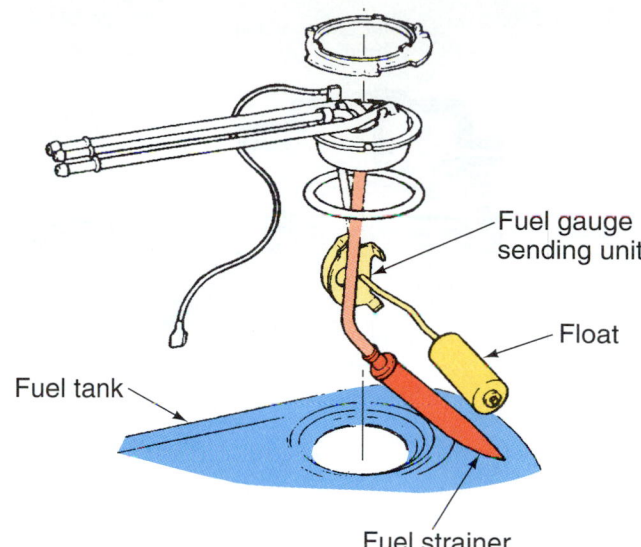

Figure 26.3 A fuel tank cluster with a filter. *(Courtesy of General Motors Corporation, Service Technology Group)*

The tank also includes expansion and overfill protection, covered in Chapter 40.

■ FUEL LINES, HOSES, AND FITTINGS

To transport fuel from the tank to the engine, steel lines made of seamless tubing run the length of the frame. Wherever there is a flexible connection, a hose is used. Fuel tubing, fittings, hoses, and clamps are covered in other chapters.

■ MECHANICAL FUEL PUMP

Earlier carbureted cars usually use a low pressure mechanical fuel pump with a *diaphragm* driven by an *eccentric* on the engine's camshaft. Operation of a fuel pump is shown in Figure 26.4. There are two *one-way check valves*: one that allows fuel to flow into the pump and the other that allows fuel to flow out of the pump.

The pump diaphragm is moved up by the movement of its *rocker arm* on the eccentric. This draws fuel into the pump past the *inlet check valve*. The diaphragm cannot be forced back down by the eccentric. A spring pushes the diaphragm back down when the float level in the carburetor drops. This forces fuel out of the pump past the *outlet check valve*. A spring on the pump rocker arm keeps it in contact with the eccentric.

■ ELECTRIC FUEL PUMP

Fuel for both mechanical and electronic fuel injection systems is supplied by an electric fuel pump (Figure 26.5). A few of the later carburetor systems used electric fuel pumps too. Electric fuel pumps are of either the external or internal (inside the fuel tank) types.

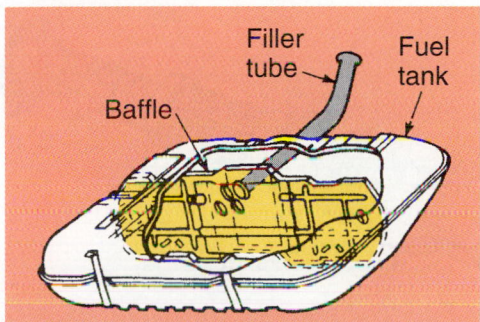

Figure 26.2 A gasoline tank with a baffle.

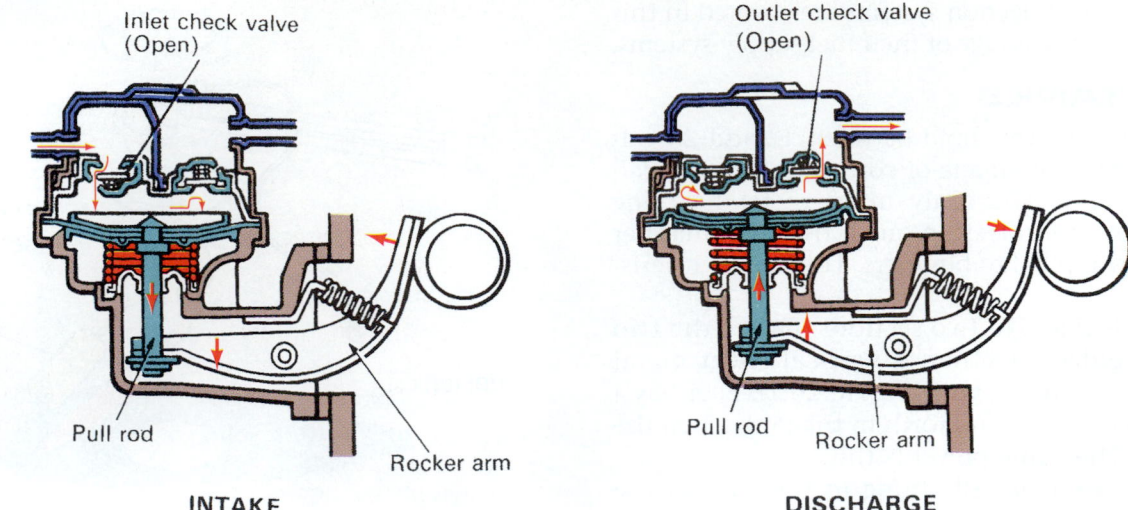

Figure 26.4 Mechanical fuel pump operation.

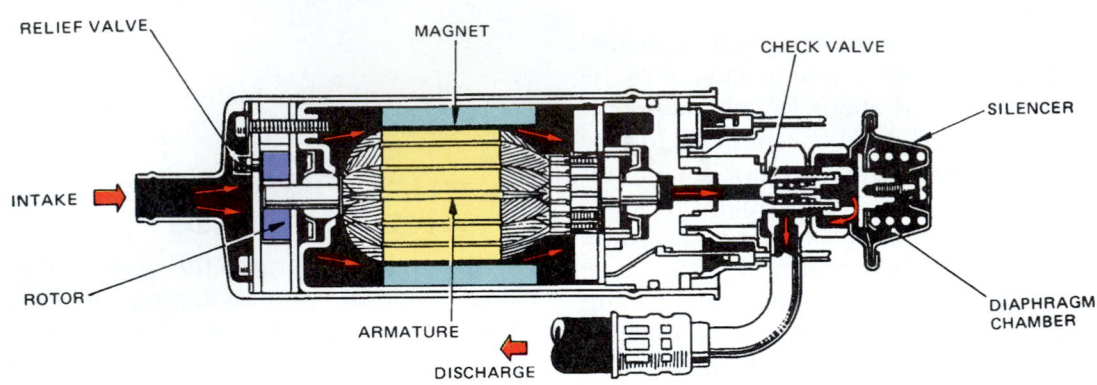

Figure 26.5 An electric fuel pump.

Fuel from the pump flows in a fuel rail loop between the engine and the fuel tank (Figure 26.6). A pressure regulator controls fuel pressure within the system so that an exact, predetermined amount of fuel will be injected each time an injector is open.

◼ FUEL FILTERS

Fuel filters can be located in a fuel line, in the tank as described earlier, in a carburetor, or in any combination of these. A carburetor often has a filter behind the fitting at its fuel inlet (Figure 26.7).

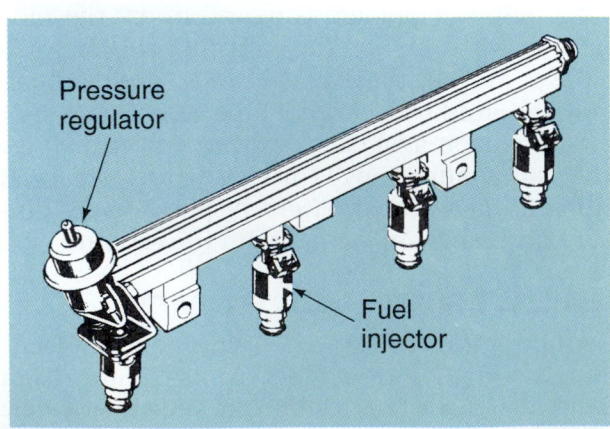

Figure 26.6 A fuel rail assembly for electronic fuel injection. [*Courtesy of General Motors Corporation, Service Technology Group*]

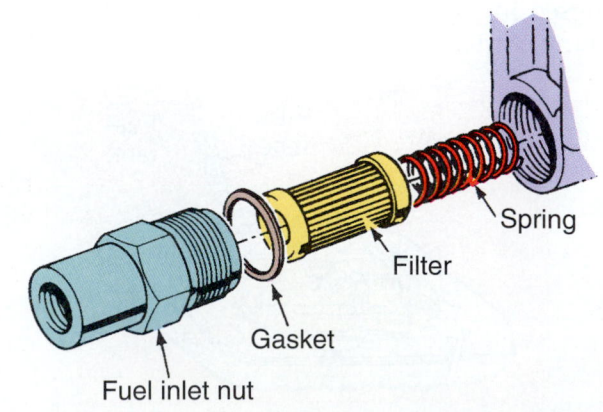

Figure 26.7 A carburetor inlet filter. [*Courtesy of General Motors Corporation, Service Technology Group*]

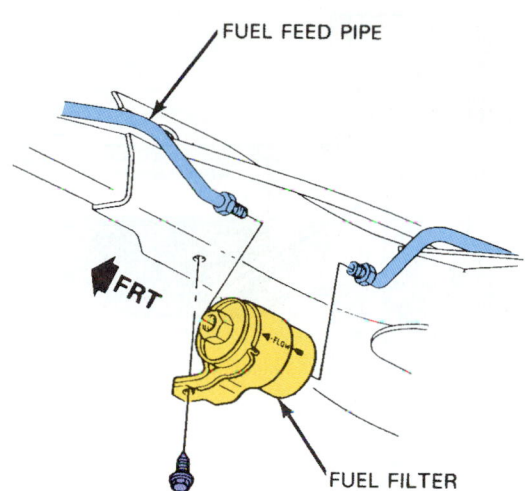

Figure 26.8 An external fuel filter. *(Courtesy of General Motors Corporation, Service Technology Group)*

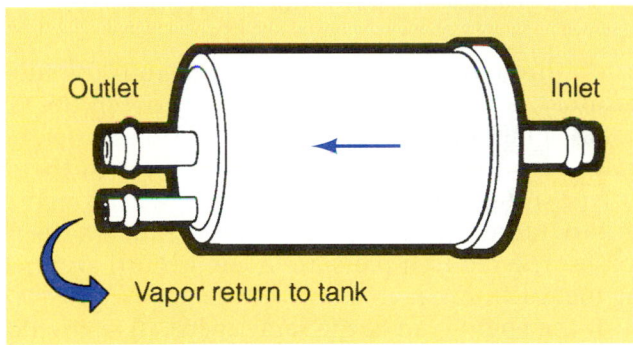

Figure 26.9 A fuel filter with a vapor return outlet. *(Courtesy of AlliedSignal Automotive Aftermarket)*

Fuel filters that are installed on the outlet side of a fuel pump are called *outlet filters* (Figure 26.8). Engine heat causes fuel to vaporize, which can result in vapor lock in the filter. Some fuel filters have vapor return lines to allow for expanding vapors and fuel to return to the tank (Figure 26.9).

Fuel injection systems require larger, heavier duty filters than carburetors. They filter out smaller particles of dirt while still allowing the pump to supply a substantial amount of fuel.

■ CARBURETORS AND FUEL INJECTION

Carburetors are found on older cars. For better control of vehicle emissions, most engines since the middle of the 1980s have been equipped with fuel injection systems. As government exhaust emissions and fuel economy standards became more stringent, the increasing complexity of carburetors made them more and more costly. Fuel injection became competitive with them and is the prevalent system today. Carburetor basic operations are covered first, followed by an overview of fuel injection systems.

■ CARBURETORS

There are still many cars on the road that have carburetors. Later model carburetors (since the start of emission controls) are quite complicated, but all carburetors operate on the same basic principles. The following material will give you a basic understanding of those principles.

There is a difference in pressure between outside air (*atmospheric pressure*) and the pressure inside of the intake manifold (**intake manifold vacuum**). As the pistons in the engine move down in their cylinders, a lower pressure area develops in the intake manifold (Figure 26.10). Outside air, because it is under higher pressure, moves into the intake manifold. Before it reaches the manifold, the air must first travel through the carburetor.

Fuel is delivered to the cylinders as a mist atomized in air so that it can vaporize and burn easily. **Atomization** is the name of the process by which the fuel is suspended in the air in a mist of tiny droplets (Figure 26.11). Atomized fuel has more exposure to air, which helps it to vaporize.

Liquid fuel must become a vapor before it can burn in the cylinders. **Vaporization** occurs when atomized fuel turns into a gas (vapor). The carburetor and intake manifold serve this function. The carburetor's job is to atomize air and fuel in the proper proportion. The carburetor is mounted on top of an intake manifold that vaporizes and distributes the mixture to the intake valve ports.

The inside of the carburetor barrel has a **venturi** that restricts air flow (Figure 26.12). The smallest area of the venturi has the least amount of pressure. This is where the main *nozzle* is positioned. The lower end of

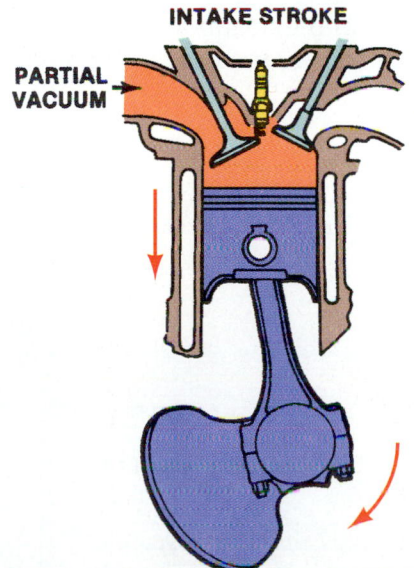

Figure 26.10 A lower pressure area develops in the intake manifold as the piston moves down. *(Courtesy of Ford Motor Company)*

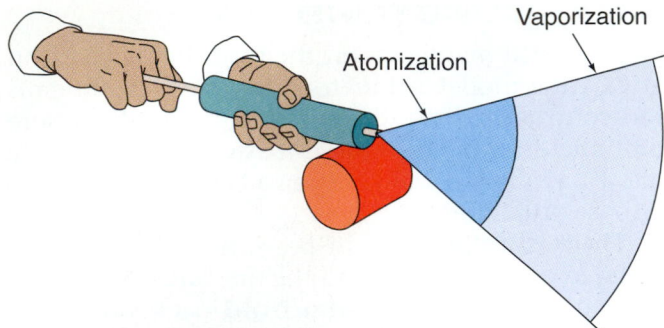

Figure 26.11 Atomization and vaporization.

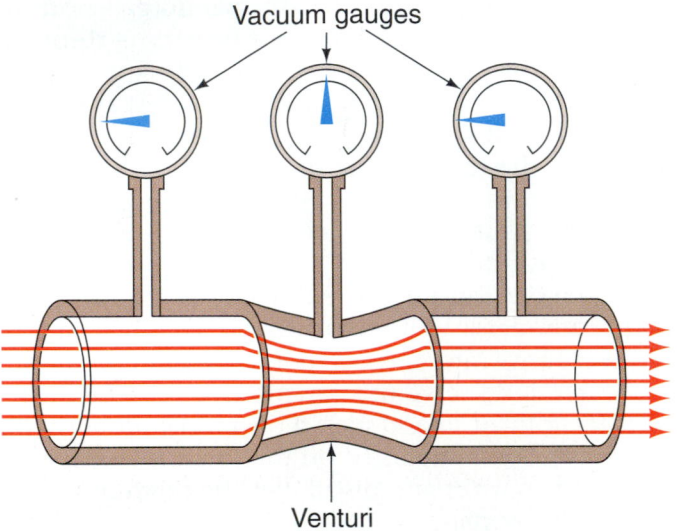

Figure 26.12 The center gauge shows more vacuum (less pressure) in the venturi.

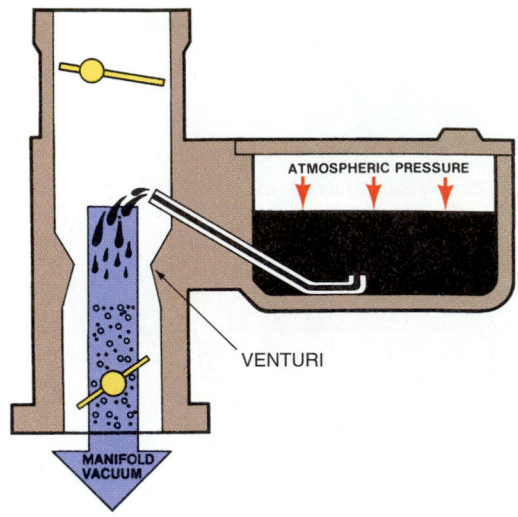

Figure 26.13 Higher pressure in the float bowl and lower pressure in the venturi results in fuel being drawn into the stream of air flowing through the carburetor.

the passage to this nozzle goes to the *float bowl*, which serves as a reservoir for the fuel.

The gasoline in the float bowl is under atmospheric pressure. When the engine is running or cranking, air rushes through the carburetor venturi into the intake manifold. The difference in pressure between the float bowl (higher) and the venturi (lower) causes fuel to be drawn into the stream of air flowing through the carburetor (Figure 26.13). As more air flows through the venturi, more vacuum is exerted on the fuel in the fuel nozzle. The reduced pressure in the venturi also helps the fuel to atomize.

If the engine ran at the same speed all of the time, no provision for changing the amount of air and fuel intake would be needed. But the flow of air and fuel needs to be changed to allow for differences in engine load and speed. Figure 26.14 shows the parts of a basic carburetor. Air flow through the carburetor is changed by opening the *throttle plate*, a butterfly valve positioned in the bottom section of the carburetor throttle body.

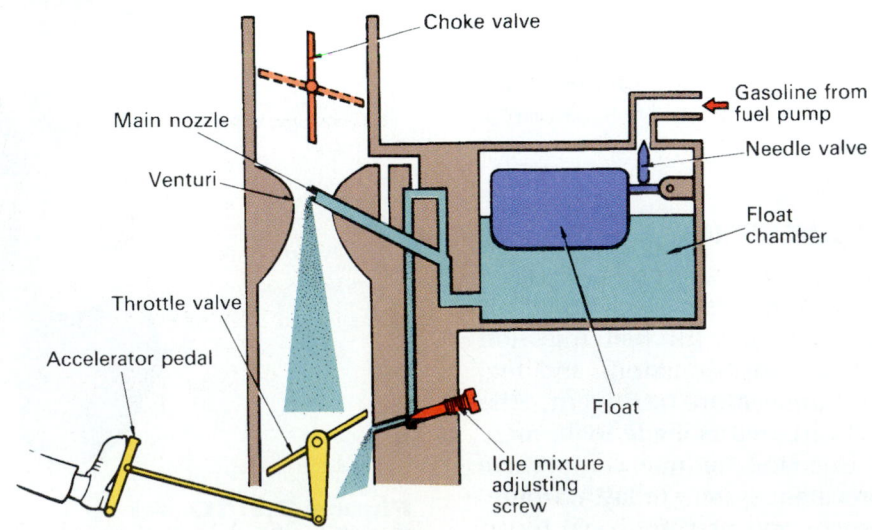

Figure 26.14 Basic parts of a carburetor.

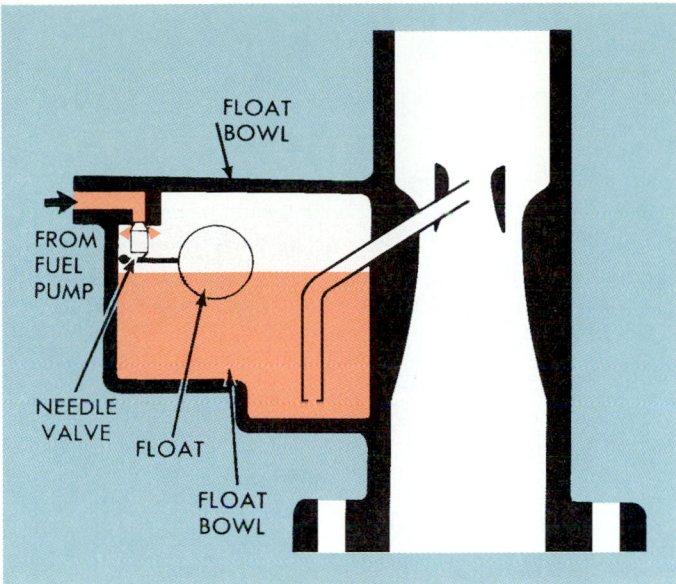

Figure 26.15 The float circuit. *(Courtesy of General Motors Corporation, Service Technology Group)*

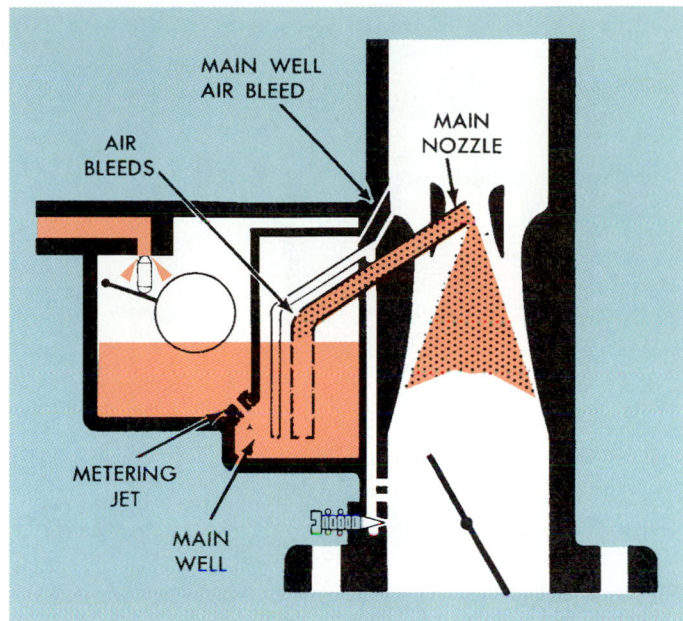

Figure 26.16 The main metering circuit. *(Courtesy of General Motors Corporation, Service Technology Group)*

When the driver depresses the accelerator pedal, the throttle plate opens. This allows more air/fuel mixture to enter the engine, raising engine rpm. Closing the throttle causes engine speed to drop.

Carburetor Circuits

Carburetors have several circuits: the float circuit, the idle and low speed circuits, the main metering circuits, the power enrichment circuit, the accelerator pump circuit, and the choke circuit.

The *float circuit* (Figure 26.15) works the same way that a toilet does; when the level of fuel rises to a predetermined level, the float closes a *needle valve* against its *seat*, stopping further flow from the fuel pump. As the engine uses fuel the float level drops, opening the needle valve and allowing fuel to flow in again from the pump.

The *float level* is important. If it is too high, the air/fuel mixture will be too rich; if too low, the mixture will be too lean The *main metering circuit* (Figure 26.16) includes a replaceable *main jet*. It has a specified opening that meters the amount of fuel that can enter the fuel nozzle from the float bowl. Flow through the jet determines how rich the air/fuel mixture will be. Fuel flow starts at the main jet (in the bottom of the float bowl) and ends up at the main nozzle, already premixed with air. At higher speeds, air needs to be added to the fuel because the air flowing through the venturi will not be as dense. *Air bleeds* add air to the fuel and also prevent siphoning action once the engine is off.

The *idle circuit* has a small tube that goes from the float bowl to a port in the base of the carburetor (Figure 26.17). The tube has an air bleed above the venturi that allows some air to be drawn in. The bottom of the tube ends at a small hole, the *idle port*, located just below the throttle plate.

The idle port allows a small amount of air and fuel to be metered into the intake manifold when the throttle is closed. When vacuum is present at the idle port, fuel (in proportion to the vacuum) is drawn from the float bowl (together with air from the idle bleed).

The *accelerator circuit* has a small, spring-operated piston (Figure 26.18) or diaphragm called an *accelerator pump*. It provides an extra squirt of fuel when the vehicle is accelerated quickly, such as when passing another car.

The *power enrichment*, or *high speed circuit* (Figure 26.19), allows more fuel into the engine than the main

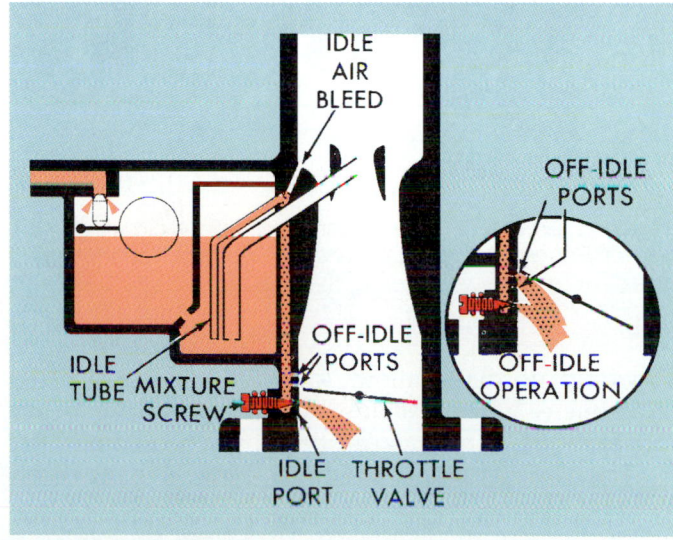

Figure 26.17 The idle circuit. *(Courtesy of General Motors Corporation, Service Technology Group)*

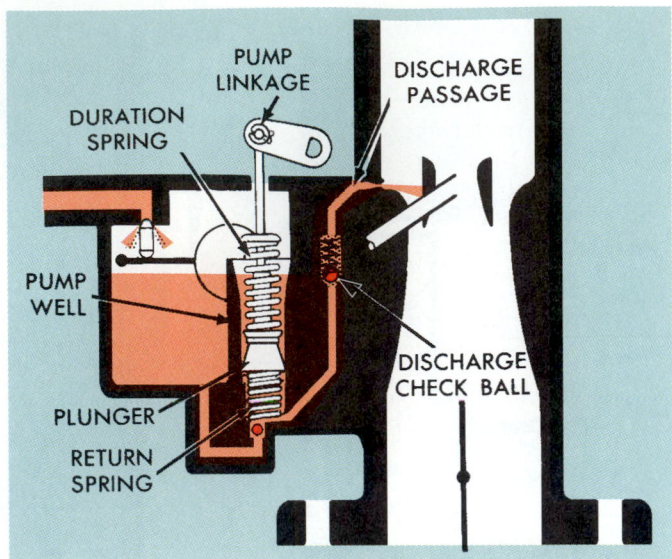

Figure 26.18   The accelerator pump circuit. *(Courtesy of General Motors Corporation, Service Technology Group)*

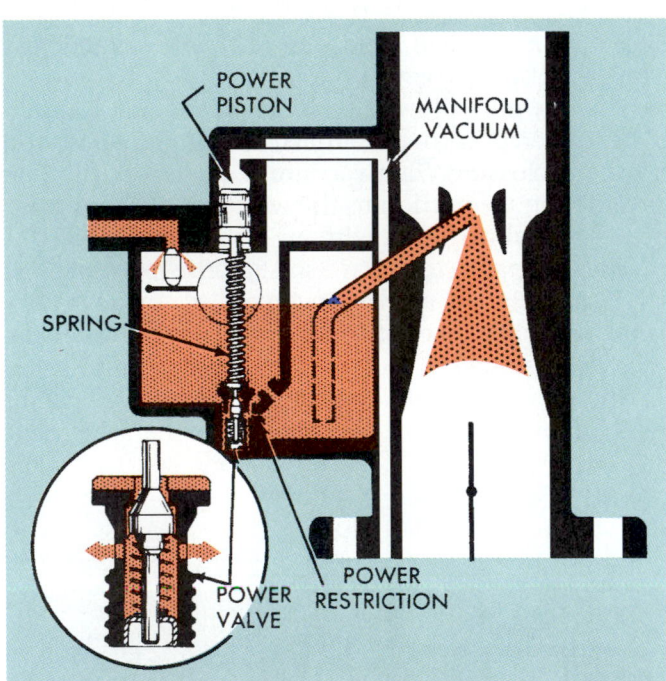

Figure 26.19   The power enrichment circuit. *(Courtesy of General Motors Corporation, Service Technology Group)*

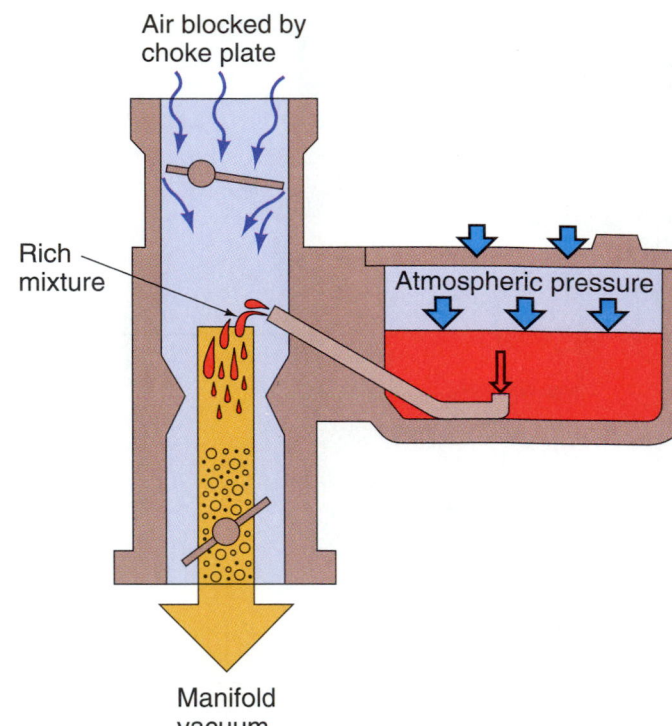

Figure 26.20   The choke is a butterfly valve at the top of the air horn. *(Courtesy of Ford Motor Company)*

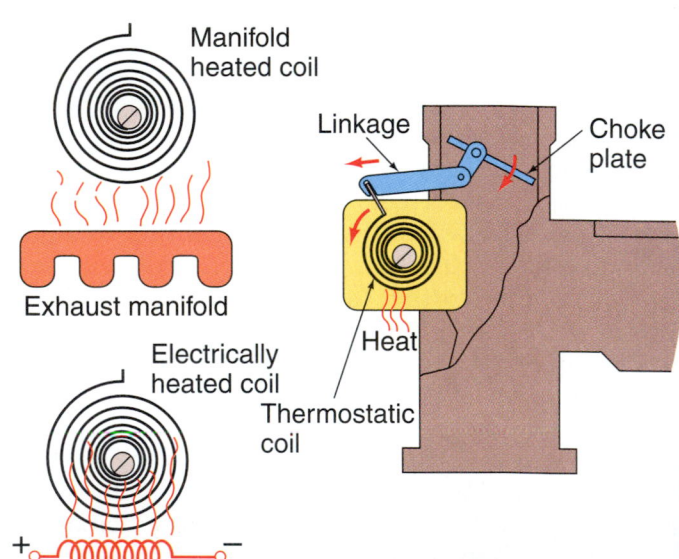

Figure 26.21   The choke bimetal spring is heated by the exhaust manifold or by an electrically heated coil. *(Courtesy of Ford Motor Company)*

jet would normally allow. A *metering rod* or a *power valve* opens when vacuum is low (as during heavy acceleration). Extra fuel can then bypass the main jet.

The *choke circuit* has a butterfly valve like the throttle valve that is located in the top of the carburetor above the venturi (Figure 26.20). When the choke is "closed," incoming air is restricted. The result is a rich air/fuel mixture that helps the engine run better when

cold. The butterfly valve shaft is offset from the center. This allows it to open with the increased air flow of higher rpm.

The position of the choke butterfly is controlled thermostatically, usually by a bimetal spring that is heated either by warm air heated by the car's exhaust or by an electric heating element (Figure 26.21). As the spring heats up, the choke gradually opens. Some carburetors use the engine's coolant to heat the choke spring.

A *fast idle cam* (Figure 26.22), with a series of steps, fits under a *fast idle adjusting screw* and raises engine idle speed when the choke is operating. It is located on the outside of the carburetor and is controlled by the position of the choke butterfly. An *unloader* causes the choke to open when the throttle is floored.

NOTES:

■ *Sometimes, idle speed will be too high. Giving the throttle a quick flooring will cause the unloader to operate. When the choke opens, the fast idle cam will lose its contact and idle speed will return to normal.*

■ *The correct way to start a cold engine with a carburetor is to push the pedal to the floor. This will allow the cold bimetal spring to close the choke and set the fast idle cam for engine starting.*

Avoiding Engine Run-On

With emission controls came leaner air/fuel mixtures and higher idle speeds. When an engine with a lean mixture is shut off, it will usually continue to run. This is because the combustion chamber is hot enough to cause the fuel to self-ignite without a spark. At idle, the throttle plate is usually slightly open to provide air flow. Closing the throttle plate all the way will prevent the engine from running on. An idle stop solenoid is used on some engines for this purpose (Figure 26.23). Another way of preventing run-on is to shut off the flow of fuel. Some manufacturers use a solenoid for this (Figure 26.24).

Feedback Carburetors

During the time that fuel injection was coming into widespread use, a new design of carburetor was developed. Late model carburetors actually meter the fuel according to how much oxygen is found in the engine's exhaust. These are called **feedback carburetors**. An oxygen sensor in the exhaust sends a signal to the computer. The air/fuel ratio is then adjusted to 14.7:1.

Multiple Barrels

Some carburetors have multiple **barrels** to provide a progressive means of adding extra air and fuel as the

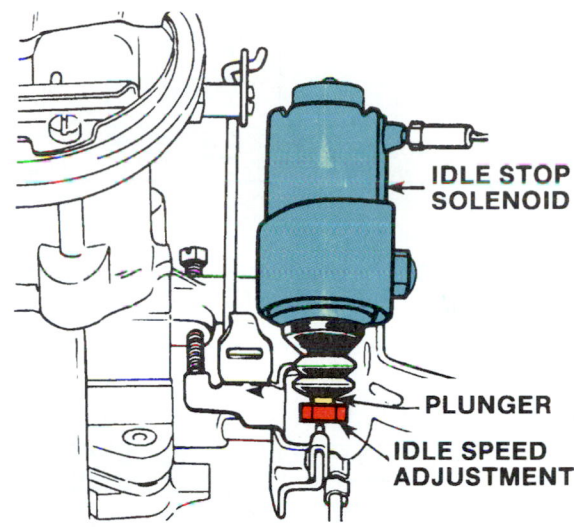

IDLE STOP SOLENOID

PLUNGER

IDLE SPEED ADJUSTMENT

Figure 26.23 An idle stop solenoid provides idle speed adjustment and allows the throttle plate to close when the engine is off. *(Courtesy of Echlin Corp.)*

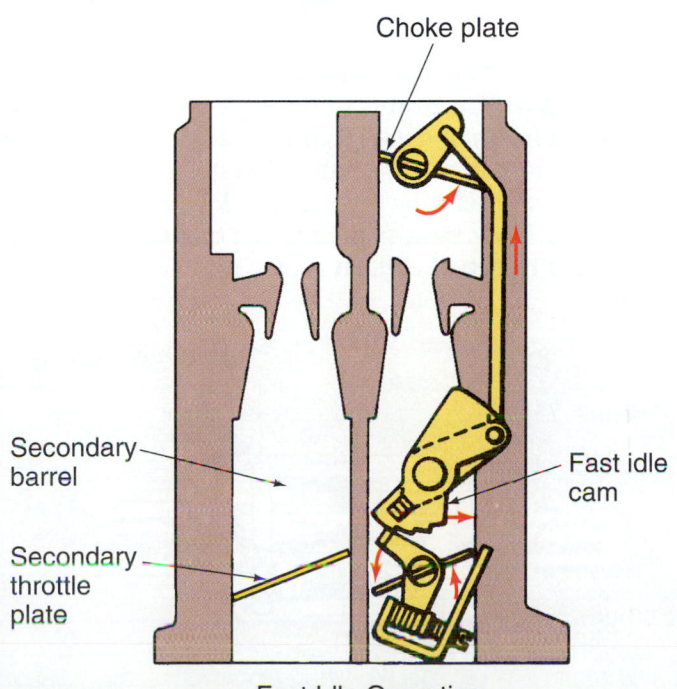

Choke plate

Secondary barrel

Secondary throttle plate

Fast idle cam

Fast Idle Operating

Figure 26.22 A fast idle cam.

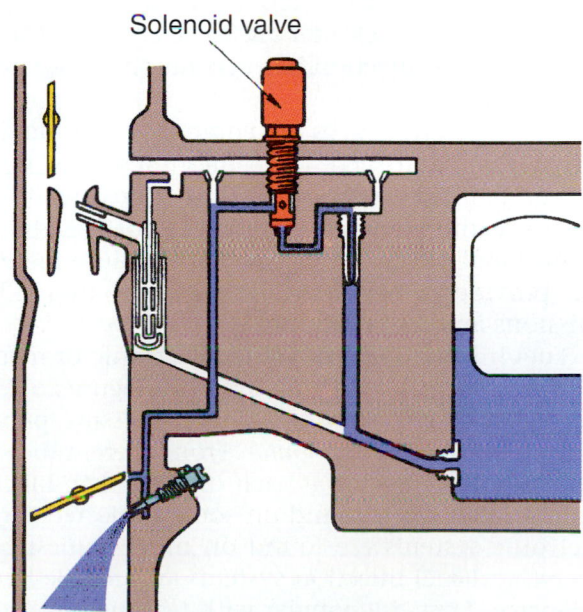

Solenoid valve

Figure 26.24 This solenoid valve closes when the engine is shut off to prevent engine run-on.

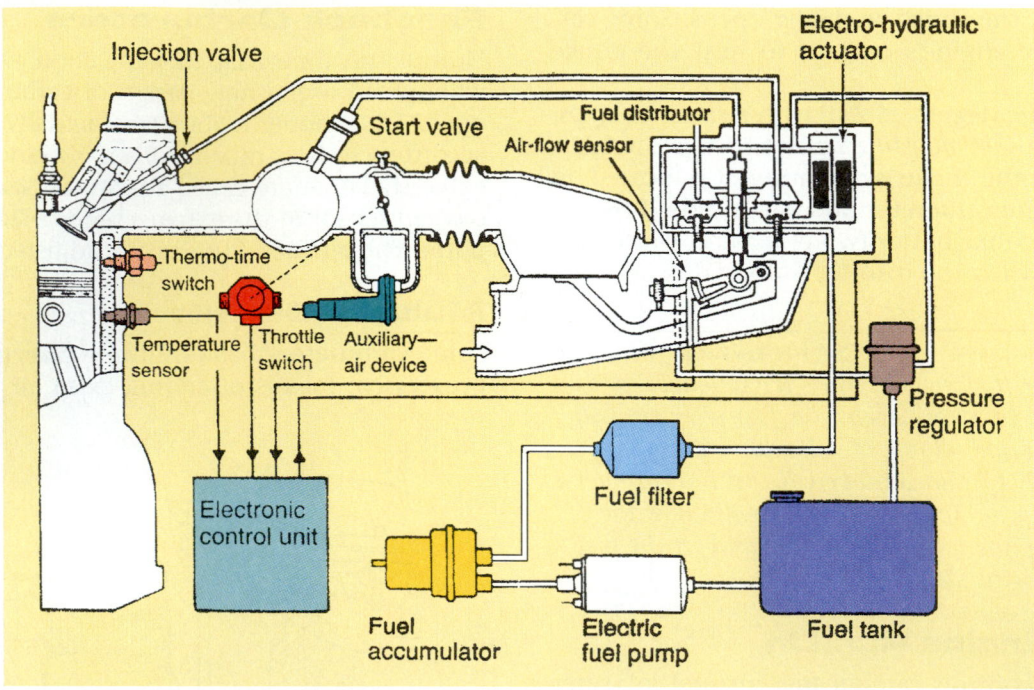

Figure 26.25 A mechanical injection system uses a fuel distributor. *[Courtesy of Society of Automotive Engineers, from October 1987 Automotive Engineering, [page 78]]*

throttle is opened further. A secondary throttle plate (see Figure 26.22) opens to give more power. A carburetor can have either one, two, three, or four barrels. Chapter 28 describes intake manifolds.

■ FUEL INJECTION

In the early 1970s the federal government established standards for fuel efficiency and pollution control. This was due to concern about excessive use of foreign oil and a desire for cleaner air. To be able to meet these standards, fuel injection has become the standard fuel system.

New cars and trucks are equipped with fuel injection systems that serve the same function as the carburetor that they replaced. While there are still millions of carbureted vehicles on the road, carburetors are no longer installed in new cars because fuel injection provides a better means of controlling exhaust emissions and fuel economy.

Fuel injection can be either electronic or mechanical. *Mechanical fuel injection* systems (Figure 26.25) are called *continuous*. They use a high pressure pump to pump fuel to a *fuel distributor*. From there, tubes carry the fuel to an injector at each intake valve. Mechanical fuel injection is found on some imported cars but electronic systems are found on most domestic cars. Mechanical fuel injection systems are trouble-free but expensive. They are popular with technicians.

Electronic fuel injection systems can be either *throttle-body* or *port* systems. Both systems use similar electro-

magnetic fuel injectors that are electrically pulsed (Figure 26.26). The amount of fuel that is injected is controlled by a computer. The computer determines the length of time that the fuel injectors remain open and turns them on and off. Because the fuel pressure in the supply rail to the injectors is closely controlled, the amount of fuel that passes through an injector can be predicted. The length of time that an injector remains open is called its **pulse width**.

The **throttle-body (TBI) system**, also called *central fuel injection (CFI)* (Figure 26.27), uses an intake manifold similar to a carbureted system. The throttle body has one fuel injector per barrel.

A **port injection** system uses a fuel rail with individual fuel injectors at each intake port (Figure 26.28).

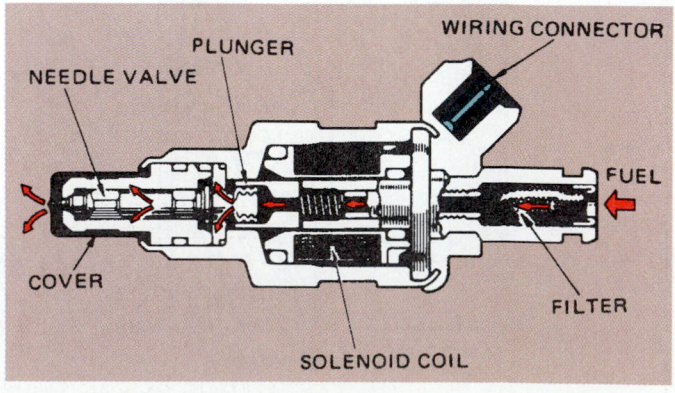

Figure 26.26 A typical electronic fuel injector.

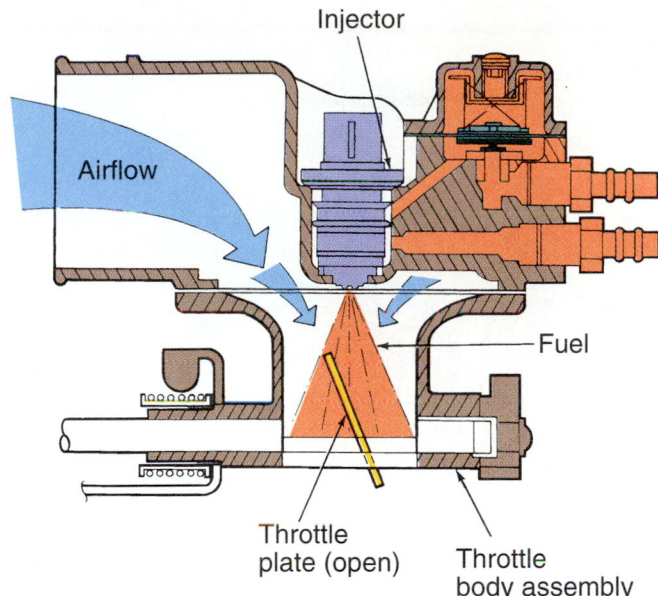

Figure 26.27 Throttle body injection. *(Courtesy of Ford Motor Company)*

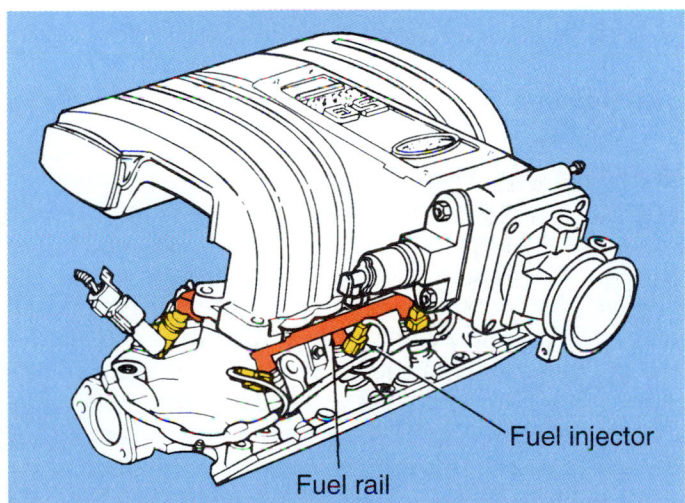

Figure 26.28 A port injection system. *(Courtesy of Ford Motor Company)*

ELECTRONIC FUEL SYSTEM OPERATION

Fuel is pushed from the fuel tank by the fuel pump. The pressurized fuel is sent to the fuel filter and to the injector rail. A *fuel pressure regulator* (Figure 26.29) controls the maximum amount of pressure that can be in the system. Excess fuel is directed by the regulator to the return line to the fuel tank. The system is shown in Figure 26.1. Some systems have a damper with a diaphragm that controls slight variations in fuel pressure from the pump. Some systems have adjustable fuel pressure regulators.

Fuel injectors are electromagnetic solenoid controlled nozzles. When electricity flows through a coil

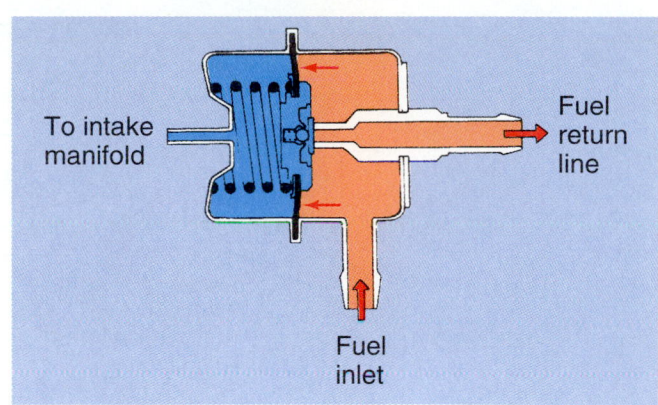

Figure 26.29 A fuel pressure regulator. *(Courtesy of Chrysler Corporation)*

of wire, a magnetic field is produced. Each fuel injector is supplied with power when the ignition is on. The computer controls the grounding of the injector to complete the circuit. When the injector is grounded the solenoid coil of the injector is energized, creating a magnetic field. The plunger of the injector is pulled against spring tension by the magnetic field. This pulls the valve from its seat (see Figure 26.26).

Figure 26.30 shows the components of a typical electronic fuel injection system. When the coolant temperature is low, a *cold start injector* operates during engine cranking. To prevent accidental flooding, a *thermal time switch* limits the maximum amount of time that the injector can operate. A bimetal spring in the switch heats up and opens the switch after the injector has been on for a predetermined length of time.

Port fuel injectors and the cold start injector are installed in a hole in the manifold or cylinder head. There is an insulating O-ring that seals vacuum and keeps fuel in the injector from becoming heated by the engine heat that surrounds it (Figure 26.31).

AIR FLOW MEASUREMENT

The fuel injection system requires a means of determining how much air is flowing into the engine. There are different ways of accomplishing this.

Speed Density (MAP) Systems

Some systems are called **speed density systems**. The computer uses a *MAP (manifold absolute pressure) sensor* and engine rpm (tach) signal to calculate the amount of air entering the engine. The MAP sensor gives an indication of the pressure in the intake manifold. Manifold pressure, density of the air, and engine speed, are all considered by the computer before it calculates the correct amount of fuel to inject into the cylinders.

Other inputs are used to fine-tune the air/fuel ratio in a speed density system. If the throttle position

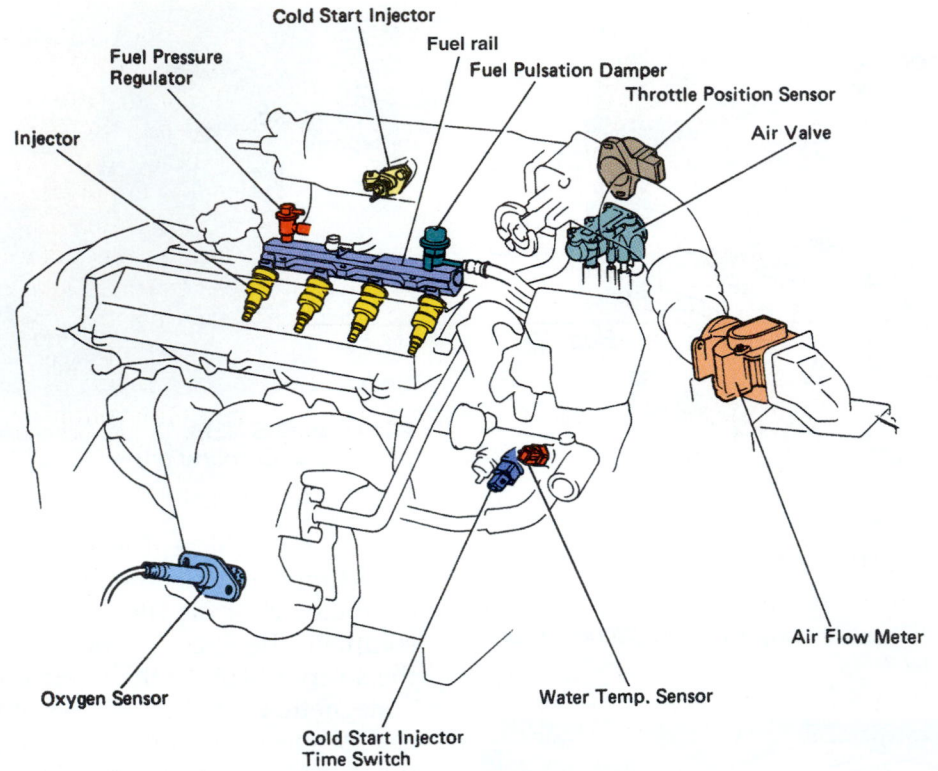

Figure 26.30 Components of a typical electronic fuel injection system.

sensor indicates sudden acceleration, for instance, the computer provides a momentarily richer air/fuel mixture.

Air Density Systems

Another system used by EFI systems is the **air density system**. An *air flow sensor* measures the volume of air entering the engine. The vane, the grid, and the hot wire are all types of air flow sensors.

Vane-Type Air Flow Sensor. A vane-type mass air flow (MAF) sensor is found on many import and domestic vehicles with EFI. All intake air must flow through the sensor. Some MAF sensors are called *volume air flow meters*. A pivoted air measuring plate is held closed with a light spring. Air movement across the plate moves it to the open position (Figure 26.32). A movable pointer that is attached to the plate wipes across a potentiometer. A variable signal is sent to the computer in response to movement of the plate.

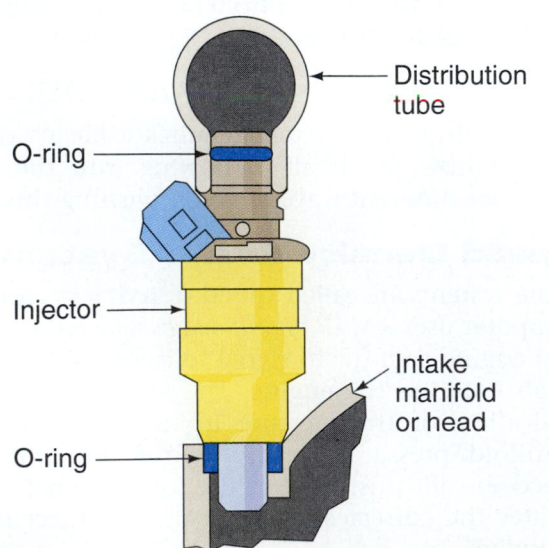

Figure 26.31 An insulating O-ring seals vacuum and keeps heat from the injector.

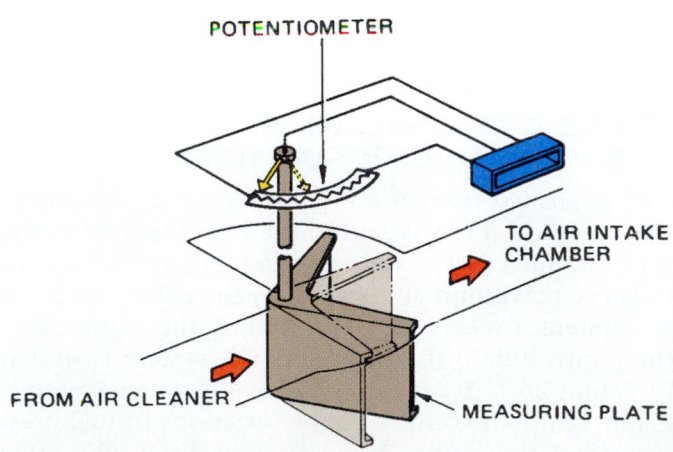

Figure 26.32 Movement of the air measuring plate changes the output from the potentiometer to the computer.

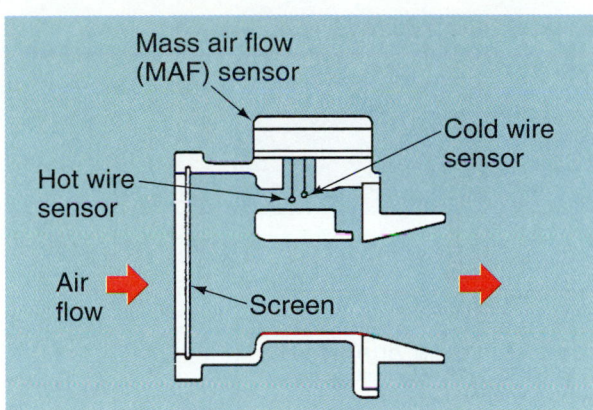

Figure 26.33 A hot wire mass air flow sensor. *(Courtesy of Ford Motor Company)*

Heated Resistor MAF Sensor. A heated resistor air flow sensor has a resistor in the air intake that is heated to a certain temperature whenever the key is on. As the engine is accelerated, more air flows across the resistor to cool it. More current is required to keep the resistor at the same temperature. The computer uses this signal to determine the amount of fuel to inject into the cylinder. Some systems use an electric grid instead of a resistor.

Hot Wire MAF Sensor. In a hot wire MAF sensor system, a sensor is located next to a hot wire (Figure 26.33). It tells the computer what the temperature in the intake system is. When the ignition is on, the computer heats the wire to a specified temperature above the temperature of the air in the intake manifold. When the engine is accelerated, fresh air cools the wire and more electrical current is provided to heat it up. The computer calculates how much air is flowing into the engine by interpreting the signal generated by this action.

NOTES: *The Robert Bosch Corporation is a major fuel injection manufacturer. The three main types of fuel injection system are characterized by Bosch as follows:*

- *D-Jetronic – this early electronic fuel injection (EFI) system is a speed density system that uses a MAP sensor.*
- *L-Jetronic – this EFI system uses a mechanical (air door) or electronic (mass air flow) sensor and is an air density system.*
- *K-Jetronic – this is the mechanical fuel injection system.*

■ IDLE SPEED CONTROL

When an engine is cold or when there are extra loads on it (air conditioning, charging, and so on) the engine's idle speed needs to be raised to compensate. Idle speed is raised by allowing more air to bypass the throttle plate. Some systems use an *auxiliary air valve*. Others use an *air bypass valve* or an *idle speed control (ISC) motor.*

A *throttle position sensor (TPS)* attached to the throttle body on the air inlet senses how many degrees the throttle plate is open.

A *coolant temperature sensor (CTS)* is installed in the cooling system to tell the computer how warm the engine is. Then the computer increases injector duration, improving cold engine driveability.

An intake *air temperature sensor* senses the temperature of the incoming air. Volume and density of air change with temperature. Even though the volume of air measured entering the engine is the same, the amount of fuel that is injected will be varied with temperature. When the temperature is above about 70°F, the computer decreases the volume of fuel coming from the injector. Below 70°F the injector volume is increased.

■ FUEL PUMP RELAY

When the engine is cranked, the fuel pump is energized. If the key is on and the engine has not been cranked for two seconds, the fuel pump relay shuts off power to the fuel pump. This prevents flooding or engine lockup (in case of leaking injectors). It also prevents the fuel pump from running with the ignition switch on, in case of an accident or broken fuel line.

Some engines are equipped with an oil pressure switch hooked in parallel with the fuel pump relay. If the fuel pump relay is defective, the *oil pressure switch* will allow the fuel pump to keep operating.

■ COMPUTER CONTROLLED FUEL SYSTEMS

Precise fuel metering has become possible with the addition of onboard computers to automobiles. These systems first became available on some cars in the late 1960s to control the fuel system. New vehicles today have computers controlling many systems, from transmission shift points to accessories. The sketch in Figure 75.35 shows a late model vehicle with ten computers.

Automotive ignition and electronics are complex specialty areas. Discussion here deals with part of the computer fuel system only. The intent is to provide a general idea of the operation of the system. The *powertrain control module (PCM)* in the sketch is the one that controls engine performance, including the fuel system.

■ FEEDBACK FUEL SYSTEMS

Three types of devices make up an automobile computer system. They are the computer, **sensors**, and **actuators**. A computer monitors input from various sensors that are part of the system. Sensors relay to the computer information about throttle position, air and coolant temperature, airflow, manifold pressure, and barometric pressure, among others. Actuators carry out the assigned change from the computer (Figure 26.34). In fuel injection systems the actuators are the fuel injectors. Electronic systems have an advantage over

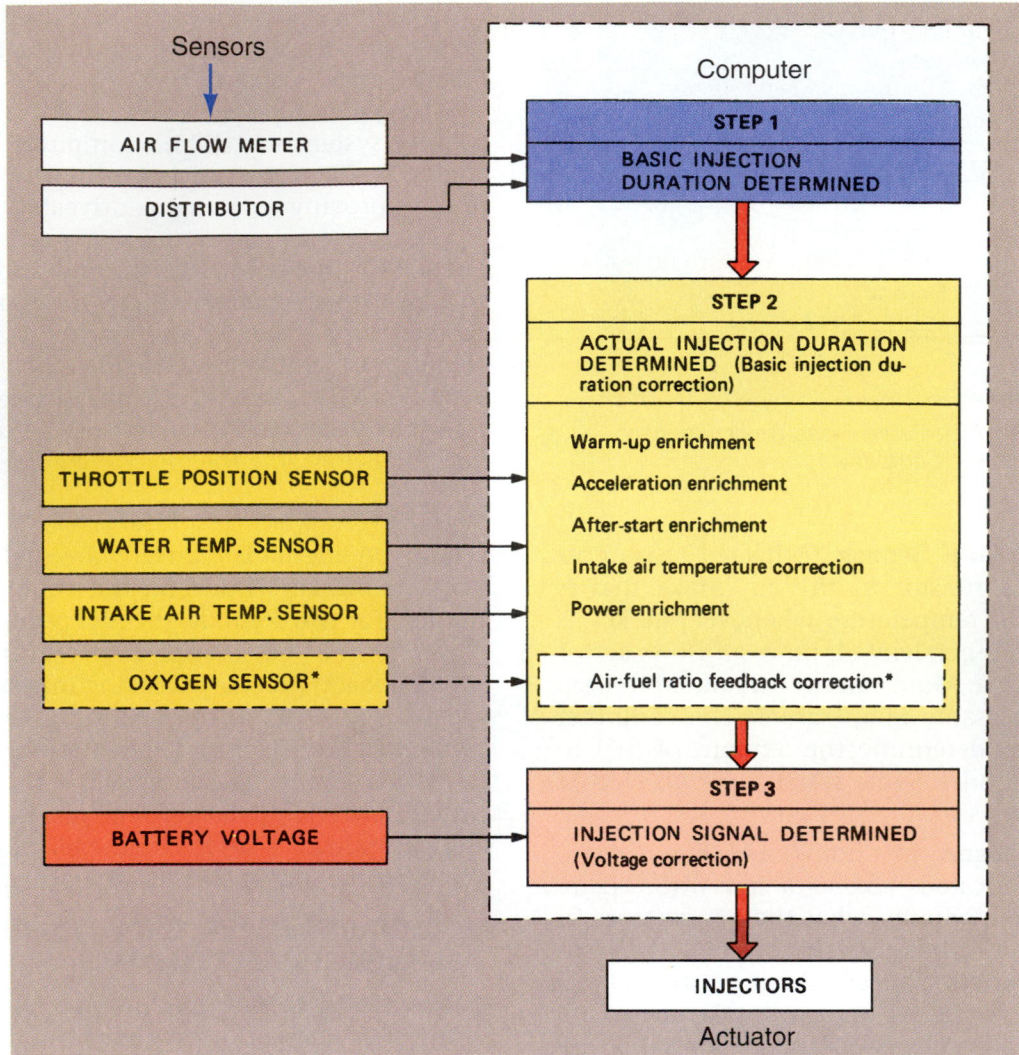

Figure 26.34 The computer receives signals from sensors and sends commands to actuators.

older carburetor systems, which did not compensate for changes in altitude and temperature. Electronic systems can vary the amount of fuel according to information received from sensors and acted upon by actuators. The mixture in a carburetor is changed by disassembling it and changing the sizes of jets.

Computer feedback cars have a sensor, usually located in the exhaust manifold, called an *oxygen sensor* (Figure 26.35). Other names for the oxygen sensor are O_2 *sensor*, *Lambda sensor*, or *heated exhaust gas oxygen (HEGO) sensor*. O_2 sensors send a voltage signal to the computer that is generated in an amount that varies according to the amount of oxygen in the engine's exhaust.

In a computer feedback fuel system, the computer acts to make corrective changes to the air/fuel mixture entering the engine's cylinders (Figure 26.36). The air/fuel mixture is varied by increasing or decreasing the pulse width in EFI systems.

Feedback carburetors are also used on some cars. These electronic carburetors change the air/fuel mixture

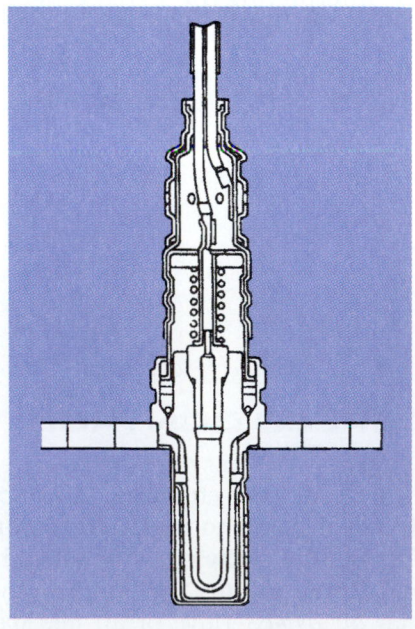

Figure 26.35 An oxygen sensor.

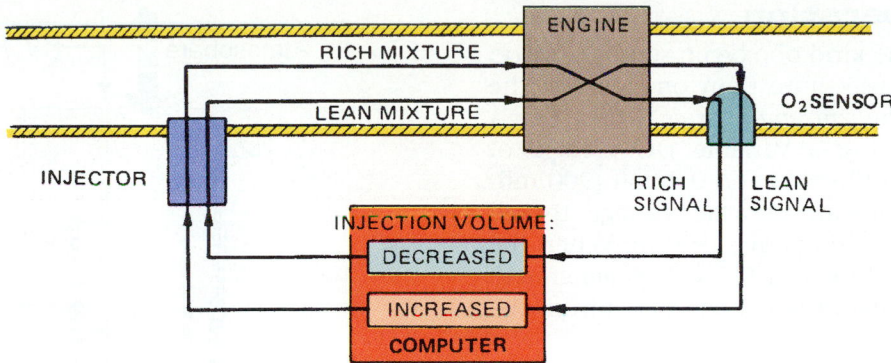

Figure 26.36 The computer changes the air/fuel mixture.

by varying the amount of pulsations a solenoid makes (dwell). The solenoid opens and closes ten times a second. When the solenoid is on, an air bleed opens or a fuel valve closes. This causes a lean mixture. When the solenoid is off, the mixture becomes rich as the air bleed closes or the fuel valve opens. A fuel specialist can interpret meter readings off of either of these feedback systems.

Open and Closed Loop

In **open loop**, the oxygen sensor does not send signals to the computer. The O_2 sensor cannot operate until it reaches about 600°F. **Closed loop fuel control** occurs when the engine reaches operating temperature and the computer starts acting upon information received from the O_2 sensor (Figure 26.37). The oxygen sensor reads and adjusts the air/fuel mixture about 10 times/second.

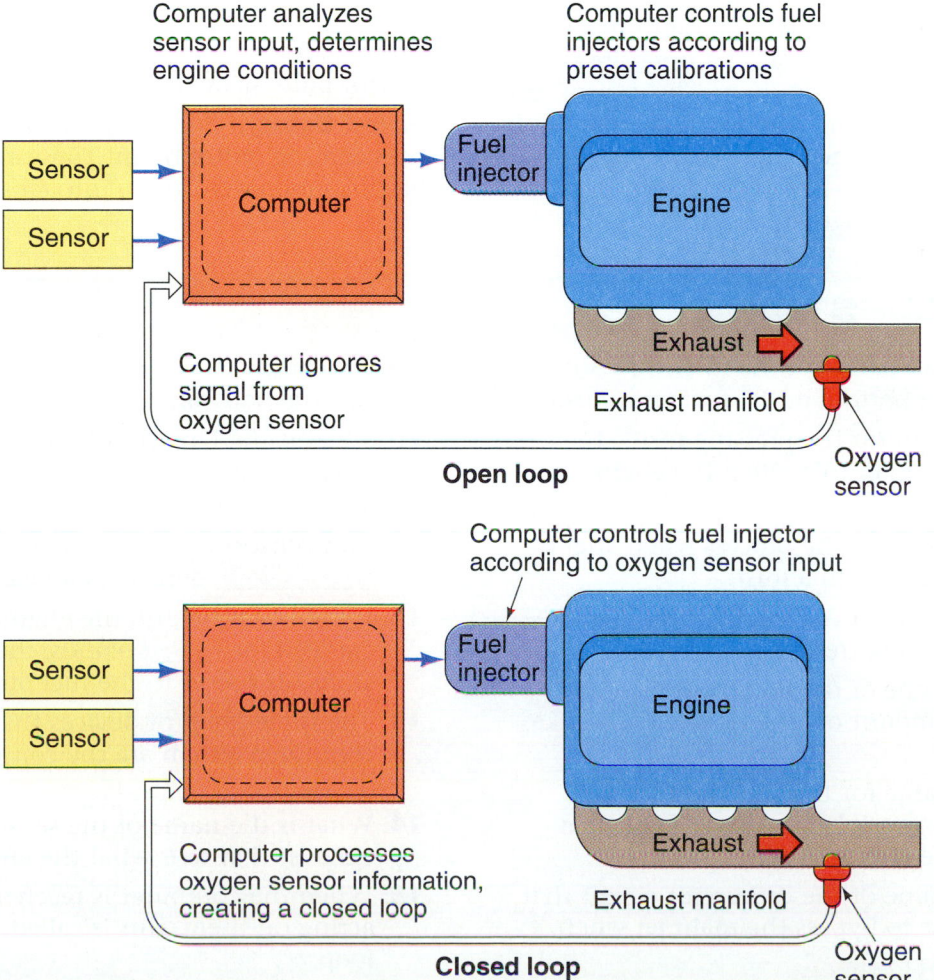

Figure 26.37 Open and closed loop. *(Courtesy of Ford Motor Company)*

O₂ Sensor Operation

There is more than one kind of oxygen sensor used on cars. Also, some cars use more than one sensor. The most popular one, the *zirconium oxide* works like a small battery, generating a variable DC voltage of between .1 volt (100 millivolts) and 0.9 volt (900 millivolts). The computer can use the voltage to tell whether the exhaust stream is rich or lean. When the mixture is rich a signal of over 0.5 volt is generated. When the mixture is lean, the voltage signal from the sensor will be less than 0.4 volt.

When heated to about 600°F, a voltage is generated on two platinum strips inside the sensor. One of these strips is exposed to outside air and the other to the exhaust stream (Figure 26.38). The two strips develop different voltages in response to the amount of O_2 ions they attract. The difference in these voltages is what makes the sensor act like a battery.

When there is a rich air/fuel mixture, there is a shortage of oxygen in the exhaust gas. The outside air side of the sensor has an abundance of oxygen. Oxygen ions, which are negatively charged atoms, move from the atmospheric side platinum strip to the exhaust side strip. This transfers a negative charge to the exhaust side of the sensor, which results in a signal generation of 0.5 volt or higher. The opposite occurs when the air/fuel mixture is lean. The sensor can change from rich to lean almost instantaneously (about $\frac{1}{10}$ of a second). When there is more oxygen in the exhaust stream, there is a lean air/fuel mixture. The computer sees a voltage of less than 0.4 volt and

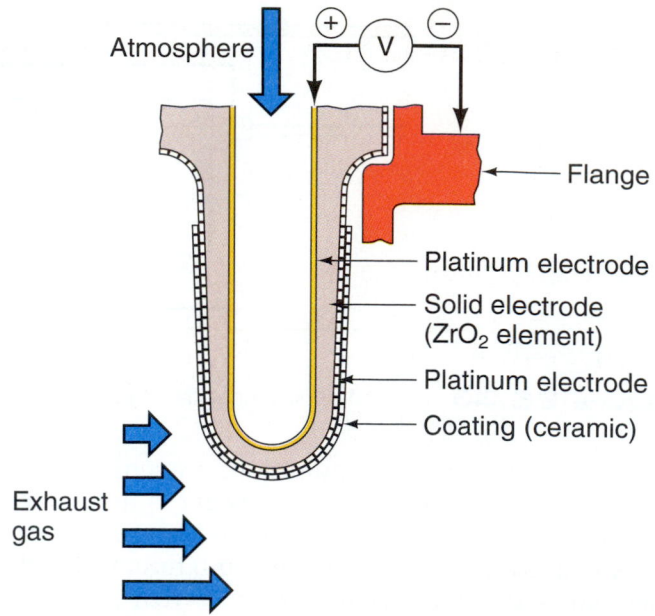

Figure 26.38 In an oxygen sensor, one platinum electrode is exposed to outside air and the other to the exhaust stream.

compensates by driving the fuel system rich (more injector pulse width).

Heated Sensor. Some O_2 sensors have three or four wires coming out of them. These are heated sensors so that the computer can go into closed loop sooner.

■ REVIEW QUESTIONS

1. What is another name for the filter that is in the fuel tank at the bottom of the fuel tank cluster?

2. What is the term for the pressure inside the engine's intake manifold when the engine is running?

3. What is the name of the process where fuel is suspended in the air in a mist?

4. What is the name of the smaller restricted area in the carburetor barrel?

5. What is the name of the butterfly valve that controls the amount of air entering the intake manifold?

6. What is the name for the little holes that prevent fuel from siphoning out of the float bowl when the engine is off?

7. What is the name of the carburetor circuit that allows gasoline to bypass the main jet when under load?

8. Fuel injection systems can be either mechanical or _____.

9. Two kinds of electronic fuel injection are throttle body and _____.

10. What is the length of time that a fuel injector is open called?

11. The two main categories of fuel injection air flow sensors are the air density and the _____ density systems.

12. If the key is on and the engine has not been cranked for _____ seconds, the fuel pump relay shuts off power to the fuel pump.

13. The three types of devices that make up a computer system are the computer, sensors, and _____.

14. What is the name of the sensor in the vehicle's exhaust that tells what the air/fuel mixture is?

15. When the computer is receiving signals and acting on them, this is called _____ loop.

ASE STYLE REVIEW QUESTIONS

1. Technician A says that carburetors require larger fuel filters than fuel injection systems. Technician B says that if carburetor float level is too high the air/fuel mixture will be richer. Who is right?

   a. Technician A   **b.** Technician B
   c. Both A and B   **d.** Neither A nor B

2. Technician A says that the idle port is located just above the throttle plate. Technician B says that shutting off air flow at the throttle plate will prevent engine run-on. Who is right?

   a. Technician A   **b.** Technician B
   c. Both A and B   **d.** Neither A nor B

3. Technician A says that when there is a rich air/fuel mixture, the oxygen sensor will give a signal that is higher than 0.5 volt. Technician B says that if there is a low voltage signal (0.4 volt or less) coming from the oxygen sensor, the computer will drive the system rich. Who is right?

   a. Technician A   **b.** Technician B
   c. Both A and B   **d.** Neither A nor B

4. Technician A says the butterfly valve that controls the amount of air entering the intake manifold is called the venturi. Technician B says the kind of fuel injection that has an injector at each intake valve port is called a throttle body injector. Who is right?

   a. Technician A   **b.** Technician B
   c. Both A and B   **d.** Neither A nor B

5. Technician A says a speed density fuel injection system uses an air flow sensor. Technician B says an air density system uses a manifold pressure sensor. Who is right?

   a. Technician A   **b.** Technician B
   c. Both A and B   **d.** Neither A nor B

CHAPTER 27

Fuel System Service

■ OBJECTIVES

Upon completion of this chapter, you will be able to:

✔ Service fuel delivery system components.

✔ Diagnose rich and lean air/fuel mixtures and their causes.

✔ Service carburetors and fuel injection systems.

■ KEY TERMS

flooding
scan tool
port fuel injected
glitch
schematic
cross counts
RTV
O$_2$ sensor safe
min/max
programmable read only
memory PROM
carbon blaster

■ INTRODUCTION

This chapter includes diagnosis and service of a number of the fuel system's parts: fuel pumps, filters, tanks, carburetors, and fuel injection systems (and their computer controls). Diagnosis and repair of problems in electronic systems can be complicated and parts are expensive (especially when they are not necessary). Information in this chapter is provided for you to gain a basic, generic understanding of related problems. Successful specialists in these systems are highly educated, experienced, and able to locate service information. Check Chapter 26 for information on how fuel system components operate.

■ FUEL SUPPLY SYSTEM SERVICE

Fuel Tank Service

It is sometimes necessary to lower or remove a fuel tank from a car when the tank is corroded or to:

■ replace fuel or vapor hoses
■ repair a fuel tank sending unit
■ replace an in-tank electric fuel pump

Fuel tank repair is commonly done by radiator shops. Possible repairs include flushing, welding, and treating with "slushing compound" to seal and protect the inside of the tank. Because of the danger associated with gasoline, many repair shops choose not to make tank repairs. The information provided here is included as a precaution for those who choose to remove fuel tanks.

Before removing a fuel tank, fuel should be pumped from the tank using commercial fuel handling equipment (Figure 27.1). The fuel is pumped out, either through the fill opening or through the outlet line to the fuel pump.

Bleed Fuel System Pressure

Fuel injection systems are designed to remain pressurized after the engine is shut off so that the engine can start quickly. If one of the clamps on the system is loosened, pressurized fuel can escape. To prevent gasoline from escaping, pinch pliers or long nose vise grips can be installed on the hose on the fuel tank side of filter (if the fuel hose is flexible).

Bleed pressure from the system before working on it. There are several methods that can be used. Check the manufacturer's manual for the recommended procedure.

Figure 27.1 Equipment for the safe handling of fuel.

CAUTION

- ■ Before removing the tank, remove the battery ground cable to avoid an accidental spark.
- ■ Gasoline weighs about 7 pounds per gallon. A full tank could easily weigh in excess of 100 pounds. Attempting to remove it while it is full could result in a dangerous gasoline spill or an injury. Remember: gasoline vapor is *extremely* flammable.
- ■ Wear safety glasses and do **not** use a drop light. Liquid falling on the hot glass can cause the bulb to explode. If it is accidentally dropped or has a bad switch or loose connections, it can cause a fire.
- ■ Work in a well-ventilated area, with no possible source of ignition, such as running cars or electrical equipment.
- ■ After draining a fuel tank, seal all openings.
- ■ The tank is filled with an inert gas such as CO_2 before welding. Even if a tank has been thoroughly flushed, it will still contain flammable vapors. Attempting to weld it can result in an explosion.

A very simple procedure for bleeding pressure from the fuel system is as follows:

- ■ Disconnect the fuel pump electrical connector or remove the fuel pump relay or fuse.
- ■ Crank the engine briefly to drop the fuel pressure or run the engine until it stalls.

There are other methods used to bleed pressure from the fuel system. One other way is to energize the fuel injector by applying positive voltage to one of the injector terminals and a ground to the other.

NOTE: *12 volts should not be applied to an injector for longer than 5 seconds.*

Another pressure bleeding procedure is to:

- ■ Disconnect the battery ground cable.
- ■ Remove the filler cap from the fuel tank.
- ■ Some systems have a *Schrader* valve that can be used to bleed off pressure from the system before disassembly. (A Shrader valve is the kind that is found on tire valve stems [see Chapter 54].) Remove the threaded cap from the fuel pressure test port on the fuel rail to find the Shrader valve.
- ■ Use a special hose that has a pressure relieving tool on the fuel rail end to drain fuel to a gas can (Figure 27.2).

Fuel Gauge Sending Unit Removal

Prior to removing the tank, the fuel gauge sending unit and float must be disconnected from it (Figure

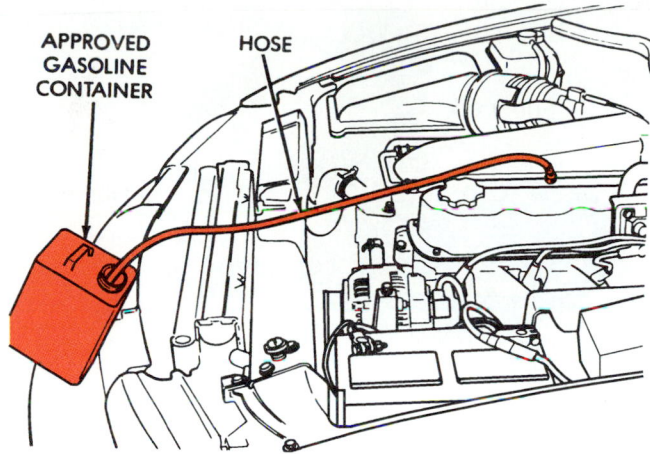

Figure 27.2 Bleed pressure from the fuel system before working on it. *(Courtesy of Chrysler Corporation)*

27.3). The fuel lines, which usually enter the tank at this point are disconnected too. The tank sending unit must be removed to get to the pickup strainer (sock).

SAFETY NOTE Use a brass drift to loosen the lock ring to prevent the danger of a spark.

Vapor Recovery

Fuel vapors need someplace to go when filling a fuel tank or when fuel expands or contracts in the tank due to temperature changes. The fuel tank is connected to a *vapor recovery device*. Operation of these devices is covered in Chapter 40. A service technician will often need to replace the fuel hoses that connect these devices. Use the correct type of fuel hose for the application (see Chapter 23). Replace hoses one at a time so that they are not accidentally reinstalled in the wrong place.

Replacing Hoses

Hoses should be inspected often as they deteriorate over time and can fail from the inside out, plugging things up. Most industry authorities recommend that hoses be replaced every three years. If a fuel hose fails, a fire can result. If a hose needs to be replaced, be sure to use one of the proper type (see Chapter 23).

SAFETY NOTE Be sure that a hose is not positioned near any part of the exhaust system or the catalytic converter.

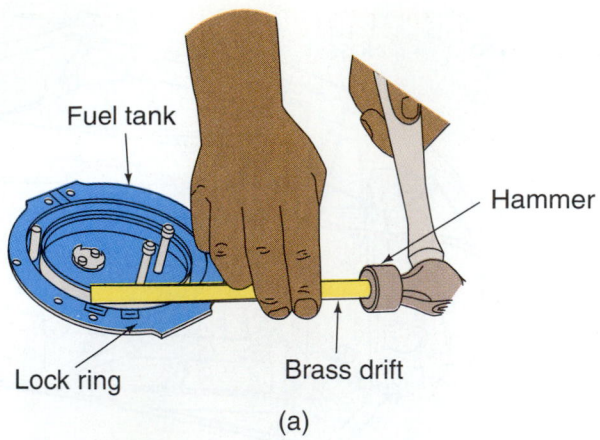

Fuel tank

Hammer

Lock ring

Brass drift

(a)

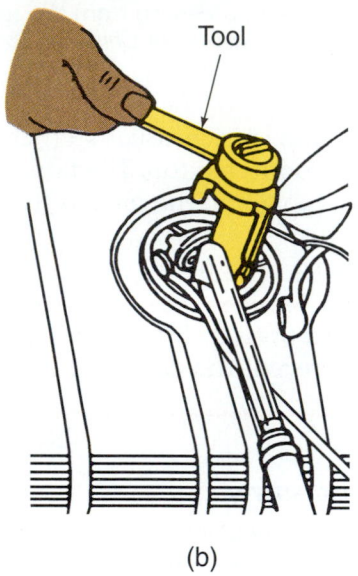

Tool

(b)

Figure 27.3 Remove the fuel sending cluster from the fuel tank before removing the tank.

To inspect a hose, flex it back and forth to see if it breaks or shows cracks. If the end of a vapor hose has gotten hard and the hose is long enough, sometimes a small section can be cut off of the end. The remaining section of the hose is reinstalled on its fitting. This is done only if it does not have a hose clamp. Clamped hoses usually contain pressurized fuel.

SAFETY NOTE Any suspected bad line should be replaced immediately so that a dangerous situation does not develop. After getting the customer's approval and replacing the line, always start the engine and check for leaks.

■ FUEL FILTER SERVICE

Fuel filters have specified intervals for replacement. They can be located in a fuel line, in the tank (see

Chapter 26), in a carburetor, or in any combination of these. Outlet fuel filters can be found on the outlet side of a fuel pump.

Carburetor Inlet Filter

A carburetor may have a filter behind the fitting at its fuel inlet. When replacing a filter, locate the proper direction to install the filter. Since 1976, there have been *rollover check valves* in fuel systems. They are often found as part of the fuel filter (Figure 27.4). The open end of the filter with the check valve in it faces toward incoming fuel. The spring goes in before the filter.

SHOP TIP When installing a fuel line, jiggling the line back and forth makes it easier to turn the nut by hand.

Be sure to hold the carburetor nut with an open end wrench while tightening the flare nut with a flare nut wrench.

➡ *Perform R & R Carburetor Inlet Filter Worksheet*

Inline Fuel Filter

Fuel filters that are installed in the fuel line are replaced at specified intervals. If the fuel filter becomes plugged in a fuel injection system, fuel to the rail will be reduced, which results in hard starting, lean running conditions, misfiring, and so on.

CAUTION Bleed off residual pressure from a fuel injection system before removing any fuel lines. To do this, disable the fuel pump. Then start the engine and let it run until it runs out of fuel.

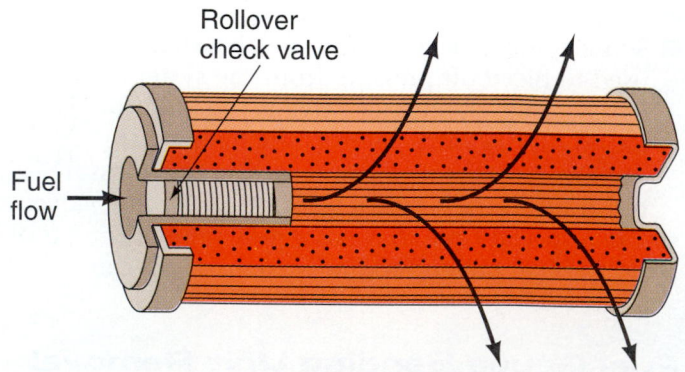

Rollover check valve

Fuel flow

Figure 27.4 A carburetor inlet filter with an anti-rollover check valve. *[Courtesy of AlliedSignal Automotive Aftermarket]*

Figure 27.5 *Various types of throttle linkage clips.*

Before disconnecting fuel lines, place a rag (or drain pan, if possible) under the filter to catch fuel that spills.

A filter will usually have an arrow to tell the proper direction of installation.

➡ *Perform* **Fuel Filter Service** *Worksheet*

◼ THROTTLE LINKAGE

Throttle control to carburetors on older cars is often by linkages and pivots. On fuel injection systems and newer carburetors, there is usually a cable. Throttle and carburetor linkages have clips of various types (Figure 27.5). One kind of clip fits around a 90° bend on the end of the linkage and then rotates to clip around the linkage. This clip can be either right or left hand. Be sure to use the right one.

◼ CARBURETOR PROBLEMS AND SERVICE

Before attempting a carburetor repair or service, perform a visual inspection. Look for obvious problems, missing parts, torn or damaged hoses. With the engine running listen for hissing, which could indicate an air (vacuum) leak. An air leak is usually accompanied by a rough, higher idle and hesitation on acceleration. Problems caused by rich and lean air/fuel mixtures are covered in detail in Chapter 25.

Carburetor Float Level

Float level is important. Too high a float level and the engine runs rich. The floats on older cars were made of brass. Later model carburetors usually have plastic (urethane) floats that can become saturated with fuel over a period of years. This makes them heavier than normal, resulting in a higher fuel level and a richer air/fuel mixture. Too low a float level will result in a lean air/fuel mixture. Weigh a urethane float with a digital scale to compare it to specifications for the vehicle. Sometimes the same carburetor will use floats of several different weights depending on the application.

Hesitation on Acceleration

A lean air/fuel mixture can cause engine surging and burned parts, as well as a *flat spot* during initial acceleration. A flat spot is when the car stumbles when first accelerated. Flat spots are a common problem that

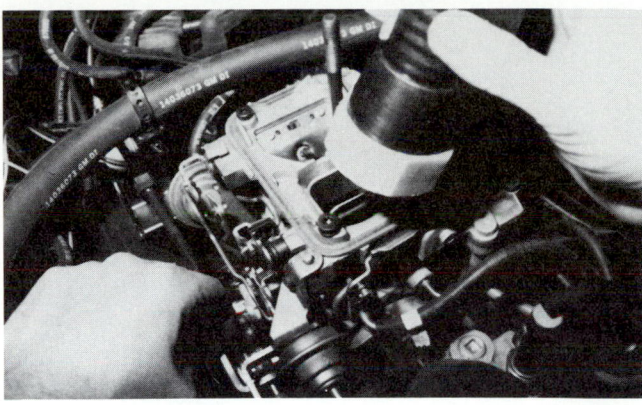

Figure 27.6 *Checking the operation of the accelerator pump.*

occurs when the *accelerator pump* wears out or its passage becomes plugged. It can also be caused by retarded ignition timing, a bad vacuum advance unit, or by a number of other problems.

To test the accelerator pump, remove the air cleaner. With the engine off and the choke plate open, look down the carburetor while opening the throttle (Figure 27.6). A strong squirt of gasoline should be visible in the venturi. If there is no fuel, or only a dribble, the accelerator pump is not working correctly. Accelerator pumps are part of a rebuild kit. Sometimes they can be purchased separately.

Idle Adjustments

Late model cars with carburetors have idle speed regulated by an idle speed control motor controlled by the computer. Earlier carburetors have an adjustment screw or nut on the throttle plate that changes the throttle opening at idle, allowing engine idle to be adjusted.

On some carburetors there is an *idle stop solenoid* that allows the throttle plate to close all of the way when the engine is shut off. This feature prevents engine run-on (dieseling) after the ignition is shut off. The idle speed is adjusted with a screw on the top of the idle stop solenoid (see Figure 26.23).

Fast Idle Cam

When the engine is cold and the choke is operating, a fast idle cam moves to raise the idle (see Figure 26.22). Sometimes, as a carburetor gets older linkages get gummed up, preventing the fast idle cam from working properly. The result can be an idle speed that is too high. The customer might complain that the throttle is sticking. Spraying carburetor spray on the linkages and the choke plate shaft that goes through the carburetor air horn will often fix the problem.

Choke Service

More recent carburetors came equipped with automatic chokes, operated by a bimetal spring. These are

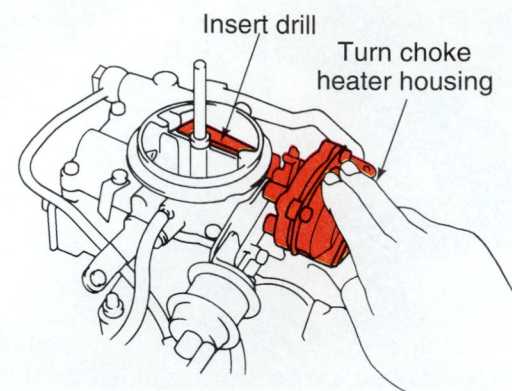

Figure 27.7 Adjusting the automatic choke.

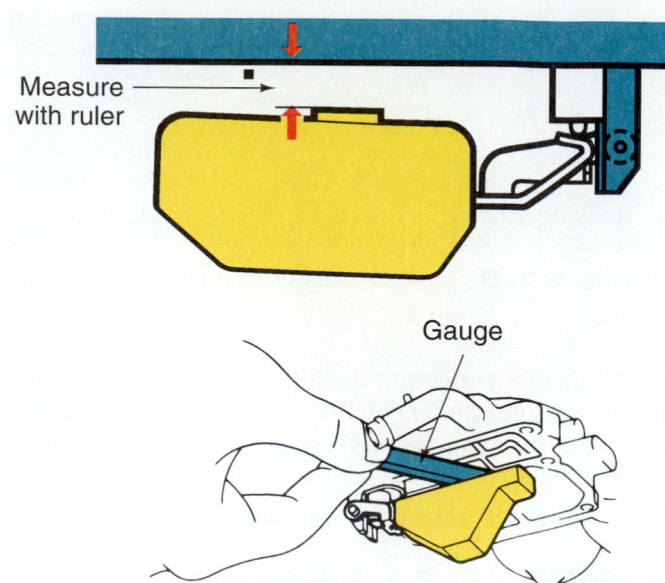

Figure 27.8 The float level is measured with a ruler or a special gauge.

adjustable. A typical adjustment would be made when the engine is cold.

■ HISTORY NOTE ■

Chokes in older vehicles were manually operated by a cable. When the engine was cold, the driver would pull on the choke cable to the carburetor to close the choke butterfly. As the engine warmed up, the driver would have to remember to open the choke, or gasoline would be wasted and exhaust emissions would skyrocket.

NOTE: *A tool called a choke checker is available from parts stores to cool the bimetal spring if the engine is hot.*

The choke spring tension is adjusted so that the choke plate is opened a specified amount (usually measured by a drill bit of that size) (Figure 27.7). Some carburetors have a choke adjustment feature on the choke heater housing and others require bending a piece of linkage. Refer to the manufacturer's service manual for instructions.

■ CAUSES OF RICH AND LEAN CONDITIONS

Float Problems

If the float system does not work properly, the engine's air/fuel mixture will not be accurate and fuel might spill out of the carburetor, causing a dangerous condition. During a carburetor rebuild, the float level is set to specifications (Figure 27.8)

Two things can cause a change in the float level. A urethane float can become saturated with fuel. Because the float is now heavier, the result is a higher than normal float level and a richer mixture. The other reason could be dirt getting into the needle and seat that control fuel entering the carburetor. When this happens there is probably dirt in the float bowl too. One of the results of this is that the main jet can become plugged, causing a lean condition.

Idle Air/Fuel Mixture

An *idle mixture screw* on some carburetors allows the amount of air and fuel entering through the idle port to be adjusted. Most later carburetors have some form of tamperproof provision for the idle mixture screw.

The idle port is very small and can become plugged with contaminants. The result is a rough idle or a car that will not run unless the accelerator is held slightly open. Study Figure 26.17 and you will see why this is true.

Power Enrichment Problems

The carburetor power circuit can affect engine operation and fuel economy. The enrichment circuit in some carburetors uses a power valve. If the diaphragm in the power valve becomes torn, fuel is pulled in causing a rich mixture. A plugged power valve can cause a lean mixture at high speed. This can result in engine overheating on the highway under load. Lean mixtures cause heat in engine parts.

Power valves are made to open at a specified engine vacuum. Power valves with several different opening values are available. The opening point can have a big effect on the vehicle's fuel economy. This is especially true with non-aerodynamic vehicles such as trucks and motorhomes.

NOTE: *Driving at 55 mph into a slight wind might cause an engine to run with 5" of intake manifold vacuum. If the power valve in this example opens at 6" of vacuum and driving at 50 mph results in 7" of vacuum, a good deal of fuel might be saved by driving at the slower speed.*

■ EXHAUST GAS ANALYSIS

The exhaust stream from a running engine can be tested using an *infrared exhaust analyzer.* Infrared light,

light that is invisible to our eyes, is used to measure these emissions. Modern exhaust gas analyzers test five gases: *hydrocarbons (HC)*, *carbon monoxide (CO)*, *carbon dioxide (CO$_2$)*, *oxides of nitrogen (NO$_x$)*, and *oxygen (O$_2$)* (see Figure 41.12). Information from an exhaust analysis can be used to diagnose incorrect air/fuel mixtures, engine and ignition system conditions, and operation of emission system components.

■ SCIENCE NOTE ■

Wavelengths of light

When white light is passed through a prism we see that it is made of many colors. Each color is due to a photon of light of a particular wavelength.

■ *White light contains wavelengths ranging to 400 nanometers (one nanometer = 10^{-9} meter)*

■ *Violet light = short wavelengths to 700 nanometers*

■ *Red light = long wavelengths*

Visible light is only a small part of the electromagnetic spectrum. At very long wavelengths are x-rays and gamma rays.

The accelerator pump can be tested with the exhaust analyzer. About 10 seconds after the accelerator pedal is pushed down, the carbon monoxide level in the exhaust should rise in response to the momentarily rich mixture created.

The air/fuel mixture can also be checked under cruise conditions. Hold the accelerator at 2500 rpm and observe the CO (carbon monoxide) reading. CO that is lower than 0.5% will result in a flat spot on acceleration. This excessively lean mixture can cause burned spark plug electrodes. Premature exhaust valve burning can also result. First, check the float level to see if it is too low. If not, carburetor jets of a larger size can be purchased.

NOTE: *An operating catalytic converter will clean the exhaust. Significant readings for these gases need to be made in front of the catalytic converter.*

Additional information on exhaust analysis is given in Chapter 41.

■ CARBURETOR REBUILDING

Some large companies specialize in rebuilding carburetors. When a rebuilt carburetors is installed, the old carburetor is returned to the rebuilder as a core for future rebuilding. When carburetors were found on most vehicles, carburetor rebuilding was commonly done in most repair shops.

A *carburetor kit* includes all of the necessary parts for a rebuild. The carburetor is disassembled and soaked in a strong chlorinated hydrocarbon, called *carburetor cleaner*. After an hour or so of soaking, the carburetor is removed, washed with hot water, and blown dry with compressed air. Instructions with the kit tell how to adjust the float and the choke. An exploded view showing all of the carburetor parts is also included.

■ ENGINE STARTING PROCEDURES

Carbureted and fuel injected cars require different treatment during starting. When a carbureted car is started in the morning, the driver depresses the accelerator pedal (one time only). This releases the choke butterfly, allowing it to spring closed during engine startup. Stepping on the accelerator pedal also results in a squirt of fuel from the accelerator pump. Both of these actions result in a rich mixture needed for starting.

Cars with fuel injection sense when a cold engine is starting. They do *not* require the pedal to be depressed as carbureted cars do. If there is an engine misfire, it is possible that EFI parts in some systems can be damaged if the throttle is depressed during startup.

Flooded Engine

Flooding of the engine can occur when an engine gets too much fuel. During flooding, spark plugs are wet and a spark cannot be generated across the plug gap. The engine will not start. On a carbureted engine, this can happen when the choke spring is adjusted so the throttle plate is too tightly closed. Flooding is also the result when the driver pumps the pedal too many times while attempting to start a car whose battery has a low state of charge or needs a tuneup.

Cars with carburetors are especially prone to flooding. Each time the pedal is depressed, the accelerator pump squirts a fresh charge of fuel into the intake manifold. This will make the flooded condition worse, so the pedal should be depressed only once.

When a carbureted engine is flooded hold the pedal all of the way to the floor during cranking, raising the pedal very slowly until the engine starts. The air rushing through the engine carries the excess fuel out the exhaust with it. The choke unloader system in the carburetor is brought into play when the throttle pedal is pushed all the way to the floor.

A service technician will often prop the butterfly valve all of the way open on a flooded car and then hold the pedal in the wide open position for a few minutes. This allows excess fuel to evaporate.

Some fuel injected cars react to a floored pedal by shutting off the flow of fuel to the fuel injectors. This is called a "clear flood" condition.

■ FUEL INJECTION DIAGNOSIS AND SERVICE

The majority of cars on the road today have electronic fuel injection systems. Diagnosis and repairs are similar among the various different EFI designs. Defects in

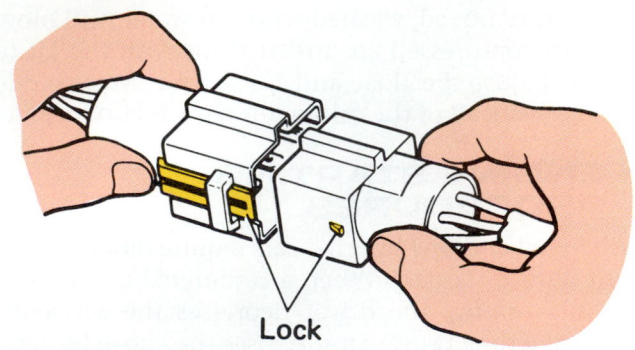

Figure 27.9 Disconnecting and reconnecting electrical connections sometimes solves the problem.

other engine systems can be mistaken for fuel injection problems. Other areas to be checked include emission controls, ignition system operation, engine compression, and battery state of charge.

Visual Check

A visual check often shows obvious problems, such as disconnected or damaged hose or wires. Electrical connections sometimes become corroded. Take the connections apart and look for corrosion (Figure 27.9). Taking the connections apart and putting them back together is sometimes enough to improve the electrical connection and solve the problem. Ford calls this a *wiggle test*.

NOTE: *Plastic electrical connections often become brittle with age. Handle them carefully.*

■ EFI COMPUTER SELF-DIAGNOSTICS

Most late model cars have self-diagnostic provisions. There are codes for many malfunctions that the computer can diagnose. The easy way to read these codes is to use a **scan tool**, a hand-held diagnostic tool. A high impedance voltmeter can also be used (see Chapter 76).

■ AIR/FUEL MIXTURE PROBLEMS

If a computer feedback car has a driveability problem that only occurs when the engine is cold, check for external causes such as an intake manifold leak or incorrect fuel system pressure.

Intake Leaks

A leak in an intake manifold gasket can allow air into the engine that is not measured by an airflow sensor. The computer pulses the injectors for less time than the amount of air warrants. This results in a lean air/fuel mixture when the engine is cold and the system is in open loop. Remember, open loop is when the engine is cold.

After the engine is warm and the computer is receiving feedback from the oxygen sensor, the computer can correct the injector pulse width to compensate for air/fuel mixture problems (if the leak is not too large).

Pressure Testing

If fuel system pressure is not at specifications, the fuel injectors will not inject the correct amount of fuel when the computer system is in open loop. If the pressure is not too far off, the computer will adjust the air/fuel mixture after the system begins closed loop operation. The vehicle will have a cold driveability problem that will disappear when it is warm.

To test fuel pressure, the pump must be operating. If it is not, check the fuel pump fuse. On Fords there is an inertia switch in the circuit that powers the fuel pump. It is located in the trunk of the vehicle. Push the reset button on it first to see if the problem goes away.

It is possible that a fuel pump produces the specified pressure when the key is turned on or at engine idle, but does not produce enough pressure under load. If the complaint is that the engine cuts out at higher speeds, the car should be road tested with a pressure gauge installed. A pressure gauge with a long hose can be taped to the windshield so it can be viewed from the passenger compartment.

NOTE: *Before a pressure gauge is installed, relieve the pressure in the fuel system.*

Installing Pressure Gauge on TBI Systems

Testing pressure in one type of throttle body fuel injection requires that the line entering the throttle body be disconnected. A pressure gauge is installed in series with the line (Figure 27.10). Another testing method

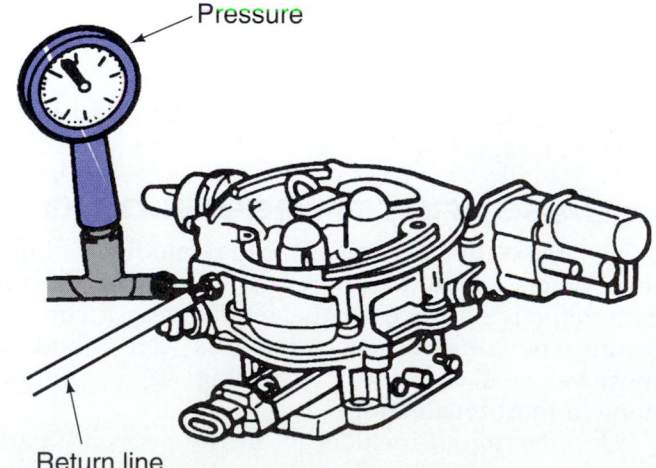

Figure 27.10 A pressure gauge is installed in the fuel line. *(Courtesy of Chrysler Corporation)*

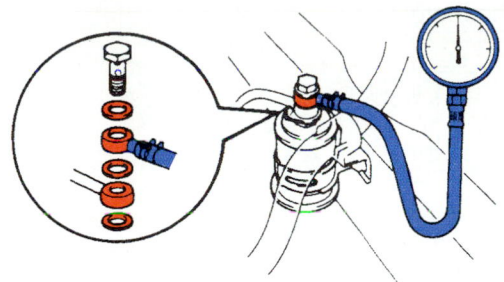

Figure 27.11 A fuel pressure gauge in series with the fuel filter inlet.

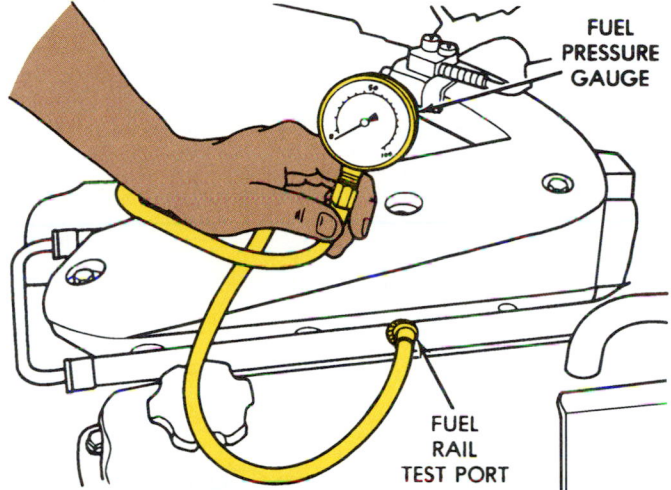

Figure 27.12 A gauge is installed in the fuel pressure test port on the fuel rail. *(Courtesy of Chrysler Corporation)*

recommended by some manufacturers is to install a gauge in series with the fuel filter inlet (Figure 27.11). After the gauge is disconnected, install new gaskets on the banjo fitting. Copper gaskets are reusable if not damaged or imprinted.

Installing Pressure Gauge on PFI Systems

In port fuel injection systems, the gauge is installed at the Schrader valve on the fuel rail (Figure 27.12) or on the fuel line to the cold start injector (Figure 27.13).

To perform the pressure test, the fuel pump is energized. One method is to run the engine at idle speed. Another way is to cycle (turn on and off) the ignition key several times. Sometimes the engine will not run and it may be necessary to energize the pump electrically. Procedures vary between manufacturers. Consult the applicable service manual.

Fuel pressure must be at least equal to manufacturer's specifications. Causes of low pressure could be a plugged fuel tank inlet sock, a kinked inlet line, or a bad auxiliary pump (some mechanical systems use an auxiliary pump in the fuel tank to push fuel to a higher pressure pump outside of the tank).

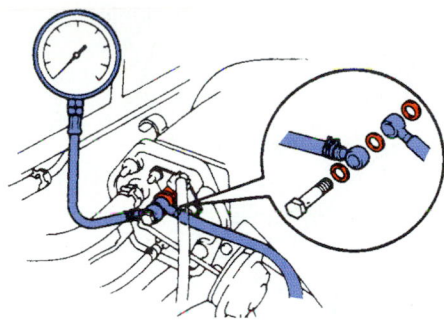

Figure 27.13 A gauge installed into the hose connection for the cold start injector.

When pressure is too low, a fuel pressure regulator could also be the cause. Fuel pressure regulators usually fail by causing pressure that is too low, rather than too high.

■ TESTING A PRESSURE REGULATOR

A pressure regulator can become defective due to foreign material or a ruined diaphragm. The result will be hard starting, poor idle quality, and lack of power. Burned spark plugs can result from operating with a lean air/fuel mixture.

High pressure is usually caused by a bad pressure regulator or when a pressure return line to the fuel tank is kinked. When there is a vacuum line to the pressure regulator, part of the procedure is to pull the vacuum hose off while the engine idles. The pressure should rise about 10 psi in most systems when the hose is pulled.

If a pressure regulator requires adjustment, something else might be wrong with the system. On nonadjustable systems, the regulator must be replaced if pressure is not correct.

■ INJECTOR PROBLEMS

Fuel injectors can be bad, leaking or dirty, shorted or open. On **port fuel injected** cars, individual injectors for each cylinder are located in the intake valve port in the cylinder head. A stethoscope can be used to listen to the opening and closing of the injector as the engine operates (Figure 27.14).

Injector electrical test lights (called *noid lights*) are available from tool companies that provide an easy way to test to see if there is power to the injectors (Figure 27.15). If there is power to the injector, the light will flash on and off as the computer cycles the injector. If the light does not light, there is no power to the injector. A wiring harness or the computer could be the cause. Sensor input to the computer could also be the cause. On many systems, if there is no tach signal, the fuel injection system will not operate.

When there is power but there is no spray from the injector, the injector is probably bad. Use an ohmmeter

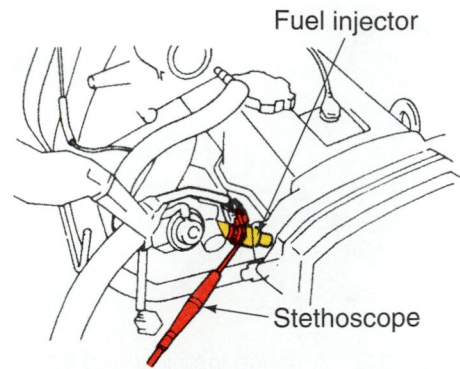

Figure 27.14 Listen to the fuel injector with a stethoscope.

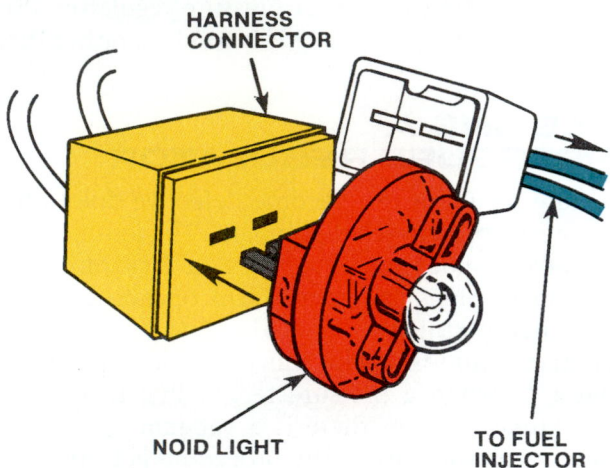

Figure 27.15 A special test light is used to check for power at the injector.

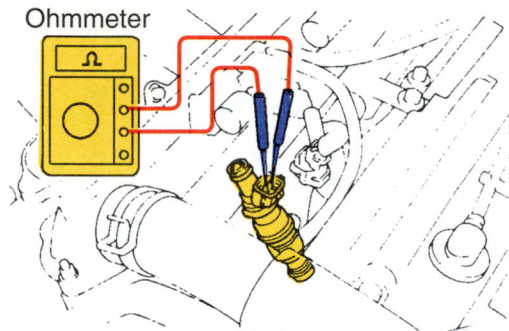

Figure 27.16 An ohmmeter is used to check for shorts and grounds.

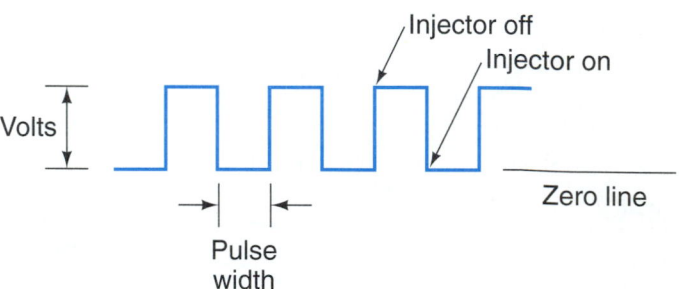

Figure 27.17 A square wave pattern of the pulse width.

to check the resistance and for shorts to ground (Figure 27.16). If the injector is open or shorted, replace the injector.

Checking Injector Pulse Width

Injector pulse width can be checked using some types of digital multimeters or a scan tool. An oscilloscope can also be used. A modified square wave pattern would be the result. The pattern on an oscilloscope shows voltage vertically (Figure 27.17). Above the zero line is positive voltage. Horizontal movement represents time.

Most injectors are powered on one side and controlled through the ground circuit. Attaching the meter feed to the ground side of one of the injectors will show the pulse width of the rest of the injectors.

Fuel injectors have two terminals. To hook up the meter or scope, backprobe one of the terminals of an easily accessible injector with a paper clip. Check either terminal on the injector for voltage. When the key is turned on, there will be power to one side of the injector. The other side is the control (ground) side. In most cases, this is the side the pattern will be able to be viewed from (ground side controlled).

Testing Fuel Injector Flow

Individual port fuel injectors are prone to plugging up from fuel deposits when low quality fuel is used. This is because they are more exposed to heat than throttle body injectors.

On some throttle body injected (TBI) cars, the fuel spray pattern can be seen coming from the end of the injector as the engine operates. This makes a visual check very simple. On other TBI cars, the injector is hidden from view.

An electronic fuel injector tester is available for testing injector balance (Figure 27.18) while reading system pressure on a gauge. With the system at the specified pressure, each injector is activated for an equal period of time. This allows fuel in the loop to escape from the injector. The drop in pressure is recorded (Figure 27.19).

The system is repressurized after each injector is bled and the next injector is operated with the tester. There should be an equal amount of pressure drop from each injector. If the pressure drop is above or below average by 1.4 psi (10 kPa), the injector is defective (Figure 27.20).

Removing and Replacing Injectors

To replace an injector, bleed pressure from the system first. Be careful that no dirt gets into the system while the fuel rail is removed. Do not soak an injector in cleaning solvent. This can damage or contaminate the injector. Be sure to replace the rubber O-rings that insulate the injectors (see Figure 26.31). Failing to do this can result in a vacuum leak, and resulting rough idle. Be sure to check for leaks after reinstalling the injectors.

Figure 27.18 A fuel injector tester.

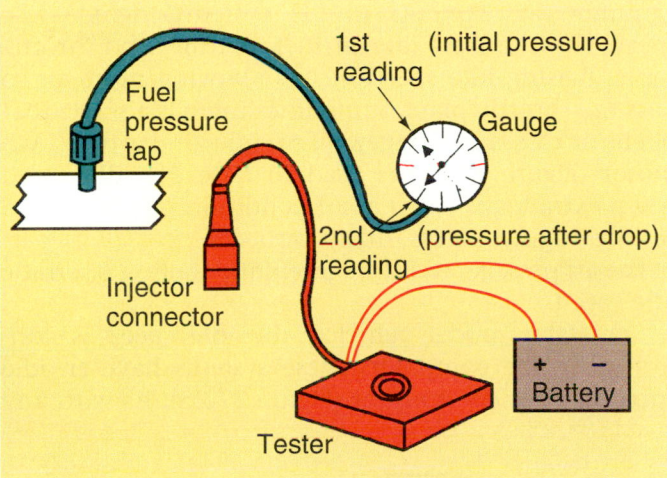

Figure 27.19 Check to see how much the pressure drops after the tester is activated. *(Courtesy of General Motors Corporation, Service Technology Group)*

Injector Flow Test

Some manufacturers use an injector flow test in which the injectors and fuel rail are removed from the engine. Each injector is energized for 15 seconds while holding it over a graduated container (Figure 27.21). The volumes of fuel flowing from each injector are compared. Variations of more than 5 cc calls for replacement of the injector.

Cold Start Injector Test

When foreign material gets into a cold start injector, it can leak. This will result in rough idle and backfire. Residual fuel pressure (pressure remaining in the line after the pump is off) can cause the injector to leak. This can cause hard starting and an overly rich mixture during starting.

Remove the cold start injector and hold it in a container (Figure 27.22). When the engine coolant is below a certain temperature, the injector should spray, but only when the engine is cranked. It is powered off of the starter solenoid circuit. If the injector does not work, check to see that it has power to it. If not, check the thermo time switch.

Leakage Test

During a system pressure test when the fuel pump is turned off, pressure in the system should remain constant for the next startup. If pressure drops while the engine is off, remove the fuel rail and repressurize it. The injectors should not leak.

- If the injectors do not leak but pressure on the fuel gauge drops, the check valve in the fuel pump could be leaking.
- Repeat the test by pressurizing the system and then pinching off the fuel line from the pump. If the pressure remains steady, the check valve is leaking.
- To test to see if pressure is leaking through the pressure regulator, plug the fuel return line and repeat the test. If the pressure remains constant, the pressure regulator valve is leaking.

If pressure leaks from the system, cycling of the ignition switch more than once is often required before starting the engine. This is because more than

Cylinder	1	2	3	4	5	6
High Reading	225	225	225	225	225	225
Low Reading	100	100	100	90	100	115
Amount of Drop	125	125	125	135	125	110
	OK	OK	OK	Faulty, Rich (Too Much) (Fuel Drop)	OK	Faulty, Lean (Too Little) (Fuel Drop)

Figure 27.20 Results of an injector balance test indicate faulty injectors. *(Courtesy of General Motors Corporation, Service Technology Group)*

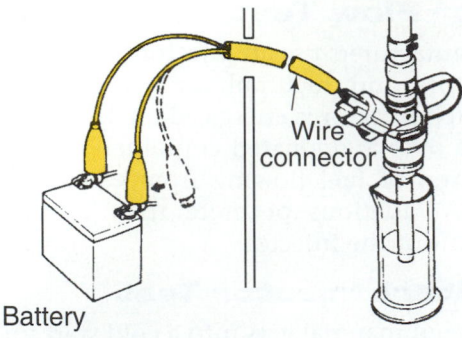

Figure 27.21 Each injector is energized for 15 seconds while holding it over a graduated container.

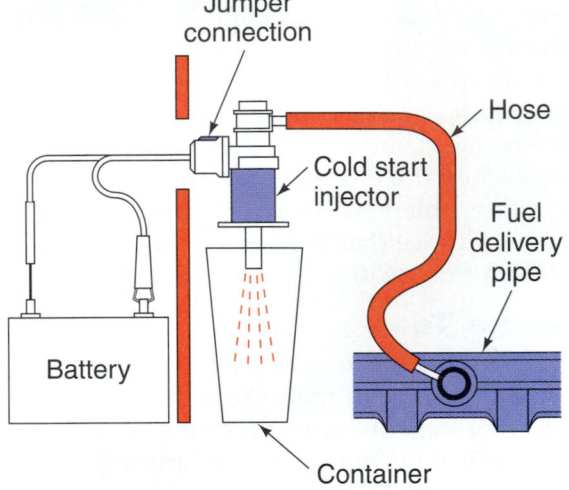

Figure 27.22 Remove the cold start injector and hold it over a container.

Figure 27.23 An injector cleaner connected to the Schrader valve on the fuel rail is pressurized by shop air. (*Courtesy of OTC-SPX Corp. Aftermarket Tool & Equipment Group*)

two seconds of pump operation might be required to fill the fuel rail.

Cleaning Injectors

Injector cleaning fluid is a mixture of the cleaner and gasoline. There are a variety of injector cleaning machines and processes. One type uses a canister, pressurized by shop air (Figure 27.23). A hose is connected from it to the Shrader valve on the fuel rail. The engine burns the pressurized fuel and cleaning solution as it runs. The fuel pump is disabled and the pressure regulator return line is blocked to prevent the solution from returning to the fuel tank.

■ THROTTLE PLATE SERVICE

Occasionally, gum and carbon can accumulate around the throttle plate on throttle body and port injection cars. The result is a rough idle. A spray can of carburetor cleaner can be used to clean this area. Be sure the spray cleaner used is safe for oxygen sensors. If spray cleaner does not work, the throttle assembly will have to be removed and soaked in a cleaner.

■ EFI ADJUSTMENTS

Common adjustments to EFI systems include idle speed, throttle stop, idle air/fuel mixture, and throttle cable. Raising idle speed means allowing more air to pass the throttle plate. Sometimes there is a screw that opens or closes a passageway. An idle air control (IAC) motor (Figure 27.24) is used to raise the idle speed when extra loads (such as air conditioning) are placed on the engine. Occasionally carbon blocks all or part of the air passage, resulting in engine stalling or erratic idle speed.

On later model vehicles, the idle speed is controlled by the computer. These systems have an idle speed control (ISC) motor (Figure 27.25). As with any

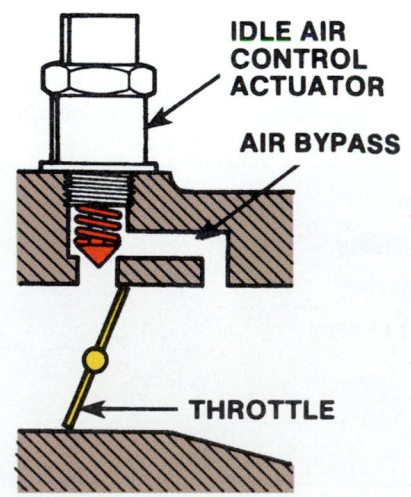

Figure 27.24 Allowing air to bypass the throttle plate raises the engine idle.

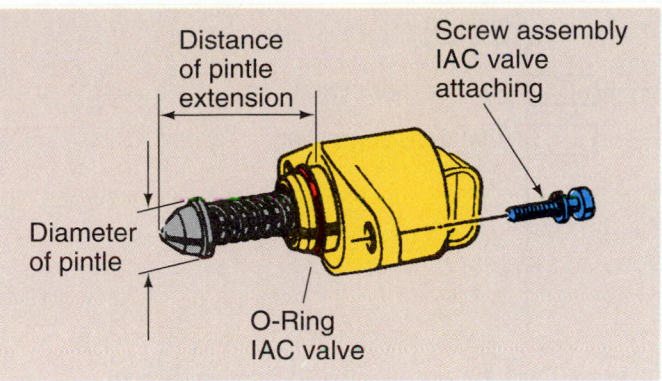

Figure 27.25 An idle speed control motor. *(Courtesy of General Motors Corporation, Service Technology Group)*

new part, when one of these motors is replaced be sure to compare the old part to the new one. On one kind of motor, the pintle on the end of it must be pushed into the motor until a specified distance is reached. If the pintle is too far out, it could be damaged during installation.

SENSOR SERVICE

Testing procedures vary for the various fuel injection system sensors. Before disconnecting a computer system component, be certain that the ignition key is turned off. Use a scan tool, a digital volt ohmmeter, or a test light. Always follow manufacturer's service manual procedures. A scan tool is a device that reads computer self-diagnosis signals (see Chapter 76).

THROTTLE POSITION SENSOR

A bad throttle position sensor (TPS) can cause a change in idle speed, a stumble on acceleration, or engine stalling. Follow manufacturer's procedures for testing. The TPS has a metal wiper arm that rubs against another metal strip (Figure 27.26). If the strip wears away, momentary interruptions of the electrical signal can occur. These interruptions are called **glitches**.

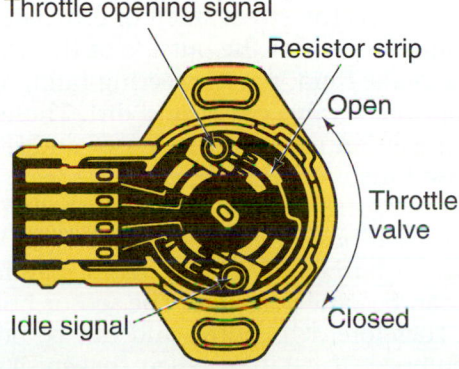

Figure 27.26 A throttle position sensor has a wiper arm that rubs against a resistor strip.

A sensor can be tested with a voltmeter while its electrical wiring is still *connected* (this is called *backprobing* a connector). With the wiring *disconnected*, the TPS can be tested with an ohmmeter. Move the throttle slowly from closed to open. At different throttle openings, varying resistances are specified. Watch for any glitches.

Some throttle position sensors have three wires and some have four. With a four-wire sensor, the fourth wire is for the idle switch. Procedures for testing are given in the service manual.

Most throttle position sensors are made so that their adjustments are tamperproof. They have screws that are either soldered or staked. To remove the switch, these might need to be drilled or filed off. After the new switch is installed and adjusted, the new mounting screws are restaked.

OXYGEN SENSOR SERVICE

The oxygen sensor is tested with the engine running at operating temperature. Check the manufacturer's service manual for the correct procedure. Use a digital voltmeter. An analog voltmeter (one with a needle dial) can damage the sensor.

The voltmeter is connected to the O_2 sensor wire and grounded to perform the test. Some sensors have only one wire coming from them. These sensors ground through the body of the sensor. To test this kind of sensor, ground one of the voltmeter wires. Attach the other lead to the wire from the sensor. If there are two wires coming from the sensor, one is for ground. Figure 27.27 shows a **schematic** of typical oxygen sensor wiring.

Heated O_2 Sensors

If there are three or four wires coming from the sensor, it is a heated sensor. The sensor is heated so the computer can go into closed loop sooner and stay in closed loop during long periods of idle. Two of the wires are for the heater. With four wires, one is for the ground signal to the computer (Figure 27.28). Probe the other wire to get the computer signal (this wire is usually a different color). Check with a voltmeter at each wire to see which one reads between 0 and 1 volt. With the key turned on (not starting), the reading will be about 0.4 volt on most cars.

The two extra wires are for the heater and will read battery voltage across their connector. The O_2 sensor heater can be tested with an ohmmeter. If it does not have the specified resistance, the heater coil is bad and the sensor is replaced.

Testing the O_2 Sensor

Run the engine and observe to see that the voltage fluctuates rapidly and repeatedly from 0.2 volt to 0.8 volt or so. This means that the sensor is working properly. Sometimes an oxygen sensor is "lazy." This means

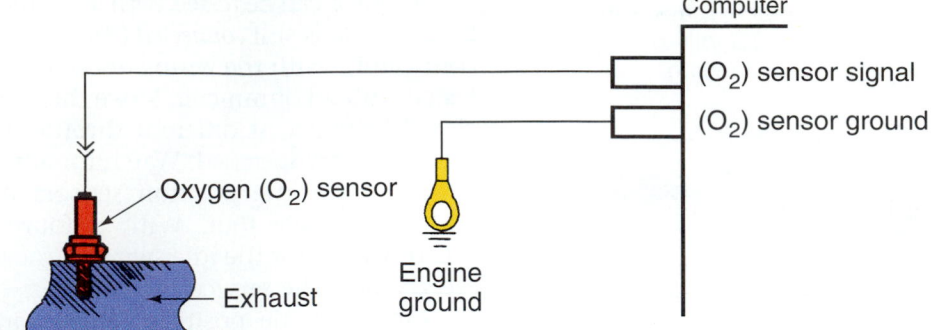

Figure 27.27 Oxygen sensor wiring. *[Courtesy of General Motors Corporation, Service Technology Group]*

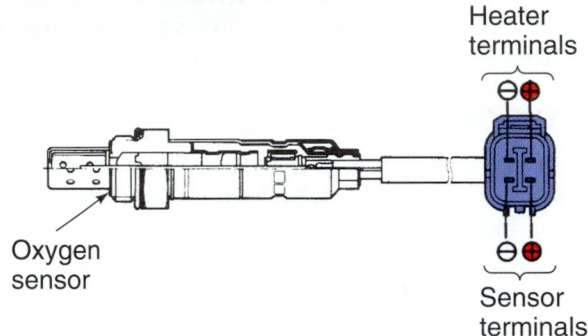

Figure 27.28 A four-wire heated O_2 sensor. *[Courtesy of American Honda Motor Co., Inc.]*

that it does not sweep back and forth fast enough between a rich and a lean signal. The speed at which this signal fluctuates is called **cross counts** and is a measurement made by scan tools.

When the O_2 sensor voltage remains low, the air/fuel ratio could be too lean, the sensor could be defective, or the wire between the sensor and computer could have high resistance.

NOTE: *The oxygen sensor only senses oxygen. A misfiring cylinder can give a false signal to the computer. Because of the abundance of oxygen in the unburned mixture, the reading sent to the computer from the oxygen sensor will be lean. The computer will compensate by increasing the injectors' pulse width, richening the air/fuel mixture.*

If the voltage remains high, the air/fuel ratio might be too rich or the sensor could be contaminated. When the platinum strip becomes insulated with a buildup of foreign material, it cannot react to oxygen ions to provide a signal to the computer.

Contaminated O_2 Sensor

The oxygen sensor can become contaminated in several ways. If the pores of the sensor become plugged, its response time and output voltage will drop. Zirconium sensors can be poisoned from either side. One side is exposed to outside air around the exhaust manifold, and the other side is exposed on the inside to the engine's exhaust stream.

Fumes from some types of **RTV** sealants used on the engine can contaminate O_2 sensors. Be sure that the RTV that you use lists **O_2 sensor safe** on its label.

Lead in fuel can also contaminate an oxygen sensor. It can come from use of some gasoline octane boosters. Another source of lead is from the rustproof coating inside some fuel tanks. It can be loosened by methanol that is added to some fuels, during the winter in particular.

Carbon buildup on the sensor can be from an overly rich air/fuel mixture, engine oil consumption, or a bad turbocharger turbine seal (see Chapter 28). Coolant leaking from a bad head gasket or cracked cylinder head can also contaminate a sensor.

Be sure to check for the source of the contamination and correct the problem before replacing the sensor. Diagnosis of these problems is as follows:

- Silicone contamination causes smooth, chalky white deposits on the tip of the sensor. Sensor output voltage can also go negative as a result of silicone contamination.
- Engine oil leaves a brown residue that causes sensor speed to slow down.
- A rich air/fuel mixture leaves a black coating that can be burned off by running the engine lean at fast idle for a few minutes. To make the engine run lean, pull a vacuum hose and unhook the oxygen sensor. This will result in a hotter exhaust.
- Coolant leaves a white flaky deposit that sometimes has the sweet smell of ethylene glycol.
- Contamination from the outside of the sensor can include brake fluid, power steering fluid, oil leaking from a valve cover, and dirt. These can block the air entrance to the sensor and slow down response time.

Sometimes, the oxygen sensor can suffer physical damage to itself or its electrical connections. Wires can be pinched or burned by a hot exhaust manifold. Connections can become corroded. The sensor housing is a ceramic. Therefore, it is brittle and can be broken like glass if bumped. If it rattles when shaken, it has been broken. Oxygen sensors are replaced on a preventative maintenance basis by many shops as a part of a tuneup.

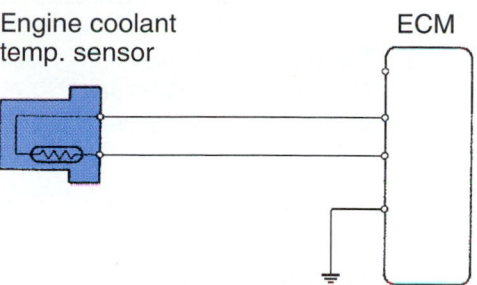

About half the O₂ sensors that are replaced are not faulty. When carbon forms on a sensor, it can be cleaned off by running the fuel system "lean" for two minutes. Pulling a large vacuum hose leans out the air/fuel mixture, which heats the exhaust and burns off the carbon. Sometimes, carbon can be burned off of the sensor with a torch during a bench test. The sensor must be removed from the car for the bench test.

Benchtesting an O₂ Sensor

When the oxygen sensor is removed from the engine, it can be tested with a propane torch. If the voltmeter is hooked up to the sensor and the torch is used to first heat the sensor and then provide a rich and lean mixture as the torch's flame is waved past the end of the sensor, the voltmeter should react. The O₂ sensor is a generator. It produces voltage in response to the richer mixture of the propane.

Replacing the O₂ Sensor

The sensor is threaded into the exhaust manifold. It can be difficult to remove unless a special anti-seize compound is coated onto its threads. Torque the sensor to 30 foot-pounds using a special socket. A sensor that is too loose or a cracked exhaust manifold can result in a lean signal to the computer.

■ COOLANT TEMPERATURE SENSOR

The resistance of the coolant temperature sensor (ECT) varies with changes in temperature. The service manual gives the resistance values at different temperatures (Figure 27.29). The sensor can be tested in hot water using a thermometer and an ohmmeter (Figure 27.30). Because of the time involved in removing a sensor to test it in hot water, a technician will usually test it on the car using a scan tool (see Chapter 76).

 **CAUTION** Do not test an ECT sensor with an open flame. The sensor will be damaged.

■ INLET AIR TEMPERATURE SENSOR

The sensor that tells the fuel system the temperature of incoming air is called an air charge temperature (ACT) sensor. It can be removed and tested in hot water like an ECT sensor (see Figure 27.30).

■ MAP SENSOR DIAGNOSIS

A defective MAP (manifold absolute pressure) sensor can cause the engine to run rich or lean. When the

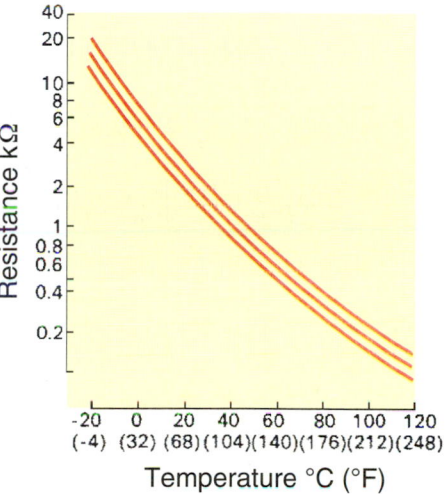

Figure 27.29 A coolant temperature sensor gives different resistance readings as its temperature changes.

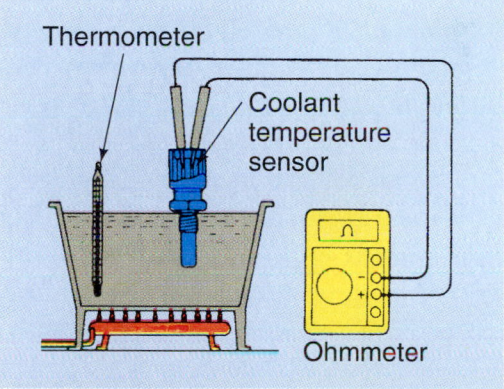

Figure 27.30 Testing a coolant temperature sensor in hot water using a thermometer and an ohmmeter.

engine is off and the key is on, MAP sensors are supposed to tell the computer what the barometric pressure is. The service manual gives the voltage specification for this test, which will vary with altitude and the weather. Voltage readings at various barometric pressure readings are given in the service manual. Figure 27.31 is an example of a chart used by one manufacturer. You will have to check with a source about the weather or have your own barometer to find out what the present barometric pressure is.

Absolute Baro Reading	Lowest Allowable Voltage at −40°F	Lowest Allowable Voltage at 257°F	Lowest Allowable Voltage at 77°F	TBI MAP Sensor Designed Output Voltage	Highest Allowable Voltage at 77°F	Highest Allowable Voltage at 257°F	Highest Allowable Voltage at −40°F
31.0"	4.548 V	4.632 V	4.716 V	4.800 V	4.884 V	4.968 V	5.052 V
30.9"	4.531 V	4.615 V	4.699 V	4.783 V	4.867 V	4.951 V	5.035 V
30.8"	4.514 V	4.598 V	4.682 V	4.766 V	4.850 V	4.934 V	5.018 V
30.7"	4.497 V	4.581 V	4.665 V	4.749 V	4.833 V	4.917 V	5.001 V
30.6"	4.480 V	4.564 V	4.648 V	4.732 V	4.816 V	4.900 V	4.984 V
30.5"	4.463 V	4.547 V	4.631 V	4.715 V	4.799 V	4.883 V	4.967 V
30.4"	4.446 V	4.530 V	4.614 V	4.698 V	4.782 V	4.866 V	4.950 V
30.3"	4.430 V	4.514 V	4.598 V	4.682 V	4.766 V	4.850 V	4.934 V
30.2"	4.413 V	4.497 V	4.581 V	4.665 V	4.749 V	4.833 V	4.917 V
30.1"	4.396 V	4.480 V	4.564 V	4.648 V	4.732 V	4.816 V	4.900 V
30.0"	4.379 V	4.463 V	4.547 V	4.631 V	4.715 V	4.799 V	4.883 V

Figure 27.31 MAP sensor voltage signals at various barometric pressures. *(Courtesy of Chrysler Corporation)*

■ SCIENCE NOTE ■

Barometric Pressure

Surrounding the earth is a blanket of air (atmosphere) approximately 500 miles thick. The atmosphere exerts a pressure on us. We can show the existence of this pressure by filling a long tube with mercury and inverting it in a dish of mercury to create a barometer. When the tube is inverted, some but not all, of the mercury runs out of it. The fact that not all of the mercury runs out of the tube shows that there must be a pressure (barometric pressure) exerted on the surface of the mercury in the dish, which is sufficient to support the amount of mercury remaining in the tube. If no pressure were exerted, there would be nothing to stop all of the mercury from running out of the upside-down tube.

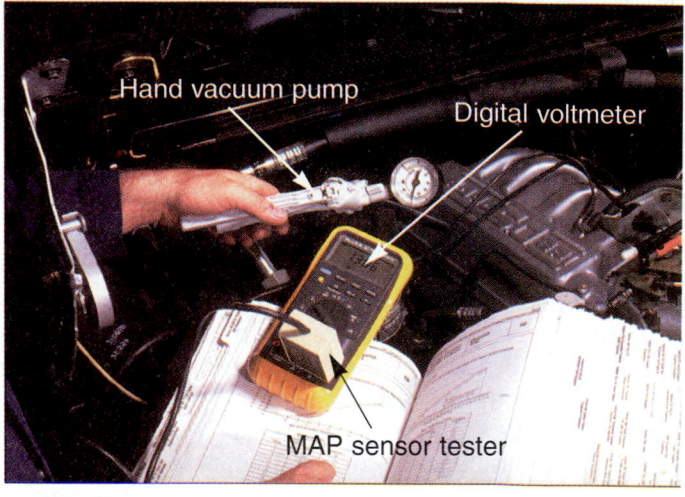

Figure 27.32 A tester is used that changes the frequency voltage to analog voltage so the voltmeter can read it.

Continue the test by applying vacuum to the MAP sensor with a hand vacuum pump. A typical test procedure is to apply 5 inches of vacuum to the MAP sensor and watch for the voltage reading to drop by a specified amount. Next the pressure is lowered (vacuum is raised) to 10 inches. The voltage should change once again. The test continues by applying vacuum in 5" intervals until 25" is reached. If the readings are out of specifications, the sensor is replaced.

Some MAP sensors produce a signal that varies in frequency. This is called a voltage frequency signal. A tester is used that changes the frequency voltage to analog voltage so the voltmeter can read it. The tester is shown in Figure 27.32.

■ MASS AIR FLOW (MAF) SENSOR DIAGNOSIS

Testing a mass air flow sensor requires a voltmeter with a **min/max** feature. With the key on, the min/max button is pressed. With a vane-type MAF sensor, the air vane is moved from closed to wide open and back to closed (Figure 27.33). Pressing the min/max button will give the maximum voltage obtained. Pressing it again will give the lowest voltage (Figure 27.34). If

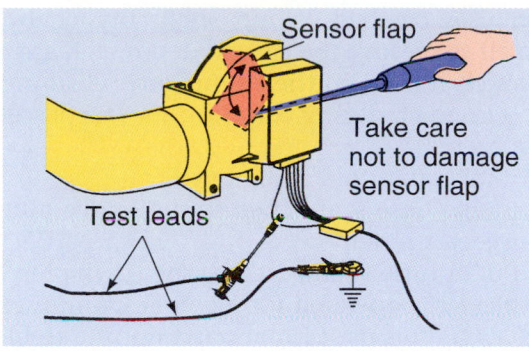

Figure 27.33 Move the MAF sensor air vane from open to closed to test it. *(Reproduced with permission from Fluke Corporation)*

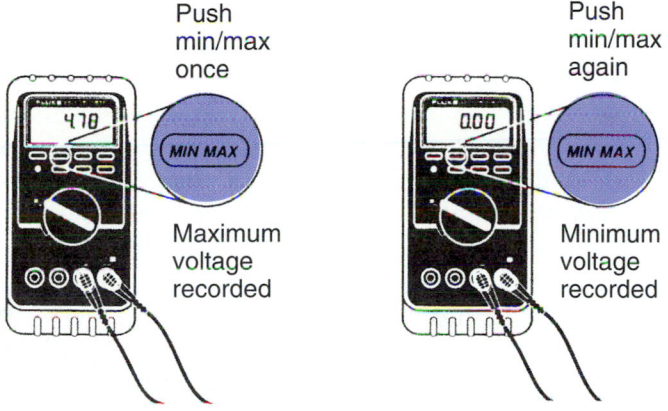

Figure 27.34 Use the min/max feature to read minimum and maximum sensor voltage signals. *(Reproduced with permission from Fluke Corporation)*

minimum voltage is zero, there might be an open circuit in the sensor's variable resistor.

With some vane-type MAF sensors, manufacturers provide ohmmeter specifications. At some terminals, a thermistor might allow temperature to affect the resistance readings. This will be stated in the service manual. The air vane is moved through its normal range of motion while observing a smooth reading on the voltmeter.

When testing a hot wire or heated resistor MAF sensor, run the engine and tap on the sensor. If the engine misfires, there is a loose internal connection requiring replacement of the sensor. There is a simple procedure for checking voltage and frequency readings with a multimeter. Follow the procedure in the service manual.

■ TESTING OTHER SENSORS

Other EFI sensors are tested in ways similar to those discussed previously. Check resistance values using a scan tool and the vehicle's computer self-diagnostic system or use an ohmmeter or voltmeter. Refer to the service manual for the procedures and values of each type of sensor.

■ COMPUTER SERVICE

The computer is rarely the cause of problems in the fuel system. If it is, be sure to locate any problem in the system that might have caused it to fail. When a computer is faulty, it is replaced. Remanufactured computers are widely available at a reduced cost for popular makes of cars. Some computers have a replaceable element called a **PROM.** This stands for **programmable read only memory**. The old PROM is installed in the replacement computer. Other later model computers have *EE Flash PROMs*. These are programmed electronically, either in the dealership or by telephone modem from the manufacturer. See Chapter 69 for more about computer systems.

■ MECHANICAL INJECTION

Servicing Continuous Injectors

Compared to electronic systems, mechanical fuel injection systems run under very high pressure. They have a spring loaded valve that requires at least 50 psi to open. When they spray, there is an even fuel pattern. To check the operation of the injectors, remove each of them using two wrenches. Pull each injector out of the head and put them in a container to catch the fuel. Activate the fuel pump and move the air flow sensor to cause fuel flow to change. Check the spray pattern. A tester is available that keeps the fuel contained for a safer and more thorough check (Figure 27.35).

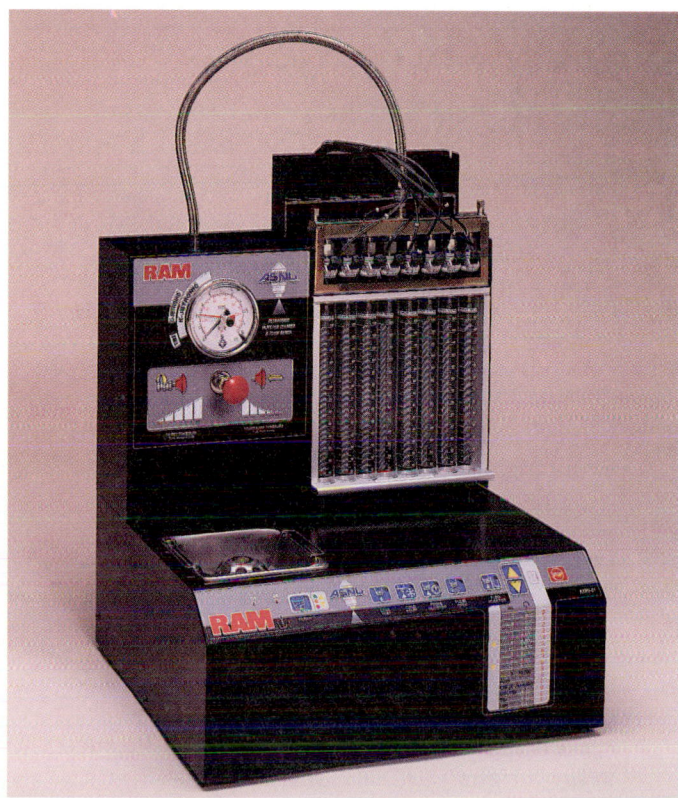

Figure 27.35 A mechanical fuel injection tester. *(Courtesy of Echlin Engine Systems)*

Volume of flow can be checked. Each injector should flow an equal amount. If not, there is a problem with either the fuel distributor or fuel injectors.

> **SAFETY NOTE**
>
> ■ There is a danger of fire from spilled gasoline when it is sprayed at higher than 50 psi from the injectors.
>
> ■ Use a plastic container. Never use glass. If it breaks, you could be badly burned and the car could be destroyed in a fire.
>
> ■ Get help when performing this test. Use two hands to hold the container.

Switch injectors between cylinders. If the problem remains at the original cylinder, the fuel distributor is at fault. Fuel distributors are replaced as a unit. Refer to a service manual for instructions.

■ CARBON DEPOSIT SERVICE

Carbon deposits cause driveability problems because fuel vapors can be absorbed into them. This results in rough idling when cold, as well as loss of power, surging, and high emissions. Carbon deposits on valves sometimes cause problems in as little as 5000 miles.

One method of removing carbon deposits is a tool that has been around for many years. A **carbon blaster** uses crushed walnut shells blasted by compressed air to remove the carbon deposits. If any pieces of the shells remain in the engine after cleaning, they will be burned up during combustion. The process is as follows:

■ The intake manifold is removed.

■ The intake hose is attached to the intake port or fuel injector hole.

■ The outlet hose is attached to the spark plug hole.

■ The blaster is operated for a couple of minutes.

■ Last, air only is used for about a minute to blow out the shells.

Sometimes carbon that has accumulated on a valve or in the combustion chamber can drop off and be crushed against the cylinder head. If this occurs during cranking, the piston can stop on its upstroke. When this happens at low rpm with the engine running, the noise resembles the sound of a bad rod bearing.

NOTE: *When carbon increases the engine's compression ratio, detonation can result. "Carbon knock" can be very noticeable when the engine is cold. Typically, engine knock occurs under load, but carbon knock can occur even when not under load.*

Carbon can be removed by using an additive such as GM Top Engine Cleaner.

NOTE: *Carbon removers can damage catalytic converters. Follow the manufacturer's instructions for their use.*

■ REVIEW QUESTIONS

1. When the float level in a carburetor is too high, the air/fuel mixture becomes rich/lean (circle one).

2. When the car stumbles when accelerated, this is called a _____ spot.

3. What is the kind of light that is invisible to your eyes called?

4. An engine is _____ when too much raw fuel has entered it and it will not run.

5. When the engine is cold, a computer feedback system is in open/closed (circle one) loop.

6. When a fuel pressure regulator fails, is the pressure usually too high or too low?

7. Most fuel injectors are powered on one side and controlled by the computer through the _____ circuit.

8. The O-ring around the base of a fuel injector does two things, insulate heat and prevent _____ leaks.

9. When testing an oxygen sensor, the voltage should fluctuate between 0.2 and _____ volt(s).

10. Some types of _____ sealer can damage oxygen sensors.

■ ASE STYLE REVIEW QUESTIONS

1. Technician A says that there is usually residual fuel pressure in fuel lines, even when the engine is off. Technician B says that a small air (vacuum) leak will result in a lower idle speed. Who is right?

   a. Technician A   b. Technician B

   c. Both A and B   d. Neither A nor B

2. Technician A says that a flat spot can be caused by retarded ignition timing. Technician B says late model cars with carburetors have idle speed regulated by an idle speed control motor controlled by the computer. Who is right?

   a. Technician A   b. Technician B

   c. Both A and B   d. Neither A nor B

3. Technician A says that to start a fuel injected car, the accelerator should be depressed once. Technician B says that to start a car with a carburetor, do not touch the accelerator pedal until the engine is running. Who is right?

a. Technician A   **b.** Technician B

c. Both A and B   **d.** Neither A nor B

4. Technician A says that when a car is flooded, the accelerator pedal is held to the floor for a few minutes to let the fuel dry out. Technician B says closing the throttle plate all of the way when the engine is shut off prevents engine run-on. Who is right?

a. Technician A   **b.** Technician B

c. Both A and B   **d.** Neither A nor B

5. A computer feedback fuel system has poor driveability when cold. Technician A says that this could be due to a bad oxygen sensor. Technician B says to check for external causes such as an intake manifold leak or incorrect fuel system pressure. Who is right?

a. Technician A   **b.** Technician B

c. Both A and B   **d.** Neither A nor B

CHAPTER 28

Intake and Exhaust Systems/ Turbochargers and Superchargers

■ KEY TERMS

plenum
siamese runners
dual-plane manifold
single-plane manifold
cross flow head
headers
resonator
catalytic converter
normally aspirated
turbocharger
draw-through
blow-through
waste gate
boost pressure
turbo lag
intercooler/charge air
    cooler/aftercooler
blowers
roots-type
plain bearings
NCFR

■ INTRODUCTION

This chapter deals with the parts, operation, and service of intake and exhaust systems, turbochargers, and superchargers. The intake system is covered first, followed by the exhaust system. Turbochargers and superchargers are covered last.

■ INTAKE SYSTEM FUNDAMENTALS

There are two sources of engine contaminants: internal contaminants generated by heat and friction within the engine, and dirt that enters through the air intake system. Every gallon of gas burned requires about 9000 gallons of air. The air that enters the engine must be filtered, but the filter must allow sufficient air flow for good engine operation. The air filter also muffles the sound of the air rushing into the engine. Another job of the air cleaner is to act as a flame arrestor in case of a popback in the intake manifold.

NOTE: *A popback is an explosion that occurs in the intake system. A backfire is an explosion that occurs in the exhaust system.*

The common types of filter in use today are the dry paper type, made of pleated paper, and the oil wetted polyurethane type. Air filters are rated for *efficiency*, *flow*, and *capacity*.
■ If 100 grams of dirt enters the filter housing and 99 grams are filtered out, the filter is said to be 99% efficient.
■ The air filter must allow enough air to pass at the engine's maximum speed to not interfere with flow.

■ Capacity is the amount of dirt an air filter can hold before it becomes restricted.
There must be a balance of efficiency, flow, and capacity for a filter to be effective.

Carburetor/TBI Air Filter

The air filter used for carburetors or throttle body injection (TBI) is usually round (Figure 28.1). It usually has an outer screen that protects the paper and holds it in place. An inner screen supports the paper and absorbs

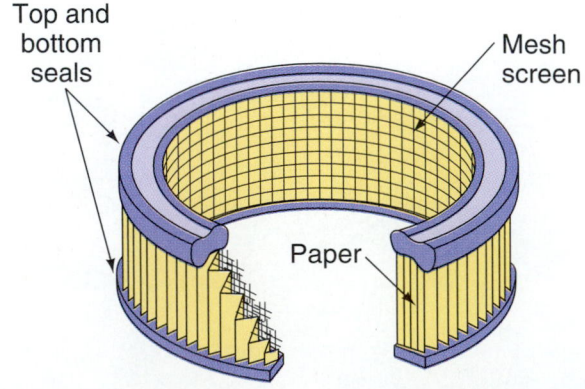

Top and bottom seals

Mesh screen

Paper

Figure 28.1 Parts of a carburetor air filter element.

heat if there is a popback in the intake manifold. The paper is *pleated* so it can offer a greater surface area in a small package. Seals at the ends of the filter keep it from leaking at the filter housing.

The filter housing is metal and has provisions for heating the incoming air during cold operation (Figure 28.2). Some air cleaners have fresh air ducted in from the front of the car to improve combustion (Figure 28.3).

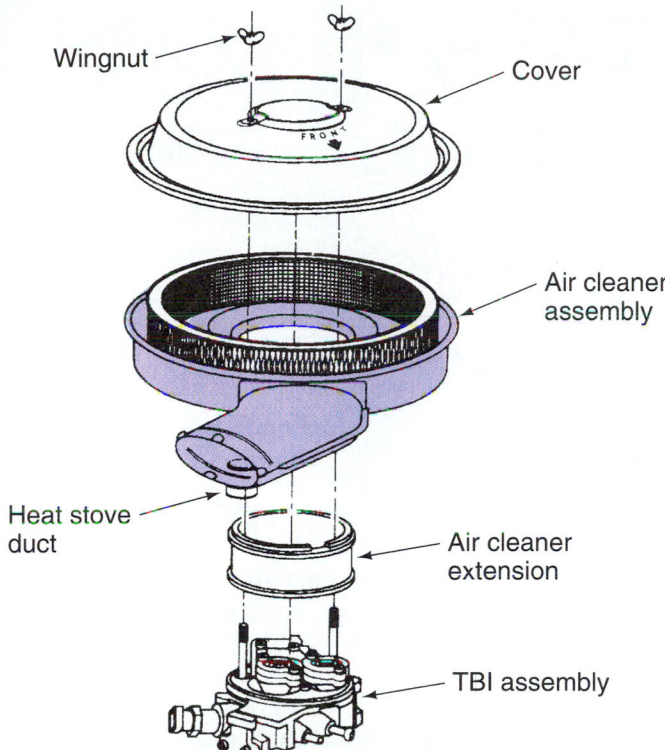

Figure 28.2 Parts of an air filter housing. *(Courtesy of General Motors Corporation, Service Technology Group)*

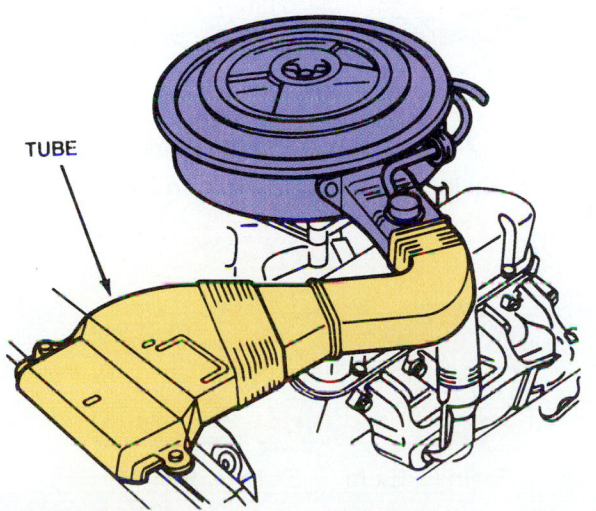

Figure 28.3 An air cleaner duct for a carburetor/TBI engine. *(Courtesy of Ford Motor Company)*

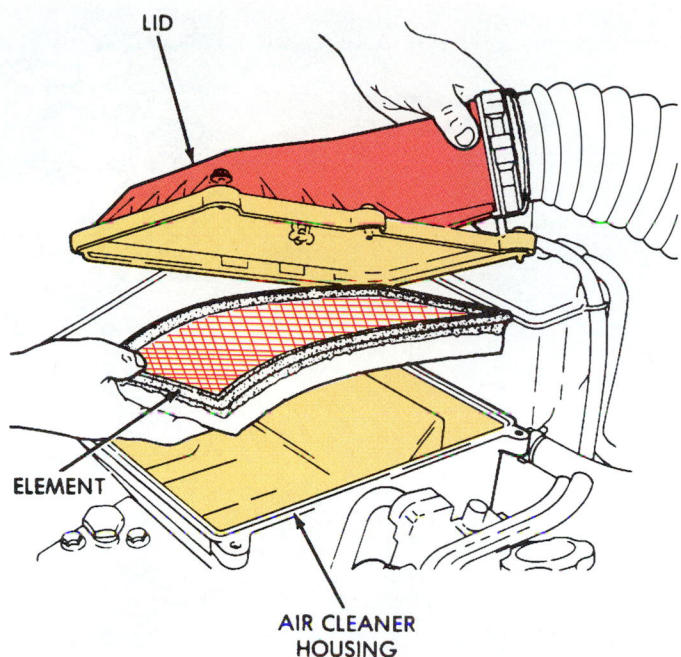

Figure 28.4 A typical fuel injection air cleaner. *(Courtesy of Chrysler Corporation)*

Fuel Injection Air Filter

Filters used with fuel injection systems are usually flat and fit into a plastic housing (Figure 28.4). Large rubber hose or ductwork carries the filtered air to the air inlet.

■ MANIFOLDS

There are two kinds of manifolds that are part of an engine's breathing system, the intake manifold and the exhaust manifold. Manifolds are carefully designed to provide a uniform air/fuel mixture to all cylinders. If they are the wrong size or design, the engine will not be able to breathe properly.

■ INTAKE MANIFOLDS

The passages in an intake manifold are known as runners. In engines that have carburetors or throttle body fuel injection (TBI), the intake manifold is designed to provide optimum flow for the *air/fuel mixture*. Engines with port fuel injection (PFI) inject the fuel directly above the intake valve, so the manifold is designed for *airflow only*.

Port injection manifolds look different than other manifolds. They can be designed with larger runners than air/fuel manifolds. When the manifold flows air only, the runners can also have sharper bends because these manifolds do not have to keep fuel suspended in air.

Intake manifolds that flow air and fuel must be designed to keep the fuel suspended in the air in fine droplets. By the time the mixture reaches the combustion chamber, most of the fuel should be evaporated so

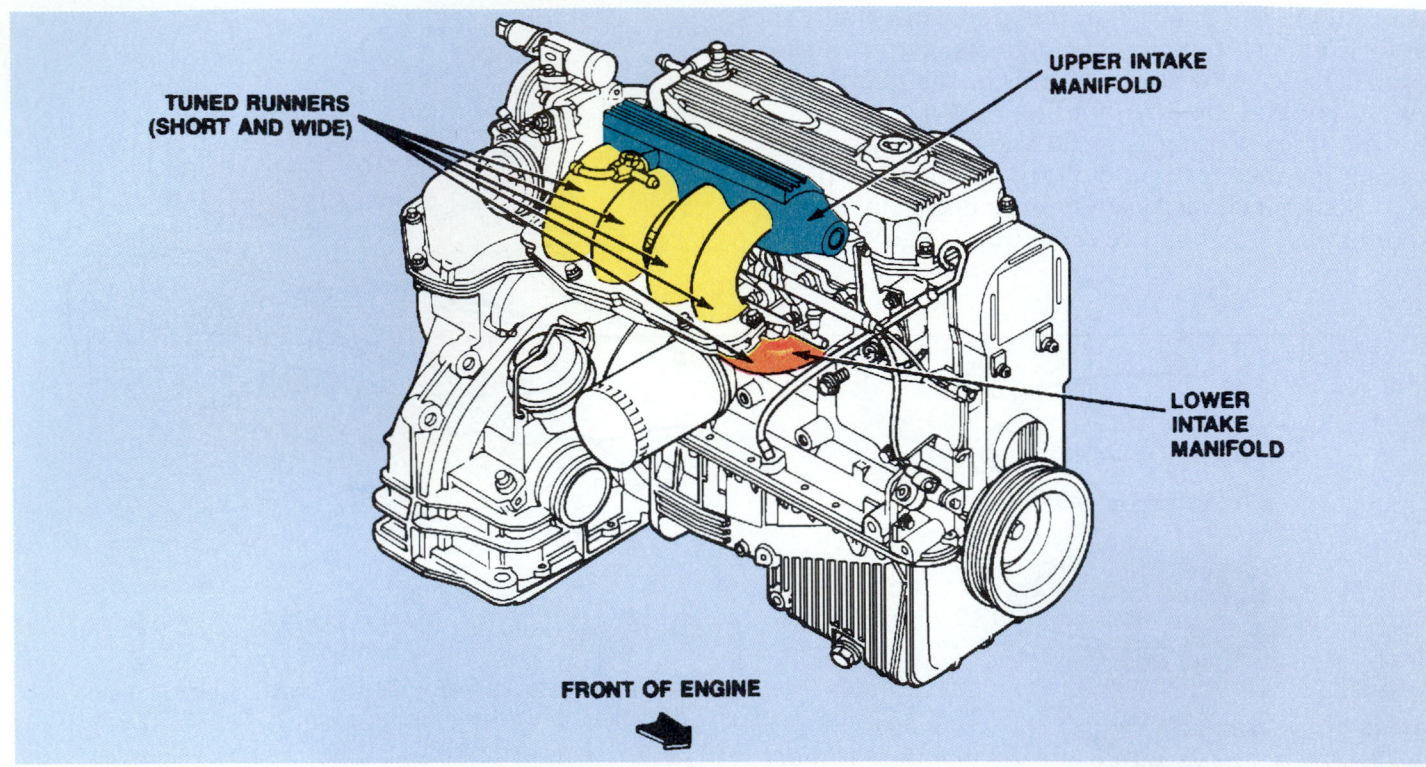

Figure 28.5 Larger diameter runners are shorter. *(Courtesy of Ford Motor Company)*

it will burn easily. If the speed of the mixture drops too low, droplets of fuel can fall out of the mixture.

Manifold runner sizes must be a compromise. Large diameter runners flow well at high speeds, but the fuel will separate from the air at lower speeds. Smaller diameter manifolds provide enough flow and will keep the fuel in suspension throughout the average rpm range of a passenger car. Usually, small diameter runners are longer, while larger diameter runners are shorter (Figure 28.5).

The air space below the carburetor is known as the **plenum**. The plenum floor is flat, and often has ridges cast into it to catch fuel that drops out of the mixture. This makes it easier for the fuel to evaporate or to rejoin the moving air/fuel mixture as it flows through the manifold.

Manifold Runner Arrangement

Manifolds used in inline engines are simple. Sometimes one runner will feed two neighboring cylinders. These are known as **siamese runners** (Figure 28.6).

The runners have as few bends as possible in order to reduce the chances of the vaporized fuel turning back into liquid fuel.

Dual-Plane Manifold. Eight-cylinder engines use a two-barrel carburetor. This means there are two openings in the bottom of it. Each of the openings is independent of the other, although they share the same float bowl. For an eight-cylinder engine with a

dual-plane two-barrel **manifold**, each barrel supplies fuel to four cylinders. To keep the runners as much the same length as possible, one barrel will serve both the inner two cylinders on the opposite side of the engine and the outer two cylinders on its own side (Figure 28.7). It is important to understand this design when troubleshooting vacuum leaks or carburetor failure. Sometimes the problem is found to be only in

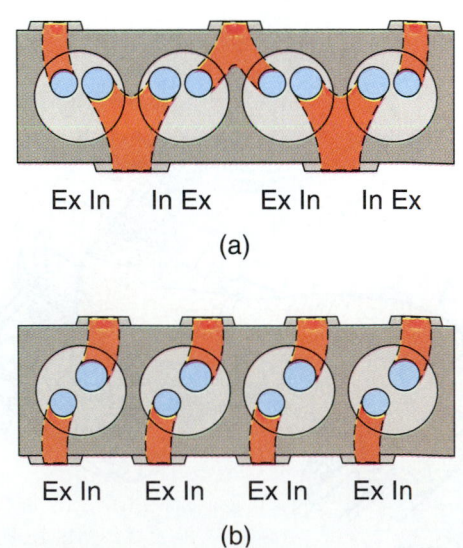

Ex In   In Ex   Ex In   In Ex

(a)

Ex In   Ex In   Ex In   Ex In

(b)

Figure 28.6 The top sketch (a) shows "siamese" valve ports that share a manifold runner. The bottom sketch (b) shows individual ports.

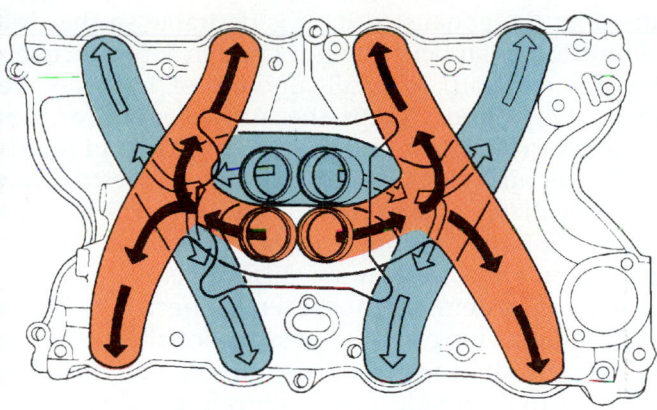

Figure 28.7 One barrel will serve both the inner two cylinders on the opposite side of the engine and the outer two cylinders on its own side. *(Courtesy of Ford Motor Company)*

EXHAUST CROSSOVER
PASSAGE FROM
CYLINDER HEAD

Figure 28.8 An exhaust gas passage in an intake manifold. *(Courtesy of Ford Motor Company)*

those cylinders served by one barrel. The dual-plane manifold has smaller runners and is better suited to lower rpm use.

Single-Plane Manifold. A manifold in which both barrels serve all eight cylinders is called a **single-plane manifold**. The single-plane manifold is more suited for high speed use. Manifold design is crucial to engine operation in much the same manner as camshaft design. The purchase of a high-performance manifold without the addition of matching components will probably hurt performance. Each design is a compromise. Breathing parts must be properly matched.

NOTE: *Better high rpm performance means worse low rpm performance.*

Manifold Heat Passage

Intake manifold heating is done on TBI and carbureted cars. There is a passage in the manifold for exhaust gas, which helps to vaporize the air/fuel mixture when the engine is cold and the manifold heat control valve (covered later) is closed (Figure 28.8).

Newer engines use port fuel injection, which does not require heating. This provides some advantages:

- Cooler air is more dense with oxygen than heated air.
- Since intake manifolds are not heated on these engines, they can be made of space age plastic materials.
- Manifolds can be made with longer curved air passages to increase air flow and power.

Coolant Passage

The intake manifold on a V-type engine has a coolant crossover passage. This passage connects the heads and provides the water outlet where the thermostat is found. A crack in this passage can cause a leak that can be difficult to diagnose.

Multiple Valve Heads

Some high-performance late model engines use three, four, or even five valves per cylinder (Figure 28.9). The use of multiple valves has become popular because this design helps higher rpm breathing. A greater amount of flow area for a given amount of valve lift is possible compared to two valve heads. A discussion of volumetric efficiency is covered in Chapter 17.

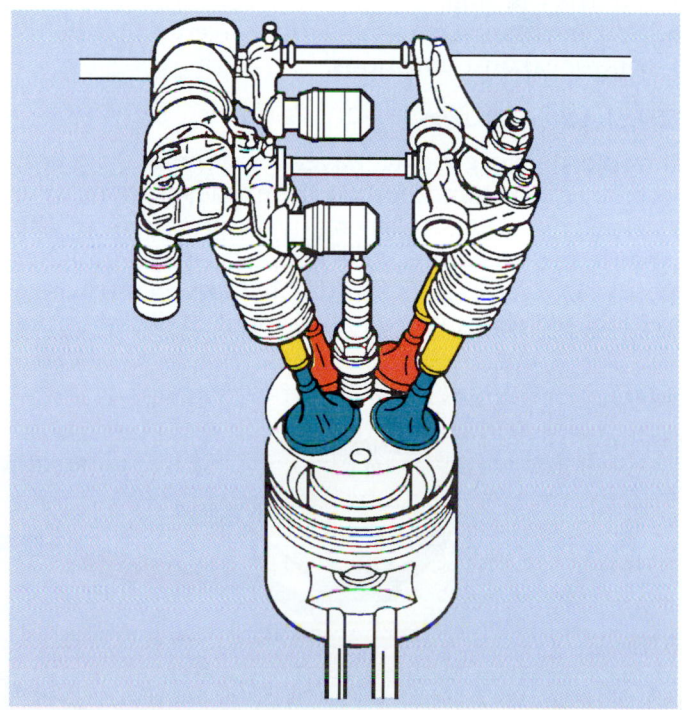

Figure 28.9 A four-valve combustion chamber. *(Courtesy of Ford Motor Company)*

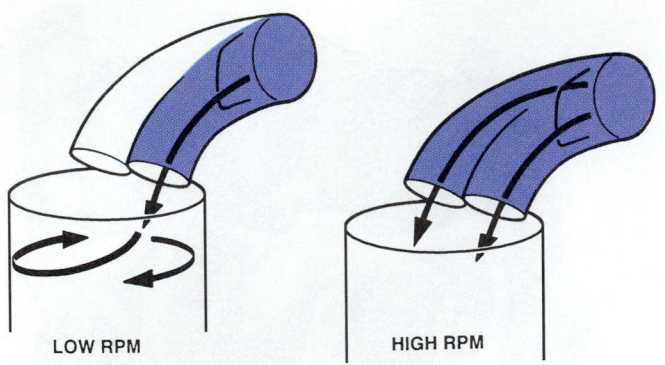

Figure 28.10 At low rpm, velocity and swirl are maintained. At high rpm, there is high flow. *(Courtesy of American Honda Motor Co., Inc.)*

Multiple Runner Manifolds

Two intake manifold runners can be used with multi-valve heads. These manifolds are variably tuned using a butterfly control valve to maintain velocity and swirl at low speed and high flow at high speed (Figure 28.10).

■ EXHAUST SYSTEM FUNDAMENTALS

Parts of the exhaust system include the exhaust manifold, exhaust pipe, the catalytic converter, the muffler (and sometimes a resonator), the tail pipe, and all of the hangers (Figure 28.11). The exhaust system serves three functions:

- ■ It carries burned exhaust gases away from the passenger compartment of the car.
- ■ It quiets the engine.
- ■ Most new cars have a catalytic converter that helps control exhaust emissions.

Backpressure

It would do no good to have a large intake system, with large valves and ports if the exhaust system were restricted. In four-stroke engines, restrictions in the exhaust can cause backpressure especially at higher speeds. Excessive backpressure reduces performance and fuel economy. The small amount of backpressure caused by the exhaust system is desirable so that the combustion chambers stay hot enough for more complete combustion. A small amount of backpressure (about 2¾ pounds) keeps the air/fuel mixture from going out the exhaust during valve overlap and results in a denser air/fuel charge in cylinder. The restrictions that the exhaust system provide also keep the exhaust system hot after the engine is shut off so that cold air cannot enter and warp an exhaust valve.

Exhaust systems are not severely affected by bends in the pipe as long as the cross-sectional area of the pipe is not diminished. There are special bending machines that provide a true radius to exhaust pipe.

■ EXHAUST MANIFOLDS

Exhaust manifolds are mounted to the cylinder head's exhaust ports (Figure 28.12). V-type engines have two exhaust manifolds and inline engines usually have one. When intake and exhaust manifolds are on opposite sides on an in-line engine, the head is called a **cross flow head** (Figure 28.13). This design improves breathing. Crossflow heads have a coolant passage to heat the intake manifold to help vaporize the fuel.

Exhaust manifolds are typically made of cast iron, although some late model cars use stainless steel manifolds. Cast iron is a good material for exhaust manifolds. Like the frying pan on your stove, it can tolerate fast, severe temperature changes.

Exhaust gas temperature is related to the amount of load on the engine. When the engine works hard, or when it has a lean air/fuel mixture, the exhaust manifold can run almost red hot. Usually, the manifold runs cooler, especially at idle.

NOTE: *At the factory, exhaust manifolds are usually bolted to the heads with no gaskets because the machined surface are perfectly flat. In service, replacement gaskets are usually used.*

Headers

Aftermarket manifolds made of tube steel are called **headers** (Figure 28.14). Headers are used when more air flow at higher speeds than stock is desired. Headers

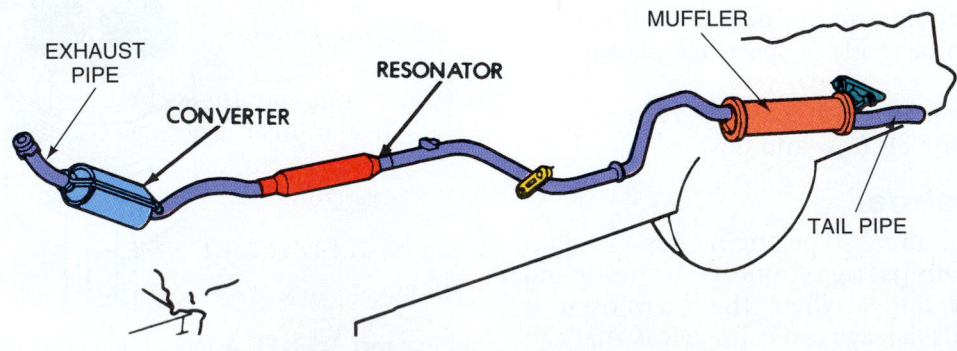

Figure 28.11 Parts of an exhaust system. *(Courtesy of Chrysler Corporation)*

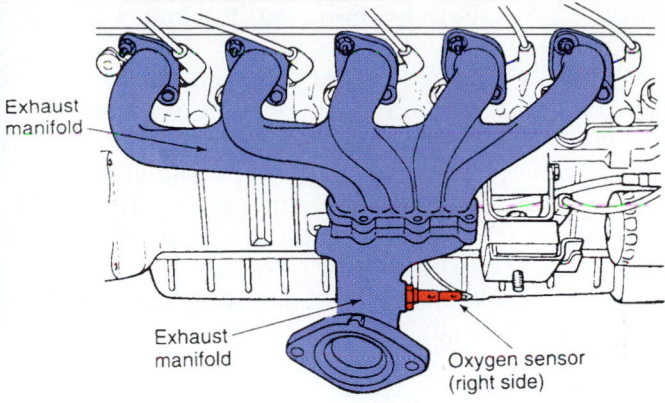

Figure 28.12 An exhaust manifold. *(Courtesy of Chrysler Corporation)*

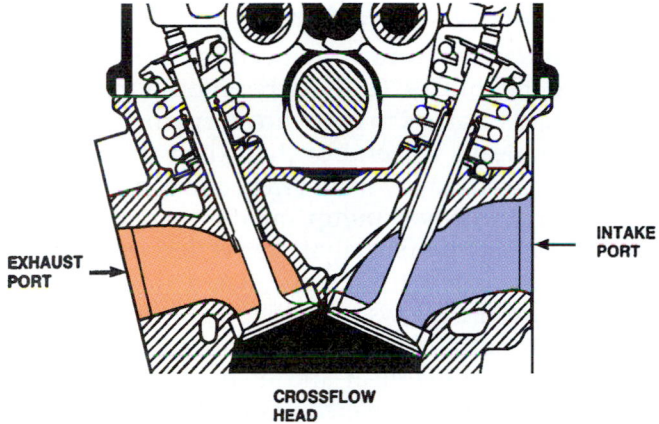

Figure 28.13 A crossflow head. *(Courtesy of Chrysler Corporation)*

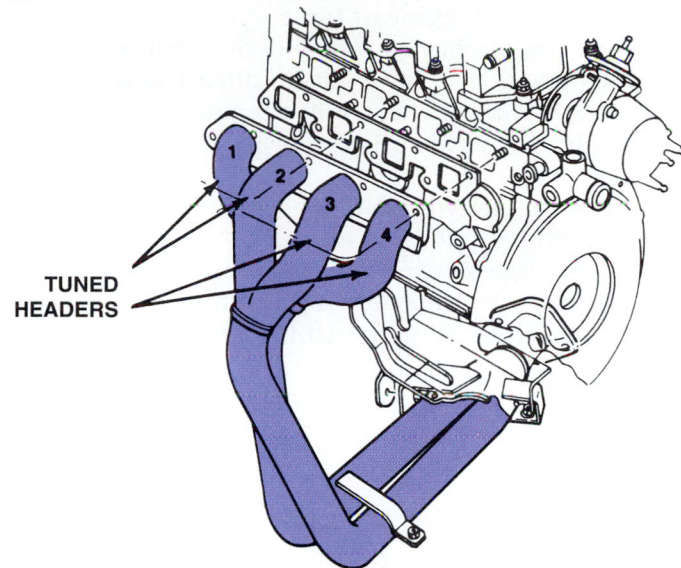

Figure 28.14 Headers. *(Courtesy of Ford Motor Company)*

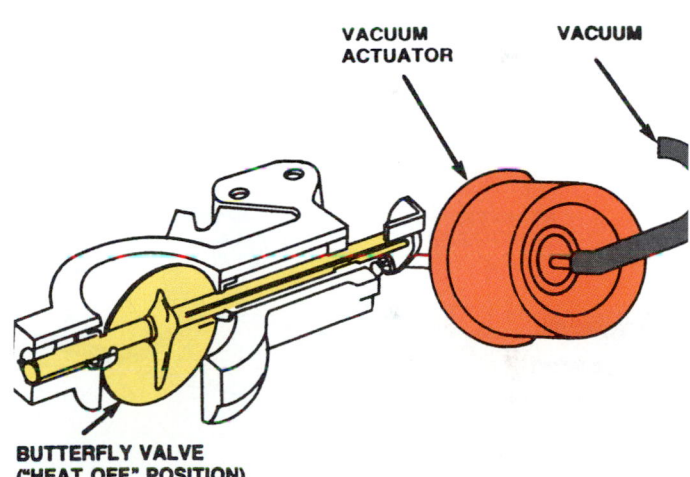

Figure 28.15 A vacuum-operated heat riser.

are short and have a large cross section to provide maximum high rpm power. The greater pressure from longer pipes tends to favor low rpm performance. Headers do not improve performance at low rpm unless they are specifically tuned for low rpm use. They also have a relatively short service life, because they can rust out.

Headers sometimes require modification to install. Emission requirements must be met or modified parts for street use cannot be installed. Any component that is substituted for an original emission part must have a BAR (Bureau of Automotive Repair) number or Federal Tampering Laws will be violated.

Manifold Heat Control

On cars with carburetors or throttle body fuel injection, the bottom of one of the exhaust manifolds can have a *manifold heat control valve*. This device, also known as a *heat riser*, is a butterfly valve that fits between the exhaust manifold and exhaust pipe. When the engine is cold, the valve blocks off part of the exhaust and sends it to the intake manifold floor under the carburetor. The hot exhaust preheats the air/fuel mixture, which helps it to vaporize.

In V-type engines, the heat riser causes a restriction in the exhaust on one side of the engine. This restriction causes exhaust to flow through the crossover passage under the carburetor to the manifold on the other side of the engine.

An older style heat riser had a large counterweight and a bimetal thermostatic spring that opened in response to heat. Another kind of heat control valve is vacuum controlled (Figure 28.15) in response to engine temperature.

■ EXHAUST PIPES

Exhaust pipes are made of steel. There are usually three exhaust pipes: the *header pipe* or *exhaust pipe*, an *intermediate pipe* between the muffler and catalytic converter, and the *tail pipe*.

Exhaust systems are of three styles:

- Inline engines have a simple design that runs down one side of the engine. Called a *single exhaust system,* there is only one pipe to the rear of the car. Single exhausts can be found on all engine sizes.
- V-type engines with *dual exhaust* have two pipes, two mufflers, and two catalytic converters (Figure 28.16). Large V-type engines use dual exhausts, especially when they are high performance. A dual exhaust allows better breathing when the engine is under load.
- A *crossover pipe* joins the two manifolds into one pipe as they leave the engine on a V-type engine's single exhaust system.

Muffler

Sound is vibration in the air. Each of the engine's exhaust valves releases a burst of pressurized exhaust every two turns of the crankshaft. The resulting noise from all of the cylinders blending together results in a loud roar. The roar is really many pressure bursts or vibrations.

The muffler has tubes and chambers that smooth vibrations out by letting the gas expand and cool (Figure 28.17). When the exhaust leaves the muffler, the

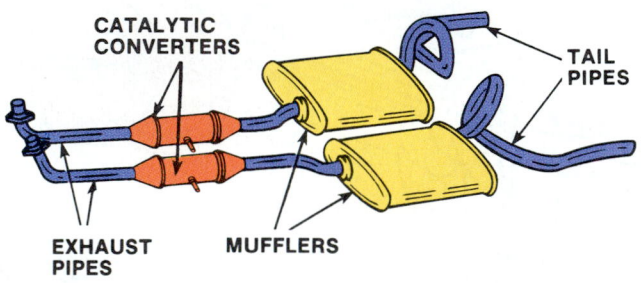

Figure 28.16 A dual exhaust system.

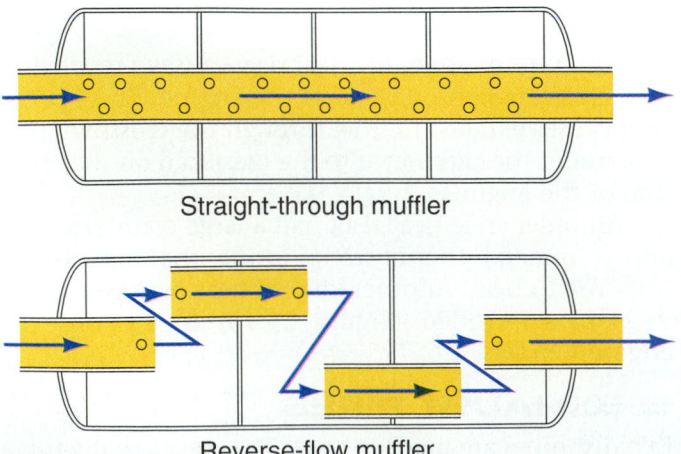

Straight-through muffler

Reverse-flow muffler

Figure 28.17 Two muffler designs. *(Reproduced by permission of Deere & Company. ©1991, Deere & Company. All rights reserved)*

pressure on it is more even, so it vibrates less. The result is less noise.

Mufflers are constructed of special materials and have different sizes and backpressures available to suit the particular need. There are two muffler designs: *straight through* and *reverse flow.* The straight through has a perforated inner pipe enclosed by the muffler housing, which is usually about three times the inner pipe's diameter. The reverse flow has short pipes and baffles that make the exhaust gas move back and forth as it travels through.

Resonator

A **resonator** is a second muffler in line with the other muffler. It further reduces the noise level.

Catalytic Converter

Catalytic converters have been installed on many cars since the mid 1970s. They resemble a muffler, but they contain catalysts to clean up an engine's emissions before they leave the end of the exhaust pipe (Figure 28.18). Some catalytic converters have plumbing to them from the smog pump on the engine. There is often a heat shield on catalytic converters because they get very hot when the engine has a misfire (Figure 28.19). A misfire allows raw gas to enter the converter because it is not burned in the cylinder. This can cause the converter to overheat and melt. Operation of the catalytic converter is covered in detail in Chapter 40.

Muffler Hangers

Muffler hangers support the muffler and pipes (Figure 28.20). A hanger is usually a piece of fabric and rubber that resembles a tire. Each end is riveted to a piece of metal that is fastened on one end to the car body or frame. The other end is clamped to the pipe or muffler. The hanger allows the exhaust to be positioned away

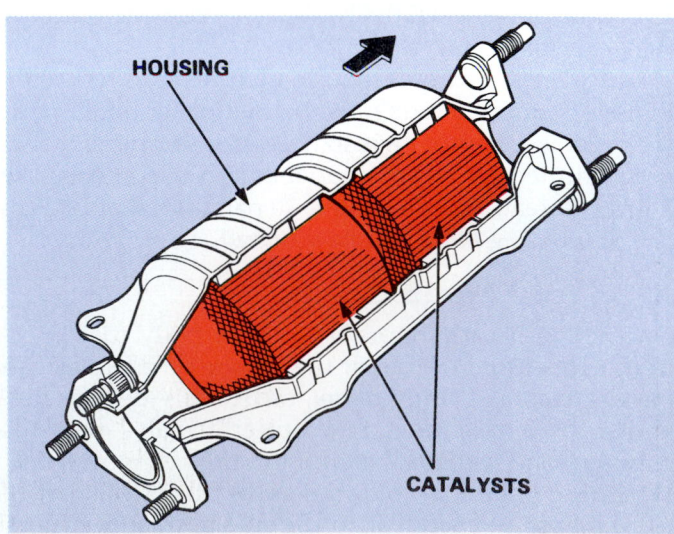

Figure 28.18 A catalytic converter. *(Courtesy of American Honda Motor Co., Inc.)*

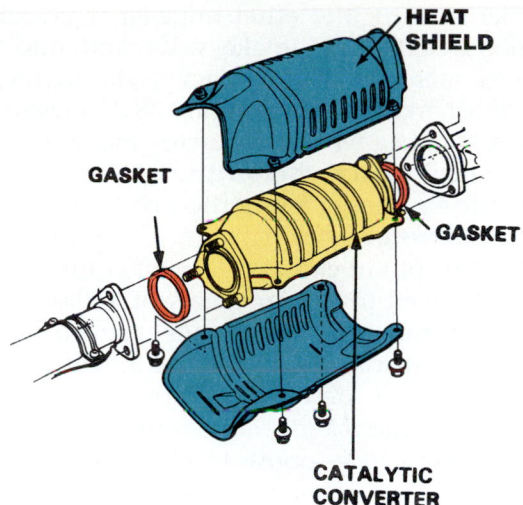

Figure 28.19 The catalytic converter has heat shields. *(Courtesy of American Honda Motor Co., Inc.)*

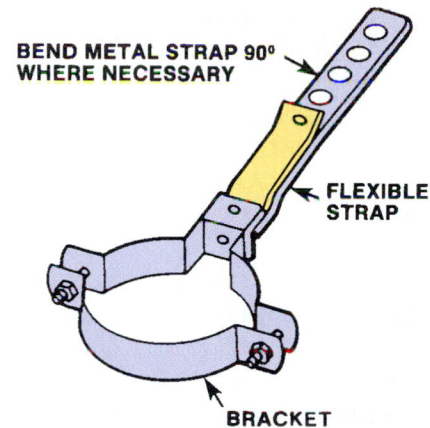

Figure 28.20 A muffler hanger.

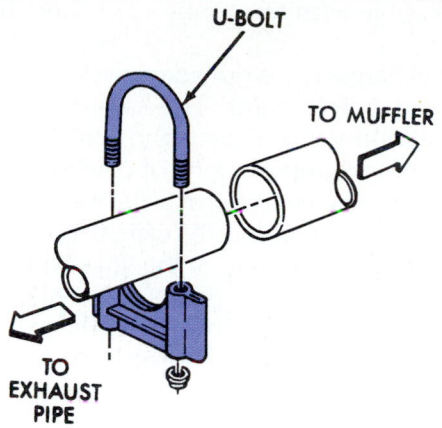

Figure 28.21 A muffler clamp. *(Courtesy of Chrysler Corporation)*

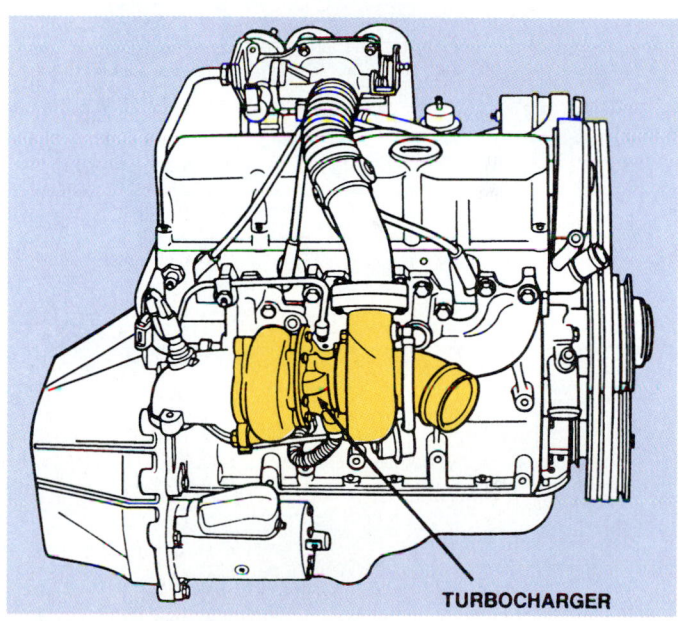

Figure 28.22 An exhaust driven turbocharger. *(Courtesy of Ford Motor Company)*

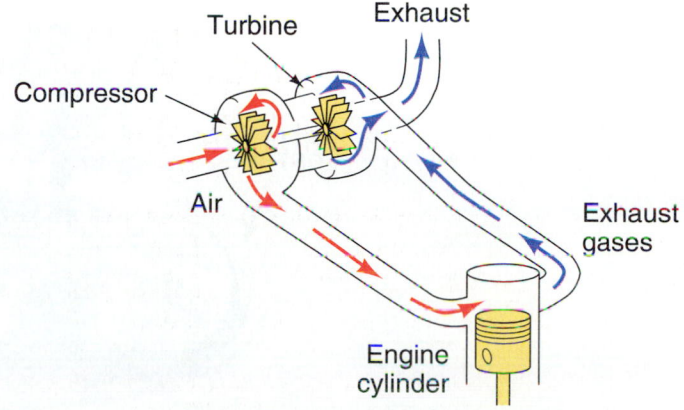

Figure 28.23 Turbocharger operation. *(Reproduced by permission of Deere & Company. ©1991, Deere & Company. All rights reserved)*

from other parts of the car and allows some flexibility from engine torque and vibration. Using a piece of rubber isolates the exhaust noise from the car body. The vibrations would also be felt in the passenger compartment.

Muffler Clamps

U-bolts of the same size as the exhaust pipe's outside diameter are used to clamp together sections of pipe. One end of one of the pipes is expanded to fit over the other piece of pipe (Figure 28.21).

■ TURBOCHARGER FUNDAMENTALS

A **normally aspirated** engine (one without a turbocharger) depends on atmospheric pressure to push air into the cylinders. Some engines have a **turbocharger** installed in the exhaust stream (Figure 28.22). A turbocharger is a small radial fan pump driven by the push of the exhaust flow. The spinning fan's energy is used to drive another fan on the intake side (Figure 28.23). This forces more air into the cylinders than would be possible with atmospheric pressure only (see Chapter 17).

The turbocharger helps provide power, when needed, from a smaller power plant. The smaller engine gets better fuel economy when the turbo is not in use than a larger, nonturbocharged engine of comparable power. Figure 28.24 shows a cutaway of a turbocharger.

A turbo (as it is commonly called) is a *centrifugal* pump. Centrifugal force takes the incoming exhaust and throws it out a snail-shaped outlet. The pump has two impellers on one shaft. The one in the exhaust side is called the *turbine*; the one that forces the air into the engine is called the *compressor*. As the exhaust forces one impeller to turn, the other impeller forces more air/fuel mixture past the intake valve and into the cylinder. This increases the efficiency of the engine.

The turbocharger shown in Figure 28.22 pressurizes the intake manifold after the carburetor and air cleaner. This is called **draw-through**. Another type pressurizes the air cleaner above the carburetor. This is called **blow-through**.

A turbo can spin in excess of 100,000 rpm. In fact, turbos are balanced to run in excess of 150,000 rpm (twenty-five times the maximum rpm of most engines). In comparison, an alternator will spin at about 20,000 rpm. A newer design turbo, called the *variable nozzle* or *variable displacement* turbo, changes its nozzle opening in response to changes in engine load.

An advantage to a turbocharger is that the vehicle will not lose power as it goes up in altitude. A normally aspirated engine will lose 3% of its horsepower with every 1000' increase in altitude.

A drawback to the turbocharger is decreased engine life for smaller engines. The smaller the engine, the larger the percentage of time the turbo is used in accelerating and climbing hills. This results in a hotter running engine. Oil must be changed more frequently in these engines.

Waste Gate

Because the system would only work well at very high rpm when the exhaust is moving fast, a higher efficiency turbo is used. The turbo works better at a lower rpm; at higher speeds, it builds excessive pressure, so a relief valve is needed. This relief valve is called a **waste gate** (Figure 28.25).

Boost

The amount of air density a turbo provides is known as **boost pressure**. A waste gate is set to open at a specific number of *pounds of boost*. Without a waste gate,

Figure 28.24 Turbocharger cutaway. *(Courtesy of AlliedSignal Automotive Aftermarket)*

WASTE GATE CONTROL

WASTE GATE

Figure 28.25 A waste gate bleeds off excessive turbo boost.

the turbo could provide the engine with so much power that it could literally destroy itself. There is a relief valve to protect the system in case the waste gate becomes stuck.

NOTE: The waste gate is the part that allows the driver of an Indy car to adjust the boost during a race. Of course, raising the boost also results in the use of more fuel so this is not done without a great deal of thought.

Turbo Lag

When the turbo is spinning at low speed, little or no boost is produced. The time required to bring the turbo up to a speed where it can function effectively is called **turbo lag**. This is a hesitation in throttle response that is noticed when coming off idle. Turbo lag, which is actually a lag in the air/fuel ratio, is more pronounced when a carburetor or throttle body is too far away from the intake valve. The problem is less noticeable with port fuel injection (when the fuel injector is located directly above the intake valve).

Intercoolers

An **intercooler**, known also as a **charge air cooler** or an **aftercooler**, is an option that may be installed with a turbo (Figure 28.26). Theoretically, a turbocharged engine operating at sea level with 7 psi boost should have about 1½ times the power of a normally aspirated engine. Unfortunately, air that is compressed becomes hotter, so fewer air molecules can enter the cylinders on each intake stroke.

An intercooler is simply a very sophisticated (and expensive) air cooler installed after the turbo. For every 10°F that the air/fuel mixture is cooled, a power gain of about 1% is achieved. Intercoolers cool the turbocharged air by about 100° before it enters the engine. With that amount of cooling an intercooler can provide about 10% more power.

Intercoolers use either air or water. They are relatively trouble free. Pressure from the turbo can cause detonation of the fuel in the cylinder. When the temperature of the incoming air is 90°F, it will become 276°F if it is compressed at 14 pounds of boost. Most production cars run 6 or 7 pounds of boost so that they can live a long life on the low octane fuels that are available. This is about half of atmospheric pressure.

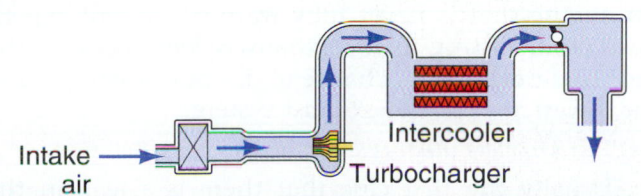

Figure 28.26 *An intercooler cools the compressed air. [Courtesy of AlliedSignal Automotive Aftermarket]*

◼ SUPERCHARGER FUNDAMENTALS

Superchargers are air pumps, often called **blowers**. A supercharger can easily produce 50% more power than a normally aspirated engine of the same size. The supercharger found on some of today's high performance computer controlled vehicles is not a new development. The *scroll-type*, used by Volkswagen, was invented in France in 1905. The **roots-type**, used by Ford and Toyota, was invented in 1854.

Like an alternator, a supercharger is driven by the crankshaft, usually with a belt but sometimes by a chain or gears. They spin at 10,000 to 15,000 rpm, much slower than turbos. Boost is limited by engine speed. Because it is belt driven, a supercharger gives stronger torque at a lower speed than a turbocharger. It also has a quicker response (no turbo lag). Older blowers made considerable noise. Modern blowers are much quieter. Unlike turbos, blowers consume horsepower as they are driven, so they are not as energy efficient under load.

NOTE: A Ford blower uses 60 horsepower at 5000 rpm but only uses ½ horsepower when cruising on the highway.

Draw-Through or Blow-Through

Air is provided in one of two manners. Fuel injected engines have air pumped directly into the intake manifold. This system, known as draw-through, is also used on most carbureted engines (Figure 28.27). Other carbureted engines use the blow-through system where the supercharger is above the carburetor (Figure 28.28). The carburetor must be enclosed in a *pressure box*, which pressurizes the air around the carburetor so the main jets will be able to supply fuel properly.

Positive Displacement and Dynamic

Two groups of superchargers are the positive displacement and the dynamic types. A positive displacement pump delivers the same amount of air with each revolution regardless of the speed. The faster it turns, the more air it pumps.

Lobe-Type Superchargers

The most popular positive displacement supercharger is the roots-type, called a lobe-type supercharger. A pair of rotors mesh to pump and compress air (Figure 28.29). The rotors are supported by bearings. Because they never touch each other, they do not wear out. Sometimes the lobes are twisted (helical) and sometimes they are straight cut.

Lobe superchargers tend to pulse at low speeds. A helical rotor tends to smooth these pulses out. Rotors are either of the two- or three-lobe design. A three-lobe rotor tends to pulse less than a two-lobe rotor. When

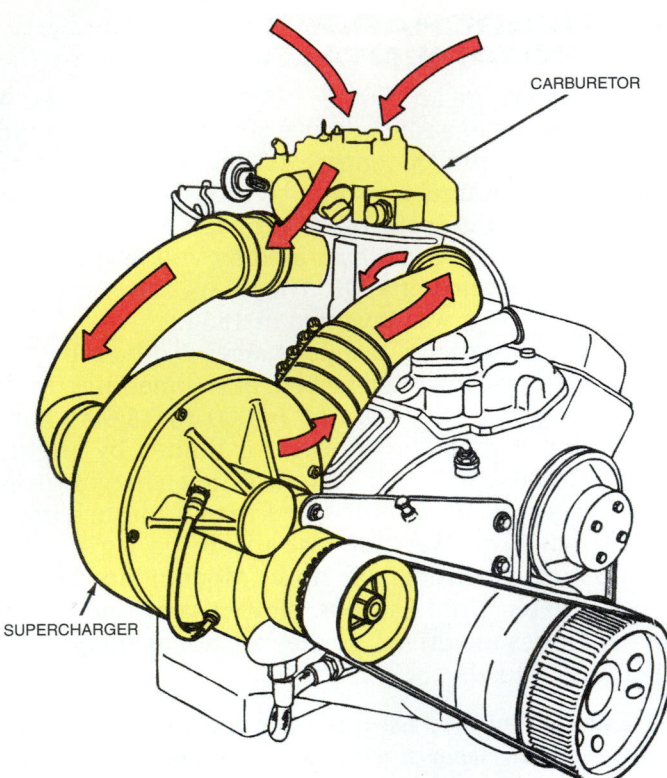

CARBURETOR

SUPERCHARGER

Figure 28.27 A draw-through supercharger. (Courtesy of Ford Motor Company)

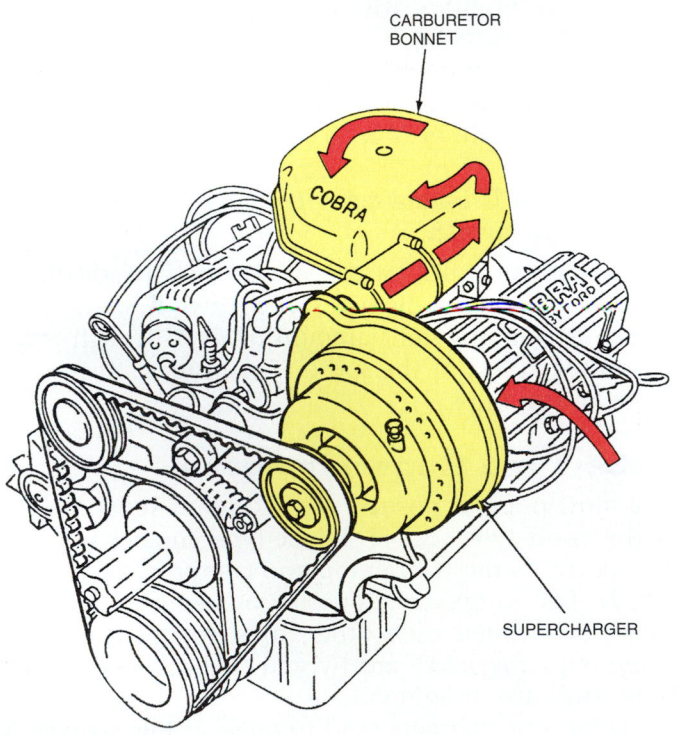

CARBURETOR BONNET

COBRA

SUPERCHARGER

Figure 28.28 A blow-through supercharger. (Courtesy of Ford Motor Company)

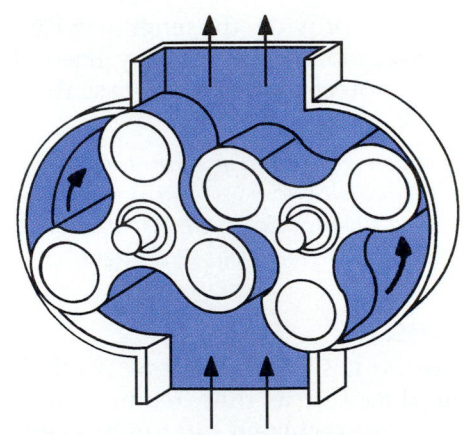

Figure 28.29 A roots-type supercharger has lobes. (Reproduced by permission of Deere & Company. ©1991, Deere & Company. All rights reserved)

the engine is not under a heavy load, intake manifold vacuum turns the rotors like a windmill, thereby using less power.

Volkswagen's *scroll-type G-lader* is so called because *lader* means "charger" and the shape of its chambers resembles the letter "G." The *displacer* inside of its housing moves in an off-center circular motion like a hula-hoop.

Another positive displacement blower is the *vane-type*, which compresses air inside its housing before forcing it through its outlet. This type of pump is used most often on small engines with high boost.

Dynamic Superchargers

A dynamic supercharger is like a turbocharger. Its output increases as the square of engine speed (if the engine turns twice as fast, boost output is quadrupled). So this pump operates best at high speeds but has less boost off idle. The three types of dynamic superchargers are the *centrifugal*, the *axial flow*, and the *pressure wave* type. The centrifugal is actually what a turbocharger is, except the turbo is exhaust-driven, rather than crankshaft-driven. The axial flow pump is not often found because of its expense to manufacture. The pressure wave pump is used mostly for two-stroke diesels.

■ EXHAUST SYSTEM SERVICE

Exhaust systems can rust out because there are acids and a good deal of moisture in the engine's exhaust. An engine that is never fully warmed up will experience rapid rusting in the exhaust system because the moisture never gets a chance to dry out. It will pool in the lowest spot in the exhaust system.

NOTE: *Mufflers often rust out on the lowest end.*

Usually the first clue that there is a leak in the exhaust system is the sound of leaking exhaust. Listen for leaks while the engine is running. Check for leaks

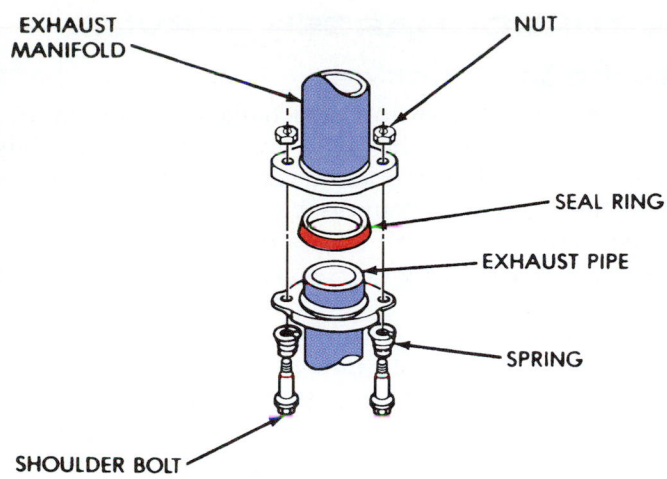

Figure 28.30 An exhaust seal, called a donut. *(Courtesy of Chrysler Corporation)*

at all pipe connections and at low spots in the system. A stethoscope can be used. Remove the metal end from the stethoscope and use just the hose to listen.

Exhaust Gaskets

Exhaust gaskets suffer a good deal of abuse. Parts expand when heated. Sometimes an aluminum part will be clamped to an iron or steel exhaust part. These metals have different expansion rates. Exhaust manifold to cylinder head gaskets are assembled with the shiny side out toward the exhaust manifold. This allows the hotter manifold to be able to slide against the gasket.

There is a round gasket, often called a *donut*, that fits between the manifold and the exhaust pipe (Figure 28.30). Donuts can be made of aluminum, steel, fiber, or ceramic. They allow a transverse engine to be able to move on its mounts without exhaust leaking from the flange.

CAUTION ■ If the car is to be run while on a wheel free, frame-contact lift, be certain that the drive tires do not contact any part of the lift.
  ■ Be careful around exhaust system parts. They are very hot.
  ■ Wear eye protection. When pounding on exhaust system parts to get them apart or put them together, rust can come off of old pipe parts and get in your eye.
  ■ Do not run the engine without adequate ventilation. Engine exhaust contains carbon monoxide.

Muffler shops specialize in this type of work. They have a large inventory of many sizes of straight pipe. Hydraulic tubing benders are used to make new pipes.

Preformed pipes are also available from aftermarket suppliers for technicians who want to perform the work themselves.

It is not unusual for the entire system to require replacement. If only the muffler is rusted out, it can be replaced with the old pipes being reused.

Header Pipe

Sometimes the section of pipe that is attached to the exhaust manifold (header pipe) is stainless steel. This is to comply with the federal law that states that any emission related component must be warranted for five years or 50,000 miles or longer, depending on the year of the vehicle. A header pipe is often two layers (laminated). The inside of these pipes can collapse, resulting in a restriction in the exhaust (Figure 28.31). A discolored stripe on the outside of the burned section of pipe often gives a clue to where the restriction is. Mufflers that have collapsed internally will often exhibit a stripe around the outside of the restricted area also.

Exhaust Part Replacement

Penetrating oil can be put on rusted fasteners and joints in the pipe before disassembly, but parts often break when they are rusted. Some of the tools used for

Figure 28.31 This collapsed laminated exhaust pipe caused breathing problems.

Figure 28.32 A chain type exhaust tubing cutter. *(Courtesy of Snap-on Tools Company. Copyright Owner)*

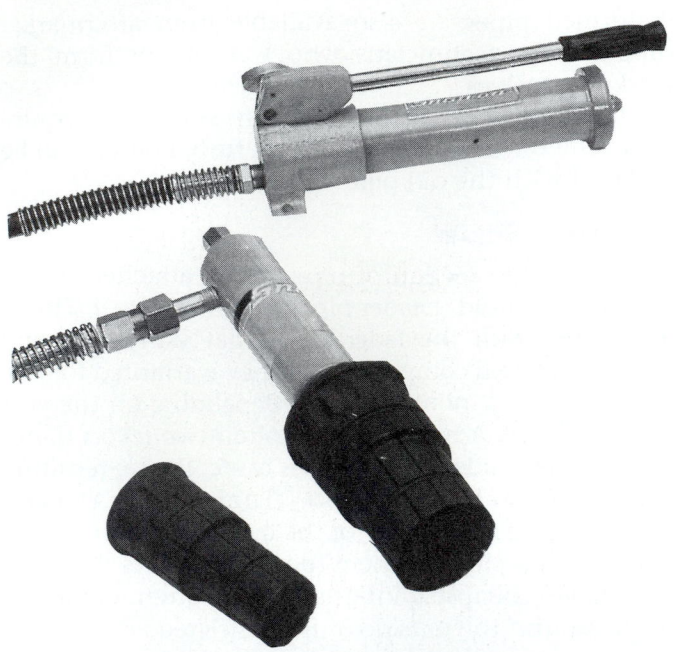

Figure 28.33 A hydraulic pipe expander. *(Courtesy of Snap-on Tools Company. Copyright Owner)*

Figure 28.34 This tool is used to slice a rusty exhaust pipe before removal.

exhaust work include a chain type pipe cutter (Figure 28.32), a roller type exhaust cutter, a pipe expander (Figure 28.33), deep sockets to clear the bolts, and a chisel cutter (Figure 28.34). The chisel cutter is often used with an air chisel.

A pipe shaper that resembles a tapered cone can be used with an air chisel, too. It is for straightening out ends of pipe that are bent.

■ HEAT RISER SERVICE

Heat risers sometimes become stuck. Most often they stick in the open position, but sometimes they stick closed. This causes carbon buildup and possible cracking of the intake manifold floor. A stuck valve can sometimes be loosened up by tapping on its shaft with a hammer. Special graphite heat riser lubricant can be applied to the shaft. To work properly, a vacuum-operated heat riser should have sufficient vacuum present at idle. If the valve is not working and there is at least 10"/hg vacuum present, there is a problem with the diaphragm, linkage, or butterfly valve.

■ TURBOCHARGER SERVICE

Turbo Lubrication

Lubrication is crucial to turbochargers (Figure 28.35). Oil changes should be more frequent. When oil is changed, the engine should be cranked for 30 seconds with the ignition system disabled to supply oil to the turbo. Failure to do this allows the turbo to run without oil when the engine is started. This can ruin an expensive turbo.

The rear bearing in a turbo can get extremely hot if the engine is shut off immediately after the turbo is used (*hot shut-down*). Motor oil burns at 430°F. The rear bearing (on the exhaust side of the turbo) can reach temperatures in excess of that when the engine is shut off after a period of acceleration. Some turbochargers have water-cooled rear bearings to help combat this problem.

Turbocharger Replacement

Rebuilding a turbo is not commonly done in the trade. There are too many different part numbers available and some turbos are balanced by grinding on the shaft nut. When this nut is removed, balance, which is critical, is upset. Unless many units are rebuilt, it is not economical for a repair shop to rebuild their own units.

A new or rebuilt unit, called a *cartridge* is obtained and installed in the old compressor housing. The compressor housing must usually be removed from the intake manifold before the turbo cartridge can be removed from it. Failure to do this can result in a bent

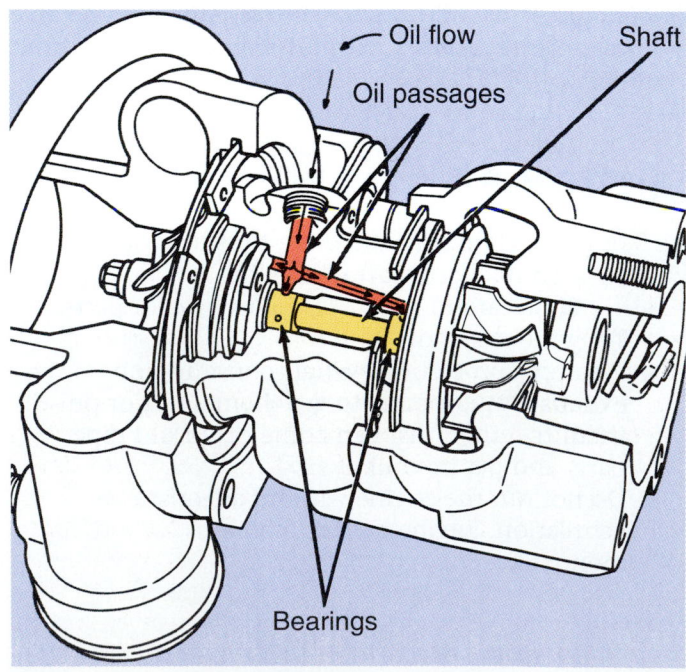

Figure 28.35 Turbocharger oil passages. *(Courtesy of Ford Motor Company)*

impeller blade. The turbo will howl and bearing failure will usually result.

Turbos, other than German ones, use full floating **plain bearings** (see Figure 28.35). Full floating bearings have oil clearance on both sides and they spin at about ⅓ shaft rpm. Because of this double oil clearance, the shaft will feel "loose." This is normal.

> **SHOP TIP** When inspecting a turbo, a rule of thumb is "if the wheel rubs the housing, the turbo needs to be replaced." Do not try to make it rub; simply see if it has been rubbing while in service.

Turbocharger Troubleshooting

The turbo runs in a difficult environment of high rpm and high heat. Exhaust gas is red hot (1800°F to 1950°F). This heat is constant, compared to that of the exhaust valve (which is cooled three-quarters of the time). The primary cause of turbo failure can be overlooked, resulting in failure of a second turbo. Blockages, leaks, and foreign object damage are common failures. Diagnose engine problems as if the engine did *not* have a turbo, before getting into turbo troubleshooting.

Manufacturers refer to unnecessary warranty returns as **NCFR** (No Cause For Removal). About half of turbo warranty returns are NCFR. The leading cause of NCFRs is blowby, which can cause oil in a turbo housing.

The rings that seal the turbo shaft are for sealing exhaust pressure out of the center housing. A plugged PCV system causes pressure to flow *into* the oil drain from the turbo. Restrictions in the oil return (Figure 28.36) will cause pressure that results in oil consumption and smoke. If a plugged drain line is not repaired, a new turbo will leak, too.

A defective compressor shaft seal can also cause oil consumption and an intake manifold air leak. The PCV valve must close when the intake side of the turbo is under boost or the crankcase can be pressurized through the PCV system. This can cause oil to enter the intake side of the turbo from the now pressurized center housing. Be sure the proper PCV valve is used and it is in good operating condition.

When the turbo has suffered foreign object damage, look at the wheel to see which direction the damage came from. Foreign object damage to the turbine (outlet) can be due to broken piston rings, valves, pistons, or any other engine failure that enters the exhaust. Damage to the compressor wheel is due to objects entering the intake (Figure 28.37). When contaminated oil damages a turbo bearing, heavy particles will tend to damage the outside of the bearing.

Figure 28.36 A turbo oil return can become plugged with carbon. *(Courtesy of AlliedSignal Automotive Aftermarket)*

Figure 28.37 Damaged compressor wheels. *(Courtesy of AlliedSignal Automotive Aftermarket)*

Turbocharger Care

Heat can be an enemy to the turbo. Oil flow to a turbo is only about 8% of an engine's total oil flow, but the turbo heats the oil 40% more than engine bearings do. Following high output use, drive slowly for 1000 yards (about a kilometer) or so to let the turbo get resupplied with cool oil before shutting the engine down.

Early turbos in the late 1970s were mounted in the rear of the engine compartment and therefore ran hotter. *After-lube* was especially important in these units. Some of today's turbos have special lubrication and cooling features that function even after shut-down.

After a period of storage, a turbo can be prelubed by disabling the ignition and cranking the engine for

15 seconds or so until oil pressure is registered on the dash gauge or light. There are special turbo oils available. Multiviscosity oils are also recommended because cold 30W can bypass the filter and the unfiltered oil will go directly to the turbo (see Chapter 12).

SUPERCHARGER SERVICE

Because superchargers are not powered by exhaust gas energy, they do not get hot. Lubrication is not a big problem like it is with turbos. In fact, the roots-type units are lubricated with their own supply of SAE 90 gear oil, which has no specified oil change interval. Blowers are very dependable, but dirt is a prime enemy to them.

Leaks are usually accompanied by a whistling sound that can be easily located by listening for its source. Vacuum leaks (intake side) suck in dust that can ruin the unit. Exhaust side air leaks (boost side) hurt performance. Vacuum leaks can also fool the computer, causing the engine to run lean. A leak on the boost side, on the other hand, will cause a rich condition. The oxygen sensor in these systems (the device that tailors the air/fuel mixture by what is coming out of the vehicle's exhaust) will only make relatively minor corrections to the mixture. It cannot compensate for major leaks.

REVIEW QUESTIONS

1. Which manifold runners are smaller in diameter, long ones or short ones?

2. What is the name of the manifold air space below the carburetor?

3. When one runner feeds two neighboring cylinders, what is this called?

4. Draw a sketch showing how a dual plane manifold distributes air and fuel to a V-8 engine.

5. What is the name of the turbocharger wheel that is on the intake side?

6. If a normally aspirated engine climbs to 3000 feet in altitude, how much power will it lose?

7. What is the name of the relief valve used to limit turbo boost pressure?

8. The time required to bring the turbo up to a speed where it can function effectively is called turbo _____ .

9. Which uses power when cruising, a supercharger or a turbocharger?

10. Another name for a lobe type supercharger is _____ -type.

ASE STYLE REVIEW QUESTIONS

1. Technician A says that one of the air filter's functions is to silence the air. Technician B says that one of the air filter's functions is to act as a flame arrestor. Who is right?
   a. Technician A   b. Technician B
   c. Both A and B   d. Neither A nor B

2. Technician A says that manifolds for throttle body fuel injection (TBI) are designed to flow air only. Technician B says that manifolds designed for carburetors are made to flow air and fuel. Who is right?
   a. Technician A   b. Technician B
   c. Both A and B   d. Neither A nor B

3. Technician A says that bends in the exhaust system do not always cause a restriction. Technician B says that headers help out low rpm performance. Who is right?
   a. Technician A   b. Technician B
   c. Both A and B   d. Neither A nor B

4. Technician A says that a car with a turbo will not operate as well in high altitude. Technician B says that a turbocharger that pressurizes the intake manifold after the carburetor is called a draw-through turbo. Who is right?
   a. Technician A   b. Technician B
   c. Both A and B   d. Neither A nor B

5. Technician A says that cooling the compressed air from the turbocharger by 100°F can increase engine power by 10%. Technician B says that turbochargers are usually called blowers. Who is right?
   a. Technician A   b. Technician B
   c. Both A and B   d. Neither A nor B

6. Technician A says that turbochargers spin slower than superchargers. Technician B says that superchargers are louder than turbochargers. Who is right?
   a. Technician A   b. Technician B
   c. Both A and B   d. Neither A nor B

7. Technician A says that exhaust systems usually rust out on top. Technician B says to take the end off of a stethoscope to listen to exhaust leaks. Who is right?

a. Technician A   b. Technician B
c. Both A and B   d. Neither A nor B

8. Technician A says that a stripe on an exhaust pipe could indicate a restriction. Technician B says that turbocharger bearings should be snug, not allowing any movement. Who is right?

a. Technician A   b. Technician B
c. Both A and B   d. Neither A nor B

9. Technician A says that if a turbocharger wheel has been rubbing the turbo housing, the turbo must be replaced. Technician B says that a bad turbo can cause engine oil consumption. Who is right?

a. Technician A   b. Technician B
c. Both A and B   d. Neither A nor B

10. Two technicians are comparing superchargers and turbochargers. Technician A says that lubrication is not as big a problem with superchargers. Technician B says that superchargers run cooler. Who is right?

a. Technician A   b. Technician B
c. Both A and B   d. Neither A nor B

Electrical System Theory and Service

THEN AND NOW: ELECTRICAL SYSTEMS

Many early cars had no electrical system whatsoever—not one wire in the entire vehicle! The engine was cranked by hand. Ignition might have been supplied by a platinum tube heated by a flame. If a driver was adventurous enough to attempt driving at night, acetylene lamps were used to light up the road.

Hand-cranking limited the use of the automobile to those with strong arms and backs. The hand crank also presented a dangerous situation when it kicked back. These were big factors in the electrification of the automobile.

In 1910, Bryon T. Carter, whose Cartercar Company had been bought by GM, stopped to help a woman whose car had stalled on a bridge. While attempting to start the engine, the hand crank kicked back, breaking his arm and jaw. He was admitted to the hospital, where he contracted pneumonia and died.

Henry Leland, head of Cadillac at that time, knew Carter. He was horrified to hear of Carter's death and directed his executives to find a way to eliminate the hand crank. Charles "Boss" Kettering was commissioned by Cadillac for this job. This ingenious inventor (one of the founders of Delco) came back with a design for the first integrated electrical system. It included not only an electric starter and battery, but also a generator. The job of the generator was to keep the battery charged and to power the system's accessories, which included headlights and a dependable point-type electrical ignition system. This landmark in automotive history appeared on the 1912 Cadillac.

As more and more electrically powered accessories appeared on cars, electrical load requirements began to exceeded the capacity of the direct current (DC) generator used until the 1960s. In 1960 Chrysler introduced the *alternator,* which is actually an alternating current (AC) generator. Diodes within the alternator convert its AC output to the DC required by the automobile electrical system. Unlike a DC generator, an alternator produces plenty of current at low engine rpm.

The ever-increasing list of electrical accessories also resulted in larger bundles of wires under the dash. Locating a short circuit or a bad ground requires an accurate wiring diagram and can be time-consuming.

Today, electronic components have been added to vehicle electrical systems. Electronic engine management systems provide increased performance, better fuel economy, and lower exhaust emissions. Many cars have a sophisticated onboard computer that controls several specialized computers. The use of fiber optics and multiplexing has made it possible to eliminate the large bundles of wiring found on older vehicles. Troubleshooting and repair of these electronic systems requires a good deal of training and experience.

The dangerous hand crank. *(Courtesy of Bob Freudenberger)*

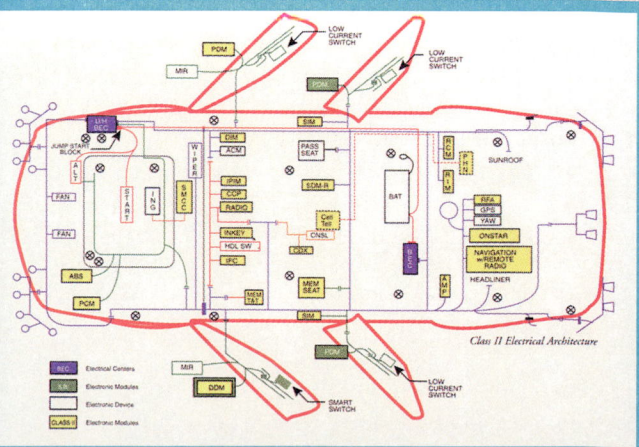

A modern electrical system. *(Courtesy of General Motors Corporation, Service Technology Group)*

Basic Electrical System Theory and Repairs

■ INTRODUCTION

Almost every system of the car uses electricity or electronics in some manner. These systems include anti-lock brakes, engine emission control devices, dash warning lights and gauges, electronic fuel injection, electrically shifted transmissions, and others. Technicians in every specialty service area must have a basic understanding of electricity to be successful. This chapter covers the basic items needed to begin learning about more advanced topics. Learning these topics now will make mastering more advanced topics easier.

■ ELECTRON FLOW

Electricity cannot be seen. It is made up of very small objects moving at the speed of light (669,600,000 miles per hour). All matter is composed of *atoms*. Atoms are composed of *protons, neutrons*, and *electrons* (Figure 29.1).

Electrons have a negative charge and orbit around the positively charged protons. Protons and neutrons (which have no charge) are located in the *nucleus* (center) of the atom. The number of protons and electrons in an atom is what determines what the atom is. Figure 29.2 compares aluminum and silver atoms.

The electrons remain in orbit around the nucleus because they have an electrical attraction to its protons. This is because opposite charges attract (Figure 29.3).

Each atom tries to remain electrically neutral. When it is neutral, it has no charge; the number of protons and electrons is equal.

Electricity is the flow of electrons from one atom to another (Figure 29.4). Protons are tightly bound within the nucleus of the atom. But electrons are free to move in their orbits at a fixed distance from the nucleus. The electrons are in orbit and stay a fixed distance from each other because they are of like charge and repel one another.

Conductors

A **conductor** is metal or another material such as carbon that wires and electrical parts are made of. Conductors have atoms that allow electricity to flow freely. Atoms

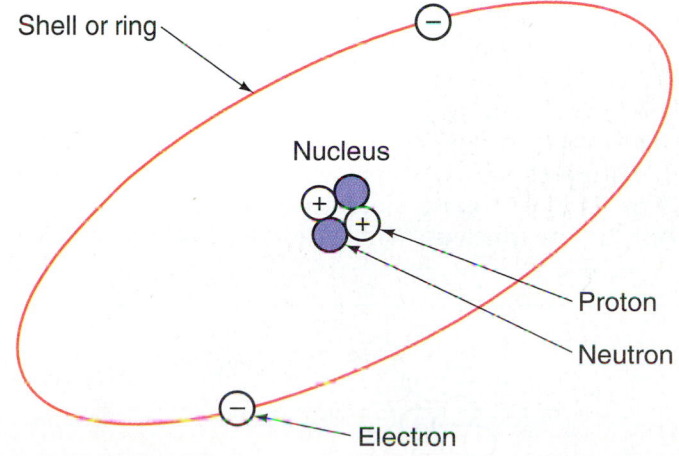

Figure 29.1 Atoms are composed of protons, neutrons, and electrons.

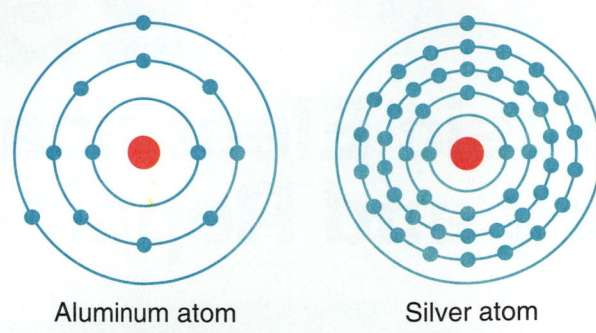

Figure 29.2 The number of protons and electrons in an atom is what determines what the atom is. *(Courtesy of General Motors Corporation, Service Technology Group)*

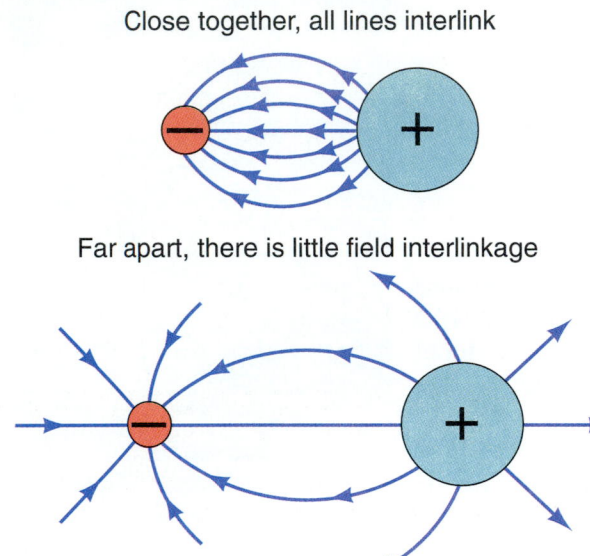

Figure 29.3 Opposite charges attract. *(Courtesy of Ford Motor Company)*

attempt to remain electrically neutral. To remain in electrical balance, an atom will shed or attract electrons from neighboring atoms. When an atom is lacking electrons and another atom has too many, electrons will flow between atoms to equalize the charges. This is what makes up electricity.

Conductors are atoms with *free electrons* (Figure 29.5). These are extra electrons not bound to the protons in the nucleus. The attraction of a negatively

charged electron for the positive charge in the atom's nucleus is less when the electron is in an outer orbit.

An atom is identified by its number of protons and its number of electrons in their orbits about the nucleus. Hydrogen is the simplest atom, with only one electron in orbit. A complex atom is compared to a hydrogen atom in Figure 29.6.

Copper is a more complex atom with twenty-nine electrons in various orbits. But it has only one electron in its valance ring (Figure 29.7). The electron can move easily from one atom to the next, which makes it a good conductor. Some good conductors used in wiring are silver, copper, and aluminum.

Insulators

An **insulator** is the opposite of a conductor. It is a material with no, or few, free electrons. Because their electrons are tightly bound in their orbits, an insulator prevents the flow of electrons between two conductors. Glass, rubber, and porcelain are good insulators.

Circuit

A complete *circuit* is when a full circle is provided for electrical flow. A light filament hooked across a battery, between its positive and negative posts, is an example of a circuit (Figure 29.8). On automobiles, the ground part of the circuit is provided by the frame and body metal (Figure 29.9).

Wiring

Electricity always takes the shortest path to ground. Without insulation, electricity can take undesired paths. With enough voltage present, electricity can jump air gaps too. Wiring from the battery's positive terminal is insulated so it cannot find ground.

Most automotive wiring is insulated with *poly vinyl chloride (PVC)*. This is a durable plastic that is designed to insulate, yet remain flexible and resistant to heat and attack from fuels and oils. The ground side of a circuit usually does not require insulation. More information on wiring is found in Chapter 36.

Control and Protection Devices

■ A *switch* can be used to open a circuit, providing a means of turning it on and off (Figure 29.10).
■ Protection devices, *fuses* or *circuit breakers,* are installed in the circuit. These are designed to open

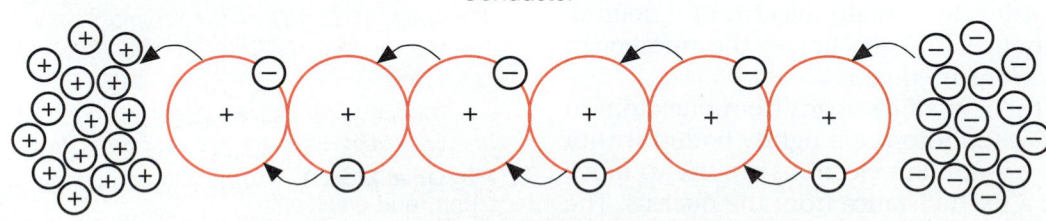

Figure 29.4 Electricity is the flow of electrons from one atom to another.

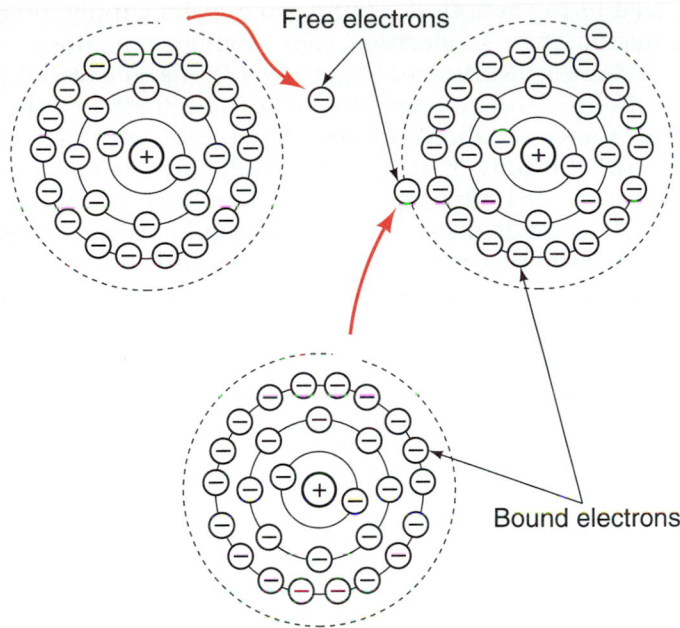

Free electrons

Bound electrons

Figure 29.5 Conductors are atoms with free electrons.

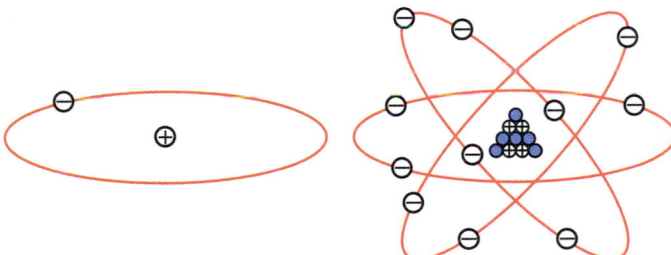

Figure 29.6 A complex atom compared to a hydrogen atom. *[Courtesy of Cooper Automotive/NAPA Belden]*

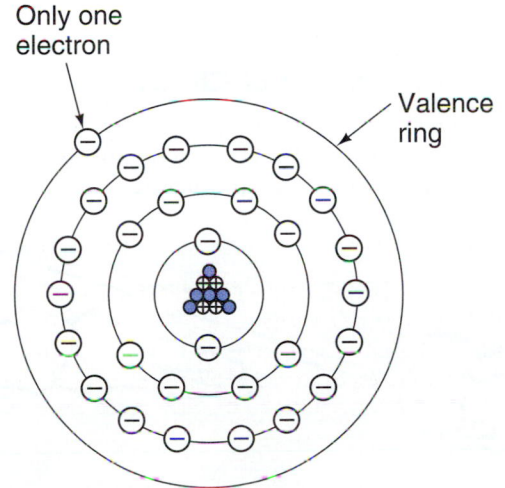

Only one electron

Valence ring

Figure 29.7 Copper has only one electron in its valance ring. *[Courtesy of Cooper Automotive/NAPA Belden]*

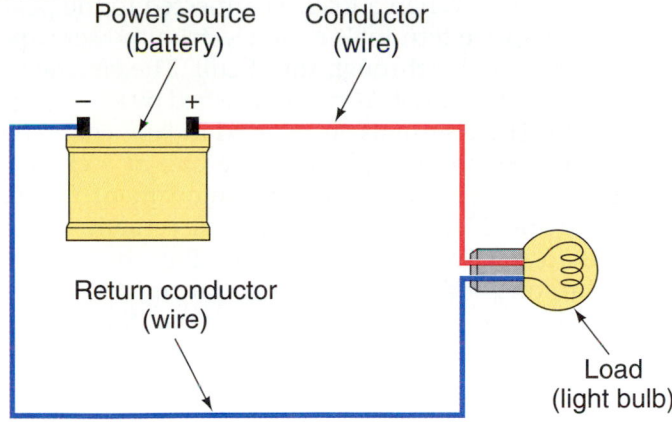

Power source (battery)

Conductor (wire)

Return conductor (wire)

Load (light bulb)

Figure 29.8 A circuit. *[Courtesy of Ford Motor Company]*

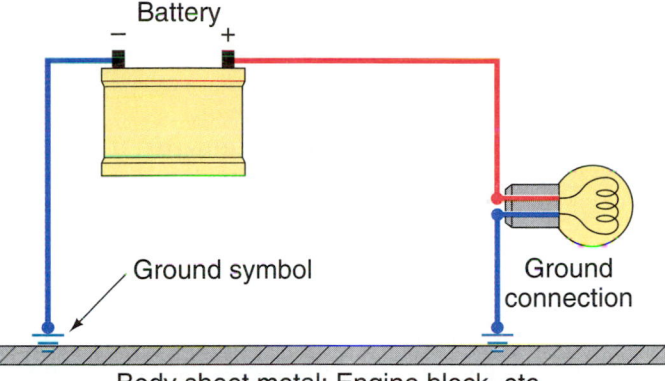

Battery

Ground symbol

Ground connection

Body sheet metal: Engine block, etc.

Figure 29.9 The ground part of the circuit is provided by the frame and body metal. *[Courtesy of Ford Motor Company]*

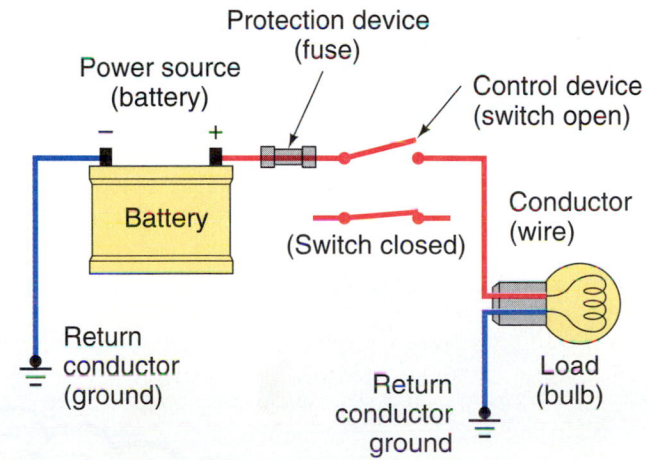

Protection device (fuse)

Power source (battery)

Control device (switch open)

Battery

(Switch closed)

Conductor (wire)

Return conductor (ground)

Return conductor ground

Load (bulb)

Figure 29.10 A switch provides a means of turning the circuit on and off. *[Courtesy of Ford Motor Company]*

BASIC AUTOMOTIVE ELECTRICAL SYSTEM

(fail) before damage can occur elsewhere in an electrical circuit. Protection devices are covered in detail in Chapter 36.

Electricity powers the *starting, ignition, lighting,* and *accessory systems* of the car. When parts of these systems operate, there is an electrical *load* on the vehicle's charging system. When a load is connected to the positive and negative terminals of an electrical power supply, electricity flows through the circuit. The electricity is supplied by a *storage battery*, replenished by the charging system. These systems are covered in later chapters.

When pressure is applied to one end of a circuit, electrons repel each other, causing movement in the circuit (Figure 29.11). Electrons do not actually move more than a few inches a second. But the energy applied to one end of the circuit is transmitted to the other end at the speed of light (186,000 miles per second). The energy transmission can be compared to the action on a row of billiard balls (Figure 29.12).

■ ELECTRICAL TERMS

Electricity is invisible. To be able to see what is happening in an electrical system, you will need to learn electrical terms and how to use a meter. Volts, ohms, and amps, terms for electrical pressure, resistance, and flow, are the common electrical terms you will need to know before using electrical meters.

The flow of electricity will be compared to the flow of water in the following discussions on voltage, current, and resistance.

Voltage

The force that is needed to push or pull an electron out of its orbit is called **electromotive force (EMF)**, which means electron moving force. EMF can be com-

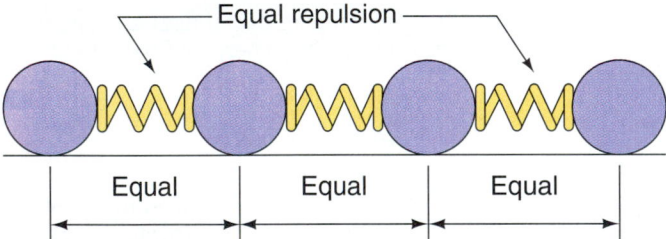

Figure 29.11 Electrons move in response to electrical pressure (volts). *(Courtesy of General Motors Corporation, Service Technology Group)*

pared to the action of a pump on a water supply. It is a measurement of electrical *pressure* and is measured in **volts** using a voltmeter (Figure 29.13). *Alessandro Volta* (1745–1827) discovered the *voltaic pile* (battery), the first source of direct electrical current. The French named the unit of electrical pressure the volt in his honor in 1881 after he had been dead for many years.

The lead acid storage battery used in automobiles produces approximately 2.1 volts per cell. There are six cells in a 12-volt battery, actually producing 12.6 volts.

Voltage controls how strongly a load in a circuit is operated. If voltage is low, a light will not be as bright or a heater defroster motor will not spin as fast. Voltage drops as various loads are placed on a battery's circuit. This means that electrical flow is directly proportional to the amount of voltage pushing it (Figure 29.14).

Voltage Differential. Electrons flow because of the difference between battery voltage at one end of a circuit and zero voltage on the ground side. Electricity will flow from a higher voltage to a lower voltage.

Current

The correct term for the flow of electricity is *current*. Current is the number of electrons flowing per second in a circuit. It can be compared to how much water flows through a pipe. *Andre Ampere* (1775-1836) was a French physicist who discovered the unit of electrical current flow, called the *ampere* or the **amp**. Water flow is measured in *gallons per minute* (Figure 29.15).

Science has calculated how many electrons can flow past a given point in one second. When 6.28 billion electrons pass a given point, this is called one *coulomb*. One coulomb per second is equal to one amp or ampere of current. Current flow is measured with an ammeter.

Alternating and Direct Current

When electrons flow in only one direction, this is called **direct current (DC)**. The automotive electrical system uses direct current because it can be stored in a

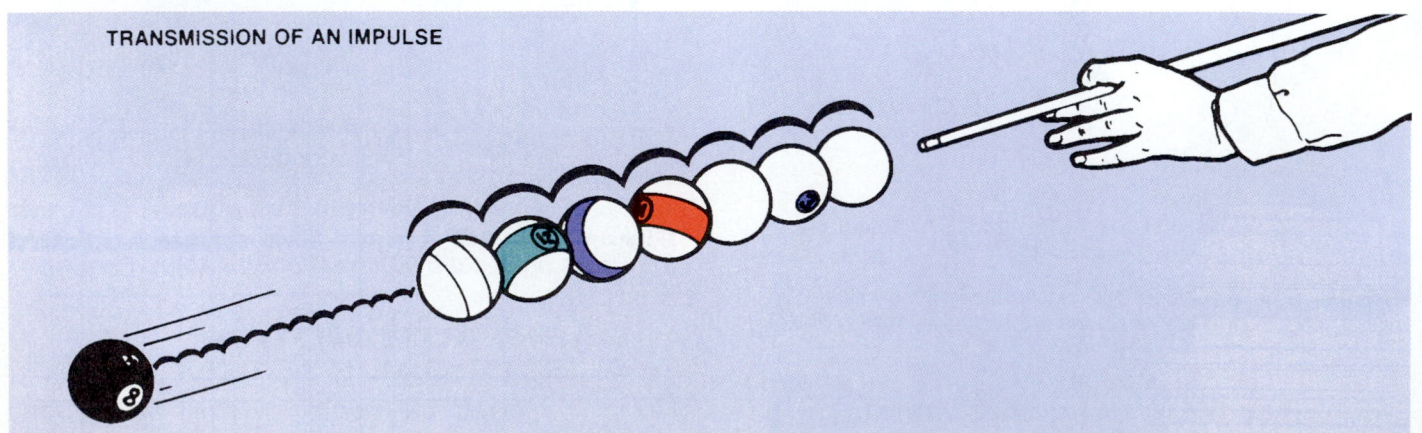

TRANSMISSION OF AN IMPULSE

Figure 29.12 Energy is transmitted almost instantaneously. *(Courtesy of Ford Motor Company)*

battery. Electricity actually flows from negative to positive. This is called the *electron theory* and is how electron flow is viewed in the electronics industry (see

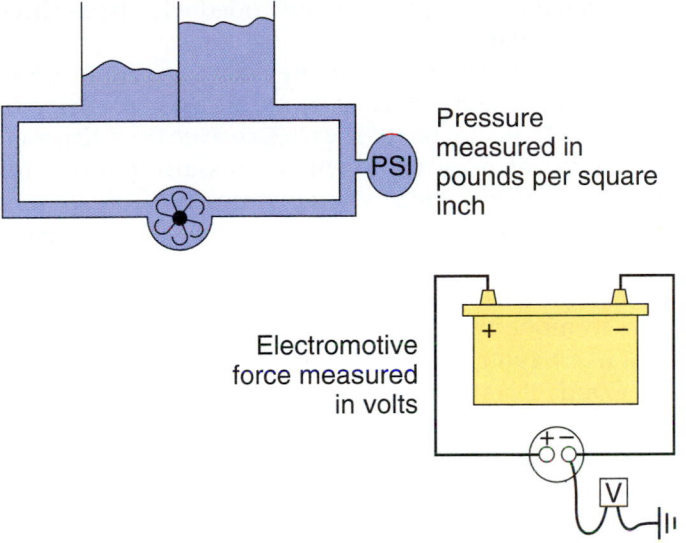

Figure 29.13 Electrical pressure is measured in volts. *(Courtesy of Cooper Automotive/NAPA Belden)*

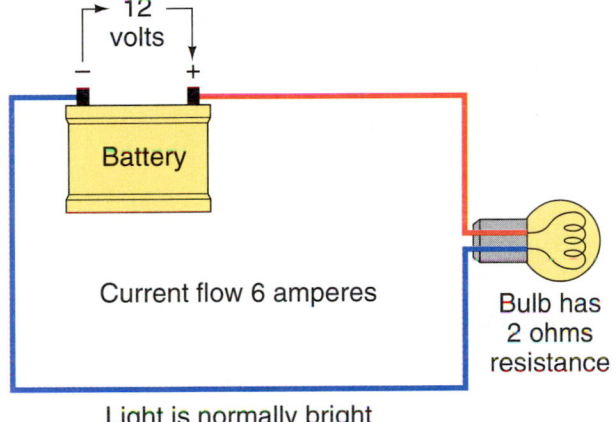

Figure 29.14 Electrical flow is directly proportional to the amount of voltage pushing it. *(Courtesy of Ford Motor Company)*

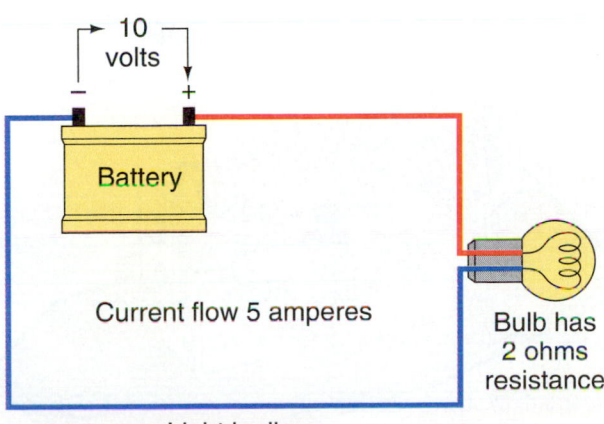

Chapter 75). In the automotive industry, electrical flow has traditionally been described as moving from positive to negative. This is called the *conventional theory*.

When electrical current surges from positive to negative and back again, this oscillation is called **alternating current (AC)**. Alternating current is used for household electricity. AC current cannot be stored in a battery. An alternator driven by a belt from the engine's crankshaft provides electrical power to the vehicle. The alternator makes alternating current, but it is rectified (changed to direct current) before it is used to recharge the battery.

Resistance

Resistance acts as an obstruction to electrical flow (such as when a smaller water pipe is used) (Figure 29.16). *Georg Ohm* (1787-1854), a German physicist, discovered

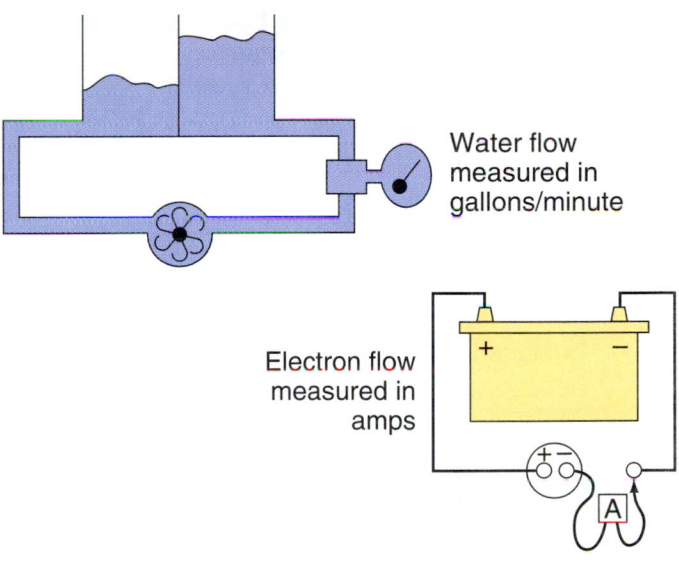

Figure 29.15 Comparing water flow to electrical current. *(Courtesy of Cooper Automotive/NAPA Belden)*

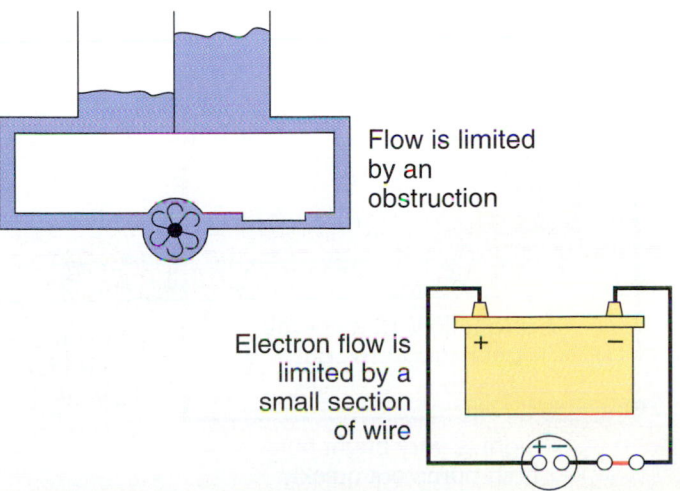

Figure 29.16 Resistance. *(Courtesy of Cooper Automotive/NAPA Belden)*

Ohm's law, which governs the relationship between volts, ohms, and amps. An **ohm** is the unit of electrical resistance measurement expressed with the Greek letter omega (Ω). One ohm is the resistance that will allow one ampere to flow when pushed by one volt.

Changes in Current Flow

You learned earlier that the amount of current flow in a circuit is directly proportional to circuit *voltage*. It is inversely proportional to the *resistance* in the circuit. This means that the more resistance there is, the less current flow there will be. Current can be increased either by increasing voltage or decreasing resistance.

The term *current draw* refers to the amount of current required to operate a load. In Figure 29.17 current draw is compared with different voltages and resistances (light bulbs). *Loads* refer to lamps, motors, solenoids, or other electrical accessories on an automobile.

Resistors are used to make heat or to control the intensity of a load. On a typical heater blower motor, two resistors control the motor's speed. Either one, two, or both resistors can be used to provide a stepped measure of control for low, medium, or high speeds (Figure 29.18).

Resistors are of different *values* (resistances). Larger resistors often have their values listed printed on them. Smaller resistors are color coded with from three to six color bands.

Variable resistors, called rheostats or potentiometers, are also used to control speed and intensity of electrical loads. A *rheostat* varies *current* flow through the circuit as a movable wiper runs along a resistor (Figure 29.19). A *potentiometer* varies the *voltage* in a circuit. A variable resistor is commonly called a "*pot*," whether it is a potentiometer or a rheostat.

Many of these devices are found in computer circuits in modern automobiles. This information is covered in Chapter 75.

Conductors offer some resistance in a circuit, but the circuit is designed so that this is not a governing factor in current flow.

■ The longer a conductor is, the more resistance it has.
■ The larger its diameter, the less resistance it has.

The longer a wire must be, the larger its diameter will be.

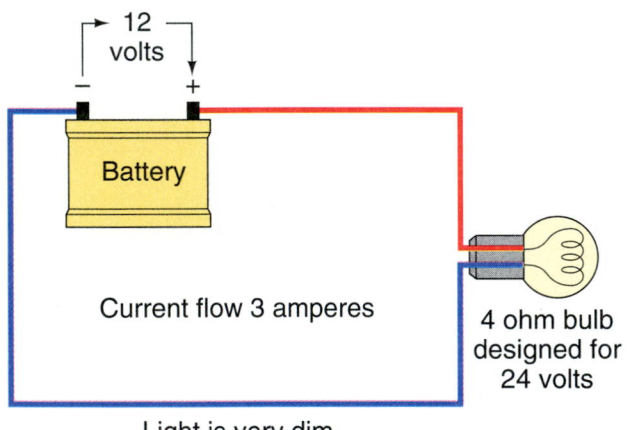

Current flow 3 amperes

4 ohm bulb designed for 24 volts

Light is very dim

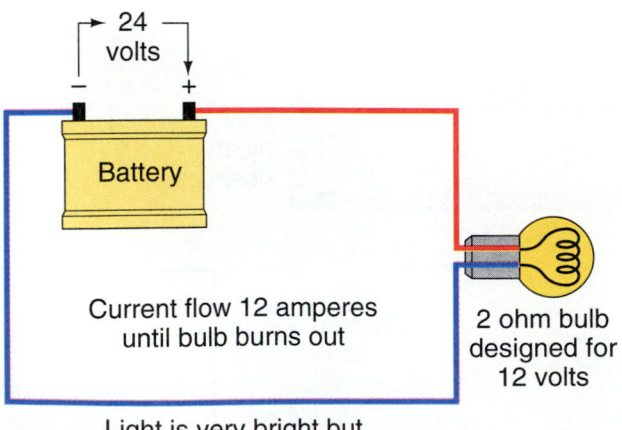

Current flow 12 amperes until bulb burns out

2 ohm bulb designed for 12 volts

Light is very bright but bulb burns out quickly

Figure 29.17 Current draw. *[Courtesy of Ford Motor Company]*

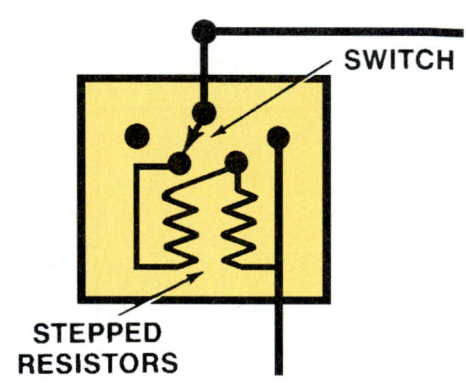

SWITCH

STEPPED RESISTORS

STEPPED RESISTOR

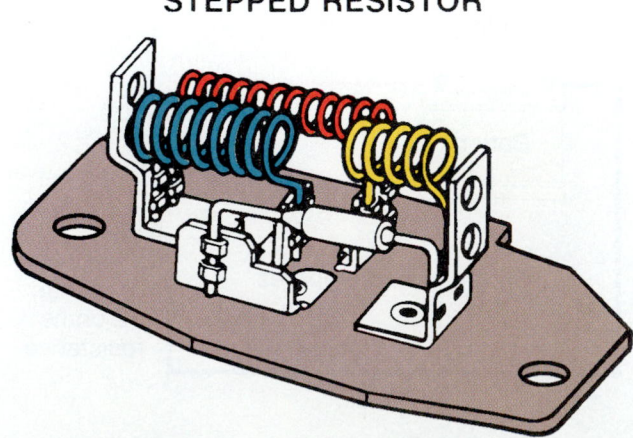

Figure 29.18 Stepped resistance controls a heater motor. *[Courtesy of Ford Motor Company]*

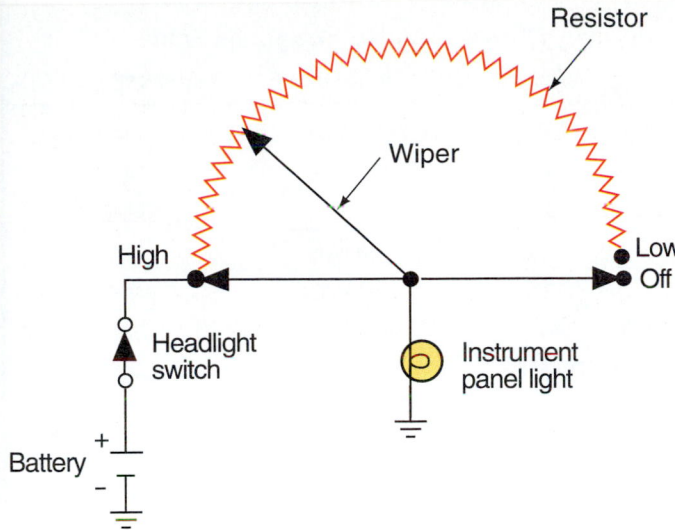

Figure 29.19 A rheostat varies current flow through the circuit.

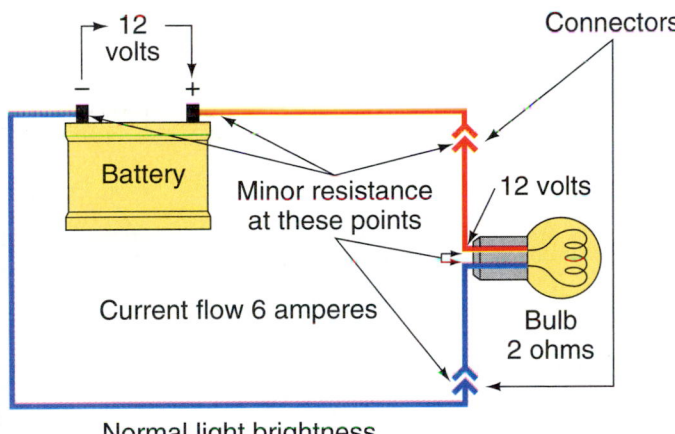

Figure 29.20 Current flow is affected by resistance. (Courtesy of Ford Motor Company)

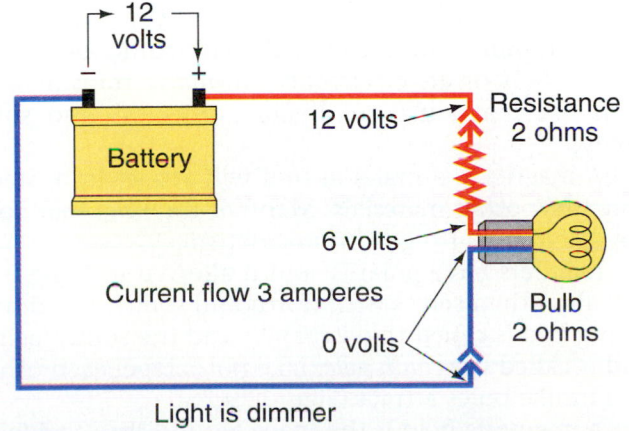

Figure 29.21 Higher resistance drops electrical flow. (Courtesy of Ford Motor Company)

A circuit is designed to have a certain amount of resistance in it. This determines the amount of current flowing in it (Figure 29.20). Problems arise when resistance becomes higher in a circuit. This happens when corrosion, grease, dirt, or another contaminant builds up in switches or on loose connections (Figure 29.21).

■ CIRCUITRY AND OHM'S LAW

Types of Circuits

Circuits can be arranged in three ways: series, parallel, and series parallel.

A **series circuit** is when current flows equally though all parts of a circuit, first to one load and then on to the next one (Figure 29.22). When current flows in the circuit, both bulbs light. If one of the bulbs fails, this acts like a switch and both bulbs go out.

In a series circuit, the resistances of all of the loads add up (Figure 29.23). Adding loads reduces current flow.

A **parallel circuit** is when a circuit starting from a common point has different branches through which electricity can flow (Figure 29.24). Each branch of the circuit has a load and a separate ground. A bulb failure in one branch will not affect the other branches of the circuit.

In a parallel circuit, the total resistance is less than the sum of the individual resistances. This is because

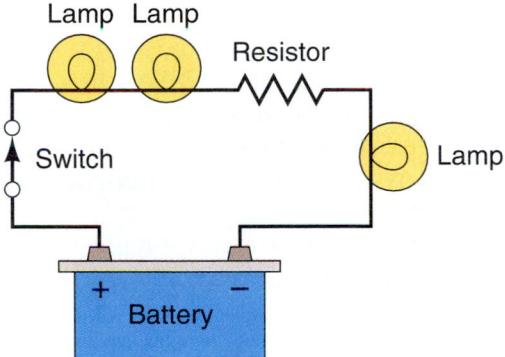

Figure 29.22 A series circuit.

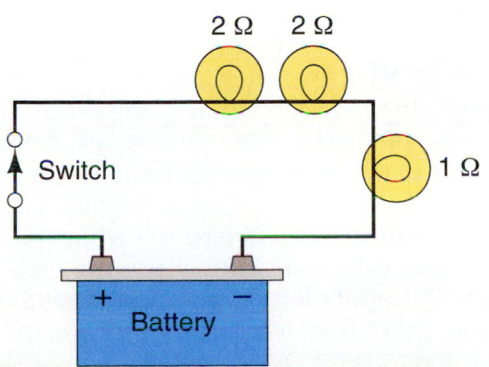

Figure 29.23 In a series circuit, the resistances of all of the loads add up.

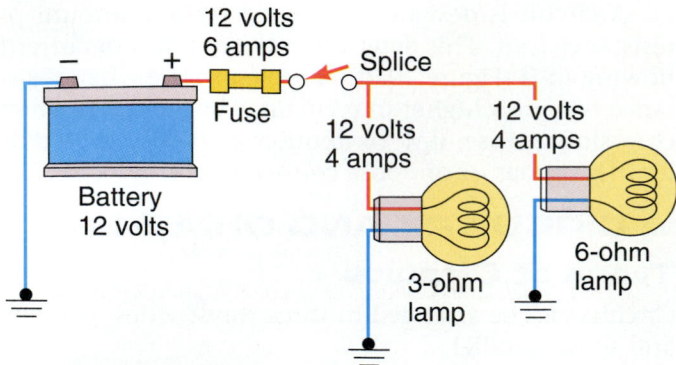

Figure 29.24 A parallel circuit. *(Courtesy of Ford Motor Company)*

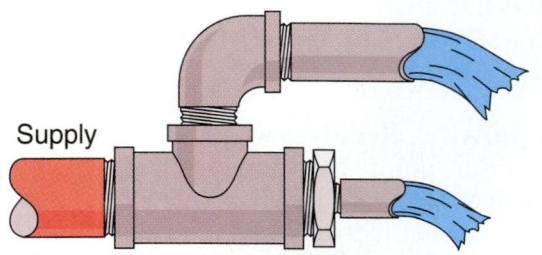

Figure 29.25 Comparing electron flow in a parallel circuit to water flow. *(Courtesy of General Motors Corporation, Service Technology Group)*

the current has two paths it can take. This can be compared to water flow as in Figure 29.25. When the same voltage is applied to the different branches, the resistance within each branch will determine the current flow. Voltage is constant in all branches until it is dropped by the load.

Remember that, in a series circuit, adding loads reduced the current. In a parallel circuit, adding loads will overload the circuit.

Series parallel circuits combine the two types of circuits. Actually the circuit shown previously in Figure 29.24 is the simplest kind of series parallel circuit. The bulbs are the parallel part of the circuit and the switch is in series with both of them.

Ohm's Law

The relationship between voltage, amperage, and resistance can be predicted using Ohm's law. Figure 29.26 shows what happens as various parts of the equation are changed.

To use Ohm's law to determine what will happen when any two values are known use the tables in Figure 29.27. The upper left corner of the sketch shows a typical pie chart that electronics technicians use. In the chart, *E* stands for volts. It is *E* instead of *V* because voltage is really electromotive force. Current is expressed as *I*. This stands for inductance.

Ohm's Law Relationship Table		
Voltage	**Resistance**	**Amperage**
UP	DOWN	UP
UP	SAME	UP
UP	UP	SAME
SAME	DOWN	UP
SAME	SAME	SAME
SAME	UP	DOWN
DOWN	DOWN	SAME
DOWN	UP	DOWN
DOWN	SAME	DOWN

Figure 29.26 The relationship between voltage, amperage, and resistance can be predicted using Ohm's law. *(Courtesy of General Motors Corporation, Service Technology Group)*

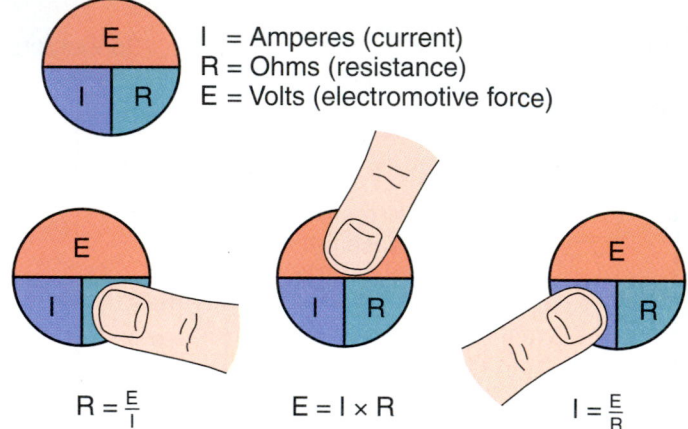

I = Amperes (current)
R = Ohms (resistance)
E = Volts (electromotive force)

$$R = \frac{E}{I} \qquad E = I \times R \qquad I = \frac{E}{R}$$

Figure 29.27 Ohm's law formulas. *(Courtesy of Ford Motor Company)*

■ MAGNETIC FIELDS

Magnetism is used in the automobile to produce electrical current in the charging system. It is also used to make the starter motor turn. An electronic fuel injection system or an electronic automatic transmission both use magnetism to operate injectors and solenoids.

A magnet is a material that will attract iron, steel, and a few other materials. Many of the laws that govern electricity also govern magnetism.

Magnets have polarity and if allowed to hang free will align themselves with north and south. The north facing end is called the *north pole* and the south facing end is called the *south pole*. Like poles repel each other and unlike poles attract (Figure 29.28).

A magnetic field is the space around the outside of the magnet that contains its lines of magnetic force. Magnetic lines of force can penetrate all objects. There

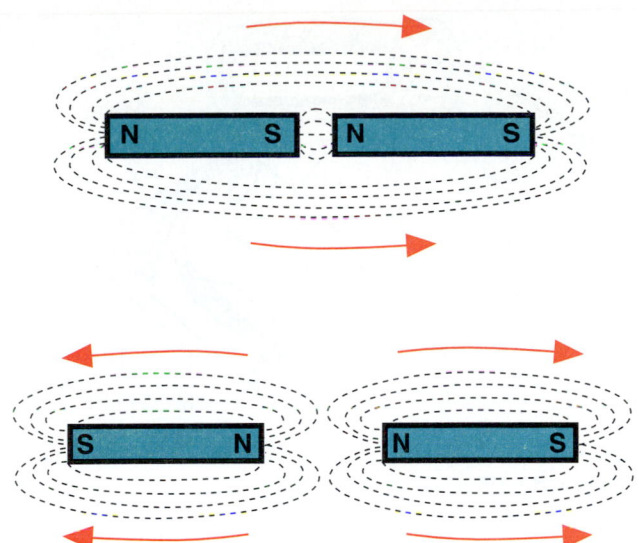

Figure 29.28 Like poles repel each other and unlike poles attract.

is no known insulator of magnetic fields. The lines of force are detected only by other magnetic materials or another magnetic field.

Electromagnetism

A magnetic field is created around the outside of a conductor with electricity flowing in it (Figure 29.29). When the wire is coiled, the magnetic field is made stronger (Figure 29.30). Adding an iron core to the inside of the coil of wire will make a still stronger magnetic field (Figure 29.31). Increasing the number of coils increases the strength of the field. The strength of

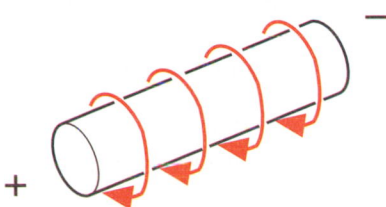

Figure 29.29 A magnetic field surrounds a conductor with current flowing in it.

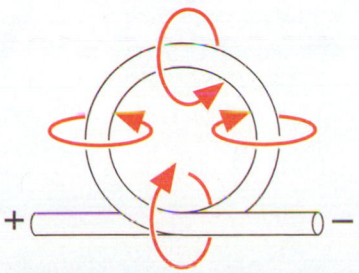

Figure 29.30 Coiling the conductor makes a stronger magnetic field.

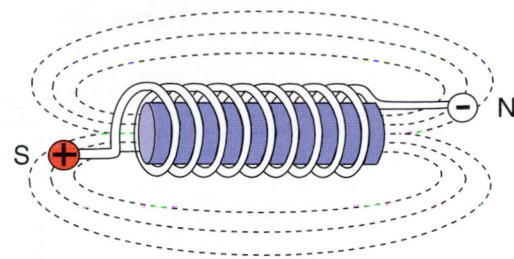

Figure 29.31 Adding an iron core concentrates the magnetic field.

the field is also related to the amount of current flowing in the conductor (Figure 29.32).

Electromagnetic Induction.   Electricity can be produced by moving a magnetic field over a conductor (Figure 29.33). This is the basic principle behind the operation of the alternator that charges the car's battery. Alternators are covered in a later chapter. **Electromagnetic induction** is also used to produce the spark used in the ignition system.

Relays

One application of magnetism is in a *relay*. A relay is a magnetically controlled switch. It is common to use a relay when a large load must be controlled by a small wire. Examples of circuits where relays are used

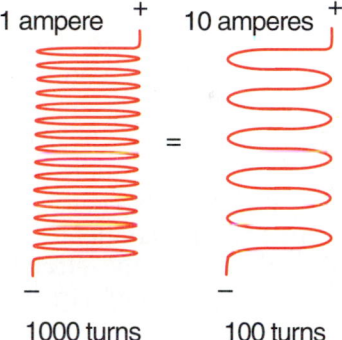

Figure 29.32 Magnetic field strength is determined by the amount of current flow and the number of coils.

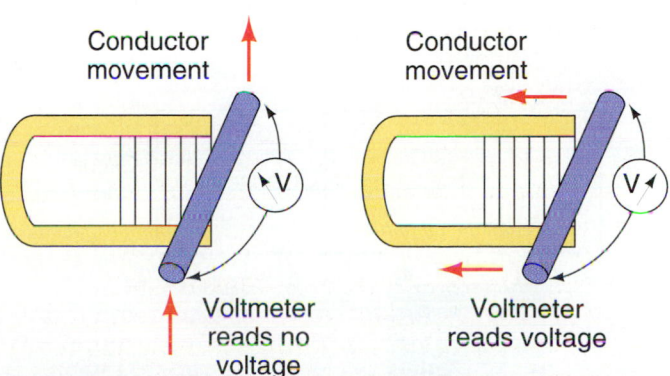

Figure 29.33 Electricity can be produced by moving a magnetic field over a conductor.

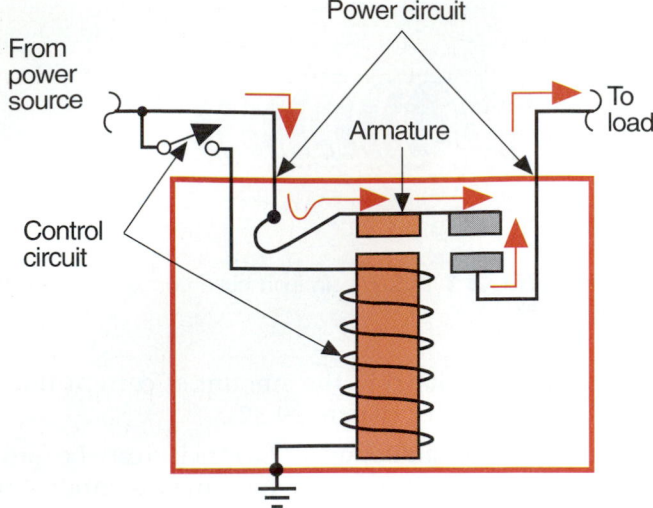

Figure 29.34 Operation of a relay.

include the starter motor and the horn. The small wire provides power to a coil of wire to create a magnetic field. The field pulls on an armature to close a switch (Figure 29.34).

■ CAPACITORS

A **capacitor (condenser)** stores electricity. Capacitors do not use energy but return all voltage to the circuit when they discharge. They are used to absorb voltage changes, called *voltage spikes,* in the system. This prevents damage to electronic components. In a DC circuit, electrons cannot flow through a capacitor.

Capacitors were used sparsely in the past in the ignition distributor or for silencing radio noise. Today's electronic cars have capacitors in most circuits. A capacitor is connected in parallel in a DC circuit (Figure 29.35). In an AC circuit, electricity will flow through a capacitor as if it were part of the wiring.

A capacitor is made up of two pieces of foil, separated by an insulator (Figure 29.36). The ground side of the capacitor (case) is connected to one piece of foil.

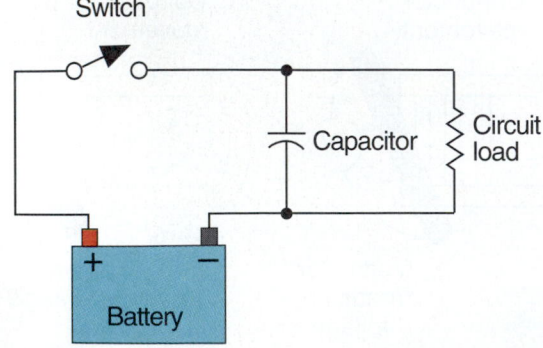

Figure 29.35 A capacitor connected in parallel in a circuit.

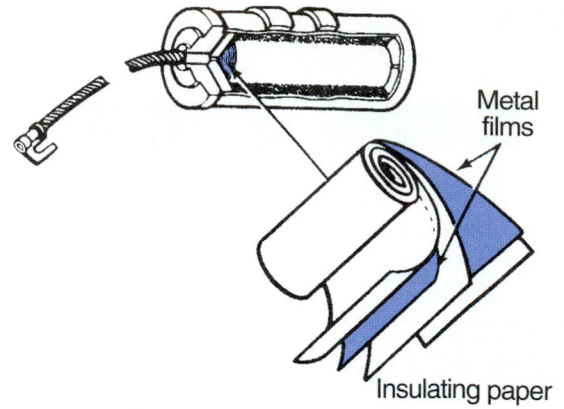

Figure 29.36 A capacitor has two layers of foil separated by an insulator.

The positive side is connected to the other piece. Electricity always looks for the shortest path to ground. The capacitor's job is to fool the electricity so that it loads one of the foil pieces with electrons. They enter the capacitor because of the attraction of the other piece of foil's opposite charge.

■ AUTOMOTIVE ELECTRONICS

Electrical components use moving or mechanical parts. These can wear, burn, or pit and are slow when compared to electronic parts. Electronic systems use *solid state* parts, which have no moving pieces.

- A **semiconductor** can act as both an insulator or a conductor. Common semiconductor materials are silicon (Si) and germanium (Ge).
- A **diode** is an electronic one-way check valve. It allows electricity to flow in only one direction.
- A **transistor** (Figure 29.37) is an electronic relay. The word is a combination of the two words *transfer* and *resist.* It is a very fast switch with no movable parts or contacts that can burn. It can resist electrical flow or allow a predetermined amount of current to flow.

Chapter 75 deals with automotive electronics in detail.

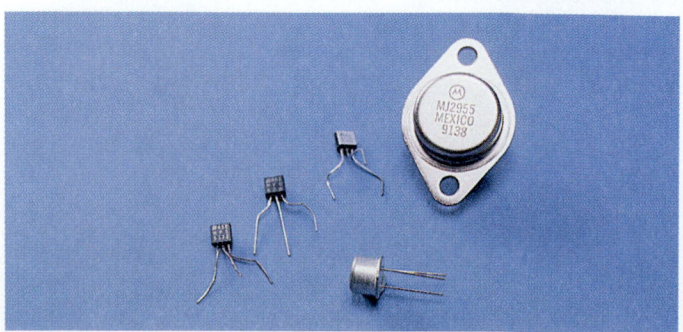

Figure 29.37 Transistors used in automotive circuits.

BASIC ELECTRICAL TESTS

Meters

A meter can be used to measure volts, ohms, and amps. Meters used for testing automotive electricity are either permanent magnet or digital. A permanent magnet meter is called an **analog meter**. This means it has a dial with a spring-loaded needle that is moved by a magnet (Figure 29.38). When three magnets are placed in a line, the center one will try to pivot into alignment by turning 180°.

To make a permanent magnet meter, the two permanent magnets are on the outside and the inside piece is an electromagnet (Figure 29.39). The electromagnet is made of several turns of wire wrapped around a rectangular aluminum frame. Having an aluminum frame (nonmagnetic) allows the magnetic field to instantly collapse when the meter is not energized. The aluminum frame floats on two jeweled bearings. Small hair springs hold the needle in the zero position.

When the coil is energized, it becomes an electromagnet, which operates on a second magnetic field produced by the permanent magnet. The electromagnet has a positive and negative pole like the permanent magnet (Figure 29.40). Because like poles repel, the meter dial moves in proportion to the electricity flowing through it.

Digital Volt Ohmmeter

A common technicians's tool is a **digital multimeter (DMM)** (Figure 29.41). When the meter has only a voltmeter and ohmmeter it is called a **digital volt ohmmeter (DVOM)**. Digital multimeters also have an ammeter (usually only up to 10 amps maximum). Because of the low amount of amperage that can be measured, a technician will need another meter for measuring amperage. Starting and charging systems can have amperage flow in excess of 100 amps. A larger volt amp tester is needed for these systems (see Figure 31.30).

The digital multimeter is more popular today than the analog meter because of its sensitivity to electronic

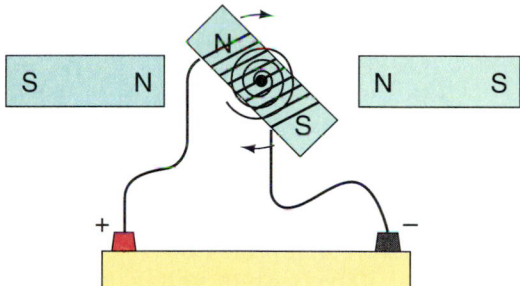

Figure 29.40 Operation of a permanent magnet meter. *(Courtesy of General Motors Corporation, Service Technology Group)*

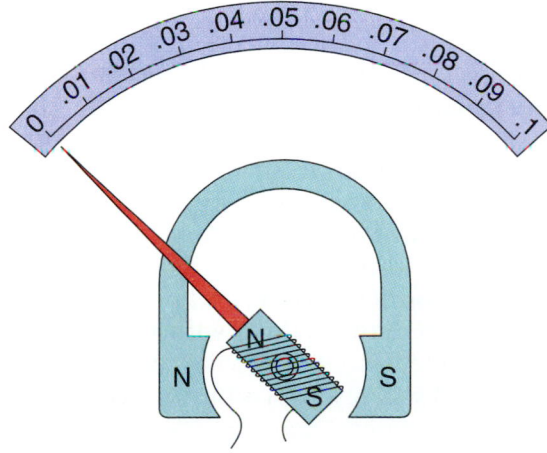

Figure 29.38 An analog meter has a dial with a needle that is moved by a magnet. *(Courtesy of General Motors Corporation, Service Technology Group)*

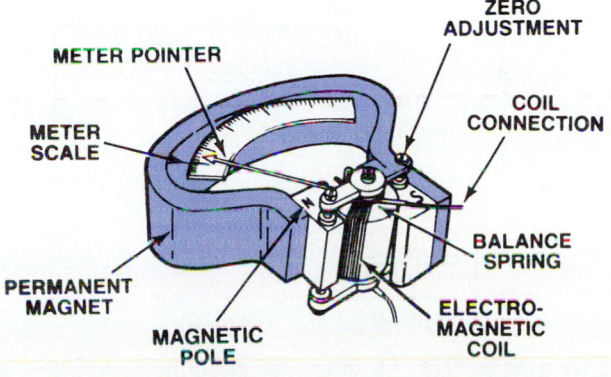

Figure 29.39 Parts of an analog meter.

Figure 29.41 A digital multimeter. *(Reproduced with permission from Fluke Corporation)*

(very small amounts of electricity) circuits and its automatic features. Because computer systems run on very small amounts of electricity, most magnetic meters will load the circuit and change the test results. Digital meters have high input resistance and have a digital display like a watch. Having high input resistance prevents the meter from drawing current when connected to the circuit. This maintains the accuracy of the meter reading and prevents damage to the circuit.

NOTE: *An analog meter usually cannot be used to measure computer circuits unless it is one of the more expensive meters having more than 10 megohms (10 million ohms)* **impedance** *(internal resistance). Most analog meters have 20,000–30,000 ohms of impedance. They can increase amperage in the circuit and damage the component or the circuit. Using a meter with resistance this low can destroy an oxygen sensor, for instance.*

Digital meters are not polarity sensitive. This means they can be hooked up in either direction. The meter will simply give a + or – reading on the dial. Most newer digital meters are self-scaling, too. This means that the dial does not have to be adjusted to read a certain level of volts, ohms, or amps like on an analog meter.

Voltmeter

A voltmeter is used for several tests:
- To measure available system voltage at the battery or alternator
- For checking the difference in voltage between two points
- For checking for excessive voltage drop due to resistance in a circuit

To read system voltage, the meter is connected to the circuit in parallel (Figure 29.42). The test leads (wires) are usually color coded, with red being positive and black negative. Hook the red lead to the positive side of the circuit and the black lead to ground. If the meter is connected backwards, the needle will deflect in the opposite direction.

The voltmeter can also be used to find resistance in a circuit. **Voltage drop** is a loss of voltage caused by current flow through a resistance. This measurement is widely used and is very popular with electrical service technicians. Voltage drop must be measured when the circuit is under load.

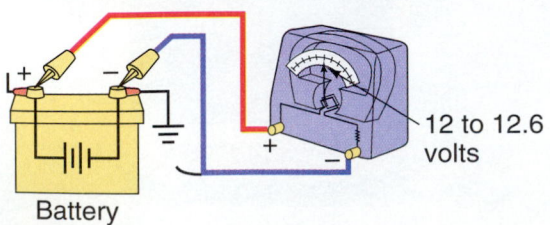

Figure 29.42 Voltmeter hookup.

Kirchoff's law states that the total voltage drop in an electrical circuit will always be equal to the available voltage at the source. All of the voltage at the source must be dropped in the circuit before it returns to the source. When resistance is excessive somewhere in a circuit, it will result in a voltage drop across that portion of the circuit. Using Ohm's law to calculate a voltage drop is done by multiplying Resistance by Current.

To measure voltage drop, the voltmeter is connected to two places on the same side of the circuit. For instance, to test a battery terminal connection, one lead of a voltmeter is connected to the battery terminal and the other lead is connected to the battery post. When the engine is cranked, any reading on the voltmeter indicates voltage that is leaking through the meter. More than a specified amount means too much resistance. This test is handy because the system does not have to be disassembled to find a problem.

NOTE: *A voltage drop test is only valid in a circuit that carries higher amounts of current. In low voltage computer circuits, there is not enough amperage to use this test.*

Ammeter

An ammeter measures amperage or current flow in a circuit. To measure water flow, a flowmeter would have to be installed in series in the water line. The same thing is true to measure electrical current flow. The ammeter must be hooked in *series* with the electrical load (Figure 29.43) or an *inductive pickup* must be used (Figure 29.44). An inductive pickup encircles the outside of a wire and measures the amount of electrical flow using the principles of electromagnetism.

Testing amp draws is done with the system under load. Using a meter that does not have enough capacity for the load will result in a blown fuse.

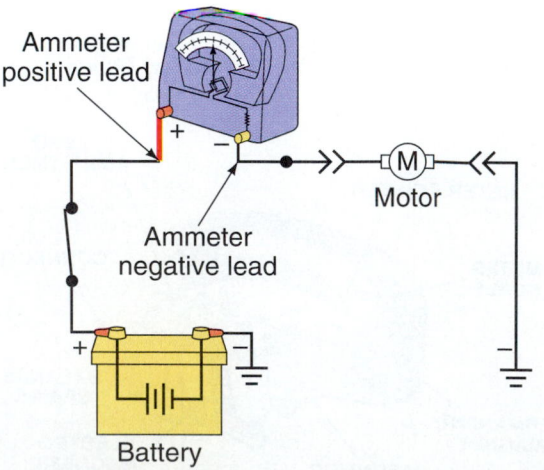

Figure 29.43 An ammeter is connected in series so electrical current can flow through it.

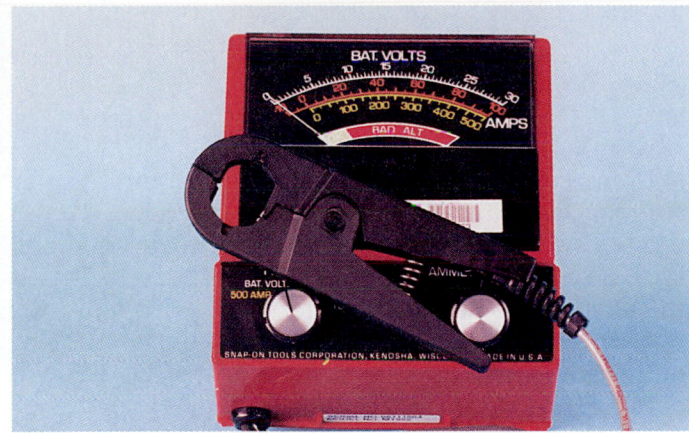

Figure 29.44 An ammeter with an inductive pickup.

Typical Amp Draw. The following list from the *BCI (Battery Council International)* gives the current (amps) loads that a typical passenger car might have.

Typical Current Loads	
Load	**Amp Draw**
Stoplights	8
Interior Lights	2
Blower (low)	6
(medium A/C)	16
(high A/C)	35
Heated back window	22
Ignition	6
Radio	0.5
Windshield Wipers	7.5
Headlights (low beam)	9
(high beam)	13
Parking Lights	7
Base load w/air cond.	
(summer)	50
(winter)	45
Summer Starting	
(gas)	150–250
(diesel)	450–550
Winter Starting (very low temperatures)	
(gas)	250–350
(diesel)	700–800

Ohmmeter

An ohmmeter measures the amount of resistance in an electrical circuit by measuring the amount of current (amperage) that can be forced to go through a circuit by a small voltage. To do this, it must have its own battery.

Because batteries do not always have the same state of charge, the ohmmeter must either be a self-calibrated type or it must be hand-calibrated to adjust its output voltage before it is used. An ohmmeter can be hand-calibrated by connecting its wires together to read 0 resistance. Some meters have an indicator that tells when the battery is low.

NOTE: *Never connect an ohmmeter across an energized circuit. No electricity can be flowing in it (Figure 29.45).*

Circuit Problems

- An *open circuit* is when there is a break in the path for electrical flow, an incomplete circle.
- A *short circuit* is when the electrical path has been shortened, when wires accidentally touch each other or grounded metal, for instance (Figure 29.46).

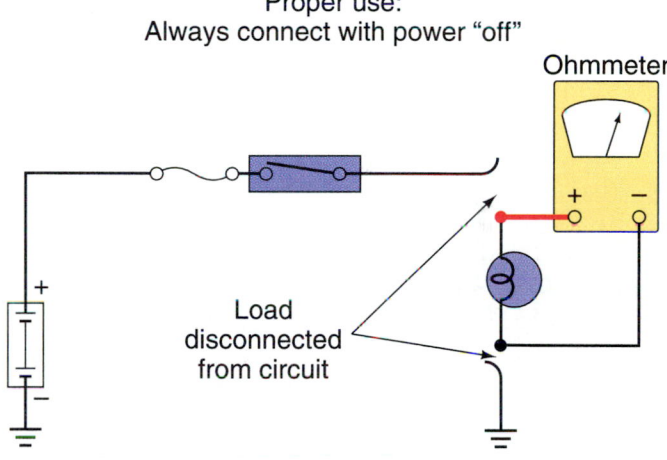

Figure 29.45 An ohmmeter is used with the power off.

Figure 29.46 A short circuit. *(Courtesy of General Motors Corporation, Service Technology Group)*

■ A *grounded circuit* is like a short circuit, but the current goes directly to ground (Figure 29.47). Notice the ground symbol in the sketch.

Jumper Wires

A jumper wire is a simple wire with alligator clips attached to its ends (Figure 29.48). It is used primarily for finding open circuits. Figure 29.49 shows how to eliminate different places in the circuit as possible causes of an open circuit.

SAFETY NOTE

■ Never use a jumper lead that is smaller in size than the circuit being tested. It could overheat and melt.
■ Never use a jumper to bypass a high resistance load (like a motor). This actually creates a short circuit, which could result in a fire.
■ Only use a jumper to temporarily bypass a component.

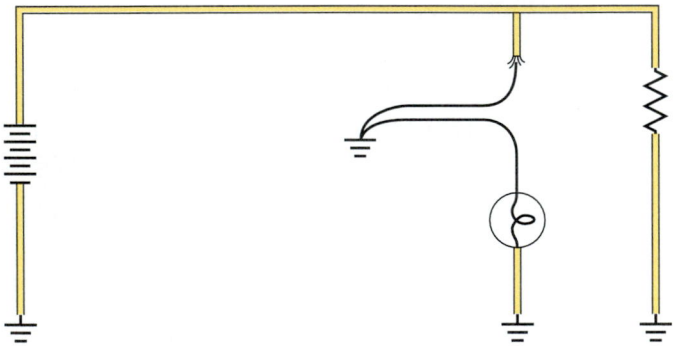

Figure 29.47 A grounded circuit. *(Courtesy of General Motors Corporation, Service Technology Group)*

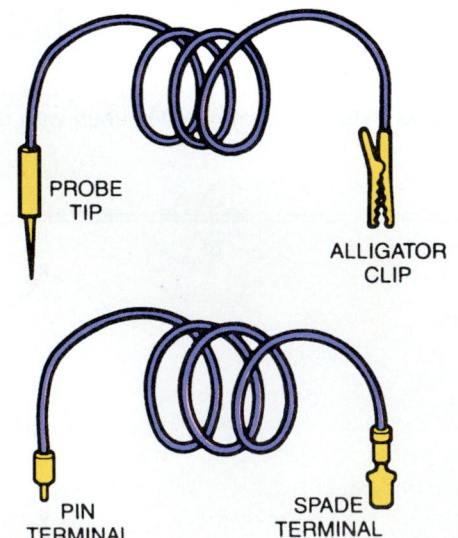

Figure 29.48 Different kinds of jumper wires. *(Courtesy of Ford Motor Company)*

Test Lights

A 12-volt test light is a handy technician's tool (Figure 29.50). To use a test light:
■ Test its operation by connecting it across the battery.
■ Connect the alligator clip to a good ground.
■ Probe where necessary with the pick end.

Testing for opens can be done with a test light. With a jumper lead, the process was to try to bypass the open to operate the circuit. With a test light, the process is to locate where voltage is in the circuit. Figure 29.51 shows how a circuit is tested at different points. This tester is not used on computer circuits.

Self-Powered Test Light. A self-powered test light is like a flashlight. It has its own battery and is used to

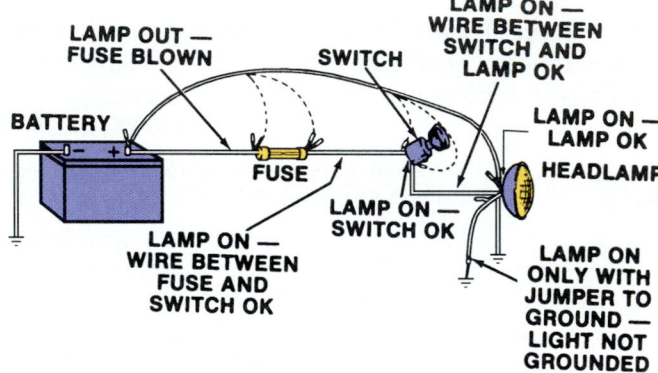

Figure 29.49 Using a jumper wire to find an open circuit. *(Courtesy of Ford Motor Company)*

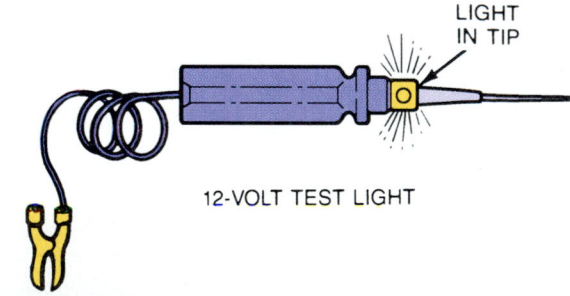

Figure 29.50 A 12-Volt test light. *(Courtesy of Ford Motor Company)*

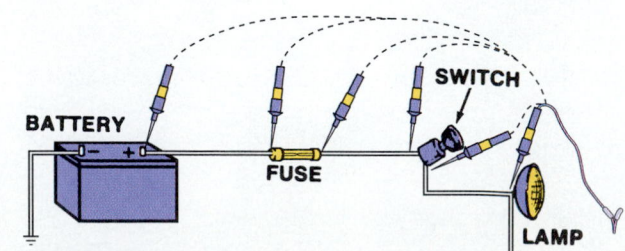

Figure 29.51 Checking for opens with a test light. *(Courtesy of Ford Motor Company)*

test for continuity much like an ohmmeter. The tester can be used for testing for open or short circuits when power is disconnected. If the light glows, the circuit or part has *continuity*. Figure 29.52 shows how a self-

powered test light can be used to test a light filament. Testing for a complete ground circuit is done as shown in Figure 29.53. Other tests and repairs to electrical wiring are covered in Chapter 37.

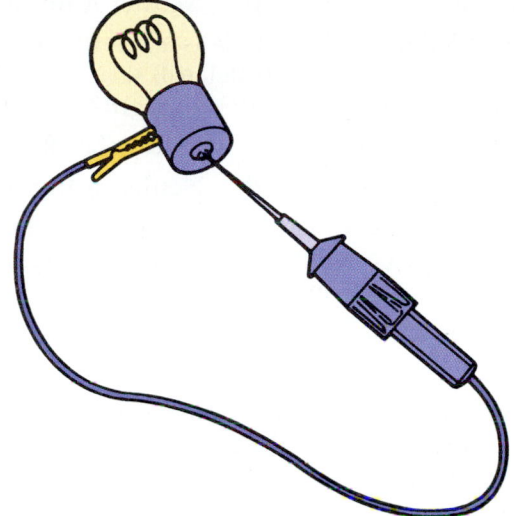

Figure 29.52 Using a self-powered test light to check a light bulb. *(Courtesy of Ford Motor Company)*

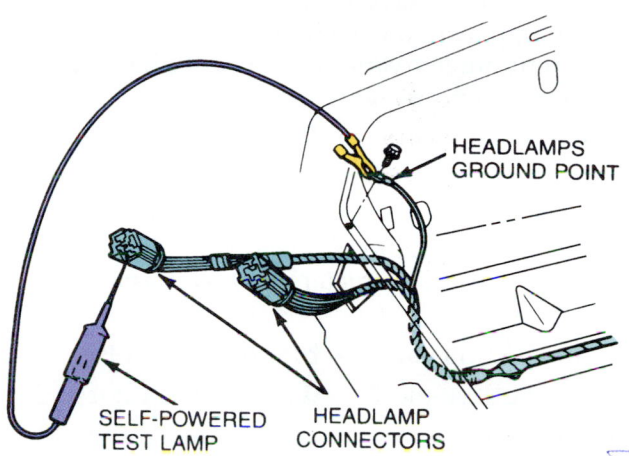

HEADLAMPS GROUND POINT

SELF-POWERED TEST LAMP

HEADLAMP CONNECTORS

Figure 29.53 If the ground circuit is complete, the self-powered test light will light up. *(Courtesy of Ford Motor Company)*

■ REVIEW QUESTIONS

1. What is the name for the materials that are made of atoms that allow electricity to flow freely?

2. An _____ prevents the flow of electrons between two conductors.

3. Name three good insulators.

4. What is the name for a complete circle provided for electrical flow?

5. What are the names of two circuit protection devices?

6. How fast does light travel?

7. What kind of current is used for houses, AC or DC?

8. What is the unit of electrical resistance measurement called?

9. If there is more resistance in a circuit, what happens to current flow?

10. What is the name of the electrical part that is used to change the amount of current that flows to a heater motor?

11. What kind of circuit has current flowing in a line, first to one load and then on to the next one?

12. What kind of circuit has different branches that current can flow through, starting from a common point?

13. What happens when a magnetic field is moved over a conductor?

14. When a large load must be controlled by a small wire, what device is used?

15. Which kind of meter is used only on a circuit that has no electrical power: voltmeter, ohmmeter, or ammeter?

■ ASE STYLE REVIEW QUESTIONS

1. Technician A says voltage is electrical pressure. Technician B says amperage is electrical flow. Who is right?

   a. Technician A   b. Technician B
   c. Both A and B   d. Neither A nor B

2. Technician A says if voltage is low, a light will not be as bright or a heater defroster motor will not spin as fast. Technician B says current flow is measured with an ammeter. Who is right?

   a. Technician A   b. Technician B
   c. Both A and B   d. Neither A nor B

3. Technician A says direct current (DC) is used in automobile electrical systems. Technician B says like magnetic poles attract each other and unlike poles repel. Who is right?

   a. Technician A **b.** Technician B
   c. Both A and B **d.** Neither A nor B

4. Technician A says a rheostat varies voltage through a circuit. Technician B says a potentiometer varies the current flow in a circuit. Who is right?

   a. Technician A **b.** Technician B
   c. Both A and B **d.** Neither A nor B

5. Technician A says the longer a conductor is, the more resistance it has. Technician B says the larger a wire's diameter, the less resistance it has. Who is right?

   a. Technician A **b.** Technician B
   c. Both A and B **d.** Neither A nor B

6. Technician A says in a series circuit, adding loads reduces current flow. Technician B says adding loads to a parallel circuit can overload the circuit. Who is right?

   a. Technician A **b.** Technician B
   c. Both A and B **d.** Neither A nor B

7. Technician A says lead is a good insulator for magnetic fields. Technician B says a magnetic field can be made stronger by coiling a current carrying wire. Who is right?

   a. Technician A **b.** Technician B
   c. Both A and B **d.** Neither A nor B

8. Technician A says a diode is a mechanical one-way check valve. Technician B says a transistor is an electronic relay. Who is right?

   a. Technician A **b.** Technician B
   c. Both A and B **d.** Neither A nor B

9. Technician A says an analog meter has a dial readout. Technician B says analog meters are better for electronic circuits. Who is right?

   a. Technician A **b.** Technician B
   c. Both A and B **d.** Neither A nor B

10. Technician A says the voltmeter can be used to check for resistance in a circuit. Technician B says the ohmmeter can be used to check for resistance in a circuit. Who is right?

    a. Technician A **b.** Technician B
    c. Both A and B **d.** Neither A nor B

CHAPTER 30

Battery Fundamentals

■ OBJECTIVES

Upon completion of this chapter, you should be able to:

✔ Describe battery parts and operation.

✔ Understand how a battery converts chemical energy to electrical energy.

✔ Explain battery capacity ratings.

✔ Select the correct battery type to use in a variety of applications.

■ KEY TERMS

element
battery terminals
cold cranking amps
  (CCA)
reserve capacity
deep-cycle
maintenance-free
  battery
low-maintenance
  battery
sulfation

■ INTRODUCTION

The car's battery is the heart of the vehicle's electrical system. It converts electrical energy received from the charging system into chemical energy that it stores. Then, it converts the chemical energy back into electrical energy.

■ When the engine is off, the battery provides power for electrical accessories that are turned on.

■ During cranking, the battery supplies current to the starter. When the engine is running and electrical demands are more than the charging system is capable of contributing, the battery serves as a reservoir, supplying electrical energy for the extra loads.

■ The battery acts like a capacitor, absorbing high voltages that happen in the electrical system. These high voltages would damage other electrical system components if the battery did not serve this function.

When a *load,* such as a light bulb, is connected to the positive and negative terminals of the battery, the bulb will light up because electrons produced by a chemical reaction in the battery are flowing through its filament (Figure 30.1). Chemical energy is being converted to electrical energy.

The largest load is the starter motor, which draws a good deal of current from the battery. The starter is the reason that the battery cables are as large as they are. If they were smaller, they would get hot as the electrons tried to flow through them to the starter. The starter draws more than 100 amperes, but the charging system is only capable of generating 35 to 100 amps (depending on its alternator's capacity).

After the starter is used, the battery voltage is momentarily low. After the engine starts, the battery needs to be recharged. It is recharged by the charging system. This is a conversion of electrical energy to chemical energy; the opposite of what took place when the battery was discharged. With all accessories running, the charging system may not be able to provide all the electricity needed. In this case, the battery will supply the extra energy needed until the load is reduced or the battery fails.

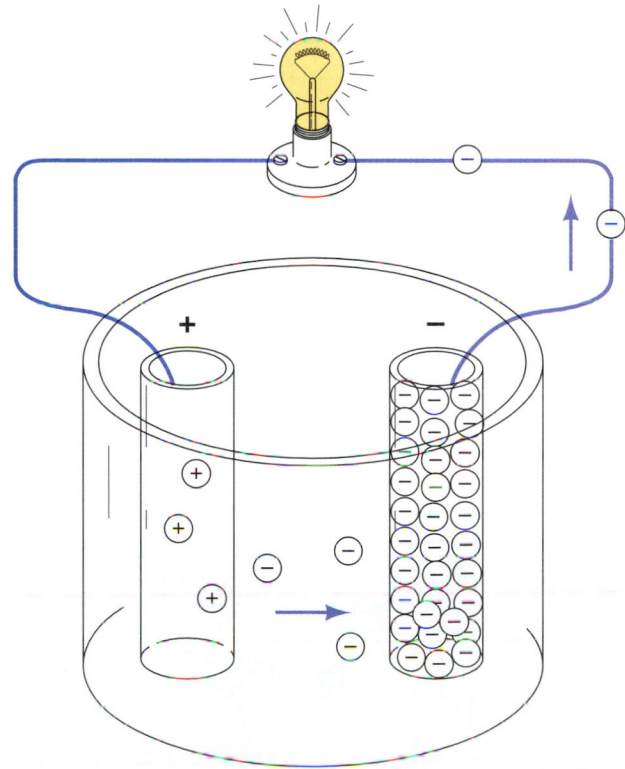

Figure 30.1 When a load is placed between the positive and negative battery terminals, electrons flow.

■ BATTERY PARTS AND OPERATION

When two different metals are immersed in an electrolyte solution, a direct current (DC) voltage is produced. You can do this using a potato, an orange, or a soft drink as the electrolyte. Poke two types of metal into the potato and use a voltmeter to read the voltage between them.

Different metals used to construct other types of batteries provide various voltages. For instance, a flashlight battery's 1.5 volts comes from the reaction of the electrolyte and carbon (the positive carbon center rod) and zinc (the negative battery case). A standard flashlight has two batteries stacked together in *series* to provide the 3 volts required to light its bulb. Notice how the positive and the negative flashlight battery terminals touch each other. That is what series is.

In automotive batteries, 2.1 volts is the amount that the two dissimilar materials of the plates provide. The typical 12-volt automotive battery has six 2.1-volt *cells* hooked to each other in series (from positive to negative to positive). This provides about 12.6 volts when the battery is fully charged (Figure 30.2).

Plates

Battery plates are constructed of grids with horizontal and vertical bars (Figure 30.3). A paste is pressed into the grid. The paste is made up of lead oxide, acid, and material expanders. After a forming charge is given to the positive plate, its lead oxide turns into lead peroxide. When the negative plate is given its forming charge, the same paste turns into sponge lead. This provides the two dissimilar metals needed to make a voltage when they are immersed in electrolyte.

Electrolyte

The battery's case is filled with an electrolyte mixture of *sulfuric acid* and *water*. Electrical current in the battery is produced by a chemical reaction between the *battery acid* (another name for electrolyte) and two different types of lead material on the battery's positive and negative *plates*.

Battery Cells

Each battery cell consists of a packet of several positive and negative plates (Figure 30.4) insulated from each other and connected in *parallel* (positive to positive and negative to negative). The positive and negative plates in the cell are held together in an **element** by plate straps (Figure 30.5). The element is submerged in electrolyte. Each cell is then hooked to its neighboring cell in series (the positive plates of one cell are hooked to the negative plates of the next cell) (see Figure 30.2).

Battery Terminals

The battery has positive and negative terminal connections made of lead. Automotive **battery terminals** will be either on the top or on the side (Figure 30.6). When battery terminals are on top they are called *battery posts* (Figure 30.6a). The positive post is larger in diameter than the negative post. The top of the battery case is usually molded with labels that say "positive" or "negative." Sometimes, the top of the larger positive post is painted red and can have a + stamped into it. The top of the smaller negative post can be painted black and have a – stamped into it.

Some batteries have *side terminals* that have reinforced internal threads (Figure 30.6b). A special terminal adapter bolt threads into the terminal.

Batteries with *L terminals* are found on marine applications, motorcycles, and some imported cars (Figure 30.6c).

Battery Case

Most battery cases are constructed of lightweight plastic (polypropylene) with a one-piece cover. Cases can also be made of hard rubber and other plastics. The case must be able to resist acid absorption and withstand extreme changes in temperature. It must be durable enough to resist vibration. The bottom of the

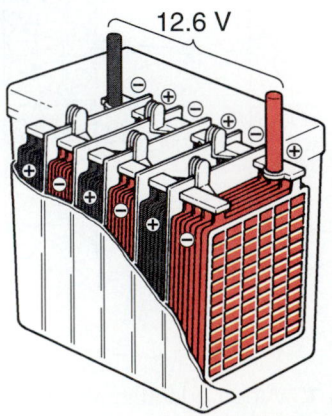

Figure 30.2 An automotive battery has six cells hooked together in series.

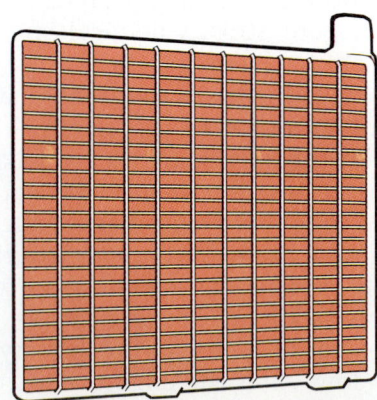

Figure 30.3 A battery grid. *(Courtesy of Ford Motor Company)*

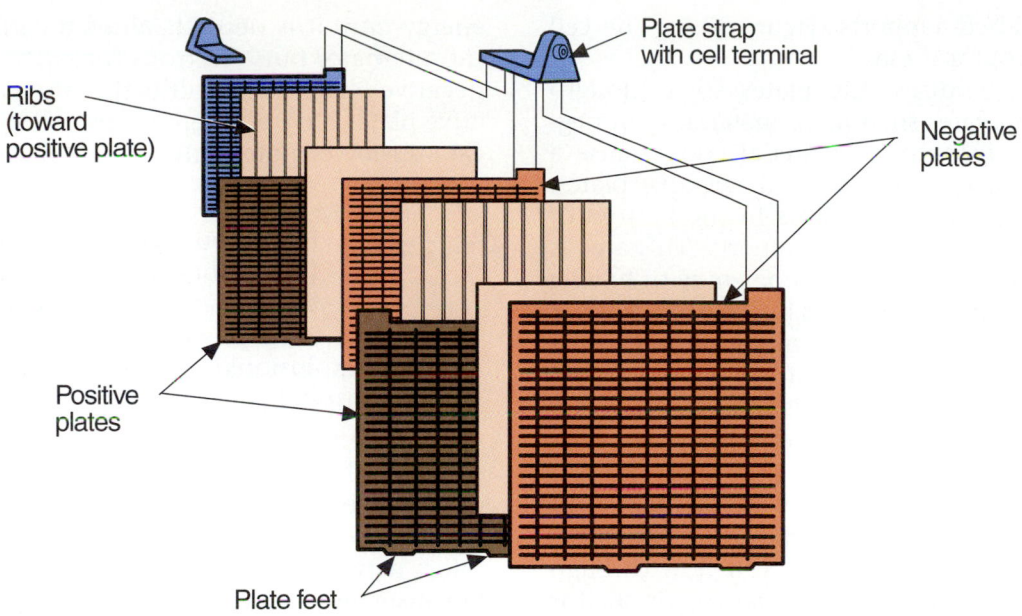

Figure 30.4 A battery cell has alternating positive and negative plates. *[Courtesy of Ford Motor Company]*

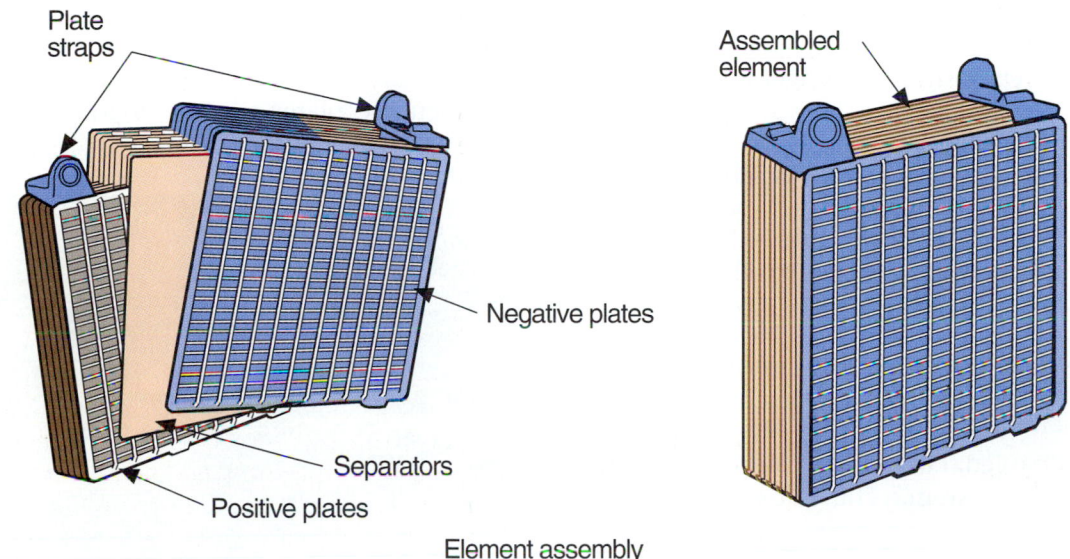

Element assembly

Figure 30.5 The cell is made up of a battery element. *[Courtesy of Ford Motor Company]*

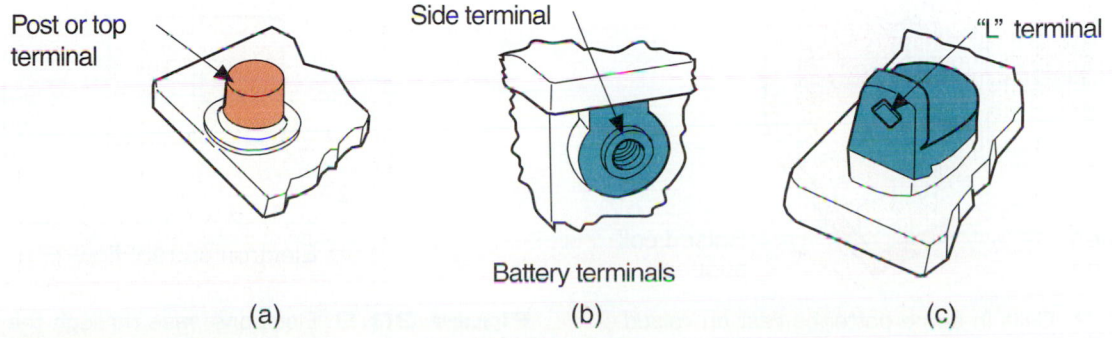

(a)   (b)   (c)

Figure 30.6 Different types of terminal connections. *[Courtesy of Ford Motor Company]*

case often has raised supports (Figure 30.7). The cell elements sit on top of these.

As a battery oxidizes, the plates shed (slough) material. Positive plates shed more material than negative plates. The loose plate material could cause a short between the cell positive and negative plates. Sloughed-off plate material collects in the bottom of the case between the raised cell supports. This area is called the *sediment chamber*. One problem with having the plates rest on the cell supports is that vibration has more of an effect on the plates. Another disadvantage is that the electrolyte that surrounds the loose material in the bottom of the battery is inactive. A bigger case is required to hold the desired amount of electrolyte.

Some cell elements have plastic envelopes that fit around the cell plates. The envelope collects material that falls off the plates. Electrolyte can flow through the plastic envelope but plate material that is shed is contained by it. The envelope prevents shed material from causing a short between the positive and negative plates. With envelopes around the plates, adding water is required less often because there is more reserve water above the plates. The envelopes also allow the battery case to be made smaller because a large sediment catch basin is no longer necessary.

Cell Caps

As a battery charges, hydrogen gas is released so the case cover must have vents. Although many batteries today are maintenance-free (without cell caps), most modern battery tops have removable cell caps (Figure 30.8). Maintenance-free batteries are covered later in this chapter. In batteries with removable cell caps, the vents are in the caps. The caps also act as spark arrestors.

Charging and Discharging

As a battery is charged, current flows into it from the vehicle's charging system. The battery stores the energy until it is needed. When a battery is charging, the alternator puts electrons (negative charges) on the negative plate. The result is that the positive and negative plates have a difference in voltage (called electrical pressure or potential).

SAFETY NOTE Hydrogen gas is very explosive. It is lighter than air so it gets trapped under the battery terminal connections and cell caps as the battery charges. Battery explosions put over 15,000 people a year in the hospital.

When the battery is discharging, current flows out of it. When this happens it is releasing stored energy. The positive and negative plates in the cell have separators between them. These allow electrolyte to flow, but insulate between the positive and negative plates and prevent them from touching.

The voltage imbalance between the positive and negative plates can only be equalized through a current path outside of the battery. When an electrical load is placed between the positive and negative terminals on the battery, a chemical reaction takes place. Current flows to the load as the voltage difference between the plates tries to neutralize. Electrons pass through the electrolyte solution as they move from

Figure 30.8 This battery has removable cell caps.

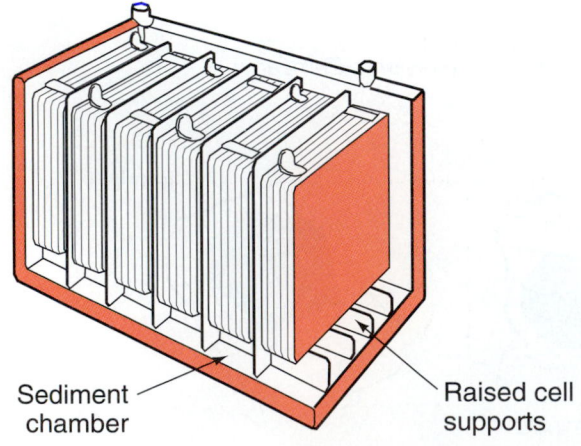

Sediment chamber

Raised cell supports

Figure 30.7 Cells in some batteries rest on raised supports in the bottom of the battery case. *(Courtesy of Ford Motor Company)*

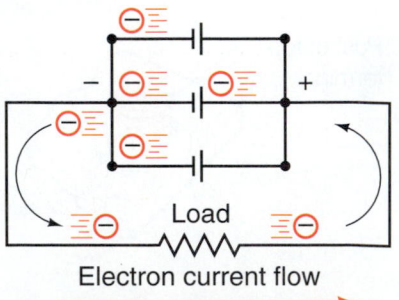

Load

Electron current flow

Figure 30.9 Electrons pass through the electrolyte solution as they move from the negative plates to the positive plates. *(Courtesy of Interstate Batteries)*

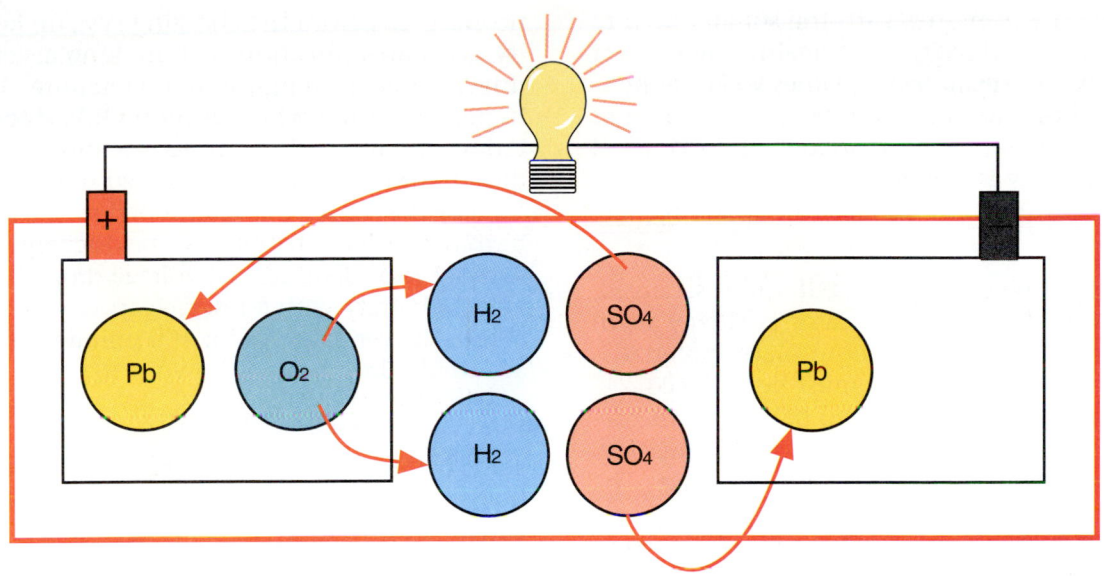

Figure 30.10 Chemical action during discharge of a battery.

the negative plates to the positive plates (Figure 30.9).

A fully charged positive plate is a combination of lead and oxygen called lead dioxide (PbO_2). The negative plate is sponge lead (Pb). The electrolyte is sulfuric acid (H_2SO_4) and water.

Discharge.   During discharge (Figure 30.10) the lead (Pb) from the positive plate combines with the sulfate (SO_4) from the battery acid. This forms lead sulfate

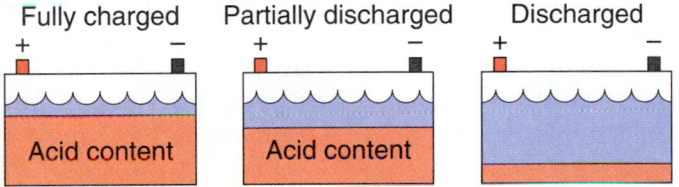

Figure 30.11 As a battery discharges, the strength of the electrolyte becomes weaker. *(Courtesy of Interstate Batteries)*

($PbSO_4$) on the plate. Oxygen in the positive plate combines with hydrogen in the electrolyte to form water (H_2O). This results in a diluted concentration of electrolyte (Figure 30.11).

Positive and negative plates are becoming alike as the battery discharges. The negative plate (Pb) combines with sulfate (SO_4) to form lead sulfate ($PbSO_4$) on the plate. When the cell is dead, the positive and negative plates are identical. As the battery discharges, additional water forms in the electrolyte to dilute it.

Charge.   During charging (Figure 30.12), the process is reversed. The lead sulfate in both the positive and negative plates is changed back to lead and sulfate as it was originally. The battery "gasses"; the water in the electrolyte is split into hydrogen and oxygen. The negative plates give off hydrogen gas and the positive plates give off oxygen; a very explosive combination.

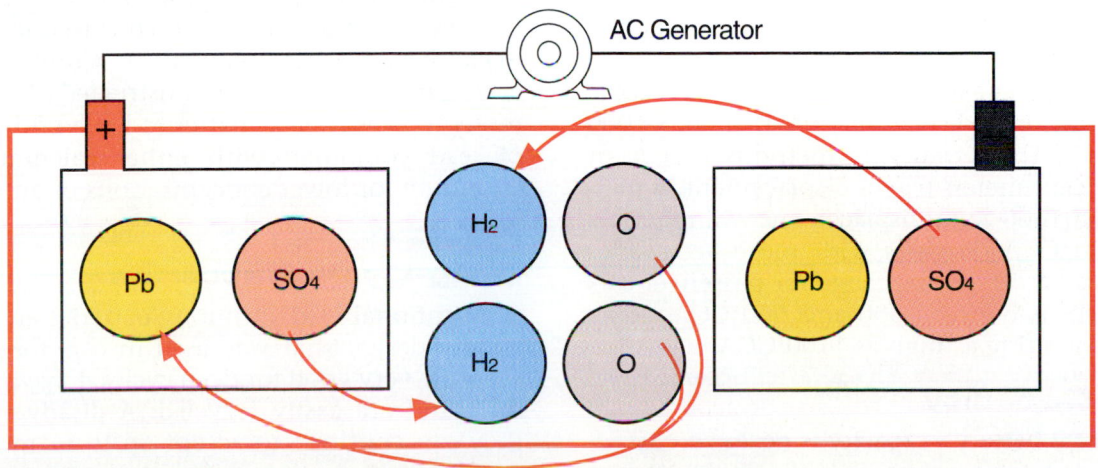

Figure 30.12 Chemical action during recharge of a battery.

The hydrogen (H_2) combines with the sulfate (SO_4) to become sulfuric acid (H_2SO_4) once again. The leftover oxygen as the water separates combines with the positive plate to form lead dioxide (PbO_2). The battery's plates and electrolyte are now back in their original form and the battery is charged.

NOTE: *The battery gasses the most as it reaches full charge.*

As a battery is repeatedly charged and discharged, the active material on the cell plates is slowly worn away. The plates oxidize and more plate material is sloughed off. Finally the battery must be replaced.

With the engine off, most good batteries can provide about 25 amperes for two hours before discharging to the point where the engine will not start. The amount of energy that the battery is capable of providing depends on the amount of plate surface material and the concentration of acid in the electrolyte.

■ BATTERY CAPACITY RATINGS

Battery capacity ratings are established by the Battery Council International (BCI) and the Society of Automotive Engineers (SAE). Current capacity is related to the following factors:

■ The surface area of the plates
■ The weight of the active materials on the plates
■ The strength of the electrolyte solution

A battery's current capacity rating reflects its ability to deliver cranking power to the starter motor and provide reserve capacity to the electrical system.

Cold Cranking Amp (CCA) Rating

The most common method of rating automotive batteries is **cold cranking amps (CCA)**. CCA is the common standard for low maintenance batteries. An older standard, called the *amp-hour* rating, is no longer widely used.

For a 12-volt battery, CCA is determined by the number of amperes a battery can deliver for 30 seconds at 0°F (–17.7°C) without terminal voltage falling below 7.2 volts.

A battery should have at least 1 CCA for each cubic inch (cu. in.) of engine displacement (61 cu. in. = 1000 cc or one liter). The formula for metric is 1 CCA for each 16 cubic centimeters (cc) of displacement. A metric engine with 1600 cc of displacement will require a battery of 100 CCA. Engines with more accessories require more CCA. The usual range for passenger cars and light trucks is between 300 and 600 CCA. Some batteries have a rating as high as 1100 CCA.

Reserve Capacity

Another test of a battery's capacity is **reserve capacity**, also called staying power. This is a measurement of the battery's ability to provide current when there is no electricity from the charging system. Reserve capacity gives an indication of how long a vehicle can be driven after a charging system failure. It is also an important rating when selecting R.V. **deep-cycle** batteries. This test is done under the most strenuous conditions and is usually not applied to car batteries.

Reserve capacity is determined by the length of time in minutes that a battery can be discharged at 25 amps without its individual cell voltage dropping below 1.75 volts. A battery with a reserve capacity of 100 would be able to deliver 25 amps for 100 minutes before the voltage would drop below 10.5 volts (6 cells × 1.75 volts).

Watt-Hour Rating

Some battery manufacturers rate their batteries in watt-hours. This is because the battery's electrical energy is converted into mechanical energy (watts).

NOTE: *A watt is the unit of electrical power that is equivalent to horsepower. One horsepower is equal to 746 watts.*

The watt-hour rating is determined at 0°F (–17.7°C) because a battery's capacity in watts changes with temperature. The rating is determined by multiplying a battery's amp-hour rating by the battery voltage.

■ BATTERY TYPES

Batteries are designed for specific uses. The state of charge of an automotive battery is kept about 90% to 100% most of the time. If the battery's state of charge falls below 75%, the battery needs to be recharged. Allowing a battery to be discharged until dead or near dead will shorten its life.

Batteries in automobiles are of three main types: **maintenance-free**, **low-maintenance** (low-water-loss) and deep-cycle (R.V.). A recreational vehicle or a boat's electric trolling motor would need to supply a relatively small amount of current for long periods of time. A passenger car, on the other hand, would require a large amount of current draw during cranking, followed by a period of recharge.

Battery grids are made of lead. But lead by itself is too soft so it must be combined with another metal to make the grid material harder. Conventional deep-cycle batteries have grids constructed of lead and antimony. Low-waterloss batteries have grids constructed of lead combined with either calcium, cadmium, strontium, or lower concentrations of antimony.

Conventional Deep-Cycle Batteries

Conventional lead antimony batteries are the original battery design that was used in cars for many years. They are very good for deep cycles, have high CCA ratings, and are easily recycled. A disadvantage is that they can continue to accept up to 5 amps of current during a recharge, even though they are fully charged. Because of this, they are no longer used in passenger cars. They use considerably more water and they gas heavily during overcharging. The gas results from electrolysis, which is when water is separated into hydrogen and oxygen. The hydrogen gas, which is dangerous, must be vented outside of the battery.

When a battery gases, corrosion of the terminals, hold-downs, and battery tray occur from the sulfuric acid fumes in the escaping gas. Lead antimony batteries also have considerable terminal corrosion and require periodic charging when stored.

These batteries can go through many deep cycles. A deep cycle is when the battery is allowed to run almost completely dead and then is recharged. Deep-cycle batteries have a thicker layer of paste on their plate grids and can stand hundreds of repeated discharges with very little damage. The thick plates are porous and can give off more electrons, but do this more slowly. Thicker plates results in less surface area in these batteries than would be the case if there were more, thinner plates.

Antimony, used in these batteries, is a stiffening material for the grids. It is not as resistant to overcharging as the other grid materials. It is temperature sensitive; as it gets hot its resistance drops. This results in an increase in the breakdown of water into hydrogen and oxygen. Antimony becomes very hot if the charging voltage is allowed to climb above 14.5 volts. As more of the water is gassed from the battery, the plates get even hotter.

Maintenance-Free Batteries

Maintenance-free batteries (Figure 30.13) have cell plates made of a slightly different material. Calcium or strontium is used to strengthen the plate grids instead of the antimony used in conventional batteries (Figure 30.14).

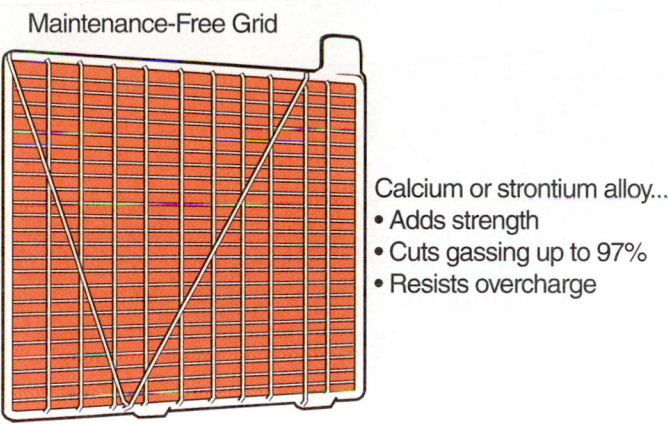

Figure 30.14 Maintenance-free battery grids are strong and thin. *(Courtesy of Ford Motor Company)*

Calcium or strontium alloy...
• Adds strength
• Cuts gassing up to 97%
• Resists overcharge

The electrical resistance of calcium does not drop when hot nearly as much as antimony. When overcharged, calcium will only use $\frac{1}{3}$ of the water that antimony does. Corrosion on the terminals and surface of the battery is minimized also.

Maintenance-free batteries are sealed except for small breather holes (see Figure 30.13). This is because the lead and calcium combination is susceptible to damage from even a small amount of dirt or grease. The battery does not require water to be added *if the charging system is operating properly*. If the voltage regulator allows the alternator to overcharge the battery, the water level will drop. If it drops too low, the battery must be replaced.

Additional advantages to the maintenance-free battery are:

■ It can be stored for long periods of time without experiencing self discharge (three times as long as lead-antimony). Eighteen months on the shelf without self discharge is common.
■ It does not normally require an activation or boost charge when new.

Disadvantages of the maintenance-free battery are:

■ It can be damaged with repeated deep discharges, reducing its plate material significantly.
■ Because there are no cell caps, hydrometer testing of the battery's state of charge is not possible (see Chapter 31).
■ It has a lower reserve capacity.
■ Its lead-calcium material is very brittle and breaks easily.

Low-Maintenance Batteries

A **low-maintenance battery** is a revision of the maintenance-free battery. Its grids contain about 3.4% antimony, and it is called a dual alloy battery. The grids intersect at a different angle for strength and better electrical conduction (Figure 30.15).

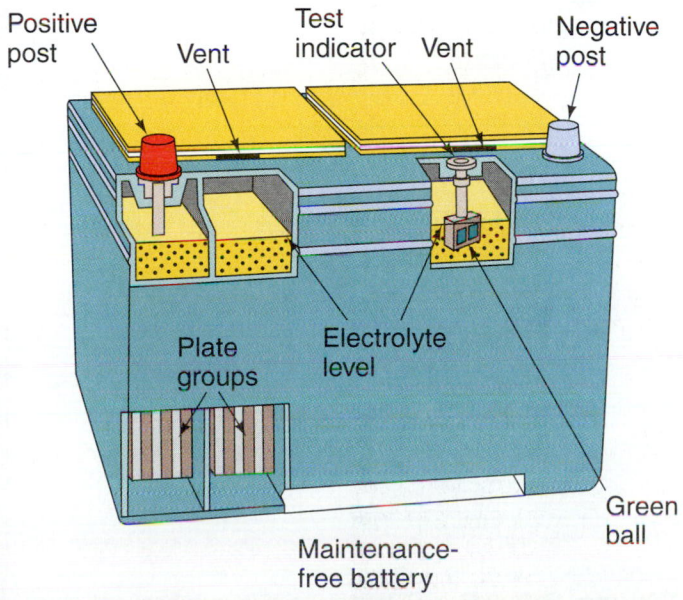

Positive post
Vent
Test indicator
Vent
Negative post
Plate groups
Electrolyte level
Maintenance-free battery
Green ball

Figure 30.13 Parts of a maintenance-free battery. *(Courtesy of Chrysler Corporation)*

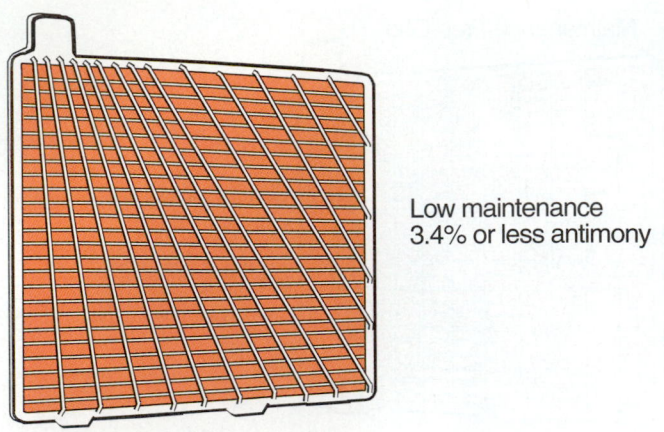

Low maintenance
3.4% or less antimony

Figure 30.15 A low-maintenance battery grid has a reduced antimony content. *[Courtesy of Ford Motor Company]*

Low-maintenance batteries are not as susceptible to damage from contaminants as calcium or strontium grid batteries. They have vent caps on top that allow hydrometer testing and the addition of water to the cells. The caps are designed to trap and condense vapors, returning them to the cell.

Low-maintenance batteries:

- Do not "gas" as much at normal voltage levels as the deep-cycle battery.
- Are the most popular battery for passenger car use
- Have a longer shelf life without charging than deep-cycle batteries (about one year)
- Have a higher cold cranking amps rating.
- Can be supplied in a dry state, ensuring that they are fresh
- Are more resistant to damage from deep discharge

■ BATTERY PLATE SIZE

The size of a battery is usually related to its reserve capacity. The amount of active plate material is what actually determines the capacity of the battery. The thicker the plates in the cells are, the more reserve capacity.

Cars with larger engines and more electrical accessories require batteries with more plate surface area (higher CCA rating). A battery with thin plates can supply a high current for a short period (such as when operating a cranking motor). A deep-cycle R.V. battery, which must supply a smaller amount of current for a longer period of time, might have fewer plates, but they are thicker.

A *series* hookup is when the positive and negative plates are hooked to each other, adding up the voltages of the cells (see Figure 30.2). Two 12-volt batteries can actually be hooked in *parallel* to provide more plate capacity. In fact, recreational vehicles often have this arrangement to provide sufficient storage capacity to supply all of their electrical needs for a night. With a parallel hookup, voltage stays the same, but the amount of plate material doubles.

■ BATTERY SELECTION

The Battery Council International (BCI) lists group numbers to indicate the physical size of batteries (Figure 30.16). Cold cranking power is also listed with the number of months warranty on the battery. The following are some considerations when selecting a battery:

Grp. Size	Vlt.	Cold Cranking power—amps for 30 secs. at 0°F*	No. of mo. warran-ted	Size of battery container in inches (incl. terminals)		
				Lgth.	Wd.	Ht.
17HF	6	400	24	7¼	6¾	9
21	12	450	60	8	6¾	8½
22F	12	430	60	9	6⅞	8⅛
	12	380	55	9	6⅞	8⅛
	12	330	40	9	6⅞	8⅛
22NF	12	330	24	9½	5½	8⅞
24	12	525	60	10¼	6⅞	8⅝
	12	450	55	10¼	6⅞	8⅝
	12	410	48	10¼	6⅞	8⅝
	12	380	40	10¼	6⅞	8⅝
	12	325	36	10¼	6⅞	8⅝
	12	290	30	10¼	6⅞	8⅝
24F	12	525	60	10¼	6⅞	8⅝
	12	450	55	10¼	6⅞	8⅝
	12	410	48	10¼	6⅞	8⅝
	12	380	40	10¼	6⅞	8⅝
	12	325	36	10¼	6⅞	8⅝
	12	290	30	10¼	6⅞	8⅝
27	12	560	60	12	6⅞	8⅝
27F	12	560	60	12	6⅞	9
41	12	525	60	11⁹⁄₁₆	6¹³⁄₁₆	6¹⁵⁄₁₆
42	12	450	60	9⅝	6⅞	6¾
	12	340	40	9⅝	6⅞	6¾
45	12	420	60	9½	5½	8⅞
46	12	460	60	10¼	6⅞	8⅝
48	12	440	60	12	6⅞	7½
49	12	600	60	14½	6⅞	7½
56	12	450	60	10	6	8¾
	12	380	48	10	6	8¾
58	12	425	60	9¼	7¼	6⅞
71	12	450	60	8	7¼	8½
	12	395	55	8	7¼	8½
	12	330	36	8	7¼	8½
72	12	490	60	9	7¼	8¼
	12	380	48	9	7¼	8¼
74	12	585	60	10¼	7¼	8¾
	12	525	60	10¼	7¼	8¾
	12	505	60	10¼	7¼	8¾
	12	450	55	10¼	7¼	8¾
	12	410	48	10¼	7¼	8¾
	12	380	40	10¼	7¼	8¾
	12	325	36	10¼	7¼	8¾

*Meets or exceeds Battery Council International rating standards.

Figure 30.16 BCI battery group numbers tell the size and power of the battery. *[Courtesy of Battery Council International]*

- The battery must fit the battery box.
- The posts must be on the correct side of the battery.
- The battery hold-down must fit the battery. Sometimes, there is a hold-down bracket cast into the bottom side of the case of the battery.
- The battery cannot be so high that it shorts out on the hood.

BATTERY SERVICE LIFE

The average battery has a service life of about three to five years. The negative plates become soft during the repeated charge and discharge cycles. A battery's life is shortened by improper (too high or low) charging system amperage or voltage. Too high a charging rate is the greatest cause of shortened battery life.

The actual length of a battery's life is determined by the amount of material that has been shed from the surface of its positive plates. Plates shed material during overcharging or vibration.

When a battery discharges, a neutral coating of lead sulfate coats the plates. The effects of this **sulfation** are usually reversible, but when it becomes too extensive the battery is scrapped (see Chapter 31).

Effect of Temperature on Batteries

Batteries do not work as well in colder weather. The following chart shows the effect that cold temperatures actually have on a battery's cranking power.

Temperature	% of Cranking Power
80°F (26.7°C)	100
32°F (0°C)	65
0°3 (–17.8°C)	40

Other battery problems caused by cold weather include:

- The engine becomes harder to crank because oil becomes thicker in colder temperatures.
- A battery that is not fully charged will freeze easier.

NOTE: Parts stores in high elevation resorts sell many batteries when tourists from areas with milder climates have batteries that are marginal. The cold weather causes battery failure.

BATTERY CABLES

Battery cables must be large enough to carry all of the current demanded by the starter and vehicle's electrical system. The large cable goes directly to the starter because it requires a great many more electrons to operate than any of the vehicle's other electrical loads. If a cable is too small, it will become hot as the engine is cranked. Battery cables for 12-volt systems are usually 4 or 6 gauge. As in the selection of other electrical wires, a larger cable will not be a problem. A cable that

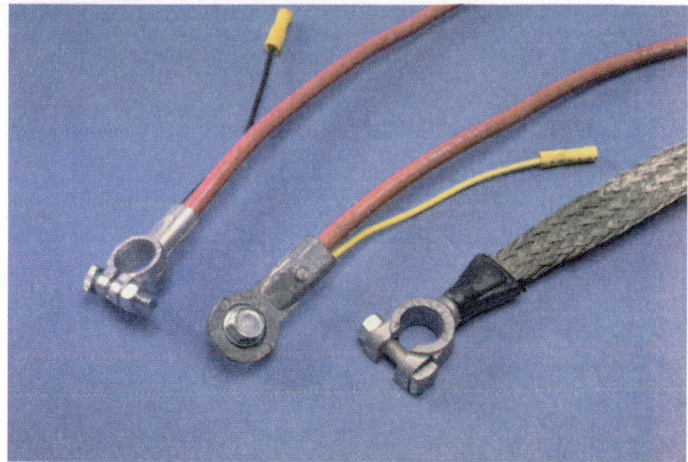

Figure 30.17 Different battery cable designs.

is too small will not deliver enough electrons to the starter motor and could possibly start a fire.

Figure 30.17 shows several designs of battery cable. The uninsulated woven cable is a ground cable. If it were the hot cable and touched against a ground, a spark would result. The positive cable is usually red. Do not take this for granted!

BATTERY HOLD-DOWNS

A battery must be held in its tray. It can fall out as the car travels over bumps. Also, excessive vibration can harm the battery, causing the plates to shed more of their material. Battery hold-downs (Figure 30.18) are made of steel or plastic. Some batteries have a molded piece near the bottom of the case that a hold-down bracket is clamped to. Others have a bracket that fits around or across the top of the battery. Long threaded

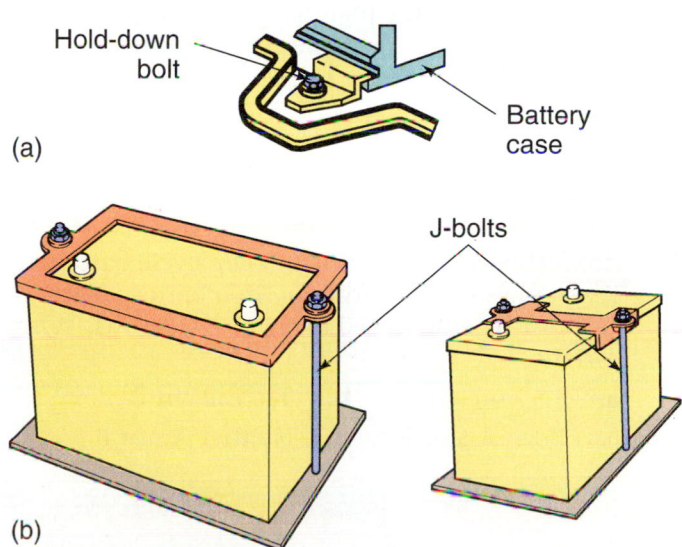

Figure 30.18 Different types of battery hold-downs. *(a, Courtesy of Chrysler Corporation; b, Courtesy of Ford Motor Company)*

rods (called *J-hooks*) are hooked into holes in the battery tray.

REASONS FOR BATTERY FAILURE

■ A damaged battery case can leak electrolyte.
■ Undercharging can result in sulfation of the battery plates, which limits electron flow.

■ Overcharging can result in buckled and warped plates from excess heat. Material will be shed from the plates.
■ Vibration can cause more material to be shed from plates.
■ A short between the plates can cause a dead cell.

REVIEW QUESTIONS

1. What is the largest load on a car battery?
2. Is the electricity produced by a battery AC or DC?
3. How many volts will a fully charged battery cell have?
4. What kind of gas is given off as a battery is charged?
5. When does a battery give off the most gas, when charging has just begun or when the battery is almost fully charged?
6. What is the most common method of rating battery capacity for automobiles?
7. What is the battery test for deep-cycle batteries called?
8. What is the electrical unit of measurement that is equivalent to one horsepower?
9. What kind of battery cannot have a hydrometer test done to it?
10. What percentage of a battery's cranking power will be available at 0°F?

ASE STYLE REVIEW QUESTIONS

1. Technician A says that when an automobile battery is fully charged a voltmeter connected to it will register 12 volts. Technician B says that the negative battery post is larger than the positive post. Who is right?
   a. Technician A   b. Technician B
   c. Both A and B   d. Neither A nor B

2. Technician A says that in a completely discharged battery, the positive and negative plates are the same material. Technician B says that in a fully charged battery, the positive and negative plates are the same material. Who is right?
   a. Technician A   b. Technician B
   c. Both A and B   d. Neither A nor B

3. Technician A says that calcium is used to strengthen the lead grids in deep-cycle batteries. Technician B says antimony in a battery plate gets hot when charged at above 14.5 volts. Who is right?
   a. Technician A   b. Technician B
   c. Both A and B   d. Neither A nor B

4. Technician A says that hooking up two batteries in series can provide more plate capacity for RV uses. Technician B says that a battery cable that is too large will become hot when the engine is cranked. Who is right?
   a. Technician A   b. Technician B
   c. Both A and B   d. Neither A nor B

5. Technician A says that the largest load on the battery is the air conditioning blower. Technician B says that a battery gives off the most hydrogen when it is almost fully charged. Who is right?
   a. Technician A   b. Technician B
   c. Both A and B   d. Neither A nor B

Battery Service

■ OBJECTIVES

Upon completion of this chapter, you should be able to:

✔ Inspect a battery and recommend the correct service for it.

✔ Service a battery.

✔ Perform a variety of tests on a battery and make a diagnosis from them.

✔ Select the best charge rate and charge a battery.

✔ Perform battery service safely.

✔ Safely and correctly jump start a car.

■ INTRODUCTION

The battery is the heart of the electrical system. The battery must be in good condition and fully charged if tests on other electrical system components are to be performed. Batteries last an average of three years and are routinely replaced when they fail. Tests will show a battery's state of charge and output voltage. Results will tell if the battery is good, needs a recharge, or must be replaced. The following case history illustrates the importance of having a good battery.

CASE HISTORY

A customer complained that his car's idle would change sometimes when he stepped on the brake pedal. The technician tested the power brake booster (a typical cause of this problem) and determined that it was good. After further testing, the technician determined that the problem was that the alternator had a bad diode. When the battery is not fully charged, the computer on modern automobiles will not work properly. On some systems, if the voltage drops below 11.6 volts at idle, the idle speed control motor will raise the idle speed to increase the charging rate. When the voltage is borderline, even the current draw requirements of the brake lights can cause the idle to raise when the brake pedal is depressed.

■ BATTERY INSPECTION

Several things are looked at when determining a battery's condition. Inspect the following items:

■ Date code on the battery label. This tells the date that the battery was installed in the vehicle.

■ Battery case condition. Electrolyte deposits on the surface of the case can conduct electricity between the positive and negative terminals, discharging the battery. Check for damage to the battery case (see Figure 13.17).

■ Level of the electrolyte. Add distilled water as needed to fill the battery to the split ring under the cell cap (see Figure 13.18). If one of the cells is low, there could be a short in that cell. Low electrolyte throughout the battery indicates the possibility of overcharging. Charging system tests are covered in the charging system chapter.

■ Condition of the cables and terminals. Clean them as needed (Figure 31.1)

■ Battery hold down and tray.

■ Built-in hydrometer. If the battery has one.

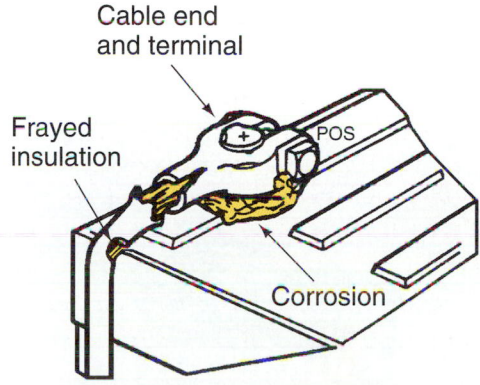

Figure 31.1 Inspect the battery cables. *(Courtesy of Chrysler Corporation)*

■ BATTERY SERVICE

Refilling a Battery with Water

Batteries lose electrolyte over a period of time due to electrolysis (using electricity to break down water into hydrogen and oxygen). If the water level is allowed to drop below the level of the plates they will rapidly become sulfated. Sulfation is discussed later in this chapter. In most batteries, water that has evaporated can be replenished through holes in the battery top.

Only water is added, not electrolyte, because the acid does not evaporate. When a battery is low on water, it has more internal resistance. This causes more heat, which is also hard on the battery.

Use clean filtered water when filling a battery. A *battery water filler* is a handy tool for filling batteries (Figure 31.2). The water filler has a nozzle that allows water to flow when it is pushed down against the split ring indicator in the cell opening (see Figure 13.18). When the cell is full, the flow of water stops automatically. A battery with an electrolyte level that is filled to the split ring will have the correct specific gravity when fully charged.

Cleaning Battery Terminals and Clamps

In a survey of auto repair shops by the *Car Care Council,* a public information agency, technicians listed battery cable maintenance as one of the most overlooked items. Battery terminal connections that have become corroded are the *most common cause of hard-starting complaints.* Terminal posts are made of lead, which forms an insulating oxide coating when exposed to air. Sometimes an oxidized terminal is black, but not always.

Lead oxidation is most often found on the positive post. This coating, which does not conduct electricity, will reduce or stop the flow of current between the battery and the vehicle's electrical system. Removing the oxide layer exposes fresh lead underneath. Use emery cloth or a terminal cleaner to clean both posts and the inside of the terminal clamps (Figure 31.3). Figure 31.4 shows several popular styles of terminal cleaners.

Corrosion, which looks like white powder on and around battery posts, sometimes happens when battery gas escapes up the terminal post from inside the case. Corrosion on both battery terminals is sometimes due to battery caps that are not installed tightly all of the way. Today's maintenance-free batteries are less prone to post corrosion because they do not gas as much. When they have *case vents* they are aimed away from the posts. Side terminal batteries are also less prone to this problem.

Battery Acid Is Corrosive. Battery electrolyte, which is sulfuric acid and water, will make holes in many types of clothing. It is not unusual to find holes in coveralls or pants after carrying a battery. The outside of some batteries is covered with a mist of acid as a battery gasses. When this gets on hands, it can be accidentally transferred to clothing.

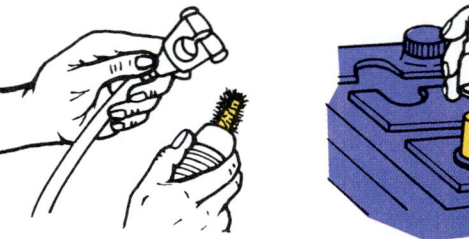

Figure 31.3 A battery terminal cleaner cleans both the inside of the clamp and the outside of the post.

Figure 31.4 Several popular styles of battery terminal cleaners. *(Photo provided by Mac Tools and is representative of products manufactured by Mac Tools or products obtained from another source.)*

Figure 31.2 A battery water filler is handy for filling batteries.

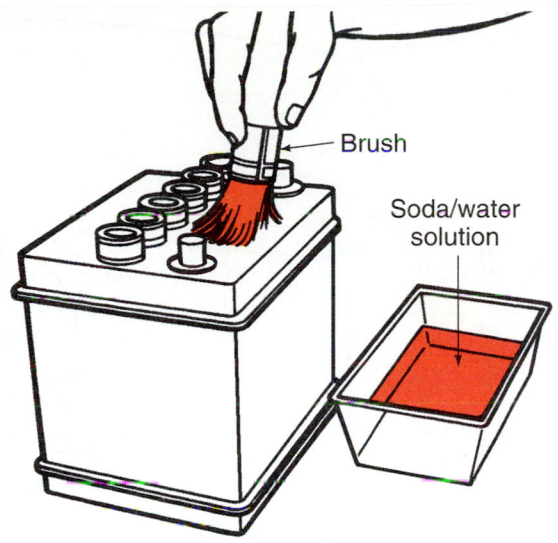

Brush

Soda/water solution

Figure 31.5 A mixture of water and baking soda or a strong detergent can be used to wash a battery. *(Courtesy of Chrysler Corporation)*

Acids can be *neutralized* with a **base**, such as baking soda. A mixture of water and baking soda or a strong detergent can be used to wash a battery (Figure 31.5). Be sure not to get any of the mixture into the cells. The battery box is cleaned with the same solution.

NOTE: *To avoid the possibility of staining a car's paint, get in the habit of always using fender covers.*

Sometimes the battery box will need to be painted after it has been thoroughly cleaned.

Dry Batteries

One of the advantages of maintenance free batteries is that they have a long shelf life. This means that they can be stored and shipped full of electrolyte. New batteries with cell caps are sometimes shipped dry, requiring the addition of electrolyte. New electrolyte, which is approximately 25% acid and 75% water (by volume), is contained in a large plastic bag inside a sturdy cardboard box. The plastic bag has a tube on it and a pinch device to shut off the flow of liquid when the plates of the cell are covered. The tube is sealed for shipping and the end needs to be cut off before the liquid can be poured.

SAFETY NOTE Be sure to wear safety goggles when performing this task. Concentrated acid will burn skin and destroy clothes. Spilled electrolyte must be neutralized immediately or it will etch a concrete floor, causing a bright white stain.

SHOP TIP Fill the battery until the electrolyte just covers the plates. Do not fill the cells to the normal fill line until *after* the battery has been charged. The electrolyte might spill over when it becomes heated during charging.

After filling a dry battery, allow it to sit for at least 15 minutes so that the plates can soak up the electrolyte. Then, charge the battery at 30 amps until the battery is charged. This is checked with a hydrometer (covered later in this chapter). Then, check the electrolyte level and add more electrolyte to fill the cells to the correct level.

■ REPLACING A BATTERY

When servicing a battery, the vehicle should be outdoors so that the battery box can be washed out. Be sure to use a fender cover on the fender nearest the battery because battery acid can ruin a paint job.

Remove the Terminal Clamps

When a battery is removed, the ground cable is always disconnected first. The ground cable is the one that is connected to the block or chassis. It is usually, but not always, the negative post. Double-check to be sure, especially on older cars (1955 and earlier and British cars before 1970) because some of them are *positive ground*.

 **SAFETY NOTE** Be certain all electrical circuits are shut off before disconnecting the battery. Otherwise a spark could cause a battery explosion (see Figure 11.27).

Check the bolt and nut used to clamp a post-type terminal clamp. Due to corrosion they will probably be smaller than they once were. This means that a wrench will not fit them. Special battery bolt pliers are available for removing these bolts (Figure 31.6).

Terminal clamps are usually easy to remove. Do not twist the battery posts. They are made of lead, which is a soft metal. The terminal clamp or the battery itself can be easily damaged. Use a battery terminal puller when a stubborn clamp is encountered (Figure 31.7).

NOTE: *When replacing a battery on computer equipped cars, memory can be lost for certain settings such as radio stations, clock, block-learned driveability, seat settings, fuel consumption rate, and temperature control. Diagnostic codes in the computer's memory will be lost, too. After the battery is reinstalled, the computer will relearn the lost information in about 50–100 miles. The owner might notice a slight difference in the driveability of the vehicle until that time.*

Figure 31.6 Battery terminal nut pliers. *(Courtesy of Snap-on Tools Company. Copyright Owner.)*

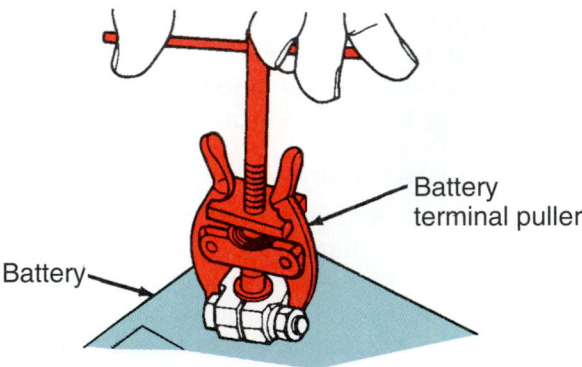

Battery terminal puller

Battery

Figure 31.7 A battery terminal puller. *(Courtesy of Chrysler Corporation)*

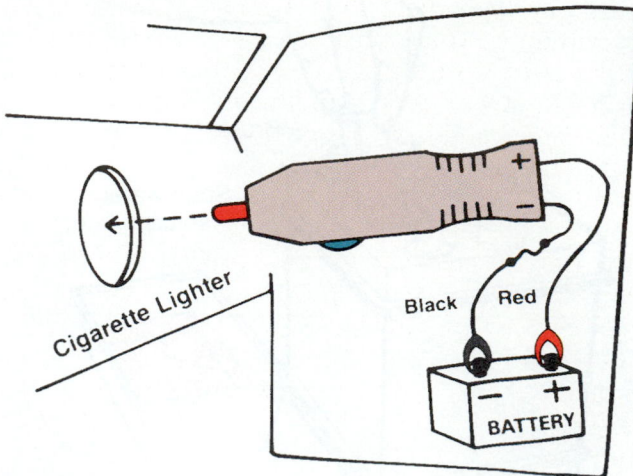

Cigarette Lighter

Black   Red

BATTERY

Figure 31.8 A tool for keeping computer memory while a battery is removed for service. *(Courtesy of Interstate Batteries)*

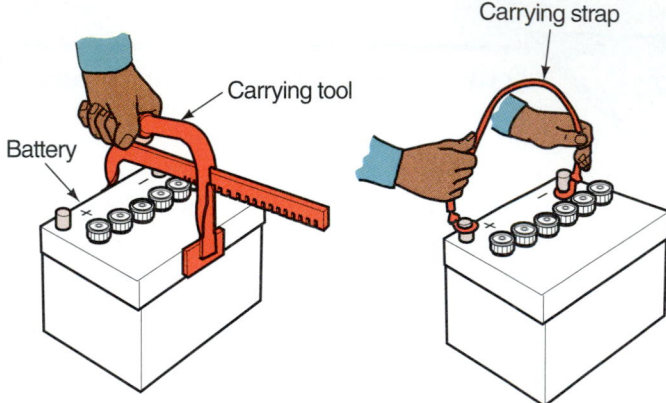

Carrying strap

Carrying tool

Battery

Figure 31.9 Two kinds of tools for carrying batteries.

➡ *Perform **Battery Service** Worksheet*

Remove the Battery

Lubricate the hold-down clamp threads. They are probably rusty. Remove the hold-down clamp and lift the battery out of its tray. Use a battery carrier to carry the battery so you can avoid dropping it (Figure 31.9). If a carrier is not available, carry the battery in a strong cardboard box. Remember to keep the battery away from clothing. Clean the battery and its tray with baking soda and water. Check and refill the battery with water as needed.

➡ *Perform **Removing the Battery** Worksheet*

Battery Terminal Clamps and Cables

Resistance at the cable connections is a major cause of starting system problems. There are two causes of increased resistance: battery acid vapors and air. The

SHOP TIP To avoid the need to reset these things, a small transistor radio battery (9 volt) can be connected to the battery cables before removing them. A special tool is available that can be connected into the cigarette lighter. The other end of it is connected to another battery to keep the circuits alive (Figure 31.8). This method also prevents battery terminal arcing when the underhood battery is disconnected.

■ The positive battery cable under the hood is hot at all times while this tool is connected. Do not let the positive battery cable touch ground or the cigarette lighter fuse will blow. This will electrically disconnect the spare battery from the vehicle. All computer memories will be lost.

■ Be sure to observe correct polarity when connecting the alligator clips to the spare battery.

■ Verify that the cigarette lighter works before inserting the tool into the cigarette lighter receptacle. This is to verify that voltage from the spare battery will actually be able to get into the electrical system to maintain memory circuits.

easiest way to prevent the resistance is to make sure that neither of these get to contact the critical areas. One fix is to install a felt washer under the terminal clamp to prevent acid from gassing out between the terminal post and battery case (Figure 31.10).

When a terminal clamp becomes corroded beyond repair, it can be replaced. After the old clamp is cut off and the insulation is removed from the end of the cable, a new clamp can be soldered to the old cable (Figure 31.11a). When the lug on the other end of a cable is replaced, it can be crimped on (Figure 31.11b).

One popular type of repair for worn cable ends is a *bolt-on type terminal clamp* (Figure 31.12). These are handy but are not recommended by manufacturers, who call them *emergency repair battery terminal clamps*. New replacement battery cables are available and should be installed as soon as possible after the emergency repair. They are better than bolt-on terminal clamps because their terminations are more thoroughly crimped or soldered to the cable. Before either type of terminal clamp is used, the end of the cable must be bright and clean. Use a sharp knife to strip back about ½" of insulation from the end of the cable. If a bolt-on type clamp is used, be sure it is sufficiently tightened to the cable so that it cannot come loose.

NOTE: *Be sure that a replacement cable has sufficient slack to allow movement of the engine on its mounts.*

➡ ***Perform Replace a Battery Cable and Terminal Clamp Worksheet***

NOTE: *Whether a cable is to be repaired or replaced, it must remain long enough after installation to permit enough slack for the engine to be able to freely rock on its mounts.*

Figure 31.10 A felt washer under the terminal clamp helps keep the battery from gassing through the terminal post.

A cable that is too short can also rub against something until its insulation wears through, causing a short to ground (Figure 31.13). A cable that is too long and is not restrained properly can burn through on an exhaust manifold. Always perform a visual check for possible electrical system problems before attempting a diagnosis.

CASE HISTORY

A customer complained that his battery kept going dead. He had already replaced his battery. His neighbor had replaced his alternator and voltage regulator but the battery continued to go dead. The customer took his car to a qualified technician in a repair shop. The technician performed a visual inspection first. She found a battery cable that had been resting against the exhaust manifold. Its insulation had burned, allowing a direct short to ground. Replacing and relocating the cable solved the problem.

Sometimes, a battery terminal does not have good contact with the terminal post or the cylinder block (in the case of the ground cable) (Figure 31.14).

Battery Hold-Downs

The life of the battery will be shortened if it is not held firmly in place. If it is allowed to bounce and vibrate against the battery box as the car travels over bumps in the road, material will be shed from the battery's plates. A battery can also fall out of its battery tray and cause damage or a fire.

CASE HISTORY

An auto shop student put a new battery in his Jeep. The battery tray is high on the bulkhead under the hood. He did not have a battery hold-down, but planned to buy one soon. After a week or so, he forgot about the hold-down. While driving to school, he heard a loud pop and oil began to leak from the bottom of his Jeep. He had recently rebuilt the engine and was very concerned. Upon opening the hood, he discovered that the battery had fallen down against the engine. Its positive post had grounded out against the oil filter and burned a hole in it (Figure 31.15). He was lucky that he stopped and did not run the engine long enough without oil to cause any damage.

Reinstall the Battery in the Vehicle

Re-install the battery and hold-down clamps.

NOTE: *Tighten the hold-down clamp only until it is snug. DO NOT OVERTIGHTEN the clamp (Figure 31.16). The battery can be damaged.*

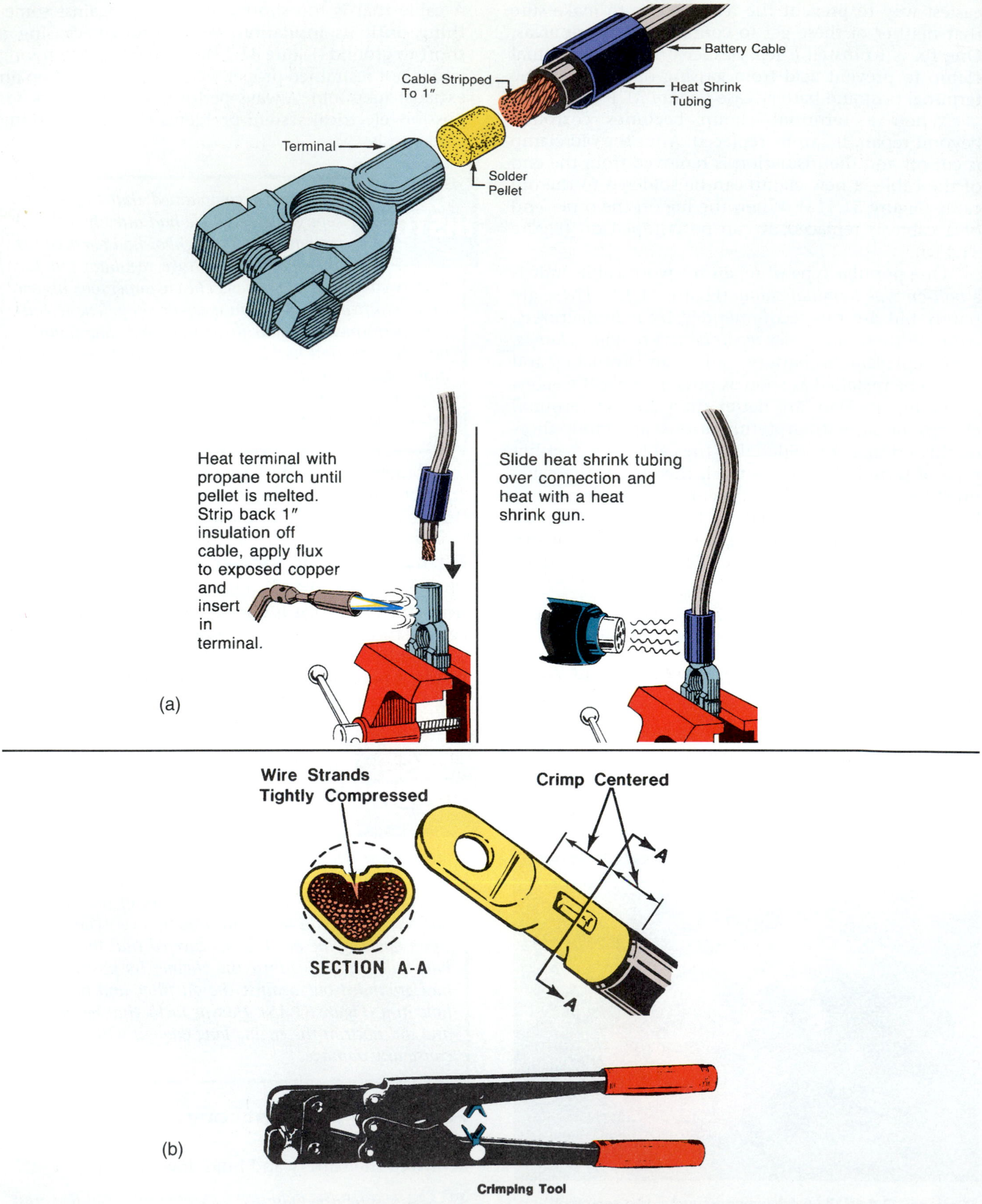

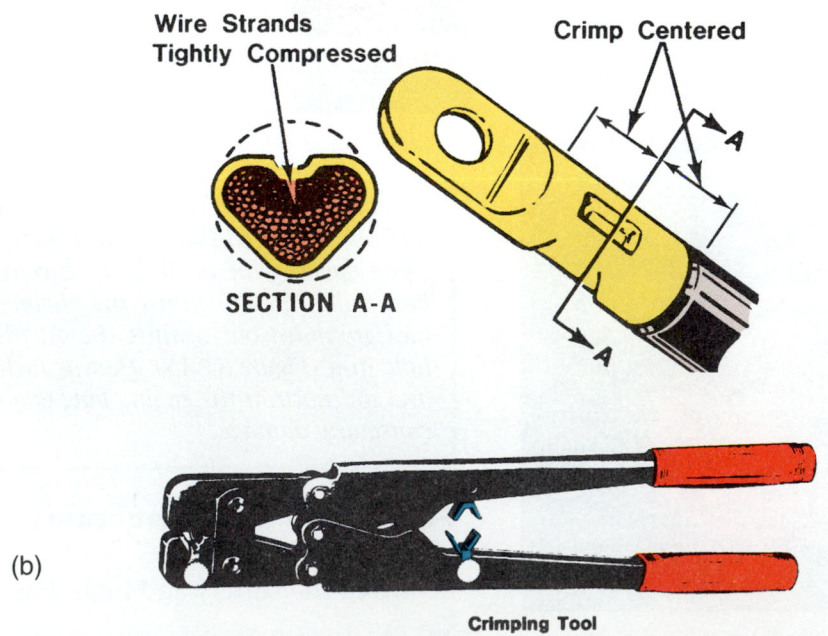

Battery Cable

Heat Shrink Tubing

Cable Stripped To 1"

Terminal

Solder Pellet

Heat terminal with propane torch until pellet is melted. Strip back 1" insulation off cable, apply flux to exposed copper and insert in terminal.

(a)

Slide heat shrink tubing over connection and heat with a heat shrink gun.

Wire Strands Tightly Compressed

Crimp Centered

A

A

SECTION A-A

(b)

Crimping Tool

Figure 31.11 Battery cable repair. (a) Soldering a terminal on a cable. (b) A new battery cable lug can be crimped to the cable. *(a, Courtesy of Cooper Automotive/Belden; b, Courtesy of Ford Motor Company)*

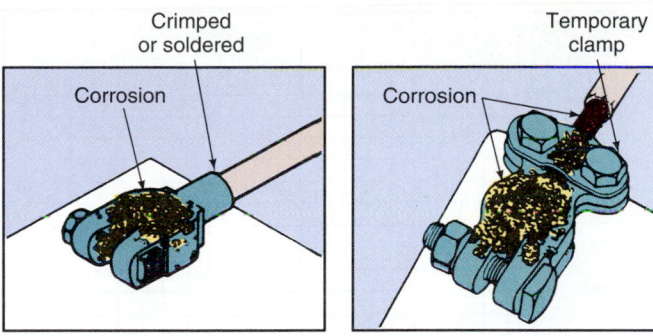

Figure 31.12 An emergency repair battery cable clamp is subject to corrosion. *(Courtesy of Cooper Automotive/Belden)*

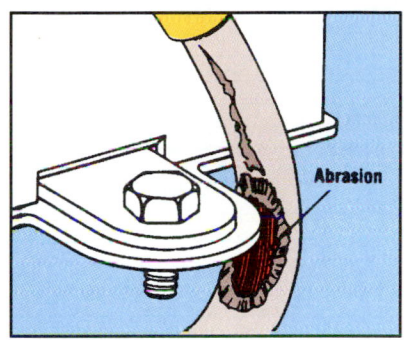

Figure 31.13 The insulation can be worn away on a cable that is too short. *(Courtesy of Cooper Automotive/Belden)*

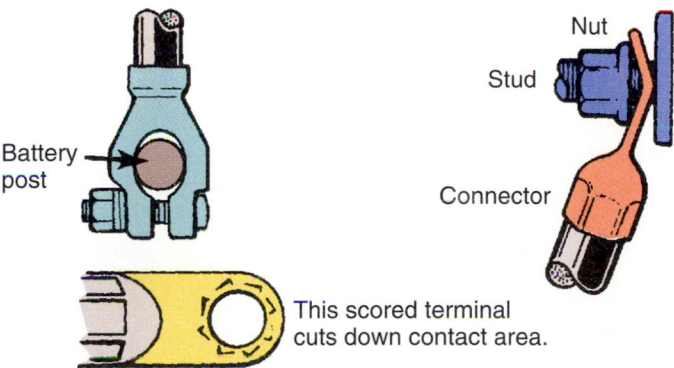

Figure 31.14 These cables do not have full contact with ground. *(Courtesy of Ford Motor Company)*

Figure 31.15 The hole in this oil filter happened when the battery fell against it.

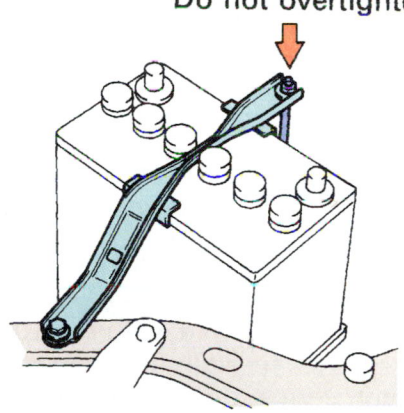

Figure 31.16 Do not overtighten the hold-down.

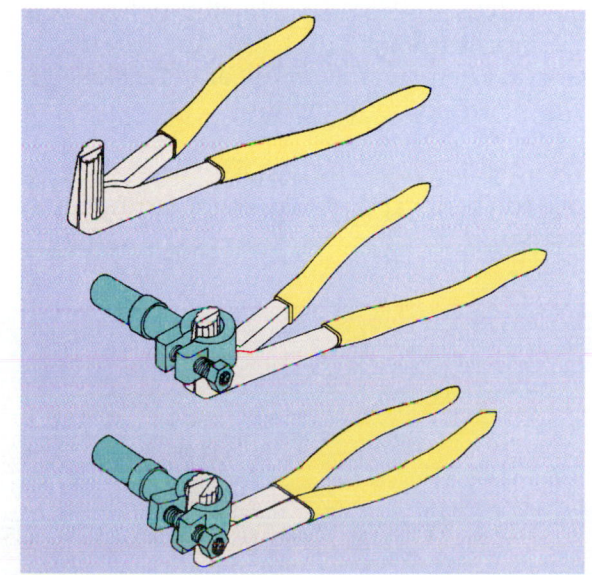

Figure 31.17 Enlarge the hole in the terminal clamp with expanding pliers. *(Courtesy of Battery Council International's Battery Service Manual)*

Before attempting to reinstall the old terminal clamp, clean the terminal clamps and enlarge their holes using expanding pliers (Figure 31.17). Reinstall the hot (usually [+]) lead. Reinstall the ground (usually [–]) lead. This is always the one connected last.

Corrosion can be prevented by not allowing air to contact the terminal posts. Smear some grease around the terminal or use battery spray or treated felt washers to prevent oxidation (needed on nonmaintenance-free only). After washing your hands, check to see that the battery starts the car.

■ BATTERY TESTING: MEASURING A BATTERY'S STATE OF CHARGE

Battery Hydrometer Testing

There are several ways of testing batteries. The best way to determine a battery's state of charge is by checking the strength of the electrolyte with a hydrometer (Figure 31.18). When a battery discharges, its **specific gravity** becomes less. This is because as the battery discharges, water dilutes the electrolyte as oxygen from the positive plate joins with hydrogen in the electrolyte.

A hydrometer compares the weight of a liquid that is drawn into it to the weight of pure water. Pure water is given the reading of 1.000. Acid weighs more than pure water (pure acid has a specific gravity of 1.840) so the more sulfuric acid in the electrolyte, the higher the specific gravity reading on the hydrometer. A fully charged battery sold in the United States usually has an electrolyte reading of 1.260–1.270 (this is the specific gravity of new electrolyte) measured at 80°F.

Batteries can freeze in cold weather. A dead battery (1.110 specific gravity) will freeze at about 19°F, while a fully charged battery will never get cold enough to freeze (Figure 31.19).

NOTE: *A battery that is at least ¾ charged (1.225 specific gravity) is in no danger of freezing.*

Electrolytes are formulated to correspond to the temperature where the battery will be used. A battery made for use in extremely cold weather has more concentrated electrolyte. In addition to preventing the electrolyte from freezing, this increases the *cold cranking power* of the battery. Higher specific gravities decrease the service life of the battery, however.

Electrolyte in tropical climates (where it never freezes) ranges from 1.210–1.230. The battery is more efficient at these temperatures and the loss of cold cranking power is not important.

Reading the Hydrometer

Remove the vent caps and lay them on the top of the battery.

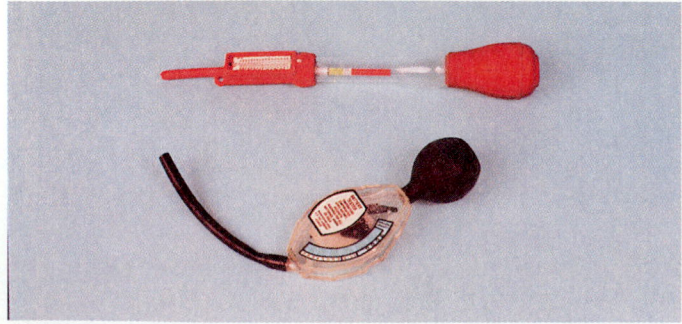

Figure 31.18 Two types of battery hydrometers.

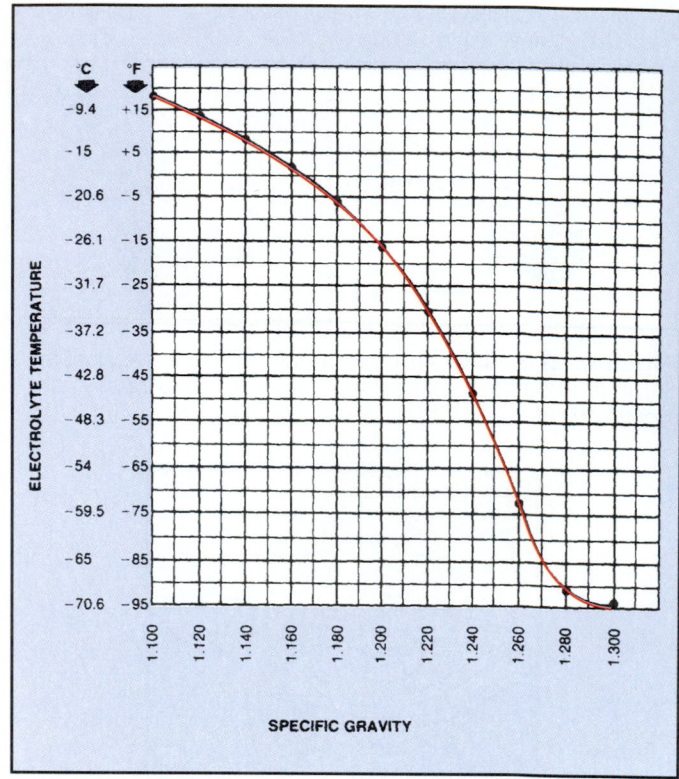

Specific Gravity Corrected to 80°F (26.7°C)	Freezing Temperature	
1.280	-92°F	(-69°C)
1.265	-71.3°F	(-57.4°C)
1.250	-62°F	(-52.2°C)
1.200	-16°F	(-26.7°C)
1.150	+5°F	(-15°C)
1.100	+19°F	(-7.2°C)

Figure 31.19 A fully charged battery will never get cold enough to freeze. *(Courtesy of Battery Council International's Battery Service Manual)*

NOTE: *Before taking a hydrometer reading, be sure that the battery water level is up to the bottom of the filler tube of each cell. If the water level is low, the specific gravity will read higher than it actually is. If the electrolyte level is too low, refill and recharge the battery.*

■ Draw electrolyte into the hydrometer (Figure 31.20) to the line on the tester bulb (if so equipped).

■ Hold the hydrometer vertically so the float can rise to its proper level. The gauge must float freely. Read the specific gravity on the float and record it on your sheet (Figure 31.21). Make a temperature correction, if necessary.

NOTE: *A hydrometer is not accurate at temperatures above or below 80°F. Most of the popular hydrometers are temperature compensated. If not, a correction of +.004 is made for each 10°F. change above 80°F. Subtract for tem-*

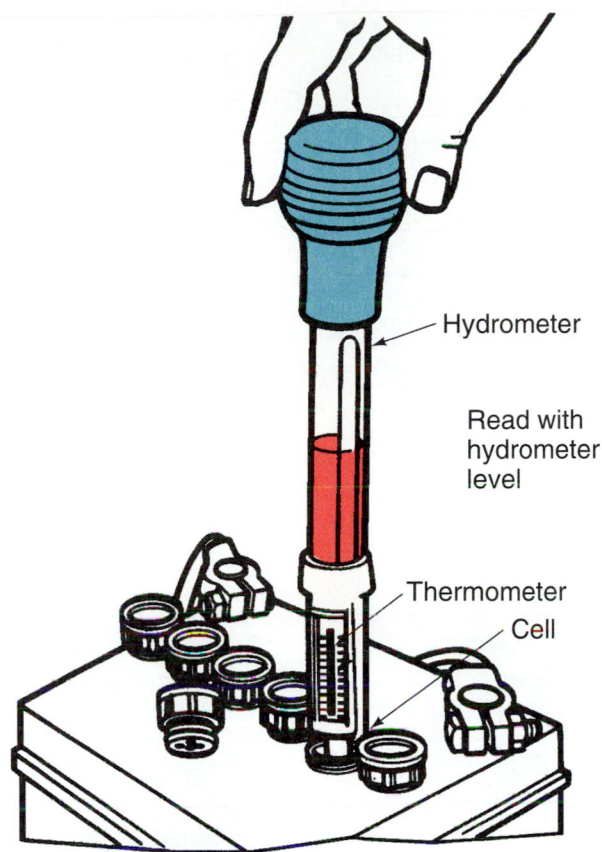

Figure 31.20 Draw electrolyte into the hydrometer. *(Courtesy of Chrysler Corporation)*

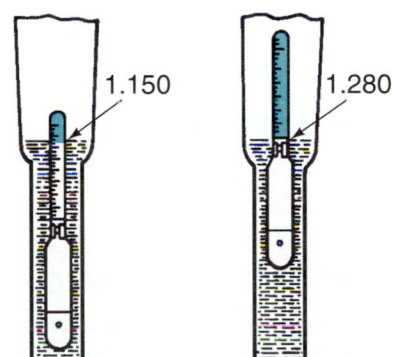

Figure 31.21 Read the specific gravity on the hydrometer float.

peratures below 80°F. The compensation factor can be important at temperature extremes.

■ Return the electrolyte to the cell and repeat the process for the other cells.

■ Differences between the cell readings should be less than 0.050 (Figure 31.22). All of the battery's cells are electrically connected to one another. If one cell is low, it will pull down the voltages of the remaining cells. While the battery might function effectively immediately after a recharge, it will discharge overnight.

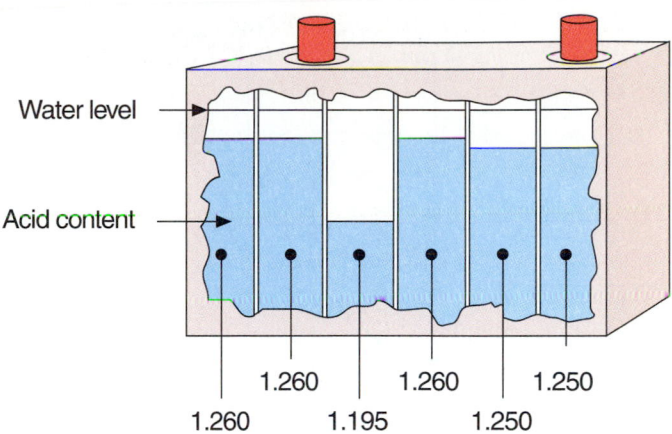

Figure 31.22 Hydrometer readings should vary no more than 0.050. This battery has a defective cell.

NOTE: *Shorted cells often are low on electrolyte while the other cells are all full.*

➡ *Perform Hydrometer Test (Specific Gravity of Electrolyte) Worksheet*

Magic Eye Batteries

Some maintenance-free batteries have a built-in hydrometer called a magic eye that tests their state of charge. These batteries have no provision for adding water. A maintenance-free battery that does not have a builtin hydrometer cannot have its specific gravity tested because it is sealed. The following are interpretations of magic eye readings (Figure 31.23).

■ If it is yellow or clear, the electrolyte is low and the battery should be discarded.

■ If it is green with a black spot in the center, recharge it.

■ If it is green and the spot shows, the battery is at least 60% charged and is fine.

NOTE: *The hydrometer checks the electrolyte strength of only the cell in which it is installed, so other cells could still have a problem.*

Open Circuit Voltage

Open circuit voltage, or *constant battery voltage*, can be checked to see if the battery has a sufficient state of charge. A battery is considered charged if its state of charge is 75% or more. The chart in Figure 31.24 shows the relationship between open circuit voltage, specific gravity, and state of charge.

Before making the test, remove the **surface charge** from the battery plates by connecting a 300-amp load across the battery for 15 seconds with a carbon pile. A battery capacity test (covered later) can also be performed to remove the surface charge.

Allow at least 10 minutes after a load test for the battery's voltage to stabilize.

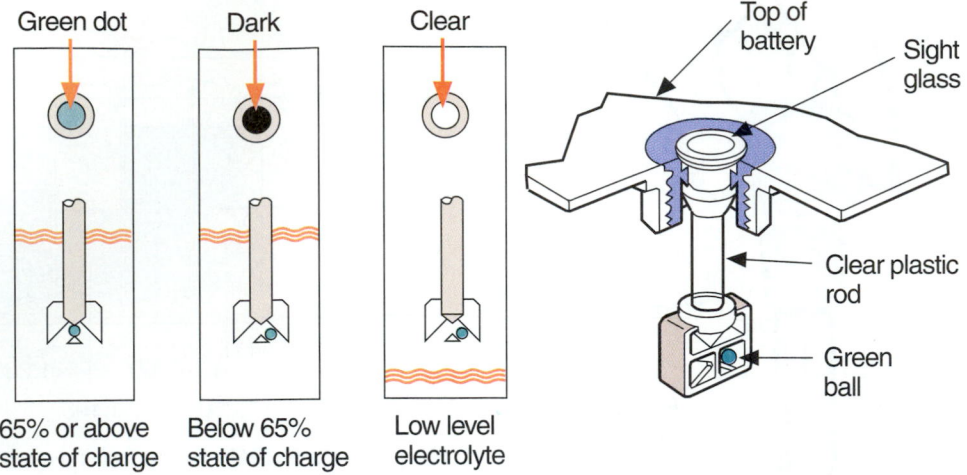

Figure 31.23 A built-in hydrometer in a sealed battery indicates the condition of the battery.

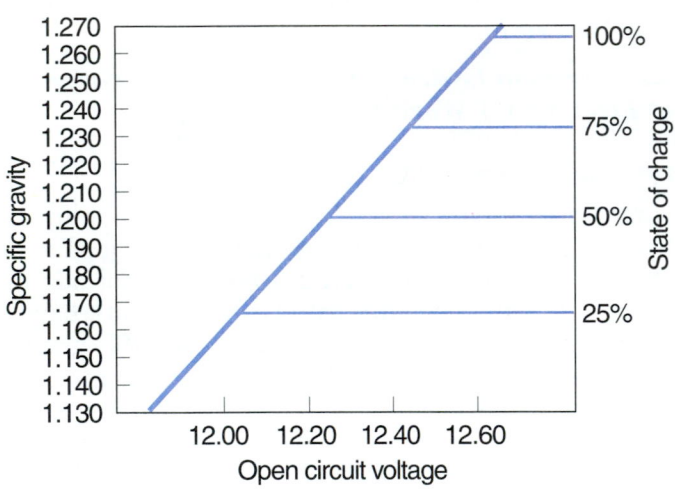

Figure 31.24 A chart comparing open circuit voltage to specific gravity. *(Courtesy of Ford Motor Company)*

Figure 31.25 Reading open circuit voltage with a digital voltmeter.

SHOP TIP The charge on the surface of the plates can actually be enough to start a car that has been recently run. But after the car sits for the night, the charge might be gone and the car will not be able to start.

NOTE: *Before reading open circuit battery voltage after charging or boosting, the battery voltage must be stabilized. Wait for at least four hours or overnight.*

With the terminals removed from the battery, check the voltage across the positive and negative terminals (Figure 31.25). A fully charged battery will have an open circuit voltage of 12.6 volts or higher after the surface charge has been removed. Open circuit battery voltage must be above 12.4 volts or the battery must be recharged.

■ BATTERY CHARGING

Most battery chargers are the *constant voltage* type. A constant voltage charger reduces the amount of charging current while maintaining the output voltage at the same level as the battery reaches full charge. This type of battery charger keeps gassing to a minimum and is easiest on the battery. Other types of battery chargers can be dangerous. When they are left on all night, voltage can climb dangerously high. Trickle chargers are often of this type.

Be sure to check the electrolyte level in the battery before attempting to charge it. Dead batteries are sometimes low on water. Fill a battery only halfway prior to fast-charging. During a fast charge, the electrolyte will become heated and expand. If the battery is full, it could overflow onto the floor. Acid etches concrete. If the battery is a sealed type and electrolyte is low, do *not* attempt to recharge it.

Some manufacturers require that the ground cable be disconnected from the battery if it is to be charged while it is in the vehicle. This is to protect the computer system.

Connect the battery charger cables to the terminals on the battery. The positive cable is usually red and the ground cable is usually black. Do not assume this. See if the negative cable is connected to the block and the positive cable is connected to the starter motor.

On top post batteries twist the clamps back and forth so that they dig into the lead on the terminals. On side terminal batteries a special adapter is required for charging batteries (Figure 31.26).

The ignition switch must be off during charging on computer controlled cars. Be sure to remove the ignition key as an extra precaution.

 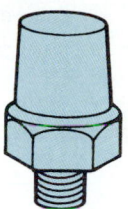 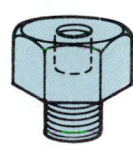

Figure 31.26 Adapters for testing and charging side terminal batteries.

Rate of Charge

Batteries are charged either fast or slow. A *fast charge* is when there is a high rate of current for a short period of time. Start charging the battery. If the battery is taking a charge that is over 30 amps, adjust its output to a lower setting.

NOTE: *If the battery will not take a charge, first check to see that the polarity of the connections is correct. Automotive battery chargers must sense some voltage in the battery or they will not work. Some battery chargers have a jump start button that will force electrons into a dead battery to give its plates a small surface charge. Depress this button for one minute and then release it. The battery should then take a charge.*

During a fast charge, the temperature of the electrolyte should not be allowed to climb higher than 125°F. If the charging voltage is kept below 15 volts, the temperature should remain low.

According to *Interstate Batteries*, 58% of battery failures are caused by overcharging. A sulfuric acid smell and/or boiling electrolyte indicates that a battery is being charged too fast. The electrolyte should never boil. It should just show a small amount of bubbling.

Check across the battery with a voltmeter to see that voltage is less than 15 volts. If it is higher, reduce the charging amperage until the voltage on the battery is less than 15 volts. Fast charge rates are listed in Figure 31.27. Do not fast charge a battery for longer than two hours.

HISTORY NOTE

Lead acid batteries were first developed in 1859. Alexander Graham Bell used one to power his first telephone call in 1876. When it was learned that they could be recharged, they were installed in automobiles. The early batteries did not hold a charge very well. They had a wooden case, made of cedar and were hung under cars so that they could drip without causing problems. These batteries were not supposed to be charged when sitting on concrete. This is not a problem with today's batteries. If a battery happens to overflow, however, the acid will etch the cement.

Battery High-Rate Charge Time Schedule

Specific Gravity Reading	Charge Rate Amperes	Battery Capacity—Ampere Hours			
		45	55	70	85
Above 1.225	5	★	★	★	★
1.200–1.255	35	30 min.	35 min.	45 min.	55 min.
1.175–1.200	35	40 min.	50 min.	60 min.	75 min.
1.150–1.175	35	50 min.	65 min.	80 min.	105 min.
1.125–1.150	35	65 min.	80 min.	100 min.	125 min.

★ Charge at 5-ampere rate until specific gravity reaches 1.250 @ 80°F.

Figure 31.27 The rate of charge is based on the state of charge and the battery capacity rating.

Sulfation

When a battery will not accept a charge, it is probably sulfated. This happens when the battery is allowed to remain in a discharged state. The lead sulfate in the battery plates becomes hard and resistant to recharging (Figure 31.28).

A dead battery that is not sulfated or damaged should readily accept a charge. If the charger is equipped with an ammeter, it should register that some amperage is flowing into the battery. If the battery still will not accept a charge, replace it. Some chargers apply a higher voltage initially to a battery. This can help a sulfated battery accept a charge.

SHOP TIP A sulfated battery can sometimes be saved. Hook up a load, such as a headlight bulb, between the positive and negative posts. Let the bulb totally drain the battery. Then recharge it slowly. Repeating the cycle of charging and discharging the battery will sometimes knock enough of the sulfation off of the plates so that the battery can charge. If after two cycles the battery still will not accept a charge, replace it.

■ SCIENCE NOTE ■

During discharge, insoluble lead sulfate forms at each electrode. The lead sulfate actually in contact with the electrode can chemically adhere to the electrode. During charging, the lead sulfate is supposed to decompose into PH, H_2SO_4, and PbO_2. However, this decomposition is never 100% due to lead sulfate's insolubility. This means that over time less and less surface area of the electrode plates is available for recharging the battery.

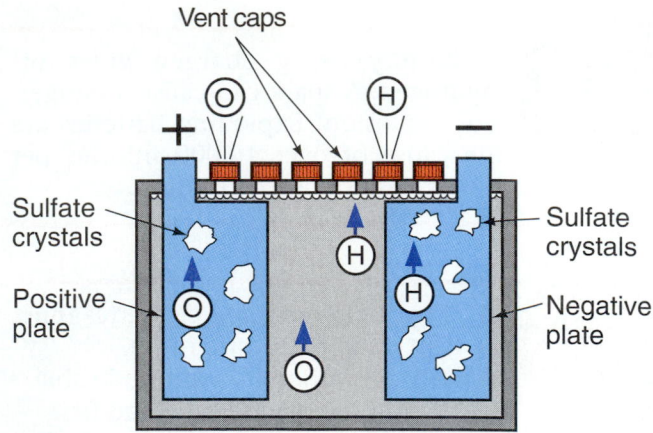

Vent caps

Sulfate crystals

Positive plate

Sulfate crystals

Negative plate

Figure 31.28 Sulfate crystals penetrate the battery plates when a battery becomes sulfated. They prevent the battery from accepting a charge.

Slow Charge

A slow charge is easiest on the battery. It is also the only way a battery can be restored to a fully recharged state. Slow charging causes the lead sulfate on the battery plates to convert to lead peroxide and sponge lead throughout the plate's total thickness. Fast charging only converts the lead sulfate on the outside of the plates.

Unfortunately, slow charging is usually impractical, unless the battery has been removed from a vehicle and can be left by the customer. If a customer needs to have a fast charge to get on the road as soon as possible, he or she should be encouraged to return the vehicle for a slow charge when time permits.

On a slow charge, the rate of charging is between 3 amps and 15 amps. A guideline for slow charging is to allow one amp for each positive plate in one cell. The time required for a slow charge is related to a battery's reserve capacity. The chart in Figure 31.29 shows the amount of time required for a slow charge at different rates of charge.

On a slow charge, a battery is fully charged when its specific gravity does not climb higher during two checks done one hour apart. The following chart gives an approximate specific gravity relationship between a fully charged battery and one that is dead.

Specific Gravity (Cold and Temperate)	State of Charge	Specific Gravity (Tropical)
1.265	100%	1.125
1.225	75%	1.185
1.190	50%	1.150
1.155	25%	1.115
1.120	Dead	1.080

Battery Capacity (Reserve Minutes)	Slow Charge
80 minutes or less	10 hours @ 5 amperes 5 hours @ 10 amperes
Above 80 to 125 minutes	15 hours @ 5 amperes 7.5 hours @ 10 amperes
Above 125 to 170 minutes	20 hours @ 5 amperes 10 hours @ 10 amperes
Above 170 to 250 minutes	30 hours @ 5 amperes 15 hours @ 10 amperes

Figure 31.29 Table showing the time and amperage for charging a battery according to its reserve capacity. *(Courtesy of Battery Council International)*

NOTE: *Following a recharge, if the difference between readings taken from different cells is more than 0.050, the battery should be replaced.*

➡ *Perform **Battery Charging (Fast Charge)** Worksheet*

■ STORING A VEHICLE

When a vehicle is left for a month at a time without being started, its battery is still being discharged by loads that are present with the key off. A battery that is only half charged can often start an engine during warm weather. This is not only hard on the battery, but presents extra challenges for the alternator that can cause it to overheat. To prevent this condition, disconnecting the battery any time it is not to be used for ten days or more is a good recommendation.

■ BATTERY CAPACITY TEST

A battery capacity test is the first test to make when testing a battery with at least a 75% state of charge. Determine the proper discharge amount to use during the battery capacity test in one of the following ways:

a. Watch the ammeter while cranking the engine to obtain a test value for battery capacity test.

b. Find the CCA of the battery (see Chapter 30) and divide by 2. Note that this may be misleading as with amp-hour rating.

c. Find the cubic inch displacement of the engine.

   1) Multiply by 2 for a 4-cylinder engine.

   2) Multiply by 1.5 for a 6-cylinder engine.

   3) Multiply by 1 for an 8-cylinder engine.

NOTE: *This will give you a CCA value (cold cranking amperage) of a suitable battery for an average vehicle. For*

heavy electrical loads or extra heavy starter draws this may have to be increased by 50 amps.

To test the battery's capacity, discharge the battery for 15 seconds at one of the above discharge amounts. Battery voltage should remain above 9.6 volts at the end of 15 seconds.

Volt-Amp Tester (VAT)

The tester most often used for the battery capacity test is a volt-amp tester or VAT (Figure 31.30). It has a voltmeter, an ammeter, and a variable **carbon pile** rheostat. The large tester leads are used for current flow (amperage). When the carbon pile is compressed, current flows through them. Voltage can be read through a small wire because voltage is only potential, not current flow.

Amperage is usually read by inserting the gauge in series, but modern ammeters have an **inductive pick-up** that wraps around one of the big leads. It senses the amount of current flowing through the wire by induction.

> ■■ **SCIENCE NOTE** ■■
>
> *Electricity can be produced by magnetic induction. Whenever a conductor is moved through magnetic lines of force or vice-versa a potential difference or voltage is created between the ends of the conductor. As soon as the conductor or magnetic field lines stop, the voltage ceases to exist. The induced voltage can be increased by increasing the speed of movement of the lines of force or increasing the number of conductors that are cut.*

NOTE: *Side terminal batteries require a special adapter if the cables are to be disconnected (see Figure 31.26).*

The carbon pile (see Figure 31.30) is made up of alternating positive and negative layers of carbon that are compressed against each other when a large knob on the face of the tester is tightened. Turning the large knob clockwise will cause a rapid discharge of the battery. As the layers of carbon under the knob are compressed, the battery is discharged at a higher rate as more current travels between positive and negative.

 Before connecting the cables to the battery, be sure that the control knob on the carbon pile (large knob) is off (counter-clockwise).

Watch the ammeter to determine the amount of current flow. Turn the control knob clockwise. The ammeter needle should move. If not, determine the cause before proceeding. Maintain the desired amperage for 15 seconds while watching the voltmeter. Read

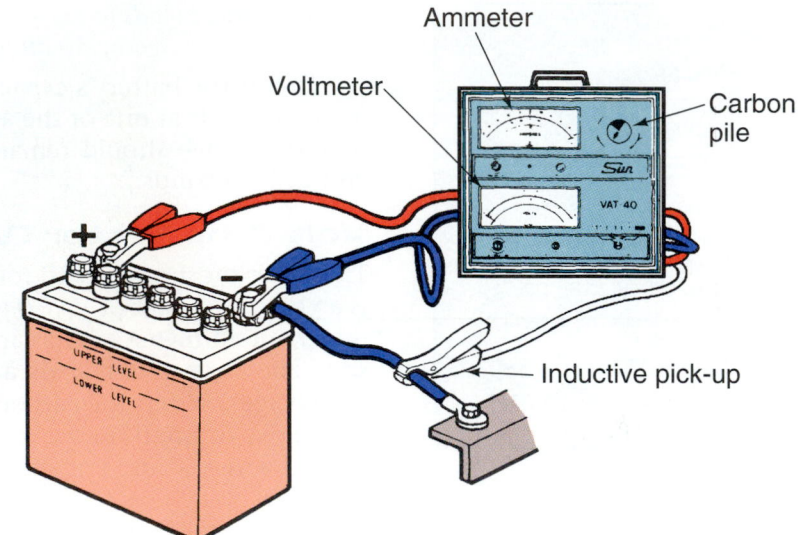

Figure 31.30 Parts of a volt-amp tester.

the voltmeter. Then, turn off the carbon pile control knob (counterclockwise). The voltmeter reading while the carbon pile is on should be above 9.6 volts at the end of the 15-second test. For every 10 degrees below 70 degrees, the voltage may be 0.1 less than 9.6.

If a battery's state of charge is 75% or higher on an open circuit voltage test (12.4 volts) and the battery fails the load test, replace the battery. If the battery is less than 75% charged and a capacity test cannot be performed, run a *3-minute charge test* (Figure 31.31). While a battery is charging, check its voltage after 3 minutes. It should not be higher than about 15.5 volts. This tests the battery for sulfation, which determines whether or not the battery will be able to be charged. If the battery passes the test, recharge it and repeat the capacity test. Discard the battery if it fails.

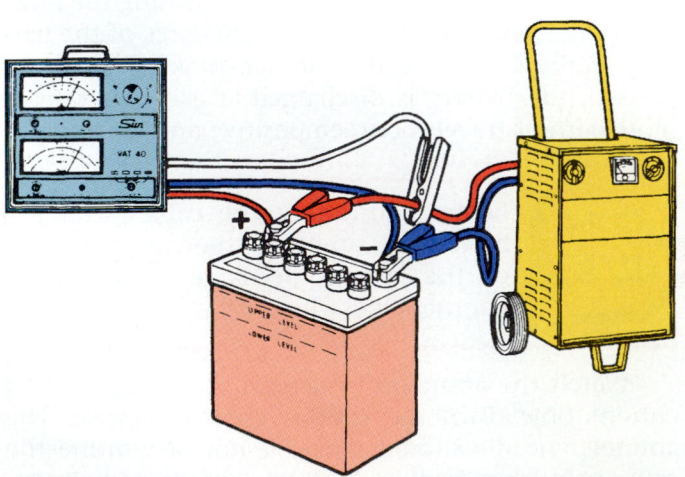

Figure 31.31 A three-minute charge test.

NOTE: *On a battery with a magic eye:*

■ *If it is yellow, the electrolyte level is low. (If this is a sealed battery it must be replaced regardless of the test results.)*
■ *A magic eye "dot" shows sufficient charge (see Figure 31.23).*

Battery capacity can be tested *without* a carbon pile volt-amp tester in the following manner:
■ Disable the engine's ignition system.
■ Remove the surface charge from the battery's plates by turning on the high beam headlights for 30 seconds.
■ Hook a voltmeter to the positive and negative terminals of the battery.
■ Crank the engine with the starter motor for 15 seconds.
■ If battery voltage is 9.5 volts or higher, the battery is good.
■ If battery voltage is below 9.5 volts, recharge the battery and repeat the test. If the result is still below 9.5 volts, check for excessive starter motor draw. If the starter is good, replace the battery.

➡ *Perform **Battery Testing** Worksheet*

■ BATTERY DRAIN TEST

If a battery that is in good condition continually goes dead, there may be a circuit that is causing the drain when everything is supposed to be off. This is called a **parasitic load** or drain. Electronic components draw small amounts of current at all times. If a voltmeter is used to check for drains, it may show battery voltage due to the parasitic loads of the vehicle's electronics.

The most correct way to test for drains is to use an ammeter that can read in tenths of amps (Figure 31.32). Less than 0.5 A (500 mA) is an acceptable amount of draw. If a sensitive ammeter is not avail-

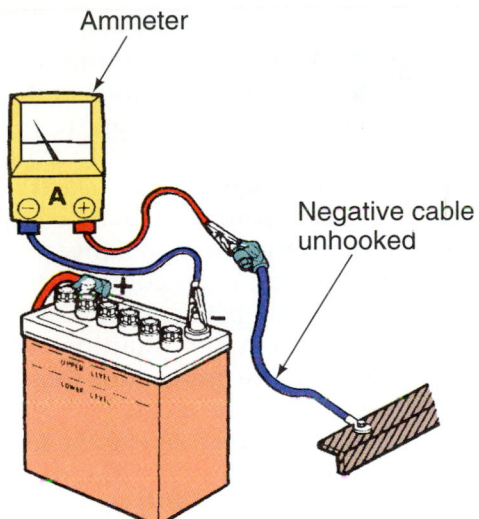

Figure 31.32 Check for parasitic drains by inserting an ammeter in series with the negative battery cable.

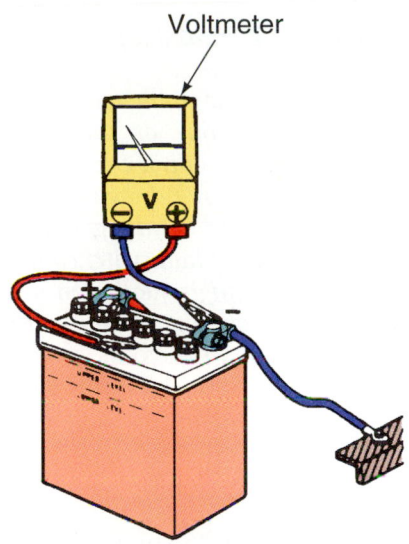

Figure 31.34 Use a voltmeter to check if dirt on the battery top is causing a drain.

able, use a test light (Figure 31.33). If a 1 candlepower or smaller bulb does not light, the drain is too small to be of concern (a 1445 bulb can be used).

■ With the key off, check all lights and accessories to see that they are off. Check all courtesy lights (those that come on when a car door is open) to see that they do not remain on at all times.

■ Disconnect the battery ground cable. Connect a test light between the battery cable end and the battery post. It should not light.

■ If the test light comes on during the last part of the test, check the lights in the trunk, ashtray, engine compartment, and glove box to see that they are not the cause of the problem. Pulling fuses one at a time can isolate the problem. Remove a fuse and see if the test light goes out. A brake light switch that stays on could also cause the problem.

A dirty battery case can also cause a drain. Test with a voltmeter between the negative battery post and the top of the case (Figure 31.34).

➡ **Perform *Battery Drain Test* Worksheet**

■ BATTERY JUMP STARTING

Dead batteries that are in good condition are common. Customers often leave lights on or crank an engine until the battery loses its charge. Jump starting from another vehicle can help to get the vehicle running again.

Jumper Cables

Use good quality braided copper jumper cables. They are flexible and will not overheat if they are large enough. Using stiff aluminum jumper cables that are too small will result in hot cables and possibly a burn to the technician.

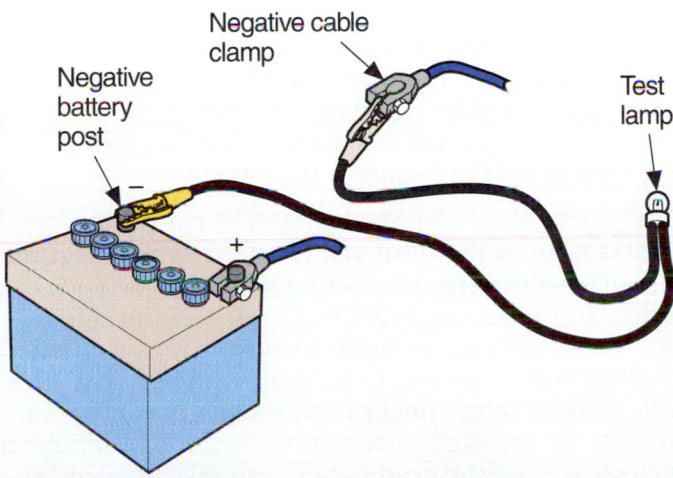

Figure 31.33 Checking for parasitic drains using a test light. *(Courtesy of Ford Motor Company)*

CASE HISTORY

A technician had completed the installation of a rebuilt engine in a motor home. When he tried to crank the engine over, it would not turn. He was concerned that maybe something in the engine was too tight. The battery was located in the rear and the cable run was quite long. The headlights were dim, so he checked the battery's specific gravity. It was about 1200 (low).

While the battery charged, he connected the shop's good jumper battery directly to the starter motor and engine block. This eliminated all of the other possible problems related to the wiring in the motor home. The engine still would not crank. He attached a ratchet and socket to the bolt on the vibration damper on the front of the crankshaft. The crankshaft turned with a reasonable amount of resistance.

Next, he connected the ammeter connection of the volt-amp tester around one of the jumper cables. Attempting once again to crank the engine, he found that the starter motor was not drawing enough current. He pulled back the insulation on one of the clamps on the jumper cable. Someone had switched jumper cables with the shop's original high quality ones. The substituted cables were of too high a gauge number (too thin), although they appeared to be good because the insulation on them was thick. When he substituted another pair of jumper cables, the engine cranked normally.

The chart in Figure 31.35 shows the size of jumper cables needed for different size engines. The best jumper cables are rope stranded (Figure 31.36). These cables are made up of wires that are much smaller. This allows the cable to be very flexible.

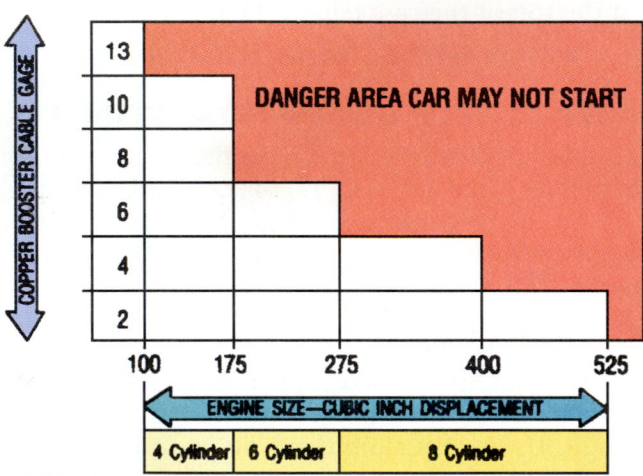

Figure 31.35 This chart shows recommended minimum jumper cable sizes. *(Courtesy of Cooper Automotive/Belden)*

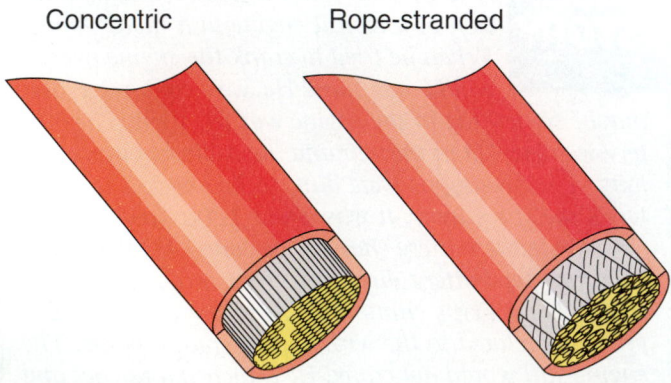

Figure 31.36 Rope stranded cables are flexible. *(Courtesy of Ford Motor Company)*

CAUTION
- Wear eye protection.
- Do not try to jump start a vehicle in freezing weather. Jump starting a frozen battery can cause it to explode.
- Serious damage to a car's electrical system can result if the cables are connected backward (with the wrong polarity).
- When possible, computer controlled vehicles should have a dead battery removed and charged, rather than jump starting. All connections and disconnections must be made with the key off on these cars.
- If a computer-equipped car is to be jump started, it is mandatory that both cars have their ignitions in the "off" position when the jumper cables are connected and disconnected.
- Check to see that the battery has electrolyte above its plates. If not, do not attempt to jump start the car.
- The two vehicles should not be touching each other. This could provide an unwanted ground path.

Jump Start Procedure

Turn on the heater blower motor in the vehicle with the dead battery. Turning on the blower motor will allow it to help absorb any damaging voltage spikes. Turn off all other switches and lights. Hook up the hot cable first and the ground cable last. When attaching the ground cable from the booster battery to the dead battery, do *not* connect the other end of the negative cable to the dead battery. Connect it instead to a ground on the engine (Figure 31.37). A metal bracket or the end of the negative battery cable that is attached to the block will do.

SAFETY NOTE
There will be a spark when the last cable is connected as the dead battery tries to equalize itself with the booster battery. Hooking up away from the battery avoids the possibility of a spark near a battery that can cause it to explode.

As soon as the dead vehicle starts, disconnect the jumper cable immediately from the block. Remove the cables in the reverse order that they were installed.

Low-maintenance batteries have a higher internal resistance than conventional lead-antimony batteries. The jumper cables may need to remain in place for a minute or so before attempting to start the disabled vehicle so that the dead battery can take on a charge.

Run the host vehicle at 2000 rpm to allow its charging system to recharge the battery.

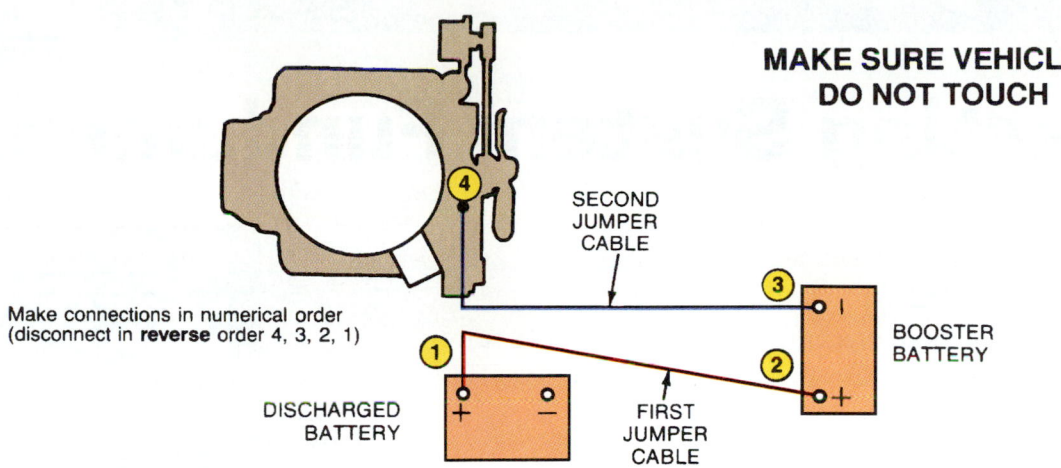

**MAKE SURE VEHICLES
DO NOT TOUCH**

SECOND
JUMPER
CABLE

③

BOOSTER
BATTERY

Make connections in numerical order
(disconnect in **reverse** order 4, 3, 2, 1)

①

②

DISCHARGED
BATTERY

FIRST
JUMPER
CABLE

Figure 31.37 Jump start procedure. *(Courtesy of Ford Motor Company)*

➡ *Perform __Battery Jump Starting__ Worksheet*

■ REVIEW QUESTIONS

1. If a battery is filled to the split ring, will it have the correct specific gravity?

2. When battery terminal posts are exposed to air, they oxidize and turn ___ in color.

3. Battery corrosion can be prevented by covering terminals with grease or installing washers under the terminals.

4. What is the name of the tester that compares the weight of pure water to the weight of electrolyte?

5. What is the specific gravity of new electrolyte?

6. What kind of gas does a battery give off as it charges?

7. What does a sulfuric acid smell or boiling electrolyte indicate about the charge rate when a battery is being charged?

8. What kind of battery charge penetrates the entire plate, fast or slow?

9. What does VAT stand for?

10. When there is a current draw on the battery even when the key is _____, this is called a parasitic load.

■ ASE STYLE REVIEW QUESTIONS

1. Technician A says that a battery that has one cell low on electrolyte could have a shorted cell. Technician B says that side terminal batteries are less prone to terminal corrosion. Who is right?
   a. Technician A   b. Technician B
   c. Both A and B   d. Neither A nor B

2. Technician A says that an acid can be neutralized with a base. Technician B says that disconnecting a battery on a computer controlled car can result in a change in a car's driveability. Who is right?
   a. Technician A   b. Technician B
   c. Both A and B   d. Neither A nor B

3. Technician A says to fill a battery to the split ring prior to charging it. Technician B says that using jumper cables that are too large can prevent the engine from starting. Who is right?
   a. Technician A   b. Technician B
   c. Both A and B   d. Neither A nor B

4. Technician A says that the large test leads are for load testing a battery. Technician B says the small test leads are for reading voltage. Who is right?
   a. Technician A   b. Technician B
   c. Both A and B   d. Neither A nor B

5. Technician A says that the battery ground cable is often black. Technician B says that lead terminal posts turn green when they oxidize. Who is right?
   a. Technician A   b. Technician B
   c. Both A and B   d. Neither A nor B

Starting System Fundamentals

■ INTRODUCTION

The starting system is an important part of the automotive electrical system. Without a starter, the car would have to be push started (most of today's cars have automatic transmissions, which prevent push starting). Henry Ford's Model T had a hand crank for the engine. This chapter deals with the operation of the starting system. If you are unclear about any of the basic principles described here, refer to Chapter 29.

■ STARTER MOTOR

The *starter circuit* includes a powerful *starter motor* and *starter drive, battery, ignition switch,* and **solenoid** (relay) (Figure 32.1). A starter operates at a high rpm. It has a gear (*drive pinion gear*) on the end of its *starter drive* that meshes with a large gear (*ring gear*) on the engine's flywheel (Figure 32.2). A gear ratio provides the starter with the leverage (torque) necessary to be able to turn the crankshaft against engine compression. The gear

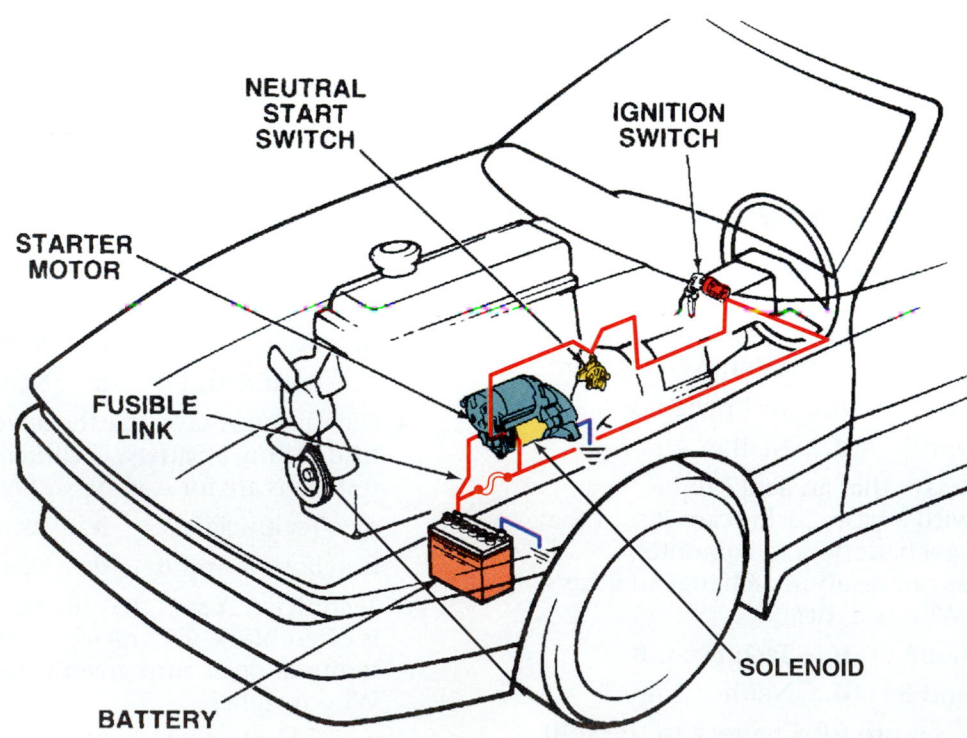

Figure 32.1 Parts of a starter circuit.

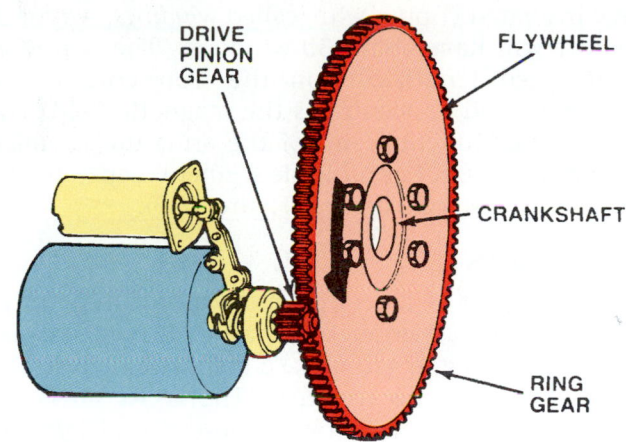

Figure 32.2 Starter drive parts. *(Courtesy of Ford Motor Company)*

ratio between the two gears is about 18 to 1. To crank the engine at normal cranking speed (200 rpm), the starter motor must be turned at 3600 rpm.

■ STARTER MOTOR FUNDAMENTALS

Starters use electromagnetism to convert electrical energy stored in the battery to mechanical power to crank the engine. Other electric motors on the car work in the same manner. Like charges repel each other and unlike charges attract. Because of this, magnetic fields can be used to cause motion. Chapter 29 gives an explanation of magnetic fields.

Figure 32.3 shows the magnetic field surrounding a conductor positioned between the north and south poles of a horseshoe magnet. There are two separate magnetic fields. One is produced by the horseshoe magnet and the other results from the current flowing through the conductor (which represents one loop in the armature).

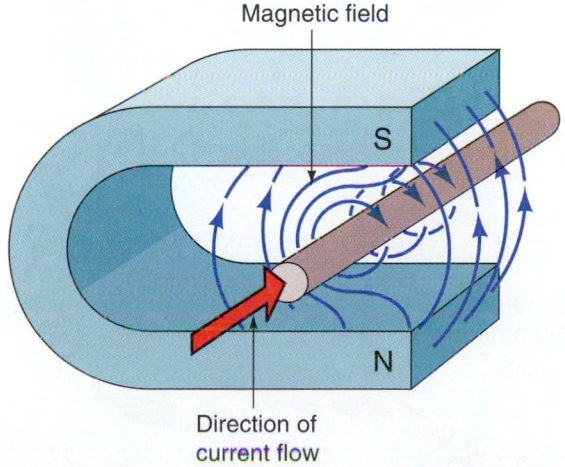

Figure 32.3 Two magnetic fields produced when current runs in the wire. *(Courtesy of General Motors Corporation, Service Technology Group)*

There is a push-pull effect on the armature (Figure 32.4) that causes the conductor to want to move from a stronger magnetic field to a weaker field (from left to right in the illustration). This action is stronger if the magnetic field is stronger or current flow in the conductor is higher.

In a motor, the conductor is formed into a loop. When electrons flow through the loop, the rotating force will be pushing in different directions on opposite sides of the loop (Figure 32.5). The two magnetic

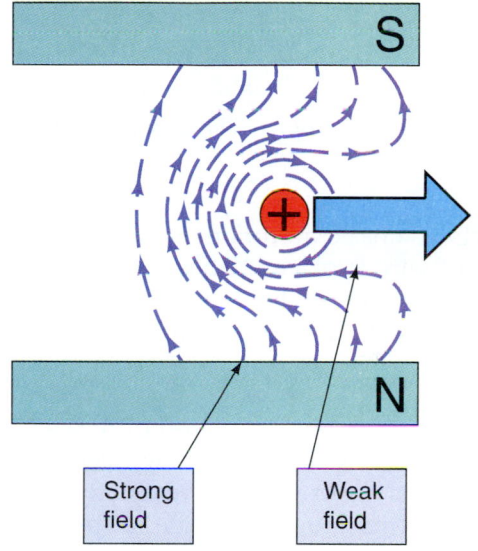

Figure 32.4 The conductor attempts to move from the strong field to the weak field. *(Courtesy of General Motors Corporation, Service Technology Group)*

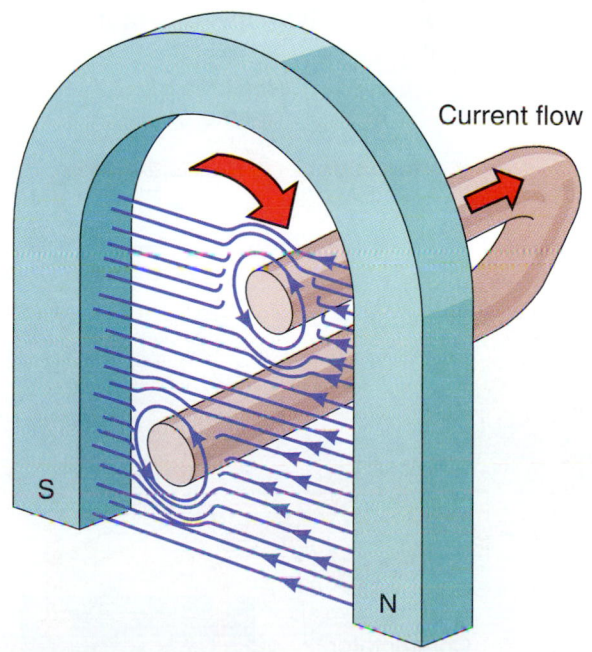

Figure 32.5 The rotating force pushes in different directions on opposite sides of the loop. *(Courtesy of Ford Motor Company)*

fields work together on one side of the armature loop to make one strong field. The other side of the loop has a weaker field. This causes the loop to turn a small amount.

Parts of a basic motor are illustrated in Figure 32.6. A loop of wire is placed between two electromagnetic **pole shoes**. The ends of the wires have *commutator* bars that electrically conducting *brushes* ride upon.

Armature

Multiple loops make up an armature (Figure 32.7). As each loop rotates slightly, a new magnetic field from the next loop reacts with the magnetic pole shoe. This causes the armature to continue to spin.

The armature has a soft iron *core* (Figure 32.8). The core is made up of soft iron laminations (thin layers of iron, each insulated from the other). If the core was solid iron (without the laminations) it would generate eddy currents. The core is wrapped with many loops of heavy insulated copper wire, called *windings*. A typical armature will have about 30 windings. The windings are arranged lengthwise on the iron core which strengthens and concentrates the magnetic field. The electrically conducting end of the armature is called the commutator. It has multiple segments separated by insulation strips, called *mica* (Figure 32.9).

Field Coils

The starter must be a very powerful motor in order to spin an engine fast enough to start it. It requires very strong electromagnets to make a magnetic field strong enough to move the armature. The **field coils** are made of heavy copper ribbons wound around soft iron cores, called pole shoes (Figure 32.10).

Permanent Magnet Starter

Some starter motors used since the mid 1980s have no field coils, but have permanent magnets instead. A special alloy magnet was developed that is ten times as strong as earlier permanent magnets. A permanent magnet starter is simpler, weighs less, and creates less heat than a conventional field coil starter. With no

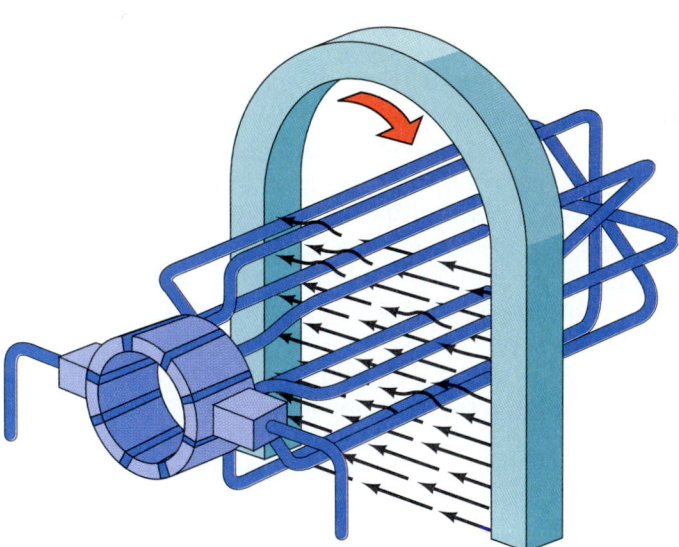

Figure 32.7 Multiple loops make up an armature. *(Courtesy of Ford Motor Company)*

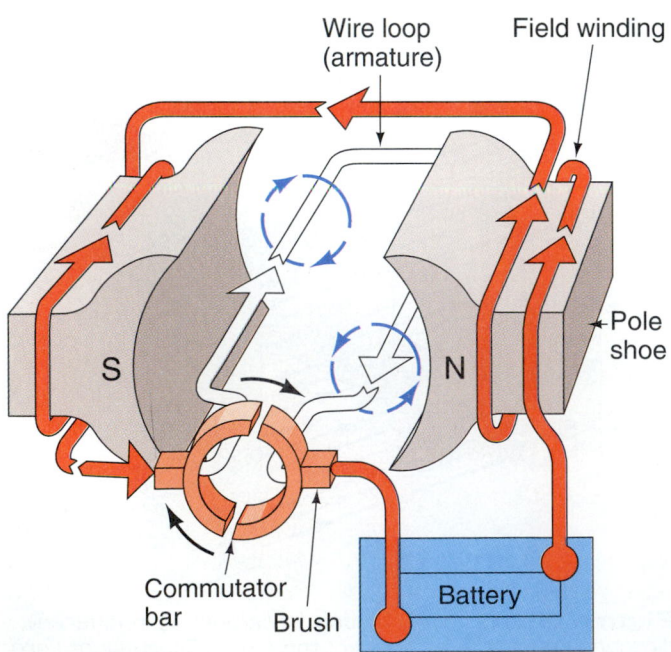

Figure 32.6 Basic parts of a motor. *(Courtesy of General Motors Corporation, Service Technology Group)*

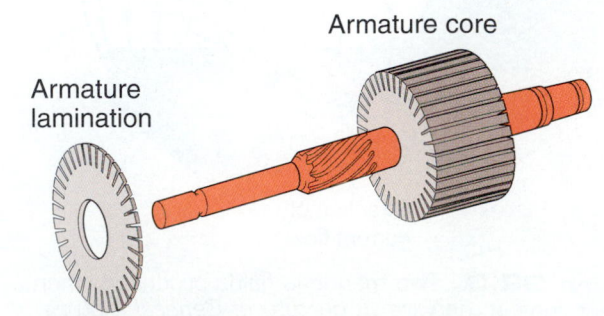

Figure 32.8 The core is made up of laminations, each insulated from the other.

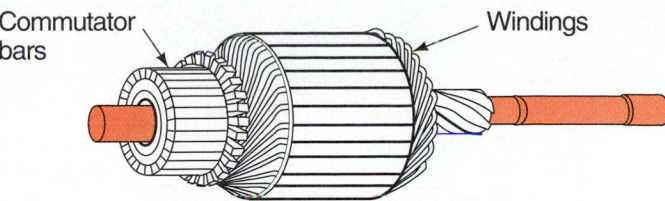

Figure 32.9 Parts of an armature.

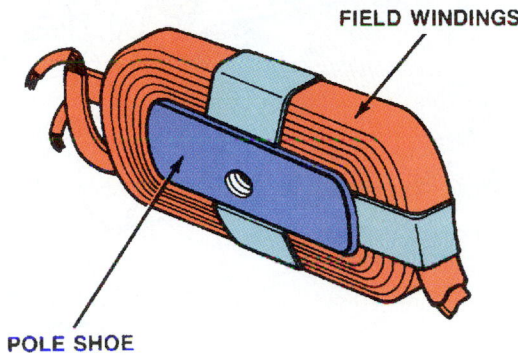

Figure 32.10 Field windings and pole shoe. *(Courtesy of Ford Motor Company)*

field coils, current goes directly to the armature through the commutator and brushes.

Brushes

Brushes, usually made of carbon, are lightly held against the commutator by springs. There are usually four brushes, together in pairs. They supply electricity to the armature windings. One brush is in contact with one commutator bar and the other is in contact with the commutator bar on the other side of the same winding (loop). One of the brushes is positive and the other is negative. The positive brush is insulated and the negative brush is connected to the starter housing (Figure 32.11).

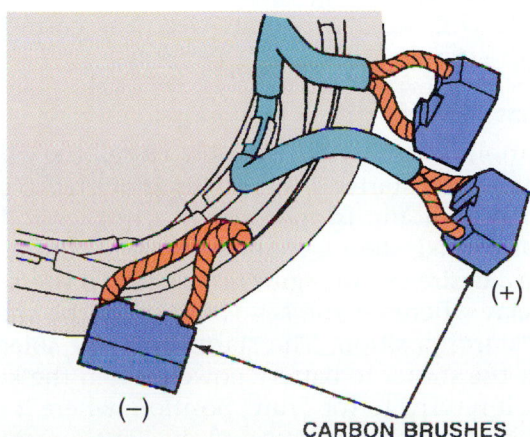

Figure 32.11 The positive brush is insulated from the frame. *(Courtesy of Ford Motor Company)*

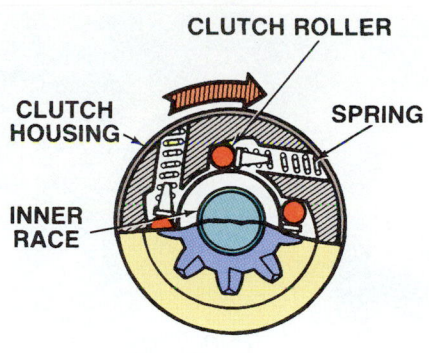

During engine starting

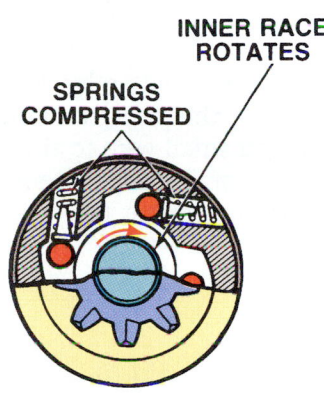

After engine starts

Figure 32.12 Overrunning clutch operation.

■ STARTER DRIVES

Overrunning Clutch Drive

Starter drives have an overrunning, or one-way clutch. The **overrunning clutch** transmits motion from the starter to the flywheel, but not from the flywheel to the starter (Figure 32.12). It disengages from the engine at startup, preventing the engine from driving the starter. If the starter were to remain engaged with the flywheel after startup and engine rpm increased to 2000 rpm (fast idle), the starter would be forced to turn at 36,000 rpm. At speeds of about 10,000 rpm, a starter can be destroyed by centrifugal force. The teeth on the starter drive gear are tapered to allow for smooth engagement with the flywheel (Figure 32.13).

■ STARTER ELECTRICAL CIRCUIT

The starter motor requires a large amount of current to operate. The battery must be in good enough condition to be able to provide substantial current for at least 15 seconds. The starter or its relay is connected directly to the top of the battery by a heavy cable. The return for the electrical circuit is through another cable (usually the negative), after flowing through the engine block.

Figure 32.13 The starter drive pinion gear has tapered teeth.

The starter is switched on by the ignition switch. A relay is required so that a high amount of current can be controlled by a small wire to the ignition switch. If the same amount of electricity went through the ignition switch and the starter, the switch would have to be very large and expensive. A starter uses a relay called a solenoid or a *magnetic switch* (Figure 32.14).

Solenoid

Most cars use a solenoid, a combination magnetic switch and mechanical device that engages the starter drive pinion with the flywheel ring gear. A relay has contact points, while a solenoid uses a movable iron core attached to a *contact disk*. Most solenoids are mounted on the top of the starter motor (Figure 32.15). Some solenoids are mounted in a remote location.

When the ignition switch is turned to the "start" position and the safety switch is closed, electricity flows from the battery to the solenoid. When a coil of wire in the solenoid is energized, a magnetic field draws a piston into the coil. On one end of the piston is a copper disk that can carry a high amount of cur-

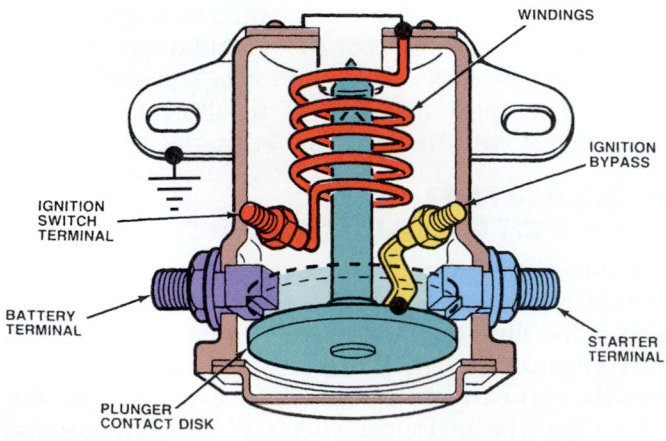

Figure 32.14 A magnetic switch. *(Courtesy of Ford Motor Company)*

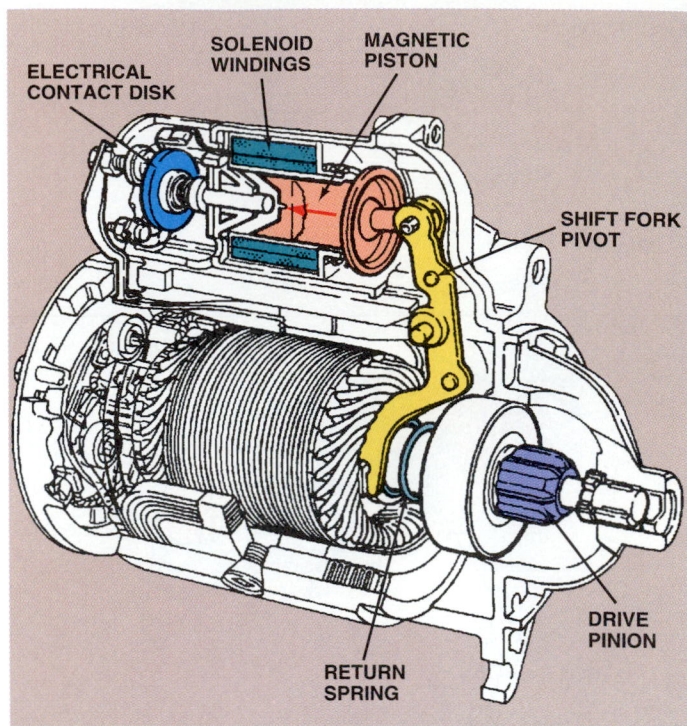

Figure 32.15 A starter motor with a solenoid. *(Courtesy of Ford Motor Company)*

rent without being damaged. It bridges the gap between two terminals in the relay to allow current to travel from the battery cable to the starter. A spring returns the piston to its original position, disengaging the starter drive from the flywheel when the key is released.

It requires more current to pull the solenoid piston against the spring than it does to keep it in engagement. There are two windings in the solenoid, a *pull-in winding* and a *hold-in winding*. Both windings are energized when the piston is pulled into the solenoid housing (Figure 32.16). When the contact disk contacts the battery electrical contacts, it shorts out the pull-in winding, leaving only the hold-in winding energized. This frees up some electrical current to operate the starter.

Ignition Switch

The ignition switch (Figure 32.17) closes the circuit that runs to the starter. It is usually mounted on the steering column and is operated by a linkage rod. As the key is turned, sliding contacts move to complete a circuit as desired. The ignition switch powers the starter relay whenever the key is turned to the spring-loaded "start" position. The starter relay or solenoid connects the starter to battery power. When the key is released, it returns to the "run" position, where it continues to provide battery power to the ignition system.

There are two paths in the starter electrical circuit that electricity can take. One of them is the large cable

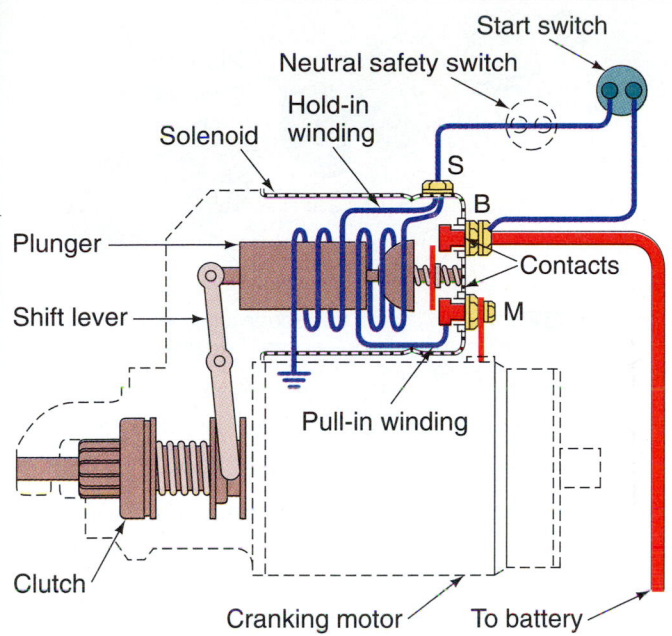

Figure 32.16 Solenoid electrical wiring. *(Courtesy of General Motors Corporation, Service Technology Group)*

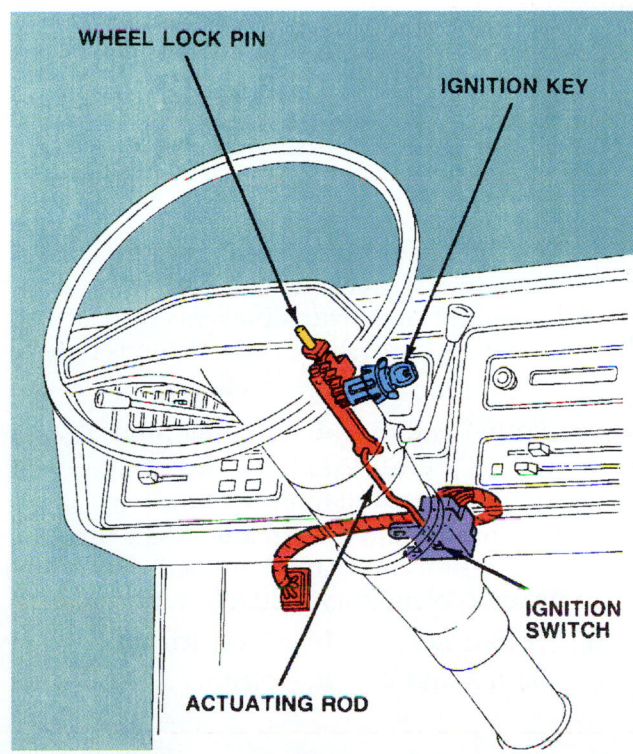

Figure 32.17 The ignition switch is operated by the key switch. *(Courtesy of Ford Motor Company)*

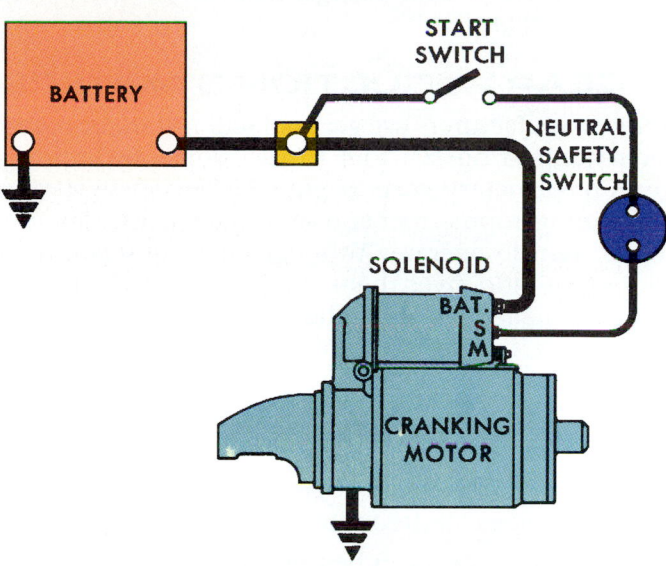

Figure 32.18 A parallel electrical path powers the solenoid. *(Courtesy of General Motors Corporation, Service Technology Group)*

Safety Switches

The circuit on newer cars with automatic transmissions has a *neutral safety*, or *neutral start switch* installed between the ignition switch and the starter. The ignition switch cannot operate the starter motor unless the transmission is in park or neutral.

Late model vehicles with manual transmissions have a *starter/clutch interlock switch* that requires the clutch to be depressed to the floor.

Starter Relay

Some cars (Fords, in particular) have a separate magnetic switch called a starter relay (see Figure 32.14). The starter relay is mounted on the fenderwell near the battery. Some cars use the relay by itself and others use a relay to reduce the load on the neutral safety and ignition switches.

Cars built before the early 1960s that were equipped with a remote starter relay had **inertia starter drive** units. These drives, also called *Bendix* drives, operate like a heavy nut on a screw. When the starter motor is turned on, the weight of the unit keeps it from turning while the armature shaft threads into it. Some British cars still had Bendix drives until the late 1960s.

Later cars with a starter relay use a *positive engagement starter* that uses the magnetic force of one of the field coils to engage the pinion with the flywheel ring gear. One of the field windings has a hollow coil, called a drive coil. A movable pole shoe (the part that fits within the field coil) is drawn into the drive coil when the starter is energized by the relay. The other end of the movable pole shoe has a lever that pushes

that goes directly from the battery to the starter. The other is the small wire that goes from the battery to the ignition switch, safety switch, and solenoid (Figure 32.18). Although the battery cable remains connected to solenoid, the solenoid breaks electrical contact to the starter motor.

the starter drive into engagement. These starters are used mostly on Fords.

■ GEAR REDUCTION STARTERS

Some manufacturers use gear reduction starters. A gear reduction of from 2:1 to 4:1 is developed by ordinary gears or planetary gears (Figure 32.19). These starters are gaining popularity because they are lighter and use less current to operate. Although they are small, their lower gear ratio gives them enough torque to turn the engine. Smaller battery cables can also be used because these starters draw less current.

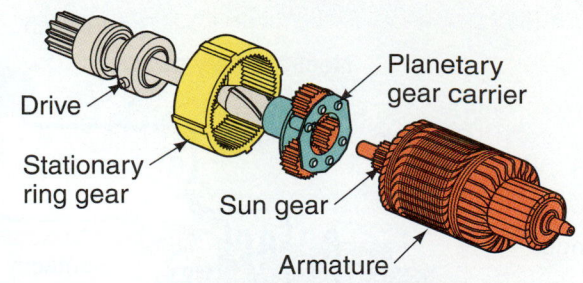

Figure 32.19   A gear reduction starter. *(Courtesy of Ford Motor Company and General Motors Corporation, Service Technology Group)*

■ REVIEW QUESTIONS

1. To crank an engine at 200 rpm, how fast must an average starter motor turn?
2. _____ fields can be used to cause motion.
3. What is the name of the parts at the ends of the armature loop that the brushes slide against?
4. The armature core is made up of soft iron _____ insulated from each other.
5. What are the loops of heavy insulated copper wire in the armature called?
6. What is another name for the one-way clutch in the starter drive?
7. If an average starter remains engaged and is turned at 2000 engine rpm, how fast would it turn?
8. How does the electrical current return from the battery to the starter?
9. What is the name of the electrical device that uses a movable iron core attached to a contact disk?
10. Which solenoid winding is shorted out when the starter motor turns?

■ ASE STYLE REVIEW QUESTIONS

1. Technician A says that motor action is stronger if the magnetic field is stronger. Technician B says that motor action is stronger if current flow in the conductor is higher. Who is right?
   - **a.** Technician A
   - **b.** Technician B
   - **c.** Both A and B
   - **d.** Neither A nor B

2. Technician A says that like magnetic charges repel and unlike charges attract. Technician B says that there is a stronger magnetic field on one side of a conductor in a motor. Who is right?
   - **a.** Technician A
   - **b.** Technician B
   - **c.** Both A and B
   - **d.** Neither A nor B

3. Technician A says that armature windings are arranged lengthwise on the iron core to keep the effect of the magnetic field minimized. Technician B says that all starter motor brushes must be insulated to keep them from touching the starter housing. Who is right?
   - **a.** Technician A
   - **b.** Technician B
   - **c.** Both A and B
   - **d.** Neither A nor B

4. Technician A says that there are two separate windings in a starter solenoid. Technician B says that a neutral start switch prevents the car from being started in park or neutral. Who is right?
   - **a.** Technician A
   - **b.** Technician B
   - **c.** Both A and B
   - **d.** Neither A nor B

5. Technician A says that on the overrunning clutch drive, the weight of the unit keeps it from turning while the armature shaft threads into it. Technician B says that Bendix is another name for a solenoid-type starter. Who is right?
   - **a.** Technician A
   - **b.** Technician B
   - **c.** Both A and B
   - **d.** Neither A nor B

CHAPTER 33

Starting System Service

■ OBJECTIVES

Upon completion of this chapter, you should be able to:

✔ Measure amperage draw on a starting system.

✔ Measure voltage drops on both the positive and ground sides of the starting circuit.

✔ Diagnose no-crank conditions with a test light.

✔ Replace a solenoid and starter drive.

■ INTRODUCTION

This chapter deals with the process for testing and repairing common problems with the starting and charging systems. Principles of operation and electrical fundamentals learned in earlier chapters will be important here. Diagnosis of failures is very important before parts are replaced. Most parts stores will not accept returns of electrical items.

NOTE: *Better techs often refer to less skilled mechanics as "parts replacers." In the long run, parts replacers cost the motoring public a great deal of unnecessary expense. In some cases, service manuals do call for testing by replacing with a known good part.*

■ STARTING SYSTEM SERVICE

Starting System Diagnosis

Examples of starting system problems include a no-crank condition and slow cranking. Methods of diagnosing those problems are covered in this chapter. Also included are on-the-vehicle tests of the starting system when the starter does not crank and preventive maintenance and diagnosis. Replacement of relays, solenoids, and entire starters is stressed. Local labor rates dictate whether a starter can be rebuilt economically by the shop or a rebuilt unit is purchased.

When testing a starter, always follow a logical procedure and do not skip steps. Some starter tests are done on an engine that still cranks over. Others must be done on a starter that will not crank at all. Be sure that the starter is the cause of the problem before removing it from the engine. Over half of the starters returned on warranty claims are not defective. Some-

times, the starter cranks normally but the engine does not start. This can be unrelated to the starter itself.

Problems can be of two types, mechanical and electrical. Mechanical problems are often identified by the noise that results. Noises that can be heard when a starter is bad include multiple clicking and single clicking, humming after cranking, and metallic noises related to the gear drive and flywheel teeth.

Electrical problems can be identified during visual and electrical tests.

Visual Check. Problems are often found during a visual check. Check the wiring connections to see that they are clean and tight. Loose or dirty connections can cause excessive voltage drop. Cables should be of the correct diameter and length. Too long a cable can cause too much resistance. The cable should not get hot during cranking, which would indicate excessive resistance.

Volt Amp Tester. The volt-amp tester (VAT) is a fundamental piece of electrical test equipment that will be used to test the starting and charging systems (see Figure 31.30). The VAT's ammeter is capable of measuring 500 amps, which could be reached in the test of a faulty starter. There is also a voltmeter and a carbon pile that is used to load the battery.

■ STARTING SYSTEM TESTS

Test the Battery First

The biggest cause of starter motor failure is low battery voltage. Remember that a battery can cause starting problems and a starter can cause battery problems. Be sure that a battery has at least 75% of full charge before attempting to test the starter. The battery must be able to keep voltage above 9.6 volts while delivering needed current to the starter.

One quick check of battery condition is done with the headlights on while cranking the engine:

- If the engine does not crank and the headlights stay off, the battery is dead.
- If the lights go out while cranking, the battery is weak.
- If the lights stay bright but the engine does not crank over, there could be high resistance or an open circuit in the ignition switch or the solenoid. If the solenoid will not click in the starter, brushes are probably worn out, not grounding the solenoid.

When the battery is weak, a solenoid can make a series of rapid clicks. This happens as the winding in the solenoid pulls the plunger and disk into contact with the contacts that operate the starter motor. When the starter begins to turn, battery voltage decreases rapidly to a point where the windings in the solenoid can no longer hold the solenoid disk against the contacts. This releases the load made by the starter, so the solenoid gets enough voltage to operate again (click).

Electrical power is measured in watts (amps × volts). When battery voltage drops, amperage decreases, cranking rpm is lower, and ignition voltage drops. The net result is poor or no starting.

NOTE: *A starter will draw twice the current (amps) if the battery voltage drops to half.*

Increased amperage results in more heat in the starter and cables. Perform battery tests as outlined in Chapter 31.

Disable the Ignition System. During starter tests, the engine must be prevented from starting. Be sure the transmission is out of gear and the parking brake is set. The ignition system must be disabled. On most cars, there is an easy way to trigger the starter motor from under the hood with a *remote starter switch* (Figure 33.1).

To disable a distributor ignition system use one of the following procedures:

- Remove the center lead from the distributor cap and attach a jumper lead from it to ground (Figure 33.2).
- Disconnect the wiring harness connector from the distributor or coil (Figure 33.3).

Do not crank the engine with the coil to distributor wire simply disconnected. This could result in damage to the coil and ignition module.

■ CRANKING VOLTAGE AND AMPERAGE TEST

The inductive cable on the VAT is installed around the battery cable (Figure 33.4). Be sure the carbon pile is off. The voltmeter should read battery voltage and the ammeter should read zero (unless there is a load such as a dome light or the ignition switch is on). When you turn on the headlights, the ammeter should move toward negative. Turn off the headlights and adjust the zero adjust knob if the machine has one.

Move the meter to where you can observe it while cranking the engine for about 5 seconds. Most V-8s

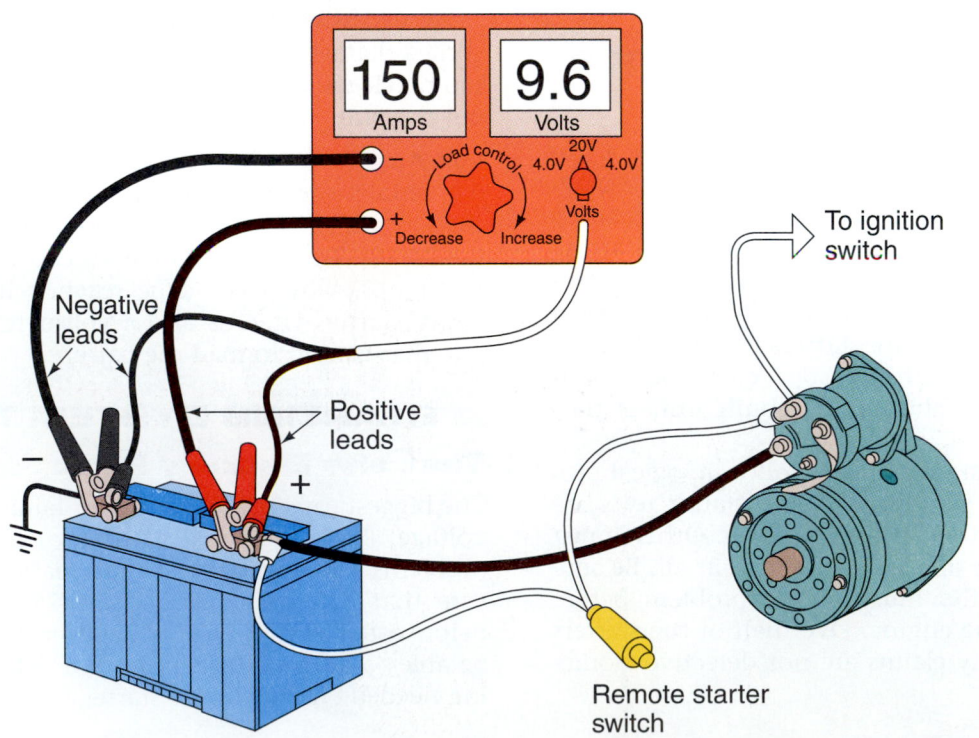

Figure 33.1 With the ignition system disabled, a remote starter is used to power the starter.

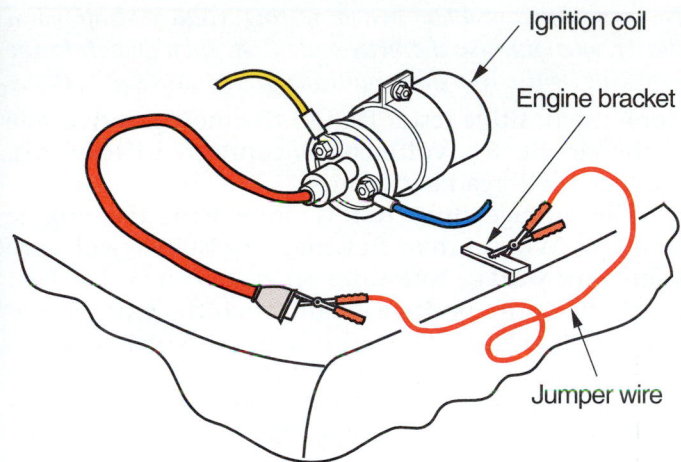

Figure 33.2 Disabling a standard ignition system.

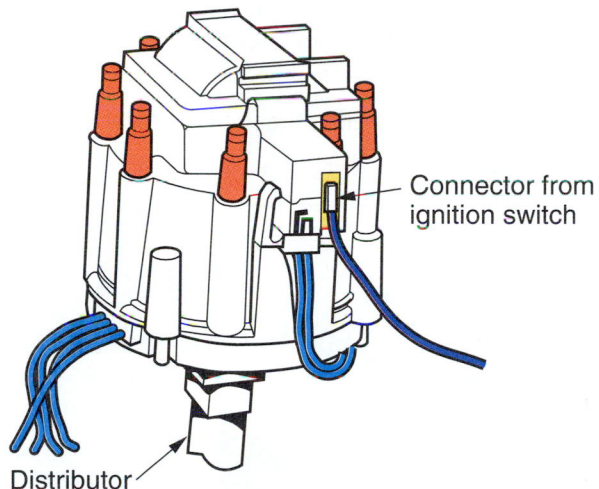

Figure 33.3 Unhook the connector at the distributor to disable an electronic ignition.

will draw about 200 amps, six cylinders about 150 amps, and four cylinders about 125 amps. These are just estimates. The service manual will list actual specifications.

NOTE: *The starter motor is designed to be used for short periods of time only. In normal use, the starter should never need to be operated longer than 15 seconds at a time. When testing, do not use the starter for more than 30 seconds. After 30 seconds of operation, allow it to cool for at least 2 minutes.*

Cranking Test Results

Normally, when the voltage drops the amperage draw goes up. If the reading was low on voltage and high on amperage, you know that more tests are required.

When there is resistance, there is a voltage drop. Cable connections or resistance in relay contacts can cause voltage to drop and amperage to become higher.

The next tests will show how to look at individual components.

■ CRANKING SPEED

Generally, 250 engine rpm is the speed for a standard starter. A gear reduction starter will turn at about 200 rpm (sometimes they turn faster). This can be measured with a tachometer but a trained ear can tell if the speed is correct.

NOTE: *When there is a combination of slow cranking speed and lower than normal cranking amperage, the starter will need to be removed. This happens only when there is too much resistance internally in the starter.*

When starter speed is low and cranking amperage is higher, this is normal. With this condition, check the positive and ground circuits.

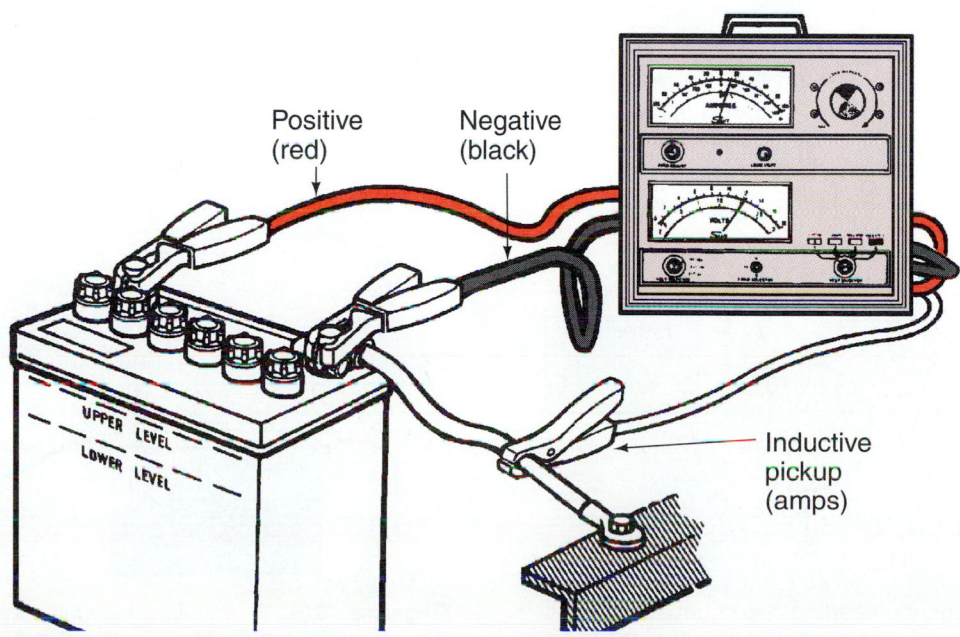

Figure 33.4 Volt-amp tester connections.

SHOP TIP When there is high resistance in a battery cable or its connections, they will become hot to the touch. When resistance is within acceptable limits, cables and connections will not become hot.

■ CIRCUIT RESISTANCE TEST

Voltage drop testing is a very popular **circuit resistance test** or way of checking for resistance in a circuit. It is a fast and simple method of testing that does not require disassembly of connections. Voltage drop is the amount that voltage drops when current is *flowing* through a resistance. What is being measured is the amount of voltage lost with current flowing from the source to the load and back to the source. Voltage drop can only be measured when the circuit is under load. Voltage is lost when current flows through a resistance. When a circuit has high voltage drop, resistance in the circuit is high.

Circuits are designed to have very little resistance. Voltage drop should be less than 0.1 volt per connection. The ground side has fewer connections. The insulated side includes the solenoid, which represents more connections. If resistance is less than 0.2 volt on the ground side of the circuit and less than 0.4 volt on the insulated side, on a *loaded test* (during cranking) voltage to the starter will be within 0.6 volt of battery voltage. To test this, connect a voltmeter across the starter positive and case ground (Figure 33.5). If the reading is less than expected, the exact location of the voltage drop (resistance) will need to be pinpointed.

For the voltage drop test, adjust the voltmeter to read on its lowest scale. This will probably be about 2 to 4 volts.

NOTE: *An analog voltmeter can be damaged if it attempts to read full battery voltage when it is set on the low scale.*

To avoid this problem when testing voltage drop on a starter solenoid, use the high scale first, then switch to the low scale or use a digital multimeter.

Hook the positive tester lead to the most positive side of the circuit. If a voltmeter is connected backwards, the meter will read backwards.

The voltage drop test is done with the engine cranking. With current flowing, a resistance will have a different voltage (pressure) on each of its sides. First check the entire positive circuit, then the entire negative circuit (Figure 33.6) to isolate on which side the excessive resistance is located. When the high voltage drop side has been isolated to either the positive or negative circuit, the circuit can be broken into smaller pieces during further voltage drop tests. Figure 33.7 shows how to check for voltage drop in the positive insulated circuit.

■ Normal voltage drop would be about 0.2 volt when normal current is flowing. A rule of thumb is 0.1

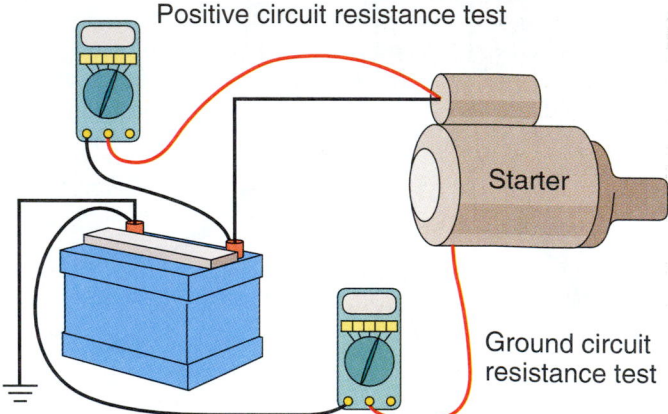

Positive circuit resistance test

Starter

Ground circuit resistance test

Figure 33.6 Isolate which side of the circuit the resistance is on.

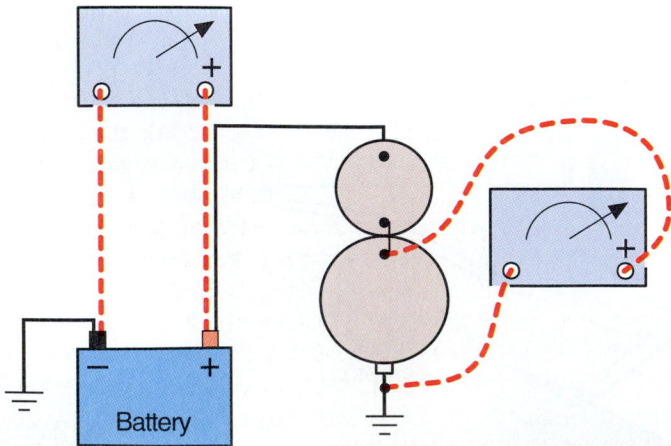

Figure 33.5 Voltage reading taken between the starter input and the starter case should be within 0.4 volt of battery voltage.

Battery

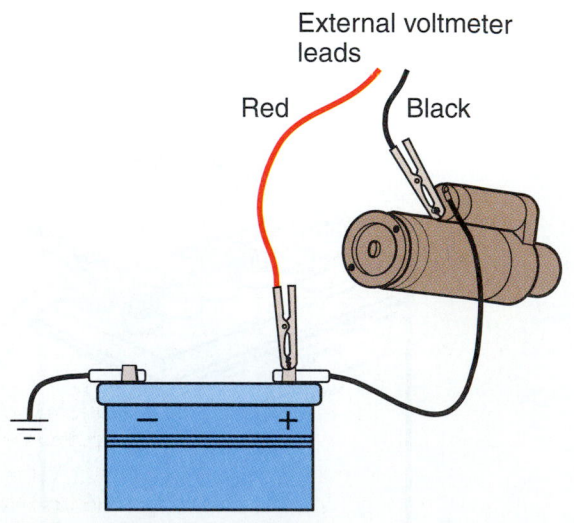

External voltmeter leads

Red   Black

Figure 33.7 Insulated circuit voltage drop check. *(Courtesy of Sun Electric Corporation)*

volt for each connection and 0.5 volt for the solenoid.

■ If 100% more current is flowing, adjust the specification up by 100% to 0.4 volt.

If the voltage drop reading is high, check each connection in the circuit. If a connection has voltage drop that is above normal, inspect it for corrosion or looseness. The solenoid is part of the overall 0.2 volt drop. The only part of the circuit that is expected to have more than 0.2 volt drop is the starter motor.

After voltage drop is restored to a normal level, cranking amperage can be used to diagnose a starter. A higher than normal amperage draw combined with no excessive resistance in the circuit points to an engine that is too tight or a starter that is bad. Remove the starter and disassemble it. Look for obvious causes of excessive current (amp) draw, like worn bushings, a rubbing armature, or shorted windings. When there are two field windings, a shorted winding results in less resistance, half the magnetic field, and twice the current flow.

■ NO-CRANK TESTS USING A TEST LIGHT

Voltage drop testing works while the engine is still able to crank. When the engine does *not* crank, this is usually because of an open circuit. A 12-volt test light is used for this diagnosis.

NOTE: *Always test voltage first with a meter (see Chapter 29).*

Start at one end of the circuit with the key in the start position. Use the 12-volt test light for a quick check to see if there is power at the outlet of the solenoid (Figure 33.8) (where power enters the starter). Because this is the end of the circuit, power here will tell you that the positive side of the circuit is complete. It will not tell you whether there is sufficient available

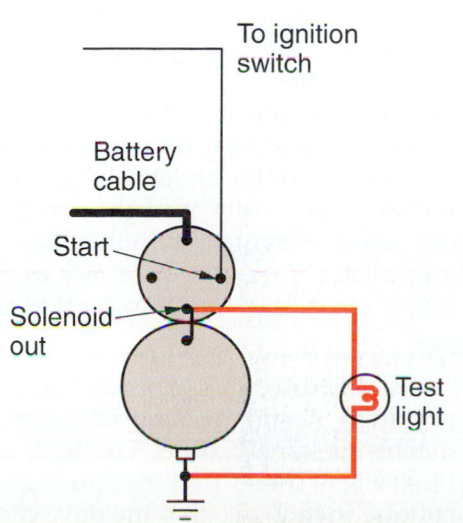

Figure 33.8 Use the 12-volt test light to check if there is power at the outlet of the solenoid.

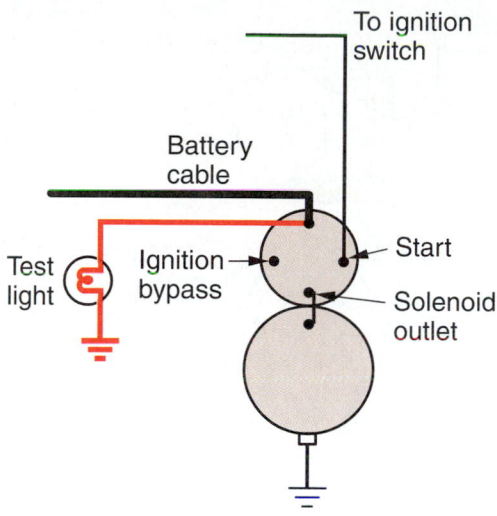

Figure 33.9 Connect the test light to the end of the battery cable at the solenoid.

voltage, however. A meter must be used for this. If there is no light, work your way back through the system until you find power (see Figure 29.51). Connect the test light to the end of the battery cable at the solenoid (Figure 33.9). The light should be on and stay on when the key is turned to the start position. If it goes out, repair the cable or connection.

The next test is to the starter's ground path. Clean off an area of the starter body and connect the test light's ground connection to it. Connect the positive probe of the test light to the known good terminal at the outlet to the solenoid. If the light comes on, the ground path is good. This means that the starter must be the problem. Always check the power and ground circuits before pulling and replacing the starter.

■ SOLENOID PROBLEMS

A rapidly clicking solenoid can be caused by several things including:
■ A weak battery
■ A corroded or loose battery cable connection
■ An open circuit in a hold-in winding

Before replacing a solenoid, check the condition of the battery. A single click when the battery is in good condition can often be traced to burned contacts in the solenoid. Sometimes these contacts can be replaced or repaired. On other solenoids, the entire solenoid is replaced. Be certain to correctly diagnose the problem before replacing the solenoid. Sometimes, they are as expensive as an entire rebuilt starter.

When the contact disk in a starter solenoid has become burned, high resistance develops in the positive side of the circuit (across the solenoid terminals). Using the voltage drop test, the resistance can be located. The symptom of resistance here is that the starter solenoid will click instead of starting the car, or the starter will turn slower than normal.

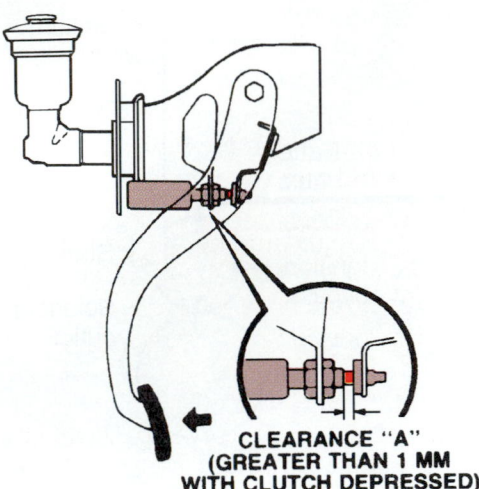

**CLEARANCE "A"
(GREATER THAN 1 MM
WITH CLUTCH DEPRESSED)**

Figure 33.10 A clutch start switch closes to allow current flow when the clutch pedal is depressed.

> **SHOP TIP** A typical pull-in winding will draw 72 amps. This will drop to 22 amps during the hold-in phase. Sometimes the hold-in winding connection becomes disconnected (open circuit). The hold-in winding has about 2500 pounds of hold-in strength. With the starter off the car and the solenoid energized, you should not be able to physically pull the piston out of the solenoid.

Neutral Start Switch

The neutral start switch is attached to either the steering column or the side of the transmission. Because it is in the relay circuit, it uses small wires. It is usually adjustable and sometimes controls the back-up lights also. Occasionally, moving the shift lever while turning the ignition key to the start position will allow the engine to crank. This means that the start switch needs to be adjusted.

Clutch Start Switch

Later model cars with manual shift transmissions have a clutch start switch (Figure 33.10). This ensures that the clutch pedal is depressed before the starter motor gets electricity. Using an ohmmeter, there should be no continuity when the clutch pedal is up. Continuity when the clutch is depressed allows current to flow through the switch to the starter.

■ STARTER REPAIR

Starters are not always economical to rebuild. This depends on the availability of competitively priced parts, wages, and whether the shop's work load is low on a particular day when a bad starter requires repair. Many shops replace starter drives and solenoids. Procedures covered here apply to replacements of those parts within the starter. The procedure for disassembly

and reassembly is also covered here so that you can disassemble a starter and inspect its working parts for a better understanding of its operation.

■ STARTER DISASSEMBLY

Get in the habit of marking parts that are disassembled so they can be reassembled in their original positions. Some starters have pins that align the parts. Others do not.

The following procedure applies to General Motors starters. Other starters are similar. To remove the solenoid, disconnect the solenoid's electrical terminal(s). Remove the two screws that hold the solenoid on the starter housing. Twist the solenoid until the locking flange is free. The spring will push the solenoid away from the housing. There are two long bolts that go through the starter frame. Remove the bolts, the end frame, and the starter body from the drive end housing. Remove the armature from the housing.

NOTE: *Solenoids and starter relays used with vehicles with computers have an internal diode to provide protection from voltage spikes that are produced when the magnetic field in the relay collapses (see electromagnetic induction in Chapter 34). A replacement solenoid with a diode may be used on applications that do not have computers. Using an early model solenoid on a computer equipped car can result in problems.*

Bearing or Bushing Service

Inspect the bearings or bushings at both ends of the housing. Inspect bearings by feeling them.

NOTE: *Commercial rebuilders have listening devices that spin a bearing and tell whether it makes too much noise.*

If a bearing is tight or feels rough, replace it. Most of today's starters use bushings because they do not rotate for long periods of time. Bushings are visually inspected. When they wear, it is usually on one side. A worn bushing is pounded or pressed out.

■ STARTER DRIVE SERVICE

Starter drives often do not last the life of the starter motor. Replacing a starter drive is commonly done by technicians. Noise that can be heard coming from a starter is usually from the starter drive. The starter whines or grinds and does not turn the engine over reliably. Sometimes, the overrunning clutch will slip in one direction when it wears internally. It might only fail to work occasionally. Test it by trying to turn it in both directions (Figure 33.11). Sometimes, it is necessary to wiggle the gear and lightly attempt to turn it to get it to slip.

> **SHOP TIP** A starter drive can be bench tested to see if it will slip. Clamp the removed starter drive in a suitable holding fixture. Locate a 12-point socket that will fit snugly over the pinion gear teeth. Using a torque wrench, see that the drive clutch will not slip when a 50-foot-pound load is applied to it.

Figure 33.11 Test the starter drive one-way clutch.

Replacing a Starter Drive

On the General Motors starter, the drive unit is held in place on the armature shaft by a split metal ring covered by a full round ring.

- Tap down on the outer ring to remove it from covering the inner ring (Figure 33.12).
- Remove the snap ring (Figure 33.13).
- Slide the starter drive from the armature.

Figure 33.12 Remove the outer ring from the retainer using a socket and a soft hammer.

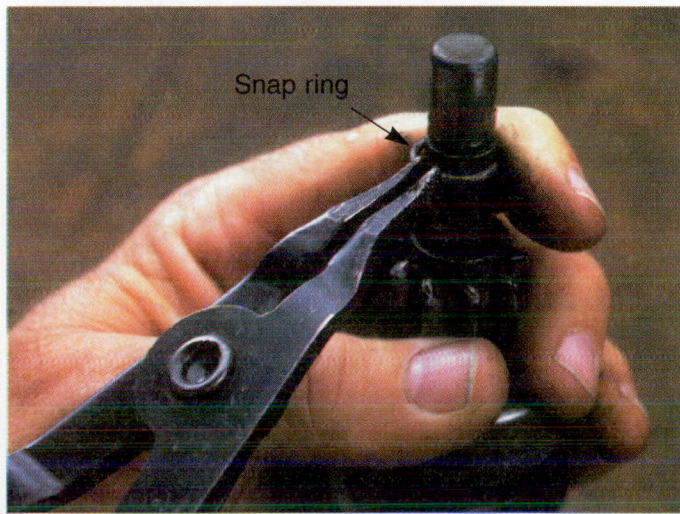

Figure 33.13 Remove the snap ring.

NOTE: *If the starter drive is to be reused, do not immerse it in cleaning solvent. The bearing is permanently lubricated and can be damaged by solvent.*

- Before replacing the starter drive with a new one, count the number of teeth on the drive pinion and match up the old and the new ones to see that they are the same.
- To reinstall the snap ring, use a hammer and a block of wood or a rubber mallet (Figure 33.14).
- Use two pairs of pliers to compress the ring assembly together (Figure 33.15).

Brush Service

Brushes occasionally wear too thin and have to be replaced. Some must be soldered in and others have connectors and can be installed with screws. If the

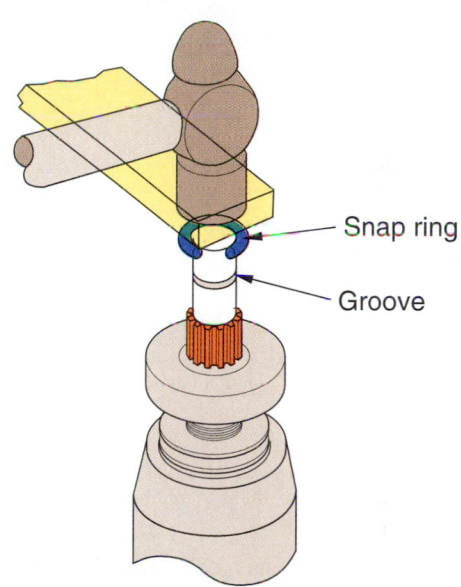

Figure 33.14 Reinstalling the snap ring.

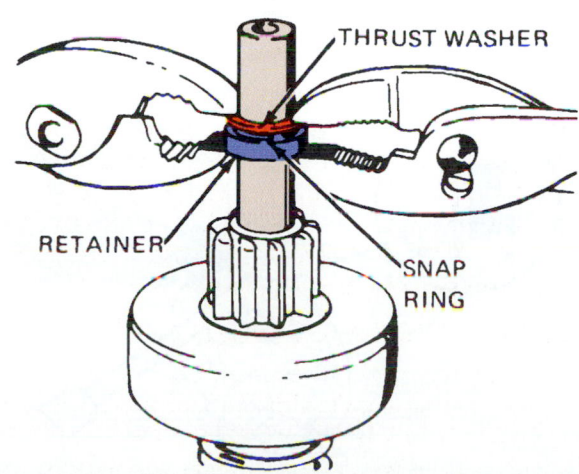

Figure 33.15 Compress the retainer assembly together. *(Courtesy of General Motors Corporation, Service Technology Group)*

commutator is in good condition it can be cleaned with crocus cloth (Figure 33.16).

STARTER REASSEMBLY

Reassemble the starter. Some starters have brushes that are on pivots and are easy to install. Others require pulling up on the springs that hold the brushes against the commutator (Figure 33.17).

PINION CLEARANCE TESTS

There are two tests that can be made. One is with the starter off the car. With the solenoid energized, push the pinion back toward the armature to remove any slack. Check the clearance with a feeler gauge (Figure 33.18). The General Motors specification is 0.010–0.140".

With the starter on the car, the pinion to flywheel ring gear clearance is checked. If the pinion clearance to the ring gear is excessive, the starter can be loud and the teeth can be damaged. With too little clearance, the starter could bind and the amp draw could be higher. The clearance is adjusted using shims (Figure 33.19). Each 0.015" shim changes the clearance by about 0.005".

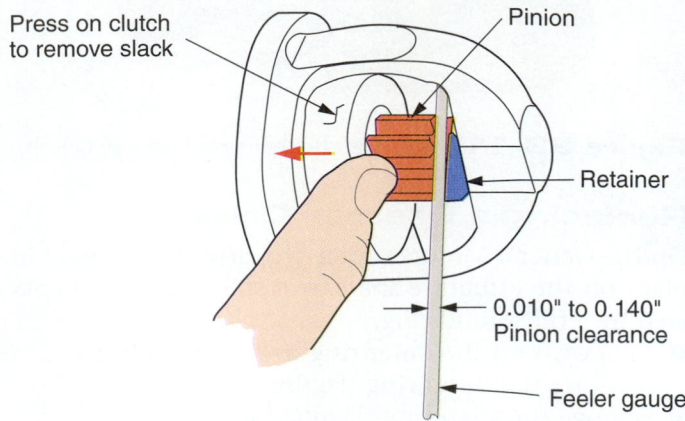

Figure 33.18 Measuring pinion gear to drive housing clearance. *(Courtesy of General Motors Corporation, Service Technology Group)*

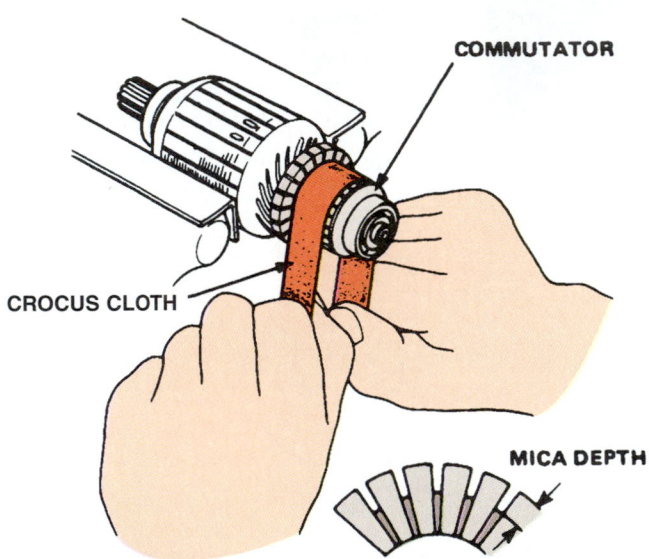

Figure 33.16 Cleaning the commutator with crocus cloth. Check the mica depth. *(Courtesy of American Honda Motor Co., Inc.)*

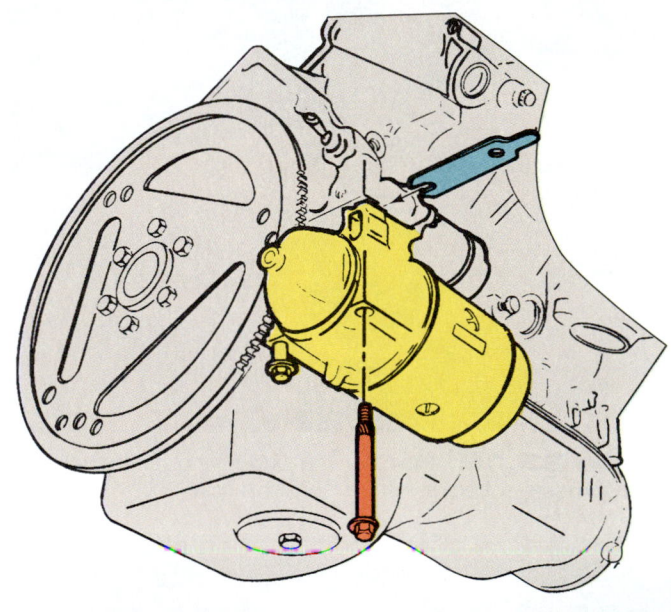

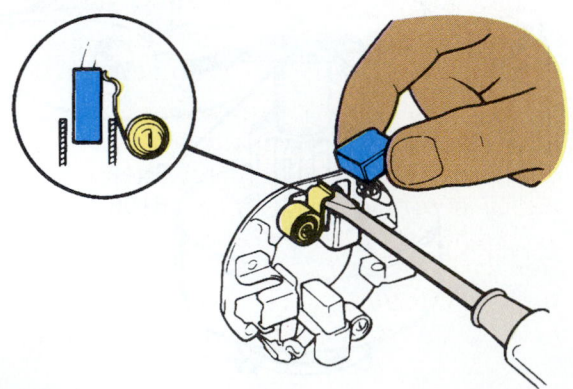

Figure 33.17 Installing brushes during starter reassembly. *(Courtesy of American Honda Motor Co., Inc.)*

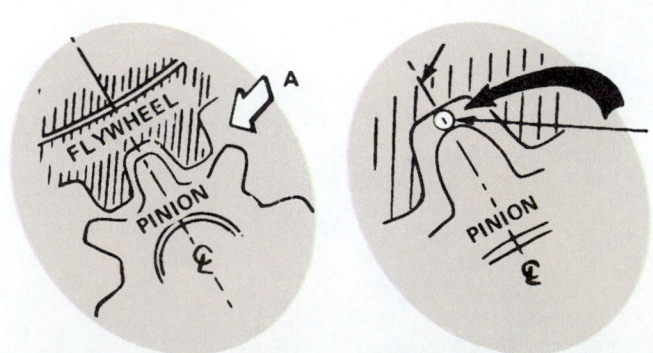

Figure 33.19 Adjusting pinion depth with shims. *(Courtesy of General Motors Corporation, Service Technology Group)*

■ REVIEW QUESTIONS

1. Problems in a starter can be either electrical or _____ .

2. What is the biggest cause of starter motor failure?

3. Be sure that a battery has at least _____ of full charge before attempting to test the starter.

4. What part makes a series of clicking sounds when the battery is low or there is a bad connection?

5. What tool is used to energize the starter motor with the ignition system disabled?

6. When testing, do not use the starter for more than _____ seconds.

7. What resistance test can be made with a voltmeter?

8. Worn bushings, a rubbing armature, or shorted windings could cause excessive _____ draw.

9. What part of the starter does not always last the life of the starter?

10. Before replacing a starter drive, what do you compare between the old gear teeth and the new gear teeth?

■ ASE STYLE REVIEW QUESTIONS

1. Two technicians are discussing an engine being cranked with the headlights on. Technician A says if the lights go out while cranking, the battery is weak. Technician B says if the lights stay bright but the engine does not crank over, there could be high resistance or an open circuit. Who is right?
   a. Technician A   **b.** Technician B
   c. Both A and B   **d.** Neither A nor B

2. Technician A says that a starter will draw half the amperage if the battery voltage drops to half. Technician B says that most V-8 engines will draw about 50 amps during cranking. Who is right?
   a. Technician A   **b.** Technician B
   c. Both A and B   **d.** Neither A nor B

3. Two technicians are discussing disabling ignition systems. Technician A says on electronic ignition systems, disconnect the wiring harness connector from the distributor. Technician B says to crank the engine with the coil to distributor wire disconnected. Who is right?
   a. Technician A   **b.** Technician B
   c. Both A and B   **d.** Neither A nor B

4. Technician A says that if a cranking reading was low on voltage and high on amperage, this is normal. Technician B says that when there is a voltage drop, there is resistance in a circuit. Who is right?
   a. Technician A   **b.** Technician B
   c. Both A and B   **d.** Neither A nor B

5. Technician A says that if the engine cranks slow and cranking amperage is lower than normal, there is too much resistance internally in the starter. Technician B says that when cranking speed is low and cranking amperage is higher, this is normal. Who is right?
   a. Technician A   **b.** Technician B
   c. Both A and B   **d.** Neither A nor B

6. Technician A says that current must not be flowing in a circuit when measuring voltage drop. Technician B says that when voltage drop through a circuit is low, resistance is high. Who is right?
   a. Technician A   **b.** Technician B
   c. Both A and B   **d.** Neither A nor B

7. Technician A says that the starter motor can have more voltage drop than the rest of the starting circuit. Technician B says that voltage drop testing is used to diagnose a no crank condition. Who is right?
   a. Technician A   **b.** Technician B
   c. Both A and B   **d.** Neither A nor B

8. A solenoid clicks repeatedly. Technician A says that a weak battery could be the cause. Technician B says that this could be caused by an open circuit in the solenoid hold-in winding. Who is right?
   a. Technician A   **b.** Technician B
   c. Both A and B   **d.** Neither A nor B

9. Technician A says that when current is flowing through a resistance, the resistance will have the same voltage on both sides of it. Technician B says that if there is excessive resistance in a circuit, a test light will be brighter. Who is right?
   a. Technician A   **b.** Technician B
   c. Both A and B   **d.** Neither A nor B

10. A starter with good circuit wiring makes a strong, *single* click when attempting to start the engine. Technician A says that the solenoid electrical contacts are probably burned. Technician B says that a low battery could cause this. Who is right?
   a. Technician A   **b.** Technician B
   c. Both A and B   **d.** Neither A nor B

Charging System Fundamentals

■ INTRODUCTION

The charging system is an important part of the automotive electrical system. The charging system allows a battery to maintain a charge and operate electrical accessories. This chapter deals with the operation of the charging system. If you are unclear about any of the basic principles described in the chapter, refer to Chapter 29.

■ CHARGING SYSTEM

The charging system consists of the alternator, the voltage regulator, the dash light or gauge, and related wiring (Figure 34.1). A battery alone can supply a vehicle's electrical needs for a period of time, but a charging system is needed to replenish the battery. The charging system output is increased whenever the load of various components causes battery voltage to drop below a certain point. The starter motor is a particularly large load on the battery. The charging system works hardest after the engine is first started and when all electrical options are operating.

The alternator is driven by a belt from the engine crankshaft pulley (Figure 34.2). From the time a car is first started until the alternator is rotating fast enough

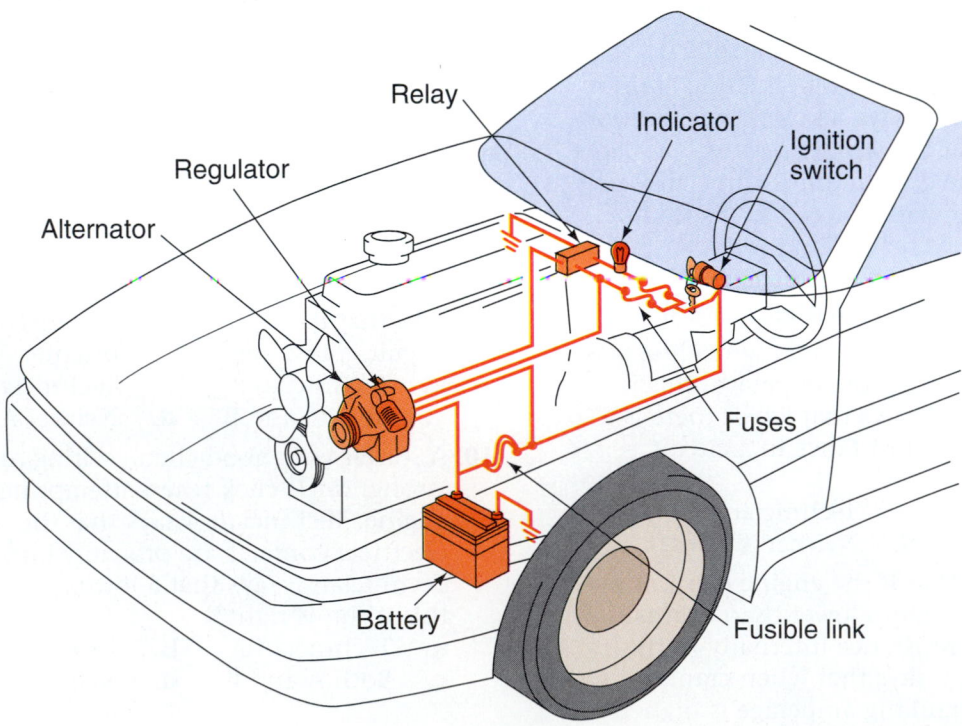

Figure 34.1 Parts of the charging system.

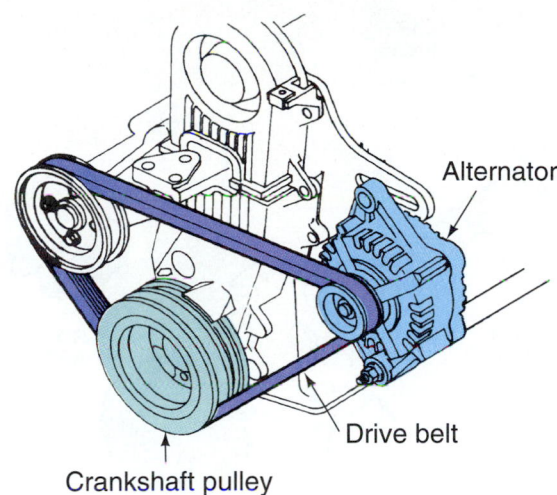

Figure 34.2 The alternator is driven by a belt from the engine crankshaft pulley.

to begin charging, the battery supplies the electrical needs of the vehicle. When the alternator starts to work, it recharges the battery.

DIRECT CURRENT (DC) GENERATORS

Older cars used DC (direct current) generators, driven by a belt on the crankshaft. A **generator** resembles an electric motor, having a stationary magnetic field with an armature output winding spinning inside. Although the generator produces alternating current, its output is direct current because its commutator has brushes on its north and south poles.

■ SCIENCE NOTE ■

Anytime a rotating member carrying current passes a stationary conductor, the current flow in the conductor changes direction as it enters the magnetic field and leaves it. The generator actually produces alternating current, which is rectified by the commutator on the generator armature.

Generators were used on automobiles until the early 1960s when they were replaced by alternators. Some drawbacks to the generator when compared to the alternator are:

■ For a generator to have high output, more current must flow through the brushes. Current flow causes brushes to wear out. If a generator were designed to put out 60–80 amps, it would have to have from six to eight brushes.

■ Generator speed is limited to about 10,000 rpm.

■ Generators do not produce enough output at low speeds (city traffic) to satisfy the demands of electrical accessories.

■ ALTERNATORS

An **alternator** (Figure 34.3) is an *AC (alternating current) generator*. It is lighter, more efficient, and more dependable than a generator. Because of its design, an alternator uses brushes that require only 3-4 amps of current flow. Brushes on alternators commonly last for the life of the original engine.

The alternator works on two basic electrical principles (see Chapter 29). When current passes through a coiled wire, a magnetic field builds up around the wire (see Figure 29.29). This is called an *electromagnet*. When a magnetic field is passed over a wire, voltage is built up. When a complete circuit is added, current flow results in the wire (see Figure 29.33). This is called *electromagnetic induction*.

In an alternator, an electromagnet passes across coils of wire to induce voltage in the coils. It is different from a generator in that the magnetic field spins and the output coils are stationary.

An alternator has two different electrical windings, each having a separate function (Figure 34.4).

■ The *stator* is a stationary conductor.

■ The *rotor* is a rotating electromagnetic field.

As the rotor spins, its magnetic field cuts across the windings of the stator. This causes current to flow in the stator windings.

Rotor Construction

The rotor is the magnetic field that rotates within the stator's wire windings. There is very little clearance (about 0.015") between the two so that the effect of the magnetic field will remain strong (Figure 34.5). The spinning rotor has a magnetic field that cuts across the windings in the stator producing voltage in them.

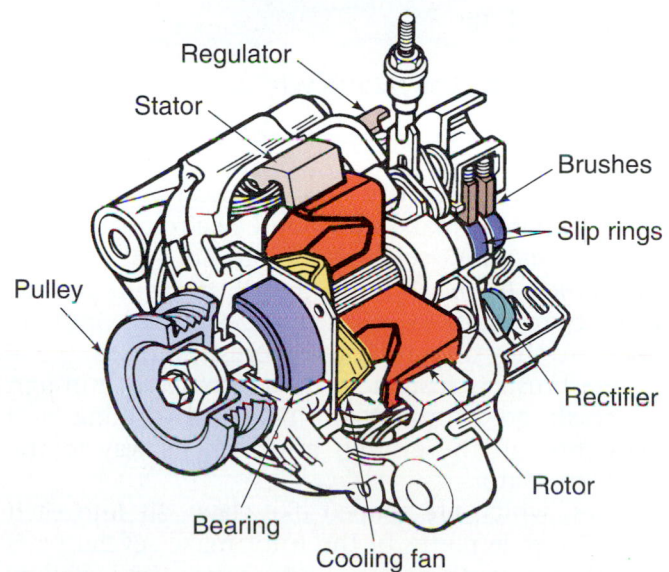

Figure 34.3 Cutaway of an AC generator (alternator).

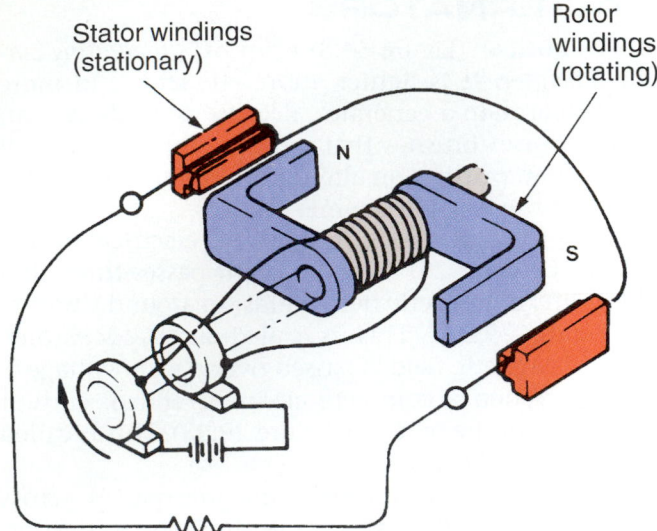

Figure 34.4 The two windings in an alternator.

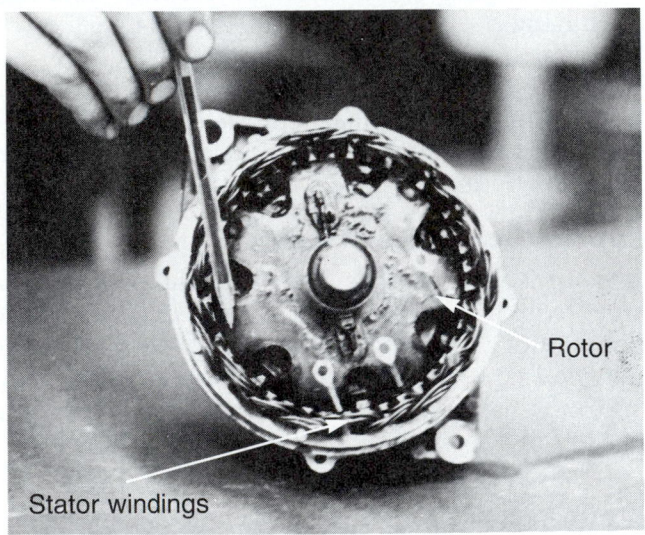

Figure 34.5 The rotor is very close to the stator windings.

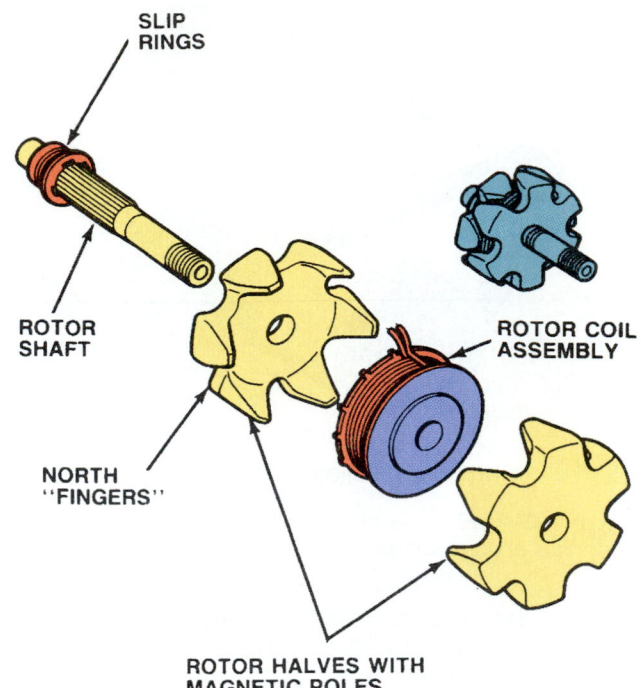

Figure 34.6 Parts of an alternator rotor. *(Courtesy of Ford Motor Company)*

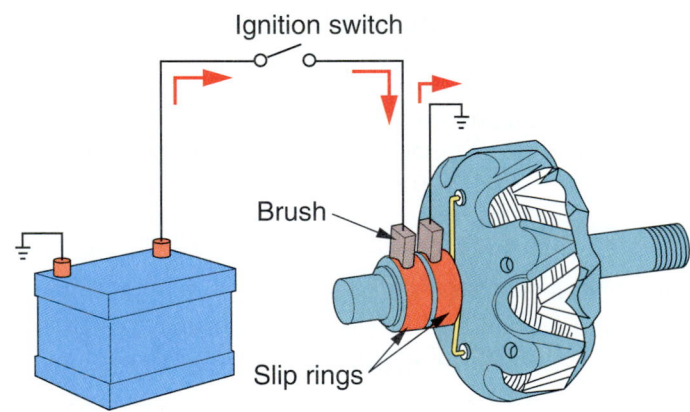

Figure 34.7 Brushes provide electricity to the field coil.

The rotor's field coil has electrical wire wound around a shaft to make a powerful electromagnet (Figure 34.6). *Slip rings* connect to both ends of the field coil. Carbon *brushes* ride on the slip rings to provide electricity to the field coil (Figure 34.7). Slip rings are smooth so they do not wear the brushes.

One brush is insulated. Power comes in through that brush, goes through the field coil winding, and leaves through the ground brush on its way to the voltage regulator.

Poles, which are shaped like claws, fit into each other. The pole pieces of the rotor make several pairs of north and south poles. This increases the magnetic flux and provides smoother electrical output. A magnetic field is formed between the pairs of north and south pole pieces (Figure 34.8). When the rotor spins,

an alternating north/south polarity is created (AC current) (Figure 34.9). This is called **sine wave voltage**. Current flow follows the same path.

An average rotor can spin at about 13,500 rpm. This is not engine rpm, but alternator rpm. The differences in size between the pulleys on the crankshaft and alternator determine the speed at which the alternator spins. The engine usually turns at about one-third the speed of the alternator.

Stator Windings

The stator fits in the frame of the alternator, between the front and rear halves of the cast aluminum housing (Figure 34.10). A stator has three sets of windings

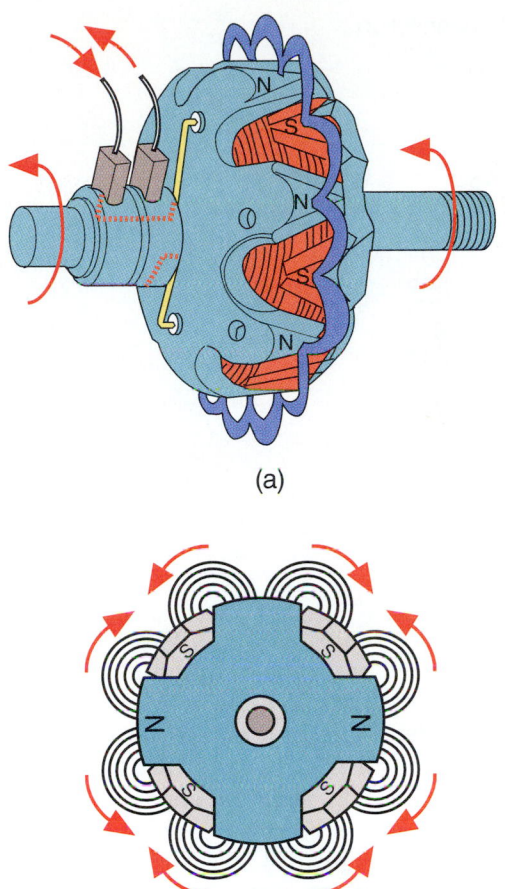

(a)

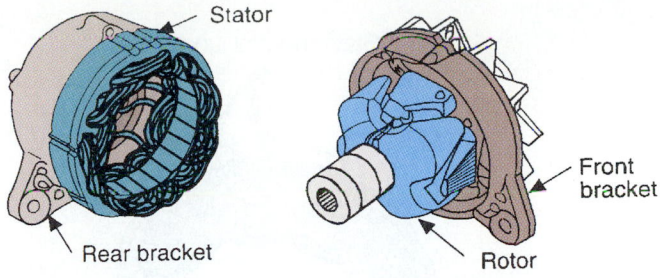

Figure 34.10 The stator fits between the front and rear halves of the alternator. *(Courtesy of Chrysler Corporation)*

Figure 34.8 A magnetic field is formed between the pairs of north and south pole pieces. *(b, Courtesy of Ford Motor Company)*

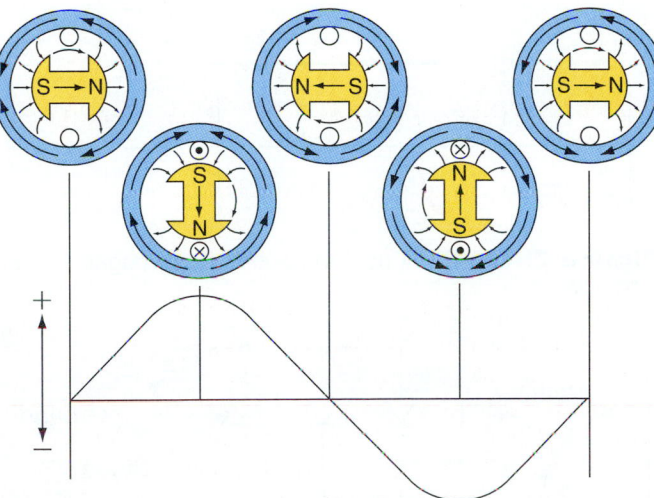

(b)

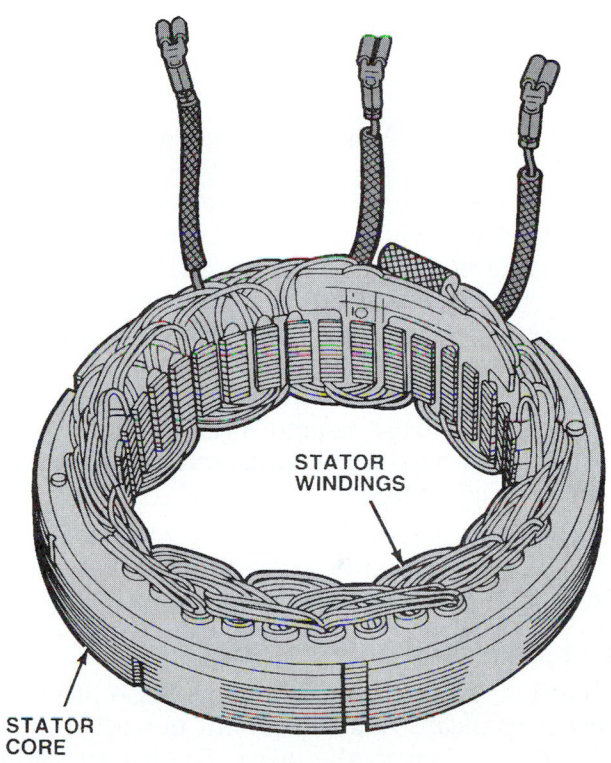

STATOR WINDINGS

STATOR CORE

Figure 34.11 A stator has three windings. *(Courtesy of Ford Motor Company)*

Figure 34.9 Alternating current is produced as the poles rotate.

and south poles. The windings are overlapped and positioned at designed places on the core.

Each winding has two leads, one for current to enter and the other for it to exit. There are two ways of connecting the leads, the wye (Y) and the delta. The wye (Y) is the most common in later vehicles. It provides higher output at lower engine speeds and is more efficient throughout the entire range of speed of alternator operation. Its magnetic field can excite (begin to work) at less than idle speed.

In the wye (Y) winding, one lead from each winding (three all together) is connected to a common connection in the middle. The remaining three leads are branched out in a "Y" pattern (Figure 34.12). Each one connects to a diode (covered later).

wrapped around slots in a laminated round iron frame, called a core (Figure 34.11). Each winding has the same number of coils as the rotor has pairs of north

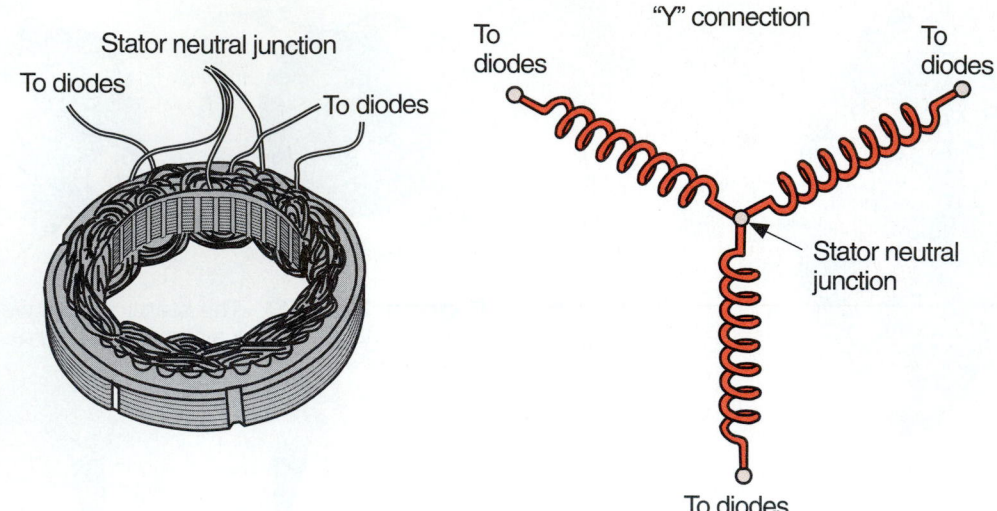

Figure 34.12 A wye stator winding.

When a stator is wound with a delta connection, the leads of the windings are connected together in series (Figure 34.13). The form resembles the Greek letter delta (Δ). A delta winding is used in high output alternators. It can put out about one-third more current at highway speeds than a Y wound stator. Because it is about one-third less efficient at idle speed, there is a problem with delta wound alternators. One complaint is the dash light will not go out until the engine is revved up enough to excite its field.

NOTE: *By cutting and changing the connections from a wye (Y) to a delta, the amp output of the alternator will go from 63 amps to 80 amps.*

Rectifier Construction

The current that an alternator produces is *alternating current.* This means that the current flows first in one direction and then in the other. The battery cannot use alternating current so alternator output must be converted to direct current. A diode rectifier bridge is used for this purpose.

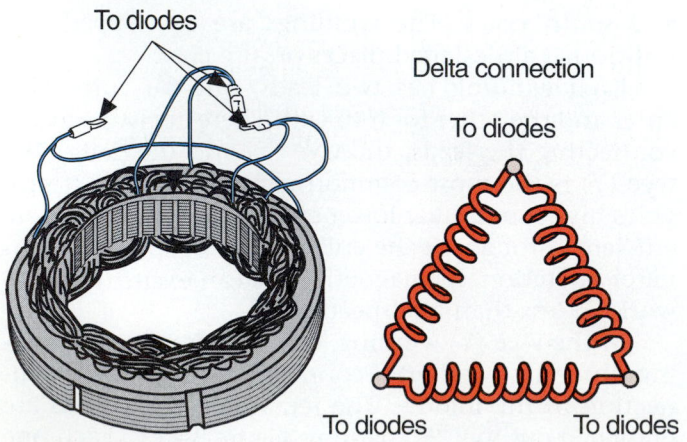

Figure 34.13 A delta stator winding.

A **diode** is a one-way check valve for electricity. There are positive and negative diodes to control the flow in either direction. When the AC current reverses itself, the diode *blocks* and no current flows. Figure 34.14 shows the pulsing DC current that results when a positive diode blocks negative current.

A pair of diodes is used for each stator winding (a total of six in all) (Figure 34.15). Three positive diodes are mounted in a heat sink (Figure 34.16). A **heat sink** helps to dissipate the heat that occurs as elec-

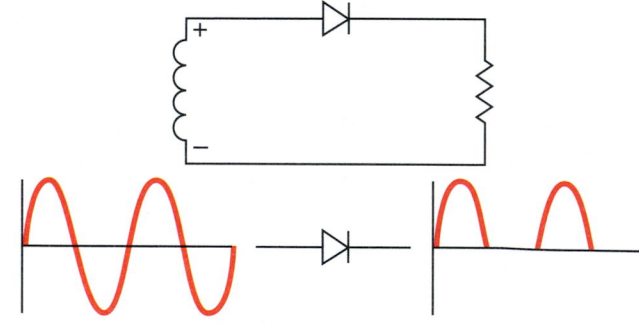

Figure 34.14 A positive diode blocks negative current flow.

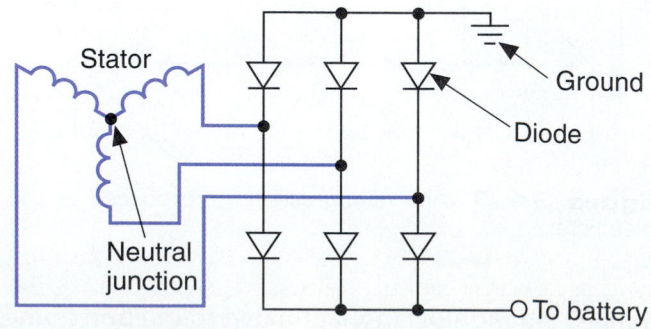

Figure 34.15 A pair of diodes is used for each stator winding.

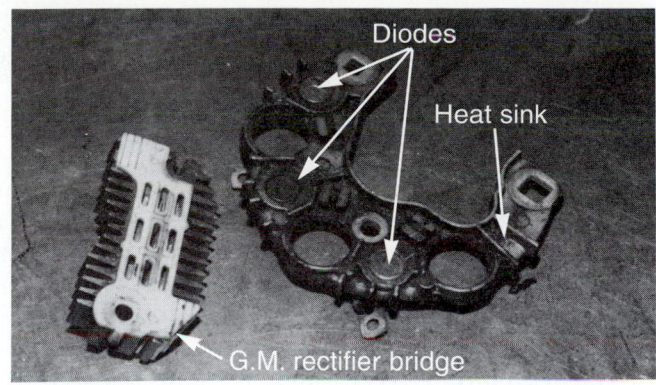

Figure 34.16 Positive diodes are mounted in a heat sink.

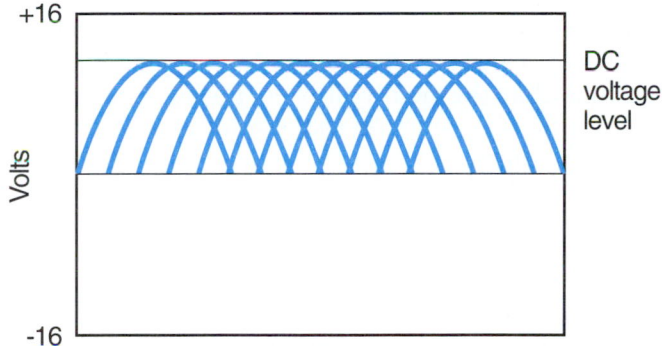

Figure 34.17 Three phase output results in very little alternator output pulsation.

trons try to pass the diode. The three negative diodes are mounted in the alternator frame.

Using a pair of diodes that are reverse biased to each other, both sides of the AC sine wave can be *rectified* to positive direct current. The three windings of the stator produce their currents in phases. The output is called **three phase**. When the three phases of alternating current are rectified, the result is almost uniform DC voltage (Figure 34.17). There is very little pulsation (about 42,000 pulses per minute at 500 engine rpm).

ALTERNATOR BEARINGS

The rotor is supported in the alternator housing using either ball or roller bearings. The bearings are usually sealed and packed with grease. The front bearing fits into an indent in the case and is retained with a small plate and screws. The rear bearing is usually press-fit into the case. On some cars, the rotor shaft slides into the rear bearing.

Alternator Fan

The cooling fan draws air into the alternator through openings at the rear of the alternator. The air leaves the alternator through openings at the front of the alternator behind the cooling fan (Figure 34.18). Some later model alternators also have an internal cooling fan.

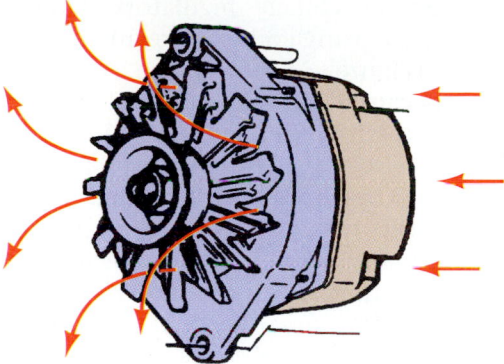

Figure 34.18 The fan draws air from the rear of the alternator for cooling. [*Courtesy of General Motors Corporation, Service Technology Group*]

VOLTAGE REGULATOR

The **voltage regulator** controls the current passing through the windings of the electromagnetic field in the rotor. This determines the amount of current produced in the stator (Figure 34.19). It senses battery voltage and decides how much current to put into the rotor (electromagnet). This determines the amount of alternator output.

When charging system voltage is low and accessories are loading the system, the regulator will increase the alternator current output. If system voltage is normal and there is a heavy accessory load, the alternator will run the accessories with little drain on the battery. When current demands are low and the battery is charged, alternator current output is low. To reduce the amount of output, the regulator puts more resistance between the battery and field coil in the rotor.

There are two main types of voltage regulators: electromechanical and electronic. Older cars had

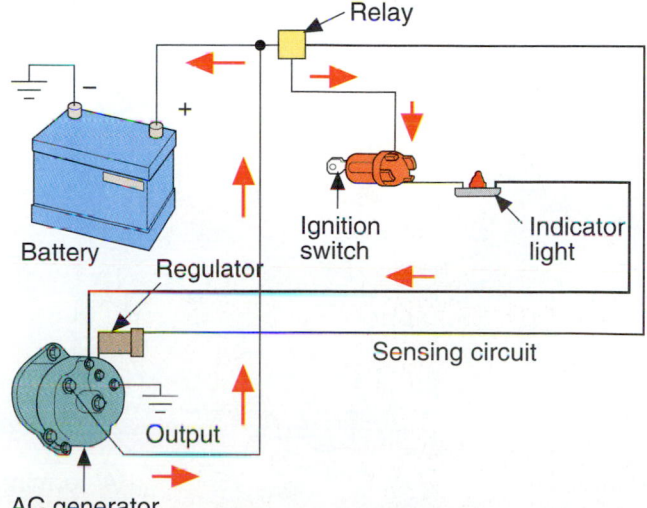

Figure 34.19 The voltage regulator controls the strength of the electromagnetic field in the rotor.

electromechanical voltage regulators, which means that they operate using coils of wire and electrical contact points (Figure 34.20). These regulators were adjustable. Because of their electrical contacts, they would wear out with extended use.

Electronic voltage regulators were phased in during the late 1970s. They have no moving parts or contacts to wear or burn out, so they are very reliable. Electronic regulators are either external or integral. *External regulators* are normally located on the fenderwell in the engine compartment. These regulators are found on cars up until the mid eighties.

Integral regulators are mounted on or in the alternator housing. They work the same way as external regulators, but are smaller with integrated circuits. Some of them are inside the alternator. Others are on the outside for easier replacement (Figure 34.21).

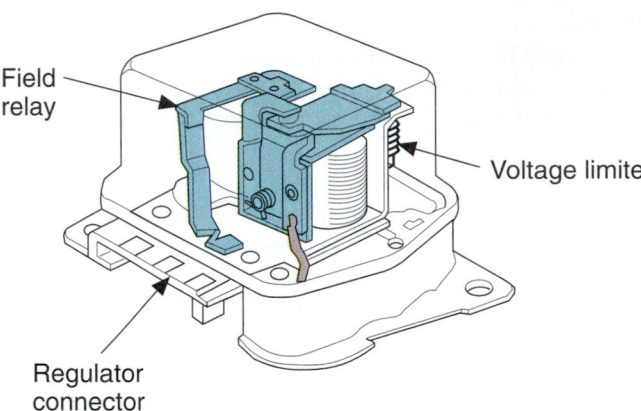

Figure 34.20 An electromechanical voltage regulator. *(Courtesy of Ford Motor Company)*

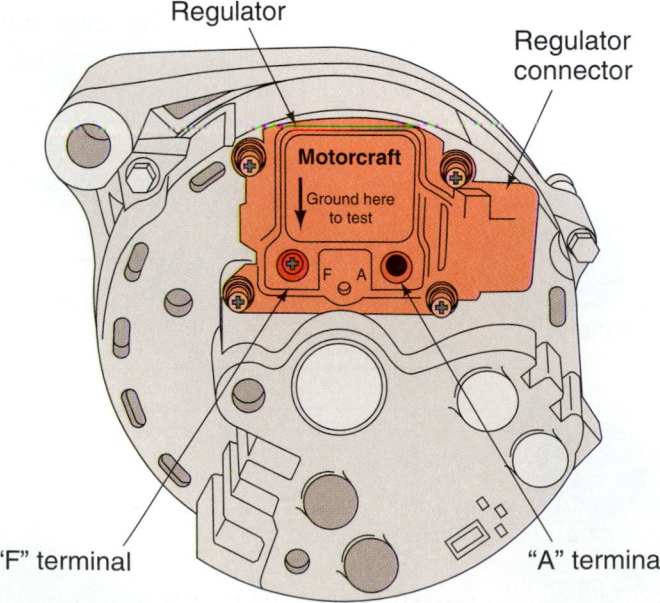

Figure 34.21 An integral regulator. *(Courtesy of Ford Motor Company)*

Voltage regulation on many late model vehicles is done by the on board computer. This has eliminated the need for an integral voltage regulator.

Electronic Regulator Operation

A **zener diode** is an electronic component in the voltage regulator that only conducts electricity when a certain voltage is reached. This capability allows an electronic regulator to keep battery voltage between two specified points. Electronic regulators are not usually adjustable.

The regulator is controlled through the ground side of the field circuit. An electronic regulator has no moving parts so it can cycle electronically between 10 and 7000 times per second. This quickness means that the regulator can control the output of the alternator more accurately. Turning the alternator rapidly on and off is called **pulse width modulation**. This varies the time that the field coil is energized. Imagine that the car has a 100 amp alternator and the system is calling for 50 amps of current. The regulator would energize the field coil for 50% of the time (Figure 34.22). If the demand increased to 75 amps, the field would be energized for 75% of the time.

■ CHARGING SYSTEM INDICATORS

When there is a problem with the charging system, the driver is warned with one of three types of charge indicators: a warning light, a voltmeter, or an ammeter. The indicator is on the car's dashboard. Most cars have an indicator light. A voltmeter or ammeter is an option found on some vehicles.

The *alternator warning light* is wired into the charging circuit and will come on when system output drops below a certain level. The light is wired in *series* with the field current. In some early systems if the light burned out, the charging system would not charge. A resistor is wired parallel to the bulb to provide a path for current flow if the bulb burns out. Figure 34.23 shows a wiring diagram for a typical wye (Y) wound alternator circuit with an indicator light. Testing the bulb is done by turning on the ignition switch without the engine running.

A *voltmeter indicator* shows system voltage when the engine is running. At 80°F, a fully charged battery is about 12.6 volts. The voltage regulator must allow higher voltage to come from the alternator to force electrons into the battery. A typical electronic voltage regulator maintains voltage at between 13.5 and 15.5 volts (check manufacturer's specifications). The voltmeter indicator can give the driver a good idea of how well the charging system is working.

An advantage to a voltmeter is that it is a parallel circuit (Figure 34.24). No current runs through the wire or the indicator so there is no danger of a fire if

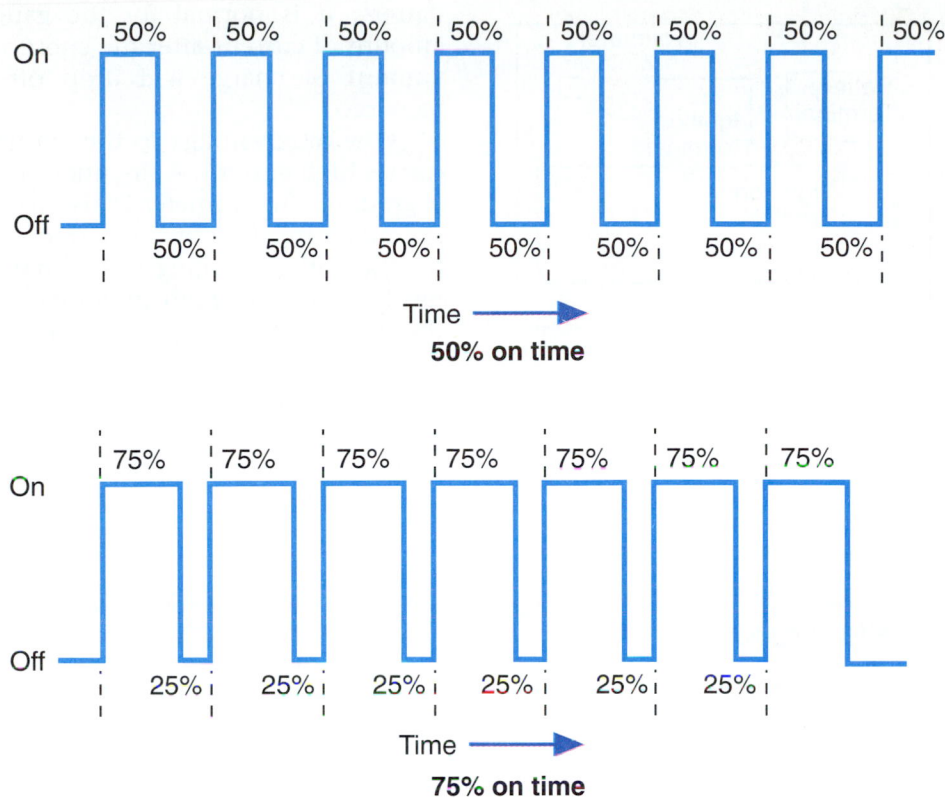

Figure 34.22 Pulse width modulation.

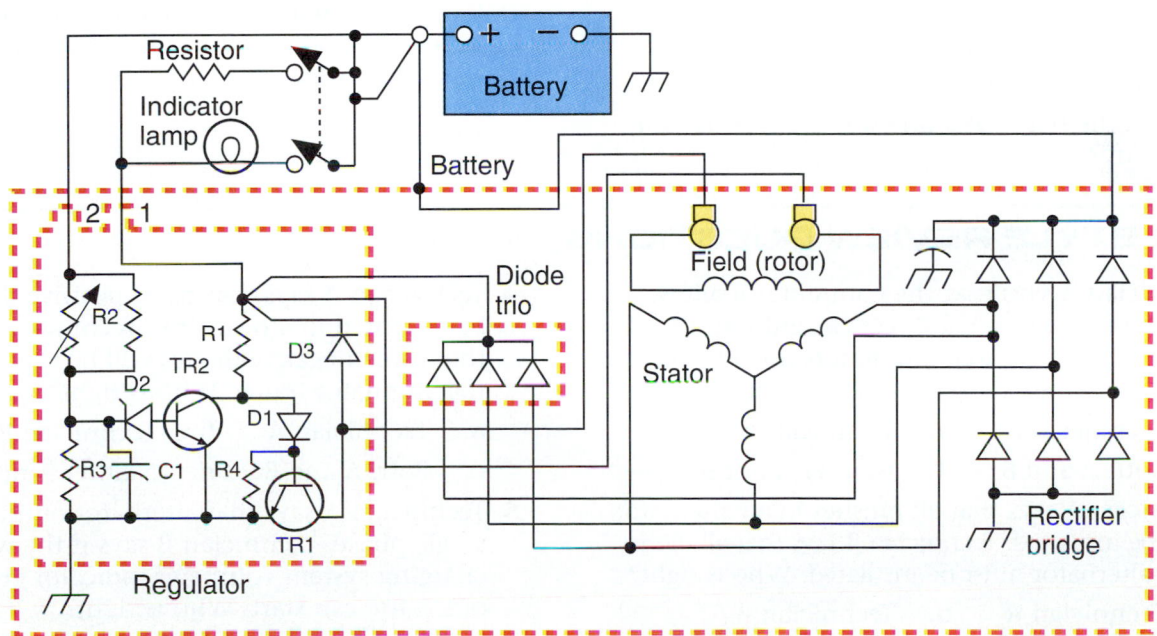

Figure 34.23 A wiring diagram for an alternator circuit with an indicator light.

the wire becomes pinched or grounded. A problem with a voltmeter is that it shows what the charging system is doing at any one moment. An uninformed driver might visit the dealership to question a nonexistent problem.

An *ammeter indicator* gives the amount of current flowing to or from the battery. The needle in the center represents zero. Usually, if the needle moves to the left side, the alternator is discharging the battery. When it moves to the right, the system is charging the

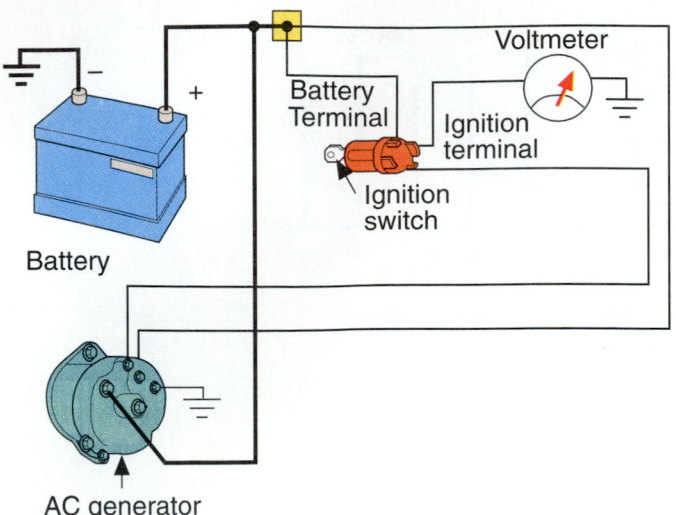

Figure 34.24   A voltmeter is wired in parallel.

battery. It is normal for the gauge to read a high amount of current after the engine is first started. The amount of charge will drop off as the battery is recharged.

One disadvantage to the ammeter is that its wire carries high current. A fire under the dash can result if it grounds. An ammeter is also too expensive because the meter must be of the correct size to carry current for the entire charging load. Charging system output has become very high in comparison to cars built in the sixties (when they had ammeters).

REVIEW QUESTIONS

1. What kind of electrical output results from the generator commutator and brushes, AC or DC?

2. What is another name for an AC generator?

3. How many amps of current flow through alternator brushes?

4. What is the name of the stationary conductor in an alternator?

5. What is the name of the rotating electromagnet in an alternator?

6. What is the name of a one-way check valve for electricity?

7. What is it called when the AC sine wave is changed to all positive pulses of current?

8. Which way does air move through an alternator, front to back or back to front?

9. What is the name of the device that controls the amount of electricity entering the rotor?

10. What is the name of the electronic component in the voltage regulator that only conducts electricity when a certain voltage is reached?

ASE STYLE REVIEW QUESTIONS

1. Technician A says that the commutator allows a generator to produce a direct current output. Technician B says that alternators use a rectifier to put out direct current. Who is right?
   a. Technician A   b. Technician B
   c. Both A and B   d. Neither A nor B

2. Technician A says that all brushes in an alternator must be insulated. Technician B says that all diodes in an alternator must be insulated. Who is right?
   a. Technician A   b. Technician B
   c. Both A and B   d. Neither A nor B

3. Technician A says that an alternator spins at engine rpm. Technician B says that the stator winding that has three wires connected to each other at a common place is called a delta winding. Who is right?
   a. Technician A   b. Technician B
   c. Both A and B   d. Neither A nor B

4. Technician A says that three positive diodes will be mounted in a heat sink. Technician B says that three negative diodes will be pressed into the alternator frame. Who is right?
   a. Technician A   b. Technician B
   c. Both A and B   d. Neither A nor B

5. Technician A says that alternator output is single phase. Technician B says if the wire to a charging system voltmeter indicator grounds out, a fire can start. Who is right?
   a. Technician A   b. Technician B
   c. Both A and B   d. Neither A nor B

CHAPTER 35

Charging System Service

■ OBJECTIVES

Upon completion of this chapter, you should be able to:

✔ Measure voltage drops on both the positive and ground sides of the charging circuit.

✔ Test and repair an alternator.

✔ Diagnose voltage regulator problems.

■ KEY TERMS

**charging system
  output test
regulator maximum
  voltage test
full-field test
circuit resistance tests
A-circuit
B-circuit
isolated field**

■ INTRODUCTION

This chapter deals with the process of testing and repairing common problems with the charging system. Principles of operation and electrical fundamentals learned in earlier chapters will be important here. Diagnosis of failures is very important before parts are replaced. Most parts stores will not accept returns of electrical items.

■ CHARGING SYSTEM SERVICE

Charging system problems usually become evident when the battery goes dead or a customer notices a noise.

Four common charging system complaints are:

■ Dead battery
■ Battery water low (overcharging)
■ Indicator light glows or incorrect voltage is indicated
■ Noises

Test Battery First

Just as with the starting system, the battery must be in good condition if its charge is to be maintained by the charging system. A complete charging system diagnosis must be performed before replacing any charging system parts. If you replace a battery just because it is old, that will not solve a charging problem caused by a loose fan belt. The car will leave the shop with a new, freshly charged battery. The customer will be upset when the battery goes dead once again and more work is required on the vehicle. Odds are that the customer will go to another shop for the work. Always test and retest your work.

Battery Condition.   First, check the battery's state of charge. A battery will not be fully charged unless the charging system is putting some current into it. If the battery state of charge is low, this could be due to a parasitic drain when the engine is off.

If the battery is at least 80% charged, perform a 15-second load test to find out if it has the capacity to stay above 9.6 volts.

If a battery fails the load test, perform the 3-minute charge test to see if the battery is sulfated. This can occur because the battery has been allowed to remain in a discharged state. Simply replacing this battery will not solve the problem. Testing the charging system is always a part of battery replacement.

■ TESTING THE CHARGING SYSTEM

First, perform a visual inspection looking for obvious problems. If the battery tests out bad, substitute a known good battery before testing the charging system.

■ Check for corroded or broken wire connections.
■ Wiggle wires while the engine runs. Have an assistant sit inside the car to see if an indicator light goes out or a gauge begins to read correctly.
■ Listen for noises as the engine runs. Be careful not to use a stethoscope close to alternator wiring.
■ Look for a loose or damaged alternator drive belt.

SHOP TIP   With the engine off, try to turn the alternator fan and pulley. If the pulley slips on the belt, it is too loose or is glazed and must be replaced.

A loose drive belt can cause two problems. The obvious problem is a low charge rate. When the regulator energizes the field in the alternator rotor, a load is put on the drive belt. If it is loose, it will slip and the charging system will not work to capacity. Today's synthetic belt materials do not squeak as badly when they slip. Older belts contained cotton fibers, which squeaked badly during slipping. Another problem caused by a slipping belt is that it can heat up the rotor shaft and cause failure of the drive end bearing.

> **CAUTION** Use a belt tension gauge to check belt tension. If a belt is too tight, wear to bearings for the water pump, alternator, and engine crankshaft can result.

> **SHOP TIP** There is a simple test to see if an alternator field is strong, done with the key on and engine off. Check the rotor shaft at the outside of the housing of the rear bearing for magnetism. If there is enough magnetism to lightly hold a screwdriver, the rotor is energized. If the rear bearing is not magnetized, the problem could be related to the regulator, the brushes, or a defective rotor.

Check Alternator Rating

Check to see what the alternator's maximum rated output is. This can be found in the specification manual. Some manufacturers identify their alternators by stamping the output on the alternator frame or by color coding. The more accessories the vehicle has, the larger the alternator will be. Figure 35.1 shows typical amp draw of accessories that the charging system might have to provide for.

Connect the volt-amp tester to the battery (see Figure 33.4). The inductive pickup should be as close to the battery as possible to avoid picking up a stray magnetic signal from the alternator or another electrical device under the hood. Set the meter range to the highest scales during starting and then adjust them to lower ranges as needed.

There are four charging system tests to perform.

- Charging system output test
- Regulator voltage settings
- Alternator full-field test (when available)
- Charging circuit resistance tests

The **charging system output test** loads the charging system and measures its current output.

The **regulator maximum voltage test** checks to see that the voltage regulator is energizing the rotor field and is not allowing the alternator to overcharge.

The **full-field test** of the alternator takes the regulator out of the circuit and causes the alternator to give full output.

Circuit resistance tests check the resistance in the ground and insulated circuits by measuring voltage drop.

NOTES:

- *Alternator ground wires are usually braided ground straps. The condition of these wires is as important as the insulated wires because they carry the same current flow.*
- *Some alternators are insulated in rubber bushings. These must have a separate ground strap.*

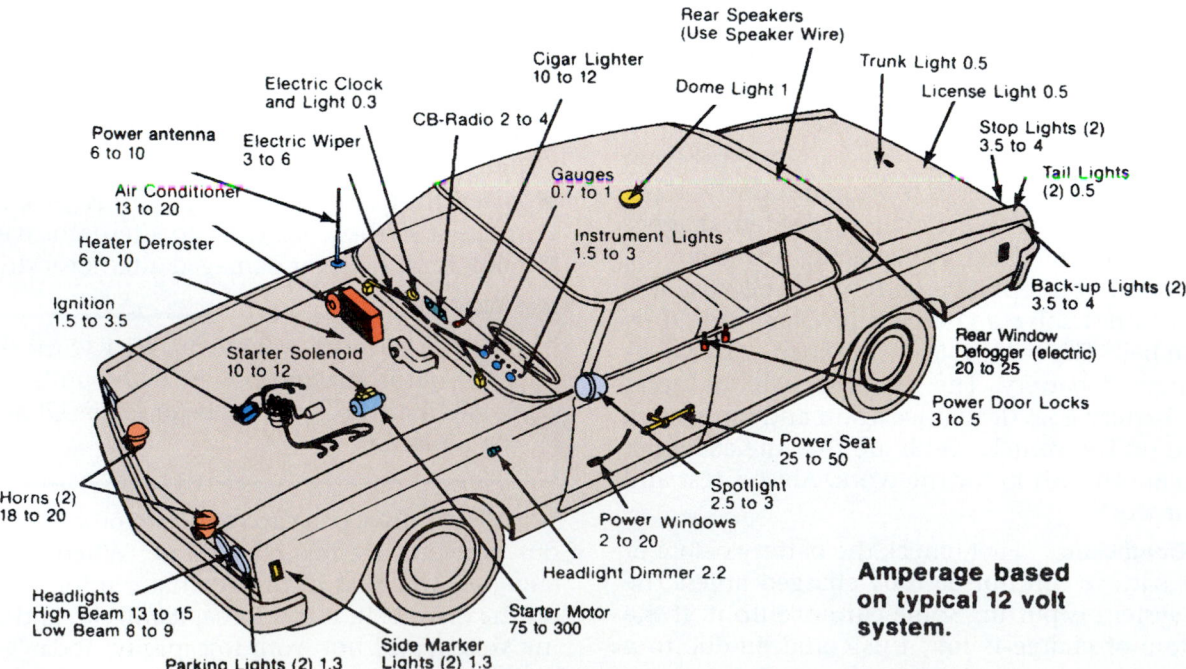

Figure 35.1 Typical amp draw for accessories. *(Courtesy of Cooper Automotive/NAPA Belden)*

CHARGING SYSTEM OUTPUT TEST

The following outlines the test of charging system output with a volt-amp tester.

- Start the engine. The speed of the alternator determines its output. At idle, the alternator is not capable of making maximum output.
- Raise the idle speed to about 2000 rpm as you turn the knob on the carbon pile to lower battery voltage. As the voltage is lowered, the voltage regulator will energize the alternator's rotor.
- Read the amount of current being put into the battery by the alternator.
- The maximum output of the alternator will be more than the amount that shows on the gauge because the vehicle was consuming some current in running. This amperage amount was never delivered to the battery where it could be read on the ammeter.
- With the engine off, turn the key to "on" and read the amount of amp draw. Add this amount to the ammeter reading to get the actual alternator output.

NOTE: *A hot alternator loses 25% of its rated output. Alternators are rated at 160° so they could have higher than the rated output when cold. This is because the electrical resistance of most conductors goes up with heat.*

REGULATOR VOLTAGE SETTING CHECK

The regulator voltage test checks to see the high and low settings of the regulator. A regulator must be able to full field an alternator immediately. It must also be able keep the system at a predetermined voltage at normal operating temperature (usually about 13.5-14.5 volts). With lower temperatures, the regulator will raise the voltage.

With the engine running, the meter should show a gradual decrease in the charging current. The voltage should remain within normal operating limits. If the regulator is below the manufacturer's specifications, the regulator must be replaced. The battery will never be fully charged and it will become sulfated. Older electromechanical regulators were adjustable. If the voltage is higher than specified, this might not be due to the alternator.

NOTE: *A voltage drop (resistance) on the ground side of the regulator will raise system voltage by the amount of the voltage drop.*

Test the regulator ground with a voltmeter (Figure 35.2). If the ground side has no resistance, check the sensing voltage input to the regulator to see that it is the same as the battery voltage. If it is, the regulator should be able to keep voltage within specifications. It must be replaced if it does not.

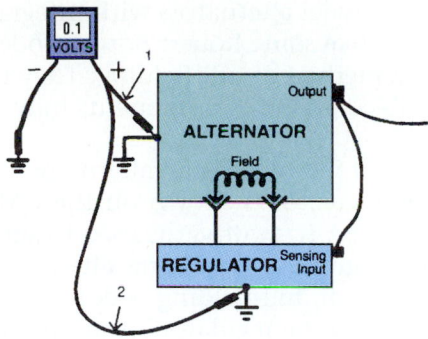

Figure 35.2 Checking the ground on a voltage regulator. *(Courtesy of Interstate Batteries)*

ALTERNATOR FULL-FIELD TEST

When the amount of output of an alternator is too low, this could be due to the alternator or the regulator. Full-fielding the alternator is a test that eliminates the regulator from the circuit and energizes the rotor fully. This will cause the alternator to produce full output.

- If the output is too low with the alternator full fielded, the alternator must be the source of the problem.
- If the alternator puts out its rated amperage, the problem is with the regulator.

There are several procedures for full fielding alternators, depending on the design. Be sure to follow manufacturers' directions.

Regulators are connected to the alternator in one of two ways, called A and B circuits. The **A-circuit** has the regulator in the ground circuit. This type is used most often with solid state (electronic) systems. A **B-circuit** regulator has its regulator between the positive feed (insulated side) and the field coil (which is grounded inside of the alternator). This type is usually found on the older electromechanical voltage regulators.

Directions for full-fielding alternators vary according to the type of circuit. They can be found in service manuals and information sheets provided by electrical test equipment manufacturers.

- To full-field an A-circuit, supply a ground to the field terminal.
- To full-field a B-circuit, supply battery voltage (B+) to the field.

NOTE: *If you are not sure whether you are working with an A type or a B type, try both power and ground to the field terminal. You will not do any damage. The alternator will just not put out current if hooked up incorrectly.*

- Some regulators are of the **isolated field** type. They have two field leads and must have one lead jumped to battery positive (B+) and the other to ground.

■ Some late model alternators with integral regulators (like some Robert Bosch models) do not provide a method of full-fielding. They must be disassembled and the components individually tested.

One way of full-fielding an alternator with an integral voltage regulator is shown on the GM alternator in Figure 35.3. A screwdriver is used to short a tab on the frame to ground. On some alternators with an external regulator, full-fielding is done by unplugging the connector to the regulator. A jumper wire is used to bypass the regulator, connecting the field directly to the battery. Figure 35.4 shows the recommended connections to jump for some Ford and GM external alternators.

Some computers have taken over the regulator function by controlling the alternator duty cycle. These are tested in the same manner as other alternators, usually by grounding the field terminal. Follow the manufacturer's recommendations. If the regulator has failed, these vehicles will require a new computer.

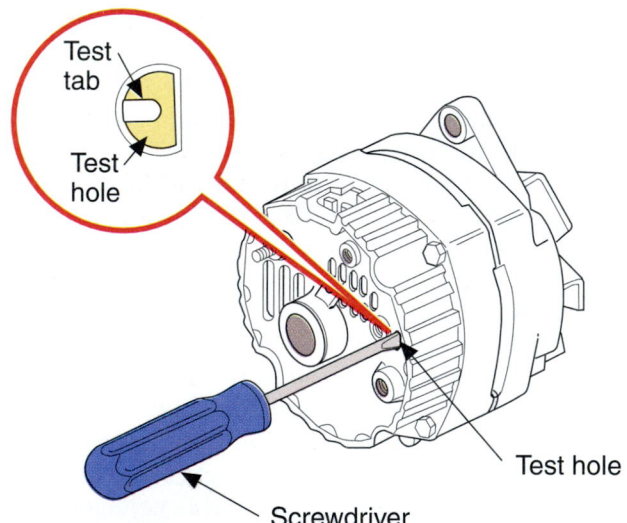

Figure 35.3 Full fielding a GM alternator.

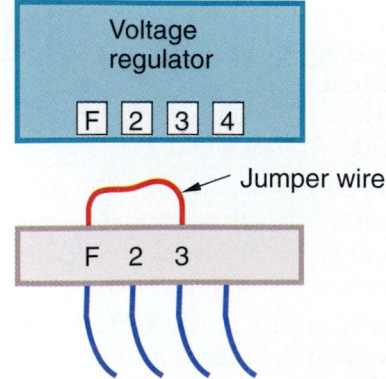

Figure 35.4 A way of full fielding alternators.

Alternator Testing Precautions

■ Connecting the wrong wire to a ground or hot lead can ruin diodes or a regulator.
■ Full-fielding should not be done for more than 10 seconds.
■ Do not connect or disconnect electrical components with the key on. They could be damaged.
■ Do not connect the regulator with the engine on.
■ Do not operate the alternator without an external load connected to it. This can cause extremely high voltage and ruin the alternator.
■ Do not run the engine with the battery disconnected. The battery helps stabilize voltage and acts like a shock absorber for damaging voltage spikes.

CAUTION Be careful not to let the voltage in the system rise above regular regulated voltage. This can be controlled with the carbon pile. If you let the voltage continue to climb without the voltage regulator in control, an electronic component might be accidentally damaged. A safe maximum is about 16 volts.

Full-Field Test Results

When you bypass the regulator, the voltage should be controlled to the same amount it was at during maximum output with the regulator in control. These figures will be compared. If the alternator puts out more at the same voltage with the regulator bypassed, the regulator is faulty. When the current output remains below specification, a bad alternator or bad drive belt is the probable cause. Before condemning the regulator, full-field the alternator at the field terminal of the alternator. This will eliminate excessive resistance in the wiring as a possible cause.

Differences between A and B Circuit Faults. If an A-circuit regulator loses battery + voltage, the alternator will overcharge if the field coils still have power. If it loses its ground, the alternator will stop working.

If a B-circuit regulator loses ground, it will overcharge. If it loses its battery + voltage, the alternator will not work.

■ CHARGING SYSTEM VOLTAGE DROPS

Checking the resistances in the charging circuit is done when locating a stubborn problem that does not appear to be related to either the alternator or regulator. Resistance due to corrosion usually takes a long time to get bad enough to result in a failure. Checking voltage drops as a preventive maintenance measure will catch potential problems before they fully develop. The tests tell if the battery, regulator, and alternator are each working at the same potential.

NOTE: Extra resistance in the regulator circuit will cause the battery to have a higher than normal charging voltage.

Perform a voltage drop test in the same way you did on the starter circuit. Current must be running through the system to be able to perform the test. Turn on the headlights or use the carbon pile on the volt-amp tester to load the system. The combined voltage drop for an entire circuit should not be more than 3% of the system voltage.
- If the system has an indicator lamp, when checking the positive circuit with 10 amps flowing, no connection should have more than 0.3 volt drop.
- With an ammeter, 0.7 volt drop is the limit.
- A small amount of voltage drop past the fuse link (which is a smaller wire than the rest of the circuit) is normal. The amount of voltage drop is very small because the fuse link is so short.

There are no fuse links in the ground circuit, so lower voltage drop (0.1 volt) is desired.

■ ALTERNATOR SERVICE AND REPAIR

Remove the Alternator

When the alternator has been determined to be the cause of a charging system problem, remove it from the vehicle.
- Always disconnect the battery ground cable first.
- Label the connections to the alternator before disconnecting them.

Loosen the drive belt and remove it. Then, unbolt the alternator. Be careful that any spacers that are used in mounting are kept with their original bolts. This will make correct reinstallation much easier.

A bad alternator could be the result of several problems. A shop that specializes in alternator rebuilding will have a test bench for alternators. This tester provides for fast, easy mounting of the alternator. The alternator can be tested for rectifier and stator grounds and for opens or shorts in the field winding.

■ ALTERNATOR DISASSEMBLY

The alternator is easy to disassemble. Although alternators are commonly replaced with a professionally rebuilt unit, the procedure for disassembly is covered here so that you can become familiar with the parts. Some shops will also replace brushes or diodes. Procedures for the electrical testing of rotors and stators are described in applicable shop manuals.

As you did when disassembling a starter, mark the side of the case for easy realignment of parts on reassembly. After removing the bolts, the rotor is separated from its housing.

Brush Service

If only the brushes are to be serviced, the rest of the alternator will not have to be disassembled. Some alternators have brushes that can be replaced from the outside, without splitting the alternator housing. Brushes are spring loaded. When they wear, spring tension becomes less. Brushes that are too short can cause an alternator to stop working intermittently (once in a while) before failing completely.

Stator Service

After removing the stator leads from the rectifier bridge (the part that holds the diodes), the stator can be removed from the alternator housing. The stator is discarded if it has any signs of burned or overheated insulation.

Rotor Service

To remove the rotor from the front frame requires removal of the pulley and fan. Use a small impact wrench to loosen the nut so you do not have to restrain the rotor from turning. Do not put the rotor in a vise to hold it. It can be easily damaged. Some rotor shafts have a means of holding the rotor from turning (Figure 35.5). This is important for reinstallation torque.

Test the continuity of the rotor winding to see that it does not have any opens or grounds (Figure 35.6). Resistance is checked with an ohmmeter. Manufacturers' specifications are different and range from 2.4 to 6.0 ohms.

Any of the following are cause for replacement of the rotor:
- A reading below specs points to a shorted rotor.
- High resistance mean connections are corroded.
- Infinite resistance means the rotor is open.

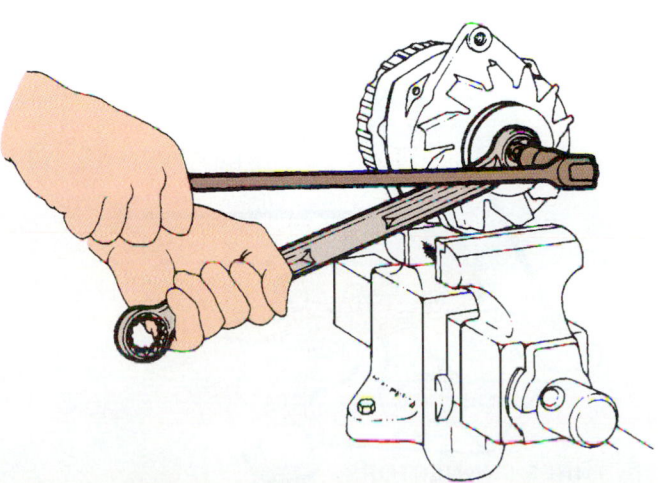

Figure 35.5 Holding the rotor shaft while tightening the pulley nut. *(Courtesy of Chrysler Corporation)*

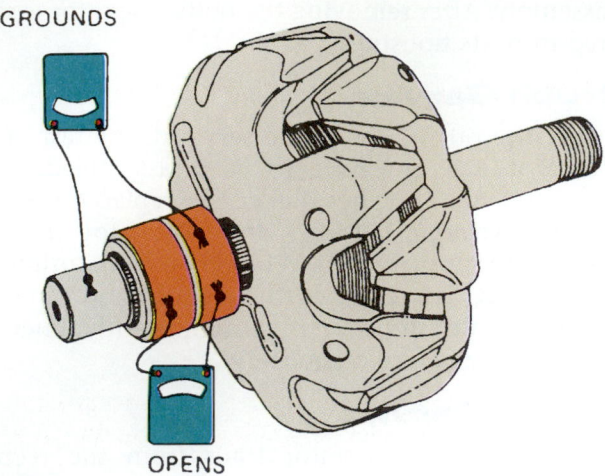

Figure 35.6 Testing the rotor for opens and grounds. *(Courtesy of General Motors Corporation, Service Technology Group)*

Diode Service

Bad diodes are a common cause of alternator charging problems. A diode must be tested under a load. There are several ways of testing diodes. A diode tester is a common way. Using an ohmmeter is another way. A diode should allow current flow in one direction only. Figure 35.7 shows a diode trio. Connect the test leads of the ohmmeter to the one of the diode trio leads and the other to the case.

■ A good diode will show high resistance in one direction and low resistance in the opposite direction.
■ A shorted diode will have low readings in both directions.

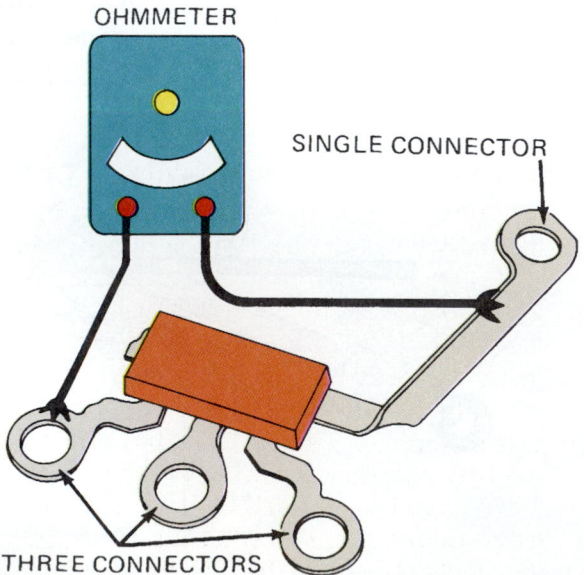

Figure 35.7 Testing a diode trio. *(Courtesy of General Motors Corporation, Service Technology Group)*

■ An open diode will show high resistance in both directions.

Individual diodes are tested in the same manner. Sometimes, diodes are soldered together and must be unsoldered to test them. They will have to be resoldered after the test. Be sure to apply the solder as soon as the wire is hot enough. Overheating a diode can ruin it. Pliers can be clamped onto the diode lead and used as a heat sink during soldering to avoid overheating the diode.

■ ALTERNATOR REASSEMBLY

Reassembling the alternator is usually not difficult. When the brushes are inside of the alternator, they must be held up against their springs so the slip rings of the rotor can be positioned under them. There is usually a hole that a pin, paper clip, or drill bit can be installed in while doing this (Figure 35.8). The drill bit is removed from a hole in the back of the alternator after the rotor is installed in the frame. Brushes that mount from the outside are simply compressed against their springs as they are installed.

Whenever an electronic part has failed, it is a good idea to check the alternator diode pattern on the oscilloscope. DC electronic components cannot be exposed to alternating current. If you simply replace an electronic component and do not fix the cause of the problem, the new part might fail.

■ DIODE NOISES

When there is excessive alternating current coming from the alternator, a car radio will sometimes make a whining sound like a siren. The whine changes with engine rpm. When this sound is loud enough for a customer complaint, a bad diode is probably the cause. Newer car radios are less likely to be susceptible to noise.

A bad diode can also cause the alternator to growl or whine. The pitch of the noise changes with engine rpm.

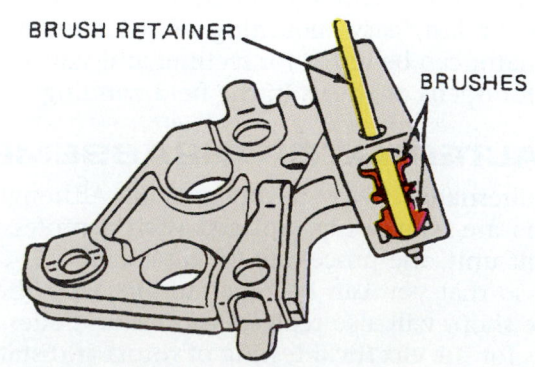

Figure 35.8 Holding the brushes against their springs for installation on the slip rings. *(Courtesy of General Motors Corporation, Service Technology Group)*

■ REVIEW QUESTIONS

1. What is the name of the test that takes the regulator out of the charging circuit and causes full alternator output?

2. What test with a voltmeter measures the resistance while current flows in the alternator circuits?

3. As the voltage is lowered during the charging system output test, what part of the alternator does the regulator give current to?

4. What happens to electrical resistance of most conductors when heated?

5. What is the normally expected voltage range of a regulator?

6. What is the maximum time that full-fielding should be done for?

7. Two components that should not be disconnected with the engine running are the alternator and the _____.

8. What is a safe voltage maximum to keep under when full fielding an alternator?

9. Performing _____ _____ tests to a charging system on a preventive maintenance basis will catch resistance problems before they become too serious.

10. When an electronic component has failed, what type of part failure should you look for on an oscilloscope pattern?

■ ASE STYLE REVIEW QUESTIONS

1. Technician A says that the inductive pickup on a volt-amp tester should be kept as far from the battery as possible. Technician B says to set the VAT meters to the lowest scales and then adjust to higher scales as needed. Who is right?
   a. Technician A   b. Technician B
   c. Both A and B   d. Neither A nor B

2. Technician A says that the alternator must be capable of producing maximum output at idle. Technician B says that an alternator will produce more output at higher temperatures. Who is right?
   a. Technician A   b. Technician B
   c. Both A and B   d. Neither A nor B

3. Technician A says if the voltage setting is too low on an electronic voltage regulator, replace the regulator. Technician B says that if the voltage is higher than specified, this might be because the regulator has a bad ground. Who is right?
   a. Technician A   b. Technician B
   c. Both A and B   d. Neither A nor B

4. Technician A says that an A-circuit has the regulator in the ground circuit. Technician B says that a B-circuit has the regulator in the insulated circuit. Who is right?
   a. Technician A   b. Technician B
   c. Both A and B   d. Neither A nor B

5. Technician A says that to full field an A-circuit, supply a ground to the field terminal. Technician B says that to full-field a B-type, supply battery voltage to the field. Who is right?
   a. Technician A   b. Technician B
   c. Both A and B   d. Neither A nor B

6. Technician A says that applying the wrong polarity (+ or -) to the *field* terminal can ruin diodes. Technician B says that connecting the wrong wire to a ground or hot lead can ruin diodes or a regulator. Who is right?
   a. Technician A   b. Technician B
   c. Both A and B   d. Neither A nor B

7. Technician A says if the alternator puts out more current at the same voltage with the regulator bypassed than it did during a VAT test, the regulator is faulty. Technician B says that extra resistance in the regulator circuit will cause the battery to have a higher than normal charging voltage. Who is right?
   a. Technician A   b. Technician B
   c. Both A and B   d. Neither A nor B

8. During a full-field test, the current output remains below the specification. Technician A says that a bad alternator could be the cause. Technician B says that a bad drive belt could be the cause. Who is right?
   a. Technician A   b. Technician B
   c. Both A and B   d. Neither A nor B

9. Technician A says that a charging system with an ammeter will have more voltage drop than one with only an indicator light. Technician B says that the fuse link in the ground circuit causes higher resistance in the circuit. Who is right?
   a. Technician A   b. Technician B
   c. Both A and B   d. Neither A nor B

10. Technician A says that a good diode will allow electricity to flow freely in both directions. Technician B says that a diode can be ruined when soldering it to a wire. Who is right?
   a. Technician A   b. Technician B
   c. Both A and B   d. Neither A nor B

Lighting and Wiring Fundamentals

■ **OBJECTIVES**

Upon completion of this chapter, you should be able to:

✔ Describe differences between wire and cable

✔ Explain the fundamentals of operation of automotive lighting and wiring.

✔ Relate when different circuit protection devices would be used.

■ **INTRODUCTION**

Lights, and the wiring to power them, make up a sometimes complicated system. The lights and wiring on older cars was simple, consisting of dash lights, the dome light, headlights, tail lights, and the license plate light. Today, there are many lights for all kinds of conveniences. This chapter deals with lighting, wiring, and circuit protection devices.

Electricity, for lights and other systems and accessories, is provided through wiring or cables. Circuit protection devices include fuses, circuit breakers, and **fusible links.** They prevent fires and damage in the event a circuit becomes shorted or grounded.

■ **WIRE AND CABLE**

Chemicals, corrosion, vibration, and heat can damage wiring. Also, aftermarket electrical accessories may be added to cars and sometimes this work is done carelessly. A service technician should know the basics of wiring to be able to perform professional service and minor repairs.

Low voltage wire and cable are made up of one or more copper conductors encased in an insulator. In most cases, this insulator is plastic and is color coded so it can be traced and/or located using a wiring diagram.

Primary and Secondary Wiring

Low voltage wiring on a car is called **primary wiring.** Ignition wiring (high voltage) is called **secondary wiring.** The terms primary and secondary wiring are used in describing transformer wiring. The ignition coil is an example of a transformer. The low voltage wiring supplying current is termed primary wiring and the output wiring is called secondary. Secondary wiring differs from primary in that it has very thick insulation so it can carry voltages of up to 100,000 volts.

Cables are large wires that allow more electrical current to flow. The size of the wire or cable used is described using either the **American Wire Gauge (AWG)** numbers (larger wire equals smaller number), or metric size designations. A chart of wire gauge sizes is included in Appendix E at the back of the book.

■ **CIRCUIT PROTECTION DEVICES**

Fuses

A fuse is a protective device designed to melt when the flow of current becomes too high for the wires or loads in the circuit (Figure 36.1). Without the fuse, the circuit would be damaged. Each fuse has an amp rating at which it is designed to fail. Each vehicle's electrical system has a fuse panel (Figure 36.2).

There are three types of replaceable fuses: the *cartridge*-type (*glass tube*-type), the *ceramic* type, and the *blade*-type (Figure 36.3).

Glass Tube Fuses. Glass tube fuses have a small wire sacrificial strip inside of a glass tube. They are usually ¼" in diameter and come in lengths increasing in ⅛" increments from ⅝" to 1⅞". There are three

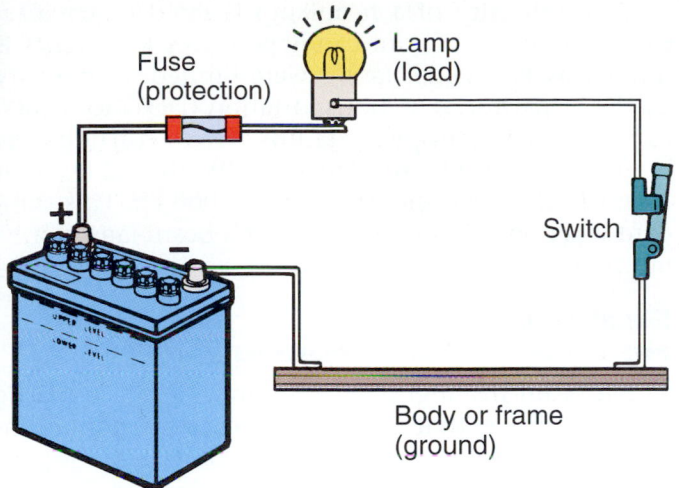

Figure 36.1 A fused circuit.

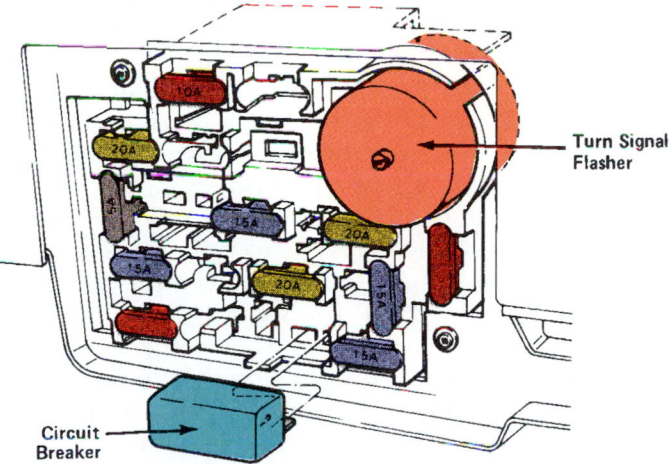

Figure 36.2 A fuse panel.

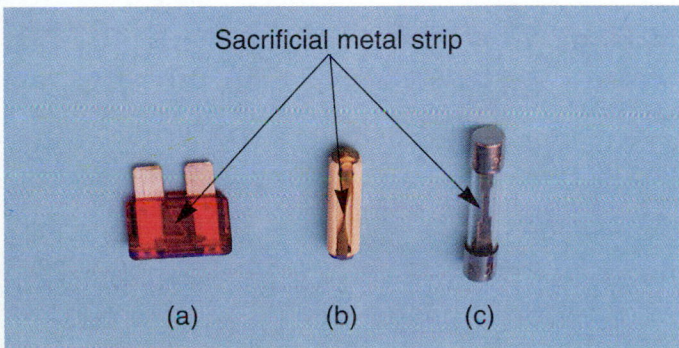

Figure 36.3 Three types of fuses: (a) blade-type, (b) ceramic, (c) glass cartridge.

common lengths used in automobiles. Usually, the longer the fuse, the higher the amperage rating.

There are two classifications for glass tube fuses.

- SFE fuses are designed to meet the standards of the **Society of Fuse Engineers (SFE)**. These fuses will be longer if their amperage rating is higher.
- AG (all glass) fuses, also called *Bussman*, or *Buss*, are manufactured by McGraw Edison Company. They have a number code to determine their application and have identification letters, such as AGA or AGC or AGW. These are simply Bussman designations, with no other meaning. Fuses within each of these classifications are all the same length no matter what the amperage. For instance, AGA fuses are ⅝", AGC fuses are 1¼", and AGW fuses are ⅞". A 5 amp and a 15 amp AGC fuse would each be the same length.

Some fuses are "slow-blow" to allow for a heavier draw when wiper or blower motors first start to operate.

Ceramic Fuses. Ceramic fuses are used in some imported cars. The fuse strip is exposed to the air and is mounted over the outside of its ceramic backing (see Figure 36.3b). Ceramic fuses are color coded with the amperage rating cast into the ceramic back of the fuse.

Blade-Type Fuses. The blade-type has been popular since the late 1970s in both import and domestic vehicles. The fuse element is cast into a clear plastic outer body (see Figure 36.3C). Blade-type fuses are color coded according to the SAE color code.

Fuse Links

In some circuits in which current control is not as critical, a fuse link will be used. A fuse link is a small length of wire that is smaller in diameter than the wire it is connected to. It is usually installed close to the power source. When the wire overheats, it will melt, opening the circuit (Figure 36.4).

Circuit Breakers

Sometimes a circuit only has temporary overloads or must have power restored if it goes out. This is true with the headlight circuit. If the dimmer switch develops a short or ground, you would not want the headlights to go out and remain out.

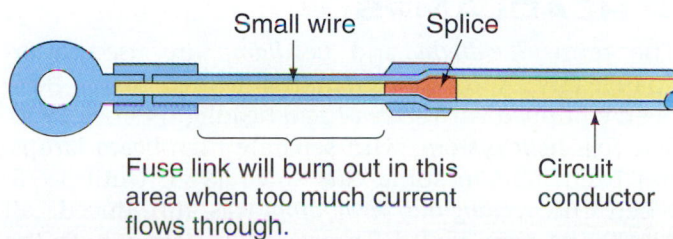

Figure 36.4 A fuse link will burn through if the wire is overloaded. *(Courtesy of Ford Motor Company)*

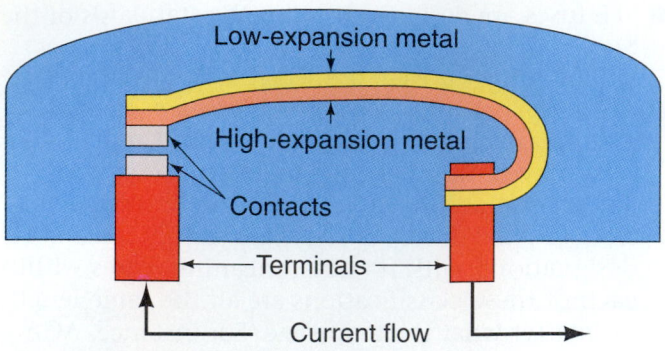

Figure 36.5 Circuit breaker operation. *(Courtesy of Ford Motor Company)*

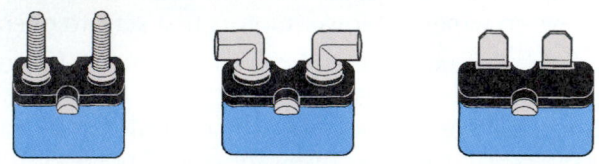

Figure 36.6 Different kinds of circuit breakers. *(Courtesy of Bussmann, Division of Cooper Industries, St. Louis, MO)*

A circuit breaker used in automobiles is usually self-resetting. In other words, when there is an overload in the circuit, the breaker "trips" and then resets. It has two metal strips with different expansion rates (Figure 36.5). This is called a **bimetal strip**. During an overload, the high expansion metal will become longer, breaking the switch contact. When it cools, contact is reestablished. Unlike the fuse and fusible link, the circuit breaker is not destroyed by overloads. Different styles of circuit breakers are shown in Figure 36.6.

When a circuit breaker is manually resettable, a button pops out on the case. It must be reset by depressing the button.

■ LIGHTING

Light bulbs, often called *lamps*, have **filaments** that electricity flows through. The filament provides a resistance to electron flow so it heats up, causing light.

■ HEADLAMPS

The terms *headlight* and *headlamp* are used interchangeably. Automobiles in the United States have been equipped with *sealed-beam* headlights since 1940. The *four-light* system, with separate high-beam lamps, has been used in some cars since 1958. Until 1975, when the *rectangular headlamp* was introduced, all headlamps were round. There are two sizes of both the round and the rectangular lamps. *High-beam* lamps are used at highway speeds with no vehicles oncoming or in front of the car. *Low beams*, used for city driving, are used much more often.

The intensity of a headlamp is rated in **candlepower** (cp). Maximum candlepower of headlights is limited by law. Older lamps were limited in intensity by the Department of Transportation (DOT) to 75,000 candlepower. European lamps were considerably brighter (300,000 cp). Since 1979 allowable light intensity has been increased to 150,000 cp. This standard was originally allowed for high beam lamps only, but now applies to all lamps.

Sealed Beam Headlamp Construction

Sealed beam headlights have an inner glass or plastic *reflector surface* that is sprayed with an aluminum reflective material (Figure 36.7). An outer glass or plastic *lens* is fused to the reflector and the lamp is then filled with *argon* gas. The beam of light from a 50 candlepower bulb's *tungsten* bar filament is aimed at the focal point of the reflector, where it is intensified to 20,000 cp. The outer lens focuses the reflected light into a beam pattern (Figure 36.8).

In a *dual beam headlight*, which has filaments for both low and high beam, the upper (high beam) filament is offset so it aims higher. A **type I lamp** has high beam only, with two lugs on its electrical connector. A **type II lamp** has both low and high beams and a three lug connector. When a type II headlamp is on low beam, only one of its filaments is lit. When the high beam is turned on, the second filament lights, too. There are only three wires on the connector because both filaments use the same ground wire.

When there are four headlamps, the single filament of the type I lamp lights on high beam, along with the two filaments in the type II lamp. A type II lamp will be located above or outside of the type-I lamp on a four-lamp system (Figure 36.9).

Halogen Lamps

Halogen headlamps say "halogen" on the center of the lens. These lamps, used in many new cars since

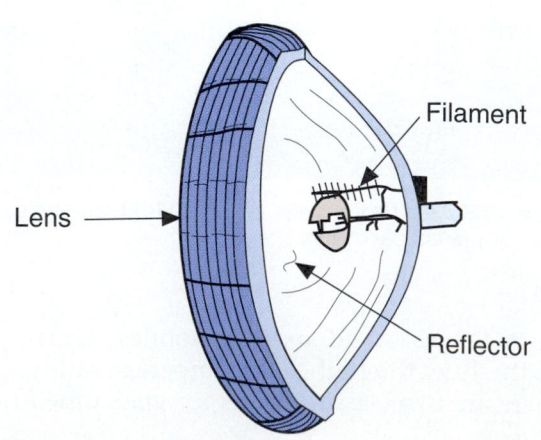

Figure 36.7 Parts of a sealed beam headlight.

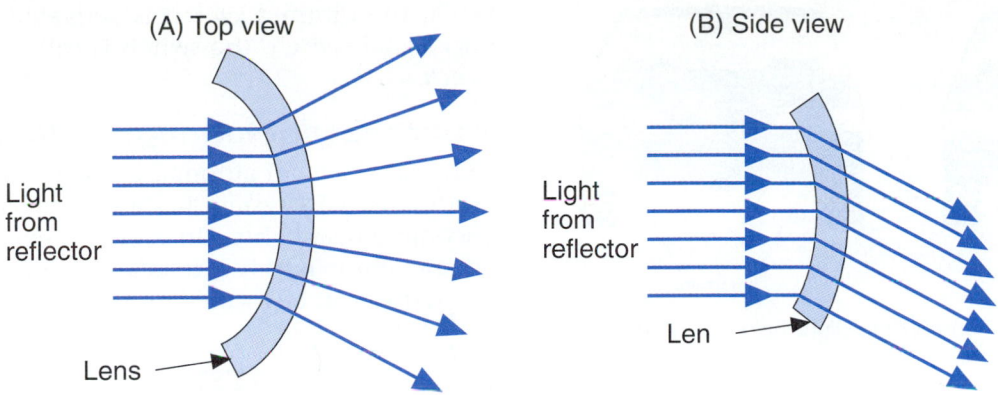

Figure 36.8 The outer glass lens focuses the reflected light into a beam pattern.

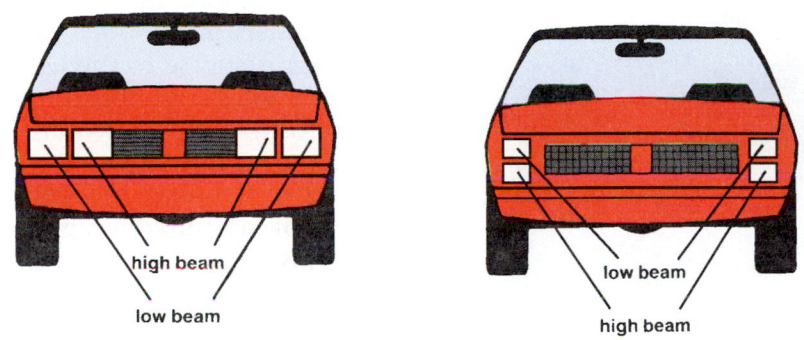

Figure 36.9 Four-light headlight system placement. *(Courtesy of GE Lighting)*

1979, produce a 25% higher output of "whiter" light on the same amount of power as a conventional sealed beam headlamp. The tungsten filaments are enclosed in pressurized halogen vapor that allows them to be heated to a higher temperature, making a higher output. Halogen lamps are so bright that, even in a four-light system, no more than two filaments will be on at once.

Halogen lamps are either of the conventional sealed beam type or they are **composite headlamps**, with a fixed lens cover and a replaceable halogen bulb (Figure 36.10). A composite headlamp has a glass balloon that the halogen lamp fits inside of. There are both two-light and four-light halogen replacement bulb systems (Figure 36.11).

Some composite headlamp designs develop condensation inside of the lens. The moisture dissipates from the heat of the lamps once the lights are turned on. Other composites are sealed and are defective if condensation develops in them.

Headlamp Switch

The headlamp switch is located in the dash or on the steering column. The dash switch is a combination on/off switch and rheostat (Figure 36.12). The rheostat is for adjusting the intensity of the dash lights. There is usually an on/off switch at the end of the range of the rheostat for turning on the dome light(s). When

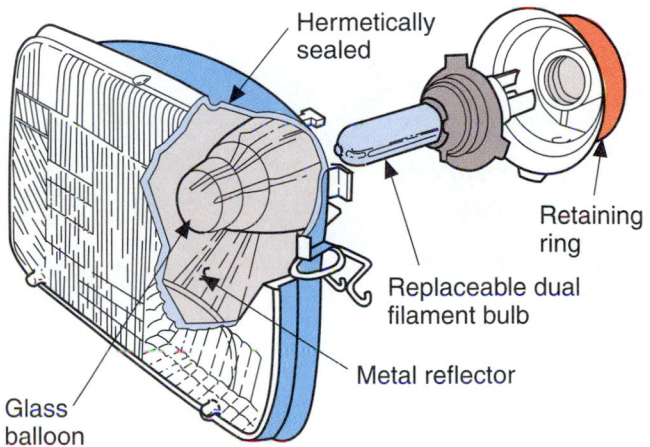

Figure 36.10 A headlight with a replaceable halogen bulb.

the switch is on the steering column, a separate rheostat will be located somewhere on the dashboard.

Dimmer Switch

For changing headlights from high beam to low beam, a dimmer switch is mounted on the floorboard or on the steering column (turn signal lever) (Figure 36.13). One position turns on the high beams (brights) and the other switches on the low beams for city driving.

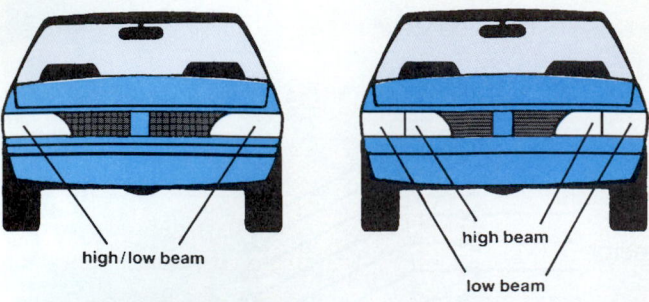

Figure 36.11 Two and four light halogen systems. *(Courtesy of GE Lighting)*

Figure 36.12 A headlight switch.

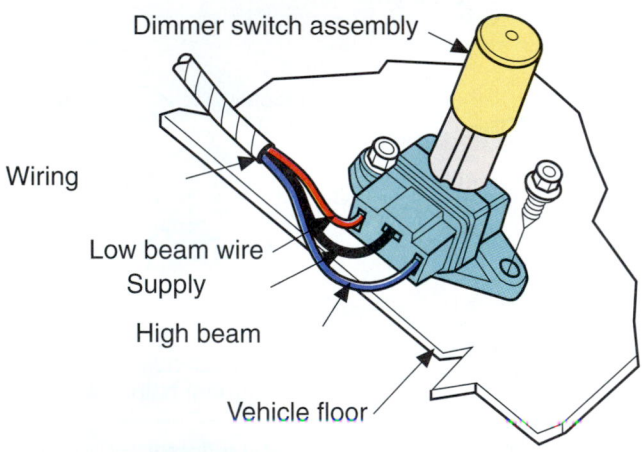

Figure 36.13 Two types of dimmer switches. *(Courtesy of Ford Motor Company)*

When the dimmer switch is activated by raising the turn signal switch, the switch is called a *multifunction switch*.

Automatic Headlight Dimmer

Some cars have an automatic system for dimming the headlights. This system has a light sensor that detects oncoming headlights. On some systems, when there is no oncoming light it switches the high beams back on. The sensor can be located in the grille, or on the top of the dash.

■ TAIL LIGHTS

Tail light bulbs come in many sizes and types. Bulbs have different numbers of filaments and different methods of grounding. Some bulbs have two terminals and others have only one. Bulbs usually ground through the case to the light socket. The light socket usually will be grounded to the frame through a wire or directly. A bulb used on a ceiling dome light or a fiberglass vehicle, such as a boat or motor home, will have two terminals for only one filament. The second terminal is to provide a ground path for the filament.

■ IDENTIFICATION OF LIGHT BULBS

It is important that the correct bulb be used or damage to a circuit could result. Automotive bulbs are numbered by the American National Standards Institute (ANSI). No matter who the manufacturer is, the bulb will still have the same number, called the **bulb trade number**.

■ When there is an *A* or an *NA* after the bulb number, that means that the bulb is an amber (yellow) color.

■ The **natural amber (NA)** bulb is colored glass while the less expensive A is painted glass.

Bulbs that are smaller than headlamps are also known as *miniature lights*.

Stoplight Switch

Brake lights (stoplights) are activated by a mechanical switch on the brake pedal (see Figure 51.46). Some older cars had hydraulically activated stoplight switches. The wire that supplies current to the brake lights is interrupted at the stoplight switch. When the brakes are applied, the two ends of the wire are connected electrically to allow current to travel to the light filaments. On some cars, the brake lights operate through the turn signal switch.

Tail and Brake Lights

A tail light bulb has two filaments. When the brakes are applied, the switch activates one filament on each side of the car. On domestic cars, this filament is also used for the turn signals and emergency flashers. The circuit is controlled through the turn signal switch. Figure 36.14 shows a typical parking and tail light circuit.

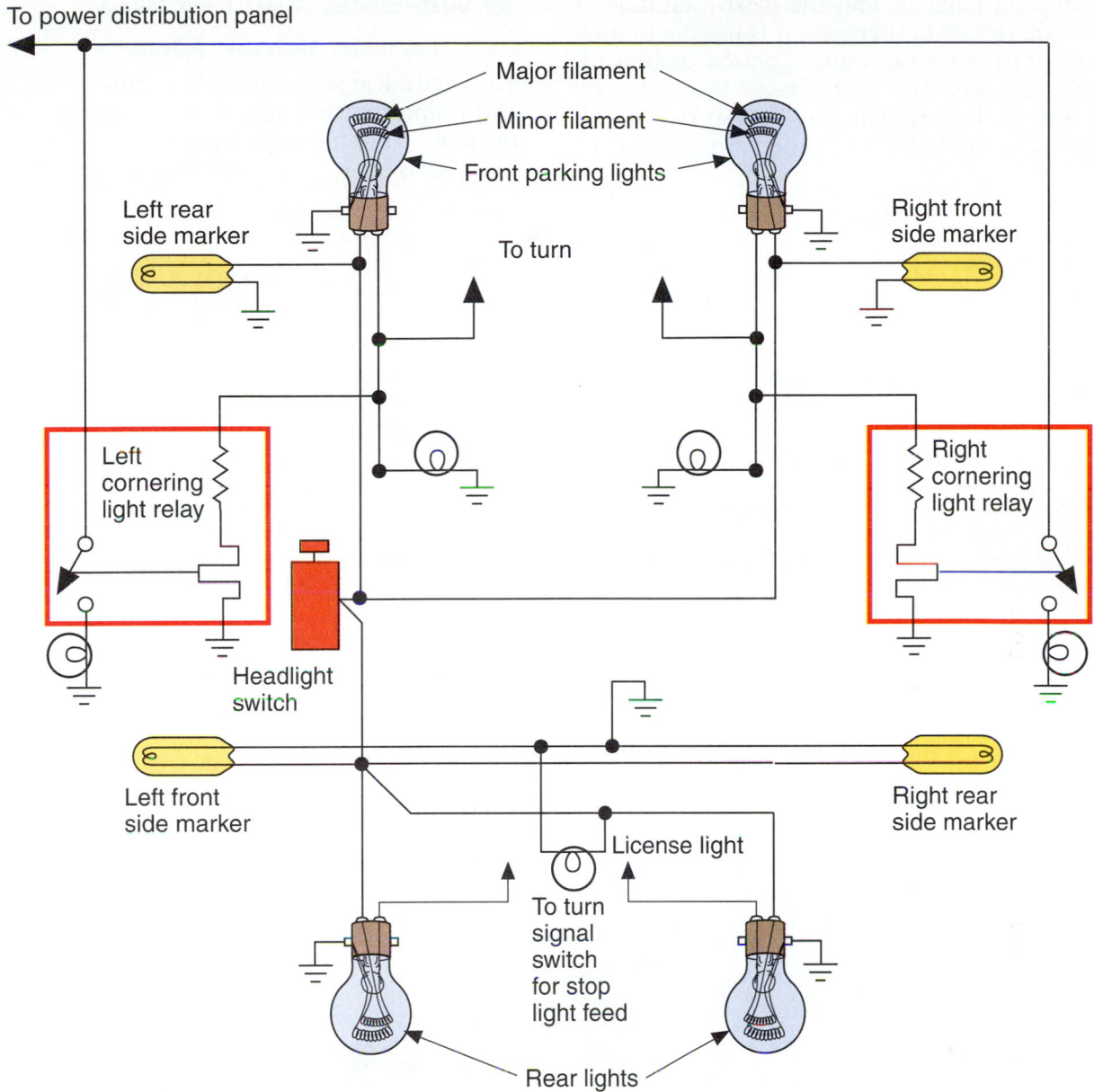

To power distribution panel

Major filament

Minor filament

Front parking lights

Left rear side marker

Right front side marker

To turn

Left cornering light relay

Right cornering light relay

Headlight switch

Left front side marker

Right rear side marker

License light

To turn signal switch for stop light feed

Rear lights

Figure 36.14 A typical brake light circuit. *(Courtesy of Ford Motor Company)*

Imported cars often use an extra bulb for the brakes and turn signals. The turn signal bulb has a lens cover that is yellow and a brake light lens cover that is red.

Late model cars also have a center brake light, mounted high for safety purposes. These lights are easier for other drivers to see in a panic stop situation.

TURN SIGNALS

The circuit for the turn signals has a switch, a **signal flasher**, two indicator bulbs in the dashboard, the stoplight filaments of the tail lights or rear stoplight bulbs, and the two bulbs in the front of the car. The turn signal switch is located in the steering column, beneath the steering wheel.

SIGNAL FLASHER OPERATION

A flasher is activated by the heat of the electricity travelling through it to the signal bulbs. It normally flashes at between 60 and 120 cycles a minute. It has a bimetal strip (Figure 36.15) and works the same way that a circuit breaker does. There are usually two flashers, one for the turn signals and one for safety hazards.

HAZARD FLASHERS

Hazard flashers are required on all cars manufactured in 1967 and later. This system is powered directly from the battery, independent of the turn signals, and turns on all of the turn signals and the dash indicators at once. The emergency flasher switch is usually located on the side of the steering column, on the dashboard, or in the glove box. The flasher used in emergency warning systems is a *variable load* one so that it will flash whether or not bulbs are burned out.

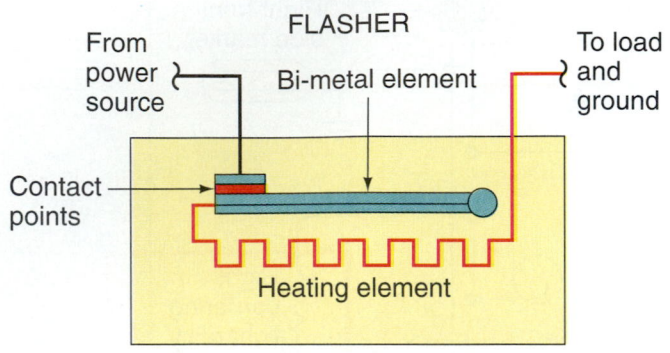

Figure 36.15 Operation of a turn signal flasher. *(Courtesy of Ford Motor Company)*

WIPERS AND HORN

Windshield Wiper Motor

The windshield wiper circuit has a fuse or circuit breaker, and a wiper switch. A relay is used to complete the circuit because of the relatively large amp draw of the motor. The relay is often located on the fuse panel (Figure 36.16).

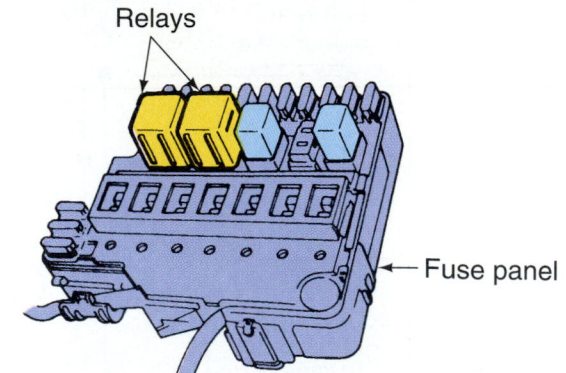

Figure 36.16 These relays are located in the fuse panel. *(Courtesy of American Honda Motor Co., Inc.)*

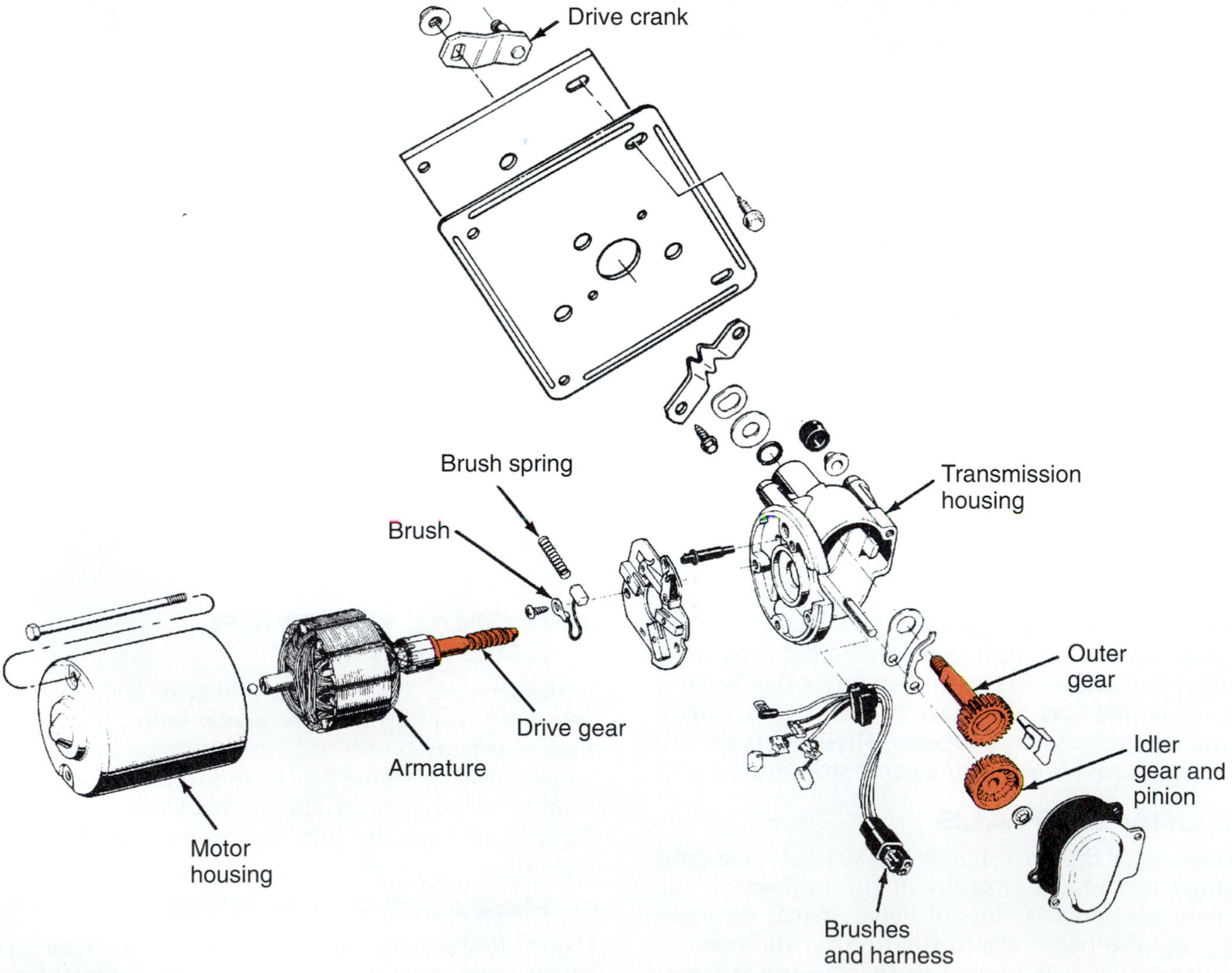

Figure 36.17 Parts of the windshield wiper motor. *(Courtesy of Chrysler Corporation)*

The motor assembly includes a permanent magnet motor, a transmission (plastic gears), a housing, and drive crank (Figure 36.17). Linkage arms connect the wipers to the crank. The parts of the entire wiper assembly are shown in Figure 36.18.

The switch on most new cars is a rheostat (see Chapter 29), which allows changes in speeds. Sometimes, the switch has a delay function that works like a signal flasher. This allows the wipers to come on periodically during fog or misty conditions.

HISTORY NOTE

The first windshield wipers had to be operated by hand. Early powered wipers operated on vacuum from the engine. This presented a unique problem. When a vehicle was under a load, such as when climbing up a hill, the wipers would lose vacuum and stop moving. When the accelerator was released an increase in vacuum caused the wipers to move rapidly again. Electric wipers solved this problem. A 1968 federal law called for all vehicles to have wipers with at least two speeds and a washer system.

Windshield Washer

The windshield washer has a reservoir, a switch, a pump, washer nozzles below the windshield, and hoses and connections. Inexpensive anti-freeze solvent is used in the reservoir to clean the windows.

There are two types of pumps used, a rotary type and a bellows (diaphragm) type. Most new cars use a rotary pump. It is located in the solvent reservoir. A small motor spins an impeller. The bellows-type pump is usually mounted on the wiper motor.

Horn

The horn circuit includes the horn, a fuse and wiring, the horn button in the steering wheel, and a relay (Figure 36.19). Pressing on the horn button closes a circuit to activate the horn relay. The relay provides the current to the horn. The horn takes quite a bit of current so most horns use a relay, like motors do.

Most horns have a diaphragm vibrated by an electromagnet (Figure 36.20). The contact points are normally closed. One of the points is attached to the armature, which moves in response to current flow through the field coil. When the armature moves, it opens the connection between the contact points. The diaphragm retracts to its original position. When the contacts close again, the cycle repeats itself. The

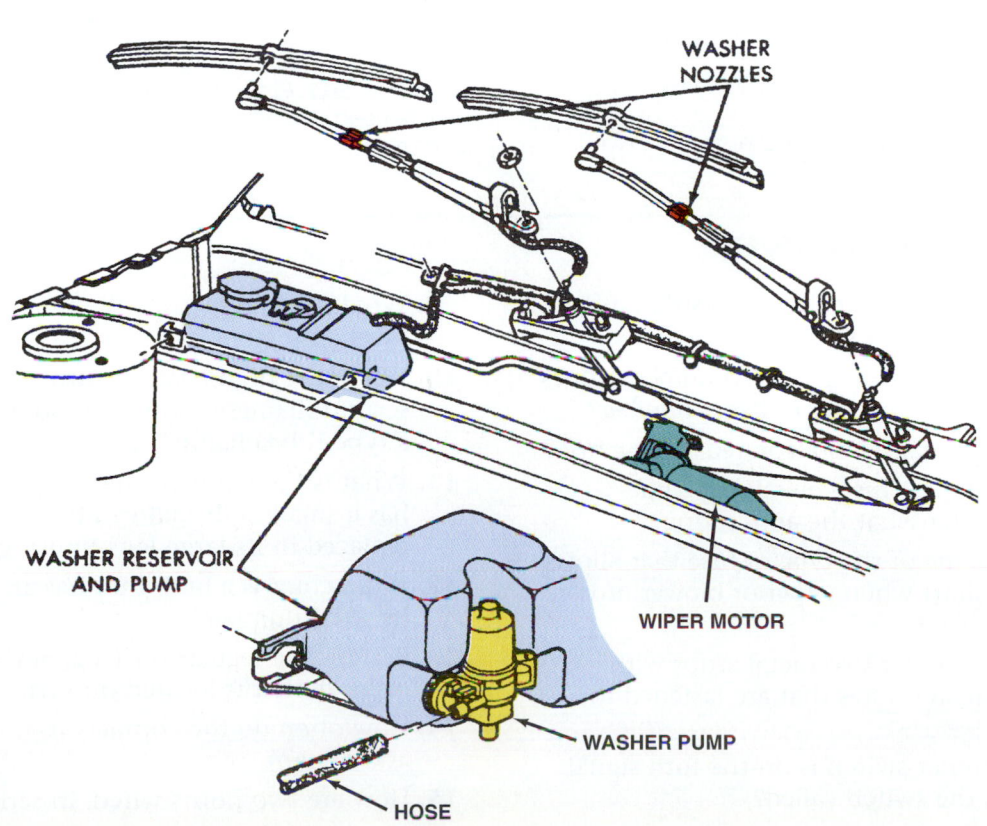

Figure 36.18 Parts of the windshield wiper assembly. *(Courtesy of Chrysler Corporation)*

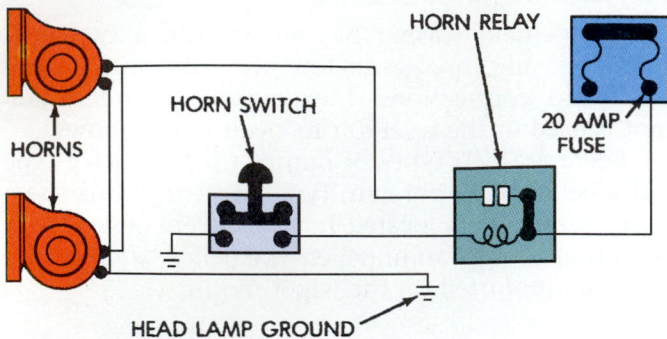

Figure 36.19 The horn circuit. *(Courtesy of Chrysler Corporation)*

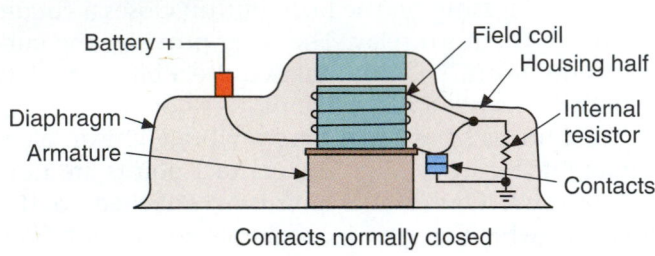

Contacts normally closed

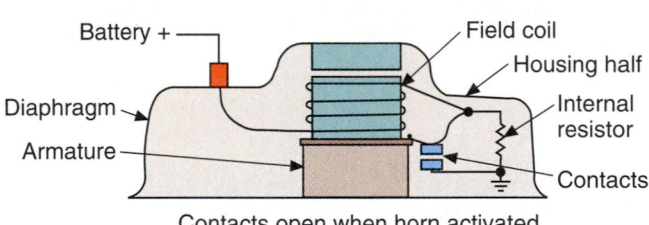

Contacts open when horn activated

Figure 36.20 Operation of a horn.

vibration of the diaphragm happens several times per second. This causes a column of air in the horn to vibrate, producing sound.

Most vehicles have two horns wired in parallel to each other and in series with the switch. One of the horns will have a slightly lower pitch than the other.

The horn switch is either a single or multiple button switch in the steering wheel. Sliding contacts must be used so that the steering wheel can turn, while maintaining contact with the switch (Figure 36.21). There is a circular contact in the steering wheel that slides against a spring-loaded contact in the steering column.

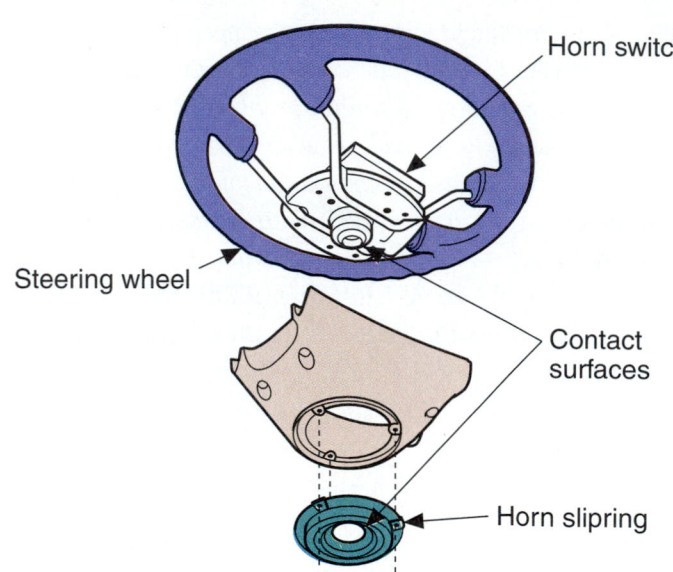

Figure 36.21 The horn has a sliding electrical contact beneath the steering wheel. *(Courtesy of Ford Motor Company)*

■ REVIEW QUESTIONS

1. What is the name for the large wires that allow more electrical current to flow?

2. With American Wire Gauge (AWG) numbers, does a larger wire mean a smaller or larger number?

3. What is another name for a colored, plastic fuse?

4. Which kind of glass fuses are all the same length, no matter what the amp rating?

5. What is the name of the type of fuse that allows for a heavier draw when wiper or blower motors first start to operate?

6. What is the name for two metal strips with different expansion rates that are fastened to each other?

7. When the dimmer switch is on the turn signal lever, what is the switch called?

8. In what year were rectangular headlamps introduced?

9. What measurement is the intensity of a headlamp rated in?

10. When the high beam is turned on and the second filament lights are too, is this a type I or a type II headlamp?

11. What is the name of the type of headlamp that has a small bulb that can be removed and replaced in its large lens housing?

12. What color is a bulb that has an A or an NA in its ANSI number?

13. Besides the regular turn signal flasher, what other flasher is located on vehicles since 1967?

14. How often do the contacts in the horn open and close?

15. How are two horns wired, in series or parallel?

■ ASE STYLE REVIEW QUESTIONS

1. Technician A says that high voltage wiring on a car is called primary wiring. Technician B says that low voltage wiring is called secondary wiring. Who is right?
   - **a.** Technician A
   - **b.** Technician B
   - **c.** Both A and B
   - **d.** Neither A nor B

2. Technician A says that each fuse has an amp rating at which it is designed to fail. Technician B says that a fuse is designed to melt when the flow of voltage in the circuit becomes too high. Who is right?
   - **a.** Technician A
   - **b.** Technician B
   - **c.** Both A and B
   - **d.** Neither A nor B

3. Technician A says that an AWG 0 wire size is smaller than an AWG size 20. Technician B says that a fuse link is a section of a smaller diameter of wire. Who is right?
   - **a.** Technician A
   - **b.** Technician B
   - **c.** Both A and B
   - **d.** Neither A nor B

4. Technician A says that ceramic fuses are color coded. Technician B says that blade-type fuses are color coded. Who is right?
   - **a.** Technician A
   - **b.** Technician B
   - **c.** Both A and B
   - **d.** Neither A nor B

5. Technician A says that some circuit breakers must be reset after they are tripped by an overload. Technician B says that some circuit breakers are self-resetting. Who is right?
   - **a.** Technician A
   - **b.** Technician B
   - **c.** Both A and B
   - **d.** Neither A nor B

6. Technician A says that a type II headlamp has two wires on the connector. Technician B says that a type I headlamp has three wires on the connector. Who is right?
   - **a.** Technician A
   - **b.** Technician B
   - **c.** Both A and B
   - **d.** Neither A nor B

7. Technician A says that a halogen headlamp burns brighter than a standard sealed beam headlamp because the tungsten filament gets hotter. Technician B says that any halogen headlamp with moisture in it is defective. Who is right?
   - **a.** Technician A
   - **b.** Technician B
   - **c.** Both A and B
   - **d.** Neither A nor B

8. Technician A says that if an ANSI light bulb has two terminals, one must be for ground. Technician B says that a bulb with an ANSI designation of NA is painted glass. Who is right?
   - **a.** Technician A
   - **b.** Technician B
   - **c.** Both A and B
   - **d.** Neither A nor B

9. Technician A says that on many imported cars, the turn signal lens is red. Technician B says that on many imported cars, the brake light lens is yellow. Who is right?
   - **a.** Technician A
   - **b.** Technician B
   - **c.** Both A and B
   - **d.** Neither A nor B

10. Technician A says that a rheostat in the wiper switch allows the wiper speeds to be changed. Technician B says that a wiper delay function works like a signal flasher. Who is right?
    - **a.** Technician A
    - **b.** Technician B
    - **c.** Both A and B
    - **d.** Neither A nor B

CHAPTER 37

Lighting and Wiring Service

■ OBJECTIVES

Upon completion of this chapter, you should be able to:

✔ Diagnose problems in lighting and wiring systems.

✔ Adjust headlamp aim.

✔ Make repairs to automotive wiring, lamps and bulbs, and protection devices.

■ KEY TERMS

flux
"tin" a wire
"cold" solder connection
heat-shrink tubing
Gauss meter or gauge

■ ANALYZING ELECTRICAL PROBLEMS

In this chapter, you will build on your knowledge of electrical system service. Wiring serves all of the electrical components in the automobile and a circuit is sometimes the cause of a problem. Before attempting to repair electrical wiring and lighting problems, you must understand how the system works and how to use electrical system tools. You need to be methodical in going about your repair. Do not just replace a starter or alternator until you have thoroughly diagnosed the cause of the problem.

■ The first step is to verify the complaint to make sure what the problem is.

■ Check to see if there are related symptoms. A related symptom might be that the lights dim when the engine is cranked. Check a wiring diagram to see if there are common problems that could be tied together by a bad ground or power feed.

■ Take time to consider the symptoms and their possible causes.

■ Check for a quick fix. Sometimes, a loose or corroded connection could be the cause of a problem.

■ WIRING SERVICE

Color Coding

Automotive wire insulation is color coded or numbered. This is useful when tracing wiring with a wiring diagram. A wiring diagram is like a roadmap that shows all of the car's wiring and loads. Figure 37.1 shows the rear lamp part of an electrical circuit. It gives color codes of the wires that supply the loads. There are several methods of marking wires. A wire might be a solid color, be striped, have hash marks, or be dotted.

SAFETY NOTE When disassembling a number of wires or vacuum hoses that are not clearly color coded, use numbered marker tape to keep them in order.

■ CRIMP TERMINALS

Crimp (solderless) terminals are a popular way of repairing wire ends. There are many varieties of terminals and connectors available (Figure 37.2). Terminals are made with different size crimping tabs to accommodate wire sizes from 22 gauge (small) to 10 gauge (larger). They can be soldered, or crimped with a special wire stripper/crimper (Figure 37.3).

To install the terminal:

■ About ¼" of the insulation is stripped from the end of the wire. If the wire is not clean and shiny, cut the insulation and wire back further until it is.

■ Insert the end of the wire into the terminal and crimp it using a crimping tool. The dimple from the crimping tool should be opposite to the seam in the connector.

■ The crimping tool will usually have labeled grooves of the proper size to crimp most sizes of wires.

Figure 37.4 shows how to crimp an open type connector end.

NOTE: *Crimp connectors are not used in computer circuits. Special weatherproof connectors are used to prevent corrosion. This type of connector prevents air from entering and corroding the connection. Corrosion causes resistance that will change the computer's sensor readings.*

Terminals are often held in a terminal block to keep several wires organized. Removing terminals from the block usually requires depressing a locking tang. Figure 37.5 shows a typical connector release.

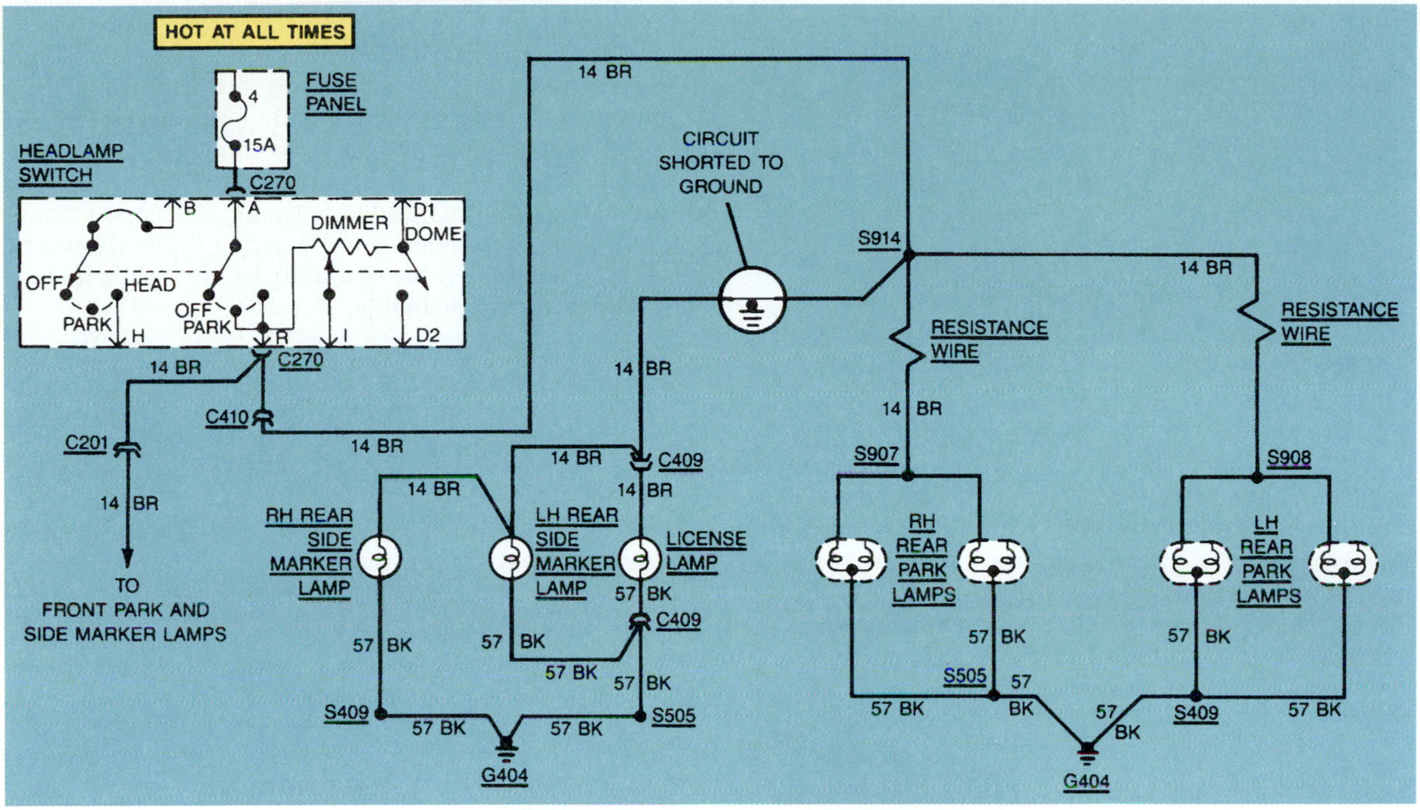

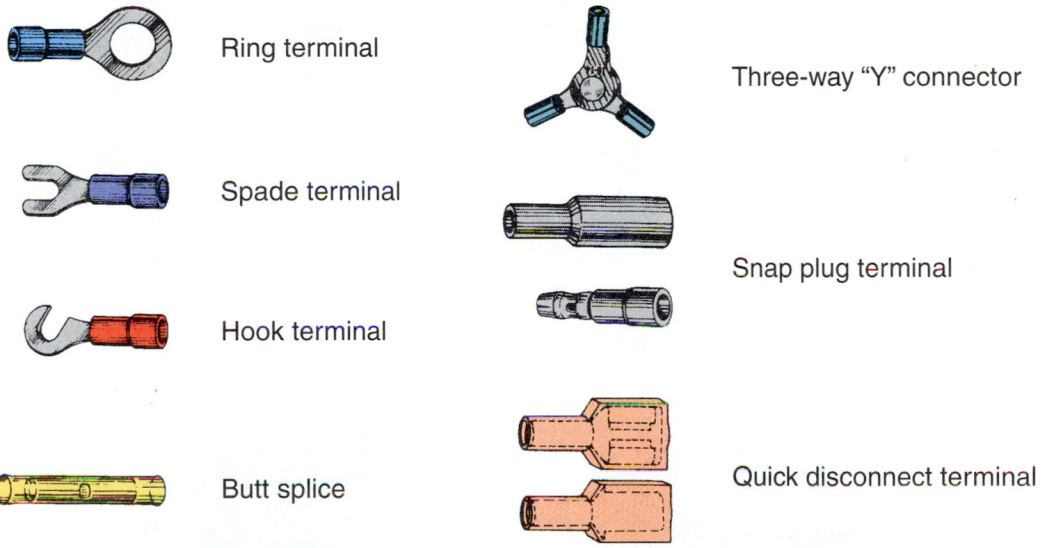

Figure 37.1 Part of a wiring diagram. [*Courtesy of Ford Motor Company*]

Ring terminal

Spade terminal

Hook terminal

Butt splice

Three-way "Y" connector

Snap plug terminal

Quick disconnect terminal

Figure 37.2 Different terminals and connectors. [*Courtesy of Cooper Automotive/NAPA Belden*]

Crimp Connectors

There are also crimp connectors available for splicing a wire together (Figure 37.6). These *butt connectors* are a fast and relatively inexpensive method of repairing wires provided the crimp is solid and is done carefully. Insert both ends of the wire into the connector so that both wires are side by side and crimp.

➡ *Perform Splice a Wire with A Crimp Connector Worksheet*

SELECTING REPLACEMENT WIRE

Replacement wire and cable comes wound in spools. Be sure to use wire of adequate size for the load. A chart in

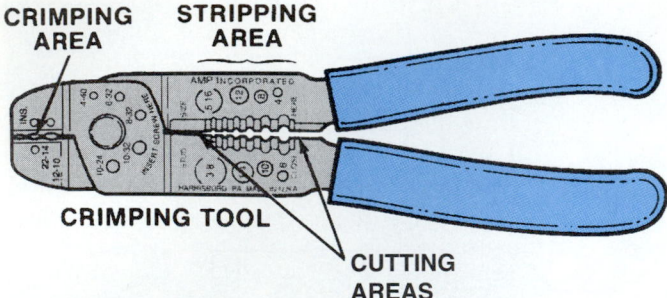

Figure 37.3 A wire crimping tool. *(Courtesy of Ford Motor Company)*

Appendix F shows the wire size needed for different length runs. Replacement wire should always be at least as big as the original wire. Wire that is too small or too long for a circuit can cause excessive resistance, resulting in a loss of lighting or accessory efficiency. Wire of double the original length has twice the resistance.

ADDING ELECTRICAL ACCESSORIES

Be sure that proper techniques are followed when adding items to a vehicle's wiring system. When an aftermarket accessory is added to the vehicle, it is usually necessary to add a separate fused circuit. Trailer lights are commonly added with an aftermarket trailer hitch. The color code for the existing lights should be followed when possible. A variety of multiple wire harness adapters are available.

NOTE: *Adding extra loads, such as stereos, c.b. radios, or fog lights can tax a vehicle's charging system so that it may no longer be adequate. Be sure to investigate this first before adding any extra load.*

Fuse Holders

When an original equipment accessory is added to a vehicle's electrical system, power can usually be taken from a fuse on the fuse panel.

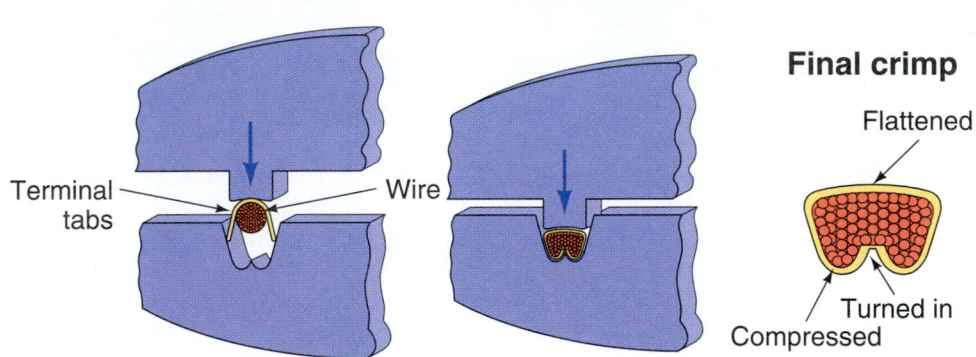

Figure 37.4 Crimping an open connector. *(Courtesy of Ford Motor Company)*

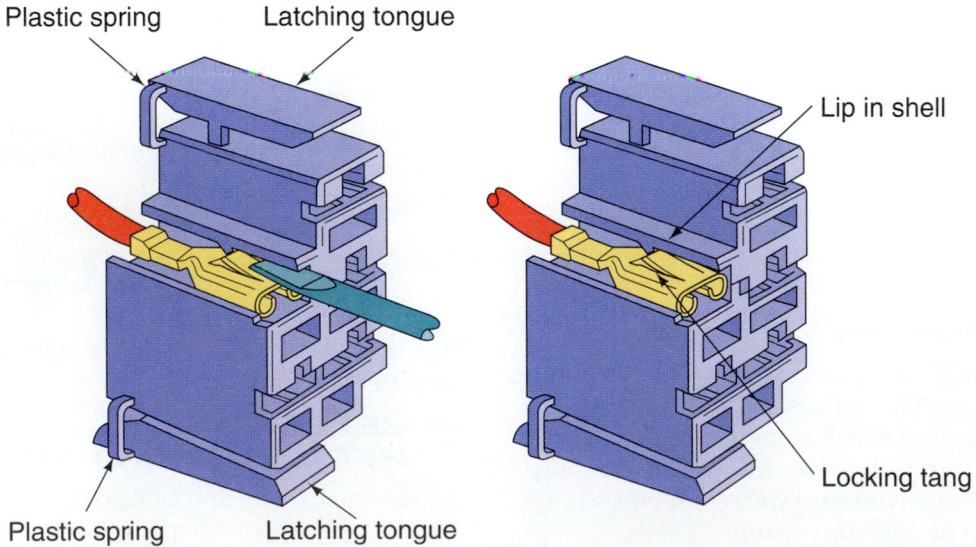

Figure 37.5 Depress the lip to remove the connector from the junction. *(Courtesy of Ford Motor Company)*

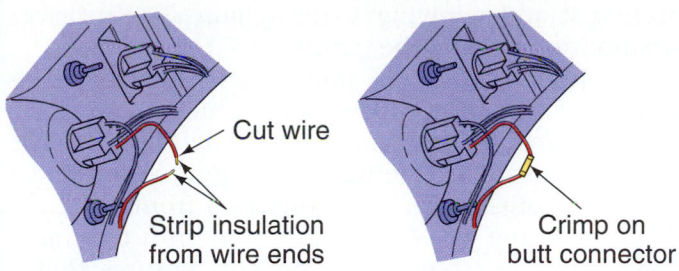

Figure 37.6 A butt connector is used to join wire ends. *(Courtesy of Ford Motor Company)*

If a fuse is not available, a fuse holder can be connected to an existing adequate power source. A radio will usually come with a fuse holder and instructions for its installation. A handy *blade-type* fuse holder can

be easily installed into the wire (Figure 37.7) to provide a safe source of power.

SAFETY NOTE To avoid damage to electrical components and wiring or a possible fire, it is very important that wire of the correct size is used. Be sure that the correct fuse size is used also.

A *tap splice* connector can be used to tap into a power wire without the need to strip or solder. This connector is used for tapping another wire into an existing wire without cutting it or stripping its insulation. Figure 37.8 shows how it is used. This kind of connector is used in electrical circuits only, not in electronic circuits.

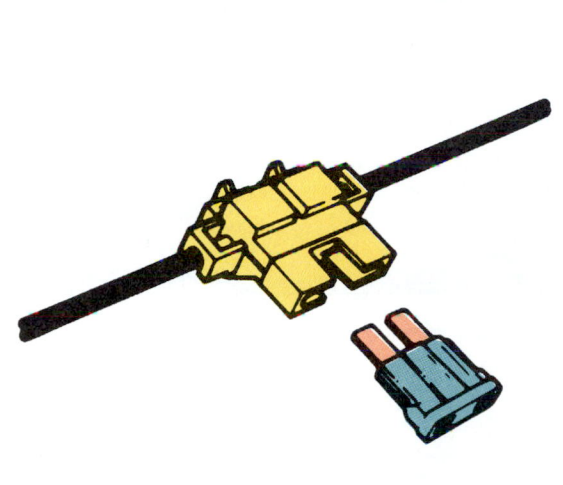

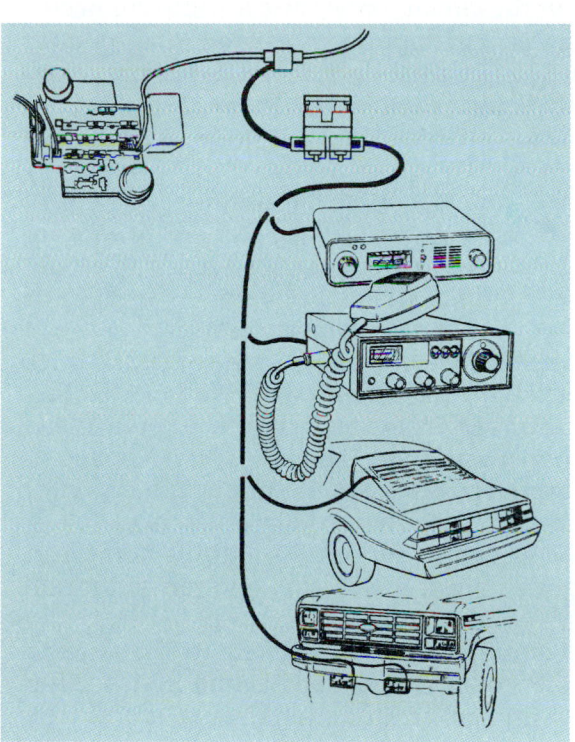

Figure 37.7 A fuse holder installed in series in a line. *(The 3M Switchlok™ Insulator Displacement Connector is printed with permission of 3M, St. Paul, MN)*

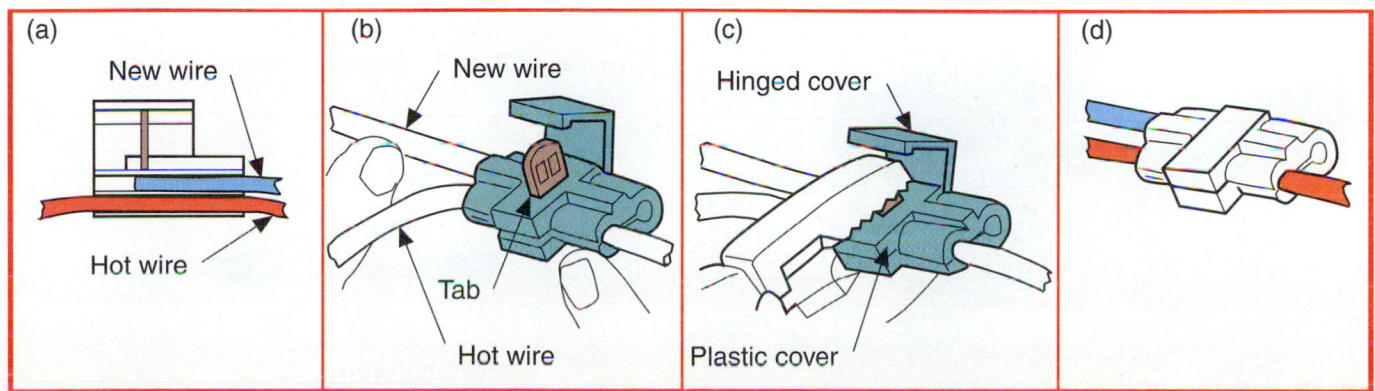

Figure 37.8 Installing a tap splice.

■ SOLDERING

Soldering is a professional method of repairing a connection. For *electronic* connections, such as those in computer circuits, soldering is especially preferred because of the small amount of electrical current that runs through them. A properly soldered connection will not suffer increased resistance due to oxidation with the passage of time.

A *soldering iron* or a *soldering gun* (see Figure 8.26) is used for connections of wiring. A small *propane flameless soldering pencil* can also be used in tight places (Figure 37.9). For radiators or larger cable terminal connections a propane, acetylene, or Mapp gas torch is used.

Sometimes, soldering is not convenient to do because of the unavailability of electricity, the need to avoid heat in the circuit, or because a crimp connector is simply faster and more convenient. A crimp connector is used to join stainless steel wire because stainless is not solderable. Stainless steel wire is used in boat trailers instead of copper because of its resistance to corrosion.

Solder will not stick to metals unless they are very clean, and not corroded. Corroded sections of wire are cut out and replaced. When making a repair to an old wire, a new section at the end of a wire is stripped of insulation to provide a clean area to solder.

Flux is used to clean metal so that solder will stick to it. Solder usually has a hollow core full of flux. Solder for *electrical repairs* is commonly a 40/60 tin/lead mix with a *rosin* flux core (Figure 37.10). *Acid-core* solder is also available, but it should not be used on electrical connections because it will corrode them.

Heat the soldering gun until the tip is hot. Clean the hot tip with a wet towel. Flux and renew the solder coat.

A wire terminal can be soldered to the wire to make a professional connection (Figure 37.11). After the wire is stripped of insulation, **"tin" a wire** by heating it and applying solder (Figure 37.12) before attempting to solder the terminal to it.

To join two wires, splice their stripped ends together. Figure 37.13 shows several methods of splicing. The wires are soldered after completing the splice.

When using a soldering iron or gun, apply solder to the opposite side of the connection from the tip of the iron (Figure 37.14). Heat the wire until the solder flows into the strands of wire. This ensures that a **"cold" solder connection** does not happen. A cold connection happens when the solder is melted by the iron, but the wire is too cold to bond to it. Also, be careful not to heat the wire too much. Overheating a wire can ruin its insulation.

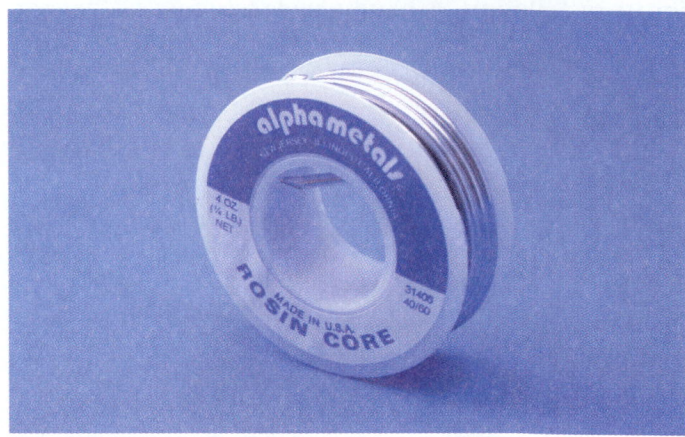

Figure 37.10   This rosin core solder is an alloy of tin and lead.

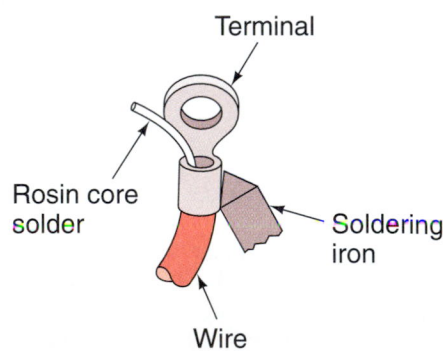

Figure 37.11   Soldering a terminal to a wire.

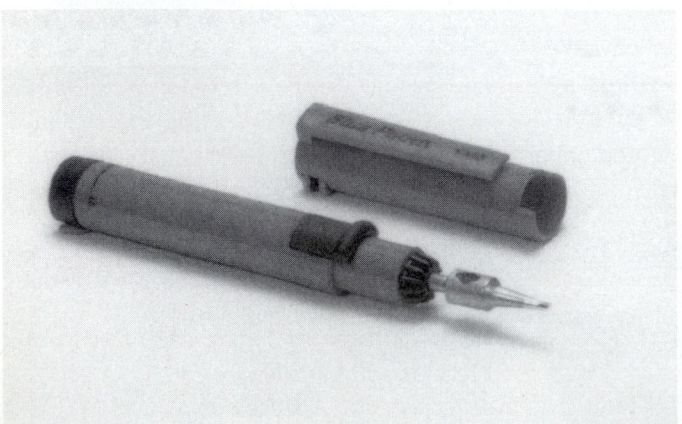

Figure 37.9   A propane soldering pencil. *(Courtesy of Snap-on Tools Company. Copyright Owner.)*

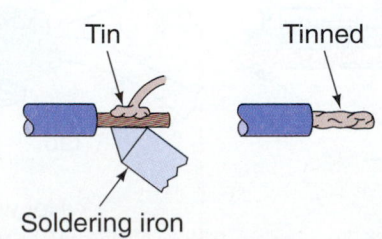

Figure 37.12   Tinning a wire.

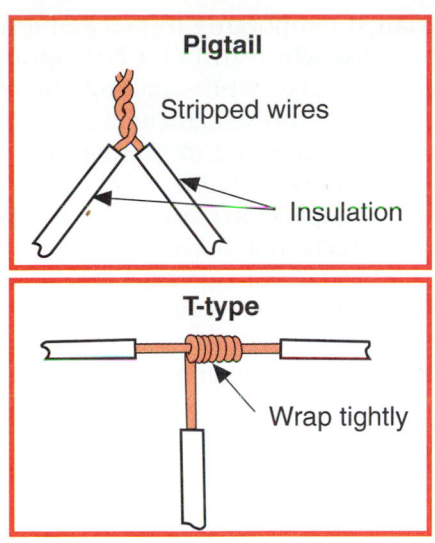

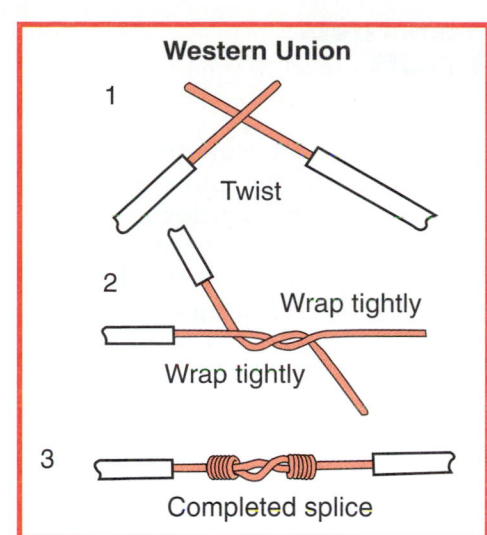

Figure 37.13 Several methods of splicing wires.

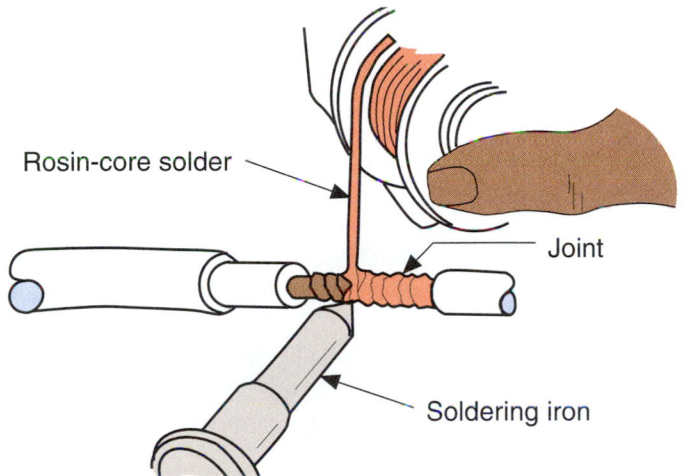

Figure 37.14 Apply the solder to the side of the wire opposite to the tip of the soldering iron or gun.

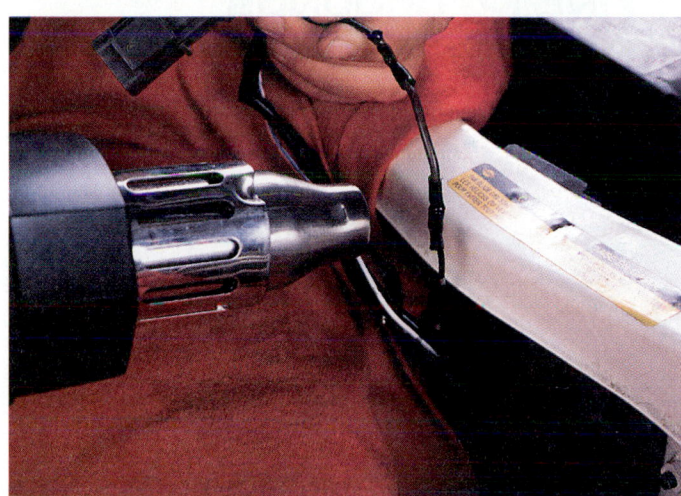

Figure 37.15 The shrink tubing is heated with a heat gun.

Heat-Shrink Tubing

Vinyl **heat-shrink tubing** can be used to insulate a solder joint and make it airtight. Cut the tubing to the correct length and slip it over one of the wires *before* they are spliced. After the soldering job is completed, the shrink tubing is slid over the soldered connection. When the tubing is heated with a flame or a heat gun, it shrinks to provide an airtight connection (Figure 37.15).

NOTE: *Heat-shrink tubing is thin. It will not withstand abrasion.*

If heat-shrink tubing is not used, insulate the connection with vinyl electrical tape.

➡ **Perform _Solder a Wire Connection_ Worksheet**

■ BROKEN OR DAMAGED GROUND STRAPS

The engine and the car body are isolated from the frame with rubber mounts. Rubber is an insulator, so ground straps are installed between the engine and chassis (see Figure 44.4). This provides electricity with a complete path back to the battery through the frame.

A broken ground strap can cause electricity to hunt for a different path to ground. When this happens, strange part failures can occur. Some of the possible results are:
- A burned transmission bushing and drive shaft yolk
- Burned emergency brake cables
- A burned carburetor return spring
- Flickering headlights
- Burned front wheel bearings or CV joints (on front wheel drive cars)

CIRCUIT TESTING AND SERVICE

Fuse Failure

A fuse failure could be a once only occurrence or there could be an electrical problem that must be fixed. Replacing a fuse or fuse link or resetting a circuit breaker does not fix the problem that caused the overload. The overload device will probably just fail again. Always fix the problem before restoring circuit protection.

Occasional fuse failure can be due to a defective fuse. Corrosion on the end of the fuse can cause failure too.

Automotive fuses are rated according to current capacity, not voltage. A 12-volt fuse will also work in a 110-volt application. Voltage does not cause a fuse to blow, current does. Fuses have a ten percent overload factor to guard against minor power surges.

FINDING GROUNDS

Locating the cause of a grounded circuit with a test light is not usually possible because the fuse blows as soon as the circuit is energized. A circuit breaker can be installed temporarily in place of a fuse for diagnosing grounded wires. Install a test light in series with the circuit breaker while making the test (Figure 37.16). Disconnect individual circuits until the light goes out. The circuit that was disconnected when the light went out is the one at fault.

A compass or **Gauss meter or gauge** (a gauge that detects magnetism) can be used to locate the exact location of a ground when a wire has been accidentally pinched (Figure 37.17).

Another method of detecting a grounded circuit is to use an ohmmeter. Connect it to the circuit side of the fuse holder and ground.

NOTE: *Be careful not to allow the ohmmeter to contact the power side of the fuse holder. When there is continuity between the load and ground, the circuit is grounded.*

FUSE TESTING AND SERVICE

The fuse panel is usually under the hood, under the instrument panel, or in a kick panel in the driver's compartment.

A visual check of the fuses in question can often show a burned fuse (Figure 37.18). Sometimes, a fuse

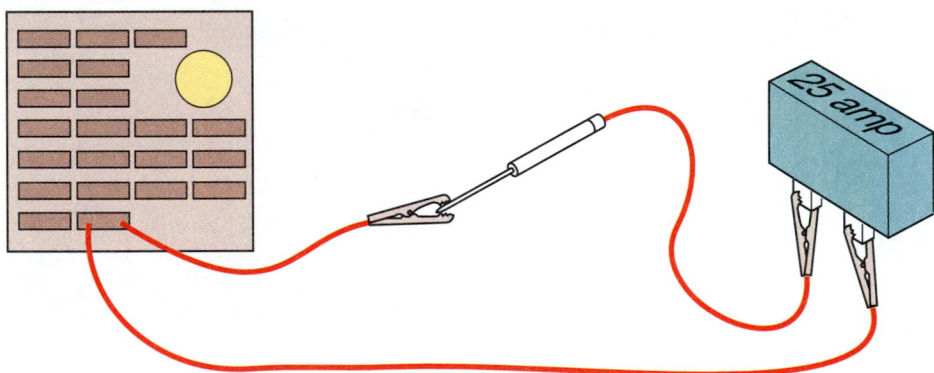

Figure 37.16 A test light installed with a circuit breaker for finding a faulty circuit.

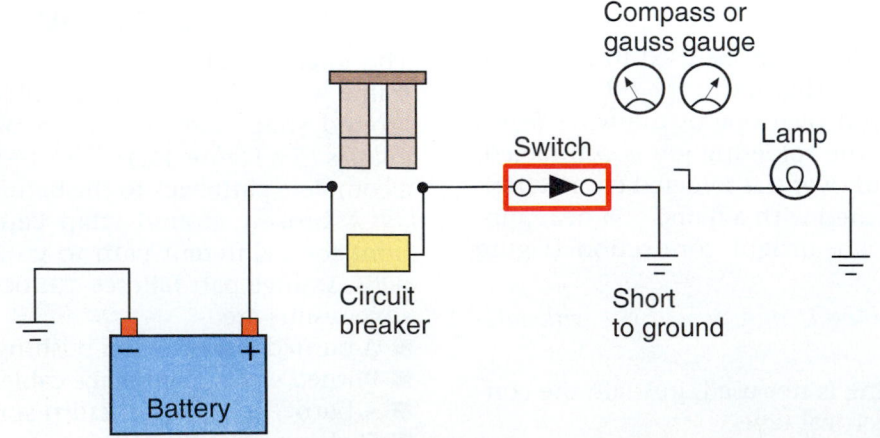

Figure 37.17 Use a compass or Gauss gauge to check for changes in the magnetism in the circuit at the location of the pinched wire.

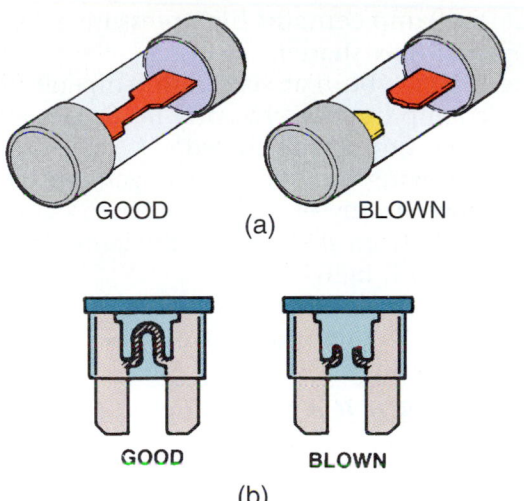

Figure 37.18 Good and blown fuses. *(a, Courtesy of Ford Motor Company)*

can blow where it cannot be seen (Figure 37.19). Remove and test a questionable fuse with an ohmmeter.

With the fuse in the socket, a test light can be used to test the fuse. With one end of the test light clipped to ground, the tester should light up when probing both sides of the fuse.

- If the tester glows when touched to either side of an installed fuse, the fuse is good and the voltage is on.
- If the tester glows only when touched to one end of the fuse, the fuse is defective. (The side that does not light is the ground side of the circuit.)
- The end of a blade-type fuse has a small hole that exposes its fuse metal to the outside of the fuse so it can be probed with a test light.

SAFETY NOTE If you use a screwdriver or a needle nose pliers to remove a glass cartridge or ceramic fuse wear eye protection. The glass tube or the ceramic body could shatter.

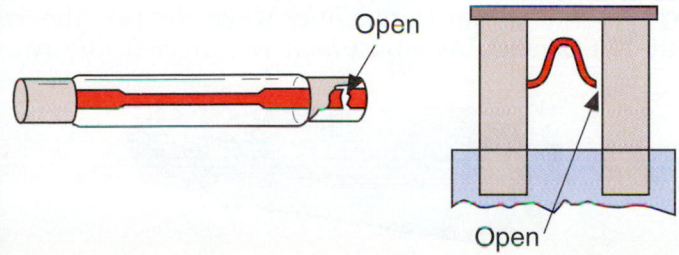

Figure 37.19 Test a questionable fuse with an ohmmeter.

Removing Fuses

With glass cartridge or blade-type fuses, a special fuse removing tool can be used to remove a bad fuse. Blade-type fuses are easy to remove by hand or with a needle nose pliers.

Testing the Fuse Socket

With the fuse removed, use the test light to determine which side of the fuse socket is the ground side. The tester will not light when attached to that side. If the tester does not glow on either side of the circuit:
- The circuit is shut off.
- The circuit is broken.
- The tester does not have a good ground connection.

Replacing the Fuse

Be sure to replace a fuse with the correct one for the application. SFE and Bussman glass cartridge fuses will interchange with one another, so care is required. Be sure that the fuse is the correct rating. Too low an amp rating will result in burned fuses. Too high a rating can cause a fire.

Check the end of the fuse for its amp rating. The correct amp rating of the fuse to be used is listed on the fuse block.

➡ *Perform **Fuse Testing and Service** Worksheet*

■ FUSE LINK SERVICE

Fuse link wire is covered with insulation that bubbles if a fuse link melts. The blisters on the insulation indicate the failure in the fuse link. To replace a burned fuse link, cut out the damaged part of the wire. Splice a new fuse link in with a butt connector. Be sure to use a fuse link of the correct size.

■ HEADLAMP SERVICE

Headlamp Replacement

The back of a headlamp has tabs that align to holes in the headlight brackets. Halogen and conventional sealed beams are both of the *standard SAE design* and will fit into the same brackets. Round type I and type II lamps will *not* interchange because they are of different sizes.

NOTE: *While standard and halogen lamps can be interchanged, this should only be done in an emergency. Conventional lamps will not illuminate equally with halogen. Also, there can be a false signal generated to the headlight warning light on the dashboard due to different electrical resistances.*

A lamp substitution guide is located in Appendix G at the back of the book.

The headlight is held in a bracket that has adjusting screws on it.

NOTE: *If the headlight is being replaced because it is burned out, the new bulb will not normally require readjustment.*

Before attempting to remove a headlight, locate and identify the aiming screws (Figure 37.20). If these screws are accidentally turned, the adjustment of the headlights will be changed. Remove only the retaining screws for the trim piece around the headlight.

Use a test light to check both the high and the low beams of the electrical plug of a suspected burned out headlight. If there is no power to the connection, consult a service manual. Use a nonconductive (**dielectric**) grease on the connection when installing the new headlight.

NOTE: *The headlamp only fits properly one way. Locate the alignment tabs on the rear of the lamp when installing the lamp in the bracket.*

Halogen Lamp Replacement

Sealed-beam halogen lamps are replaced as a unit and are more costly than conventional sealed beam lamps.

- When the outer housing breaks on a halogen bulb, the inner housing (which could still be intact) will still light up. The quality of the light will be poor, however.

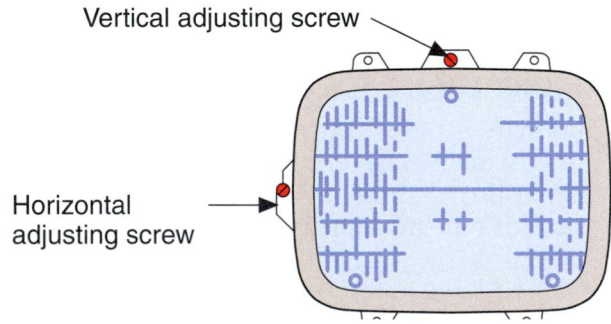

Vertical adjusting screw

Horizontal adjusting screw

Figure 37.20 Headlight aiming screws.

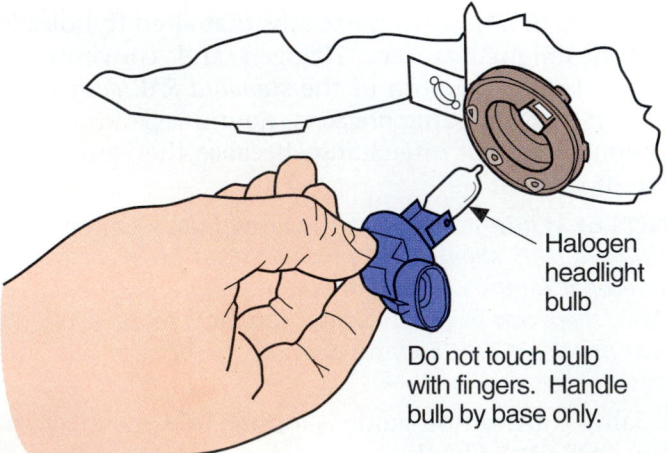

Halogen headlight bulb

Do not touch bulb with fingers. Handle bulb by base only.

Figure 37.21 Halogen bulb removal. *(Courtesy of Ford Motor Company)*

- A halogen lamp contains high pressure gas. If it is dropped, it may shatter.
- Halogen lamps burn at very high temperatures. When a lamp has been recently lit, do not attempt to replace it until it has cooled.
- When removing a composite halogen replaceable lamp, touch it only on its plastic base (Figure 37.21). If oil from skin gets on the lamp, it will shatter when it lights.
- Remove the old bulb only when ready to install the new one so dirt and moisture does not get into the lamp housing.

➡ **Perform** *R & R a Sealed Beam Headlight Worksheet*

■ HEADLIGHT AIMING

When properly aimed, low-beam headlights face down and to the right. This is so they do not shine in the eyes of oncoming drivers. Headlights are adjusted to a point that is 25 feet in front of the car. At that distance, the allowable margin of error is 4". From 300 feet, a driver is supposed to be able to see obstacles. But if a headlight is out of adjustment by the allowable amount of 4" at 25', the amount of the error is 4' at 300'. This can be dangerous, especially at high speeds.

Mechanical headlight aiming equipment is lightweight and is usually stored in a small suitcase. It is accurate and simple to use. Vertical and horizontal adjustments are made by turning two screws on the headlight bracket (see Figure 37.20). A headlight aiming set consists of two headlight aimers, mounting adapters, and level compensating adapters. The two adapters are mounted on the headlights using suction cups (Figure 37.22).

Before adjusting headlights, several things must be checked. The vehicle should be carrying its usual load. The trunk should contain its typical amount of material. Ideally, the gas tank would be half-full and the driver's weight in the front seat would be taken into account. Tire pressures should be adjusted.

Headlight aiming equipment uses a *bubble level* to calibrate vertical alignment. Horizontal alignment is made by comparing one headlight to the other to see if they are parallel. A target (sight bar) on one aimer is viewed through mirrors inside a window on the other aimer (Figure 37.23). The target on the aimer being viewed will appear as two lines when the two aimers are not parallel. As adjustment is corrected, the two

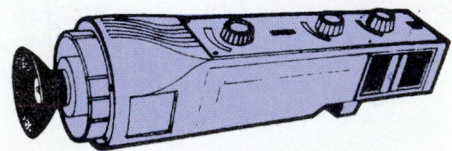

Figure 37.22 A headlight aimer. *(Courtesy of Hopkins Mfg. Co.)*

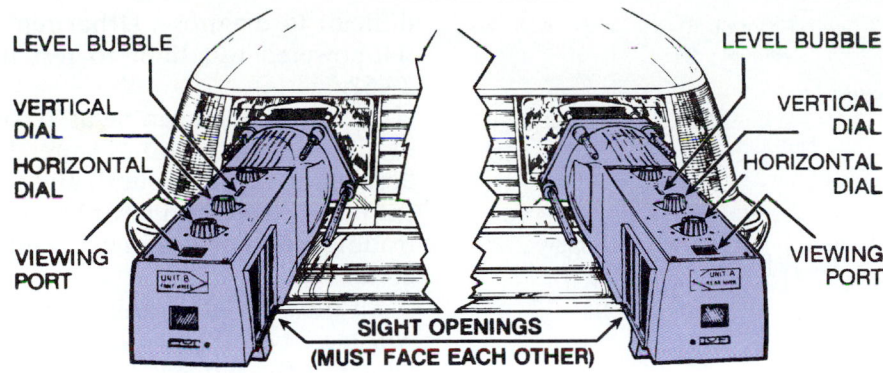

Figure 37.23 Headlight aimers read off of each other. *[Courtesy of Hopkins Mfg. Co.]*

lines merge to become one (Figure 37.24). Access to the adjusting screws is often possible without having to remove trim pieces or headlight doors.

The aimers must be adjusted to compensate for the level of the floor that the vehicle is on. When one area of the shop is always used for adjusting headlights, the compensation step can be skipped.

Mounting the Aimers

The aimers are attached to the headlights using their suction cups. The headlight must be clean for the suction cup to stick to it. Some lamps require a suction cup extension that is screwed into threads on the inside of the aimer's suction cup. Otherwise the suction cup will not reach the lamp. Different styles of headlights require different adapters to allow the aimer to fit the light.

Some new vehicles have composite lamps with a curved surface. Mounting the aimers to these lamps requires different length rods that are part of the adapter kit. American style headlamps have three guide pins on the front outside of the lens that locate the adapter. Each guide pin has a number next to it. Each rod is adjusted to the number on its corresponding headlight guide pin.

Prior to adjusting headlamps, each aimer can be adjusted using either its vertical or horizontal knob (see Figure 37.23). These knobs can also be used to determine the amount that a headlight is out of adjustment before aiming it. With the aimer mounted on the headlight, the amount of out-of-adjustment is determined by reading the gauge dial. Prior to reading the gauge dial, center the level bubble using the vertical knob and turn the horizontal knob until the split image is aligned.

NOTE: *If the gauge reads 6", the aim is off calibration 6" at a distance of 25' in front of the headlight.*

Some states have requirements other than zero for proper headlight adjusting.

■ TAIL AND PARK LIGHT BULB SERVICE

Bulbs in some tail lights are easy to remove and can be accessed from inside the trunk. The light socket is turned slightly to remove it from the tail light housing (Figure 37.25).

Other lights require removal of the light lens (Figure 37.26). Be careful when removing a light lens. The lens sometimes gets stuck to the rubber molding around it. If it is forced, an expensive lens can be broken. Be careful when reinstalling the lens that the gasket is in place in its groove.

To remove a bulb from its socket:
- ■ Push it in and turn it approximately ⅛ turn counter-clockwise.
- ■ It will spring out of the socket when it is released (see Figure 13.20).

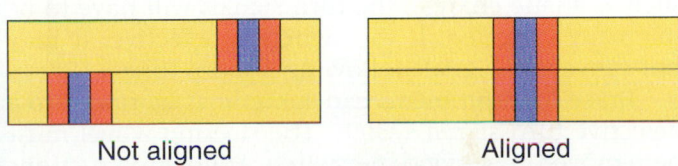

| Not aligned | Aligned |

Figure 37.24 When the headlights are aligned, both halves of the screen align. *[Courtesy of Hopkins Mfg. Co.]*

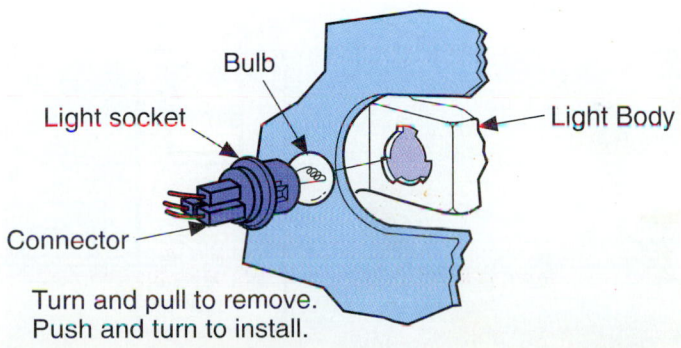

Turn and pull to remove.
Push and turn to install.

Figure 37.25 A removable light socket. *[Courtesy of Nissan Motors]*

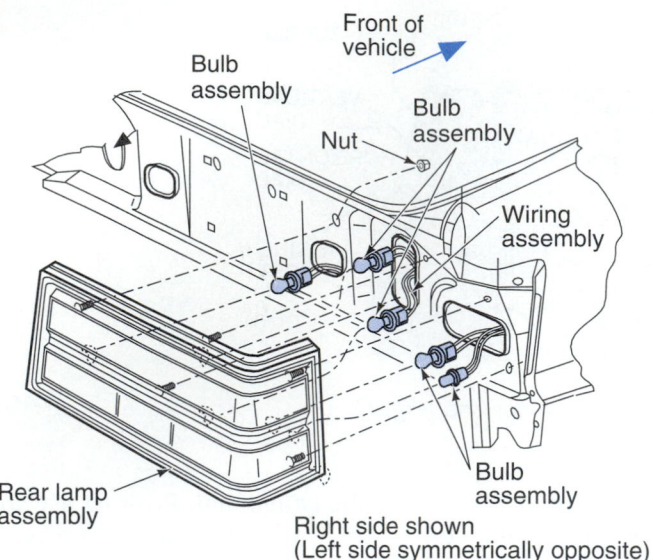

Figure 37.26 Access to this light requires removal of the lens. *(Courtesy of American Honda Motor Co., Inc.)*

Excessive voltage shortens the life of a light bulb. Voltage as little as 5% higher can reduce a bulb's life by half. Be sure the bulb socket is clean of oxidation, which would prevent electron transfer.

Check the part number (called the bulb trade number) printed on the base of the bulb to be sure it is the correct one. A wrong bulb could burn with a different intensity compared to the one on the other side of the vehicle. The bulb trade number is the same for a specified bulb, no matter who the manufacturer is.

You should be able to verify the condition of most bulbs by visually checking the filaments. A bad filament is usually obvious. If a problem with the lights is

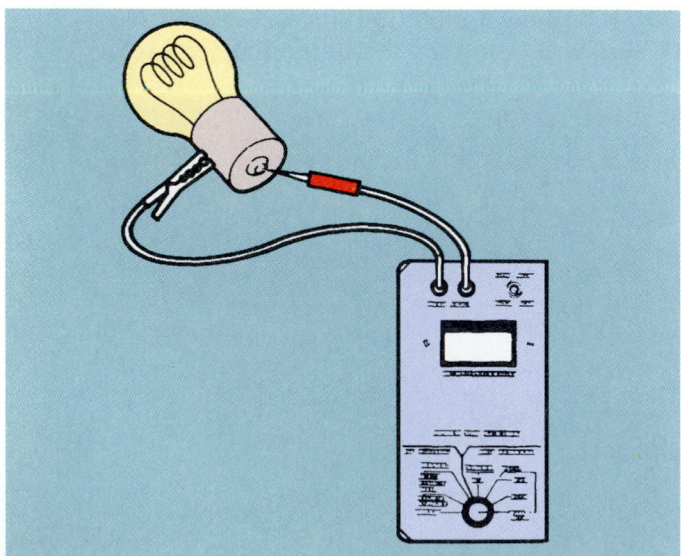

Figure 37.27 Checking a bulb with an ohmmeter. *(Courtesy of Ford Motor Company)*

difficult to diagnose, either replace the bulb or use a self-powered test light to test it as shown in Figure 29.52.

An ohmmeter can be used to check the continuity of the bulb filament (Figure 37.27). If the reading between the case (ground) and the terminal is infinity, the bulb is open and must be replaced. A relatively low reading on the ohmmeter gauge means the bulb is good.

If the bulb is a combination brake and tail light, it will have two filaments and two terminals. To check which of the wires is hot during a particular application, use a test light.

NOTE: *The socket is usually the ground for most bulbs. If you short the tester between the hot wire and the light socket housing, a fuse will blow.*

➡ *Perform* **R & R a Tail Light Bulb** *Worksheet*

STOPLIGHT SWITCH SERVICE

To test a suspected bad stoplight switch, bypass it with a jumper wire. The stoplights should light when the wires are connected. Use the test light to see if there is power at one of the wires.
- If the lights come on when the wires are connected but not when the brakes are applied, replace the switch.
- If the stoplights operate without the key on, remove the stoplight fuse before removing the switch.
- If the new switch is adjustable, adjust it so it is open when the pedal is released. The lights should come on about ¼" after the pedal is applied.

➡ *Perform* **Stop Light Switch Service** *Worksheet*

BACK-UP LIGHTS

The back-up light circuit includes a fuse, shift lever or transmission mounted switch, wiring, and lights. Sometimes the shift lever switch is adjustable. It is possible that the back-up lights could come on in a gear range other than reverse. Check the service manual for adjustment procedures.

TURN SIGNAL SWITCH

When a turn signal switch is operating properly, as the steering wheel returns to straight ahead after a turn, the signals should cancel. Following a minor turn, such as a lane change, the turn signals will have to be manually canceled. If the switch is defective, it may not cancel the signals following a sharp turn.

Brake light problems can sometimes be traced to a defective turn signal switch. The steering wheel must be removed to service the switch. Follow instructions in the service manual for steering wheel removal. Be especially careful if there is an air bag in the steering wheel (see Chapter 59).

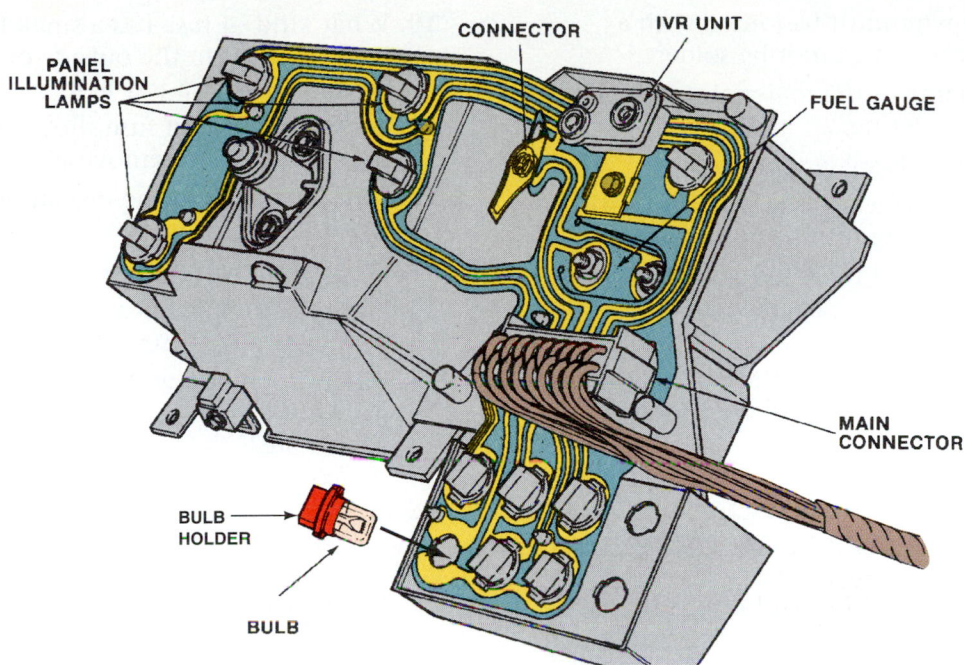

PANEL ILLUMINATION LAMPS

CONNECTOR

IVR UNIT

FUEL GAUGE

MAIN CONNECTOR

BULB HOLDER

BULB

Figure 37.28 Dash bulbs located in the rear of a printed circuit dashboard. *(Courtesy of Ford Motor Company)*

■ SIGNAL FLASHER

If a bulb burns out, there is less current travelling through the flasher and it will flash slower, or not at all. The speed of some flashers is dependent upon the current flowing through the circuit. The use of the wrong bulb can cause the flasher to flash too fast or too slow. Turn signals on American cars use either an 1157 or a 1034 bulb, which are interchangeable. An 1157 bulb draws more current than a 1034, however, and if it is used in place of a 1034, the flasher will flash faster. If a 1034 bulb is used in place of an 1157 bulb, it will flash slower.

There are different flashers available to match the load of the bulbs in the circuit. For instance, if a trailer is being towed and its lights are wired into the turn signal circuit, the extra current of the trailer lights will cause the flasher to flash too fast. A heavy-duty (variable load) flasher will solve the problem. If the heavy-duty flasher is used when the trailer is disconnected, the signals would normally flash more slowly. But a heavy-duty flasher always flashes at the same speed, no matter what the load (even if a bulb burns out).

To check the filaments of all of the turn signal bulbs and the brake lights, turn on the hazard flashers and walk around the car.

■ LOCATING A SIGNAL FLASHER

A flasher can be located in the fuse panel (see Figure 36.2), under the dash in a wiring loom, in a glove box, under the hood, or somewhere else. Most flashers are located under the dashboard on the driver's side. When working properly, a flasher is easy to locate because it makes a ticking sound. When it is not functioning, a service manual may be needed to help locate it. Lighting company catalogs generally have pages listing the locations of flashers.

➡ *Perform R & R Turn Signal Flasher Worksheet*

■ DASH LIGHT BULBS

Dash lights are usually very small bulbs that resemble small photographic flash bulbs (Figure 37.28). In printed circuit dashboards, they are housed in a plastic connector. Turning the connector ¼ counterclockwise removes it from the dash. The bulb is removed by pulling it straight out of the connector.

■ HORN SERVICE

The horn pitch can be adjusted by changing the spring tension on the armature. Turning a screw on the horn housing changes the rate of vibration in the horn.

■ REVIEW QUESTIONS

1. What is the name of the roadmap that shows all of the car's wiring and loads?

2. Besides color coding, striping, or hash marks, how might a wire be marked from the factory?

3. What is the name of the type of splice that can be used to join one wire to another without having to strip the first wire?

4. What kind of wire must be joined with a connector because it cannot be soldered?

5. What is the name of the material used to clean wire prior to soldering?

6. What two metals is solder made of?

7. What is it called when a wire is coated with solder before joining it to a connection?

8. What is the name of the vinyl material that is used with a heat gun to insulate an electrical connection?

9. What can be used to locate the exact location of a ground when a wire has been accidentally pinched?

10. What kind of fuse has a small hole that exposes its fuse metal to the outside of the fuse so it can be probed with a test light?

11. For which kind of fuse should you wear eye protection when removing?

12. What happens to the insulation on a fuse link when it fails?

13. What is nonconductive grease called?

14. Where is the target located that the mirror on a headlight aimer "sees"?

15. What do you do with the two wires on a stoplight switch to make the rear brake lights illuminate?

◼ ASE STYLE REVIEW QUESTIONS

1. Technician A says that wire of double the length has twice the resistance. Technician B says that wire that is too small for a circuit can cause excessive resistance. Who is right?
   a. Technician A **b.** Technician B
   c. Both A and B **d.** Neither A nor B

2. Technician A says that acid core solder is used for electrical connections. Technician B says that rosin core solder is used for electrical connections. Who is right?
   a. Technician A **b.** Technician B
   c. Both A and B **d.** Neither A nor B

3. Technician A says that a burned transmission bushing and drive shaft yolk could be caused by a broken ground strap. Technician B says that burned emergency brake cables could be caused by a broken ground strap. Who is right?
   a. Technician A **b.** Technician B
   c. Both A and B **d.** Neither A nor B

4. Technician A says that a 12-volt fuse will also work in a 110-volt application. Technician B says that voltage does not cause a fuse to blow. Who is right?
   a. Technician A **b.** Technician B
   c. Both A and B **d.** Neither A nor B

5. Technician A says that an ohmmeter is used on a circuit that is powered. Technician B says to remove and test a questionable fuse with an ohmmeter. Who is right?
   a. Technician A **b.** Technician B
   c. Both A and B **d.** Neither A nor B

6. Technician A says that installing a fuse with too high an amp rating will result in a burned fuse. Technician B says that installing a fuse with too low an amp rating can cause a fire. Who is right?
   a. Technician A **b.** Technician B
   c. Both A and B **d.** Neither A nor B

7. Technician A says that halogen and conventional bulbs of the standard SAE design are not interchangeable. Technician B says that excess voltage shortens bulb life. Who is right?
   a. Technician A **b.** Technician B
   c. Both A and B **d.** Neither A nor B

8. Technician A says that headlight aim must be readjusted after replacing a headlight. Technician B says that some halogen headlamps can be ruined if they are handled by bare hands. Who is right?
   a. Technician A **b.** Technician B
   c. Both A and B **d.** Neither A nor B

9. Two technicians are using ohmmeters to check a lightbulb whose ground is its case. Technician A says that if the reading between the case and the terminal is infinity, the bulb is good. Technician B says that a high reading means the bulb is good. Who is right?
   a. Technician A **b.** Technician B
   c. Both A and B **d.** Neither A nor B

10. Technician A says the speed of some flashers is dependent upon the voltage flowing through the circuit. Technician B says using the wrong bulb can cause the flasher to flash too slow. Who is right?
    a. Technician A **b.** Technician B
    c. Both A and B **d.** Neither A nor B

Engine Performance Diagnosis: Theory and Service

THEN AND NOW: IGNITION SYSTEMS

Believe it or not, some of the earliest cars used *hot-tube ignition*—a platinum tube extended through the cylinder wall into the combustion chamber. A flame from an alcohol lamp outside kept the tube red hot. Even more unbelievable was the *flint and steel* idea: a rough piece of steel mounted on the top of the piston rubbed against a spring-loaded flint to make a spark.

A Frenchman named Etienne Lenoir invented the spark plug in 1860, 26 years before the first automobile appeared. But most early cars did not use spark plugs. Instead, they had *low-tension magnetos*, which produced electricity from magnets and coils of wire. A spark resulted when two contacts inside the cylinder were pulled apart by a complex mechanism.

The *high-tension magneto* appeared in 1903. It added an induction coil to produce enough voltage to jump the gap of a spark plug. This ignition system was popular in Europe into the 1930s. The *trembler coil* was first used on the 1896 Benz. It made a steady stream of sparks from a vibrating blade that controlled current through the coil. The trembler coil was very successful in the United States, sparking the 15 million Ford Model Ts built between 1909 and 1927.

Charles "Boss" Kettering, one of the founders of Delco, developed the dependable *breaker point and coil ignition system*, which was in use for over half a century. A drawback to this type of ignition was that the rubbing block of the movable point slowly wore down against the distributor cam. This resulted in spark timing occurring later and later. Also, firepower fell off dramatically at high rpm.

Electronic ignition solved this problem. Chrysler used it on all models in 1973. By 1975, most cars sold in the United States had electronic ignition systems. With nothing to wear out, spark timing remained constant throughout the life of the car. The coil also got its current so fast there was no danger of running out of spark during hard acceleration.

Modern *distributorless ignition systems (DIS)* came on the scene with the 1984 Buick. Camshaft and crankshaft position sensors inform an electronic ignition module of the position of each piston in its cylinder in terms of the combustion cycle. The ignition module on this system fires multiple coils, which have spark plug wires running directly to the spark plugs. This design eliminated the need for the potentially troublesome distributor rotor and cap of the earlier distributor systems. Another DIS version called *direct ignition*, eliminates the need for spark plug wires by mounting the coils in an assembly that snaps onto the spark plugs.

A Ford Model A ignition system. *(Courtesy of Bob Freudenberger)*

A modern platinum-tipped spark plug. *(Courtesy of Bob Freudenberger)*

CHAPTER 38

Ignition System Fundamentals

■ OBJECTIVES

Upon completion of this chapter, you should be able to:

✔ Describe the functions of ignition system parts.

✔ Explain the operation of points, electronic, and computer ignition systems.

✔ Give an overview of the different spark advance methods.

✔ Describe distributorless and conventional ignition distributor variations and operation.

✔ Make a wiring diagram showing the primary and secondary ignition systems.

■ KEY TERMS

coil saturation
dwell
vacuum advance
distributorless ignition systems (DIS)
ground electrode
reach
heat range
television/radio suppression (TVRS)
trigger
electronic ignition systems
Hall-effect switch
light emitting diode (LED)
crank angle sensor
current limiting system
variable dwell
ignition timing setting
centrifugal advance
electronic computer advance
ported vacuum
base timing
waste spark method

■ INTRODUCTION

The ignition system's job is to create a timed spark and distribute it to the engine's cylinders. The ignition system also serves the function of turning the engine on and off. The spark is distributed to the spark plugs where it jumps the gap and ignites the air/fuel mixture just before the piston reaches the top of its compression stroke. The ignition timing must change with engine speed because the amount of time it takes for the fuel to burn in the cylinder is relatively constant.

There are several different types of ignition systems. New cars have computer controlled ignition systems. The specifics of those systems are covered in Chapter 75. This chapter covers the parts and operation of mechanical and electronic ignition systems.

■ BASIC IGNITION SYSTEM

Parts of the ignition system include the battery, ignition switch, coil, points or transistor switching device, distributor, cap and rotor, plug wires, and plugs (Figure 38.1). An ignition system has two circuits called primary and secondary. The primary is the low voltage (battery) part of the system and the secondary is the high voltage (spark) side of the circuit.

■ PRIMARY CIRCUIT

The primary circuit carries current from the battery. When the flow of current in the primary circuit in a coil is switched on, a magnetic field is energized. This creates a high voltage spark from the ignition coil to the spark plug.

Parts of the primary ignition system include the:

■ Battery
■ Charging system
■ Ignition switch
■ Resistor bypass circuit
■ Ignition coil primary windings
■ Breaker points or electronic switching device
■ Distributor cam lobes or crank/cam sensor
■ Ground return path

Battery voltage is converted to high voltage by the ignition coil. An electrical spark jumps across a gap at the end of a spark plug. Electricity for the ignition system is supplied with battery voltage when the engine is cranking. Once the engine is running, the alternator supplies the necessary electrical flow. Spark timing is critical to power output. It cannot be too early or too late. An ignition system has spark advance systems to control this.

Ignition Switch

A multiposition switch turns power to the ignition circuit on and off. In the *start* position, it triggers the starter motor and provides a parallel path for electrical flow to the ignition system. After starting, in the *run* position it continues to allow current to flow to the ignition system and the entire electrical system. The

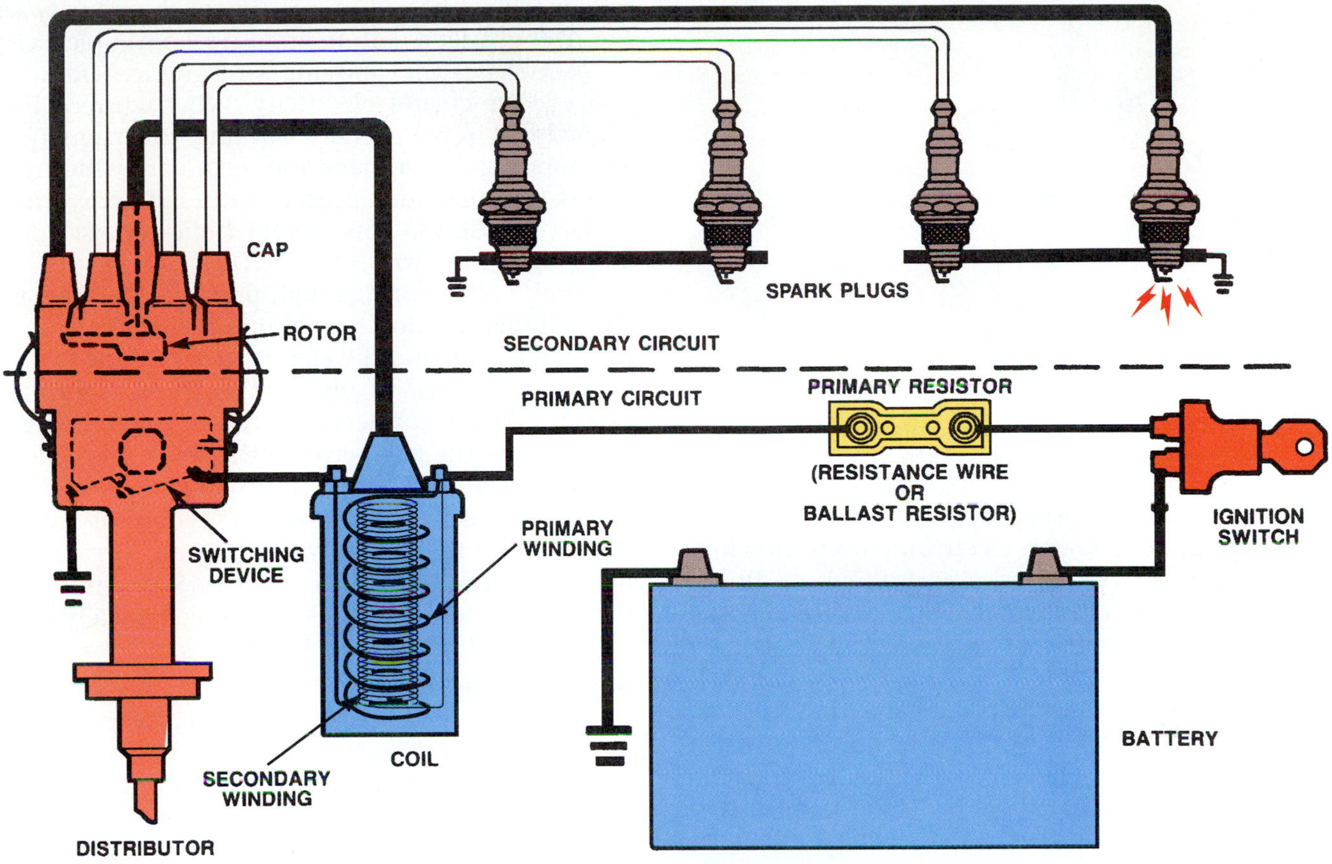

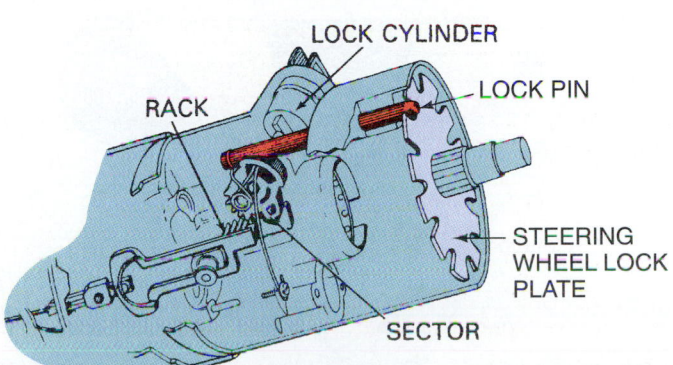

Figure 38.1   Parts of an ignition system. *[Courtesy of Cooper Automotive/NAPA Belden]*

accessory position powers electrical accessories when the engine is off. This key position is sometimes when the key is turned counterclockwise. Other times, the key is turned to the first position in clockwise rotation.

The ignition switch operates the steering wheel lock (Figure 38.2) and a buzzer or light. If the key is in the ignition when the engine is off, a door is ajar or open, or seat belts are not buckled, the buzzer and/or light will operate. Some fuel pumps are connected to electrical power through the ignition switch. More information on the ignition switch is found in Chapter 32.

Figure 38.2   The ignition switch also operates the steering wheel lock. *[Courtesy of General Motors Corporation, Service Technology Group]*

CASE HISTORY

A student had a daily commute several miles down a mountain road to get to school. He had an idea that he could save gas by coasting down the hill. One day he tried it. He started driving down the road and then shut the ignition switch off. He turned the switch all the way off and when he came to the first turn in the road, the steering wheel lock pin engaged the slot in the steering wheel lock. He couldn't steer the car so it went off of the road. Luckily, the hill he went over wasn't too steep. The car was heavily damaged but no one was hurt.

Ignition Coil

The coil is the heart of the ignition system. Battery voltage (12.6 V) is not enough electrical pressure to push a spark across the spark plug gap. The coil is a transformer that transforms battery voltage into many thousand volts.

A coil has a low voltage *primary winding* and a high voltage *secondary winding* (Figure 38.3). Because it has both low and high voltage windings, part of the coil is in the primary system and part of it is in the secondary system.

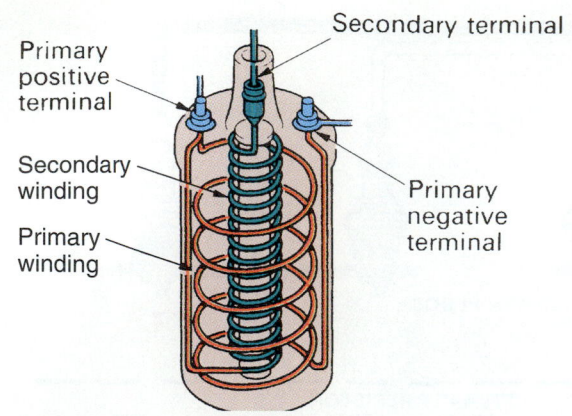

Figure 38.3 A coil has a low voltage primary winding and a high voltage secondary winding.

In Chapter 29 you learned that a magnetic field surrounds a conductor with electrical current flowing in it (see Figure 29.31). When electrical current flows through a coil it creates a magnetic field (see Figure 29.33). Coiling the conductor around an iron core creates a stronger field. The high voltage spark is created by the ignition coil when the electrical current supply is discontinued and the magnetic field breaks down.

The more windings there are in a coil, the stronger the magnetic field. When there are fewer coils in the primary part of a transformer, it is a *step-up transformer* (Figure 38.4). Fewer coils in the secondary would make a *step-down transformer*. An ignition coil is a step-up transformer.

The coil's primary winding is relatively large wire, coiled 150 to 250 times. It carries several amps of current flow at battery voltage. Within the center of the primary winding is a secondary winding. It is made up of several thousand turns of very small, hairlike insulated wire. The secondary winding is wrapped around strips of iron, layered together to make up a laminated iron core.

One end of the secondary winding goes to the center terminal that leads to the distributor cap. This is where the high voltage pulse will leave the coil on its way to the spark plug gap. The other end of the secondary

winding is connected to one side of the primary winding. This gives it a path to ground for a complete circuit.

Remember that moving a magnetic field across a coil of wire creates electricity. When current flow is interrupted in the primary winding, the magnetic field surrounding the primary and secondary windings collapses. The magnetic lines of force cut across the secondary windings. This creates high voltage and low amperage in the winding.

High voltage means high pressure. The pressure at the coil high tension terminal in the center of the coil housing is high and the electricity is looking for a path to ground. Electricity always takes the shortest path to ground. Wiring connects the coil to a spark plug. The spark jumps the gap between the spark plug's positive and negative electrodes.

When did the spark happen? When the switching device broke the circuit in the primary system allowing the magnetic field to collapse. Switching the primary current on and off is done mechanically in older cars and electronically in newer cars.

Saturation and Dwell. When the magnetic field has finished its buildup inside of the coil, the coil is said to be *saturated*. **Coil saturation** time depends on the amount of current flowing in the primary winding. The older contact point systems had slower saturation times because they were limited to about 2-3 amps of current flow. Some of the newer electronic ignition systems can flow current of up to 20 amps. This allows much faster electron flow.

Dwell is the length of time in degrees that primary current is flowing in the primary winding. It is measured in degrees of distributor rotation (½ crankshaft speed). In a contact point system, dwell is the length of time that the points are closed (Figure 38.5).

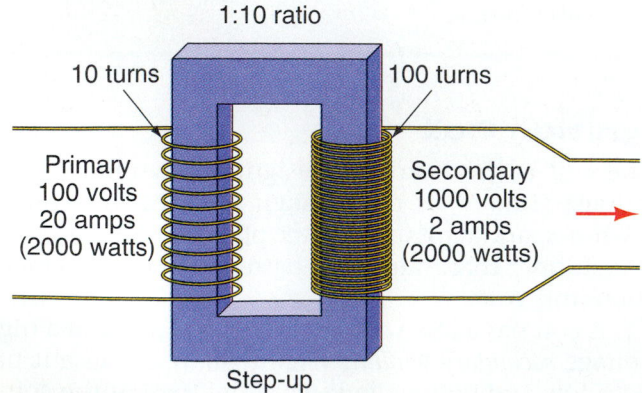

Figure 38.4 A step-up transformer has fewer windings in its primary winding. *(Courtesy of Ford Motor Company)*

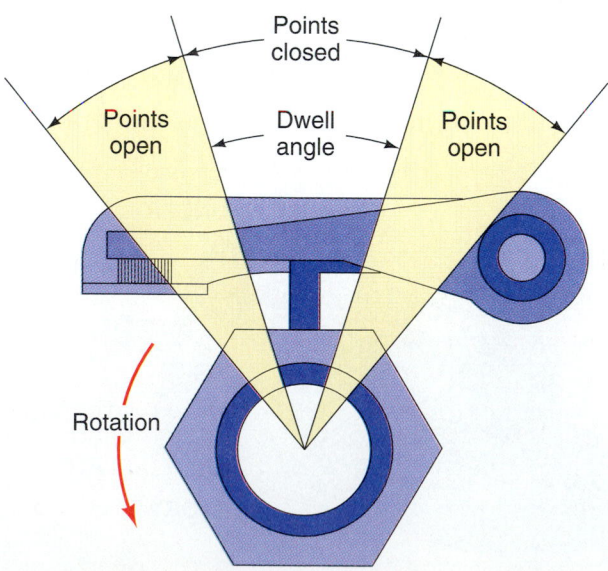

Figure 38.5 Dwell is the length of time that the points are closed. *(Courtesy of Sun Electric Corporation)*

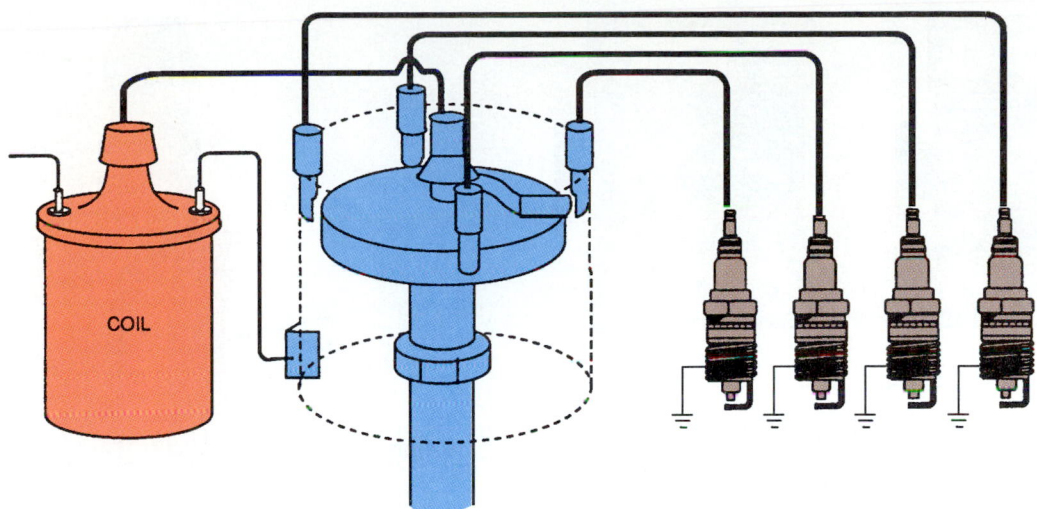

Figure 38.6 The secondary circuit is the high voltage side of the ignition system. *(Courtesy of SPX Corporation, Aftermarket Tool and Equipment Group)*

In an electronic system, it is the length of time that the ignition module allows current to flow through the coil primary winding.

SECONDARY CIRCUIT

The secondary circuit is the high voltage side of the ignition system. It delivers high voltage from the coil to the spark plugs (Figure 38.6).

DISTRIBUTOR

Contact point systems and older electronic systems have a distributor that includes a *cam* that triggers the points or an electronic *module*, also called an *igniter*. The *distributor cap*, *rotor*, and *mechanical* and **vacuum advance** mechanisms are also part of the distributor. Many engines on new cars have no distributor. These **distributorless ignition systems (DIS)** have multiple ignition coils (usually one for every two cylinders). They fire spark plugs based on a signal received from the computer. Plugs are sometimes fired through a plug wire and sometimes directly (with no plug wire). In a distributorless system, the crankshaft has a pickup that takes the place of the distributor.

SECONDARY IGNITION PARTS

Distributor Cap and Rotor

In conventional systems, electricity flows from the coil to a distributor cap and rotor (Figure 38.7). Wires plugged into holes in the top of the distributor cap connect the distributor cap and spark plugs. The ends of the plug wires fit snugly into the distributor cap.

The distributor shaft is driven by the engine's camshaft. Remember that the camshaft turns once for every two turns of the crankshaft, so the distributor rotates at one-half crankshaft speed. A spark occurs

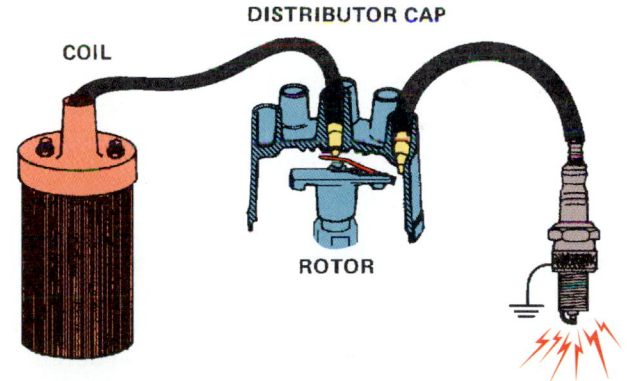

Figure 38.7 Electricity flows from the coil to a distributor cap and rotor. *(Courtesy of Sun Electric Corporation)*

just before the power stroke. Then, the crankshaft rotates through four more strokes (two turns) before another spark is required in that cylinder.

The rotor is located on top of the distributor shaft. As it turns, it points to a different terminal inside of the distributor cap. The center of the distributor cap has a button made of carbon. Electricity is fed into the center of the cap through a *high tension lead* from the coil. From there, it goes to a tab on the center of the rotor. Electricity flows through the rotor to a different terminal on the cap each time the primary circuit magnetic field breaks down. The spark plug cables are arranged around the distributor cap in the engine's firing order (Figure 38.8). For more information on an engine's firing order refer to Chapter 16.

SPARK PLUGS

In gasoline engines, spark plugs provide a gap for the spark to jump. The spark ignites the compressed air/fuel mixture in the cylinders. Figure 38.9 shows the

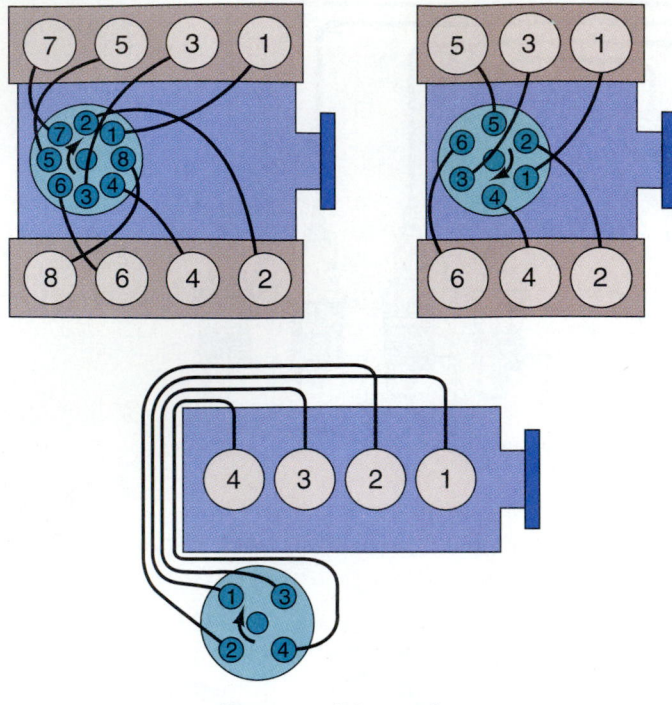

Common firing orders

Figure 38.8 The spark plug cables are arranged around the cap following the engine's firing order.

parts of a spark plug. A spark plug has a metal shell with threads that mesh with threads in the cylinder head. A metal conductor runs down the center of the plug, surrounded by a ceramic insulator. The end of the conductor forms the *positive electrode*. On the end of the metal case is the **ground electrode**.

The ground electrode can be bent toward or away from the center electrode to make the correct spark plug gap. The gap varies depending on the kind of ignition system used. Spark plugs in older point-type ignition systems specified small gaps of around 0.035". The gaps on newer electronic ignition cars can vary up to 0.080". A gap this large produces a very hot spark.

Spark Plug Threads

Spark plugs have different thread diameters (10, 12, 14, and 18 mm). The length of the threaded area of a spark plug is called its **reach**. A plug with too long a reach will extend into the combustion chamber (Figure 38.10). If the reach is too short, the end of the plug will not reach the ends of the threads in the head.

Heat Range

A spark plug's **heat range** is an indication of how fast heat can travel away from the spark plug center electrode to the cooling system's water jackets in the cylinder head. There are several different spark plug heat ranges available. The length of the ceramic insulator is what determines heat range. Figure 38.11 shows how

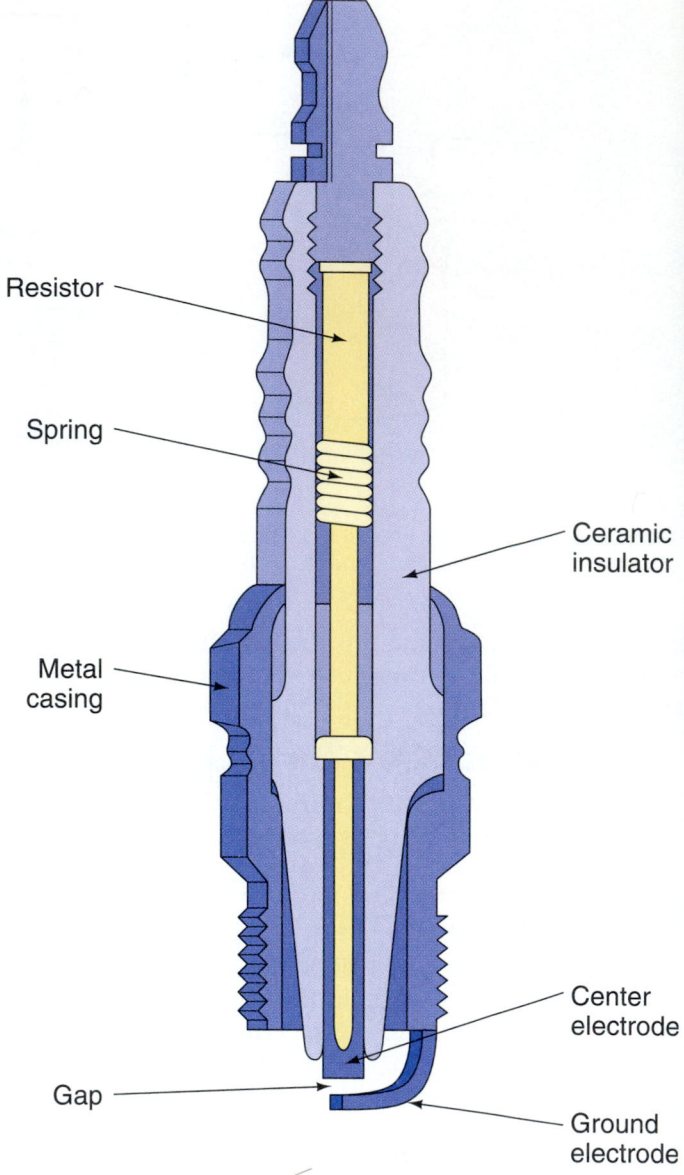

Figure 38.9 Parts of a spark plug. *(Courtesy of Cooper Automotive/Champion Spark Plug)*

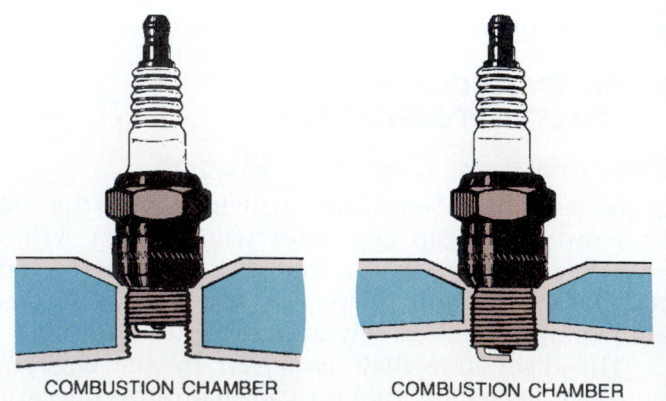

COMBUSTION CHAMBER   COMBUSTION CHAMBER

Figure 38.10 Incorrect reaches. *(Courtesy of AlliedSignal Automotive Aftermarket/Autolite)*

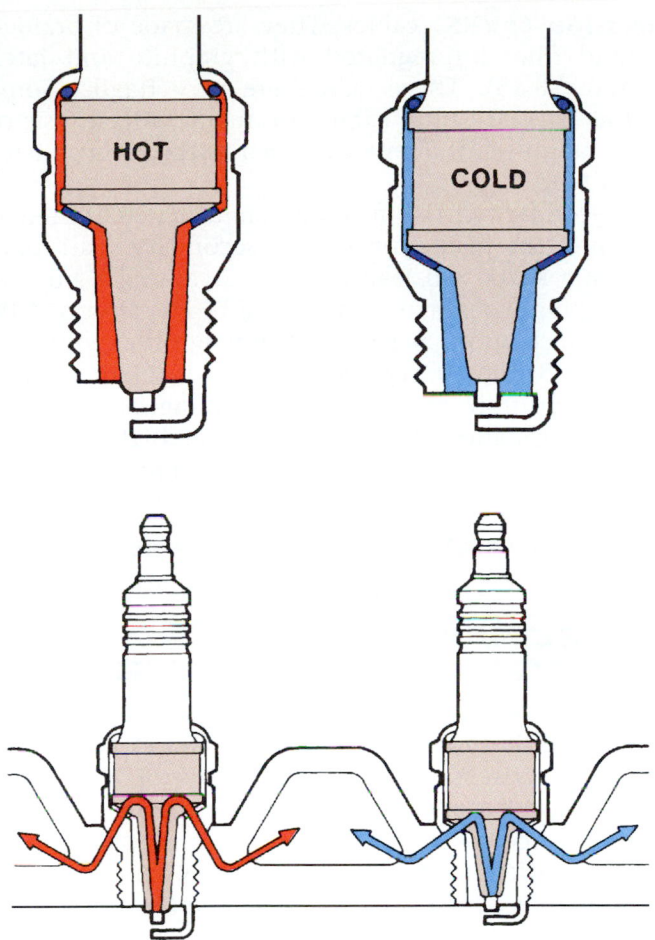

Figure 38.11 The length of the ceramic insulator determines heat range. *(Courtesy of General Motors Corporation, Service Technology Group)*

the path of heat transfer determines the temperature at which the plug will run. According to Champion Spark Plugs, over 90% of a spark plug's heat leaves it at the edge of the threads. The length of the ceramic insulator above the metal makes very little difference.

The use of spark plugs of the wrong heat range can cause major engine damage (see Figures 25.15 and 25.16). If the heat path is long, the plug will run hot. A short path means a cold plug. If the plug is too hot, it will burn the plug electrode. It can also cause pre-ignition and can burn a piston. If a plug is too cold, deposits will build up on it and it will foul. Spark plug fouling means that it gets covered with material and a spark will not jump its gap.

The amount of what the plug projects out below the threads has very little to do with the heat range. The reason for the projection is to get the spark deeper into the combustion chamber. A flush tip plug is more likely to foul when used under low-speed, light-load conditions. The extended tip acts like a hotter plug during those conditions. At higher speeds there is a charge cooling effect, which lets it perform like a cold plug. A

projected tip maintains a more stable temperature range. The only reason a flush tip plug is used is to avoid contact with a piston. Most modern pistons have enough clearance to be able to use an extended tip plug.

Taper and Gasket Type Plugs

To seal it against the cylinder head, a spark plug has either a tapered seat or a flat seat with a gasket (Figure 38.12). Tapered seats are used in all sizes of spark plugs, but especially in the smaller space-saving spark plugs. Spark plugs with gaskets are still popular also. The two types are not interchangeable.

Platinum Plugs

Platinum is used for the electrode in long life plugs because it is resistant to chemical erosion and corrosion. When these plugs are used in some new vehicles, the manufacturer might even say that the engine can go 100,000 miles without changing them.

Resistor Plugs and Wires

Resistance is added to the secondary ignition system, either in the spark plugs or spark plug cables. This increases the required firing voltage, which raises the spark plug firing frequency out of the range of radio interference.

■ SCIENCE NOTE ■

As the plug is fired, the plug voltage rises and falls rapidly. The plug is actually refired several times during about 1/1000 of a second. This produces an alternating current, which the resistance prevents. Dropping the voltage with a resistance changes the intensity of the magnetic field that happens along the length of the wire, which also lowers radio noise.

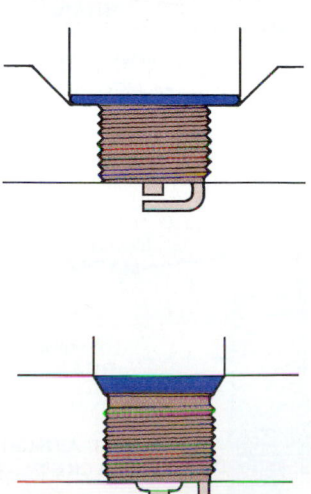

Figure 38.12 Gasketed and tapered seat plugs. *(Courtesy of General Motors Corporation, Service Technology Group)*

The wireless telegraph, invented by Guglielmo Marconi in 1895, was a spark transmitter similar to the ignition system on a car. Pulsed messages could be sent using the long and short dashes and dots of the Morse code. The transmitter had a wire antenna attached to its gap. It could be tuned by changing the width of the gap or the shape of the electrodes. This allowed each transmitter to have its own characteristic sound that could be identified by radio operators. This was important if signals were being received by more than one transmitter at a time.

When radios began to be installed in automobiles, it became apparent that ignition systems with metal spark plug wires were actually transmitting signals similar to the telegraph. Adding resistance to the circuit changed the frequency of the ignition pulses so that they would not interfere with radio reception.

Spark plugs often have resistors. A resistor inside the spark plug raises the firing voltage required by the coil (see Figure 38.9). Some spark plugs have carbon resistors. Others have a semiconductor suppressor that is not energized until 1000 volts pass through it. This kind of resistor plug cannot be checked with an ohmmeter.

Spark Plug Cables

Most spark plug cables (wires) provide a resistance, too. Resistor cables are called **television/radio sup-**

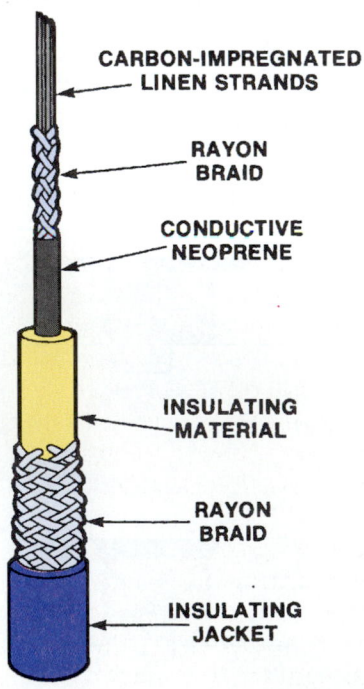

Figure 38.13 Static suppression ignition cable. *(Courtesy of General Motors Corporation, Service Technology Group)*

pression (TVRS) cables. They are made of braided aramid fiber impregnated with graphite and latex (Figure 38.13). These cables are very fragile. Some European spark plug cables are metal with a resistor boot at the end. These are used with a nonresistor spark plug.

Secondary wiring must be well insulated. If there is a leak in the insulation on the secondary spark plug cables, a spark will leak to ground causing a misfire. Older plug wires had 7 mm (0.282") Hypalon or CPE rubber insulation. Newer electronic ignition equipped cars use 8 mm (0.312") silicone for insulation. Electronic ignition creates higher voltages, requiring thicker insulation. Silicone has better high temperature resistance than Hypalon. It is rated for 400°F continuous use.

➡ *Perform* ***Ignition Wiring Diagram*** *Worksheet*

■ MECHANICAL AND ELECTRONIC IGNITIONS

Electromechanical contact point systems and electronic ignition systems are similar. Newer and better electronic systems have taken over control of the basic mechanical functions of the contact points system. Maintenance has been reduced tremendously.

The difference between the two systems is in the **trigger** mechanism that controls the flow of current in the *primary* winding. Both types of systems control the flow of current on the ground side of the coil. Refer to Figure 38.14 and compare the primary circuits of electronic and point-type ignitions.

Breaker Points

The earlier electromechanical systems have a contact switch called *contact points*, or *breaker points*. The points alternatively energize, and then open, the primary ignition circuit. The points are opened by lobes on a *distributor cam* that has as many lobes as the engine has cylinders (Figure 38.15). When the contact points separate, the magnetic field breaks down. This results in a spark from the secondary winding of the coil.

When the engine is running at 3000 rpm, a coil on an eight-cylinder engine must do remarkable things. Its magnetic field must be able to build up and collapse 200 times in one second.

3000 revolutions per minute divided by 60 seconds = 50 revolutions per second. Divide this by two (= 25) because each spark plug only fires once every two turns of the crank. Multiply that by the number of cylinders (8) and you have 200 times per second.

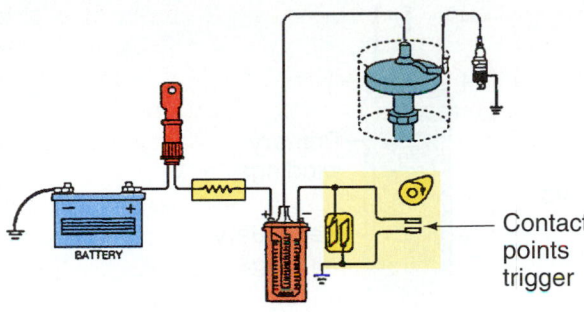

Mechanical system

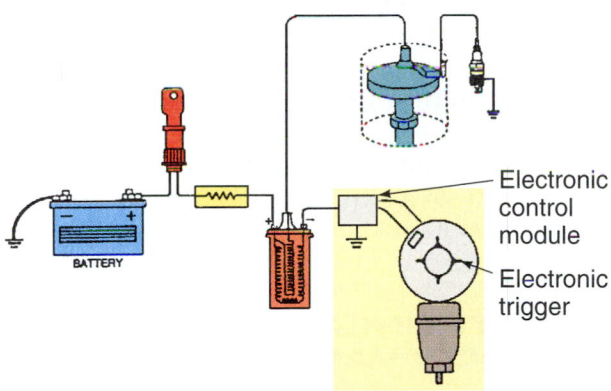

Electronic system

Figure 38.14 A comparison of primary triggers. *(Courtesy of SPX Corporation, Aftermarket Tool and Equipment Group)*

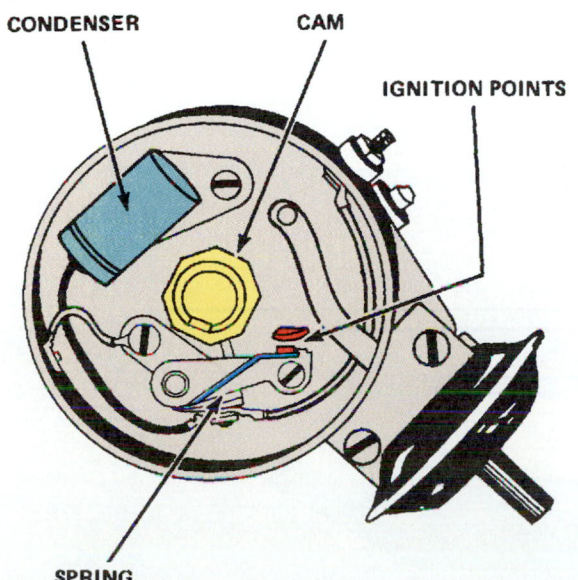

Figure 38.15 A distributor cam has as many lobes as the engine has cylinders. *(Courtesy of Sun Electric Corporation)*

This means that at ordinary highway speeds, a single set of contact points must open and close 200 times per second. Imagine what an 8-cylinder Indy car's ignition system must do at 15,000 rpm. It has 1000 spark plug firings per second.

Contact points are a mechanical switch, so they suffer from wear and pitting over a period of operation. Burned points cannot conduct enough electricity to create the rapid sparks that are needed at normal engine rpm. Also, points have a *rubbing block* that rides against the distributor cam. When this rubbing block wears down, the point gap becomes narrower. This retards the ignition timing. Points are usually replaced at about 10,000 mile intervals as part of a tuneup. An additional disadvantage to points is that they bounce at high speeds.

Condenser

A condenser, or capacitor, is installed with contact points. After the primary winding of the coil has become saturated, the points separate. When this happens, the magnetic field in the coil breaks down. As the contact points separate, current still attempts to flow across them. This results is arcing, or burning away of the tungsten material from the contact surfaces. The metal from one contact deposits on the other side until the surfaces are too pitted and burned to conduct electricity. The condenser protects against this wear. It also allows the magnetic field to collapse more quickly.

Condenser Operation. When the contact points begin to open, electricity sees an easier parallel path for electrical flow into the condenser (see Figure 29.35). By the time the electrons have filled the positive side of the condenser, the points have separated sufficiently to prevent further current flow.

■ RESISTOR BYPASS

When a distributor has mechanical breaker points, a resistor is needed to keep the points from burning prematurely. With the key in the run position, electrical power to the ignition system flows through a resistor to lower the current flow to acceptable levels. Point-type ignition systems limit current flow to under 3 amps with a maximum voltage output of 20,000 to 25,000 volts. The coil can produce this amount of voltage if it has about 9 to 10 volts available to it. So that is what the resistor limits voltage to.

During cranking, the amount of current flowing through the coil primary winding drops off as battery voltage drops. To provide a hot spark, full battery voltage is supplied during cranking. It is provided through a wire from the starter solenoid or ignition switch (Figure 38.16). This circuit is called the resistor bypass. After cranking, the ignition switch returns to the run position and ignition power is supplied through the resistor. The resistor can be within the coil or outside of it in the form of a *resistor wire* or *ballast resistor*.

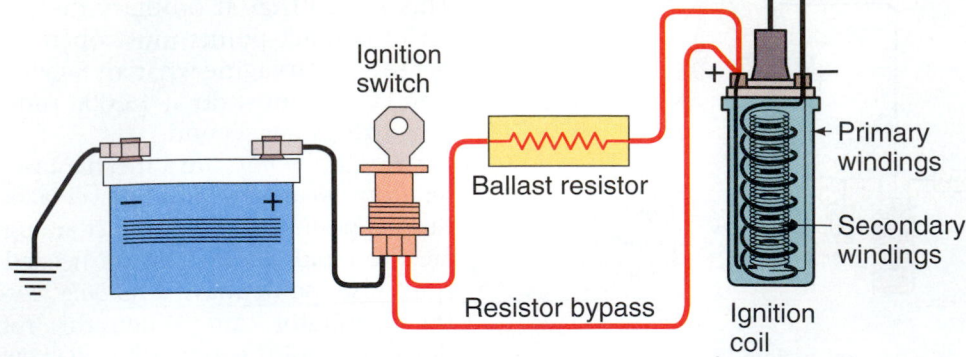

Figure 38.16 Resistor and bypass circuits. *(Courtesy of Sun Electric Corporation)*

■ ELECTRONIC IGNITION

Electronic ignition systems are more reliable and have improved emissions, performance, and fuel economy. The primary and secondary systems perform in the same way as points, but a transistor triggers the buildup and collapse of the magnetic field.

No primary current runs through the distributor. The distributor simply houses the trigger that controls the functioning of the transistor. The transistor is housed in an ignition module (Figure 38.17).

A transistor is an electronic switch, or relay. A transistor is a semiconductor, like a diode. The operation of semiconductors is covered in Chapter 75. Most ignition systems have one or more *power transistors* that switch the coil on and off. Power transistors can regularly carry 10 amps of current; far more than contact points could. The power transistor(s) are controlled by a *driver transistor* that gets its signal from a triggering device in the distributor.

Because there is no mechanical switch or electrical arcing with electronic ignition systems, regular maintenance is not a normal short mileage occurrence (as it was with breaker points).

In an electronic ignition system, the only parts that suffer wear are parts of the advance mechanisms and bushings in the distributor. In later electronic systems, advance mechanisms are functions controlled by a computer chip (see Chapter 75). The distributor consists simply of a switching mechanism, rotor, and shaft. Other systems have no distributor. These are called DIS, or distributorless ignition systems.

■ ELECTRONIC IGNITION VARIATIONS

The trigger in electronic ignitions differs between manufacturers. The trigger is either in the distributor or on the crankshaft. It can be a magnetic AC generator, a Hall switch, or an optical (photoelectric) sensor.

AC Generator Systems

A popular type of electronic trigger is the magnetic-type AC generator pickup located in the distributor

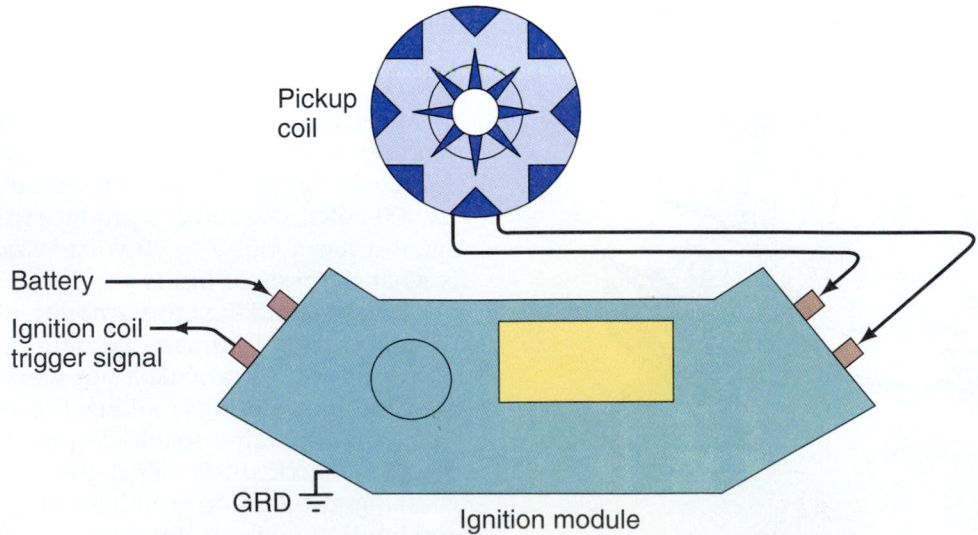

Figure 38.17 The transistor is housed in an ignition module. *(Courtesy of General Motors Corporation, Service Technology Group)*

(Figure 38.18). The system works like an alternator. Different manufacturers call the parts by different names, but they all work the same way. A stationary sensor called a *pickup coil* sends signals to a transistor located in the *ignition module.*

The pickup coil is wrapped around an iron *pole piece* near a permanent magnet. The magnetic field of the permanent magnet surrounds the pickup coil.

A nonmagnetic *trigger wheel*, also called a *reluctor, pulse ring,* or *armature,* is attached to the distributor shaft. It replaces the distributor cam on the contact point system. The trigger wheel has as many teeth on it as the engine has cylinders. Every time a reluctor tooth passes across the windings in the pickup coil, a small voltage is generated. The ignition module senses this voltage and uses it to control the buildup of the magnetic field in the coil primary winding.

The trigger wheel has low magnetic reluctance. Low reluctance can be compared to low resistance. Magnetism is easily absorbed by a material with low reluctance. As the tooth of the trigger wheel is rotated into the magnetic field around the pole piece and pickup coil, magnetic flux becomes concentrated in the tooth (Figure 38.19). The magnetic field around the pole piece changes and a small voltage is induced in the pickup coil. The amount of voltage changes with the rate of the magnetic flux. As the reluctor tooth moves away from the pole piece, the magnetic field becomes weaker (Figure 38.20).

The output voltage signal varies between positive and negative (alternating current). As the reluctor approaches the pole piece, the voltage increases toward positive. When the reluctor teeth align with

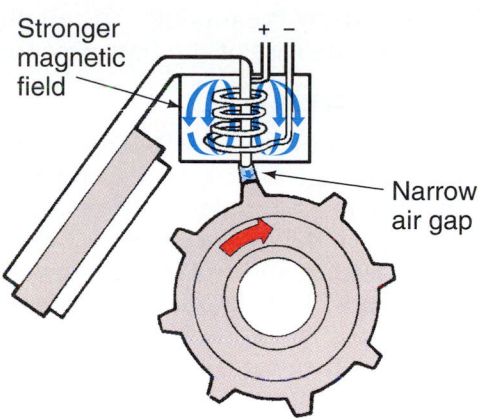

Figure 38.19 As the tooth of the trigger wheel lines up, magnetic flux concentrates in the tooth. *[Courtesy of Chrysler Corporation]*

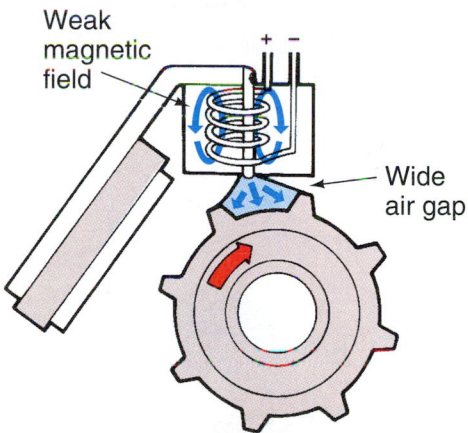

Figure 38.20 As the reluctor tooth moves away from the pole piece, the magnetic field becomes weaker. *[Courtesy of Chrysler Corporation]*

the pole piece, voltage is at zero. The voltage increases in a negative direction when the reluctor moves further away from the pole piece.

Most electronic ignition systems open the circuit as the polarity changes from positive to negative. This is the same as when breaker points open, resulting in a spark at the plug. As the tooth moves out of the magnetic field of the pole piece, the next tooth on the trigger wheel comes into the field and the cycle repeats itself. Once again current flows to create another magnetic field.

The transistor in the module shuts off current flow in the primary circuit. The AC signal voltage is made

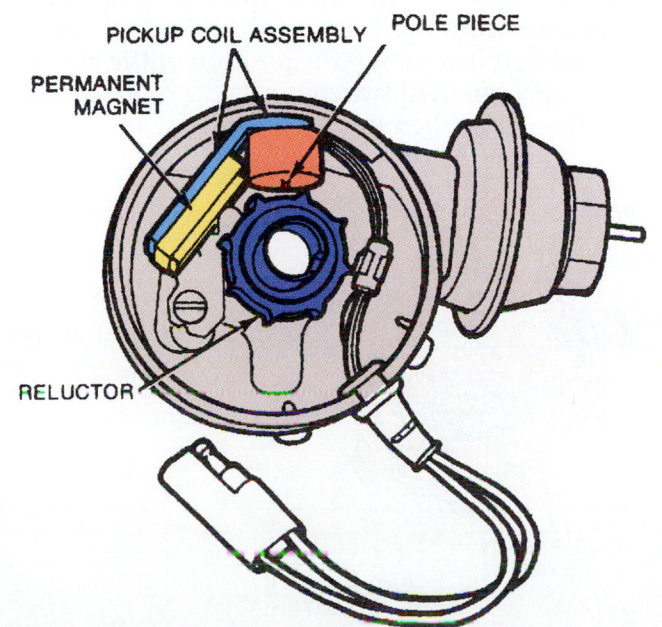

Figure 38.18 A magnetic type AC generator pickup. *[Courtesy of Chrysler Corporation]*

PICKUP COIL ASSEMBLY

POLE PIECE

PERMANENT MAGNET

RELUCTOR

greater (amplified) by the module. The amplified voltage is applied to the base of the transistor to trigger it.

Hall-Effect Pickups

In the mid-seventies, the **Hall-effect switch** was introduced. The Hall switch, named for its inventor, has become the most popular electronic ignition triggering device. Like the AC generator, a Hall switch has a stationary sensor and rotating trigger wheel. It requires an input voltage to operate. Because the input voltage it controls is consistent, it works well at any speed. This is an advantage over the AC generator type of trigger, which must turn faster to generate a stronger signal.

Parts of a Hall switch include a permanent magnet, a *semiconductor Hall element*, and a cupped metal ring with vanes and windows (Figure 38.21). The permanent magnet is mounted with a small gap separating it

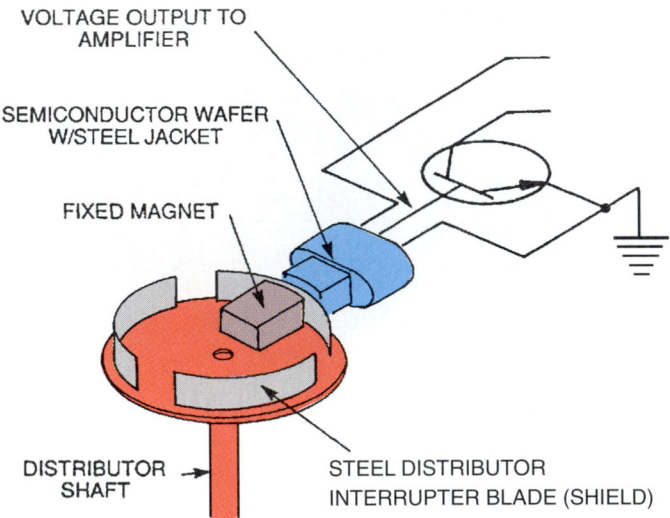

Figure 38.21 Parts of a Hall-effect distributor. *[Courtesy of SPX Corporation, Aftermarket Tool and Equipment Group]*

from the Hall element. As the distributor shaft rotates, the cupped metal ring passes through the space between the element and the magnet. There are as many windows and vanes as the engine has cylinders.

When the ring rotates, a window opens (Figure 38.22). This allows the magnetic field to surround the Hall element. In the Hall-effect process, current is passed through a thin semiconductor material while a magnetic field passes through it. This produces a small voltage (about 0.4 V) in the semiconductor. With the window open, the magnetic field causes a small voltage to be produced by the Hall element (Figure 38.23).

When the next metal tab on the shutter wheel enters the gap between the magnet and Hall element, the magnetic field is shielded from the Hall element. This stops the Hall voltage from being produced by the element. The module reacts to this by switching on the primary circuit, allowing the coil primary winding to become saturated.

The Hall switch makes its own small analog voltage signal. An amplifier strengthens the signal and converts it to a square wave (pulsing direct current) before it goes to a switching transistor. The square wave signal is more compatible with computer systems than the analog wave.

When the shutter tab moves out of the air gap, the voltage produced by the Hall switch returns. The computer chip determines what the correct spark advance should be. Then, it has the ignition module open the primary circuit. This causes the magnetic field in the coil to collapse, resulting in a spark at the spark plug.

Hall switches are also used to generate rpm signals and cause sequential fuel injection systems to pulse. They can be found in the distributor or on the crankshaft. The computer monitors the rate at which the voltage level from the switch rises and falls. From this information it determines the position of the piston in the cylinder. The Hall switch is called a *crankshaft position sensor (CPS)* when it is used this way.

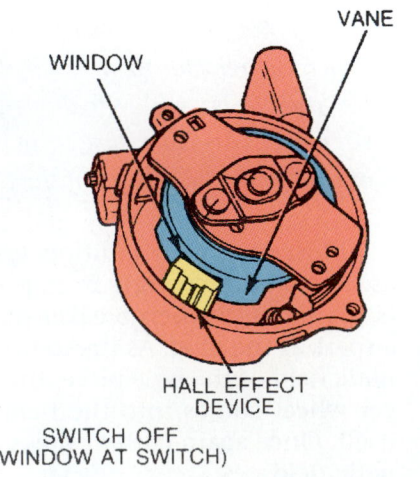

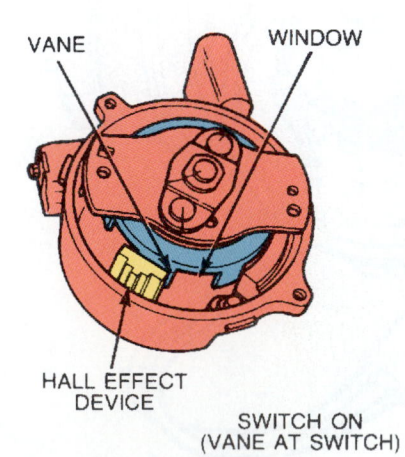

Figure 38.22 The windows and vanes move in and out of the magnetic field that acts on the Hall-effect device. *[Courtesy of Ford Motor Company]*

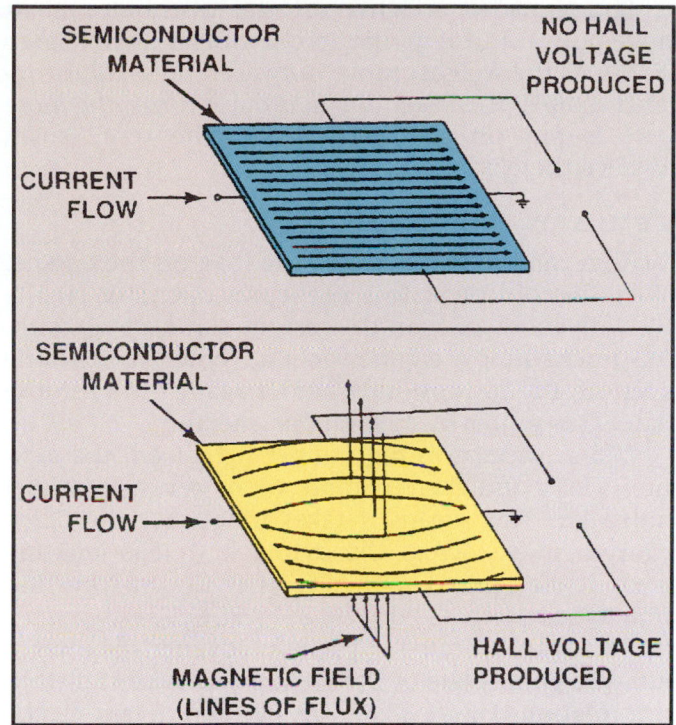

Figure 38.23 The Hall effect. When current is passed through a thin semiconductor material while a magnetic field passes through it, a small Hall voltage is produced. *(Courtesy of Ford Motor Company)*

Optical Sensors

Optical sensors, also called *photoelectric sensors*, use a beam of light to control the primary circuit. Parts of the system include a **light emitting diode (LED)**, a photo cell (a light sensitive phototransistor), and a slotted disc to interrupt the light beam (Figure 38.24). The LED and the photo cell are located across from one another, with the slotted disc in between them. Light from the LED is changed into voltage pulses by

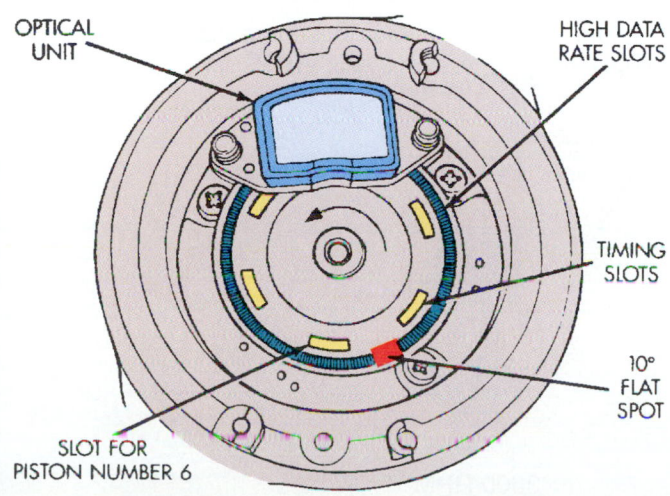

Figure 38.24 A photoelectric sensor. *(Courtesy of Chrysler Corporation)*

the photo cell. When there is voltage present, the control unit provides current to the ignition coil primary winding. When the disc interrupts the light beam, the voltage stops. The control unit shuts off current flow, collapsing the magnetic field and causing a spark at the plug.

Photoelectric sensors in cars are called **crank angle sensors** because they actually sense two things. One signal is generated by the sensor every degree of rotation by the slots in the disc (high data rate slots). There are also low data rate slots: four slots spaced every 90° for four cylinder engines and 6 slots at 60° intervals for six-cylinder engines. The computer determines crankshaft position from these openings and a section of the high data rate ring that does not have any slots.

An optical sensor makes an AC (analog) sine wave signal. Remember, an analog signal is not square like a digital computer needs so it is converted to a square on and off signal by a processor.

Crankshaft Triggered Ignition

Some engines before distributorless ignitions used a crankshaft position sensor instead of a distributor-based triggering setup. A distributor in this engine consists of only a shaft, rotor, and cap for providing secondary high voltage distribution.

■ RESISTOR CIRCUIT

Most ignition systems have a resistance somewhere in the primary circuit. It can be within the coil or outside of it in the form of a resistor wire or ballast resistor. The most common form of resistance is a variable resistance within an ignition module. This is called a **current limiting system**. The ignition module shuts back on current flow as soon as the coil primary winding is saturated. This increases the life of the coil and yet allows voltage to be higher when needed.

To find out whether an ignition system has a current limiting module, check the electrical wiring diagram in the service manual. If there is no ignition resistor bypass circuit for starting, the system has it internally within the module. The bypass circuit is therefore unnecessary.

Variable dwell is another thing that newer ignition systems have. At low engine rpm, dwell will be shorter and at higher rpm it will be longer.

There are different types of electronic ignition modules. Most of the newer ones do four things: turn primary current on, turn primary current off, limit current, and vary dwell. These functions can all be seen on an oscilloscope.

Ignition Timing

The primary ignition circuit controls the point at which the secondary ignition circuit fires the plugs. This point, known as the **ignition timing setting**, is adjustable on cars with distributors. The spark from the spark plug is supposed to ignite the mixture at just the right moment so that the flame front gently pushes the piston down in the cylinder.

The spark is timed to occur just before the piston reaches the top of the compression stroke so peak combustion pressures will occur just after the piston starts down on the power stroke. The spark must ignite all of the air/fuel mixture before the piston has reached 23° ATDC.

■ ADVANCE MECHANISMS

Spark advance mechanisms compensate for piston speed and the amount of air and fuel in the cylinders. They advance or retard the ignition timing in response to engine speed and load changes. Some systems also compensate for altitude and engine temperature.

Timing is changed in one of three ways:

■ *Mechanical* or **centrifugal advance** senses the speed of the engine.

■ Vacuum advance senses engine load and changes the timing to compensate.

■ **Electronic computer advance** uses information available to the computer for the fuel and emission systems to control spark advance. Electronic systems perform the same functions as mechanical systems, but do them more accurately.

Centrifugal (Mechanical) Advance

Whether the engine is running at low speed or at high speed, the same amount of time is required to burn the air/fuel mixture (.003 second). Spark timing must be advanced at higher engine speeds if the mixture is to be ignited soon enough to complete its burn (Figure 38.25). As engine speed increases, centrifugal advance moves the rotor and distributor cam lobe or reluctor tooth forward in its normal direction of rotation. This advances the timing.

Spring-loaded weights on the distributor shaft move outward in response to centrifugal force (Figure 38.26). As the weights move outward, the spark is triggered sooner. The faster the distributor turns, the more force is put on the weights and the more spark advance there is.

Vacuum Advance

Vacuum can be used to sense the load on the engine. When the throttle plate is all the way open, the engine fills with more air than the pistons can keep up with. This results in low engine vacuum. When the throttle is let off, the pistons draw harder against the throttle plate. This results in high engine vacuum.

When vacuum is high under light load, the mixture is lean and there is more room between its fuel molecules. This means it takes longer for the flame front to move across the cylinder. A lean mixture needs to be ignited sooner so that the mixture has time to complete its burn by 23° ATDC.

To advance timing, a vacuum advance diaphragm pulls the point plate or the pickup coil against distributor rotation (Figure 38.27). As vacuum decreases, the plate is returned to its original position by a spring. Vacuum is supplied to the diaphragm from a port above the carburetor or fuel injection throttle plate. This keeps vacuum from acting on the diaphragm at idle. This is called **ported vacuum** because the vacuum port is above the throttle plate. Vacuum lines often run from the vacuum port to a thermal vacuum switch (TVS). The TVS only allows vacuum when the engine coolant has reached a high enough temperature.

Some older engines had vacuum spark retard also. It was usually a part of a unit called a dual advance diaphragm. This system can be identified by its two vacuum ports. The retard diaphragm is smaller than the advance diaphragm and is used only at idle for

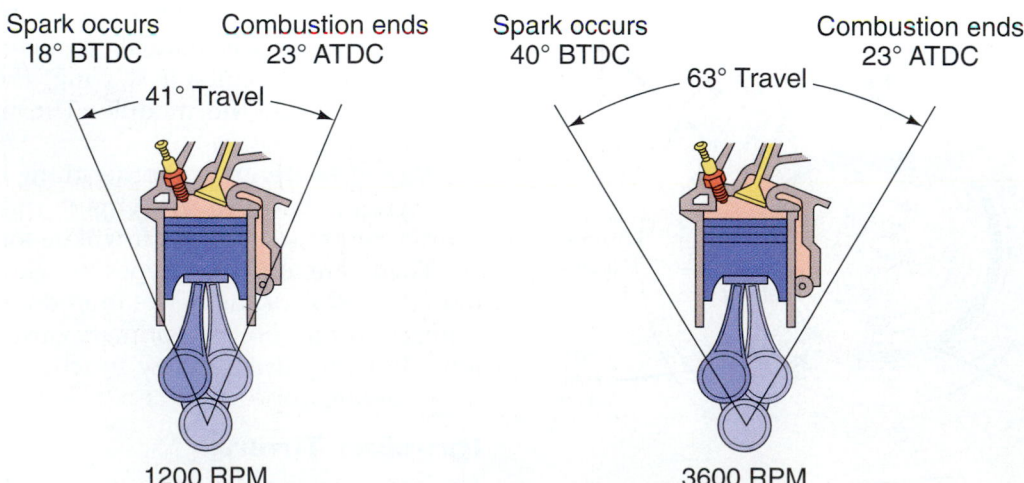

Figure 38.25 The spark must occur sooner at faster engine speeds. *(Courtesy of Cooper Automotive/Champion Spark Plug)*

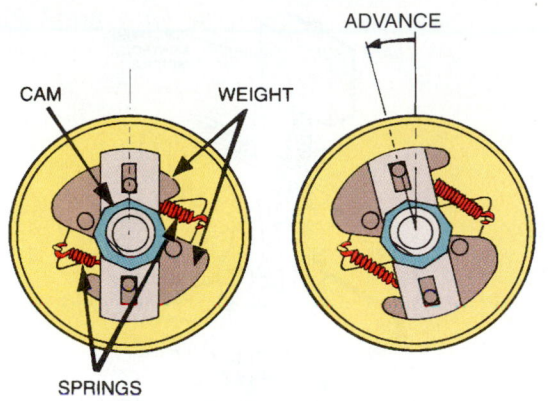

Figure 38.26 Centrifugal advance operation. *[Courtesy of AlliedSignal Automotive Aftermarket/Autolite]*

emission reasons. Some electronic ignition modules also allow for retarded timing during engine cranking. Retarding the timing allows the engine to crank faster for easier starting.

Electronic Spark Advance

Many newer cars have computer controlled spark timing (Figure 38.28). Distributor vacuum and mechanical advance are no longer needed in these cars. A computer chip receives input signals of *vacuum* (engine load), *tachometer* (engine rpm), and *road speed*. It responds to any changes in these inputs and changes timing in a fraction of a second.

The distributor only houses the primary switching trigger and distributes secondary high voltage through

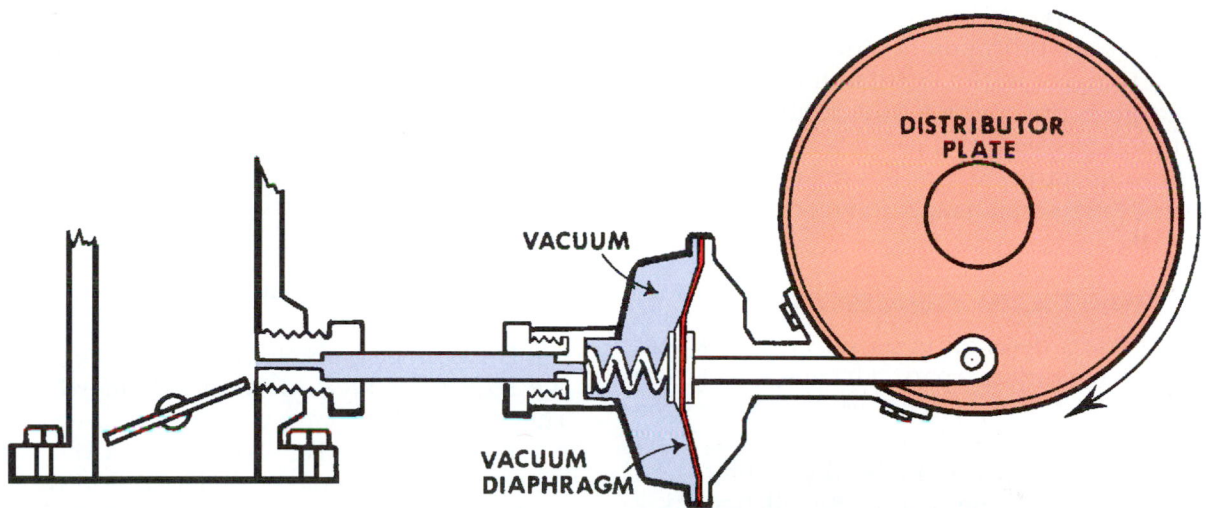

Figure 38.27 The vacuum advance diaphragm pulls the point plate or the pickup coil against distributor rotation. *[Courtesy of Echlin Inc.]*

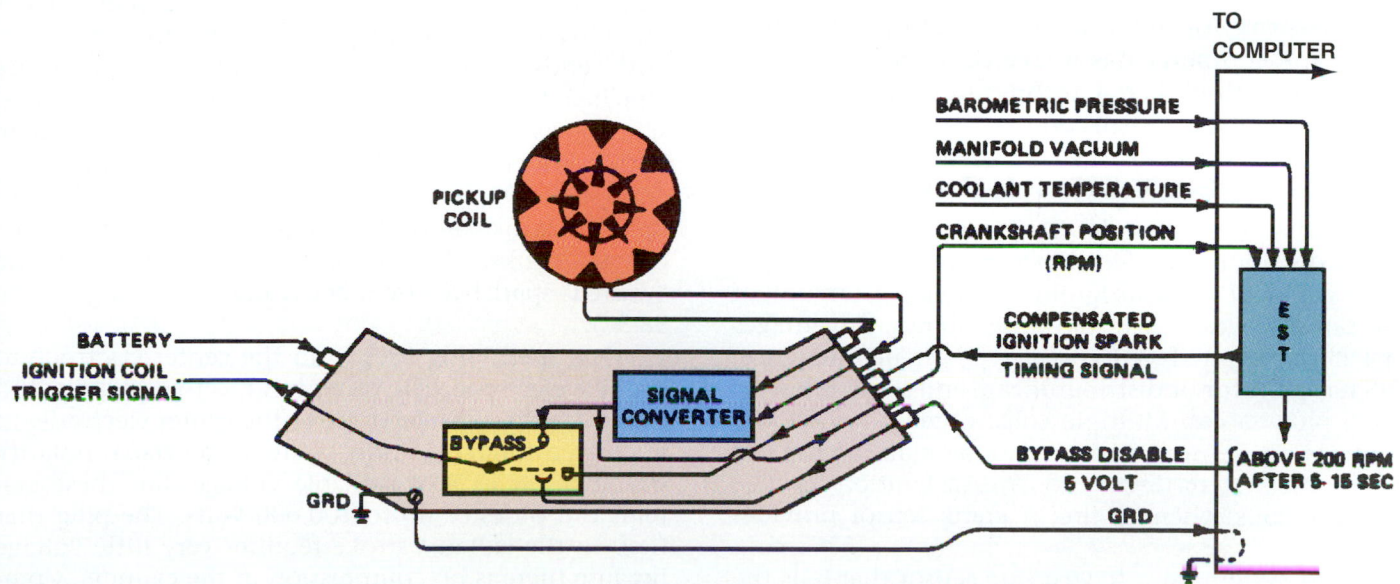

Figure 38.28 A computer controlled ignition system acts on signals received from sensors. *[Courtesy of General Motors Corporation, Service Technology Group]*

the rotor and distributor cap to the spark plug wires. Its shaft is simply that—a shaft with a gear drive on the bottom and a rotor on the top.

Late model computer systems continuously adjust spark timing to provide the best engine power and lowest emissions throughout all operating conditions. The powertrain computer receives input signals on throttle position, manifold and barometric pressures, engine rpm, crankshaft position, and coolant temperature. Some of these extra functions were not possible with mechanical distributors.

- A throttle position sensor (TPS) sends a voltage signal to the computer that tells it what position the throttle is in.
- A manifold absolute pressure sensor (MAP) sends the computer data on the pressure in the intake manifold. This replaces the vacuum advance unit and also gives an indication of altitude.
- The primary trigger in the distributor or on the crankshaft gives the computer the ability to interpret engine speed.
- A coolant temperature sensor lets the computer make adjustments for changes in engine temperature.

DETONATION SENSOR

Some engines use a detonation sensor, or knock sensor, to control maximum spark advance (see Figure 38.30). The sensor, a piezo-electric crystal, detects the frequency of spark knock and retards the timing. Later model systems continuously adjust the timing until they learn the best timing setting for all operating conditions (see Chapter 75).

Initial Timing

Base timing, or *initial timing,* is the timing setting before the computer has a chance to make changes. This is a setting that a technician will verify in a tuneup or emission inspection.

DISTRIBUTORLESS IGNITION

Ignition systems that do not have a distributor are called distributorless ignition systems, or direct ignition systems (DIS). This ignition system performs all of the same functions without a distributor. Advantages of DIS include reduced cost and lower maintenance. DIS has no rotor or distributor cap, and sometimes, no spark plug cables. Multiple coils, called a *coil pack,* a control unit, sensors, and a computer make up the rest of the DIS system (Figure 38.29). An ignition module tells the coils when to fire. A crank sensor provides input.

Most systems also have a cam sensor that tells the computer when the number 1 cylinder is on the compression stroke (Figure 38.30). The crank and/or cam

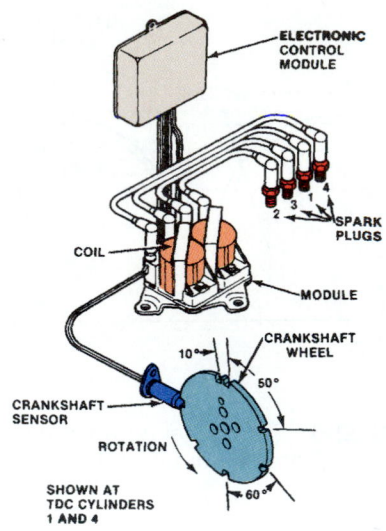

Figure 38.29 Parts of a distributorless ignition system. *(Courtesy of Cooper Automotive/Champion Spark Plug)*

will have a spaced ring with one unevenly spaced tooth. By counting the time between impulses, the ignition module can tell when the unevenly spaced slot passes. This information is used to tell the coils the correct time to spark.

The cam sensor is also used on some systems to tell the sequential fuel injection system when to pulse. There are other variations that also tell when the number 1 piston is at TDC. Some have a Hall switch and others use a magnetic AC generator.

Some DIS systems use one coil per plug. Systems that use one coil for every two spark plugs use the **waste spark method**. In other ignition systems, one end of the coil secondary grounds through an attachment to an end of the primary winding. In this system, both ends of each coil's secondary winding are attached to one of a pair of companion cylinders' spark plugs (Figure 38.31). This is a series system where both cylinders' spark plugs fire every revolution of the crankshaft. One cylinder will be at the top of its compression stroke and the companion will be on its exhaust stroke. The one on the exhaust stroke is the "wasted" spark because it does not do anything (Figure 38.32).

One spark plug fires from the center electrode to the side electrode. The other has reversed polarity and fires from the side electrode to the center electrode. In a conventional ignition system, reversed polarity would result in less available voltage. But these systems can produce up to 100,000 volts. The plug that fires on the exhaust stroke requires very little voltage because there is no compression in the cylinder. Compression makes it harder for a spark to jump a gap.

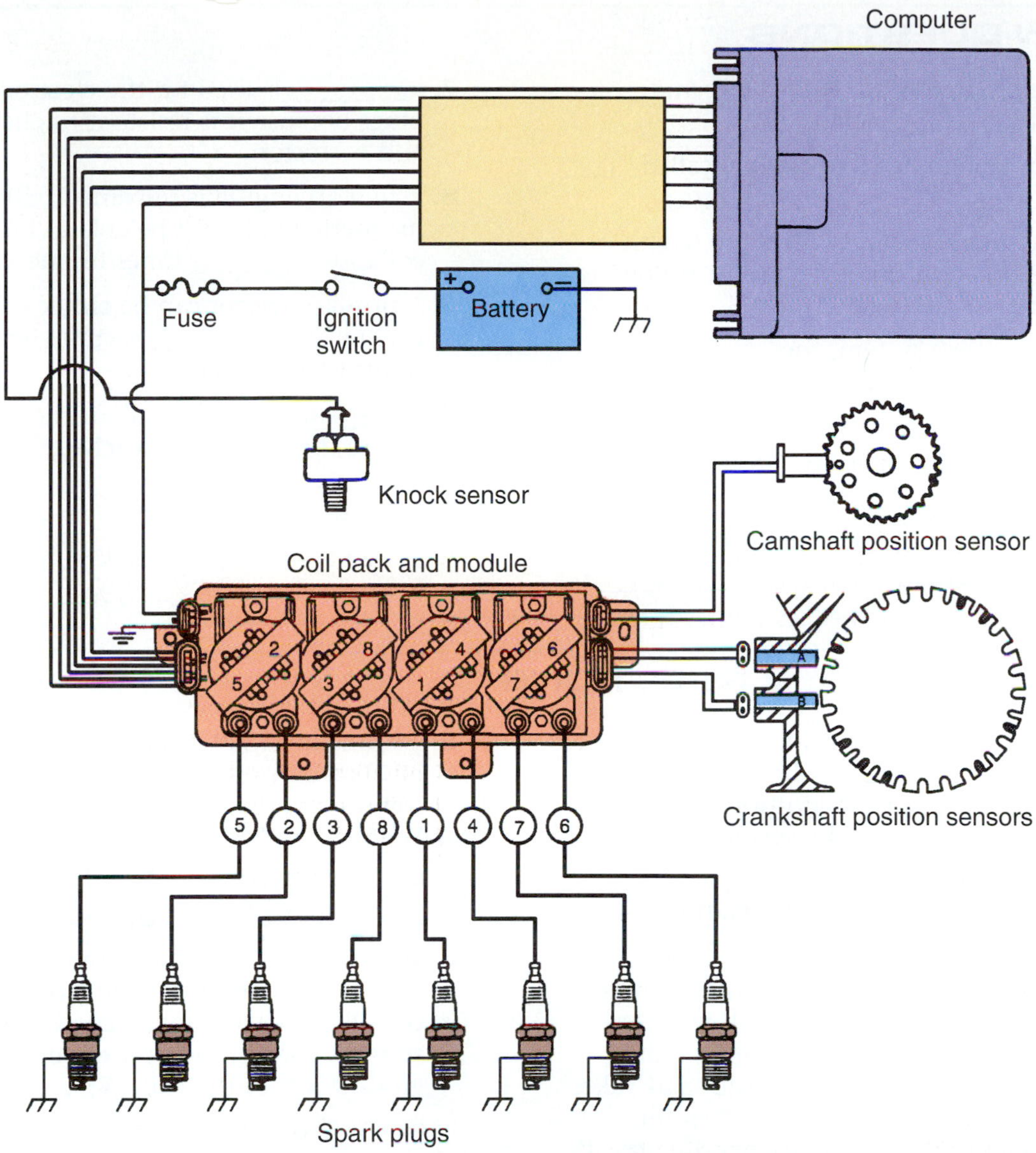

Computer

Fuse

Ignition switch

Battery

Knock sensor

Camshaft position sensor

Coil pack and module

2 8 4 6
5 3 1 7

Crankshaft position sensors

5 2 3 8 1 4 7 6

Spark plugs

Figure 38.30 This ignition system has camshaft and crankshaft position sensors, as well as a knock sensor. *(Courtesy of General Motors Corporation, Service Technology Group)*

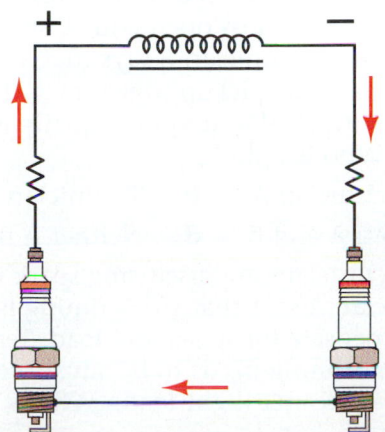

Figure 38.31 The waste spark method. Both ends of each coil's secondary winding are attached to one of a pair of companion cylinders' spark plugs. *(Courtesy of General Motors Corporation, Service Technology Group)*

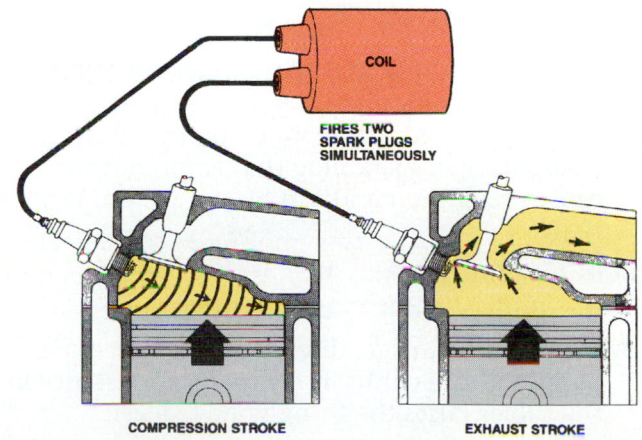

COIL

FIRES TWO SPARK PLUGS SIMULTANEOUSLY

COMPRESSION STROKE   EXHAUST STROKE

Figure 38.32 The spark in the exhaust cylinder is "wasted". *(Courtesy of Cooper Automotive/Champion Spark Plugs)*

■ REVIEW QUESTIONS

1. The ignition system's job is to create and distribute a _____ spark.

2. An ignition system has two circuits called_____ and_____.

3. What is the name that describes the length of time in degrees that primary current is flowing in the primary winding?

4. In a distributorless system, the_____ has a pickup that takes the place of the distributor.

5. After a spark plug fires, the crankshaft rotates___ turns before another spark is required in that cylinder.

6. Over 90% of a spark plug's heat leaves it at the edge of the_____.

7. _____ is used for the electrode in long life plugs because it is resistant to chemical erosion and corrosion.

8. When an engine is running at 3000 rpm, its magnetic field must be able to build up and collapse_____ times in one second.

9. A primary trigger can be either a magnetic AC generator, a_____ switch, or an optical (photoelectric) sensor.

10. The spark must ignite all of the air/fuel mixture before the piston has reached_____ ATDC.

■ ASE STYLE REVIEW QUESTIONS

1. Technician A says that the primary side of the ignition system is the low voltage side. Technician B says that the ignition spark comes from the primary part of the system. Who is right?
   - **a.** Technician A
   - **b.** Technician B
   - **c.** Both A and B
   - **d.** Neither A nor B

2. Technician A says that the high voltage surge to the spark plug is created in the coil when electricity stops flowing in the coil primary winding. Technician B says that the spark takes place from the primary winding when the magnetic field in the secondary winding breaks down. Who is right?
   - **a.** Technician A
   - **b.** Technician B
   - **c.** Both A and B
   - **d.** Neither A nor B

3. Technician A says that spark plugs used with electronic ignition use a narrower spark plug gap than those used with contact points. Technician B says using a spark plug that is too hot can cause preignition and can burn a piston. Who is right?
   - **a.** Technician A
   - **b.** Technician B
   - **c.** Both A and B
   - **d.** Neither A nor B

4. Two technicians are discussing resistance spark plug cables. Technician A says that resistance in the cables raises the firing voltage level. Technician B says that raising the resistance lowers radio interference. Who is right?
   - **a.** Technician A
   - **b.** Technician B
   - **c.** Both A and B
   - **d.** Neither A nor B

5. Technician A says that some resistor spark plugs have carbon resistors. Technician B says that some spark plugs have semiconductor resistors. Who is right?
   - **a.** Technician A
   - **b.** Technician B
   - **c.** Both A and B
   - **d.** Neither A nor B

6. Technician A says that contact points control the flow of current on the ground side of the coil. Technician B says that electronic ignition systems control the flow of current on the ground side of the coil. Who is right?
   - **a.** Technician A
   - **b.** Technician B
   - **c.** Both A and B
   - **d.** Neither A nor B

7. Technician A says that on an eight-cylinder engine running at 3000 rpm, a single set of contact points must open and close 200 times per second. Technician B says that an advantage of the magnetic pickup trigger over the Hall-effect switch is that it works equally well at any speed. Who is right?
   - **a.** Technician A
   - **b.** Technician B
   - **c.** Both A and B
   - **d.** Neither A nor B

8. Two technicians are discussing spark timing. Technician A says that spark timing is adjusted to compensate for speed and load. Technician B says that timing needs to be advanced when the mixture is leaner (light load). Who is right?
   - **a.** Technician A
   - **b.** Technician B
   - **c.** Both A and B
   - **d.** Neither A nor B

9. Technician A says that mechanical advance senses engine load. Technician B says that vacuum advance changes to compensate for lean air/fuel mixtures. Who is right?

   a. Technician A   **b.** Technician B

   c. Both A and B   **d.** Neither A nor B

10. Technician A says that on a conventional ignition system, one end of the coil secondary grounds through an attachment to an end of the primary winding. Technician B says that in a DIS system, both ends of each coil's secondary winding are attached to one of a pair of companion cylinders' spark plugs. Who is right?

   a. Technician A   **b.** Technician B

   c. Both A and B   **d.** Neither A nor B

Ignition System Service

■ **OBJECTIVES**

Upon completion of this chapter, you should be able to:

✔ Diagnose common ignition system problems.

✔ Service ignition systems and distributors properly.

✔ Install a distributor and adjust ignition timing.

✔ Operate an oscilloscope and interpret scope patterns.

■ IGNITION SYSTEM SERVICE AND REPAIRS

Tuneup is a term that originated during the days when cars had ignition points and required periodic service. The term *minor tuneup* refers to a service that includes replacement of the spark plugs, points, condenser, distributor cap, and rotor. A *major tuneup* could also include a carburetor overhaul and replacement of the spark plug cables.

Today, a tune up is usually referred to as a *30,000 or 60,000 mile service*. This service can include replacement of all filters, belts, hoses, and fluids. The thermostat is replaced and the cooling system flushed, too. Some engines also require mechanical adjustment of the valves. A timing belt replacement can also be part of a major service.

■ SPARK PLUG SERVICE

Some of today's engines use expensive spark plugs designed to last for up to 100,000 miles, but most spark plugs need to be changed periodically. During a tuneup, a complete set of new spark plugs is commonly installed. When a spark plug is dirty or has worn electrodes, a misfire can result. Symptoms of a misfire probably will not be noticeable except possibly at idle. It is possible for a spark plug to misfire half of the time with the only indication being that the engine suffers from worse fuel economy or less power.

NOTE: *In an engine running on the highway at 3000 rpm, each cylinder's spark plug will be firing 25 times every* **second**. *How could you possibly feel this?*

Resistor Plugs

Some spark plugs have carbon resistors. Others have a semiconductor suppressor that is not energized until 1000 volts. This kind of resistor plug *cannot* be checked with an ohmmeter.

Replacement Plugs

Be sure to use the correct replacement plugs. The code number on the spark plug tells about its heat range, thread size, type of seat, whether its tip is extended, and whether or not it has a resistor (Figure 39.1).

If you use a cross-reference chart on spark plugs, you will probably put in the wrong plug. Many spark plug

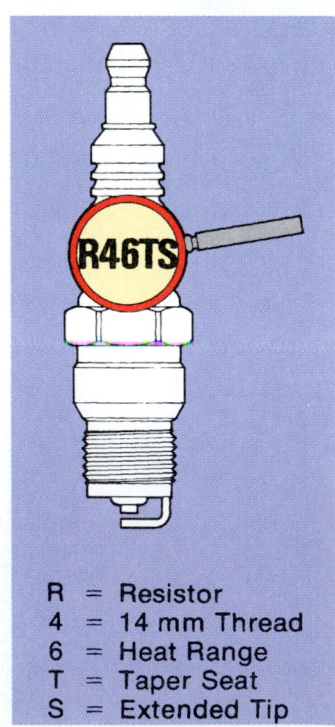

R = Resistor
4 = 14 mm Thread
6 = Heat Range
T = Taper Seat
S = Extended Tip

Figure 39.1 Spark plug code numbers tell about the plug. *(Courtesy of General Motors Corporation, Service Technology Group)*

problems are caused by misapplication. As an example, the Champion Spark Plug company has 267 different OE spark plug applications at the time of this writing.

Be sure to use the correct spark plug heat range. Too hot a plug can cause serious engine damage. Also, be sure to replace a resistor plug with a resistor plug and non-resistor with non-resistor. Check the size of the spark plug threads; 14 mm and 10 mm are the most popular. Check also to see whether the old plug has a gasket or a tapered seat without a gasket. Some square seat plugs are used without the gasket. Installing a spark plug without a gasket when one is required will allow the threads of the plug to hang into the combustion chamber. This can cause it to overheat, resulting in preignition.

Check the reach or thread length (see Figure 38.10). The use of a plug with too short a reach can result in carbon buildup in the end of the spark plug hole. Too long a plug might possibly interfere with the piston. In either case, carbon on the threads can result in a ruined spark plug hole.

Removing Spark Plugs

Tapered seat plugs in particular are often very tight and can be difficult to remove. The body of the spark plug is ceramic. It is fragile and can easily be cracked. A cracked insulator will allow the spark to go to ground, rather than jumping the gap (Figure 39.2).

Spark plugs are removed and installed with a special socket. Spark plug sockets have either a rubber insert or a magnet inside to hold the plug in the socket so it does not fall out during installation. When removing and installing spark plugs, be careful not to allow the spark plug socket to cock to the side.

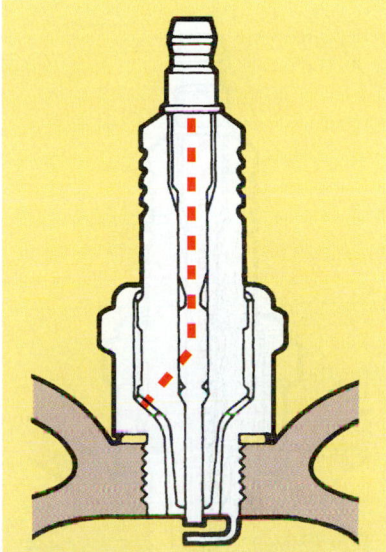

Figure 39.2 A cracked insulator will allow the plug to short to ground. (Courtesy of General Motors Corporation, Service Technology Group)

NOTE: *Be careful not to drop a good plug on the ground. A plug that has been dropped on the ground could be cracked and should be replaced.*

As you remove the spark plugs, keep them in order for later comparison.

■ SPARK PLUG DIAGNOSIS

Two things determine the life of a spark plug: the temperature of combustion and the quality of the electrical system. The temperature of a spark plug relates directly to combustion chamber temperature. There are many other factors that govern the temperature of a spark plug. These include ignition timing, work load, compression ratio, and air/fuel ratio. With a gasoline engine, 1600°F is the maximum temperature that should ever occur.

The minimum spark plug temperature of 850°F is where carbon forms. If the plug can operate at 1200°F it will have very little wear and a life expectancy of 25–35,000 miles can be expected. The plug does need to get hot enough to burn off normal deposits, however. In normal driving, the load on a spark plug is very light until about 30 mph. Plugs are more apt to foul if the car is never driven over 30 mph. Spark plug fouling does not occur when cruising at 50 mph.

Reading Spark Plugs

Reading the condition of used spark plugs will give an idea of the condition of the engine (see Figure 43.9) and fuel system. Compare the plugs to see if there is a difference between cylinders. This can tell you if the engine can be tuned or will require repair.

When engines had leaded gas, plugs that were operating properly were tan to tan/grey in color. With today's fuels, the color of correctly operating plugs has changed. They should be light in color. If there is a deposit on the insulator, it should be soft and flaky.

Blistering of the insulator tip or specks that resemble pepper indicate that a plug has been too hot. When a plug has been overheated, the center electrode's color resembles the burnt head of a match. Heat dissipation is critical. When heat results, a hole gets burned in a piston and/or the spark plug center electrode melts. Figures 25.15 and 25.16 show the results of preignition and detonation damage.

An electrode that has pointed wear often indicates a fuel mixture that is too lean. Lean mixtures burn hotter. In fact, when a drag racer runs out of fuel during a quarter mile run, a burned piston is a common result. This kind of wear can also be due to ignition timing that is too advanced or an inefficient cooling system.

Spark plugs must be properly torqued to avoid misfires. Look for evidence of etching on all threads that could indicate incorrect tightening. An overheated plug can be due to dirty threads or a tapered plug seat that is dirty. The thread and thread seat are where over 90% of the plug's heat dissipation occurs. The threads must be clean and undamaged. If the spark plug does not unscrew easily with finger pressure only, clean the threads with a thread chaser. If all of the plugs appear to be overheated and the threads and plug seats are clean, try a colder heat range plug.

On distributorless ignition systems (DIS), every other plug fires toward the ground electrode. The rest of the plugs fire toward the center electrode. The positively fired plug wears on its ground electrode and the negatively fired plug wears on its center electrode. After 40,000 miles or so, the positive electrodes are the ones that tend to show wear first. Some manufacturers recommend rotating the plugs for this reason. Ford has a platinum pad on a different electrode depending on whether it is positively or negatively fired.

There is a blue line that forms on the ground electrode of a clean plug that will tell if the ignition timing is correct. If the line is positioned at the bend in the plug, the timing is correct (Figure 39.3). If the line is up the electrode the timing is too advanced. Too close to the tip means the timing is too retarded.

Oil and ash deposits can be found on some plugs. A plug that is wet with fuel or oil usually indicates a misfiring cylinder.

Dirt on the spark plug's insulator can also cause misfiring by providing an electrical path for the spark to jump to ground.

Fouled Plugs

A spark plug that has a buildup of carbon that shorts it out is called a **fouled spark plug**. The spark plug insulator looks smooth, but under a microscope it is porous. When a spark plug has been fouled from flooding, it fills with tiny metallic additives from the fuel. Because the insulator now conducts electricity, it will not work any longer (Figure 39.4). Once a plug has fouled once, it makes sense to replace it, rather than to clean it. If an engine problem is suspected, perform a spark plug deposit test as described in Chapter 43.

SHOP TIP   After a tuneup when new plugs were installed, the car is taken on a test drive. Hard acceleration can loosen carbon deposits and they can stick to the new plugs. Some technicians like to use a combustion chamber cleaner and allow a 20 to 30 minute soak period. Another technique is to do a hard acceleration test drive with old plugs to clean out the combustion chamber deposits before replacing the plugs. The problem with this is that the spark plugs and exhaust manifold will be hot when you want to remove the spark plugs.

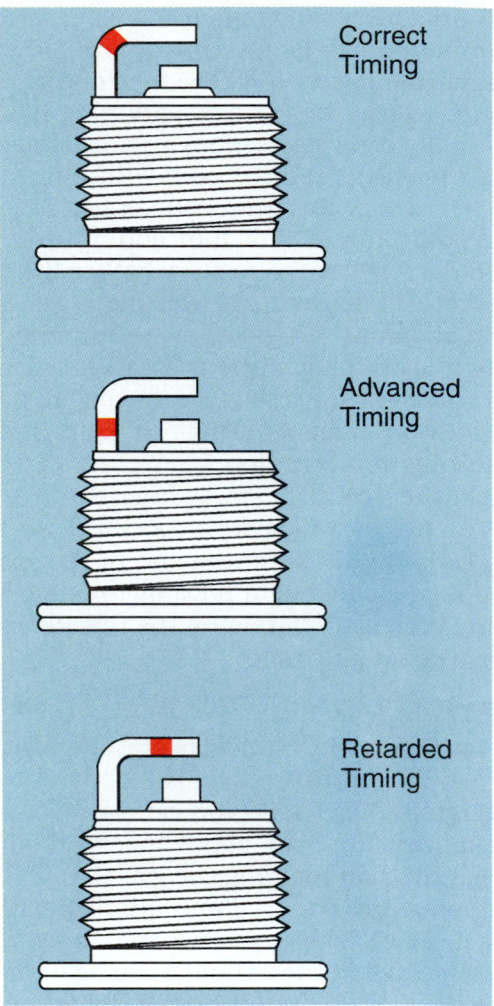

Figure 39.3 Stripes on the ground electrode indicate correct or incorrect ignition timing. *(Courtesy of Cooper Automotive/Champion Spark Plug)*

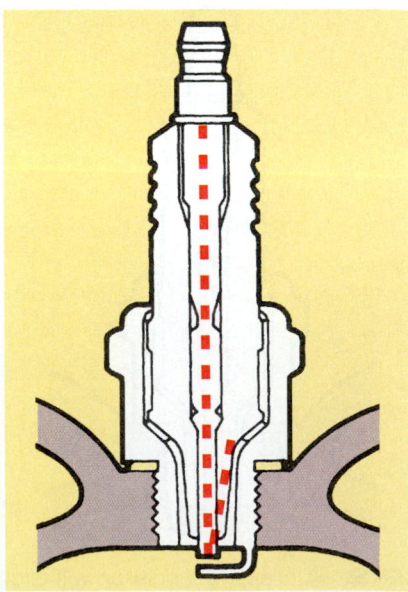

Figure 39.4 A fouled plug shorts to ground, rather than firing normally. *(Courtesy of General Motors Corporation, Service Technology Group)*

Figure 39.5   Normal and worn spark plugs. *(Courtesy of Cooper Automotive/Champion Spark Plug)*

Spark Plug Wear

The spark occurs at the outside edges of the electrode (Figure 39.5). A worn rounded electrode has more surface area that will need to be charged up with electrons during sparking. In the distant past, spark plugs were cleaned, filed, and regapped.

NOTE: *Cleaning plugs is not cost effective. This procedure used to be more common, but spark plugs are so inexpensive compared to the amount of labor required.*

Checking and Adjusting the Spark Plug Gap

The spark plug gap is usually set from the factory. Always double-check the gaps before installation. It is not unusual to find closed gaps or gaps that have changed due to rough handling. Also, spark plug companies often market the same plug for several different applications. It would be too costly to number them differently just because their gaps are of different sizes.

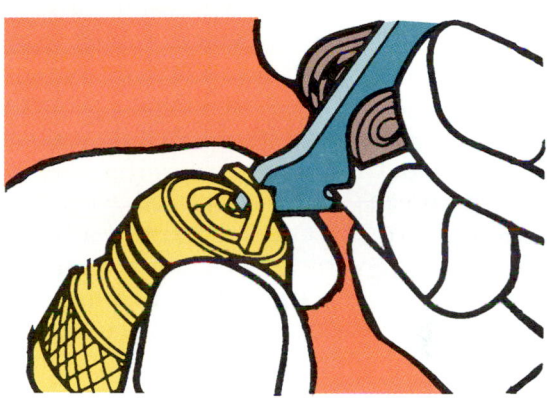

Figure 39.7   This tool is used to bend the ground electrode to set the gap. *(Courtesy of Sun Electric Corporation)*

Spark plugs in older point-type ignition systems specified small gaps of around 0.035". The gaps on newer electronic ignition cars can vary up to 0.080". Be certain to look up the specification for the gap.

A wire gauge can be used to check the gap (Figure 39.6). It has an arm that is designed to be used to reposition the bendable electrode on the plug (Figure 39.7).

■ INSTALLING SPARK PLUGS

Put a little bit of anti-seize compound on the front two threads of the plug before installation. Anti-seize is especially needed on aluminum heads to prevent electrolytic action between the steel spark plug body and aluminum head. Do not use too much.

When tightening the plug, the seal of the plug can be ruined by overtightening. Tighten a gasketed plug finger tight, then ¼ turn further. A taper seat plug is installed finger tight, then ¹⁄₁₆ turn further. Small "peanut" plugs will twist off at 50 to 60 pounds of torque. They are usually only torqued to about 10 or 15 foot-pounds. Check the specifications until you are used to the type of car you are working on.

Indexing a Spark Plug

Indexing a plug is a high-performance procedure usually used only in super high compression engines for racing. The idea is to have the back side of the ground

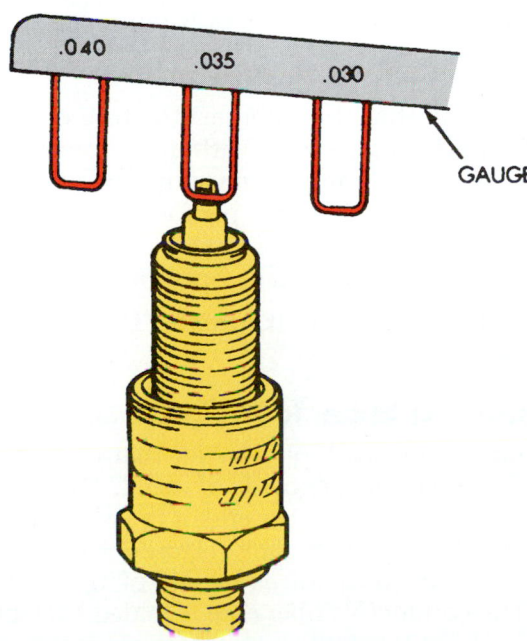

Figure 39.6   A wire gauge used to measure the spark plug gap. *(Courtesy of Chrysler Corporation)*

electrode face the exhaust valve so the fuel coming in through the intake valve can blow directly into the gap. Copper shims are used to get the indexing right. There are 0.060", 0.080", and 0.135" shims.

➡️ *Perform R & R Spark Plugs Worksheet*

■ REPAIRING DAMAGED SPARK PLUG THREADS

Spark plug threads are sometimes stripped during a careless installation. Stripping threads can be avoided if a ratchet is never used to install a plug, only to tighten it. A piece of vacuum hose can be installed on the end of the spark plug or a special rubber spark plug tool can be used. The socket can be used without the ratchet also. If the threads are aligned properly, the plug should go all the way in to touch its seat without the aid of a ratchet.

Stripped threads can be replaced with a thread insert (see Figure 6.28). On aluminum heads, this can sometimes be done with the head on the car. The aluminum shavings that come out of the plug hole during the reaming operation should prove to be harmless if some of them happen to get into the combustion chamber.

NOTE: Tapered seat spark plug ports cannot be repaired with thread inserts. The tap used to prepare the hole for the thread insert will ruin the taper in the hole, resulting in a leaking spark plug.

■ SPARK PLUG CABLE SERVICE

Replacing Spark Plug Cables

Removing spark plug cables must be a careful operation if the cables are to be reused. The rubber boots usually become formed to the ridges on the spark plugs. They must be twisted to loosen them (see Figure 42.11). TVRS (television/radio suppression) cables are made of braided fiber and graphite and are very fragile. Handle them only by the plug boots so they do not suffer internal breaks.

SHOP TIP Changing cables one at a time will avoid accidentally mixing them up in the firing order. You will also be able to put them in their correct places in their existing loom.

NOTES:
- ■ *Cables should be held in looms away from the exhaust manifold and engine drive belts.*
- ■ *Do not allow any cables to run over the computer wiring loom where it could set up radio interference for the computer signals. This can cause an engine to run very poorly.*

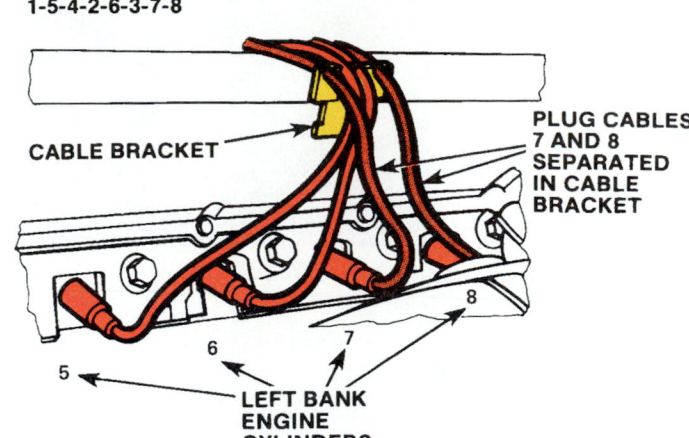

FIRING ORDER 1-5-4-2-6-3-7-8

CABLE BRACKET

PLUG CABLES 7 AND 8 SEPARATED IN CABLE BRACKET

LEFT BANK ENGINE CYLINDERS

Figure 39.8 Avoiding induced sparking. *(Courtesy of Chrysler Corporation)*

- ■ *When installing the cables on the engine, do not allow cables from cylinders that fire after one another in the firing order to run parallel to each other (Figure 39.8). The first one to fire can induce a spark in the next one, causing it to fire before its time.*

Induced spark is called **crossfire induction**. It usually happens on V-8s. In GM V-8s, cylinders 5 and 7 fire after each other (18436572). In Fords, cylinders 7 and 8 fire after each other (15426378). These cylinders are next to each other in the block.

Be sure that the cable ends fit snugly into the distributor cap. Some manufacturers recommend putting some silicone dielectric compound on the inside of each boot. You should feel the metal clip on the end of the plug cable snap into the distributor cap socket.

Testing Cable Resistance

To determine that the internal structure of a cable is sound, check its resistance with an ohmmeter (Figure 39.9). The normal range of resistance in a new cable is about 5000 to 10,000 ohms per foot. A defective cable can have 50,000 (50K) or 100,000 (100K) ohms of resistance or it can be totally open (*infinite* resistance).

➡️ *Perform Inspecting Spark Plug Cables Worksheet*

Repairing Spark Plug Cable Ends

The ends of the spark plug cables that connect to the inner fiber material are made of metal. The most common kind is crimped to the end of the plug cable. They can be replaced individually (Figure 39.10). The insulation is stripped off of about ⅜" of the end of the wire. The conductive fiber core is folded back over the outside of the insulation. The terminal is slipped over the outside of the conductor and insulation, then it is crimped tightly to the cable. A rubber boot is installed on the outside of the cable end to provide insulation.

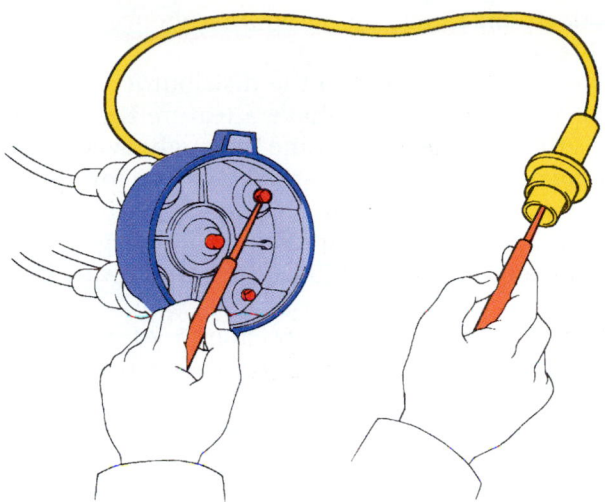

Figure 39.9 Checking the resistance of a spark plug wire and distributor cap.

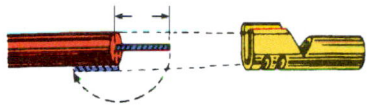

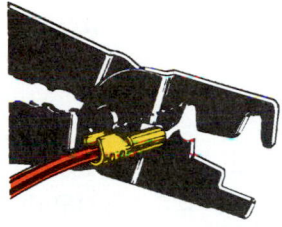

Figure 39.10 Replacing a spark plug terminal. *(Courtesy of Cooper Automotive/NAPA Belden)*

The boot is slipped over the metal end before it is installed. Special crimping pliers are used to crimp the terminal to the cable.

Sometimes, there is a resistor installed in the end of a spark plug cable. In this case, the cable will be solid cable. The spark plug might also *not* be a resistor plug.

Spark Plug Cable Installation

Spark plug cables are installed in the holes around the outside of the distributor cap in the engine's firing order. The sketch in Figure 39.11 shows a distributor with counterclockwise rotation and a firing order of 1 5 4 2 6 3 7 8. Check the book! The firing order for a 302 (5-liter) Ford can be either 15426378 or 13726548, depending on the model year.

Direction of Distributor Rotation. The direction of distributor and oil pump rotation usually follows the direction toward which the vacuum advance points. Figure 39.12 shows two distributors that turn in oppo-

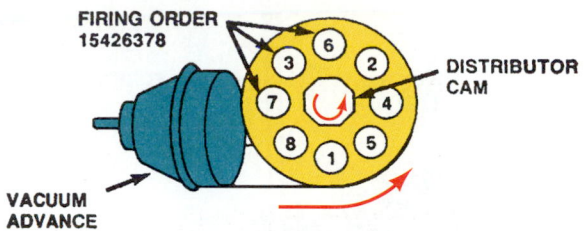

Figure 39.11 Spark plug cables are installed in the holes around the outside of the distributor cap in the engine's firing order.

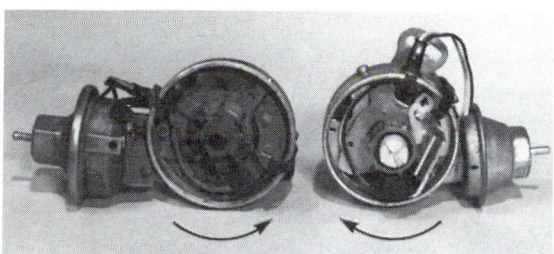

Figure 39.12 The distributor rotates the same direction that the vacuum advance points.

site directions. On distributors that have mechanical spark advance, another way to check for direction of rotation is to snap the rotor in one direction while the distributor shaft is held stationary. The rotor will be spring loaded only in one direction. That is the direction in which the distributor turns.

CASE HISTORY

A customer took her Bronco into a service station for a tuneup. When she returned to get it, she was informed that the cam timing had skipped and the engine would no longer run. The service station did not do that kind of work so she had to have the vehicle towed to another repair facility. The technician at the new facility performed a diagnosis on the engine before removing the timing cover. He discovered that the spark plug cables had been installed in the distributor cap in the wrong firing order. The engine ran fine after the order of the cables was corrected.

■ DISTRIBUTOR SERVICE

Distributor Cap and Rotor Service

Some caps are held on with screws and others have clips that snap into place (Figure 39.13). A distributor cap or rotor sometimes wears, cracks, or becomes corroded, requiring replacement (Figure 39.14). When a distributor cap is cracked cap a **carbon trail** often forms along the crack. The carbon will conduct electricity, usually to a wrong cylinder or to ground.

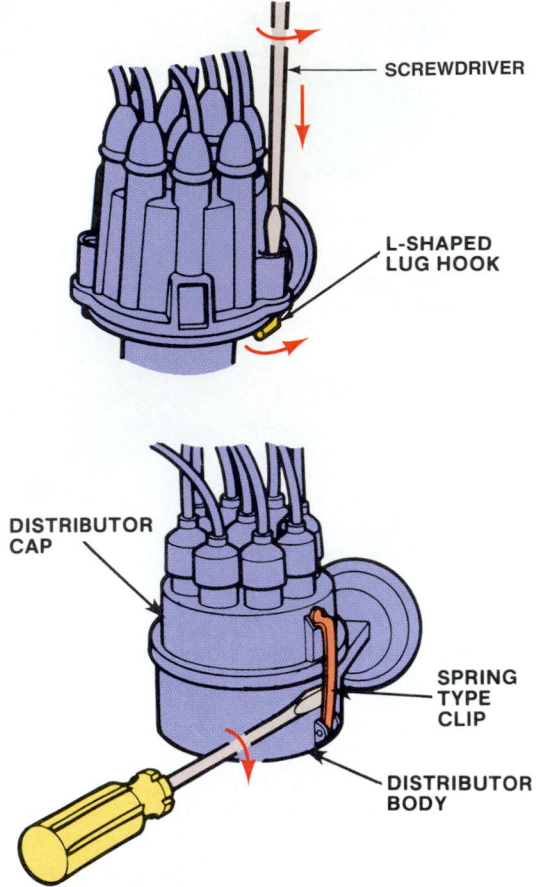

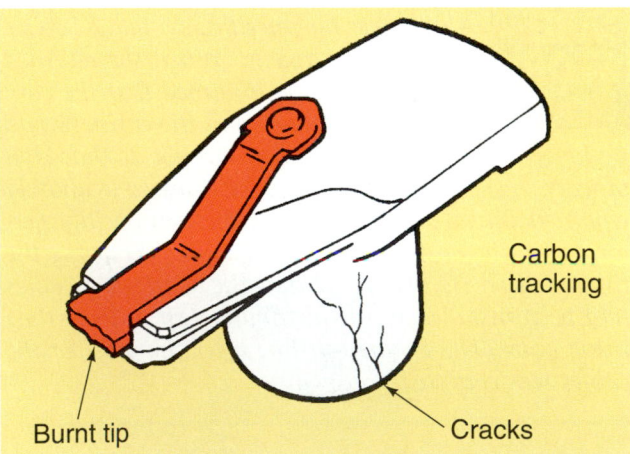

SCREWDRIVER

L-SHAPED
LUG HOOK

DISTRIBUTOR
CAP

SPRING
TYPE
CLIP

DISTRIBUTOR
BODY

Figure 39.13 Methods of retaining the distributor cap. *(Courtesy of Chrysler Corporation)*

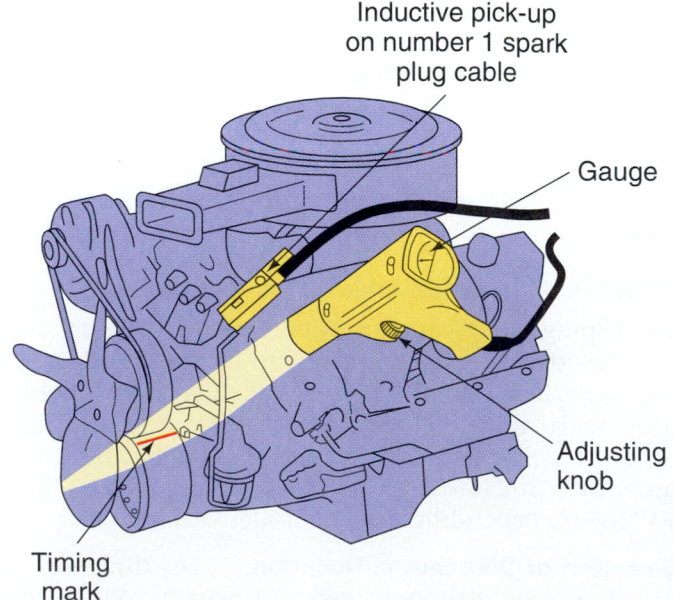

Carbon
tracking

Burnt tip

Cracks

Figure 39.14 Places where a rotor can be damaged or worn. *(Courtesy of Chrysler Corporation)*

When there is excessive resistance in the secondary system, the tip of the rotor can sometimes develop a hole. This *puncture* causes unusual symptoms. Replacement of the rotor is usually a part of a maintenance tuneup. The cap is always inspected. Some manufacturers recommend smearing some silicone dielectric compound on the tip of the rotor. This

cuts down on arcing and radio interference, extending the life of the rotor and cap.

Rotors are attached to the distributor shaft in different ways. All of them have a feature that correctly aligns them to the shaft. Some fit snugly over the shaft and are aligned by a plastic tab that fits into a notch in the shaft. Others screw on and align with a square or round peg that fits into a corresponding notch. Be sure the aligning feature goes the correct way. Do not force it!

The distributor cap has an aligning feature. There is a tab or notch somewhere on the distributor body that must align with a corresponding tab or notch in the cap. Once again, do not force something.

➡ *Perform **Replace a Distributor Cap and Rotor** Worksheet*

■ IGNITION TIMING

Measuring Timing

With the engine running, the timing can be checked using a **timing light** (Figure 39.15). A timing light is a strobe light. The light turns on in response to an inductive trigger signal picked up from the number one spark plug cable. When the spark plug fires, the timing light strobes and "freezes" the crankshaft's position.

> **CAUTION** The strobe makes the crankshaft look as though it is standing still. The parts are still moving and they are dangerous.

Inductive pick-up
on number 1 spark
plug cable

Gauge

Adjusting
knob

Timing
mark

Figure 39.15 The ignition timing is checked using a timing light. *(Courtesy of Sun Electric Corporation)*

Changing Ignition Timing

The ignition timing is changed by loosening the distributor body and rotating it in one direction or the other. The distributor hold-down clamp is sometimes difficult to get to. There are special wrenches available that can bend around the distributor body to access the bolt (Figure 39.16). Loosen it only enough to allow the distributor to be turned. On most cars, the distributor is turned until a timing mark on the front of the engine crankshaft pulley lines up with another mark on a tab on the timing cover (Figure 39.17).

NOTES:

■ *As the timing is advanced, engine idle rpm will usually increase slightly.*

■ *When the timing is retarded, idle speed will decrease.*

To advance the timing, turn the distributor in a direction opposite to that which the vacuum advance on the distributor points (see Figure 39.12). This is opposite the direction the rotor normally turns.

When the timing is too far *advanced*, the spark occurs too soon and abnormal combustion can occur

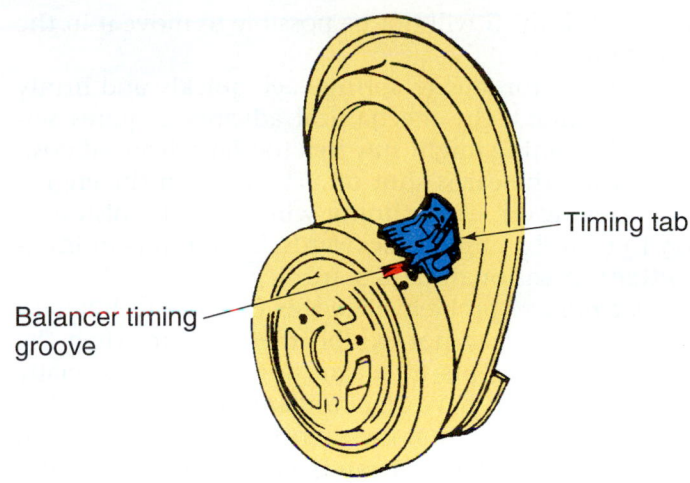

Timing tab

Balancer timing groove

Figure 39.17 A groove in the crankshaft pulley is aligned with a stationary timing tab. *(Courtesy of General Motors Corporation, Service Technology Group)*

(see Chapter 25). Pressure rises in the cylinder, causing an explosion instead of controlled burning. This can result in engine damage.

Retarded timing will cause engine coolant temperature to rise, resulting in overheating. When timing is either too advanced or too retarded, fuel economy and power will suffer. Proper timing is also very important for *exhaust emissions*.

Ignition Timing Specifications

Cars produced since 1972 have an **underhood emission label** that lists the timing specification. In order to be able to check base timing, or initial timing, on computer controlled cars, a wire that gives information to the computer is usually disconnected. The wire is called a *bypass wire*. Ford calls it a *spout* wire. The required procedure is described on the underhood label. On older cars, it is sometimes necessary to disconnect the vacuum advance hose before checking the timing. Check the service manual.

➡ **Perform _Check Ignition Timing Using a Timing Light_ Worksheet**

■ DISTRIBUTOR ADVANCE SERVICE

During a tuneup, the technician will check initial timing and total spark advance to see if the mechanical and vacuum advances are working properly.

Checking Mechanical Advance Operation

Mechanical advance can be checked with the engine on or off. With the engine off, remove the distributor cap from the distributor. Grasp the rotor and attempt to twist it. If the mechanical advance is operating properly, it will be possible to move the rotor smoothly and easily against spring pressure in one

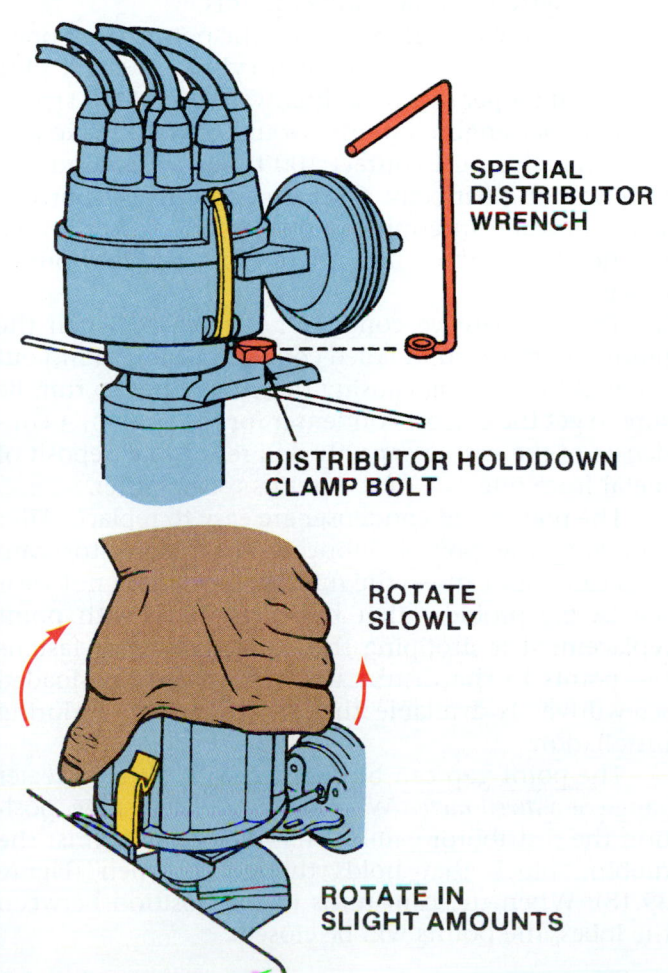

SPECIAL DISTRIBUTOR WRENCH

DISTRIBUTOR HOLDDOWN CLAMP BOLT

ROTATE SLOWLY

ROTATE IN SLIGHT AMOUNTS

Figure 39.16 Adjust the timing by loosening the distributor slightly and turning it. *(Courtesy of Chrysler Corporation)*

direction only. It will not be possible to move it in the other direction.

If the rotor fails to spring back quickly and firmly when released, the mechanical advance requires service. The timing might stay in a too-far-advanced position when the car is shut off. Then, when the engine is started again, early timing will cause the piston to try to turn the engine backwards. This results in intermittent (occasional) hard starting.

Intermittent problems like this frustrate technicians because a car may not exhibit the problem when the technician attempts to diagnose it. This is especially true on cars that have an abundance of on board electronics. A problem that is currently occurring is known as a **hard fault**. Hard faults can be diagnosed and repaired. **Intermittent faults** often require guesswork. Intermittent fault diagnosis can be done on computer controlled cars using scan tools (see Chapter 76).

To test mechanical advance operation with the engine running, aim a timing light at the crankshaft pulley while raising and lowering engine rpm. Watch for the timing to change, first advancing and then returning to its initial setting at idle.

➡ *Perform Check Centrifugal Advance Operation Worksheet*

Part of a tuneup includes lubrication of the mechanical advance. This is especially easy on GM cars.

Checking Vacuum Advance

For vacuum advance to operate correctly, two conditions must be met.

■ Sufficient vacuum must be present at the vacuum advance at the correct times.

■ A good diaphragm will hold vacuum.

Engine vacuum can be measured by tapping into the intake manifold at any point above the intake port. This is called intake manifold vacuum. Vacuum advance is usually not present at idle because the vacuum tap is at a point above the throttle plate. This is called ported vacuum, as opposed to intake manifold vacuum.

The diaphragm in a vacuum advance can sometimes fail. Its condition can be checked with a vacuum pump. Remove the hose from the diaphragm. Connect a hose from the vacuum pump to the hose bib and pump until about 15" Hg of vacuum is reached. Putting vacuum on the diaphragm should cause the advance plate in the distributor to move. The diaphragm should hold constant vacuum if it is good.

➡ *Perform Check a Vacuum Advance Unit Worksheet*

Checking Total Spark Advance

The operation of the vacuum and mechanical advance is tested by watching the timing mark on the crankshaft while aiming a timing light at it. A high quality timing light has an adjustable dial so that the amount of total advance can be measured. With the engine running under no load, off idle, if the hose to the vacuum advance is pulled off, the advance should drop back.

To check total advance, the engine is typically run at 2500 rpm with a specified amount of vacuum applied by a pump to the diaphragm. Initial, mechanical, and vacuum advance combine to make total advance. Specifications for both advances are readily available. In a tuneup shop, a distributor tester can be used for measuring and adjusting spark advance when the distributor is out of the engine.

■ CONTACT POINT DISTRIBUTOR SERVICE

Most cars built since the mid-seventies have electronic ignition systems. There are still many older cars on the road with contact points so that service is discussed briefly here. Regular service at about 10,000 mile intervals is important in these cars. The major wear point in these systems is the contact points.

Every time a spark plug fires, the points must open and close. Remember that each cylinder's spark plug fires 25 times per second at highway speed (3000 rpm). A 4-cylinder engine's points would have to make and break the electrical contact 100 times per second if it were working perfectly. Over time, points get burned. This can cause intermittent misfiring that will probably not be noticed until there is a complete breakdown.

The inexpensive condenser is replaced when the points are replaced. A failed condenser can ground out the ignition system, causing the engine not to run. Be sure to get the correct condenser for the system. A condenser of the wrong capacity will result in a deposit of metal from one side of the points to the other.

The points and condenser are easy to replace. After removing the points, lubricate the distributor cam with cam lubricant so the rubbing block does not wear out. A big problem that beginners have with point replacement is dropping the small screw that fastens the points to the distributor. A special spring-loaded screwdriver is available that holds the screw during installation.

The point gap can be checked with either a feeler gauge or a *dwell meter*. When using a feeler gauge, position the distributor cam so that its lobe is against the rubbing block that holds the points open (Figure 39.18). When the cam turns to the position between the lobes, the points will be closed.

NOTE: *Changing the point gap changes the ignition timing because it changes the dwell. If the new points are reinstalled with the correct gap, the timing should remain the same.*

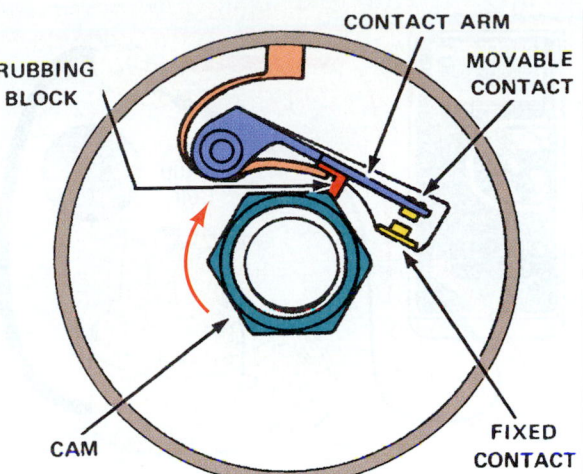

Figure 39.18 When adjusting the point gap, the rubbing block is against the lobe on the cam. *(Courtesy of Sun Electric Corporation)*

The dwell meter measures the length of time in distributor degrees that the points are closed in one revolution. Dwell is less the more cylinders there are.

◾ ELECTRONIC IGNITION DISTRIBUTOR SERVICE

Normal electronic ignition service is limited to replacement of the cap and rotor. Lubrication of the mechanical advance weights is done when the distributor is so equipped. If the distributor has vacuum advance, its diaphragm can be tested in the same way as with contact point ignitions.

The term *ignition trigger* refers to either contact points or the electronic device that signals the module to make and break the primary circuit in the coil. A triggering mechanism and module take the place of the breaker points. Diagnosis of problems in those systems are covered here.

Electronic Ignition Tests

When there is a no-spark condition, the trigger might not be opening the primary circuit. Be sure that the distributor turns when the engine is cranked. A timing chain or belt could be broken or the distributor drive gear could be damaged. Another possibility is that the trigger is good but the module is not working.

Checking the Ignition Module

When there is a no-start, no-spark condition, a module is a possible cause. The best test for a bad module is to test the other parts of the system first. If they are all functioning properly and within specifications, the module is replaced. There are expensive testers available for modules but they only work on specific systems.

First, check for power in the primary circuit. Check at the coil negative terminal with a digital logic probe or an electronic circuit tester as the engine cranks. The test light should flash.

CAUTION If you use a regular 12 V test light on Hall circuit, too much current will flow through the parallel circuit the light provides. Too much current flowing throughout the system can damage the electronic circuitry. A regular test light can be used with a magnetic trigger but the clamp is connected to battery positive instead of ground. A logic probe is the best way to test a Hall switch. Photoelectric sensors can be tested with an oscilloscope.

If there is no power in any of the following conditions, there is an open circuit that must be repaired. Use a test light to see if there is power in the run and crank positions. A wiring diagram will tell you the correct wires to check. To test for power in the run position, with the key off disconnect the connections at the module.

CAUTION Do not disconnect an ignition module with the ignition switch on. A voltage spike can ruin the module.

Turn the key on and probe the power connection. The test light should light brightly. The + (battery) side of the coil should also light the test light when the key is in the run position.

Next, test for power in the start position. First, remove the connector from the module. Probe the connector that the wiring diagram says is for the start position while holding the key in the start position. The engine will crank unless you disconnect the wire to the S terminal on the starter solenoid. There should also be power at the battery side of the coil. If the system has a resistor, check its continuity.

NOTE: *Be sure to test the ignition coil. Often a defective coil causes the failure of the module. A shorted primary winding or open secondary winding can be the cause. Check the resistance in each winding (Figure 39.19). When primary winding resistance is less than specified, there is probably a short that would allow too much current flow. This is what damages the module.*

Pickup Coil Testing and Replacement

Use an ohmmeter to test the continuity of the pickup coil (Figure 39.20). There are only two leads so this is

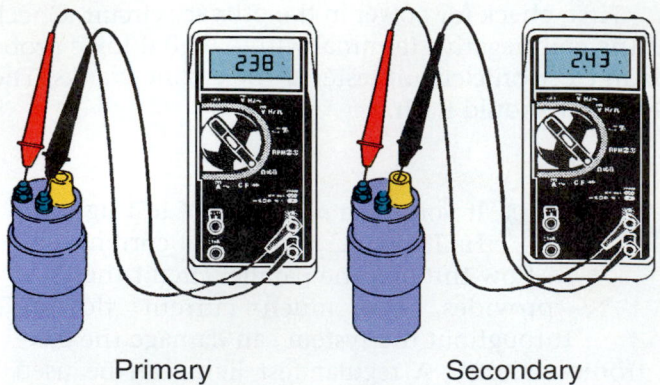

Figure 39.19 Checking the resistance in both coil windings. *(Reproduced with permission of Fluke Corporation)*

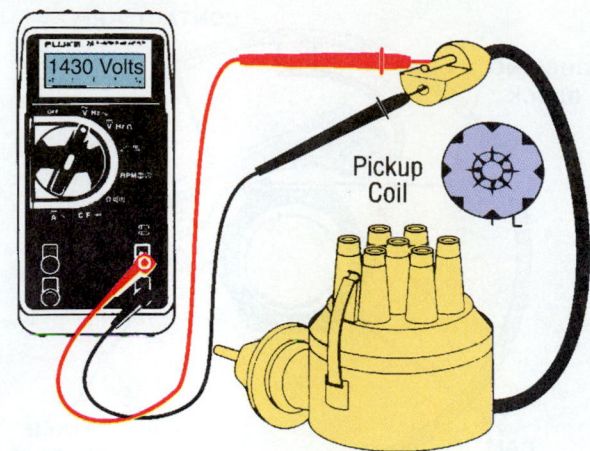

Figure 39.21 Checking a pickup coil with a voltmeter. *(Reproduced with permission of Fluke Corporation)*

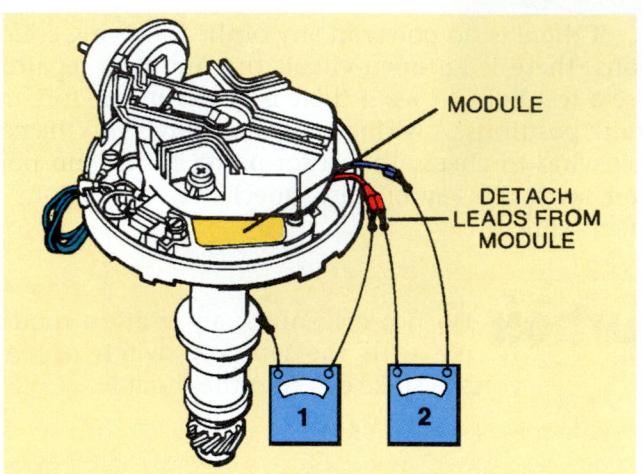

Figure 39.20 Checking a pickup coil with an ohmmeter. *(Courtesy of AlliedSignal Automotive Aftermarket/Autolite)*

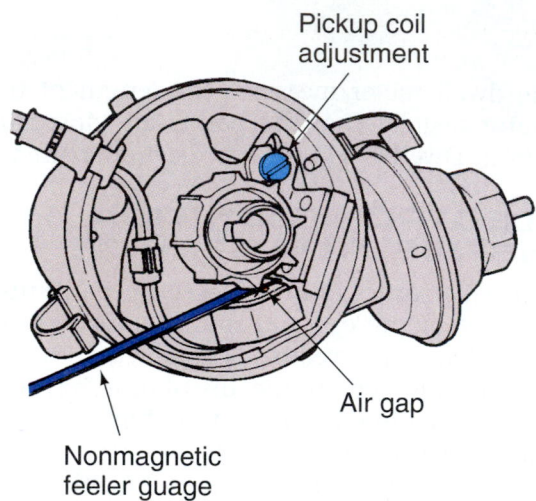

Figure 39.22 Measuring a pickup coil air gap. *(Courtesy of Chrysler Corporation)*

a simple procedure. A typical specification would be 400–800 ohms. You can also test the pickup coil by cranking the engine with a multimeter connected (Figure 39.21). You are looking for an AC signal of less than .250 volt. An analog meter (one with a needle) works better for this test because you can see the sweep of the needle from positive to negative. If there is an AC signal, it will trigger the module. Using an oscilloscope will show a sine wave.

When the pickup coil is replaced, the air gap between the reluctor teeth and pickup coil is adjusted to specifications (Figure 39.22).

Hall Switch Testing

In order to be able to generate a signal, a Hall switch requires voltage and ground. Disconnect the wiring harness and make connections to battery voltage (Figure 39.23). Connect a voltmeter to the output wire.

Insert a metal feeler gauge between the Hall device and the magnet. On a digital voltmeter, watch the bar graph. Usually, full battery voltage is shown when the shutter blocks the Hall effect. Zero should be shown when the shutter window is open. If there is no voltmeter reading, there is a problem in either the positive or ground circuit. On an oscilloscope, a square wave will be seen if the switch is operating.

Replacing the Ignition Module

Ignition modules are usually dependable. High underhood temperatures contribute to electronic failure so modules are more prone to failure during very hot weather. Failing to correctly mount a module can result in its premature failure. They must be mounted securely and have a layer of silicone dielectric grease under them to help keep them cool.

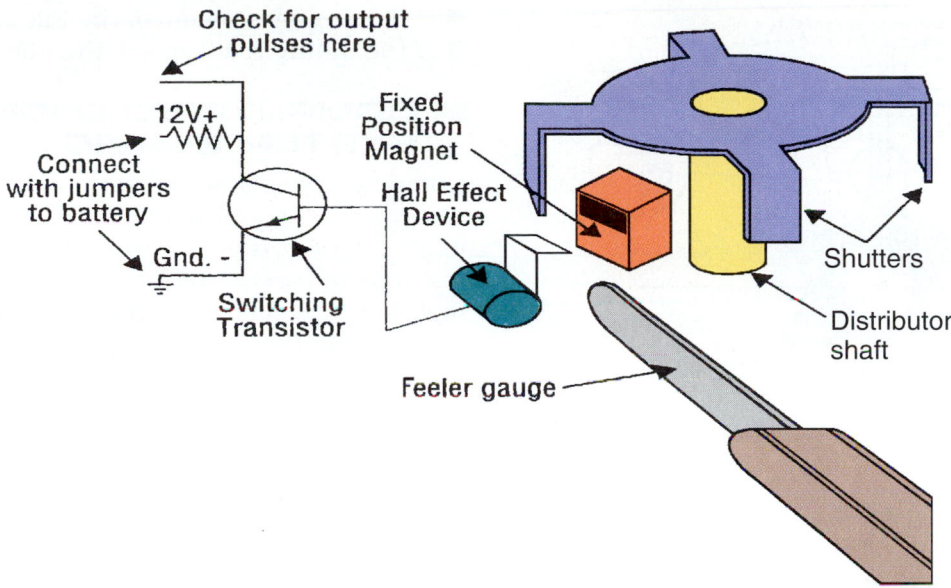

Check for output pulses here

12V+

Connect with jumpers to battery

Gnd. -

Switching Transistor

Fixed Position Magnet

Hall Effect Device

Feeler gauge

Shutters

Distributor shaft

Figure 39.23 Check the operation of a Hall switch by interrupting the magnetic field strength with a steel feeler gauge. *(Reproduced with permission of Fluke Corporation)*

OTHER DISTRIBUTOR SERVICE

Distributor Drive Gear Service

Occasionally, the distributor drive gear can become stripped. It can be replaced by driving out a roll pin that is pressed fit into the bottom of the distributor shaft. The shaft bushing and cam lobes can wear (on distributors with points). There are two ways this can be tested other than visually. Bushing wear will show up when checking ignition timing. A worn bushing will cause the timing mark to move around under the strobe. On the oscilloscope, the end of the dwell section of the pattern will change. This *dwell variation* can also be tested on a distributor tester.

Rebuilding a Distributor

The cost of parts and labor has become too high for custom rebuilding of distributors in the service industry. Rebuilt distributors are commonly available from large rebuilders through local parts stores. Occasionally, a part of a distributor will be broken. This can be ordered through a dealer and replaced without buying an entire rebuilt unit.

DISTRIBUTOR INSTALLATION

Align the timing mark on the damper with the pointer on the timing cover (see Figure 39.17). When the marks are aligned, either the number one cylinder or its companion cylinder is at the top of its compression stroke. The spark plug should be ready to fire. If the distributor is installed 180° off, backfiring will occur and the engine will not run.

SHOP TIP To be sure that the number one cylinder is on its compression stroke, observe the rocker arms to see that they are both moving. Then, rotate the crankshaft one revolution until the rocker arms for the *companion* to cylinder number one are both moving at TDC. Align the timing mark and install the distributor with its rotor pointing to the number one spark plug cable in the distributor cap.

Often, the distributor will not drop down all the way into the block. This is because it is not aligned with the drive lug on the oil pump. The engine can be cranked or turned by hand until the distributor and oil pump are aligned (see Figure 19.41). When they are aligned, the distributor will drop the last ¼" or so until it is flush with the block.

Make sure the spark plug cables are correctly installed on the distributor cap. They must be installed in the right firing order and in the direction of the distributor's rotation.

STATIC TIMING

Timing the distributor with the engine stopped is referred to as **static timing**. Turn the distributor against its direction of normal rotation until the points just begin to open, or until the electronic ignition module pulses. The electronic ignition pulse occurs when the armature pole, or trigger, lines up with the magnetic pickup in the distributor (Figure 39.24). The tooth on the armature must align perfectly

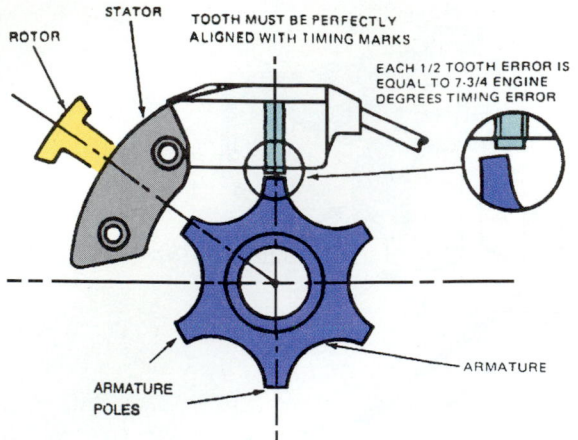

Figure 39.24 Setting ignition timing. *(Courtesy of Ford Motor Company)*

with the pickup, or an error in timing will result. When the key is on in a distributor with contact points, a spark would occur at the instant that the points separate.

The separation of the points can be confirmed in this manner:

- Connect a test light to the distributor side of the coil and to ground.
- Turn the ignition key on.
- Rotate the distributor slightly in the direction it normally rotates, until the test light goes out.
- Then, slowly turn the distributor in the opposite direction until the test light comes on (the points are open).

After reviewing the previous section on static timing, carry out the following procedure. On either type of ignition system, the ignition system can be static timed by causing the ignition system to make a spark.

- Align the timing indicator at the crank pulley to the desired timing specification.
- Install the distributor with the rotor pointing to the number one plug cable.
- Rotate the distributor body in the direction of normal distributor rotation until the rubbing block, or pole piece, is between cylinder number one and the cylinder that fires before it in the firing order.
- Hold the spark plug end of the number one plug cable near a bare metal spot on the block.
- With the ignition switch on, rotate the distributor slowly in the direction *opposite* to its normal rotation. When the spark occurs, the distributor is properly timed. Be sure to turn off the key.
- Tighten the distributor hold-down.

The timing should now be set closely enough so that the engine will start right up. If the idle speed is not controlled by the computer, the engine idle will be off if the timing is wrong. Remember this before changing the idle adjustment screw.

- If the timing is retarded, the idle will be too low.
- If the timing is advanced, the idle will be too high.

■ COMPUTERIZED IGNITION SYSTEM SERVICE

Computers have brought many advantages to the ignition system. One of the disadvantages is that components can be costly. In the old days, a mechanic could just replace parts until the problem was fixed. Today, replacing a defective electronic component without determining the cause of the problem can result in a repeat failure of an expensive part. Determine the cause of the problem before replacing the part! Computer system service is covered in Chapter 76.

■ ENGINE ANALYZER

Large engine analyzers are used by many shops, especially those who do emission certification. An infrared exhaust analyzer (see Chapter 41) is usually included in these testers. Engine analyzers can be used to check the battery, charging, starting, and fuel systems. An oscilloscope is used to test the ignition system. The tester also has functions for testing engine condition.

An engine analyzer includes many meters and gauges that would also be available as separate units. Meters include a voltmeter, ohmmeter, ammeter, and a tachometer. A vacuum gauge and cylinder leakage tester are usually included. Also included is an automatic cylinder power balance tester and a timing light. There is usually a printer that prints out results of tests.

■ OSCILLOSCOPE

The oscilloscope, commonly called a scope, is a helpful tool for pinpointing ignition and engine problems. Early scopes were analog but today's scopes show a simulated pattern on a digital display screen. A scope has wires that attach to the battery, the number one spark plug, and the ignition coil.

Scope Pattern

The image on the screen is called a scope pattern. The patterns are all the same whether the engine has points or an electronic trigger.

The scope is a voltmeter with a builtin time frame. Vertical *spikes* on the screen indicate voltage and the horizontal pattern represents time (Figure 39.25). The pattern shows the firing of the spark plug, followed by the rest of the ignition cycle leading up to the next firing. By analyzing the pattern and making comparisons between cylinders, a technician can determine which cylinder(s) are causing a problem. Electrical problems are easily pinpointed. Combustion problems and some engine problems can also be diagnosed.

There are different types of scope patterns that can be selected for viewing by the technician. There is

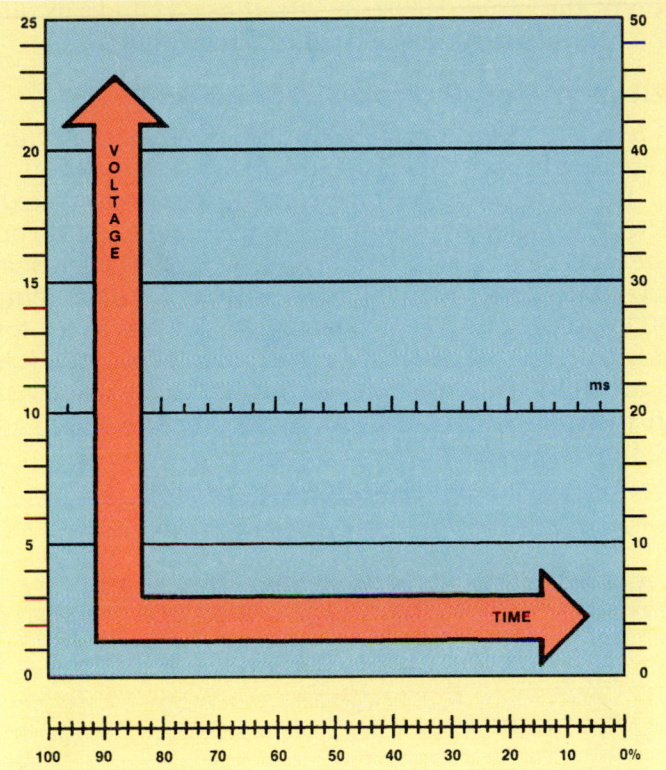

Figure 39.25 A scope is a voltmeter that also measures time. *(Courtesy of Sun Electric Corporation)*

spark across the spark gap. The firing line will be somewhere between 5 kV and 12 kV.

NOTES:

■ *It does not matter if the system is mechanical or electronic. Resistance in the secondary is what causes the firing voltage requirement to rise.*

■ *Higher spikes indicate more resistance to spark. This can be due to a lean air/fuel mixture, a wide spark plug gap, high resistance in a spark plug cable, or high compression due to carbon buildup.*

■ *Lower spikes indicate lower resistance. This can be due to a spark plug gap that is too narrow, the use of an incorrect type spark plug or cable (no resistance), a rich air/fuel mixture, or compression that is too low.*

Spark Line

The **spark line** is a horizontal line that begins at the voltage level where electrons start to flow across the gap. This horizontal pattern, made during the combustion process, indicates the continuous firing voltage requirement of the spark plug. A technician can use this to determine conditions in the combustion chamber.

The length of the spark line shows how long the spark actually lasts. The spark drains the voltage off of the coil secondary winding. When the voltage lowers to a certain level where it can no longer flow, this part of the pattern stops. The next part of the pattern has oscillations of the remaining voltage in the coil as it dissipates back to zero. Notice how the pattern's oscillations become lower and lower.

Dwell Section

The dwell section of the pattern represents the time that the primary current is switched on (Figure 39.26). A short downward line shows where current starts to flow into the coil. As current is entering the coil, the line is below the zero line. The small oscillations show the beginning of magnetic field buildup. As saturation occurs, the horizontal line comes up to zero. An upward line is the point where dwell ends. What happens when dwell ends is that the magnetic field in the

a selector position for either a primary or secondary pattern. The secondary pattern is the one most often used for ignition system diagnosis because it also gives an indication of what is occurring in the primary side of the system.

Firing Line

The secondary *firing section* is the first part of the scope pattern. The upward line that starts the firing section is called the **firing line**. Voltage builds up in the ignition coil until it is greater than the secondary resistance against it. This is the voltage required to start the

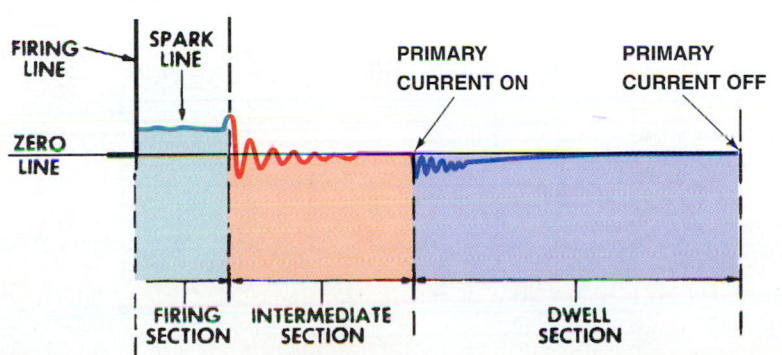

Figure 39.26 The dwell section of the pattern represents the time that the primary current is switched on. *(Courtesy of Sun Electric Corporation)*

coil collapses, resulting in a spark at the plug (the beginning of the next pattern).

■ TYPES OF SCOPE PATTERNS

Raster Pattern

A **raster**, or *stacked*, **pattern** displays all of the cylinders, one above the next (Figure 39.27). The voltage firing line is difficult to see when using this pattern. The scope patterns are arranged in the firing order from bottom to top. The pattern shown is for an eight-cylinder engine with the firing order 18436572. The scale on the bottom represents percent of dwell. Some scopes measure this in degrees.

Superimposed Pattern

A **superimposed pattern** is used to compare all of the cylinders while their patterns are displayed one on

top of the other (Figure 39.28). If one cylinder is different from the others, it will be easy to see.

Display or Parade Pattern

A *parade* or **display pattern** displays all of the cylinders next to each other, side by side so that the heights of the firing lines (voltage spikes) can be compared. The scope selector is switched to read a lower scale on the left side or a higher scale on the right. The left side represents 25 kV (25,000 volts) and the left represents 50 kV (see Figure 39.25). The higher side is selected for this pattern. The heights of the spikes are slightly less than 10 kV. Patterns from each cylinder are displayed in the order of the engine's firing order, from left to right (Figure 39.29).

NOTE: *In a display pattern, most scopes display the number one spark plug firing spike at the far right, with the rest*

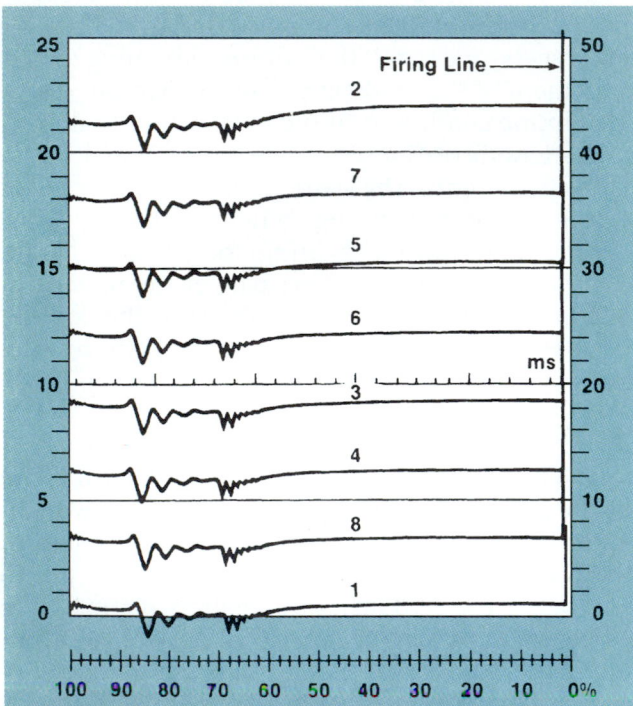

Figure 39.27 A raster pattern shows each individual pattern arranged from bottom to top. *(Courtesy of Sun Electric Corporation)*

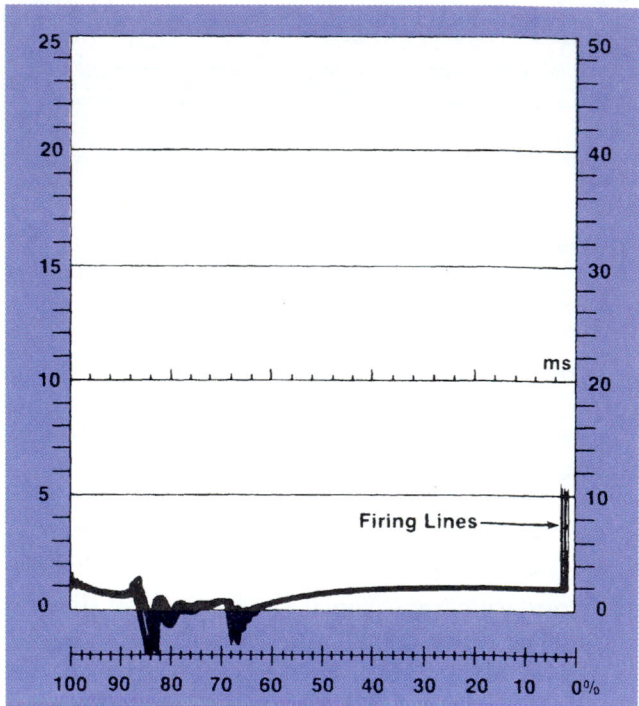

Figure 39.28 A superimposed pattern shows all of the cylinders' patterns on top of each other. *(Courtesy of Sun Electric Corporation)*

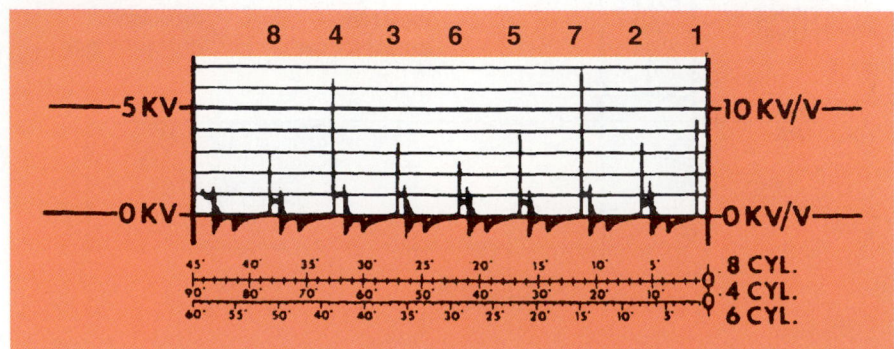

Figure 39.29 These uneven firing voltages indicate problems. *(Courtesy of Sun Electric Corporation)*

of the number one pattern at the left. The firing order shown here is 18436572.

➡ **Perform <u>Hook up and Read an Oscillo-</u><u>scope</u> Worksheet**

■ OSCILLOSCOPE TESTS

Available Voltage Test

With older contact point systems, a popular test was the available voltage test. A plug cable was removed using insulated pliers while the engine idled. A coil could put out somewhere around 20,000 volts. On modern electronic systems, this open circuit test is no longer advisable. It can result in a breakdown of the insulation of the coil. Today, the special spark plug tester is used (see Figure 42.1). It requires about 35 kV to fire this special plug.

Connect the spark plug tester to a spark plug cable. Then, crank the engine and watch the scope pattern. If the spark jumps the gap, the system has enough available voltage. If it does not, the voltage can be read as less on the scope. Battery voltage can be too low, the module could have excessive resistance, or the coil could have a problem.

Reading a Parade Pattern

Run the engine at 1200–1500 rpm for oscilloscope tests. Select the parade pattern and look at all the firing lines to see that they are within 3 kV of the height of each other (see Figure 39.29). With new spark plugs, the firing voltage will be lower. As the gap becomes larger with wear, the firing voltage requirement will increase.

A leaner air/fuel mixture causes a higher resistance to spark, requiring more voltage to jump the plug gap. Watch the height of the firing lines. A poorly performing cylinder with a high firing line indicates a lean air/fuel mixture; a low firing line indicates low compression.

Acceleration Load Test. Rapidly snap the throttle open and closed. RPM does not need to climb above 2000 for this test. This puts a load on the ignition system so you can see if it misfires. The firing lines will all rise equally as more load is put on the system. Again, the difference between the cylinders should be no more than 3 kV.

Reading a Raster Pattern

The *spark section* of the pattern is viewed from the raster or superimposed pattern selection. The spark sections should all be the same length and have a slight downward slope. The length of this line gives an indication of the amount of voltage in reserve. This part of the pattern gives an indication of what is going on in the combustion chamber. Further investigation

is called for when not all cylinders are the same. Examples are:

- ■ A steep slope in the spark line can indicate high resistance in the spark plug wires (Figure 39.30).
- ■ When the slope starts at the top of the firing line, a fouled plug is indicated.
- ■ An upward slope in this section can indicate that the combustion chamber is receiving less fuel or there is high plug resistance.
- ■ Normal turbulence in the air/fuel mixture can show oscillations in this line.

The *coil condenser oscillations* are the next part of the pattern. There should be five to seven gradually diminishing oscillations. If there are fewer, there could be shorted windings in the coil.

The *dwell section* is the next part of the pattern. The length of the dwell section will vary with design. To observe dwell variation, look at the dwell section of the pattern (Figure 39.31). A worn distributor shaft bushing can cause dwell to jump around. When the throttle is snapped, the dwell will often stabilize.

On later systems, dwell varies with engine speed. If the engine is running faster, the coil needs more time to saturate so dwell is longer. Raising engine rpm will result in a lengthening of the dwell section on the scope. When an ignition system is supposed to have

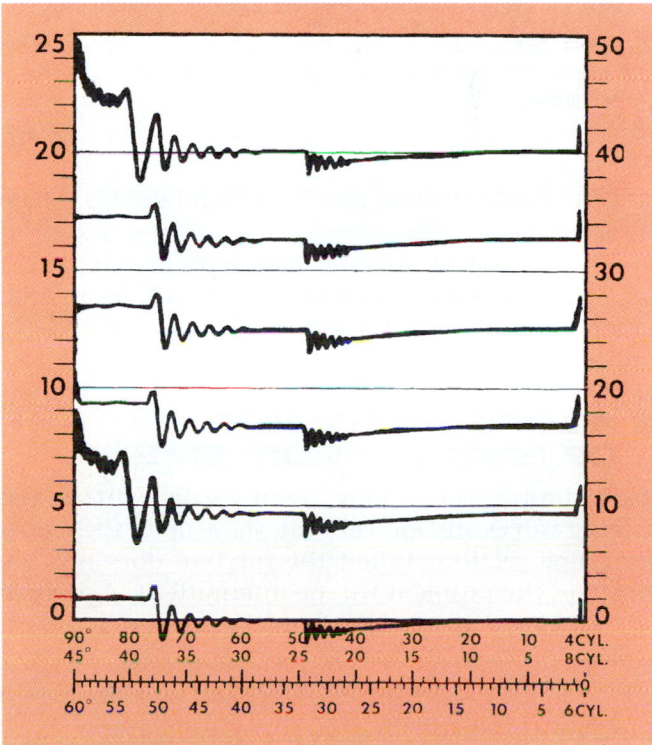

Figure 39.30 Two cylinders have problems. If the firing order is 153624, then cylinders number 5 and 4 are the ones to check. *(Courtesy of Sun Electric Corporation)*

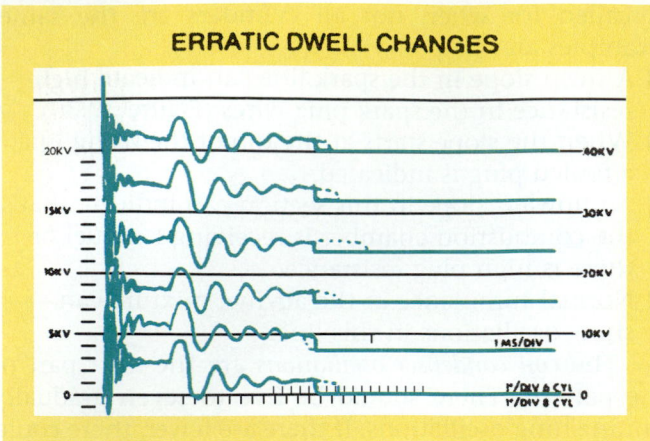

Figure 39.31 Dwell variation. *(Courtesy of Automotive Diagnostics)*

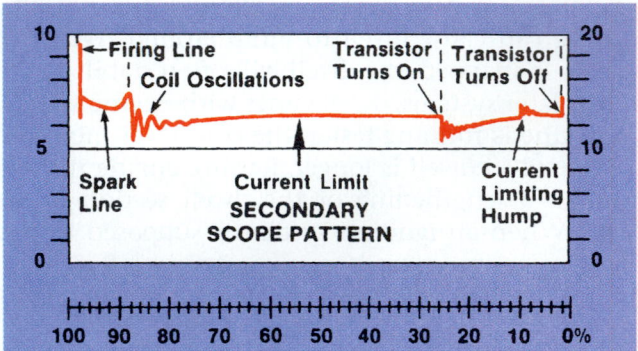

Figure 39.32 This pattern is for an ignition system with current limiting. *(Courtesy of Sun Electric Corporation)*

variable dwell but does not, a faulty module is usually the cause.

When a system does not have a primary resistor, it has a current limiting function. There will be an additional hump in the dwell section of these patterns (Figure 39.32).

■ SCOPE DIAGNOSIS OF ENGINE PROBLEMS

Engine problems can show up in a scope pattern, too. Sticking valves and air leaks can show up in the scope's firing line section. When the mixture does not stay constant, the problem will be intermittent. A problem that is related to electrical resistance would tend to remain constant.

NOTE: *In most cases, an oil change will solve the problem of sticking valves.*

A scope reading can also reveal a *worn timing chain* when there are erratic changes in the dwell section. This problem can appear at idle when slack develops in the chain. If the engine is momentarily accelerated,

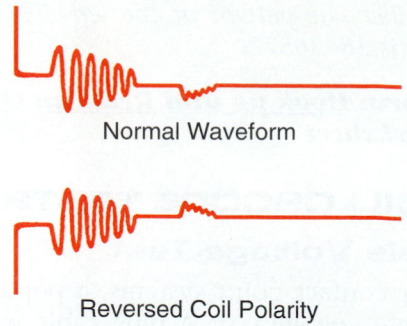

Figure 39.33 If the pattern is upside down, the coil is connected backwards. *(Courtesy of AlliedSignal Automotive Aftermarket/Autolite)*

the problem disappears. Further investigation will be required as this can also be due to a worn distributor.

■ SCOPE DIAGNOSIS OF IGNITION PROBLEMS

When the engine is off, use a ground probe to ground the spark plug end of a plug cable. The firing voltage requirement will drop, leaving the height of the spike only as the amount remaining for the rotor air gap. If the pattern is upside down, the coil is connected backwards (reversed polarity) (Figure 39.33).

■ STRESS TEST

Temperature can cause an electronic ignition to fail. To **stress test** parts like a module or pickup coil, parts can be cooled or heated. Direct compressed air at the base of the distributor to cool it. Technicians used to use freon refrigerant to cool parts but this is illegal now. With the scope set on raster pattern, watch for changes in the dwell zone.

Watch the firing line and spark line while spraying water on the distributor cap and plug cables. Spark plug cable insulation can break down under moisture or load. To create a simulated load, you can put the engine under a light load with the transmission in gear and the brakes applied. Then bring the rpm up to approximately 2000.

Problems related to heat will usually be found in the dwell section of the pattern. You can use a heat gun to warm up a module, trigger pickup, and electronic ignition connectors.

■ OTHER SCOPE TESTS

Tests of individual components can be done with the scope. When checking a trigger on an oscilloscope, a Hall switch, contact points, and an LED all give a square wave. The pickup coil gives an AC sine wave that increases with speed (Figure 39.34). An alternator diode can also be tested. EFI injectors and knock sensors can also be diagnosed with an oscilloscope.

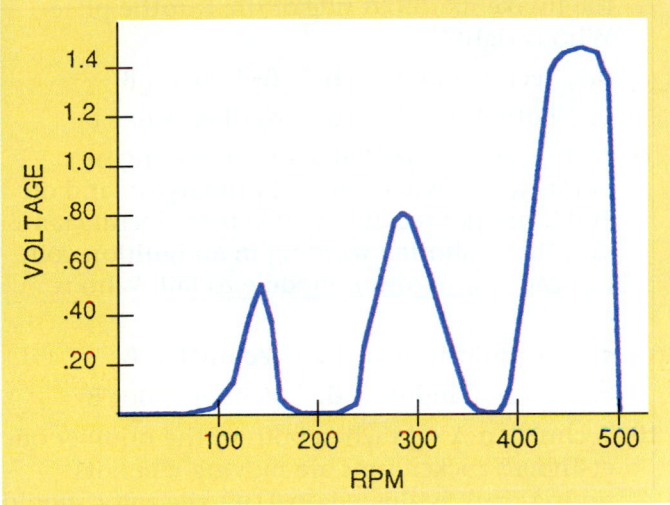

Figure 39.34 The pickup coil gives an AC sine wave that increases with speed. *(Courtesy of SPX Corporation, Aftermarket Tool and Equipment Group)*

■ OTHER DIAGNOSTIC INSTRUMENTS

With computer controlled cars, other test instruments have become common. Computer systems store trouble codes when an electrical malfunction occurs. Handheld scan tools are used to read codes and interpret computer data. A digital storage oscilloscope (DSO) is a popular piece of equipment used to diagnose electronic problems in newer vehicles. These instruments are covered in Chapter 76.

■ REVIEW QUESTIONS

1. In an engine running on the highway at 3000 rpm, each cylinder's spark plug will be firing _____ times every second.

2. The spark plug thread and thread seat is where over _____ of the plug's heat dissipation occurs.

3. A spark plug that has a buildup of carbon that shorts it out is called _____.

4. Where should anti-seize compound be installed on the spark plug?

5. Tighten a gasketed plug finger tight, then _____ turn further.

6. The normal range of resistance in a new spark plug cable is about _____ to _____ ohms per foot.

7. Initial, mechanical, and vacuum advance combine to make _____ advance.

8. Vertical spikes on the screen indicate _____ and the horizontal pattern represents time.

9. A _____ or parade pattern displays all of the cylinders next to each other, side by side.

10. If the scope pattern is upside down, what is wrong?

■ ASE STYLE REVIEW QUESTIONS

1. Technician A says that using a spark plug with too hot a heat range can damage the engine. Technician B says that only half of the negative electrodes wear on spark plugs on DIS engines. Who is right?

   a. Technician A   b. Technician B
   c. Both A and B   d. Neither A nor B

2. An engine has one overheated spark plug. Technician A says this can be due to dirty threads or a tapered plug seat that is dirty. Technician B says that the fuel injection system could be running too lean. Who is right?

   a. Technician A   b. Technician B
   c. Both A and B   d. Neither A nor B

3. Technician A says that a blue line at the bend in a spark plug's ground electrode is normal. Technician B says that a blue line on the ground electrode that is positioned near the plug gap indicates incorrect ignition timing. Who is right?

   a. Technician A   b. Technician B
   c. Both A and B   d. Neither A nor B

4. Technician A says gasketed seat spark plug ports cannot be repaired with thread inserts. Technician B says that some manufacturers recommend smearing some silicone dielectric compound on the tip of the distributor rotor. Who is right?

   a. Technician A   b. Technician B
   c. Both A and B   d. Neither A nor B

5. Technician A says if a distributor's rotor can be turned against spring pressure in a clockwise direction, the distributor rotates counterclockwise. Technician B says that on

most distributors, the direction the rotor turns can be determined by looking at the vacuum advance housing. Who is right?

   a.   Technician A   **b.**   Technician B
   c.   Both A and B   **d.**   Neither A nor B

6. Technician A says that distributor rotors must be aligned in the correct position on the distributor shaft. Technician B says that distributor caps must be installed on the distributor in only one position. Who is right?

   a.   Technician A   **b.**   Technician B
   c.   Both A and B   **d.**   Neither A nor B

7. Technician A says that advanced ignition timing can cause the engine's coolant temperature to rise. Technician B says that advancing the ignition timing can raise the engine's idle speed on some cars. Who is right?

   a.   Technician A   **b.**   Technician B
   c.   Both A and B   **d.**   Neither A nor B

8. Technician A says that a stuck mechanical advance can cause hard starting. Technician B says that ported vacuum is taken from a tap on the intake manifold side of the throttle plate. Who is right?

   a.   Technician A   **b.**   Technician B
   c.   Both A and B   **d.**   Neither A nor B

9. Technician A says that a 4-cylinder engine's points must switch the coil primary on and off 100 times per second at 3000 rpm. Technician B says that a shorted winding in an ignition coil can cause an ignition module to fail. Who is right?

   a.   Technician A   **b.**   Technician B
   c.   Both A and B   **d.**   Neither A nor B

10. Technician A says when both of the number one cylinder's rocker arms are moving when its timing mark is aligned on TDC, the rotor should be facing the number one terminal on the distributor cap. Technician B says no voltage spikes are shown in a raster pattern. Who is right?

   a.   Technician A   **b.**   Technician B
   c.   Both A and B   **d.**   Neither A nor B

CHAPTER 40

Emission Control System Fundamentals

■ OBJECTIVES

Upon completion of this chapter, you should be able to:

✔ Describe the different types of air pollution caused by motor vehicles.
✔ Explain the fundamentals of the major emission control systems.
✔ Label the parts of emission control systems.
✔ Explain the operation of computer controlled emission systems.

■ KEY TERMS

photochemical smog
inversion layer
Environmental Protection Agency (EPA)
oxides of nitrogen (NO$_x$)
skin effect
particulates
closed ventilation system
air injection system
smog pump
inert gas
catalyst
monolithic
reduction catalyst
oxidation catalyst
charcoal canister
hot soak
liquid/vapor separator
purge
malfunction indicator light
surface-to-volume ratio

■ INTRODUCTION

Emission controls have been included on cars since the 1960s. The area of emission control and tuneup has become a complicated specialty area. Emission specialists are required to be licensed to perform repairs. The emphasis in this section is to cover the basic theory of emission devices. Understanding this information is necessary for you to learn to diagnose and repair emission systems.

■ AIR POLLUTION

The air is polluted by many different things: oil exploration, industry, nature, home fireplaces, furnaces, paints, and transportation (cars, trucks, trains, boats, and airplanes). **Photochemical smog** is produced when hydrocarbons and oxides of nitrogen react with sunlight to form a dirty yellow-brown haze in the air. Warm air close to the ground rises, where it is cooled by the cooler air found at higher altitudes. This process usually rids the air of smog. But when there is a warm air **inversion layer**, the warm air becomes trapped within 1000 feet of the ground. This trapped air causes high concentrations of smog, resulting in health hazards.

Beginning in the early 1960s, the U.S. Congress began to pass laws related to air pollution. These laws are administered by the **Environmental Protection Agency (EPA)**. Vehicles have been designed to lessen the amount of pollutants produced when they burn fuel. Engine controls and emission devices have reduced vehicle emissions to the point that a car manufactured today will produce less than 5% of the air pollution that a 1960 model did.

■ AUTOMOTIVE EMISSIONS

There are three sources of emissions: crankcase, evaporation, and exhaust. The exhaust emission control system is designed to minimize byproducts of combustion leaving the engine in its exhaust. The main byproducts of combustion that parts of the emission control system are designed to control or eliminate are unburned *hydrocarbons* (*HC*), *carbon monoxide* (*CO*), and **oxides of nitrogen (NO$_x$)**.

Hydrocarbons are used as automotive fuels. When they are not completely burned in the engine and enter the air during daylight, smog can result. Emissions can come from several places on the car (Figure 40.1):

■ The exhaust pipe—all three emissions
■ The crankcase—mostly hydrocarbons
■ Vapors—hydrocarbons that evaporate from the fuel tank or vents in the fuel system

The percentages of various emissions in each of the above categories are listed in Figure 40.2. All of these emissions are defined together as *exhaust emissions*.

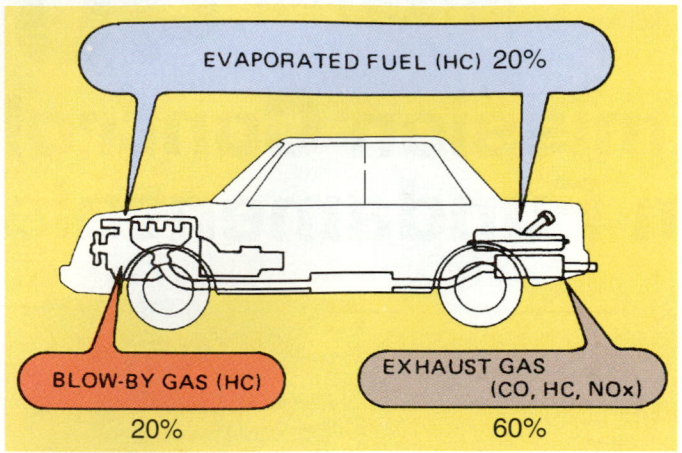

Figure 40.1 Sources of automotive air pollution.

Composition Type of Gas	CO	HC	NOx
Exhaust gas	100%	55%	100%
Blow-by gas	—	25%	—
Evaporated fuel	—	20%	—

Figure 40.2 Composition of various emissions.

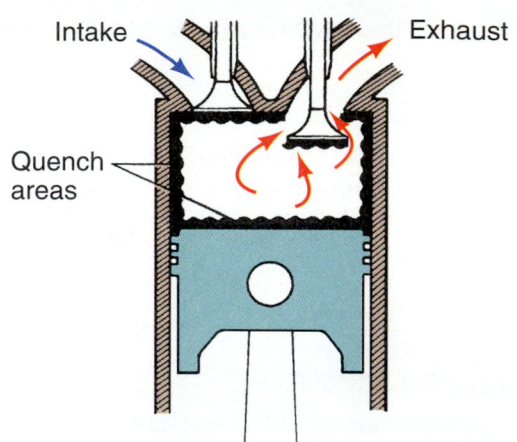

Figure 40.3 Unburned fuel is left on all surface areas.

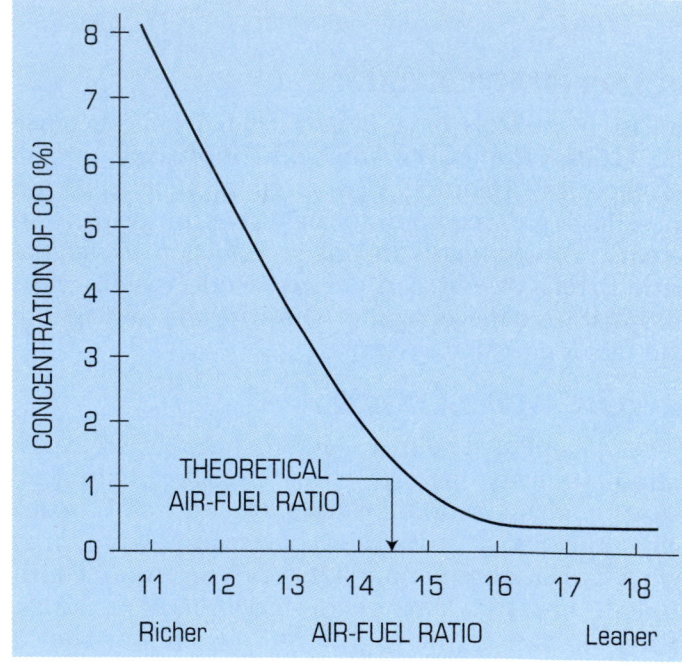

Figure 40.4 The amount of carbon monoxide (CO) in the exhaust varies with the air/fuel ratio.

HC forms from several sources:

■ Blow-by gases are those that leak past the piston rings (see Figure 15.4). Crankcase ventilation systems, explained later, remove these vapors from the crankcase.

■ After each four-stroke cycle is completed, raw gas is found on all areas of the combustion chamber. This happens because the surface area of the combustion chamber is cold so fuel burning is *quenched* (Figure 40.3). This is called the **skin effect**. HC emissions are worse when there is more surface area.

■ When the engine is coasting or decelerating, unburned gasoline (hydrocarbon) is left in the cylinder.

■ When combustion is incomplete because of too rich an air/fuel mixture, raw gas results in the exhaust.

■ If a cylinder does not produce sufficient compression to burn the mixture, unburned hydrocarbons will exit the cylinder.

Carbon monoxide (CO) emissions result when gasoline is not completely burned. The amount of CO in the exhaust varies with the air/fuel ratio (Figure 40.4). When an air/fuel mixture is too rich, there will be more carbon monoxide. Uneven air/fuel distribution in the combustion chamber and too much quench from low cylinder wall temperatures can also cause CO to be high.

Oxides of Nitrogen

Oxides of nitrogen (NO_x) are produced when combustion temperatures are too high (above 2500°F), causing nitrogen to react with oxygen. High compression and excess cooling system temperatures can be causes.

Particulates

Particulates are airborne microscopic particles of things like dust and soot. These are a source of secondary air pollution that contains lead and carbon. Particulates can be harmful to your health.

CO₂ and O₂

There are actually two other gases that are measured in vehicle exhaust. These two, *carbon dioxide* (CO_2) and *oxygen* (O_2) are not harmful but can be used to diagnose combustion problems. A five-gas emission analyzer, covered in the next chapter reads these two gases and the other three polluting gases.

> ### ■ SCIENCE NOTE ■
>
> *Most scientists agree that gradual global warming is occurring, although they are not all in agreement as to how fast it is occurring. Regardless of the rate that it is occurring, scientists do agree that if it is allowed to continue for a sufficient amount of time, even a 1° to 3°F change in temperature could dramatically affect coastlines and crop growing areas.*

■ POLLUTION CONTROL

In the early 1960s, California became the leader in emission control legislation. This was due to the smog problem of the Los Angeles basin, caused in part by the large vehicle population. Studies during this time showed that 25% of hydrocarbon emissions were contributed by the open automotive crankcases of that day.

Beginning with the 1961 model year, all new cars sold in California were required to have a crankcase emission system. In the six years following the introduction of this system, 1 million more cars were registered in California, but emission levels remained the same. In 1963, this system was required on all cars sold in the United States.

By 1966, new California cars were required to have exhaust emission systems. Some early systems had an air pump for reducing HC and CO emissions. Adding air to the exhaust ports helped the air/fuel mixture to continue to burn in the vehicle's exhaust system. Burning pollutants more completely lowers air pollution.

In 1970, the U.S. Congress passed the *Clean Air Act*. Maximum levels of pollution were set nationwide for HC, CO, and NO_x for automobiles and industry. The Arab Oil Embargo of 1973 caused oil prices to rise significantly. The U.S Congress passed a law that required improved fuel economy on vehicles. This is the Corporate Average Fuel Economy (CAFE) standard (see Chapter 1). Manufacturers had to design cleaner, more fuel-efficient engines. Engines since 1973 have had many improvements made on them. Some of these are discussed in this chapter.

> ### ■ SCIENCE NOTE ■
>
> *Burning a pollutant more completely reduces pollution by one of the following methods.*
> - *Providing sufficient oxygen so that all carbon in the compound is oxidized to CO_2*
> - *Providing enough heat so that large, tightly packed complex compounds can relax and stretch out enough so that oxygen can penetrate the inside of the compound for efficient burning*
> - *Breaking toxic compounds down into nontoxic ones*
>
> *One method of disposal for toxic materials is incineration in furnaces on ships 200 miles off shore. In the automotive industry, parts cleaned by thermal cleaning have very little toxic waste residue.*

■ AUTOMOBILE EMISSION CONTROL SYSTEMS

To make exhaust cleaner in today's cars, a combination of things has been done. These include:
- Engine design
- Fuel and ignition system controls
- Devices designed specifically for the control of emissions
- Modified fuel blending techniques

Computers have taken over the management of emission devices on today's cars. The computer uses an engine load (vacuum) signal, a road speed (speedometer) signal, temperature signals, and when the engine is warm, the oxygen sensor reading from the vehicle's exhaust. It then decides when to control various devices.

Older carbureted cars often have many vacuum hoses that control or supply various emission devices. Vacuum spark timing, controlled by temperature sensitive switches, is also popular. Most new vehicles have electronic controls for these items.

Emission control devices used on each car are listed on the underhood emission label (see Chapter 41). Some of those devices are:
- Positive crankcase ventilation
- Air injection system
- Exhaust gas recirculation
- Catalytic converter
- Fuel evaporative controls
- Heated air inlet

■ CRANKCASE VENTILATION

The first emission control device, installed on all new cars since the early 1960s was the PCV system. PCV stands for positive crankcase ventilation. It prevents the emission of hydrocarbons (blow-by gases) from the crankcase by scavenging them and reintroducing them into the combustion chambers.

NOTE: *Some people mistakenly call this system a PVC system. PVC stands for poly vinyl chloride, a common plastic used for wire insulation and plastic pipe.*

Early engines were equipped with a *road draft tube,* the end of which was positioned below the bottom of the engine to help purge the crankcase of harmful vapors. Since the early 1960s, cars have had crankcase ventilation systems instead of road draft tubes (Figure 40.5). The PCV system allows a small amount of intake manifold vacuum to leak to the crankcase (Figure 40.6).

A PCV valve, or a metering device, has a major advantage over a road draft tube. The road draft tube depended on air movement to operate; it did not work when the vehicle was not moving. The PCV system works at idle and part throttle.

The modern crankcase ventilation system (since the mid-sixties) is a **closed ventilation system** (see Figure 40.5). Atmospheric pressure must enter the crankcase for the ventilation system to operate. Filtered intake air is supplied through a hose from the air cleaner. The fresh air in the hose (under atmospheric pressure) flows toward the negative pressure created by the PCV valve (which is an engineered vacuum leak).

The fresh air hose also provides an escape route for excessive crankcase pressure when blow-by is more than the PCV valve can handle or the PCV valve is plugged. Excess pressure will go up the hose into the air cleaner, where it will be drawn once again into the intake system with incoming fresh air. On the older open systems, excess blow-by could escape through a breather into the outside air.

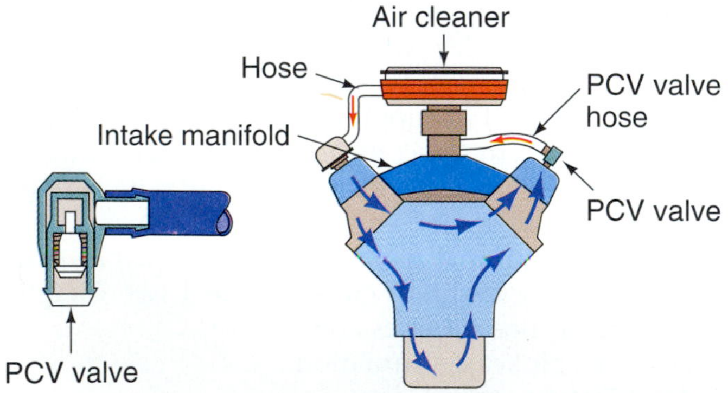

The orifice in the crankcase ventilation system (usually part of the PCV valve) is a specific size that matches the size of the engine. Therefore, it is very important that the correct valve be used. Performance problems, as well as oil leaks, can be traced to the use of a wrong-sized PCV valve.

Because it operates on intake manifold vacuum, the PCV valve does not function as well under a load (when there is less vacuum). Therefore, it opens up further to allow more blow-by to pass through.

Besides eliminating one source of air pollution, the PCV system has other benefits.

- It has cleaned up the insides of engines enormously. Acids are not formed in stagnant air like with a road draft tube.
- It reduces sludge by removing moisture from the crankcase.
- Reduced oil leakage caused by excessive crankcase pressure is another major benefit.

■ AIR INJECTION SYSTEM

An **air injection system** has two basic functions. It provides a low pressure air supply needed to lower HC and CO in the exhaust. It also provides air to the catalytic converter to help heat it up when it is cold. The first air injection systems appeared in California in 1966.

Some unburned gases remain in the exhaust after combustion in the cylinder. In early emission controlled engines, engine modifications were used to control these emissions. Another method for reducing these emissions is to keep them burning in the exhaust pipe. An air injection system is used to feed these hot gases and keep them burning. Air is provided by either a belt-driven air pump or by a non-pump pulse air system.

The air injection system uses an air pump (commonly called a **smog pump**), control valves, and lines to the manifolds. Parts of the system are shown in Figure 40.7.

- A low pressure *vane-type pump* driven by a crankshaft belt provides air (Figure 40.8). Operation of a vane-type pump is shown in Figure 40.9.
- A *diverter valve* (Figure 40.10), or a similar type of device, prevents a backfire during deceleration on carbureted cars. Carburetors allow a rich charge of fuel to enter the cylinder during deceleration. Without this valve, this rich mixture would be ignited when the fresh air from the pump contacts it. The diverter sends the air to the intake manifold or air cleaner. On computer cars the diverter directs air pump flow to the correct place.
- A one-way *check valve* keeps the exhaust from

Figure 40.5 A closed crankcase ventilation system. *(Courtesy of Cooper Automotive/NAPA Belden)*

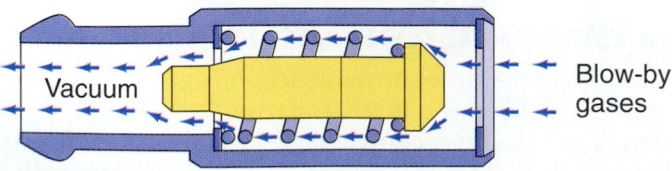

Figure 40.6 A PCV valve allows a small amount of intake manifold vacuum to leak to the crankcase. *(Courtesy of Ford Motor Company)*

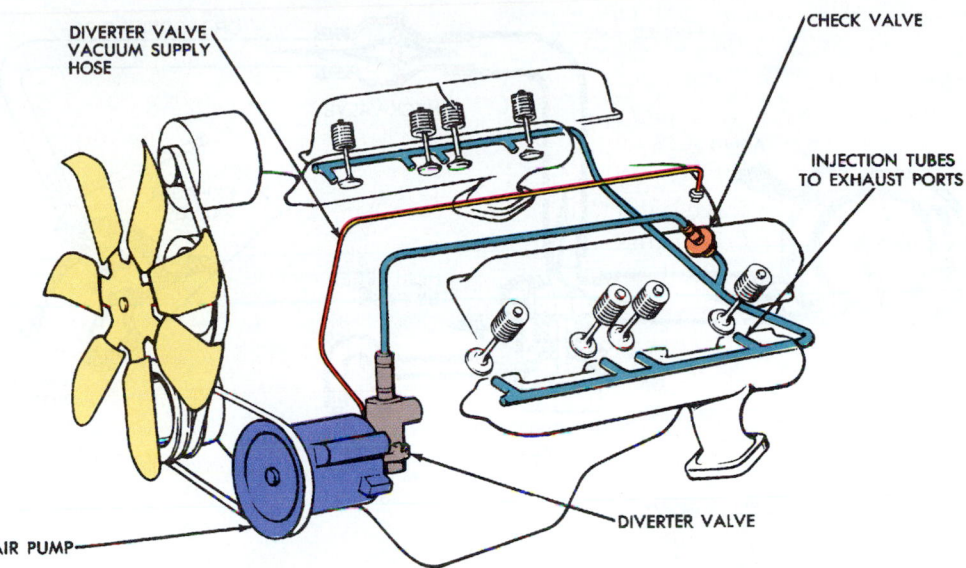

Figure 40.7 Parts of an air injection system. *(Courtesy of Chrysler Corporation)*

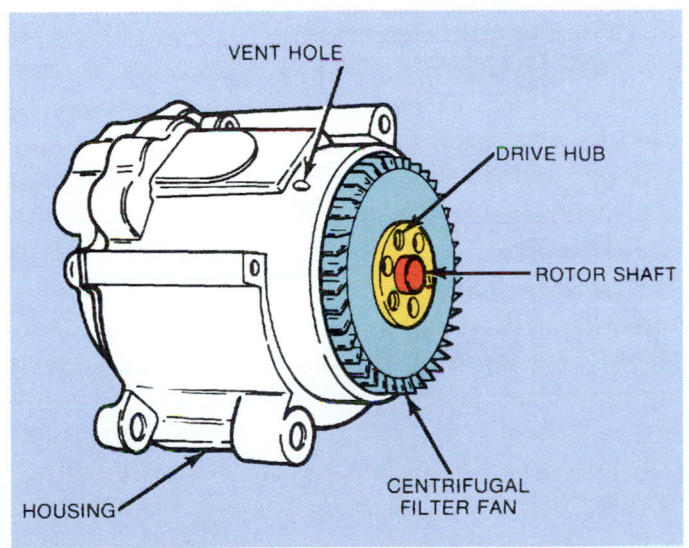

Figure 40.8 Parts of a smog pump. *(Courtesy of Chrysler Corporation)*

backing up into the hoses and pump.

- An air distribution manifold or cast in passageways in the head distribute the air to the exhaust ports.
- Computer controlled three-way catalyst systems (covered later in this chapter) include an air switching valve that moves the injected air into the exhaust manifold or into part of the catalytic converter.

ASPIRATOR VALVE OR PULSE AIR SYSTEM

At the end of the exhaust stroke when the intake valve opens, a momentary low pressure condition (pulse) occurs. The *aspirator valve* or *pulse-air system* uses these pulses to blow fresh air into the exhaust (Figure 40.11). This system saves the cost of the air pump. It works well at idle and low speed but is not efficient at high speeds. It is mainly found on small engines. One type is used mainly to heat the catalytic converter.

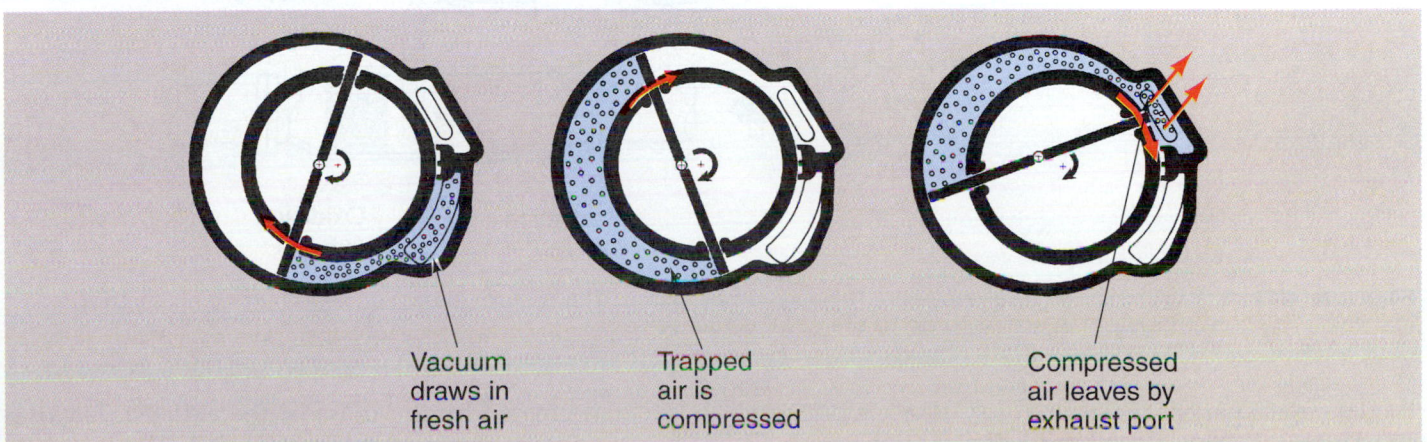

Vacuum draws in fresh air

Trapped air is compressed

Compressed air leaves by exhaust port

Figure 40.9 Operation of a vane-type smog pump. *(Courtesy of General Motors Corporation, Service Technology Group)*

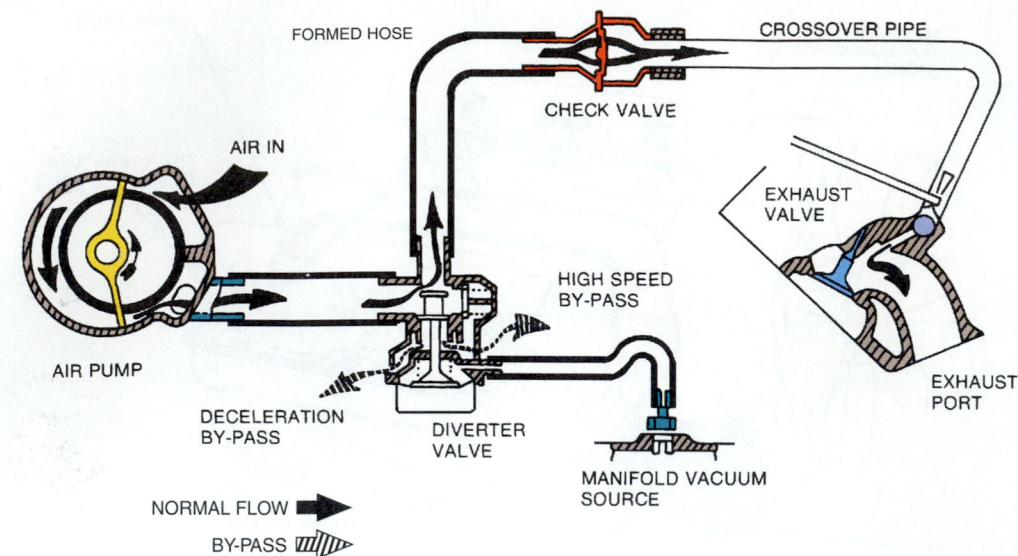

Figure 40.10 A diverter valve prevents backfire on deceleration on carbureted cars. *(Courtesy of General Motors Corporation, Service Technology Group)*

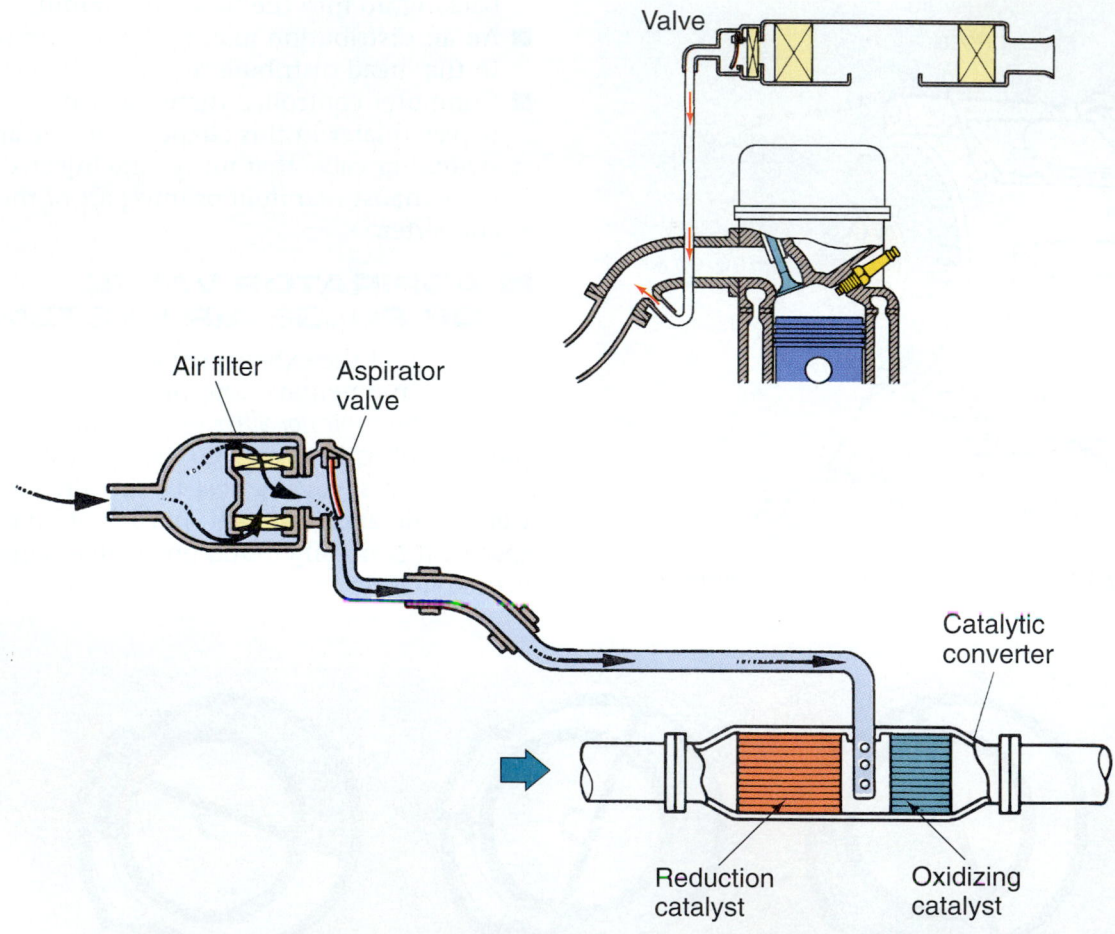

Figure 40.11 An aspirator valve system replaces an air pump. The system shown on top is providing air to the manifold. The bottom system is providing air to the catalytic converter.

The main part of the aspirator system is a one-way check valve, called an aspirator valve. It allows fresh air from the air cleaner to flow through it when there is a vacuum created by the pulse of the exhaust. It closes when exhaust pressure begins to build. There can be one aspirator for an engine or one for each cylinder.

■ EXHAUST GAS RECIRCULATION

Scientific studies of the late 1960s, discovered that NO_x was adding to smog production. The EPA added NO_x to its list of regulated items. In the early 1970s manufacturers were designing their engines for lower HC and CO emissions. Unfortunately, the resulting lean air/fuel mixtures and higher engine operating temperatures raised NO_x in the exhaust. Adding an **inert gas** (one that does nothing but take up space) into the incoming air and fuel was found to cut NO_x emissions. Because an inert gas will not burn, it does not contribute to emissions at the tailpipe.

The *exhaust gas recirculation (EGR)* system allows a small amount of exhaust gas (less than 10% of the total) to be routed into the incoming air/fuel mixture (Figure 40.12). Diluting the air/fuel mixture with this already burned gas, lowers combustion temperatures by about 300°F because there is very little free oxygen

left in exhaust to support combustion. Remember from the earlier discussion that oxides of nitrogen (NO_x) are formed when combustion temperatures are above 2500°F. The purpose of EGR is to reduce NO_x emissions at specified times.

EGR is also used to improve fuel economy. That is why some of these systems are so expensive. Metering the valve provides a means of varying the engine's compression ratio. Under light load conditions, the compression ratio can be lowered so the engine can operate without having to compress mixtures to needlessly high levels.

By 1973 most cars used an EGR system. Early systems developed a bad reputation so mechanics were disabling them to improve fuel economy and driveability. Later systems are vastly improved.

■ EGR SYSTEM OPERATION

A simple EGR system has an EGR valve that is operated by engine vacuum. The valve is located on the intake manifold or on a spacer plate mounted under the carburetor. There is either an internal passageway in the manifold or cylinder head, or a small pipe from the exhaust manifold that carries exhaust to the valve.

Very little NO_x is formed at idle or when the engine is cold, so the EGR valve is closed at those times. EGR is needed to reduce combustion temperatures when accelerating under load or during part throttle driving. As the throttle is opened, vacuum from a port above the throttle plate is applied to the valve. The valve opens to allow the passage of exhaust gas into the intake manifold (Figure 40.13).

Having the valve open is like having an air leak (vacuum leak). It is only open when the throttle is open so it will not have an effect on engine idle quality. It cannot be open during engine cranking either. This would act like an air leak and make the engine hard to start.

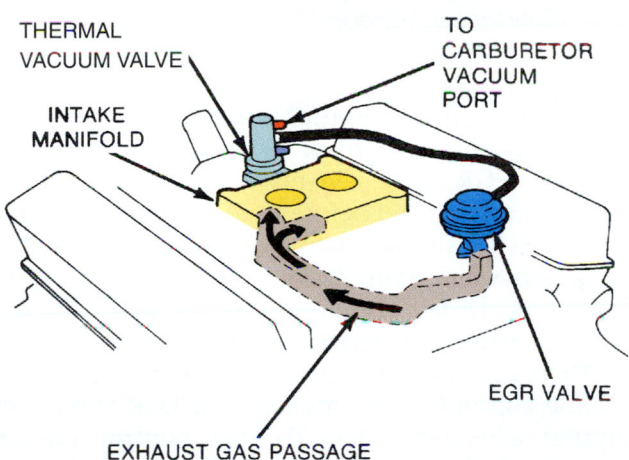

Figure 40.12 The EGR system allows a small amount of exhaust gas to leak into the intake stream. *(Courtesy of General Motors Corporation, Service Technology Group)*

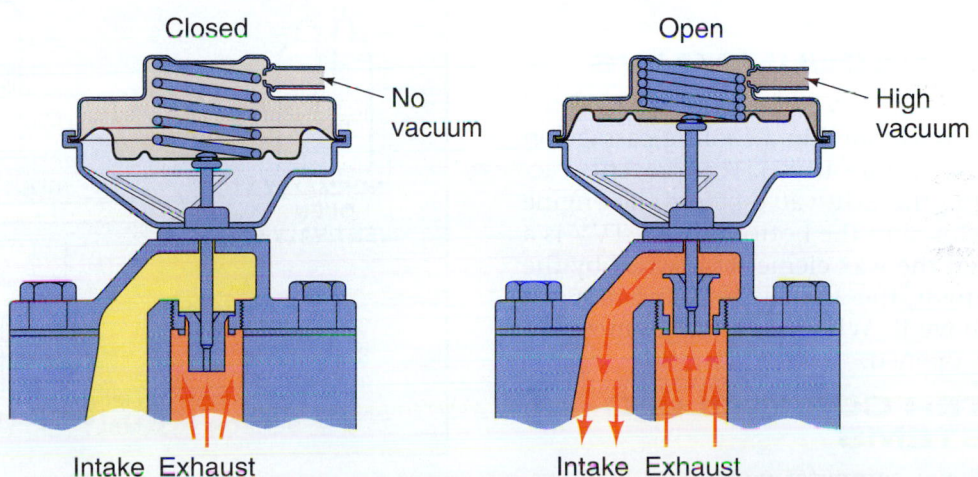

Figure 40.13 A simple EGR valve allows exhaust to flow into the intake when it is opened by vacuum. *(Courtesy of General Motors Corporation, Service Technology Group)*

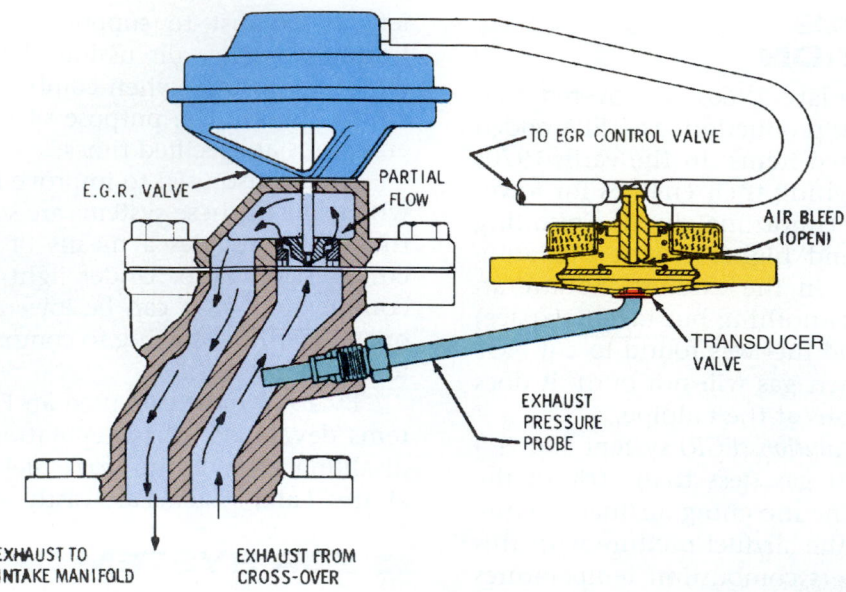

Figure 40.14 A backpressure transducer controlled EGR valve allows different amounts of flow. When backpressure is low, the air bleed is open, which allows only partial exhaust flow. *(Courtesy of General Motors Corporation, Service Technology Group)*

Backpressure EGR Valve

Exhaust backpressure is a good indicator of engine load. When an engine is under heavy load, more NO_x is formed. At this time more EGR flow is needed. A *backpressure transducer valve* can either be inside the valve or within its own housing. A diaphragm in the valve senses exhaust pressure and closes a vacuum bleed hole in response to it. Opening the bleed hole reduces the vacuum available to the EGR valve so it does not open as far (Figure 40.14). When the engine is under load and backpressure is high, the bleed hole is covered. This allows full exhaust gas recirculation.

There are two types of backpressure transducers, positive and negative. The one described above is the positive type. The negative type does the same thing, but reacts to vacuum in the exhaust (decreasing back pressure) to regulate valve flow.

■ THERMAL VACUUM VALVE

A thermal vacuum valve (TVV), also called a ported vacuum switch (PVS), is found in a cooling passage on precomputer cars (see Figure 40.12). This prevents vacuum from operating the EGR valve before the engine is warmed up. Enclosed in the bottom of the TVV is a wax element. When the wax element is heated by the coolant surrounding it, the wax expands and moves a plunger located above it. When the plunger moves, a vacuum passage is opened.

■ COMPUTER CONTROLLED EGR SYSTEMS

Most newer cars have computer controlled EGR systems. Computer systems do not need a thermal valve (TVV) for vacuum control. Also, they use the throttle position sensor and computer activation of the EGR to control a vacuum switch, or switches, in the vacuum line to the EGR valve. On many cars, the computer must see vehicle speed before the EGR valve operates.

Many EGR valves on late model engines have *position sensors*. Some manufacturers have used these for many years. Different manufacturers have different ways of sensing the position. Some use temperature and others use actual physical position of the pintel.

Some computer systems have an EGR valve with computer activated solenoids that control vacuum (Figure 40.15). These valves control vacuum coming to

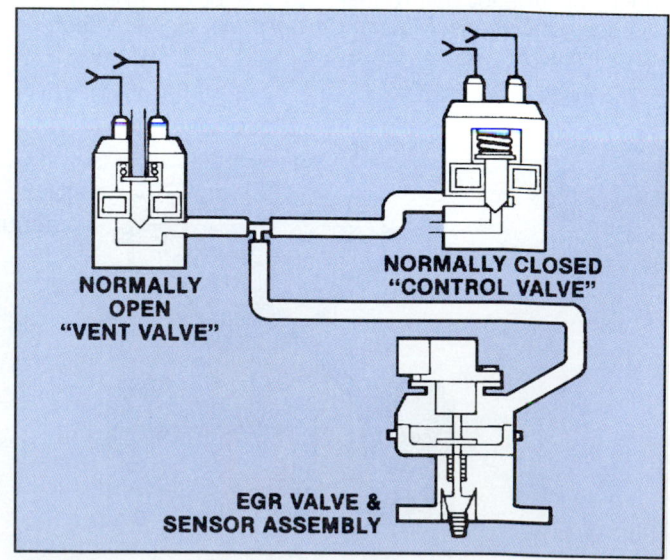

Figure 40.15 Computer activated solenoids control the vacuum to the EGR valve. *(Courtesy of Ford Motor Company)*

the solenoid and bleed off vacuum when it is not wanted in the same manner as the backpressure transducer.

Some computer systems use pulse width modulation, like on fuel injection systems, to rapidly cycle an EGR vacuum control solenoid on and off. This regulates EGR more closely.

Digital EGR valves are used on some cars. EGR flow is regulated by the computer by controlling a series of solenoids. There are three metered orifices opened and closed by solenoids (Figure 40.16).

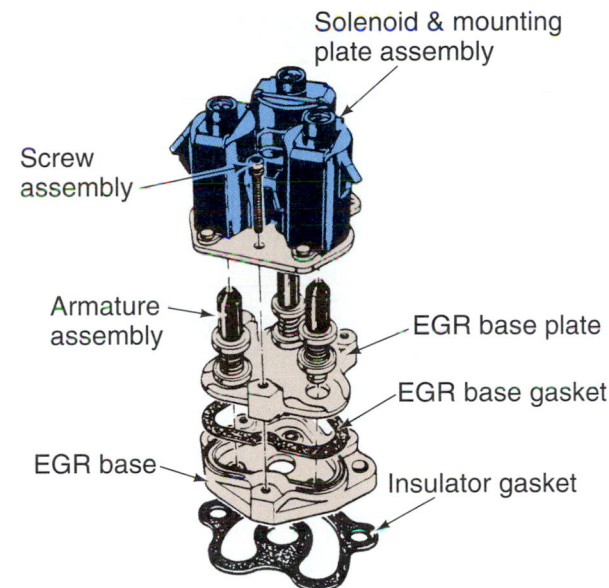

Figure 40.16 A digital EGR valve with three electric solenoids. *(Courtesy of General Motors Corporation, Service Technology Group)*

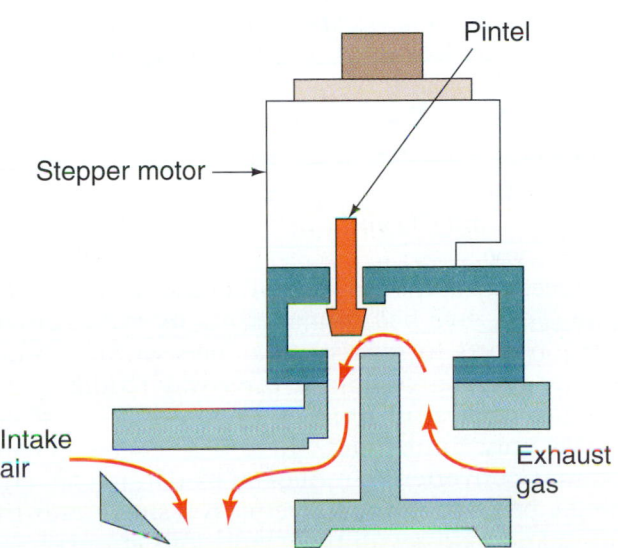

Figure 40.17 A linear EGR valve has a stepper motor. *(Courtesy of General Motors Corporation, Service Technology Group)*

The latest development in computer controlled EGR systems is a *linear EGR valve* (Figure 40.17). Instead of being opened by engine vacuum, this valve has a stepper motor. This means that the valve does not depend on engine vacuum and can be partly or fully opened in increments.

■ CATALYTIC CONVERTER

A **catalyst** is a substance that causes a chemical reaction to occur without undergoing any change to itself. The chemical reaction is one that normally would not occur at all or one that occurs at a much faster rate than normal due to the catalyst. A catalytic converter is an emission device that has been installed on many cars since the middle 1970s (the same time unleaded fuel came into use). The *catalytic converter,* often called a *cat,* is located in front of the muffler in the exhaust system (see Figure 28.11). It looks very much like a heavy muffler.

A catalytic converter must be hot to operate. It is mounted closer to the engine than a muffler so less exhaust heat is lost before reaching the cat. The point where the converter begins to work is called its *light off temperature.* The light off temperature of the catalyst is about 500°F.

Prior to the invention of the catalytic converter, engine emissions were controlled by engine systems, which included leaner air/fuel mixtures than were used previously, and altered ignition timing. Controlling emissions in this fashion resulted in lower performance and fuel economy. The catalytic converter allowed automotive manufacturers to improve engine performance and fuel economy and let the converter take care of the emissions.

Catalytic Converter Construction

The catalyst is either **monolithic** (see Figure 28.8) or of the *pellet design* (Figure 40.18). The pellet design uses small aluminum pellets coated with the same catalyst material used in a monolithic catalyst. A monolithic catalyst is like a big honeycomb. It has a thin coating of *platinum* and *palladium* applied to either a ceramic or metal *monolith* coated with alumina. Alumina is an oxide of aluminum that is very porous. The catalyst's metals (platinum and palladium) fill the holes in the alumina.

Monolithic catalysts are more widely used than pellets because they warm up more quickly. They also produce less exhaust backpressure. A disadvantage is their cost because they require more platinum and palladium.

Both types of converters are contained within a stainless steel shell. In order to provide a long life to the expensive components within the converter, stainless steel is required. It resists corrosion better than the galvanized steel used in mufflers.

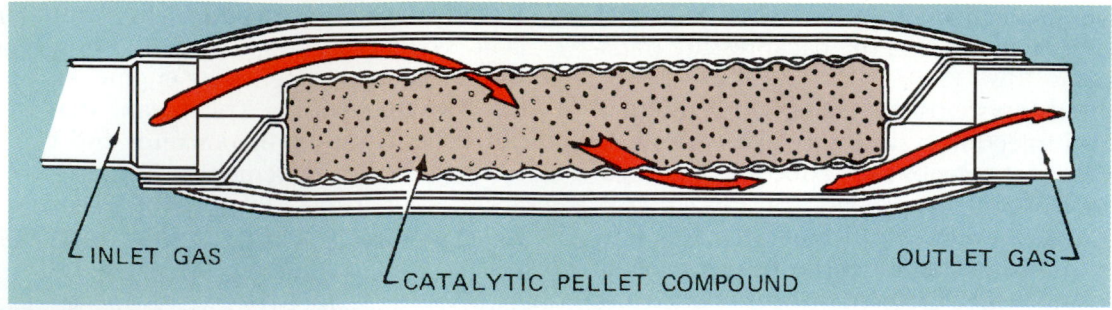

Figure 40.18 A pellet-type converter. *[Courtesy of General Motors Corporation, Service Technology Group]*

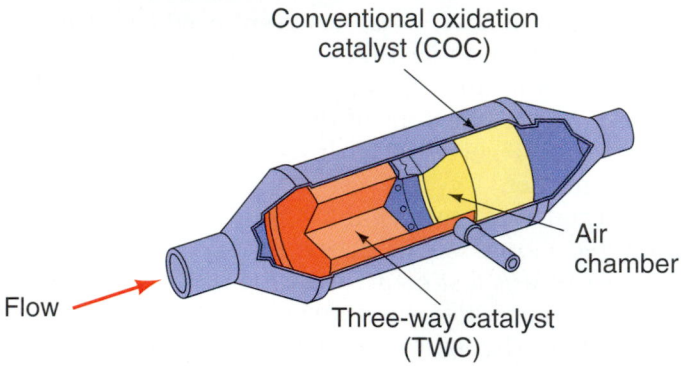

Figure 40.19 A three-way oxidation catalyst. *(Courtesy of Ford Motor Company)*

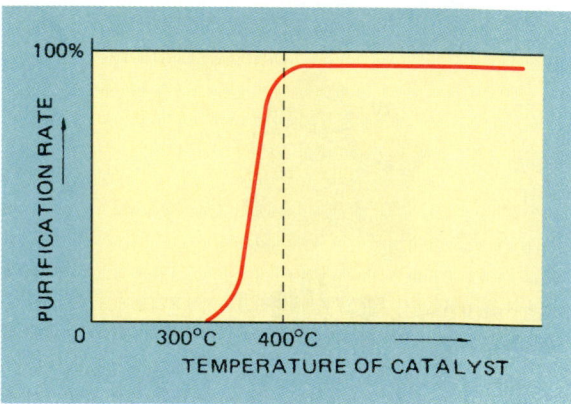

Figure 40.20 An oxidation catalyst does not work until it is hot.

■ TYPES OF CATALYTIC CONVERTERS

There are three types of catalytic converters:

■ The earliest catalytic converter, called a *two-way converter*, was designed to oxidize HC and CO, converting them into H_2O and CO_2.

■ The *three-way single bed converter* oxidizes HC and CO, and also reduces harmful NO_x into harmless nitrogen and oxygen. The NO_x portion of the converter is called a **reduction catalyst**. The oxygen produced in the NO_x portion of the converter aids in oxidizing HC and CO.

■ The *three-way dual bed converter* has two chambers. It has a tube between the two chambers to provide oxygen from the emission system's air pump to its rear oxidation chamber (Figure 40.19).

Two-Way Oxidation Converters

A *two-way catalyst*, also called an **oxidation catalyst**, changes harmful HC and CO into harmless carbon dioxide (CO_2) and water vapor (H_2O). It is often used with air injection to help it operate more efficiently. It is found mostly on cars built prior to 1980 without oxygen sensors.

An oxidizing catalyst begins working at temperatures between 500° to 600°F, but works most effectively when heated to 750°F (400°C) or higher (Figure 40.20). Converters often operate at temperatures near those in the combustion chamber. The graph shows a purification rate, which is a measure of the proportion of pollutants in the exhaust that can be converted to nonpollutants.

NOTE: *When a catalytic converter works, heat is a byproduct. The gas coming out of the converter can be from 50° to 200°F higher than the exhaust coming into it. Insulating or shielding material keeps the heat inside of the converter and away from the floor of the car (see Figure 28.19).*

Three-Way Catalytic Converters

A three-way catalytic converter is used on cars with oxygen sensor feedback systems. It reduces NO_x and oxidizes HC and CO and can be of either the monolith or pellet type.

Three-way converters are either a *single bed* or *dual bed design*. A dual bed converter has its reduction and oxidation parts in separate chambers within a single housing. It has a combined three-way oxidizing and reduction catalyst in front of a two-way catalyst that oxidizes only. Sometimes, there is a separate smaller three-way converter in front of an oxidation converter. A single bed three-way converter is a single unit that reduces and oxidizes all three emissions at once.

A three-way catalytic converter uses *rhodium* (or sometimes palladium) as a catalyst for its reduction portion (the part that controls NO_x). It only works properly when operated at the correct air/fuel mixture. Extra oxygen hinders the conversion of NO_x. This is why the tube from the air pump enters between the chambers in a dual bed cat. When the front part of the converter catalyzes NO_x, it creates more oxygen. The extra oxygen leaving the reduction part of the converter can also be used to assist the rear oxidation catalyst in its operation. In fact, a substantial amount of CO is oxidized to CO_2 before leaving the reduction part of the converter. The reduction part of the catalyst is therefore always located in front of its oxidation part. The reduction process does not produce enough oxygen for the oxidation catalyst to work at its best, however. The air added at the center of the converter assists in the continuation of the oxidation process.

A three-way converter needs heat and a reasonable air/fuel mixture to work effectively. It works far less efficiently when the mixture is too rich or too lean. Fuel mixtures on computer controlled cars are tailored to 14.7 to 1 to help the cat to function efficiently (Figure 40.21). The oxidizing part of the catalyst does not remove CO and HC very well at mixtures richer than 14.7 to 1. NO_x reduction in the catalyst does not work well at mixtures leaner than 14.7 to 1. The air/fuel mixture in an oxygen feedback system constantly fluctuates between rich and lean. This allows the oxidation part to operate when the mixture is lean and the reduction part to operate when the mixture is rich. The reduction part needs the extra CO to scavenge the oxygen it produces.

Unleaded gasoline became required for cars when catalytic converters first began to appear on cars in 1975. Lead can coat the catalyst, preventing it from functioning properly. A common result of a coated catalyst is a restricted converter. This reduces engine power and sometimes causes overheating. Lead was

previously used by refiners to raise the octane level of fuels, but lead is no longer an additive in gasolines sold in the United States.

■ AIR SWITCHING VALVE

Dual bed catalytic converters include an air switching valve, also called a diverter, that directs the air *upstream*, *downstream*, or to the air cleaner (*bypassing*) (Figure 40.22). The valve sends the air upstream to the exhaust manifold when the engine is cold. This helps heat the converter and the oxygen sensor (Figure 40.23a). On an engine with an oxygen sensor, if air was injected into the manifold when the engine was in closed loop, the sensor would think that the air/fuel mixture was too lean (see Chapter 26). The result would be that the computer would cause the air/fuel mixture to become too rich.

Injecting air in front of the *oxidation* (HC/CO) part of the cat helps that part of the cat to work more efficiently. When the engine is warm, the air is directed downstream in the exhaust to the oxidation chamber of the catalytic converter (Figure 40.23b). *Preventing* the injection of air into the reduction (NO_x) chamber of the cat helps reduce oxides of nitrogen. A small amount of CO (rich mixture) is required for the *reduction* part of the catalyst to function.

When the engine is cold, the air pump can blow upstream to the manifold and flow through the entire converter. At lower engine temperatures, not as much NO_x is produced and the computer is not reacting to the O_2 sensor. The air provided by the air pump helps to heat the O_2 sensor and catalytic converter more quickly to operating temperature. The switching valve also sends air to the atmosphere on some cars during deceleration to prevent backfire during open loop

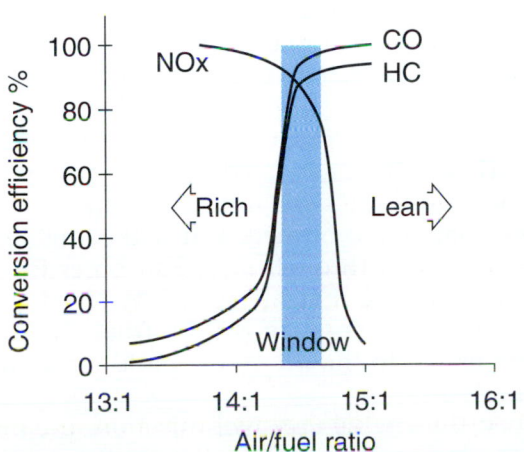

Figure 40.21 The air/fuel mixture needs to be close to 14.7:1 for the three-way catalytic converter to be effective. [*Courtesy of Chrysler Corporation*]

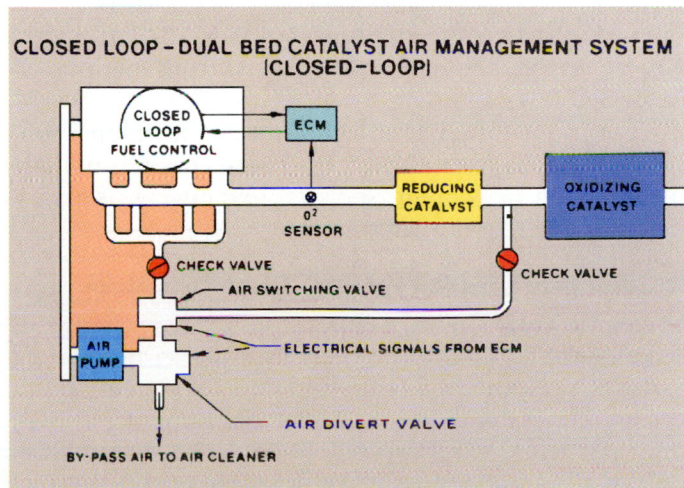

Figure 40.22 An air switching valve directs the air from the pump to go upstream, downstream, or bypassing to the air cleaner. [*Courtesy of General Motors Corporation, Service Technology Group*]

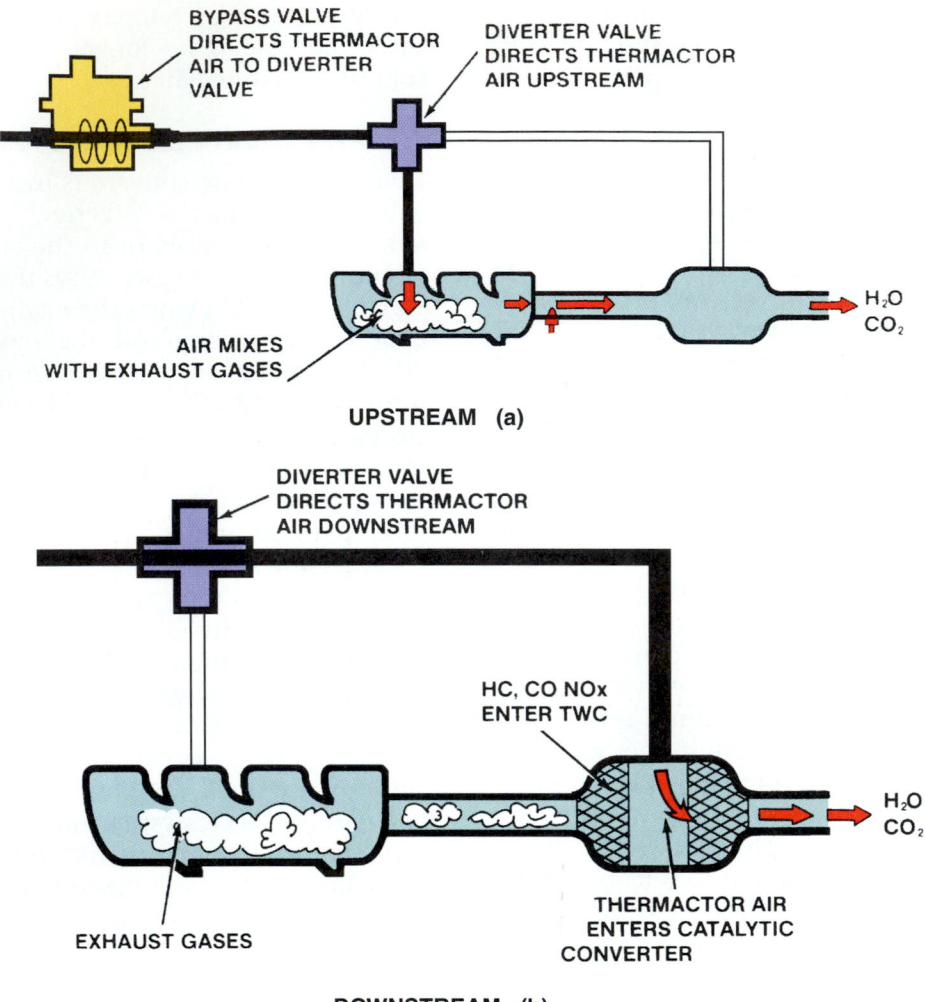

BYPASS VALVE
DIRECTS THERMACTOR
AIR TO DIVERTER
VALVE

DIVERTER VALVE
DIRECTS THERMACTOR
AIR UPSTREAM

H₂O
CO₂

AIR MIXES
WITH EXHAUST GASES

UPSTREAM (a)

DIVERTER VALVE
DIRECTS THERMACTOR
AIR DOWNSTREAM

HC, CO NOx
ENTER TWC

H₂O
CO₂

EXHAUST GASES

THERMACTOR AIR
ENTERS CATALYTIC
CONVERTER

DOWNSTREAM (b)

Figure 40.23 (a) The valve sends the air upstream to the exhaust manifold when the engine is cold. (b) When the engine is warm, the air is directed downstream in the exhaust to the catalytic converter to help it operate. *(Courtesy of Ford Motor Company)*

operation (when the engine is cold). It also can divert air from the catalytic converter during periods of extended idle.

For more complete information on computer controls, read Chapters 75 and 76. These chapters include information on engine sensors, exhaust gas sensors, the computer, and related items.

■ EVAPORATIVE CONTROLS

Since the early 1970s, evaporative emission controls have been required on all vehicles sold in the United States. Check the ECS (emission control system) application manual for the emission equipment required on a particular car. Evaporative control systems (ECS) reduce the emission of gasoline vapors from the fuel tank and carburetor vents (Figure 40.24). In warm weather, gasoline evaporates more easily. The vapor pressure of gasoline is more closely controlled, which is helping in the control of these emissions also (see

Chapter 25). Evaporative emissions are said to account for about 20% of a car's emissions (see Figure 40.2). A parked car can emit substantial emissions if it does not have one of these systems or if the system is not working properly.

The heart of the evaporative emission system is a **charcoal canister**. Activated charcoal can store gasoline vapors until the right time for them to be drawn into the engine and burned. When the engine is off, vapors are routed through lines and hoses to a charcoal canister where they are stored. When the engine is running, the vapors are purged from the canister. They are drawn to the intake manifold for burning in the cylinders.

When the engine uses fuel injection, the fuel system is sealed so there are no evaporative emissions from the engine. The fuel tank must still be sealed to prevent vapor loss.

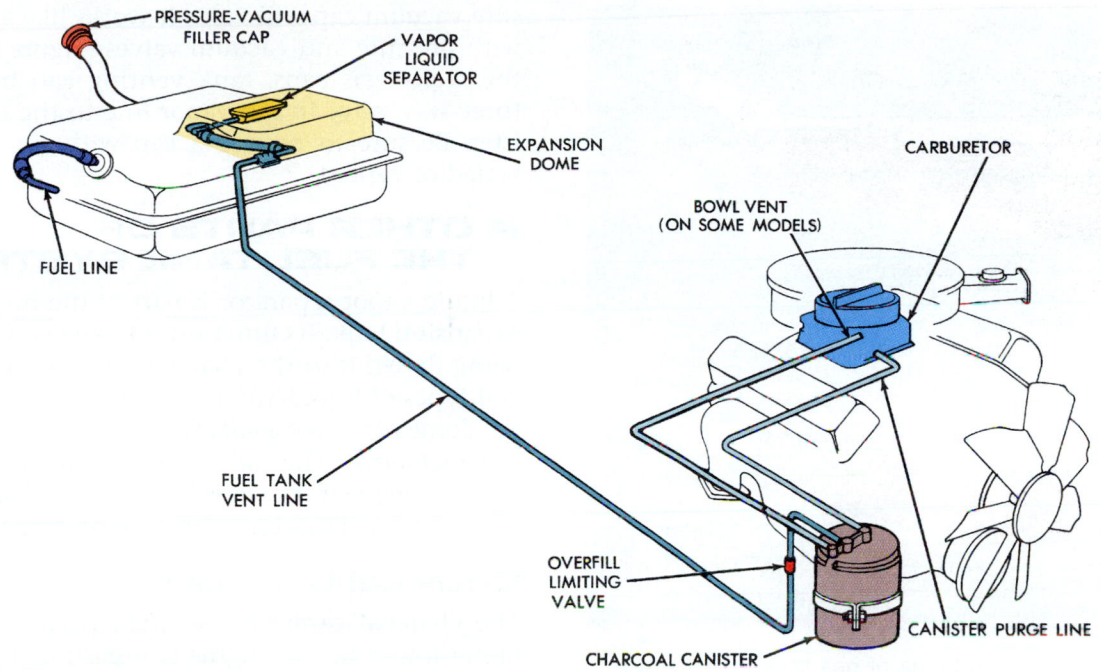

Figure 40.24 An evaporative emission system. *(Courtesy of Chrysler Corporation)*

■ SEALING THE SYSTEM

Emission of fuel vapors from the fuel tank or carburetor float bowl can be controlled by sealing the fuel system from the atmosphere. Air must be allowed into the fuel tank to take the place of fuel as it is consumed. Otherwise, the fuel tank would collapse as the fuel pump continued to create a vacuum within it. Flow of fuel would also be restricted to the engine. The fuel tank is usually vented by the gas cap.

On a hot day, gasoline in the tank expands. A tank filled on a cool morning must have sufficient reserve capacity to hold the expanded fuel.

The carburetor float bowl must have means of venting also. Without a vent, fuel could not flow and the air/fuel mixture would become lean. In fact, if a gas cap check valve becomes restricted, a car will surge and then stop running when driven for a few miles. Loosening the gas cap results in a large rush of air, and then the car will run again. The same effect would happen if the float bowl were not vented.

As the temperature of the engine and air under the hood increases, the evaporation rate of the fuel increases. With the engine running, fresh fuel is always coming into the float bowl. This keeps it cool. Once the engine is shut off, a condition known as **hot soak** occurs. On hot days, gasoline can boil out of the float bowl. The evaporative emission system must be able to store fumes for a long time.

Fuel Tank

Gasoline tanks in today's cars are designed to allow for expansion of the fuel by around 10% of the total vol-

ume. Major parts of the different designs of fuel tank systems include a *pressure-vacuum filler cap*, an *expansion area*, and a **liquid/vapor separator**. An expansion area's function is to prevent the overfilling of fuel, which could allow liquid into the vapor recovery system. During a tank fill, the filler nozzle will shut off before the expansion area can be filled. There are several ways of providing a reservoir for expansion of excess fuel:

Expansion Dome. Some cars have a dome above the fuel tank for excess fuel (see Figure 40.24).

Expansion Tank. An expansion tank is sometimes located inside of the main gas tank (Figure 40.25a). When the gasoline in the main tank expands on a hot day, it can slowly fill the expansion tank. There are small holes drilled in the expansion tank that allow the expanding fuel to enter it.

Some manufacturers install an extra tank that holds one or two gallons of fuel to cover expansion.

Filler Neck Design. The filler neck can be designed to prevent overfilling. On one type, when the gasoline reaches a certain level during a tank fill it flows through a tube back into the filler neck. This shuts off the gas nozzle. If the customer does not try to further top off the tank, it will not be overfilled. On another type, the lower end of the tank filler tube is positioned so its end is low enough to prevent the tank from overfilling (Figure 40.25b).

Gas Cap

Gas caps have either a *pressure vacuum valve* or they are *sealed.* If pressure in the tank exceeds 1 psi, the pres-

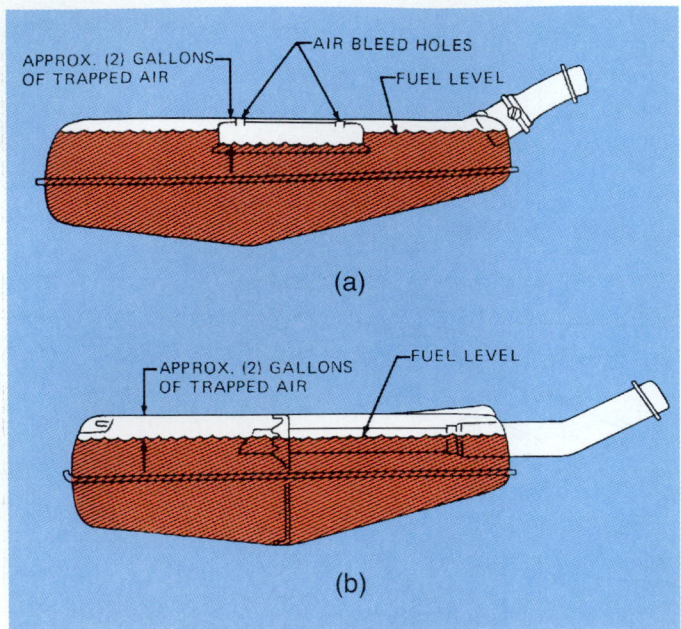

Figure 40.25 Two types of gas tank overflow protection. (Courtesy of General Motors Corporation, Service Technology Group)

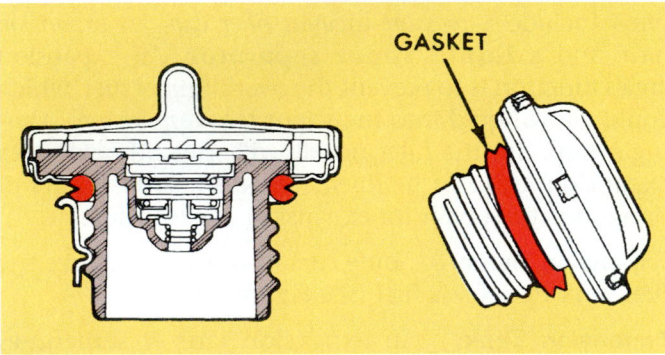

Figure 40.26 A pressure vacuum gas cap. (Courtesy of Chrysler Corporation)

sure vacuum cap will vent. It works like a radiator cap with pressure and vacuum valves (Figure 40.26). With the sealed gas caps, tank venting can be done by a three-way valve in the vapor line to the charcoal canister. Be sure to replace a cap with the correct one, sealed or vented.

■ OTHER PARTS OF THE FUEL TANK SYSTEM

A liquid vapor separator is part of the fuel tank or the expansion tank. Its function is to keep liquid fuel from being drawn into the charcoal canister. There are several types of liquid/vapor separators.

Some cars have a *roll over valve* in the vent line from the fuel tank. The valve has a ball or plunger that blocks the line if the car is in an accident and rolls over. This prevents liquid fuel from being able to leak out.

Charcoal Canister

The charcoal canister is a small plastic or steel container found in the engine compartment or in a front fenderwell. The canister stores vapors from the gas tank, and sometimes the carburetor float bowl, until they can be safely burned in the engine, rather than allowing them to escape to the atmosphere.

The canister has about 1.5 pounds of activated charcoal that can soak up twice its weight in fuel. It stores vapors until the engine is running. The vapors are then drawn into a port on the side of the carburetor or through the PCV system. Air on some systems enters the canister through a filter in the bottom. On other systems, fresh air enters through the top of the canister. Air is drawn through the charcoal and leaves through the purge line.

A vapor canister is only allowed to **purge** (empty) itself of vapor storage when the engine is running (Figure 40.27). It collects vapors when the engine is off. Control is usually done by either a purge control valve or an electric solenoid (when there is computer control).

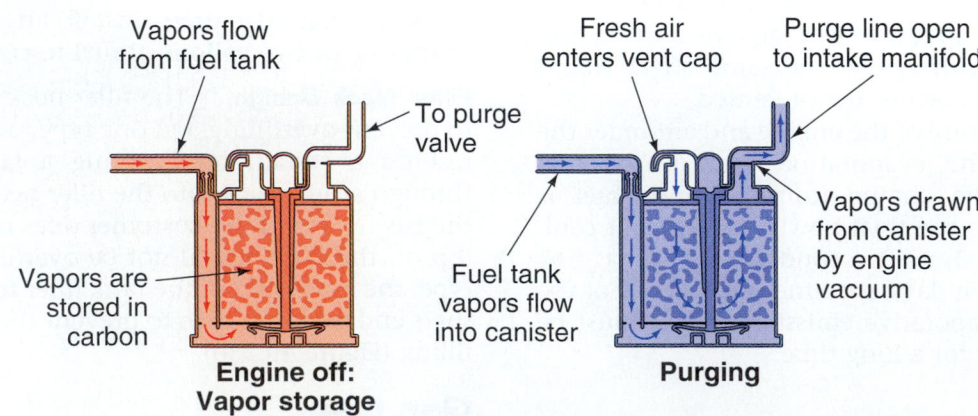

Figure 40.27 Vapors are stored in the charcoal canister until they can be purged with the engine running. (Courtesy of Ford Motor Company)

HEATED AIR INLET

Carbureted and throttle body injection engines have heated air controls in the air cleaner (Figure 40.28). This is called a *thermostatic air cleaner (TAC)*. The density of air changes with temperature. Fuel remains atomized better in warm air. Consistent temperatures of air entering the carburetor make it possible to maintain a more consistent air/fuel mixture. Cold air is more dense than hot air. Heating incoming air when the engine is cold prevents the mixture from becoming too lean when the engine is hot. A door in the air cleaner is opened to heat radiated up a tube called a *stove* from the exhaust manifold. A temperature sensor inside of the air cleaner routes engine vacuum to a vacuum motor that shuts off the hot air when the engine is warm.

Manifold Heat Control Valve

Port fuel injected cars do not require preheat to vaporize the fuel. A manifold heat control valve is included on many cars with carburetors or throttle body fuel injection. Its purpose is to reduce exhaust emissions and improve *driveability* during engine warmup. Some systems include a butterfly valve, located at the bottom of the exhaust manifold. When the engine is cold, exhaust gas is directed under the floor of the intake manifold (Figure 40.29). The manifold heat control valve routes exhaust gas under the floor of the intake manifold when the engine is cold. This improves the vaporization of the cold fuel.

Some heat control valves are operated by engine vacuum and others have a bimetal spring that expands when heated to open the valve. This type of valve is known as a heat riser. When its temperature sensitive spring is cold, the valve is closed. This diverts exhaust gas under the floor of the manifold. When the spring heats up, the valve opens.

The heat control valve shown in Figure 28.15 is controlled by engine vacuum. The vacuum signal can be controlled either by a computer or by a thermal vacuum switch (TVS). This kind of valve is called an *EFE (early fuel evaporation)* valve.

In V-type engines, the heat riser causes a restriction in the exhaust on one side of the engine. This restriction causes exhaust to flow through the crossover passage under the carburetor (see Figure 28.8) to the manifold on the other side of the engine.

Some EFE systems use an *electrically heated grid* installed under the base of the carburetor or throttle body (Figure 40.30). It serves the same function as the manifold heat control valve, which is to improve vaporization of the fuel during the first few minutes of engine operation.

ON BOARD DIAGNOSTICS

The California Air Resources Board (CARB) developed the first on board diagnostics regulations (OBD-I) in 1985. The regulations became law in 1988. The regulations require the computer to monitor the engine's oxygen sensor, EGR valve, and charcoal canister purge solenoid to see that all of these systems continue operating properly. A **malfunction indicator light** was also required by the regulations. Newer regulations were later developed that were more thorough on emission problem detection.

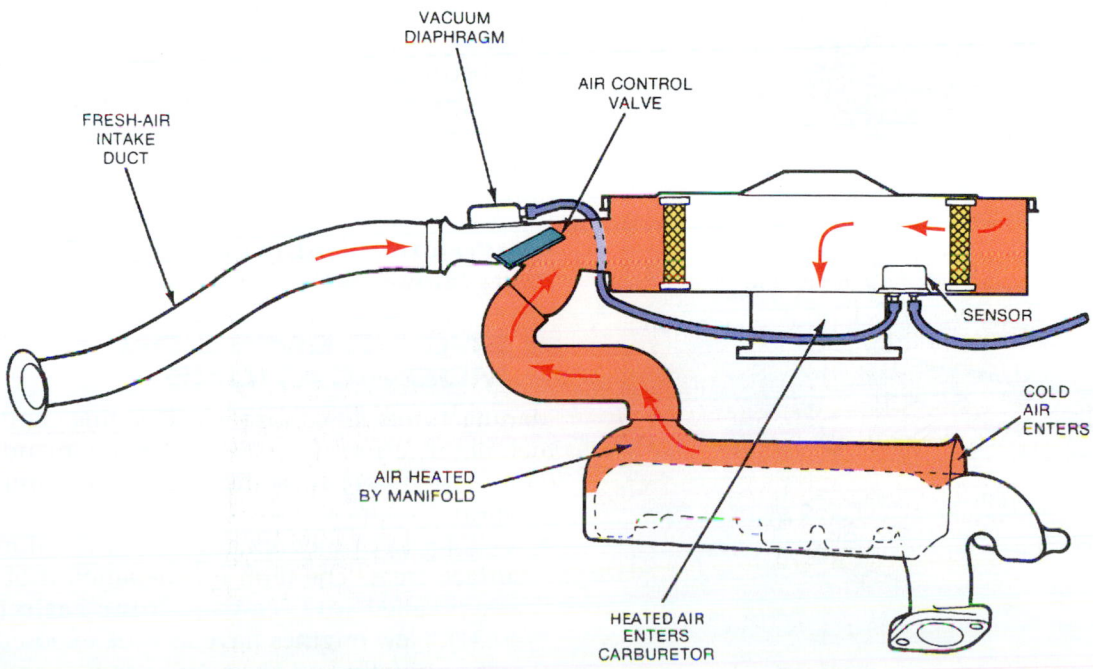

Figure 40.28 Hot air comes into the air cleaner when the outside air is cold. *(Courtesy of Chrysler Corporation)*

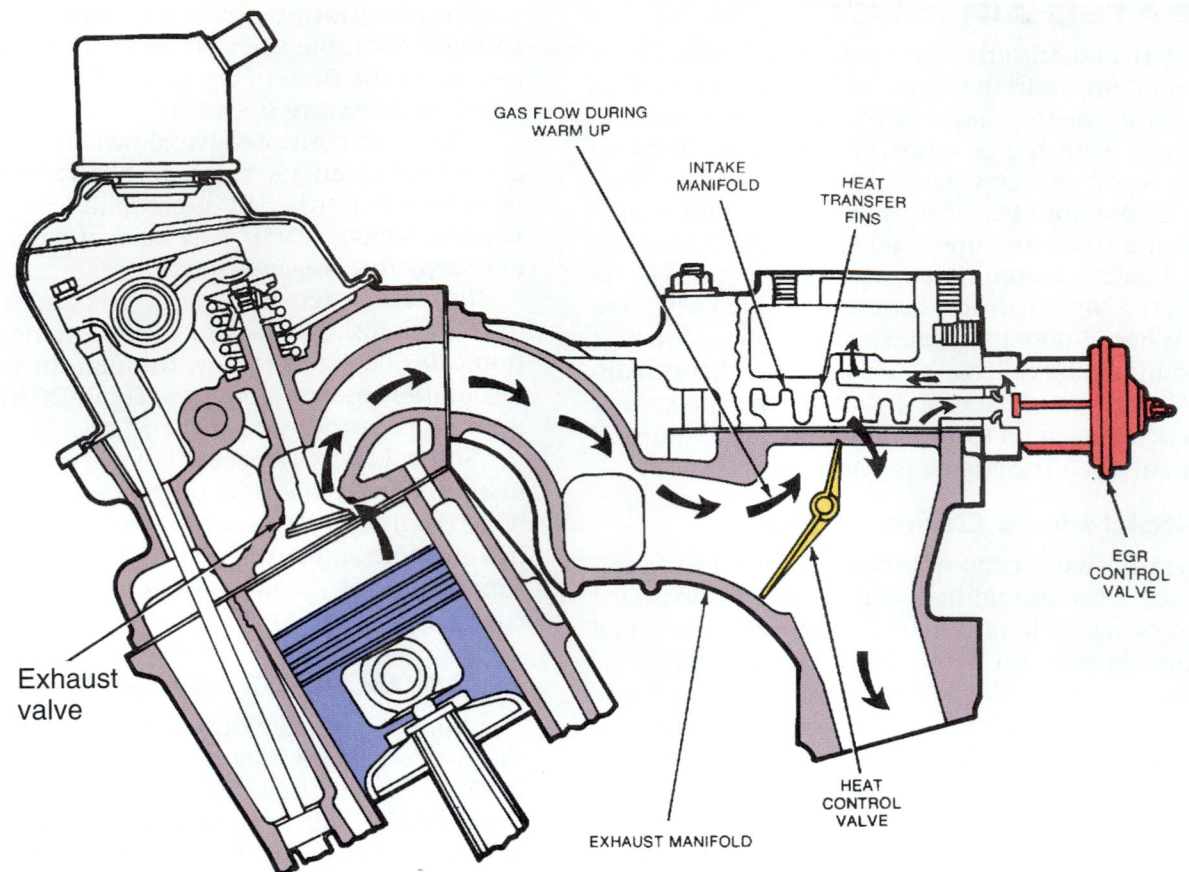

GAS FLOW DURING
WARM UP

INTAKE
MANIFOLD

HEAT
TRANSFER
FINS

EGR
CONTROL
VALVE

Exhaust
valve

HEAT
CONTROL
VALVE

EXHAUST MANIFOLD

Figure 40.29 When the engine is cold, exhaust gas is directed under the floor of the intake manifold by the manifold heat control valve. *(Courtesy of Chrysler Corporation)*

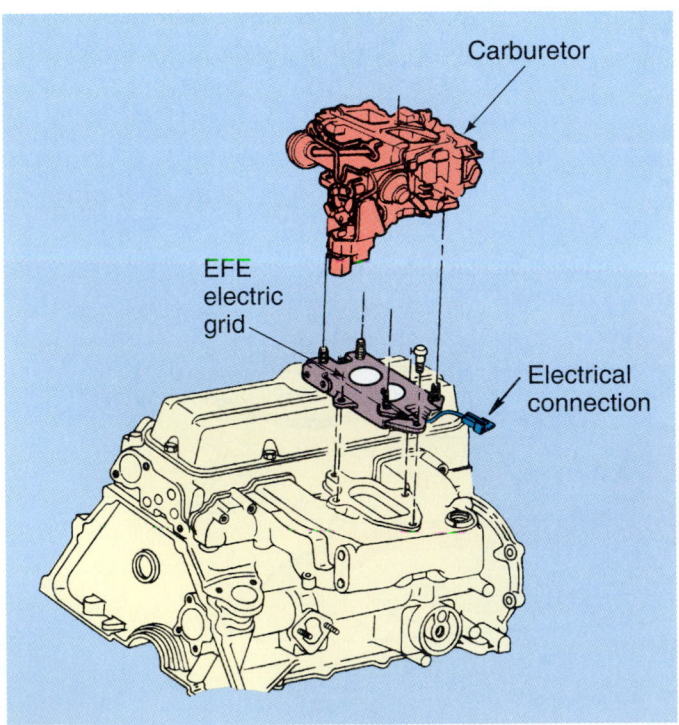

Carburetor

EFE
electric
grid

Electrical
connection

Figure 40.30 An EFE electric grid. *(Courtesy of General Motors Corporation, Service Technology Group)*

There have been many types of emission controls and also different names for the same part, depending on the manufacturer. On late model cars, many of the major systems that have been covered in this chapter have become somewhat standardized. Names of emission parts and connections for test equipment were also standardized by the Society of Automotive Engineers. These changes are part of the OBD-II regulations.

➡ *Perform Identify and Inspect Emission Parts Worksheet*

■ ENGINE EMISSION MODIFICATIONS

Manufacturers have also used engine, ignition, and fuel modifications to achieve federally mandated standards. Listed here are some of the more common modifications.

- A cool area of fuel remains unburned along all cool surface areas. The term for the amount of surface area is called **surface-to-volume ratio** (Figure 40.31). New engines have as little exposed surface area as possible on the combustion chamber and the top of the piston. The top piston ring has been

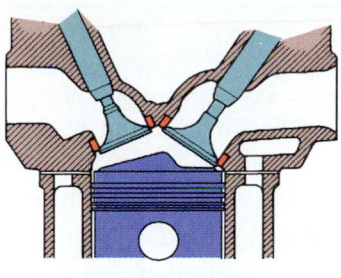

S/V RATIO LARGE

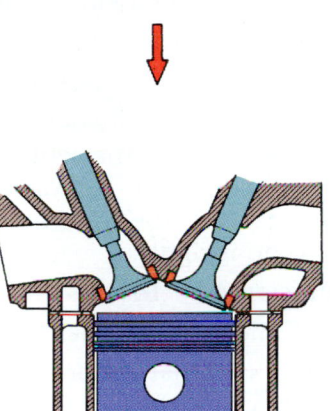

S/V RATIO SMALL

Figure 40.31 Surface-to-volume ratio is kept small for control of hydrocarbons.

moved as close as possible to the top of the piston so there is not much surface area of the piston exposed to flame.

- Engines are running with *higher cooling system temperatures*. This improves thermal efficiency. Older cars ran at 160° to 180°F while newer cars operate at from 190° to 218°F.
- *Advancing ignition timing* will increase HC and NO_x but if done properly increases fuel economy. More efficient use of fuel helps reduce emissions overall. The computer is programmed to provide the best spark advance for the condition at hand.
- *Changing the cam design specifications* can result in different emissions, especially at idle. Increasing valve overlap increases HC emissions, but lowers NO_x emissions at low engine rpm. The effect is the same as exhaust gas recirculation. On the exhaust valve side, exhaust will be drawn back into the incoming air/fuel mixture when the intake stroke begins. On the intake side, exhaust moves up into the intake port because the intake valve is opened during the exhaust stroke. This was done on engines in the mid-1970s. Their fuel economy and performance suffered greatly. See Chapter 18 for more information on valve overlap.

Today's engines have many refinements that have made their engine design, emission, and fuel systems operate very well with each other.

REVIEW QUESTIONS

1. Smog is produced when hydrocarbons and oxides of nitrogen react with_____to form a dirty yellow-brown haze in the air.

2. A warm air_____ layer traps air pollution within 1000 feet of the ground to produce smog.

3. What are the three main byproducts of combustion that parts of the emission control system are designed to control or eliminate?

4. _____ gases are those that leak past the piston rings.

5. What are two other nontoxic gases included in the vehicle exhaust?

6. What does PCV stand for?

7. What part prevents a backfire during deceleration on carbureted cars?

8. A one-way_____ _____ keeps the exhaust from backing up into the hoses and pump.

9. What type of emission does the EGR valve diminish?

10. When the engine is under load, is the exhaust backpressure high or low?

11. A_____ is a substance that causes a chemical reaction to occur faster without any measurable change to itself.

12. What is the name of the catalytic converter type that is a honeycomb style?

13. When is the evaporation rate higher in the fuel system, when the engine is running or after it is shut off?

14. How much extra tank capacity do fuel tanks have after filling?

15. When does the purge valve allow the canister to get rid of vapors in storage, when the engine is running or when it is off?

ASE STYLE REVIEW QUESTIONS

1. Technician A says that when an air/fuel mixture is too rich, there will be more carbon monoxide. Technician B says that PVC means poly vinyl chloride. Who is right?
   - **a.** Technician A
   - **b.** Technician B
   - **c.** Both A and B
   - **d.** Neither A nor B

2. Technician A says that a crankcase ventilation system includes a metered vacuum leak. Technician B says that a positive crankcase ventilation system will not work when the vehicle is not moving. Who is right?
   - **a.** Technician A
   - **b.** Technician B
   - **c.** Both A and B
   - **d.** Neither A nor B

3. Technician A says that a properly operating PCV system reduces oil leakage. Technician B says that an aspirator valve air injection system works better at highway speeds. Who is right?
   - **a.** Technician A
   - **b.** Technician B
   - **c.** Both A and B
   - **d.** Neither A nor B

4. Technician A says that air injection lowers HC and NO_x in the exhaust. Technician B says that it also provides air to the catalytic converter to help heat it up when it is cold. Who is right?
   - **a.** Technician A
   - **b.** Technician B
   - **c.** Both A and B
   - **d.** Neither A nor B

5. Technician A says that high combustion temperatures produce more hydrocarbons. Technician B says that higher combustion temperatures produce more oxides of nitrogen. Who is right?
   - **a.** Technician A
   - **b.** Technician B
   - **c.** Both A and B
   - **d.** Neither A nor B

6. Technician A says that more NO_x forms under load. Technician B says that more NO_x forms when the engine is hot. Who is right?
   - **a.** Technician A
   - **b.** Technician B
   - **c.** Both A and B
   - **d.** Neither A nor B

7. Technician A says that the EGR valve is open at idle. Technician B says that the EGR valve is open during engine cranking. Who is right?
   - **a.** Technician A
   - **b.** Technician B
   - **c.** Both A and B
   - **d.** Neither A nor B

8. Technician A says that a two-way catalytic converter has a tube in the center that goes to the air injection system. Technician B says that a reduction catalyst is for reducing oxides of nitrogen. Who is right?
   - **a.** Technician A
   - **b.** Technician B
   - **c.** Both A and B
   - **d.** Neither A nor B

9. Technician A says that a monolithic converter has platinum coated beads. Technician B says that a converter with an air tube is called a dual bed catalyst. Who is right?
   - **a.** Technician A
   - **b.** Technician B
   - **c.** Both A and B
   - **d.** Neither A nor B

10. Technician A says that port fuel injected cars must have manifold preheat to help vaporize the fuel in the manifold. Technician B says that if air were directed upstream during closed loop, the computer would drive the fuel system rich. Who is right?
    - **a.** Technician A
    - **b.** Technician B
    - **c.** Both A and B
    - **d.** Neither A nor B

CHAPTER 41

Emission Control System Service

■ **OBJECTIVES**

Upon completion of this chapter, you should be able to:

✔ Determine the causes of various emission system problems.

✔ Perform service on the emission control system.

✔ Repair or replace defective emission control parts.

✔ Diagnose emission problems with an exhaust gas analyzer.

■ **KEY TERMS**

**converter light off
high temperature
    pyrometer
infrared thermometer
oxidizing**

■ **INTRODUCTION**

The area of emission control and tuneup has become a complicated specialty. In the United States, the Federal Environmental Protection Agency (EPA) is involved in the regulation of air quality. Emission specialists are required to be licensed to perform repairs to the emission system.

The aim of this chapter is to provide an apprentice with sufficient understanding to be able to service emission systems, diagnose problems, or to remove and replace components when performing other types of repairs. Diagnosing emission problems using the five-gas exhaust emission analyzer is also stressed in this chapter.

The text will provide you with basic, generic service information. Learning the specifics of a particular vehicle will require use of the manufacturer's or an aftermarket service manual. Shop practice will be required in order to become familiar with the different vehicles' systems.

■ **INSPECTING EMISSION CONTROL SYSTEMS**

Periodic emission inspections are required in many states.

Vehicles produced since 1972 have an underhood label (see Figure 4.3). It describes such things as:

■ The size of the engine

■ Ignition timing specifications

■ Idle speed

■ Valve lash clearance adjustment

■ The emission devices that are included on the engine

A vacuum hose routing label is usually found under the hood also (Figure 41.1).

➡ *Perform Underhood Emission Label* **Worksheet**

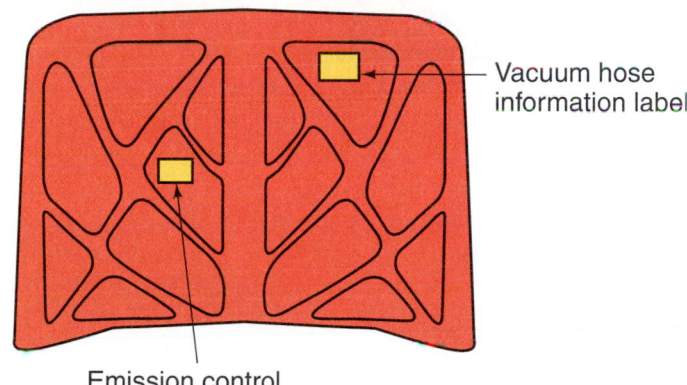

Underside of hood

Vacuum hose information label

Emission control information label

Figure 41.1 Location of vacuum and emission labels.

Aftermarket *Emission Control Applications manuals* provide such information as EGR testing procedures, reminder light instructions, ignition timing procedures, and emission equipment requirements.

Like other engine service areas, a visual inspection is performed first. Look at hoses and wires and check to see that all parts are in place. Run the engine and listen for air leaks with a length of hose or stethoscope with the metal probe removed from the end. Check the pump belt for condition and tension. Inspect the air cleaner to see that it is clean and unplugged.

■ **CRANKCASE VENTILATION SYSTEM SERVICE**

Maintenance of the crankcase ventilation system includes inspecting the operation of the PCV valve, checking hoses and passages for condition and internal deposits, and inspection and/or replacement of the PCV air filter (Figure 41.2).

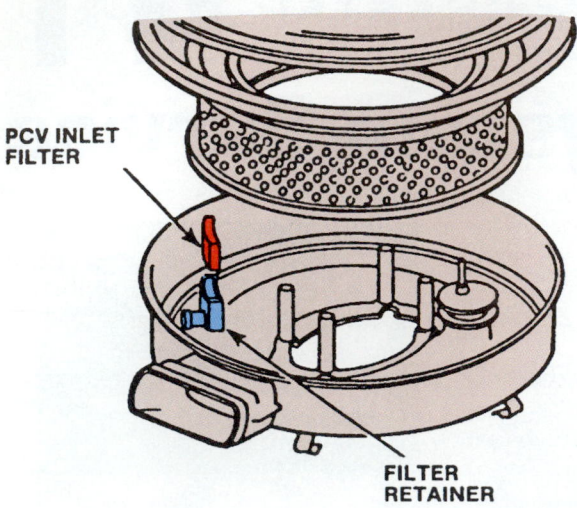

PCV INLET FILTER

FILTER RETAINER

Figure 41.2 Some air cleaners have a crankcase ventilation inlet filter. *(Courtesy of Ford Motor Company)*

Checking Crankcase Ventilation System Operation

The crankcase system must be airtight (Figure 41.3). A leaking or misplaced gasket to a valve cover or intake manifold will cause air leakage in amounts that are sufficient to cause failure of the PCV system. Instead of creating a vacuum in the crankcase, the suction from the PCV valve is not sufficient because the engine is not airtight.

NOTES:

■ *Gasket failures that cause failure of the ventilation system are often a major cause of oil leaks. With the proper amount of suction in the crankcase, slight seal leakages should actually result in outside air leaking* **into** *the crankcase, rather than oil leaking out.*

■ *Leaking vacuum hoses, vacuum control units, vacuum accessories, or manifold leaks allow dirt to enter the engine unfiltered. Unfiltered air entering the engine can cause engine wear.*

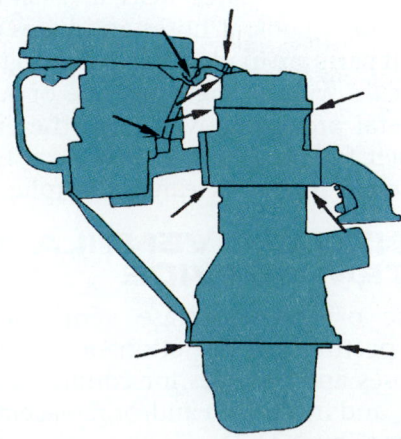

Figure 41.3 The crankcase system must be airtight for the ventilation system to operate efficiently.

■ *A system that is not airtight can also cause driveability problems with computerized fuel systems that measure air density.*

■ TESTING THE PCV SYSTEM

With the engine running and the PCV valve installed, there should be a vacuum at the rocker cover oil filler opening. The tester shown in Figure 41.4 can be used to see if there is a vacuum in the crankcase. Another simple test to see if there are any leaks in the crankcase is to:

■ Pull one end of the hose from the valve cover to the air cleaner (at the air cleaner side).

■ Put your thumb over the end of the hose and wait for a couple of seconds. Vacuum should be felt if the system is working properly.

To locate a leak in the crankcase ventilation system:

■ Seal the breather and PCV valve.

■ Blow (lightly) into the dipstick tube with a rubber-tipped blow-gun.

■ Listen for leaks using a piece of hose or a stethoscope with the metal end pulled off.

A leak is often not readily apparent, especially at the top side of a valve cover gasket or where the intake manifold meets the block at the front or back. Oil might not leak out because of gravity and suction from crankcase vacuum of the PCV system.

Testing the PCV Valve

Test the PCV valve as follows:

■ Pull the PCV valve from its mounting. There are several places the PCV valve can be located. It is usually in a grommet in the valve cover, intake manifold, or a hose from the crankcase (see Figure 40.5).

Figure 41.4 A tool for testing the crankcase ventilation system.

- With the engine stopped, the PCV valve can be removed and shaken. It should rattle (some smaller valves will not).
- With the engine running, cover the end of the valve with your thumb or finger (Figure 41.5) or pinch the line that leads to it. Engine rpm should drop 50 to 80 rpm, because the slight air leak that the PCV valve causes has been closed off.
- Pull the fresh air hose from the air cleaner. With the engine idling, a piece of paper placed over the end of the hose should remain suspended there.

Blocking the flow of air to the PCV valve results in an enriched air/fuel mixture. Remember, a leaner air/fuel mixture means a higher idle speed.

If a PCV valve is replaced, be sure to get the correct one for the vehicle. Valves with different rates of flow are available. Installing the wrong one can result in poor engine operation.

Check the PCV Breather Hose

Make sure that the breather hose running to the air cleaner is not kinked or its filter plugged. During a heavy load, this hose must allow blow-by to escape from the crankcase; otherwise, excessive pressure can build up and cause oil leakage and oil consumption.

NOTES:

- *Any blow-by that can cause crankcase pressure **must** be able to escape through the hose to the air cleaner. An oily air cleaner is often a clue to a crankcase pressure problem. Excess pressure in the crankcase causes blow-by to escape up the hose to the air cleaner.*
- *Crankcase pressure can cause oil to migrate up the distributor shaft and into the distributor.*

■ EVAPORATIVE CONTROL SYSTEM SERVICE

One of the major causes of air pollution from motor vehicles is evaporation of fuel. Evaporative emission controls are installed on gasoline filler nozzles at filling stations (Figure 41.6). These pump nozzles will not pump unless the pleated rubber end is compressed against the vehicle fill opening (or manually held in the compressed position during pumping). This device is responsible for preventing massive amounts of air pollution.

Vehicles have their own evaporative systems. These systems are often neglected until a driveability problem occurs. Evaporative fuel controls are included in an emission inspection.

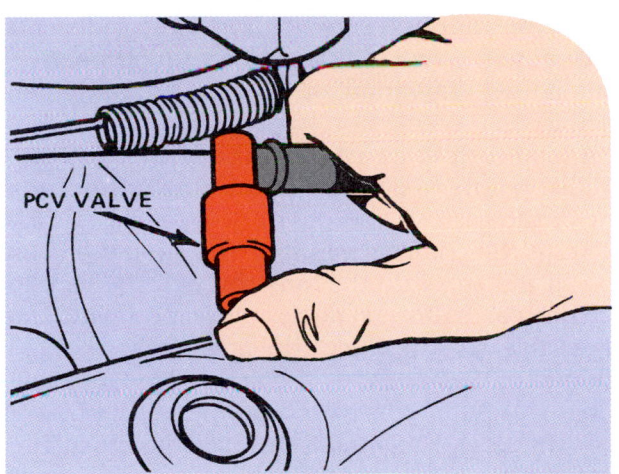

Figure 41.5 Engine rpm should drop when the valve is sealed off at idle. *(Courtesy of Chrysler Corporation)*

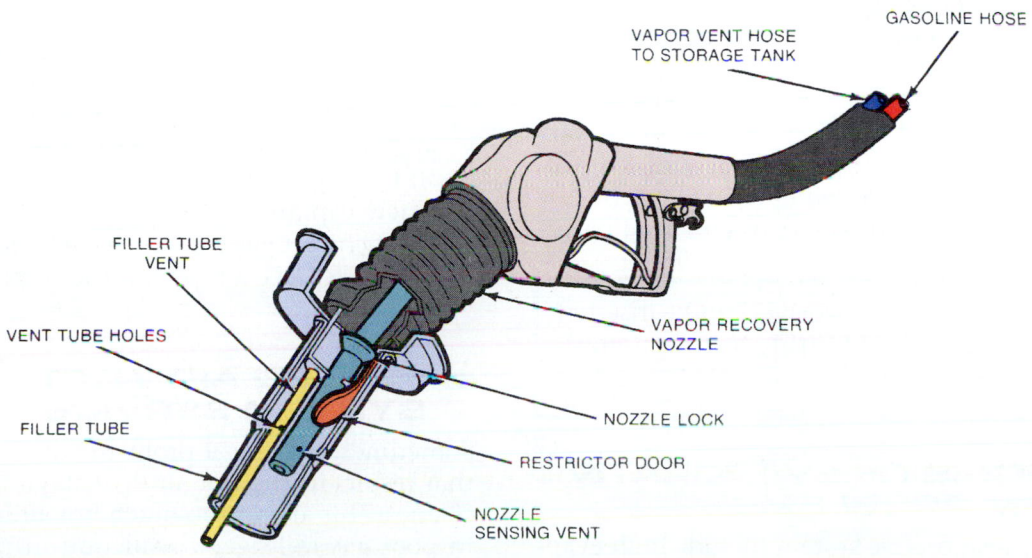

Figure 41.6 An evaporative control filling nozzle prevents HC emissions at gasoline stations. *(Courtesy of Chrysler Corporation)*

Figure 41.7 Some vapor canisters have a replaceable filter on the bottom. *(Courtesy of Chrysler Corporation)*

◼ EVAPORATIVE SYSTEM MAINTENANCE

Filter Replacement

Fresh air is drawn into the canister during purging. Because the charcoal canister breathes fresh air, some canisters have a replaceable filter (Figure 41.7). The filter is either foam or fiberglass. Check the manufacturer's specifications for the recommended service interval, which is usually every two years. In dusty conditions, the filter will require service more often. The filter will be held in with a retaining ring or a bar. If the canister must be removed to service the filter, be sure to mark the hoses so that they can be properly replaced.

Check Condition of Hoses

Check the condition of all system hoses. When replacing old or damaged hoses, use fuel hose or hose that is designed to be used with fuel vapors. Ordinary vacuum hoses will be damaged by the fuel vapors.

Care should be used when disconnecting a vacuum hose from a plastic fitting on a vacuum tee connection or a thermal vacuum switch. These can be easily broken, causing unnecessary replacement of parts.

SHOP TIP A small pick or screwdriver can be inserted between the hose and fitting before attempting to remove the hose.

◼ DIAGNOSIS OF EVAPORATIVE SYSTEM PROBLEMS

Problems in the evaporative system include high evaporative emissions and a rich mixture at idle, if the purge valve is stuck open. It should only purge on a warm engine off idle. Sometimes, a novice has attached lines to the wrong locations or a hose has fallen off. The customer might complain of the smell of raw gasoline from inside the car.

If you find liquid fuel in a charcoal canister, check the liquid/vapor separator. Its return line to the tank can become plugged with rust from inside the gas tank. The main vent line can become plugged or kinked. A hole can develop in the vent line from rust. The liquid/vapor separator can become full when a customer keeps topping off the tank during a refill. Gasoline is forced into the liquid/vapor separator. This can cause poor starting on a hot engine along with black smoke. After you clear the flooded condition, run the engine at fast idle for a few minutes. Air running through the canister during this time will dry it out.

There can be a blockage in the liquid/vapor separator or the vent line between it and the canister. This could cause the fuel tank not to breathe properly. The result is fuel starvation or a collapsed gas tank (on sealed gas cap type).

A loud rush of air entering the tank when the cap is pulled points to a venting problem. To test for this, remove the gas cap and disconnect the gas tank vent line from the charcoal canister. You should be able to blow all the way into the tank through the vent line.

Charcoal canisters rarely cause problems, but a purge valve can fail. Check the purge valve with a vacuum pump. Vacuum should hold for 20 seconds. If the car has computer controls for a solenoid purge valve, check the service manual for the testing procedure.

Check the condition of the gas cap. This is where the majority of the problems are and there are no tools available for checking it. See that it fits snugly and that its gasket is in good condition.

To test a pressure vacuum-type gas cap, clean it and blow it dry. Try to blow through the inner vent hole with your mouth. It only takes 1.5 pounds of pressure to open the valve. You should feel the valve open when you blow lightly. Sucking on the cap vent should easily overcome the vacuum relief valve, too.

If the cap is suspect, check it against the operation of a new cap and replace it if needed. It is important that the correct gas cap for the vehicle always be used. Be sure to replace a cap with the correct one, sealed or vented.

◼ HEATED AIR INLET SYSTEM SERVICE

Sometimes, there is a problem with a heated air inlet that results in the hot air duct staying open to the air cleaner. This allows too much hot air in, and can result in poor gas mileage, a reduction in power, and detonation (this also increases NO_x). The opposite problem occurs when hot air is not sent to the air cleaner when

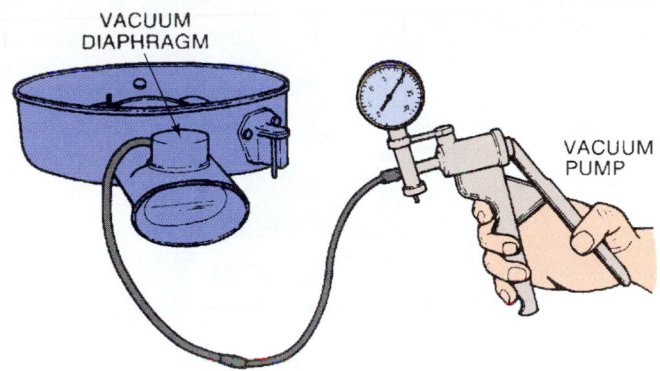

Figure 41.8 Testing the vacuum motor with a vacuum pump. [*Courtesy of Chrysler Corporation*]

the engine is cold. This causes hesitation and stalling when the car is cold.

When the vacuum motor does not get vacuum, this can be because of a leak, misrouted hose, bad temperature sensor, or bad diaphragm in the motor. The result is that the air door closes off heated air. This is the most common problem. The air door only sticks in the hot air position when something physical is wrong, like binding pivots.

Testing the Heated Air Inlet System

Be sure that the heat stove is intact. This is a common emission test failure. The stove is often either missing or rusted and damaged. Aftermarket replacement stove kits made of flexible metal hose are available when a sheet metal original stove fails.

Make sure the hose is good and the valve is operating. Operate the vacuum motor with a vacuum pump (Figure 41.8). Use a mirror so you can see it move. Otherwise a screwdriver can be used to feel the position of the door.

■ EARLY FUEL EVAPORATION SERVICE

Manifold heat control valves sometimes become stuck. Usually they stick open, but sometimes they can stick closed. The excess heat that results can crack the manifold or cylinder head. Sometimes, a stuck valve can be loosened by tapping the end of its butterfly shaft with a hammer. Graphite heat riser lubricant can be applied to the shaft. In fact, this is sometimes part of a lubrication safety service.

To work properly, a vacuum-operated heat control valve should have sufficient vacuum present at idle. If the valve is not working and there is at least 10"/Hg vacuum present, there is a problem with the diaphragm, linkage, or butterfly valve.

Some heat-control valves are built into the manifold, and others are replaceable.

On engines equipped with an EFE (early fuel evaporation) electric grid under the carburetor or throttle body injection, current is supplied to the grid when the engine is cold and running. If there is no current at the grid, check the wiring and relay that control it. The grid should have a specified amount of resistance. Check this with an ammeter while there is power to the grid. No current flow indicates an open circuit. Checking this device with an ohmmeter while it is disconnected will also show an open circuit, but if the grid is barely connected and opens under load, the test will not detect a problem. Computer controlled cars will set a fault code when there is a problem with this system.

■ EGR SYSTEM SERVICE

EGR valves are the most often repaired emission devices. A faulty EGR valve is a common occurrence. A computer operated EGR valve can experience the same mechanical symptoms as a vacuum switch operated valve. The computer system will set a fault code if it experiences an electronic failure or sees a sensor reading that is out of specification.

There are three typical symptoms for EGR system problems: pinging, stalling or rough idle, and hesitation when cold.

Does the engine ping under load? If the EGR valve is inoperative, combustion temperatures become too high. The result is often abnormal combustion (pinging or spark knock) on older cars if there is no knock sensor. Higher NO_x emissions also result.

Rule out other causes of spark knock first.
■ If the engine pings under load, check the initial ignition timing first.
■ Check the cooling system temperature. A failed cooling fan switch, a stuck thermostat, or a plugged radiator can also cause pinging.
■ Spark plugs of too high a heat range will cause pinging.
■ Fuel of too low an octane for a high compression engine can cause pinging.

■ DIAGNOSING EGR PROBLEMS

Testing of all types of valves requires these three things:
■ Test valve operation
■ Clear passageways
■ Check sources of vacuum

Check the service manual for individual service procedures. There are many different procedures for the different types of valves and controls that have been used over the years.

Testing EGR Valve Operation

If the EGR valve is stuck open, the engine will stumble and have a rough idle. An open EGR valve causes hard starting because it acts like a vacuum leak.

If the EGR valve is of the type where its stem can be seen, the engine can be run while watching the valve stem move with a mirror. The stem can be hot so do not touch it. If the stem does not move in response to engine revving to about 2500 rpm when there is a good source of vacuum present, something is wrong with the valve. You can check for vacuum by pulling the hose and feeling the vacuum at the end of it as the engine is revved. An accurate reading can be taken by teeing a vacuum gauge into the hose to the EGR valve.

On some types of EGR valves, you can apply vacuum to the valve with a pump while the engine idles. This causes a vacuum leak that will make the engine stumble. This test will not work with a positive back-pressure valve.

Clear Passageways

It is very common to have an EGR valve that appears to operate properly, but does not control NO_x or prevent detonation under load. This can be due to an exhaust passageway that has become plugged with carbon.

> **SHOP TIP** A plugged passageway can sometimes be cleaned with a homemade tool. Use a short length of parking brake cable or speedometer cable on a drill. Spread the end of it open, and spin it through the passageway.

Check Vacuum Source

Check for a vacuum source problem. Run the engine at above 2000 rpm at normal operating temperature and check to see that there is a good source of vacuum to the valve. Vacuum problems can be caused by thermal valves, delay valves, or improper hose routing. Improper hose routing is a common occurrence when a novice removes a cylinder head for repair and does not label the vacuum lines during disassembly.

If the EGR valve works before the engine is warmed up, the engine can run rough or stumble. A defective thermal vacuum valve (TVV) could be causing this. These valves can stick in the open or closed position. A valve that is stuck closed will keep the EGR valve from operating, causing fuel knock and excess NO_x. Figure 41.9 shows how to test a TVV.

Most new cars have computer controlled EGR valves (see Chapter 40). On many of these systems, there will be no vacuum to the EGR valve during park or neutral operation.

■ AIR INJECTION SYSTEM SERVICE

Inspect Air System Plumbing

Look for rusty check valves and air manifolds on air pump and pulse air systems. Hoses used on the air

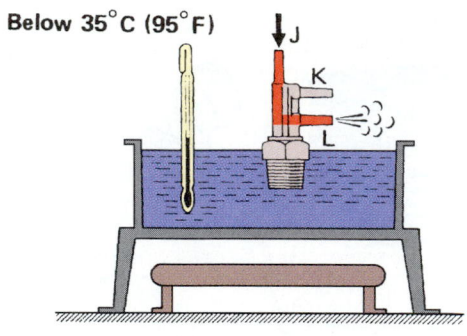

Below 35°C (95°F)

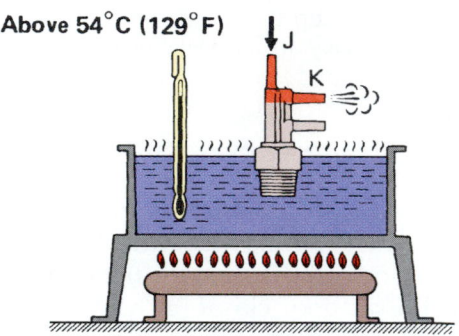

Above 54°C (129°F)

Figure 41.9 Testing a thermal vacuum valve.

injection (smog pump) system can be exposed to exhaust gas if a part in the system fails. When there is a burned hose near a check valve, the check valve could be allowing exhaust to flow backwards.

Smog hoses that are mounted near exhaust parts can be exposed to excessive heat. These hoses are often candidates for replacement and many of them are formed hoses. Smog hose with the wire in it is a good choice for these applications.

Diagnosing Problems

There are several problems that can happen with an air injection system.

Exhaust Noise Under the Hood. This could be from either a leaking check valve or manifold pipe. These can sometimes rust through.

Backfiring on Deceleration. When a car is backfiring on deceleration, check the diverter valve with a vacuum pump. Vacuum should cause the pump air to be diverted to the muffled exhaust port. This is a typical early system. Check the service manual to see what kind of system you have.

Air Pump Noisy or Frozen. Remove the belt and spin the pulley by hand. It is normal for the pump to make some noise, but it should not feel hard to turn or squeal loudly. If the pump is bad, check for a leaking check valve that has allowed exhaust to enter the pump.

Emission Test Failure. When an engine does not pass an emission analyzer test, check the pump to see if it is pumping air. Pinch the hose and listen to the pressure relief valve start releasing air. You can pressure test for pump capacity with a pressure gauge.

To test to see that passageways are clear, you can check the O_2 content of the exhaust with the air pump working. Pinch the hose and compare the reading to the amount of O_2 with the system inoperative.

■ CATALYTIC CONVERTER SERVICE

Catalytic converters can become plugged or inoperative or inefficient to the point of not being able to pass an emissions test. The most common failure to a converter is overheating caused by a misfiring cylinder. This is anything that allows unburned fuel to enter the exhaust. If a spark plug wire falls off, the cat can get really hot. This can result in the inside of the converter getting so hot that it actually melts.

A plugged converter restricts the engine's exhaust flow. The symptom of this is that the engine might start and accelerate briefly, but then has trouble breathing and will not produce power. A hissing sound can come from the exhaust, accompanied by a roaring sound or spitting from the fuel intake in the engine compartment.

Do Not Just Replace the Converter

A catalytic converter is designed to reduce normal exhaust emissions. It can mask problems that are related to the fuel or ignition systems, or mechanical problems. This will overwork the converter and it probably will not last as long. An overworked converter will lose efficiency. Converters do not usually just stop working. They just work less and less effectively. Before replacing a converter, determine whether there is another problem that has contributed to its failure. Installing a new converter might fix the problem for a while, but it is sure to come back when the converter fails once again.

■ CONVERTER TESTING

The cat must be heated to at least 600°F to work. As it begins to oxidize HC and CO, it will become even hotter. When the converter begins to oxidize, this is called **converter light off**. An oxidizing converter that is working will have a hotter exhaust at its outlet than at its inlet. A temperature probe is used to take temperature measurements of the pipe at the inlet and the outlet. This is one of the easiest tests to see if the converter is operating:

■ The engine is run at 2000 rpm until the converter is at normal operating temperature. A converter will cool off at idle speed and might stop working.

■ A good converter should measure 5–10% hotter at its outlet than its inlet. This is only true when the converter is working. If the engine is running with very clean exhaust, the converter is dormant.

■ Shut off the engine. Create a rich mixture or pull a spark plug wire and ground it. This will result in higher HC and oxygen to the converter.

NOTE: *Do this test quickly (30 seconds should be sufficient). A converter will overheat in a very short time (2–3 minutes).*

The temperature is measured with a **high temperature pyrometer** that is touched against the surface of the pipe. Another good tool is an **infrared thermometer**. It is not held against the surface, just near it. The surface pyrometer is slower than the infrared thermometer.

A problem with the temperature test for three-way cats is that this test does not tell whether they are working well in both halves.

The best test for a converter is to test its operation with an emission analyzer. This and other emission analyzer tests of the converter are covered later in this chapter.

■ CONVERTER REPLACEMENT

A replacement converter is often bolted on with new gaskets. Some of them require welding for installation. Be sure to get an EPA approved equivalent replacement. Installation of a used catalytic converter is illegal for an emission technician unless it has been tested and certified (a process that costs as much as a new one).

NOTE: *When a customer complains of a rotten egg smell coming from the exhaust, this could be due to an excessively rich or lean air/fuel mixture, or it could be due to fuel with too much sulfur in it. In the latter case, recommend that the customer change brands of fuel.*

■ COMPUTER CONTROLLED EMISSION SERVICE

Today's new cars have sophisticated emission systems that are controlled by a computer, sensors, and actuators. When problems occur, many of them are recorded in the computer's memory. A fault code is set and a warning light on the dash will come on. A technician will use a scan tool to read the codes in the computer. This information is covered in Chapter 76.

Some cars have emission maintenance reminder lights that indicate service needed to emission control devices based on time or mileage. There are many ways of testing the devices and resetting the light once it has been illuminated. Refer to an emission control applications manual or a manufacturer's service manual for instructions.

■ ANALYZING EXHAUST EMISSIONS

Air is mostly nitrogen (N) and oxygen (O_2). Perfect combustion would leave only water (H_2O) and carbon dioxide (CO_2) in the exhaust (Figure 41.10). Nitrogen would be in the exhaust stream, but it would not have

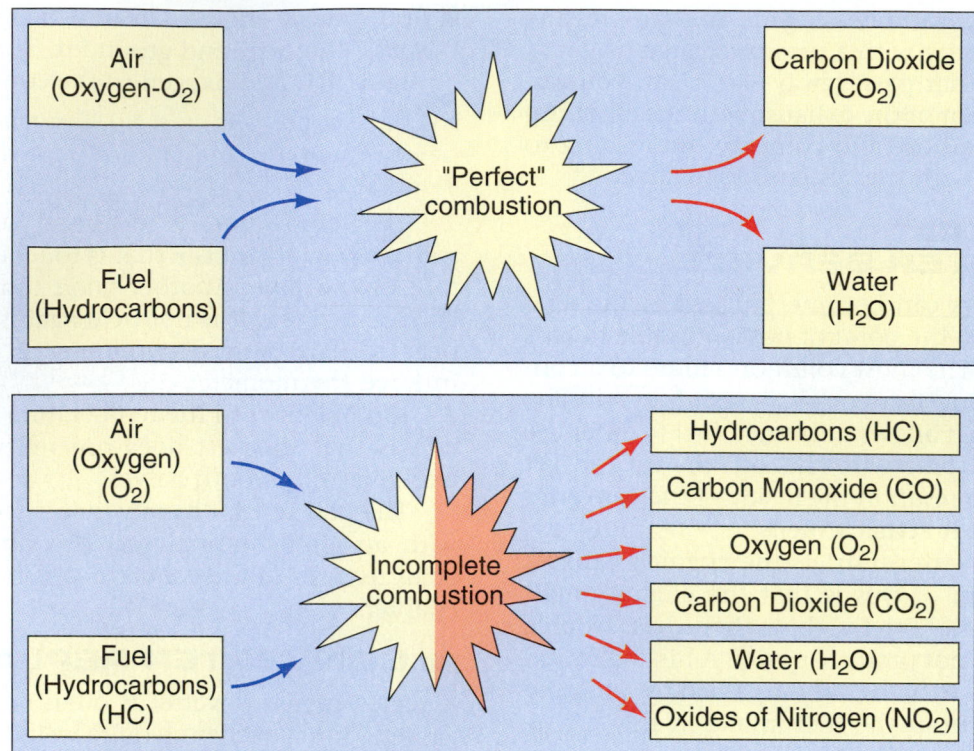

Figure 41.10 Byproducts of combustion. *[Courtesy of Environmental Systems Products, Inc.]*

changed throughout the burning process. If it did, this was caused by high temperatures. Certain things vary in the combustion process that make it less than perfect. These include variations in ignition, air and fuel, combustion temperatures, and physical limitations such as the skin effect and combustion chamber shape.

Fuel Burning

To understand what the readings in the exhaust mean, you will need to understand what happens when fuel burns. Hydrocarbons (HCs) are what gasoline is made up of. In analyzing exhaust readings, the lower the HC's the better because HC is unburned fuel. Burning of the fuel is called **oxidizing**. To be able to burn, fuel requires oxygen and heat. About 21% of the air we breathe is oxygen.

Atomized fuel in the correct proportion with air is compressed in the combustion chamber. Compressing the air/fuel mixture pushes the oxygen molecules and fuel molecules closer together. When some of the molecules are ignited by the heat of the spark from the spark plug, they burn. During combustion, heat is given off. The heat ignites molecules next to the burning molecules and a chain reaction takes place. This is called the *flame front*. Any of the mixture that does not burn must go out the tailpipe. That is what is measured in the exhaust as HC.

Some HC in the exhaust is normal. If the reading is excessive, the cause could be anything that inter-

feres with the normal compression, ignition, or burning of the fuel.

Compression. Compression must be enough to push the molecules close together. If not, burning could start but not continue because the chain reaction is broken. HC levels will be higher.

Ignition. With an ignition misfire, the entire charge of raw fuel in the cylinder will go out the tailpipe and be measured as HC.

Air/Fuel. If the air/fuel mixture is too rich, there will not be enough oxygen in the mixture to combine with all of the fuel so it can be burned. HC emissions are lowest at an air/fuel mixture of about 14.7:1, called stoichiometric (Figure 41.11). Computer controlled

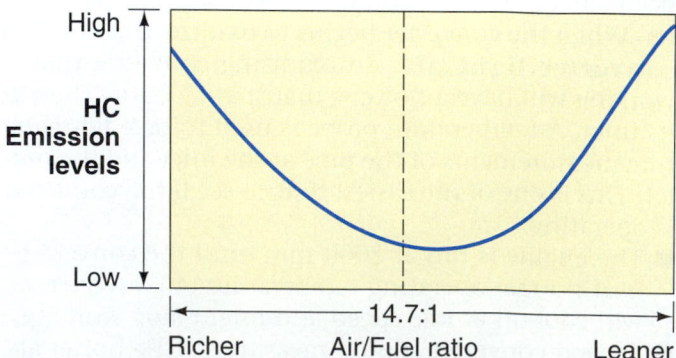

Figure 41.11 HC emissions are lowest at an air/fuel mixture of about 14.7:1. *[Courtesy of NAPA Institute of Automotive Technology]*

emission systems are set to run at this ratio. As the mixture gets leaner than stoichiometric, there is not enough fuel to continue the chain reaction and the engine misfires. With either a rich or lean misfire, HC in the exhaust rises.

■ SCIENCE NOTE ■

Gasoline is actually a mixture of over 100 different hydrocarbons in varying proportions. Any particular hydrocarbon will have from 4 to 14 carbon atoms. A four carbon hydrocarbon will require far less oxygen for complete combustion than a 14 carbon atom will. Therefore, the stoichiometric air/fuel ratio will depend on the percentage of each hydrocarbon in the mixture. This is the reason you may sometimes encounter a stoichiometric air/fuel ratio of other than 14.7:1.

■ EMISSION ANALYZER

The exhaust emission analyzer measures the content of gases in the exhaust through a probe installed in the exhaust pipe of a running engine.

SAFETY NOTE CO is colorless, odorless, and dangerous. It deprives the body of oxygen. Headaches, coordination problems, unconsciousness, and death can result from exposure to too much CO. Be sure to use adequate ventilation when running an engine during testing.

The emission analyzer has a pump that pulls an exhaust sample through a hose connected to the exhaust pipe probe. The gas sample is moved through filters to trap solids and remove moisture. There are sampling tubes inside the tester that the exhaust sample flows through. An infrared light beam is focused through the exhaust gas onto a detector. The energy required to penetrate the exhaust sample varies with the amount of gas in the exhaust.

A two-gas analyzer is the older type that measures HC and CO. Some of these machines required a long warmup period. Later model computerized machines have automatic features for warmup and calibration and tell the operator when to accelerate the engine. These machines also read four gases or five gases, rather than two. A four-gas analyzer measures HC, CO, CO_2, and O_2. A five-gas analyzer (the most modern) measures five gases, HC, CO, CO_2, O_2, and NO_x (Figure 41.12).

An exhaust analyzer is helpful in diagnosing problems with the fuel system, catalytic converter, evaporative emission system, and air injection system.

Figure 41.12 A portable five-gas analyzer. *(Courtesy of WPI Micro Processor Systems, Inc.)*

NOTE: *When testing for emissions in the exhaust, remember that the air pump, catalytic converter and its temperature, and exhaust leaks will all affect test results.*

■ HYDROCARBONS

Hydrocarbons are oxidized by the catalytic converter and air injection system. HCs are diluted by exhaust leaks. HCs are measured by the analyzer in parts per million (ppm). On modern dynamometer (loaded mode) tests, ppm is converted to grams per mile (gpm).

Hydrocarbons are measured in parts per million. 1 ppm means that out of every million parts of an exhaust sample, one part is HC. Parts per million gives a precise way of measuring percentage. Because the amount of HC we are dealing with is so small, the percentage would be too minute to be read on a percent scale. A reading of 500 ppm is only 0.05%, five hundredths of one percent. Specifications can range from a high of 700 ppm for an early model car down to 100 or less for later models.

Hydrocarbon Readings

The most common cause of high HC is unburned fuel. This can be caused by any misfire, mechanical problem, or advanced ignition timing. Hydrocarbons indicate unburned gasoline resulting from a misfire (Figure 41.13). High HC could be because of too much fuel or not enough air. Too much fuel could be due to high fuel pressure, a bad fuel injector, oil saturated with

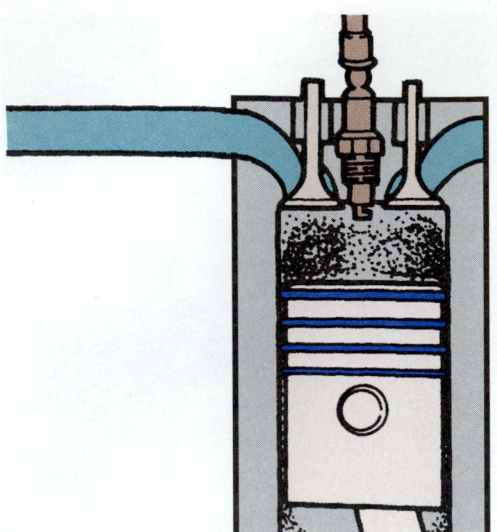

Figure 41.13 Weak or broken piston rings can cause a misfire, higher HC emissions from the exhaust, and increased crankcase pressure. *[Courtesy of Chrysler Corporation]*

fuel, canister purge, or problems with electronic engine controls.

Restricted air flow is usually caused by a mechanical malfunction. These include a burned or misadjusted valve, a broken piston ring, a flat cam lobe, cam timing off, a restricted air cleaner, or a restricted exhaust system. These problems all result in less air flowing through the engine. If there is a restriction to intake or exhaust, the engine cannot bring in enough fresh air and fuel. The distance between the hydrocarbon molecules is increased and burning is hampered. Whatever remains in the cylinder after a misfire ends up in the exhaust.

Other causes of high HC include:

■ Advanced ignition timing. When the timing is advanced, the flame starts to burn the mixture before the molecules are fully compressed. This results in a break in the chain reaction and the flame front goes out before all of the fuel is burned.
■ A defective spark plug or wiring.
■ An air/fuel mixture that is too rich or too lean can cause a misfire. This causes high HC.
■ A manifold vacuum leak that results in a mixture that is too lean to burn.
■ An excessively rich or lean air/fuel mixture.

Analyzer Readings

A lean mixture can cause fluctuating HC readings (CO remains mostly fixed). With a working catalytic converter, a reading of over 100 ppm can indicate a rich or lean condition or a misfire due to advanced timing, low compression, a fouled spark plug, inoperative air injection, and so on.

Excess carbon on a valve stem can restrict intake flow. The carbon can also absorb fuel like a sponge.

This promotes a misfire when the mixture cannot stay stable because of the fuel going into or leaving the carbon at uncontrolled times.

■ CARBON DIOXIDE

Carbon dioxide (CO_2) is a byproduct of complete combustion, giving an indication of how thoroughly fuel is burning in the cylinder. The closer combustion is to being complete, the higher the CO_2 reading will be. CO_2 is a good indicator of the efficiency of the engine.

The air/fuel ratio affects CO_2. CO_2 is highest at 14.7:1 air/fuel ratio. You can begin to see why the stoichiometric ratio (14.7:1) is used for emission control. CO_2 levels drop when lower or higher than this level (Figure 41.14).

Unlike HC, CO_2 only results from combustion of the fuel:

■ If there is no flame, no CO_2 is in the exhaust.
■ When the combustion process is incomplete, the CO_2 level will be lower. A rich or lean mixture will cause a *drop* in the CO_2 reading.

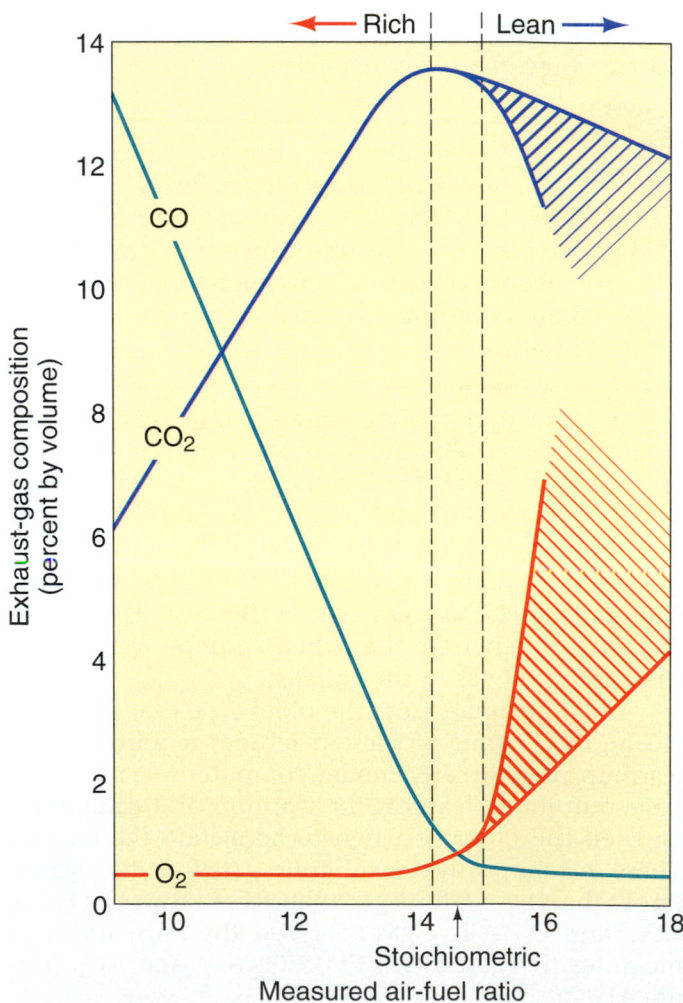

Figure 41.14 If CO_2 is low and O_2 is high, the air/fuel mixture is lean. *[Courtesy of Sun Electric Corporation]*

- A high CO_2 level means that the air/fuel mixture must be near stoichiometric.
- CO_2 is not a good rich lean indicator. If CO_2 is low, the air/fuel ratio could still be correct.
- If CO_2 is low and O_2 is low, the air/fuel mixture is rich.
- If CO_2 is low and O_2 is high, the air/fuel mixture is lean (see Figure 41.14).
- A typical good running engine with a catalyst will have from 13% to 16% CO_2. The amount of CO_2 produced between different vehicles is not consistent and rises as the catalytic converter heats up.

Reading CO_2 is different from HC. It is read as a percent of the total, rather than parts per million. If the reading is 13.2%, then 13.2% of the total volume of the exhaust is CO_2.

If the air pump is running during a test reading, the volume will be changed and the reading will not be accurate. Also, leaks in the exhaust system can change CO_2 readings.

Figure 41.15 shows the effect extra air in the exhaust has on each of the five gases. Notice that the percentage of CO_2 is lower because the volume of the exhaust gas sample has become larger. O_2 (oxygen) is an exception to this because adding outside air adds oxygen to the exhaust.

NOTES:
- *Be sure to disable the smog pump or pulse air system during exhaust gas analysis.*
- *Readings are taken after the catalytic converter.*

An engine that is running will always have at least 4% CO_2. Some state emission tests use a combination of CO and CO_2 to check the condition of the exhaust system. A combined reading of *less* than 7–8% (depending on the car) would fail the test. When a reading is below that level, look for a source of air entering the exhaust.

A test of catalytic converter efficiency using CO_2 is covered later in this chapter.

CARBON MONOXIDE (CO)

CO should be CO_2 but there is not enough oxygen. A rich air/fuel mixture (not enough oxygen) causes incomplete combustion. Carbon monoxide gives a

EXHAUST DILUTION	
Hc ppm	Lower than actual
CO%	Lower than actual
CO_2%	Lower than actual
O_2%	Higher than actual
NO_x ppm	Lower than actual

Figure 41.15 The effect extra air in the exhaust has on each of the five gases. *(Courtesy of NAPA Institute of Automotive Technology)*

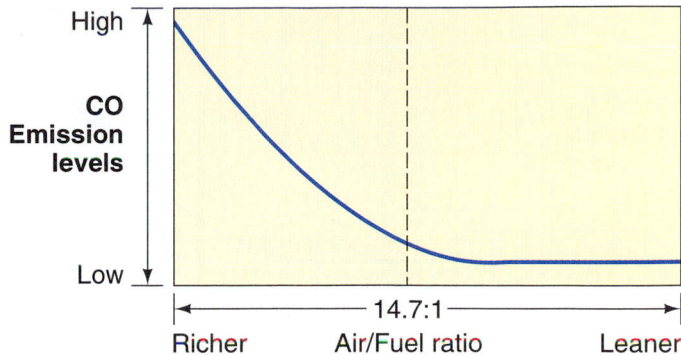

Figure 41.16 Carbon monoxide readings give a good indication of the air/fuel ratio. *(Courtesy of NAPA Institute of Automotive Technology)*

good indication of the air/fuel ratio (Figure 41.16). With a rich mixture, levels of CO are high. They are low with a lean mixture (plenty of air). When there is high CO, you know you do not have a lean mixture.

CO is measured as a percentage of the total exhaust. With a catalytic converter, a reading of over 1% indicates a rich mixture. Without a converter, the reading can be over 3%. Before the cat, you could expect to read 0.6% CO with a 14.7:1 mixture. CO is not lowest at stoichiometric, but gets lower beyond that point.

High readings can be caused by a rich mixture, retarded ignition timing, or a worn engine. CO is oxidized into CO_2 by the catalytic converter and air injection system. It is diluted by exhaust leaks.

When working on older cars with carburetors, a two-gas analyzer was handy for reading and adjusting air/fuel ratios using the CO reading. This procedure is covered in Chapter 27.

CO, like CO_2, is a byproduct of combustion. If there is no flame, there is no CO. If a spark plug is disabled, HC readings will be high, but CO will not register for that cylinder.

SHOP TIP The exhaust analyzer can be used to test to see if a fuel injected engine with a no-start condition is getting fuel. Crank the engine over and watch the tester. If HCs are coming out of the exhaust, there is fuel.

OXYGEN (O_2)

When the air/fuel mixture is richer than stoichiometric, the level of oxygen will be low (Figure 41.17). This is because a rich mixture provides enough fuel to use up all of the oxygen in the cylinder. When the O_2 reading is low, a worn engine could also be the cause.

As the mixture becomes leaner, O_2 levels increase quickly. When the mixture is lean, O_2 rises because there is not enough fuel to use up all of the oxygen.

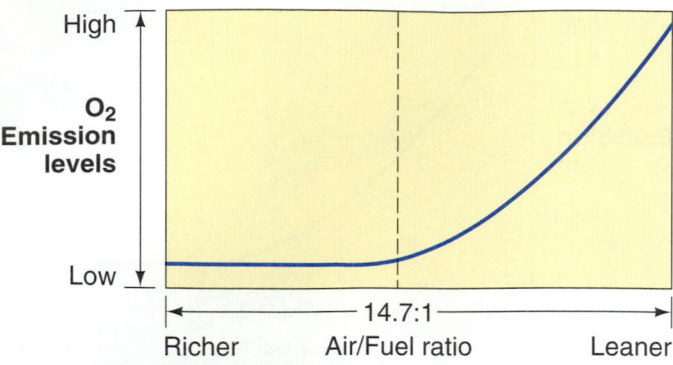

Figure 41.17 When the air/fuel mixture is richer than stoichiometric, the level of oxygen will be low. *(Courtesy of NAPA Institute of Automotive Technology)*

Diagnosis with O_2

Like HC, O_2 is not formed during combustion so it should be consumed during each combustion cycle. Before combustion, there is 21% O_2 in the air. The amount left over after combustion subtracted from 21% leaves the amount of oxygen used during combustion. Without air injection, a normal reading going *into* the catalytic converter is 1–2% (14.7:1 air/fuel ratio).

- Lean mixture—Higher O_2 means less oxygen was used during combustion.
- Rich mixture—Lower O_2 means more oxygen was used during combustion.
- Over 5% O_2 usually indicates a misfire.
- O_2 can be higher because of an air leak (vacuum leak).

Oxygen is consumed by a catalytic converter. On a late model car running at 2000 rpm with a hot converter, normal readings from the exhaust pipe should be below 0.1% with the air injection system disabled. A higher reading can be because the smog pump has not been disabled or because of dilution from an exhaust leak.

Oxygen is used in combustion, so anything that ends combustion or prevents it from occurring will raise O_2 levels. Look for an ignition, fuel, or mechanical problem.

■ OXIDES OF NITROGEN

Air is made up of 78% nitrogen and 21% oxygen. Oxides of nitrogen (NO_x) form when the two bond under high heat and pressure in the engine.

■ SCIENCE NOTE ■

NO (nitrogen oxide) is one part of nitrogen and one part of oxygen. NO_2 (nitrogen dioxide) means that there are two oxygen atoms. It is five times as toxic as NO. Most of the NO that leaves the exhaust will become NO_2 eventually.

Oxides of nitrogen include the entire family of nitrogen oxides. The "x" in NO_x means a variable number.

Controlling NO_x

- Rich mixture—A rich mixture will cool combustion and it has less oxygen, so NO_x emissions are lower (Figure 41.18).
- Lean mixture—NO_x emissions are also less when the mixture is leaner than 16:1. This is because combustion is slower, which keeps temperatures lower.

NO_x emissions are highest at a point near stoichiometric. This is obviously not the best air/fuel ratio if you want to control NO_x. The catalytic converter is more efficient at controlling HC and CO when the air/fuel mixture is stoichiometric.

NO_x is kept low in two ways:

- NO_x emissions are prevented from forming by the EGR valve.
- NO_x emissions are reduced after they are formed, by the reduction catalytic converter.

Oxides of nitrogen will not form under normal atmospheric conditions. It takes heat to make them form. The EGR valve lowers combustion temperatures when the engine is cruising and under load. Adding inert exhaust gas spreads out the hydrocarbons, slowing down their chain reaction burn. The inert exhaust gas also takes up space, preventing the normal amount of fresh air and fuel from entering the cylinder. The gases also absorb some of the heat formed during combustion.

Diagnosing High NO_x

The automotive service industry has not been required to test NO_x emissions until recently. Emission analyzers had provisions for testing only four gases, the ones discussed to this point. NO_x only forms under load and heat conditions, so a dynamometer or a portable tester is required. Figure 41.19 shows a comparison of NO_x emissions from the same engine run at 40 mph with different horsepower loads on it.

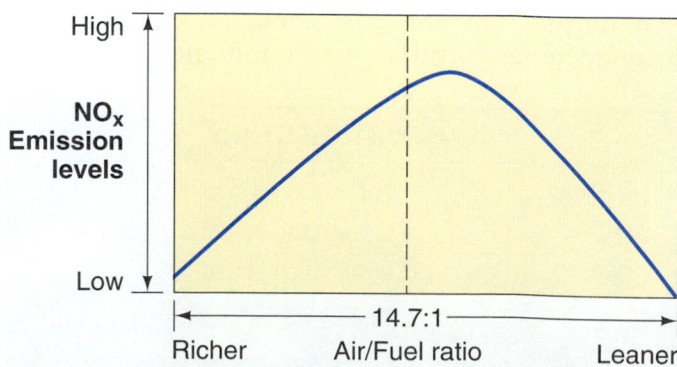

Figure 41.18 A rich mixture will cool combustion and it has less oxygen, so NO_x emissions are lower. Slower burning of lean mixtures results in less NO_x, too. *(Courtesy of NAPA Institute of Automotive Technology)*

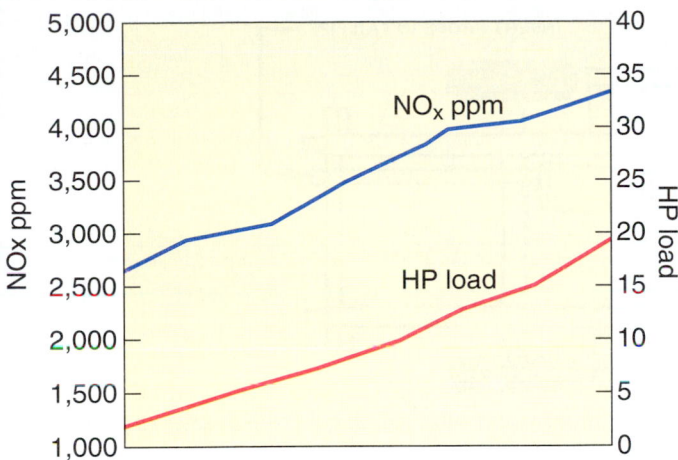

Figure 41.19 The line on the top represents NO_x emissions with the same engine under a higher load. *[Courtesy of NAPA Institute of Automotive Technology]*

EGR problems are the biggest contributor to high NO_x emissions. During one dynamometer test done at an emission testing lab, a 1988 car with a stoichiometric air/fuel ratio was run at 50 mph under light load. The five gas readings are shown in Figure 41.20. When the vacuum hose to the EGR valve was disconnected and plugged, the NO_x reading increased dramatically (Figure 41.21). Driveability did not change.

If you do not have a dyno but have a portable five-gas analyzer, you can still test EGR function. Have an assistant drive the car while you watch the emission levels on the tester. Disable the EGR valve and try the same road test to see if the EGR valve is functioning. The NO_x on second reading after the EGR valve has been disabled should be much higher than the first. If

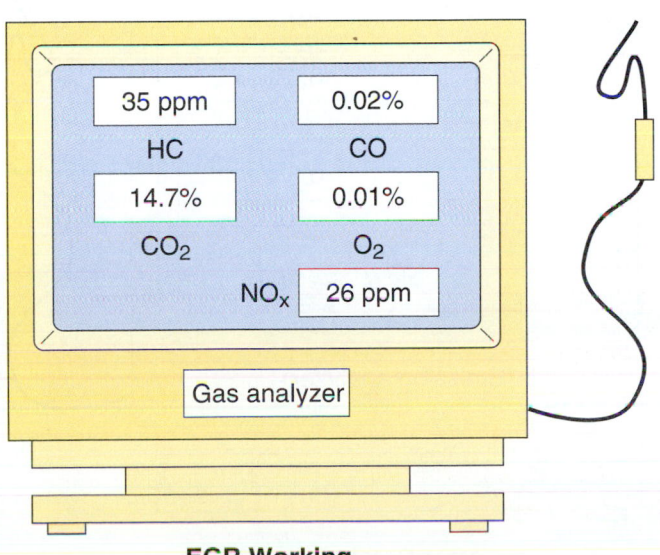

EGR Working

Figure 41.20 Results of a dynamometer test with all emission systems operating. *[Courtesy of NAPA Institute of Automotive Technology]*

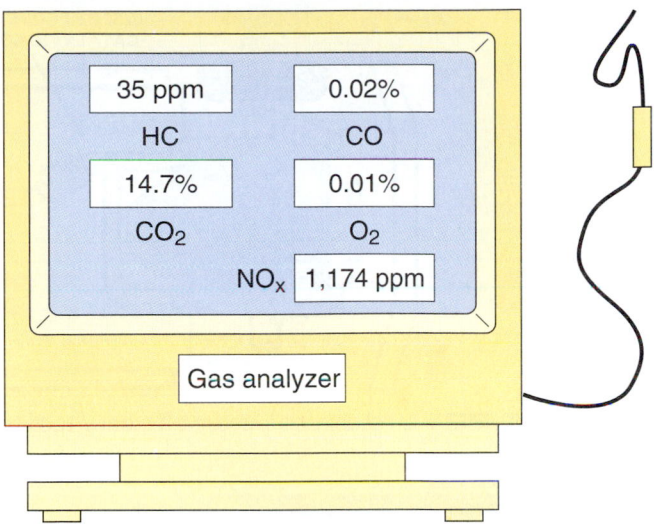

EGR Disabled

Figure 41.21 Results of the same test with the EGR system not working. Notice the level of NOx emissions. *[Courtesy of NAPA Institute of Automotive Technology]*

the shop's five-gas analyzer is not portable, a stall test (see Chapter 67) will create NO_x.

Anything that increases temperatures can be a possible cause of NO_x levels that are too high. Problems with the cooling system, heated air intake, and compression raised by carbon deposits can all cause NO_x to increase. Ignition timing also affects NO_x levels. Advanced timing raises pressure, causing increased combustion temperatures. Always check the initial timing and spark advance when there is high NO_x.

■ CATALYTIC CONVERTER TEST WITH ANALYZER

An oxidizing converter produces CO_2 and water. One of the reasons a converter must have oxygen is to be able to catalyze CO; CO + oxygen = CO_2. Oxygen levels in the exhaust vary, so air injection is used. Single bed three-way converters work on only one catalyst and do not have an air pipe. They are popular on newer vehicles without air injection. The air/fuel ratio is important to these catalysts. They depend on the air/fuel mixture fluctuating to both sides of 14.7:1 to be efficient.

■ A rich mixture will oxidize HC and CO but will not reduce NO_x.

■ A lean mixture will lower reduction, but will not do oxidation efficiently.

The best test for catalytic converter efficiency is when readings are taken, first in front of the converter and then at the tailpipe (Figure 41.22). The two readings are compared to tell if the cat is efficient. A hole is drilled in the exhaust pipe in front of the cat. The probe of the analyzer is inserted into the hole. Tool

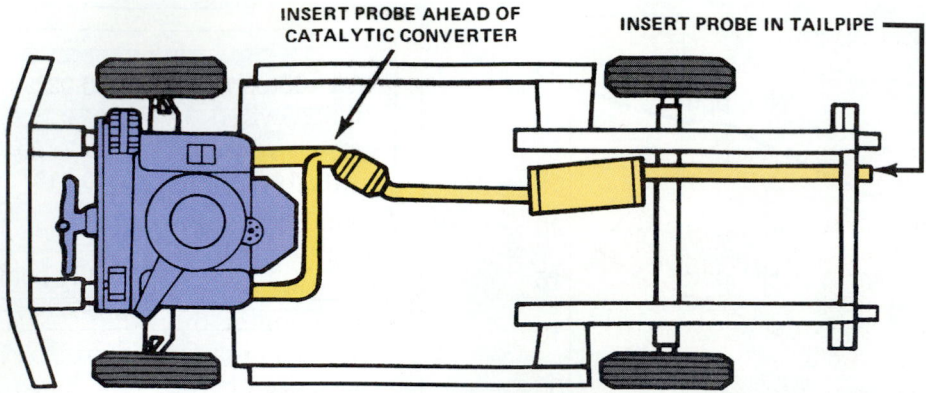

Figure 41.22 A reliable converter efficiency test takes readings before and after the cat. *(Courtesy of Chrysler Corporation)*

GAS	BEFORE	AFTER
HC	69 ppm	4 ppm
CO	0.77%	0.06%
CO_2	13.6%	14.5%
O_2	0.6%	0.01%
NO_x	2502 ppm	814 ppm

Figure 41.23 Dynamometer test results showing exhaust analysis done before and after the converter. *(Courtesy of NAPA Institute of Automotive Technology)*

kits are available that make it easy to permanently seal the hole after taking the readings.

A converter was tested in this manner while operating the car on a dynamometer. The speed was 50 mph with a slight uphill load. Tests were taken before and after the converter. Figure 41.23 shows that the converter was working with 94.2% efficiency for HC oxidation and 92.2% efficiency for CO reduction. These tests can be done without the dynamometer, but NO_x results will not be significant without the load. Results of the converter efficiency test at idle are shown in Figure 41.24.

NOTES:
- *Be sure that the cat is hot. Driving the car on the road or on a dyno is the best way to heat up the cat. Running the engine in a stall is not as effective.*
- *Be sure there is no air leaking in when taking the post-cat readings.*

■ CO_2 CONVERTER TEST

A cranking test for CO_2 can be used to tell the condition of a catalytic converter. With a hot catalytic converter, crank the engine for 10 seconds with the ignition system disabled. Observe the CO_2 reading. A reading below 12% indicates that the converter is weak or defective. Be sure to disable the secondary ignition system and not the primary. Otherwise you will disable the fuel injection system.

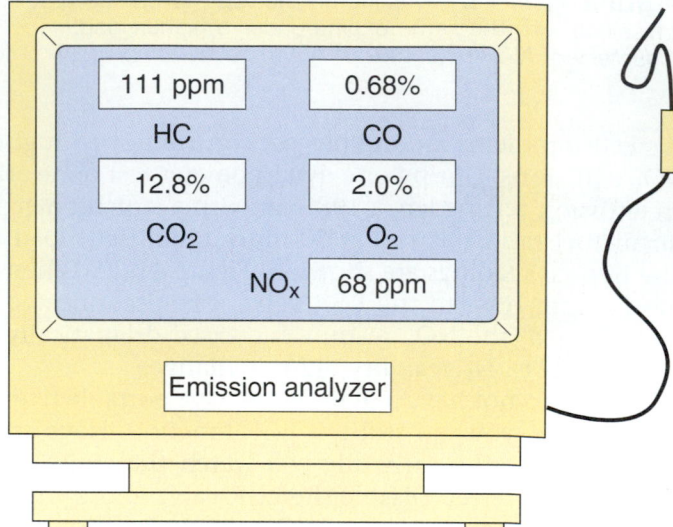

Before Catalyst

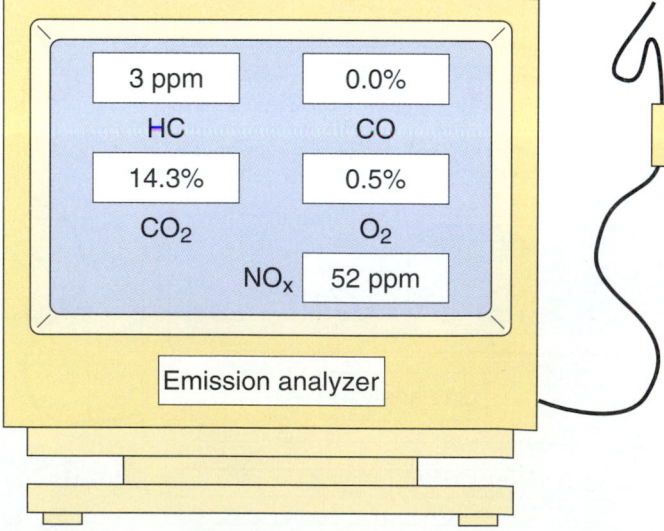

After Catalyst

Figure 41.24 When readings are taken at idle, NO_x does not change much. *(Courtesy of NAPA Institute of Automotive Technology)*

■ REVIEW QUESTIONS

1. What will happen to engine idle when the PCV valve is plugged off?
2. What electrical test instrument is used to check an electric EFE grid?
3. A defective temperature_____switch (TVS) can cause the EGR valve to work when the engine is cold.
4. The catalytic converter must be heated to at least_____ °F before it starts to work.
5. What does oxidizing mean?
6. Hydrocarbons are measured in_____ per million.
7. CO_2 is a good indicator of the_____ of the engine.
8. A typical good running engine with a catalyst will have from_____to 16% CO_2.
9. Is CO read as a percentage, or parts per million?
10. Which of the five exhaust gas readings becomes higher when the exhaust is diluted with air?
11. Which smog device must be disabled during emission analyzer tests?
12. What is the lowest amount of CO_2 a running engine can have without any outside air entering the exhaust system?
13. If an engine that will not start is being cranked with an emission analyzer connected, the presence of what gas tells you that there is a fuel supply?
14. When an O_2 reading is from 2–5%, what is suspected?
15. During a cranking CO_2 test, a reading below_____% indicates that the converter is weak or defective.

■ ASE STYLE REVIEW QUESTIONS

1. Technician A says that a valve cover gasket leak can hamper crankcase ventilation system operation. Technician B says that when the PCV valve is plugged off while the engine idles, the rpm should remain constant. Who is right?
   a. Technician A   b. Technician B
   c. Both A and B   d. Neither A nor B

2. Technician A says that a purge valve should open when the engine is cold. Technician B says that a purge valve should open at idle. Who is right?
   a. Technician A   b. Technician B
   c. Both A and B   d. Neither A nor B

3. Technician A says that when a heated air cleaner fails, usually the air door closes off heated air. Technician B says that a bad vacuum motor in a heated air cleaner causes hesitation and stalling when the car is cold. Who is right?
   a. Technician A   b. Technician B
   c. Both A and B   d. Neither A nor B

4. Technician A says that a faulty EGR valve can cause spark knock. Technician B says that a faulty EGR valve can cause a rough idle. Who is right?
   a. Technician A   b. Technician B
   c. Both A and B   d. Neither A nor B

5. Technician A says that an EGR valve that is stuck closed can cause hard starting. Technician B says that applying vacuum to an EGR valve while the engine idles will cause it to stumble. Who is right?
   a. Technician A   b. Technician B
   c. Both A and B   d. Neither A nor B

6. Technician A says that a backfire on deceleration can be due to a faulty air injection system valve. Technician B says a catalytic converter can melt if a spark plug wire falls off. Who is right?
   a. Technician A   b. Technician B
   c. Both A and B   d. Neither A nor B

7. Technician A says that an engine that has more than 14.5% CO_2 probably has a functioning converter. Technician B says that when there is a misfire, there is no CO_2 in the exhaust. Who is right?
   a. Technician A   b. Technician B
   c. Both A and B   d. Neither A nor B

8. Technician A says that CO_2 is a good rich lean indicator. Technician B says that HC is a good rich lean indicator. Who is right?
   a. Technician A   b. Technician B
   c. Both A and B   d. Neither A nor B

9. Technician A says that high combustion temperatures and load cause high NO_x. Technician B says that anything that increases temperatures can be a possible cause of HC levels that are too high. Who is right?
   a. Technician A   b. Technician B
   c. Both A and B   d. Neither A nor B

10. Technician A says that retarded timing raises NO_x emissions. Technician B says that testing before and after the converter with a five-gas analyzer does not work as well for NO_x without a dyno. Who is right?
   a. Technician A   b. Technician B
   c. Both A and B   d. Neither A nor B

Diagnosing Engine Performance Problems

■ INTRODUCTION

Problems with the engine and ignition system can result in poor fuel economy, high emissions, and poor driveability. Being able to correctly tune an engine requires related knowledge from material in other related chapters. Chapter 38 covered ignition system theory. Firing orders and companion cylinders are covered in Chapter 16. Engine mechanical problems are covered in detail in Chapter 43. Refer to those chapters as needed.

For an engine to start and run properly, three things must be present:

■ Sufficient compression
■ A timed spark
■ Fuel and air in the correct mixture

The spark created by the ignition system must be delivered to the correct spark plug at a specified point before TDC. This is called *ignition timing*. Vaporized air and fuel in the correct quantity and ratio must be drawn into the cylinder. If it is not compressed enough, it will not burn when the spark occurs at the spark plug gap.

■ VISUAL CHECKS

The first thing to do when an engine is running poorly, or not at all, is a visual check. A broken vacuum line or an electrical wire that has fallen off is often the cause of a problem. Check the fuel gauge to see if there is sufficient fuel. Believe it or not, sometimes the car is just out of gas. This is comparable to an appliance repair person making a house call only to discover that an appliance does not work because it is not plugged in.

■ IGNITION SYSTEM CHECKS

For an engine to run, the ignition system must produce a strong spark at the correct moment. Spark is the quickest thing to check for. If an engine will not start, check for spark at any of the spark plugs. A tester is available that provides a visible gap (Figure 42.1). You can also make a tester (Figure 42.2). If there is a strong hot spark, the next tests are for fuel and compression.

With a fuel injected car, it is usually easier to test compression than check for fuel. With a spark plug removed, hold your thumb over the spark plug hole while cranking the engine over. Be sure to disable the ignition system during cranking. If the engine has good compression, the piston will blow air hard on your thumb. A compression tester can also be used, but you are just trying to determine that compression is present at this point.

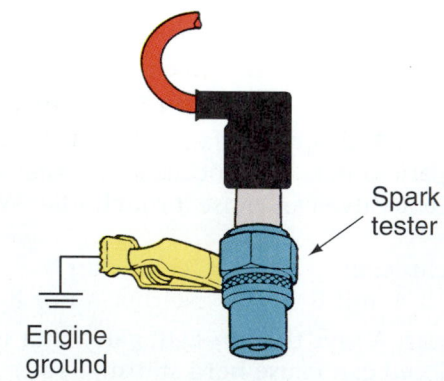

Spark tester

Engine ground

Figure 42.1 A spark tester. *(Courtesy of Ford Motor Company)*

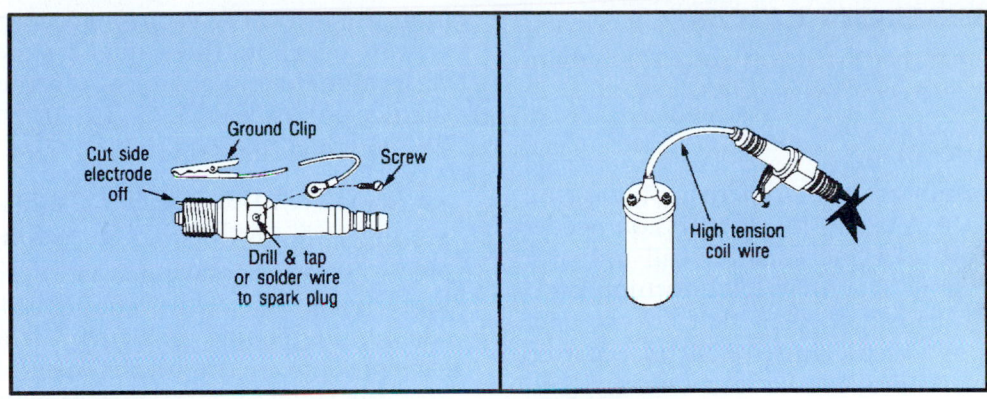

Figure 42.2   How to make a spark plug tester. *(Courtesy of Cooper Automotive/NAPA Belden)*

Sometimes an engine runs but one cylinder has a *dead misfire*. This could be due to ignition or mechanical problems. Further testing will be required.

ENGINE PERFORMANCE TESTING

When looking for reasons why an engine runs poorly, first rule out other causes.

Fuel Problems

- Dry black soot at the exhaust pipe often indicates an overly rich air/fuel mixture; not oil consumption.
- A bad fuel pump can show up as low fuel pump pressure and an oil dipstick that smells of gasoline. The latter may result in the car running out of fuel when part way up a grade. It can also cause vapor lock (when the fuel boils in the line, temporarily depriving a carbureted engine of gas).
- Fuel in the crankcase causes a dangerous condition to develop. A faulty fuel pump will allow the crankcase to collect fuel. Also, fuel enters the crankcase when an engine has become flooded. Flooding occurs when the owner of a vehicle that needs a tuneup continues to crank the engine after it fails to start.

 **SAFETY NOTE** If you are working on a vehicle and your clothing should catch fire, fall on the ground and roll around to smother the flames. Do not run.

The air/fuel mixture can also be affected by air leaks in a fuel injection intake system. An oxygen sensor can also be defective. They sometimes slow down as they get older. Scan tools used in diagnosing computer control problems (see Chapter 76) show whether the oxygen sensor responds quickly to control the air/fuel ratio.

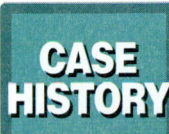 **CASE HISTORY** *A vehicle was brought into a shop for a tuneup. The technician decided that it would be easier to service the distributor if it was removed from the engine. Somehow, the ignition switch was left in the "on" position. When the distributor was removed a spark happened, igniting the air/fuel mixture in the crankcase and causing an explosion. The explosion caused the burning mixture to come out the ignition distributor hole, burning the technician and catching his clothing on fire.*

Sometimes on a throttle body injection car, idle speed is different each time it returns to idle. A buildup of soot on the idle air control motor pintel can cause this. Remove it and clean it. Also, when tuning TBI cars always spray carburetor cleaner around the throttle plate to keep it clean.

On port fuel injected cars, blow-by gases and EGR gases come up through the manifold and cause a buildup behind the throttle plate. This first causes a gummy deposit. After many miles, the deposit gets hard and restricts air travel around the plate. The idle air control motor reaches its limit of movement and shuts off the engine. To clean the area, wrap a rag around the end of a screwdriver and soak it with carburetor cleaner.

Dirty injectors can cause driveability problems. When fuel injectors get dirty, this can cause the engine to ping at full throttle. Dirty injectors do not always make the clicking sound associated with a normal injector. Use a stethoscope to listen for different sounds between the injectors. After cleaning the injectors (see Figure 27.23), the normal click can return. The new ball-type injectors are self-cleaning.

A plugged fuel filter will act just like vapor lock, causing high speed starvation. Some technicians recommend a fuel injection cleaning and fuel filter change at each tuneup.

■ COMPRESSION LOSS

Low compression can be traced to two causes: engine *breathing problems* and *compression leaks*.

Engine Breathing

An engine that cannot breathe properly is suffocating. It will not be able to develop the compression needed to function properly. Engine vacuum will drop off, lowering compression. Examples of breathing problems include worn camshaft lobes that do not open the valve far enough (see Figure 18.34) or incorrect valve timing. Valve timing can become retarded (late) when a timing chain becomes so worn that it skips a tooth (see Chapter 46).

NOTE: *Late valve timing will cause poor low rpm performance. At higher rpm, engine performance might be acceptable.*

If the timing chain has skipped and valve timing is retarded, suction will be felt at the exhaust pipe. This happens because the exhaust valve is still open during the piston's intake stroke. Procedures for checking timing chain wear are covered in the section on oscilloscope tests in this chapter and in Chapter 46.

Breathing problems can also be traced to carbon buildup around the neck of the valve (see Figure 18.6) or to restrictions such as a dirty air cleaner or a blocked exhaust. A blocked exhaust will be evident when the engine rpm is raised quickly. A roar will be heard through the carburetor or fuel injection air intake.

NOTE: *A collapsed laminated (two-layer) exhaust pipe (see Figure 28.32) will often cause a stripe of discoloration on the outside of the pipe at the point of restriction.*

Catalytic converter mufflers can become plugged after running for a prolonged period of time with an ignition system defect. A rich air/fuel mixture can also result in a plugged converter when it overheats and melts internally.

> **SAFETY NOTE** A catalytic converter can actually become so hot that it can start a fire. Some converters have ceramic pellets inside of them and when the converter fails, the hot pellets fly out and can start a fire also.

A car with an exhaust restriction might also experience automatic transmission harsh or late shifts due to the resulting faulty vacuum signal or increased throttle pressure in the transmission.

Exhaust backpressure can be tested using a fuel pump vacuum/pressure tester hooked either to the AIR (smog pump) lines into the intake manifold, or an adapter can be substituted in place of the EGR valve. According to TRW, removing the oxygen sensor to per-

form the test can give an inaccurate reading due to a venturi effect in the exhaust system. Specifications vary among manufacturers. As a general rule, pressure should not exceed 1.75 psi at wide open throttle (WOT) under full load.

■ VACUUM TESTING

A vacuum/pressure gauge (Figure 42.3) is commonly used to measure intake manifold vacuum or fuel pressure. Intake manifold vacuum readings are useful in determining engine problems. Vacuum readings compare pressure in the intake manifold to atmospheric pressure. Hook up a vacuum gauge to a manifold vacuum source. For an unmodified engine, vacuum should range between 16 and 22 inches of mercury ("Hg) at idle, and the needle should be steady (Figure 42.4).

NOTE: *Vacuum readings will drop approximately 1" for each 1000 feet above sea level.*

Rough Idle

A *leaking intake manifold* gasket can cause a rough idle.

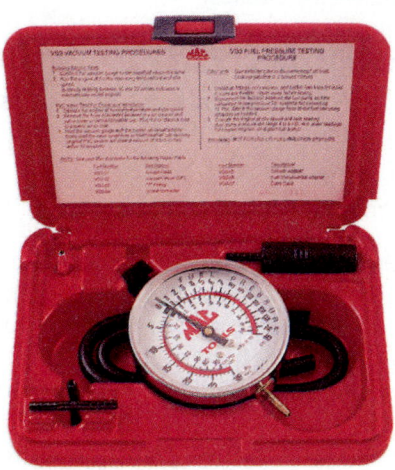

Figure 42.3 A vacuum/pressure tester used to measure intake manifold vacuum. *(Photo provided by Mac Tools and is representative of products manufactured by Mac Tools or products obtained from another source.)*

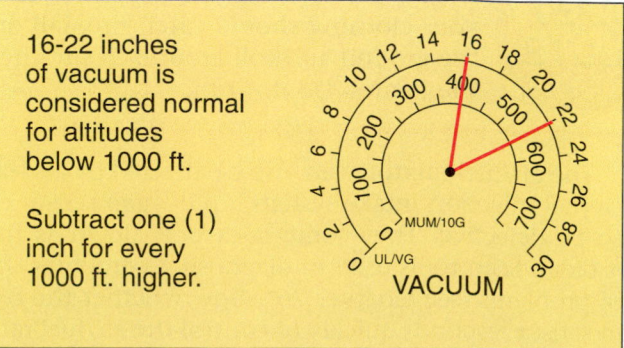

16-22 inches of vacuum is considered normal for altitudes below 1000 ft.

Subtract one (1) inch for every 1000 ft. higher.

Figure 42.4 Normal engine vacuum. *(Courtesy of SPX Corporation, Aftermarket Tool and Equipment Group)*

NOTE: *At speeds above idle, symptoms of a leaking manifold diminish and do not show up on the vacuum gauge. This is because the size of the leak is proportionally less as the engine breathes more air.*

Sometimes, an engine will idle rough for no apparent reason. Port fuel injected engines use o-rings to seal each individual fuel injector where it enters an intake port (see Figure 26.31). A leaking o-ring, which causes a vacuum leak, will result in a lean air/fuel mixture for that cylinder. A rough idle is the result.

With a vacuum leak, an engine with an oxygen sensor feedback fuel system can run rough when cold, yet run fine when hot. This is because the oxygen sensor (which does not work until the engine is warmed up) tells the computer to compensate with a richer air/fuel mixture (see Chapter 27).

Low Vacuum

If low, steady vacuum is found, check the ignition timing.

NOTE: *Late (retarded) ignition timing will often cause an engine to run hot.*

Bad struts bottoming out can cause a knock sensor to retard the timing. A bad AC bracket can do this, too.

NOTE: *Vacuum readings for an idling engine with a performance camshaft will be much lower. These engines are inefficient at low rpm.*

If the timing is retarded a great deal, check the cam timing too (see Chapter 46). Excessive timing chain slack can allow the timing chain to skip a tooth on its sprocket. Because the distributor is usually driven by the cam, this also results in late ignition timing.

NOTE: *A high-performance camshaft will have low vacuum at idle, but the vacuum will go up as rpm is raised. This is caused by valve overlap.*

Weak piston rings can also result in low vacuum. To test for this with a vacuum gauge, raise the engine speed to about 2000 rpm. Then, snap the throttle closed and watch for an increase of from 2" to 6" of vacuum above normal. Worn rings will not increase vacuum sufficiently during deceleration. Generally, the higher the rise, the better the condition of the rings.

■ VACUUM TESTS FOR VALVES PROBLEMS

Sticky valves can be indicated when the needle drops quickly, or drifts. The movement will have no apparent rhythm. Leaking valves are indicated when the needle drops at regular intervals. A power balance test will pinpoint the low cylinder.

Restricted Exhaust

You can hear cars with restricted exhaust systems if you ever have occasion to be stopped on a hill or grade. A restriction in the exhaust causes a hissing sound from the tailpipe when under load. To test for a restricted exhaust, raise the engine rpm quickly to 2000 to cause a vacuum reading that is momentarily low. Then, release the throttle quickly. Vacuum should return smoothly and quickly to higher-than-normal levels. A slow, hesitating return usually indicates a restriction.

Cranking Vacuum

A **cranking vacuum test** can be performed to check for internal air leaks. If the test is to be effective, the engine must be sealed off. The throttle plate must be all the way closed and the crankcase ventilation system must be plugged off.

To perform a cranking vacuum test, close the throttle all the way and plug the breather hose to the air cleaner. The ignition system must be disabled. During cranking, the needle on the gauge should be steady. This is more important than the amount of vacuum on the gauge. The engine should produce a minimum of 3" of vacuum while cranking but most engines will produce far more.

Next, pull the PCV (positive crankcase ventilation) valve and cover its opening with your thumb. Cranking vacuum should rise. If it does not, there is a problem with the PCV system.

NOTE: *If an engine is equipped with an air injection smog pump, be sure that it does not have a defective antibackfire valve, which can cause low vacuum.*

■ OTHER VACUUM TESTS

Sometimes, a V-type engine can have a vacuum leak from the underside of the intake manifold to the crankcase. To test for this, pinch off the PCV valve hose to the manifold and the breather hose to the air cleaner. When the engine is running, if there is vacuum at the oil filler opening, an internal vacuum leak is indicated.

Another test for an internal leak is the **propane enrichment test**. All PCV system hoses are also closed off for this test. If the engine idle changes when propane or acetylene gas is put into the oil filler opening, the gas is getting to the combustion chambers by way of an intake manifold leak.

■ COMPRESSION PROBLEMS

Compression can leak due to several causes such as a blown head gasket (Figure 42.5), burned valves (Figure 42.6), worn or broken piston rings (Figure 42.7), or a broken valve spring.

When valve clearances are adjusted too tightly they cannot properly seal the cylinder. This is a common problem after an overhaul or servicing.

■ DIAGNOSING ENGINE COMPRESSION PROBLEMS

A quick judgment about engine compression can often be made simply by cranking the engine with the

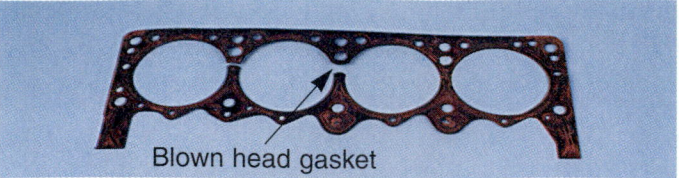

Figure 42.5 A blown head gasket.

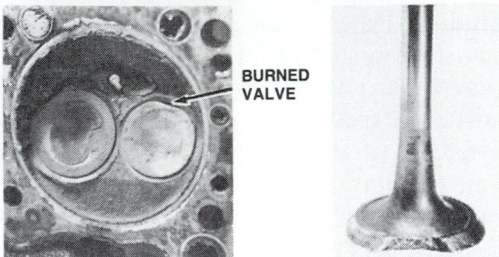

Figure 42.6 A burned valve.

Figure 42.7 A damaged piston resulted in a broken ring.

ignition system disabled and listening for an uneven rhythm. An engine whose cylinders have equal compression will have an even cranking rhythm.

To quickly locate a weak cylinder, perform a **cylinder power balance test**. A **compression test** or leakage test can be performed to pinpoint compression problems located by the power balance test.

NOTE: *If any of these tests indicate a possible valve sealing problem, be sure to check the valves for proper adjustment before proceeding with repair.*

Cylinder Power Balance Test

Shorting out the spark to a cylinder should result in a drop in engine rpm. A cylinder that does not drop as much as the others is not pulling its full load. Variations in the amount that rpm drops between cylinders should be less than 5%. The problem could be in the ignition system or fuel system, or the engine could have vacuum leaks or compression problems. A com-

pression test or leakage test (covered later) can be performed to pinpoint compression problems located by the power balance test.

To perform a power balance test on cars equipped with electronic ignitions, an *electronic power balance tester* is necessary. It allows cylinders to be electronically grounded out without removing plug cables. When cars had points, simply removing a spark plug cable with the engine idling was all it took to perform this test. Pulling a plug cable that is situated above the control module in an electronic ignition distributor can result in a blown control module.

Most scopes have an electronic cylinder balance tester built into them. The cylinder selector follows the firing order (Figure 42.8). Handheld scan tools used with computer controlled engines (see Chapter 76) also have power balance test capability.

NOTES:
- *To perform a power balance test on cars equipped with computer controlled fuel systems, the computer system must be disabled. These systems have idle speed control motors that automatically raise the engine's idle to compensate as each cylinder is grounded out. Consult a service manual for the proper procedure.*
- *Do not allow cylinders to be shorted out for a long period of time (more than 30 seconds) on catalytic converter-equipped cars. Raw fuel entering the converter can cause overheating. Converter overheating can be minimized by disconnecting the smog pump during the test.*

Occasionally, rpm might *rise* as a cylinder is shorted out due to exhaust gas entering the intake manifold. This can be caused by an exhaust gas recirculation

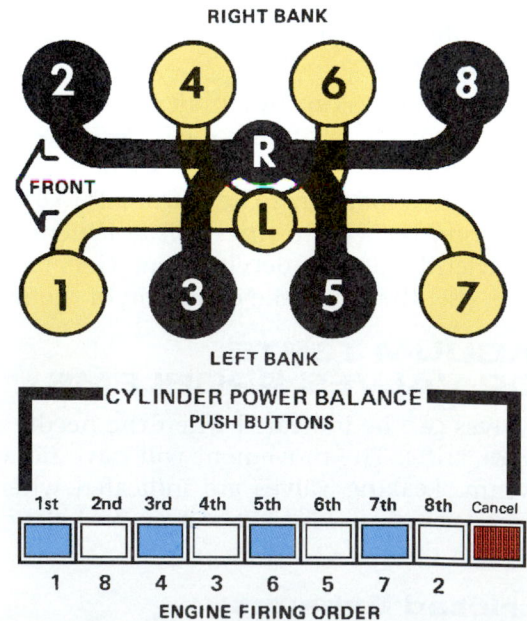

Figure 42.8 A cylinder power balance tester checks cylinders in the engine's firing order. *(Courtesy of Sun Electric Corporation)*

(EGR) valve that is open. EGR should be closed at idle (see Chapter 41). The cylinder that causes the rpm to rise is the one from which the exhaust gas for the EGR valve was picked up. Retest at cruise rpm and the problem will disappear.

Compression Test

One of the most common and least expensive pieces of test equipment is the compression tester. A compression tester is simply a pressure gauge that is inserted into a spark plug hole and registers a reading as the engine is cranked. Differences in pressure between the cylinders pinpoints problem areas. If all cylinders are performing equally and engine performance is acceptable, the engine passes the test.

NOTE: *A compression test will* **not** *tell the condition of oil control rings, only compression rings.*

Compression Testers. There are two styles of compression testers. One is held in place while cranking the engine (Figure 42.9). It is handy on inline engines because it is fast and easy. The other type is the screw-in tester. The tester shown in Figure 42.10 has two sizes of spark plug threads on it. It has a shrader valve (a tire valve). This is for releasing the pressure that is saved in the gauge so that the technician can read the compression after the engine is cranked.

The advantage of the screw-in tester is that it can be threaded into the plug hole while the engine is cranked with the ignition switch. Ideally, the test is done with the engine at normal operating temperature. The battery must be fully charged for the test to be effective.

■ COMPRESSION TEST PROCEDURE

The spark plugs are removed to perform a compression test. Twist the rubber boots on the spark plug cables to loosen them from the spark plugs (Figure 42.11). The boots stick to the plugs because of heat radiated from the exhaust manifold. The inside material in the cables can easily be damaged if the cables are handled roughly.

Before removing the spark plugs, blow dirt away from their outsides with compressed air. The specified way to do a compression test calls for the removal of *all* of the spark plugs. This is so the starter can crank the engine easily. Keep the spark plugs in order so after they are removed, you can compare them. Spark plug diagnosis is covered later in the chapter.

Block the throttle in the wide open position. This can be done with a throttle depressor.

(a)

Figure 42.9 This compression tester is held in place while cranking the engine. *(Courtesy of Snap-on Tools Company. Copyright Owner.)*

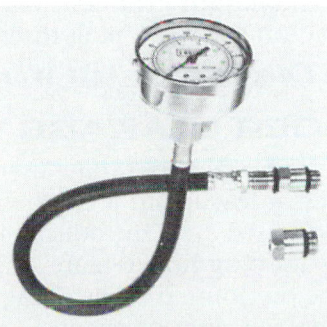

Figure 42.10 This tester can fit two different spark plug thread sizes. *(Courtesy of K-line)*

INCORRECT WAY

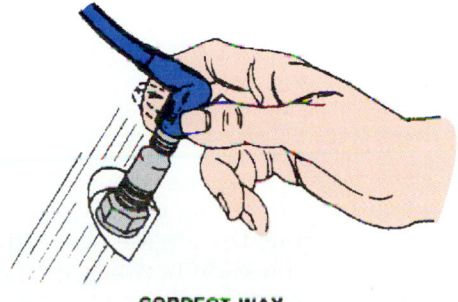

CORRECT WAY

Figure 42.11 Twist the rubber boots on the spark plug cables to loosen them from the spark plugs. *(Courtesy of Cooper Automotive/NAPA Belden)*

SHOP TIP On vehicles equipped with cruise control, throttle position can be controlled by connecting a hand held vacuum pump to the vacuum diaphragm.

Blocking the throttle open during a compression test prevents an engine with a carburetor from sucking fuel into the cylinders and lets the engine breathe air more easily.

CASE HISTORY *A group of students were performing a compression test on the teacher's car. The compression was checked repeatedly by several students (the throttle was not blocked open). After completing the compression test, they tried repeatedly to get the engine started. Finally, a spark happened, fuel in the muffler of the flooded engine was ignited, and a deafening explosion ripped the muffler apart.*

Hook up a remote starter switch between the S terminal on the starter solenoid and the ungrounded battery post.

SHOP TIP On some vehicles, when the starter is not easily accessible, a wire in an easy-to-reach location under the hood can often provide access to the starter circuit. Study the wiring diagram for the vehicle.

Disable the ignition system. Pull the battery-to-distributor wire on electronic ignition-equipped cars. Follow the manufacturer's instructions. For most cars, the compression test should be done with the ignition switch in the "on" position to prevent damage to the electrical system.

Insert the compression gauge into a spark plug hole and crank the engine through at least *four* compression strokes. The gauge will move four times, or with all the plugs removed, you can hear each compression stroke as the compression in the cylinder being tested slows the engine. Check and record each reading.

Interpreting Compression Test Results

If all cylinders are performing equally and engine performance is acceptable, the engine passes the test.
- Variations in compression between cylinders should be no more than 20%.
- When two cylinders next to each other have low compression, a blown head gasket is usually indicated (see Figure 42.5).

- One or several cylinders with low compression and no apparent pattern of loss often indicates burned exhaust valves. Rough idling is a symptom. At higher rpm, the rough running from the burned valves disappears.

SHOP TIP On an engine at idle, a burned exhaust valve will cause a dollar bill to be sucked against the end of the exhaust pipe every time the bad cylinder's piston has an intake stroke.

■ MATH NOTE ■

When compression test specifications are available, they are only an estimate. If specifications are not available, locate the compression ratio in the specification manual and use the following formula:

$$\text{Compression Ratio} \times \text{Atmospheric Pressure} +$$

$$\text{Atmospheric Pressure} + 5 \text{ (Volumetric Efficiency)}$$

For example, to figure out the approximate compression on an 8:1 engine at sea level (14.7 psi atmospheric pressure): 8.0 × 14.7 + 14.7 + 5 = 137.3 psi.

Wet Compression Test

If any cylinders show poor results, perform a **wet compression test**. Squirt about a tablespoon of oil into each low cylinder. The oil makes a seal around worn rings, boosting the compression reading. When cylinder readings that were low increase to normal during a wet test, a piston ring problem is indicated.

NOTE: *Adding too much oil takes up volume and raises compression.*

When the compression test is completed, install the spark plugs. Be especially careful not to strip the threads on an aluminum cylinder head and use anti-seize compound on the spark plug threads.

➡ *Perform **Compression Test** Worksheet*

■ CYLINDER LEAKAGE TEST

The **cylinder leakage test** can be used to accurately pinpoint causes of leakage in a combustion chamber. Regulated compressed air is introduced into a cylinder through the spark plug hole (Figure 42.12). If there is a leak, it can be pinpointed by listening at the:
- Oil filler = leaking rings or piston
- Carburetor or fuel injection intake = leaking intake valve

SHOP AIR

REGULATED AIR TO SPARK PLUG HOLE

Figure 42.12 A cylinder leakage tester is connected to the cylinder through the spark plug hole.

- Exhaust = leaking exhaust valve
- Bubbles in the radiator = blown head gasket, or a crack in the head or block, which allows the regulated air to enter the cooling system.

The leakage tester offers several advantages over a compression test.

- The test can be performed on an engine that is removed from a car (such as an engine purchased at a junkyard).
- A performance camshaft will not affect the results of the test. It would cause lower readings on the compression test because engine vacuum is lower at cranking speeds with a racing cam.
- The exact source of leakage can be pinpointed before engine disassembly.

To perform a leakage test, the piston is positioned at TDC on the compression stroke. This ensures that both valves are completely closed. It is best to perform a cylinder leakage test when the engine is warm and the rings are sealed with oil. Otherwise, a small amount of movement at TDC can allow the piston ring to move off its ring land, allowing leakage (see Chapter 19).

NOTE: *When a cylinder bore is worn to a considerable taper, the reading can be improved if the piston is moved slightly past TDC into a less worn area.*

Acceptable leakage on the face of the tester's gauge is usually less than 10 to 15%, although vehicles with up to 30% leakage might still be performing to the owner's satisfaction. The owner does not notice the power difference until it is restored after an overhaul because the loss in power has happened gradually.

NOTE: *Any time there is leakage past the piston rings, the PCV valve can allow air to travel into the intake manifold where it can be mistaken for a leaking intake valve. To avoid this situation, remove the oil filler cap, or unhook the vacuum line to the PCV valve or pinch it with pliers.*

■ CARBON RELATED PROBLEMS

Carbon can cause performance problems in an engine. It can cause the engine to have increased compression, which can cause fuel to detonate in the cylinder. Fuel can also become saturated in carbon deposits on valves. This causes driveability problems when the fuel goes into or comes out of the carbon at the wrong times. The carbon deposits can also block the flow of air and fuel into the cylinder.

There are two kinds of deposits: oil based and carbonaceous. The oil-based deposits are the traditional gummy, black ones like the deposit on the valve in Figure 18.6. They are caused when oil and heat come together.

Carbonaceous deposits are ones that result from fuel. They are called cauliflower deposits because of their resemblance to the vegetable. These deposits are not as thick as oil deposits. They are also hard, dry, and tougher to remove. They cause driveability problems because fuel vapors can be absorbed into them. Driveability problems include rough idling when cold, loss of power, surging, and high exhaust emissions. Valve deposits sometimes cause problems in as little as 5000 miles. Additives are put into fuels by oil companies to minimize this problem. Removal of carbon deposits is covered in Chapter 27.

■ REVIEW QUESTIONS

1. An engine will not start. If there is a strong hot spark, the next tests are for _____ and compression.

2. Some technicians recommend a fuel injection cleaning and ____ ____ ____ change at each tuneup.

3. A roar will be heard through the carburetor or fuel injection air intake when an _____ system is restricted.

4. What happens to the outside of an exhaust pipe at the point of an internal restriction?

5. Vacuum readings will drop approximately 1" for each _____ feet above sea level.

6. An engine should produce a minimum of ___ of vacuum while cranking but most engines will produce far more.

7. When engine idle rises during a cylinder power balance test, what could be the cause?

8. During a compression test, remove all of the ____ ____ so the starter can crank the engine easily.

9. How many compression strokes should the engine crank a minimum of during a compression test?

10. If a dollar bill gets sucked against an exhaust pipe when the engine is idling, what could this indicate?

■ ASE STYLE REVIEW QUESTIONS

1. An engine will not start. Technician A says that spark timing might be off. Technician B says that compression could be too weak. Who is right?
   - **a.** Technician A
   - **b.** Technician B
   - **c.** Both A and B
   - **d.** Neither A nor B

2. A car has a buildup of dry black ash (soot) in the exhaust pipe. Technician A says this could be due to an overly rich air/fuel mixture. Technician B says this could be due to oil consumption. Who is right?
   - **a.** Technician A
   - **b.** Technician B
   - **c.** Both A and B
   - **d.** Neither A nor B

3. Technician A says that dirty fuel injectors can cause an engine to ping at full throttle. Technician B says that a dirty injector might not sound the same as a clean one. Who is right?
   - **a.** Technician A
   - **b.** Technician B
   - **c.** Both A and B
   - **d.** Neither A nor B

4. An engine has late valve timing. Technician A says this will cause poor low rpm performance. Technician B says that at higher rpm, engine performance might be acceptable. Who is right?
   - **a.** Technician A
   - **b.** Technician B
   - **c.** Both A and B
   - **d.** Neither A nor B

5. Technician A says that, with a vacuum leak, an engine with an oxygen sensor feedback fuel system can run fine when cold, yet run rough when hot. Technician B says that advanced ignition timing will often cause an engine to run hot. Who is right?
   - **a.** Technician A
   - **b.** Technician B
   - **c.** Both A and B
   - **d.** Neither A nor B

6. Technician A says that vacuum readings for an engine with a performance camshaft will be lower at idle. Technician B says that during a cranking vacuum test, when the PCV valve is covered, the vacuum reading will go down. Who is right?
   - **a.** Technician A
   - **b.** Technician B
   - **c.** Both A and B
   - **d.** Neither A nor B

7. Technician A says that a compression test can be used to tell the condition of the oil control rings. Technician B says that several cylinders with low compression test readings are likely due to burned or tight valves. Who is right?
   - **a.** Technician A
   - **b.** Technician B
   - **c.** Both A and B
   - **d.** Neither A nor B

8. Two technicians are discussing compression test results. Technician A says that low compression in adjacent (neighboring) cylinders indicates a bad head gasket. Technician B says that if the reading on the compression tester goes up after adding oil to the cylinder, the compression rings are probably worn. Who is right?
   - **a.** Technician A
   - **b.** Technician B
   - **c.** Both A and B
   - **d.** Neither A nor B

9. Technician A says that a cylinder leakage test is better than a compression test for testing the condition of a junkyard engine before installing it in a car. Technician B says that a compression tester is better than a cylinder leakage tester for testing the condition of an engine with a performance camshaft? Who is right?

a. Technician A   **b.** Technician B

c. Both A and B   **d.** Neither A nor B

10. Technician A says that an engine that idles rough due to burned valves will probably run fine at speeds above idle. Technician B says that carbon buildup in a combustion chamber can cause an engine to have higher compression. Who is right?

a. Technician A   **b.** Technician B

c. Both A and B   **d.** Neither A nor B

Diagnosing Engine Mechanical Problems

■ **KEY TERMS**

black light test
spark plug deposit test
engine knocks
piston slap
flat cam
oil pressure sending unit
pressure relief valve
oil analysis
cross fluid contamination
seized engine
oil wash

■ **INTRODUCTION**

This chapter focuses on the diagnosis of problems before engine disassembly. Internal problem diagnosis after disassembly is covered in more detail in later chapters. Those mechanical problems that are related to engine driveability and performance are covered in Chapter 42.

It is very important to diagnose the cause of the problem before performing a repair. It is not unusual for an inexperienced technician to spend many hours of work only to discover that the repair was unnecessary. Four major diagnosis areas will be covered:

1. Oil consumption
2. Engine noises
3. Oil pressure problems
4. Cooling system problems

Problems related to engine performance, such as rough idle or low compression, are covered in Chapter 42.

There are many causes of engine problems. Some are the result of normal wear and tear or due to a lack of maintenance. Others might be due to work previously done to the engine. Additionally, problems that appear to be engine related can be from other automotive specialty areas, such as transmission repair or emission controls. Sometimes a problem with a system causes an engine to fail. If the problem is not taken care of, the failure will recur.

■ **DIAGNOSING PROBLEMS BEFORE A REPAIR**

An engine should be properly diagnosed before disassembly for two reasons. First, it should be determined that a repair is really necessary. Second, the exact location of a problem should be determined while the engine is running (before disassembly). A thorough discussion of the problem with the vehicle's owner is also helpful. Many times an owner's driving habits or maintenance procedures can be the cause of the problem.

■ **OIL CONSUMPTION**

A vehicle owner will usually blame piston rings for oil consumption even though oil can be lost through a variety of other conditions. Oil can be lost through either *external leakage* or *internal oil consumption*. Internal oil consumption can sometimes be spotted as an oily coating on the inside of the exhaust pipe. Oil consumption of this magnitude will also be accompanied by blue smoke. Black soot at the exhaust pipe often indicates an overly rich air/fuel mixture, not oil consumption. Smoke from an overly rich mixture will be black.

The rate of normal oil consumption depends on the size of the engine, the weight and shape of the vehicle, the viscosity (thickness) and service rating of the oil, engine rpm during use, engine temperature, and the amount of oxidation and dilution of the oil. Information on oil is included in Chapter 12.

A customer might complain of a unique instance of rapid oil use. Sometimes this happens when the car was driven for 1000 or more miles of city driving, followed by a highway trip. During the city driving, the engine might have consumed a normal amount of oil, but because of fuel and water dilution from city driving, the dipstick registers "full." When the customer goes on a highway trip, the oil thoroughly heats up and the fuel and water diluting it evaporates. This gives the appearance of rapid oil consumption.

■ CAUSES OF OIL CONSUMPTION

Bad Valve Guides or Seals

When an engine has relatively low mileage (under 60,000), the cause of internal oil consumption is often the valve guides. The car might smoke during deceleration (high engine vacuum) because of oil leaking into the combustion chamber through the intake valve guides. Smoke at other times might be from exhaust valve guides (see Chapter 46).

When only one side of a spark plug is fouled with carbon, leaking guide seals are indicated. The carbon will be on the side of the spark plug that was facing the intake valve. Leaking valve guide seals also result in carbon deposits in the "neck" area of the intake valves (see Figure 18.7). Watch for carbon deposits when disassembling a cylinder head.

Different kinds of valve guide seals are described in detail in Chapter 18. When the valve cover is removed, the standard umbrella guide seal can be checked to see if it has become brittle. Insert a pocket steel rule between the coils of the valve spring, and push on the seal to see if it is still soft. The seal should also be checked to see that it still fits the valve stem tightly. Valve guide seals are replaced during a valve job while the heads are disassembled. Chapter 46 includes a procedure for replacing valve guide seals *without* removing the heads from the engine.

Oil Consumption from Piston Rings

Smoke during deceleration can also be due to worn or broken piston rings. Worn rings are found more often on a high mileage engine. The worn rings allow oil on the cylinder walls to escape into the combustion chamber during the high vacuum conditions associated with engine deceleration.

Piston rings are usually the first thing a customer suspects when a car starts to use oil. When oil is consumed past piston rings, the cause is usually high engine mileage or a poor maintenance schedule. Engines that suffer from too few oil changes can have plugged oil control rings (Figure 43.1). Oil rings scrape oil from the cylinder walls and return it to the crankcase through the inside of the piston (see Figure 15.27).

Figure 43.1 Plugged oil control ring. *(Courtesy of Dana Corp.-Perfect Circle Division)*

NOTE: *According to the Ford Motor Company, there are approximately 36,500 drops of oil in one quart. If the engine consumed as little as 1/1100 of a drop of oil on each stroke, it would use one quart of oil in 1000 miles. So you can see the fantastic job the oil rings do, and the major problem created when they are plugged up.*

When looking for a cause for oil consumption, technicians will often check an engine's compression (see Chapter 42). While this procedure might locate worn or broken compression rings, it does not check the condition of oil control rings.

Increased Oil Consumption after a Valve Job

The condition of the entire engine must be considered before performing a valve job on a high mileage engine. A valve job increases *compression*, but it also increases engine *vacuum*, which can cause more oil to be sucked past the worn piston rings into the combustion chamber.

If cylinder heads are removed for a valve job, look at the tops of the pistons to see if oil consumption caused by worn or stuck piston rings is indicated (Figure 43.2). During the valve job, cleaning carbon from the tops of the pistons can also lead to oil consumption if pieces of carbon get wedged between piston rings and ring grooves.

Oil Consumption from Excessive Rod Bearing Clearance

An engine with high mileage can have worn connecting rod bearings. Results of this are:

■ Low oil pressure when the engine idles (with normal pressure off idle)

■ Oil consumption when excessive oil leaks out between the rod journals and bearings

The oil is thrown onto the cylinder walls at high speed (Figure 43.3). The oil rings do not have enough capacity to return all of this oil to the crankcase. What-

Figure 43.2 A clean area around the outside edge of a piston indicates excessive oil consumption past the rings.

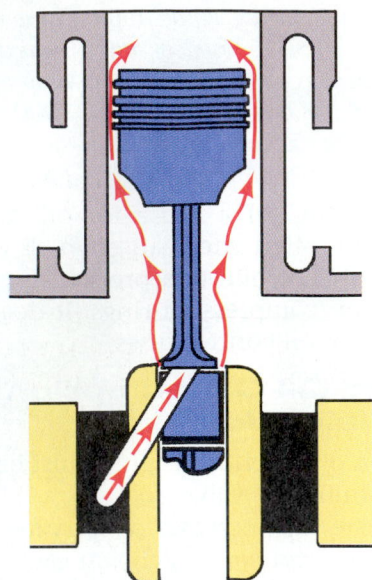

Figure 43.3 Effect of excessive bearing clearance on oil throw-off.

ever oil enters the combustion chamber will be burned with the air/fuel mixture.

High-speed driving also causes increased oil consumption because of the extra oil thrown from the rods. In one test, an engine run at 70 mph used seven times the oil that it used at 40 mph.

Vacuum Modulator

Some automatic transmissions have a vacuum modulator (Figure 43.4). It senses engine load by using intake mani-

fold vacuum to tell the transmission when to shift. If the diaphragm in the modulator leaks, transmission fluid will be sucked into the intake manifold. This can produce smoke that can be confused with engine oil smoke, even though the engine may be in good condition.

The leaking diaphragm can also cause a rough idle (air leak into the intake manifold) and harsh, late shifts of the transmission. Spark plugs near the vacuum tap on the manifold often become oil fouled, too.

Be sure to question the vehicle owner thoroughly. The combination of any or all of these symptoms can lead an owner to believe that an engine overhaul is needed. The key to diagnosing a faulty vacuum modulator is that the transmission is probably using automatic transmission fluid (ATF), without leaking externally.

Oil Consumption from an Incorrect Dipstick

The same engine is often installed in several different model cars. Each engine can be equipped with an oil pan of a different shape, which requires a different length oil dipstick. Sometimes oil consumption is blamed when the only problem is that the car has the wrong dipstick (Figure 43.5). Every time the owner mistakenly adds a quart of oil to the crankcase, the crankshaft whips up the oil like an egg beater whips cream. Excessive oil is thrown onto the cylinder walls. The oil rings cannot handle this amount of oil so it migrates into the cylinders and is burned off. It is especially

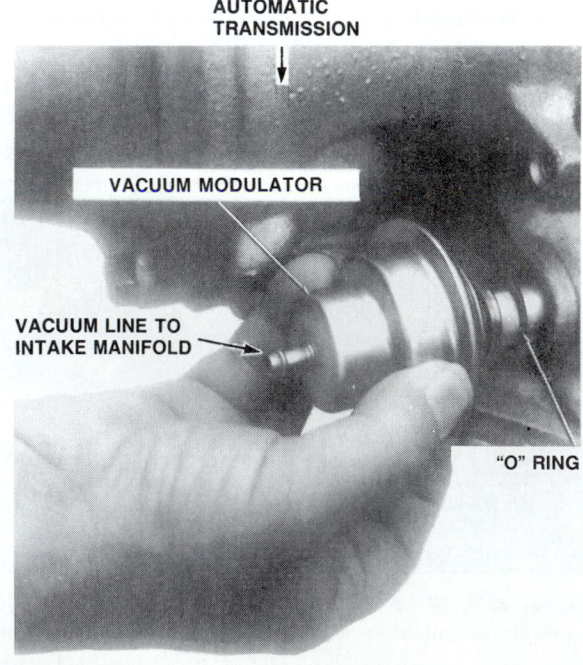

Figure 43.4 A vacuum modulator. *(Courtesy of General Motors Corporation, Service Technology Group)*

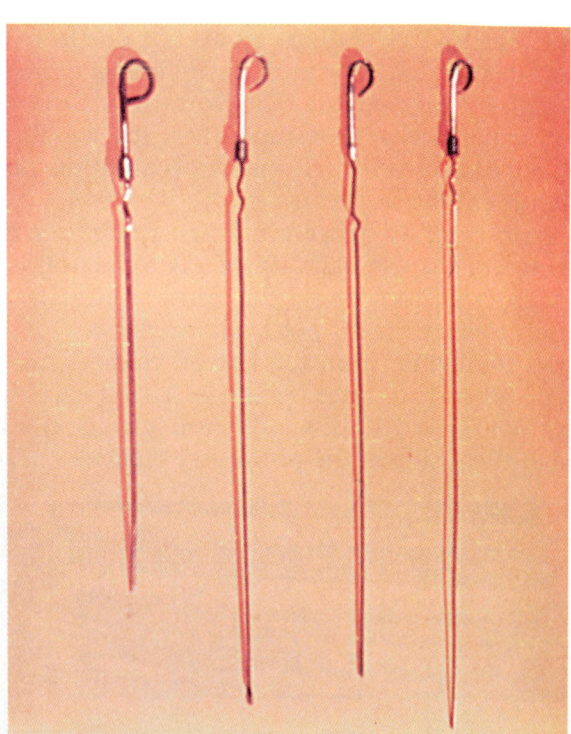

Figure 43.5 These dipsticks are the ones available for one size of engine that comes with different oil pans. *(Courtesy of Ford Motor Company)*

important to check that the correct dipstick is used after an engine change or short-block installation.

Excessive oil can also result in aeration of the oil, noisy lifters, unstable oil pressure, gurgling sounds coming from the crankcase, and oil leakage.

Oil Consumption from Plugged Cylinder Head Drainback Holes

Be sure to check for plugged drainback holes before removing an engine for a rebuild. When engine oil is not changed often enough, thick, dirty oil can plug the holes in the cylinder head. These holes are there to allow upper end oil to drain back to the crankcase (Figure 43.6). The problem can be temporarily solved by cleaning out the holes, but it is a symptom of a poorly

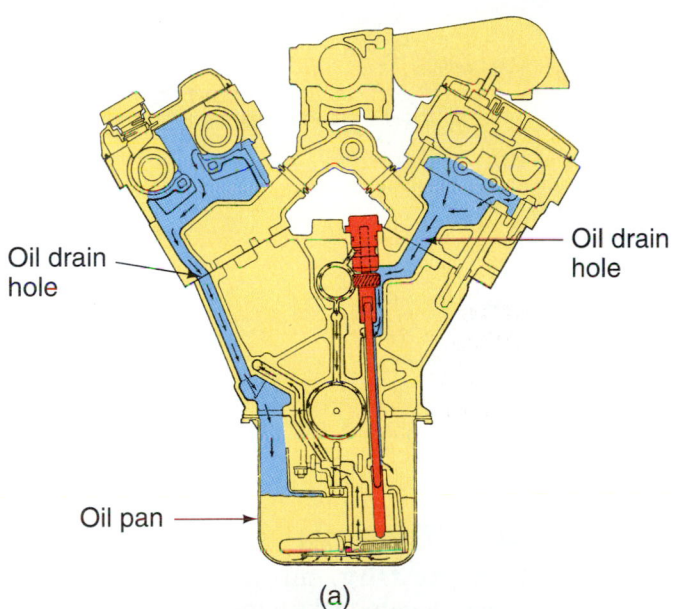

(a)

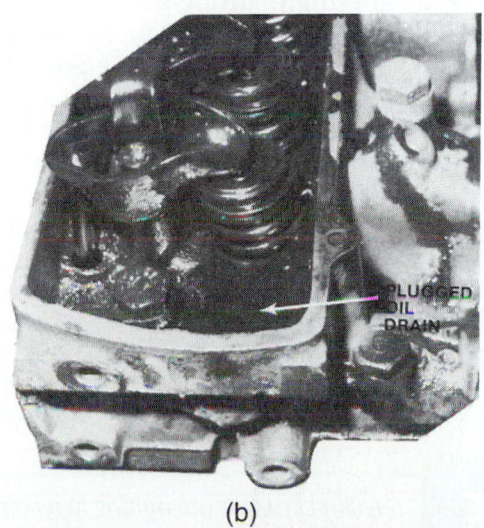

(b)

Figure 43.6 (a) The cylinder head has holes that allow oil to drain back to the oil pan. (b) Plugged oil drainback holes caused this engine to smoke. *(a, Courtesy of General Motors Corporation, Service Technology Group)*

maintained engine that will soon require major service. The oil stays up in the valve cover area instead of returning to the crankcase. It floods the valve guide, making the valve stem seal ineffective.

Oil Consumption from a Leaking V-type Intake Manifold Gasket

Intake manifold vacuum can draw oil into the intake ports from the lifter valley area under the intake manifold (Figure 43.7). This is a tough problem to find. A cranking vacuum check or a propane test can be performed to check for internal air leaks before the engine is disassembled. The procedure is covered in Chapter 42. When removing the manifold, always inspect visually for the possibility of previous intake gasket leakage.

V-type engines equipped with an exhaust gas recirculation (EGR) valve (see Chapter 40) on the intake manifold often experience oil fouling of the spark plugs that are closest to the EGR valve. This is caused when the intake manifold warps or the manifold gasket fails. A replacement gasket designed for high temperature applications is available.

Oil Consumption Due to Crankcase Pressure

One possible reason for excessive oil leakage is a plugged PCV valve, which can cause pressure to build up in the crankcase. This pushes on the gaskets and seals, allowing oil to leak out. Operation of the positive crankcase ventilation (PCV) system is covered in Chapter 40. If oil is leaking from the breather hole of a mechanical fuel pump, be sure to inspect for excessive crankcase pressure. Crankcase pressure can result in increased internal oil consumption, too.

Oil Consumption Due to a Faulty Mechanical Fuel Pump

Engines with carburetors usually have mechanical fuel pumps. A faulty mechanical fuel pump can be responsible for oil consumption as well as poor fuel economy. A break in the fuel pump diaphragm can cause fuel to dilute the oil in the crankcase, making it thinner (Figure 43.8). Thinner oil is more easily consumed by the engine.

■ OIL LEAK TESTING

Oil that leaks through gaskets and seals is a common cause of oil consumption.

NOTE: *A loss of one drop of oil every 30 feet results in a loss of about three quarts of oil every 1000 miles.*

A rear main bearing seal leak can be identified by the presence of oil on the engine side of the flywheel or torque converter. Oil on the transmission side of the torque converter indicates a leakage of a front transmission seal.

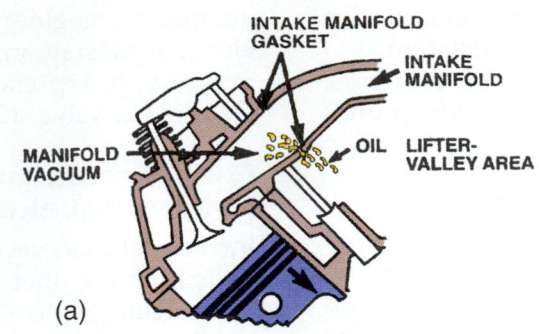

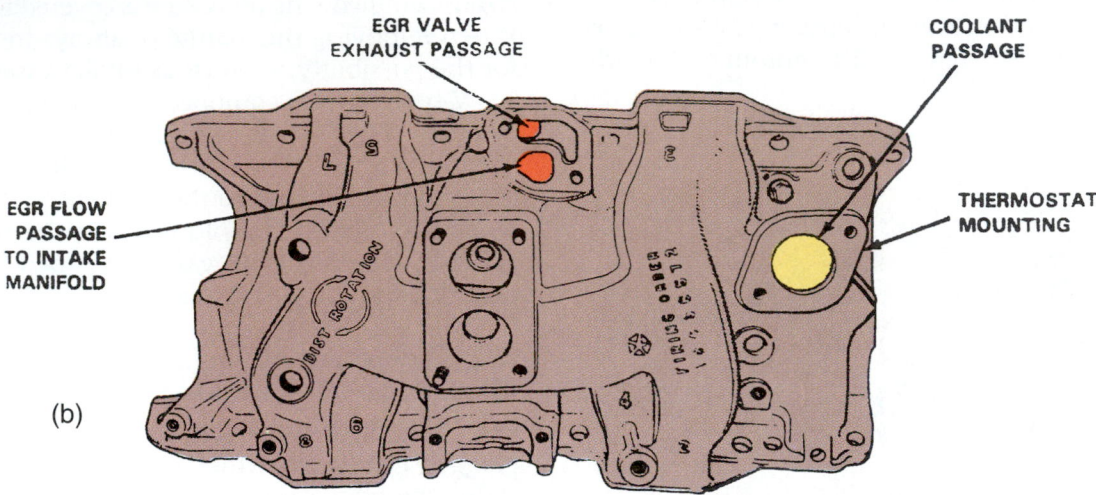

Figure 43.7 (a) Oil can be drawn into the intake manifold past a faulty manifold gasket. (b) The EGR valve passage in the intake manifold carries hot exhaust gases. *(Courtesy of Chrysler Corporation)*

A mechanical fuel pump has an oil seal above its diaphragm spring (see Figure 43.8). If oil leaks past this seal, it can result in an external leak as it comes out of the vent hole in the dry chamber above the diaphragm. The problem will be worse if crankcase pressure is excessive.

If the engine's oil pressure sending unit leaks, a great deal of oil can be lost in a short amount of time. If the sending unit shows any signs of seeping oil, it should be replaced as soon as possible.

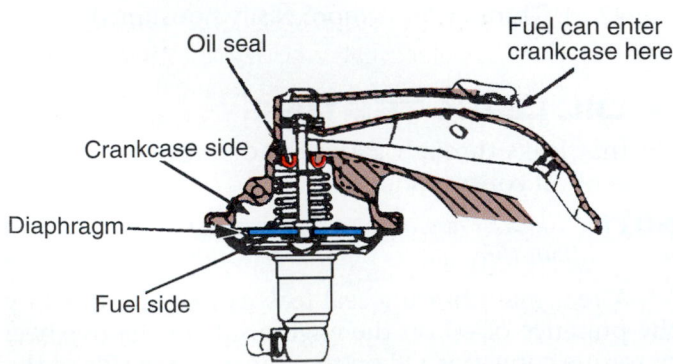

Figure 43.8 A broken fuel pump diaphragm can cause dilution of oil in the crankcase.

Black Light Testing

In **black light testing**, an additive is added to engine oil to help locate leaks. One product shows up under a black light, highlighting the source of the leak in bright yellow streaks (Figure 43.9). A 1-ounce bottle of dye is added to the engine oil. A very powerful black light is used so the leak will be apparent even under normal shop lighting. A mirror can be used to bounce the black light into hard-to-see areas. Washing the block first is helpful, but not necessary.

Sometimes when a leak is minor, it does not show up after just a short time, so the car might need to be driven for a day or so. After repairing the leak, the engine is cleaned and rechecked with the black light. The fluorescent dye stays in the oil. The dye is not harmful and the manufacturer says that it dissipates within 300 miles of driving.

CASE HISTORY *A technician was attempting to repair an oil leak on a dual overhead cam six-cylinder engine. He replaced the valve cover gasket but the leak continued and the unhappy customer returned to the shop. After pouring some leak*

detection fluid into the crankcase and taking the car on a test drive, he aimed the black light at the engine. Bright streaks of yellow oil were visible coming from behind one of the camshaft sprockets. He removed the valve cover, timing belt, and sprocket, and replaced the circular rubber seal behind the timing sprocket (see Figure 46.43).

There are special formulations of leak detection fluid for automatic transmissions, fuel systems, air conditioning systems, and power steering leaks.

To check for internal oil consumption, a **spark plug deposit test** can be performed. During this test:
- The engine's spark timing is retarded to 10° after top dead center (ATDC).
- The engine is allowed to idle at 500 rpm for 15 minutes.

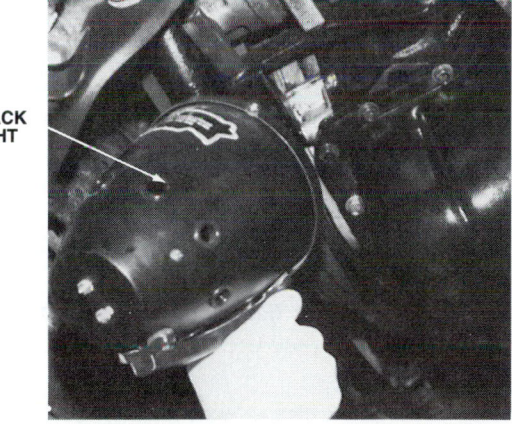

BLACK LIGHT

Figure 43.9 A black light oil leak detector.

(a) (b)

(c)

Figure 43.10 Spark plug deposits. (a) Normal deposits. (b) Oil wetting deposits. (c) Oil ash deposits. *(Courtesy of Champion Spark Plug)*

- Next, the spark plugs are examined for oil wetting.
- The engine should not be run above idle speed until the timing is properly reset. Retarded timing can cause exhaust valves to run dangerously hot.

Spark plugs can be a good indication of what is taking place in a cylinder. Figure 43.10 shows normal and abnormal spark plug conditions associated with oil consumption.

Miscellaneous Oil Leak Detection Methods

Another product used to help determine the source of a leak is red dye that is added to oil. Yet another way of checking for leaks is to spray foot powder on a cleaned off area. When the oil leaks again it will stain the white powder. Refer to Chapter 45 for more information on gaskets and seals.

ENGINE PERFORMANCE AND COMPRESSION LOSS

Compression loss can be a result of such things as a blown head gasket, burned valves, or broken piston rings. Engine compression testing and mechanical problems that result in compression loss are covered in detail in Chapter 42.

ENGINE NOISES

It is important to try to determine the location of noises before disassembling the engine. There have been cases where engines have been disassembled, inspected, and rebuilt, but when reinstalled in the car still had the same problem. Pinpointing the origin of the noise might have resulted in more careful scrutiny of the offending part while the engine was apart.

Noises are often transmitted from their origins to other locations, and can be difficult to isolate. Listening through a stethoscope (Figure 43.11), or listening at the end of a large screwdriver, a piece of hose, or a long wooden dowel are some of the ways to help pinpoint noises.

CAUTION Both the stethoscope and screwdriver are electrical conductors. They should be used with caution around electrical connections such as those located on the back of the alternator.

Accessory Noises

Accessories often cause noises that can be mistaken for other problems. Be sure to check alternators, smog pumps, air-conditioning compressors, and coolant pumps carefully.

Belts are a very common source of noise. If you suspect a belt of making noise, disconnect the belt and

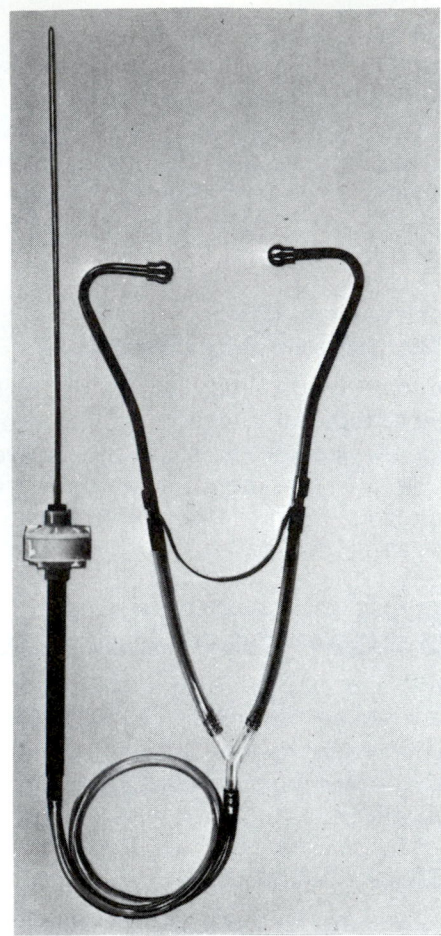

Figure 43.11 A stethoscope. *(Courtesy of Lisle Corporation)*

run the engine for a short time, or spray soapy water on the belt while the engine runs.

A fan clutch on the coolant pump can cause a noise that sounds serious and is hard to pinpoint. A stethoscope cannot be used to listen to the noise because the fan is spinning when the engine runs. Sometimes, the extra looseness (bearing play) in the fan clutch can be felt when the engine is off.

> **SHOP TIP** To isolate noises from defective belt driven parts or accessories, unhook the fan belt and run the engine at idle for a short time.

■ ENGINE KNOCKS

Crank Noises

Engine knocks such as crankshaft noises can be caused by a variety of things. They are generally deeper in pitch than other engine noises. It is important to isolate the source of the noises so that an accurate diagnosis can be made.

Front Main Bearing Knock. Excessive front main bearing clearance results in a heavy knock when the engine warms up. The knock is generally most pronounced at 1500–2500 rpm. Loosening accessory belts will often reduce the intensity of the knock. Upper front main bearing wear can result from accessory belts that are adjusted too tightly.

Thrust Bearing Knock. The crankshaft surface that controls end thrust can be worn, allowing the crank to move back and forth. End thrust is movement of the crankshaft in a forward and backward direction. Excessive end thrust will cause a clunk to happen when the vehicle leaves a stop sign. This is usually more pronounced on vehicles with standard transmissions.

Rod Knock. Excessive clearance at a connecting rod journal commonly results in a rod knock. During a cylinder power balance test, the intensity of the knock will diminish or disappear altogether as the offending cylinder's spark plug is grounded out. This condition is sometimes accompanied by low oil pressure, especially at idle.

Related Noises. A loose flywheel, torque converter, or vibration damper can cause a very serious sounding noise. Torque converter flex plates sometimes crack, causing a serious sounding knock. To test for a cracked flex plate:
- Run the engine at about 2000 rpm.
- Turn the key off and then on, listening for a clunk as the engine restarts.
- Shut off the engine and use an inspection mirror (Figure 43.12) to try to see the crack in the flex plate.

NOTE: *Shining a flashlight into the mirror will bounce the light onto the flex plate so you will be able to look at hard-to-see locations.*

Figure 43.13 shows cracks in a flex plate made visible by magnetic crack inspection. Magnetic crack inspection is covered in Chapter 46.

Bent Oil Pan. A bent oil pan can cause a deep knocking sound when the connecting rod hits the dented spot. Complete engines can weigh several hundred pounds. It is not uncommon for engine sheet metal, such as an oil pan or valve cover, to become dented when the engine is removed or replaced in the vehicle.

> **CASE HISTORY** *A student did a complete rebuild of an inline six-cylinder engine. After installing the engine and priming the lubrication system, he started it up. The engine had a very serious knock that could be heard with a stethoscope as originating from the oil pan. The student was very discouraged and wanted to remove the engine from the vehicle to repair the problem. The instructor noticed an indentation in the oil pan and suggested that the student remove the pan to inspect it. Examination of the inside of the oil pan showed a shiny spot where the crankshaft had been hitting it. Knocking out the dent fixed the problem.*

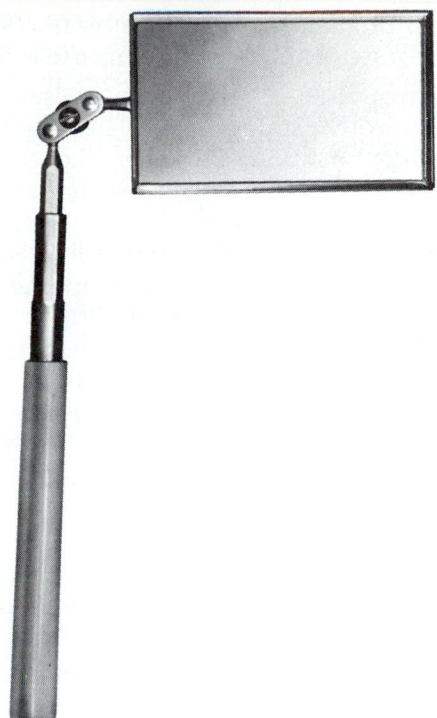

Figure 43.12 An inspection mirror. This one telescopes and has a replaceable mirror on a pivoting head. *(Courtesy of K-D Tools)*

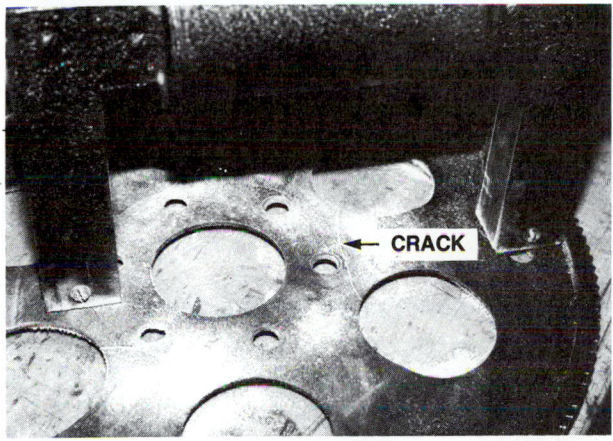

Figure 43.13 A cracked flex plate.

Rod Side Clearance. Excessive clearance on the sides of connecting rods can sometimes cause a ticking sound that resembles a valve train noise, but coming from the crankcase. The noise can go away during a power balance test. Spark knock (see Chapter 25) can intensify noise caused by excessive rod side clearance.

Piston Noises

There are several different types of piston noise resulting from cracked pistons, piston slap, excess piston pin clearance, and other causes.

■ Retarding the spark timing will generally reduce the intensity of a piston noise.

■ Grounding out the plug during the power balance test can *increase* a piston noise. This is the opposite result from the way a bad connecting rod reacts to ignition grounding.

Cracked Pistons. Cracked pistons are often the source of noise (Figure 43.14).

■ The noise is sometimes higher pitched than a crank related noise and could be confused with a valve train noise, except that it occurs at a faster rate than a valve noise.

■ Cracked pistons are often the result of a broken timing chain or improper valve timing, which can allow a valve to strike a piston on some engines.

Piston Slap. **Piston slap** is caused by excessive clearance between the piston skirt and the cylinder wall.

■ Sometimes the noise gets louder during acceleration, often clearing up when the engine warms up.

■ Piston slap at TDC and BDC causes oval wear as the piston rings scrub the sides of the cylinder.

■ Look for worn or collapsed piston skirts when excess oval cylinder wall wear is found.

Piston Pin Noise. Noise from excessive piston pin clearance makes a "double click" sound at idle or fast idle.

■ Pin noise often becomes more intense after the installation of new piston rings. The noise will gradually become less as the rings wear.

■ With the engine running at the speed where the most noise occurs, grounding the plug wire will sometimes increase the noise even more. The noise level might not increase, but it will not become less. Piston inertia causes the noise, which is why the noise does not go away when the spark plug is shorted out.

■ The noise usually becomes less or goes away when the engine warms up.

Other Piston Sounds. Another piston noise is caused by a *broken ring*, which can rattle during acceleration.

Figure 43.14 A cracked piston.

A cylinder ring ridge (see Chapter 47) that is not removed when new rings are installed can cause a clicking sound because the square edge of the new ring hits against the rounded edge of the ring ridge (see Figure 47.13). This can also force a ring land down on the second ring. The result is a stuck second compression ring or a broken ring or ring land.

Valve Train Noises

Valve train noises make a loud "ticking" sound and are the most common of engine noises. A finger placed on the valve spring retainer while the engine idles will feel the shock each time the loose valve hits its seat.

NOTE: *Valve train noises occur at half the speed of engine crankshaft rpm. This is because the cam only turns once for every two turns of the crankshaft.*

Noise from excessive valve guide-to-stem clearance can be pinpointed with the engine running. Squirt oil on the suspected guide to take up the clearance and stop the noise.

Sticking Valve. A sticking intake valve often results in a popping noise through the carburetor as burning gases escape into the intake manifold past the leaking intake valve. To find a sticking valve on a pushrod engine, hook up a timing light to one spark plug wire at a time with the valve cover off. The strobe action of the light will catch the offending valve in the open position.

Worn or Flat Cam Lobe. A similar situation arises when an engine has a smooth idle but runs rough under acceleration, "popping back" through the intake valve. This can often be traced to an exhaust lobe on the cam that has "gone flat," resulting in a **flat cam**. Pressure builds up on the power stroke and cannot escape during the exhaust stroke, so it goes back up the intake port. The lobe can go bad fairly quickly once it starts to wear, so the condition seems to occur overnight.

The pop back can be more severe with port fuel injected engines because each individual injector is near an intake valve port. The drop in suction from the offending cylinder does not cut the flow of fuel as it would in a carbureted engine or an engine with throttle body injection.

> **SHOP TIP** If you hold your hand over the carburetor of an engine with a bad cam, it might get wet with fuel.

Noisy Fuel Pump. A defective mechanical fuel pump makes a noise resembling a bad hydraulic lifter. The noise is loudest at idle speed when little fuel is being used by the engine.

Rocker Arms. Lack of lubrication to rocker arms can cause a loud, "squeaky" sound in some older engines. This problem is not as common as it used to be, partly because of the better oils in use today.

Timing Components. Valve train noises can also come from inside the timing cover. They can be caused by a bad timing chain or a loose sprocket or gear. The noise usually is a rattle or knock that becomes louder when decelerating. For engines with a timing chain tensioner, a worn chain can become loose enough on the sprocket to rattle whenever the engine floats or cruises between load and coast conditions. In severe cases, a chain can actually wear a hole in the timing cover resulting in an oil leak (Figure 43.15). Severely worn cam bearings can also be the cause of excessive timing chain slack. Depending on the design of the lubrication system, this problem can also be accompanied by low oil pressure at idle.

Lifter Noises

A very common valve noise is caused by a noisy lifter. Sometimes this occurs when the engine is first started because a lifter has lost its oil while the engine was off. Whenever an engine is shut off, there will be some valves held open, putting spring pressure on lifters to bleed them down. When pressurized oil reaches the lifter, the noise goes away. If the noise goes away in less than 15 seconds, this is considered normal. Some other lifter noises and their causes are:

Intermittent Noise at Idle or Low Speed. This can often be traced to dirt in the lifter check valve or wear (see Chapter 46).

Noise at Idle That Goes Away at Higher Speeds. This usually indicates excessive wear between the lifter body and its plunger. This noise could also be caused by low oil pressure or too thin an oil.

Quiet at Idle but Noisy at High Speed. The oil could be full of air. This occurs when the oil level is so high that the crank whips the oil, filling it with air. It

WEAR CAUSED BY LOOSE TIMING CHAIN

Figure 43.15 A worn timing chain rubbed on this timing cover, causing noise and an oil leak.

could also happen because there is air leaking into the suction side of the oil pump.

Lifter Noise at All Engine Speeds

Lifter noise can be due to several things:
- Dirt or varnish buildup inside the lifter

> **SHOP TIP** Lifters that stick because of varnish buildup can sometimes be loosened up by squirting carburetor spray cleaner down the hollow pushrod lubricating channel, where it contacts the rocker arm.

- Worn parts such as worn rocker arms or a cam lobe that is going flat
- Insufficient oil supply
- Oil too thin
- Oil pressure too low

Spark Knock Noise

Abnormal combustion can cause engine noise. Sometimes it can be serious enough that engine damage can result. Spark knock can be a result of carbon buildup, incorrect ignition timing, fuel of too low an octane, an engine with a compression ratio that is too high, or cooling system problems. Chapter 25 deals with abnormal combustion in detail.

Excessive Carbon Buildup. In a car with relatively low mileage, carbon buildup in the combustion chambers can cause an increase in the compression ratio (Figure 43.16). This problem is not as common since the introduction of unleaded gas, which leaves fewer deposits, but some late model engines develop carbon problems in less than 10,000 miles.

Broken Motor Mount

The engine is attached to the frame with a rubber and metal motor mount. A broken motor mount can cause vibrations, sticking throttle and shift linkages, torn radiator hoses, interference between the fan and radi-ator, and broken ground straps from the engine to the firewall or frame. Late model motor mounts are designed with safety provisions so that, although they may fail, the engine is restrained.

To check for a broken motor mount, have another person put the transmission in both forward and reverse ranges while keeping the vehicle braked. This will cause the engine to lift on one side and then on the other.

A broken ground strap (see Figure 44.4) can cause several problems that result as the electricity hunts for its shortest path to ground. These include an etched transmission bushing and drive shaft yoke, burned emergency brake cables, a burned carburetor return spring, flickering headlights, failure of the engine to start, burned front wheel bearings or CV-joints (on front wheel drive cars), and other seemingly mysterious conditions. Because the problem is not obvious (as it would be if the engine did not run, for instance), a defective ground strap is sometimes ignored.

> **SHOP TIP** When the engine runs and all lights and accessories are on, check to see if a voltage can be measured between the transmission housing and the frame. If so, attach a ground strap from it to the frame.

■ OIL PRESSURE PROBLEMS

Oil pressure can be too low or too high. Low oil pressure is far more common. Engines with high odometer mileage begin to use oil between oil changes. People who use self-serve gasoline stations tend to neglect to check the engine's oil level at each fill-up. Allowing the oil level to drop too low can cause major engine damage. Usually the first thing to happen is lower main bearing wear (Figure 43.17). After this happens, oil pressure will remain permanently low at idle speed.

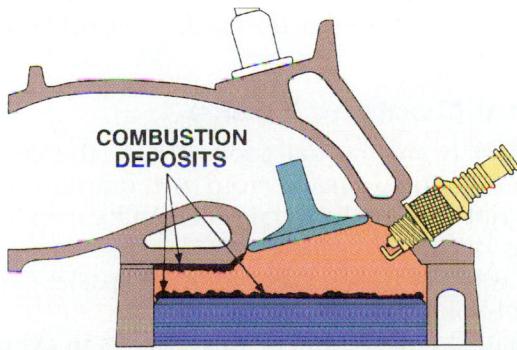

Figure 43.16 Carbon in the combustion chamber can raise the engine's compression ratio. *(Courtesy of Chrysler Corporation)*

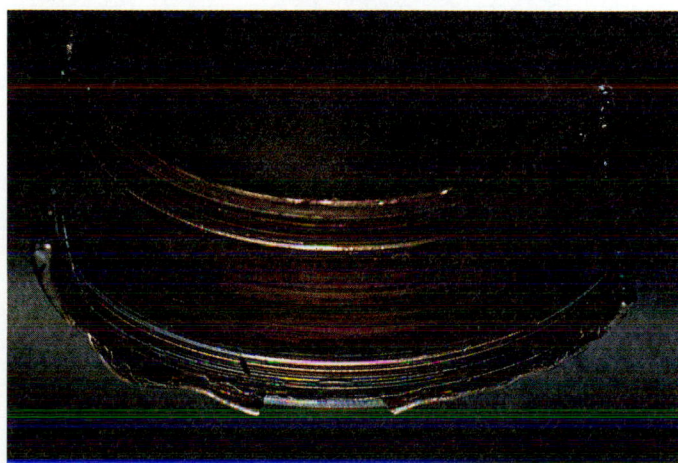

Figure 43.17 Lower main bearing wear resulting from lack of oil. *(Courtesy of AE Clevite Engine Parts)*

Because the pump turns faster when the engine rpm is increased, clearance problems do not usually result in low oil pressure at off-idle speeds.

Low Oil Pressure

Low oil pressure can ruin an otherwise good engine in a short time. Often the cause of a low oil pressure reading on an electric dash gauge or light is a faulty **oil pressure sending unit**.

■ Test the gauge by grounding the wire that leads to the sending unit on the block. When the wire is grounded with the key switch on, the gauge should show maximum oil pressure, or the light should go on. If either happens, the gauge and wiring are good and the sending unit is at fault.

■ Pressure can be tested by temporarily installing an oil pressure gauge in place of the sending unit (Figure 43.18). A special *oil sending unit socket* is used to remove the sender (Figure 43.19). Using pliers can damage the unit. A static oil clearance test can be performed using a pressure primer with the oil pan removed (see Figure 50.7).

The oil pump can also cause low oil pressure. It might be worn excessively or have a sticking relief valve.

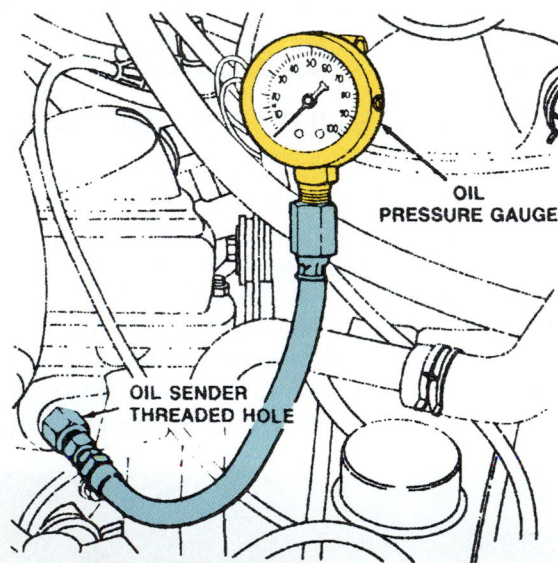

Figure 43.18 An oil pressure gauge is installed in the oil sending unit hole in the block. *(Courtesy of Chrysler Corporation)*

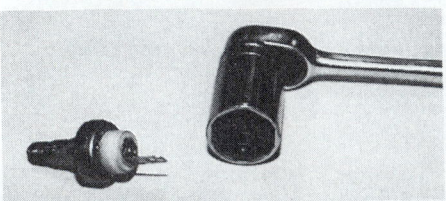

Figure 43.19 An oil pressure sending unit is removed with a special socket. *(Courtesy of Chrysler Corporation)*

The intake sump (oil pickup) can have an improperly located or partially plugged screen, or the pump body could be loose on the block (see Chapter 47).

High Oil Pressure

Occasionally, an engine can have oil pressure that is too high. This can result in a bursting oil filter or cause oil consumption or bearing material to be washed from the bearings. The problem can be caused by a stuck **pressure relief valve** in the oil pump or by a severe blockage in an oil gallery near the cam or crank (close to the oil pump). In either case, the oil pump relief valve would be unable to bypass enough oil at high speeds.

Oil Analysis

A common test done by fleets (large companies with many vehicles) is **oil analysis**. A sample of crankcase oil is sent to a laboratory to be tested. The results of this test can point to mechanical problems before they have become serious. Some of the things that are measured are:

■ Coolant or moisture (which can indicate internal coolant leaks or insufficient maintenance intervals)

■ Metals (which can indicate which part of an engine is wearing)

■ Dirt

◼ COOLING SYSTEM PROBLEMS

When a cooling system has been neglected, expensive engine damage often results. When a radiator has become plugged or its fins have become corroded and are flaking away, it cannot conduct enough heat away and the engine will overheat on the highway. Also, the water jackets in the engine can develop a buildup of minerals and scale, which prevents the transfer of heat to the cooling system. Material will continually flake off, resulting in a plugged radiator. Testing for cooling system problems is covered in detail in Chapter 21.

◼ INTERNAL ENGINE LEAKAGE

Internal leaks are those that are within the engine. They can be either combustion leaks, coolant leaks, or oil leaks.

Internal Coolant Leaks

When there is an internal coolant leak, the coolant level often drops. Leakage from the outside of the engine is not evident. Internal leaks can happen in the following locations:

■ In the water crossover passage of the intake manifold on V-type engines

■ In threaded plugs inside of valve covers in cylinder heads

■ In combustion areas, such as the head gasket

■ In a cracked head or block

A leaking thermal vacuum switch can cause a leak into the intake manifold when the engine is warm only. A vacuum operated heater valve can cause a similar problem.

Internal leaks are diagnosed using the block tester, pressure tester, or infrared analyzer. Water in the exhaust is commonly caused by condensation. Steam in the morning from the exhaust is a normal condition. Excessive water from the exhaust can be caused by a crack in the combustion chamber or a head gasket leak. A coolant leak in a combustion chamber will be evident when the cylinder head is removed from the engine. A normal combustion chamber will be coated with a small amount of carbon. A cylinder with the leak will not have any carbon in it (Figure 43.20). Checking for cracks after disassembly is covered in Chapter 47.

Cross Fluid Contamination

Water leaking into the crankcase will contaminate the oil. This is called **cross fluid contamination**. The condition will be very evident when a valve cover is removed (Figure 43.21). An internal water leak can occur when a loose timing chain wears a hole in the inside of a timing cover that has a water pump mounted to it. An additional problem that can occur with an internal leak is failure of a computerized fuel system's oxygen sensor. This happens when silicates that are a part of the additive package of some coolants coat the sensor. This is much the same as O_2 sensor failure caused by some silicone RTV sealants (see Chapter 45).

Figure 43.21 A blown head gasket caused the oil to take on this appearance.

Internal Oil to Coolant Leaks

When a leak occurs between an oil and water passageway, pressurized oil (approximately 30 psi) will force its way into the cooling system (approximately 15 psi). The engine will overheat and pour a messy oil and water mixture from the radiator overflow. This can also be a symptom of a leaking automatic transmission cooler (see Chapter 21). The transmission will also fill with coolant when the engine is turned off.

Internal leakage can be spotted by installing a pressure tester on the radiator filler neck of a warmed-up engine (see Figure 21.12). The pressure tester can also pinpoint the location of external leaks. Experienced technicians will not begin a repair until they are positive of the locations of leaks.

Exhaust Gas in the Coolant

A leaking head gasket will not always show up on a pressure test. A block check tester or an infrared exhaust analyzer can also be used to check to see if there is exhaust gas in the coolant (see Figures 21.19 and 21.20). Bubbles in the coolant can also indicate a leak.

▨ SEIZED ENGINE

A **seized engine** occurs when a starter motor will not crank the engine over and the engine cannot be turned over by hand. Loosen all fan belts. A frozen smog pump, power steering pump, water pump, or other belt-driven accessory can actually keep an engine from turning over (Figure 43.22). A drive belt on a frozen accessory can become so hot it melts. When the engine is shut off, it vulcanizes to the pulley, preventing the engine from rotating during the next morning's startup attempt.

If the engine will not turn in either direction with the belts loose, the cause could be coolant thermoplastic seizure (resulting from coolant mixing with engine oil) or other serious engine damage, such as

NO CARBON
DEPOSITS ON
COMBUSTION CHAMBER

NO CARBON
DEPOSITS
ON PISTON

Figure 43.20 A cracked combustion chamber or leaking head gasket that resulted in a coolant leak will remove all carbon from one piston and combustion chamber.

Figure 43.22 This damaged starter motor was the cause of a "frozen" engine.

piston seizure, seized bearings, a broken crank, or a seized rod or valve.

Sometimes pressurized coolant from the radiator fills a cylinder after an engine is shut off. If both of that cylinder's valves are closed, the engine will not be able to turn completely over when cranked. This is called hydrolock (see Chapter 21). If the spark plugs are removed, the engine will be able to crank. Water will pour out of the offending plug hole.

> ## CASE HISTORY
>
> *A technician bought an old pickup with the intention of rebuilding the engine. The previous owner said that the engine was seized up (wouldn't turn over). After towing the vehicle to his garage, he began to diagnose the engine before disassembling it. He discovered that the engine wasn't really seized but was hydrolocked. He knew that head gaskets on long inline six cylinders are more likely to fail in the middle, so he took out the center two spark plugs and cranked the engine. After water came out the spark plug holes, he was able to replace the spark plugs and run the engine. Surfacing the cylinder head, replacing the head gasket, and changing the oil were the only things required to put the truck in good working order.*

> ## CAUTION
>
> If the vehicle has an electric fuel pump, *be sure the liquid in the cylinder is not gasoline before cranking the engine.* A fire could result. See the related case history in Chapter 21.

◼ ELECTRONIC FAILURES/ ENGINE DAMAGE

Engine damage such as burned valves, scuffed pistons, worn bearings, and damaged cylinder heads can sometimes be traced to electronic component failures. Today's fuel and emission control systems are computer controlled. Computers receive input from various engine sensors.

An EGR valve that is controlled by a computer can become inoperative if two of its input sensor signals are interrupted. When the EGR valve does not operate properly, detonation can be the result (see Chapter 25).

A failure of an electric cooling fan can be due to an inoperative sensor.

An overly rich air/fuel mixture because of a failed sensor or electronic part can cause oil dilution, resulting in piston or crankshaft bearing damage. A tip-off here is that the catalytic converter in the vehicle's exhaust system may overheat and melt, causing exhaust back pressure. Always trace a problem to its root cause.

◼ ENGINE PERFORMANCE AND FUEL MIXTURE PROBLEMS

Emission control and fuel system malfunctions can sometimes mimic problems related to the engine. Sometimes important items are neglected during an engine job. A fuel or emission problem that caused an engine problem will probably cause it to happen again if it is not repaired. Larger engine shops often employ a specialist capable of diagnosing these complicated problems.

An air/fuel mixture that is too *lean* (too much air/too little fuel) can increase heat in the combustion chamber and result in detonation and/or burned internal engine parts. An overly *rich* mixture (too much fuel/too little air) can cause **oil wash** (when oil is washed from cylinder walls resulting in cylinder wall wear). Leaking fuel injectors can be a cause of cylinder wall oil wash. Intake valve deposits, which will affect engine idle and emissions, can also result.

◼ REVIEW QUESTIONS

1. Carbon has built up in the neck area of a valve. What is the probable cause?

2. If a spark plug has a carbon deposit on only one side, what is a probable cause?

3. What is one problem that can result from both of the following: worn rod bearings and high-speed driving?

4. At the end of a spark plug deposit test, what must be done before the engine is accelerated off idle?

5. A noise goes away when the spark plug wire for one cylinder is grounded out. What could this noise be?

6. In the above condition, the noise becomes *louder* when the cylinder is grounded out. What could this be?

7. What is a common cause of a cracked piston?

8. When a cylinder is worn excessively oval, what should be looked for?

9. When an engine runs for a long period with an excessively lean air/fuel mixture, what type of engine damage can result?

10. What possible problem can be pinpointed by looking at the tops of these pistons (Figure 43.23)?

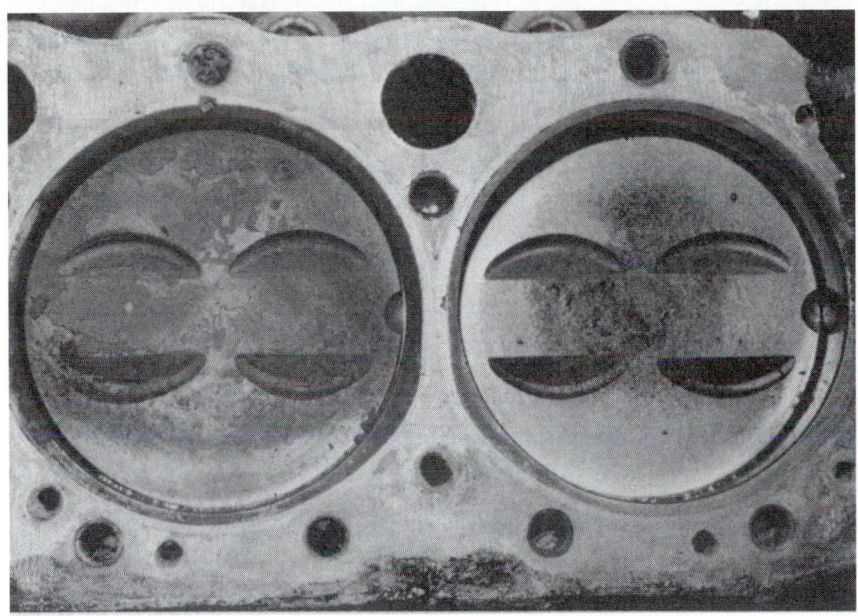

Figure 43.23   What could have caused this?

ASE STYLE REVIEW QUESTIONS

1. A car with 40,000 miles on it is using oil. Technician A says that the most probable cause is bad valve guide seals. Technician B says that the cause is probably piston rings. Who is right?

  a. Technician A   **b.** Technician B

  c. Both A and B   **d.** Neither A nor B

2. Technician A says that if valve guide seals need to be replaced, the cylinder heads must be removed from the engine. Technician B says an engine that has worn lower main bearings will have lower oil pressure at freeway speeds. Who is right?

  a. Technician A   **b.** Technician B

  c. Both A and B   **d.** Neither A nor B

3. A noise occurs at one-half engine rpm. Technician A says this could be a bad lifter. Technician B says this could be a valve that needs to be adjusted. Who is right?

  a. Technician A   **b.** Technician B

  c. Both A and B   **d.** Neither A nor B

4. An engine has a crack between a water jacket and an oil gallery. Technician A says that oil will go into the cooling system when the engine is running. Technician B says that coolant will go into the oil when the engine is off. Who is right?

  a. Technician A   **b.** Technician B

  c. Both A and B   **d.** Neither A nor B

5. A car has oil smoke from its exhaust during deceleration. Technician A says that bad valve guide seals could be the cause. Technician B says that worn piston rings could be the cause. Who is right?

  a. Technician A   **b.** Technician B

  c. Both A and B   **d.** Neither A nor B

Automotive Engine Service and Repair

THEN AND NOW: GASKETS

Man has been presented with the challenge of sealing things up since prehistoric times. Early examples include stuffing mud and leaves into the gaps between the logs of a hut and, according to the Bible, Noah smearing pitch on the hull of the Ark. With the coming of the industrial revolution in the mid-1700s, sealing jobs became more precise. But the basic idea was still to fill up the imperfections in mated surfaces, thus preventing leakage through the seam.

It was not until the beginning of the automotive age, over a century ago, that this challenge began to be faced often by so many people. Engineers and mechanics used all kinds of materials and substances to contain and separate engine oil, coolant, vacuum, and compression. Early vehicles used many different types of sealing materials, including soft metals, leather, rubber, paper, and even beef fat. It was a tough fight, however, and leaks of every kind were a common type of mechanical failure.

The engine gasket with the most difficult job to do is the head gasket, located between the head and the cylinder block. Sealing this area was a very big problem in early engines. Some makers eliminated this seam altogether by casting the head and block in one piece (not a practical arrangement from a service standpoint).

In the old days, mechanics commonly made gaskets by laying a piece of paper-like material against the casting. Then they would tap along the edges with the rounded end of a ball peen hammer, which would cut the material.

Early on, bolts were tightened by feel. This method of tightening could only be learned through experience, overstretching bolts and sometimes breaking them off. The tool that solved this problem was the torque wrench. It told the mechanic exactly how many foot-pounds of twisting power was being exerted on the fastener. A recent tightening method, called "torque turn" or "angle torquing" reduces the effect of friction on the torque measurement. After tightening the bolt a small amount with a torque wrench, the bolt is further tightened a specified number of degrees.

Now, gaskets have to face higher temperatures, increased pressures, fewer and thinner bolts, and the differing expansion rates in aluminum head/iron block engines. Gaskets have become highly engineered (and expensive) parts. A head gasket might be constructed of heat-proof graphite on an expanded core or a multilayer steel gasket might be an engineer's choice to stabilize the joint. Cork/rubber is still being used for valve cover or oil pan seams, but sometimes molded silicone rubber is the material of choice.

Cutting a gasket with a ball peen. *(Courtesy of Bob Freudenberger)*

A modern head gasket. *(Courtesy of Fel-Pro)*

Engine Removal and Disassembly

■ INTRODUCTION

This chapter deals with removal and disassembly of the engine. It is important that procedures be followed carefully. It is very easy to damage the car body or paint when removing the engine. The engine is also very heavy and your personal safety is at stake.

Engine parts must be removed and inspected in an orderly manner. Signs of wear noticed during engine disassembly can be clues to problem areas. The correct repair can prevent the problem from occurring again.

Be sure to consult the applicable repair manual before beginning the repair job. Procedures differ slightly between manufacturers.

■ ENGINE REMOVAL

Install fender covers on both fenders and over the grill to protect the vehicle's paint. Some shops like to protect the windshield with a piece of corrugated cardboard.

Battery Cables

Disconnect the battery cables. The ground cable (usually the negative) should be disconnected first. This is a good habit to follow whenever doing major repairs. Before disconnecting the battery, make a note of which radio stations the customer has set on the radio so they can be reprogrammed after the engine is reinstalled.

The outside of the battery is often covered with a thin film of battery acid. After handling the battery, be sure to clean your hands. Battery acid can ruin clothing that you touch later.

NOTE: *Sometimes, there is a small ground wire that is connected from the battery terminal to the fenderwell. Be sure to remove the battery terminal connection from cars that have this or a complete circuit could still be present through the ground strap from the block to the bulkhead or firewall (panel that separates the engine compartment from the passenger compartment).*

Hood

Remove the hood before working on the engine. Before loosening the hood bolts, mark the location of the hood to the hood hinges so that it can be properly reinstalled. One effective method is to use a pencil to mark the exact hood positions with an outline.

Be careful not to damage the paint on the hood. Set it down on a fender cover or cardboard.

Label Vacuum Lines

A carburetor can have as many as fourteen hoses connected to it. Later models usually have some of the hoses grouped together with male and female connectors on the ends to make it less likely to make an error in a hose connection. Sometimes the hoses are numbered, color coded, or are different sizes.

Technicians sometimes have several narrow rolls of different colors of tape to label hoses when the hoses are unmarked. Put a piece of one color on the end of the vacuum hose, and another piece of the same color on the connection the hose was removed from to make reassembly simple.

Late model vehicles often have a vacuum diagram label under the hood. When there is not, draw a map of the vacuum hoses. Assign a number to each hose on the map. Then use masking tape and a pen to number each line as it is removed. Rolls of numbered tape are available from automotive tool suppliers. Electrical supply houses also sell books of numbered tape, which electricians use to keep the wires in order (Figure 44.1).

Figure 44.1 Tape is numbered to label matching vacuum and electrical lines.

NOTE: *Later-model cars are usually equipped with dedicated electrical and vacuum connectors that can only be attached to their counterparts. Even if this is true on the vehicle you are working on, label the connection with information as to which accessory it connects to.*

Drain Coolant and Oil

Drain all coolant from the radiator and block. If the coolant will not be reused, be sure to comply with local regulations for its disposal. If the block is equipped with a coolant drain plug (Figure 44.2), the engine block should also be drained.

Drain engine oil and remove the oil filter. The oil filter is made from thin sheet metal that is easily crushed or torn if the filter wrench is not held as close to the filter base as possible (see Chapter 12).

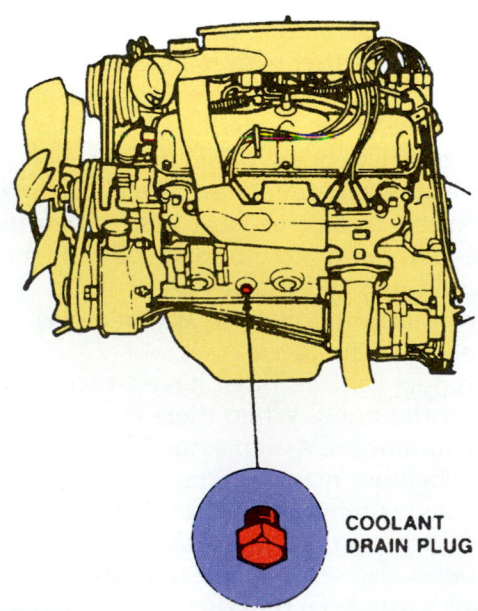

COOLANT DRAIN PLUG

Figure 44.2 A coolant drain plug in a cylinder block. *(Courtesy of Chrysler Corporation)*

Radiator

Remove the radiator. Disconnect the radiator hoses from the engine. If the hoses are left attached to the radiator, they do not need to be marked for reassembly. Remove the radiator from the car. It can be stored in the vehicle's trunk with its hoses facing upward and the radiator cap installed so any remaining coolant does not leak out. The radiator is fragile so be sure that other parts are not stored on it.

If the car has an automatic transmission, it is probably equipped with a heat exchanger (cooler) on the bottom or side of the radiator (see Chapter 20). The two lines leading to the radiator from the transmission must be removed and plugged. Use a flare nut (tubing) and open end wrench to disconnect the line (see Figure 7.7b).

SHOP TIP A short length of hose plugged at one end with a bolt can be installed over a disconnected line to keep fluid from leaking from it.

NOTE: *A broken or damaged line can be repaired using a union (see Chapter 24).*

Distributor and Spark Plug Wires

Remove the distributor and spark plug wires. It is a good idea to remove the distributor and spark plug wiring before removing the engine to prevent damage caused by interference with the lifting sling.

Do not remove the spark plug wires from the distributor cap. They are already in the correct firing order and need not be disturbed. Mark the location of the number one plug wire on the distributor cap before removing any wires. If the wires are to be replaced, replace them one at a time so they can be measured easily and kept in order.

Alternator, Fan, and Accessory Drives

Remove the alternator. If possible, leave the wires hooked to the alternator; simply unbolt it and use wire to fasten it to something out of the way. Use masking tape to label any electrical wiring that must be disconnected. Taking the time to do this will save the time that would be wasted in trying to determine wires during reassembly.

The fan blades, belts, and smog pump can also be removed before removing the engine from the car to prevent possible damage.

If the engine is to be cleaned before removal, protect the alternator. Some soaps can damage alternator bearings. After engine cleaning, be sure to clearly label the electrical connections to the starter and sending units for oil pressure and coolant temperature (Figure 44.3).

Figure 44.3 Label the wires to the sending units.

Heater Hoses and Ground Strap

Remove the heater hoses and ground strap. Some heaters have a control valve in one of the heater hoses. On some engines, it matters which way the heater hoses go. Label the hose that comes from the coolant pump so it can be reattached correctly later. Use care when removing the heater hose. Cut the hoses and replace them if necessary to avoid damaging the heater core.

There is usually a ground strap from the engine to the **firewall** (Figure 44.4). Be sure to disconnect it before the engine is removed.

SHOP TIP A missing or broken ground strap can cause a transmission bushing, wheel bearing, or emergency brake cable to burn or fail. This is because electricity follows the path of least resistance to ground, causing etching and heat. A bushing, bearing, or cable may become an alternate "ground" circuit if the strap is not attached.

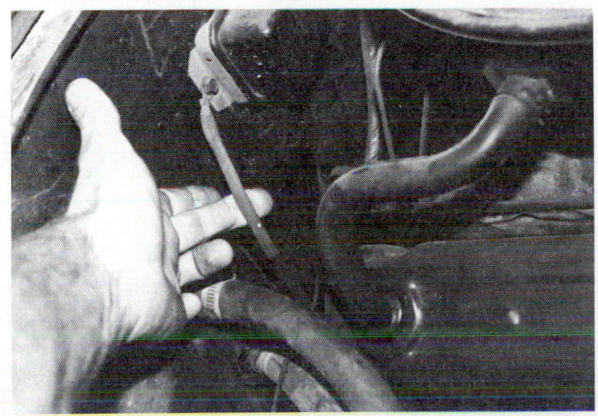

Figure 44.4 Be sure to remove the ground strap.

Cooling System Switches

Most engines use a coolant temperature-operated device to control vacuum to emission control devices. These switches, often called thermal vacuum valves (TVV), are very easily damaged and can be expensive to replace. Use caution when working around these devices. If it is necessary to remove them, use a special socket (Figure 44.5).

Carburetor and Intake Manifold

Remove the carburetor and intake manifold. If the carburetor will be removed, use a flare-nut wrench and an open end wrench to remove the fuel line (see Figure 7.7b). Failure to use both wrenches can result in a ruined fuel line.

Remove the throttle linkage or cable. Then remove the carburetor, valve covers, and motor mount bolts. Before the engine is removed from the car, the intake manifold can also be removed. Store the carburetor in an upright position. Do not turn it over unless it is going to be rebuilt.

Accessory Brackets and Accessories

Mark accessory brackets and remove accessories. Any accessory brackets (such as air conditioning) that are attached to the head or block may be removed. If the vehicle has many accessories, label the brackets to show their location on the head or block.

SHOP TIP Some accessories have multiple-piece brackets. After removal, reassemble them to the accessory so you will not forget how to reassemble them later. Repair manuals do not usually show bracket locations.

Because of the unusual lengths of many accessory bracket bolts, it is a good idea to put them in labeled bags for easier reassembly. An air conditioning compressor bolt that is too long can damage an expensive compressor housing. When there are bolts of several

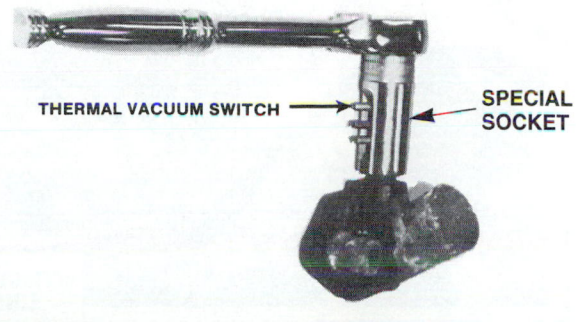

THERMAL VACUUM SWITCH   SPECIAL SOCKET

Figure 44.5 A special thermal vacuum switch socket.

different lengths, it saves time to assemble each one back into its accessory bracket as it is removed.

CAUTION *Do not unhook air conditioning hoses* as they contain pressurized refrigerant. If a refrigerant line must be disconnected, *use extreme caution and wear eye protection.* Refrigerant vaporizes so fast that it freezes anything it touches. It can cause blindness and turns to deadly phosgene gas (nerve gas) if allowed to contact an open flame or an extremely hot metal surface.

It is unlawful to release R12 refrigerant into the atmosphere. It must be reclaimed and can be recycled.

NOTE: *R12 (Freon) refrigerant, used in most vehicles until 1994, has been implicated in the depletion of the earth's ozone layer. In the Montreal Protocol of 1987, twenty-three countries (including the United States and Canada) signed an agreement setting limits on the production of ozone-depleting chemicals. Since 1992 the use of refrigerant recycling machines has been required by law. Persons using the equipment must be certified. Certification is available from several industry associations.*

The *air conditioning compressor* can usually be wired up out of the way with the lines still attached (Figure 44.6). When air conditioning lines must be disconnected, be sure to plug all openings immediately. Moisture must not be allowed to enter the system.

Remove the *power steering pump without* disconnecting the lines and wire it in a position so that fluid cannot leak out.

Exhaust Components

Remove exhaust components. Bolts holding the exhaust manifold and exhaust pipe are often rusted and will be difficult to remove and have a tendency to break. Spray penetrating oil on them and use a six-

point socket. See Chapter 6 for information on how to remove broken fasteners.

SHOP TIP
■ A rusted fastener can sometimes be tightened slightly to help penetrating oil get into the threads.
■ Rusted manifold bolts are easier to remove with an impact wrench (Figure 44.7).
■ Use a six-point socket to remove most engine items, especially exhaust bolts and nuts. A twelve-point socket will be more likely to slip, rounding off the corners of a nut.

A car with a computer controlled fuel system will use an oxygen sensor to determine the proper air/fuel mixture (Figure 44.8). Unhook the wire to the sensor. Leave the sensor in the exhaust system.

Fuel Line

Remove and plug the fuel line. The fuel line from the tank must be disconnected. To prevent fuel leakage

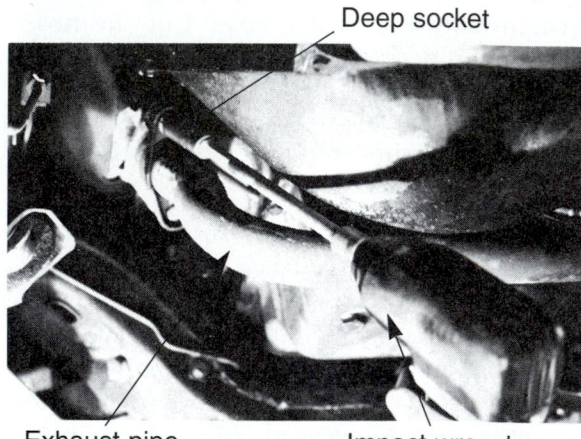

Figure 44.7 Use an impact wrench and impact socket to remove exhaust manifold-to-pipe bolts.

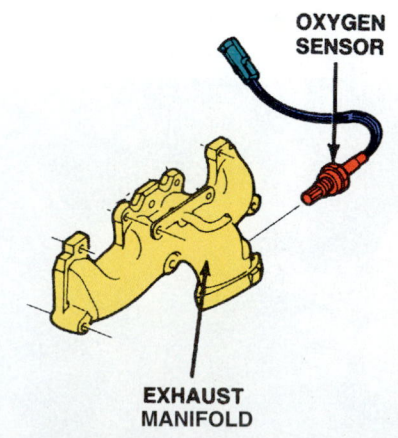

Figure 44.8 An oxygen sensor located in an exhaust manifold. *(Courtesy of General Motors Corporation, Service Technology Group)*

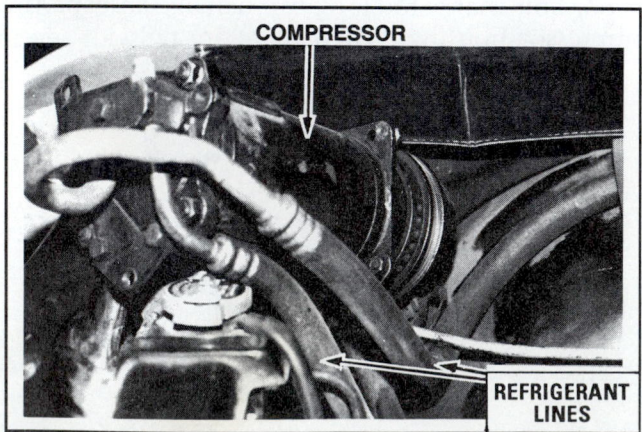

Figure 44.6 Wire the air conditioning compressor out of the way with both refrigerant lines still attached.

from the line, plug it with a bolt and hose clamp (Figure 44.9) or pinch pliers (Figure 44.10).

■ Even though fuel may not pour out of a fuel line when it is first disconnected, a hot day can cause expansion of the fuel in the tank. This can result in the fuel in the tank siphoning out all over the ground. Be sure to plug all fuel lines including those to the smog-control vapor canister. Also remember to loosen the fuel cap to prevent pressure from building in the fuel tank on a hot day.

■ Fuel injected engines have residual pressure in the fuel lines when the engine is off. Relieve this pressure before loosening a fuel line connection (see Chapter 27).

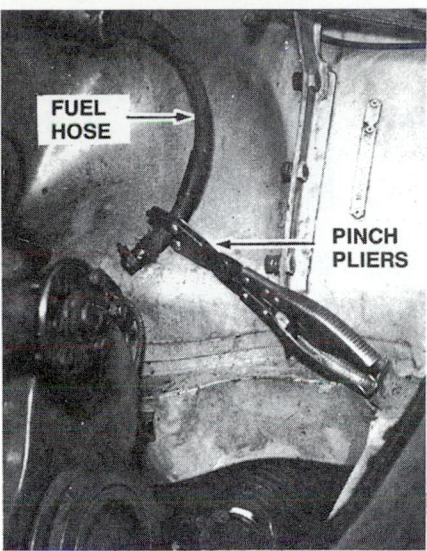

Figure 44.10 A fuel line closed off with pinch pliers.

Fuel Pump

Remove the fuel pump. Cars with carburetors usually have a mechanical fuel pump operated by an eccentric on the camshaft. Sometimes, it is better to remove the fuel pump before removing the engine from the car. Loosen the bolts that hold the fuel pump to the block. Then turn the engine until all spring tension is removed from the pump rocker arm. The cam eccentric will then be in its low position (see Figure 18.28), so the fuel pump can be easily removed.

Transmission

Determine whether the transmission must be removed with the engine. Before engine removal, check for recommendations in the repair manual. Front wheel

drive engines are often removed with the transmission. The axles are disconnected from the front wheels first. The procedure for removing front axles is covered in Chapter 71.

On rear wheel drive cars, usually it is easier to leave an automatic transmission in the car. If the transmission is to be removed, make center punch marks on the converter and the flexplate so that they can be correctly aligned on reassembly.

Remove the torque converter attaching bolts from the flywheel flexplate (Figure 44.11). To gain access to each bolt, rotate the engine by turning the crankshaft with a special flywheel turning tool, or by using a large socket and ratchet on the damper bolt at the front of the engine crankshaft (Figure 44.12).

SHOP TIP

■ If an impact wrench is used, the crankshaft will not have to be held to keep it from turning.

■ Be sure always to turn the engine in its direction of normal rotation. Some OHC engines will skip cam timing if they are turned backwards.

Pry the torque converter toward the transmission, away from the flexplate. Leave the converter installed in the front of the transmission during engine removal so transmission fluid will not pour out of it.

If the transmission dipstick tube is attached to the cylinder head, unbolt it from the head.

Remove the engine-to-transmission bolts. On rear wheel drive vehicles, these bolts are easily loosened from underneath the car using a very long extension and a universal socket (Figure 44.13).

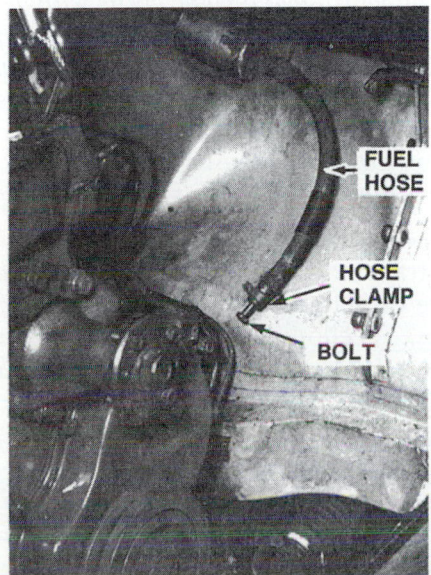

Figure 44.9 A fuel hose plugged with a bolt and hose clamp.

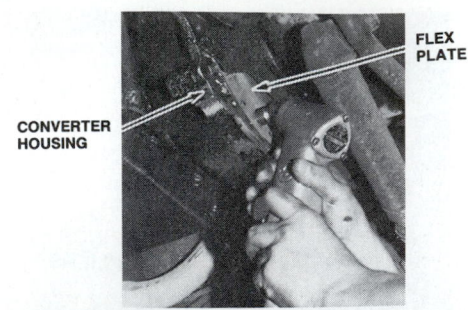

Figure 44.11 Use an impact wrench when removing torque converter bolts or nuts so that you will not have to hold the crank to keep it from turning.

Figure 44.12 Turning the crank using the damper bolt.

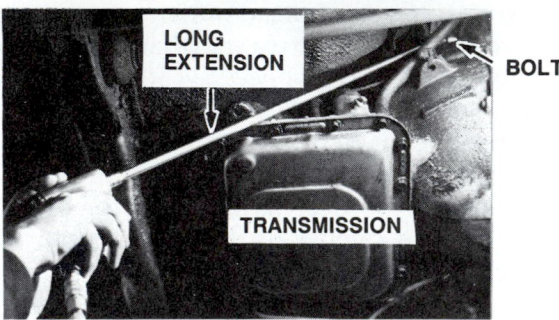

Figure 44.13 This tool setup is good for removing top transmission-to-engine bolts.

NOTE: *When a long extension is used, a larger impact wrench might be needed because some of the impact is absorbed by the long tool.*

Sometimes, it is necessary to unbolt the rear transmission crossmember and allow the rear of the transmission to drop. This provides easier access to the top engine-to-transmission attaching bolts and also to the transmission cooler lines located high on some transmissions.

A C-clamp can be clamped on the transmission housing to to keep the converter in place in the front pump of the transmission. The converter can also be held into the transmission with safety wire or a bar bolted across the front of the converter housing.

If the converter slides too far forward, it will come out of the transmission front pump. It must be realigned with the front-pump drive gear to prevent damage to the pump and flexplate when the engine is reinstalled.

SHOP TIP When the engine and transmission are first separated, measure and record the distance from the edge of the converter housing to the front of the converter (see Figure 67.5). Although performing this step is not necessary, it can give peace of mind that the converter is installed all the way during reassembly.

Ideally, converters without studs should be only ⅛" from the flexplate when pushed as far into the transmission as possible. This ensures that the transmission front-pump drive gear is sufficiently engaged by the torque converter drive lug. If the distance is more than ⅛", shims can be installed.

The transmission *must* be supported during and after engine removal. Support the transmission with a jack until it can be wired up so that it will not hang (Figure 44.14).

It is a good practice to remove the torque converter from the transmission and replace the front transmission seal during an engine rebuild. This is sometimes necessary after an engine rebuild because an old transmission front seal that was not replaced can begin to leak. Front seal replacement is covered in Chapter 67.

CASE HISTORY *An engine with an automatic transmission was replaced with a rebuilt unit. While the engine was out of the car, the torque converter, which was no longer supported by the rear of the engine crankshaft, hung on the old seal, which had become brittle with age. When the new engine was started, oil came pouring from the front of the transmission. The transmission had to be removed to make the repair. Removing the engine would have been more difficult. The customer blamed this leak on the shop because the transmission was not leaking when he brought the car in for the engine.*

If the transmission is to be removed from a rear wheel drive vehicle, disconnect the shift linkages, electrical wires, speedometer cable, and the driveshaft. Tape the rear U-joint cups with masking tape so they will not accidentally fall off the U-joint (Figure 44.15).

NOTE: *Some two-piece driveshafts are splined in the middle. The U-joints on both sections must be in "phase" (in the same plane) or serious vibration will occur. If the halves are to be separated, mark them for easier reassembly.*

Plug the end of the transmission after removing the driveshaft so transmission fluid will not leak out (Figure 44.16).

Figure 44.14 The front of an automatic transmission that has been wired up.

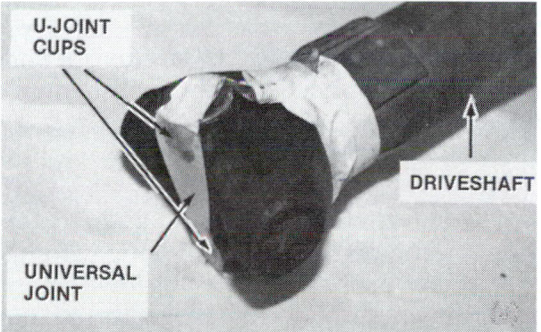

Figure 44.15 Taped U-joints.

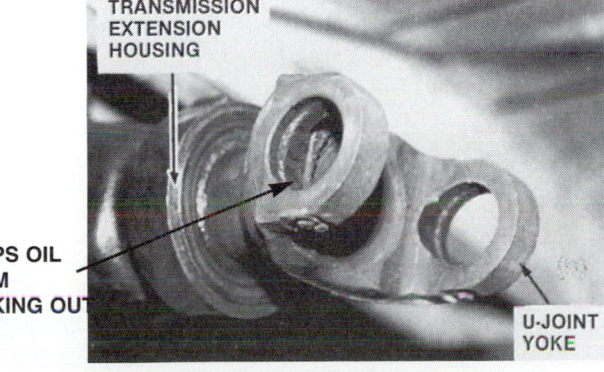

Figure 44.16 The end of this transmission is plugged with an old slip yoke after the drive shaft was removed.

Unhook the speedometer cable from the transmission end (Figure 44.17).

Before removing the transmission from a standard transmission car, the clutch activating fork and gearshift linkages must be disconnected.

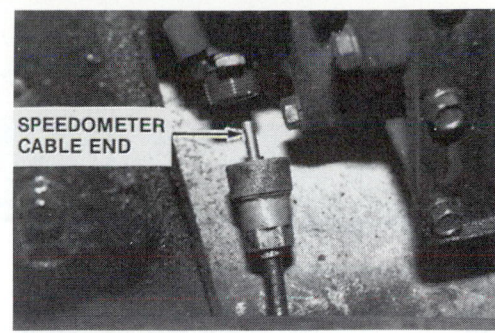

Figure 44.17 Remove the speedometer cable from the transmission.

Engine Mounts

Unbolt the engine mounts. Mark them with a center-punch or marker to show which side of the mount is the front and which side is left or right.

An apprentice was installing an engine in an Oldsmobile and was having a very difficult time. Further investigation showed that the engine mounts were installed on the wrong side of the engine, which caused the engine to be at least one inch too far forward in the chassis.

In some vehicles, the engine can actually be installed with the mounts reversed. The rear transmission mount stretches until there is metal-to-metal contact with the mount and the frame. This results in noticeable engine vibration as the vehicle is driven.

Removing the Engine

Remove the engine from the vehicle. Attach a cable sling, a chain, or a special lifting tool to the heads or block. Some engines are equipped with lifting brackets (Figure 44.18). Make sure the bolts are tightened all the way up against the sling brackets to protect them from excessive stress that can break them (Figure 44.19). Spacers can be made from old piston pins or pieces of pipe cut to different lengths.

Be careful of the car's paint when using a chain hoist to remove an engine.

An apprentice technician was using a chain hoist to remove an engine from a car that had just been painted. As the chain was being pulled, it was rubbing against the paint on a fender, scraping the paint as the chain links moved (Figure 44.20).

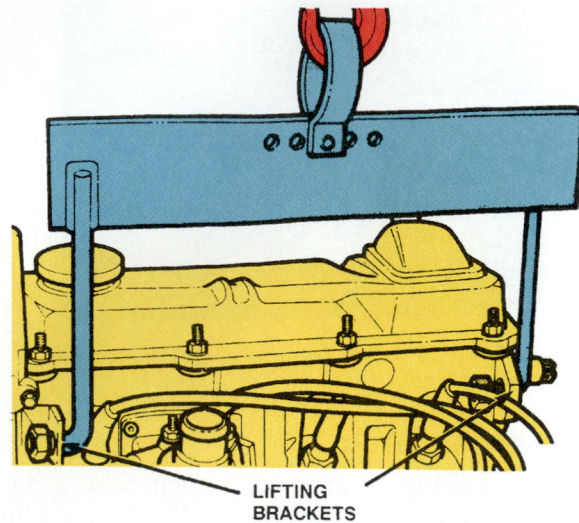

Figure 44.18 An engine with lifting brackets. *(Courtesy of Chrysler Corporation)*

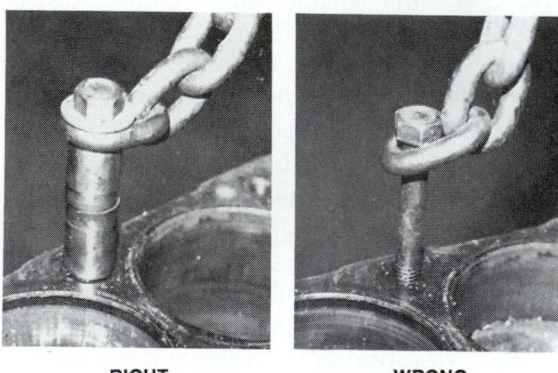

RIGHT   WRONG

Figure 44.19 Avoid excessive stress on sling bolts.

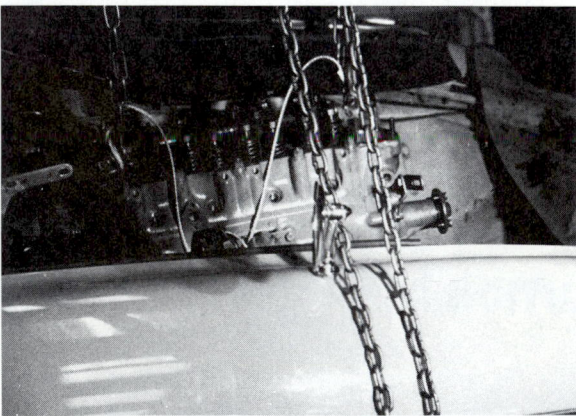

Figure 44.20 Be careful not to let the chain hoist damage the car's paint.

The grill in front of the radiator is also an area where paint can be damaged by a chainfall. Even if the engine rebuild is perfect in every other respect, you can be sure that paint or grill damage is the one thing the customer will notice.

On *front wheel drive vehicles* if the transmission is to be removed with the engine as a unit, the drive axles are removed first (Figure 44.21). This will require removal of the lower ball joints from both sides of the vehicle so that the wheel hubs can be moved out far enough to allow the splines on the ends of the axles to be disengaged from the transaxle. Be ready with a drain pan. When the axles are removed, oil will spill from the transaxle. Remove both axles and store them for later reassembly.

CASE HISTORY *An apprentice removed an engine and transmission from a front wheel drive vehicle. After the new engine was installed and tuned, the customer left with the car. Later that day, the car was towed into the shop with the complaint that it would not go into fifth gear. Further investigation showed that the transmission did not have any oil in it. The apprentice forgot to replenish the oil that had drained out when the axles were pulled. The transmission required extensive and costly repair.*

When lifting the engine, be sure to double-check for anything left connected between the engine and the vehicle.

SAFETY NOTE Be careful when using a cherry picker (engine hoist) for moving an engine. Let the engine down as low as possible in order to keep the center of gravity low. If the center of gravity is too high, the cherry picker can tip over.

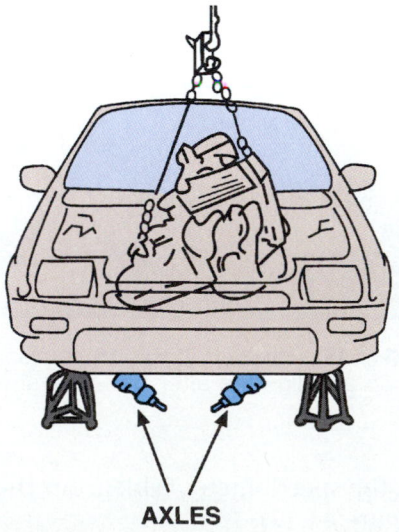

AXLES

Figure 44.21 Before removing a front wheel drive engine and transaxle, the axles are removed. Do not allow them to hang.

NOTE: *When removing engines from vans and motor homes special solutions are often required. Sometimes, it is easier to remove an engine from a van by dropping it out of the bottom. Other times, the cylinder heads might need to be removed first.*

■ ENGINE DISASSEMBLY

Before and during an engine disassembly, inspect for problems that might add to the cost or feasibility of the rebuild. Items like broken castings, stripped threads, broken studs, and damaged sending units or vacuum switches should be noted on the work order and included in the estimate to the customer. Noting such items now can eliminate controversy later, should the customer dispute whether such items were damaged when the vehicle or engine entered the shop.

Save all old parts, including gaskets, until the engine is reassembled and running again. Parts might be needed for comparison. Sometimes, on the assembly line a damaged engine will be salvaged by machining a valve lifter or main bearing bore oversize or a crankshaft undersize.

SHOP TIP The tops of used oil containers can be cut off and used to store nuts and bolts. Plastic baggies are also handy for storing small parts. Make notes during disassembly regarding special parts or service needs.

CAUTION Before beginning engine disassembly, make sure the engine is cold. Tearing down a hot engine can cause warped cylinder heads, especially on engines with aluminum heads.

The following information is necessary before performing the checklist in the lab manual. Pushrod and overhead cam engines have slightly different procedures that are outlined here also.

Clutch Parts

Remove clutch parts. If the engine is equipped with a standard transmission, remove the transmission, *bell housing*, and clutch before installing the engine stand mounting head. Remove the screws on the pressure plate evenly to prevent it from being bent by the clutch apply spring. It is possible to install some clutch discs backwards. Mark the clutch disc on the flywheel side so that it (or its replacement) can be reinstalled in the same direction. Clutch parts and installation are

covered in detail in Chapters 62 and 63. Leave the flywheel or flexplate bolted to the crankshaft.

Engine Mount

Mount the engine on a stand. While the engine is still on the floor, mount the universal mounting head to the engine (see Figure 8.14). Then, install the engine and mounting head on the engine stand. Use four bolts of the proper length with washers. The bolts should have a minimum amount of thread contact into the block of at least 1½ times the diameter of the bolt.

SAFETY NOTE
■ Do not work on the engine while it hangs in the air. It should be mounted on an engine stand or rested on the floor.
■ The mounting head should be mounted so that its center of gravity when it is on the stand will not force the engine to rotate, possibly causing injury to a technician.

Coolant Pump

Remove the coolant pump. Inspect the impeller to see that it is undamaged. Also, feel the bearing for roughness and check to see that there is no evidence of leakage from the pump's vent hole.

Oil Pan

Remove the oil pan. Loosen all bolts to the oil pan. Keep the engine upright and disassemble the top end. It is better to remove the pan before turning the engine over on the stand. This step will help prevent any oil accumulated in the bottom of the pan from pouring back into the engine.

Sometimes the pan will loosen by itself. If not, wedge a gasket scraper or a rolling-head prybar between the pan and block, and loosen the pan being careful not to bend or distort it. If necessary, break the seal by tapping a scraper blade all along the gasket.

Valve Covers

Remove the valve cover(s). On V-type engines, label one of the valve covers "left" or "right" before removing it. Sheet metal parts on engines assembled at the factory are usually vulcanized to the head or block with *RTV sealants* (see Chapter 45). Valve covers and oil pans sealed in this manner can be difficult to remove. Here are two ways:
■ Slip a knife blade between the head and sheet metal valve cover to break the seal.
■ Sheet metal parts can be tapped with a rubber mallet to loosen them. Tap on a curved (strong) area to avoid damaging a part.

■ ENGINES WITH PUSHRODS

Stud-Mounted Rockers

Loosen the nuts on their studs before disassembly and cleaning and turn the rocker arms to the side to remove the pushrods. After the heads are cleaned, they can be removed one at a time to inspect the parts for unusual wear. Keep the pushrods in order so they can be reassembled to their original places.

Shaft-Mounted Rockers

Shaft-mounted rockers should be loosened slowly and evenly. If all the bolts but one are loosened, the pressure of multiple valve springs will be exerted on only one rocker tower, and damage can result. Next, remove the pushrods.

SHOP TIP

Keep all parts in order:
- Inspecting for worn parts will be easier when organized. It is very important to find the root cause of any problem and repair it so it will not happen again.
- Parts become "wear-mated" to each other; they should be returned to their original positions if they are to be reused.

Pushrods can be pushed through holes made in a piece of cardboard. Some engines use pushrods of varying lengths. These with *must* be kept in order.

Valve Lifters

Remove valve lifters. Wipe oil off of the bottoms of valve lifters and label them with a felt marker to keep them in order for reassembly. If they are to be reused, lifters must be used on the cam lobe on which they originally ran (see Chapter 42). Inspect the lifters for unusual signs of wear. The cam and lifters are usually replaced during a major engine overhaul because of wear factors.

The bottoms of lifters often have varnish built up on them in the area that extends out of the lifter bore (Figure 44.22). This makes it difficult to remove them. Varnish can be softened with lacquer thinner or spray carburetor cleaner. If the lifters are hard to remove, wait until the camshaft is removed and try to push them out from the bottom or remove them with a special lifter puller. If the puller is not available, or if the lifters will not be reused, tap them out from the top with a drift punch after the cam is removed. Be careful not to nick the lifter bores. Excessively dirty lifters that are forced out from the bottom can damage lifter bores.

■ OVERHEAD CAM ENGINES

- Before the cam belt or chain is removed, position the number one piston at TDC and note the

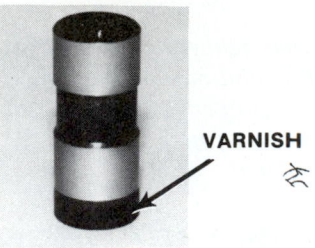

Figure 44.22 Varnish buildup on the bottom of a lifter.

location of the timing marks on the cam and crank sprockets (Figure 44.23).
- Locate a sketch in the repair manual that shows how the camshaft and crankshaft are timed and compare the sketch to the marks on the engine before disassembling it.
- Draw a sketch of the cam timing to keep for future reference.

CAUTION When an OHC head is removed, the camshaft will be holding some of the valves open. Be careful not to set the head face down or the open valves can be bent.

Vibration Damper

Remove the vibration damper. Some engines have a bolt that holds the vibration damper on the crankshaft. Some vibration dampers will slip off after the bolt is removed. Others are pressed on. A vibration damper puller is required to remove these (Figure 44.24). Using the wrong puller can ruin the damper. Grabbing the damper by the outer ring can pull the ring off the damper or pull it off-center (Figure 44.25).

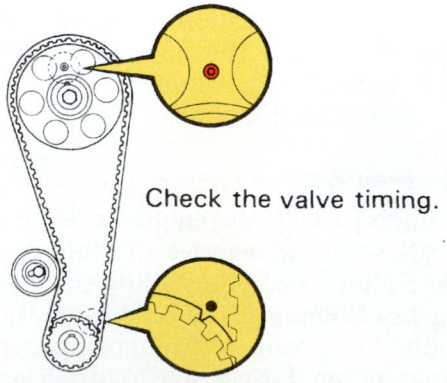

Check the valve timing.

Figure 44.23 Align the timing marks for the camshaft and crankshaft sprockets before removing the timing belt or chain.

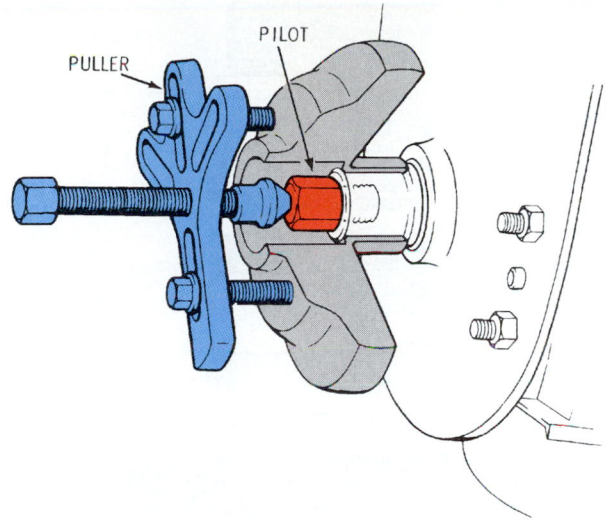

Figure 44.24 A damper puller in use. The threads in the crankshaft are protected. *(Courtesy of General Motors Corporation, Service Technology Group)*

Rubber piece

Figure 44.25 Use of the wrong type puller ruined this damper.

> **SHOP TIP** It is easiest to remove the damper using an impact wrench on the puller screw. This will eliminate the need to hold the crankshaft to keep it from turning. Wear safety goggles during this operation. The damper bolt is also easiest removed with an impact wrench.

On some engines, the timing cover seal can be removed before removing the timing cover. Many timing cover seals are removed and replaced from the inside of the timing cover. These will need to be removed after taking off the timing cover (see Chapter 45).

Timing Cover

Remove the timing cover. Measure timing chain slack as described in Chapter 46.

Cam Drive Assembly

Remove the cam drive assembly.
- Pushrod engines. Unbolt the cam sprocket and slide or pry it off the cam. Then remove the chain. Reinstall the sprocket and tighten one bolt fingertight. The sprocket will be used to help remove the cam later.
- Overhead Cam. Remove the chain or belt tensioner to remove the cam drive (Figure 44.26).

Cylinder Head(s)

Remove the cylinder head(s). Mark one of the cylinder heads (if there are more than one) "left" or "right" and remove them. Remember, left is when viewed from the flywheel end. Most repair manuals now give a head bolt *removal* sequence.

Some engines have cast-in pry points. Be careful not to break a casting by using excessive force when prying. *During the entire engine disassembly process, be absolutely certain that all bolts have been removed before using force.* If a part cannot be pried loose easily, recheck the repair manual.

During removal of the head bolts, look to see if any head bolts are of different lengths or sizes. If they are all of the same lengths and sizes, they do not need to be kept in order. Sometimes, a special bolt is designed with a smaller shank area to allow for the passage of oil to the rocker arm area of the head.

Head Gasket

Inspect the head gasket. After removing the head, look for evidence of coolant or oil leakage (see Chapter 42). Save the head gasket(s) until the job is completed. It will be handy for diagnostic purposes.

NOTE: *In many areas, laws give the customer the right to inspect all old parts.*

Inspect the head gasket for signs of detonation damage and compression leakage. If the gasket was sealing properly, there will be a well-defined line of

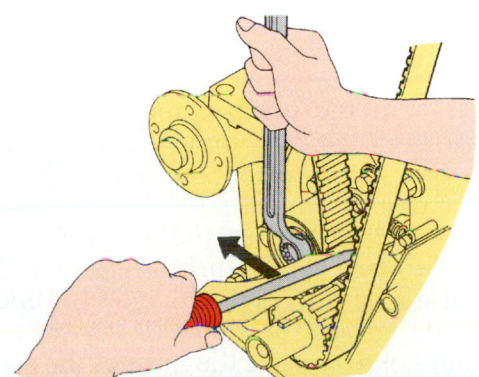

Figure 44.26 This timing belt has a spring-loaded tensioner. Loosen the tensioner bolt, pry it away from the belt, and retighten the bolt.

thin carbon around the combustion chamber on both the head and block. Carbon deposits on the metal rings of the gasket or a poorly defined combustion chamber seal indicate possible compression leakage.

■ CYLINDER BLOCK DISASSEMBLY

Ring Ridges

Remove the ring ridges. A **ring ridge** sometimes forms at the top of the cylinder (see Figure 19.3). Move each piston to BDC and remove the ring ridge before removing the piston. The pistons can only be removed through the top of the block because main-bearing webs are in the way at the bottom.

The ridge is caused by two things:

■ The pressure of combustion forcing the piston ring against the cylinder wall (especially the top one),

■ A lack of lubrication at the top of the cylinder.

Sometimes, the ridge is simply an edge of carbon that can be easily removed with a scraper. Other times (especially on older engines) it can be quite deep. The ridge is removed so that the new square ring will not strike the rounded edge of the ring ridge and possibly break a piston land (Figure 44.27). When in doubt about whether a ring ridge is large enough to require removal, use the following rule of thumb: If a ridge is large enough to catch a fingernail, it will probably catch a piston ring.

A ridge reamer is used to remove the ring ridges. There are several types. One that is easy to use is shown in Figure 44.28.

■ First the reamer is centered in the cylinder. During ridge removal it is turned, rising on a thread, and cutting away the ridge.

■ Be careful not to pull the tool off-center. A T-handle works well for this.

■ After removing the ridge, move the piston to TDC and wipe the chips from the top of the cylinder. Then, do the next cylinder that is at BDC.

SHOP TIP If both the damper and the flywheel have been removed from the engine, turn the crank by adjusting an adjustable-end wrench (crescent wrench) to the size of the crank. When the jaw contacts the woodruff key on the crank, the crank can be turned (Figure 44.29). There are also special tools that fit over the end of the crank and engage the woodruff key. Be especially careful not to damage the front of the crankshaft. A damaged snout will make the crankshaft unacceptable as a core return.

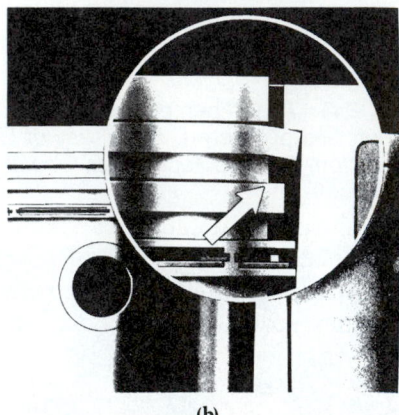

Figure 44.27 (a) If the ring ridge is not removed, the new ring can hit the ring ridge. (b) The ring land above the second ring can be forced down. *(a, Courtesy of Dana Corp., Perfect Circle Division; b, Courtesy of Federal-Mogul Corporation)*

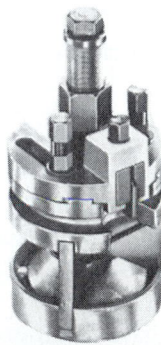

Figure 44.28 A ridge reamer. *(Courtesy of K-line Industries, Inc.)*

NOTE: *The cutter should be run over the ridge only once, or it might cut away too much of the cylinder wall.*

Main Caps

Mark main caps. Turn the engine over so that the crankshaft is facing up. Mark the main caps and rod caps, if they have not been previously marked. Main bearings are numbered from the damper side of the

Figure 44.29 An adjustable wrench grasps the woodruff key to turn the crank. Is the wrench in this example being used properly?

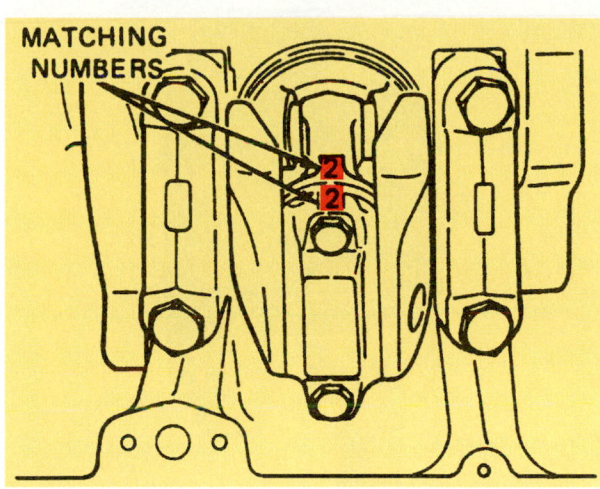

Figure 44.31 Mark rod caps before disassembly. *(Courtesy of General Motors Corporation, Service Technology Group)*

block (#1) to the flywheel end. They can be marked with number stamps. If these are not available, use a centerpunch (Figure 44.30).

Connecting Rods and Caps

Mark connecting rods and caps. *Rod caps are not interchangeable and must be reinstalled on their respective connecting rods.* Rod caps and connecting rods should be marked while they are still tightened to the crank. Be certain to inspect connecting rods previously marked (at the factory or during a previous rebuild) for correct location (Figure 44.31).

NOTE: *Do not file notches on the rod beam. This can cause stress raisers, which weaken the rod.*

When marking rods on a V-type engine, mark them according to the cylinder's number. The only side of the rod that is easily accessible is the side that faces the outside of the engine (away from the cam on a pushrod engine) (Figure 44.32). Figure 38.8 shows many of the firing orders used by various manufactur-

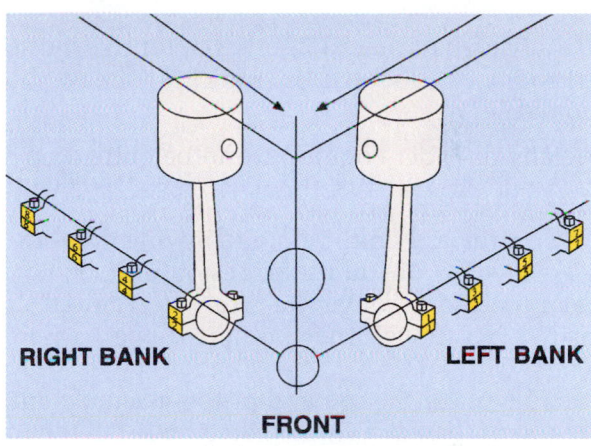

Figure 44.32 Mark the connecting rods on the correct side.

ers. On V-type engines, the number one cylinder will almost always be the one that is the farthest forward on the block. Because cylinders on V-type engines share crankpins with cylinders on the opposite side of the block, the cylinders must be staggered.

Connecting rods must not be reinstalled backwards. Figure 44.33 shows two V-8 connecting rods. Notice that they are machined more on one side than the other.

Piston and Rod Assembly

Remove and inspect the piston and rod assembly.

■ Move each piston to BDC so the rods will clear the crank during piston removal.

■ Loosen each rod nut. The amount of torque required to loosen the bolts can be determined using a dial indicator torque wrench. This will tell you whether nuts were properly torqued before.

NOTE: *The click-type torque wrench should not be used for loosening.*

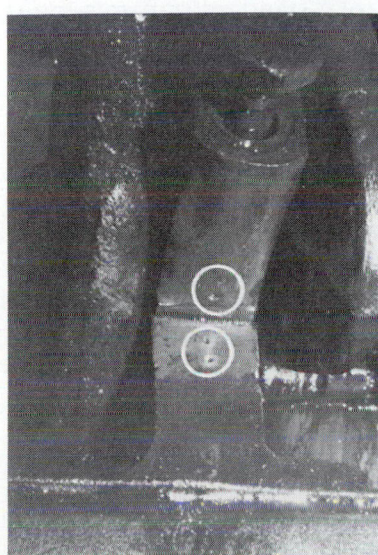

Figure 44.30 Main-bearing cap marked with a center punch.

Figure 44.33 Offset connecting rods from a V-type engine.

- Use a brass hammer to lightly tap on the ends of the rod bolts to loosen the rod caps for easy removal.
- The crank is soft and nicks easily. To protect the crank, install **rod bolt protectors** (available from your parts source) or pieces of fuel hose 3" to 4" long on the rod bolts before removing the rods (Figure 44.34). If you use pieces of hose, do not cut them too short. Longer ones can be removed easily, even though they become slippery when coated with oil.
- Carefully push the rod and piston assembly out of the top of the bore. Immediately reinstall the rod cap on the rod. Remember, connecting rod caps are not interchangeable.

NOTE: *Keep the old bearing in place in the rod for future diagnosis.*

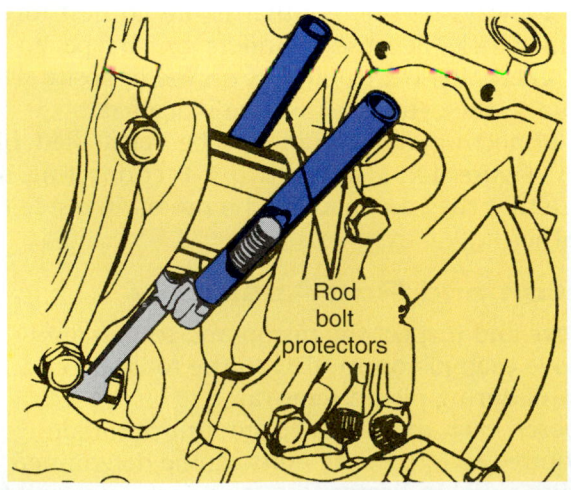

Figure 44.34 Install hoses or rod bolt protectors on the rod bolts before removing the piston/rod assembly. *(Courtesy of General Motors Corporation, Service Technology Group)*

Piston, Rings, Rod, and Bearings

Inspect the Piston, Rings, Rod, and Bearings for the following:
- Detonation (see Chapter 25) can affect the rod bearing. If the rod bearing falls out as the piston and rod are removed from the cylinder, this means the bearing has lost its crush. Check for detonation damage, especially on the upper rod bearing.
- Inspect the pistons for obvious wear and breakage.
- This is also a good time to visually inspect each cylinder for corresponding wear.
- Inspect the oil rings to see if they are plugged.
- If compression rings were working properly, the piston land between the top and second rings should be relatively clean. Chapter 48 describes a procedure for checking piston ring wear.

Crankshaft

Remove the crankshaft and inspect for wear.
- Remove the main cap bolts. Use a dial indicator torque wrench to see how tight they were.
- Remove the main caps. They fit tightly in the block.

NOTE: *Main caps are not interchangeable, and must be returned to their original positions.*

- Normally, bearings are replaced. If bearings are reused, they *must* be returned to their original positions, as they have been wear-mated to the crank journals.
- Carefully lift out the crankshaft. It has a flywheel or flexplate bolted to it (see Figure 15.28).
- The crank should be rested on end (vertically) during storage or hung in a rack to prevent warping. Leave the flywheel or flexplate bolted to the crank to help hold it upright during storage.
- Check to see if there is a pilot bearing or bushing in the rear of a standard transmission crankshaft. It can be removed with a puller as shown in Figure 7.39.
- Check the condition of the crankshaft surface where the rear main seal rides.
- Inspect the bearing surfaces of the crankshaft for wear. Measure the main and connecting rod journals with a micrometer and compare them to manufacturer's specifications (see Chapter 47).
- Inspect the thrust bearing surfaces. These surfaces control fore and aft movement of the crankshaft. Wear, which is unusual, is usually greatest on the rear side (see Figure 45.32).
- Inspect the front upper bearing to see if there is any sign of wear resulting from belts that were too tight.
- Label the backs of the main bearings with a felt marker with their positions. The number one *upper* main bearing is marked 1U, the number one *lower* is marked 1L, and so on. Place them in a line to

inspect them for wear. Bearing wear problems are discussed in Chapter 47.

Camshaft

Remove the camshaft.

■ Pushrod Engines. Some engines use a bolt-on cam thrust plate (see Figure 18.29). Gear-drive cam bolts can be accessed through holes in the cam gear to remove the cam.

An impact screwdriver is sometimes necessary to remove Phillips head screws (see Figure 7.16).

Varnish often builds up on the edges of the cam journals, making it difficult to remove the cam (Figure 44.35). Squirt some penetrating oil on the varnish. Reinstall the cam sprocket and use it as a "handle" to remove the cam. It is easier to remove the camshaft without damaging the lobes if the engine is stood on end.

NOTE: *If the engine is mounted on an engine stand and the cam will not come out, try supporting the cam gear end of the block. Sometimes the block will sag just enough to bind the cam. This is especially true on inline six-cylinder engines.*

■ Overhead Cam Engines. The camshaft is removed with the cylinder head. Follow manufacturer's instructions for removal of the cam from the head. This varies with the design of the head.

Inspect the cam lobes and journals for visible wear. When several cam lobes are excessively worn, look for a fuel leak or an internal coolant leak causing a loss of critical lubrication.

If the lifters on a pushrod engine are too worn to be reused, the following procedure will save time in removing the cam:

■ With the engine in the upside down position, move the lifters to the highest positions in their bores by turning the camshaft one complete revolution.

■ Use a piece of wooden dowel to finish pushing the lifters into their bores so that they will clear the cam (Figure 44.36).

■ Now the cam can be carefully removed. This is a *delicate* operation. Be careful not to chip any cam lobes.

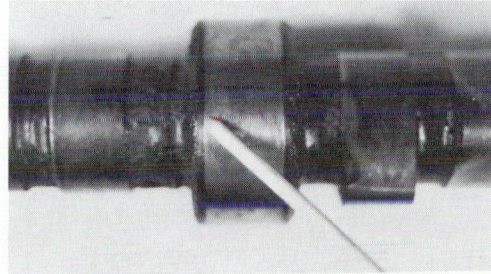

Figure 44.35 Varnish buildup on the edge of the cam journal makes camshaft removal difficult. *[Courtesy of Dana Corp., Perfect Circle Division]*

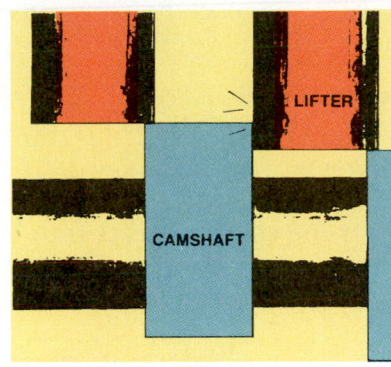

Figure 44.36 The lifters must be removed high enough to allow the cam to be removed. *[Courtesy of Dana Corp., Perfect Circle Division]*

Cam Bearings

Remove and label cam bearings. Cam bearings are inexpensive and are usually replaced.

■ Remove the cam bearings with a cam bearing tool (see Chapter 47).

■ The rear cam plug can be knocked out with the cam bearing tool.

■ Label the bearings with a felt marker, or put them in labeled baggies. They may need to be referred to when determining block positions of the new bearings.

Core Plugs

Remove core plugs. Remove core plugs before hot-tanking the block (see Figure 21.21). Use a large punch to knock them sideways. They can then be removed easily with a roll-head punch or a large pair of pliers.

NOTE: *Be careful not to pound them into the side of a cylinder wall (see Chapter 21).*

Hot-Tanking

Hot-tank the parts. Use a magnet to see whether or not a metal is ferrous (iron or steel). Remove anything left on the block that is nonferrous. It is permissible to hot tank brass fittings or copper, but do *not* hot-tank aluminum. Refer to Chapter 9 for more information on cleaning.

NOTE: *Following cleaning in a caustic tank, oil galleries **must** be cleaned with a brush after the plugs sealing the ends of the galleries are removed. Over many miles of driving under various maintenance conditions, grime builds up in the oil galleries. This material is loosened during the cleaning process. If it is not physically removed, it will end up ruining new engine bearings and possibly more. Chapter 47 covers the procedure for removing oil gallery core plugs from the block before using a long rifle brush to clean the galleries.*

Torque Main Caps

Torque main caps to the block. Following hot-tank cleaning, replace the main caps and torque them down to help keep the block properly stressed during any machining.

Crank Sprocket

Remove the crank sprocket. The crank gear, or sprocket, is driven by a woodruff key or bar key on the front of the crankshaft. Some crank sprockets or gears are pressed-fit. If the crank sprocket is to be replaced, it can be removed as shown in Figure 49.3. The woodruff key can be removed from a crank or cam by

striking it on its rear end with a brass punch (see Figure 49.4). Then strike the underside of the front to roll it out of its channel. Be certain to install a woodruff key flat in its groove or the crank sprocket may be damaged (see Figure 49.10).

Finish Diagnosis and Repair

Finish diagnosis and repair of the engine assembly. The procedures for cleaning, measuring, and repairing parts are found in subsequent chapters.

■ REVIEW QUESTIONS

1. What could cause a transmission bushing, wheel bearing, or emergency brake cable to burn or fail?

2. What is the name of the refrigerant that is responsible for depletion of the earth's ozone layer?

3. Name two belt-driven accessories that can be wired out of the way instead of disconnecting them.

4. What kind of transmission is usually best left in the vehicle during engine removal?

5. Are both engine mounts always interchangeable?

6. When moving an engine that is suspended on an engine hoist, what position should it be in?

7. How long should bolts be that hold the engine to the engine stand?

8. Which parts *must* be kept in order if they are to be reused?

9. Which tool can be used to assist when pulling a damper so that the crankshaft will not need to be held from turning?

10. What is the name of the ledge that forms at the top of a cylinder as the engine wears?

■ ASE STYLE REVIEW QUESTIONS

1. Technician A says refrigerants can instantly freeze anything that they come into contact with. Technician B says that it is illegal to allow R12 refrigerant to escape into the air. Who is right?

   a. Technician A   **b.** Technician B
   c. Both A and B   **d.** Neither A nor B

2. Technician A says that all vibration dampers are pressed-fit on the crankshaft and require a special puller to remove. Technician B says that main and rod caps are interchangeable. Who is right?

   a. Technician A   **b.** Technician B
   c. Both A and B   **d.** Neither A nor B

3. Technician A says that V-8 connecting rods are sometimes machined more on one side than the other. Technician B says to use a click-type torque wrench to check torque when loosening rod nuts. Who is right?

   a. Technician A   **b.** Technician B
   c. Both A and B   **d.** Neither A nor B

4. Technician A says that a rod bearing that has lost its crush could be due to detonation. Technician B says that wear on a thrust bearing is usually on the rear side. Who is right?

   a. Technician A   **b.** Technician B
   c. Both A and B   **d.** Neither A nor B

5. Technician A says using an impact wrench sometimes makes the removal of rusted bolts easier. Technician B says that a broken ground strap can cause transmission bushing failure. Who is right?

   a. Technician A   **b.** Technician B
   c. Both A and B   **d.** Neither A nor B

Engine Sealing, Gaskets, Fastener Torque

■ OBJECTIVES

Upon completion of this chapter, you should be able to:

✔ Order the correct gaskets for a variety of engine repair jobs.

✔ Explain the theory related to torque and clamping force.

✔ Understand the role that friction plays when tightening parts.

✔ Describe the differences between different types of head gaskets.

✔ Understand the theory behind torque-to-yield fasteners.

✔ Select the correct sealer for use in a variety of applications.

■ INTRODUCTION

This chapter deals with the very important area of sealing lubricants and coolants in the engine. Leaks are not only unprofessional, but they run the risk of serious mechanical part failures. Cleanliness is very important when assembling parts. If a leak is to be avoided, there must be **clamping force** on two parts that are held together. Clamping force results from the tension or stretch on bolts securing the two pieces together (Figure 45.1). It also results from the springiness of a gasket between parts. Knowledge of the techniques in this chapter will help make you a better technician and avoid costly errors.

■ TORQUE AND FRICTION

Torque is the measurement of the twisting effort required to tighten a fastener. The amount of clamping force actually applied by a fastener changes with the use of lubricants. Heavy lubrication of threads can result in overstretched bolts.

■ Approximately 90% of the torque applied in tightening a fastener is used to overcome friction. About half of that 90% is lost to friction between the head of the bolt and the work surface; the other half is lost between the threads.

■ The final 10% provides the clamping force necessary to hold the assembly tight.

■ Fastener threads must be sealed and lightly oiled.

Using the wrong torque can distort an engine's cylinder block. Follow the OE (original equipment) manufacturer's torque and fastener recommendations. In some cases, the manufacturer calls for the use of a sealer or thread lock compound prior to installation and torquing.

■ TORQUE WRENCHES

The amount of torque applied to a fastener is measured with a torque wrench (Figure 45.2). There are three styles of torque wrenches.

■ The inexpensive *beam wrench* is very common (Figure 45.2a).

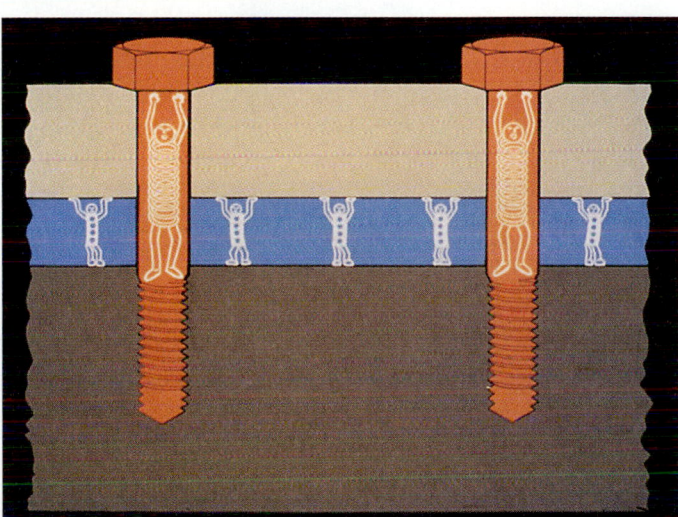

Figure 45.1 Clamping force results from bolt stretch. *(Courtesy of Fel-Pro Incorporated)*

- The *"click" (spring) wrench* is available with a ratchet (Figure 45.2b). It is the most versatile, because it can be used in hard-to-reach places in the engine compartment. The "click" can be felt, even if the gauge cannot be seen.

NOTE: *Reset this torque wrench to zero after each use; this will unload the spring and maintain its accuracy.*

- The *dial torque wrench* is the most accurate (Figure 45.2c). Some are available with a light or a buzzer for use when the dial face cannot be seen.

Torque wrenches used in automobile repair commonly come with ½" or ⅜" drives.

To get an accurate reading with a torque wrench:

- Hold it at 90° to the fastener being tightened.
- Torque in three or four steps.
- Tighten first to about one-third of the torque specification.
- Then, tighten it two-thirds.

- The third step should be to within 10 foot-pounds of the final specification. When the last step is within 10 foot-pounds, the final torque will be more accurate.
- The final step is to double-check all torque readings. During the recheck, if a bolt still moves it should be removed and examined for water or oil in the hole. Liquid can cause initial resistance to torquing and eventually "bleed" up the threads, causing the bolt to loosen.

A bolt can be overstretched (see Figure 6.1a) because it is too poor a grade of fastener for the torque being applied or because a thread lubricant is used without a corresponding reduction in torque (see Chapter 6). Some technicians depend on "feel" when torquing. An experiment that tests a special tension gauge against a group of experienced technicians will show how inaccurate this method of torquing really is.

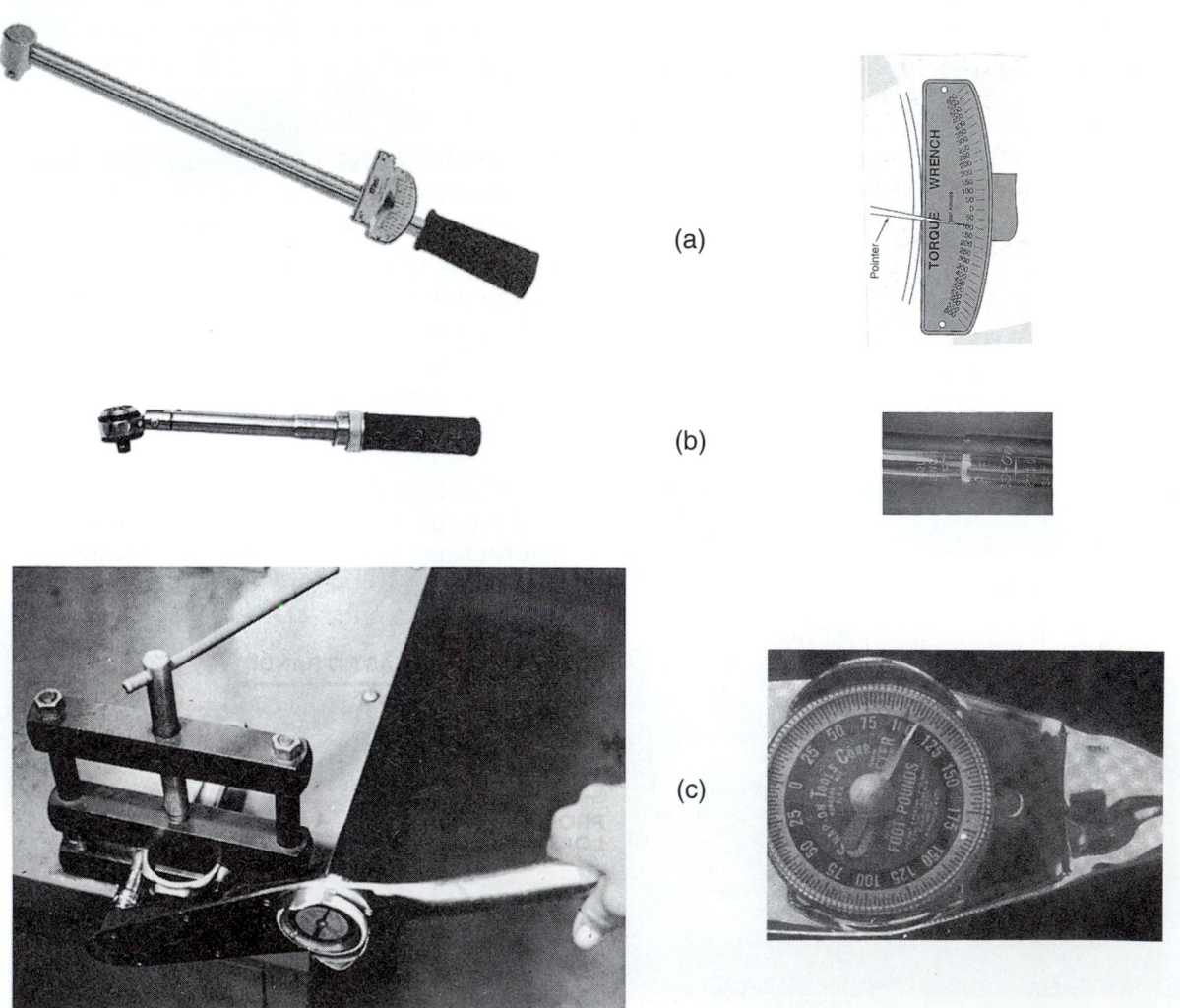

Figure 45.2 Torque wrenches: (a) A beam scale torque wrench. (b) A "click" torque wrench. (c) A dial indicator torque wrench. Hold the torque wrench at 90° to the fastener being tightened. *(a, Courtesy of K-Line Industries, Inc.; b, Courtesy of K-D Tools)*

Torque is measured in foot-pounds or inch-pounds. The metric measurement of torque is the newton-meter. Torque specifications for metric cars are generally given in foot-pounds in English-language manuals.

■ One inch-pound equals $\frac{1}{12}$ of one foot-pound.
■ A foot-pound torque wrench is not accurate below 15 foot-pounds; use an inch-pound torque wrench and multiply the torque figure by 12.

■ TORQUE-TO-YIELD

A common misconception is that when a fastener is torqued to its yield point (see Chapter 6), it will no longer exert the necessary clamping force on the fastened joint. Figure 45.3 shows the relationship between clamping force, bolt stretch, and yield point. Normal head bolt torque values have a calculated safety factor of about 25% less than the maximum amount a bolt can be torqued without becoming permanently stretched. So clamping force is only 75% of its potential for the bolt size. Further study of the chart in Figure 45.3 will show that a small amount of installation error in one or more of these head bolts can result in a wide variation in clamping force.

Because torque is a function of friction, head bolts torqued to the same specification can produce variables in clamping force of up to ± 200%. Some of the bolts might be called upon to do more than their share of the clamping, rather than sharing the load equally. The result can include distortion of engine cylinders. Some manufacturers have even used different amounts of torque on head bolts of the same engine in order to stress the block in a desired manner.

In some late model engines, head bolts are purposely torqued to within 2% of their yield point. This method, known as **torque-to-yield**, provides a more consistent clamping force and the most fatigue resistant joint possible. Notice in Figure 45.3 that the bolt may be elongated considerably at its yield point before it reaches it failure point. Notice also that the clamp load through the entire yield section on the chart is more constant, especially when compared to the large changes that occur in clamping load when the bolts are torqued to different values that are within the elastic range of the fastener.

NOTE: *Bolts torqued into yield have been stretched beyond their elastic limit and must be replaced whenever the head is removed. Gasket sets often come with a new set of head bolts when their replacement is called for.*

■ TORQUE BY DEGREES

On the assembly line, equipment tightens bolts until they are stretched a prescribed amount. **Torque turn**, or *torque angle*, is one method used by manufacturers to be able to ensure equal tightness of fasteners. To duplicate this method of tightening when a torque turn specification is available:

■ First, tighten the bolt to the specified torque.
■ Then, turn it an additional 35° to 180°, depending on the specification.

Tightening in this manner lessens the possibility that variations in friction will have an effect on torque

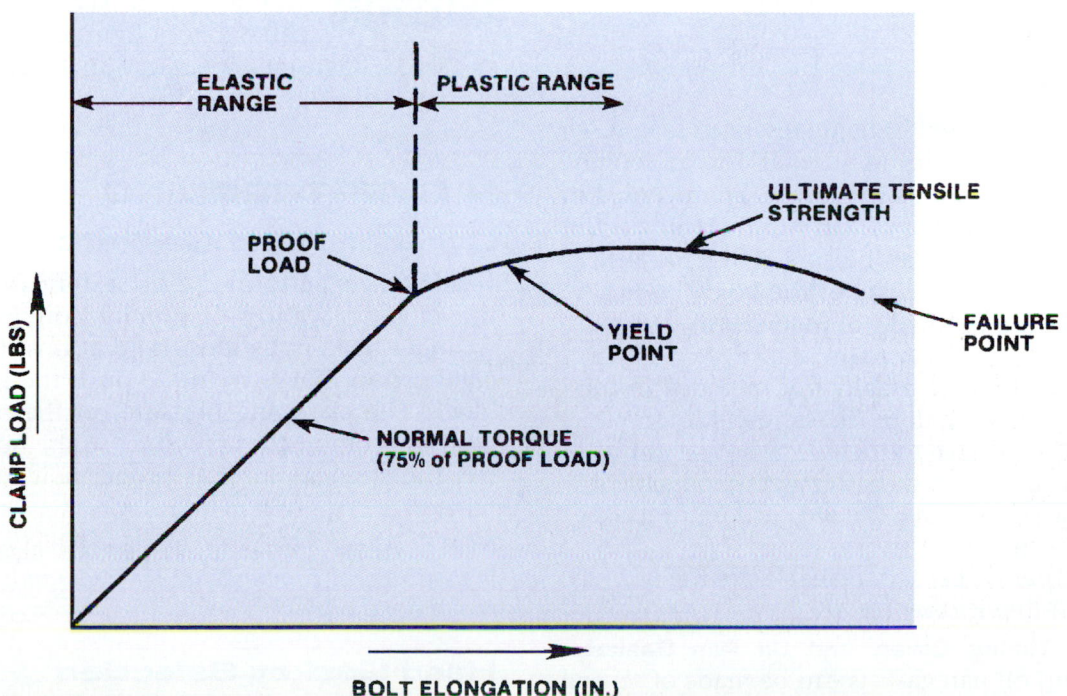

Figure 45.3   Relationship between proper clamp load and bolt failure.

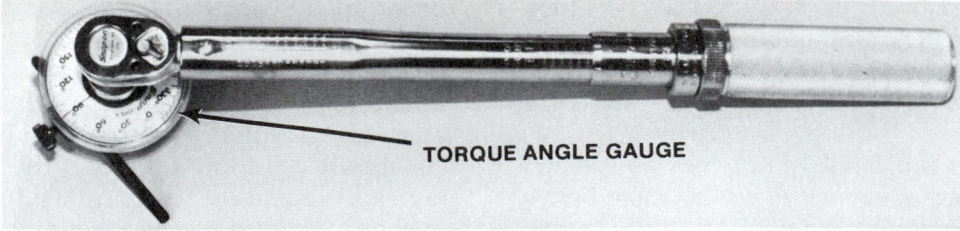

TORQUE ANGLE GAUGE

Figure 45.4 A torque angle gauge and torque wrench.

readings. A *torque angle gauge* that helps keep track of how far a fastener is turned is available from tool manufacturers (Figure 45.4).

FASTENER CLAMPING LOAD

The clamping load on a fastener is known as **preload**. There are three ways of measuring the preload on a fastener.

- The most common is with a torque wrench.
- A torque angle gauge is another method.
- The most accurate method is to measure the actual amount of bolt stretch. This is not practical, however, unless both ends of the bolt are accessible. High-performance engine builders measure bolt stretch on connecting rod bolts.

GASKETS

Gaskets are installed between two surfaces to prevent leakage. Some gaskets seal low pressure fluids; head gaskets seal the high pressures of combustion. Seals are used to seal in lubricants around a rotating shaft. Figure 45.5 shows the locations of many gaskets and seals used in an engine.

Gasket Sets

Gaskets can be purchased individually or in sets. A *full set* (FS) contains everything needed for an engine rebuild. Often, extra unneeded gaskets are included in the set so that the number of part numbers stocked in a parts house inventory can be kept down. Be sure to read the contents on the cover of the box:

- Sometimes a full set does not include valve guide seals or a crankshaft rear main seal.
- Carburetor gaskets are usually not included in the set. A 1955–1980, small block General Motors gasket set would have to include twenty-eight different carburetor gaskets to meet all possible needs. So be sure to save the old carburetor gasket for comparison.

Some of the other gasket sets available are the *head set*, *oil pan set*, and *timing cover set*.

Valve Cover, Timing Cover, and Oil Pan Gaskets.

Valve cover and oil pan gaskets can be made of several different materials. Valve cover gaskets are molded to fit the shape of the cover. They often have tabs that fit

into slots in the valve cover to hold them in place (Figure 45.6). Cast aluminum valve covers on overhead cam engines often have rubber semi-circular plugs that fill holes in the head left from manufacturing (Figure 45.7). These harden with age and begin to leak. They are not included with a valve cover gasket and must be ordered separately.

Oil pan gaskets used on sheet metal oil pans are usually made up of several pieces (Figure 45.8). Rubber seals on the ends join with side gasket strips. Sometimes a timing cover gasket is replaced without removing the oil pan. Figure 45.9 shows how to cut a pan gasket and install a new piece of gasket.

Sheet metal parts often become distorted when bolts are tightened against them during assembly. Always check to see that the valve cover or oil pan is straight before assembly (Figure 45.10). Use a hammer to carefully pound down areas around the bolt holes (Figure 45.11).

> ⚠️ **CAUTION** Torque on gaskets to sheet metal parts is very light. Check the manufacturer's specification. Overtorquing can damage the gasket.

OVERTORQUING

Cylinder Head Gaskets

Head gaskets perform the most difficult sealing job in the engine. Sealing of combustion is very critical. Cylinder head bolts must hold against the pressure of combustion that is trying to push the head off of the block. The clamping pressure on the bolts is greater than the combustion pressure, so the head bolts never feel the stress as long as proper bolt torque is maintained.

NOTE: *About 75% of the clamping load of the head bolts is used to seal combustion. The remaining load is to seal coolant and oil.*

Head Gasket Selection

Carefully match a new gasket to the old one or to the block and head to be sure it is the right one.

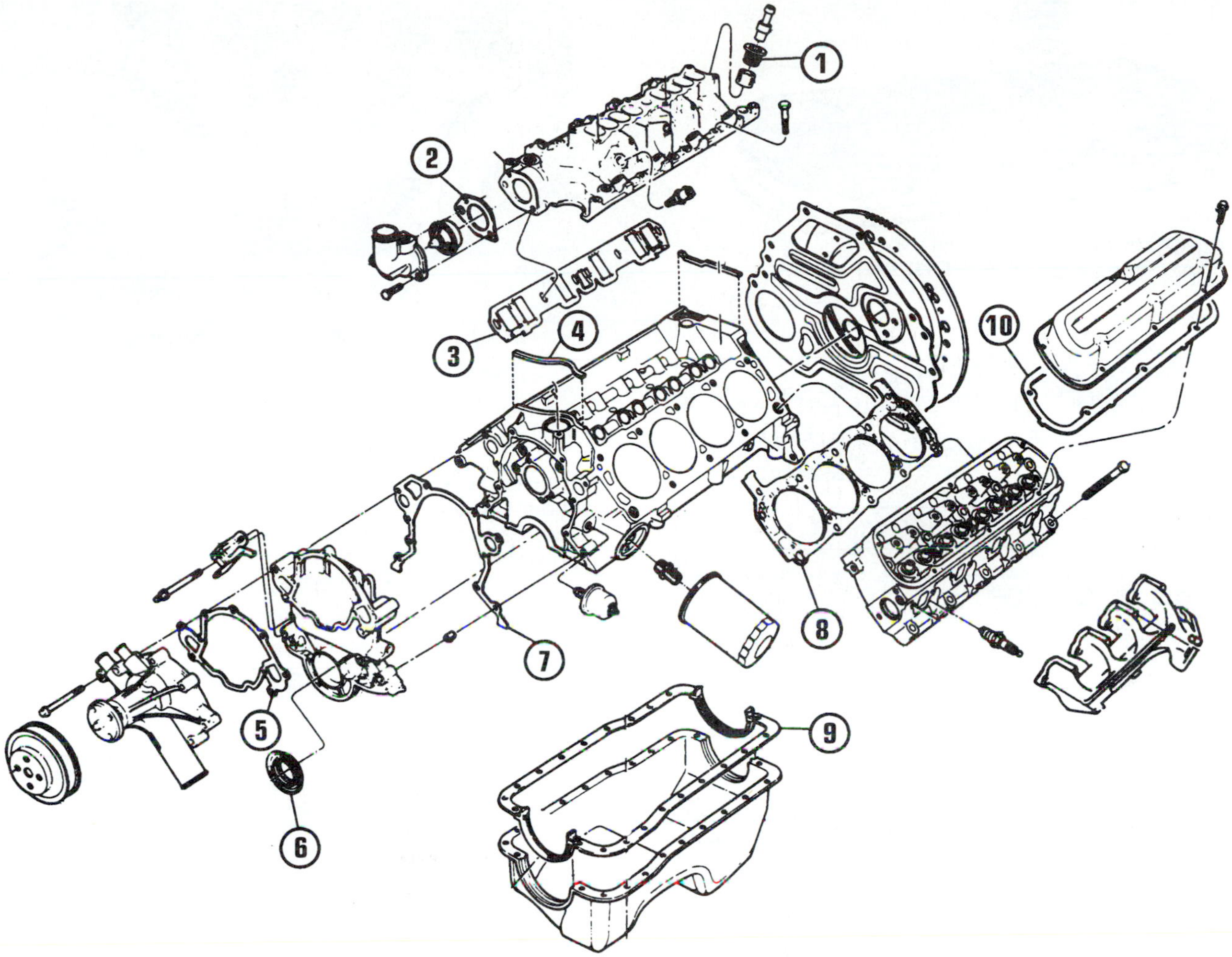

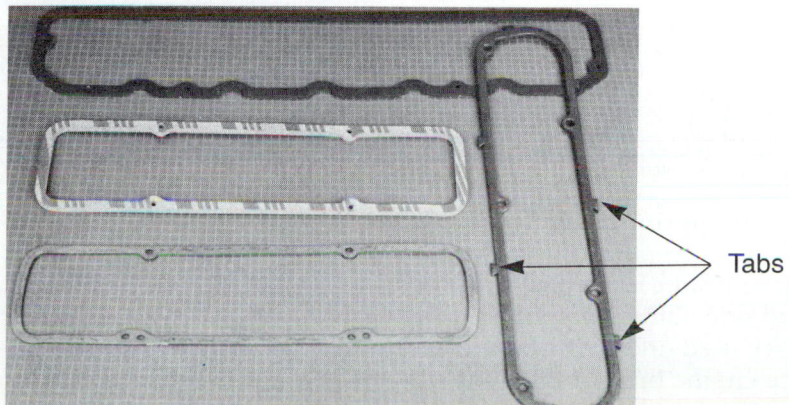

1	PCV valve grommet	6	front crankshaft seal
2	thermostat housing gasket	7	timing cover gasket
3	intake manifold gasket	8	cylinder head gasket
4	intake manifold end seal	9	oil pan gasket
5	water pump gasket	10	valve cover gasket

Figure 45.5 Typical engine gasket and seal locations.

Tabs

Figure 45.6 Valve cover gaskets are molded to fit the valve cover. *(Courtesy of Fel-Pro Incorporated)*

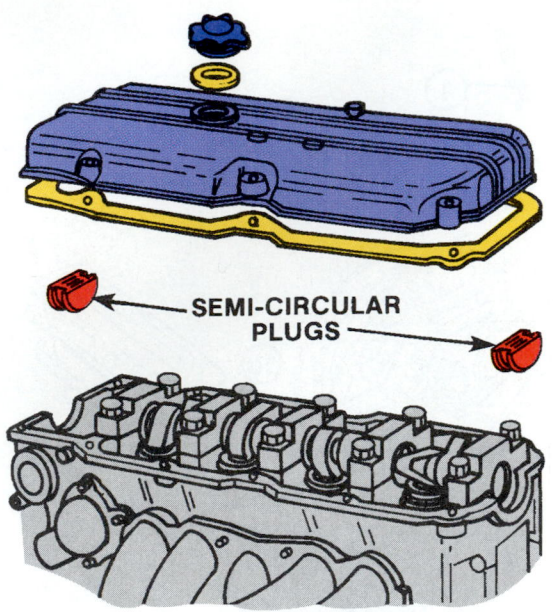

Figure 45.7 Some OHC heads have semicircular plugs at the ends.

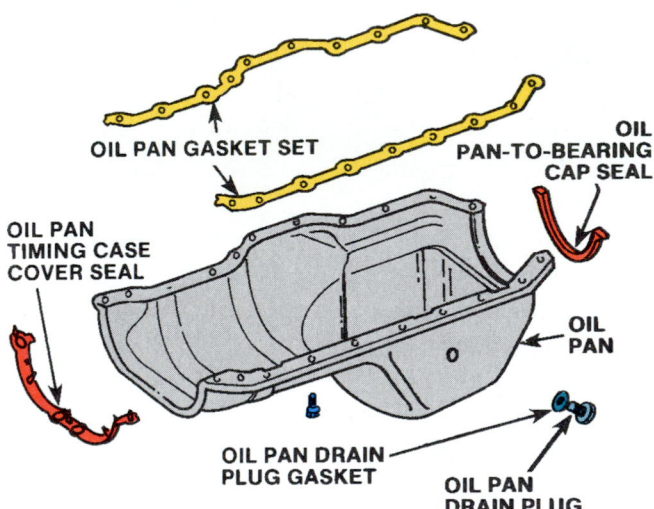

Figure 45.8 Oil pan gaskets and end seals.

Figure 45.9 When the timing cover is removed without removing the oil pan, the front part of the old pan gasket is removed and replaced. *(Courtesy of Fel-Pro Incorporated)*

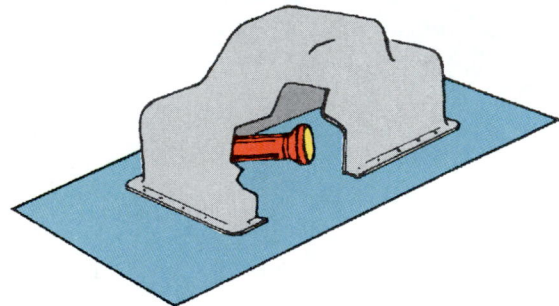

Figure 45.10 Check to see that the oil pan is straight before installing it. *(Courtesy of Fel-Pro Incorporated)*

- Head gaskets have small holes that control the flow of coolant in the head (Figure 45.12).
- These should *not* be cut out to match the old gasket. If the cooling and oil holes do not match up, serious lubrication or cooling problems can result.
- Sometimes gaskets have an imprint marked *top* or *front* (Figure 45.13).
- Some V-type engines have head gaskets that interchange from side to side. Others do not.
- Some manufacturers have used different head gaskets for the same size engine built in different years. Use of the wrong gasket in one of these engines can result in cooling problems.

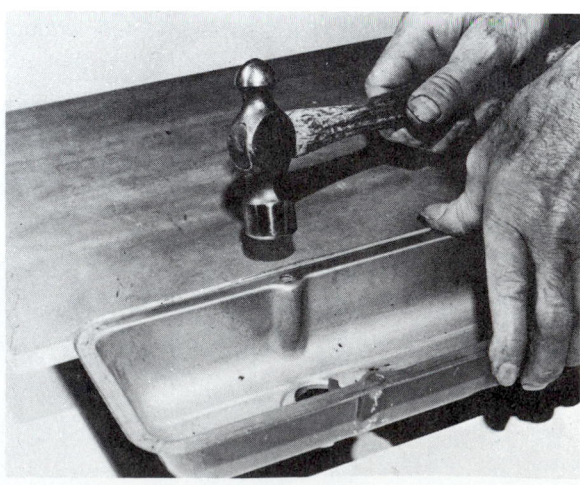

Figure 45.11 Flatten out sheet metal bolt holes. *(Courtesy of Fel-Pro Incorporated)*

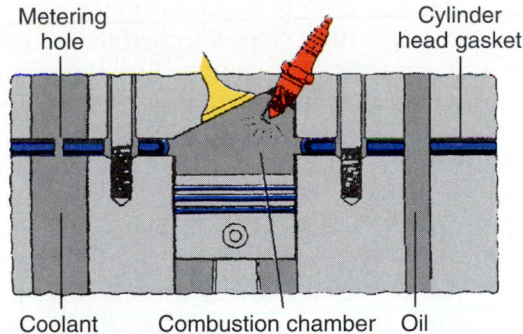

Figure 45.12 Head gaskets have coolant metering holes. *(Courtesy of Fel Pro Incorporated)*

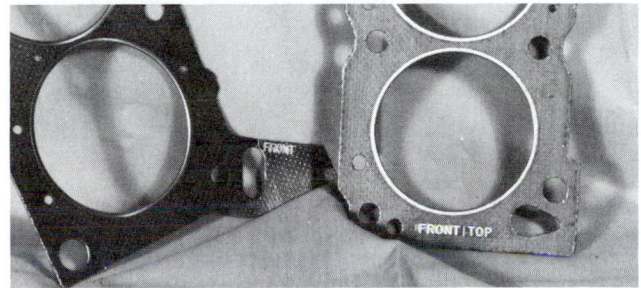

Figure 45.13 Some head gaskets are printed with installation directions. *(Courtesy of AE Clevite Engine Parts)*

Head Gasket Installation

Installing a head gasket correctly will help guard against combustion leakage.

- Chase the bolt holes with a tap.
- Wire-brush the bolt threads and the underside of the bolt head.
- Then oil each bolt with 10W-30 oil and wipe off the oil with a rag.

Head bolts should be torqued in the sequence prescribed in the manual (Figure 45.14). This torque pattern ensures that the head will be pulled evenly against the head. Remember to torque the head in three or four stages as described earlier.

■ EMBOSSED STEEL HEAD GASKET

In the past, an inexpensive embossed steel shim head gasket was used extensively. It is no longer used by

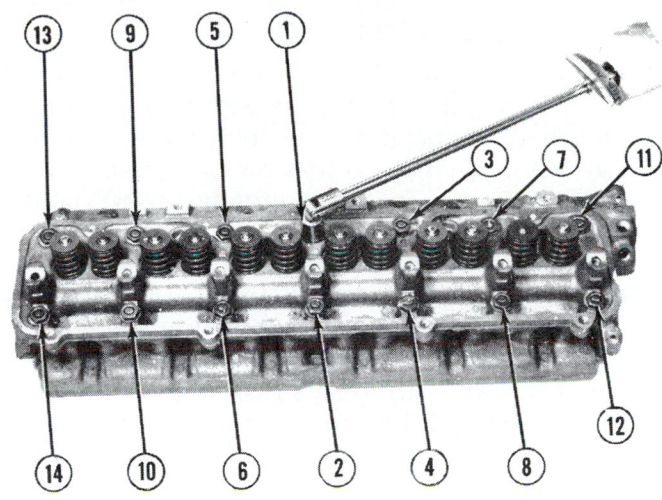

Figure 45.14 Head bolts are torqued in a prescribed sequence. *(Courtesy of Chrysler Corporation)*

most manufacturers. Its use is not recommended in the aftermarket. In the past it was used on new vehicles only because mating surfaces between new parts are nearly perfect. Replacement gaskets are about twice as thick as the steel shim. Their extra thickness compensates for the amount they will compress and also makes up for metal removed during routine head resurfacing.

SAFETY NOTE Older gaskets often contained asbestos. Asbestos is being phased out as a gasket material. Manufacturers use either aramid, expanded graphite, or all-metal as substitute materials. When cleaning a gasket surface, remember that the gasket might contain asbestos.

Today's head gaskets have a more difficult job in sealing. Blocks are lighter and more flexible. A head gasket has an especially difficult job sealing engines with an aluminum head and an iron block, or vice versa. The expansion rate of aluminum is twice that of iron (Figure 45.15).

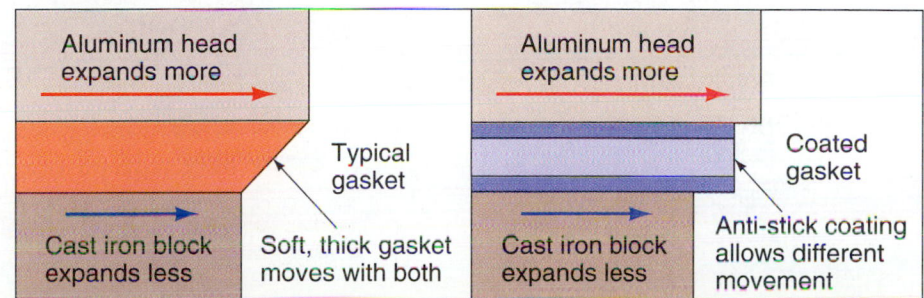

Figure 45.15 An aluminum head expands more than a cast iron block. *(Courtesy of Fel-Pro Incorporated)*

■ NON-RETORQUE HEAD GASKETS

A gasket must be able to maintain a good seal over its lifetime. Small changes in the gasket's thickness can result in a substantial loss of clamping force. Continuous elongation of the bolt so that clamping force is maintained is necessary. The greatest torque loss will occur in the first hour of operation of a rebuilt engine.

Today's **non-retorque gaskets** will compress less than conventional gaskets, but all gaskets will relax some in use. Older gaskets required retorquing after about 500 miles. Retorquing cylinder heads is not commonly done with today's gaskets. Many of today's cars have air conditioning or turbochargers, which make head bolts hard to reach. Non-retorque gaskets can save 20 minutes to two hours of the time usually required for retorquing.

NOTE: Whether the gasket is a permanent torque-type or not, always follow the gasket manufacturer's torque and retorque recommendations.

Although it is not required, modern non-retorque gaskets can benefit from retorquing after several heating and cooling cycles. The recommendation is usually for a gasket retorque at 500 miles because this is when the car should come back for its first oil change and a thorough inspection.

A modern non-retorque gasket has a coating, facing, and core (Figure 45.16). The core is either solid or clinched steel. The core is faced with either expanded graphite or rubber fiber. These materials are dense so they do not compress too much. But they can compress enough to conform to minor irregularities of a sealing surface. Non-retorque gaskets retain their resistance to compression better than regular gaskets; therefore, bolt stretch during torquing does not diminish their clamping force as much (Figure 45.17).

A Teflon or silicone-based anti-friction coating helps seal minor surface imperfections. These gasket coatings do not stick to parts so they have the added advantage of easy disassembly. Teflon-coated gaskets should be used *without* gasket sealers. Some premium Teflon gaskets have silk-screened silicone beads that help them to seal around crucial areas (Figure 45.18). Some technicians use lacquer as a sealer. Lacquer can eat silicone sealing material, although it will not harm Teflon.

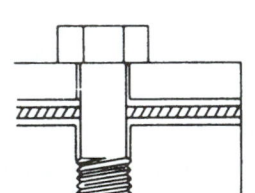

BOLT STRETCH @ .015" @ CLAMP UP
GASKET SET .005"

BOLT STRETCH = .010" @ END

THEREFORE $\frac{.010"}{.015"}$ OR 2/3 OF
CLAMP UP TORQUE RETAINED

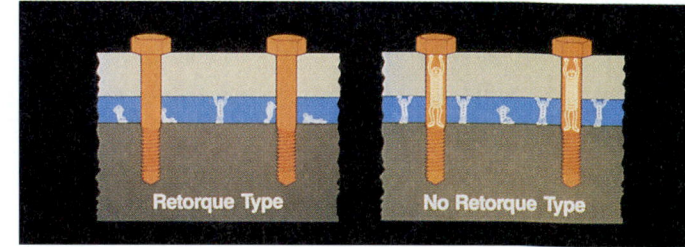

Figure 45.17 Clamping force on a non-retorque gasket. *(Courtesy of Fel-Pro Incorporated)*

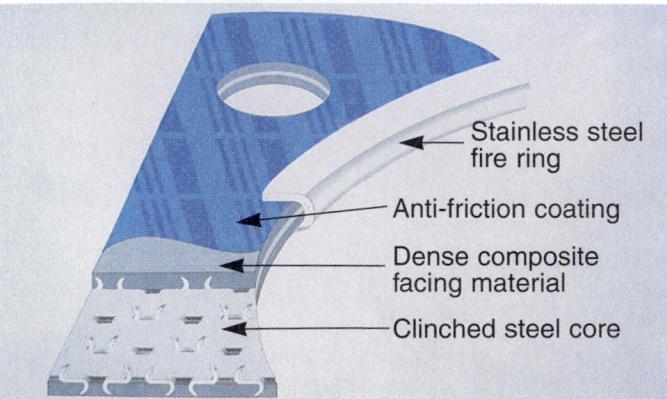

Figure 45.16 Construction of modern non-retorque head gaskets. *(Courtesy of Fel-Pro Incorporated)*

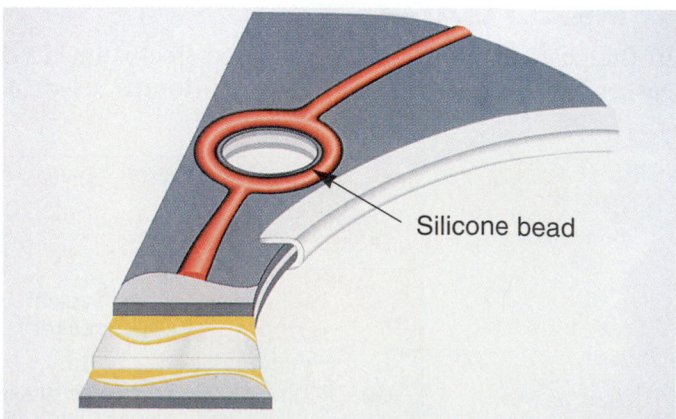

Figure 45.18 A non-retorque gasket with a silicone bead. *(Courtesy of Fel-Pro Incorporated)*

■ SURFACE TEXTURE

Metal surfaces in the engine are designed to have various **surface textures**. If a gasket is used with too smooth or rough a surface, it might leak. Measurements of surface texture are rated on the ANSI (American National Standard Institute) scale. There are several types of measurements, but roughness measured in **microinches** is the most important.

NOTE: *A microinch is one millionth of an inch.*

Roughness above 63 microinches can be checked visually. Visual scales are available from gasket manufacturers. Machinists use manufacturer's recommendations. The following are some guidelines:

Block and cylinder head	60–125 microinches
Main bearing bores	63–200 microinches
Cylinder bores	13–25 microinches
Camshaft	16–25 microinches
Crankshaft	8–25 microinches

■ CLEANING THE HEAD

Make sure that the cylinder head and block surfaces are clean before assembly. Be careful not to nick the surface when scraping gaskets, especially on aluminum heads.

SHOP TIP When cleaning aluminum heads, it is better to use a chemical gasket remover, lacquer thinner, or alcohol with a plastic scraper to prevent damage to the sealing surface. Spray gasket cleaners can be left to penetrate for 5 to 10 minutes.

SAFETY NOTE Spray cleaners can damage skin and eyes. Be sure to follow the instructions printed on the can and check the MSDS (Material Safety Data Sheet) for hazards.

■ OTHER ENGINE GASKETS

There are several materials used to seal oil and coolant in the engine (Figure 45.19).

Cork Gaskets

Cork is a common gasket material. Because it shrinks in the presence of air, it should come in a vacuum-sealed package.

SHOP TIP A cork gasket that has shrunk can be soaked in water for five minutes or so to help it regain its size.

NOTE: *Cork becomes brittle with heat. A brittle gasket can crack if screws are tightened further in an attempt to correct a leak.*

The two most common gasket materials used for valve cover gaskets are **rubberized cork** and *synthetic rubber*. Modern cork gaskets are "rubberized." Rubberized cork is a superior material to regular cork because it does not shrink. All cork gaskets are somewhat resilient—that is, they are "springy"—but they should only be *lightly* torqued. Follow the torque recommendations to guard against breaking the oil pan or valve cover gasket screws

■ Recommendations range from 5 to 15 foot-pounds. This is about the same amount the bolts could

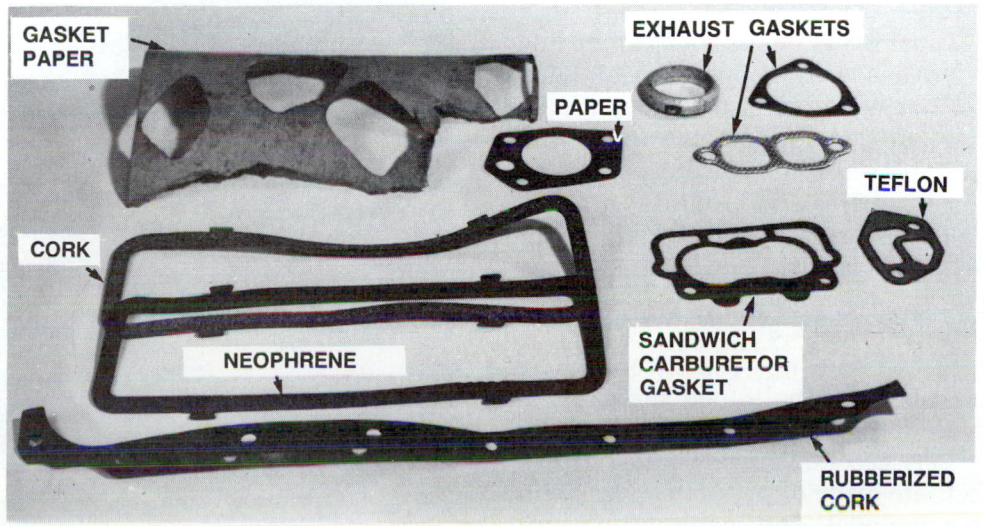

Figure 45.19 Various gasket materials.

comfortably be tightened with a screwdriver or nutdriver.

■ Typical torque on a valve cover or oil pan is about 60 inch-pounds. Use the inch-pound torque wrench because foot-pound torque wrenches are not accurate at less than 15 foot-pounds of torque. To convert foot-pounds to inch-pounds, multiply by 12.

Neoprene Gaskets

Neoprene gaskets (artificial rubber) are the best, but they are expensive. Because they are reusable, these gaskets are especially useful for valve covers, which must be periodically removed (to adjust mechanical valve clearance, for instance).

SHOP TIP Neoprene gaskets are used without sealers. If a neoprene gasket must be held in position, use an adhesive on one side of the gasket.

Paper Gaskets

Paper gaskets are generally the only gaskets, other than retorqueable head gaskets, that require the use of a sealer. Paper gaskets are found on timing covers, water pumps, water outlets, fuel pumps, and some carburetors. Several manufacturers make tool sets for cutting the holes in paper gaskets.

SHOP TIP Gasket paper can be purchased in sheets. A gasket can be roughed out by holding the paper against the part and tapping on it with a ball-peen hammer.

Intake Manifold Gaskets

Intake manifold gaskets must be resistant to air and water leaks. They are made of embossed steel shim, asbestos faced steel, or the non-retorque materials. On V-type engines, there are two side gaskets and two end gaskets. Silicone RTV (covered later in this chapter) is used to join the gasket pieces and around water passages (Figure 45.20). Be sure to read any instructions that come with the gasket set.

Exhaust Manifold Gaskets

Exhaust manifold gaskets are not usually found on new, cast-iron heads because the joined surfaces seal perfectly. Gaskets are used once the parts have been separated during service, however (Figure 45.21). Replacement gaskets are made of perforated steel with asbestos or another material. The manifold expands much more than the head.

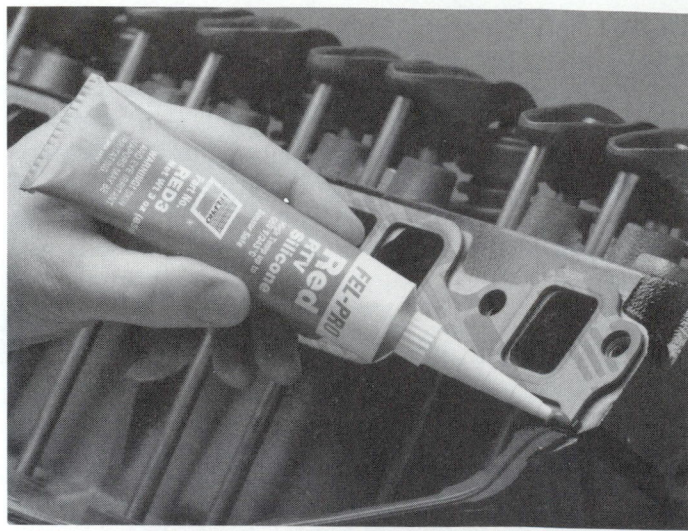

Figure 45.20 Use silicone where gaskets join or to seal water passages. *(Courtesy of Fel-Pro Incorporated)*

■ The steel side of the gasket faces toward the manifold.
■ Do *not* torque the manifold at the outside ends first; it may crack (Figure 45.22).
■ When there is one hole smaller than the others, its bolt goes in first.

NOTE: *When a manifold cracks or a manifold bolt breaks, it is not unusual to find a burned exhaust valve in a port close to the exhaust leak. This happens because of the shock of the relatively cold air (from the exhaust leak) on the red hot valve.*

The bottom of the manifold uses a "donut" O-ring gasket (see Figure 45.19). In the past, these were made of readily available and inexpensive asbestos. The

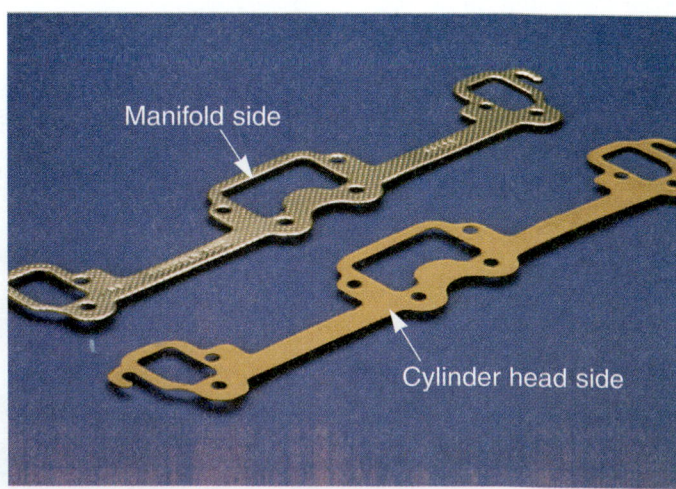

Manifold side

Cylinder head side

Figure 45.21 One side of an exhaust manifold gasket has a steel facing that goes against the manifold. *(Courtesy of Fel-Pro Incorporated)*

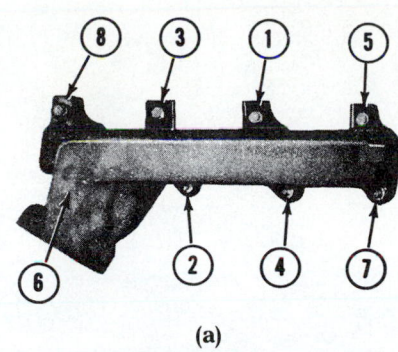

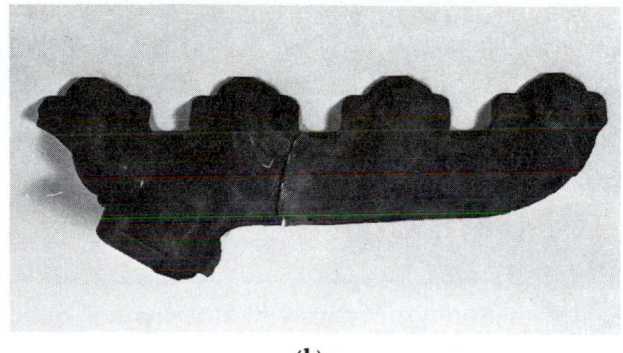

Figure 45.22 Exhaust manifold torque: (a) Exhaust manifold torque sequence. (b) An exhaust manifold that cracked because it was torqued first at the outside ends. *(a, Courtesy of Ford Motor Company)*

trend today is toward ceramic O-rings, which are also reusable.

In front wheel drive vehicles with transverse (sideways mounted) engines, acceleration and deceleration forces movement of ± 4° (8° total possible) between the manifold and exhaust pipe. Consequently, the seal for this joint must be flexible as well as durable. Stainless steel wire mesh or expanded graphite are used for these gaskets. The gaskets wear out. When they leak, they must be replaced. Simply loosening and retightening will only result in a future failure.

Chemical Gaskets

Silicone (covered later in this chapter) and other chemical sealers are sometimes used instead of gaskets (Figure 45.23). These "formed-in-place" gaskets have been used since the middle 1970s in new cars. Chemical gaskets are used for everything except the head gasket and the carburetor-to-manifold gasket.

Chemical gaskets are sometimes difficult to use in the repair trade, so aftermarket anaerobic materials are available for these uses.

NOTE: *Anaerobic means that the chemical works only in the absence of air. That is why the container it comes in is not full; if all the air were removed, the sealer would*

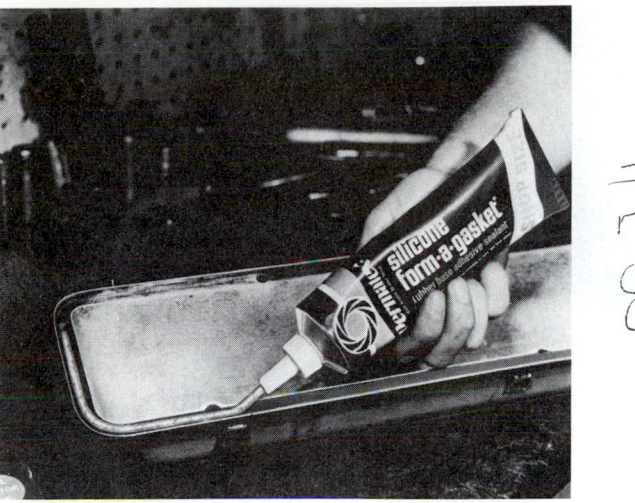

Figure 45.23 Silicone is sometimes used instead of a gasket. *(Courtesy of Loctite Corporation)*

*harden. Silicone RTV, an **aerobic** sealer, comes in a full tube because it cures when exposed to air.*

Anaerobic sealers are used only on precision machine parts. The closer the fit, the faster the sealer cures.

■ GASKET SEALERS

Gasket sealers come in several configurations. Some come in tubes that are squeezed to apply the sealer to the part. Others are spread on from a can with a brush (Figure 45.24). Sealers also come in spray cans for easy application.

Gasket sealers serve three purposes:
■ They help gaskets to seal.
■ They hold gaskets in place.
■ Some gasket sealers also take the place of gaskets.

SHOP TIP Gasket sealers are not always necessary. Some major gasket manufacturers suggest that sealers are necessary only on paper gaskets, or to hold a gasket in position. No sealers should be used on the non-retorque head gaskets.

Neoprene intake manifold end gaskets on V-type engines sometimes "squish" out. Neoprene will usually stay in place if it is completely dry.

Do not use *hardening gasket cements* as sealers in automotive applications; they are inflexible and can begin to leak after vibration or heating and cooling. A small amount of hardening cement can be used on the corners of some gaskets, simply to hold them in place during assembly.

Some technicians use grease to hold gaskets in place for assembly, but the greased parts can shift during assembly. A better choice would be an adhesive or an adhesive sealant (Figure 45.25).

Figure 45.24 Some sealers are spread on from a can with a brush. *(Courtesy of Fel-Pro Incorporated)*

Figure 45.25 An adhesive sealant is often used to hold a gasket in position. *(Courtesy of Fel-Pro Incorporated)*

Locking Thread Sealers

These "hard glue" sealers, often called *loctite* (after one of several manufacturers of these products), are anaerobic. That is, they harden to their maximum strength when tightened between two metals without the presence of air (Figure 45.26). Some adhesives have very little shear strength and their effectiveness is affected by elevated temperatures.

NOTE: *Chemical adhesives are not recommended for bolts over ⅜" in diameter.*

There are several kinds of thread sealers, which are differentiated by their colors. The *blue* sealer is used when parts may be disassembled again. It is said that the torque required to fasten parts with the blue sealer will have to be doubled for disassembly. This sealer hardens only when torqued. Its use is advisable when reusing connecting rod nuts.

The other popular color is *red stud and bearing mount*. This sealer is used to fasten parts that are not intended to be disassembled in the future.

Figure 45.26 Anaerobic sealers harden in the absence of air. *(Courtesy of Loctite Corporation)*

SHOP TIP Although loctite will not harden in its container, it will harden between the container and its cap. Be sure to wipe off the threads on the container before installing the cap.

Teflon Tape

Threads can be sealed with **Teflon tape** to prevent liquid leakage. Figure 45.27 shows Teflon tape stretched around an oil gallery plug.

SHOP TIP
■ Use caution when Teflon tape is used on tapered pipe thread. The tape acts as a lubricant and overtightening can result in a cracked casting.
■ Apply Teflon tape starting with the second thread to prevent some of it from entering and contaminating the system being sealed.

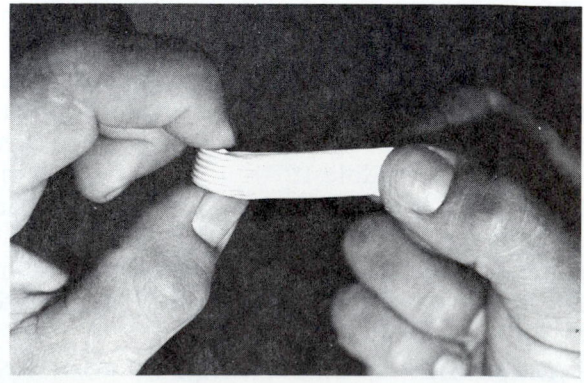

Figure 45.27 Sealing an oil gallery threaded plug with Teflon tape.

Rubber Cement

Rubber cement sealers are very popular and easy to use. Be sure to keep the lid on the can as much as possible, because it will thicken when exposed to air.

■ RTV SEALANT

Room temperature vulcanizing (RTV) has become the most popular sealer in recent years. It is commonly known as *silicone rubber* (see Figure 45.23). Because it cures when exposed to air, silicone is an aerobic sealer. It comes in several colors.

Much of the assembly of modern engines is done by robots. With the exception of cylinder head and intake manifold gaskets, many gaskets can be eliminated by substituting a bead of RTV that can be applied by a robot. Aftermarket gaskets are available as a replacement for most factory RTV applications.

SHOP TIP
■ Because RTV sealed parts are actually vulcanized together, disassembly can be very difficult. Sometimes a sharp knife can be inserted between the parts to cut the bead of RTV.
■ Oil pans and valve covers can be soaked in solvent for an hour or so to help soften the RTV for easier removal.

RTV sets up faster in warm temperatures and high humidity. It usually begins to cure ("skin over") in about 15 minutes, but it does not set completely for about a day. This is a good feature if a part is forgotten, or parts must be disassembled for any other reason. The gasket will not be ruined.

RTV silicone can be used to make a gasket anywhere but on head gaskets, on carburetors (it is eaten by gasoline) or on exhaust systems (because of the very high temperatures). Surfaces must be clean and dry before using RTV or anaerobic sealers. Use a nonpetroleum cleaner such as alcohol to avoid leaving an oil film that can prevent the sealer from bonding.

When using RTV sealer in place of a gasket:
■ Be sure to apply the bead of sealer to the sheet metal surface on the *inside* of screw holes to prevent leakage.
■ Put the sealer in place and finger-tighten the fasteners.
■ Give the material a chance to vulcanize so that it will not squish out.
■ After it sets, the screws can be tightened further.

If the gasket film is too thin, heat and motion can cause leakage. For that reason, grooves and stopper devices are installed to ensure a thick enough gasket (0.020"–0.060" minimum).

NOTE: *Excessive RTV can squeeze out and get into the oil, where it can clog the oil pump pickup screen and ruin the engine.*

Silicone is recommended for use where gaskets join and for sealing water passages in intake manifolds (see Figure 45.20).

NOTE: *RTV will not vulcanize or bond to a gasket, so its use as a gasket sealer is not recommended.*

Many new engines use preformed one-piece silicone gaskets that have steel washers around the bolt holes to prevent overtightening.

Low-Volatile RTV

The *acetic acid* in household silicone gives it the smell of vinegar as it sets up. General Motors has reported that acetic acid, which gives off formaldehyde fumes as it sets up, was responsible for coating the oxygen sensors of its computer controlled fuel system. Low-volatile RTV has been developed by substituting an alcohol-related chemical for acetic acid. Be sure to use a low-volatile silicone on vehicles equipped with oxygen sensors. Low-volatile silicones take somewhat longer to cure than regular silicones.

High-Temperature RTV

Adding iron oxides to RTV can increase its temperature rating by 50–100°F. This silicone is *red* in color. Because silicone is not used in high-temperature areas of the engine, it may not be worth the extra cost to buy high temperature RTV.

■ SEALS

Dynamic seals, also called *chevron* seals, are used at the front and the rear of the crankcase. They are known as dynamic seals because they seal moving parts. Gaskets are *static seals* because they seal nonmoving parts.

Seals are generally made of butyl rubber or neoprene. Neoprene seals should always be lubricated during installation. This is to prevent damage due to overheating during initial startup of the engine.

■ FRONT SEALS

Front seals are usually *lip-type seals* with a garter spring (see Figure 53.27).

SHOP TIP
The seal lip, or open side of the seal, faces toward the oil.

When a seal is exposed to outside air and dust, such as at the ends of the oil pan, it also has a *dust lip*, which deflects dirt from the outside of the seal.

The front seal is installed in the timing cover and seals against the hub on the vibration damper. The friction of the seal against the vibration damper hub sometimes causes wear. A *harmonic balancer repair sleeve* kit will repair a worn damper hub at minimal cost (Figure 45.28). Some replacement seals are designed so that the

Figure 45.28 Worn damper: (a) This damper has a groove worn in its hub. (b) The sleeve installed on the damper hub.

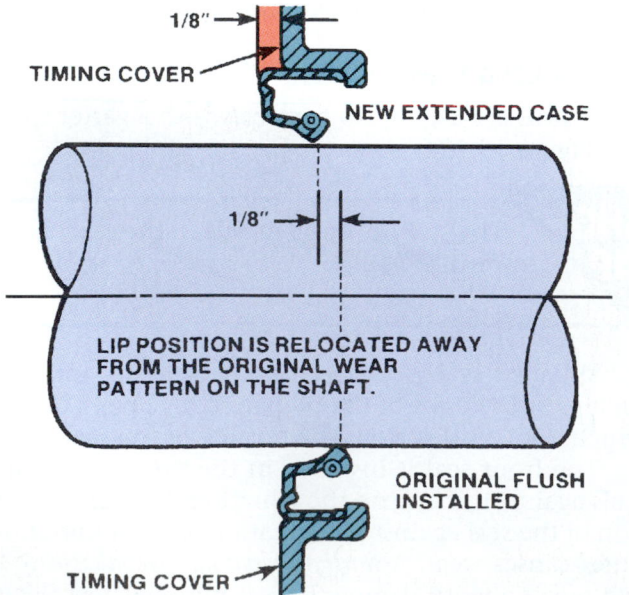

Figure 45.29 A seal with a relocated lip.

lip of the seal will meet the damper hub at a different place than the original seal (Figure 45.29). Front seals can be replaced with the timing cover on or off the engine (Figure 45.30). With the cover off the engine, the seal can be removed with a seal puller (Figure 45.30a). Be careful not to ruin the timing cover by prying on the sheet metal lip that the seal seats against (Figure 45.30b). Pound the new seal into place (Figure 45.30c). The seal can also be installed without removing the timing cover (Figure 45.30d).

OHC Camshaft Seals

On belt-driven overhead cam engines, there is a round lip seal at the front of the camshaft (Figure 45.31).

■ REAR MAIN SEALS

The crankshaft rear seal is called the rear main seal because it is located behind the rear main bearing. A leaking rear seal can destroy a clutch. There are two types of rear seals: the neoprene lip seal and the rope seal.

NOTE: *Rear main seals are often needlessly replaced when an intake manifold, valve cover, or oil pan gasket is the true source of the leak.*

Neoprene Rear Seal

The most common rear seal is a neoprene lip seal. Some of these are split in half, and others are a single piece. The split seal should be installed with its parting line offset, as in Figure 45.32, and with its lip facing toward the oil.

NOTE: *Be sure to lubricate all neoprene seals with grease during assembly. When the engine is started for the first time, there is a short interval before the seal gets any oil; greasing the seal will protect it from failure during this interval.*

Single-piece neoprene rear seals have become popular in recent years. Manufacturers can reduce engine weight and length by not having a separate flywheel flange at the rear of the crankshaft.

NOTE: *Care must be taken that flywheel bolts are carefully torqued, or the crankcase rear main seal surface can become distorted, causing a leak.*

CASE HISTORY *A technician replaced a clutch on an engine with a single-piece rear main seal. When he installed the flywheel, he used an impact wrench. The customer returned later with a rear main bearing seal leak because the crankshaft sealing surface was distorted by the uneven torque.*

Flywheel bolts on some of these engines protrude into the crankcase (Figure 45.33). Sealer must be

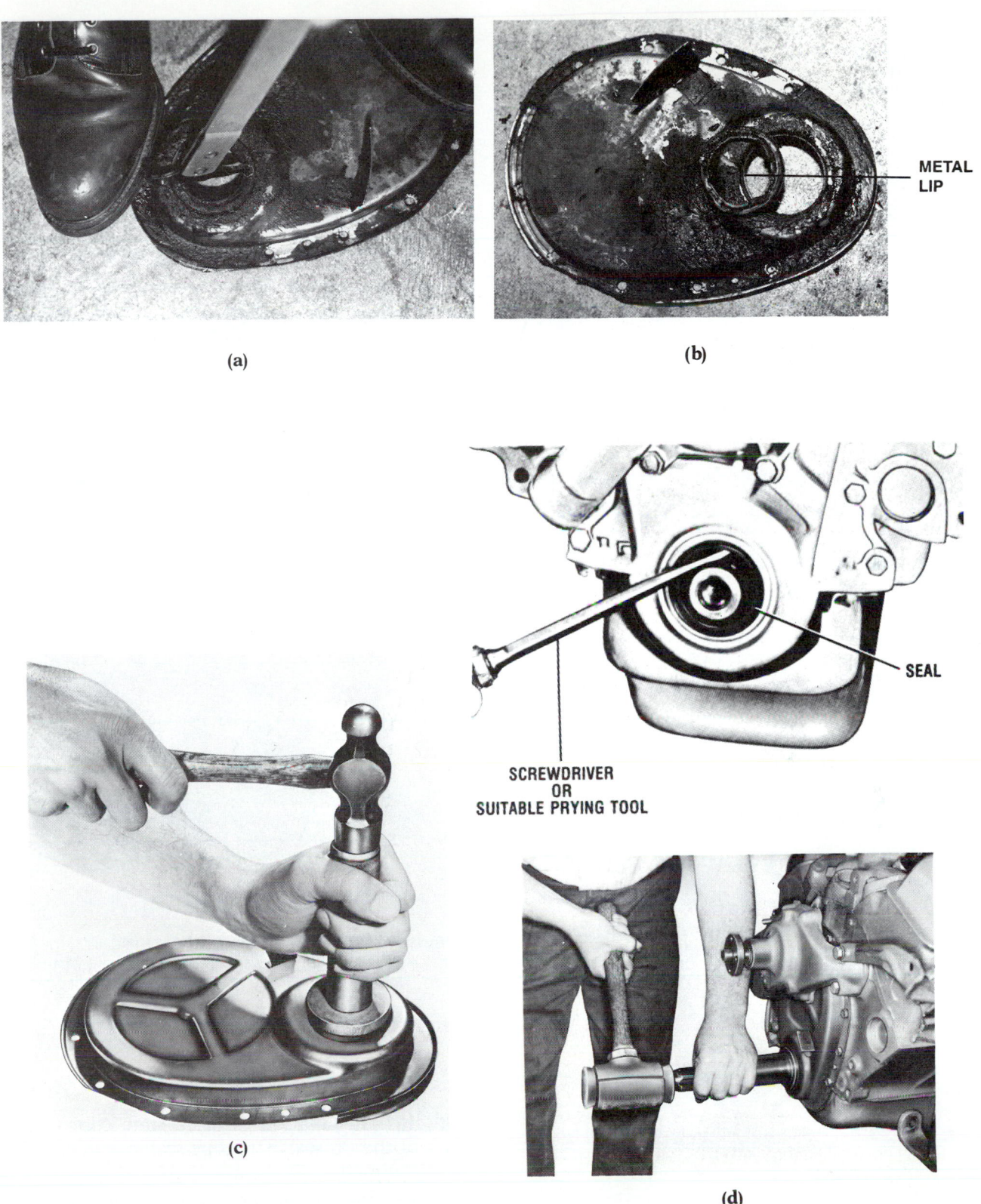

Figure 45.30 Timing cover seal replacement: (a) Removing the timing cover seal. (b) The seal is removed. Notice the metal lip in the timing cover. (c) The new seal is pounded into place. (d) Installing the front seal with the timing cover on the engine. *(Courtesy of General Motors Corporation, Service Technology Group; d, Courtesy of Chrysler Corporation and General Motors Corporation, Service Technology Group)*

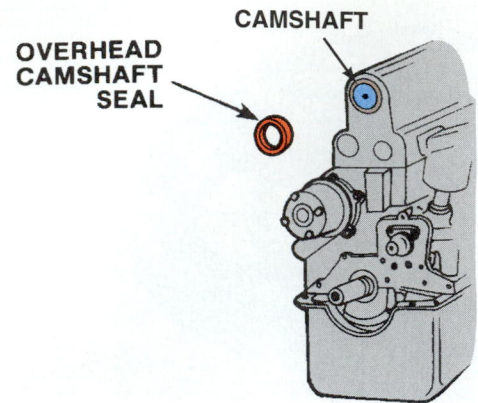

Figure 45.31 A belt-driven OHC engine has a seal at the front of the camshaft.

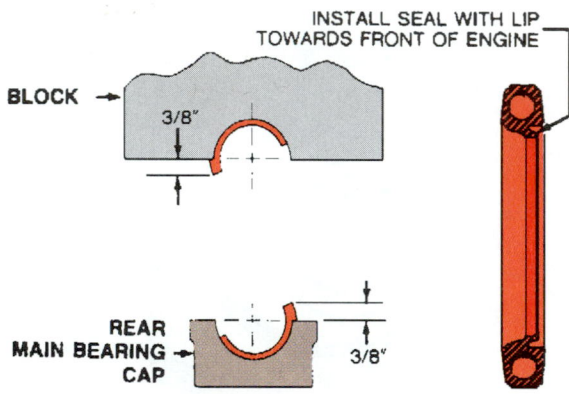

Figure 45.32 Neoprene rear seal installation. *(Courtesy of Ford Motor Company)*

Figure 45.33 These bolt holes go all the way through the crankshaft flange into the crankcase. *(Courtesy of Ford Motor Company)*

applied to them to prevent oil leakage from the interior of the oil pan to the clutch surface.

Rope Seal

Almost all earlier engines used another type of rear seal, called the rope, or *wick*, seal. These seals are inexpensive,

(a)

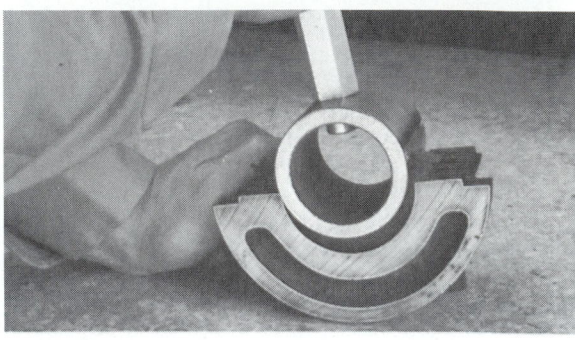

(b)

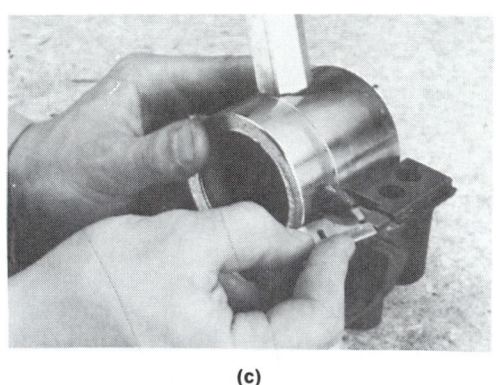

(c)

Figure 45.34 Rope rear seal installation: (a) Install the seal. (b) Form the seal to the channel of the block. (c) Trim the seal. *(Courtesy of Fel-Pro Incorporated)*

but their installation is more time consuming than that of the lip seal. The rope seal is formed into place and trimmed to fit (Figure 45.34).

NOTE: *A conversion from a rope to a lip seal is often possible.*

When a rope seal is used, the crankshaft is knurled on the rear sealing surface to help return oil to the crankcase. The rope seal should not be used if there is no knurl on the crank. Some blocks also use side seals between the rear main cap and the block (Figure 45.35).

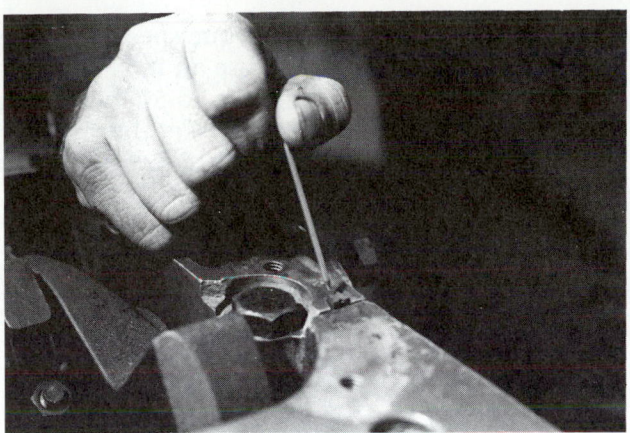

Figure 45.35   Some engines have seals on the sides of the rear bearing cap. *(Courtesy of Fel-Pro Incorporated)*

■ TRANSMISSION FRONT PUMP SEAL

Another seal found on automobiles is the front transmission seal. This seal keeps transmission oil from leaking out between the torque converter and transmission oil pump. It suffers the abuse of severe heat and often becomes brittle with age. If the seal fails, a substantial amount of transmission fluid will leak from between the engine and transmission. As a preventative measure, the seal is often replaced when the engine is out of the car for an overhaul. The procedure for replacing the seal is covered in Chapter 67.

■ REVIEW QUESTIONS

1. Approximately how much of the torque applied to a fastener is used to overcome friction?

2. What is the name of the metric measurement for torque?

3. Refer to a bolt torque chart. What is the proper torque for a grade five, ⅜" bolt?

4. How much of the clamping load on a head gasket is used for sealing combustion?

5. Does the lip of an oil seal face toward or away from the oil?

6. When installing half of a rear main lip seal in a block, the parting line is positioned offset/even (choose one) with the parting line of the main cap.

7. What color is the locking sealer that is used to glue parts that will *not* be disassembled again?

8. What does RTV stand for?

9. Low _____ RTV silicone sealant that does not use acetic acid to cure is used with oxygen sensor systems.

10. Before putting a click-type torque wrench aside after using it, what should be done?

■ ASE STYLE REVIEW QUESTIONS

1. Technician A says that head gaskets have small metering holes to direct coolant flow. Technician B says occasionally one manufacturer's engine might use different head gaskets for different years. Who is right?
   a. Technician A   b. Technician B
   c. Both A and B   d. Neither A nor B

2. Technician A says that non-retorque head gaskets do not compress so they maintain all of the original clamping force. Technician B says that a microinch is one millionth of an inch. Who is right?
   a. Technician A   b. Technician B
   c. Both A and B   d. Neither A nor B

3. Technician A says that sealers are used with all gaskets but paper ones. Technician B says that a broken exhaust manifold bolt can result in a burned exhaust valve. Who is right?
   a. Technician A   b. Technician B
   c. Both A and B   d. Neither A nor B

4. Technician A says that blue loctite is used when parts might be disassembled again. Technician B says that using some types of RTV for engine sealing can damage an oxygen sensor. Who is right?

   a. Technician A   **b.** Technician B
   c. Both A and B   **d.** Neither A nor B

5. Technician A says if a torque specification is 5 foot-pounds, he would use an inch-pound torque wrench and tighten to 50 inch-pounds. Technician B says a cork gasket that has shrunk can be made large again by soaking it in water. Who is right?

   a. Technician A   **b.** Technician B
   c. Both A and B   **d.** Neither A nor B

CHAPTER 46

Engine Diagnosis and Service: Cylinder Head and Valve Train

■ INTRODUCTION

This chapter deals with service to the cylinder head and valve train. When a cylinder head or heads are removed for valve grinding, this is called a *valve job*. When a head gasket has been leaking, sometimes the head is removed for resurfacing and gasket replacement only. This chapter also deals with timing chain or timing belt service.

■ HEAD DISASSEMBLY

Heads are easier to work on if they are clean. After cleaning, rinse heads thoroughly and lubricate all machined surfaces immediately to prevent rust of ferrous parts. Be sure to also lubricate valve springs that have been hot-tanked. Springs will lose their protective coating during hot-tanking and will rust quickly. Rust will increase the tension of the spring.

⚠ **CAUTION** Be careful when handling a cylinder head with an overhead cam. While the cam is still installed in the head, some of the valves will be held open. Do not set the head down on the combustion chamber side or some of the valves can be bent.

Some OHC camshafts act directly on the valve, others use rocker arms. It is necessary to remove rocker arms from OHC heads first in order to remove the valve and spring. Pry against the spring retainer and remove the rocker arm as shown in Figure 46.1. Check the manufacturer's recommendations for camshaft removal. Then the camshaft can be removed.

Varnish buildup often results in the valve retainer being stuck to the valve locks (keepers). Before using a spring compressor, strike each spring retainer with a brass hammer and a short length of pipe, an old piston pin, or a special tool (Figure 46.2). If this is not done, the jaws of the spring compressor can be bent or broken.

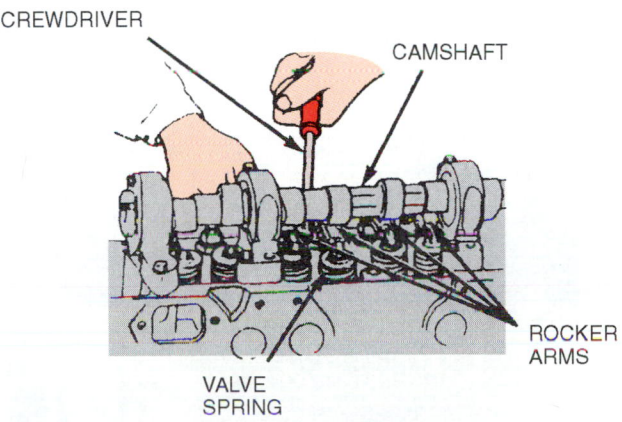

Figure 46.1 Remove OHC rocker arms before disassembling the valve and spring. [*Courtesy of Nissan Motors*]

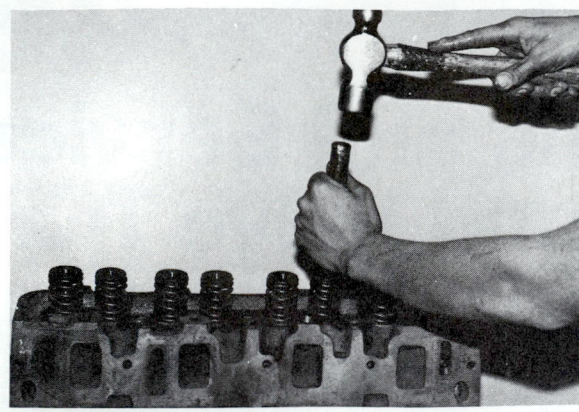

Figure 46.2 Strike the retainer with a piece of pipe, an old piston pin, or a special tool.

CAUTION Be especially careful not to bend a valve when striking the retainer. Sometimes the retainer is so firmly stuck that it moves the valve down against the spring when struck. If the cylinder head is face down on the bench, the valve head will strike the bench and be bent. The solution to this problem is to mount the heads on the head stands. Head stands can be purchased, or they can be made by welding a large starting punch to a piece of steel (Figure 46.3).

Spring-Removal Tools

For protection from flying parts *be sure to wear face protection* when removing valve springs. There are several types of spring-removal tools. The most common type is the manually operated *spring compressor* (Figure 46.4). The jaws of this type of compressor must be adjusted to fit the spring retainer (Figure 46.5). Adjust the compressor so that the spring will be compressed just enough to allow the keepers to be removed.

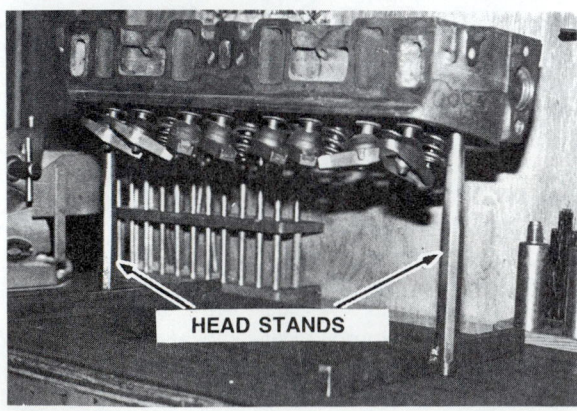

Figure 46.3 These head stands were made by welding a punch to a piece of steel.

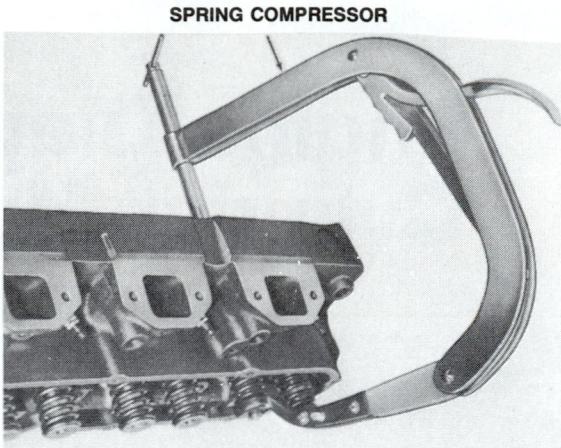

Figure 46.4 A common type of valve spring compressor in use. *(Courtesy of General Motors Corporation, Service Technology Group)*

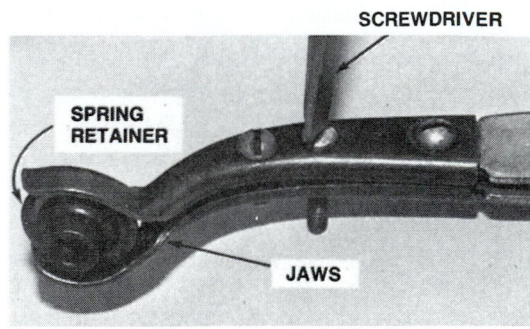

Figure 46.5 Adjust the jaws to fit the spring retainer.

Another type of spring compressor is air operated (Figure 46.6). Be careful that fingers are not accidentally pinched when using this tool as the plunger moves very fast.

Another type of compressor is the small *screw-type compressor* (Figure 46.7). This compressor is especially handy for compressing springs when the heads are on the engine. Another type of compressor, shown in Figure 46.8, is used on OHC engines. It can be used with the head on the engine to replace valve guide seals or with the head off the engine.

Keep Valves in Order. Keep all valves in the numerical order in which they were removed (Figure 46.9).

- One or more oversized valve stems might have been used in a previous head repair procedure.
- If the valve guides are not to be serviced, the valves should be removed in numerical sequence and returned to their original guides.
- Valves can be placed through holes in a piece of wood, metal, or cardboard to ensure their respective order positions.
- Some technicians like to use an electric engraver to number valves before disassembly.

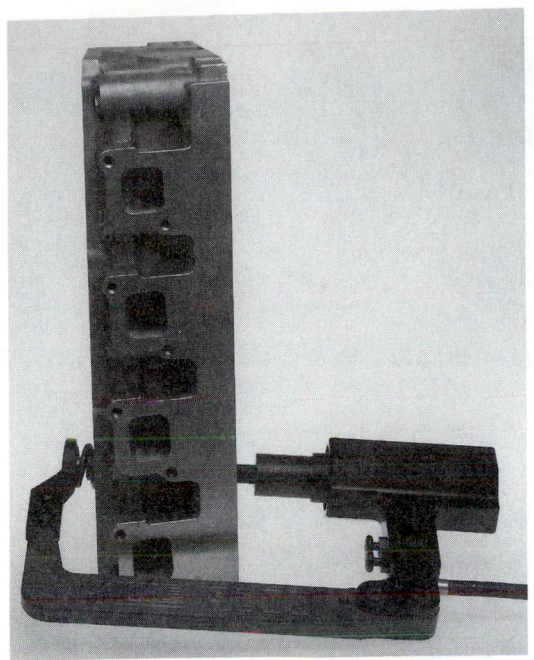

Figure 46.6 An air-operated spring compressor. *(Courtesy of K-line Industries, Inc.)*

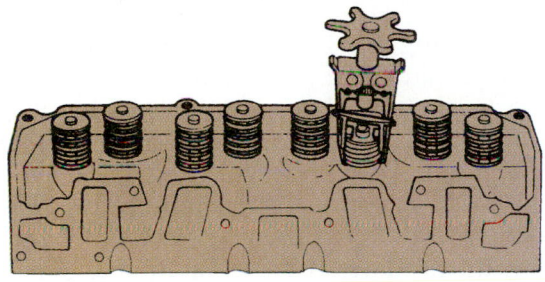

Figure 46.7 This valve spring compressor can be used with the heads on or off the car. *(Courtesy of K-D Tools)*

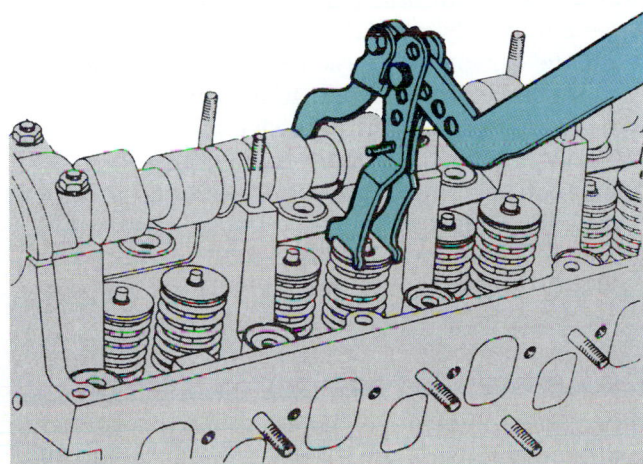

Figure 46.8 A valve spring compressor for overhead cam engines. *(Courtesy of Chrysler Corporation)*

Be especially careful not to lose any keepers. They are very small and easily lost. Put all small valve components in a sealable storage container, such as a coffee can.

Figure 46.9 Keep all valves in order.

Sometimes the tips of valve stems become mushroomed from the pounding of running with excessive clearance (Figure 46.10). This **mushroomed valve tip** area must be dressed with a file before the valve can be removed from the guide.

 **CAUTION** Do *not* try to drive out mushroomed valves with a punch and hammer. The valve or guide can be damaged.

Measuring Stem Height

Measure and record **valve stem height** during disassembly of the head. After springs are removed, measure the height of the stem tip prior to removal of the valves (Figure 46.11). Record these figures on the shop

(a)

MUSHROOMED TIP

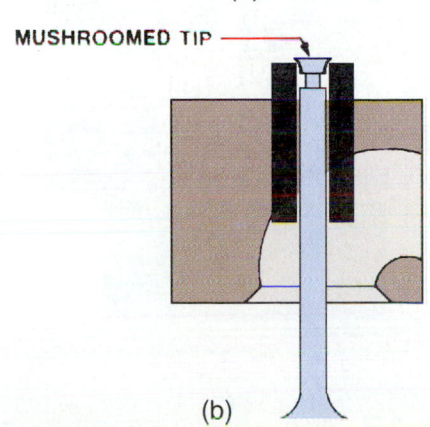

(b)

Figure 46.10 A mushroomed valve tip is filed (a) before removing the valve from its guide (b).

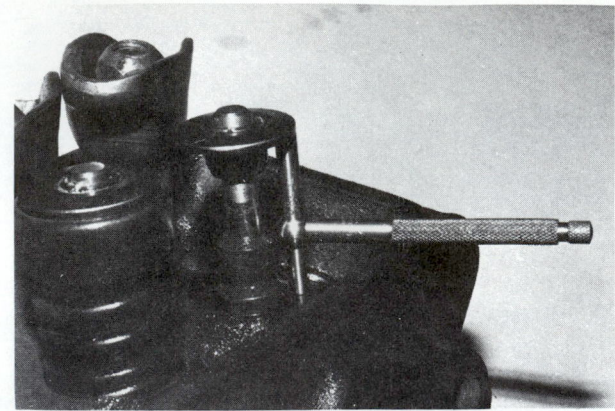

Figure 46.11 *Measure valve stem height prior to disassembly of the head.*

work order. If extensive seat or valve work becomes necessary, obtaining this measurement now will be an important step in the process. Factory specifications are not readily available for stem height, although many of these can be found in the AERA *Cylinder Head and Block I.D. Manual* (Figure 46.12). The stem height measurements should be nearly the same from valve to valve. Prior to disassembly, the heights of the valve springs can also be checked and recorded (Figure 46.13). This can be a handy reference.

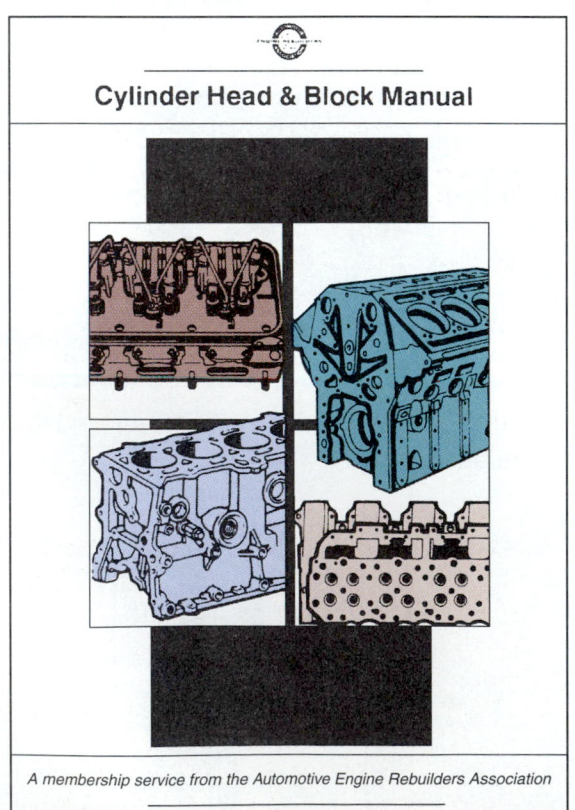

Cylinder Head & Block Manual

A membership service from the Automotive Engine Rebuilders Association

Figure 46.12 *Automotive Engine Rebuilders Association book of head and block casting numbers. (Courtesy of AERA)*

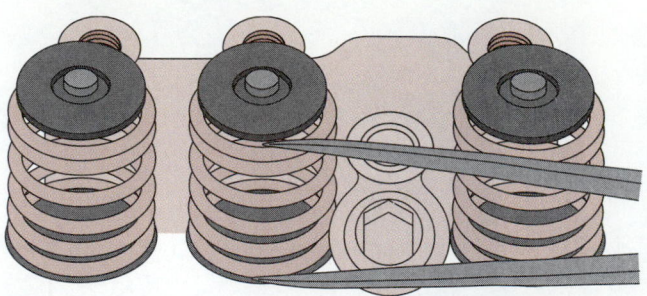

Figure 46.13 *Measure valve spring height prior to disassembly of the head. (Courtesy of Ford Motor Company)*

■ CARBON REMOVAL

Removing Carbon from Combustion Chambers

After disassembly, machine shops sometimes glass bead blast heads that *do not* have oil galleries (most OHC heads *do* have oil galleries). The beads can become trapped in the galleries and cause engine damage later.

Bead blasting removes carbon very quickly; it will also highlight any cracks in the combustion chambers or valve ports (see Chapter 9). Bead blasting is an excellent method of removing carbon from aluminum heads but it will *not* work on greasy surfaces. The heads must be clean and dry. After blasting, use water to thoroughly wash any remaining beads from the head and blow the head dry. Another method to remove the beads is with a tumbling machine.

A portable wire wheel with a drill motor, an air drill, or a die grinder are other available carbon removal tools commonly used (see Figure 7.50)
- These methods can damage an aluminum head. Remove carbon and gasket material by hand, using a gasket scraper or carbon removing tool.
- Be careful not to exert excessive pressure on the narrow head surface between combustion chambers; even the harder surface of an iron head could be damaged.
- Chemical carbon removers can be applied to aluminum heads to help soften the carbon and gasket materials stuck to the surface.

Removing Carbon from Valves

Carbon can be removed from valves by glass bead blasting, or by using a wire wheel buffer on a grinder (Figure 46.14). In case of extremely hard carbon deposits, use an old valve to gently break off carbon before cleaning on the wire wheel. Be careful not to nick the good valve (Figure 46.15). Sometimes, soaking carbon deposits in solvent before wire wheel cleaning will soften the deposits and speed cleaning time.

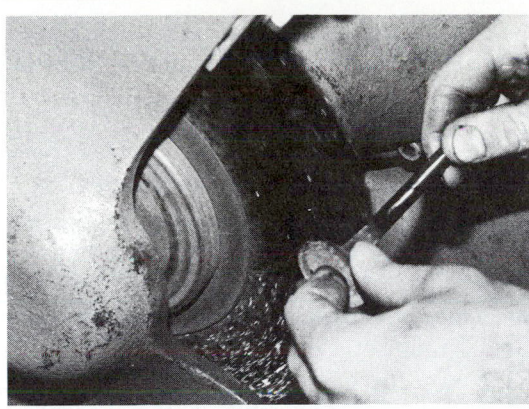

Figure 46.14 Using a wire wheel to remove carbon from a valve.

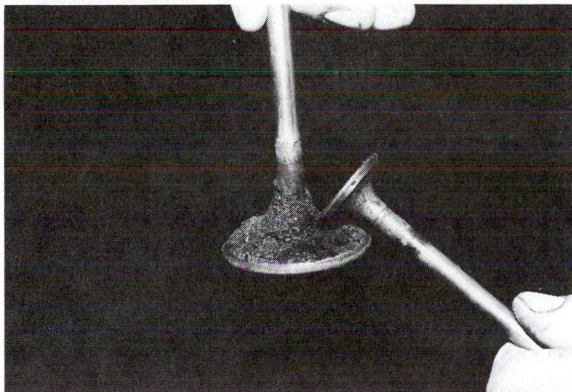

Figure 46.15 Chipping off heavy carbon deposits with an old valve.

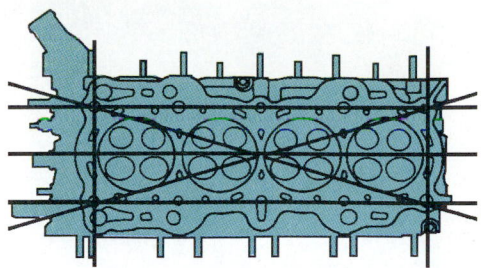

Figure 46.16 Check in several directions for excessive warpage. *(Courtesy of American Honda Motor Co., Inc.)*

■ *Resurface heads warped more than 0.006" on inline sixes, 0.004" on four or eight cylinders, and 0.003" on three cylinders or V-6s.*

■ *Heads should not be warped more than 0.003" in any 6" length.*

■ *Flatness across the width of a head should not vary by more than 0.004".*

When checking head flatness on a four cylinder, try to fit a 0.004" feeler gauge under the straightedge. If it fits, but a 0.005" does not, then it is fine. Also, aluminum heads with corrosion around water passages must be resurfaced. Also check the deck of the block for flatness (see Figure 45.12).

■ RESURFACING BY GRINDING, CUTTING, OR SANDING

Resurfacing is accomplished either by *"fly-cutting"* the head on a milling machine, grinding the head on a head grinder, or sanding the head on a belt sanding machine (Figure 46.17). Most heads clean up with less than 0.010" of metal removed. Head resurfacing can increase compression, so remove as little metal as possible. Following machining, the surface should not be too smooth. It must be rough enough to bite into the gasket, but not enough to leak (see Chapter 45). In the event more metal is removed than is desirable, gasket companies make 0.020" copper shims (for some applications only) that can be installed on reinstallation of the head.

■ CYLINDER HEAD INSPECTION

Checking for Flatness

Cylinder heads sometimes warp, especially when a head gasket has failed when an engine was overheated (see Figure 42.5). Heads that are excessively warped are resurfaced to ensure proper head gasket sealing. Clean the head surface before checking for flatness. Use a straightedge and feeler gauge diagonally, vertically, and horizontally on the head (Figure 46.16). When checking for warpage on the ends of a head, be sure to rock the straightedge so one edge of it rests against the opposite side of the head. This will leave an opening for the feeler gauge on the other side of the head. If you do not do this, you will only get half of the warp reading.

NOTES:

■ *Warpage on both sides of an OHC head is not always equal. Check the valve cover rail for straightness, too. The spark plug side of the valve cover rail often has more warpage. Maximum warpage on the cam side is 0.002". If it is more, straighten the head as explained later. Otherwise the camshaft will not be able to turn without resistance and could break.*

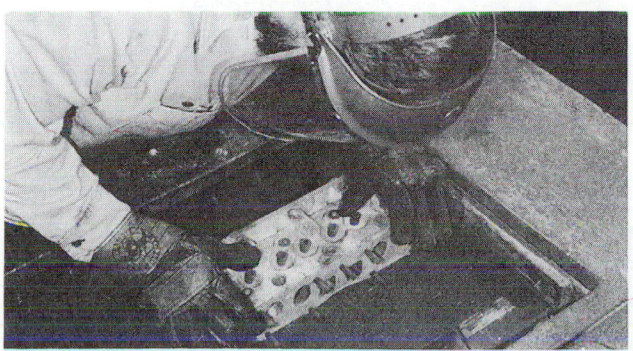

Figure 46.17 Surfacing a head on a belt surfacer.

Figure 46.18 Straightening a head in an oven. *(Courtesy of AMPRO)*

■ STRAIGHTENING CYLINDER HEADS

Warped aluminum OHC heads are commonly straightened. Besides measuring the top surface of the head for flatness, check to see if the camshaft turns easily.

NOTE: *If the camshaft does not turn easily and the machine shop only resurfaces the bottom side of the head, a broken camshaft can result.*

There are several methods of straightening cylinder heads. The one considered the best is to use a heating oven (Figure 46.18). The head is bolted to a thick plate with shims positioned under the outside edges of the head. The head is heated in the oven for five to six hours and allowed to cool slowly.

An added advantage to straightening the head prior to surfacing is that combustion chamber volumes will remain equal. Normally during surfacing of the warped head, the outer cylinders' combustion chambers will end up with reduced volume.

NOTE: *Removal of 0.020" from OHC head surfaces results in about 1° of retard in valve timing. OHC engines have chain tensioners, but they only control slack on one side of the chain (see Figure 18.46). Slack on the drive side of the chain is taken up by rotation of the crank. This results in valve timing that is retarded after surfacing or chain stretch.*

■ CRACK INSPECTION

Cracks are sometimes found in combustion chambers, between adjacent combustion chambers, and also, rarely, on the valve spring side of the head.

SHOP TIP When water has been entering the combustion chamber from a leaking head gasket or a crack in the head, there will be no carbon on the surfaces of the combustion chamber and piston (see Figure 43.20).

There are a number of ways to detect cracks. Sometimes after the head is cleaned, cracks will be apparent to the eye. Glass bead blasting is especially helpful in highlighting cracks.

Magnetic Crack Inspection

Crack inspection on *iron* heads can be performed using a simple magnetic crack detector as shown in Figure 46.19. The electromagnet sets up a magnetic field. The head is dusted with iron powder and the powder lines up with the magnetic lines of force. A crack interrupts these lines of force, causing the powder to gather around the crack. *Magnaflux* is the common trade name for a more sophisticated type of magnetic crack detection.

Dye Penetrant

A **dye penetrant** kit can be used to detect cracks in any metal (Figure 46.20). It is usually not used on iron because of its expense compared to magnetic detection. After the surface is cleaned, a red dye is sprayed on and allowed to dry for about 3 to 5 minutes. The surface is then wiped clean. Sometimes a dye remover is used to help clean up the surface. A light spray of a white *developer* makes any cracks visible to the eye. Cracks show up as red or pink lines on a white background. A black light crack detector works in a similar fashion. The black light is shined on the surface after spraying a special solution on the head. The flourescent dye that penetrates the crack is illuminated by the black light.

Pressure Testing

Heads that are especially susceptible to cracks are often inspected by pressure testing. This test is very effective, but can be time consuming. All openings in the head

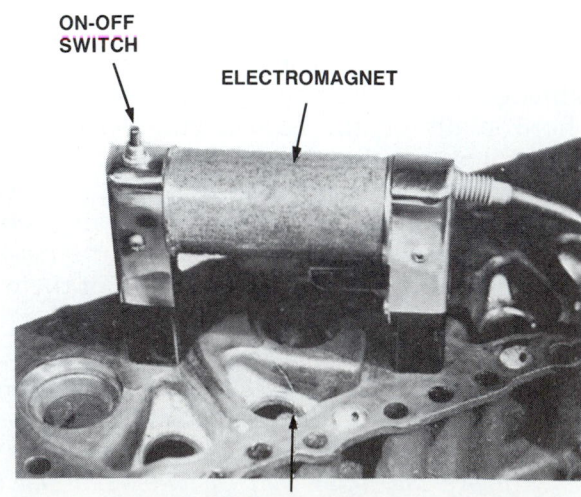

ON-OFF SWITCH

ELECTROMAGNET

Crack made visible

Figure 46.19 A portable magnetic crack detector. *(Courtesy of Geo. Olcott Co.)*

(a)

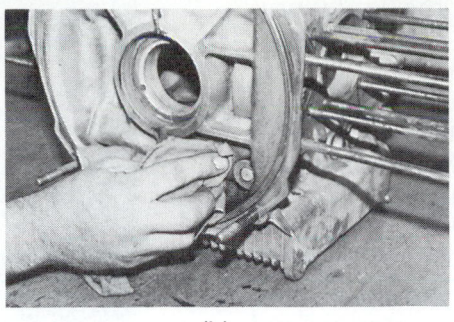
(b)

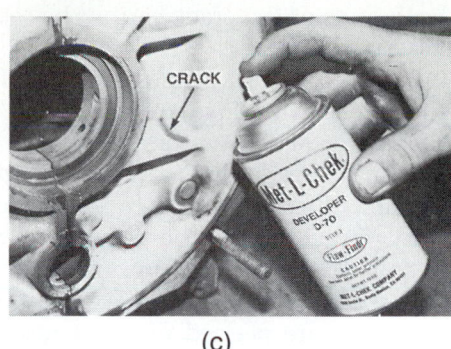
(c)

Figure 46.20 Checking for cracks with a dye penetrant: (a) Spray on penetrant. (b) After five minutes, clean the surface. (c) Spray on developer to highlight the crack.

are plugged and the head is filled with water or air. If air is used, soapy water is sprayed over the head surface, or the head is submerged in water to check for air bubbles.

CRACK REPAIR

Cracks are sometimes repairable, but the repair is only practical if the cost of a bare head is more than twice the cost of the crack repair. If there is any question whatsoever as to the effectiveness of a crack repair, it is not worth taking a chance. A new or used head should be obtained.

Tapered Plugs

Cracks in iron heads are commonly repaired with tapered, threaded plugs. This process is known in the industry as **pinning or stitching** a crack. The plugs are usually iron, but sometimes brass plugs are used. Both ends of the crack must be drilled to prevent the crack from spreading any further (Figure 46.21a). If

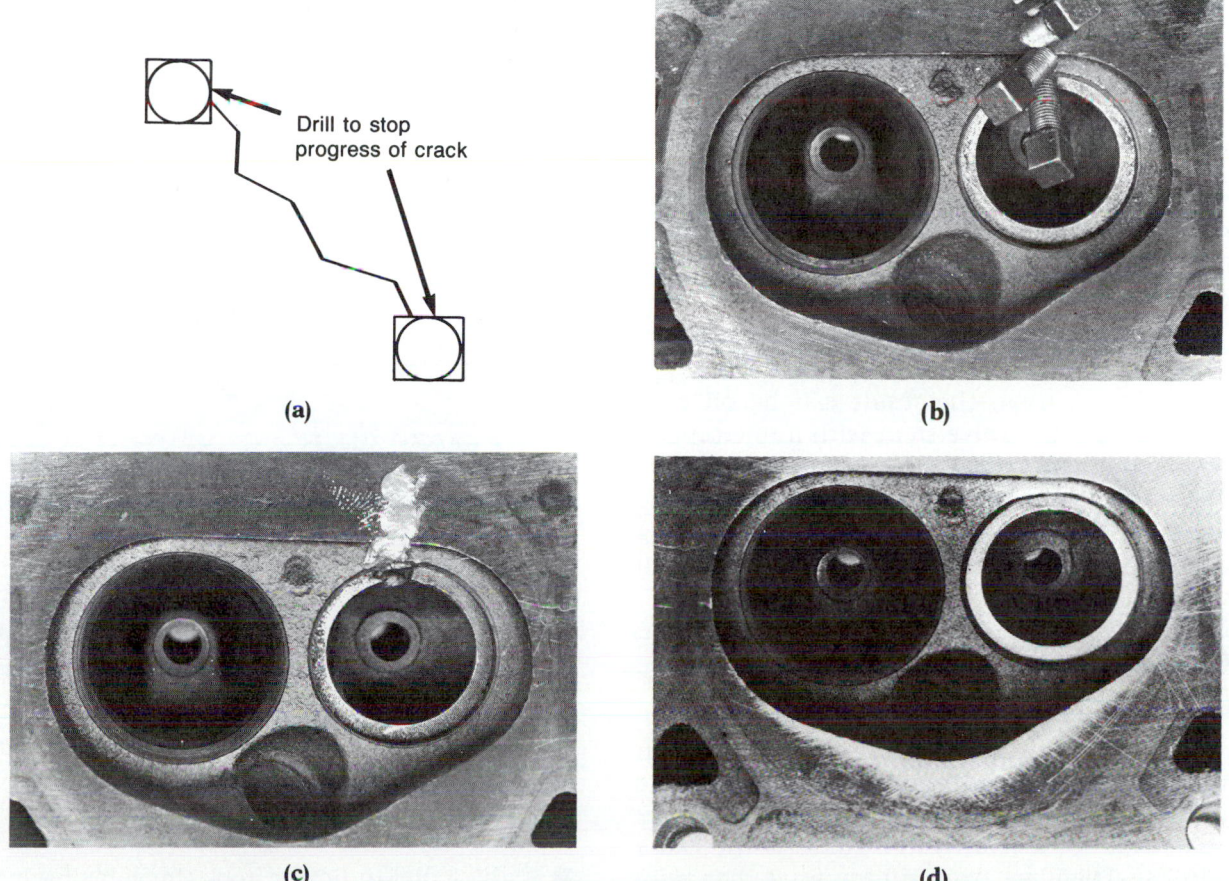

(a) Drill to stop progress of crack

(b)

(c)

(d)

Figure 46.21 Pinning a crack: (a) Drill both ends of the crack. (b) Install the pins one at a time and then cut them off. For illustration purposes, the plugs were not cut off as each one was installed. (c) Peen the cutoff pins to seat them. (d) Grind and clean the chamber. Install an insert seat. *(Courtesy of Irontite Products Co.)*

both ends of the crack cannot be seen, the head is usually scrapped.

To pin a crack, holes are drilled and tapped along the crack. The plugs are dipped in ceramic sealer and then threaded into the holes and broken off (Figure 46.21b). More plugs are installed so that they overlap the first plugs. Each plug overlaps about one-third of the next plug all along the crack (Figure 46.21c). They are installed at different angles so that they will interlock above and below the surface. To help close the crack, the plugs are peened outward toward the casting and the casting is peened inward toward the plugs. The surface is ground flush to finish the surface (Figure 41.21d).

Welding Heads

Welding is a common method of repairing aluminum head cracks. The aluminum is preheated and then welded using an inert gas shield. The process is known in the industry as *heli-arc*, or *TIG* (*tungsten inert gas*) welding.

Iron heads are not welded as often. Welding an iron head is a very specialized job. An iron head must first be heated in an oven to about 700°F, or the head will crack as it is welded. Specialty shops usually weld iron heads using an inert gas welder. The head is resurfaced to complete the repair.

■ CHECKING VALVE SPRINGS

Springs are tested for tension, squareness, and height. Spring tension is tested on a spring tester (see Figure 46.37). A scale on the side of the tester provides a place to check for squareness as well as spring height. Specifications are available in the service manual and in booklets published by spring manufacturers. A spring is allowed to be 10% less than specifications.

■ CHECKING VALVE STEMS

When valve stems wear, the result can be oil consumption. Measure the valve stem with a micrometer at an area of the valve stem above where it normally rides in the valve guide (Figure 46.22). Then measure on the worn area at the top of the stem. Comparing the two will give the amount of wear. Typical wear will be less than 0.001".

■ VALVE GUIDE SERVICE

Checking Valve Guides

Valve guides wear in a bellmouth fashion, which can result in an increase in oil consumption (see Figure 18.5). When a valve seat has worn and is much wider than usual, look for a worn valve guide as the cause (Figure 46.23). There are two primary ways that *valve stem-to-guide clearance* is checked.

■ Use a split ball gauge and a micrometer (see Figure 5.24b). Measure at the top, bottom, and center.

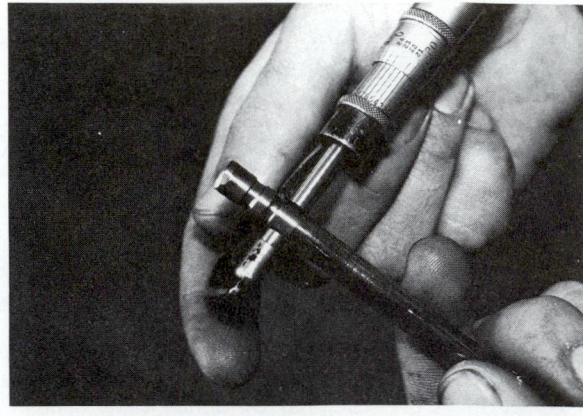
(a)

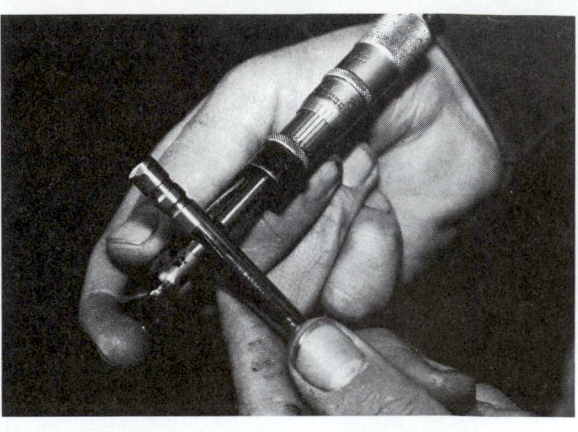
(b)

Figure 46.22 Inspecting the valve stem for wear: (a) Measuring the unworn portion. (b) Measuring the worn portion.

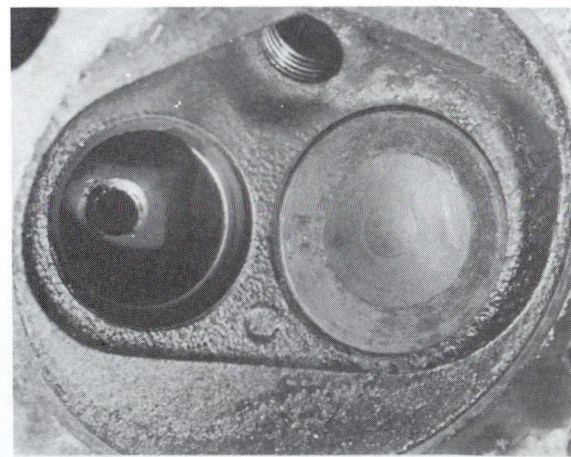

Figure 46.23 This exhaust valve seat is badly worn and is considerably wider on one side than the other. Look for a badly worn valve guide as the cause.

■ Using a dial indicator, rock the valve back and forth. When measuring with an indicator, divide the measurement by two for the amount of valve-to-guide clearance.

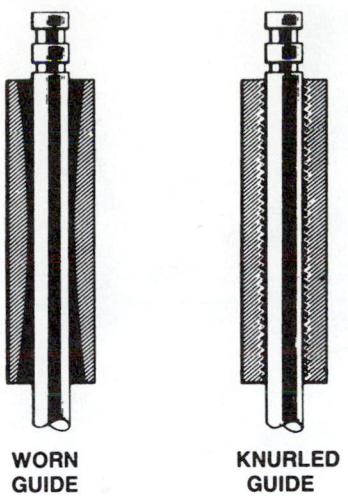

WORN
GUIDE

KNURLED
GUIDE

Figure 46.24 A valve guide before and after the knurling process. *[Courtesy of Kwik-Way Mfg. Co.]*

Figure 46.25 A thinwall guide liner (L) and a solid guide liner (R).

■ GUIDE REPAIR

Guides can be repaired in several ways:
- A worn integral guide can be bored out to accept a pressed-fit insert guide.
- A worn insert guide can be pressed out and replaced with a new one.
- Guides can be repaired using a process called **knurling.** Knurling makes the inside of the guide smaller. Then it is reamed out to fit the valve stem (Figure 46.24).
- A **thinwall** insert can be installed in a valve guide (Figure 46.25). First the guide is bored oversize and the insert is pounded in. Next it is broached to form it to the guide bore. Finally it is reamed to provide the correct clearance for the valve stem.

A pressed-fit guide can also be repaired by knurling (only on cast iron guides) or installing a thinwall insert. This practice is sometimes easier or less expensive than replacing the valve guide.

■ GRINDING VALVES

Valves are refinished on the face angle using a valve grinder (Figure 46.26). The stem tip is reground flat

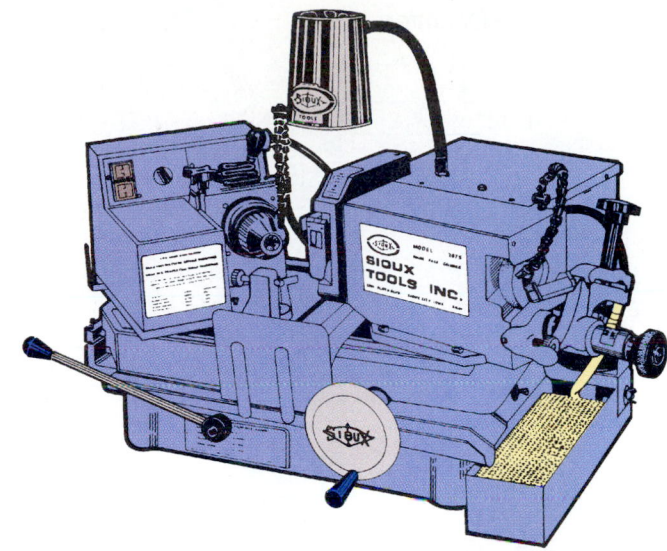

Figure 46.26 A valve grinder. *[Courtesy of Sioux Tools, Inc., Sioux City, IA]*

during valve grinding (Figure 46.27). It is also ground to a chamfer (Figure 46.28). Be sure to wear eye protection when working around grinding wheels. Dust comes off of the grinding wheel and grinding wheels have been known to explode. The grinding wheel is dressed with an industrial diamond that wears very little as it removes the outer surface of the grinding wheel (Figure 46.29).

The angle of the chuck is adjustable for grinding valve faces to different angles. Forty-five degrees is the most common angle. Some machinists grind an **interference angle** between the valve and the seat (Figure 46.30). The valve is usually ground to 44° and the seat is ground to 45°. This helps the newly reground seats and valves to seat into each other.

The valve grinder has an adjustable *stop* (Figure 46.31) that prevents the neck of the valve from acci-

Figure 46.27 Grinding the stem tip.

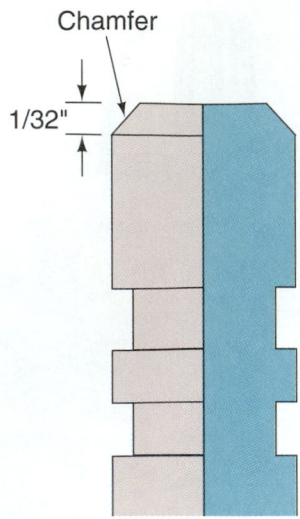

Figure 46.28 The stem tip is ground to a chamfer. *(Courtesy of Federal-Mogul Corporation)*

Figure 46.29 The grinding wheel is dressed with an industrial diamond.

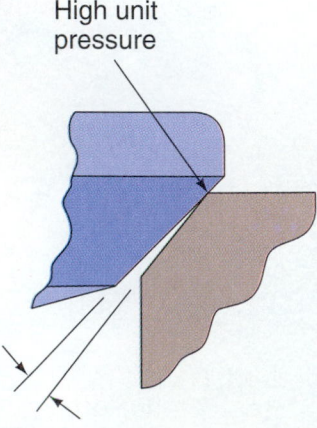

Figure 46.30 An interference angle. *(Courtesy of Federal-Mogul Corporation)*

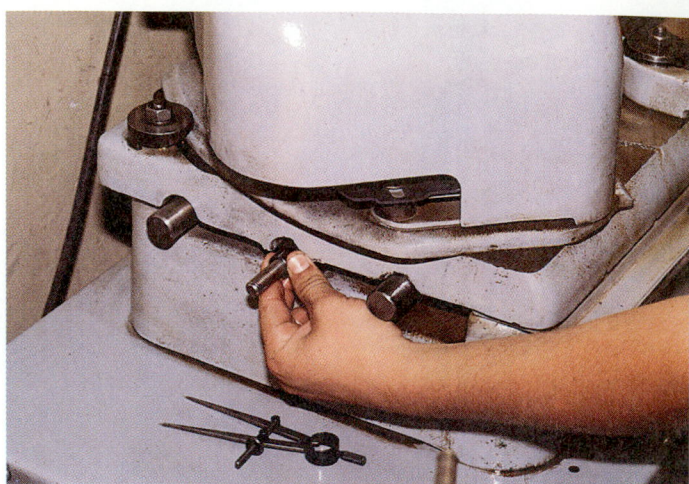

Figure 46.31 Adjust the carriage stop so that the neck of the valve will not come into contact with the stone.

Figure 46.32 This valve has been ruined because the neck area was ground.

dentally contacting the grinding wheel, which would ruin the valve (Figure 46.32).

Grinding oil is pumped to a nozzle that flows it onto the face of the valve during grinding. The valve can be observed as it turns in the chuck. If it wobbles, it is either bent or it is not chucked properly.

Very little metal is removed from the surface of the valve face during grinding. If too much metal is removed from the valve face, the valve face margin will be too thin and the valve could burn. The face is ground until all "pits" are removed. Pits are small, dark indentations in the surface of the metal on the valve face.

▉ GRINDING VALVE SEATS

Valve seats are refinished with a grinding stone (Figure 46.33) or a seat cutter (Figure 46.34). Grinding three angles makes it possible to control the width of the seat (Figure 46.35). The intake seat is usually ground to

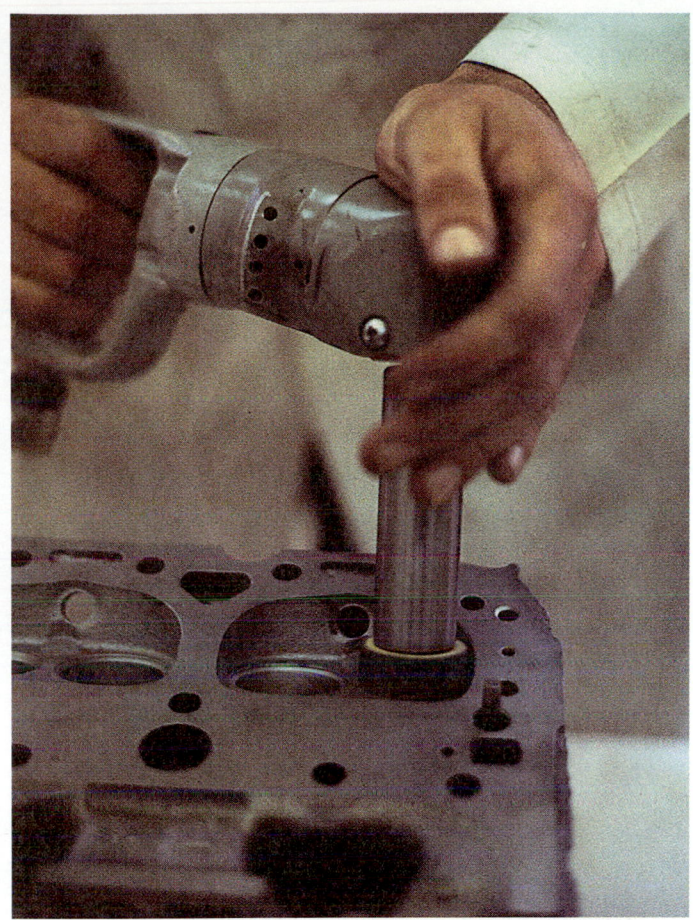

Figure 46.33 Grinding a valve seat.

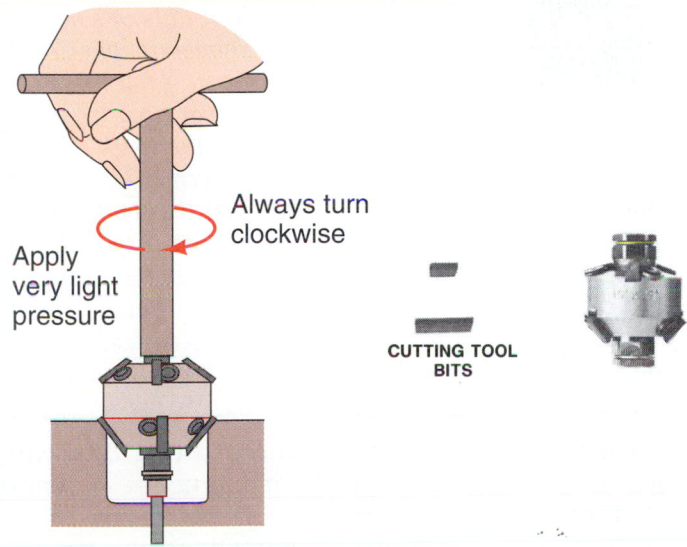

Figure 46.34 Seats are sometimes refinished with a carbide seat cutter. *(Courtesy of Neway and K-Line Industries, Inc.)*

a width of about ¹⁄₁₆" and the exhaust valve is ground to about ³⁄₃₂" (Figure 46.36). Exhaust seats are wider than intake seats because valve cooling is critical. If the seat is too narrow or the valve clearance adjust-

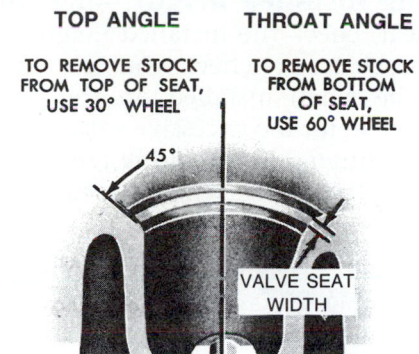

Figure 46.35 A three angle valve seat. *(Courtesy of Ford Motor Company)*

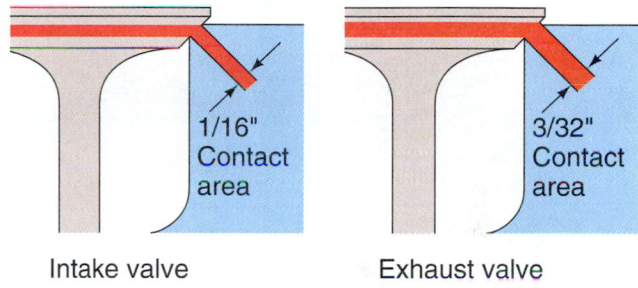

Figure 46.36 Intake and exhaust seat width. *(Courtesy of Federal-Mogul Corporation)*

ment is too tight, the valve will not be able to dissipate enough heat to keep it from burning.

The 45° angle that contacts the valve face is ground or cut until the entire seat area is cleaned up and is free of pits. Sometimes the seat will polish up with very little metal removed. When this happens, machining of the other two angles may not be necessary.

The 60° angle in the bottom of the seat, called the throat angle, is cut *very lightly*. This angle can be done by hand (with no grinding motor). Cut the throat until a ring shows up around the entire bottom of the 45° area. Be careful not to encroach into the seat area with this angle.

> **CAUTION** Removing too much of the seat with the 60° cutter or grinder will result in a need to replace the seat.

The top angle is cut until the width of the seat is correct. The head must be thoroughly cleaned of all grit before beginning assembly.

■ CHECKING INSTALLED HEIGHT OF THE VALVE STEM

When the seat and valve are reground, the stem moves further into the cylinder head. This results in increased

valve spring installed height. After grinding the valve and seat, check the installed height. That measurement can be used to check what the tension of the spring will be when installed on the head (Figure 46.37). To correct for too excessive valve stem installed height and maintain correct spring tension, shims are installed under the springs when a head is reassembled (Figure 46.38).

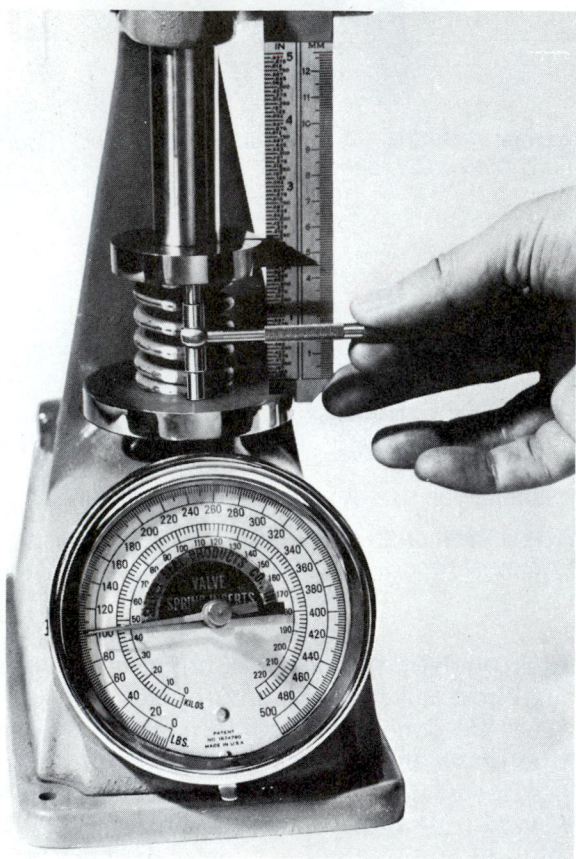

Figure 46.37 Use the installed height measurement to check what the actual tension will be. *[Courtesy of Silver Seal Products]*

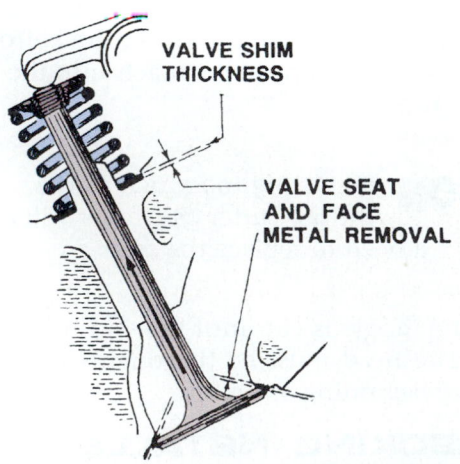

VALVE SHIM THICKNESS

VALVE SEAT AND FACE METAL REMOVAL

Figure 46.38 The installed height of the spring changes due to wear or metal removal during machining. *[Courtesy of Hastings Manufacturing Company]*

■ SOLVENT TESTING THE VALVE AND SEAT

After the valve and seat have been ground:
- Turn the head over so that the combustion chambers face up.
- Place the head on head stands and put it on a shelf in the solvent tank.
- Install the valves in their ports.
- Install the spark plugs in their holes.
- Fill the combustion chambers with solvent and check for leaks.

The solvent test is a very important test. If the valves seal against leakage without springs, they will seal when the engine is started. If any of the valves leak, they can be repaired now while it is easy to do.

■ REASSEMBLING THE HEAD

Clean the head thoroughly before reassembly. Be sure that the guides have been thoroughly cleaned too. Lubricate all valve stems, faces, and seats.

■ VALVE GUIDE SEAL INSTALLATION

Install the guide seals before installing the springs on all but O-ring seals.

NOTE: *O-ring valve guide seals must be installed **after** the spring is compressed, or they will be ruined during assembly.*

CASE HISTORY *A student was restoring a Camaro that had 90,000 miles on it. It did not smoke or use oil, but he wanted to put the engine in as-new condition. After a careful and complete rebuild, he took the car on a trip. It used a quart of oil every 200 miles. All the spark plugs had carbon deposits on one side of the center insulator only. With compressed air injected into the spark plug holes, the valve springs were removed to inspect the valve guide seals. It was discovered that the O-ring seals had been put onto the valve stem **before** the spring was installed. Each seal was twisted. Excessive oil consumption was the result.*

Oil the seals before installation so that they will slip easily into the spring retainer and not be accidentally torn.

On a worn engine, valve guide seals can be replaced with the head on the engine (see Figures 46.7 and 46.8). Air is put into the cylinder to hold the valve while the spring is compressed (Figure 46.39). The piston for each cylinder is positioned at TDC with both valves closed. If both rocker arms are removed, piston position is unimportant (unless the air is disconnected while the valve keepers and spring are removed from the valve).

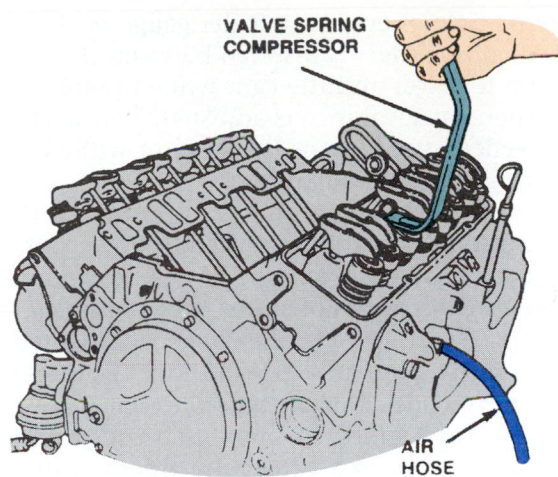

Figure 46.39 The valve is held closed with air pressure while the spring is compressed. *[Courtesy of General Motors Corporation, Service Technology Group]*

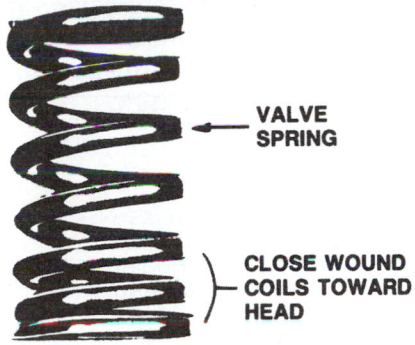

Figure 46.40 When a spring has coils that are more tightly wound at one end, that end is installed toward the head. *[Courtesy of General Motors Corporation, Service Technology Group]*

INSTALL THE VALVE AND SPRING ASSEMBLY

Some springs have coils more closely spaced at one end than at the other. During assembly, be sure that the end that is more tightly coiled is positioned against the cylinder head (Figure 46.40). To prevent damage to the valve guide seal, compress the spring just enough to install the keepers. Inspect each keeper for wear before installing it. Use grease to help hold keepers in place during reassembly. After assembly, tap the top of each valve tip with a soft face hammer to check that the keepers are seated.

ROCKER ARM SERVICE

Engines that do not have hydraulic clearance adjustment sometimes experience wear between the rocker arm and valve stem tip. This makes it difficult to get a proper valve clearance adjustment. Cast rocker arms that are shaft-mounted can be reground. Stud-mounted rocker arms are not serviceable and are replaced when worn.

Sometimes rocker studs pull out of the head, or the rocker wears against the rocker stud. An oversize stud can be installed by a machine shop. A threaded replacement stud, called a *screw-in stud*, is also available.

INSPECT PUSHRODS

Inspect the ends of the pushrod and the surface of the socket where it pivots on the rocker arm. Look for pitting or other unusual wear. Roll the pushrods on a bench to check if they are bent.

INSPECT OHC CAMSHAFT

Overhead camshafts often have oil galleries and holes drilled in cam lobes for direct lubrication. On some camshafts, the oil holes are small and are prone to plugging (Figure 46.41). Check that the oil holes are clear before installing a used camshaft in the head. This is a good practice to follow when installing a new cam too.

REASSEMBLING OHC HEADS

On overhead cam engines, reinstall the camshaft in the head.
- Turn the cam until a cam lobe faces away from the valve.
- Reinstall the rocker arm with a screwdriver or prybar to compress the valve spring a small amount, as shown in Figure 46.1.
- Adjust the **valve clearance (lash)** before installing the head on the engine.

Valve Lash

Valve lash must be enough to allow heat to dissipate from the valve to the valve seat. As pushrods, rocker arms, and other valve train parts wear, lash *increases*. Lash *decreases* as the valve and seat wear, allowing the valve stem to move further into the valve guide (Figure 46.42).

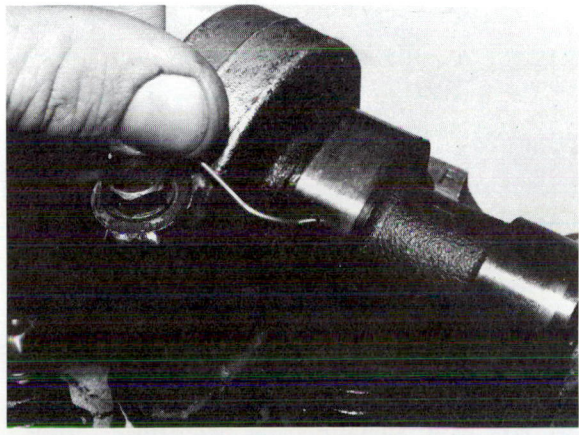

Figure 46.41 Some OHC engines have drilled camshafts to provide lubrication to the cam lobes.

NOTE: *If the lash is too tight the valve will not contact the valve seat for a long enough time to cool. This will cause the valve to burn.*

Overhead cam engine valve lash is usually easy to adjust. When an OHC cylinder head has been reassembled, the valve lash can be adjusted before installing the head on the engine.

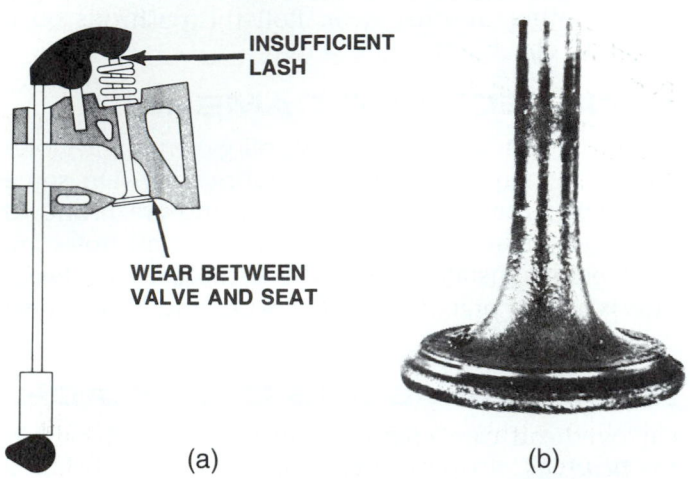

(a)   (b)

Figure 46.42 Problems related to valve lash: (a) Lash decreases because of head expansion or valve and seat wear. (b) The wear on this valve would result in decreased lash. *(Courtesy of Federal-Mogul Corporation)*

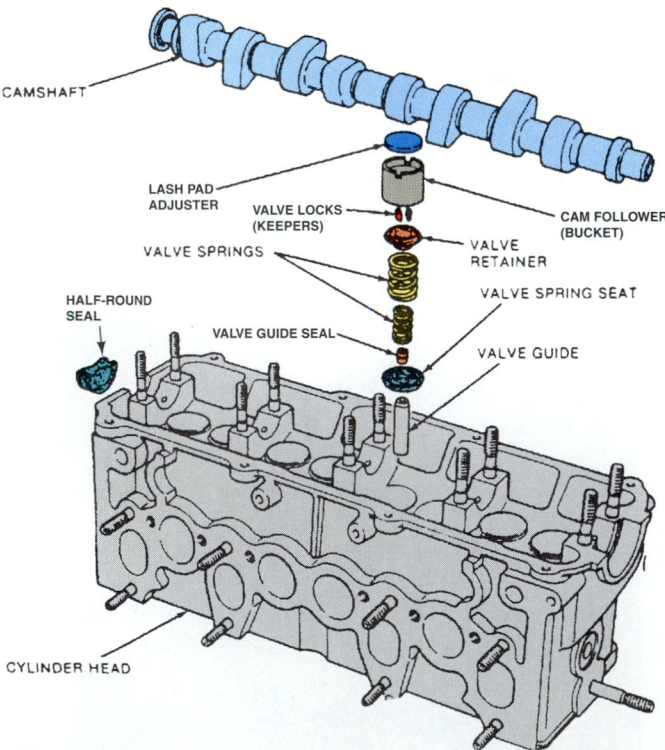

Figure 46.43 Some engines use different thicknesses of disks to adjust valve clearance. *(Courtesy of Chrysler Corporation)*

- On one type of engine, a feeler gauge of the specified thickness is inserted between the rocker or cam follower and the cam while the lobe faces up. Then the clearance is adjusted. Two wrenches are needed: one loosens the locknut while the other makes the adjustment.
- On some OHC engines, the cam pushes on the valve through an adjustable cam follower (often called a "bucket"). One type of head design uses different thicknesses of shims called *lash pad adjusters* (Figure 46.43). Clearance is measured between the shim and the cam lobe (Figure 46.44)

  There is a special tool available that forces the bucket away from the cam so that the shim can be replaced during a valve adjustment. Shims can be removed with a blast of compressed air, a screwdriver and magnet, or a pick. The shim is measured (Figure 46.45), and then a new one of the proper thickness is installed.
- Another kind of cam follower has a tapered adjusting screw. Each turn of the screw changes clearance by 0.003". Adjustments must be done in one turn increments (0.003").

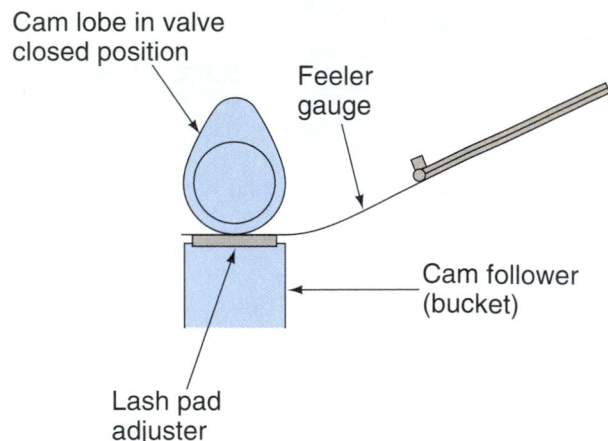

Figure 46.44 Clearance is measured between the shim (lash pad adjuster) and the cam lobe. *(Courtesy of Ford Motor Company)*

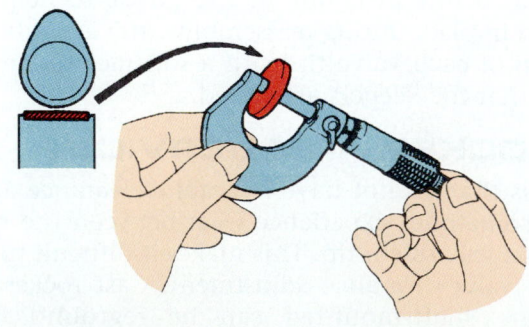

Figure 46.45 The shim is measured and a new one of the correct thickness is installed.

■ CAMSHAFT SERVICE

Measuring the Camshaft for Wear

Specifications are not readily available for the height of cam lobes. A comparison measurement can be made by measuring each lobe and seeing if they are all close to the same. A visual check of the cam lobes is the standard way of looking for wear (see Figure 18.36). The cam and lifters become wear-mated during the first few minutes the engine is run.

NOTE: *If the cam is replaced, replace the lifters also. Used lifters will rapidly wear out a new cam. If a cam is to be reused, it is important that the lifters be kept in order for replacement on their respective lobes.*

Roller Cam Wear

Lobes on roller cams are polished to a fine matte finish. During engine break-in, the lifter burnishes the lobe to a smooth mirror finish. The visual wear pattern is not as important as it is with conventional cam lobes. To inspect for wear, measure the lobe and look for a wear ledge greater than 0.005".

■ LIFTER SERVICE

Hydraulic lifters can fail for several reasons:
- Dirt lodged in the check valve can allow the lifter to leak. Too much wear between the lifter and the body can cause excessive leak-down. This results in a noisy lifter.
- A lifter might be noisy because of an oil pressure problem.

> **SHOP TIP** An oil pressure problem will be evident on hollow pushrod engines if no oil is reaching the rocker arms.

- A lifter can become stuck because of varnish that accumulates between its plunger and body due to a lack of periodic oil changes.

> **SHOP TIP** A stuck lifter can sometimes be freed up by the addition of an additive to the oil, or by spraying fuel system spray cleaner down a hollow pushrod oil channel.

Hydraulic lifters are not rebuilt because they are relatively inexpensive. It is possible to disassemble and clean lifters but they *must* be reassembled with their mated parts. Because of the labor time involved, this procedure is usually impractical. Sometimes a piece of foreign material becomes lodged in the lifter check valve. If the lifter is removed and disassembled before serious wear between the lifter and cam lobe results, it may be cleaned.

Worn mechanical lifters can be reground. The base of the lifter is ground to a radius on a special lifter grinder.

Roller lifters have a longer service life than standard lifters. Wear can occur on the roller pin. Some of these pins are serviceable.

■ CAM AND LIFTER BREAK-IN

Lubrication and break-in are critical to the life of a cam. On a pushrod engine, the cam is usually lubricated by oil that is thrown from the crank, so long periods of idle are hard on the cam. A cam that survives the first half hour of use without wear should last the life of the vehicle with minimal additional wear. Chapter 50 covers the correct way to break in a camshaft and lifters.

■ TIMING GEAR, CHAIN, AND BELT SERVICE

Timing Gear Service

Cam gears are either pressed or bolted to the camshaft. Although it is uncommon, cam gears sometimes break. Inspect them for cracks. To install a pressed-fit camshaft gear:
- The rear of the cam gear must be supported while the cam is pressed from its center.
- After the new cam gear is pressed on, end play is checked with a feeler gauge (Figure 46.46).

The clearance between a cam thrust plate and the sprocket or gear varies from 0.0015" to 0.005". If a spacer is used, it must be from 0.0015" to 0.005" *thicker* than the thrust plate. The cam will press against the spacer, which fits inside the thrust plate hole, controlling end thrust. Aluminum gears can be heated for easier installation.

NOTE: *Do not forget the spacer on a cam with a bolt-on sprocket. The result could be a cam that will not turn after the bolt is tightened.*

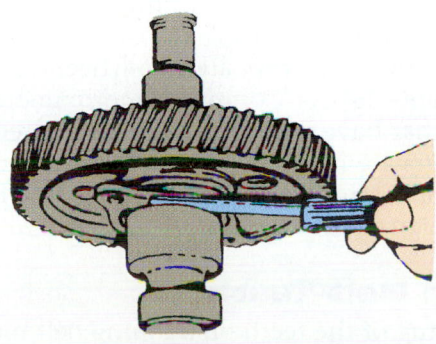

Figure 46.46 Checking clearance behind the thrust plate to determine camshaft end play. *(Courtesy of General Motors Corporation, Service Technology Group)*

CAUTION Aluminum must not be heated to more than 200°F. If it is, it will not return to its original size when it cools. The result will be a gear that is loose on the camshaft.

■ TIMING BELT SERVICE

Because timing belts are not lubricated, there are no sealing gaskets on the timing cover. A belt's service life can be affected by contact with foreign material: oil, mud, smog, or even ice. Timing covers for belts are designed to *thoroughly* cover the belt. If a foreign object gets into the belt area, damage to the tensile cord can result in belt failure. A damaged flange behind the sprocket can cut the belt.

Timing belts have a very strong fiberglass cord structure, but they are delicate. When overflexed, the cords can break. Do not twist them more than 90° and do not coil them up or hang them for storage.

To inspect a timing belt:
- Look for fraying, cracks at the base of a tooth, or loose fibers.
- Rotate the engine slowly by hand while checking it.
- Do not twist the belt.
- Wear on one side of a tooth indicates a misalignment problem.
- Check for oil leaks, which could damage the belt. An OHC head with a timing belt has a camshaft oil seal and some have an O-ring behind the cam sprocket. It is a good idea to replace these during a timing belt replacement.

See Chapter 22 for more information on timing belts.

■ TIMING BELT REPLACEMENT

Follow manufacturer's recommendations for the replacement interval for the belt. This varies from 20,000 to 60,000 miles. Some manufacturers have longer intervals. Goodyear recommends replacing all timing belts at 40,000 to 50,000 miles. When no recommendation is made, timing belts should be changed every four years.

Most American cars are free-wheeling. According to the Gates Rubber Company, approximately 40% of engines that have timing belts are interference. Japanese imports are more likely to be interference engines. Those American cars that are interference most often have foreign-made engines.

Timing Belt Teeth

The spacing of the teeth on a timing belt must exactly match the cogs on the cam and crank sprockets. Belts have different tooth rib profiles. Some have round ribs and others are square. Be certain of an exact match.

Compare an old belt to its replacement for belt width and for tooth shape and spacing.

When replacing a timing belt:
- Sprockets should be cleaned with a nonpetroleum-based solvent.
- Inspect the sprockets to see that wear on them will not cause wear on the timing belt teeth. Replace a sprocket that shows any kind of wear.
- Put a sticker on the valve cover or door record listing the mileage when the belt was installed.
- The belt tension is adjusted according to manufacturer's specifications.

Before installing the belt on the sprockets, put the crankshaft at TDC and be sure that the camshaft and crankshaft timing marks are aligned correctly. Check the manual for the correct positions of the marks. They vary among manufacturers.

Many engines have spring-loaded belt tensioners. On these engines, install the belt and make sure that there is tension on the drive side of the belt (Figure 46.47a).

Next, loosen the tensioner to allow the spring to force the tensioner roller against the belt (Figure 46.47b). Tighten the tensioner bolt and crank the engine over at least ten times. Tension on the belt will drop as it seats into the timing sprocket. Loosen and retighten the tensioner adjusting bolt to complete the job.

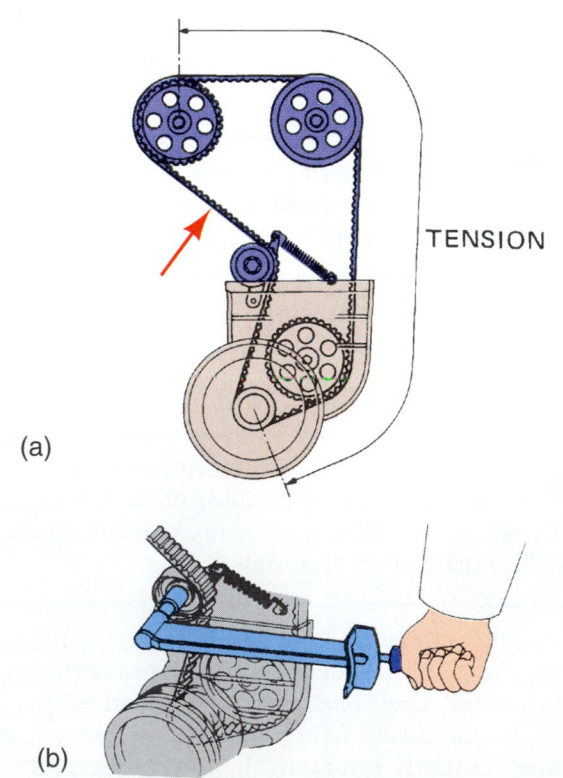

Figure 46.47 (a) Loosen the tensioner and let the spring pull the roller against the belt. (b) Before tightening the tensioner see that the belt is tight in the area noted here.

- Some manufacturers require the use of a belt tension gauge to set timing belt tension.
- An OHC head with a timing belt has a camshaft oil seal and some have an O-ring behind the cam sprocket (see Figure 46.32). It is good insurance to replace these during routine replacement of a timing belt.
- If a camshaft must be relocated to set cam timing on a nonfree-wheeling, belt-driven OHC engine, turn the crank to about 45° BTDC. Then all pistons will be slightly down in their bores and valve-to-piston interference will not occur.

NOTE: *Do not adjust timing belt tension on a hot engine. The belt will be too tight, which can result in a broken camshaft.*

TIMING CHAIN SERVICE

Excessive timing chain stretch can be checked in one of the following ways:

1. On engines without chain tensioners that have distributors driven by the camshaft, watch the timing mark while checking ignition timing at idle. If the mark moves back and forth, the chain is loose.

2. To check a pushrod engine for chain slack while the engine is in the car, remove the distributor cap and turn the vibration damper in one direction until the distributor rotor begins to turn. Then, turn the crank in the opposite direction until the rotor moves again. Observe how far the damper moves before the rotor turns. Because most distributors are driven by the cam, this is a good way to check for chain slack on these engines. The engine may still run acceptably with 10° to 15° of chain slack. But ideally, movement should be less than 5° at the crank.

3. The most obvious way to check chain slack is to measure the slack during engine disassembly, after removing the timing cover. Be sure to turn the crank in one direction first, in order to tighten up one side of the chain. Then measure the amount of slack in the chain.

NOTES:
- *Large pin silent chains are so flexible that they appear excessively sloppy when new. Typical movement on the slack side is around 0.200" in and out (0.400" total). Replacement is suggested when total movement is 1.00".*
- *Some timing chain tensioners are spring loaded and are locked by oil pressure when the engine runs. Do not turn these engines over backward by hand. Some types of tensioners will allow the chain to skip off correct valve timing.*
- *Long chains have chain guides with rubber surfaces that wear. The chain guides are usually replaced when the chain is replaced.*

Before installing a new chain, soak it in oil.

Timing the Cam to the Crank

There are several ways to time the cam to the crank. Be sure to check the manual before you install the timing chain or belt.

- Some timing sprockets are properly timed when the marks face each other.
- On others, there must be a certain number of chain links between the marks.
- Sometimes chains have colored links that must be aligned with the marks on the sprockets.
- Some overhead cams have a mark on the cam gear that lines up with a mark on the cylinder head when the timing mark on the damper is at the TDC mark.

If the old parts are available, carefully compare the new gears or sprockets with the old ones. Check the keyway and timing marks just in case there might have been an error made in manufacturing. Sprockets are sometimes stamped backward. Its much better to find the problem during assembly, rather than waiting until problems show up after reassembly and reinstallation.

VALVE COVER SERVICE

Most valve covers are made of sheet metal or cast aluminum. Some engines have been manufactured with plastic valve covers. These are bonded to the cylinder head at the factory with RTV (silicone) sealant. They must be clean of all oil and grease so that the new silicone will bond or stick to the valve cover.

Before installing a sheet metal or cast valve cover, check it for straightness. Put it on a flat surface with a flashlight under it. Look for light leaking through.

When bolts are tightened against a gasket, the bolt holes tend to pull down into the gasket. Flatten the bolt holes with a hammer before installing a valve cover (see Figure 46.10).

A valve cover gasket that is compressed a *small* amount will spring back against the torque of the valve cover screws. If the screw is tightened too much, the gasket is overly flattened and does not hold tension back against the screw. Apprentices often overtighten valve cover screws when a cork or rubber gasket is used. Check bolt tightness with an inch-pound torque wrench. Torque specifications are usually about 50 to 60 inch-pounds. You will be surprised at how little this is.

Besides the cylinder heads and camshaft, other parts of the breathing system include the intake manifold and exhaust manifold. Those topics, as well as turbochargers and superchargers are covered in Chapter 28.

REVIEW QUESTIONS

1. After removing the valve spring, what measurement is done before removing the valve?

2. What is the maximum amount of warp that a V-6 cylinder head can have?

3. If 0.020" is removed from a cylinder head, approximately how much will valve timing change?

4. If a crack in a head is to be repaired with tapered plugs, what is done to both ends of the crack?

5. Three things that a valve spring is checked for are tension, squareness, and _____.

6. The guide repair process that displaces metal inside of the valve guide to make it smaller before reaming it for stem-to-guide clearance is called _____.

7. An _____ angle is when a valve face is ground to an angle that is slightly flatter than the valve seat.

8. A _____ is installed under a valve spring to correct the installed spring height.

9. A _____ test can be done after a valve grind to ensure that the seats do not leak.

10. When _____ are to be reused, they must be kept in order during engine disassembly.

ASE STYLE REVIEW QUESTIONS

1. Technician A says that bead blasting is a good method for cleaning OHC heads. Technician B says that an OHC head that has a stuck camshaft will probably require straightening. Who is right?
   a. Technician A   b. Technician B
   c. Both A and B   d. Neither A nor B

2. Technician A says that surfacing a cylinder head results in an advance in valve timing. Technician B says that a head that has no carbon in only one of its combustion chambers probably has a crack. Who is right?
   a. Technician A   b. Technician B
   c. Both A and B   d. Neither A nor B

3. Technician A says that magnaflux is a crack detection method that is commonly used on aluminum heads. Technician B says that an exhaust valve seat should be wider than an intake seat Who is right?
   a. Technician A   b. Technician B
   c. Both A and B   d. Neither A nor B

4. Technician A says that valve guide seals can often be replaced while the head is installed on the engine. Technician B says that pushrod engines can have the valve lash adjusted before installing the head on the engine. Who is right?
   a. Technician A   b. Technician B
   c. Both A and B   d. Neither A nor B

5. Technician A says that timing chain stretch is sometimes noticed when using a timing light. Technician B says to always be sure that timing marks are facing each other when timing the camshaft to the crankshaft. Who is right?
   a. Technician A   b. Technician B
   c. Both A and B   d. Neither A nor B

6. Technician A says if a spring has rusted, its tension will change. Technician B says if the cam side of an OHC head is warped excessively, the head can be straightened. Who is right?
   a. Technician A   b. Technician B
   c. Both A and B   d. Neither A nor B

7. A valve has burned. Technician A says this could be the result of a valve seat that is too wide. Technician B says this can be the result of a valve margin that is too thin after regrinding a valve. Who is right?
   a. Technician A   b. Technician B
   c. Both A and B   d. Neither A nor B

8. Technician A says that if a head is warped, running an engine without the thermostat is a likely cause. Technician B says that an overheated engine can result in a warped cylinder head. Who is right?
   a. Technician A   b. Technician B
   c. Both A and B   d. Neither A nor B

9. Technician A says that OHC cylinder heads often have oil galleries in them. Technician B says that pushrod engine cylinder heads often have oil galleries in them. Who is right?
   a. Technician A   b. Technician B
   c. Both A and B   d. Neither A nor B

10. Technician A says the torque recommendation for cork gaskets is 60 foot-pounds. Technician B says that if more metal is removed from a head surface than is desirable, a shim can be installed to compensate. Who is right?
    a. Technician A   b. Technician B
    c. Both A and B   d. Neither A nor B

Engine Diagnosis and Service: Block, Crankshaft, Bearings, and Lubrication System

■ OBJECTIVES

Upon completion of this chapter, you should be able to:

✔ Analyze wear and damage to the cylinder block.

✔ Select and perform the most appropriate repairs to the block, crankshaft, and bearings.

✔ Analyze wear and damage to the crankshaft and bearings.

✔ Analyze wear and damage to lubrication system parts.

✔ Select and perform the most appropriate repairs to the lubrication system.

■ KEY TERMS

line honing or line boring
cylinder glaze
crosshatch
piston slap
lugging
tolerance
crank polishing
plastigage
sending unit

■ INTRODUCTION

Usually, a cylinder block can be reused after certain service procedures are performed. Some blocks with excessive wear will have to be bored oversize to accept new, larger pistons; others will need only cleaning and some minor service operations. Others will require major service operations, such as align boring of main bearing bores, or sleeving of cracked or damaged cylinders. As you read this chapter, you should become familiar with the practice of diagnosing the cause of problems before attempting a repair. This is the skill that separates the successful technicians from the average ones.

■ CLEANING THE BLOCK

First, the block must be thoroughly cleaned. All core plugs, oil gallery plugs, cam bearings, and any other removable parts must be removed.

■ OIL AND WATER PLUG REMOVAL

Oil gallery plugs and threaded heater hose connections have tapered pipe threads. Pipe threads wedge together, making a very tight seal (see Chapter 6). There are three types of gallery plugs. The female plug is removed with a special plug driver that resembles a ⅜" to ¼" socket adapter, except that it is solid (Figure 47.1a).

SHOP TIP Using a socket adapter to remove these plugs will usually result in a broken tool.

Male plugs can be removed with an 8-point or square drive socket (Figure 47.1b). Use an allen wrench with a socket drive to remove an allen type of plug (Figure 47.1c). An air impact wrench usually makes removal of these plugs easier. Following removal, oil passages can be cleaned with a cleaning brush.

The front of the block often has small core plugs that seal the ends of the oil galleries. After the rear threaded plugs are removed, the front core plugs can be knocked out from the rear with a long metal rod.

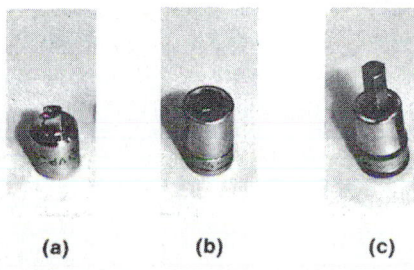

(a)   (b)   (c)

Figure 47.1 Three types of gallery plug tools: (a) Square oil-plug tool. (b) An 8-point socket. (c) An Allen wrench with socket drive.

SHOP TIP Occasionally, plugs will not come out because they are very tightly rusted in place. A rusted plug can usually be removed by heating it with an oxyacetylene torch (Figure 47.2a) and then applying paraffin wax (a door-ease wax stick is handy) to the drive hole of the plug (Figure 47.2b). The wax acts like a heat sink, shrinking the plug. The plug is then removed (Figure 47.2c). If all else fails, an extractor set may be required (see Chapter 6).

Clean the block with one of the methods discussed in Chapter 9. Lubricate all machined areas *immediately* to prevent rusting.

Check for Cracks

If the engine had a coolant consumption problem, check the block for cracks in the cylinder bores after hot-tanking (Figure 47.3). Some late model V-type blocks also have a tendency to crack in the lifter valley area.

Clean Oil Galleries

Deposits from engine oil galleries and supply holes must be removed after block cleaning to prevent loos-ened particles in the passageways from damaging a newly rebuilt engine. Oil galleries are filled with sludge that can trap metal shavings and grinding grit produced during block machining. Use an inexpensive stiff bristle brush (called a "rifle brush") with hot soapy water to clean galleries that run the full length of the block (Figure 47.4). Gallery cleaning brush sets are available from engine parts suppliers.

NOTE: *Failure to clean galleries is an invitation to engine failure.*

CASE HISTORY *An apprentice was rebuilding an engine that had burned a rod bearing. He neglected to carefully clean the oil galleries. After the engine was installed and started, metal from the galleries mixed with the engine oil. The result was severe damage to the crankshaft and bearings.*

■ OIL AND WATER PLUG INSTALLATION

After cleaning the galleries, reinstall the plugs. Threaded oil gallery plugs are coated with sealer or

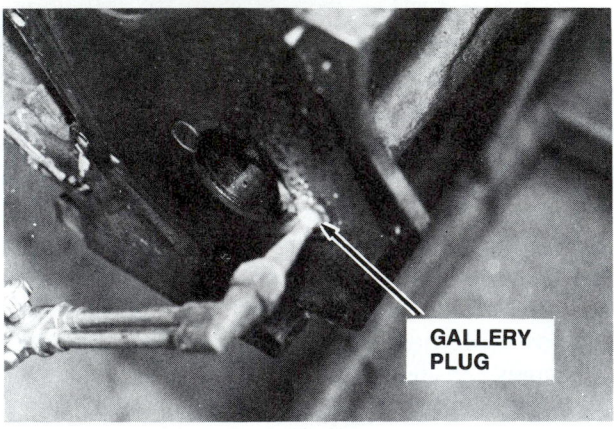

(a)

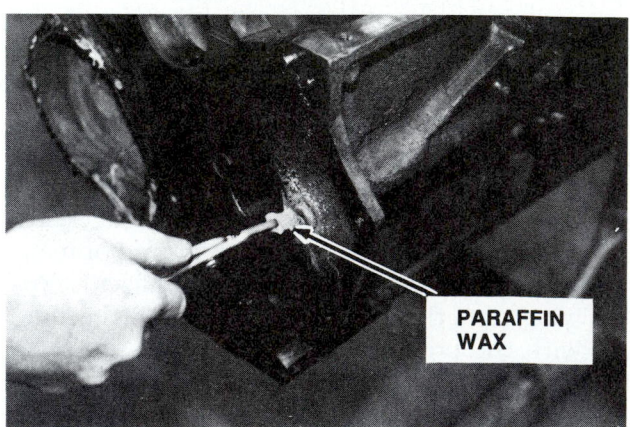

(b)

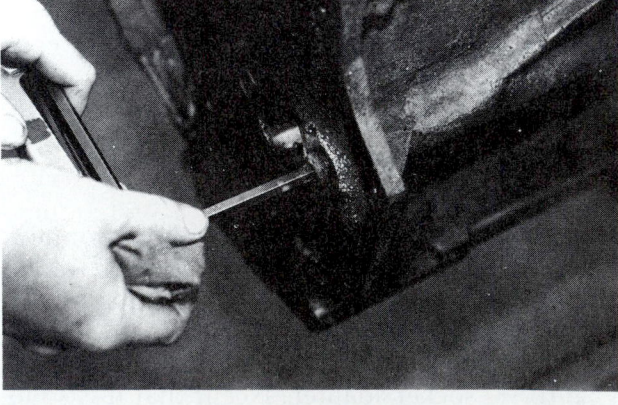

(c)

Figure 47.2 Removing stuck oil gallery plugs: (a) Heat the plug with a torch. (b) Apply paraffin to the plug. (c) Remove the plug.

Figure 47.3 A crack in a cylinder wall. *(Courtesy of Geo. Olcott Co.)*

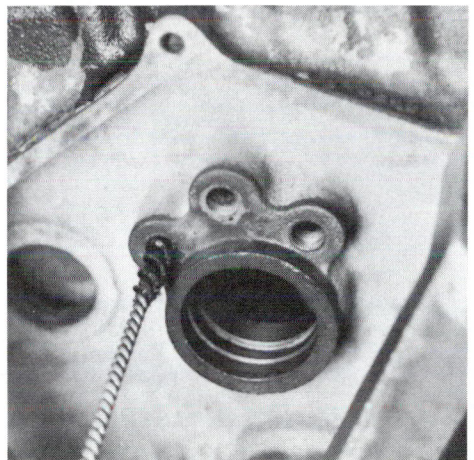

Figure 47.4 Clean all oil passages with a brush.

teflon tape and tightened into the block (see Chapter 6). Be careful not to overtighten threaded plugs. Pipe threads are tapered and wedge tightly to the block as the diameter expands.

Oil gallery pressed-fit core plugs are installed with red loctite. *Cross stake* the outside of the core holes with a chisel to prevent high oil pressure from blowing the plugs out (Figure 47.5). If the plugs come out, the engine will lose oil pressure.

■ INSPECT AND CLEAN LIFTER BORES

Lifter bores must be clean to ensure that the lifters will spin. The bores can be cleaned with a brake hone turned by hand. Be careful not to enlarge the lifter bore by excessive hone contact during the cleaning operation.

■ CHECKING MAIN BEARING BORE ALIGNMENT

Repeated heating and cooling of an engine block can result in misalignment of the main bearing bores. Main bearing bore alignment can be checked with a

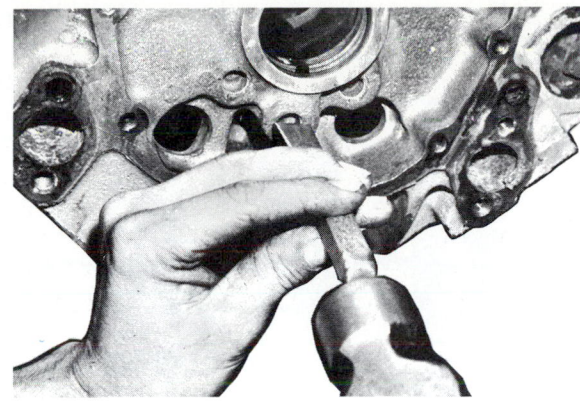

Figure 47.5 Oil-gallery core holes are cross-staked with a cold chisel. If the plugs come out, the engine will lose oil pressure.

0.0015" feeler gauge and a straight bar or a straight-edge (Figure 47.6).

SHOP TIP Do not mount the block on a universal engine stand during the alignment check. Hanging the block from one end could produce a false reading because of the flexibility of the casting (Figure 47.7). Set it on the floor or a bench top.

NOTE: *When a main bearing has become hot enough to burn or the block has changed color the bore will usually shrink (Figure 47.8), which means that the main bearing bores must be remachined. When a crankshaft breaks, check for excessive main bearing bore stretch. Any main bearing bore that shows discoloration should also be checked for cracks before doing any repair work on the block.*

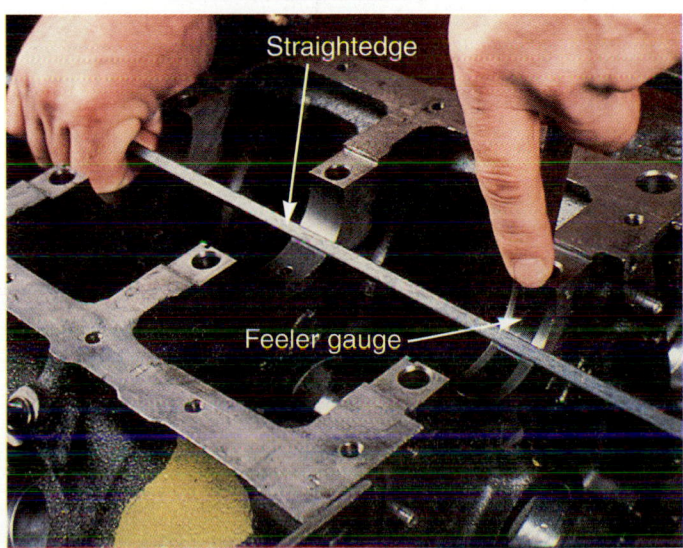

Straightedge

Feeler gauge

Figure 47.6 Checking main bearing bore alignment. *(Courtesy of Federal-Mogul Corporation)*

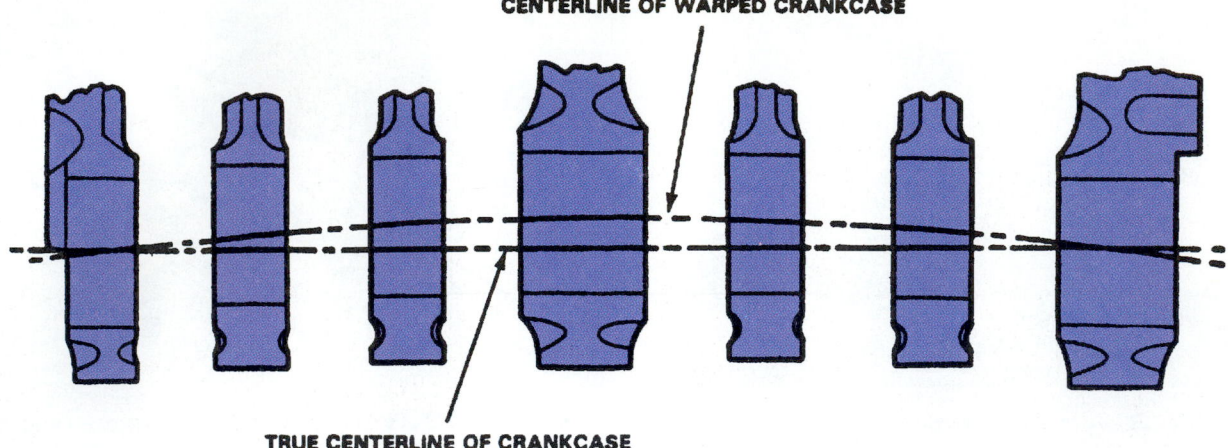

CENTERLINE OF WARPED CRANKCASE

TRUE CENTERLINE OF CRANKCASE

Figure 47.7 The crankcase can be warped. *(Courtesy of Federal-Mogul Corporation)*

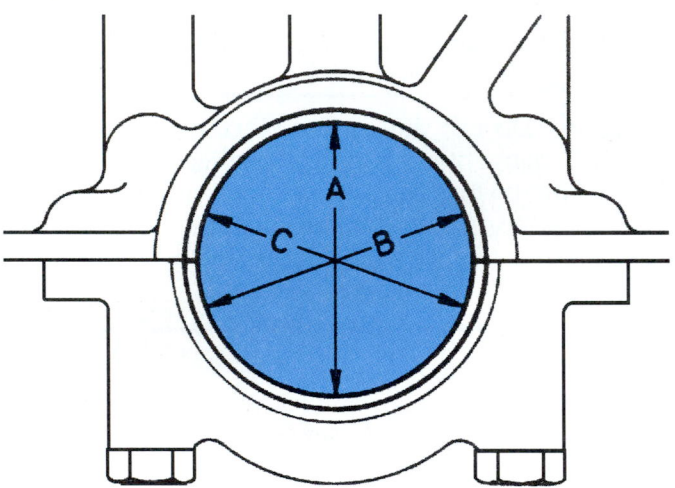

Figure 47.8 If a main bearing has become hot, the measurement at C and B might be less than the measurement at A. *(Courtesy of Mercedes-Benz of North America, Inc.)*

Figure 47.9 Checking main bearing bores with a precision gauge. *(Courtesy of Sunnen Products Company, St. Louis, Missouri)*

Figure 47.10 Honing main bearing bores with a line honing machine. *(Courtesy of Sunnen Products Company, St. Louis, Missouri)*

Main bearing bores can be checked with a dial bore gauge as shown in Figure 47.9. The vertical measurement should not be larger than the horizontal; if it is, stretch has occurred. An out-of-round measurement of less than 0.001" is acceptable if the horizontal reading is the largest.

Realigning Main Bores

Realignment of the main bores is accomplished either by **line honing or line boring** (Figure 47.10). First, the main caps are ground or milled on their parting faces (where they meet with the block). Then, the bores are aligned by honing or boring to the original main bearing bore size. The limitation to this procedure is that removing too much metal during this procedure can move the crank up too far into the block. This results in excessive timing chain slack (or retarded cam timing in engines with a chain tensioner). It also means that the piston will come up higher in the cylinder, which causes higher compression.

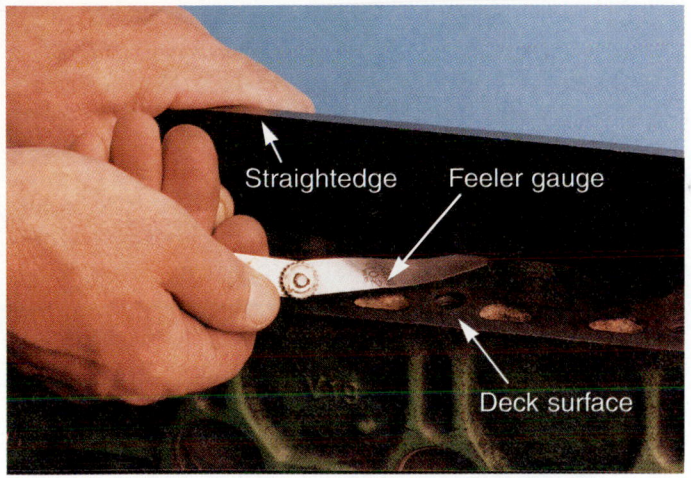

Figure 47.11 Check the deck surface for flatness.

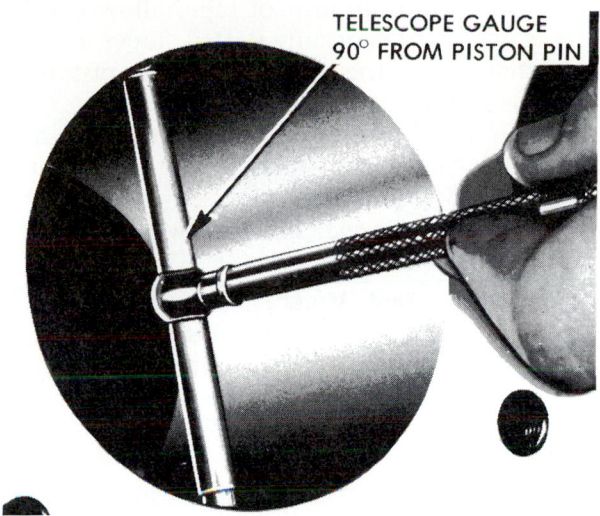

Figure 47.12 Maximum wear is 90° from the piston pin just below the ring ridge. *(Courtesy of General Motors Corporation, Service Technology Group)*

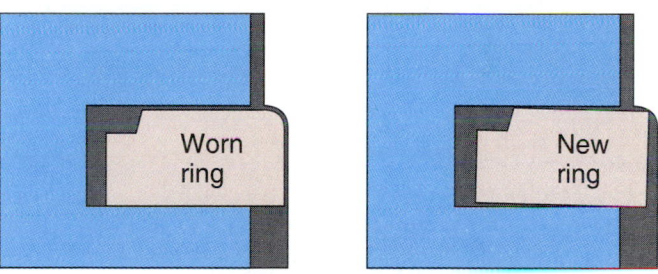

Figure 47.13 A new square ring will strike the worn round area. *(Courtesy of Dana Corp.-Perfect Circle Division)*

CHECK THE DECK SURFACE FOR FLATNESS

A whetstone or a file can be used to clean the deck surface of the block. Do not make the surface too smooth; head gasket sealing problems may result (see Chapter 45). Just remove any nicks or burrs that might give false readings when checking for surface warpage.

Check the deck surface for flatness (Figure 47.11). Blocks do not warp nearly as often as heads. When the block surfaces are warped, or are not parallel to the main bearing bores, they can be surfaced (*decked*) with a milling machine or grinder in the same manner as a cylinder head.

CLEAN BOLT HOLES

Threads in the block must be clean so that correct clamping load is achieved. When a block is hot-tanked, a film develops on the threads. Chase them with a tap. Failure to chase the threads can result in improper torque, which can lead to a leaking head gasket. Head bolt holes often run into water jackets, so they can rust on the bottom. Rust is very hard so be very careful not to break a tap while cleaning head bolt threads.

INSPECTING CYLINDER BORES

Cylinder Bore Taper

Cylinder bores wear in a taper and *out-of-round* fashion. Maximum wear is at 90° to the wrist pin, just under the ring ridge (Figure 47.12). The top of the cylinder receives less lubrication, and it is subjected to the high pressure of the piston rings against the cylinder wall as the air/fuel mixture is ignited. This causes taper wear, which forms the ring ridge at the top of the ring travel. The ridge must be removed or square edges of the new rings can come into contact with the

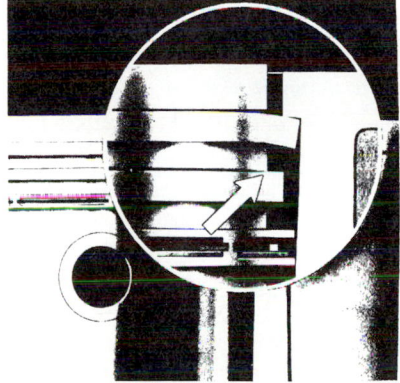

Figure 47.14 The ring land above the second ring can be forced down. *(Courtesy of Federal-Mogul Corporation)*

rounded ring ridge (Figure 47.13). There is only about 0.002" to 0.004" clearance between the side of the ring and the ring groove. The top ring land can be forced down against the second land, jamming the second ring in its groove (Figure 47.14). Ridge removal is covered in Chapter 44.

The maximum amount of taper allowed before a rebore is necessary is usually 0.010". Taper in the bore causes changes in the end gap of the piston rings. Modern engines use low tension rings (see Chapter 19), which must have less bore taper. Low-tension rings are less "forgiving" than standard rings and are less apt to follow tapered and eccentric cylinders. *For low-tension rings, the maximum allowable cylinder wall taper is 0.005" to 0.006".*

Cylinder Bore Oval Wear. Out-of-round wear is caused by the rocking of the piston on its wrist pin at TDC and BDC. A major cause of out-of-round cylinder wear is gasoline washing oil from the cylinder walls when the engine is cold. Excessive piston-to-cylinder wall clearance will cause out-of-round wear. It is more difficult for new rings to seal against an out-of-round bore than to a tapered one. Maximum permissible out-of-round wear is 0.005".

■ MEASURING THE BORE

The cylinder bore can be measured in several ways.

> ### ■ MATH NOTE ■
>
> *A simple way to check for taper without any special measuring instruments is to use a feeler gauge and an old piston ring:*
>
> - *Square up the ring just below the ring ridge and measure the ring end gap with a feeler gauge.*
> - *Compare this measurement to the end gap measurement at the bottom of the cylinder.*
> - *To determine the amount of taper, divide the difference between the two gaps by 3 (Figure 47.15). This is done because the change in the gap is really a measurement of change in cylinder circumference, not diameter. So, actually the number 3.14 (π) would be used.*

The bore can be measured using a telescoping gauge and micrometer, an inside micrometer, or a *cylinder dial bore gauge* (Figure 47.16). The dial bore gauge is usually much more accurate than the other instruments for determining the bore size. To measure the bore using a dial bore gauge, rock the gauge back and forth in the cylinder. The correct size is the smallest diameter that the gauge reads. This is because when the tool moves past the smallest point in either direction the reading will get larger (Figure 47.17).

Figure 47.16 Using a cylinder dial bore gauge to measure a cylinder bore. *(Courtesy of Sunnen Products Company, St. Louis, Missouri)*

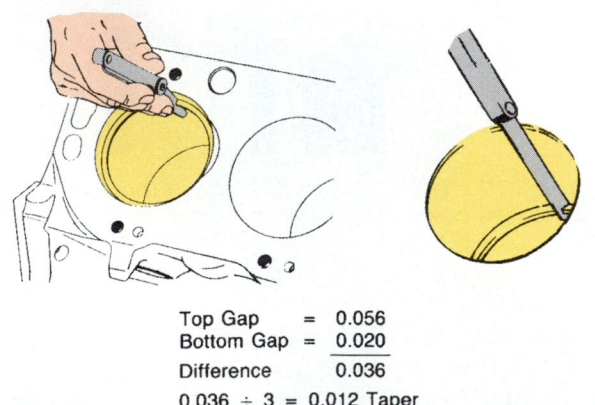

Top Gap	=	0.056
Bottom Gap	=	0.020
Difference		0.036

0.036 ÷ 3 = 0.012 Taper

Figure 47.15 Divide the difference between the top and bottom ring gaps by 3. *(Courtesy of General Motors Corporation, Service Technology Group)*

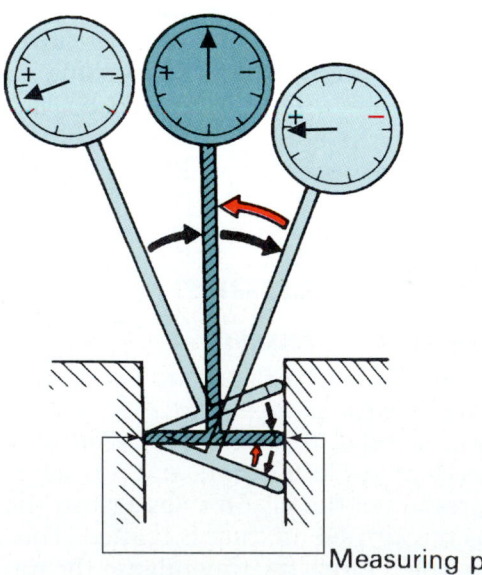

Measuring point

Figure 47.17 Rock the dial bore gauge back and forth. Select the smallest measurement.

■ DEGLAZING THE CYLINDER BORE

Cylinders become glazed where the piston rings contact the cylinder wall. The **cylinder glaze** can be removed with lacquer thinner or carburetor cleaner or with a glaze breaker that leaves a honed, **crosshatch** appearance (Figure 47.18). The purpose of the crosshatch is to provide channels to hold oil while new piston rings wear into the cylinder walls. If the factory crosshatch is still visible, glaze breaking is not necessary.

The angle of the crosshatch should be between 20° and 60°. Crosshatching that is too steep (more than 60°) will allow oil to run down the cylinder wall; too flat (less than 20°) causes too thick an oil film. The rings will skate over the film, causing oil consumption.

Stroking about once per second will provide the proper crosshatch (depending on the speed of the drill motor). If the crosshatch is flatter than desired, either slow down the drill speed or speed up the stroke speed. A drill with a rotation speed of approximately 450 rpm is recommended for deglazing cylinders.

SHOP TIP Deglaze the ring contact area of the cylinder only (Figure 47.19). There is no need to travel through the bottom of the cylinder.

NOTE: *Dana Corporation recommends that engines that are rerung while they are still in the vehicle should* **not** *have the glaze broken with a glaze breaker unless the block is to be thoroughly cleaned afterwards. They feel that glaze breaking is desirable but that the damage done by leaving grit in the engine outweighs the benefits gained by glaze breaking.*

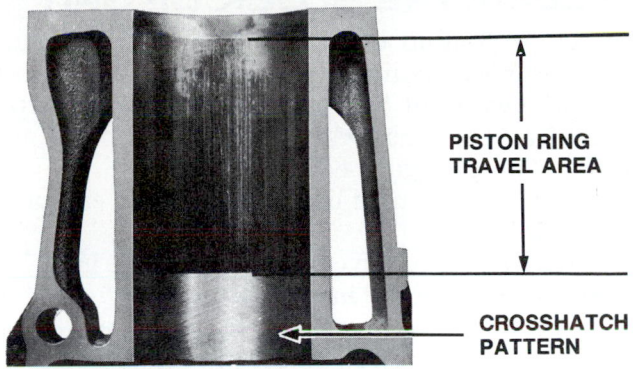

Figure 47.19 The area below ring travel still shows the factory crosshatch. *(Courtesy of Mercedes-Benz of North America, Inc.)*

Glaze Breakers

There are two principal types of glaze breakers: the spring-loaded glaze breaker and the ball-type glaze breaker or *flex hone* (Figure 47.20). Both are driven by drill motors and cooled with honing oil or cleaning solvent. Glaze breakers remove metal at a much slower rate than the rigid hones used to finish a cylinder to the correct size after boring.

The stones on a spring-loaded glaze breaker can be accidentally broken in two ways:
■ This happens when the tool is removed from the cylinder while the drill motor is still spinning.
■ Stones also break when they are allowed to spin below the bottom of the cylinder bore and come into contact with main bearing webs.

The flex hone can be rotating as it enters and leaves the cylinder. This will leave a proper crosshatch finish. It is not as susceptible to stone breaking, although the balls tend to wear off on the bottom end of the hone from repeatedly striking the main bearing webs.

Figure 47.18 Crosshatch appearance of a honed cylinder wall. *(Courtesy of Federal-Mogul Corporation)*

Figure 47.20 A flex hone is easy to use. *(Courtesy of Brush Research Mfg. Co., Inc.)*

■ CLEAN THE BLOCK OF GRIT

A thorough cleanup of the block is necessary after glaze breaking or honing. Any grit left on the block will rapidly wear out new parts. After all block preparation, clean the block with a stiff bristle brush and hot, soapy water (Figure 47.21).

■ The brush can be used by hand or on the end of an air drill.

■ Ordinary cleaning solvent will not lift the grit from the pores of the metal.

■ Check cylinder walls and the crankcase for cleanliness with a clean cloth. Grit remaining in the bore will be deposited on the cloth.

■ Following cleaning, grit can often be found in the crankcase area, just under the cylinder bores (Figure 47.22).

Ferrous parts (iron or steel) should be thoroughly coated with oil to prevent rusting, which begins immediately after cleaning.

Figure 47.21 Clean the cylinder bores with hot, soapy water. *(Courtesy of Federal-Mogul Corporation)*

Figure 47.22 Be certain this area is cleaned thoroughly of honing grit.

■ BORING FOR OVERSIZED PISTONS

Cylinders are deglazed for use with new rings *only* if they do not have excessive bore taper. Damaged cylinders or cylinders with too much taper should be rebored and honed by a machine shop to fit new oversized pistons.

Piston Oversizes

Pistons are available in standard, 0.020" (0.50 mm), 0.030" (0.75 mm), 0.040" (1.0 mm), and 0.060" (1.5 mm) oversizes. Some later model blocks do not have enough cylinder wall thickness to accommodate reboring to 0.060" oversize. Pistons of 0.030" oversize are most commonly used. Using a 0.030" cut for each rebore (0.015" of metal removal from each cylinder side) allows for two overhauls (0.030" and 0.060") per block, which is usually more than enough.

■ MATH NOTE ■

The compression ratio will change slightly with a rebore. A 0.030" oversize increases compression by about $\frac{1}{10}$ of a ratio; 0.060" will result in a change of about $\frac{1}{4}$ of a ratio.

Example:

4.000" compression ratio	*9:1*
4.030" compression ratio	*9.1:1*
4.060" compression ratio	*9.25:1*

■ BLOCK DISTORTION

Block castings typically distort when heads and main cap bolts are torqued. Cylinder distortion can be from 0.001" to 0.005". Piston clearance is usually only 0.001" to 0.003".

In extreme cases, this distortion can result in piston scuffing (momentary welding to the cylinder wall). Most often, distortion results in poor piston ring sealing or **piston slap**. Piston slap occurs when excessive piston-to-cylinder wall clearance allows the piston skirt to tap against the cylinder wall.

A boring stand is desirable for reboring and honing blocks (Figure 47.23). It supports the block at the main bearing bores.

NOTE: *Cylinders can be up to 0.002" out-of-round when the block is hung from one end on a conventional engine service stand. Boring cylinders with a portable boring bar while the block is mounted to one of these stands can produce oval-shaped cylinders.*

Torque Plates

Because a cylinder block is so flexible when the heads are not holding it in place, a torque plate made of 1¾"

Figure 47.23 Boring a block with a boring bar and stand. *(Courtesy of Rottler Mfg. Co.)*

thick cast iron is sometimes torqued to the top of the block. It stresses the block to simulate assembled conditions when refinishing cylinders (Figure 47.24). Main caps should be torqued in place also. A torque plate is also called a *deck plate*, *honing plate*, or *stress plate*.

■ HONING AFTER BORING

The machine shop will bore the cylinders to within 0.0025" to 0.003" of the desired finished bore size. Then they are finished to the proper size by honing with a *honing machine* (Figure 47.25).

Honing after boring provides a much more satisfactory surface for new rings and promotes longer ring life. The ring and bore do not fit exactly to each other, so a crosshatched honed surface provides an area where they can wear into each other.

Manufacturers of honing equipment recommend honing to increase the cylinder's diameter by at least 0.025" to 0.003". The boring cutter leaves a fine thread and microscopic fractures in the cylinder wall (Figure 47.26). Small chips break away from the cylinder surface as it is cut. The cavities left behind are about 0.001" deep depending on the condition of the tool bit.

NOTE: *Honing the cylinder to remove 0.001" of metal from the surface of its wall will make the cylinder 0.002" larger in diameter.*

Piston Clearance

The machine shop will want to have the new pistons on hand before honing, so that they can be properly fitted to the bores. There is some variation in sizes of pistons within a set. Typical clearance is 0.001" to 0.002" for cast automotive pistons. They come about 0.0015" undersize to the standard bore sizes. For example, a piston for a 4.030" bore will actually measure about 4.0285".

■ MATH NOTE ■	
Standard Bore Size	4.000"
Replacement Piston Size	4.0285"
Piston Clearance	+ 0.0015"
Rebore Finished size	4.030"

Figure 47.24 A torque plate. *(Courtesy of Federal-Mogul Corporation)*

Torque Plate

Figure 47.25 An automatic cylinder bore honing machine. *(Courtesy of Sunnen Products Company, St. Louis, Missouri)*

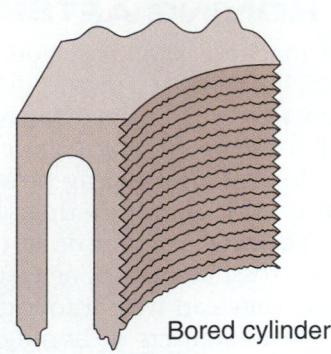
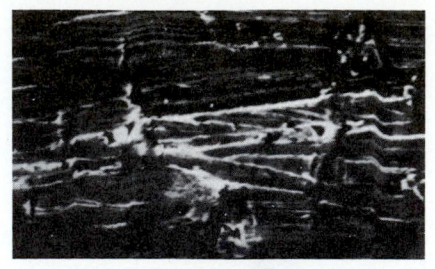

Bored cylinder

Figure 47.26 A bored cylinder wall is torn and has folded metal, which will cause excessive ring face wear. *(Courtesy of Federal-Mogul Corporation)*

The machine shop will measure the bore with a dial bore gauge and compare it to the size of the pistons. The dial bore gauge is set with a very precise micrometer setting fixture (see Figure 5.28).

Chamfering the Cylinder

After boring and honing, the top of the bore is chamfered about $\frac{1}{16}$". This is so that during installation of the pistons in the block, the new rings can easily enter the cylinder without being chipped.

■ SLEEVES

When a sleeve is used to repair a damaged cylinder, it is installed with an interference fit of 0.0005" per inch of cylinder bore. Thus, a 4" bore would require the block to be bored 0.002" smaller than the outside diameter of the sleeve.

■ MATH NOTE ■

Measure the top of the sleeve at three places 120° apart.

4.285"
4.287"
4.286"

Repeat this at the bottom.

4.284"
4.288"
4.286"

Add the six measurements together and divide by 6 to find the average outside diameter of the above.

4.285"
4.287"
4.286"
4.284"
4.288"
+4.286"
25.716" *divided by 6 = 4.286" (average diameter)*

After the sleeve is pressed into the bore, the top of it is finished flush with the block. Then the inside diameter (I.D.) of the sleeve is bored to the finished size.

NOTE: *The inside diameter (I.D.) of the sleeve is usually less than a standard bore, so it can be bored to original standard diameter.*

One popular sleeving method is to make a step in the bottom of the bore by cutting the cylinder only to within $\frac{1}{8}$" or $\frac{1}{4}$" of its bottom. The bottom of the sleeve will rest on this step. When the final boring is completed on the sleeve, the line between the sleeve and step in the block will be almost impossible to see. Many machine shops use this procedure.

■ CAM BEARING INSTALLATION

Cam bearings are press-fitted into the block. (The outside diameter of the bearing is larger than the bearing bores in the block.) Before installing cam bearings, emery cloth is used to clean the cam bearing bores. The cam bearings in a pushrod engine are installed before the crankshaft is installed. This makes it easier to align the oil holes in the bearings to the corresponding holes in the block. It will also be easier to shave off any high spots on the bearing that occurred during bearing installation.

Sometimes cam journals are different sizes. The smallest is at the rear of the block. New bearings can be checked for approximate fit by putting them on cam bearing journals before installation.

Cam Bearing Removal and Installation Tools

There are several types of cam bearing removal and installation tools. The most popular one is the universal type (Figure 47.27). Its use is covered here.

1. A special washer must be used with the driver, or the tool's mandrel pieces may be ruined. Washers of two different sizes are provided with the tool. Use the largest washer possible. Too small a washer can result in broken segments. Line up the segments with the spaces in the mandrel.

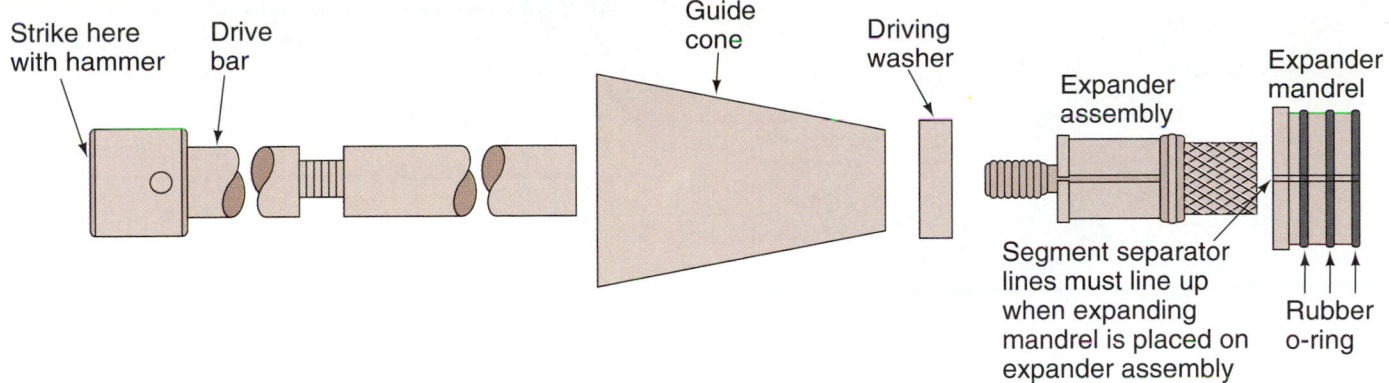

Figure 47.27 A universal cam bearing driver. *(Courtesy of Lisle Corporation)*

2. When starting to press a bearing into a bore, tighten the tool as much as possible by hand. Give the tool one sharp rap with a large hammer, and then loosen the mandrel approximately ⅛ turn before continuing. Be sure to re-check this adjustment occasionally when driving the bearing into place because the tool will sometimes loosen and damage the bearing.

3. Hold the guide cone tightly against the front of the block. Hold the cam bearing driver snugly against the bearing during installation so it does not bounce, or the end of the bearing can be damaged.

Although cam bearings are usually chamfered on the outside edges of both sides, some bearings are chamfered only on one side. Make sure that the chamfer faces the bore before installation. Use a large hammer because of the interference fit of approximately 0.002" to 0.003" that prevents rotation of the bearings in the block while the engine is running. The use of 90- or 140-weight oil makes installation much easier. Another helpful tip is to chamfer the inside edge of the bearing with a scraper if it is not already chamfered.

Positioning the Oil Hole. Follow the manufacturer's recommendations when positioning the oil hole. When there is an oil groove around the total cam bore in the block, positioning the oil hole toward the top also allows oil to enter the bearing easily. It will be carried under the cam as it rotates (Figure 47.28). The following recommendation from Federal-Mogul Corporation can be followed:

> Install the bearings with their oil holes at approximately the 2 o'clock to 4 o'clock position, viewed from the front, so that oil hole alignment can be observed easily. Avoid positioning the oil hole between the 5 o'clock and 8 o'clock positions.

Double-check oil hole alignment with a piece of bent wire or welding rod (Figure 47.29). Some engines, such as the small block Chevrolet V-8, have a rear cam

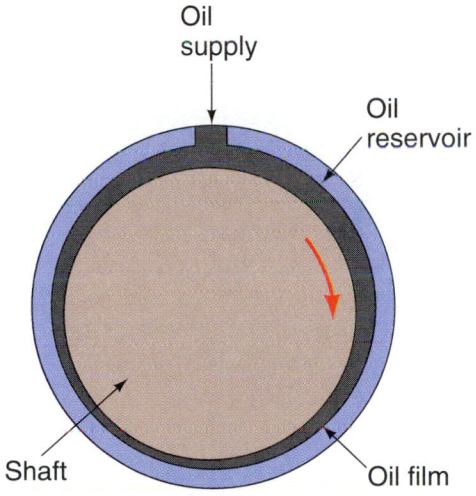

Figure 47.28 Oil is carried under the cam as it rotates. *(Courtesy of Federal-Mogul Corporation)*

bearing bore that is wider than the bearing. Failure to align the bearing oil hole to the block oil groove can result in a total loss of oil pressure in this engine.

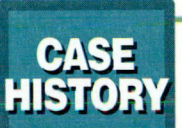 *A student rebuilt a small block Chevrolet and installed it in the vehicle. When the engine was started, it was discovered that it did not have any oil pressure at idle. Raising the engine speed produced normal oil pressure. The engine was removed and disassembled for inspection. The rear camshaft bearing had been installed flush with the front of the bearing bore, but the alignment of the oil hole was not confirmed with a piece of wire. The oil groove that is supposed to line up with the oil hole in the bearing was partially uncovered because the bearing had not been installed all of the way into the bore. The result was a loss of engine oil pressure that returned to normal when the engine was revved high enough for the oil pump to fill the leak.*

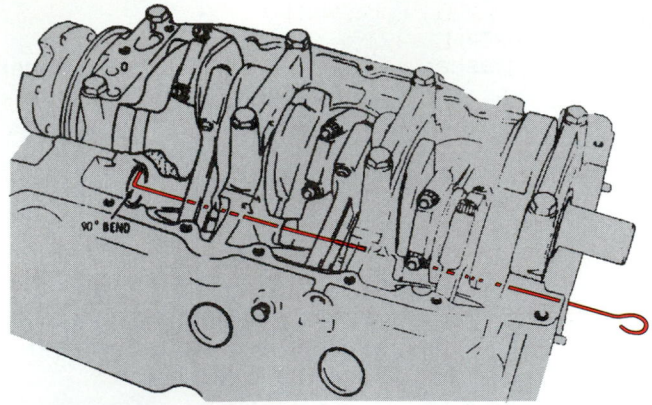

Figure 47.29 Checking oil hole alignment. (Courtesy of General Motors Corporation, Service Technology Group)

■ FRONT CAM BEARING INSTALLATION

On pushrod engines, the timing sprockets and chain are lubricated by oil from the front cam bearing. Many front bearings are pounded in past the block surface (Figure 47.30). This leaves an oil passage in front of the bearing that provides lubrication to the cam timing sprocket and chain. Other bearings have a notch cut into them. The bearing is installed with the notch facing the cam gear.

Check Fit of Bearing

After installing the bearing, install the cam and turn it. If it does not turn easily, remove it and look for any high spots that can be scraped with a bearing scraper. High spots on the bearing will become polished as the cam is rotated after it is first installed.

NOTE: *Bearings should **not** be sanded with emery cloth because pieces of emery grit can become embedded in the bearing surface.*

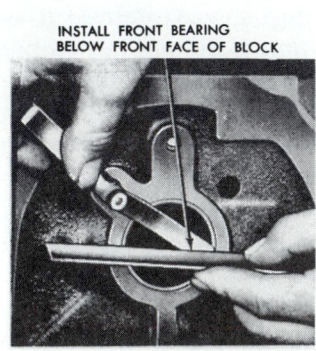

Figure 47.30 The front cam bearing is installed past the front of the block surface to allow for timing chain oiling. (Courtesy of Ford Motor Company)

SHOP TIP A cam bearing reamer can be made by grinding grooves into the journals of an old cam core. As the tool is turned, it shaves off any high spots from the bearing.

- A special flex hone is available for honing small amounts off of cam bearings.
- If necessary, Scotch Brite™ can be used to polish cam bearing surfaces.

■ CHECKING CRANKSHAFT CONDITION

Checking a Crankshaft for Straightness

Keep bearings in respective position order during disassembly so that the cause of unusual wear conditions can be diagnosed. A bent crank can be indicated when one bearing wears more than the others (Figure 47.31). The resulting bearing wear is usually worse at the center.

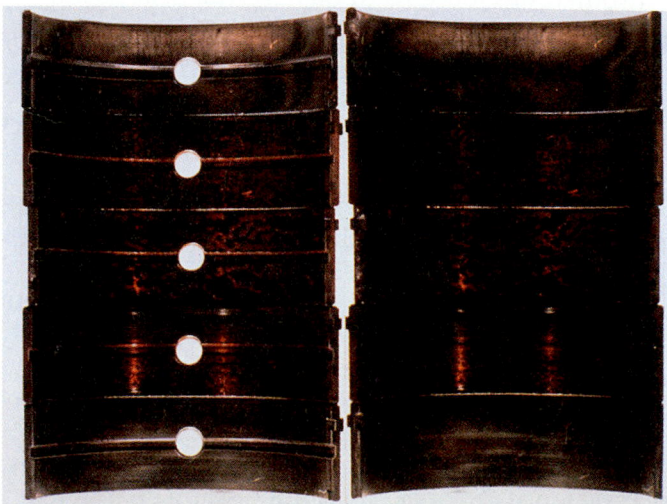

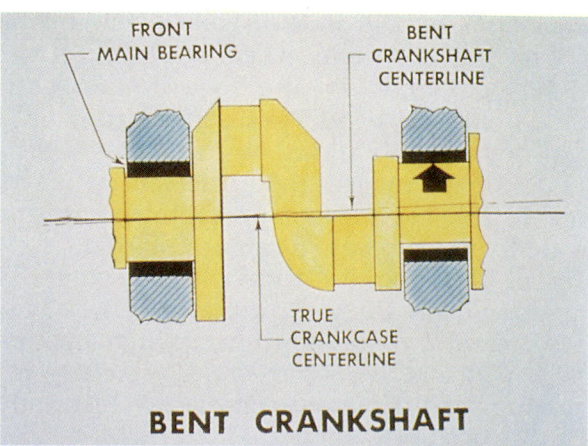

Figure 47.31 Crankshaft misalignment. Main bearing wear caused by a bent crankshaft or a misaligned crankcase. (Courtesy of AE Clevite Engine Parts)

Checking for Cracks

The crankshaft can be checked for obvious cracks by "ringing" the counterweights with a light tap of a hammer. A dull sound indicates the presence of a crack. Machine shops also have more accurate magnetic crack detection methods.

Checking the Vibration Damper

Whenever a crankshaft is broken, be sure to check the vibration damper for obvious signs of damage. The following are some additional vibration damper considerations:

- The outer ring on the damper can slip, causing an out-of-balance condition. With the cylinder head removed and the piston at TDC, make a visual comparison. Do the timing marks for TDC on the timing cover and damper appear to be accurate?
- Be sure the damper is the correct one for the vehicle. The wrong one can be worse than none at all, resulting in a broken crankshaft.
- Sometimes a damper can become loose on the crank. Check the keyway for wear that can result from improper damper installation, the wrong size key, or a loose damper hub-to-crank fit.
- Be sure all pulleys are straight. A damaged pulley can force a crankshaft to bend during engine operation.

■ CRANKSHAFT AND BEARING WEAR

Bearings have *loaded* and *unloaded* bearing halves. The upper rod bearing and the lower main bearing are the loaded halves. Upper rod bearing wear is often found in combination with lower main bearing wear. A *little* of this kind of wear is considered normal.

It is a good idea to keep all bearings in numerical position order for inspection. The main cause of short bearing life is dirt including grit, metallic particles, and other abrasives. Sometimes sandy dirt enters the engine through the air cleaner or the dipstick opening. A crankshaft sometimes wears excessively because of abrasives in the oil.

Journals can wear out-of-round or become tapered. Figure 47.32 shows how to evaluate crankshaft journal measurements.

Out-of-Round Journal Wear

When the engine is first turned over after the engine has not been run for a period of time, there is little or no lubrication between the crank and the lower main bearings. The result is that the lower main bearing wears excessively and the main journals wear out-of-round. When the main bearing that is farthest from the oil pump shows more wear than the other main bearings, a *dry start* condition is indicated. This means that the engine was probably revved before oil had

A vs. B = vertical taper
C vs. D = horizontal taper
A vs. C and B vs. D = out-of-round

Check for out-of-round
at the end of each journal

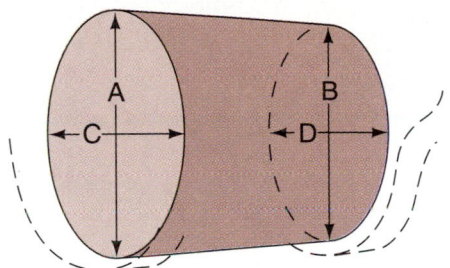

Figure 47.32 Crankshaft journal measurements. *(Courtesy of Ford Motor Company)*

filled the system. This problem occurs most often in cold weather.

Sometimes, all the lower main bearings will be worn except for the front bearing. This bearing usually wears less on the bottom because of the upward tension of the fan belt. Excessive belt tension can cause wear on the *upper* front main bearing.

Connecting rod journals also wear out-of-round, wearing on their top sides because of excessive loads during the power stroke. Crankshaft main bearing journals wear out-of-round. Excessive loads cause the oil film to break down, resulting in wear. Excess loads can be caused by **lugging** the engine or by abnormal combustion.

NOTE: *Lugging occurs when the load on the engine is greater than the rpm needed to develop enough horsepower to pull the load.*

Measure the rod journal in a horizontal and vertical direction to check for out-of-round wear. Crank journals are *miked* (measured with a micrometer) at 90° angles to check for out-of-round wear, which should be less than 0.0005".

Tapered Wear

Rod journals sometimes suffer taper wear due to misalignment of the connecting rod (see Figure 48.23). The presence of uneven rod bearing wear, and sometimes piston skirt wear, usually indicates taper. Connecting rods should be checked for misalignment whenever uneven wear is found (see Figure 48.24).

Thrust Bearing Wear

The thrust bearing surface that faces the rear of the engine sometimes shows excessive wear (Figure 47.33). Most thrust bearings have concaved reliefs cut into them to provide lubrication. Under normal conditions, the thrust surface is only under load when the clutch pedal is depressed or if the automatic transmission torque converter is under a load. Thrust bearing

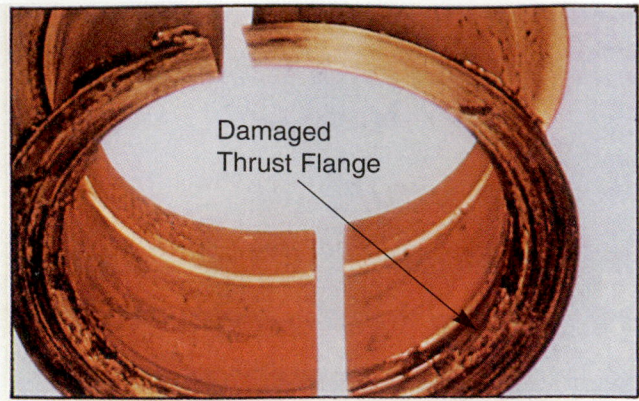

Figure 47.33 One side of this thrust bearing is burned. *[Courtesy of Federal-Mogul Corporation]*

wear and failure occurs when the load is continuous, such as when there is an improper clutch adjustment (too little free play), or when the driver has been "riding the clutch."

■ CRANKSHAFT JOURNAL TOLERANCE

Wear specifications for crankshaft journals are found in repair manuals. An example specification might be 2.2592–2.5600; this range is called the **tolerance**. When the crank is worn only slightly and is still within tolerance, it can be reused.

Crank polishing with an *emery belt* by a machine shop is sometimes desirable. The crankshaft is turned in a lathe or crank grinder while the emery belt is turned against its surface. This is a relatively inexpensive operation. Check the crank bearing surface for excessive roughness by rubbing a penny on it. If the journal picks up copper, the shaft is too rough.

■ REGRINDING THE CRANKSHAFT

A machine shop can regrind a crankshaft with a *crankshaft grinder* (Figure 47.34). Crankshafts are usually reground to 0.010", 0.020", or 0.030" undersize.

NOTE: *A 0.020" undersized crank actually has only 0.010" of metal removed from its surface. Because metal is removed from both sides when the crank is ground, the crank will be 0.020" undersize.*

Sometimes, rod journals and main journals are ground to different undersizes. In this case, the main journal size is listed first—for example, 0.010–0.020". The main bearing size is 0.010". Burned or severely worn crank journals can be built up on a crank welder before regrinding. After grinding, crank journals are polished approximately 0.0002" with the emery belt.

NOTE: *Do not hand-polish a crank journal with emery cloth or sandpaper. The shiny, worn crank surface will cause less bearing wear if left alone.*

Figure 47.34 A crankshaft grinder. *[Courtesy of Peterson Machine Tools, Inc.]*

■ MEASURING BEARING CLEARANCE WITH PLASTIGAGE

Bearing clearance can be checked with **plastigage** (Figure 47.35) or with a micrometer. Plastigage, a special thickness of plastic string, is applied to the bearing journal between the bearing and the crank. When the cap is torqued, the plastic string flattens. The bearing cap is removed to inspect the plastigage. The wider the flattened string, the less clearance there is. Plastigage is oil soluble and is readily cleaned from the bearing or crankshaft.

NOTE: *An engine with double oil clearance will throw off five times as much oil onto the cylinder walls, causing increased oil consumption (Figure 47.36).*

Be careful not to allow too much clearance:
■ A front bearing with excessive clearance will knock.
■ Excessive clearance can also cause low oil pressure.

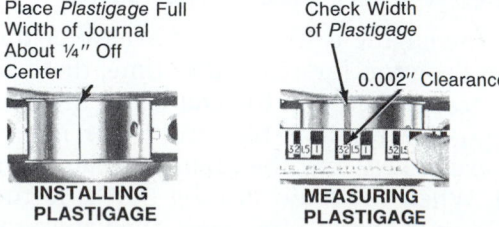

Figure 47.35 Checking bearing oil clearance with plastigage. *[Courtesy of Ford Motor Company]*

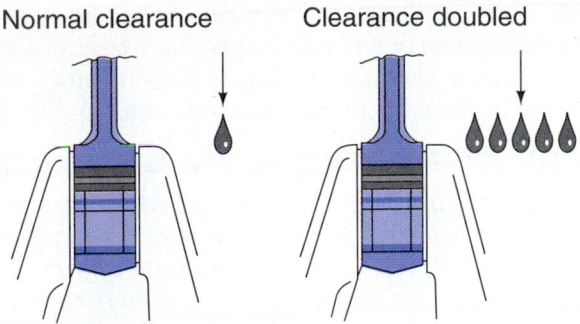

Normal clearance   Clearance doubled

Figure 47.36 Doubling oil clearance results in five times as much oil throw-off from connecting rods. *(Courtesy of AE Clevite Engine Parts)*

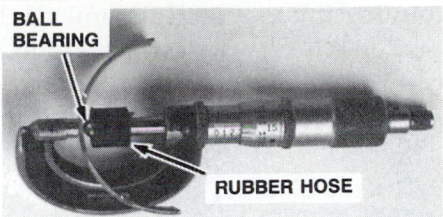

Figure 47.37 Measuring bearing wall thickness.

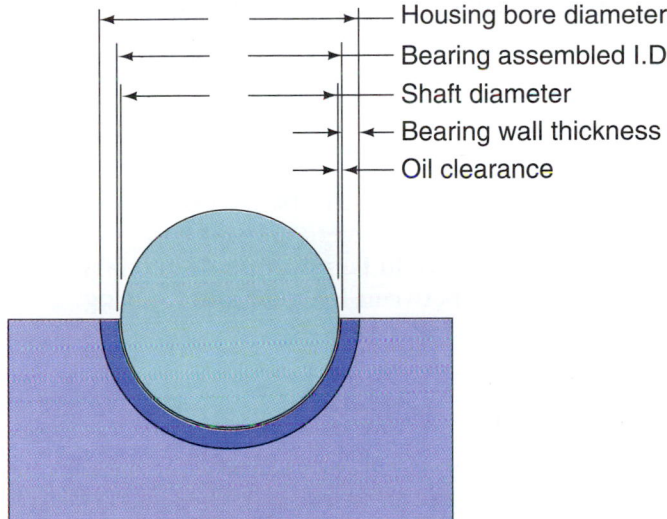

- Housing bore diameter
- Bearing assembled I.D.
- Shaft diameter
- Bearing wall thickness
- Oil clearance

Figure 47.38 Determining oil clearance. *(Courtesy of Federal-Mogul Corporation)*

CASE HISTORY

A technician working in a dealership repaired an engine that had a worn crankshaft. The crankshaft was removed and sent to a machine shop for regrinding. After the job was completed, the engine had a minor knock that was diagnosed as a front main bearing. The technician pulled the oil pan and measured all of the crankshaft clearances with plastigage. The front crankshaft bearing had 0.0025" clearance, enough to make noise. He cleaned the main bearing cap and lightly sanded the parting surface with very fine sandpaper. When he rechecked the clearance it was 0.0015", within specifications. After reassembly the noise was gone. After that, he always checked crankshaft clearances carefully during initial assembly.

Actual clearance can also be determined by using a micrometer to measure the crankshaft journal, the housing bore, and the bearing insert. When you measure the bearing insert, use a ball bearing and subtract its thickness from your measurement (Figure 47.37). Total oil clearance can be determined by subtracting the sizes of the bearing journal and bearing from the housing bore.

MATH NOTE	
(see Figure 47.38)	
Housing Bore	*2.124"*
Bearing Wall Thickness (0.061) × 2	*-0.122"*
Bearing Assembled Inside Diameter	*2.002"*
Shaft diameter	*-2.000"*
Oil Clearance	*0.002"*

LUBRICATION SYSTEM SERVICE

Oil Pressure

An engine's oil pressure provides a good indication of the condition of its lower end. Low oil pressure at idle can indicate excessive oil clearance at a bearing or a worn oil pump. The engine might take too long to build oil pressure when it is first started. Low oil pressure at idle can also result if the oil pump pressure relief valve is stuck in the "open" position. The relief valve opening is so large that the pump cannot fill the system at low speed, although pressure will usually build to normal levels at higher speeds.

Long periods of idling with low oil pressure can lead to cylinder wall lubrication problems. Throw-off from the connecting rods cannot supply the cylinder walls with enough oil for adequate lubrication.

Testing Oil Pressure

To test oil pressure against manufacturer's specifications, remove the indicator light **sending unit** and install an external gauge in its place (see Figure 43.18). An engine idling at 800 rpm should have a minimum of 8 psi oil pressure.

An engine with too much oil pressure can have oil consumption problems and bearing lining material can be washed off the bearing. High oil pressure results when the relief valve sticks in the "closed" position.

■ The oil pressure can be normal at idle, but will increase beyond desired levels as engine rpm increases.

■ High pressure can cause oil leaks and can blow out oil gallery core plugs, which will result in no oil pressure at low rpm.

■ Sometimes an oil filter can burst under too much oil pressure.

■ CHECKING OIL PUMPS FOR WEAR

Use a feeler gauge to check an oil pump for wear. The clearances here are approximate and are used only in the absence of specifications in a manual.

Rotor pump clearance tolerances are:

■ 0.010" between the inside and outside rotors (Figure 47.39a).

■ 0.014" between the outside rotor and housing (Figure 47.39b).

Gear pump clearance can be checked by inserting plastigage between the cover and gear ends.

■ End clearance should be no more than 0.003".

■ Side clearance between the gear and housing should be less than 0.005".

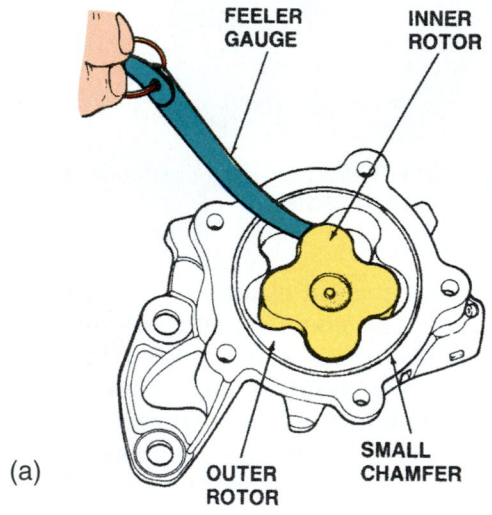

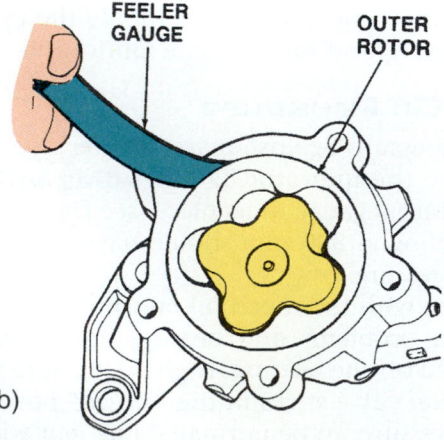

Figure 47.39 Checking a rotor oil pump for wear using a feeler gauge. *(Courtesy of Chrysler Corporation)*

NOTE: *The pump housing can be easily bruised or distorted during installation of pressed-fit oil pickup screens or because it was carelessly clamped in a vise. This can cause interference between the gears or rotor and the pump housing.*

■ OIL PUMP SCREEN SERVICE

The oil pump sump screen, or pickup, must be checked to be sure it is clean. A new oil pump does not come with a new screen. Some rebuilders replace every screen with a new one. A plugged screen can ruin a new engine in a very short time (see Figure 19.45).

On a newly cleaned screen, make sure that the bypass is not struck permanently open (Figure 47.40). Be certain that a screen does not have any loose or damaged wire mesh that could break loose and cause oil pump failure.

■ OIL PUMP FAILURE

Oil pumps usually wear or seize because of improper engine maintenance or because pieces of a metal part, such as a failed camshaft or bearing, get into the pump (Figure 47.41). Sometimes, an oil pump hex driveshaft will twist off when a foreign object becomes wedged in the pump (Figure 47.42).

Another major cause of pump failure is when it draws in plastic material, such as nylon timing sprockets (see Figure 18.43) or deteriorated valve guide seals. Seals become brittle with age and can find their way into the

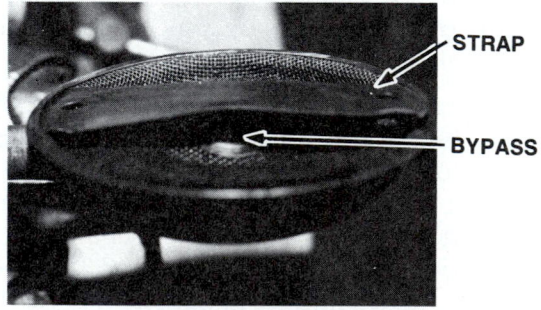

Figure 47.40 This pump screen is defective; the bypass should be seated against the strap.

Figure 47.41 This gear-type oil pump was destroyed when metal particles entered the pump from a failed part elsewhere in the engine. *(Courtesy of Federal-Mogul Corporation)*

oil system (Figure 47.43). Aged or damaged umbrella seals can break up during normal engine operation, but O-ring seals usually enter the system during a careless seal replacement job. Plastic pieces can ruin the oil pump or cause a pump relief valve to stick (Figure 47.44).

CASE HISTORY *A technician was replacing a timing set in a small block Chevrolet. The flat-rate manual includes time on this job for removal of the oil pan. From previous experience, the technician knew that he could complete the job without removing the pan. He simply loosened the pan bolts and pried the timing cover out. This cut his time on the job. A short time after the job was completed, the car returned to the shop once again; this time for a failed oil pump and resulting severe engine damage. Pieces of the old timing sprocket were found in the pump screen and oil pump.*

Air in the System

The bottom of an oil pump does not require a gasket unless it is mounted on the outside of the block. A slight amount of oil leakage on the output side of the pump is harmless. Leakage on the input side of the pump will allow air to be drawn into the oil.

Check the tube that connects the screen to the oil pump to be sure that there are no holes that can allow the entry of air into the system. Bolt-on oil pickup screens

Figure 47.42 The hex drive on this rotor pump was ruined when foreign material jammed the pump. *(Courtesy of Federal-Mogul Corporation)*

Figure 47.43 Broken umbrella valve guide seals.

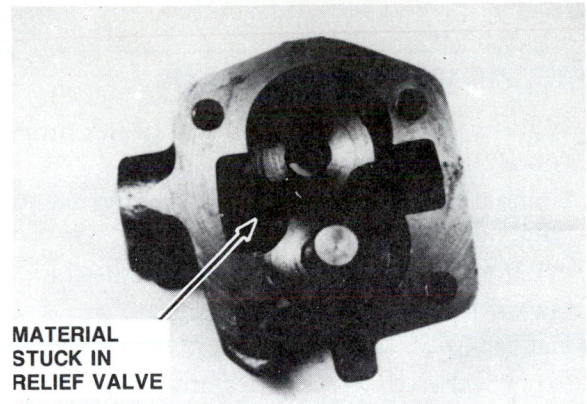

MATERIAL STUCK IN RELIEF VALVE

Figure 47.44 The relief valve has a piece of nylon timing sprocket jammed in it. *(Courtesy of Federal-Mogul Corporation)*

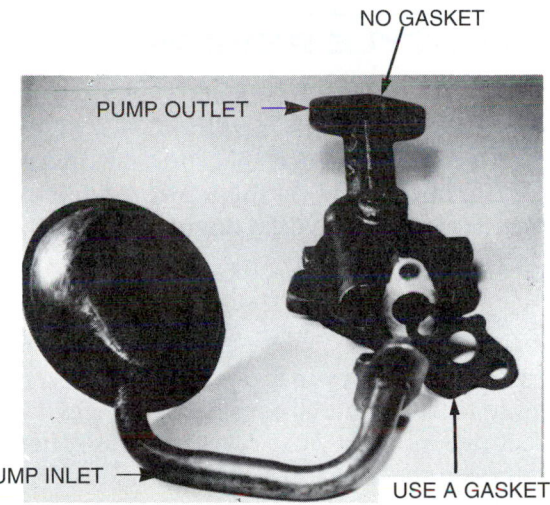

NO GASKET
PUMP OUTLET
PUMP INLET
USE A GASKET

Figure 47.45 A bolt-on pump pickup requires a gasket. *(Courtesy of Federal-Mogul Corporation)*

usually require a gasket (Figure 47.45). Be sure to remove the old gasket carefully. Some screens are pressed-fit.

NOTE: *Do not pound on the pump when installing the screen. The pump housing can be damaged. Clearances within the pump are so small that if the pump is improperly clamped in a vise or pounded on, interference between the gears or rotors and housing can result.*

- If the screen does not fit tightly, is damaged, or is dirty, it should be replaced.
- Position the screen ¼" to ½" from the bottom of the oil pan. This will prevent it from drawing in sediment that has accumulated in the bottom of the oil pan.

■ INSTALLING THE OIL PUMP

Oil pumps are filled and tested at the factory before shipment. Before installation, a new or rebuilt pump should be turned by hand and filled with oil to prime it and to make sure that it has not been damaged by previous mishandling. Priming the lubrication system is covered in detail in Chapter 50.

■ REVIEW QUESTIONS

1. What type of threads do oil and heater fitting connections have?

2. When a main bearing bore has changed color, what should be looked for?

3. What are two processes that can be used to realign main bearing bores?

4. List two ways that a cylinder bore can wear.

5. What is the crisscross appearance called that results from honing a cylinder?

6. Before measuring a block that is mounted on its end by a universal engine stand, what should be done?

7. What is typical piston-to-cylinder wall clearance?

8. When the main bearing that is farthest from the oil pump shows excessive wear, what is indicated?

9. What is the material called that is used for measuring bearing clearance?

10. Besides metal from worn bearings or a failed camshaft, what are two types of foreign material that can find their way into the oil pump?

■ ASE STYLE REVIEW QUESTIONS

1. Technician A says that tapered oil gallery plugs can usually be removed without heating them. Technician B says that inaccurate cylinder bore measurements will be the result when a block is mounted on a universal engine stand. Who is right?
   a. Technician A   b. Technician B
   c. Both A and B   d. Neither A nor B

2. Technician A says that block deck surfaces warp more often than cylinder head surfaces. Technician B says that rusted steel is softer than unrusted metal. Who is right?
   a. Technician A   b. Technician B
   c. Both A and B   d. Neither A nor B

3. Technician A says that changing the speed that the drill is stroked at will change the angle of the crosshatch pattern in the cylinder. Technician B says that it is not necessary to stroke a glaze breaker through the entire cylinder. Who is right?
   a. Technician A   b. Technician B
   c. Both A and B   d. Neither A nor B

4. Technician A says to position the oil hole in a cam bearing directly under the camshaft. Technician B says when a crankshaft is broken, check if the vibration damper is defective or loose. Who is right?
   a. Technician A   b. Technician B
   c. Both A and B   d. Neither A nor B

5. Technician A says that excessive belt tension can cause main bearing wear. Technician B says riding the clutch can cause wear to the rod bearings. Who is right?
   a. Technician A   b. Technician B
   c. Both A and B   d. Neither A nor B

6. Technician A says that a gasket must be installed under an oil pump that is mounted inside of the oil pan. Technician B says that a gasket must be installed under a bolt-on pump screen. Who is right?
   a. Technician A   b. Technician B
   c. Both A and B   d. Neither A nor B

7. Technician A says that the loaded rod bearing is the upper one. Technician B says that when a crankshaft is ground to 0.030" undersize, 0.030" of material is removed from the journal surface. Who is right?
   a. Technician A   b. Technician B
   c. Both A and B   d. Neither A nor B

8. Technician A says when measuring cylinder bore taper using a piston ring and feeler gauge, divide the sum of the difference between the end gap readings by 3 to get the correct cylinder bore diameter. Technician B says when the engine speed is too low for the load on the engine, this is called lugging. Who is right?
   a. Technician A   b. Technician B
   c. Both A and B   d. Neither A nor B

9. Technician A says to hone at least 0.001" after boring. Technician B says to use solvent to clean the block after honing. Who is right?
   a. Technician A   b. Technician B
   c. Both A and B   d. Neither A nor B

10. A cylinder bore is damaged. Technician A says it can be repaired with a sleeve. Technician B says boring it oversize might clean it up. Who is right?
    a. Technician A   b. Technician B
    c. Both A and B   d. Neither A nor B

Engine Diagnosis and Service: Piston, Piston Rings, Connecting Rod, Engine Balancing

■ OBJECTIVES

Upon completion of this chapter, you should be able to:

✔ Analyze wear and damage to the piston, piston rings, and connecting rod.

✔ Select and perform the most appropriate repairs to the piston, piston rings, and connecting rod.

✔ Explain the theory of engine balancing.

■ KEY TERMS

stress raiser
scuffing
four corners scuffing
end gap
connecting rod resizing
internal balancing
bob weights
primary vibration
rocking couple
force, static, or kinetic
    imbalance
dynamic and couple
    imbalance

■ INTRODUCTION

This chapter deals with diagnosis and service of the piston, piston rings, and connecting rod. Engine balancing service is also covered.

Complete rebuilt engine assemblies are commonly installed in the industry. Information in this chapter deals with actual engine repairs. Some repair shops will remove an engine for rebuilding or perform an overhaul while the engine remains in the chassis. Those procedures are covered in Chapter 50.

Occasionally, a piston ring will break or an engine will be overheated (which causes rings to lose their spring tension). If a cylinder head is removed for repairs and the oil pan is readily accessible, the customer might opt for the installation of new piston rings. Information learned in this chapter will help you decide on the best repair option.

Piston rings are replaced whenever an engine is disassembled. Usually the old pistons are reused unless the cylinder bores are worn badly enough to require reboring. Connecting rods do not usually require service, but you need to know what to look for to avoid costly engine failures. As in previous service chapters, as you read this chapter you should become familiar with the practice of diagnosing the cause of problems before attempting a repair. This is the skill that separates the most successful technicians from the average ones.

■ PISTON SERVICE

It is easier to clean the piston ring grooves if the piston and rod assembly is held in a vise. Be careful not to damage a slipper piston skirt when clamping the rod in the vise. Figure 48.1 shows the right way to clamp the rod on a slipper piston in a vise.

NOTE: *The jaws of the vise must be soft metal. Steel-toothed vise jaws can mark the connecting rod and weaken it. Damaging a metal surface raises stress, weakening the part. This is called a "**stress raiser**."*

Removing Piston Rings

Compression rings are removed with a ring expander (Figure 48.2). If a ring expander is not available, a shop towel can be used in the manner shown in Figure 48.3. Oil rings are easily removed by rolling off the rails and removing the expander spacer.

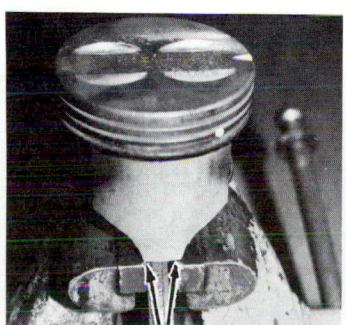

Figure 48.1 The right way to mount a slipper skirt piston and rod assembly in a vise.

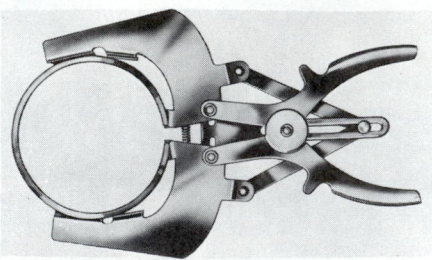

Figure 48.2 One type of ring expander. *(Courtesy of Ammco Tools)*

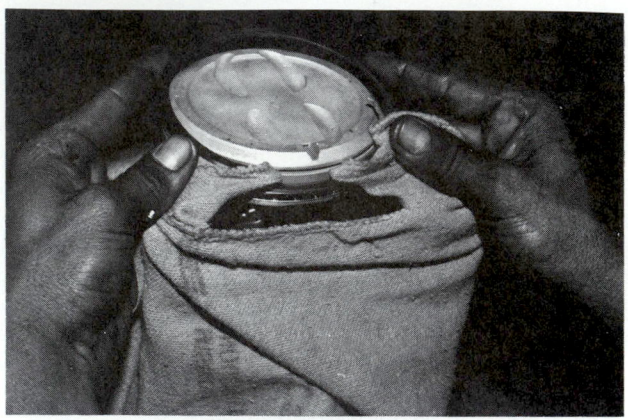

Figure 48.3 Using a shop rag to remove or install a compression ring.

Cleaning the Piston

The top of the piston should be cleaned with a scraper (Figure 48.4). The *top* (and only the top) of the piston can also be cleaned on a wire wheel, but be especially careful not to round off the edges of the piston head.

> ⚠️ **CAUTION** Do not use the wire wheel to clean the skirt of the piston or the ring groove area.

Figure 48.4 Use a scraper to clean the top of a piston.

Supervisors often advise against cleaning any part of the piston with a wire wheel because it requires skill and care. A beginner can accidentally ruin an otherwise good piston by trying to clean the skirt and ring grooves using the wire wheel. It is a good idea to clean the piston top before removing the piston rings to avoid accidental ring groove damage.

Cleaning Piston Ring Grooves

During engine operation, carbon forms in the back of the compression ring grooves. The carbon must be removed; otherwise, the carbon deposits in the ring grooves might prevent the new rings from compressing enough to enter the cylinder during piston installation. Clean the carbon from the ring grooves with a ring groove cleaner (Figure 48.5).

- When using a ring groove cleaner, be careful not to remove aluminum from the rear of the ring groove after all of the carbon has been removed.
- The ring groove cleaning tool works very well on compression ring grooves but can easily nick oil ring grooves, which do not usually get as much hard carbon buildup anyway.
- Oil return holes sometimes become plugged and must be cleaned, or the engine will continue to use oil. Drilled oil holes can be cleaned with a drill bit that correctly fits the oil hole.

If a ring groove cleaner is not available, an old ring can be broken off and ground sharp to clean out carbon (Figure 48.6). Tape the ring so it does not cut you. Use a ring to double-check for correct groove depth. Roll the ring around the entire groove to ensure that the ring does not bind up on the edges of the ring land.

Ring Groove Wear/ Side Clearance Check

The top ring groove wears the most. Rings with excessive side clearance can break. Also, the old ring wears into the groove. New rings cannot seal against a worn ring groove.

Figure 48.5 Cleaning carbon from the ring grooves with a ring groove cleaner.

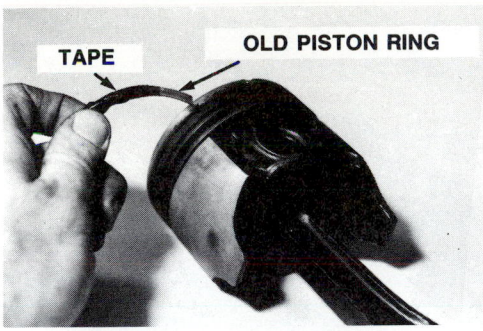

Figure 48.6 Cleaning a piston ring groove using a broken piston ring.

NOTE: *The cast iron piston ring often wears more than the aluminum piston.*

Before cleaning the piston, check the top ring groove for excessive wear (Figure 48.7). Normal ring-to-groove side clearance is from 0.002" to 0.004". A new ring is placed in the groove. If a 0.006" feeler gauge can be inserted under it, the groove is worn excessively. This check also ensures that a ring of the proper width and depth is used. A worn ring can be used for this test if the ring is held in the groove. Only part of the ring wears so the unworn part is good enough for performing this measurement (Figure 48.8)

NOTE: *Engines that use low-tension rings usually clean up at 0.020" oversize when boring. They do not wear the cylinder wall as much as standard tension piston rings because they do not push as hard against it.*

Low-Tension Ring Depth. Be sure the rings used are the proper ones for the piston. Pistons designed for use with low-tension rings have ring grooves that are shallower than standard ones.

Measuring a Piston

The place to measure a piston varies among manufacturers, so check the service manual. Figure 48.9 shows

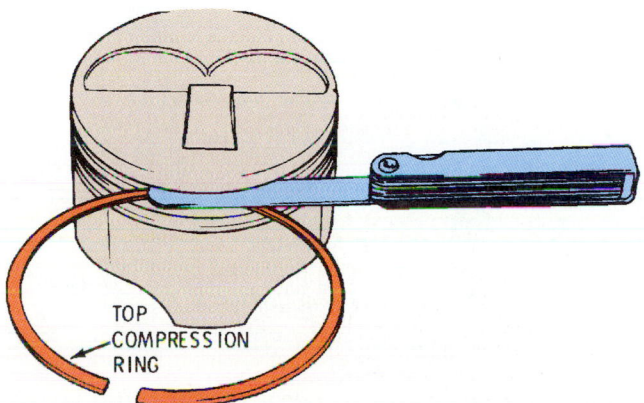

Figure 48.7 Checking a ring groove for wear. *(Courtesy of General Motors Corporation, Service Technology Group)*

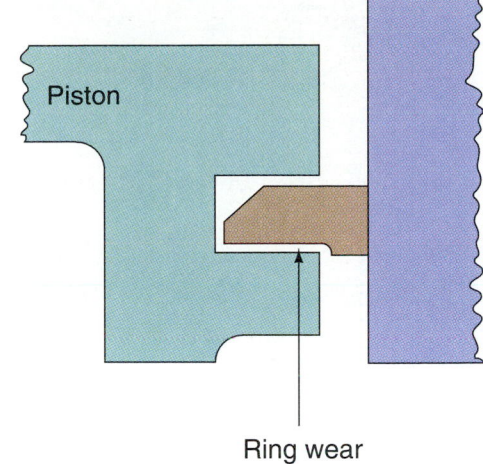

Ring wear

Figure 48.8 When ring wear is evident, it is found on part of the bottom edge.

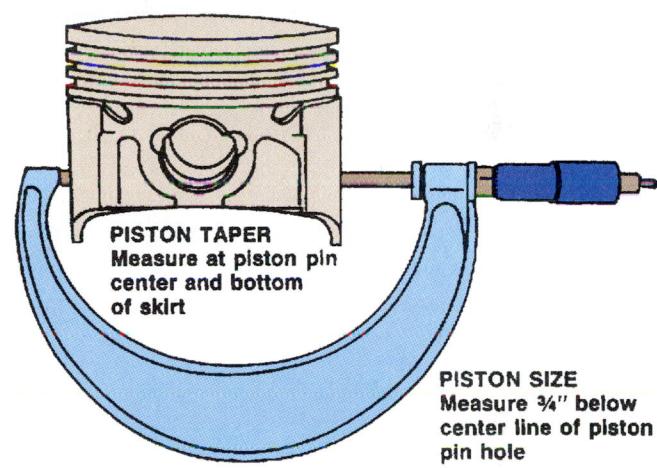

Figure 48.9 Measuring maximum piston skirt diameter. *(Courtesy of General Motors Corporation, Service Technology Group)*

where most manufacturers recommend a piston skirt be measured.

Knurling Pistons

To compensate for worn piston skirts and cylinder walls, piston skirt diameter can be enlarged by knurling (Figure 48.10). A knurled piston requires only half the clearance of a stock piston.

NOTE: *A piston skirt that has collapsed (become smaller because of heat) should be replaced, rather than knurled.*

Piston Weight

To maintain engine balance, it is important that a new replacement piston's weight match that of the original piston. Replacement pistons are designed to weigh the same as originals, even if they are oversize.

When a replacement piston is heavier than the other pistons it can be lightened at the *balance pads*

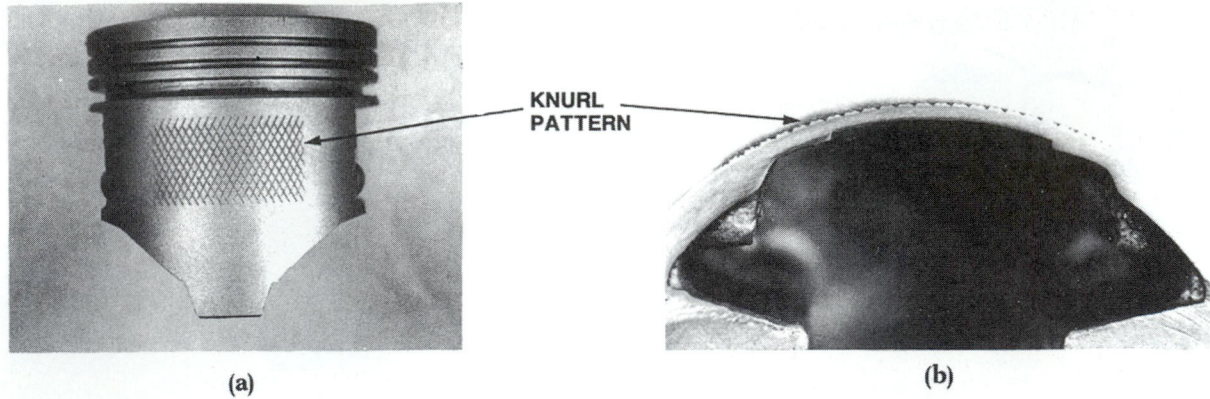

Figure 48.10 Knurling a piston skirt: (a) A knurled piston skirt. (b) Cutaway of a knurled piston skirt.

(see Figure 19.20). Balance pads are usually located on the edge of the skirt under the wrist pin.

NOTE: *Replacement pistons for passenger cars and light trucks are destroked 0.020" to compensate for block and head resurfacing. If 0.020" of resurfacing is not done, the compression ratio will be diminished by about 0.25 (9:1 reduced to 8.75:1).*

■ MATH NOTE ■

Compression Ratio

Stock block and stock piston	8:1
Stock block and new replacement piston	7.75:1
Head or block resurfaced 0.020" + new piston	8:1

Diagnosing Piston Wear

Scuffing results from excessive heat. It occurs when the cylinder wall and piston momentarily weld to each other as the piston stops at TDC. The welds are constantly made and broken.

■ Scuffing on both skirts is a problem usually caused by insufficient clearance between new pistons and cylinder walls.

■ Scuffing on only one skirt can be caused by excessive idling at too low an rpm or by lugging the engine. In either case, there is not enough oil thrown from the rods to provide adequate cylinder wall lubrication (Figure 48.11).

■ Scuffing can also be caused by cylinder wall hot spots that are the result of poor cooling system maintenance.

■ Tight wrist pins cause what is commonly called **four corners scuffing**. Both skirts are scuffed on the edges next to the piston pin (Figure 48.12). This is usually a result of an external cause such as too lean an air/fuel mixture, which causes the top of the piston to run too hot and expand against the cylinder bore.

Figure 48.11 A scuffed piston skirt. *(Courtesy of Federal-Mogul Corporation)*

Figure 48.12 Scuffing near the wrist pin. *(Courtesy of Federal-Mogul Corporation)*

■ A piston that overheats because of cooling system problems or abnormal combustion will expand excessively near the piston pin. This can cause scuffing of the piston skirt near the pin.

■ PISTON RING SERVICE

Ring Wear Diagnosis

The following are some of the things that can cause piston ring wear:
- Leftover honing grit from a careless block cleanup
- Running the engine with a missing or damaged air cleaner or broken vacuum lines
- Using a contaminated oil fill spout or funnel

Figure 48.13 shows compression and oil rings worn by abrasives.

When inspecting rings for wear, look for the following:
- When wear is due to dirty air getting in, the top ring will show more wear and vertical abrasive lines will be visible. The cylinder wall will show scratches in the area where the rings ride.
- When ring wear is due to abrasives in the oil, the lower rings and cylinder wall will have more wear and the top ring will have less wear.
- When abrasives cause wear, the bottom side of a ring will wear, leaving a lip on the outside edge (see Figure 48.8).

Ring Oversizes

When engines are rebored, oversized pistons and rings are used. Rings are made for standard, 0.020" (0.50 mm), 0.030" (0.75 mm), 0.040" (1.0 mm), and 0.060" (1.5 mm) oversizes.

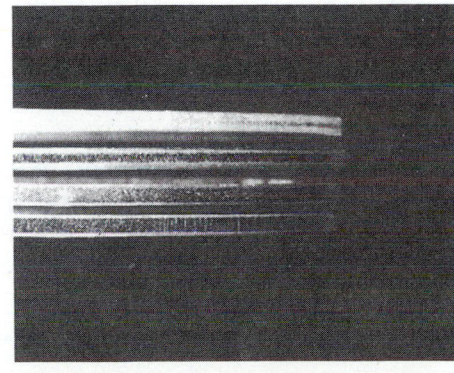

NEW RING   WORN RING

Figure 48.13 Compression and oil rings worn by abrasive particles. *(Courtesy of Caterpillar and Dana Corp., Perfect Circle Division)*

Compression Ring End Gap Clearance

Before installing rings in a cylinder bore, check the ring **end gap**. To measure gap clearance, install the ring in the cylinder and square it up with a piston. Then measure the gap with a feeler gauge. The ring must be positioned in the unworn portion of the bore, below ring travel. The end gap is tapered, so be sure to measure at the outside edge of the ring for an accurate measurement.

To give themselves more latitude during production, manufacturers produce rings that have 0.005–0.010" more gap clearance than the minimum specification. The ring gap should be at least 0.003" to 0.004" for each inch of cylinder bore diameter unless otherwise specified in the repair manual. The ring end gap can be filed to fit if it is not wide enough. One manufacturer states that maximum gap clearance is not as critical and can actually be as much as 0.030" *more* than the minimum specifications without causing blow-by.

The important thing about ring gap is the minimum specification. Too small an end gap can cause the rings to lock up in the bore as they heat and expand. Ring end gap lockup results in scuffing and ring failure. The ends of the ring will appear polished if this happens.

NOTE: *An increase of 0.002" in the bore size would increase the gap by about 0.006". Installing standard rings in a 0.030" oversized cylinder bore would result in an end gap increase of approximately 0.090".*

■ MATH NOTE ■

As an interesting experiment, check the gap on a worn ring to see how much metal it has lost.

An approximation of wear may be made as follows: If the gap of an old ring is 0.050" and the gap of a new ring is 0.020", divide the difference by 3 (or π) and then by 2 (wear from both sides).

> 0.050" Old Ring Gap
>
> −0.020" New Ring Gap
>
> 0.030" Difference

$$\frac{0.030"}{3} = 0.010" \text{ (Conversion of Circumference to Diameter)}$$

$$\frac{0.010"}{2} = 0.005" \text{ Ring Face Wear}$$

Too much gap clearance can point to a bore too large or the use of rings that are too small.

NOTE: *The gap will change by about 0.030" for each 0.010" error in size.*

Figure 48.14 Rings that are too small for the cylinder will have the appearance of carbon deposits near the gap. *[Courtesy of Dana Corp., Perfect Circle Division]*

A common customer error is to want to install oversized rings in badly worn cylinders. An oversized ring might fit into the top of a tapered cylinder. Because of the tapered cylinder wear, the gap on an oversized ring would lock up as the piston moved down the cylinder wall.

Figure 48.14 shows rings that were too small for the cylinders they were run in. Notice the dark area (carbon deposits) near the gap.

■ INSTALLATION OF PISTONS AND RINGS

Installing Pins to Connecting Rods

A machine shop will separate pressed-fit pins and pistons from their connecting rods using a machine shop tool called a pin press (Figure 48.15). Installing pressed-fit pins into connecting rods is accomplished using a

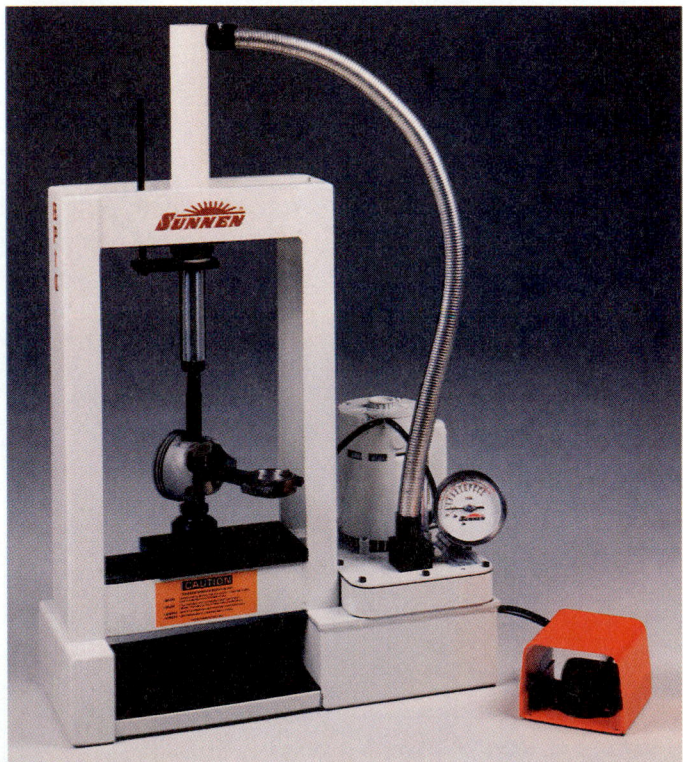

Figure 48.15 Using a piston pin press to remove a pin from a pressed-fit connecting rod. *[Courtesy of Sunnen Products Company, St. Louis, Missouri]*

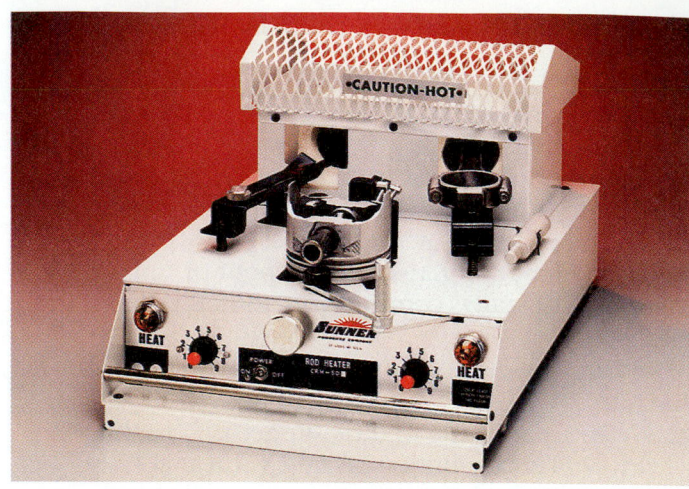

Figure 48.16 A deluxe rod furnace. *[Courtesy of Sunnen Products Company, St. Louis, Missouri]*

rod heater to heat the eye of the rod (Figure 48.16). While the rod is hot, a new pin will slip into place easily without the risk of ruining a piston.

SHOP TIP
■ When original pistons are to be reused during an overhaul, it is best to simply leave the pistons and rods assembled. Pressing them apart serves no useful purpose and risks ruining an otherwise good piston.
■ Do not glass bead blast pistons while pistons are assembled to the connecting rods.
■ Do not soak the piston and rod assembly in carburetor cleaner. The pin could seize on the piston.
■ Be sure to keep track of the direction the connecting rod faces in relation to the top of the piston. Pistons have a notch on the side of the piston head that faces the front of the engine.

■ INSTALLING RINGS ON PISTONS

Oil rings are installed first. Then, the second compression ring is installed. Finally, the top ring is installed.

■ OIL RING INSTALLATION

Most automobiles use three-piece oil control rings, consisting of an expander spacer in the center with two outside rails. Install them as shown in Figure 48.17.
■ Install the expander spacer, being careful not to overlap its ends. The ends are usually painted different colors to make it obvious to the installer if they are accidentally overlapped. Some expanders are filled with a Teflon button on their

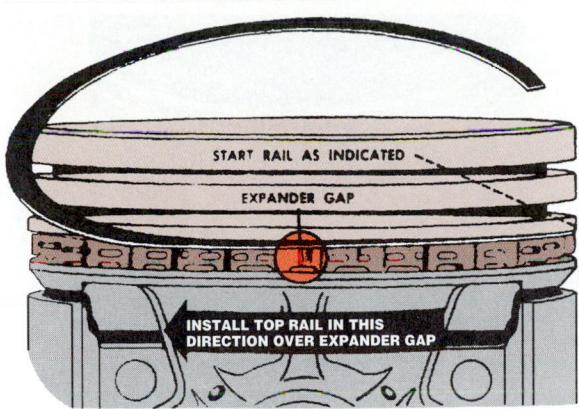

Figure 48.17 Installing the oil ring rail. *(Courtesy of Federal-Mogul Corporation)*

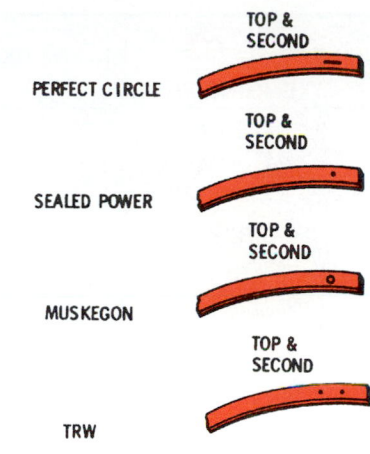

Figure 48.19 Ring I.D. marks face up. *(Courtesy of General Motors Corporation, Service Technology Group)*

ends to prevent improper assembly. In the absence of a recommendation, position the expander gap above one end of the wrist pin.

- Next, install the rails. Installing the top rail first is easiest. While holding your finger over the butted ends of the expander, install the top rail. Position its gap above the skirt on one side of the expander spacer gap.
- Install the lower rail with its gap placed above the skirt on the opposite side of the expander gap (Figure 48.18).

Compression Ring Installation

Remember from Chapter 19 that piston rings are often tapered or have chamfers or reliefs to cause them to twist. These piston rings must be installed with their identification marks facing up (Figure 48.19). Installing them upside down will result in severe oil consumption. In fact, one compression ring installed upside down can double an engine's oil consumption.

The second compression ring actually controls more oil than compression.

Use a ring expander to install the compression rings (see Figure 48.2). If a ring expander is not available, a shop rag can be used as shown in Figure 48.3. It is important not to "spiral" or roll the rings on; they can become distorted to resemble a lock washer (Figure 48.20). Overexpanding plain cast iron rings during installation can very easily result in a broken ring.

Compression Ring Gap Position

The gaps are placed at different locations around the piston. Manufacturers specify different gap positions in their service manuals. According to information published by engineers in the Perfect Circle Division of Dana Corporation, the reason for the practice of staggering ring end gaps is to guard against scuffing when an engine is started for the first time. As the engine operates, rings will rotate from the position they were first installed in.

- End gap position is *not* a cause of oil consumption.
- There are many differing opinions on ring gap placement. The most prudent policy is to follow recommendations of the vehicle manufacturer, when available.

The positions shown in Figure 48.21 are popular in the aftermarket.

Figure 48.18 Oil ring gap positions. *(Courtesy of Federal-Mogul Corporation)*

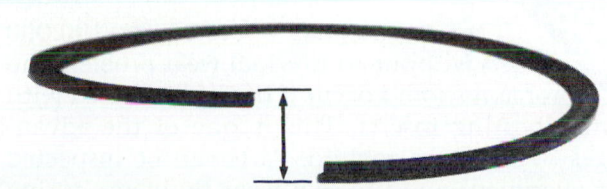

Figure 48.20 Improper installation can ruin a piston ring.

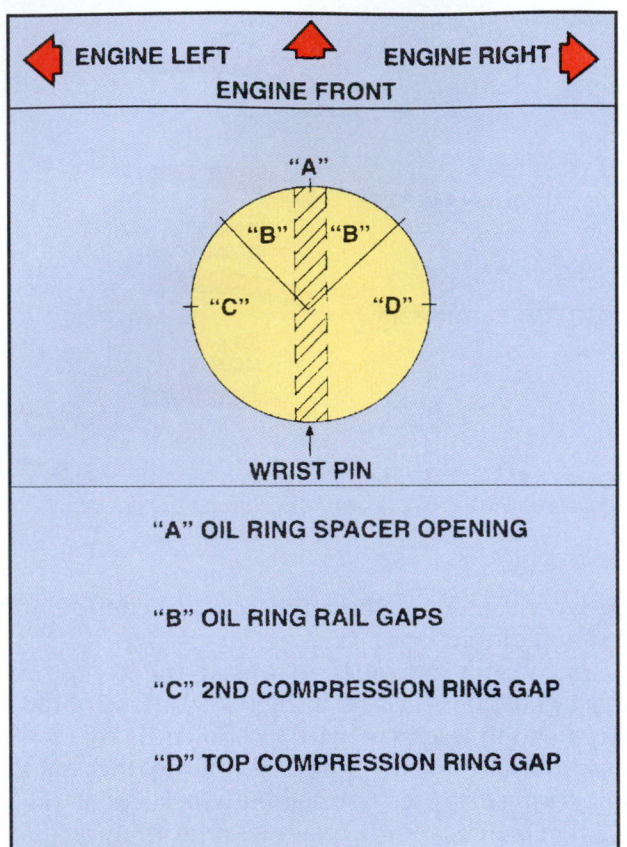

ENGINE LEFT   ENGINE FRONT   ENGINE RIGHT

"A"
"B"   "B"
"C"   "D"

WRIST PIN

"A" OIL RING SPACER OPENING

"B" OIL RING RAIL GAPS

"C" 2ND COMPRESSION RING GAP

"D" TOP COMPRESSION RING GAP

Figure 48.21 Chart showing popular ring gap placements.

■ CONNECTING ROD SERVICE

Be sure to keep rod caps on the rod that they came on. Both the upper and lower pieces should be numbered to ensure accurate pairing on reassembly.

CAUTION If the cap is installed backwards, the rod bore will not be round.

Rod Alignment

Closely examine all piston skirts for unusual wear patterns that can indicate a *twisted rod* (Figure 48.22). A *bent rod* may show wear on opposite sides of the rod bearings (Figure 48.23).

SHOP TIP If an engine has operated for 100,000 miles without an unusual wear problem, no problem should occur if the rod is reused with new bearing inserts. This is one of the advantages of a custom rebuild. Parts can be inspected for alignment and unusual wear problems during disassembly.

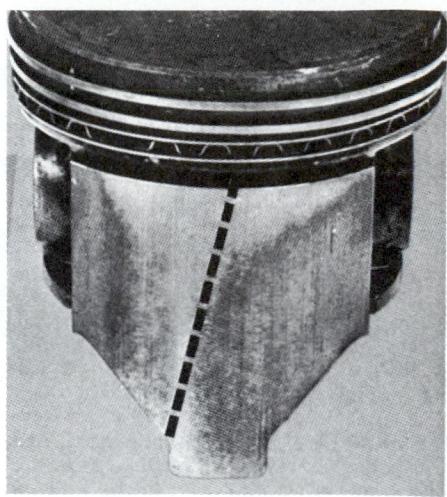

Figure 48.22 A twisted connecting rod caused this wear pattern on the piston skirt. *(Courtesy of Dana Corp., Perfect Circle Division)*

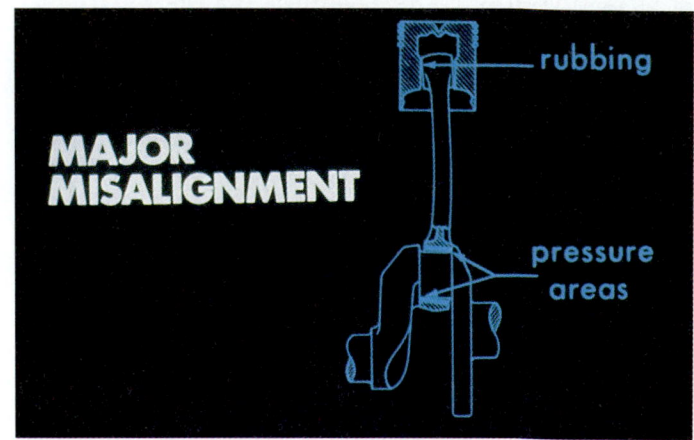

MAJOR MISALIGNMENT

rubbing

pressure areas

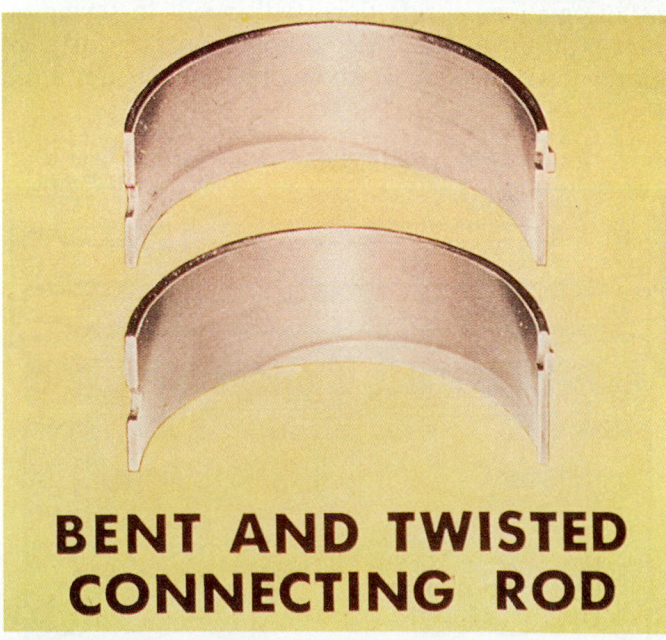

BENT AND TWISTED CONNECTING ROD

Figure 48.23 Bearing wear caused by a bent or twisted connecting rod. *(Courtesy of AE Clevite Engine Parts)*

Rods suspected of being twisted can be checked on a rod aligning fixture (Figure 48.24) and straightened as needed. Figure 48.25 shows the difference between bent and twisted rods and how each can be corrected.

Rods can also be checked for alignment when they are installed on the crankshaft. A feeler gauge is used to measure rod side clearance at several points around the rod (Figure 48.26).

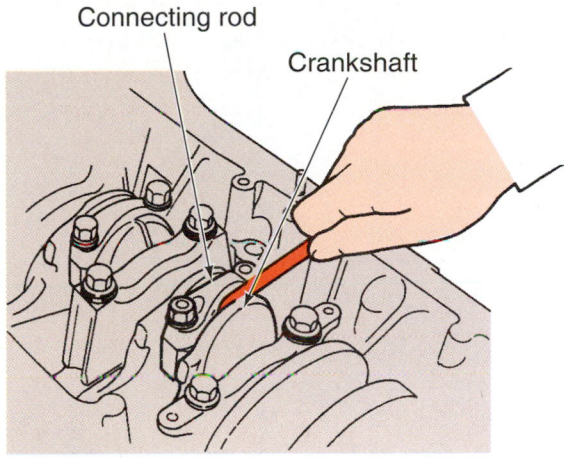

Figure 48.26 Check rod bearing side clearance at several points next to the rod. *(Courtesy of American Honda Motor Co., Inc.)*

Connecting Rod Rebuilding

When a rod bearing has "spun" or burned or if the big end has "stretched," the rod can be resized by a machine shop. This is known as **connecting rod resizing**. After machining, the bore of the rod is still the same size it originally was.

Rod stretch sometimes causes the rod to draw closer together at the rod cap parting line. This sometimes causes bearing wear at the ends of the inserts (Figure 48.27). Rods are measured for "out-of-round" with a special gauge (Figure 48.28). The gauge measures in tenths of thousandths. Usually, rods can be up to 0.001" out-of-round before resizing is necessary.

To resize a rod:
- The pressed-fit rod bolts are pressed or pounded out.
- A small amount of metal (usually less than 0.002") will be ground off of both the rod and cap mating surface.

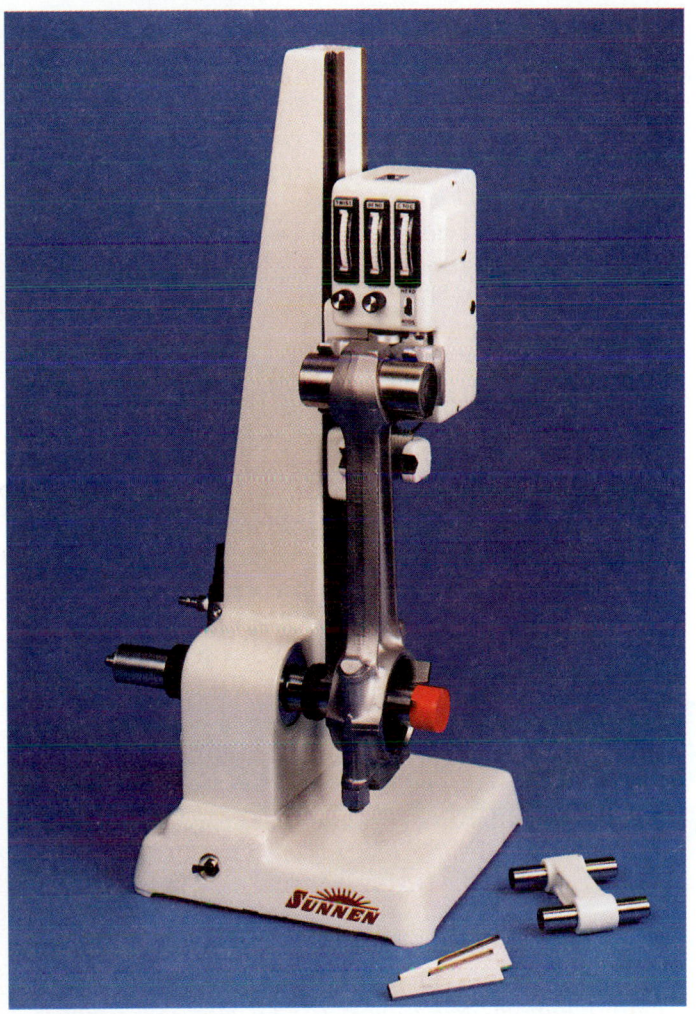

Figure 48.24 Checking connecting rod alignment. *(Courtesy of Sunnen Products Company, St. Louis, Missouri)*

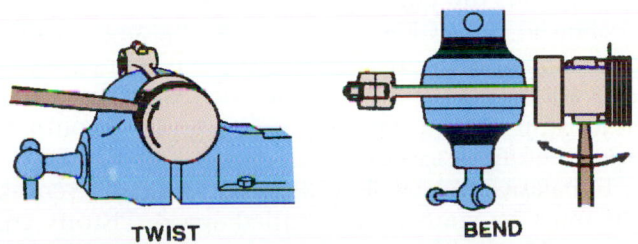

TWIST   BEND

Figure 48.25 Correcting rod twist and bend. *(Courtesy of Sunnen Products Company, St. Louis, Missouri)*

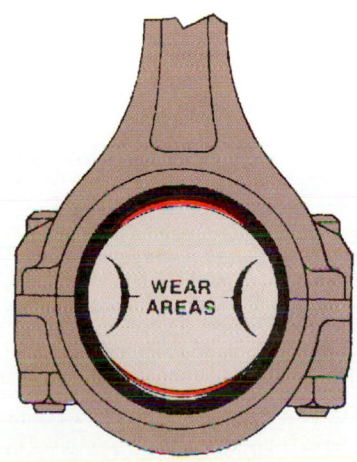

WEAR AREAS

Figure 48.27 Rod stretch causes bearing wear at the parting line. *(Courtesy of AE Clevite Engine Parts)*

Figure 48.28 Checking for rod stretch. *(Courtesy of Sunnen Products Company, St. Louis, Missouri)*

- The rod cap is reinstalled on the rod and the rod nuts are torqued.
- The rod bore (which is now smaller as a result of grinding) is honed out with a rod hone (Figure 48.29) until the original diameter of the rod bore is reached.

NOTE: *Overtorque of rod bolts can cause them to fail during deceleration (when the load is on the bolt instead of on the rod).*

To repair the small end of a full-floating type connecting rod, a bronze bushing is installed and the bore is then honed to the proper size.

Figure 48.29 Honing connecting rods to size. *(Courtesy of Sunnen Products Company, St. Louis, Missouri)*

ENGINE BALANCING SERVICE

Balancing is done by a machine shop or a balancing specialist. Reciprocating parts (the piston assembly, including rings, wrist pins, and the pin end of the rod) are balanced to weigh approximately the same amount. Figure 48.30 shows a connecting rod being weighed separately on each end. All of the rods are weighed and then lightened to the weight of the lightest one.

Rotating parts are balanced by spinning them on a balancing machine (Figure 48.31) to determine the location of heavy spots. Balancers operate at low speed (400 rpm) for safety. This rpm is sufficient to calculate what the amount of imbalance will be at higher speeds.

Heavy counterweights can be lightened by drilling (Figure 48.32). Figure 48.33 shows balance pads on a connecting rod; weight can be removed from them during the balancing procedure. Counterweights that are too light can be drilled for the addition of a heavier metal, like lead or tungsten, which is twice as heavy as steel. Weight can be added to a wrist pin if a piston is too light.

Cast cranks sometimes use a balanced vibration damper and torque converter flexplate. On these cranks, there might not be any balancing done on the crank counterweights. **Internal balancing** of forged cranks is usually achieved by drilling holes on the counterweights.

CASE HISTORY

A technician rebuilt a 440 Dodge engine for a motor home using spare parts from the shop inventory. He used the existing engine's torque converter and vibration balancer. When the engine was started, a serious vibration was evident. The crankshaft used in the rebuilt engine was forged and the one in the engine that was removed was cast. One of the crankshafts was externally balanced, requiring a special vibration damper and weights on the torque converter.

When balancing an engine used with a standard transmission, the clutch cover (pressure plate) should be balanced too. It is usually balanced along with the flywheel. Some engines have vibration dampers that are balanced with the belt pulley bolted to them. The relationship of the damper to the pulley should be marked before disassembly.

To balance the crank, rods, and pistons, the crankshaft must be spun at a specified speed. Pistons and rods would not be able to be spun while attached to the crankshaft unless the parts were assembled in the block. **Bob weights** (Figure 48.34) are used when

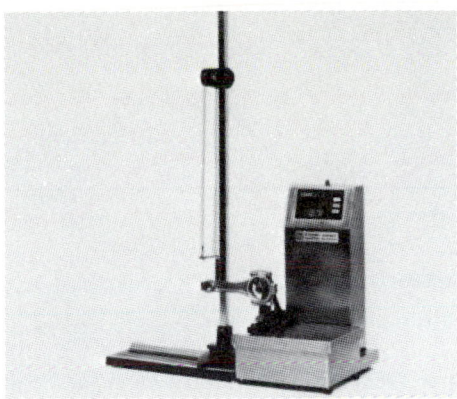

Figure 48.30 A connecting rod is weighed on each end using an accurate scale. *(Courtesy of Pro-Bal Industrial Balancers)*

Figure 48.31 A crankshaft in an engine balancer. *(Courtesy of Pro-Bal Industrial Balancers)*

Figure 48.32 Holes are drilled in crankshaft counterweights.

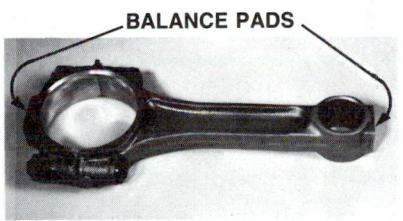

Figure 48.33 Connecting rod balance pads.

spinning the crankshaft to simulate the correct weight. Inline engines do not use bob weights.

Replacement Piston Balance

Balance is not as important with inline engines, but it is critical in V-type engines. When replacing a piston on a V-type engine, match the weight of the new piston to the stock piston to maintain correct balance. Pistons in a new set should not have a spread of more than 5 grams from the heaviest to the lightest.

NOTE: *A dollar bill weighs about 1 gram.*

SHOP TIP When there is a complaint of engine vibration, start your investigation with the engine mounts. They could be worn out or stiff ones could have been installed by mistake.

Figure 48.34 A bob weight mounted on a crankshaft throw. *(Courtesy of Pro-Bal Industrial Balancers)*

Balance Shafts

Engines with balance shafts run very smoothly. The timing of the shafts is critical. They must be replaced in the proper manner to maintain balance.

> ### CASE HISTORY
>
> *After replacing a timing belt on a Mitsubishi 1.8L four cylinder with dual balance shafts, the engine had a serious vibration problem. The timing marks on the front balance shaft and oil pump drive were inspected and found to be aligned properly. After investigation in the repair manual, the technician discovered that the oil pump drives a rear balance shaft that was actually installed 180° out of position. When putting on the timing belt, installation instructions called for installing a shaft through a hole in the rear of the block to correctly align the rear balance shaft to the oil pump. This corrected the problem.*

■ ADVANCED BALANCING INFORMATION

The following section describes the theory of different types of imbalance. It is more advanced information for the serious student. More information on engine balancing theory is included in Chapter 19.

Types of Vibration

When the piston slows down as it approaches TDC, its force pulls the engine up. When it approaches BDC, it pulls the engine down. This is known as **primary vibration**. Engines that have two rod throws 180° apart counterbalance each other to give perfect primary balance.

If there are only two cylinders that fire 180° apart instead of 360° apart, another form of vibration called a **rocking couple** takes place. This results in the engine rocking from end to end. A four-cylinder engine eliminates the rocking couple problem because one pair of couples cancels out the other pair. Cylinders number one and four move in the opposite direction of the inner two cylinders, number two and three (Figure 48.35).

Another form of vibration called *secondary vibration* occurs in inline four-cylinder engines. Secondary vibrations are only about one-fourth of the strength of primary vibrations but can become quite severe at the higher rpms at which four-cylinder engines quite

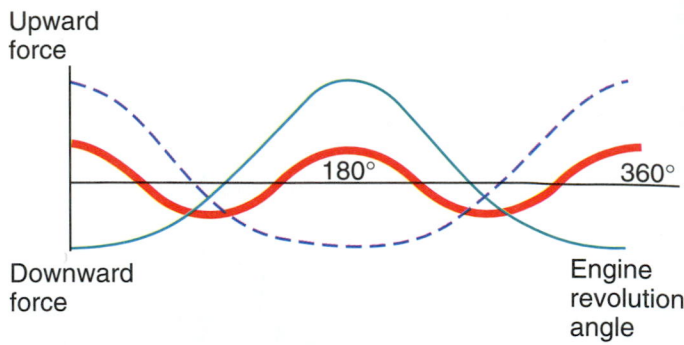

Figure 48.35 The dotted line represents the force generated by cylinders one and four. The fine line is cylinders two and three. *(Courtesy of Ford Motor Company)*

often operate. Secondary vibration is represented in Figure 48.35 by the solid line. Balance shafts (see Figure 19.39) are used to cancel out secondary vibration. They are driven by the crankshaft in opposite directions at twice crankshaft speed.

Types of Imbalance

Force, also called **static** or **kinetic**, **imbalance** can be compared to the type of imbalance that is corrected when balancing tires with a bubble (level) balancer. As the crankshaft spins, the balancer senses the vibration whenever a heavy area is forcing down. Removing this heavy spot, or counterbalancing it by adding an equal amount of weight to the light side corrects the force imbalance. Force imbalance is smallest at 90° of rod angle and most at TDC or BDC.

Two other types of imbalance that must be corrected are called **dynamic and couple imbalance**. Both require adding or removing metal at two different places on the part. Correcting dynamic imbalance can result in the correction of force imbalance at the same time. Computer balancers compute the combined amount of dynamic and force imbalance and tell where to remove metal to correct them.

Ninety-degree V-6 engines have especially strong primary (up and down) imbalance. These engines are installed transversely in front wheel drive cars. The vibration can be felt inside the car on the steering wheel. Soft engine mounts are used that allow the engine to move from side to side. Engine mounts and exhaust connections suffer increased wear and failure because of this, but this is an effective way of smoothing out engine vibrations for front wheel drive cars.

■ REVIEW QUESTIONS

1. What is it called when a metal surface is damaged, weakening it?
2. When the compression ring grooves are cleaned, what substance is removed?
3. What causes scuffing?
4. When both piston skirts are scuffed near the wrist pin, what is this called?

5. What oversizes (in inches) are pistons and rings made for?

6. What minimum size should the ring gap be per inch of cylinder bore?

7. If the bore size is increased by 0.010", how much does the ring gap increase by?

8. If a worn ring's end gap is 0.060" and a new ring's end gap is 0.015", how much is the old ring face worn?

9. What is it called when the big end of a connecting rod is remachined?

10. Where is the extra weight added when an engine is externally balanced?

■ ASE STYLE REVIEW QUESTIONS

1. Technician A says that the ring belt area of a piston (just above the piston skirt) is best cleaned with a wire wheel. Technician B says that ring grooves used with low-tension rings are deeper that standard ring grooves. Who is right?

   a. Technician A   **b.** Technician B

   c. Both A and B   **d.** Neither A nor B

2. Technician A says that a piston skirt that has collapsed can be repaired by knurling it. Technician B says that a piston pin can become tight on the piston if a piston and rod are cleaned in carburetor cleaner. Who is right?

   a. Technician A   **b.** Technician B

   c. Both A and B   **d.** Neither A nor B

3. Technician A says that pressed-fit pins are best installed on connecting rods using a rod heater. Technician B says that pressed-fit piston pins are best removed from connecting rods using a press. Who is right?

   a. Technician A   **b.** Technician B

   c. Both A and B   **d.** Neither A nor B

4. Technician A says that if the piston ring face wears by 0.005", the end gap change will change by 0.010". Technician B says that a 0.030" oversize piston should weigh the same as a standard piston. Who is right?

   a. Technician A   **b.** Technician B

   c. Both A and B   **d.** Neither A nor B

5. Technician A says that the ring gap can be up to 0.003" too small without causing blow-by. Technician B says that the second ring wears the most. Who is right?

   a. Technician A   **b.** Technician B

   c. Both A and B   **d.** Neither A nor B

6. A cylinder wall has cylinder wall scratches only in the area where the piston rings ride. Technician A says this can happen if unfiltered air has been entering the engine's air intake. Technician B says this is due to dirty engine oil. Who is right?

   a. Technician A   **b.** Technician B

   c. Both A and B   **d.** Neither A nor B

7. Technician A says a connecting rod bolt is under the most load during deceleration. Technician B says that if the ring gap at the top of the cylinder is larger by 0.010" than the gap at the bottom, the cylinder has 0.010" of bore taper. Who is right?

   a. Technician A   **b.** Technician B

   c. Both A and B   **d.** Neither A nor B

CHAPTER 49

Ordering Parts, Short and Long Blocks, Engine Assembly

■ OBJECTIVES

Upon completion of this chapter, you should be able to:

✔ Order the correct parts for an engine.

✔ Decide the best course of action for engine repair, rebuild, or replacement.

✔ Reassemble an engine in a proper and organized manner.

■ KEY TERMS

engine kit
cores
short blocks
custom rebuild
long block
spin tests

■ INTRODUCTION

Engine Assembly

Before assembly of the engine:
■ Look through the service manual for special instructions.
■ Have all tightening specifications handy.
■ Be sure all parts are thoroughly cleaned.
■ Obtain replacement parts.

■ ORDERING PARTS

After the block has been stripped, inspect all parts for wear and damage, and compile a list of new parts that are needed (Figure 49.1).

■ ENGINE KITS

Parts kits are available at wholesale prices for most of the more common foreign and domestic engines. The **engine kit** contains all or most of the parts necessary to completely rebuild an engine. Many of the parts are not individually boxed and a kit usually costs far less than the individual parts when purchased separately. There are kits that contain various groups of parts—for example:
■ a crank kit
■ a timing chain set
■ an overhaul kit
■ a master kit

A crankshaft kit includes a reground crankshaft and bearings. A timing chain set includes the chain and sprockets, or chain guides and a tensioner for OHC engines. An overhaul kit includes gaskets, piston rings, crankshaft bearings, and sometimes, a timing set.

The parts normally found in an engine "master kit" are pistons, rings, reground crank and bearings,

reground cam and bearings, new or rebuilt oil pump, timing chain set, and a complete gasket set (Figure 49.2). Lifters may or may not be included. Many different types and qualities of kits are available. Be sure to compile the parts list carefully. It is not uncommon to spend twice as much on separate engine parts and not get as many new parts as would come in a packaged engine kit.

Part Cores

When an engine or an engine kit is purchased from a parts source, the crankshaft, camshaft, and oil pump are usually returned to them as **cores**. The "core charge" that is paid will be refunded when these parts are returned (provided the core is rebuildable).

Before returning a crankshaft as a core, remove the timing gear or sprocket. If the gear does not have threaded holes, use the puller setup shown in Figure 49.3, or a bearing separator and puller like the one shown in Figure 7.41. Save the crankshaft *woodruff key* and pilot bushing. They are not always easy to locate in the correct size. The procedure for removing a woodruff key is shown in Figure 49.4.

■ DETERMINING PART SIZES

The engine size can be determined in several ways. Engine rebuilders use books that list casting numbers (Figure 49.5). These numbers identify blocks, crankshafts, cylinder heads, and so forth, by groups as they were cast at the foundry. The block usually has numbers stamped somewhere on it that can be used to identify the engine before ordering parts.

The cylinder bore diameter and the diameter of the crankshaft journals can be measured and compared to the specifications in the repair manual.

NOTE: Be sure to save all old engine parts.

Engine year and size _____

Piston and ring size: STD _____ .020" over _____ .030" over _____ .040" over _____ .060" over _____
(Check one)

Crank and bearing sizes: Mains STD _____ .010" under _____ .020" under _____   Rods STD _____ .010" under _____ .020" under _____
(Check one)

Core plug size _____

✓ Those items needed

	Quantity	Price
*1. Reground crankshaft and bearings (Crankshaft kit) (be sure to save old woodruff key)		
**2. Piston, wrist pins		
*3. Piston rings		
**4. Reground cam		
*5. Cam bearings		
**6. Valve lifters		
NOTE: Sometimes these are not included in a master kit.		
**7. Cam and crank sprockets (Sometimes not in kit)		
*8. Timing Chain		
**9. Oil pump		
A. New		
B. Rebuilt		
**10. Core plugs (brass is recommended, at extra cost) including rear cam plug and oil gallery plugs		
**11. Overhaul gasket set (includes the following)		
A. Head gaskets		
B. Valve guide seals		
C. Valve cover gaskets		
D. Intake manifold gaskets		
E. Timing cover gasket		
F. Oil pan gaskets		
G. Rear main seal		
H. Front cover seal		
I. Exhaust manifold gaskets		
J. Water outlet gasket (gooseneck)		
NOTE: Carburetor gaskets are usually not included, save your old gasket so you know which to use		
12. Oil		
13. Oil filter		
14. Spray paint		
15. Large trash bag (to keep engine clean during assembly)		

	Quantity	Price
The following items are inspected during engine repair and may require replacement at extra cost to the vehicle owner.		
16. Hoses		
A. Radiator		
B. Heater hoses		
C. Thermostat bypass (molded hose)		
D. Vacuum hose		
E. Fuel hose		
F. Smog hose		
17. Hose clamps		
A. Heater		
B. Fuel		
C. Radiator		
18. Fuel filter		
19. Thermostat		
20. Water outlet housing (gooseneck)		
21. Radiator cap		
22. Coolant (recommended)		
23. Motor mounts		
NOTE: If kit is purchased, old crank, cam, and oil pump must be returned to the parts house or a core charge must be paid.		
24. Ignition parts		
A. Spark plugs (recommended)		
B. Points, condenser		
C. Rotor and distributor cap		
D. Plug wires		
25. Air filter		
26. PCV valve		
27. Fan belts		
28. Ground straps, battery cables or clamps		
29. Valves		
A. Intake		
B. Exhaust		
30. Valve springs		
31. Push rods		
32. Clutch parts		
33. Automatic transmission front pump seal (recommended)		
34. Automatic transmission fluid		
35. Fuel pump		

Engine Master Kit (parts marked * or ** are included in a high quality kit)
Engine overhaul kit (parts marked * only are included in this kit)

Figure 49.1 An engine parts checklist.

Figure 49.2 An engine master kit.

Figure 49.3 A puller is used to remove the crankshaft sprocket. *(Courtesy of General Motors Corporation, Service Technology Group)*

During inspection of engine parts, watch out for unusual oversized or undersized parts. When errors are made at the factory, an engine block is often salvaged. An original engine might have a crankshaft that has been ground undersized on its main and/or rod journals. Cylinder bores, core plug holes, and valve guides and stems can be oversize. Sometimes one or more lifter bores will be machined oversize and fitted with oversize lifters.

Manufacturers have codes (usually in their service manuals) to indicate the use of nonstandard parts. Numbers or letters may be stamped on the parts or on

Figure 49.4 Removing a woodruff key.

Cylinder Head & Block Manual

A membership service from the Automotive Engine Rebuilders Association

Figure 49.5 Automotive Engine Rebuilders Association book of block and head casting numbers. *(Courtesy of AERA)*

the oil pan rail or crankshaft or there may be paint on the part (green is a favorite color). Do not leave this to chance. Measure all parts.

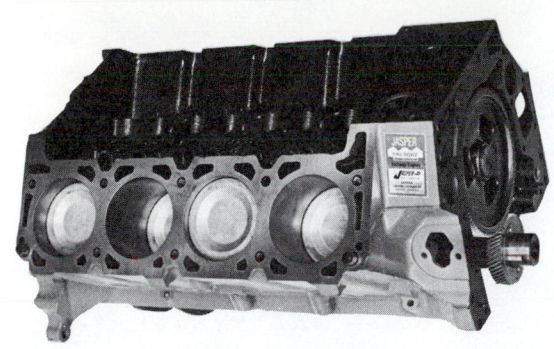

Figure 49.6 A short block. *(Courtesy of Jasper Engines)*

■ SHORT BLOCKS

Short blocks are sometimes used by independent shops and dealerships. Short blocks are completely assembled rebuilt blocks purchased from automotive machine shops and engine rebuilding companies (Figure 49.6). They do not include any external parts such as mounting brackets, sheet metal, pumps, or accessories. Some larger companies rebuild engines on an assembly line. Some of these rebuilders are factory-authorized by automakers such as Chrysler or Ford, and they rebuild engines to the authorizing manufacturers' standards.

The repair shop will often do the head reconditioning, install the heads on the short block, and install the engine in the car. In this way, they earn the money for a valve job and engine R&R, as well as earning the markup price from wholesale to retail on the short block assembly.

An advantage to the shop owner over buying an engine kit is that a short block carries a time and mileage guarantee from the rebuilder. Another consideration is that many rebuilt short block assemblies use lower quality cores. They may have crankshafts that are 0.030" undersize and pistons and blocks that are 0.060" oversize. On a **custom rebuild** of a core, a first time overhaul would usually have a 0.010" undersize crank and 0.030" oversize pistons and cylinders. A custom rebuild is when a customer's engine is rebuilt for use in the same vehicle from which it was removed.

Lifters, gaskets, and/or an oil pump may have to be purchased in addition to the short block. An assembly that needs these parts is known as a "short" short block.

■ LONG BLOCKS

Many shops prefer **long blocks** to short blocks. A long block, which includes the cylinder heads assembled on a short block, might be a brand new factory engine or a rebuilt engine. Long blocks are tested in run-in stands or spin testers so more problems can be spotted before installation.

Engine Cores

When buying a short block or a long block, the old block assembly is returned to the rebuilder as a core. If the block core is unacceptable or if the crankshaft, camshaft, cylinder block, or heads are found to be defective, a core charge is charged. Typical practice is for the core charge to be returned only after the rebuilder inspects the parts.

■ SELECTING THE CORRECT REPLACEMENT ENGINE

When you swap one engine for another, be sure that you have the correct replacement. Some of the engine parts can be different. Check where the dipstick is located. On the same manufacturer's engine, it can be in a different place from one vehicle model to another. Sometimes crankshafts are different, too. Check the forging numbers on them.

Engines can also have different mounting points and fittings. If a customer asks for an engine with a different displacement than was originally installed in the vehicle, he or she could have problems with emission certification. Also, the engine could have more power than the drive line was designed to manage.

NOTE: *Federal law prohibits anyone from substituting one engine with another engine of a different year. The engine's emission components must all be the same.*

■ WARRANTY

Some rebuilt engines have warranties of as little as ninety days or 4000 miles. A longer guarantee usually means a higher cost. When a rebuilt engine fails while under warranty, some rebuilders pay for repairs according to the time listed in the flat-rate manual (see Chapter 4). The percentage of flat rate paid is usually such that the installation shop will not make a profit on the warranty repair, but the technician will be compensated.

Proper diagnosis is very important. If something that is not the fault of the rebuilder is causing the problem, under the terms of the warranty the rebuilder will not be liable for the cost of the repair. A needlessly replaced long block can be a very costly expense to the installation shop. If an engine failed due to an overheating problem, it is possible that the sensors for coolant temperature and the fan relay could be defective. Sometimes a piston will *seize* in the cylinder because of an air leak in the intake manifold or a fuel system problem causing an air/fuel mixture that is too lean.

All fuel and cooling system hoses that are not in excellent condition must be replaced. The preliminary estimate for the customer should include these items. If the customer does not agree to their replacement, the shop cannot guarantee the job. Also, make sure that the water pump and radiator are in good condition. All the work that has been done will be wasted if a careful and methodical engine reassembly is not done.

■ REASSEMBLY

Before beginning reassembly, inspect and count all new parts. It is possible that too many parts could be packaged in a master parts kit. For instance, if there were too many oil gallery plugs or compression rings in a set, the extra one might appear to be a leftover. It would be frustrating to disassemble the engine just to see if a part had been forgotten.

Prepare the cylinder block for reassembly as described in Chapter 47. Clean the block with soap and water and then oil it after drying. Be sure all oil galleries have been cleaned with a rifle brush before installing gallery plugs. This is extremely important. Failure to do a thorough job on this can result in an engine failure. Install the oil gallery and coolant core plugs in the block (see Chapter 21). On pushrod engines, install the cam bearings as described in Chapter 47. Chase all threads in the block with a tap.

■ BEGIN REASSEMBLY

After the block has been thoroughly cleaned and the cam bearings and core plugs have been installed, you can begin to reassemble the engine.

SHOP TIP
- Once an engine is assembled, the only way for dirt to get in is past the air cleaner, so cleanliness during assembly is a must. If work is stopped at any time during the reassembly, cover the engine completely with a large trash bag to keep dirt out (Figure 49.7).
- If a bolt cannot be turned into a hole by finger pressure only, something is wrong. *Never force threads together with hand tools or an impact wrench.*
- Finger start *all* bolts that fasten a particular part before proceeding to torque any of them down. This will allow a part to be shifted around so that the threads of all the bolts can be started more easily.
- Bolts or studs that are threaded into aluminum should be coated with anti-seize compound to prevent the aluminum from oxidizing to the bolt (see Figure 6.3).

Assembly Lubricants

Lubricate all possible wear areas generously during assembly. Assembly lubricants are used in the assembly of high-load parts such as the cam lobes (Figure 49.8). Two of these black, gummy solutions are *moly-disulfide* and *graphite*. Both are popular in greases and gear lubes.

Figure 49.7 Cover the engine with a large trash bag to keep out dirt.

Figure 49.8 A special assembly lube is used on cam lobes. *(Courtesy of Dana Corp., Perfect Circle Division)*

SHOP TIP Some assembly lubes should be used only between the cam lobes and lifters, not on engine bearing surfaces. The melting temperature of these lubricants is higher than the melting temperature of the engine's bearing materials.

There are other assembly products that have low melting temperatures and are soluble in oil. An assembly lube with the consistency of grease at room temperature is desirable in case the engine is left sitting for a long period of time before it is installed in a vehicle.

■ INSTALL THE CAM

- Install the cam and make sure that it turns easily in the newly installed bearings. Be careful not to nick a lobe.
- Install the sprocket or gear and use it as a temporary handle to turn the cam during installation.

- If the block is not installed on the engine stand, it can be rested upright (vertically) to make cam installation easier.

    After installation, make sure that the cam can be turned by hand! Cylinder blocks are flexible. If the block is not supported, it is possible that the camshaft will not enter it easily.

An apprentice was installing cam bearings in an inline six cylinder mounted on a universal engine stand. After all of the bearings were installed, he attempted to install the camshaft in the block. It would not go in and light was visible between part of the journal and the bearing. The machinist was consulted. He suggested removing the block from the engine stand and resting it on a workbench. After this was done, the camshaft went easily into the block. On the engine stand, the block had been sagging under its own weight.

Once the head and oil pan are installed on the block, the block will become more rigid.

◗ PREPARE THE CRANKSHAFT FOR INSTALLATION

The following items should be observed when preparing the crankshaft for installation:

- Remove the old crankshaft sprocket. If the new one is pressed-fit, install it before installing the crank in the block (Figure 49.9). This will prevent damage to the thrust main bearing surface when the sprocket is pounded on.
- Make certain that the woodruff key is perfectly flat in its groove in the crank. An improperly installed

woodruff key can cause a cracked sprocket (Figure 49.10).
- The sprocket can be heated for easier installation if the crank is already in the block.
- Some crank sprockets slide easily into place.
- Install the sprocket with the timing mark facing outward. The inside edge of the sprocket has a chamfer that corresponds to the fillet on the crank; if the sprocket is backward it cannot be installed all the way (Figure 49.11).

SHOP TIP Be sure that all oil passages in the crank are clean (Figure 49.12). The machinist is not paid to clean the crankshaft after grinding; this is the assembler's responsibility.

- Be sure that the surface the rear seal rides on is clean. If it is not, clean it with very fine emery or crocus cloth (Figure 49.13).

Figure 49.10 Careless installation of the crank sprocket can ruin it. *(Courtesy of Federal-Mogul Corporation)*

Figure 49.9 Replace the crankshaft sprocket before installing the crankshaft in the block. *(Courtesy of General Motors Corporation, Service Technology Group)*

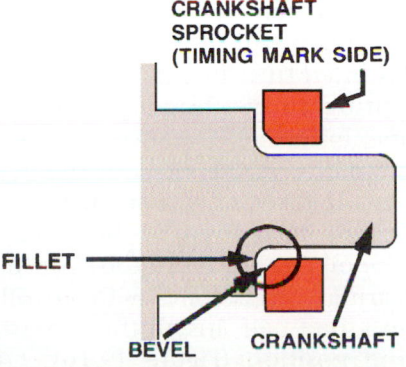

CRANKSHAFT SPROCKET (TIMING MARK SIDE)

FILLET

BEVEL   CRANKSHAFT

Figure 49.11 The inside edge of a crankshaft sprocket is beveled to clear the fillet on the crankshaft.

Figure 49.12 Make sure that all crankshaft oil holes are clean. *(Courtesy of Federal-Mogul Corporation)*

Figure 49.13 Clean the crankshaft rear sealing surface with crocus cloth.

■ INSTALL THE CRANK

Clean the main bearing bores to prevent oil clearance and heat transfer problems (Figure 49.14). If dirt is left on the bearing bore, the bearing can be forced against the crankshaft (Figure 49.15).

Be especially careful to clean the recesses where the bearing lock tab will fit. Failure to clean this area thoroughly is often the cause of a tight crankshaft.

Often, during an oil clearance check on a newly rebuilt engine, excessive clearance is discovered. This could be due to dirt on the block or the main bearing parting halves. It takes only a small film of dirt left from the hot tank, bake oven, or shot-peening process to increase the oil clearance by 0.001" or more.

Main bearings often come with an oil hole and groove in only one half; install these bearings in the upper bearing position (Figure 49.16). Lower main bearings (especially for heavy-duty use) are sometimes solid with no oil hole or groove. Other times, the same

Figure 49.14 Clean the main bearing bore. *(Courtesy of Federal-Mogul Corporation)*

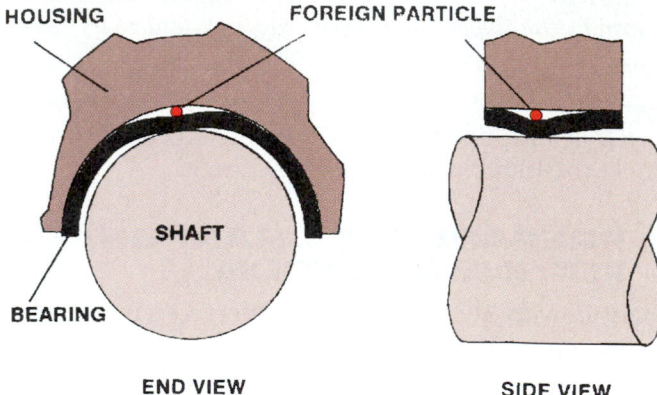

Figure 49.15 Results of careless cleaning. *(Courtesy of AE Clevite Engine Parts)*

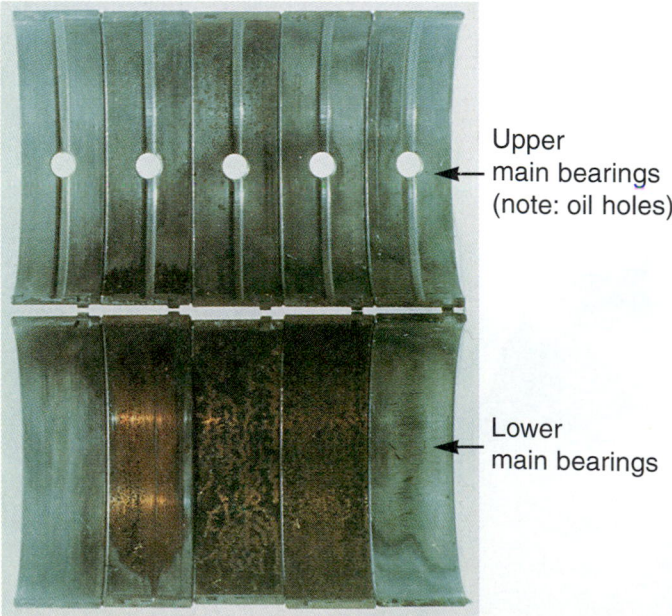

Upper main bearings (note: oil holes)

Lower main bearings

Figure 49.16 These upper main bearings have oil holes and the lower ones do not. These lower bearings are extremely worn. *(Courtesy of AE Clevite Engine Parts)*

bearing style with a hole and groove is used in both the upper and lower positions.

- Install the upper main bearings by pushing them into the main bearing bores (Figure 49.17).
- Be sure that the lubricating holes in the bearings line up properly with the corresponding holes in the block (Figure 49.18).
- Do not touch the bearing surface with your hands.
- Lubricate the bearings only on the surface that is *toward* the crankshaft.

NOTE: *Do not oil the bearing backs. They are not a bearing surface.*

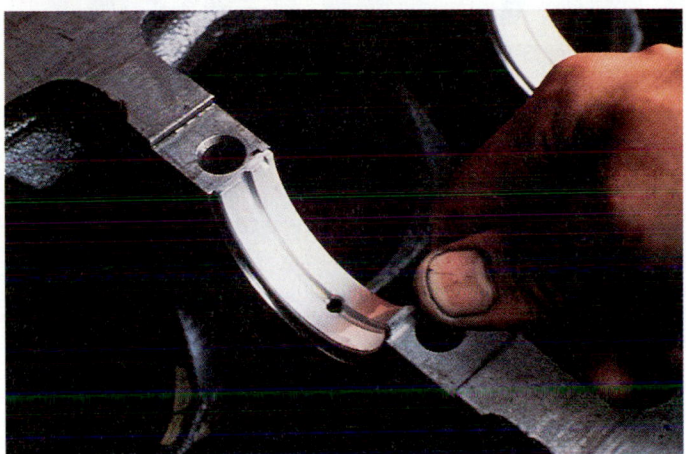

Figure 49.17 Snap the bearing into place. *(Courtesy of Federal-Mogul Corporation)*

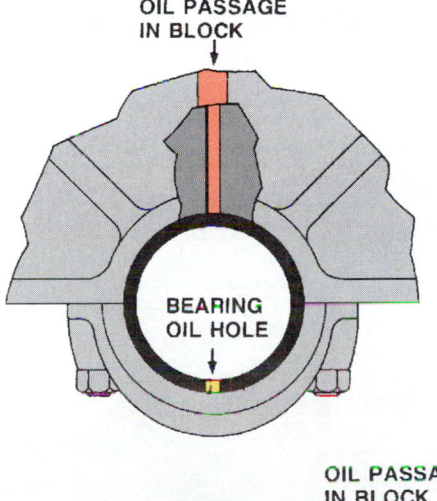

OIL PASSAGE IN BLOCK

BEARING OIL HOLE

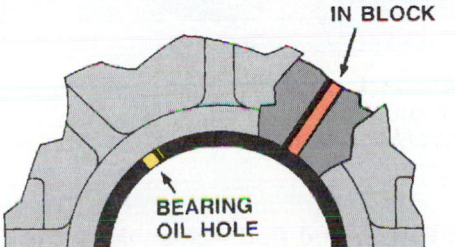

OIL PASSAGE IN BLOCK

BEARING OIL HOLE

Figure 49.18 Installing the main bearings in the wrong positions can block the passage of oil to the journal. *(Courtesy of AE Clevite Engine Parts)*

Check Bearing Oil Clearance

Install the crank and check the clearance with plastigage as described in the section on bearing oil clearance in Chapter 47. If the crank and/or the bearings are being reused, it is advisable to check each bearing's clearance.

- When using a reground crank and new bearings, check the main and rod bearings for proper clearance and to be sure that the right bearings are being installed.
- Remove the crankshaft and install the rear main seal according to the instructions in Chapter 45. Be sure to offset the parting lines on a two-piece seal.

NOTE: *If a rope seal is installed too tightly, heat can cause the rear main bearing to seize. The seizure will occur on the rear part of the bearing.*

Although it was commonly recommended in the past, Fel-Pro Inc. recommends that rope rear seals *not* be soaked in oil before installation. Soaking of the wick can cause it to disintegrate. Some of the seal's graphite impregnation might be lost. *Lightly* lubricate the seal with chassis grease to avoid damage from a dry startup.

Tighten the Main Caps

For a five-main bearing block, the torque sequence is 1-4-3-2-5.

NOTE: *As each main cap is torqued down, check to see that the crank continues to turn easily.*

After the rear cap is removed, check the rear seal drag. Some manufacturers give a torque specification for the amount of effort required to turn the crank with the damper bolt in an assembled engine.

Align the Thrust Bearing Halves

- Torque all bearing caps except the thrust main. Its halves should be aligned (Figure 49.19) before torquing. Misaligned thrust halves could eliminate end play (Figure 49.20). This is done by prying on the crankshaft while the thrust main is still loose.
- After aligning the thrust halves, check crankshaft end play (Figure 49.21) with a feeler gauge or a dial indicator.

NOTE: *End play tolerance is usually from 0.004"–0.006" for a crankshaft with a 2" to 2¾" main bearing diameter.*

Assemble the Piston and Rings

- Assemble the new or serviced pistons to the connecting rods. Be sure that the rods have been installed on the pistons in the right direction (see Chapter 48).

NOTE: *Be sure to lubricate the wrist pin. This is often forgotten.*

- Check the butt gap of the compression rings (see Chapter 48). Then, install them according to the

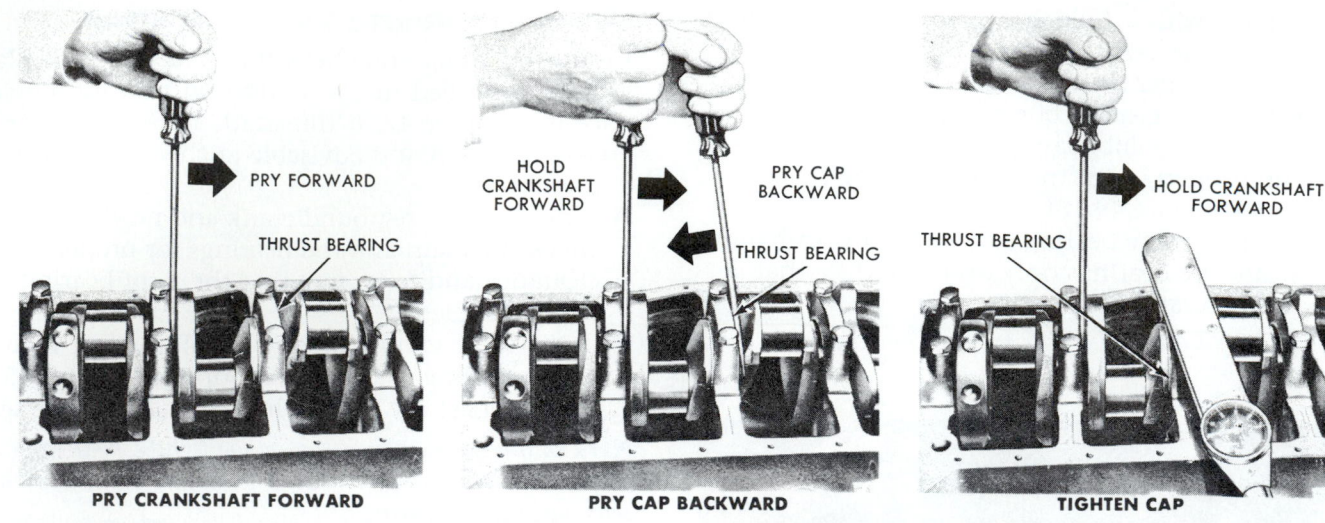

Figure 49.19 Align the thrust surfaces as shown. *(Courtesy of Ford Motor Company)*

Figure 49.20 Misaligned thrust flanges would prevent end play. This upper thrust bearing is shifted to the left.

instructions in the ring package with the dots facing up and the gaps properly placed.

Install the Rod Bearings.

■ Install the new bearing inserts in the rod and cap.

■ Bearing spread keeps the bearings in the rod while the pistons are being installed (see Chapter 19). A good bearing that is going to be reused might have lost its spread. According to Federal Mogul Corporation, bearings can be respread by placing the bearing on a hardwood surface with the parting face down and gently tapping the back with a soft-faced mallet.

NOTE: *When a bearing has lost its spread, inspect the piston for evidence of abnormal combustion.*

■ Make sure that the parting surface of the rod cap does not have any burrs or foreign material that might prevent proper mating when torquing (Figure 49.22).

■ Make sure the bearing lock tabs are aligned in the cap (Figure 49.23).

(a)

(b)

Figure 49.21 Checking crankshaft end play: (a) With a feeler gauge. (b) With a dial indicator. *[a, Courtesy of Federal-Mogul Corporation; b, Courtesy of Ford Motor Company]*

Install the Piston and Rod Assembly in the Block.

■ Oil the piston rings and rod bearings thoroughly (Figure 49.24).

■ Install a short length of fuel hose or rod bolt

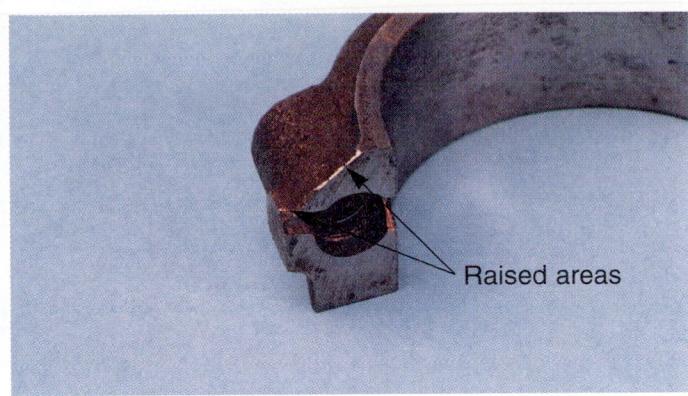

Figure 49.22 A burred cap parting surface.

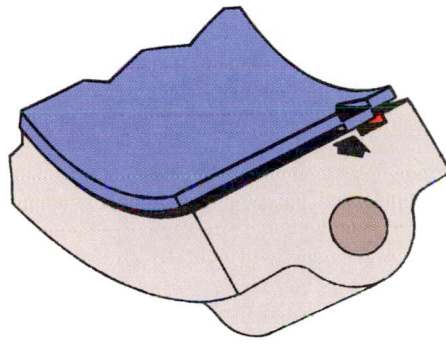

Figure 49.23 Make sure that the bearing tang is properly located before torquing the rod cap. *(Courtesy of AE Clevite Engine Parts)*

Figure 49.24 Oil the piston rings and rod bearings thoroughly. *(Courtesy of Federal-Mogul Corporation)*

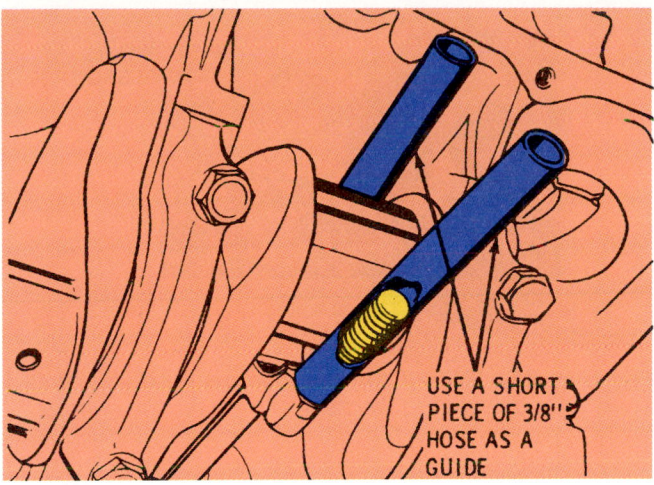

USE A SHORT PIECE OF 3/8" HOSE AS A GUIDE

Figure 49.25 Install pieces of hose or a special tool on the connecting rod bolts to protect the crankshaft from accidental nicks. *(Courtesy of General Motors Corporation, Service Technology Group)*

Figure 49.26 A nick on the rod journal will cause a line all around the bearing insert. *(Courtesy of AE Clevite Engine Parts)*

protectors on each rod bolt (Figure 49.25). This will guard against nicking the crank, which would result in a damaged bearing insert (Figure 49.26).

■ Face the notch on the piston head to the front of the engine (Figure 49.27).

■ Using a ring compressor, hold a rubber hammer against the top of the piston and tap the hammer lightly with the palm of your hand while holding the ring compressor firmly against the block until all the rings have entered the cylinder.

SHOP TIP Installation of the three-piece oil ring occasionally presents some difficulty. Sometimes, there is a small space between the bottom of the ring compressor and the top of the block. Try to quickly push the oil ring past the end of the ring compressor and into the cylinder bore before one of the narrow oil ring rails can spring out into this space.

■ If the piston does not go in easily, something might be wrong. Do not force it! You can damage a ring or a ring land (Figure 49.28).

■ Once all the rings have entered the cylinder, the rod can be pulled down against the rod journal by hand from the crankcase side of the block. Twist the rod if necessary to align it properly with the journal.

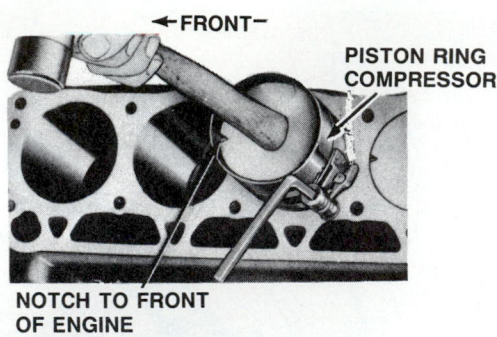

Figure 49.27 Install the piston and rod assembly in the right direction. *(Courtesy of Ford Motor Company)*

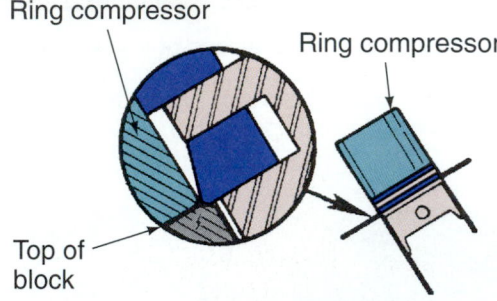

Figure 49.28 Do not force a piston into the cylinder. *(Courtesy of Ford Motor Company)*

- Be certain that the connecting rod faces in the right direction. On a V-type engine, with the notches on the piston facing forward, the left cylinder bank's rods should face the opposite direction from the right bank's (Figure 49.29). If they are facing the wrong way, the rods might have been improperly installed on the pistons. Do not continue with assembly until the rods are correctly installed.
- Remove the hoses or rod bolt protectors from the rod bolts and install the rod caps. Make sure that the numbers on the rod caps correspond to the numbers on the rods (Figure 49.30) and that they face in the right direction. The numbers on each rod and cap should be on the same side with the lock tabs facing each other.
- Rod nuts are square and flat on the bottom side and often are curved on the top. Be sure that the flat side of each rod nut is faced against the cap. Use some loctite, or similar adhesive, when rod nuts are reused (see Chapter 45). Torque the nuts to specifications. Ideally, nuts should not be reused (see Chapter 6). In the industry, rod nuts are commonly reused without adhesives. Repeated reuse of nuts is asking for trouble.

Check Installation of Rod and Bearings.
- After installing each piston, rotate the crank one complete turn to check for unacceptable drag (Figure 49.31).

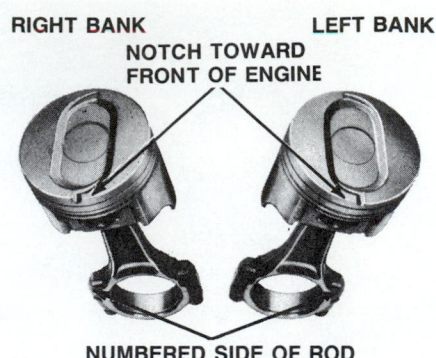

Figure 49.29 On V-type engines, the rods are installed with their numbers facing away from the crankshaft. *(Courtesy of Ford Motor Company)*

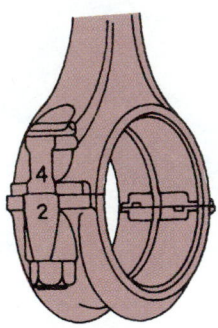

Figure 49.30 Bearing cap on wrong connecting rod. *(Courtesy of AE Clevite Engine Parts)*

Figure 49.31 Rotate the crankshaft after installation of *each* piston and rod. *(Courtesy of Federal-Mogul Corporation)*

- On V-type engines, check the side clearance between the connecting rods with a feeler gauge and compare it to specifications. Use the low side of the tolerance when selecting a feeler gauge. Check at two or three locations around the circumference of the rod. Variations in clearance indicate a bent or twisted rod.

■ INSTALL THE CYLINDER HEADS

- ■ Be sure to install the cylinder head gaskets in the right direction (see Figure 45.13).
- ■ Most engines have dowels to align the heads. If there are no dowels, make alignment pilots by cutting the heads off of two long bolts and hacksawing screwdriver grooves in their tops. These aligning studs will help protect the head gasket from damage during installation and can be removed easily after the head is installed.
- ■ Be sure to compare the head gasket to the head to see that it is the correct one. Check that the fire rings around the combustion chambers fit. Also see that oil and water passageways line up.

■ INSTALL OHC HEADS

Before installing overhead cam heads on the block, the number one piston must be at TDC and the camshaft must be turned in the head until its timing mark is properly located. Otherwise valves held open by the cam might be forced against a piston and become bent as head bolts are tightened.

Install the Head Bolts

After the head bolt threads have been cleaned on the wire wheel, install the head on the block and bolt it on.
- ■ Any bolts with rusted shanks should be replaced.
- ■ Torque the bolts in the proper sequence (see Figure 45.14).
- ■ Apply sealer to any bolts that go into holes that go all the way through into water jackets (Figure 49.32), so that coolant cannot migrate into the oil. A long, narrow screwdriver can be used to probe the head bolt holes to see if they are "blind" holes (holes that have bottoms).
- ■ Align the TDC timing mark found on the cam or cam sprocket with the mark on the cylinder head (Figure 49.33).
- ■ Position the crank-pulley timing mark at TDC. To double-check that the cam is actually positioned at TDC, either the number one cylinder or its companion cylinder should have its cam lobes facing as shown in Figure 49.34. The companion cylinder's cam lobes will face down in the valve overlap position.
- ■ When the cam and crank sprockets are in their proper positions, install the timing chain.

Figure 49.32 This head bolt came from a hole that was threaded into a water jacket.

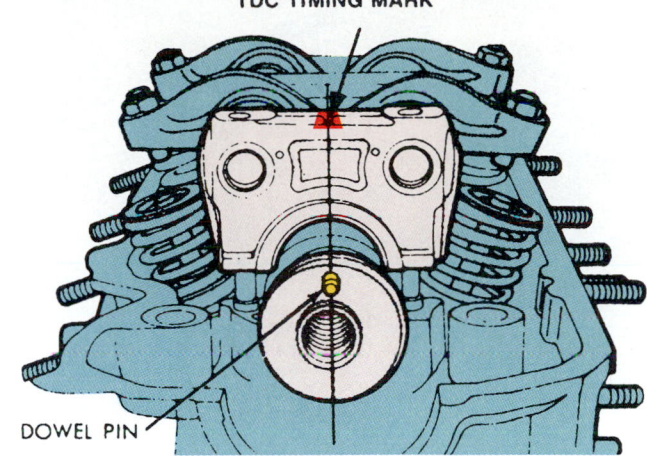

Figure 49.33 Position the cam correctly before installing the cylinder head. *(Courtesy of Chrysler Corporation)*

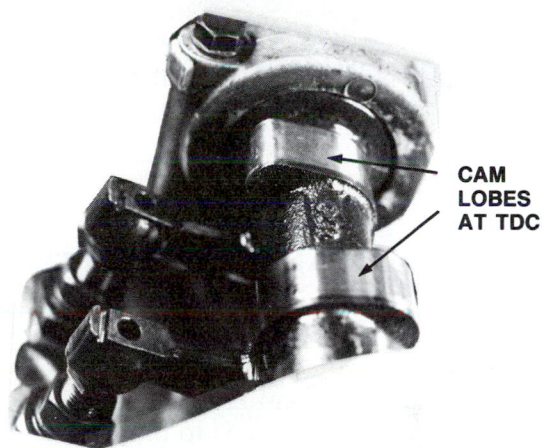

Figure 49.34 When the cam lobes face away from the valves, the cam is positioned at TDC before the power stroke.

Sometimes the chain has plated links that align with marks on the cam and crank sprockets. Be sure to check the repair manual.

■ INSTALL THE CAM DRIVE (PUSHROD ENGINES)

Install the chain and cam sprocket. Be sure to set the cam timing properly (see Chapter 42).
- ■ If the sprocket was not previously installed on the crank, lay the timing chain and sprockets on the bench and align their marks carefully.
- ■ Hold both sprockets in the chain while installing the crank gear.
- ■ Turn the crankshaft until the cam sprocket aligns with its pin or key.
- ■ Install the cam sprocket on the cam. The marks should now be in perfect alignment (Figure 49.35).
- ■ Double-check cam timing by rotating the crankshaft until the marks line up again.

Figure 49.35 Be certain timing marks are aligned according to specifications. *(Courtesy of General Motors Corporation, Service Technology Group)*

> **CAUTION** Tightening down an OHC camshaft sprocket with an impact wrench can sometimes result in a broken camshaft snout.

INSTALL VALVE TRAIN PARTS

- Oil and install the lifters; make sure they turn freely in their bores.
- Coat the bottom of each lifter with an assembly lube (see Figure 49.8), an EP 90-weight oil, or an engine oil supplement (EOS) to protect them against wear when the engine is first started. EOS is available from all major auto dealers. It contains a higher percentage of zinc dithiophosphate than normal motor oil. EOS helps during cam-to-lifter break-in, but is not necessary (or recommended) at subsequent oil changes.

NOTE: *The Automotive Engine Rebuilders Association recommends against prefilling hydraulic lifters. If they are overfilled, the valves can be held off their seats, which will make starting the engine difficult.*

- Install the pushrods and rocker arms. Lubricate all wear areas thoroughly.
- On shaft-mounted rocker arms, tighten the bolts on the rocker shaft that are closest to the center first. Then pull all the rest of the bolts slowly tight.

INSTALL THE OIL PUMP

- Fill the pump with oil so that it will not be run while dry and will "prime" right away.
- Install the pump as described in Chapter 49. Oil pumps do not usually require a gasket if they are housed inside the oil pan.

- Make sure that the pick-up screen is properly positioned so it will be about ¼" from the bottom of the oil pan.

INSTALL THE TIMING COVER

- If the engine has a bolt-on fuel pump eccentric, use loctite when installing it.
- Install the crankshaft woodruff key and the oil slinger (if the engine uses one) before installing the timing cover (Figure 49.36).

NOTE: *Be sure to install the oil slinger in the correct direction or it may rub and make noise. Tearing down a newly rebuilt engine to replace a backwards or forgotten oil slinger is time consuming and frustrating.*

- Install the timing cover seal as described in Chapter 45. Be sure to grease the lip of the seal. Timing covers are usually installed before installing the oil pan because the oil pan and timing cover often share a gasket surface.
- If the engine does not have timing cover aligning pins, use a special tool to align the damper to the timing cover seal. If this tool is not available, temporarily install the damper. After tightening a pair of the timing cover screws, the damper can be removed to permit tightening of the remaining screws.

NOTE: *Timing cover screws are small. Torque the screws to specs (usually 50–60 inch-pounds) with an inch-pound torque wrench to prevent breaking the screws.*

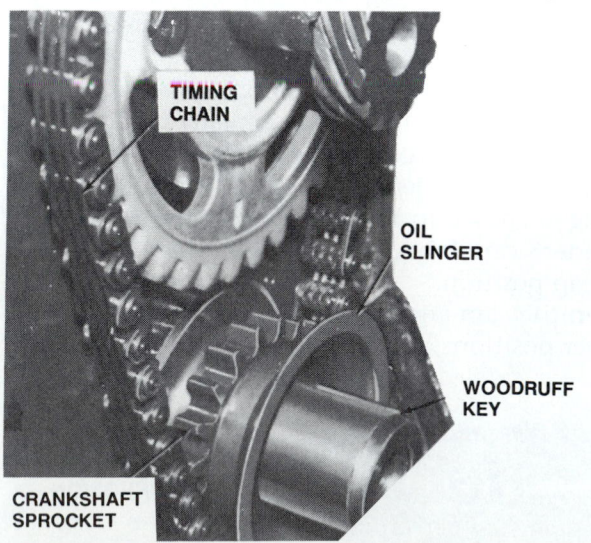

Figure 49.36 Install the woodruff key and oil slinger before installing the timing cover. *(Courtesy of General Motors Corporation, Service Technology Group)*

◼ INSTALL THE DAMPER

Install the vibration damper by pounding it on, or use a special installation tool. Generally, the damper is installed until it bottoms out against the oil slinger and the timing sprocket.

> **SHOP TIP** If you are unsure whether the damper is all the way on, install the water pump and pulley. Then, install the damper until its pulley aligns with the water pump pulley.

It is advisable to use the installation tool because pounding on the damper can cause stress on the crank thrust bearing surface. It is best to stand the engine block on end and support the crankshaft if the damper must be pounded on.

Some vibration dampers are not pressed-fit on the crankshaft. Be sure to install a large washer behind the damper retaining bolt on these engines; otherwise, the damper might fly off, causing damage and a safety hazard.

Gasket Installation

Before installing any gaskets, read the instructions included in the gasket set. They often contain special tips that can be important in assembly of the engine.

◼ INSTALL THE OIL PAN

The installation of the oil pan is especially important. An otherwise perfect engine overhaul will appear amateurish to a car owner if the pan leaks oil. Be careful here. Replacing a pan gasket is a much more difficult repair job when the engine is in the car.
◼ Be sure that the oil pan is flat.
◼ Sheet metal oil pan or valve cover bolt holes are often distorted. Straighten them by carefully flattening them with an anvil and hammer (as shown in Figure 45.11).
◼ Make sure that the pan was not dented when it was removed from the car. A dented pan that is hit by a rotating crank counterweight can sound like a rod knock.

>
> **CASE HISTORY** *An apprentice technician was installing a rebuilt long block in a vehicle. When the engine was started, a loud knock was heard. The disheartened apprentice sought help from the shop owner. Further investigation showed that the oil pan had been damaged during installation of the engine. The pan was removed from the engine and a shiny spot was visible where the crankshaft had been contacting it. After the pan was straightened and reinstalled, the engine ran fine and the car was returned to the customer.*

LIFTERS IN VALVE-OPEN POSITION

Figure 49.37 If hydraulic lifter lash is adjusted before installing the intake manifold, it is easy to see which lifters are on the heel of the cam.

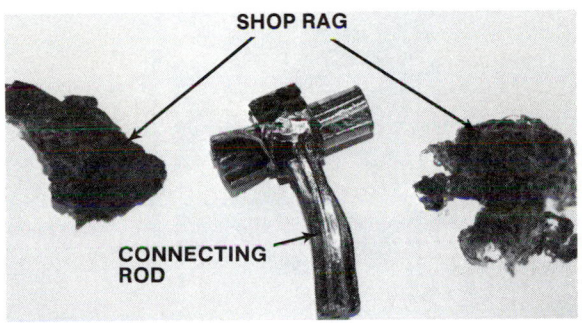

SHOP RAG

CONNECTING ROD

Figure 49.38 A shop rag was left inside of this engine during a careless reassembly.

◼ Apply a small amount of RTV where the cork gaskets join the rubber gasket on the front of the pan. Do not use too much; the excess silicone can break off and plug the oil pump screen.
◼ Sometimes the same engine is used in different sizes and types of vehicles. It is not uncommon for different oil pans to be used on the same engine block. Make sure the pan used is the right one. If the type of oil pan is changed during an engine swap, usually the oil pump screen must be changed too.
◼ When installing the oil pan screws, start each one into its threads before tightening any of them.

During engine assembly, it is best to adjust *hydraulic lifter lash* on V-type engines before the intake manifold is installed because you will be able to see that the lifters are on the heel of the cam lobes and are positioned low in their bores (Figure 49.37). Valve adjustment procedure is covered in Chapter 46.

Be sure that nothing is accidentally left in the engine before the oil pan or intake manifold are installed. Figure 49.38 shows the result of a careless engine assembly where a shop towel was left inside the engine.

◼ INSTALL THE INTAKE MANIFOLD

The intake manifold can be installed now. Chapter 45 describes the precautions to take when sealing an

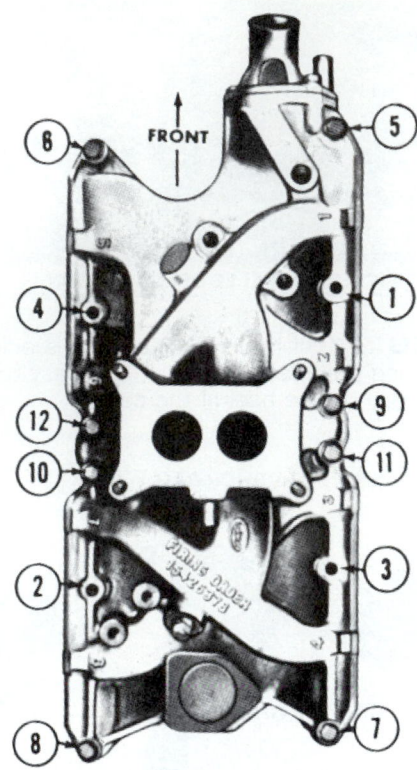

Figure 49.39 Intake manifold torque sequence. Notice the firing order and cylinder numbers cast on the manifold. *(Courtesy of Ford Motor Company)*

intake manifold. Torque the manifold according to specifications found in the repair manual (Figure 49.39).

■ INSTALL THE THERMOSTAT AND WATER OUTLET HOUSING

Install the thermostat with the temperature sensor facing into the block. If the thermostat is installed upside down, the engine will overheat.

CASE HISTORY

An apprentice was finishing the installation of a rebuilt engine by installing the thermostat and all of the belts and hoses. After the engine was started, he took the car on a test drive. After a very short distance, the engine overheated. Further investigation by the shop foreman showed that the thermostat had been installed backwards. With the temperature sensing bulb facing toward the radiator, the thermostat could not sense the heated coolant in the block. Correctly installing the thermostat corrected the overheating problem.

■ INSTALL THE FUEL PUMP

If the engine has a mechanical fuel pump, rotate the crankshaft until the fuel pump eccentric is at its low-

est position in relation to the pump (see Chapter 18). Failure to do this makes it difficult to install the pump.

■ COMPLETION OF ASSEMBLY

Paint the Engine

A good paint job is an important part of a professional engine rebuild. It is easier to paint the engine after it is completely assembled. Exhaust manifolds should not be painted, so paint the engine before they are installed.

SHOP TIP Apply a thin layer of grease to exhaust gasket surfaces and carburetor gasket surfaces. This makes it easier to remove paint from these areas.

■ Screw old spark plugs into the plug holes during painting.
■ Spray one light coat of paint (*tack coat*). Wait until it becomes tacky to the touch.
■ Spray the second coat. Spraying the second coat after the first coat becomes tacky will prevent runs and promote better paint coverage.

Spin Testing

High-volume engine rebuilders use a *run-in stand* to **spin test** rebuilt engines following assembly. The engine is rotated at 600 rpm by a machine while the lubrication system has full oil pressure. Major problems can be spotted using this test before a defective engine leaves the rebuilder. One of the advantages of a long block over a short block is that certain tests and adjustments are possible during spin testing.
■ Compression and oil pressure can be checked.
■ Cam, rod, and main bearings can be visually checked for excessive bearing oil leakage.
■ Valve adjustment can be performed.
■ Lifter rotation can be checked.

■ INSTALL EXHAUST MANIFOLD(S)

After the block assembly is painted, install the exhaust manifold(s). Some inline engines use a combined intake and exhaust manifold gasket.
■ Tighten the bolts in the center of the manifold first to prevent cracking it.
■ If there are dowel holes in the exhaust manifold that align with dowels in the cylinder head, make sure that these holes are larger than the dowels. If the dowels do not have enough clearance because of the buildup of foreign material, the manifold will not be able to expand properly and might crack.
■ On engines with shared bolts between the intake and exhaust manifolds, tighten the individual manifold-to-engine bolts first. Then, tighten the bolts where the two parts meet.

■ REVIEW QUESTIONS

1. Use a _____ to clean bolt holes in the block before reassembly.

2. Pry the crank forward and rearward before torquing the thrust bearing cap to align the halves of the _____ bearing.

3. Crankshaft end play can be checked with a dial indicator or with a _____ gauge.

4. Install pieces of _____ or rod bolt protectors on the rod bolts to protect the crank against accidental nicks during piston installation.

5. The crankshaft should be _____ after installing each piston to be sure that nothing is too tight.

6. When installing an OHC cylinder head, be sure that the camshaft and piston are correctly _____ so they do not hit each other.

7. If the block does not have aligning pins for the timing cover, what can be used to align the timing cover?

8. Distorted valve cover or oil pan bolt _____ are flattened with a hammer.

9. When installing the fuel pump, the fuel pump _____ should be in the down or away position so that the pump arm is not activated.

10. Before painting, _____ is applied to areas that are not to be painted.

■ ASE STYLE REVIEW QUESTIONS

1. Technician A says to apply moly assembly lube to all internal lubricated engine surfaces. Technician B says to prefill all hydraulic lifters before installing them in an engine. Who is right?

   a. Technician A   **b.** Technician B
   c. Both A and B   **d.** Neither A nor B

2. Technician A says that oil pumps enclosed in the oil pan usually require a gasket. Technician B says that exhaust manifold bolts are tightened in a pattern that has the outside bolts tightened first. Who is right?

   a. Technician A   **b.** Technician B
   c. Both A and B   **d.** Neither A nor B

3. Technician A says when installing an oil pan to torque each screw as it is installed. Technician B says a cylinder block is flexible until the head and other engine parts are installed on it. Who is right?

   a. Technician A   **b.** Technician B
   c. Both A and B   **d.** Neither A nor B

4. Two technicians are discussing main bearings. Technician A says that both upper and lower main bearings can have oil holes. Technician B says that upper main bearings often do not have an oil hole. Who is right?

   a. Technician A   **b.** Technician B
   c. Both A and B   **d.** Neither A nor B

5. Technician A says that crankshaft end play can be checked with a dial indicator. Technician B says that crankshaft end play can be checked with a feeler gauge. Who is right?

   a. Technician A   **b.** Technician B
   c. Both A and B   **d.** Neither A nor B

CHAPTER 50

Engine Installation, Break-In, and In-Chassis Repairs

■ KEY TERMS

guide pins
pressure primer
scuffing
ether starting fluid
wide open throttle (WOT)
overhaul
steering linkage tapers
tempilstick

■ INTRODUCTION

The content of this chapter deals with installing an engine in the vehicle. After installation, certain procedures are followed to break it in. This chapter also deals with some of the repairs that can be done to an engine while it is in the vehicle.

■ ENGINE INSTALLATION

Installing a completely assembled engine in a chassis is an important part of engine service. Before installing the engine in the vehicle, be sure that fender covers are installed on the fenders. Be especially careful when using a chainfall. The chain can nick the car's paint.

■ INSTALL ENGINE MOUNTS

Install the engine mount bolts *loosely* on the block. Leave them loose during engine installation so that the mounts can be more easily aligned with the frame mount brackets.

■ INSTALL THE ENGINE

Raise the engine and position it in the engine compartment. A *rolling head prybar* (Figure 50.1) is handy when aligning mounts. Use the pointed end to line up the mount bolt holes before installing the mount bolts.

NOTE: *Do not use the engine mount bolts to pull a V-type engine into place. Use shims if necessary to fill any gaps between the mount and block. The Automotive Engine Rebuilders Association reports numerous cases of block distortion, resulting in scuffed pistons in cylinders, near mounts that were forced tight.*

USE END TO ALIGN
BOLT HOLES

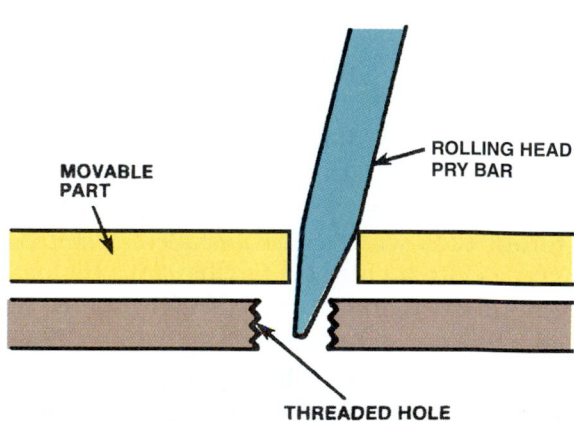

MOVABLE
PART

ROLLING HEAD
PRY BAR

THREADED HOLE

Figure 50.1 A rolling head prybar is helpful when aligning motor mounts.

If the transmission was left installed in the car, a floor jack or a transmission jack under the transmission will help to align it with the engine. **Guide pins** will be especially helpful when a standard transmission is still in the vehicle (Figure 50.2). They can be made by cutting the heads off of some bolts and using a hacksaw to cut a screwdriver slot in them. A standard

Figure 50.2 Guide pins.

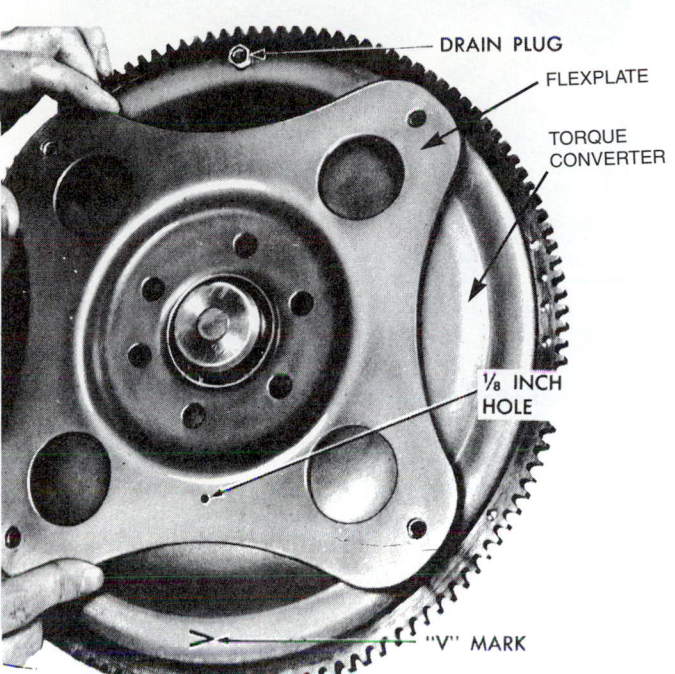

Figure 50.3 Make sure that the flexplate and converter are properly aligned. *(Courtesy of Chrysler Corporation)*

transmission is often left in the vehicle when a heavy truck transmission is involved.

NOTE: *Engines also have dowel pins attached to the engine block. These are important for engine-to-flywheel alignment and must not be left off the vehicle. Engine-to-transmission housing screws do not provide alignment. Without dowels, damage, especially to an automatic transmission flexplate, can occur in a short period of time.*

Hook Up Accessories

Bolt the engine to the transmission housing, if the engine and transmission were not installed as a unit. On automatic transmissions, slide the torque converter forward and bolt it to the flexplate (Figure 50.3). Be sure that it is in the right position. Some converters have drain plugs that must align with holes in the flexplate. Align the converter with the factory marks or with marks made during disassembly.

Install all previously disconnected parts:

- Bolt on the exhaust pipes. Use new nuts (brass nuts are preferable, if they are available).
- Install all pulleys, accessories, and belts. Use a belt tension gauge to adjust belt tension (see Chapter 22). When adjusting power steering belt tension, be careful not to pry against the sheet metal or plastic portion of the pump; the pump may be damaged. Manufacturers usually provide a suitable place to pry on the pump housing (Figure 50.4). Check the repair manual.
- Install the temperature and oil pressure sending units.
- Install the radiator and hoses and fill the cooling system with coolant. Make sure that heater hoses are installed with the return line to the block coming from the heater core, *not* from the heater valve (see Figure 21.26). These should have been marked during disassembly.

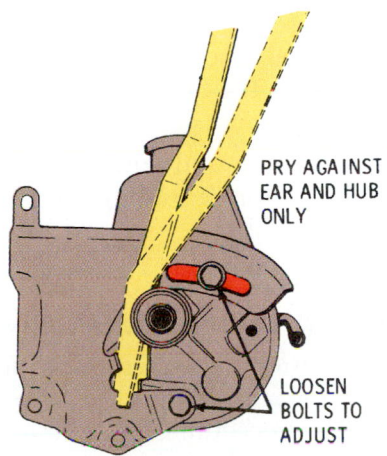

Figure 50.4 When tightening the power steering belt, pry only in the designated area. *(Courtesy of General Motors Corporation, Service Technology Group)*

- All electrical wires and vacuum lines should be reattached as labeled.

■ INSTALL THE CARBURETOR

If the vehicle is equipped with a carburetor, it must be installed with the right gasket. Compare it to the old gasket. Unplug and install the fuel line to the fuel

Figure 50.5 Priming the lubrication system with a drill. When the distributor for an engine rotates counter-clockwise, a reversible drill is used.

pump. Fill the float bowl with gasoline through the vent opening.

SHOP TIP In use, a fuel pump diaphragm accumulates dirt particles. When a carburetor is dry before an engine is first started, the fuel pump diaphragm moves farther than it would in normal use. The dirt breaks off of the diaphragm, clogging the fuel system. Purge the fuel line, prefill the carburetor, or crank some fuel out of the pump before hooking up the line to the carburetor.

Install the oil filter and add oil to the crankcase. Prime the system before installing the distributor (Figure 50.5).

PRIMING THE LUBRICATION SYSTEM

The lubrication system can often be primed by removing the distributor and driving the oil pump with an electric drill. This procedure is often easier to do before engine installation, while the engine is mounted on the engine stand. A priming tool can be made from an old distributor with the gear removed. Commercially made pump priming tools are also available. To prime the system by driving the oil pump, note the following items:

- Drive the tool with a slow drill in the normal direction of distributor rotation.

NOTE: *Because the oil pump is usually driven by the distributor, the pump must be turned in the same direction that the distributor rotates if it is to suck oil. The pump will not fill with oil if it is turned in the wrong direction. Chapter 39 explains how to determine the direction of the distributor's rotation.*

- After pressure builds in the system, rotate the crankshaft one complete revolution by hand on engines that have hydraulic lifters.

- Prime the system once more. This will allow all hydraulic lifters to fill up with oil.
- Sometimes pressure will not be sufficient to pump oil to the rocker arms unless the distributor shaft used in priming the system is installed in a distributor housing. Check the lubrication diagram in the service manual to see if oil travels around the distributor housing in the engine being primed.

The distributor must have a gear on the bottom of its shaft or it will not be possible to prime the system by driving the oil pump (Figure 50.6). If there is no gear on the bottom of the distributor, the oil pump has a gear that is always in mesh with the cam gear, making oil pump priming impossible.

PRESSURE PRIMING

If it is not possible to prime the system with the distributor removed, either use a **pressure primer** or remove all spark plugs and crank the engine over. Continue cranking until oil is distributed throughout the engine or until the gauge has registered oil pressure for 30 seconds. This procedure can also fill the carburetor with gasoline as the fuel pump works. It is much easier on the engine to prime the system first than to start the engine and wait for the pump to prime itself while the engine runs without oil.

CASE HISTORY *A group of students rebuilt a small block Chevrolet engine and installed it in a boat. When the engine was started, no oil pressure was displayed on the dash gauge. The oil sending unit was removed and an oil pressure gauge installed in its place (see Figure 43.19). The engine was started and the gauge showed low oil pressure that did not increase with engine rpm. The engine was removed and the oil pan was taken off it. An oil pressure primer (Figure 50.7) was attached to the hole where the sending unit was previously threaded. When pressurized oil was introduced to the system, oil could be seen gushing out near the front camshaft bearing. The timing cover and cam sprocket were removed. The oil gallery plugs had been left out of the front of the engine. Replacing them solved the problem.*

Another group of students was not so lucky. They continued to run their newly rebuilt engine while it did not have oil pressure and burned all of the pistons and cylinder bores. Pressure priming would have saved both of these groups of students a good deal of disappointment.

A pressure primer can be used to:
- Prime the system.
- Check for excessive bearing clearance.
- Check for sufficient oil pressure.
- Flush oil galleries during an in-car engine repair job.

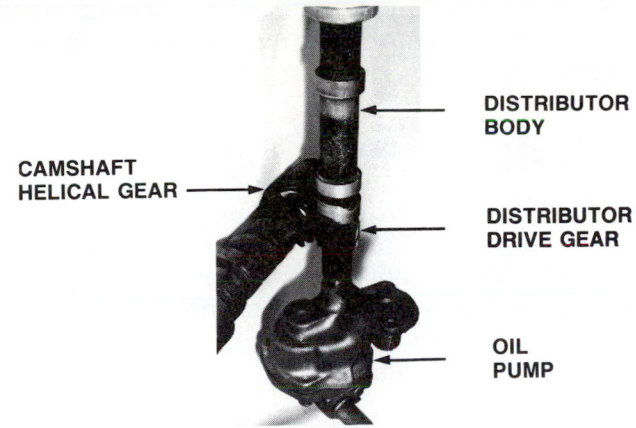

Figure 50.6 This distributor has a gear on its bottom that meshes with the gear on the cam.

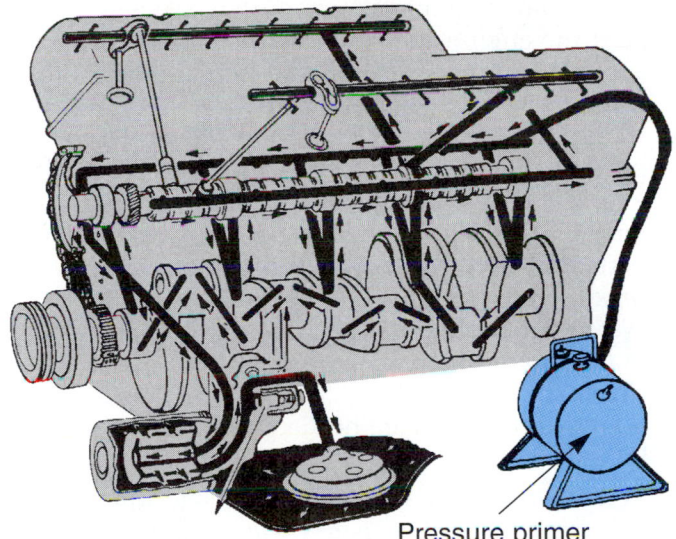

Figure 50.7 A pressure primer is attached to the oil sending unit hole in the block. *(Courtesy of Hastings Manufacturing Company)*

The supply of oil in the primer tank should be enough to fill the new oil filter as well as the oil lines in the block.

- With the oil pan removed, block off the oil pump inlet to perform the test.
- Be sure to remove the gasket when finished.
- Acceptable leakage from bearings can be from 20 to 150 drops in a minute. A more rapid flow indicates excessive clearance (Figure 50.8).
- Rotate the crank one-half turn and test again before condemning the bearing. If the oil holes in the crank and block were indexed, the appearance of excessive leakage could result.
- An absence of oil flow indicates a blocked oil line or insufficient clearance.

Some manufacturers recommend filling the pump cavity with petroleum jelly; others say that this will

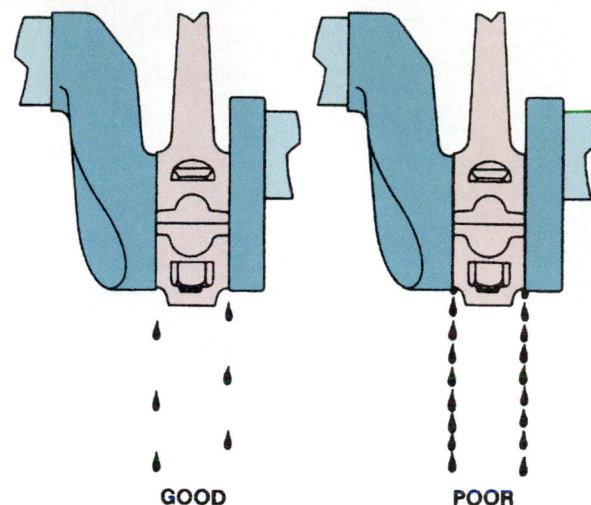

Figure 50.8 The amount of oil leakage depends on the amount of oil clearance. *(Courtesy of AE Clevite Engine Parts)*

cause an air bubble in the pump, which will keep it from filling with oil. Some pumps (the internal/external in particular) are not self-priming, and *must* be filled with petroleum jelly or assembly lube before installation. Follow the manufacturer's directions.

■ INSTALL VALVE COVERS

When oil is apparent at some of the rocker arms, the lubrication system is primed and the valve covers can be installed. If oil is not reaching the valve area during priming, double-check to see that an internal oil leak does not exist. It is easier to position the engine at TDC on #1 now, before installing the valve covers (see Figure 49.35).

NOTE: *On V-type engines, be sure to install the valve covers on their proper sides to ensure that breather and oil filler openings are in the right positions.*

If all the valve cover screws cannot be screwed in easily by hand, turn the valve cover around and try installing it in the opposite direction. Do not force the screws.

■ INSTALL THE DISTRIBUTOR

Align the timing mark on the damper with the pointer on the timing cover (Figure 50.9). When the marks are aligned, either the number one cylinder or its companion cylinder is at the top of its compression stroke. The spark plug should be ready to fire. If the distributor is installed 180° off, backfiring will occur and the engine will not run. Directions for distributor installation are included in Chapter 39. Tighten the distributor hold-down. Following installation of the distributor, the ignition timing should be set closely enough so that the engine will start right up.

- If the timing is retarded, the idle will be too low.
- If the timing is advanced, the idle will be too high.

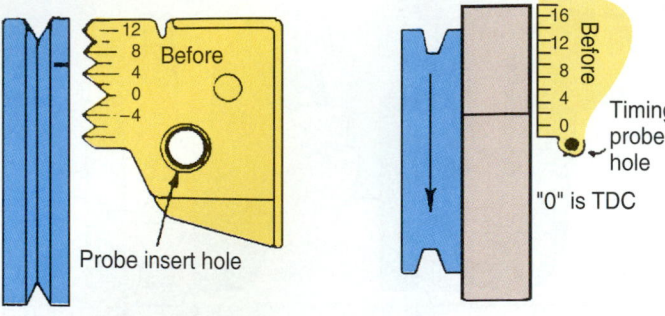

Figure 50.9 Typical timing marks. *(Courtesy of SPX Corporation Aftermarket Tool & Equipment Group)*

SHOP TIP If the idle speed is not controlled by the computer, the engine idle will be off if the timing is wrong. Remember this before changing the idle adjustment screw.

■ ENGINE STARTING AND INITIAL BREAK-IN OF THE CAMSHAFT

The first few minutes of an engine's operation are critical to its long life. A new part might look quite smooth to the naked eye, but compared to a used, worn-in part, it is quite rough. Moving parts are continuously separated by a film of oil; only the peaks on rough surfaces touch. If this contact occurs without **scuffing** (welding), the parts will wear in conforming to each other. Temperatures must be controlled so that the oil film will not become too thin or the clearances become too little. Heavy engine loads should also be avoided when the engine is new.

SAFETY NOTE Make sure that everything has been hooked up properly. Block the front wheels and apply the parking brake before starting the engine. If the transmission linkage was disturbed during the repair, it is possible that the car might start in gear.

SHOP TIP Sometimes it is difficult to get fuel from the gas tank to a mechanical fuel pump on a carbureted vehicle. Try jacking up the rear of the car or hooking up a temporary electric fuel pump at the fuel hose on the carburetor to help prime the vehicle's mechanical fuel pump. A siphon spray gun can be hooked to the fuel line also (see Figure 9.7).

A favorite starting aid is **ether starting fluid**, which lowers the temperature required for combustion to begin.

CAUTION Excessive use of starting fluid can cause an explosion. This can cause damage to a piston, piston rings, connecting rod, or gaskets. Use a little instead of a lot. Do not spray starting fluid into the engine when it is cranking or running. Because of this danger, most manufacturers discourage the use of *any* starting fluid.

Camshaft Break-In

It is especially important that the engine starts up immediately to avoid excessive loading between the cam lobes and the lifters.
- Start the engine and run it at 1500–2000 rpm for about 20 minutes to allow the cam and lifters to begin to wear in to each other.
- Idling should be avoided during this period to prevent cam and lifter failure and because, during idle, less oil is thrown off the connecting rod journals onto the cylinder walls, cam, and other parts that need this critical lubrication.
- Make sure the engine has oil pressure.
- Watch the coolant temperature to see that it does not climb too high.
- Check for oil leaks.
- If any adjustments are needed, the engine should be shut off immediately.

NOTE: *With the idle set high, a carburetor will probably have to be choked off to stop the engine from dieseling (running on).*

- After the initial 20-minute period, the idle can be reset and the ignition timing can be checked with a timing light.

NOTE: *Proper timing is essential:*
- *Just a few degrees of late timing can lower power output and cause overheating.*
- *Advanced timing can result in abnormal combustion, which can result in piston damage. Also, for every 3° of extra ignition advance, a gasoline octane increase of about four is required to avoid spark knock.*

Reinstall the hood. Squeeze the top radiator hose to see if it is hard (indicating a full cooling system).

CAUTION Do not open the radiator cap unless the hose collapses when squeezed. Escaping coolant will boil when the pressure is released. Someone could be burned, and coolant will be wasted.

■ VALVE CLEARANCE ADJUSTMENT

Adjust Mechanical Valve Lash—Engine Running

To locate and silence hard-to-find valve clearance noises, clearance adjustment on some engines can be done with the engine idling. Mechanical lifter clearance fine tuning can be done at low idle speed by inserting a feeler gauge and adjusting the screw until a slight drag is felt. Check the condition of the feeler gauge because it can easily be pounded out of shape. This procedure can be messy because of the oil that is being supplied to the rocker arms. Figure 50.10 shows a tool that will cut down on the oil mess. Attempting an *OHC adjustment* with the engine running results in oil being thrown from the timing chain and cam.

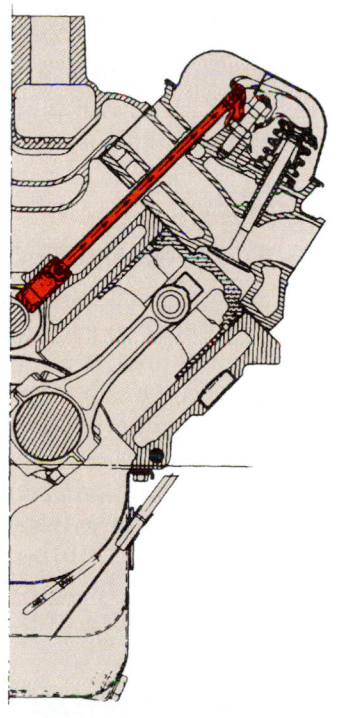

(a)

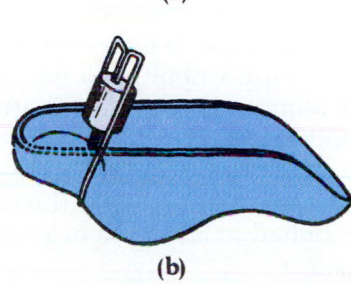

(b)

Figure 50.10 (a) Oil is supplied to the rocker arm and valve through the pushrod. (b) An inexpensive way to keep oil from spraying all over during a valve adjustment with the engine running. *(a, Courtesy of General Motors Corporation, Service Technology Group; b, Courtesy of Lisle Corporation)*

NOTE: Try to minimize oil spillage. When running an engine with the valve cover(s) removed, engine oil can run onto a hot exhaust manifold where it might catch fire. Oil is not flammable at room temperature but will ignite when it reaches its flash point, in the same way that diesel oil self-ignites when heated under high compression.

Adjusting Hydraulic Valve Lash—Engine Running

With the engine running (idling), loosen the rocker arm adjusting nut until a clicking sound is heard. Then, tighten the nut just until the noise stops. This point is called zero lash. Tightening past this point is done slowly (usually ¼ turn at a time) to give the lifter time to leak down. When the engine idle smooths out, the lifter has adjusted itself, and the adjusting nut can be turned farther.

This procedure may not be performed on diesel engines because there is not enough clearance between the piston and valve. Some technicians simply tighten the adjusting nut from ¼ to ½ turn past zero lash. This works well, but it does not allow as much margin for wear and temperature differences as the specified procedure allows.

SHOP TIP A popular method of engine running adjustment is to loosen and adjust each adjuster until zero lash is reached. Then, shut the engine off and tighten each adjuster the specified amount. Wait twenty minutes for the lifters to bleed down before starting the engine.

■ ROAD TEST AND BREAK-IN

Road Test

Take safety precautions before taking a test drive:
- Double-check all hose connections and fluid levels.
- With the key *off*, push the accelerator to **wide open throttle (WOT)**. Be certain there are no binds or obstructions in the accelerator linkage. When released, the throttle pedal should return quickly to the idle position.

NOTE: Be sure there are no loose parts that could cause a rattle. Many late model cars have knock (detonation) sensors that will react to the frequency of the vibration of spark knock, retarding spark timing. A knock sensor can be triggered by the rattle of a bad fuel pump, a loose valve adjustment, or a muffler vibrating against the frame. When the sensor retards timing, the engine idle will drop and performance will be erratic.

Piston Ring Seating

Because they are pre-lapped during manufacture, most piston ring companies now say that their rings will seat in a very short period of time. If in doubt about whether ring seating is necessary, use the following procedure to help seat the piston rings:

- Drive the car on the freeway.
- In high gear, accelerate from 35 mph to 50 mph and coast back to 35 mph several times. This helps to seat the rings under pressure and vacuum conditions. The deceleration (vacuum) phase also helps in sucking extra oil up into the cylinder walls to prevent scuffing.

It may take 2000 to 3000 miles to completely seat the rings. Drive easily (no high rpm) during this period. Engine bearings are soft and will also conform to irregularities in the bearing surfaces during this break-in time.

■ FINAL INSPECTION

After the road test and break-in procedures have been completed:

- Double-check the engine for oil leaks.
- Make certain that all wires and lines have been correctly reinstalled.
- Check that all warning lights or gauges are operating properly.

Vapor lock occurs when fuel vaporizes in a line and a bubble blocks fuel flow. Pump pressure will then cause the bubble to contract each time the pump pulses. When the line cools and the bubble condenses, the problem disappears. To avoid vapor lock, be sure all fuel lines are positioned at least ½" from any heat source and fuel filters are positioned where they can be cooled by air. A common cause of vapor lock is when a fuel line runs too close to an exhaust manifold or pipe.

■ RETURNING THE CAR TO THE CUSTOMER

The vehicle should be returned to the owner in a clean condition. Grease on the fenders, steering wheel, seats, or carpet will contribute to a poor first impression of the job. An owner who is favorably impressed will be the best source of free advertisement.

The customer should use the following break-in procedures suggested by Clevite Engine Parts:

1. Do not allow excessive engine idle for the first three hours of rebuilt engine operation.
2. Keep in the normal rpm range at about 75% load for the first two to three hours.
3. Engine speed should be varied as much as possible.
4. Full load or high-speed operation should be limited to less than 2 to 3 minutes at a time.
5. After high-load operation, allow the engine to return to a stable operating temperature by running at light load before shutting it off.

500-Mile Checkup

The customer should return to the shop after 500 miles for an oil and filter change and a head retorque (if required). A phone call reminder is a good public relations measure and will offer a chance to check up on any complaints. The following engine items can be checked:

- If head bolts are easily accessible, it is a good practice to retorque them (even when the engine has non-retorque gaskets).
- Mechanical valve lifter lash can also be readjusted after the valves have seated.
- After a head retorque, valve adjustments might also change on pushrod engines.

NOTE: *Head retorque should not affect valve lash on OHC engines.*

- A V-type engine with an iron intake manifold should have the manifold retorqued when *hot* to help guard against air leaks.

Oil consumption may not stabilize until all parts have broken in (seated). The customer should not be concerned until the vehicle has been driven at least 4000 miles.

■ ENGINE REPAIR— ENGINE IN THE VEHICLE

Technicians often perform major repair to an engine while it is still in the vehicle. In addition to rebuilding the cylinder heads (valve job), they might do lower end work. This could be the replacement of a single defective piston, or it might be a piston-ring-and-crank-bearing-replacement, called an engine **overhaul**. This job often includes replacing the timing chain and sprockets. Many OHC engines require a new chain tensioner (see Chapter 18). In-the-car repair is often less expensive for the vehicle owner. An engine that is to be completely rebuilt must be removed from the car.

■ VALVE JOB OR HEAD GASKET REPAIRS

- The engine should be cold before removing the head. When the head is bolted to the block, it forms a rigid unit. Unbolting a hot cylinder head (especially aluminum) can cause parts to warp.
- It is a good habit to unbolt the head in a direction opposite to the normal tightening sequence. A flat-rate technician will often leave the carburetor and manifolds bolted to an inline head during a head gasket repair.
- Check the cleanliness of head bolt threads as each one is removed. A flat-rate technician will use a tap to chase only the bolt holes in the block that correspond to dirty head bolt threads. A small wire brush on a die grinder is also effective in cleaning threads.

■ When reinstalling the head, be careful that nothing is accidentally pinched between the head and the block (Figure 50.11).

⚠ CAUTION Before reinstalling a cylinder head, use a suction gun or air nozzle to remove any oil or water from blind head bolt holes. Failure to do this can result in a cracked block next to the bolt hole. The water or oil will not compress during torquing.

Maintaining Valve Timing

During a valve job, it is essential to keep the timing chain or belt in place to maintain correct valve timing. Position the number one cylinder at TDC. Some overhead cam engines use a single long chain for a cam drive. The chain can be wedged against its guides with a tapered block of wood (Figure 50.12). If the crankshaft is turned, an unsecured chain tensioner can fall out, causing much extra work (Figure 50.13). Some engines have a lower and upper chain. These engines do not require special attention to wedging the chain. Be sure to look for hidden head bolts and check the repair manual before removing the OHC head.

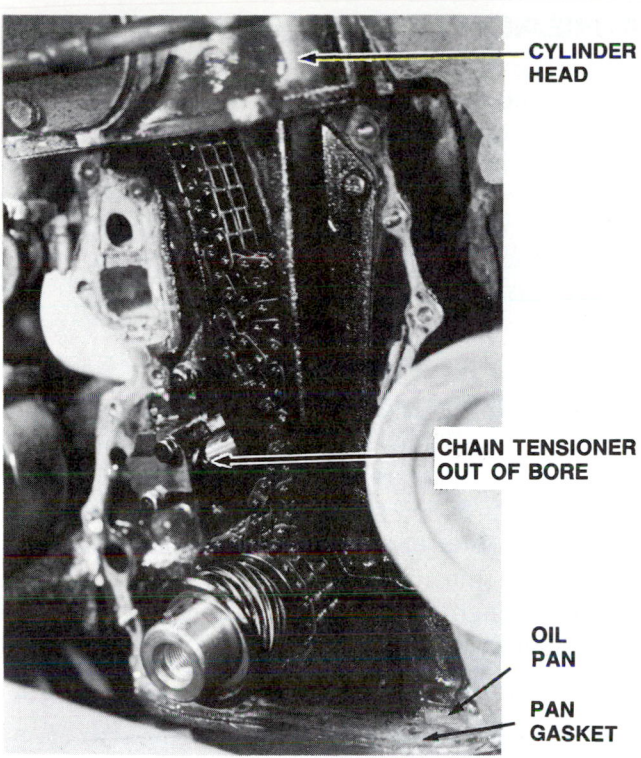

Figure 50.13 This chain tensioner came out of its bore when the timing chain was allowed to drop.

Figure 50.11 This ground strap was clamped between the head and the block during a careless installation.

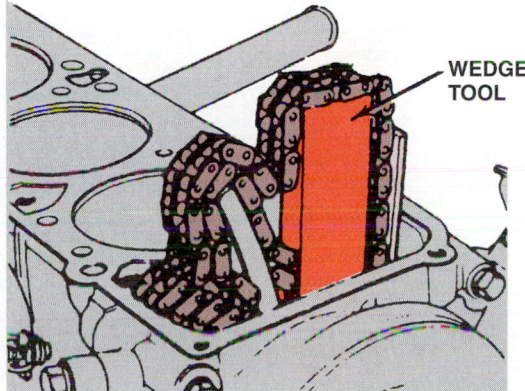

Figure 50.12 The chain tensioner on some OHC engines must be wedged to keep the chain in position during cylinder head removal. *(Courtesy of Nissan Motors)*

■ VALVE JOB OR COMPLETE ENGINE OVERHAUL

The question of whether to do a valve job only or an engine overhaul is an important one. Valve guide seals are responsible for a good many oil consumption complaints. Figure 50.14 shows what pistons look like when rings have not been controlling oil. There is a band around the top of each piston where no carbon is present. A valve job on this engine would not solve the oil consumption problem.

Figure 50.14 A clean area around the outside edge of a piston indicates excessive oil consumption past the rings.

■ HEAD GASKET PROBLEMS

Inspect the head gasket for damage. A faulty head gasket often points to a need for head surfacing, cooling system service, or repair to the engine's fuel or ignition system. Excessive temperatures can turn a metal head gasket blue or black. A teflon head gasket can turn brown. Look for signs of coolant leaks and damage from abnormal combustion. Carbon buildup around the combustion seal can indicate the use of a gasket of the wrong size (too large).

■ IN-CHASSIS LOWER END REPAIRS

An engine that is to be completely rebuilt must be removed from the car. A low-mileage engine is often repaired without removing it from the car.

The cherry picker (engine hoist) can be used to raise the engine in order to remove the engine's oil pan. The engine is blocked up under the engine mounts or hung from a fixture from above during repair (see Figure 50.19).

■ REMOVING THE OIL PAN

Removing the oil pan could require the removal of some steering linkage. Often, simply unbolting the idler arm bracket (Figure 50.15) will provide enough clearance for the oil pan to be removed. If not, one or more of the **steering-linkage tapers** (Figure 50.16) might have to be broken loose. Many technicians use a *"pickle fork"* with an air chisel but this procedure often tears an otherwise good tie-rod seal. A better way to break a taper is to use a special tie-rod puller (Figure 50.17) or two large hammers as shown in Figure 50.18. Sometimes the engine mounts must be loosened and the engine raised a few inches in order to remove the pan.

NOTE: *Be careful not to damage the radiator or the fan shroud when raising the engine off its mounts. Sometimes the radiator hoses must be removed or the radiator unbolted*

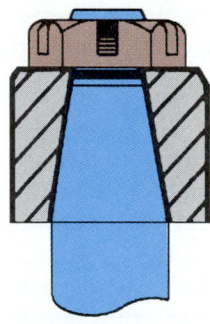

Figure 50.16 A steering-linkage taper.

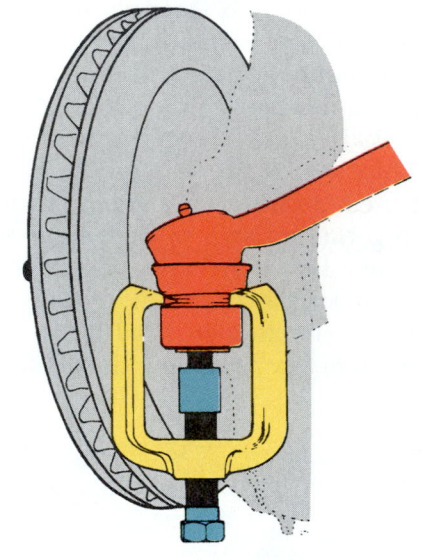

Figure 50.17 A tie-rod puller. *(Courtesy of General Motors Corporation, Service Technology Group)*

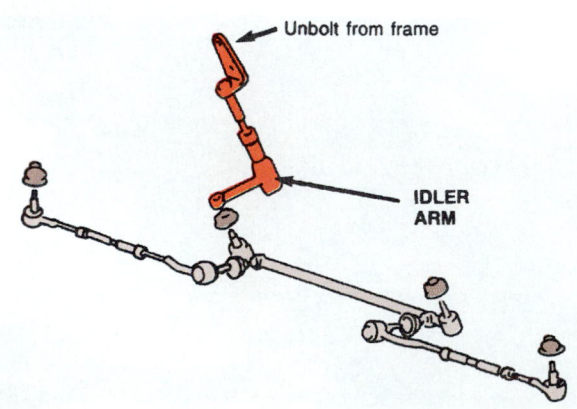

Figure 50.15 Unbolt the idler arm bracket to allow the steering linkage to drop. *(Courtesy of Chrysler Corporation)*

Unbolt from frame

IDLER ARM

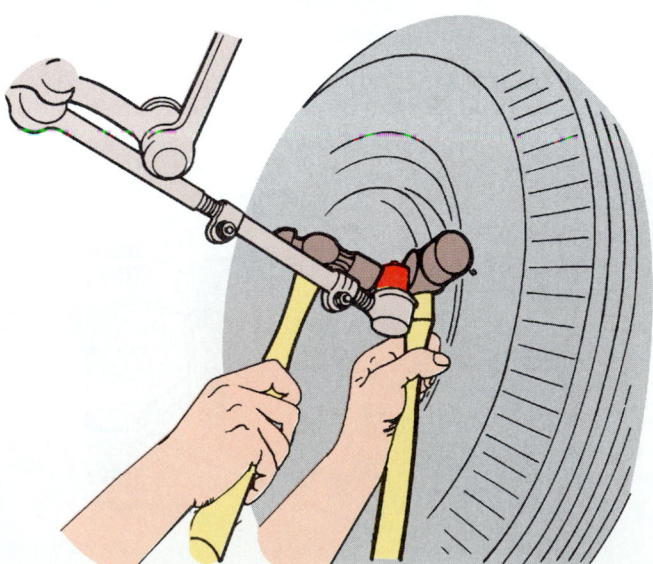

Figure 50.18 Using two large hammers to break steering-linkage tapers. *(Courtesy of General Motors Corporation, Service Technology Group)*

before the engine is raised. Use a piece of plywood to protect the oil pan, and jack the engine up with a hydraulic jack. The engine can then be blocked up at the engine mounts.

SAFETY NOTE

■ Be sure to use plywood instead of ordinary lumber when jacking or supporting something. The many layers of wood that make up plywood are laminated with the grain running in different directions. This makes plywood unlikely to split under pressure like ordinary lumber would.

■ If the oil pan is to be removed with the car on a lift, the safest choice is a drive-on lift that raises the car by the wheels (see Chapter 10). Raising the engine from the bottom can upset the balance of the car on a frame-contact hoist.

If a frame-contact hoist is to be used, there are special fixtures that can be used to raise the engine from its top side (Figure 50.19), or from the bottom (Figure 50.20).

SHOP TIP

Sometimes it is necessary to unbolt the oil pump after the pan is loosened to be able to get enough clearance to remove the pan. The pump is dropped into the pan and removed along with it. When reinstalling the pump, wrap a rubber band around the screws to prevent them from falling out of the holes in the pump while you reinstall it on the engine through the small opening between the pan and the block.

Figure 50.19 The engine can be lifted from the top. *(Courtesy of SPX Corporation Aftermarket Tool & Equipment Group)*

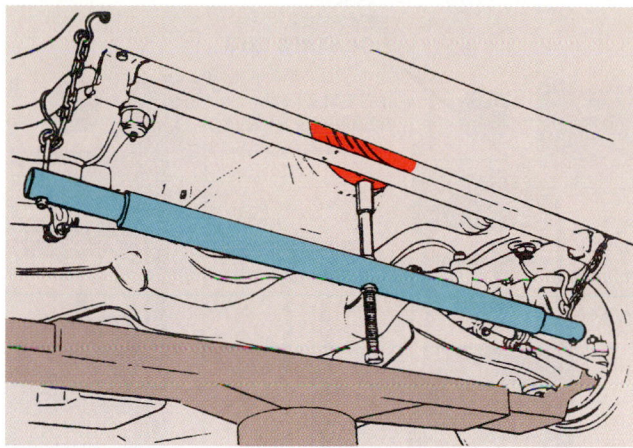

Figure 50.20 The engine can be raised from the bottom. *(Courtesy of General Motors Corporation, Service Technology Group)*

■ REMOVE THE PISTON AND ROD ASSEMBLY

If the lower end is to be repaired, remove the cylinder head(s). If there is a ring ridge in the cylinder, remove it (see Chapter 44). First, move the piston to BDC and place a rag in the cylinder to catch the metal chips. With the pan removed, the connecting rods can be unbolted and the pistons removed according to the procedure described earlier in this chapter.

Rod Bearing Replacement

Although bearings are usually replaced, an in-car repair might call for the replacement of only one defective piston. In this case, an unworn rod bearing might be reused. The piston shown in Figure 50.21 is just one example of such a case. If a rod bearing is to be reused, mark the back of the bearing with a felt marker so it can be returned to its original position. If a bearing has lost its spread (see Chapter 19) it may have to be spread slightly before reinstallation.

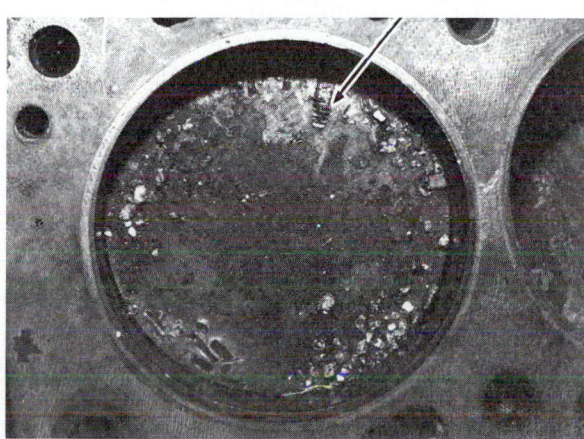

Figure 50.21 This piston was damaged when a bolt accidentally fell into the intake manifold before the carburetor was installed.

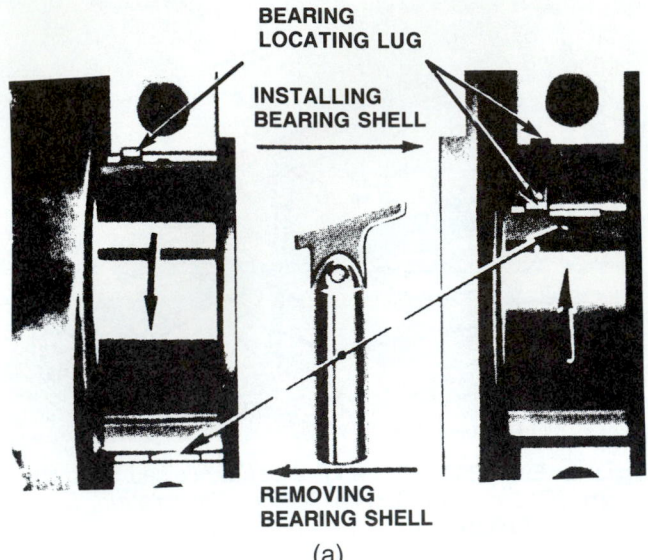

BEARING LOCATING LUG

INSTALLING BEARING SHELL

REMOVING BEARING SHELL

(a)

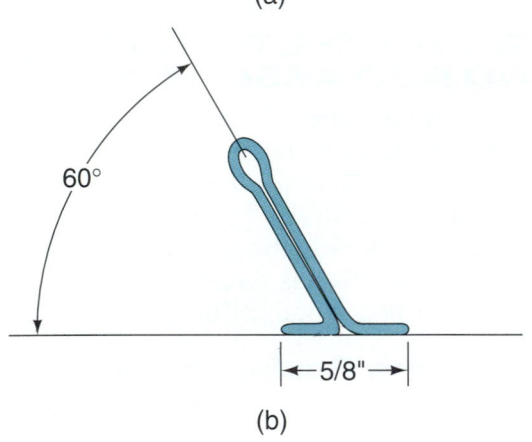

60°

5/8"

(b)

Figure 50.22 Main bearings can be rolled out and new bearings rolled back in. If the special tools are not available, you can make one out of a cotter pin. (a, Courtesy of General Motors Corporation, Service Technology Group)

Main Bearing Replacement

Main bearings are replaced with the crankshaft in the engine using a tool installed in the oil feed hole in the journal (Figure 50.22). The bearings must be rolled out on the side opposite the bearing locating lug, or tang.

■ REMOVE THE TIMING COVER

To remove the timing cover, remove the radiator, fan belts, and vibration damper first. The damper bolt can be loosened with a large socket and impact wrench.

SHOP TIP An angle attachment is available for use with an impact wrench in case an air conditioning condenser (in front of the radiator) is too difficult to move. However, using the angle attachment lowers the amount of torque available to the damper bolt.

Some OHC engines use a chain drive; others use a belt drive. Service on belt drives is relatively simple because no oil is sealed by the timing cover (as in the chain drives). Removing the timing cover on some OHC engines that have timing chains is more difficult because the cover often fits between the oil pan and the cylinder head (see Figure 50.13). There are special procedures for replacing cam timing components in these engines. Sometimes the head and pan might have to be loosened, which can result in leaks after the repair. When the cover intersects the pan, the front part of the pan gasket needs to be cut. Next, part of a new pan gasket is cut and installed to match the intersection of the oil pan and timing cover (see Chapter 45).

■ FREEWHEELING AND INTERFERENCE ENGINES

An engine that has enough piston-to-valve clearance to prevent contact is known as a freewheeling engine. Whenever timing chain service is performed because the chain has broken or skipped, it is possible that valves have come into contact with pistons. Cracked pistons or bent valves can result (see Figure 18.48). Before a chain repair job, perform a leakage test on non-freewheeling (interference) engines to check for bent valves. This step will help ensure that an accurate repair estimate can be made (Figure 50.23).

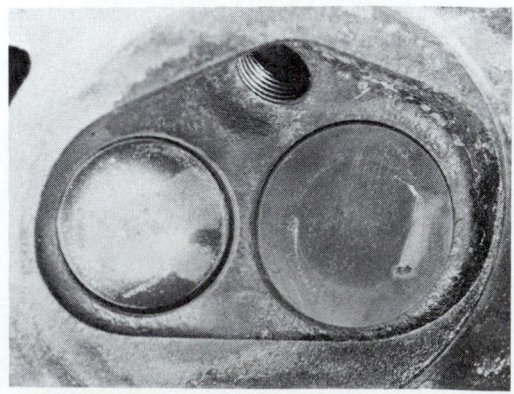

Figure 50.23 This exhaust valve was bent when a timing chain broke. After the chain was replaced, the valve burned because of uneven cooling.

■ REPLACE THE TIMING COMPONENTS

When reinstalling the head on an OHC engine, it is very important that the #1 piston be at TDC and the cam in the head be timed properly. Otherwise, it is possible that a valve could be open and be forced against a piston. If the head is tightened in that position, a valve will be bent. The valves are adjusted before installing the head.

If the camshaft and crankshaft drive gears require replacement, this can be done with the engine in the car. Figure 50.24 shows a crank gear being removed while the crankshaft is in the block.

When replacing a pressed-fit camshaft gear, use a puller to remove it from the cam. When reinstalling the gear, take care that the cam does not move to the rear; it could knock loose the core plug in the rear of the block or chip cam lobes on neighboring lifters. Manufacturers of engines with pressed-fit camshaft gears make special tools that can be used to hold the cam forward during gear installation.

NOTE: *If the cam sprocket is the nylon type and chunks of gear teeth are missing, the oil pan must be removed in order to extract teeth from the oil pump screen and oil pan (Figure 50.25). Failure to do this will result in oil pump failure later when the pieces get sucked into the pump and block oil flow (Figures 50.26 and 50.27).*

■ CRANKSHAFT SEAL REPLACEMENT

Crankshaft front and rear seal replacements can be performed with the engine in the car. The procedures are covered in Chapter 45.

Figure 50.24 The crank sprocket can be removed with the crankshaft still in the engine. *(Courtesy of General Motors Corporation, Service Technology Group)*

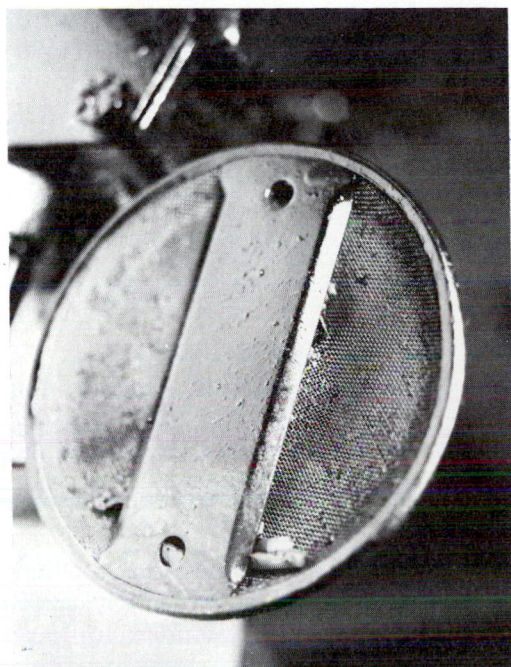

Figure 50.25 This oil pump screen is filled with the teeth of a nylon timing sprocket that failed.

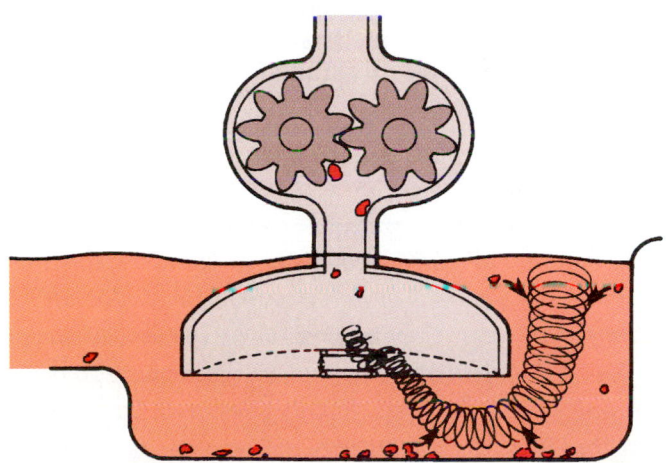

Figure 50.26 When the bypass valve opens on a cold morning, foreign material will be sucked in by the oil pump. *(Courtesy of Federal-Mogul Corporation)*

Figure 50.27 The hex drive on this rotor pump was ruined when foreign material jammed the pump. *(Courtesy of Federal-Mogul Corporation)*

SHOP TIP For single-piece (full round) seals that are in a casting bolted to the rear of the engine block, remove the seal by prying between the crankshaft and seal. Do this before removing the casting. The casting is fragile and is easily broken during seal removal unless it is bolted to the block.

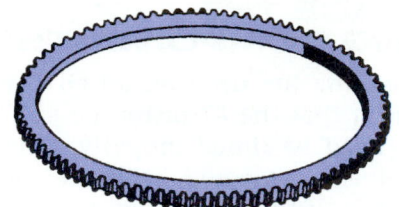

Figure 50.28 A replaceable ring gear. *(Courtesy of Mercedes Benz of North America, Inc.)*

■ FLYWHEEL RING GEAR SERVICE

Most vehicles with standard transmissions have a replaceable ring gear on the flywheel (Figure 50.28). Sometimes, the flywheel ring gear has been worn by the starter motor (Figure 50.29). To remove the worn ring gear from the flywheel, drill a hole between the teeth (Figure 50.30) and break the ring with a chisel.

Heating a Ring Gear

Heat the new ring gear evenly around its circumference during installation. It should not be heated to more than about 400°; too much heat can remove the hardness from the gear.

Figure 50.29 Damaged ring gear teeth.

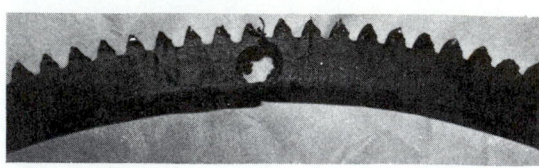

Figure 50.30 Drill a hole between the teeth and strike with a chisel to remove the gear.

> **■ SCIENCE NOTE ■**
>
> *Pure iron is a silvery-white soft metal that rusts rapidly. The addition of small amounts of carbon strengthens it greatly. The hardness of iron is also related to cooling it at different rates. Cooling iron slowly allows carbon to separate from iron almost completely giving large smooth crystals of iron. Cooling it rapidly gives small jagged crystals in which the carbon does not separate but remains combined to iron as iron carbide (Fe_3C).*

The temperature of the ring gear can be checked with a **tempilstick**. Tempilsticks are pencil-shaped sticks that melt at different temperatures. A 400° tempilstick will leave a melted film on the ring gear when it is stroked across a surface hotter than 400°.

Temperature can also be checked by polishing several spots on the ring gear using emery cloth or sandpaper. Heat the ring gear until the spots turn blue.

Solder can also be used to check the temperature of a ring gear. When the solder melts, the ring gear is hot enough and can be positioned onto the flywheel.

NOTE: *The chamfered side of the teeth should be on the same side as they were on the old ring gear.*

■ REVIEW QUESTIONS

1. A tool made of a _____ shaft with the gear removed is used for driving the oil pump when priming the lubrication system.

2. During lubrication system priming, the tool is turned in the same direction the _____ rotates.

3. When the crankshaft timing marks are aligned, number one cylinder and its companion are at _____.

4. When an engine is first started, it should be run at 1500–2000 rpm for _____ minutes.

5. The point where valve clearance is eliminated is called zero_____.

6. What does WOT mean?

7. _____ _____ is when fuel boils in a fuel line.

8. At what mileage should the first service and checkup be performed after an engine rebuild?

9. If a head gasket needs to be replaced, what machine work is probably necessary?

10. What kind of wood should be used when jacking or supporting something because it is less likely to break than standard lumber?

■ ASE STYLE REVIEW QUESTIONS

1. Technician A says that forcing an engine into alignment with its mounts can result in a distorted cylinder bore. Technician B says to pry on the sheet metal part of a power steering pump to adjust the belt tension. Who is right?

   a. Technician A   b. Technician B
   c. Both A and B   d. Neither A nor B

2. Technician A says that only engines without a gear on the distributor can have the engine primed by driving the oil pump. Technician B says that if the crankshaft timing mark on a GM V-8 (firing order 18436572) is aligned with the mark on the timing cover, the number 6 cylinder could be at TDC. Who is right?

   a. Technician A   b. Technician B
   c. Both A and B   d. Neither A nor B

3. Technician A says that the engine should only run at idle during the first 20 minutes of operation after initial startup. Technician B says that valve clearance adjustment can be done on some engines with the engine idling. Who is right?

   a. Technician A   b. Technician B
   c. Both A and B   d. Neither A nor B

4. Technician A says that on an engine with a detonation sensor, the engine's idle can change due to a rattle such as a muffler vibrating against the car's frame. Technician B says a detonation sensor can cause the computer to change ignition timing. Who is right?

   a. Technician A   b. Technician B
   c. Both A and B   d. Neither A nor B

5. Technician A says that if ignition timing is retarded, engine idle will be higher. Technician B says when installing an engine in a chassis, the engine mount bolts on one side must be torqued before installing the mount on the other side. Who is right?

   a. Technician A   b. Technician B
   c. Both A and B   d. Neither A nor B

Miscellaneous Chassis Theory and Service

THEN AND NOW: BRAKING SYSTEMS

As self-propelled vehicles came on the scene, it became apparent that an effective means of stopping them would be required. Borrowing from wagons, the first cars used *"spoon" brakes*. This was simply a block of wood pressed against the wheel by a lever. Next came the *contracting brake,* a steel strap or cable wrapped around the outside of the wheel's hub that could be tightened by the driver.

In 1903, a Frenchman named Louis Renault dramatically improved stopping power with the *internal expanding drum brake*. This brake design has curved friction material-lined "shoes" that are pushed against the inside of a drum to cause drag. Internal expanding drum brakes are still in wide use today.

A brake drum is a lot like a cast iron cooking pot. Getting rid of heat has always been a big problem with this brake design. The development of the disc brake did much to help solve that problem. A disc brake uses friction elements that "pinch" a flat rotor that is exposed to outside air. One of the earliest disc brake versions appeared on the front wheels of an electric car designed by Elmer Ambrose Sperry in 1898. Then came the 1949 Crosley with its *"spot" brakes* (round pads). Front wheel disc brakes became popular on domestic cars in the mid-sixties.

Another early challenge was how to transmit the driver's signal to stop from his foot to the brake. The pedal replaced the hand lever early on, but mechanical means of applying the brakes (cams, cables, and levers) remained for some time. These systems were dangerous and could cause skidding, however, because they were impossible to equalize perfectly. They also required constant adjustment. The idea of using an enclosed liquid to do the job was proposed back in 1897. This would solve the problem because hydraulic systems have equal pressure throughout the area in which the brake fluid is enclosed.

It took many years to develop dependable hydraulic designs, however. The first American car with hydraulically actuated brakes was the 1921 Dusenberg, followed by Chrysler in 1924. In 1967, a federal law stated that all cars sold in the United States must have two separate hydraulic circuits.

Anti-lock braking systems (ABS) are the "hot" brake topic today. ABS pulses the brakes on hard stops or slippery surfaces preventing a wheel from locking up causing a skid. The idea is far from new. Patent applications were made for mechanical versions in the mid-1920s. Electronic systems were offered for a while in the early 1970s. Neither of these systems were dependable or affordable enough to be acceptable. Today, most vehicles come with ABS as standard equipment.

The 1910 Buick had the pedal for its transmission brake on the right. *(Courtesy of Bob Freudenberger)*

A modern disc brake caliper being bled. *(Courtesy of Bob Freudenberger)*

Brake Fundamentals

■ INTRODUCTION

When a car is travelling at freeway speed, a large amount of **kinetic energy** is stored. This is energy that wants to remain in motion (*inertia*). Energy cannot be consumed, but its form can be changed. When you apply the brakes to stop the car, *dry friction* is used to change energy of motion (kinetic energy) to *heat energy*.

The amount of horsepower changed to heat during stopping can amount to several times the power developed by the engine during acceleration. The amount of kinetic energy to be changed into heat depends on the weight and speed of the vehicle. The temperature in the brake linings during a stop can approach 600°F. The amount of dry friction varies, depending on the force applied, the material that the friction surfaces are made of, and the roughness or finish of the friction surface.

■ Friction is the force that resists movement between any two contacting surfaces.

■ The ratio of the force holding two surfaces in contact to the force required to slide one over the other is known as the **coefficient of friction**. If it takes 70 pounds of force to drag one piece of material across another, the coefficient of friction is 70 divided by 100, or 0.70.

■ The coefficient of friction varies with the temperature of the surfaces in contact, the rubbing speed, and the condition of the surfaces.

■ BRAKE LININGS

The friction materials used in cars and trucks are called brake linings. Linings are either **bonded** (glued) or **riveted** to the disc backing or shoe (on drum brakes). Some newer pads are *integrally molded*. From the back of the pad, you observe holes that are either full or partially full of lining material. The linings are actually molded to the metal disc backing plate to become one unit.

Linings are either asbestos, nonmetallic organic, semi-metallic, or metallic. *Asbestos linings* are a health hazard and have been phased out as a brake material. The materials used to replace asbestos are organic materials bonded together with a resin binder.

Semi-metallic linings are organic linings with sponge iron and steel fibers mixed into them to add strength and temperature resistance. They are fast at taking heat away from the rotor and putting it into the lining. This heat transfer does *not* affect the service life of the lining. The hotter they get, the better they work.

Metallic linings are used in very heavy-duty and racing conditions. They work poorly when cold.

Weight Transfer

During a stop, the weight of the vehicle shifts onto the front brakes. Because of this weight transfer, rear brakes do not usually wear out as fast as front brakes. Front brakes must be able to absorb more heat than rear brakes, so linings with more surface area are required. Heavier vehicles require wider linings with more surface area to carry off the increased heat.

The ratio between the front and the rear brakes is about 60/40 for rear wheel drive vehicles. When the vehicle has front wheel drive, the braking ratio is about 80% for the front and 20% for the rear. This is because of the added weight of the power train components in the front.

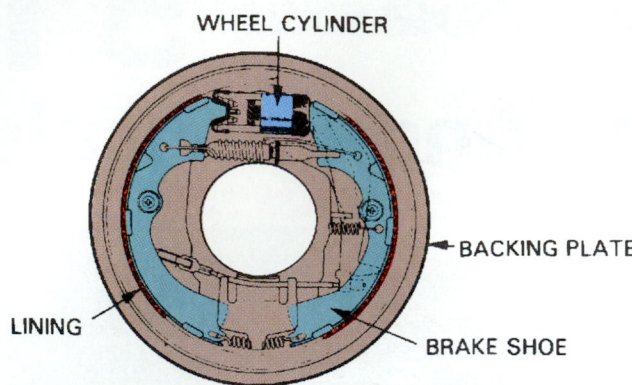

Figure 51.1 Drum brakes are mounted on a backing plate.

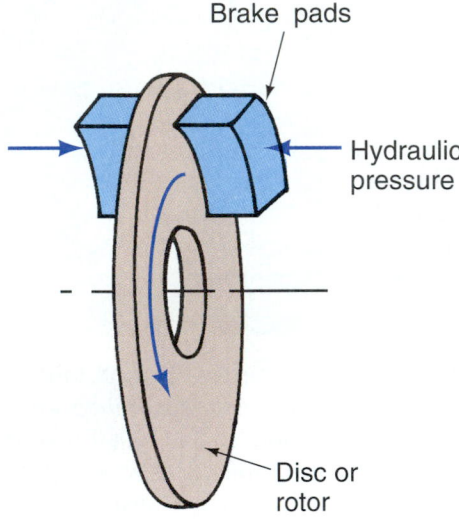

Figure 51.2 Disc brake systems have a rotor and caliper with brake pads, similar to that used on a bicycle.

There are two main types of brakes: *drum* and *disc*. Disc brakes are usually found in the front, with drum brakes in the rear. Drum brake systems have metal brake drums that are bolted to the wheels. Drums are made in several materials and styles. The linings and braking components are mounted on a fixed backing plate (Figure 51.1). Disc brake systems have a rotor and caliper, similar to that used on a bicycle (Figure 51.2). On drum brakes, the friction materials are called linings or *shoes*. On disc brakes, they are called linings or *pads*.

■ BRAKE HYDRAULICS

Liquid under pressure can be used to transfer motion or multiply and apply force. This is called **hydraulics**. Depressing the brake pedal moves a piston in a *master cylinder* (Figure 51.3). The piston pushes fluid under pressure through brake lines and hoses to hydraulic cylinders at each wheel. The pressure acts on hydraulic cylinders at each wheel to produce force (Figure 51.4).

■ The force applied when the master cylinder operates is increased with a larger diameter *wheel cylinder* (Figure 51.5a).

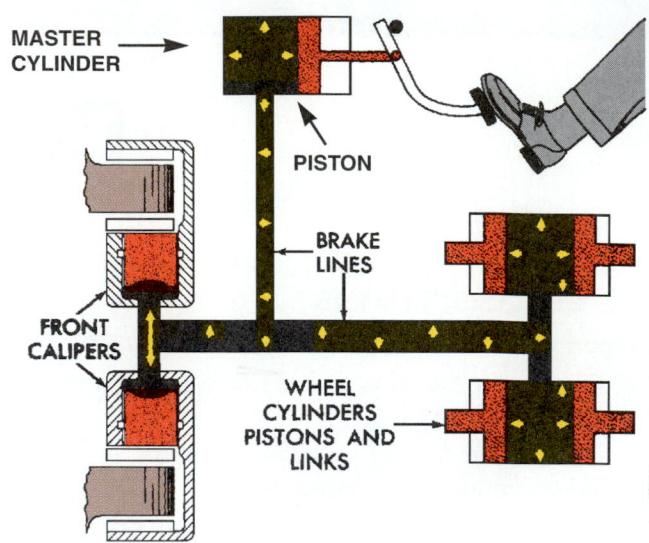

Figure 51.3 Depressing the brake pedal moves a piston in a master cylinder. *(Courtesy of General Motors Corporation, Service Technology Group)*

■ When larger wheel cylinders are used, the distance that the pedal has to travel before building up pressure increases.

■ In order to increase the pressure coming from the master cylinder, the diameter of its bore must be *smaller*.

Pascal's Law

Pascal's law states that "pressure in an enclosed system is equal and undiminished in all directions." This means that the fluid is not moving. The following hydraulic formulas apply (Figure 51.5b):

Pressure = force ÷ area
Force = pressure × area

■ ■ **HISTORY NOTE** ■ ■

At the age of 23, Pascal began a series of experiments with atmospheric pressure, proving a year later that vacuum did in fact exist. He observed that atmospheric pressure decreases with height and theorized that a vacuum existed above the atmosphere. At the age of 30 he explained "Pascal's Law of Pressure."

Pascal's law is fundamental to the development of your automotive diagnostic ability. It applies to such items as the hydraulic brake system, the lubrication system, engine clearances, the pressurized cooling system, engine compression, suspension and steering, and air conditioning. Laws of hydraulic pressure can also be applied to electrical fundamentals.

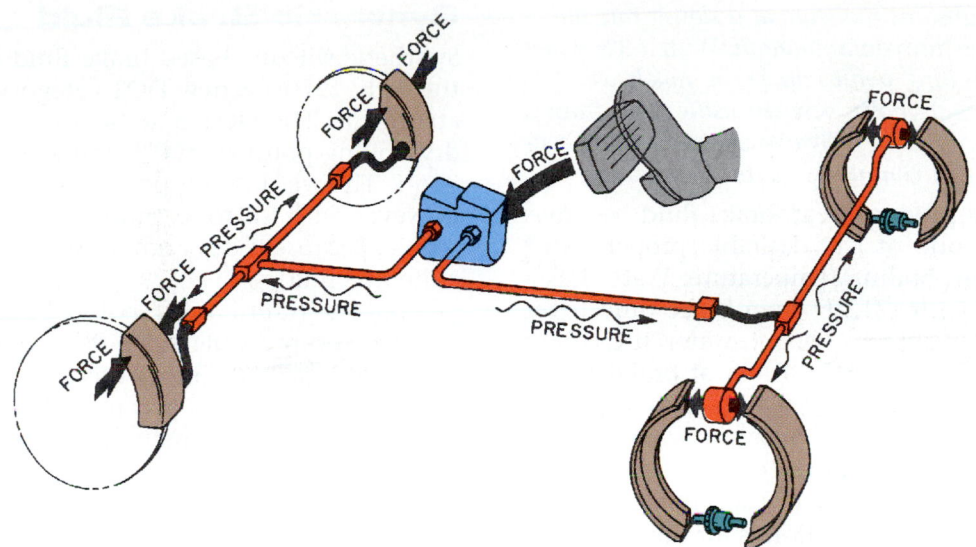

Figure 51.4 Pressure acts on hydraulic cylinders at each wheel to produce force. *(Courtesy of EIS Brake Parts)*

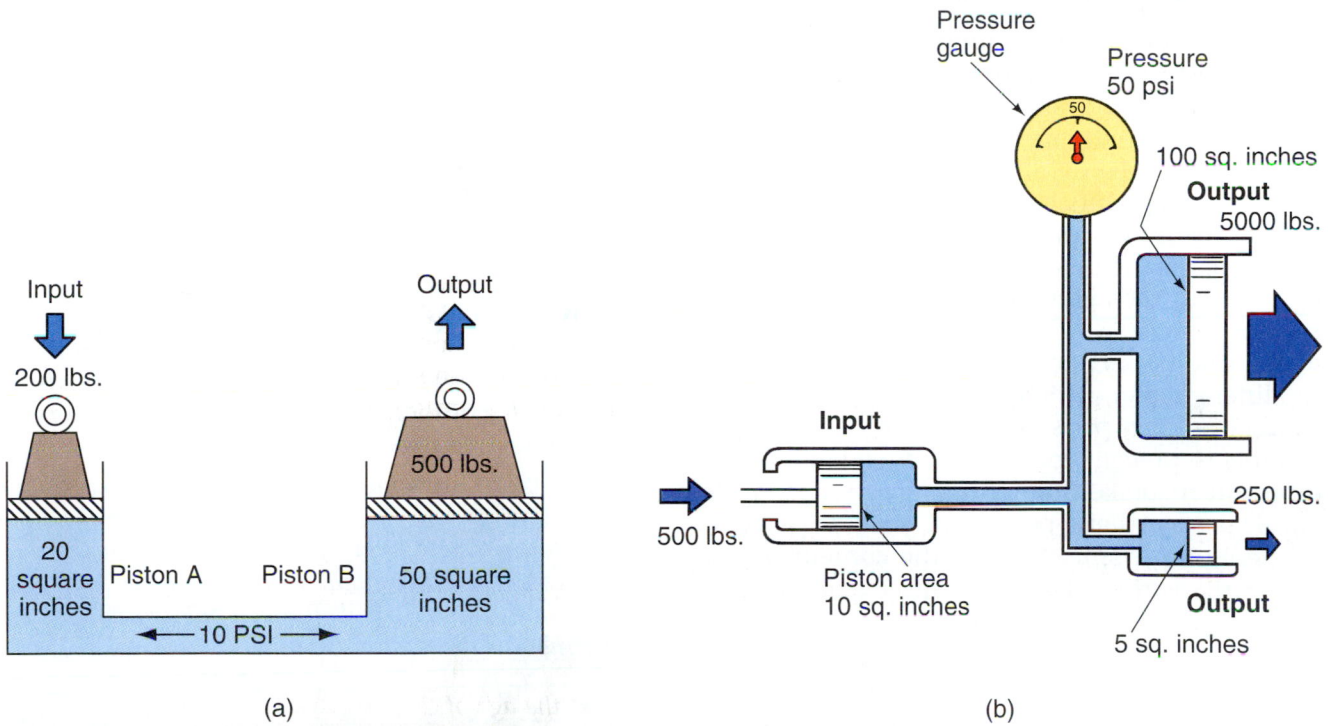

(a)   (b)

Figure 51.5 (a) The force applied when the master cylinder operates is increased with a larger diameter wheel cylinder. (b) Application of formulas for hydraulic pressure. *(Courtesy of Nissan Motors)*

▪ BRAKE FLUID

The fluid used in the brake hydraulic system meets standards set by the Society of Automotive Engineers (SAE) and the Department of Transportation (DOT). There are three main categories of brake fluid: DOT 3, 4, and 5. DOT 5 is *synthetic brake fluid*. DOT 3 and 4 are fluids made from a *polyglycol* base, similar to that of engine coolant.

Glycol based fluids are **hygroscopic**, which means that they absorb water. Being hygroscopic is an advantage for brake fluid because it can absorb mois-

ture that enters the system. This prevents the formation of water drops that could boil or freeze. Dispersing the moisture throughout the fluid also prevents localized corrosion that could cause holes to form in brake lines.

NOTE: *Brake fluid can absorb a large enough quantity of water to be ruined in as little as one hour if left uncovered. According to the EIS Brake Company, a typical vehicle's brake fluid that has been changed in the last eighteen months will have accumulated 2 to 3% water. A water content of 3% lowers the boiling point of DOT 3 brake fluid by*

25%. DOT 4 fluid absorbs moisture at a slower rate but is more affected by moisture contamination. With a 3% water concentration, its boiling point can be as much as 50% lower. When there is a choice between using fluid from a large, partly empty container or two unopened smaller containers, the two smaller containers are the best choice.

Because brakes produce heat, brake fluid becomes hot. That is why one of the desirable properties of brake fluid is a high boiling temperature. Water has a far lower boiling point (212°F) than brake fluid (over 400°F). If brake fluid absorbs enough water, it can boil in the lines. This can result in a loss of braking efficiency, a brake pedal that feels "spongy" during application, and even a total loss of brakes.

There are DOT specifications for both *dry and wet boiling points* (Figure 51.6). The dry specification is for new fluid and the wet specification is for fluid that has absorbed 2% water.

■ DOT 3 fluid has a dry boiling point of 401°F and a wet boiling point of 284°F.

■ DOT 4 fluid has a higher boiling point (446°F wet). Fluid that meets *DOT 3* or *DOT 4* specifications can be used in drum and disc systems.

Glycol-based fluids are *not generic* (not all the same). They are a mixture of various substances. Up to ten different ingredients can be blended to make up the fluid. Some fluids perform better than others. All brake fluids are made up of four things:

■ A lubricant to keep parts sliding freely

■ A solvent diluent, which determines the fluid's viscosity and boiling point

■ A modifier-coupler, which changes the amount of swelling of rubber parts exposed to it

■ Inhibitors to prevent corrosion and oxidation

The lubricant in brake fluid makes up 20–40% of its content. It is a synthetic polyglycol, such as polyethylene or poly propylene. Most of the content of the fluid (50–80%) is the solvent-diluent, glycol ether. In some fluids, ethylene glycol (the same compound as automotive coolant) is also used for this purpose because it also controls rubber swelling. A wide range of chemicals make up the small remainder of the fluid (0.5 to 3%).

NOTE: *Brake fluid prices often vary with the price of engine coolant, due to the amount of glycol used.*

STANDARDS			
	DOT 3	**DOT 4**	**DOT 5**
Boiling Point	min. 205°C/401°F	min. 230°C/446°F	min. 260°C/500°F
Wet Boiling Point*	min. 140°C/284°F	min. 155°C/401°F	min. 180°C/356°F

* Indicates decline in boiling point caused by increasing water content.

Figure 51.6 DOT specifications for both dry and wet boiling points.

Synthetic Brake Fluid

Synthetic silicone-based brake fluid was developed in the early 1970s. A new DOT category, DOT 5, was created for it. The DOT 5 standard requires a minimum dry boiling point of 500°F and a wet boiling point of 356°F. The wet boiling point does not come into play however with silicone fluid because it is *not* hygroscopic. Because it does not absorb moisture, it is called a lifetime fluid.

Silicone fluid has a "spongy" pedal feel because it is more compressible than ordinary brake fluid. The pedal will also have more travel before the brakes are fully applied. This is because silicone fluid contains about three times as much dissolved air as glycol-based fluid. Glycol has 5 percent dissolved air and silicone has 15 percent.

NOTE: *Silicone fluid tends to develop air bubbles when it is cycled rapidly. Most manufacturers recommend it **not** be used in ABS applications. Check the manufacturer's recommendation for the correct fluid. Anti-lock brake systems (ABS) usually call for DOT 3 fluid.*

■ BRAKE HOSES

Hydraulic brake tubing that runs the length of the vehicle frame is made of steel. Rubber hoses are used on both front wheels and above the rear axle to provide a flexible connection from tubing to brake system components (see Figure 23.7). A flexible rubber connection is needed because the front wheels pivot during steering and the rear axle moves up and down when the springs deflect. When the car has an independent rear suspension, each rear wheel is served by its own flexible hose.

NOTE: *Water molecules are smaller than brake fluid molecules. Over a long period of time, since brake fluid is hydrophilic, (water seeking) water vapor from the outside migrates to the brake fluid. This can happen even though brake fluid does not leak through the woven fibers and outer rubber layer of a brake hose.*

■ SCIENCE NOTE ■

Gore-Tex® is a membrane that is laminated to a variety of fabrics. Its expanded PTFE (similar to teflon) membrane is hydrophobic (water hating) and contains 9 billion pores per square inch. Each pore is 20,000 times smaller than a drop of water, yet 700 times larger than a molecule of water vapor. Therefore, it allows moisture from perspiration to escape during exercise, yet liquid water cannot penetrate the pores of the material. An oleophobic (oil hating) substance is also integrated into the membrane to prevent the entry of contaminants (such as bug spray, body oil, suntan lotions, and so forth).

Brake lines are usually made of double-walled steel tubing coated with a rust preventative material. Replacement brake lines are available in several precut lengths with flared ends already on. Sometimes a loop or bend can be put into a line to shorten it to the desired length without having to cut it and reflare one of its ends. When replacing a brake line, be careful to copy the original line as closely as possible. Complete information on brake tubing theory and service is covered in Chapter 24.

■ MASTER CYLINDER OPERATION

Sometimes a service technician who does not understand the operation of the hydraulic system will perform brake repairs just by replacing parts. This is unfortunate because it often means that some repairs will be needlessly performed while searching for the solution to a problem. To be able to efficiently diagnose brake system problems, an understanding of the operation of the master cylinder and the complete hydraulic system is necessary. Pay close attention to the following explanation of master cylinder operation.

The master cylinder, connected to the foot pedal, supplies hydraulic pressure to operate the wheel cylinders during braking (see Figure 51.3). Figure 51.7 shows a simple single-piston master cylinder with one seal, called the *primary cup*. The primary cup compresses fluid when the pedal is depressed. Another seal, called the *secondary cup*, keeps fluid from leaking out the back of the master cylinder bore. This seal does not seal against pressure.

Fluid from the reservoir fills the master cylinder bore through the compensating port (Figure 51.8). When the pedal is applied, the primary cup moves forward in the master cylinder bore. It moves past the *compensating port*, allowing pressure to build in the

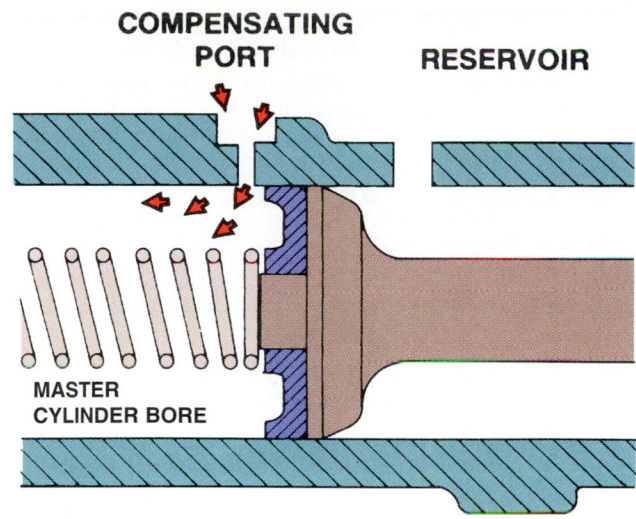

Figure 51.8 Fluid from the reservoir fills the master cylinder bore through the compensating port. *(Courtesy of Ford Motor Company)*

line. As the pedal is forced further down, fluid pressure builds in the system. Remember Pascal's law.

Each wheel cylinder has a pair of rubber cup lip seals (Figure 51.9). The lips of the cup seals face inward against the pressurized fluid created by the master cylinder. The lip on the master cylinder primary cup

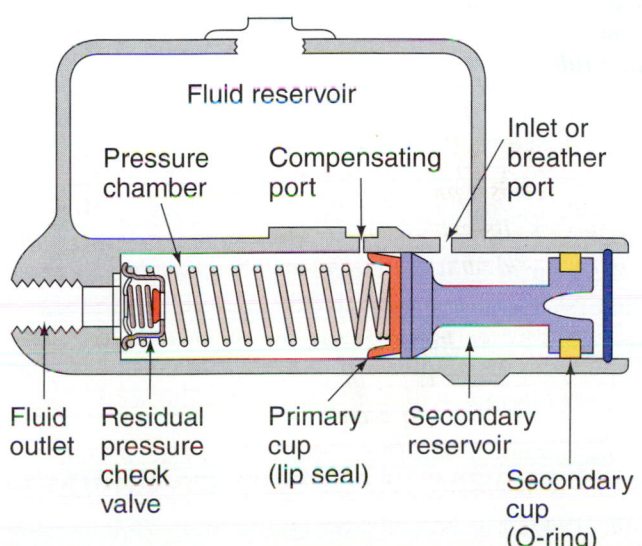

Figure 51.7 A single piston master cylinder.

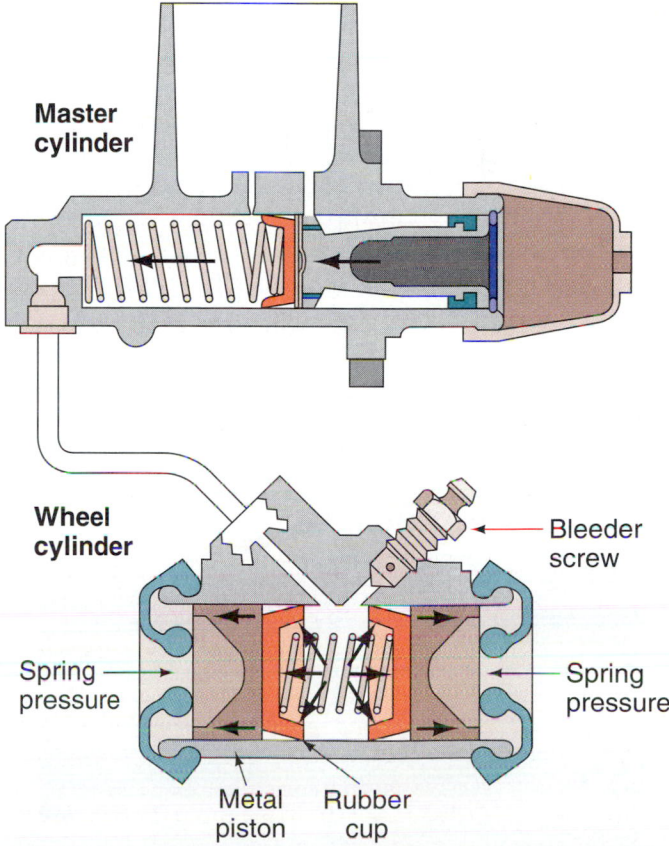

Figure 51.9 The lips of the rubber cups must face toward the brake fluid.

also faces toward the fluid. Notice the direction of the primary cup lip seal in the master cylinder. When a seal is installed in this direction, it *seals*. A seal installed backwards will leak.

■ LOW PEDAL MASTER CYLINDER ACTION

As brake linings wear, clearance between the friction surfaces becomes greater. Disc brakes and most drum brakes are *self-adjusting*. When there are no self-adjusters or if a self-adjusting mechanism fails to work, a drum brake will develop extra clearance. The result is that the brake pedal moves closer to the floor before the brakes are applied. This is called a *low pedal*.

When excessive wear of the friction materials occurs, the pedal might be able to travel all the way to the floor. When this situation develops, a second application of the pedal will result in a higher pedal that will stop the car. What makes this happen?

When the brakes are properly adjusted, only a small volume of fluid in the master cylinder is moved to operate the brakes. When a low pedal is released and then quickly reapplied, fluid becomes momentarily trapped in the cylinders at the wheels. The fluid cannot return quickly to the master cylinder because the inside diameter of the brake line is very small. This small size prevents the quick movement of a large volume of fluid. Large diameter tubing would allow fluid to move more quickly.

When the pedal is released, the primary cup moves back quickly in the master cylinder bore. An area in front of the primary cup, called the *pressure chamber*, is momentarily empty of fluid. Stored just behind the primary cup is a reservoir of fluid. Because the sealing lips on the primary cup lip are facing away from the fluid pressure, fluid can leak past it from the back to the front. This refills the temporarily empty pressure chamber in the master cylinder bore (Figure 51.10).

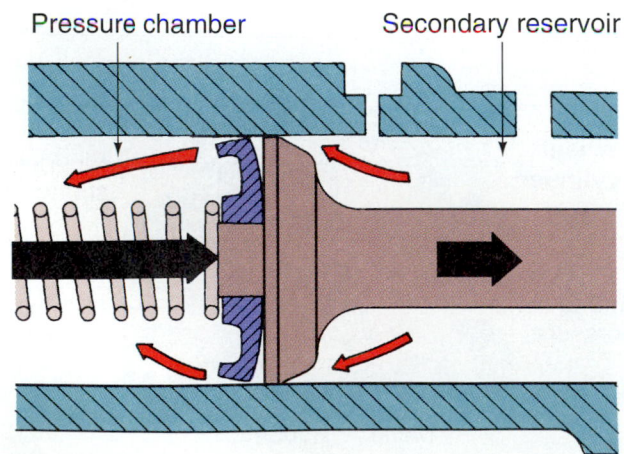

Figure 51.10 Fluid moves from the secondary reservoir to the pressure chamber when the brakes need adjusting. [Courtesy of Ford Motor Company]

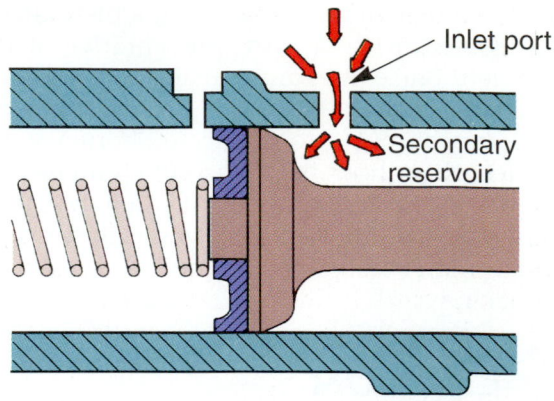

Figure 51.11 Fluid is replenished to the secondary area through the inlet port. [Courtesy of Ford Motor Company]

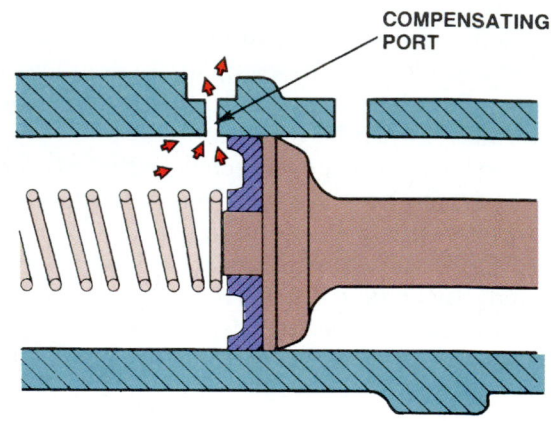

Figure 51.12 Excess fluid bleeds back to the reservoir through the compensating port. [Courtesy of Ford Motor Company]

The secondary area is refilled through the inlet port (Figure 51.11).

When the pedal is quickly reapplied, the linings will contact the drum and pressure will build on the fluid in the wheel cylinders. When the pedal is released again, the brake shoe return springs *slowly* force the excess fluid back through the brake line from the wheel cylinder. The excess fluid bleeds off to the reservoir through the now-uncovered compensating port (Figure 51.12). Until the brakes are properly adjusted, the pedal height will increase on the second pedal application. Low pedal height can also be caused if one part of the master cylinder is leaking internally but pedal height will not increase on the second application.

■ TANDEM MASTER CYLINDER

Since 1967, tandem master cylinders have been mandated by law. Older cars were equipped with only a single master cylinder. When a pre-1967 car had a hydraulic system failure, it would experience a total brake system failure.

NOTE: *Remember Pascal's Law: Pressure is equal everywhere in an enclosed system.*

A tandem master cylinder has one cylinder bore with two separate pistons and chambers (Figure 51.13). During normal stopping, the primary cup on the rear (secondary) piston pushes fluid to the brakes it serves. It also pushes fluid forward against a rearward facing lip seal on the front (primary) piston. When the friction materials begin to apply force at the wheels, pressure begins to build in both front and rear systems at the same time.

When one-half of this system fails, the pedal will be lower but the remaining half of the hydraulic system should have enough braking capacity to stop the car. During a failure in the part of the brakes served by the rear piston, a stub on the front of that piston bottoms out against the back of the front piston (Figure 51.14a). Thus, the primary piston is applied mechanically instead of hydraulically.

When the front part of the system experiences a failure, it cannot build up pressure. A stub on the end of the piston bottoms out at the end of the bore (Figure 51.14b). This allows the rear piston to build up pressure as it forces fluid against the rearward facing seal on the back of the front piston.

Tandem systems are either *longitudinally split* or *diagonally split*. The longitudinal system operates the front and rear brakes separately. The diagonal system operates the brakes on opposite corners of the vehicle (Figure 51.15). They are popular on front wheel drive vehicles. Weight transfer on one of these vehicles can

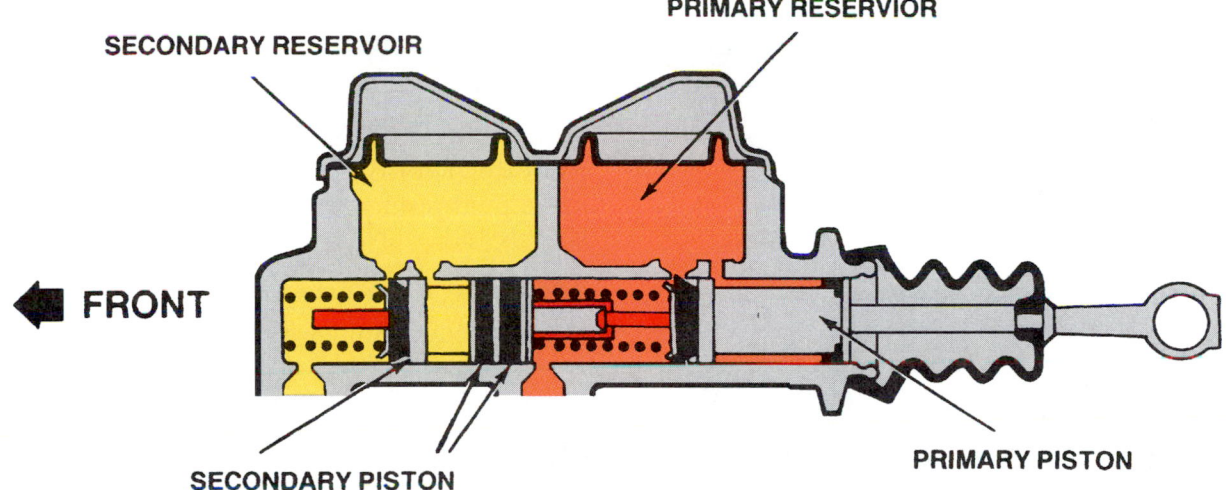

Figure 51.13 A tandem master cylinder. *[Courtesy of Ford Motor Company]*

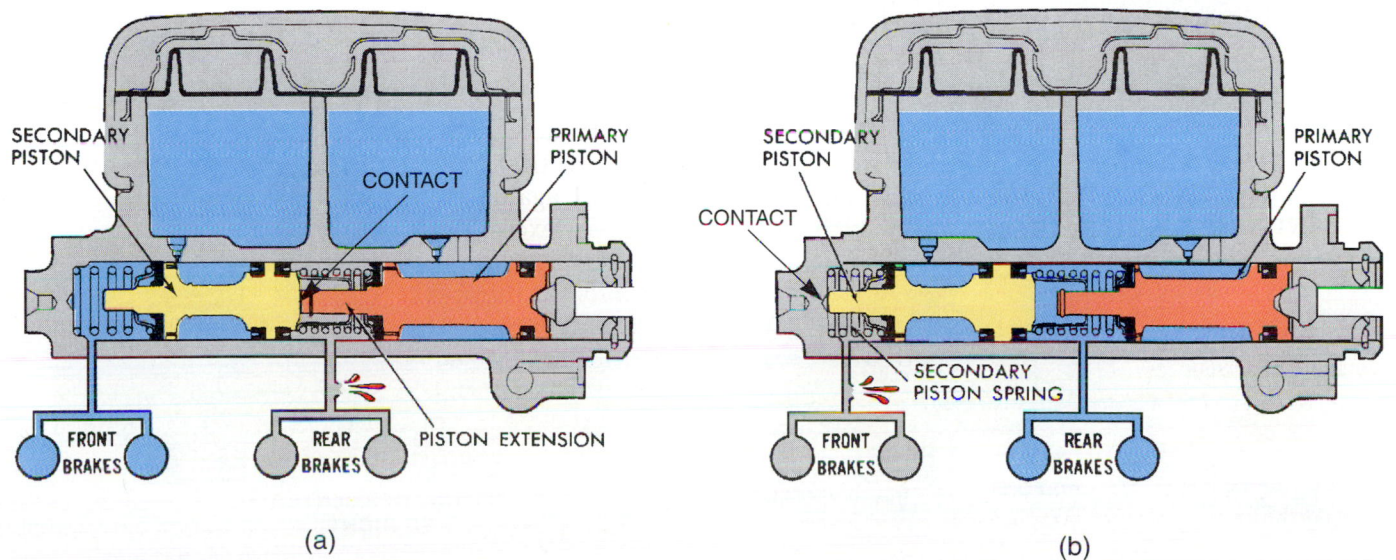

Figure 51.14 (a) The secondary piston hits the primary piston during rear system failure. (b) The primary piston bottoms out in the bore when there is failure in the front system. *[Courtesy of General Motors Corporation, Service Technology Group]*

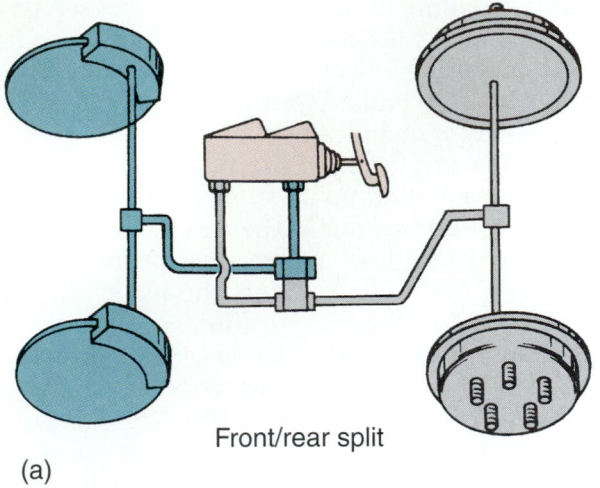

Front/rear split

(a)

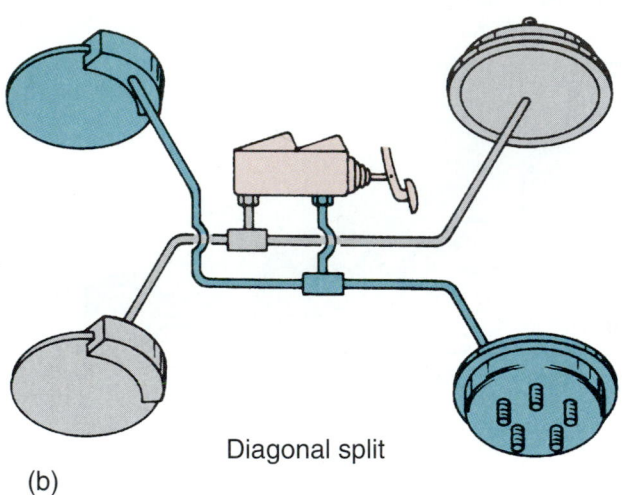

Diagonal split

(b)

Figure 51.15 Tandem brake cylinders are hydraulically split in different ways. *(Courtesy of Ford Motor Company)*

shift to 80% at the front during a hard stop. If the front brakes were longitudinally split and the front failed, the rear brakes would not be effective in stopping the vehicle.

Some master cylinders are designed with a step bore. These are called *quick take-up master cylinders*. They are designed to move a large amount of fluid when the pedal is first applied, using the larger part of the bore. When the friction materials contact braking surfaces, the smaller part of the bore comes into play to give a large pressure boost during stopping. The result is less pedal travel.

Brake Warning Light

Tandem systems are equipped with a brake warning light that tells the driver when half of the hydraulic system has failed. When pressure on one side of the system drops, a piston inside the *hydraulic safety switch* moves off-center, completing the circuit to a light on the driver's compartment instrument panel (Figure 51.16).

Other conditions can also illuminate the brake warning lamp. Applying the emergency brake illuminates this light on some cars. Many cars have a low brake fluid level warning system. The brake warning lamp is part of this system.

Master Cylinder Vent

Master cylinder fluid level fluctuates through the compensating port as the brakes are used because the fluid heats up or cools down (see Figure 51.12). Therefore, the cover to the master cylinder must be vented. The reservoir cover has a rubber diaphragm or a plastic float (Figure 51.17). This feature allows atmospheric pressure to act on the fluid in the reservoir without the fluid becoming contaminated with moisture from the outside air.

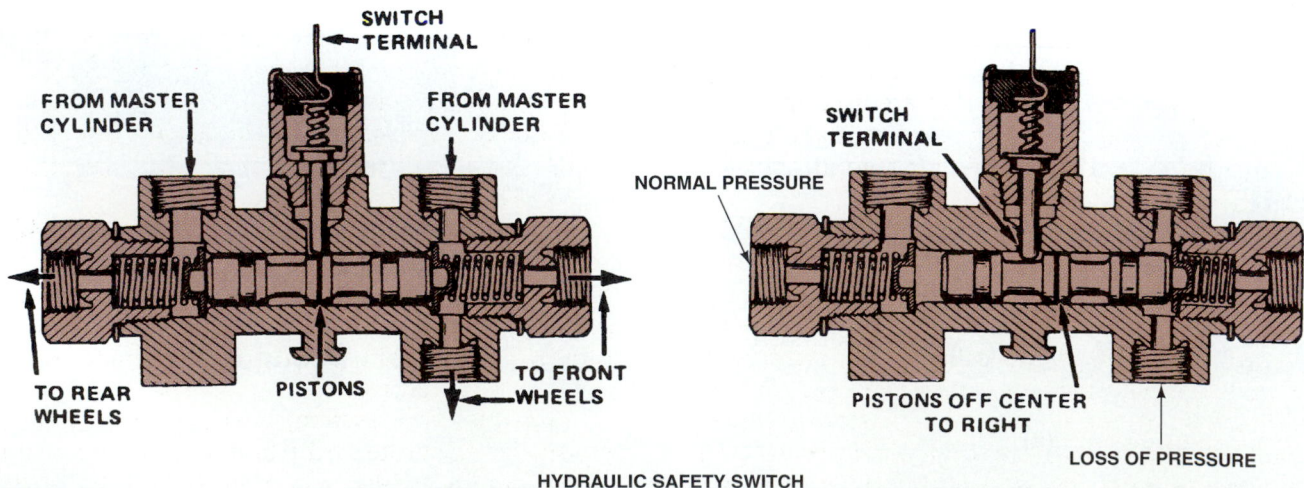

Figure 51.16 During a hydraulic system failure, the piston moves off-center to complete an electrical circuit to the dash light. *(Courtesy of AlliedSignal Bendix)*

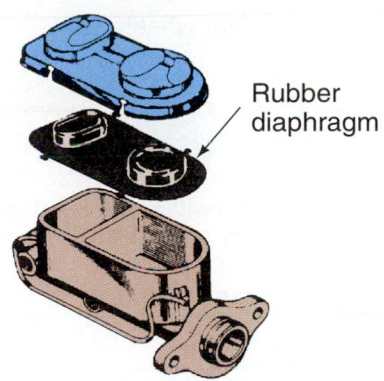

Figure 51.17 The reservoir cover has a rubber diaphragm or a plastic float. *(Courtesy of AlliedSignal Bendix)*

Master Cylinder Check Valve

In some drum systems, there is a *residual check valve* at the outlet to the master cylinder. The purpose of this check valve is to keep a small amount of *residual pressure* (6–25 psi) in the brake system when the brakes are not applied. This helps keep the wheel cylinder rubber cup sealed against the wall of the wheel cylinder to prevent air from entering.

Not all drum brake systems have a residual check valve. Cup expanders on the ends of the center spring in the wheel cylinder make the valve unnecessary. Disc brake systems do *not* have a check valve either. If they did, the pressure would overcome the return action of the disc brake seal (covered later in this chapter).

In a tandem master cylinder, the residual check valve is a **flapper valve**, also called a *duck bill valve*. A flapper valve allows fluid to flow through its center in one direction only. It blocks fluid attempting to flow in the opposite direction. Figure 51.18 shows the operation of a flapper-type residual check valve.

■ DRUM BRAKES

In the past, almost all vehicles were equipped with drum brakes. Today, drum brakes are found in some rear brake applications. Disc brakes are used mostly in the front. Drum brakes provide good initial stopping ability and also provide a means for a good, inexpensive, mechanical parking brake.

There are several different drum brake designs. One popular style is known as the *Bendix brake* or *duo-servo brake* in which both shoes are self-energizing. During stopping, the leading shoe on a drum brake is forced into the brake drum. This is called *self-energization* or **servo action**. There is no anchor at the bottom between the front and rear shoes. The front (*primary*) shoe floats during braking and applies additional pressure against the rear (*secondary*) shoe (Figure 51.19). During a stop in reverse, the secondary shoe is energized and pushes on the primary shoe.

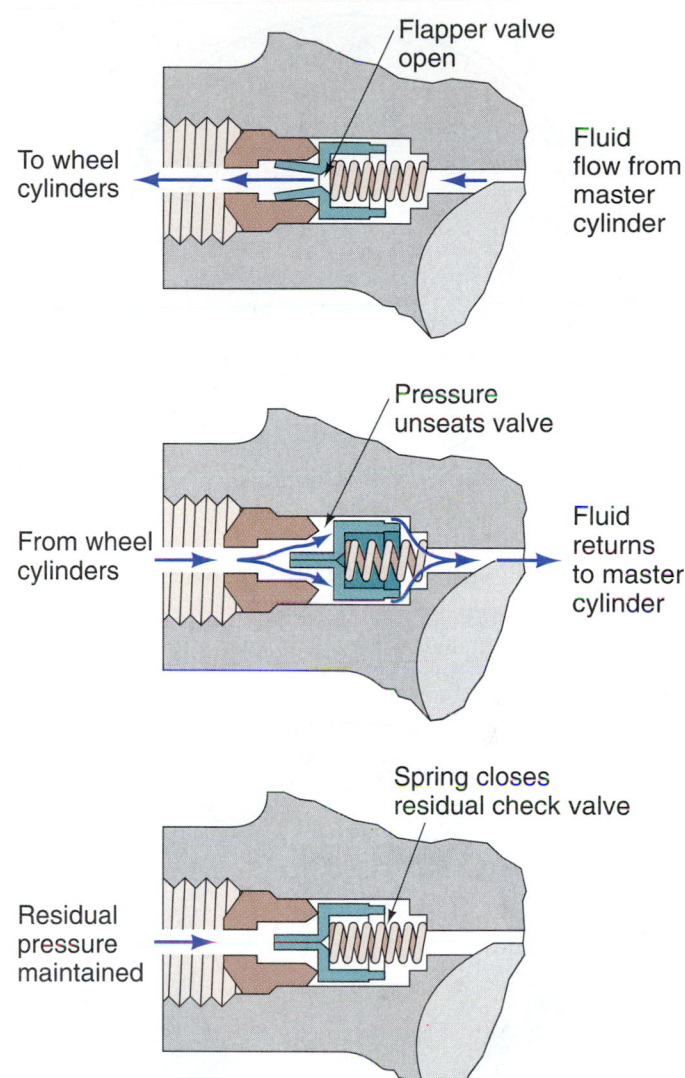

Figure 51.18 Operation of a master cylinder residual check valve.

Another popular style of brake used mostly on rear brakes in front wheel drive cars is a non-servo brake called the *leading-trailing* brake (Figure 51.20). Non-servo brakes have an anchor at the end of each shoe. The front brake shoe, called the leading shoe, is energized, but the trailing shoe is not. As a result, the front shoe wears more than the rear.

When each shoe has its own single-piston wheel cylinder at opposite ends of the brake shoe, this is called a *leading-leading brake*. Each shoe is energized during forward stopping, but is not energized in reverse. This design was used in some older domestic and imported vehicles.

In a duo-servo system, because the rear lining is responsible for more of the braking, its shoe will have a longer lining, which will often be made of different material than the front lining. Leading-trailing shoes are the same length.

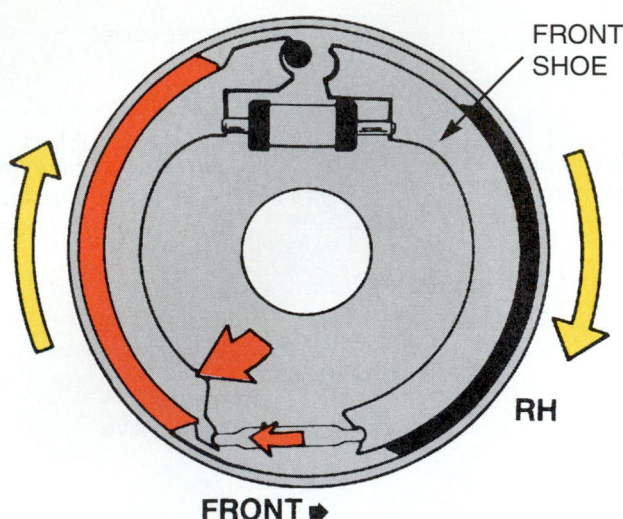

SELF-ENERGIZING BRAKE

Figure 51.19 In a self-energizing brake, the energized front shoe pushes against the rear shoe. *(Courtesy of Ford Motor Company)*

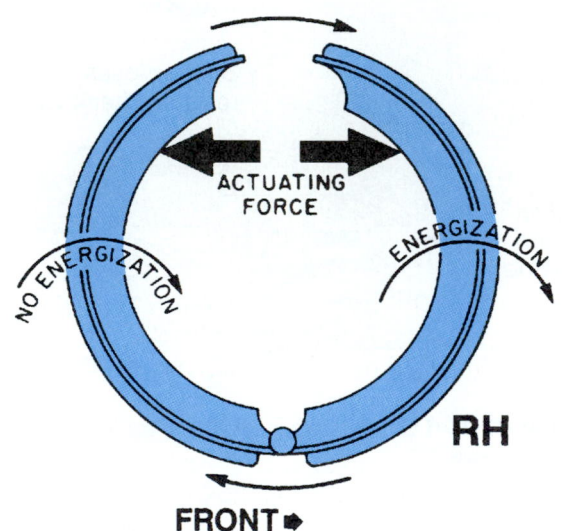

LEADING/TRAILING BRAKE

Figure 51.20 In a leading-trailing brake, only the front lining is energized. *(Courtesy of Ford Motor Company)*

Self-Adjusting Drum Brakes

Springs are used in both types of drum brakes to return the shoes after braking. A starwheel is turned to adjust the brakes (Figure 51.21). Figure 51.22 compares adjusters for leading-trailing and duo-servo brakes. Leading-trailing brakes have one of two kinds of self-adjusting mechanisms. When the linings have worn a certain amount, self-adjustment takes place whether traveling forward or in reverse.

Duo-servo drum brakes usually have self-adjusters that operate only when the brakes are applied with the car backing up. When there is excessive clearance, the

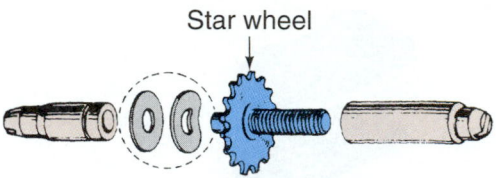

Figure 51.21 A starwheel is turned to adjust the brake clearance. *(Courtesy of EIS Brake Parts)*

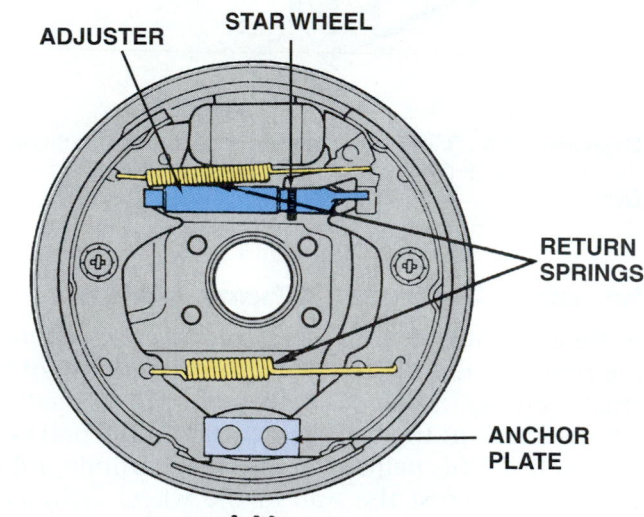

LH
LEADING-TRAILING
◀ FRONT

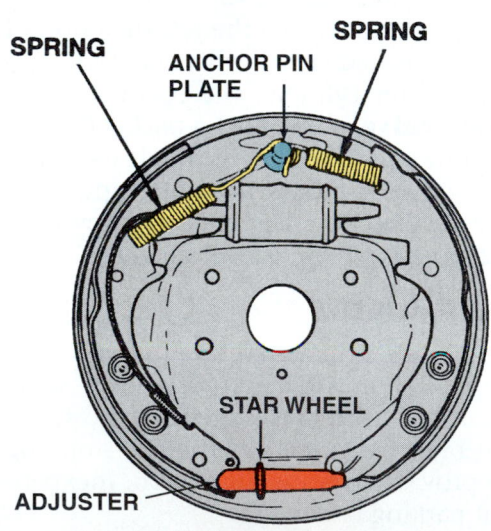

RH
DUO-SERVO
FRONT ▶

Figure 51.22 Adjusters for leading-trailing and duo-servo brakes. *(Courtesy of Ford Motor Company)*

top of the rear brake shoe moves away from the anchor pin (Figure 51.23). This causes the adjusting cable to pull up on the adjusting lever against spring

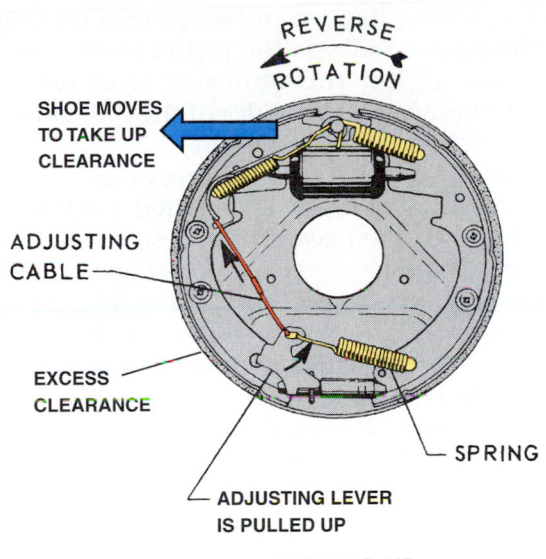

REVERSE
ROTATION

SHOE MOVES
TO TAKE UP
CLEARANCE

ADJUSTING
CABLE

EXCESS
CLEARANCE

SPRING

ADJUSTING LEVER
IS PULLED UP

BACKING UP

Figure 51.23 When there is excessive clearance, the top of the rear brake shoe moves away from the anchor pin, pulling the adjuster lever up. *(Courtesy of General Motors Corporation, Service Technology Group)*

tension. When the brake is released, the spring pulls the adjusting lever back down. This turns the starwheel to adjust the brakes (Figure 51.24). The adjusting lever and starwheel parts are labeled left or right and can only be used on one side of the car.

Leading-trailing brakes have either an expanding strut adjuster or an incremental adjuster. The *expanding strut* works off the parking brake. A rod and strut assembly fits between the two brake shoes. When the parking brake is applied, if too large a gap exists between the shoes, the strut and rod assembly will expand when the brake is released.

An *incremental adjuster* has a starwheel. It operates in either forward or reverse, whenever the gap between the lining and shoe becomes too large. It is named an incremental adjuster because the starwheel only moves one tooth at a time to compensate for wear.

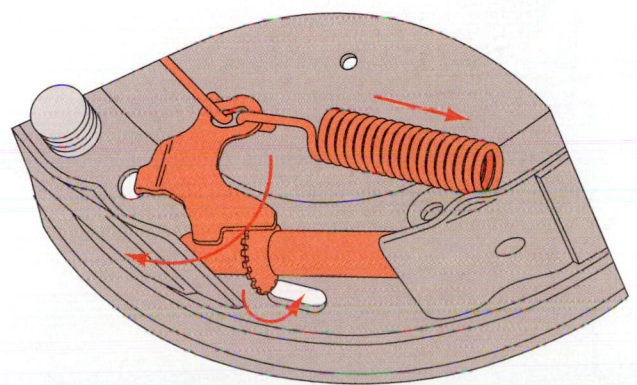

Figure 51.24 The starwheel turns to adjust the brakes. *(Courtesy of Ford Motor Company)*

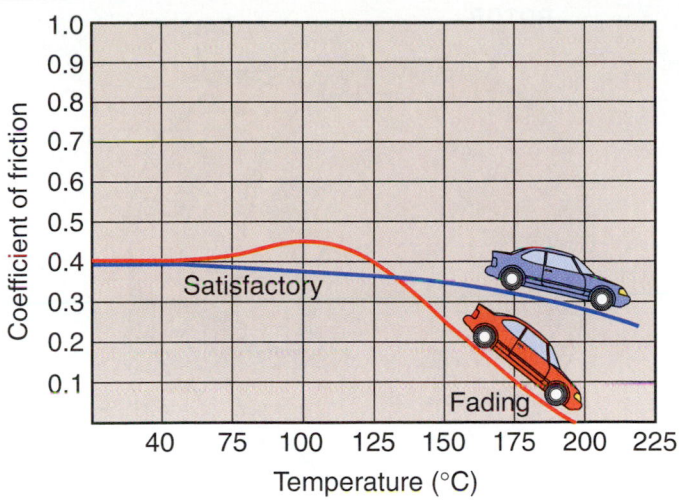

Figure 51.25 Coefficient of friction drops off as brakes fade. *(Courtesy of Ford Motor Company)*

Brake Fade

Brake fade results from excessive brake heat. One drawback to drum brakes is that they do not dissipate heat as well as disc brakes. Disc brakes are exposed to abundant air. In drum brakes, heat is trapped within the drum. With increased heat, the drum expands, causing the pedal to move closer to the floor before the brakes are applied. The coefficient of friction of the lining also drops off (Figure 51.25). This means that more effort is required to stop the car. As fade becomes worse with the addition of more heat, stopping ability will drop off and may disappear altogether. Brake drums often have fins to help dissipate the heat.

■ DISC BRAKES

Many American cars since the mid-1960s have been equipped with front disc brakes. More expensive and performance cars sometimes have four wheel disc brake systems. A disc brake system has a *rotor* (disc) and a *caliper* (like a bicycle brake) (Figure 51.26). Figure 51.27 shows the parts of a typical disc brake system.

Brake Rotors

Disc brake rotors are either *solid* or *ventilated* (see Figure 51.28a). Disc brake parts represent **unsprung weight**, which affects vehicle handling. It is desirable for weight to be kept as low as possible. Lightweight solid rotors are used in lighter cars. Ventilated rotors are heavier but have abundant surface area, so they can dissipate the extra heat generated by the brakes of heavier vehicles.

Brake Calipers

There are two types of disc brake calipers: the *fixed caliper* and the *floating (sliding) caliper* (Figure 51.28). Fixed calipers have pistons on both sides of the caliper.

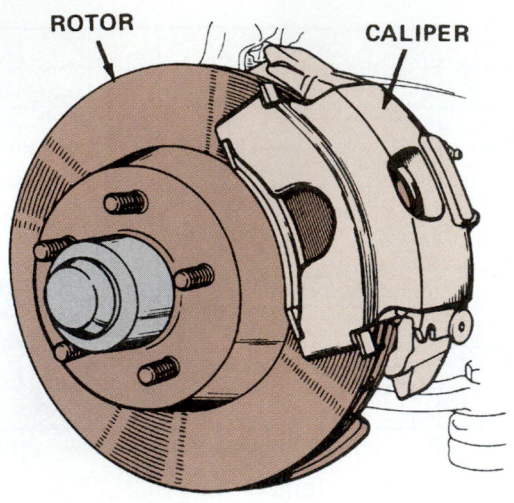

Figure 51.26 A disc brake rotor and caliper resembles a bicycle brake. *(Courtesy of AlliedSignal Bendix)*

A floating caliper has one or two pistons on only one side. The caliper slides as the piston moves out of its bore (Figure 51.29). This results in equal force being applied to linings on both sides of the caliper from the hydraulic action on only one side.

One of the drawbacks to floating caliper disc brakes is that a power assist is usually required. Early American disc brakes used the fixed caliper design because more piston area could be used, so a power brake would not be necessary. Most later designs use a floating caliper, which has fewer parts and is easy to repair.

A rubber, square-cut O-ring seals the disc brake piston to its bore. Disc brakes do not require a return spring like drum brakes do. When the brakes are applied, the seal distorts (Figure 51.30). When the brakes are released, the seal retracts to its original position, pulling the piston back and allowing the linings to release the rotor.

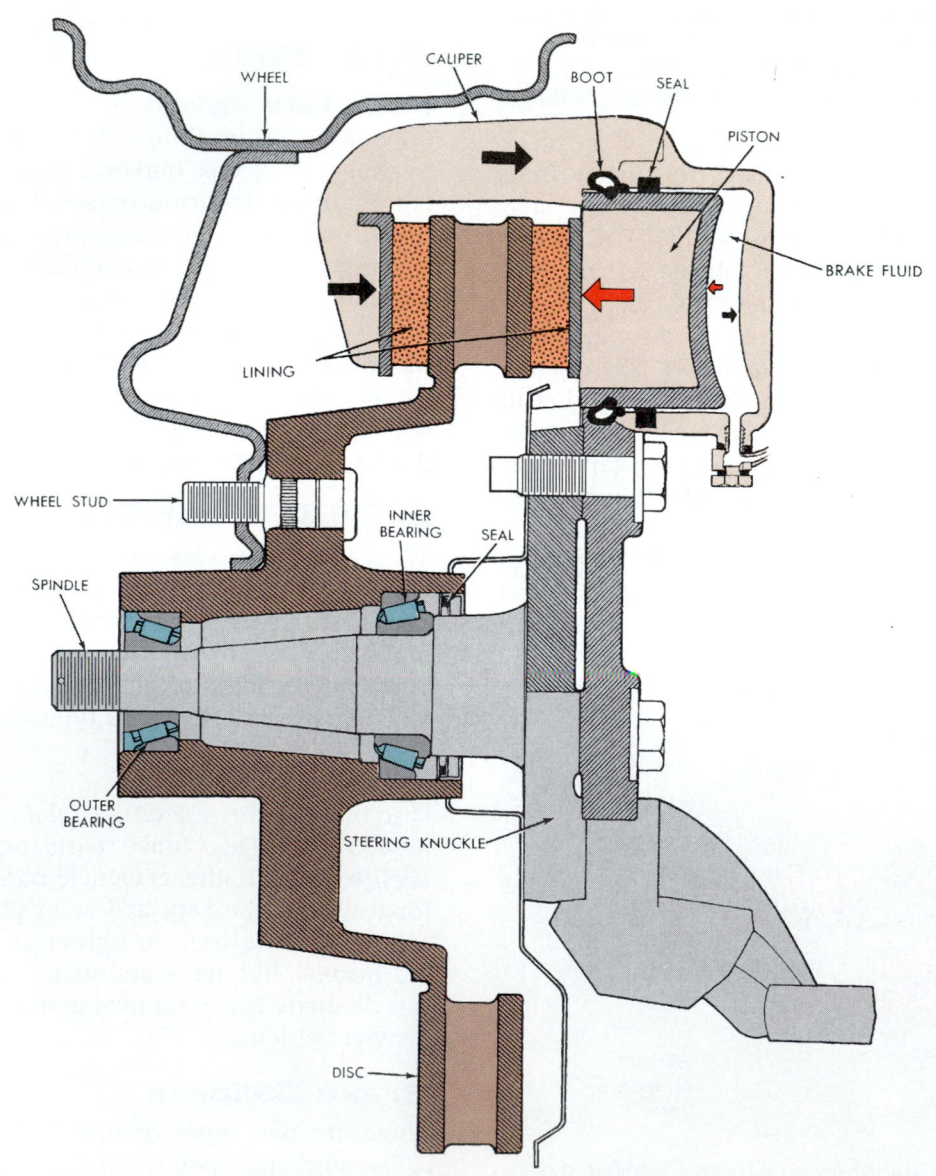

Figure 51.27 A typical disc brake system shown here with the wheel bearing assembly. *(Courtesy of Chrysler Corporation)*

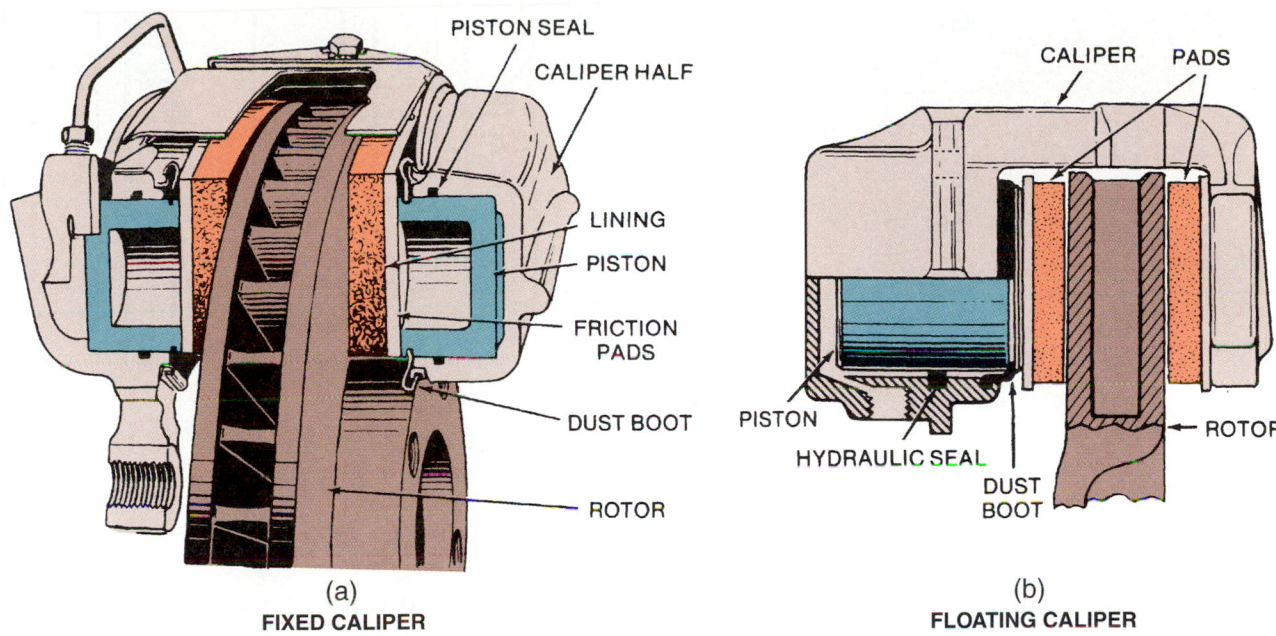

(a)
FIXED CALIPER

(b)
FLOATING CALIPER

Figure 51.28 A fixed and a floating caliper. *[Courtesy of EIS Brake Parts]*

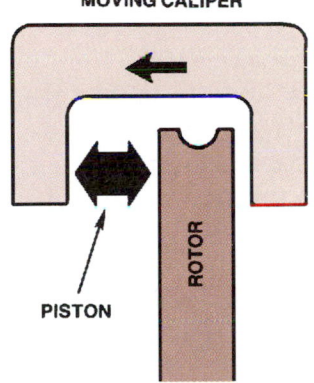

Figure 51.29 A floating caliper moves as the brakes are applied. *[Courtesy of Ford Motor Company]*

Front disc brakes and rear disc brakes without an integral emergency brake are *self-adjusting*. As floating caliper linings wear, the caliper moves on its mount and the piston slides forward in its bore to take up the slack (Figure 51.31). As this occurs, the caliper requires more fluid. On a master cylinder with reservoirs of different sizes, the larger one is for the disc brakes and the smaller one is for the drum brakes.

When a rear disc brake includes a mechanism for the parking brake, there will be an automatic adjuster (Figure 51.32). Operating the parking brake causes the piston to thread outward to adjust the clearance.

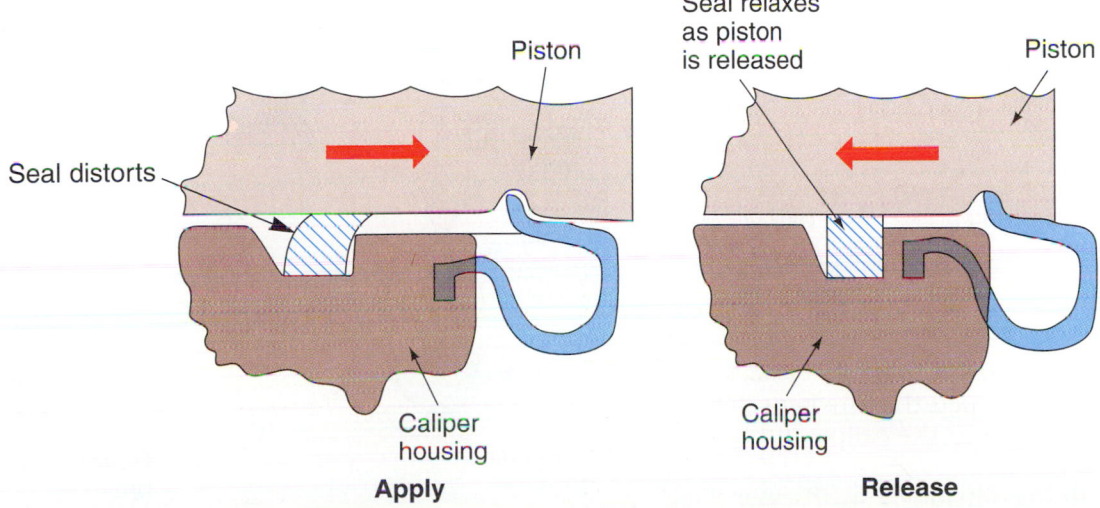

Apply

Release

Figure 51.30 When the brakes are applied, the seal distorts. When the brakes are released the seal retracts pulling back the piston.

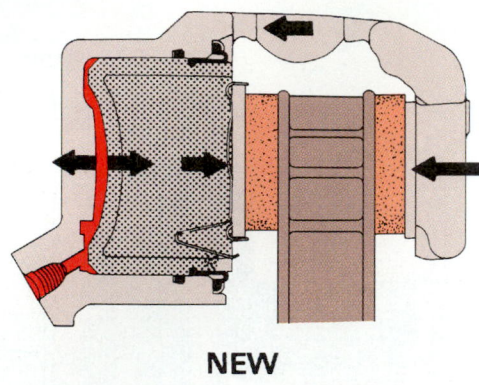

NEW

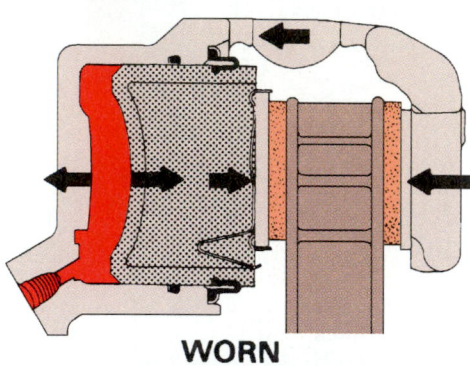

WORN

Figure 51.31 As the disc linings wear, the piston moves out to self-adjust. *(Courtesy of General Motors Corporation, Service Technology Group)*

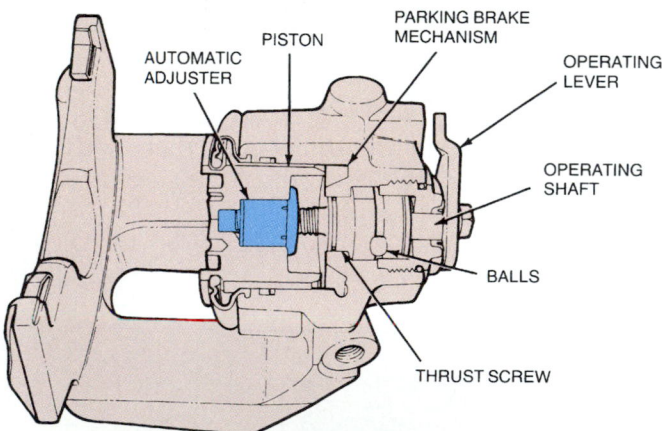

AUTOMATIC ADJUSTER — PISTON — PARKING BRAKE MECHANISM — OPERATING LEVER — OPERATING SHAFT — BALLS — THRUST SCREW

Figure 51.32 A rear disc brake caliper with a self-adjuster operated by the parking brake. *(Courtesy of Ford Motor Company)*

Disc Linings

Disc brake linings are fastened to a metal back. Some have tabs on the back of the pad that are bent during installation to hold it snugly to the caliper. If the pad vibrates, there will be noise.

Some disc brake linings include a *wear sensor* (Figure 51.33), which is a small metal tab that rubs against the rotor when the lining wears thin. The resulting

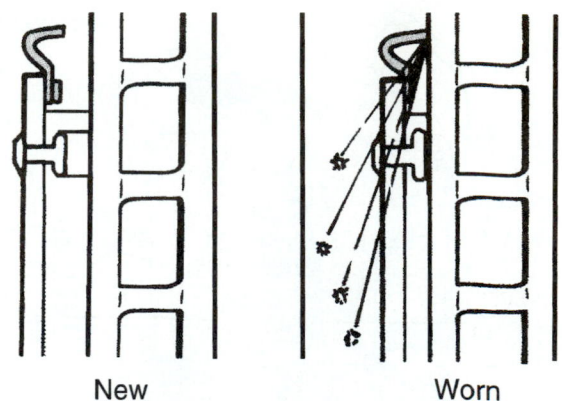

New   Worn

Figure 51.33 A wear sensor rubs against the rotor to create a noise that warns the driver. *(Courtesy of General Motors Corporation, Service Technology Group)*

noise alerts the driver that brake work is needed before serious damage to the rotor occurs.

Metering and Proportioning Valves

Metering and *proportioning* **valves** are also used in disc brake systems to keep the front discs and the rear drums in balance during stopping (Figure 51.34).

Metering Valve. The metering valve (Figure 51.35) is used on front disc brakes. It does not operate until 10 psi is reached in the system. Then it shuts off pressure to the front brakes until about 125 psi builds up in the rear brakes. When the metering valve operates, the stem can be seen moving out away from the body of the

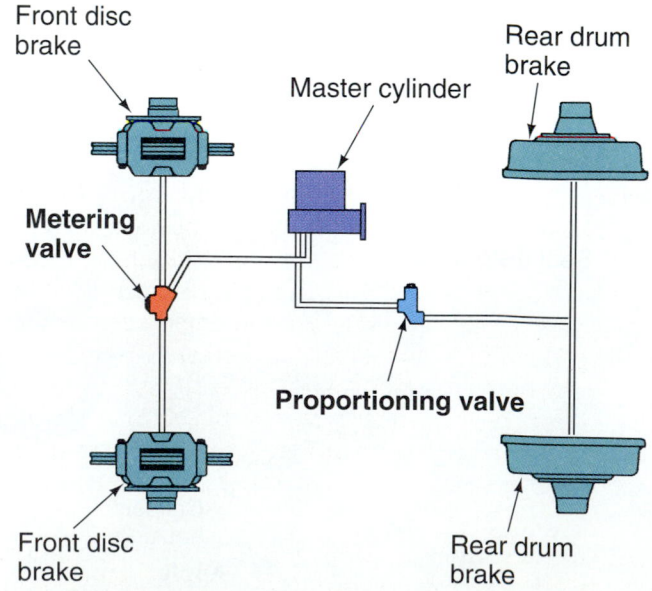

Front disc brake — Rear drum brake — Master cylinder — **Metering valve** — **Proportioning valve** — Front disc brake — Rear drum brake

Figure 51.34 A metering valve controls the front brakes and a proportioning valve controls the rear brakes.

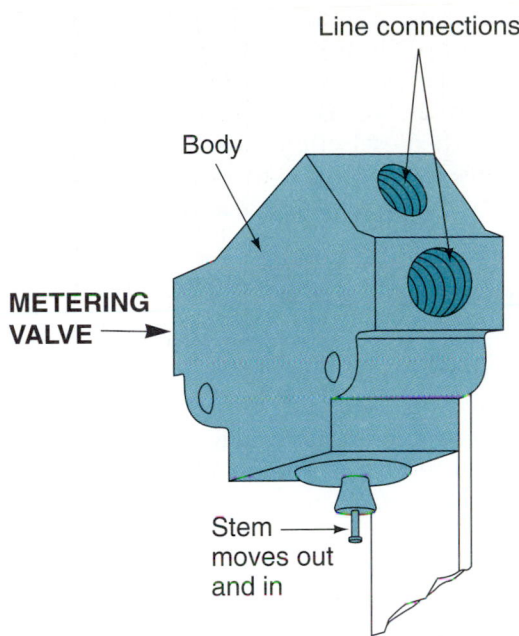

Line connections

Body

METERING VALVE →

Stem moves out and in

Figure 51.35 A metering valve keeps the front disc brakes from operating until rear drum linings are in contact with the brake drum. *(Courtesy of Chrysler Corporation)*

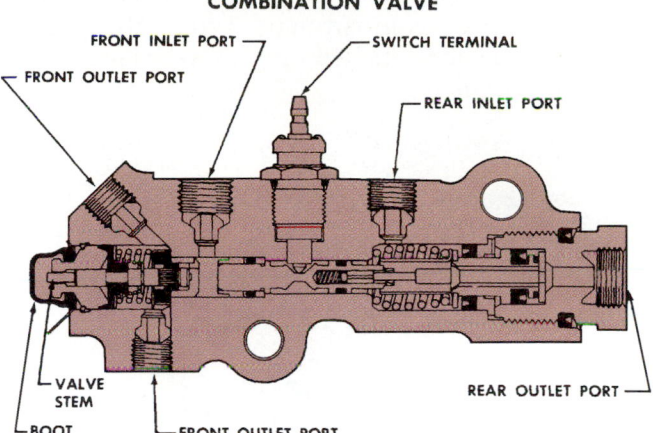

COMBINATION VALVE

FRONT INLET PORT — — SWITCH TERMINAL

FRONT OUTLET PORT — — REAR INLET PORT

VALVE STEM — BOOT — FRONT OUTLET PORT — REAR OUTLET PORT

Figure 51.36 Cutaway of a combination metering and proportioning valve. *(Courtesy of General Motors Corporation, Service Technology Group)*

valve. Keeping the front brakes from applying until the rear brakes start to build pressure serves two purposes:

- It keeps front pads from doing too much of the light braking, preventing them from wearing out too soon.
- It helps prevent dangerous skids that could result on slick surfaces if the front brakes were to apply before the rears.

Without a metering valve, the rear brake cylinders would have to overcome the drum brake return spring pressure before the rear brakes could begin to operate. Front disc brakes do not have return springs so they would be able to start to apply before the rears, even though both sides of the system had equal pressures.

Proportioning Valve. The ability of the brakes to do their job is limited by the grip of the tires to the road surface. During hard stopping, weight shifts forward away from the rear wheels. The proportioning valve helps prevent the rear wheels from locking up during a panic stop or when brakes are applied hard. When pressure reaches a predetermined level, the proportioning valve prevents a further increase in pressure.

Rear wheel lockup is a problem on some trucks and cars. Some proportioning valves are load sensitive. They compensate to control stopping pressure to the rear brakes according to how much weight is on the rear of the vehicle.

Combination Valve. The metering and proportioning valves are usually combined with a hydraulic safety switch in one valve called a combination valve (Figure 51.36). Combination valves can also be just a

metering and proportioning valve or a safety switch and metering valve.

■ POWER BRAKES

The difference between atmospheric pressure and vacuum can be used to apply force (Figure 51.37). Most modern vehicles have *vacuum brake boosters*. The booster is located behind the master cylinder on the firewall (also called the bulkhead), which is the metal wall between the engine and passenger compartments. The pedal apply rod goes into the back of the unit and the master cylinder is bolted to the front of it (Figure 51.38)

The common kind of power brake booster is called the *vacuum suspended power brake*. When the brakes are not in use, vacuum is present on both sides of a diaphragm inside of the booster (Figure 51.39). During a stop, air is drawn into the rear chamber of the booster (Figure 51.40). There is a filter in the back of

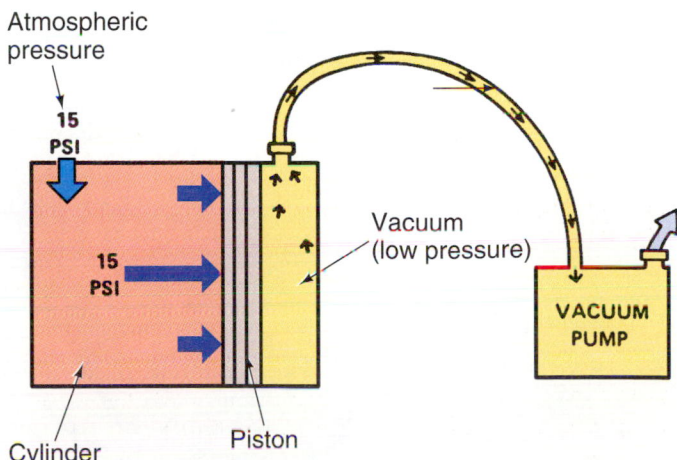

Atmospheric pressure

15 PSI

15 PSI

Vacuum (low pressure)

VACUUM PUMP

Cylinder

Piston

Figure 51.37 The difference between atmospheric pressure and vacuum can be used to apply force. *(Courtesy of Nissan Motors)*

the unit that filters the air that comes into the booster from the driver's compartment. When the pedal is lightly held, a spring-loaded valve prevents more pedal assist from occurring.

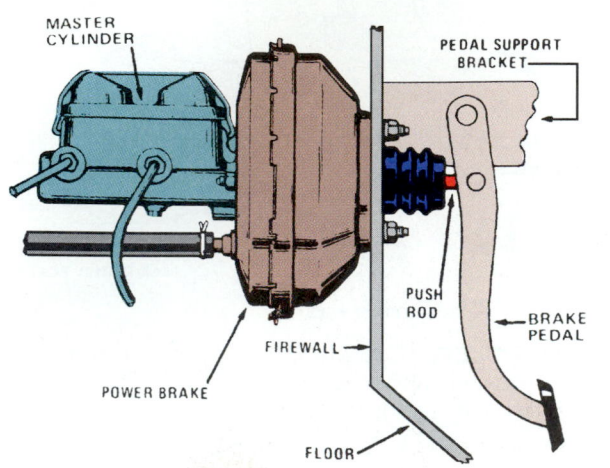

Figure 51.38 A typical power brake combined with a master cylinder. *(Courtesy of AlliedSignal Bendix)*

Some power brake systems are powered off of the power steering pump. These systems, called *hydro-boost brakes* (Figure 51.41), use fluid pressure instead of vacuum to produce the power assist. These are commonly found on cars with diesel engines because diesels do not produce sufficient vacuum to operate a vacuum booster. Hydro-boost systems are also found on some gasoline engine vehicles.

■ PARKING BRAKE

The parking brake, also called the *emergency brake*, must operate independently of the *service brakes* (the normal hydraulic brake system). On *integral-type parking brakes*, a cable-actuated bar applies the drum-type parking brake (Figure 51.42). The cable is connected to either a hand brake or foot pedal in the driver's compartment and to an *equalizer* (Figure 51.43). The equalizer has a cable from each rear wheel attached to both of its sides. Because it pivots in the center, the equalizer applies each rear emergency brake equally.

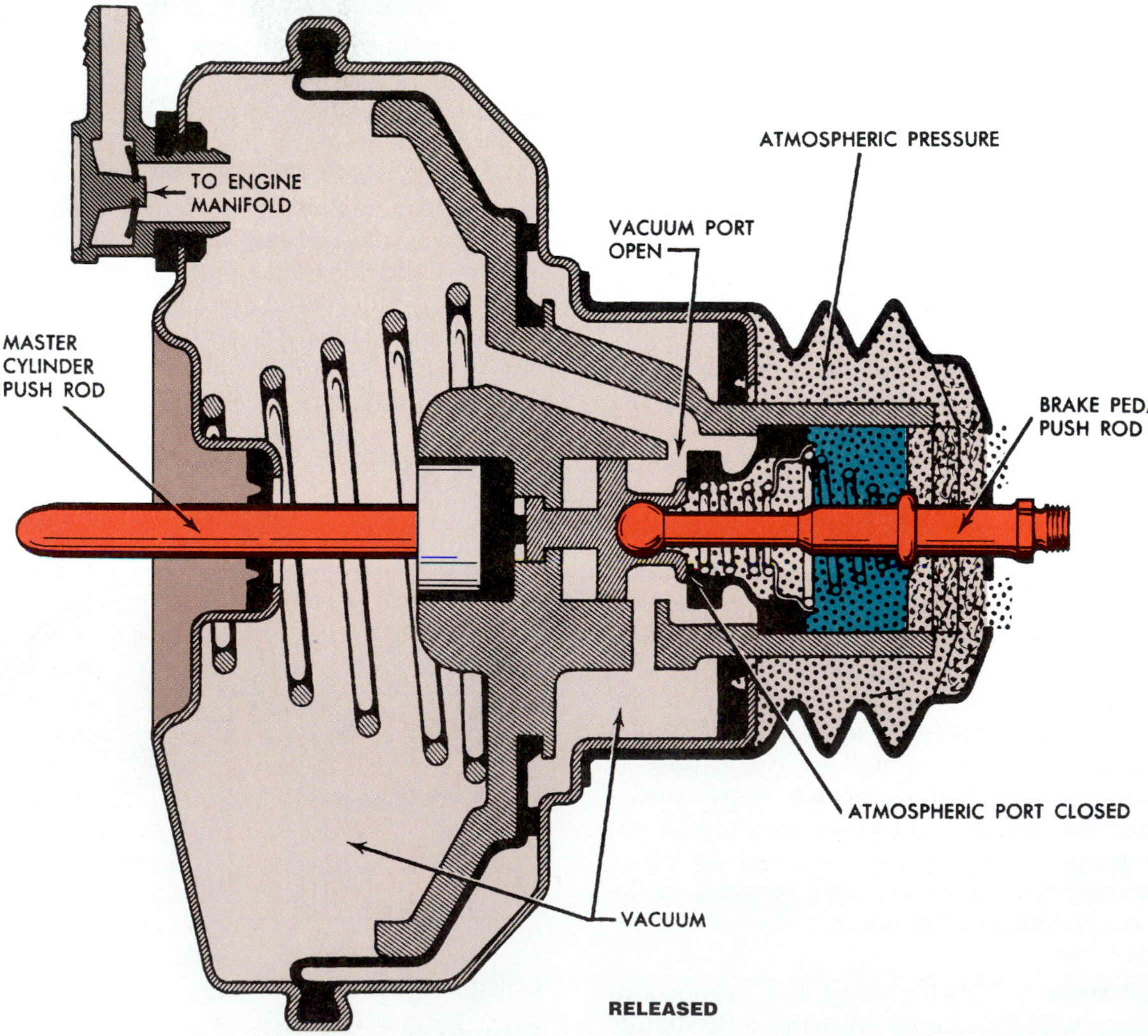

Figure 51.39 A power brake in the release position. *(Courtesy of General Motors Corporation, Service Technology Group)*

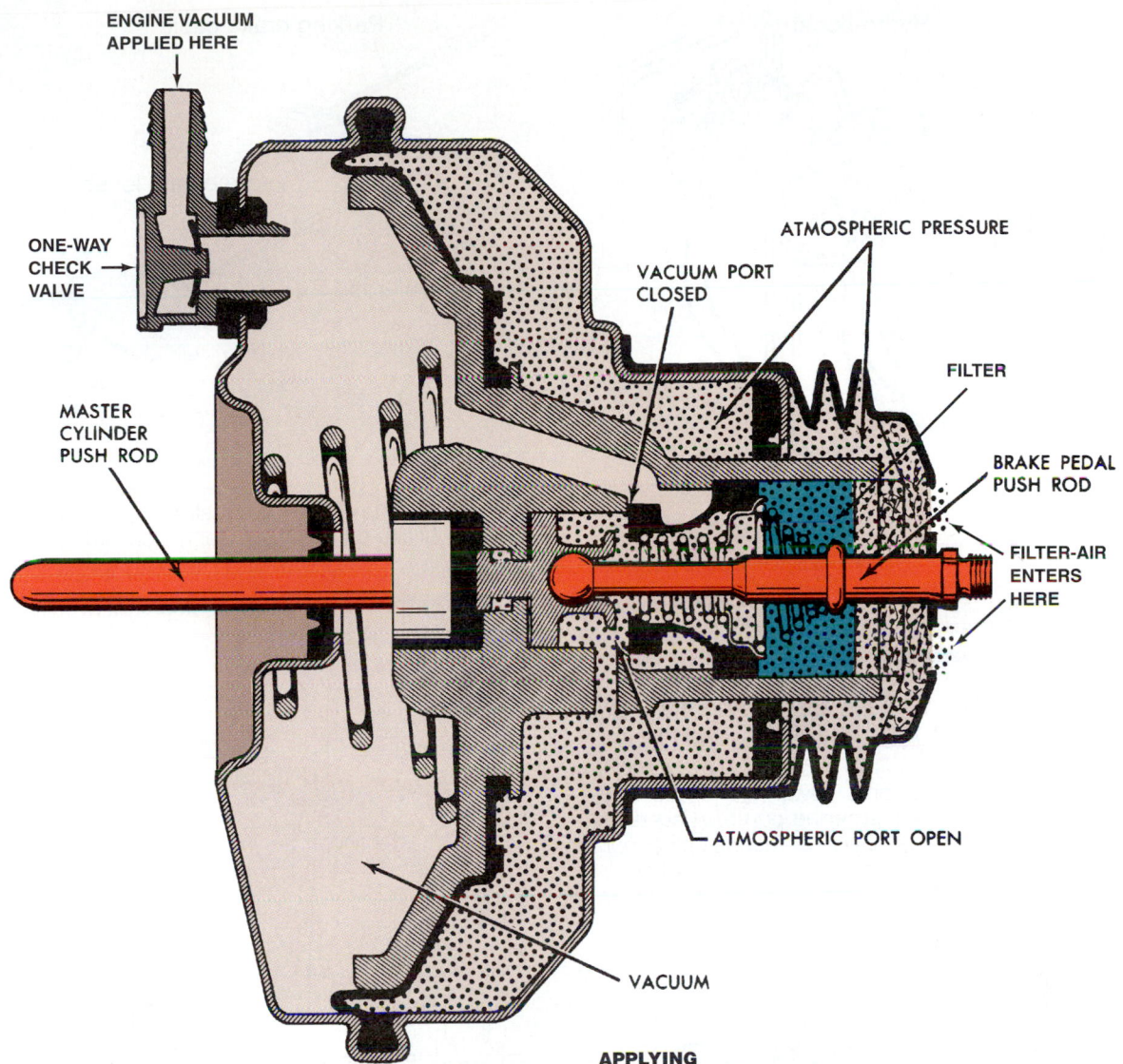

ENGINE VACUUM
APPLIED HERE

ONE-WAY
CHECK
VALVE

MASTER
CYLINDER
PUSH ROD

ATMOSPHERIC PRESSURE

VACUUM PORT
CLOSED

FILTER

BRAKE PEDAL
PUSH ROD

FILTER-AIR
ENTERS
HERE

ATMOSPHERIC PORT OPEN

VACUUM

APPLYING

Figure 51.40 A power brake in the applying position. *(Courtesy of General Motors Corporation, Service Technology Group)*

An *independent-type emergency brake* is found on heavier vehicles and RVs. It is mounted behind the transmission on the front of the driveshaft. It can be either an internal-expanding or an external-contracting type of brake.

■ STOPLIGHT SWITCHES

The stoplights are turned on by a stoplight switch, usually mounted on the brake pedal arm (Figure 51.44). When the pedal is depressed, contacts in the switch are closed to complete the circuit to the lights. Some older cars' stoplight switches are operated by hydraulic pressure.

■ ANTILOCK BRAKES

The ability of the brakes to do their job is limited by the grip of the tires to the road surface. Even with the best quality tires, if the car skids there will be a loss of

stopping ability and control. If the driver could release pressure on the brake pedal just before the wheel locked up, the skid could be avoided.

Most newer cars are equipped with a computerized **antilock brake system (ABS)** to keep the wheels from locking up (Figure 51.45). ABS was first used on B-47 jet bombers in 1947 for better stopping on slick runways. Most large passenger and military planes have ABS today. In the late 1970s, many European luxury cars imported to the United States came equipped with ABS. Antilock brakes have been found on some American cars since 1985 when Ford introduced the Teves system. Other domestic manufacturers soon had ABS.

During normal operation, the ABS acts like conventional brakes. It only functions differently when a wheel locks up. It has electronic sensors at each wheel. Each sensor monitors the speed of the rotation of the wheel by measuring how fast a toothed ring moves

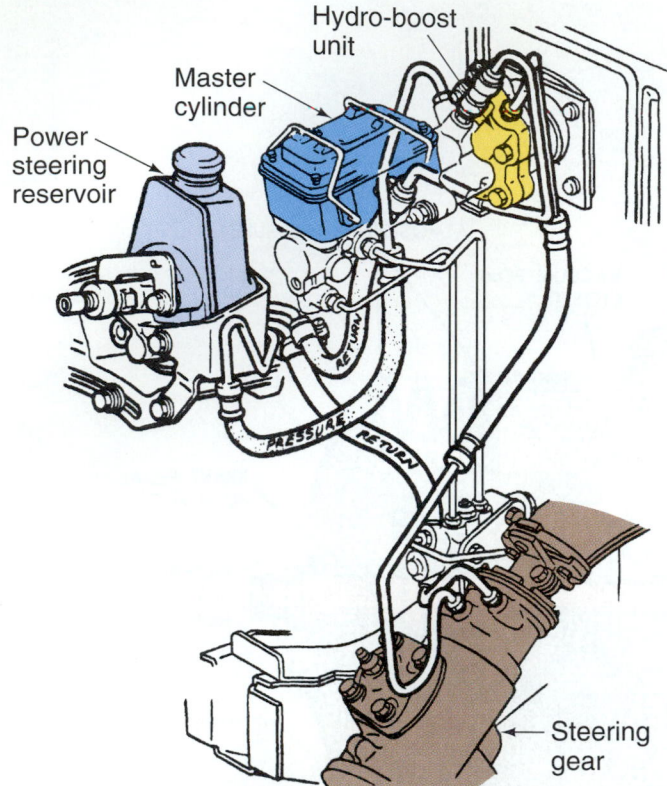

Figure 51.41 A hydro-boost power brake operates off of pressure from the power steering pump. *(Courtesy of General Motors Corporation, Service Technology Group)*

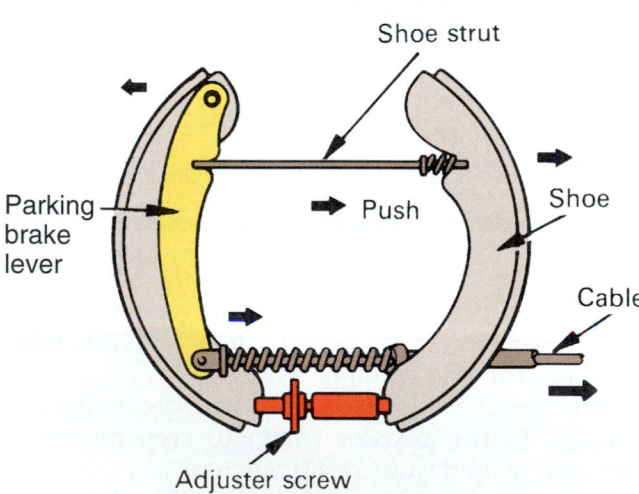

Figure 51.42 A parking brake used with a duo-servo brake.

past the sensor (Figure 51.46). If a wheel locks up, the ABS pulsates (modulates) the pressure to that wheel in much the same manner as a race car driver pumps the brakes during a turn to avoid a skid. ABS can do this pulsation much faster than a human, however. An ABS can pulsate the pressure to the brake system up to 15 times per second.

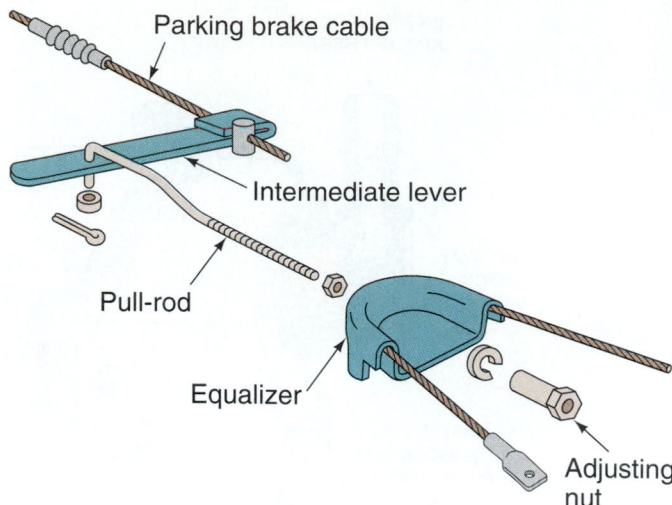

Figure 51.43 An equalizer applies pressure to parking brake cables from each side of the vehicle.

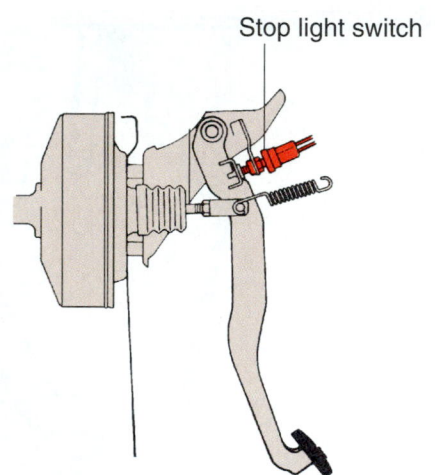

Figure 51.44 A stoplight switch.

NOTE: *At stock car races, a pulsating chirp can be heard coming from ABS-equipped cars as the tires lock and unlock during turns.*

Pedal Feel

An ABS-equipped car has a different pedal feel than a conventional braking system. When the ABS becomes active, a small bump followed by rapid pulsation is felt in the brake pedal. The feel of pulsation is noticed in some systems more than in others. When the road is bumpy, there is loose gravel, or the road is slick, less pedal force is needed to activate the ABS. During normal stops, there is no pedal pulsation.

■ TYPES OF ABS

Several manufacturers produce ABS. Bendix, Bosch, Delco (GM), ITT/Teves, Kelsey-Hayes, and Lucas Girling are some of the companies. Be sure you know which system you are working on before starting any repairs.

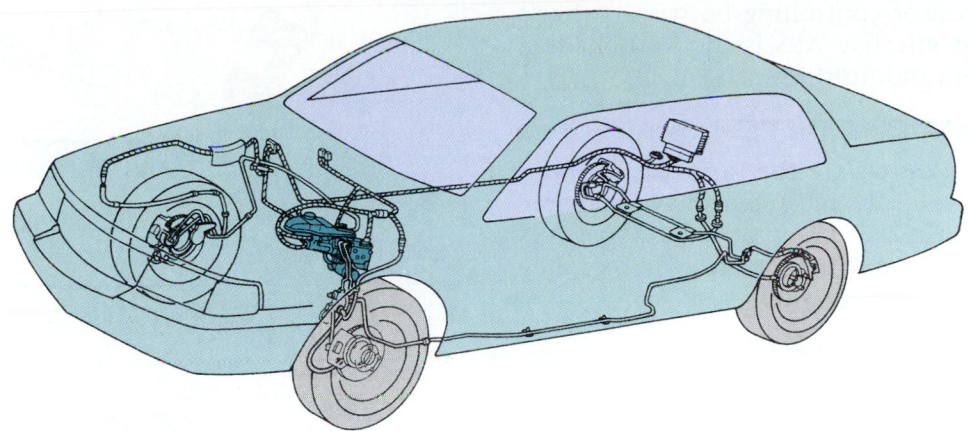

Figure 51.45 An anti-lock brake system. *[Courtesy of Ford Motor Company]*

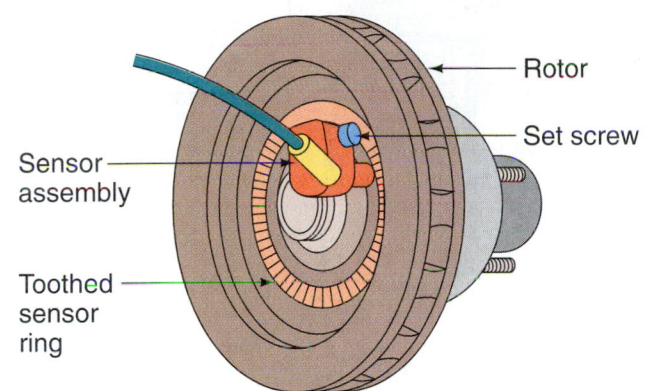

Figure 51.46 An ABS sensor and sensor ring. *[Courtesy of Ford Motor Company]*

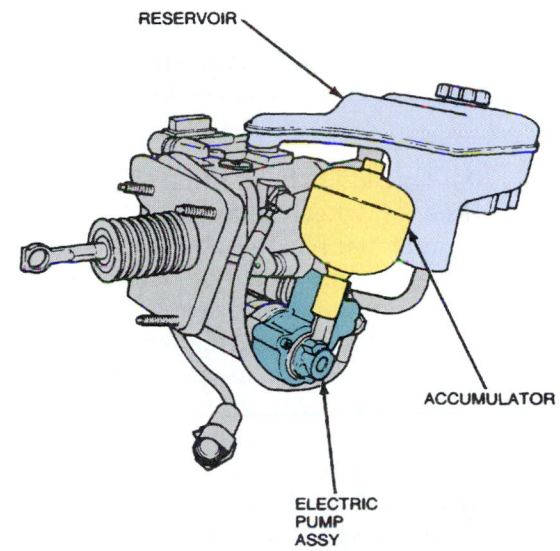

Figure 51.47 An integral ABS. *[Courtesy of Ford Motor Company]*

Most systems are *integral* or *integrated systems* (Figure 51.47). This means that they combine the master cylinder, power brake booster, and ABS hydraulic circuitry in one single hydraulic assembly. *Non-integral* systems have a conventional power brake and master cylinder with the ABS unit separate (Figure 51.48).

The method by which fluid pressure is controlled depends on the design of the system. An ABS can be either a *two-wheel* or *four-wheel* system. It can be a *one-*, *two-*, *three-*, or *four-channel* system.

Two-Wheel Systems

A two-wheel ABS only works on the rear wheels. They are found most often on sport utility vehicles (SUVs) and light trucks. These systems are either one or two channel. To control skids, one-channel systems modulate the rear brakes at the same time. There is a centrally located speed sensor, normally on the top of the differential or in the transmission or four-wheel drive transfer case. Two-channel systems pulsate the pressure independently according to signals received from speed sensors at each rear wheel.

Some manufacturers offer an off/on switch for ABS brakes. It may be desirable to shut off ABS. When dri-

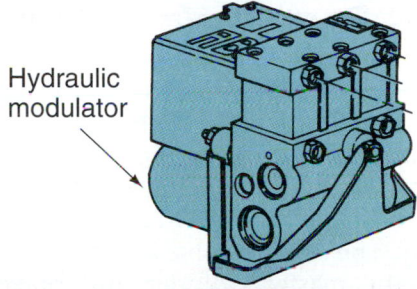

Figure 51.48 A non-integral ABS. *[Courtesy of General Motors Corporation, Service Technology Group]*

ving in snow, for instance, a locked-up wheel builds a wedge of snow in front of it. This aids in stopping.

Four-Wheel ABS

These systems can either be three channel or four channel. A three-channel system has a sensor at each front wheel. The rear is like the single-channel, two-wheel system with one sensor controlling both rear wheels together. The most effective ABS is the four-channel system with a sensor monitoring each wheel.

◾ ABS COMPONENTS

Brake assemblies and lines are the same for ABS and non-ABS vehicles. Non-integral systems have the same master cylinder and power brake booster. Some ABS components are common to all systems, whether they are one-, two-, three-, or four-channel.

Electronic Control Unit

Sensor inputs, such as the wheel speed sensors and brake pedal sensor, provide data to the controller. The controller acts on those inputs to correct differences in wheel speed during a skid. There are solenoid-operated valves in the brake lines. These valves are normally open. When the controller senses a wheel locking up, a current is directed to the solenoid, energizing a magnetic field and closing the valve.

Wheel Sensors

Each wheel that has skid control has a wheel sensor to detect the speed of wheel rotation. Wheel sensors operate in the same manner that the magnetic trigger in the distributors of some electronic ignition systems do. Most sensors are permanent magnet types with a coil of wire wound around a permanent magnet core (Figure 51.49a). The sensor is positioned near a toothed ring that spins with the wheel. When the rotating ring passes the magnetic sensor, changes in the magnetic lines of flux around the sensor cause an alternating current voltage. The frequency of the voltage changes with the speed of wheel rotation (Figure 51.49b). The electronic control unit uses this information to calculate the speed of the wheel and compare it to the speed of the other wheels.

Hydraulic Control Valve Assembly

A hydraulic control valve assembly has mechanical and electrical parts that cause hydraulic pressure to pulsate or modulate. It works when a wheel's speed drops a certain amount below the speed of the wheel on the opposite side of the vehicle. When the wheels are rotating normally, the ABS is bypassed and operates normally. The control valve assembly can be combined with the master cylinder and brake booster (integrated) or it can be separate (non-integrated).

Pump and Motors

An electric pump is usually used as a source of hydraulic pressure because the ABS is independent of the normal braking system. Fluid is pumped under pressure to a relief valve, accumulator, and hydraulic controls.

Accumulator

Many integral ABS have accumulators (see Figure 51.47). An accumulator stores brake fluid under very high pres-

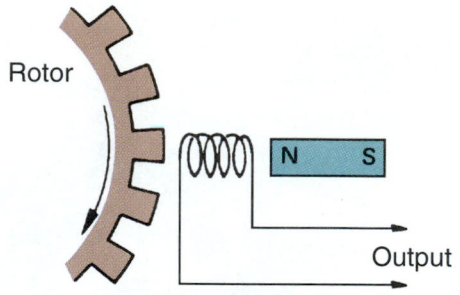

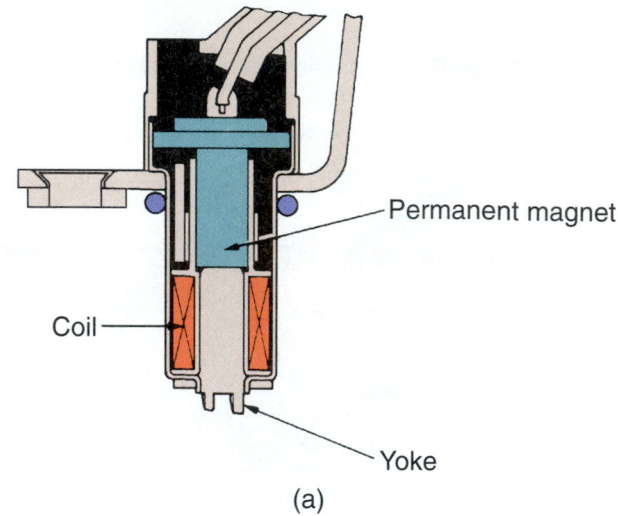

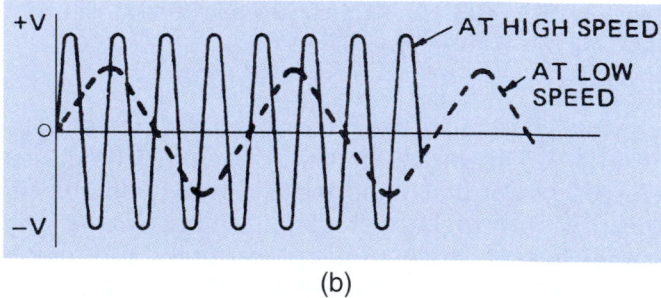

Figure 51.49 (a) Most sensors are a coil of wire wrapped around a permanent magnet core. (b) The frequency of the voltage changes with the speed of wheel rotation.

sure. This keeps a ready source of constant pressure to operate the ABS when needed. The accumulator is filled with a charge of high pressure nitrogen gas.

CAUTION Because of the high pressure, extra care is necessary when servicing these systems. Up to 2000 psi can be available at the wheels even when the engine is off. Pressure this high is about ten times as high as shop air pressure. It can force brake fluid to go right through your hand.

Some non-integral systems, such as the Delco ABS-VI, do not use an accumulator or single electric pump. These systems use a motor pack instead (Figure 51.50). Small screw-drive electric motors increase and decrease fluid pressure in each wheel circuit. An electromagnetic brake in the pump provides precise control. It stops the pump immediately as it alternately raises or decreases pressure. System pressure in this ABS is close to normal brake system pressure.

Reservoir

The brake system reservoir is usually much larger than a normal brake system reservoir. Some systems use a second reservoir. This reservoir, which acts like a holding tank, collects pressurized fluid as it returns from the brakes at the wheels. Its place in the hydraulic circuit is before the pressure pump.

Pump Control Switches

Systems with a pump have a switch to turn the pump on and off. Some systems have a pressure sensitive switch that monitors accumulator pressure, turning on the pump when pressure drops to a certain level. If pressure drops below a certain level, an amber ABS warning light on the dash will come on. This light is in addition to the red brake hydraulic system warning light.

Some systems, Fords in particular, have an antilock brake pedal switch. The switch is normally closed until the pedal travel exceeds a certain point. When this point is reached, extra pedal pressure has been called for. The switch opens, turning on the pump motor. The pressure produced by the pump pressurizes the system, pushing the brake pedal up far enough to close the switch once again.

ABS OPERATION

Signals from each of the wheel sensors are pulsed to the *ABS control module*. When a wheel starts to lock up, the signal from that wheel's sensor drops off rapidly. The control module will shut off hydraulic fluid pressure to that wheel to prevent lockup. If the frequency of the signal continues to drop off, pressure will be *released* to that wheel. As soon as the wheel starts to spin freely again, pressure is reapplied to stop the vehicle. If the wheel locks again, the cycle repeats itself up to 15 times a second.

In most ABS, solenoid valves control the holding and releasing of hydraulic system pressure (Figure 51.51). Some systems have separate solenoid valves for holding and releasing pressure and others have combination valves. In separate valve systems, a *hold*, or inlet, solenoid controls fluid flow to the wheel cylinders. During a lockup stop, extra pedal pressure is prevented from acting on the wheel cylinder. When action by the hold solenoid is not sufficient to prevent wheel lockup, a *release*, or outlet, solenoid bleeds pressure back to the reservoir.

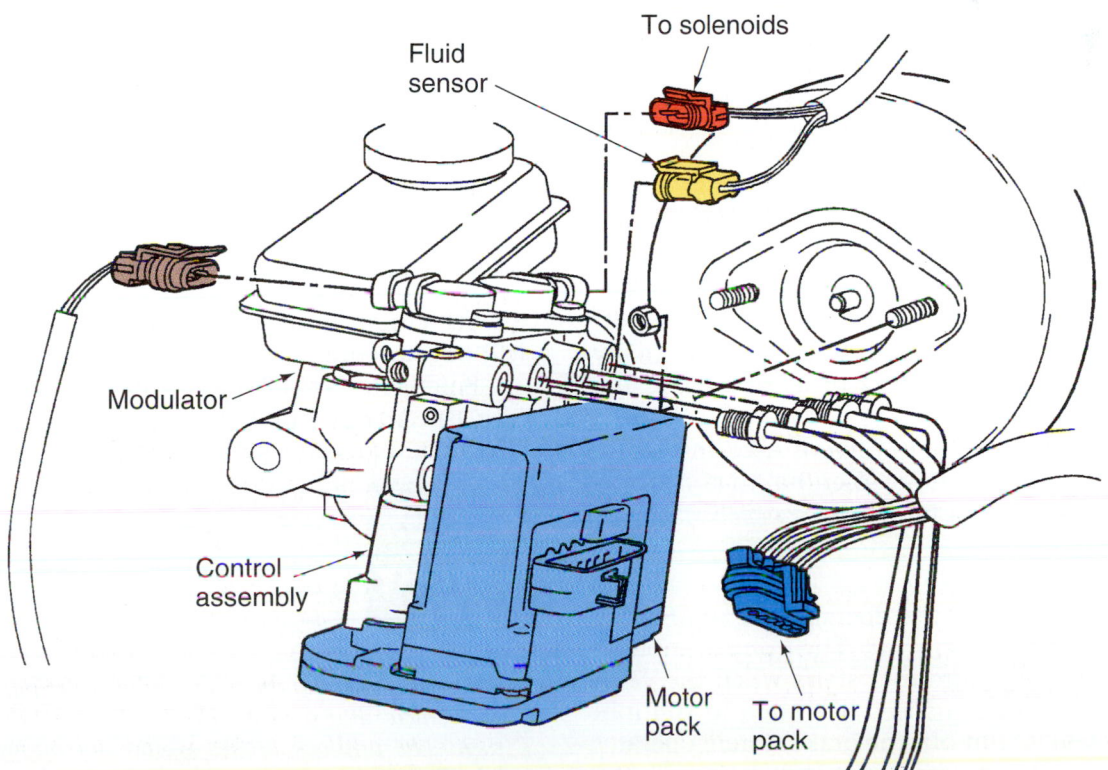

Figure 51.50   An ABS motor pack. *(Courtesy of General Motors Corporation, Service Technology Group)*

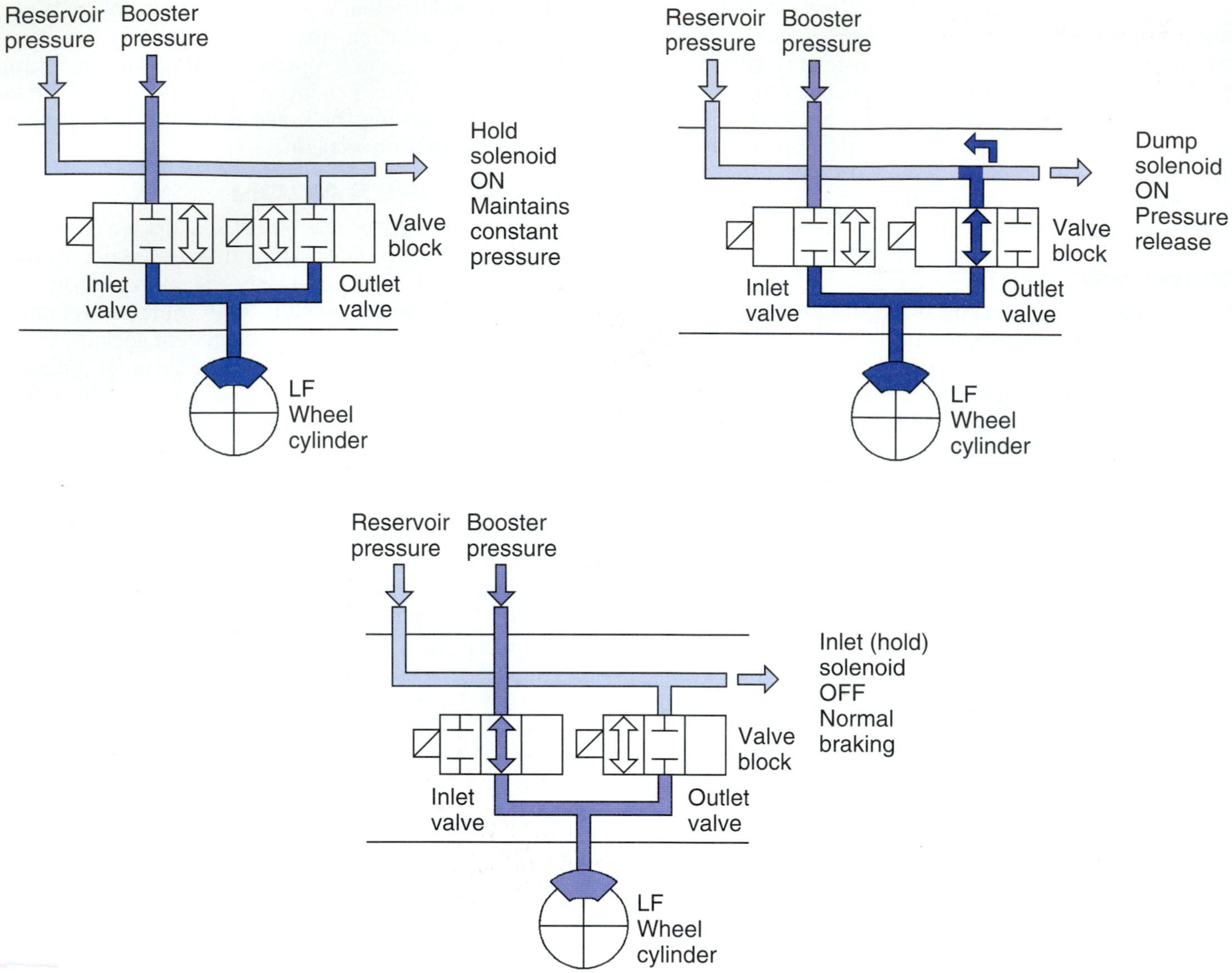

Reservoir pressure   Booster pressure

Hold solenoid ON Maintains constant pressure

Valve block

Inlet valve   Outlet valve

LF Wheel cylinder

Reservoir pressure   Booster pressure

Dump solenoid ON Pressure release

Valve block

Inlet valve   Outlet valve

LF Wheel cylinder

Reservoir pressure   Booster pressure

Inlet (hold) solenoid OFF Normal braking

Valve block

Inlet valve   Outlet valve

LF Wheel cylinder

Figure 51.51 Solenoid valves control the holding and releasing of hydraulic system pressure. *(Courtesy of General Motors Corporation, Service Technology Group)*

The hold solenoid is normally open, allowing normal brake function. The release solenoid is normally closed, which prevents fluid from leaking back to the fluid reservoir.

In some systems, a single valve is used instead of two. It has a three-position solenoid that controls hold, release, and normal fluid flow.

ABS Safeguards

If a malfunction occurs in the operation of an ABS, the computer shuts the system off. The control module has a diagnostic procedure that starts when the vehicle starts and finishes at a speed of from 4 to 10 mph. When the system is shut off, the brake system operates as a normal system without antilock brakes. If an emergency spare is used, for instance, the differences

in wheel rotation speed on that axle is sensed. The ABS will not function.

Emergency vehicles are equipped with a switch to turn the ABS on and off. A skid is sometimes desired when attempting to get a car to spin.

CASE HISTORY *A highway patrol officer received notice of a burglary in progress. As he sped toward the business address, the suspect's speeding vehicle passed him going in the opposite direction on the street. From habit, the officer slammed on the brakes while attempting to throw his vehicle into a skid in order to make a quick 180° turn. The antilock brakes would not allow the inside wheel to stop turning, and he lost control of the vehicle.*

■ REVIEW QUESTIONS

1. If it takes 50 pounds of force to drag one material across another one, what is the coefficient of friction?

2. On disc brakes, the friction linings are called _____ .

3. _____'s law states that "pressure in an enclosed system is equal and undiminished in all directions."

4. What is the DOT number for synthetic brake fluid?

5. What is the term for a material that absorbs water?

6. What material is part of brake fluid that is also part of automotive coolant?

7. When the leading shoe on a drum brake is forced into the brake drum as it rotates, this is called _____ action or self-energization.

8. The two main kinds of drum brake designs are the duo-servo and the leading-_____ shoe.

9. What do most power brake booster diaphragms use to provide an assist to pedal effort?

10. What is the name of the ABS part that is a magnetic pickup positioned near a toothed ring that spins with the wheel?

■ ASE STYLE REVIEW QUESTIONS

1. Technician A says that silicone brake fluid has a "spongy" pedal feel. Technician B says that with silicone brake fluid, the pedal will have more travel before the brakes are fully applied. Who is right?
   - **a.** Technician A
   - **b.** Technician B
   - **c.** Both A and B
   - **d.** Neither A nor B

2. Technician A says that when drum brakes fade, brake pedal travel increases during a stop. Technician B says that the coefficient of friction becomes lower during brake fade. Who is right?
   - **a.** Technician A
   - **b.** Technician B
   - **c.** Both A and B
   - **d.** Neither A nor B

3. Technician A says that disc brakes do not require return springs. Technician B says that front disc brakes are self adjusting. Who is right?
   - **a.** Technician A
   - **b.** Technician B
   - **c.** Both A and B
   - **d.** Neither A nor B

4. The brake pedal on a disc/drum car becomes higher on the second application. Technician A says that the rear brakes might be in need of adjustment. Technician B says that the front brakes might be in need of adjustment. Who is right?
   - **a.** Technician A
   - **b.** Technician B
   - **c.** Both A and B
   - **d.** Neither A nor B

5. Technician A says that drum brakes would usually be found on the front of a vehicle. Technician B says that pressure from the master cylinder is increased if the diameter of its bore is smaller. Who is right?
   - **a.** Technician A
   - **b.** Technician B
   - **c.** Both A and B
   - **d.** Neither A nor B

6. Technician A says pressure equals force divided by area. Technician B says the lips on a brake cylinder cup face toward the fluid. Who is right?
   - **a.** Technician A
   - **b.** Technician B
   - **c.** Both A and B
   - **d.** Neither A nor B

7. Technician A says when a second application of the brake pedal produces a higher pedal, the brakes need bleeding. Technician B says when a second application of the brake pedal produces a higher pedal, the brake pads need to be replaced. Who is right?
   - **a.** Technician A
   - **b.** Technician B
   - **c.** Both A and B
   - **d.** Neither A nor B

8. Technician A says the fixed caliper design has a piston or pistons on only one side. Technician B says the floating caliper design must be able to slide during and after the brakes are applied. Who is right?
   - **a.** Technician A
   - **b.** Technician B
   - **c.** Both A and B
   - **d.** Neither A nor B

9. Technician A says that tandem brake systems can be split front to rear. Technician B says that tandem brake systems can be split diagonally. Who is right?
   - **a.** Technician A
   - **b.** Technician B
   - **c.** Both A and B
   - **d.** Neither A nor B

10. Technician A says that linings are bonded to the metal brake backing. Technician B says that linings are riveted to the metal brake backing. Who is right?
    - **a.** Technician A
    - **b.** Technician B
    - **c.** Both A and B
    - **d.** Neither A nor B

Brake Service

■ OBJECTIVES

Upon completion of this chapter, you should be able to:

✔ Inspect brake systems and recommend needed repairs.

✔ Diagnose brake system problems.

✔ Perform brake repairs and adjustments using the correct materials and procedures.

■ BRAKE INSPECTION

A thorough inspection of the braking system should be performed before attempting any repairs. Start with the pedal and master cylinder.

■ CHECK BRAKE PEDAL FEEL

Apply the foot brakes and check the travel of the brake pedal. There should be an ample amount of *pedal reserve* (above the floor) after the brakes begin to stop the car. When there is no pedal, this is a serious problem indicating failure in both halves of the brake system (see Chapter 51). A low pedal can indicate a leak in one-half of the system. Check the fluid level first. The leak can be either external or internal (no fluid leaking out).

The pedal should feel firm, not "**spongy**." A spongy pedal indicates air in the system, calling for a brake bleed.

NOTE: *Cars with ABS brakes will have a feeling of pedal pulsation during hard stops. Some of these systems will also experience a rise in pedal height during hard stops.*

■ MASTER CYLINDER INSPECTION

Hold the pedal down very lightly. It should not fade away toward the floor. If the pedal drifts toward the floor and there is no visible sign of fluid leakage anywhere in the system, the master cylinder is leaking internally past the primary cup. Usually, pushing harder on the pedal will result in the brakes working properly as the seal distorts against the wall of the cylinder, cutting off the internal leak.

Check the level of the fluid in both halves of the master cylinder reservoir. Disc brake lining wear will result in a lower level of fluid. This is because the pistons move out in their bores to compensate for the lin-ing wear. If one of the chambers is larger and it is lower in fluid than the other one, the disc linings served by that chamber are probably worn.

If a secondary piston seal is defective, fluid can be pumped from one of the fluid chambers into the other. The result is that one chamber is low, while the other is overflowing.

Look for external leaks from the master cylinder. Leakage can occur past the master cylinder cover when the metal bails are bent and loose. Fluid can also escape past the secondary cup on the rear piston. If there is no power assist, the fluid will be visible on the carpet or on the inside of the firewall or bulkhead. If there is a power booster, fluid will leak from the front of the power booster. A small amount of seepage is acceptable.

Check the Compensating Port

Fluid movement past the compensating port should always be visible when the pedal is applied (Figure 52.1). It is a very small opening and can become plugged. The result is brakes that do not release all the way after the fluid absorbs heat resulting from braking action. This can cause a pedal that is high and very hard. As the driver attempts to get the vehicle to move after a stop, the brakes might continue to hold.

Sometimes, brake pedal travel is adjusted improperly. A lack of pedal free play can cause the primary cup to be positioned past the compensating port even when the brakes are released. This will cause the same symptoms as a plugged compensating port.

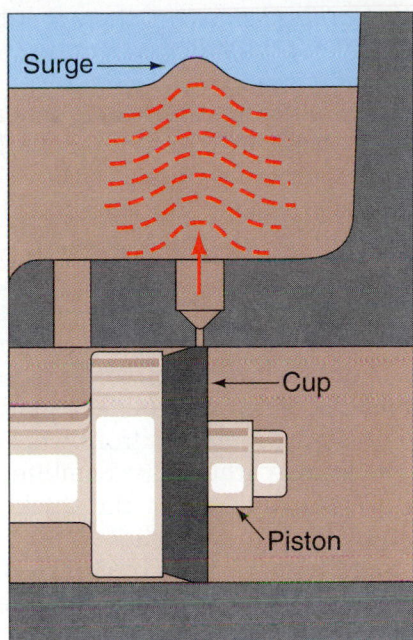

Figure 52.1 Fluid movement past the compensating port should always be visible when the pedal is applied. *(Courtesy of EIS Brake Parts)*

Brake drag is the term given when the brakes stay on after a stop.

Checking for Air in the System

To check for air in the system, have a helper apply the brakes about ten times, holding it the last time. With the cover removed from the master cylinder, watch the movement of fluid while the pedal is released. A heavy surge from the compensating port (Figure 52.2) indicates air in the system. Note which chamber has the high fluid surge. This is the side of the system that has the trapped air.

NOTE: *Be sure fender covers are installed on the fender. Splashed brake fluid can ruin paint. Wear eye protection.*

Power Brake Checks

A power brake booster can experience problems such as a hole in a diaphragm or a sticking valve. To test the operation of a power brake booster, exhaust all vacuum reserve from the power booster by applying the foot pedal several times with the engine off. The sound of the air rushing into the reservoir should be heard. Next, hold your foot on the pedal while starting the engine. If the power booster is operating correctly, the pedal will move about an inch closer to the floor after the engine starts.

Stoplight Switch

Apply the brakes and test the stoplights. If they do not come on, check to see if the key must be on first. Then,

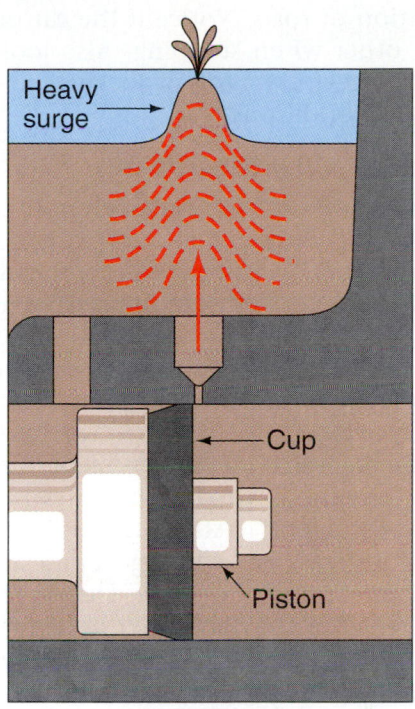

Figure 52.2 A heavy surge from the compensating port indicates air in the system. *(Courtesy of EIS Brake Parts)*

check the fuse. There are two wires to the stoplight switch. Remove the connector from the switch and use a jumper wire to connect them together. If the stoplights work with the wires hooked together, the stoplight switch is faulty.

Hydraulic Safety Switch Service

In some early systems from the sixties, the bulb indicating a hydraulic system failure would light and remain on until it was re-centered by a technician after repair. On modern systems, the light only operates during braking. Some systems share the same bulb to indicate when the parking brake is applied. Testing the bulb in these systems can be done with the parking brake applied and the key turned on. Low fluid level is also indicated by the same bulb.

Chassis Problems

Problems related to the chassis can cause symptoms felt in the brakes, such as brake pull. Chassis parts involved include loose wheel bearings, worn ball joints or bushings, and worn steering linkage parts.

Road Test

Before a road test, be sure there is fluid in both reservoir sections of the master cylinder. Check to see that pedal height is normal. If not, do not road test the car. During the road test, stop the car on a deserted, straight section of road. Notice if the car pulls to one side or the other when stopping. Also look for pedal pulsation, a wheel grabbing or locking up, noises, or other unusual conditions.

■ BRAKE DIAGNOSIS

Brake pedal pulsation results when hydraulic pistons are moving during a stop. This can be caused from rear drum brakes that have become out-of-round. This happens when overheated brakes have the parking brake set firmly against them.

Pedal pulsation also results when disc brake rotors have become overheated. Sometimes rotors that are only slightly warped can be corrected by machining them on a brake lathe, but the problem often returns.

Brakes sometimes **grab**. This means that they apply quickly and tend to stick on. This happens when there is oil or grease on a lining. It can also happen when metallic linings are used. Their coefficient of friction increases as they become hot (Figure 52.3).

Extra braking effort can be required when a power booster is not operating properly or when the wrong linings have been installed.

Brake pull happens due to several reasons.
■ When there is a difference in friction between sides of a vehicle it will pull.
■ Linings can have different friction due to oil or grease.
■ A hose can be restricted or a caliper can be frozen.
■ Changes in wheel alignment resulting from loose parts can also cause pull.

NOTE: *A low tire will probably pull all of the time, not just during stopping.*

When the *warning light* is on, there is either a problem in the brake hydraulic system or a problem ground in the switch's electrical circuit.

Noises during stopping can include squeaks, metal-to-metal sounds from excessive wear, rattles due to loose parts, and rubbing from a distorted backing plate. Noise usually results from vibration. This is covered in more detail later in this chapter.

■ UNDER CAR CHECKS

With the car in the air, visible checks can be made of the brake system. Disc brakes are inspected to see that a sufficient amount of lining material remains. Usually it is possible to simply remove a front wheel and visually inspect the lining without disassembling the brake caliper. You can usually inspect the thickness of the linings through a hole in the caliper.

Linings that are worn to approximately the same thickness as the metal shoe back should be replaced. After a complete brake inspection, the customer should be informed of the need for brake work. Write on the repair order that the inspection was performed. A service writer will usually inform the customer by telephone in case the customer wants to have the work done while the car is in the shop for the day.

Inspect the Rotor

Inspect the rotor. Scoring and unusual wear will be visible as the rotor is turned. Feel it with your hand.

CAUTION If a car has been recently driven, the rotor will hold heat for some time. You could be burned when you touch a hot rotor.

Disc brake warpage can be checked with a dial indicator (Figure 52.4). A quick check can be made by rotating the rotor by hand while holding a screwdriver or pencil near it as a reference point.

Inspect the Caliper

There should be no signs of leakage at the caliper piston seals. Check the dust boots to see if they are torn. Damaged dust boots will allow in moisture that will ruin the caliper and the brake fluid.

Inspect Wheel Seals

Look at backing plates to see that there is no sign of a fluid leak at the axles or grease leaking from the wheel bearings.

Parking Brake Inspection

Apply the parking brake, checking to see that it engages before it is at half travel. Check for damage or rust on the cable.

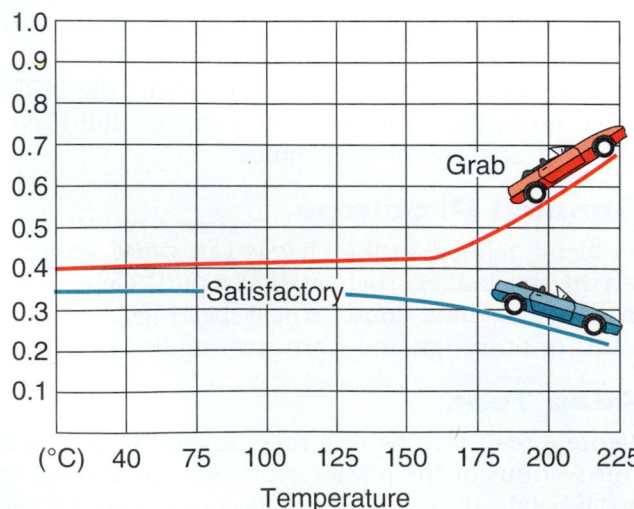

Coefficient of friction

Figure 52.3 Linings of various materials perform differently with changes in temperature. *(Courtesy of Ford Motor Company)*

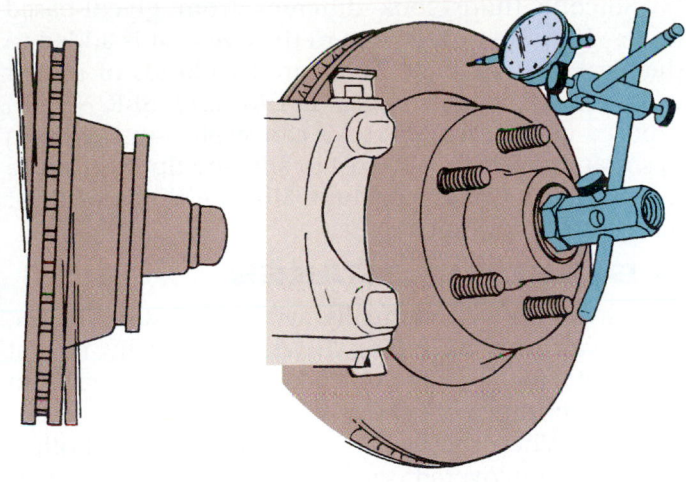

Figure 52.4 Checking disc brake warpage with a dial indicator. *(Courtesy of EIS Brake Parts)*

Inspect Tubing and Hoses

Look for physical damage to hoses, such as cracking, swelling, or softening from exposure to oil (Figure 52.5). Woven hoses sometimes experience internal failure, especially when a careless technician allows a disk brake caliper to hang from a hose during service to brakes, suspension, or wheel bearings. Even though a hose appears to be good from the outside, the inside of the hose can be collapsed trapping fluid inside the front brake and causing excessive wear or brake pull to one side. Excessive brake pad wear results when the brakes do not fully release.

A partially plugged brake hose will cause the car to pull at the beginning of a stop. When the pressures equalize, the pull goes away.

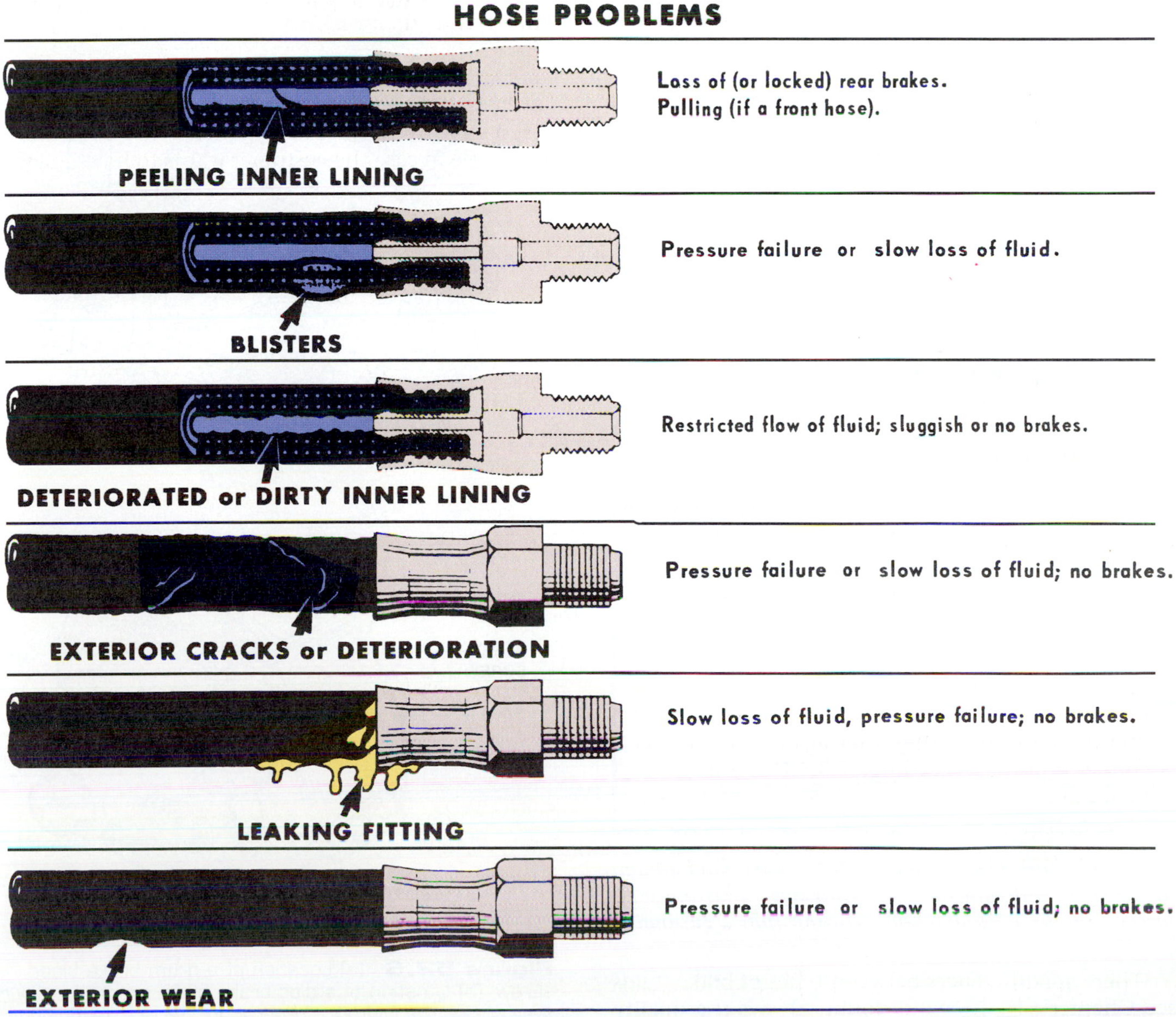

HOSE PROBLEMS

PEELING INNER LINING — Loss of (or locked) rear brakes. Pulling (if a front hose).

BLISTERS — Pressure failure or slow loss of fluid.

DETERIORATED or DIRTY INNER LINING — Restricted flow of fluid; sluggish or no brakes.

EXTERIOR CRACKS or DETERIORATION — Pressure failure or slow loss of fluid; no brakes.

LEAKING FITTING — Slow loss of fluid, pressure failure; no brakes.

EXTERIOR WEAR — Pressure failure or slow loss of fluid; no brakes.

Figure 52.5 Various types of hose damage. *(Courtesy of EIS Brake Parts)*

Checking for Fluid Leaks

When the master cylinder fluid level is low, look for an external leak. If the seal at the rear of the master cylinder bore (secondary cup) leaks, fluid can escape. When a wheel cylinder is leaking, the insides of the tires will often be streaked with fluid and dirt. Occasionally the outside of the tire can even show signs of fluid leakage. A wet rear tire can be because of a leaking axle seal too.

> **SHOP TIP** To tell the difference between an axle leak (gear oil) and brake fluid leak, try washing the fluid off with water. Brake fluid is water soluble and gear oil is not. Also, brake fluid causes paint damage. The paint on the brake backing plate can become bubbled, indicating a brake fluid leak.

■ BRAKE FLUID SERVICE

Periodic fluid changing and flushing is recommended. Also, after a brake job the system will need to be purged of air that becomes trapped. Water that is absorbed into the hydraulic fluid corrodes the inside of the brake system. To avoid moisture contamination, *brake system flushing* is done whenever a brake job is performed and sometimes more often.

Brake linings usually wear out fast enough in average driving so that brake fluid will remain in good condition between brake jobs. Brake manufacturers recommend changing the brake fluid every two years or 30,000 miles on a preventive maintenance basis. With ABS more frequent changes (as often as once a year) are sometimes recommended.

Most of today's brake systems are bimetal. There are aluminum master and wheel cylinders and pistons and steel lines. Moisture in this environment results in galvanic (battery) action and corrosion. Most manufacturers recommend a flush at least every two years.

> **SHOP TIP** A voltmeter can be used to measure the conductivity of the brake fluid in the master cylinder reservoir. Place one meter probe in the fluid and hold the other against the master cylinder housing. A reading of 0.3 DC volts is the maximum.

NOTE: *Use care when handling glycol brake fluid around car paint. It will ruin the paint. Silicone fluids are non-corrosive so an accidental spill will not ruin a customer's paint.*

There are differences between types of brake fluids (see Chapter 51). To avoid doubt, always use quality brake fluid of a type specified by the manufacturer.

Silicone fluids look different from glycol-based fluids. They are purple, due to the dye that is added to the clear silicone base. There are two kinds of rubber used for brake part seals, EPDM and SBR. When exposed to silicone, SBR seals can swell, causing them to soften or to leak. The cost of silicone fluid is four to five times that of glycol fluid. Silicone fluids will not mix with glycol fluids.

■ BLEEDING BRAKES

Brakes can be bled manually, with a pressure bleeder, or by vacuum. **Bleeding brakes** manually is best done with two people. Air and fluid are released from the system from a bleed screw on the back of the wheel cylinder. When the bleed screw is loosened, a hollow opening is uncovered (see Figure 51.18). Fluid escapes out of the center of the screw. Figure 52.6a shows a typical location of a bleed screw for a drum brake. Figure 52.6b shows a typical location for a disc brake

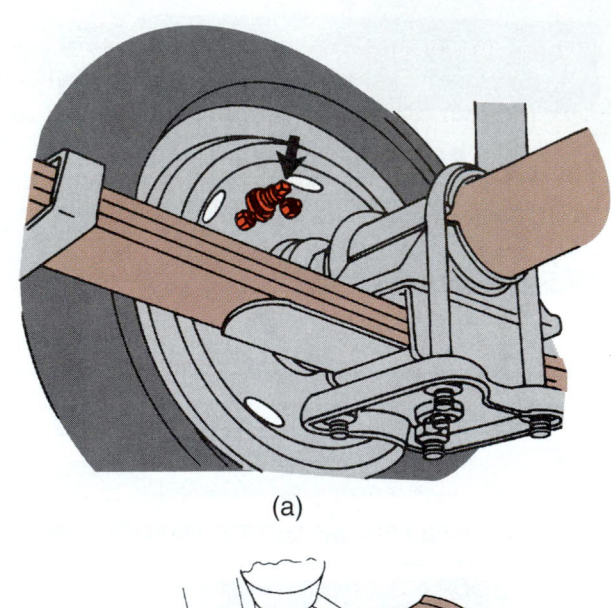

(a)

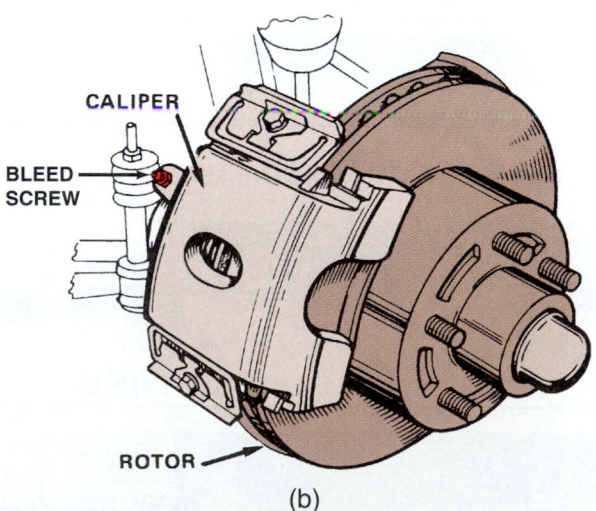

(b)

Figure 52.6 (a) Location of a drum brake bleed screw. (b) Location of a disc brake bleed screw. *(a, Courtesy of General Motors Corporation, Service Technology Group; b, Courtesy of AlliedSignal Bendix)*

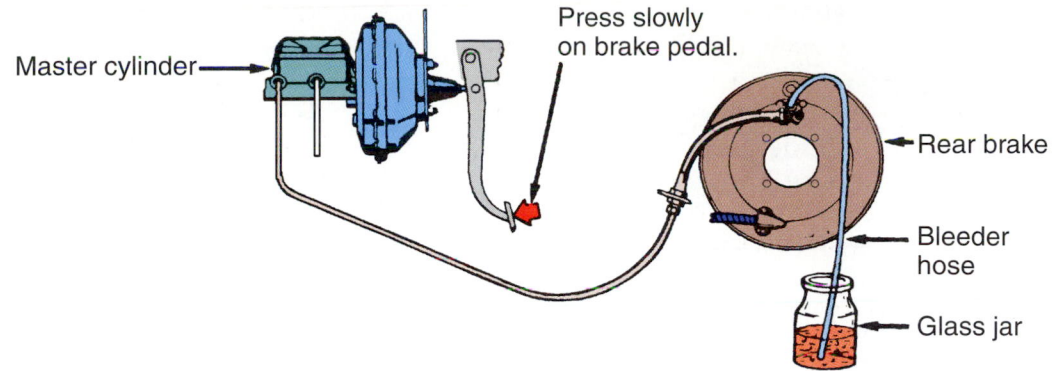

Figure 52.7 Manual brake bleeding procedure. *(Courtesy of AlliedSignal Bendix)*

bleed screw. Notice the bleed screw is at the top of the caliper. Disc brake calipers are different on the left and right sides of the vehicle.

NOTE: When a caliper is installed on the wrong side of the car, the bleed screw will be on the bottom and it will be difficult to purge air from the caliper.

There are different bleeding sequences, depending on the design of the brake system. Be sure to check the service manual for the correct procedure. Some import calipers have two or three bleed screws per caliper. These must be bled at the same time, with the lowest closed first and the highest closed last.

A typical procedure for bleeding brakes is:
- With the bleed screw loosened about one turn, depress the brake pedal to the floor.
- Close the bleed screw.
- Have your helper release the brake pedal.
- Repeat this procedure until all air is gone from the fluid coming out of the bleed screw.
- Be sure to check the master cylinder fluid level frequently and replenish it as it fluid is used.

One of the methods recommended by manufacturers includes installing a hose from the open bleed screw into a jar with brake fluid in it (Figure 52.7). Depress and release the pedal several times. When there are no more bubbles showing in the jar, bleeding is complete.

Most front wheel drive vehicles have diagonally split hydraulic systems. These require a special bleeding sequence, described in the service manual. Without the correct sequence the pedal may never get a solid feel.

A **pressure bleeder** is a canister with two chambers separated by a rubber diaphragm. There is an air chamber and a fluid chamber (Figure 52.8). The air chamber is filled with compressed air, which forces the diaphragm to push against the fluid. Compressed air is full of moisture, which is bad for brake fluid so the diaphragm keeps the two separated. The pressurized fluid is directed to an adapter on the top of the master cylinder (Figure 52.9). Opening a bleed screw releases fluid and air to bleed the brakes.

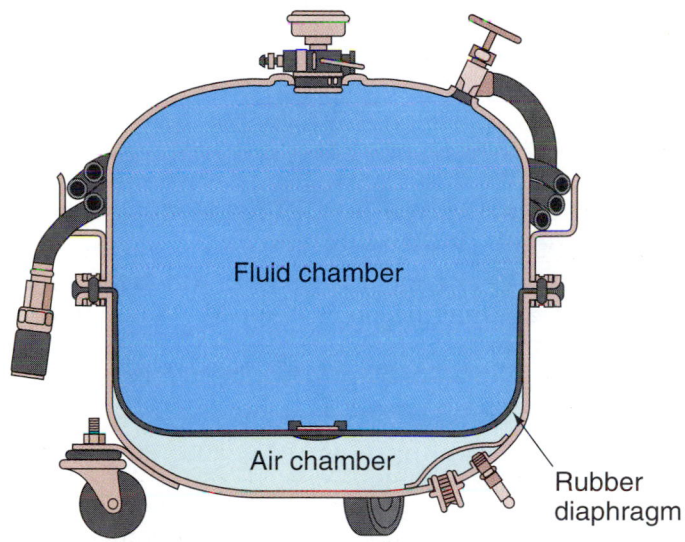

Figure 52.8 A pressure brake bleeder. *(Courtesy of EIS Brake Parts)*

Air in the pressure chamber should be 15–20 psi. It is important that the adapter be securely fastened to the top of the master cylinder. First, fill the master cylinder to within ½" of the top. On diagonal split systems, the master cylinder must be bled first. The manufacturer's manual will list the correct sequence in which to bleed the wheel cylinders. A typical sequence has the wheels furthest from the master cylinder bled first. With a hose over the bleed screw leading into a clear container, open the bleed screw. Allow fluid to escape until it is completely clean and has no air bubbles.

When bleeding brakes with a pressure bleeder, the metering valve must be disabled so it will not prevent fluid from reaching the wheel cylinders. When manually bleeding brakes using the brake pedal, the metering valve can be opened with pedal pressure.

Another way of bleeding brakes is to use a *vacuum bleeder*, or *suction bleeder*, to pull fluid through the system. One system uses a tank that has a hookup for compressed air (Figure 52.10). The air creates a vacuum as it moves past an opening in the tank. A hose from

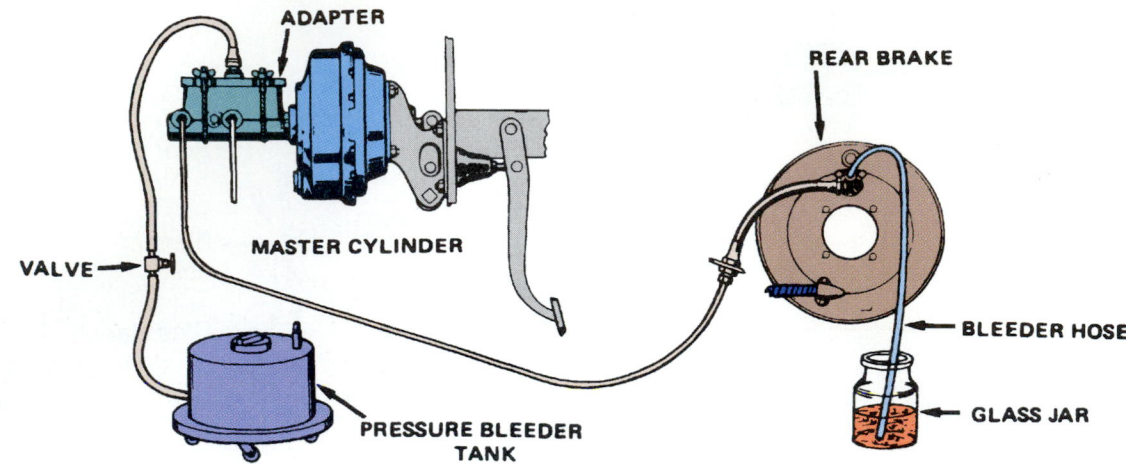

Figure 52.9 Pressure bleeder setup. *(Courtesy of AlliedSignal Bendix)*

the vacuum tank is connected to the open bleeder screw and fluid is collected in the vacuum tank as bleeding is completed. This method is clean because fluid is collected in the tank.

NOTE: *Avoid accidental contamination of the brake hydraulic system with petroleum products. Rubber parts are used in the brake system. Rubber is not resistant to attack by petroleum.*

As an experiment, put a rubber dust boot and wheel cylinder cup into a jar of engine oil. Contact with oil will cause rubber parts to swell. When this happens, the following are necessary:
- Completely disassemble the entire hydraulic system.
- Thoroughly flush the system with alcohol.
- Install new rubber parts.

Alcohol is the solvent of choice in cleaning the brake system. It can be used to clean petroleum, leaving no residue. It also absorbs water. Bleeding ABS brakes is covered later in this chapter.

Figure 52.10 A vacuum brake bleeder. *(Photo provided by Mac Tools and is representative of products manufactured by Mac Tools or products obtained from another source.)*

■ SCIENCE NOTE ■

Rubber is a polymer (a molecule formed by joining many smaller molecules together) made up of smaller units of isoprene molecules. Natural rubber is tacky because the isoprene polymer chains are not connected to one another. When rubber is **vulcanized**, *sulfur bridges, or cross links, are formed between different polymer chains. These cross links allow the polymer chains to be stretched to several times their original length without breaking. This elastic behavior makes synthetic rubber an elastomer. Synthetic rubbers are made by joining different molecules together, resulting in rubbers with different characteristics. Oil is soluble in rubber. It dissolves in and around the polymer chains, unlinking them and stretching them out. The result is that the rubber swells when it comes into contact with oil.*

➡ *Perform **Bleed Brakes** Worksheet*

■ ADJUSTING BRAKES

Brake fluid is not compressible. That is why it works to transmit pedal pressure to the wheel cylinders. What happens when air is in the fluid of the hydraulic system? Air *is* compressible, so a "spongy pedal" results. Air in the system will *not* cause a low, firm pedal, but improper brake adjustment will.

To make the correct diagnosis, pump the pedal twice quickly. If the pedal height rises higher on the second application, the brakes need adjustment. To understand why, an understanding of the operation of a simple master cylinder is needed (see Chapter 51).

Most brakes today have self-adjusting mechanisms, but some small cars and light trucks still require manual adjustment. Brake adjustment is covered later in this chapter.

An understanding of hydraulics is valuable when it is applied to the theory of brake adjustment. Pascal's law states that pressure in an enclosed system is equal in all parts of that system. The key to this is the word "enclosed," which also means "not moving." When fluids are moving, different laws apply.

Consider this example. Both drum brakes on a rear axle are served by one of the master cylinder's chambers. They are adjusted to different lining-to-drum clearances. Will the car pull to one side? If both wheel cylinders have the same pressure applied to their pistons when the fluid is not moving, then the drum with the excessive clearance will be contacted by its linings before the cylinder on the other side of the car will allow pressure to build. In wet weather, the correctly adjusted wheel might grab before the loose one. But as soon as fluid movement in the system stops, both sides will begin to receive equal pressure.

■ BRAKE LINING INSPECTION

Inspecting Drum Brake Assemblies

Removing the Brake Drum. Before inspection, a wheel and drum must be pulled. First, mark the drum and axle so it can be reinstalled in the same location. Remove the front hubs and drums and inspect the condition of the wheel bearings and grease. Rear drums are centered off the axle flange and often become rusted to it. Hammering between the wheel studs will usually loosen the drum from the axle. Be careful not to damage a wheel stud. Do not try to pound on the rear lip of the drum.

NOTE: *When a car owner continues to drive a car after the linings have worn down to the shoe, the shoe will wear a channel in the drum. Then, the self-adjusters operate, adjusting the shoe into the worn out area of the drum. When this has happened, the drum cannot be removed, unless the brake adjustment is backed off. The self-adjuster lever must be moved out of the way while performing the adjustment (Figure 52.11).*

Inspecting Drum Brake Cylinders

Pull back the rubber boots on the wheel cylinders to check for leakage and rust (Figure 52.12). A small amount of dampness from brake fluid is acceptable but fluid should not be dripping. A cast iron wheel cylin-

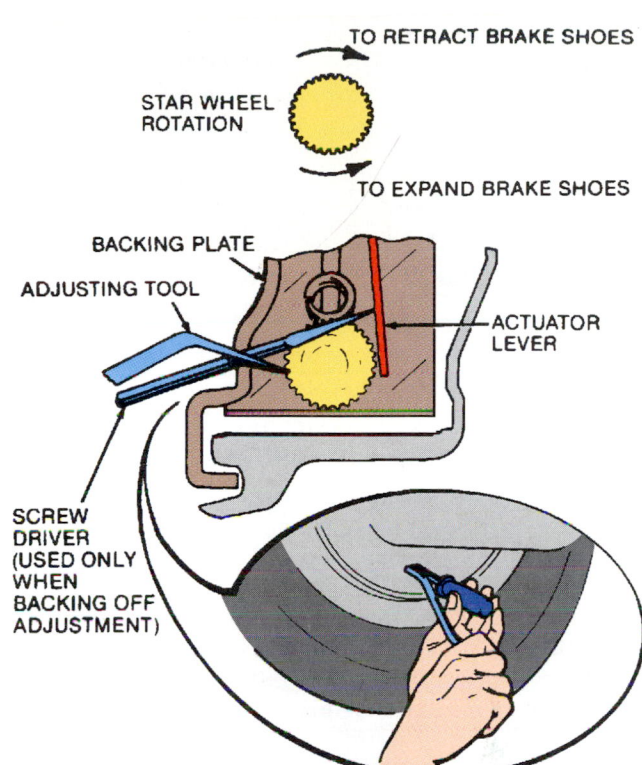

Figure 52.11 Move the self-adjusting actuator lever out of the way to loosen the brake adjustment. *(Courtesy of General Motors Corporation, Service Technology Group)*

Figure 52.12 Inspect under the wheel cylinder boot for leakage and rust. *(Courtesy of General Motors Corporation, Service Technology Group)*

der bore should not have rust and an aluminum piston should not be oxidized.

Rust (oxidation of iron or steel) is a serious problem. About one-seventh of the iron produced each year by industry is used to replace iron that has rusted. The exact process that occurs during rusting is not known. What is known is that oxygen and water are necessary for rusting to take place. Rusting occurs more rapidly in the presence of acid. (continued)

A sacrificial metal can be used to protect iron or steel from rusting. Oxidation requires the removal of one or more electrons from an atom. The oxidation potential of one metal is higher than for another if it requires less energy to remove an electron from the former than the latter. For example, the oxidation potential of magnesium is higher than the oxidation potential of iron or steel. This is because it is easier to remove an electron from magnesium than from steel. This is why blocks of magnesium are strapped to steel ships to prevent rusting. Not only does the magnesium oxidize instead of the iron, but the electrons stripped from the magnesium migrate to the steel. This happens because electrons associated with steel are held more strongly, and if electrons are given a choice, they prefer to be held more strongly. The transferring of electrons from magnesium to steel means that the steel cannot rust until it loses all of its excess electrons. The magnesium will prevent this electron loss by continually supplying the steel with electrons, giving the steel even more protection. Application of this principle can also be found in the magnesium rod screwed into your household water heater to prevent rusting of the tank.

Inspect Drum Brake Linings

Examine the condition of the friction lining surfaces. The front brakes wear more than the rears, so check one of them first.

- A bonded lining should be at least the same thickness as its metal backing. Figure 52.13 shows the lining (friction material) and the brake shoe.
- Riveted linings should be above the rivet heads.

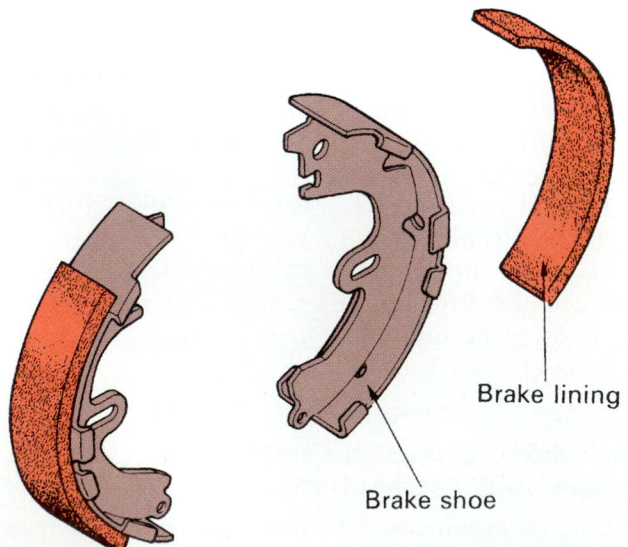

Figure 52.13 The brake lining and shoe.

Check Self-Adjuster Operation

Self-adjusters can become rusty, requiring disassembly so that their threads can be cleaned. Operate the self-adjusters to see that they move freely.

Inspect Disc Brakes

To inspect disc brakes, remove a wheel. Removal of the caliper is usually not necessary. Use a flashlight to look through the top of the caliper or around its sides. Check to see that the brake pad backing plate is thinner than the lining material, just as you did with drum brakes. Inspect the rotor to see that it does not have visible wear. While rotating the rotor, hold a fixed object near it and watch for visible runout. This can be more accurately checked with a dial indicator.

➡ **Perform _Brake Inspection_ Worksheet**

■ MASTER CYLINDER SERVICE

Check Master Cylinder Vent

When a master cylinder cover vent is obstructed, air can be drawn in at the back of the master cylinder, aerating the fluid. When there is air in the fluid, it becomes compressible and causes a spongy pedal rather than a firm one. When checking the fluid level, the vent can be easily checked for an obstruction.

■ MASTER CYLINDER REMOVAL

Remove the master cylinder from the vehicle. Be sure to use fender covers to protect the paint from accidental spills. Loosen and remove the two flared metal lines using a flare-nut wrench. If there is no vacuum power brake, unhook the pedal pushrod from under the dash. This is not necessary on power brake cars. Remove the nuts that hold the master cylinder in place (Figure 52.14). Remove the master cylinder and dump the brake fluid in a suitable container for disposal.

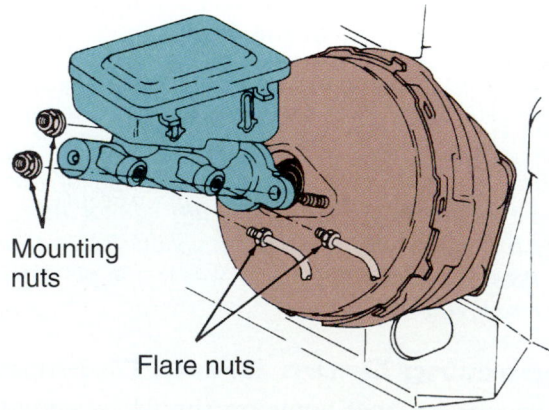

Figure 52.14 Remove the master cylinder. *(Courtesy of General Motors Corporation, Service Technology Group)*

MASTER CYLINDER DISASSEMBLY

When a master cylinder is defective, take it apart and use what you know about master cylinder operation to locate the source of the problem. Master cylinders can be purchased as new or rebuilt units. Rebuild kits are available for use when the cylinder bore is not damaged by corrosion. Liability issues have made the practice of a shop rebuilding a unit less popular than it was in the past.

To disassemble a master cylinder, push the piston deeper into the bore using a large Phillips head screwdriver or metal rod. While holding the piston in against piston return spring pressure, remove the circlip at the outside of the bore. Remove the piston from the bore. Sometimes, there is a stop screw in the bottom of the center of the bore that must be removed before the front piston can be removed (Figure 52.15).

Clean the inside of the cylinder bore and use a flashlight to inspect it. If the bore of a cylinder is corroded or pitted, the cylinder must be replaced.

Quick take-up master cylinders are serviced in the same manner as other master cylinders. They are usually made of aluminum, however. Aluminum cylinders cannot be cleaned with abrasives, such as hone stones. They are anodized to provide a protective coating to the aluminum.

SCIENCE NOTE

Anodizing is a chemical process that accelerates and controls the formation of an oxide coating on aluminum. The oxide coating provides corrosion protection, abrasion resistance, and improves the metal's appearance. During the anodizing process, the part is first dipped in a series of cleaners to prepare its surface. Next, it is dipped in an anodizing tank full of an electrolyte (acid and water). When electrical current is applied to the electrolyte, oxygen atoms are attracted to the aluminum. This happens because the electrolyte is negatively charged (cathode) and the aluminum is positively charged (anode). An aluminum oxide begins to grow on the part like roots on a human hair. The oxide grows until the coating rises above the surface of the part.

Anodizing prevents the conduction of electricity. This is important when making such things as aluminum ladders.

If the reservoir is plastic (Figure 52.16), remove it and clean it with hot water. After drying it thoroughly, replace it. Be careful not to install it backwards.

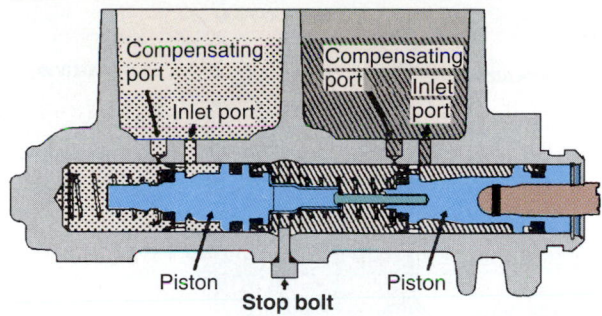

Figure 52.15 This master cylinder has a stop bolt at the front of the rear master cylinder piston. *(Courtesy of EIS Brake Parts)*

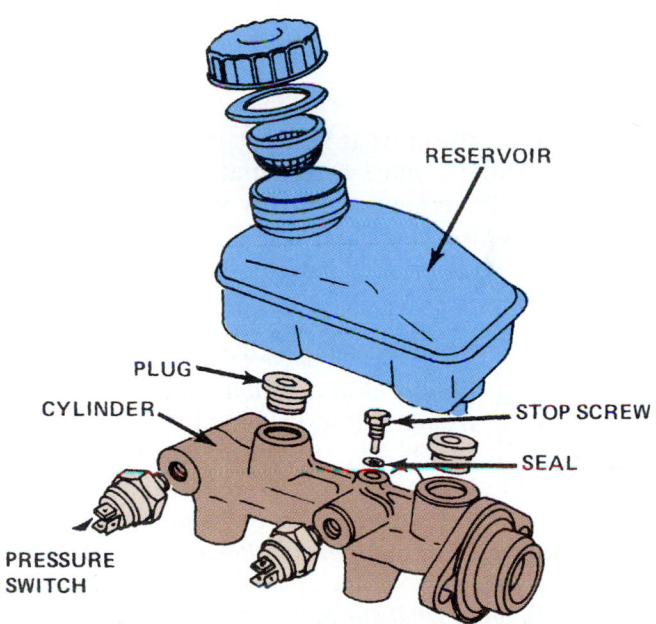

Figure 52.16 A master cylinder with a removable plastic reservoir. *(Courtesy of General Motors Corporation, Service Technology Group)*

SHOP TIP To avoid crossthreading when reinstalling flared lines, turn the fitting counterclockwise first to help seat the threads. Then, turn the fitting carefully clockwise to tighten it. Jiggle the metal tubing while turning the fitting so you can turn it all the way against its seat using only your fingers. Then use a flare-nut wrench to tighten it securely. Be sure to use another wrench to hold the flare fitting from turning.

RESIDUAL CHECK VALVE SERVICE

Some master cylinders for drum brake systems have a residual pressure check valve in the fluid outlet under the flare nut. If a check valve is used in a disc brake

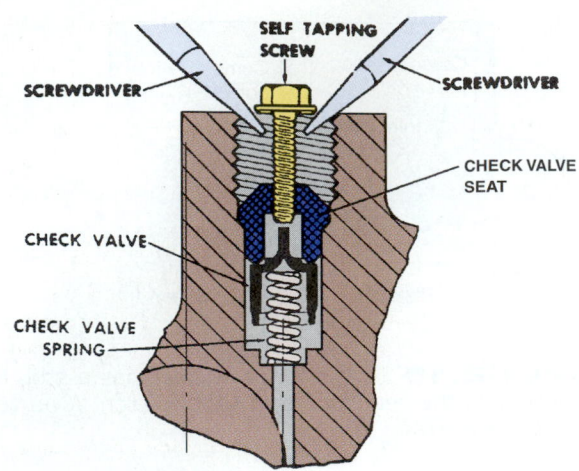

Figure 52.17 Removing a residual pressure check valve. *(Courtesy of Brake Parts Inc.)*

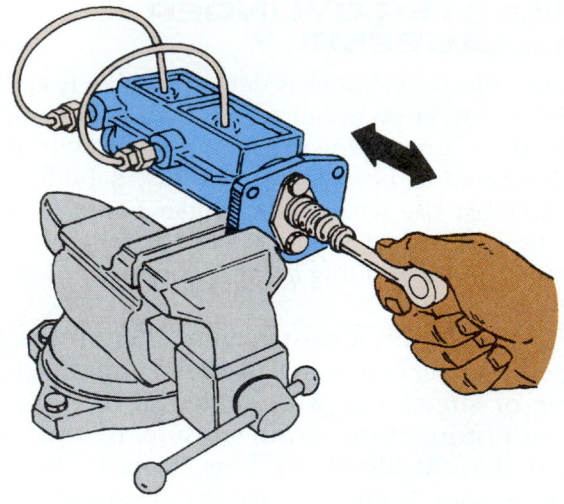

Figure 52.18 Bench bleeding a master cylinder. *(Courtesy of EIS Brake Parts)*

system, rapid lining wear will occur. The check valve is located under a brass insert that fits tightly in the fluid outlet. To remove the check valve, install a self-tapping screw in the hole in the middle of the insert. Then, pry it out with a pair of screwdrivers (Figure 52.17).

> **CASE HISTORY**
>
> *A student in an automotive brakes class installed disc brakes in place of drum brakes on a 1968 Camaro. He used the original master cylinder. About a year later, his girlfriend enrolled in the same brake class. She brought his car in for installation of new disc brake linings (for the third time since the disc brakes were installed). From what she had learned in class, she had a hunch that the residual pressure check valve might still be in the master cylinder. She removed it and solved her boyfriend's problem.*

A restricted check valve can cause excessive pedal travel before the brakes apply. To test the operation of the check valve, apply and release the brakes. Then, open a wheel cylinder bleed screw. A brief spurt of fluid is normal as the residual pressure is released.

■ BENCH BLEEDING THE MASTER CYLINDER

Before installing a master cylinder, it can be filled with fluid and bled of air. Figure 52.18 shows a master cylinder in a vise with two fabricated tubes returning fluid to the reservoir. Pumping the apply rod forces air and fluid out of the piston chambers. New, airless fluid enters the master cylinder through the reservoir. A variation of this process is to hold your fingers over the fluid outlets while releasing the pedal rod. This prevents air from being sucked back into the piston chamber. When the pedal rod is applied, fluid and air

escape into a drain pan. The process is repeated until all air is expelled from the master cylinder.

Quick take-up master cylinders have a step bore in which one piston is larger than the other. The large piston provides low pressure but high volume to take up clearance between the linings and rotor with less pedal travel. Feel the bottom of the master cylinder. If the casting is larger near the rear, it is a quick take-up cylinder. These are bench bled with a suction device. Without that, use a large Phillips screwdriver or wooden dowel to slowly depress the rear piston ¾ of the way into the bore. Do not bottom out the piston.

Adjusting Brake Pedal Free Travel

Adjustment of pedal free travel is not usually required, but should be checked when the master cylinder is replaced. There should be less than ⅛" of free play between the brake pushrod and the back of the primary master cylinder piston or power booster. This translates to free pedal travel of about ¼ to ½ inch.

■ BRAKE JOB

Front or rear linings are replaced in pairs. Parts are purchased in *axle sets* (both wheels on the front or the rear of the car).

> **SHOP TIP** It is a good idea to disassemble only one side of an axle at a time. The other side can then be used for comparison until the job has been repeated often enough so that a reference is no longer needed.

Be sure to consult the appropriate service manual for the vehicle being worked on. The major suppliers of brake parts have brake service manuals available to users of their products.

A front wheel *bearing repack* is usually a part of a front axle brake job on a rear wheel drive car (see Chapter 53). A complete brake job also includes all of the internal hydraulic parts of the wheel cylinders and disk calipers as well as new brake fluid throughout. New hardware and springs are often included. Drums or rotors may be turned on a lathe. A master cylinder and new hoses are usually *not* included in a brake job. Figure 52.19 shows what one manufacturer recommends for a complete brake job.

Ethics in Brake Work

A common practice in recent years was for large merchandisers to advertise incomplete brake jobs at very low prices. Once the car was on the lift, the cost of the job often rose rapidly as needed items were sold to the customer. When the job was finished, the price was often much higher than it would have been if the customer had patronized his or her regular repair shop. Consumer affairs divisions in some states have determined this type of selling to be unfair advertising. When an incomplete brake job is advertised in those states, a disclaimer must be included that says something to the effect that the job may cost substantially more when other necessary are parts are added.

■ SELECTING LININGS

Various lining materials are available, depending on the application and the friction characteristics desired. Some brakes stop well when they are cold, but do not work well when hot. Others work very well when hot, but will not stop the car until after a few stops (after they have had a chance to become hot) (see Figure 52.3). Both of the previous situations can be very dangerous. The best policy is to use only the type of lining material that came on the vehicle as original equipment and use premium quality materials. Besides taking a chance with someone's safety, there could be major liability for a repair shop if an accident results from the use of the inferior linings.

The Society of Automotive Engineers (SAE) has a standard for brake linings of which the intent is to provide a uniform means of identification, which may be used to describe initial friction characteristics of various linings. Friction coefficient ratings are done on a special machine in a test laboratory. The test is done on a one square inch piece of friction material. The standard is printed on the edge of the lining. It does not indicate quality. A lining with good stopping characteristics could wear out quickly or cause excessive wear to a rotor or drum.

THE CONDITIONS OF THE BRAKE SYSTEM ON THIS VEHICLE IS AS FOLLOWS:

BRAKE SYSTEM COMPONENTS	REQUIRED SERVICES	$ ESTIMATE PARTS	LABOR
Diagnosis ABS	❑ Recondition ❑ Replace		
Disc Pads*	❑ Replace Front ❑ Replace Rear		
Disc Caliper*	❑ Recondition ❑ Replace		
Disc Hardware*	❑ Replace		
Disc Rotor*	❑ Resurface ❑ Replace		
Grease Seals*	❑ Replace		
Front Wheel Bearings*	❑ Repack ❑ Replace		
Brake Shoes*	❑ Replace Front ❑ Replace Rear		
Brake Drums*	❑ Resurface ❑ Replace		
Wheel Cylinders*	❑ Recondition ❑ Replace		
Brake Hardware*	❑ Replace		
Parking Brakes	❑ Adjust/Lubricate ❑ Replace		
Power Brake Booster	❑ Service ❑ Replace		
Master Cylinder	❑ Recondition ❑ Replace		
Brake Fluid*	❑ Flush System, Add New Fluid		
Lines, Hoses, Combination Valve	❑ Replace		
Stop Light	❑ Replace Bulb ❑ Replace Switch		
Other			
	SUB-TOTAL		

*Total brake service includes reconditioning or replacement of these items.

ADDITIONAL SERVICES RECOMMENDED:

SYSTEM	REQUIRED SERVICES	$ ESTIMATE PARTS	LABOR
Lighting			
Lube/Oil/Filter			
Exhaust			
Shock Absorbers			
Tires			
Alignment			
Tune-up			
Spark Plug Wires			
Emission Control			
Battery			
Battery Cables			
V-belts			
Hoses			
Radiator Coolant			
Wipers			
Fluids			
Other			
	SUB-TOTAL		
	GRAND TOTAL:		

Figure 52.19 Items in a complete brake job. *[Courtesy of Cooper Automotive/Wagner Brake]*

Figure 52.20 A brakes parts washer.

DRUM BRAKE LINING REMOVAL

Before disassembling brakes, clean the entire assembly using either a HEPA vacuum (see Figure 11.15) or a brakes parts washer (Figure 52.20).

 **SAFETY NOTE** Brake dust is hazardous to breathe. Repair shops are required by law to have a brake parts washer or a HEPA vacuum when servicing brakes.

The removal of brake linings is accomplished using special tools. Figure 52.21 shows a typical tool used to remove and reinstall brake springs. Before disassembly, be sure to carefully inspect the linings being replaced. Sometimes there are differences in materials or lengths of the primary and secondary linings.

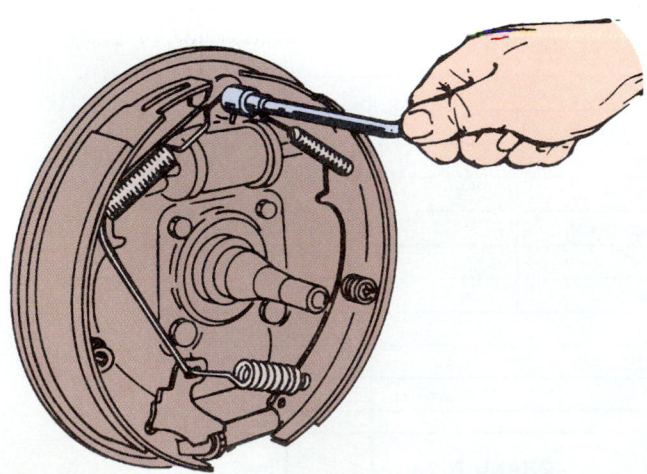

Figure 52.21 A special tool is used to remove and replace brake springs. *(Courtesy of General Motors Corporation, Service Technology Group)*

REBUILDING HYDRAULIC CYLINDERS

Hydraulic wheel cylinders and disc brake calipers can be rebuilt if they are made of cast iron and are not corroded. Cylinders can often be rebuilt on the car while they are still bolted to the backing plate. After removing the brake linings, remove the dust covers from the wheel cylinders. Push on the piston on one side of the cylinder to force the parts from the cylinder (Figure 52.22).

Two kinds of hone are available: one has two or three stones and the other is a flex hone similar to those used in engine rebuilding (Figure 52.23). A two-stone hone can go into smaller bores. It is used for rebuilding master cylinders and smaller import car cylinders.

After the wheel cylinder is lightly honed to clean it up (Figure 52.24), new rubber parts are installed from a **wheel cylinder kit** (Figure 52.25).

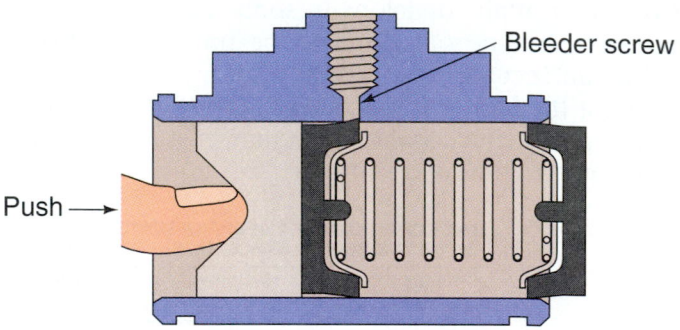

Figure 52.22 Push on the piston on one side of the cylinder to force the parts from the cylinder. *(Courtesy of EIS Brake Parts)*

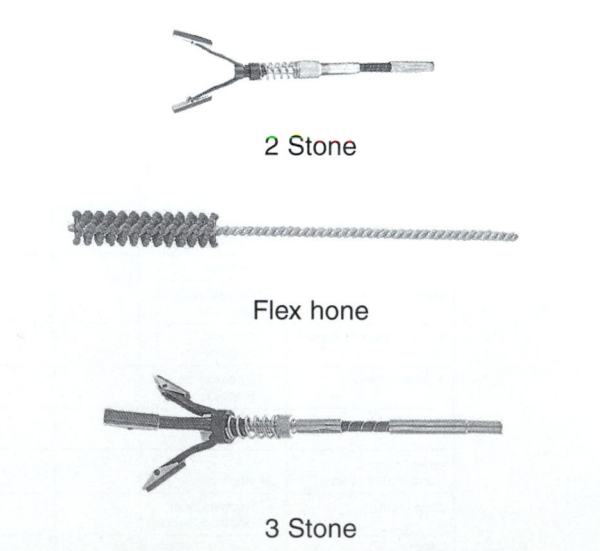

2 Stone

Flex hone

3 Stone

Figure 52.23 Various brake cylinder hones. *(Photo provided by Mac Tools and is representative of products manufactured by Mac Tools or products obtained from another source.)*

Figure 52.24   Honing a wheel cylinder. *(Courtesy of EIS Brake Parts)*

NOTES:

■ *If the size of the wheel cylinder is increased during honing by more than 0.005", the cylinder must be replaced.*

■ ***Anodized aluminum*** *cylinders, found on many late model cars, should* not *be honed because their hard surface layer will be removed and they will corrode.*

■ *If in doubt as to whether a wheel cylinder is cast iron or aluminum, check with a magnet. A magnet will not be attracted to an aluminum cylinder.*

■ *If an aluminum cylinder's bore has pit marks it must be replaced, rather than rebuilt.*

■ REASSEMBLING A WHEEL CYLINDER

Reassemble the wheel cylinder as shown in Figure 52.26. The lips on the wheel cylinder cups must face toward the fluid. A seal lip installed backwards will

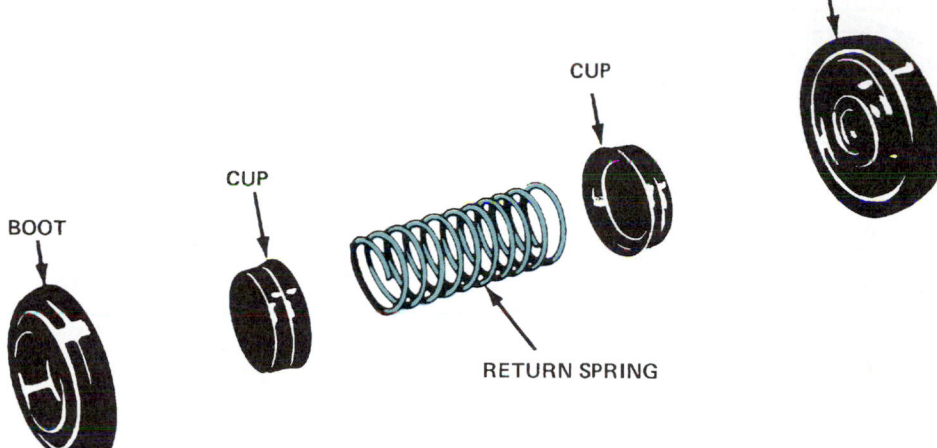

Figure 52.25   Parts of a wheel cylinder kit. *(Courtesy of General Motors Corporation, Service Technology Group)*

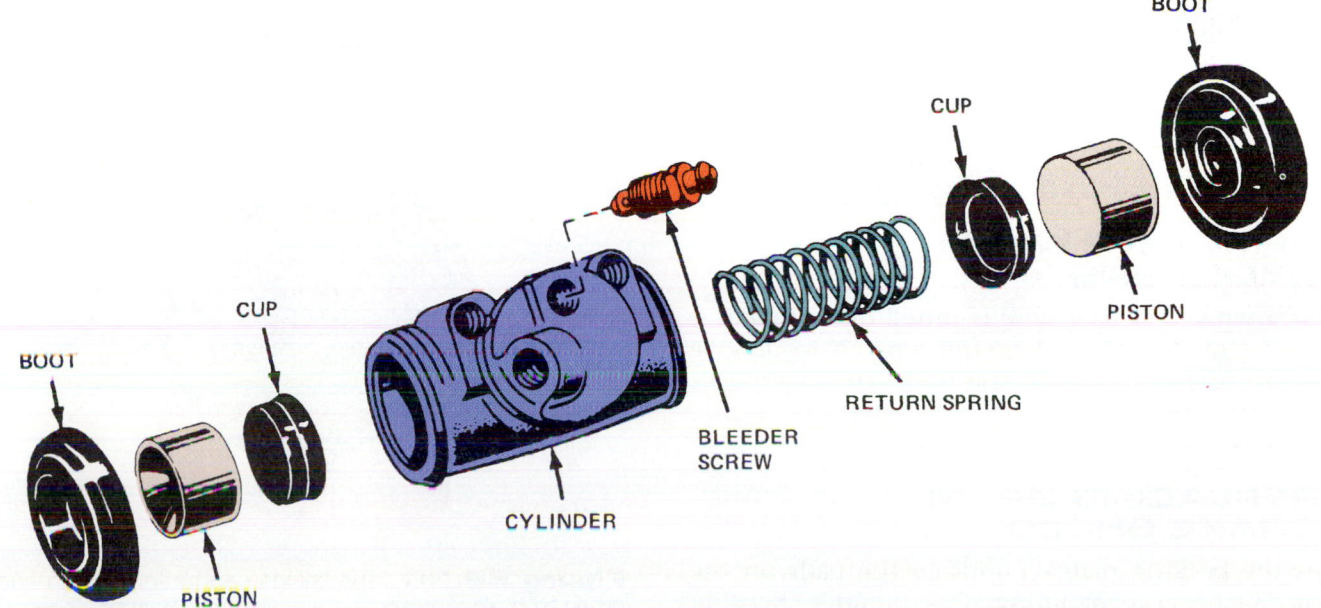

Figure 52.26   Position of parts in a wheel cylinder. *(Courtesy of General Motors Corporation, Service Technology Group)*

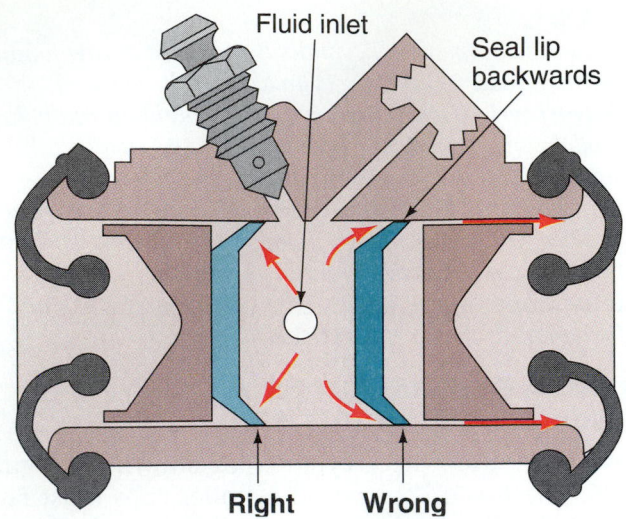

Figure 52.27 A lip seal installed backwards will leak.

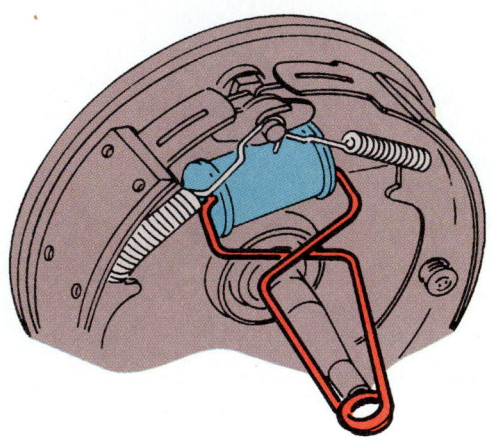

Figure 52.28 A brake cylinder clamp. *(Courtesy of General Motors Corporation, Service Technology Group)*

result in fluid leaking from the wheel cylinder when the brakes are applied (Figure 52.27).

When reassembling brake hydraulic parts, use brake fluid liberally as an assembly lubricant. After assembling a wheel cylinder, use a clamp to hold the assembly together while installing the linings (Figure 52.28).

■ REMOVING WHEEL CYLINDERS

When a wheel cylinder bore is pitted or corroded, it must be replaced. Use a flare-nut wrench to remove the brake tubing fitting on the back. The cylinder is held to the backing plate by two screws or a clip, which requires a special service tool.

■ REPLACING DRUM BRAKE SHOES

Clean the backing plates. Lubricate the pads on the backing plates that the linings slide on after checking them for grooves (Figure 52.29). When reinstalling

brake shoes on the backing plates, it is sometimes easier to assemble the springs to the brake linings first. Then, the linings can be folded into each other making installation easier (Figure 52.30).

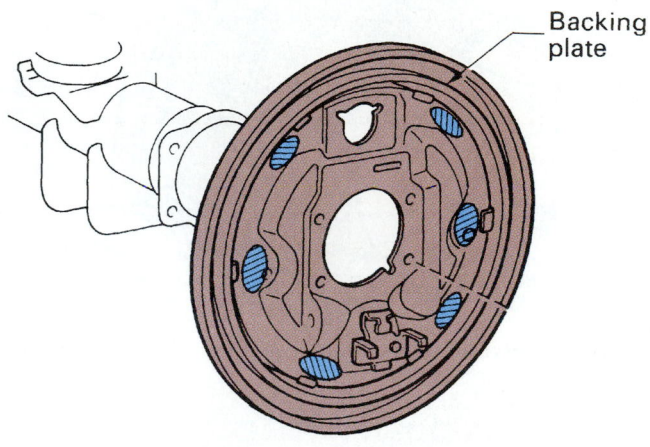

⬭ : Brake shoe contact surface

Figure 52.29 Lubricate the pads on the backing plates that the linings slide on after checking them for grooves.

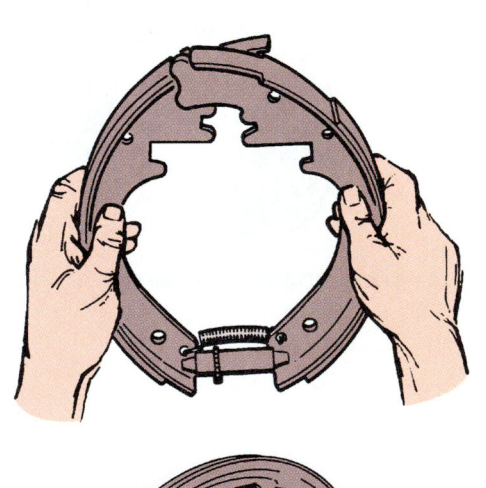

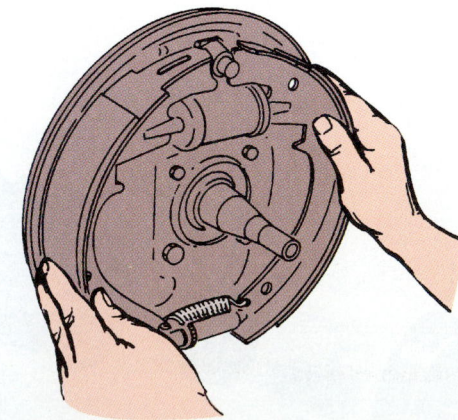

Figure 52.30 The linings can be folded into each other to make installation easier. *(Courtesy of General Motors Corporation, Service Technology Group)*

■ CHECKING RETURN SPRING CONDITION

Check the return springs visually. Shoe return springs should not be loose or broken. The paint on the springs should be in good condition, indicating that they have not been overheated. When a return spring has been removed from the brake assembly, see if a piece of paper can be inserted between its coils. If so, replace the spring. Another test is to hold the spring up to light. If light is visible between the coils of the spring, replace it.

 **SHOP TIP** If a spring is dropped on a hard surface, it should not ring.

■ INSTALLING SELF-ADJUSTERS

Self-adjusters must be reinstalled on the correct side of the car because they have either right or left hand threads and cannot be interchanged (Figure 52.31).

■ ADJUSTING DRUM BRAKE CLEARANCE

After brake linings are installed on the backing plates, an initial clearance adjustment is made before the drums are installed. A brake adjusting gauge is adjusted to the size of the drum (Figure 52.32). Then, the starwheel of the brake adjuster is turned until the shoes expand to the size of the adjusting gauge. This will provide about 0.010" of clearance between the lining and drum. If additional adjustment is necessary after the brake job is completed, drive the car slowly in reverse while applying the brakes repeatedly. This completes the adjustment.

To do an adjustment while the drum is installed requires accessing the starwheel through a hole in

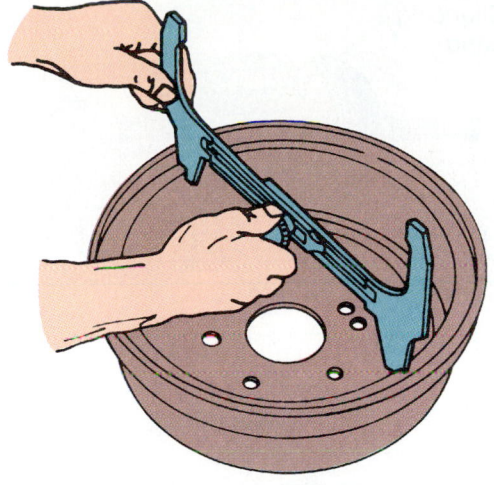

(a) SETTING TOOL TO DRUM

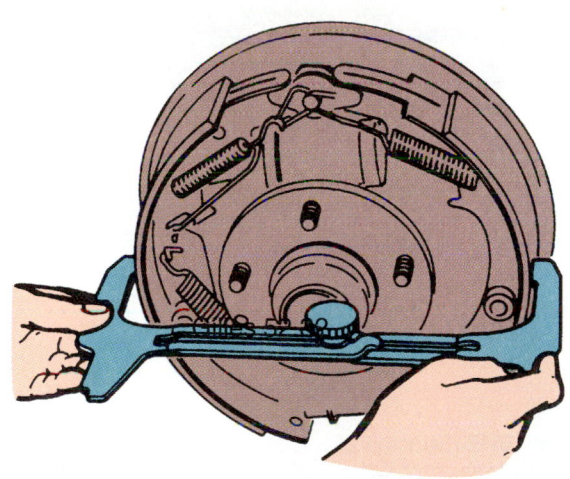

(b) SETTING BRAKE SHOES TO TOOL

Figure 52.32 A brake adjusting gauge. *(Courtesy of General Motors Corporation, Service Technology Group)*

the backing plate or front of drum. While turning the wheel by hand an adjusting tool, called a **brake spoon**, is used to turn the starwheel until the wheel will no longer turn. Then, the self-adjusting mechanism is held out away from the starwheel (see Figure 52.11) while loosening it about five to ten teeth until the wheel turns freely once again. During tightening, the self-adjuster does not have to be held out of the way. The starwheel will move easily in that direction. Older cars and trucks without self-adjusters use the same procedure, but the self-adjusting mechanism does not have to be held out of the way during loosening.

➡ ***Perform Brake Adjustment** Worksheet*

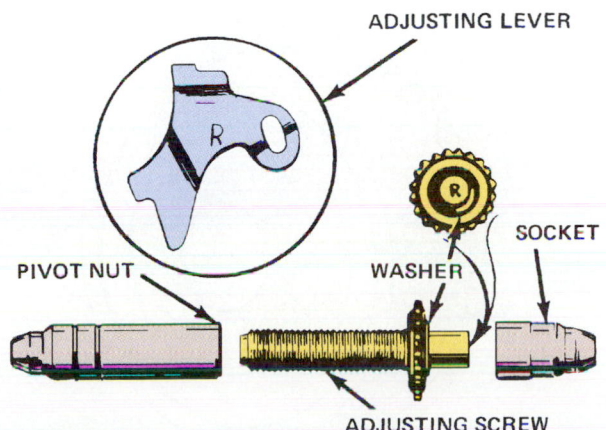

Figure 52.31 Brake adjusters have either right hand or left hand threads. *(Courtesy of General Motors Corporation, Service Technology Group)*

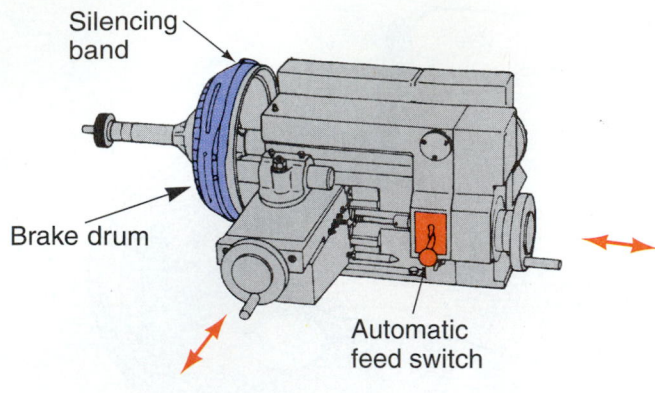

Silencing band

Brake drum

Automatic feed switch

Drum lathe

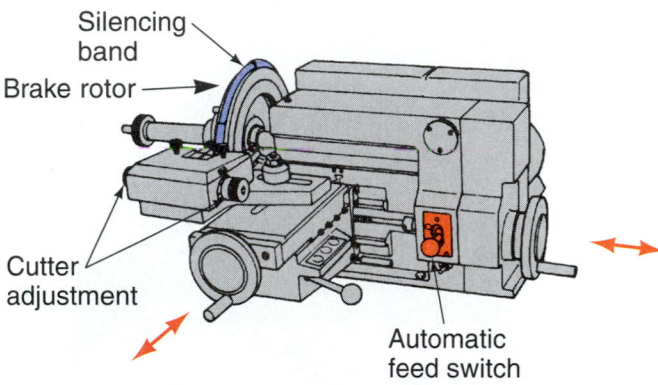

Silencing band

Brake rotor

Cutter adjustment

Automatic feed switch

Rotor lathe

Figure 52.33 Brake lathes. *(Courtesy of EIS Brake Parts and Van Norman)*

■ DRUM AND ROTOR SERVICE

When drum or disk linings are replaced, it is a common practice to resurface (*turn*) the drums or rotors using a drum or rotor lathe (Figure 52.33). If too much metal needs to be removed to clean up its surface, the drum or rotor must be replaced. Service manuals (available from brake parts manufacturers) give the maximum discard diameter for brake drums and the minimum thickness of disk rotors.

> **CAUTION** When using a brake lathe to machine a disk or a drum, do not wear loose clothing. Hanging jewelry or long hair should be secured out of the way.

■ DRUM SERVICE

A drum is measured with a special caliper. The one in Figure 52.34 has a dial indicator. Since 1972, the maximum allowable size for a brake drum has been cast

Figure 52.34 This tool is used for measuring drum brakes. *(Courtesy of General Motors Corporation, Service Technology Group)*

into the outside or inside of brake drums (Figure 52.35). Each brake job usually results in a drum turned to 0.030" oversize. The maximum amount that can be cut from most drums is 0.060". A typical drum will list a **discard diameter** that is 0.090" larger than stock. This means that the drum can be machined 0.060", leaving 0.030" for future wear. The size listed on a 10" drum will be 10.090", leaving 0.030" for wear after turning to 0.060".

NOTE: *A 0.030" increase in the diameter means that 0.015" of surface wear has occurred (Figure 52.36).*

Inspect drums to see that they are not out-of-round or scored. Be sure there is no grease or oil on the drum. Cast iron absorbs oil. It is very difficult to remove all of the oil from the pores of the metal when an axle seal has leaked. Soaking the drum in a caustic tank is sufficient to remove the oil from the iron.

Some drums have lug studs that are swaged to hold the drum to the wheel hub (Figure 52.37). If a drum with **swaged lugs** must be replaced, it must first be removed from the wheel hub. A cutter is used to assist in removing the studs. When the new stud is installed, a tool is used to deform it to hold it tightly to the hub.

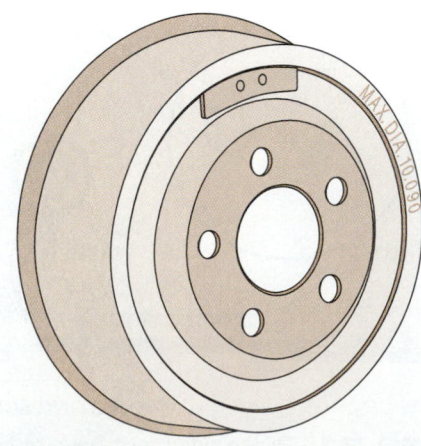

Figure 52.35 The maximum diameter is cast into a brake drum. *(Courtesy of Chrysler Corporation)*

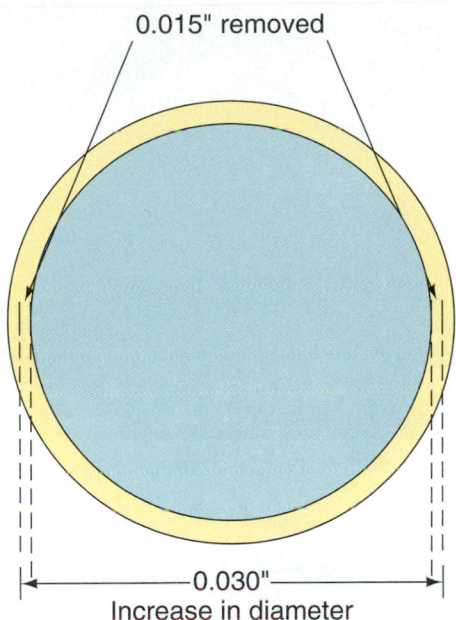

Figure 52.36 Surface wear of 0.015" results in a 0.030" change in diameter.

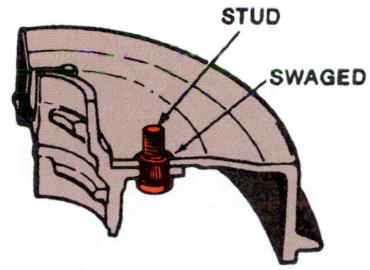

Figure 52.37 Some wheel studs are swaged. *[Courtesy of EIS Brake Parts]*

Turning Drums and Rotors

Drums and rotors are mounted on a lathe using adapters that fit into the wheel bearing races (Figure 52.38). Check the adapters for nicks and scratches. If you find a flaw, remove it with a fine stone before use. Rear drums and FWD rotors are usually mounted on tapered cones and clamped between adapter cups.

Runout of the arbor should be no more than 0.001". Turn on the motor and watch to see that the drum or rotor turns without any visible runout (wobble). If the part is not mounted properly, disassemble it and remount it.

NOTE: *Noise is the result of vibration. Vibration can cause a very rough cut when turning parts on a lathe. To prevent this, a silencing band is used (see Figure 52.33).*

The end of the cutter bit has a **carbide** *insert* in it. Carbide is a harder metal than ordinary tool steel. It has a longer life if vibration is kept to a minimum. When cutting with carbide, a minimum cut of over 0.005" is recommended. A typical cut will increase the diameter of a drum by 0.015".

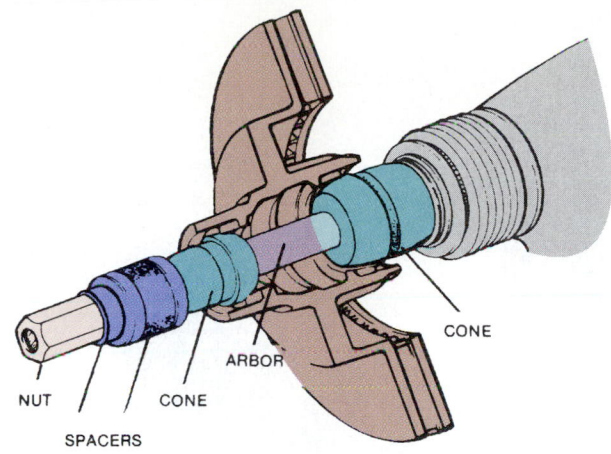

Figure 52.38 Cone adapters fit into the wheel bearing races. *[Courtesy of EIS Brake Parts]*

Hard Spots

Sometimes, **hard spots** appear in the surface of a drum or rotor. This condition happens as a result of excessive heat. The spots, which are blue in color, are places where the metallurgical composition has been changed from cast iron to steel. Attempting to cut them on a lathe results in high spots in the finish. The carbide cutter cuts the surrounding material but skips over the hard spots. Grinding can be used to remove hard spots, but they may return.

■ ROTOR SERVICE

The thickness of a rotor must be measured to see if it is still usable. There are special micrometers available that have a pointed tip so the depth of a groove can be measured. This will save machining a rotor that will turn out to be too thin.

NOTE: *According to Wagner Brake, turning the rotors is only necessary if there is pedal pulsation due to warpage. Score marks up to 0.050" deep are said not to affect brake operation.*

Resurfacing drums or rotors any time the linings are replaced is a good idea because it provides a smooth surface. This results in less break-in time for the new lining material. Check rotors for parallelism (Figure 52.39) and runout (see Figure 52.4)

Rotor lathes have speed controls. It is important to take a rough and a finish cut. The speed of the cut is the key to providing the correct surface finish. Feel the surface of a new rotor with your fingernail. The surface of a rotor turned on the lathe should be at least as smooth as a new rotor. Figure 52.40 lists recommended speeds for rotor turning.

A rough cut will cause vibration and noise during braking. As the cutter goes around the surface of the rotor (Figure 52.41), it leaves a microscopic thread. This must be removed. Some machines have an attachment for this purpose. If not, use an orbital sander to

Figure 52.39 Checking a rotor for parallelism. *(Courtesy of EIS Brake Parts)*

	Rough Cut	**Finish Cut**
Spindle Speed		
10" & under	150–170 RPM	150–170 RPM
11"–16"	100 RPM	100 RPM
17" & larger	60 RPM	60 RPM
Depth of Cut		
(Per Side)	0.005"–0.010"	0.002"
Tool Cross Feed		
(Per Rev.)	0.006"–0.010"	0.002" max
Vibration Dampener	Yes	Yes
Sand Rotors	No	Yes
Final Finish		

Figure 52.40 Recommendations for rotor refinishing. *(Courtesy of Cooper Automotive/Wagner Brake)*

Figure 52.41 Threaded appearance as a rotor is refinished.

Figure 52.42 Use an orbital sander to sand off the threaded finish while the lathe turns.

sand off the threaded finish while the lathe turns (Figure 52.42). Sand with fine grit paper for a minimum of 60 seconds with the rotor turning about 150 rpm. After sanding, clean with brake cleaner, denatured alcohol, or a damp rag to remove metal dust.

NOTE: *When cleaning up a disc or drum lathe, do not use compressed air. Metal chips can get into the inside of the machine, causing damage.*

Many new rotors allow for only 0.030" of machining. Some allow as little as 0.020". Specification charts are available from all of the brake manufacturers. It is not necessary to refinish new rotors.

On-the-Car Lathes

Lathes are available that allow the rotor to be machined without removing it from the car (Figure 52.43). These

Figure 52.43 An on-the-car lathe.

machines are popular in high volume shops. If the front bearing must be disassembled on a front wheel drive car, removal of some rotors can be a time-consuming proposition. On-the-car lathes save this labor requirement. Some manufacturers require the use of an on-the-car lathe to prevent damage to a bearing that is disassembled.

Installing a Rotor

Before reinstalling a brake rotor, be sure that any rust or dirt is removed from the rotor and hub mating surfaces. Failure to do this can result in lateral runout of the rotor. Some rotors are designed to be installed on only one side of the car (Figure 52.44). The fins on these rotors slant in one direction.

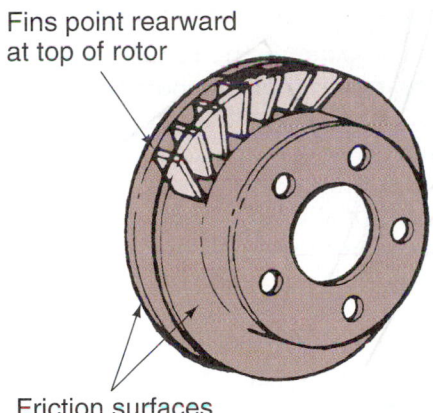

Fins point rearward at top of rotor

Friction surfaces

Figure 52.44 Some rotors are designed to be installed on only one side of the car. *(Courtesy of Cooper Automotive/Wagner Brake)*

When tightening lug nuts, be sure to use a torque wrench and torque in a star pattern to avoid warping the rotor.

■ DISC BRAKE SERVICE
Replacing Disc Linings

Disc linings are usually easy to replace. Sometimes the linings can even be replaced without removing the caliper from its mount. This type of caliper will have an access hole in the top through which the pads can be removed and installed. Some caliper designs require removal of the caliper to remove the pads (Figure 52.45). When the caliper is unbolted, use wire or a hook to hold it so its weight is not allowed to hang on the hose (Figure 52.46). The hose can be damaged internally.

Before replacing linings, spin the rotor and inspect it for roughness on the front *and back* that could result from a worn-out brake pad.

NOTE: *Be sure to check both sides of the rotor. Sometimes one pad wears out and not the other.*

Before installing new pads, make sure that a floating caliper can slide freely on its mounts. Slide it back and forth *before* installing the pads.

When replacing linings on floating calipers, a C-clamp or a large pair of pliers is used to move the piston back in its bore (Figure 52.47). This is so that the caliper can be easily removed and the new linings (which are considerably thicker than the worn ones) will fit during installation.

NOTE: *Before attempting to push the piston back in its bore, open the bleed screw on the back of the caliper. Then, move the piston all of the way back into its bore. Tighten*

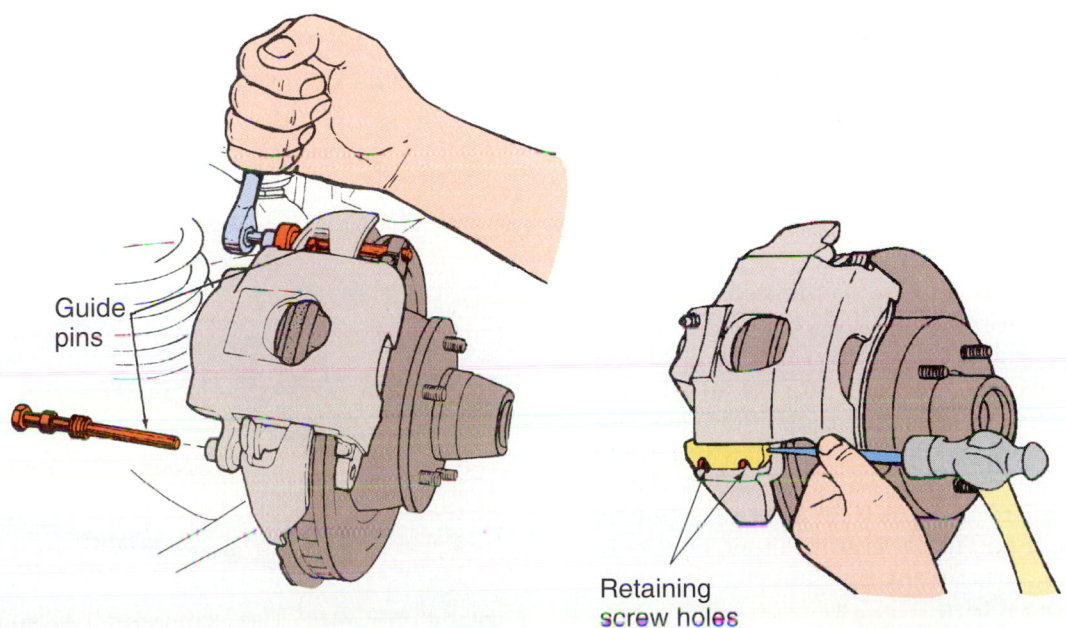

Guide pins

Retaining screw holes

Figure 52.45 This caliper design must be removed to install new brake pads. *(Courtesy of EIS Brake Parts)*

Figure 52.46 Do not allow the caliper to hang from a brake hose.

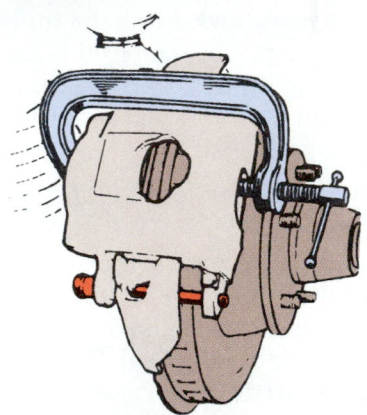

Figure 52.47 A C-clamp or a large pair of pliers is used to move the piston back in its bore. *(Courtesy of EIS Brake Parts)*

the bleed screw immediately so it is not accidentally left loose. If left loose, the master cylinder will empty of fluid.

■ REAR DISC PARKING BRAKE CAUTION

Rear wheel disc brakes have a parking brake built in. There is either a miniature drum brake that is inside of the disc rotor or there is a mechanism to clamp the disc brake pads to the rotor when the emergency brake is set (see Figure 51.32). The latter will have either a screw or a cam attached to a lever. The lever on the back of the caliper is the clue. If there is a lever, do not try to force the piston into the bore. Check the service manual for instructions on how to proceed.

Sometimes rust forms in the self-adjusting mechanism when the parking brake has not been used regularly. This freezes up the caliper. Like other self-adjusting mechanisms, left and right sides are not interchangeable. Because of the complexity of these calipers and the chance that parts are not reusable, many shops install rebuilt units instead of servicing them themselves. If you use a rebuilt caliper, be sure to keep all of the old parts on the old caliper so it will be acceptable as a core return.

Pedal travel on four wheel disc brakes tends to be further than with disc/drum combinations. This is because the rear pads have to travel further during brake application. When there is a problem with the self-adjuster, pedal travel will become excessive more quickly. Also, after replacing a caliper, the initial clearance must be adjusted to within $\frac{1}{16}$" or less. Adjustment can usually be done by repeatedly applying the parking brake. If the adjustment is too tight to install the caliper, the piston must be screwed back into its bore. This requires a special tool.

The rear disc parking brake that has a miniature drum brake will usually last for the life of the vehicle. The thickness of the parking brake lining needs only to be sufficient to hold the car. It does not have to dissipate heat or resist wear because it is only used to hold the vehicle when parked.

■ DISC BRAKE CALIPER REBUILDING

In disc brake systems, the caliper pistons are the sealing surface, rather than the bore. There is a piston seal and a boot that keeps contaminants away from the sealing area (see Figure 51.27). A caliper *rubber parts kit* contains a new boot, a piston seal, and sometimes rubber O-rings (for sliding caliper bolts). To disassemble a caliper, use compressed air and a rubber tipped blowgun (Figure 52.48).

CAUTION Pressure behind the piston will become equal to shop air pressure and output force is multiplied by the area of the piston (remember Pascal's law). Apply pressure in short bursts to gently push the piston from the bore. Full pressure can shoot the piston out with dangerous force.

To protect the piston from damage, put a stack of shop rags below the piston before applying air pressure. If a caliper has pistons on both sides, place a

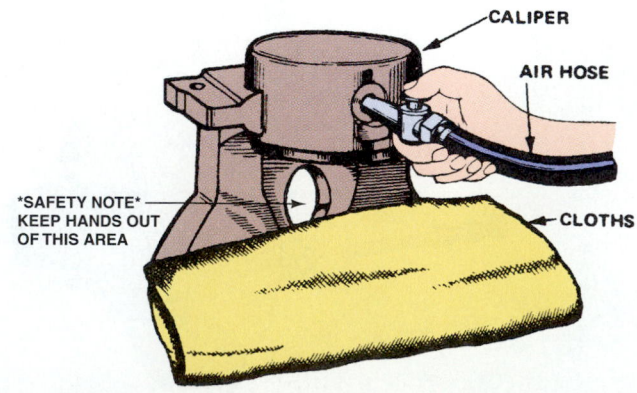

Figure 52.48 Use compressed air and a rubber tipped blowgun to remove a caliper piston. *(Courtesy of AlliedSignal Bendix)*

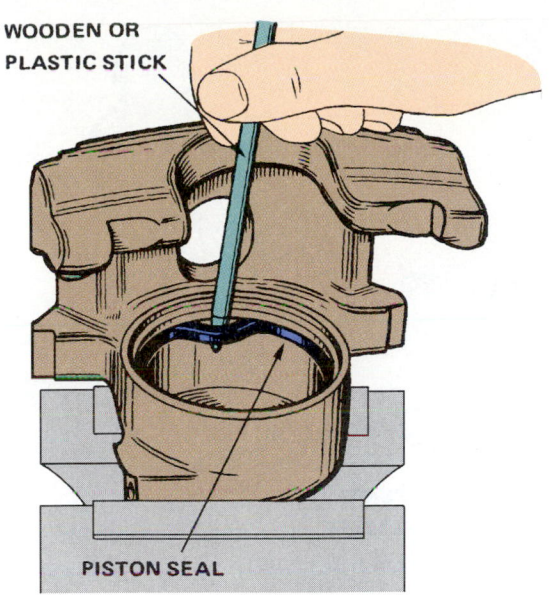

WOODEN OR PLASTIC STICK

PISTON SEAL

Figure 52.49 Removing the old seal from the caliper bore. *(Courtesy of AlliedSignal Bendix)*

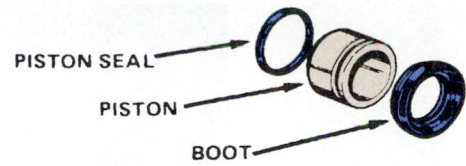

PISTON SEAL
PISTON
BOOT

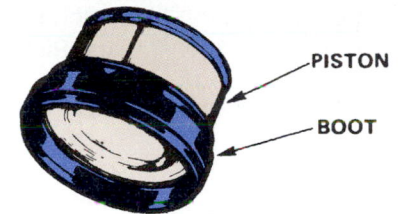

PISTON
BOOT

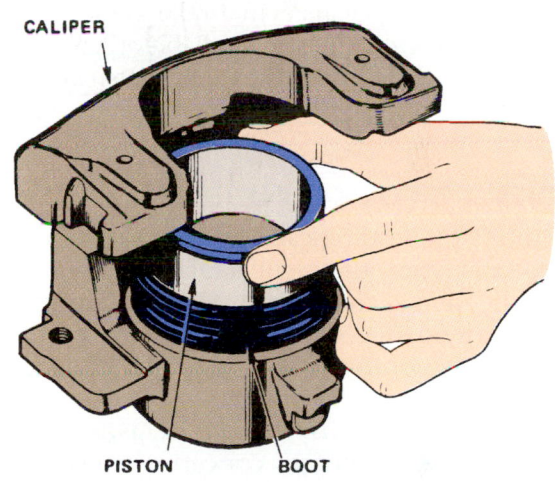

CALIPER

PISTON   BOOT

Figure 52.50 Caliper reassembly. *(Courtesy of AlliedSignal Bendix)*

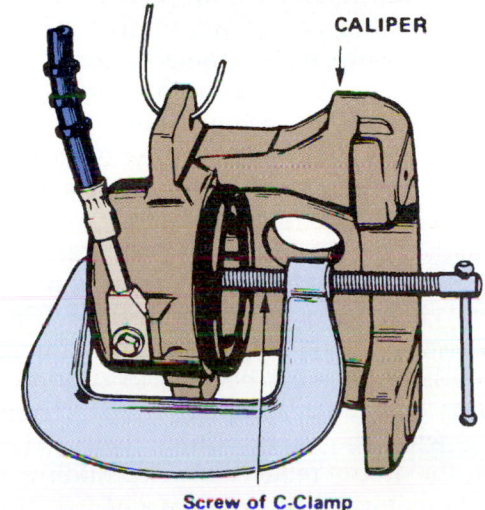

CALIPER

Screw of C-Clamp inside piston.

Figure 52.51 Push the piston to the bottom of its bore. *(Courtesy of AlliedSignal Bendix)*

piece of wood between the pistons so that one piston cannot come all the way out of its bore before the other one does. This would prevent air pressure from being used to remove the other one.

Remove the old piston seal from the bore (Figure 52.49). Thoroughly clean and inspect for scratches, rust, and corrosion. Varnish can be cleaned from the piston using spray brake cleaner.

NOTE: *Do not use emery cloth to clean pistons.*

Stubborn buildup can be cleaned with crocus cloth. *Crocus cloth* is very fine polishing cloth. It does not leave scratches or grit. The bore is not a sealing surface so it does not need to be honed like a drum brake wheel cylinder bore.

■ ASSEMBLE THE CALIPER

There are commercial brake assembly lubricants that make assembling disc brake calipers easier. Apply a liberal amount of lubricant to the piston seal and install it in the bore. Install the boot on the piston. Figure 52.50 shows the parts and procedure for caliper assembly. Sometimes the outer lip on the dust boot must be installed in its place on the caliper before pushing the piston into the bore.

Carefully push the piston into the bore, shimmying it past the seal.

NOTE: *Once a caliper has been disassembled, a new seal must be used. The old seal will usually expand, making reassembly impossible.*

After the piston is installed in the bore, push it all the way to the bottom of the bore using a C-clamp (Figure 52.51). This is to make enough room for the pads when the caliper is installed.

**Loaded Caliper
Single Piston**

**Unloaded Caliper
Dual Piston**

Figure 52.52 Rebuilt calipers. *(Courtesy of Cooper Automotive/Wagner Brake)*

Rebuilt Calipers

Some rebuilders specialize in brake calipers. Loaded and unloaded calipers are available (Figure 52.52). **Unloaded calipers** are also called *bare* calipers. The **loaded caliper** comes with new friction pads, hardware, and shims.

■ DISC BRAKE NOISE

Vibration is the leading cause of disc brake noise. The noise occurs when the metal brake pad back vibrates against the metal caliper piston. Harder lining materials have a tendency to make noise, especially when cold. But these linings last longer and provide better hot stopping than softer linings. Also, asbestos, a common (and excellent) lining material, is being phased out due to environmental concerns. Newer vehicles come from the factory with semi-metallic linings. These are more prone to low frequency vibration. A small amount of noise from these linings is considered normal.

It is important that disc linings be firmly attached to the apply piston or caliper so that vibration is avoided. On floating calipers, there are sometimes tabs on the pads that hold them tightly to the caliper. These are adjusted to fit, using either a hammer (Figure 52.53) or pliers (Figure 52.54). Disc brake hardware kits include new *anti-rattle clips* that are attached to the backs of the brake pads (Figure 52.55).

Aftermarket glue/insulator materials can be spread on the metal back of the shoe before it is installed in the caliper. When it sets up, it helps prevent the vibration from occurring and helps dampen noise.

Floating calipers must be able to slide freely on their mounts. Caliper kits usually include O-ring bushings along with the other rubber parts. These are important for noise reduction. The sliding caliper mounting bolts must also be clean and free of rust or they must be replaced. Bushings and sleeves can become gummed up to the point that the caliper cannot slide easily. A result of this can be that one lining

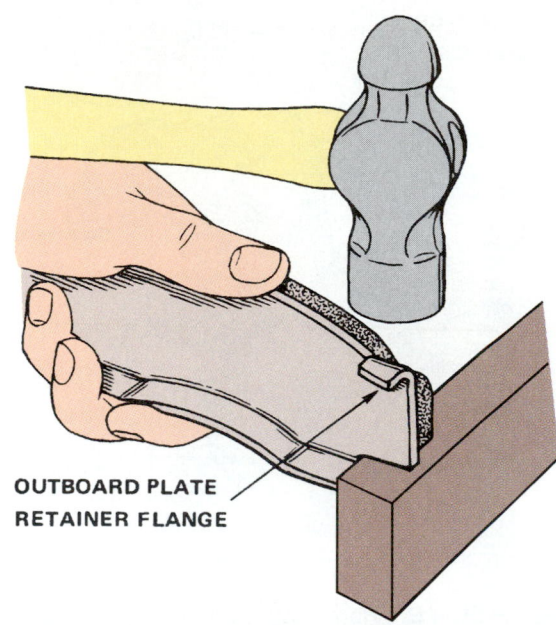

**OUTBOARD PLATE
RETAINER FLANGE**

Figure 52.53 Using a hammer to fit the disc pad to the caliper. *(Courtesy of AlliedSignal Bendix)*

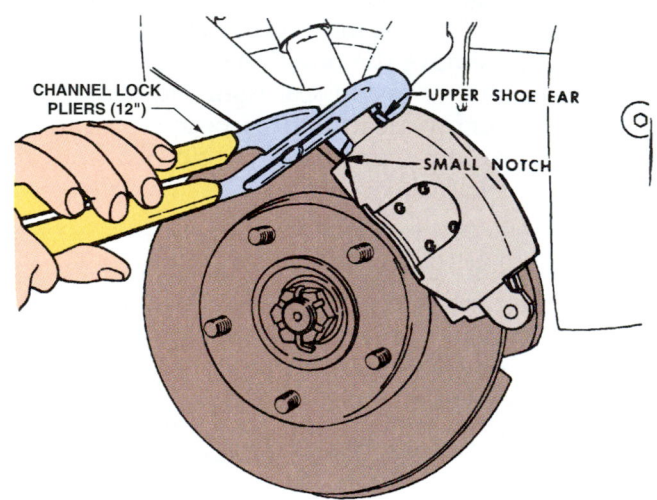

**CHANNEL LOCK
PLIERS (12")**
UPPER SHOE EAR
SMALL NOTCH

Figure 52.54 Tightening a disc pad to the caliper. *(Courtesy of General Motors Corporation, Service Technology Group)*

wears more than the other. One other result is that a brake does not release all of the way after a stop. This can result in an intermittent annoying squeal that goes away whenever the brakes are applied. Be attentive to the caliper design and be sure that all sliding surfaces are rust-free and in good shape.

■ INSTALL THE CALIPER

Install the caliper and check that it can move freely. Torque the bolts that hold it to the steering knuckle.

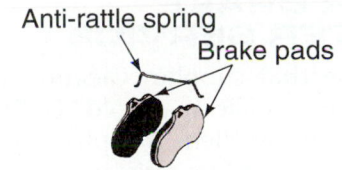

Anti-rattle spring
Brake pads

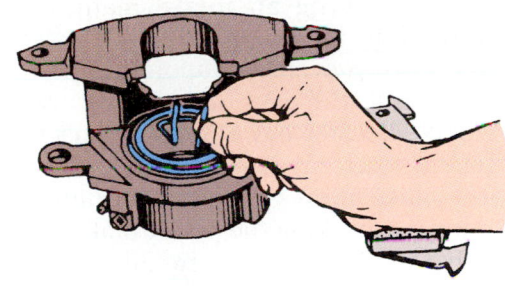

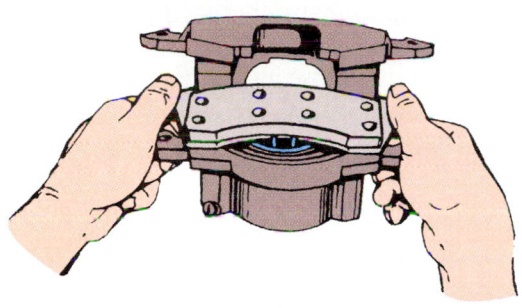

Figure 52.55 Various disc brake clips and springs. *(Courtesy of EIS Brake Parts)*

CAUTION Immediately after installing the caliper, be sure to apply the brake pedal. Only hand pressure is necessary so if the car is on the lift, this can still be done. Sometimes more than one application of the pedal is required to push the linings against the rotor.

BREAK-IN NEW LININGS

One disadvantage to disc brake pads is that they tend to squeal if the rotor does not have the correct surface finish. The pads also have to be broken in correctly. This means do not stop too fast for the first few hundred miles. To break in linings, make ten to twenty gentle stops from 30–15 mph.

Brake linings are not completely cured until after several hundred miles of driving. Overheating them when they are new damages the resins in the lining. The resin softens and solidifies on the surface of the linings, resulting in a glaze that causes brake squeal. Caution the owner not to use the brakes hard for the first 100 miles of city driving.

SERVICE THE PARKING BRAKE CABLE

During a rear brake job, the emergency brake cables are disconnected from the brake linings. This provides a good opportunity to lubricate the cables. From the underside of the vehicle, pull the cables as far as possible. Wipe them off and inspect them. Apply clean grease to them and push them back into their sheaths. Another way of lubricating a cable is to apply penetrating oil to its sheath (the metal cover that surrounds it). Usually, you can only do this on the ends where the cable has no plastic insulating cover.

The parking brake cable usually has a spring-loaded clip that fits into the brake backing plate. If a parking brake cable requires replacement, use a small hose clamp to help remove it from the backing plate (Figure 52.56).

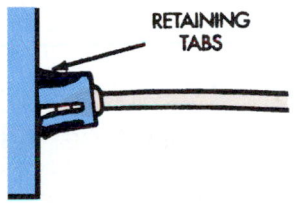

RETAINING TABS

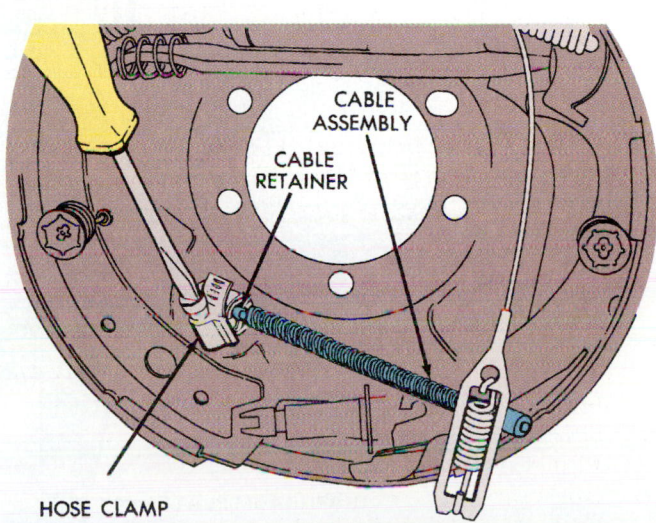

CABLE ASSEMBLY
CABLE RETAINER
HOSE CLAMP

Figure 52.56 Use a small hose clamp to depress the retaining tabs. *(Courtesy of Chrysler Corporation)*

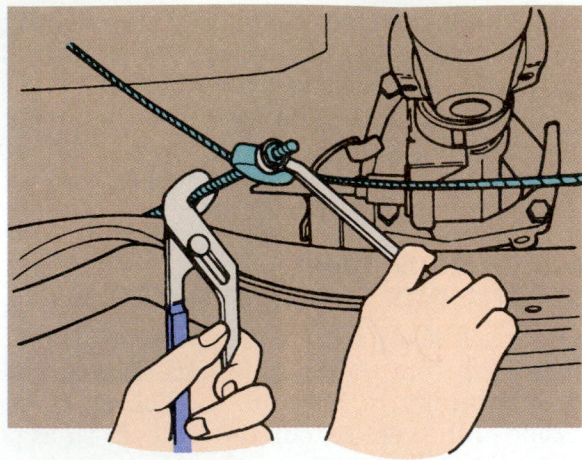

Figure 52.57 Typical parking brake adjustment. *(Courtesy of General Motors Corporation, Service Technology Group)*

Parking Brake Travel

Service manuals often describe parking brake adjustment as the number of clicks that the brake handle or pedal makes before being fully applied. The specification might say "less than nine clicks." A good rule of thumb is that the brake should be fully applied at half travel. Figure 52.57 shows a typical undercar adjustment.

*NOTE: Changing the service brake adjustment will affect the parking brake adjustment (the **service brakes** are the brakes that stop the car). Do not adjust the parking brake unless the service brakes have been adjusted first.*

➡ *Perform **Parking Brake Adjustment** Worksheet*

■ POWER BRAKE BOOSTER SERVICE

Check the hose that supplies vacuum to the power booster from the intake manifold. It should not be damaged, hard, or swollen. A replacement hose to a power booster must be fuel resistant. Some hoses have a fuel filter in them to prevent fuel from entering the power booster during an intake manifold popback. Fuel will damage the diaphragm in a power booster. The filter also prevents evaporating hydrocarbons from leaking into the passenger compartment.

The brake booster has a one-way check valve that traps vacuum when the engine is off. If the check valve is bad, the brake effort will vary according to the load on the engine. Checking the power brake check valve is done with the engine off. Carefully bend the valve against its rubber grommet (Figure 52.58). If the check valve is holding, air will rush into the front part of the booster. Some manufacturers use a check valve in the vacuum line. To test check valve operation in these cars, remove the vacuum hose at the booster side and apply vacuum. If vacuum holds, the check valve is good.

Leaking Power Booster Seal

A leak in the seal in the front of a power booster can allow fluid to be drawn into the power booster and burned in the intake manifold. The rear chamber of the master cylinder will continually use fluid. The key to diagnosing this is that there is no evidence of external fluid leakage.

Replacing a Power Brake Booster

Most shops do not rebuild power brake boosters. Rebuilt units are available for most cars and trucks.

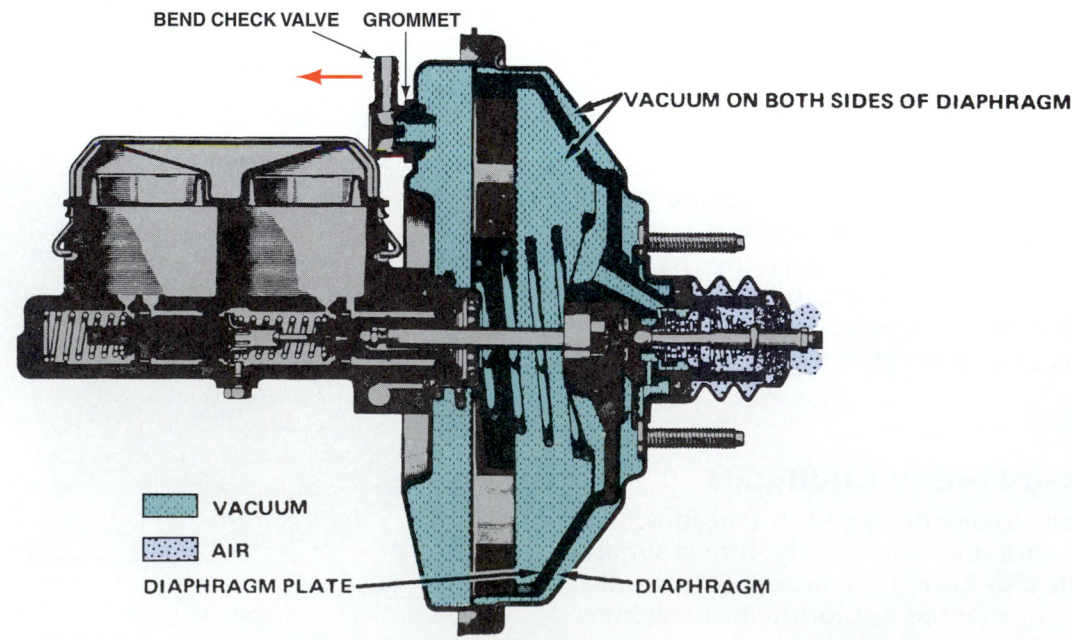

BEND CHECK VALVE   GROMMET

VACUUM ON BOTH SIDES OF DIAPHRAGM

▨ VACUUM
▧ AIR

DIAPHRAGM PLATE ─────   ───── DIAPHRAGM

Figure 52.58 Testing the check valve. *(Courtesy of AlliedSignal Bendix)*

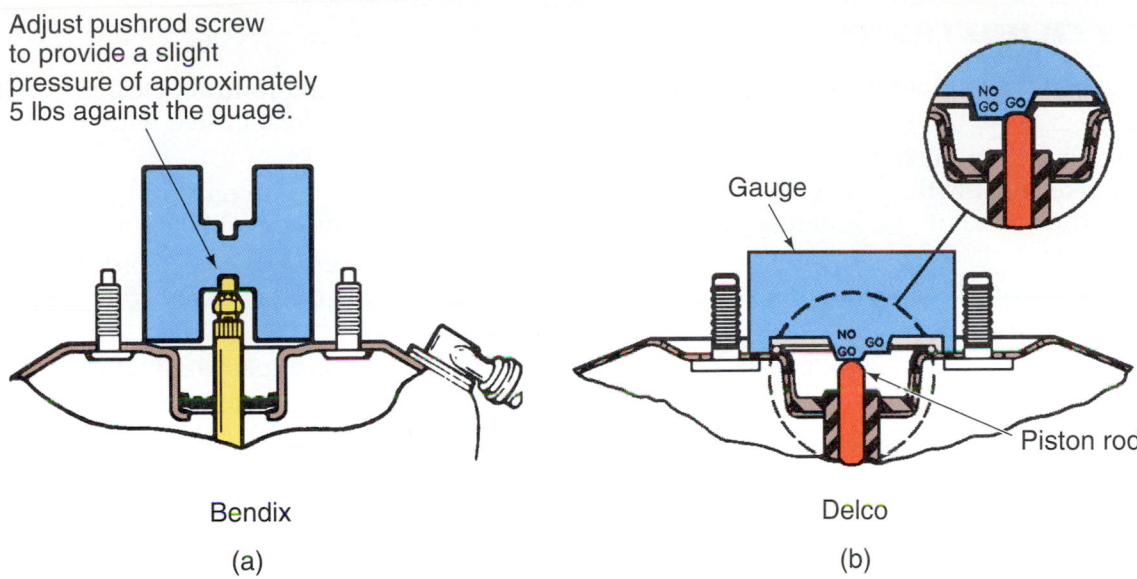

Adjust pushrod screw to provide a slight pressure of approximately 5 lbs against the guage.

Gauge

NO GO | GO

Piston rod

Bendix

(a)

Delco

(b)

Figure 52.59 Tools for checking power booster pushrod depth. *(a, Courtesy of Brake Parts Inc.; b, Courtesy of General Motors Corporation, Service Technology Group)*

When replacing a power booster, it is important that the pushrod depth is correct. There are special tools and different methods specified for adjusting this (Figure 52.59). Complete rebuilt units are available that include a rebuilt master cylinder assembled to them.

■ ANTILOCK BRAKE SYSTEM (ABS) SERVICE

Antilock brake systems have proven to be very trouble-free. In fact, less than 1% of problems in the brake system come from the ABS. Most of these problems result from damage to a wheel sensor. The output of a wheel sensor is checked while spinning a wheel and measuring sensor output. The sensor has a specified air gap that is checked, too.

Electrical problems are diagnosed through the vehicle's onboard diagnostics. Detailed ABS service manuals are available from brake manufacturers.

■ An annual fluid flush is a good recommendation with antilock brake systems.

■ Silicone brake fluids are not hygroscopic (do not accumulate water) like ordinary brake fluid. Water can accumulate in some areas of the system, causing problems.

Depressurizing ABS Pressure

Some ABS operate off of extremely high pressure. These systems have an **accumulator** that must be bled of pressure before attempting to service the system.

Before removing a caliper or master cylinder, be sure the system is depressurized. Most systems are depressurized by applying the brake pedal at least forty times while the key is off. The pedal should feel like a

normal power brake with no vacuum left in the reservoir. After restarting the vehicle, the ABS indicator light will go out when the system is repressurized.

⚠️ **CAUTION** Failing to bleed off pressure before opening a brake line or bleed fitting could result in serious injury to the technician.

Follow the correct procedure for inspecting brake fluid level. Check the service manual for the car or the underhood label. Some systems must be depressurized before checking. Others require the key to be on with the accumulator pressurized.

Bleeding ABS

Before bleeding ABS, always check the manual. There are many different procedures. Some require a dedicated scan tool. Without this, you will not be able to actuate the pump and valves to bleed air from the system. If you do not need a special tool to retrieve codes on the ABS, the system should be easily bled following procedures in the ABS repair manual.

A good rule to follow is not to let the ABS or master cylinder run dry. Then special bleeding procedures will not be required.

After bleeding the system, put the ABS into operation with a couple of hard stops during a test drive. This will help to get any remaining air out of the hydraulic control unit (HCU). It will not completely bleed the air from the device, but can help after the correct procedures are followed.

REVIEW QUESTIONS

1. A soft brake pedal due to air in the system is called a _____ pedal.

2. With a vacuum power brake booster, when you apply the brake pedal with the engine off you should hear the sound of ____ entering the booster.

3. What color is silicone brake fluid?

4. Does alcohol absorb water?

5. What do you do to a master cylinder before installing it on a car?

6. What is the name of the type of vacuum cleaner filter that is used to clean asbestos?

7. How much lining-to-drum clearance will there be after brakes are adjusted with a brake adjustment gauge?

8. What can you use to remove the threaded finish left after machining a brake rotor?

9. A _____ is used to push a piston to the bottom of the caliper bore before installing new brake linings.

10. A rebuilt brake caliper that includes all of the parts is called a _____ caliper.

ASE STYLE REVIEW QUESTIONS

1. Technician A says that a brake pedal that drops toward the floor under light pressure is probably due to a leaking master cylinder piston seal. Technician B says that worn drum brake linings will result in a lower master cylinder fluid level in one chamber. Who is right?
   a. Technician A   b. Technician B
   c. Both A and B   d. Neither A nor B

2. A vehicle has a high, hard pedal and sometimes the brakes remain applied. Technician A says that the compensating port in the master cylinder could be plugged. Technician B says that the brake pedal rod clearance adjustment might be too high. Who is right?
   a. Technician A   b. Technician B
   c. Both A and B   d. Neither A nor B

3. Technician A says that silicone brake fluid can be mixed with glycol fluid. Technician B says that air in a pressure bleeder acts directly on the brake fluid. Who is right?
   a. Technician A   b. Technician B
   c. Both A and B   d. Neither A nor B

4. A car has a low, firm brake pedal. Technician A says this can be due to brakes that need to be bled. Technician B says this could be due to brakes in need of adjustment. Who is right?
   a. Technician A   b. Technician B
   c. Both A and B   d. Neither A nor B

5. Technician A says that brake linings are purchased separately for each wheel. Technician B says the master cylinder is usually replaced or rebuilt as part of a brake job. Who is right?
   a. Technician A   b. Technician B
   c. Both A and B   d. Neither A nor B

6. Technician A says wheel cylinder cups must face away from fluid pressure. Technician B says a brake return spring should ring if dropped on a hard surface. Who is right?
   a. Technician A   b. Technician B
   c. Both A and B   d. Neither A nor B

7. Two technicians are discussing checks that are made on a disc brake rotor. Technician A says to check it for parallelism. Technician B says to check it for runout. Who is right?
   a. Technician A   b. Technician B
   c. Both A and B   d. Neither A nor B

8. Technician A says to put noise suppressor material on the friction surface of a disc brake lining before installing pads. Technician B says that master cylinder fluid consumption from the rear chamber could be due to a leaking power brake booster seal. Who is right?
   a. Technician A   b. Technician B
   c. Both A and B   d. Neither A nor B

9. Technician A says when there is a large spurt from the master cylinder compensating port when the brakes are released, there is air in the system. Technician B says that it is not necessary to hone a disc brake caliper during a rebuild because the bore is not the sealing surface. Who is right?
   a. Technician A   b. Technician B
   c. Both A and B   d. Neither A nor B

10. Technician A says emery cloth can be used to polish disc brake pistons. Technician B says to hone an aluminum master cylinder before installing a wheel cylinder kit. Who is right?
   a. Technician A   b. Technician B
   c. Both A and B   d. Neither A nor B

Bearings, Seals, and Greases

■ **KEY TERMS**

frictionless or
    anti-friction bearings
needle bearings
race
bearing cage
radial load
thrust load
wheel bearings
live axles
axle bearings
semi-floating axle
full-floating axles
grease
National Lubricating
    Grease Institute (NLGI)
dropping point
copolymer
spalling
brinelling
press fit
push fit

■ INTRODUCTION

Bearings of different types are found throughout the automobile. This chapter will first deal with fundamentals of bearings, seals, and lubricants. The last part of the chapter will cover bearing service. Bearing replacement for non-serviceable bearings found on driving axles is covered in Chapters 64 and 66.

■ PLAIN BEARINGS

Plain bearings are the kind used as engine crankshaft bearings. They do not use rolling parts and they provide sliding contact between two mating surfaces. **Frictionless, or anti-friction, bearings** provide a rolling contact. They have balls, rollers, or needles that allow easy rotation when any part moves.

■ FRICTIONLESS BEARINGS

Frictionless bearings are either *ball, roller,* or **needle bearings** (Figure 53.1). Bearings are made of hardened steel alloys. They are ground to a precise finish and size.

Frictionless bearings must be lubricated. Some bearings are sealed, having a seal on either or both sides of the bearing. The function of the seal is to keep grease in the bearing and dirt or oil out.

In a ball or roller bearing, the balls or rollers ride between an inner **race** (*cone*), or raceway, and an outer race (*cup*) (Figure 53.2). The balls or rollers are usually held in position in a **bearing cage** or *separator*. The cage, usually made of stamped steel or plastic, keeps them from bunching up and rubbing against each other. It also serves to keep the bearings properly

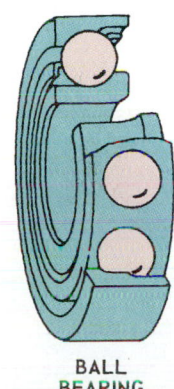
BALL
BEARING

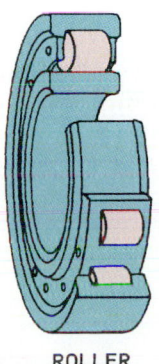
ROLLER
BEARING

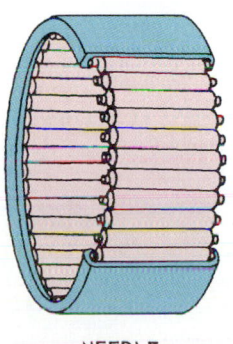
NEEDLE
BEARING

Figure 53.1 Types of frictionless bearings. *(Reproduced by permission of Deere & Company ©1992, Deere & Company. All rights reserved.)*

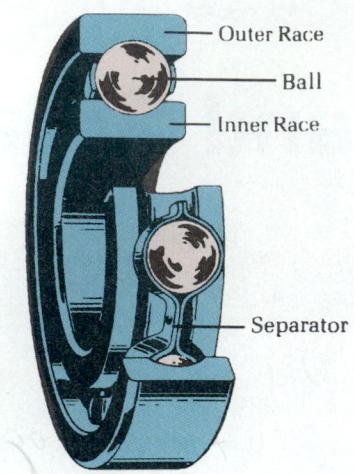

Figure 53.2 Parts of a ball bearing. *[Courtesy of Chicago Rawhide]*

located so that the load on them is evenly distributed. When the bearing is the type that can be disassembled, the cage prevents the loss of the parts.

◼ DIRECTION OF BEARING LOAD

Bearing manufacturers make bearings to handle different types of loads. A bearing load that is in an up-and-down direction, is called a **radial load** (Figure 53.3). When the load is in a front-to-rear direction, this is called a **thrust load** or *axial thrust* (Figure 53.4).

◼ BALL BEARINGS

Ball bearings ride in grooves that are ground into the surfaces of the inner and outer races (see Figure 53.2).

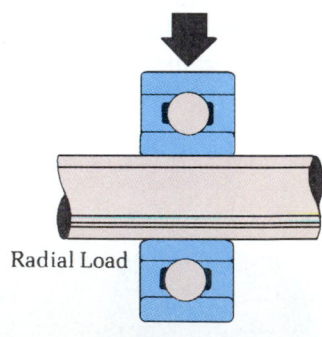

Figure 53.3 Direction of radial loads on a bearing. *[Courtesy of Chicago Rawhide]*

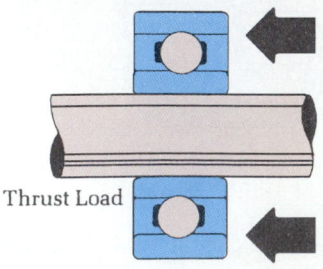

Figure 53.4 Direction of thrust loads on a bearing. *[Courtesy of Chicago Rawhide]*

A big disadvantage to a ball bearing is that most of the load of the vehicle is exerted on the bottom of the bearing, on a very small surface area of the ball and race. The rest of the balls do not do much in supporting the load. This can be an advantage though. Ball bearings have less friction than roller bearings and can operate at higher speeds. Figure 53.5 shows the difference between ball and roller contact areas.

When a ball bearing is at rest, its load is distributed equally wherever the balls and races are in contact with each other. When there is motion and the ball begins to roll, material in the race actually bulges out in front of the ball (Figure 53.6). Then, it flattens out behind the ball. Metal-to-metal contact cannot be allowed, so lubrication is critical.

Ball bearings control both end thrust (side movement) and radial movement (up and down) (Figure 53.7). Single row ball bearings are popular in transmissions, alternators, steering gears, and differentials. They are designed primarily for radial loads but can withstand thrust loads also. When more load capacity is needed, extra balls are added to the bearing. This reduces the bearing's thrust capacity, however.

When a ball bearing must control thrust, the groove in the bearing race will be offset to one side. Figure 53.8 shows a typical *radial thrust*, or *angular contact*, ball wheel bearing setup where two radial thrust control bearings face each other. This controls thrust in both directions.

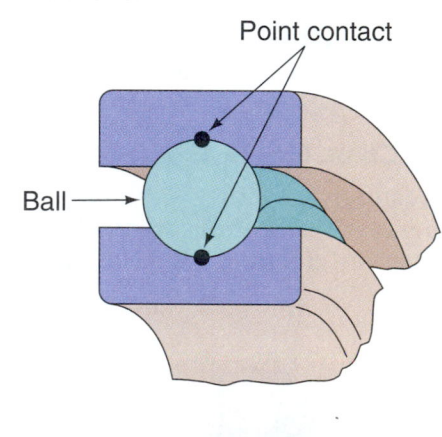

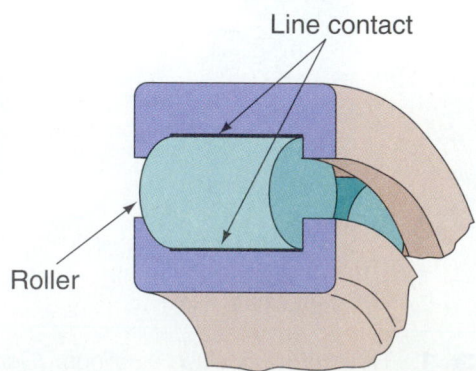

Figure 53.5 Ball and roller bearings have different contact areas. *[Courtesy of Ford Motor Company]*

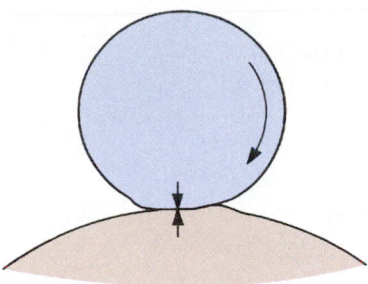

Figure 53.6 Metal bulges out in front of the ball when a load is applied to the bearing. *[Courtesy of Chicago Rawhide]*

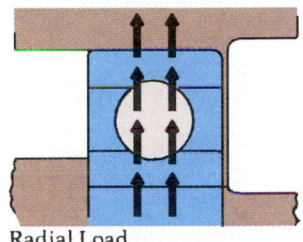

Radial Load

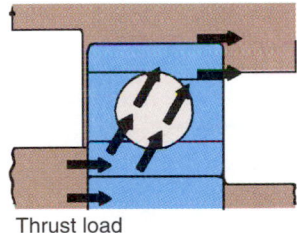

Thrust load

Figure 53.7 Ball bearings control both end thrust and radial movement. *[Courtesy of Chicago Rawhide]*

Nut

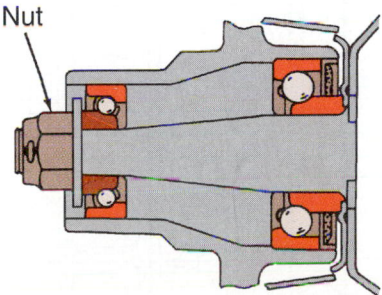

Figure 53.8 A radial thrust, or angular contact, ball wheel bearing set controls thrust in both directions. *[Reproduced by permission of Deere & Company. ©1992, Deere & Company. All rights reserved.]*

Single row bearings are susceptible to damage when a shaft is misaligned. When a ball bearing is required to have more thrust capacity or when misalignment is likely to be a problem, a *double row bearing* is used. This bearing is typically found in air conditioning clutches and front wheel drive front bearing hubs (Figure 53.9).

Figure 53.9 A double row front wheel ball bearing. *(Courtesy of SKF USA Inc.)*

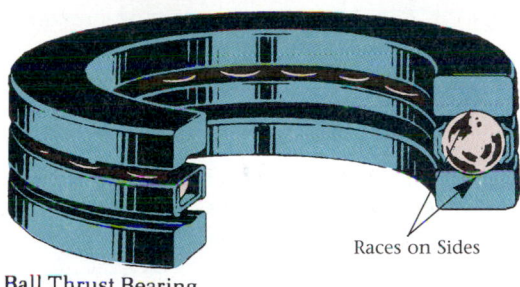

Ball Thrust Bearing   Races on Sides

Figure 53.10 In a ball thrust bearing, the races are faced sideways, rather than up and down. *[Courtesy of Chicago Rawhide]*

Ball bearings can also be designed primarily to control thrust. This is done by facing the races sideways, rather than up and down (Figure 53.10).

■ ROLLER BEARINGS

When a greater load carrying capacity is need, roller bearings are used instead of ball bearings. They provide more surface area of contact with the race and can carry a greater load (see Figure 53.5). There are several types of roller bearings. Some roller bearings have only an outer or inner bearing race. The shaft or bore will be precision ground and hardened to serve as the other race. This is often the case with RWD rear axles in light vehicles (Figure 53.11). Rear wheel drive cars have straight roller bearings on the rear axle. These bearings support radial loads well but cannot control thrust. Axles must be retained or keyed so that end thrust cannot be applied to the bearing.

Roller bearings do not control end thrust. A thrust bearing (or a pair of tapered roller bearings) must be used when end thrust is to be controlled. A thrust bearing or bushing is one that is mounted at 90° to the wheel bearing's load.

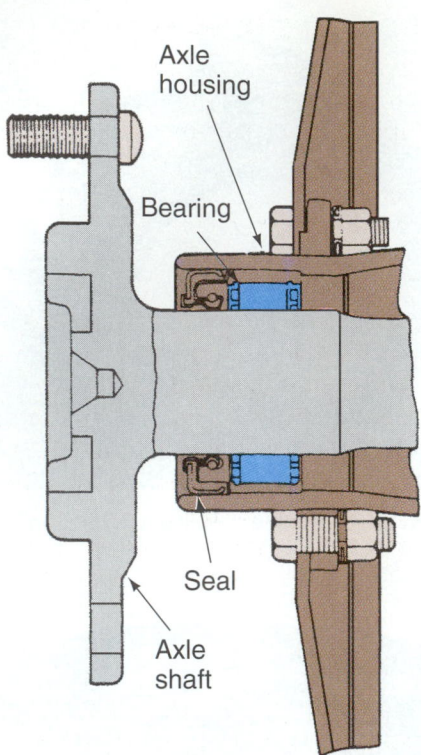

Figure 53.11 A straight roller rear axle bearing for a rear wheel drive car. *(Courtesy of General Motors Corporation, Service Technology Group)*

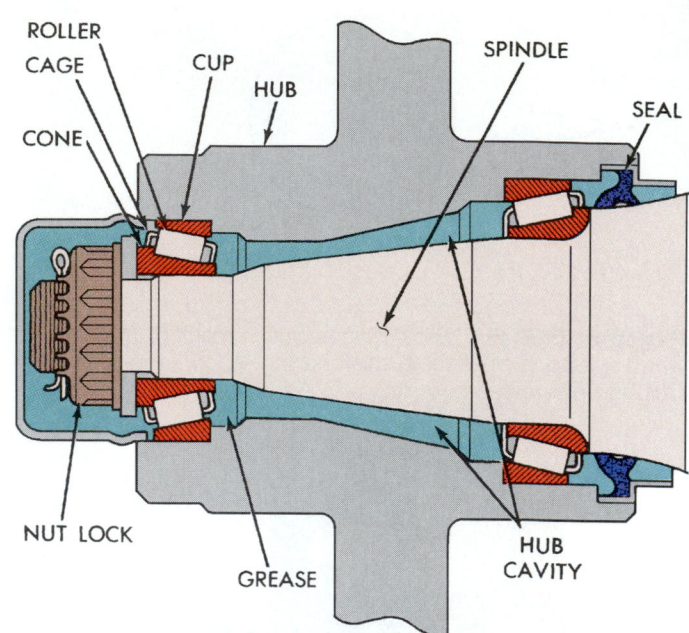

Figure 53.12 Tapered front wheel bearings are installed with their tapers facing in opposite directions. *(Courtesy of Chrysler Corporation)*

Tapered Roller Bearings

The most popular type of roller bearing for use in automobiles is the tapered roller bearing. Tapered roller bearings are used for front wheel bearings because they can control end thrust when two of them are installed with their tapers facing in opposite directions. The small diameters of the tapers face toward each other (Figure 53.12). In front wheel bearings, the larger of the two bearings is installed on the inside (where it supports the weight of the vehicle). The outer (smaller) bearing simply keeps the wheel aligned and provides end thrust control. Two outside bearing cups (races) are pressed into the *hub*.

Sometimes, two tapered roller bearings are used in a single bearing assembly. These are found where thrust loads are greater, such as on front wheel hubs on front wheel drive cars (Figure 53.13).

Tapered roller bearings tend to be self-aligning. The angles of the tapers of the bearing rollers are such that all of the angles meet at a common point (Figure 53.14). This makes each bearing roller align itself with the shaft.

Tapered roller bearings assembled with the inner and outer races as a single unit have lips on the inside and outside edges of the inner bearing race (Figure 53.15). This prevents the rollers from being able to slide out of the bearing.

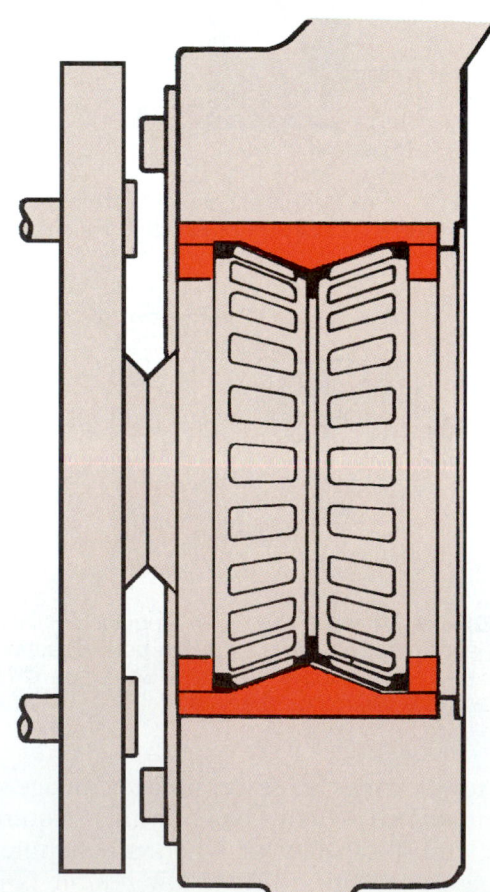

Figure 53.13 This FWD drive axle bearing has two tapered roller bearings. *(Courtesy of Chrysler Corporation)*

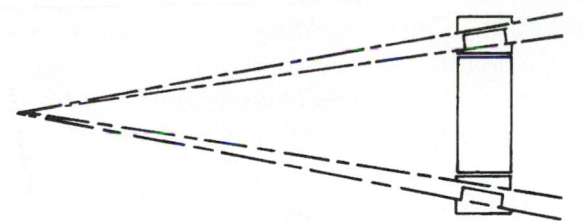

Figure 53.14 The angles of the tapers of the bearing rollers meet at a common point. *[Courtesy of Chicago Rawhide]*

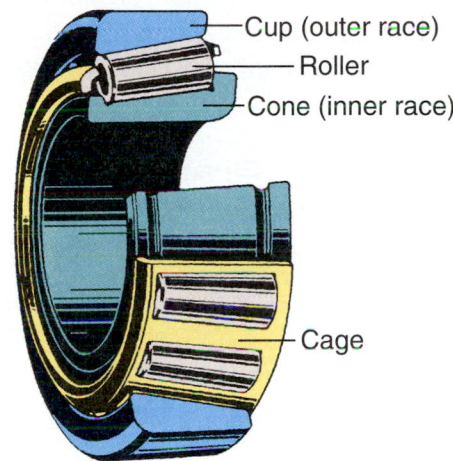

Cup (outer race)
Roller
Cone (inner race)
Cage

Figure 53.15 A tapered roller bearing has lips on the inside and outside edges of the inner bearing race. *[Courtesy of Chicago Rawhide]*

Needle Bearings

When a roller bearing is very small, it is called a needle bearing. Needle bearings can be used to control thrust or radial loads (Figure 53.16). They are most often used where space is limited and are not as good for high speeds as roller or ball bearings. Needle bearings do not tolerate misalignment. One example of a common use of needle bearings is in universal joints (see Chapter 68).

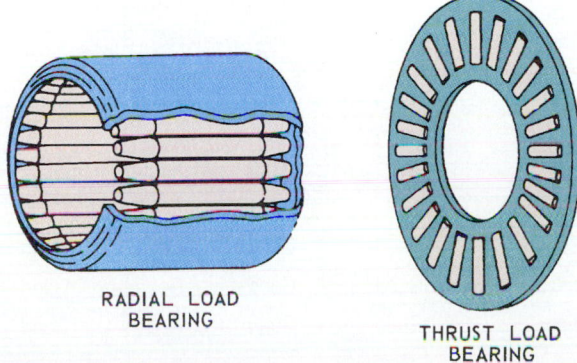

RADIAL LOAD
BEARING

THRUST LOAD
BEARING

Figure 53.16 Needle bearings can be used to control thrust or radial loads. *[Reproduced by permission of Deere & Company, ©1992. Deere & Company. All rights reserved.]*

■ WHEEL BEARINGS

Wheel bearings or axle bearings are found on all of the wheels of a vehicle. Some of them require service and others are simply replaced when they fail. The term, **wheel bearing**, refers to non-drive front and rear wheel (serviceable) bearings. Bearings on **live axles** (those that drive wheels) are referred to as **axle bearings** (non-serviceable).

Drive Axle Bearings

Drive axle bearings are located at the ends of the rear axle housing on a rear wheel drive car or on the hub on a front wheel drive car. Passenger car rear axles use a **semi-floating axle** design (Figure 53.17). This design has a bearing that rides on the axle. This design is unsafe for heavy loads. If the axle breaks, the wheel can fall off. This axle design is found on ½ ton pickup trucks. That is why they are not often used with large campers.

Full-Floating Axles

Full-floating axles are found on some light duty and larger trucks (usually ¾ ton and larger). In this axle design, the bearings do not touch the axle. They are located on the outside of the axle housing (Figure 53.18). If the axle breaks, the rear brake drum and wheel will still be supported. This axle design is for heavy loads.

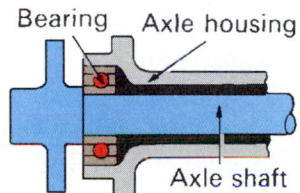

Bearing   Axle housing

Axle shaft

SEMI-FLOATING TYPE

Figure 53.17 A semi-floating axle has the bearing mounted on the axle.

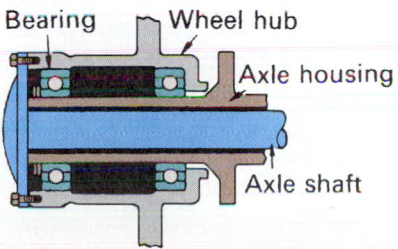

Bearing   Wheel hub
Axle housing
Axle shaft

FULL-FLOATING TYPE

Figure 53.18 A full-floating axle has the bearing mounted on the axle housing.

Front Wheel Drive Bearings

Front wheel drive bearings are compact because they must fit into a tight space. There are either a pair of ball or tapered roller bearings housed in the steering knuckle (see Figure 53.13). The small stub axle that the bearing supports is usually the end of the CV-joint. It has splines that attach it to the front hub. A nut holds the hub to the CV-joint.

■ GREASES

Lubricants have several purposes. They reduce friction and wear. They dissipate heat and protect metal surfaces from rusting. Seals are lubricated by lubricants also. This helps them keep the lubricant in and dirt out.

Greases are used in wheel bearings, chassis joints, universal joints, and gearboxes. Some components are equipped with grease fittings for replenishing grease. Others must be disassembled to replace the grease.

Grease is a combination of oil and a thickening agent. In some lubricating situations, grease has certain advantages over oil. Oil circulation systems are not needed, the tendency for leakage is reduced, and the lubricant remains in place after a long shutdown.

Grease does not leak or flow out of a bearing like an oil would. It acts like a liquid lubricant when a shearing force is applied to it, but the direction of fluid flow is only in the direction that the bearing turns. It does not flow out of the bearing because it is not fluid at right angles to the direction that the bearing is turning.

How Grease is Made

The properties of a grease are limited by the quality of oil that it is made of. Greases are usually 70–90% liquid petroleum oils made semi-fluid or solid by adding a thickening agent. The thickening agent, at a concentration of from 5–25%, is usually a soap such as *aluminum*, *lithium*, *sodium*, or *calcium*. Calcium soap is often used for pressure gun lubricant. Sodium is found in wheel bearing grease, and lithium is used for multipurpose grease. Additives make up 0–10% of a grease.

Soaps used in greases are similar to soaps made for household uses except they are soluble in oil and not soluble in water. Soaps have more ability to stick to metals than oils do. The lubricating oil in grease is held tightly against a metal surface where it can do a better job of lubricating and of protecting against rust than oil does. Lubricating greases are made to suit a particular application. Figure 53.19 show some of the different characteristics of greases.

■ SCIENCE NOTE ■

Aluminum, lithium, sodium, and calcium are not actually soaps, but soaps can be made from them by incorporating these atoms in place of the acidic hydrogen in an organic acid. Soaps used in greases are similar to soaps made for household uses. Household soaps, which are metallic salts of long-chain organic acids, appear to dissolve in water. They do not actually dissolve in water but form micelles in which the long-chain organic parts of the soap congregate together leaving the metal salt on the outside. It is the metal salt that is dissolved in water.

Greases are usually long-chain organic compounds. "Like dissolves like" so they dissolve and thus are held together by the long-chain portion of the soap.

Consistency of Grease

Greases are fibrous. Different sizes of fibers are available. Some are so large that they are visible to the naked eye, while others appear to be smooth.

The **National Lubricating Grease Institute (NLGI)** defines grease consistency in *NLGI grade numbers*. Numbers begin with grade #000 and increase in whole numbers through grade #6. The grade is determined during a penetration test in which a standard cone is penetrated into the grease under controlled circumstances of weight, time, and temperature. The depth in millimeters that the cone penetrates the grease is what determines the NLGI number of the grease. This standard denotes the consistency of a grease only and does *not* indicate quality or characteristics. A lower NLGI number relates to a softer grease (Figure 53.20).

When a grease is softer (lower NLGI number), it has a greater tendency to leak out. A grease that is too

Soap Type Comparison				
Soap Type	**Calcium Sulphonate**	**Polyurea**	**Lithium Complex**	**Lithium**
Texture	Smooth	Smooth	Smooth	Smooth
Dropping Point,°F	550	530	340/375	500+
Mechanical Stability	Excellent	Good	Good	Excellent
Shock Resistance	Excellent	Good	Good	Good
Water Resistance	Good	Good	Good	Good
Rust Protection	Good	Good	Good	Good
Adhesion	Good	Good	Good	Good
Temperature Resistance	Excellent	Excellent	Good	Excellent

Figure 53.19 Characteristics of different types of greases. *(Courtesy of Kendall Motor Oil)*

NLGI Consistency Grades

NLGI Grade	Worked Penetration at 25°C (77°F) mm/10
000	445 to 475
00	400 to 430
0	355 to 385
1	310 to 340
2	265 to 295
3	220 to 250
4	175 to 205
5	130 to 160
6	85 to 115

Figure 53.20 A lower NLGI number is a softer grease. *(Courtesy of Shell Lubricants)*

Thickener	Structure	Dropping Point (°F)	Max Service Temperature (°F)	Other Properties
Sodium Soap	Fibrous	350	200-275	Natural rust resistance, high water resistance, poor low-temperature properties
Calcium Soap Simple	Smooth	270-290	250	Excellent water resistance
Complex	Smooth, buttery	>450	300	Inherent extreme pressure load-carrying properties, good water resistance
Lithium Soap Simple	Smooth	390	325	Good water resistance, good mechanical stability
Complex	Smooth, slightly stringy	>450	350	Same as above
Aluminum Complex	Smooth gel	>450	300	Excellent water resistance, shear stability and pumpability
Clay	Smooth	>500	350	Nonmelting, very water resistant, marginal shear stability
Polyurea	Opaque, Slightly mealy	>450	350	Good oxidation resistance and water resistance
150				

Figure 53.21 Characteristics of various greases. *(Courtesy of The Lubrizol Corporation)*

thick can cause friction losses, become channelled during use, and not perform satisfactorily. An *all-season lubricant* is used in most climates, but thicker and thinner greases are available for severe situations.

Temperature Characteristics

A grease does not have a sharp (definite) melting point. As it is heated, it will become softer until it becomes fluid. A general idea of the safe temperature range of a grease can be gotten by referring to its **dropping point** (melting temperature). This is the temperature at which the grease turns into a liquid.

To test for its dropping point, a sample of grease is heated in a special test cup that has a small orifice (hole) of a specific size. The dropping point is the temperature when the first drop of material falls from the orifice. This is *not* the temperature that it would be safe to use the grease. According to Quaker State, greases should generally be used at temperatures no higher than 100° below their dropping points. Figure 53.21 shows dropping points of various thickeners.

Viscosity

The viscosity of the oil used in making a grease is important to the grease's *apparent viscosity*. A lubricant must remain fluid in cold weather while still having enough body to protect parts. The ability of a lubricant to maintain its viscosity across a wide variation in temperatures is called its viscosity index.

Lubricant manufacturers make up their greases to fit many applications. Figure 53.22 is a photo of two samples of the same base oil taken at –40°C. The oil on the left is frozen on a popsicle stick. At the right is the

Figure 53.22 The frozen sample of oil on the left is the same viscosity of oil as the grease on the right. The grease has had a thickening agent added and is still able to flow. *(Courtesy of Imperial Oil Limited)*

same oil with a thickening agent added to make a grease. The grease is still able to flow at that low temperature, while the oil is solidified.

Because the rate of oil oxidation doubles with every 20°F increase in temperature, when a bearing runs hot its grease will require more frequent changing. A grease with a life expectancy of 1000 hours at 100°F would have a 500-hour life expectancy if used at 120°F. Grease made from petroleum should not be used at temperatures in excess of 325°F unless it is being changed every hour or two.

Grease Performance Classification

Automotive grease is classified by NLGI for performance. This standard is based on an ASTM standard grease specification approved in May 1990. Automotive grease is classified in two general groups. If the standard has an "L" prefix, the grease is for the lubrication of suspension components such as ball joints and steering pivots. If the prefix is "G," the grease is intended for wheel bearings. Figure 53.23 shows a further breakdown of the classifications.

Some greases might overlap and be suitable for both chassis and wheel bearings. The NLGI has a logo, which can be displayed on the product (Figure 53.24).

Only the highest of each category of a combination of the two can be included in the logo.

Extreme Pressure Lubricants

Extreme pressure (EP) lubricants, which are added to some greases, are the same as those found in gear lubricants (see Chapter 68). Metallic soap thickening agents add a small amount of EP effect. Graphite or molybdenum do not add EP properties, but they increase anti-wear and frictional modification characteristics.

Chassis Lubricants

A chassis lubricant is a grease of a consistency that allows it to be applied through a zerk fitting with a grease gun. It must adhere to the bearing surface and seal out dirt and water. Chassis lubricant is highly resistant to being washed away with water.

Chassis parts move up and down repeatedly, which can break down the structure of the grease. Grease that does not have adequate *shear resistance* will break down, become like oil, and flow away from the surfaces where it is needed.

LA — Service typical of chassis components and universal joints in passenger cars, trucks, and other vehicles under mild duty only. Mild duty will be encountered in vehicles operated with frequent relubrication in noncritical applications.

LB — Service typical of chassis components and universal joints in passenger cars, trucks and other vehicles under mild to severe duty. Severe duty will be encountered in vehicles operated under conditions which may include pro-longed relubrication intervals, or high loads, severe vibration, exposure to water or other contaminants, etc.

GA — Service typical of wheel bearings operating in passenger cars, trucks and other vehicles under mild duty. Mild duty will be encountered in vehicles operated with frequent relubrication in noncritical applications.

GB — Service typical of wheel bearings operating in passenger cars, trucks, and other vehicles under mild to moderate duty. Moderate duty will be encountered in most vehicles operated under normal urban, highway, and off-highway service.

GC — Service typical of wheel bearings operating in passenger cars, trucks, and other vehicles under mild to severe duty. Severe duty will be encountered in certain vehicles operated under conditions resulting in high bearing temperatures (disc brakes). This includes vehicles operated under frequent stop-and-go service (buses, taxis, urban police cars, etc.), or under severe braking service (trailer towing, heavy loading, mountain driving, etc.).

Reference: ASTM Designations D 4950

Figure 53.23 Automotive grease classifications. *(Reprinted with permission from the National Lubricating Grease Institute)*

Figure 53.24 NLGI logos. *(Reprinted with permission from the National Lubricating Grease Institute)*

Wheel Bearing Grease

Wheel bearing lubricant is grease with a high resistance to heat. It is usually *thicker* and specially formulated for use with various types of bearings. Because the grease is on a spinning part, it must resist the tendency to be thrown off. A poor quality bearing grease can cause a safety hazard by leaking onto the brake linings. It can also cause bearing failure due to lack of lubrication.

Universal Joint Grease

Universal joint grease is made specifically for universal joints. Some universal joint designs require special lubricants. Be sure to follow manufacturer's recommendations.

Multipurpose Grease

A multipurpose grease satisfies the requirements of chassis, wheel bearing, and universal joint lubricants. It is the most common type of grease used in service shops. But multipurpose does *not* mean all-purpose. It meets only certain requirements. The lubricant most often used in chassis lubrication is a multipurpose *lithium-based grease*.

Solid Lubricant Greases

Some greases contain solid lubricant materials such as *molybdenum (moly)* or *graphite*. These are often used to lubricate speedometer cables, emergency brake cables, splines, and leaf springs.

▩ WHEEL BEARING SEALS

Automobiles and equipment can use seals for the following reasons:
- ▪ To seal in lubricants
- ▪ To keep different lubricants separated
- ▪ To keep out dirt
- ▪ To maintain vacuum or pressure

Seals are either *dynamic* (for sealing moving parts) or *static* (for sealing fixed parts) (Figure 53.25).

Wheel seals are usually lip seals (Figure 53.26). There is usually a garter spring behind the seal lip to further increase the pressure of the seal on the shaft. The spring can be separate or cast into the seal itself. The open side of the sealing lip always faces the lubricant (Figure 53.27). This is so that any pressure from the lubricant will increase the wiping pressure of the seal on its shaft.

Some seals have more than one lip; one is the *sealing lip* and the other is the *auxiliary lip* for sealing dust (Figure 53.28). Auxiliary seals are only used when there is a large amount of dirt because the auxiliary lip causes increased frictional heat.

Some bearings are sealed on one or both sides. When there is a seal on both sides, the bearing is not serviceable, but must be replaced if there is a problem. If lubricant leaks from one of the seals, replace the

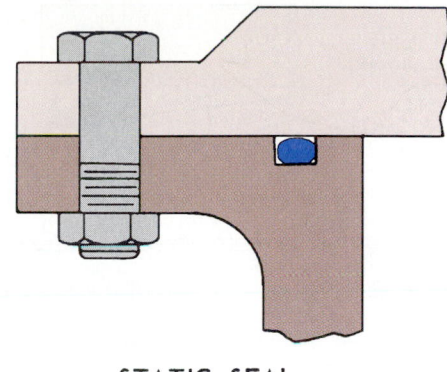

STATIC SEAL

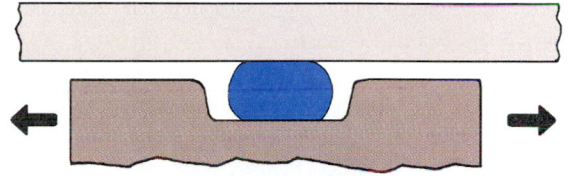

DYNAMIC SEAL

Figure 53.25 Static seals do not move. Dynamic seals work on moving parts. *(Reproduced by permission of Deere & Company, ©1992. Deere & Company. All rights reserved.)*

Figure 53.26 A lip seal. *(Reproduced by permission of Deere & Company, ©1992. Deere & Company. All rights reserved.)*

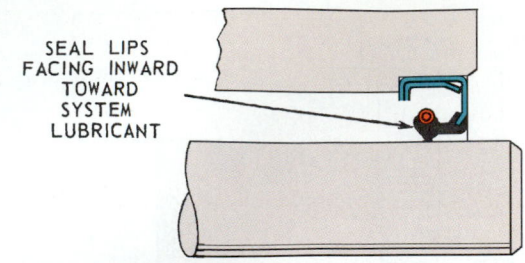

Figure 53.27 The open side of the sealing lip faces the lubricant. *[Reproduced by permission of Deere & Company, ©1992. Deere & Company. All rights reserved.]*

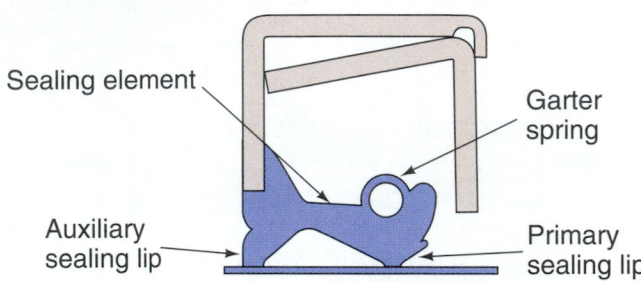

Figure 53.28 This seal has an auxiliary dust sealing lip. *[Courtesy of Chicago Rawhide]*

bearing. Be sure not to immerse the bearing in solvent. If solvent gets into the bearing, it will dilute the lubricant. This can lead to bearing failure.

Often, there is a seal on only one side of a bearing. This type of bearing is commonly found on the rear axle of RWD vehicles. These bearings are lubricated by oil from the differential.

■ SEAL MATERIALS

Seals are made of different types of materials depending on their intended use. The most popular seals are made of synthetic materials. These materials include nitriles, polyacrylates, silicones, and fluoroelastomers.

Leather used to be a common material for lip seals. Today most lip seals are made of *nitrile*. Nitrile is a grey-black or shiny black mixture of two synthetic rubbers: the polymers Buna and acrylonitrile. The combination of the two polymers is called a **copolymer**.

> ### ■ SCIENCE NOTE ■
>
> *A polymer is defined as a substance consisting of giant molecules formed from smaller molecules of the same kind. A continuous chain results from the repeated addition of small molecules. This chain is the polymer. Polymers are generally known as plastics.*

The disadvantage to nitrile seals is that they are not resistant to some synthetic oils. They are good with most mineral oils and greases. Their operating range is between –65°F to 225°F.

Polyacrylate seals are *elastomers* designed for higher temperatures (to 300°F–350°F depending on the lubricant). They are also resistant to extreme pressure lubricants. Polyacrylate is often used for differential pinion bearing seals. They look the same as nitrile seals.

Silicone seals are often used in engines and transmissions. They are usually colored gray, red, orange, or blue. Silicone is a polymer that is very flexible and absorbs lubricant. Silicone seals are resistant to friction and wear. They can operate at temperatures of from 325°F to 350°F depending on the lubricant used.

Fluoroelastomers are used with special lubricants and chemicals, which the other seal materials cannot handle. Viton is one popular fluoroelastomer. An example of its use is for carburetor inlet needle valves and other fuel system applications because of its resistance to methanol fuels. The color can be brown, black, blue, or green.

Non-Synthetic Seals

Leather used to be the main seal material used. It is expensive, though, and has been mostly replaced by synthetics. Some truck seals are still made of leather, however. Leather does not do a good job around moisture or higher temperatures (above 200°F). To tell if a used seal is synthetic rubber or leather, scratch its surface. It will turn to a rougher, lighter color like brown or beige if it is leather. Rubber will appear smooth and black.

Felt is a material that is sometimes used to keep out dirt. It is made of wool, which absorbs oil well and keeps out dirt. It does not do a good job of keeping oil confined and can also absorb water, contributing to rusting.

Other Seals. When a lubricant is to be directed back to its source, the lip of the seal is sometimes fluted (Figure 53.29). A newer seal design is the *wave*

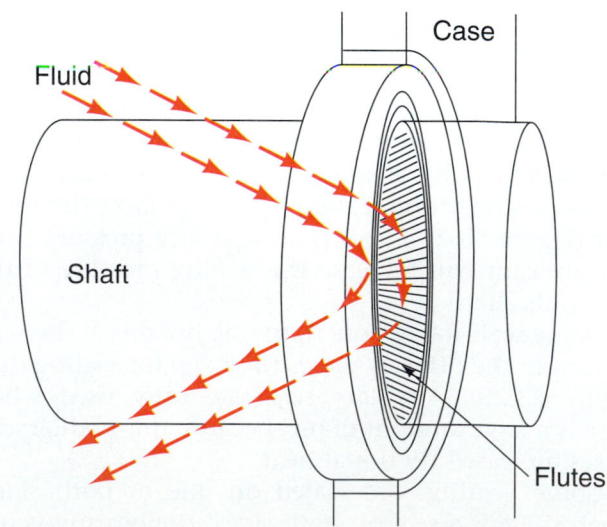

Figure 53.29 This fluted lip seal directs oil back to its source. *[Courtesy of Chrysler Corporation]*

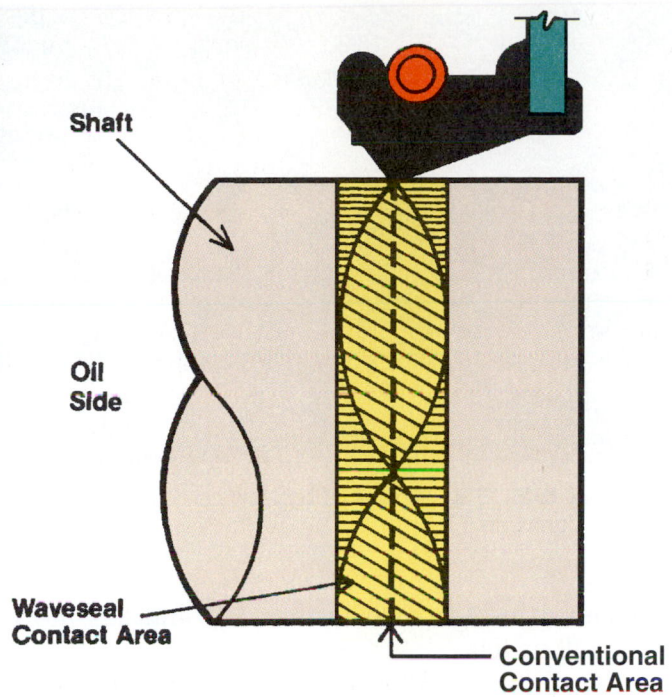

Figure 53.30 A wave seal oscillates, pumping oil back toward its source. *[Courtesy of Chicago Rawhide]*

seal. This seal does not ride on the shaft in a straight line like a conventional seal. It oscillates, pumping oil back toward its source (Figure 53.30). It has less friction, lasts longer, and works equally well no matter which direction the shaft turns.

■ SEAL TOLERANCE

Seals can usually accommodate a shaft that is undersized up to 1/64" (0.016") but only if all parts are in per-fect alignment. The Rubber Manufacturer's Association (RMA) recommends that runout tolerance be held to + or −0.003" for shafts up to 4 inches in diameter. The surface finish should be smooth (10–20 microinches).

■ WHEEL BEARING DIAGNOSIS AND SERVICE

Seal and Bearing Failure

If the lubricant can leak out, moisture can leak in. Also, once grease has leaked out, the bearing will soon suffer shock loads due to the lack of lubrication. Pieces will come off the bearing and races. The pieces will circulate around the bearing, resulting in heat buildup and failure of the bearing.

Periodic maintenance to bearings on non-drive axles consists of cleaning the bearings, repacking them with grease, and adjusting the clearance or preload after reinstallation. Parts of the wheel bearing assembly are shown in Figure 53.31. On drive axles, the only service done to a bearing is to press it off and replace it if it is bad. That service is covered in Chapter 69.

Boat Trailer Bearing Failures. Boat trailer bearings are well sealed and often have abundant high-quality grease. Boat trailers on smaller boats use tires that are of a smaller diameter than normal too. This causes the bearings to rotate at a higher than normal rpm.

One common problem happens when the boat trailer has just been towed for a long distance. Bearings are at operating temperature when the boat trailer is backed into the water. As the warm air inside of the bearing shrinks, cold lake or ocean water is drawn into the bearing. If the trailer is parked for a long period of time before it is used again, the balls or rollers in the bearings can rust.

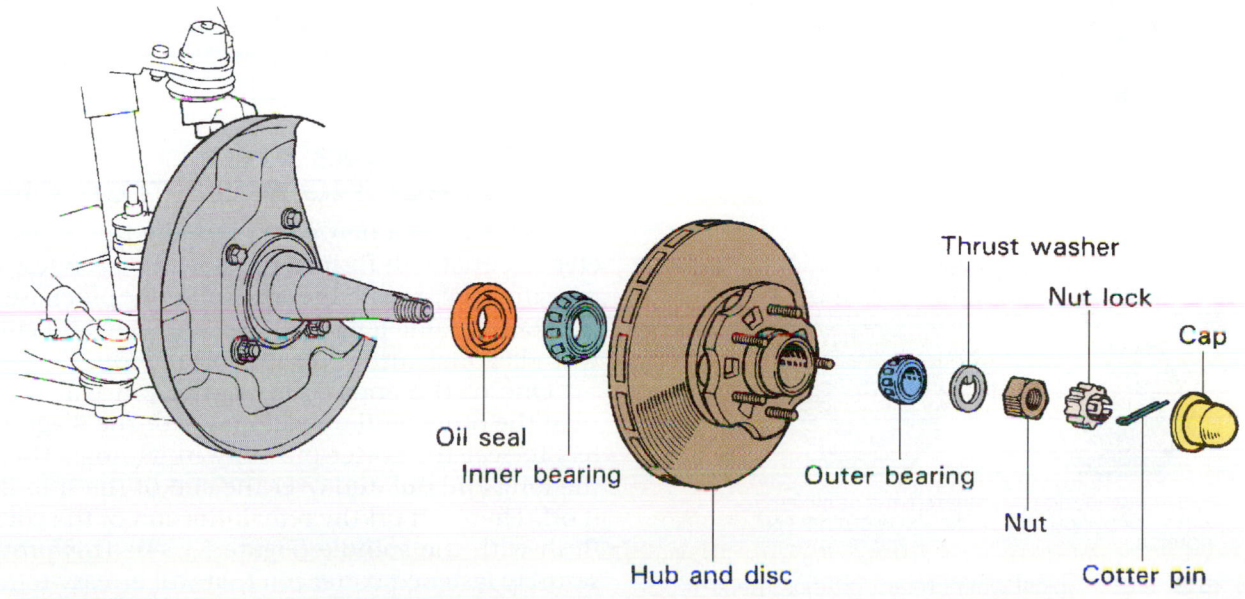

Figure 53.31 Parts of the wheel bearing assembly.

■ WHEEL BEARING ADJUSTMENT

Wheel bearings must be properly adjusted. If they are too loose, the wheel will be able to move. This can mimic problems that result from worn parts. Some of these problems are shake, noise, wander, steering wheel play, cupped tire wear, and an intermittent low brake pedal on disc brake cars.

The bearing is designed to operate with little or no clearance. This is called "0-lash." As a bearing rotates under a load, some heat is developed and the bearing expands. A bearing adjusted too tightly will develop more friction, use more power, and may ultimately fail.

Manufacturers list methods of adjusting their bearings (Figure 53.32). A common adjustment will result in the bearing having from about 0.001"–0.005" of running clearance. This is somewhere around the thickness of a hair.

An easy method of adjusting a loose bearing can be done with the tire raised off the ground. There is a *dust cap* that is pressed into the hub at the end of the spindle. The dust cap is removed with a large slip-joint pliers or a special dust cover tool (Figure 53.33).

Some spindle nuts have lock tabs or lock nuts, but most are kept in place with a cotter pin. Remove the cotter pin so the spindle nut can be tightened. Using both hands, grasp the top and bottom of the tire and try to rock it back and forth (see Figure 60.2). As the

Figure 53.33 A special tool for removing a wheel bearing dust cap.

spindle nut is tightened further, less and less movement will be felt.

The washer under the spindle nut has a tab that fits into a groove on the spindle (see Figure 53.31). Its function is to keep the bearing from trying to turn the spindle nut. It fits the spindle loosely so it should be easy to move.

SHOP TIP As a test to see that the bearing is not too tight, insert a screwdriver under the tabbed washer that is under the spindle nut and pry the washer up and down. There is enough slack between the washer and the spindle that it should move freely. If the washer is hard to move, the adjustment is too tight.

NOTE: *The above shop tip works on most cars; however, be certain to always check the service manual for the correct procedure when you are working on a make of automobile for the first time.*

■ SELECTING AND INSTALLING A COTTER PIN

When selecting a new cotter pin, use the largest diameter one that will fit into the hole. To keep less inventory in stock, repair shops commonly purchase cotter pins of the same length (2" for instance) and cut them with diagonal cutters during installation.

One of the ends of the cotter pin will be longer than the other. Pull on this end with the diagonal cutters to seat the cotter pin fully in its hole. Then, pull the long end out and over the end of the spindle. Cut it off. Then, cut off the remaining end of the cotter pin flush with the spindle (Figure 53.34). This provides a securely fastened cotter pin that will be easy to remove during the next service.

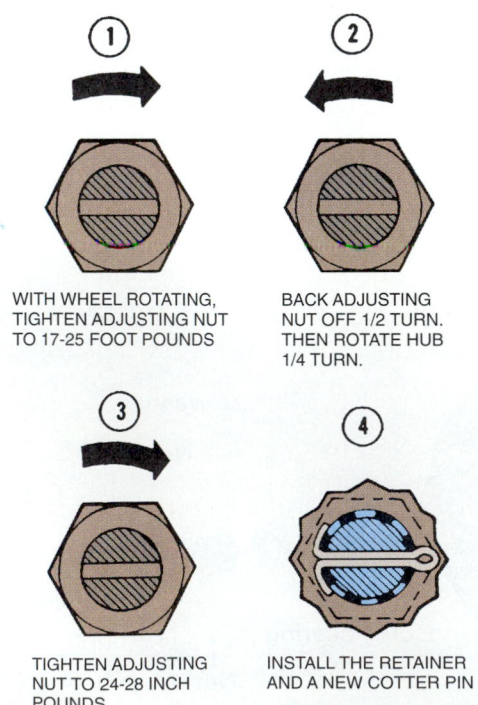

① WITH WHEEL ROTATING, TIGHTEN ADJUSTING NUT TO 17-25 FOOT POUNDS

② BACK ADJUSTING NUT OFF 1/2 TURN. THEN ROTATE HUB 1/4 TURN.

③ TIGHTEN ADJUSTING NUT TO 24-28 INCH POUNDS

④ INSTALL THE RETAINER AND A NEW COTTER PIN

Figure 53.32 Typical wheel bearing adjustment directions. *(Courtesy of Ford Motor Company)*

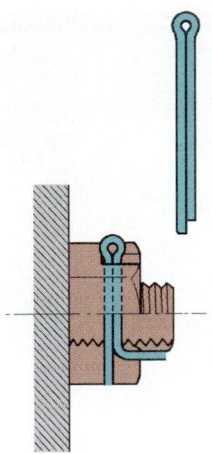

Figure 53.34 Cotter pin installation.

➡ *Perform Adjust a Tapered Roller Bearing Worksheet*

■ REPACKING WHEEL BEARINGS

Tapered wheel bearings are lubricated with grease, which also protects the metal from corrosion and helps carry away heat. Front wheel bearings are sealed from the elements, but over a period of time, moisture, brake, and road dust can accumulate. It is customary to clean and repack the wheel bearings with grease whenever the front brakes are relined or at 30,000 mile intervals.

On FWD (front wheel drive) cars, the procedure for repacking the rear axle bearings is similar to the procedure for repacking the front axle bearings on RWD cars. On drum brake cars, leave the lug nuts tight. It is not necessary to remove the wheel and tire assembly from the hub. Do one side at a time so that parts are not accidentally interchanged from side to side.

After removing the dust cap and cotter pin, remove the spindle nut and the tabbed washer that is under it. To easily remove the outer bearing, rock the tire back and forth at the top. The bearing will usually pop out on the spindle (Figure 53.35).

Remove the Seal

Pull the wheel and hub from the spindle. When working with the wheel bearings, be sure to keep grease and solvent off of braking surfaces.

Seals are usually replaced during a bearing repack. Using the old seal over is taking a chance on grease leaking out and contaminants leaking into the bearing. To remove the seal, use a long dowel or drift to pound on the bearing from the inside (Figure 53.36). The seal can also be removed with a screwdriver (Figure 53.37).

NOTE: *Keep the bearing parts together so they can be reassembled in their original bearing races. Bearing parts become wear-mated to each other.*

Figure 53.35 Remove the outer bearing. *(Courtesy of The Timken Company)*

Figure 53.36 Use a long dowel or drift to pound on the bearing from the inside. *(Courtesy of Chicago Rawhide)*

Clean out the old bearing; do not just add grease. Use a paper towel to wipe all of the old bearing grease from the spindle, the bearing, the bearing cups, and the hub. Do not waste a shop rag for this. Laundry service is expensive. Clean the bearing with solvent and let it air dry on a paper towel or blow it dry from the ends (parallel to the rollers) using compressed air (Figure 53.38). Be careful not to blow the old dirty grease into the bearing. Rewash the bearing and blow dry it again.

NOTES:
- ■ *Do not spin the bearing with air. Spinning a dry bearing can damage it and can also be dangerous if bearings come loose from the bearing cage.*
- ■ *When blowing off parts, blow into the solvent tank to avoid making a mess.*

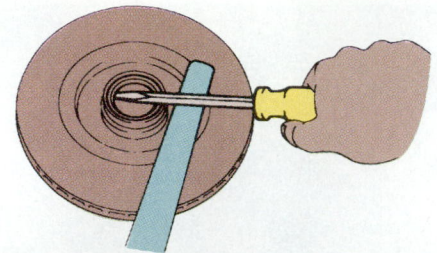

Figure 53.37 Using a screwdriver to remove a bearing.

Figure 53.38 Hold the bearing to keep it from spinning as you blow it off. *(Courtesy of The Timken Company)*

Figure 53.39 A spalled bearing and race. *(Courtesy of Chicago Rawhide)*

■ BEARING INSPECTION AND DIAGNOSIS

After the bearings are cleaned, inspect them for damage. If a bearing is damaged, the cause should be determined so the problem does not happen again. Check the bearing and bearing race for pitting and other signs of damage. If any damage is apparent, replace the bearing and its race. Save the old bearing to compare it with the new one. Bearing replacement is covered later in this chapter.

Bearings usually fail slowly. Noise is the clue that something is wrong. As metal is deformed or comes off the bearing races, the noise will become more pronounced. There are two main kinds of damage to a bearing, spalling and brinelling.

Spalling is when pieces break off of the bearing metal (Figure 53.39). The noise from spalling is a random, high-pitched sound. **Brinelling** is when the bearing or race has indentations from shock loads (Figure 53.40). The noise that results from brinelling damage is a regular, low-pitched sound.

The biggest shock load on a bearing is often during installation. Pressing should be done directly on the race that is a pressed-fit only. Another cause of bearing damage can be due to an improperly grounded arc welder.

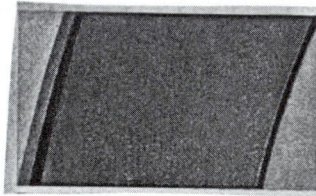

Figure 53.40 Dents in the bearing race, called brinelling, result from the roller hammering against the race. *(Courtesy of Chicago Rawhide)*

CASE HISTORY

A vehicle had a trailer hitch welded to its frame using an electric arc welder. The owner went on a vacation pulling his new trailer. Shortly into the trip, a front wheel bearing failed. This happened because the negative cable to the welder was not clamped directly to the frame but to the energy absorbing bumper mount. The vehicle had a broken ground strap between the engine and chassis. Electricity always takes the shortest path to ground. This time it went through the wheel bearing (Figure 53.41).

Figure 53.41   This bearing was damaged by electrical current. *(Courtesy of Chicago Rawhide)*

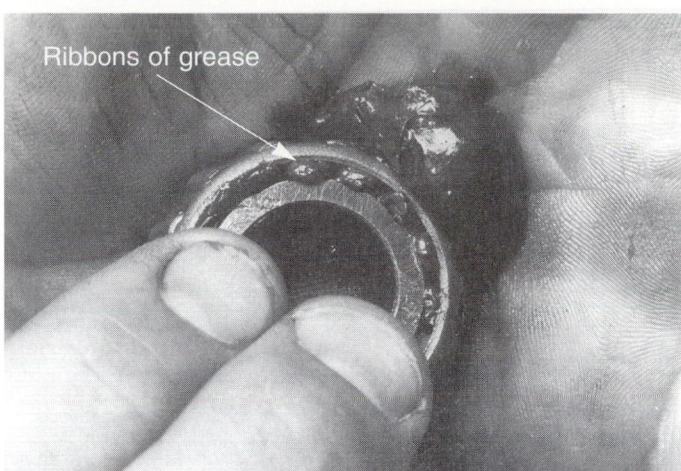

Figure 53.42   Apply grease until ribbons of grease start to appear at the opposite end of the bearing. *(Courtesy of The Timken Company)*

Figure 53.43   A pressure bearing packer. *(Courtesy of The Timken Company)*

Adding Grease to Bearings

To pack bearings by hand, put a small amount of grease in the palm of your hand. Stroke the large open end of the bearing cage against the grease until ribbons of grease start to appear at the opposite ends of the bearing (Figure 53.42). Smear a fresh layer of grease all around the bearing and on the clean bearing races.

A *pressure bearing packer* is owned by most shops. A grease gun is applied to a grease fitting to pack the bearing (Figure 53.43).

After the bearings are packed with fresh grease, place them on a paper towel. Remember to reinstall them into the same bearing races. Leave a ring of grease below the bearing race to help keep the fresh grease inside the bearing area after it heats up and begins to flow.

Grease the Inside of the Hub

Put a small amount of grease in the cavity of the hub. Do not fill it up. According to bearing engineers, the amount of grease that is packed into a bearing determines how well it will work. If the cavity is full, there is no place for the excess grease or pressurized air to go when the bearing heats up. If the hub is full, there is no place for excess grease in the bearing to go. This results in excess heat and fluid friction, which can cause the bearing or seal to fail.

High-speed bearings, like those on race cars, are supposed to be filled to only 25% of their free space. When a bearing is totally full, it should be used at low speeds only. Figure 53.44 shows the recommended fill for passenger cars.

Bearing Race Installation

Anti-friction bearings usually have one race that is **press fit**. The other race is a **push fit**. A push-fit race slides into place by hand. The pressed-fit race is usually pressed onto or into the rotating part (outer race in brake hub). The push-fit race is usually pushed onto or into the stationary part (inner race on spindle).

Inspect the Spindle

Just behind the area where the bearing rides on the spindle, there is a raised area that the seal rides on

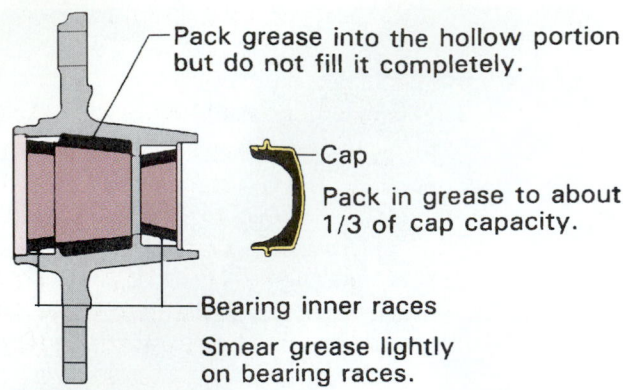

Figure 53.44 Partly fill the hub with grease.

Figure 53.45 Clean the sealing area on the spindle so that the new seal is not accidentally ruined. *(Courtesy of The Timken Company)*

Figure 53.46 Inspect the bearing races in the hub. *(Courtesy of The Timken Company)*

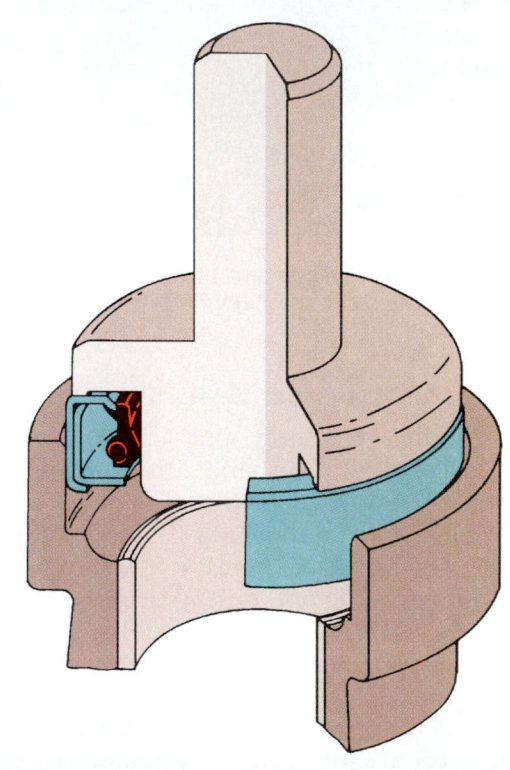

Figure 53.47 A special tool for installing seals. *(Reproduced by permission of Deere & Company, ©1992. Deere & Company. All rights reserved.)*

(Figure 53.45). Clean this area so that the new seal is not accidentally ruined. Inspect the condition of the spindle and check the fit of the large bearing on it.

When a wheel bearing fit is too loose on the spindle, an old trick is to use a punch to raise the level of the spindle so that the bearing fits snugly. This is not recommended as it will cause misalignment problems with the bearing. Also, if the spindle is worn and breaks later, the shop can have a liability.

NOTE: *A worn spindle is usually caused by a bearing seizure that forces the inner race of the bearing to spin on the spindle. The resulting heat softens the hardened spindle, which requires its replacement.*

A bearing is designed to *creep* on the spindle when loaded, so sometimes there are marks on the spindle. This is normal as long as the bearing feels snug and yet moves freely on the spindle. Inspect the bearing cups in the hub (Figure 53.46).

After the inner bearing is installed in the hub, the seal can be installed. Be careful to install it straight. A

special tool is helpful for this (Figure 53.47). Install the seal with its open end, or lip, facing in toward the bearing (see Figure 53.27). Be sure to lubricate the seal lip so it does not burn up (Figure 53.48). Adjust the wheel bearing and install the cotter pin according to the instructions provided earlier.

Pack some grease into the dust cap. Do not fill it all the way though. This keeps out contaminants and provides a reservoir for fresh oil. As the grease in the bear-

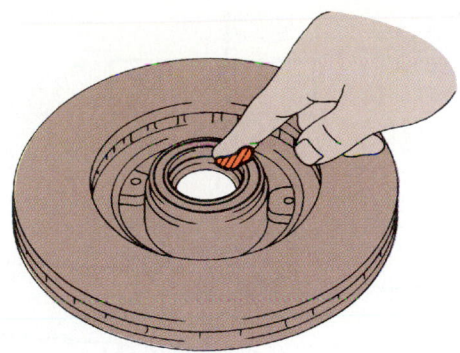

Figure 53.48 Lubricate the seal lip before installing the bearings on the spindle.

ing oxidizes, it can dry out. Oil in the fresh grease in the cap can replenish the old grease through *capillary action*.

➡ *Perform Repack Tapered Wheel Bearings Worksheet*

◼ REPACKING DISC BRAKE WHEEL BEARINGS

The procedure for repacking disc brake wheel bearings is the same as that followed for drum brakes, except that the disc *caliper* must be removed in order to gain access to the inside wheel bearing. The caliper must be supported or wired to the steering knuckle support. Do not let calipers hang on brake hoses. The inside of the brake hose can be damaged. When reinstalling the caliper, torque the caliper bolt to specifications. When the caliper bolts have an allen head, a socket drive allen head tool should be used with a torque wrench.

➡ *Perform Repack Wheel Bearings (Disc Brake) Worksheet*

◼ DIAGNOSING WHEEL BEARING NOISE

If a possible wheel bearing groan is heard, driving the car can sometimes help pinpoint the problem. First, check the tires for damage and be sure they are properly inflated. Find an empty parking lot or deserted road and make slow left and right turns. This shifts the weight of the vehicle from one side to the other. When the weight is increased on the bearing, the noise increases. The inside tire always turns at a lower rpm than the outside tire.

◼ The noise from a wheel bearing that is bad will change pitch as the wheel speeds up and slows down when turning from one side to the other.

◼ When the outside wheel has the bad bearing, the noise will become worse because that wheel is turning faster.

◼ Applying the brakes can also cause the noise level from a bad bearing to become less as they contact the drum or rotor with the bad bearing.

◼ Spinning the wheels can be done using an on-the-car wheel balancer for non-drive wheels.

◼ REPLACING BEARING RACES

When a damaged wheel bearing is replaced, the pressed-fit race must be removed from the bearing hub. The race might still look good to the eye but has been subjected to loads for just as long as the failed bearing. Leaving the old race to be used with a new bearing is an invitation for failure. Manufacturers plan on all but 10% of a given set of bearings lasting about 150,000 miles.

The old bearing race is removed by pounding it with a drift punch or a special tool. When using a drift punch there are recesses in at least two places in the side of the hub that allow for hammering on the back of the race. Be sure to hammer a little bit from one side then the other, so the race is not distorted during removal, damaging the hub. A special tool or a flat bar can be used to remove the race (Figure 53.49).

The new race must fit the hub tightly. If not, replacement of the hub is usually required. A loose bearing race is often the result of a bearing seizure, which can be caused by a bent spindle. Drive the race into the hub, being careful not to cock it off to one side. If using a soft punch (Figure 53.50), hit on one side of the cup and then on the other. Drive it all of the way into the bore until it seats solidly against the ridge at the bottom of the hub. When the race bottoms out, the sound from the pounding will change.

A special tool is handy for installing a bearing race (Figure 53.51). A new race can be chilled in a refrigerator to make it easier to install. If a bearing is allowed to run in a misaligned position, it will be overloaded and fail (Figure 53.52).

➡ *Perform Replace Wheel Bearing Cup Worksheet*

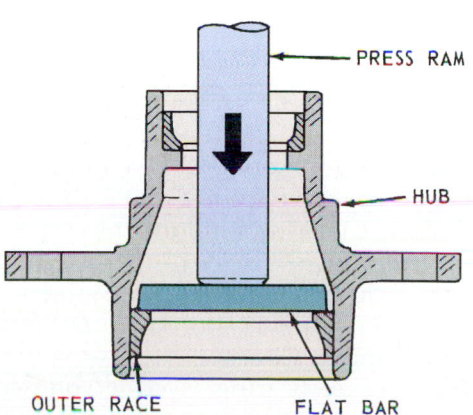

Figure 53.49 Removing a bearing cup. *(Reproduced by permission of Deere & Company, ©1992. Deere & Company. All rights reserved.)*

Figure 53.50 Drive on one side and then the other to avoid cocking the bearing race. *(Courtesy of The Timken Company)*

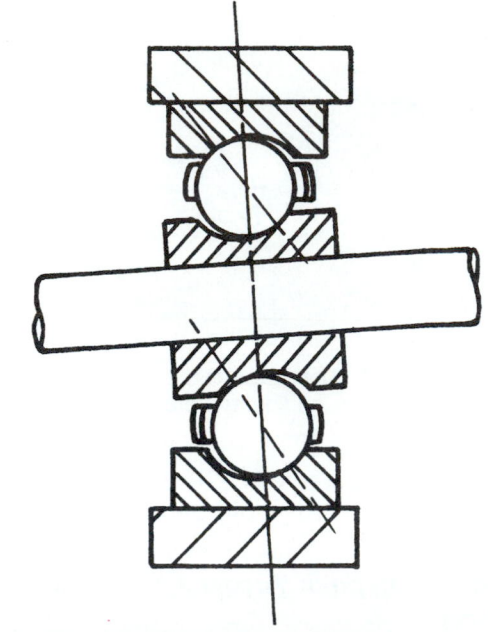

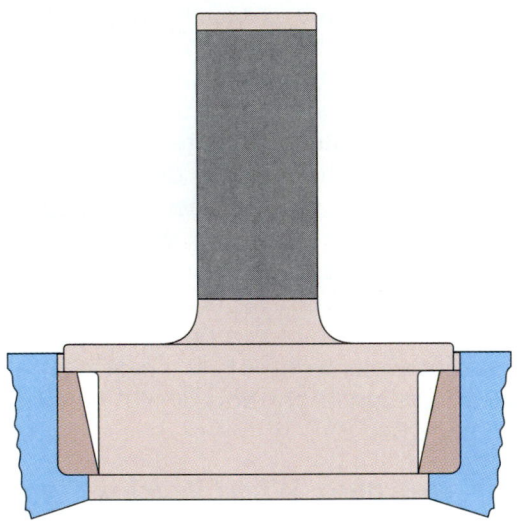

Figure 53.51 A special tool for installing a bearing race. *(Courtesy of Chicago Rawhide)*

■ SERVICING FRONT WHEEL DRIVE BEARINGS

Some bearings on these cars are serviceable but most are sealed, requiring no service. The end of front wheel drive axle shaft has drive splines that fit into splines in the rotor hub. The bearing supports the end of the axle just behind the splines. To get to the bearing requires removing the axle (half-shaft and CV joints). A puller is often required (Figure 53.53). The procedure is covered in more detail in Chapter 71.

The front wheel bearing is either pressed on or bolted onto the steering knuckle. Special tools are available for removing pressed-on bearings without having to remove the steering knuckle from the vehicle. Check the service manual for the correct procedure.

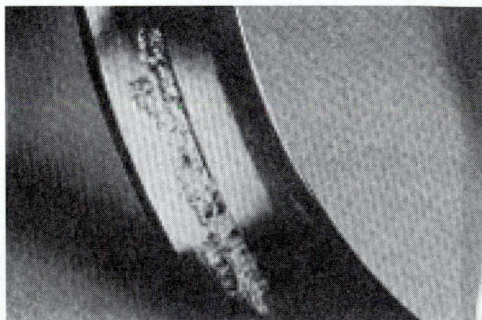

Figure 53.52 Misalignment damage. *(Courtesy of Chicago Rawhide)*

Some manufacturers recommend that the bearing be replaced any time a front wheel drive assembly is disassembled (when the front hub is removed from the spindle). End play is controlled by the size of the parts. When they are tightened together, the end play should be correct.

Special care is required during reassembly. When a puller is required to disassemble an axle from the hub, a certain degree of force will be required to reinstall it in the hub. Be certain that the parts are aligned before using any force. There are usually two rows of ball bearings with their races tapered toward each other to avoid end thrust. Be sure that the bearing assembly is held together as a unit before forcing it onto the axle shaft. The bearing can be ruined during this process.

NOTE: *The vehicle should not be rolled on its wheels with the axle removed or the bearings can be damaged.*

On ball bearings, which is what most front wheel drive cars are now equipped with, the bearings are preloaded. Torque specifications range up to 200 pounds. Do not torque with the vehicle on the ground. Have an assistant hold pressure on the brakes while torquing. Then, stake the nut into the groove in the spindle (Figure 53.54). Do not overtighten and then back off.

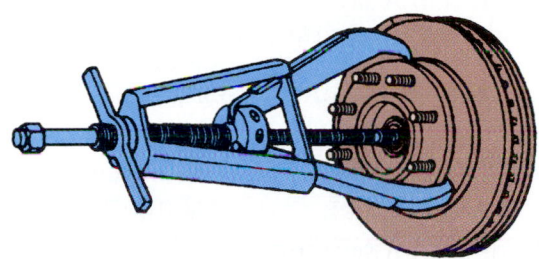

Figure 53.53 A puller used for removing a front hub from a front wheel drive car. *(Courtesy of General Motors Corporation, Service Technology Group)*

Figure 53.54 A front wheel drive bearing retaining nut is torqued and staked to keep it in place. *(Courtesy of Chicago Rawhide)*

■ REVIEW QUESTIONS

1. What kind of bearing is a ball or roller, anti-friction or plain?

2. The separator that holds roller or ball bearings properly spaced in the bearing assembly is called a _____ .

3. An up-and-down load on a bearing is called a _____ load.

4. A front-to-rear or side-to-side load on a bearing is called a _____ load.

5. Which bearing design has greater load carrying capacity, roller or ball?

6. A _____ bearing is a very small roller bearing.

7. What kind of soap would probably be used as a thickening agent for wheel bearing grease?

8. If a grease has a lower NLGI number, is it thicker or softer?

9. What is the term for the temperature at which the grease turns into a liquid?

10. Which grease would be intended for use in wheel bearings, L or G?

11. Which kind of thickener is most often used for multipurpose grease?

12. What kind of oil can be damaging to a nitrile seal?

13. How much running clearance would a correctly adjusted tapered roller front wheel bearing have?

14. What is it called when the bearing or race has indentations from shock loads?

15. On a race car, how far should the wheel bearing hub cavity be filled with grease?

■ ASE STYLE REVIEW QUESTIONS

1. Two technicians are discussing non-drive tapered front wheel bearings. Technician A says that the outside wheel bearing supports the load of the vehicle. Technician B says that the inside bearing holds the wheel in alignment. Who is right?
   a. Technician A   b. Technician B
   c. Both A and B   d. Neither A nor B

2. Technician A says that tapered roller bearings tend to be self-aligning. Technician B says that ball bearings tend to be self-aligning. Who is right?
   a. Technician A   b. Technician B
   c. Both A and B   d. Neither A nor B

3. Technician A says that tapered roller bearings have lips on the edges of the inside and outside

races. Technician B says that the angles on tapered rollers are what make them self-aligning. Who is right?

a. Technician A   b. Technician B
c. Both A and B   d. Neither A nor B

4. Technician A says that "axle bearing" is the term for those bearings found on non-drive front and rear wheels. Technician B says that "wheel bearing" is the term for those bearings found on live axles. Who is right?

a. Technician A   b. Technician B
c. Both A and B   d. Neither A nor B

5. Technician A says that heavy-duty trucks use semi-floating axles. Technician B says that if an axle breaks on a full-floating axle, the rear brake drum and wheel will still be supported. Who is right?

a. Technician A   b. Technician B
c. Both A and B   d. Neither A nor B

6. Technician A says that grease is made up mostly of lubricating oil. Technician B says that soaps are what make a grease thick. Who is right?

a. Technician A   b. Technician B
c. Both A and B   d. Neither A nor B

7. Technician A says that the open side of the sealing lip always faces the lubricant. Technician B says to clean a sealed bearing before reassembly by soaking it in solvent. Who is right?

a. Technician A   b. Technician B
c. Both A and B   d. Neither A nor B

8. Technician A says during a bearing repack to fill up the hub cavity with grease. Technician B says that a wheel bearing should fit the spindle tightly. Who is right?

a. Technician A   b. Technician B
c. Both A and B   d. Neither A nor B

9. Technician A says when turning a corner, the inside tire always turns at a lower rpm than the outside tire. Technician B says when selecting a new cotter pin, use the largest diameter one that will fit into the hole. Who is right?

a. Technician A   b. Technician B
c. Both A and B   d. Neither A nor B

10. Technician A says that the rate of oil oxidation doubles with every 20°F increase in temperature. Technician B says that even if a used bearing race looks good to the eye, it should not be reused with a new bearing. Who is right?

a. Technician A   b. Technician B
c. Both A and B   d. Neither A nor B

Tire and Wheel Theory

■ OBJECTIVES

Upon completion of this chapter, you should be able to:

✔ Describe how a tire is constructed.

✔ Understand the various size designations of tires.

✔ Tell the design differences between radial and bias tires.

✔ Be able to select the best replacement tire for a car.

■ INTRODUCTION

A service technician should be able to advise customers about tires, discuss aspects of tire design, and help the customer to make the safest (and best) choice when purchasing new tires and/or wheels. Tires and wheels are an important automotive safety and service specialty area. In-depth information about them is presented in this and the next chapter.

■ TIRE CONSTRUCTION

Tires are constructed of several layers of rubber materials, cords, and two rings of wire, called beads (Figure 54.1). The *casing* or *carcass* is the internal structure of the tire. A *ply* is metal or fabric *cord* that is rubberized (covered with a layer of rubber). The plies provide strength to the tire to support the load of the vehicle.

The ends of the plies wrap around the steel *bead* before being bonded to the side of the tire. The beads are coils of wires at the side edges of the tire. These give the tire the strength to stay firmly attached to the wheel. *Chafing strips* are hard strips of rubber that protect the beads from damage that could result from chafing against the rim.

A **belt** is cord structure made up of plies. It is only in the area of a tire under the tread and does not extend under the sidewalls.

The *tread* is the section of the tire that rides on the road. A sidewall covering of rubber protects the casing plies between the tire tread and the tire bead.

■ TUBELESS TIRES

Because of safety considerations and their ease of servicing, car manufacturers in the 1950s began to put tubeless tires on all of their cars. Almost all passenger car tires sold since the early 1960s are of the tubeless design. Some imported cars still had tube-type tires until the middle 1970s, and wire wheels have tubes to prevent leakage from the spoke holes.

The inside of a tubeless tire found on a passenger car has an *inner liner* bonded to it that seals air into the tire. The liner is thicker than the liner on a *tube-type tire*. Tubeless tires are actually safer than tube tires. When a tubeless tire is punctured, it will usually not go flat immediately. A nail tends to be held in the tire by

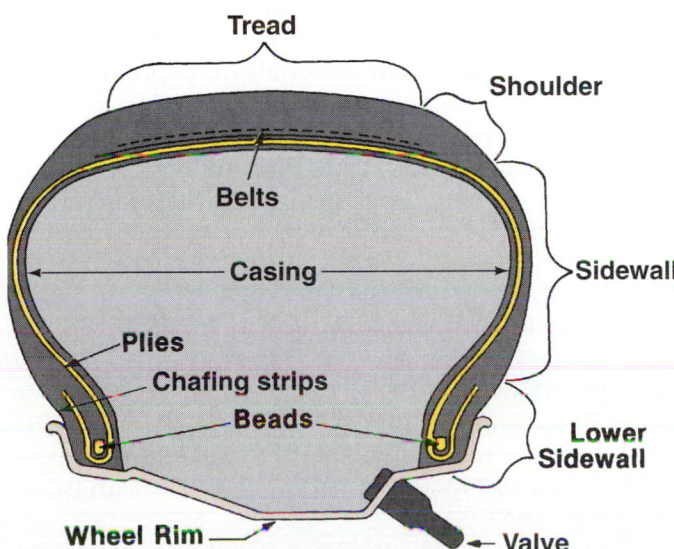

Figure 54.1 Construction of a tire. *(Courtesy of The Goodyear Tire & Rubber Company)*

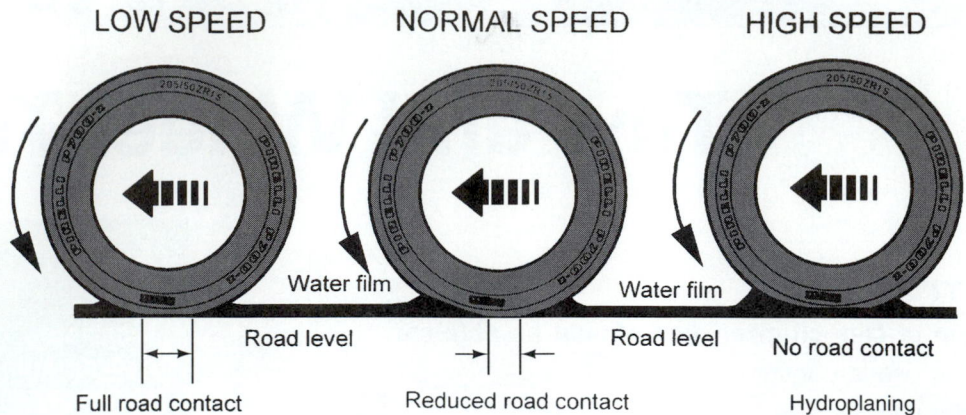

LOW SPEED   NORMAL SPEED   HIGH SPEED

Water film   Water film   No road contact
Road level   Road level
Full road contact   Reduced road contact   Hydroplaning

Figure 54.2 Hydroplaning. *[Courtesy of Pirelli Tire Corporation]*

the inner liner, allowing air to escape more slowly. A tube-type tire tends to go flat instantly when punctured because the walls of the inner tube tend to tear.

TRACTION

A tire's **traction** is defined as how well it grips the road. Traction is affected by the road surface and contaminants such as water, ice, or debris. It is also affected by the tire's tread, the tread material, inflation pressure, width of the tread, cord ply design, wheel alignment, and other things.

TIRE TREAD

The tread is a band made of a rubber compound designed to have various traction and wear characteristics. A federal grading standard (discussed later in this chapter) that is cast into the sidewall of the tire describes a tire's traction and wear characteristics. Grooves in the tread allow traction on wet surfaces, giving the water a place to go. They also allow the tire to flex without squirming, which would cause wear.

Treads are designed for specific types of weather and conditions. The design selected is always a compromise. The best traction on a dry paved road would be with a racing *slick*, or a bald tire. That same tire would be dangerous in the rain. Water forms a wedge under a tire that can actually float the car. This is called **hydroplaning** or *aquaplaning*. A deep tread pattern will break through a water film and grip the road at low speeds, but at high speed the tire can hydroplane (Figure 54.2).

Tires with large grooves are designed for use in mud and snow. But the large tread pattern can result in noise on the highway. Treads are often spaced at random intervals to minimize noise.

Sipes are small grooves in the tire tread that look like knife cuts (Figure 54.3). They allow extra gripping as the tire flexes. Sipes also clear water off of the road, wiping the contact area to provide a better grip. *Ribs* in the tire tread are designed to pump water from the road through the grooves to the back of the tire, where it is thrown out onto the road.

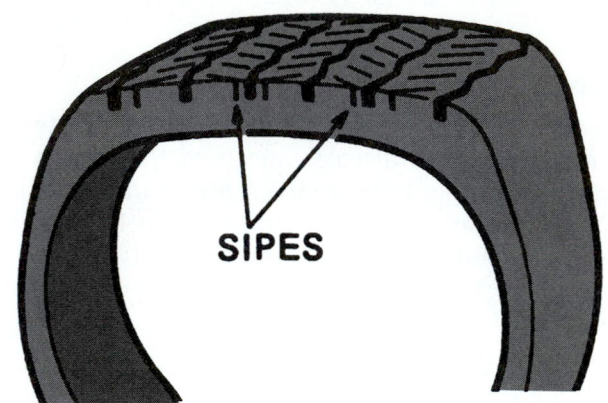

SIPES

Figure 54.3 Tread sipes.

TIRE TREAD MATERIAL

The tread material calls for compromise. Hard materials might wear longer but not provide sufficient traction. Materials for mud and snow tires must remain soft in cold weather. Soft materials must provide sufficient wear. Natural rubber is compounded in different proportions with synthetic rubber to achieve the desired characteristics. Synthetic rubbers are more resistant to heat and solvents, while natural rubbers are better in other areas.

Rubber

Pure rubber is a hydrocarbon grown in all of the subtropical areas of the world. It freezes at only 40°F and becomes sticky at 86°F. It swells when contacted by many liquids and is damaged by sunlight. For rubber to be useful, it must be vulcanized (heated) to make it stable. Charles Goodyear patented the vulcanization process in 1842.

Isoprene, the core substance of natural rubber, was synthesized in 1910 in Germany. Natural and/or synthetic rubbers have chemicals such as carbon black and antioxidants added to them to improve grip, abrasion resistance, flexibility, and oxidation resistance.

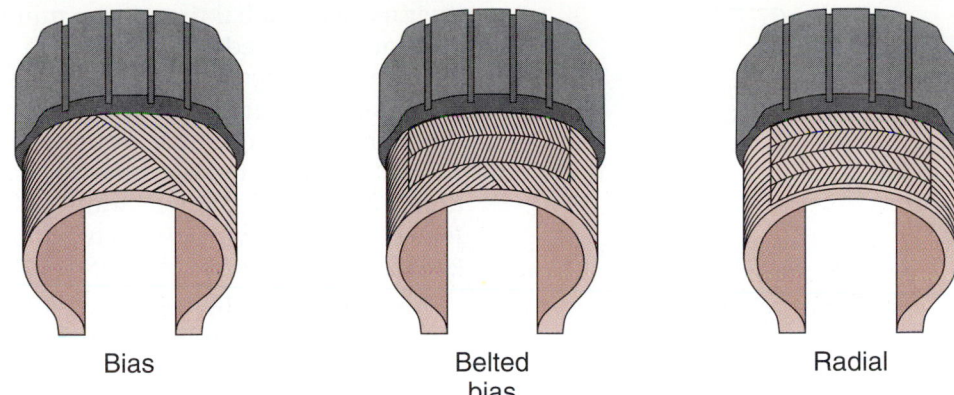

Figure 54.4 Comparison of radial and bias tire construction. *[Courtesy of General Motors Corporation, Service Technology Group]*

Hysteresis is a term used by chemical engineers to describe a rubber's energy absorption characteristics. A high hysteresis compound results in quiet running, a comfortable ride, and better wet and dry grip. A low hysteresis compound has good lateral stability, low rolling resistance, and minimized tread wear.

◼ TIRE CORD

Because rubber is elastic and not very strong, it must be reinforced with material such as fabric, fiber, or steel cords. Without these materials, a tire would blow up like a balloon. The most common cord material until World War II was cotton. Today, cord material in the casing is made of either rayon, nylon, or polyester. Cord material for belts can be steel, rayon, nylon, fiberglass, or aramid (Kevlar), which was developed specifically for the tire industry.

◼ TIRE PLY DESIGN

Most of today's tires are **radial-ply** tires, although some light trucks and RVs still use **bias-ply** tires. Figure 54.4 shows the difference between radial and bias tire construction.

Radial-Ply Tires

Radial tires have casing plies that run across the tire from bead seat to bead seat in the "radial" direction of the wheel. The outside circumference of the tire is held together by reinforcing belt rings of slightly angled cord material (Figure 54.5).

Bias-Ply Tires

Bias-ply, diagonal, or cross-ply tires have casing plies that cross each other at angles of 35–45°. They ride softer than radials, but their tread tends to squirm when rolling. This results in tire wear. Belts beneath the tread give the tire stability. Bias tires with belts under the tread last longer than unbelted bias tires because the belts keep the tread from squirming (Figure 54.6).

Radial tires give longer tread life, better grip to the road surface, and improved fuel economy. They ride rougher at low speeds than bias tires, but can actually be smoother at faster speeds while travelling over highway expansion joints. A radial is more expensive to construct than a bias tire because during manufacturing it requires more labor.

A bias tire is made in a *full circle mold* that has two halves, like a bagel cut in half. The parting line for a tire made in this type of mold will run down

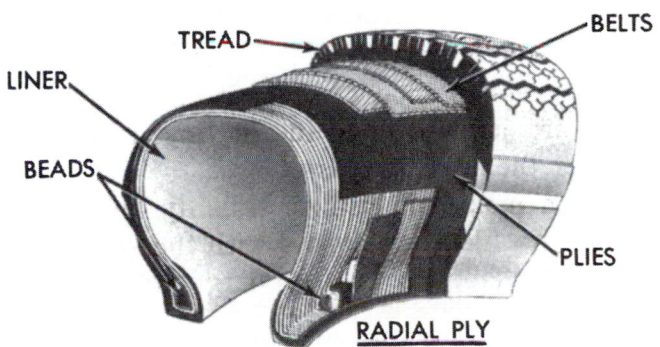

Figure 54.5 The radial cords are held together under the tread by belts. *[Courtesy of Chrysler Corporation]*

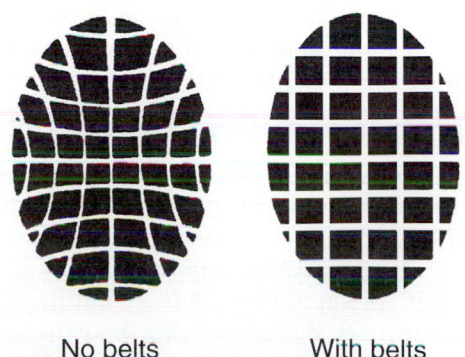

No belts　　　　With belts

Figure 54.6 Belts stabilize the tread in the tread contact area.

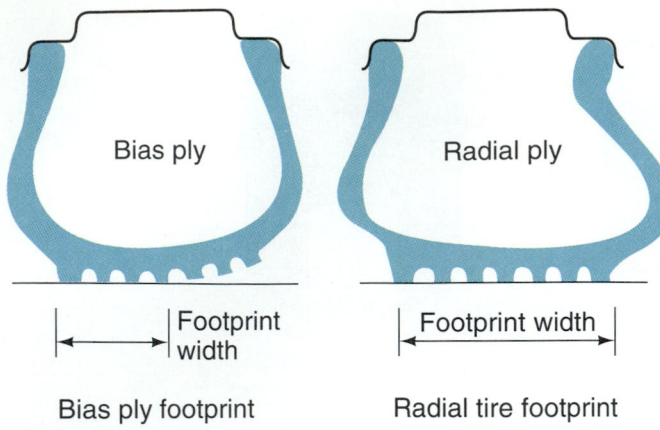

Bias ply

Radial ply

Footprint width

Footprint width

Bias ply footprint

Radial tire footprint

Figure 54.7 The flexible radial sidewall allows more of the tread to remain in contact with the road.

the center of the tire's tread. A segmented mold is often used for constructing radial tires. A radial tire that has been made in a segmented mold will have several radial parting lines running from bead to bead across its tire tread.

A tire acts like a part of the suspension system as it supports the load of the car, isolating the passengers from road shock as its sidewall deflects. *Sidewall deflection* allows more of the tread to actually be in contact with the road surface. The larger area of contact, called the tire's **footprint**, allows the load on the tire to be spread across a wider area of the tire. A larger footprint also causes the tire to grip better so it can transmit forces of the engine and brakes to the road surface.

A radial tire flexes on its sidewall and is more resistant to wear because its tread surface stays flat on the ground (Figure 54.7). Because of the bulging sidewall, a properly inflated radial will appear to be low on air.

Radial tires have less resistance to rolling, which improves a car's gas mileage. Fuel economy standards have been mandated by the United States Congress for several years. Each manufacturer must meet a *corporate average fuel economy (CAFE)* standard or pay a *"gas-guzzler"* penalty to the government. Because radial tires help the vehicle to achieve better fuel economy, they are included on all new cars as original equipment.

■ SWITCHING TO RADIALS

Customers with bias tires often wish to upgrade to radial tires. Cars built prior to 1972 usually do not have suspension systems designed for radial tires. Installing radials on these cars can result in a somewhat harsher ride at slow speeds. Installing radials on wheels that were designed only for bias tires can result in a dangerous wheel failure because radials exert more pressure against the sides of the rim. Cars produced after 1975 have numbers on the rims that designate

their use with radial tires. If the number includes an R, the rim is designed to be used with radials.

The difference in handling characteristics between radials and diagonal bias tires makes it best not to mix them on the same vehicle. Sometimes, a customer's car will have two belted-bias tires that are in good condition. The customer might want to begin to make the switch to radials without buying all four new tires. The *Rubber Manufacturers Association (RMA)* recommends that the two new radial tires be installed on the rear. See Chapter 55 for more information on radial tire service.

■ TIRE SIDEWALL MARKINGS

The U.S. Department of Transportation (DOT) requires the listing of certain information on the tire (Figure 54.8). Included on the tire sidewall for a typical passenger car are:

- The tire size
- Maximum permissible cold air pressure
- Load rating—an indication of the load limit for each of the vehicle's tires under cold inflation.
- The name of the material that the cords of the tire are made of
- The number of plies in the tread and sidewall areas
- If the tire is a radial tire
- Whether the tire is tube-type or tubeless
- The DOT manufacturing code
- M+S—this indicates that the tire meets the RMA definition for a mud and snow tire
- Uniform Tire Quality Grade Standard (UTQG)— DOT grading for traction, treadwear, and temperature

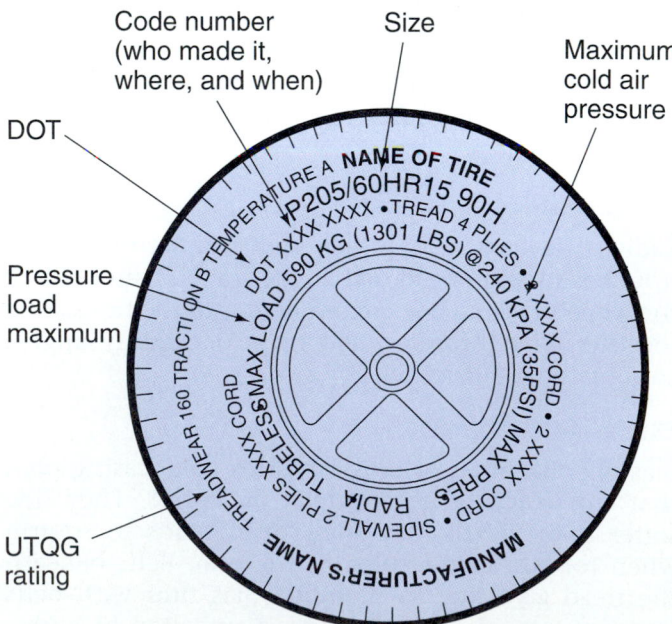

Figure 54.8 Information found on the tire sidewall. *(Courtesy of Rubber Manufacturers Association)*

Tire Size

A tire information sticker called a **placard** (see Figure 13.30) has been required on cars sold in the United States since 1968. It is located on the door post, the edge of the door, the gas filler door, or on the glove box door. The placard indicates the correct OE tire size, the cold inflation pressure, and the gross axle weight (for commercial vehicles). If there is no placard, check the owner's manual for the information.

The tire's size is listed on the sidewall, using one of several ratings (Figure 54.9).
- P-Metric (P205-75 R15)
- European Metric (185/70 R14)
- Numeric (6.70-15)
- Alpha/numeric (FR78-15)
- Light trucks (LT 215/85R-16)

The numeric rating was commonly used until the early 1970s, but metric cross section measurement is universal today. Figure 54.10 shows how the tire size designation is interpreted for the P-metric radial tire, the most common tire in use today. The first letter tells the type of tire it is:
- P means passenger car
- C means commercial (for light trucks)
- T—temporary spare (covered in more detail later)

The tire's cross-section width (205 mm) is listed next. With each size increase (from 205 to 215, for instance) the width of the tire increases by 10 millimeters.

There is usually a letter in the size designation. Some of the possible letters are:
- R (radial)
- B (belted bias construction)—this is sometimes left blank
- D (diagonal bias construction)

A tire's height is called its **profile**. A low profile tire is shorter than a normal tire. The number that comes after the cross-sectional width of the tire is the **aspect ratio**, which is a measurement of the height-to-width ratio (Figure 54.11). In Figure 54.10, aspect ratio is expressed as the number 75. An example of a lower profile tire would be a 70. When a designation

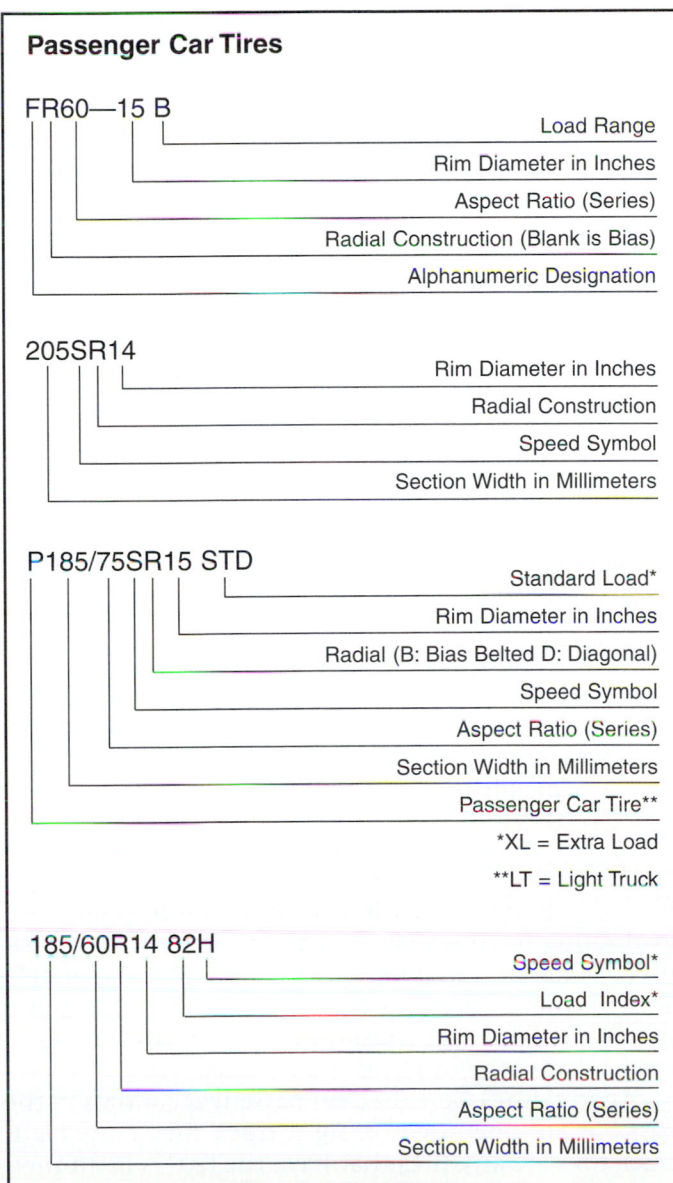

Figure 54.9 Different ways of measuring tire size. *(Courtesy of Bridgestone/Firestone, Inc.)*

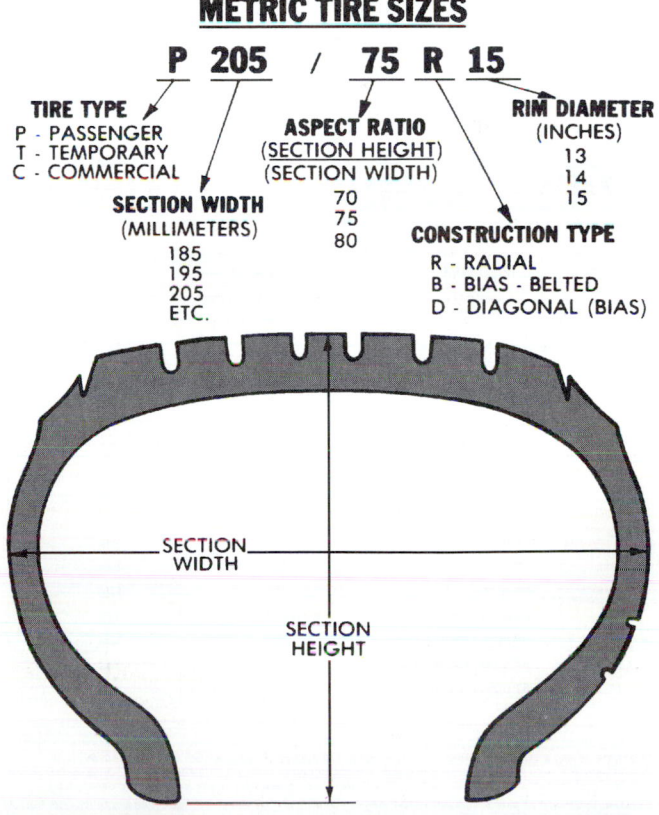

Figure 54.10 Tire size designation on a P-metric radial. *(Courtesy of General Motors Corporation, Service Technology Group)*

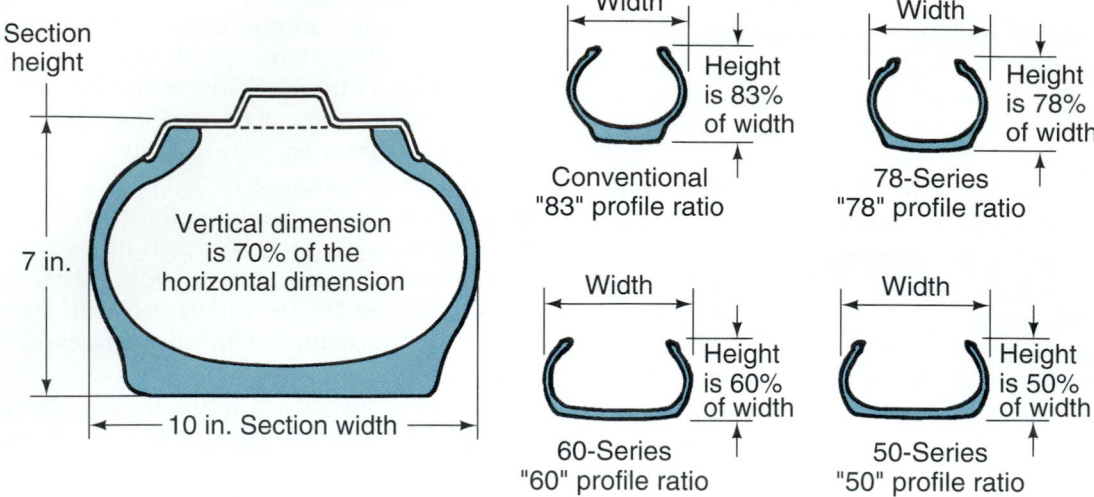

Figure 54.11 Aspect ratio compares the tire's height and width.

does not include a designation for the aspect ratio, the tire is an 80 series (P205 R15).

The last part of the sequence (R15) tells that the tire is a radial tire to be mounted on a 15" diameter rim. This is sometimes followed by LT (light truck) and/or 90H (load/speed index).

Speed Rating

Sometimes a **speed rating** (Figure 54.12) is listed as part of the size designation. If the tire used in the earlier example included a speed rating in its label, it would be listed as P 205-75 HR15. The H is the speed designation for 130 mph. This means that a properly

inflated tire with an H designation has been designed to operate at up to 130 mph for short periods of time (such as when passing cars). Sustained high speeds can damage the tire.

Newer tires have the speed rating and load index listed separately, after the size designation. The new designation is a two- or three-digit **load index** followed by the speed symbol. The load index is the maximum load at the designated speed rating. This designation might read P 205-75 R15 90H. Individual load capacities are listed in Appendix H at the back of the book. A tire's load carrying ability is related to the strength of its sidewall plies. A tire with a higher load capacity will also have a higher inflation pressure.

Speed symbols for passenger cars range from the L rating (74.5 mph/120 km/h), to ZR (over 149 mph/240 km/h). Examples of the most common ratings are as follows:

> S—112 mph (180 km/h)
> T—118 mph (190 km/H)
> H—130 mph (210 km/H)

Tires with speed designations of Z or V are for higher speeds. Consult the manufacturer for the exact speed rating of those tires.

SPEED SYMBOLS/RATINGS

SPEED SYMBOL	MAXIMUM SPEED	APPLIES TO Passenger Car Tires	APPLIES TO Light Truck Tires
ZR *	*above 149 mph (240km/h)	Yes	No
Y **	186 mph (300 km/h)	Yes	No
W **	168 mph (270 km/h)	Yes	No
V	(with service Description)		
	149 mph (240 km/h)	Yes	Yes
H	130 mph (210 km/h)	Yes	Yes
U	124 mph (200km/h)	Yes	Yes
T	118 mph (190 km/h)	Yes	Yes
S	112 mph (180 km/h)	Yes	Yes
R	106 mph (170 km/h	No	Yes
Q	99 mph (160 km/h)	No (winter tires Yes)	Yes
P	93 mph (150 km/h)	No	Yes
N	87 mph (140 km/h)	No	Yes
M	81 mph (130 km/h	Temporary Spare Tires	No

* For tires having a maximum speed capability above 149 mph (240 km/h), a "ZR" may appear in the size designation. For tires having a maximum speed capability above 186 mph (300 km/h), a "ZR" must appear in the size designation. Consult the tire manufacturer for maximum speed when there is no service description.

Example: P275/40 R17 93W at 168 mph (270 km/h) or P275/40ZR17 at above 149 mph (240 km/h). Consult the tire manufacturer.

Figure 54.12 Speed ratings. *(Courtesy of Pirelli Tire Corporation)*

Load Rating

It is very important *not* to use a tire that has too low a load rating for the weight of the vehicle. Tire companies recommend that heavy RVs be weighed separately at each wheel. This is to determine the load rating and the air pressure that should be used to safely support the load at each corner of the vehicle.

Load ratings were used on passenger car tires in the past but are now used on light truck tires only. Light truck tires are any tires that have the letters LT in their size callout. The load rating tells how much weight that tire can safely support at a specified air pressure. In the past, load range B was a typical passenger tire rating. As the letters progress through the alphabet, the load rating becomes higher. A load range C or D tire would be found on a light truck or van.

P-metric radial tires found on today's passenger cars are all of a uniform standard load rating or an extra load rating. The amount of load one of these tires can support is determined by the area of the tire and the amount of air pressure in it.

DOT Codes

The DOT symbol signifies that the tire meets DOT safety standards. The first two letters of the DOT code tell what manufacturer made the tire. There is a separate code for the same manufacturer in different countries, so the country of origin can be determined. The remaining characters in the code tell the size and type of the tire and the week and year it was made (026 = second week of 1996, 374 = 37th week of 1994).

■ ALL-SEASON TIRES

Since the mid 1970s, all-season tires have been available for customers who previously needed to change to specialized tires twice a year. Any combination of the letters M and S (*M+S, MS, M&S, M/S*) on the tire sidewall means that the tire meets *snow tire* definitions set by the Rubber Manufacturer's Association (Figure 54.13). These tires have treads with specially designed pockets and slots in at least one tread edge.

All-season tires are also available as original equipment. Vehicles equipped with these tires generally have a narrower range of acceptable wheel alignment settings. Some all-season tires are susceptible to uneven wear when alignment angles are not set correctly. More frequent tire rotation is recommended.

Snow Tires

Snow tires have deeper tread grooves designed to provide a better grip when driving on snow covered roads. Most manufacturers recommend that they be installed

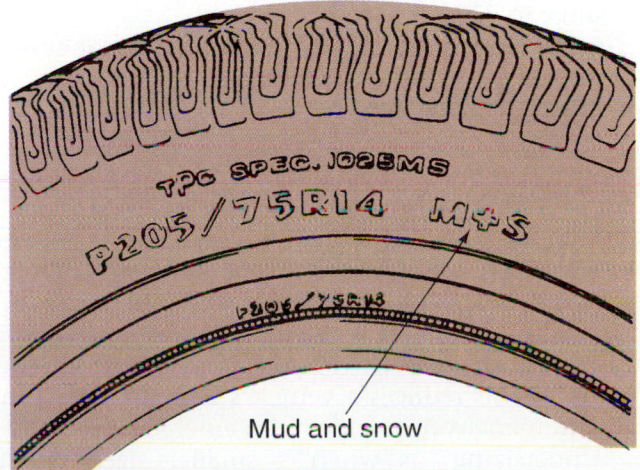

Figure 54.13 This tire is rated for snow.

on all four wheels to prevent handling problems. They are fitted in the original equipment size.

■ TIRE QUALITY GRADING

American manufacturers use the **UTQG (uniform tire quality grade)** system, which rates tread wear, traction, and *temperature resistance*. It is printed on the sidewall of the tire (see Figure 54.6). An example of this rating is *200AA*, a number and two letters.

Tread Wear

Tread wear is a rating that compares tires tested on a government test course. The number 100 represents a standard tire. A 200 would be expected to last twice as long on the government test course, and a 150 would last about one and one-half times as long. The actual life of a tire can vary due to road conditions, climate, air pressure, alignment, driving habits, vehicle loading, and other factors.

Traction Grade

When interpreting the letter ratings, A is the highest rating, while C is the lowest. The first letter is the traction grade, which indicates resistance to skidding on government test surfaces of wet asphalt pavement and concrete. This rating covers braking only in a straight ahead direction, and not cornering.

Temperature Grade

The second letter is the temperature grade. It indicates a properly inflated tire's resistance to generating heat and its ability to dissipate heat at highway speeds. Grade C is the minimum standard required by law. Standards B and A exceed this standard. Continuous high speed driving can result in deterioration of the tire's material. This and excessive temperatures can lead to sudden tire failure.

A compact spare tire is what most new cars come equipped with. The compact tire is considerably smaller than a regular tire and is to be used temporarily only. Many have a limited speed of 50 kilometers per hour (31 mph) and a distance of 50 kilometers (31 miles). The speed and distance warning will be printed on the sidewall of the tire.

➡ *Perform **Tire Identification** Worksheet*

■ CHANGING TIRE SIZE

Customers often want to make a change in the size of an original equipment tire. Selecting a replacement tire that is the exact same size as shown on the placard is not always possible. If tire size is changed, be sure to substitute a tire that has an equal or greater load-carrying capacity. Tire companies provide charts that give the maximum load that various sizes of tires will provide at listed cold pressures (see appendix). A change in tire size can usually be accomplished without sacrificing safety or design considerations. For instance, the following five tire sizes all have the same diameter and load capacity. These are called *size equivalents*.

175/70SR13

205/60R13 H

185/60R14 H

205/55VR14

195/50VR15

When changing tire sizes, three things need to be considered in the replacement tire:

- Its overall diameter must not be changed by more than +2% or −3% from the original tire's diameter. Changing tire diameter can affect anti-lock brakes, the speedometer, gear ratios, and four wheel drive.
- Its speed rating must be equal to or greater than that of the original tire.
- It must have an overall load carrying capacity that is equal to or greater than the load index number listed on the original tire.

SAFETY NOTE Changing from a P185/65R14 (85 load index/1124 pound maximum carrying capacity) to a P185/60R14 (82 load index/1047 pound capacity) will result in almost 7% less load capacity.

Tire manufacturers publish application handbooks that give information such as tire dimensions, revolutions per mile, diameter, acceptable rim sizes for each tire size, and the cross section of the tire. When unsure about a possible change in tire size, technical assistance is available through all of the major tire manufacturers.

The three considerations when choosing tires include the tire's diameter and load capacity, the width of the rim, and the intended use of the tire. DOT standards require the overall diameter of a replacement tire to be within −3% to +2% of what the tire was as original equipment. Usually, as the diameter of the tire increases, the load capacity of the tire increases also. Notice how light trucks and recreational vehicles often use taller tires.

Since the early 1970s, tires with lower profiles have become increasingly popular. When a tire with a lower profile is being installed, a wheel with a larger inside diameter will be used to make up the difference in height. A recent trend is to a 1" larger wheel diameter with ½" less section height. This tire, which has the same overall diameter of the original equipment tire (Figure 54.14) has less sidewall flexibility but has the same load capacity. It also has a larger footprint.

Lower profile tires grip better and they are more responsive. A high profile tire provides better loose

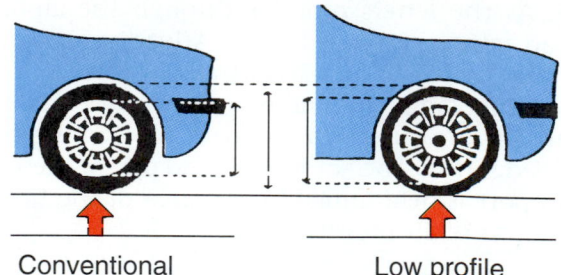

Conventional   Low profile

Figure 54.14 Same tire diameter with a low profile tire on a larger wheel rim. *(Courtesy of Pirelli Tire Corporation)*

snow or mud traction because it does a better job of cutting through these materials. Narrow wheels and tires with the proper load capacity can be a better choice for a customer who finds their snow traction to be inadequate.

Wheel rim width must also be considered when changing tire size. A tire must not be installed on a wheel that is narrower or wider than the manufacturer's recommended wheel width. Installing wider tires on a vehicle will probably mean that wider wheels will have to be installed. A wider wheel provides more support to the tire sidewall. A narrower wheel allows the sidewall to flex more easily, providing a softer ride.

It is very important that the intended use of the tire be known before making a change from original equipment tires. On a computer controlled car, changing the tire size will require a change in the computer's program. Also, the speed rating of the replacement tire must meet or exceed the speed capability of the vehicle. The load range must also be sufficient for the weight of the vehicle.

When changing tire size, tire load capacity and diameter can be maintained by using the following formula:

- When changing to a 5% lower profile (reducing the aspect ratio by 5), choose a tire with a 10 mm wider cross section. For instance: From 215/75 R15 change to 225/70 R15
- Tire manufacturers recommend against changing from a lower profile to a higher profile tire, which results is reduced vehicle performance.
- When making a change in the size of tires on a vehicle, the recommendation is to replace tires in sets of four.

When the diameter of a tire is changed, front end geometry is altered and the speedometer will need to be recalibrated. When higher diameter tires are installed to provide increased load, this is called "oversizing." Higher diameter tires raise the vehicle's center of gravity. This reduces a vehicle's ability to hold the road and maneuver quickly in an emergency.

"Undersizing" is when a smaller sized tire is installed, often because they are less expensive. Never

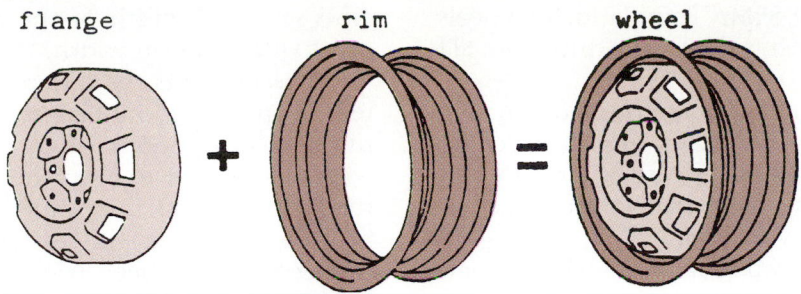

flange   rim   wheel

Figure 54.15 Parts of a wheel rim. *(Courtesy of Pirelli Tire Corporation)*

install undersized tires on a vehicle. They will wear faster and have less load carrying ability. The vehicle will be lower, the speedometer will no longer be accurate, and the increase in engine rpm for a given speed will result in a decrease in fuel economy.

NOTE: *Many newer vehicles are equipped with anti-lock brake systems (ABS) (see Chapter 51). A small pickup coil at each wheel measures wheel rotation speed and generates a signal that is sent to the ABS computer. Tires of a different height than original can cause excessive tire chirping, or erratic system operation during a panic stop. When the size difference is large enough, the ABS warning light can come on. A small amount of tire height difference between tires on different sides of the vehicle is enough to cause the computer to set a trouble code and disable the ABS function. In this case, brakes will continue to operate normally.*

■ WHEELS

Wheels are made either of steel, aluminum, or an alloy of aluminum and magnesium (commonly called *mags*). Original equipment wheels on less expensive passenger cars are made of steel. Wheels have two parts, the *center* or *flange*, and the *rim* (Figure 54.15). The center flange of steel wheels is stamped because this is the least expensive production method. A strip of steel is rolled and butt welded at the ends to form the rim, which is then spot welded to the center flange.

The raised sections on either side of the drop center area are called *bead seats*. This is where the tire actually seals. There are raised sections on the inside edges of the bead seats called *safety beads* (Figure 54.16). These help to keep the tire bead on the bead seat in case of an "air out," until the car can be safely stopped. Rims for tubeless tires must have safety beads in order to be DOT approved.

A **drop center** or *rim well* provides a means of removing and installing a tire from the wheel. The tire bead is reinforced with wire and it will not stretch. When a tire is installed on the wheel, one side of the bead is pushed into the drop center so that the other side of the bead can be pulled over the edge of the rim (see Chapter 55).

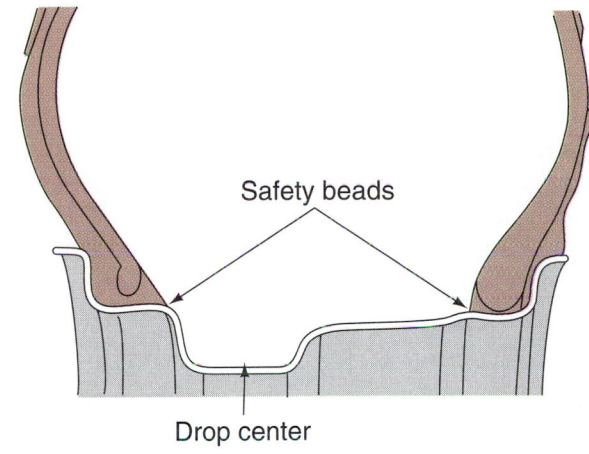

Safety beads

Drop center

Figure 54.16 Safety beads keep the tire on the wheel in case of a flat tire.

Wheels are centered on the hub by one of two methods:
- **Hub-centric**—This means that the center of the wheel has a machined counterbore that pilots on a machined area of the hub. This is the most precise method of centering the wheel to the axle. Most original equipment wheels are hub-centric.
- **Stud-centric**, or *lug-centric*, wheels locate on the wheel studs. Most aftermarket or custom wheels are stud-centric. Custom wheels that are made for a specific model of vehicle can be hub-centric, but they are more expensive because they are not universal to many makes of vehicle. When stud-centric wheels, which have a bigger center hole, are installed on hub-centric vehicles the result can be an improperly centered wheel, which causes vibration.

Service information on hub and stud centering is found in Chapter 55, under the heading of wheel balancing.

■ CUSTOM WHEELS

Customers sometimes purchase custom wheels when they want to have a cosmetic change in the appearance of the vehicle or when different sized tires are installed. Aftermarket wheel quality is rated by the **Specialty Equipment Manufacturer's Association (SEMA)** by their affiliate *SFI* (The SEMA Foundation). Wheels carrying their certification are manufactured to SFI standards.

"Mag wheels" are made of aluminum or an aluminum alloy. Aluminum can be cast, forged, or rolled. Mags can be either a single piece casting or they can have lighter rolled rim halves bolted to a cast center section. Race cars use alloy wheels.

An *alloy* occurs when two or more metals are combined to make one. The alloy used for mags is a combination of magnesium and silicon. These wheels are strong and light but are not practical for passenger cars because they are expensive and do not resist corrosion.

Custom wheels for street use are single-piece castings of light alloy aluminum with a weather resistant coating. The more costly custom wheels fit a single application only. Less expensive wheels are made of weaker materials. They are stud-centric (center off of the wheel lugs) and are made to fit a variety of applications.

Rim width is the measurement from bead seat to bead seat (Figure 54.17). It is usually about 80% of the cross-sectional width of the tire. The centerline is at one-half of the rim width.

Wheel offset is the difference between the rim centerline and the mounting surface of the wheel. Sometimes a certain amount of offset is designed into a wheel to allow it to clear the fenderwell. The offset of the wheel is also important to proper brake cooling because it affects the distance between the brake caliper and the wheel.

Wheel clearance is *not* included in application handbooks. This is something that must be carefully checked on the vehicle. The use of wider tires and wheels, or wheels that are offset a different amount than stock, can result in interference between the tire and fenderwell or suspension components.

The way that offset is described varies. The most common method of offset classification is given here. When the wheel is offset in such a way as to increase the track width of the tires, this is called **negative offset**. The opposite, which is often found on FWD cars, is called **positive offset**. Wheels should be replaced with wheels of the same offset to maintain the proper **scrub radius**.

> **CAUTION** *Scrub radius, which is shown in Figure 60.12, has an effect on the handling and steering effort of the vehicle. Changing the height or centerline of a wheel from that designed by the manufacturer can result in a change in scrub radius from positive to negative or vice versa. The result can seriously effect vehicle handling.*

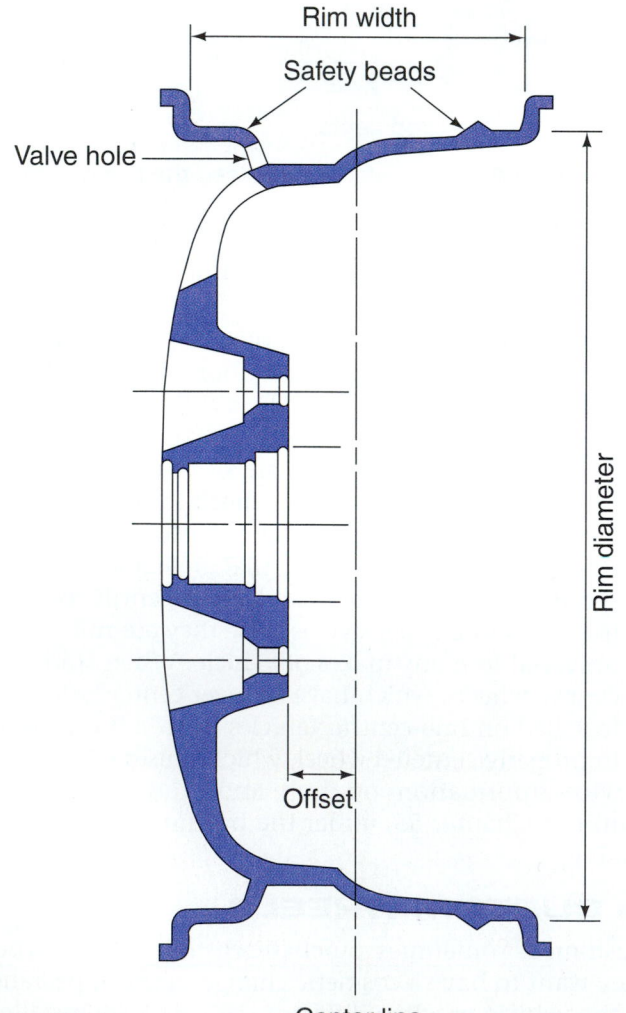

Figure 54.17 Wheel terminology. *(Courtesy of Pirelli Tire Corporation)*

Labels in figure: Rim width, Safety beads, Valve hole, Rim diameter, Offset, Center line

■ LUG STUDS

Wheels have different numbers of lug studs (between three and eight), depending on the load on the vehicle. Most passenger cars use four or five lug studs, while light trucks usually use six or eight. Heavier trucks and some RVs sometimes use fewer lug bolts or studs, but they are larger in diameter and are tightened to a much higher torque.

There are also various bolt patterns used. A *bolt pattern* that is listed in a catalog as being 6–5½, is a six-bolt pattern spaced around a 5½ inch circle. Bolt patterns with an even number of bolt holes are easy to measure. Simply measure the distance from the center of one bolt to the center of the one across from it. Five bolt patterns are more difficult to measure. Templates are available to help determine the size of a bolt pattern.

■ LUG NUTS

Most lug nuts use metric screw threads. Lug studs have a serrated shank so that they will remain tight in the hole in the hub during tightening. Lug nuts for cast wheels (mags) are long and thick and must fit into a large, deep hole. These lug nuts must be used with a washer to avoid damaging the wheel. Most lug nuts are made with the washer permanently installed on them (see Chapter 55).

■ TIRE VALVE STEMS

Passenger car tire valve stems are usually rubber and are designed to be used at pressures of less than 4.2 bar (62 psi). A spring-loaded valve core is screwed into the valve stem (Figure 54.18). Valve stems have a screw-on dust cap, some of which have a gasket that prevents air loss past the valve core.

NOTE: *According to the Pirelli Armstrong Tire Corporation, an imperceptible leak of one bubble per minute can result in the loss of .1 bar (1.5 psi) per month.*

There are two common valve stem lengths available. A short stem is used when there is a hubcap and a long stem accommodates the use of full wheel covers (Figure 54.19). Rubber valve stems come in common diameters of 11.3 mm or 15.7 mm to match the two common wheel rim hole sizes. When a longer stem is required, tire valve extensions can be screwed onto an existing valve stem. Extensions have a spring-loaded check valve so that air pressure can be checked and adjusted as needed.

Light trucks and custom wheels are usually equipped with threaded metal stems that have a rubber bushing and are fastened to the wheel with a nut (Figure 54.20). The common wheel hole size for metal stems is 11.5 mm.

■ RETREADS

Retreads were very popular in the past, but are not very common any longer except for the rear tires of large trucks. A retread is a worn tire that has had its entire tread removed. A new tread area is vulcanized (bonded) to the tire carcass.

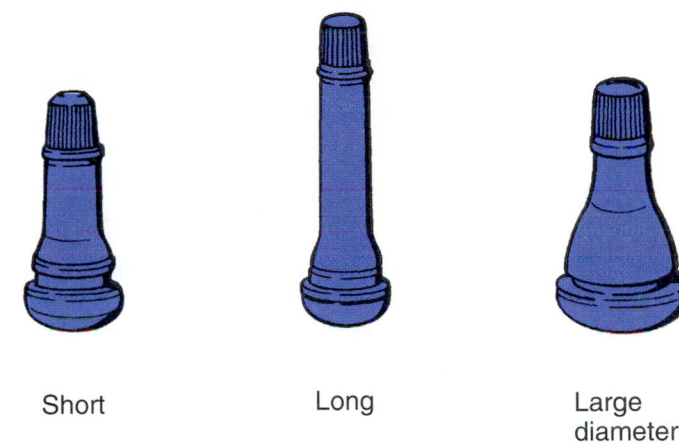

Short   Long   Large diameter

Figure 54.19 *Different sizes of rubber valve stems. [Courtesy of Plews/Edelmann Division, Stant Corporation]*

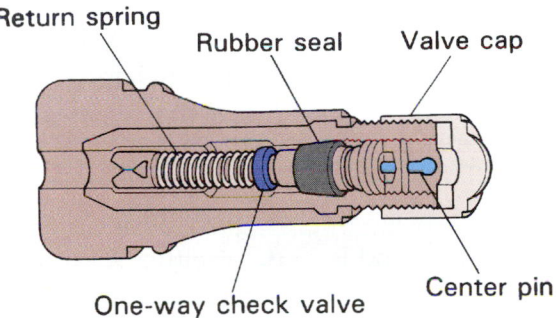

Figure 54.18 Parts of a rubber valve stem.

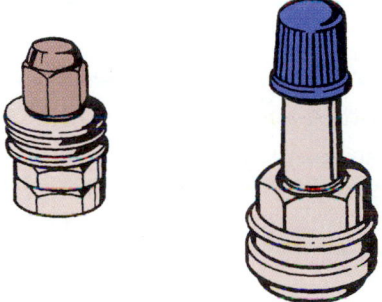

Figure 54.20 *Metal valve stems. [Courtesy of Plews/Edelmann Division, Stant Corporation]*

■ REVIEW QUESTIONS

1. What is it called when water forms a wedge under the tire, causing it to lose traction?
2. Which tire allows better fuel economy, radial or bias?
3. What is the area of tread contact called?
4. What is another name for a tire's height?
5. What is the miles-per-hour limit for a speed rating of H?
6. What are the two letters that are found on snow or all-season tires?

7. When changing a tire sized 195/75 R15 to a 5% lower profile, the correct tire size would be _____.
8. What is the lower area between the bead seats of a wheel rim called?
9. When a wheel is offset to increase the track width of the front tires, this is called_____offset.
10. A small leak of one bubble per minute can result in the loss of _____ psi per month.

■ ASE STYLE REVIEW QUESTIONS

1. Technician A says that a properly inflated bias tire will have a bulging tire sidewall. Technician B says when installing two radials with two bias tires, install the radials on the front. Who is right?
   - **a.** Technician A
   - **b.** Technician B
   - **c.** Both A and B
   - **d.** Neither A nor B

2. Technician A says that a tire with a tread wear rating of 200 would be expected to last twice as long under the same conditions as one with a 100 rating. Technician B says when the diameter of a tire is changed, front end geometry is altered. Who is right?
   - **a.** Technician A
   - **b.** Technician B
   - **c.** Both A and B
   - **d.** Neither A nor B

3. Technician A says that an alloy occurs when two or more metals are combined to make one. Technician B says that custom wheels are usually hub-centric. Who is right?
   - **a.** Technician A
   - **b.** Technician B
   - **c.** Both A and B
   - **d.** Neither A nor B

4. Technician A says that the tire ply design that has the cords intersecting at an angle is called a radial tire. Technician B says that the P in a P-metric radial's size means performance. Who is right?
   - **a.** Technician A
   - **b.** Technician B
   - **c.** Both A and B
   - **d.** Neither A nor B

5. Technician A says that the UTQG standard rates a tire's traction. Technician B says that the UTQG standard rates a tire's tread wear rate. Who is right?
   - **a.** Technician A
   - **b.** Technician B
   - **c.** Both A and B
   - **d.** Neither A nor B

6. Technician A says that a drop center is when the center of the wheel has a machined counterbore that pilots on a machined area of the hub. Technician B says that a tubeless tire will deflate faster than a tube-type tire. Who is right?
   - **a.** Technician A
   - **b.** Technician B
   - **c.** Both A and B
   - **d.** Neither A nor B

7. Technician A says that radial tires ride smoother at low speeds than bias tires. Technician B says that radial tires ride smoother at faster speeds while travelling over highway expansion joints. Who is right?
   - **a.** Technician A
   - **b.** Technician B
   - **c.** Both A and B
   - **d.** Neither A nor B

8. Technician A says the amount of load a tire can support is determined by the area of the tire. Technician B says the amount of load a tire can support is determined by the amount of air pressure in it. Who is right?
   - **a.** Technician A
   - **b.** Technician B
   - **c.** Both A and B
   - **d.** Neither A nor B

9. Technician A says that a 70 is a lower profile tire than a 75. Technician B says that an H rating is for higher speeds than a Z rating. Who is right?
   - **a.** Technician A
   - **b.** Technician B
   - **c.** Both A and B
   - **d.** Neither A nor B

10. Technician A says that a tire's height is called its footprint. Technician B says that a leak of one bubble per minute from a tire can cause the tire to go flat overnight. Who is right?
    - **a.** Technician A
    - **b.** Technician B
    - **c.** Both A and B
    - **d.** Neither A nor B

Tire and Wheel Service

■ INTRODUCTION

Tire service is a large area of automobile repair. The average owner can expect to replace at least one set of tires on his or her car. Tire life depends on tire quality, air pressure, vehicle weight, driving conditions, suspension condition, and wheel alignment.

■ TIRE INFLATION

Maintaining correct air pressure is the most important factor in the safety, performance, and life expectancy of a tire. Results of low pressure are:

■ Temperature of the tire rises.

■ Load carrying capacity of the tire is lowered.

■ Life of the tire is reduced.

■ Fuel consumption increases.

■ Outside edges of the tire wear excessively.

When a bias tire has low air pressure, a driver will often feel it in the handling of the vehicle. A car will pull to the side that has the low tire, especially when it is a front tire.

Low tire pressure in a radial tire changes the normal deflection of its sidewall. This raises the amount of heat generated within the tire's carcass. Underinflation is the most common cause of radial tire failures. Bias-type tires experience a strong pull to the side of a tire with lower inflation. A radial tire can often be run at too low a tire pressure without exhibiting handling symptoms.

Tire pressure that is too high can wear the center of the tire tread and cause a rough ride. The tire is actually the first part of the suspension and spring system.

■ CHECKING AIR PRESSURE

NOTE: *An air pressure gauge must be used to check radial tire pressures. Do not try to check them visually. Radial tires*
that are normally inflated have a bulging sidewall and often appear underinflated. Figure 55.1 compares the appearance of bias and radial tires when fully inflated.

Vehicle owners should be encouraged to check tire air pressure at least once a month.

The following items refer to the effect of temperature on the tire's pressure:

■ As the air in the tire expands due to heat, pressure normally increases. Air pressure should be checked when the tires are cold.

■ Pressure normally increases 4 to 8 psi from cold to hot. It takes less than 3 minutes, or one mile of driving at moderate speeds, to make tires too hot to check accurately. This increases air pressure by about 4 psi. According to Michelin Tires, when adjusting pressure in a hot tire, add 4 psi to the maximum gauge reading desired. For instance, if the recommended cold pressure is 24 psi and the gauge reads 26 psi, fill the tire to 28 psi. Be sure to recheck it when cold the next day.

■ Air should *not* be let out of a tire when it is hot.

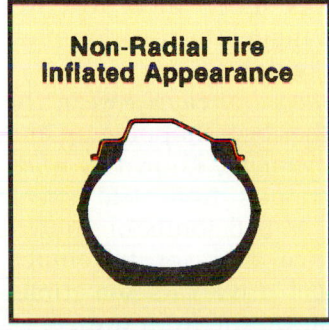

Figure 55.1 The bulging sidewall of a fully inflated radial tire gives it the appearance of being underinflated. *(Courtesy of The Goodyear Tire & Rubber Company)*

- Each change in outside temperature of 10°F will result in about 1 psi change in tire pressure.
- A hot tire that has lower pressure than the recommended cold pressure is seriously underinflated.
- Cold tire pressure should never be higher than the maximum pressure molded into the tire sidewall.

Tire inflation pressures should always be the same for both tires on one axle so ride and handling are not affected. A high quality *tire gauge* should be used. Inexpensive gauges are often inaccurate. One type of gauge is a part of the *air chuck*. These air gauges are often abused and become inaccurate when left installed on an air hose.

All valve stems should have screw caps on them. These keep out dirt and moisture and provide a backup in case the valve core leaks. Before adding air to a tire, blow air through the air chuck to clear it so dirt is not forced into the valve core.

TIRE WEAR

According to Goodyear, a 4 psi decrease in pressure below the recommended amount can result in a 10% loss of tread life. Additional losses of air pressure can result in even more wear and the possibility of serious damage to the tire. Underinflation can also cause the edges of the tire to wear (see Figures 14.1–14.3).

The fastest tire wear occurs during hard cornering, braking, and acceleration. Rough pavement also contributes to accelerated tire wear. Slow speed sharp cornering wears the front tires while high speed cornering will remove tread from the tires on the side of the vehicle to which the weight is transferred.

When a tire wears to within 1/32" of the bottom of its tread, *wear bars* show up around the tire (see Figure 14.4). The wear bars are raised areas cast into the bottom of the tire tread area.

Belted bias tires often experience *step wear*, which is wear on the second row of tread from the outside. This is a normal condition with these tires. It occurs because the inside and outside rows of tread on these cars carry most of the weight, while the second and sixth ribs carry the least. This allows them to scrub slightly as the tire rotates. Changing the inflation pressure of the tire will not affect this wear.

When a tire exhibits *scalloped* or *cupped* wear, the cause is usually that the tire has been hopping up and down on the road. This movement, known as *wheel tramp* (see Figure 55.36), results from bad shock absorbers, worn parts (such as ball joints or control arm bushings), out-of-balance tires, or too much runout of a tire. Front wheel drive cars sometimes experience abnormal tire wear on the front tires.

Inspect the tire for physical damage. When a car is driven with a tire that is flat or underinflated, the tire can become damaged. Look for evidence of tread or sidewall separation. Sometimes, damage from

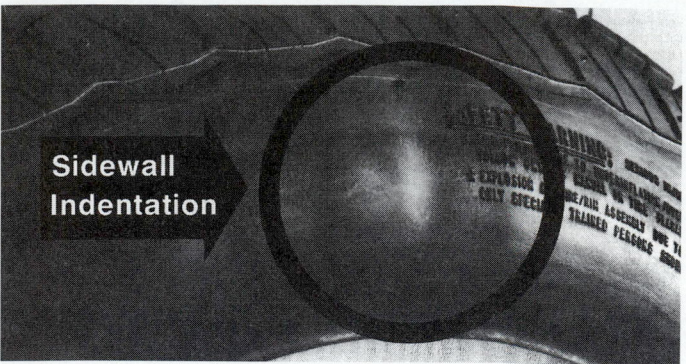

Figure 55.2 A slight indentation in the sidewall of a radial tire is normal. *(Courtesy of Rubber Manufacturers Association)*

underinflation might not be visible from the outside of the tire.

SIDEWALL CHECKS

Sidewall cracks are usually caused by years of exposure to the sun. Cracks on the outside of sidewalls often develop on recreational vehicles because RV driving is seasonal. Motor homes are used mostly on the highway (very low tire wear). Tires can be many years old before the tread wears out. When an RV is parked in the same spot for long periods of time and the tires have not been regularly rotated, only the tires on one side of the vehicle might be cracked.

According to the Rubber Manufacturer's Association, a slight sidewall indentation is common on radial tires (Figure 55.2). These are due to the construction of the tire and are normal. Outward bulges, on the other hand, are cause for the tire not to be used unless approved by the tire manufacturer or its representative. Cuts or cracks in the sidewall that allow cords to be exposed are cause for replacement of the tire.

➡ *Perform Tire Inspection Worksheet*

Recreational vehicles and light trucks with dual rear tires often experience uneven wear on the rear tires. The diameters of dual wheel tires must be within 1/4" of each other.

TIRE ROTATION

Front wheels on all cars experience the most wear because they are used for steering and also because weight transfers forward during a stop. The need for rotation is basically to keep the tires all wearing equally for safety and so that they can be replaced in complete sets. Most manufacturers specify regular rotation intervals. When there is no recommendation, rotation is recommended at 6000 to 8000 miles and then again at 18,000 to 24,000 miles.

Vehicle manufacturers provide a recommended rotation pattern for tires on their cars. Some patterns include a crisscross pattern. Others rotate front to rear (Figure 55.3). Some include rotation of the spare tire.

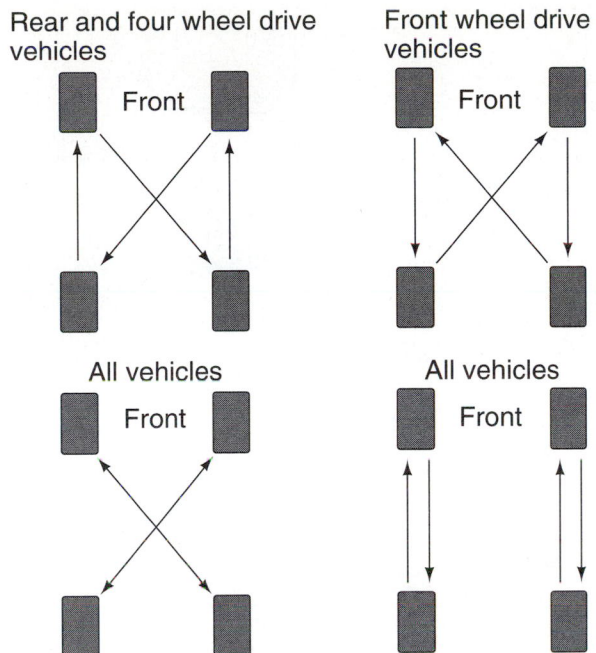

Figure 55.3 diagram labels:
Rear and four wheel drive vehicles — Front
Front wheel drive vehicles — Front
All vehicles — Front
All vehicles — Front

Figure 55.3 Recommended tire rotation patterns. *(Courtesy of Rubber Manufacturers Association)*

Other cars are equipped with only an emergency compact spare, which is not rotated onto the vehicle.

NOTE: *Some vehicles have tires of a smaller size on the front. These should not be rotated to the rear.*

Some tires are designed to be mounted and run in a specified direction of rotation only. These tires are rotated in the front-to-rear pattern and are cross-rotated only if they are removed from the rim and remounted so that they will still rotate in the designated direction.

- In the past, moving radial tires to the other side of the car was not recommended, but this is no longer true *as long as there is not a specified rotation pattern to the tire.* If one front tire shows uneven wear, it can be switched with the other front tire.
- Before rotating tires, check to see that radials are not being mixed with bias tires. This is an unsafe practice. When bias tires and radial tires are mixed on passenger cars, the radials should be installed on the rear.
- When towing a trailer, the tire types on the trailer and the tow vehicle should be the same.
- According to the RMA (Rubber Manufacturer's Association), when a recreational vehicle or light truck has dual rear wheels, radials may be used on either the front or rear axle.

Front wheel drive cars experience far more wear on the front tires. Rear tires on these cars last much longer. One popular rotation pattern for front wheel drive cars involves moving the front tires to the rear on the same side of the car. The rear tires are then

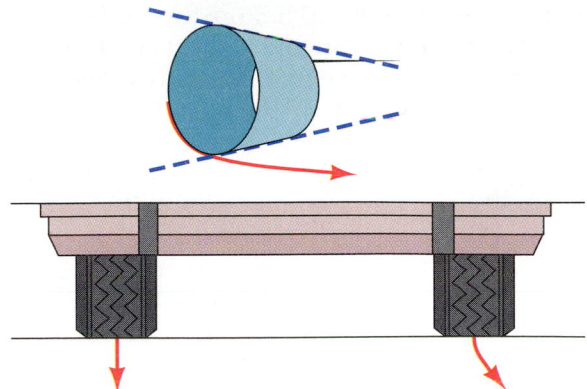

Figure 55.4 Tire conicity results in a pull. *(Courtesy of Chrysler Corporation)*

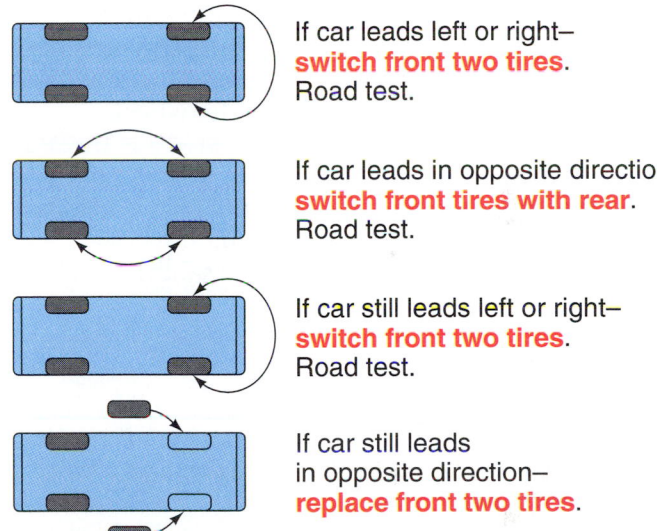

If car leads left or right— **switch front two tires**. Road test.

If car leads in opposite direction— **switch front tires with rear**. Road test.

If car still leads left or right— **switch front two tires**. Road test.

If car still leads in opposite direction— **replace front two tires**.

Figure 55.5 Diagnosing tire pull caused by conicity. *(Courtesy of Chrysler Corporation)*

moved to the front, but to opposite sides of the vehicle. This is said to even out and minimize "drive cornering scrub" wear.

Sometimes after a tire rotation, the car can exhibit a pull to one side or the other. This can be because of an inherent pull within a radial tire caused by tire conicity (when the tread is tapered like a cone), or off-center belts (Figure 55.4). If the offending tire was previously installed on the rear of the car (non-steering wheels), it can cause a problem when installed on the front of the car. Figure 55.5 shows some recommendations for isolating the problem. Off-center belts can also result in outside shoulder wear on the tire.

■ REMOVING AND TIGHTENING LUG NUTS

Most lug nuts have right hand threads and are loosened by turning counterclockwise. A few vehicles have left hand threads on the lug nuts one side of the car.

These are labeled with an "L" on the end of the lug stud. Loosening and removing lug nuts is easiest when done with an impact wrench. When the wheel is in the air, it will not have to be held from turning.

CAUTION Do not loosen the lug nuts when the wheel is on the ground. The car can fall down. Have someone hold the brakes applied while you loosen them with the wheel off the ground.

Tighten lug nuts evenly, in a crisscross pattern. Use a torque wrench to avoid warping a disc brake rotor (Figure 55.6). Specifications are available for the different makes of cars. Just as with other fasteners, lug bolts are stretched when they are properly installed so that they maintain clamping force. Lug nuts that are loose allow the wheel to exert all of the weight of the vehicle on the lug bolts, rather than on the wheel.

Lug nuts for steel wheels are tapered on the side that faces the wheel (Figure 55.7a). Installing these backwards can ruin a rim and result in a dangerous accident. Some wheels have special lug nuts that cover the ends of the wheel lugs, so these can only be installed in one direction.

CASE HISTORY *A student complained of a clunking sound and poor vehicle handling. He asked the teacher to inspect the vehicle. The teacher was amazed to find all of the lug nuts on inside out. All of the holes in the rims were damaged so badly that new wheels were required.*

Anti-theft lug nuts are popular on custom wheels. They require a special key (Figure 55.7b). Sometimes, the customer loses the key or forgets to bring it to the

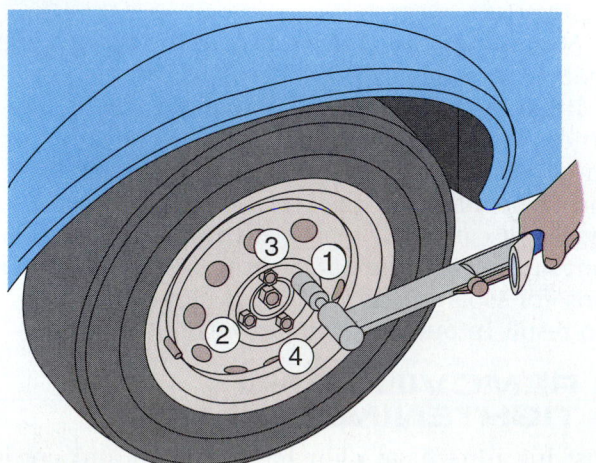

Figure 55.6 Lug nuts are tightened with a torque wrench.

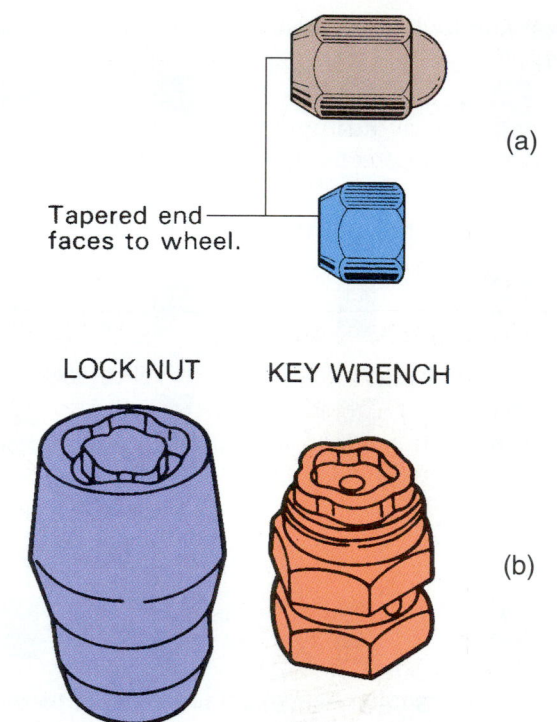

Figure 55.7 Lug nuts for steel wheel rims: (a) The tapered side faces the rim. (b) Lock nut and key. *(Courtesy of General Motors Corporation, Service Technology Group)*

shop for a tire or brake repair. Most shops that do tire repair have a special tool kit for removing these without the key. It has several adapters that are wedged against the outside of the lug nut. The tool is driven rapidly by an impact wrench and forced against the lug nut to loosen it.

A special type of lug nut is used with aluminum wheels. It is usually equipped with a washer that cannot be removed. The shank on the nut must not be too long or it can bottom out, leaving the wheel loose (Figure 55.8). Also, the lug nut must have a small amount of clearance so it is free to turn inside of the hole in the wheel.

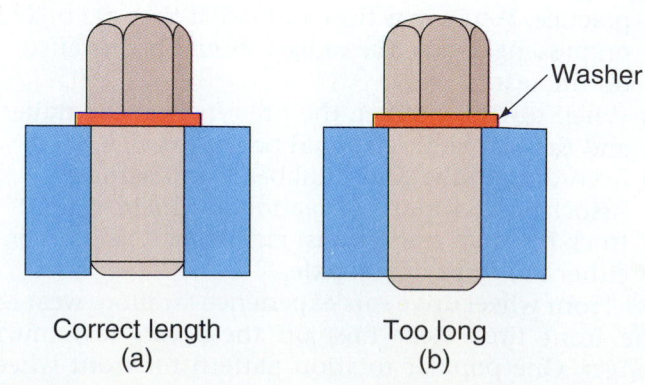

Figure 55.8 Lug nuts in cast aluminum wheels: (a) is correct and (b) would allow the wheel to be loose.

→ *Perform Tire Maintenance Worksheet*

REPAIRING WHEEL STUDS

Occasionally, a lug bolt will be stripped or broken. If only a couple of threads are damaged, they can be cleaned up with a thread chaser. Broken lug bolts can be replaced. On drum brake cars, when the drum is removed with the hub, the studs are **swaged**, which means that they are deformed to keep them tight. The swaged area can be cut off with a special tool to make them easier to remove (see Figure 52.37). Drum brakes are no longer found on the fronts of newer vehicles.

If the rotor or drum separates easily from the hub, the lugs can usually be driven from the hub with a brass hammer or punch. If much effort is needed, a tie rod press can be used to force the lug bolt out of its hole. The new lug stud can be installed with an inverted lug nut and washers as shown in Figure 55.9.

REMOVING AND MOUNTING TIRES ON RIMS

There are some important points to be aware of before attempting to remove and install tires on wheel rims. *Tires can explode and fingers can be cut off if proper caution is not observed.*

- Be sure to use the proper size and construction of tire to match the wheel rating.
- Be sure that the rim diameter matches the diameter molded on the tire sidewall. Be especially careful that a tire or wheel is not a metric size that was manufactured for the overseas market. Although you might be able to mount a metric tire on a standard wheel, it will not fit properly and can come off.

Tire problems such as leaks and vibrations can sometimes be traced to improperly mounted tires.

The following information applies to either of the two most common tire changing machines.

Deflate the Tire

Before removing the tire from the wheel, remove the valve core to deflate the tire completely. Figure 55.10 shows a tool used to remove the valve core. If the tire is to be patched and reinstalled, be sure to mark the locations of the valve stem and any wheel weights with a marking crayon. If another tire is to be installed on the rim, remove any balance weights that are on it (Figure 55.11).

Tire Machines

There are two types of tire changers in common use, the rim clamp and the center post (Figure 55.12). The procedures for mounting and dismounting tires are similar and will be covered here.

Unseating the Beads

Before the tire can be removed from the wheel, the beads must be unseated from the bead seats (Figure 55.13). This is called "breaking the beads." Lubricate the tire bead with rubber lube. Rubber lube reduces friction between the tire and wheel, preventing damage to the tire and making removal easier.

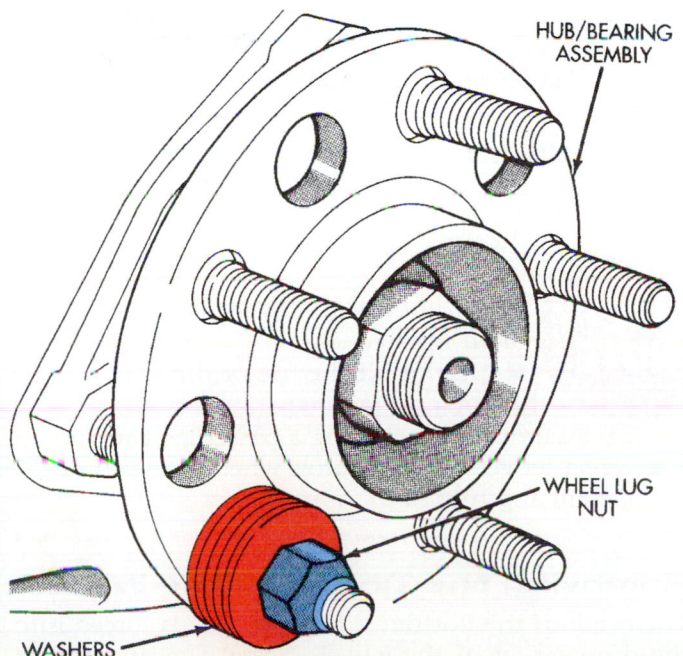

Figure 55.9 Installing a new stud. *(Courtesy of Chrysler Corporation)*

HUB/BEARING ASSEMBLY

WHEEL LUG NUT

WASHERS

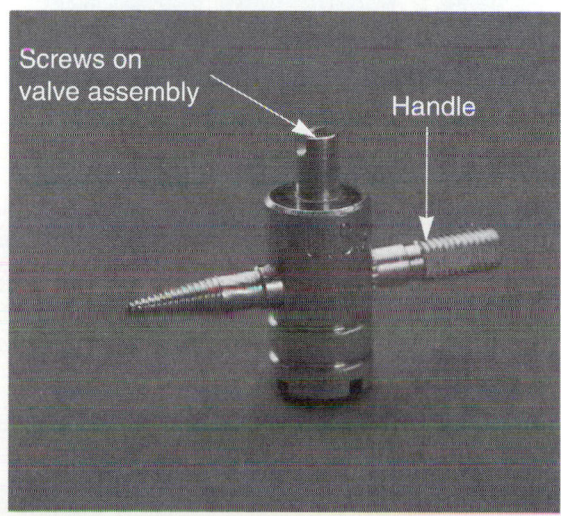

Figure 55.10 A tool for removing and replacing valve cores.

Screws on valve assembly

Handle

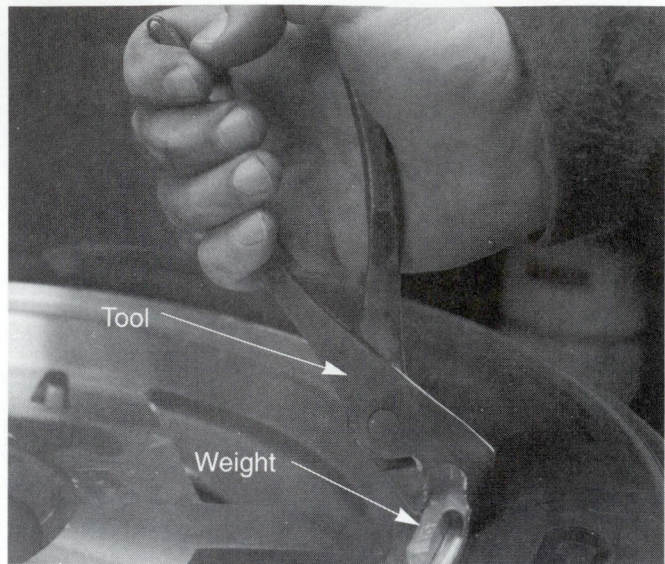

Figure 55.11 Removing wheel weights from the rim.

(a) Rim Clamp

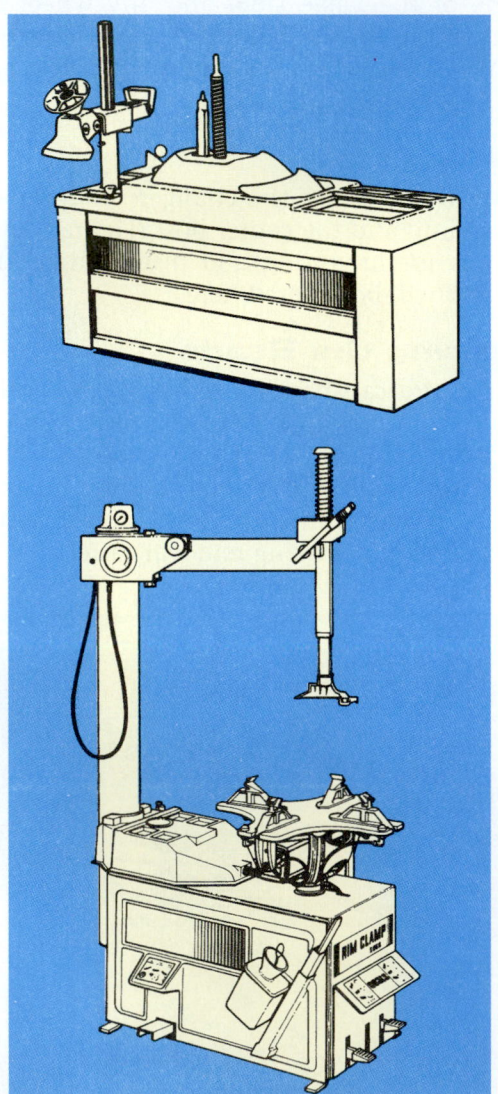

Figure 55.12 A center post tire change (top) and a rim clamp tire changer (bottom). *(Courtesy of Hennessy Industries, Inc.)*

(b) Center Post

Figure 55.13 Breaking the bead on each type of tire changer. *(b, Courtesy of Rubber Manufacturers Association)*

SAFETY NOTE
- Do not attempt to unseat the beads of an inflated tire.
- Do not hit the tire or rim with a hammer. A hammer should not be necessary when removing a passenger car tire.
- Never let go of the tire iron when in use. It can flip up and hit you.

Removing the Tire from the Wheel

The bead on the bottom side of the tire is forced into the drop center of the wheel so that the upper bead can be pulled over the edge of the rim (Figure 55.14). On either type of tire machine, removal of the tire from the wheel requires the use of a bar called a *tire iron* (Figure 55.15). First, the upper bead is pulled over the outside of the rim. Then, the lower one is. Do not try to remove both beads at the same time.

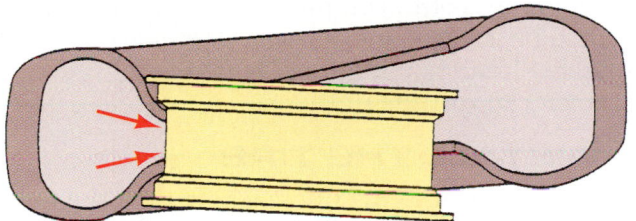

Figure 55.14 Both beads are in the drop center of the rim while the bead is pulled over the edge of the tire. *(Courtesy of Pirelli Tire Corporation)*

(a) Rim Clamp

Bead is in drop center of rim

(b) Center Post

Figure 55.15 A tire iron is used to pull the top bead over the edge of the rim. Be sure that the other side of the tire bead is in the drop center of the rim. *(b. Courtesy of Rubber Manufacturers Association)*

NOTE: *Be careful not to tear the bead. This will ruin the tire. If the bead does not come off the rim without binding:*

■ *Be sure there is enough rubber lube on the bead.*
■ *Double check to see that the lower bead is totally into the drop center of the wheel.*

■ INSPECTING THE TIRE AND WHEEL

After the tire is removed from the rim, inspect its inside for cuts, carcass damage, penetrating objects, loose cords, dirt and liquid. Inspect the condition of the bead by pulling out on it in several places around its circumference. If there is a sharp bend in the bead, do *not* mount the tire. The bead seat could be broken. It is not worth taking a chance with someone's safety.

 **SAFETY NOTE** Never mount a tire or wheel that is damaged. This could result in injury or death to the driver of the vehicle and the shop may be held liable in court for negligence.

Inspect the condition of the wheel rim for sharp edges, dents, cracks, and other damage. Small dents in the rim flange can be straightened. When there is a larger dent in the wheel, it must be checked for **radial** and **lateral runout** before it can be used. Excessive runout can cause the car to shake at highway speeds. Checking runout is best done with a dial indicator, but spinning the wheel while holding something (a screwdriver or a pencil) near it for reference will give a good indication of whether a wheel is bent.

It is quite common for rust to damage the bead seat on a wheel. If the rust is on the surface only, it can be removed with a wire brush. If the bead seat is not smooth, the tire will have a slow leak and the wheel will need to be replaced.

■ VALVE STEM REPLACEMENT

Rubber valve stems are customarily replaced when new tires are installed. Removal of the old stem is usually done while the tire is off the rim. The stem is cut off with a knife (Figure 55.16a), or it is forced through the hole in the rim using the valve stem installing tool.

To install the new stem, thread it into the installing tool, lubricate it with rubber lube, and pull it into the hole (Figure 55.16b). Be sure that it is pulled all of the way into place and is properly seated in the hole in the rim. A new valve stem comes with a new valve core. Remove the valve core before attempting to inflate the tire.

➡ *Perform* **Replace a Rubber Valve Stem Worksheet**

(a)

(b)

Figure 55.16 (a) Cut the bottom of the old valve stem. (b) The valve stem is installed with a special tool.

Some tires are designed to be run in only one direction. Check the sidewall of the tire to see if there are direction arrows.

■ RUBBER LUBE

Lubricate both tire beads with an appropriate rubber lube. Good quality rubber lube is slippery and fast drying. Lubricate the bead seats on the rim with rubber lube also. Using rubber lube provides the following advantages:

- Rubber lube reduces friction between the tire beads and the edge of the rim during mounting.
- Rubber lube helps to seal around the bead during initial inflation of the tire.
- Friction between the bead seats and the tire bead will be reduced when inflating the tire. This is important so that the beads will be all of the way seated and the tire tread will not be distorted.
- Safety glasses should be worn when inflating the tire.

Radial tire sidewalls are more flexible and their tire beads are not forced into position as readily as bias-type tires. If the beads are not seated properly, the tire can have an out-of-round condition that will make it difficult to balance.

Cautions with rubber lube:

- Do not use petroleum products, which will damage the rubber in the tire.
- Rubber lube should not be diluted with water, which can rust the rim.

■ INSTALL THE TIRE

Clamp the wheel to the tire machine with the narrow bead ledge up and install the inside bead of the tire over the flange of the rim (Figure 55.17). As more and more of the bead is passed over the edge of the rim flange, force the bead down into the drop center well of the wheel. This is important! If one side of the bead is not in the drop center, the other side will not be able to be stretched over the flange. When the first bead is totally installed and positioned in the drop center of the wheel, install the other bead over the flange of the rim (Figure 55.18).

Narrow bead ledge

Wide bead ledge

Figure 55.17 The narrow bead ledge is up and the lower bead is in the drop center of the rim. *(Courtesy of Rubber Manufacturers Association)*

Tool rotation

Bead is in drop center

Figure 55.18 As the bead is drawn over the edge of the wheel rim, it is important that the other side of the bead stay in the drop center. *(Courtesy of Hennessy Industries, Inc.)*

Inflating the Tire

Many tire machines have a provision for tire inflation.

NOTE: *Inflation can be the most dangerous part of tire installation as tires can explode during inflation.*

If a machine is not equipped with a lock-down clamp to hold the wheel, inflate the tire in a safety cage. Use an extension hose with a clip-on air valve. This is so the technician can stand away from the tire during inflation.

 **CAUTION** Do not leave tools on the tire side-wall when inflating the tire.

Seating the Beads

Tubeless tires require a substantial volume of air flow for the beads to start to seat on the rim. Occasionally, getting the tire to take on air can be quite difficult.

SHOP TIP Removing the valve core from the valve stem during filling allows a greater volume of air to enter the tire.

Most tire machines have an inflation chamber with a sealing ring. The sealing ring forces a good deal of air from the air chamber into the area surrounding the lower bead of the tire (Figure 55.19). The air chamber is required so that there will be a tremendous amount of air flow at once. If air was simply pumped through the normal shop air lines, the volume of air could not be maintained.

CASE HISTORY *A shop purchased a new glass bead blaster and a technician was attempting to use it. The pressure gauge showed 120 psi. When the pedal was first applied, the glass beads and air from the blast nozzle cleaned effectively for only about one second. Then, the pressure would drop to about 60 psi. Further investigation showed that the machine had been fitted with a small regulator with ⅜" pipe nipples on its inlet and outlet sides. The air supply line from the air compressor was ¾" pipe. Installing a larger ¾" regulator solved the problem. The smaller regulator was acting as a restriction or orifice in the line. This is an application of Pascal's law.*

Pressure after an orifice will be lower than the pressure before it. When trying to fill a tire with air from a remote air compressor, the air line acts as a similar orifice. That's why an inflation chamber is used. If the tire is pulled against the upper bead seat while inflating it, both beads will usually seat.

What to do when the beads will not seat:
■ Sometimes, the outer bead can be pushed against the inner bead using the bead breaking attachment to force it down over the safety hump.
■ Another trick is to bounce the tire vertically on the floor so that the air in it can force the beads outward.

Radial tires are more difficult to seat than bias tires. An inflatable rubber tube can be used when filling a bias tire to distort the tire tread, which forces the beads outward. The inflatable tube does not work on steel belted tires such as radials, because the tire is too stiff in the tread area.

When inflating a tire on a center post tire machine, the hold-down clamp should be in place. When the tire begins to inflate, loosen the hold-down clamp one turn. Otherwise as the tire expands, the lower sidewall can wedge against the tire machine, making it difficult to loosen the hold-down clamp.

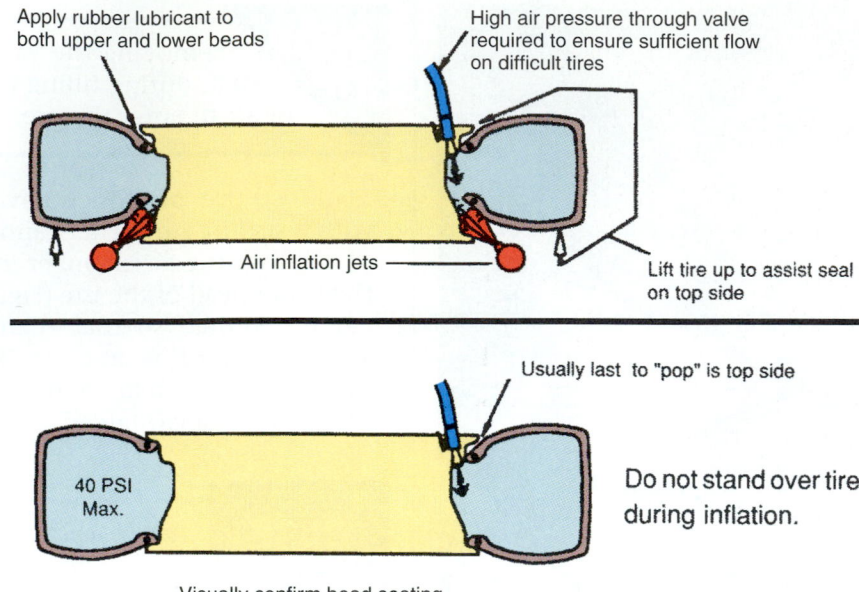

Apply rubber lubricant to both upper and lower beads

High air pressure through valve required to ensure sufficient flow on difficult tires

— Air inflation jets —

Lift tire up to assist seal on top side

Usually last to "pop" is top side

40 PSI Max.

Do not stand over tire during inflation.

Visually confirm bead seating

Figure 55.19 Hold the tire against the top bead seat and use the air jets to seal the bottom bead. *(Courtesy of Hennessy Industries, Inc.)*

The last bead to seat is usually the top one. As the tire beads slip over the safety humps on the bead seats of the rim, a loud pop is usually heard. This is normal.

Many tire machines include an inline dial indicator-type tire gauge. Do not put more pressure than 40 psi (or the maximum pressure listed on the tire sidewall) when attempting to seat the beads. If the beads will not seat with 40 psi, break the beads again and thoroughly lubricate the tire and rim again. Reposition the tire on the rim and reinflate it. When a tire is inflated in a safety cage, it can be inflated to a maximum of about 50 psi (3.5 bar) to seat the beads.

 **SAFETY NOTE**
■ Do not inflate a tire higher than 50 psi. The metal bead wire might break.
■ Be careful in which direction the valve stem is pointing when inflating a tire. It can be shot out of the wheel.

CASE HISTORY
A trick used by some for seating the beads on tires is very dangerous. The procedure consists of putting a flammable substance into the tire and lighting it on fire. The tire in Figure 55.20 is one example of the force of an explosion caused by this practice. The explosion can cause serious injury or death, not to mention internal damage to the tire that could result in a later accident.

For especially difficult tire inflations, many tire specialty shops have special *bead seaters* (Figure 55.21).

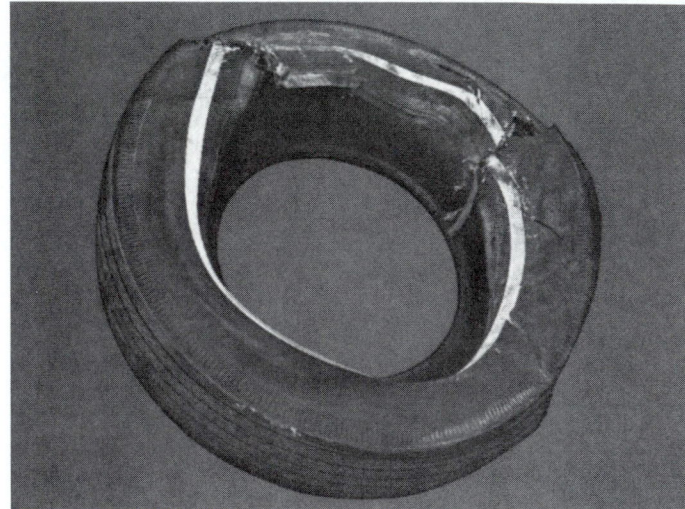

Figure 55.20 This tire exploded when a flammable substance was ignited in it. *(Courtesy of Rubber Manufacturers Association)*

These represent an extra cost to the shop because a different size bead seater is needed for each rim diameter. On wider wheels, a bead seater might be necessary.

Install the Valve Core

Install the valve core. The valve core should not extend above the top of the valve stem. If it does, the valve cap could contact the top of the valve core and let air out. A shorter valve core should be installed. After the valve core is installed, inflate the tire to the amount recommended on the vehicle placard. Check

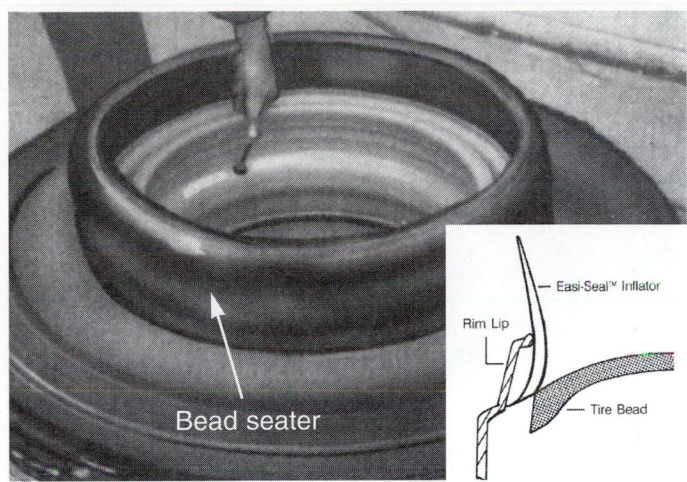

Figure **55.21** A bead seater helps to inflate a tire. *(Courtesy of Hardline International, Inc.)*

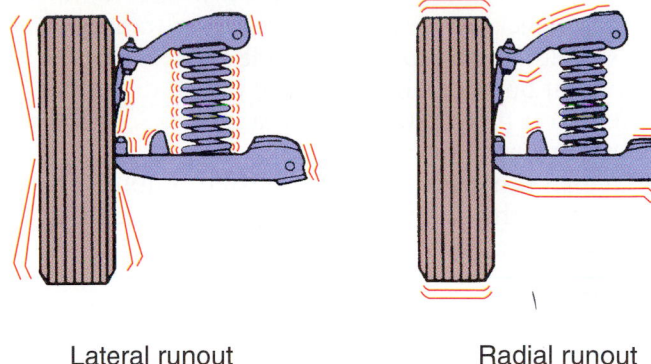

Lateral runout   Radial runout

Figure **55.23** Tire runout can be either radial or lateral. *(Courtesy of Ford Motor Company)*

to see that the valve core is sealing. Then, install the valve stem cap.

▰ TIRE RUNOUT

When a tire is not correctly mounted on a rim or is out of round it will have runout. Runout is when the tire is not round as it spins. If the tire is higher on one side than the other, handling problems will occur.

Pirelli Tires recommends that tires be inflated initially to 40 psi to thoroughly seat the beads. Sometimes, a tire is not properly seated on the rim. The sidewall has *centering ribs* or *locating rings* (Figure 55.22) that can be inspected to see that they are even with the edge of the rim flange all the way around the tire. If not, the beads need to be broken down again to center the tire. Dismounting and remounting the tire on the rim 180° from its original position can sometimes result in improvement.

Another possible cause of runout is that a second-quality or blemished tire might be out-of-round. Runout can be either lateral (sidewall wobble) or radial (tread up and down) (Figure 55.23). Runout can be

checked with a dial indicator. Some shops have *tire-truers* for correcting radial runout.

➡ *Perform **Dismounting and Mounting Tires Worksheet***

▰ TIRE REPAIR

Checking for Leaks

To check for leaks, a water tank is commonly used. Roll the inflated tire slowly around in the tank while looking at the area where the tread meets the water (Figure 55.24). Small bubbles from a slow leak can usually be found in this manner. If a leak is not evident, push the valve stem from side to side. Also, check carefully where the tire bead meets the rim. If a soak tank is not available, apply soapy water to the outside of the tire.

Inspecting the Tire

When inspecting a tire before repairing it, remove it from the wheel. A special tool called a bead spreader can be used to hold the beads apart during the tire inspection and repair. Check the entire outside and inside of the tire carefully to see that it is not damaged from having been run flat. Use a bright light. The tire must be dry prior to inspection. Inspect the outside of

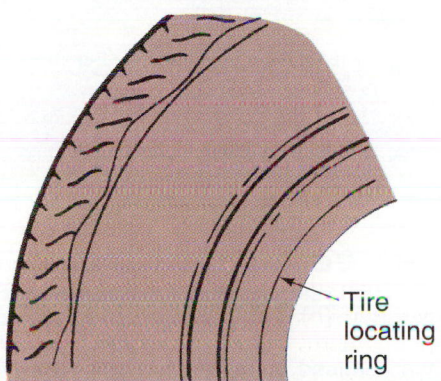

Figure **55.22** Check the locating ring to see that the tire is evenly mounted.

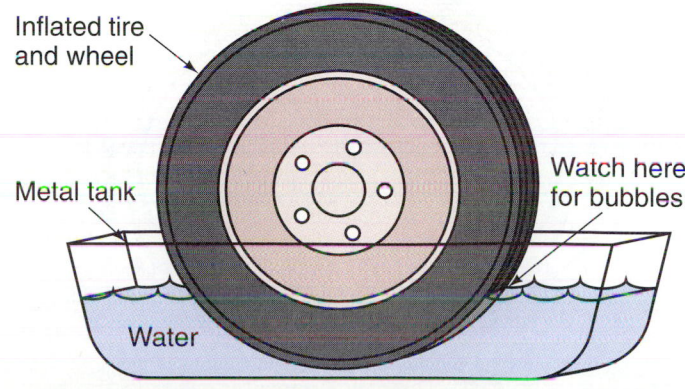

Figure **55.24** Check for bubbles at the water line as you spin the tire slowly in the tank.

the tire for anything that can allow moisture in, like weather checking, cracks, or tread separation. Tires that should *not* be repaired include those that:

■ Smell of burned rubber or have bluish discolored rubber on the sidewall flex area
■ Have evidence of loose cords
■ Have an inner liner that has blisters, bubbles, cracks, or the casing cord pattern showing through
■ Are worn beyond the wear bars

■ REPAIRING A TIRE

Tire puncture repairs can be done with a rubber **tire plug**, a **patch**, or both. Pinhole repairs can be made using a patch only. A pinhole is a very small hole that closes up visually when the nail is removed.

Repairs can be made when they are within the tread area (crown) (Figure 55.25) as long as the puncture is less than ¼" in diameter. Tire sidewalls continually flex, so repairs in that area may not last long because the patch can come off. Repairs to holes larger than ¹⁄₁₆" in the sidewall area or ¼" in the tread area can often be made, but they must be done by a full service tire facility. Repairs to the bead area cannot be made.

The Rubber Manufacturer's Association publishes guidelines for the proper repair of all types of tires. Tires are sometimes repaired improperly. Rubber plugs are often installed without removing the tire from the wheel. This is a substandard repair because the tire should be removed from the wheel and carefully inspected to see that it is in good shape. Damage can happen when a repair is done to a tire that has been run flat, has a hole that is too large to be repaired, or has suffered damage from an impact or a nail. Figure 55.26 shows types of tire damage that are not seen from the outside.

There are differing opinions as to the repairability of some tires. Each tire manufacturer publishes

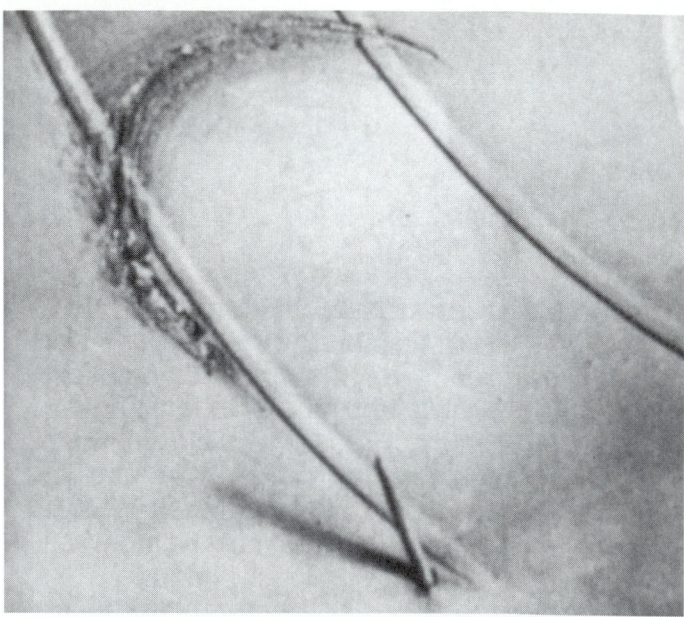

(a) Interior damage caused by nail

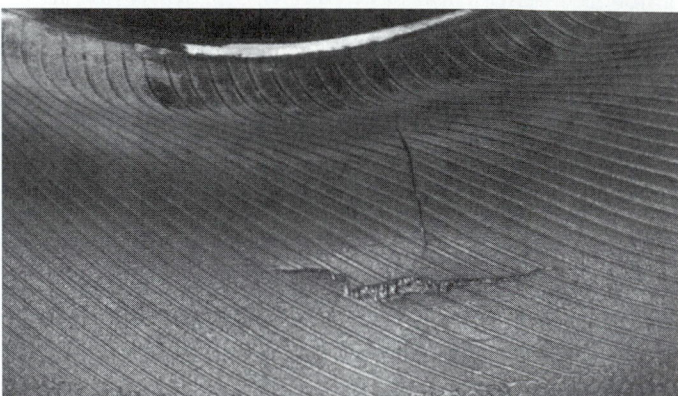

(b) Impact damage

(c) Rim bruise break

Figure 55.26 Tire damage unseen from the outside: (a) The damage to the interior of this tire was caused by a nail. (b) A rim bruise break is caused by impact against something like a curb or pothole. (c) This damage was caused by driving for a short distance on a severely underinflated tire. *(a, Used with permission of Michelin North America, Inc. All rights reserved.; b and c, Courtesy of Rubber Manufacturers Association)*

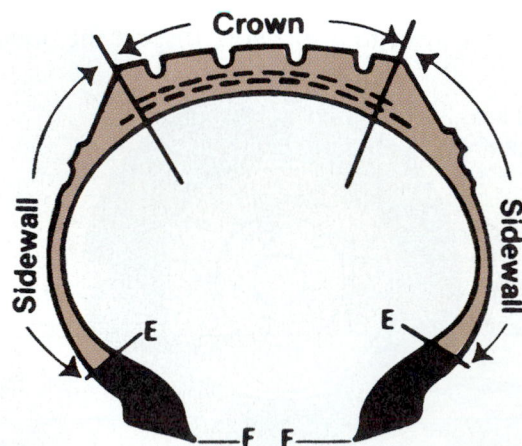

Figure 55.25 Repairs cannot be made to the area between E and F. *(Used with permission of Michelin North America, Inc. All rights reserved.)*

repairability standards that they distribute to their dealers. Some manufacturers say that high speed tires are not repairable because these tires are used on powerful cars at high rates of speed. Others say that a high speed tire will lose its speed rating if repaired. Still others give repair procedures for these tires. In any event, the tire should be removed from the rim if a repair is to be made.

A proper repair consists of:

- Removing the tire from the wheel
- Inspecting the tire carefully
- Running a reamer or drill through the hole
- Installing a tire plug
- Smoothing off the inside of the plug
- Installing a patch on the inside of the tire

PREPARING THE TIRE FOR REPAIR

Before attempting to repair a tire, the inside of the tire liner must be cleaned. The rough inner surface of a tire is covered with a material that will prevent the adhesion of a patch unless the surface is properly prepared. Using rubber cleaning solvent, clean the area where the new patch will be installed. A scraper is used to clean the liner because it is more effective in removing oils, grease, or silicone (Figure 55.27).

Plugging the Hole

A steel belted radial tire that has a puncture through a steel belt probably has steel exposed. Moisture can damage the inside of the belt area of the tire. It is advisable that this tire be filled with a rubber plug as well as patched.

Ream the Hole

The hole must be drilled or reamed first so that the plug will fit into the it. Holes that go through steel belts beneath the tread can have sharp pieces of steel sticking into them. The result can be a cut plug. This is another reason that the hole is drilled or reamed before installing a plug. If a drill is used, it should have a speed of less than 5000 rpm to avoid excess heat. Before drilling, probe the hole with an awl to determine the direction of the puncture.

A rubber plug is inserted into the hole in the tire. First, vulcanizing cement is applied to the plug (Figure 55.28). The cement is the same as that which is used to vulcanize a tire patch to the tire liner. Next, the plug is forced into the hole until ½" of the plug extends above the tread surface of the tire. Last, the end of the plug is cut off with a flexible knife to ⅛" to 1/16" above the inner liner. Cut the plug flush with the tread surface on the outside of the tire. Be careful not to stretch the plug while cutting it.

The liner is lightly buffed with a fine buffing stone and a low speed buffer motor (less than 5000 rpm) to complete the preparation of the surface (Figure 55.29).

Figure 55.27 Clean the area to be repaired with a scraper. *(Courtesy of Patch Rubber)*

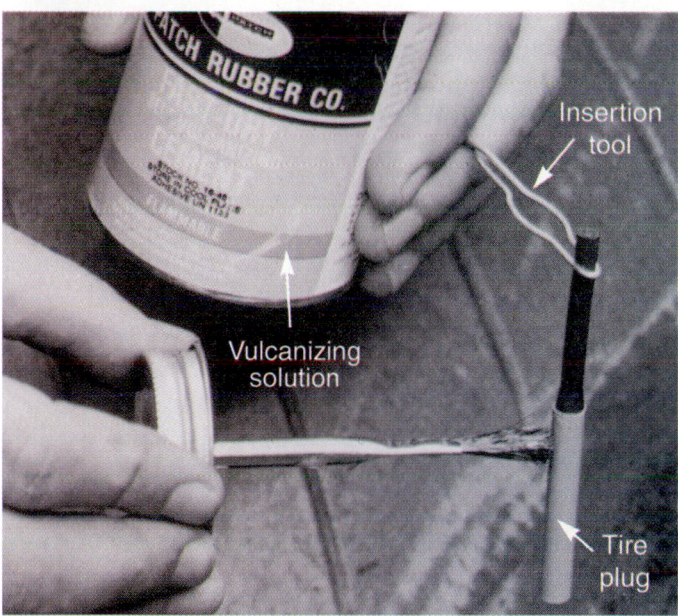

Figure 55.28 Apply vulcanizing cement to the plug and install it in the hole. *(Courtesy of Patch Rubber)*

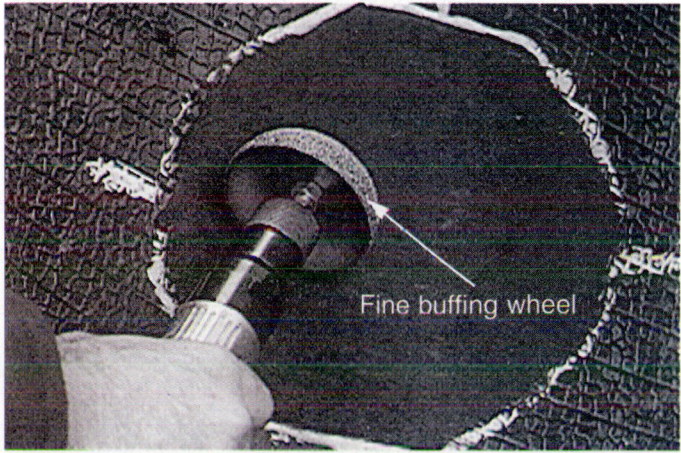

Figure 55.29 Buff the area where the patch will be. *(Courtesy of Patch Rubber)*

First, outline an area ½" around the outside of the patch with a marking crayon. Be careful not to buff through the liner into the casing plies. Just clean the surface of the rubber. If a portion of steel belt extends into the inside of the tire, it is carefully ground away. Because compressed air contains oils and moisture, it should not be used to remove the dust. The dust from the buffing operation is vacuumed away.

■ PATCHING THE TIRE

After the plug is cut flush with the inside of the tire liner, clean and scrape the area with rubber cleaning solvent and a scraper. After the cleaned area dries, apply an even amount of the vulcanizing cement to the buffed area where the patch will be installed (Figure 55.30). Vulcanizing cement is liquid when it is in good condition. If the cement has solidified or is like jelly, it should be discarded. It is a good idea that the cement be the same brand as the patch so that the two are compatible.

Allow the vulcanizing cement to *completely* dry before applying the patch. Drying time varies with temperature and humidity.

SHOP TIP Moving the puncture to the up position will allow the cement to dry quicker and more evenly. Solvents are heavier than air.

Install the Patch

Carefully remove the backing from the adhesive side of the patch. Be careful not to touch the sticky adhesive.

NOTE: *The patch should be installed while the tire is in its natural shape. A bead spreading tool is commonly used by many shops when affixing a patch to the inside of the tire. This is not recommended for radial tires when applying the patch.*

Center the patch over the hole, making sure that the bead arrows on the patch point to the beads of the tire. Roll the patch into place with a corrugated tire stitcher (Figure 55.31). Roll firmly from the center outward, using as much hand pressure as possible. After stitching is complete, remove the thin plastic covering from the top of the patch.

Combination Plug/Patches

Some plugs are a combination of plug and patch. These can only be used when the hole goes straight into the tire. If the hole is slanted, use a plug and patch. Some combination plug/patches are mushroom-shaped and must be installed using a special plug installation tool. Be sure to dip a blunt probe in vulcanizing cement and push it through the hole to thoroughly coat the walls of the hole. After reinstalling the tire on the wheel and inflating it, double-check with water to see that the leak has been fixed.

➡ *Perform* ___Patching a Tire___ *Worksheet*

Liquid Puncture Sealants. Liquid puncture sealants are commonly available in automotive stores. While these products may be handy when a flat tire occurs, they are not recommended by tire manufacturers.

SAFETY NOTE Some puncture sealants are flammable and can result in an explosion from a spark caused during an external plug repair. The spark results when drilling through the steel belts of the tire. Removing the tire from the wheel lessens the danger of an explosion.

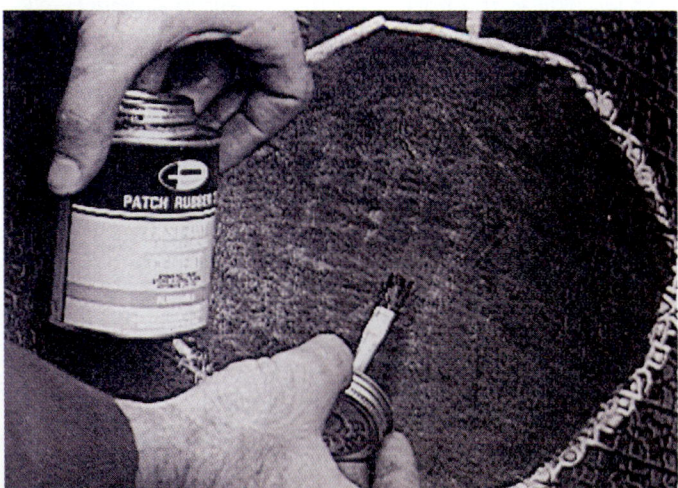

Figure 55.30 Apply vulcanizing cement to the buffed area. *(Courtesy of Patch Rubber)*

Figure 55.31 Use the stitcher to seat the patch to the tire casing. *(Courtesy of Patch Rubber)*

TIRE AND WHEEL BALANCE

Tire imbalance is one of several possible causes of vehicle vibration. Some other possibilities are a bent axle or wheel rim, an out-of-phase driveshaft, bad universal joints, or misaligned drive train components. Tire imbalance can result in a cupped tire, a loss of traction, and premature wear to steering and suspension parts.

While tire imbalance is not always the cause of vehicle vibration, it is a major cause. Front tires are the most prone to exhibit symptoms from imbalance. Vibration from rear tire imbalance may not be as obvious, but tire wear can still result.

Imbalance results when the weight of a tire's materials is not equally distributed around the tire. The result is that one side of the tire is heavier than the other. This can be due to manufacturing or it can result from tire wear. Sometimes new tires can have a minor imbalance that is correctable by adding specific amounts of weight to the wheel rim at specified points.

Wheel Weights

Wheel weights are attached to the rim to correct imbalance (Figure 55.32). Clip-on wheel weights are attached to the rim using a special wheel weight hammer (see Figure 55.11).

Several styles of lead clip-on weights are available in increments of ¼ ounce. There are different types of clips available. Some have an extra long clip for fitting under wheel covers. Another style has a wider, larger clip to fit on pickup truck wheels. Aluminum wheels require alloy or coated clips if they have a flange that will accept a clip-on weight. The coated weight will not corrode the wheel.

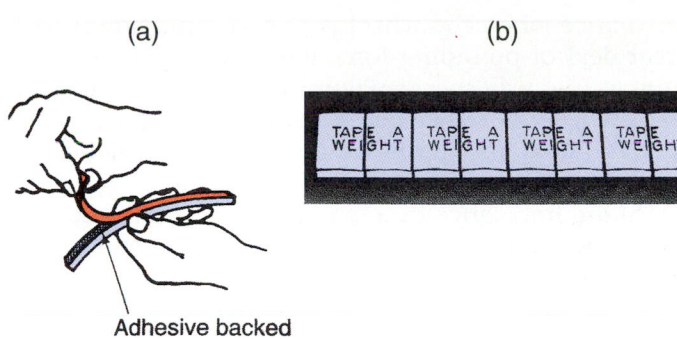

(a)   **(b)**

Adhesive backed

Figure 55.33 Lead tape weights. *(b, Courtesy of Hennessy Industries, Inc.)*

Lead strips that are cut to length are also available for aluminum wheels (Figure 55.33). These are attached to the wheel with double-sided tape after cleaning a place on the inside of the wheel with sandpaper. Sometimes they are installed on the inside and outside of the wheel, but this is unsightly. Chrome weight strips are available also.

NOTE: *Be careful when working around aluminum wheels. They are easily scratched. Plastic shields are available that cover tire tools to prevent damage.*

TYPES OF WHEEL BALANCE

The two types of wheel balancing methods used on cars are static and dynamic. Static (which means an object is stationary) imbalance is measured with the wheel at rest. If a wheel with static imbalance were mounted on a spindle with the heavy spot at the top, the heavy spot would rotate to the lowest possible position on its own. If the wheel were statically balanced, it would *not* have a tendency to rotate by itself.

Static imbalance subjects the wheel to vertical impacts that become worse with higher speed. The impacts occur because the tire has a heavy spot on one end of its tread (Figure 55.34). A small amount of

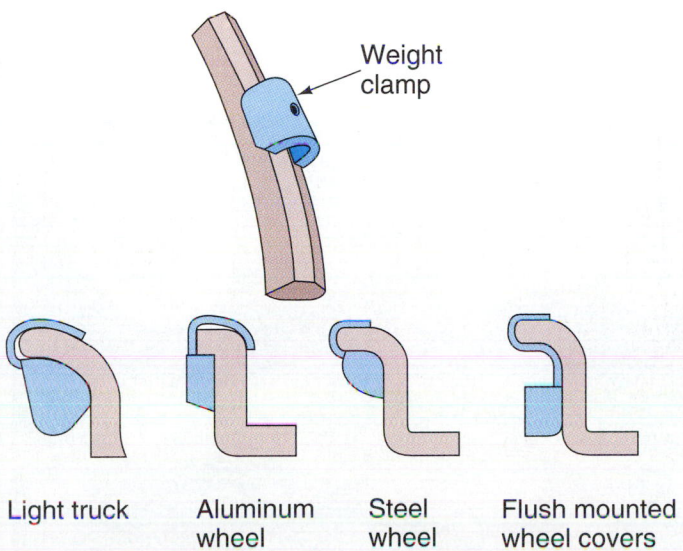

Weight clamp

Light truck   Aluminum wheel   Steel wheel   Flush mounted wheel covers

Figure 55.32 Lead wheel weights are installed on the wheel to counterbalance a tire. There are several styles for different applications. *(Courtesy of Hennessy Industries, Inc.)*

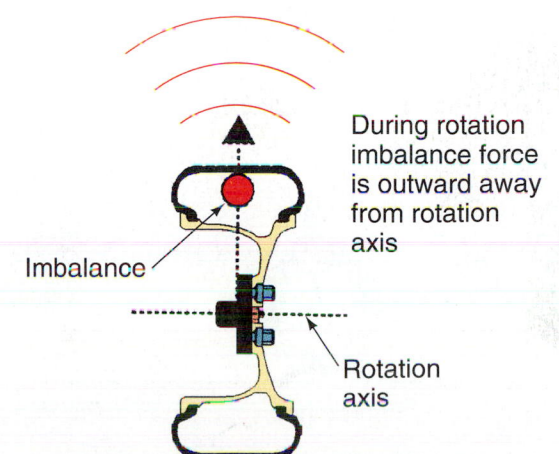

During rotation imbalance force is outward away from rotation axis

Imbalance

Rotation axis

Figure 55.34 When the heavy spot is rotated, force is directed out away from the center of the wheel. *(Courtesy of John Bean Company)*

imbalance when the wheel is at rest can amount to a great deal of pounding force when the wheel is spinning at highway speeds (Figure 55.35). For instance, if a 15" wheel is 1 ounce out of balance, when the wheel rotates at 60 mph, the pounding force will be 4.6 pounds.

Static imbalance causes wear to mechanical parts, vibration, and gouged tire tread wear, called *cupping* (see Figure 14.5). In severe cases, especially when accompanied by a bad shock absorber, a tire can actually hop so badly that it leaves the road surface. This is called wheel tramp (Figure 55.36).

The part of the tire that is contacting the ground is actually travelling at 0 mph. This is called *static friction* or *static contact*. During wheel tramp when the tire recontacts the road, a small amount of the tire's rubber is scrubbed off. This is because the tire was travelling at the same speed as the vehicle when it left the road surface. This scrubbing happens all around the tire, accounting for the cupped wear all around the tire's tread surface.

Static balancing is done off the car on a bubble balancer. To balance a tire in a single plane, compensating weight is added on the opposite side of the wheel (Figure 55.37). If it were possible to put the compen-

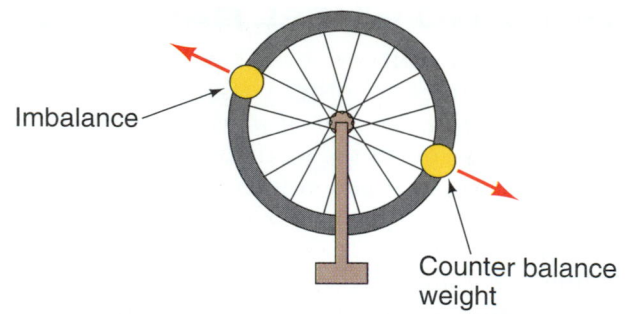

Figure 55.37 To balance a tire in a single plane, compensating weight is added on the opposite side of the wheel. *(Courtesy of John Bean Company)*

sating weight at exactly the same place on the opposite side of the tire, the amount of the weight would be exactly the same as the imbalance. Because this is not possible, a greater amount of weight is added at the edge of the wheel rim.

■ ON-THE-CAR BALANCERS

Wheels can be balanced on the car using a spin balancer (Figure 55.38). When a wheel is spun with a balancer, static balance is not an accurate term. The correct term is *single plane balance* or *kinetic balance*. This method is popular for use on exotic and high speed cars because the tire, wheel, and spinning brake parts are all balanced as a unit. No special adapters are required for the special wheels that some of these cars use.

NOTE: *If the wheel is removed from the car, the balance can be changed if it is reinstalled in a different position. It should be marked for both the wheel location on the car and the lug position.*

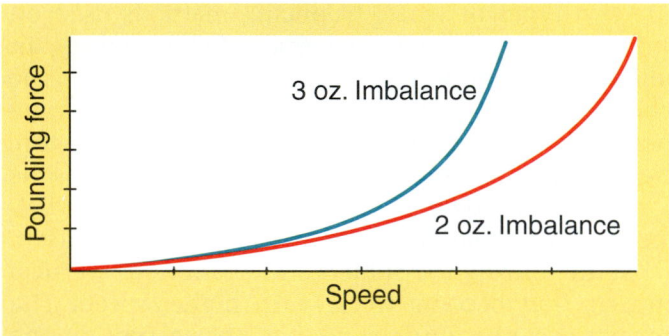

Figure 55.35 Pounding force increases as speed and the amount of imbalance increase. *(Courtesy of John Bean Company)*

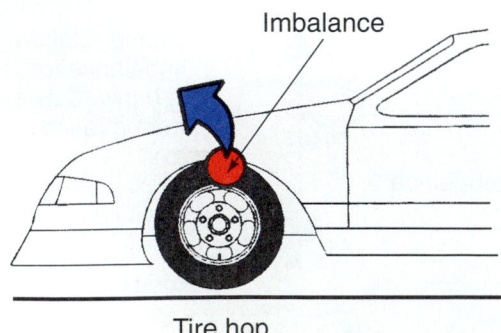

Figure 55.36 Wheel tramp happens when speed and force of the imbalance become large enough to cause the tire to leave the road surface. *(Courtesy of John Bean Company)*

Figure 55.38 An on-the-car spin balancer. *(Courtesy of Hunter Engineering Company)*

■ COUPLE IMBALANCE

When a tire tread is lopsided, with more of its weight on the outside or the inside of its tread, this is known as couple imbalance. Couple imbalance only shows up when the wheel is spinning. The wheel will have a tendency to shimmy, which also results in faster tire wear. Figure 55.39 shows how a spinning heavy spot will seek the centerline of the tire. This is why the tire shimmies (wobbles from side to side). When the heavy spot is in the front, it causes the wheel to move one way. After the tire rotates one-half revolution, the heavy spot causes the wheel to try to turn the opposite direction (Figure 55.40).

The previous example is of an imbalance on only one side of the tire. Usually there are imbalances on both sides of the tire, requiring counterbalancing weights to be installed on both sides of the wheel.

A wheel may be in static balance but not couple balance. If a tire has been statically balanced by adding all of the weight to one side of the offset wheel rim, that weight will be thrown side-to-side in a different direction as the tire spins, resulting in shimmy. The proper way to balance a wheel statically, is to split the amount of weight to be used in half, placing an equal amount of weight on both sides of the rim (Figure 55.41). Pirelli Tires recommends this procedure anytime the amount of weight to be added is in excess of 20 grams (0.71 oz.). With tape weights, weights are often added in two places on the inside of the wheel (Figure 55.42)

■ DYNAMIC BALANCE

It is not likely that a tire and wheel will have only static imbalance, and not couple imbalance, too. Dynamic imbalance is the combination of both static and couple imbalance. Dynamic balance means balance in motion. It is also called *two-plane balance* because it measures side-to-side (lateral) force as well as up-and-down (axial) force. Lateral forces are felt when a steering wheel moves back and forth.

Dynamic wheel balancers spin the wheel and locate vibration in it. The computer splits the tire into two halves and measures lateral and radial (axial) forces on each side of the tire's center. Weights are added to the proper side of the rim to correct the imbalance. A tire that is dynamically balanced will be statically balanced, too. Dynamic balancing can be done with a computer balancer or with an on-the-car spin balancer.

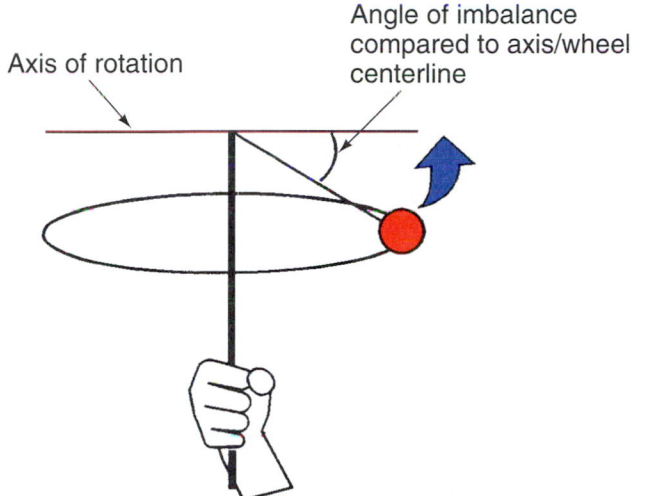

Axis of rotation

Angle of imbalance compared to axis/wheel centerline

Figure 55.39 As the speed of rotation increases, the imbalance moves toward the axis of rotation. *[Courtesy of John Bean Company]*

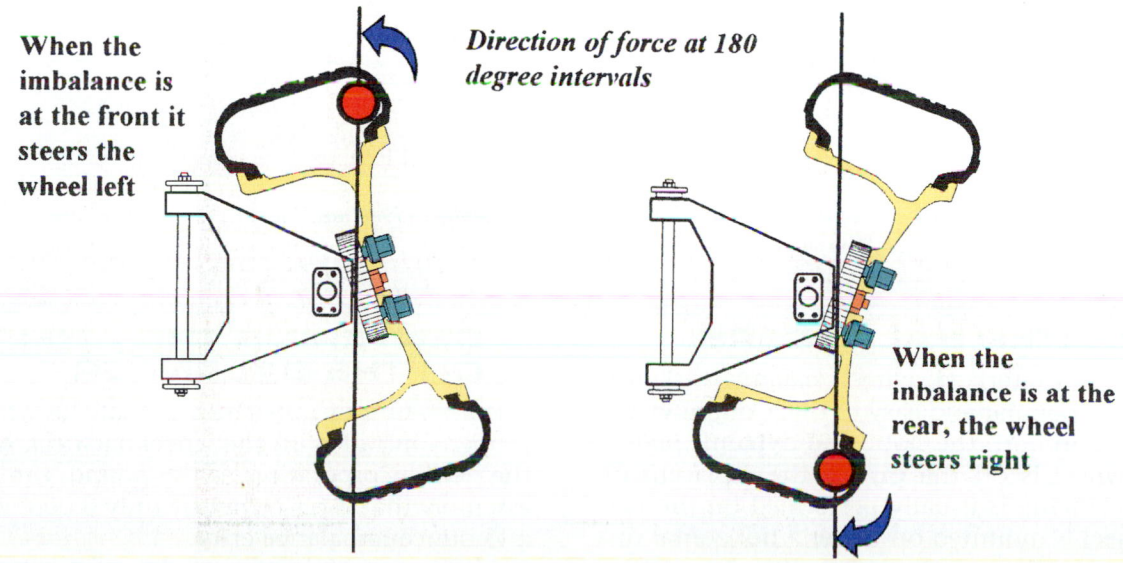

When the imbalance is at the front it steers the wheel left

Direction of force at 180 degree intervals

When the inbalance is at the rear, the wheel steers right

Figure 55.40 Action of dynamic or couple imbalance. *[Courtesy of John Bean Company]*

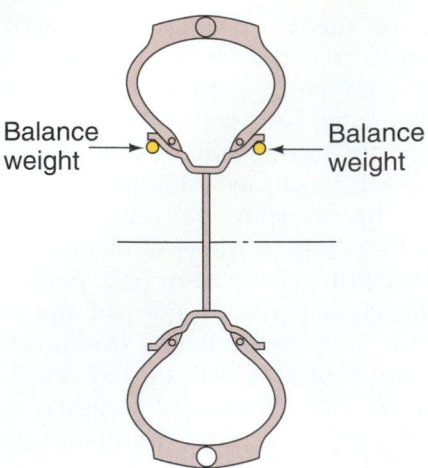

Figure 55.41 The amount of weight is split in half and put on both sides of the rim to avoid creating a couple imbalance.

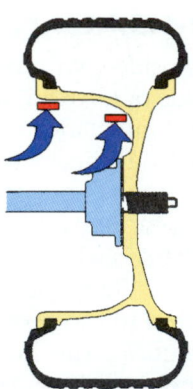

Figure 55.42 Tape weights added to the inside of the wheel for cosmetic purposes. *(Courtesy of John Bean Company)*

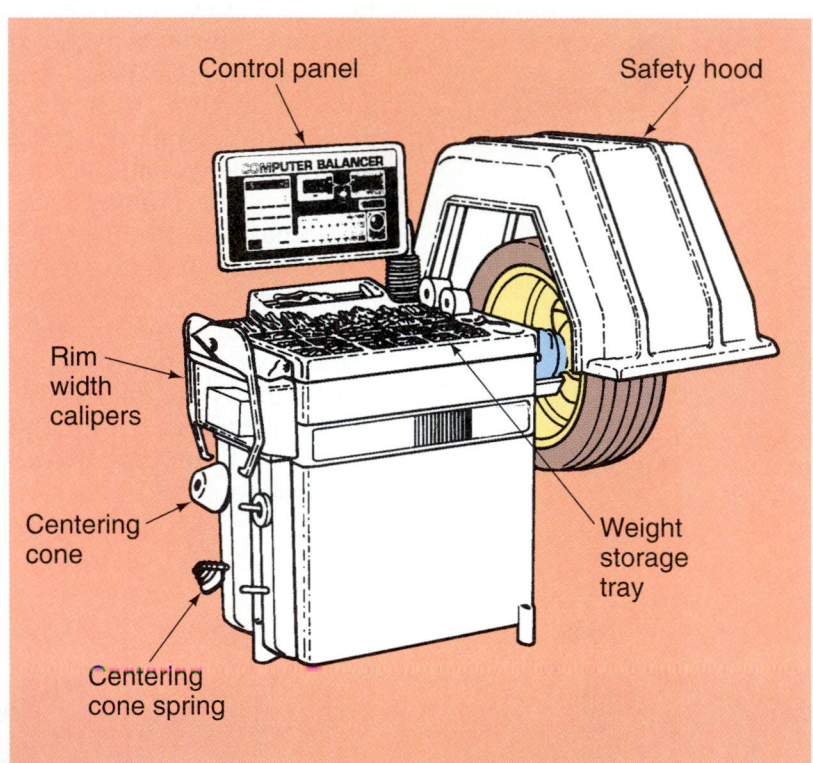

Figure 55.43 A computer wheel balancer. *(Courtesy of Hennessy Industries, Inc.)*

■ COMPUTER BALANCERS

Off-the-car computerized wheel balancers (Figure 55.43) are very popular and easy to use. Computer balancers balance in both the static and dynamic planes. When the wheel is off the car for tire replacement, computer balancing is usually performed on the new tire. The wheel is mounted on either a horizontal or a vertical threaded shaft using adapters that are supplied with the machine.

■ CENTERING THE WHEEL ON THE BALANCER

Another cause of imbalance is that the wheel was not properly installed on the wheel balancer. According to the FMC Corporation, a 36-pound tire and wheel assembly that is off-center by only 0.006" will result in a ½-ounce imbalance error.

When mounting a wheel on the wheel balancer, the best method is to center the wheel by the same

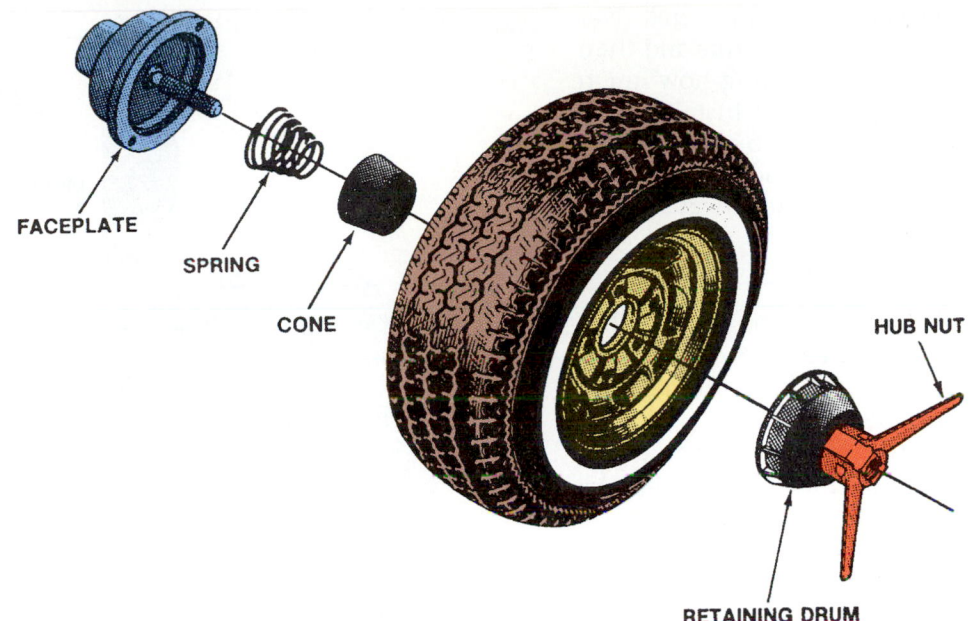

FACEPLATE

SPRING

CONE

HUB NUT

RETAINING DRUM

Figure 55.44 Recommended computer balancer mounting arrangement for hub-centric wheels. *(Courtesy of Sun Electric Corporation)*

was method as designed by the manufacturer. There are two ways that wheels are centered (see Chapter 54):

- Hub-centric—this is when the hole at the center of the wheel locates the wheel.
- Lug-centric—this is when the lug nuts center the wheel.

To determine whether or not a wheel is hub-centric, see if the wheel fits the hub snugly with the lug nuts removed.

When mounting a hub-centric stamped steel wheel, use a centering cone. The most accurate centering system is to install it from the backside of the wheel. This is because the center hole was originally stamped from the back. Figure 55.44 shows the recommended mounting arrangement.

It is important that the centering cone fit the shaft snugly. When new, it has about 0.001" clearance. Expandable collets are available that provide the most accuracy in centering. This is because they expand when tightened, eliminating all clearance between the collet and the shaft. To mount lug-centric wheels, use a special lug centering adapter.

The backing plate and wheel lug flange must be clean and undamaged. Be sure that the wheel and tire have all foreign objects removed from them before attempting tire balance. When the wheel and tire assembly is first spun, look for signs of obvious runout and double-check the mounting.

After the wheel is mounted, the technician needs to program three references into the computer:

- The width of the rim, measured by a rim caliper (Figure 55.45)
- The location of the flange on the wheel, measured by a gauge (Figure 55.46)
- The diameter of the wheel (for instance: 13", 14", 15")

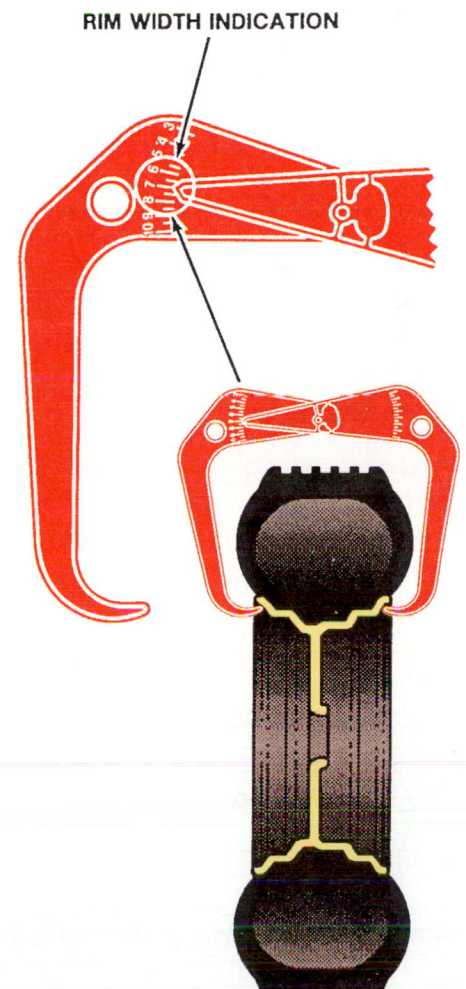

RIM WIDTH INDICATION

Figure 55.45 A rim caliper is used to measure the width of the rim. *(Courtesy of Sun Electric Corporation)*

To balance a tire, the safety hood is lowered over the wheel. The wheel is spun for a short time and then stops. The machine gives a readout telling how much weight to use and where to put it. After installing the weights, the wheel is spun once again to check for the accuracy of the balance job. A reading of "zero" on both sides of the wheel means that the wheel is ready for installation on the car.

■ MATCH MOUNTING

Some computer balancers have an extra feature that matches a tire's imbalance to a wheel's imbalance. This procedure is used when a weight of more than 2 ounces is required on one side of a wheel. The tire is given an initial spin on the balancer. Then, it is deflated and installed on a tire machine so that the beads can be broken down. After rotating the tire on the rim 180°, the tire is reinflated and rebalanced. The computer can now tell how much the wheel is out of balance and how much the tire is out of balance. If neither is unacceptable, the tire is once again rotated on the rim to complete the match mount.

➡ *Perform Computer Tire Balance Worksheet*

■ MOUNTING THE WHEEL ON THE CAR

Before installing a wheel on a car, double-check to see that the bolt holes in the wheel center are in good condition. It is not uncommon to find steel wheels where the bolt holes are worn from a wheel that was loose, lug nuts that were installed upside down, or lug nuts that were overtightened. Finger-tighten the lug nuts. Shake the wheel to center it over the centering area on the hub. Cross-tighten the lug nuts with a torque wrench.

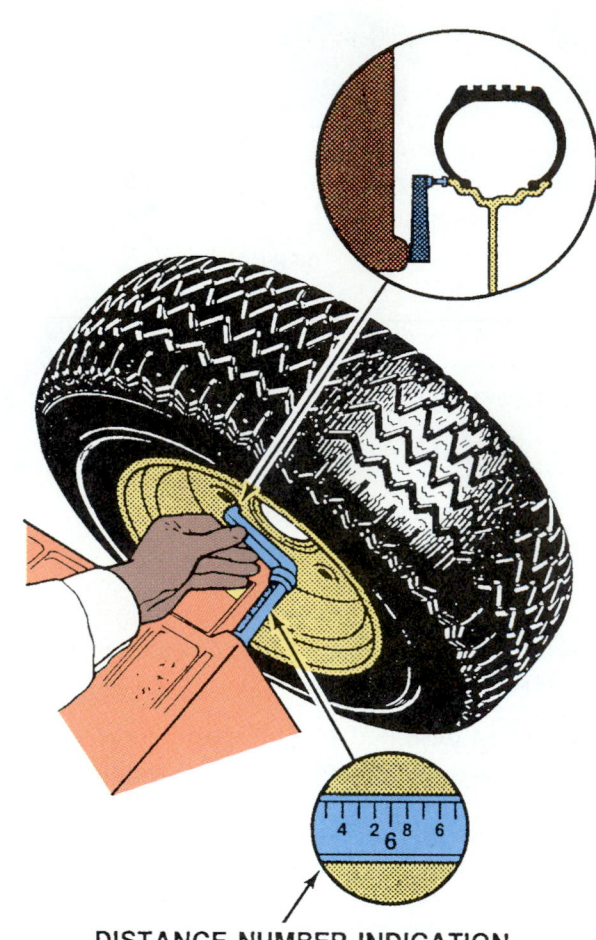

DISTANCE NUMBER INDICATION

Figure 55.46 This measurement tells the computer where the wheel is located on the balance shaft. *(Courtesy of Sun Electric Corporation)*

■ REVIEW QUESTIONS

1. What kind of tire has a bulging sidewall when properly inflated, radial or bias?

2. About how much does pressure in a tire increase as the tire warms up?

3. Wear bars show up around the tire tread at what tread depth?

4. What is the name for the up-and-down action of the tire that results in scalloped tire wear?

5. What is applied to the tire bead before installing a tire on a rim?

6. Should the valve core be in the valve stem or out of it when filling the tire with air to seat the bead?

7. What kind of hole can be repaired using only a patch?

8. A hole in the tire tread area of a steel belted tire must be _____ or _____ before installing a plug in it.

9. What kind of tool is used to clean the inside of a tire liner before installing a patch?

10. What type of wheel balance is measured with the wheel stationary?

11. If a 15" wheel is 1 ounce out-of-balance, when the wheel rotates at 60 mph, the pounding force will be _____ pounds.

12. How fast is the part of the tire that is contacting the ground travelling?

13. When a tire is lopsided, with more weight on one side than the other, what kind of imbalance results?

14. What is the combination of both static and couple imbalance called?

15. To mount _____ centric wheels, use a special lug centering adapter.

■ ASE STYLE REVIEW QUESTIONS

1. Technician A says when static balancing a wheel, it is better to place the weight on the inside of the wheel. Technician B says tire balance problems become less as speed increases. Who is right?
   - **a.** Technician A
   - **b.** Technician B
   - **c.** Both A and B
   - **d.** Neither A nor B

2. Technician A says that a computer tire balancer balances the tire statically. Technician B says that the computer tire balancer balances a tire dynamically. Who is right?
   - **a.** Technician A
   - **b.** Technician B
   - **c.** Both A and B
   - **d.** Neither A nor B

3. Wheel lug nuts on front wheels are being installed using an impact wrench. Technician A says that the lug nuts can be loosened while the car is off the ground without holding the wheel from turning. Technician B says that tightness must be double-checked with a torque wrench. Who is right?
   - **a.** Technician A
   - **b.** Technician B
   - **c.** Both A and B
   - **d.** Neither A nor B

4. Technician A says that wheel lugs on most cars are loosened by turning them clockwise. Technician B says that lug nuts are installed with the tapered side away from the rim. Who is right?
   - **a.** Technician A
   - **b.** Technician B
   - **c.** Both A and B
   - **d.** Neither A nor B

5. Technician A says a tire's inflation pressure drops as the car is driven. Technician B says the tire bead must not be lubricated for assembly or the rubber will rot. Who is right?
   - **a.** Technician A
   - **b.** Technician B
   - **c.** Both A and B
   - **d.** Neither A nor B

6. Technician A says the valve core should be installed before inflating a tire to seat its beads. Technician B says to use a shop towel and solvent to clean a tire's inner liner before applying cement. Who is right?
   - **a.** Technician A
   - **b.** Technician B
   - **c.** Both A and B
   - **d.** Neither A nor B

7. Technician A says holes larger than a pinhole must be plugged. Technician B says holes larger than a pinhole must be patched. Who is right?
   - **a.** Technician A
   - **b.** Technician B
   - **c.** Both A and B
   - **d.** Neither A nor B

8. Technician A says the part of the tire that is in contact with the ground is travelling at the same speed as the car. Technician B says the combination of static and couple balance is called dynamic balance. Who is right?
   - **a.** Technician A
   - **b.** Technician B
   - **c.** Both A and B
   - **d.** Neither A nor B

9. Technician A says to unseat the tire beads while the tire is inflated. Technician B says to pound on the beads of a passenger car tire with a special tire hammer. Who is right?
   - **a.** Technician A
   - **b.** Technician B
   - **c.** Both A and B
   - **d.** Neither A nor B

10. Technician A says not to inflate a tire over 25 pounds when seating a bead. Technician B says if a hot tire has a lower pressure than the recommended cold pressure, it is seriously underinflated. Who is right?
    - **a.** Technician A
    - **b.** Technician B
    - **c.** Both A and B
    - **d.** Neither A nor B

Suspension, Steering, Alignment

THEN AND NOW: STEERING

Some of the early automobiles could reach surprisingly high speeds. From a safety standpoint, the steering systems used in these vehicles were very primitive. Direct linkage from a tiller to the drag link could result in a broken arm when you hit a bump. The first domestic vehicle equipped with a steering wheel was the 1901 Packard.

Even after the tiller evolved to the less lethal steering wheel, automobile steering systems were not satisfactory. Various types of steering mechanisms were designed to reduce kickback and improve mechanical advantage. These included a *chain drive* and the *worm-and-sector steering gear* found on some cars still on the road today. The best idea was the *recirculating ball steering gear*, which made its debut on V-16 Cadillacs. Migrating ball bearings on the steering shaft reduced the high friction that occurred between the internal gears of the earlier gear boxes. The recirculating ball steering gear also resists the shock of running over a pothole.

There were earlier attempts at *power steering* but quality systems did not appear until after World War II. The war effort resulted in much research on hydraulics for aircraft servos and tank controls. In 1951, Chrysler applied this new technology to the production of the first automotive power steering system.

The *rack and pinion* steering gear used on many vehicles today uses a round gear on the steering wheel shaft. It meshes with a long, flat gear cut into the top of a long shaft that connects between the front wheels. Rack and pinion steering is not a new idea. In fact, it was used on huge steam tractors in the mid-19th century. The 1886 Daimler use a curved-rack version to control its center-pivoted axle. The planetary steering used on Ford's Model T was based on the same principle.

Rack and pinion steering was abandoned on early vehicles due to severe kickback and its limited mechanical advantage. It was not until the 1950s that European engineers began to rediscover its potential and improve its design to eliminate these problems. In 1971, Ford showed its European connection by producing the Pinto, the first volume domestic car to come with rack and pinion steering.

An electrically powered, electronically controlled rack and pinion steering gear is found on the Acura NSX. An electric motor assists the rotation of the pinion, giving fast response and eliminating the pump and hoses. This system has not come into widespread general use due to the excessive amount of electrical energy it consumes.

The 1901 Mercedes had a steering wheel instead of a tiller. *(Courtesy of Mercedes-Benz of N.A., Inc.)*

The Acura NSX has electrical power steering. *(Courtesy of Acura)*

CHAPTER 56

Suspension Fundamentals

■ OBJECTIVES

Upon completion of this chapter, you should be able to:

✔ Identify parts of typical suspension systems.

✔ Describe the function of each suspension system component.

✔ Compare the various types of suspension systems.

■ INTRODUCTION

The vehicle chassis includes the frame, shocks and springs, steering parts, tires, brakes, and wheels (Figure 56.1). Part of the chassis is the **suspension** system. There are several suspension designs in use today. Over the years, there have been many non-standardized part names to go with these designs. The names used in this chapter are the ones that have become standard. As you read the chapter, become familiar with the names of the parts.

■ SUSPENSION

The suspension supports the vehicle, cushioning the ride while holding the tire and wheel correctly positioned in relation to the road. Suspension system parts include the springs, shock absorbers, control arms, ball joints, steering knuckle, and spindle or axle.

Sprung and Unsprung Weight

Sprung weight is the weight that is supported by the car springs. Vehicle control increases with the reduction of unsprung weight. Anything *not* supported by the springs is unsprung weight. This includes the tires and wheels, bearings, axles, and differential.

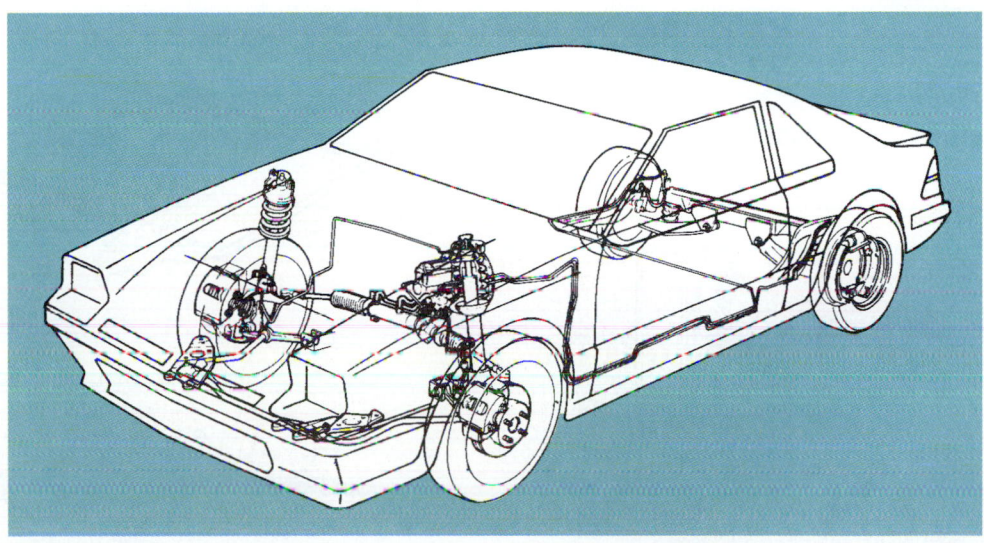

Figure 56.1 Parts of a vehicle chassis. *(Courtesy of Ford Motor Company)*

■ FRAME AND SUSPENSION DESIGNS

There are various types of frame designs used in cars and trucks. Cars are designed to be as lightweight as possible to improve fuel economy. Many newer cars have front wheel drive. Almost all older, heavier cars had rear wheel drive.

A front wheel drive car is made with a **sub-frame** in the front that includes the engine, transaxle, and steering/suspension system (see Figure 58.5). These cars, as well as modern rear wheel drive cars, rarely have a frame running the entire length of the car. Instead they have a sheet metal floor pan with small sections of frame at the front and rear.

■ INDEPENDENT AND SOLID AXLE SUSPENSIONS

A **rigid axle**, called a *straight axle* (Figure 56.2), is found on most rear ends and some heavy truck front ends (called an *I-beam*). I-beam axles are very strong and can support a great deal of weight.

Independent suspensions are found on most passenger car front ends and some rear ends. When a car with independent suspension goes over a bump, only that wheel will deflect (move up or down). When a wheel attached to a rigid axle goes over a bump, the wheel on the other side is affected by that movement too. Independent suspension systems have less unsprung weight. Because the rigid axles on I-beam front ends are heavy, they increase the vehicle's

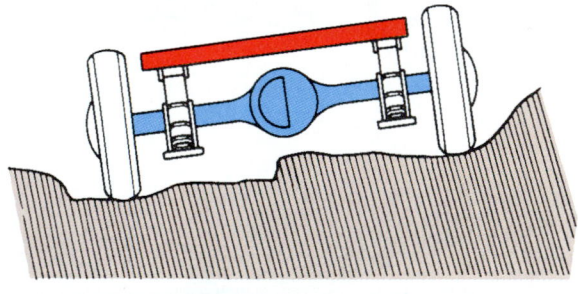

RIGID AXLE SUSPENSION

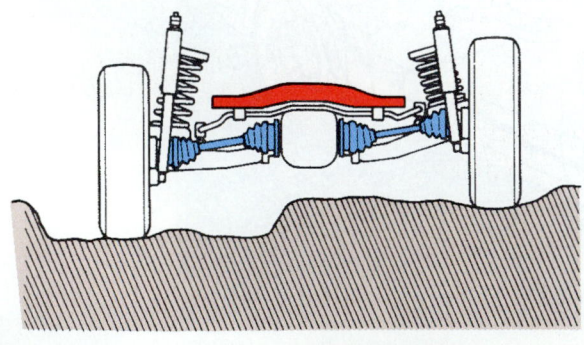

INDEPENDENT SUSPENSION

Figure 56.2 When an independent suspension goes over a bump, only one wheel deflects.

unsprung weight. A rougher ride is usually associated with a rigid axle.

■ SPRINGS

Springs support the load of the car, absorbing the up and down motion of the wheels that would normally be transmitted directly to the frame and the car body. There are three types of springs in use: the coil spring, the torsion bar, and the leaf spring. Springs are painted or coated with epoxy to prevent rust or nicks. These would raise stress in the spring, which could result in breakage.

Coil Spring

The *coil spring* is the most common type of spring in passenger cars. It is simply a spring steel rod wound into a coil (Figure 56.3). The coil spring is dependable, lightweight, and relatively inexpensive.

There are different ways to make variable rate springs. One way is to use a tapered rod. The ends of the spring do not work as hard as the center coils. As the spring is being compressed, it becomes stiffer. This makes for a smoother ride when going over smaller bumps and still allows for heavier carrying capacity.

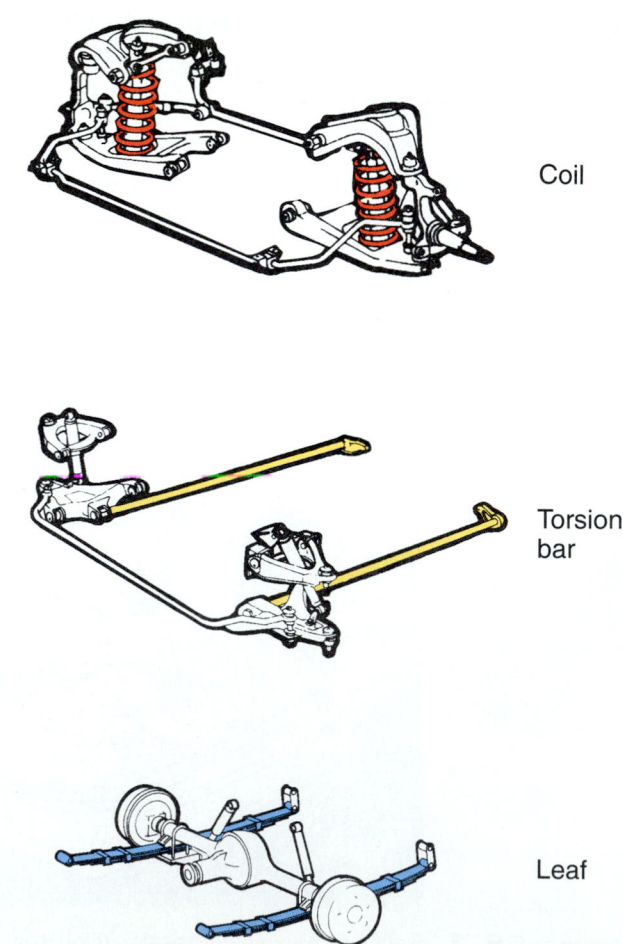

Coil

Torsion bar

Leaf

Figure 56.3 Three types of springs. *(Courtesy of Hunter Engineering Company)*

To keep a rear straight axle in the correct position when used with a coil spring, a fore and aft control arm is used. When a car has leaf springs, the leaf spring maintains the position of the axle.

Torsion Bar Spring

A **torsion bar** is a straight rod that twists when working as a spring (Figure 56.4). Torsion bars have a hex heads or splines on both ends that fit into a mating surface at the frame and movable suspension parts of the vehicle. A torsion bar can be mounted in the chassis to run either front to rear or side to side.

Some vehicles use torsion bars because they do not require much vertical space and the car can be designed to be lower on the front. One advantage of a torsion bar is that spring height can be adjusted by turning a screw against a bracket mounted on one of the torsion bar ends. The vehicle's correct ride height can be restored prior to an alignment.

Leaf Spring

A leaf spring is made of a long, flat strip of spring steel rolled at both ends to accept a rubber bushing. The front end of the spring is attached to the frame and the rear of the spring is attached to the frame with a *spring shackle* (Figure 56.5). The spring shortens and lengthens as it compresses and rebounds. The shackle is needed to compensate for changes in spring length.

To provide a **variable rate spring**, extra springs, called *leaves*, of varying lengths can be added to the *master* leaf. A center bolt and metal clips keep the leaves centered. Each leaf is curved more than the one above it. Only the master leaf is rolled at the ends.

One advantage of multiple leaf springs is that there is friction between the leaves during spring deflection. This friction dampens spring oscillations. It also causes a rougher ride because the spring cannot flex as easily as a coil spring. Passenger cars have plas-

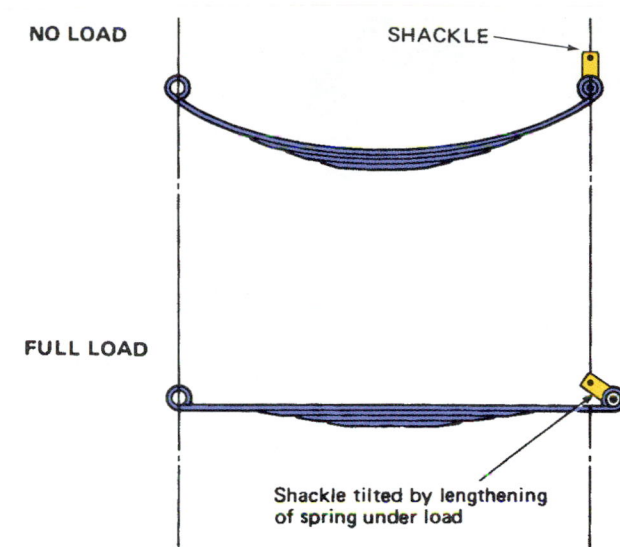

Figure 56.5 Action of a leaf spring.

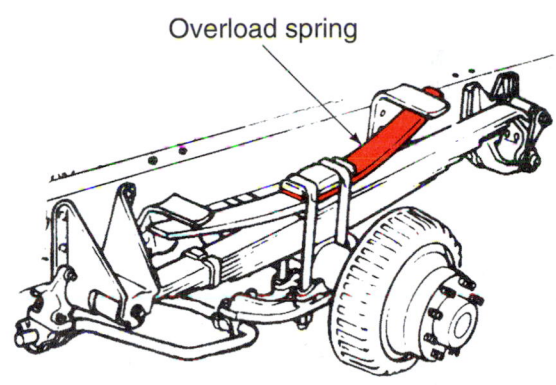

Figure 56.6 An overload spring does not work until the other leaves have deflected enough under load to contact it. [*Courtesy of General Motors Corporation, Service Technology Group*]

tic or synthetic rubber insulating strips between the leaves to provide a softer ride and minimize squeaks.

Some vehicles made since the early 1980s have composite (fiberglass) springs. They cut weight by about 30 pounds per spring.

Some trucks have **overload springs** that do not work until the other leaves have deflected enough under load to allow contact with them (Figure 56.6).

■ SUSPENSION CONSTRUCTION

Control Arms

Control arms allow the springs to deflect (move up or down) (Figure 56.7). When they are A-shaped (like the ones shown in Figure 56.7), they are called *A-arms*. Another type of control arm commonly found on lighter cars has a single bushing with a strut rod for stability (Figure 56.8).

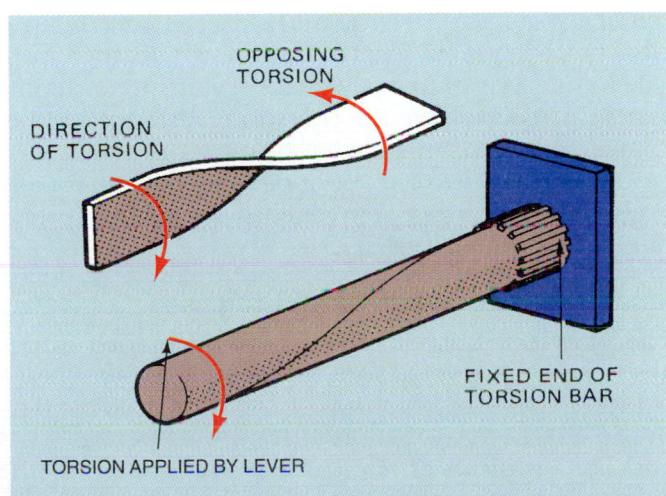

Figure 56.4 A torsion bar is a straight rod that twists when working as a spring.

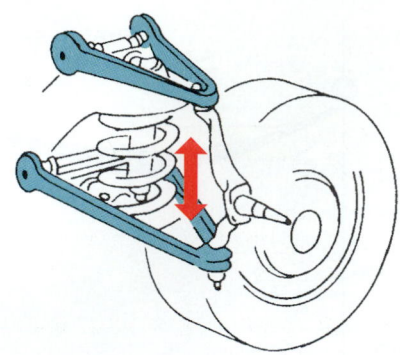

Figure 56.7 Control arms allow the springs to deflect. *(Courtesy of Federal-Mogul Corporation)*

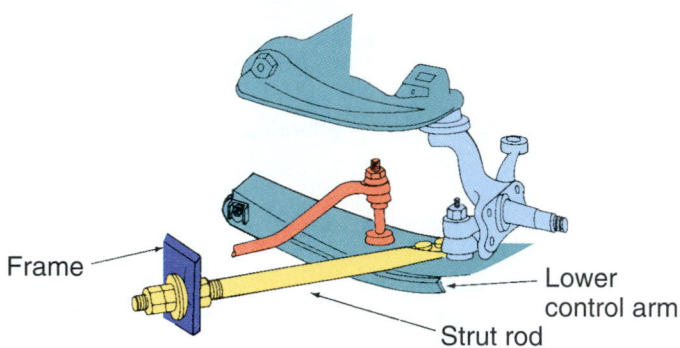

Figure 56.8 The strut rod provides stability to a narrow control arm. *(Courtesy of Federal-Mogul Corporation)*

Bushings

Rubber bushings are used to keep many suspension parts separated. The bushing has an outer and inner metal shell (Figure 56.9). The outer part of the bushing is pressed into the control arm and the inner part fits against a pivot shaft (Figure 56.10).

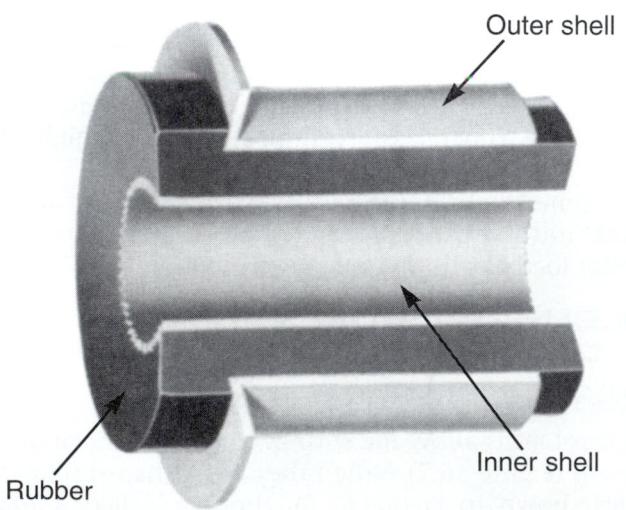

Figure 56.9 Rubber bushings keep suspension parts separated. *(Courtesy of Moog Automotive, Inc.)*

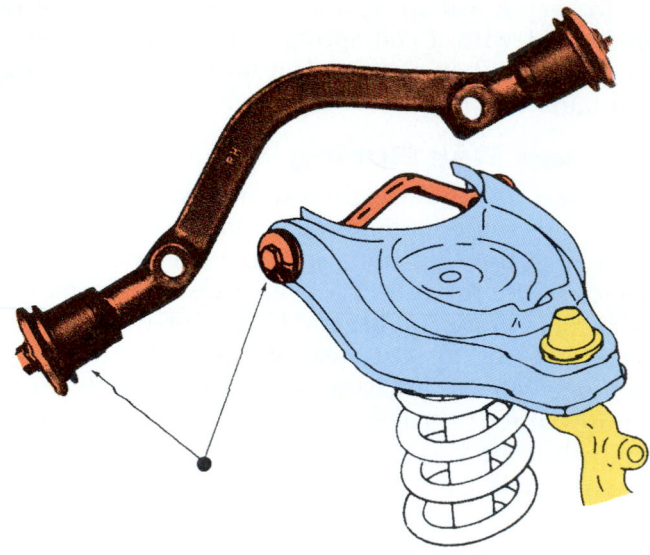

Figure 56.10 A control arm, bushings, and pivot shaft. *(Courtesy of Federal-Mogul Corporation)*

When the wheel moves up as the spring compresses, this is called **compression or jounce**. When the wheel moves back down, it is called **rebound**. During compression or rebound, the control arm moves up or down, twisting the rubber bushing and allowing parts to pivot without any metal-to-metal contact (somewhat like a vulcanized motor mount). Some of the suspension's resistance to body roll comes from the resistance of the bushings to twisting.

■ SUSPENSION TYPES

Automotive front suspensions are of two main designs: the *short/long arm* suspension *(SLA)* and the *Macpherson strut* (Figure 56.11). The SLA suspension uses two control arms of unequal length so that when the spring is compressed, the wheel will tilt *inward* at the top. This allows the tread width of the front tires to remain constant (Figure 56.12). If the control arms were both the same length, the tire would slide from side to side as it went over bumps (Figure 56.13). Some SLA suspensions have the coil spring above the upper control arm and some have it below.

Rear suspensions can be independent, but usually they have rigid axles. Some are used with leaf springs but most have coil springs.

■ SHOCK ABSORBERS

The tires on the car are actually the first shock absorbers. The springs absorb shocks also. But when a compressed spring rebounds, it begins to oscillate (Figure 56.14). Each oscillation of the spring becomes less and less, until all of the compression energy in the spring has been used up. Cars have four **shock absorbers**, commonly called shocks, one at each wheel. Shock absorbers do not actually absorb shock;

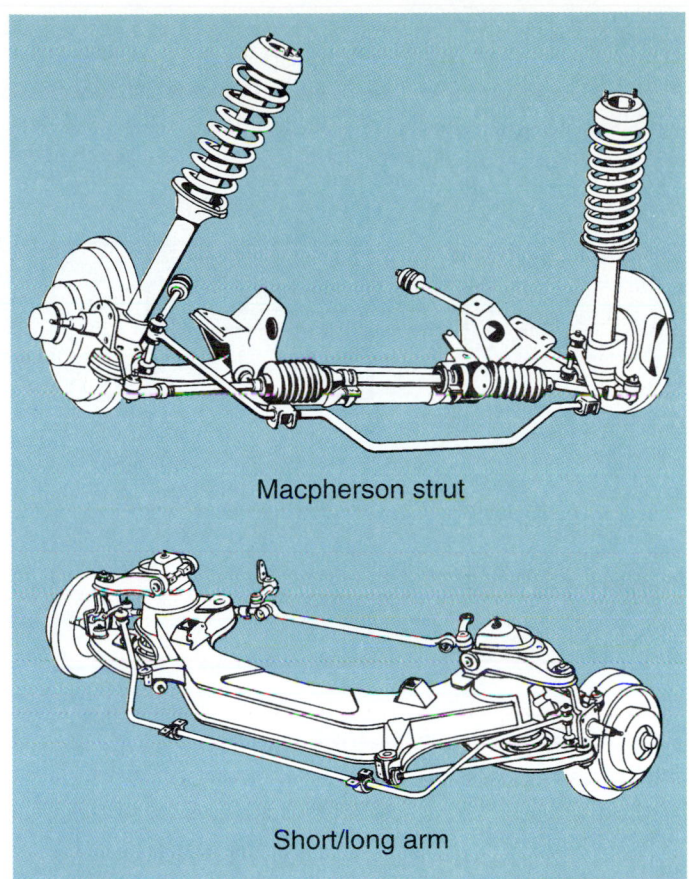

Macpherson strut

Short/long arm

Figure 56.11 Two popular suspension designs. *(Courtesy of Moog Automotive, Inc.)*

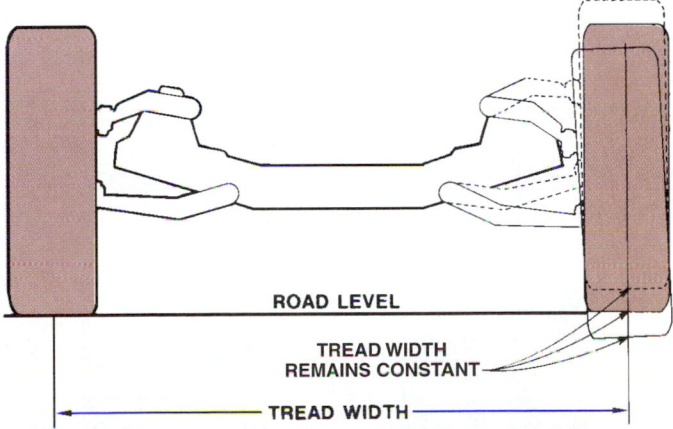

ROAD LEVEL

TREAD WIDTH
REMAINS CONSTANT

TREAD WIDTH

Figure 56.12 The SLA suspension allows the tread width to remain constant. *(Courtesy of Moog Automotive, Inc.)*

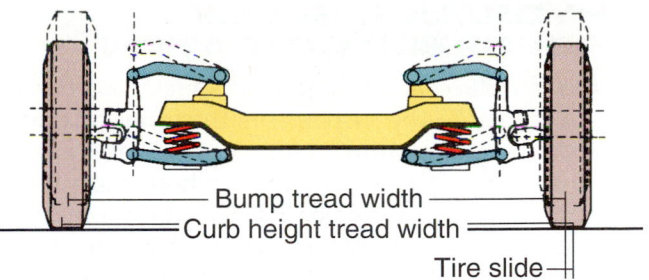

Bump tread width
Curb height tread width
Tire slide

Figure 56.13 With control arms of equal length, the tire slides as it goes over a bump. *(Courtesy of Federal-Mogul Corporation)*

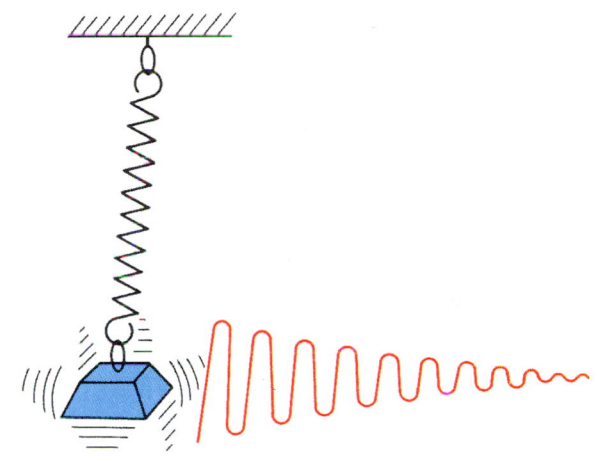

Figure 56.14 When a compressed spring rebounds, it begins to oscillate. *(Courtesy of General Motors Corporation, Service Technology Group)*

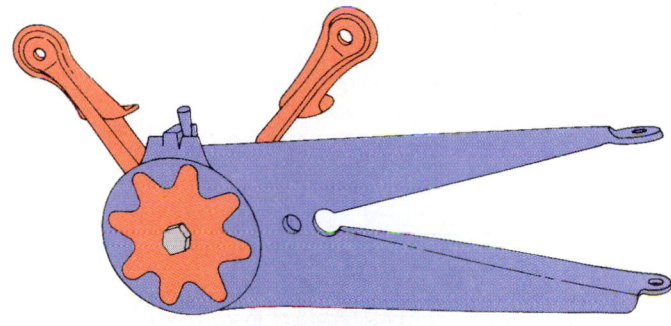

Figure 56.15 An early friction shock absorber. *(Courtesy of Tenneco Automotive, Inc.)*

their function is to *dampen spring oscillations* by converting the energy from spring movement into heat energy.

The shock absorber affects the speed at which the suspension moves. For instance, a "firm" shock resists spring movement as the spring is compressed. Normal motion of the springs, which can be controlled using shock absorbers, is uncomfortable to passengers, causes cupped tire wear, and is unsafe.

Early shock absorbers were mechanical friction devices. Friction material similar to that found on a clutch or brake was fitted between two levers. One of the levers was attached to the frame and the other to the spring mount or spring (Figure 56.15). They required frequent adjustment because they were prone to wear.

■ HYDRAULIC SHOCK ABSORBER OPERATION

In a modern *direct acting* hydraulic shock absorber, one end is attached to the suspension; the other is attached to the car body or frame (Figure 56.16). There are rubber bumpers on the frame to protect body parts in case of a pothole or worn shocks. The shock is mounted on rubber bushings to allow for slight changes in its angle of installation as it is compressed and extended.

To dampen the action of a spring, a hydraulic shock absorber forces oil through a small opening, called a *valve*, like water through a squirt gun (Figure 56.17). Using energy in this fashion lessens the oscillations of the springs.

A *double-tube shock* has two chambers, or tubes, with a piston that forces fluid through the valves from one chamber to the other. The faster the piston moves, the more resistance it encounters. During compression, the fluid is forced from the *pressure chamber* (the inner chamber under the piston) into the outer and upper chambers (Figure 56.18). When the shock extends, oil returns once again to the pressure chamber.

Front and rear shocks are not the same, usually having chambers and valves of different sizes to control the differences in weight between the front and rear of the car. Also, the center of gravity shifts when the vehicle stops or is thrown into a turn, causing the front shocks to do more of the work.

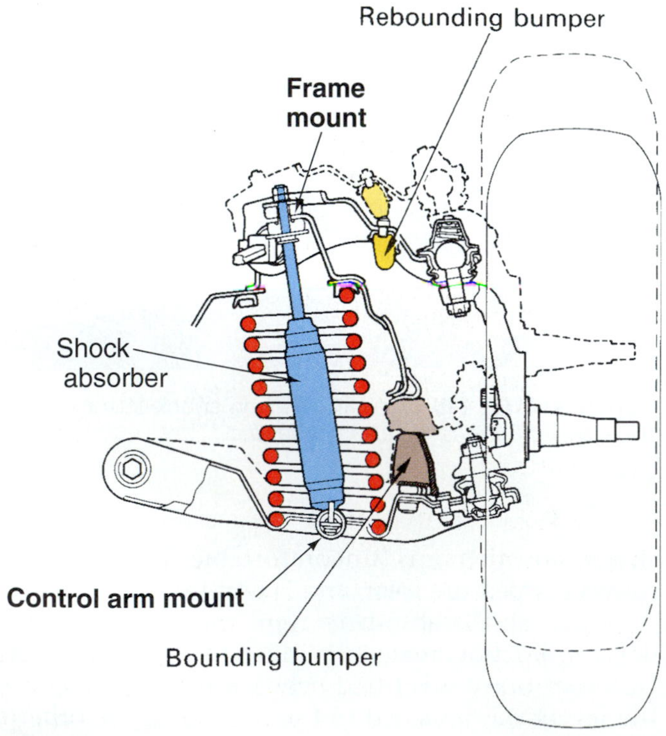

The up and down movements of a shock are called compression (or jounce) and *extension* (or rebound). Shock absorbers are called **double-acting** because they control motion when moving both up and down. Early friction shocks had a *ratio of 50–50*, controlling

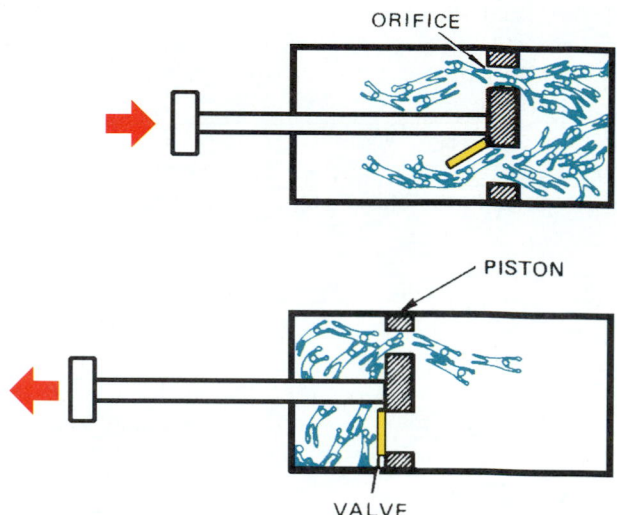

Figure 56.17 Fluid is forced through an orifice to dampen spring action.

Figure 56.16 One end of the shock is attached to the control arm and the other end is attached to the frame.

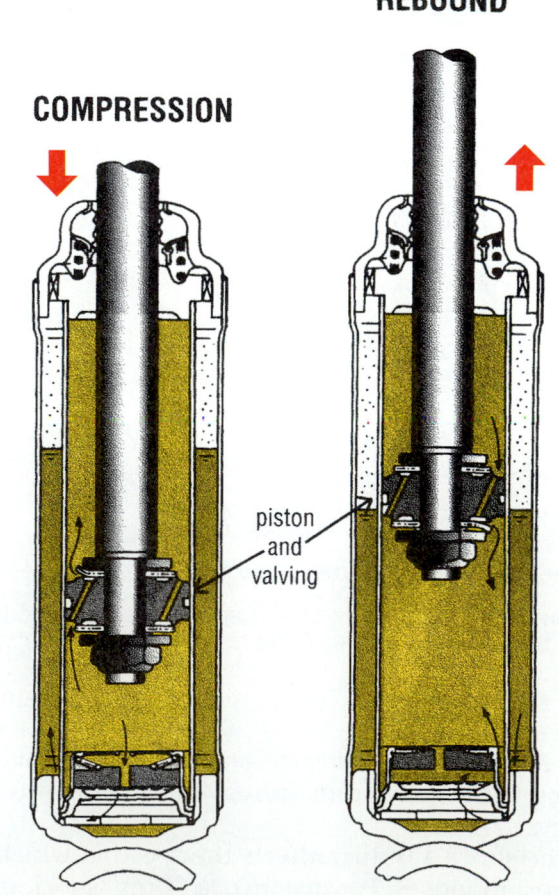

Figure 56.18 Shock action during the compression and rebound cycles. *[Courtesy of Tokico [USA], Inc.]*

the motion of the springs equally on compression and rebound.

◼ SHOCK RATIO

Hydraulic shocks have different valves that control fluid movement in one direction or the other, providing different resistances on compression and extension. The difference between the amount of control on compression and extension is called **shock ratio**. A typical shock will compress more easily (when controlling relatively light unsprung weight). It will provide more aggressive control on extension, when the heavier weight of the car body is in motion.

Shock absorber valves have different sized openings, depending on whether the shock is moving up or down, and also, on how fast the shock is moving. When a shock has *multiple stage valving*, the speed at which shock fluid is allowed to go through the valves differs. This depends on whether the car hits a hole (fast fluid movement), or whether the car leans over (called *body roll*) as it goes through a turn (slow fluid movement). The initial valve (first stage) is a small hole. As the piston moves faster, an additional valve opens.

Information on the ratio of stock shock absorbers is not readily available, even in parts catalogs. The information is presented here to provide a background in the operation of shocks. Also, shock absorbers are a big area of sales in the service industry and a good service technician will be able to provide intelligent answers to questions from customers.

When listing the ratio of an original equipment (OE) shock absorber, design engineers first refer to the percentage of dampening action during extension. For example, a 70/30 shock gives 70% of its resistance during extension and 30% on compression. In the high performance market, the opposite is true; the compression percentage is listed first.

Shock Control Force

Shock absorber ratio is actually just a comparison between the amounts of control that a particular shock provides in each direction. The **control force** of a shock absorber is actually a more important factor than ratio in determining the performance of a shock. It is what determines the softness or harshness of the ride. In the chart in Figure 56.19, both of the shock absorbers have the same ratio. But when installed on the same vehicle, the one with the higher control force (400#) will give a harsher ride. The one with the least control force (200#) provides a softer ride with more body roll.

◼ AERATION OF FLUID

The shock must be installed in a nearly vertical position so when it extends, air will not be drawn in place of fluid from the outer reservoir. This would result in a

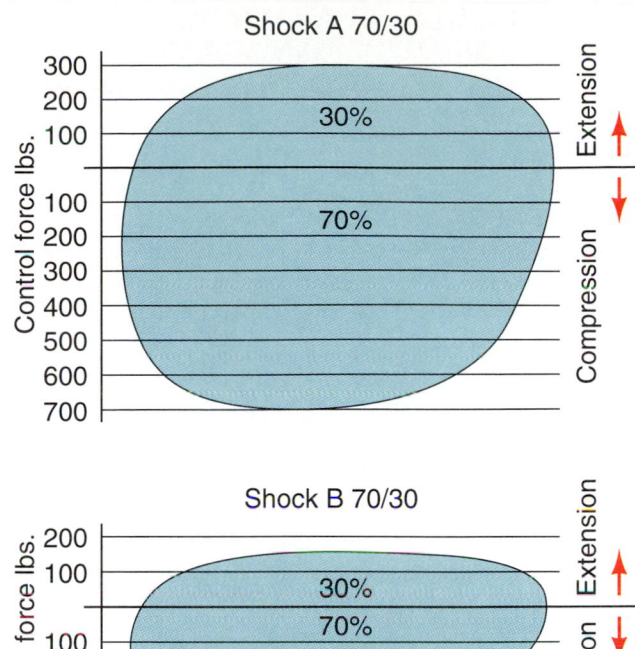

Figure 56.19 The control force of the shock determines the softness or hardness of the ride. *(Courtesy of Tenneco Automotive, Inc.)*

"skip" as the shock moves through its range of motion. **Aeration, or cavitation**, is when hydraulic fluid becomes mixed with air.

As the shock piston moves, pressure builds up in the fluid in front of it. A drop in pressure happens behind the piston, causing air bubbles and making the fluid foamy. When driving on rough roads fluid is rapidly forced through the check valves. The entire area of fluid around the piston becomes aerated and shock operation suffers (Figure 56.20). During normal operation, the air will usually work its way back to the air chamber.

There are two ways that designers avoid the tendency toward aeration. One of them is to use spiral grooves or a flat spiral ring around the outside of the reservoir tube.

◼ GAS SHOCKS

Another way to avoid aeration of the shock fluid is to use **gas shocks** for off-road and bumpy conditions. In the early 1950s a French physicist, DeCarbon, designed a gas pressurized shock. Pressurizing the oil column in the shock absorber keeps the bubbles in the solution. Compare this to what happens when a carbonated beverage is first opened. When the top is removed, the pressure drops and the bubbles come out of the soft drink. Pressurizing the shock keeps the

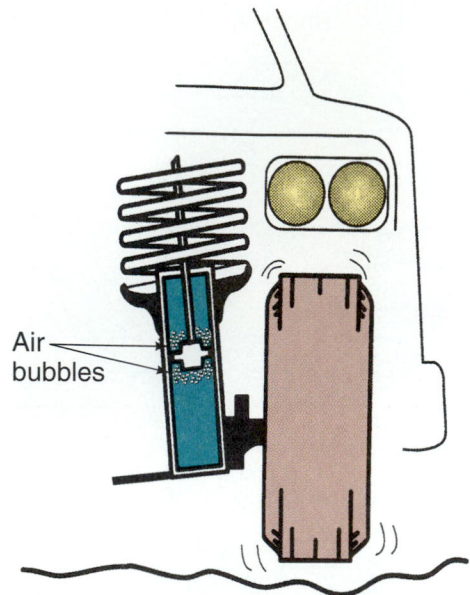

Figure 56.20 Air bubbles form around the valves during bumpy rides. *(Courtesy of Tokico [USA], Inc.)*

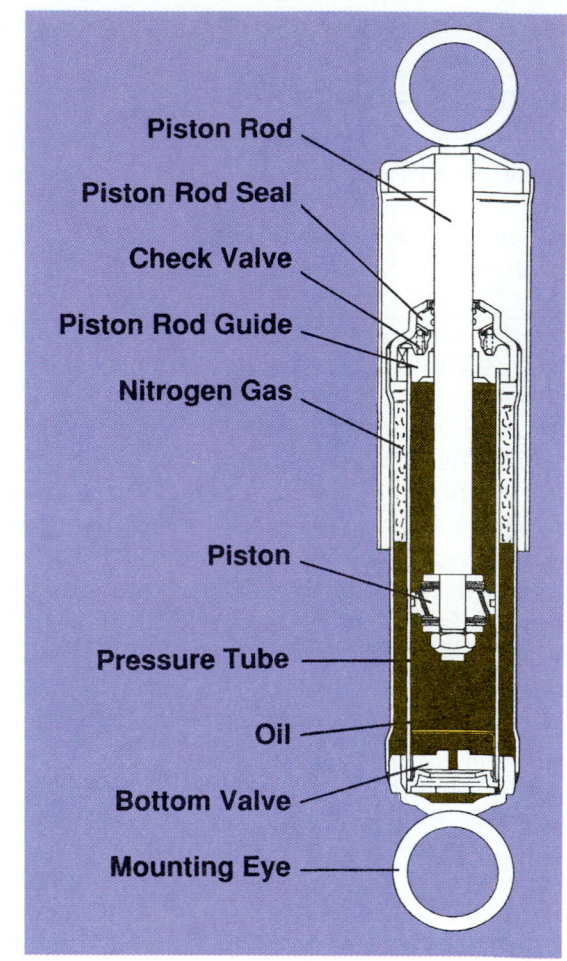

Figure 56.21 Parts of a pressurized gas shock. *(Courtesy of Tokico [USA], Inc.)*

shock oil clear and eliminates the skips that occur in the action of a regular shock absorber.

Some gas shocks have a pressurized gas-filled cell. This is a plastic bag that takes the place of the free air in a normal shock. These bags usually have low pressure (10–20 psi of refrigerant gas) in them. True gas shocks have a reservoir pressurized with nitrogen gas at 100–200 psi (Figure 56.21).

NOTE: *Gas shocks are packaged with a strap to hold them in the compressed position. They extend on their own when the strap is removed for installation.*

Single tube gas shock absorbers are also available. They offer better cooling of the shock fluid and can be mounted upside down. Single tube shocks must be pressurized with twice as much gas pressure to operate properly. They are also more susceptible to physical damage than double tube shocks.

■ AIR SHOCKS/ LEVELING DEVICES

Shock absorbers are not normally designed to carry the weight of the vehicle. If they were, the height of the vehicle would be affected when they were removed or when they wore out.

There are aftermarket devices that use the shock absorber as a means of correcting or adjusting the height of the vehicle. The two common ones are air shocks and coil springs that are mounted on the outside of the shock body. There is a disadvantage to leveling the vehicle in either of these ways. When shocks support the weight of the vehicle, the shocks and shock mounts are prone to breakage. Air shocks (Figure 56.22) have a rubber bladder that, when filled with

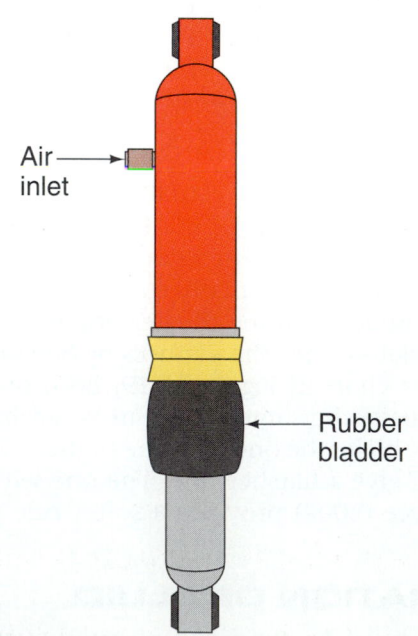

Figure 56.22 An air shock. *(Courtesy of Tenneco Automotive, Inc.)*

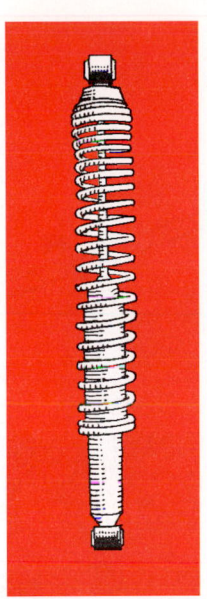

Figure 56.23 A coil spring shock. *(Courtesy of Tenneco Automotive, Inc.)*

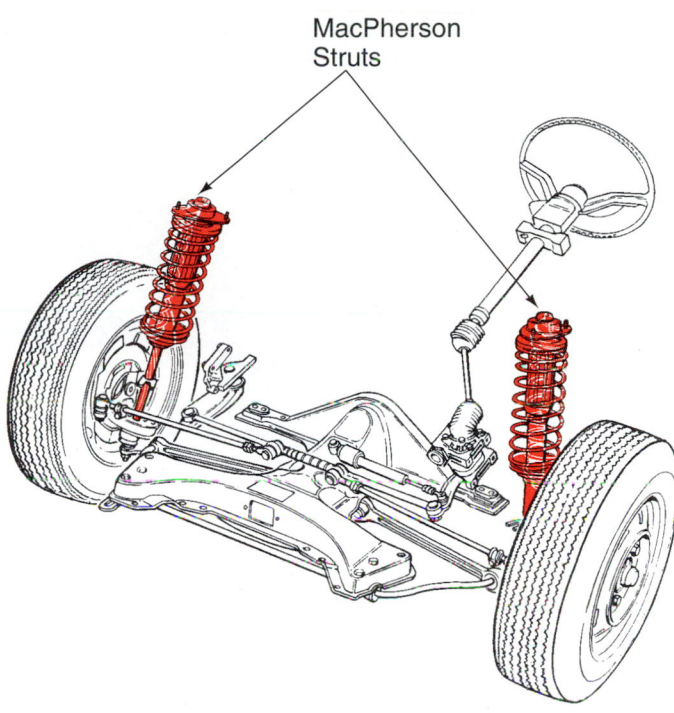

MacPherson Struts

Figure 56.24 A MacPherson strut suspension system. *(Courtesy of McQuay Norris)*

varying amounts of air, raises the rear of the vehicle to compensate for a heavier than normal load.

NOTE: *A small amount of air must be kept in the bladder at all times. If the vehicle is driven while the bladder is empty, the bladder can be folded into the shock and become torn.*

A coil spring shock (Figure 56.23) has a constant rate spring that works the same during both extension and compression. Its action in preventing body roll is similar to that of a stabilizer bar (covered later in the chapter).

■ MACPHERSON STRUT SHOCKS

Many smaller cars use the **MacPherson strut** design. It incorporates the shock absorber into the front suspension (Figure 56.24) using only a single control arm on the bottom. Most consider the short and long control arm (SLA) suspension to be a better design, but weight and cost savings account for the MacPherson strut's popularity.

When a shock absorber, also called a damper, is serviced in a MacPherson strut system, the strut is usually removed from the car. The coil spring is removed from the strut assembly and the strut is disassembled. The old shock absorber is discarded and a sealed replacement cartridge (Figure 56.25) is installed. The procedure is covered in Chapter 57.

■ OTHER FRONT END PARTS

Other parts are attached to the suspension to help control the ride. Parts like stabilizers and strut rods are insulated from front suspension parts and the frame with rubber bushings (Figure 56.26).

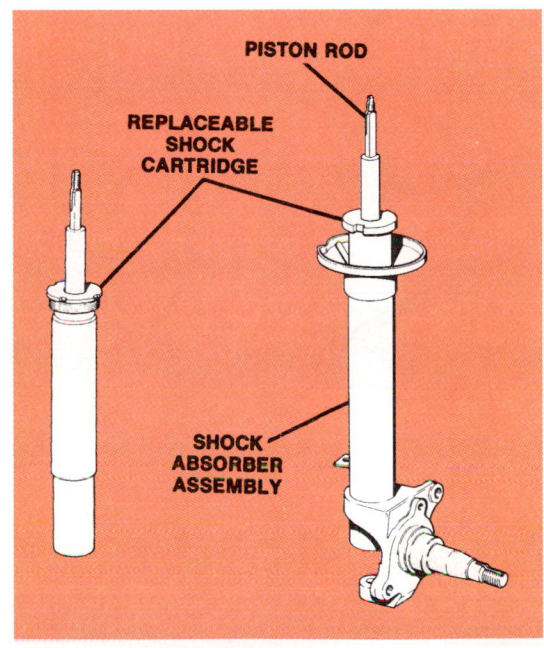

PISTON ROD

REPLACEABLE SHOCK CARTRIDGE

SHOCK ABSORBER ASSEMBLY

Figure 56.25 A replacement cartridge is installed in the old strut. *(Courtesy of Ford Motor Company)*

■ STABILIZER BAR

A stabilizer bar, also called a *sway bar* or an *anti-roll bar*, is used on the front or rear of many suspensions (Figure 56.27). It connects the lower control arms on both sides of the vehicle together, reducing sway and functioning as a spring when the car leans to one side. When both tires move up or down an equal amount,

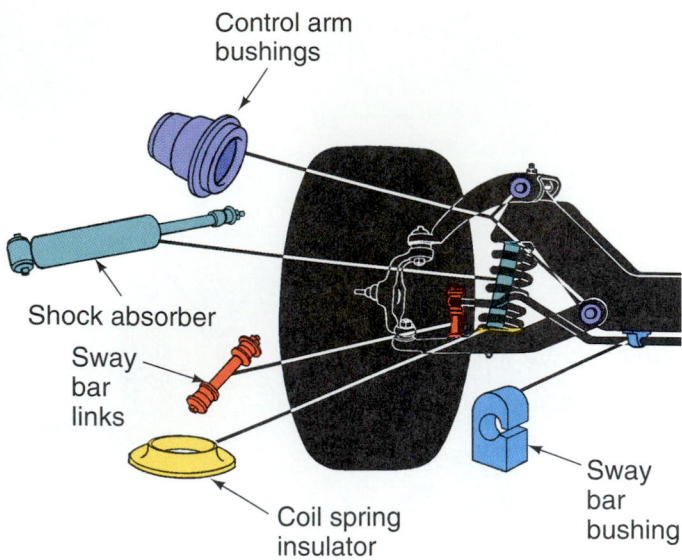

Figure 56.26 labels: Control arm bushings, Shock absorber, Sway bar links, Coil spring insulator, Sway bar bushing

Figure 56.26 Parts are insulated from each other with rubber bushings. *(Courtesy of Moog Automotive, Inc.)*

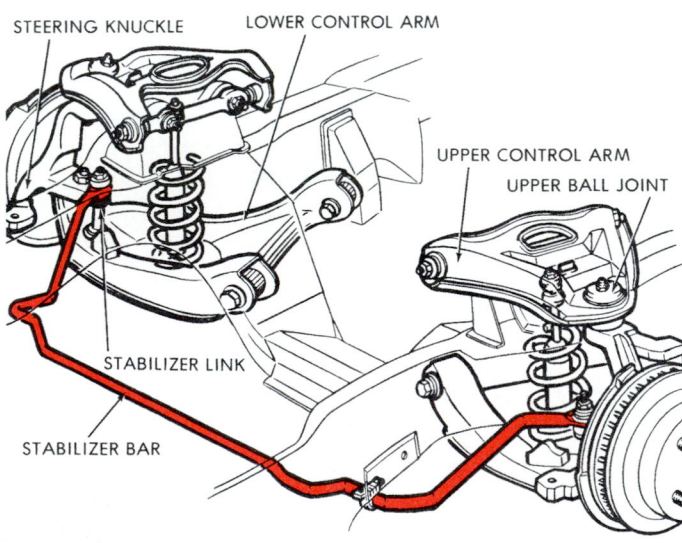

Figure 56.27 labels: STEERING KNUCKLE, LOWER CONTROL ARM, UPPER CONTROL ARM, UPPER BALL JOINT, STABILIZER LINK, STABILIZER BAR

Figure 56.27 A stabilizer bar and links. *(Courtesy of General Motors Corporation, Service Technology Group)*

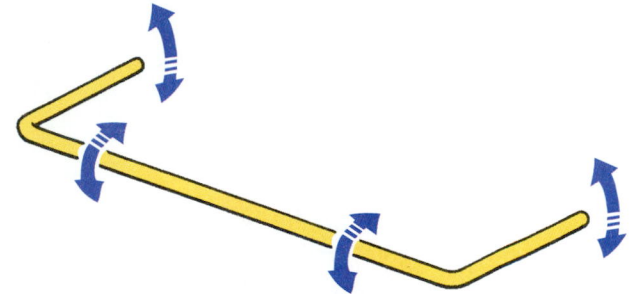

Figure 56.28 Action of a stabilizer.

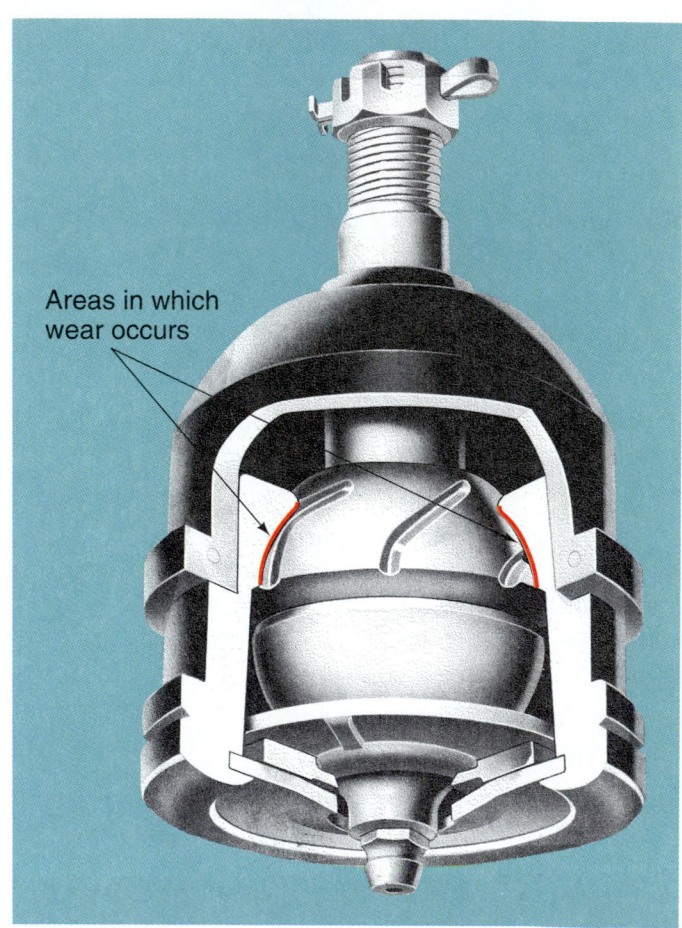

Figure 56.29 label: Areas in which wear occurs

Figure 56.29 A ball joint cutaway. *(Courtesy of Moog Automotive, Inc.)*

the stabilizer simply rotates in its bushings (Figure 56.28). If one of the wheels moves up, the bar twists as it tries to move the other wheel along with it.

Stabilizer links and bushings provide some flexibility and softness to the sway control so the suspension can still operate somewhat independently during minor bumps.

Spindle and Ball Joints

The *spindle support arm* (also called the *steering knuckle*) includes the axle that the wheel bearing is mounted on. Ball joints (Figure 56.29) are used to attach the control arm to the spindle. A ball joint allows motion in two directions, moving with the same up and down

motion as the bushings on the other end of the control arm. A ball joint also allows the spindle to pivot for steering. Depending on the design, ball joints will be located on top of or under the control arm.

On suspensions with two control arms, ball joints function as either a **load carrier** or a **follower ball joint** (Figure 56.30). The ball joint on the control arm that has the spring mounted on it is the load carrier. The function of the follower ball joint is to maintain parts in the proper position.

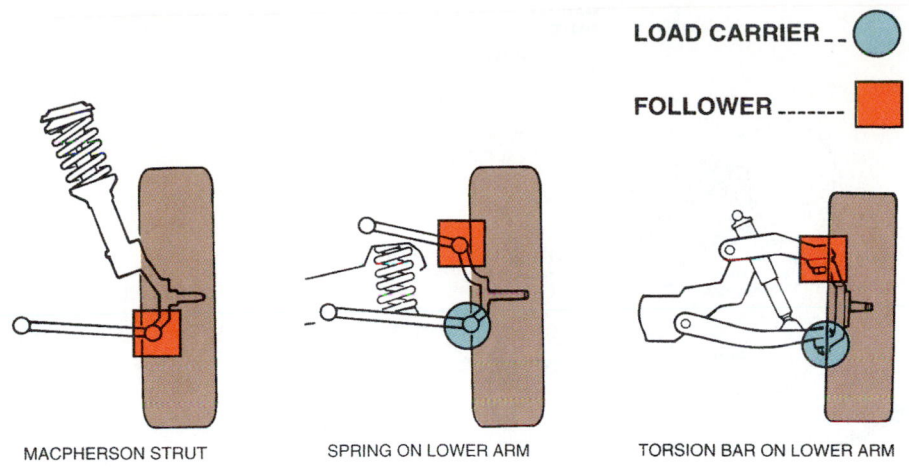

LOAD CARRIER ——

FOLLOWER ———

MACPHERSON STRUT   SPRING ON LOWER ARM   TORSION BAR ON LOWER ARM

Figure 56.30 Location of load carrier and follower ball joints. *[Courtesy of Moog Automotive, Inc.]*

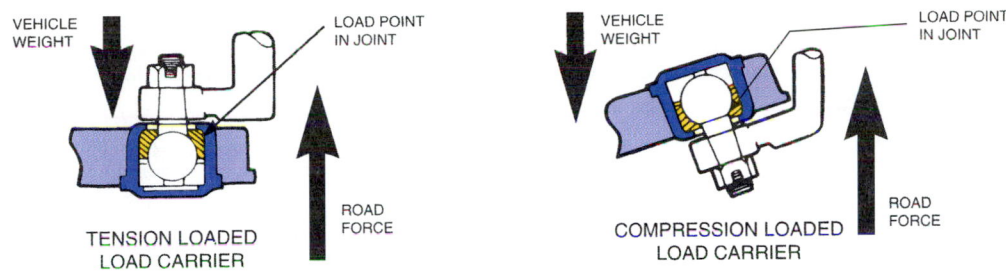

VEHICLE WEIGHT   LOAD POINT IN JOINT

TENSION LOADED LOAD CARRIER   ROAD FORCE

VEHICLE WEIGHT   LOAD POINT IN JOINT

COMPRESSION LOADED LOAD CARRIER   ROAD FORCE

Figure 56.31 Tension and compression loaded ball joints. *[Courtesy of Moog Automotive, Inc.]*

There are two styles of ball joints. They are either compressed all the time *(compression type)*, or always pulling apart *(tension type)* (Figure 56.31).

The MacPherson strut suspension system uses only one ball joint because it has only one control arm. It has a pivot bearing at the top of the strut that allows the strut to rotate for steering (Figure 56.32). On a strut

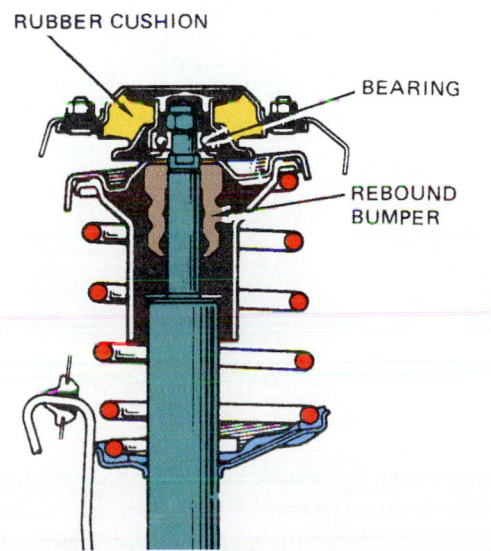

RUBBER CUSHION

BEARING

REBOUND BUMPER

Figure 56.32 A pivot bearing allows the strut to rotate.

suspension, the pivot bearing carries the load and the ball joint is a follower.

■ SUSPENSION LEVELING SYSTEMS

Suspension leveling systems are used on some luxury vehicles to keep the vehicle at the same height when weight is added to parts of the car. These systems are called **adaptive suspension systems**. Early leveling systems were manual, using air shocks and a compressor. A manual switch was used to change the height of the car body. Today's systems are automatic.

■ AUTOMATIC SUSPENSION LEVELING

Automatic leveling systems use air shocks or air springs filled by air from a compressor. An air dryer is usually attached to the pump to condition the air before it enters the shocks. A height sensor connected to the frame and axle housing is used for vehicle height input (Figure 56.33). It can turn on the compressor or bleed air to correct changes in height. These systems do not require a computer. On computerized systems, a module acts on signals from sensors at all four wheels to change the amount of air in air springs at each wheel (Figure 56.34).

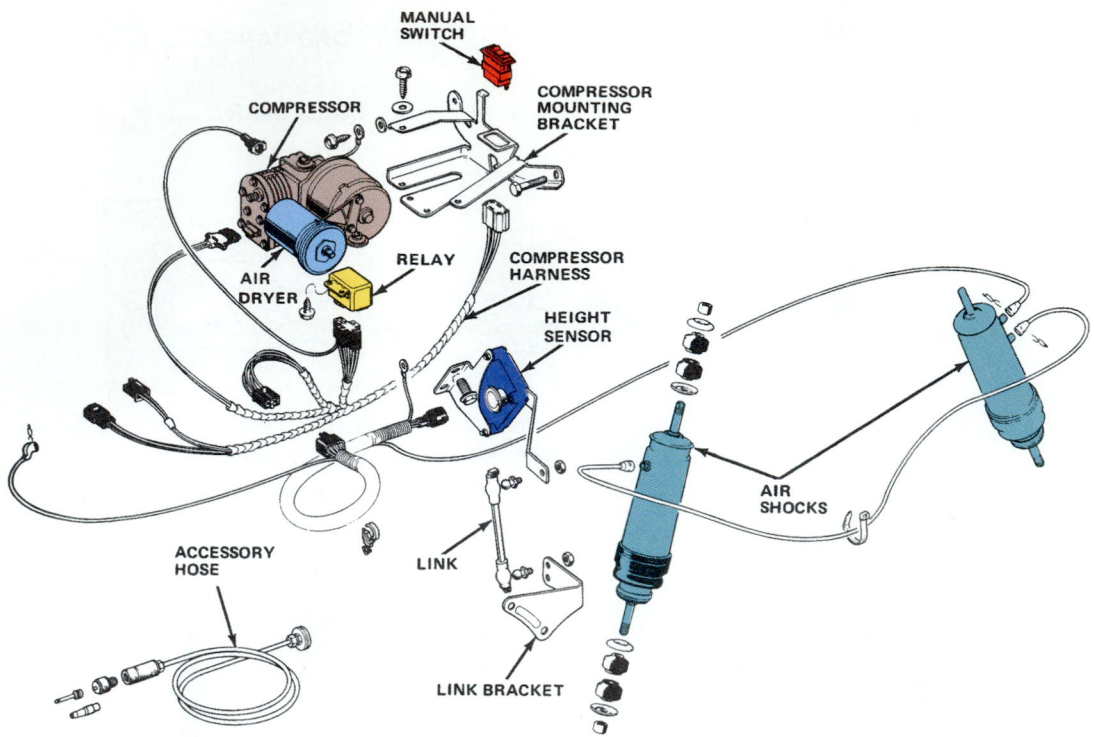

Figure 56.33 An automatic leveling system that uses air shocks. *(Courtesy of Chrysler Corporation)*

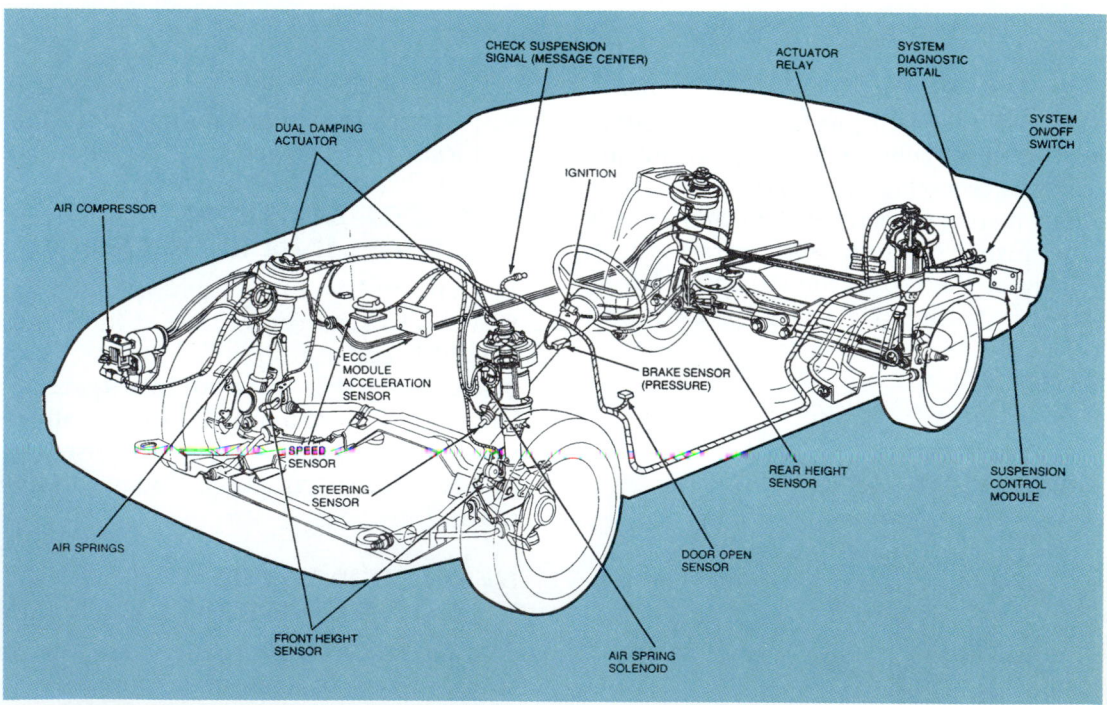

Figure 56.34 A computer controls the air springs at each wheel. *(Courtesy of Ford Motor Company)*

■ ACTIVE SUSPENSION

Active suspensions were first developed by Lotus in England. Each wheel has a hydraulic cylinder (high speed actuator) to keep the car body level during all driving conditions. Springs and shock absorbers are not necessary on these vehicles because the actuators replace them. These systems provide a smooth ride and yet give the handling characteristics of a vehicle with a stiff suspension. The suspension can be programmed by the driver as to the type of ride desired.

The system is powered by a hydraulic pump driven by the engine. The system requires about the same amount of power as a typical power steering pump during a turn (3–5 horsepower). The high speed actuators can raise or lower the vehicle in 0.003/second. As soon as a bump has been absorbed by an actuator, pressure is reestablished to keep the wheel in contact with the road and maintain ride height. During hard stops, the system reduces the tendency for the vehicle to dive or lean. Power consumption from the system is least when the vehicle is riding flat.

The computer uses signals received from sensors to track the position of each actuator. It can sense whether a wheel is in jounce or rebound. It also senses how heavily each wheel is loaded and whether the wheel is turning or is straight ahead.

NOTE: *In case of a flat tire, the system can be told to raise the tire so a jack is not needed.*

◾ REVIEW QUESTIONS

1. Which kind of spring is adjustable?
2. What is the name of the suspension type that uses two different length control arms?
3. What is the name of the part that is used to control oscillations of a spring?
4. What is the name of the difference between the amount of shock control on compression and extension called?
5. An original equipment 70/30 shock gives ___% of its resistance during extension and ___% on compression.
6. What is the name for the term that describes the softness or harshness of the ride?
7. What kind of shock can be mounted upside down?
8. If a small amount of air is not kept in an air shock, what can happen?
9. What is the name of the kind of front suspension that has a built-in shock absorber and only one ball joint?
10. What is the name of the ball joint design that is always trying to pull apart?

◾ ASE STYLE REVIEW QUESTIONS

1. Technician A says that when a rigid axle goes over a bump, the wheel on the other side of the vehicle moves too. Technician B says that on a MacPherson strut, the ball joint is a follower. Who is right?

    a. Technician A   **b.** Technician B
    c. Both A and B   **d.** Neither A nor B

2. Technician A says that a typical shock absorber will compress easier than it extends. Technician B says that some shocks have gas under pressure to prevent cavitation of the fluid. Who is right?

    a. Technician A   **b.** Technician B
    c. Both A and B   **d.** Neither A nor B

3. Technician A says that shock absorbers are designed to carry the weight of the vehicle. Technician B says that the MacPherson strut is a superior front suspension design than the SLA. Who is right?

    a. Technician A   **b.** Technician B
    c. Both A and B   **d.** Neither A nor B

4. Technician A says that some types of ball joints carry a load and others do not. Technician B says that the load carrying ball joint is the one that has the spring attached to it. Who is right?

    a. Technician A   **b.** Technician B
    c. Both A and B   **d.** Neither A nor B

5. Technician A says that unsprung weight is desirable for better handling. Technician B says that a MacPherson strut suspension has only one control arm. Who is right?

    a. Technician A   **b.** Technician B
    c. Both A and B   **d.** Neither A nor B

Suspension System Service

■ INTRODUCTION

Emphasis in this chapter is on the diagnosis and service of suspension system parts. The reader will become familiar with commonly performed chassis diagnosis and repair procedures. When suspension parts are in good condition and properly aligned, they are subjected to two forces. These forces are the weight of the car on the springs, and the force of the road on the tires (Figure 57.1).

Suspension parts take a large amount of abuse as the car is regularly driven over potholes, speed bumps, and concrete expansion joints. Forces are transmitted through the suspension parts, causing wear over time (Figure 57.2).

■ DIAGNOSING SUSPENSION SYSTEM PROBLEMS

When diagnosing suspension problems, carefully question a customer about the symptoms. Problems with the suspension system usually come to light when the customer complains of noises such as a clunk or squeak, vibration, or a handling problem. All of the suspension system parts depend on the others

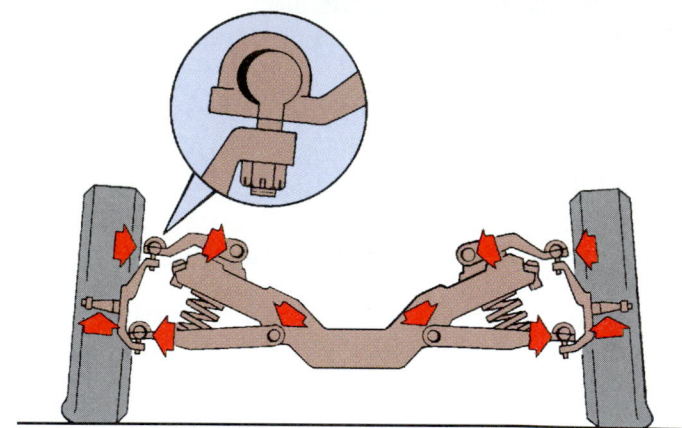

Figure 57.2 Locations of wear. *(Courtesy of McQuay Norris)*

for their performance. The rest of the system is affected when one part becomes worn.

Wheel alignment settings can change when a suspension component wears. Bushing wear can cause suspension parts to change position (see Figure 57.11). This can cause a car to pull, either all the time or just when braking.

Tire wear can result when springs sag. This is because the camber angle will change on most cars when the ride height drops (Figure 57.3). Ride height measurement is covered in Chapter 61.

Check all ball joints and bushings for looseness. Check rubber bushings and the sway bar links for wear or cracking.

■ SHOCK ABSORBER SERVICE

When there is a bad shock absorber, a tire will "hop." Feel the tire tread around the total circumference of the tire. When a scalloped or gouged wear pattern

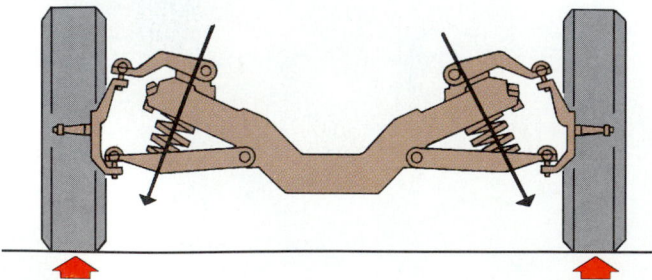

Figure 57.1 The weight of the car and the force of the road act on the suspension system. *(Courtesy of McQuay Norris)*

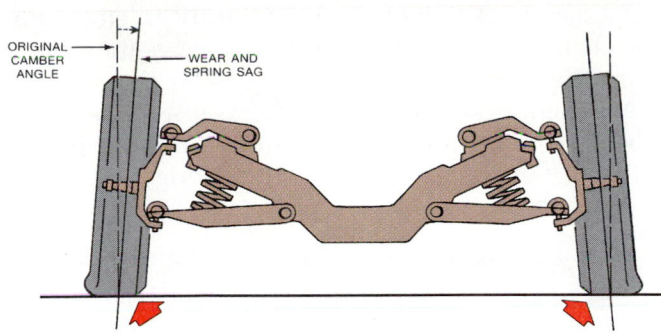

Figure 57.3 The camber angle will go toward negative on most cars when the ride height drops. *(Courtesy of McQuay Norris)*

develops on a tire (see Figure 14.5), look for a bad shock absorber. Tire imbalance could also be a contributing factor.

A shock can be defective because it is leaking or physically damaged. When one damaged shock is found, both shocks on that end of the vehicle (front or rear) are replaced.

Original equipment shock absorbers are designed to give a satisfactory (soft) ride when used with tight, new suspension bushings. By the time a vehicle has accumulated over 20,000 miles and the suspension has loosened up, some owners choose to replace the shocks simply to stiffen the ride on the highway. Internal parts can also experience wear. Shocks should be checked to see that they still provide good control.

■ TESTING A SHOCK

To test a shock with the car on the ground, perform a *bounce test*. Push down hard two or three times on the fender at each corner of the car. After the fender is pushed down, it should oscillate only about 1.5 cycles and then settle.

Shocks do not often fail at the same rate. Usually one shock has a problem, requiring the replacement of both. Compare the amount of movement side to side in the front or rear to see that they are the same. Do not compare front operation to rear operation.

Sometimes there is a complaint of noise, the location of which can be determined during the bounce test. The sound of the fluid being forced through the valves in the shock is normal.

Perform a visual inspection of the shock:
■ Inspect the condition of the shock mounts and rubber cushions.
■ Look to see if any fluid has leaked out of the shock, indicating a bad seal. Fluid cannot be replenished and the shock will have to be replaced. It is normal for a slight amount of moisture to be on the seal.

■ If the outside of the shock body is damaged, replace both shocks.

When the operation of the shock is in question, it can be unbolted from its lower mount. Then, the shock can be moved to see that it moves slowly and with equal resistance through its normal range of motion. A skip or lag in a shock as its direction of motion is changed at midstroke indicates a defective shock.

A shock must have a means of limiting its travel. If the shock is allowed to reach full travel, the valves in the shock could be ruined. Shocks sometimes limit spring travel also. In fact, on some coil spring cars, if the shocks are removed, the spring can actually fall out. When the shock is removed, the control arm can drop quickly downward in response to spring pressure. If the shocks are to be replaced, the car should be raised on a hoist that supports the wheels.

Shock Mounts

Check shock mounts to see that the shock is secured to the vehicle. A common problem with a shock absorber is for it to become loose. Shocks are mounted to the chassis with rubber cushions. Shock mounts for passenger cars are usually the *single stud (bayonet)* type or the *ring mounting type*. Some of the ring mountings have a bar (cross-pin), stud, or bolt pressed into them (Figure 57.4).

Rubber cushions allow the shock to have some flexibility in the mount. On some single stud type shocks, the tightness of the nut can determine how well the shock cushion functions. Some stud mount nuts are properly tightened when torqued against a shoulder on the stud. Others are tightened only until the rubber bulges out even with the end of its metal retainer (Figure 57.5).

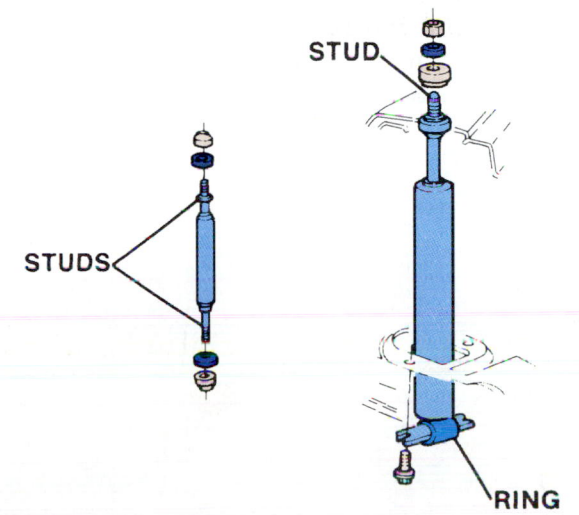

Figure 57.4 Different shock mounts. *(Courtesy of Ford Motor Company)*

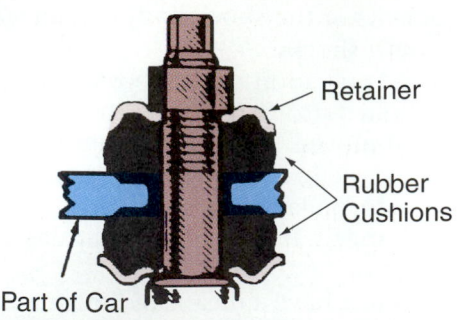

Figure 57.5 Tighten the nut until the cushions bulge almost to the outer edge of the retainers. *[Courtesy of Tenneco Automotive, Inc.]*

Bleeding/Purging Shocks

When a conventional shock absorber has been stored on a shelf in a horizontal position, an air space can develop within it. This will cause a skip in the middle of the shock's travel. If the air is not purged or bled, the shock can be misdiagnosed as defective. **Bleeding/purging shocks** of air is done as follows (Figure 57.6):

■ Extend the shock while it is in its normal vertical position.

■ Turn the shock over so that the top is down and fully collapse the shock.

■ Repeat the process four or five times while working out any skips caused by air.

➡ *Perform R & R Shock Absorbers Worksheet*

Air and Gas Shocks

An air shock can sometimes fail because of a hole in its rubber bladder. Sometimes the rubber will rot if oil comes into contact with it. If the shock is installed in a tight location, the rubber can rub on suspension parts when the air shock is inflated. Air shocks can be checked for leaks using a soapy water solution. Rub the solution over the lines, fittings, and bladder while the shocks are inflated. Look for bubbles, which would indicate a leak.

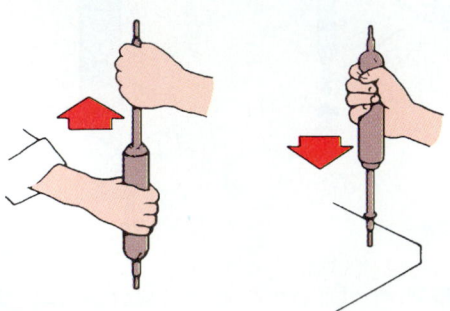

Figure 57.6 Purging the shock of air. *[Courtesy of Ford Motor Company]*

A gas shock will expand to its fully extended position if not restrained. When gas shocks are removed from the box, there is a band around them to hold them compressed. If a gas shock has lost its gas charge, it will no longer expand on its own. It may still have oil in it, but the gas charge will no longer be there to prevent the oil from foaming.

■ MACPHERSON STRUT SERVICE

A large majority of vehicles have MacPherson struts. When a MacPherson strut shock fails, sometimes an entire strut assembly must be replaced. The usual repair is to install a **strut cartridge** into the original shock housing (Figure 57.7).

The entire strut assembly is removed from the car (Figure 57.8). First, mark one of the bolts and its location at the top of the strut tower (Figure 57.9). A spring compressor is used to compress the coil spring (Figure 57.10). Removal of the retainer at the top of the strut tube allows the old shock to be removed.

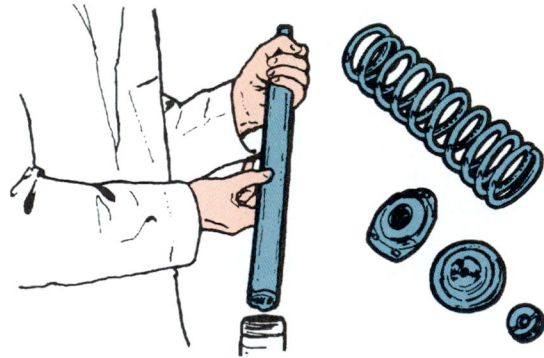

Figure 57.7 A strut cartridge fits into the old strut housing. *[Courtesy of Tenneco Automotive, Inc.]*

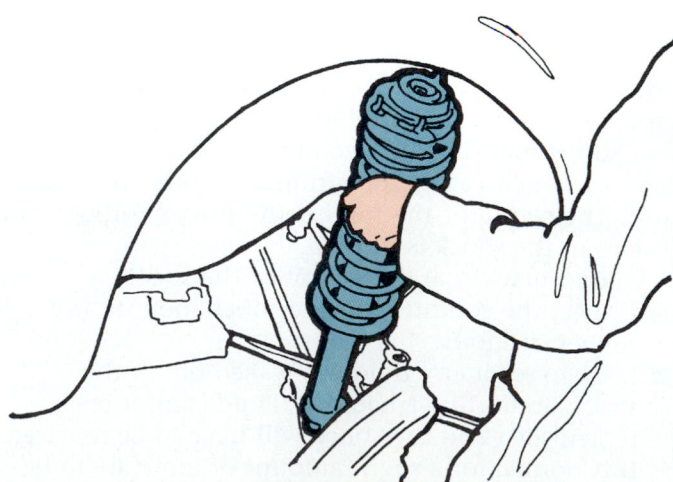

Figure 57.8 The entire strut assembly is removed from the car. *[Courtesy of Tenneco Automotive, Inc.]*

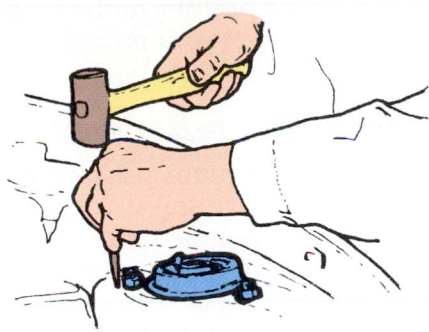

Figure 57.9 Mark the location of one of the bolts at the top of the strut tower. *(Courtesy of Tenneco Automotive, Inc.)*

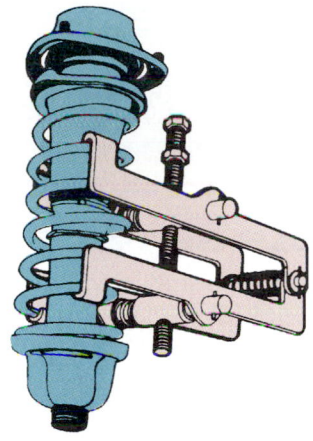

Figure 57.10 A MacPherson strut compressor. *(Courtesy of Federal-Mogul Corporation)*

SAFETY NOTE

■ The center nut must not be removed from the top of the strut before the strut assembly is removed from the vehicle. This would allow the spring to be decompressed during strut removal.

■ Be careful when handling a coil spring while it is compressed. If it is accidentally dropped, it can fly out of the compressor causing a serious injury.

ADD OIL TO STRUT BODY

Unlike a replacement cartridge, an original equipment strut is not enclosed in a cartridge. The old shock fluid will need to be poured out of the strut. Some lightweight oil is poured into the strut body to help conduct heat between the new strut cartridge and the outside of the strut housing. The old shock oil can be used for this. Two or three ounces of oil is usually enough. The amount is not important, but it should almost reach the top of the strut tube with the cartridge replaced.

SHOP TIP ATF is a good choice of oil to add to the strut because it is red in color. If the shock leaks later on, you will be able to determine if the oil is from the strut or is ATF.

Center the strut cartridge in the strut body and install the locking nut. Install the coil spring. Be sure both ends of the spring are correctly seated before removing the compressor.

Reinstall the strut assembly on the car in the same position it was in before. Perform a wheel alignment and/or bleed brakes as needed.

BUSHING SERVICE

Bushings are made of synthetic rubber. They insulate suspension parts from noise and road shock. Rubber bushings deteriorate with age. They are also susceptible to heat damage. Upper control arm bushings are especially prone to heat damage because of their close proximity to engine exhaust manifolds. Driving on bumpy roads can also result in heat and fatigue of the bushing's rubber.

Bushings should not be lubricated because petroleum attacks rubber and can ruin it. Rubber lubricant (the kind used to aid in mounting tires) can be used to help quiet a squeaky, hardened bushing but the fix is only temporary.

CONTROL ARM BUSHINGS

Damage or distortion to a control arm bushing can result in changes in wheel alignment settings (Figure 57.11). Inspect the control arm bushings for deterioration and splits in the rubber (Figure 57.12). See if the bushing is off-center. Sometimes bushings are in a position where visual inspection is difficult. Using a flashlight and mirror will sometimes help.

Push on the fenders of the vehicle while listening for noise. To inspect bushings for looseness, use a prybar to see if the control arm can be moved.

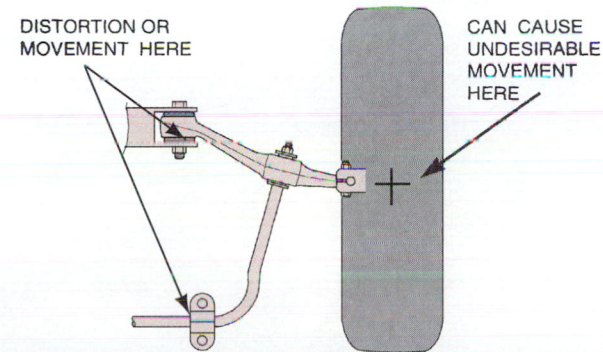

Figure 57.11 Defective bushings can change alignment settings. *(Courtesy of Moog Automotive, Inc.)*

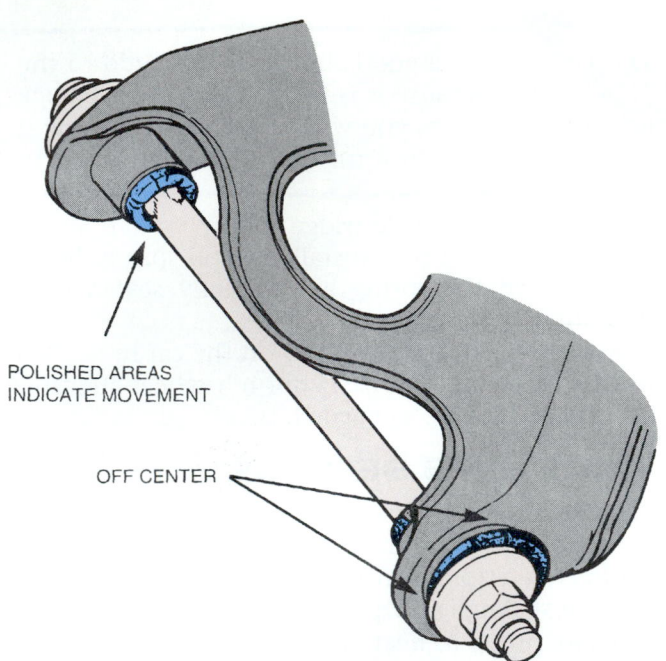

Figure 57.12 Inspect control arm bushings. *(Courtesy of Moog Automotive, Inc.)*

Special offset replacement control and bushing sets are available that compensate for sagged frame components that can cause front end alignment problems.

Control arm bushings are pressed or driven out. It is important that the control arm holes not be damaged by the process. If an air chisel is used (Figure 57.13), be sure that the chisel bit is wide and dull. If it

Figure 57.13 Using an air chisel to remove a control arm bushing. *(Courtesy of Moog Automotive, Inc.)*

is sharp, it will cut the bushing rather than push it out. The new bushing is pressed or pounded in with a special driving tool.

◼ STABILIZER BAR SERVICE

Inspect the bushings at both ends of the stabilizer bar (Figure 57.14). Stabilizer links do not have to be removed to replace the inner bushings. Replacement stabilizer bushings have a split in them. Some original bushings do not have a split. These can be cut off with a razor blade. Then install the new bushing.

⚠ CAUTION When replacing stabilizer bar bushings, be sure that *both* front wheels are either on the ground or in the air. Otherwise, the sway bar will be spring loaded. With one wheel jacked up, removal of the nut on the top of the stabilizer link could result in a serious injury.

◼ SPINDLE SERVICE

The steering knuckle and spindle do not require service. If the spindle is damaged, it is replaced. Damage from a collision will be noticed during the SAI alignment angle check. If a wheel bearing fails it can wear or heat the spindle. This changes the metallurgy of the spindle and it must be replaced.

CASE HISTORY *A front wheel bearing failed on a 1965 Mustang. There was wear on the spindle that the technician failed to notice. He replaced the wheel bearing and returned the car to the customer. Several months later, the customer's daughter was driving the car when she drove over a small pothole in the road. The spindle broke off just behind the area where the old bearing had burned (Figure 57.15). She was not travelling fast and, luckily, was not hurt when the front wheel fell off of the car.*

◼ BALL JOINT SERVICE

Ball joints are relatively trouble-free, but occasionally they wear out. The ball joint is sealed within a rubber boot filled with grease. A small bleed hole in the boot allows for grease movement during lubrication. Other cuts or tears in the boot will allow water and dirt into the joint, causing it to wear.

Feel around the outside of the boot, looking for tears. If the boot is torn, the joint will probably fail soon and should be replaced. Inspect also for signs of rust or cracks on the control arm near the joint.

Manufacturers list specifications for movement of ball joints. Always check the specifications and be sure

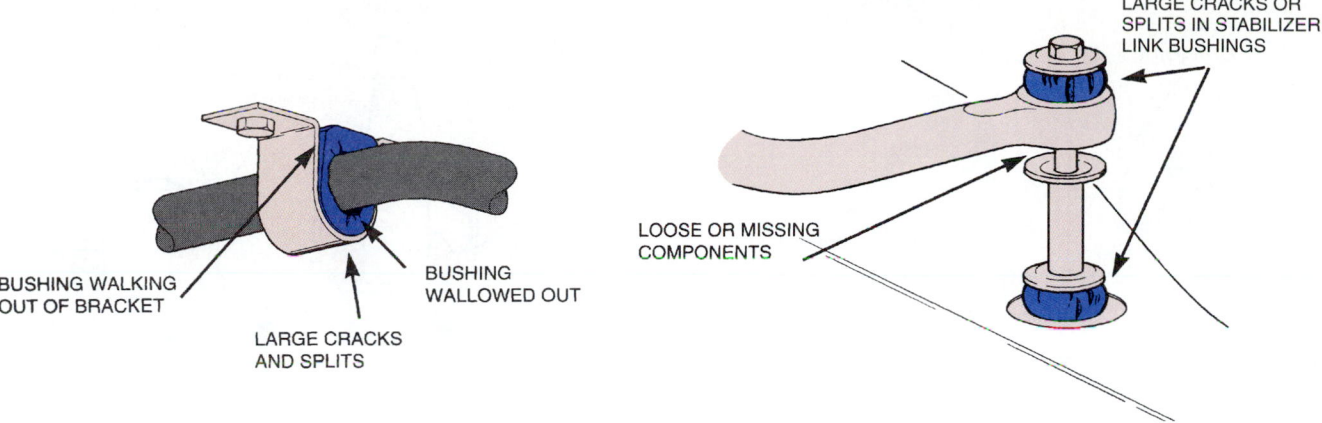

Figure 57.14 Inspect stabilizer bushings. *(Courtesy of Moog Automotive, Inc.)*

Figure 57.15 This spindle broke after a burned wheel bearing was replaced.

that the proper procedure is followed. Vertical or **axial movement** is usually specified, although some manufacturers specify horizontal or **radial movement** (Figure 57.16).

A load carrying ball joint usually has some movement when unloaded. In the past, the unnecessary replacement of good ball joints was a common occurrence. As a result, some states now require measurement and documentation of the amount of clearance found on a worn ball joint before it can be replaced. Most load carrying ball joints have a wear limit of 0.060" of vertical movement. But some joints can have as much as 0.200" movement and still be within specified limits.

Before checking ball joint wear, determine whether the ball joint is a load carrier or follower (see Figure 56.30). It is important to know this when testing ball joints for wear because the load carrying joint must be unloaded to test it. Load carrying ball joints are either tension or compression types (see Figure 56.31). Tension loaded ball joints are more common.

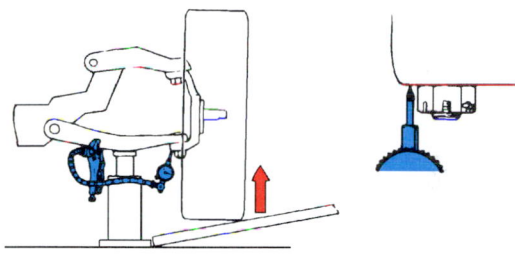

Axial check

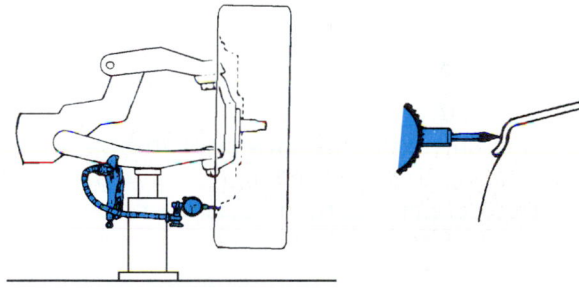

Radial check

Figure 57.16 Ball joint checks. *(Courtesy of Moog Automotive, Inc.)*

- The ball joint on the control arm that has the spring mounted on it is the load carrier.
- The ball joint on a strut suspension is a load carrier.

MEASURING BALL JOINT WEAR

For an accurate wear check, the ball joint must not support the weight of the vehicle. Some SLA suspension systems have the spring above the upper control arm. Figure 57.17 shows how to unload the joint for both the spring-above and the spring-below types.

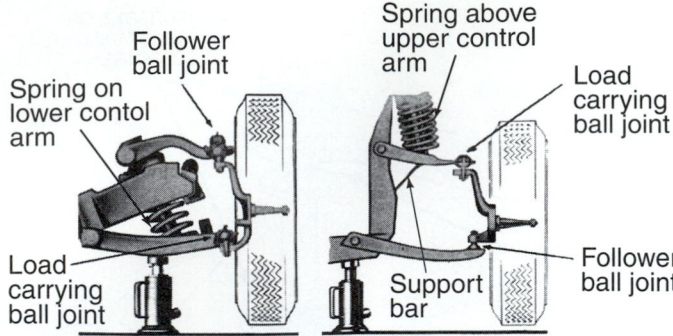

Figure 57.17 Methods of checking ball joints. *(Courtesy of Federal-Mogul Corporation)*

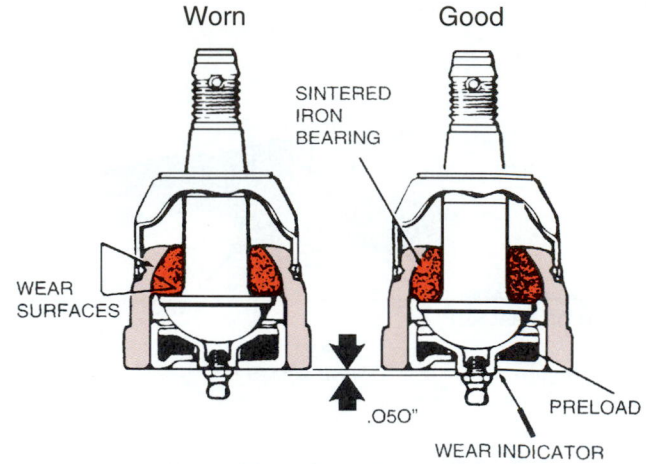

Figure 57.18 Checking a wear indicator ball joint. *(Courtesy of Moog Automotive, Inc.)*

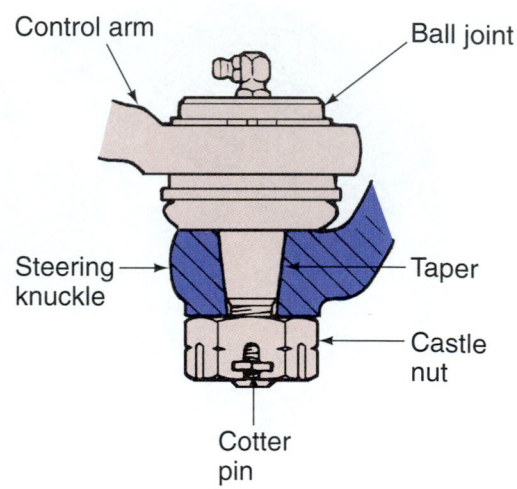

Figure 57.19 A ball joint taper. *(Courtesy of Moog Automotive, Inc.)*

The follower joint should be checked at the same time that the loaded ball joint is unloaded for testing. Follower ball joints hold the steering knuckle in the correct position and allow it to pivot during bumps and turns. Recommendations for most follower ball joints call for replacement if there is "any perceptible movement" in the joint. Always check specifications. They are not all that way.

To check a follower ball joint for movement, unload the joint and try to move the tire back and forth while looking for movement. For this test to be accurate, wheel bearing clearance cannot allow the wheel to move.

SHOP TIP
■ A quick way to eliminate wheel bearing clearance from the measurement is to apply the brakes during the test. When a helper is not available, brakes can be applied using a pedal depressor.
■ To measure ball joint clearance without a dial indicator, measure the length of the ball joint with a caliper and then measure it again when the joint has been unloaded.

Wear Indicator Ball Joints

Some ball joints have a **wear indicator** built into them. The most common type of wear indicator has a shoulder that sticks out of the bottom of the joint about 0.050" when it is new (Figure 57.18). If the ball joint has worn, this shoulder will recede into the ball joint housing. When it is flush, the ball joint should be replaced. A wear indicator ball joint must be loaded and at normal ride height to read the indicator.

■ SEPARATING BALL JOINT TAPERS

The ball joint is connected to the steering knuckle with a tapered connection, called a **steering taper** (Figure 57.19), which is also used in other steering connections. To remove a ball joint requires the use of a spe-

cial tool. Another way of "breaking" the taper is to use a large hammer.

■ First, remove the cotter pin from the ball joint nut and loosen the nut several turns.
■ Position the vehicle so that the coil spring is pushing on the ball joint. This could require lifting the vehicle or allowing its weight to rest on the wheels.
■ Use a hammer to pound sharply on the steering knuckle on the outside of the taper. This will deform the taper and spring pressure will separate the ball joint from the steering knuckle.

■ REPLACING THE BALL JOINT

A ball joint can be retained by one of several methods. Figure 57.20 shows three of the most common ones. Some ball joints are held in with rivets from the factory. The rivets are drilled and punched to remove them. Bolts and nuts are used to hold the replacement joint in place (Figure 57.21).

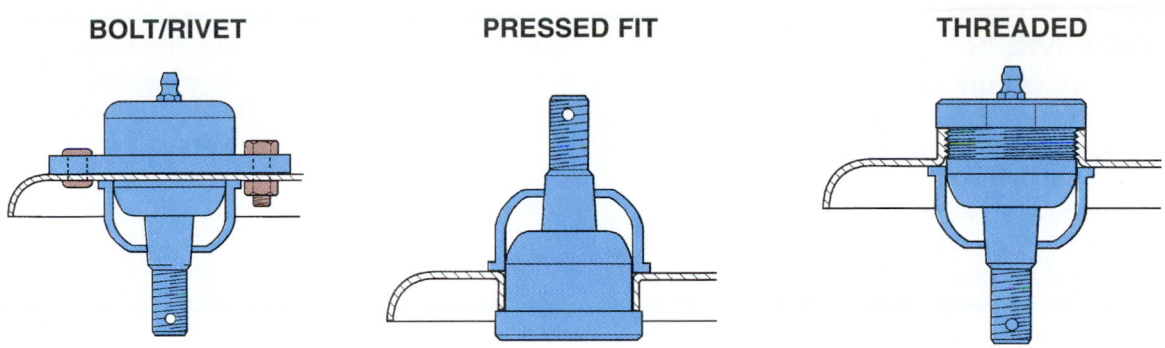

BOLT/RIVET **PRESSED FIT** **THREADED**

Figure 57.20 Three common ball joint retaining methods. *(Courtesy of Moog Automotive, Inc.)*

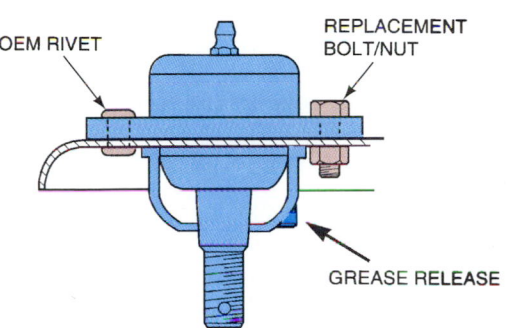

Figure 57.21 Original equipment rivets are replaced with nuts and bolts. *(Courtesy of Moog Automotive, Inc.)*

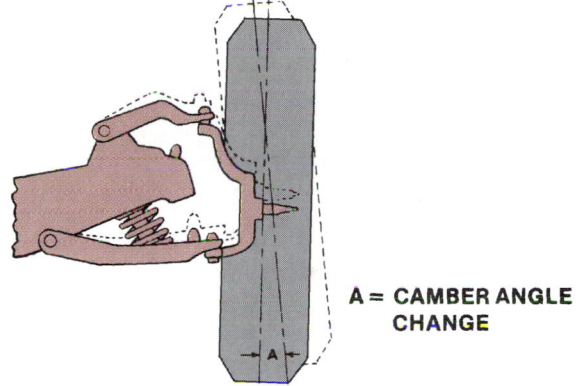

A = CAMBER ANGLE CHANGE

Figure 57.22 The position of the control arm changes in its arc due to spring sag. *(Courtesy of McQuay Norris)*

Ball joints can also be pressed or threaded into the hole in the control arm. If the control arm has been removed from the vehicle, a pressed-fit ball joint can be removed using a standard hydraulic press. When the control arm is still on the car, a special press set allows the removal and replacement of these ball joints.

Some pressed-fit ball joints have a spot weld that holds them in place. This must be carefully removed. When a replacement ball joint is installed, a snap ring often takes the place of the weld.

NOTE: *Be sure the grease release hole in the new ball joint's rubber boot is aimed away from the brakes.*

■ COIL SPRING SERVICE

Over the years, the weight of the car causes the coil spring to lose some of its tension. This results in a lowered ride height. A ride height check can be performed to see if the height of the car conforms to specifications.

Incorrect ride height affects the camber and toe alignment angles. With an SLA suspension system, the car is designed to ride at a height that minimizes changes in tread width. When the suspension system drops due to sagged springs, the upper control arm is

in a different place in its arc of travel. This means that there will be even more camber change during bumps (Figure 57.22). Tires will wear faster and handling is affected.

A vehicle that is too low cannot be aligned properly, so this test should be done before attempting a wheel alignment. The ride height check is covered in Chapter 61.

■ ADJUSTING SPRING HEIGHT

An advantage of the torsion bar spring used on a few vehicles is that spring height can be adjusted. A screw is turned against a bracket mounted on one of the torsion bar ends. The vehicle's correct ride height can be restored prior to an alignment. Coil springs must be replaced when they have sagged beyond specifications.

Air shocks or shock absorbers with coil springs around them are designed to be used only for temporary overload conditions. The weight of the vehicle will rest on the shock mounts instead of the spring seat. Shock mounts are not designed to continually support the vehicle.

■ SLA COIL SPRING REPLACEMENT

Most passenger cars require the use of a coil spring compressor for coil spring removal and replacement. When a spring is installed in a vehicle, it is held in a compressed position. When it is fully extended, it is much longer. Before removing a spring, the wheel, shock absorber, and stabilizer links are removed. The outer tie rod is disconnected from the steering arm.

During a coil spring replacement, only the lower ball joint needs to be removed. The upper ball joint can remain in place. The upper control arm and steering knuckle will be moved out of the way during the coil spring replacement. Sometimes, both upper and lower ball joint tapers must be broken to be able to get the disc brake splash shield and spindle to clear the lower control arm. Remove the cotter pin from the ball joint nut and back the nut off several turns.

NOTE: A pickle fork (see Figure 59.6) will probably ruin the rubber seal on the ball joint. If the ball joints are not going to be replaced, this will be a costly subtraction from the profit of the job.

When using the special ball joint press to loosen a taper connection, unload the ball joint with a floor jack. For maximum leverage, position the jack so it is as close to the outer ball joint as possible (Figure 57.23). A piece of wood will help get more leverage and keep the jack level.

 Reach across to jack up the control arm from the opposite side of the vehicle. This will allow the jack to roll toward the inside of the vehicle as the control arm is lowered, releasing spring pressure.

Figure 57.23 Position the jack so it is as close to the outer ball joint as possible. *(Courtesy of Moog Automotive, Inc.)*

Raise the jack to compress the spring until the vehicle begins to lift. Remove the lower ball joint nut and lift the steering knuckle assembly away from the ball joint stud. Support it out of the way.

Use two clips (see Figure 57.24) to hold the spring compressed. The clips are either short or long. Use the one that fits best. Short clips are used to hold four rungs of the spring. Long clips are used to clip five rungs. Use two clips, positioned side by side on the same spring rungs. Position the clips as low as possible on the inboard side of the spring.

Pry down on the lower control arm and pry the coil spring from its place (Figure 57.24). Mark the locations of the clips. Use a coil spring compressor to compress the spring and remove the clips.

 Never compress the spring far enough for the coils to stack up.

Lay the old spring next to the new spring. Align the dominant ends (Figure 57.25). The **dominant end** is the end that aligns in the coil spring seat in the frame or lower control arm. Mark the new spring in the same location of the old spring. Compress the spring and install the spring clips (Figure 57.26).

Because of the spring clips, the spring will be bowed. This will make aligning the upper and lower spring seats easier during installation. If there is an insulator on the bottom of the spring, tape it onto the spring to make installation easier.

The spring seats must be accurately aligned. During installation, align the upper end first. Then, push the bottom of the spring in and align the lower seat. Jack the control arm to compress the spring and install the ball joint stud in the tapered hole (Figure 57.27). Be sure to align the cotter pin hole so it is parallel to the brake backing plate. Otherwise, it will be difficult

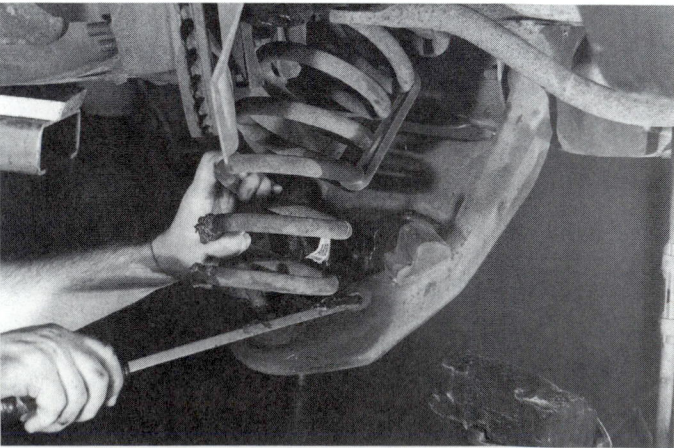

Figure 57.24 Pry down the control arm and remove the spring. *(Courtesy of Moog Automotive, Inc.)*

Figure 57.25 Align the dominant ends of the old and new springs. *(Courtesy of Moog Automotive, Inc.)*

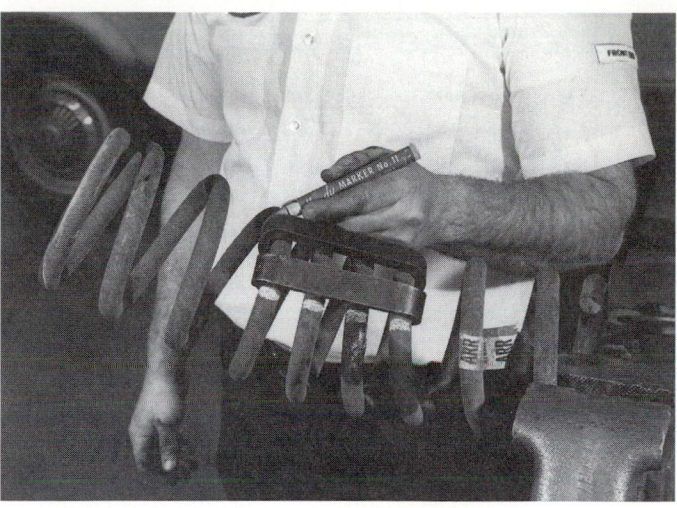

Figure 57.26 Compress the spring and install the spring clips. *(Courtesy of Moog Automotive, Inc.)*

to install the new cotter pin in the castle nut. Torque the castle nut to specifications.

NOTE: *If the cotter pin hole does not line up, tighten until the next hole does. Do not back off on the nut to align the hole.*

After connecting the ball joint(s), remove the spring clips from the spring. A small prybar may be needed for this.

Complete the assembly of all of the components. Before lowering the wheels onto the ground, loosen the bolts that compress the control arm bushings. Drive the vehicle a short distance and jounce the suspension several times. Retorque all bushings to specifications before doing a wheel alignment.

■ WHEEL ALIGNMENT

Following completion of suspension repair jobs, a wheel alignment will be required. This will reposition suspension components so that the car will be safe to drive and will go straight without unusual tire wear.

Figure 57.27 Align the cotter pin hole and install the ball joint in the steering knuckle. *(Courtesy of Moog Automotive, Inc.)*

■ REVIEW QUESTIONS

1. A _____ shock expands to its full travel when not installed.
2. What is added when installing a MacPherson strut cartridge to help it conduct heat to the outside of the strut?
3. If a wheel bearing burns and damages the spindle, what part should be replaced?
4. Are there any holes in a good ball joint boot?
5. What are two directions ball joints are checked for wear?
6. Which kind of ball joint is more common, tension or compression?
7. When a wear indicator ball joint wears, does the shoulder move into or out of the ball joint?
8. When a replacement ball joint is held in place with nuts and bolts, what held the previous ball joint in position?
9. What is used with a spring compressor to aid in the removal and replacement of coil springs?
10. What is the name of the end of a coil spring that aligns in the coil spring seat in the frame or lower control arm?

ASE STYLE REVIEW QUESTIONS

1. Technician A says that a leaking shock should have the fluid refilled. Technician B says it is normal for a small amount of moisture to be on the outside of the shock body. Who is right?
   - **a.** Technician A
   - **b.** Technician B
   - **c.** Both A and B
   - **d.** Neither A nor B

2. Technician A says that a shock mount should be tightened until the rubber bumper is totally compressed. Technician B says to lubricate shock absorber bushings with light oil. Who is right?
   - **a.** Technician A
   - **b.** Technician B
   - **c.** Both A and B
   - **d.** Neither A nor B

3. Technician A says when replacing control arm bushings with an air chisel, the tool must be sharp. Technician B says when replacing stabilizer bar bushings, be sure that both front wheels are either on or off the ground. Who is right?
   - **a.** Technician A
   - **b.** Technician B
   - **c.** Both A and B
   - **d.** Neither A nor B

4. Technician A says a load carrying ball joint must be unloaded to test it. Technician B says load carrying ball joints must be replaced if there is any perceptible movement. Who is right?
   - **a.** Technician A
   - **b.** Technician B
   - **c.** Both A and B
   - **d.** Neither A nor B

5. Technician A says applying the brakes will temporarily eliminate wheel bearing clearance. Technician B says to install a new ball joint with its rubber boot grease hole facing against the brake backing plate. Who is right?
   - **a.** Technician A
   - **b.** Technician B
   - **c.** Both A and B
   - **d.** Neither A nor B

6. Technician A says air shocks are designed to permanently raise a vehicle's height. Technician B says when jacking on a control arm to compress a coil spring, the jack should be 90° (perpendicular) to the tires. Who is right?
   - **a.** Technician A
   - **b.** Technician B
   - **c.** Both A and B
   - **d.** Neither A nor B

7. Technician A says to align the cotter pin hole in a ball joint so it is parallel to the brake backing plate. Technician B says after reaching torque specifications, back off on the ball joint lock nut to align the cotter pin hole. Who is right?
   - **a.** Technician A
   - **b.** Technician B
   - **c.** Both A and B
   - **d.** Neither A nor B

8. Technician A says that wear on one side of a tire could result from a bad shock absorber. Technician B says that when one shock absorber is bad, all four shocks should be replaced. Who is right?
   - **a.** Technician A
   - **b.** Technician B
   - **c.** Both A and B
   - **d.** Neither A nor B

9. Technician A says that worn bushings can change alignment settings. Technician B says that the sound of the fluid being forced through the valves in the shock is normal. Who is right?
   - **a.** Technician A
   - **b.** Technician B
   - **c.** Both A and B
   - **d.** Neither A nor B

10. A MacPherson strut shock absorber is leaking badly. Technician A says that sometimes an entire strut assembly must be replaced. Technician B says that a strut cartridge is often installed into the original shock housing. Who is right?
    - **a.** Technician A
    - **b.** Technician B
    - **c.** Both A and B
    - **d.** Neither A nor B

CHAPTER **58**

Steering Fundamentals

■ **OBJECTIVES**

Upon completion of this chapter, you should be able to:

✔ List the parts of steering systems.
✔ Describe the principles of operation of steering systems.
✔ Compare linkage systems to rack and pinion.
✔ Describe how power steering systems operate.
✔ Understand the operation of four wheel steering systems.

■ **KEY TERMS**

recirculating ball and nut
 steering gear
rack and pinion steering
steering ratio
lock to lock
steering damper
parallelogram steering
ball sockets
turnbuckle
toe out on turns
spool valve

■ STEERING SYSTEMS

The steering system works with the suspension system. It allows the driver to steer the car while providing a comfortable amount of steering effort. Steering system parts include the *steering gear,* the *steering linkage,* the *steering wheel,* and the *steering column.* There are two styles of steering. One has a gear box and parallelogram linkage (Figure 58.1). The other is a simple long rack with linkage extending from its ends (Figure 58.2).

■ STEERING GEARS

The two common types of automotive steering gears are the conventional **recirculating ball and nut steering gear** and the **rack and pinion steering**. There are other types of steering gear box designs, but these are the primary ones used in automobiles and light trucks.

The number of teeth on the driving gear compared to the number of teeth on the driven gear help determine the **steering ratio**. The length of the steering arms, pitman arms, and idler arms (parts of the con-ventional steering linkage) also play a part in determining steering ratio.

When the steering wheel is turned all the way in one direction, it stops against a *lock.* Turning the wheel all the way from one lock to the other is called **lock to lock**. The steering ratio refers to the amount of space it takes for a vehicle to be able to turn around. A "fast" steering ratio is about three turns lock to lock. Slower ratios require the wheel to be turned about four times. The ratio of a steering gear varies, depending on whether or not the car has a power assist. Power steering cars usually have faster ratios. A 15:1 ratio means that when the steering wheel is turned 15° the front wheels will turn 1°.

■ RECIRCULATING BALL STEERING GEAR

In a recirculating ball steering gear (Figure 58.3), a *sector gear* (part of the *pitman* or *sector shaft*) meshes with

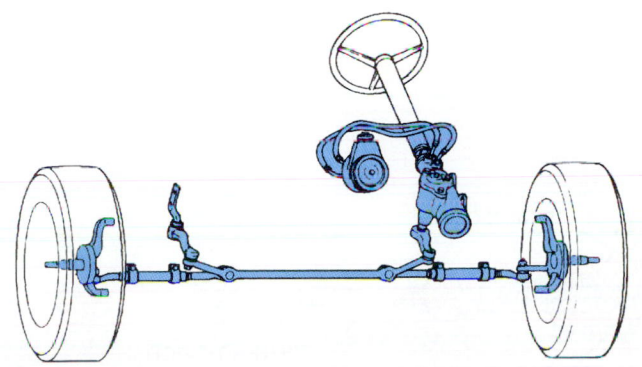

Figure 58.1 A steering gear box and parallelogram linkage. *(Courtesy of Moog Automotive, Inc.)*

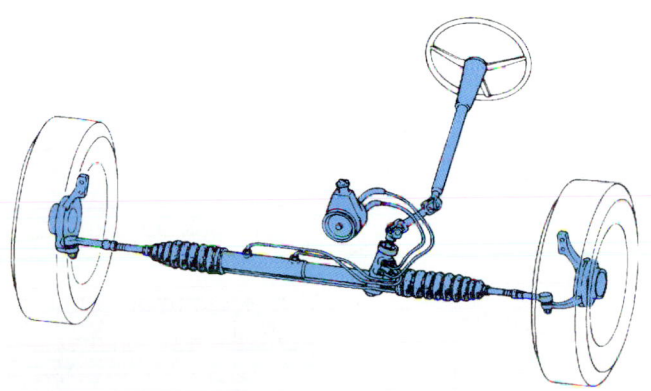

Figure 58.2 Rack steering with linkage. *(Courtesy of Moog Automotive, Inc.)*

811

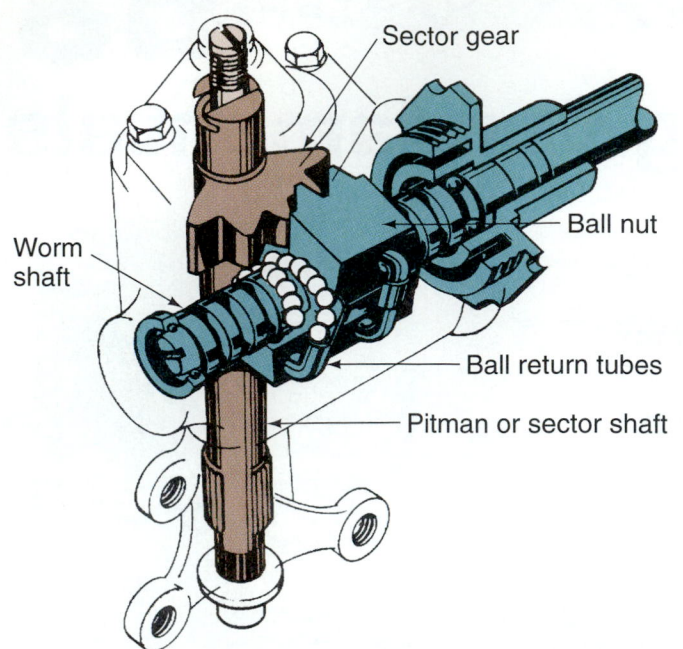

Figure 58.3 A recirculating ball steering gear. *(Courtesy of Moog Automotive, Inc.)*

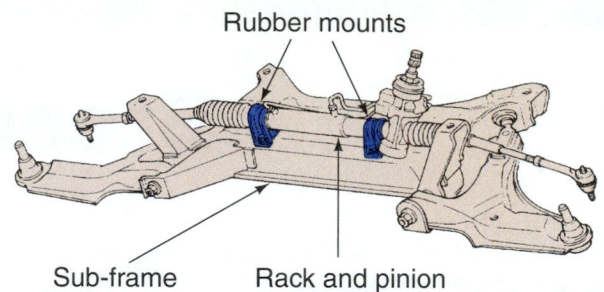

Figure 58.5 The rack is mounted on rubber bushings. *(Courtesy of Moog Automotive, Inc.)*

a *ball nut* that rides on bearings on the *worm (steering) shaft* to provide a smooth steering feel. The ball nut has curved channels for ball bearings to ride in. The steering shaft also has bearing channels. The balls rotate and recirculate through tubes (ball returns).

RACK AND PINION STEERING

On a rack and pinion steering system, the end of the steering shaft has a *pinion gear* that meshes with the *rack gear* (Figure 58.4). They are used on many cars because they are lighter than standard steering gears and are easier to assemble to the vehicle at the factory. A typical rack and pinion often has a faster ratio.

Rack and pinion steering systems are more easily damaged when the front wheels hit a curb or rock. They transmit more road shock that can be felt back

through the steering wheel. The rack is mounted on rubber bushings to help cushion shocks (Figure 58.5). Some rack and pinions use a **steering damper**, a horizontal shock absorber for road shocks.

A spring-loaded damper preloads the rack gear to prevent the rack from flexing, which would cause gear backlash (Figure 58.6). An adjustment screw or shims are used to adjust the amount of tension against the rack. The pinion shaft is usually supported by needle bearings. The rack is most often supported by plain bushings.

The rack and pinion unit can be mounted in several locations. Some are on the sub-frame and others are on the firewall.

STEERING LINKAGE

The steering gear is connected to the wheels by the steering linkage. Steering linkage parts vary depending on the design used, but all designs include *tie rods*, *steering arms*, and a *steering-knuckle*. A comparison of the steering linkage parts of the two major steering systems is shown in Figure 58.7.

When a conventional steering gear box is used, there can be a number of different linkage designs used, depending on the suspension design. The most popular steering design in use with the long and short arm suspension is the parallelogram.

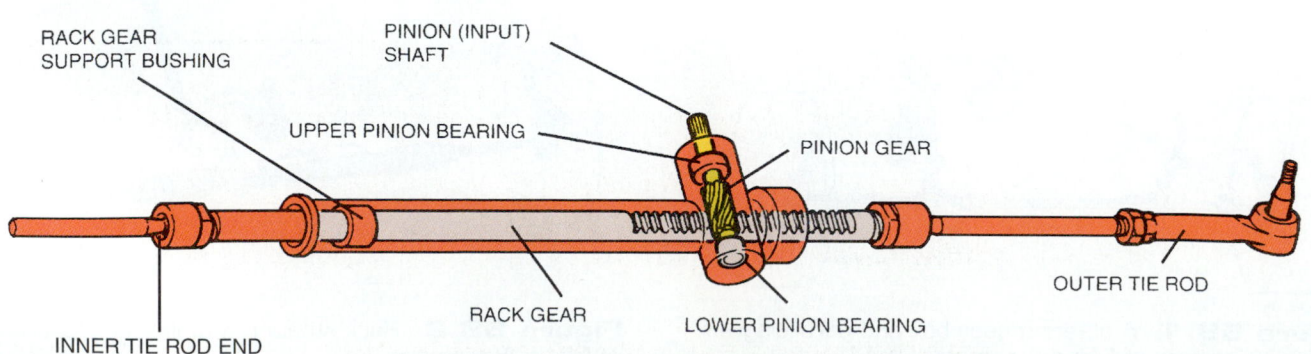

Figure 58.4 Parts of a rack and pinion steering gear. *(Courtesy of Moog Automotive, Inc.)*

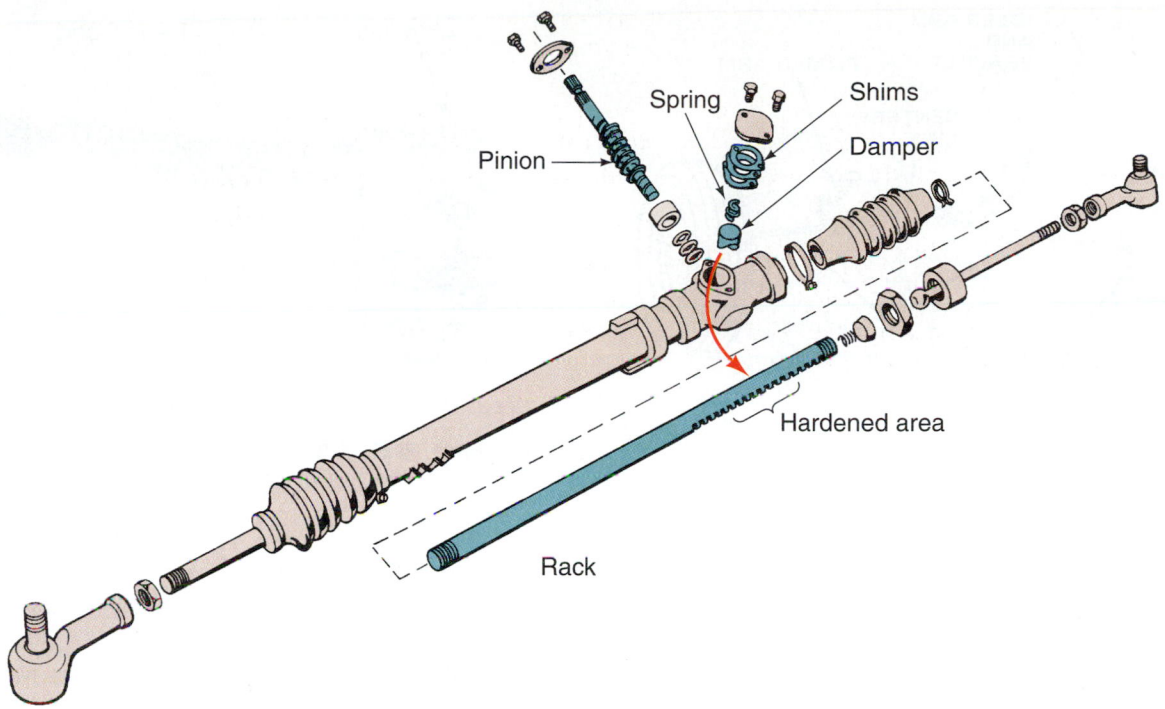

Figure 58.6 A damper holds tension against the rack. *[Courtesy of Moog Automotive, Inc.]*

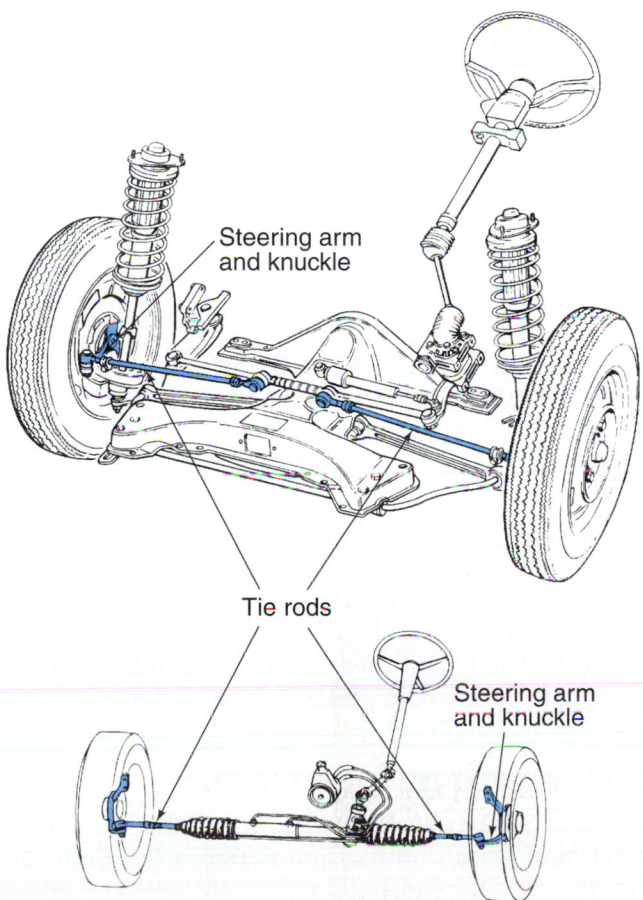

Figure 58.7 Both types of steering have tie rods, steering arms, and a steering knuckle. *[Courtesy of McQuay Norris and Moog Automotive, Inc.]*

■ PARALLELOGRAM STEERING LINKAGE

On passenger cars, a recirculating ball gear usually uses a **parallelogram steering** system (Figure 58.8). The name comes from the parallelogram shape made by the steering linkage during a turn. Parallelogram steering parts are shown in Figure 58.9. *Tie rods* on each side are connected by the *center link*. The *pitman arm* is the part that connects the steering box to the center link. An *idler arm* supports the center link on the passenger side (Figure 58.10).

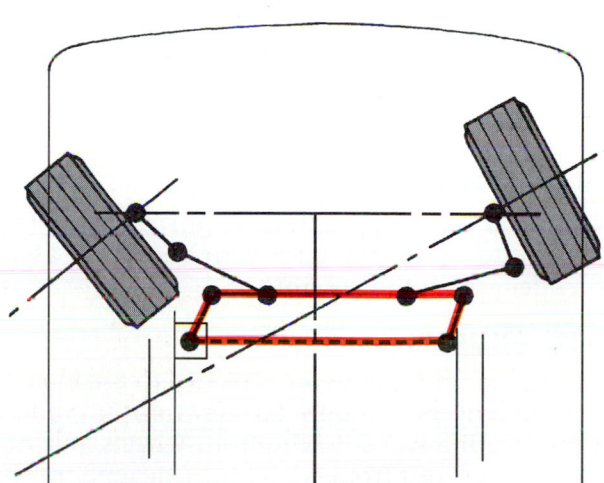

Figure 58.8 The shape of a parallelogram is made by the steering linkage during a turn. *[Courtesy of General Motors Corporation, Service Technology Group]*

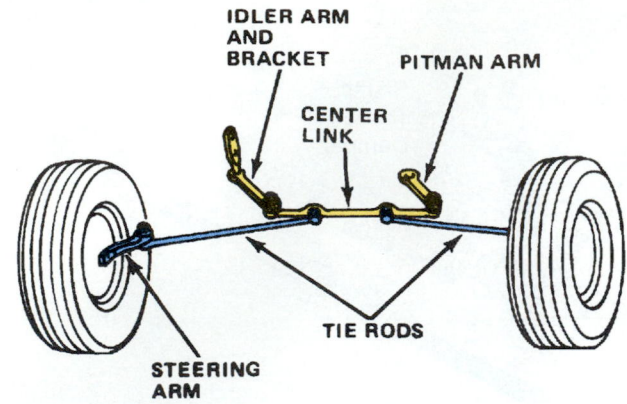

Figure 58.9 Parts of a parallelogram steering during a turn. *(Courtesy of Chrysler Corporation)*

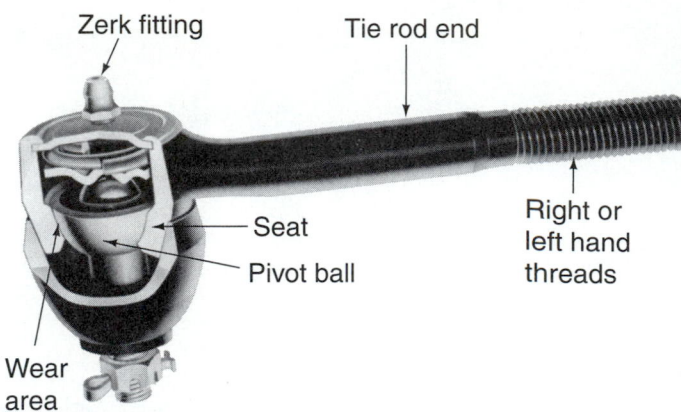

Figure 58.11 Cutaway of a tie rod end. *(Courtesy of Federal-Mogul Corporation)*

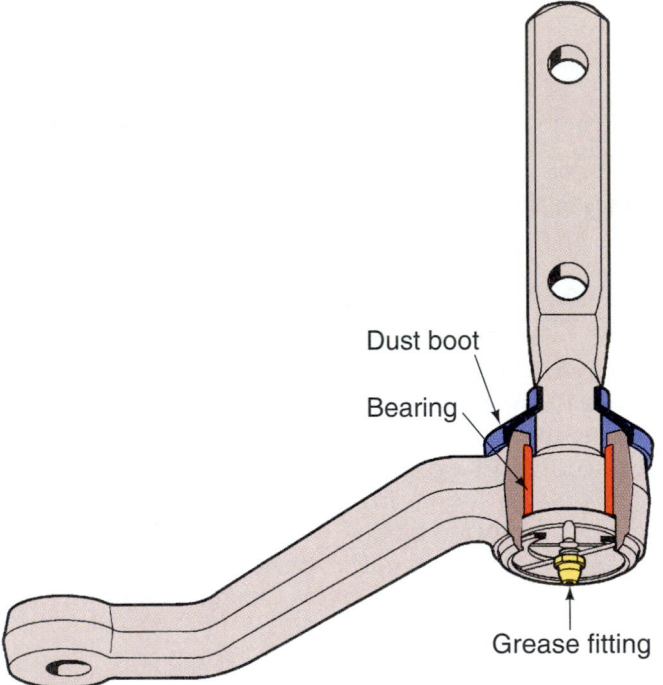

Figure 58.10 A cutaway of an idler arm. *(Courtesy of Moog Automotive, Inc.)*

■ BALL SOCKETS

Ball sockets connect the steering linkage parts. They allow parts to rotate during a turn and pivot as the steering deflects during a bump.

■ TIE RODS

Tie rod (Figure 58.11) ends are attached to pivot points at the front wheels. Not only do they transmit motion from the steering wheel to the front wheels, but they maintain the correct front wheel toe (the amount that the front tires are aimed inward or outward at the front). There is a threaded adjusting sleeve that connects the inner and outer tie rods (Figure 58.12). It has

a right hand thread on one end and a left hand thread on the other end. During an adjustment, when it is turned it acts like a **turnbuckle**. Turning it one way shortens the tie rod assembly. Turning it the other way lengthens it.

■ STEERING ARM

The tie rods attach to the front wheels at the steering arms (Figure 58.13). The steering arm is attached to the *steering knuckle,* which includes the *spindle* (Figure 58.14)

During a turn, the inside wheel must turn sharper than the outside wheel because they follow different circular paths. The steering arms are angled inward, rather than being parallel to the frame. This angle provides an important steering angle, **toe out on turns** (Figure 58.15).

■ RACK AND PINION STEERING LINKAGE

Rack and pinion does not have the complicated steering linkage used with a conventional steering box. In most systems, two tie rods come out of the steering rack (see Figure 58.6). They have conventional tie rod end ball sockets only on the outer ends. These attach to the steering knuckles. The inner pivots are ball sockets, enclosed within *rubber bellows* or *boots* on the rack (Figure 58.16). In another type of rack and pinion system, the inner tie rods attach to the center of the rack gear (Figure 58.17).

■ STEERING COLUMN

Included in the steering column is the turn signal switch and horn control. Sometimes a headlight and dimmer switch, windshield wiper and washer controls, transmission shift selector, cruise control, and ignition switch and lock (between the shift selector and the steering wheel) are built into the steering column. This

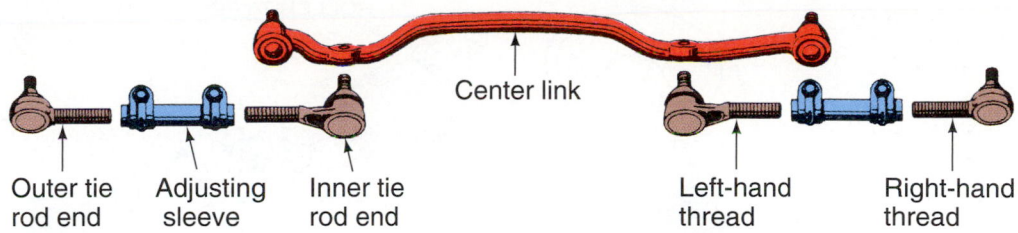

Figure 58.12 A threaded sleeve connects the inner and outer tie rods. *[Courtesy of Federal-Mogul Corporation]*

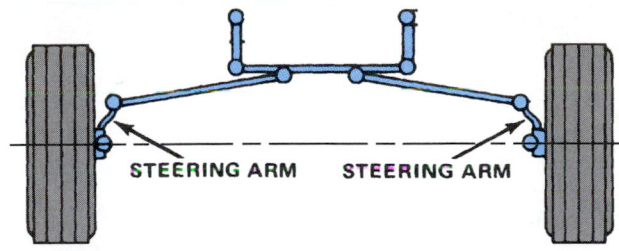

Figure 58.13 The tie rods attach to the front wheels at the steering arms. *[Courtesy of Chrysler Corporation]*

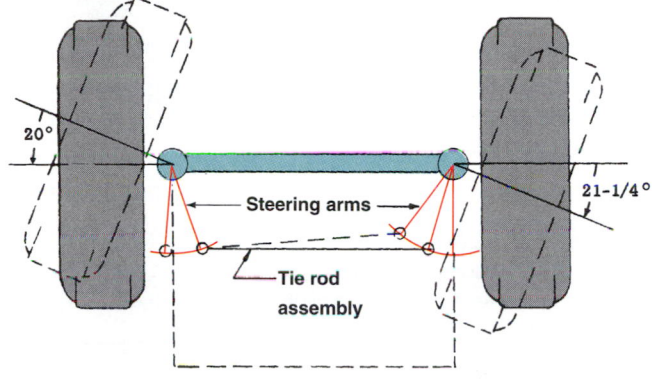

Figure 58.15 The steering arms are bent at an angle to allow the fronts of the tires to toe out during a turn. *[Courtesy of General Motors Corporation, Service Technology Group]*

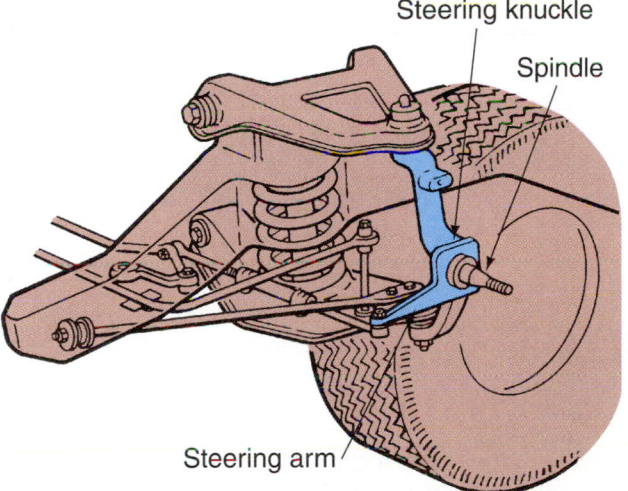

Figure 58.14 Steering arm and knuckle. *[Courtesy of McQuay Norris]*

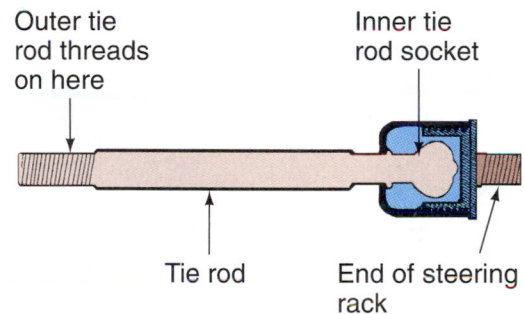

Figure 58.16 The inner pivots on a rack tie rod are ball sockets. *[Courtesy of McQuay Norris]*

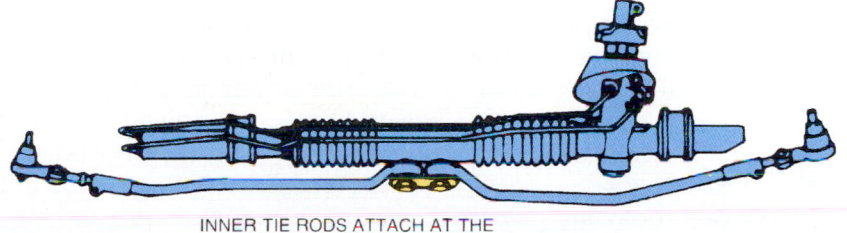

INNER TIE RODS ATTACH AT THE
CENTER OF THE RACK GEAR

Figure 58.17 Inner tie rods attached at the center of the rack. *[Courtesy of Moog Automotive, Inc.]*

is called the *European style* of controls. Tilt steering wheels and collapsible columns are also available.

■ STEERING SHAFT

The steering shaft connects the steering wheel to the steering gear. Because the steering column is mounted to the body of the car and the steering gear box is mounted to the frame, there is movement or flex between the two. There is a *flexible coupling* or a *universal joint* between the steering shaft and the splined input shaft of the steering gear (Figure 58.18). It allows a small amount of misalignment between the steering

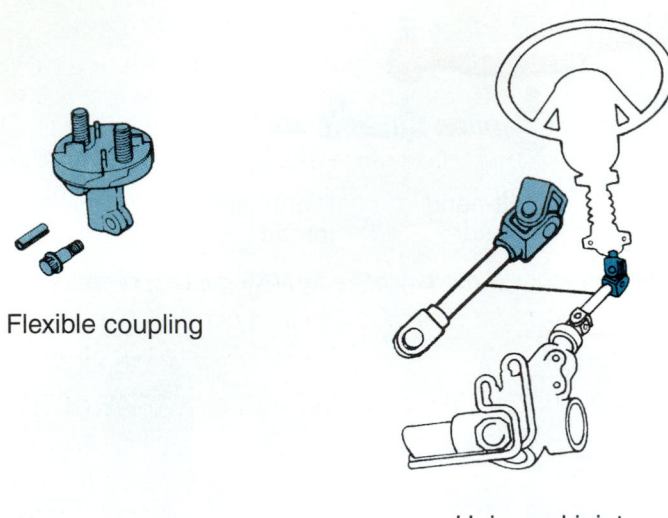

Flexible coupling

Universal joint

Figure 58.18 A flexible coupling or a universal joint provides the attachment between the steering shaft and the splined input shaft of the steering gear. *(Courtesy of Moog Automotive, Inc.)*

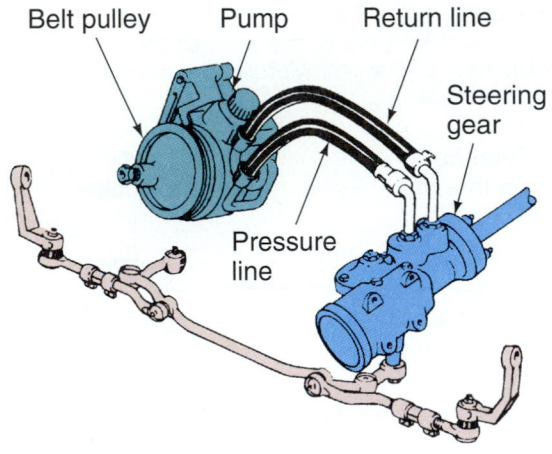

Belt pulley   Pump   Return line

Steering gear

Pressure line

Figure 58.19 Parts of a hydraulic power steering system. *(Courtesy of Moog Automotive, Inc.)*

ROLLER TYPE

ROLLER

VANE TYPE

VANE

SLIPPER TYPE

SLIPPER

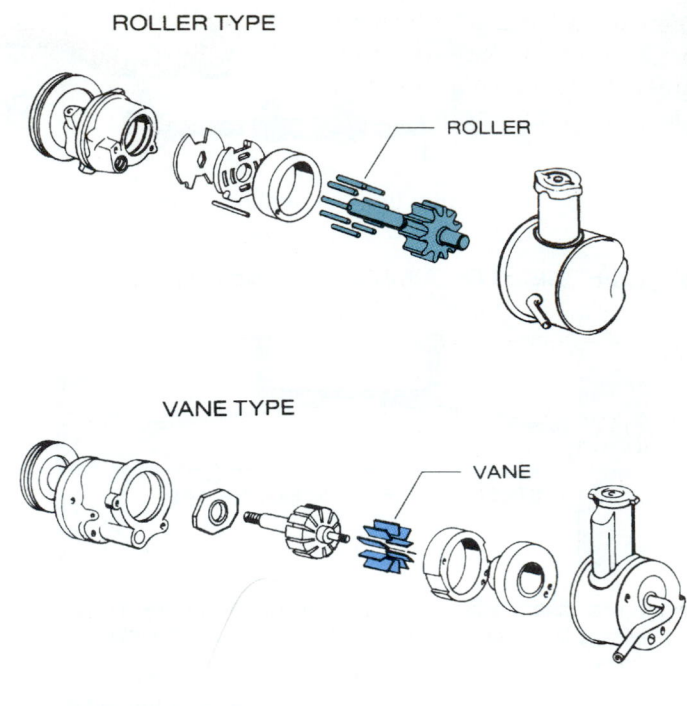

Figure 58.20 Types of power steering pumps. *(Courtesy of Moog Automotive, Inc.)*

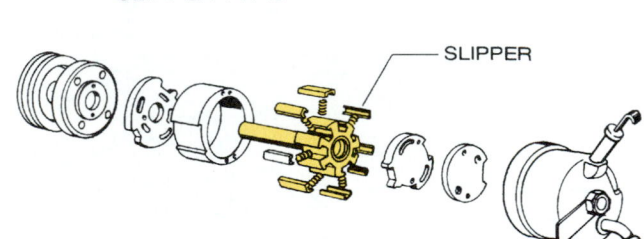

column and the steering gear. The flex coupling also keeps shock from transferring from the road to the steering wheel. When there is a greater angle, a universal joint is used.

■ POWER STEERING

Steering systems found on most cars today are power assisted, although there are still some manual units made. Most power steering is hydraulic, with pressure supplied from the crankshaft by a belt-driven pump (Figure 58.19).

■ POWER STEERING PUMP

When the driver turns the wheel, a steering pump driven by a crankshaft belt supplies hydraulic power to assist steering effort. Three main types of power steering

pumps have been used on cars (Figure 58.20). They are the *roller, vane,* and *slipper* types. Of these, the vane design is the most used. There are several types of power steering pumps. All of them work in the same way. As the pump shaft is turned, oil is drawn into the pump. The oil is squished into a smaller area, which traps it and pressurizes it for delivery to the steering gear.

The power steering pump works under difficult conditions. Steering assist is needed most when the car is stopped or nearly stopped. So a steering pump must deliver sufficient flow to be able to provide steering assist at low engine rpm and idle. Pumps develop more flow as they are turned faster. Therefore, when the pump is turned at cruising rpm, it must divert its fluid back to the inlet side of the pump.

There is a two-stage relief valve in the pump. A *control valve* monitors the turning effort on the wheel to provide the correct amount of assist (Figure 58.21). When the pump turns at low speed, fluid action is as shown in the drawing. At higher speeds, the pump can flow too much fluid. The bypass port opens, allowing the excess fluid to return to the pump intake. When pressure in the steering system becomes too high, the pressure relief valve opens to allow the pressure to bleed off to the fluid intake side of the pump.

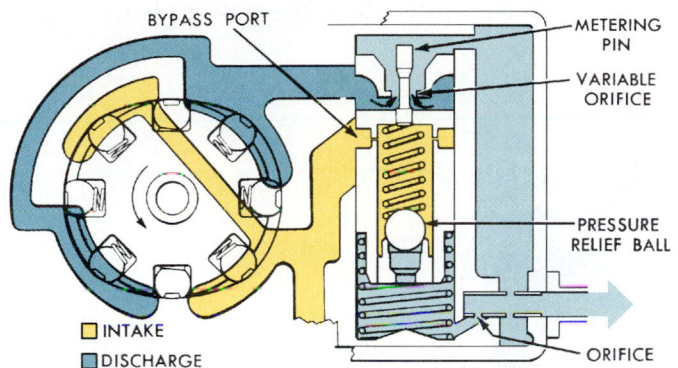

Figure 58.21 A control valve regulates flow in the power steering pump. *[Courtesy of Ford Motor Company]*

Steering Power Consumption

Power steering pumps require a considerable amount of horsepower to operate. Sometimes, on small vehicles, the air conditioning compressor will be shut off at low engine speeds to compensate for the draw of the power steering system. Many late model cars have computer controlled charging systems that also shut off the alternator at idle.

Power assist on very few later model rack and pinion steering units is supplied by an electric motor. These use the battery to supply power to the steering unit.

■ TYPES OF POWER STEERING

Power steering systems are either rack and pinion or conventional recirculating ball units with a hydraulic control system added. Most power steering systems are of the *integral type* (see Figure 58.19). Integral means "part of." This means the power steering components are within the steering gear. Integral units can be either recirculating ball or rack and pinion (Figure 58.22).

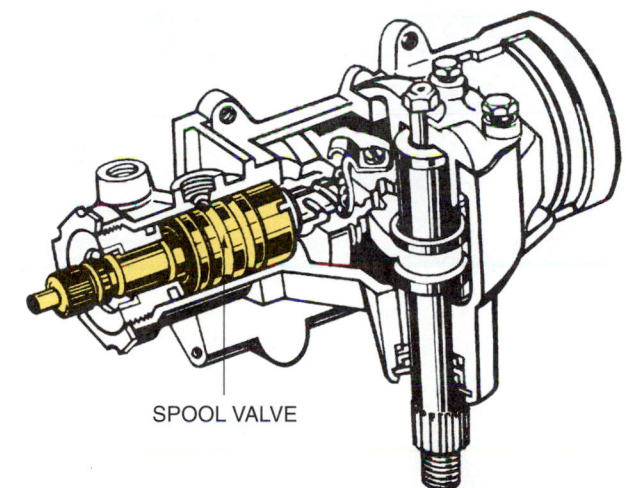

INTEGRAL POWER STEERING GEAR

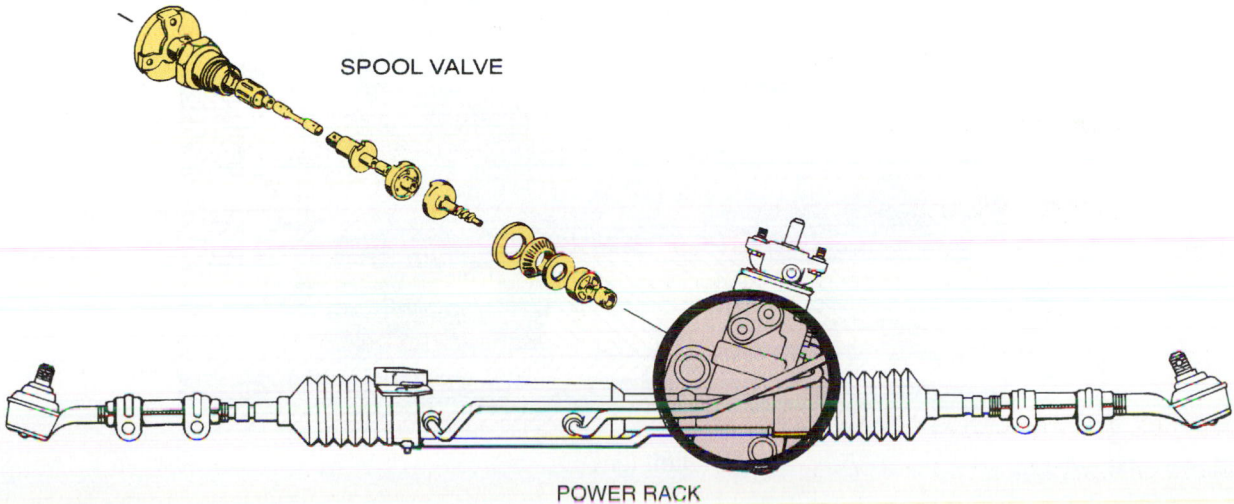

Figure 58.22 Types of integral power steering. *[Courtesy of Moog Automotive, Inc.]*

Integral recirculating ball gear boxes have a ball nut and sector gear, just like a manual gear box. The ball nut is housed within a *power piston.* Pressurized oil enters a chamber on either side of the power piston to provide steering assist (Figure 58.23).

To sense and control power steering assist, gear boxes use either a pivot lever or a torsion bar acting on a **spool valve**. On the *pivot lever type,* turning effort causes

a spool valve to be moved by a pivot lever (Figure 58.24). When the spool valve moves, fluid is directed to one side or the other of the power piston to provide the assist.

On the *torsion bar type,* also called a *rotary valve type,* a sensitive, small torsion bar twists in response to steering effort. This turns a rotary spool valve to direct pressure to the correct side of the power piston (Figure 58.25).

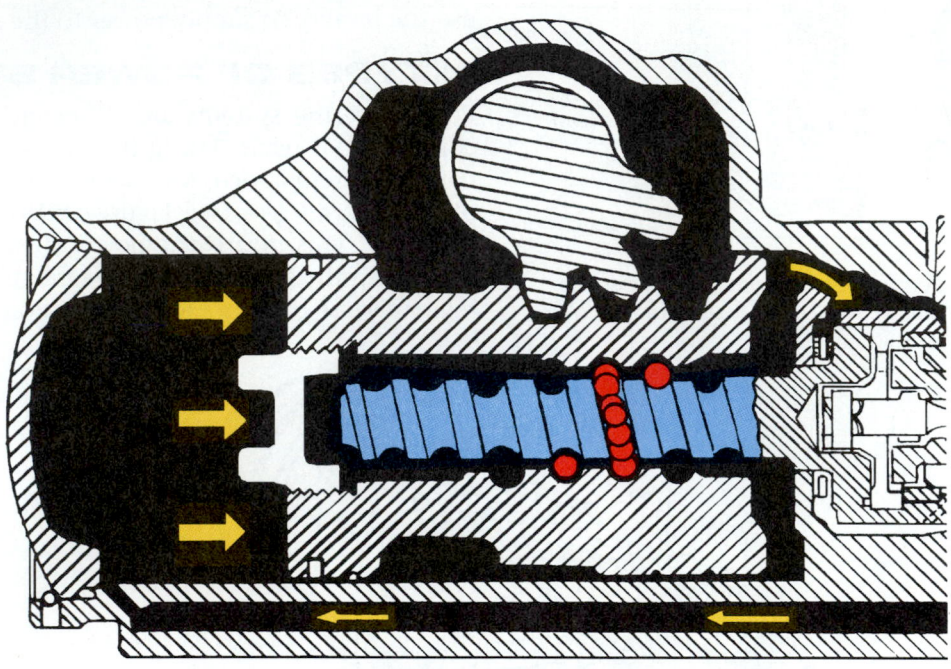

Right turn

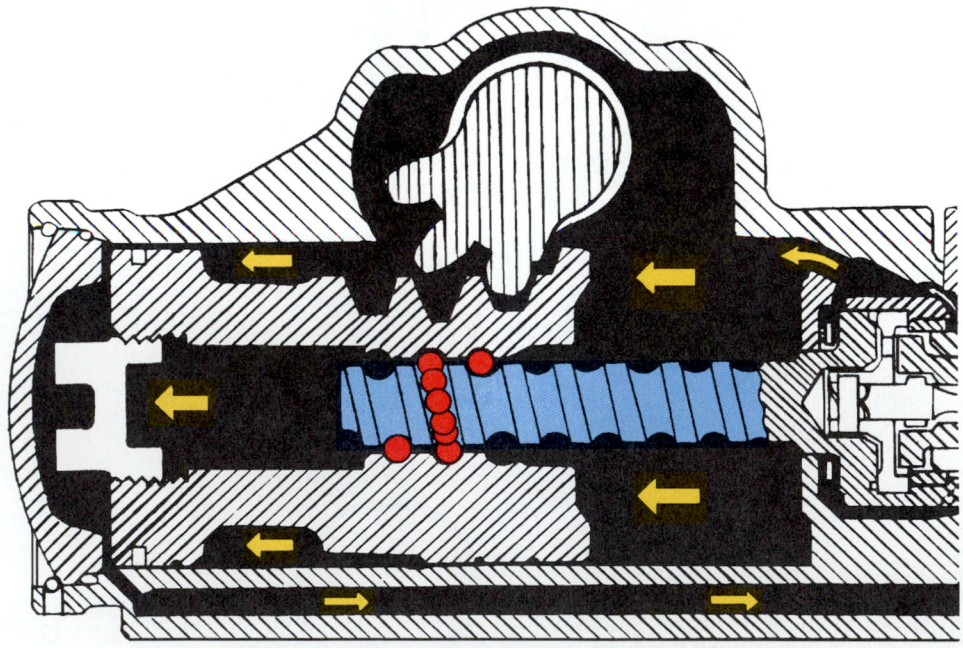

Left turn

Figure 58.23 Pressurized oil enters a chamber on either side of the power piston to provide steering assist. *(Courtesy of Chrysler Corporation)*

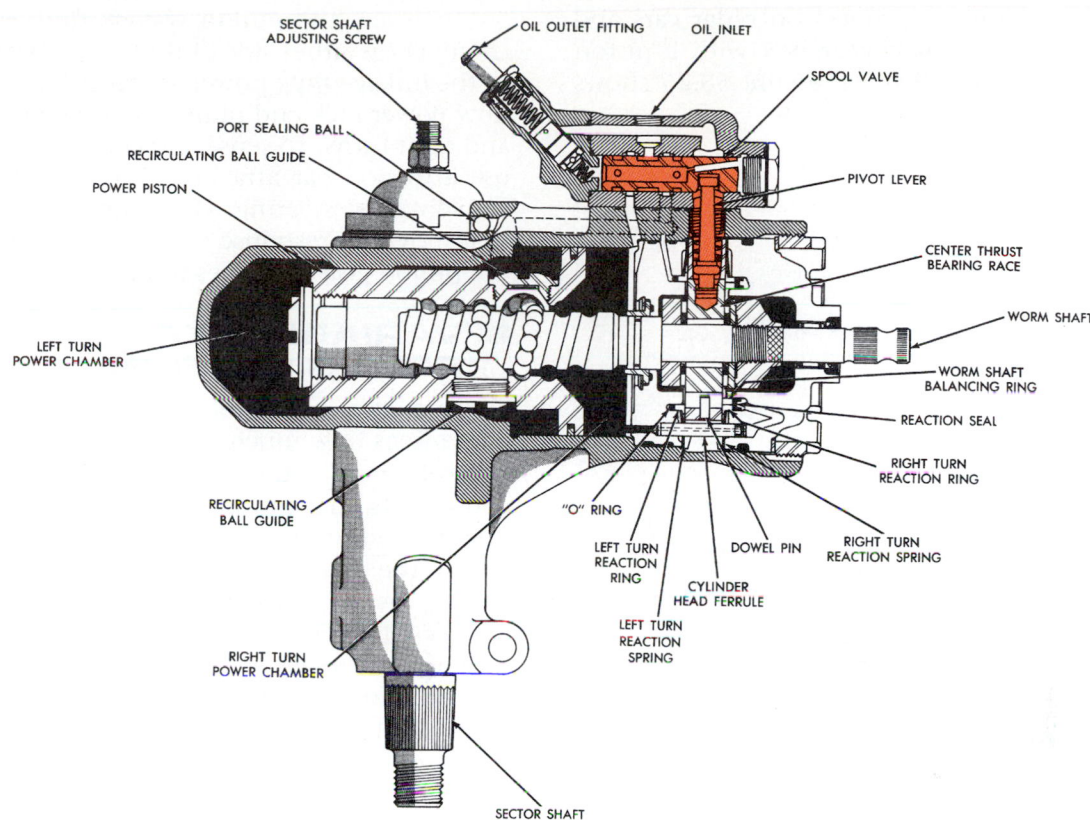

Figure 58.24   A pivot lever type power steering gear. *(Courtesy of Chrysler Corporation)*

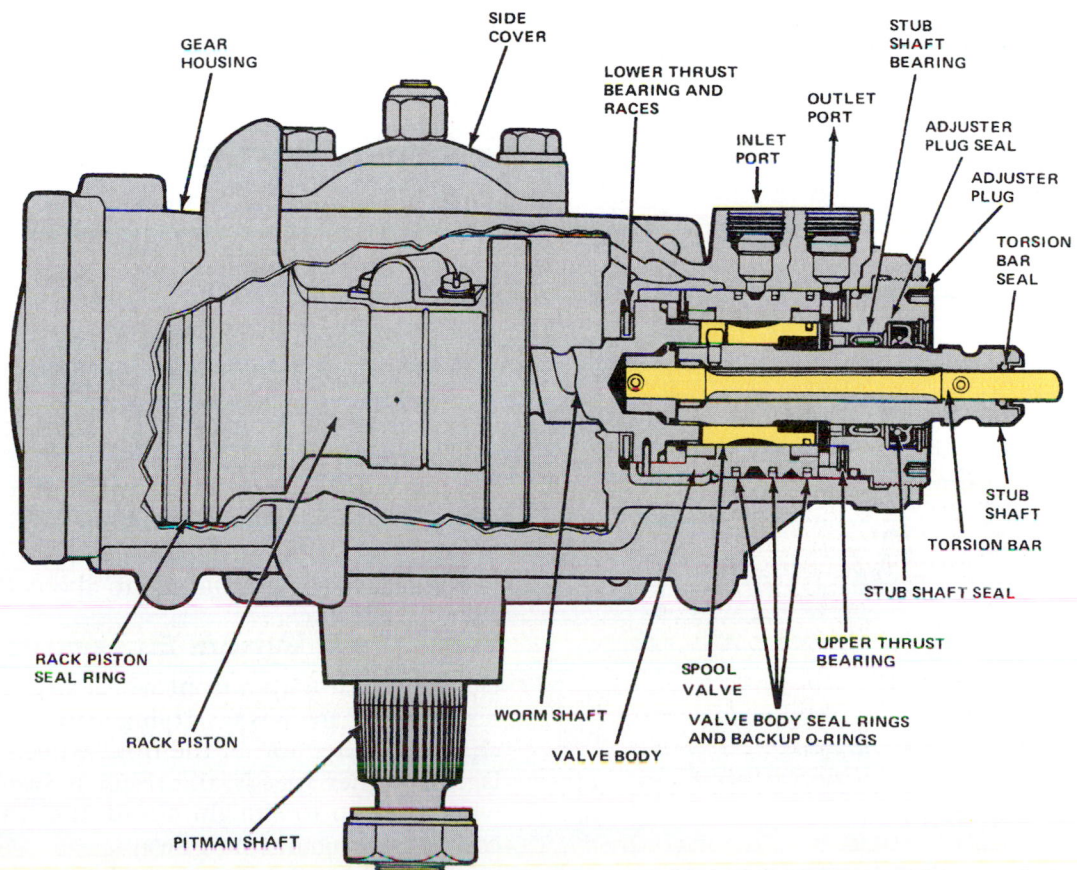

Figure 58.25   A torsion bar-type power steering gear. *(Courtesy of Chrysler Corporation)*

A *linkage-type* gear box, found on older cars and light trucks, uses a standard gear box with a piston attached to the steering linkage. Figure 58.26 shows fluid assist during left and right turns.

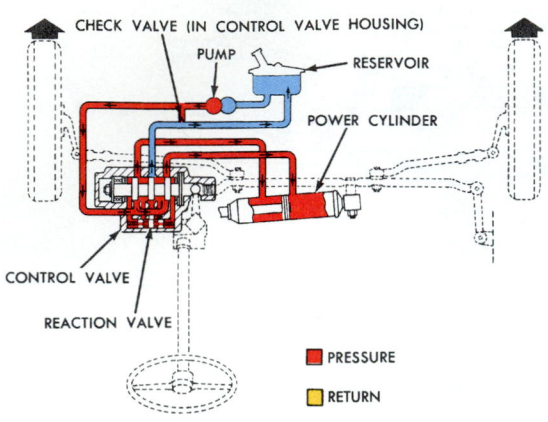

Straight-Ahead

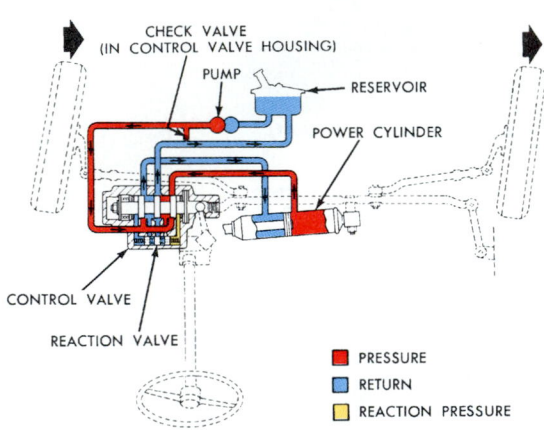

Right Turn

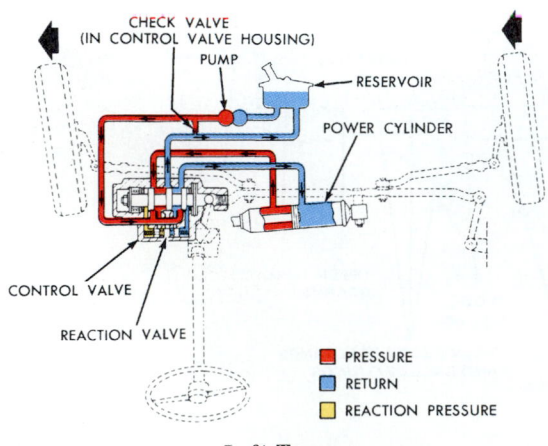

Left Turn

Figure 58.26 Fluid flow in a linkage power steering system. *[Courtesy of Ford Motor Company]*

In a rack and pinion system, fluid is directed to a chamber on either side of the rack in a manner similar to the linkage-type power steering. Figure 58.27 shows how power rack and pinion steering works during left and right turns. To sense and control assist, some racks use a torsion bar attached to the input shaft like in conventional steering. Other units use a spool valve that moves in response to up and down movement of the pinion gear as it tries to drive the rack.

■ VARIABLE EFFORT POWER STEERING

On some late model vehicles, the speed of the vehicle determines how much power assist is given. One type controls fluid output from the pump. Another type controls the amount of fluid pressure available in the power steering gear.

In the pump controlled type, a solenoid controlled by a power steering module (computer) changes fluid flow in the pump control valve. Maximum power assist is at 1500 rpm (fast idle) and the vehicle not moving. As the speed of the vehicle increases, the amount of pump flow is decreased. This increases the steering effort and gives the driver a better feel of the road.

In the steering gear controlled units, the amount of the boost available at the power steering gear is controlled by a solenoid control valve that sends signals to a module (Figure 58.28). The module responds by changing fluid flow in the pump control valve to provide the correct amount of assist.

■ FOUR WHEEL STEERING

Some manufacturers produce four wheel steering systems. In four wheel steering all four wheels turn, improving handling and helping the vehicle make tighter turns.

Generally, at low to medium speeds the rear wheels steer backwards to reduce turning radius, helping with parking. At high speeds they turn in the same direction as the front wheels to improve maneuverability during lane changes.

The front wheels do most of the steering. Rear wheel turning is generally limited to 5°–6° during an opposite direction turn. During a same direction turn, rear wheel steering is limited to about 1°–1.5°.

Why Four Wheel Steering

During a turn with a front wheel steering vehicle, the rear wheels are always trying to catch up with the change in direction of the front wheels. This is called lag. At higher speeds, the result is *sway*. As the front wheels return to straight ahead, the rear wheels must compensate again (more sway).

Centrifugal force moves the rear of the vehicle sideways, causing the tires to slip. This can cause the

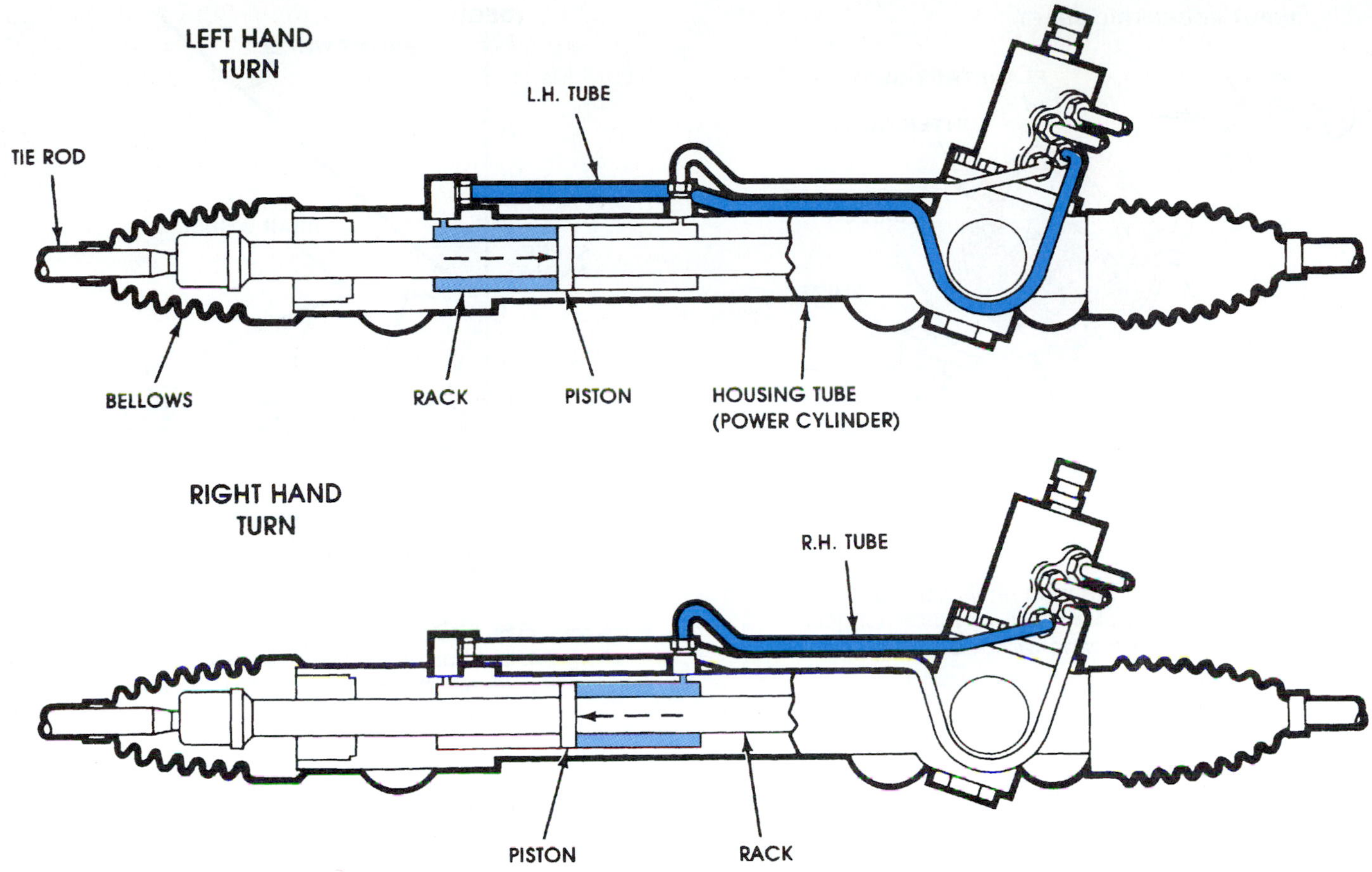

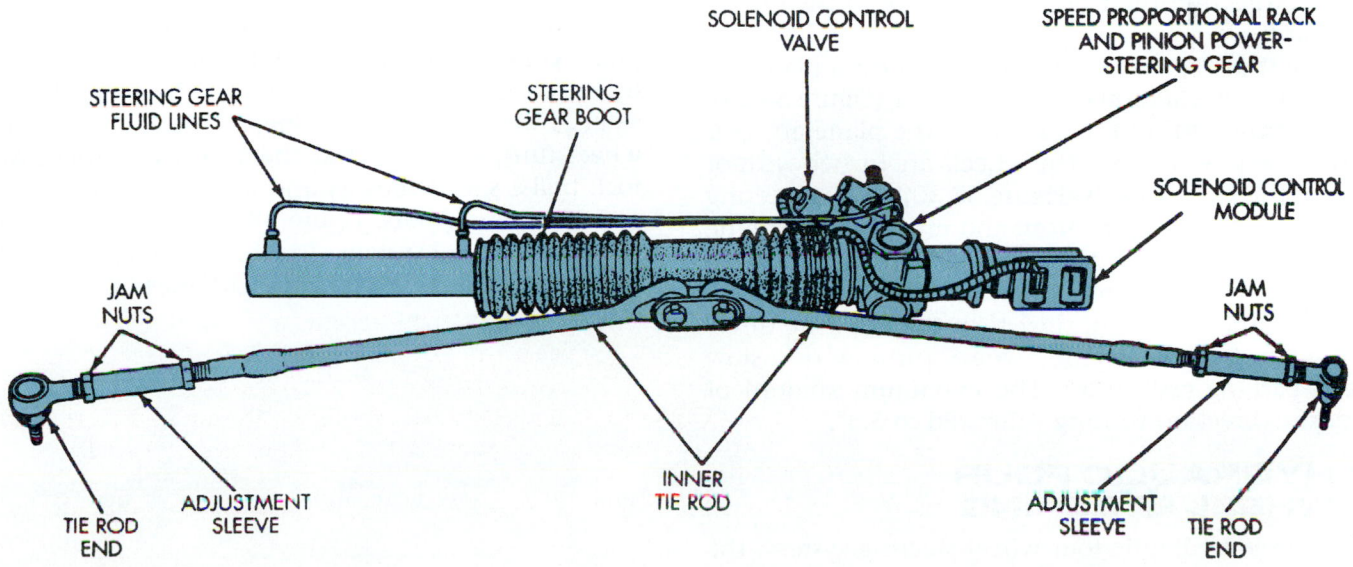

LEFT HAND TURN

TIE ROD

L.H. TUBE

BELLOWS   RACK   PISTON   HOUSING TUBE (POWER CYLINDER)

RIGHT HAND TURN

R.H. TUBE

PISTON   RACK

Figure 58.27 Fluid power assist in a rack and pinion unit. *(Courtesy of Hunter Engineering Company)*

STEERING GEAR FLUID LINES   STEERING GEAR BOOT   SOLENOID CONTROL VALVE   SPEED PROPORTIONAL RACK AND PINION POWER-STEERING GEAR

SOLENOID CONTROL MODULE

JAM NUTS   JAM NUTS

TIE ROD END   ADJUSTMENT SLEEVE   INNER TIE ROD   ADJUSTMENT SLEEVE   TIE ROD END

Figure 58.28 A speed proportional power steering unit. *(Courtesy of Chrysler Corporation)*

car to spin out. During a high speed turn, when the rear wheels are turned the same direction as the front wheels, the entire vehicle moves one way. *Side slip* is reduced and stability is improved. Steering response is faster because the lag between the front and rear wheels is eliminated.

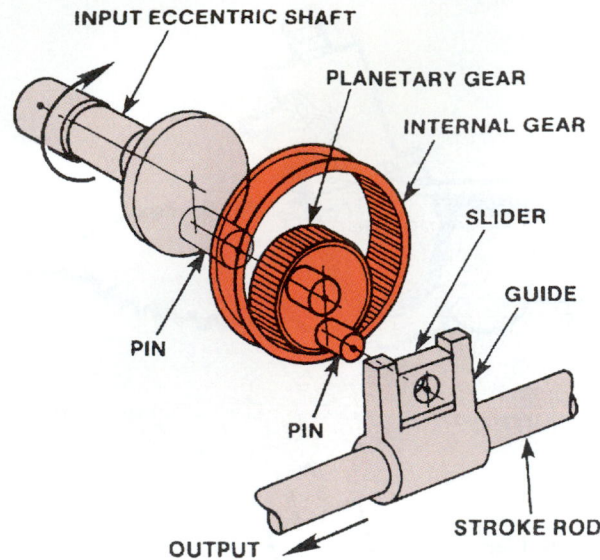

Figure 58.29 A steering shaft from the front turns a planetary gear inside of a stationary internal gear. *(Courtesy of American Honda Motor Co., Inc.)*

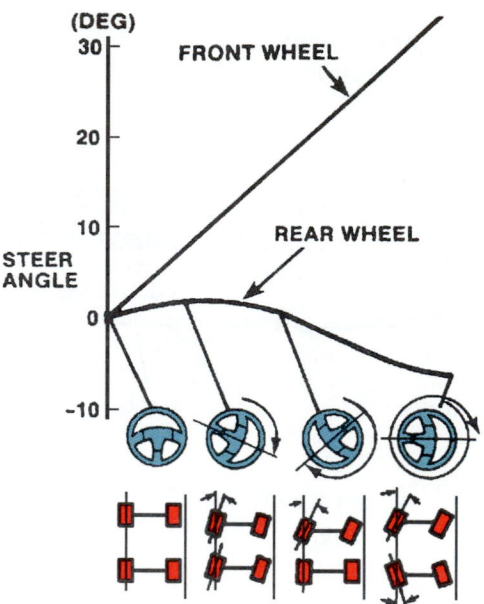

Figure 58.30 As the steering wheel is turned, the rear wheels are turned with or against the front wheels. *(Courtesy of American Honda Motor Co., Inc.)*

◼ TYPES OF FOUR WHEEL STEERING

There are several types of four wheel steering: mechanical, hydraulic, and electric/hydraulic. The *mechanical four wheel steering* system uses two steering gears, one in the front and the other in the rear. A conventional steering arrangement is used in the front. There is a shaft connecting the front steering box to the one in the rear.

The steering shaft from the front turns a planetary gear inside of a stationary internal gear (Figure 58.29). An eccentric pin on the outside of the planetary gear moves the steering rod. The wheels are turned with or against the front wheels (Figure 58.30). As the steering wheel is first turned, the front and rear wheels turn the same direction. After the steering wheel has turned past 120°, they start to straighten out again. At 240° they are straight. Then, they turn the opposite direction with further steering wheel turning (for slow speed parking assistance). The maximum amount of opposite direction turning is limited to 5.3°.

◼ HYDRAULIC FOUR WHEEL STEERING

In a simple hydraulic four wheel steering system, the rear wheels are turned in only the same direction as the front wheels. Turning is limited to 1.5°. The system only operates at speeds in excess of 30 mph in forward gears. A two-way hydraulic cylinder mounted in the rear turns the wheels. Fluid pressure comes from a pump driven by the differential. Fluid is supplied by a fluid storage reservoir in the engine compartment. As the front wheels are turned, pressure is increased to the rear wheel hydraulic cylinder to turn the rear wheels.

◼ ELECTRONICALLY CONTROLLED FOUR WHEEL STEERING

Electronically controlled four wheel steering systems use computer controls to control a hydraulic steering phase control unit (Figure 58.31) or an electric motor driven steering gear (Figure 58.32). The computer considers three factors: road speed, amount of steering wheel turn, and how fast the wheel is turned. Antilock brake speed sensors and a steering angle sensor provide these inputs. Figure 58.33 shows the components of a typical system. During hydraulic failure, the rear wheels are locked in a straight ahead direction for safety.

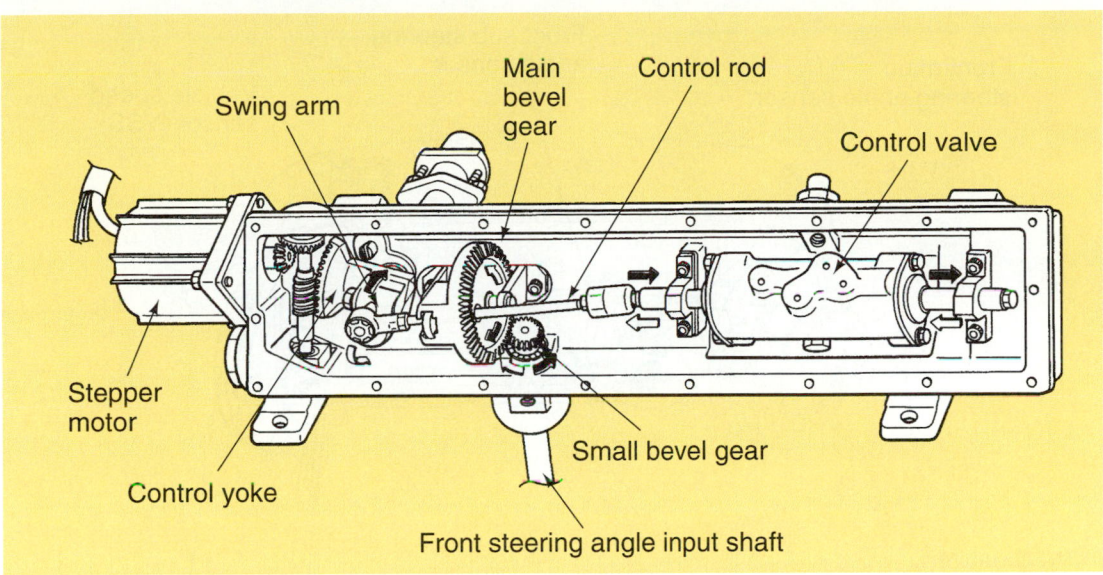

Figure 58.31 A hydraulic steering phase control unit. *(Courtesy of Society of Automotive Engineers, figure from SAE Technical Paper 820253, 1982.)*

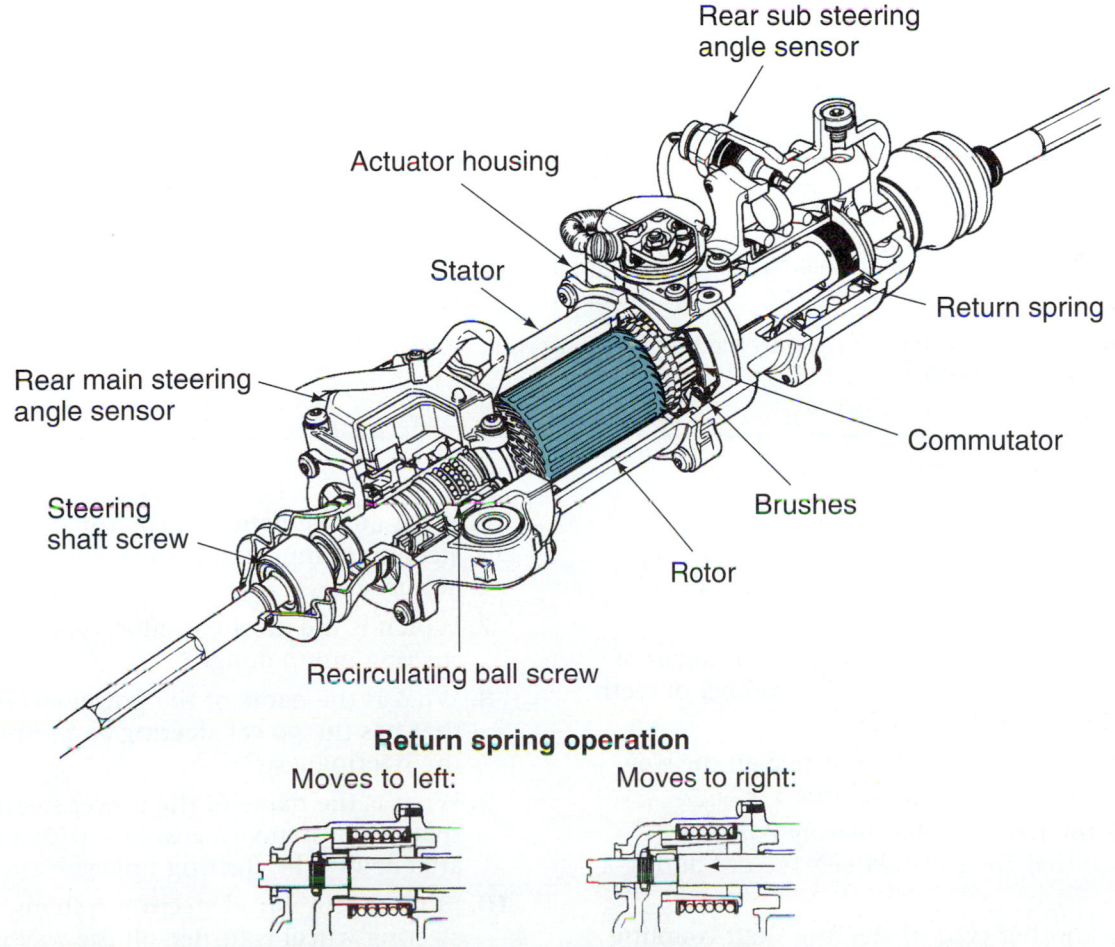

Figure 58.32 An electronically controlled electric motor-driven recirculating ball rear steering rack. *(Courtesy of American Honda Motor Co., Inc.)*

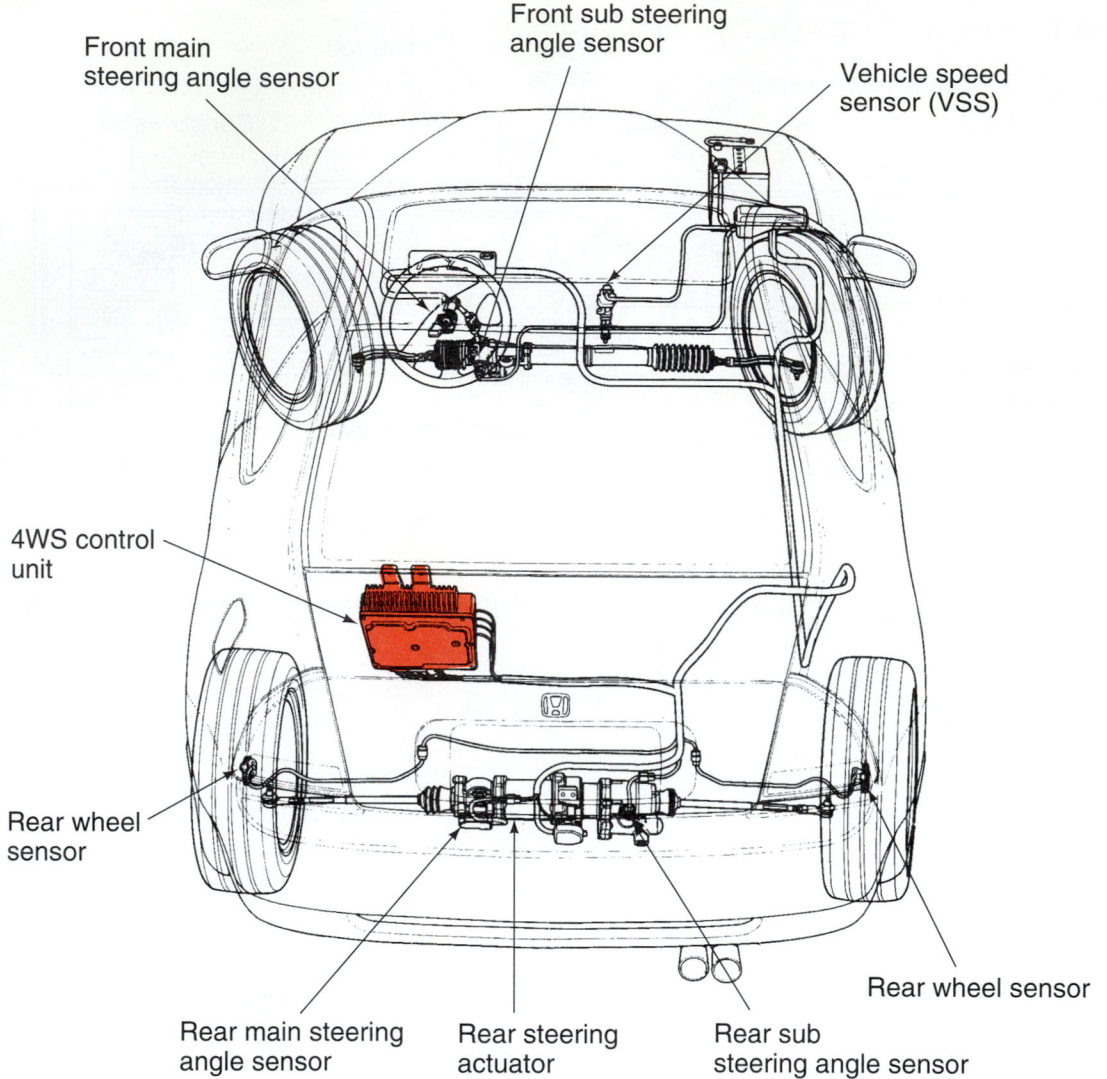

Front main
steering angle sensor

Front sub steering
angle sensor

Vehicle speed
sensor (VSS)

4WS control
unit

Rear wheel
sensor

Rear wheel sensor

Rear main steering
angle sensor

Rear steering
actuator

Rear sub
steering angle sensor

Figure 58.33 An electronically controlled four wheel steering system. The computer control unit is in the trunk. *(Courtesy of American Honda Motor Co., Inc.)*

■ REVIEW QUESTIONS

1. One of the two most common types of automotive steering gears is the recirculating ball and nut. What is the name of the other one?

2. What is the term that compares the number of teeth on the driving gear to the number of teeth on the driven gear?

3. When the steering wheel is turned all the way in one direction, it stops against a _____.

4. What is the name of the steering part that is angled so that the front wheels toe out during a turn?

5. What is another type of steering shaft coupling besides a flex coupling?

6. What are the names of the three main types of steering pumps that have been used on automobiles?

7. Which is the most common type of power steering pump design?

8. What is the name of the power steering design that has the power steering components within the steering gear?

9. What is the name of the power steering design that uses a standard gear box with a piston attached to the steering linkage?

10. In most four wheel steering systems, when the steering wheel is turned all the way in one direction, do the rear wheels turn the same way?

■ ASE STYLE REVIEW QUESTIONS

1. Technician A says that the outer wheel must turn sharper during a turn. Technician B says that front tires toe outward during a turn. Who is right?

   a. Technician A   **b.** Technician B

   c. Both A and B   **d.** Neither A nor B

2. Technician A says that steering assist is needed most when the car is stopped or nearly stopped. Technician B says that the power steering must be able to provide more steering assist at low engine rpm and idle. Who is right?

   a. Technician A   **b.** Technician B

   c. Both A and B   **d.** Neither A nor B

3. Technician A says that on some power steering systems, the faster the vehicle is driven, the less power assist is given. Technician B says that in four wheel steering systems, the front and rear wheels turn the same amount. Who is right?

   a. Technician A   **b.** Technician B

   c. Both A and B   **d.** Neither A nor B

4. Technician A says that on most four wheel steering systems, at low to medium speeds the rear wheels steer backwards to reduce turning radius. Technician B says that during hydraulic failure in a four wheel steering system, the rear wheels are locked in a straight ahead direction. Who is right?

   a. Technician A   **b.** Technician B

   c. Both A and B   **d.** Neither A nor B

5. Technician A says that the most popular steering design in use with the long and short arm suspension is the rack and pinion. Technician B says that a "fast" steering ratio is about three turns lock to lock. Who is right?

   a. Technician A   **b.** Technician B

   c. Both A and B   **d.** Neither A nor B

Steering Service

■ INTRODUCTION

In normal use, with good lubricants and seals, steering parts remain separated by lubricants. When seals are damaged and a lubricant leaks out, parts can rub against each other. Also, road hazards such as potholes can cause hard impacts to steering parts. This can result in immediate damage or damage that will result in later wear. Belts and hoses can fail, resulting in operating problems with the steering system.

■ FLUID LEVEL CHECKS

Fluid should be hot when checking its level. With the engine idling, turn the wheel several times in each direction to raise the temperature. Sometimes a system has a reservoir located in a remote location from the pump.

■ TYPE OF FLUID

Most manufacturers recommend either special power steering fluid or automatic transmission fluid in their power steering systems. Honda systems must use a special type of fluid. Installing the wrong type of fluid can damage seals within the system. For further information on belts and hoses refer to Chapters 22 and 23).

■ DIAGNOSING STEERING PROBLEMS

Steering and suspension problems can be caused by play, looseness, incorrect height, and wheels not aligned to specifications. Wheel alignment theory is discussed in a later chapter, but some of the terms will be discussed here when they apply to a particular problem.

■ NOISE DIAGNOSIS

Noise from the steering system can be due to several things. A loose power steering belt can result in a jerky feeling during a turn. It can also squeal when the wheel is turned. Belt squeal will be especially evident when the wheel is turned all the way to the end of its travel against a steering lock. If allowed to continue slipping without a belt tension adjustment, the belt will become glazed. The pulley can also wear out.

When the steering wheel is turned against a lock, the pump must bypass the extra pressure that develops. A "whirring" sound is normal as bypass occurs.

Noise can also happen due to a lack of fluid in the power steering reservoir. This will cause a whine under load when the steering wheel is turned.

Loose parts will cause a clunk when going over bumps or when turning the steering wheel from side to side.

■ STEERING PART INSPECTION

The condition of the steering system must be inspected before doing a wheel alignment. Excessive vibration of front end parts can allow the wheels to move on their own. This can cause vibration, noise, tire wear, and unsafe driving conditions.

Remember to inspect the wheel bearings to see that they are not too loose or too tight. Also, check to see if they feel rough before performing other tests on the front end. If a wheel bearing is loose, the wheel alignment equipment will not be able to make accurate readings.

HARD STEERING

Hard steering can be caused by binding in the steering linkage, steering box problems, low tires, or incorrect wheel alignment settings. Always check power steering belt tension and fluid level. These are common causes of hard steering.

TIRE WEAR

Steering and suspension problems often result in unusual tire wear (see Figure 54.3). Worn parts can cause scalloped or gouged tire wear (see Figure 14.5). Tire wear can also be due to improper alignment adjustments (see Chapter 61).

STEERING LINKAGE INSPECTION

The fastest way to discover obvious looseness is by performing a "dry park check." Have an assistant turn the wheel from side to side with the tires on the ground or supported by a wheel contact lift. Look for loose parts while observing the steering and suspension as the wheels turn. A vehicle with power steering must have its engine running for this test. Otherwise, there can be looseness in the steering linkage at the steering gear.

While the car is in the air during a lube inspection, the steering linkage can be inspected. With the tires raised and hanging free, turn the steering wheel through its full range of travel. Check to see that there is no binding in the steering linkage or gear box.

STEERING GEAR LOOSENESS

Excessive steering wheel free play can be caused by a worn steering column flexible coupling, worn steering linkage, or a worn or misadjusted steering gear. Steering gear box looseness is usually adjustable.

Check steering wheel free play with the engine off. Power steering cars should have the engine on. Move the wheel back while checking to see how far it moves before the wheels start to move.

NOTE: *Check the steering gear box to see if it feels smooth throughout its entire travel. Check it in the straight ahead position for binding at the center position and from lock to lock to see that there is no looseness or rough feel.*

PARALLELOGRAM INSPECTION

Tie rods are held in position by the pitman arm, idler arm, and center link. Parts should allow pivoting and turning, but prevent movement. They are tested by attempting to move them.

Pivot sockets are filled with lubricant and sealed with rubber seals. Check sockets for looseness, or damage to the seal.

NOTE: *When checking for steering looseness, be sure that the steering wheel is in the straight ahead position. This is where the most wear would tend to be.*

Idler Arm Inspection

An idler arm can have either one or two pivoting connections. If it does not have a pivot socket at the outside end, the socket will be a part of the center link. To check the idler arm, grasp the center link at a point that is as near to the idler arm as possible. Push it firmly up and down while looking for movement (Figure 59.1). Movement here can result in a change in toe-in of the front wheels, causing tire wear.

Pitman Arm Inspection

The pitman arm is the lever that attaches the steering linkage to the steering gear box. Some pitman arms do not have a pivot socket at the center link end. These are hardly ever replaced unless they have been bent or damaged in a collision.

Tie Rod Inspection

Inspect tie rods for wear by moving them vigorously to see if they are loose (Figure 59.2). There are two types of tie rod sockets. One is preloaded, with no clearance. The other is spring loaded and can be compressed vertically. One test for a tie rod is to compress it (Figure 59.3). Spring-loaded sockets should have firm spring pressure when compressed all the way. Preloaded sockets should not give at all.

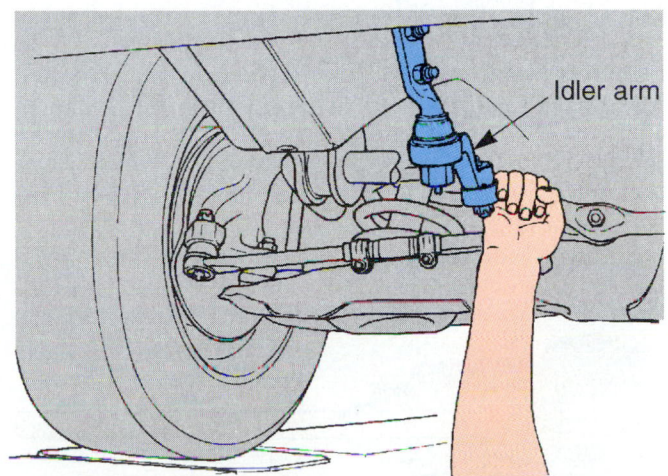

Figure 59.1 Checking an idler arm. *(Courtesy of Moog Automotive, Inc.)*

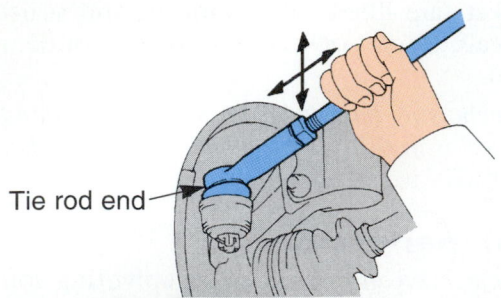

Figure 59.2   Checking for tie rod end wear.

Figure 59.3   Testing a compressible tie rod end.
[Courtesy of McQuay Norris]

■ RACK AND PINION STEERING LINKAGE INSPECTION

The inside tie rod is housed within the bellows boot (Figure 59.4). Feel for looseness by grasping the tie rod through the bellows boot. There should not be any looseness between the stud and the housing. The steering linkage must be in its usual position for this test. If the tie rod is hanging down in its socket, it could be bound up, preventing an accurate check.

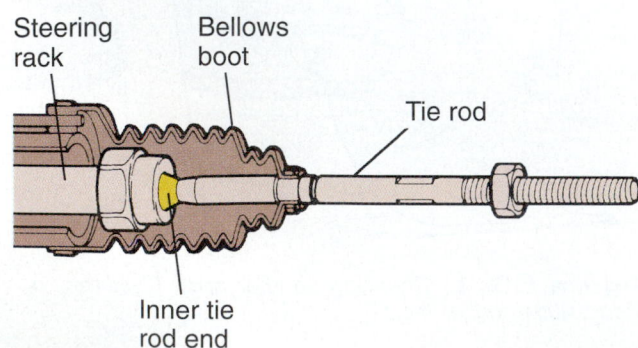

Figure 59.4   Location of inner tie rod end. *[Courtesy of Moog Automotive, Inc.]*

■ STEERING LINKAGE REPAIRS

Steering linkage parts are locked by wedging them against each other with beveled connections called "tapers" (Figure 59.5). Locknuts, usually castle nuts with cotter pins, provide additional protection. When disassembling tapers to service or replace parts, there are several ways to get them apart:

■ A *pickle fork* can be used if the part will not be reused (Figure 59.6). The pickle fork will probably cut the seal and ruin the part.

■ Tapers can usually be separated using a puller (Figure 59.7).

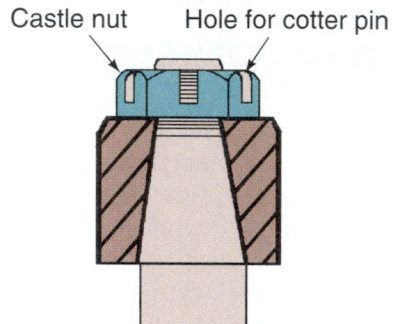

Figure 59.5   A steering linkage taper connection.

Figure 59.6   A pickle fork adapter for an air chisel.
[Courtesy of Chicago Pneumatic]

Figure 59.7   Using a tie rod end puller. *[Courtesy of Moog Automotive, Inc.]*

■ Linkage parts can also be separated by deforming the outside of the taper using a hammer (Figure 59.8). This is the best way if the part is to be reused.

■ IDLER ARM REPLACEMENT

During replacement of an idler arm, check to see if it is one of the adjustable types (Figure 59.9). If the idler arm and pitman arm are not level to each other, the car will experience bump steer. Bump steer causes toe to change when going over bumps, which will cause tire wear (see Figure 61.8).

■ PITMAN ARM REPLACEMENT

Removal of the pitman arm requires a special puller (Figure 59.10). It must be replaced in the same position on the splines of the steering gear. Be sure to check its position before removing it. Pitman arms often have a "blind spline." One wide spline ensures that the pitman arm can only be installed in one position.

Figure 59.10 A pitman arm puller. *(Courtesy of Moog Automotive, Inc.)*

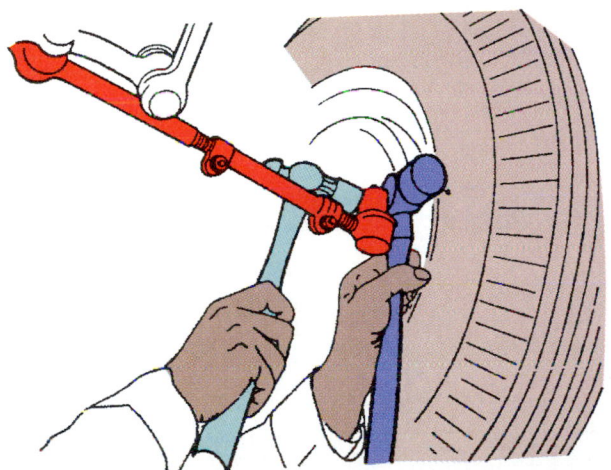

Figure 59.8 Using two hammers to loosen a steering taper connection. *(Courtesy of General Motors Corporation, Service Technology Group)*

■ TIE ROD END REPLACEMENT

Tie rods ends are replaceable, whether the steering system is a rack and pinion or parallelogram steering system. Tie rod ends are threaded to provide a means of adjusting toe. When a tie rod is replaced, measure the old tie rod assembly before disassembling it. An approximate toe-in adjustment can be made to the new one prior to installation (Figure 59.11). When removing the old tie rod end, count the number of turns it takes to remove it. Then, turn the new tie rod end onto its threads the same number of turns.

Some vehicles have tie rods that appear to be the same on both ends. The difference is in the threads. One end has left hand threads and the other has right hand threads. It is possible to install the entire shaft backwards

Figure 59.9 These slotted holes are for mounting an adjustable idler arm. *(Courtesy of Moog Automotive, Inc.)*

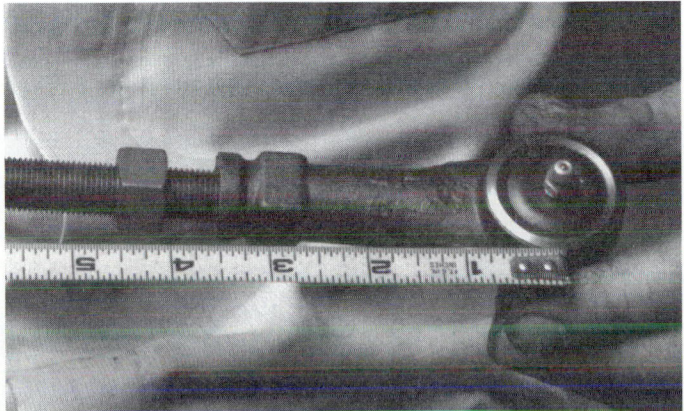

Figure 59.11 Measure the tie rod before disassembling it. *(Courtesy of Moog Automotive, Inc.)*

(which will work). Mark the tie rod to identify the inner or outer end before removing it.

CASE HISTORY *A technician ordered a tie rod end for a small pickup truck. He specified that the tie rod end was for the outside left. When the new tie rod was delivered from the parts store, he attempted to install it. The threads did not match up. Further investigation showed that the tie rod was for the opposite side. The parts store double-checked and found that they had sent the correct part. The tie rod had been previously installed backwards. This had no effect on the operation of the steering but resulted in some confusion during service.*

Before tightening tie rod clamps, check to see that they are in good condition and are positioned properly so they can be clamped tightly (see Figures 61.23 and 61.24). Before doing a front end adjustment, spray penetrating oil on the threads of the tie rods. Do this during the steering linkage inspection so the lubricant has time to soak in.

■ RACK AND PINION TIE RODS

Outer rack and pinion tie rod ends are serviced the same way as other tie rod ends. There are several ways that the inner tie rod socket is held to the ends of the rack. Be sure to follow the instructions that come with the replacement part.

A typical Asian import uses a *claw washer retainer*. The washer has inner tabs that fit into a recessed area of the rack (Figure 59.12). Then, deform the outside of the washer with a hammer and punch so that it conforms to the flats on the outside of the tie rod socket.

Three other types of retaining methods are shown in Figure 59.13.

■ One method uses a **jamb nut**. Some of these types use a thread adhesive. The rack must be held from turning while tightening the jamb nut (Figure 59.14).

■ Another method uses a special washer that is staked into a flat area in the end of the rack.

■ When a roll pin is used to hold the tie rod to the

rack, it must be drilled to remove it. After installing the new tie rod end, a new hole is drilled for a new pin 90° to the original one.

New bellows boot kits are available. Urethane boots are used by many domestic manufacturers. These are superior to neoprene boots.

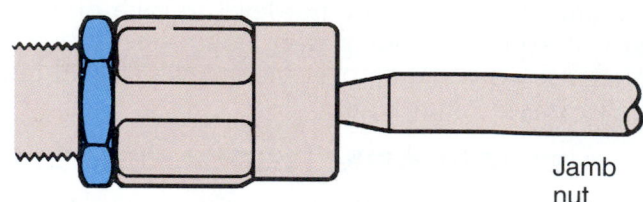

Jamb nut

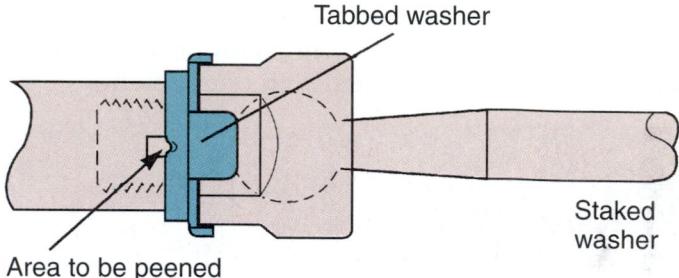

Tabbed washer

Area to be peened

Staked washer

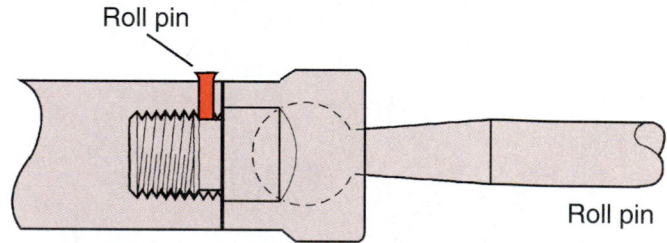

Roll pin

Roll pin

Figure 59.13 Three types of tie rod retaining methods. *[Courtesy of Moog Automotive, Inc.]*

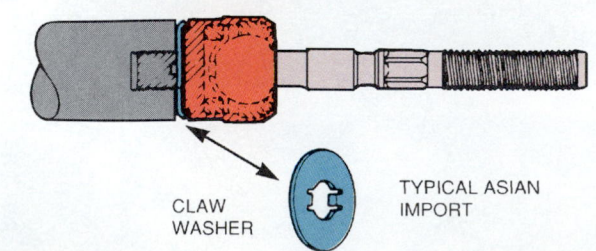

CLAW WASHER

TYPICAL ASIAN IMPORT

Figure 59.12 An inner tie rod end held by a claw washer. *[Courtesy of Moog Automotive, Inc.]*

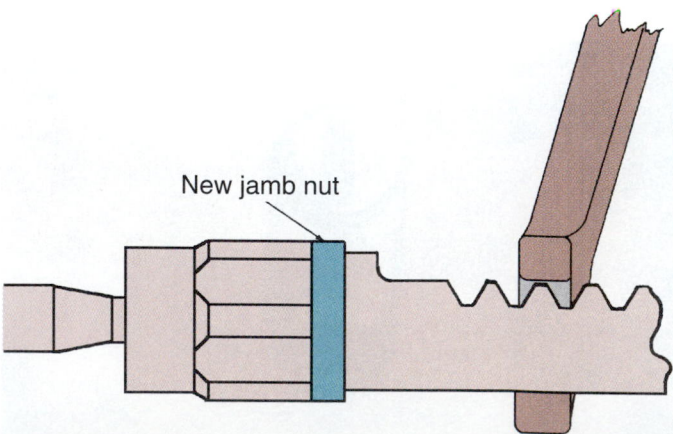

New jamb nut

Figure 59.14 Hold the rack with a wrench while tightening the jamb nut. *[Courtesy of Federal-Mogul Corporation]*

STEERING WHEEL, COLUMN, AND AIR BAG SERVICE

Removal of the steering wheel is sometimes necessary. This could be because a turn signal switch requires replacement or a horn does not work.

NOTE: *On most steering systems, when the steering wheel is off-center adjustment is usually made by turning the tie rods. It is **not** done by removing the steering wheel.*

AIR BAGS

Many newer vehicles come equipped with air bags, also called **supplemental inflatable restraints (SIR)** to protect occupants of a vehicle during a crash (Figure 59.15). Before attempting to remove a steering wheel on one of these vehicles, consult a service manual. Accidentally deploying (activating) an air bag can result in an injury or death.

NOTES:

- *Air bags cannot be reused and they can be relatively expensive, depending on what parts the manufacturer stipulates should be replaced with them.*
- *Air bags contain rocket propellant. They are shipped in cartons that say "**pyrotechnics**" on their label.*
- *When an air bag deploys, it opens so fast that only a very good video camera can freeze the action enough to slow it down for frame by frame investigation.*
- *It is not unusual for a windshield to be broken by the passenger side air bag.*
- *A driver who survives a crash with an air bag often has burn marks on his or her arms from holding the steering wheel when the bag went off.*

The driver's side air bag is located in the steering wheel. It is usually easily removable by loosening two or three screws beneath the steering wheel.

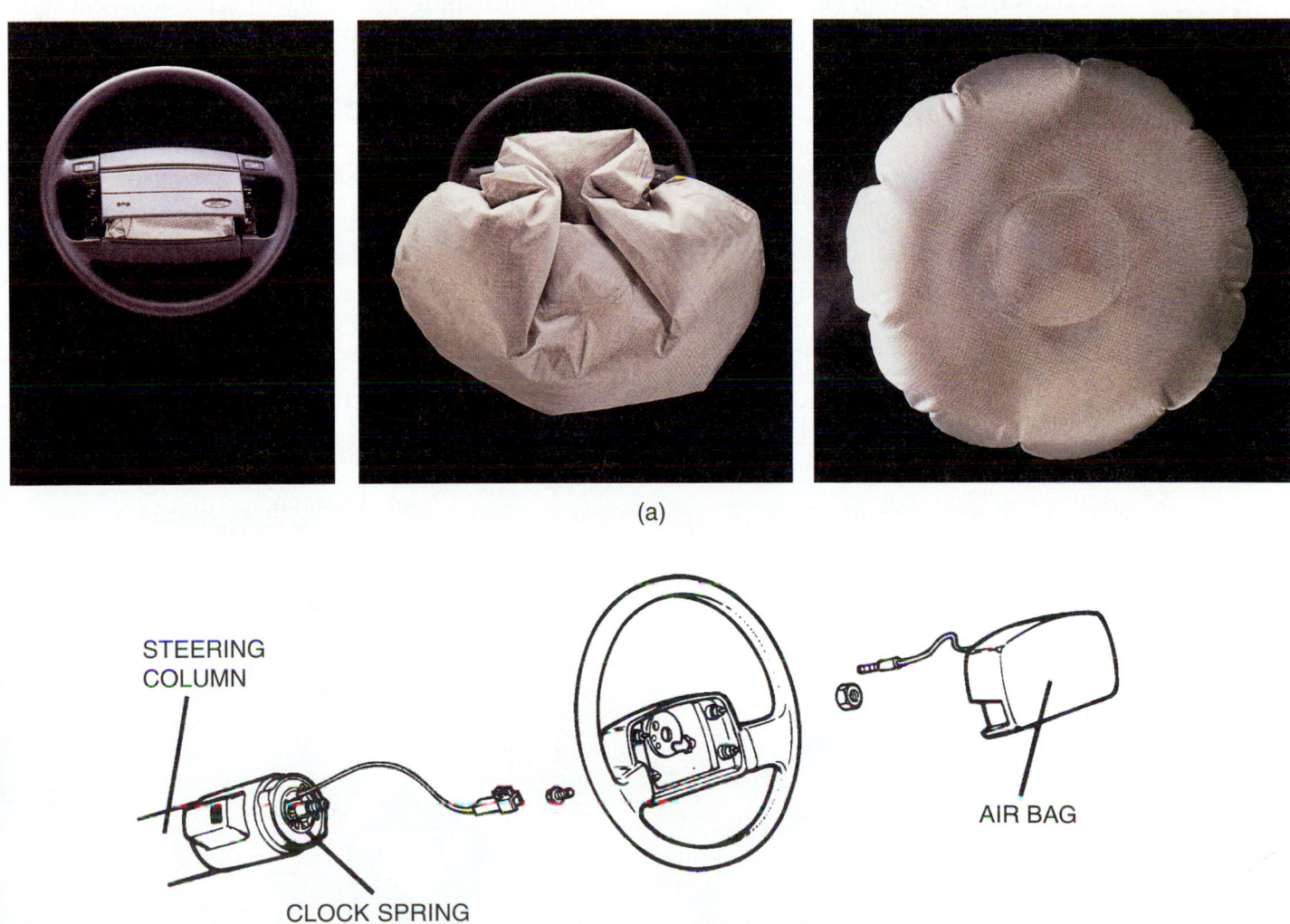

(a)

(b)

Figure 59.15 (a) Various stages of air bag inflation. *(Courtesy of TRW, Inc.)* (b) Steering column parts of an air bag. *(Courtesy of Moog Automotive, Inc.)*

Air Bag Clock Spring

The driver's side air bag has a **clock spring** or **spiral cable** (Figure 59.16) that sends an electrical signal from the air bag sensors to the steering wheel in case of an accident. The spring is called a clock spring because of its resemblance to the spring in a mechanical clock.

SHOP TIP When removing a steering column assembly, turn the wheel all of the way against the steering lock and remove the ignition key, locking the steering column. During reinstallation, be sure the wheels are positioned in the same full travel position.

SAFETY NOTE
- When removing a steering wheel on a car with an air bag, be sure that the air bag wiring harness is disconnected. Simply disconnecting the battery will not ensure that the bag will not deploy. There is a backup capacitor that will deploy the air bag in the event the battery becomes disconnected during a crash. After a battery has been disconnected for a few minutes, the air bag can be removed. *Disconnect* the wiring harness to be safe.
- Always store an air bag facing up on the workbench. If it accidentally deploys, an entire steering column could be shot into the ceiling. If the air bag is facing upward, the chance of injury to bystanders is less.

The clock spring has a certain amount of travel from fully wound to fully unwound. It is in its midway position when the steering wheel is straight ahead. When the wheel is turned one way against a steering lock, it unwinds. Turning the wheel the other way winds it up.

If the spring is not installed in the correct position, it will break when the wheel is turned in both directions. The air bag dash light will come on and the spring will have to be replaced. A new clock spring comes with an aligning mark so it can be installed properly.

■ STEERING WHEEL SERVICE

Pulling the Steering Wheel

Do not hammer on the end of the steering shaft when removing the steering wheel. This can damage the column. Install a puller in the threaded holes in the steering wheel (Figure 59.17). If there are no alignment marks, make some with a centerpunch and hammer. When reinstalling a steering wheel, double-check to see that it is in the neutral position (half way between the steering locks).

■ STEERING COLUMN SERVICE

Steering column service can be complicated and the proper service manual should be consulted before repairs are attempted.

Flexible Joint Replacement

A worn flex coupling (see Figure 58.18) can cause looseness in the steering wheel. To replace a flex coupling, loosen its bolt at the steering gear. Figure 59.18 shows the locations of the parts. Remove the steering column mounting bolts where the column mounts to the instrument panel inside the passenger compartment.

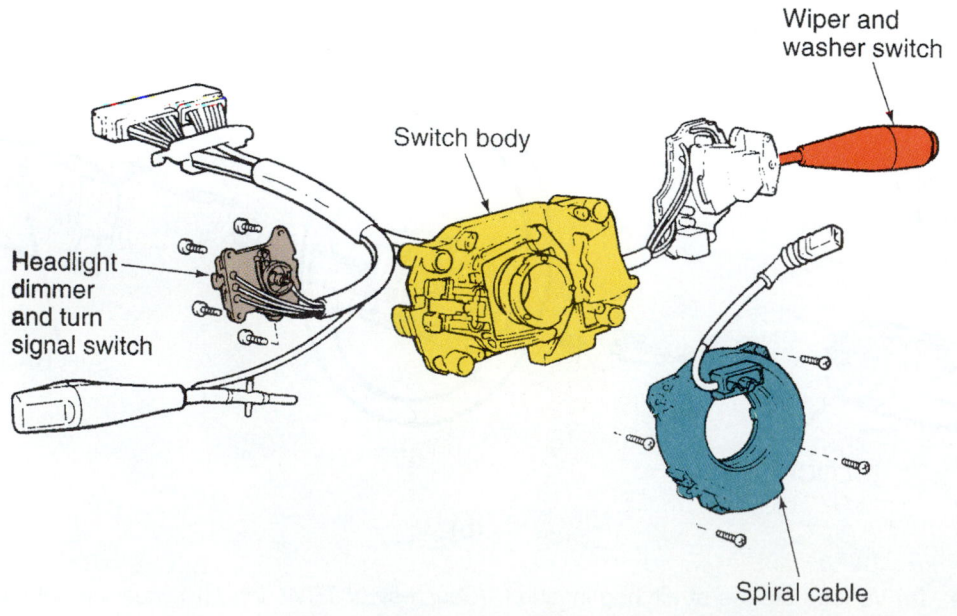

Wiper and washer switch

Switch body

Headlight dimmer and turn signal switch

Spiral cable

Figure 59.16 The spiral cable is commonly called a clock spring.

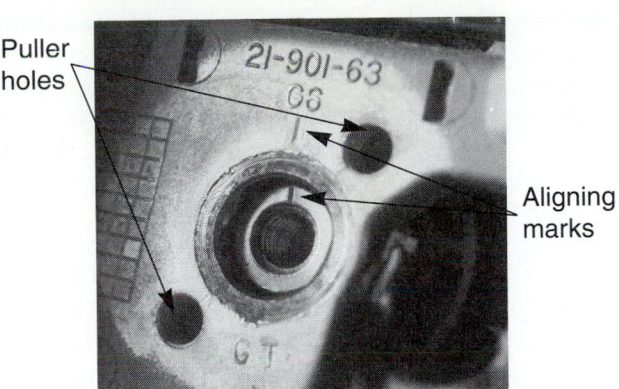

Puller holes

Aligning marks

Figure 59.17 Removing a steering wheel with a steering wheel puller.

Move the column far enough toward the rear of the car for the coupling to be removed from the steering gear.

STEERING GEAR SERVICE

Steering gears are generally reliable. Problems can be caused by collisions or deep potholes. Rebuilt steering gears are available.

Steering gears are sometimes rebuilt by the technician in the repair shop. Recirculating ball and nut

manual steering gears are simple to rebuild. There are two adjustments to make (Figure 59.19). One of them is a **preload adjustment** to the bearings at the ends of the steering shaft (worm shaft). These are ball bearings, which call for a small amount of load to be on them. The load adjustment is measured with an inch-pound torque wrench (Figure 59.20).

The second adjustment is to the **high point**. The sector gear teeth in a steering gear have a high point where the gear teeth come closer together when the car is travelling straight ahead (Figure 59.21). The high point must be centered when the vehicle travels straight ahead so the steering response is not faster in one direction than the other.

MANUAL RACK SERVICE

It is important that the steering wheel on a rack and pinion be properly centered. This is for two reasons:
- The center of the rack is hardened.
- If the tie rods are of different lengths, the car can steer to one side farther than it can to the other.

RACK AND PINION LOOSENESS

A steering damper holds tension against the steering rack. A rack gear preload adjustment can be made (see Figure 58.6), but be sure to check the service manual for the correct procedure and specification. If the rack is worn in the center, tightening this adjustment can result in binding steering at the outsides during turns.

POWER STEERING SYSTEM SERVICE

Reservoir and Hose Service

The fluid reservoir is usually made of sheet metal. Sometimes it is attached to the pump (integral) and sometimes it is in a remote location, attached by hoses. When the reservoir is attached, O-rings seal it to

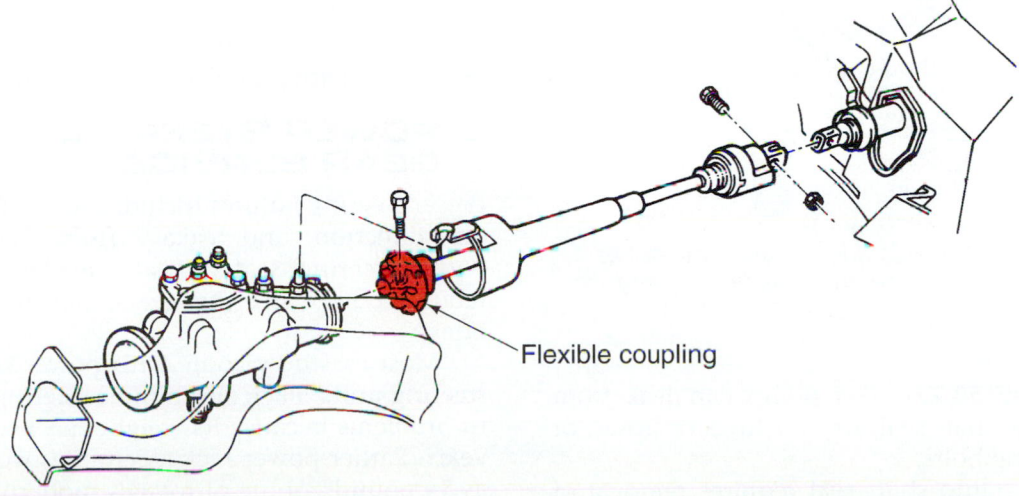

Flexible coupling

Figure 59.18 Parts of a steering column. *(Courtesy of General Motors Corporation, Service Technology Group)*

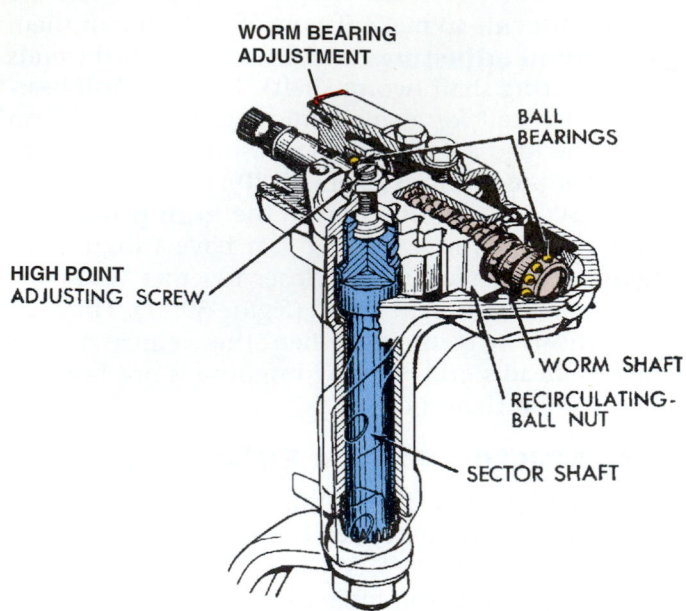

Figure 59.19 High point and worm bearing preload adjustments on a recirculating ball nut steering gear. *(Courtesy of Chrysler Corporation)*

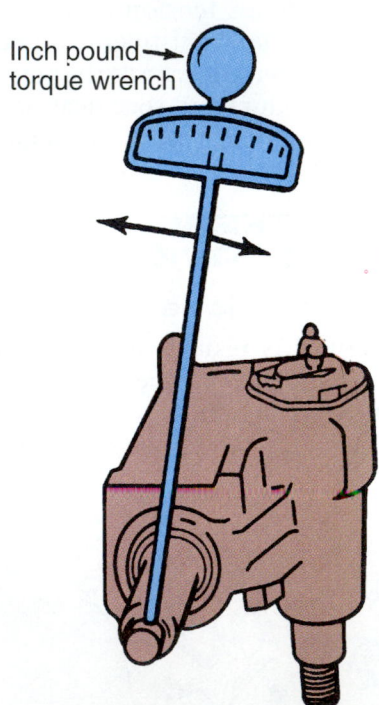

Figure 59.20 The load adjustment is measured with an inch-pound torque wrench. *(Courtesy of Chrysler Corporation)*

the pump (Figure 59.22). The pump can leak from these O-rings, the shaft seal, from fittings or hoses, or through mounting bolts.

To replace a pump shaft seal requires removal of the pulley. The pulley can be easily damaged with a

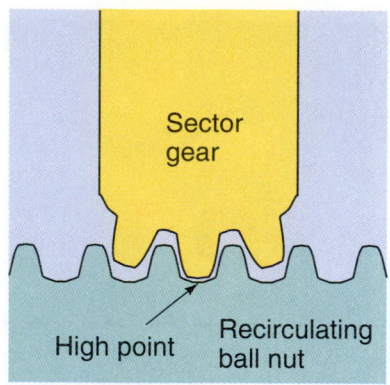

Figure 59.21 The high point is where the closest mesh between the teeth happens when the steering wheel is centered. *(Courtesy of General Motors Corporation, Service Technology Group)*

conventional puller. Special pullers and installers are used for this purpose (Figure 59.23).

■ POWER STEERING HOSES

The system has two hoses: a pressure hose and a return hose (see Figure 23.6). The pressure hose is made of reinforced oil resistant synthetic rubber that is crimped to metal tubing. It must be able to withstand temperatures of 300°F and handle pressures up to 1500 psi.

Sometimes the pressure hose has two different ends. The larger end fits on the pump side and acts as an "accumulator," which absorbs pulsations in the hose from variations in pressure. The ends of the pressure hose have flared connections. The return hose is usually attached only with hose clamps because it is not under high pressure.

When checking a hose, look for signs of leakage or dampness at the connections. Also, look for signs of deterioration such as cracks, signs of rubbing, or swelling. A failed pressure hose can leak, blowing oil onto an exhaust manifold where it can cause a fire. More information on hoses can be found in Chapter 23.

■ POWER STEERING GEAR SERVICE

Power steering failures include a lack of power assist in one direction, and leakage from the steering gear. Power steering seal kits are available. Some shops rebuild power steering gears, but most buy rebuilt ones.

Most cars now come with power rack and pinion steering units. Rack and pinion steering is more prone to problems because its weight has been cut over the years. Earlier power rack assemblies weighed as much as 54 pounds. Some of today's modern racks weigh as little as 8–10 pounds. Weight reductions of this mag-

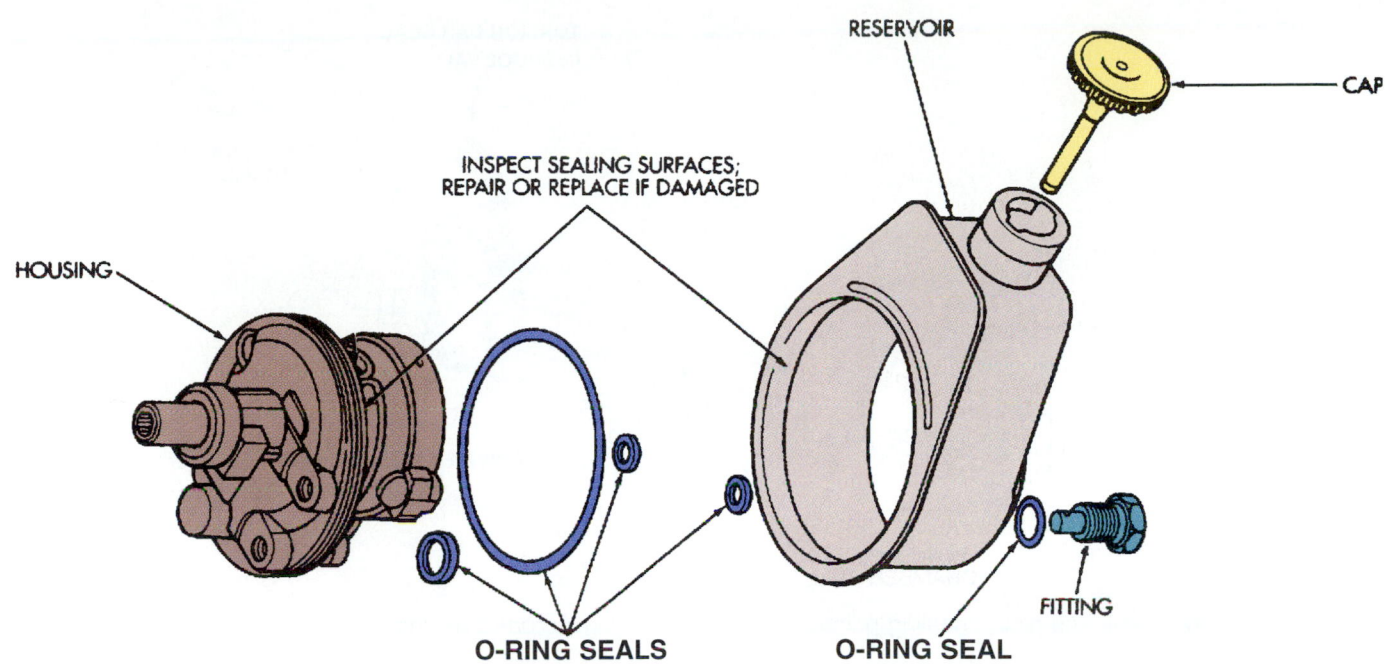

Figure 59.22 O-rings seal the pump reservoir to the pump body. *(Courtesy of Chrysler Corporation)*

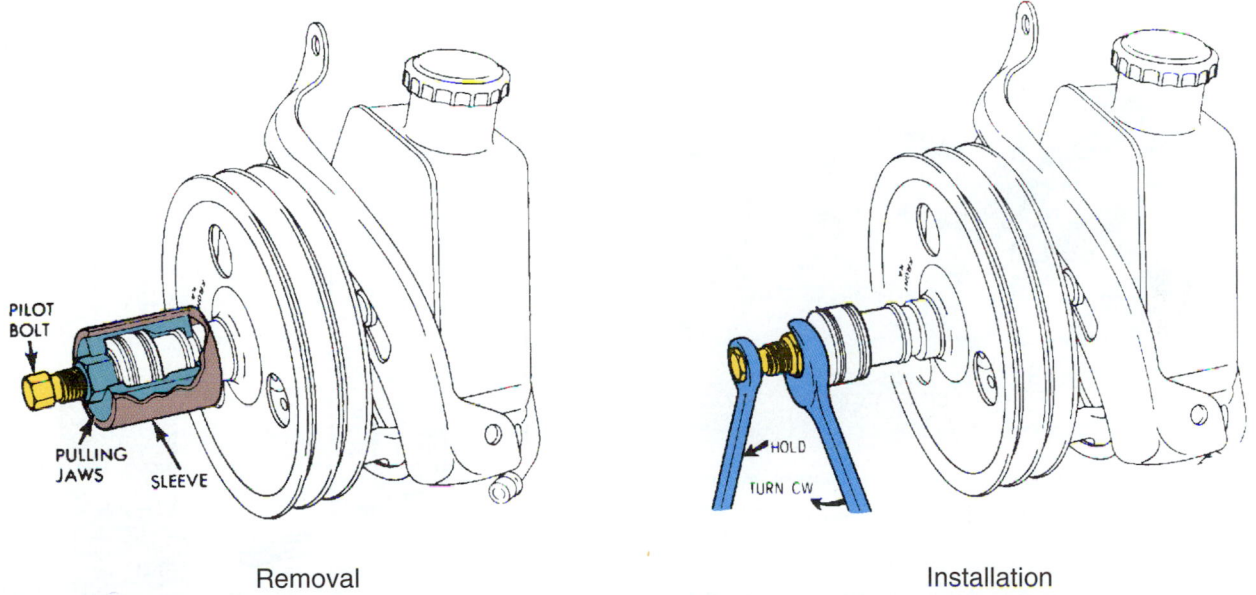

Removal   Installation

Figure 59.23 Tools for removing and installing a power steering pump pulley. *(Courtesy of General Motors Corporation, Service Technology Group)*

nitude have had an effect on the durability of the steering gears. Some of these lighter units have not been very durable, requiring replacement while the vehicle still has relatively low mileage.

A rack and pinion steering gear can have an external leak at the pinion shaft (attached to the steering column) or from the rubber bellows at either end of the rack (Figure 59.24). Rack failures can be due to a torsion bar bent from a hard impact or serious internal or external leaks.

■ REPLACING RACK AND PINION UNITS

Rack and pinion gears are often replaced with rebuilt units. Some units have a good record of success when rebuilt by technicians and others do not. Sometimes grooves wear in the control valve housing (Figure 59.25). The symptom of this problem is usually a loss of power assist on cold startups (called "**morning sickness**"). Rebuilt units usually have a nickel-plated sleeve installed to correct this and prevent it from

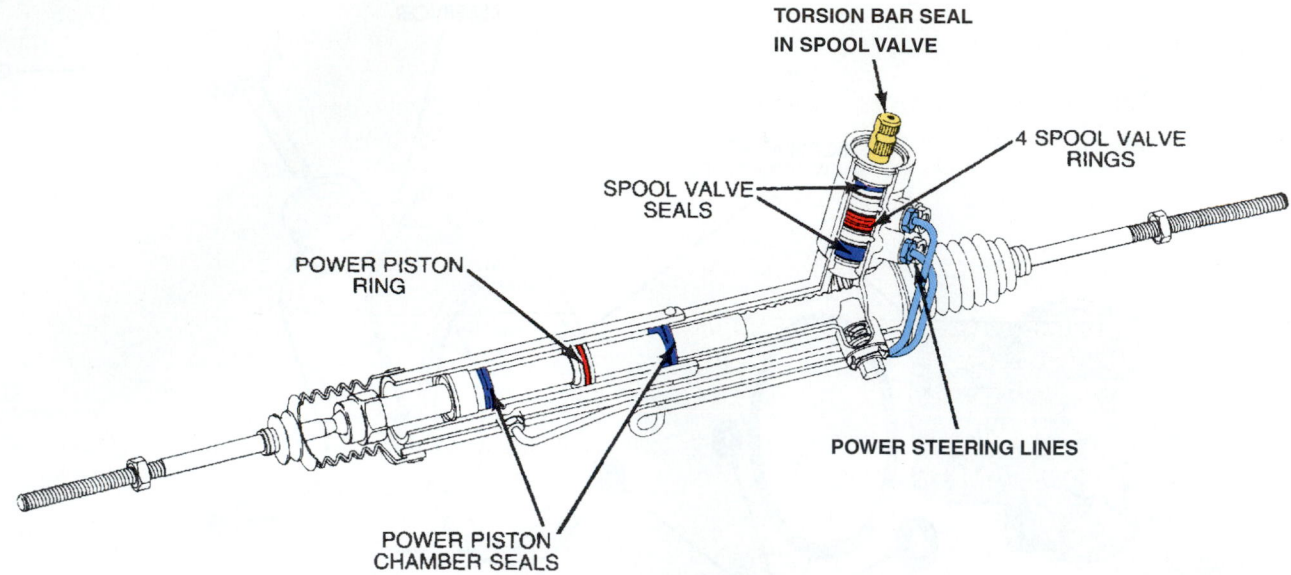

Figure 59.24 Rack and pinion sealing points. *(Courtesy of Moog Automotive, Inc.)*

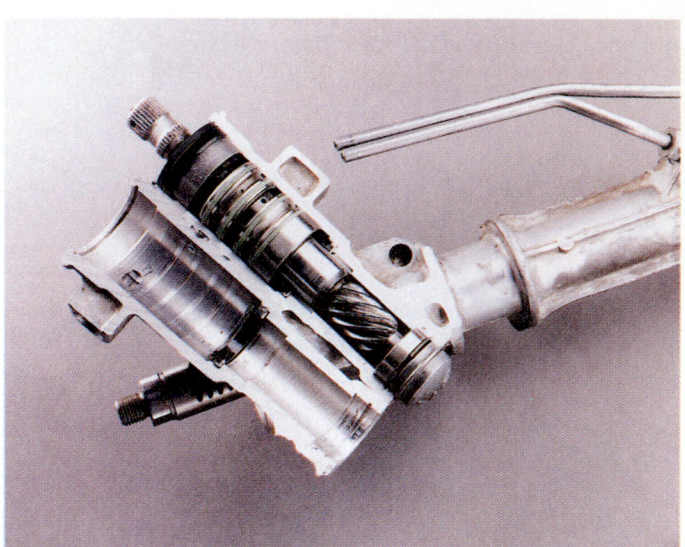

Figure 59.25 Grooves worn in the control valve housing. *(Courtesy of Moog Automotive, Inc.)*

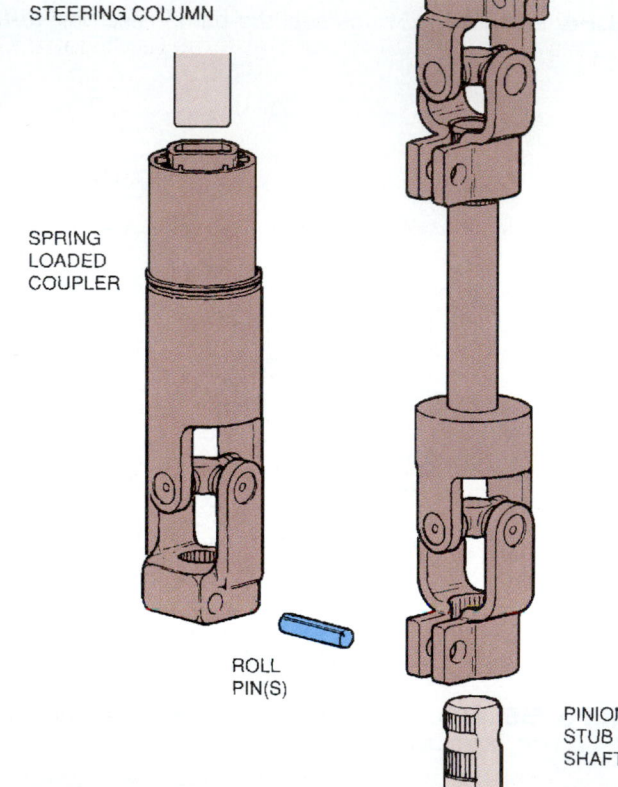

Figure 59.26 Two designs of steering shaft couplers. *(Courtesy of Moog Automotive, Inc.)*

happening again. When deciding whether to rebuild a unit or buy a replacement, an auto parts supplier is a good reference source.

Before removing a rack and pinion unit from the car, the steering shaft coupler must first be removed. There are two main types of coupler designs. Both types use a roll pin or threaded shoulder bolt that fits into a groove on the pinion stub shaft (Figure 59.26). One type uses a roll pin or bolt at the top too. The other uses a "D" shaped spring loaded slip fit coupler. The spring clip can fall out on the interior carpet of the car as the steering gear is lowered for removal. It must be reused.

Roll pins must be removed with a punch and chisel. On the type that has two roll pins, the lower

roll pin is usually corroded and difficult to remove while the rack is on the car. Once the steering rack is on the workbench, this pin can more easily be removed.

There are usually four tubing connections into the top of the rack and pinion unit. The two that are from the hoses to the power steering pump must be removed. Disconnecting and reconnecting the hoses is often not possible with ordinary flare-nut wrenches. A special crowfoot flare socket can be used. Some power steering lines have swivel ends. When these are fully tightened, they can still rotate. Be sure to tighten them to the correct torque specification and replace the two seals that seal them to the tube fitting. Sometimes, it is necessary to lower the crossmember under the engine to gain access to the rack and pinion.

When replacing an entire rack and pinion unit, be sure that the brackets are installed properly. If the rack is not level, bump steer can result.

FLUSHING THE HYDRAULIC SYSTEM

Be sure to check the condition of the hoses. Power steering pumps or gears often fail because the inside of the power steering hose has deteriorated.

When replacing a rack and pinion steering unit, the system must be thoroughly flushed. Most failures from morning sickness result in contaminants being circulated throughout the fluid system. These contam-

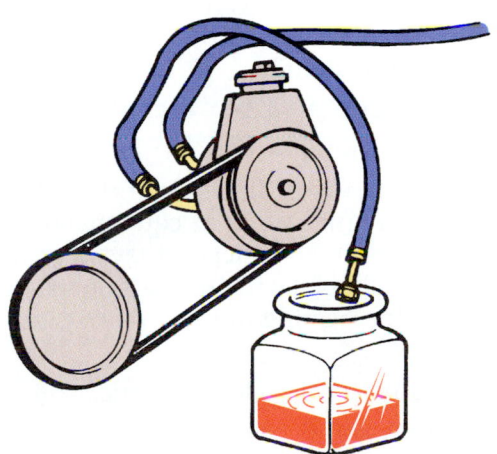

Figure 59.27 Crank the engine to pump out the fluid. *(Courtesy of Moog Automotive, Inc.)*

inants can ruin a new rack if they are not flushed from the system. Some manufacturers recommend the installation of a filter when a rack is replaced.

To flush the system, put the pressure line from the pump into a container. Disable the engine's ignition system and crank the engine to pump out the old fluid (Figure 59.27). While the rack is being replaced, let the hoses drain by gravity.

To flush the pump, put the pressure line in a container. Fill the system with clean fluid and operate the pump for two or three seconds. Then refill the reservoir. Repeat this process until the fluid in the reservoir is clean.

BLEEDING THE SYSTEM

The following procedure is recommended for Ford vehicles but can be applied to other makes as well.
- Reinstall the rack and refill the system.
- Run the engine and turn the wheel from lock to lock 20 to 25 times. Refill the reservoir as needed.
- When the fluid is warm, turn the wheel slowly from lock to lock.
- Sometimes vacuum is required to remove air from the fluid. A rubber stopper is installed in the reservoir filler neck. Vacuum is applied through the stopper for about 15 minutes.
- Test drive the vehicle. If there is a groan from the power steering system, air is still in the system.

More information on power steering fluid and bleeding is included in Chapter 23.

VARIABLE POWER STEERING SERVICE

Problems with variable power steering found in some newer cars can be either mechanical or electrical. When there is an electrical system failure, the power steering system goes back to operating on full time power assist. In other words, steering becomes easier at higher speeds. If this is the case, refer to the electrical trouble-shooting procedure in the vehicle service manual. When there is a mechanical problem, there will almost always be noise or a loss of power assist.

REVIEW QUESTIONS

1. What happens to power steering temperature when the wheel is turned several times?
2. What is the name of the test for looseness where an assistant turns the wheel from side to side with the tires on the ground?
3. How much looseness should there be between a rack and pinion inner tie stud and housing?
4. What is used to remove a pitman arm from a steering gear?
5. What measuring tool is used to make an approximate toe setting when replacing a tie rod end?
6. What is another name for the spiral cable that sends an electrical signal from the steering column to the steering wheel?
7. What can happen if the clock spring is not centered before steering linkage is reconnected to a steering gear?

8. What is the name of the point in a steering gear where the gear teeth come closer together when the car is travelling straight ahead?

9. Which kind of power steering system is more prone to problems, rack and pinion or conventional?

10. What is it called when there is a loss of steering power assist on cold startups?

■ ASE STYLE REVIEW QUESTIONS

1. Technician A says that automatic transmission fluid can be used in all power steering systems. Technician B says that the engine must be on when checking steering looseness on a car with power steering. Who is right?
   a. Technician A   b. Technician B
   c. Both A and B   d. Neither A nor B

2. Technician A says that a loose belt can cause the steering wheel to jerk during a turn. Technician B says that a loose belt can make a loud squeal, especially when going straight. Who is right?
   a. Technician A   b. Technician B
   c. Both A and B   d. Neither A nor B

3. A car has been driven for a long time with a loose belt. Technician A says that the belt has probably become glazed. Technician B says that the pulley groove could be worn out. Who is right?
   a. Technician A   b. Technician B
   c. Both A and B   d. Neither A nor B

4. Technician A says when testing looseness on a rack and pinion tie rod, the wheels should be hanging free. Technician B says that inner and outer parallelogram tie rod ends are identical. Who is right?
   a. Technician A   b. Technician B
   c. Both A and B   d. Neither A nor B

5. Technician A says if the steering wheel is off-center when driving on the highway, remove the steering wheel using a special puller and replace it in the straight ahead position. Technician B says that air bags can be reused after they have been deployed. Who is right?
   a. Technician A   b. Technician B
   c. Both A and B   d. Neither A nor B

6. Technician A says to store an air bag face down on a workbench. Technician B says the wiring harness must be disconnected before removing an air bag. Who is right?
   a. Technician A   b. Technician B
   c. Both A and B   d. Neither A nor B

7. Technician A says that the center of a rack and pinion steering rack is hardened. Technician B says if a rack is not centered, the car can steer further to one side than to the other. Who is right?
   a. Technician A   b. Technician B
   c. Both A and B   d. Neither A nor B

8. Technician A says to use a conventional pulley puller to remove a power steering pump pulley. Technician B says that a recirculating ball steering gear must be centered so the steering response is not faster in one direction than the other. Who is right?
   a. Technician A   b. Technician B
   c. Both A and B   d. Neither A nor B

9. Technician A says if the ends of the idler arm and pitman arm are not at equal height, bump steer will result. Technician B says that the wheels should be facing to the side when checking for steering linkage wear? Who is right?
   a. Technician A   b. Technician B
   c. Both A and B   d. Neither A nor B

10. Technician A says when a power steering hose has a hose clamp on its connection to the pump, this is the pressure hose. Technician B says on a variable power steering system with an electrical malfunction, steering effort on the highway becomes easier. Who is right?
    a. Technician A   b. Technician B
    c. Both A and B   d. Neither A nor B

CHAPTER 60

Wheel Alignment Fundamentals

■ **OBJECTIVES**

Upon completion of this chapter, you should be able to:

✔ Describe each wheel alignment angle.

✔ Tell which alignment angles cause wear or pull.

■ INTRODUCTION

Correct alignment of the wheels provides a vehicle with the ability to run straight on the highway with very little steering effort. This chapter deals with the principles of the different alignment angles.

■ ALIGNMENT ANGLES

There are five wheel alignment angles:

■ Toe
■ Camber
■ Caster
■ Steering axis inclination (SAI)
■ Turning radius

■ TOE

The alignment angle most responsible for tire wear is **toe**. Toe is a comparison of the distances between the fronts and the rears of a pair of tires (Figure 60.1). When the tires are closer together at the front, this is called *toe-in*. When the tires are further apart at the front, this is *toe-out*.

Every ¹⁄₁₆" of toe-in results in 11 feet per mile of **scuff**. This means that the tires actually move side-ways for 11 feet out of every mile travelled. If a car has 1" of toe, the tire is dragged 182 feet sideways every mile. The result is severe tire wear. Incorrect toe can occur because of:

■ An improper adjustment
■ Looseness in the steering linkage due to wear
■ A collision with a curb
■ A change in either the caster or camber adjustment

Tie rods and other steering linkage parts are built to be flexible if there is an impact. If steering linkage parts were brittle they might break during an impact, causing a dangerous loss of vehicle control. Because of their flexibility, steering linkage parts can bend during an impact. This will change the toe setting. When this happens, the steering wheel will usually be off-center when travelling straight down the road.

Toe is adjustable on all vehicles. When the car is rolling, the amount of toe changes. Whether the tires deflect inward or outward when rolling depends on if the car has front or rear wheel drive. On rear wheel drive cars, rolling tires tend to toe out when driving. This happens as the steering linkage deflects due to the rolling resistance of the tires.

During an alignment, compensation is made so that front tire toe will be as close as possible to zero while rolling. The specification for adjustment for rear wheel drive (RWD) cars usually calls for the tires to be slightly toed-in (closer together at the front) while on the alignment rack.

Front wheel drive cars react in an opposite fashion. Wheels on these cars tend to toe inward as they are pushed by engine torque. Specifications for front

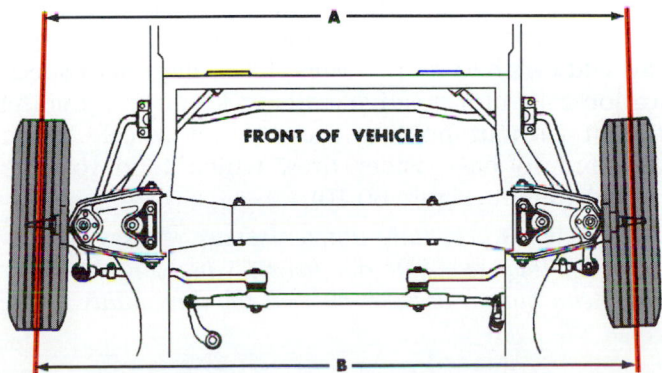

Figure 60.1 Toe is a comparison of the distances between the fronts and the rears of a pair of tires. *(Courtesy of Ford Motor Company)*

wheel drive cars usually call for a slight toe-out setting or zero.

■ CAMBER

Camber, an adjustable angle on most vehicles, is the inward or outward tilt of a tire at the top. A tire that tilts out at the top has *positive camber* (Figure 60.2). A tire tilted in at the top has *negative camber*.

Camber is a *tire wearing* angle. If the tire leans to one side, the tread will wear on that side as the smaller diameter of the tire squirms against the road surface (Figure 60.3).

The camber angle is controlled by the position of the control arms or struts. The control arms are quite strong and are not usually affected by an impact, like running into a curb. On long and short arm suspension systems, negative camber typically develops as springs sag with age. Sagged springs are often accompanied by wear on the inside edges of the front tires from the resulting negative camber.

Camber is a *directional control* angle also. A cambered tire tends to roll in a circle, as if it were at the large end of a cone (Figure 60.4). This is called **camber roll**. Unequal camber can cause a vehicle to pull to one side.

■ CASTER

Caster is the angle that describes the forward or rearward tilt of the spindle support arm (Figure 60.5). When the top is tilted to the rear as shown in Figure 60.6, the wheel has *positive caster*. Its *lead point*, or *point of load*, is in front of true vertical. Negative caster is the forward tilt of the steering axis. Figure 60.7 compares positive and negative caster.

Caster is a *directional control* angle but does not cause tire wear when the car is going straight. When riding a bicycle (which has positive caster), it tends to continue in a straight-ahead direction without holding the handle bars. If the handle bars are aimed in the wrong direction (negative caster), the bicycle loses its stability.

When the front wheels on a car have different caster settings, the car will pull toward the side with the most negative caster. That wheel will have its point of load behind the wheel on the other side of the car. Caster causes the spindle to move either toward the ground or away from the ground during a turn. Because the tire cannot move closer to the ground, the vehicle must lift (Figure 60.8). When turning effort is removed from the steering wheel after the turn, the tire attempts to move back to a straight-ahead position.

Positive caster results in road shock being transmitted through the steering column (Figure 60.9). Too much positive caster can result in a front wheel shimmy. Some cars that have a large amount of positive caster use a steering damper to prevent this. When the damper is worn out, the front wheels will shimmy uncontrollably when a front tire hits a small bump.

NOTE: *Excessive positive caster can actually cause cupped tire wear if the resulting shimmy is allowed to continue.*

Negative caster makes a vehicle easier to steer. It can also cause a car to wander and weave on the highway. Old cars had narrow tires. Their alignment specification called for positive caster to keep the car going straight without holding the steering wheel. Newer cars tend to have wider tires, which tend to keep rolling straight (Figure 60.10).

NOTE: *Light trucks have different caster settings depending on the their rear axle ride height. Changing the caster setting can have a drastic effect on the driveability of the vehicle.*

Caster and Tire Wear

Caster is actually a measurement of the camber angle as it changes during a turn. A high amount of caster can result in tire wear if the car is driven excessively on

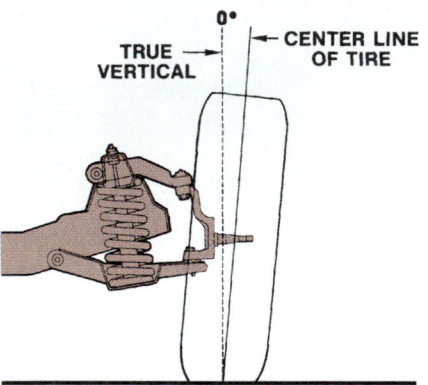

Figure 60.2 This wheel has positive camber. *(Courtesy of Moog Automotive, Inc.)*

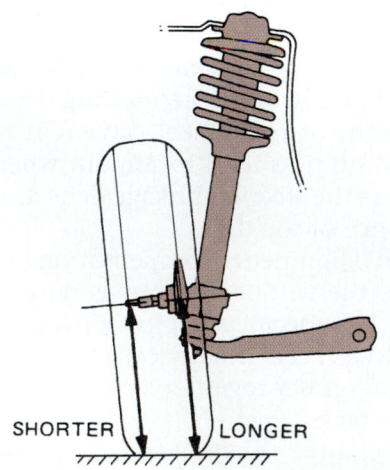

Figure 60.3 The tread will wear on the shorter side as the smaller diameter of the tire squirms against the road surface.

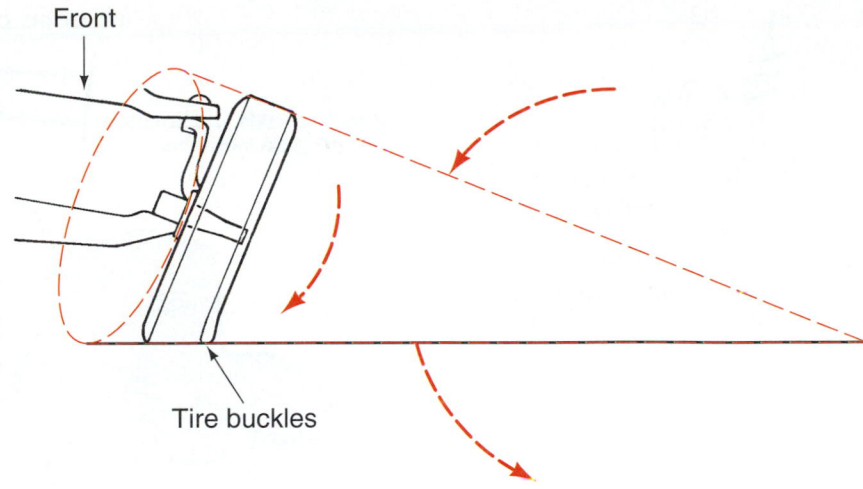

Figure 60.4 A cambered tire tends to roll in a circle.

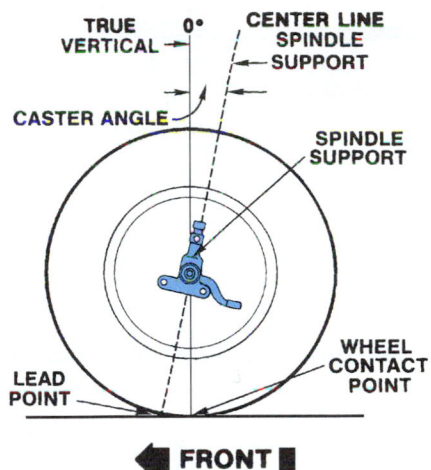

Figure 60.5 Caster. *(Courtesy of Moog Automotive, Inc.)*

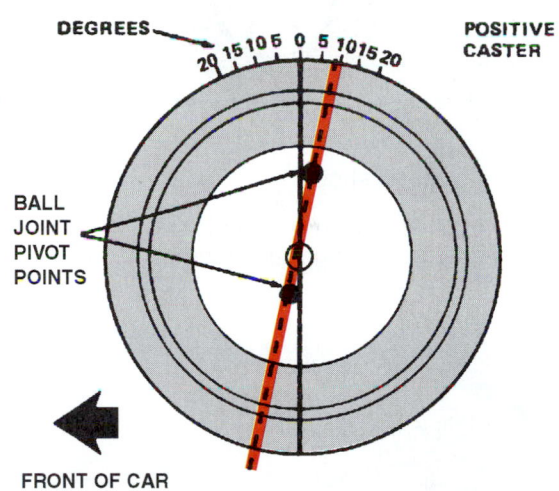

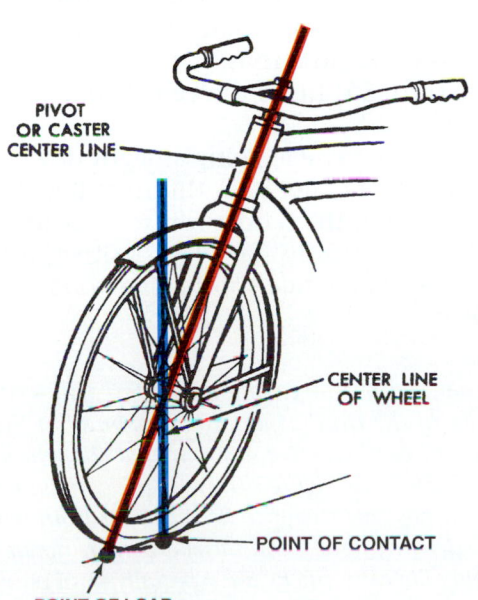

Figure 60.6 A bicycle has positive caster. *(Courtesy of Ford Motor Company)*

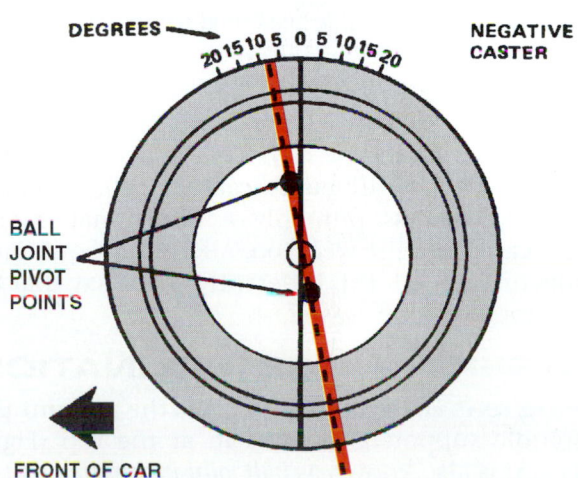

Figure 60.7 Comparison between positive and negative caster. *(Courtesy of Ford Motor Company)*

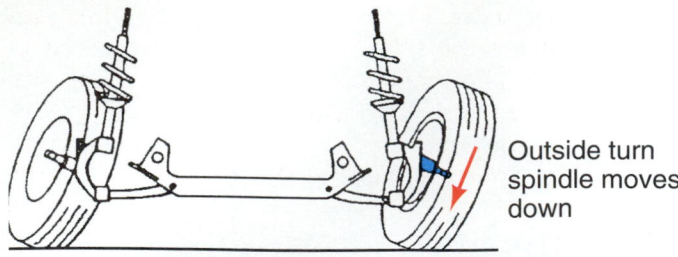

Figure 60.8 Because the tire cannot move closer to the ground, the vehicle must lift. *(Courtesy of Hunter Engineering Company)*

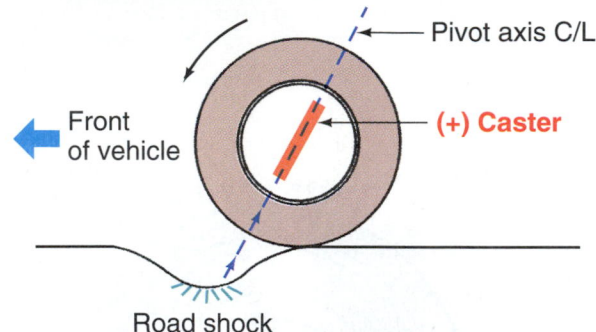

Figure 60.9 Positive caster results in road shock being transmitted through the steering column.

Figure 60.10 Wider tires tend to roll straight. *(Courtesy of Ford Motor Company)*

winding roads or in the city. This is because a high amount of caster influences camber roll. Although caster is known as primarily a directional control angle, it can cause tire wear too. Alignment equipment manufacturers teach this differently. Some say that tire wear is not a result of caster.

■ STEERING AXIS INCLINATION

Steering axis inclination (SAI) is the amount that the spindle support arm leans in at the top (Figure 60.11). SAI is also known as *ball joint inclination (BJI)* or *king pin inclination (KPI)*. It is *not* a tire wearing angle. SAI basically helps the vehicle steer in the straight-ahead direction.

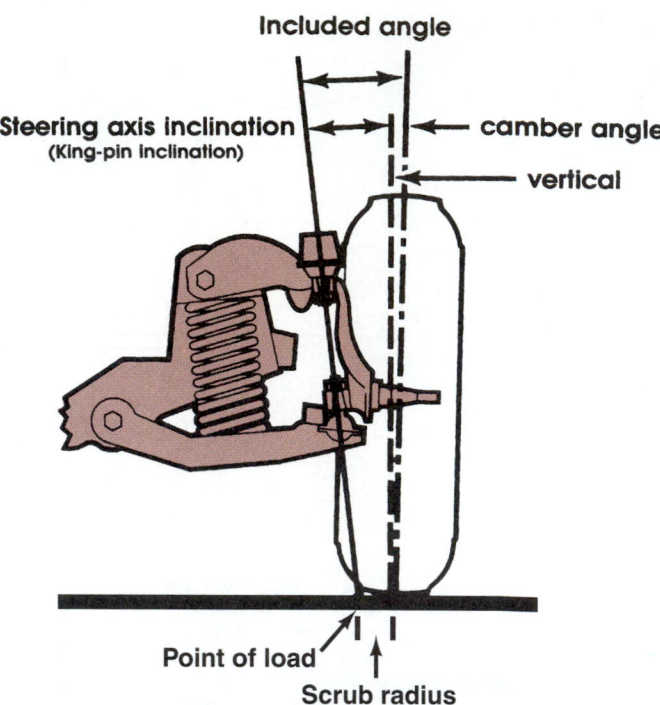

Figure 60.11 Steering axis inclination puts the pivot point under the tire. *(Courtesy of Hunter Engineering Company)*

How SAI Works

Although the spindle is horizontal to the ground (except for the camber angle) when the wheels are pointed straight ahead, it attempts to move closer to the ground during a turn (see Figure 60.8). The spindle cannot move any closer to the ground because of the tire, so the car must lift. During a turn, the weight of the car puts pressure on the wheel, causing it to want to return it to the straight-ahead position.

SAI has three functions:

■ After a turn, SAI helps the vehicle return to straight ahead.

■ SAI keeps the vehicle going straight down the road. (Positive caster does this, too. But the more positive caster, the harder it is to steer the car.)

■ SAI allows the car to have less positive caster (for easier steering), while still having good directional stability.

NOTES:

■ *Some cars with a large amount of SAI will wear the outsides of the tires (especially on nonradial tires) because of the excessive amount that the wheel cambers during turns (see Figure 60.8). These cars do not require as much positive caster to keep the car from wandering.*

■ *In city driving, right hand turns are made about 80% of the time. Delivery trucks are especially hard on right hand springs, bearings, and tires (which must turn sharper).*

The combination of SAI and camber is called the **included angle**.

■ SCRUB RADIUS

Scrub radius is a factor of steering axis inclination. It is the pivot point for the front tire's footprint. Scrub radius is the distance at the road surface between the centerline of true vertical at the center of the tire tread and the steering axis pivot centerline.

The junction of the steering axis and centerline pivot point is normally below the surface of the road (see Figure 60.11). How far below the surface of the road determines how much scrub radius there is.

The more scrub radius there is, the harder it is to steer the car. Positive camber reduces scrub radius. But the best tire life happens when the running camber angle is 0°.

■ Scrub radius on rear wheel drive cars is called *positive scrub radius*. The front wheels toe out when rolling (Figure 60.12).

■ Front wheel drive cars usually have *negative scrub radius*. The tires toe in when rolling (Figure 60.13).

NOTE: *If a left front tire on a front wheel drive vehicle with positive scrub radius blows out, the tire will pull hard to the left. The still-inflated right tire would pull inward. To prevent this dangerous situation, SAI has been increased on FWD cars. This results in negative scrub radius, which causes the blown tire to pull inward. The car will continue to go straight because this direction of motion counteracts the motion of the right front tire.*

Incorrect Scrub Radius

Installing lower profile tires and offset wheel rims on a vehicle will result in a big increase in scrub radius (Figure 60.14). The pivot point can actually move outside of the tires' footprint area. The result is much harder steering, wheel shimmy, and a tendency to wander.

Negative scrub radius can result when tires and wheels that are too tall are installed on a rear wheel drive vehicle. The result can be instability. When a tire on an SLA suspension system goes over a bump, the scrub radius can change from positive to negative as the camber changes. Vehicle handling can become dangerous.

NOTE: *When changing to tires and wheels of a different size, modifications to the suspension must be made. Wheel alignment cannot compensate for this.*

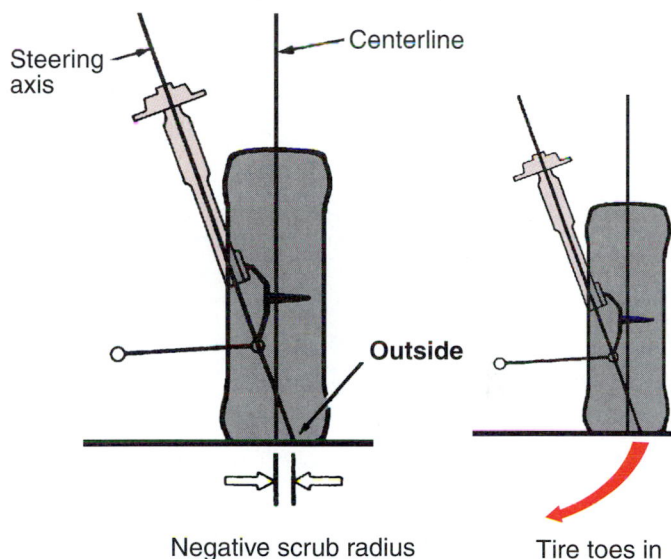

Negative scrub radius   Tire toes in

Figure 60.13 With negative scrub radius, the tire toes in when rolling.

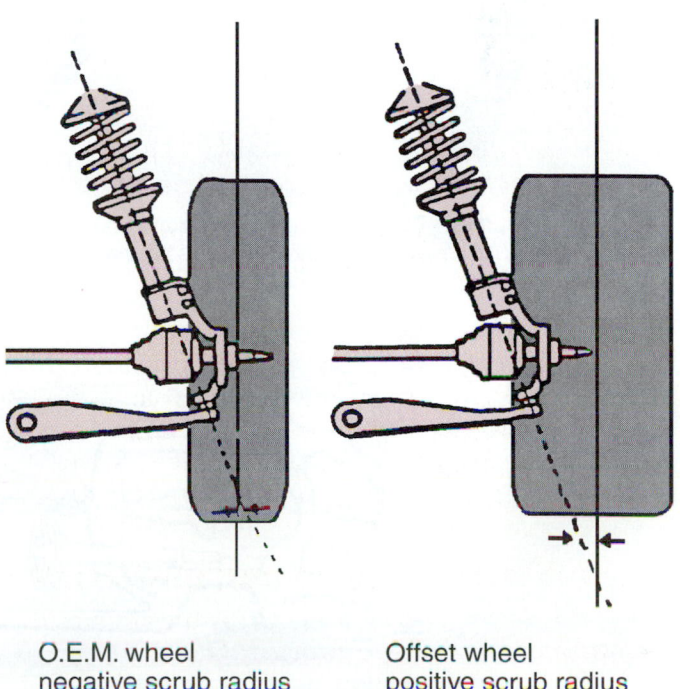

O.E.M. wheel negative scrub radius   Offset wheel positive scrub radius

Figure 60.14 Installing lower profile tires and offset wheel rims on a vehicle will result in a change in scrub radius. (*Courtesy of SPX Corporation*)

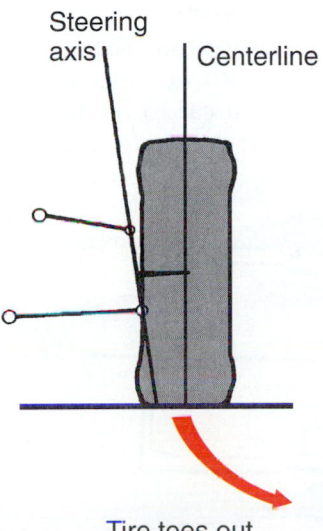

Tire toes out

Figure 60.12 With positive scrub radius, the tire toes out when rolling.

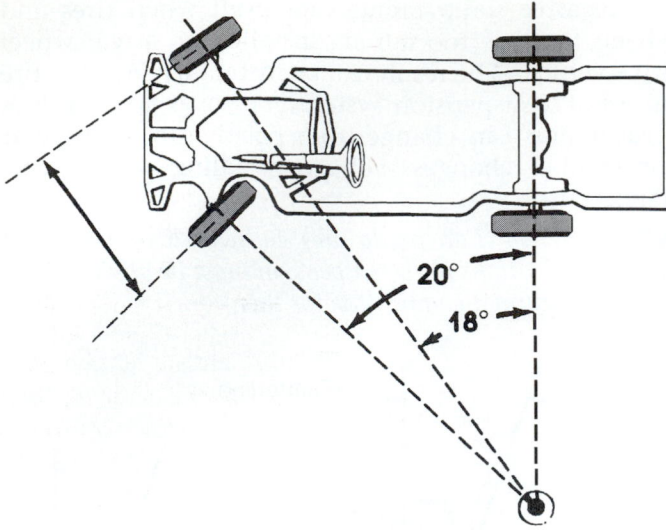

Figure 60.15 Front tires must toe out during a turn. *(Courtesy of Hunter Engineering Company)*

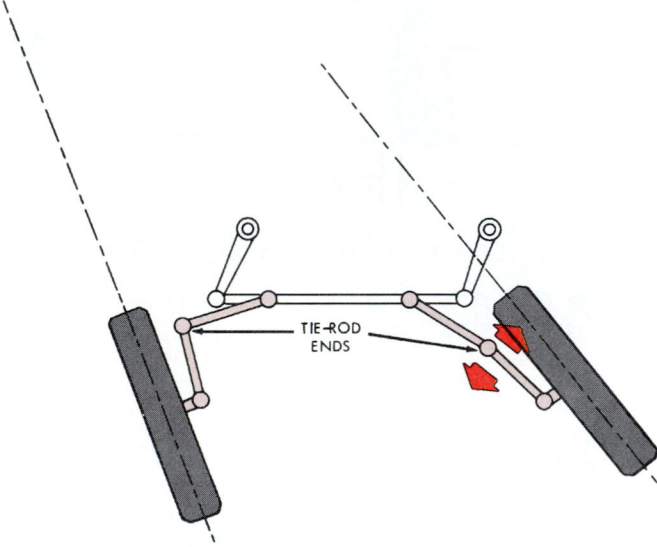

Figure 60.16 The tires turn different amounts as the steering arms move through their arcs of travel. *(Courtesy of Moog Automotive, Inc.)*

Other causes of incorrect scrub radius are a bent front suspension part or damage to the frame at the **crossmember** or *strut tower*. The crossmember, or **cradle**, is the large steel part of the frame beneath the engine and between the front wheels. The strut tower is the area inside of the front fenders that supports a MacPherson strut.

■ TURNING RADIUS

When a car makes a turn, the outside wheel must travel in a wider arc than the inside wheel (Figure 60.15). The alignment angle that controls this is called **turning radius**, toe-out-on-turns, or the **Ackerman angle**. The tires actually toe out during a turn because the steering arms are bent at an angle (see Figure 58.13). When the wheels are turned, they move at different amounts as they move through their arc of travel (Figure 60.16).

■ HISTORY NOTE

Horse drawn wagons used a steering system that pivoted at the center. In 1884 Rudolf Ackerman patented a unique steering system with both of the wheels pivoting at the outside of the axle. The angle he designed into the steering arms caused the front wheels to toe out when turning.

■ TRACKING

The distance between the front and rear tires is called the **wheel base**. The side to side distance between an axle's tires is called the *track*. For minimum tire wear and good fuel economy, the wheels must run on track. The rear tires are supposed to follow in the tracks of the front tires.

Tracking is a term that refers to the relationship between the average direction that the rear tires point, when compared to the front tires. When tracking is off, the front wheels do not follow the rears (Figure 60.17). This causes the car to try to steer to the side, increasing tire wear and upsetting handling.

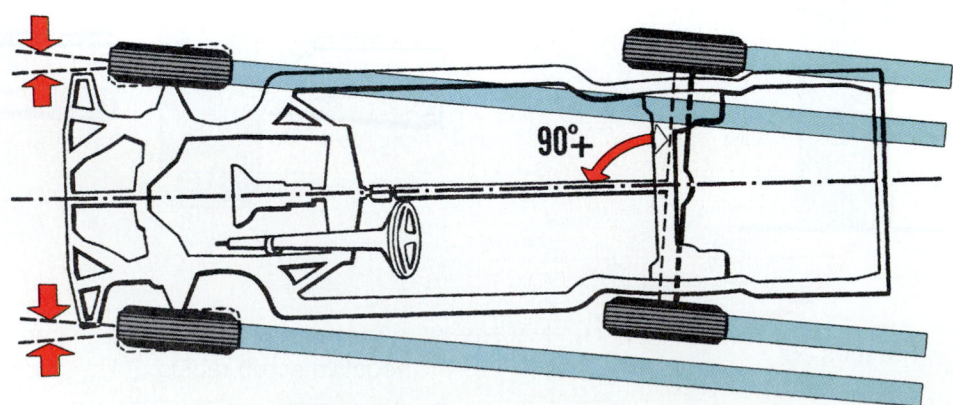

Figure 60.17 When tracking is off, the front wheels do not follow the rears. *(Courtesy of Hunter Engineering Company)*

If the rear axle is out of line to the right, it will cause the steering wheel to be aimed to the right. This will cause right front toe-out and left front toe-in. The result will be inside or outside toe wear. Toe wear is described in the pre-alignment section of this chapter.

■ SET-BACK

Set-back is the amount that one front wheel is behind the one on the other side of the car (Figure 60.18). It is measured in degrees. A negative angle indicates that the wheel on the left side is set back. This will cause the vehicle to steer to the left. It can also cause a brake pull. When set-back is incorrect, a collision could be the cause.

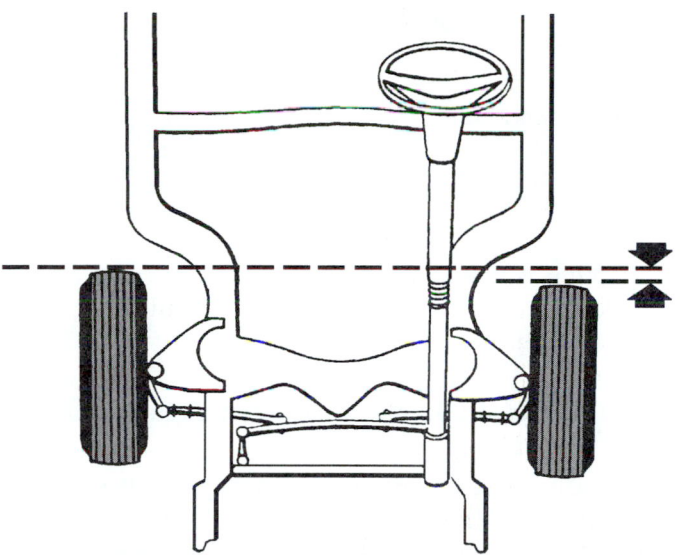

Figure 60.18 Set-back is the amount that one front wheel is behind the one on the other side of the car. *(Courtesy of Hunter Engineering Company)*

■ SPECIAL HANDLING CHARACTERISTICS

Slip Angle

Tire manufacturers sometimes recommend different inflation pressures or alignment angles to change a vehicle's handling characteristics. A tire with lower inflation has a greater **slip angle**. Slip angle is the tendency during a turn for a tire to continue to go in the direction it was going, even though the rim has turned in response to steering wheel movement (Figure 60.19). Besides tire pressure, the amount of slip angle a tire has depends on the weight exerted vertically on it, the structure and type of tire, and the wheel alignment setting on the wheel. Positive camber causes a tire to have a greater slip angle.

Understeer and Oversteer

When a car does not seem to respond to movement of the steering wheel during a hard turn, this is called **understeer**. When a car turns too far in response to steering wheel movement, this is called **oversteer** (Figure 60.20). Desirable steering is said to be *neutral*.

Rear wheel drive cars sometimes have lower inflation pressures specified for the rear tires to prevent the car from understeering. That way, when a car begins to slide and the driver lets up on the accelerator, the driver gains control again. If the car had oversteer, letting off of the gas would result in a spin-out.

When a car has four wheel alignment capability, setting rear wheel camber more negative than the front tires results in less tendency to oversteer. When tires of different profiles are installed on the same vehicle, the ones of the lower profile are installed on the rear to prevent oversteer.

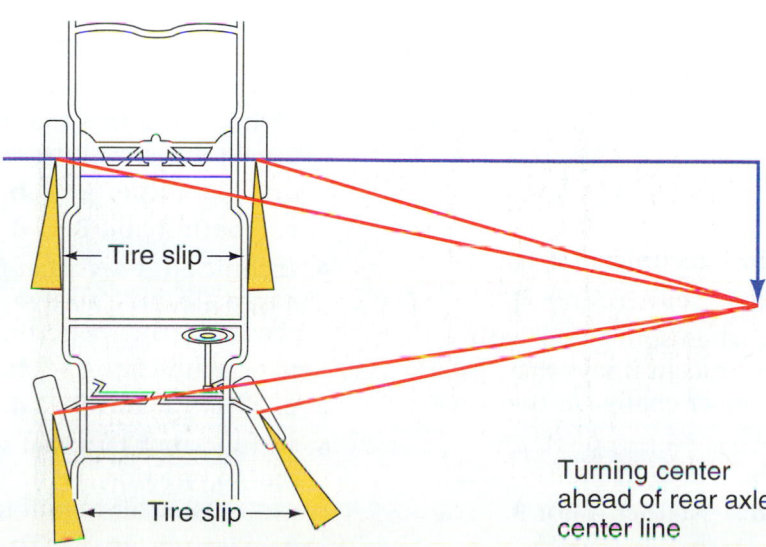

← Tire slip →

← Tire slip →

Turning center ahead of rear axle center line

Figure 60.19   Slip angle. *(Courtesy of AMMCO® Wheel Alignment Parts I & II by Hennessy Industries, Inc.)*

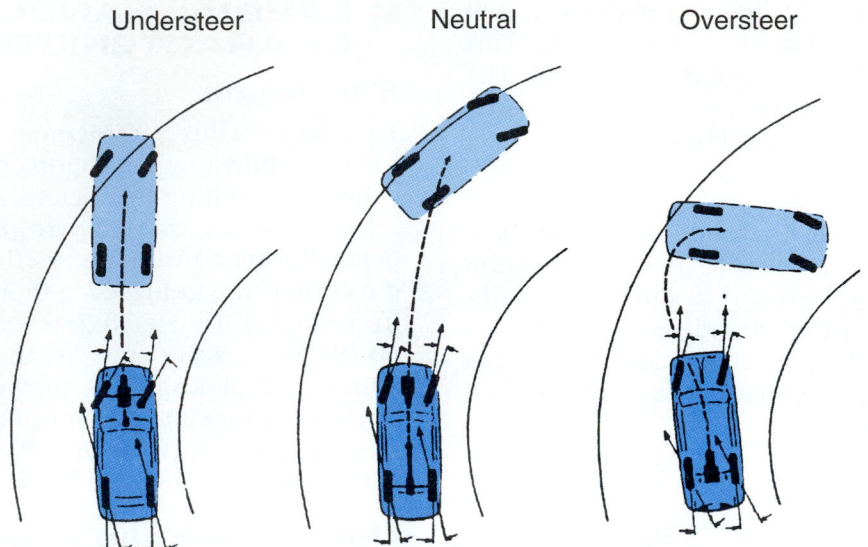

Understeer   Neutral   Oversteer

Figure 60.20 Understeer and oversteer. *(Courtesy of Pirelli Tire Corporation)*

■ REVIEW QUESTIONS

1. List the five wheel alignment angles.
2. Which alignment angle is most responsible for excessive tire wear?
3. If a tire has 1" of toe-in, how far is it dragged sideways in one mile of driving?
4. Which kind of car has its toe set inward when not rolling, front wheel drive or rear wheel drive?
5. When the top of a tire tilts inward (toward the center of the vehicle), what positive or negative angle is this?
6. The angle that keeps a bicycle going straight when your hands are off the handle bars is called _____.

7. What is the name of the alignment angle that keeps the car going straight while allowing easy steering?
8. When tires of a different height or wheel offset are installed on a vehicle, what alignment factor changes?
9. During a turn, which front wheel turns sharper, the outside or the inside?
10. When a car does not seem to respond to movement of the steering wheel during a hard turn, what is this called?

■ ASE STYLE REVIEW QUESTIONS

1. Technician A says that camber can cause a car to pull to one side. Technician B says that camber can cause a tire to wear on one side of its tread. Who is right?

   a. Technician A   b. Technician B
   c. Both A and B   d. Neither A nor B

2. Technician A says that when springs sag with age, caster changes. Technician B says that caster is a measurement of camber change as the wheel is turned. Who is right?

   a. Technician A   b. Technician B
   c. Both A and B   d. Neither A nor B

3. Technician A says that scrub radius is what causes the tires to toe in or out when rolling.

Technician B says that front wheel drive cars usually have positive scrub radius. Who is right?

   a. Technician A   b. Technician B
   c. Both A and B   d. Neither A nor B

4. Technician A says that front tires toe out during a turn. Technician B says that turning radius is caused by the toe setting. Who is right?

   a. Technician A   b. Technician B
   c. Both A and B   d. Neither A nor B

5. Technician A says that set-back can cause a car to pull to one side. Technician B says that tracking can cause a car to pull to one side. Who is right?

   a. Technician A   b. Technician B
   c. Both A and B   d. Neither A nor B

Wheel Alignment Service

■ INTRODUCTION

Before performing a wheel alignment, a thorough inspection of the steering and suspension systems must be performed. Loose parts will prevent an accurate and lasting adjustment. Looseness of suspension or steering parts can result in slack in the steering wheel, shimmy, or an intermittent pull to one side or the other. For an alignment to be successful, suspension and steering components must be in good working order. A vehicle with worn or loose parts cannot be aligned.

Incorrect alignment settings can result in pull, instability, and tire wear. Unusual tire wear or vehicle handling problems are usually what direct a driver to a shop for a wheel alignment. Front axles experience far more stress than rear axles because they support the weight of the engine and provide for steering. Although the rear axle is usually not the cause of problems, the technician must take all of the frame, steering, and suspension into consideration to be able to properly align a car.

This chapter deals with suspension and steering system wear; inspection procedures are covered next. The last part of the chapter deals with wheel alignment measuring and adjusting procedures.

■ PRE-ALIGNMENT INSPECTION

Prior to a wheel alignment, the suspension and steering systems must be inspected. If parts are loose or worn, an alignment will not be successful:

■ Tire pressures must be adjusted to specifications.

■ The frame must be at the correct height.

■ Worn bushings and pivots parts must not allow movement of suspension and steering parts.

■ Tires must be new or be worn evenly for the vehicle to be level during alignment measurement.

■ TIRE WEAR INSPECTION

Tire wear can be caused by worn parts, incorrect inflation, hard cornering, or incorrect wheel alignment.

Camber or toe are the two adjustable wheel alignment angles that cause wear. It is better if the alignment technician can look at worn tires before they are replaced. If tires are evenly worn and there is no pull or wandering, caster and camber should not require adjustment.

Tire Wear from Camber

Camber causes the inside and outside of the tire to have different diameters (see Figure 60.3). Wear from incorrect camber shows up on either the outside or inside of the tire tread (Figure 61.1). Outside wear is due to positive camber. It usually results from incorrect settings or a vehicle with a high amount of steering axis inclination driven with a high amount of cornering.

NOTE: *Camber wear results as springs sag over time, changing the height of suspension components.*

Tire Wear from Toe

Driving a vehicle with excessive toe is dangerous because the front tires are sliding. When toe is incor-

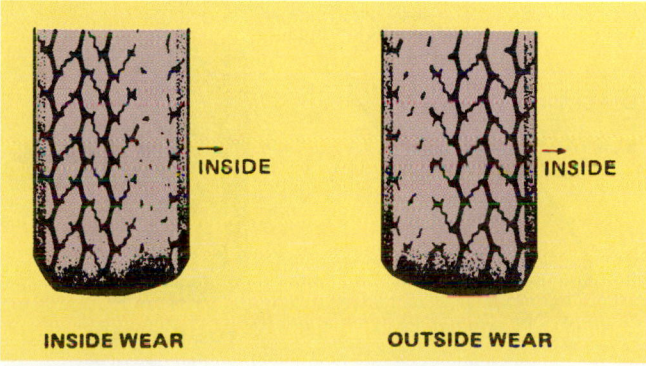

INSIDE   INSIDE

INSIDE WEAR   OUTSIDE WEAR

Figure 61.1 Effect of camber on tire diameter.

rect, the tire will develop a feathered edge (Figure 61.2). The feathered edge can be especially severe on bias tires, but also occurs on radial tires. To diagnose excessive toe on a worn tire:

■ Move your hand across the tire's footprint area from the outside to the inside. If you feel a feathered edge, the car has excessive toe-out (see Figure 61.2).

■ If you feel a feathered edge when moving from the inside to the outside, the car has excessive toe-in (Figure 61.3).

Other Toe Wear Factors

■ On a rear wheel drive car, the right front tire tends to toe in more than the left. The outside edge wears because the tire is rolling under.

■ The left wheel will roll under if toed out. Toe-out results in left front wheel wear to the inside edge.

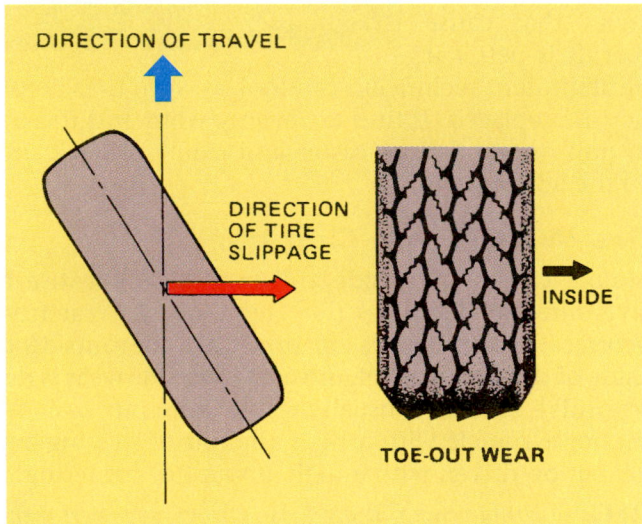

Figure 61.2 Toe-out wear.

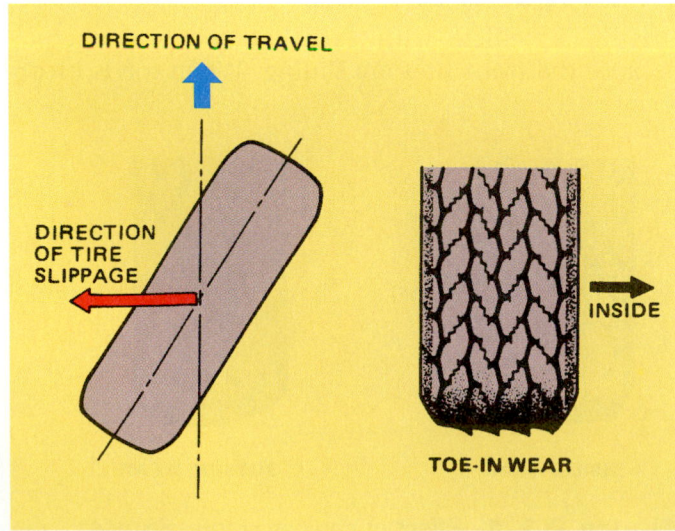

Figure 61.3 Toe-in wear.

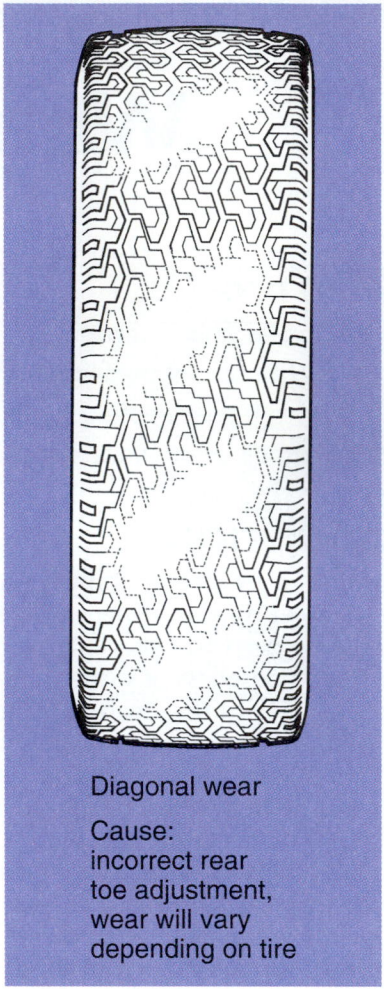

Diagonal wear

Cause:
incorrect rear
toe adjustment,
wear will vary
depending on tire

Figure 61.4 Wear from incorrect rear toe. *(Courtesy of Moog Automotive, Inc.)*

■ Radial tires with toe-in will both roll under resulting in wear that looks like positive camber.

■ Toe on the rear of a car does not equalize like toe on the front does. If the tires are not rotated, they will develop diagonal wear (Figure 61.4).

■ RIDE HEIGHT CHECK

Because the weight of the car is always resting on them, springs tend to sag as a car ages. Alignment specifications are based on the assumption that the ride height of the car is correct. If alignment is adjusted when the springs have sagged beyond specifications, tire wear and unusual handling can result. Prior to making any wheel alignment adjustments, ride height must be measured.

A car with a short/long arm (SLA) suspension with springs that have sagged will not function within its desired alignment range as the springs deflect. This can cause changes in the way the car reacts to bumps. Figure 61.5 shows where typical ride height measurements would be checked.

➡ *Perform Ride Height Check Worksheet*

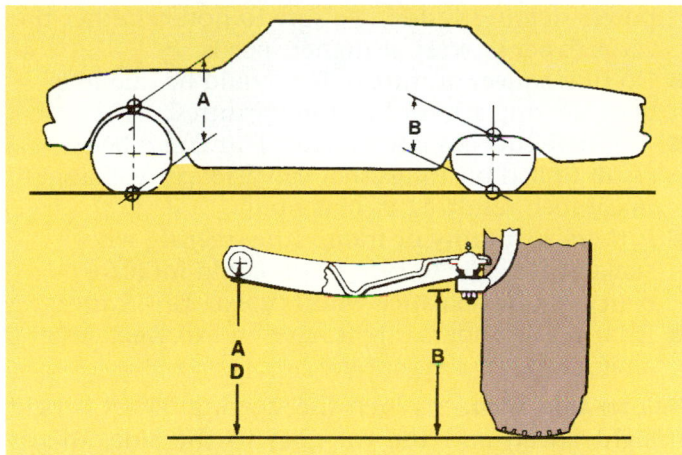

Figure 61.5 Typical ride height measurement. The lower sketch shows the typical measurement locations on an SLA system. *(Courtesy of Moog Automotive, Inc.)*

TOE CHANGE

As suspension height changes, toe measurement can change with it (Figure 61.6). The tie rods are designed to remain parallel to the lower control arms when they pivot (Figure 61.7). When springs have sagged, **toe change** can result. Toe change causes the tire to move on the road surface, scrubbing away tread. A small amount of toe change can be absorbed by the flexing

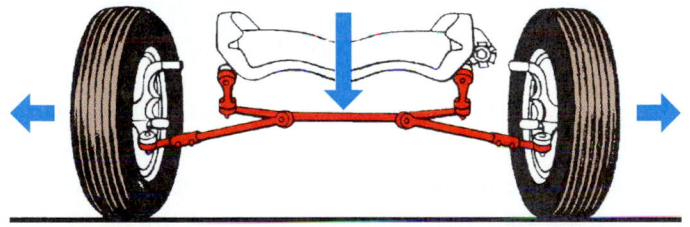

Figure 61.6 Toe change during changes in ride height. *(Courtesy of Hunter Engineering Company)*

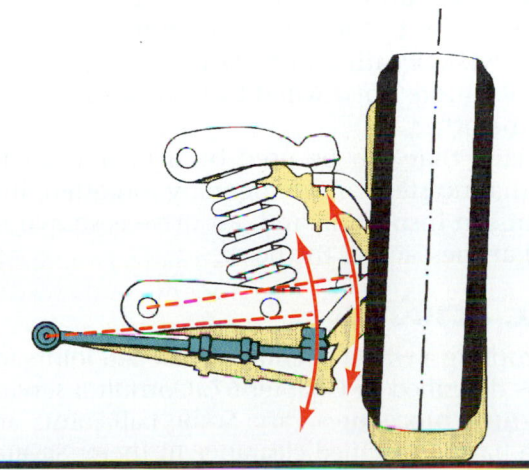

Figure 61.7 The tie rods are designed to remain parallel to the lower control arms when they pivot. *(Courtesy of Hunter Engineering Company)*

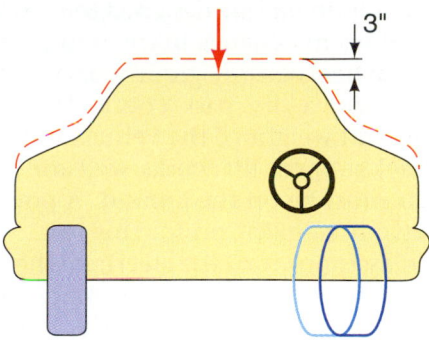

Figure 61.8 Bump steer. *(Courtesy of AMMCO®, Registered trademark of Hennessy Industries, Inc.)*

of a properly inflated tire. Loose parts or incorrect wheel alignment can move the amount of allowable toe change out of normal limits.

When toe change is confined to one side of the vehicle, **bump steer**, or *orbital steer,* can be the result. Bump steer happens when a wheel with tie rods at unequal heights goes over a bump. The car momentarily steers in the direction that the wheel turns as the toe changes (Figure 61.8).

NOTE: *The height of the steering arm changes as caster changes. Toe change can be caused by having different caster angles from side to side.*

TORQUE STEER

Sometimes a vehicle will turn abruptly to the side during initial acceleration. This action, called **torque steer**, is usually found on a front wheel drive car with axles of unequal lengths. It can also be caused by anything that causes the axles to be at different heights. This results in unequal CV joint angles. The height difference could be due to a loose sub-frame or a problem with unequal spring height.

HISTORY NOTE

On cars with small engines, torque steer wasn't a problem. When front wheel drive became popular and horsepower increased, torque steer was more noticeable. Older front wheel drive cars sometimes had front driving axles of different lengths. During acceleration the long axle would twist, delaying power transfer to its front wheel. During quick deceleration it would recoil. Some of the solutions attempted by auto makers were to use a thicker axle, change scrub radius, or use an intermediate axle shaft so that drive axles would both be the same length.

SUSPENSION LOOSENESS

A motto of one aftermarket suspension and steering part manufacturer is "you can't align looseness." Steering and suspension components are designed to be able to pivot, without allowing any change in the positions of parts.

With the wheels on the ground or on a wheel support lift, perform a dry park check. When checking any part, the full weight of the vehicle must be on the tires. A wheel support lift works well for this. Otherwise, the car must be on the ground. A power steering car must have the engine on for this test.

Have an assistant turn the steering wheel back and forth a short distance while you look for looseness in the steering linkage or suspension. The loose parts will be evident if there is any up or down movement in a linkage or movement of a control arm. Check Chapter 59 for more information on steering system testing and diagnosis. Check the adjustment of the wheel bearings to see that they are not loose before attempting a wheel alignment.

■ TEST DRIVE

Unless a vehicle is unsafe to drive, a test drive is always done before performing repairs.

Before Driving the Car

Before driving the car, a visual inspection is performed to check the following items:
- Inspect suspension bushings visually and with a prybar.
- Inspect steering linkage pivot connections. Firmly grasp the part and rock it to check for looseness.
- Inspect rubber grease boots on tie rods and ball joints.
- Inspect shock absorbers.
- Check to see if the vehicle has any signs of collision damage.

Tire Checks.
- Always adjust tire pressures to specifications. Check the condition of the valve stems and look for signs of impact damage that might have resulted in a bent rim.
- Check for signs of damage to the sidewalls or tread area.
- Be sure that tires of the correct size are used.
- Radial and bias tires should not be mixed. Front tires should be of the same brand and tread pattern.

Power Steering Checks.
- Check power steering fluid level. Look for evidence of fluid leaking.
- Check drive belt tension.
- Check the power assist to see that it works with equal ease in both directions.

During the Test Drive

During the test drive, the car is checked for several conditions:
- Hard steering can be caused by binding parts, incorrect alignment, low tires, or a failure in the power steering system. Be sure to note whether the car is easier to steer at higher speeds.
- Do tires squeal on turns? This could be due to a bent steering arm or low tire pressures.
- Are there squeaks and clunks? This could be the result of bad bushings that can also cause changes in camber, resulting in brake pull.
- Is there a shimmy or tramp (the steering wheel shakes from side to side)? This could indicate a bent or out-of-balance wheel or excessive caster.
- A car that wanders might have an incorrect caster angle setting.

Pull to One Side. When the steering wheel is held straight ahead and the car goes to the side, this is called **pull**.
- Does the car pull to one side or the other? Does it pull always or only during braking? This could be due to either alignment or brakes.
- If the car pulls, does the direction or amount of pull change if the tires are rotated? A defective or damaged tire can cause a pull.
- Is it possible that the power steering is the cause of the pull? Some cars have adjustable spool valves. With others, the steering box might require service. While driving in a straight line in a parking lot, shut off the key and see if the pull goes away. If it does, the power steering is at fault.

Rough Ride.
- Does the owner complain of a rough ride? This could be due to tire pressures that are too high, a bent or frozen shock absorber, or the installation of radial tires on an older car.

Body Roll.
- Does the car lean excessively to one side or the other during fast turns? The shock absorbers could be worn out.

Wheel Bearings.
- Is there a noise during turns that changes in pitch as the car weaves to the left and then to the right? The outer wheel bearing turns faster during a turn. It will make more noise when turning to one side than to the other.

A checklist that can be used by a technician to make sure that no steps are accidentally forgotten during a suspension inspection and test drive is shown in Appendix I at the back of the book.

■ BALL JOINTS

Before attempting a wheel alignment, test ball joints for looseness as described in Chapter 57. Consult a service manual because procedures vary. Some ball joints are designed to have a specified clearance in them. Several part manufacturers make specification tables available.

➡ *Perform **Inspect Front Suspension and Steering Linkage** Worksheet*

WHEEL ALIGNMENT PROCEDURES

The front suspension is designed to keep the wheels in the best possible position when rolling. Wheels must roll freely with as little tire scuff (side to side wear) as possible. Alignment settings can change according to vehicle speed, roughness of the road surface, acceleration, braking, weight distribution, or cornering. Specifications are developed by manufacturers so that alignment can be adjusted with the vehicle at rest on a level alignment rack (Figure 61.9). If tire wear is experienced when the settings are within specifications, the technician can make adjustments to compensate.

Adjustments to original alignment settings might be needed because of wear on the inside of pivoting parts, bad road conditions, collision damage, spring sag, or unusual loads. A new car might need an alignment because it came from the factory with adjustments outside of the normal range of specifications. Design engineers list a range of adjustment limits for wheel alignment angles. If the adjustments fall within those limits and the car "tracks" straight (goes straight on a level road), little tire wear should occur.

Only three of the five alignment angles are normally adjustable:
- Caster
- Camber
- Toe

The other two angles are measured to check for damage to steering parts.

NOTE: *Of the two adjustable angles, tire wear when travelling straight only results from incorrect camber and toe. Directional pull can be caused by camber and caster.*

NOTE: *On some MacPherson strut systems, only toe is adjustable.*

MEASURING ALIGNMENT

Ball bearing supported plates are placed beneath the tires. These allow the tires to assume a relaxed position (see Figure 61.9).
- On the alignment rack, the front wheels are positioned on **radius plates** (Figure 61.10). These have a gauge that measures in degrees how far a wheel is turned to the right or left.
- On four wheel alignment racks, *slip plates* are under the rear tires.

Pins hold the upper plate from moving on the bearings. After the vehicle is driven onto the plates, the pins are pulled out of the plates. Pushing down on the bumpers lets the wheels creep into a relaxed position as they would be in when rolling on the road. Air or hydraulic jacks are used to raise the car off the lift during alignment adjustments or to reposition the radius plates.

Sophisticated computerized alignment machines (Figure 61.11) are used to do **four wheel alignments** on many newer cars that are equipped with rear wheel adjustment capability. Mechanical measuring systems are covered here first.

The tools used for measuring caster and camber in older and portable mechanical alignment measuring systems have bubble levels. These are used for making comparisons to an exact level position. To get accurate measurements, the car must be level.

NOTE: *If tires are to be replaced, the new tires should be installed before attempting an alignment. An alignment*

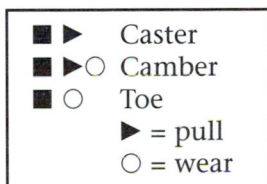

■ ▶	Caster
■ ▶○	Camber
■ ○	Toe
▶ = pull
○ = wear

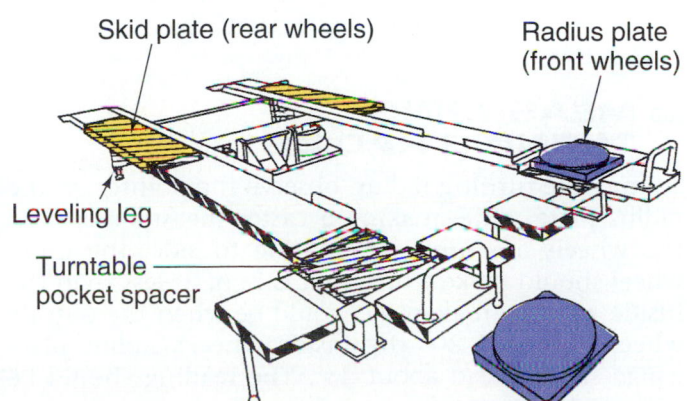

Figure 61.9 A wheel alignment rack with ball bearing supported plates for the front and rear wheels. *(Courtesy of Snap-on Tools Company, Copyright Owner.)*

Labels: Skid plate (rear wheels), Radius plate (front wheels), Leveling leg, Turntable pocket spacer

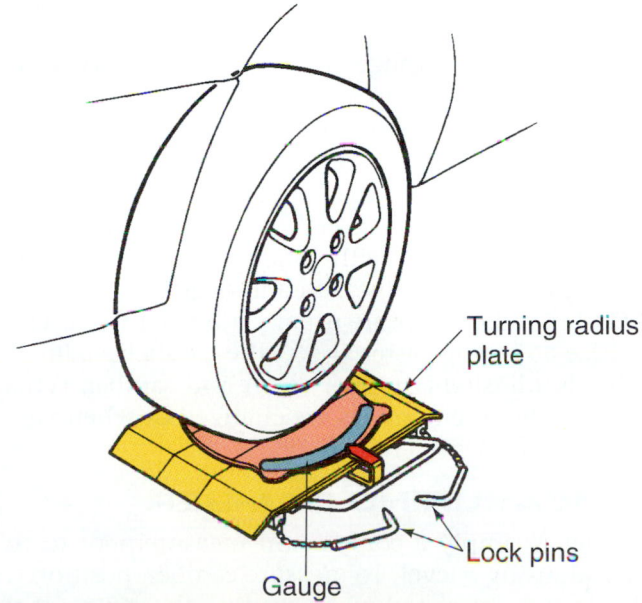

Figure 61.10 A radius plate. *(Courtesy of American Honda Motor Co., Inc.)*

Labels: Turning radius plate, Lock pins, Gauge

Figure 61.11 A modern four wheel alignment machine. *(Courtesy of Hunter Engineering Company)*

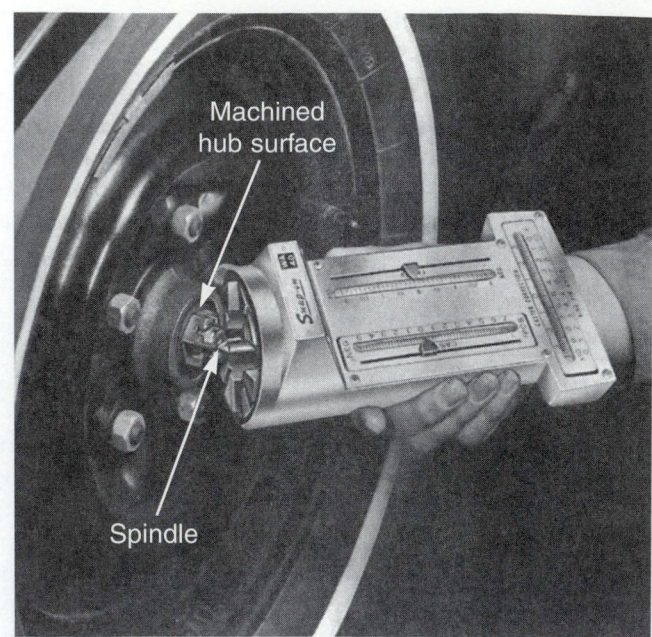

Machined hub surface

Spindle

Figure 61.12 A universal wheel rim adapter. *(Courtesy of Snap-on Tools Company, Copyright Owner)*

rack provides a level location for checking suspension angles. The heights of worn tires will affect alignment measurements.

To attach the gauges to the wheels, a wheel rim adapter clamp is used or a large magnet holds the gauge to the wheel hub (Figure 61.12). The tool is mounted against a machined area of the hub. The wheel bearing dust cap is removed so that a pilot can center the gauge on the center hole in the end of the spindle.

Toe is measured in either inches or millimeters. The remaining angles are measured in degrees of a circle.
- Caster and camber are usually adjusted together since adjusting caster affects the camber reading.
- Toe is adjusted last, after caster and camber. When other alignment angles are changed or when parts are replaced, toe changes.

■ MEASURING CAMBER

Camber is simply a comparison measurement to true vertical, using a level. To measure camber, position the wheels straight ahead while reading the gauge. If the wheels are not straight ahead, the caster angle can cause an incorrect camber reading.

■ CASTER MEASUREMENT

Caster is measured using the same gauge as camber. Caster causes the wheel's camber angle to change during a turn.

NOTE: *The caster measurement is actually a reading of camber while turning.*

To measure the caster setting:
- The wheel is turned either inward or outward 20° (depending on the design of the equipment).
- There are two levels on the gauge. One of the levels has a thumbscrew that allows it to be leveled. At the 20° point on the radius plate, set the level to 0°.
- Turn the wheel through a 40° sweep until the wheel is 20° in the opposite direction. At this point, the reading on the gauge is the caster reading. It will read in degrees, either positive or negative.

■ MEASURING TURNING RADIUS

To measure turning radius, observe the pointer on the radius plate while making a caster measurement. As the wheels are turned from side to side, the outer wheel should make a turn that is 2° or 3° less than the inside wheel. An example would be when the outside wheel is turned 20° the inside wheel's radius plate gauge should read about 23°. The reading should be within 1½° of specifications found in the service manual. Turning radius is *not* an adjustable angle, but if a steering arm becomes bent the turning radius changes and the tires will squeal on turns.

■ ADJUSTING CAMBER AND CASTER

Determining Correct Alignment Settings

Alignment settings for caster and camber are usually within a range. Just having the settings fall within the range will not ensure that the car will go straight.

■ ROAD CROWN AND PULL

Roads are **crowned** (higher at the center than the outside) so that rain will run off (Figure 61.13). A car aligned with equal settings from side to side will drift to the right on a crowned road.

To compensate for road crown, two methods can be used:
■ Camber can be set slightly more positive on the driver's side.
■ Caster can be set so that it is slightly more negative on the driver's side of the car.

There are many methods of adjustment for caster and camber. Alignment equipment manufacturers provide detailed charts and catalogs that describe specific adjustment methods. On SLA suspensions, the camber adjustment is done with shims or eccentrics. Figure 61.14 shows two common SLA adjustments in which *shims* are removed or installed to reposition the upper control arm.

> **SHOP TIP** Before making an alignment adjustment, unload the suspension components so the weight of the car is not resting on the wheels. This makes removing and replacing shims, or turning an adjustment eccentric easier.

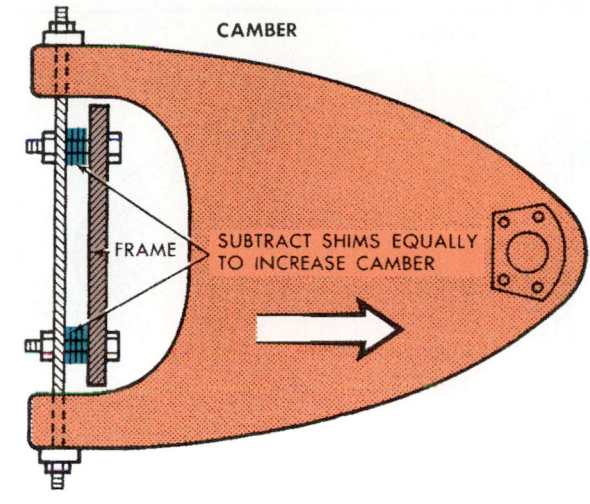

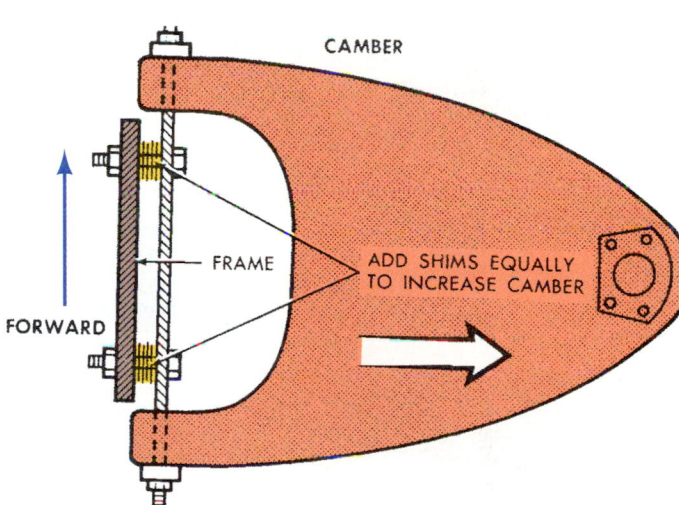

Figure 61.14 Changing camber with shims on inboard and outboard pivot shafts. *(Courtesy of General Motors Corporation, Service Technology Group)*

■ PLAN AHEAD

When there are shims, caster and camber are changed together. Removing and replacing shims is sometimes a fair amount of work. Plan ahead. Think about the effect the shim will have. Look at the control arm inner shaft to see what the effect of adding or removing shims will be. On some vehicles (usually light trucks) the pivot shaft is outboard of the frame. Shims have the opposite effect as the normal control arm with the pivot shaft inboard of the frame. Figure 61.14 shows how to change camber on inboard and outboard pivot shaft locations.

To change caster with shims requires removing or adding shims at either end of the control arm pivot shaft (Figure 61.15). This will change camber because it moves the control arm out or in on one side. To keep camber from changing much during a caster

Figure 61.13 Roads are crowned so water can run off them.

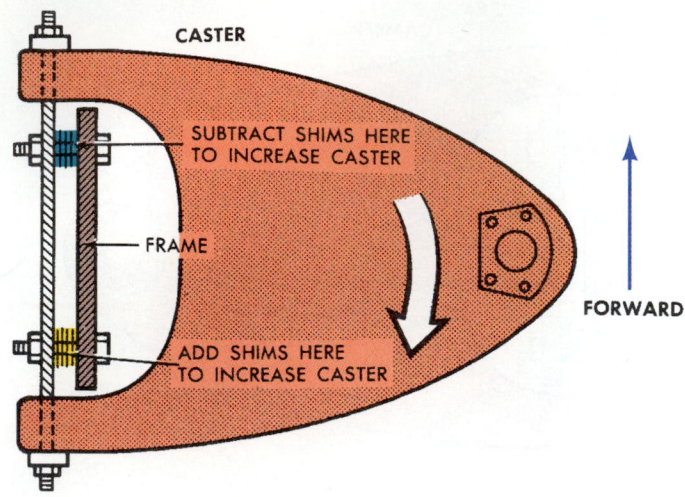

CASTER

SUBTRACT SHIMS HERE
TO INCREASE CASTER

FRAME

ADD SHIMS HERE
TO INCREASE CASTER

FORWARD

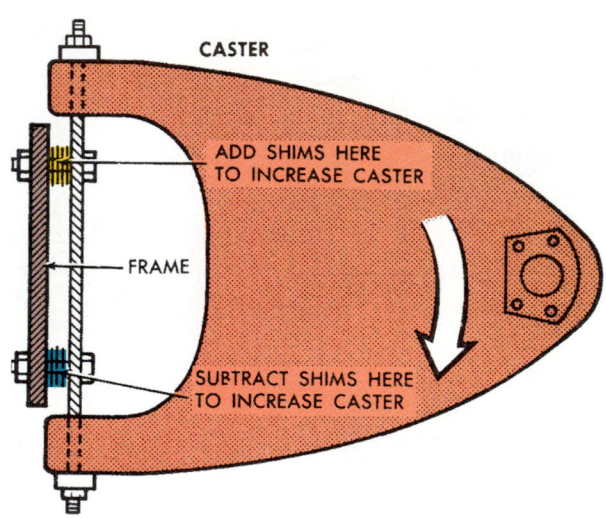

CASTER

ADD SHIMS HERE
TO INCREASE CASTER

FRAME

SUBTRACT SHIMS HERE
TO INCREASE CASTER

Figure 61.15 Adding shims from one side and/or subtracting them from the other moves the position of the upper ball joint fore or aft. *(Courtesy of General Motors Corporation, Service Technology Group)*

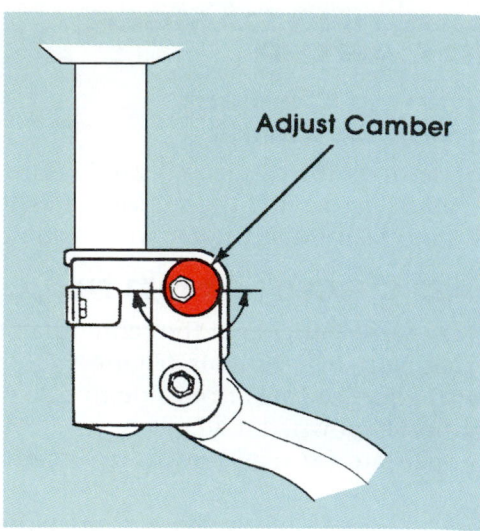

Adjust Camber

Figure 61.16 Camber can be adjusted at the bottom of this strut by turning the eccentric. *(Courtesy of Hunter Engineering Company)*

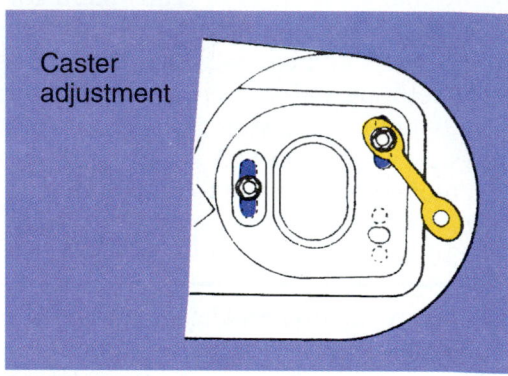

Caster adjustment

Figure 61.17 These slots at the top of the MacPherson strut tower provide a means of moving the strut to adjust alignment. *(Courtesy of Hunter Engineering Company)*

adjustment, remove a shim from one side of the pivot shaft and install at the other end.

Service manuals often state the effect each 1/64" of shim thickness will have on both the caster and camber readings. Often this information is not available. When a technician works repeatedly on one make of car, experimenting with one size of shim will show the effect that shim has on alignment settings.

An approximation to use with shims is:

■ 1/16" shim = 1/2° caster, or
■ Remove 1/32" from one and add to the other for 1/2° caster change.

Service manuals sometimes give a chart showing the amount of shim change to make to get a desired setting.

Some cars use an *eccentric cam* adjustment on the upper or lower control arm or strut (Figure 61.16).

Turning the adjustment repositions the camber angle.

When MacPherson struts are adjustable (not all of them are) there are slots in the upper bearing bracket for adjusting caster (Figure 61.17).

When a car has a narrow lower control arm, there is a strut rod running from it to the frame for strength. There are often threads on one end of the strut. Repositioning the nuts on the threaded area moves the control arm in an arc (Figure 61.18). The result is that caster changes. Moving the lower control arm toward the rear makes caster more negative.

■ MEASURING STEERING AXIS INCLINATION

Older vehicles used caster as the primary directional control angle. Steering axis inclination was only secondary for directional control. On front wheel drive MacPherson strut cars, SAI is now the primary direc-

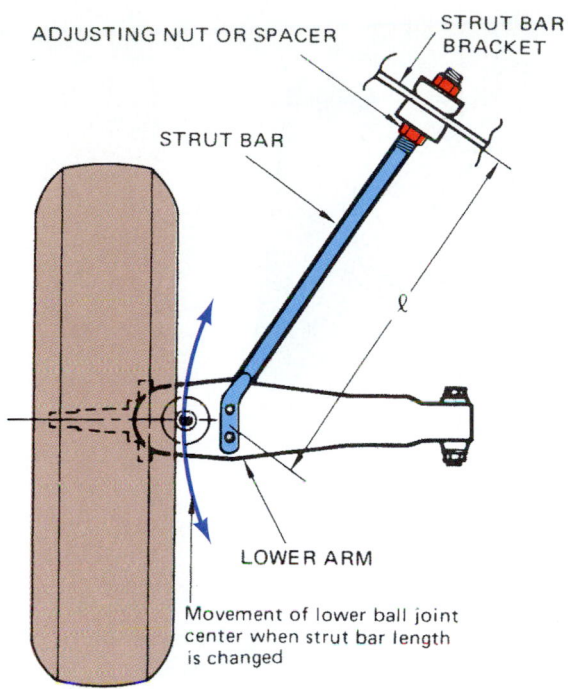

Figure 61.18 Adjusting the length of the strut bar changes the caster.

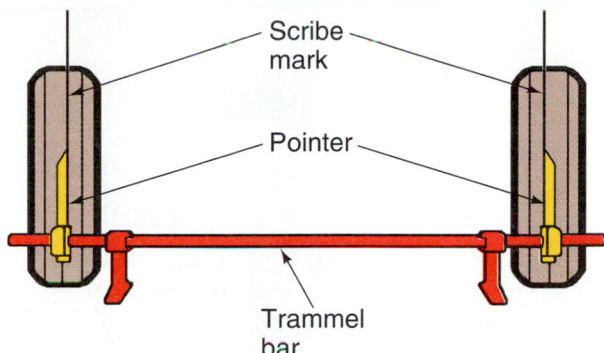

Figure 61.19 A trammel bar for measuring toe.

tional control angle and must be checked on every alignment. The newer cars have lighter suspensions that are more easily damaged.

SAI is not adjustable. It is not measured directly and some equipment does not measure SAI. SAI does not change unless the spindle has been bent. If this happens, the camber angle will also change. If the included angle is correct but the camber angle is not, the spindle is not bent. Replacing it will not correct the problem. If the included angle is not correct, parts must be replaced.

NOTE: *Look for a bent spindle when a front wheel bearing wears out. Front wheel bearings do not wear out very often, but a bent spindle can cause the bearings to be misaligned, increasing the load on it.*

SAI specifications are not usually provided by the manufacturer. The typical range of SAI specifications is:
- SLA/RWD—6–8°
- MacPherson/FWD—12–16°
- Maximum side to side variation is 2°

A typical car with an SLA suspension might have 7.5° SAI and 2.8° caster. A MacPherson strut front wheel drive car might have 14° SAI and 1.6° caster.

■ MEASURING INCLUDED ANGLE

When camber is out of specs, check the included angle. When the included angle is off, the spindle is bent. This requires replacement of the part.

The included angle is the amount of SAI minus camber. If camber is negative, subtract it from the SAI reading. If camber is positive, add it to the SAI reading. Included angle is usually within ½° from side to side. Double-check the specification.

■ MEASURING TOE

Checking and adjusting toe after replacing a tie rod end or other steering linkage component is an important part of the job. When measuring toe, the distances between the fronts and the rears of the front tires are compared (see Figure 60.1). Rear wheel drive cars are usually set slightly toed in (about ¹⁄₁₆" on new cars). Front wheel drive cars are set with 0 toe or a slight amount of toe out.

To measure toe, a *trammel bar* or *tram gauge* (Figure 61.19) or an optical toe measuring setup is used. The optical toe gauge projects an image to a gauge on the opposite side of the car (see Figure 61.22).

When measuring toe *without* an optical device, an accurate line must be scribed on the tread's footprint. With the tire raised off the ground, spin it while applying chalk to the center of its tread. Then, scribe a narrow line in the middle of the chalked area. Lower the car to the ground and compare the distances between the scribed lines on the tread surfaces at the front and rear of the tire. Use the trammel bar adjusted to measure at the same height as the spindles.

■ ADJUSTING TOE

Steering linkages on cars have two tie rods. On conventional parallelogram steering, the toe adjustment is made by turning the threaded "turnbuckle" sleeves on the tie rods (Figure 61.20). Because one of the tie rod ends has a left-hand thread and the other has a right-hand thread, as the sleeve is turned the tie rod will become either longer or shorter. This changes the toe setting.

First, center the steering wheel and hold it in place with a steering wheel clamp (Figure 61.21). Then make the adjustment. When making a mechanical measurement with a trammel bar, sight down the side of the

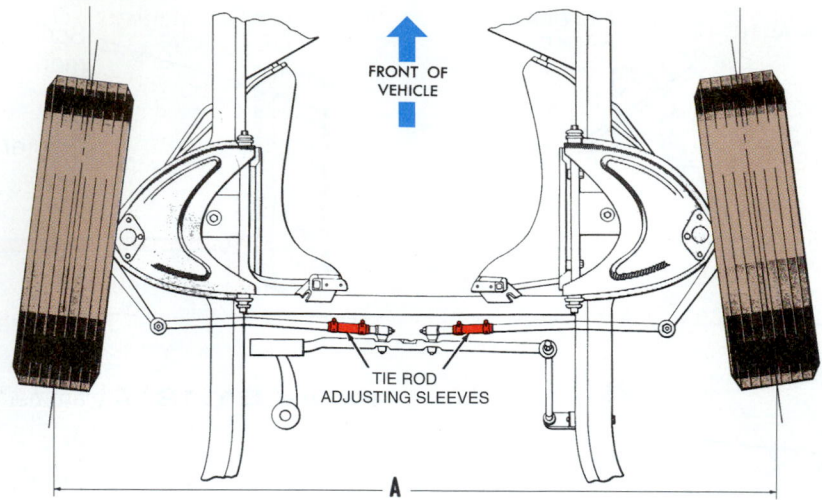

Figure 61.20 Turn the adjusting sleeves to adjust toe. *(Courtesy of SPX Corporation)*

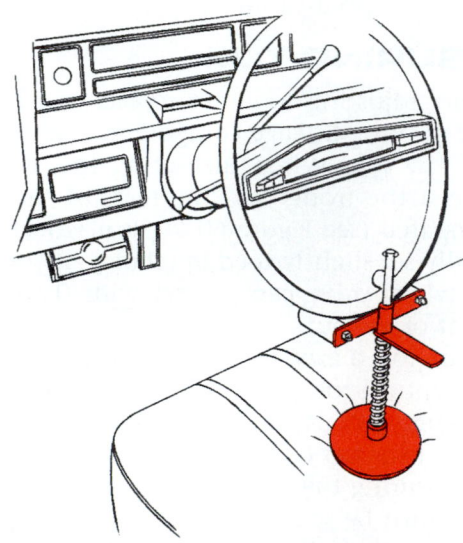

Figure 61.21 Center the steering wheel and hold it in place with a steering wheel clamp. *(Courtesy of Hunter Engineering Company)*

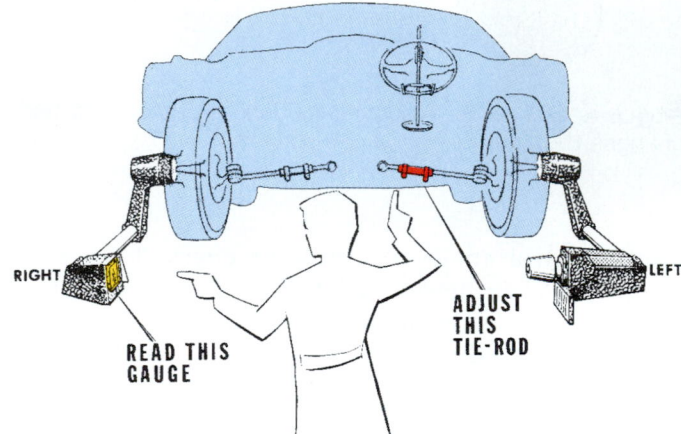

Figure 61.22 Adjust the tie rod on the opposite side of the gauge. *(Courtesy of Snap-on Tools Company, Copyright Owner)*

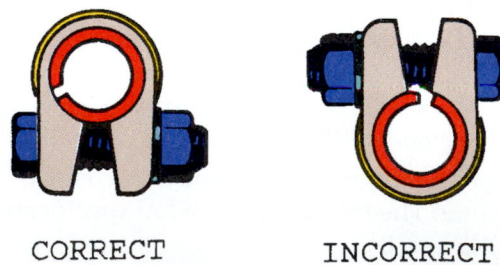

CORRECT   INCORRECT

Figure 61.23 Position the clamp correctly.

front tire to the rear tire and make an adjustment so that each front wheel looks to be in line with its rear one. With an optical toe gauge, read the gauge on one side while adjusting the tie rod on the other (Figure 61.22).

After turning the adjusting sleeves, the clamp must be properly positioned before tightening it. It must not come into contact with anything when the wheels are turned from side to side. Be sure that the opening of the clamp is not positioned over the split in the adjusting sleeve (Figure 61.23). On parallelogram systems if one tie rod happens to be tilted one way and the other is tilted the other way, the tie rod will not be able to pivot (Figure 61.24). The tie rods are positioned so that they are not binding up. This is most easily

done by turning both tie rods in the same direction before tightening the clamp.

Rack and pinion steering systems have a single tie rod end with a jamb nut on each side (Figure 61.25). Hold the tie rod end with pliers while loosening and tightening the jamb nut. The inner end of the tie rod can be rotated with pliers to adjust toe.

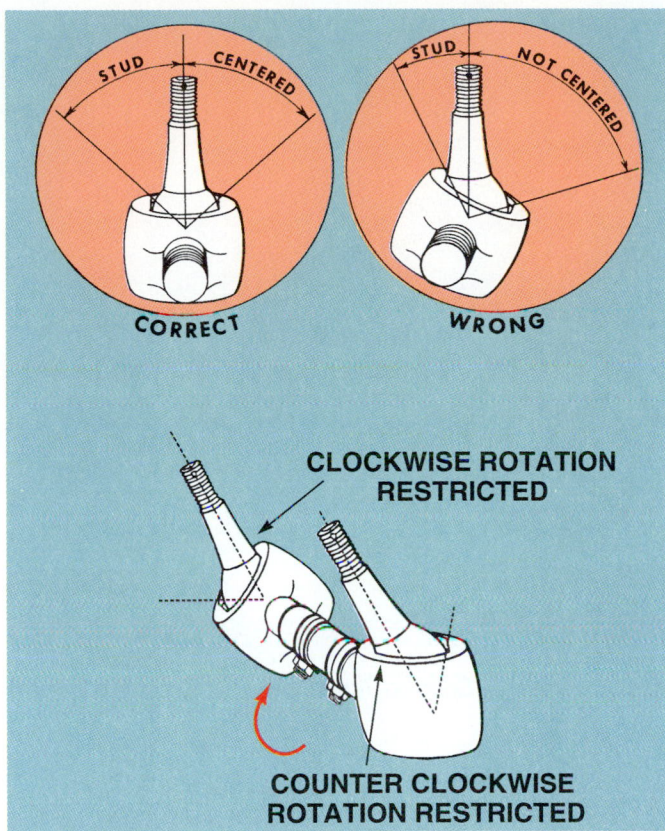

Figure 61.24 Position the tie rods so that they will not bind. *(Courtesy of Moog Automotive, Inc.)*

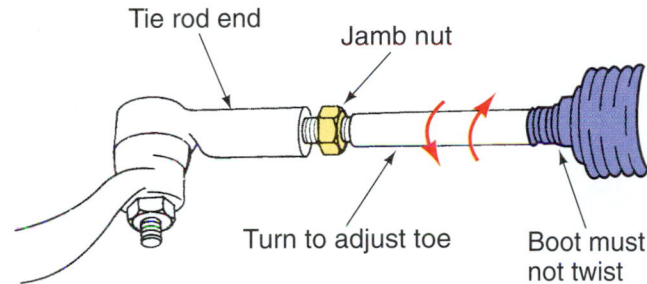

Figure 61.25 Rack and pinion tie rod end. *(Courtesy of Federal-Mogul Corporation)*

■ CENTERING THE STEERING WHEEL

Adjusting one of the tie rods more than the other will cause the steering wheel to be off-center.

NOTE: *In most cars it is important that the steering wheel be centered using the tie rods, not by removing the steering wheel and putting it back on straight.*

If the steering wheel is not straight ahead when driving, follow this procedure to straighten it:

■ Count the number of turns of the steering wheel while turning it from lock to lock.

■ Position the steering wheel so that it is half-way between the locks. It should be straight. If not, remove it and put it back on straight.

■ Use a steering wheel lock to hold the steering wheel in the centered position (see Figure 61.21).

SHOP TIP From the front of the car, sight down the tires on each side of the car to see if the front one aligns with the rear one. This will give a rough estimate of what direction the tie rods need to be turned.

■ When making a rough adjustment, turn each tie rod an equal amount in opposite directions. This will approximately maintain the current toe setting.

■ Test drive the car on a straight, level road. If the steering wheel is off-center and points to the left, adjust the tie rods so that the tires also point to the left (Figure 61.26). Adjust the wheels to face the opposite direction if the wheel points to the right.

■ TOE CHANGE CHECK

Because toe can change with bumps, toe is accurate when the vehicle is at the correct ride height only. Use the jack on the alignment rack to raise the vehicle an equal amount on each side (Figure 61.27). Check to see that toe changes equally on each wheel. If not, one end of the steering linkage is at an incorrect height.

Sometimes, an idler arm will have slotted mounting holes so that its height can be adjusted. Rather than bending a steering part (which can present a liability problem), try to find the reason for the problem. The vehicle might have been in an accident and require the services of a frame straightening shop.

■ FOUR WHEEL ALIGNMENT

Modern wheel alignment equipment measures wheel alignment of all four wheels. *Rear wheel toe* is often adjustable. The **geometric centerline** of the vehicle is a line drawn between the center of the front axle and the center of the rear axle (Figure 61.28). When individual toe is measured at all four wheels, the geometric centerline is used.

When the rear wheels are held in a fixed position, the **thrustline** defines the wheels' true straight-ahead position. **Thrust angle** is the angle formed by the thrustline and the geometric centerline. Correct rear wheel alignment involves adjusting rear wheel toe to factory specifications with the thrust angle at or near zero. This means that the left rear might be toed out ¼" and the right rear toed in ¼". The result is that the steering wheel will be straight ahead when the car is driving on a level highway.

NOTE: *Be sure there are no heavy loads in the vehicle. The rear axles of smaller front wheel drive cars are so light that they can flex under load, resulting in camber change.*

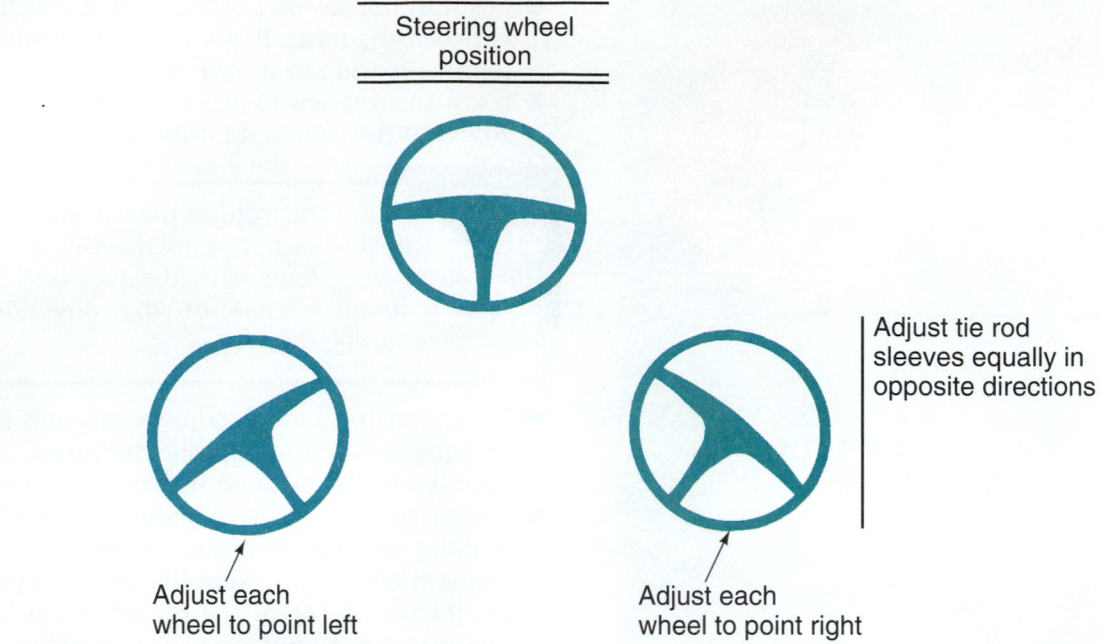

Figure 61.26 Adjusting steering wheel center using the tie rods. *(Courtesy of Snap-on Tools Company, Copyright Owner)*

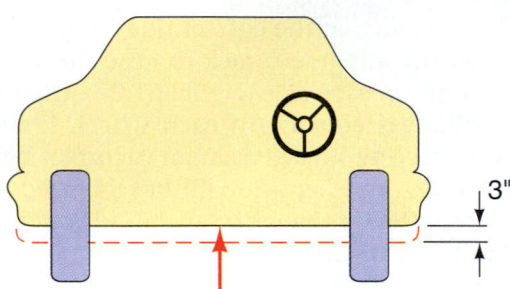

Figure 61.27 Raise the vehicle equally from side to side to check for toe change. *(Courtesy of AMMCO®, Registered trademark of Hennessy Industries, Inc.)*

During a computer wheel alignment:
- The steering wheel is held in the straight-ahead position.
- Alignment gauges are installed on all four wheels.
- The rear wheels are aligned first, then the fronts.
- The front wheels receive signals from the rears telling them the thrust angle. Rear toe must be read in order to calculate this.
- When toe on all four wheels has been adjusted correctly, the steering wheel will be correctly centered.

■ GENERAL WHEEL ALIGNMENT RULES

A suspension/steering specialist will experiment with different vehicles to see what changes result from various adjustments. The following general rules are provided as a starting point. Due to variations in design, they will not always be true.

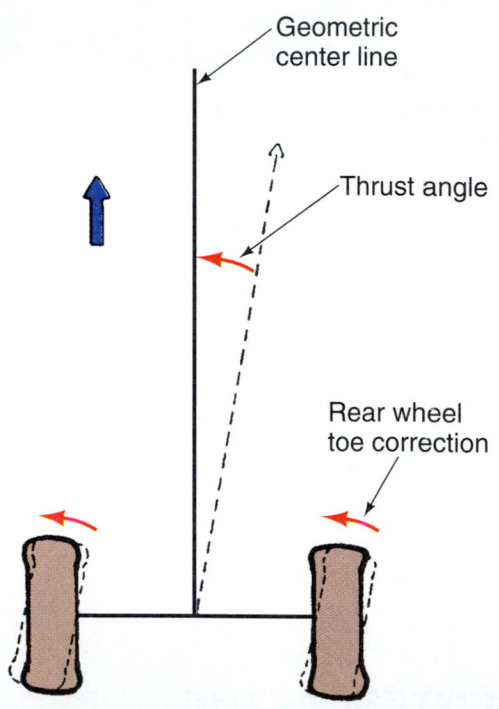

Figure 61.28 Factors in rear wheel toe adjustment. *(Courtesy of Hunter Engineering Company)*

Caster/Camber

- The car will pull to the side with the least (most negative) caster.
- The car will pull to the side with the most (most positive) camber.
- Negative caster results in easier steering.

- The driver's weight will usually cause camber to increase on the left front wheel and decrease on the right front wheel.
- A ¹⁄₁₆" shim will cause approximately ½° change in camber.
- On shim type cars, adjust caster first, then camber.
- Caster for both wheels should be set either positive or negative; not one positive and one negative.
- Caster **spread** between the front wheel settings should not be more than ½°. Spread is the difference between the measurements from side to side.
- Make caster equal from side to side. Use camber to compensate for road crown. Set the left side with ¼° more positive than the right (if caster is equal).
- On manual steering cars, caster is 0 to 1° negative. Power steering cars can have caster as high as 10° (Mercedes).

- On a MacPherson strut car, jounce the car while measuring camber. If camber changes drastically on either front wheel, the strut is bent.

Toe-in

- Before driving on the alignment rack, be sure the pins are in the radius plates.
- Every ¼ turn of the adjusting sleeve results in about ¹⁄₁₆" of change in toe.
- Loosen both adjusting sleeves while adjusting toe.
- Changes in caster and camber affect toe, so it is adjusted last.
- Toe is measured in either inches or millimeters.

■ REVIEW QUESTIONS

1. When a front wheel drive car has two drive axles of different lengths, what can happen during acceleration?

2. If you move your hand across the tire's footprint area from the outside to the inside and feel a _____ edge, there is excessive toe-in.

3. On a rear wheel drive car with excessive toe-in, the _____ edge of the right front tire tends to wear.

4. What is the name of the test where an assistant turns the steering wheel back and forth a short distance while you look for looseness in the steering linkage?

5. Which of the alignment angles are normally adjustable?

6. Which alignment angle is usually measured in inches or millimeters, rather than degrees of a circle?

7. _____ caster results in easier steering.

8. When the road is higher in the center than on the outside edges, what is this called?

9. Is SAI adjustable?

10. The included angle is the amount of SAI _____ camber.

■ ASE STYLE REVIEW QUESTIONS

1. Technician A says that a car with the steering wheel off-center needs a toe adjustment. Technician B says that a bent steering arm will cause a change in turning radius. Who is right?
   a. Technician A   b. Technician B
   c. Both A and B   d. Neither A nor B

2. Technician A says that a change in camber can cause a change in toe. Technician B says that a change in toe will cause a change in camber. Who is right?
   a. Technician A   b. Technician B
   c. Both A and B   d. Neither A nor B

3. Technician A says that toe can change when springs sag. Technician B says that centering the steering wheel is done by adjusting the tie rods. Who is right?
   a. Technician A   b. Technician B
   c. Both A and B   d. Neither A nor B

4. Technician A says that caster is read with the wheels facing straight ahead. Technician B says that caster spread between the front wheel settings can be up to 2°. Who is right?
   a. Technician A   b. Technician B
   c. Both A and B   d. Neither A nor B

5. Technician A says that the car will pull to the side with the least (most negative) camber. Technician B says that the car will pull to the side with the most (positive) caster. Who is right?
   a. Technician A   b. Technician B
   c. Both A and B   d. Neither A nor B

Drive Train

THEN AND NOW: FRONT WHEEL DRIVE

Although *front wheel drive (FWD)* did not appear as a regular design in American cars until the mid-1980s, it is not a new idea. It dates back to 1877, when New York attorney George B. Selden filed a patent on a "road engine" with a transversely mounted engine driving its front wheels. The car, which was not actually built until 1905, could not really be called a usable vehicle.

The advantages of pulling with the front wheels instead of pushing with the rears have long been an attraction to vehicle designers. French automotive pioneer Andre Citroen is credited with saying, "driving the front wheels is a natural idea. After all, the horse doesn't push the cart." The idea was fine in theory, but front wheels are required to steer while rising and falling with bumps in the road. The ordinary universal joint used in rear wheel drive cars simply could not handle the sharp angles required for directing torque to front wheel drive wheels. Front wheel drive cars required a component that had not been invented yet.

In 1928, Alfred Rzeppa invented a new type of joint capable of transmitting twisting power through very sharp bends. This device was given the name *constant velocity* or *CV joint*. There were serious durability problems with early CV joints. They needed more development and better lubricants before they could be considered dependable, so most car makers stayed with RWD. There were some notable exceptions, however, such as the beautiful Cord of the 1930s.

Improved technology and the genius of English engineer Sir Alec Issigonis came together in 1960 with the introduction of the first truly successful FWD automobile. The Austin-Morris Mini stands as a landmark in automotive history. In the United States, the 1966 Oldsmobile Toronado was the first modern-day FWD car.

Today, well over half of the automobiles on our roads are equipped with front wheel drive. This layout is popular because it makes more efficient use of space, improves traction, and helps avoid spin-outs. But FWD also provides more complicated and expensive repair and service opportunities.

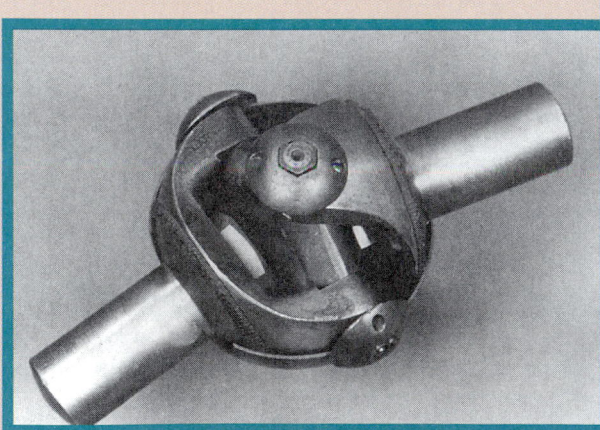

Plain universal joints, such as Clarence Spicer's original 1901 version, cannot cope with the angles FWD requires. *(Courtesy of Dana-Perfect Circle)*

Inside a modern CV joint. *(Courtesy of Bob Freudenberger)*

Clutch Fundamentals

■ KEY TERMS

clutch hub
dampened hub
pressure plate assembly
release levers
release bearing
diaphragm spring
quill
clutch fork
overcenter assist spring
slave cylinder
clutch free play

■ INTRODUCTION

A clutch is found on cars with manually shifted transmissions. The clutch disengages the engine from the transmission, allowing it to continue running while the car is at rest (Figure 62.1). It also allows for releasing the engine from the transmission during gear shifts. The driver controls the application of a clutch from inside the vehicle, using the clutch pedal.

Driver control allows the clutch to gradually be applied. This is important because the internal combustion engine does not make sufficient torque at lower engine rpm to be able to move the car. The clutch must be "slipped" so it gradually connects the rear wheels to the engine. There are two *driving* members of the clutch. An additional part, a friction disc that acts as a *driven* member, is positioned between the driving parts.

■ CLUTCH PARTS AND OPERATION

The clutch has several parts: a flywheel, pressure plate, disc, and release mechanism (Figure 62.2). If the clutch disc is pushed against the flywheel with enough force, the disc will rotate with the flywheel.

■ CLUTCH DISC

The clutch disc has splines that fit over splines on the transmission input shaft. The splines allow the disc to slide back and forth on the input shaft as it is first clamped to the flywheel and then released. It has several main parts: a hub and damper, facings, and plate.

Hub and Damper

The **clutch hub** is the inner part of the disc. It is the part that has the splines that slide over the input shaft. The number of splines differs from make to make. Some hubs have as many as 26 splines. Others have as few as 10.

Most clutch disc hubs are **dampened hubs** (Figure 62.3). *Torsional dampers* that are either coil springs or rubber are positioned between the disc plate and the clutch hub to absorb shock during engagement. Some hubs used on smaller cars are rigid, with no dampening.

A series of large rivets, called *stop pins*, prevent the hub from compressing the springs or rubber too far. The springs cushion the application. Engineers can tune

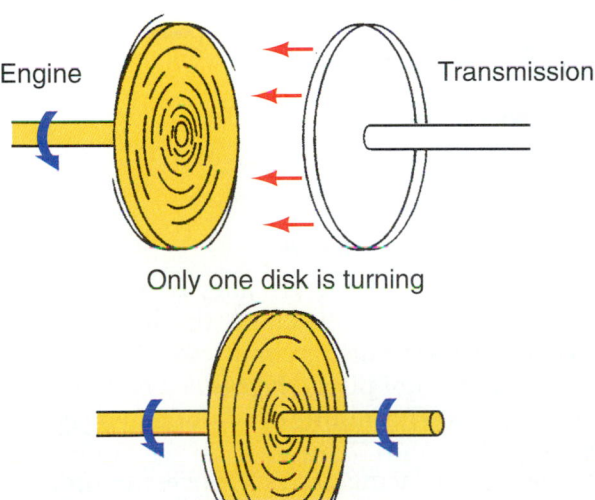

Engine — Transmission

Only one disk is turning

Both disks are turning

Figure 62.1 The clutch disengages and engages the engine to the transmission. *(Reproduced by permission of Deere & Company. ©1991, Deere & Company. All rights reserved.)*

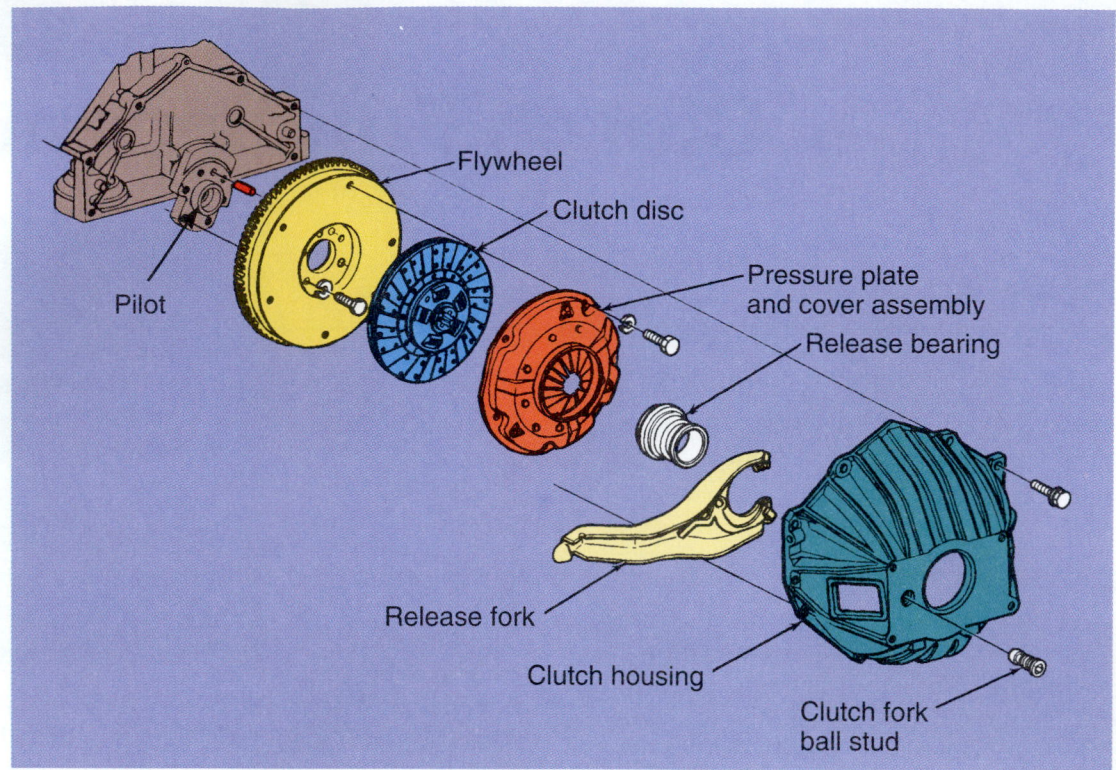

Figure 62.2 Parts of the clutch assembly. *(Courtesy of General Motors Corporation, Service Technology Group)*

Figure 62.3 A dampened clutch hub. *(Courtesy of LuK, AS, Automotive Systems)*

the design of the springs and windows to give different characteristics to the clutch. Torsional dampers keep engine vibration from causing noise and wear in the transmission.

Clutch Disc Facing

The clutch disc has facings made of *friction material,* traditionally containing *molded* or *woven asbestos.* Facing material must be able to withstand the extreme heat that is generated as the clutch is slipped during engagement. Fibers are visible in a woven disc. Woven asbestos transfers heat better than molded does.

Asbestos is an excellent friction material for brakes and clutches but it is regulated because it is dangerous to breathe. It can cause lung problems and a cancer of the chest cavity. Non-asbestos materials using fiberglass or Kevlar are being substituted for asbestos.

Clutch Disc Plate

Facings are riveted to both sides of a cushion plate. The *cushion plate* is riveted to the *disc plate* (Figure 62.4). As the clutch disc is compressed between the flywheel and pressure plate, the *cushion plate* lets the facings compress. This results in smoother engagement of the clutch.

The clutch facings have grooves so they do not stick to the flywheel. Air is trapped in the grooves when the clutch is engaged. When the clutch is released, the trapped air has centrifugal force that pushes the disc away from the pressure plate and flywheel.

■ PRESSURE PLATE

The **pressure plate assembly** is also called the *clutch cover assembly.* The pressure plate is a cast iron plate that is actually part of the cover assembly. The cover

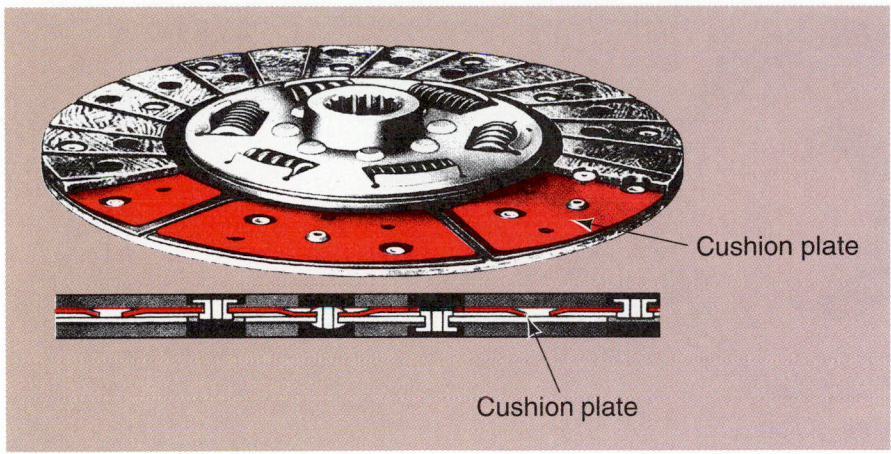

Cushion plate

Cushion plate

Figure 62.4 Between the two clutch facings is a cushion plate. *(Courtesy of LuK, AS, Automotive Systems)*

assembly is bolted to (and rotates with) the flywheel. The clutch disc is wedged between the pressure plate and the flywheel. There is spring pressure on the disc during normal operation (Figure 62.5). This is because

Clearance with foot off of pedal

Release bearing

Clutch fork

Clutch disc squeezed against flywheel

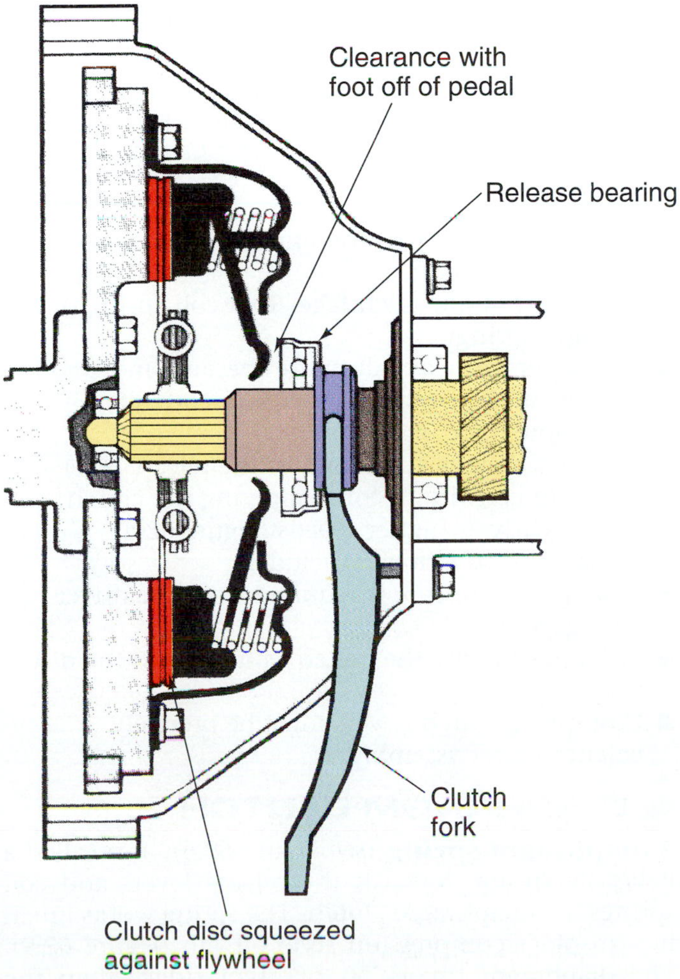

Figure 62.5 The clutch disc is squeezed against the flywheel by spring pressure when the pedal is up. *(Courtesy of Federal-Mogul Corporation)*

the space between the pressure plate and the flywheel is less than the thickness of the clutch disc. Remember, the inside of the clutch disc hub has splines that connect it to the input shaft on the transmission. This means that the engine and transmission are physically connected when the clutch pedal is released.

■ TYPES OF CLUTCH COVERS

There are two main types of clutch covers: coil spring and diaphragm (Figure 62.6). There are other designs of clutches but these are what most cars have.

■ COIL SPRING CLUTCH

A coil spring clutch cover is shown in Figure 62.5. Pressure plate springs are preloaded when the clutch cover assembly is put together at the factory. More compression of the springs takes place when the cover assembly is bolted to the flywheel. When the clutch is engaged (foot off of the pedal), the pressure plate exerts a force of between 1000 and 3000 pounds on the disc. Heavy-duty clutches exert force nearer 3000 pounds and light duty is closer to 1000 pounds. The pressure plate is designed so that when the disc is totally worn out, there will be about 10% more torque carrying capacity left in the clutch than the engine can deliver.

NOTE: *There is often confusion regarding the terms that apply to clutch application and release.*
■ *When you **apply** the pedal, you **release** the clutch.*
■ *When you **release** the pedal, you **apply** the clutch.*

■ RELEASE LEVERS

Release levers, also called *fingers*, are attached to the cover assembly at pivot points. One end of the release lever contacts the **release bearing**. The other end pulls or pushes on the pressure plate. Pushing on the clutch pedal moves the pivot lever, which pulls the pressure plate away from the flywheel (Figure 62.7).

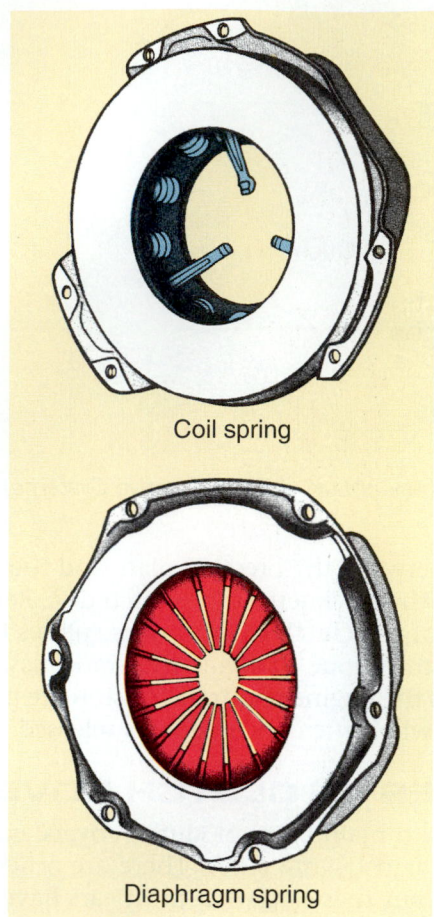

Coil spring

Diaphragm spring

Figure 62.6 Coil and diaphragm spring clutch covers. *(Courtesy of Federal-Mogul Corporation)*

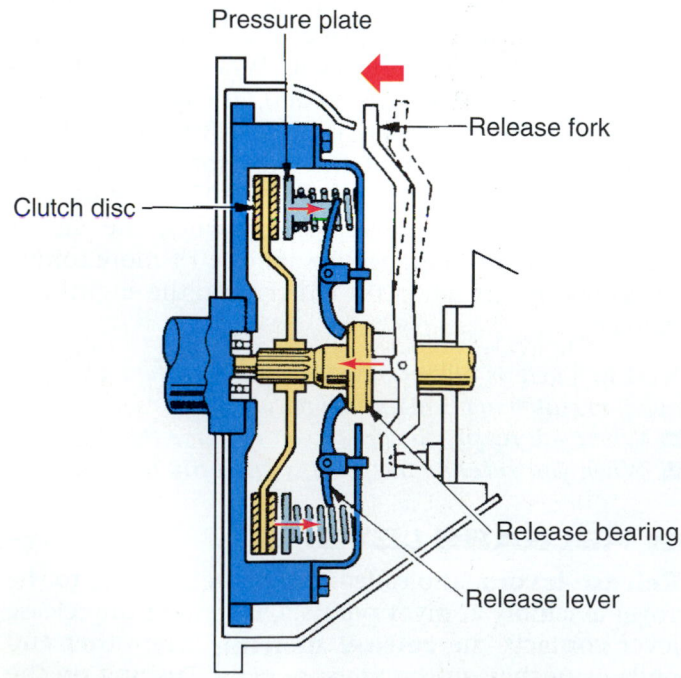

Pressure plate

Release fork

Clutch disc

Release bearing

Release lever

Figure 62.7 Pushing on the pedal moves the pivot lever to pull the pressure plate away from the disc.

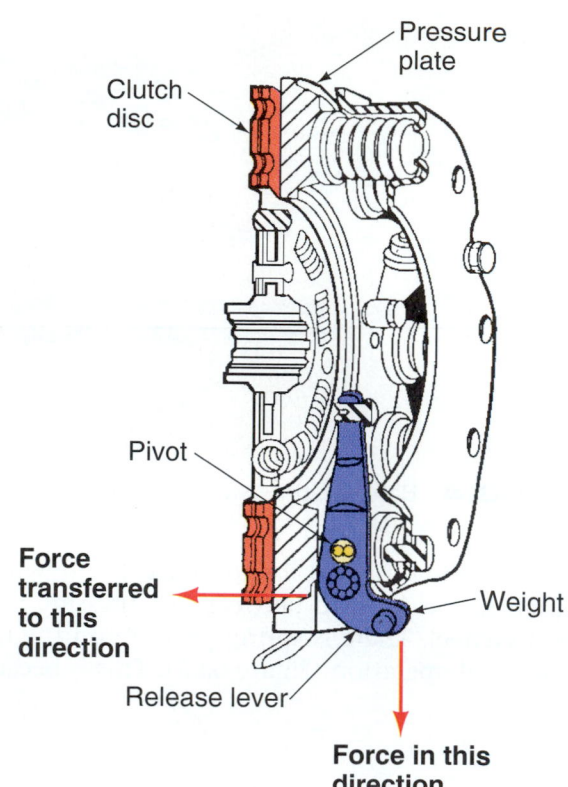

Pressure plate

Clutch disc

Pivot

Force transferred to this direction

Release lever

Weight

Force in this direction

Figure 62.8 A weight on the release lever causes centrifugal force to apply more force against the disc at higher speeds. *(Courtesy of Ford Motor Company)*

This action releases the disc's link between the engine and the transmission.

There are some advantages to a coil spring over a diaphragm spring.

■ It is better for heavy-duty uses because more coil springs can be installed to make a clutch apply with more force.

■ Putting a weight at the end of the release lever results in centrifugal force applying the clutch more tightly at higher speeds (Figure 62.8).

Coil spring disadvantages include:

■ More pedal pressure is required from the driver to disengage it.

■ It does not apply the clutch as heavily as the disc wears.

■ Coil spring clutch covers must be precisely balanced after assembly.

■ DIAPHRAGM CLUTCH

A **diaphragm spring** (see Figure 62.6), also called a Belleville spring, replaces the release levers and coil springs in a diaphragm clutch. The spring works much like an older compression style oil can (Figure 62.9). The diaphragm pivots off of pivot rings when the clutch pedal is depressed.

■ A diaphragm clutch requires lower pedal operating pressure.

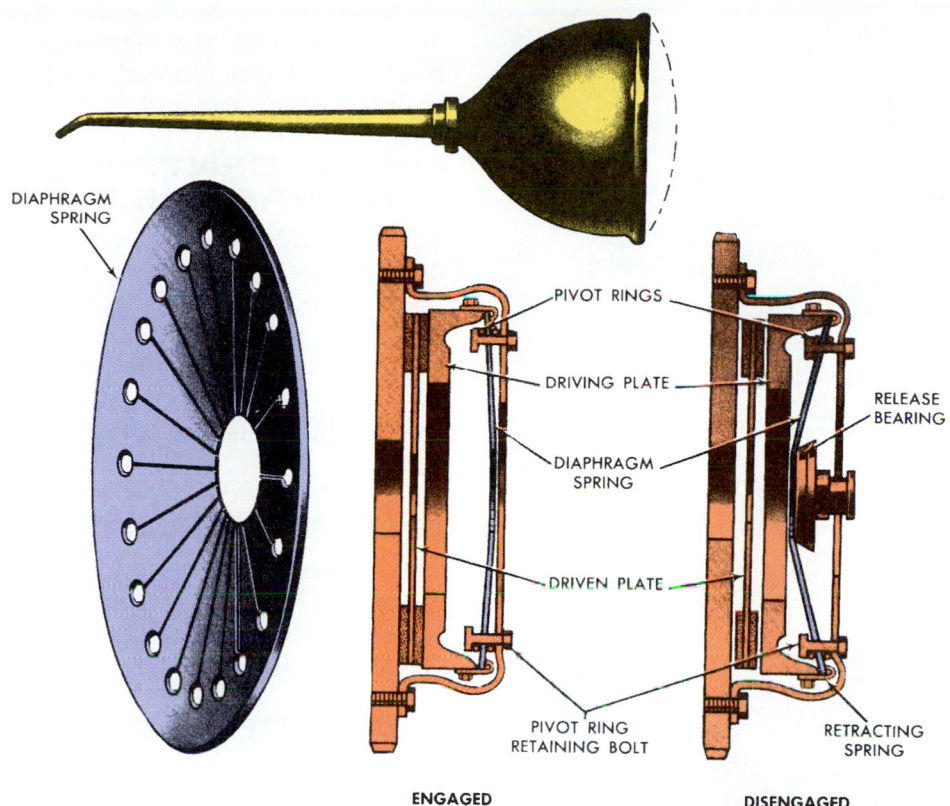

DIAPHRAGM
SPRING

PIVOT RINGS

DRIVING PLATE

DIAPHRAGM
SPRING

RELEASE
BEARING

DRIVEN PLATE

PIVOT RING
RETAINING BOLT

RETRACTING
SPRING

ENGAGED

DISENGAGED

Figure 62.9 The diaphragm spring works like an oil can. *(Courtesy of General Motors Corporation, Service Technology Group)*

- It also takes up less space. Sometimes, this is helpful when a clutch is being installed in a tight spot in a front wheel drive car.
- As the clutch disc wears, spring pressure exerted on the disc actually becomes greater due to the overcenter design of the diaphragm.
- Diaphragm clutches are well balanced when compared to coil spring clutches because of their type of construction.

PILOT BEARING OR BUSHING

The engine side of the transmission input shaft must be supported. It is supported by a sealed pilot bearing or sintered bronze bushing pressed into the end of the crankshaft (see Figure 62.2). Figure 62.10 shows three styles of pilots. Some front wheel drive transaxles do not use a pilot bearing because a roller or two ball bearings support the front of the transmission shaft (Figure 62.11).

RELEASE BEARING

The release bearing, also called a *throwout bearing* (see Figure 62.5) allows the pressure plate release mechanism to operate as the crankshaft rotates. The release bearing slides on the front transmission bearing retainer, often called a **quill** (Figure 62.12). Conventional release bearings are used in rear wheel drive

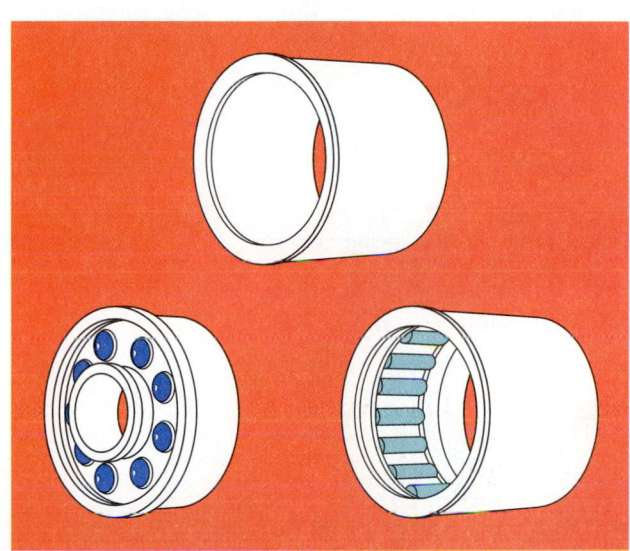

Figure 62.10 Three styles of pilots. *(Courtesy of Federal-Mogul Corporation)*

cars. A special kind of release bearing is used in some front wheel drive cars.

The release bearing is lubricated and then sealed at the factory. A *hub*, or *collar*, is sometimes part of the bearing (Figure 62.13). Other times the hub is pressed fit and is replaceable.

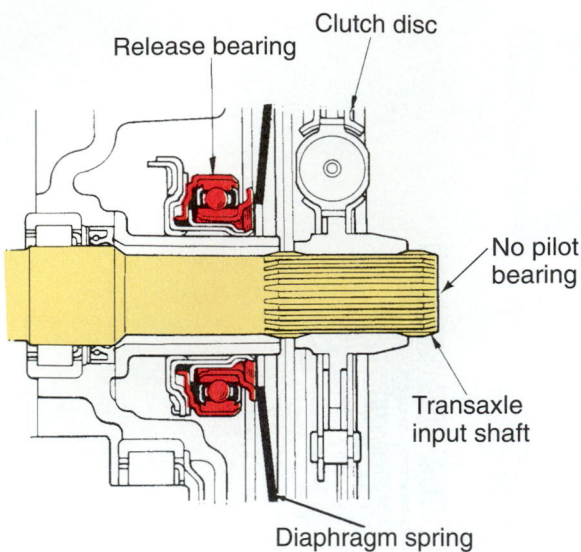

Figure 62.11 This front wheel drive transaxle does not use a pilot bearing.

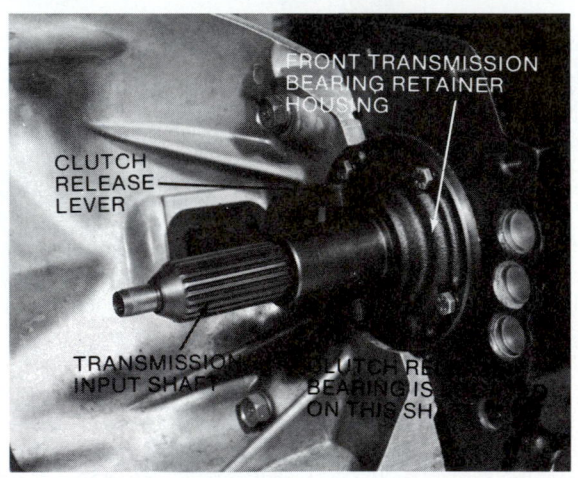

Figure 62.12 The release bearing slides on the transmission front bearing retainer.

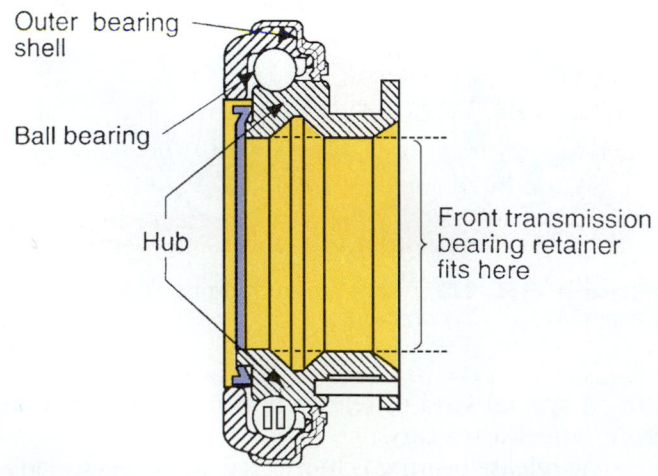

Figure 62.13 The hub or collar is part of this release bearing. *(Courtesy of General Motors Corporation, Service Technology Group)*

Some release bearings are flat on their face and others are curved. This depends on whether the release levers are flat or curved. Curved release levers go against flat bearings and vice versa.

■ SELF-CENTERING RELEASE BEARING

Self-centering release bearings are used on front wheel drive cars because they do not use a pilot bearing in the crankshaft. Without a pilot bearing, the transmission input shaft might not be correctly aligned with the rear of the crankshaft. If there is misalignment during clutch application, the bearing can apply the clutch while it is not correctly aligned. The result is noise.

■ OTHER RELEASE BEARINGS

Some release bearings found on older European cars use a round carbon ring that slowly wears away with each use. There are also special release bearings found on cars that have pressure plates that pull to release.

■ CLUTCH FORK

The release bearing hub has a provision to attach it to the **clutch fork** (Figure 62.14). The clutch fork, also called a *throwout lever* or *release arm*, fits between the release bearing and the clutch cable or linkage. The clutch fork has a pivot shaft (sometimes called a *cross shaft*) or a pivot ball or raised area in the bell housing that it pivots off of.

■ CLUTCH RELEASE METHODS

The clutch pedal operates the clutch fork using either linkage, a cable, or hydraulic cylinders. A clutch start switch is included on the clutch pedal on late model vehicles (see Figure 33.10).

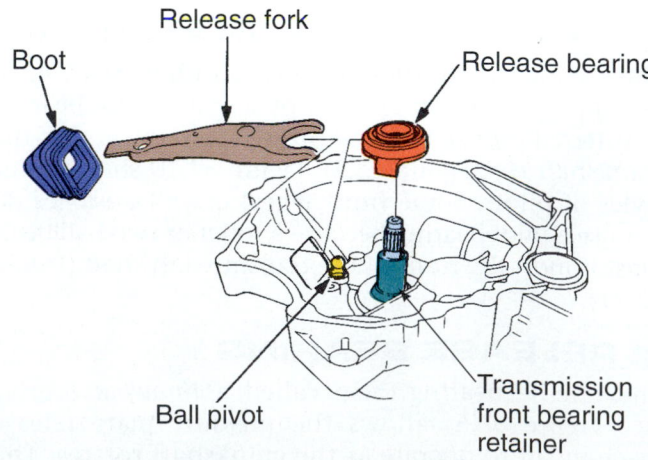

Figure 62.14 The clutch fork fits between the release bearing and the clutch linkage. *(Courtesy of American Honda Motor Co., Inc.)*

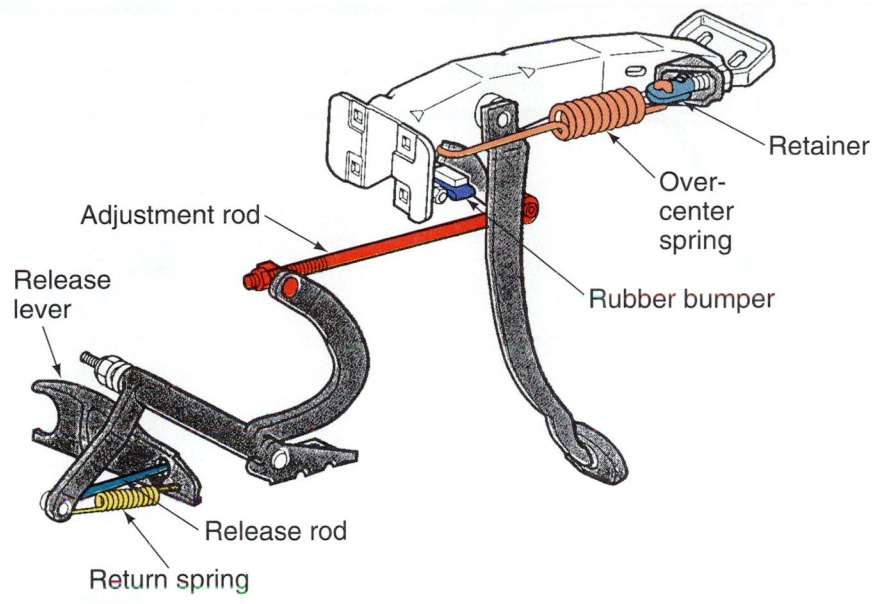

Figure 62.15 Mechanical clutch linkage. *(Courtesy of Ford Motor Company)*

■ MECHANICAL LINKAGE

Mechanical linkage found on some older cars and trucks has rods and pivot arms that carry motion from the clutch pedal to the release fork (Figure 62.15). Clutch free play adjustment can be made on a threaded rod. An **overcenter assist spring** pulls the pedal upward during the first half of travel and toward the floor during the second half of travel. This makes releasing the clutch (applying the pedal) easier. Using an overcenter spring is the same idea as used on hood springs.

NOTE: *To see how an overcenter spring works, carefully operate the clutch pedal with the linkage disconnected. The pedal will fall strongly toward the floor during the second half of travel. Keep your fingers out of the way.*

A drawback to mechanical linkage systems is that when an engine or transmission moves on its rubber mounts, clutch adjustment and amount of application changes.

■ CLUTCH CABLE

Newer cars use a less expensive cable to operate the clutch (Figure 62.16). A cable is flexible so its adjustment remains the same as the engine moves. This is especially important on four cylinder, front wheel drive cars. The engine moves quite a bit on these cars. Drawbacks to cables are that they must be lubricated and they develop friction and wear with repeated use.

Linkage can push on a clutch arm, while a cable can only pull on it. With a cable, the pivot point of the fork must be on the outside of the input shaft away from the cable end. The amount of available adjustment to a cable system is less than with linkage.

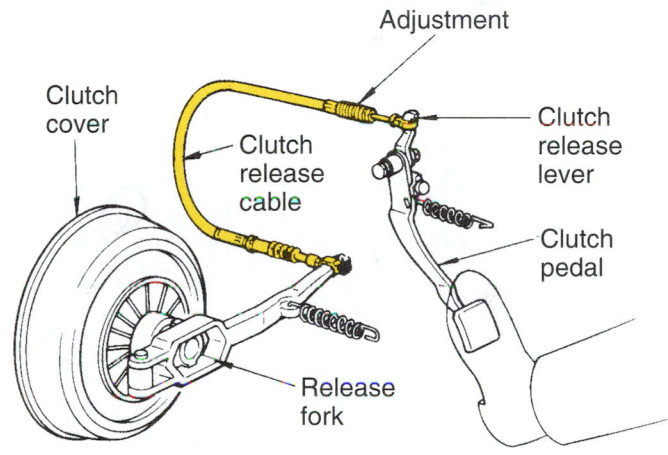

Figure 62.16 A cable-operated clutch release.

■ HYDRAULIC CLUTCH OPERATION

Hydraulic clutches are found on many types of vehicles. Hydraulic operation of clutch linkage is similar to the way brakes operate. Liquids cannot be compressed, so they can transmit motion and increase torque. Brake master cylinder operation is covered in detail in Chapter 51.

The input piston is located in the *master cylinder* (action cylinder) connected to the clutch pedal (Figure 62.17). It is located next to the brake master cylinder. The output piston is located in the reaction or actuator cylinder, commonly called a **slave cylinder**. It is attached to the release lever at the clutch. The two cylinders are attached hydraulically by tubing and hose, just like in the brake system. Hydraulic systems

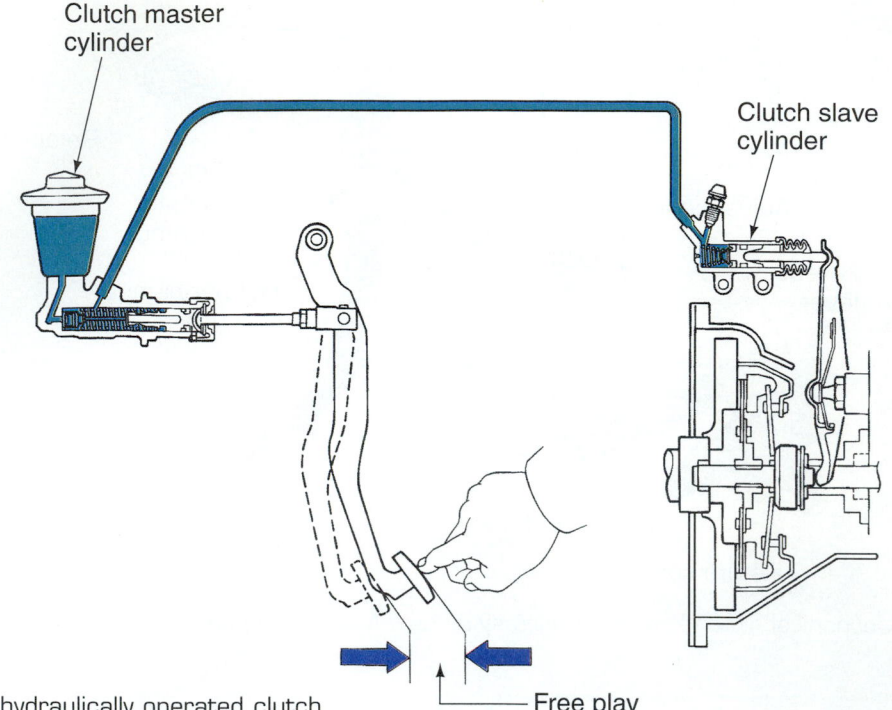

Figure 62.17 A hydraulically operated clutch.

are popular on custom vehicles because adaptation of the clutch release mechanism is easy.

The only difference between a clutch master cylinder and a single brake master cylinder is that the brake master cylinder has a residual check valve if it is for drum brakes. A clutch master cylinder does not have one because the clutch would stay applied as if a foot were always resting on the pedal. This would result in failure of a standard release bearing.

Some newer hydraulic systems have the slave cylinder connected to the release bearing. This eliminates the need for a release fork or cross shaft.

■ CLUTCH FREE PLAY

Often, the clutch linkage has a spring that pulls the release bearing away from the pressure plate after the clutch pedal is released. **Clutch free play**, or free travel, at the pedal is the result. Free travel is usually adjusted to about 1" at the pedal, which translates to about ⅛" at the release lever on the clutch.

Newer cars that have self-adjusting clutches maintain contact between the release levers and release bearing. Figure 62.18 shows a clutch slave cylinder with a return spring to keep the release bearing in con-

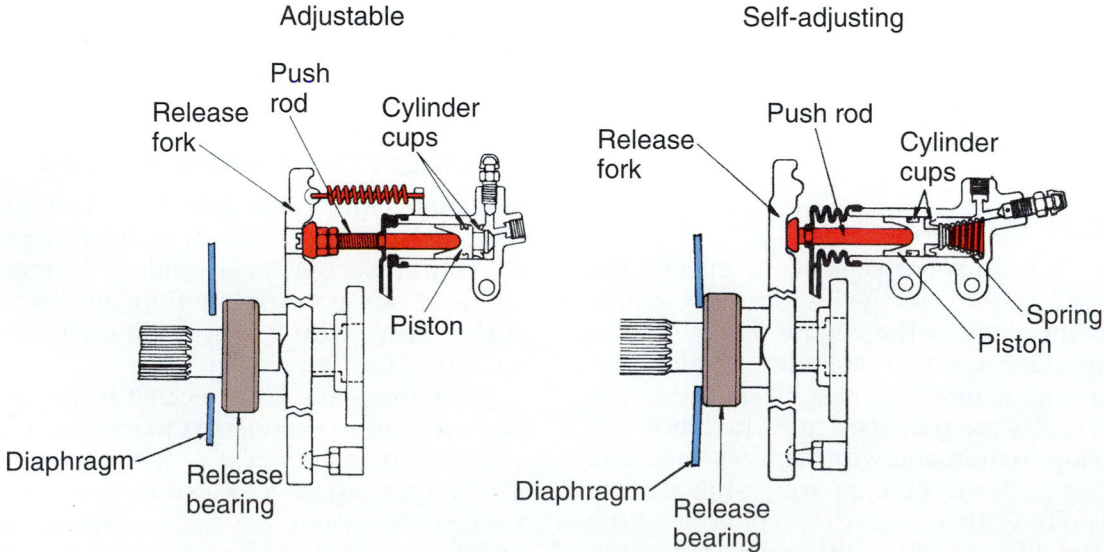

Figure 62.18 Comparison of adjustable and self-adjusting slave cylinders. Note the return spring at the right side of the self-adjusting slave cylinder.

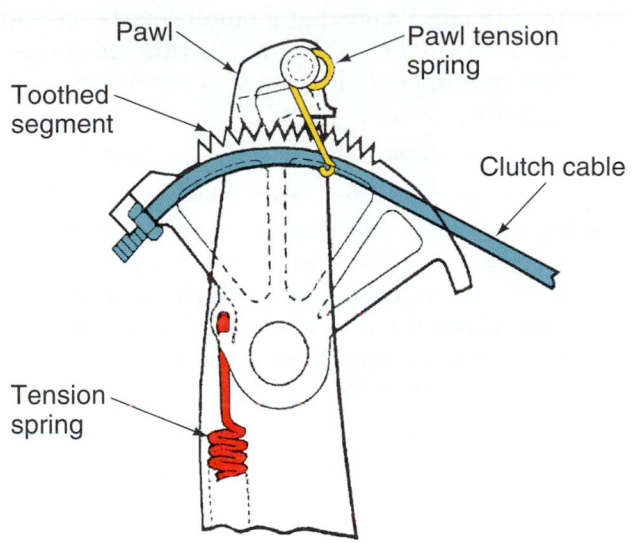

Pawl

Pawl tension spring

Toothed segment

Clutch cable

Tension spring

Figure 62.19 A clutch cable self-adjuster. *(Courtesy of Ford Motor Company)*

tact with the clutch cover. This requires a different style of release bearing. Standard release bearings do not remain in constant contact with the clutch cover like the ones on self-adjusting systems do.

Many newer cars have self-adjusting cables. A spring-loaded toothed sector gear is pinned to the pedal arm (Figure 62.19). Whenever the clutch is released, a *pawl* is lifted and released. If there is any slack in the adjustment, the gears spring causes it to rotate. This takes up the slack and the pawl slips into its new position on the gear to complete the adjustment.

■ REVIEW QUESTIONS

1. Where do the clutch disc splines ride?
2. Clutch torsional dampers are either rubber or _____ springs.
3. What is the name of the part of the clutch disc that is between the two facings and lets them compress?
4. What is another name for the pressure plate assembly?
5. When the clutch disc is totally worn out, about how much more torque carrying capacity is left in the clutch than the engine can deliver?
6. What is another name for clutch release levers?
7. The front of the transmission input shaft is supported in the crankshaft by a _____ bearing or bushing.
8. What is another name for a release bearing?
9. Which clutch release system changes its adjustment as the engine moves on its mounts: linkage, cable, or hydraulic?
10. What is the name of the hydraulic cylinder that the clutch master cylinder sends fluid pressure to?

■ ASE STYLE REVIEW QUESTIONS

1. Technician A says that some clutch hubs are rigid with no dampening. Technician B says that air gets in the clutch facing grooves and helps release the clutch. Who is right?
   a. Technician A   b. Technician B
   c. Both A and B   d. Neither A nor B

2. Technician A says that a clutch cover can have a diaphragm spring. Technician B says that a clutch cover can have coil springs. Who is right?
   a. Technician A   b. Technician B
   c. Both A and B   d. Neither A nor B

3. Technician A says that when you apply the pedal, you release the clutch. Technician B says that when you release the pedal, you apply the clutch. Who is right?
   a. Technician A   b. Technician B
   c. Both A and B   d. Neither A nor B

4. Technician A says that diaphragm clutches are easier to apply than coil spring clutches. Technician B says that coil spring clutches are better balanced than diaphragm clutches. Who is right?
   a. Technician A   b. Technician B
   c. Both A and B   d. Neither A nor B

5. Technician A says that a special kind of release bearing is used in rear wheel drive cars.

Technician B says that some transaxles do not use a pilot bearing. Who is right?

   a.   Technician A   **b.**   Technician B
   c.   Both A and B   **d.**   Neither A nor B

6. Technician A says that curved release levers must be installed against flat bearings. Technician B says that some release bearings slowly wear away with each use. Who is right?

   a.   Technician A   **b.**   Technician B
   c.   Both A and B   **d.**   Neither A nor B

7. Two technicians are discussing an overcenter assist spring. Technician A says that it pulls the pedal downward during the first half of travel. Technician B says that it pulls the pedal upward during the second half of travel. Who is right?

   a.   Technician A   **b.**   Technician B
   c.   Both A and B   **d.**   Neither A nor B

8. Technician A says that liquids can be compressed. Technician B says that liquids cannot be used to increase torque. Who is right?

   a.   Technician A   **b.**   Technician B
   c.   Both A and B   **d.**   Neither A nor B

9. Technician A says that a clutch master cylinder has a residual check valve. Technician B says that manual clutch linkage is often self adjusting. Who is right?

   a.   Technician A   **b.**   Technician B
   c.   Both A and B   **d.**   Neither A nor B

10. Technician A says that some clutch designs maintain constant contact between the release levers and release bearing. Technician B says that some clutch designs keep the release bearing away from the release levers except during shifting. Who is right?

   a.   Technician A   **b.**   Technician B
   c.   Both A and B   **d.**   Neither A nor B

CHAPTER 63

Clutch Diagnosis and Service

■ **OBJECTIVES**

Upon completion of this chapter, you should be able to:

✔ Diagnose clutch problems before disassembly.

✔ Adjust a clutch.

✔ Install a replacement clutch.

✔ Inspect worn or damaged clutch parts and determine the cause of the problem.

■ **KEY TERMS**

dragging clutch
chattering clutch
grabbing clutch
gravity bleeding
engine lugging
riding the clutch

■ INTRODUCTION

Many cars manufactured today have clutches. Because the clutch wears with use, it is not unusual for a car to have at least one new clutch installed during its lifetime. Clutch work is not that difficult to learn and can be a lucrative service area for technicians. As in other automotive service areas, the important thing to learn to be a true professional is to be able to diagnose problems. Problems are diagnosed before disassembly and after disassembly.

SAFETY NOTE Many clutches contain asbestos. Breathing asbestos can be dangerous to your health. Raw asbestos fibers are very small in diameter and resemble long, curly hair. Your lungs cannot rid themselves of these fibers as they normally do with other dusts. Before working on a clutch assembly, it is a good idea to wash it with a brake parts cleaner or vacuum it with a HEPA vacuum (see Figures 11.15 and 48.39).

■ DIAGNOSIS OF CLUTCH PROBLEMS

A clutch will normally last in excess of 100,000 miles. City driving in heavy traffic reduces clutch life. Extra heavy-duty uses can actually damage a clutch before its normal life expectancy is reached. Such abuse includes excessive slipping of the clutch in stop-and-go hilly traffic. Pulling a boat up a launching ramp or carrying a heavy load also represent potential abuse

situations. When pulling a heavy load, the objective is to have the clutch fully engaged (pedal up) as soon as possible. Pulling heavy loads calls for a transmission with a much lower gear ratio in first gear.

Diagnosing a problem before disassembly is very important. You must first remove a transmission or transaxle before you can remove a clutch. Be sure to verify the problem. The labor in a clutch job is time consuming and costly.

■ CLUTCH NOISES

Listen for unusual noises and try to isolate them to the clutch only. Sometimes a transmission noise can be mistaken for a clutch noise. When the noise only happens when the pedal is moved, try moving the pedal while the engine is off.

With the engine running, use your knowledge of clutch operation while determining the cause.

■ A noise when the clutch is first engaged (pedal let out and disc being wedged against the flywheel and pressure plate) is due to a problem with the friction lining (Figure 63.1).

■ The release bearing on cars without self-adjusting features is not designed to contact the clutch release levers *except* when the pedal is applied. If the noise only happens when your foot is resting lightly on the pedal, the release bearing is probably at fault.

■ The pilot bearing is pressed into the rear of the crankshaft. For it to make noise, the input shaft must be held still. Put the transmission in gear, depress the clutch pedal, and start the engine. The input shaft will be held still while the crankshaft and pilot bearing rotate.

Figure 63.1 The disc and pressure plate will wear into one another when the friction material is worn too thin. The rivets will be the first thing to touch. *(Courtesy of LuK, AS, Automotive Service)*

■ TRANSMISSION NOISE

Sometimes a car makes a noise only when the engine is idling in neutral with the clutch engaged (pedal all the way up). The transmission input shaft is turning at this time. The front transmission bearing on the input shaft and the mainshaft pilot on the back of the input shaft are the only bearings that are rotating. One of these bearings is probably the source of the noise.

■ PEDAL PROBLEMS

Some diagnosis can be done on the basis of pedal feel:
- A pulsating pedal is due to something internal in the clutch. The clutch will probably have to be disassembled, but looseness or misalignment somewhere could also be the culprit.
- Sometimes a pedal is hard to depress. A binding linkage or cable could be the problem.

Further exploration of these problems is covered in the next section. Problems are often accompanied by photos of the worn parts that resulted.

■ SLIPPING CLUTCH

A clutch is tested for slipping by putting it in the highest gear range (this would be fifth gear on a five-speed). Set the parking brake firmly and attempt to slip the clutch as if trying to make the vehicle move. If the clutch is slipping, engine rpm will rise. If the clutch holds, the engine will die. There are several possible reasons for a clutch to slip.

Partial Engagement

Partial engagement is a common cause of a slipping clutch. In a manually adjusted system as a clutch disc wears, free play becomes *less*. Because of the multiplication effect of the clutch linkage parts, a small amount of wear on the clutch disc can result in a total loss of original free travel. If the wear progresses beyond this point, it will be just as if the driver is always "riding the clutch" (pushing a small amount on the clutch pedal). Slipping is the result. This ruins the clutch disc (if it is not worn out already).

When a clutch slips, temperatures inside the clutch housing can reach 500°F in just a few seconds. Burn marks on the pressure plate are one of the results.

Partial Disengagement

A cause of partial disengagement is when there is a problem with a hydraulic clutch release system. Look for a low fluid level first. Then check for a failed master or slave cylinder. Replacement of the leaking part fixes the problem. A leaking clutch or brake master cylinder might be leaking into the passenger compartment. Check the carpet or floormat for evidence of leakage.

It is possible that there might be no evidence of leakage. This is because a master cylinder might have an *internal* leak. The test for an internal leak is to pump the clutch pedal and hold it to the floor while starting the engine. If the clutch slowly begins to engage, there is an internal leak.

Air in the lines can cause a spongy pedal on hydraulically released cars. Bleed the system to solve this problem.

■ DRAGGING CLUTCH

When the clutch *drags*, it does not release properly. The disc stays attached to the flywheel. There will be often too much pedal free travel, either some or all of the time. The transmission is often difficult to shift between gears or to put in low or reverse after starting.

To test for a **dragging clutch**:
- Start the engine and run it until the engine and transmission are at normal operating temperatures.
- With the engine at idle and the transmission in neutral, depress the clutch pedal to the floor.
- Wait 10 seconds.
- Shift the transmission into reverse.

Ten seconds should have been enough time for the clutch to come to a complete stop, allowing a shift into reverse without gear clashing. If there is a gear clash, the clutch is dragging (not releasing completely). Most cars will take only three or four seconds before allowing a smooth, quiet shift into reverse.

A dragging clutch can be due to a bent disc or clutch bearing retainer or to a rusted input shaft. Discs are checked at the factory for runout. They can be

Figure 63.2 This disc failed because it was bent during installation. *(Courtesy of LuK, AS, Automotive Service)*

damaged in shipping. A more likely cause of a bent clutch disc is that the transmission was allowed to hang on the disc during installation.

Figure 63.2 shows a disc that failed after the transmission sagged during installation. The symptom of the problem was that it was impossible to disengage the clutch. The thin piece of sheet metal that makes up the center portion of the disc is called the *retainer plate*. This thin sheet metal is very strong in a radial direction (perpendicular to the input shaft). But sideways, the disc is very lightweight and flimsy. It can be easily bent, resulting in later damage.

Sometimes, there is too much friction between the splines in the clutch hub and input shaft. The splines can be rusty because they were never lubricated. Rust can also happen from a leaky core plug on the back of the engine, driving through stream beds, or when the transmission fills with water through its vent hole when driving during a flood.

Clutch linkage can become stiff. This is an easy repair. Simply lube the linkage and free it up. Any worn or bent linkage part should be replaced or straightened.

OILY CLUTCH FACINGS

Oil leaking onto the clutch disc can result in clutch slipping. The disc will often overheat from the slipping, resulting in a burned disc. The disc and pressure plate must be replaced and the other parts thoroughly cleaned. If any oil remains in the pores of the metal surfaces of the flywheel, the clutch will probably begin to chatter soon after the disc is replaced.

The source of the oil leak must be determined and repaired or the problem will return. Oil leakage that is evident on the front side of the flywheel is from either a crankshaft rear seal or external engine leaks such as a valve cover or oil pan gasket. Oil on the rear side of the flywheel is often due to a bad front transmission seal or gasket or an overfilled transmission.

DAMAGED FRICTION SURFACES

Damage to friction surfaces of the clutch cover or flywheel can cause a clutch to slip or give inconsistent performance. When a clutch slips, heat is generated. A flywheel or pressure plate can be warped from the heat. This can cause uneven or incomplete contact with the disc. The flywheel, pressure plate, or disc can all become glazed from excess heat. A poor friction coefficient between the disc and metal friction surfaces results. The flywheel is commonly surfaced to ensure proper operation as part of the clutch job.

CHATTERING OR GRABBING CLUTCH

A **chattering clutch** occurs when the pedal shakes as the clutch is engaged (pedal raised from the floor). Figure 63.3 shows a pressure plate with marks on it that resulted from clutch chatter. A **grabbing clutch** is one where the friction disc does not slip normally but grabs all at once. There are several causes for these conditions. One common cause is oil on the friction

Figure 63.3 The marks on this pressure plate are from chattering. *(Courtesy of LuK, AS, Automotive Service)*

lining. This calls for replacement of the disc and repairing of the oil leak that caused the problem.

Worn or broken motor mounts often cause clutch chatter. Worn motor mounts are especially noticeable on four-cylinder engines. When there are only four cylinders, each firing impulse is more clearly defined. These engines will vibrate noticeably when resting at idle at a signal light. Coupled with clutch chattering, this points to mounts that, while not broken, might have become too soft. Oil leaking from a longtime leak onto a motor mount will soften the rubber too.

A worn or bent clutch fork can cause clutch chatter because it does not allow the release bearing to contact the pressure plate evenly (Figure 63.4). The clutch does not fully disengage when stepping on the pedal. This can also cause hard shifting, especially into first or reverse.

Uneven release fingers will cause the same problems as a worn or bent clutch fork. The force on the clutch disc will be diminished when the clutch pedal is released (clutch engaged). One release lever remaining in contact with the release bearing will diminish clutch clamping force by one-third. This problem can be caused by shipping damage, improper installation, uneven bolting to the flywheel, or a poor machine job on the flywheel. Be sure to resurface the flywheel when replacing the pressure plate.

Misalignment of parts or a bent friction disc can also cause uneven clutch operation. Misaligned parts can happen when pieces of dirt or debris become trapped between the transmission and its housing or the housing and the engine block. A bent friction disc or even a bent transmission input shaft can happen during a careless installation.

■ CLUTCH SERVICE

Clutch Adjustment

Some clutches require adjustment. As the disc wears, it becomes thinner. This results in the release levers moving *closer* to the release bearing, which means less clearance (Figure 63.5). If the disc wears too far, the release bearing would release the clutch disc at all times. Adjust the clearance to provide about 1" of free travel at the pedal.

When a car has a cable, adjustment is made by adjusting the length of the outside of the cable housing (Figure 63.6). Adjusting a linkage type requires a similar adjustment to a threaded rod (see Figure 62.15). Hydraulic systems are often self adjusting and have a full-time contact release bearing. On the systems that require adjustment, there is an adjustment on the slave cylinder pushrod.

■ SERVICING HYDRAULIC COMPONENTS

Leaks in the hydraulic system can be in hoses or lines but usually result from age of internal rubber sealing parts. The system has a master cylinder and a slave cylinder, either of which can develop leaks. The master cylinder (Figure 63.7) has a primary cup that can

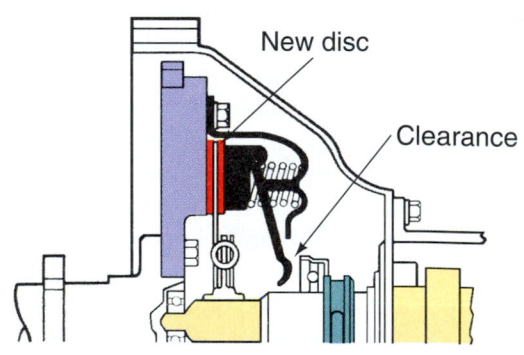

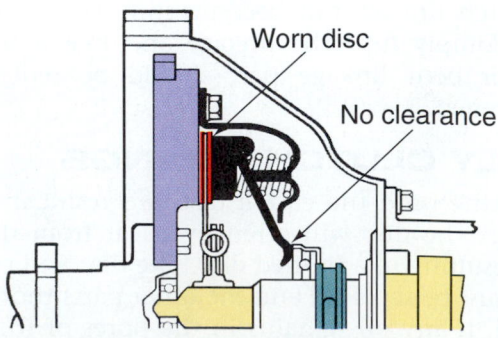

Figure 63.4 The release bearing was not contacting this diaphragm spring evenly. *(Courtesy of LuK, AS, Automotive Service)*

Figure 63.5 As the clutch disc wears, clearance becomes less. *(Courtesy of Federal-Mogul Corporation)*

cause internal leaks. The secondary cup can cause external leaks onto the floor in the passenger compartment.

The slave cylinder has a dust boot and one or two sealing rings that fluid pressure from the master cylinder acts upon (Figure 63.8). A slave cylinder will usually leak fluid onto the ground when it fails. Sometimes they leak slowly so adding fluid can put off the repair.

Bleeding the Hydraulic System

Air is removed from the hydraulic system by using a **gravity bleeding** procedure:

■ Open the bleeder screw on the slave cylinder.
■ Remove the lid from the master cylinder reservoir and let the cylinder drain. Refill it with fresh fluid.

■ Occasionally close the bleeder screw and refill the master cylinder with fresh fluid.
■ Repeat until all of the air is out of the system.
■ Sometimes it is necessary to have an assistant push on the pedal while the bleeder screw is open (Figure 63.9) and let up on the pedal when the screw is closed.

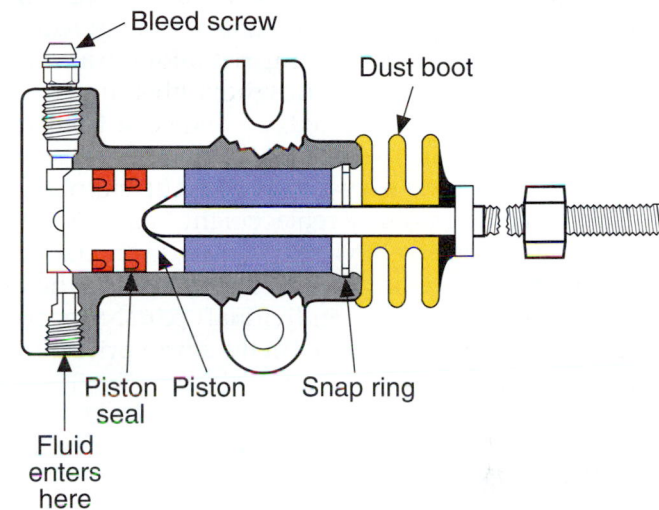

Figure 63.8 Parts of a slave cylinder. Replaceable rubber parts are highlighted. *(Courtesy of General Motors Corporation, Service Technology Group)*

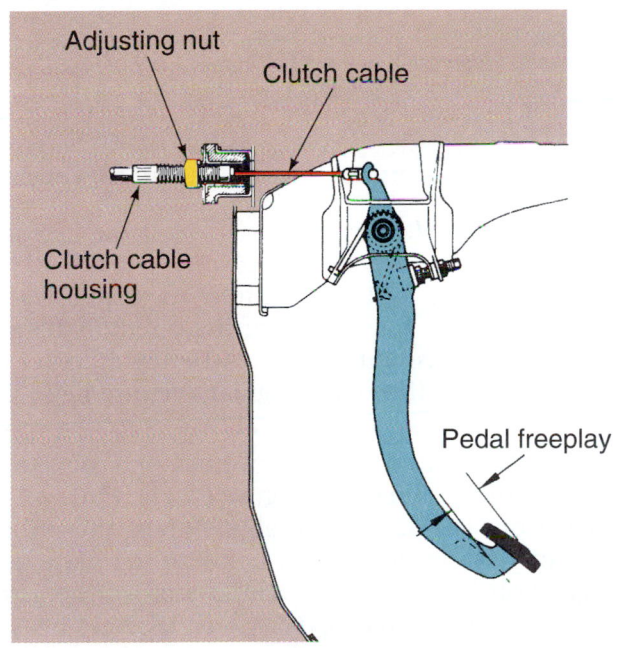

Figure 63.6 Cable adjustment.

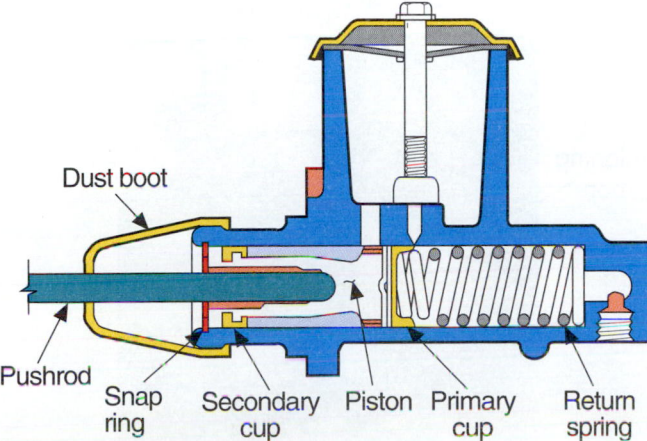

Figure 63.7 A clutch master cylinder has two rubber cup seals that can fail. *(Courtesy of General Motors Corporation, Service Technology Group)*

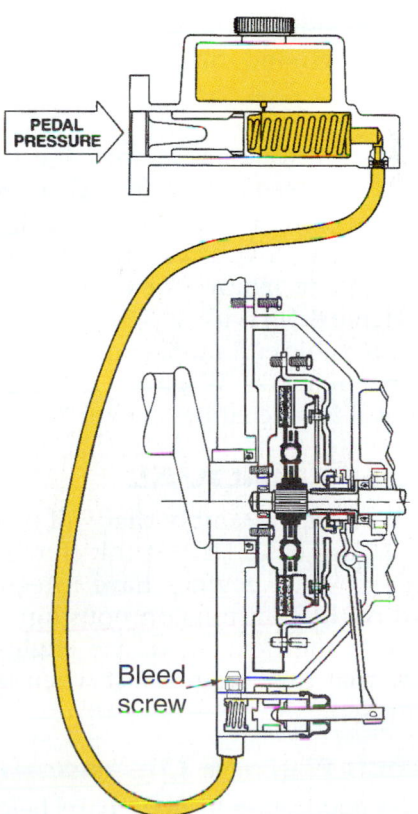

Figure 63.9 Fluid is moved from the master cylinder to the slave cylinder and exits from the bleed screw. *(Courtesy of LuK, AS, Automotive Service)*

The system can also be bled using a pressure bleeder (see Figure 52.8). Use fender covers. Be careful that fluid does not spill on paint.

■ CLUTCH REPLACEMENT

When a clutch job is done, it is common to replace the disc, release bearing, clutch cover, and pilot bearing. These often come in a clutch parts kit. The pressure plate and disc are often damaged and usually require replacement. The release bearing and pilot bushing are usually replaced on a preventive maintenance basis. With all of the parts replaced, a comeback from the failure of an old part two months later will not occur. If a new part fails, it is customary for the supplier to pay labor on the warranty replacement.

One advantage to buying an entire kit is that the shop will get a price break on volume, making the kit less expensive than its individual parts. Another advantage is that the shop can make a firm price quote to the customer.

■ REMOVE THE TRANSMISSION OR TRANSAXLE

Remove the transmission or transaxle. Follow the instructions in the service manual.

CAUTION Before attempting to remove a transmission or transaxle, disconnect the battery ground cable.

On rear wheel drive cars, remove the driveshaft, release mechanism, speedometer drive gear, crossmember, and transmission. On front wheel drive cars, remove the halfshafts (drive axles) and transaxle. Sometimes it is necessary to remove the engine. A transmission jack (see Figure 8.16) is used to lift the transmission in and out of the vehicle. A more complete description of transmission removal is given in Chapter 65. Transaxle removal is covered in Chapter 71.

■ CLUTCH REMOVAL

The parts of the clutch assembly that will be removed are shown in Figure 62.2. Unbolt the clutch housing from the engine. Some engines have integral clutch housings (part of the transmission housing). Note the dowels that align the housing to the engine (Figure 63.10). Be sure that they are not lost when the transmission is removed.

Mark Parts Before Disassembly

It always makes good sense to mark parts before disassembling them. Clutch parts are balanced at the factory while on the engine. If the clutch cover is to be used again, it should be replaced in its original posi-

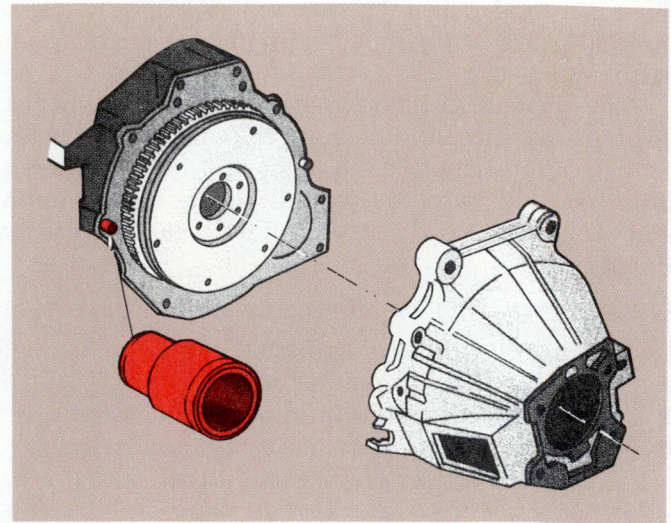

Figure 63.10 Dowels align the clutch housing with the engine. *(Courtesy of Federal-Mogul Corporation)*

tion. Make a centerpunch mark on the flywheel and on the clutch.

SAFETY NOTE Use a HEPA vacuum to vacuum any asbestos dust from the assembly before disassembling it. Asbestos fibers can pass through a regular vacuum bag.

Use an aligning arbor to hold the disc while removing the clutch cover (Figure 63.11). This will prevent it from falling accidentally. Unbolt the pressure plate bolts evenly to prevent bending the plate when the last bolt is removed. Carefully remove the assembly. Be careful not to drop the parts; they are heavy.

Inspect the clutch parts for damage. You will want to determine the cause before replacing the parts. A repeat failure must be avoided. The disc shown in Fig-

Aligning arbor

Figure 63.11 Use an aligning arbor to hold the disc while removing the clutch cover.

Figure 63.12 This clutch hub was damaged by a driver lugging the engine. *(Courtesy of LuK, AS, Automotive Service)*

ure 63.12 was damaged because the engine was pulling a load at too low an engine rpm. This situation, called **engine lugging**, is caused by driver error.

Clean Parts

Use hot soapy water, brake parts cleaner, or alcohol to clean off any grease from clutch parts. Metal surfaces can be deglazed with sandpaper.

> **CAUTION**
> ■ Do not wash the release bearing in solvent. It is packed with grease and will be ruined if solvent enters it.
> ■ Do not wash clutch parts in solvent. Solvent contains oil that will be absorbed into the friction surfaces of the metal.
> ■ Do not blow off parts with compressed air because of the danger of blowing asbestos around.

■ FLYWHEEL REMOVAL

Before removing the flywheel, mark it to make realignment with the crankshaft easier. Usually the bolt pattern on the flywheel is such that the flywheel will only bolt on one way.

Inspect the flywheel for damage. If the surface is flat, a sanding disc can be used to deglaze the surface. Heat from a slipping clutch causes cracks and warpage.

A flywheel with this type of damage can be resurfaced by a machine shop. This is necessary to validate most clutch disc warranties.

■ FLYWHEEL STARTER RING GEAR REPLACEMENT

Check the starter ring gear teeth to see that they are not damaged. Standard transmission flywheels have a pressed-fit ring gear. Sometimes a ring gear will be damaged due to starter motor problems (Figure 63.13). It can be removed and replaced with a new one. To remove a worn ring gear, drill a hole in it between two teeth (Figure 63.14) and break the ring gear with a chisel.

To install the new ring gear, it must first be heated to expand it. It must be heated evenly all around its circumference. If it gets hotter on one side than the other, it will become smaller in diameter on the cold side. Do not heat it to more than 400°F.

> **SHOP TIP** Methods of testing heated temperature:
> ■ Check ring gear temperature with a tempilstick. Tempilsticks are pencil-shaped temperature indicating sticks that melt different temperatures. A 400° tempilstick will leave a melted film on the ring gear when it is stroked across a surface hotter than 400°.
> ■ Polish several places on the ring gear with emery cloth or sandpaper. Heat the ring gear until the shiny places turn blue.
> ■ As the gear is being heated, touch solder to its surface. If the solder melts, the ring gear is hot enough to be slipped over the outside of the flywheel.

The chamfered side of the teeth should be aimed to the same side of the flywheel as the old teeth were.

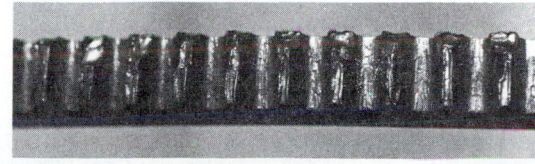

Figure 63.13 Damaged ring gear teeth due to starter motor problems.

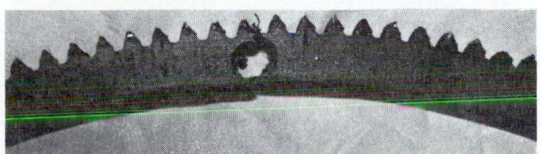

Figure 63.14 To remove a ring gear, drill a hole between two teeth and strike it with a chisel.

■ FLYWHEEL INSTALLATION

Use the correct quality screws on the flywheel and the clutch cover. These screws are usually shoulder screws (screws that do not have thread all the way to the top). When lock washers are used, they are usually serrated (especially on the flywheel).

CASE HISTORY *A technician did a complete clutch job that included flywheel resurfacing. When he installed the flywheel on the engine, he couldn't find the original screws and lock washers so he bought some new ones. He completed the reassembly, put the transmission in neutral, and started the engine. He depressed the clutch pedal and attempted to shift the transmission into gear. The clutch would not release because split lock washers are too tall for this application. They caused the flywheel screws to stick out too far and contact the hub on the clutch disc, locking it to the flywheel.*

When installing the flywheel, use the correct grade of screw (usually SAE Grade 8, see Chapter 6). Torque the screws to their correct specification. This is especially important on crankshafts that have a full round rear seal. The sealing area can be distorted if not evenly torqued.

CASE HISTORY *An apprentice did a clutch job during which he installed the flywheel with his impact wrench. He test drove the car upon completion of the job and returned it to the customer. The customer returned the next week complaining of clutch slippage and oil leaking from the bottom of the clutch housing. The rear engine seal had begun to leak when the flywheel bolts were not properly torqued.*

Inspect the rear main engine seal and the front transmission seal for leaks while the flywheel is removed.

■ INSPECT NEW PARTS

Before attempting to replace parts, always compare the new ones with the old ones (Figure 63.15). If the clutch was disassembled before the new parts were ordered, take the old parts to the parts store with you for comparison. There are often different sizes of release bearings and different diameters of clutches available within the same make of car.

■ PILOT BUSHING SERVICE

The pilot bushing or bearing in the crankshaft is often replaced as part of a clutch job. When the bearing is not to be replaced, inspect it by pushing on it while

Figure 63.15 Compare the old and new parts. *(Courtesy of Federal Mogul Corporation)*

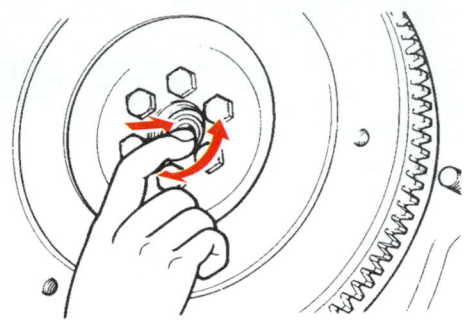

Figure 63.16 Checking the pilot bearing.

attempting to rotate it (Figure 63.16). It should spin smoothly without binding.

If the bushing must be removed, use a puller. A slide hammer or a bridge yoke bearing puller can be used (Figure 63.17). A screw is turned to expand a puller head inside of the pilot bushing (see Figure 7.39).

SHOP TIP A trick that sometimes works is to pack the cavity behind the pilot bearing with grease. Then, insert the largest bolt that will fit into the ID of the bearing. Pound on the bolt to force out the bearing.

Check the fit of the new bearing or bushing on the transmission input shaft before installing it in the crankshaft. The new bushing or bearing is installed with a bushing driver (Figure 63.18) or an old input shaft with a washer installed over the end of it.

■ CLUTCH DISC SERVICE

The clutch disc is normally replaced any time the clutch is apart. A disc is relatively inexpensive and is the one part that must wear with use. If the pressure

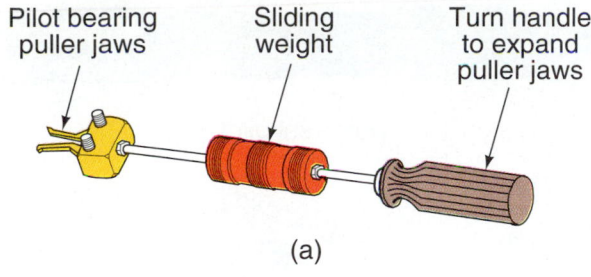

(a)

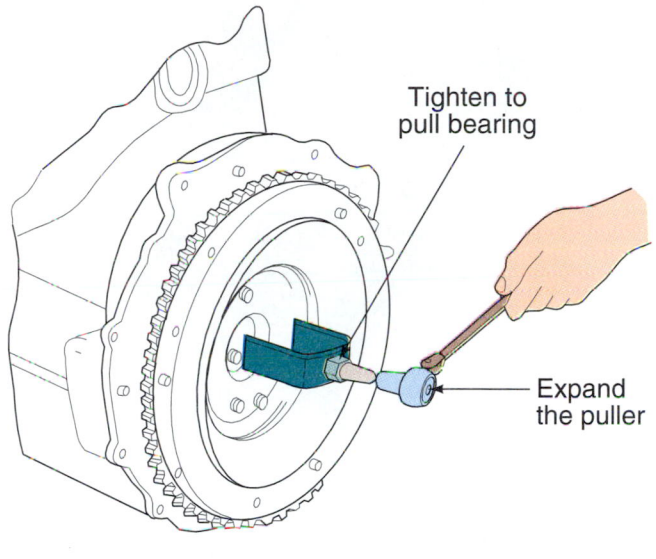

(b)

Figure 63.17 Pullers for pilot bushings and bearings: (a) Turning the handle on the slide hammer expands the jaws of the puller. (b) After expanding the puller to fit the I.D. of the bearing, turn the puller nut to remove it. *(a, Courtesy of Kent-Moore Division, SPX Corp; b, Courtesy of General Motors Corporation, Service Technology Group)*

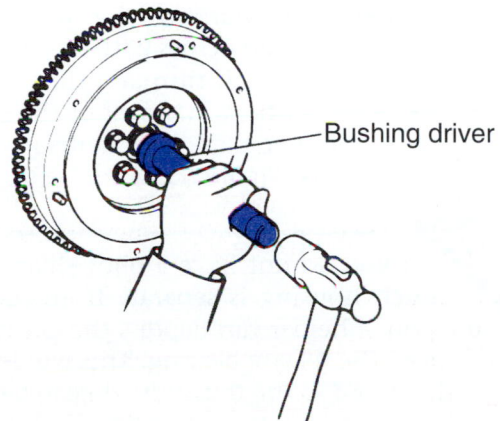

Figure 63.18 Use a bushing driver to install the new bearing or bushing.

plate and flywheel have not been overheated, replacing the disc will often result in like-new clutch performance.

Figure 63.19 Check the pressure plate for warpage.

■ CLUTCH COVER/ PRESSURE PLATE SERVICE

A clutch cover assembly is usually replaced when a disc is. If it is unworn, it may be reused. If you choose to reuse a clutch cover that appears to be in good condition, inspect the pressure plate to see that it is flat (Figure 63.19).

■ CLUTCH INSTALLATION

Before installing a clutch disc, be sure to clean your hands. When handling the disc, hold it on its edges like you would hold a photograph. Clutch discs often get a small amount of grease on them from careless handling.

A thin film of grease or anti-seize compound is applied to the input shaft splines. If too much grease is applied, it can be thrown outward until it is trapped on the friction lining (Figure 63.20). Even this small amount of grease on the friction lining results in clutch chatter.

Be careful to install the disc in the right direction. There should be a marking on it that says *"flywheel side."* If in doubt, check the repair manual. Figure 63.21 shows the difference between the flywheel and transmission sides of the clutch disc.

Attach the disc and clutch cover to the flywheel. Align the punch marks on the flywheel and clutch cover if the parts are being reused.

Using a clutch-aligning tool or an old transmission input shaft, align the clutch disc using the pilot (see Figure 63.23). Injection molded plastic clutch pilots are available for each make of vehicle (Figure 63.22). There are different sizes of splines and pilot ends to match the different sizes of clutch pilots available. One company makes a kit that includes thirty-five different sized pilots. A universal aligning tool is also available. Check the pilot shaft to see that it fits the pilot and the disc splines.

Figure 63.20 Excess grease on the input shaft was thrown from the center of this clutch hub to the friction material. *(Courtesy of LuK, AS, Automotive Service)*

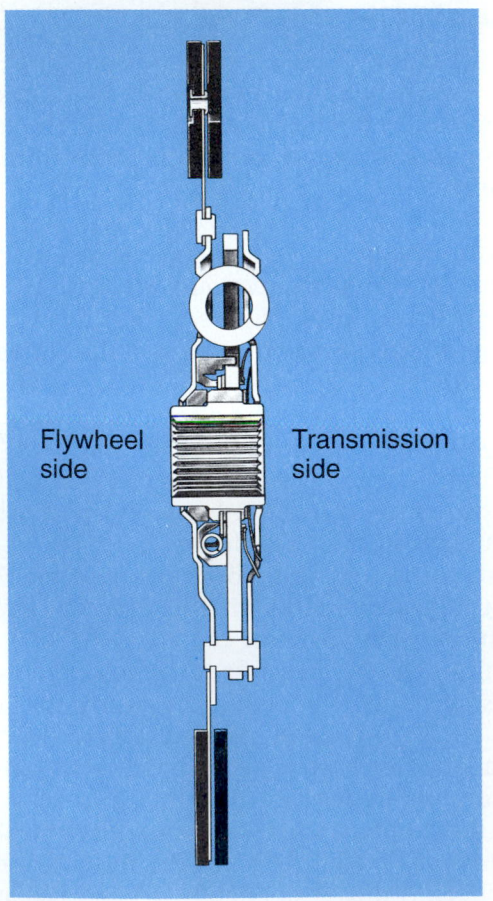

Figure 63.21 Install the clutch disc in the correct direction. *(Courtesy of Luk, AS, Automotive Service)*

Flywheel side

Transmission side

Universal pilot

Injection molded plastic pilots

Figure 63.22 Clutch pilots. *(Courtesy of LuK, AS, Automotive Service)*

With the pilot holding the disc, tighten all of the clutch cover screws finger tight (Figure 63.23). To avoid any chance of warping the pressure plate, tighten the screws in a crisscross pattern. Tighten each one no more than one-half turn at a time. Then turn the next one one-half turn. After all of the screws are tight, torque them in gradual increments in a star pattern to specifications (Figure 63.24).

SHOP TIP If you do not have a pilot shaft and the clutch housing is separate from the transmission, a helper can depress the clutch pedal to activate the release bearing. This will leave the clutch disc loose so the transmission can be moved around until the pilot can be aligned and slipped into the bearing or bushing.

Sometimes, the temptation is to try to lift a manual transmission into place without the aid of a jack. While this is often a necessity, especially when a job is being done at home, the weight of the transmission

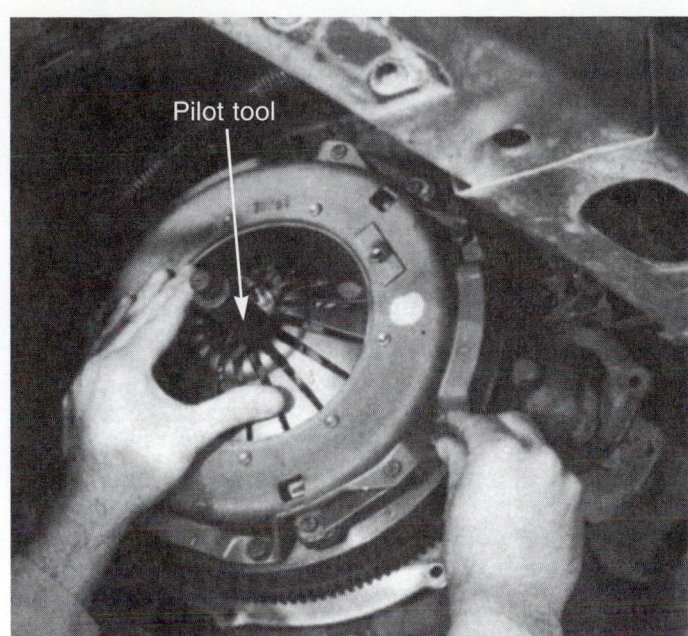

Figure 63.23 With the pilot holding the disc, tighten all of the clutch cover screws finger tight. *(Courtesy of Federal-Mogul Corporation)*

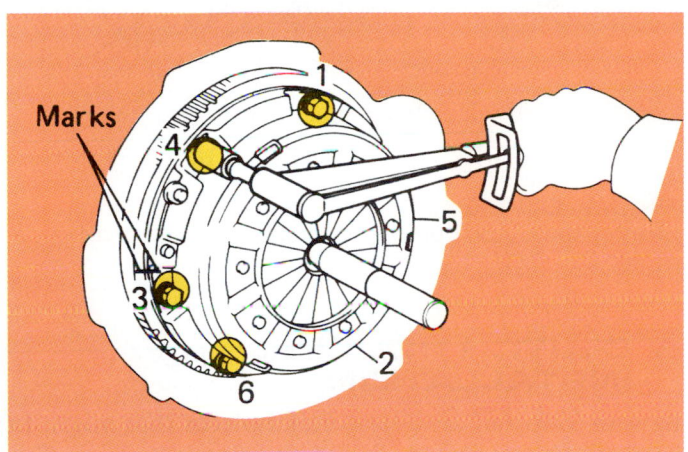

Figure 63.24 Torque in a star pattern.

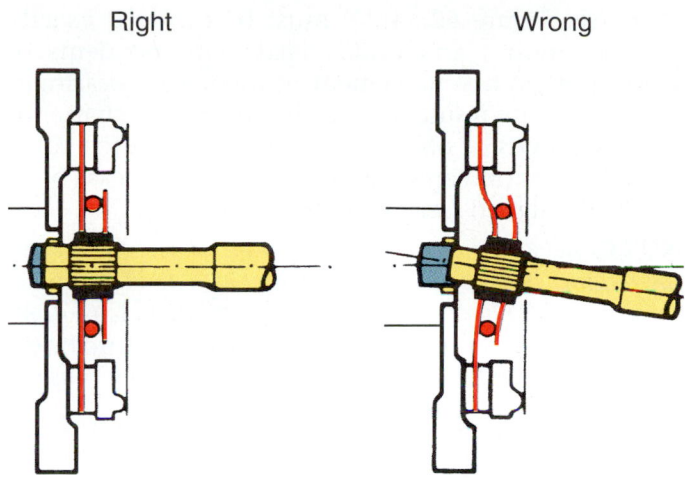

Figure 63.25 Allowing the transmission to hang on the clutch disc can bend the disc. *(Courtesy of LuK, AS, Automotive Service)*

A release bearing is usually replaced on a preventive maintenance basis during a clutch job. If the rest of the clutch parts are not being replaced, check the bearing for roughness or excessive looseness.

Sometimes when a new release bearing is purchased, the collar (hub) is reused. The fit between the bearing and collar is snug but not very tight. Tap it off with a hammer or use a press. Install the collar in the new bearing in the same manner (Figure 63.26).

Inspect the end of the input shaft. The splines must be clean, unworn, and free of rust. Inspect the pilot area.

Check the *front bearing retainer*, often called a quill. It fits around the outside of the transmission input should never be allowed to hang on the clutch disc. This can easily bend the clutch disc, ruining the job (Figure 63.25).

■ RELEASE BEARING SERVICE

The release bearing sometimes fails. This happens often on clearance type release bearings when the driver has a habit of **riding the clutch** (resting a foot on the clutch pedal when driving). With the engine running and the transmission in neutral, if resting your foot on the pedal produces a noise the release bearing is probably bad.

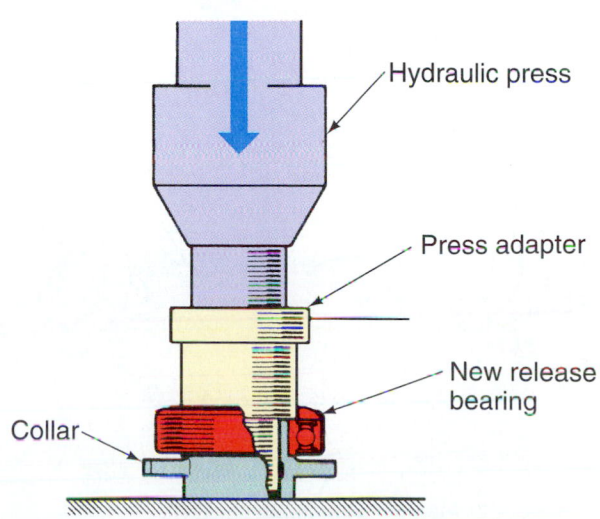

Figure 63.26 Installing a used release bearing collar in a new bearing.

shaft (see Figure 62.14). It must be centered exactly and be smooth and undamaged, with no dents or worn spots. If free movement of the release bearing is prevented, the result will be clutch grab or chatter. If there are signs of off-center wear on the clutch cover release levers or diaphragm fingertips, the release bearing is misaligned (see Figure 63.4).

NOTE: *Be sure that the oil return hole on the front bearing retainer is correctly lined up with the hole in the transmission case. Otherwise, the oil that lubricates the front bearing will not be able to return to the transmission case. Clutch failure will be the likely result when the oil runs out the input shaft onto the clutch disc.*

Lubricate Release Mechanism Contacts

Various parts of the release mechanism have sliding contacts that require application of a small amount of grease (Figure 63.27). Install the release bearing. Be sure that the clutch fork is not damaged. Locations of parts that attach to the clutch fork are shown in Figure 63.28. Attach the clutch fork to the release bearing and clutch housing (Figure 63.29).

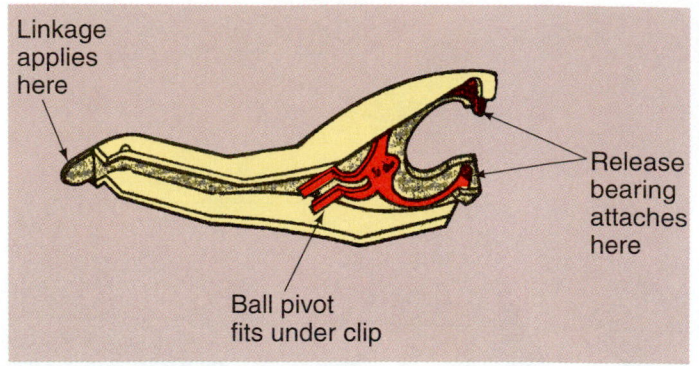

Figure 63.28 — Linkage applies here — Release bearing attaches here — Ball pivot fits under clip

Figure 63.28 Locations of parts that attach to the clutch fork. *(Courtesy of Federal-Mogul Corporation)*

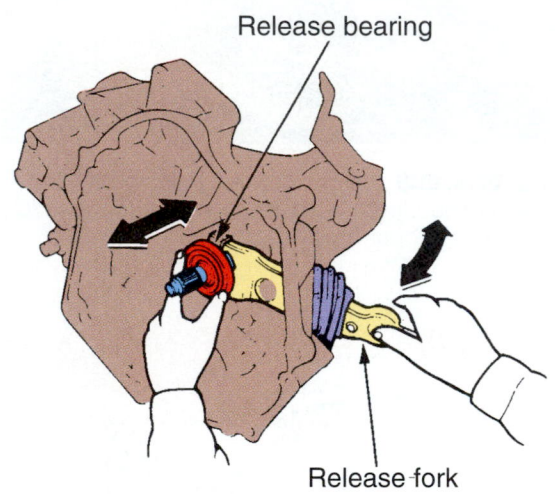

Release bearing — Release fork

Figure 63.29 Attach the clutch fork to the release bearing and clutch housing. *(Courtesy of American Honda Motor Co., Inc.)*

■ CLUTCH HOUSING INSTALLATION

Be sure to install any sheet metal pieces that might fit between the engine block and the clutch housing (Figure 63.30). These are commonly left out by mistake. Install the clutch housing or the entire transmission if there is no separate clutch housing. Be sure all of the dowels are in place. Missing dowels can result in misalignment problems (see Figure 63.10).

Be careful that nothing becomes accidentally trapped between the surfaces of the clutch housing and the engine block. Check to see that there are no nicks or pieces of debris on the engine and clutch housing mating surfaces. Misalignment will cause clutch problems. If the clutch assembly was removed because of chattering and a previous misalignment is suspected, inspection of the clutch housing surfaces will be done.

To complete the job, install the transmission and remaining parts. Perform any needed clutch adjustment, fluid level checks, and test drive the car.

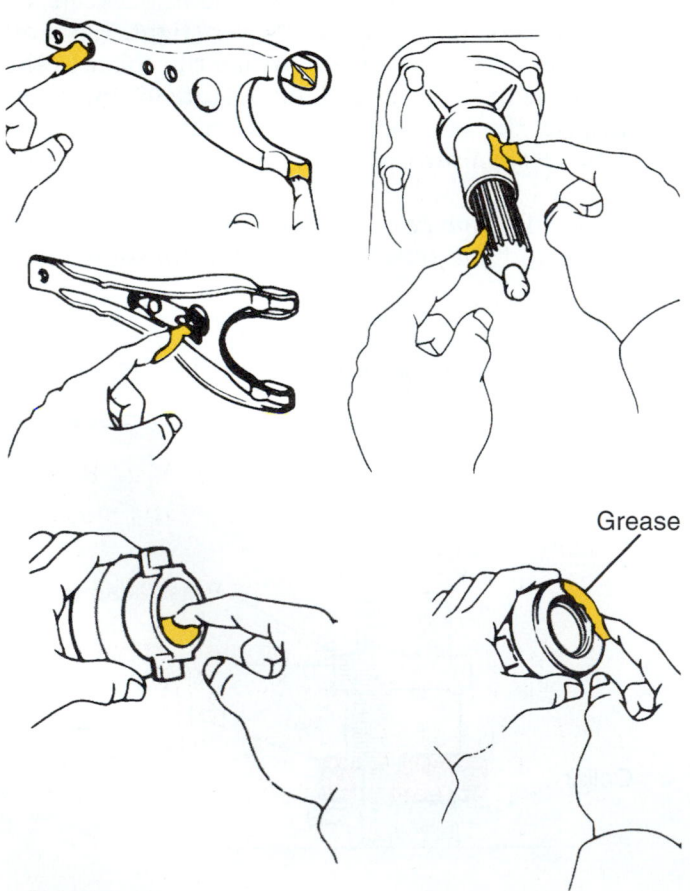

Grease

Figure 63.27 Lightly lubricate the various contacts in the release mechanism.

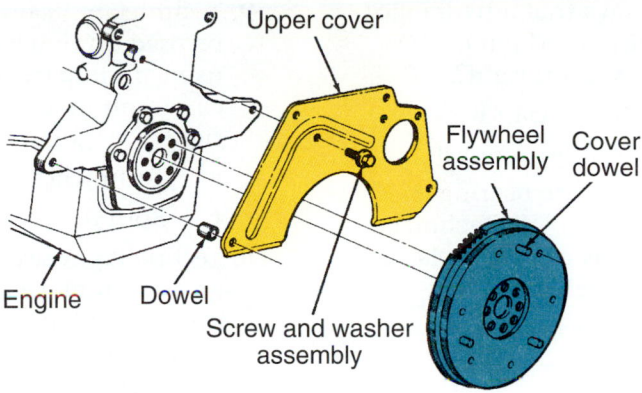

Upper cover

Flywheel assembly

Cover dowel

Engine

Dowel

Screw and washer assembly

Figure 63.30 Be sure to install any sheet metal pieces that might fit between the engine block and the clutch housing. *(Courtesy of Chrysler Corporation)*

■ REVIEW QUESTIONS

1. When the clutch pedal is let up from the floor, is the clutch being engaged or disengaged?

2. Should the transmission be in gear or in neutral when checking for pilot bearing noise?

3. If there is no sign of a fluid leak but the clutch slowly engages with your foot on the pedal, what kind of leak does a clutch master cylinder have?

4. If the transmission grinds when shifting into reverse after 10 seconds in neutral with the clutch in, what is the problem?

5. When the pedal shakes as the clutch is engaged, what is this called?

6. When something becomes trapped between the engine and the clutch housing during installation, what condition results?

7. How much free travel at the pedal should you adjust a clutch to?

8. The names of the cylinders in a hydraulic clutch system are the master cylinder and the _____ cylinder.

9. What is the name of the puller that has a weight on it that is slid against a stop to pull a pilot bearing?

10. What is another name for the front transmission bearing retainer?

■ ASE STYLE REVIEW QUESTIONS

1. Technician A says that if a clutch noise only happens when the pedal is all the way to the floor, the release bearing is probably at fault. Technician B says that noise that only happens when the engine is idling in neutral with the clutch engaged is probably an input shaft bearing. Who is right?
   a. Technician A   b. Technician B
   c. Both A and B   d. Neither A nor B

2. Technician A says to put a transmission in low gear to test for a slipping clutch. Technician B says that as a manually adjusted clutch disc wears, free play becomes less. Who is right?
   a. Technician A   b. Technician B
   c. Both A and B   d. Neither A nor B

3. Technician A says that a clutch disc is very tough and not easily bent. Technician B says that an oil leak that makes the front of the flywheel wet is probably due to a bad transmission seal. Who is right?
   a. Technician A   b. Technician B
   c. Both A and B   d. Neither A nor B

4. Technician A says that worn motor mounts can cause clutch chatter. Technician B says that worn motor mounts can cause clutch slippage. Who is right?
   a. Technician A   b. Technician B
   c. Both A and B   d. Neither A nor B

5. Technician A says that some hydraulic clutches are self adjusting. Technician B says that some mechanical clutches are self adjusting. Who is right?
   a. Technician A   b. Technician B
   c. Both A and B   d. Neither A nor B

6. Technician A says that a hydraulic system can be bled by putting fluid into the slave cylinder

bleed screw. Technician B says that a hydraulic system can be bled by letting fluid out of the slave cylinder bleed screw. Who is right?

a. Technician A   b. Technician B
c. Both A and B   d. Neither A nor B

7. Technician A says that the release bearing is usually replaced during a clutch job. Technician B says that the pilot bearing is usually replaced during a clutch job. Who is right?

a. Technician A   b. Technician B
c. Both A and B   d. Neither A nor B

8. Technician A says that clutch parts should be cleaned thoroughly in cleaning solvent. Technician B says that when a new starter ring gear is heated and turns blue, it has been heated too much. Who is right?

a. Technician A   b. Technician B
c. Both A and B   d. Neither A nor B

9. Technician A says that split lock washers must be used on flywheel screws. Technician B says using an impact wrench to tighten a flywheel can result in a leaking rear engine seal. Who is right?

a. Technician A   b. Technician B
c. Both A and B   d. Neither A nor B

10. Technician A says that even a small amount of grease can make a clutch chatter. Technician B says to tighten clutch cover screws with a torque wrench in a circular pattern. Who is right?

a. Technician A   b. Technician B
c. Both A and B   d. Neither A nor B

CHAPTER 64

Manual Transmission Fundamentals

■ OBJECTIVES

Upon completion of this chapter, you should be able to:

✔ Describe the relationship between gears and torque.

✔ Understand the basic types of gears.

✔ Calculate gear ratios.

✔ Trace the power flow through three-, four-, and five-speed transmissions.

✔ Name all of the transmission parts.

■ KEY TERMS

manual transmission
low gear
gear radius
gear ratio
overdrive
wide ratio transmission
close ratio transmission
granny gear
final drive ratio
mesh
pitch diameter
spur gears
helical gears
idler gear
synchronizer
blocker ring synchronizer
countergear
dog teeth

■ INTRODUCTION

A **manual transmission** is used with a clutch. It must be shifted between gears manually (Figure 64.1). This kind of transmission is also called a *stick shift* or a *standard transmission*. Standard transmission is not an accurate name anymore because more cars today are built with automatic transmissions. The name became popular when automatic transmissions first appeared as an option on cars.

A transmission is used in rear wheel drive cars. A similar unit, called a transaxle, is used in front wheel drive cars. It is covered in another chapter, along with drive axles and CV (constant velocity) joints.

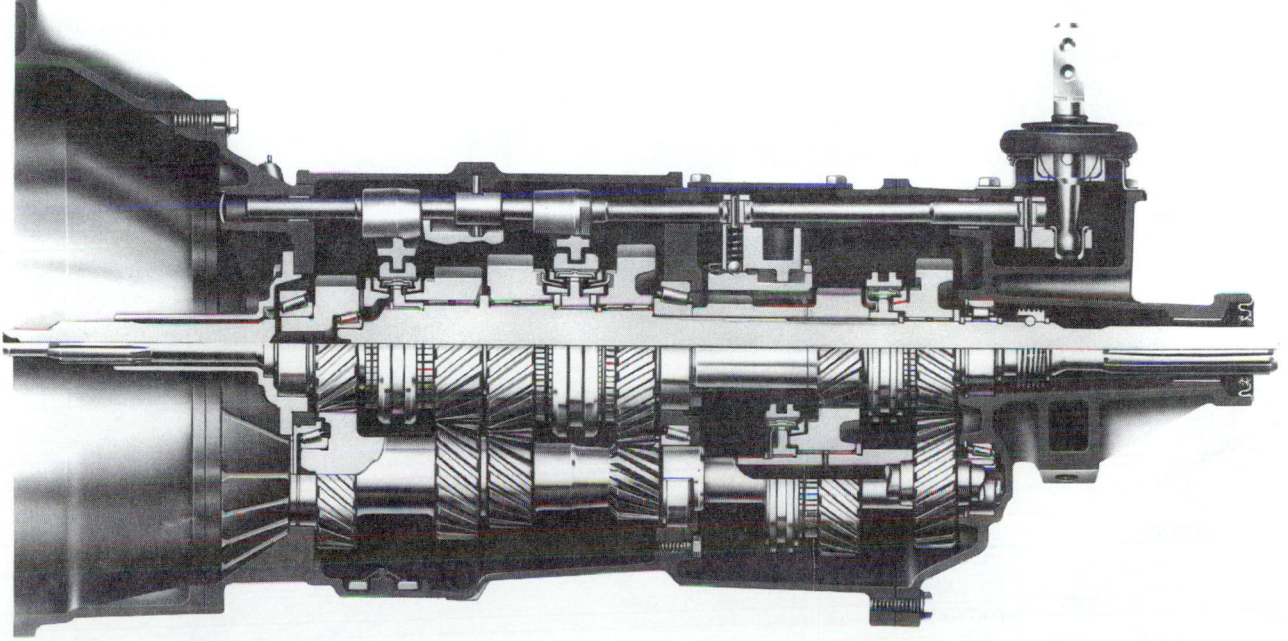

Figure 64.1 Cutaway of a manual transmission. *(Courtesy of Borg-Warner Automotive, Inc.)*

■ PURPOSE OF A TRANSMISSION

A four-stroke engine does not produce equal torque at all times. It has lower torque when starting from a stop. That is why you have to slip the clutch during take off to keep the engine from stalling. A transmission provides a means of changing torque to fit an engine's operating requirements.

When the engine is operated in low gear, the transmission turns approximately three times to one turn of the engine's crankshaft. This is called a 3:1 *ratio*. Without the extra leverage that this gear reduction provides, the engine would stall or lug during take-off. Figure 64.2 compares the torque an engine produces in various gear ranges. Compare this to riding a ten-speed bicycle. When you shift into a lower gear, the pedal crank revolves faster and you climb the hill easier. In a **lower gear**, a small gear drives a larger gear. The gears provide leverage like that of a light person lifting a heavy person on a teeter-totter (Figure 64.3).

NOTE: *If you ever drive an electric golf cart, notice how much torque the electric motor makes when you first take off. Electric motors produce excellent torque from a standing start.*

■ USING GEARS TO INCREASE TORQUE

To measure torque, the force applied is multiplied by the distance from the centerline of rotation. Twenty pounds of force applied to the end of a 1-foot rod produces 20 foot-pounds of torque. Ten pounds of force applied to a 2-foot rod also produces 20 foot-pounds of torque.

When the driving gear is smaller than the driven gear, its speed decreases and the output torque increases. The distance from the center of a gear to its outside edge is called the **gear radius**. The radius is where the torque is measured. In Figure 64.4, the radius of the smaller gear is 1 foot, and 25 pounds of force is applied to the gear. The output of the gear is 25 foot-pounds of torque. That torque is applied to the second gear, which has a 2-foot radius. The resulting output is 50 foot-pounds of torque.

NOTE: *The amount of torque increase is proportional to the speed decrease. The output speed is half, so the torque is doubled.*

■ GEAR RATIO

A **gear ratio** can be calculated by dividing the number of teeth on the driven gear by the number of teeth on the driving gear. The input gear is called the *driving* gear. The output gear is called the *driven* gear. If there are 12 teeth on the driving gear and 24 teeth on the driven gear, the resulting gear ratio is two to

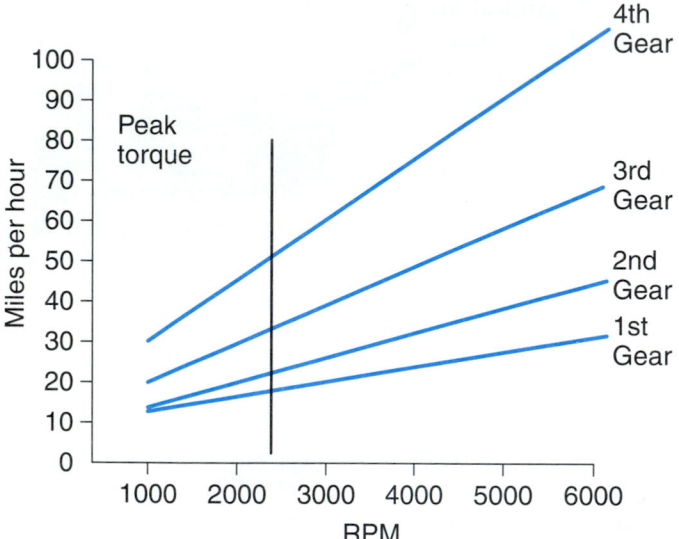

Figure 64.2 The vehicle must be travelling faster before the transmission is shifted into fourth gear.

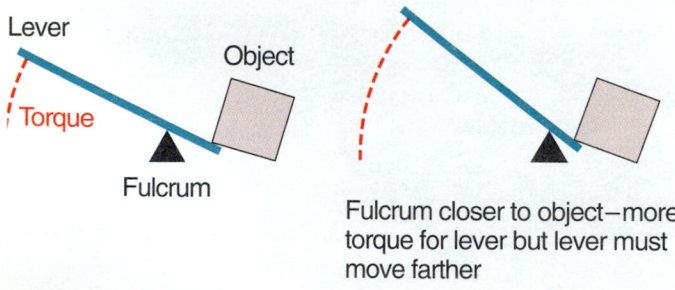

Figure 64.3 Gears provide leverage like that of a light person lifting a heavy person on a teeter-totter. *(Reproduced by permission of Deere & Company, ©1991. Deere & Company. All rights reserved.)*

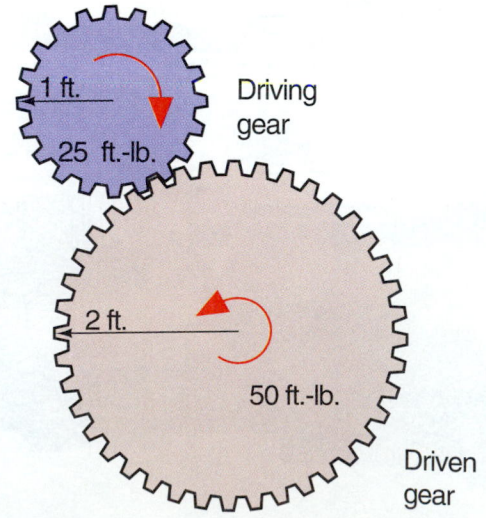

Figure 64.4 A driving gear of half the size results in double the torque output.

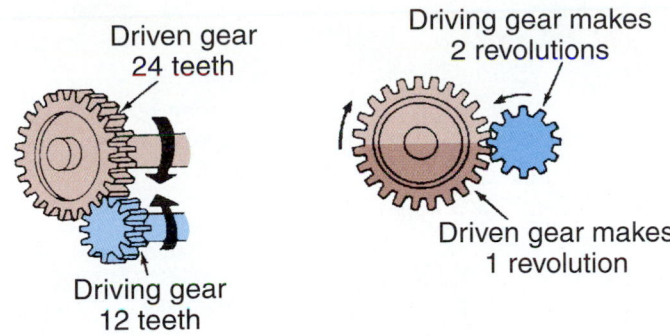

Driven gear
24 teeth

Driving gear makes
2 revolutions

Driven gear makes
1 revolution

Driving gear
12 teeth

Figure 64.5 A 2 to 1 gear ratio. *(Reproduced by permission of Deere & Company, ©1991. Deere & Company. All rights reserved.)*

one (written 2:1) (Figure 64.5). The driving gear must turn two times to one turn of the driven (larger) gear. The torque on the larger gear would be twice that on the input shaft (smaller) gear.

When figuring a gear ratio when there are more than two gears, number the gears first and be certain which is driving and which is driven. Then, multiply driving gears together and driven gears together before dividing. This would be done when figuring reverse gear ratio because it always has an extra gear.

TRANSMISSION GEAR RANGES

Transmissions in cars and light trucks can have either three, four, five, or six forward gear ranges, called speeds. A reverse gear range allows the vehicle to be backed up. A neutral gear range is useful for long periods of idle and when servicing a vehicle. Lower power engines require more gear ranges to keep the engine at maximum torque.

Approximate transmission gear ratios for a three-speed transmission are:
- 3:1 first
- 2:1 second
- 1:1 high
- 3:1 reverse

Typical ratios with a four-speed transmission are:
- 3:1 first
- 2.5:1 second
- 1.5:1 third
- 1:1 fourth
- 3:1 reverse

When there is a fifth gear, its ratios are usually the same as for a four speed but with an added overdrive gear:
- 0.75:1 fifth gear (overdrive)

OVERDRIVE

Overdrive is the opposite of gear reduction. The output shaft turns *faster* than the input shaft. It provides

a ratio that is another step beyond the 1:1 ratio of high gear, 0.75:1 for instance (a 25% change in speed). With a modern five-speed transmission, it is when the large gear drives a smaller gear. There are also other types of overdrives.

Besides the common overdrive found in five-speed transmissions, there are planetary gear types. Using planetary gears is the way overdrive is achieved in an automatic transmission with a lock-up torque converter (except for Honda and Acura transmissions). Planetary overdrive manual transmissions were popular in the 1950s and 1960s before automatic transmissions became popular. A planetary gear set was attached to the rear of a manual transmission output shaft. The overdrive unit was located in the extension housing.

CLOSE AND WIDE RATIO

In the muscle car days of the 1960s and early 1970s, a four-speed manual transmission could sometimes be ordered with a close ratio or a **wide ratio transmission**. When there is a larger difference between the ratios of the gears, the transmission is called a wide ratio transmission. This transmission might have a first gear ratio close to 3:1 while a **close ratio transmission** would have a first gear ratio closer to 2.2:1. The wide ratio transmission provides more of a torque increase in first gear.

Shifting between gears in a close ratio transmission would result in less drop of engine rpm, keeping the engine within its torque band. It would be more applicable to racing. Shifting while driving in traffic could feel almost like staying in the same gear and letting out the clutch. The close ratio transmission was most commonly used with a low (i.e., 4.11:1) final drive gear ratio. This made up for the 2.2 first gear ratio, which would make accelerating from a standing stop more difficult.

Some light trucks come equipped with a four-speed transmission that has a very low first gear, sometimes called a **granny gear**. These transmissions usually have a very long floor shift lever. When driving these trucks, it is customary to start from a stop in second gear. This is because the low gear ratio is in the 7:1 range. This ratio is handy for idling a boat up a boat ramp or pulling a very heavy load. The second gear ratio is comparable to first gear in a three-speed transmission.

FINAL DRIVE RATIO

The **final drive ratio** is the ratio between the transmission output shaft and the differential ring gear. Let us say a differential has a ratio between its pinion gear and ring gear of 3:1. If the transmission was shifted into a 3:1 first gear, the final drive ratio to the rear wheels would be 9:1.

■ MATH NOTE ■

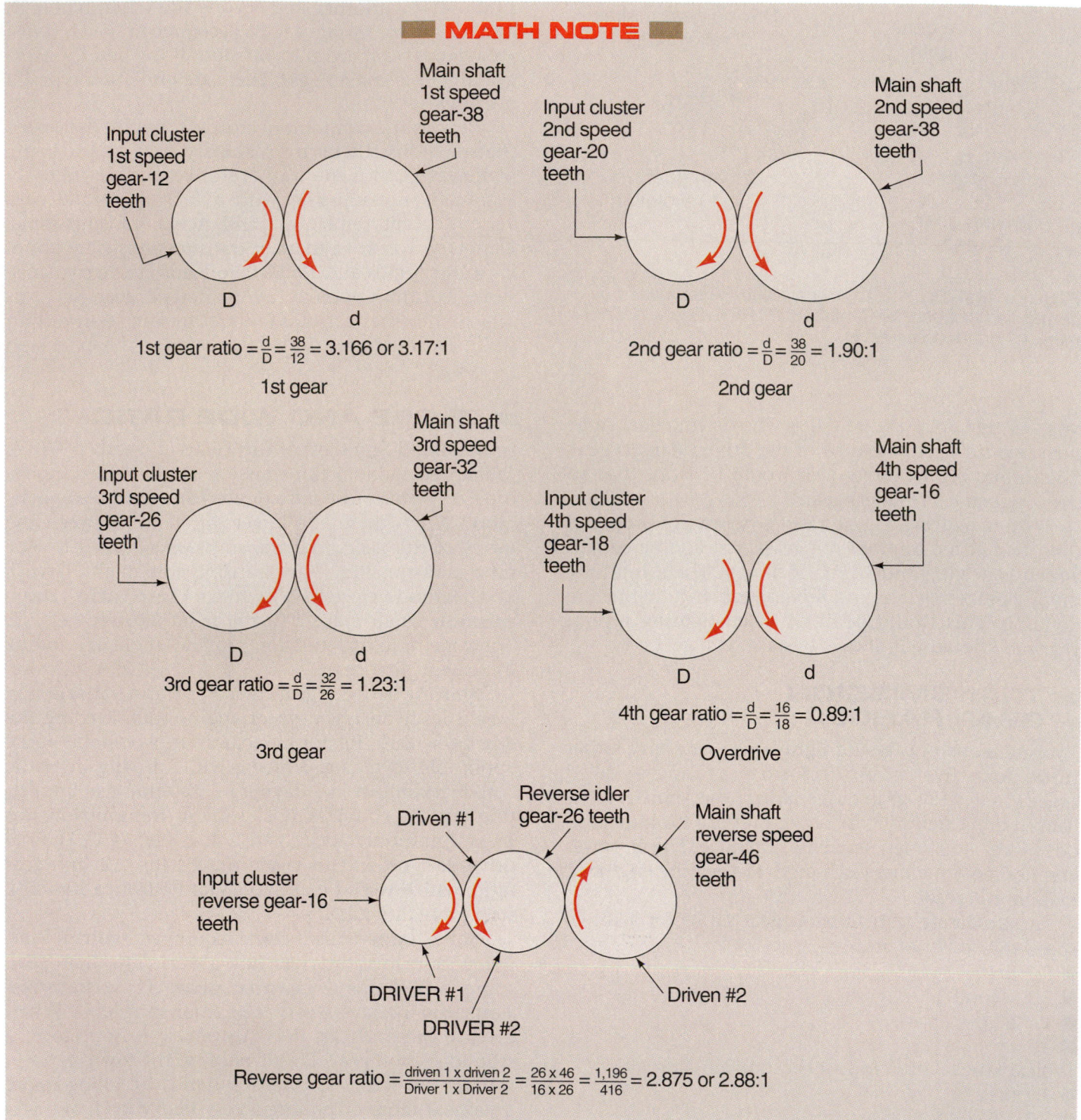

Input cluster
1st speed
gear-12
teeth

Main shaft
1st speed
gear-38
teeth

D   d

1st gear ratio $= \frac{d}{D} = \frac{38}{12} = 3.166$ or 3.17:1

1st gear

Input cluster
2nd speed
gear-20
teeth

Main shaft
2nd speed
gear-38
teeth

D   d

2nd gear ratio $= \frac{d}{D} = \frac{38}{20} = 1.90:1$

2nd gear

Input cluster
3rd speed
gear-26
teeth

Main shaft
3rd speed
gear-32
teeth

D   d

3rd gear ratio $= \frac{d}{D} = \frac{32}{26} = 1.23:1$

3rd gear

Input cluster
4th speed
gear-18
teeth

Main shaft
4th speed
gear-16
teeth

D   d

4th gear ratio $= \frac{d}{D} = \frac{16}{18} = 0.89:1$

Overdrive

Driven #1

Reverse idler
gear-26 teeth

Main shaft
reverse speed
gear-46
teeth

Input cluster
reverse gear-16
teeth

DRIVER #1

DRIVER #2

Driven #2

Reverse gear ratio $= \frac{\text{driven 1} \times \text{driven 2}}{\text{Driver 1} \times \text{Driver 2}} = \frac{26 \times 46}{16 \times 26} = \frac{1,196}{416} = 2.875$ or 2.88:1

■ GEAR TYPES AND OPERATION

The shape of a gear tooth is designed to allow the teeth to roll into and out of **mesh** with a minimum of friction. As gears contact each other, the load rolls across the gear teeth. The contact pattern (patch) is where the teeth of the two gears meet. The effective diameter of the meshed gear (used for determining its circumference) is called the **pitch diameter** (Figure 64.6).

Gear teeth can also slide against each other, depending on the design. Manual transmissions use two kinds of gears, a spur gear and a helical gear.

■ SPUR GEARS

Spur gears are a simple gear design with straight cut teeth (Figure 64.7). Spur gears have only one tooth in contact at a time. With this design, there is no end thrust and the transmission will not attempt to pop out of gear during acceleration or deceleration. That is one of the reasons why spur gears are often used in reverse. Reverse often does not have a synchronizer clutch (covered later) as forward gears do. Spur gears can also slide together easily.

Backlash is the clearance between meshing gear teeth (Figure 64.8). It allows for expansion and lubrication oil to get between the gear teeth. Because there is only one tooth in contact at a time, there is a clicking sound that happens as a gear tooth rolls out of

contact and a new one takes its place. As this noise gains speed, it turns into *gear whine*.

■ HELICAL GEARS

Helical gears have replaced spur gears in transmissions because they are much quieter. They are machined at an angle which gives them a continuous flow of power across the gear teeth (Figure 64.9). Engagement starts at the tip of one end of the gear tooth and rolls down the tooth. This keeps backlash to a minimum and makes them quieter. It also results in more gear strength because there is more area of tooth contact. A problem with helical gears is that they cause end thrust under load.

■ IDLER GEARS

When gears rotate against each other, the driven gear will be turned in the opposite direction. An **idler gear** is one that is used between two other gears. Its purpose is to change the direction of rotation (Figure 64.10).

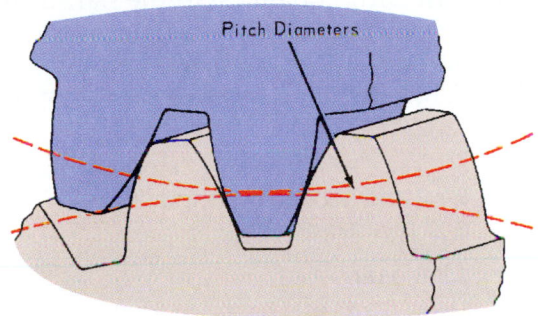

Figure 64.6 The point where the teeth of the two gears meet is called the pitch diameter. *(Reproduced by permission of Deere & Company, ©1991. Deere & Company. All rights reserved.)*

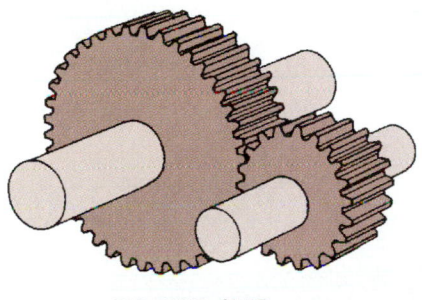

STRAIGHT SPUR

Figure 64.7 Spur gears have square cut gear teeth. *(Reproduced by permission of Deere & Company, ©1991. Deere & Company. All rights reserved.)*

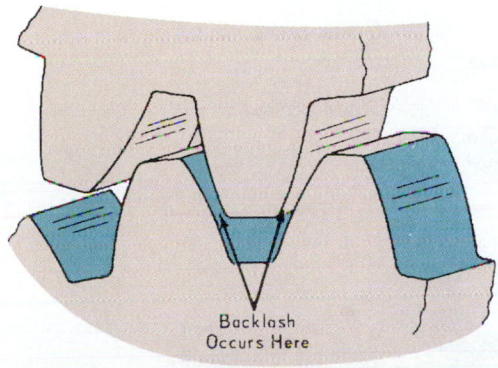

Backlash Occurs Here

Figure 64.8 Backlash is the clearance between meshing gear teeth. *(Reproduced by permission of Deere & Company, ©1991. Deere & Company. All rights reserved.)*

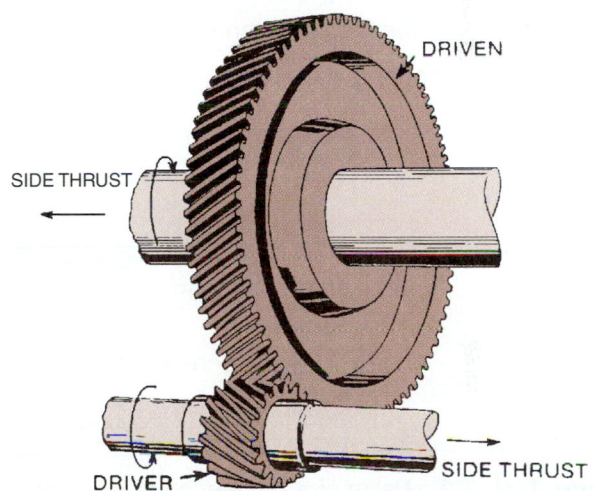

DRIVEN

SIDE THRUST

SIDE THRUST

DRIVER

Figure 64.9 Helical gears have slanted gear teeth and move away from each other under load (thrust).

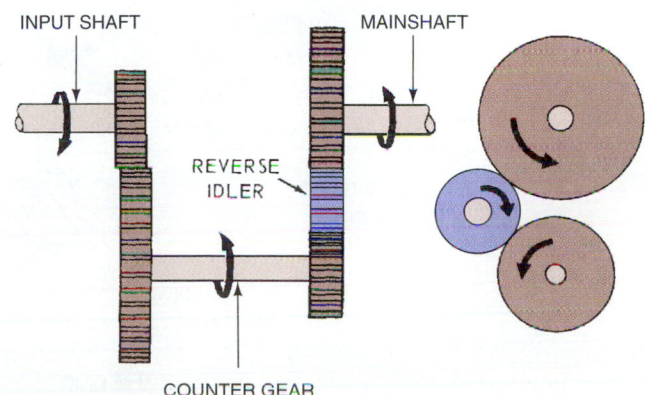

INPUT SHAFT

MAINSHAFT

REVERSE IDLER

COUNTER GEAR

Figure 64.10 A reverse idler gear changes the direction of rotation. *(Reproduced by permission of Deere & Company, ©1991. Deere & Company. All rights reserved.)*

■ TRANSMISSION PARTS

Figure 64.11 shows the basic parts of a four-speed manual transmission. Gear flow is from the clutch disc to the input shaft. The various gears provide a means of changing torque and speed of the output shaft. Each forward gear has a **synchronizer**, sometimes called a synchro, that keeps two meshing gears from clashing during a shift. Shift linkage acts on shift forks within the transmission to select a gear. Power flows from the input shaft to a countergear and then to the mainshaft or output shaft.

Parts are housed in a transmission *gear case* made of iron or aluminum (Figure 64.12). The case is usually made of aluminum because of its lighter weight. Bolted to the back part of the case is the *extension hous-ing*, or tailshaft housing. Sometimes a case is iron and the extension housing is aluminum. The extension housing or case often has threaded flange holes for the transmission-to-crossmember mount. A *front bearing retainer* holds the input shaft bearing against the case. It also acts as a sleeve for the throwout bearing.

The case has drain and fill plugs for adding and draining oil. The drain plug often has a magnet attached to collect particles that might break or wear off of gear teeth. If these pieces attach themselves to the magnet, they will not become lodged between gear teeth and cause further damage to the transmission.

There is a seal at each end of the transmission, at the input and output shafts. Gaskets seal between the parts of the case. A breather is provided at the top of the case.

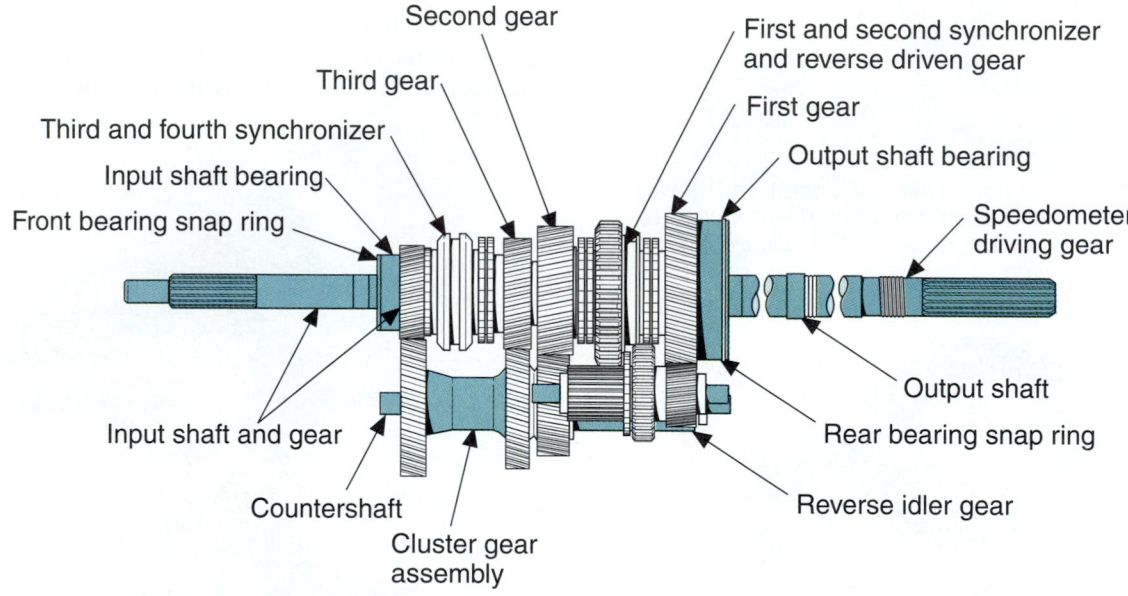

Figure 64.11 Parts of a four-speed manual transmission. *(Courtesy of Ford Motor Company)*

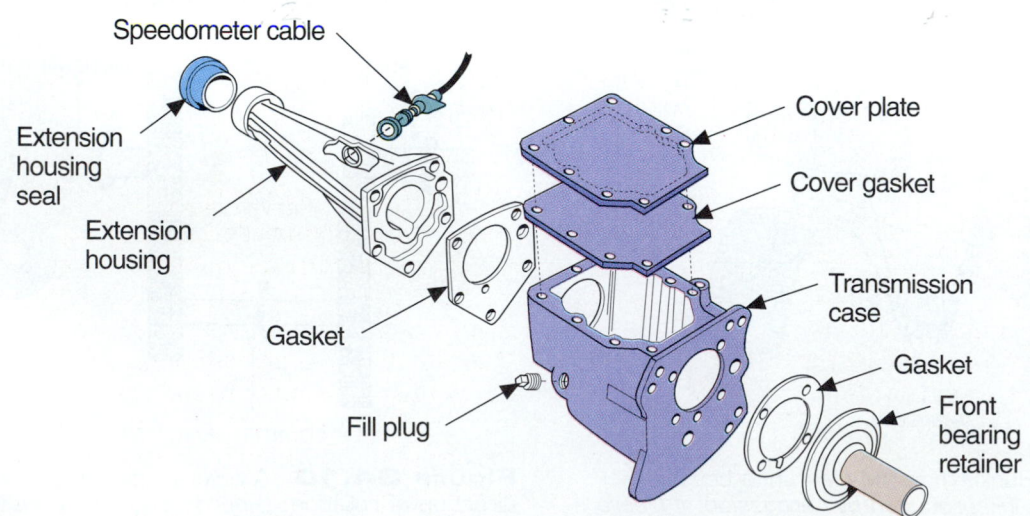

Figure 64.12 Parts attached to a transmission case. *(Courtesy of Ford Motor Company)*

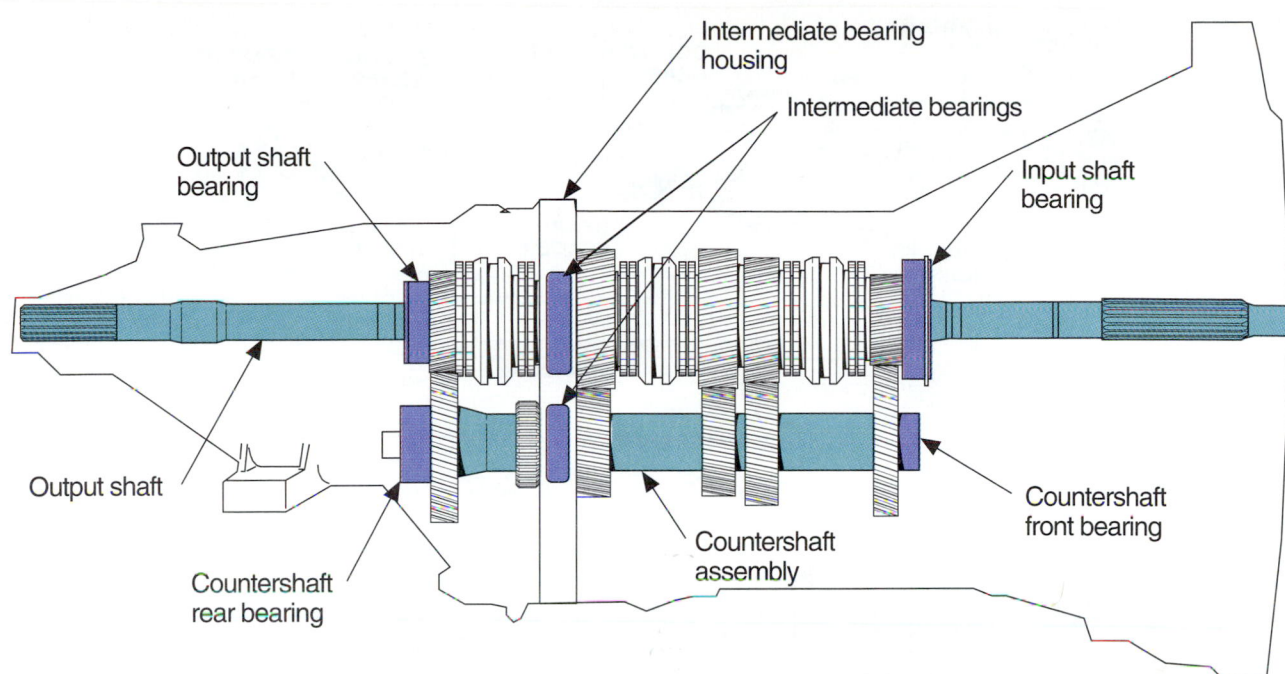

Figure 64.13 Locations of various bearings in a manual transmission. *[Courtesy of Ford Motor Company]*

■ TRANSMISSION LUBRICATION

Transmission parts are separated by oil at all times. Oil is moved throughout the case by the rotating gears. This is called *splash lubrication*. The lubricant varies by manufacturer. Some use SAE 80 or 90 gear oil, the same as is used in differentials. Others use SAE 30 motor oil or automatic transmission fluid (ATF). Using the incorrect oil can drastically affect the transmission's operation. Be sure to follow the manufacturer's recommendations.

■ TRANSMISSION BEARINGS

Bearings support the ends of almost all rotating parts within a transmission. They allow parts to rotate with very little friction. Several types of bearings are found in manual transmissions (Figure 64.13). Reverse idler shafts and gears are sometimes supported by bushings, but the bearings are either ball, roller, or needle.

■ TRANSMISSION GEARS AND SHAFTS

Different parts of the gear train are commonly called "shafts." These include the input shaft, countershaft, and output shaft or mainshaft.

The *input shaft* is often called a *clutch shaft* (Figure 64.14). Its bearing is called an input shaft bearing or *clutch bearing*. You should be aware of this in case you hear a front bearing retainer called a clutch bearing retainer.

The *countershaft* is usually one gear made up of a series of gears that mesh with the various gears on the

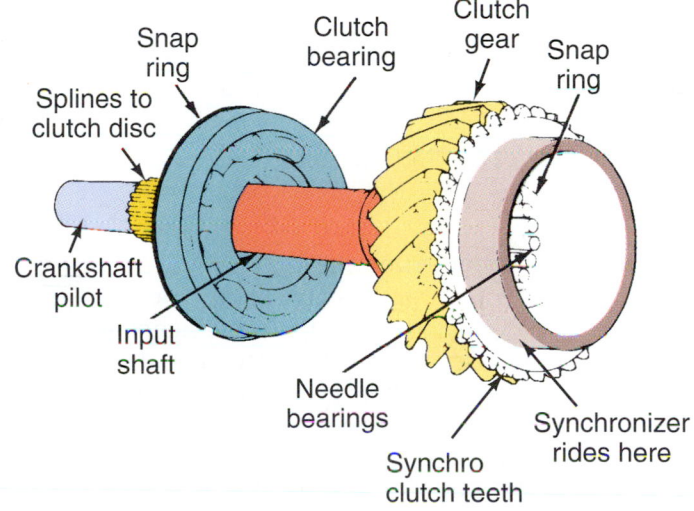

Figure 64.14 Parts of an input shaft. *[Courtesy of General Motors Corporation, Service Technology Group]*

mainshaft (Figure 64.15). It is often called a *cluster gear*. Older transmission designs have short plain steel shafts that ride inside of the countergear and reverse idler gear. Rows of small needle bearings separate the shaft and the inside of the gear to support the load. The ends of the shafts are pressed fit in the case. Some heavy load transmissions have two rows of bearings on each end. Later model transmissions often have the countergear supported by a large tapered roller bearing at the ends.

The *output shaft* or *mainshaft* has all of the gears and synchronizers mounted on it (Figure 64.16). On manual

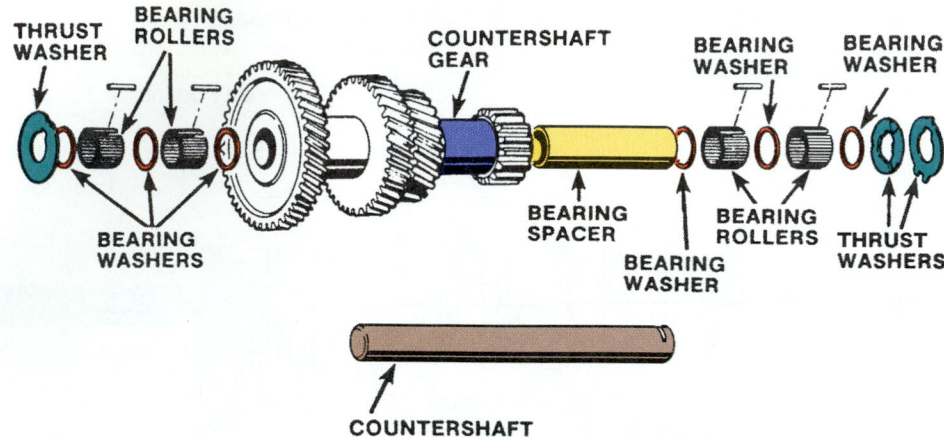

Figure 64.15 Parts of a countershaft assembly. *(Courtesy of Chrysler Corporation)*

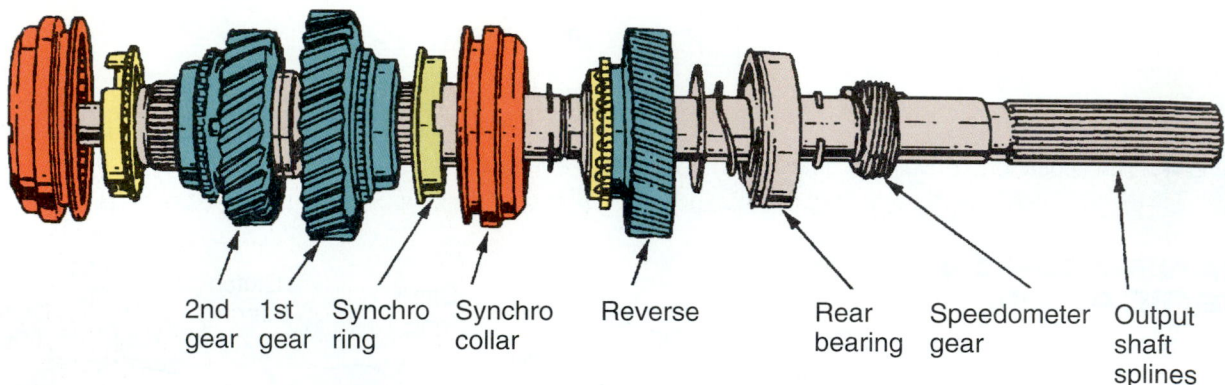

Figure 64.16 Parts of a mainshaft on a three-speed transmission. *(Courtesy of General Motors Corporation, Service Technology Group)*

transmissions made since the mid 1960s, all forward gears are in *constant mesh*. This means that only shift collars move to engage each gear to the mainshaft.

The *reverse idler gear* is the only gear that moves into mesh with another gear (Figure 64.17). The reverse gear set is often made up of spur gears, rather than helical gears. That is why a transmission will often howl when in reverse gear.

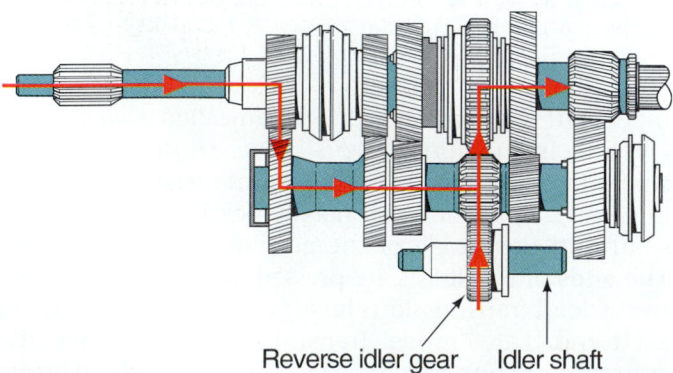

Figure 64.17 A reverse idler gear slides into contact with another gear. The one in this transmission is a spur gear. *(Courtesy of Ford Motor Company)*

■ SYNCHRONIZER ASSEMBLY

The job of the synchronizer is to help two gears spinning at different speeds mesh without clashing. The synchro blocks the shift and brakes the two parts together using a cone clutch type of action.

Blocker ring synchronizers are the common type used in automobiles. There are other kinds, but they are rare. Parts of a typical synchro are shown in Figure 64.18. Locate the *hub* in the sketch. It is splined to the mainshaft (Figure 64.19). A *shift collar*, or *synchronizer sleeve*, fits around the outside of the hub (Figure 64.20). The gears are in constant mesh with their counterparts on the **countergear** and rotate freely on bearing areas on the mainshaft. To make a gear shift, the splines on the outside of the hub attach to a gear.

Figure 64.21a shows the parts of a synchro, assembled as a unit. The assembly is sometimes called a *dog clutch*. Each gear has **dog teeth**, which are little teeth around the circumference of the edge of the gear (Figure 64.21b). When a gear shift is made, the sleeve is moved by shift linkage to select the desired gear. The synchro sleeve has splines around its inside diameter. As it is moved toward the gear, it engages its dog teeth.

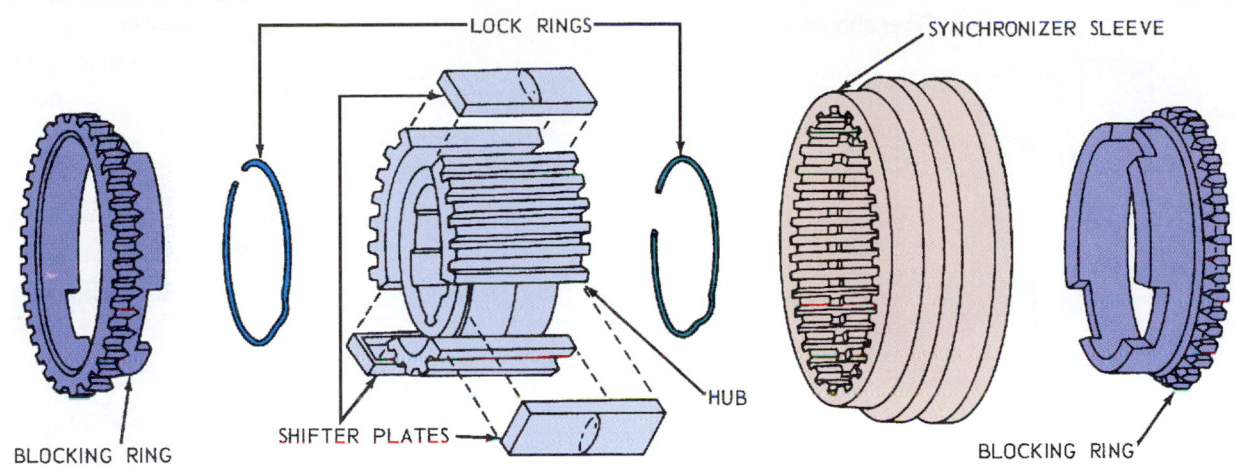

Figure 64.18 Parts of a synchronizer assembly. *(Reproduced by permission of Deere & Company, ©1991. Deere & Company. All rights reserved.)*

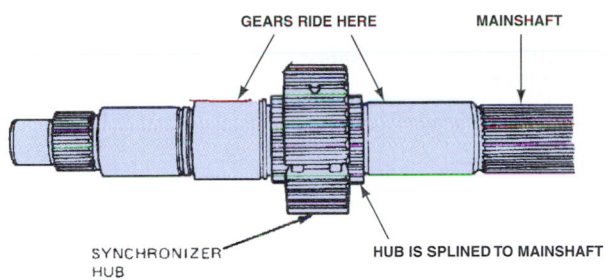

Figure 64.19 The hub is splined to the mainshaft and the gears are free to rotate. *(Courtesy of Ford Motor Company)*

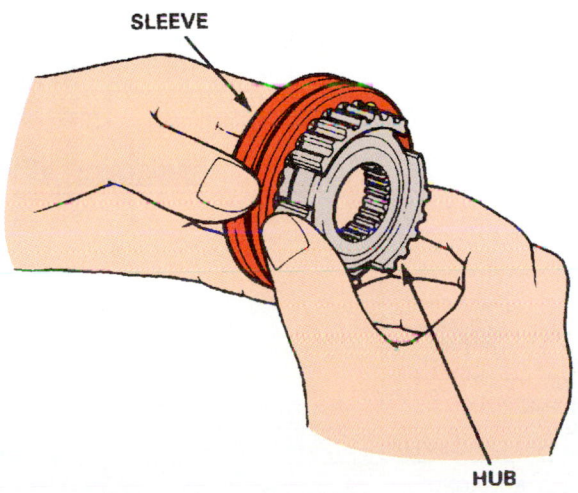

Figure 64.20 A sleeve is splined to the outside of the hub. *(Courtesy of American Honda Motor Co., Inc.)*

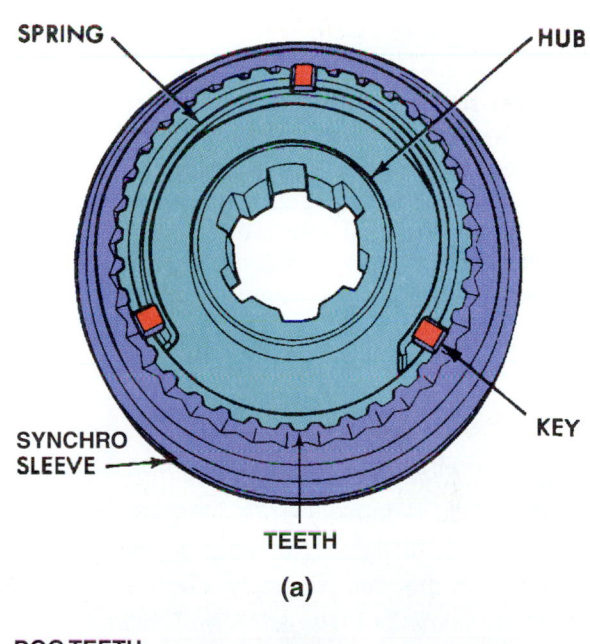

(a)

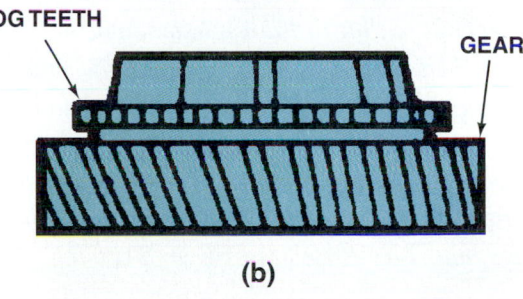

(b)

Figure 64.21 (a) An assembled synchro. (b) Each constant mesh gear has dog teeth. *(Courtesy of General Motors Corporation, Service Technology Group)*

Within each shift collar is a blocking ring synchronizer called a *synchro ring* or blocker ring (Figure 64.22). Blocker rings are usually made of brass. Some newer blocker rings are paper lined. Automatic transmission fluid must be used with these.

Besides preventing the clashing of gears, the synchro's purpose is to lock the input shaft gear to the output shaft gear. Keys fit into notches in the synchro rings to tie the hub to the synchro ring and keep the two spinning together (Figure 64.23). The keys prevent

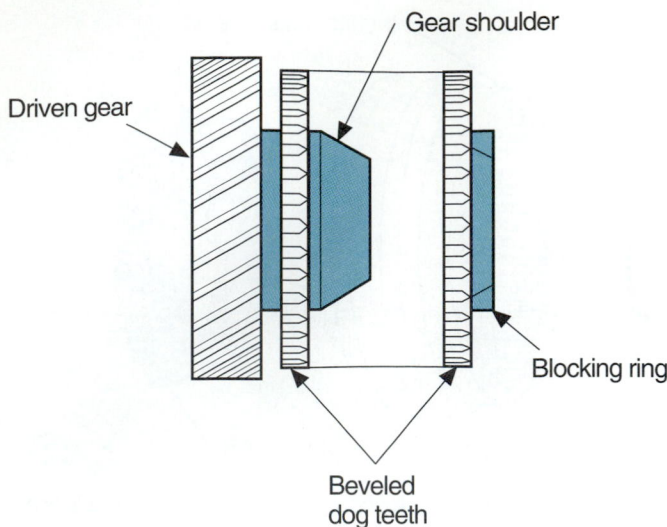

Figure 64.22 A blocking ring wedges against the shoulder of the gear.

the blocker ring teeth from moving more than half of a tooth's width in either direction.

As the shift progresses, the tapered ends of the dog teeth on the blocker ring attempt to mesh with the tapered ends of the sleeve teeth (Figure 64.23a). As the shift is completed, the teeth on the sleeve overlap the dog teeth on the gear (Figure 64.23b). Figure 64.24 shows how the blocker ring wedges against the cone of the gear to stop it from spinning.

■ HISTORY NOTE ■

A fully synchronized transmission is a transmission that has synchros in all forward gears. Some automobiles produced in the first half of this century had sliding gear transmissions that had no synchronizers. To shift gears in these transmissions, the driver had to double clutch. This involves putting the transmission in neutral between shifts and letting the clutch out. While the clutch pedal is up, the engine is accelerated to an estimate of how much engine rpm will be after the shift. Then the clutch is depressed and the shifter moved to the desired gear. For the transmission to shift without clashing, the entire process must take place in less than a second.

■ GEAR SHIFT MECHANISMS

Shift forks fit into grooves cut in the outside of the synchro collar. Shift linkage can be either the internal shift rail type (Figure 64.25) or the external rod type (Figure 64.26). There are also mechanisms that use cables instead of linkage. Each type has features for keeping the transmission in gear and for keeping it out of two gears at once. The internal linkage type will be used here for illustration purposes.

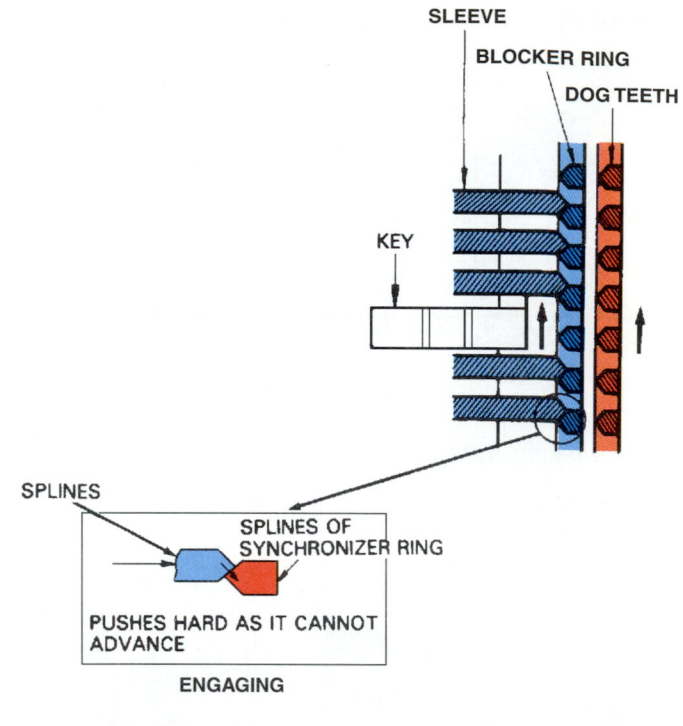

(a)

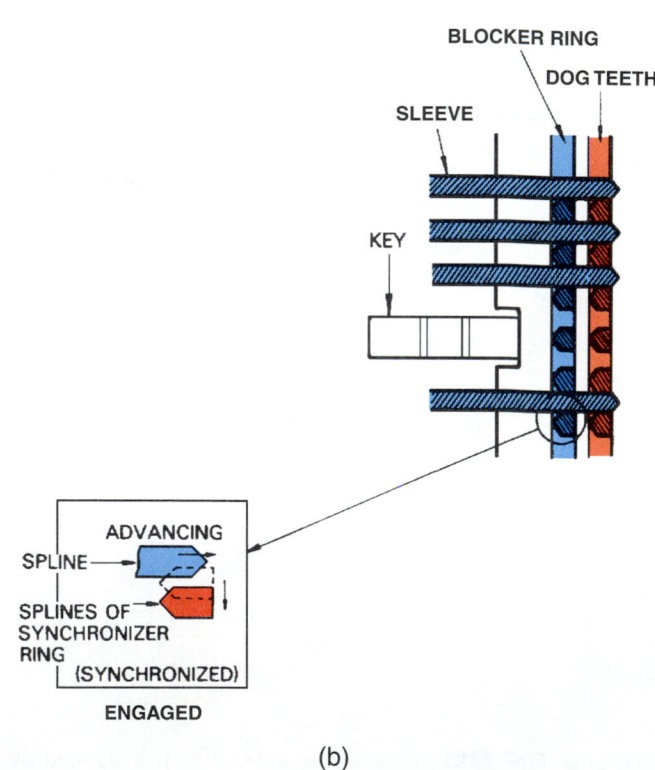

(b)

Figure 64.23 Action of a synchronizer assembly.

A *detent mechanism* holds the transmission in gear (provided there are no worn parts in the synchronizer shift collar). Spring tension holds the detent balls into the detent notches in the shift rail (see Figure 64.25).

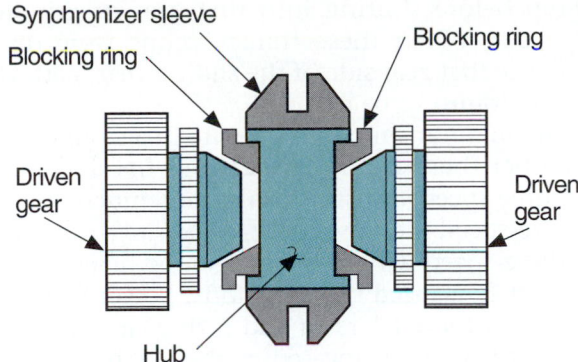

Synchronizer sleeve

Blocking ring

Blocking ring

Driven gear

Driven gear

Hub

Synchronizer in neutral position before shift

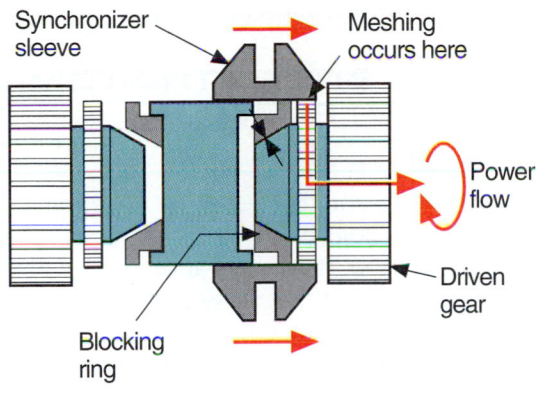

Synchronizer sleeve

Meshing occurs here

Power flow

Driven gear

Blocking ring

Shift completed—collar locks driven gear to hub and shaft

Figure 64.24 The blocker ring in neutral position (above) and wedging against the cone of the gear (bottom).

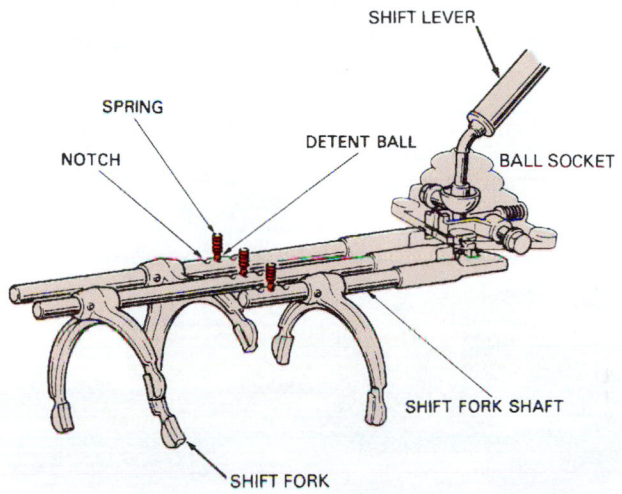

SHIFT LEVER

SPRING

DETENT BALL

BALL SOCKET

NOTCH

SHIFT FORK SHAFT

SHIFT FORK

Figure 64.25 Typical internal shift mechanism.

This keeps the shift rails in place, preventing the transmission from popping out of gear.

An interlock mechanism prevents the selection of two gears at once (which would destroy the transmission). When one of the shift shafts is moved during a shift, a set of interlock pins holds the other shafts in

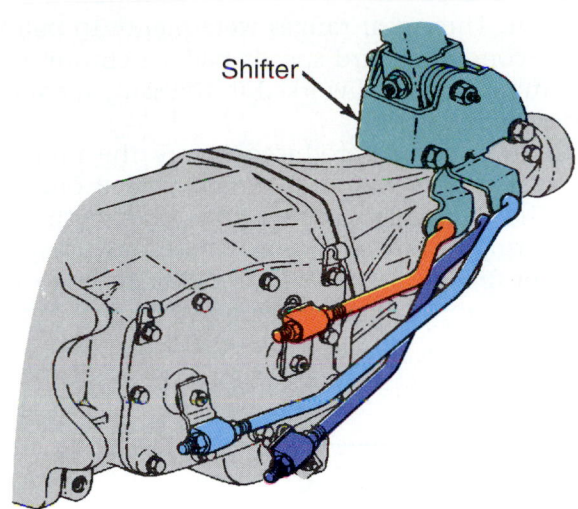

Shifter

Figure 64.26 External shift mechanism on a four-speed manual transmission. [*Courtesy of General Motors Corporation, Service Technology Group*]

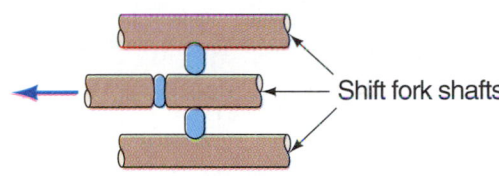

Shift fork shafts

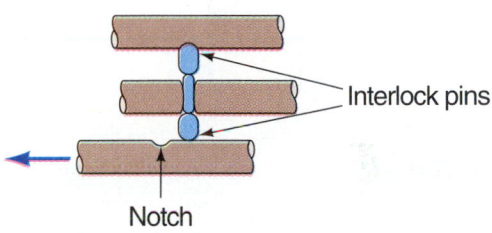

Interlock pins

Notch

Figure 64.27 Interlock pins prevent shifting into two gears at the same time.

their neutral positions (Figure 64.27). There is also a shift restricting pin so that a shift cannot be made from fifth to reverse during a downshift.

■ SHIFT PATTERN

There are various shift patterns used for different transmissions, but most of them use a standard one (Figure 64.28). With four or five speeds, sometimes reverse is in a different position. Other times, the handle is pushed down to locate the reverse gate.

■ TRANSMISSION POWER FLOW

Old transmissions had sliding spur gears. Today's transmissions are constant mesh. The only thing that moves is the synchro collar, which selects the gear. Until the mid 1960s most transmissions were three-speed manuals. Engines of the day were large and powerful. Three-gear ranges were plenty to pull these cars. Second and third speeds had synchronized constant mesh gears and worked in the same manner as a modern transmission.

In three-speeds produced before the mid 1960s, first gear (called low) and reverse shared one sliding gear. The sliding gear came into contact with an idler gear during a shift to reverse. There was no synchronizer for first gear. The vehicle had to come to a com-plete stop before shifting into first gear or grinding would occur. When these transmissions were disas-sembled, the first gear side of the sliding gear teeth was commonly found to be chipped.

All manual transmissions operate in a similar fash-ion, whether there are three speeds or five speeds. In fact, a three-speed transmission is not much smaller than a five-speed. Because three-speeds are from older cars, a three-speed will probably even be heavier.

In most five-speed transmissions, power in fourth gear is direct, for a 1:1 ratio and fifth gear is an over-drive. The fifth gear is located in the extension hous-ing (Figure 64.29). A four-speed is simply this kind of five-speed without the overdrive. It has the same gears in the case, but no extra gear in the extension housing.

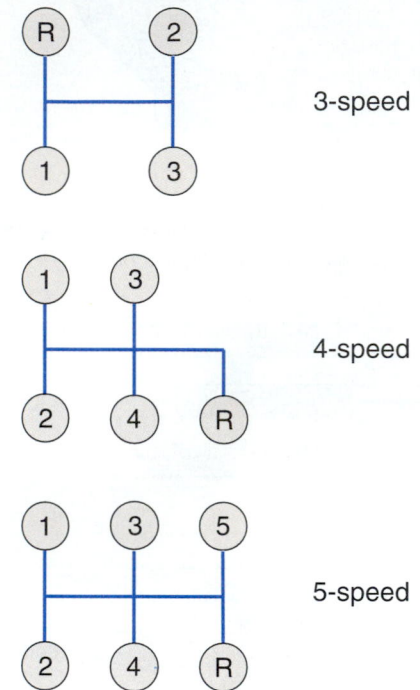

Figure 64.28 Typical gear shift patterns.

■ FOUR-SPEED TRANSMISSION POWER FLOW

Neutral

In neutral, the synchro sleeves are all centered and do not mesh with the dog clutch teeth of any gear (Figure 64.30). In the illustration, the spur gear on the outside of the syn-chronizer sleeve is the reverse gear. This is a typical setup in transmissions. Notice that there is a blocking ring on each side of the synchro assembly. If this is a three or four synchronizer, one ring is for stopping third gear on the mainshaft, while the other is for stopping fourth gear. The other synchro assembly is for 1 to 2 shifts.

High Gear

The easiest gear is high gear, with a 1:1 ratio. Its power flow runs straight through the transmission from the input shaft to the output shaft. The synchro sleeve

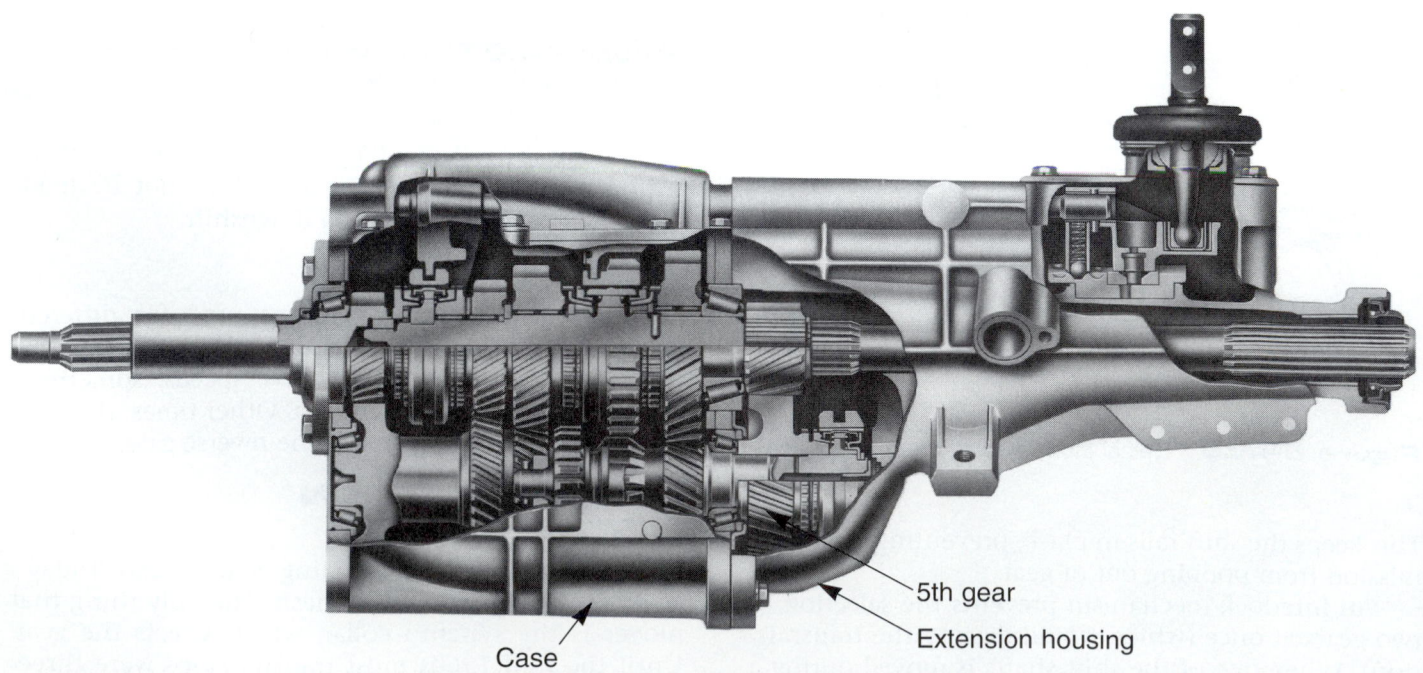

Figure 64.29 Fifth gear is located in the extension housing. [*Courtesy of Borg-Warner Automotive, Inc.*]

Dog teeth showing synchronizers centered

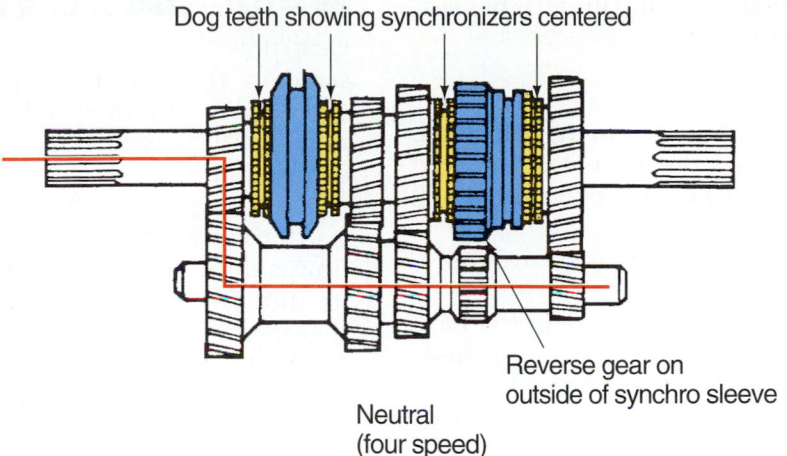

Reverse gear on
outside of synchro sleeve

Neutral
(four speed)

Figure 64.30 In neutral, none of the dog teeth are engaged. *(Courtesy of Volvo North America Corporation)*

moves to the front of the transmission to engage the dog teeth on the back of the input shaft. This attaches the input shaft to the output shaft. Figure 64.31 shows high gear in a three-, four-, or five-speed transmission.

Third Gear

In third gear power comes in through the input shaft, which is engaged to the countergear. The synchro

sleeve has moved to the right and the dog clutch teeth of third gear are engaged (Figure 64.32). This locks third gear to the mainshaft. In a three speed transmission, this would be the position of the second gear.

Second Gear

In second gear, the synchro sleeve in the rear of the transmission is moved toward the left to engage the

Sleeve moves left (forward)

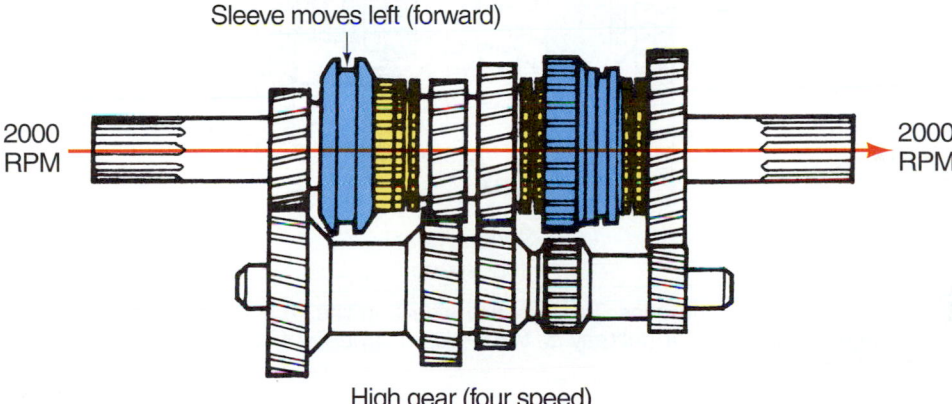

2000 RPM

2000 RPM

High gear (four speed)

Figure 64.31 In high gear, the synchro sleeve engages the dog teeth of the input shaft (clutch gear). *(Courtesy of Volvo North America Corporation)*

Sleeve moves right (rearward) 3rd gear

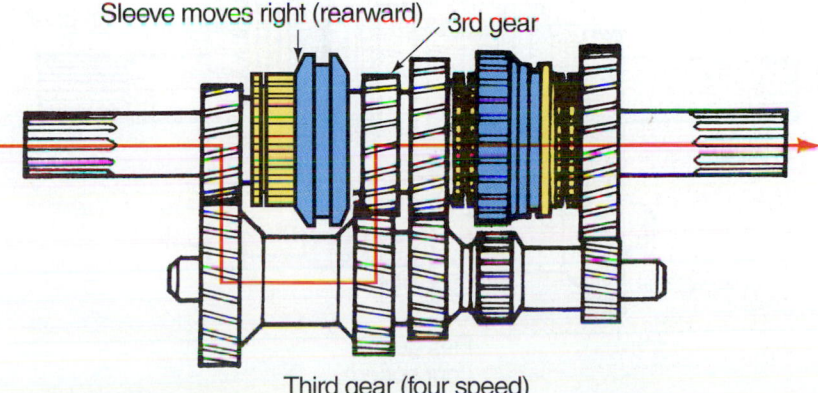

Third gear (four speed)

Figure 64.32 Power flow in third gear. The synchro sleeve has moved to the right to engage the third gear dog teeth. *(Courtesy of Volvo North America Corporation)*

second gear dog teeth (Figure 64.33). This attaches second gear to the output shaft.

First Gear

In first or low gear, the synchro sleeve in the rear of the transmission is moved toward the right to engage the first gear dog teeth. This attaches first gear to the output shaft (Figure 64.34).

Reverse

In reverse, both of the synchro sleeves are in the neutral position. The reverse idler gear is slid into mesh between the spur gear on the outside of the rear synchro sleeve and the reverse gear on the countershaft (Figure 64.35). The hub inside of the synchro sleeve is splined to the mainshaft. Because the idler gear is one more gear than is normally used, the direction of rotation is in reverse.

■ FIVE-SPEED TRANSMISSION

Gear flow in a typical five-speed transmission is the same in the first four speeds and reverse as that described for a four-speed transmission. As previously noted, a five-speed has an extra set of gears in the extension housing.

In fifth gear, both of the synchro sleeves in the transmission case are in the neutral position. Power flow is through the end of the countergear to a gear at its end. The gear has dog teeth and a synchro sleeve that meshes with it during fifth gear operation (Figure 64.36). In some five-speeds, both reverse and fifth gear are in the extension housing or rear section of the case. These transmissions have a dog clutch that is between the fifth and reverse gears. Figure 64.37 shows power flow in fifth and reverse in one of these transmissions.

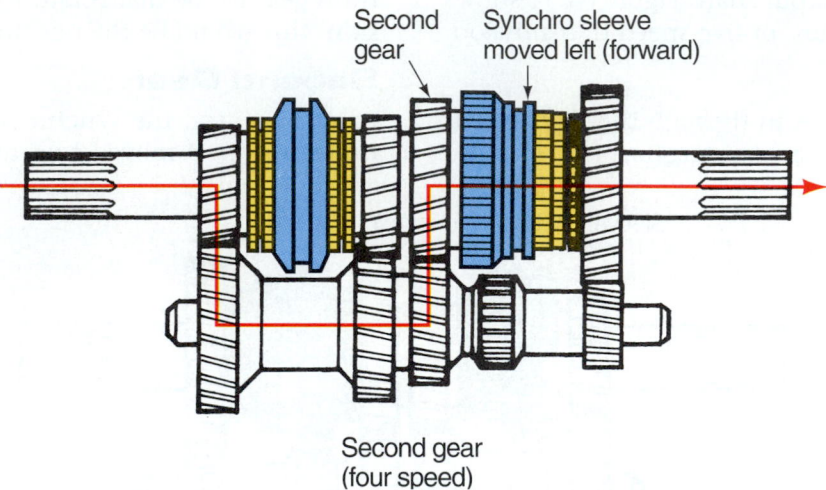

Figure 64.33 Power flow in second gear. The synchro sleeve in the rear of the transmission has moved toward the left to engage the second gear dog teeth. *[Courtesy of Volvo North America Corporation]*

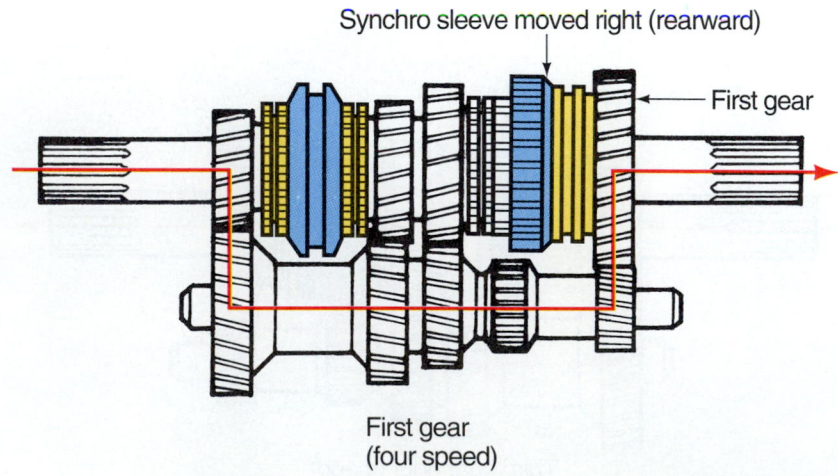

Figure 64.34 Power flow in first gear. The synchro sleeve in the rear of the transmission is moved toward the right to engage the first gear dog teeth. *[Courtesy of Volvo North America Corporation]*

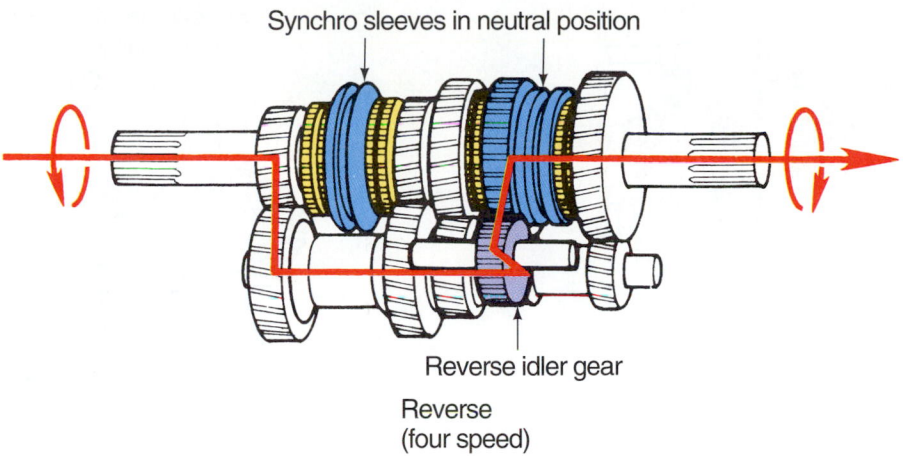

Synchro sleeves in neutral position

Reverse idler gear

Reverse
(four speed)

Figure 64.35 Power flow in reverse. The reverse idler gear is slid into mesh between the spur gear on the outside of the rear synchro sleeve and the reverse gear on the countershaft. *(Courtesy of Volvo North America Corporation)*

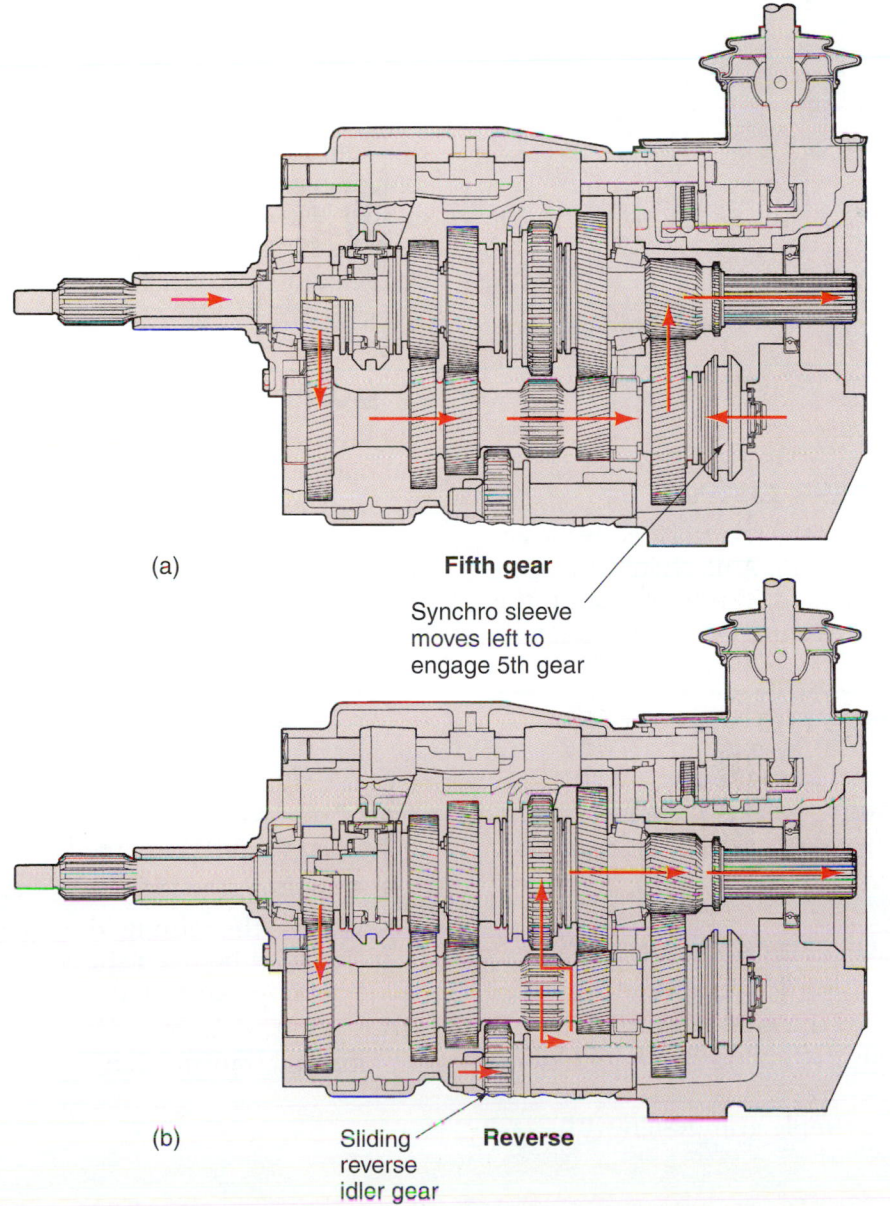

(a)

Fifth gear

Synchro sleeve
moves left to
engage 5th gear

(b)

Sliding
reverse
idler gear

Reverse

Figure 64.36 Power flow in fifth and reverse in some five-speeds. *(Courtesy of Borg-Warner Automotive, Inc.)*

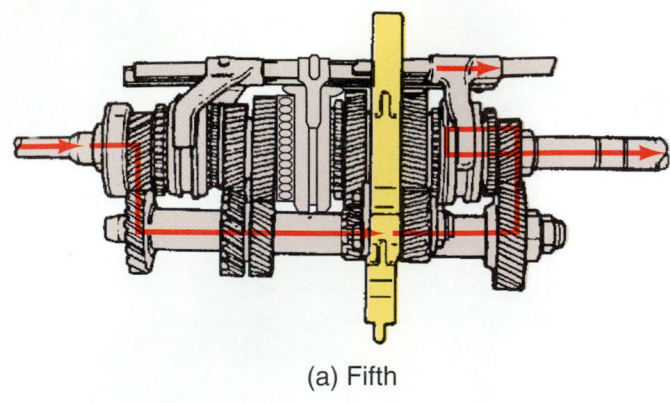

(a) Fifth

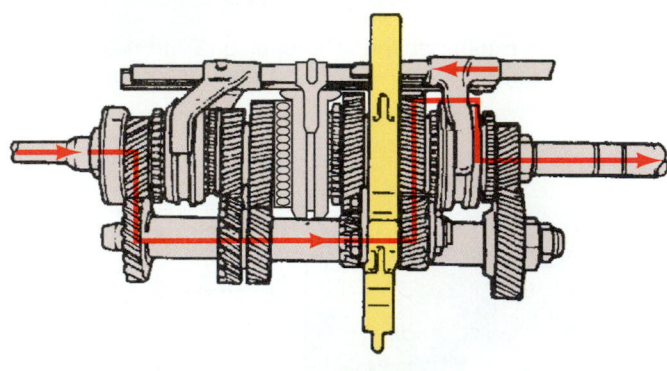

(b) Reverse

Figure 64.37 (a) Power flow in fifth gear in some five-speed transmissions. (b) In reverse a sliding idler gear moves into mesh. *(Reprinted with permission by American Isuzu Motors, Inc.)*

■ SPEEDOMETER DRIVE

Some cars use electric speedometers that receive a signal from a vehicle speed sensor. A mechanically operated speedometer has a cable driven by a gear on the transmission output shaft (Figure 64.38). The gear on

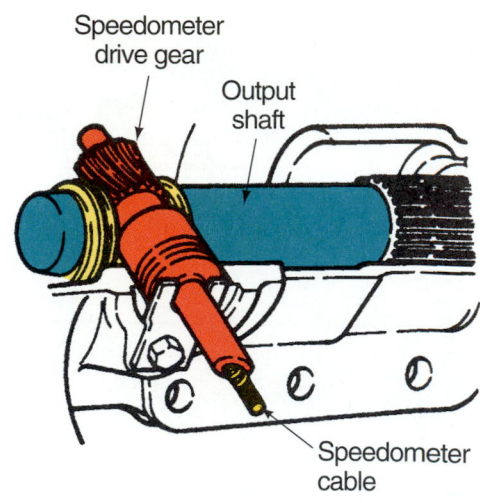

Speedometer drive gear
Output shaft
Speedometer cable

Figure 64.38 A speedometer gear driven by a gear on the transmission output shaft. *(Courtesy of Ford Motor Company)*

the output shaft is either part of the shaft or it is a replaceable part made of metal or plastic. The gear that meshes with it is usually plastic. Speedometer gears come with different numbers of teeth so that they can be replaced to adjust for differences in tire diameter. Some makes of cars come with different wheel sizes and use the same transmission.

■ SWITCHES

A transmission will often have switches that tell the computer or a mechanical system what gear range it is in. Switches are used for backup lights or ignition advance control. Shifting into reverse will close a switch to light the backup lights. Shifting into a particular gear can result in a change in ignition timing to assist in controlling emissions.

■ REVIEW QUESTIONS

1. A manual or standard transmission can also be called a _____ shift.

2. If there are 16 teeth on the driving gear and 48 teeth on the driven gear, the resulting gear ratio is _____ to one.

3. What is the approximate gear ratio in low gear in a transmission with a granny gear?

4. What is the name for the ratio between the transmission output shaft and the differential ring gear?

5. A _____ gear is a simple gear design with straight cut teeth.

6. The clearance between meshing gear teeth is called _____

7. What is another name for the input shaft?

8. A countergear can also be called a _____ gear.

9. What is the term used to describe the method of clutching between shifts when there are no synchronizers in a transmission?

10. What is a common gear ratio for high gear in a manual transmission?

■ ASE STYLE REVIEW QUESTIONS

1. Technician A says that when speed goes down, torque goes up. Technician B says that in overdrive, the input shaft turns faster than the output shaft. Who is right?

   a. Technician A **b.** Technician B
   c. Both A and B **d.** Neither A nor B

2. Two transmissions have different ratios. One has a 3:1 gear ratio in first gear and the other has a 2.2:1 gear ratio in first gear. Technician A says the 2.2:1 transmission is a close ratio. Technician B says the 3:1 ratio is a wide ratio. Who is right?

   a. Technician A **b.** Technician B
   c. Both A and B **d.** Neither A nor B

3. Technician A says that reverse gears are often spur gears. Technician B says that spur gears have replaced helical gears in transmissions because they are quieter. Who is right?

   a. Technician A **b.** Technician B
   c. Both A and B **d.** Neither A nor B

4. Technician A says that some manual transmissions use SAE 80 or 90 gear oil. Technician B says that some manual transmissions use automatic transmission fluid. Who is right?

   a. Technician A **b.** Technician B
   c. Both A and B **d.** Neither A nor B

5. Technician A says all of the gears are always in mesh with each other in a constant mesh transmission. Technician B says transmissions often howl in reverse gear because of the gear tooth design. Who is right?

   a. Technician A **b.** Technician B
   c. Both A and B **d.** Neither A nor B

6. Technician A says the synchronizer hub is splined to the mainshaft. Technician B says that the synchronizer wedges against the gear teeth to stop the gear from spinning. Who is right?

   a. Technician A **b.** Technician B
   c. Both A and B **d.** Neither A nor B

7. Technician A says a typical five-speed transmission has an extra set of gears located in the extension housing. Technician B says some five-speed transmissions have a dog clutch that is between the fifth and reverse gears. Who is right?

   a. Technician A **b.** Technician B
   c. Both A and B **d.** Neither A nor B

8. Two technicians are comparing spur and helical gear designs. Technician A says a spur gear has more backlash. Technician B says a spur gear is stronger. Who is right?

   a. Technician A **b.** Technician B
   c. Both A and B **d.** Neither A nor B

9. Technician A says that a transmission changes torque to fit an engine's operating requirements. Technician B says that in a higher gear, a small gear drives a larger gear. Who is right?

   a. Technician A **b.** Technician B
   c. Both A and B **d.** Neither A nor B

10. Technician A says an idler is a gear that changes the direction of rotation. Technician B says a synchro keeps two meshing gears from clashing during a shift. Who is right?

    a. Technician A **b.** Technician B
    c. Both A and B **d.** Neither A nor B

Manual Transmission Diagnosis and Repair

■ OBJECTIVES

Upon completion of this chapter, you should be able to:

✔ Diagnose problems related to transmissions.

✔ Remove transmissions in a safe and professional manner.

✔ Disassemble transmissions correctly.

✔ Use presses and pullers to disassemble and reassemble transmissions.

✔ Diagnose and service synchronizers.

✔ Reassemble and install a transmission.

■ INTRODUCTION

This chapter deals with the diagnosis and repair of manual transmissions. Emphasis is on generic diagnosis and repair procedures. Knowing these procedures will make diagnosing and repairing transmissions an orderly process.

Learning how to use a press and pullers properly is a very important part of transmission repair. There are many different ways that transmissions are disassembled and reassembled. Be sure to locate service information for the transmission being worked on. Disassembling transmissions usually requires the use of controlled force. Do not use a press, puller, or hammer until you are sure what you are doing is correct.

NOTE: *Unless you are an apprentice or are working on the transmission with someone who has repaired that particular transmission before, service information will be necessary to avoid damage to the transmission.*

Instructions specific to a particular transmission will not be covered in this chapter. Each transmission has its own specific peculiarities when it comes to disassembly, but the basic operation and diagnostic procedures are similar. Gears require a certain amount of clearance between each other. Synchronizers are mostly of the same design, and procedures for testing them are similar.

■ TRANSMISSION DIAGNOSIS

Quite often complaints of hard shifting or a transmission that does not go into gear are due to a clutch that is out of adjustment or defective. Always check clutch adjustment and operation of linkage or the hydraulic assist. See that the clutch fork moves the expected distance before removing the transmission from the car.

Transmission failure can result from abuse or misapplication of shocks to the transmission. This can be because of hot rodding or due to a wet foot slipping off of the clutch during a downshift. Worn linkage, driving with a defective clutch, or wear resulting from extra high mileage can also cause transmission problems.

Before removing the transmission, try to determine the cause of the failure. Ask the driver about the symptoms. A test drive will often help confirm your diagnosis if the car is driveable.

Some typical symptoms a transmission can have include gear clashing during a shift, hard shifting, jumping out of gear, or unusual noise. Noises can be due to a failed front or rear bearing, or defective needle bearings in the countergear or input shaft. Bad linkage often contributes to a transmission becoming locked in gear. This can be because of the internal failure that results from someone forcing a faulty linkage to shift gears.

NOTE *Driver error can result in low mileage manual transmission damage. This occurs when the overdrive ratio is selected at speeds below approximately 50 miles per hour. The result is that too much load without enough leverage is exerted on the helical gear teeth. Helical gear teeth wipe against each other creating end thrust. This can cause the transmission to begin to jump out of gear, something that will become worse and worse each time it happens again.*

Selecting overdrive at too low a speed can also result in engine damage due to lugging.

LUBRICANT CHECKS

Oil leaks often contribute to transmission failure. If the transmission runs low on lubricant, check for leaks at the front and rear seals any switches, plugs or vents, and gaskets. Lubricant level is checked by removing the fill plug and inserting your little finger inside the hole. You should feel oil within ½" of the bottom of the hole (see Figure 14.22).

TRANSMISSION REMOVAL

This section is a generic description of a rear wheel drive transmission removal. Removing a transmission can be dangerous if you do not understand exactly what you are doing. Seek assistance before pulling your first transmission. Use a hoist and transmission jack or jack stands and a hydraulic jack if doing the job on the ground.

Before installing the jack under the transmission, drain the gear oil. Be certain that the bolt being removed is the correct one for draining the fluid.

CASE HISTORY

An apprentice at a lube and oil service outlet was told to drain the transmission oil out of a Ford Mustang with a five-speed transmission. After he drained the transmission, he loosened and removed a bolt in the side of the transmission that he figured was the fill plug. Unfortunately, this was not the fill plug. The transmission had to be removed and disassembled for repair (Figure 65.1).

Sometimes, it may be necessary to remove exhaust pipes before the transmission can be removed. Be sure to use penetrating oil on the rusty studs. Use an impact wrench with a deep socket to remove the nuts.

Disconnect the battery. Before the transmission can be removed, the driveshaft must be removed. Mark the rear universal joint on the driveshaft and its corresponding yoke for proper balance upon reinstallation. When the gear oil is not to be drained ahead of time, a yoke or a seal installation tool can be installed on the output shaft splines to keep oil from running out (Figure 65.2).

Remove the speedometer cable from the extension housing. Remove and label any wires to electrical switches (backup lights, transmission controlled spark switches, and so forth).

Disconnect external shift linkage at the transmission. This can involve pulling the handle from the top of a shift rail type of transmission, or it can mean removing the shift linkage from the shift levers on the side of the transmission.

SHOP TIP To gain easier access to the nuts that hold a shifter handle in place, remove the cross-member bolts (Figure 65.3) and slowly lower the jack a small amount. This will usually let the transmission hang low enough to gain access to the shifter nuts. Be careful not to put too much pressure on the front motor mounts when you lower the jack.

If the vehicle is equipped with one, remove the crossmember. Support the rear of the engine.

Figure 65.1 Removing this bolt to check the fluid level resulted in having to remove and partially disassemble the transmission.

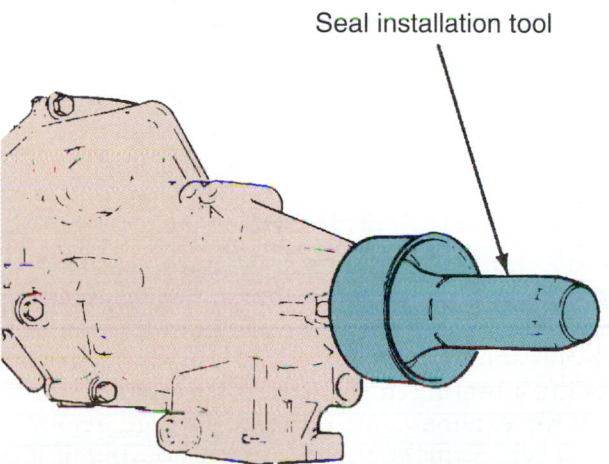

Seal installation tool

Figure 65.2 A seal installation tool or slip yoke installed on the output shaft will prevent lubricant leakage. *(Courtesy of Chrysler Corporation)*

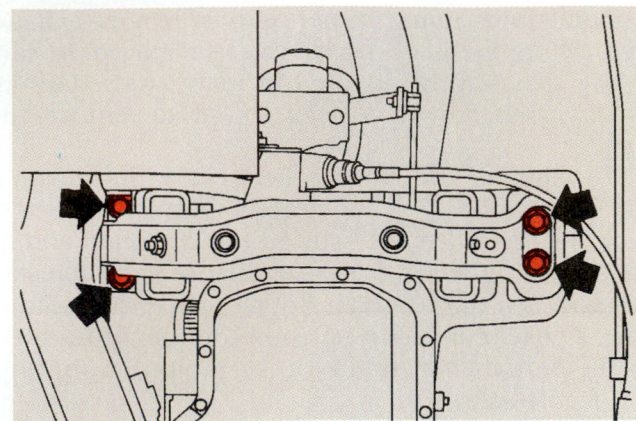

Figure 65.3 Support the transmission and remove the crossmember bolts. *(Courtesy of Nissan Motors)*

 **SHOP TIP** If lifting beneath the engine from the oil pan with a jack, make sure to use a piece of plywood to protect the oil pan.

Use a jack under the transmission to support it. Remove the transmission. Slide it straight back away from the flywheel until the input shaft is clear of the clutch. Do not let the engine hang only by its two mounts. Be aware of the cooling fan-to-radiator shroud clearance.

SAFETY NOTE Use a jack to remove the transmission or be sure to have someone help you so you do not hurt your back.

Remove the transmission and clean it. After cleaning the transmission, put it on a workbench for disassembly.

Identify the Transmission

When ordering parts from a dealer or parts store, you will need to be able to identify the transmission. There will be either a transmission I.D. tag under one of the bolts or numbers stamped on the transmission case.

■ TRANSMISSION DISASSEMBLY

Before disassembling a transmission, always check the service manual. Procedures vary from transmission to transmission. Sometimes a case can be damaged by pressing a bearing in the wrong direction.

With external shift rail type linkage, remove the shift levers. Remove the extension housing. If it does not come off easily after unbolting it, use a soft mallet or brass hammer to knock it loose (Figure 65.4). On some transmissions with internal shift rails, the linkage will have to be disconnected before the extension housing can be removed. The extension housing on

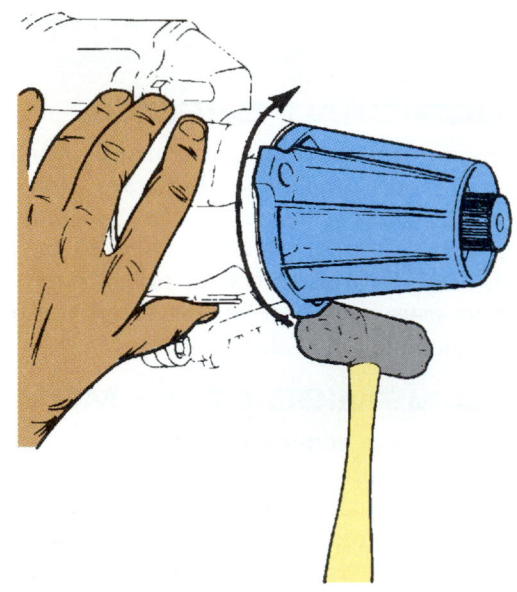

Figure 65.4 If the extension housing does not come loose easily, tap it lightly with a brass hammer. *(Courtesy of Chrysler Corporation)*

these types must be removed before the cover can be removed from the transmission. Then, remove the shift fork assembly and cover (Figure 65.5).

After removing the side cover or top cover, try to shift the gears and rotate the input shaft. Inspect the gears for any obvious signs of damage. As you disassemble the transmission, look for unusual signs of wear on parts that rub against each other. Gears will develop a normal wear pattern with use. They should not have any damage on the face of the gear in the tooth contact area. Damage to the edges of the gear tooth can be ground down if there is a high spot (Figure 65.6).

Remove the front bearing retainer (see Figure 64.12). Note the position of its oil return hole so it can be correctly reinstalled. Save the gasket. There are different thicknesses available, and when it is replaced,

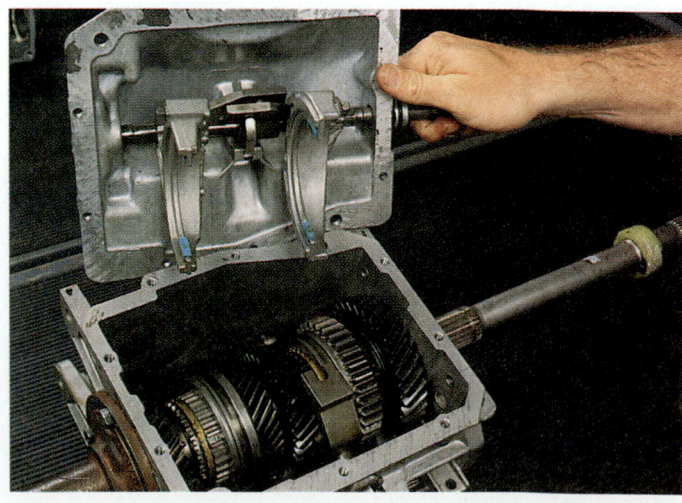

Figure 65.5 Remove the cover and shift forks.

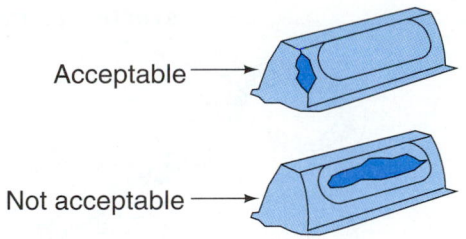

Figure 65.6 Acceptable and unacceptable gear tooth wear. *(Courtesy of Ford Motor Company)*

one of the same size must be used. They control how tightly the bearing is held by the bearing retainer. Next, remove the snap rings on the outside and inside of the bearing.

On many transmissions, before the mainshaft assembly can be removed from the case, the countershaft must be removed to allow the countergear to drop away from the gears on the mainshaft.

Knock the countershaft and reverse idler shafts out of the case. Check the manual to see which direction to drive the shafts. Sometimes they can only go one way or the case will be damaged. A countershaft might be tapered or it might have a pin through it on one side.

SHOP TIP
■ The end of a shaft can easily be damaged by pounding on it with a steel hammer. Develop the habit of using brass hammers or drifts on hard steel parts.
■ A short shaft can be used to keep the needle bearings in the countershaft. This shaft, called a **dummy shaft** (Figure 65.7), is slightly shorter than the case. This will allow the gears to drop free of the mainshaft gears.

The mainshaft is removed from the rear or the top of the transmission, depending on transmission design. The countergear can be removed next. On some transmissions, the countergear bearing must be

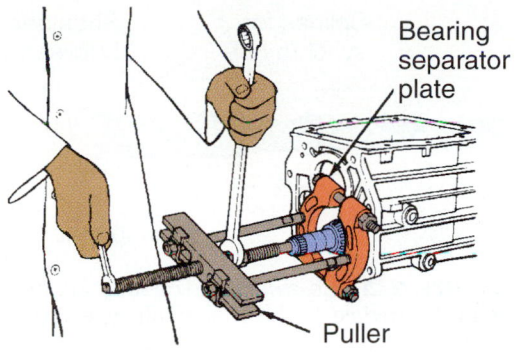

Figure 65.8 Removing the countergear bearing. *(Courtesy of Ford Motor Company)*

removed with a puller before the countergear can be removed (Figure 65.8).

Remove the input shaft from the case. Be careful not to damage the gears. Depending on the type of transmission, sometimes the input shaft is removed before removing the mainshaft. Other times, the countershaft must be removed first. Check the recommendation in the repair manual.

■ DISASSEMBLE THE MAINSHAFT ASSEMBLY

When disassembling the parts, remember to lay them out as they are disassembled and keep them in order. The high gear synchro clutch hub is usually held in place on the mainshaft with a snap ring. It is removed with snap ring pliers (Figure 65.9). The synchronizer clutch hub fits snugly on splines on the mainshaft. A press or puller will probably be required to pull it from the mainshaft. It must be removed to release the gear behind it.

CAUTION There is often a shoulder on the mainshaft that separates the gears (Figure 65.10). Be aware of where this shoulder is when pressing or using a puller. Sometimes, a removable collar has the shoulder on it.

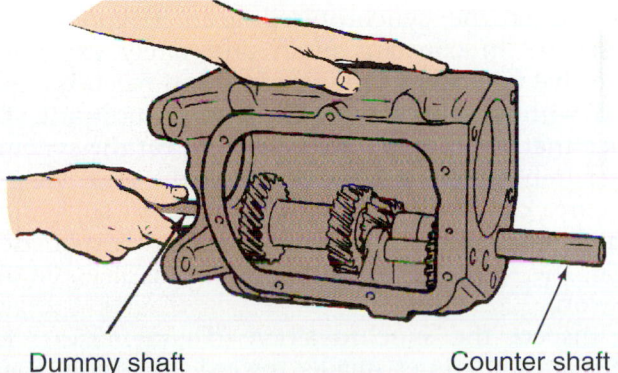

Figure 65.7 Using a dummy shaft to remove the countershaft. *(Courtesy of General Motors Corporation, Service Technology Group)*

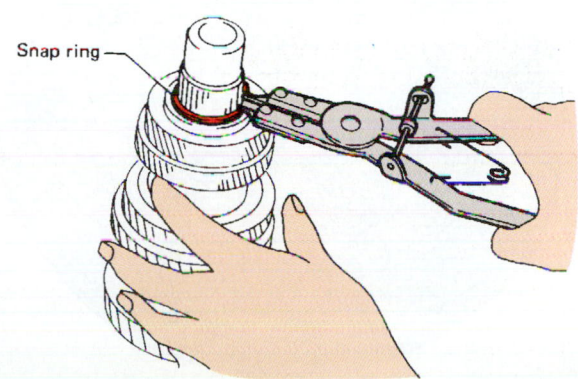

Figure 65.9 Removing a snap ring on the front of the mainshaft. *(Courtesy of Nissan Motors)*

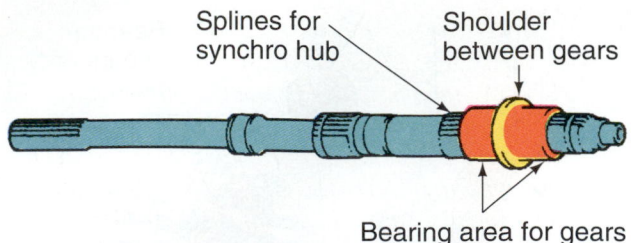

Splines for synchro hub

Shoulder between gears

Bearing area for gears

Figure 65.10 Be aware of the location of the shoulder in the middle of the mainshaft when pressing or pulling gears.

■ SYNCHRONIZER SERVICE

If synchronizers are to be reused, scribe a line on the outside so that they can be reassembled in the same position (Figure 65.11). Some synchronizers are marked from the factory.

After the gears and parts are disassembled from the shaft and cleaned, inspect them. Check the bottom of the transmission case for metal. If brass dust is found in the oil, this is due to synchronizer or thrust washer wear. You can check for gear wear by dragging a magnet through any remaining oil in the case. This will not work for brass synchros or thrust washers because brass is not magnetic. Thoroughly clean the oil from the case with solvent and clean all the transmission parts.

CAUTION When blowing bearings dry, do not allow them to spin. They can fly apart, possibly causing an injury.

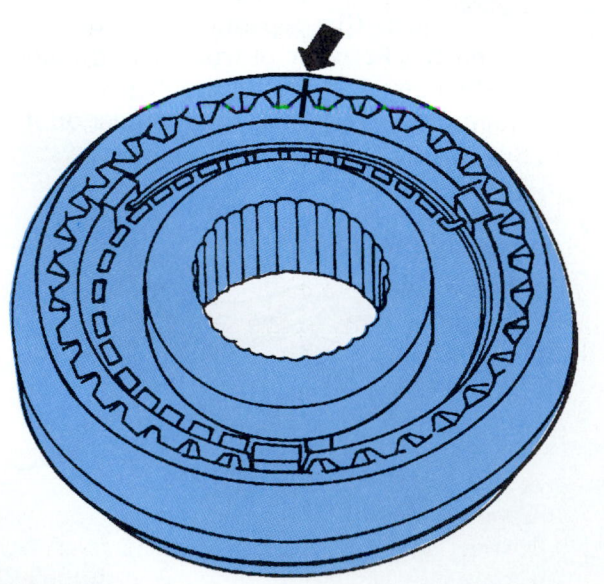

Figure 65.11 Mark the synchro before disassembly. *(Courtesy of Volkswagen of America)*

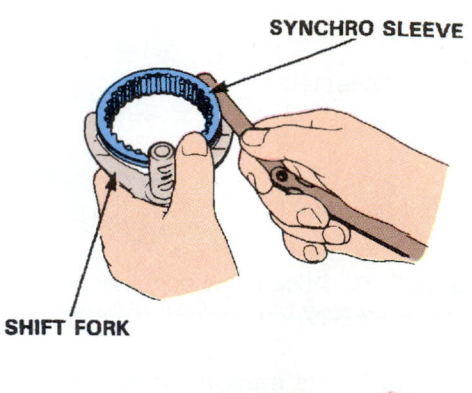

SYNCHRO SLEEVE

SHIFT FORK

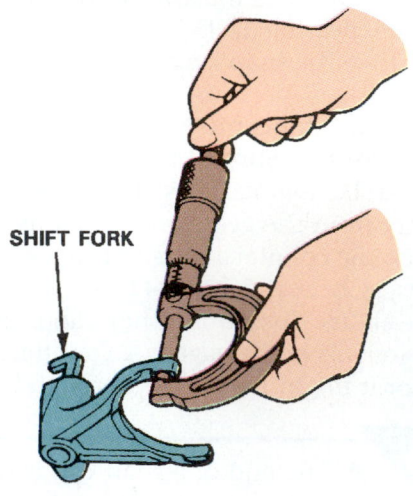

SHIFT FORK

Figure 65.12 Checking for wear on the synchro sleeve and shift fork. *(Courtesy of American Honda Motor Co., Inc.)*

Look for wear on the gears and shift sleeves. Check the shift mechanism for wear. Wear can occur between the shift fork and synchro sleeve (Figure 65.12) or between parts of the linkage.

■ SYNCHRONIZER INSPECTION

The best way to tell if a synchro is not working is to test drive the car before you remove the transmission. A worn synchro will cause gear clashing, especially on downshifts (from third to second, for instance).

Inspect the synchronized gears (Figure 65.13). Gears are in constant mesh and rarely experience excessive wear unless the transmission has been operated without oil or with the wrong lubricant. The place that often wears is where teeth on the synchronizer hub rub against the teeth on the gear. Wear in this area can result in a transmission that slips out of gear or sticks in gear. When synchros have experienced wear, the teeth on the corresponding gear could also be worn.

Inspect the synchro sleeve (Figure 65.14). The teeth are sometimes smaller toward the inside (Figure 65.15). Their width should be even across their length, not worn more on one side. When synchro sleeves wear, they either wear on the outside edge of the tooth

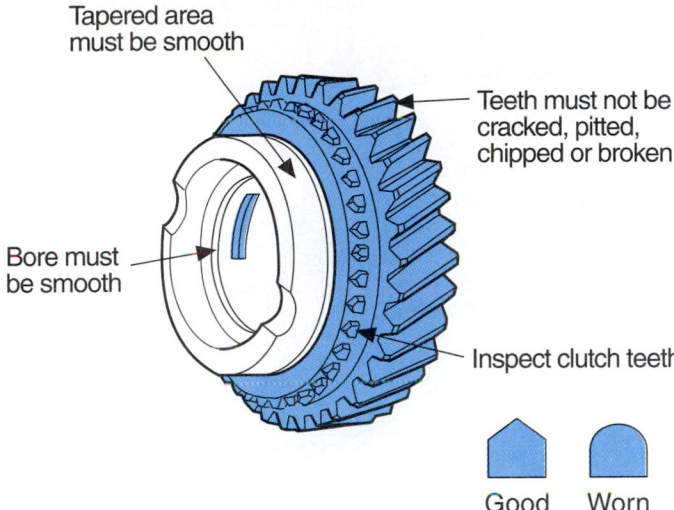

Tapered area
must be smooth

Teeth must not be
cracked, pitted,
chipped or broken

Bore must
be smooth

Inspect clutch teeth

Good   Worn

Figure 65.13 Inspect the gears. *(Courtesy of Chrysler Corporation)*

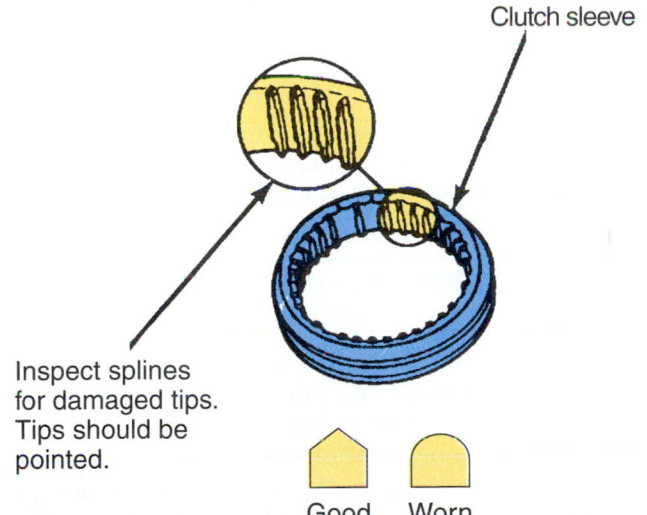

Clutch sleeve

Inspect splines
for damaged tips.
Tips should be
pointed.

Good   Worn

Figure 65.14 Inspect the synchro clutch sleeve. *(Courtesy of Chrysler Corporation)*

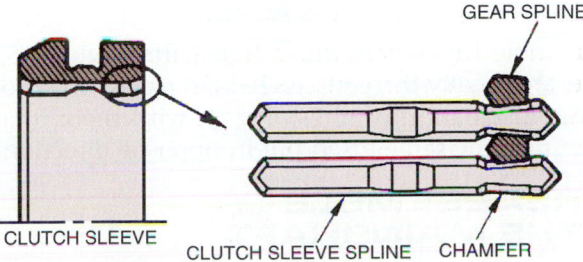

GEAR SPLINE

CLUTCH SLEEVE

CLUTCH SLEEVE SPLINE   CHAMFER

Figure 65.15 Shape of the synchro clutch sleeve teeth.

or on the inside, part of the way into the sleeve. If you replace a clutch sleeve, look for problems with the shift linkage that could have caused the problem.

CASE HISTORY

A customer had her Jeep towed into a repair shop because it jumped and made terrible grinding sounds when driven. She had loaned it to a friend and it had "stopped working." The customer related that she had been having difficulty shifting the transmission into fifth gear for some time and sometimes it would get stuck in fifth.

When the transmission was disassembled, a stripped fifth gear was discovered. The synchro hub was also severely worn. Part of the shift mechanism that keeps the transmission from shifting into two gears at once was also broken. Third gear was damaged badly on the countergear and mainshaft. Before reading the next paragraph see if you can figure out why these parts were broken.

Diagnosis:

The transmission became stuck in fifth gear and the friend forced it with enough pressure to break the shift rail. With the shift lockout mechanism broken, the transmission was able to shift into third gear while it remained locked in fifth. The result was a very costly transmission overhaul. Gears, especially the countergear, are very expensive.

■ INSPECT BLOCKER RINGS

Carefully inspect the brass blocker (synchro) rings if they are to be reused. The inside surface of the blocker rings should have sharp edges to the thread cut into them. Because removing and disassembling a transmission is difficult and time consuming, synchros are often replaced when a transmission is apart. This will depend on the cost of the new parts. The test drive is especially important for this part of transmission diagnosis.

Push the blocker ring against the polished tapered surface on the gear that it rides on. It should grab the chrome surface when pushed against it. If it is too worn, it will slip. Manufacturers often give a minimum specification for clearance between the blocker ring and the gear (Figure 65.16).

■ INSPECT INPUT SHAFT AND MAINSHAFT

Inspect the input shaft. It has several areas that could require service. Inspect the mainshaft splines. They occasionally get twisted from abusive driving. A twisted spline will prevent the driveshaft slip yoke from sliding in and out of the transmission when the car goes over bumps.

The speedometer gear is also located on the output shaft. Inspect it for damage. This gear rarely suffers damage because the gear from the speedometer that meshes with it is usually made of a softer material. Sometimes, the speedometer gear on the mainshaft is

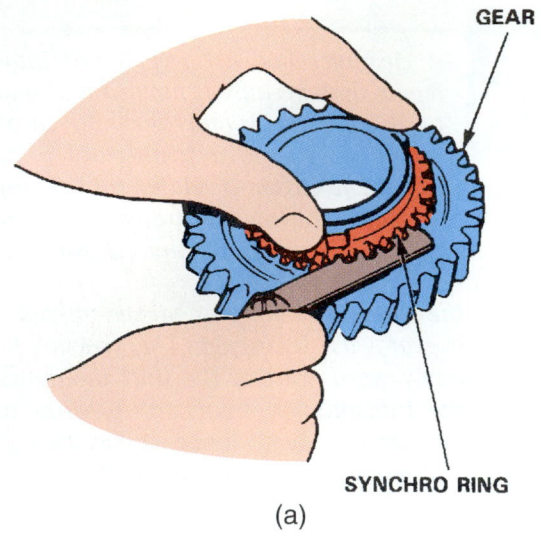

(a)

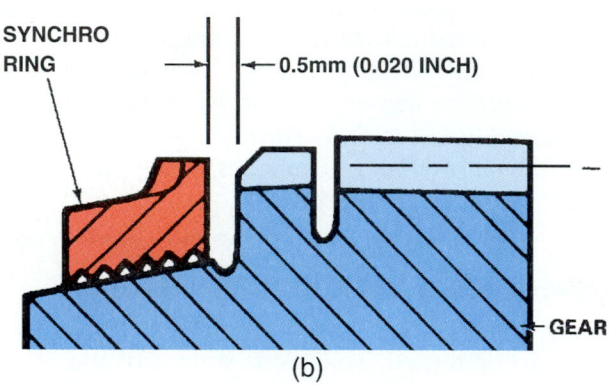

(b)

Figure 65.16 Checking synchro wear. *(Courtesy of American Honda Motor Co., Inc. and Ford Motor Company)*

replaceable. A clip can hold it in place or it can be a pressed fit. A special puller is available to pull these gears on the car.

■ REPLACE WORN BEARINGS

Replace any worn bearings. Use a press and support the inside of the bearing with a bearing separator (Figure 65.17). When reassembling the new bearing to the

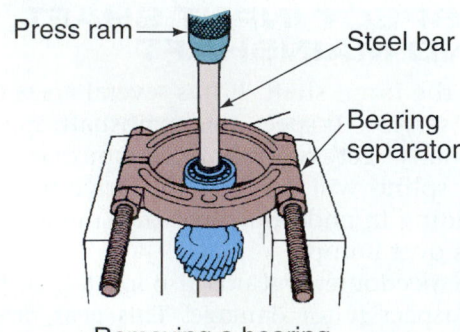

Removing a bearing

Figure 65.17 Using a press to remove a bearing from a shaft.

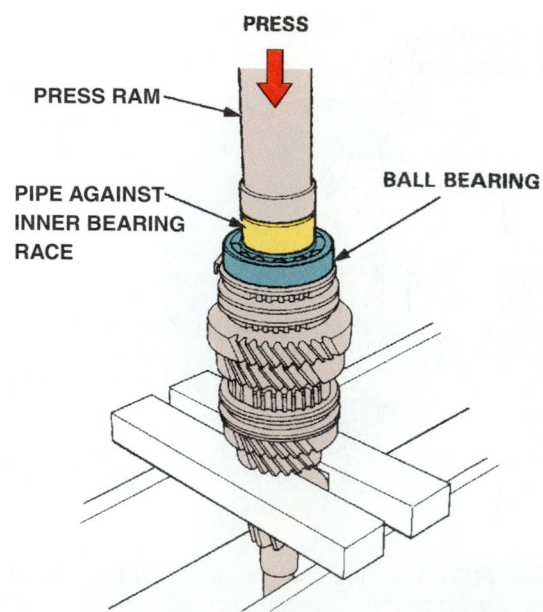

Figure 65.18 When installing a bearing on a shaft, press against the inner bearing race only. *(Courtesy of American Honda Motor Co., Inc.)*

shaft, always put pressure on the inside of the bearing only (Figure 65.18). Pressing on the outside of the bearing can result in damage to the bearing. It is important that the bearing be installed all of the way onto the shaft or clearance problems can result.

■ REASSEMBLE THE TRANSMISSION

Before assembling the transmission, locate and assemble any new replacement parts. Two gears operating in mesh create a wear pattern between them. When a gear is replaced, the corresponding gear that rides on it is customarily replaced too. Otherwise, using a new gear with an old gear can result in unacceptable gear noise. The cost of replacement gears often makes it advantageous to consider the cost of a guaranteed rebuilt unit.

■ REASSEMBLE THE SYNCHRONIZERS

Reassemble the synchronizer hub parts (Figure 65.19). There are usually three inserts held in place with a round spring. Installation of the springs is with them rotating away from the same insert but in opposite directions.

■ REASSEMBLE THE MAINSHAFT

Install the gears on the mainshaft. Lubricate all parts liberally. Install the synchro hub with a press or pipe and hammer. Check end play of assembled parts as described in the service manual. On some transmissions selective snap rings are available. The thickness of the snap ring determines the clearance between the gears (Figure 65.20).

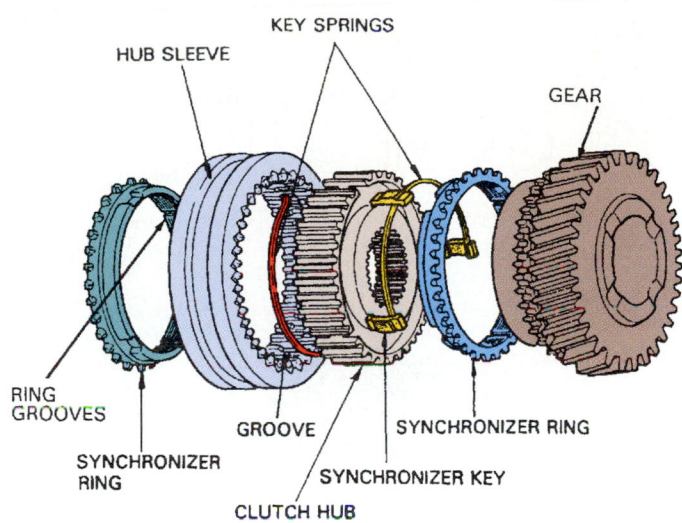

Figure 65.19 Parts of a synchronizer hub.

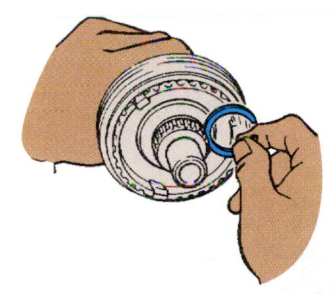

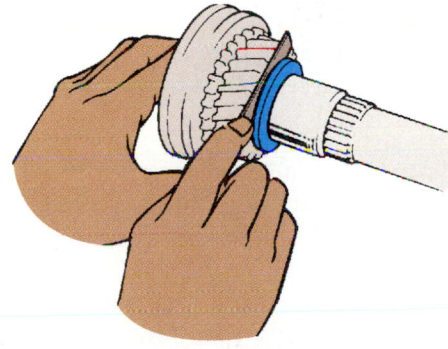

Figure 65.20 The thickness of the snap ring determines the clearance between the gears.

■ END PLAY

The thrust washers at the ends of the countergear determine where the gear will be positioned in the case and also how much clearance there will be. There might be a thicker washer on the rear of the countergear or vice-versa.

NOTE: *Positioning the washers on the wrong ends of the gear can result in interference between the countergear and other gear(s) on the mainshaft.*

Clearance is checked on the mainshaft after all of the gears are assembled on it (Figure 65.21). Normal

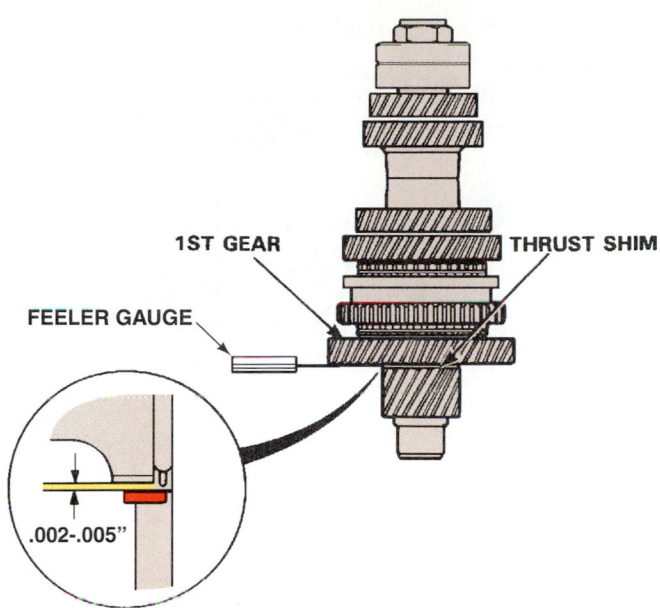

Figure 65.21 Measuring clearance between the gears on the mainshaft. *(Courtesy of American Honda Motor Co., Inc.)*

clearance is usually 0.002" to 0.005". This clearance is adjusted by changing the thickness of the snap ring or by installing a thrust washer at the center of the shaft. Clearance is checked at each of the gears.

■ NEEDLE BEARING INSTALLATION

Needle bearings are sometimes *caged* (they are held in place by a cage). They can also be loose and are installed one at a time (Figure 65.22). If this is the case, use very heavy grease to hold them in place during assembly. Be sure that all needle bearings have been installed. The service manual often tells how many of each needle bearing there are. Experienced technicians will also count the needle bearings during disassembly.

SHOP TIP

■ A grease that is not fibrous (sticky) works best (a **fibrous grease** leaves no strings of grease when you put a small amount between two fingers and then spread them apart).

■ A dummy shaft can be used during transmission assembly to make reinstallation of the countershaft much easier. The short shaft is installed and the bearings are inserted around it. The countergear is installed in the bottom of the transmission case. When the countershaft is pounded into the case and into the end of the countergear, it displaces the short shaft. The short shaft comes out the other end of the transmission case during assembly.

Countergear

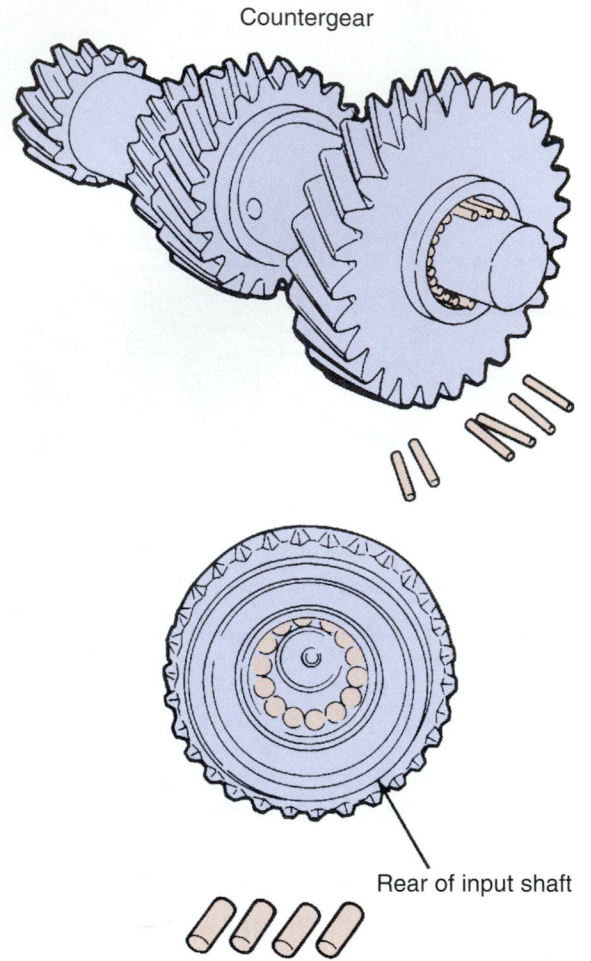

Rear of input shaft

Figure 65.22 Installing loose needle bearings in the countergear and input shaft. *(Courtesy of General Motors Corporation, Service Technology Group)*

BUSHING

BUSHING REMOVER

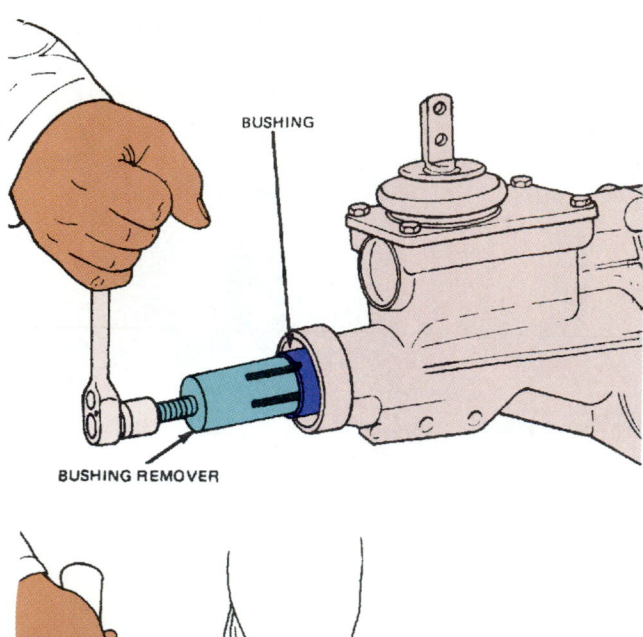

BUSHING

BUSHING INSTALLER

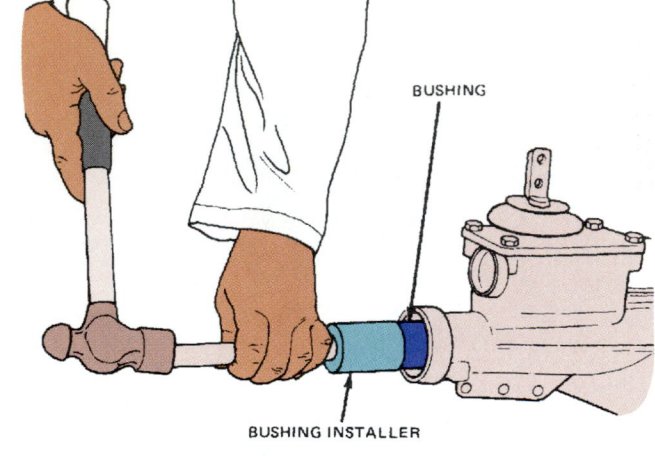

Figure 65.23 Removing and installing a transmission bushing. *(Courtesy of Ford Motor Company)*

◼ INSTALL NEW GASKETS

Purchase a new gasket set for the transmission. Gaskets are not reusable. A gasket that was working fine previously could leak after reassembly. Replacing this gasket would be a complicated repair that would be done at no charge to the customer.

Install the shift mechanism.

NOTE: Check to see if the transmission shifts properly before completing the assembly. On some transmissions, if the shift forks are installed backwards, the transmission will not shift properly. Fixing the problem could require removal of the transmission once again.

⚠ **CAUTION** Be sure that metric and English fasteners are not interchanged by accident.

On rear wheel drive cars, replace the extension housing bushing and seal. The bushing rides against the slip yoke on the front of the driveshaft. It can be

replaced with a hydraulic press or bearing driver before the extension housing is installed. After the extension housing is installed, the bushing can be removed and replaced with special tools (Figure 65.23). A seal installer is used to install a seal without distorting it (Figure 65.24).

NOTE: When installing rubber seals and O-rings, be sure to oil them. Failure to do this can result in damage to the seal.

SEAL INSTALLER

EXTENSION HOUSING SEAL

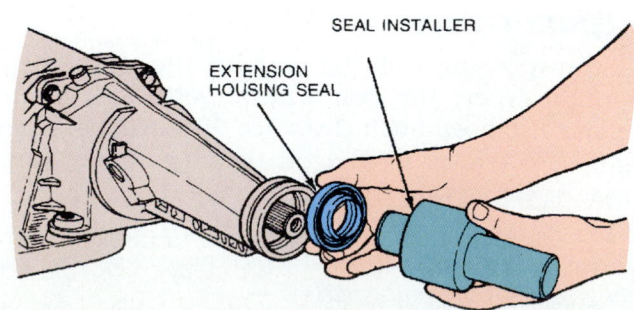

Figure 65.24 Installing a seal with a seal installer. *(Courtesy of Ford Motor Company)*

■ INSTALL THE TRANSMISSION

Before attempting to install the transmission, be sure that the clutch fork is properly seated on its pivot. Align the transmission with the engine. The transmission must be at the same angle so it is able to slip into the splines in the clutch disc. If the clutch has been disassembled, see the instructions on clutch disc alignment in Chapter 63. Shake the transmission while rotating it slightly back and forth to help the input shaft to slip into the clutch disc and pilot bushing.

CAUTION Do not use the transmission bolts to draw the transmission into the clutch. The transmission should slide all the way in by hand.

CASE HISTORY *An apprentice technician was installing a clutch in a rear wheel drive car. During reinstallation of the transmission, he found that it would only slide in to about ½" of the housing. He installed the bolts and tightened them lightly and evenly until the transmission was flush with the housing. After completion of the job, he started the car in neutral. When he attempted to shift it into gear, it would not go. He shut off the engine and put the transmission in gear. When he attempted to start it this time it lurched forward. The clutch would not release because the pilot bushing was deformed and held the transmission input shaft tightly.*

Install the bolts in the transmission. Install the crossmember and its vulcanized mount, if applicable. Reinstall any clutch or transmission linkage that was removed previously.

■ TRANSMISSION LINKAGE ADJUSTMENT

Most external linkages include a provision for aligning the shifter to neutral. With the linkages disconnected from the shifter, place the shifter in neutral and put a pin through the hole (Figure 65.25). Then, adjust the lengths of the rods until they can be attached to the arms of the shifter. There is usually an adjustable swivel that is threaded up or down the shift rod to adjust it.

■ ADD LUBRICANT

Remember to add the recommended lubricant before driving the vehicle. The lubricant can be added prior to installation in the vehicle. This can result in oil leaking out of the extension housing as the transmission is installed if a yoke is not installed on the output

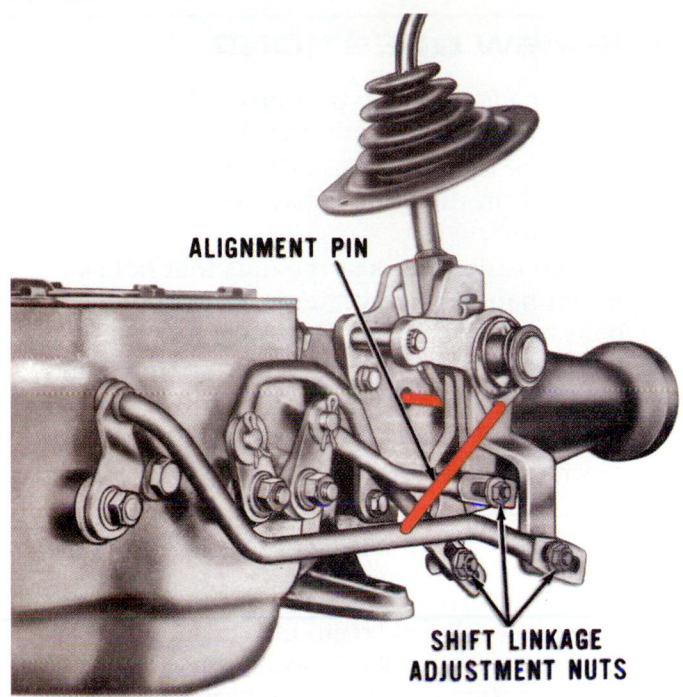

Figure 65.25 A pin is used to align the shift gate while adjusting the transmission linkage pieces to the correct length. *(Courtesy of Ford Motor Company)*

shaft. The lubricant level should come to the bottom of the fill hole (see Figure 14.22).

CASE HISTORY *An apprentice finished installing a recently rebuilt transmission in a car. He asked the master technician how much gear lube to put in the transmission. The technician said to just fill it until the gear lube begins to come out of the fill hole in the side of the transmission. The drain plug on this transmission happened to be on the side, rather than the bottom. Not seeing the fill plug, he mistakenly pumped some oil through the drain opening. Of course, it began to come out in a very short time. Had he checked the service manual for the system's capacity, he would not have destroyed the second gear synchronizer early in the test drive.*

NOTE: *Be sure to check the shop manual for the correct type of lubricant to use. In some manual transmissions, ATF or engine oil is specified. Using the incorrect oil can result in transmission damage.*

■ TEST DRIVE

During the test drive, check for proper operation of the transmission. Be sure that shifts are smooth. The shifter should feel firm and the neutral gate should be accurately aligned. After returning to the shop, check the outside of the transmission for leaks.

REVIEW QUESTIONS

1. Before condemning a transmission that shifts hard, check the operation of the _____.

2. A yoke or a _____ installation tool can be installed on the output shaft splines to keep oil from running out.

3. To gain easier access to the nuts that hold a shifter handle in place, remove the _____ bolts and slowly lower the jack a small amount.

4. What is the name of the short shaft used to drive out a countershaft?

5. There is often a _____ on the mainshaft that separates the gears.

6. What should you do to synchronizer rings before disassembling them from the hub and shift collar?

7. What is the best way to determine if synchronizers are working?

8. How many inserts are there usually on a synchronizer assembly?

9. What will an experienced technician count during transmission disassembly?

10. What gear position is the transmission shifter placed in to perform a transmission linkage adjustment?

ASE STYLE REVIEW QUESTIONS

1. Technician A says that damage to the edges of the gear tooth can be ground down if there is a high spot. Technician B says that gears should not have any damage on the face of the gear in the tooth contact area. Who is right?

   a. Technician A   **b.** Technician B
   c. Both A and B   **d.** Neither A nor B

2. Technician A says that the mainshaft is always removed from the top of a transmission. Technician B says that you can check for damage to synchro rings by dragging a magnet through the transmission oil. Who is right?

   a. Technician A   **b.** Technician B
   c. Both A and B   **d.** Neither A nor B

3. Technician A says that a transmission that jumps out of gear could have a worn synchronizer hub. Technician B says that a worn synchronizer hub can cause a transmission to stick in gear. Who is right?

   a. Technician A   **b.** Technician B
   c. Both A and B   **d.** Neither A nor B

4. Technician A says to spin bearings dry with compressed air. Technician B says that blocker rings should have sharp threads. Who is right?

   a. Technician A   **b.** Technician B
   c. Both A and B   **d.** Neither A nor B

5. Technician A says that there is often a specification for blocker ring to gear clearance. Technician B says when reassembling the new bearing to the shaft, always put pressure on the outside of the bearing only. Who is right?

   a. Technician A   **b.** Technician B
   c. Both A and B   **d.** Neither A nor B

6. Technician A says that sometimes the speedometer gear on the mainshaft is

replaceable. Technician B says that the speedometer gear on the end of the cable must be replaced with the cable. Who is right?

   a. Technician A   **b.** Technician B
   c. Both A and B   **d.** Neither A nor B

7. Technician A says that on some transmissions, the thickness of the snap ring determines the clearance between the gears. Technician B says that clearance is checked between the gears after they are assembled on the countergear. Who is right?

   a. Technician A   **b.** Technician B
   c. Both A and B   **d.** Neither A nor B

8. Technician A says to use oil on loose needle bearings to keep them in place during assembly. Technician B says that some paper transmission gaskets are reusable. Who is right?

   a. Technician A   **b.** Technician B
   c. Both A and B   **d.** Neither A nor B

9. Technician A says that the extension housing bushing can be replaced when the extension housing is off the transmission. Technician B says that the extension housing bushing can be replaced when the extension housing is on the transmission. Who is right?

   a. Technician A   **b.** Technician B
   c. Both A and B   **d.** Neither A nor B

10. Technician A says to use the transmission bolts to draw the transmission into the clutch disc. Technician B says that lubricant can be added to a transmission prior to installing it. Who is right?

   a. Technician A   **b.** Technician B
   c. Both A and B   **d.** Neither A nor B

Automatic Transmission Fundamentals

◼ INTRODUCTION

An automatic transmission shifts gears automatically and does not require a manual clutch. Its purpose is the same as that of a standard transmission: to match the torque of the engine to the rear wheels. Automatic transmissions can be found in both front and rear wheel drive vehicles. In front wheel drive vehicles, they are combined with a differential in a single unit called an automatic **transaxle**. The operation of the automatic transmission part of the transaxle is the same. Transaxles are covered in another chapter.

Today's automatic transmission is a *torque converter automatic*. There were several designs of automatic transmissions before the first torque converter automatic was introduced by General Motors in the 1948 Buick. The first automatic transmission was actually developed in Boston in 1904 by the Sturdevant brothers, but its design was not the same as today's reliable unit.

◼ AUTOMATIC TRANSMISSION PARTS

The automatic transmission consists of several parts. Refer to Figure 66.1 for the names and locations of the basic parts.

◼ A torque converter takes the place of the clutch in a manual transmission.

◼ An *input shaft* connects the torque converter to the transmission section.

◼ The transmission *pump* provides hydraulic fluid to a valve body, which makes shift decisions and acts on them. Fluid pressure from the pump is directed to perform work in various places in the transmission.

◼ *Bands* and clutches, called *planetary holding members*, are applied by hydraulic pressure to drive or hold parts of a *planetary gear system* to make the gear shifts.

◼ The *transmission case* includes the torque converter housing, fluid pan, and extension housing.

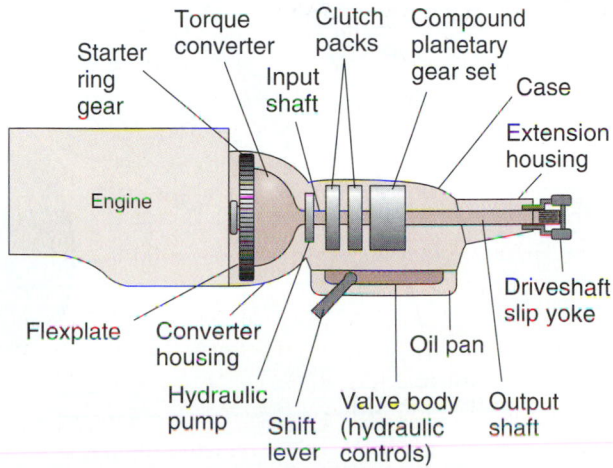

Figure 66.1 Basic parts of an automatic transmission.

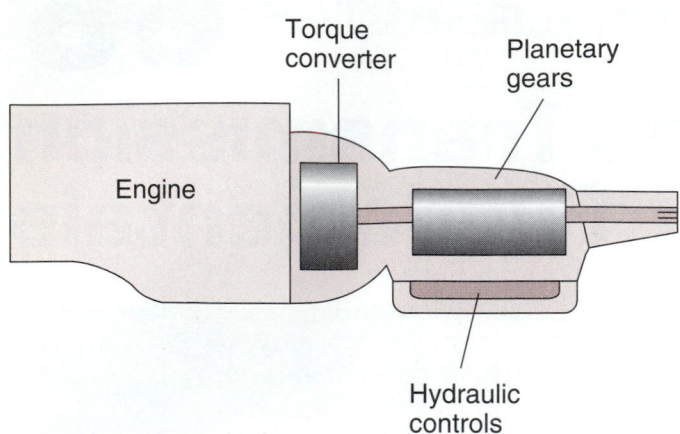

Figure 66.2 Basic areas of an automatic transmission.

POWER TRANSMISSION

Automatic transmissions use three methods of transmitting power: fluid, friction, and gears. The torque converter transmits power using fluid. The planetary holding members use fluid and friction. Gears in the transmission transmit power and also change speed and torque.

The basic areas of the transmission are the torque converter, planetary gears, and hydraulic controls (Figure 66.2). Automatic transmissions sense road speed and engine load to determine when to make a correct shift. In conventional automatic transmissions, gear shifts are made by hydraulic pressure. Electronically controlled transmissions are shifted by electric solenoids triggered by the onboard computer.

FLEXPLATE

A standard transmission has a flywheel. A *flexplate* and torque converter are used with automatic transmissions instead of a flywheel (Figure 66.3). It is a thin stamped version of the heavy steel flywheel, usually with a starter ring gear on its outside circumference. Some torque converters have their starter ring gears on the outside of the housing.

The flexplate is bolted to the engine's crankshaft. Bolts or studs on the outside of the torque converter housing are used to fasten the converter to the flexplate. A torque converter is large and holds 3 or 4 quarts of fluid depending on its size. This weight allows the torque converter to take the place of a normal flywheel's function of smoothing the engine's power impulses.

TORQUE CONVERTER

The torque converter takes the place of the clutch. It allows the vehicle to idle at a stop sign. It also slips during initial acceleration so that the engine will not stall.

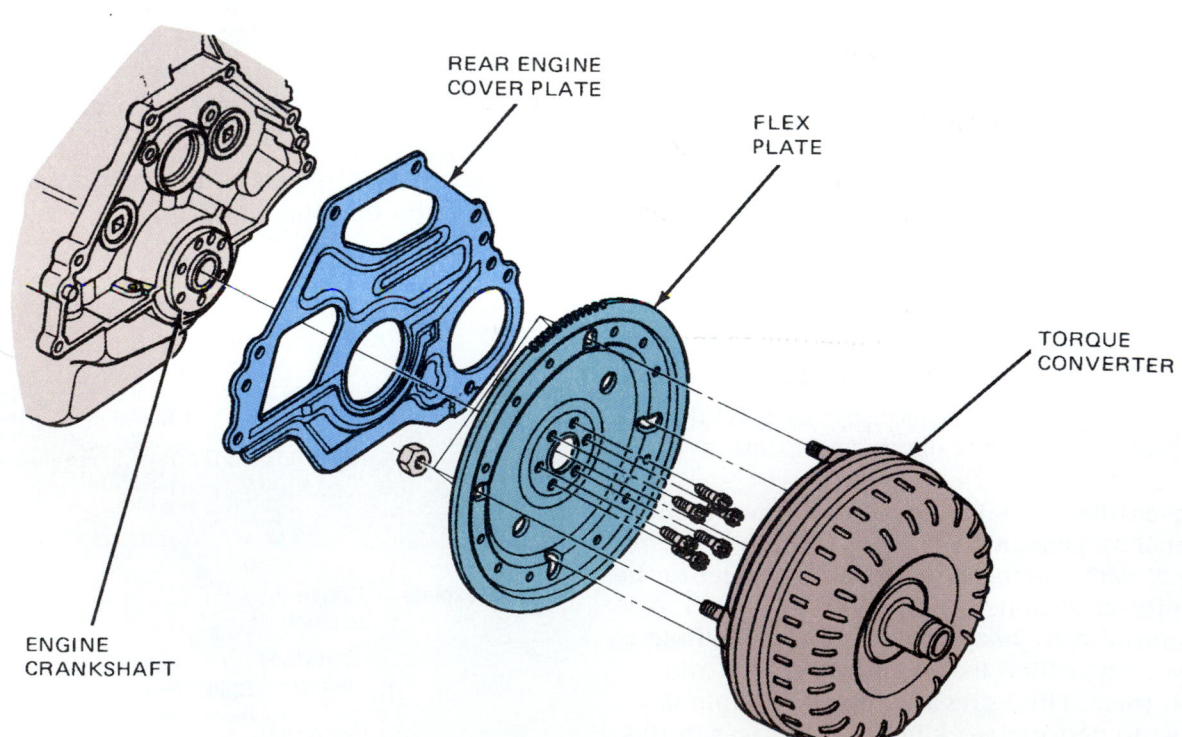

Figure 66.3 A flexplate is used with automatic transmissions instead of a flywheel. *(Courtesy of Ford Motor Company)*

Driving

Driven

ONE PART CAN DRIVE
ANOTHER BY FORCE
OF AIR – OR OIL

Figure 66.4 Operation of the fluid coupling can be compared to two fans. *(Reproduced by permission of Deere & Company, ©1992. Deere & Company. All rights reserved.)*

■ HISTORY NOTE ■

Chrysler developed the predecessor to the torque converter in 1937. Called a fluid coupling, it was used with a manual transmission that also had a clutch. With the clutch pedal all of the way up and the vehicle stopped, the engine would not stall.

Operation of the **fluid coupling** can be compared to two fans (Figure 66.4) with fluid used to transmit motion instead of air. As the first fan is turned faster, the second fan picks up the energy and is turned. There are two parts contained within the welded fluid coupling housing. The driving member is called the *impeller* or *pump* (not to be confused with the hydraulic pump). The driven member is called the *turbine*. There is another part of a transmission that is also called a pump. That pump produces fluid flow and develops the pressure necessary to operate the transmission. To avoid confusion when referring to the names of transmission parts, impeller will be used in the following discussion when referring to the converter pump.

■ TORQUE MULTIPLICATION

A torque converter has the same parts as a fluid coupling but also has an additional member, a *stator*. The difference between a fluid coupling and a torque converter is that the torque converter can actually *increase* torque. Some industrial torque converters can increase torque to the equivalent of 9:1 gear reduction when the pump first starts to drive the turbine. The amount of torque increase in an automotive torque converter when starting from a stop is about 2:1.

Torque is multiplied whenever the impeller spins faster than the turbine. When an automatic transmission equipped vehicle is used to pull a boat up a

launching ramp, the following ratios are combined to increase torque available to the drive wheels:

- Transmission low gear ratio = about 3:1
- Torque converter ratio = about 2:1
- Differential ratio = about 2.5:1

Total ratio = about 15:1

■ TORQUE CONVERTER OPERATION

The turbine has splines in its center that mesh with splines on the end of the transmission input shaft (Figure 66.5). The converter impeller (pump), is actually part of the torque converter housing. It is the part that is closest to the transmission. When the converter housing (impeller) rotates at idle speed, fluid is thrown from the impeller toward the turbine (the driven member of the torque converter). The centrifugal force of the rotating torque converter also throws fluid to the outside of its housing (Figure 66.6).

■ TORQUE CONVERTER STATOR

A fluid coupling has no stator and has *straight blades*. A torque converter has a stator and has *curved blades*. The stator is installed between the impeller and the turbine to redirect fluid flow. This makes the torque increase possible. The stator is mounted on a one-way clutch supported by the outside of the stationary front pump housing.

When the engine is idling, the converter housing (impeller) spins slowly at engine speed. A small amount of fluid is thrown from the impeller to the turbine, but

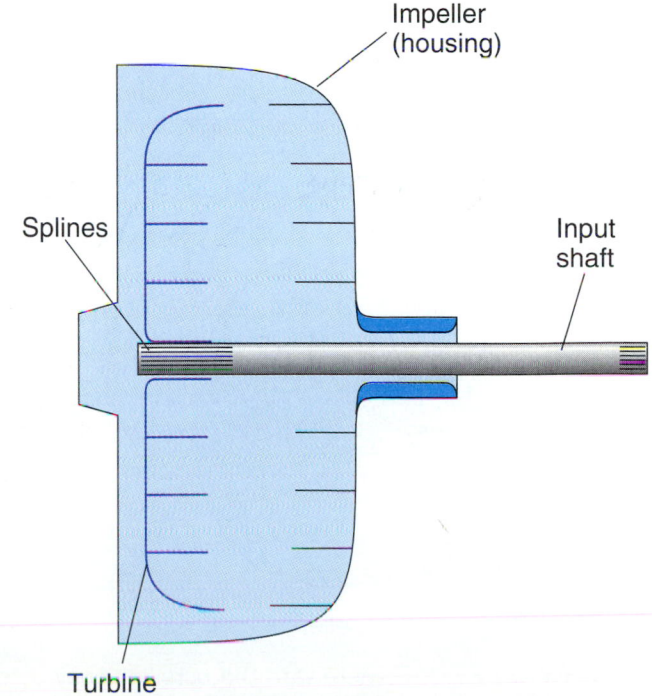

Impeller
(housing)

Splines

Input
shaft

Turbine

Figure 66.5 Parts of a fluid coupling.

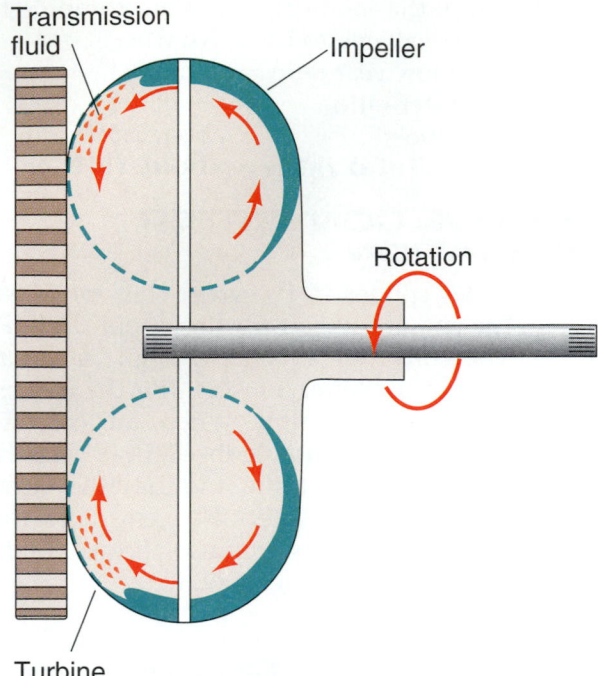

Transmission fluid

Impeller

Rotation

Turbine

Figure 66.6 The centrifugal force of the rotating torque converter throws oil to the outside of its housing.

not enough to lock the engine to the transmission. Unless the engine's idle is too high, the car will remain at rest when the brakes are released.

When the engine is accelerated, more fluid is thrown from the impeller to the turbine. Fluid flows toward the outside, away from the center of the impeller (Figure 66.7). When the fluid hits the turbine, it is redirected inside toward the center of the converter (Figure 66.8).

As the turbine is driven by the impeller, the transmission input shaft turns and the vehicle begins to move. There is still some slippage between the impeller and turbine. Because they are not turning at the same speed, torque is being increased.

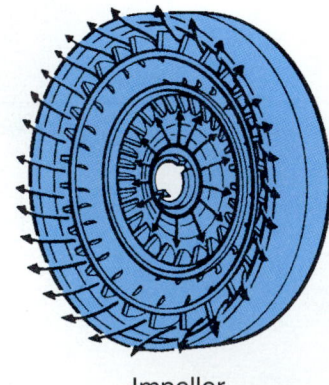

Impeller

Figure 66.7 Oil flows toward the outside, away from the center of the impeller. (*Courtesy of Ford Motor Company*)

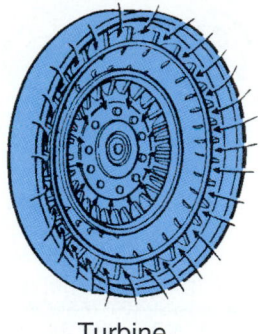

Turbine

Figure 66.8 When the oil hits the turbine, it is directed inside toward the center of the converter. (*Courtesy of Ford Motor Company*)

When the vehicle reaches cruising speed, the impeller and turbine are turning at almost the same speed with very little slippage. There is no further torque multiplication because the parts are turning at the same speed.

■ STATOR OPERATION

The stator is installed between the impeller and the turbine to redirect fluid flow. In a torque converter, the blades are curved so that transmission fluid is thrown at an angle against the turbine. This is what increases torque. A fluid coupling has straight blades and does not provide a torque increase.

As it rotates, the engine's flywheel turns the transmission pump, creating pressure in the transmission and converter. Fluid is *circulated* in a pattern throughout the torque converter. This spiral pattern of fluid flow is called **vortex** flow (Figure 66.9).

From a standing start, the fluid is thrown against the turbine at a sharp angle. The turbine can only be driven clockwise but fluid exits from it moving in a counterclockwise direction. If this fluid was allowed to hit the impeller, it would try to stop it from turning.

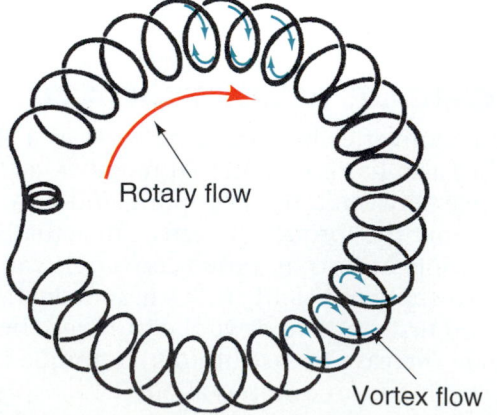

Rotary flow

Vortex flow

Figure 66.9 The spiral pattern of oil flow is called vortex.

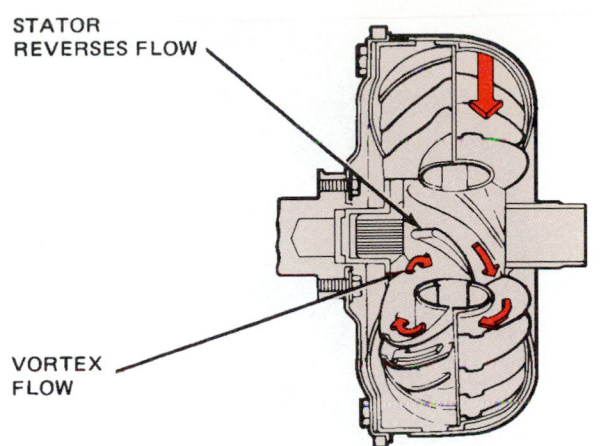

Figure 66.10 The stator catches the fluid coming out of the turbine and redirects fluid flow. *[Courtesy of Ford Motor Company]*

The stator catches the fluid coming out of the turbine and redirects fluid flow (Figure 66.10). When fluid leaves the stator it strikes the impeller, flowing in the same direction. This actually accelerates the impeller.

The torque converter has split rings attached to the blades of the turbine and impeller (Figure 66.11). These direct the fluid in a smooth pattern. Without the split rings, flow in the center of the converter would be turbulent during periods of the most torque multiplication.

■ STATOR CLUTCH OPERATION

The stator has a **one-way clutch**, also called an overrunning clutch, that locks in one direction and freewheels in the other (Figure 66.12). When the fluid strikes the stator at a high angle, the stator's one-way clutch locks up, preventing rotation of the stator (Figure 66.13). When the speed of the turbine starts to

catch up with the speed of the impeller, the stator does not need to redirect fluid flow any longer. In fact, redirecting the fluid at this point in time would actually work against the turbine. This would cause the stator to act as a brake to the engine. To prevent this, the stator's one-way clutch freewheels. This allows the stator, impeller, and turbine to turn freely with the mass of fluid that is now turning with rotary flow (without vortex). There is little vortex flow because the impeller and turbine have reached **coupling speed**. At this point, all of the converter parts and ATF all turn as a unit (Figure 66.14).

The converter does not become efficient at transferring power from the engine until coupling speed is reached at about 2300 rpm. When the turbine rpm reaches about $^9/_{10}$ that of the impeller, there is no more torque multiplication.

■ STALL SPEED

Stall speed is when the vehicle is being prevented from moving while the engine is accelerated. The torque converter slips, allowing engine rpm to increase. When the engine rpm can no longer rise, this is referred to as the stall speed. Stall speed is the point of maximum torque multiplication.

NOTE: A high stall speed converter is used in drag racing. This is a torque converter that is too small for a normal vehicle. This allows the engine to accelerate to a higher rpm before reaching stall speed. The camshaft in the engine is such that its maximum torque is not reached until higher engine speeds. A higher stall speed converter allows the engine to be at a higher rpm during engine braking at the start of the drag race for a faster take-off.

Lower stall speed converters are used for motor homes and for pulling trailers. This is because they produce less heat and are more efficient.

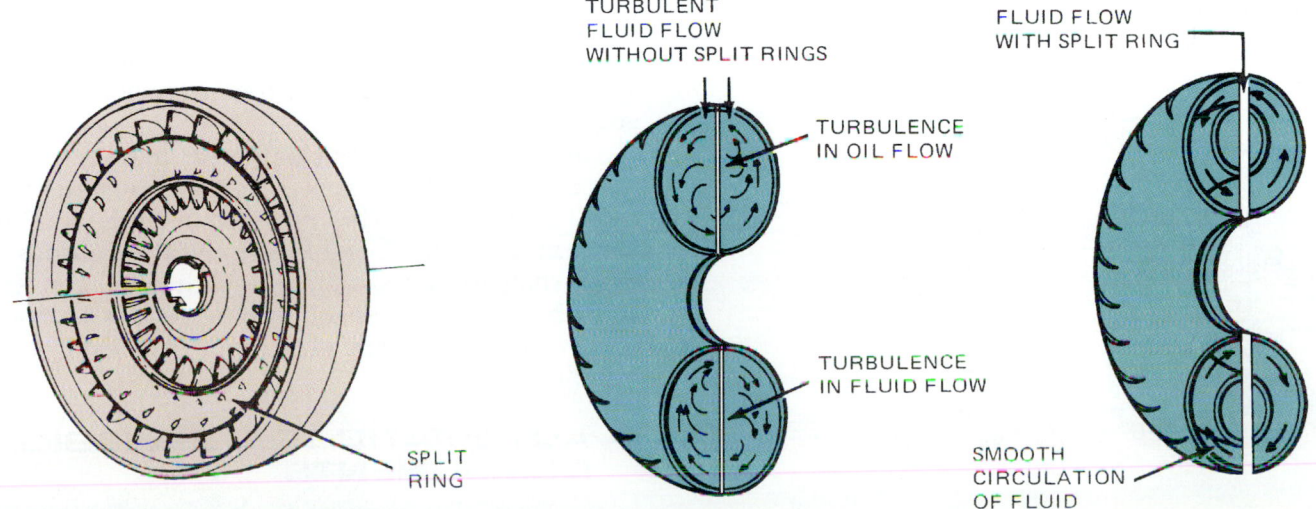

Figure 66.11 The torque converter has split rings attached to the blades of the turbine and impeller. *[Courtesy of Ford Motor Company]*

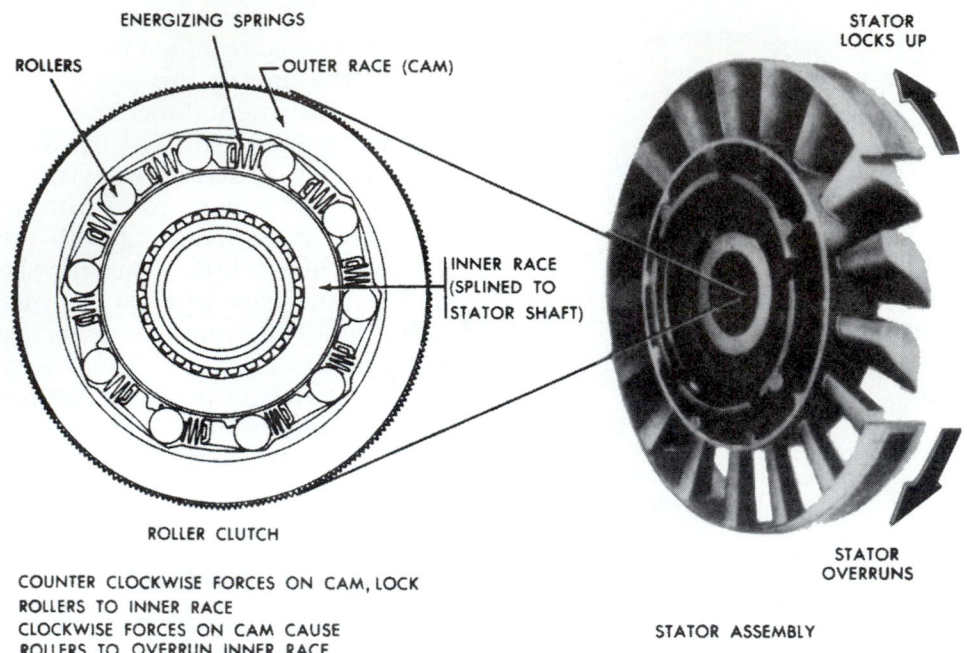

ENERGIZING SPRINGS

ROLLERS

OUTER RACE (CAM)

INNER RACE (SPLINED TO STATOR SHAFT)

ROLLER CLUTCH

COUNTER CLOCKWISE FORCES ON CAM, LOCK
ROLLERS TO INNER RACE
CLOCKWISE FORCES ON CAM CAUSE
ROLLERS TO OVERRUN INNER RACE

STATOR LOCKS UP

STATOR OVERRUNS

STATOR ASSEMBLY

Figure 66.12 The stator clutch locks in one direction and freewheels in the other. *(Courtesy of General Motors Corporation, Service Technology Group)*

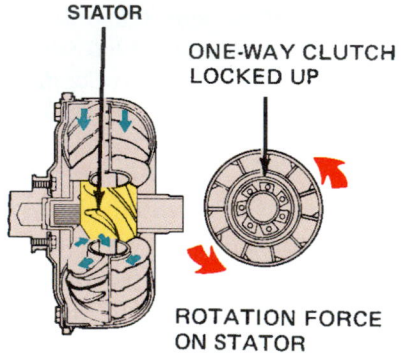

STATOR

ONE-WAY CLUTCH LOCKED UP

ROTATION FORCE ON STATOR

Figure 66.13 Stator clutch locked up. *(Courtesy of Ford Motor Company)*

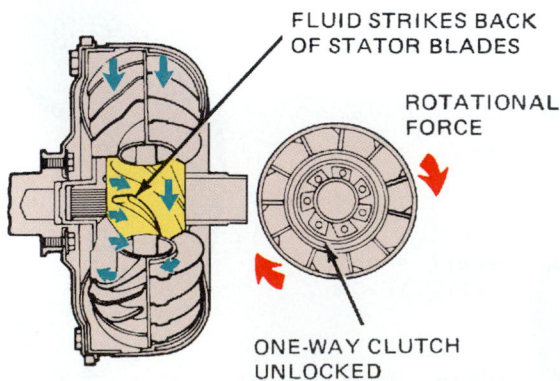

FLUID STRIKES BACK OF STATOR BLADES

ROTATIONAL FORCE

ONE-WAY CLUTCH UNLOCKED

Figure 66.14 When the stator clutch unlocks at coupling speed, all of the converter parts and ATF turn as a unit. *(Courtesy of Ford Motor Company)*

■ LOCK-UP CONVERTERS

Newer vehicles have **lock-up torque converters**. With a standard torque converter, actual coupling speed only happens when the vehicle crests a hill. Coupling is at that point between where the engine stops driving the wheels and the wheels start to drive the engine. The rest of the time, there is up to about 10% slippage between the impeller and the turbine. Slippage causes a cutting action on the transmission fluid. The result is worse fuel economy and heat buildup.

Some transmissions do not lock up the converter until high gear or overdrive. In overdrive the gear ratio is 0.7 to 1. With this ratio and lower engine speed, a standard converter would have more of a tendency to slip. Newer transmissions often have lockup in all gears.

In a lock-up converter, a pressure plate behind the turbine locks it to the back part of the converter housing (Figure 66.15). This provides a mechanical link between the engine's crankshaft and the transmission input shaft to prevent slippage within the converter. To lock up the impeller and turbine, fluid is directed to one side of the pressure plate and exhausted from the other side.

■ AUTOMATIC TRANSMISSION COMPONENTS

Like a manual transmission, an automatic transmission has a *mainshaft* or *output shaft*. Output of the transmission is directed to the differential. An *input*

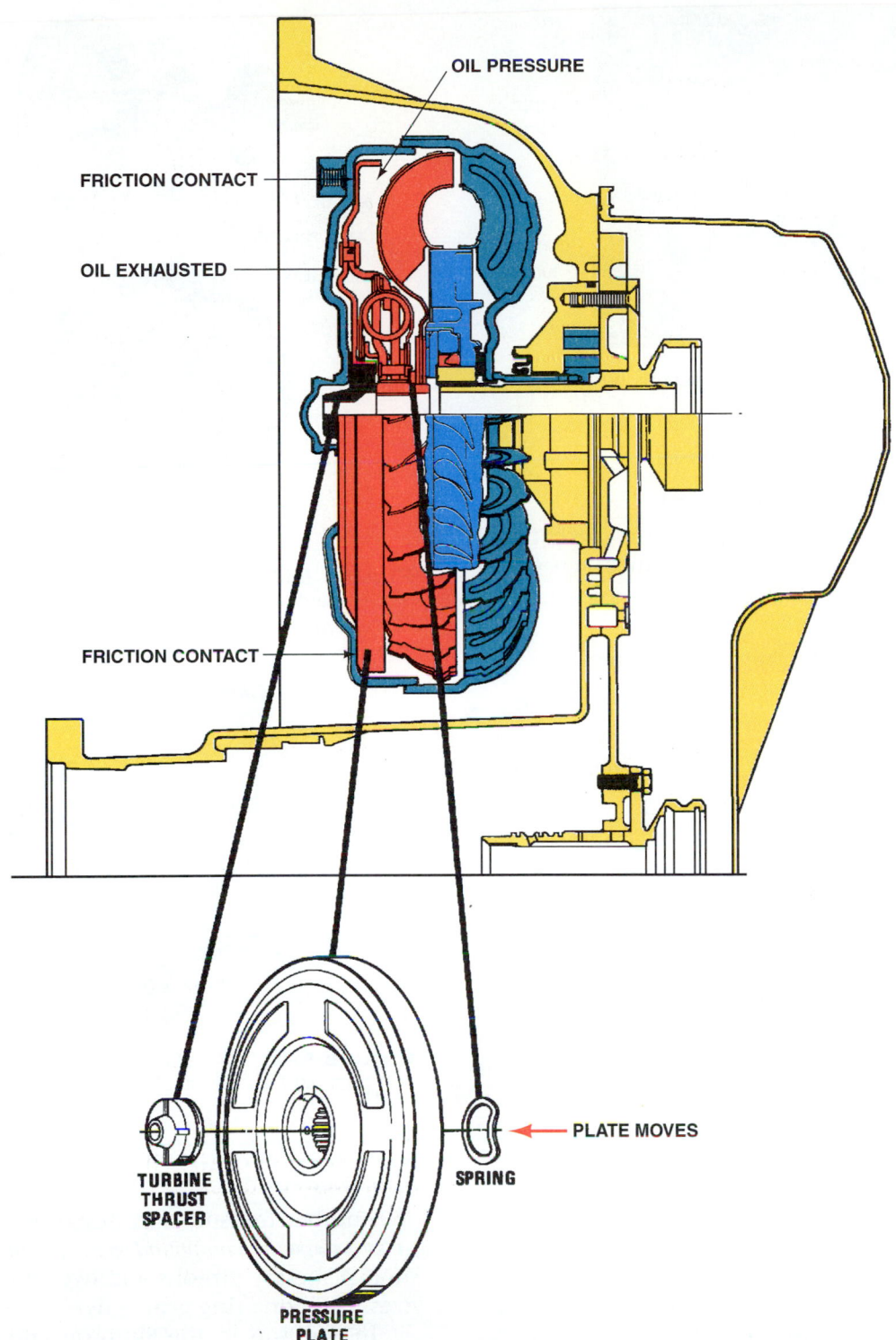

OIL PRESSURE

FRICTION CONTACT

OIL EXHAUSTED

FRICTION CONTACT

TURBINE
THRUST
SPACER

SPRING

PLATE MOVES

PRESSURE
PLATE

Figure 66.15   When oil is applied to the front of this pressure plate and exhausted from the other, the converter is locked up. *(Courtesy of General Motors Corporation, Service Technology Group)*

shaft or *turbine shaft* fits between the torque converter and front of the transmission.

The *stator support* is a large splined support that extends out of the transmission's front pump body

(Figure 66.16). It surrounds the input shaft. The splines mesh with splines on the inside of the stator's overrunning clutch. The transmission input shaft extends through it and is supported by bushings.

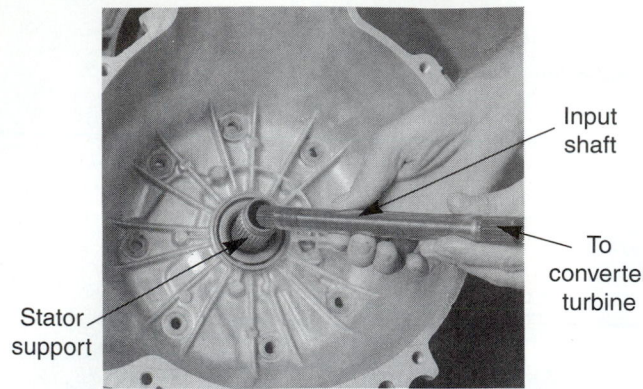

Figure 66.16 The torque converter is mounted on the stator support.

PLANETARY GEARS

The planetary gear set provides a means of changing gear ratios by coupling and releasing members to get different results.

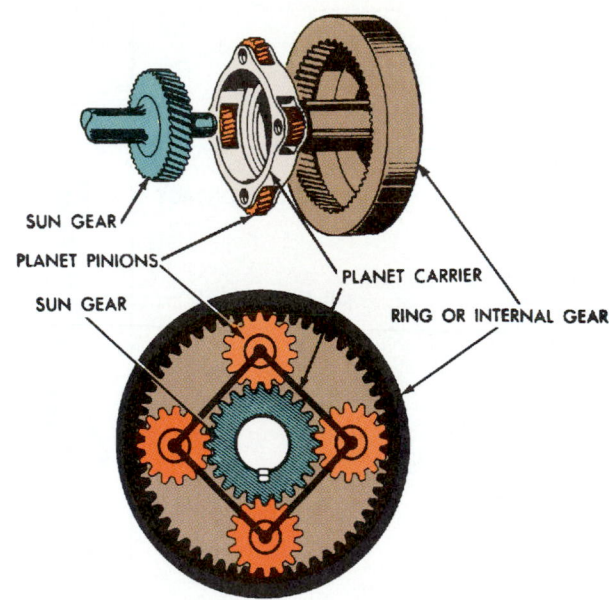

Figure 66.17 Parts of a simple planetary gear set. *(Courtesy of General Motors Corporation, Service Technology Group)*

> ### HISTORY NOTE
>
> *James Watt improved the steam engine in many ways. The reciprocating motion of the beam attached to the piston needed to be changed into rotary motion in order to be able to harness the engine's energy. Watt didn't want to pay royalties to the person who had patented one device that could be used to attach the connecting beam to a crankshaft, so he devised the planetary gear system consisting of a sun and planets to do the conversion.*

One major advantage to a planetary gear set is that all of the gears are in *constant mesh*. This means no gear clashing or wear as gears turning at different speeds are engaged and disengaged with each other. Another advantage is that the load is distributed over several gears instead of only two.

SIMPLE PLANETARY GEAR SET

In a simple planetary gear set there is a *sun gear, planetary pinions,* a *carrier,* and a *ring gear* (Figure 66.17). The ring gear is also called an *internal gear* because of its internal teeth. The planetary carrier is also called a *cage*. A **compound planetary gear set** combines two planetary gear sets to provide more gear ratio possibilities.

One popular design is the **Simpson gear train**. It has one long sun gear that operates two cages with pinions and two ring gears. Invented by a retired transmission designer, it is a design that uses interchangeable gears in each gear set. This lowers manufacturing costs. The Simpson gear train is often called a "salad

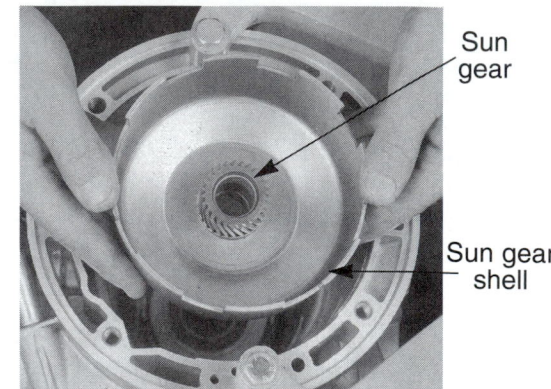

Figure 66.18 The Simpson gear set showing the sun gear and sun gear shell.

bowl" because of the shape of its outside metal shell (Figure 66.18).

Another transmission design is the **Ravigneaux gear train** or *compound gear design*. It has long and short planetary pinions and two sun gears of different sizes. Only one ring gear is used.

In addition to the Simpson and Ravigneaux gear designs, some General Motors and Ford transmissions use two simple planetary gear sets in series, and many overdrive transmissions use simple and Simpson planetary gear sets in series.

SIMPLE PLANETARY OPERATION

Several shift options are available from one simple gear set. The chart in Figure 66.19 shows the six possible conditions with a simple planetary gear set.

Conditions	1	2	3	4	5	6
Ring	D	T	H	H	T	D
Carrier	T	D	D	T	H	H
Sun	H	H	T	D	D	T
Speed	I	L	L	I	IR	LR

D = driven (output)　　　L = reduction
H = hold　　　　　　　　R = reverse
I = increase (overdrive)　T = turn (input)

Figure 66.19 The six possible conditions with a simple planetary gear set.

Basic gear rules are:
- Two gears with external teeth in mesh will rotate in opposite directions.
- Two gears in mesh, one with internal and one with external teeth (ring and planet pinions) will rotate in the same direction.

When pinion gears move in an automatic transmission, this is referred to as *"walking"* or *"idling."*

- When walking inside of the ring gear, pinions are moving in a direction that is opposite to their own rotation.
- When the carrier is held from turning, the pinions idle. This produces a change in direction (reverse).

Forward gear reduction can be accomplished in a simple planetary set in two ways. Maximum gear reduction is accomplished by turning the sun gear, while holding the ring gear (Figure 66.20). Output is through the planet carrier. This is condition number 3 on the chart:

Sun　　　=　input　　(T)
Ring　　　=　held　　　(H)
Carrier　=　output　(D)
Output　=　normal rotation, gear reduction

If the sun gear is held while turning the ring gear, output is through the carrier for a lesser gear reduction (Figure 66.21). This condition (number 2 on the chart) is used in the Simpson gear train for second gear operation. The Simpson gear train will be used here to illustrate how an automatic transmission operates.

Ring　　　=　input　　(T)
Sun　　　=　held　　　(H)
Carrier　=　output　(D)
Output　=　normal rotation, gear reduction

Reverse is accomplished by using only the rear gear set (Figure 66.22). Input power is through the sun gear shell to the sun gear. Holding the carrier turns the pinions into idler gears. This results in reverse operation of the ring gear for output.

Sun　　　=　input　　(T)
Carrier　=　held　　　(H)
Ring　　　=　output　(D)
Output　=　reverse

Overdrive is when the output shaft turns faster than the input shaft. It is the result when the carrier is turned. The sun gear is held and output is through the ring gear. This is condition number 1 in the chart (Figure 66.23).

Carrier　=　input　　(T)
Sun　　　=　held　　　(H)
Ring　　　=　output　(D)
Output　=　normal rotation, overdrive

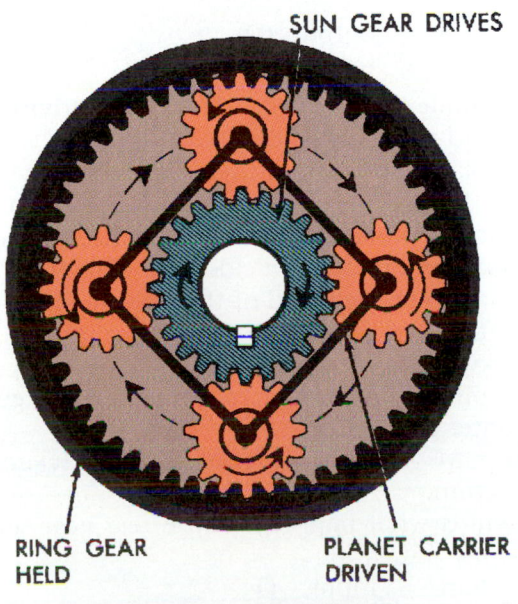

MAXIMUM REDUCTION

Figure 66.20 Maximum gear reduction is accomplished by turning the sun gear, while holding the ring gear. *(Courtesy of General Motors Corporation, Service Technology Group)*

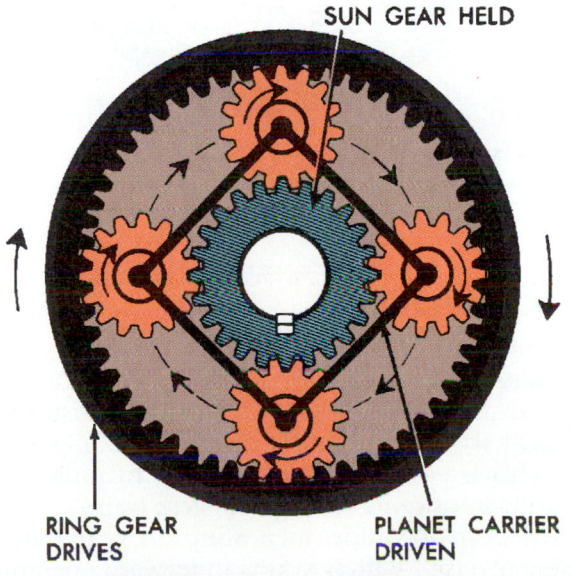

MINIMUM REDUCTION

Figure 66.21 If the sun gear is held while turning the ring gear, output is through the carrier for a lesser gear reduction. *(Courtesy of General Motors Corporation, Service Technology Group)*

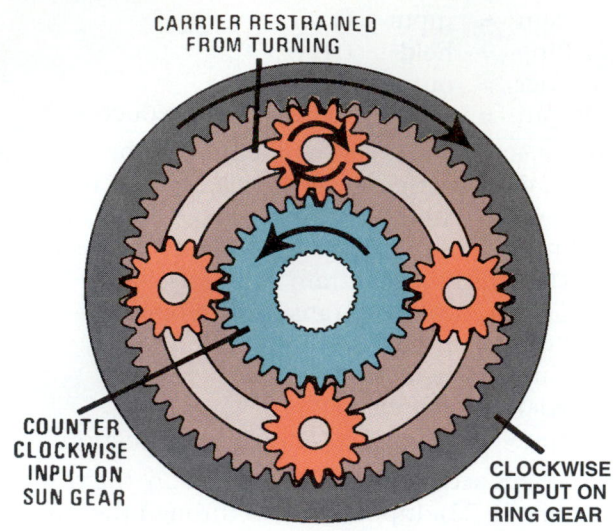

Figure 66.22 In reverse, the pinions act as idlers. *(Courtesy of General Motors Corporation, Service Technology Group)*

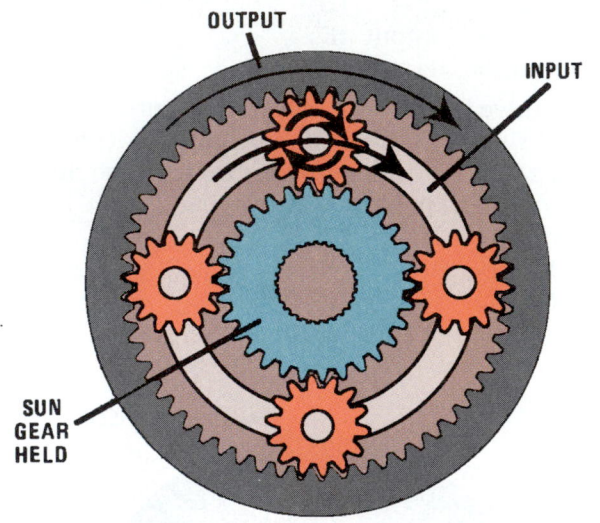

Figure 66.23 Overdrive is when the output shaft turns faster than the input shaft. *(Courtesy of General Motors Corporation, Service Technology Group)*

■ COMPOUND PLANETARY OPERATION

In compound planetary gear sets, two sets of planets are used together. In non-overdrive transmissions, high gear (third or drive) is direct drive, giving a 1:1 ratio. This is easily accomplished by turning on two gear units at once to lock up the gear train.

Low gear operation in a Simpson gear train is a "double reverse," which results in forward operation (a combination of conditions number 5 and 6).

Here is what happens in the front gear set (condition number 5) (Figure 66.24a):

Ring = input (T)
Carrier = held (H)

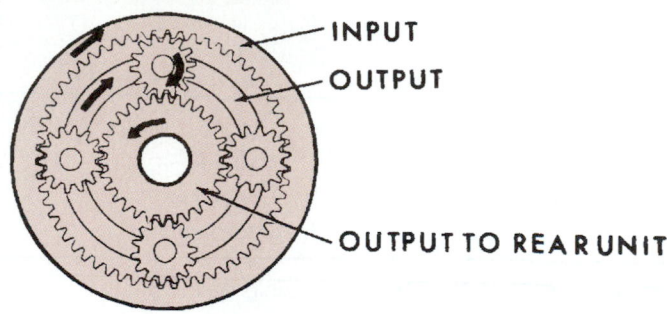

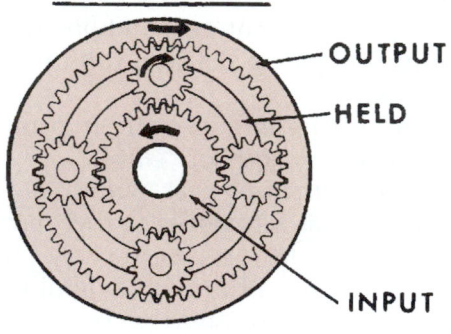

Figure 66.24 Low gear uses both planetary gear sets. *(Courtesy of General Motors Corporation, Service Technology Group)*

Sun = output (D)
Output = reverse

■ The input shaft is turned clockwise, driving the forward ring gear clockwise.

■ The planetary carrier is held by the rear wheels to turn the pinions into idlers. This is because the carrier is splined to the output shaft. Although the carrier does rotate in a clockwise direction, it can only turn at the speed of the output shaft.

■ The ring gear drives the planetary pinions clockwise.

■ The pinions act as idlers to drive the sun gear counter-clockwise.

■ Output from the front gear set is in a reverse direction.

Here is what happens in the rear gear set (Figure 66.24b):

Sun = input (T)
Carrier = held (H)
Ring = output (D)
Output = reverse

■ Input is through the sun gear in a counterclockwise direction.

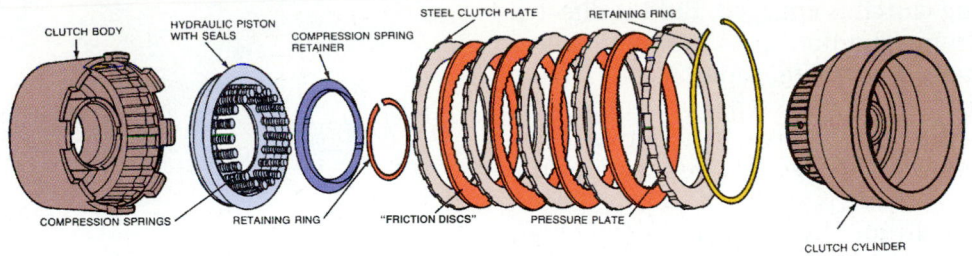

Figure 66.25 Parts of a multiple disc clutch pack. *(Courtesy of Ford Motor Company)*

- The rear carrier is held by a one-way clutch or band. This causes the pinions to be driven clockwise.
- The ring gear is driven clockwise. It is splined to the output shaft.

Combining the two reverse actions results in forward operation.

In neutral, all holding devices are released, allowing engine freewheeling when the transmission gear shift selector is in neutral or park.

To achieve overdrive, some transmissions have an overdrive planetary set added to either the front or the rear of the existing Simpson gear train.

DRIVING AND HOLDING DEVICES

To operate a planetary gear set, one member must be held while another is driven. A driving device, a *fluid clutch*, is used to lock a rotating planetary member to the input shaft. *Bands* and locking one-way clutches are holding members used to hold a part of the planetary set. Clutches and bands are friction devices. They operate any time fluid pressure is applied to them.

CLUTCHES

Multiple disc clutches can be used for *holding* or *driving* parts of the planetary gear set. They contain steel plates (called *"steels"*) and friction discs (Figure 66.25). The steels are splined against one element of the clutch pack, while the friction discs are splined to the corresponding part. Half of the clutch plates in a clutch pack can be splined to the case of the transmission or to another component of a planetary gear set. When clutches have half of their plates splined to the transmission case, they are used as holding devices (Figure 66.26).

CLUTCH OPERATION

There is an apply piston in the clutch pack that is applied by hydraulic pressure, and released by spring pressure (Figure 66.27). This means that whenever hydraulic pressure is not directed at the clutch, the clutch will release and the friction discs and steels will be free to turn independently of each other.

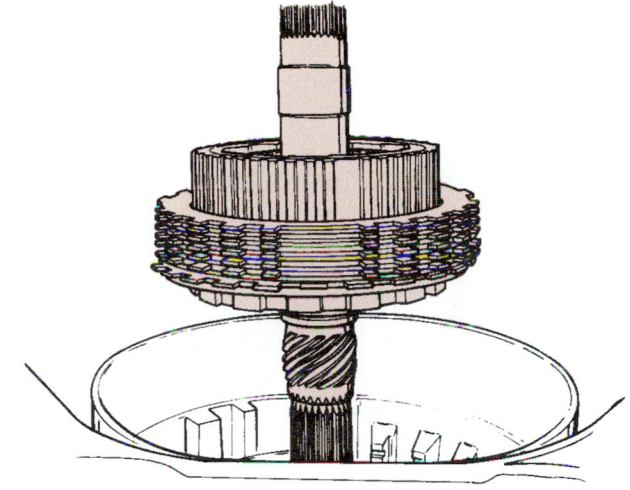

Figure 66.26 A clutch that is splined to the case. *(Courtesy of General Motors Corporation, Service Technology Group)*

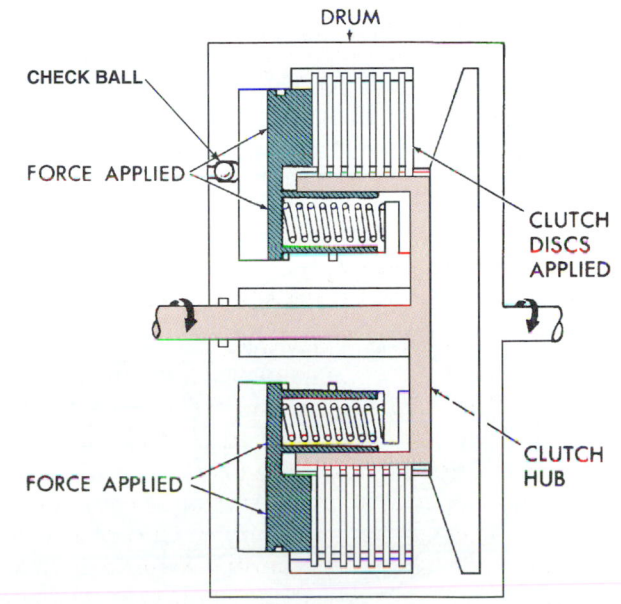

Figure 66.27 A clutch pack is applied by hydraulic pressure and released by spring pressure. *(Courtesy of General Motors Corporation, Service Technology Group)*

When a driving clutch is engaged, fluid is directed into the clutch drum. The fluid passageway is drilled into the mainshaft. There are round seals (Figure 66.28) that isolate the fluid and keep it under pressure. The fluid travels through the passages between the seals into a drilled fluid passage on the inside of the clutch drum. Fluid pressure is applied to a large piston on the inside of the drum. The piston has rubber seals on its inside and outside diameters (Figure 66.29).

The piston is applied against the discs to compress the return springs and lock up the clutch through the pressure plate. When the pressure is released, the piston is pushed away by the return springs. It is important that this occur quickly because another action will be taking place in another clutch pack during a gear shift. The piston will often have a check ball that allows fluid to be released from the clutch when pressure is released (Figure 66.30). The check ball also releases air when the clutch refills with fluid on the next application.

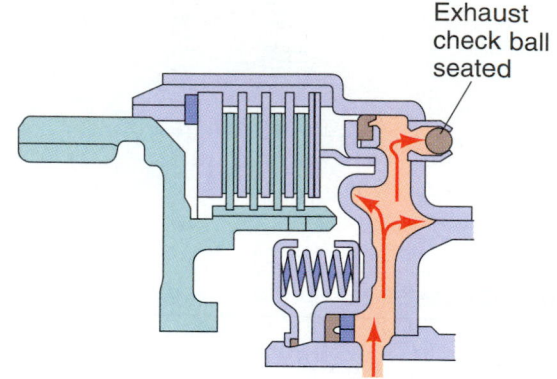

Clutch applied

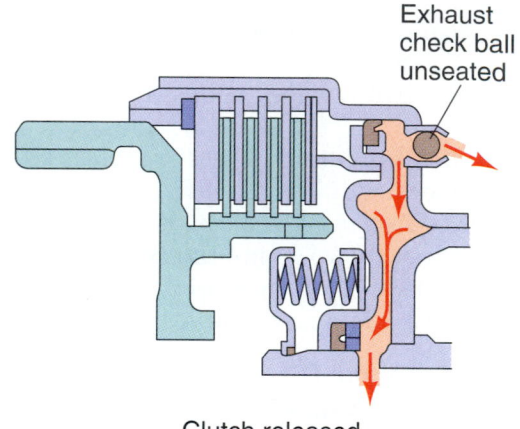

Clutch released

Figure 66.30 A check ball allows fluid to be released from the clutch when pressure is released. *(Courtesy of General Motors Corporation, Service Technology Group)*

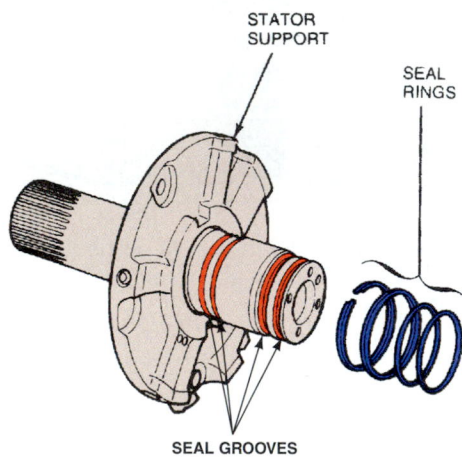

Figure 66.28 Sealing rings. *(Courtesy of Ford Motor Company)*

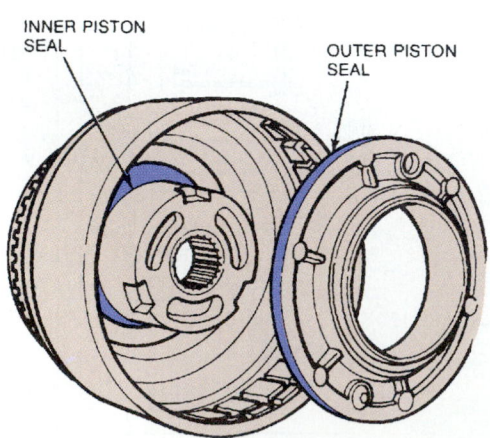

Figure 66.29 The piston has rubber seals on its inside and outside diameters. *(Courtesy of Ford Motor Company)*

NOTE: *A major advantage of the hydraulic clutch is that it does not require periodic adjustment like a brake band does.*

■ ONE-WAY CLUTCHES

In order for a planetary gear set to operate, various holding devices must be used. An overrunning, or one-way, clutch can be used to hold a part of a planetary gear set from turning. It locks in one direction and freewheels in the other. It locks instantly when an attached planetary member tries to reverse itself. An overrunning clutch is commonly used in *drive low* gear (when the gear shift is positioned in drive). The car freewheels when coasting.

NOTE: *If a driver wants the engine to slow the car down during deceleration, he or she must select the manual low position. In that position, a band or clutch holds the planetary device from turning.*

Overrunning clutches have an inner and outer race and a set of springs and rollers, just like the over-

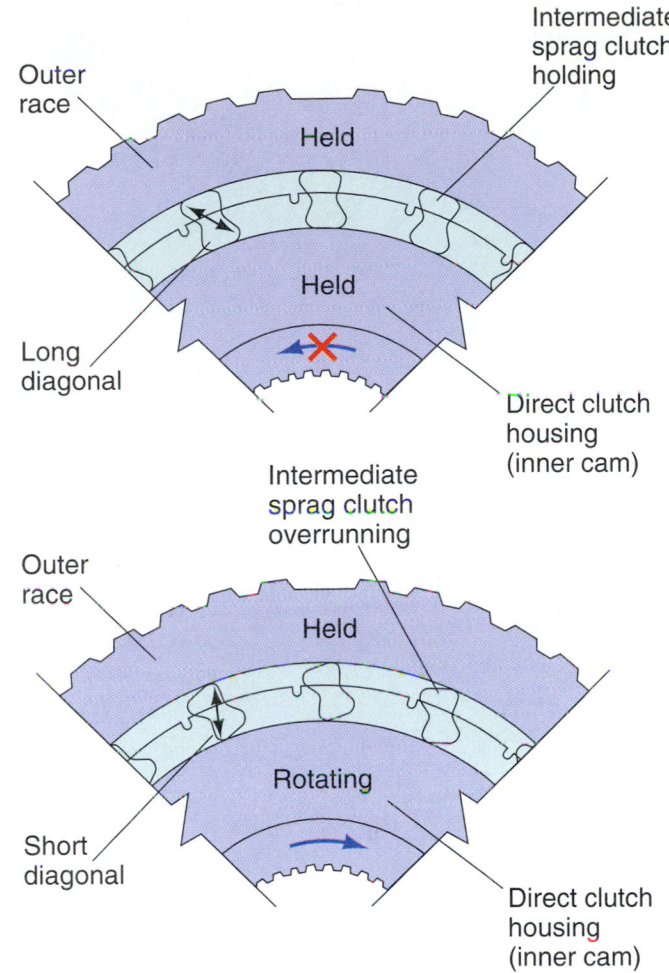

Figure 66.31 labels: Outer race, Held, Intermediate sprag clutch holding, Long diagonal, Direct clutch housing (inner cam), Intermediate sprag clutch overrunning, Outer race, Held, Rotating, Short diagonal, Direct clutch housing (inner cam)

Figure 66.31 Operation of a sprag clutch. *(Courtesy of General Motors Corporation, Service Technology Group)*

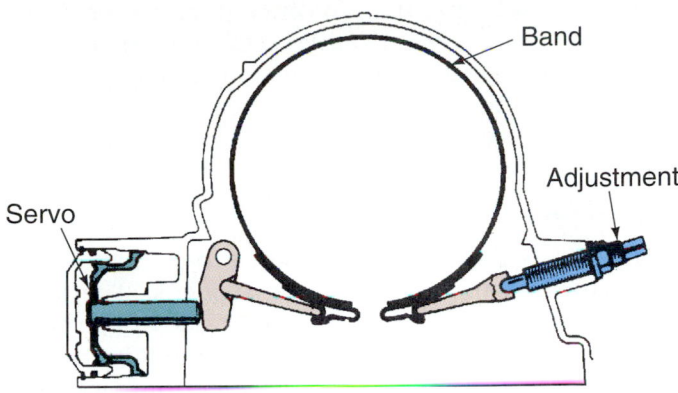

Figure 66.32 labels: Band, Adjustment, Servo

Figure 66.32 A band and servo. *(Courtesy of Ford Motor Company)*

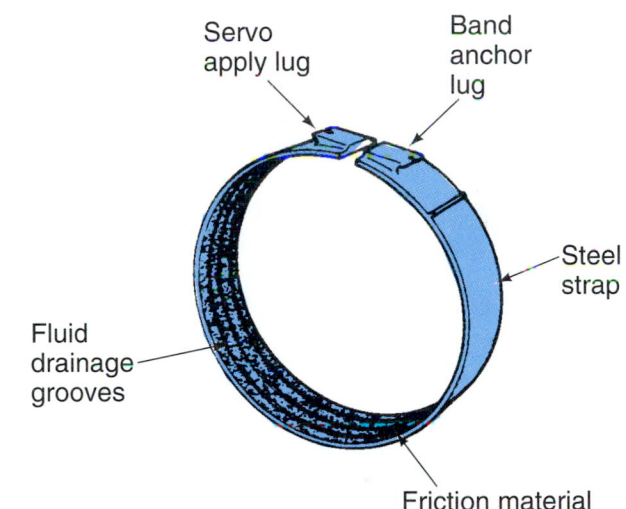

Figure 66.33 (top) labels: Servo apply lug, Band anchor lug, Steel strap, Fluid drainage grooves, Friction material

Single wrap

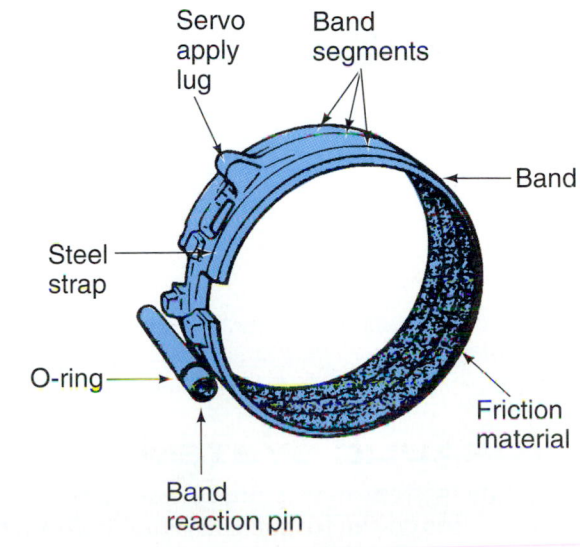

Figure 66.33 (bottom) labels: Servo apply lug, Band segments, Band, Steel strap, Friction material, O-ring, Band reaction pin

Double wrap

Figure 66.33 Single wrap and double wrap bands. *(Courtesy of Chrysler Corporation)*

running clutch used in a starter motor drive (see Figure 66.12). Another type of overrunning clutch is called a *sprag* clutch (Figure 66.31). It works in the same manner but has a different shaped locking device between the inner and outer races. Sprag clutches are strong because more sprags are possible in a smaller area. A drawback to a sprag is that machining tolerances must be held quite close. Otherwise, a sprag can "flip" under heavy torque load, such as when shifting from reverse to low while the vehicle is rolling.

■ BANDS

An *external brake band*, commonly called a band, is clamped around the outside of a drum to hold it from turning (Figure 66.32). An external band can only hold a component of the transmission stationary to its case. A clutch pack can hold two rotating planetary components together. A band is a steel strap with friction lining material on its inside diameter. A band can be either flexible or rigid. Some bands are *single wrap* and others are *double wrap* (Figure 66.33). Torque in low and reverse is higher than in other gears so more holding power is necessary. Double wrap bands are used for low and reverse gears because they can grip tighter with a lesser amount of force applied to them. Double wrap bands are also more flexible for better friction material contact.

Servos

A servo is used to operate a band. It has an apply piston with a seal on its outside diameter. The servo pushes on a rod that applies to one side of the band. There is a band adjustment screw on the other end of the band (see Figure 66.32). Bands require periodic adjustment as they wear, so they are not often found in late model transmissions.

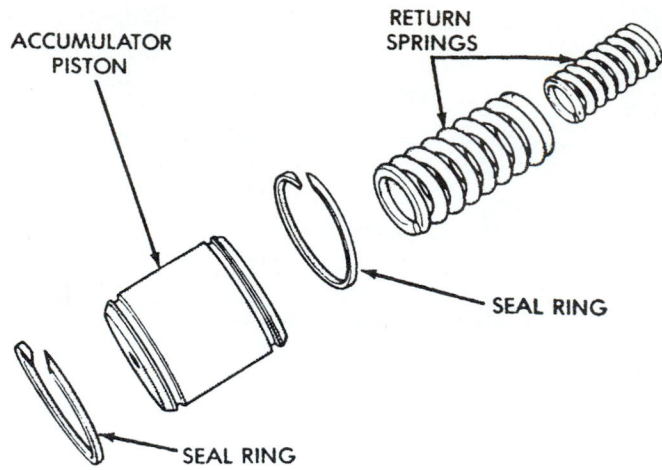

■ BAND OPERATION

When the band is applied, fluid pressure is directed into the servo's cylinder. The piston applies pressure to one end of the band. The other end of the band is fixed in place, so the band tightens around the outside of the drum. The friction material on the inside of the band keeps the drum from turning. A servo includes a spring or springs to return the piston after fluid pressure is exhausted. One disadvantage to bands is that they require periodic adjustment to compensate for wear of the friction material.

Note: Some transmissions have bands that operate only when the shift selector is placed in the D1 or D2 position for grade retard or engine braking. In normal drive range, a one-way clutch is the holding member in lower gear ranges. During deceleration, the clutch freewheels so engine braking is not possible without a band.

■ ACCUMULATOR

During shifts, new parts are held from turning while other parts are driven. Shifts must happen at a specified time. Shuddering or damage can occur as the transmission is momentarily locked up when two of the wrong components are applied at the same time. An accumulator works like an adjustable shutter on a camera. It has a piston and reservoir that must fill before pressure can actually be applied to a driving or holding device (Figure 66.34).

NOTE: *Pressure results from a resistance to flow. If there is a reservoir (accumulator) that is empty, it must fill before pressure can be used to apply force to a clutch or band.*

■ HYDRAULIC SYSTEM

The hydraulic system makes fluid pressure that transmits power through the torque converter. It also senses vehicle speed and engine load, directing fluid to reaction members at the correct time to cause shifts. Parts of the hydraulic system include a pump, pressure regulator, manual valve, vacuum modulator, governor,

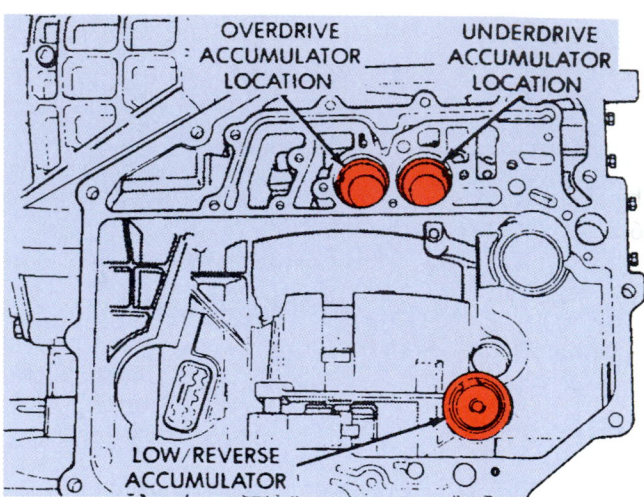

Figure 66.34 An accumulator has a reservoir that must fill before pressure can actually be applied to a driving or holding device. *(Courtesy of Chrysler Corporation)*

shift valves, servos, pistons, and control valve body. A transmission technician will use a service manual chart of the hydraulic system to solve transmission problems.

■ FLUID PUMP

Fluid is distributed throughout the transmission to control shifts and provide lubrication to critical parts. The transmission fluid pump, located just behind the torque converter at the front of the transmission, is driven by lugs on the outside of the snout of the torque converter (Figure 66.35). It is often called the *front pump* because old transmissions often had a front and rear pump.

The fluid pump does several things:

■ It creates the hydraulic pressure to apply the clutches and bands.

■ It provides lubrication to the parts in the transmission.

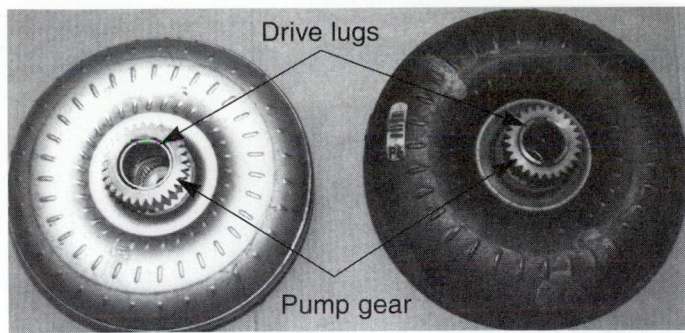

Figure 66.35 The pump is driven by lugs on the torque converter.

- It fills the torque converter.
- It circulates fluid throughout the transmission and to the heat exchanger in the radiator in the front of the vehicle to cool the transmission.
- Its pressure operates valves in the hydraulic valve body.

TYPES OF PUMPS

There are three types of pumps: the rotor type (Figure 66.36), the internal/external gear crescent type (see Figure 66.39), which is the most common, and the vane type (Figure 66.37).

PUMP OPERATION

Fluid is drawn through a sump screen or *filter* in the bottom of the fluid pan (Figure 66.38). The filter can be a simple fine mesh metal screen or it can be made of paper or felt. A filter only keeps the larger particles out because it must allow sufficient fluid flow for transmission operation. Paper filters are made of cellulose or Dacron fabric. Cellulose is the resin-based material that paper is made out of. These filters keep out smaller particles and can plug more easily. Some transmissions have a bypass circuit in case the filter becomes plugged.

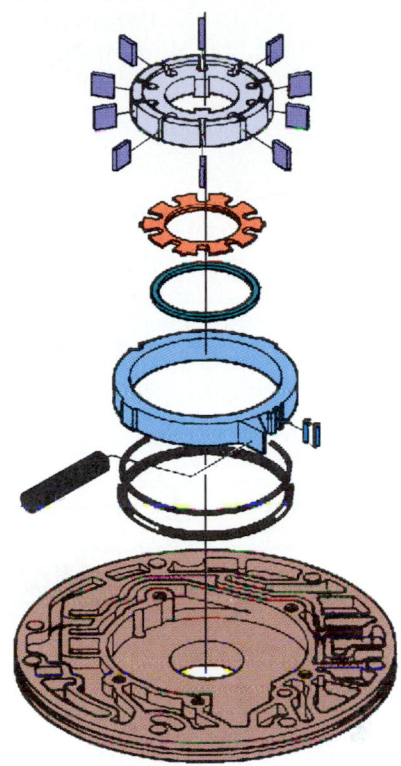

Vane pump

Figure 66.37 A vane pump. *[Courtesy of General Motors Corporation, Service Technology Group]*

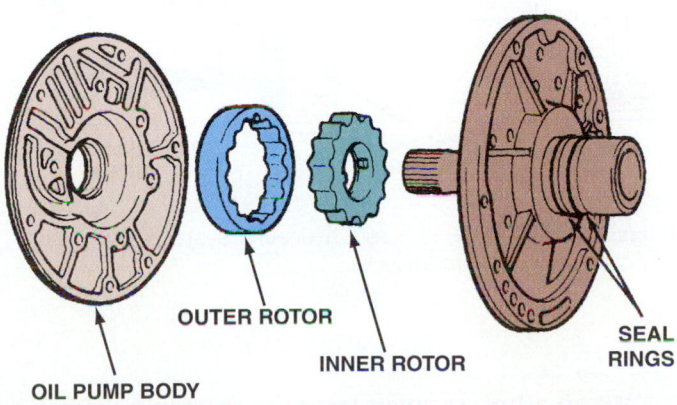

OIL PUMP BODY · OUTER ROTOR · INNER ROTOR · SEAL RINGS

ROTOR PUMP

Figure 66.36 A rotor-type pump. *[Courtesy of Chrysler Corporation]*

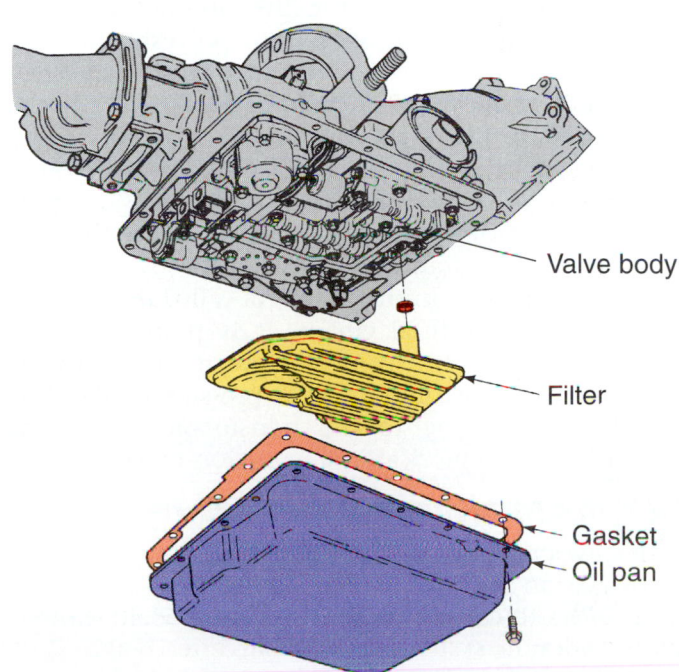

Valve body

Filter

Gasket

Oil pan

Figure 66.38 The filter is located within the oil pan. *[Courtesy of General Motors Corporation, Service Technology Group]*

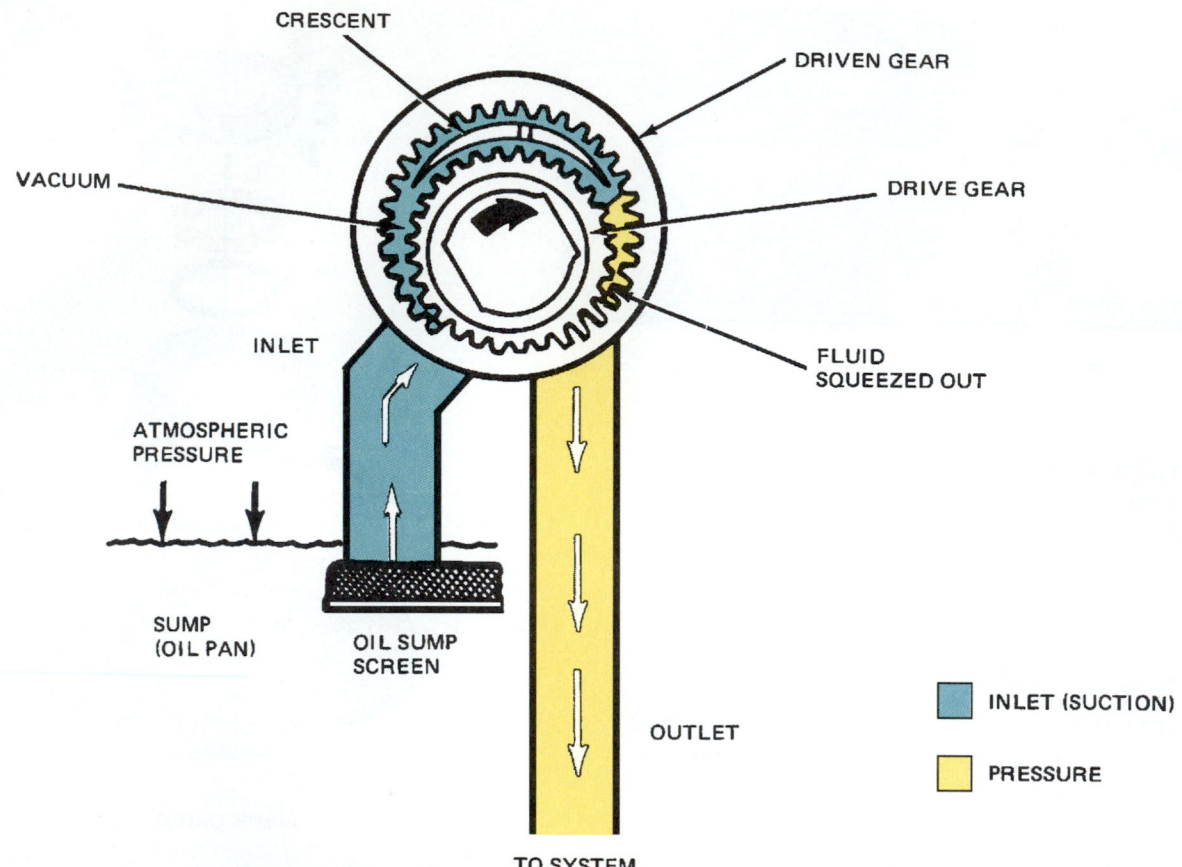

CRESCENT

DRIVEN GEAR

VACUUM

DRIVE GEAR

INLET

FLUID
SQUEEZED OUT

ATMOSPHERIC
PRESSURE

SUMP
(OIL PAN)

OIL SUMP
SCREEN

OUTLET

INLET (SUCTION)

PRESSURE

TO SYSTEM

Figure 66.39 Operation of an oil pump. *(Courtesy of Ford Motor Company)*

Felt-type filters are the most common. They trap contaminants throughout the filter and not just on its surface. Felt filters are made from polyester material that is randomly spaced to be able to trap smaller particles and yet allow sufficient fluid flow.

Secondary filters are found in transmissions too. They are simply screens located in fluid passageways.

The fluid, called automatic transmission fluid (ATF), is sent from the pump (Figure 66.39) to the pressure regulator before being distributed by the valve body. The pressure in this circuit is called *line pressure*. In the converter fluid circuit, it is pumped to the torque converter and cooler before returning to the transmission. This fluid is under pressure of about 35 psi. Fluid returning to the transmission from the cooler is used to lubricate transmission parts.

■ TRANSMISSION VALVES

The valves used to control fluid flow in an automatic transmission are called spool valves. They have *lands* and *valleys* to control fluid flow. Figure 66.40 shows a basic hydraulic system.

Valves can be moved in three ways:
■ By spring
■ By a lever or rod
■ By hydraulic pressure

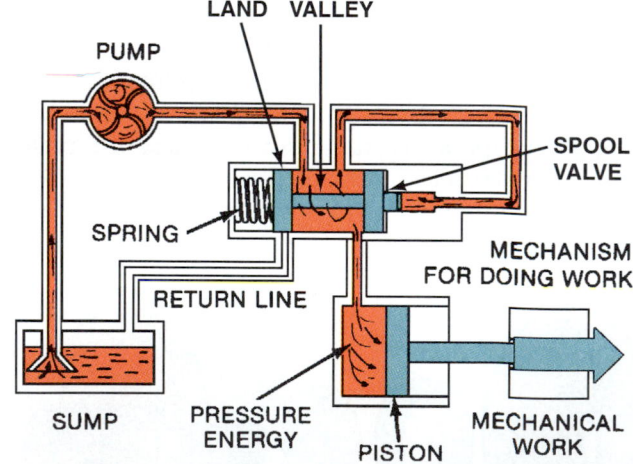

LAND VALLEY

PUMP

SPOOL
VALVE

SPRING

MECHANISM
FOR DOING WORK

RETURN LINE

SUMP

PRESSURE
ENERGY

PISTON

MECHANICAL
WORK

Figure 66.40 A basic hydraulic system. *(Courtesy of Ford Motor Company)*

When all of a valve's faces are the same diameter, this is called a balanced valve. If one face is larger than the other, fluid directed between them will result in movement of the valve in the direction of the larger face (Figure 66.41).

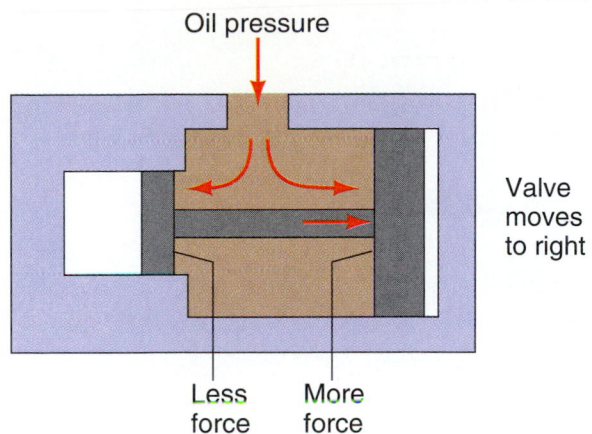

Figure 66.41 Fluid pressure will move the piston in the direction of the larger land.

■ PRESSURE REGULATOR

The pressure regulator valve determines the pressure in the system. As engine rpm increases, the pump turns faster and puts out more volume, which can result in excess pressure. The excess pressure is relieved by allowing fluid to flow back to the fluid pan (Figure 66.42) instead of forcing it through galleries, which would increase pressure (remember Pascal's law).

The pressure regulator is a variable pressure regulator. When higher pressures are required, fluid is diverted to the back side of the valve. This works with the relief valve spring to increase the pressures throughout the transmission.

An **orifice** restricts fluid flow. As fluid tries to move through the orifice, pressure on the other side of it drops (Figure 66.43a). When fluid movement is restricted, pressure equalizes on both sides of the ori-

fice (Figure 66.43b). Orifices can be used to reduce pressure of moving fluid such as when a delay in a shift is desired.

■ HYDRAULIC VALVE BODY

The **valve body** is the brains of the transmission (Figure 66.44). It senses engine load and vehicle speed and adjusts the transmission's shift points and fluid pressures to match. The valve body contains many valves, springs, and orifices to control the shifting of the transmission. It is usually bolted to the bottom of the transmission, inside of the pan. The pump inlet filter or screen is usually attached to the bottom of the valve body.

A *valve body separator plate*, also called a spacer plate or channel plate (Figure 66.45) fits between the valve body and the transmission. It has holes of many different sizes to regulate fluid flow.

The *manual control valve* is the longest valve in the valve body. It is attached by linkage to the shift lever. When the driver moves the shift lever, the manual valve moves to direct fluid through the valve body to the correct driving or holding devices in the transmission.

The **shift quadrant** is the readout on the gear selector that tells what gear the transmission is in. For safety reasons, the positions of automatic transmission gears are controlled by federal law. The shift order is always PRNDL, which sounds like "prindle" when spoken. Later model transmissions have several lower gear ranges (PRNDD2L). On some vehicles (notably those with Hydramatic transmissions) up until the mid-fifties, reverse was positioned on the opposite end of the selector from normal (NDLR). An owner of two vehicles with different shift positions could accidentally back up instead of going forward, or vice versa.

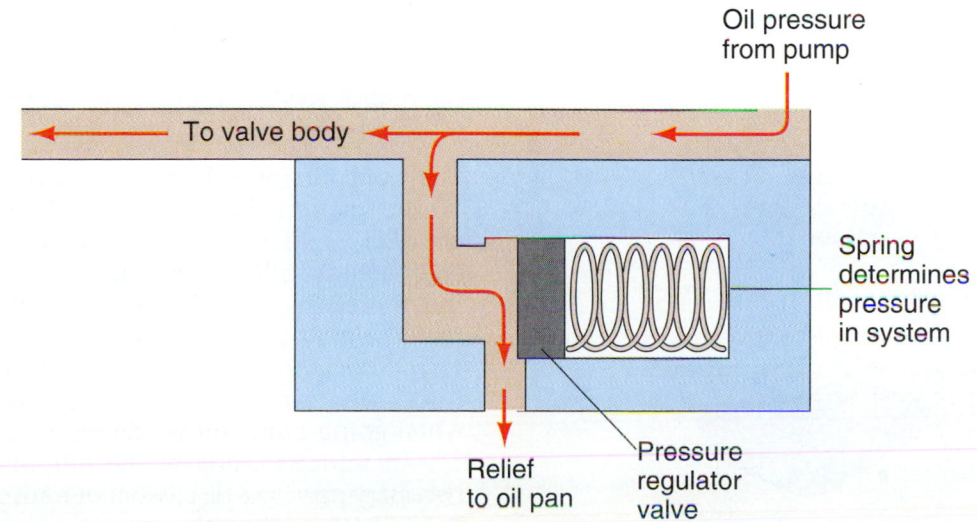

Figure 66.42 Excess pressure is relieved by allowing oil to flow back to the oil pan.

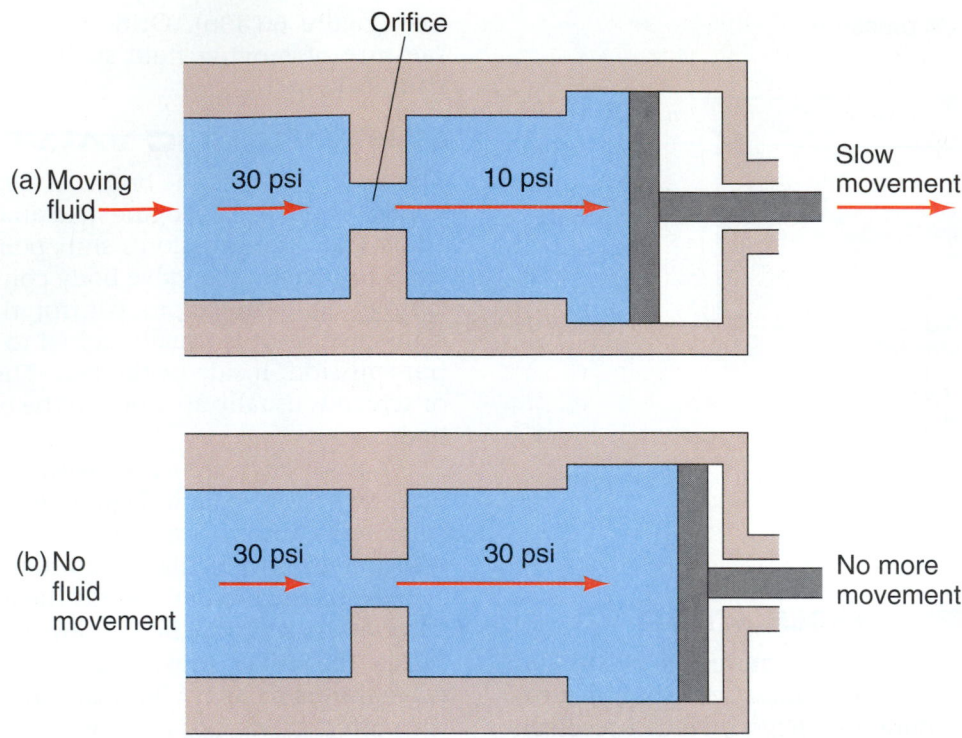

Figure 66.43 As fluid tries to move through the orifice, pressure on the other side of it drops.

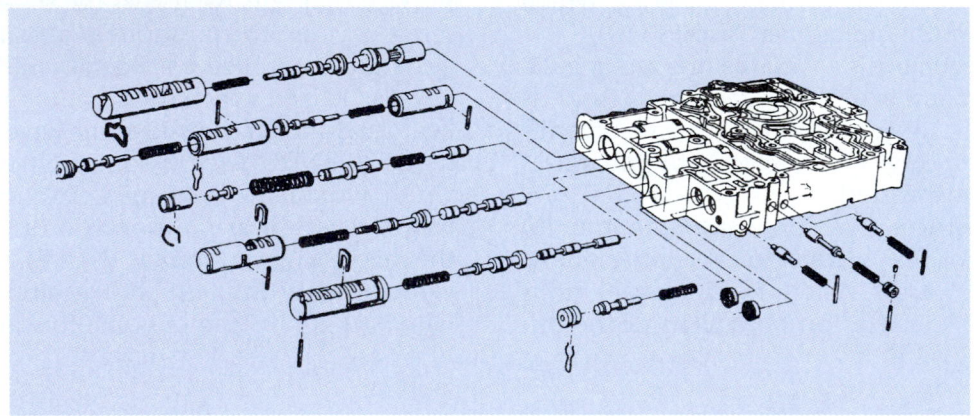

Figure 66.44 Parts of a valve body. *(Courtesy of General Motors Corporation, Service Technology Group)*

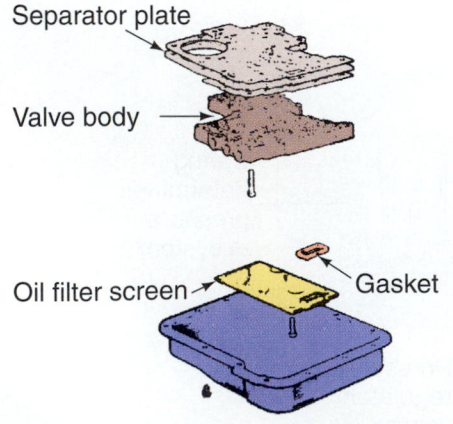

Figure 66.45 Location of a separator plate. *(Courtesy of Ford Motor Company)*

■ TRANSMISSION AUTOMATIC SHIFT SELECTION

The transmission selects the correct gear range based on the load on the engine and the speed of the vehicle. For instance, as a vehicle on the highway approaches a hill, the engine starts to work harder and the vehicle begins to slow down. Pressing harder on the accelerator pedal results in lower engine vacuum. This is because the engine cannot breathe enough air to be able to produce maximum vacuum as it would when going downhill or idling.

An understanding of the following terminology is necessary prior to a discussion of transmission operation:

■ **Upshift** is when the transmission shifts from a low gear to a higher gear; second to third, for instance.

■ **Downshift** is when the transmission shifts to a lower gear; second to first, for instance.

Shift valves respond to engine speed and load by moving back and forth in their *shift cylinders* to create gear shifts. There are shift valves for each of the gear shift positions. For instance, in a three-speed automatic transmission, there is a *1–2 shift valve* and a *2–3 shift valve*. The shift valve has two forces working on it: throttle pressure and governor pressure.

Throttle pressure is the kind of pressure that results when engine vacuum changes. Low vacuum, due to an open throttle or heavy load, results in high throttle pressure. High throttle pressure will tell the shift valve to move into a lower gear position. For instance, a 2–3 shift valve will attempt to move to the second gear position due to the increased load on the engine (Figure 66.46a).

Governor pressure is the kind of pressure that results from increases in vehicle speed. When a car is travelling at 75 miles per hour and the throttle is moved to the wide open position (called W.O.T. or wide open throttle), the 3–2 shift valve wants to move to the second gear position. High governor pressure prevents this shift until the speed drops (Figure 66.46b). The amount of drop in speed would be whatever the engineers who designed the valve body determined would be acceptable for such a shift.

■ **GOVERNOR**

The governor is located on the output shaft of the transmission. Some governors are gear-driven by the speedometer gear. Other governors are pinned to the output shaft (Figure 66.47).

The governor is actually a variable pressure relief valve. Governor pressure can be equal to its source (line pressure) but it cannot be greater than that. When the transmission's output shaft is moving at high speeds, weights in the governor are thrown outward. They are attached to a spool valve that closes off a fluid passageway causing governor fluid pressure to climb. When the speed is slow, the weights are in and the fluid passage is open. This results in lower governor fluid pressure.

■ **VACUUM MODULATOR**

To control throttle pressure, some transmissions have a vacuum modulator valve (Figure 66.48). Others have a cable that moves the *throttle valve* based on the position of the accelerator (gas pedal). Throttle pressure times upshifts and also increases pump pressure under load. It aids the spring on the shift valve to work against governor pressure.

A vacuum modulator is similar in appearance to a distributor vacuum advance. It has a diaphragm inside it and a hose fitting that is attached to a manifold vacuum source (a hose to the intake manifold).

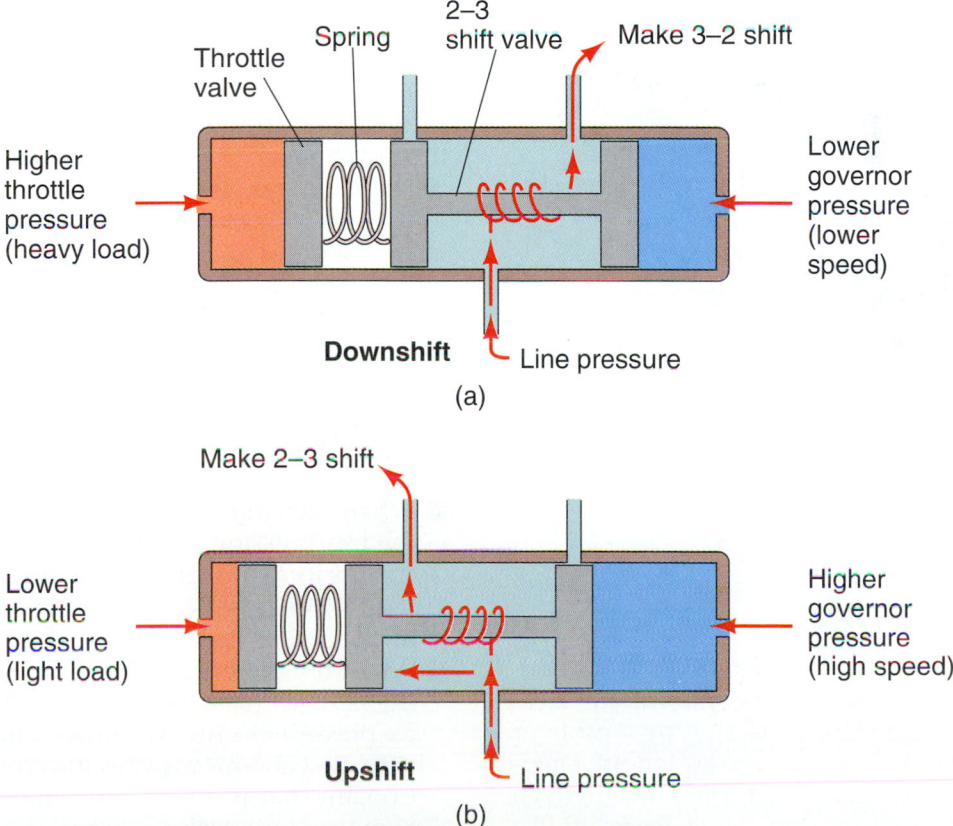

Figure 66.46 (a) The 2–3 shift valve will attempt to move to the second gear position due to the increased load on the engine. (b) Under light load, governor pressure causes a 2–3 shift.

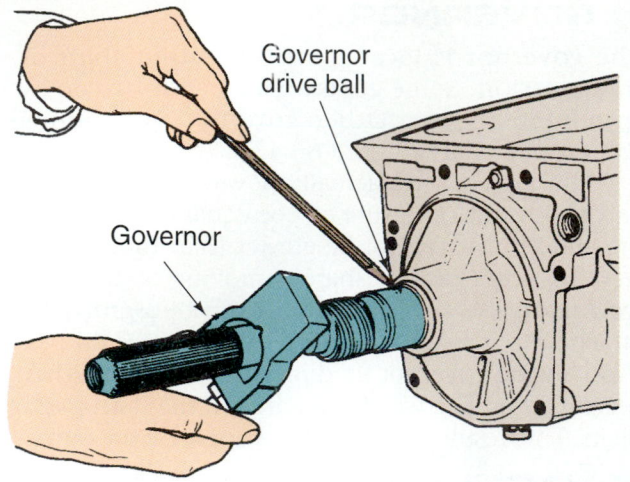

Figure 66.47 A governor that is pinned to the output shaft. *(Courtesy of Ford Motor Company)*

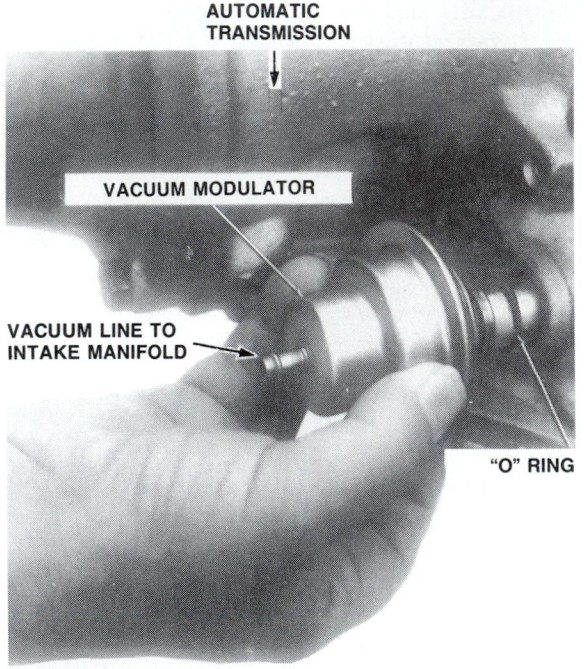

Figure 66.48 A vacuum modulator. *(Courtesy of General Motors Corporation, Service Technology Group)*

■ KICKDOWN VALVE

At wide open throttle, a kickdown (also called downshift or **detent**) valve comes into operation. The kickdown valve is either manually operated by a cable or linkage, or it is an electrically operated solenoid. Operating the kickdown valve causes throttle pressure to go to its highest point. Attempting a *forced downshift* like this will still not result in a kickdown until governor pressure is low enough to allow it.

It is easy to find a manual kickdown valve when inspecting a valve body. It is the one that is spring loaded. It extends out of the side of the valve body where it contacts the lever that applies it from outside of the transmission.

■ AUTOMATIC TRANSMISSION FLUID (ATF)

Automatic transmission fluid is oil that is specially formulated for use in automatic transmissions. Facts about the different types of fluids and their correct uses are covered in Chapter 13.

■ AUTOMATIC TRANSMISSION COOLING

The transmission develops a great deal of heat during operation. This is especially true during starting and stopping, in hot weather, and when pulling a load. Whenever there is a difference in the speeds of the torque converter impeller and turbine, the fluid is heated as it is cut.

Heat damages transmission fluid. Under normal conditions, transmission fluid will last for over 50,000 miles without being damaged. When fluid is heated to excessive temperatures, however, it oxidizes. Ruined fluid can damage a transmission. Oxidized fluid gets gummy or varnishy. This causes clutch pistons and valves to stick throughout the transmission. When this happens the transmission will need to be disassembled and rebuilt.

Most transmissions have an external *fluid cooler*, called a heat exchanger. It is located in the bottom or side tank of the radiator. If your transmission has a heat exchanger, there will be two lines between the transmission and radiator (see Figure 20.10). It is called a heat exchanger because in cold weather, the fluid must reach operating temperature as quickly as possible. The engine's cooling system has a thermostat that results in quick warming of the engine's coolant. The warm coolant heats the transmission fluid by way of the heat exchanger in the radiator.

One cause of transmission failure is leakage of the cooler. If a radiator heat exchanger leaks, the results are:

■ When the engine is operating: Fluid pressure in the transmission is higher than that in the cooling system so ATF migrates into the radiator. This results in a lower cooling ability for the radiator because fluid does not cool as well as coolant.

■ When the engine is turned off: Pressure remains in the radiator (usually about 14–17 pounds). There is no pressure on the ATF unless the engine is running, so coolant migrates into the transmission. Coolant that is mixed with fluid becomes very gummy. If the leak is serious, the transmission will require rebuilding after the internal coolant leak is fixed.

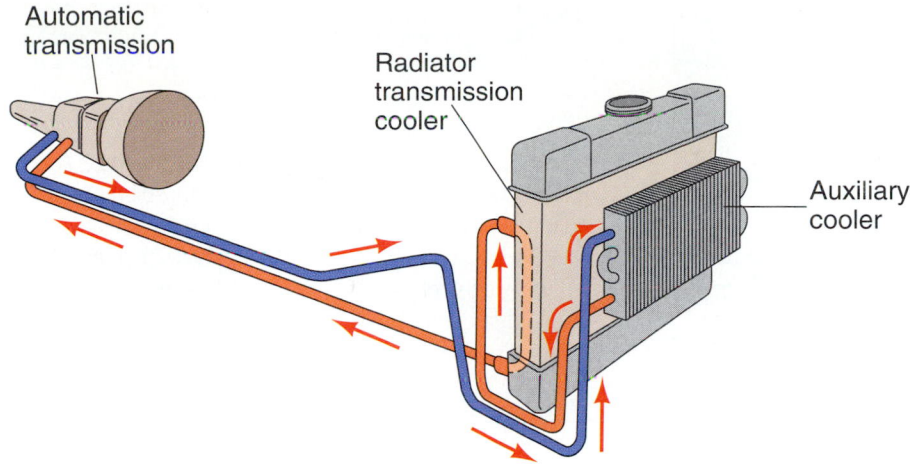

Figure 66.49 An auxiliary cooler is installed in series before the radiator transmission cooler.

AUXILIARY COOLER/ HEAT EXCHANGER

An auxiliary cooler/heat exchanger is often added to motor homes and vehicles that pull trailers. It resembles a small radiator and is hooked into the cooler line in series (Figure 66.49). Some manufacturers of auxiliary coolers recommend that they be installed in the line *before* the radiator cooler so that flash heat is removed by the cooler before transmission heat is exchanged with the engine's coolant. If it is installed after the radiator, engine coolant can actually become hotter during heavy-duty operation.

PARK PAWL

The **park pawl** is a lever that locks the output shaft of the transmission when the shift lever is placed in park (Figure 66.50).

SAFETY NOTE The park pawl is not supposed to be used in place of an emergency brake. In fact, it is dangerous to trust this device. If you study Figure 66.50, you will see that there are high spots between the notches in the park pawl gear.

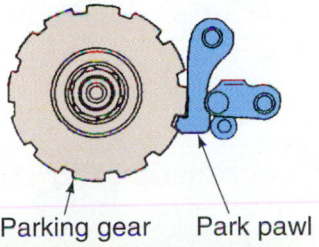

Parking gear   Park pawl

Figure 66.50 A park pawl. *(Courtesy of American Honda Motor Co., Inc.)*

CASE HISTORY *A driver left his car idling with the gear shift in the park position when he went back into his house for something he had forgotten. The park pawl was not fully engaged and engine vibration caused the transmission to slip into reverse. The car went down his driveway and ran over a chicken.*

ELECTRONIC AUTOMATIC TRANSMISSIONS

Shifts controlled by the car's onboard computer are used by manufacturers on some of their transmissions. Electronic controls can provide more precise shift control. They also replace part of the relatively costly valve body with less expensive solenoids.

In a conventional hydraulically controlled transmission, shift valves are moved by hydraulic pressure from the governor. In an electronically controlled transmission, the upshift is controlled by a solenoid.

Electronic transmissions vary by manufacturer, but they all operate in a similar fashion using basically the same parts. The computer is the heart of the electronic transmission shift control. It receives signals from various sensors and acts on them through one or more fluid control solenoids (Figure 66.51).

Shift solenoids used in an automatic transmission are similar to solenoids that have been used in previous automatic transmissions for forced downshifts and for engaging the torque converter clutch. A solenoid can open or close a passageway. It is similar to a fuel injector used with electronic fuel injection. A solenoid operates when its coil of wire is energized. It is returned to its original position by a spring when current flow stops and the magnetic field in its coil breaks down. Solenoids can be either normally open

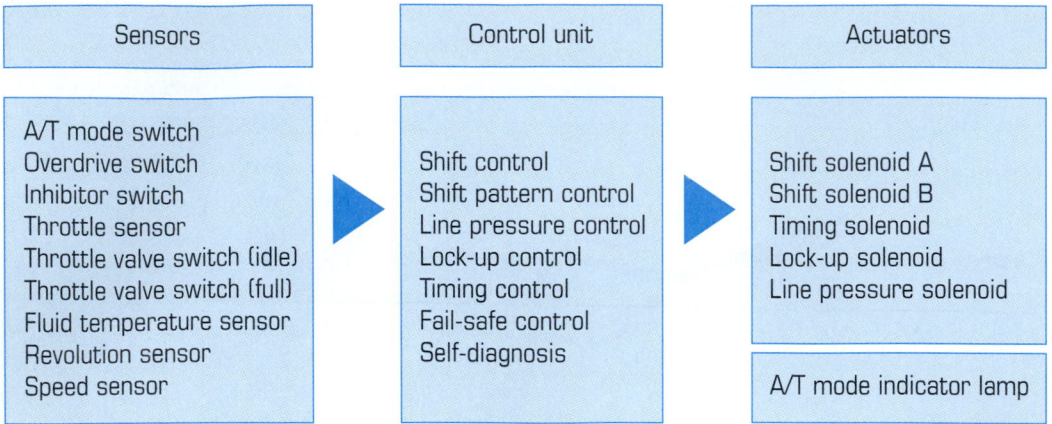

Sensors	Control unit	Actuators
A/T mode switch Overdrive switch Inhibitor switch Throttle sensor Throttle valve switch (idle) Throttle valve switch (full) Fluid temperature sensor Revolution sensor Speed sensor	Shift control Shift pattern control Line pressure control Lock-up control Timing control Fail-safe control Self-diagnosis	Shift solenoid A Shift solenoid B Timing solenoid Lock-up solenoid Line pressure solenoid
		A/T mode indicator lamp

Figure 66.51 The control unit (computer) receives signals from various sensors and acts on them through fluid control solenoids. *(Courtesy of Nissan Motors)*

or normally closed when they have no electricity flowing through them.

In some transmissions, fluid flows to a clutch or servo (direct control). Cycling the solenoid rapidly on and off (like pulse width in a fuel injector) can control how fast the clutch applies and thus the shift feel. In other transmissions, the solenoid controls a shift valve that sends fluid to the clutch or servo. Various solenoids are shown in Figure 66.52.

Conventional transmissions use road speed and engine load information to make shift decisions. These same inputs, along with others, are used to make electronic transmissions shift.

- A vehicle speed sensor (Figure 66.53) takes the place of the governor. It sends an output voltage to the computer that varies with road speed. This

information is also used for the electronic speedometer.

- A throttle position sensor gives the computer information that was previously made available by a throttle cable. A voltage signal changes in response to throttle opening.
- Engine load signals are already known to the computer through the existing devices used to operate the engine's fuel and emission systems.
- A neutral start switch mounted on the transmission keeps the engine from being able to start in any gear but park or neutral. On electronically controlled transmissions, it is not just for park and neutral. It also operates the back-up lights and gives information to the computer as to other gear range positions as selected by the driver.

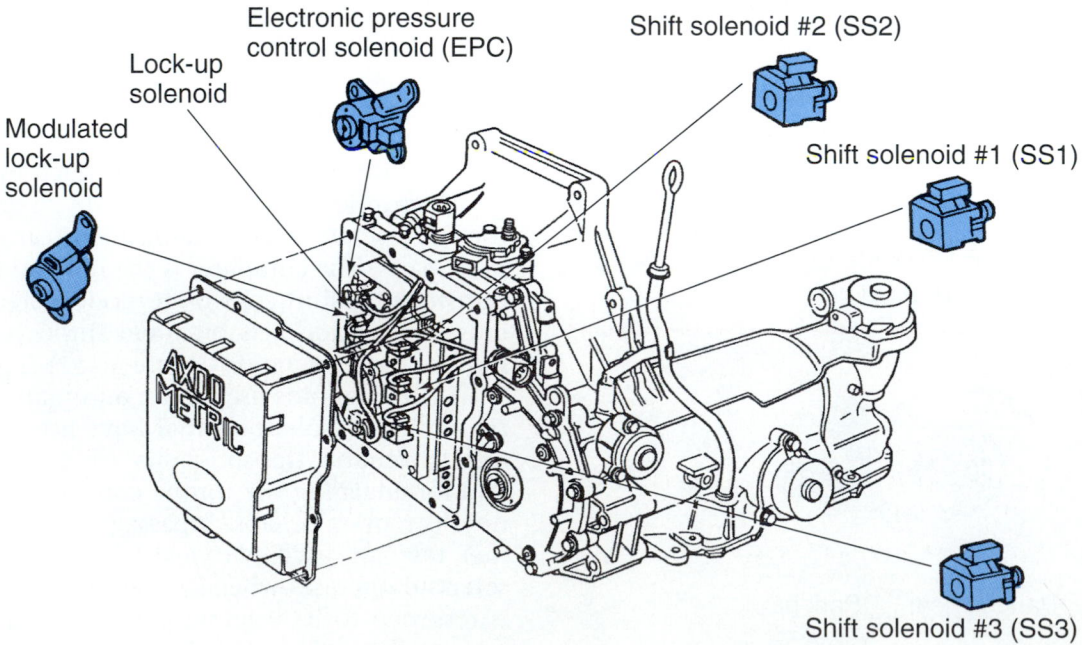

Figure 66.52 Various shift solenoids. *(Courtesy of Ford Motor Company)*

- A power or sport range selection switch on some cars is controlled by the driver. It allows the transmission to make shifts at a higher rpm.
- Torque converter lockup is also controlled by the computer.
- Engine coolant temperature is monitored by the computer. It prevents the torque converter clutch from operating when the engine is cold. Some transmissions also have later upshifts when the engine is cold.
- A brake switch functions as part of the cruise control and stoplight switch. It tells the computer

to release the torque converter clutch during a stop to prevent the engine from stalling.

■ CONTINUOUSLY VARIABLE TRANSMISSION

Continuously variable transmissions (CVT) are being used by some manufacturers and are under development by others. They work in a manner similar to a variable speed drill press (see Figure 8.18b). Where a conventional transmission has two, three, four, or more speeds, the CVT has infinite driving ratios. The aim of these transmissions is to increase fuel economy in the range of about 25%. This is possible because the engine can be run with a relatively constant rpm. Because the engine does not have to accelerate through each gear, a smooth increase in vehicle speed is also possible. A Honda CVT is shown in Figure 66.54.

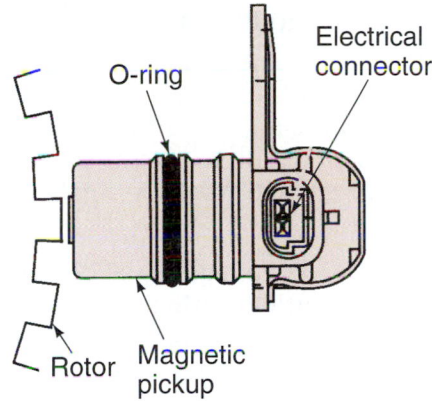

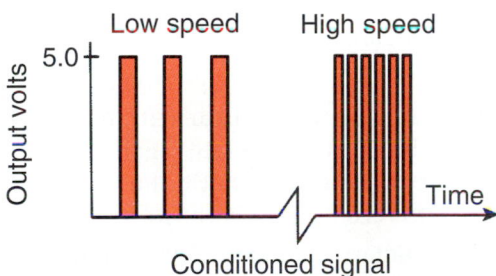

Figure 66.53 A speed sensor takes the place of the governor. *(Courtesy of General Motors Corporation, Service Technology Group)*

Figure 66.54 Drive assembly from a continuously variable transmission. *(Courtesy of Tony Jewell)*

■ REVIEW QUESTIONS

1. What is the name of the part that is actually part of the converter housing, the impeller or the turbine?
2. What is it called when the impeller and turbine turn at the same speed?
3. What is the name of the part that redirects fluid flow at the center of the converter?
4. In a simple planetary gear set there is a sun gear, planets, a carrier, and a _____ gear.
5. In a planetary gear set with input in a clockwise direction, if the planet carrier is held from turning, what is the direction of output?

6. _____ pressure is in the main line of the transmission.
7. What is the order of gear shifts on the shift quadrant?
8. A vacuum _____ is a vacuum-operated diaphragm that controls shift points on some automatic transmissions.
9. What is a detent valve for?
10. What is another name for a transmission cooler?

■ ASE STYLE REVIEW QUESTIONS

1. Technician A says that the torque converter stator clutch freewheels when coupling speed is reached. Technician B says that there is less than 10% slippage in a lock-up converter when its clutch is not applied. Who is right?

   a. Technician A   **b.** Technician B
   c. Both A and B   **d.** Neither A nor B

2. Technician A says that a clutch can be used to drive planetary gear parts. Technician B says a clutch can be used to hold planetary parts from turning. Who is right?

   a. Technician A   **b.** Technician B
   c. Both A and B   **d.** Neither A nor B

3. Technician A says that clutch packs require periodic clearance adjustment. Technician B says that some cars must be manually shifted to a specified gear range in order to get engine braking during deceleration. Who is right?

   a. Technician A   **b.** Technician B
   c. Both A and B   **d.** Neither A nor B

4. Technician A says that as a vehicle climbs a hill, engine vacuum becomes lower. Technician B says that as a vehicle slows down, governor pressure drops. Who is right?

   a. Technician A   **b.** Technician B
   c. Both A and B   **d.** Neither A nor B

5. Technician A says when starting from a stop, the approximate torque increase in an automotive torque converter is 2:1. Technician B says that the type of one-way clutch that uses rollers is called a sprag. Who is right?

   a. Technician A   **b.** Technician B
   c. Both A and B   **d.** Neither A nor B

6. Technician A says that a band can be used to hold or drive a planetary gear set. Technician B says that the name for the maximum engine rpm that can be reached when the brakes are on with the transmission in gear is vortex. Who is right?

   a. Technician A   **b.** Technician B
   c. Both A and B   **d.** Neither A nor B

7. Technician A says that a band is a holding member. Technician B says that a roller clutch is a holding member. Who is right?

   a. Technician A   **b.** Technician B
   c. Both A and B   **d.** Neither A nor B

8. Hydraulic pressure is directed to the inside of a spool valve with one face smaller than the other. Technician A says the valve will move toward the larger side. Technician B says the valve will move toward the smaller side. Who is right?

   a. Technician A   **b.** Technician B
   c. Both A and B   **d.** Neither A nor B

9. Technician A says the device that controls shift points on an automatic transmission is the valve body. Technician B says the device that controls shift points on an automatic transmission is the computer. Who is right?

   a. Technician A   **b.** Technician B
   c. Both A and B   **d.** Neither A nor B

10. Technician A says that a car with an automatic transmission has a flexplate. Technician B says that a car with an automatic transmission has a torque converter. Who is right?

    a. Technician A   **b.** Technician B
    c. Both A and B   **d.** Neither A nor B

CHAPTER 67

Automatic Transmission Diagnosis and Service

■ **KEY TERMS**

Dexron III/Mercon
severe service
varnish
union
mushy shift
harsh shift
stall test
pressure test
air test
soft parts
hard parts

■ OBJECTIVES

Upon completion of this chapter, you should be able to:

✔ Perform maintenance service on automatic transmissions.

✔ Diagnose automatic transmission problems and recommend repairs.

✔ Change the transmission fluid and filter.

✔ Repair automatic transmission leaks.

✔ Perform basic automatic transmission tests.

✔ Understand what is involved in a transmission overhaul.

■ INTRODUCTION

Major automatic transmission repairs are typically done by an automatic transmission specialty shop. But a knowledge of transmission service and diagnosis is important for all technicians. Problems with throttle pressure, a leaking vacuum modulator, or a locked-up torque converter stator can all cause problems that can mimic engine problems. A technician lacking basic knowledge of transmission operation could do extensive repairs to the fuel system or replace an engine because he or she lacks an understanding of transmission operation.

CASE HISTORY

A customer called a repair shop to make an appointment to have a rebuilt engine installed. Before ordering an engine, the shop owner suggested that the car be properly diagnosed. The technician test drove the car and found that it had a rough idle, white smoke from the exhaust, and harsh, late shifts. When he returned to the repair shop, he pulled the hose from the vacuum modulator and found that there was fluid in it. The transmission fluid level was low too. The spark plugs closest to the vacuum tap on the intake manifold were oil-fouled. He replaced the modulator, cleaned the spark plugs, and returned the car to the very happy customer.

When a transmission has been tested and proves to need major repairs, it must be removed from the vehicle. Manufacturers' service manuals give detailed instructions on re-building procedures. Although this kind of work is not any more difficult than other types of automotive work, it is a complicated specialty area. Special schooling is recommended before attempting to perform major repairs. Some special tools are also required to perform transmission work.

In a specialty automatic transmission repair shop, there are two levels of employees: an R&R (remove and replace) specialist and a *rebuilder*. The R&R specialist removes the transmission and reinstalls it after it is rebuilt. The rebuilder typically performs rebuilds at the workbench.

■ AUTOMATIC TRANSMISSION MAINTENANCE

Transmission service is most often preventive maintenance, rather than correction of a problem. Automatic transmissions generally require very little maintenance. A transmission that has not been run under severe service conditions should easily last for 100,000 miles or more.

■ TRANSMISSION FLUID SERVICE

The oil in an automatic transmission is called automatic transmission fluid (ATF). Most types of ATF are dyed red in color to help differentiate them from engine oil. ATF has a low viscosity, about like an SAE 10 engine oil. Additives include viscosity index improvers, oxidation and corrosion inhibitors, extreme pressure lubricants, anti-foam additives, detergents and dispersants, friction modifiers, and pour point depressants.

There are several types of automatic transmission fluids. **Dexron III/Mercon** is the latest and most used. It is designed for use with electronic transmissions, which require fluid of low viscosity in low temperatures. It is used in all late model domestic and imported automatic transmissions. Dexron III/Mercon can be used as a replacement fluid for all automatic transmissions except some 1982 and earlier Ford and Borg-Warner products, which use Type F fluid.

ATF is long lasting and does not collect byproducts of combustion like engine oil does. Manufacturers have different recommendations about the length of service before a fluid change is called for. Automatic transmission fluid additives can be depleted under severe service.

Anything that causes heat is **severe service**. This includes trailer towing and stop and go city driving in hot weather. Repeated starts from a stop cause the torque converter to shear the fluid. A transmission in a vehicle pulling a trailer can easily develop 300°F of fluid temperature. Fluid starts oxidizing rapidly at about 250°F, and the rate of oxidation doubles every 20°F after that point. Older transmissions were operated at about 175°F and had a fluid life expectancy of about 100,000 miles. At 195°F, fluid lasts about 50,000 miles and at 215°F about 25,000 miles. It is not unusual for a modern transmission to operate at 215°F.

■ FLUID LEVEL

The dipstick is located in a tube that goes into the transmission pan. The top end of the dipstick is in the engine compartment. The fluid level should be between the add and full marks on the dipstick (see Figure 13.29). Correct fluid level is very important to proper transmission operation. Too low a fluid level can result in transmission damage.

A low fluid level is usually caused by a leak. A *leaking vacuum modulator* will also result in an internal leak when ATF is pulled into the intake manifold where it is burned. When burned, ATF causes white smoke out of the exhaust.

A high fluid level can result in foamed fluid when the transmission gears churn up the fluid. With a normal fluid level the gears do not dip into the fluid sump. A high fluid level can result when too much fluid is added after a fluid change and the excess is not removed.

A high fluid level can also happen because a customer adds fluid mistakenly after improperly checking the fluid level. When checking engine oil level, the add line means that the crankcase is one quart low. It only takes one *pint* of fluid to raise the level from add to full on a hot transmission.

Checking Fluid Level

The following conditions must be met when checking ATF level:
- Car is on level ground
- Engine and transmission warm (dipstick should be uncomfortable to touch)
- Parking brake set
- Gear selector in park position

NOTE: *Look for special fluid checking information stamped on the dipstick. For instance, Chrysler rear wheel drive vehicles specify neutral as the correct position for checking fluid level.*

■ CHECK FLUID CONDITION

Fluid condition can tell a good deal about the overall condition of a transmission. The fluid should be clean and red. It should not be pink and foamy or milky from contamination with water or coolant. Fluid should be easy to wipe from the dipstick and should not smell burnt.

The following are some of the conditions that might be found:
- *Milky fluid* is usually found when coolant and fluid mix together when the transmission cooler in the radiator leaks.
- **Varnish** is indicated when the dipstick will not wipe clean. Internal parts in the transmission cannot operate properly when the fluid has become oxidized like this. The transmission will need to be removed, disassembled, and rebuilt.
- *Burned fluid* will be dark and have a burnt smell.

NOTE: *Some fluids have a burnt smell when they are in good condition, so be sure this is not the normal condition. Burned fluid is usually caused by failed friction parts in the clutch packs or bands. Friction material might be evident on the dipstick. The transmission will probably require rebuilding soon.*

When adding fluid, be sure to use the correct type. Add fluid through the dipstick hole with a funnel with a small opening that will fit into the dipstick tube. Be sure the funnel is clean.

SHOP TIP   A funnel that has had oil in it will collect shop dust. After use, rinse it and keep it in a plastic bag.

■ CHANGING TRANSMISSION FLUID

When changing transmission fluid, check the service manual for special instructions. Some manufacturers call for draining the converter also. In this case, the converter has drain plugs. Most transmissions do not have drain plugs in the converter, so they do not require a complete change of the fluid. Only the ATF in the pan is removed. Adding new fluid after the pan is reinstalled replenishes any additives that have been depleted during service.

■ TRANSMISSION FILTER SERVICE

The pan must be removed to clean or replace the filter. Newer transmissions work harder, so they leave more clutch material in the fluid. While ATF might not need to be changed, the filter should be.

Some filters are replaceable. Some older cars had reusable brass screens. It is customary to replace either the filter or the screen. They usually come in a filter kit with a new gasket. Some filters have a drainback check valve to prevent fluid from draining from the valve body when the engine is off.

Transmission filters are *unlike* engine oil filters, which have a bypass valve in case they become plugged. Polyester felt filters are the most popular filters in late model transmissions. Felt will trap materials as small as 60 microns (0.0025"). This is about the size of a human hair. A clogged transmission filter can cause shifting problems, pressure loss, and possibly transmission failure. A plugged filter will sometimes cause a whining sound like an empty power steering pump.

Removing the Pan

Most transmission pans do not have drain plugs. To drain the fluid requires removal of the pan.

> ⚠ **CAUTION** There are usually 3–4 quarts of fluid in the pan. This weighs about 6–8 pounds. On the first transmission fluid change, beginners often make a large mess. Be careful! The fluid is often quite hot and if the pan is not carefully removed, the fluid will spill out fast.

> **SHOP TIP** When removing the pan, leave a row of screws in the front of the pan. Remove all of the other screws. Then, loosen the remaining screws in the front of the pan (Figure 67.1). Be sure to have a large fluid catch tray ready. Let the fluid spill into the catch tray and then carefully remove the pan.

Inspect the Pan

Inspect the inside of the pan for pieces of friction material or metal particles. These might indicate serious transmission problems. A small amount of debris from friction material is normal. Clean the pan with a lint-free rag.

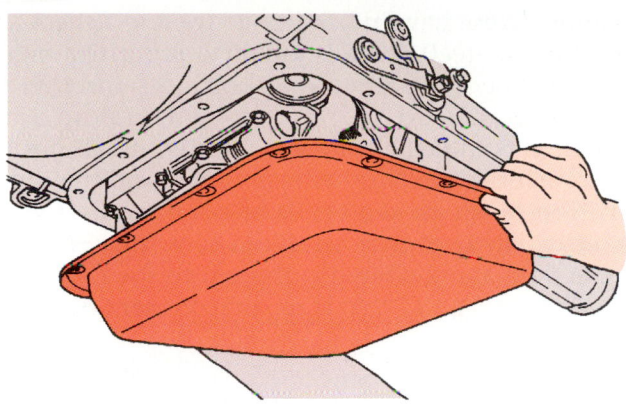

Figure 67.1 Slightly loosen the screws in the front of the pan and let the oil flow out of the back. *(Courtesy of Ford Motor Company)*

NOTE: *Occasionally, lint from shop rags is found plugging up the transmission screen or filter.*

Check the screw holes in the pan to see that they are not raised up from previous overtightening. Use a hammer to flatten them against an anvil.

■ REMOVE AND REPLACE THE FILTER

It is important that a pump suck only fluid and not air. Some filters have a gasket (Figure 67.2). Other filters plug into the transmission case or pump and are sealed with only a neoprene O-ring or sleeve. Be sure to remove the old O-ring or gasket and replace it with a new one. Lubricate the new O-ring with ATF. Do not use sealer. Install the filter and check to see that its tube fits snugly in its bore. Be sure to replace any parts that come off with the filter.

NOTE: *Most transmissions are aluminum. They are easy to strip. Be sure to use a torque wrench on filter and valve body screws.*

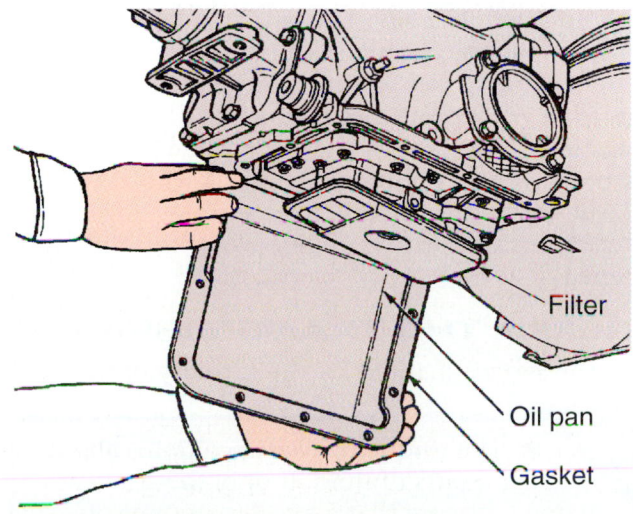

Filter

Oil pan

Gasket

Figure 67.2 A pan, gasket, and filter. *(Courtesy of Ford Motor Company)*

Some filters have nylon housings and others are aluminum or steel. There is usually a protective metal grommet where the mounting bolt goes through to the valve body.

If air gets into the torque converter, it will not be mixed with the fluid but will gather around the input shaft. Too much air can result in a burned up torque converter. The converter acts like a centrifuge, separating the fluid and air. Fluid is thrown to the outside of the converter. The air that is separated is sent on to the cooler.

NOTE: *The bottom of anything spinning is always the outside.*

SHOP TIP To check for aeration from the converter:
- Run a hose from the cooler line to the transmission dipstick tube.
- Start the engine.
- If there is an air leak, for a few seconds air will be visible in the fluid.

Replace the Pan

Gasket sealer is not normally used on pan gaskets except to hold a gasket in place. Rubber cement or gasket positioning sealer can be used if necessary.

NOTES:
- *Usually, the screw holes in the gasket are smaller than the screw diameter. When the screws are installed through the pan and the gasket, they will remain there during installation to position the gasket.*
- *Do not use silicone on the pan. ATF attacks it. It is not uncommon to find silicone plugging a transmission filter or screen.*

Tighten the screws evenly. Do not tighten any of the screws all of the way until all of the screws are started in their threads. Use an inch-pound torque wrench to torque the screws to specifications. Look this up in the service manual. Specifications usually range from 60 to 150 inch-pounds. This is not very much, but it compresses the gasket and allows it to work properly. Overtightening a gasket will take it past its elastic limit and cause it to lose its sealing ability. Gaskets often split in the middle when they are over-tightened.

■ REFILL THE TRANSMISSION

Check the service manual for the capacity of the pan.

SHOP TIP ■ The torque converter contains about 3 to 4 quarts of fluid. If only the pan was drained, be sure that a smaller amount of fluid than transmission capacity is added.

- If in doubt about the fluid capacity, add 3 quarts (the approximate amount contained in the pan). Then, run the engine, step on the brake, put the transmission in each gear, and recheck the fluid level. Putting the transmission in each gear range fills the cavities in each circuit. If you have a helper, you can run the engine with the transmission in low gear while you fill the transmission.
- If the fluid does not touch the bottom of the dipstick, add one quart. Then, start the engine and recheck the fluid level.
- When fluid is on the dipstick, remember that only ½ quart will move the fluid level from *add* to *full*.
- Cold fluid will expand by about 1 pint when heated to operating temperature so leave the fluid level at the add mark until the car can be test driven to heat the fluid.

Do not overfill the transmission. Foaming and siphoning can result. Some older Ford C6 and GM TH400 transmissions had shallow pans and were prone to problems from overfilling. The fluid would siphon out the dipstick tube. Later models of those transmissions had deeper pans to solve that problem.

SHOP TIP Excess fluid can be removed by one of the following methods:
- Loosen and remove one of the transmission cooler lines from the radiator. Run the engine to pump ATF out into a pan. Be sure to use two wrenches (one of them a flare-nut wrench) to prevent damage to the radiator.
- Through the transmission dipstick hole:
  - Use a hand suction gun.
  - Use a solvent siphon gun hooked to shop air.

■ DIAGNOSIS AND REPAIR OF LEAKS

Leaks can happen from the front pump seal, the rear seal to the drive shaft yoke (rear wheel drive cars), the shift lever shaft seal, the pan gasket, or the extension housing gasket. Sometimes, a leak happens on one of the transmission cooler lines or the dipstick tube.

External leaks are obvious. Locating the source of a leak can be as easy as cleaning the outside of the transmission and watching for new fluid. ATF is usually red in color, which helps differentiate it from engine oil. A fresh leak will usually clean the dirt off of the outside of the transmission or leave a washed area that looks like water running through dirt.

If a leak is particularly stubborn, fluorescent dye and a black light can be used. The leaking fluid will glow under the black light. Most leaks (except for the front pump seal) can be repaired without removing the transmission.

LEAKS FROM THE CONVERTER HOUSING

A leak from the converter housing can be due to several causes (Figure 67.3). To check the converter for leaks, remove the access cover on the housing. Some housings do not have an access cover. In this case, disconnect the battery ground cable and remove the starter. An oil leak from the engine crankshaft seal could be the cause. This will usually leave a film on the front side of the flexplate. Engine oil is not red, so it is usually easy to tell the difference.

TRANSMISSION COOLER LINE LEAK

Pressure in the transmission cooler lines is usually about 35 psi. Some technicians attempt to repair leaks in these lines by cutting out the bad section of line and replacing it with a length of fuel hose. Ordinary fuel hose is not recommended for use with transmission fluid. Use transmission oil cooler (TOC) hose (see

SAFETY NOTE Leaking fluid can cause a fire. Although fluid is not flammable at room temperature, it ignites just like gasoline does when heated in excess of about 300°F. Leak from a pressurized line can vaporize more easily. If transmission fluid is sprayed onto a hot exhaust manifold, it could cause a fire.

Chapter 23). The hose can easily blow off, even though there are hose clamps on the line so both ends of the line must be flared.

NOTE: *The ends of the lines must be double-flared (see Chapter 24) so that the flare is not so sharp that it cuts its way through the line. The body and engine move and vibrate on different mounts. In time, the hose could suffer a cut.*

A better repair is to cut out the old section of line and replace it with a **union** (see Figure 24.15) or install flare fittings and install a short piece of reinforced hose that has threaded ends crimped onto it.

SPEEDOMETER DRIVE GEAR LEAK

Many late model vehicles have an electronic speedometer pickup, which has no mechanical connection.

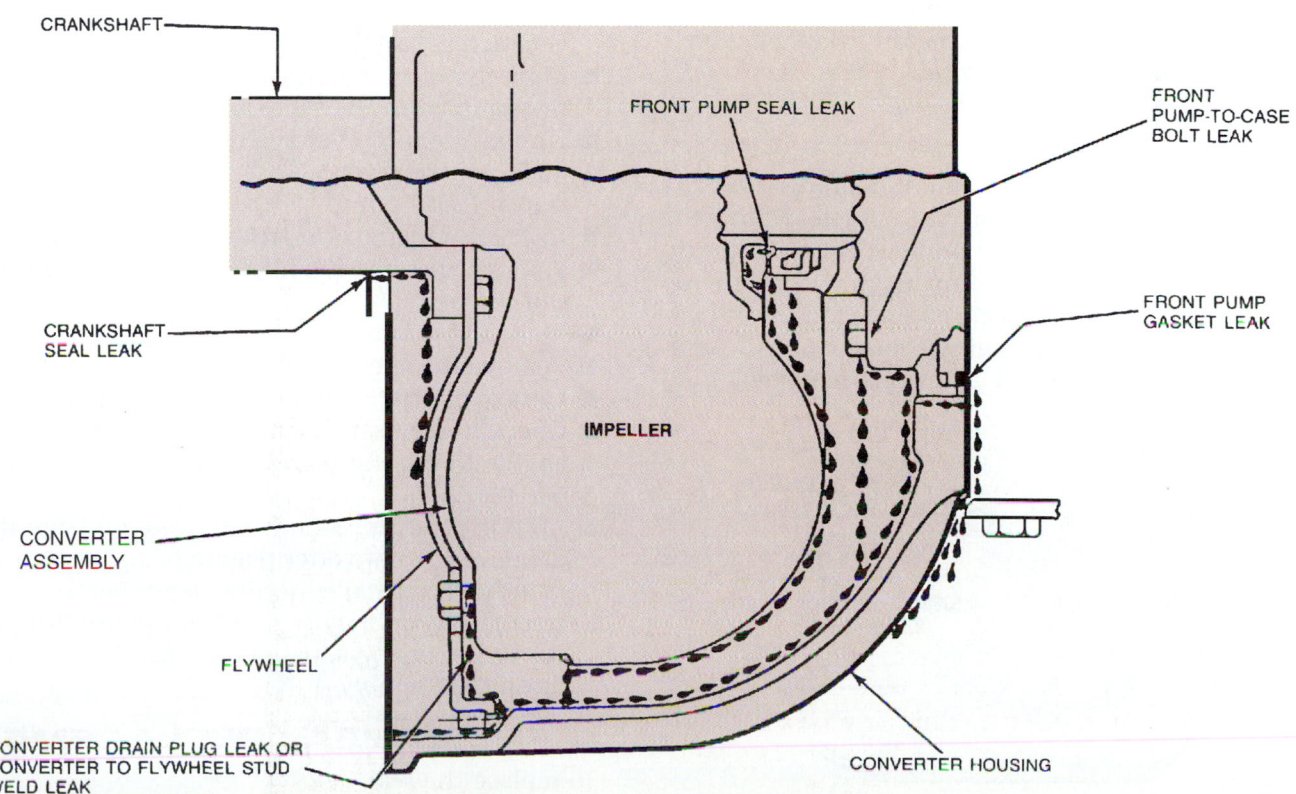

Figure 67.3 Leaks from the converter housing can be due to several causes. *(Courtesy of Ford Motor Company)*

Mechanical speedometers have an O-ring seal that can leak from the extension housing.

■ SHIFT LEVER SEAL REPLACEMENT

The shift lever to the manual valve is sealed with a lip seal or O-ring. It is sometimes necessary to remove the pan and valve body to be able to remove the shift lever shaft. Check the service manual for procedures.

■ PUMP SEAL REPLACEMENT

The transmission has a full round seal that rides on the front of the torque converter where it enters the transmission pump (Figure 67.4). This seal suffers from a good deal of heat because the torque converter operates at a temperature higher than the rest of the transmission. When this seal fails, transmission fluid *pours* out of it. A front pump seal often fails after an engine R&R.

CASE HISTORY *An engine in a van was worn out. The engine was pulled and sent to the machine shop to be rebuilt. While the engine was out of the vehicle, the torque converter was left hanging in the transmission, unsupported on its front end.*

NOTE: *When installed between the engine and the transmission, the front end of the converter is supported in the rear of the engine's crankshaft.*

The old transmission seal, which had lost some of its elasticity, became distorted and began to leak after the engine was reinstalled and started. The transmission had to be removed to replace the seal.

A *front pump seal* replacement requires that the transmission be removed from the vehicle. Follow the manufacturer's instructions. Front wheel drive procedures vary. Be sure to support the engine from the top (see Figure 50.19). Sometimes, engine cradles, suspension parts, or complete engine/transmissions assemblies must be removed. If axles are removed, fluid

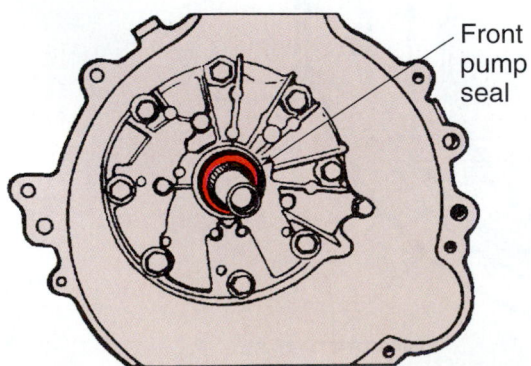

Figure 67.4 A front pump seal inside of the converter housing. *(Courtesy of Ford Motor Company)*

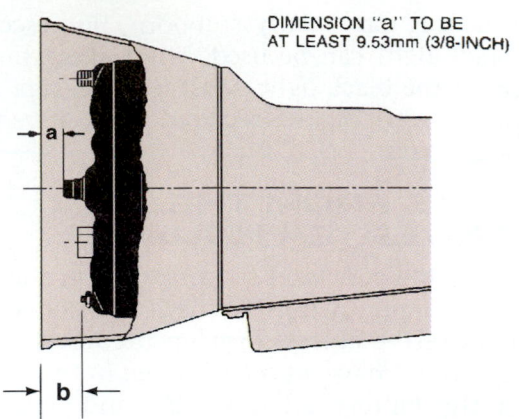

DIMENSION "a" TO BE AT LEAST 9.53mm (3/8-INCH)

Figure 67.5 Measure the depth of the converter. Be sure the converter (a) is installed to this depth before reassembly, and the studs (b) are deep enough to clear the flexplate. *(Courtesy of Ford Motor Company)*

drains from the transaxle. Save the fluid in a clean can for refilling later. If the container is not completely clean or is not kept covered, the fluid must be replaced to prevent contamination. When reinstalling the transmission, try to put the transmission back in the same position it was in before. Be sure that the CV joint boots are not torn.

Rear wheel drive procedures are more straight forward.

These general instructions apply to all transmissions.
- Remove the inspection cover or starter motor to gain access to the engine side of the flexplate.
- Mark the converter and flexplate so they can be reassembled in the same position.
- Unbolt the torque converter from the flexplate and push the converter all the way into the transmission pump.
- Install a transmission jack under the transmission.
- Remove any crossmembers or braces supporting the engine.
- Mark and remove any linkage.
- Remove the speedometer cable or wire.
- Unbolt the transmission from the engine.
- Check to see that any dowel pins that align the engine to the transmission are in place (see Figure 63.10).
- After the transmission is removed, measure the depth of the converter (Figure 67.5). Be sure the converter is installed to this depth before reassembly. Otherwise, the front pump, torque converter, or transmission case can be damaged during reassembly.

■ FRONT SEAL REPLACEMENT

To replace the seal:
- Place a drain pan under the torque converter and remove it from the transmission. (Figure 67.6). If

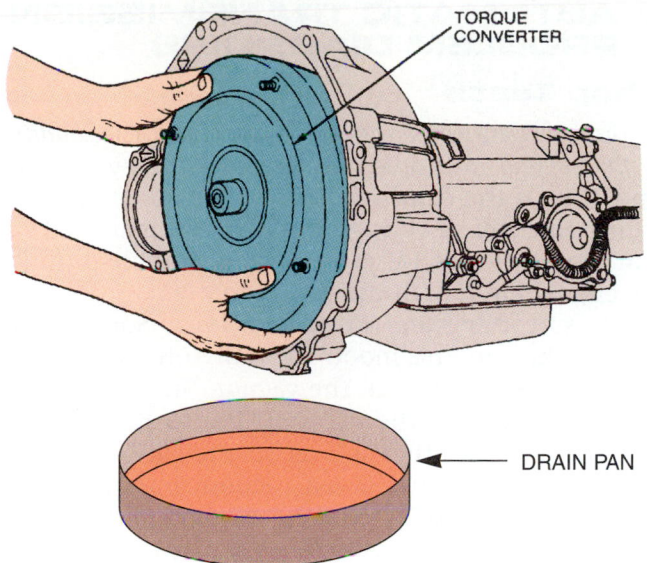

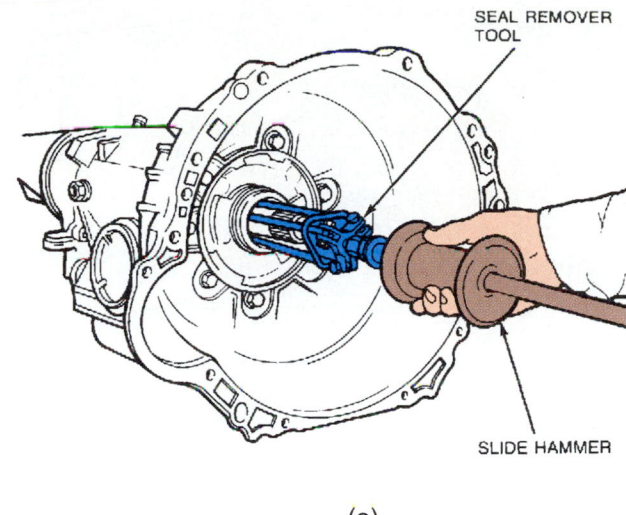

(a)

Figure 67.6 Remove the torque converter from the transmission. *(Courtesy of Ford Motor Company)*

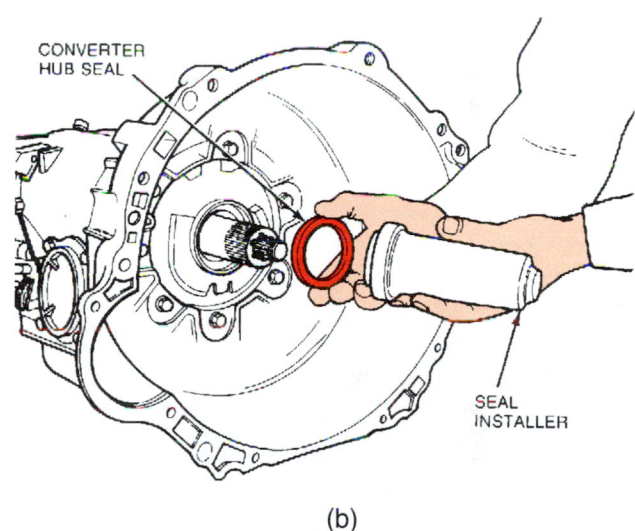

(b)

Figure 67.7 (a) Remove the old seal from the pump housing. (b) Install the seal on the installer and drive the seal into the pump. *(Courtesy of Ford Motor Company)*

you quickly rotate it so the transmission end faces up, very little fluid will leak out.

- Remove the old seal with a chisel, hook puller, or slide hammer seal puller (Figure 67.7a).
- Install the seal on the installer and drive the seal into the pump (Figure 67.7b).

PUMP BUSHING REPLACEMENT

There is a bushing in the transmission pump that supports the snout of the torque converter. When a pump seal is replaced, the pump is often removed and the bushing replaced too.

Check the service manual for the procedure on bushing replacement. This often requires removal and disassembly of the pump. Removing the pump sometimes requires the use of a pair of slide hammers threaded into two of the bolt holes in the pump (Figure 67.8).

Mark the gears or rotors so they can be put back in the same position in relation to each other. The pump gear must be reinstalled with the chamfer down so that the converter can be installed into it (Figure 67.9). Some pumps require the use of a special clamp that fits around the outside of the pump body to align the halves.

REINSTALLING THE TRANSMISSION

When reinstalling the transmission, be sure the converter is all the way into the pump. Install the transmission and bolt it to the engine. Be sure that the engine and transmission are touching each other com-

pletely before installing any bolts. If the converter is installed properly in the pump, there will be enough room to rotate it to align it with the bolt holes in the flexplate. Then, slide it forward and bolt the converter to the flexplate.

NOTE: When installing the torque converter to the flexplate, be sure to use the correct bolts. Bolts that are too long can be forced through the converter housing or they can leave the flexplate loose. This can result in elongated holes and a cracked flexplate due to excessive vibration. The result is a clunk when shifting from neutral into gear. Bolts too long can also deform the converter housing and lock up the converter clutch.

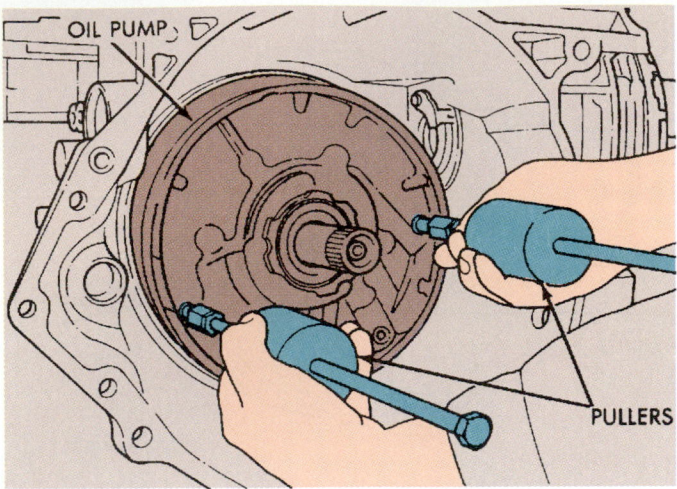

Figure 67.8 Slide hammer pullers for removing the pump from the transmission body. *(Courtesy of Chrysler Corporation)*

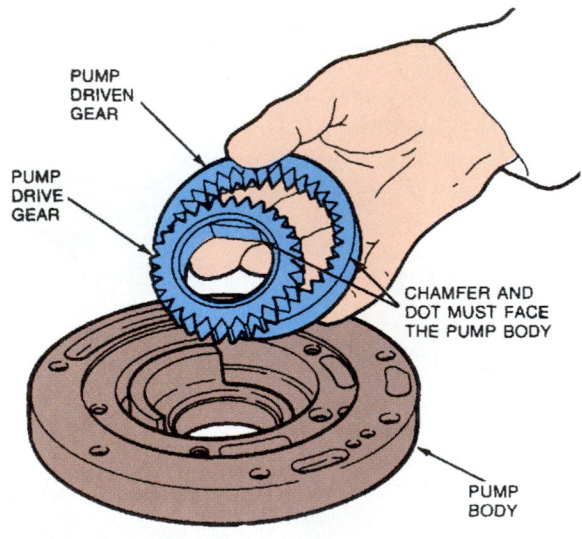

Figure 67.9 Install the pump gear with the chamfer down so the torque converter can fit easily into it. *(Courtesy of Ford Motor Company)*

◼ REAR OIL SEAL AND BUSHING REPLACEMENT

Replacement of the extension housing seal and bushing on rear wheel drive cars requires removal of the driveshaft (see Chapter 63). The old seal can be removed with a chisel or a special seal puller. Special drivers are available for installing the new seal (see Figure 65.24).

Using a special puller, the old bushing can be removed without removing the extension housing (see Figure 65.23). The bushing sometimes has an oil return hole that must be aligned with a hole in the extension housing during installation. The puller set includes an installation tool too.

◼ AUTOMATIC TRANSMISSION PROBLEM DIAGNOSIS

Shop Tests

Before condemning a transmission, check the *condition of the engine*. It should accelerate normally and idle smoothly. If the cam timing has skipped, engine vacuum at idle will be below normal. Intake manifold vacuum that is below 15" can indicate a problem (unless the engine has a high-overlap cam).

Be sure there are no *vacuum leaks*. Sometimes a vacuum line to the modulator falls off, resulting in harsh, late shifts. Check the vacuum modulator, if so equipped. Remove the vacuum line to it. If there is fluid in the line, that is a good indication that the diaphragm is leaking. Use a vacuum pump to put a vacuum of at least 24" on the diaphragm. It should hold vacuum. If the needle drops slowly or quickly the diaphragm is leaking. Be sure that the hose connection is good.

Check the operation of the *kickdown cable* or *linkage*. If the linkage has fallen off, forced kickdown will not work properly. If the linkage binds or is forcing the kickdown valve to stay applied, the transmission will stay in lower gears.

The *quadrant indicator* might not be properly aligned. As you move the shifter, count the number of clicks from park to drive. Perform the test drive with the shifter in the correct position and make any needed adjustment to the indicator later.

Perform a visual check of *electrical connections*. Look for corrosion, bad insulation, and disconnections.

Test Drive

Before taking a test drive, talk to the customer and get a clear idea of the complaint. Road testing is usually done on a warm vehicle. Ask the customer how the transmission operates when it is cold.

Be sure there is at least some fluid on the dipstick before driving the car. Operation without fluid can damage a transmission.

Some technicians suggest that the car be driven before adjusting the fluid level. This way, if adding fluid solves the problem, that can be noted during a retest.

Use a *road test checklist* to help diagnose a problem. Be sure to turn off the radio during the test drive so unusual noises can be heard.

Start your test drive by checking *minimum throttle upshifts*. The specified speed varies between manufacturers. Minimum throttle is at a speed just above idle. A typical 1–2 shift should happen no later than about 22 to 24 mph (this would be under heavy load). The rest of the shifts should happen a few miles per hour apart and not be stacked on top of each other. Note shift quality to see that they are not too soft and the rpm does not slip up. Try forced detent downshifts at WOT.

■ SLIPPAGE

Slippage can be caused by a low fluid level, incorrect linkage adjustment, leakage in a clutch pack or servo, a plugged pump inlet screen or filter, or a problem with the valve body.

■ MORNING SICKNESS

As transmissions age, heat takes a toll on the synthetic rubber seals within it. As seals age, they harden. This was especially true with the lower quality seals used on older vehicles. A hard seal will sometimes leak when the transmission is first put in gear on a cold morning. The transmission will hesitate for anywhere from a few seconds to a minute, leaving the vehicle standing still as if in neutral while the engine races. Usually the seals soften sufficiently to quit leaking and the vehicle begins to move. The transmission might operate perfectly fine after this initial problem. A transmission such as this is said to be "sleepy" or have "morning sickness."

A similar condition results all of the time when a transmission is slightly low on fluid. Once the fluid expands, the transmission will work in a normal manner.

■ TRANSMISSION DRAINBACK

Another problem resembling morning sickness could be worn input shaft bushings. The complaint might be that the transmission slips on initial take-off for about 5 to 10 seconds. When the engine is shut off, the torque converter is supposed to stay full and ready for the next time the engine is started. There is a one-way check valve in the line that provides fluid to the converter circuit. Worn input shaft bushings can allow fluid to run out of the converter and overfill the pan (Figure 67.10). The owner might complain of an occasional puddle of ATF under the vehicle. This is because of fluid escaping from the vent hole high on the side of the transmission case.

■ SHIFT FEEL

Another symptom of leaking seals occurs as the transmission shifts into a higher gear. During the shift, the engine rpm will increase in speed briefly until the clutch pack or servo piston finally seals and pressure builds up. This is called a **mushy shift**.

A **harsh shift**, the opposite of a mushy shift, is when the transmission makes a gear change that is too fast. It might shudder due to actually having clutches applied for two planetary operations at once.

From the factory, shift feel is usually soft so that the customer cannot notice it. A properly timed harsh shift is actually better for transmission life.

NOTE: A shift kit is something that is often added to a transmission to make it shift more aggressively. The kit includes springs for the valve body and instructions for drilling enlarged holes in the valve body separator plate to

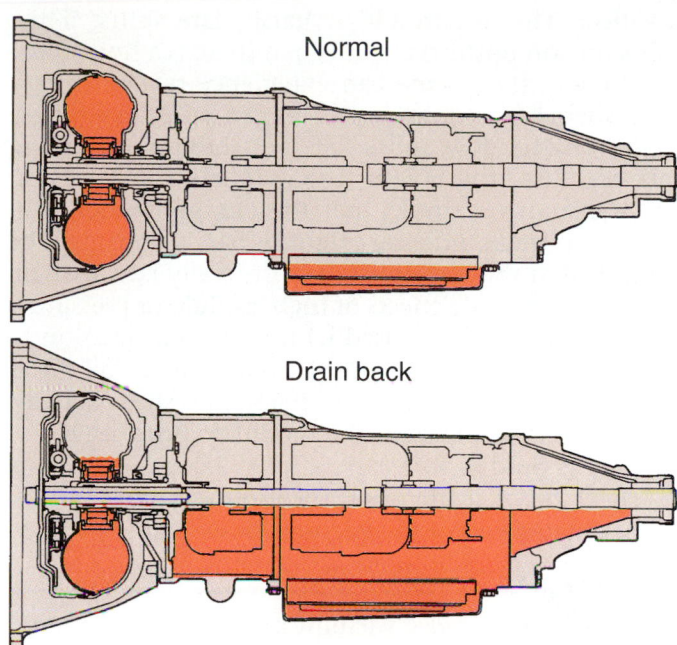

Figure 67.10 The torque converter can empty past worn input shaft bushings. *(Courtesy of Nissan Motors)*

allow fluid to get to a clutch sooner. Aggressive shifts are easier on clutch facing materials, but allow shocks to damage transmission hard parts in response to driver abuse.

■ IMPROPER SHIFT POINTS

Shifts can be either too late, with the engine rpm too high, or too early, with the engine lugging. Transmission shifts at the wrong time could be caused by a pressure problem or by a sticking valve. An engine problem can also cause the transmission to shift later because more throttle is necessary to get the car to move. Examples of pressure-related problems include a leaking or misadjusted vacuum modulator, throttle valve cable, or a sticking governor. Adjustments are covered later in the chapter.

Governor Sticking

If the governor sticks in the "out" position, the valve body is mistakenly receiving information that says the vehicle is travelling at high speed. In response, the shift valve will try to move toward high gear. The 3–2 shift valve will be in the third gear position during a start, so the car will start out in high gear. If the governor sticks in the "in" position, the vehicle would want to stay in a lower gear range.

■ VACUUM MODULATOR DIAGNOSIS AND ADJUSTMENT

If the vacuum modulator has a hole in its diaphragm, the valve body thinks that the throttle is wide open, and the shift valve will move into the lower gear position. This will result in harsh, late shifts, if the transmission upshifts at all. When there is a hole in the diaphragm, the engine can experience several things. First, there is a vacuum leak. This can result in a rough, slightly higher idle and a flat spot (hesitation) during take-off. Also, the hole in the diaphragm means that the transmission thinks that the engine is under full load, even though the car might be barely cruising. The result is that the transmission shifts really late, if at all.

One of the side effects of high modulator pressure is that pressures are increased within the transmission to provide a firm shift under the heavier load of full throttle. Because of this, the transmission also shifts harshly.

■ First, pull the hose on the modulator and see if there is fluid in it.

■ Check to see if there is sufficient vacuum at the modulator end of the vacuum hose (Figure 67.11).

> **SHOP TIP** *Aftermarket* vacuum modulators are often adjustable. There is an allen head fitting accessible through the vacuum inlet to the valve. Because the modulator controls transmission pressure, this adjustment should be accompanied by a transmission pressure check, too. Turn the screw only one revolution at a time. Tightening the screw makes the shift point later. Keep track of the amount of adjustment. Loosening the adjustment can result in low pressures and transmission damage.

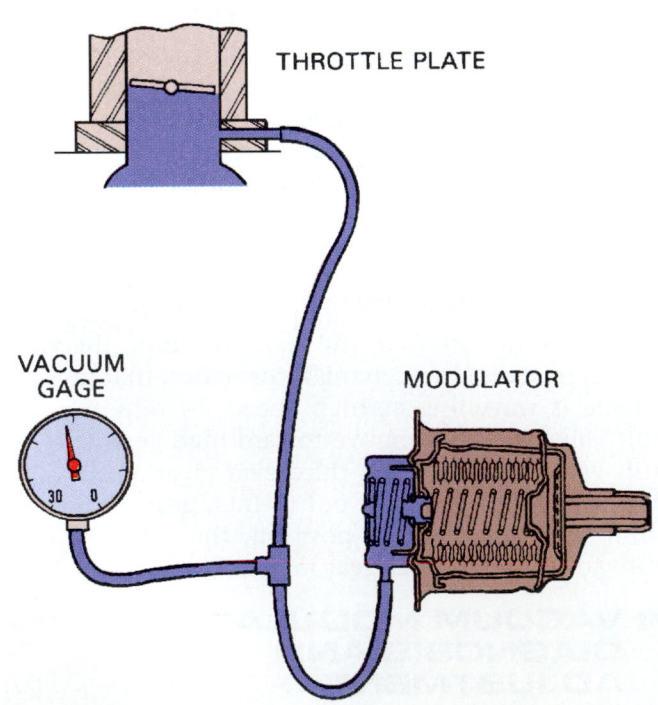

Figure 67.11 Check to see that the engine vacuum signal to the modulator is strong. *[Courtesy of General Motors Corporation, Service Technology Group]*

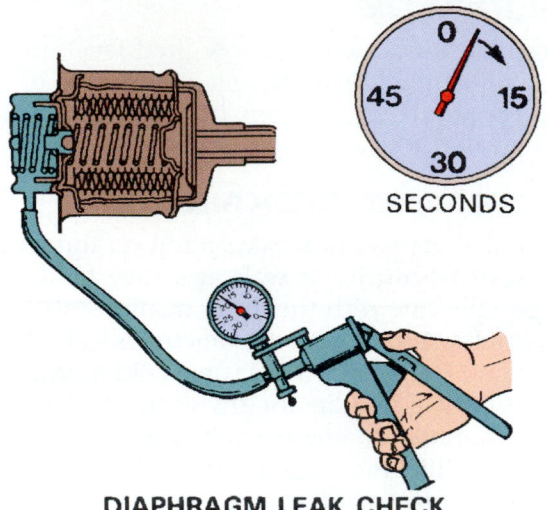

DIAPHRAGM LEAK CHECK

Figure 67.12 The diaphragm should hold steady vacuum for at least 30 seconds. *[Courtesy of General Motors Corporation, Service Technology Group]*

Run the engine and see that the vacuum signal responds quickly to changes in throttle opening.

■ Check the diaphragm for leaks. It should hold vacuum for at least 30 seconds (Figure 67.12).

■ NOISES

Transmission noises can result from a high or low fluid level, gear or bearing wear or damage, or a bad torque converter bushing or one-way clutch. Noises can be transmitted through other parts.

Engine accessories such as the power steering pump, smog pump, timing belt idler bearing, or bad fan belts can also mimic transmission noise. A bad universal joint or axle bearing can also be mistaken for transmission noise.

■ Use a stethoscope to pinpoint the source of a noise.

■ Check to see if the noise occurs in all gear ranges. This could be a torque converter or pump; something that operates in all gears.

■ A defective pump will cause humming and buzzing that gets louder with increased engine rpm. It will occur in any gear position.

■ A grinding noise when the vehicle is moving that increases under load could be a failure in a planetary gear set, a needle bearing, or a bushing.

■ FLEXPLATE

The flexplate is so named because it flexes inward toward the engine. The converter normally slides forward during deceleration. When the engine is under load, the converter is driving its turbine and there is no forward load on the crankshaft. During deceleration the rear wheels drive the converter, reversing the direction of thrust. When the converter moves forward, the pilot hub on the nose of the torque converter (Figure 67.13) moves into the bore in the rear of the crankshaft. The flexplate will allow movement of up to 0.080" to 0.100". Rust, paint, or damage to the crankshaft pilot bore can result in damaging loads to the engine's crankshaft thrust surface.

A cracked flexplate (Figure 67.14) can cause a knocking sound. A badly cracked flexplate will cause a sound similar to a fan hitting the radiator. The noise changes with speed and load.

SHOP TIP
- One test for an intermittent flexplate sound is to run the engine at about 2000 rpm (fast idle). Shut off the ignition and quickly turn it back on. A loud rattle will be heard when the engine is restarted.
- Another test is to shift the transmission from neutral into gear while listening for a clunk.

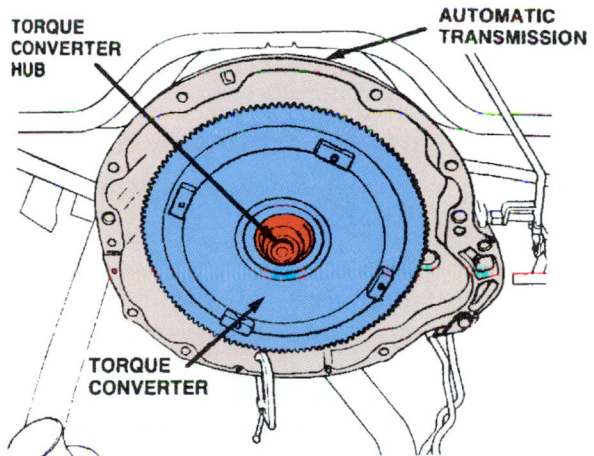

Figure 67.13 The torque converter hub moves in and out of the crankshaft hub pilot hole. *(Courtesy of Chrysler Corporation)*

TORQUE CONVERTER HUB

AUTOMATIC TRANSMISSION

TORQUE CONVERTER

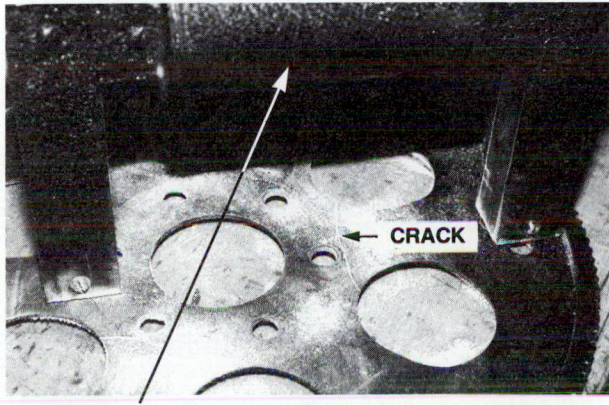

Magnetic crack detector

CRACK

Figure 67.14 A cracked flexplate can cause a knock.

CASE HISTORY

An automotive student installed a factory rebuilt engine in a vehicle. A few weeks after the vehicle was returned to the customer, it returned to the shop. The complaint was of a noise that sounded like a cracked flexplate. After the transmission was removed, it was discovered that the alignment dowels from the engine to the transmission were missing from the factory rebuilt engine. The clunk was in response to the engine shifting back and forth on the engine to transmission attaching screws that had worked themselves loose.

■ TRANSMISSION TESTS

Stall Test

A **stall test** is used to tell if a transmission is slipping or a torque converter is defective.

To perform a stall test:
- Connect a tachometer to the engine's ignition to measure engine rpm.
- Set the emergency brake firmly.
- Apply the *service brakes* (foot brake).
- Start the engine.
- Put the gear selector in low gear.
- Accelerate the engine until speed no longer increases. Note engine rpm at this point.

NOTE: *Be sure that you do not do the stall test for longer than 5 seconds because of the excess heat that the test causes. Also, if rpm climbs beyond specifications, stop the test immediately. Always check the service manual for manufacturer's specifications and special instructions prior to performing the test.*

Stall Test Results.
- If the rpm is lower than specifications, the torque converter stator clutch could be slipping.
- If the rpm is higher than specifications, the transmission is slipping. A band adjustment might correct this (if the transmission has bands), but the transmission will probably require major repair.

Pressure Test

A **pressure test** will tell if the transmission is experiencing internal leakage or requires adjustment to linkages or the vacuum modulator. There are pipe plugs installed in the outside of the transmission case that tap into various circuits in the transmission. The service manual will tell where each pressure tap point is located as well as what the specifications are.

The pressure gauge that is used should have at least a 300 psi capacity. It should have a long hose so it can be stretched through the window into the driver's compartment.

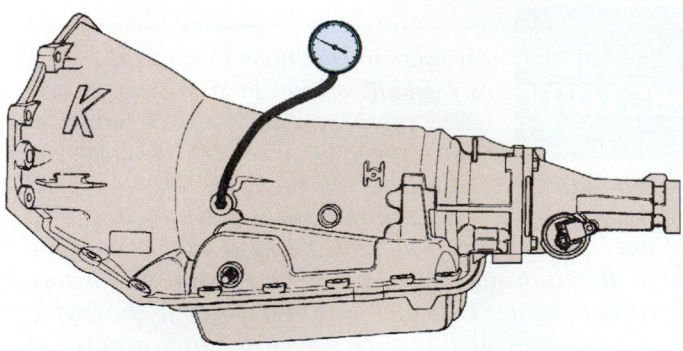

Figure 67.15 Install a pressure gauge in the correct pressure tap. *(Courtesy of General Motors Corporation, Service Technology Group)*

To perform a pressure test:
- Connect a pressure gauge to the correct pressure tap (Figure 67.15).
- Run the engine until it reaches operating temperature.
- Shift through all of the gear ranges and note the pressures that result.
- Test drive the car to note shift points and pressures.

Air Test

An **air test** can be done to test the operation of various clutch packs and servos before disassembling a transmission fully. The pan and valve body are removed and air is directed into the fluid passages that supply the clutch or servo. The service manual tells where each passage leads.

A rubber tipped blowgun is used for performing an air test. When pressure is introduced into the passageway, you will be able to hear a clunk as the clutch applies. If it is leaking, you will hear air leaking out of the leaking seal.

⚠️ **CAUTION** A rubber-tipped blowgun does not have a pressure relief feature like a normal blowgun. Pressure at the tip will equal shop air pressure. Use caution with this high air pressure. If you have an air pressure regulator, set it for no more than 35 psi. If the clutch or servo seals work at 35 psi, it will work at normal transmission pressure.

NOTE: *An air test will not always pinpoint a small leak in a seal. Transmission fluid must first pass through an orifice in the separator plate. A higher volume of air can pass through this orifice, enough to compensate for the leak.*

On some transmissions, the gear train is visible when the valve body is removed. You can feel a clutch apply or watch a band apply (Figure 67.16). When air

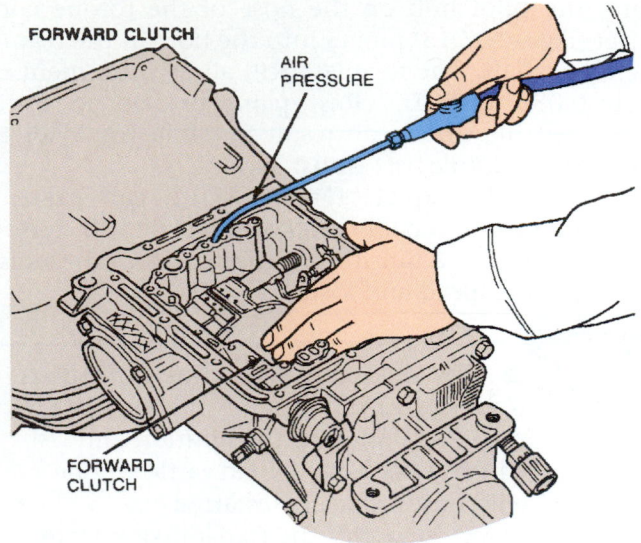

Figure 67.16 When the valve body is removed, you can feel a clutch apply or watch a band apply. *(Courtesy of Ford Motor Company)*

testing a governor, sometimes a vibrating or whistling sound is heard, depending on the style of governor.

■ VALVE BODY REMOVAL

If the valve body is being removed while the transmission is in the car, be sure it is cold.

NOTE: *If valve body is removed when hot, it can warp.*

When reinstalling the valve body, the torque sequence is important to avoid warping the valve body.

When removing a valve body, there are often *check balls* that must be replaced in the same position on reassembly (Figure 67.17). Failure to put these in the correct position will result in faulty transmission operation. With the transmission in the car, reassembly

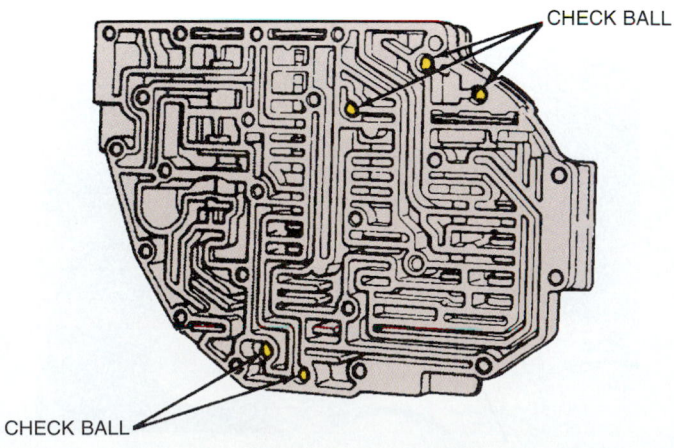

Figure 67.17 Check balls must be replaced in the correct location in the valve body. *(Courtesy of Chrysler Corporation)*

can be difficult. A good trick is to put the check balls in their correct positions on top of the separator plate and put it into place first. Petroleum jelly can also be used to hold the check balls in place.

■ TRANSMISSION ADJUSTMENTS

Adjustments that can be made on a transmission include those to the transmission linkage, the kickdown lever or switch, the neutral safety switch, and band clearance.

■ LINKAGE ADJUSTMENT

Check to see if the shift linkage has any slack or looseness in it. This can cause a transmission manual valve to be in the wrong position (Figure 67.18). Adjustment procedures vary between manufacturers. Check the service manual for the exact procedure. On most shift linkages, there is a locknut on the shift rod. Loosening it allows the adjustment to be shortened or lengthened as needed.

SHOP TIP There is usually no reason for an adjustment to change. Whenever a linkage part is removed, it should be marked or removed in such a manner that it can be easily replaced in its original position.

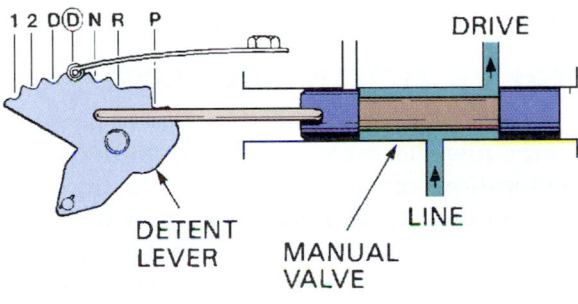

DRIVE

DETENT LEVER MANUAL VALVE LINE

CORRECT

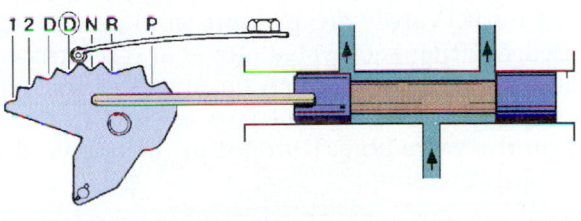

INCORRECT

Figure 67.18 If the shift linkage adjustment is off, fluid can be diverted. *(Courtesy of General Motors Corporation, Service Technology Group)*

The quadrant indicator might not be properly aligned. To check this, move the shifter as you count the number of clicks from park to drive.

■ NEUTRAL SAFETY SWITCH

If the engine does not crank when the shift selector is in the park position, the neutral safety switch might need adjustment. Moving the shift selector while turning the key often results in engagement of the starter motor.

To readjust the neutral safety switch:
■ Put the selector in the park position.
■ Loosen the screws on the neutral safety switch.
■ With the ignition switch held in the start position, move the safety switch until the engine cranks.
■ Tighten the switch.
■ Double-check the adjustment by attempting to start the engine in the park and neutral positions.
■ The switch sometimes controls the operation of the back-up lights also. With the engine off and the key on, put the selector in the reverse position. Check the back-up lights to see that they are on. Then, check in the drive position to see that they are out.

Sometimes, there is a pin hole to align while adjusting the neutral safety switch.

To check a neutral safety switch to see if it is bad:
■ Connect an ohmmeter across the terminals of the switch.
■ With the switch in park and neutral, the reading should be zero ohms. This means the switch is "closed."
■ With the switch in the other gear positions, the reading should be infinite. This means the switch is "open."

■ THROTTLE VALVE ADJUSTMENT

Throttle valves give the transmission valve body information on how far the throttle is open. When there is a throttle valve cable it is usually adjustable. Incorrect adjustment of a throttle valve cable can result in incorrect shift points and transmission damage from pressures that are too low for the condition at hand. Check the repair manual for the correct procedure.

■ BAND CLEARANCE ADJUSTMENT

A brake band is used on some transmissions as a holding device to keep a planetary gear set drum from turning. The friction material on a band is on its inside surface. When this surface wears, the timing of a shift can be affected. Slippage might occur also.

Some transmissions have bands that are adjustable. Others have bands that only operate occasionally when the shift selector is placed in manual second or

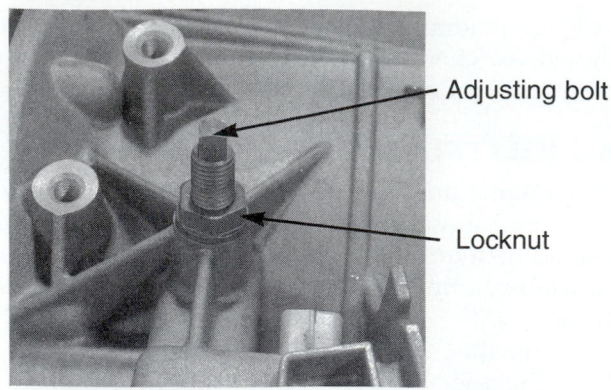

Figure 67.19 A typical band adjusting bolt and locknut.

Figure 67.20 A transmission rebuild kit. *(Printed with permission of Transtar® Industries, Inc.)*

low for engine braking. These bands are not usually adjustable.

To adjust a band:

■ Locate the specification in the service manual.
■ Locate the band adjustment bolt. It has a square head and a large locknut (Figure 67.19). Use an 8-point socket and an inch-pound torque wrench.
■ Tighten the band adjustment bolt (turn clockwise) by the specified amount of torque to seat the friction material against the drum.
■ Loosen the bolt by the specified amount to provide the correct clearance.
■ Tighten the locknut, while holding the adjustment bolt from turning.

The locknut has a sealing surface to keep transmission fluid from leaking around the adjusting screw. These nuts are usually replaced.

Some transmissions have more than one band. Adjustment specifications on these often differ because the bands are often of different designs. One band adjustment specification might call for loosening the adjustment screw by 1¼ turns and the other might call for loosening 2¾ turns.

■ TRANSMISSION REBUILDING

Procedures vary for rebuilding various transmissions. Some things are typical to almost all of them. Parts can be purchased separately or in kits (Figure 67.20). Parts contained in a rebuild kit are called **soft parts**. These include all rubber seals and gaskets, bands, clutch friction discs, modulator, thrust washers, and bushings. A rebuilt converter will be installed too.

When internal damage has occurred, **hard parts** will be necessary. This raises the cost of the rebuild considerably. Hard parts include metal parts such as gears, valves, pump bodies, clutch drums, and other things that would not normally be replaced in a rebuild.

NOTE: *When soft or hard parts have failed in a transmission, be sure to flush the cooler thoroughly with com-*

pressed air from a rubber-tipped blowgun. A new rebuild can be ruined when the particles in the cooler enter the transmission. Be careful of the transmission fluid exiting the cooler when blowing it out.

■ SEAL KIT OVERHAUL

Sometimes, a transmission is overhauled simply to replace hard seals that are causing morning sickness. If all of the transmission soft parts are still in good condition, a very inexpensive seal kit (under $20) might be all that is needed to put the transmission back in good order.

■ VALVE BODY SERVICE

If a valve body is disassembled, be sure to lay out each piece on a towel as it comes from the valve body. The correct locations of valves and springs are important to the proper operation of the transmission. Service manuals give a layout picture of a valve body, but changes are often made in the middle of model years to some of the valves.

When there has been no internal failure, there is a safer means of servicing the valve body. Use carburetor cleaner spray to spray each valve. Carefully move each valve with a plastic screwdriver to avoid accidentally scratching it. Valves are pressure applied and spring released. See that each valve moves and returns easily. If valve movement seems to be sluggish, disassemble that valve and clean its bore. Crocus cloth can be used to clean the valve bore. Unroll it in the bore and twist it to polish the bore.

■ CLUTCH PACK SERVICE

Remove the large snap ring from the clutch pack and remove the clutch discs (Figure 67.21). Some clutch pistons require a spring compressor to disassemble them. The compressor pushes the piston return spring

Figure 67.21 Remove the large snap ring from the clutch pack and remove the clutch discs.

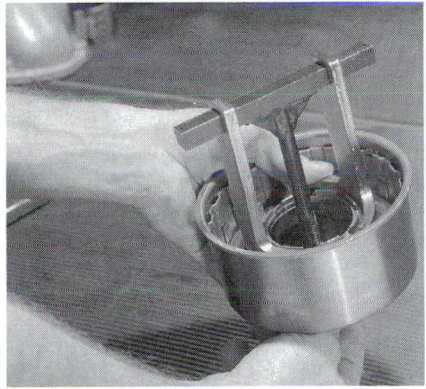

Figure 67.22 Compress the piston return spring and remove the snap ring.

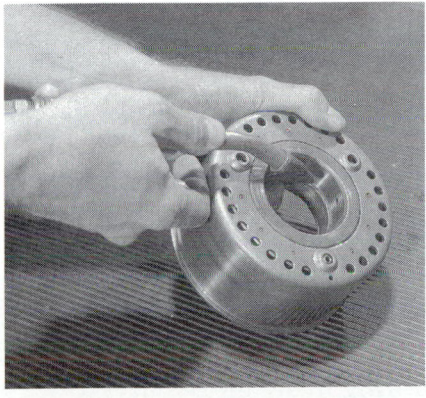

Figure 67.23 The piston can be removed with shop air.

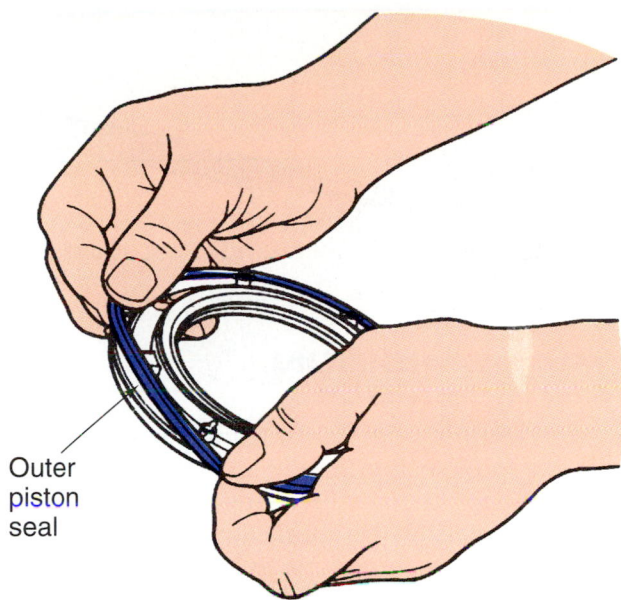

Outer piston seal

Figure 67.24 Replace the piston seals. *(Courtesy of Ford Motor Company)*

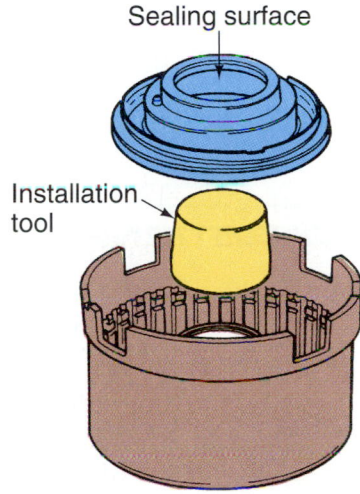

Sealing surface

Installation tool

Figure 67.25 A tool that helps install the inside piston seal over the drum. *(Courtesy of General Motors Corporation, Service Technology Group)*

or springs down while the snap ring is removed (Figure 67.22). The piston can be removed with shop air (Figure 67.23).

Replace the old seals on the piston and inside the clutch pack (Figure 67.24). Lubricate the seals and install the piston in the clutch pack. Sometimes, there is a special tool that is installed over the inside of the clutch drum during piston installation (Figure 67.25). Some piston seals are O-ring seals and some are lip seals. Lip seals are more delicate and require more care during installation so they are not cut. A special "piano wire" type tool has rounded edges to help install the seals (Figure 67.26). Before clutch discs are reinstalled, be sure to soak them in automatic transmission fluid.

■ BUSHING SERVICE

Check all bushings for visible signs of wear. Bearings often have a soft grey metal surface when new. A worn bushing will appear shiny.

Figure 67.26   This tool is used to help install lip piston seals. *(Courtesy of General Motors Corporation, Service Technology Group)*

■ TRANSMISSION REASSEMBLY

When reassembling the transmission parts in the case, be sure not to force anything. If any thrust washers are worn, replacing them will change end play in the transmission. End play is checked with a dial indicator (Figure 67.27). It is important to keep end play to the lower side of specifications when possible. Otherwise, a clutch drum will move fore and aft too much, wearing out sealing rings.

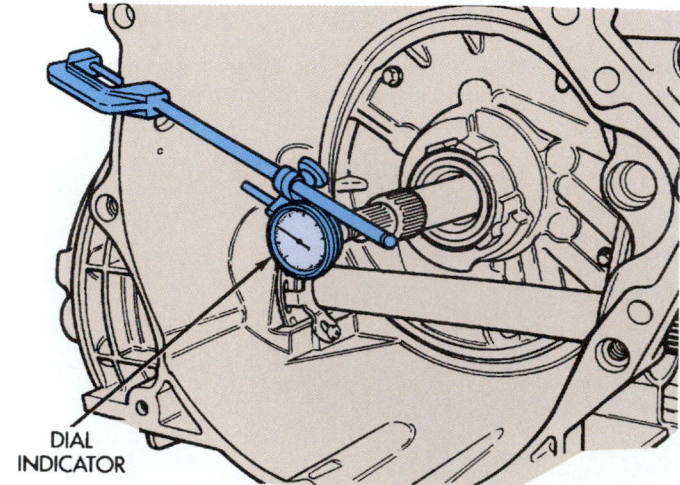

DIAL
INDICATOR

Figure 67.27   End play is checked with a dial indicator. *(Courtesy of Chrysler Corporation)*

■ REVIEW QUESTIONS

1. _____ service is the name for the condition during stop-and-go hot weather driving and trailer towing.

2. What color smoke does ATF cause when burned in the engine?

3. What is indicated if the dipstick will not easily wipe clean?

4. If there is fluid in a vacuum line to the modulator, what could be wrong?

5. What is better for transmission life, a harsh shift or a mushy shift?

6. If the governor sticks in the "out" position, what gear would the transmission want to be in, second or third?

7. If the vacuum modulator had a hole in its diaphragm, what gear would the transmission want to be in, second or third?

8. How much is a flexplate designed to be able to move?

9. How much pressure capacity should a transmission fluid pressure gauge have?

10. What is the name of the test that is done by pressurizing hydraulic passages to see if clutches are working?

■ ASE STYLE REVIEW QUESTIONS

1. Technician A says that all automatic transmission vehicles have their fluid checked with the engine running in park. Technician B says that most torque converters have a drain plug. Who is right?

   a. Technician A   **b.** Technician B
   c. Both A and B   **d.** Neither A nor B

2. Technician A says that automatic transmission filters are designed to last the life of the car. Technician B says when checking ATF level, the fluid on the transmission dipstick should be uncomfortable to touch. Who is right?

   a. Technician A   **b.** Technician B
   c. Both A and B   **d.** Neither A nor B

3. Technician A says that transmission fluid filters have a bypass valve in case they become plugged. Technician B says the fluid level will climb on the dipstick about 1 quart from cold to hot. Who is right?

   a. Technician A   **b.** Technician B

   c. Both A and B   **d.** Neither A nor B

4. Technician A says that if the governor sticks in the "in" position, the transmission will shift late or stay in low gear. Technician B says that if the hose to a vacuum modulator falls off, the transmission will shift late and harsh if it shifts at all. Who is right?

   a. Technician A   **b.** Technician B

   c. Both A and B   **d.** Neither A nor B

5. Technician A says that a stall test is when the brakes are applied and the engine is accelerated to its highest possible rpm. Technician B says an inch-pound torque wrench is used when adjusting bands. Who is right?

   a. Technician A   **b.** Technician B

   c. Both A and B   **d.** Neither A nor B

Driveline Operation

■ INTRODUCTION

Driveline is a term that describes the parts that transfer power from the transmission to the rear wheels. On rear wheel drive vehicles, this includes a long driveshaft between the transmission and differential (see Figure 1.20). Front wheel drive vehicles have a transaxle with two **halfshafts** that deliver power to the front wheels. Front wheel drive operation is covered in Chapter 70, but many of the basic concepts are covered here. Four wheel drive operation is covered later in the chapter.

■ DRIVESHAFT

The open driveshaft, or **Hotchkiss drive** design, has been in use since the 1950s and is the only one found on most vehicles since that time. The older design that it replaced was an enclosed type called the *torque tube*. The driveshaft not only transfers power, but it allows for changes in driveline length as a car goes over bumps.

The driveshaft, or *propeller shaft,* is usually made of steel tubing, although some late model driveshafts are aluminum. Driveshafts are strong and light. They must be balanced and straight. Universal joints on both ends attach the driveshaft to other components. Yokes to accept the universal joints are welded onto the shaft at both ends. A typical driveshaft (Figure 68.1) includes two universal joints, a slip yoke, and sometimes a rear yoke that bolts to a flange on the differential.

■ SLIP YOKE

As the vehicle goes over bumps, the rear springs allow the rear axle assembly to go up and down (Figure 68.2). The distance between the differential and the transmission changes so the driveshaft must be able to move in and out of the transmission. A slip yoke is attached to a universal joint on the front end of the driveshaft. The other end of the slip yoke fits over splines on the output shaft or mainshaft of the transmission (Figure 68.3). It slides in and out of the transmission as the distance between the transmission and differential changes.

The slip yoke is machined smooth on its outside diameter. This provides a sealing surface for the extension housing seal. It also provides a bearing surface for the extension housing bushing to act upon.

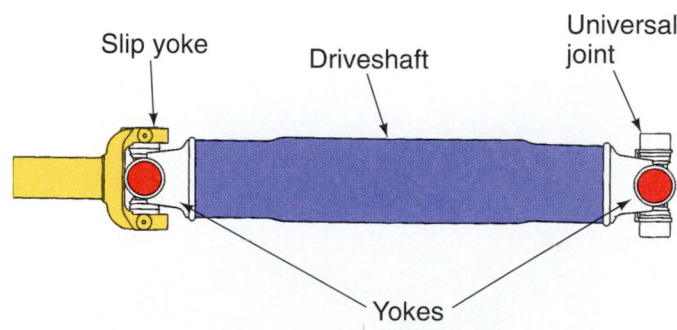

Figure 68.1 A typical driveshaft. *(Courtesy of General Motors Corporation, Service Technology Group)*

Slip yoke Driveshaft Universal joint

Yokes

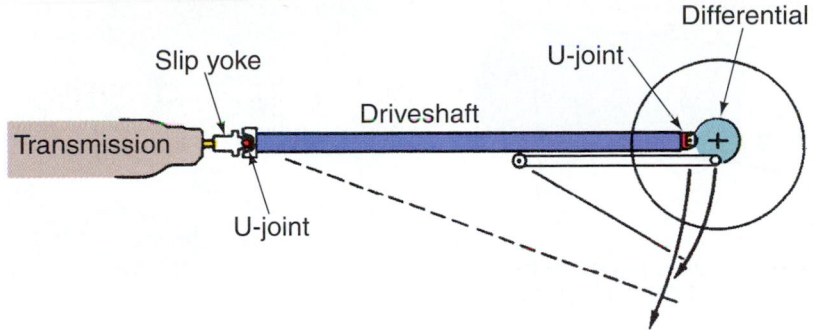

Figure 68.2 A slip yoke allows the driveshaft length to change as the car goes over bumps. *(Courtesy of Moog Automotive, Inc.)*

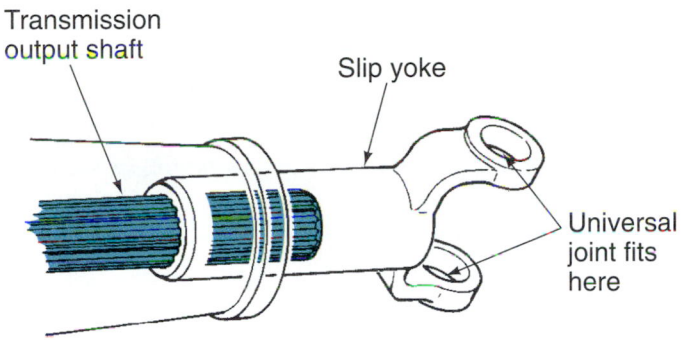

Figure 68.3 The slip yoke fits over splines on the output shaft. *(Courtesy of Moog Automotive, Inc.)*

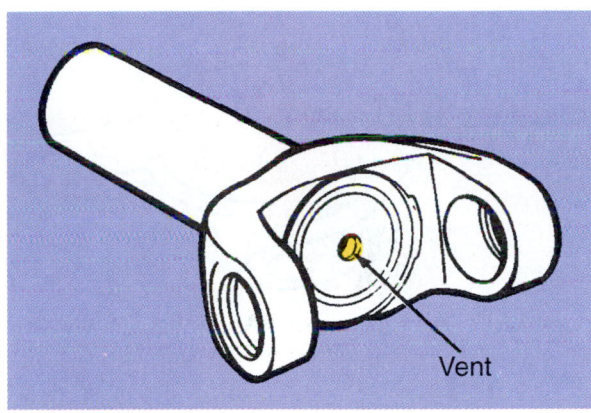

Figure 68.5 A slip yoke with a vent hole. *(Courtesy of Ford Motor Company)*

When a slip yoke is used with an automatic transmission, there is sometimes a seal that goes over the output shaft. It rides against the inside of the slip yoke (Figure 68.4). Its purpose is to keep ATF, which is as thin as SAE 10 engine oil, from leaking out of the slip yoke through its vent hole (Figure 68.5). The splines on the output shaft are lubricated by the ATF. There must be a vent hole to allow the slip yoke to move in

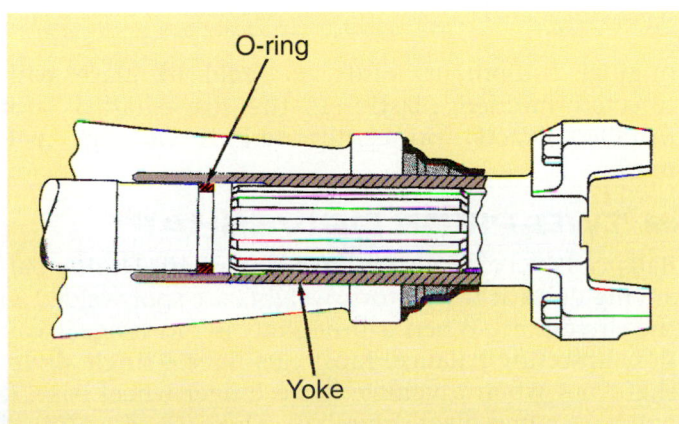

Figure 68.4 Automatic transmission slip yokes sometimes have seals. *(Courtesy of Ford Motor Company)*

and out. Some yokes are greased and have a grease fitting. They are sealed at the end of the yoke.

If the rear yoke is not attached to the driveshaft, there will be a *flange* that is bolted to the front of the *differential pinion shaft,* the splined shaft that comes out of the front of the differential (covered later).

■ UNIVERSAL JOINTS

Universal joints, called U-joints, are located at both ends of the driveshaft. They transmit power at an angle (Figure 68.6). When the axle moves up or down, the universal joint allows the changes in angle at the ends of the driveshaft to take place.

The most popular universal joint design is called a **cross and yoke**, or *Cardan.* It is two Y-shaped yokes connected by a cross, called a *spider* (Figure 68.7). Most U-joints are made of forged, carburized steel. At the ends of the universal joints are four bearing caps with needle bearings (Figure 68.8). The needle bearings ride on *trunnions,* which are bearing areas ground on the ends of the cross or spider. The caps and bearings allow the joint to swivel as its angle changes. The caps and the ends of the trunnions must have a groove for grease.

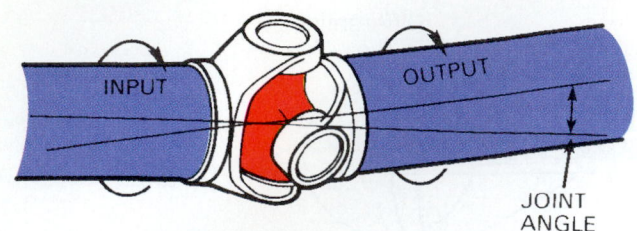

Figure 68.6 Universal joints allow power to be transmitted at an angle. *[Courtesy of Dana Corporation, Spicer Driveshaft Division]*

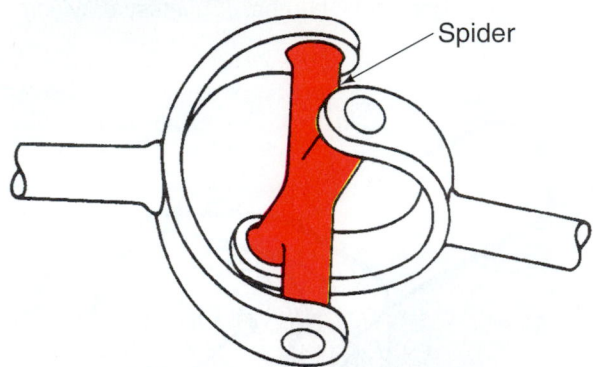

Figure 68.7 A universal joint is two Y-shaped yokes connected by a cross called a spider. *[Courtesy of General Motors Corporation, Service Technology Group]*

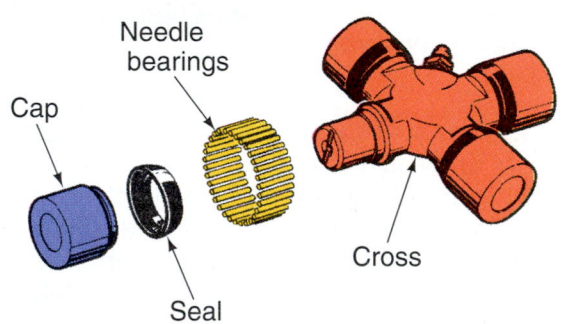

Figure 68.8 Parts of a universal joint. *[Courtesy of Moog Automotive, Inc.]*

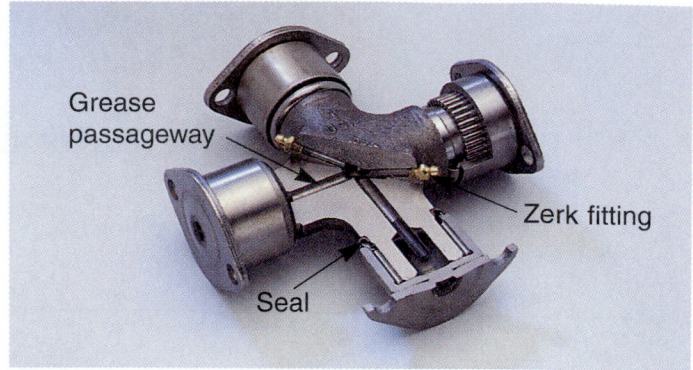

Figure 68.9 Replacement joints usually have a zerk fitting for lubrication during service. *[Courtesy of Moog Automotive, Inc.]*

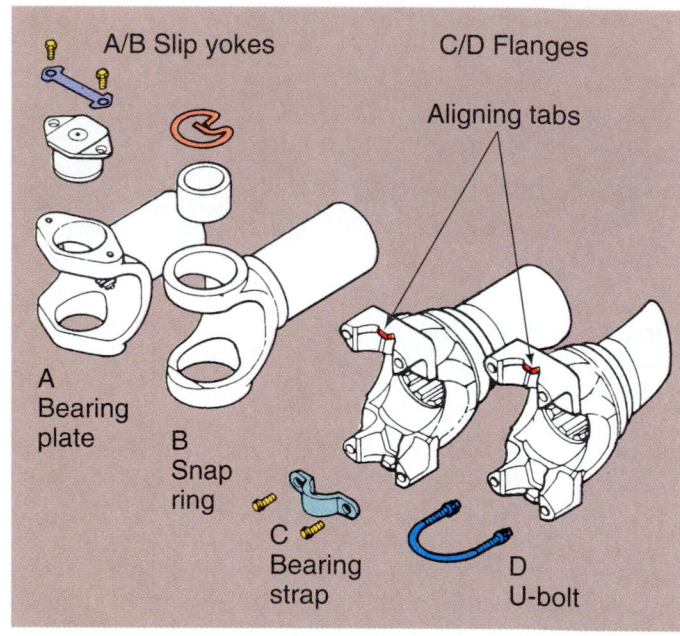

Figure 68.10 Several methods of holding universal joints to yokes. *[Courtesy of Universal Joint Division/Dana Corporation]*

Grease seals fit onto the ends of the bearing caps to keep dirt out and lubricant in. Replacement joints usually have a zerk fitting for lubrication during service (Figure 68.9). Original equipment joints do not have this feature.

There are several methods of holding the U-joints to the yokes (Figure 68.10). Universal joints are usually pressed fit into the yokes on the driveshaft. Snap rings fit into grooves in the yoke to keep the joint centered and retained. The most popular mounting found on cars is one where U-bolts hold the U-joint in place between tabs in the rear flange. On the rear flange, there must be a feature that aligns the joint on center. Alignment is done by either tabs on the outsides or snap rings on the ends. Instead of a snap ring, some

original equipment joints are held in place with injected molded plastic. If the driveshaft is not installed exactly on center, serious vibration will result.

■ TWO-PIECE DRIVESHAFT

Balance is very important on a driveshaft. During balancing done at the factory, weights are spot-welded to the driveshaft. When a driveshaft is too long, it can flex, upsetting balance. Most cars have a single driveshaft, but when a vehicle has a longer wheel base, it will have a two-piece driveshaft. These are often found on pickup trucks.

On a two-piece driveshaft, a *center support bearing* is used to hold the center of the shaft where the two shafts attach to each other. It bolts to the frame, crossmember, or underside of the vehicle. The sealed bearing is supported in a rubber mount that dampens noise and vibration (see Figure 14.27).

■ DRIVESHAFT ANGLE

Cardan universal joints have a problem that limits their use. When they are operated at an angle, the

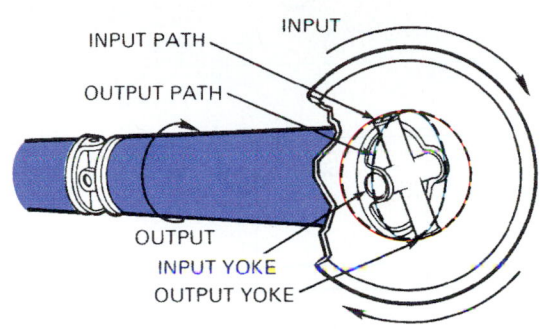

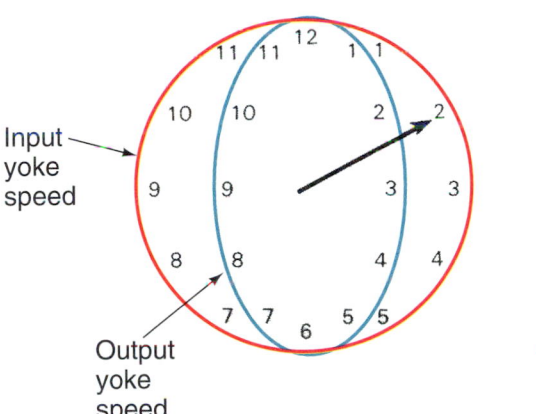

Figure 68.11 The elliptical path that the input and output yokes take can be compared to the face on a clock. *(Courtesy of Perfect Circle Div., Dana Corporation)*

speed of the driven shaft varies as it revolves. The driven shaft turns at the same average rpm as the driving shaft, but its speed increases and decreases four times during every revolution (see Figure 68.12).

The driving yoke rotates at a constant speed, but the driven yoke speeds up and slows down every time it makes a turn. The elliptical path that the input and output yokes take can be compared to the face on a clock (Figure 68.11). At 12 and 6 o'clock the speeds of the input and output yokes are the same. In between, the output yoke turns at a speed that is not constant. The arrow in the sketch shows that the output yoke has travelled to approximately the 2.2 o'clock position compared to the 2 o'clock position of the input yoke.

If the angle is increased, the change in *velocity* during each revolution increases. This can cause the driveshaft to whip like a jump rope. Angles of greater than 3° are not used with single universal joints because they cause excessive vibration. Figure 68.12 compares the speed change of the universal joints with the shaft at both 10° and 30° angles.

At 10° of driveshaft angle, there is a change in velocity of 3%. According to Chevrolet, if the driveshaft angle were set at 30°, at a speed of 1000 rpm, the speed of the driven yoke would change from 856 rpm to 1155 rpm in one quarter of a revolution. Then it would change back. Imagine the vibration that would result from this.

With a one-piece driveshaft, arranging the universal joints so that they can cancel each other's angle out solves the problem (Figure 68.13). The angle is canceled out by the universal joint at the opposite end of the driveshaft. Both ends of the driveshaft must be in **driveshaft phase** (Figure 68.14) for the canceling action to take place. This means that the trunnions of the front and rear U-joints are in the same plane (parallel). On a two-piece driveshaft, the front and rear halves can be assembled out of phase, resulting in vibration.

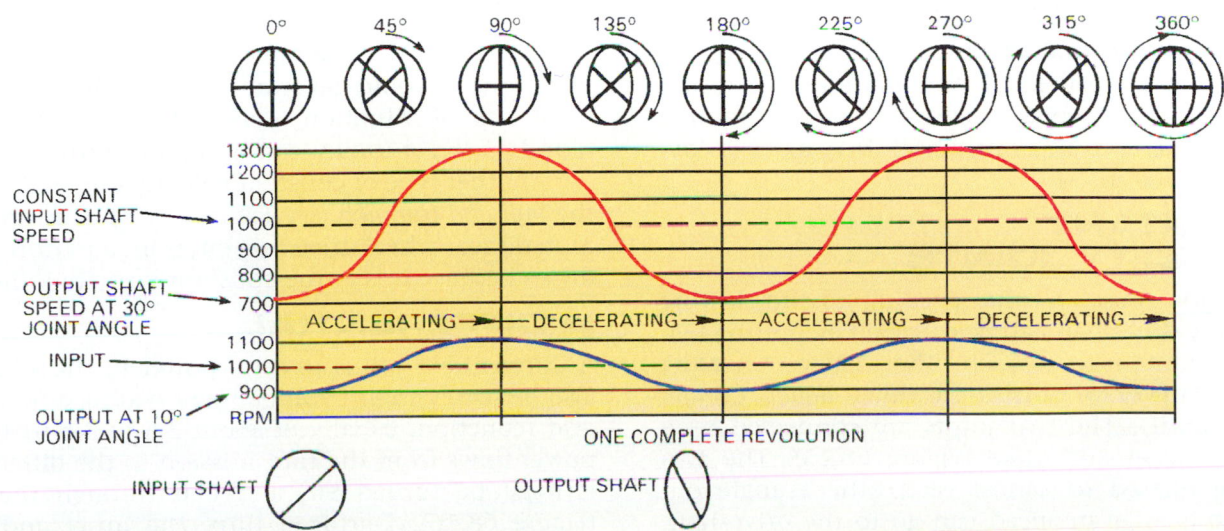

Figure 68.12 A comparison of the speed change of universal joints with the shaft at both 10° and 30° angles. *(Courtesy of Perfect Circle Div., Dana Corporation)*

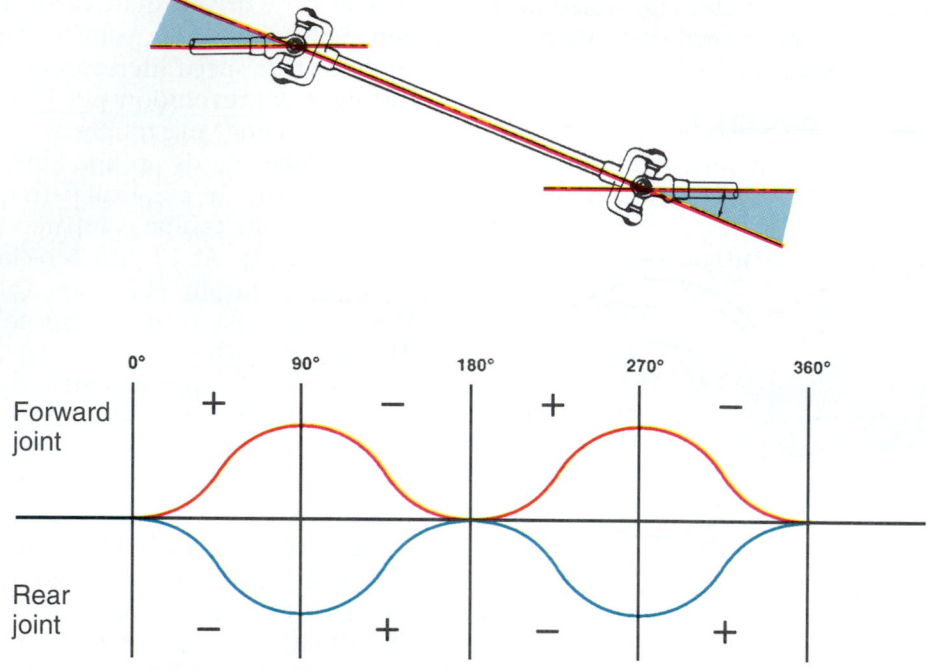

Figure 68.13 Universal joints can be positioned to cancel each other's angle. *(Courtesy of General Motors Corporation, Service Technology Group and Dana Corporation, Spicer Driveshaft Div.)*

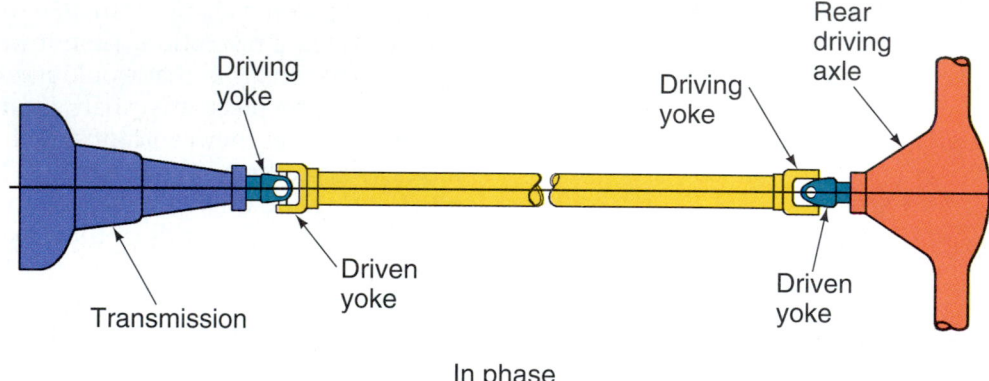

Figure 68.14 Both universal joints are in phase. *(Courtesy of Ford Motor Company)*

It is important that the angles at each end of the driveshaft be almost equal or vibration will result. When an angle is steeper, a constant velocity joint is used. The type used in front wheel drive axles is covered in Chapter 70.

CONSTANT VELOCITY JOINTS

The vibration caused by the speeding up and slowing down of the driveshaft can be canceled by putting two Cardan U-joints next to each other. These **constant velocity universal joints** are called *double Cardan* universal joints. The two joints are connected by a centering socket and yoke (Figure 68.15). The two joints are phased to cancel each other's angle out before the change in speed can go to the driveshaft. This is not a true constant velocity joint, but the speed change never leaves the joint. Its output is at a con-

stant speed so the driveshaft does not get any of the vibrations it would get from a single joint. Because of their lack of vibration, constant velocity joints are found on larger luxury cars and pickup trucks.

Another type of constant velocity universal joint is the *ball and trunnion* (see Figure 70.25). It is not common on rear wheel drive cars but is used in front wheel drives because it is a true constant velocity joint.

DIFFERENTIAL

A differential's job is to send power to the wheels. It also increases engine torque by providing a final drive gear reduction, usually of about 2.5 to 3.5 to 1. After power flows from the transmission to the differential, it must be turned 90° and exit through the axles (Figure 68.16). During a turn, the inner and outer wheels rotate at different speeds (Figure 68.17). The differential must also accommodate this.

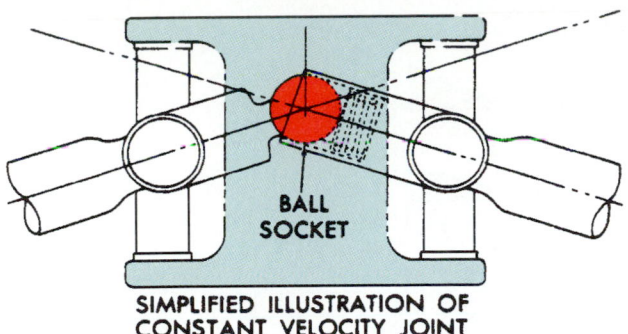

Figure 68.15 A double Cardan U-joint has two single cardan U-joints connected by a centering socket and yoke. *(Courtesy of General Motors Corporation, Service Technology Group)*

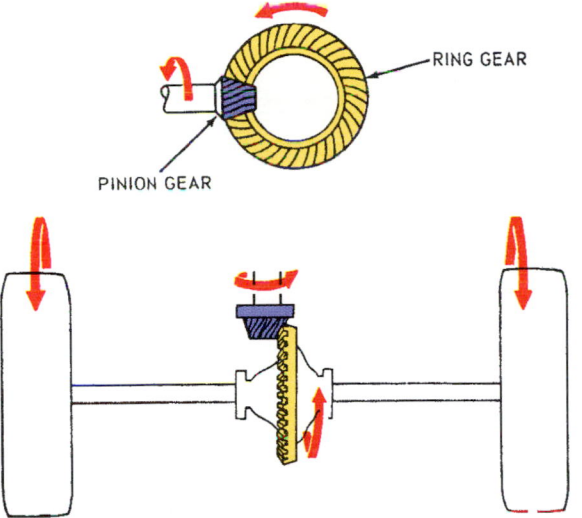

Figure 68.16 Power is turned 90° and exits through the axles. *(Reproduced by permission of Deere & Company, ©1992. Deere & Company. All rights reserved.)*

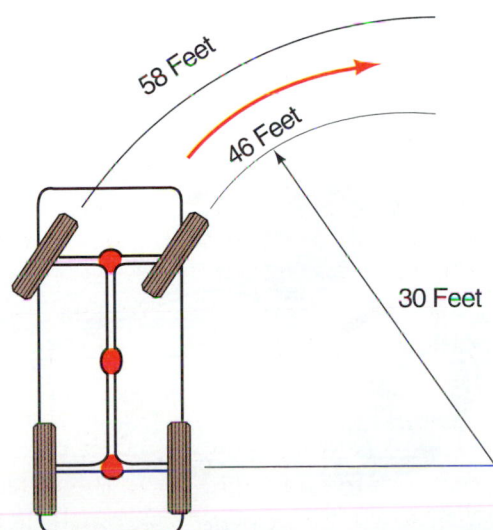

Figure 68.17 During a turn, the inner wheel does not travel as far as the outer wheel.

NOTE: *The differential is called a* **third member**. *The engine is the first member; the transmission is the second member.*

■ DIFFERENTIAL CONSTRUCTION

Parts of the differential include the differential pinion gear, the ring gear, the case, two drive axles, and rear bearings, all housed within the axle housing (Figure 68.18). The pinion gear is splined to a *flange*, or *yoke*, that attaches to the driveshaft. When the flange is flat (Figure 68.19) it is called a *companion flange*. Compare the two flanges in Figures 68.18 and 68.19.

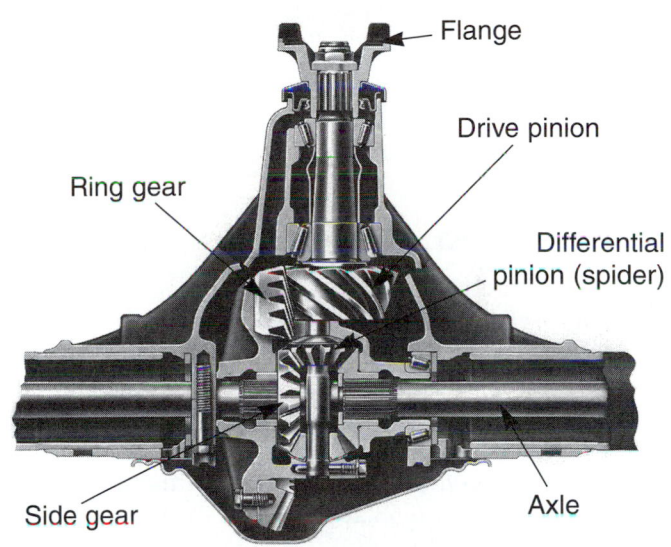

Figure 68.18 Parts of a differential assembly. *(Courtesy of General Motors Corporation, Service Technology Group)*

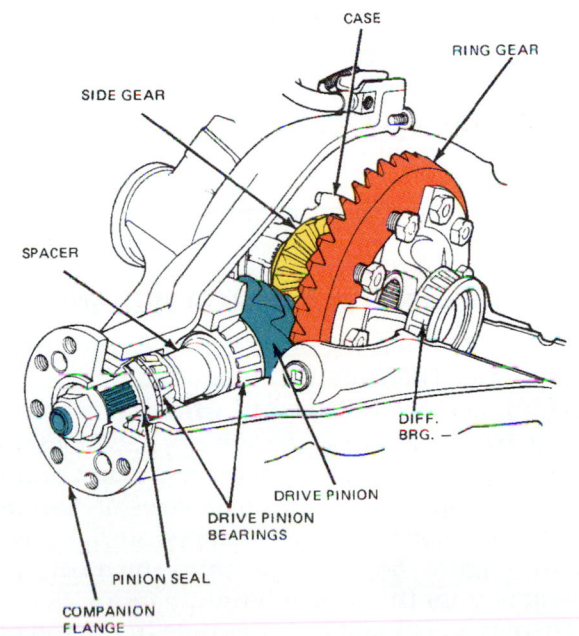

Figure 68.19 The flat flange on the front of this differential is called a companion flange. *(Courtesy of Ford Motor Company)*

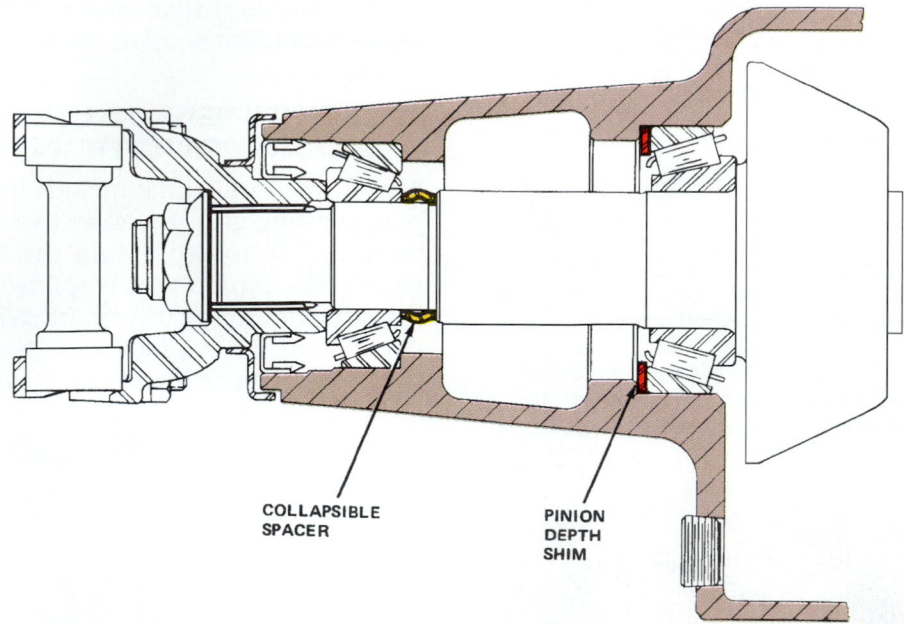

Figure 68.20 A collapsible spacer separates the two pinion bearings. *(Courtesy of Chrysler Corporation)*

The *pinion gear* drives the *ring gear*. The ring has more teeth than the pinion. The *differential case* that the ring gear is bolted to is mounted between two tapered roller bearings. The tapers of these bearings face each other. There is either a *crush sleeve*, sometimes called a *collapsible spacer*, or shims between the two bearings (Figure 68.20). The crush sleeve has three functions:

■ It keeps the bearings separated from their bearing races.

■ It maintains the preload (tension) on the bearings. The bearing must be held close against the bearing races to prevent the pinion from moving back and forth on the ring gear. This would cause damage, looseness, or noise.

■ It keeps the front bearing race from spinning with the bearing.

A shim controls how deep the pinion is positioned into the ring gear.

■ DIFFERENTIAL HOUSING

The differential housing or carrier holds the drive pinion and case. There are two types. One has a removable third member, called a **pumpkin**. The housing is called a **banjo housing** because it resembles a two-sided banjo when the third member is removed (Figure 68.21). The other is called an *integral* or **Salisbury axle**. It has a cover of either stamped steel or cast aluminum (Figure 68.22). The third member is not removable from this axle as a unit.

There is a vent and breather tube in the top of the axle housing. As lubricant warms up during use it expands. The breather makes room for this. The tube usually runs up into a fenderwell. It prevents water

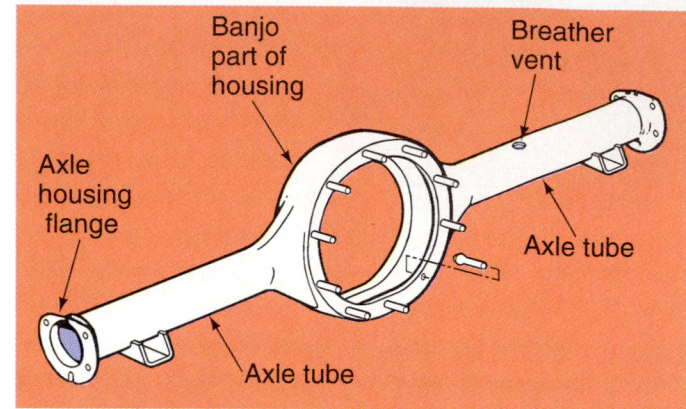

Figure 68.21 A banjo housing. *(Courtesy of Ford Motor Company)*

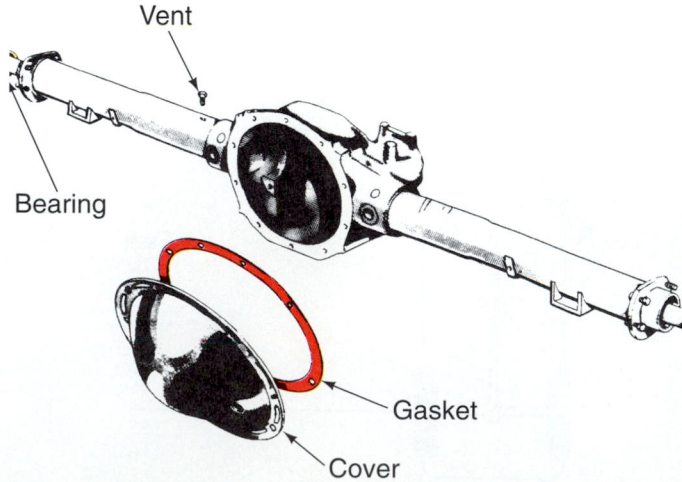

Figure 68.22 A Salisbury or integral axle has a cover on the back. *(Courtesy of Chrysler Corporation)*

from entering the axle when the car is driven through puddles or deep water. The vent often has a one-way check valve in it that allows only pressurized air to escape from the valve. Air can leak back in past the wheel seals as the lubricant cools.

■ DIFFERENTIAL OPERATION

During a turn, the outside wheel must be able to roll farther than the inside one. On the end of each axle is a set of splines. The *side gears* have splines that mesh with the axle splines (Figure 68.23). *Differential pinions* mesh with the side gears (Figure 68.24).

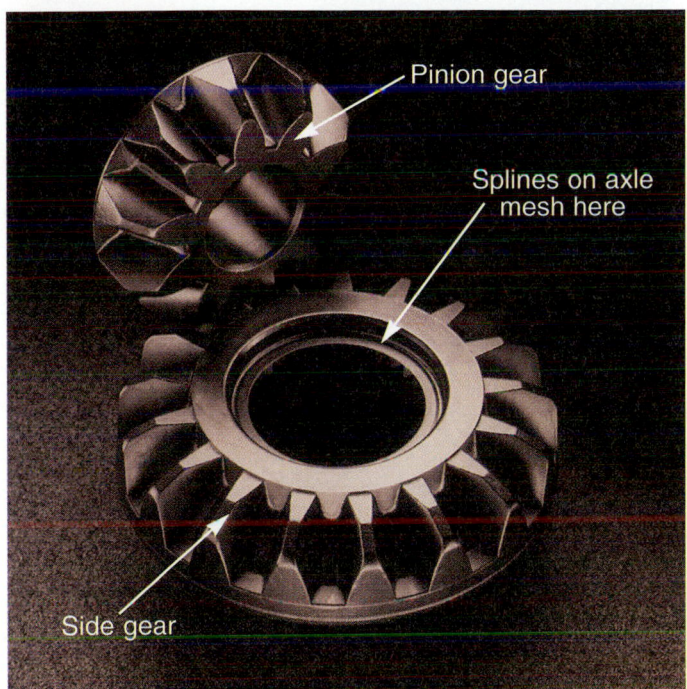

Figure 68.23 The side gears have splines that mesh with the axle splines. *(Courtesy of Eaton Corp.)*

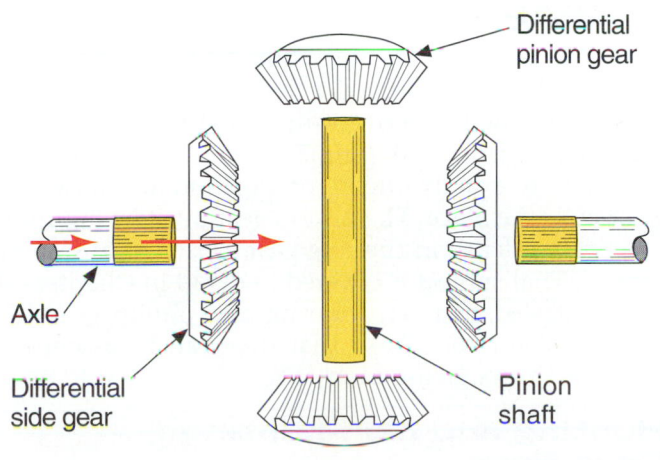

Spider gears

Figure 68.24 Differential pinions and side gears mesh with each other.

NOTE: *These are called differential pinions, not to be confused with the drive pinion, which meshes with the ring gear.*

The differential pinions are mounted on a *pinion shaft*. It must be removed to get the axles out of some types of cars. This procedure is covered in Chapter 69.

The side gears and differential pinions are called **spider gears**. The differential case has the ring gear mounted on it. It has the spur bevel spider gears inside of it. There are always two side gears, but sometimes there are more than two differential pinion gears that mesh between the side gears.

Power Flow

When the car is traveling straight ahead, the pinion gears do not rotate against each other. They turn with the case. The pinion shaft is mounted in the differential case and rotates with it (Figure 68.25). During a corner, when one of the wheels turns faster than the other, the side gears (splined to the axles) rotate against the differential pinions (Figure 68.26).

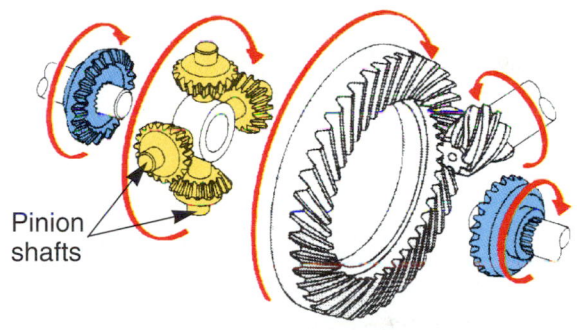

Pinion shafts

Figure 68.25 When traveling straight ahead, the pinion shaft is mounted in the differential case and rotates with it. *(Reproduced by permission of Deere & Company, ©1992. Deere & Company. All rights reserved.)*

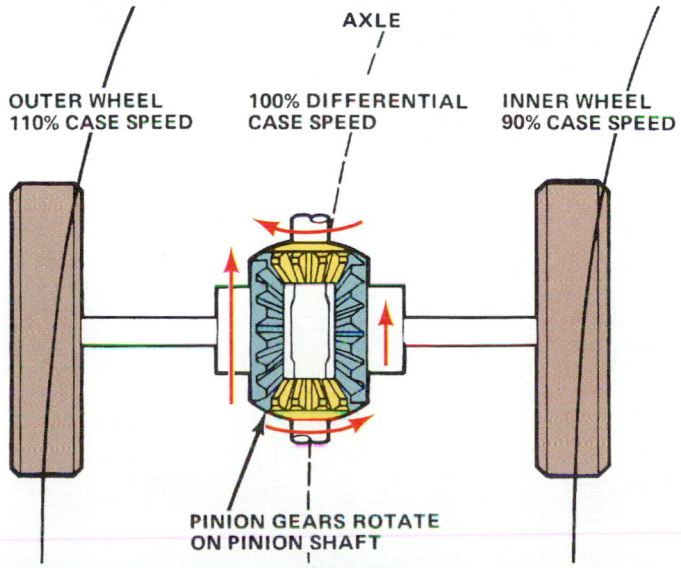

Figure 68.26 When one of the wheels turns faster than the other, the side gears rotate against the differential pinions. *(Courtesy of Chrysler Corporation)*

■ DIFFERENTIAL GEARS

Differential gears have progressed from plain bevel gears to spiral bevel, and finally to hypoid gears (Figure 68.27). In a hypoid gear set, the pinion gear is lower than the centerline of the ring gear. This is so that the transmission hump can be lower. The teeth of a hypoid gear are curved in a spiral shape. Sometimes, a drive pinion gear has a pilot bearing that supports the front of it (Figure 68.28).

NOTE: When gears are machined, they are made by a process called hobbing, a rough machine process that leaves heavy machine marks on the surface of the gear tooth (Figure 68.29a). Next, the gear is finish-machined by a shaving process. The root of the gear tooth remains as hobbed without the finish-machining (Figure 68.29b).

Because hypoid gears are curved, each tooth has a concave and a convex side. The convex side is the drive side and the concave side is the coast side (Figure 68.30).

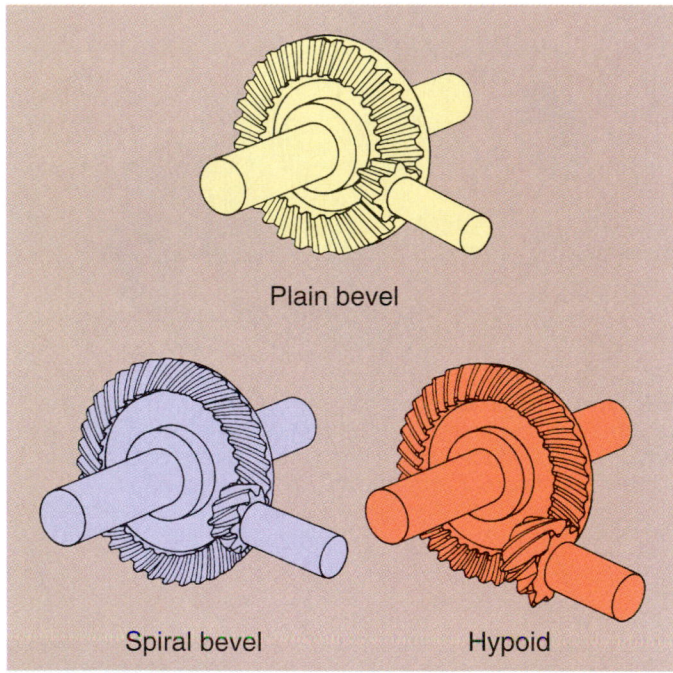

Figure 68.27 Types of differential gears. *(Reproduced by permission of Deere & Company, ©1992. Deere & Company. All rights reserved.)*

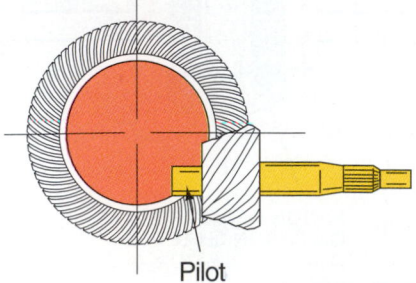

Figure 68.28 Sometimes the drive pinion has a pilot to support the front of it. *(Courtesy of Nissan Motors)*

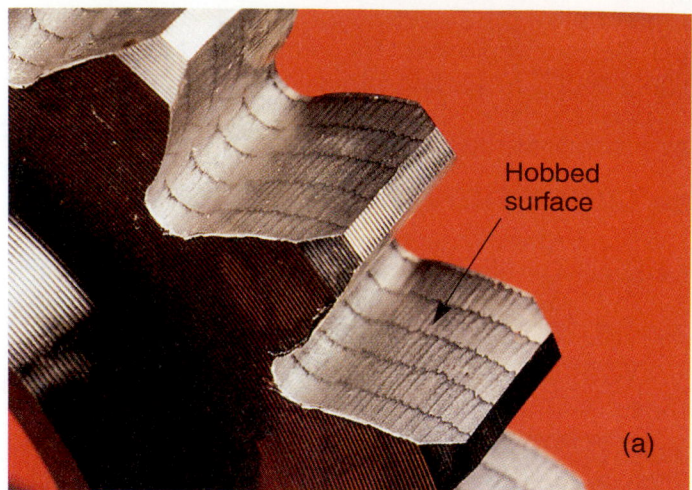

Figure 68.29 (a) Gears are made by a process called hobbing, a rough machine process that leaves noticeable machine marks on the surface of the gear tooth. (b) When the tooth is finish-machined (shaved), the root of the tooth remains rough. *(Courtesy of Caterpillar Inc.)*

■ GEAR SETS

Ring and pinion gears are produced as a *matched set.* They are like the parts of a hydraulic lifter. They are either matched electronically or they are lapped together. A ring and pinion set must be mounted together in exactly the right position or noise and wear will take place. The pinion gear must be mounted at a correct depth in the ring gear. Differential setup is critical. That subject is covered in detail in Chapter 69. In a matched gear set, the ring and pinion gears are marked (Figure 68.31) so that they can be assembled with the marks facing each other.

Hunting and Non-hunting Gear Sets

Gear sets are either hunting or non-hunting. Most ring and pinion sets are **hunting**. With a hunting gear set, a pinion gear tooth will move around until it contacts

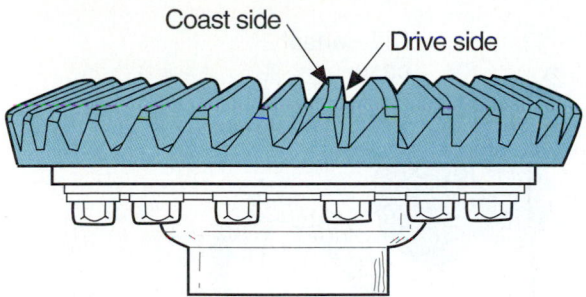

Figure 68.30 The convex side is the drive side and the concave side is the coast side.

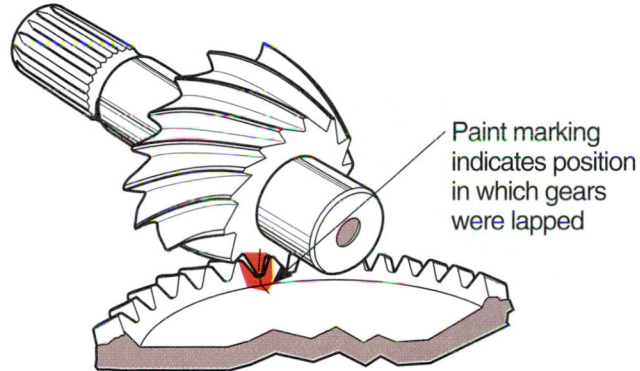

Figure 68.31 In a matched gear set, the ring and pinion gears are marked. [Courtesy of Ford Motor Company]

all of the ring gear teeth. This gear set will not have to be timed.

When a gear set is called **non-hunting** one tooth on the pinion gear will mesh with the same tooth on the ring gear again and again, rather than "hunting" for all of the rest of the teeth. This will have an even gear ratio, such as 4:1. One pinion tooth would contact four different ring gear teeth every revolution. A non-hunting gear set has a timing mark that needs to be aligned.

A **partial non-hunting** gear set is one with a ratio like 2.5:1 that allows one pinion tooth to contact two or three different ring gear teeth during one revolution and two or three different ones the next revolution. At the third revolution it will be back to the same teeth on the ring gear.

■ AXLE RATIO

To find the ratio in the differential, divide the number of teeth on the pinion gear into the number of teeth on the ring gear. If there are 10 teeth on the pinion gear and 33 teeth on the ring gear, this would be a 3.30:1 ratio.

NOTE: *A higher ratio means a lower number; 3.5 is a lower ratio than 3.3. For racing, differential ratios can go as low as 5:1 or 6:1.*

When a car has a manual transmission, it will have a lower ratio, i.e., 3.50:1. With an automatic transmission, the ratio will be higher 3.0:1, for instance. For better fuel economy, axle ratios tend to be higher (lower number).

■ LIMITED SLIP DIFFERENTIAL

A standard differential allows one wheel to rotate if it begins to slip. The other wheel will remain at rest and the vehicle will be stuck. A **limited slip** differential prevents this from happening. A limited slip differential locks up the spider gears when one wheel starts to lose traction. This puts traction to the still wheel. The spider gears are not locked up during normal operation.

There are other names for locking differentials. Some of them are *Positraction* (Chevrolet), *Equa-lok* or *Traction-lok* (Ford), *Sure-Grip* (Chrysler), *Power-Loc* (Toyota), *Trac-loc* (Dana), and *Super Brute* (GM).

■ TYPES OF LIMITED SLIP DIFFERENTIALS

There are several designs of locking differentials. The most popular one has clutch packs like in an automatic transmission (Figure 68.32). When torque is applied to the clutch pack, the side gear locks to the case. This causes the wheel on that side to be driven (Figure 68.33).

There are different ways that the clutch pack is applied to the side gear. Usually a spring (either a coil, multiple coils, an S-shaped spring, or a belleville spring) pushes the side gear against the clutch pack. When the side gear on the side that is spinning tries to climb up out of the pinion gears it is in mesh with, the pinions push against the clutch pack splined to the side gear on the opposite side. This is called torque loading.

On some older units, there are four pinion gears and two pinion shafts. The pinion shafts are positioned in V-shaped ramps, instead of a round hole (Figure 68.34). This allows the pinion gears to move to one side or the other and apply pressure to the clutch packs.

The other main type of limited slip differential is the *cone-type* (Figure 68.35). Cones are forced against the case by springs. Because the axles are splined to the cones, they will want to rotate with the differential case.

One other design used in some applications is the *viscous coupling* differential. In the past, it was used in the Jenson Interceptor. It had an expensive patent that is now expired so it is expected to be seen in more and more vehicles. Figure 68.36 shows a viscous coupling used in an all wheel drive. It is located in the center of the driveshaft that runs between the front and rear differentials.

A viscous coupling is a sealed unit that contains silicone fluid. The clutch plates in the coupling are all

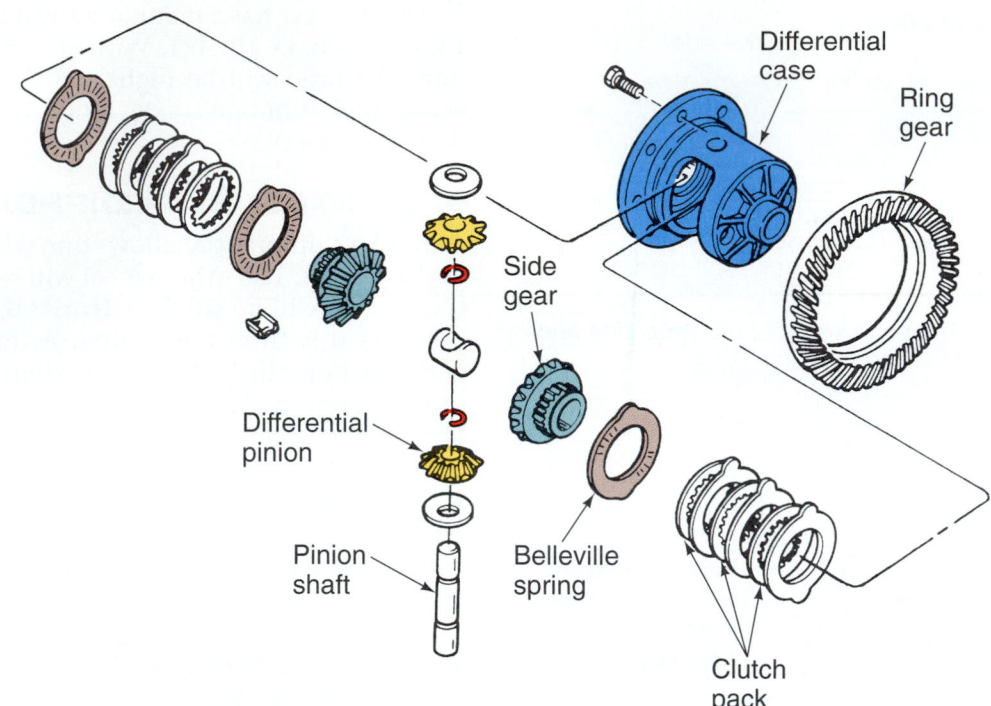

Figure 68.32 A limited slip differential with clutch packs. *(Courtesy of Chrysler Corporation)*

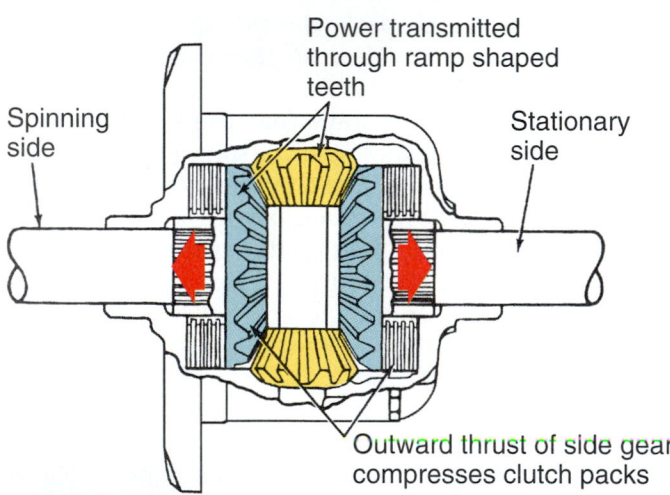

Figure 68.33 When torque is applied to the pinion gears, pressure on the side gears compresses the clutch packs. *(Courtesy of General Motors Corporation, Service Technology Group)*

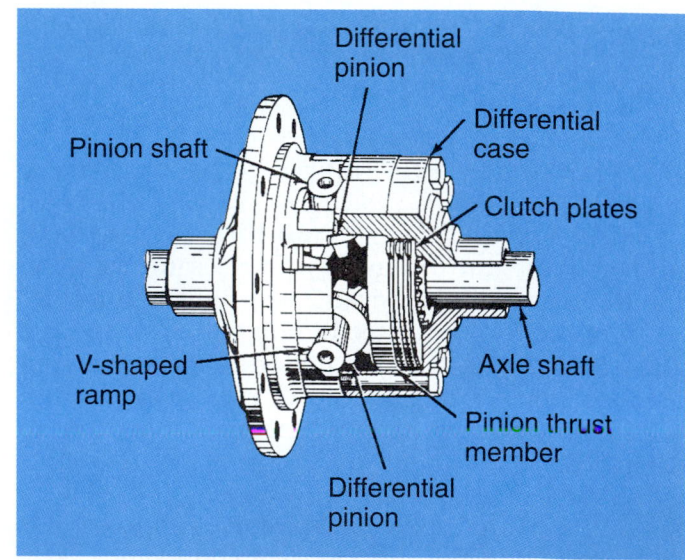

Figure 68.34 These pinion shafts are positioned in V-shaped ramps, instead of a round hole. *(Courtesy of Chrysler Corporation)*

steel with many holes drilled in them to dissipate heat. There is no spring to apply the clutches. Silicone does not change in viscosity as it heats up. The viscous coupling is not totally full of fluid. When there is slippage between the spider gears, air bubbles form in the silicone. The fluid is sticky and when it gets hot, the bubbles create pressure that locks the plates. When they cool down, they release once again. This cycling can take place several times a minute. This is called hys-

teresis. The viscous coupling is not serviceable and is expensive if it fails. Be careful not to mismatch tires or the unit will work all of the time.

NOTE: *It is important that tires be the same size on all limited slip differentials. Otherwise, the clutches will always be working against each other and cause premature wear or failure.*

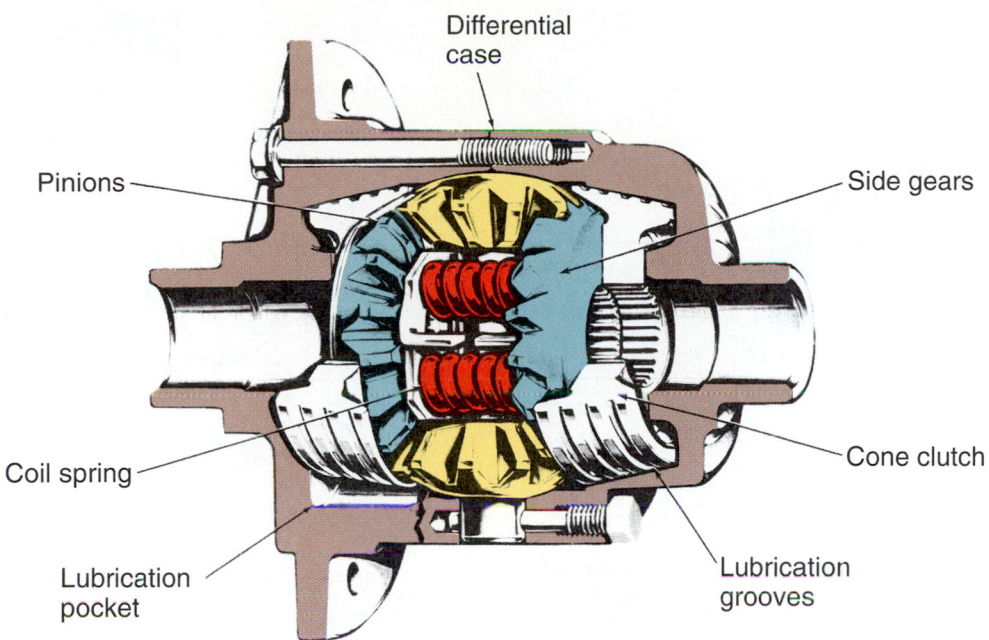

Figure 68.35   A cone-type limited slip differential. *(Courtesy of Chrysler Corporation)*

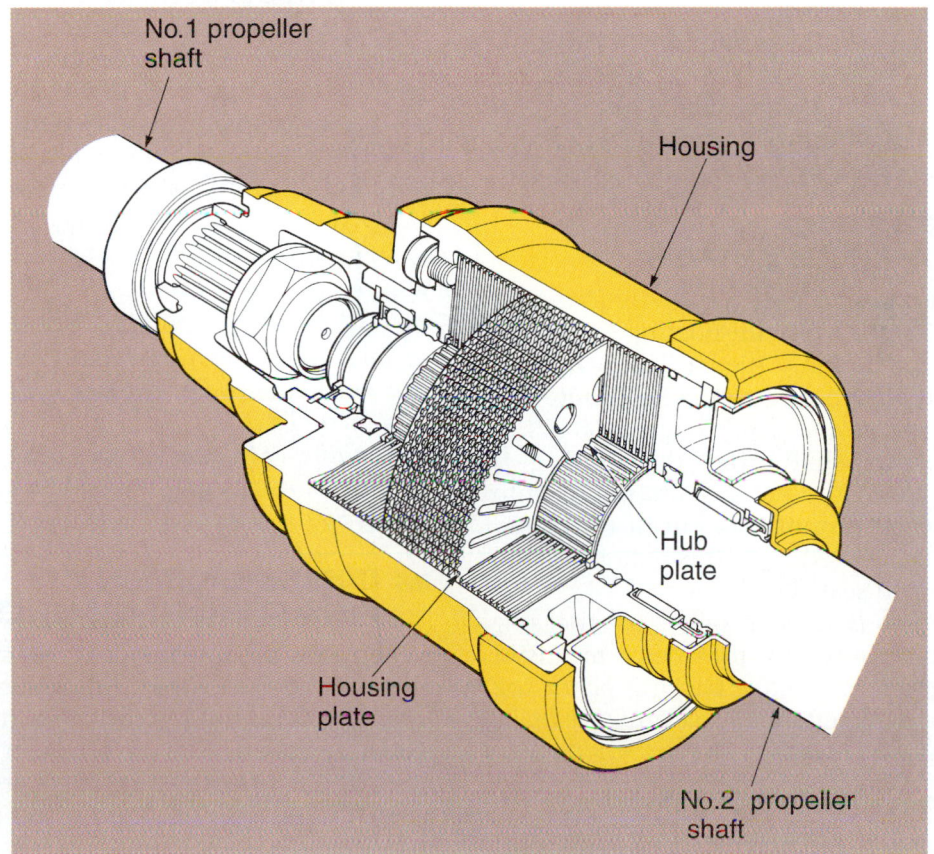

Figure 68.36   A viscous coupling is sealed and is almost full of silicone. *(Courtesy of American Honda Motor Co., Inc.)*

Other differential designs are used only for racing or off-road applications. A popular one is the Detroit Locker. It has a ratcheting pair of clutch packs that physically lock the unit up. A drawback to it is that it makes a ratcheting noise when it releases. Off-road people recommend that this differential not be used on the front of a four wheel drive because it affects steering.

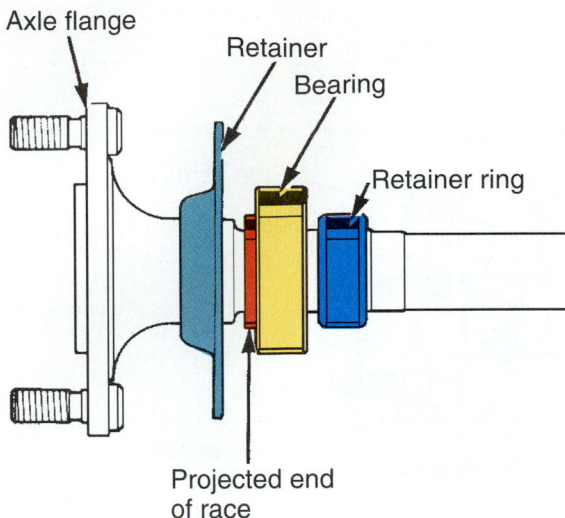

Axle flange

Retainer

Bearing

Retainer ring

Projected end
of race

Figure 68.37 A retaining ring is pressed onto the shaft after the bearing. *(Courtesy of American Honda Motor Co., Inc.)*

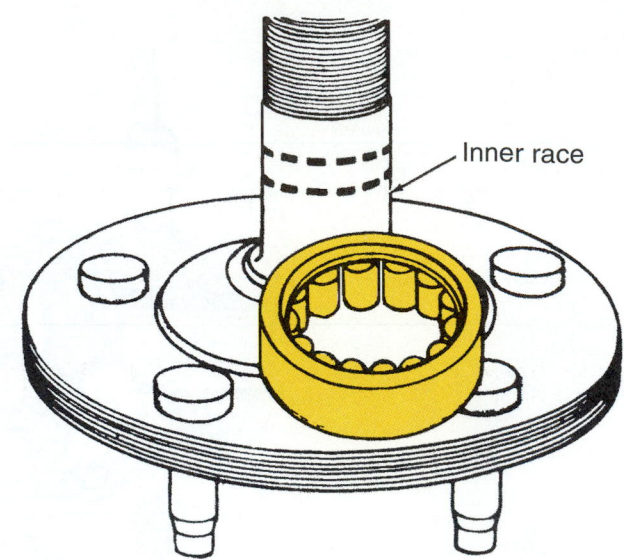

Inner race

Figure 68.38 On a C-lock axle, the bearing rides on a hardened area of the axle. *(Courtesy of Ford Motor Company)*

■ DRIVE AXLES AND BEARINGS

Drive axles usually support the weight of the car. Bearings are covered in Chapter 53. Refer to that chapter for more information about particular types of bearings. The coverage here is limited to specific rear wheel bearing designs.

■ SEMI-FLOATING AXLE BEARING TYPES

Semi-floating bearings (see Figure 53.17) are found on passenger cars. There are two semi-floating designs in common use. On one design, called a *bearing retained axle,* a bearing with an inner race is pressed onto the axle shaft. A bearing *retainer ring* (Figure 68.37) is pressed onto the axle shaft after the bearing is installed. The outside of the bearing fits tightly into the axle housing so oil cannot leak out of the differential. The axle bearing used with this axle is usually packed with grease and sealed.

A bearing retainer that is installed on the axle before the bearing is pressed on provides a means of bolting the assembly to the brake backing plate. The bolts holding the bearing retainer also clamp the brake backing plate to the axle housing.

■ C-LOCK AXLE

The other style of axle is called a *C-lock* or *C-clip axle.* On most axles of this type, the bearing rides on a hardened area of the axle rather than having an inner bearing race (Figure 68.38). The outer bearing race and its bearing rollers fit tightly into the end of the axle housing. The axle bearing is lubricated by a mist of oil from the differential. A lip seal keeps the oil from leaking out onto the brake shoes.

This type of axle retaining is found only on Salisbury axle types with a cover on the rear. A C-lock at the inside end of the axle holds the axle into the differential (see Figure 69.18). The outside of the C-lock fits into a recess in the differential side gear (the gear that is splined to the rear axle and drives it). When the differential pinion shaft is installed, the C-locks are trapped within the side gears.

NOTE: *With either of the semi-floating axle designs, a broken axle can result in the wheel moving out away from the axle housing. With the C-lock axle, there will be nothing to retain it. With the bearing retained axle, the bearing can wear through the retainer if the car is driven in this condition. That is why light-duty trucks and heavier use full floating axle bearings (see Figure 53.18). The full floating axle can break and the wheel and hub will still be supported by the bearings.*

■ INDEPENDENT REAR SUSPENSION

When a car has independent rear suspension, the axles must be able to pivot independently as each wheel goes over a bump. Instead of solid axles, there are two **swing axles** (Figure 68.39). Each axle has a universal joint at one or both ends. These are similar to the ones used for front axles in front wheel drive cars (see Chapter 70).

■ GEAR OILS

Gear oils are special heavy liquid lubricants used to lubricate gears and bearings in manual transmissions and differentials. A hypoid axle set is the most difficult of gear applications to lubricate. The gear teeth can come under pressures as high as 400,000 psi. Because

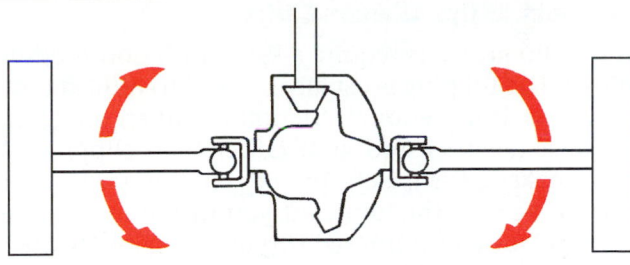

FLEXIBLE AXLE SHAFT

Figure 68.39 An independent rear suspension has two swing axles. [*Reproduced by permission of Deere & Company, ©1992. Deere & Company. All rights reserved.*]

of the shape of the gear tooth, there is also a high amount of sliding between the gear teeth. Gears must remain separated by a film of oil, even when under heavy loads. A good oil prevents high temperatures and scoring of the parts. As one gear tooth moves against another, oil is forced from between them. This oil must be continuously replenished.

When the lubricant film fails, wear results. Gear oils have additives similar to the ones found in engine oils. Additives prevent oxidation and corrosion, reduce friction, limit wear, and prevent foaming. **EP additives** (extreme pressure) are added to prevent welding of the gear teeth under extreme load conditions. These are chemicals that cause the formation of compounds on the gear teeth that will not weld.

There are many varieties of gear oils. Some gear oils are thin, as in automatic transmissions. Other are thick, as in differentials. Oil used in manual transmissions must remain fluid enough to allow easy shifting in cold weather. Automatic transmission fluids must remain fluid to as low as –40°F and remain at almost the same viscosity at over 300°F.

Viscosities for gear oil range from 75 to 250. There are also multigrades like 80W–140. The W is for winter grade, like in engine oil. Gear oil has a viscosity rating that is different than that for engine oil. Viscosity is rated in centiStokes. The chart in Figure 68.40 compares viscosities of SAE gear and engine lubricants.

NOTE: *An SAE 90 gear lube is comparable in viscosity to an SAE 40 or 50 engine oil.*

API Gear Oil Ratings

The American Petroleum Institute (API) and *Coordinating Research Council (CRC)* have established a system for classifying gear lubricants. The Society of Automotive Engineers (SAE) lists these as shown in Figure 68.41. The GL in the number stands for gear lube. GL1 is the lowest grade only for light duty and low speed. Automotive gear lubes are GL4, GL5, and GL6. GL6 is a synthetic classification.

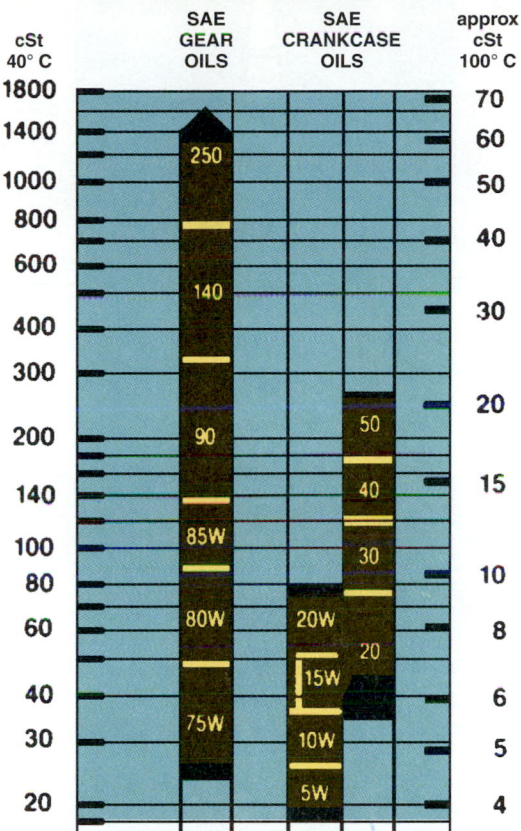

Viscosity Equivalent Chart

Figure 68.40 A comparison of viscosities of SAE gear and engine lubricants. [*Courtesy of Shell Lubricants*]

API Classification	Amount of EP Additives	Type of Service
GL-1	none	Light duty only. Spiral bevel and worm gear axles. Some manual transmissions.
GL-2	small amount	Worm gear axles under more severe conditions than GL-1.
GL-3	small amount	Spiral bevel axles and manual transmissions under more severe conditions than GL-1.
GL-4	medium amount	Hypoid gears—light service. Spur and bevel gears—medium service.
GL-5	high amount	Hypoid gears—medium service. Spur and bevel gears—heavy duty service.
GL-6	highest amount	Most gears—extremely heavy duty service.

Figure 68.41 Gear lubricant ratings.

Figure 68.42 Limited slip differentials require a special lubricant. *(Courtesy of Quaker State Corporation)*

Limited Slip Gear Oils

Limited slip gear oils require a special friction modifier additive. The fill plug is usually marked to indicate this. When there is not enough friction modifier, the clutch discs stick together and grab rather than slipping. As the car makes a turn, a chirping or clunking sound will be heard. Be sure the lubricant you install in a limited slip axle is suited for limited slip uses (Figure 68.42).

NOTE: *Limited slip gear oil will work in standard axles.*

■ FOUR WHEEL DRIVE

There are different types of four wheel drives. One type is the conventional four wheel drive that operates like a rear wheel drive vehicle with an extra differential in the front (Figure 68.43). Another type is like a front wheel drive vehicle with an extra differential in the rear (Figure 68.44). Still another is called **all wheel drive (AWD)**. It remains in four wheel drive all of the time.

■ FOUR WHEEL DRIVE AXLE ASSEMBLY

An axle assembly for a four wheel drive is very similar to the two wheel drive rear axle. A driveshaft and universal joints transfer power from the transmission by way of a transfer case. The ends of the axle housing on front axles are different because they have to pivot to allow the front wheels to be turned during steering (Figure 68.45).

■ TRANSFER CASE

On four wheel drive vehicles there is a transfer case on the rear of the transmission that provides for power

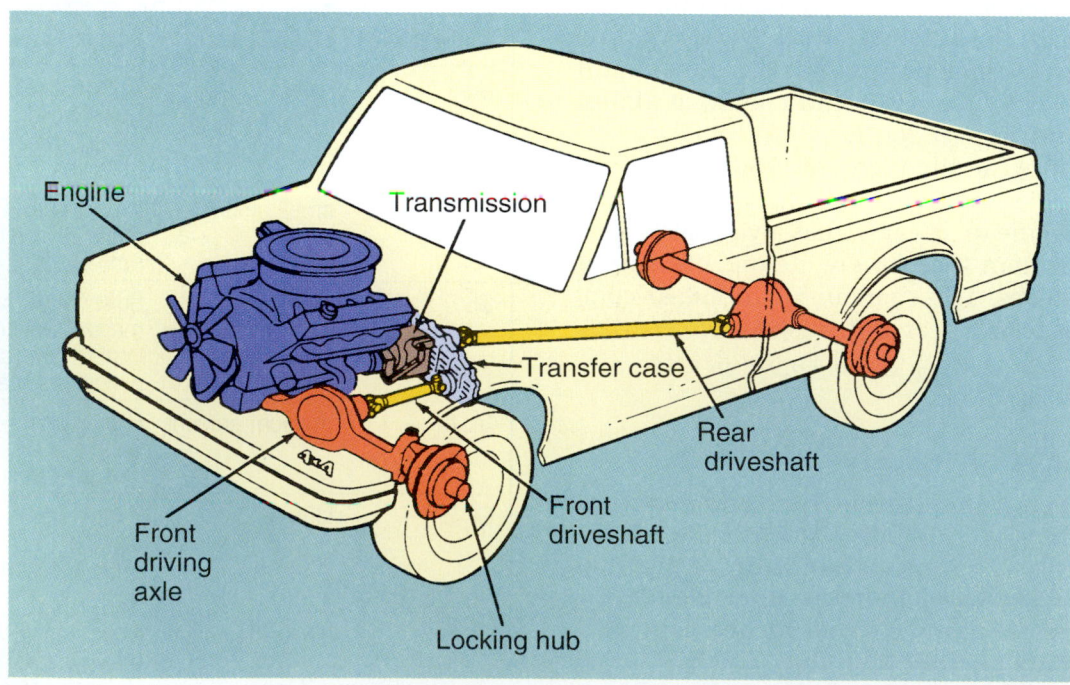

Figure 68.43 Parts of a four wheel drive. *(Courtesy of Ford Motor Company)*

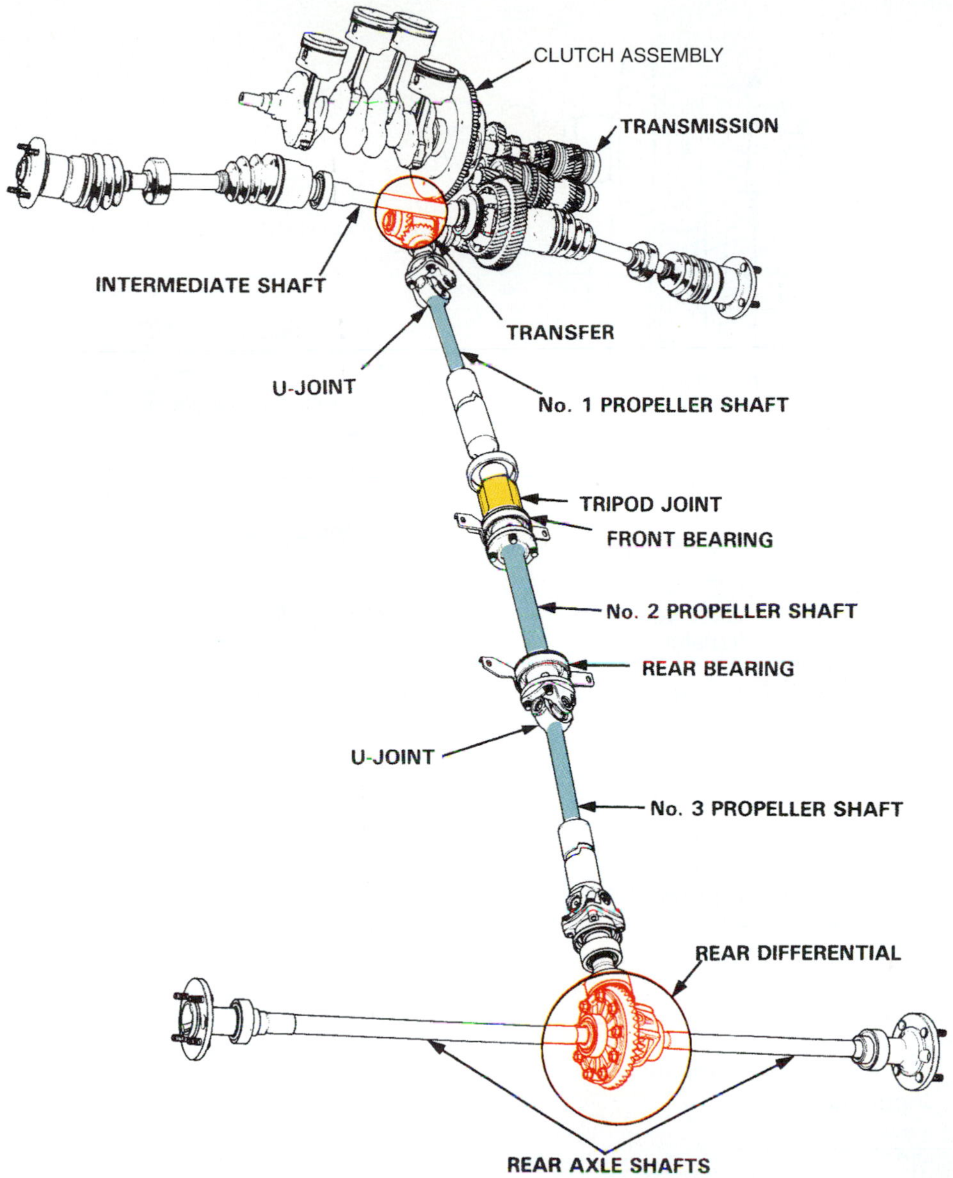

CLUTCH ASSEMBLY

TRANSMISSION

INTERMEDIATE SHAFT

TRANSFER

U-JOINT

No. 1 PROPELLER SHAFT

TRIPOD JOINT
FRONT BEARING

No. 2 PROPELLER SHAFT

REAR BEARING

U-JOINT

No. 3 PROPELLER SHAFT

REAR DIFFERENTIAL

REAR AXLE SHAFTS

Figure 68.44 A four wheel drive train with a front transaxle. *(Courtesy of American Honda Motor Co., Inc.)*

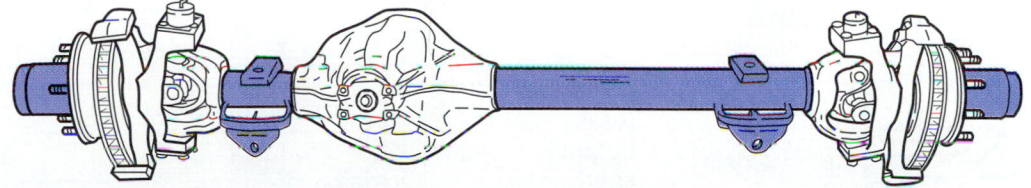

Figure 68.45 The ends of the four wheel drive front axle must pivot for steering. *(Courtesy of Ford Motor Company)*

transmission to the front axle assembly (see Figure 68.43). Two driveshafts come out of the transfer case; one goes to the front and the other goes to the rear.

Most four wheel drives provide the options of high and low ranges for steep climbing or for highway driving. The transfer case allows for shifting between low range and high range. Low range usually provides an additional 2:1 reduction. High range is 1:1. A transfer case also allows for shifting between two wheel and four wheel drive.

The transfer case can consist of either a gear drive or a planetary gear set with a chain drive (Figure 68.46). Power flow is covered here using a gear drive transfer case.

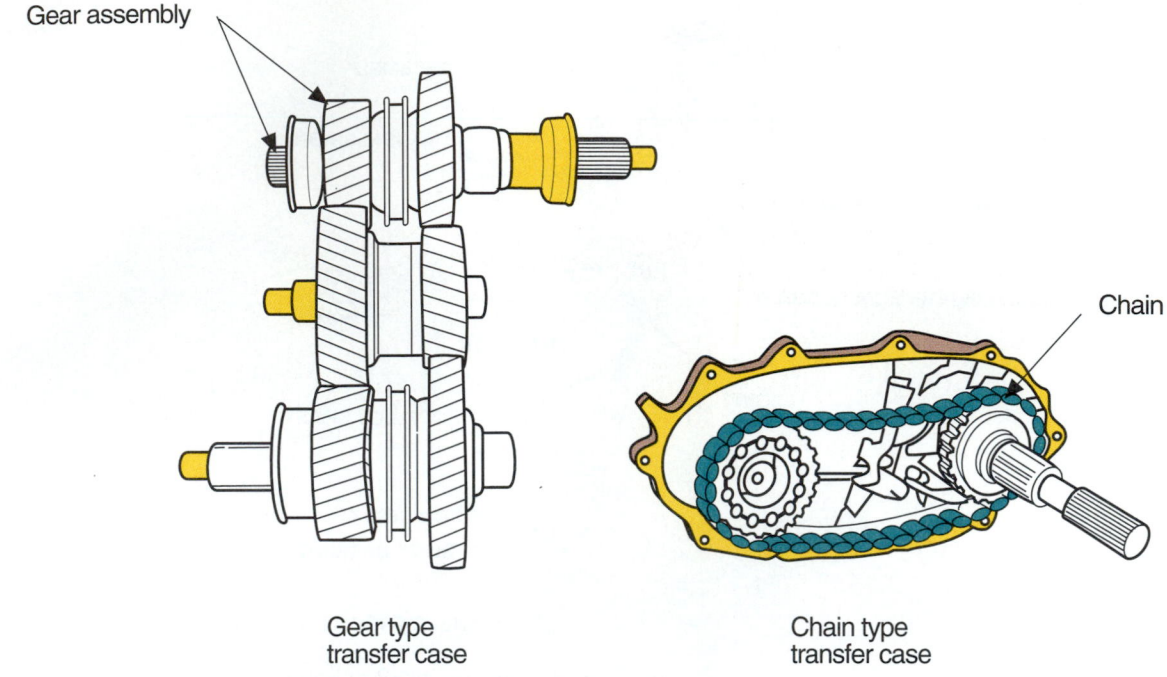

Gear assembly

Chain

Gear type
transfer case

Chain type
transfer case

Figure 68.46 Gear drive and chain drive transfer cases. *(Courtesy of Ford Motor Company)*

■ In *two wheel drive high range (2H)*, power flows through the locked planetary gear set to the mainshaft and to the rear differential.

■ In *four wheel drive high range (4H)*, power flows from the input shaft to the locked planetary gear set in the same way it does in 2WD (Figure 68.47).

A *sliding clutch*, called a *range clutch*, shifts into the mainshaft clutch gear. Power flow is through the drive chain and front output yoke to the front driveshaft. Power also flows through the rear driveshaft so that both axles drive the vehicle.

■ In *four wheel drive low range (4L)*, everything is the same as in high range except that the shifter moves the range clutch rearward to engage low gear (Figure 68.48).

Input
drive gear

Range
clutch

Input shaft

Differential

Idler
gear

Low
speed
gear

Front output
shaft

Chain

Rear
output
shaft

Front of
vehicle

High range

Figure 68.47 Power flow through the transfer case in four wheel drive high range. *(Courtesy of Ford Motor Company)*

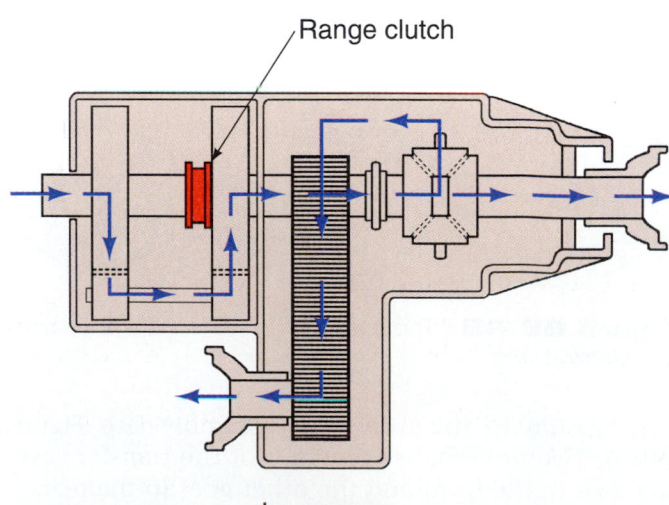

Range clutch

Low range

Figure 68.48 Power flow through the transfer case in four wheel drive low range. *(Courtesy of Ford Motor Company)*

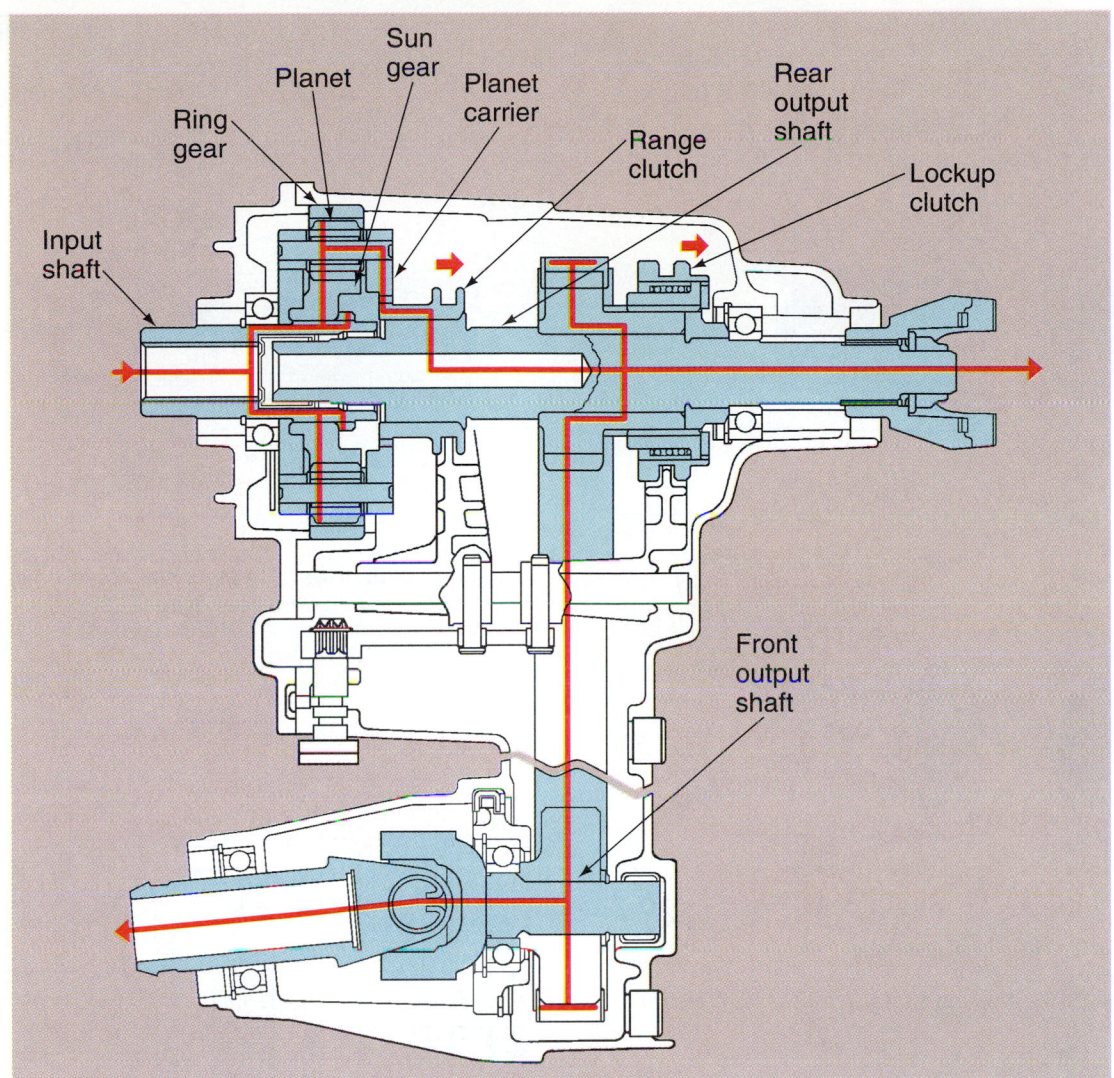

Figure 68.49 Power flow through a planetary gear transfer case in low range. *(Courtesy of Ford Motor Company)*

Planetary Low Range

In a planetary transfer case, the ring gear is pressed into the housing, so it is held stationary. Power flows into the sun gear and out through the cage or carrier (Figure 68.49). A gear reduction is produced that is transferred to both the front and rear output shafts.

■ LOCKING HUBS

Many axles also have **locking hubs**. There are three types of locking hubs. Their purpose is to allow a four-wheel drive vehicle to be used in two wheel drive without the axles and differential being attached to the front wheels.

With a *manual locking hub,* the hubs have on and off locks on the front spindles (Figure 68.50). These must be turned to lock and unlock the axles from the wheels. When the hubs are unlocked, the front wheels are not driven.

- The vehicle should not be operated in four wheel drive with the hubs disengaged or driveline wear will occur.
- When shifting the hubs back out of locking range, the transfer case must be shifted to two wheel drive. This releases the pressure from the hubs and makes them easy to turn.

Automatic locking hubs come on whenever the vehicle is shifted into four wheel drive. *Full time hubs* are always locked to the front axles and differential.

■ ALL WHEEL DRIVE

In all wheel drive (AWD), all four wheels are driven. These vehicles are not designed for off-road use but are to improve poor traction in icy or snowy driving. Control is improved by transferring engine power to the wheel with the most traction. Most AWD systems have a center differential to split the power between the

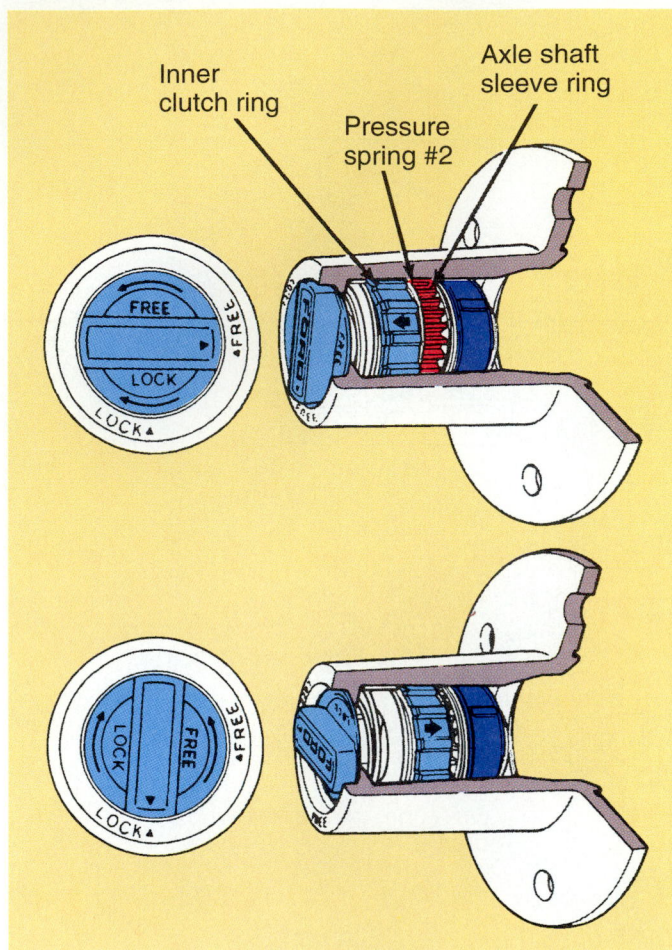

Figure 68.50 Locking hubs. *(Courtesy of Ford Motor Company)*

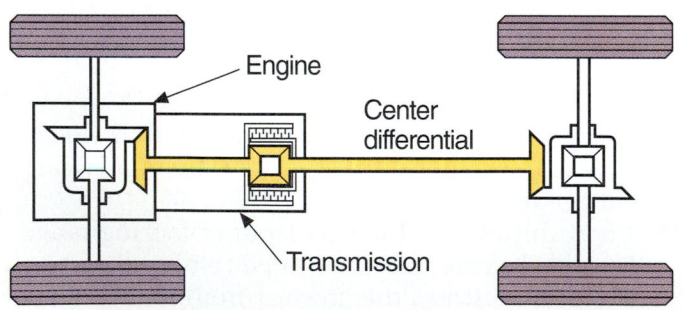

Figure 68.51 A center differential splits the power between the front and rear axles. *(Courtesy of Chrysler Corporation)*

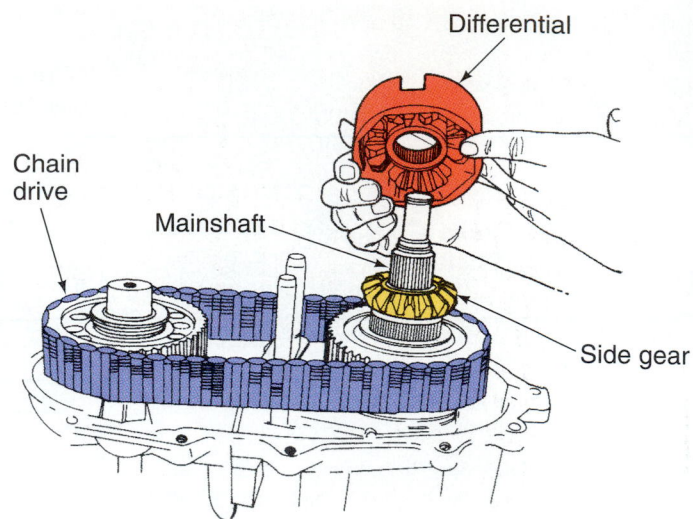

Figure 68.52 This transfer case has a differential in it for full time four wheel drive. *(Courtesy of Chrysler Corporation)*

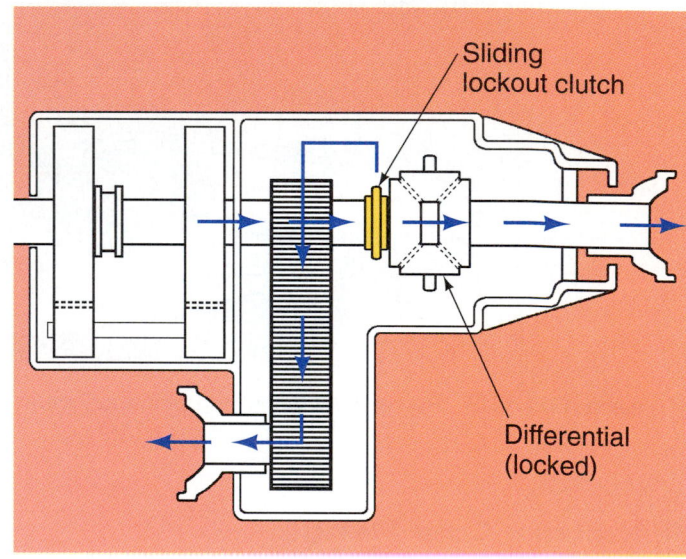

Figure 68.53 A lockout for the center differential. *(Courtesy of Ford Motor Company)*

front and rear axles (Figure 68.51). The transfer case shown in Figure 68.52 has the differential within the transfer case. Four wheel drive transfer cases do not have this differential.

When there is a center differential, if one wheel is lifted free of the ground, it is possible for all of the torque to go to that wheel. This would mean no power goes to the other wheels. A lockout for the center differential must be included in these types (Figure 68.53). A popular new way of connecting the front and rear axles is the viscous coupling. It allows the front and rear wheels to revolve at different speeds (Figure 68.54).

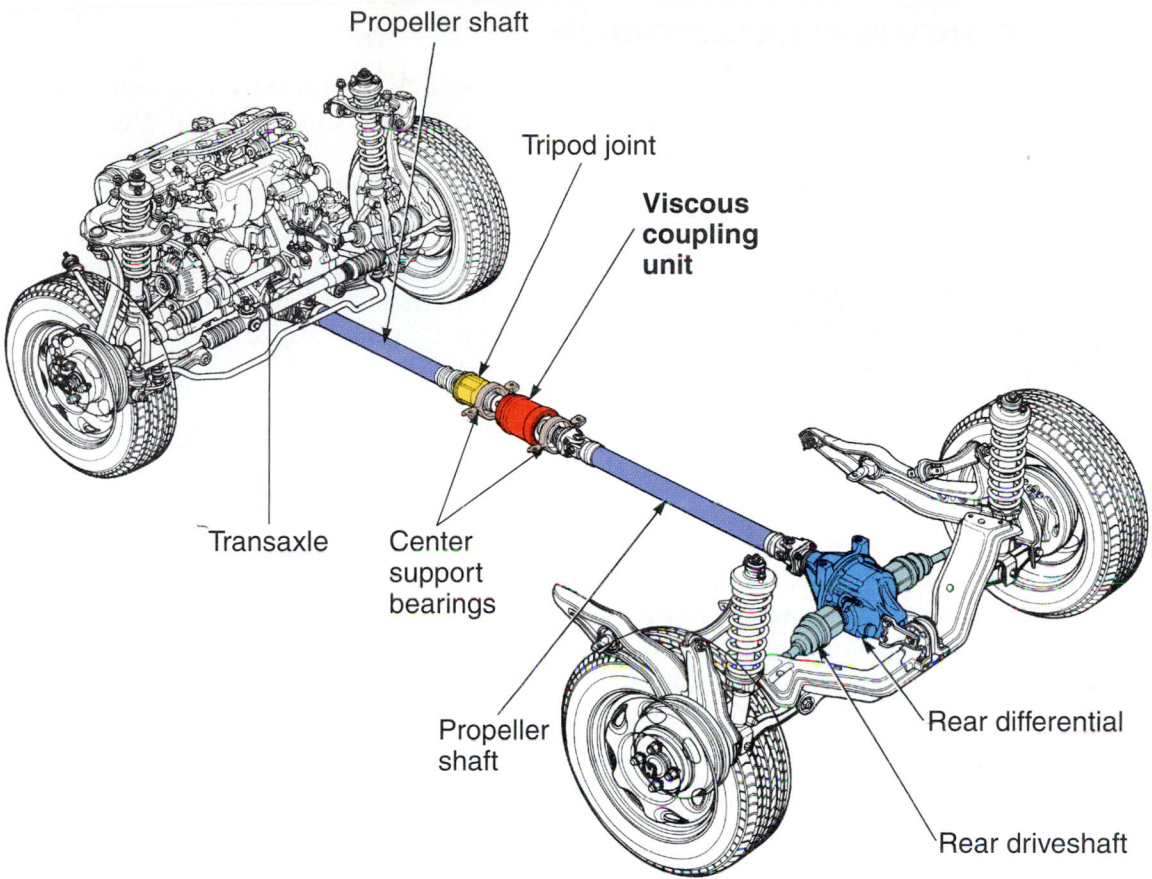

Propeller shaft

Tripod joint

Viscous coupling unit

Transaxle

Center support bearings

Propeller shaft

Rear differential

Rear driveshaft

Figure 68.54   All wheel drive with a viscous coupling. *(Courtesy of American Honda Motor Co., Inc.)*

■ REVIEW QUESTIONS

1. What is the name of the open driveshaft system?

2. What is the name of the part that slides in and out of the transmission as the distance between the transmission and differential changes.

3. How many times in one revolution does speed increase and decrease when a U-joint is operated at an angle?

4. Angles of greater than _____° are not used with single universal joints because they cause excessive vibration.

5. At 30° of driveshaft angle, when a shaft is turned at 1000 rpm, the speed of the driven yoke will change from _____ rpm to 1155 rpm in one-quarter of a revolution.

6. What is the name of the constant velocity joint that uses two universal joints?

7. The differential is called the _____ member.

8. Which has more teeth, the ring gear or the pinion gear?

9. The name of the part that separates the two pinion bearings is called a crush sleeve or collapsible _____.

10. During a turn, which wheel must move farther than the other, the inside or the outside?

11. The side gears and differential pinions are called _____ gears.

12. To find the ratio in the differential, divide the number of teeth on the _____ gear by the number of teeth on the _____ gear.

13. Which car will have a higher numerical axle ratio, one with an automatic or a manual transmission?

14. An SAE 90 gear lube is comparable in viscosity to an SAE 40 or 50 _____ oil.

15. What does the GL stand for in the gear lube rating?

ASE STYLE REVIEW QUESTIONS

1. Technician A says original equipment universal joints have a grease fitting. Technician B says that when a Cardan universal joint is operated in a straight line, the speed of the output member is different than the speed of the input member. Who is right?

   a. Technician A **b.** Technician B
   c. Both A and B **d.** Neither A nor B

2. Technician A says universal joints are angled to cancel each other out. Technician B says that on a one-piece driveshaft, the front and rear yokes can be out of phase, resulting in vibration. Who is right?

   a. Technician A **b.** Technician B
   c. Both A and B **d.** Neither A nor B

3. Technician A says constant velocity joints are found on smaller rear wheel drive cars. Technician B says that the speed of the input yoke is constantly changing as it revolves. Who is right?

   a. Technician A **b.** Technician B
   c. Both A and B **d.** Neither A nor B

4. Technician A says that a differential allows its two wheels to turn at different speeds. Technician B says that a differential increases the torque from the engine. Who is right?

   a. Technician A **b.** Technician B
   c. Both A and B **d.** Neither A nor B

5. Technician A says that the differential is removed from a Salisbury axle as a complete unit. Technician B says a removable third member is called a pumpkin. Who is right?

   a. Technician A **b.** Technician B
   c. Both A and B **d.** Neither A nor B

6. Technician A says that the differential pinion gears have splines that mesh with the axle splines. Technician B says when the car is travelling straight ahead, the differential pinion gears rotate. Who is right?

   a. Technician A **b.** Technician B
   c. Both A and B **d.** Neither A nor B

7. Technician A says the convex side of a gear tooth is the drive side. Technician B says that with a hunting gear set, one pinion gear tooth will contact all of the ring gear teeth. Who is right?

   a. Technician A **b.** Technician B
   c. Both A and B **d.** Neither A nor B

8. Technician A says a standard differential will not allow one wheel to spin without putting power to the other. Technician B says that a limited slip differential requires a different lubricant than a standard differential. Who is right?

   a. Technician A **b.** Technician B
   c. Both A and B **d.** Neither A nor B

9. Technician A says silicone in a viscous coupling changes its viscosity to apply the clutches. Technician B says C-lock axles are sometimes installed on differentials with removable third members. Who is right?

   a. Technician A **b.** Technician B
   c. Both A and B **d.** Neither A nor B

10. Technician A says gear lube viscosity is different than motor oil viscosity. Technician B says that a four wheel drive vehicle requires a differential or viscous coupling between the front and rear axles. Who is right?

    a. Technician A **b.** Technician B
    c. Both A and B **d.** Neither A nor B

CHAPTER 69

Driveline Diagnosis and Service

OBJECTIVES

Upon completion of this chapter, you should be able to:

✔ Diagnose and repair universal joints.

✔ Remove and replace axle bearings and seals.

✔ Disassemble and reassemble a differential.

✔ Run a gear pattern and adjust a differential.

KEY TERMS

galling
drive
cruise
coast
float
howl
breakaway torque
gear pattern
drive side
coast side
heel
toe
pitch line
face
flank
preload

INTRODUCTION

In this chapter, service to the driveline is covered. The driveline includes the driveshaft and U-joints, axles and axle bearings, and differential. Live axles are covered in this chapter. They are ones that turn with the wheels. Transaxles, a later chapter, have similar service to their differentials. Your understanding of differential service procedures will apply later. Differentials are also referred to as front or rear *axles*.

In this chapter, you will also learn more than just driveline service and repair procedures. Pay attention to the methods used in pressing bearings and other parts. Press skills are used in many other areas of automotive repair.

DRIVESHAFT DIAGNOSIS

Driveshaft problems can result in noise or vibration from worn or rusted U-joints, a worn slip yoke, or a bad center support bearing. Worn U-joints can cause *squeaking* or *grinding* sounds.

SHOP TIP Before disassembling a universal joint, check to see that a wheel cover or hub cap is not making the noise. Remove all of them and drive the car to see that the noise has not gone away.

Sometimes a car will have a *clunking* sound when changing from acceleration to deceleration. This can be due to worn slip yoke splines or a bad extension housing bushing. It can also be because of problems in the differential or a very worn U-joint. Sometimes leaf springs can be loose at the differential allowing the housing to wind up.

A *ringing sound* is sometimes a complaint. This often results from a bad clutch disc damper. Replacing the clutch disc usually solves the problem. If the car has an automatic transmission, the problem can be due to a bad lock-up converter.

A worn center support bearing can cause a *whining sound* that varies with vehicle speed. The noise is constant in pitch, rather than changing or intermittent like U-joint noise. A U-joint noise changes pitch because of the changing angle of the U-joint.

Driveshaft Balance

Driveshafts are a possible source of vibration. In high gear, the driveshaft spins at engine rpm. If the shaft is bent or a universal joint is worn, a vibration can occur. Driveshaft balance is covered in Chapter 72.

Some driveshafts are built in two pieces with rubber dampening rings inside of them. Another driveshaft style has a damper like a crankshaft vibration damper mounted on its outside. This absorbs torsional vibration.

Driveshaft Inspection

Several checks are made when inspecting a driveshaft.
- Look for undercoating, missing balance weights, or obvious physical damage.
- Move the driveshaft up and down while watching the U-joints for looseness.
- Look for a rusty appearance around U-joint cups.
- Move the slip yoke up and down against the extension housing bushing while looking for excessive movement.
- Check to see that the motor mount on the transmission crossmember is in good condition.

A dent in the driveshaft tubing will weaken it, which can cause it to kink easily under load. The strength of a driveshaft is longitudinal. Try standing on a soda can and quickly touch both sides of the can. It will immediately collapse when you touch it. Cracks in a driveshaft result from physical damage. They always start on the surface, never at the inside.

■ DRIVESHAFT SERVICE

To remove the driveshaft:

■ Mark the driveshaft so that it can be replaced in the same position. Use a crayon to mark the rear differential yoke and the companion flange (Figure 69.1).

■ Unbolt the rear U-joint from the differential companion flange.

■ Pry the rear U-joint forward away from the differential.

■ Wrap tape around the U-joint cups so that they cannot fall off of the U-joint cross.

NOTE: *If one of the cups falls off the U-joint, one or more of the small needle bearings might fall out. If one gets lost, the entire U-joint must be replaced.*

■ On a two-piece driveshaft, unbolt the center support bearing.

NOTE: *Be sure to mark both halves of the shaft in the center where the splines are. The shaft will be out of phase if it comes apart and is not reassembled correctly (see Figure 68.14). Serious vibration will result.*

■ The driveshaft will now slip out of the transmission. When it is removed, oil will probably come out of the transmission.

SHOP TIP To prevent oil from leaking out of the transmission, install an old slip yoke onto the mainshaft splines. If an old yoke is not available, a plastic plug or a bushing installation tool can be used.

■ UNIVERSAL JOINT DIAGNOSIS AND SERVICE

When a univeral joint begins to fail, a squeaking sound is often noticed just when the car begins to go forward. The most common cause of U-joint failure is when its grease dries out. This often happens because the seal on the U-joint has failed allowing moisture in. A vibration can also occur when a U-joint starts to fail. With a worn U-joint, a sharp, one-time click sound often occurs when the vehicle direction is changed from forward to reverse or when the vehicle first takes off.

■ UNIVERSAL JOINT DISASSEMBLY

The procedure described here is for single cross and yoke (cardan) universal joints. If the U-joint has any snap rings, remove them (Figure 69.2). Some snap rings are on the inside of the yoke. Others are on the outside. When the snap ring is on the outside a sturdy pair of pliers can be used.

NOTE: *Smaller needle nose pliers can become damaged because they are not sturdy enough.*

Sometimes snap rings are on the inside (Figure 69.3). A punch or a special tool can be used to remove them. If the U-joint is retained by plastic resin, follow the manufacturer's service manual instructions. A small tube of resin is usually used.

There are three ways U-joints are commonly disassembled. The most common method is to use a bench vise because every shop has one. When pressing a universal joint out of its yoke, the bearing cap on one side must be pushed into a socket or pipe that is slightly larger in diameter (Figure 69.4). After the cap is removed from one side, the cross is forced against the other cap to force it out of the other side. The cross can be removed at this time. Then the cap is pounded out with a punch (Figure 69.5). The process is repeated on

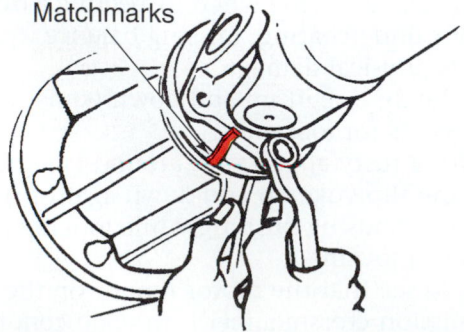

Figure 69.1 Mark the rear differential yoke and the companion flange before removing the driveshaft.

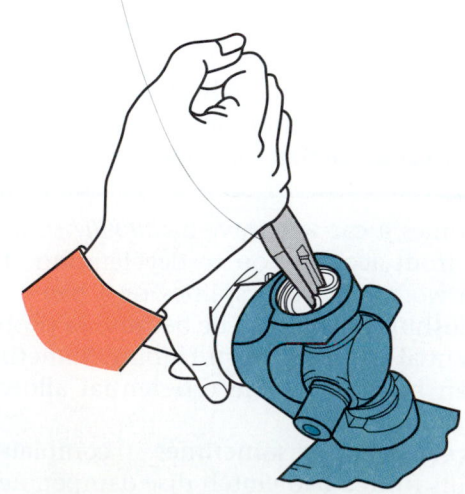

Figure 69.2 Removing a snap ring from a U-joint. *(Courtesy of Ford Motor Company)*

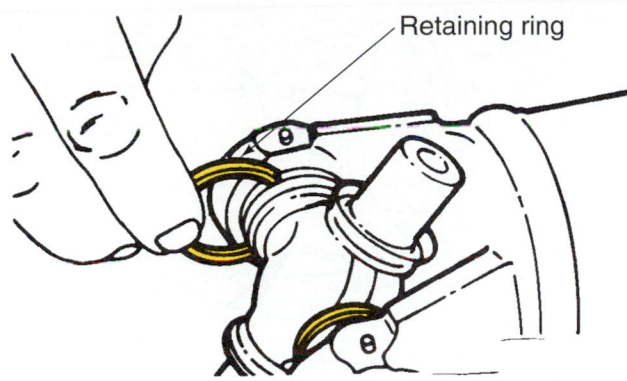

Figure 69.3 Sometimes snap rings go on the inside of the yoke. *(Courtesy of General Motors Corporation, Service Technology Group)*

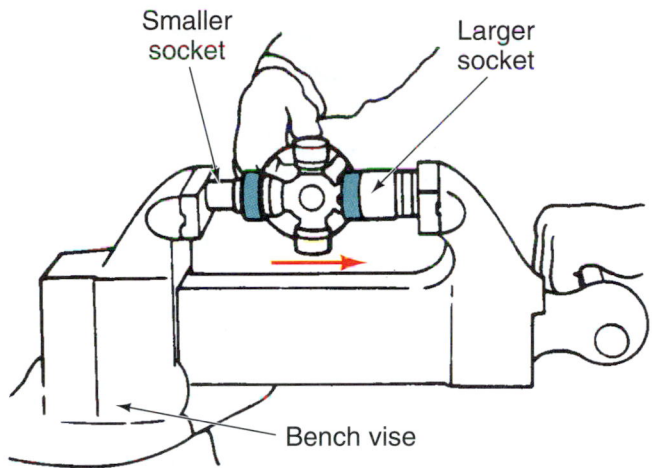

Figure 69.4 Removing a U-joint with sockets and a vise. *(Courtesy of General Motors Corporation, Service Technology Group)*

the other side of the cross to complete the removal of the U-joint.

NOTE: *One problem with using a vise is that it is weak when opened as far as it takes to accommodate the U-joint. A vise can be broken if excessive force is used when its jaws are far apart.*

There is also a special universal joint tool that many shops have. It is used in the same manner (Figure 69.6). The hole in the opposite side of the tool is larger than the U-joint bearing cup that will be pressed into it during removal. A third method of universal joint removal is to use a hydraulic press with a special tool.

Some shops pound the U-joint out with a hammer. This process works, but you must be careful not to bend a yoke. A very small amount of misalignment is enough to cause a vibration. Also, if you install the tube part of the driveshaft in a vise during installation, be careful not to damage the tube.

Universal Joint Inspection

One problem caused during installation in a vise is that driveshaft yoke ears commonly become sprung inward. This results in brinelling. Brinelling is when small indentations wear into the bearing surface (Figure 69.7). Brinelling is often the result of a faulty U-joint installation. The joint should always feel loose and relaxed (not binding up) after a correct installation.

A problem with driveshaft angles can cause **galling**, which often happens on the ends of the trunnions. Usually galling will be found on a trunnion 180° to brinelling (Figure 69.8).

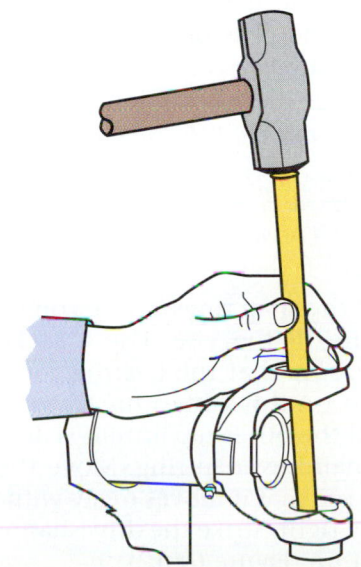

Figure 69.5 Pound out the bearing cap with a punch.

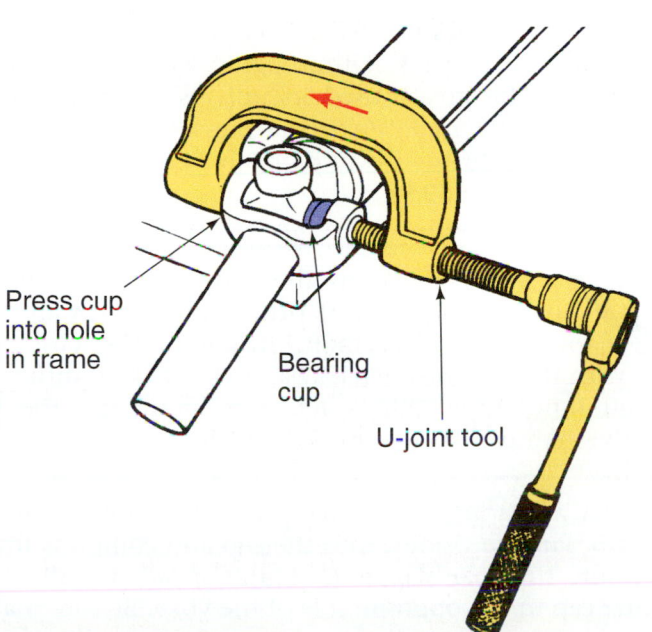

Figure 69.6 A special universal joint tool. *(Courtesy of Ford Motor Company)*

Figure 69.7 Brinelling of the trunnion. *(Courtesy of Rockwell Automotive)*

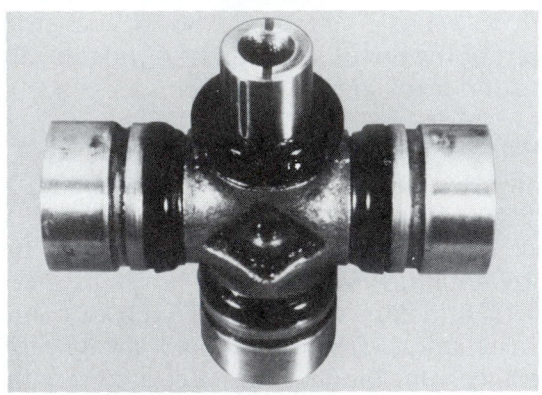

Figure 69.8 Galling on the trunnion. *(Courtesy of Rockwell Automotive)*

■ UNIVERSAL JOINT REASSEMBLY

If the U-joint does not have a zerk fitting for lubrication, be sure to pack it thoroughly with the correct lubricant. The grease used in universal joints is an NLGI #1, with an EP (extreme pressure) additive. If there is a zerk fitting it will probably be slanted in one direction (Figure 69.9).

SHOP TIP When installing the cross into the yoke be sure that the zerk fitting is angled inward toward the driveshaft. Installing the cross with the zerk fitting backwards will make it difficult, if not impossible, to lube the U-joint after the driveshaft is installed on the vehicle.

Put a bearing cap partially into the yoke (Figure 69.10). Put the U-joint into the cap and compress the cap into the yoke (Figure 69.11a). *Carefully* install the other cap in the opposite side of the yoke. Be sure that you do not accidentally knock one of the needle bearings out of position. A very common occurrence is for one bearing to be knocked into the bottom of the cap.

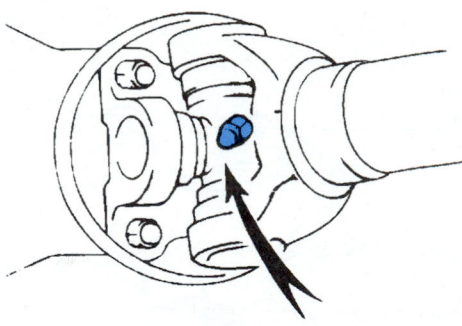

Figure 69.9 The zerk fitting is slanted toward the driveshaft.

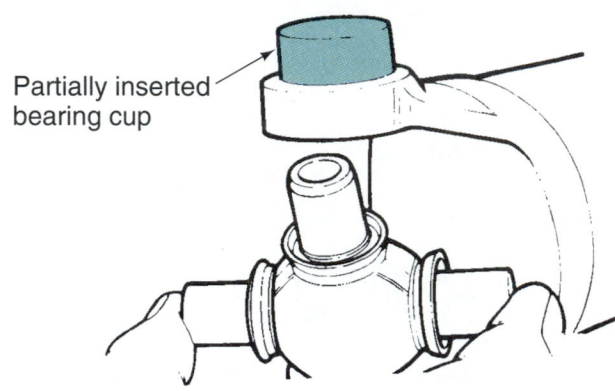

Partially inserted bearing cup

Figure 69.10 *Put a bearing cap partially into the yoke. (Courtesy of General Motors Corporation, Service Technology Group)*

When the caps are pressed together all the way, the joint locks up. When you disassemble it to fix the problem, you might find a damaged needle bearing. This would require the purchase of an entire new U-joint.

Move the joint back and forth to see that it is free and nonbinding in both caps (Figure 69.11b). Place the yoke between the jaws of the vise and compress the remaining cap into the yoke (Figure 69.11c). While you are pressing it in, move the joint back and forth and watch for binding. Do not force anything.

NOTE: *If the bearing cap turns in the hole in the yoke, the yoke is defective. The cap is hardened and the ear of the yoke is not. A new yoke will need to be installed on the driveshaft by a machine shop.*

Install one of the snap rings before completing the pressing procedure in the vise. Use a socket that is smaller than the hole that the bearing cap goes into and press the cap until it comes up against the snap ring. Then, install the other cap until it is deep enough to install the remaining snap ring. Move the joint in each direction to see that it moves freely without binding. If it is slightly tight, strike the driveshaft yoke with a punch to free it up (Figure 69.12).

NOTE: *Sometimes, universal joint caps are held in place with plastic material that is injected into a hole in the yoke.*

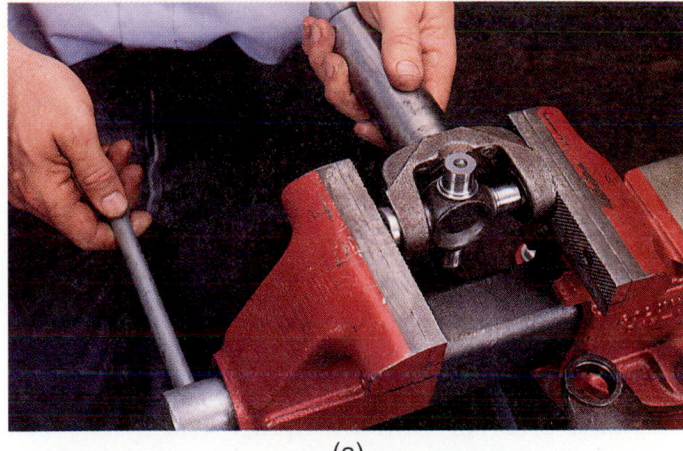

(a)

(b)

(c)

Figure 69.11 (a) Put the U-joint into the cap and compress the cap into the yoke. (b) Move the U-joint to see that it is not binding up. (c) Compress the remaining cap into the yoke.

Follow manufacturers' instructions when dealing with one of these U-joints.

Some replacement U-joints use internal snap rings when replacing injection molded plastic retainers (see Figure 69.3).

Strike tube yoke ear in this area

Figure 69.12 Striking the yoke realigns the needle bearings in the bearing cap. *(Courtesy of General Motors Corporation, Service Technology Group)*

Figure 69.13 Install the other half of the cross onto the driveshaft.

After two bearing caps are installed in the slip yoke, install the other half of the cross onto the driveshaft (Figure 69.13).

■ DRIVESHAFT INSTALLATION

When installing a driveshaft that bolts to a yoked flange, be sure that the universal joints fit exactly between the tabs on the flange. If the driveshaft is not installed exactly on center, there will be serious vibration. Be sure that all of the contact surfaces are clean before installing the driveshaft. Dirt or a burr on the companion flange where the U-joint cap will fit can cause the shaft to vibrate after installation. With the transmission in neutral, slide the slip yoke into the transmission. Align the marks that you made during removal and install the rear U-joint onto the companion flange. Install the retaining bolts to complete the job.

NOTE: *If a vehicle experiences vibration after the installation of new universal joints, sometimes it can be corrected by removing the driveshaft and reinstalling it, turning it 180° in the companion flange.*

■ TWO-PIECE DRIVESHAFT SERVICE

On a two-piece driveshaft, the center support bearing sometimes fails. The seized bearing can tear away the rubber mount that supports the outside of it allowing the bearing to rotate with the driveshaft. This allows the driveshaft to wobble up and down causing a vibration. The bearing is pressed off and a new one is pressed on. Be sure to press on the inside bearing race so you do not damage the bearing. Some bearings are designed to be installed in one direction only. Installing them backwards will result in damage to the rubber ring on the bearing.

If the two pieces of the driveshaft are separated, be sure that it is reassembled in phase. Assembling the front and rear halves out of phase can cause extreme vibration. When the shafts were not marked before disassembly, use a tape measure or a steel rod to check the alignment of the halves.

■ DIFFERENTIAL AND AXLE DIAGNOSIS AND SERVICE

Differential gears rarely wear out. In fact, they often last the life of the vehicle on the original lubricant. Causes of damage to the gears include moisture and dirt getting into the differential and lubricant leaking out. Abuse by a driver is another prime cause of damage to a differential. Sometimes, a gear ratio change is desired and this is the reason why a differential is disassembled.

■ PROBLEM DIAGNOSIS

When diagnosing a noise or vibration, isolate the problem. Be sure it is not from the transmission. The differential and transmission are connected by the driveshaft. Noises can transmit from one to the other. When two gears mesh, some noise is normal. Also, with a van or station wagon, the vehicle can act like a big speaker box and amplify noises.

Road Test

Does the noise change under different driving conditions? There are several driving conditions to be aware of when listening for noises or feeling for vibration. The following is the terminology you need to become familiar with:

■ **Drive**—under acceleration, power is on the convex side of the gear tooth.
■ **Cruise**—the car is maintaining its speed.

■ **Coast**—deceleration, power is on the concave side of the gear tooth.
■ **Float**—car speed is slowly dropping.

Noise that resembles a **howl** or *whine* can be due to adjustment of the ring and pinion or due to bearings that are worn. Incorrect differential gear adjustment can result in a howl that occurs only under drive or only under coast conditions. Worn bearings will make a constant sound that changes in relation to road speed. Clunking noises can be due to damaged gears or bearings. When a gear is badly damaged, a shudder can sometimes be felt along with the noise.

Noise that happens only during a turn is probably due to a problem with the spider gears. They can become damaged when a wheel is allowed to spin in a puddle and then gets traction. The differential pinions are very small gears. They cannot withstand the punishment of a heavy load that the large ring and pinion can. Remember that the pinion gears (spiders) are only turning during a turn. When they come to an abrupt halt, they can easily lose teeth. Damaged side gears are usually on the side that received the stress.

Other problems related to the spider gears include:
■ Pinion gears too tight on the shaft
■ Side gears too tight or too loose in the differential case
■ Excessive backlash between the spider gears

■ AXLE BEARING DIAGNOSIS

A worn or damaged axle bearing can result in a fluid leak or bearing noise. If a groan is heard that could be due to a wheel bearing, driving the car can sometimes help pinpoint the problem. First, check the tires for damage and be sure they are properly inflated. Find an empty parking lot or deserted road and make slow left and right turns. The inside tire always turns at a lower rpm than the outside tire. The noise from a wheel bearing that is bad will change pitch as the wheel speeds up and slows down when turning from one side to the other. When the outside wheel has the bad bearing, the noise will become worse because that wheel is turning faster.

Applying the brakes can also cause the noise level from a bad bearing to become less as the brakes contact the drum or rotor on the axle that has the bad bearing.

After the road test, raise the car on a hoist and try to move the tires and wheels by hand. The only kind of bearing that has any normal movement is the one used with a C-lock axle. These sometimes have end play that can be felt. The other kinds of wheel and axle bearings should have less than 0.005" end play. Up and down movement is not normal. When pushing in and out on the top and bottom of a tire on a wheel with a tapered roller bearing, a small amount of movement is normal. Testing an axle bearing in this manner should result in no movement.

Non-serviceable axle bearings have no service interval but they sometimes fail. When this happens they are replaced. The axle must be removed from the housing to repair it. Axle removal is commonly done by service establishments, but the presswork required to install the new bearing on the axle is usually done by parts stores, many of which have machine shops. Axle seals are sometimes part of the axle bearing. Other times they are separate from the axle bearing. Repairing an axle seal requires removal of the axle from the differential.

Listening for Noises

With the car on the lift, have a helper run the engine with the transmission engaged. Sometimes, the vibration from a bad bearing can be felt by placing your hand on a part close to it. Listen for the noise with a stethoscope. Put the end of the tool against the brake backing plates to locate bad axle bearings. Listen below the pinion gear for a bad pinion bearing.

CAUTION Be careful of spinning wheels during this test. Be especially careful that tires do not accidentally touch parts of the lift. This could result in a very serious accident!

If the noise is related to tires, it will go away when the car is in the air. An axle bearing noise will also be diminished because the load has been removed.

Lubricant Leaks

Gear oil can leak out of the differential at the pinion seals, from the cover, from the banjo housing, or from the axle seals. When axle seals leak, a rear brake could become covered with oil and not work properly. Because rear wheels do not steer, this is not usually an obvious problem to the driver. Because the oil is thrown by the spinning brake drum, the entire brake backing plate might be stained. Oil should fill the axle housing as shown in Figure 14.22.

SHOP TIP To tell the difference between a brake cylinder leak and an axle seal leak, spray the area with water. Brake fluid will wash off and gear oil will require solvent to remove.

Bearing Problems

A differential can experience bad carrier bearings, or a bad pinion bearing. Carrier bearings and a pinion bearing will make a constant sound because the carrier is always driven at the same speed as the ring and pinion. The sound will usually become louder with an increase in speed.

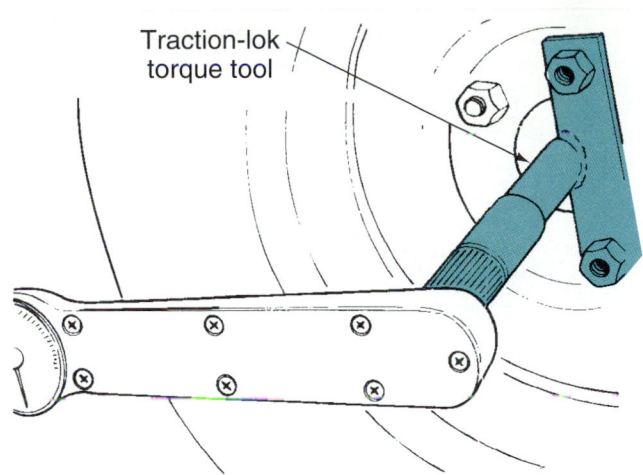

Figure 69.14 Checking a limited slip axle. *(Courtesy of Ford Motor Company)*

Limited Slip Problems

Limited slip differentials can make a chattering sound during turns. The recommended fix is to drain the differential and refill it with fluid with a friction modifier (see Figure 68.42). There are three kinds of differential lubricants. Experienced technicians like to put limited slip fluid in all differentials to avoid problems. The lubricant can be tested by mixing some alcohol with it. If it turns blue it is the correct lubricant. If it turns yellow it is the wrong lubricant.

A limited slip axle can be tested for **breakaway torque** (Figure 69.14). This is the amount of torque needed to make a side gear rotate the clutches. A special tool is used to center the torque wrench on the axle. Typical turning torque on a clutch disc type axle with used clutch plates is 75 foot-pounds. With new discs, torque can range from 155 to 195 foot-pounds. If breakaway torque is too low or too high, there is a problem in the limited slip part of the axle.

■ AXLE BEARING SERVICE

Removing an Axle

Axles are supported on the inside end by the side gears in the differential. Axle on pumpkin-type differentials are called bearing retained axles. The bearing is pressed fit on the axle shaft and fits tightly in the axle housing. There is a retaining plate that must be removed before pulling the axle. Its bolts are accessed through holes in the axle flange (Figure 69.15). A socket and extension can be used (usually with an air impact wrench) to remove the bolts and nuts. After each nut and bolt is removed, the axle is turned to allow access to the next bolt. Some axles have four bolt holes, but most have only one. The outside of the bearing is often tight in the hole in the axle housing. A slide hammer is often required to remove it (Figure 69.16).

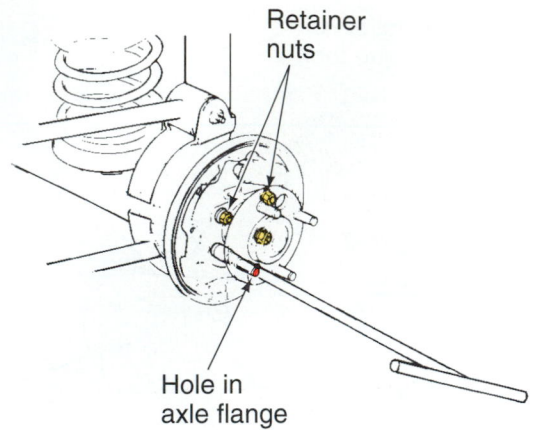

Figure 69.15 A hole in the axle flange allows for removal of the retainer nuts. *(Courtesy of American Honda Motor Co., Inc.)*

Figure 69.17 Remove the bolt and pinion shaft.

Figure 69.16 Removing an axle with a slide hammer.

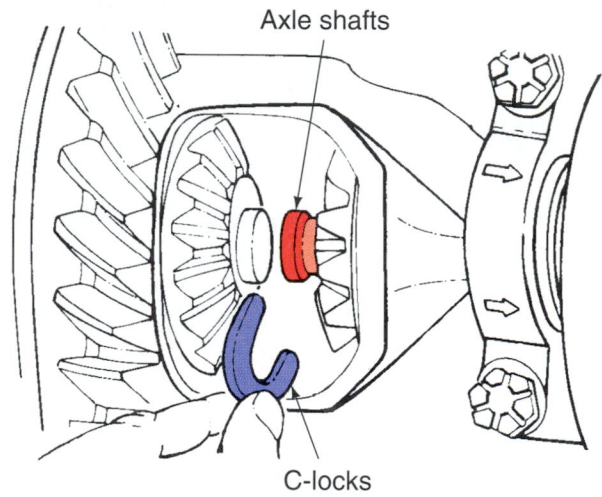

Figure 69.18 When the axle is pushed inward, the C-lock becomes free of the inside of the side gear. *(Courtesy of Ford Motor Company)*

■ C-LOCK AXLE REMOVAL AND REPLACEMENT

The end of a C-lock axle has a machined groove that a C-lock fits into (see Figure 69.18). The cover on the back of the differential must be removed to allow access to the differential pinion gears. The differential pinions must be removed to allow access to the C-locks. The pinion shaft has a lock bolt. After it is removed, the shaft is pulled out (Figure 69.17). After the pinion lock bolt, shaft, and pinions are removed, the axles can be pushed inward, allowing the C-locks to be removed (Figure 69.18).

SHOP TIP Occasionally, a lock bolt can be broken off. There are special tools available for removing these.

On most C-lock axles, the bearing rides on a hardened area of the axle, rather than having an inner bearing race (see Figure 68.38). The outer race and bearing rollers are held in the axle housing. The hardened area of the axle can be damaged where the seal rides on it. Some replacement seals are designed so that the lip will ride in a different area of the axle providing a new sealing surface. During reinstallation, support the axle and do not let it drag on the surface of the new seal.

Axle Seal Service

After the axle is removed, pry out the old seal (Figure 69.19). Check the surface of the axle shaft to see that it is clean and undamaged. Apply lubricant to the lip of the new seal before installing it. Use a seal installing tool to avoid cocking it to the side and damaging it.

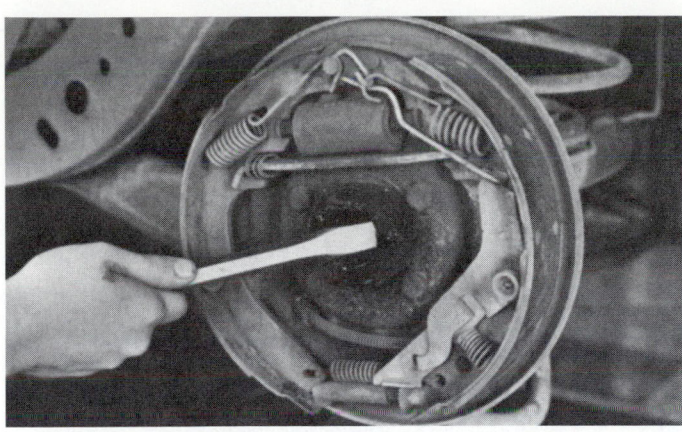

Figure 69.19 Removing the old axle seal. *(Courtesy of Chicago Rawhide)*

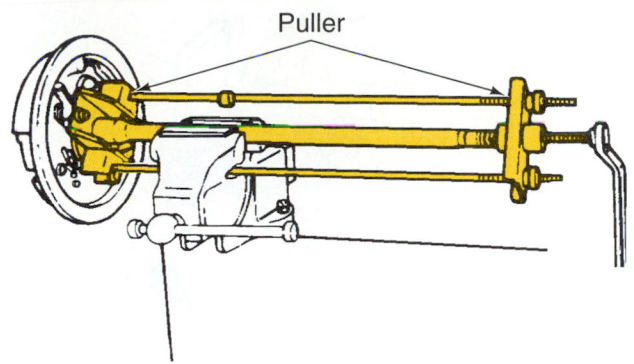

Figure 69.21 A puller setup for removing an axle bearing. *(Courtesy of Mazda Motor Corporation)*

■ AXLE BEARING REPLACEMENT

Some axle bearings are pressed on after the brake backing plate. When removing the axle on these types, the brake line is first unhooked. The backing plate is unbolted and the still assembled brakes are removed along with the axle (Figure 69.20). A press or a special puller (Figure 69.21) can be used to remove the bearing. Other types of axles with pressed-fit bearings come out independent of the backing plate (Figure 69.22).

On C-lock type axles, the bearing remains in the housing after the axle is removed. The bearing is pulled with a slide hammer and special attachment (Figure 69.23).

Pressed-Fit Bearing Replacement

Do not try to press the bearing and retaining ring off at the same time. Grind a notch on the retaining ring

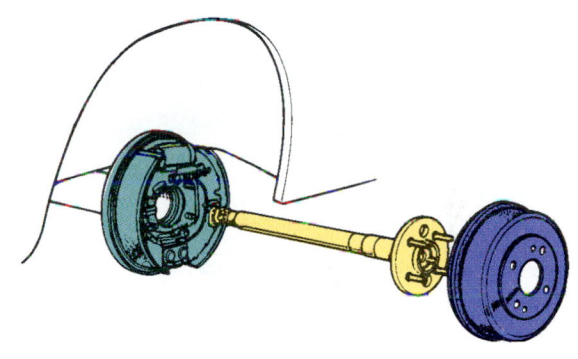

Figure 69.22 Some axles are removed without the backing plate.

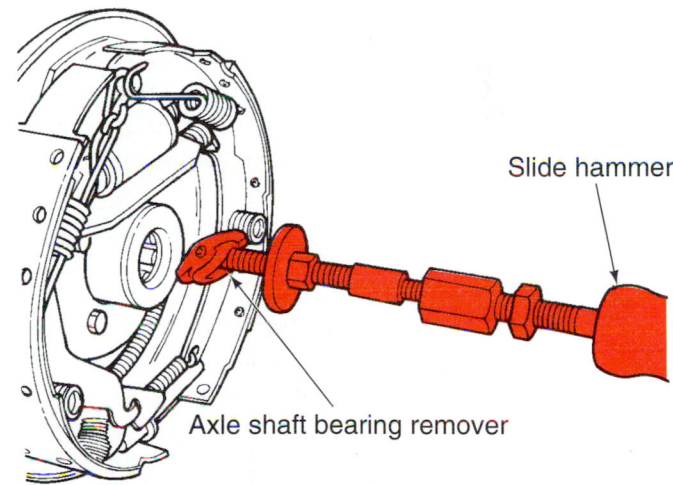

Figure 69.23 Removing an axle bearing with a slide hammer. *(Courtesy of General Motors Corporation, Service Technology Group)*

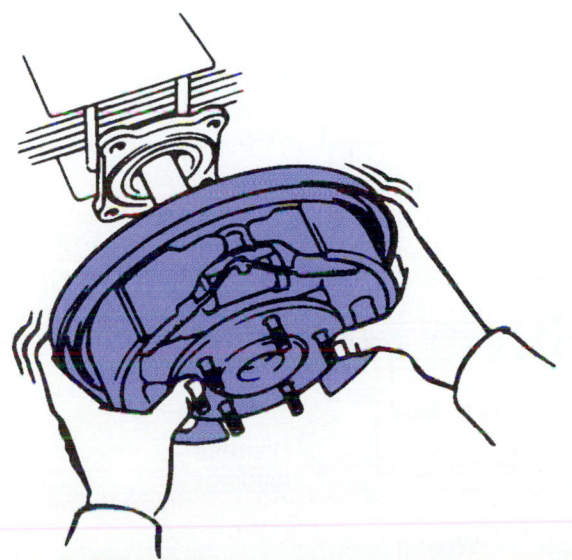

Figure 69.20 Some axles must be pulled together with the brake backing plate.

and strike it with a chisel to weaken it prior to pressing it off (Figure 69.24). Presses usually have a large safe cage made of large diameter steel pipe that can be installed around the bearing when pressing it off. It is not unusual for a bearing to require in excess of 20 tons of pressure to remove it.

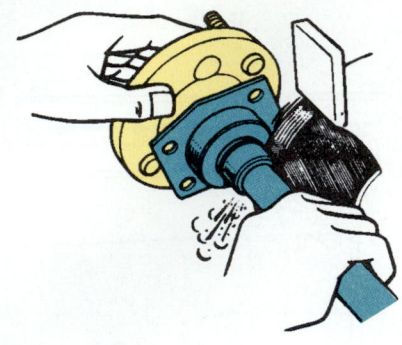

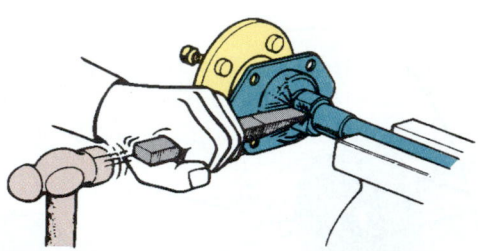

Figure 69.24 Removing a bearing retaining ring.

> ⚠ **CAUTION** If a bearing explodes during removal (and they quite often do), this can be comparable to a grenade exploding!

If the gauge on the press reads over 15 tons, something could be wrong. Be certain that you are pressing parts apart and not pressing against something.

■ AXLE BEARING INSTALLATION

When pressing a bearing onto the axle it is supported by the inner race (Figure 69.25). The cage and rollers should always be able to be turned during the installation process. That ensures that there is no load on them. First the bearing is installed on the axle. The retaining ring is often heated (150°C/300°F) for easier assembly. Then, the retaining ring is pressed into place against the axle bearing (Figure 69.26).

> ⚠ **CAUTION** Do not attempt to press the bearing and the retaining ring onto the axle at the same time.

■ REINSTALL THE AXLE

Reinstall the axle in the housing and bolt on the retainer. Be sure that the oil return slot in the retainer lines up with the return hole in the axle housing (Figure 69.27). This is similar to the oil return on the clutch bearing retainer on a manual transmission.

■ FULL FLOATING AXLE SERVICE

Full floating axles are found on ¾ ton and larger trucks. Axle bearings are located in the hub and brake drum assembly (see Figure 53.18). To remove the drum on a full floating system, the axle must be removed first (Figure 69.28). There are soft metal beveled washers that surround the studs that hold the axle to the hub. Remove them by striking the outside of the axle flange with a brass hammer. Do not force a chisel in between the axle and hub or you could create a future seal leak.

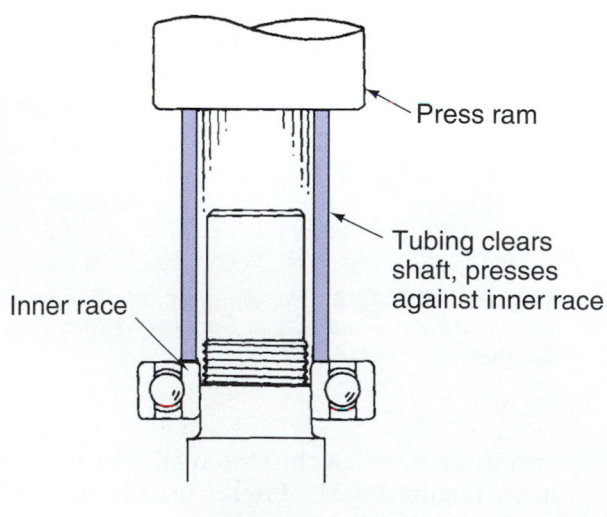

Right

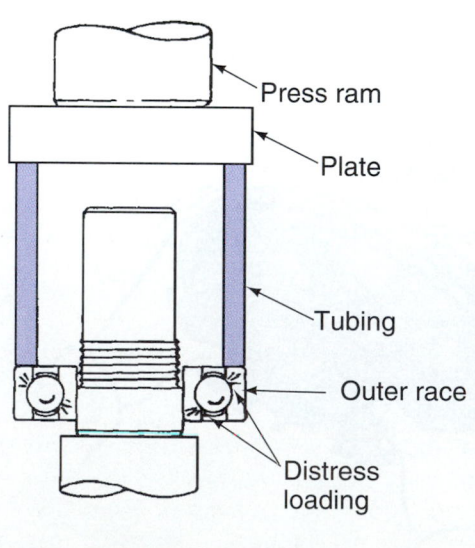

Wrong

Figure 69.25 Press bearings off by putting pressure on the inner race. *(Reproduced by permission of Deere & Company, ©1992. Deere & Company. All rights reserved.)*

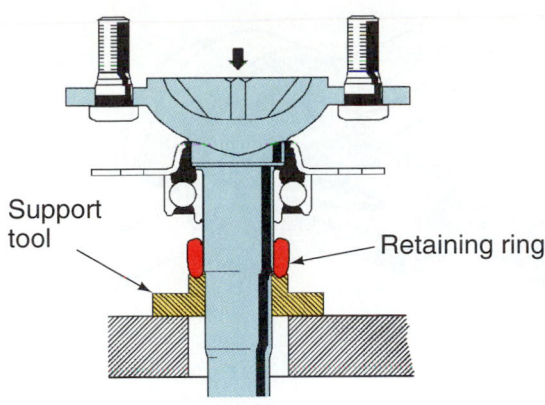

Figure 69.26 Pressing the retaining ring into place against the axle bearing.

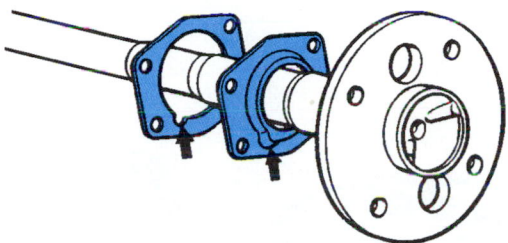

Figure 69.27 Be sure that the oil return slot in the retainer lines up with the return hole in the axle housing.

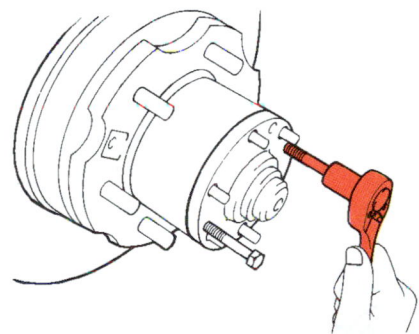

Figure 69.28 Removing a full floating axle.

After the axle is removed, the locknut that holds the brake drum can be removed (Figure 69.29). There are usually two nuts. The outer nut is a locknut. Lift off the brake drum. Catch the outer bearing so it does not fall on the ground.

An apprentice did a brake job on a ¾-ton pickup truck. When he reinstalled the rear drum and hub, he didn't pay attention to how the locking nut was supposed to work and assembled it improperly. The owner was driving his truck later that day and the axle, hub, and wheel came off during a turn. The result was that expensive repairs were required on the truck. Luckily, there was no accident and no one was hurt.

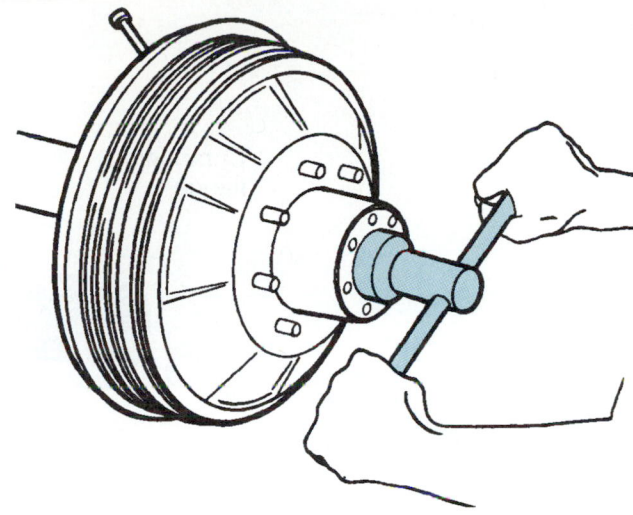

Figure 69.29 Removing the axle locking nut. *(Courtesy of General Motors Corporation, Service Technology Group)*

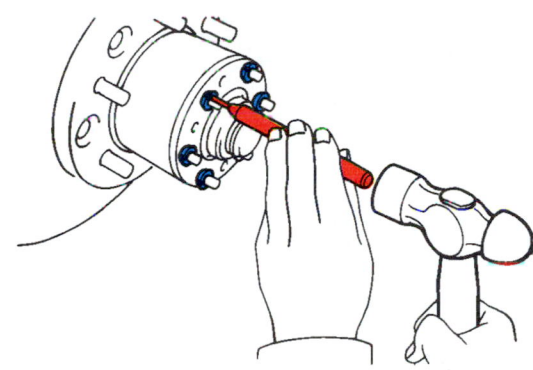

Figure 69.30 Installing the beveled washers.

During reassembly, the hub and drum are reinstalled on the outside of the axle housing. The retaining nut is tightened until it has little or no clearance.

After the hub is correctly reinstalled, install the axle and beveled washers. Pound the beveled washers in until the whole axle is seated (Figure 69.30).

■ DIFFERENTIAL PINION SEAL REPLACEMENT

A common repair to a differential is the replacement of the pinion seal. Parts involved in a typical pinion seal replacement are shown in Figure 69.31. A very important part of this job is to maintain the tension on the pinion bearing crush sleeve. One way of maintaining crush sleeve tension is to mark the nut and pinion shaft before disassembly (Figure 69.32). After reassembly, tighten an additional ⅛ turn past the mark.

Another method of maintaining pinion bearing crush sleeve tension is to use a dial indicator torque wrench when removing the nut. Note the torque

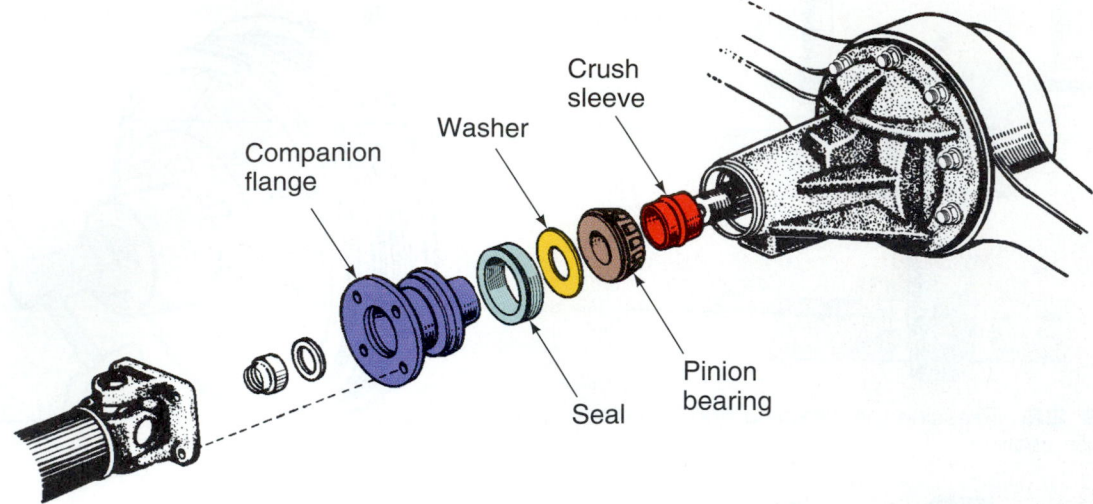

Figure 69.31 Parts involved in a pinion seal replacement.

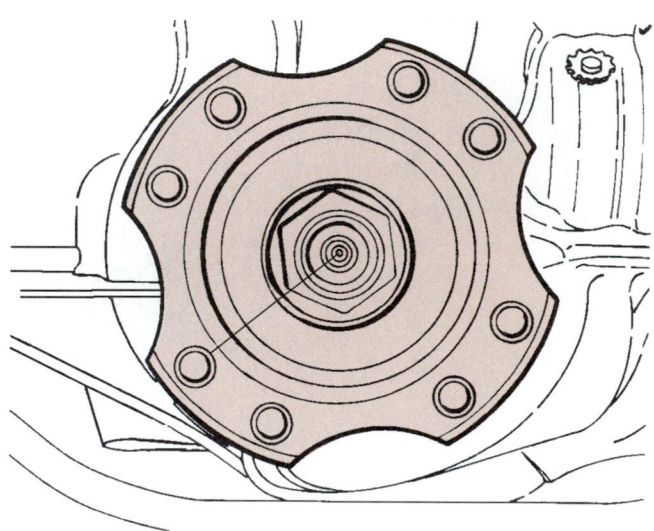

Figure 69.32 Mark the flange, nut, and pinion shaft before disassembly. *(Courtesy of General Motors Corporation, Service Technology Group)*

required to loosen it. During reinstallation of the nut, add 5 pounds to the former torque to recompress the old crush sleeve.

A student's mother's car had a leaking pinion seal. The student removed the driveshaft and pinion flange. He used a hammer to remove the flange. Then he replaced the seal and reassembled all of the parts. He tightened the pinion nut securely, without a torque wrench. Within 3000 miles, the pinion bearing failed because the bearing adjustment was too tight (Figure 69.33). The small end of the bearing will experience wear first.

Figure 69.33 This pinion bearing failed because it was too tight. The small ends of the rollers failed first. *(Courtesy of Quaker State Corporation)*

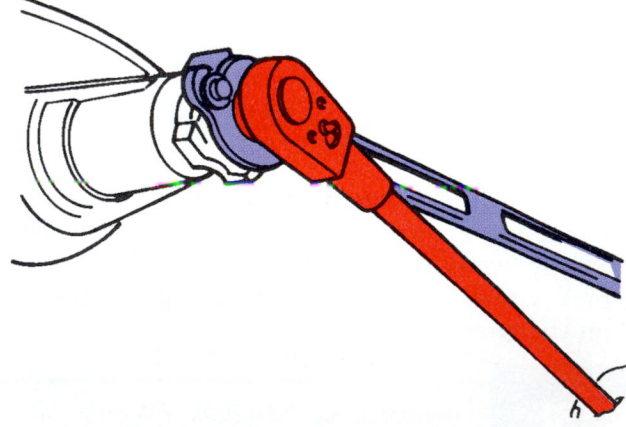

Figure 69.34 Use a bar to restrain the flange while loosening the pinion nut. *(Courtesy of General Motors Corporation, Service Technology Group)*

Always use a long bar to hold the yoke from turning while loosening the pinion nut (Figure 69.34). Using an impact wrench can damage the ring and pinion or the pinion bearings.

Figure 69.35 Use a puller to remove the flange.

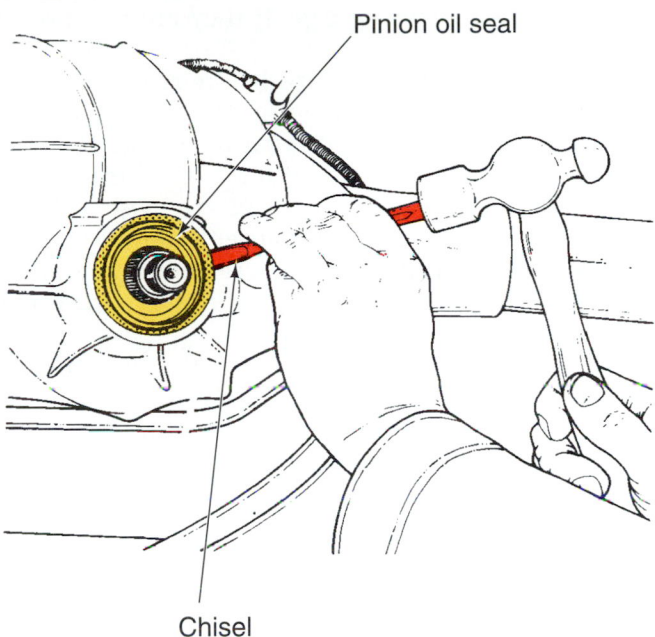

Figure 69.36 Removing a pinion seal. *[Courtesy of General Motors Corporation, Service Technology Group]*

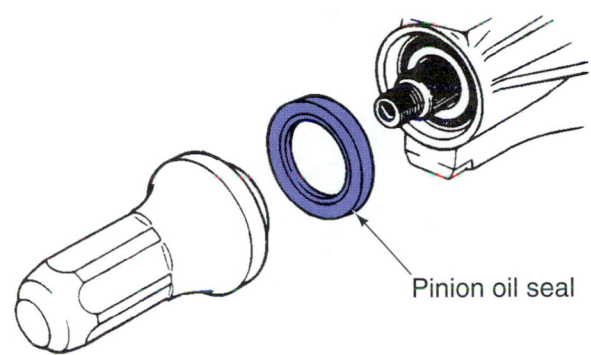

Figure 69.37 To avoid distorting the seal, install it using a seal installer. *[Courtesy of General Motors Corporation, Service Technology Group]*

Use a puller to remove the flange (Figure 69.35). Do not use a hammer. A flange can be easily bent, which will cause a driveline vibration. The student's car in the case history always had a driveline vibration after the pinion seal job because he bent the flange.

Remove the seal with a hammer and chisel (Figure 69.36). Install the new seal with a seal installer to avoid cocking or damaging it (Figure 69.37). Lube the lip of the new seal and the mating surface on the companion flange before reinstalling it.

■ DIFFERENTIAL REPAIR

Identify the Differential

Look for some sort of identification tag somewhere on the differential. It is usually located under one of the nuts holding the third member or one of the bolts holding the cover.

■ REMOVING A THIRD MEMBER

To remove a separable third member from a banjo housing, first pull the axles. If there is a drain plug, drain the housing of gear oil. Then remove all of the nuts from the studs around the outside of the third member. Each stud will usually have a copper washer under its nut. These are sometimes hard to see because of an accumulation of grease. Use a sharp scraper to remove each of these washers. If they are not removed,

the third member will not be able to come out of the housing after the nuts are removed.

A gasket will occasionally hold the third member to the housing if a previous technician has glued it in place on both of its sides. Usually the third member can be easily removed. Be careful not to drop it. It is heavier on one side than the other and must be carefully supported when lifting it away from the housing. Use a hydraulic jack if the vehicle is raised on jackstands (Figure 69.38).

■ DISASSEMBLING A SALISBURY AXLE

To remove a carrier from a Salisbury axle requires removal of the cover. Be sure to have a drain pan under it to catch the gear oil as it flows from the housing. Remove the C-locks and the axles as described earlier.

Mark the carrier housing caps if they are not already marked.

Mark the side bearing caps before removing them (Figure 69.39). After removing the bolts, use the pointed end of a roll head prybar in the bolt holes to pry the caps off. Pry back and forth, first on one side and then the other. Once the caps are loose, you can pry against a solid part of the carrier to remove it from the axle housing (Figure 69.40). Use wire to attach any shims to the caps so they can be returned to their original positions.

On some types of differentials, the drive pinion gear can be removed by unbolting it from the housing (Figure 69.41). On other types, the pinion nut and flange are removed as described earlier. To loosen a pinion nut, use a holding bar as described earlier or clamp the yoke in a vise if working on a replaceable third member or a replaceable pinion.

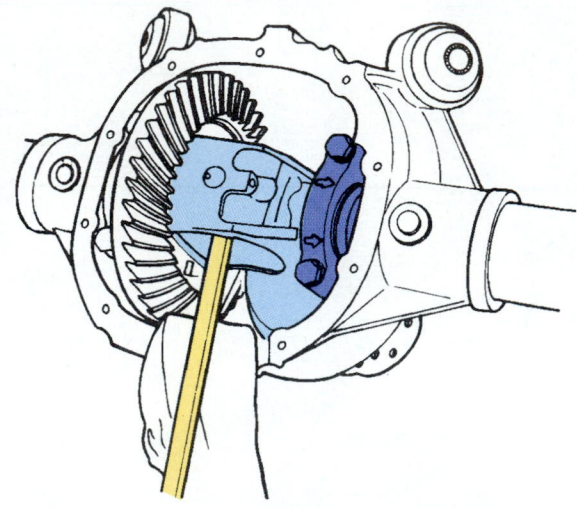

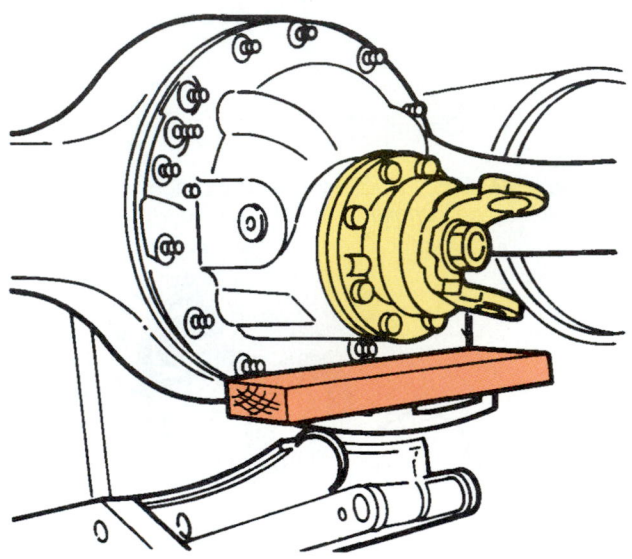

Figure 69.38 Use a jack to help lift out a differential. *(Courtesy of General Motors Corporation, Service Technology Group)*

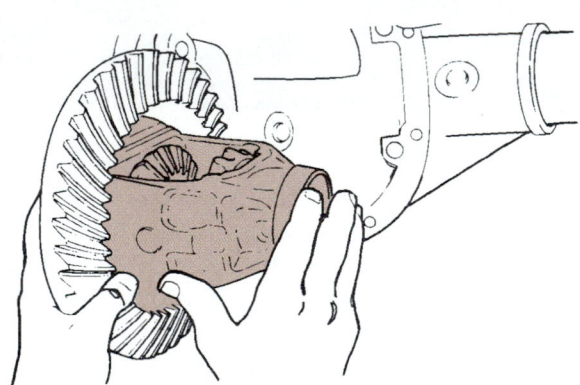

Figure 69.40 Removing an integral carrier from the housing. *(Courtesy of Ford Motor Company)*

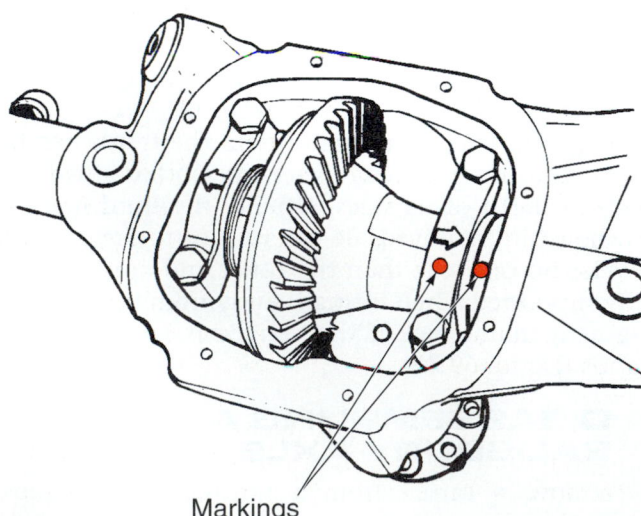

Markings

Figure 69.39 Mark the side bearing caps before removing them. *(Courtesy of Ford Motor Company)*

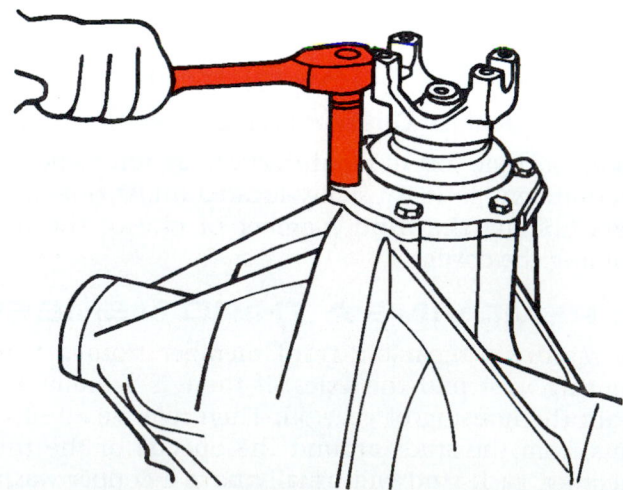

Figure 69.41 This differential has a removable pinion housing. *(Courtesy of General Motors Corporation, Service Technology Group)*

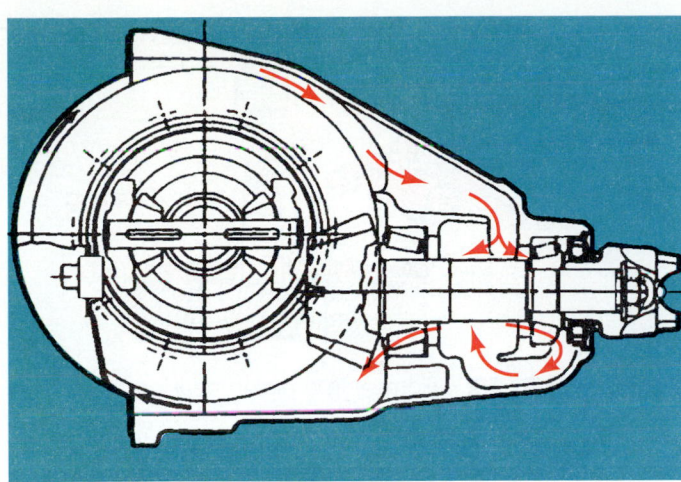

Figure 69.42 Be sure that the passageway that provides lubrication to the pinion bearings is clear. *(Courtesy of General Motors Corporation, Service Technology Group)*

■ CLEAN AND INSPECT PARTS

As you disassemble a differential, look for signs of wear or misalignment on bearing cups. Bearings can also fail due to lack of lubricant, the use of the wrong lubricant, or an incorrect bearing adjustment.

Be sure to keep bearing cups with their bearings in case they will be reused. Only use a puller to remove bearings that are worn or damaged. Good bearings can remain in place on the case or pinion gear. When removing a carrier bearing from the case, be sure to use the correct pullers to avoid damage. During reinstallation, be sure the case bearings are seated all the way against the shoulder on the case.

Inspect the ring and pinion gears for damaged teeth. Check the case to see that the lubricant passageway to the pinion bearings is clear (Figure 69.42). If gears are worn, adjusting the ring and pinion clearances will not solve the problem. The gear set will have to be replaced.

■ DIFFERENTIAL REASSEMBLY

If a ring and pinion are being replaced, replace them as a matched set. If one gear is damaged, they must both be replaced. Before reassembling any components in the differential, clean all parts in solvent and blow them dry. Remember to blow solvent back into the solvent tank instead of onto the floor or a workbench. Coat all parts with differential oil prior to reassembly. Install the side gears and pinions in the case and torque all of the case bolts.

If a ring gear is being replaced, it can be heated in oil to make it easier to install on the case. It must be installed perfectly flat. Pilot studs made from cutoff screws with a hacksaw slot for a screwdriver are a good idea (Figure 69.43). Be sure to torque the ring gear retaining screws after installation (Figure 69.44).

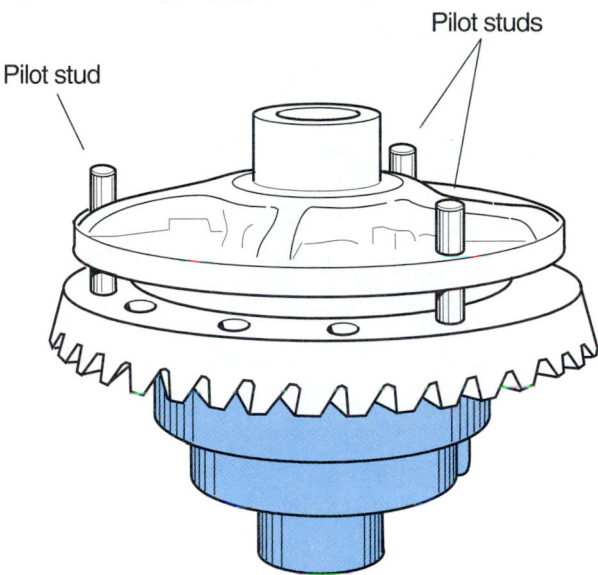

Figure 69.43 Pilot studs assist in installation of a ring gear.

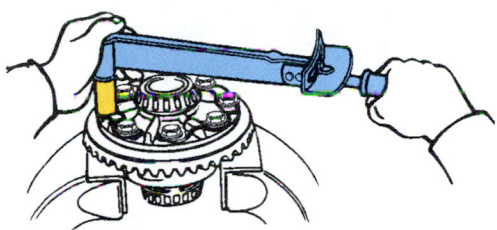

Figure 69.44 Torque the ring gear screws.

■ ADJUSTING A DIFFERENTIAL

Ring and Pinion Adjustments

A ring and pinion must mesh with each other in the correct position or noise and rapid wear can result. The two adjustments that are made are *pinion gear depth* and *ring and pinion backlash* clearance.

■ PINION GEAR DEPTH

Pinion gear depth is a measurement of how far the pinion gear extends into the differential housing. The measurement can vary according to where the pinion bearing race is positioned in the case. During manufacture, if more was machined from the case or if the ledge on the pinion gear had more or less metal on it, pinion depth will vary. This adjustment requires a special tool when starting from scratch without existing gears (Figure 69.45).

If a gear set is produced that will result in a depth that is different than stock, there will be a number on it that will tell how much compensation to make when setting up the clearance between the two (Figure 69.46). A +2 on the pinion gear means that the shoulder for the shims should be 0.002" too large for stock.

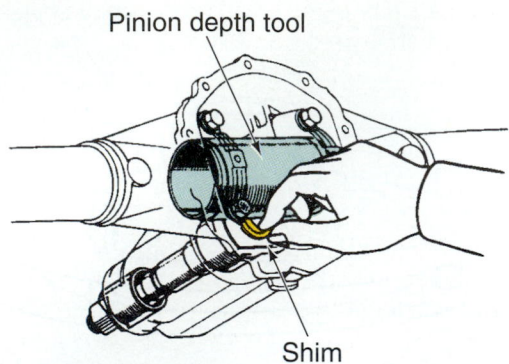

Figure 69.45 A special tool is used to adjust pinion depth.

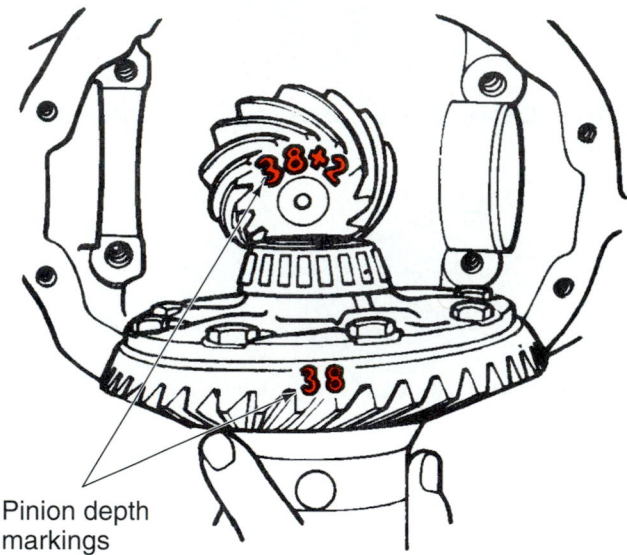

Figure 69.46 Pinion depth shim mark. *[Courtesy of Chrysler Corporation]*

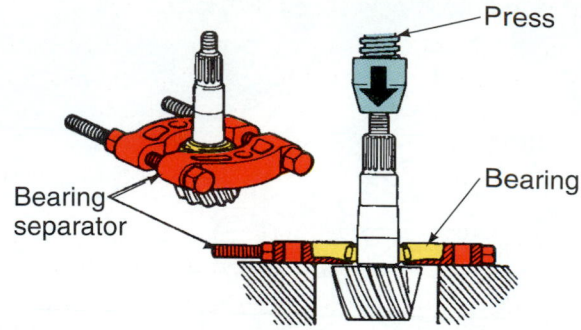

Figure 69.47 Removing a pinion bearing.

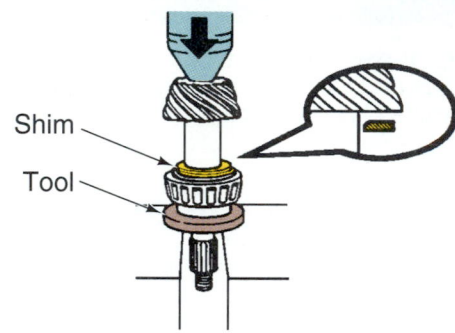

Figure 69.48 Installing a pinion bearing.

This calls for a shim that is 0.002" less than the shim that was on the old gear (if that gear had a "0" mark).

Shims vary in thickness. Use a micrometer to double-check the size of labeled shims. If a pinion depth gauge is not available, try checking the gear tooth pattern (covered later) while using a shim of half the thickness of the maximum shim specification. Tighten the pinion nut to 25 foot-pounds for the preliminary test so you do not ruin the crush sleeve.

To remove a pinion bearing, use a bearing separator and press (Figure 69.47). When pressing the shaft back into the bearing, use a tool or piece of pipe that is larger than the diameter of the pinion shaft (Figure 69.48). Do not forget to install the pinion depth shim under the bearing.

■ RING GEAR BACKLASH

Check gear backlash with a dial indicator (Figure 69.49). Backlash specifications vary by manufacturer.

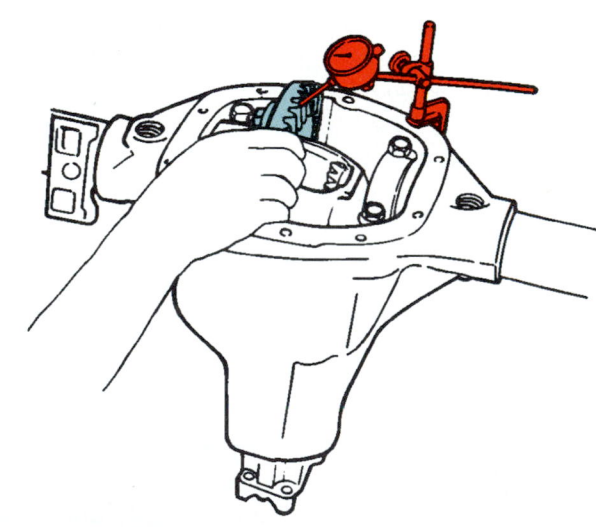

Figure 69.49 Checking ring gear backlash. *[Courtesy of Chrysler Corporation]*

Try to adjust backlash to the middle of the specifications. A setting that will work on any hypoid gear set is 0.007". Backlash that is too little can result in a broken gear set after expansion takes place. Also, the lubricant will not have enough room to be able to separate the metal of the gear teeth so they might become burned.

To check runout of the ring and pinion, read the backlash at several points around the gear. The maximum variance should be no more than 0.002" to 0.003".

SIDE BEARING PRELOAD

Case side bearings are adjusted to a specified preload. If bearings are too tight, they can fail and if they are too loose the ring gear can move in the case, causing noise and wear. A good indication of preload is that your little finger should strain when trying to turn the ring gear.

On a Salisbury axle, the preload and backlash adjustments are made with shims (Figure 69.50). Hold the ring gear away from the pinion and measure the clearance between the bearing and the case with feeler gauges (Figure 69.51). Adding a 0.010" shim will change the backlash by about 0.005". Add an additional 0.004" thickness to the shim pack on each side of the gear (0.008" total) to preload the bearings.

Shims are installed by pounding them in, either with a punch or with a special tool. Some shops have a case spreader that can be used instead (Figure 69.52).

A dial indicator is used with most spreaders. Be sure not to spread the case more than 0.020".

NOTE: *A case spreader is an example of one of the more expensive tools that a service manual will list for repair of a component. In the real world, many dealerships and independent shops do not purchase many of the more costly tools. The job can be done without this tool, but it is more time consuming to perform.*

In a separable third member, the side bearing adjustment is made in combination with the backlash adjustment too (Figure 69.53). Adjusting nuts with

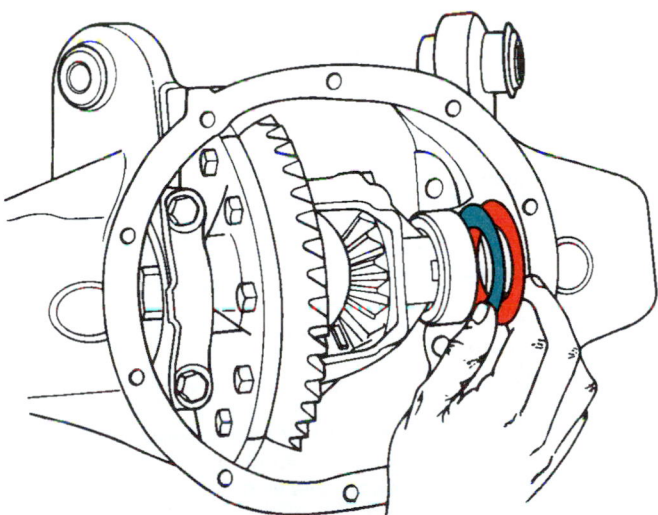

Figure 69.50 Preload and backlash settings made with shims. *(Courtesy of General Motors Corporation, Service Technology Group)*

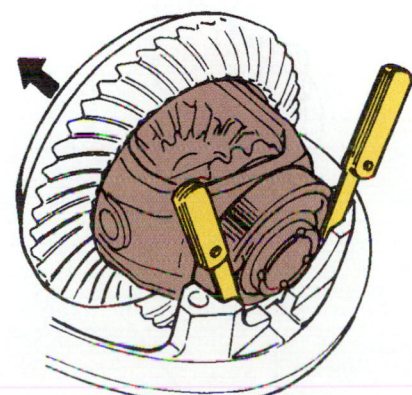

Figure 69.51 Measure clearance before determining the shim size. *(Courtesy of American Honda Motor Co., Inc.)*

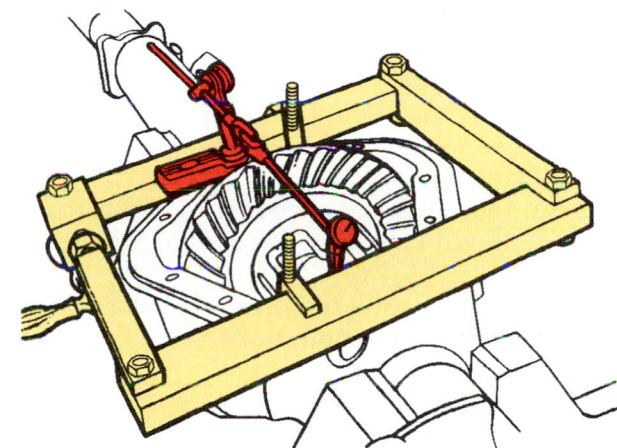

Figure 69.52 A case spreader. *(Courtesy of General Motors Corporation, Service Technology Group)*

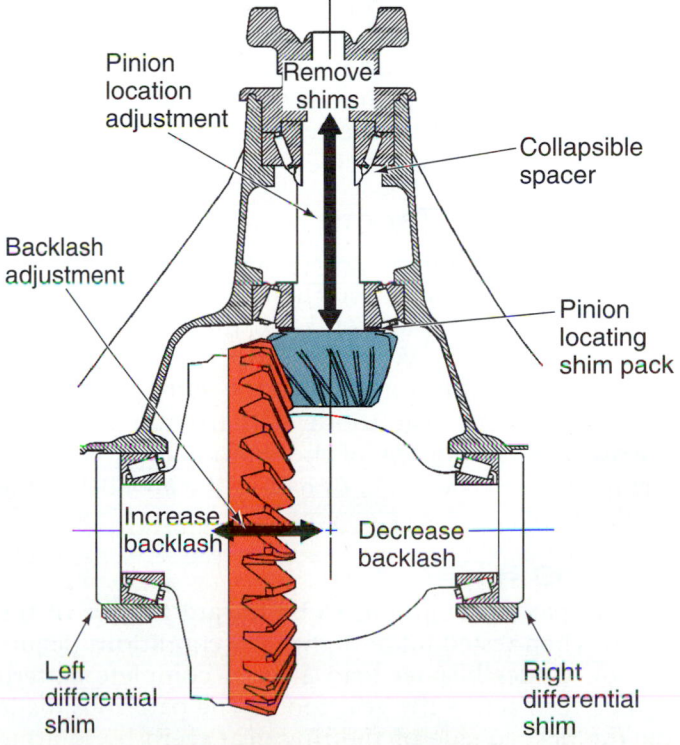

Figure 69.53 Adjustments to the ring and pinion gear positions. *(Courtesy of Ford Motor Company)*

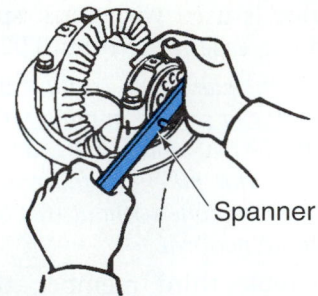

Figure 69.54 Adjusting the side bearing preload with a spanner.

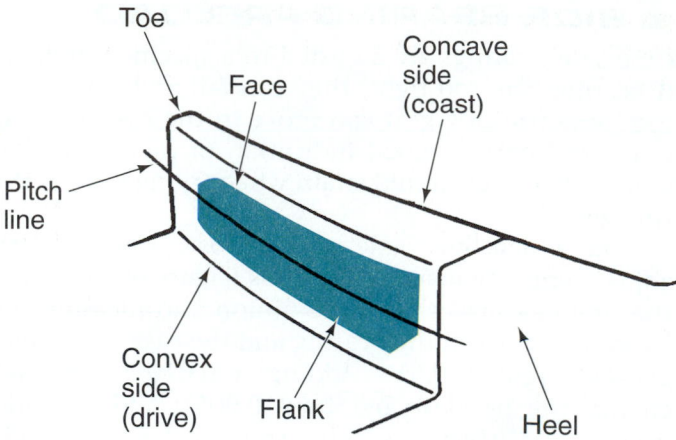

Figure 69.55 Gear tooth terms. *(Courtesy of General Motors Corporation, Service Technology Group)*

holes in them are turned with a spanner wrench (Figure 69.54). With the caps torqued to 25 foot-pounds, the adjusting nuts should be easy to turn in their threads. Be sure to keep the adjusting nut in its original bearing cap.

- Tighten the adjusters to move the ring gear against the pinion until all of the backlash clearance is removed.
- Tighten the adjusting nut on the side facing the ring gear teeth until there is the correct amount of backlash.
- Tighten each side an additional 1 to 2 notches to preload the side bearings. Check the repair manual for the specification.

CONTACT PATTERN

When a differential is adjusted, a **gear pattern** is taken off of the ring gear teeth. A coating of colored paste is painted onto the gear teeth. Manufacturers sell gear marking compound in different colors. Yellow is popular because the pattern is easier to see. A desirable pattern is located near the center of the gear teeth on both the drive and coast sides.

Gear Tooth Terms

The **drive side** of the gear teeth is convex and the **coast side** is concave (see Figure 68.30). The **heel** of a gear tooth is the outer end and the **toe** is the inner end (Figure 69.55). A **pitch line** runs through the center of the tooth from heel to toe. Other terms used when analyzing gear tooth patterns are **face** and **flank**. The face, or *top*, of the tooth is the area above the pitch line. The flank, or *root*, is the area below the pitch line.

Making a Pattern

A good pattern will tend to be toward the toe of the tooth when tested under light load conditions (Figure 69.56). Under heavier load a more complete pattern can be read across the gear tooth. The pattern is made on the desired side of the ring gear teeth by rotating the ring gear while holding the pinion gear lightly (Figure 69.57). A rag wrapped around the yoke is a

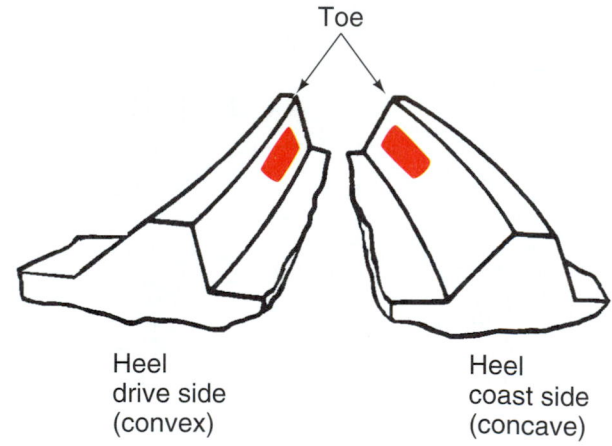

Figure 69.56 A pattern will be toward the toe of the gear tooth under light load conditions. *(Courtesy of General Motors Corporation, Service Technology Group)*

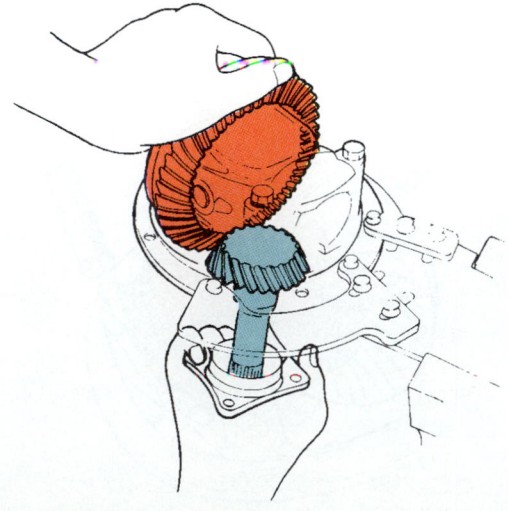

Figure 69.57 Hold resistance against one of the gears while rotating the other in both directions. *(Courtesy of American Honda Motor Co., Inc.)*

good way to put resistance on the pinion gear. Another method is to put a prybar between the side of the ring gear and the differential housing. Install a ½" ratchet and socket on the pinion gear nut to drive it.

Changing backlash or pinion depth changes both patterns. The change in a pattern when changing pinion depth is opposite of what you would expect. Pulling the pinion away from the ring moves the pattern reading toward the toe instead of the heel.

When changing pinion depth, leave the pinion seal out until the correct depth has been established. Then install the seal and the crush sleeve and complete the assembly procedure. When the patterns are at opposite ends of the tooth on the drive and coast sides, this is a pinion depth problem.

■ When the drive pinion is too far away from the centerline of the ring gear, the pattern on the drive side will be too far toward the heel and too far toward the toe on the coast side. This means the shim is too thin. Increasing the shim thickness will

cause the patterns to move back toward the center of the gear faces (Figure 69.58a).

■ When the shim is too thick, the drive pinion will be too close to the centerline of the ring gear. The result is a pattern that is too far toward the toe on the drive side of the tooth. It will be too far toward the heel on the coast side of the tooth. Putting in a thinner shim will move the pattern closer to center (Figure 69.58b).

■ PINION BEARING PRELOAD

The pinion bearings must be **preloaded** because of the tremendous amount of force on the gears during acceleration and deceleration. If the drive pinion gear were allowed to move, clearance between it and the ring gear would change. To preload the bearings a collapsible spacer is used (see Figure 68.20). The preload is checked by seeing how difficult it is to turn the drive pinion in its bearings. The more the nut on the pinion shaft is tightened, the more preload there will be on the bearings. The nut is tightened in small increments while checking repeatedly for drag with an inch-pound torque wrench (Figure 69.59). A dial-type indicator is best. Do not overtighten because you cannot back off the adjustment. Once the crush sleeve is compressed, you will have to replace it.

Check for the amount of effort it takes to turn the pinion nut. A typical reading is 15 to 25 inch-pounds, not very much. Do not take the reading right at the start of turning, but during turning. Be sure the bearings have been lubricated first if you want an accurate reading. Pound on both ends of the pinion shaft to seat the bearings.

When reusing pinion bearings, the preload is set to one-half the new preload specification. Always replace the crush sleeve when a differential is apart.

NOTE: *A new crush sleeve is difficult to collapse at first. Start to collapse it with an impact wrench and then use the bar and a breaker bar (see Figure 69.34).*

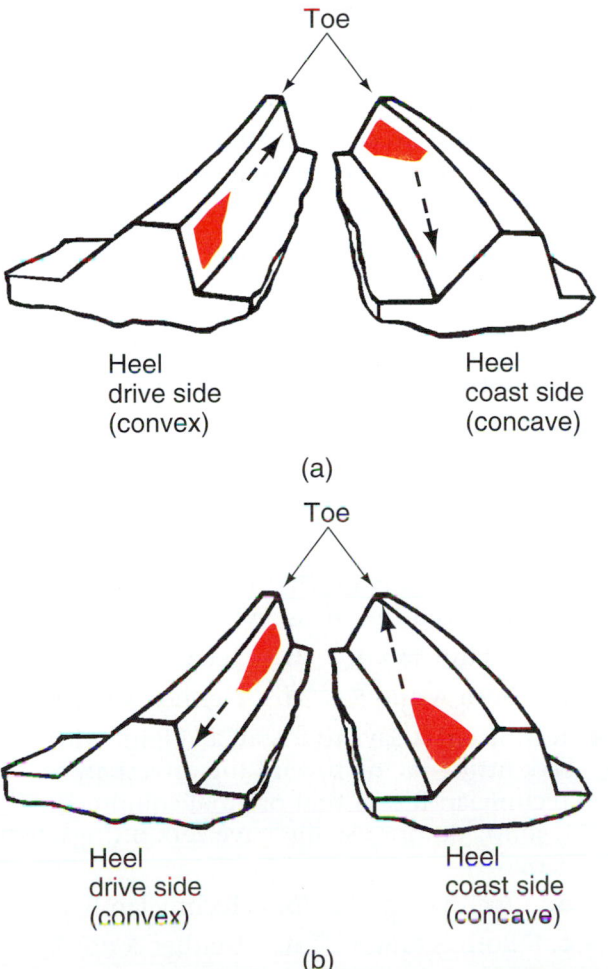

Toe

Heel   Heel
drive side   coast side
(convex)   (concave)

(a)

Toe

Heel   Heel
drive side   coast side
(convex)   (concave)

(b)

Figure 69.58 (a) Increasing the shim thickness will cause the patterns to move back toward the center of the gear faces. (b) Putting in a thinner shim will move the patterns closer to center. *(Courtesy of General Motors Corporation, Service Technology Group)*

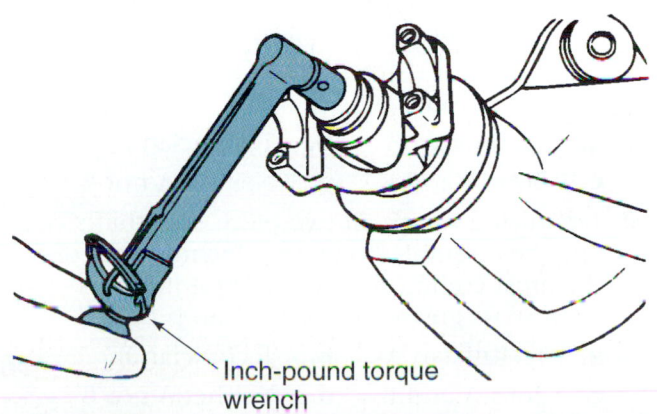

Inch-pound torque wrench

Figure 69.59 Checking pinion bearing preload. *(Courtesy of General Motors Corporation, Service Technology Group)*

Some differentials have solid spacers, rather than collapsible ones. These use shims to adjust the preload. The torque is adjusted to a specified amount.

RING AND PINION NOISE

If the pattern is good but there is a gear howl, replace the ring and pinion. Incorrect backlash can be isolated during a test drive. Noise during acceleration points to heavy heel contact. The ring gear needs to be moved closer to the pinion. The pattern is usually set slightly toward the toe because during acceleration it will move toward the heel.

If the noise happens when coasting in gear, the pattern will be too heavy on the toe of the gear tooth. Move the ring gear away from the pinion. Noise does not usually occur during float.

REVIEW QUESTIONS

Driveshaft and Axle

1. What is the term for an axle that turns with the wheel?

2. What can you wrap around the U-joint cups to keep them from falling off?

3. What is it called when a driveshaft is assembled in a position where the front and rear U-joints are not parallel?

4. What is it called when small indentations wear into the bearing surface?

5. Which part is hardened, the U-joint cap or the ear of the driveshaft yoke?

6. The four test drive conditions for the driveline are drive, cruise, coast, and _____.

7. Which gears in a differential are the ones that tend to break when a spinning rear wheel gets traction?

8. A limited slip axle can be tested for ____ torque.

9. After a pinion nut is retightened, how much further should you tighten it to recrush the sleeve?

10. When a tapered bearing is too tight, which end will fail?

Differential

1. What part is held with a bar while loosening the drive pinion nut?

2. If a ring gear is worn, what other gear must also be replaced?

3. The two adjustments that are made are pinion gear depth and ring and pinion ____ clearance.

4. Which setting does the number found on a ring and pinion affect?

5. When testing pinion depth setting, tighten the pinion nut to 25 foot-pounds for the preliminary test so you do not ruin the _____ sleeve.

6. On a Salisbury axle, adding a 0.____ " shim will change the backlash by about 0.005".

7. What is the name of the outer end of the gear tooth, heel or toe?

8. Does pulling the pinion away from the ring move the pattern reading toward the toe or heel of the tooth?

9. When changing pinion depth, leave the ____ ____ out until the correct depth has been established.

10. A typical pinion bearing preload reading is ____ inch-pounds.

ASE STYLE REVIEW QUESTIONS

Driveshaft and Axle

1. Technician A says that a dent in the driveshaft can weaken it. Technician B says that cracks in the surface of a driveshaft almost always come from the inside. Who is right?
   - **a.** Technician A
   - **b.** Technician B
   - **c.** Both A and B
   - **d.** Neither A nor B

2. Technician A says all two-piece driveshafts can only be assembled one way. Technician B says the most common reason a U-joint fails is because its grease dries out. Who is right?
   - **a.** Technician A
   - **b.** Technician B
   - **c.** Both A and B
   - **d.** Neither A nor B

3. Technician A says that a very small amount of runout of a driveshaft is enough to cause vibration. Technician B says that damaged side gears are usually on the side opposite to the side that received the stress. Who is right?
   - **a.** Technician A
   - **b.** Technician B
   - **c.** Both A and B
   - **d.** Neither A nor B

4. Technician A says to install a U-joint with the zerk fitting facing toward the driveshaft. Technician B says that original equipment U-joints do not usually have zerk fittings. Who is right?
   - **a.** Technician A
   - **b.** Technician B
   - **c.** Both A and B
   - **d.** Neither A nor B

5. Technician A says that during a turn when the outside wheel has a bad bearing, the noise will become worse. Technician B says applying the brakes can cause the noise level from a bad bearing to become higher as the brakes contact

the drum or rotor that has the bad bearing. Who is right?

a. Technician A   b. Technician B
c. Both A and B   d. Neither A nor B

6. Technician A says axle seals are sometimes part of the axle bearing. Technician B says axle seals are sometimes separate from the axle bearing. Who is right?

a. Technician A   b. Technician B
c. Both A and B   d. Neither A nor B

7. Technician A says that a limited slip axle that chatters during turns could have the wrong lubricant. Technician B says that a Salisbury axle must have the cover removed from the back to remove the axles. Who is right?

a. Technician A   b. Technician B
c. Both A and B   d. Neither A nor B

8. Technician A says that on most pumpkin-type differentials, the bearing rides on a hardened area of the axle rather than on an inner bearing race. Technician B says to press an axle bearing and retaining ring onto the axle at the same time. Who is right?

a. Technician A   b. Technician B
c. Both A and B   d. Neither A nor B

Differential

1. Technician A says a bent pinion flange will result in driveline vibration. Technician B says to remove a pinion flange using a brass hammer. Who is right?

a. Technician A   b. Technician B
c. Both A and B   d. Neither A nor B

2. Technician A says to remove both carrier bearings from the case for inspection. Technician B says if the ring and pinion are worn, to readjust them for proper clearance. Who is right?

a. Technician A   b. Technician B
c. Both A and B   d. Neither A nor B

3. Technician A says a +2 on a pinion gear means that a shim that is 0.002" thicker must be used. Technician B says that a backlash setting of

0.007" will work on any differential. Who is right?

a. Technician A   b. Technician B
c. Both A and B   d. Neither A nor B

4. Technician A says on a pumpkin-type differential, the side bearing preload and gear backlash settings are made with shims. Technician B says a pattern taken under light load will tend to be toward the heel of the tooth. Who is right?

a. Technician A   b. Technician B
c. Both A and B   d. Neither A nor B

5. Technician A says to be sure there is no resistance against the ring or pinion gears when rolling a contact pattern. Technician B says changing backlash or pinion depth results in a change in both patterns. Who is right?

a. Technician A   b. Technician B
c. Both A and B   d. Neither A nor B

6. Technician A says when the patterns are at opposite ends of the tooth on the drive and coast sides, backlash is the problem. Technician B says a collapsible spacer is used to preload the pinion bearings. Who is right?

a. Technician A   b. Technician B
c. Both A and B   d. Neither A nor B

7. Technician A says if you overtighten pinion bearing preload, back off on the adjustment ⅛ turn. Technician B says if there is noise during acceleration, heel contact is too much. Who is right?

a. Technician A   b. Technician B
c. Both A and B   d. Neither A nor B

8. Technician A says to pound side bearing shims into the differential housing with a hammer. Technician B says to check gear ring backlash with a dial indicator. Who is right?

a. Technician A   b. Technician B
c. Both A and B   d. Neither A nor B

CHAPTER 70

Front Wheel Drive (Transaxle and CV Joint) Fundamentals

■ OBJECTIVES

Upon completion of this chapter, you should be able to:

✔ Describe differences between front and rear wheel drive trains.

✔ Tell the names of parts of a transaxle.

✔ Trace the power flow through four- and five-speed transaxles.

■ INTRODUCTION

A front wheel drive car has a transaxle, which combines a transmission and differential into one unit. Drive axles extend to the front wheels out of each side of the transaxle. At each end of the drive axle is a constant velocity universal joint, called a CV joint.

A transaxle can be either manual (Figure 70.1) or automatic (Figure 70.2). The theory of operation of the clutch, transmission, and differential sections of a transaxle is almost identical to those components on

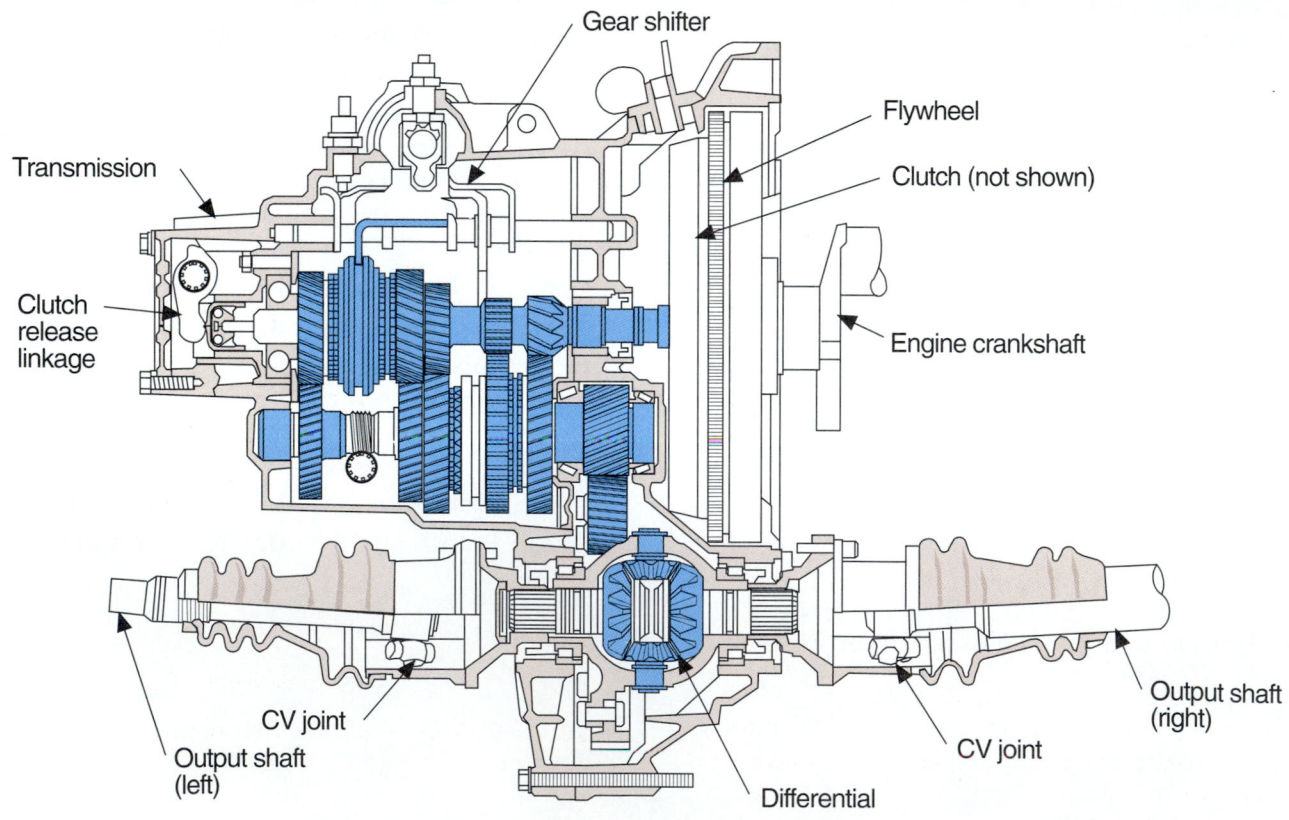

Gear shifter

Transmission

Clutch release linkage

Flywheel

Clutch (not shown)

Engine crankshaft

CV joint

Output shaft (left)

Output shaft (right)

CV joint

Differential

Manual transaxle

Figure 70.1 A manual transaxle. *(Courtesy of Chrysler Corporation)*

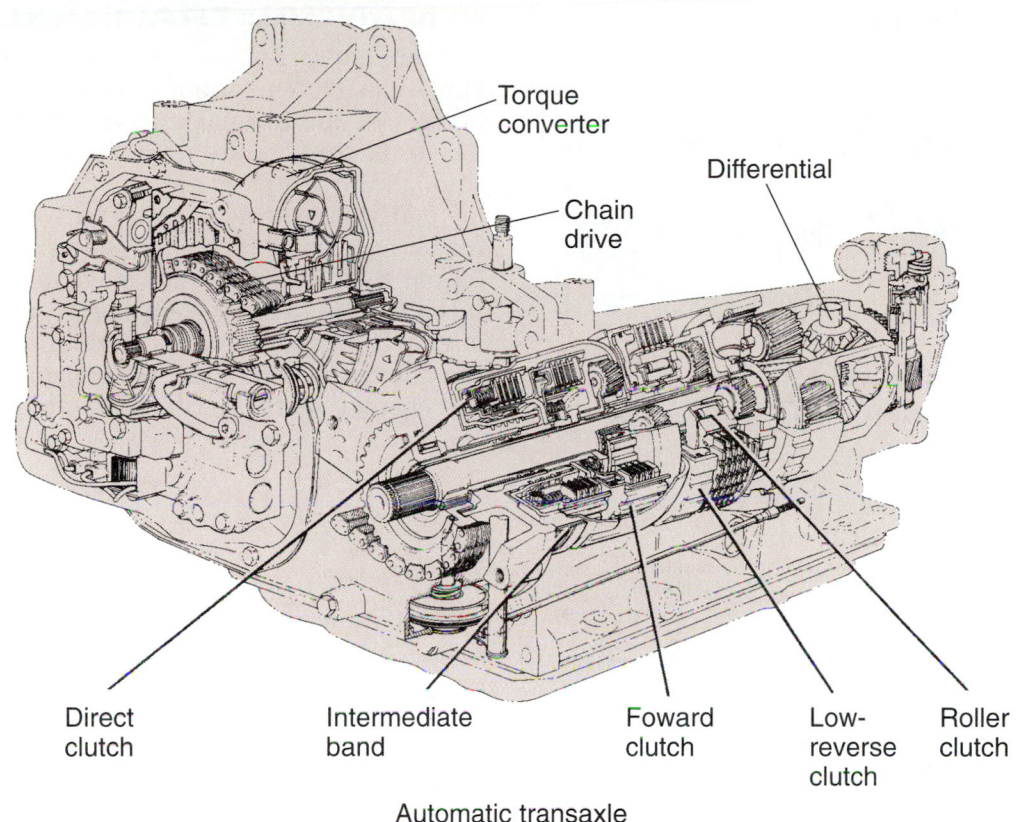

Automatic transaxle

Figure 70.2 An automatic transaxle. *(Courtesy of General Motors Corporation, Service Technology Group)*

rear wheel drive vehicles. Those fundamentals are covered in previous chapters. This chapter will emphasize things that are unique to transaxles.

■ FRONT WHEEL DRIVE

Most cars until the last twenty years came equipped with rear wheel drive. Front wheel drive (FWD) first appeared on an American car in 1930, on the Auburn. The Cord, built later in the 1930s, also used front wheel drive. The next American-made car to be built with front wheel drive was the 1966 Oldsmobile Toronado. Front wheel drive showed up on a few more vehicles but remained limited until the mid 1980s. Today, a majority of cars come with front wheel drive.

Advantages of front wheel drive include reduced weight and a more efficient drive train, resulting in better fuel economy. When combined with MacPherson struts there is less unsprung weight for better handling too. The transmission hump necessary in rear wheel drives is also eliminated.

A few front wheel drive engines have been mounted in the engine compartment longitudinally, from front to rear with the transaxle (Figure 70.3). Most transaxles are mounted in a sideways (transverse)

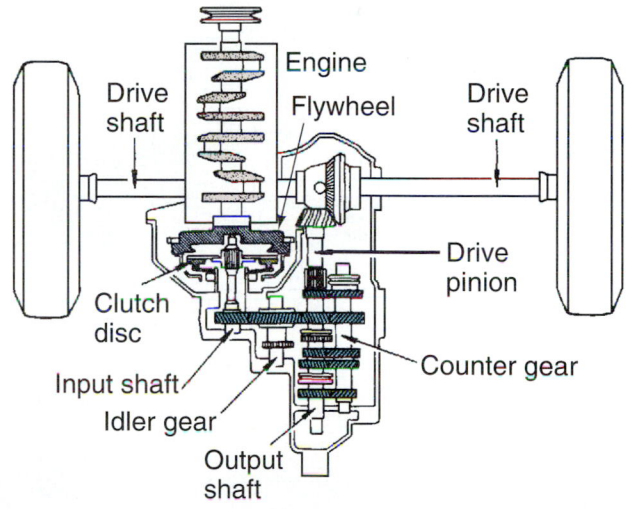

Front to rear

Figure 70.3 A front to rear engine and transaxle.

fashion (Figure 70.4). The transaxle is bolted to the engine. A clutch or torque converter connects it to the crankshaft.

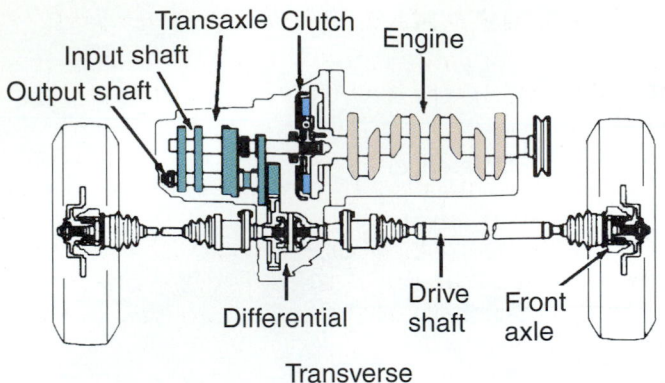

Figure 70.4 A side-to-side engine and transaxle.

■ MANUAL TRANSAXLE

Manual transaxles and transmissions use the same kind of a clutch. Some clutch release mechanisms operate in the normal manner described in Chapter 62. Other clutches are released by a pushrod running through the center of the input shaft (Figure 70.5).

Transaxle part names vary with different manufacturers. There is an input shaft and an output shaft. The output shaft is commonly called an *intermediate shaft*. It can also be called a mainshaft, countershaft, or pinion shaft because it drives the pinion gear that turns the differential ring gear. For explanation purposes in this chapter, it will be called an intermediate shaft.

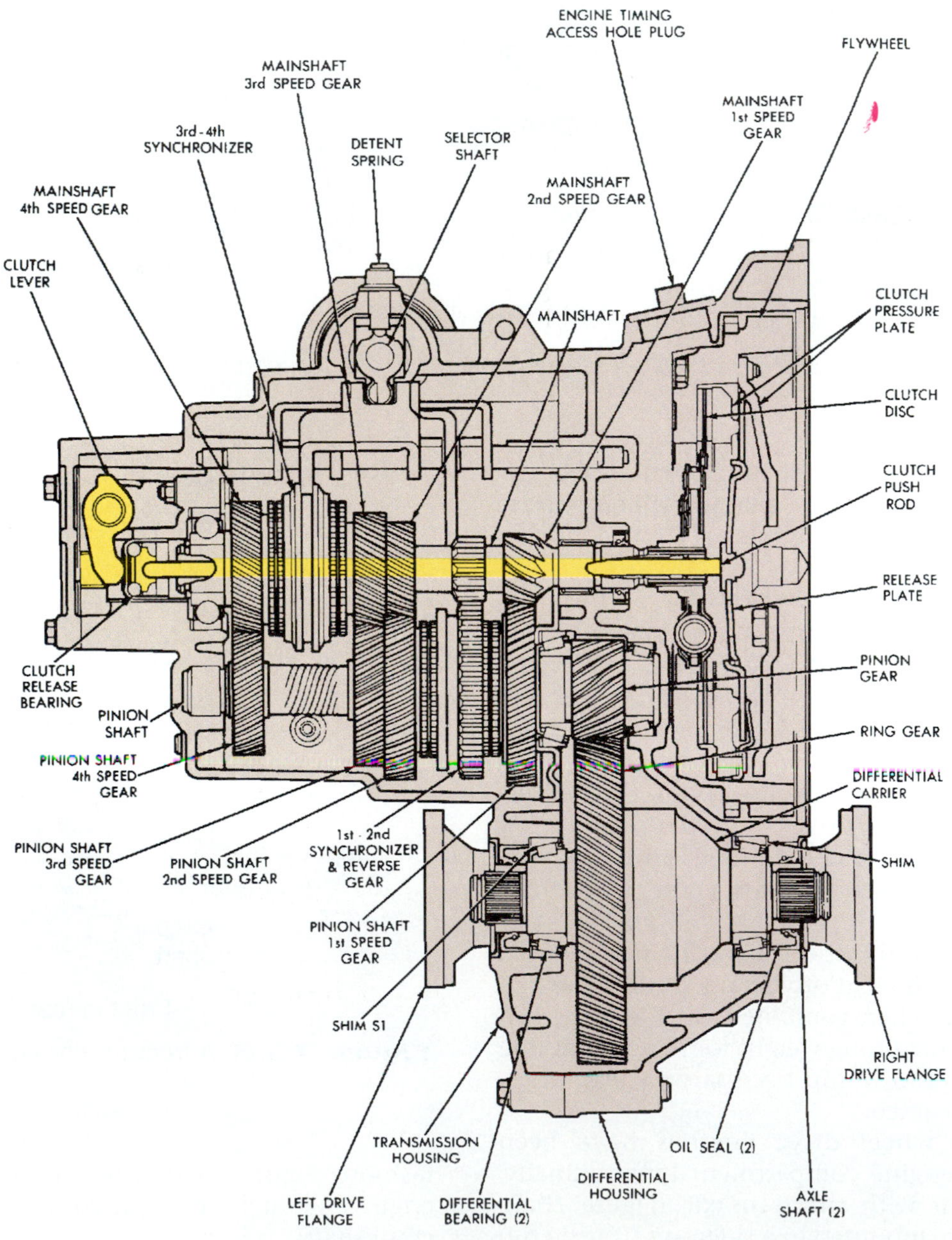

Figure 70.5 This transaxle has a clutch apply rod that runs through the inside of the input shaft. *(Courtesy of Chrysler Corporation)*

The gears and synchronizers are the same idea as a manual transmission (see Figure 70.1). All of the gears are in constant mesh except for a sliding reverse idler. To save space, narrower gears and pressed-fit synchro hubs are sometimes used.

Transaxles have three parallel shafts. The input shaft is usually located above the intermediate shaft (called that because it is the shaft in the middle of power flow). The third parallel shaft is the differential assembly. Input shaft gears directly drive output shaft gears. Remember in the rear wheel drive transmission, the input shaft drove a countergear that powered gears on the mainshaft. In the transaxle, one of the shafts has been eliminated. Some transaxles have an additional shaft when the design calls for even more space saving, but most have the kind of gear train described here.

The gear shafts are primarily supported by larger ball, roller, or tapered roller bearings rather than shafts and needle bearings. End play between the gears is controlled by thrust washers or spacers between the bearing end plate and case.

■ SHIFT LINKAGE

Transverse transaxles are sometimes shifted by cables (Figure 70.6). Other times, shift linkage is used (Figure 70.7). There are two shift cables or rods. One of them moves a selector on the transaxle that determines which of the shift forks will move. The other one moves the shift fork back and forth. An advantage to cables is that engine shake is not transmitted back to the driver's hand on the shift lever.

■ TRANSAXLE DIFFERENTIAL

The purpose of a differential is to allow the wheels to turn at different speeds when rounding corners. This is accomplished the same way as in the rear wheel drive differential, with pinion gears and side gears (Figure 70.8). There is no need to turn power 90° as in rear wheel drive so an ordinary helical gear set (like the kind used in the transmission section) is used instead of bevel gears. Power from the differential side gears is transmitted to the front drive axles through axle shafts (Figure 70.9).

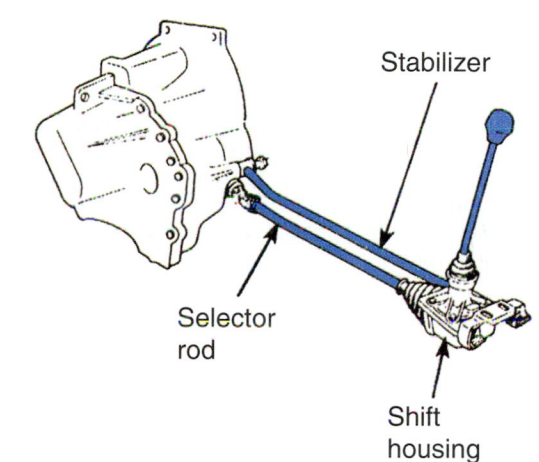

Figure 70.7 A transaxle shifted by linkage. *(Courtesy of Ford Motor Company)*

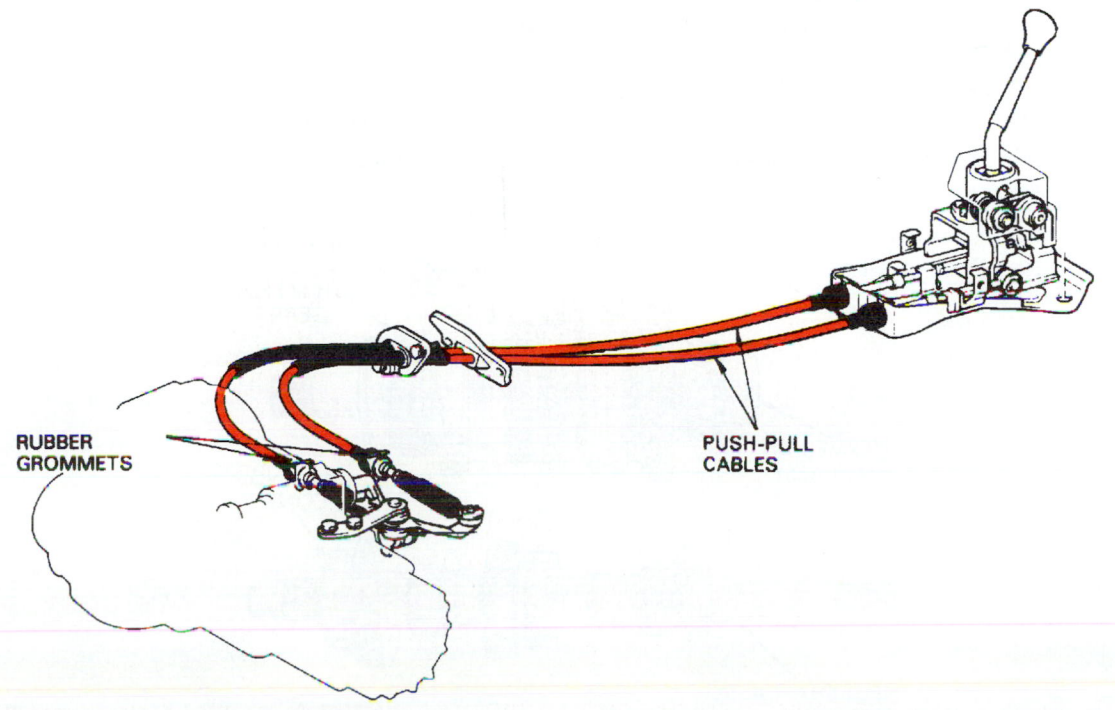

Figure 70.6 A transaxle shifted by cables.

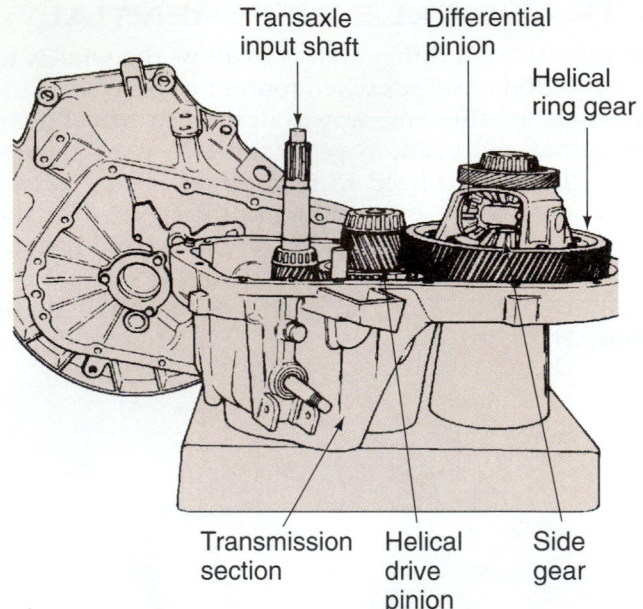

Figure 70.8 A transaxle has a differential section and helical drive gears. *(Courtesy of General Motors Corporation, Service Technology Group)*

■ **TRANSAXLE POWER FLOW**

There are different designs in transaxles. Some of them have gears clustered (two or more fixed gears) on the intermediate shaft, and others have them clustered on the input shaft. The four-speed transaxle in Figure 70.10 has an input shaft that has all of the fixed gears on it. This is a true cluster gear. Compare that to the four-speed transaxle in Figure 70.11. It has two fixed gears and two synchronized gears on the input shaft.

In the five-speed transmission used in the following power flow explanation, the fixed gears for first, second, and reverse are on the input shaft. The fixed gears for third, fourth, and fifth are on the intermediate shaft. In all gear ranges, power flow leaves the transmission intermediate shaft and continues through the drive pinion to the axles. Because the engine is mounted sideways, the axles run parallel to the input shaft.

■ Low gear—power comes in through the input shaft first gear. The first and second synchro clutch selects first gear on the intermediate shaft providing a gear reduction (Figure 70.12).

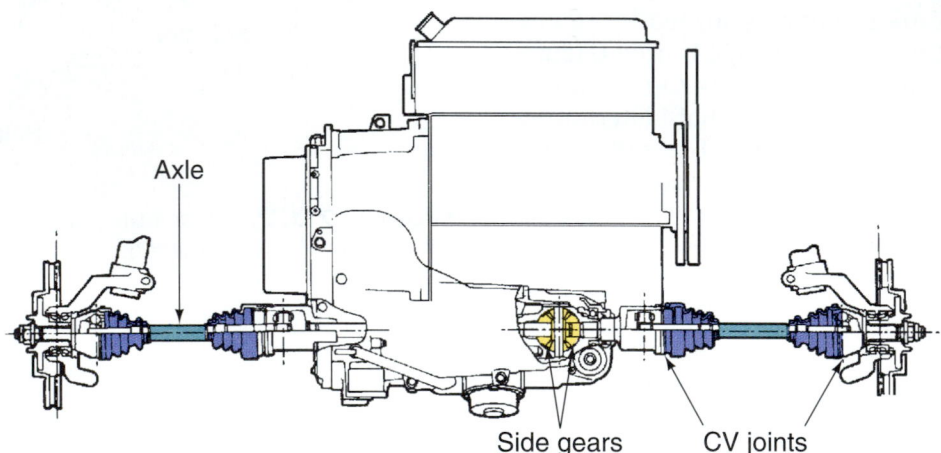

Figure 70.9 Power from the differential side gears is transmitted to the front drive axles through axle shafts. *(Courtesy of Ford Motor Company)*

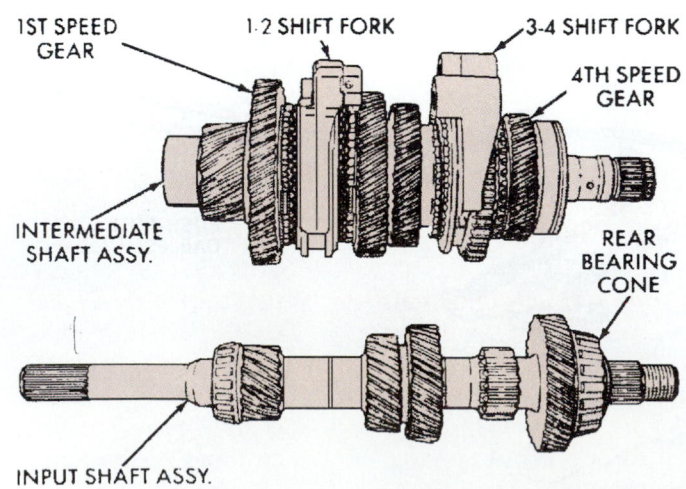

Figure 70.10 A transaxle with all of the fixed gears on the input shaft. *(Courtesy of Chrysler Corporation)*

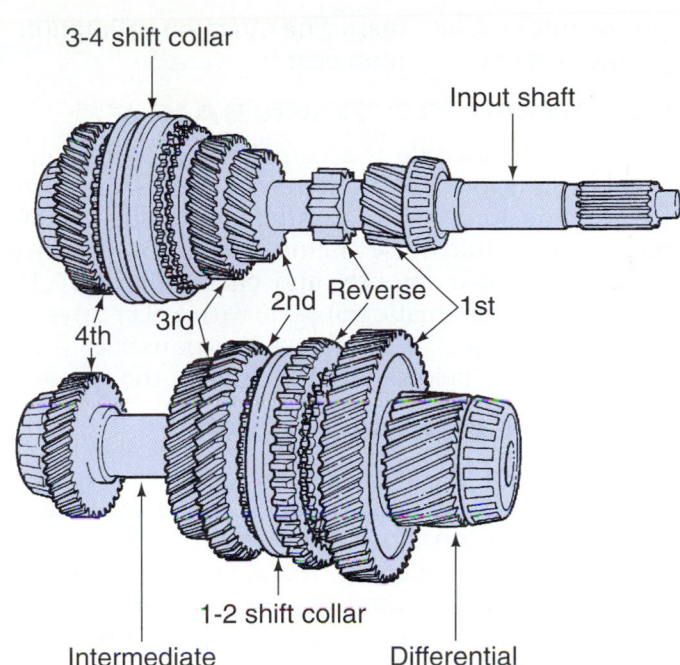

Figure 70.11 A transaxle with two fixed gears and two synchronized gears on the input shaft. *[Courtesy of General Motors Corporation, Service Technology Group]*

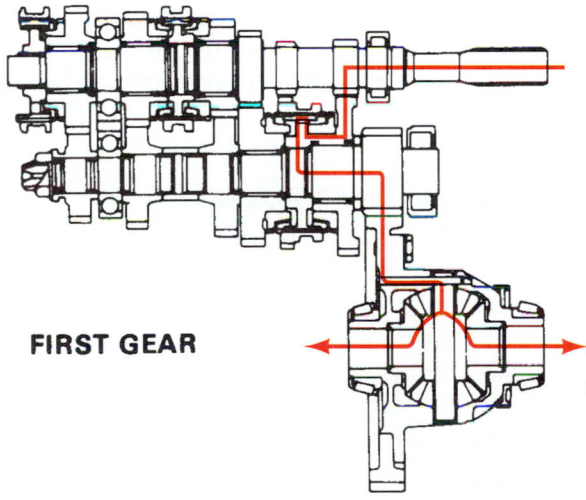

Figure 70.12 Power flow in first gear.

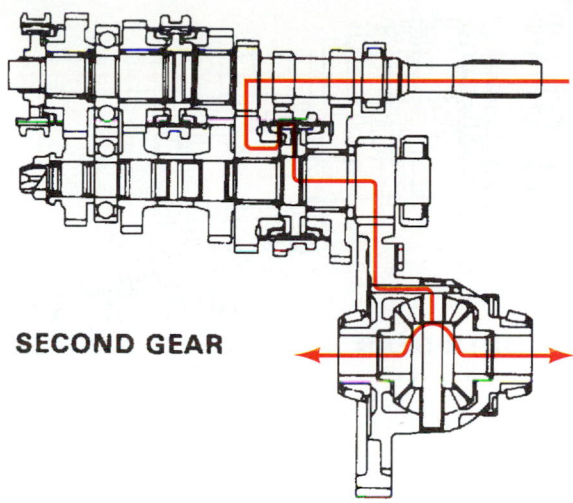

Figure 70.13 Power flow in second gear.

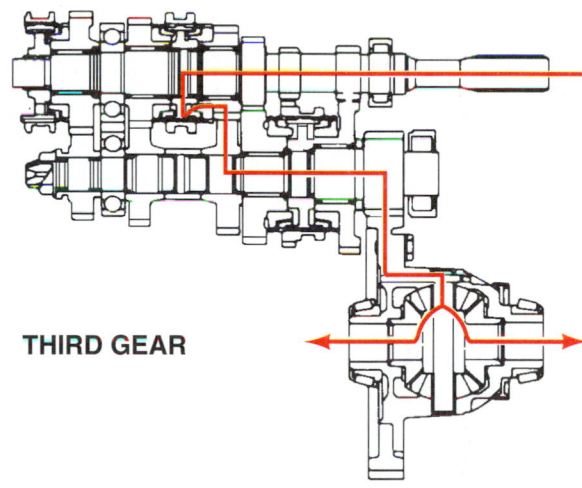

Figure 70.14 Power flow in third gear.

- Second gear—power comes in through the input shaft second gear. The first and second synchro clutch selects second gear on the intermediate shaft providing a gear reduction (Figure 70.13).
- Third gear—power comes in through the input shaft third gear, which is connected to the input shaft by the third and fourth synchro clutch. The fixed third gear on the intermediate shaft receives the power, providing a gear reduction (Figure 70.14).
- Fourth gear—power comes in through the input shaft fourth gear, which is connected to the input shaft by the third and fourth synchro clutch. The

fixed fourth gear on the intermediate shaft receives the power, providing a gear reduction (Figure 70.15).

NOTE: *In a typical rear wheel drive transmission, fourth gear would be accomplished by locking the input shaft to the output shaft for a 1:1 ratio. Notice that in a transaxle, power flow is always through gears in each gear range.*

- Fifth gear—power comes in through the input shaft fifth gear, which is connected to the input shaft by the fifth gear synchro clutch. The fixed fifth gear on the intermediate shaft receives the power (Figure 70.16). It is smaller than the gear on the input shaft so it provides an overdrive.
- Reverse gear—in reverse, power comes in through the input shaft where it leaves through a fixed spur gear (Figure 70.17). With all of the other synchro clutches in the neutral range, a sliding spur idler gear is moved into contact with a spur gear on the outside of the 1–2 synchro clutch, which is splined

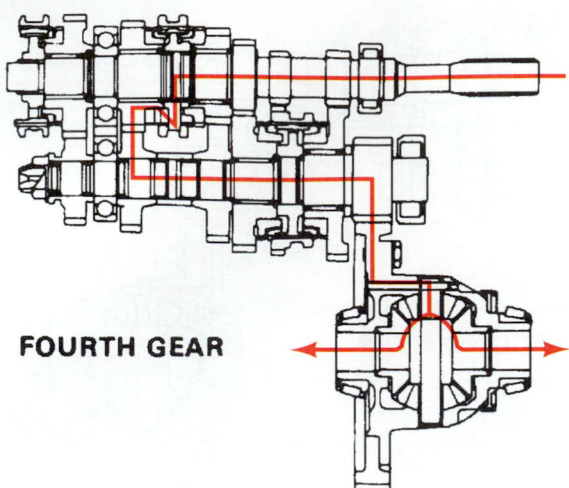

Figure 70.15 Power flow in fourth gear.

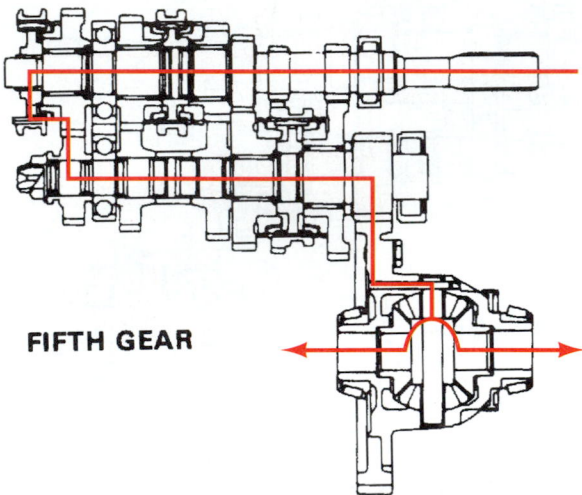

Figure 70.16 Power flow in fifth gear.

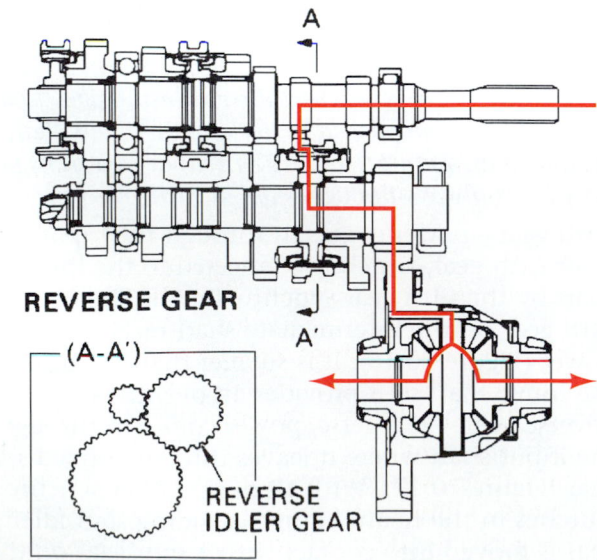

Figure 70.17 Power flow in reverse gear.

to the intermediate shaft. The direction of rotation of the differential pinion gear is reversed.

AUTOMATIC TRANSAXLE

An automatic transaxle is a combination of an automatic transmission and a differential. The same parts and operation apply, with the differential being the difference. With a transverse engine, power flow is either through gears or a sprocket and chain. Figure 70.18 shows a typical automatic transaxle with a gear drive. A chain drive (Figure 70.19) allows the transaxle to be mounted slightly below and to the side of the engine.

FRONT DRIVE AXLES

There is a major difference between rear wheel drive axles and front wheel drive axles. The front wheel drive axles have constant velocity (CV) joints at each end. CV joints are necessary because the axles are driven at sharper angles than can be tolerated by ordinary universal joints (see Chapter 68). They also allow steering of the front wheels to take place during power transmission (Figure 70.20).

An ordinary universal joint changes its output speed twice in every revolution when run at an angle. A true constant velocity joint provides a constant output speed no matter what the shaft angle.

In a rear wheel drive car, the driveshaft turns very fast because it is positioned before the gear reduction of the differential. Front wheel drive axles turn slower than a rear wheel driveshaft by the difference of the axle ratio (about one-third as slow).

AXLE SHAFT PARTS

The drive axle is called a half shaft or an *axle shaft*. There is a short shaft at the outside end called a **stub shaft** or *stub axle*. It is splined to the front hub so it can drive the front wheels. Some axle shafts have a stub shaft on the inside too. Others have a female-type splined yoke that slides over male splines similar to a rear wheel drive transmission output shaft.

CV Joints

At each end of a front wheel drive axle is a constant velocity joint, called a **CV joint** (see Figure 70.9). There are three ways of classifying CV joints.
- Inboard and outboard
- Fixed and plunge
- Ball and tripod

The inside end of the axle shaft is commonly called the **inboard side** and the outside end nearest the wheel is called the *outboard side*.

Plunge CV Joints

The inboard CV joint is called a **plunge joint**. When the suspension system deflects, the axle's length must change (Figure 70.21). The joint that takes care of this change in length is the plunge joint. It acts like a slip

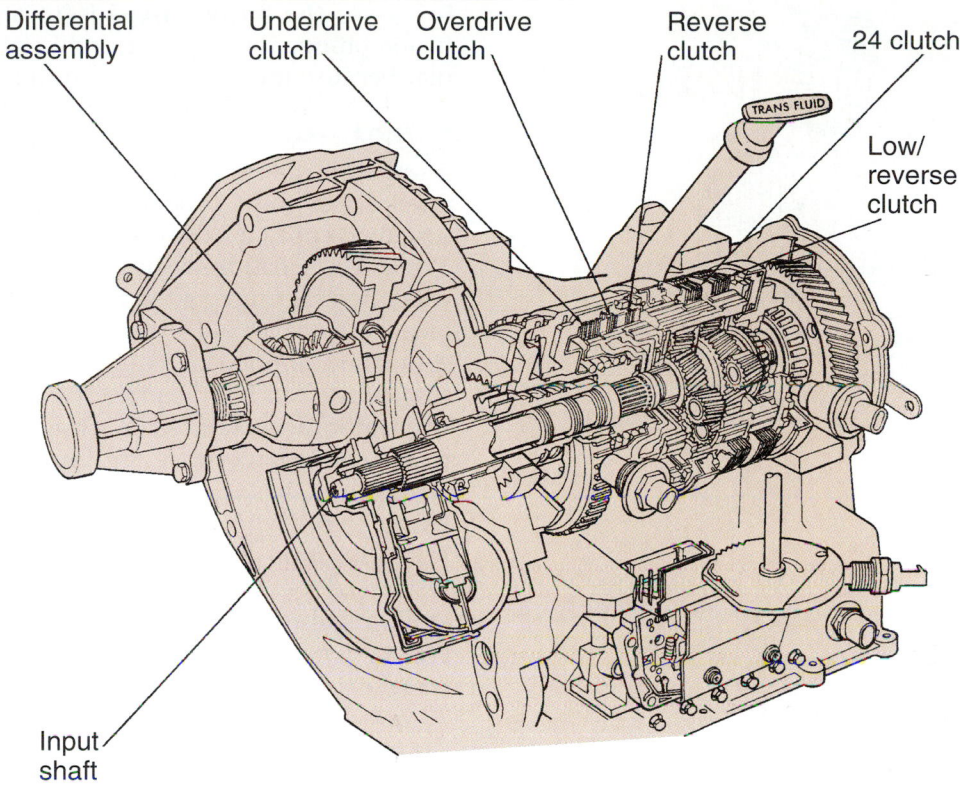

Figure 70.18 An automatic transaxle with a gear drive. *(Courtesy of Chrysler Corporation)*

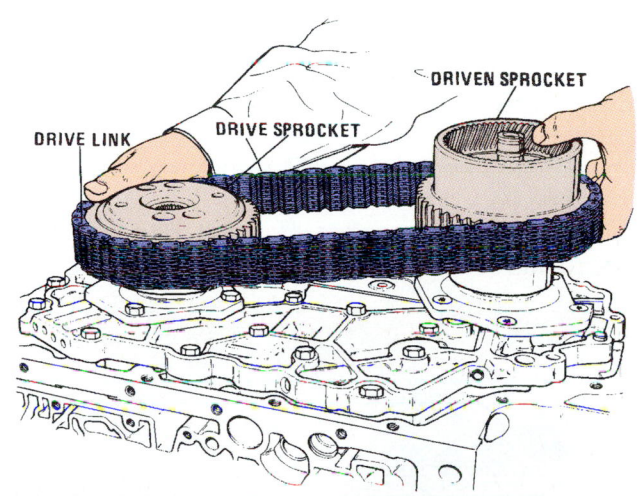

Figure 70.19 An automatic transaxle with a chain drive. *(Courtesy of General Motors Corporation, Service Technology Group)*

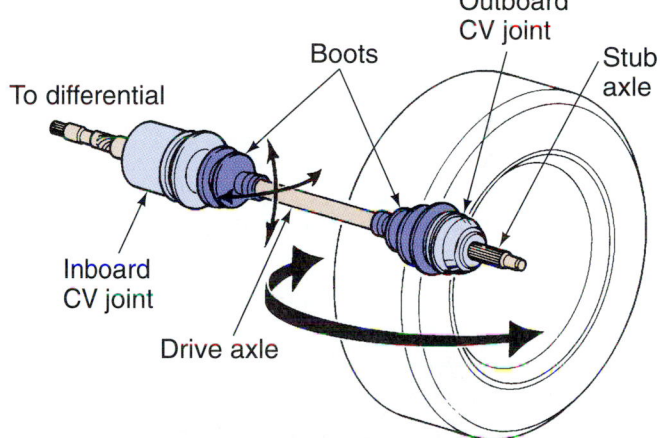

Figure 70.20 CV joints pivot during steering. *(Courtesy of Moog Automotive, Inc.)*

yoke does in a rear wheel drive car. As its name implies, it plunges in and out to compensate for changes in axle length. Because of its design, it can also transmit motion through an angle.

Fixed CV Joints

The outboard joint is called a **fixed joint**. This means that it does not allow for a change in length like the inboard joint. It allows for a change in the angle of the

axle as the suspension moves over bumps in the road. A fixed joint also allows for severe pivoting during steering. It can accommodate angles up to 50°, although the maximum amount of travel is usually limited to 46°. The large angle that the outer joint is subjected to is why they wear out more often than inner ones. Plunge joints cannot change angles as effectively as a fixed joint (Figure 70.22). They are limited to about 20°.

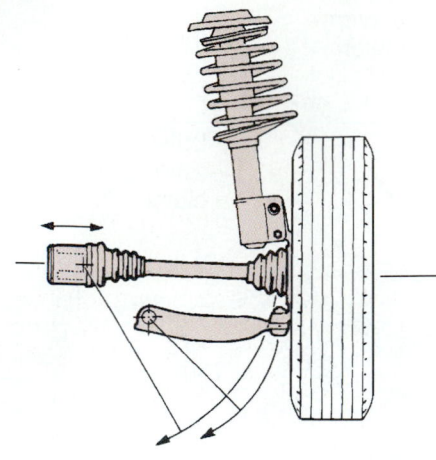

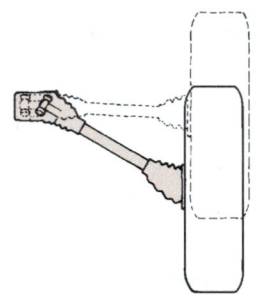

Figure 70.21 A plunge joint allows the axle to change length. *[Courtesy of Moog Automotive, Inc.]*

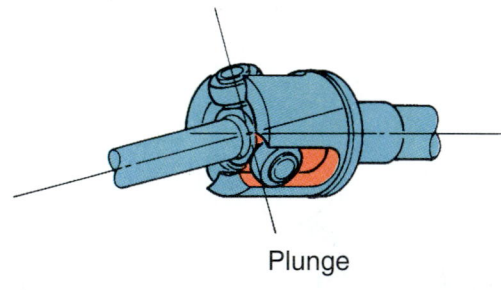

Plunge

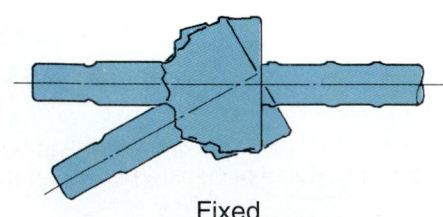

Fixed

Figure 70.22 A plunge joint does not allow as much angle change as a fixed joint. *[Courtesy of Moog Automotive, Inc.]*

■ CV JOINT CONSTRUCTION

There are different types of plunge and fixed joints. Different combinations of joints are found on axles. The most common combination uses a **Rzeppa CV**

joint for the outside fixed joint and a *tripod* for the inside plunge joint. The tripod has become more popular because it costs less to manufacture.

Fixed Joints

Almost all newer vehicles use a Rzeppa joint for the fixed outer joint. The Rzeppa joint is named after Alfred Rzeppa, a Ford engineer who invented it in 1929. It is a ball and socket-type joint (Figure 70.23) with the inner race splined to the axle shaft. The outer race is either part of the stub shaft or it is splined to the stub shaft. Splined parts are held to their shafts with snap rings. Six ball bearings fit in between the inner and outer races, housed in a cage. The joint flexes in tracks in the inner race and outer housing. The balls act as both bearings and a medium for transferring power to the wheels.

A Rzeppa joint is like a bevel gear with balls instead of teeth. It provides even power transmission at all angles. As the inner and outer races move back and forth compensating for the angle change, the balls remain in a constant perpendicular plane (Figure 70.24). The cage holds the balls in this position while the inner and outer races move.

Types of Plunge Joints

There are three types of plunge joints. A *tripod tulip plunge joint* (Figure 70.25a) has a spider assembly with three spherical rollers that ride on needle bearings, similar to a cross and yoke universal joint. The yoke or housing, often called a **tulip**, has three grooves that the rollers can roll in and out on as the axle length changes. The housing is usually spline-mounted, but can also be bolted to a flange.

A **double offset plunge joint** (Figure 70.25b) uses six balls evenly spaced by a cage. Most double off-

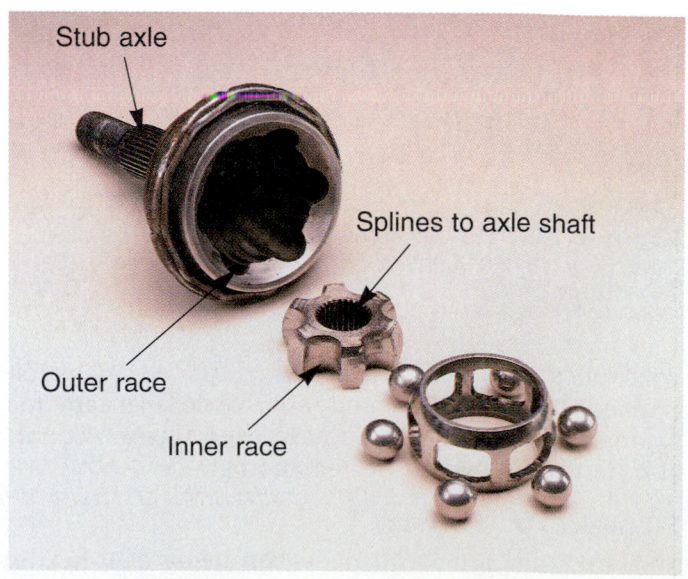

Figure 70.23 Parts of a Rzeppa CV joint. *[Courtesy of Moog Automotive Inc.]*

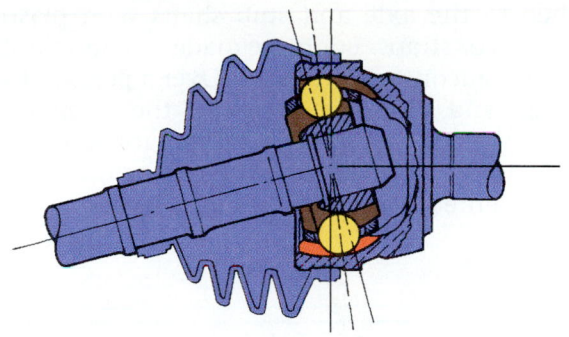

Figure 70.24 As the inner and outer races move back and forth compensating for the angle change, the balls remain in a constant perpendicular plane. *(Courtesy of Moog Automotive, Inc.)*

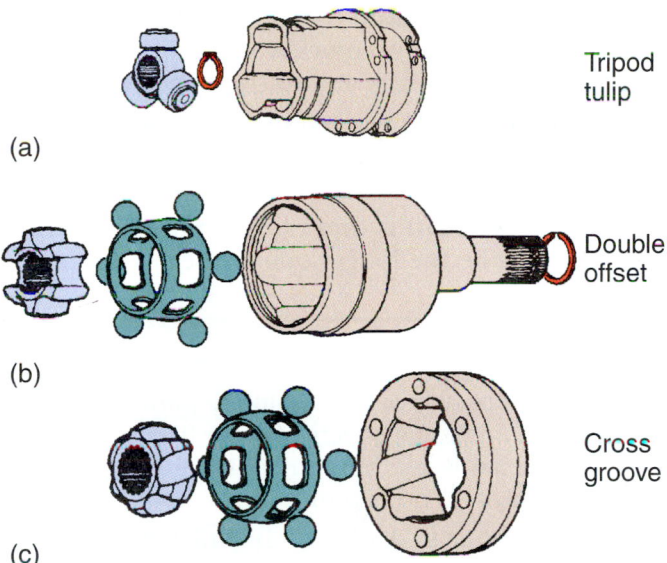

(a)

(b)

(c)

Tripod tulip

Double offset

Cross groove

Figure 70.25 Plunge joints: (a) a tripod tulip joint; (b) a double offset joint; (c) a cross groove joint. *(Courtesy of Moog Automotive, Inc.)*

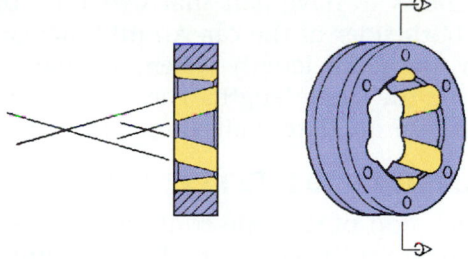

Figure 70.26 In a cross groove joint, grooves in the bearing races would cross each other if they extended far enough. *(Courtesy of Moog Automotive, Inc.)*

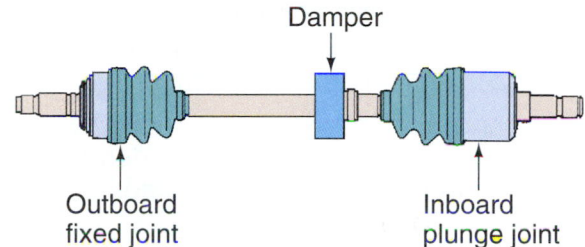

Damper

Outboard fixed joint

Inboard plunge joint

Figure 70.27 A damper on some axles absorbs vibrations. *(Courtesy of American Honda Motor Co., Inc.)*

LONGER SHAFT IS THICKER

TYPICAL FRONT WHEEL DRIVE DRIVELINES

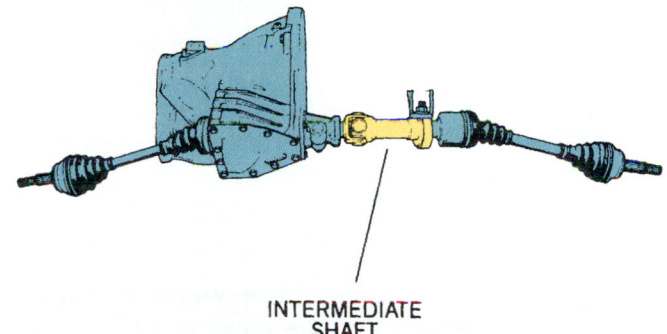

INTERMEDIATE SHAFT

Figure 70.28 When axle shafts are the same length there is an intermediate shaft. *(Courtesy of Moog Automotive, Inc.)*

set joints are spline-mounted to the transaxle. The outer housing is considerably deeper than a cross groove so it can plunge further. The races have straight grooves.

A **cross groove plunge joint** (Figure 70.25c), also called a *pancake joint,* has grooves in the bearing races that would cross each other if they extended far enough (Figure 70.26). There are six balls and a cage, in addition to the inner and outer races.

■ AXLE SHAFTS

Axle shafts can be either solid or hollow. Some of them have damper weights to absorb vibration (Figure 70.27). Because the axle shafts turn so much slower than a rear wheel drive driveshaft, balance is not as important.

Axle shafts can be the same length or they can be different lengths (Figure 70.28). Earlier cars had unequal length axle shafts because the transaxle is mounted off to one side of the engine. When there are different length shafts, the long one wraps up and there is a slight lag before it puts its torque to the wheel. The more horsepower the engine has, the greater the effect of this torque steer. One way of preventing torque steer is to make the longer axle shaft of larger diameter tubing.

The trend is to have half shafts that are of equal length on both sides of the car. An intermediate shaft is used on the equal length system so that the axle shafts can be the same length. The Ford Taurus actually has the intermediate shaft as part of the transaxle.

■ CV JOINT BOOTS

Boots at each end of the axle contain grease and protect the joint from the elements. The CV joint boot is attached to the axle and stub shafts with plastic or steel bands or straps. Boots are made of natural rubber, neoprene, silicone, or urethane. Over a period of time, boots age and suffer the effects of the elements and repeated flexing. When a boot fails, grease is thrown from it. If the boot is not replaced before contaminants can enter it, the entire joint will fail.

■ REVIEW QUESTIONS

1. How are most cars produced today, with front wheel drive or rear wheel drive?

2. When a transaxle is connected to the shifter by cables, how many cables will there be?

3. The parts of the transaxle that allow the front wheels to turn at different speeds during a turn are the differential pinion gears and _____ gears.

4. A FWD axle is also called a _____ shaft.

5. What is the name of the short splined shaft at the outside end of a FWD axle.

6. The two kinds of CV joints are fixed and _____.

7. A _____ joint is like a bevel gear with balls instead of teeth.

8. What kind of sound does a worn Rzeppa joint make if its cage becomes worn, allowing the races to move back and forth in their planes?

9. What is it called when a car with unequal length half shafts steers one way during acceleration?

10. If one half shaft is thicker than the other, will it probably be the longer one or the shorter one?

■ ASE STYLE REVIEW QUESTIONS

1. Technician A says that transaxles have all of the fixed gears on the input shaft. Technician B says that FWD axles run parallel to the input shaft. Who is right?

   a. Technician A   **b.** Technician B
   c. Both A and B   **d.** Neither A nor B

2. Technician A says that in high gear in a five-speed transaxle, the input shaft and output shaft are locked together. Technician B says that in fifth gear the input shaft turns faster than the output shaft. Who is right?

   a. Technician A   **b.** Technician B
   c. Both A and B   **d.** Neither A nor B

3. Technician A says that ordinary universal joints cannot handle as steep an angle as a CV joint. Technician B says that a RWD driveshaft turns at a faster rpm than a FWD axle. Who is right?

   a. Technician A   **b.** Technician B
   c. Both A and B   **d.** Neither A nor B

4. Technician A says that a plunge joint cannot change angles as effectively as a fixed joint. Technician B says that the outboard joint is a fixed joint. Who is right?

   a. Technician A   **b.** Technician B
   c. Both A and B   **d.** Neither A nor B

5. Technician A says that outboard CV joints wear out more often than inboard ones. Technician B says that almost all front wheel drive cars use a Rzeppa inboard joint. Who is right?

   a. Technician A   **b.** Technician B
   c. Both A and B   **d.** Neither A nor B

Front Wheel Drive (Transaxle and CV Joint) Service

■ OBJECTIVES

Upon completion of this chapter, you should be able to:

✔ Diagnose CV joint problems.

✔ Service CV joints.

✔ Replace CV joint boots.

✔ Disassemble and repair transaxles.

■ KEY TERMS

bridge-type clamp
earless clamp
knock-off type CV joint
rebuilt half chaft

■ INTRODUCTION

The emphasis in this chapter is the removal, service, and replacement of front drive axles and CV joints. Not only is this a lucrative service area, but transaxle removal is relatively straightforward, once axle removal is accomplished. A related area, vibration analysis, is covered in Chapter 72.

■ TRANSAXLE AND FRONT WHEEL DRIVE SERVICE AND REPAIR

Inspecting Axle Assemblies

Some checks of the axle assemblies can be made with the axle on the car.

■ LEAKING CV JOINT BOOT

Failure of the boot is the most common problem with CV joints. They are designed to last about 75,000 miles. Boot replacement is an often required service. The outboard boot is the one that fails most of the time. Also, the one on the driver's side fails most often because it is the outside wheel in most turns (right hand).

Check the boots for damage. Grease will begin to spray from a tear in the boot. Check to see if a part has been rubbing on the boot. If a boot is torn and you can see evidence of contamination, the joint will probably have to be replaced.

NOTE: *CV joints are lubed for life. If a boot tears or the clamp falls off, the joint will fail if not serviced immediately. The estimated life of a joint is eight hours without lubricant.*

The action of an operating CV joint is like a vacuum cleaner, causing suction of water into the joint. Storm runoff water in a gutter carries a good deal of mud and grit that will destroy the bearing areas inside

of a CV joint. When a boot has failed, the correct repair is to disassemble and inspect the CV joint before replacing the boot and grease. Be sure to check the seal at the transaxle and plunge joint for leakage, too.

■ BOOT KITS

Replacement boot kits are available that come with new clamps and the correct amount of high temperature grease to refill it with. A new nut is required too, but that does not come in the boot kit because there are too many different nuts used. The half shaft is removed to replace the boot. Just removing the stub shaft from the hub is all that is required to replace the outer boot.

There are a few axle shafts where the fixed joint cannot be disassembled. These can have the boot installed from the plunge joint end.

■ CV JOINT BOOT CLAMPS

There are several types of clamps (Figure 71.1). The **bridge-type clamp** is the most common. It requires a special crimping tool. Urethane boots are stiffer. When the boot is urethane, a tighter crimp is necessary so a heavy-duty crimping tool is needed. The tool is used with a torque wrench to ensure the correct clamping load on the boot (Figure 71.2). Bridge-type universal clamps are available that can be cut to length.

An **earless clamp**, also called a *low profile* or *locking clamp*, is used when there is a problem with the axle clearing other parts. Another type of clamp might accidentally get knocked off.

One type of universal clamp is a continuous strap made of stainless steel. It comes in two sizes. A special tool pulls it tight around the boot. A universal clamp will work on any boot except a urethane boot.

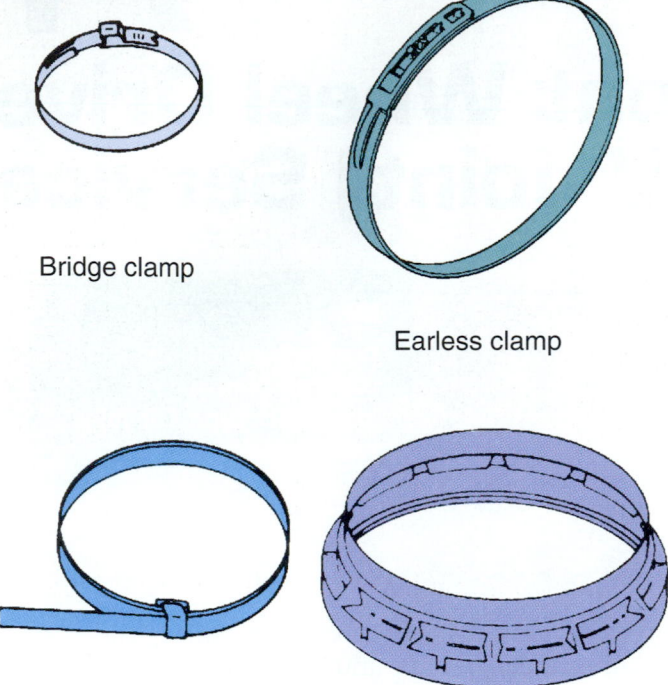

Bridge clamp

Earless clamp

Universal clamp

Press-on ring clamp

Figure 71.1 Types of boot clamps. *(Courtesy of Moog Automotive, Inc.)*

Torque wrench

Bridge clamp

Figure 71.2 Crimping a bridge clamp on a urethane boot using a torque wrench. *(Courtesy of Moog Automotive, Inc.)*

SHOP TIP Spray the clamp with penetrating oil to reduce friction during installation.

Another type of clamp is a pressed-fit ring. It must be installed in perfect alignment or it will become overexpanded. These clamps are not reusable.

NOTE: *Do not try to use a regular hose clamp to replace a boot clamp. It will probably ruin the boot. In a pinch, a nylon tie strap can be used for a day or two until a clamp can be purchased.*

■ AXLE INSPECTION AND DIAGNOSIS

Check the half shaft for obvious looseness. Noises coming from the front axle can be listened for during a test drive. To test for noise, weight must be on the suspension. Otherwise the joint is not in the same position where all of the wear occurs. Just feeling the joint does not work for diagnosing noise unless the joint is extremely damaged. Ask the customer about the noise. Ask what it sounds like and whether the noise is only on acceleration or at all speeds.

■ CV JOINT DIAGNOSIS

Fixed Joint Problems

Test the car during turns like you did when diagnosing axle bearings. A clicking sound during a turn could be because of a bad outboard joint. The clicking will increase and decrease with wheel speed. If the cage becomes worn, the result is a clicking sound when turning as the races move back and forth in their planes (Figure 71.3). Opposing balls in a Rzeppa joint always work together in a pair. When one ball wears its track, the opposite side will have identical wear.

Always check the splines on a shaft. They could be the cause of a clicking sound, too.

Plunge Joint Problems

A bad plunge joint will make a clunking sound when starting from a stop, or during deceleration and when braking. If an accurate diagnosis cannot be made with the axle on the car, remove it and disassemble it to find the cause of the problem.

■ AXLE SHAFT REMOVAL

There are some differences in the methods that axle shafts are removed and serviced. Check the service manual for the procedure before attempting a job for the first time.

Loosen the Axle Nut

To remove the axle, first loosen the stub shaft nut in the center of the wheel while the wheel is still on the ground. Most manuals advise against using an impact wrench on the nut with the vehicle raised on a lift because of possible damage to parts.

When the car has custom wheels or if the wheel is already off the car, you can block a ventilated rotor

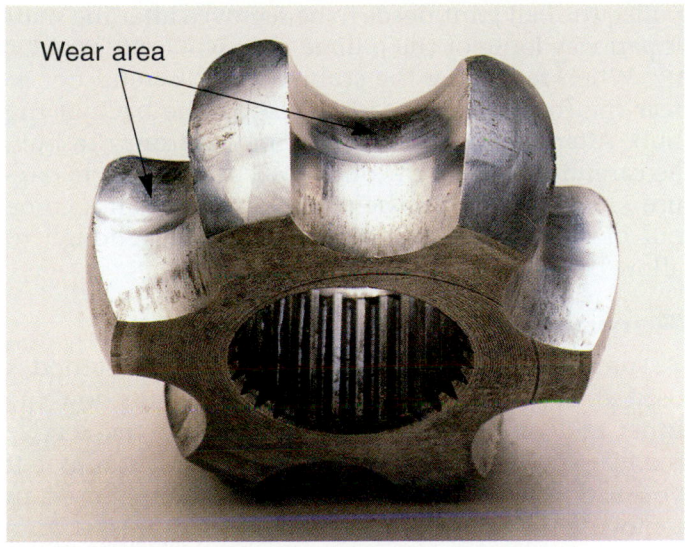

Wear area

Figure 71.4 Most axle nuts are staked to lock them in place. *(Courtesy of Moog Automotive, Inc.)*

joint is disconnected, removing the sway bar from the other side will make it easier to pry down on the control arm to separate it (Figure 71.5).

SHOP TIP If you remove a strut, mark it either at the top or the bottom so that it can be reinstalled in the same position. Otherwise, the wheel alignment will be off after the job is completed.

Remove the Stub Shaft

Removing the stub shaft from the hub can sometimes be done by tapping it with a brass hammer. Other times a puller is required (Figure 71.6). When using the

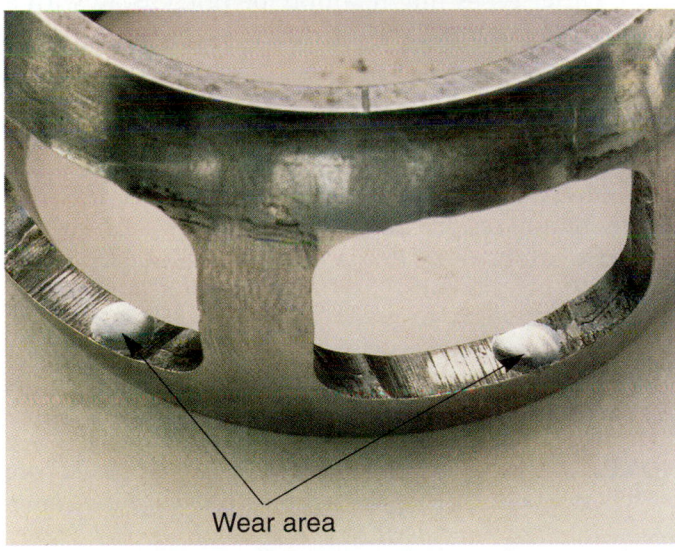

Wear area

Figure 71.3 A worn cage or race can cause a clicking sound during a turn. *(Courtesy of Moog Automotive, Inc.)*

from turning. Wedge a large screwdriver in one of the vent holes. If you have a helper, you can just have them apply the foot brake tightly while you loosen the nut. Most axle nuts are staked to lock them in place (Figure 71.4). A new nut will be required when these are disassembled.

Remove the Stub Shaft from the Hub

On some cars it is easier to move the hub away from the stub shaft if the tie rod is removed first (see Figures 59.7 and 59.8).

Front wheel drive cars usually have MacPherson struts. The ball joint or the strut will have to be disconnected on these cars to be able to move the hub away from the stub shaft. The plunge joint will not be able to move in enough to allow this. When the ball

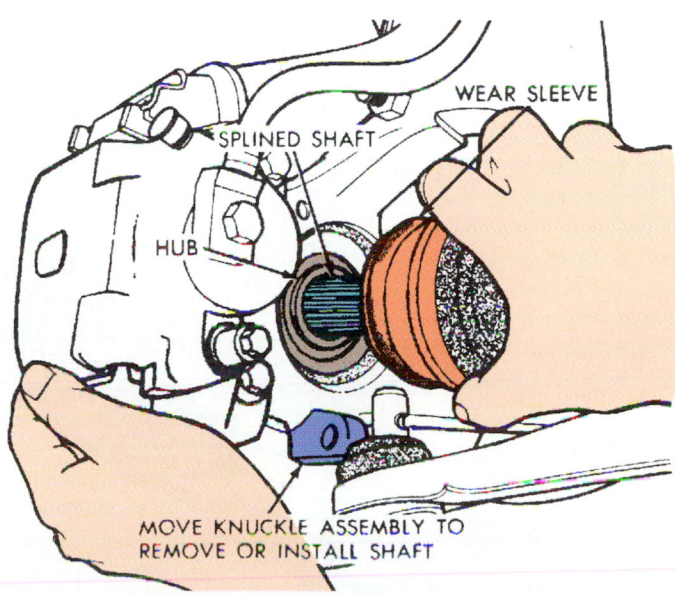

WEAR SLEEVE
SPLINED SHAFT
HUB
MOVE KNUCKLE ASSEMBLY TO REMOVE OR INSTALL SHAFT

Figure 71.5 Lifting the steering knuckle off the ball joint allows enough room to remove the stub shaft from the hub. *(Courtesy of Chrysler Corporation)*

Figure 71.6 Removing a stub shaft. On some cars a puller is needed. *[Courtesy of Moog Automotive, Inc.]*

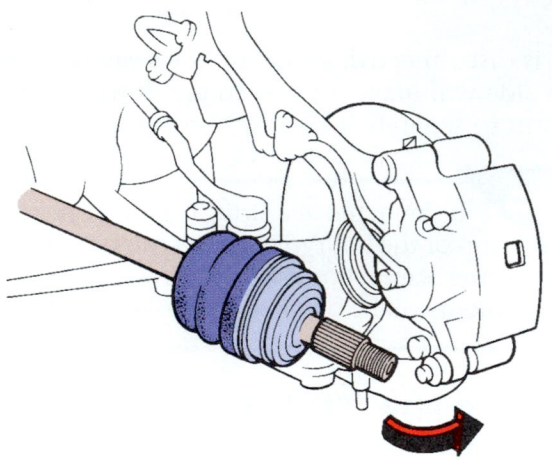

Figure 71.7 Move the stub shaft away from the hub and be careful not to tear the boot. *[Courtesy of American Honda Motor Co., Inc.]*

puller, the ball joint needs to be removed after the shaft is part way loose or the plunge joint will bottom out.

When removing the stub shaft, be careful not to tear the front wheel bearing seal on the back of the hub. After the outer joint is removed from the hub, be careful not to tear it on the suspension parts (Figure 71.7). Watch that the plunge joint does not come out of its housing. The boot can stretch enough to allow this.

Remove the Axle

Be sure to note which kind of axle retention method is used at the transaxle. If you are not certain how the inner end of the axle is retained, check the service manual. You could damage an aluminum transaxle if you use excessive force. A slide hammer is used to remove some types of splined inner joints (Figure 71.8). Other types might simply be pried away with a screwdriver (Figure 71.9) or unscrewed from a flange (Figure 71.10).

If the plunge joint has a spline, oil will leak from the differential when the axle shafts are pulled from the housing of a manual transaxle. If you did not drain the transaxle first, place a drain pan under the axle before removing it. Replacing the axle seal in the transaxle can be done if it is hard or was leaking.

Boot Removal

Position the axle in a vise. Cut off the boot clamps and remove the boot. Wipe off the joint and inspect it to see what kind of retaining method is used on it.

NOTE: Always mark the location of the end of the old boot on the axle with tape so the new one can have the clamp installed in the same position.

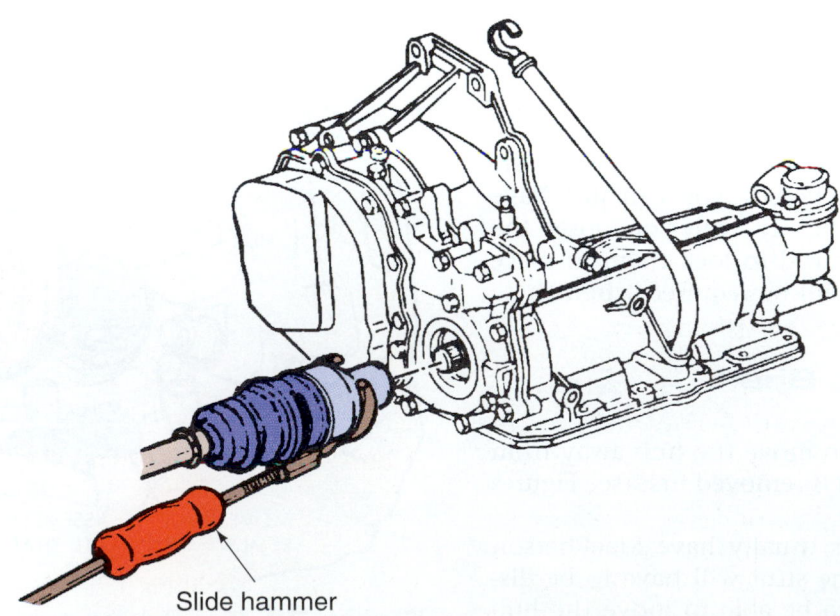

Slide hammer

Figure 71.8 A slide hammer is used to remove some types of splined inner joints. *[Courtesy of General Motors Corporation, Service Technology Group]*

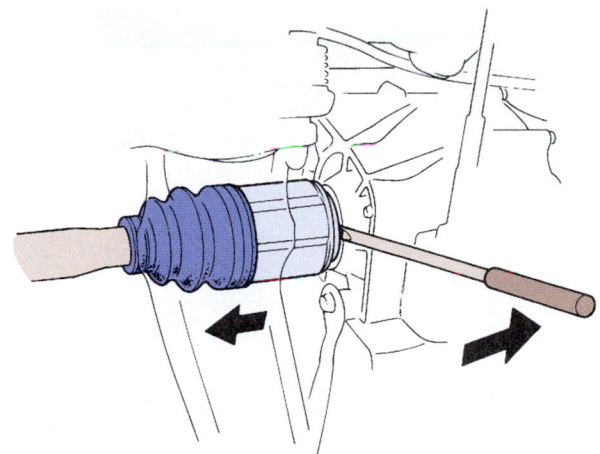

Figure 71.9   Prying a plunge joint out of a transaxle. (*Courtesy of American Honda Motor Co., Inc.*)

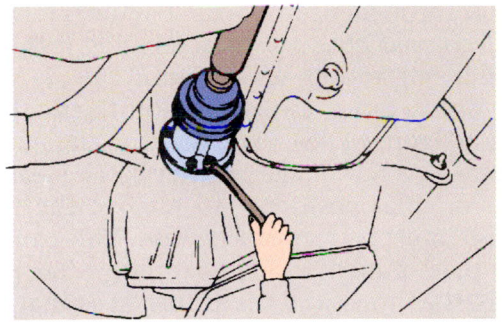

Figure 71.10   Sometimes a plunge joint is unbolted from a flange.

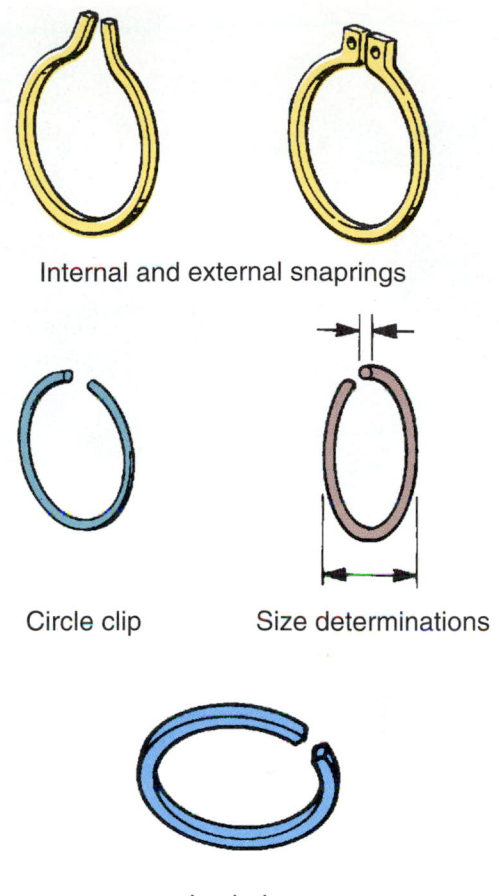

Internal and external snaprings

Circle clip   Size determinations

Lock ring

Figure 71.11   Several types of retaining rings. (*Courtesy of Moog Automotive, Inc.*)

Clips and Snap Rings.   There are several types of clips used to retain CV joints on the axle shafts or to the transaxle (Figure 71.11). These are not reused but are replaced.

Circlips are commonly used. They come in various thicknesses and sizes. Measure the wire diameter before selecting a replacement. *Internal snap rings* and *external snap rings* are also common. They require snap ring pliers for removal. *Lock rings* are sometimes installed in a groove to locate or limit movement of a part on a shaft.

■ CV JOINT REPLACEMENT

Fixed Rzeppa joints are held to their shafts by one of two methods. On one of them, when the joint is wiped off during an inspection an external snap ring will be visible. There is a recess in the inner race where the tabs on the snap ring fit (Figure 71.12). The snap ring is removed with snap ring pliers.

An internal circlip is the second way of retaining a fixed joint to the shaft. The clip rides in a groove in the shaft. Its outer diameter goes against the inner race (Figure 71.13). At the other side of the inner race there is often a lock ring or a thrust washer. It will be the only thing visible when this type of joint is wiped off

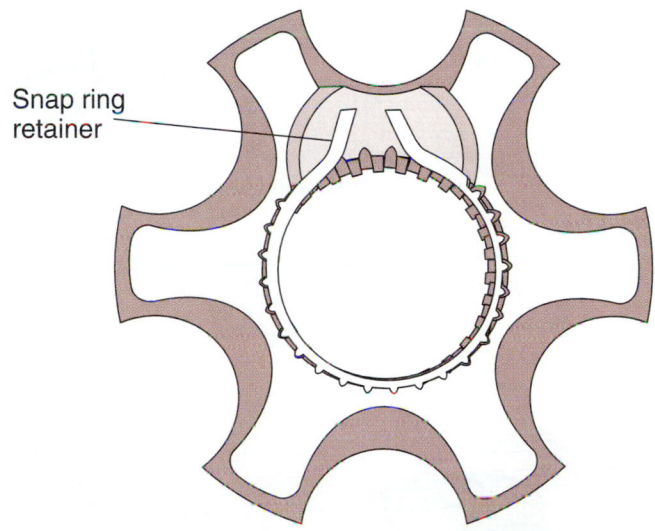

Snap ring retainer

Figure 71.12   The tabs of the snap ring fit into a recess in the inner race. (*Courtesy of Chrysler Corporation*)

during an inspection. This joint design is called the **knock-off type CV joint**. It is removed using a brass punch and hammer. There is a chamfer on the inner race that helps compress the clip during removal.

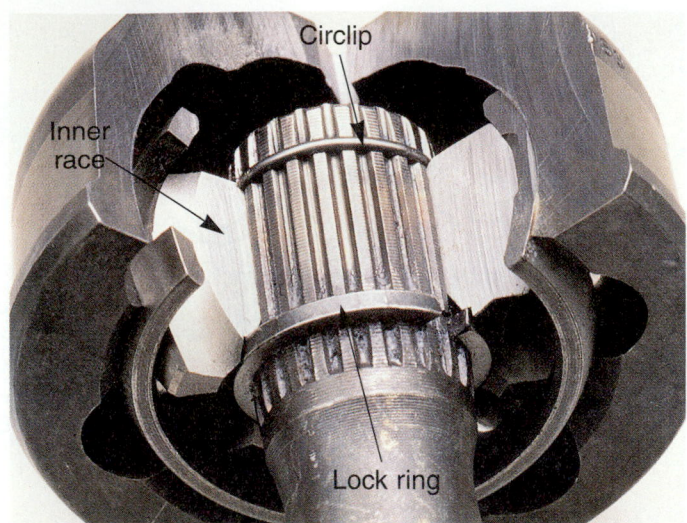

Figure 71.13 This inner race has a lock ring on one side of it and a snap ring on the other. *[Courtesy of Moog Automotive, Inc.]*

Some Japanese cars use another version of the knock-off type joint. A circlip rides in a groove in the axle and fits into a corresponding groove inside the inner race (Figure 71.14). To remove this joint, pressure must be applied to the inner race. There is a special tool for this (Figure 71.15). If you use this tool to knock off another type of fixed joint that has a lock ring, be sure you are not driving against the lock ring.

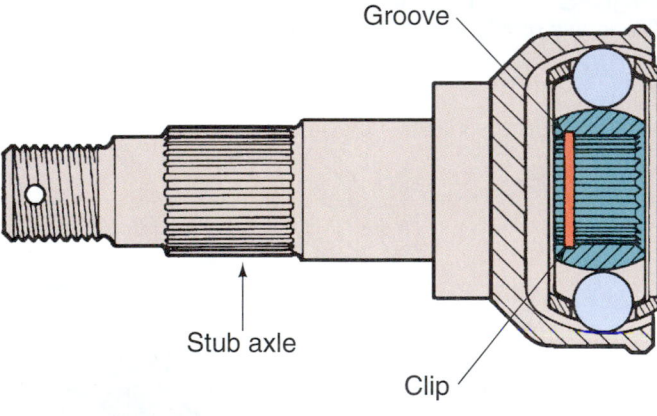

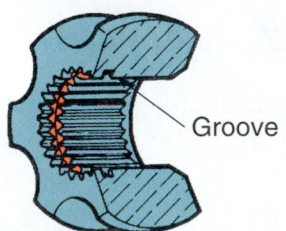

Figure 71.14 Some CV joints have a circlip that fits into a groove inside of the inner race. *[Courtesy of Moog Automotive, Inc.]*

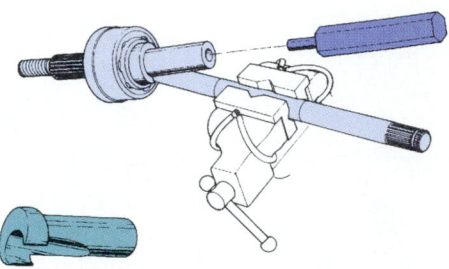

Figure 71.15 A special tool for removing knock-off type CV joints. *[Courtesy of Moog Automotive, Inc.]*

■ FIXED JOINT DISASSEMBLY AND INSPECTION

After the boot is removed, use a drift punch or a special tool to move the inner race to the side for a more complete inspection (Figure 71.16). Wipe the parts off and look for evidence of wear (Figure 71.17). If the metal is colored blue, this is acceptable if the bearing surfaces are all smooth and unworn.

To disassemble a Rzeppa joint, the balls are removed (Figure 71.18). First, mark all the major components with a felt marker so they can be reassembled in their original positions. To remove the balls, the inner race must be turned sideways. This can be done using a brass punch or a special tool. After removing the balls, the inner race can be turned so that it can be removed from the stub shaft (Figure 71.19). All of the parts are cleaned and inspected once again for wear.

Grease in a CV joint has a very difficult job to do. Parts of the CV joint should not be cleaned in solvent.

Figure 71.16 Move the CV joint to the side for cleaning and inspection. *[Courtesy of Moog Automotive, Inc.]*

Figure 71.17 These CV joint balls are pitted. The joint must be replaced. *(Courtesy of Moog Automotive, Inc.)*

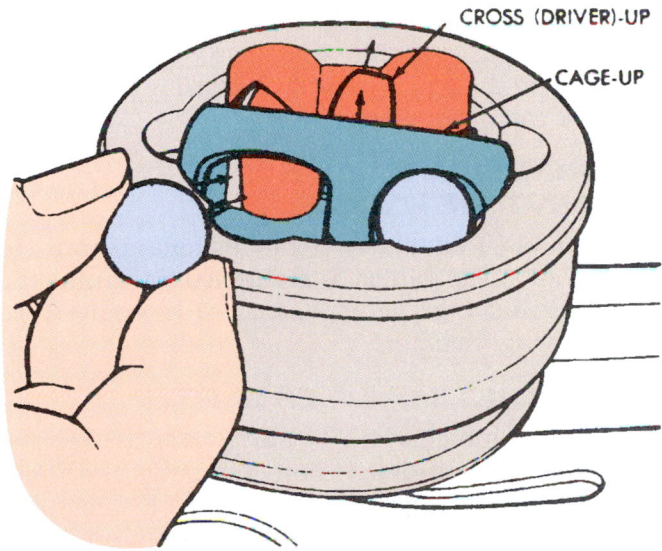

Figure 71.18 With the cage turned to the side, the balls can be removed. *(Courtesy of Chrysler Corporation)*

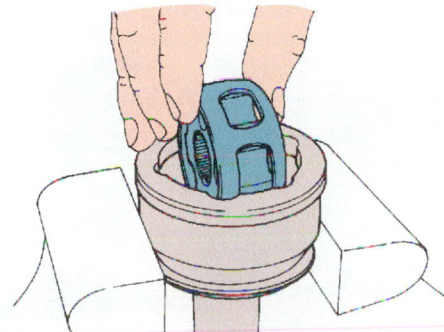

Figure 71.19 After removing the balls, the inner race can be turned so that it can be removed from the stub shaft. *(Courtesy of Moog Automotive, Inc.)*

Figure 71.20 After installing the large clamp on the boot, let the air out of the small end before installing the clamp. *(Courtesy of Moog Automotive, Inc.)*

The solvent film left behind can ruin the grease. Some technicians like to remove the balls one at a time. Then they clean and inspect them and put them back in. This way they do not have to label them. After wiping the joint clean, lightly grease all of the parts. Reassemble the joint and install it on the axle using a new clip.

Install the Boot and Joint

Included in a boot kit is a tube that contains the correct amount of CV joint grease. Half of the grease is applied to the CV joint itself. The other half goes into the boot.

The lock ring can be temporarily removed from the shaft to make installation of the boot easier. After the boot is slipped over the end of the axle, install a new circlip. Be careful not to overstretch the circlip. After the joint is pounded onto the shaft, pull on it to see that its snap ring or circlip is seated in its groove.

Pull the big end of the boot over the CV joint and install its retaining clamp. Next, vent the small end of the boot on the axle (Figure 71.20). Slide the small end of the boot until it is even with the mark you made on the shaft before disassembly and install the clamp in that position.

■ SERVICING AN INNER TRIPOD JOINT

Before removing the inner boot, tape the shaft and joint to show where the clamps should be reinstalled (Figure 71.21). This is especially important on plunge joints because they must be able to move in and out. Mark the location of the housing to the shaft so parts can be reassembled in their original locations. Cut away the clamps and boot. Some tripods will simply come out of the tulip. Others will have retaining tabs, which must be bent out of the way, or a wire ring retainer.

Figure 71.21 Put tape next to the ends of the boot and make marks to make reassembly easier. *(Courtesy of Moog Automotive, Inc.)*

Check to see if the tripod rollers can come off of the spider. Put tape around a tripod so the balls and rollers do not come out (Figure 71.22). It can be difficult to get all of the rollers back in. Mark the tape to locate the tripod to the shaft on reassembly. It is not necessary to disassemble these. Simply roll them to feel that they are free and smooth.

Inspect the condition of the tulip. Feel its grooves for depressions in the metal. A polished appearance in the center of travel is normal. The tripod will have to be removed to install the boot.

SHOP TIP On some axles, the boot can be installed from the other side while the fixed joint is off.

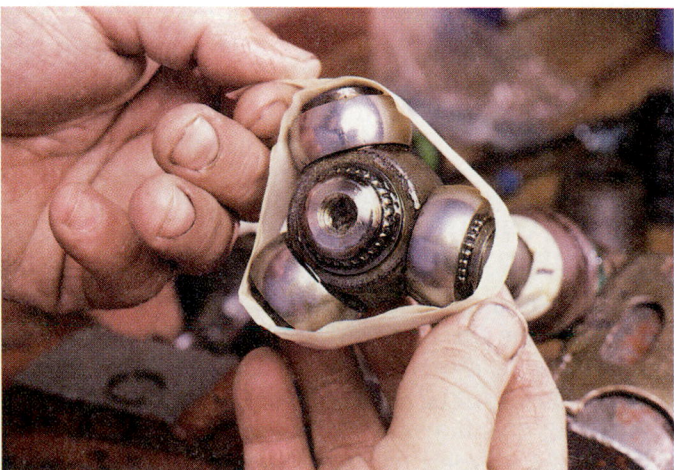

Figure 71.22 Put tape around the rollers so they cannot come off and spill out their needle bearings. *(Courtesy of Moog Automotive, Inc.)*

When reinstalling the joint onto the axle shaft, be sure to align any marks made during disassembly. Use a soft-faced hammer to tap the joint back onto the shaft. If it has a circlip, you will hear it click when it reaches the correct position.

After the boot and tripod are reinstalled, grease the rollers and fill the boot with the rest of the grease. Put the tulip back over the tripod and bend any retaining tabs as needed. Position the joint at mid-travel and crimp the clamps.

Remove the retaining ring that holds the tripod to the shaft and remove the tripod. Some have different types of circlips and others have to be driven off.

■ DOUBLE OFFSET PLUNGE JOINTS

Double offset plunge joints vary in how they are connected to the axle shaft. Most of them are held by an internal circlip. To remove the joint requires it to be disassembled first. This kind of joint is identified by a large round retaining ring in a groove in the joint's outer housing (Figure 71.23). The other kind of double offset joint has a circlip that can be removed by pounding on the outer housing. A metal retaining ring holds the parts in the housing.

■ CROSS GROOVE JOINT SERVICE

A cross groove joint is serviced in a similar manner to Rzeppa joints. The joint is disassembled by turning the inner race and cage perpendicular to the outer cage and removing them.

■ REBUILT HALF SHAFTS

Installing a complete **rebuilt half shaft** has become a popular repair. Although both the outer and inner joints are replaced, an entire rebuilt shaft is sometimes less expensive than just one joint kit. This is especially

Retaining ring

Figure 71.23 This double offset plunge joint is disassembled by removing a retaining ring. *(Courtesy of Moog Automotive, Inc.)*

true with some imports. The determining factor in whether a technician will rebuild a shaft or buy a complete rebuilt shaft is often whether there is other work to be done. If the technician has finished the day's work and has nothing else to do, he or she will often rebuild the shaft to be able to earn the labor.

■ INSTALLING THE AXLE

To install a spline-type joint into the transaxle, push in on the axle while turning it to align the splines. Push the splines all of the way in. If there is a circlip, it will click when it is in its groove. Pull on the housing of the joint to see if the snap ring is seated. Do not pull on the axle shaft. The plunge joint could come apart.

After the inner end is installed, position the stub shaft into the hub as far as it can go. Install the washer and a new axle nut and tighten the nut until it is snug. Reinstall the front end components. Do not forget to torque the ball joint stud bolt if that was reassembled. A new pinch bolt is recommended. Leaving these loose can cause a serious accident. Some of the pinch bolts have a knurled area under the head to keep them from turning. Be sure only to turn the nut.

CASE HISTORY *A staff member brought her car into a school shop to have the students replace her CV joint boots. After the job was completed, she drove the car home with no problem. But when she backed down her driveway the next morning the wheel fell off to one side and turned 90° to the other. She left a big skid mark on her driveway and ruined her tire before she realized there was a problem. The car was towed to the shop where it was discovered that the pinch bolt had come loose and the ball joint had fallen out of its hole. Luckily no one was killed in an accident that surely would have happened if the wheel had fallen off while on the highway.*

When the car is back on the ground, torque the new stub shaft nut. The torque on the spindle nut can be substantial. A planetary torque multiplier works well for this (see Figure 7.46). Use a dull chisel to *stake* the nut into the slot in the spindle (see Figure 71.4). Always check the clearance of the boot clamps before test driving the car.

■ TRANSAXLE REPAIR

Transaxle repair is similar to transmission repair on rear wheel drive cars. On transaxles used with transverse engines, the differential gears are helical gears, like the rest of the gears in the transmission section (see Figure 70.8). Because these are not hypoid gears, there are no pinion depth or backlash adjustments. The only adjustment to the differential is side bearing preload. Clutch adjustment is the same as for rear wheel drive.

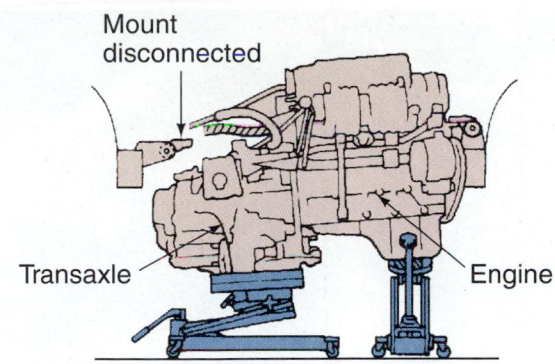

Figure 71.24 Removing the transaxle using a pair of jacks.

Figure 71.25 Lowering the transaxle to the ground. *(Courtesy of Nissan Motors)*

■ TRANSAXLE REMOVAL

On front wheel drive cars, it is often easier to pull the engine with the transaxle. If it is easier to leave the engine in the car, be sure to support the engine. The engine can sometimes be supported from the top with a special fixture (see Figure 48.19). Most shops that do this type of work will have one of these.

A pair of jacks can be used to perform the job with the car supported on jackstands (Figure 71.24). The transaxle is unbolted from the engine and lowered to the ground (Figure 71.25).

The axles are pulled in the manner described previously. After the axles are removed, removing the transaxle is the same as with rear wheel drive cars.

■ TRANSAXLE SERVICE

Separate the halves of the case on a transaxle (Figure 71.26). As you did with a RWD transmission, check the service manual for any specifics prior to disassembly. Service and assembly of the components are similar to that of RWD transmissions. Those items and press and puller procedures are covered in detail in Chapter 65.

End play of the gears is checked with a feeler gauge or dial indicator. Selective thrust washers and snap

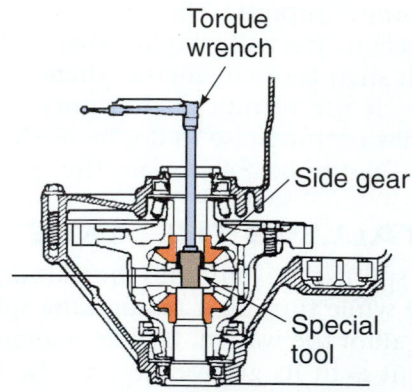

Figure 71.27 A torque wrench and special tool are used to check to see how difficult it is to turn the side bearings.

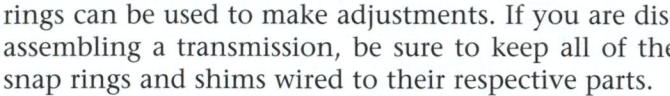

Figure 71.26 Separate the halves of the transaxle.

rings can be used to make adjustments. If you are disassembling a transmission, be sure to keep all of the snap rings and shims wired to their respective parts.

When the case halves are reassembled, the differential side bearing preload is adjusted. Check the manufacturer's instructions. Some of them use special tools. Figure 71.27 shows a torque wrench and special tool used to check to see how difficult it is to turn the differential.

◼ INSTALL THE TRANSAXLE

Installation of the transaxle is the reverse of the removal sequence. After the transaxle is reinstalled be sure to check the transaxle lubricant level. Some transaxles have a fill plug on the side like an ordinary RWD transmission. Checking the lubricant on others is sometimes different. Some transaxles have a dipstick. Other transmissions require the removal of the speedometer gear to check the fluid level.

◼ REVIEW QUESTIONS

1. What is the only adjustment to a transaxle differential?
2. How many hours is the estimated life of a CV joint that has lost its lubricant?
3. What is the most common repair to CV joints?
4. What material is a universal clamp made out of?
5. Before pulling a splined axle from a manual transaxle, what do you position under it?
6. Before removing a boot from a CV joint what do you do to the axle?
7. When there is no snap ring visible when a CV joint is wiped off for inspection, this kind of joint is called a _____- off joint.
8. Which end of a CV joint boot is installed first, the big end or the little end?
9. What do you do to a plunge joint's tripod rollers to keep them from coming apart?
10. What do you do to the stub shaft nut after it is torqued to specification?

◼ ASE STYLE REVIEW QUESTIONS

1. Technician A says that transverse engine transaxles use helical differential gears. Technician B says that most transaxles do not have pinion depth and backlash adjustments. Who is right?
   a. Technician A **b.** Technician B
   c. Both A and B **d.** Neither A nor B
2. Technician A says that the outboard boot is the one that fails most often. Technician B says that the outboard boot on the passenger side fails more often than the one on the driver's side. Who is right?
   a. Technician A **b.** Technician B
   c. Both A and B **d.** Neither A nor B
3. Technician A says that it is not necessary to remove an axle and disassemble a CV joint just because the boot has failed. Technician B says to spray the CV joint boot with penetrating oil

to reduce friction during installation. Who is right?

a. Technician A   **b.** Technician B

c. Both A and B   **d.** Neither A nor B

4. Technician A says that to test an axle shaft for noise, the weight must be off of it. Technician B says that a clicking sound during a turn could be due to a bad plunge joint. Who is right?

a. Technician A   **b.** Technician B

c. Both A and B   **d.** Neither A nor B

5. Technician A says that a bad plunge joint will make a clunking sound when starting from a stop. Technician B says that a bad plunge joint will make a clunking sound when braking. Who is right?

a. Technician A   **b.** Technician B

c. Both A and B   **d.** Neither A nor B

6. Technician A says to use an impact wrench to remove the stub shaft nut with the car in the air. Technician B says that a new stub shaft nut is required when replacing CV joint boots. Who is right?

a. Technician A   **b.** Technician B

c. Both A and B   **d.** Neither A nor B

7. Technician A says that removing a stub shaft from a hub always requires a puller. Technician B says that sometimes a stub shaft can be

removed from the hub with a brass hammer. Who is right?

a. Technician A   **b.** Technician B

c. Both A and B   **d.** Neither A nor B

8. Technician A says that if the metal parts of a CV joint are colored blue, the joint must be replaced. Technician B says to clean all CV joint metal parts in solvent. Who is right?

a. Technician A   **b.** Technician B

c. Both A and B   **d.** Neither A nor B

9. Technician A says that a polished appearance in the center of travel of a plunge joint tulip is normal. Technician B says to bottom out the tripod in its bore before clamping the boot to the axle shaft. Who is right?

a. Technician A   **b.** Technician B

c. Both A and B   **d.** Neither A nor B

10. Technician A says that a transaxle must have the correct drive pinion depth and backlash. Technician B says to roll a gear contact pattern before assembling a transaxle. Who is right?

a. Technician A   **b.** Technician B

c. Both A and B   **d.** Neither A nor B

CHAPTER 72

Driveline Vibration Service

■ **OBJECTIVES**
Upon completion of this chapter, you should be able to:
✔ Understand terms related to vibration.
✔ Describe the different types of vibration.
✔ Test for vibration using test instruments.
✔ Check driveshaft runout.
✔ Balance a driveshaft.
✔ Check driveshaft angle.

■ **KEY TERMS**
vibration
frequency
hertz
amplitude
velocity
natural frequency
first order vibration
second order vibration
beat/boom vibration
electronic vibration
    analyzer (EVA)
snapshot
moan
whine
match mount
transducer
launch shudder

■ VIBRATION ANALYSIS

Tire and wheel imbalance is the most common cause of vibration complaints. Those procedures were covered in Chapter 55. After tires and wheels have been eliminated as a cause of vibration, the items dealt with in this chapter come into play.

Vibration is something that does not usually shut the car down. It does annoy the driver and can result in his or her unhappiness with the repair job or the car in general.

To analyze vibration problems, a process of elimination is used. The following are terms used in analyzing vibrations:

■ **Vibration** describes a part that is in motion in waves, or *cycles*. Figure 72.1 is an example of vibration as it cycles through a material.

■ **Frequency** is how many cycles take place in a period of time. The number of cycles in one second is measured in **hertz** (Figure 72.2).

■ **Amplitude** measures how intense a vibration is (Figure 72.3). An intense vibration has more amplitude.

■ **Velocity** is a combination of amplitude and frequency (Figure 72.4). Velocity is not constant. It is an average, measured in inches per second. If amplitude is 4" and frequency is 10 Hz, this is 40"/sec velocity.

Resonant or **natural frequency** is the frequency at which a body vibrates. A tuning fork can be used to illustrate this. Unibody vehicles often have inherent vibration characteristics that cause them to vibrate at a predictable road speed. This was a problem especially with older unibody vehicles. More recently engineers have designed these vehicles so that their vibration range is outside of the normal driving range. Engines, suspension systems, and driveshafts are some of the things that vibrate at different frequencies. Stiffer materials tend to have higher natural vibrating frequencies. When you tighten a string on a guitar, it becomes stiffer and the sound from its vibration occurs at a higher frequency.

The severity (amplitude) of a vibration will be greatest at its *point of resonance*. An automotive suspension system has a natural frequency in the range of 10 to 15 Hz. The suspension's frequency is always the same, but a driveshaft that is out of balance will have a frequency that can become equal to the suspension frequency at a particular speed (Figure 72.5). This is the point of resonance, the speed where the vibration

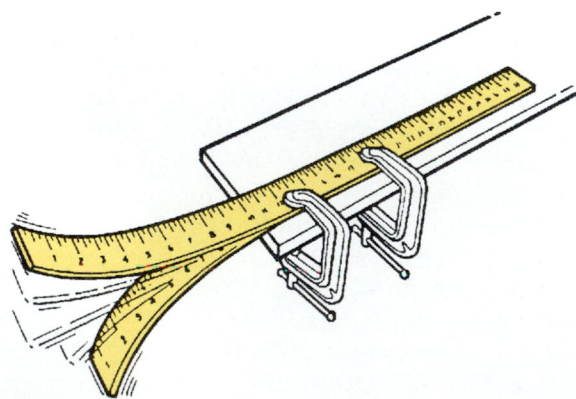

Figure 72.1 Vibration happens in cycles. *(Courtesy of General Motors Corporation, Service Technology Group)*

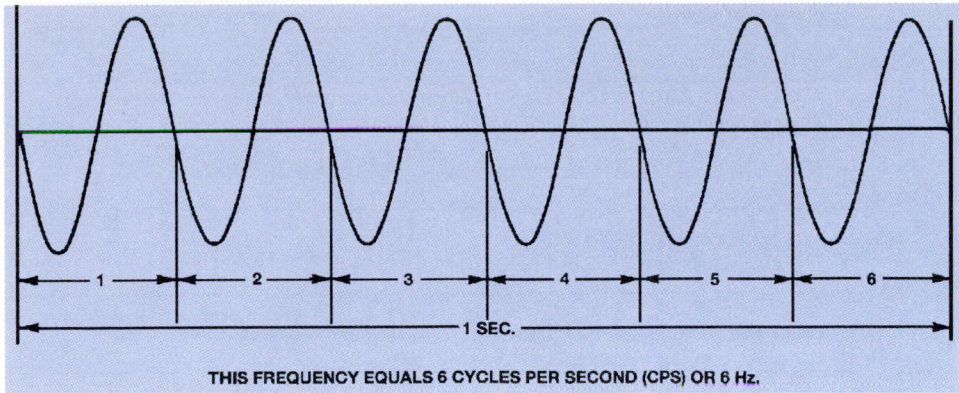

THIS FREQUENCY EQUALS 6 CYCLES PER SECOND (CPS) OR 6 Hz.

Figure 72.2 Frequency is how many cycles take place in a period of time. *(Courtesy of General Motors Corporation, Service Technology Group)*

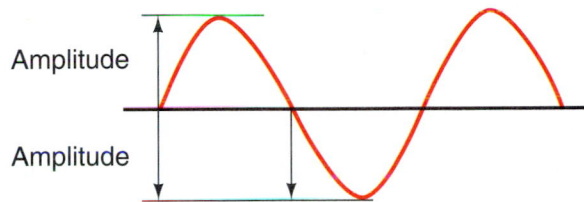

Figure 72.3 Amplitude is a measurement of the intensity of a vibration. *(Courtesy of General Motors Corporation, Service Technology Group)*

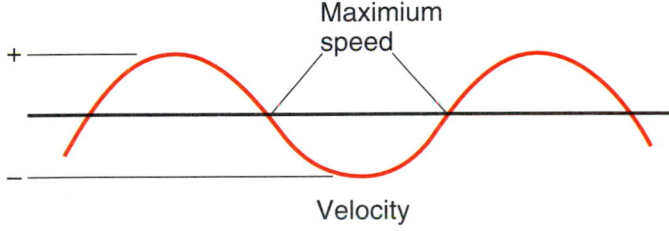

Figure 72.4 Velocity is a combination of amplitude and frequency. Maximum speed (velocity) is reached when the vibrating wave is at half travel.

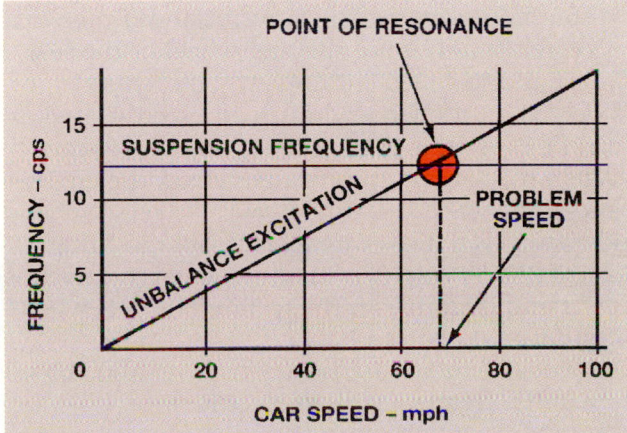

Figure 72.5 A driveshaft that is out of balance will have a frequency that can become the same as the suspension frequency at a particular speed. *(Courtesy of General Motors Corporation, Service Technology Group)*

will be greatest. Front and rear suspension systems are designed with frequencies that are different by 3 Hz to avoid a combination of their vibrations.

■ TYPES OF VIBRATIONS

First order vibration is anything that spins at driveshaft speed and vibrates only once every revolution.

Second order vibration is a universal joint. Remember, as the driveshaft rotates, the U-joints speed up and slow down twice during each revolution. Two vibrations per revolution is second order vibration. This vibration is always related to road speed and is usually worse under torque.

A **beat/boom vibration** occurs when one vibration interacts with another. Canceling one of the vibrations can sometimes solve the problem.

■ VIBRATION TEST INSTRUMENTS

Because every vibration has its own frequency, vibrations can be analyzed using test instruments. One type of mechanical vibration analyzer that was invented in the 1940s senses the frequency of vibrations from 10 to 80 Hz. This tool is called a *reed tachometer* (Figure 72.6). It is a tachometer because it measures in hertz (cycles per second).

The reed tach has two rows of reeds of different lengths. The reeds vibrate at different frequencies, lower or higher as you move from left to right or vice versa. The vibration shown in the illustration is strongest at 16 Hz. The tool must be placed near the source of the vibration or it cannot react.

A newer and better version of the reed tachometer is the **electronic vibration analyzer (EVA)** (Figure 72.7). It has a 20-foot cord and an electronic pickup. The pickup is held in place with putty, Velcro, or a magnet allowing the car to be driven while checking for vibrations. The EVA measures acceleration of a part. The bigger the reading the worse the vibration is. One side of the gauge shows the frequency of a vibration in

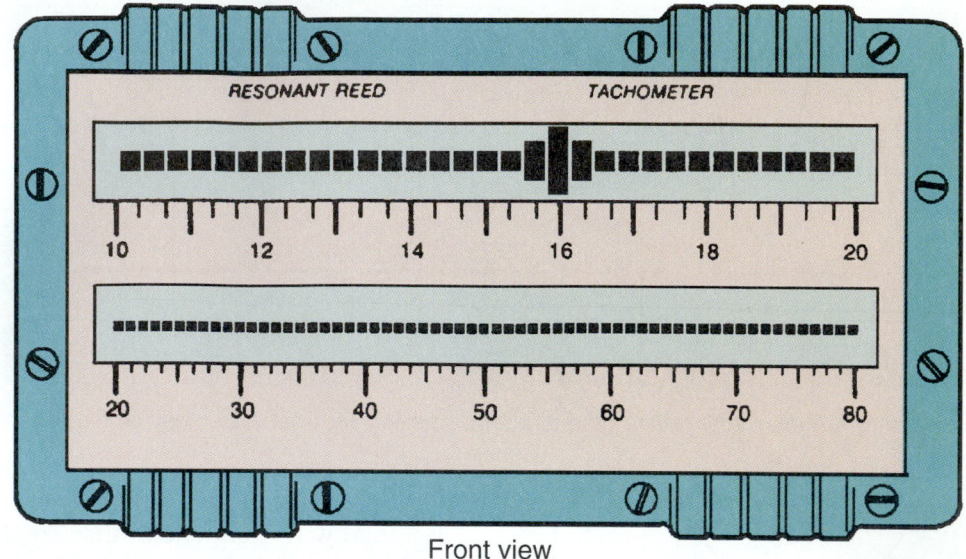

RESONANT REED   TACHOMETER

Front view

Figure 72.6   A reed tach. *(Courtesy of General Motors Corporation, Service Technology Group)*

Figure 72.7   An electronic vibration analyzer (EVA) *(Courtesy of Kent Moore Div.-SPX Corp.)*

rpm or hertz. The other side of the gauge shows the amplitude of the vibration in *Gs (gravity)*.

Like other electronic instruments, the EVA has the capability of *freezing* an event. This is handy when a vibration occurs quickly and then goes away. A recorded event is called a **snapshot**. A snapshot has ten frames. During playback, the technician can freeze and unfreeze the display to isolate the event.

The EVA is handy for balancing driveshafts, too. It has a coil of wire that the inductive pickup on an ignition timing light can hook to.

■ VIBRATION AND FREQUENCY

General Motors has done extensive vibration testing. Earlier front wheel drive cars had unibodies with resonant frequencies that came into play in the normal driving range. This has been controlled by stiffening the floor pans until their frequency is beyond the normal range of use. Different types of vibrations and their frequencies are shown in Figure 72.8.

Some vibrations can be felt and others can be heard. The ones that are heard go into the higher frequencies above 100 Hz. With vibrations that cannot be heard, work on the problem with the highest amplitude and ignore the rest until the worst are fixed. The following definitions can be used to isolate vibrations even if you do not have access to a vibration analyzer. The following are some vibrations that can be felt:

■ *Shake* is a low frequency vibration of 5 to 20 Hz. It is commonly the feeling an unbalanced tire would give, felt through the steering wheel or the seat. Shake is also called shimmy, wobble, shudder, or hop. When the vibration is road speed-related, it can be caused by tires, wheels, or brake drums. When it is engine rpm-related, check the engine and accessories.

■ *Roughness* is a higher frequency vibration than shake (20–50 Hz). It is usually related to driveline parts and is similar to the feeling one gets when cutting wood with a jigsaw.

■ *Buzz* comes in at a frequency of 50 to 100 Hz, about the frequency of an electric razor. It is often caused by a vibration in the exhaust system or an engine accessory like the air conditioning compressor. If the buzz is road speed-sensitive, a driveline problem can be the cause.

■ *Tingling* is vibration above 100 Hz. It feels like pins and needles in the steering wheel or through the

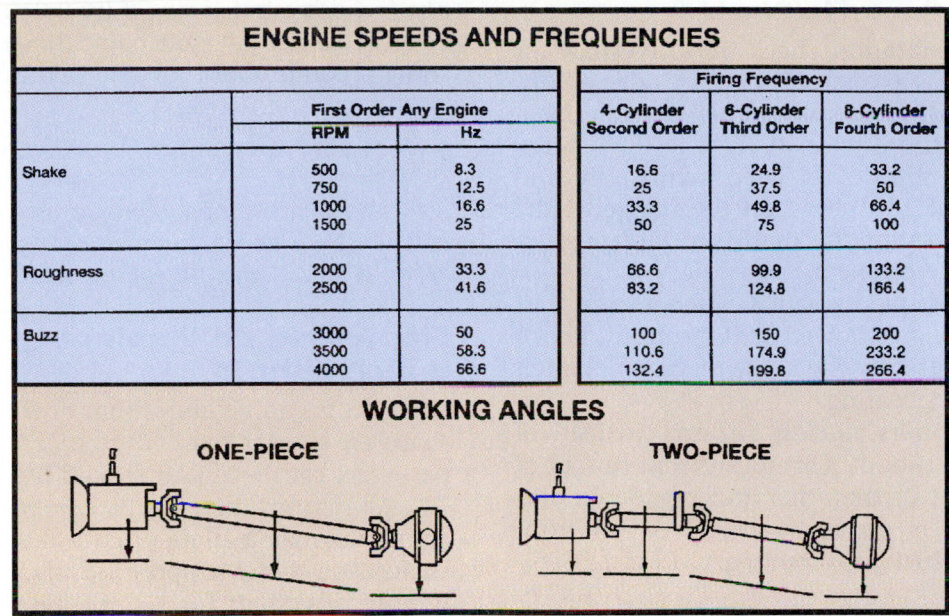

Figure 72.8 *Different types of vibrations and frequencies. (Courtesy of General Motors Corporation, Service Technology Group)*

floor. The reed tachometer does not measure above 100 Hz. The EVA can pick up tingling problems. Again, road speed-related is drive train, and engine rpm-related is engine.

The following vibrations can be heard:

- *Boom* is a low frequency noise in the range of 20 to 60 Hertz. It is described as droning, moaning, or humming. Pressure can be felt in the ears. There can be a feeling of roughness with it. Look at the chart. Boom is normally caused by the driveline, like shake.

- **Moan** or *drone* is a noise that is compared to the hum of a bumblebee. Its range is from 60 to 120 Hz. Powertrain mounts or exhaust system problems are possible causes.

- *Howl* is a noise like the wind across the top of a soda bottle. Its range is from 120 to 300 Hz.

- **Whine** sounds like a turbine or mosquitos. The range is from 300 to 500 Hz. It is usually related to gear noise.

Road Test

During the road test, use a process of elimination. Driveline vibration is slightly different than the feeling that results from tire imbalance. It is not normally identified as a possible problem until wheel balance has been done first and a vibration still remains. The vibration usually occurs at a specific rpm and goes away at other speeds, just like wheel balance.

When test driving for drive shaft vibration, drive in high gear at the engine rpm where the vibration is worst. Then, shift into a different gear or put the transmission into neutral to see if the vibration changes or goes away entirely. When there is no change in the vibration, the driveline is suspected as a possible source of the problem. If the problem does change, one of the components whose rpm changed with the gear change is responsible. These include the engine, clutch or torque converter, or the transmission.

Be sure you understand the difference between amplitude and frequency when attempting to diagnose vibration problems. When road testing, note the speed at which a vibration occurs. A vibration should always occur at the same speed or frequency. If a second road test is less in intensity but occurs at the same speed, the amplitude has changed. Drive the vehicle and note the speed at which the amplitude of the vibration is the worst. Then, repair the suspected problem and retest at the same speed.

Modern test instruments for vibration are so sensitive that you could be attempting to fix a problem that cannot be felt. If you cannot feel the vibration, do not try to fix the problem.

Testing in the Service Bay

To test for a first order vibration, raise the tires in the air and run the engine in gear to check the driveline. Move the pickup of the vibration analyzer around and try to locate where the energy from the vibration is located. If there is no vibration without a load on the tires, this is not a balance problem. The problem is probably related to misalignment of parts. Angles are covered later in this chapter.

While the driveshaft is spinning, check for vibrations at the front and rear of it. A torque sensitive problem is usually in the differential. This is often caused by a bearing cup that is not seated squarely in its bore.

■ DRIVESHAFT RUNOUT

After checking for vibration check the driveshaft for runout. When there is excessive runout in a driveshaft, suspect the driveshaft or the pinion flange. First, measure the runout of the driveshaft (Figure 72.9). Mark the high spot on the shaft. Then remove it and remount it at a spot 180° away on the flange. If the runout remains the same, the shaft is at fault. Otherwise the flange is the problem.

One part of a two-piece shaft is called a stub shaft (Figure 72.10). These are allowed 0.003" runout. Excessive runout here is usually the cause of vibrations felt at the center support bearing.

Some manufacturers **match mount** instead of balancing to correct runout. This means that runout of one part and runout of the other are considered during assembly. This is especially prevalent on front wheel drive axles, which are often marked for reassem-

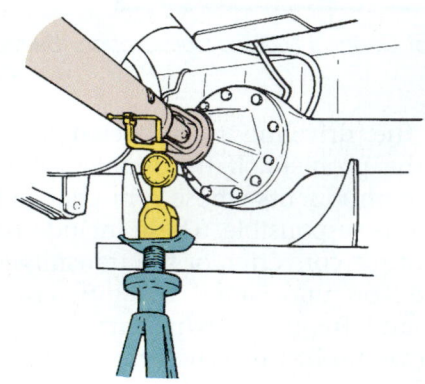

Figure 72.9 Measuring driveshaft runout. *(Courtesy of Ford Motor Company)*

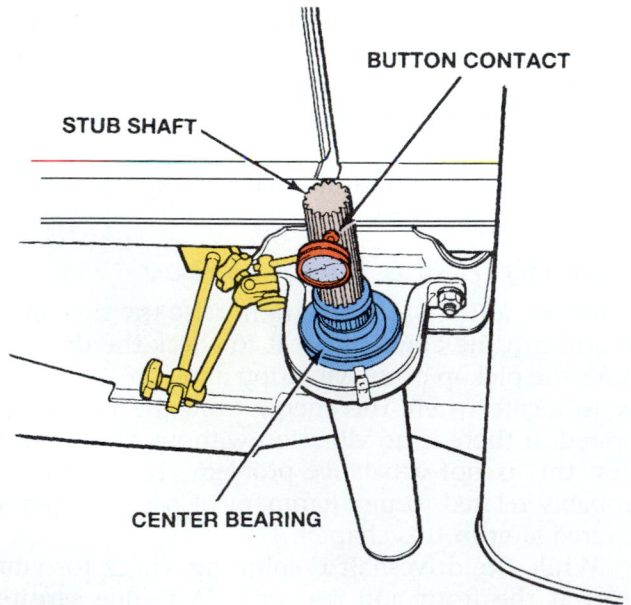

Figure 72.10 Measuring stub shaft runout. *(Courtesy of General Motors Corporation, Service Technology Group)*

bly. If a driveshaft was off by 0.010" and a flange was also off by 0.010" you could have either a perfectly straight combination or one that is 0.020" off.

NOTE: *A part with runout cannot be balanced if it causes parts to bind up once per revolution. Consider that driveshaft parts that are square can be balanced. This is not true with tires because they touch against the road surface so they must not have runout. You cannot balance a box and expect it to run smoothly on the road.*

Repairing or Replacing a Driveshaft

Driveshafts can be aluminum or steel. Most steel drive shafts are shorter than 60" to 65". A 3" to 3.5" diameter passenger car driveshaft should be shorter than 60" to 70". Aluminum drive shafts can be 85" to 90" because aluminum has a higher "critical speed." Longer steel shafts use a center support bearing. Do not swap an aluminum driveshaft for a steel one unless the length is less than 60" to 65".

Driveshafts can be straightened by a specialty machine shop. They heat and quench it until it is straight within 0.010". Longer shafts are allowed less runout in the center.

■ OTHER CAUSES OF VIBRATION

Other things besides the driveline can cause vibrations. Engine accessories can also cause vibrations. Test to eliminate outside causes of vibrations in the repair bay. If a vibration is still evident when the car is not moving, how could it be driveline-related? Some engines are especially prone to vibrations. Engine accessory brackets have been designed to *excite* at 400 Hz or higher. This is equivalent to 6000 engine rpm on a V-8, a driving condition that the average driver should not encounter.

The engine, torque converter, and driveshaft change in frequency as rpm changes. A rebuilt torque converter is often the cause of a vibration that is rpm-related.

Check to see that a muffler is not vibrating against the frame. All muffler hangers should be in place and undamaged.

■ DRIVESHAFT BALANCE

Driveshafts are balanced on the ends, not in the middle. The driveshaft can be compared to a very wide, low profile tire. At the factory, weights are welded onto the ends of the driveshaft when the shaft is balanced. The balance weights should be at least 1" down the shaft from the weld so the integrity of the weld is not disturbed. PTOs (power take-offs) and shafts that spin at under 1000 rpm do not need to be balanced. Specialty shops in the aftermarket shorten or lengthen driveshafts and do balancing.

When the driveshaft is out of balance, a technician can correct it using an on-the-car strobe-type wheel balancer. Heavy spots are counterbalanced by installing hose clamps with the screw positioned on the opposite side of the heavy spot (Figure 72.11). The pinion nose end of the driveshaft is where most of the first order driveline vibrations come from. Driveshaft flanges are sometimes balanced. This is to compensate for runout created during manufacturing. Weight will either be on the front of a flange or the outside depending on whether the correction was for flange runout or differential balance (Figure 72.12). Runout of the flange is checked with a dial indicator (Figure 72.13). It is limited to 0.006" when there is no weight on the flange.

Checking Driveshaft Balance

One end of the driveshaft is balanced first. After the other end is balanced, the first end is rechecked. A **transducer** is a good tool for testing for driveshaft

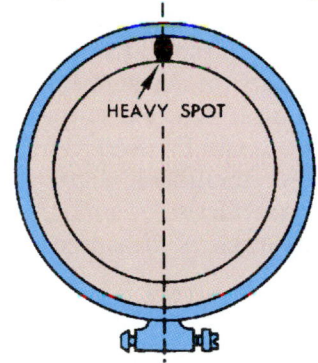

Figure 72.11 A heavy spot is counterbalanced with a hose clamp. *(Courtesy of General Motors Corporation, Service Technology Group)*

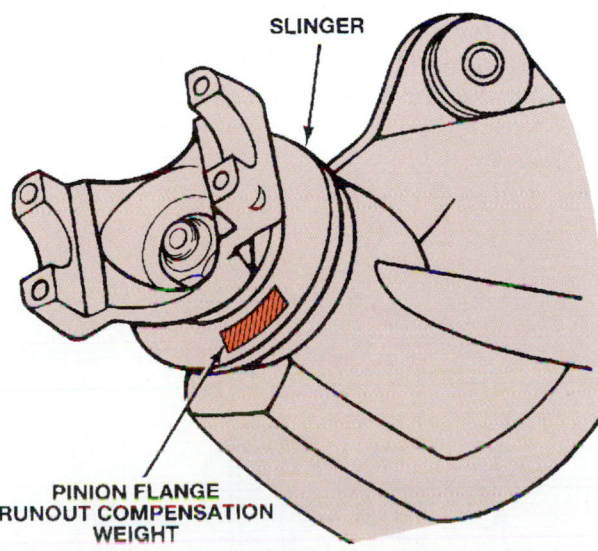

Figure 72.12 When there is weight on a pinion flange, it was put there to correct for runout. *(Courtesy of General Motors Corporation, Service Technology Group)*

imbalance. This is part of an on-the-car wheel balancer. It is mounted on the differential case, under the pinion gear (Figure 72.14).

A transducer is a magnet that is spring loaded on both sides. It moves back and forth in a coil of wire, sending a signal to a meter in response to vibration. The drive shaft is marked anywhere with a piece of chalk. When the driveshaft is spun at the speed where the vibration occurs, the transducer triggers a strobe light. When the light illuminates the mark on the driveshaft, make a mental note of its position. When the shaft is stopped, position the mark at the same place (6 o'clock, for instance). At this position, the heavy spot on the driveshaft is on the bottom of the shaft.

Large hose clamps are positioned on the shaft opposite the heavy spot. Then, the shaft is spun again to see whether more clamps are needed to correct the problem. You can move the hose clamp to one side to see if the vibration improves. If it gets worse, try it the other direction. The EVA has a pickup wire that can be used with an induction timing light that will determine the heavy spot.

Driveshaft balance can also be checked without special instruments. First, put four numbers around the circumference of the driveshaft (Figure 72.15).

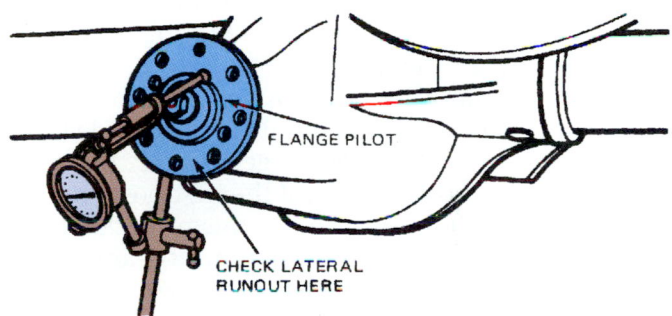

Figure 72.13 Checking flange runout. *(Courtesy of Ford Motor Company)*

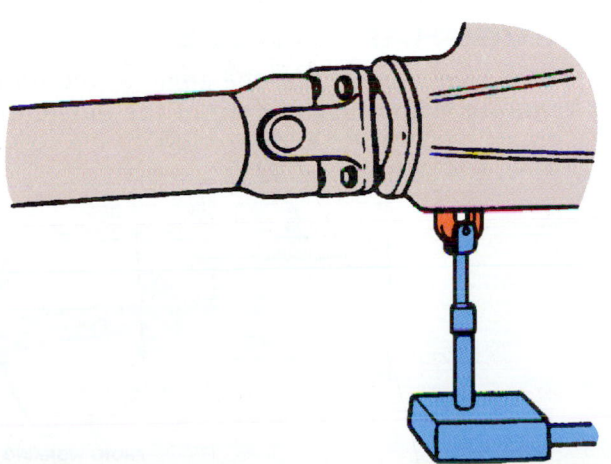

Figure 72.14 A transducer is used to pick up vibrations.

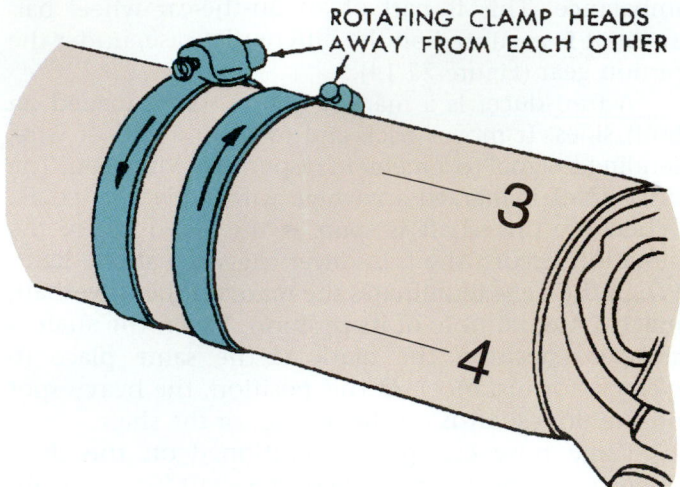

Figure 72.15 Two hose clamps can be separated or brought closer together to change the amount of weight. *(Courtesy of General Motors Corporation, Service Technology Group)*

Because imbalance can be related to runout, put the #1 mark at a point opposite to the amount of greatest runout. Put the clamp with the weight at the #1 position and run the shaft. Check to see if it feels better or worse. Then, move it to the #2 position and try it again. Try all four positions. You are trying to find out where to put the weight.

When you find the best position, move the hose clamp halfway between the numbers to one side to see if the vibration improves. If it gets worse, try it the other direction. When you find the best position, add another hose clamp and repeat the procedure. The clamps can be separated to fine-tune the amount of weight needed (see Figure 72.15).

Another procedure is to run the engine to turn the shaft while holding a piece of chalk near the shaft. The heavy spot will tend to be marked by the chalk. The hose clamp will go opposite to the chalk mark.

■ DRIVESHAFT ANGLE

To extend U-joint service life, the angle of the transmission output shaft and the front of the differential should be within ½° of each other. If not, either the transmission or differential should be shimmed with tapered wheel alignment shims. A ¼" shim will result in a change of about ¾°. Leaf springs can be shimmed on the rear axle housing or the transmission can be shimmed on the crossmember mount. A faster shaft rpm requires a lower angle.

The *working angle* is the difference between the driveshaft angle and the angle of the transmission or differential (Figure 72.16). For passenger cars, the working angle should not be more than 4°. There must always be at least a small angle (at least ½°) to keep the needle bearings rolling or brinelling will occur.

The trim height (ride height) measurement (see Figure 61.5) is important to universal joint angle, and also to front wheel drive tripod joints (see Chapter 70). Four wheel drives have a front shaft that is short and has a sharp angle. Riser kits that make room for tall tires cause the angle to be even more. Slip yokes and splines tend to wear out in these drive shafts.

Launch shudder is a common second order vibration complaint in raised trucks with a high driveshaft angle. It occurs on acceleration from a standing start to about 25 mph before disappearing.

Several methods can be used to measure the driveshaft angle. These include a digital measuring tool (Figure 72.17), a protractor, or a magnetically attached inclinometer. The angle is measured in both the front and the rear.

Suggested maximum driveshaft angles are as follows:

5000 rpm–3¼°	3000 rpm–5⅚°
4500 rpm–3⅔°	2500 rpm–7°
4000 rpm–4¼°	2000 rpm–8⅔°
3500 rpm–5°	1500 rpm–11½°

You can see from the chart that a four wheel drive truck with high driveshaft angles is better operated at low shaft rpm. If the transmission and differential angles are not the same a double Cardan U-joint will be needed.

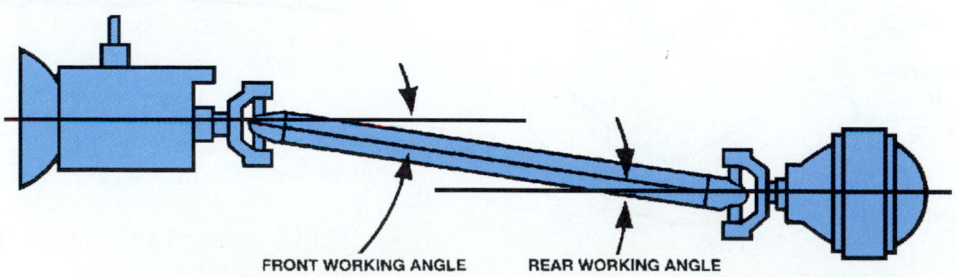

FRONT WORKING ANGLE   REAR WORKING ANGLE

Figure 72.16 The working angle is the difference between the angle of the driveshaft and the angles of the transmission and differential. *(Courtesy of General Motors Corporation, Service Technology Group)*

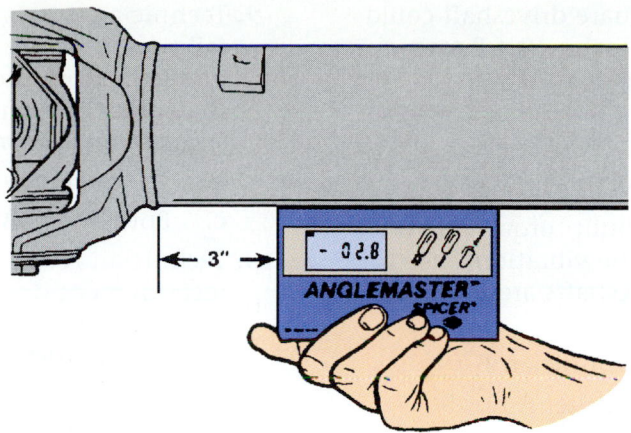

Figure 72.17   A digital driveshaft alignment tool. *(Courtesy of Spicer/Universal Joint Division/Dana Corporation)*

■ REVIEW QUESTIONS

1. The number of cycles in one second is measured in _____.

2. A _____ fork can be used to illustrate resonant frequency.

3. A vibration that interacts with another is called a _____ boom vibration.

4. What is the name for the vibration that is about the same frequency as an electric razor?

5. When road testing, note the _____ at which a vibration occurs.

6. A differential vibration is often caused by a _____ cup that is not seated squarely in its bore.

7. Excessive runout of the stub shaft is usually the cause of vibrations felt at the _____ support bearing.

8. When the runout of one part and the runout of the other are considered during assembly, this is called _____ mounting.

9. The angle of the transmission output shaft and the front of the differential should be within _____° of each other.

10. The difference between the driveshaft angle and the angle of the transmission or differential is called the _____ angle.

■ ASE STYLE REVIEW QUESTIONS

1. Technician A says that front and rear suspension systems are designed to have different resonant frequencies to avoid wheel shimmy. Technician B says that a vibration that occurs twice every revolution is related to a universal joint. Who is right?
   - **a.** Technician A
   - **b.** Technician B
   - **c.** Both A and B
   - **d.** Neither A nor B

2. Technician A says that an electronic vibration analyzer can measure acceleration of a part. Technician B says that an electronic vibration analyzer has a memory function. Who is right?
   - **a.** Technician A
   - **b.** Technician B
   - **c.** Both A and B
   - **d.** Neither A nor B

3. Technician A says that an electronic vibration analyzer can be used with a timing light to balance a driveshaft. Technician B says that vibrations above 100 Hz are usually felt, rather than heard. Who is right?
   - **a.** Technician A
   - **b.** Technician B
   - **c.** Both A and B
   - **d.** Neither A nor B

4. Technician A says that a vibration should always occur at the same speed or frequency. Technician B says that if you cannot feel the vibration, do not try to fix the problem. Who is right?
   - **a.** Technician A
   - **b.** Technician B
   - **c.** Both A and B
   - **d.** Neither A nor B

5. Technician A says that excessive runout in a driveshaft could be because of the pinion flange. Technician B says that excessive runout in a driveshaft could be because of a bent driveshaft. Who is right?
   - **a.** Technician A
   - **b.** Technician B
   - **c.** Both A and B
   - **d.** Neither A nor B

6. Technician A says that a square driveshaft could theoretically, be balanced. Technician B says that a square tire could, theoretically, be balanced. Who is right?

   a. Technician A   **b.** Technician B

   c. Both A and B   **d.** Neither A nor B

7. Technician A says that a rebuilt torque converter is often the cause of driveline vibrations. Technician B says that driveshafts are balanced in the middle. Who is right?

   a. Technician A   **b.** Technician B

   c. Both A and B   **d.** Neither A nor B

8. Technician A says that heavy spots on a drive shaft can be counterbalanced by installing hose clamps with the screw positioned even with the heavy spot. Technician B says that the transmission end of the driveshaft is where most of the first order driveline vibrations come from. Who is right?

   a. Technician A   **b.** Technician B

   c. Both A and B   **d.** Neither A nor B

9. Technician A says that four wheel drives with tall tires are more likely to have vibration problems from the driveshaft. Technician B says that shuddering in a raised truck will occur at speeds above 25 mph. Who is right?

   a. Technician A   **b.** Technician B

   c. Both A and B   **d.** Neither A nor B

10. Technician A says that the ride height measurement does not affect universal joint angle. Technician B says that shafts that spin at under 1000 rpm do not need to be balanced. Who is right?

    a. Technician A   **b.** Technician B

    c. Both A and B   **d.** Neither A nor B

Comfort Systems and Vehicle Electronics

THEN AND NOW: CLIMATE CONTROL

The dangers and discomforts in the early days of motoring would probably discourage even the toughest motocross racer today. Headlights were so dim you could barely see a locomotive in your path. Primitive suspensions were rough enough to chip your teeth. Climate control systems (air conditioning and heating) were simply nonexistent.

The first heater simply used hot bricks lined up on the floorboards. The next step was a footwarmer that burned blocks of charcoal. Those ideas were obviously inefficient since internal combustion engines create a good amount of waste heat anyway. Early car makers developed ducts that directed air over the exhaust manifold into the passenger compartment, but this was never very satisfactory.

Using the liquid from water-cooled engines to provide heat made good sense. This idea was tried as early as 1897 on the Canstatt-Daimler. It was not until 1931, however, that Lincoln produced a heater of the type we know today. It had finned tubes in a housing (heater core) coupled with an electric fan to move air. Flaps were used to control the volume and direction of heat.

The *gasoline heater* was made popular by Ford in the 1930s. A blowtorch-like flame and a fuel line inside the passenger compartment would seem dangerous today, but this system was very effective in dealing with frost.

A car's interior is very successful in collecting solar energy. There were attempts to chill the passenger compartment long before refrigeration—or even the automobile—was invented. In 1884, William Whiteley placed blocks of ice in trays under horse carriages. A fan attached to a wheel forced air to the interior of the carriage. A bucket of ice in front of a floor vent was the automotive equivalent to this early comfort control system.

The first car with real air conditioning was the 1939 Packard. A huge evaporator was mounted in the trunk, leaving very little space for luggage. Cadillac was the next car with air conditioning. In 1941, 300 Cadillacs were built with the air conditioning option.

A recent change in modern air conditioning systems has been the switch from R-12 Freon (which damages the earth's ozone layer) to a more environmentally friendly refrigerant, R134A. Freon production has been outlawed, and older cars are being converted to the new refrigerant.

A gauge set for servicing A/C systems with the environmentally-friendly R-134a refrigerant. *(Courtesy of Bob Freudenberger)*

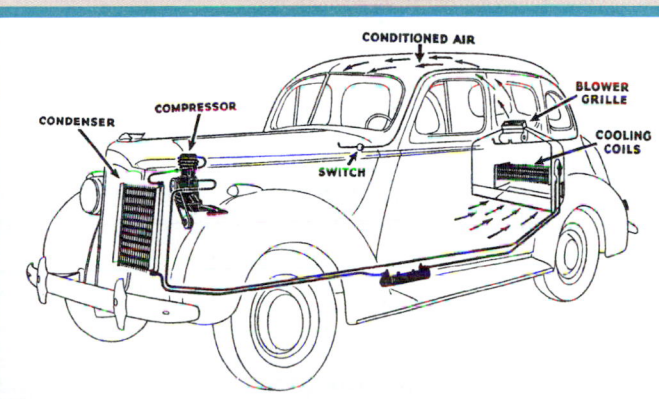

A 1939 Packard with A/C. *(Public Domain)*

Heating and Air Conditioning Fundamentals

■ OBJECTIVES

Upon completion of this chapter, you should be able to:

✔ Explain refrigeration fundamentals.

✔ Describe the difference between the high and low sides of the system.

✔ List the major heating and air conditioning parts and describe their operation.

■ INTRODUCTION

Comfort systems of a car's interior include heating, ventilation, and air conditioning. Many automotive service establishments perform heating and air conditioning service work. In this chapter, the basic operation and service of these systems will be covered. As in previous chapters, the emphasis is on learning how the system operates and how to perform minor service and diagnostic procedures.

■ SOURCES OF HEAT

Like systems for buildings, the automobile heating and air conditioning system is called a heating, ventilation, and air conditioning system (HVAC). In the winter, the system will need to add heat to the inside of the vehicle.

In the summer, it will need to remove interior heat. Heat comes from several sources (Figure 73.1).

■ Each passenger adds heat to the inside of the car, mostly from their breath. Each passenger's temperature is approximately 98.6°F. The more passengers, the more heat.

■ Heat from the outside air adds only 15% of the total.

■ Heat that comes off of the road, the engine, and the catalytic converter accounts for 20%.

■ Most of the heat comes from sunlight, which radiates through glass or shines on the painted surfaces of the car. Insulation of the interior helps to keep the heat inside the car.

■ VENTILATION

Fresh air is introduced into the vehicle interior to replace stale air and to prevent entry of carbon monoxide from the vehicle's exhaust. Ventilation of an automobile can be provided by blowing or ducting outside air into its interior. Some cars have air ducts, or vents, that can be opened allowing outside air to ventilate the inside of the car. Ventilation using outside air does not work when the car is moving at slow speeds or is stopped in traffic. The ventilation system's electrically driven blower takes care of these needs. The blower motor has a **squirrel cage fan** (Figure 73.2).

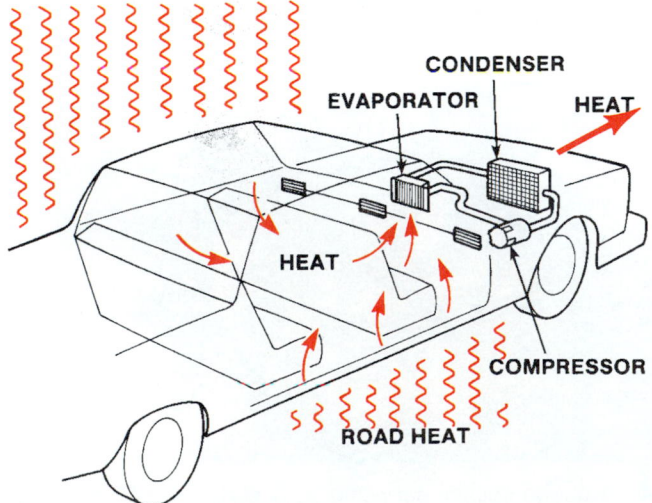

HEAT FROM SUN AND OUTSIDE AIR

CONDENSER
EVAPORATOR
HEAT
HEAT
COMPRESSOR
ROAD HEAT

Figure 73.1   Sources of heat. *(Courtesy of Everco Temp Control)*

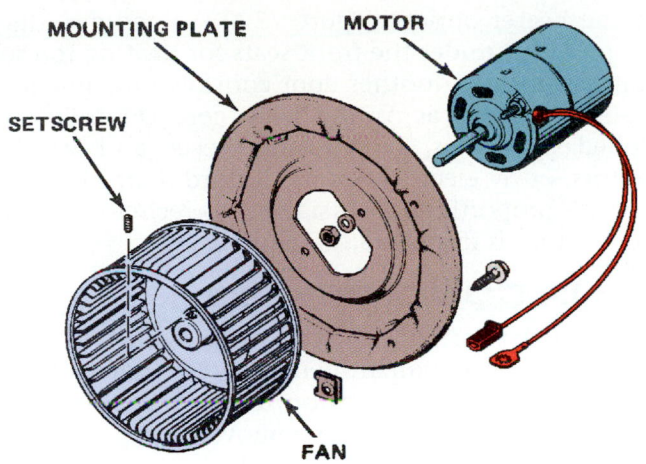

Figure 73.2 A blower motor and squirrel cage fan. *(Courtesy of Chrysler Corporation)*

Heating, ventilation, and air conditioning systems share the same blower motor. The motor usually has multiple speed settings. On many cars, instead of an "off" setting, the blower runs at low speed when the ignition key is on. This maintains a fresh flow of air into the interior of the car.

■ HEATING

Hot air is provided by routing engine coolant to a heater core in the passenger compartment (Figure 73.3). Heat from the engine coolant is radiated to the passenger compartment. When the heater operates, outside air or air from the car's interior passes over the fins of the heater core.

NOTE: *An engine cooling system operating at the correct temperature with a properly operating thermostat is necessary for the heater to be effective.*

Some systems have a control valve (Figure 73.4) operated by a cable or vacuum. Other systems allow coolant to constantly flow through the heater core. Heat to the passenger compartment in these systems is controlled by whether or not air is passed over the heater core. This system is more responsive to automatic controls and is found on most new cars. The *air mix damper*, or **blend air door**, can be opened and closed to close off hot air immediately (Figure 73.5). If the system has a heater control valve, it is used during periods of maximum cooling only.

A heater has doors for three purposes. The air flow doors control whether air from the heater goes to the windshield defroster (on the top of the dashboard) or

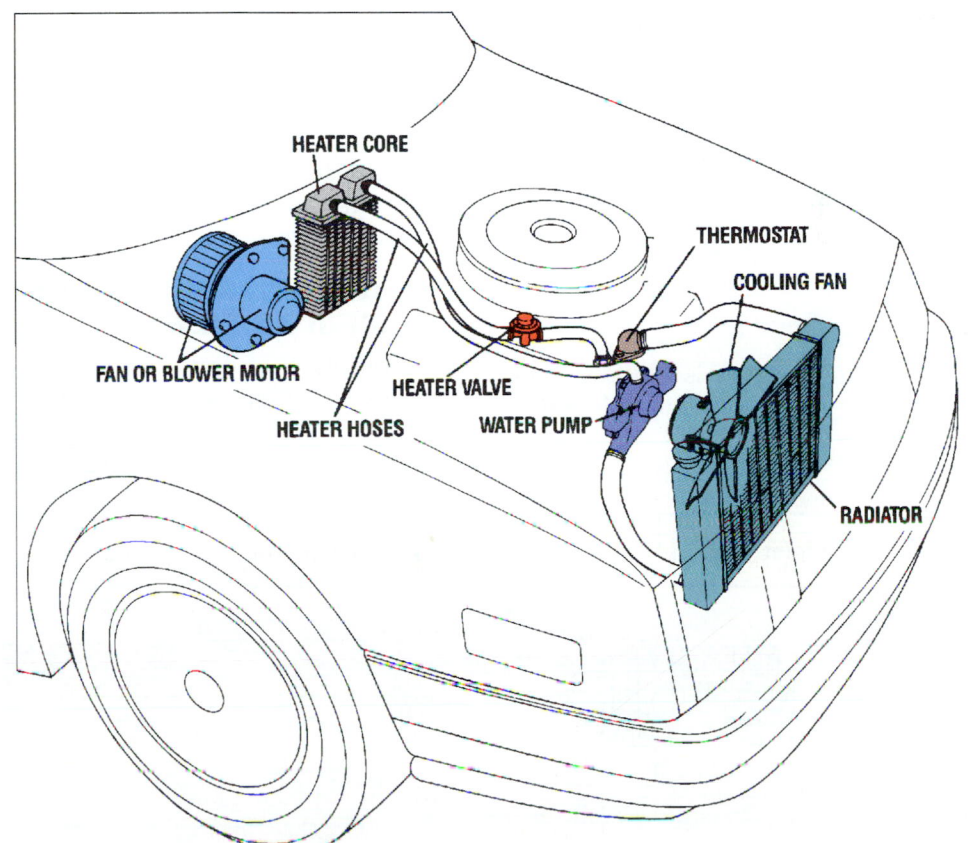

Figure 73.3 Coolant flows through the heater core to warm the passenger compartment. *(Courtesy of Everco Temp Control)*

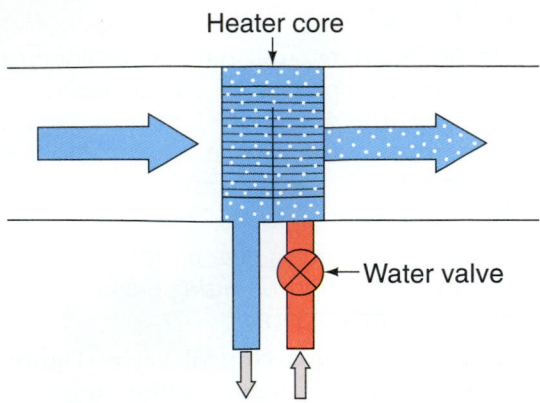

Figure 73.4 This heater uses a control valve.

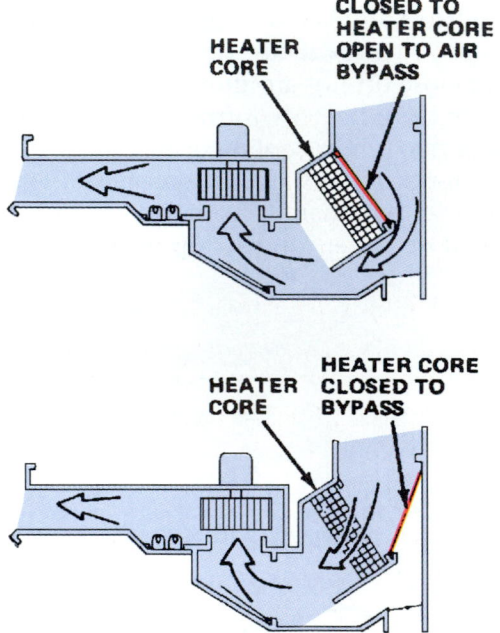

Figure 73.5 The air damper door is closed to control the flow of air across the heater core. *(Courtesy of Chrysler Corporation)*

to the heater outlets (Figure 73.6). Sometimes, ducts are provided under the front seats for heating the rear seat passengers. Another door controls how much air is allowed to pass across the heater core. Doors are controlled by cables, by vacuum hoses and vacuum motors, or by electric motors. A third door may regulate the proportion of outside air to recirculated interior air that is ducted to the blower.

■ AIR CONDITIONING

Air conditioning is the process in which air inside of the passenger compartment is cooled, dried, and circulated. Comfort depends on temperature, humidity, and air movement. Heat is removed from inside the vehicle and transferred to the outside air.

In the 1920s, household refrigerators were first introduced. The first refrigeration unit on an automobile was available as a luxury extra on the Packard automobile in 1939, but air conditioning did not become popular with the motoring public until the 1960s. Today, over 80% of cars sold have air conditioning. Refrigeration systems are also used on off-road and farm machinery.

Air conditioning systems originally were manually controlled by the passengers. Many of today's systems are automatic. At one time it was thought that air conditioning was not energy efficient because it used power. But today's aerodynamic cars are designed to be run on freeways with the windows up for the least wind resistance. A car gets better freeway fuel economy with the air conditioning on and the windows up than it does with the windows down and the air conditioning off. In fact, at 40 mph more gas is used with the windows down.

■ AIR CONDITIONING PRINCIPLES

Automotive air conditioning works on the same principles on which household refrigerators and air conditioners work. A liquid refrigerant is changed to a gas and then back to a liquid again (Figure 73.7). If a

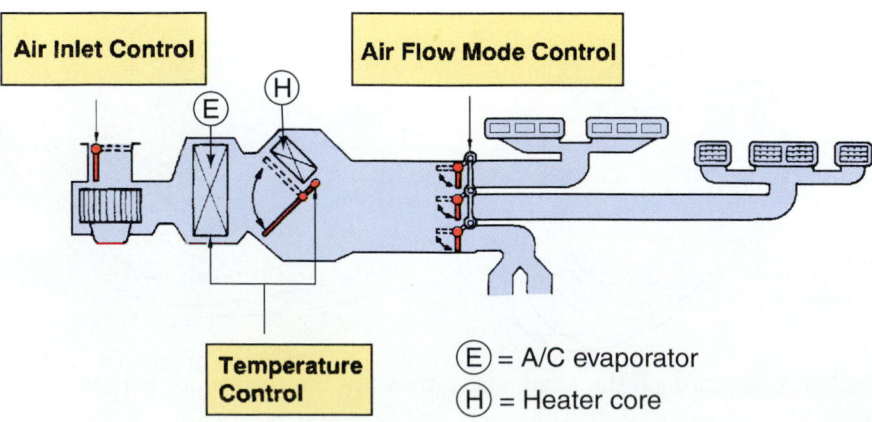

Figure 73.6 Doors control whether the air goes to defrost, the floor or the dash.

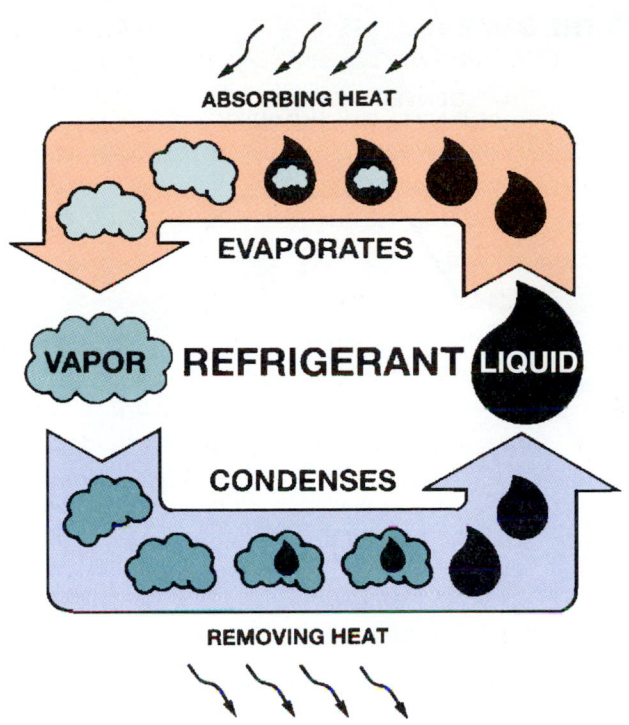

Figure 73.7 The refrigeration process. *(Courtesy of General Motors Corporation, Service Technology Group)*

change of state of the refrigerant is to take place, there must be a transfer of heat. Two principles apply here:

■ For a liquid to change to a gas, it must absorb heat.
■ For a vapor to change to a liquid, it must release heat.

■ HEAT TRANSFER

Whenever there is a difference in the temperatures of two objects, heat can be transferred. Heat will flow to anything that has less heat. Heat transfer occurs by **convection**, **radiation**, or **evaporation**.

Convection

When a body gives off heat, the surrounding air becomes warmer and moves upward. Air that contains less heat takes its place. This is convection, one of the processes of removing heat. Two principles of convection are:

■ Heat rises (notice how steam or smoke rises)
■ Heat always flows from hot to colder

   In a car air conditioning system, air is not actually cooled. Heat is taken away. Have you ever walked past the outside of a house air conditioner installed in a window? Did you feel the hot air coming off the back of it? Heat is being removed from the house.

Radiation

Another way heat is transferred is by radiation. The effects of the sun's radiation are felt by the human body when moving from the shade into bright sun-

light. Dark car colors absorb and radiate heat better than light colors.

Evaporation

Heat is also transferred by evaporation. As moisture is vaporized it absorbs heat, cooling the surface. Your body feels the effect of evaporation with the cooling effect of wind.

■ HUMIDITY

Air conditioning systems provide **humidity** control too. One hundred percent (100%) humidity is when the air is totally saturated with moisture. When humidity is 50%, the air is holding half the amount of moisture that it is capable of holding at a given temperature. Low humidity (dry air) permits heat to be taken away from the human body by evaporation of perspiration. Notice how after a shower if a breeze hits you, you feel cool from evaporation.

   High humidity makes evaporation more difficult. The human body is most comfortable at between 72° to 80°F. (22.2°C to 26.6°C) with humidity at 45% to 50% (in street clothes while riding in an enclosed vehicle). People feel just as cool at 79°F with 30% humidity as they do at 72°F and 90% humidity (Figure 73.8).

■ STATES OF MATTER

All matter can exist in three different states, depending on its temperature. The three states of matter are *solid*, *liquid*, and *gas*. When a solid is heated above its freezing point, it begins to melt (become a liquid).

■ LATENT HEAT

The heat that goes into matter that results in a temperature increase is called **sensible heat** because it is easy to understand (sensible). Before matter can actually change its state, extra heat is required. This additional heat is called **latent heat** (hidden heat). Automotive air conditioning operates on this principle.

NOTE: *After the refrigerant reaches its boiling temperature (–22°F at atmospheric pressure) it absorbs more and more heat, yet its temperature does not increase. It is simply absorbing heat as it attempts to change its state from a liquid to a gas.*

   One candle's temperature is measured at 500°F. The temperature of two candles at 500°F is still 500°F. Latent heat cannot be recorded on a thermometer. Instead, it is measured in British thermal units (Btus). One Btu is the amount of energy required to raise the temperature of one pound of water by 1°F.

Vaporization

The boiling point of water is 212°F at sea level. One pound of water that has already been heated to 212°F will require an additional 970 Btus of heat to make it boil. None of this heat can be recorded on a

COMFORT LEVEL IS THE SAME

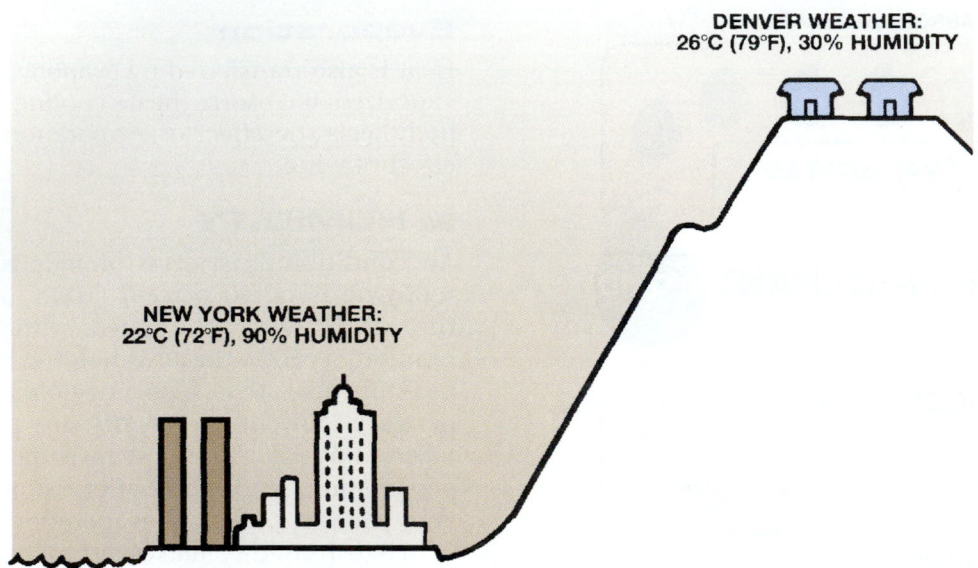

DENVER WEATHER:
26°C (79°F), 30% HUMIDITY

NEW YORK WEATHER:
22°C (72°F), 90% HUMIDITY

Figure 73.8 Humidity affects the level of comfort. *(Courtesy of General Motors Corporation, Service Technology Group)*

thermometer, so the temperature of the boiling water and steam will remain at 212°F. This is called the **latent heat of vaporization** (Figure 73.9). Figure 73.10 shows the relationship between latent heat and sensible heat.

NOTE: *The amount of energy it takes to raise water's temperature from freezing to boiling is much less than the amount of latent heat of vaporization (the amount of heat needed to make the water boil).*

Condensation

When moisture from steam condenses on a cool bathroom mirror, a vapor changes to a liquid. This is called **condensation**. As a vapor condenses, it releases its latent heat. When steam condenses back to water, it releases 970 Btus of heat per pound. This heat released during condensation is called **latent heat of condensation**.

Cooling means "taking away heat." During evaporation, heat is absorbed. The lower the boiling temperature of a liquid, the easier it evaporates. Alcohol feels cooler on your hands than water does. This is because alcohol has a lower boiling point. As it evaporates, it pulls heat from your hands.

A household window air conditioning unit is rated in how many Btus it removes (20k Btu). A typical GM vehicle air conditioner is rated at 21,000 Btu. It needs to be so much greater because of the heat load mentioned before. The way that the heat load is figured is with the **tons rating**. The tons rating is how much heat needs to be added in 24 hours to turn 1 ton of ice at 32° into water; 1 ton = 12,000 Btu/hr.

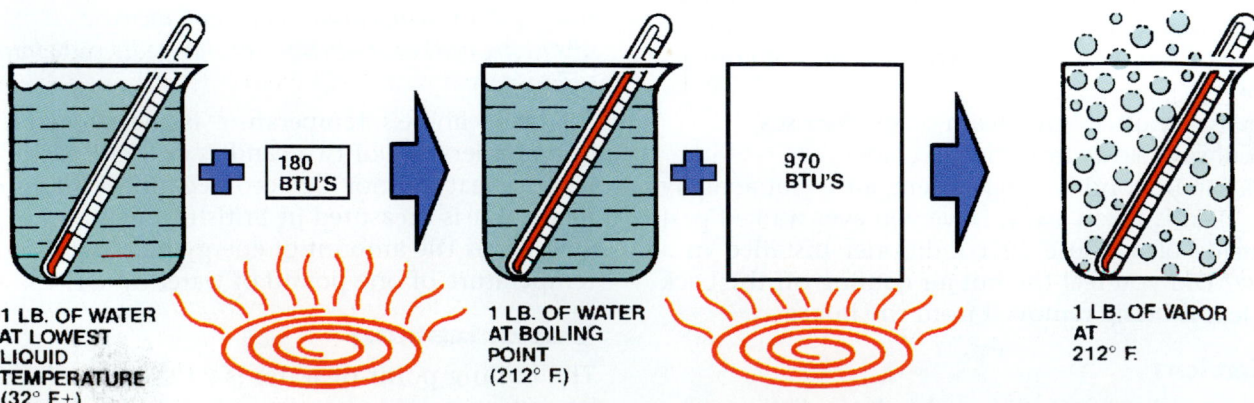

+ 180 BTU'S + 970 BTU'S

1 LB. OF WATER AT LOWEST LIQUID TEMPERATURE (32° F+)

1 LB. OF WATER AT BOILING POINT (212° F.)

1 LB. OF VAPOR AT 212° F.

Figure 73.9 It takes 970 BTU's to make water boil without raising its temperature. *(Courtesy of Ford Motor Company)*

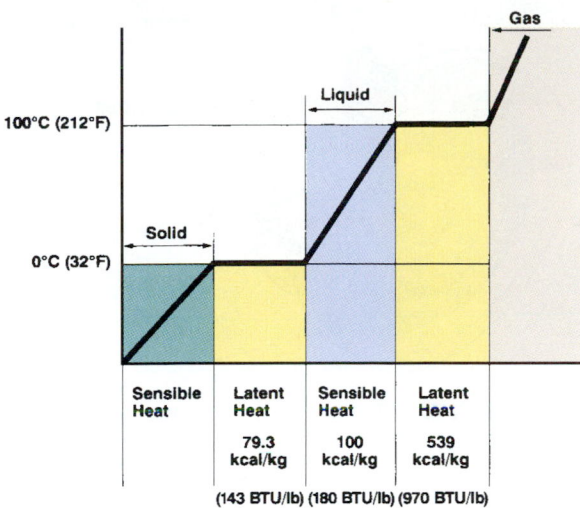

Figure 73.10 The relationship between latent heat and sensible heat.

AIR CONDITIONING SYSTEM OPERATION

An A/C system consists of three major devices in a closed system: the *compressor*, the *condenser*, and the *evaporator*. Refrigerant circulates among these devices. A flow control device regulates the flow of refrigerant among them (Figure 73.11). For an air conditioning system to operate, there must be large differences in pressure within the system. Changing pressure and the gas or liquid state of the refrigerant regulates the oper-

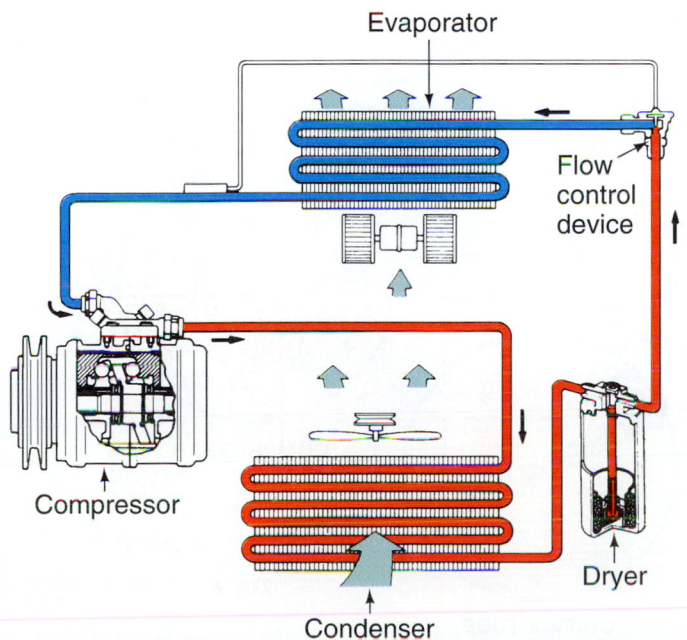

Figure 73.11 A flow control device regulates the flow of refrigerant.

ation of the cooling cycle. The air conditioning cycle has four stages: *compression*, condensation, *expansion*, and vaporization.

ABSORBING HEAT

Located inside of the car's interior is a small, radiator-like device called the evaporator (see Figure 72.13). *Liquid* refrigerant is circulated to the evaporator. As the refrigerant enters the evaporator it loses pressure. There is a slight amount of flash gas at the metering device. Then liquid flows into the evaporator.

The refrigerant is a **saturated vapor**. A saturated vapor is a liquid that is in contact with its vapor within an enclosed space.

The liquid refrigerant absorbs heat from the inside of the car. This is the same as when alcohol or another liquid evaporates from skin, leaving it cool. When the refrigerant absorbs enough heat from the air flowing across the evaporator, it boils and changes to a vapor.

When the refrigerant is pressurized again in the engine compartment, it gives off the heat to the surrounding outside air. When the air conditioning system is first turned on, each cycle through the evaporator absorbs at least 25° of heat from the air blowing across it. Typical temperature readings would be 90° air entering the evaporator and 65° air exiting it. As the temperature in the car goes down, there is not as much heat to remove, so the temperature difference will become less.

REDUCING HUMIDITY

Humidity can enter the car's interior from moisture from the outside air or from the breathing of passengers. When moisture condenses on the cool window surfaces, visibility problems can result. One of the air conditioning system's duties includes dehumidifying the air inside the car. Warm air holds more moisture than cold air. When warm air passes across the cool fins of the evaporator, it loses heat. Moisture in the air tends to condense on the evaporator fins, like it does on the cool glass of a mirror.

The moisture that deposits on the cold evaporator core collects in a drain pan under it and is drained off through the floor as water. This accounts for why a car recently parked after the air conditioning has been operating will often have a stream of water draining onto the ground from the bottom of the front/right side of the passenger compartment.

Because of the principle of latent heat, as the vapor turns to a liquid more heat is released into the surrounding air. This means that the process of dehumidification consumes energy that would normally be used in cooling the passenger compartment. Therefore, when the humidity is high, the air conditioning system does not cool the air as much. You will get more comfortable, however, because once the humidity is lowered your body can lose its heat more effectively.

COMPRESSING THE REFRIGERANT

The vaporized refrigerant is pumped through the suction line from the evaporator to the *compressor* in the engine compartment. The compressor is driven by a belt from the engine crankshaft. It pressurizes the heated refrigerant, further increasing its temperature. Raising the temperature of the refrigerant before it goes to the condenser makes the condenser more efficient at removing heat.

Compressor Clutch

The compressor has a clutch that connects and disconnects it from the crankshaft pulley. The clutch is an electromagnet. It is energized when refrigerant pressure is required (Figure 73.12). In some systems, clutches are only released during periods of heavy loads. Other systems use the clutch to cycle the compressor on and off.

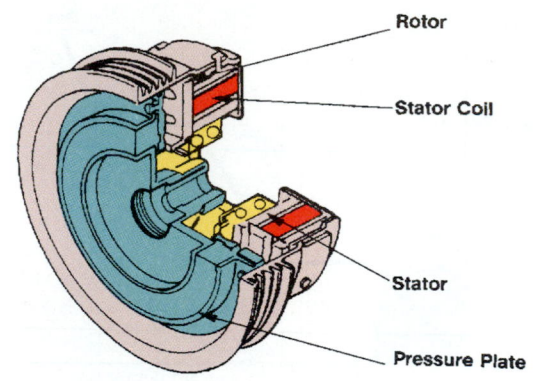

Figure 73.12 Cutaway of a magnetic air conditioner clutch.

TRANSFERRING REFRIGERANT HEAT TO OUTSIDE AIR

From the compressor the refrigerant is pumped to the condenser, located in front of the engine's radiator. The condenser has metal tubes/fins. The condenser is a radiator for refrigerant (see Figure 72.13). Its job is to

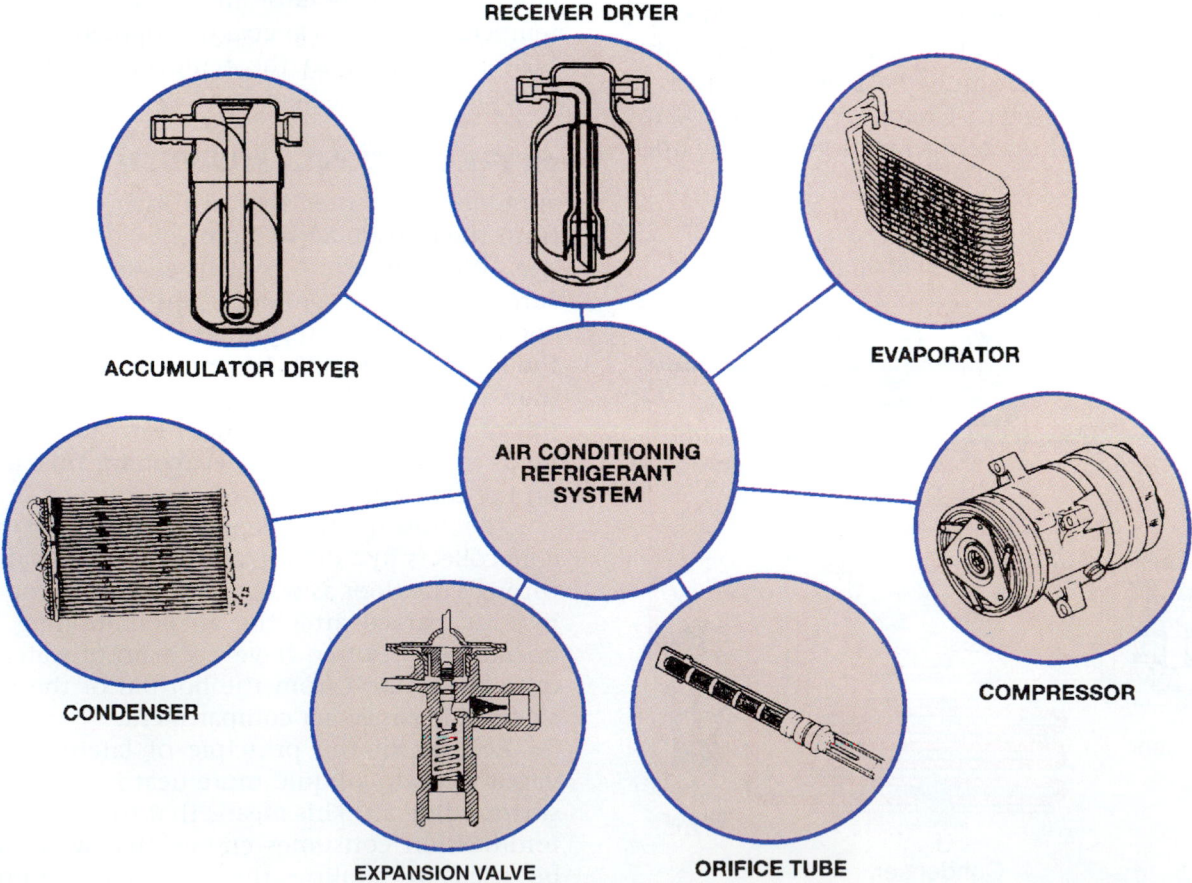

Figure 73.13 Major parts of the air conditioning system. *[Courtesy of General Motors Corporation, Service Technology Group]*

transfer the heat that was absorbed in the passenger compartment to the cooler air blowing through it. The refrigerant's boiling temperature becomes higher when it is pressurized. This is the same idea as pressurizing radiator coolant with a radiator cap to raise its boiling point. Cooling the pressurized refrigerant in the condenser causes it to change from a gas to a slightly cooled but still warm liquid.

Pressure in the condenser must be high enough to operate the flow control device and raise the refrigerant temperature well above that of the **ambient (surrounding) air**. The refrigerant becomes concentrated and very hot when it is compressed. It must be hotter than the air coming across the condenser or heat transfer cannot take place. This allows rejection of the heat that was absorbed as the refrigerant was flowing through the evaporator.

■ AIR CONDITIONING SYSTEM PARTS

All air conditioning systems use an evaporator, condenser, and compressor. Those and other major air conditioning system parts are shown in Figure 73.13. Two of the parts shown at the bottom of the sketch are the *expansion valve* and *orifice tube*. The system will use one or the other of these flow control devices. The system will use a dryer also. The expansion tube system uses a receiver dryer and the orifice tube system uses an accumulator dryer.

■ FLOW CONTROL DEVICES

To raise pressure, there must be a restriction in the air conditioning system. The restriction divides the system into the **high side** and the **low side**. The terms high side and low side refer to high pressure and low pressure within the system (Figure 73.14). The flow control device, located in the inlet pipe of the evaporator, lets the high pressure off of the refrigerant as it "trickles" into the evaporator. When the pressure on the refrigerant is lowered it can evaporate at a lower temperature. During evaporation, heat is absorbed from the passenger compartment. The low pressure side is after the flow control device.

The refrigerant flowing through the evaporator has boiled, and yet it continues to absorb more heat and increase in temperature. Even though the boiling point is cold (–22°F) the refrigerant is still superheated past the evaporator.

Superheat refers to temperatures that are above a liquid's boiling point. Superheat in the air conditioning system is the temperature difference of the refrigerant between the inlet and outlet of the evaporator. It is the difference between the refrigerant's boiling point at system pressure at that moment and the outlet temperature of the refrigerant.

Expansion Valve

The flow control device can be either an orifice tube or a metering valve called an expansion valve (Figure 73.15).

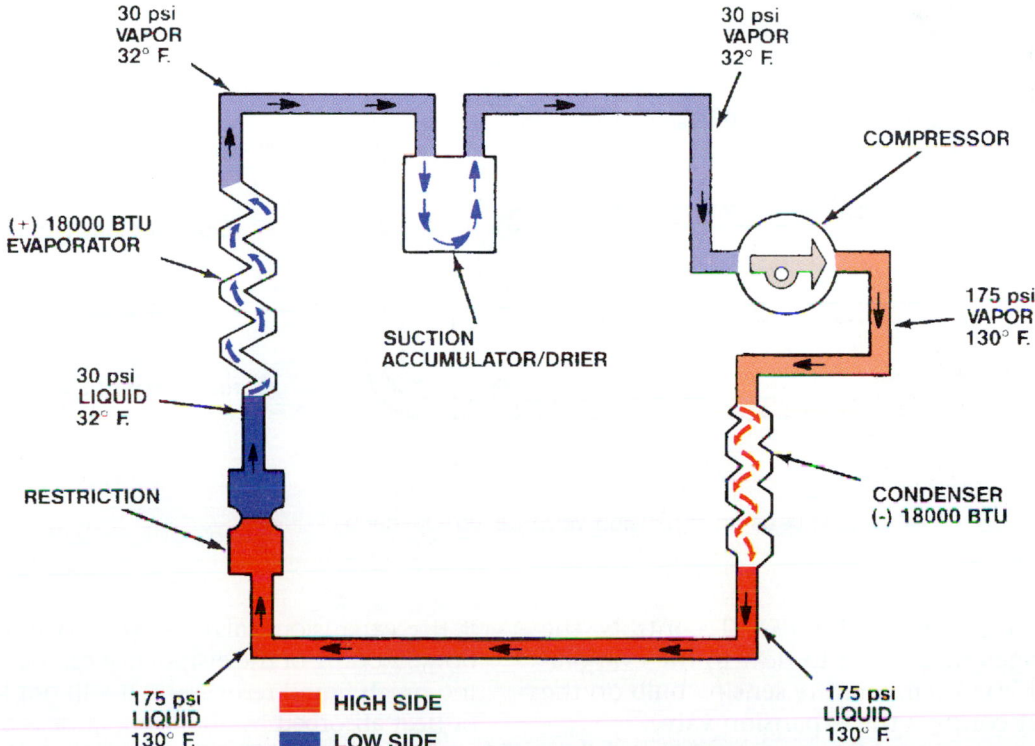

Figure 73.14 The high pressure side and the low pressure side of the air conditioning system. *(Courtesy of Ford Motor Company)*

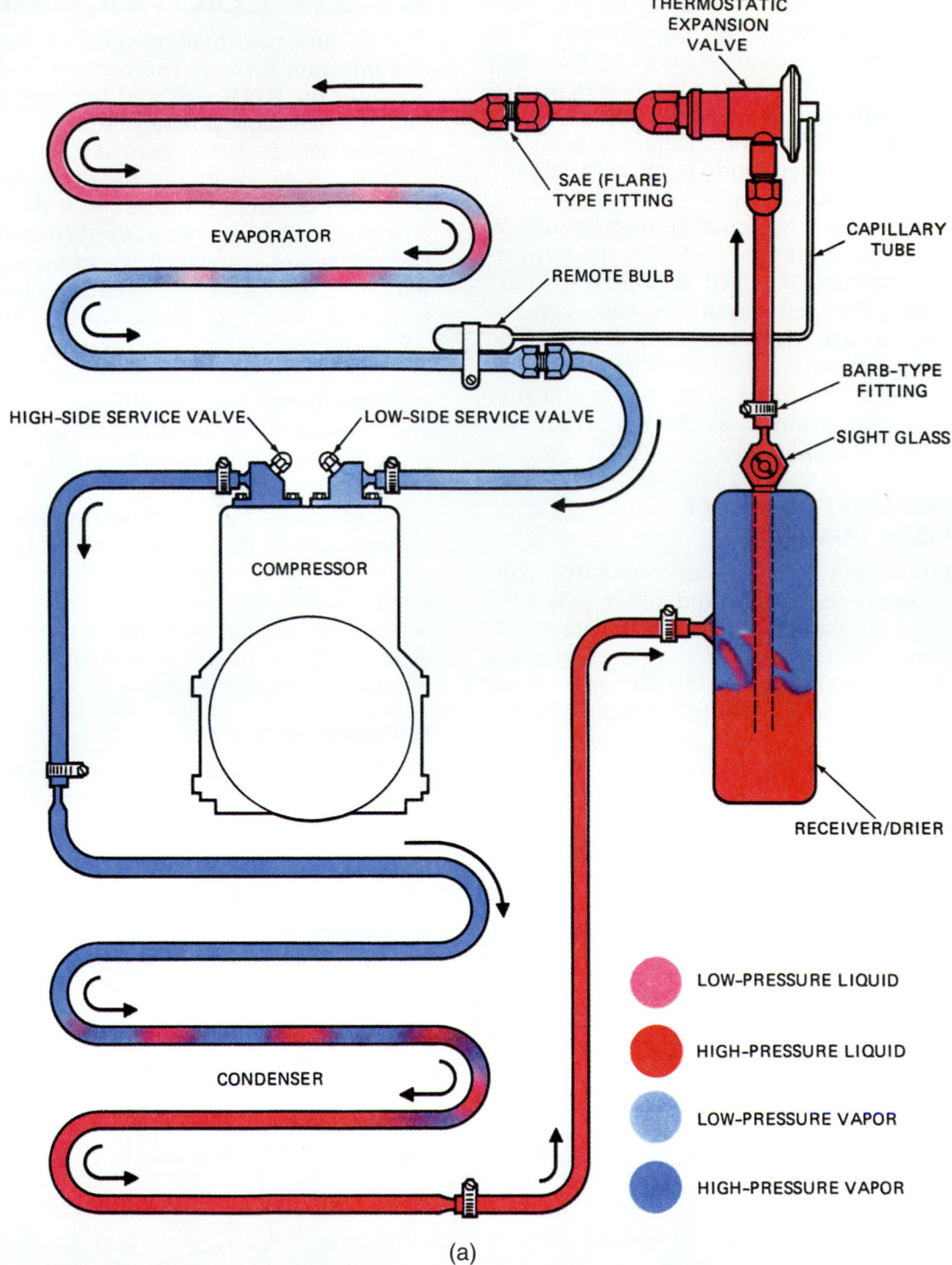

THERMOSTATIC
EXPANSION
VALVE

SAE (FLARE)
TYPE FITTING

CAPILLARY
TUBE

REMOTE BULB

EVAPORATOR

HIGH-SIDE SERVICE VALVE

LOW-SIDE SERVICE VALVE

BARB-TYPE
FITTING

SIGHT GLASS

COMPRESSOR

RECEIVER/DRIER

CONDENSER

● LOW-PRESSURE LIQUID

● HIGH-PRESSURE LIQUID

● LOW-PRESSURE VAPOR

● HIGH-PRESSURE VAPOR

(a)

Figure 73.15a This system uses an expansion valve as its flow control device.

A *thermostatic expansion valve (TXV)* controls the amount of refrigerant allowed to flow to the evaporator (Figure 72.15a). A temperature sensing bulb on the evaporator inlet controls the expansion valve.

A primary purpose of the flow control device is to control the amount of refrigerant flowing into the evaporator. A colder evaporator removes more humidity. But

if the expansion valve allows too much refrigerant to flow, flooding of the evaporator can occur. When there is too much liquid refrigerant, it will not boil. The amount of heat absorbed by the evaporator is directly related to how much of the liquid refrigerant inside of it boils.

If enough refrigerant does not flow into the evaporator, a starving condition occurs. This causes the

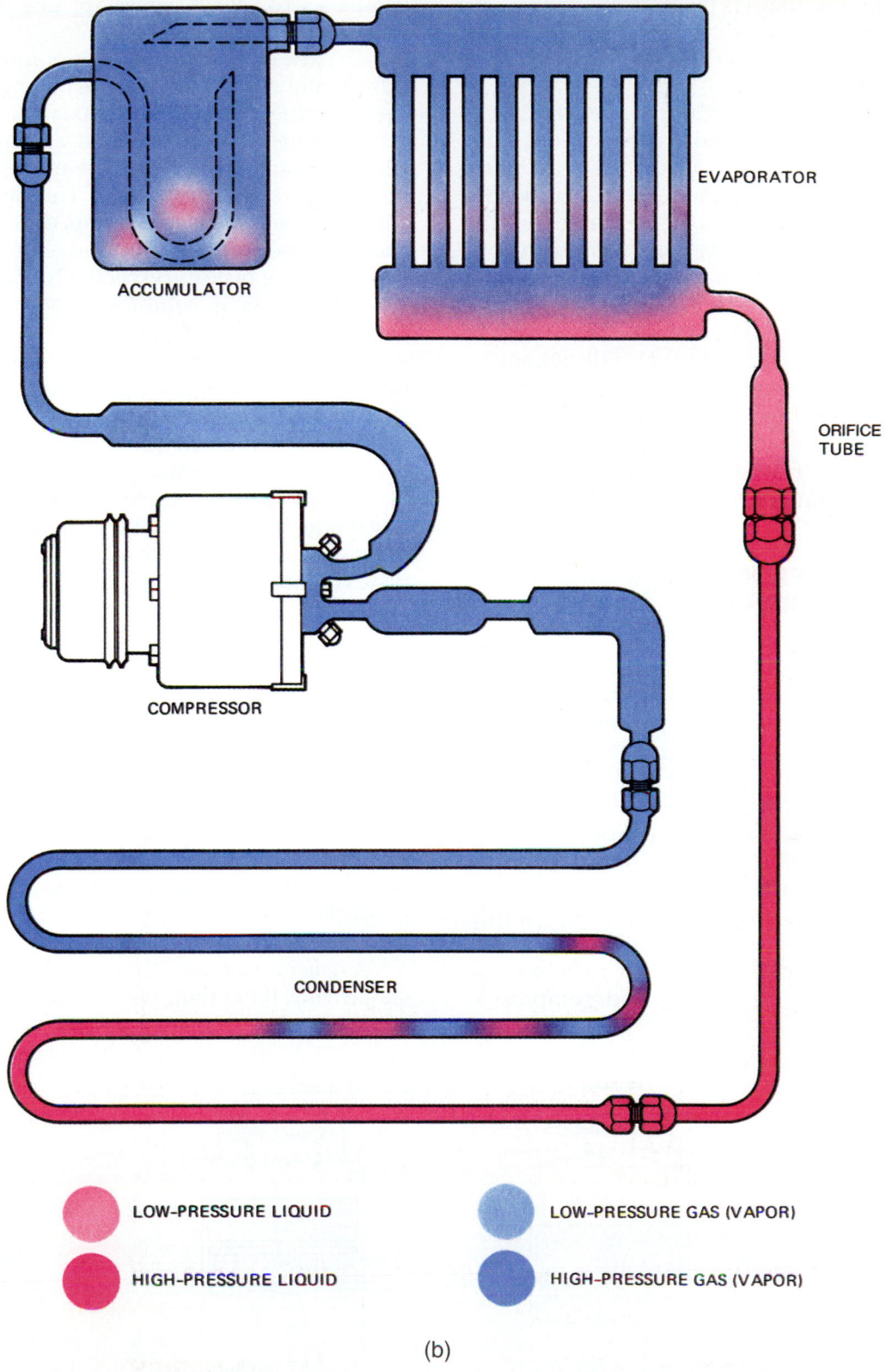

EVAPORATOR

ACCUMULATOR

ORIFICE
TUBE

COMPRESSOR

CONDENSER

LOW-PRESSURE LIQUID

HIGH-PRESSURE LIQUID

LOW-PRESSURE GAS (VAPOR)

HIGH-PRESSURE GAS (VAPOR)

(b)

Figure 73.15b This system uses an orifice tube as its flow control device.

evaporator to freeze up. Once again no cooling takes place. This can be due to a low charge of refrigerant or a restriction in the expansion valve or the line to it.

When the evaporator becomes too cold, moisture that has accumulated on the fins freezes, preventing air flow and hampering heat dissipation. The evaporator can also freeze up if its drain is plugged, allowing water to build up in the evaporator case.

Orifice Tube

An orifice tube system cycles the compressor clutch, instead of using an expansion valve (see Figure 72.15b). It has a *fixed orifice* just before the evaporator, which is simply a hole that cannot change size like the expansion valve does. The magnetic clutch on the compressor cycles the compressor on and off to control temperatures in the evaporator. This is called a *CCOT (cycling clutch orifice tube)* system.

There are other orifice tube system designs that do not cycle the clutch. These have variable displacement or variable output compressors. The clutch remains engaged except during high load conditions, like WOT or during high power steering pressure requirements. This system is used with some smaller engines because it allows steadier engine operation without the clutch cycling on and off.

■ AIR CONDITIONING COMPRESSORS

Earlier compressors were large and heavy. Today's compressors are small and light. There are several types of compressor designs. Three of the designs use pistons and reed valves. Earlier *crankshaft-type compressors* have two cylinders and resemble a gas engine. Crank compressors can be either inline or V-type. They have cast iron piston rings. Later compressors use teflon rings. These have low friction so they can be used in an aluminum bore while causing very little wear.

The compressor works the same way an air compressor does (see Chapter 10). The *reed valve* is a thin piece of sheet metal that bends to act as a one-way check valve. It is opened and closed to allow the compressor to draw in refrigerant and compress it

(Figure 73.16). Reed valves are like the ones used in some two stroke engines.

An *axial compressor* has four or more cylinders. The pistons move lengthwise in the compressor body. Connected to the driveshaft is an axial plate, usually called a *swash plate*. It is mounted at an angle and wobbles when it rotates. This pivots the pistons back and forth in their bores (Figure 73.17). The pistons are double ended, so both sides can pump. Six- and ten-cylinder axial compressors are the most common. They have three and five pistons respectively. Figure 73.18 shows a 6-cylinder axial compressor body and reed valve plate.

The larger cast iron A-6 compressor used by GM up until the early 1980s weighed nearly 35 pounds. Its aluminum replacement weighs only 12 pounds.

Axial compressors sometimes have a *wobble plate* instead of a swash plate (Figure 73.19). Where a swash plate rotates with the driveshaft, the wobble plate does not rotate. It simply wobbles in place. The pistons are only one sided and have connecting rods. Wobble plate compressors usually have five or seven cylinders.

Several manufacturers make newer wobble plate compressors that are *variable displacement compressors* (Figure 73.20). A wobble valve moves to shorten the stroke. This reduces the load on the compressor when it is not needed or wanted. These dependable compressors work well because you do not feel the compressor kicking in and out when it is used on a low power engine. With cycling clutch compressors, the clutch shuts off the compressor to prevent icing of the evaporator. With a variable displacement compressor, no clutch cycling is used. The compressor just changes its displacement. Figure 73.21 shows a variable displacement compressor without its housing.

A *radial compressor* (Figure 73.22) is like a radial aircraft engine. It has multiple cylinders with double-ended

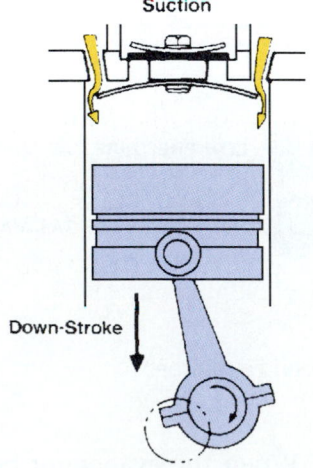
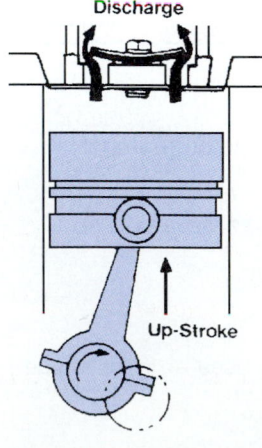

Figure 73.16 Operation of a reed valve.

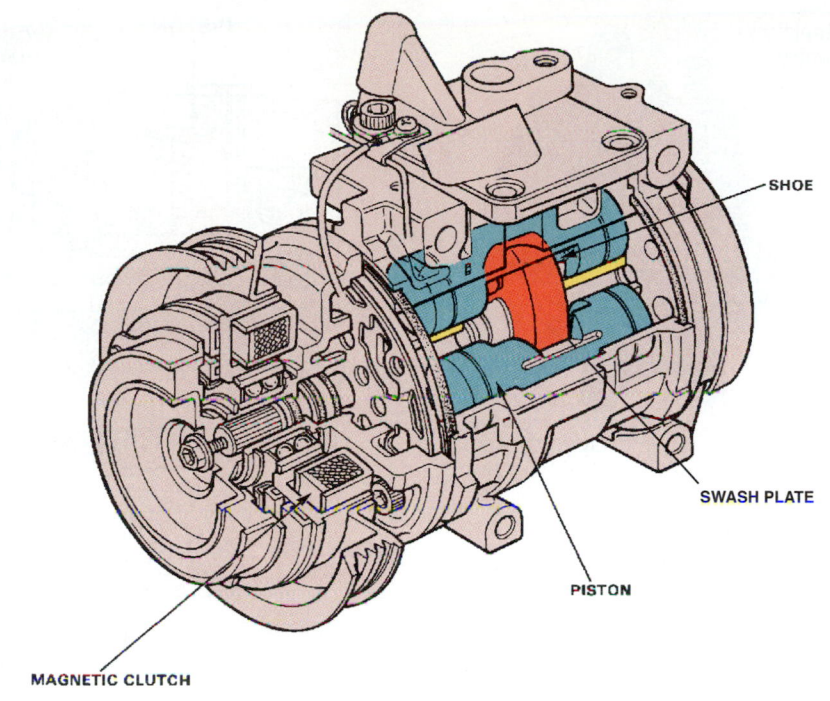

SHOE

SWASH PLATE

PISTON

MAGNETIC CLUTCH

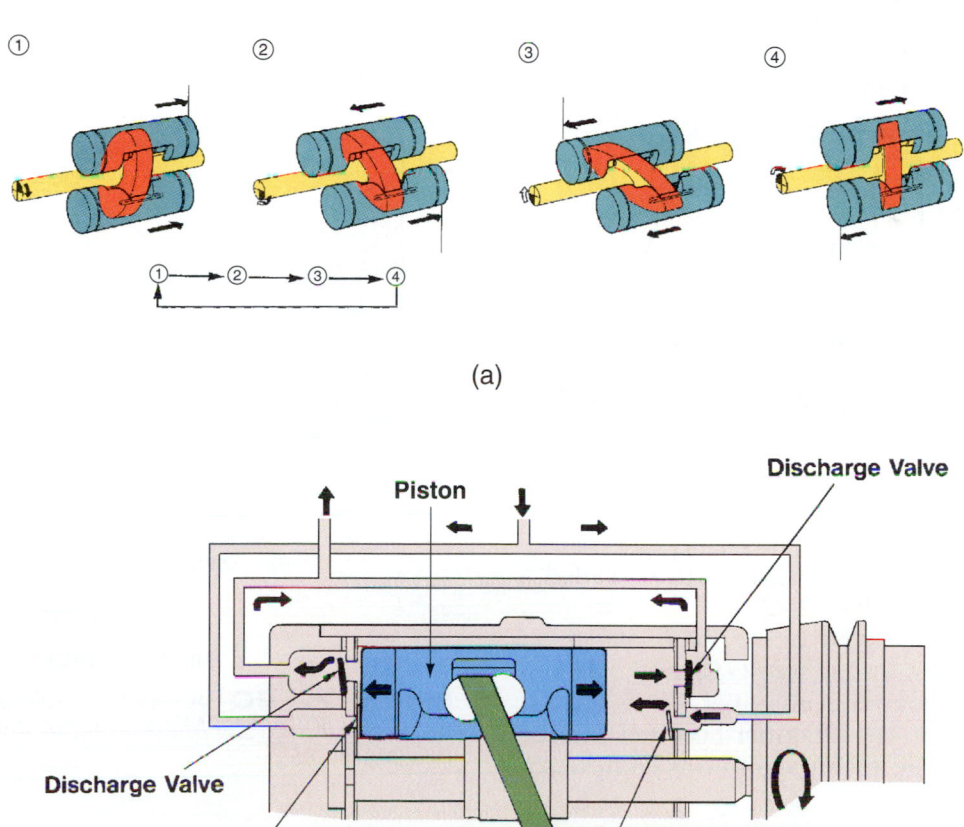

(a)

(b)

Figure 73.17 Operation of a swash plate and double sided piston. *(a, Courtesy of American Honda Motor Co., Inc.)*

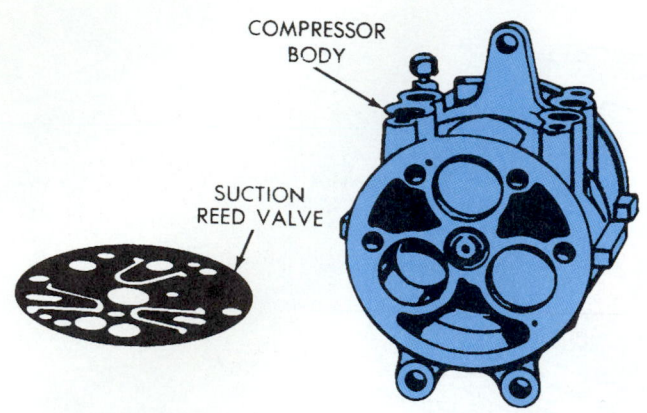

Figure 73.18 An axial compressor body and reed valve. *(Courtesy of Chrysler Corporation)*

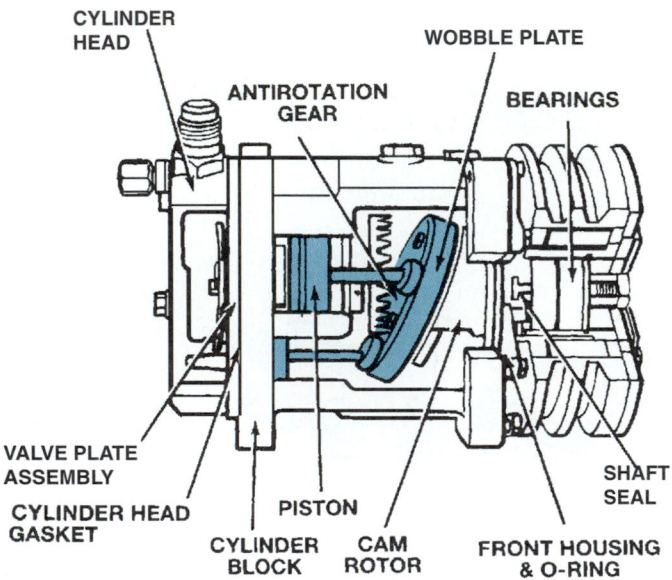

Figure 73.19 A wobble plate compressor. *(Courtesy of General Motors Corporation, Service Technology Group)*

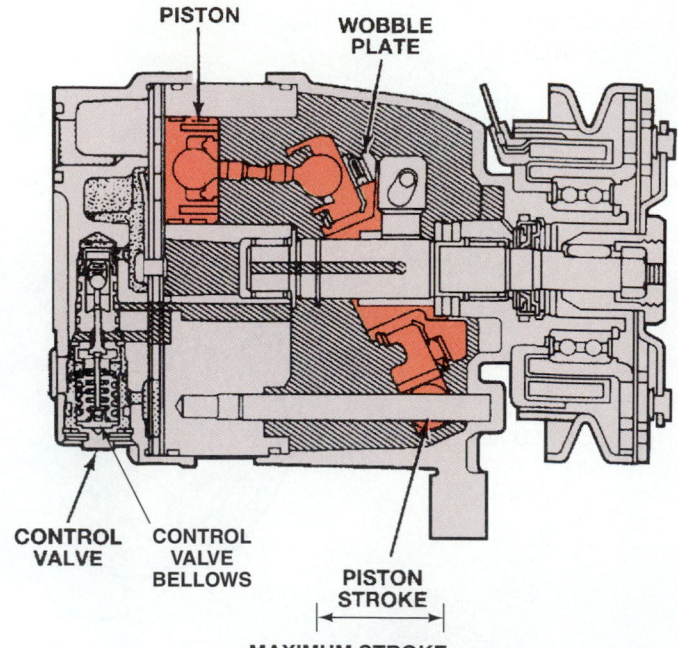

MAXIMUM STROKE

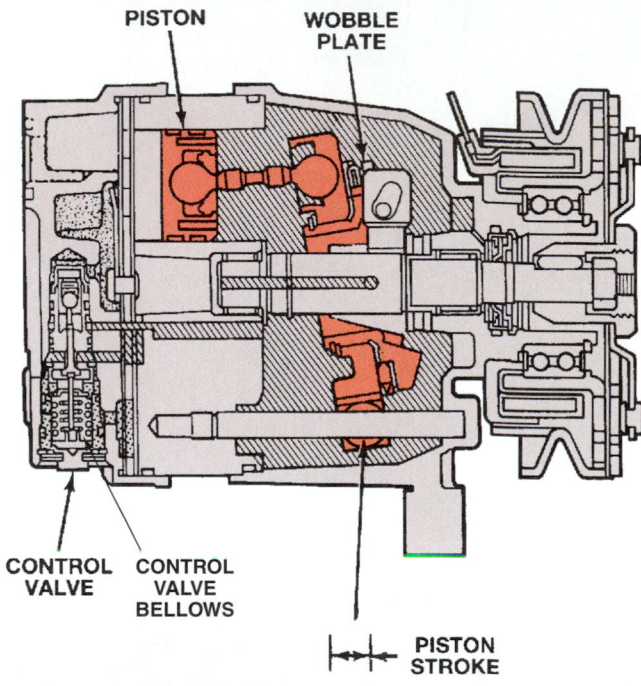

MINIMUM STROKE

Figure 73.20 A variable displacement compressor. *(Courtesy of General Motors Corporation, Service Technology Group)*

pistons and one eccentric crankshaft throw called a *Scotch yoke*. It uses reed valves, too.

Another compressor type is the *scroll compressor* that has a fixed and a movable scroll (Figure 73.23). As the scroll orbits, a pumping chamber forms that is open at the outer end. The chamber becomes smaller as the scroll rotates. The advantage to this compressor is its smooth operation.

One other compressor style, the *rotary vane*, has blades like a power steering pump or smog pump (Figure 73.24).

■ COMPRESSOR LUBRICATION

Another way that compressors are similar to two-stroke gasoline engines is the way they are lubricated. Most compressors are lubricated by oil that is carried in the refrigerant. The old, heavy A-6 has a sump and pump.

■ MUFFLER

Some air conditioning systems have a muffler installed on the outlet of the compressor (Figure 73.25). This is

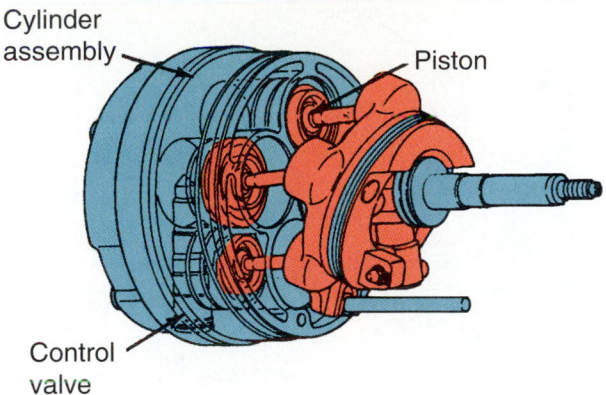

Cylinder assembly

Piston

Control valve

Figure 73.21 Parts of a variable displacement compressor. *[Courtesy of General Motors Corporation, Service Technology Group]*

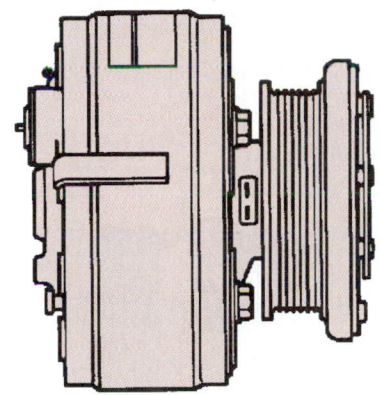

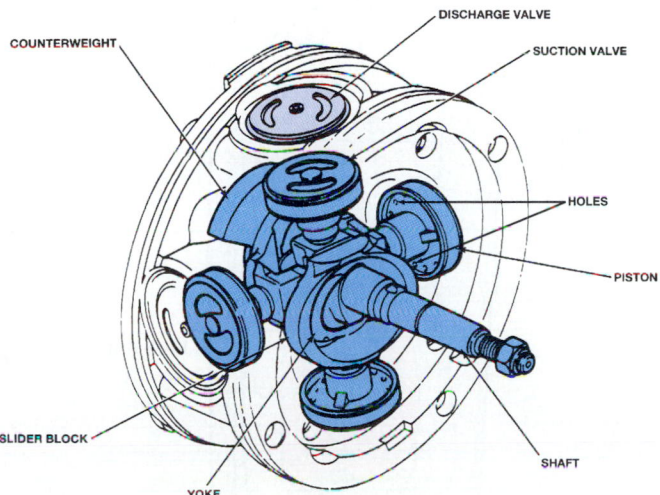

COUNTERWEIGHT

DISCHARGE VALVE

SUCTION VALVE

HOLES

PISTON

SLIDER BLOCK

SHAFT

YOKE

Figure 73.22 A radial compressor. *[Courtesy of General Motors Corporation, Service Technology Group]*

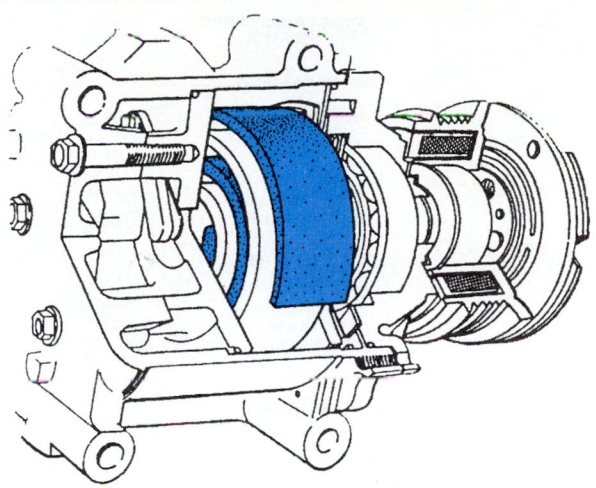

Figure 73.23 A scroll type compressor.

■ ACCUMULATOR OR RECEIVER DRYER

Air conditioners use either a *receiver dryer* (Figure 73.26) or an *accumulator* (Figure 73.27). The two share two basic functions:

■ Both devices have a **desiccant** that removes moisture from the system. Moisture is an enemy to air conditioning systems because it can react with refrigerant and corrode the inside of the system (see Chapter 74).
■ Providing a reservoir for extra system capacity on a cool day is also a function of these devices. The amount of refrigerant needed by the system varies according to heat load and temperature.

These devices also vary slightly as to function and differ as to where they are installed in the system (see Figure 73.15):

■ A receiver dryer is installed in the *high side*. One of its functions is to be sure that pure liquid refrigerant is supplied to the expansion valve.
■ An accumulator is installed in the *low side*. One of its functions is to be sure that pure refrigerant vapor is supplied to the compressor.

When a receiver dryer is used, it is installed after the condenser where it provides a storage place for excess liquid refrigerant until it is needed again by the evaporator. Remember, refrigerant is a high pressure liquid after the condenser.

Some systems have a sight glass on the receiver dryer or in the high pressure liquid line so that unwanted vapor bubbles can be spotted as the system operates. This can give a quick check of whether the refrigerant level is too low. A sight glass is only found on the high pressure side of the system (see Chapter 74).

If the system has an accumulator instead of a receiver/dryer it will be found in the low side at the outlet to the evaporator. All of the refrigerant does not

because some systems make pumping noises due to high or low side pressure vibrations. Mufflers are occasionally wrapped with insulation to further reduce noise.

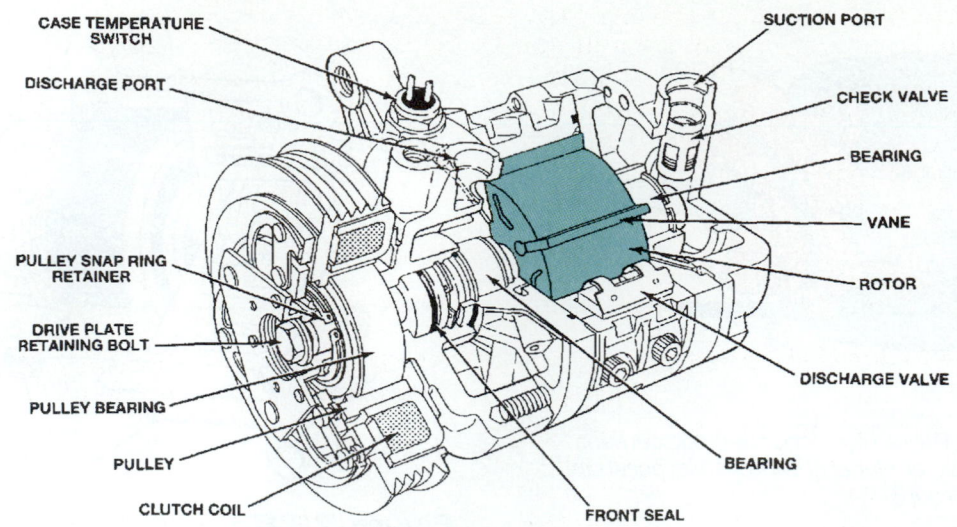

CASE TEMPERATURE SWITCH
DISCHARGE PORT
SUCTION PORT
CHECK VALVE
BEARING
VANE
ROTOR
PULLEY SNAP RING RETAINER
DRIVE PLATE RETAINING BOLT
PULLEY BEARING
DISCHARGE VALVE
PULLEY
BEARING
CLUTCH COIL
FRONT SEAL

Figure 73.24 A rotary vane compressor. *(Courtesy of General Motors Corporation, Service Technology Group)*

Figure 73.25 Some systems have a muffler for the compressor. *(Courtesy of General Motors Corporation, Service Technology Group)*

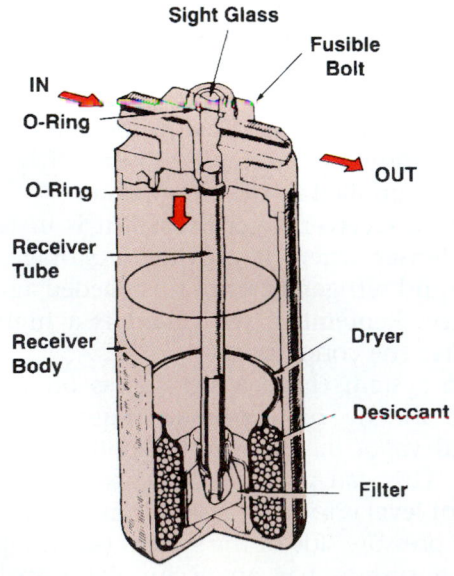

Sight Glass
Fusible Bolt
IN
O-Ring
O-Ring
OUT
Receiver Tube
Receiver Body
Dryer
Desiccant
Filter

Figure 73.26 A receiver dryer fills and siphons liquid refrigerant from the bottom.

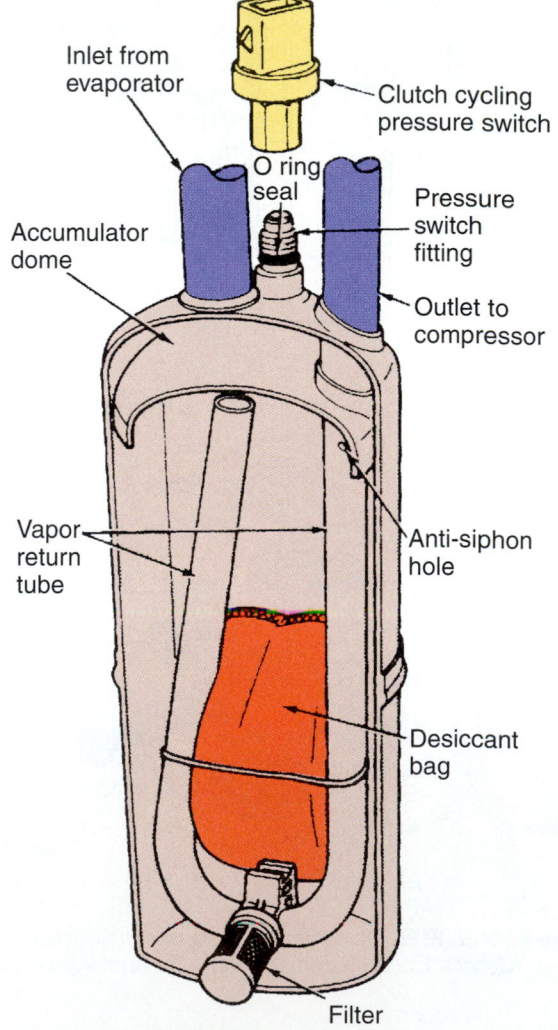

Inlet from evaporator
Clutch cycling pressure switch
O ring seal
Accumulator dome
Pressure switch fitting
Outlet to compressor
Vapor return tube
Anti-siphon hole
Desiccant bag
Filter

Figure 73.27 An accumulator separates vapor and liquid. It pulls off vapor from the top. *(Courtesy of Ford Motor Company)*

vaporize in the evaporator core so the accumulator must capture any liquid to keep it from going to the compressor. Liquid refrigerant would "slug" the compressor, ruining it. The compressor is only supposed to pump vapor. A sight glass will not work on the low side because it is full of vapor, not liquid.

An accumulator is used with an orifice tube. The hole in the orifice tube is only 0.072". This sounds small, but 200 psi will blow a substantial amount of refrigerant through that size hole.

EVAPORATOR ICING CONTROL

There are several methods of shutting off the clutch to keep the evaporator from freezing on an orifice tube system. A *thermostatic switch* on the evaporator is one method (Figure 73.28). It has a mercury tube that senses when the temperature of the evaporator falls to about 32°F. The switch turns off the compressor clutch. Remember, the evaporator accumulates moisture as it dehumidifies the air in the passenger compartment. This water would freeze if the air conditioning system were to continue to pump a fresh supply of refrigerant to the evaporator core. Air from the blower would not be able to carry heat away from the passenger compartment and the system would cease operation.

Sometimes a *pressure cycling switch* is used instead of a thermostatic switch (Figure 73.29). The temperature in the evaporator can be predicted by the pressure of the refrigerant inside. The pressure cycling switch is mounted on the accumulator at the outlet of the evaporator.

Some systems, mostly in the 1960s and 1970s, use an *STV (suction throttling valve)* installed between the evaporator and compressor. These systems did not

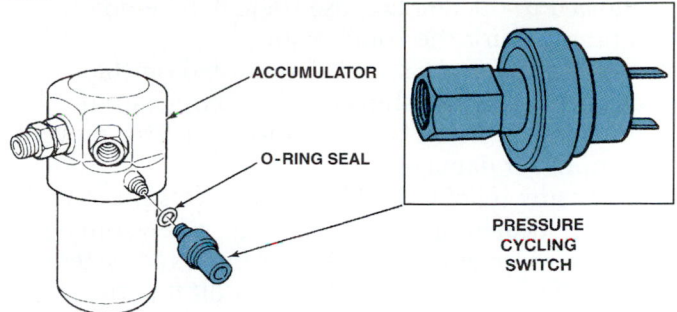

Figure 73.29 A pressure cycling switch on the accumulator. *[Courtesy of General Motors Corporation, Service Technology Group]*

cycle the clutch. Lower pressure means that the system is getting colder. An STV monitors pressure in the evaporator and shuts off refrigerant flow when the pressure drops below 30 psi with R12 refrigerant. An STV can also be called a *POA (pilot-operated absolute)* valve, or an *EPR (evaporator pressure regulator)* valve.

SYSTEM SWITCHES

There are many different switches and controls used to protect the system. Some of them are listed here.

■ Some air conditioning systems run in the defrost position to dry the air going to the windshield. An *ambient temperature switch* keeps the compressor from working when outside temperatures are cold (about 35°F). This protects the compressor from damage.

■ When pressure in the system drops too low, usually because the system is low on refrigerant, the *low pressure cutout switch* will open the circuit to the magnetic clutch (Figure 73.30). This switch is mounted in the compressor inlet. When there is little or no refrigerant, the system must not be

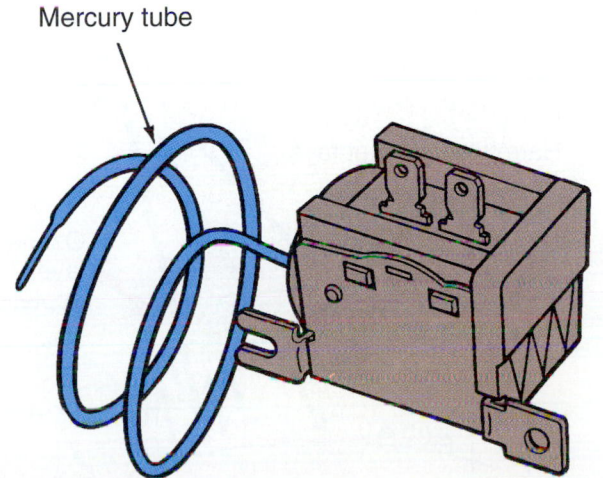

Mercury tube

Figure 73.28 A thermostatic switch on the evaporator shuts off the clutch when the evaporator starts to freeze. *[Courtesy of General Motors Corporation, Service Technology Group]*

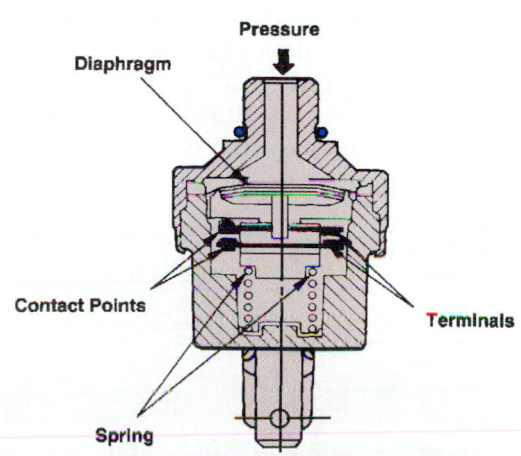

Pressure
Diaphragm
Contact Points
Terminals
Spring

Figure 73.30 A low pressure cutout switch will shut off the clutch if the system becomes too low on refrigerant.

allowed to operate because there will be no lubrication for the compressor.

- A *high pressure cutout switch* mounted on the compressor outlet shuts off the compressor if discharge pressure becomes too high. This prevents compressor damage.
- A pressure relief valve bleeds off excess pressure. It can be located on the compressor, receiver dryer, or anywhere else on the high side of the system.
- A *cutoff switch* can be used to shut off the clutch during WOT operation on small cars. It can also shut off operation during high power steering pressure.
- An air conditioning control switch can be turned on or off from the passenger compartment.

HEATING AND AIR CONDITIONING CONTROLS

The heating and air conditioning system is controlled either manually or automatically. The driver has control over whether to turn the system on or off.

In a manual control system, the driver controls the blower speed, position of air doors, and temperature. A typical manual dash control is shown in Figure 73.31. Electric switches, cables, or vacuum motors are used to activate the selected positions. Automatic temperature controls have sensors and a computer that control the temperature in the passenger compartment. Setting the temperature on the dash control is all the driver has to do (Figure 73.32).

The normal position brings in fresh air. When the selector switch on the dashboard is in the "max," or recirculating position, only about 7% outside air is brought into the vehicle.

REFRIGERANTS AND THE ENVIRONMENT

The refrigerant that was widely used until recently is called R-12 or Freon. It has been around for over 60 years. It is nonpoisonous, stable, and inexpensive to manufacture. Because it has been outlawed, its cost today is very high. The major problem with this refrigerant was not discovered until recently.

Located in the earth's stratosphere (10–30 miles above the surface of the earth) is its protective ozone

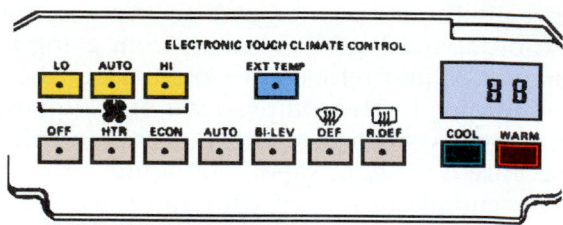

Figure 73.32 A typical automatic air temperature control. *[Courtesy of General Motors Corporation, Service Technology Group]*

layer, which filters out most of the sun's harmful ultraviolet rays (Figure 73.33). The amount of ozone, which is what makes the sky appear blue, has remained about the same in the stratosphere for centuries. Freon (R-12) is a **chlorofluorocarbon (CFC)**. CFCs are depleting this protective ozone layer through a chemical reaction. When CFCs are released into the atmosphere, they slowly travel into the stratosphere where they can remain for 100 years or more (Figure 73.34).

At ground level, ozone is a part of photochemical smog. A lack of ozone in the stratosphere actually increases ozone at ground level. Because of centrifugal force, the ozone layer is thickest at the equator and thinnest at the north and south poles.

CFCs are made of chlorine, fluorine, and carbon. They have been used for such items as fire extinguishers, dry cleaning fluid, solvents, and aerosol can propellants. In 1978, due to concerns about CFCs contributing to global warming, they were banned by the United States and other countries for use as aerosol propellants and dry cleaning solvents.

The concentration of chlorine in the stratosphere was measured at 80 parts per trillion (ppt) in 1978. In 1985, scientists confirmed that the ozone layer was

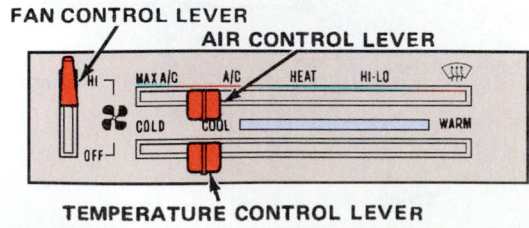

Figure 73.31 A typical manual dash control for heating and air conditioning. *[Courtesy of Chrysler Corporation]*

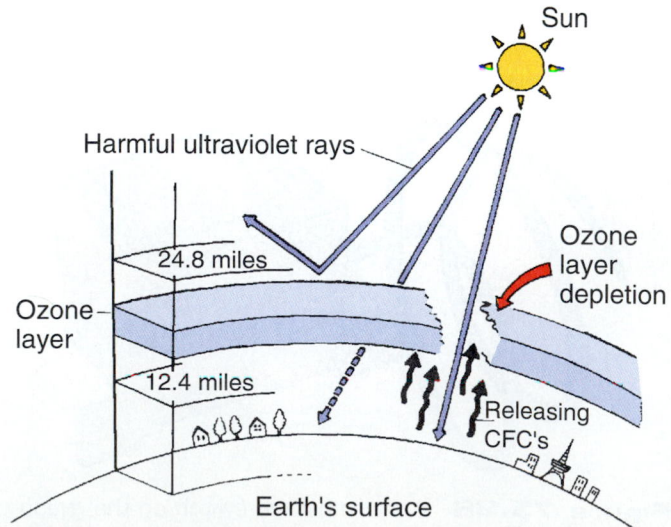

Figure 73.33 The ozone layer filters out the suns harmful rays.

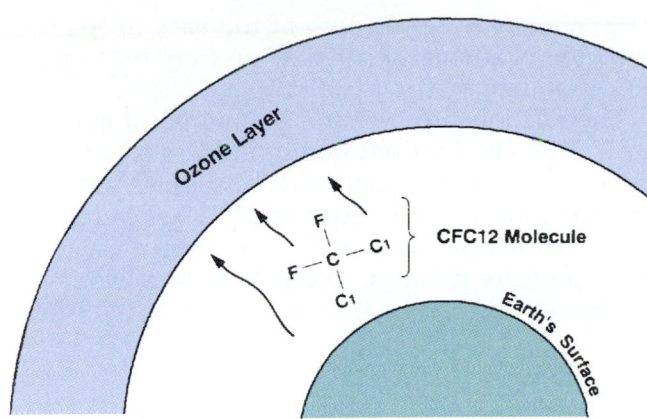

Figure 73.34 CFC migration to the stratosphere.

depleted to the point that there was a large hole in it over the south pole. This was attributed to CFCs. Chlorine in the stratosphere was measured in 1990 to be 499 ppt. Chlorine attacks the ozone in the stratosphere and destroys it. Without ozone in the stratosphere, ultraviolet light reaches the earth's surface (Figure 73.35).

In 1987, the United States and twenty-two other countries signed an agreement known as the Montreal Protocol. The agreement set limits on the production of ozone-depleting chemicals, with a total phase out by the year 2000. In Europe and the United States the date for the total ban was accelerated to 1996.

The estimate is that about 30% of released CFCs are from mobile air conditioning sources. Some leaks out naturally but most of the problem comes during repair and service. By 1993, service technicians were required to be licensed to work on air conditioning systems or be able to purchase CFC refrigerants.

There are only two refrigerants used in cars, R-12 and R-134A. R-22 is used in stationary air conditioners and refrigerators. In the past, inexpensive R-12 refrigerants were commonly released into the air. Today

they are recovered and recycled. Recovering means to capture the refrigerant and store it for reuse. Recycling is when a special machine is used to remove water, air, and oil from it so it can be reused.

Beginning in 1994, almost all cars were equipped with R-134A refrigerant. This refrigerant is a *hydrofluorocarbon*, rather than a CFC. It is simply a greenhouse gas and does not damage the ozone layer. There have been changes made to the systems to accommodate the different characteristics of the refrigerants. The most notable is that the condenser is larger. The system will cool effectively with the new changes.

■ TEMPERATURE AND PRESSURE

When dealing with refrigerants and diagnosing problems with the air conditioning system, you will need an understanding of the temperature and pressure and their relationship to one another.

Temperature Measurement

The *Fahrenheit* system of temperature measurement (most often used in air conditioning work) was developed in the early 1700s by Gabriel D. Fahrenheit. Zero in the Fahrenheit system was arrived at by marking a mercury tube on what felt like the coldest day ever in Iceland. When the weather became warmer and ice began to melt, the mercury had expanded to 32 thousandths of its original volume. At the point at which the water boiled, the mercury had expanded by 212 thousandths of its original volume. Thus, the freezing point of water was designated to be 32°F and the boiling point 212°F.

The *Celsius* system of temperature measurement was also developed in the early 1700s by Anders Celsius, a Swedish astronomer. In the Celsius system, ice melts at 0°C and water boils at 100°C. This system is the scale of metric temperature measurement. Figure 73.36 gives a comparison of temperature readings between different temperature scales.

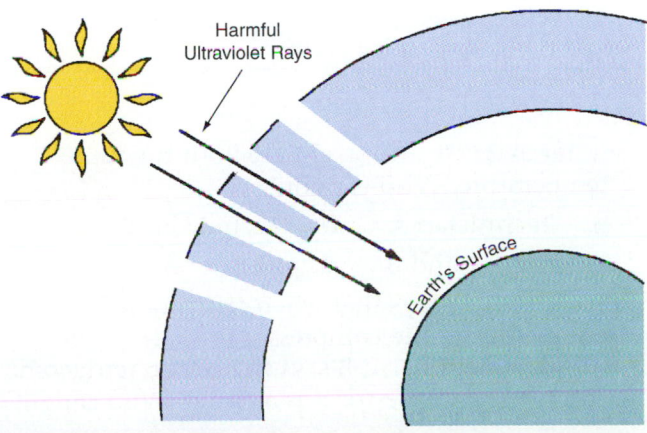

Figure 73.35 Without ozone in the stratosphere, ultraviolet light reaches the earth's surface.

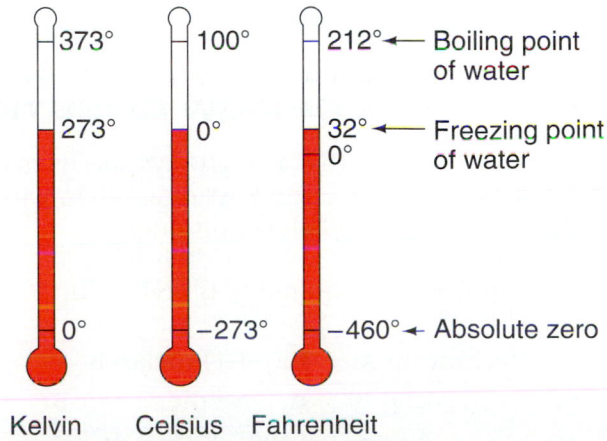

Figure 73.36 Comparison of different temperature scales.

■ PRESSURE

The atmosphere, consisting primarily of the gases, oxygen (21%) and nitrogen (78%), extends about 600 miles above the earth's surface. The pressure that the weight of atmospheric gases exerts on the earth's surface is about 14.7 pounds per square inch.

On a pressure gauge, pressures above atmospheric are called pressure, while pressures below atmospheric are called *vacuum*. That is because the gauge is designed to read 0 psi and 0 vacuum regardless of actual atmospheric pressure. Pressure above atmospheric is actually *PSIG*, which means *pounds per square inch gauge*. With this system 0 on the gauge represents 0 at sea level. Air conditioning pressures are measured in this way.

Pressure below atmospheric is called vacuum. It is measured with a gauge that lifts a column of the liquid heavy metal mercury (Hg) up a small tube. Vacuum is measured in **inches of mercury (in.Hg)**. Hg is the scientific abbreviation for mercury. The strongest vacuum that can be achieved in the earth's gravity is 30 in.Hg. A chart in Appendix J at the back of the book shows the relationship between pressure measured in PSIG, BAR (barometric), and in.Hg.

The relationship between boiling point and pressure is different for each liquid. Because it is a liquid that everyone is familiar with, examples are given here using *water*. Water boils at 212°F at sea level. Its boiling point is lower above sea level because atmospheric pressure is lower. Every 1000 feet increase in altitude results in a boiling point that is 1.1°F lower than 212°F.

Example:
at 8900' altitude:
8.9 (8900 divided by 1000) × 1.1° = 9.8°
212°F – 9.8°F = 202.2°F

Increasing the pressure on water will raise its boiling point by about 3°. Cooking food in a pressure cooker raises the boiling point of water so that the temperature of the water can be above 212°F. The radiator cap on the engine's cooling system raises the boiling point of coolant.

■ REVIEW QUESTIONS

1. What is the name of the device that engine coolant flows through to heat the car's interior?

2. Comfort depends on temperature, _____, and air movement.

3. Heat transfer occurs by _____, radiation, or evaporation.

4. When the air is totally saturated with moisture, this is called 100% _____.

5. What is the name of the type of heat that is required before matter can actually change its state?

6. How is latent heat measured?

7. What is the name of the small, radiator-like device that absorbs heat from the car's interior?

8. Each cycle through the evaporator absorbs at least ____° of heat from the air blowing across it.

9. What is the name of the part in front of the radiator that carries heat away from the refrigerant?

10. The flow control device can be either an orifice _____ or an expansion valve.

11. What is the name of the one-way check valve used in an air conditioning compressor?

12. The _____ compressor with a swash plate uses pistons that are double ended so they can pump in each direction.

13. Which compressor type resembles an aircraft engine?

14. Which refrigerant is a CFC, R-12 or R-134A?

15. What is the name of the substance that is a pollutant at ground level, but when in the stratosphere, protects the earth's surface from ultraviolet rays?

■ ASE STYLE REVIEW QUESTIONS

1. Technician A says that at highways speeds more fuel is used with the windows down than with the windows up and air conditioner on. Technician B says that the body is more comfortable when humidity is high. Who is right?
   a. Technician A   b. Technician B
   c. Both A and B   d. Neither A nor B

2. Technician A says that the "normal AC" position is better for controlling humidity than "MAX". Technician B says that at sea level, water at 212°F and steam are both the same temperature. Who is right?
   a. Technician A   b. Technician B
   c. Both A and B   d. Neither A nor B

3. Technician A says that when refrigerant is pumped from the compressor it is a high pressure gas. Technician B says that when refrigerant enters the compressor it is a liquid. Who is right?
   a. Technician A   b. Technician B
   c. Both A and B   d. Neither A nor B

4. Technician A says that refrigerant cools off when it is compressed. Technician B says that too little refrigerant flow in the evaporator will result in a frozen evaporator core. Who is right?

   a. Technician A   **b.** Technician B
   c. Both A and B   **d.** Neither A nor B

5. Technician A says that the expansion valve keeps the evaporator from freezing. Technician B says that cycling the clutch keeps the evaporator from freezing. Who is right?

   a. Technician A   **b.** Technician B
   c. Both A and B   **d.** Neither A nor B

6. Technician A says that with a variable displacement compressor, no clutch cycling is used. Technician B says that the compressor is only supposed to pump liquid. Who is right?

   a. Technician A   **b.** Technician B
   c. Both A and B   **d.** Neither A nor B

7. Technician A says that a receiver dryer is installed in the high side of the system. Technician B says that a receiver dryer supplies refrigerant vapor to the expansion valve. Who is right?

   a. Technician A   **b.** Technician B
   c. Both A and B   **d.** Neither A nor B

8. Technician A says that an accumulator is installed in the low side of the system. Technician B says that an accumulator supplies refrigerant vapor to the compressor. Who is right?

   a. Technician A   **b.** Technician B
   c. Both A and B   **d.** Neither A nor B

9. Technician A says that refrigerant is a high pressure liquid when it leaves the condenser. Technician B says that most compressors have an oil sump. Who is right?

   a. Technician A   **b.** Technician B
   c. Both A and B   **d.** Neither A nor B

10. Technician A says that the temperature in the evaporator can be predicted by the pressure of the refrigerant inside. Technician B says that a low pressure cutout switch will open the circuit to the magnetic clutch if the refrigerant level is low. Who is right?

    a. Technician A   **b.** Technician B
    c. Both A and B   **d.** Neither A nor B

Heating and Air Conditioning Service

■ OBJECTIVES

Upon completion of this chapter, you should be able to:

✔ Locate obvious problems in heating and air conditioning systems with a visual inspection.

✔ Test air conditioner efficiency and pressures.

✔ Locate leaks in the refrigeration system.

✔ Diagnose and repair problem components.

✔ Evacuate and recharge a refrigeration system in a safe and legal manner.

■ KEY TERMS

discharge service valve
suction service valve
compound gauge
static pressure
molecular sieve
polyalkylene glycol (PAG)
polyalester (ester)
dual pass machine
single pass machine

■ HEATER SERVICE

Complaints in the heating system are usually related to coolant leaks or inappropriate temperatures for the season. With modern automobiles and the necessary practice of maintaining the coolant concentration, many of the old problems related to coolant flow have disappeared. If a heater is not often used, rust can accumulate in it.

The heater flow can be checked by feeling the hoses at the inlet and outlet to the heater core. If the heater has a control valve and it is open, both hoses should be warm. This indicates that coolant is flowing through the heater core. If the outlet hose is cold, there must be an obstruction in the heater core or control valve.

NOTE: *Be sure to check the coolant level first. If the coolant level is low, the heater might not have sufficient coolant to operate.*

A plugged heater core can often be flushed with pressurized water and chemical compounds designed for flushing a cooling system. Unhook both hoses at the most accessible locations and force water through one hose while it comes out the other.

Heater Core Replacement

When a heater core leaks, a new heater core is installed or the old one repaired. The heater housing is usually under the dash and must be removed to gain access to the heater core (Figure 74.1). Once the housing is removed, it can be dismantled to gain access to the heater core. Some housings have two halves, held together by clips (Figure 74.2).

■ AIR CONDITIONING SERVICE

Air conditioning service is a challenging and rewarding specialty area. Because of recent discoveries regarding the hole in the ozone layer (see Chapter 73), technicians are now required to be licensed to service these systems. This has resulted in the work that was formerly done by do-it-yourself people at their homes now being done by repair shops.

There is an abundance of work in this area, but there are also several different types of refrigerants now on the market. A technician must have a good understanding of the entire air conditioning system before attempting to work on it.

Air Conditioning Safety

Safety is always a top priority. When working on air conditioning, be careful around fans and hot coolant. Refrigerant requires other safety precaution.

CAUTION ■ Be sure to wear eye protection when attaching pressure gauges or servicing the system. Have you ever seen a demonstration in a science class where a rubber band is frozen with liquid nitrogen? When it is dropped on the table it shatters. Refrigerant will instantly freeze anything it comes into contact with. If it touches your eyes it will cause blindness.

■ R-12 refrigerants produce nerve gas (phosgene) when exposed to an open flame.

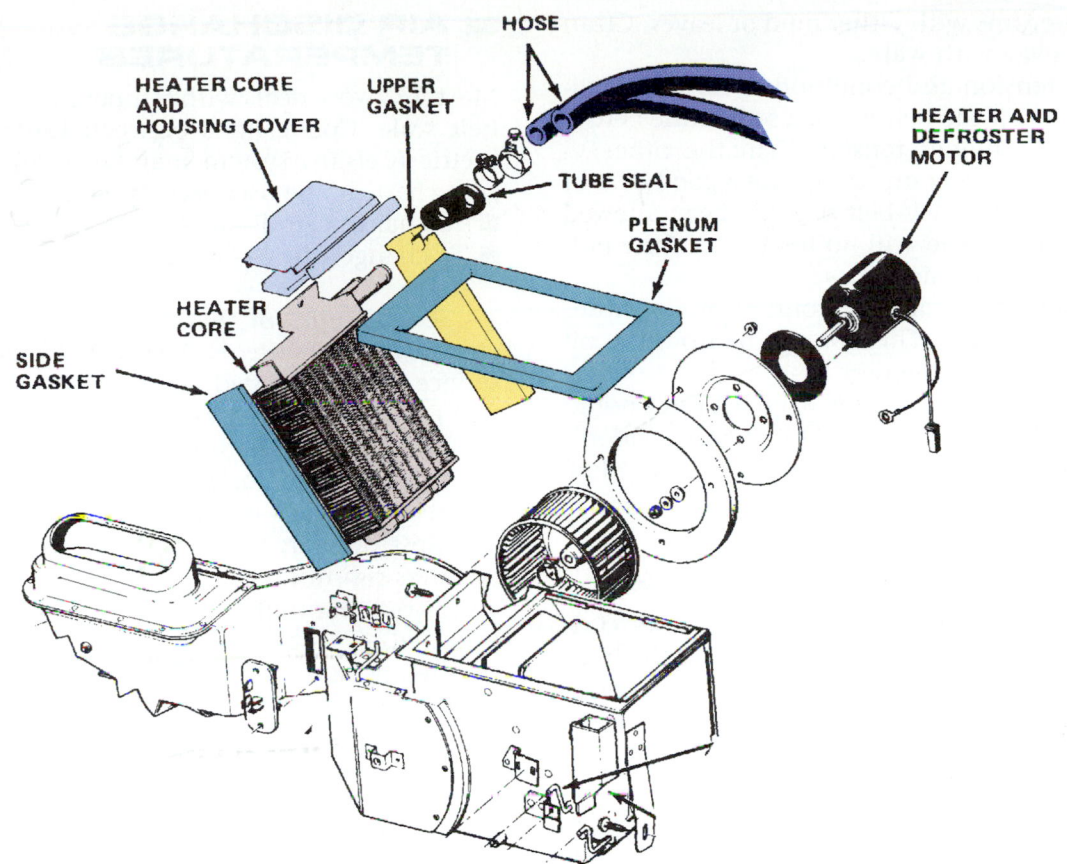

HOSE

HEATER CORE AND HOUSING COVER

UPPER GASKET

TUBE SEAL

PLENUM GASKET

HEATER AND DEFROSTER MOTOR

HEATER CORE

SIDE GASKET

Figure 74.1 The heater housing is disassembled to get to the heater core. *(Courtesy of Chrysler Corporation)*

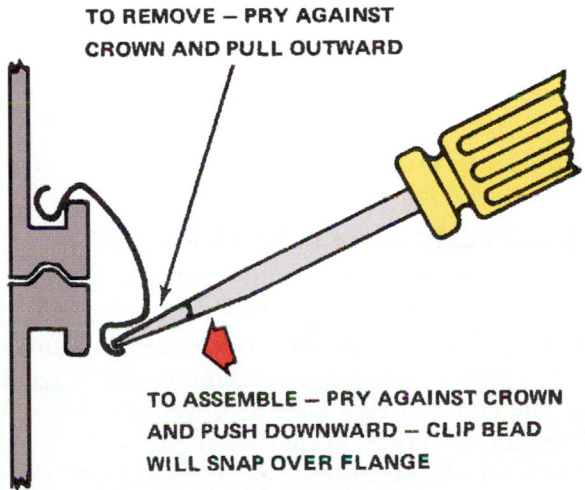

TO REMOVE – PRY AGAINST CROWN AND PULL OUTWARD

TO ASSEMBLE – PRY AGAINST CROWN AND PUSH DOWNWARD – CLIP BEAD WILL SNAP OVER FLANGE

Figure 74.2 To separate the halves of the heater housing, use a screwdriver on the clip. *(Courtesy of Chrysler Corporation)*

■ AIR CONDITIONING SYSTEM TESTING

When diagnosing the air conditioning system, you will have a chance to use the basic understanding you gained of system theory while reading Chapter 73.

Check simple things first to try to form a good idea of where the problem is. Review the main parts of the system in Figure 73.13.

Before the hole in the ozone layer was discovered, R-12 refrigerant cost less than one dollar a pound and was used wastefully. During an air conditioning service, it was a common practice to let the refrigerant escape into the atmosphere. Not only is this now against the law, but it can be expensive. Some large systems with rear air can take up to 9 pounds of refrigerant. At over $20 per pound and up for R-12, making a mistake could be quite costly.

Another error is to allow an air conditioning system to remain empty. This allows moisture to enter the system, ruining a receiver dryer or an accumulator. Replacing these parts is costly, especially when combined with the cost of lost refrigerant. Replacement of these devices is covered later in the chapter.

Visual Inspection

Always perform a quick visual inspection first, looking for obvious causes of problems. Check the coolant level in the radiator. The engine's cooling system has an effect on the operation of both the heating and refrigeration systems. Look for blockage of the condenser fins with either mud or leaves. Clean mud and dirt away with water.

Check the tension and condition of the air conditioning drive belt. If the belt is not a serpentine belt, it probably requires higher tension than the other V-belts. Air conditioning compressors use a good deal of power when they run. A V-belt that has been allowed to slip will be glazed and will no longer hold the pulley, even if properly tensioned.

Look for wet stains around connections of hoses and parts of the system. These would be from the oil that circulates in the refrigerant.

Check the operation of the clutch. You should hear the loud click of the clutch as it engages periodically. Electric cooling fans should begin to run at low speed when the clutch comes on. Check wiring and hoses for obvious problems.

System Operation Quick Check

Run the engine for 5 to 10 minutes with the air conditioning system operating. Feel the tubes on the high and low sides of the system. If the line does not feel hot or cold where it is supposed to, look for a restriction in the line at that point. This is commonly a frost ring on a line or part.

The *high side* should be warm or *hot*. This is because it is under pressure and wants to lose heat to the condenser. If frost accumulates on the receiver dryer, it could be restricted.

The *low side* should be *cool*. On accumulator systems, the accumulator should feel like ice water is running through it. This is because the refrigerant in it has come through the evaporator and is still mostly vapor. The boiling point of the refrigerant is lower at low pressure, so the line is cooler because the vapor is absorbing heat.

When the low side is cool but the system does not cool the passenger compartment, the problem could be in an air door or panel controls. Check their operation next. If the high side is not hot and the low side is not cool, you will know the problem is in the refrigeration section of the system.

■ TESTING AIR CONDITIONING EFFICIENCY

Before temperature and pressure tests, the following list of items should be met:
- Set the temperature control to maximum cool and normal AC. Some manufacturers call for MAX AC so that recirculating air is used. Be sure to check the specifications.
- Set the blower position to high.
- The temperature inside of the car must be stabilized (not getting cooler).
- Engine speed must be at least 1500 rpm to ensure adequate refrigerant flow.
- The compressor clutch must be engaged.

■ AIR DISCHARGE TEMPERATURES

Most A/C work deals with temperatures on the Fahrenheit scale. Conversions between Fahrenheit and the metric (Celsius) system may be necessary. A conversion chart can be used (see Appendix K) or:
- To change C to F—multiply C × 1.8 and add 32°
- To change F to C—subtract 32° and multiply by 0.556

Several types of thermometers are available. The *stem/dial thermometer* is the most popular one for A/C service. There are two common ranges, from 0°F to 220°F and −40°F to +160°F (this one is the most popular for A/C work). Both ranges can work on either the high side (in the condenser) or the low side (in the outlets in the car's interior) (Figure 74.3).

Most discharge temperatures specified by manufacturers run from 35°F to 55°F depending on temperature and humidity. Figure 74.4 shows a chart of expected air discharge temperatures at different outside temperatures. The vertical temperature readings refer to actual duct readings. The horizontal readings refer to the day's outside temperature. The drop in the air's temperature as it passes through the evaporator should be at least 20°F. It is very difficult to tell how well A/C works on a cool day.

Rules of thumb:
- With 70°F air entering the evaporator, expect about a 20°F drop in air temperature (50°F).
- With 80°F air entering the evaporator, expect about a 25°F drop in air temperature (55°F).
- With 90°F air entering the evaporator, expect about a 30°F drop in air temperature (60°F).

It is important that the hood be closed during this test. The hardest work an air conditioner does is on "normal" because it is cooling outside air.

■ CLUTCH CYCLING TIMES

A majority of vehicles built after 1980 have an orifice tube system. Total cycling time is the length of time between clutch engagement and re-engagement after going through a complete engagement/disengagement cycle. The chart in Figure 74.5 shows a typical clutch cycling time. A typical cycle would be:
- Clutch on—60 seconds
- Clutch off—15 seconds

In this example, the total clutch cycle would be 75 seconds.

■ PRESSURE TESTING

A pressure gauge set is used to measure pressures in the system. One of the gauges is for high side pressure and the other is for low side pressure. The gauges and hoses are of a standard color code (Figure 74.6).
- *Red is the high side.*
- *Blue is the low side.*
- *Yellow is for charging and discharging equipment.*

Figure 74.3 A thermometer used for checking condenser and evaporator temperatures.

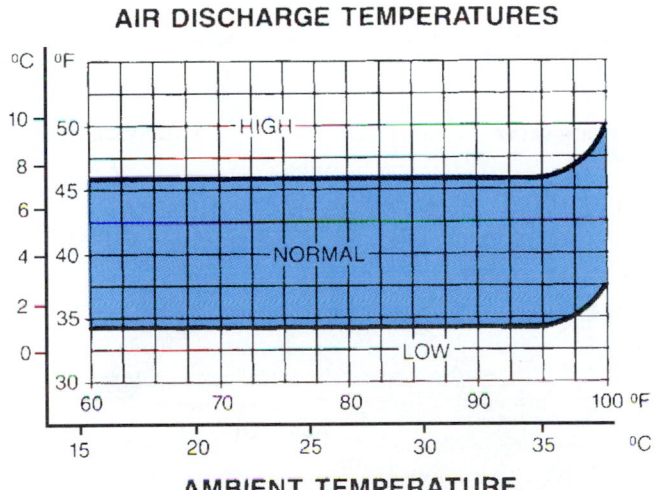

Figure 74.4 A chart showing expected air discharge temperatures at different outside temperatures. *(Courtesy of Ford Motor Company)*

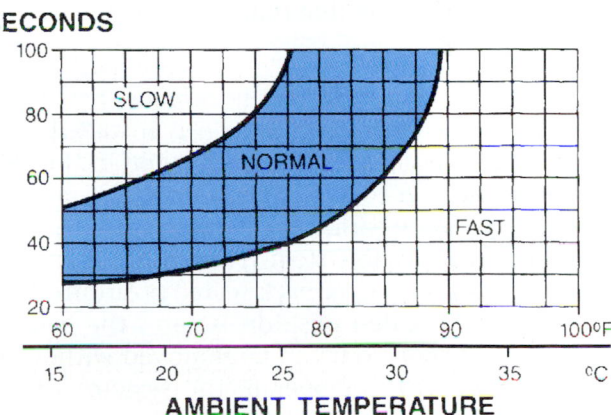

Figure 74.5 A chart of typical clutch cycling times. *(Courtesy of Ford Motor Company)*

This is the same color code that plumbing follows for hot and cold water. The gauges are connected to one manifold with another hose fitting for performing service procedures. Evacuation and recharging of the system can be done through this port.

Service Valves

Service valves are found in the high and low sides of the system to provide for refrigerant installation.

- The **discharge service valve** is on the output side of the compressor on the *high side*.
- The **suction service valve** is on the inlet to the compressor on the *low side*.

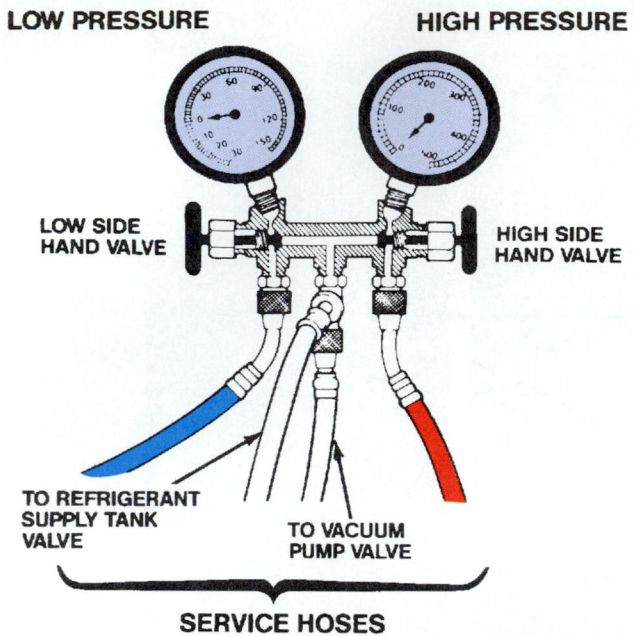

Figure 74.6 A pressure gauge set is used to measure pressures in the system. *(Courtesy of Ford Motor Company)*

> **CAUTION** Be careful that the connections are made to the correct sides of the system. Older R-12 systems (before 1976) can have fittings that are the same size. On later systems, the high side fitting is always in a smaller diameter line than the low side fitting.

Most R-12 valves are of the *Schrader-type* (Figure 74.7). When the gauge connection is installed over a Schrader valve, it depresses the center pin in the valve. This opens the port to allow refrigerant pressure to act on the gauge. Some earlier R-12 systems used a service valve with an adjustable stem. This valve can be operated at the front-, mid-, or back-seated positions (Figure 74.8). The front-seated position isolates the compressor from the system so it can be removed without losing refrigerant. Do not operate the system with the valve in this position. See the service manual.

R-134A service ports use a different type of connector so they cannot accidentally be mixed with R-12 service equipment. A quick coupler connects the hoses to the system. When the hoses are disconnected, they seal off to keep refrigerant trapped in the hoses. The high side fitting is larger than the low side.

Sealed caps are installed over the ports to prevent refrigerant leakage when they are not attached to the gauges. They have an O-ring and are to be fingertightened only. Adapters are often required to attach the gauges to the fittings (Figure 74.9).

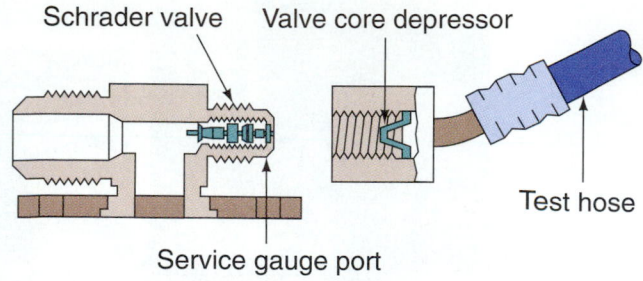

Figure 74.7 A Schrader valve service port.

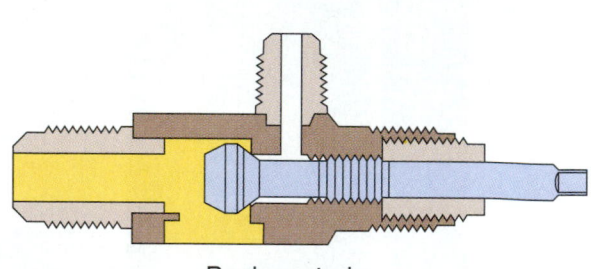

Figure 74.8 Service valve positions.

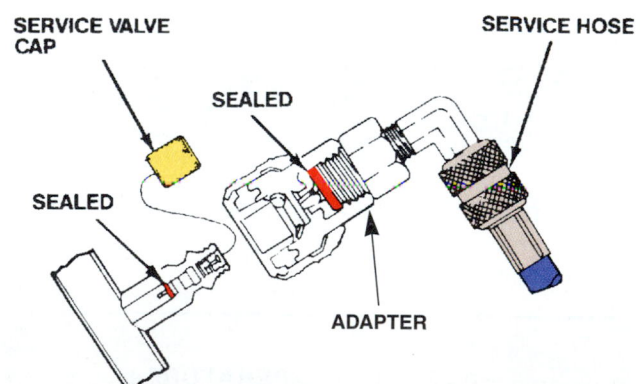

Figure 74.9 Adapters are used to attach the gauges to the fittings. *(Courtesy of Ford Motor Company)*

Connecting the Gauges

Before connecting the gauges to the service ports, close the manifold hand valves and any valves on the service hoses at a refrigerant tank or vacuum pump. Be careful that the hoses do not hang in the fan or on an exhaust manifold. Then, connect the gauges (Figure 74.10).

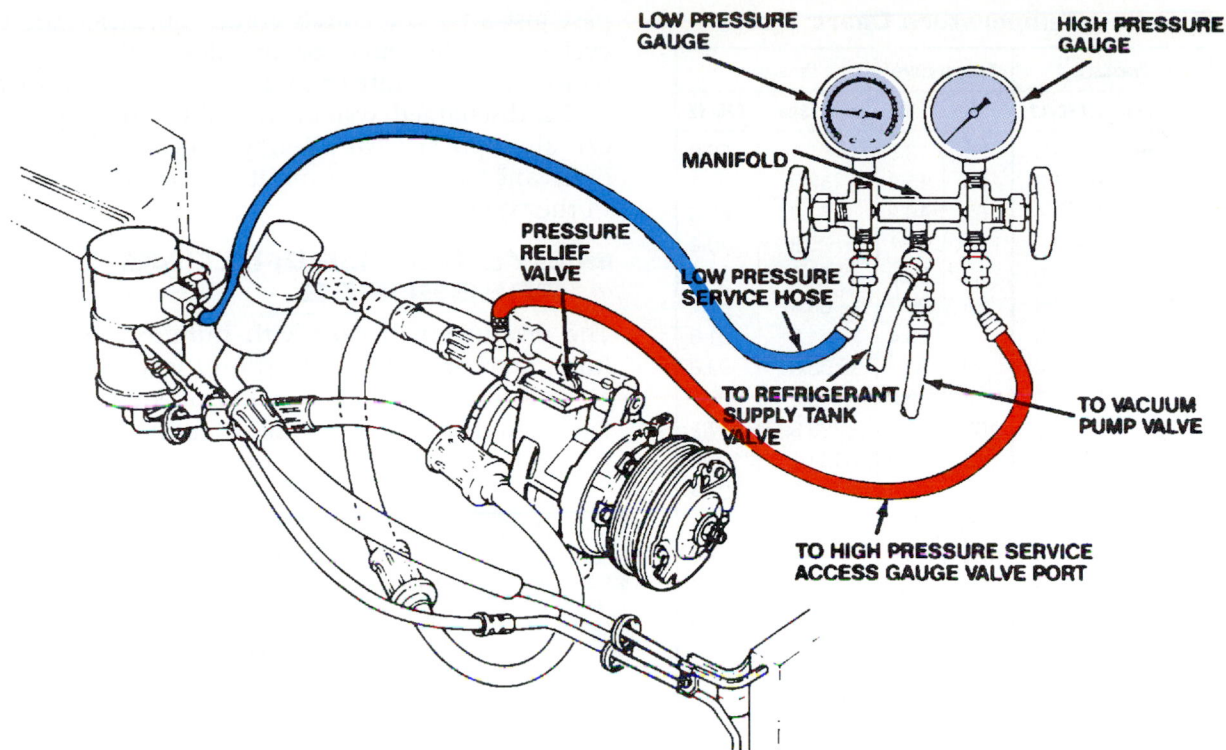

Figure 74.10 Connect the gauges to the service ports. *(Courtesy of Ford Motor Company)*

Start the threads onto the service port threads a couple of turns. Hold the knurled knob against the fitting while you tighten it to avoid refrigerant escaping before the valve is tight. Only fingertighten; do not use pliers.

The recommendation before the hole in the ozone layer was discovered was that hoses be purged of air by leaving one of the hoses off while allowing some refrigerant to leak out. This process is no longer allowed. Newer hoses have a check valve that keeps refrigerant in the line (see Figure 74.9). Older hoses are not approved for use in licensed shops.

The low pressure gauge is a **compound gauge**. It reads in either low pressure or vacuum. This is because it is in the suction side of the system. There is a hand-operated valve that opens and closes the suction line to the service hose used to charge the system. The gauge is always open to the suction line and registers that side's pressure.

The high side gauge registers *pressure* only. The gauge usually reads to 500 psi. Most high side gauges have a small orifice in the inlet. This keeps the needle from fluctuating rapidly as the system pulses. Many gauges have a zero calibration adjustment screw on the gauge face also. Sometimes gauges have a scale to convert pressure readings to relative temperature.

The hand control valve for the high side opens and closes the line to the high pressure service hose, used for discharging the system. The gauge always measures pressure in the high pressure side.

With manual service valves, use a special wrench to mid-seat the valve by turning the stem one or two turns. It must be back-seated before removing the hoses.

Modern service hoses have a check valve in the end of the line to prevent refrigerant from escaping when the line is disconnected. A small amount can still get out. When removing a hose from a connection, put a shop rag around it to prevent oil and refrigerant from spraying. Disconnecting the low side while the system is running results in the lowest pressure at that connection.

■ STATIC PRESSURE READING

The pressure reading in the system when it is not operating is called the **static pressure**. This will be at least 50 psi if there is enough refrigerant in the system. High and low side pressures will be equal when there is no movement through the system. Remember Pascal's law? *Pressure is equal when in a confined area.* The pressure should be the same as the temperature/pressure relationship for the refrigerant (Figure 74.11). At 70° in an R-12 system pressure should be about 70 psi.

High pressure = high temperature
Low pressure = low temperature

Pressure Temperature Chart

Temperature		Pressure		Temperature		Pressure	
°F	°C	HFC-134a	CFC-12	°F	°C	HFC-134a	CFC-12
-60	-51.1	21.8	19.0	55	12.8	51.1	52.0
-55	-48.3	20.4	17.3	60	15.6	57.3	57.7
-50	-48.6	18.7	15.4	65	18.3	63.9	63.8
-45	-42.8	16.9	13.3	70	21.1	70.9	70.2
-40	-40.0	14.8	11.0	75	23.9	78.4	77.0
-35	-37.2	12.5	8.4	80	26.7	86.4	84.2
-30	-34.4	9.8	5.5	85	29.4	94.9	91.8
-25	-31.7	6.9	2.3	90	32.2	103.9	99.8
-20	-28.9	3.7	0.6	95	35.0	113.5	108.3
-15	-26.1	0.0	2.4	100	37.8	123.6	117.2
-10	-23.3	1.9	4.5	105	40.6	134.3	126.6
-5	-20.6	4.1	6.7	110	43.3	145.6	136.4
0	-17.8	6.5	9.2	115	46.1	157.6	146.8
5	-15.0	9.1	11.8	120*	48.9	170.3	157.7
10	-12.2	12.0	14.6	125	51.7	183.6	169.1
15	-9.4	15.0	17.7	130	54.4	197.6	181.0
20	-6.7	18.4	21.0	135	57.2	212.4	193.5
25	-3.9	22.1	24.8	140	60.0	227.9	206.8
30	-1.1	26.1	28.5	145	62.8	244.3	220.3
35	1.7	30.4	32.6	150	65.6	261.4	234.6
40	4.4	35.0	37.0	155	68.3	279.5	249.5
45	7.2	40.0	41.7	160	71.1	298.4	265.1
50	10.0	45.3	46.7	165	73.9	318.3	261.4

Red figures — in.Hg Vacuum   * Do not heat can above 120°F
Black figures — PSIG

Figure 74.11 A temperature/pressure relationship chart for refrigerant. (Courtesy of International Mobile Air Conditioning Association [IMACA])

■ OPERATING PERFORMANCE TEST

When you start the engine and turn on the AC system, the low side pressure will drop and the high side will increase. Pressure on the low side will change according to the temperature in the evaporator. The high side will reflect the temperature of the liquid refrigerant leaving the condenser. The pressures will continue to change until the system stabilizes (reaches its normal operating temperature). The system pressures are stabilized when maximum high side pressure when the system cycles off does not change from cycle to cycle.

Check the Sight Glass

Inspect the sight glass, if the system has one. This is done in the shop, out of the sun. The sight glass will be located on top of the receiver dryer or in a refrigerant line. When the sight glass is on the receiver dryer, you are looking to see if the bottom of the pickup tube is uncovered. This would tell you there is not enough reserve refrigerant in the system.

Normal appearance in the sight glass is for the refrigerant to be clear, after a burst of bubbles flowing past just after the clutch comes on. After the clutch cycles off, the burst of bubbles will appear again within half a minute or so. Excessive bubbles can indicate a discharged system. A badly discharged system can also appear clear. Cloudy appearance means that desiccant has escaped from its bag or there is moisture in the system.

■ SYSTEM OPERATING PRESSURES

The pressures will vary with humidity because high humidity puts a higher load on the system. Adequate air flow over the condenser is also required. A portable fan might be necessary to blow additional air through the grill on a car with a belt driven fan.

SHOP TIP A rule of thumb for normal high side pressure is with the temperatures of the condenser and compressor at 130°F, the high side pressure should equal the outside (ambient) air temperature + 100 (+ or –20°).

Expected low side pressures differ between expansion valve and orifice tube systems (see chart in Appendix L). Troubleshooting charts in manufacturers' service manuals will help pinpoint the causes of pressure problems.

The following are rules of thumb for the pressure test:
- The higher the high side pressure, the more heat there is.
- The lower the low side pressure is, the lower the temperature will be at the outlet.
- High side pressure affects low side. When you squirt water on the condenser, the high and low pressures *both* drop at once.

■ LEAK DETECTION

Leaks can be found using several means. These include soapy water, internal dyes, and several types of special leak detectors. *Soapy water* is made by mixing water and soap to a thick solution that can be applied with a small brush. If the leak is large enough, soap bubbles will appear (Figure 74.12).

An internal charge detection method includes injecting a *coloring agent* into the low side of the system. Some red dyes are available in pound-size refrigerant containers. The dye is harmless and will be left in the system after the test. If dye leaks out, it will be visible at the source of the leak. A problem with the red dyes is that they can leak onto interior carpet and cause a damaging stain. If there is a leak, the dyed refrigerant will have to be recovered for reuse.

Another dye method uses fluorescent dye. A black light is used to locate the leak (Figure 74.13). This method is very effective on difficult leaks that only

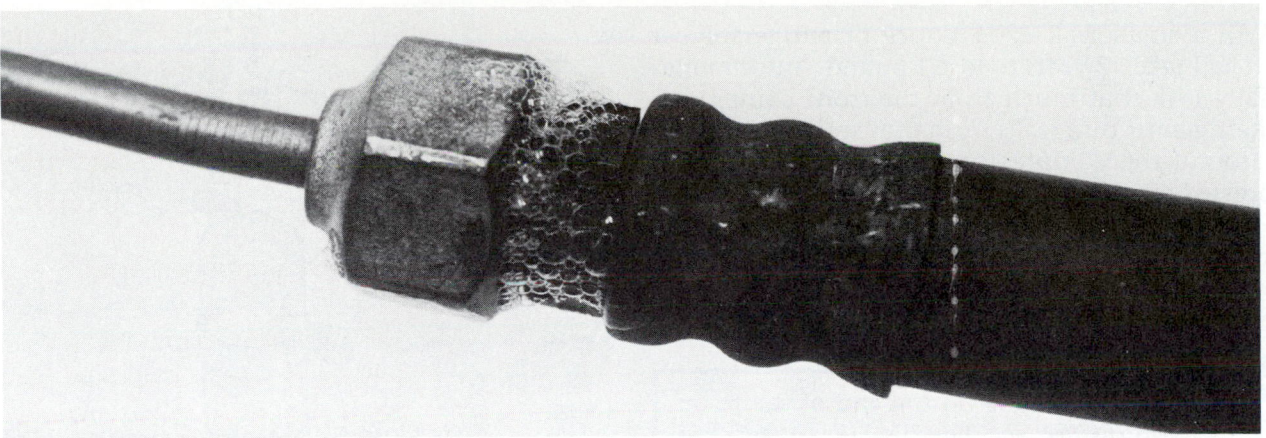

Figure 74.12 Soap bubbles showing a leak in a hose. *[Courtesy of Best Educational Trainers, Inc.]*

Figure 74.13 A fluorescent dye is added to the system before checking for leaks with a black light. *[Courtesy of Tracer Products, Westbury, NY]*

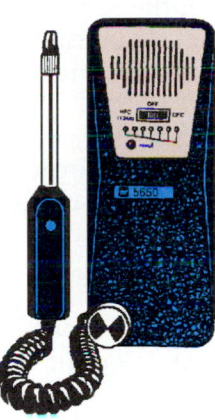

Figure 74.14 An electronic leak detector that works on both R-12 and R-134A refrigerants.

happen during the vibration that results from driving the vehicle. The dye will pinpoint where it leaked. The system is charged to at least 10% capacity for either test. This is so the system can develop normal operating pressure. There should be at least 50 psi static pressure (when the engine is off).

There are other types of leak detectors. A *flame-type detector* used to be used with R-12 systems. It would cause a change in the color of the flame. It will not cause R-134A to change colors, however. A major problem with this method is that it exposes refrigerant to flame. When R-12 is burned it creates mustard gas. If the leak were large enough it could be dangerous. The tester had to be used in an area with lots of ventilation.

Electronic leak detectors are the most popular ones used today. Many of them work on both R-12 and R-134A refrigerants (Figure 74.14). The tester nozzle is moved around the various joints in the system. The probe is held at a distance within ¼" of the joint being checked. When a leak is found, the tester lights up and its slowly pulsing, high-pitched horn or clicking sound becomes more rapid.

The tester has a sensitivity adjustment. You can adjust it so that it just stops ticking. When it starts to tick you have found a leak. The "gross leak" position can find leaks of 4 ounces a year. The sensitivity in this position is less so you can locate the exact source of the leak.

An electronic halogen detector can detect other leaks too. Because it senses halogen, you can "check for leaks in your headlights." It is so sensitive that windshield washer solvent or heavy perfume will set it off.

■ LOCATION OF LEAKS

Leaks are most often found at connections in the system. Check around all the possible leak points. Be sure to check all around the low points in the system. On older cars, connections are sometimes held with hose clamps. Sometimes, simply tightening a fitting a small amount will eliminate the leak. Be careful not to overtighten a fitting. If an O-ring is damaged, tightening the fitting will not correct the problem.

With a small leak of ½ ounce of refrigerant per year, it will take 32 years to lose 1 pound. Automobiles normally leak that much from the front compressor seal. Refrigerant that leaks from that seal can be pocketed around the compressor. Blow off the area with compressed air and test again before condemning the seal. If it really is a leak, it will still be there. The tester is very sensitive. Check the evaporator core at its lowest point where the water from condensation comes out.

> **SHOP TIP** Use a clear tube on the end of the tester so you do not accidentally suck water into the tester.

Test the bottom of the accumulator during an A/C leak test in case its metal can leaks. The accumulator is spin welded during manufacture so this is a possibility.

O-rings and Seals

Threaded fittings and block fittings are often sealed with O-rings. O-rings are used because the system must be able to tolerate vibrations. A rigid pipe joint would fail over time. O-rings used since the late 1980s are oval shaped (Figure 74.15). The newer O-rings do not fill the O-ring groove completely. They are said to last longer. The new style O-ring can also be used on older cars.

The oil used with R-134A will damage conventional rubber seals. Be sure to use compatible O-rings. Newer O-rings are usually color coded. R-134A O-rings are also thicker.

Block fittings (Figure 74.16) are used at connections to parts because they allow sealing surfaces to be more accurately positioned.

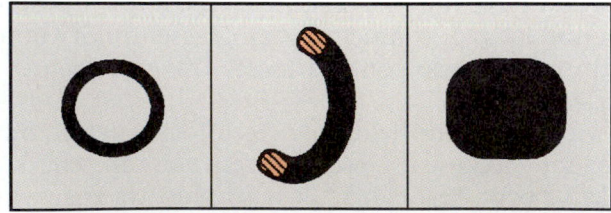

OVAL

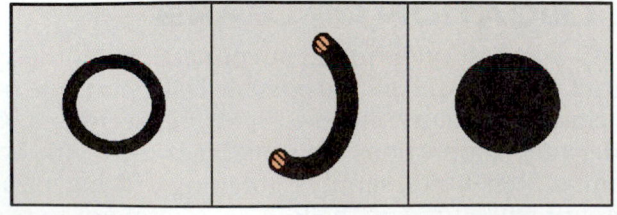

ROUND

Figure 74.15 Different types of O-rings. *(Courtesy of General Motors Corporation, Service Technology Group)*

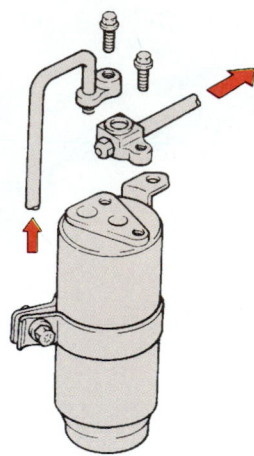

Figure 74.16 Block fittings are used at connections to parts because they allow sealing surfaces to be more accurately positioned.

Always coat O-rings with refrigerant oil before installing them. When they are moistened, they seal better. Newer style sealing washers do not require lubrication and, in fact, will leak if lubricated.

■ COMMON AIR CONDITIONING COMPONENT PROBLEMS

Whenever a component is removed and replaced in the system, be sure to capture all of the refrigerant in the system. Remember that the system does not like being exposed to moisture, even for a short time. Be ready to cap off all components and seal the system while you work on a part. Do not leave the system open or it will take on excess moisture causing the receiver dryer to require replacement!

Accumulator or Receiver Dryer Service

Moisture is an enemy to air conditioning systems because it can react with refrigerant and corrode the inside of the system.

> **■ SCIENCE NOTE ■**
>
> *When moisture gets into an R-12 system, three types of acids are formed: carbonic, hydrofluoric, and hydrochloric. These acids destroy aluminum. When they mix with the refrigerant oil, sludge forms. The majority of the sludge accumulates in the evaporator, causing corrosion and leaks. When moisture combines with R-134A, acids are not formed. The only result is corrosion.*

To remove moisture from the system, air conditioners use either an accumulator or a receiver dryer (see Figures 73.26 and 73.27).
- A receiver dryer is installed in the high side.
- An accumulator is installed in the low side.

NOTE: *Both systems contain a desiccant that requires periodic replacement of the receiver or accumulator.*

Desiccants are in the form of pellets. They are held in place by a screen or a bag. A desiccant removes moisture from refrigerant. It is a chemical that forms a molecular bond to water (H_2O). Refrigerant can pass through the desiccant but moisture bonds permanently to the pellets. The desiccant changes into a different molecule so there is no way that the water can be broken from the pellets.

NOTE: *The desiccant will become saturated when it becomes full of water. On a day with 80% humidity, a desiccant can become completely saturated if left fully open to the atmosphere for as little as 10 minutes.*

The receiver dryer or accumulator should be replaced whenever the system has been leaking, left empty, or left open for a long time, or when a compressor or its reed valve has failed.

The material used for desiccants since the early eighties is a **molecular sieve** that will not crumble when it is wet. The old silica/aluminum desiccant would absorb 11 grams of water. Today's sieve material will absorb 45 grams of water. The desiccant is said to be good for ten years of normal use (if the system has never been without pressure).

Many manufacturers will not warranty an accumulator because they do not believe it is cost effective. The accumulator is an insurance policy for the technician especially if the system has been left without pressure for a while.

■ REFRIGERANT OIL

Air conditioning systems use a special non-foaming refrigeration oil to lubricate the compressor, gaskets and seals, and expansion valve. When the air conditioning system is operating, there is always a small amount of oil located in each of the system's components. The amount of oil in a system can vary from 6 to 9 ounces. The only compressor that uses more oil is the GM A-6 compressor. It uses 12 ounces and has its own oil sump.

All of the system's components contain oil. When a part is replaced, a certain amount of oil is added to the part before replacing it. Oil requirements for an orifice tube (expansion tube) system are shown in Figure 74.17.

Oil collects in the bottom of the compressor during use. When the old compressor is replaced, drain out the oil and measure it so a like amount of oil can be added to the new compressor. The viscosity of refrigerant oil ranges from 300 to as high as 1000.

NOTE: *Like brake fluid, refrigerant oil should be stored with the cap on. It can absorb moisture, which is damaging to the A/C system.*

Replacement compressors are shipped with a full charge (5 to 9 ounces) of oil in the crankcase. This oil must be drained or an overcharge will result. A common amount of oil to be drained from the old compressor is about 4 ounces. Never add less than 2 ounces when you replace a compressor. When there has been an abrupt loss of oil, figure 3 ounces of oil was lost (plus the oil required for the failed part). Too much oil in the system takes up space that would normally be used for the refrigerant. This reduces cooling system capacity. If an oil overcharge is suspected, try to drain the accumulator. A two-bag accumulator should hold 3 ounces of oil. Some aftermarket accumulators may have only one bag. The desiccant will hold the oil, so there should not be very much oil that pours out. If you suspect the system is too full, pour some oil out of the accumulator, refill, and test the system once again.

Choosing the Right Oil

The oil must be compatible with the refrigerant. R-12 uses *mineral oil* and R-134A uses **polyalkylene glycol (PAG)** or **polyalester (ester)** oil. The correct oil is very important in R134A systems. Oils used in different systems are not the same. Check the compressor

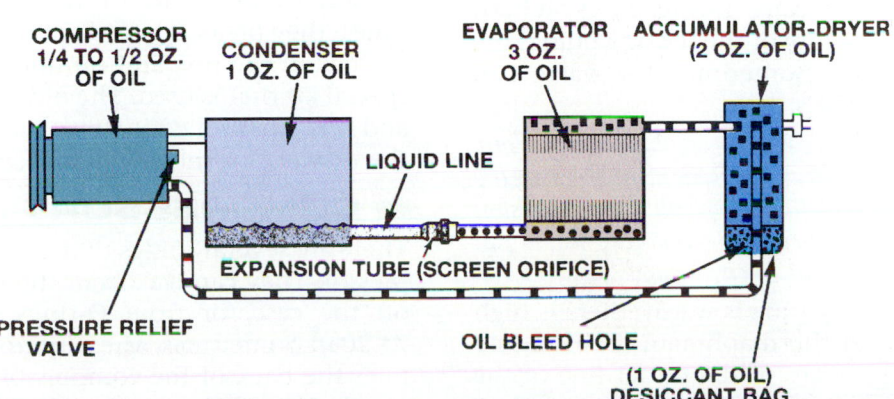

Figure 74.17 Oil requirements for different parts of an orifice tube system. *(Courtesy of General Motors Corporation, Service Technology Group)*

label or the service manual to see what the manufacturer recommends. The mineral oil used with R-12 systems is not soluble in R-134A or PAG oil so it will separate and form sludge in recesses in the system. PAG oil does not tolerate chlorine, which is part of CFC gas. It will turn to a gelatin-like substance if it is accidentally put into an R-12 system.

CAUTION
- PAG oil is nontoxic. R-134A is also nontoxic. But when R-134A and PAG oil are mixed together, both are toxic.
- PAG oil can damage some paints.
- Use regular mineral oil as a lubricant. Do *not* use PAG oil as a lubricant because it absorbs water more easily than mineral oil. If PAG oil leaks, it will corrode fittings. One-half of every connection is exposed to air. This can cause corrosion on the exposed area of an R-134A system.

■ EVAPORATOR PROBLEMS

Some manufacturers put a layer of foam on the outside of the passenger side of an evaporator. This is to prevent water from being blown on the passengers. Mildew is a problem with air conditioning systems.

SHOP TIP Some technicians spray Lysol on the foam if it mildews. There are also commercial evaporator cleaning kits available.

■ EXPANSION VALVE OR ORIFICE TUBE PROBLEMS

An expansion valve or orifice tube can become plugged with debris. An expansion valve can have other problems too.

Expansion Valve

A bad expansion valve will cause low readings on both the low and high side. Failure of the sensing bulb is the most common problem. Sometimes the valve can become plugged with debris.

NOTE: *If the expansion valve requires replacement for any reason, replace the receiver dryer. It is right in front of the expansion valve and should have prevented the corrosion that ruined the expansion valve. The evaporator would fail next if the expansion valve was open.*

If the sensing capillary tube is warm, there is high pressure that pushes on the diaphragm. The sensing bulb checks the temperature of the outlet line of the evaporator. If it is warm, it opens the valve. Pressure on top of the diaphragm tries to push the valve open. The operation of the expansion valve can be checked.

SHOP TIP It used to be the custom to test an expansion valve by releasing R-12 refrigerant onto the sensing bulb to freeze it. This is no longer allowed or cost effective. A CO_2 fire extinguisher will work to test an expansion valve.

Some expansion valves are wrapped with insulation. When the sensing bulb is in the evaporator case, it does not need to be wrapped because it stays cold. When an expansion valve is packed in that type of system, it is only to silence the noise that the evaporator makes.

Some vans have a rear evaporator that has an expansion valve. There is always some flow in the rear unit to keep the expansion valve bulb cold. If you can pinch the hose to the rear unit while the rear unit is off and this changes pressure drastically, then the expansion valve is stuck open.

Orifice Tube Service

With an orifice tube system, also called an expansion tube system, the accumulator (see Figure 73.13) should be cold and sweaty. The same thing occurs in it as the evaporator. Like a bad expansion valve, a plugged orifice tube will cause low system pressures. There is a filter screen in the inlet to the orifice tube that can become plugged. A saturated desiccant can rupture, sending its material throughout the system. This can result in a plugged orifice tube screen.

The orifice tube is easily checked and replaced once the refrigerant has been recovered. An orifice tube can be installed in either direction, but there is one right way. It has an arrow that points in the direction of refrigerant flow. If you put it in backwards, the filter screen will be on the wrong end. It might plug up and be difficult to remove. When it has removal tabs on one end, simply grasp it by that end when you reinstall it. A special tool makes installation the correct way easier (Figure 74.18).

Occasionally, an orifice tube gets stuck in its pipe. A special removal tool can help to remove it. Sometimes, they break up inside of the pipe. When the tube is stuck in the line and cannot be removed, there is a special kit that is used. The old section of line is cut off and it is replaced with a new section of line that holds the new orifice tube (Figure 74.19).

■ COMPRESSOR PROBLEMS

There are several things that can go wrong with a compressor. They can leak from the front shaft, from seals on the case, or from O-rings on the back (Figure 74.20a). Sometimes, the aluminum tubing that supplies the back of the compressor develops a leak (Figure 74.20b). There are hundreds of configurations of these. Aftermarket companies produce them and have catalogs listing them by number. Some mufflers (see

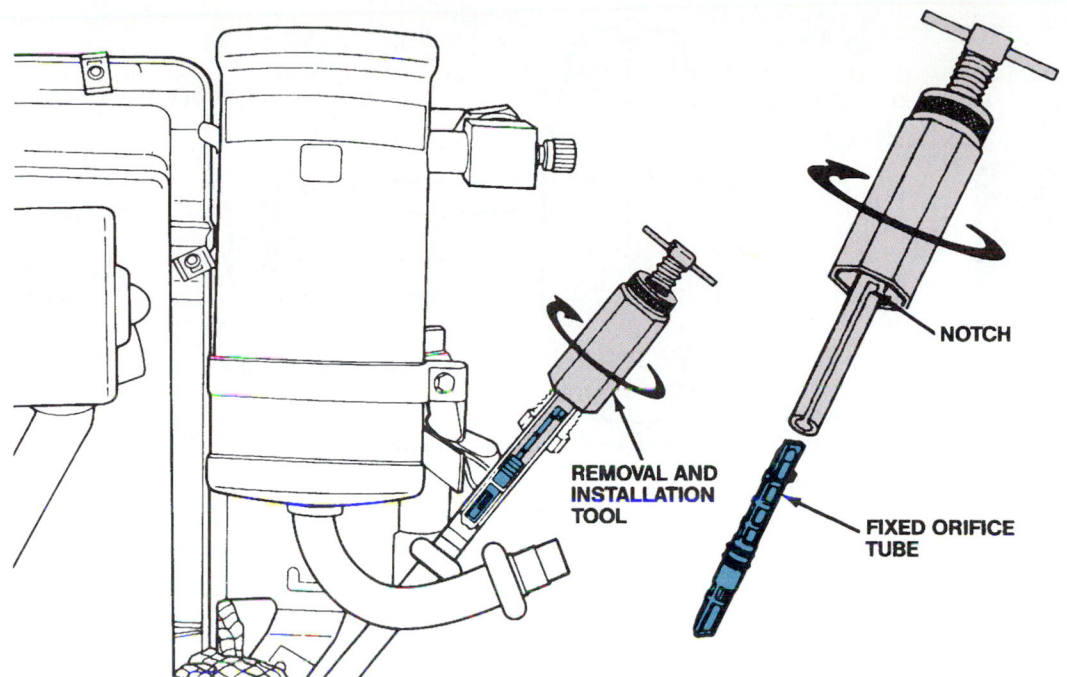

Figure 74.18 A special tool for removing and replacing orifice tubes. *(Courtesy of Ford Motor Company)*

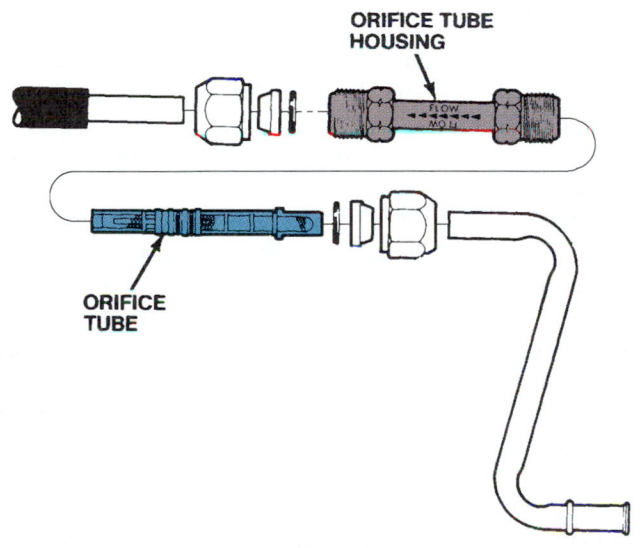

Figure 74.19 An orifice tube kit for replacing a damaged orifice tube. *(Courtesy of Ford Motor Company)*

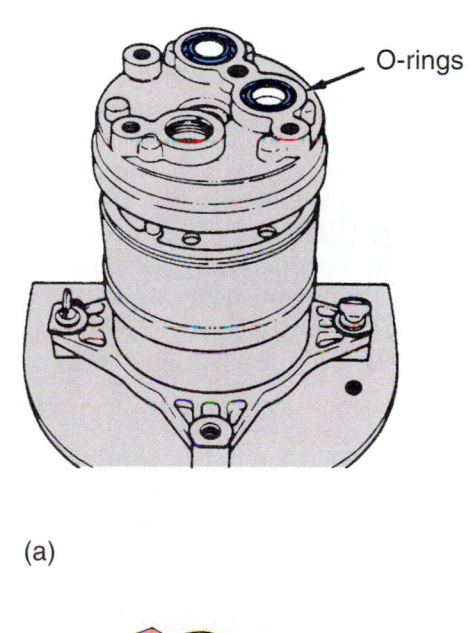

(a)

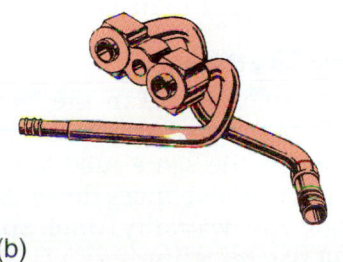

(b)

Figure 74.20 (a) O-rings on the back of a compressor. (b) An aluminum tubing manifold. *(a, Courtesy of General Motors Corporation, Service Technology Group; b, Courtesy of Everco Temp Control)*

Figure 73.25) have a filter that must be replaced in the event of a compressor failure.

When you replace one compressor with another, it is important that the correct one be used. Be sure you have the correct part number. On the GM radial six-cylinder compressor there are twelve different part numbers. The difference is in where the compressor is mounted. It is important that the correct mounting be used so the oil passage is in the right place.

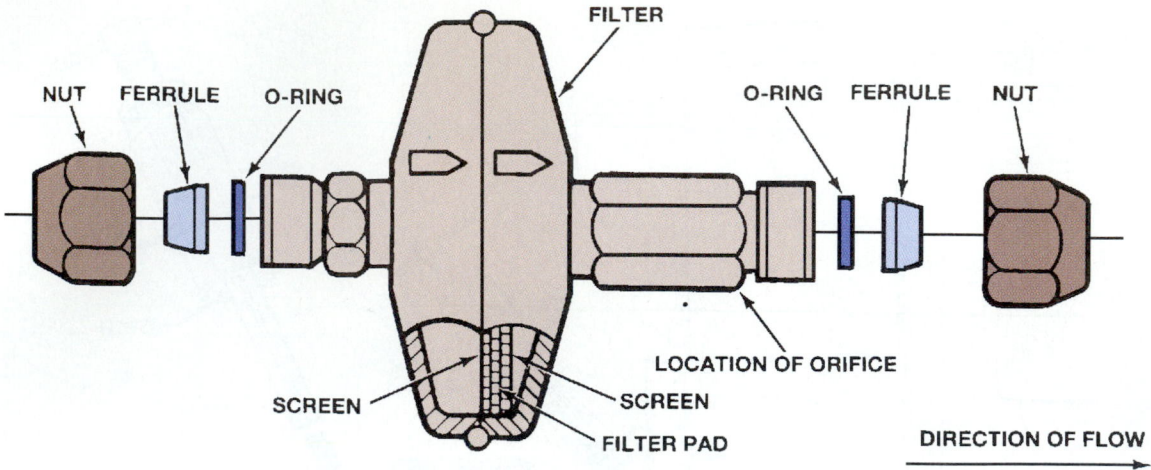

Figure 74.21 An air conditioning filter. *(Courtesy of General Motors Corporation, Service Technology Group)*

Compressor Oil

Because oil flows throughout the system, rather than collecting in the compressor, the level of the oil is only checked when there is evidence of a major oil leak. This will be visible at the source of the leak. The oil is also checked if the compressor is repaired or replaced.

NOTE: *To check the oil level in the compressor requires the evacuation of the system. Procedures for checking the oil vary with the manufacturer.*

On orifice tube systems, oil is pulled up from the bottom of the receiver dryer. If the refrigerant level gets too low or if the oil hole gets plugged, the compressor can fail. Rebuilders recommend replacing the accumulator with every compressor replacement. This is because they do not want to take a chance that the oil hole is plugged and the newly rebuilt compressor will fail due to lack of lubrication.

Filters

A filter can be added to an air conditioning system (Figure 74.21). When a compressor has failed, be sure to put a filter after the outlet of the compressor. A filter is available either with or without an orifice tube. Put the new filter in the line out of the condenser where it is most convenient to install it.

Flushing the System

A *manifolded condenser* came out in the late 1980's. It cannot be flushed like the *dual parallel condenser*. It has its inlet and outlet on the same tube but has seams that separate about every five tubes down. Some manufacturers will not pay warranty time on a system flush but they will pay for a filter.

Flushing must be done with a liquid so grit is carried out. Do not use a vapor (shop air). Using 15 pounds of R-11 refrigerant was the old way that systems were flushed, before the hole in the ozone layer was discovered. Today, mineral spirits are often used for flushing. Mineral spirits are chemically related to mineral oil so they do not present a problem with the refrigerant on R-12 systems. They should not be used on R-134A systems.

Compressor Shaft Seal

Compressor shaft seals occasionally require replacement. On rear wheel drive cars, this can often be done using special tools while the compressor remains on the engine. There are special tools available for removing and installing the clutch seal (Figure 74.22).

Clutch Problems

When clutch engagement problems occur, check for voltage from the source. Then use a jumper wire to apply power and ground to the terminals on the clutch. This will bypass the protection devices and determine whether the problem is the clutch or the circuitry. Parts of the clutch and compressor are shown in Figure 74.23.

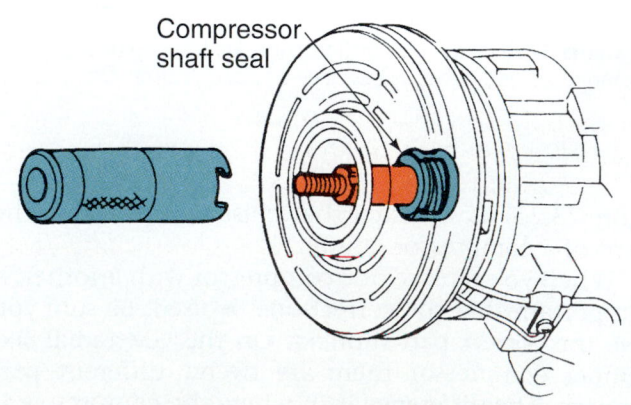

Figure 74.22 Installing a compressor clutch seal. *(Courtesy of American Honda Motor Co., Inc.)*

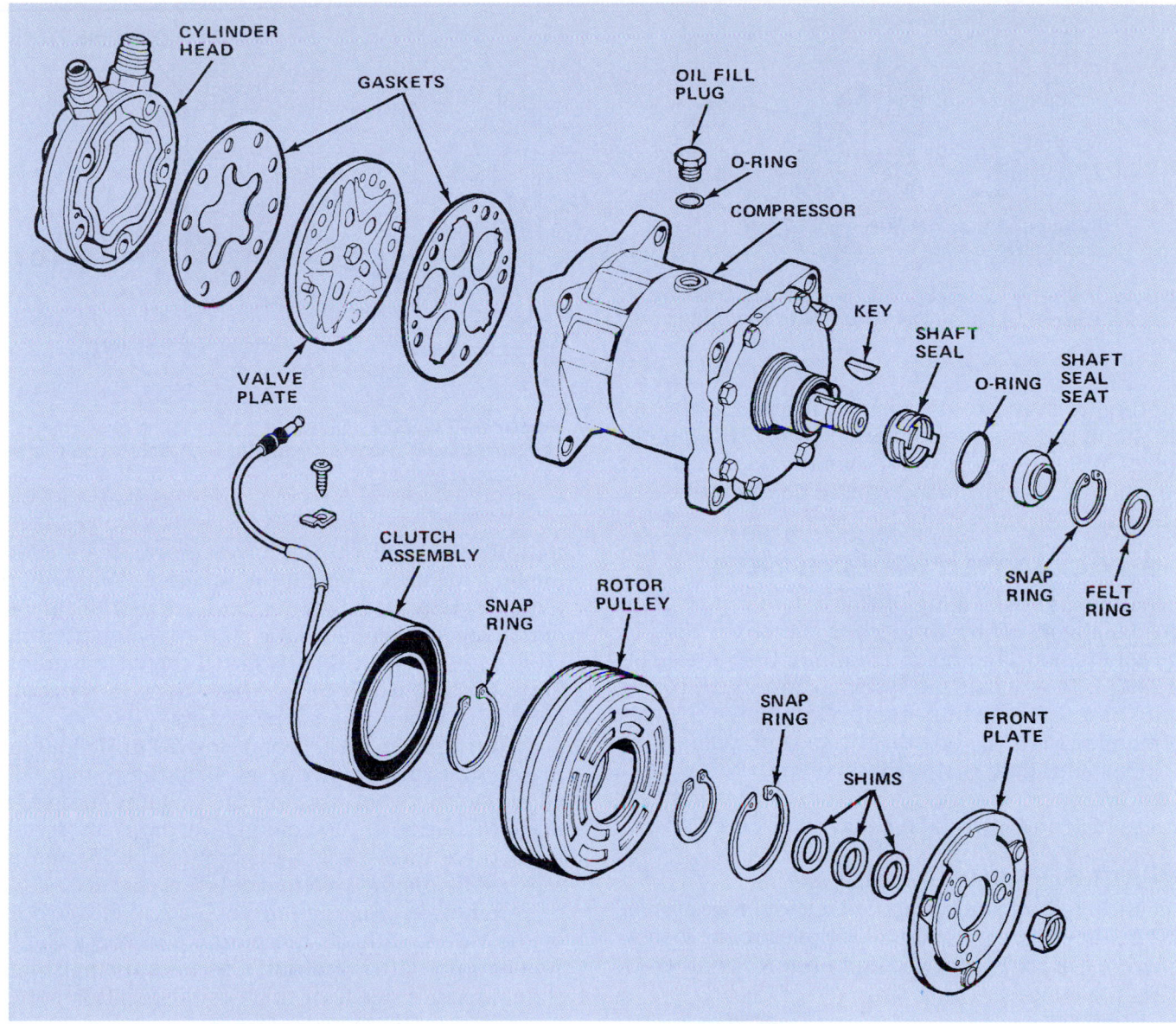

Figure 74.23 Parts of the clutch and compressor. *(Courtesy of Chrysler Corporation)*

■ EVACUATING AND RECHARGING

Two additional hoses can be attached to the center connection on the manifold set. These are for a refrigerant supply tank hookup and for a remote vacuum pump. Yellow or white are the preferred hose colors for R-12 and solid yellow with a black stripe is the R-134A color. Service ports use color-coded quick couplers that automatically close the hose end when disconnected.

■ VACUUMING A SYSTEM

When a system is empty or has been repaired and drained, all of the air and moisture must be removed. They are removed with a vacuum pump. The vacuum pump removes the air right away. To remove the moisture, it must be boiled out. The system does not have

to be heated to make moisture boil. It simply has the pressure removed from it to at least 28 in.Hg. Pressure this low will boil water in a glass at room temperature (Figure 74.24).

Vacuuming a system is done not only to dry out moisture but to help the efficiency of the system. Air builds up in the top of the condenser. Cooling will not take place until the refrigerant gets below the air level so you lose condenser efficiency. The result is that when the refrigerant comes into the expansion valve or tube, it carries more heat. Vacuuming is said to drop the outlet temperature 6°F to 8°F.

Vacuum Pump Rating

The micron rating of the vacuum pump tells how deep a vacuum it will pull. Too small a pump will take a

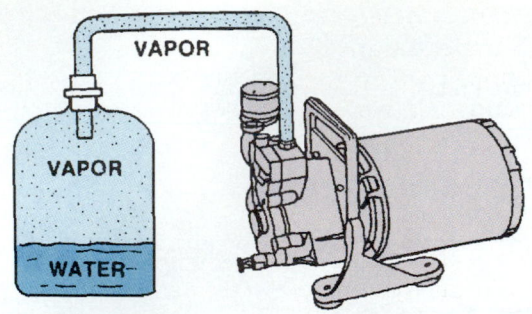

Figure 74.24 A vacuum pump reduces pressure within the system to cause water to boil at room temperature.

long time to evacuate the system. A really good vacuum pump will pull to about 50 microns. This is more than 29 in.Hg and drops water's boiling point to below zero. Oil in a vacuum pump should be changed every 10 hours.

■ EVACUATE THE SYSTEM

Evacuation and recharging of the system can be done with the engine off by connecting the center hose to both service ports and opening both of the valves (Figure 74.25). To evacuate the system, connect the center service hose to the vacuum pump. Open both valves on the manifold and the one on the vacuum pump (if it has one) and start the vacuum pump. When both valves are open, the gauges do not read accurately because high and low side pressures affect each other.

Pulling a Vacuum

After the vacuum pump starts, you should see system pressure drop. Vacuum is usually applied to the system for ½ hour or so. Keep a vacuum on it for another 15

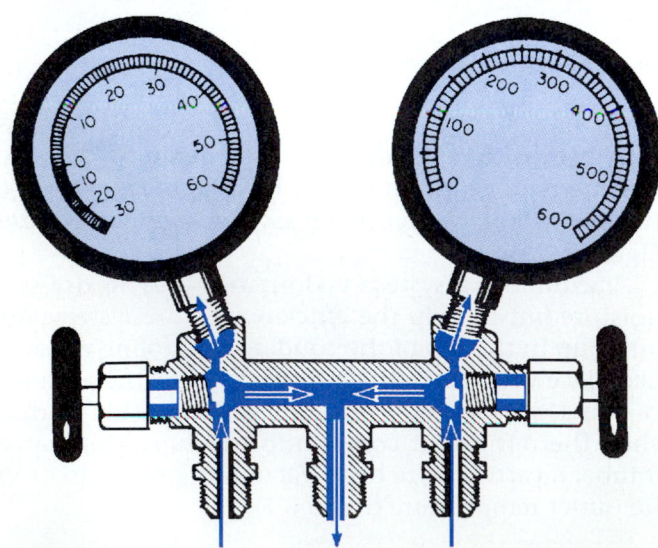

Figure 74.25 For evacuating and recharging the system, connect the center hose to both service ports and open both of the valves.

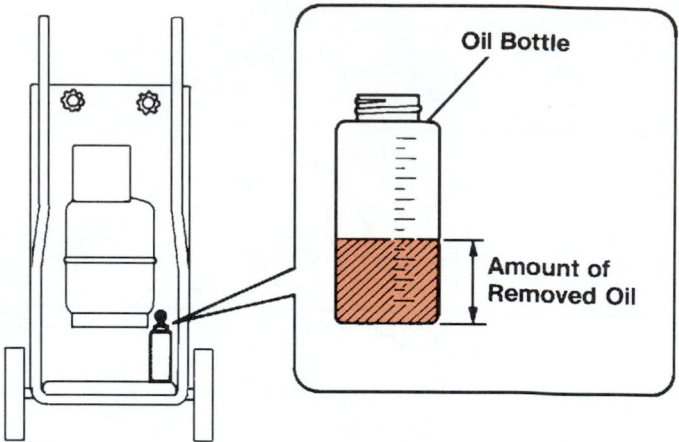

Figure 74.26 Measure the amount of oil that is removed during recovery and add that amount to the system during recharge.

minutes after it reaches its lowest point. As water boils in the system, it takes up more space. Watching the low side pressure will give an indication of the amount of moisture to be boiled away. If the system drops right away to 28 in.Hg, it is fairly dry. If the system drops to 26 in.Hg and holds for a while, there is probably a good deal of moisture in the system.

If vacuum does not drop below 20 in.Hg in 5 minutes or so, there is probably a leak in the system. Close all of the valves and shut off the vacuum pump to see if there is a leak. If the vacuum level drops off (pressure rises) there must be a leak. It must be located and repaired before the system can be evacuated.

After the evacuation process is completed, close all of the valves, shut off the pump, and check the low side pressure. After 5 minutes, recheck the pressure. It should remain steady or there could be a leak.

NOTE: *Vacuum leak checking is not a dependable way of locating tough leaks. An O-ring could leak under pressure, but not under vacuum.*

During recovery, oil is removed with the refrigerant. It collects in a bottle in the recovery station. Remember to empty the oil reservoir first. After recovery, measure the amount that is removed and add new oil in that amount when you recharge the system (Figure 74.26). During recovery of R-134A you will usually lose 1 ounce of oil at least. This is much more than is lost with R-12.

NOTE: *Some commercial desiccants will give up moisture under vacuum, but not those used in automobiles.*

■ MODERN AIR CONDITIONING SERVICE EQUIPMENT

A modern charging station (Figure 74.27) will usually have a switch to control the process. Electric solenoids open the valves.

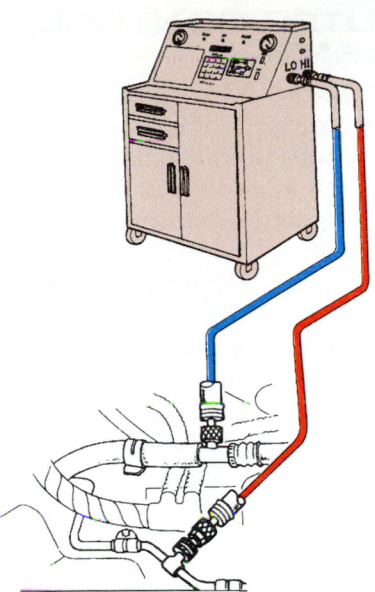

Figure 74.27 A modern charging station. *(Courtesy of American Honda Motor Co., Inc.)*

Some machines do everything from one place. Two machines will be needed for dealing with R-134A and R-12 refrigerants. They cannot be mixed or contamination results.

With a **dual pass machine**, first recover, then recycle. With a **single pass machine** (one motion) there is no need to ever charge with vapor (from the low side). The machine removes the refrigerant and recycles it during the 15-minute minimum vacuum cycle. The refrigerant bottle will be heated sufficiently by that time so there is enough heat to blow the refrigerant into the high side without starting the vehicle. The machine has a switch that allows changing between recycled and new refrigerants.

■ RECHARGING THE SYSTEM

After the evacuation process is complete, refill the system with the correct amount of refrigerant. The amount of refrigerant is important. Each system is different. Putting too much or not enough refrigerant in the system will hamper efficiency. Filling the system until the sight glass is clear is not good enough.

The amount of refrigerant required is normally found on an underhood label. Put in the recommended amount on the label to ensure that the receiver dryer is a reservoir (when the expansion valve opens, it calls on the receiver dryer for more refrigerant). If the label is missing, check the manufacturer's service manual for the vehicle. Older systems had larger receiver dryers or accumulators. These systems are more forgiving if the recharge amount is excessive.

Refilling Through the Low Side

Recharging the system from a large bulk tank installed in a charging station is pretty much the rule, now that

refrigerant has become so expensive. You will have to weigh the can during the refill. You can approximate the amount left in the can by feeling the can. Liquid refrigerant at the bottom of the can will be colder than the vapor in the top.

> **SHOP TIP** A good rule of thumb is if the system calls for 3½ pounds of refrigerant and you install the whole 4 pounds, for every 5 psi of unwanted high side pressure it will cost ⅒ mpg of fuel economy on a 4-cylinder vehicle.

During the fill, the refrigerant boils as it is leaving the container. The first part goes in quickly because the system is under vacuum. The boiling refrigerant generates pressure and the process slows down. Larger stations have heaters to heat and raise pressure in the refrigerant container. Be sure to keep the can upright when filling the system so only gas enters. Liquid refrigerant entering the low side can damage the compressor.

Completing the Refill

After the refrigerant starts to go in more slowly, the air conditioning system can be run so that refrigerant can be sucked into the low side. Be sure that the high side valve is always closed when the system is working. If the system has a low pressure cutout it will not operate. These can be jumpered to bypass them temporarily.

The system should begin to operate normally when the fill level is about ½ pound less than full. The remaining part is to fill the accumulator or receiver dryer. When the system is being refilled with a charging station, the amount of the fill is entered into the machine and it takes care of the rest.

■ RECYCLING AND RETROFIT

Since CFC refrigerants have been phased out, their cost has skyrocketed. R-134A has been the refrigerant that new vehicles are equipped with since 1994. Look under the hood for a label that states what kind of refrigerant the vehicle has.

Retrofitting an R-12 system is possible, although there are some drawbacks. These are different refrigerants with different properties. Many things must be considered. With the cost of R-12 refrigerant increasing rapidly, do not make the mistake of substituting the wrong refrigerant. R-134A is the only refrigerant any manufacturer is recommending.

Manufacturers have kits that generally go back ten model years for coverage. Follow manufacturers recommendations for retrofitting.

Some of the items a retrofit might require are:
■ A compressor front seal.
■ Some vehicles will require condenser replacement.
■ Some will need an extra cooling fan.
■ Hoses.

The larger condenser required for an R-134A system is the biggest hardware change.

There are many differences between the way R-134A and R-12 react in the air conditioning system:

- R-134A has a 30–60 psi increase on the high side and it is 6°–7° warmer.
- The low side pressure with R-134A is typically 3 pounds less than with R-12. This means the pressure sensing switch will probably have to be changed during a retrofit.
- R-134A requires less weight of refrigerant during a refill because R-134A is a different density. At 79°F, R-12 weighs 81.5 pounds per cubic foot and R-134A weighs 75 pounds per cubic foot. That is why an R-134A 30-pound container is larger than an R-12 30-pound container.
- R-134A is about 15% less efficient than R-12. That is why the condenser is larger on cars that come equipped with it.

■ COMPUTER CONTROL PROBLEMS

Computer problems with the air conditioning are limited to those that are electromechanical in nature. The computer makes 600,000 calculations/second and can change the engine's idle speed in 0.4 seconds.

The HVAC programmer controls the clutch circuit, cutting it out when:

- Refrigerant temperatures are above 240°.
- Engine vacuum and throttle opening, read by the throttle position sensor and manifold pressure sensor, shut off the clutch when a heavy load is sensed. The amount of vacuum where shutoff occurs varies from one car to the next.
- When power steering pressure is above 500 psi.

The computer's idle air control seeks a desired idle. If it cannot get it, the compressor clutch shuts off. With some systems, try putting your foot on the gas pedal and raise the rpm to see if you can bring the clutch on.

■ REVIEW QUESTIONS

1. If the heater has a control valve and it is open, the hoses should both be _____.

2. What gas is produced when R-12 is exposed to flame?

3. When there is a restriction in a line or part, what will commonly appear downstream of that spot?

4. Most air discharge specifications are between 35°F and ____°F.

5. If the high and low side fittings are not the same size, which one will be bigger?

6. What is the name of the material that absorbs water in the AC system?

7. When a part of the system is replaced, a measured quantity of _____ is added.

8. After a compressor failure, a _____ is usually installed in a line after the condenser.

9. What is used for flushing a system?

10. When refilling from the low side, should the can be upside down or right side up?

■ ASE STYLE REVIEW QUESTIONS

1. Technician A says that electric cooling fans should begin to run at low speed when the compressor clutch comes on. Technician B says that a high side line should feel cold. Who is right?

   a. Technician A   **b.** Technician B
   c. Both A and B   **d.** Neither A nor B

2. Technician A says that the drop in the air's temperature as it passes through the evaporator should be at least 20°F. Technician B says that the higher the ambient air temperature, the higher the amount of temperature drop to the air through the evaporator. Who is right?

   a. Technician A   **b.** Technician B
   c. Both A and B   **d.** Neither A nor B

3. Technician A says that high pressure means high temperature. Technician B says that low pressure means low temperature. Who is right?

   a. Technician A   **b.** Technician B
   c. Both A and B   **d.** Neither A nor B

4. Technician A says that refrigerant oil will absorb moisture if exposed to the air. Technician B says that too much oil in the system can reduce cooling capacity. Who is right?

   a. Technician A   **b.** Technician B
   c. Both A and B   **d.** Neither A nor B

5. Technician A says that a bad expansion valve can cause low pressures on both sides of the system. Technician B says that if the expansion

valve is replaced, the accumulator should be replaced, too. Who is right?

a. Technician A   b. Technician B
c. Both A and B   d. Neither A nor B

6. Technician A says evacuating a system will make it cool more efficiently. Technician B says that water can be boiled at room temperature by pressurizing it. Who is right?

a. Technician A   b. Technician B
c. Both A and B   d. Neither A nor B

7. Technician A says that vacuum checking is a good way to locate difficult leaks. Technician B says to refill an evacuated air conditioning system until the sight glass is clear. Who is right?

a. Technician A   b. Technician B
c. Both A and B   d. Neither A nor B

8. Technician A says that the sight glass can be located on the accumulator. Technician B says

that high side pressure affects the low side. Who is right?

a. Technician A   b. Technician B
c. Both A and B   d. Neither A nor B

9. Technician A says that an accumulator should feel cold and sweaty. Technician B says when both the high and low side valves are open, the gauges do not read accurately. Who is right?

a. Technician A   b. Technician B
c. Both A and B   d. Neither A nor B

10. Technician A says that disconnecting the low side while the system is running results in the lower pressure at that connection than with the engine off. Technician B says that the suction service valve is on the inlet to the compressor on the high side. Who is right?

a. Technician A   b. Technician B
c. Both A and B   d. Neither A nor B

CHAPTER 75

Automotive Computers and Electronics Fundamentals

■ INTRODUCTION

Electronics is the science of using very small amounts of electricity to control larger amounts of electricity. All of the laws of basic electricity still apply. Those basic laws were covered in Chapter 29. The operation of semiconductors, computers, sensors, and actuators is the emphasis in this chapter.

■ SEMICONDUCTORS

Semiconductors, used to make diodes and transistors, are materials that can be either a conductor or an insulator. Silicon and germanium are common semiconductor materials. These materials have a crystalline structure. This means they share outer electrons with each other (Figure 75.1) so they do not gain and lose electrons like conductors do.

Valence rings were discussed in Chapter 29. Refer to that chapter to refresh your knowledge. When a material has fewer than four electrons in its outer valence ring it is a conductor. If it has more than four, it is an insulator. Crystals are made of materials that have four electrons in their outer valence rings. A pure crystal has its atoms linked in a way that leaves no *holes* that would allow electron movement. A pure crystal is therefore a good insulator.

The insulating property of a crystal can be changed by *doping* the crystal to make a semiconductor. This process adds a very small amount of *impurity* (one atom in 10 million). Yet, enough holes are added so that the material will conduct electricity when a voltage is applied to its base. This acts on the semiconductor material to change it from an insulator to a conductor. Current in the circuit can then pass through it.

Doping a semiconductor creates one of two types of semiconductor materials, depending on the type of impurity added. The negative type is called an *N-type*. These crystals are doped with an atom like phosphorous, with extra electrons. N-type semiconductors have five or more electrons in their outer valence ring. The extra electron is free, which gives the material a negative charge (Figure 75.2).

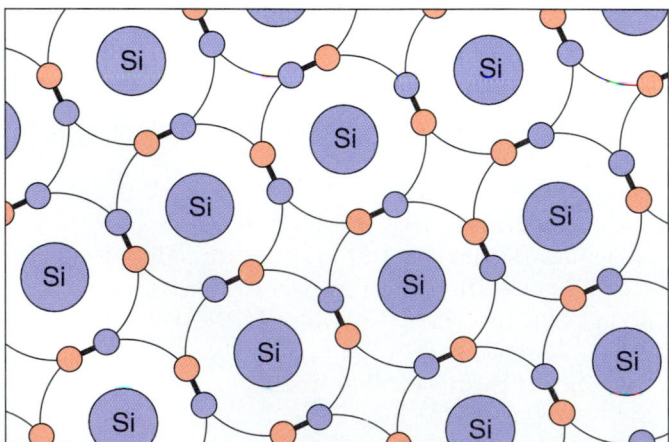

Figure 75.1 Semiconductor materials share outer electrons with each other. *(Courtesy of General Motors Corporation, Service Technology Group)*

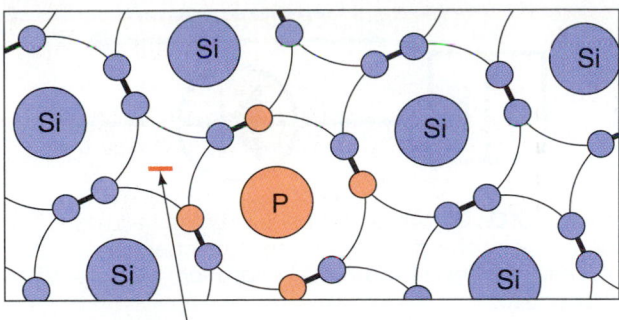

Figure 75.2 N-type semiconductors have extra electrons in their outer ring, which gives the material a negative charge. *(Courtesy of General Motors Corporation, Service Technology Group)*

A *P-type* semiconductor is positively charged and can carry electrical current. An impurity with a three-electron outer ring, such as aluminum or boron, is added to the crystal. Wherever this element fits into the crystal there is a hole for a fourth electron to fit into (Figure 75.3). The hole is a positively charged space. It carries the current in a P-type semiconductor when a voltage is applied. Because the holes are positively charged, they attract electrons. The electrons cannot become free of their atoms, but the crystals can rearrange their patterns to fill a nearby hole. This leaves a hole where the electron came from. That hole is filled by another atom's electron and so forth, resulting in electron flow (Figure 75.4). Hole movement only occurs in the semiconductor, while electron movement occurs in the entire circuit.

Electron Flow

Conventional electrical theory has been understood as electricity flows from positive to negative. In the field of electronics, electron theory says that electron flow

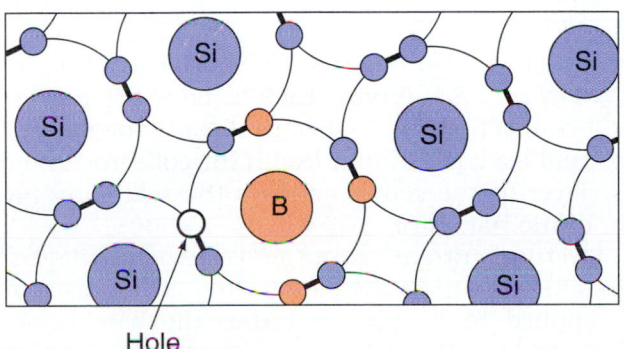

Hole

Figure 75.3 A P-type semiconductor, which can carry electrical current, is doped with a material that leaves a hole for a fourth electron to fit into. *(Courtesy of General Motors Corporation, Service Technology Group)*

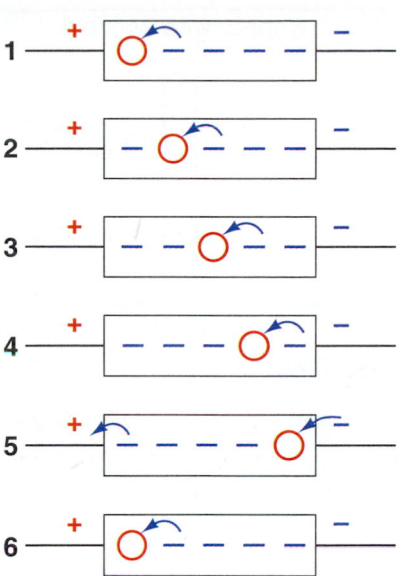

Figure 75.4 Hole movement in a semiconductor. *(Courtesy of General Motors Corporation, Service Technology Group)*

is from − to +. Think about how traffic flows. As cars move forward after a stoplight to fill holes left from previous cars, did the cars move forward or did the holes move backward? Holes flow from + to −; electrons flow − to +. In a diode, holes flow in the direction of the arrow, while electrons go against the arrow.

Semiconductors are designed to handle a limited amount of current. If too much current is applied in a reverse direction, it can force through and ruin the diode or transistor.

◼ DIODES

A diode (Figure 75.5) is a one-way electrical check valve. It is made by placing P-type and N-type crystals back to back. This *P-N junction* will only allow electrons to pass when a voltage of more than 0.5 V to 0.7 V (in silicon) pushes holes in the positive material toward the extra electrons in the negative material (Figure 75.6).

When the polarity is such that the P-N junction conducts current, this is called **forward bias**. Forward bias diodes allow current flow. When current tries to flow in the opposite direction, it cannot. Reverse bias diodes do not allow current to flow. The P and N sides of the diode are connected to opposite charges, which attract (Figure 75.7). The P material is drawn toward the negative part of the circuit and the N material is drawn toward the positive part. With the P-N junction empty, current flow stops.

The electrical symbol for a diode is an arrow with a bar at its point (Figure 75.8). The point indicates the

Copper wire terminal

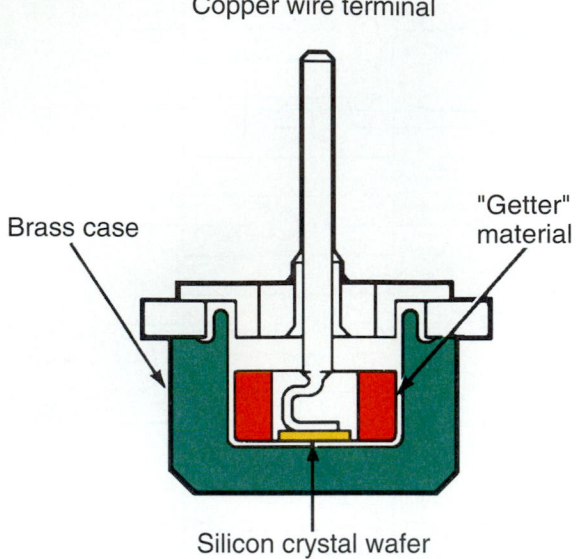

Brass case

"Getter" material

Silicon crystal wafer

Figure 75.5 Parts of a diode. *(Courtesy of Ford Motor Company)*

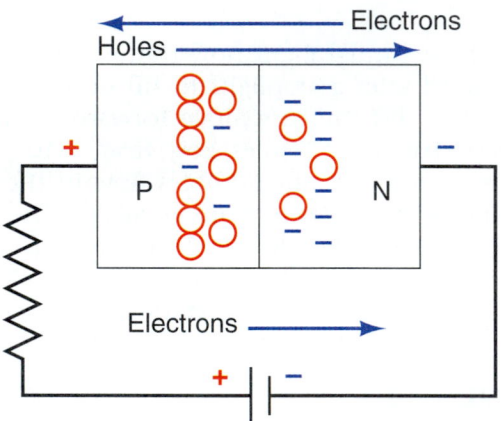

Figure 75.6 A voltage pushes holes in the positive material toward the extra electrons in the negative material. *(Courtesy of General Motors Corporation, Service Technology Group)*

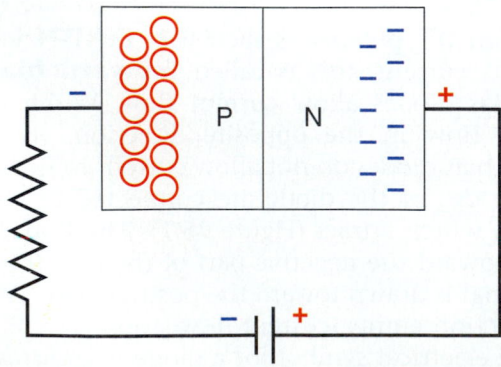

Figure 75.7 A reverse bias diode does not allow current flow. *(Courtesy of General Motors Corporation, Service Technology Group)*

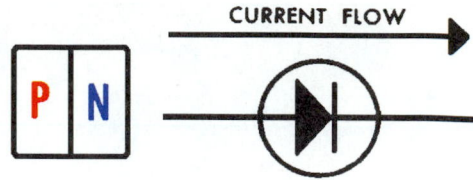

CURRENT FLOW

Figure 75.8 An electrical symbol for a diode. The point indicates the direction of conventional current flow. *(Courtesy of General Motors Corporation, Service Technology Group)*

direction of conventional current flow. Electron flow is in the opposite direction. The wide end of the arrow indicates the P side. It is often called an *anode* and is like the positive terminal on a battery. The bar end of the anode is the N side. Often called a *cathode*, it is like the negative terminal on a battery.

■ TRANSISTORS

Electronics was first used in an automobile in the 1930s when a radio was installed in a car. Radios used *vacuum tubes*, which work like a transistor but are large and relatively unreliable. The transistor was invented in 1948 by Texas Instruments. In the late fifties, the transistor (see Figure 29.37) made smaller and better car radios a reality.

Transistors turn electrical circuits on and off. They are controlled by another electrical circuit. A transistor is an electronically controlled relay or switch. A small amount of current applied to the transistor causes it to relay larger amounts of electricity through it. The difference between a diode and a transistor is that the transistor has two P-N junctions, while the diode has one. It is like a diode with an extra side.

There are three semiconductor crystal layers in a transistor. They are called the *emitter*, the *base*, and the *collector* (Figure 75.9). The base is always in between the emitter and the collector. The transistors used in automobiles are called *bipolar transistors*. They have two polarities, *electrons* and *holes*. A transistor is either an *N-P-N* or a *P-N-P* type. Each layer of the transistor has a connection for an electrical lead. The emitter is the input lead, the output lead is the collector, and the base layer in the center controls the switching function of the transistor.

Electrical current cannot move across the layers of the transistor unless the base (center layer) has a voltage applied to it. Voltage causes the base layer to reverse its charge and become a conductor. Extra electrons or holes are added to the base by the controlling current. The transistor can be controlled by whatever amount of voltage is desired by putting a resistor on its base (input). This will control at how many volts it opens or closes. It also keeps current flow low through that part of the transistor.

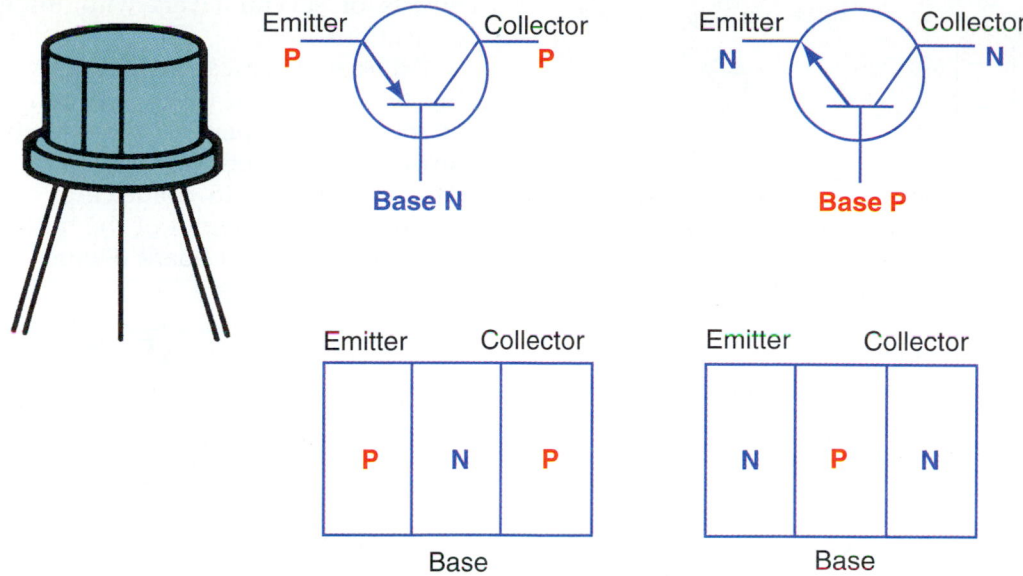

Figure 75.9 P-N-P and N-P-N transistors. *(Courtesy of Chrysler Corporation)*

A transistor is said to be forward biased when it is allowing current to flow. When the voltage is removed from the base, the circuit shuts off. This is called reverse bias.

The schematic for a transistor has the arrow on the emitter pointing in the direction of current flow. The N-P-N transistor is normally off. When the base is forward biased with a more positive voltage, the emitter-to-collector circuit is turned on (Figure 75.10). The amount of output current is proportional to the amount of current through the base leg. This type of transistor can also be used as a variable resistor by varying the current applied to the base. More current at the base results in more current passing to the collector. This kind of variable resistance can be found in a power transistor for an automatic climate system blower motor.

A P-N-P transistor is controlled by its ground. A more negative voltage than is in the emitter must be applied to the base of a P-N-P transistor to turn it on.

This results in current flow between the emitter and collector. A small amount of current flows from the source of the control voltage through the base and is grounded through the emitter when the base has voltage applied to it.

A transistor actually never shuts off. Current either flows in from the base or from the collector. It never flows in from the emitter so it is a "dual diode." When a contact point closes, it takes $\frac{1}{100}$th of a second to accelerate from 0 to 4 amps. A transistor can regulate at 10,000 times a minute because current always continues to flow.

Zener Diodes

A **zener diode** (Figure 75.11) is one whose crystals are more heavily doped during manufacturing. Zeners allow current flow at a certain voltage but halt current flow below that. They are used to control transistors and electronic voltage regulators. When a transistor is switched off and on several thousand times a minute, a zener is needed to control the backwash or double bounce of the voltage.

Figure 75.10 When the base is forward biased with a more positive voltage, the emitter to collector circuit is turned on.

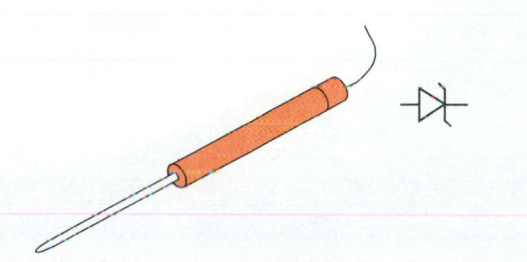

Figure 75.11 A zener diode and its electrical wiring diagram symbol.

Figure 75.12 An LED and its electrical wiring diagram symbol.

■ LEDS

Light emitting diodes (LED) (Figure 75.12) are found in digital displays on the dashboard and in test instruments. They are also used in some center high-mounted brake lights and as a trigger in some ignition and fuel systems. They have a crystal that operates like a light bulb and glows (usually red) when current flows through it. They consume less power and do not have a filament so they have a longer life than a light bulb.

■ AUTOMOTIVE COMPUTER SYSTEMS

Integrated Circuits

An integrated circuit (IC) is a complete miniaturized electric circuit. It consumes considerably less electrical current than a large-scale circuit. Transistors, diodes, and resistors are included in the *chip*. The chip consists of tiny sandwiched silicon wafers of P-type or N-type material. One transistor has limited ability when it comes to performing complicated tasks. When many semiconductors are used in a circuit, the functions they can perform are amazing. As many as 30,000 transistors can be placed on a chip that is only ¼" square (Figure 75.13). The circuit is constructed by photographically reproducing circuit patterns onto a silicon wafer. An electrical circuit that would normally fill a large room can be put in this small area. A chip

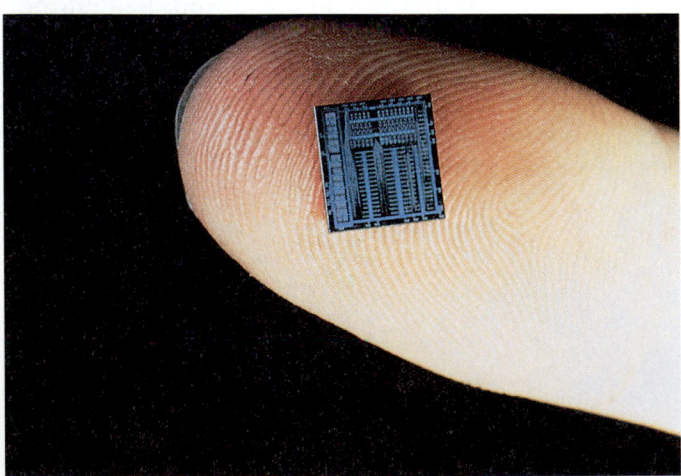

Figure 75.13 The actual size of this integrated circuit (chip) is less than ¼" square. *(Courtesy of Texas Instruments)*.

consists of several layers with underlying electrical connections.

External electrical connections come off of the chip (Figure 75.14). Chips can be mounted on a flat ceramic or metal plate or they can be surrounded by plastic. Several methods of chip packaging are illustrated in Figure 75.15. Some chips are replaceable and have pins that come out of the bottom that plug into sockets in the circuit board (Figure 75.16).

Figure 75.14 This chip from a fuel injection module is enlarged 12 times here. *(Reprinted with permission from Robert Bosch Corporation)*

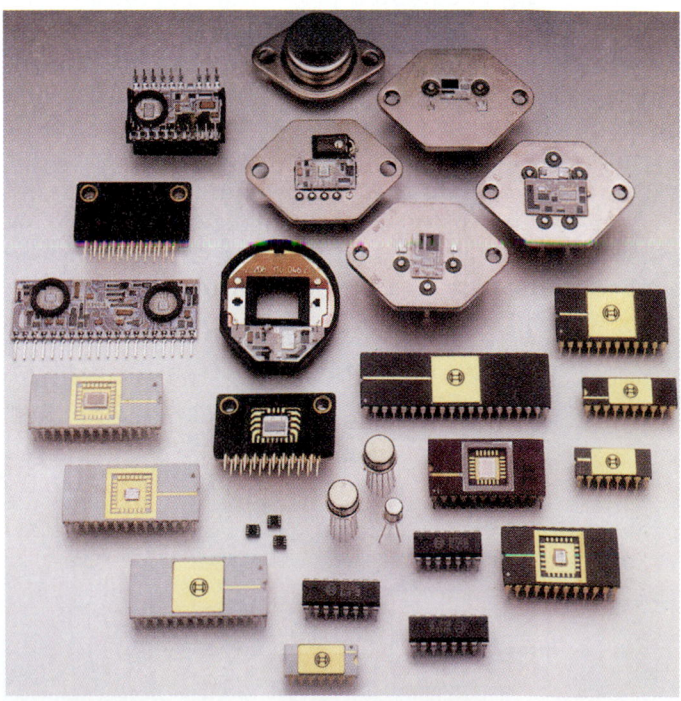

Figure 75.15 Several methods of chip packaging. *(Reprinted with permission from Robert Bosch Corporation)*

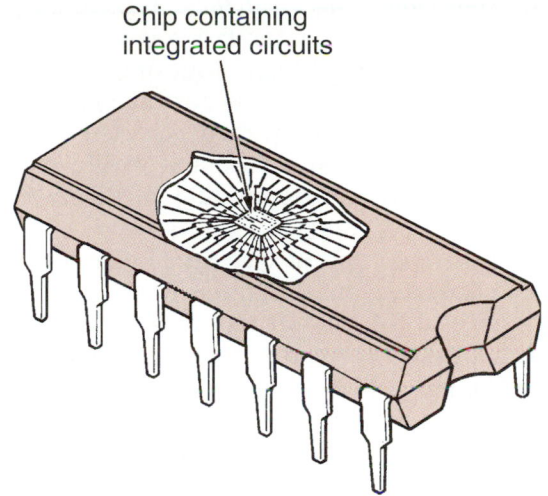

Figure 75.16 Some chips are replaceable and have pins that come out of the bottom that plug into sockets in the circuit board.

ICs are limited in the amount of electrical current that they can carry. Too much electrical current causes heat that can damage the circuit.

PARTS OF A COMPUTER SYSTEM

There are three main parts to automotive computer systems: the *computer,* sensors, and actuators. The many sensors and actuators control engine and other vehicle functions (Figure 75.17). Sensors relay information to the computer on air or coolant temperature, air flow, manifold pressure, and barometric pressure. The computer processes the information and sends command signals to actuators.

ONBOARD COMPUTER

An automotive onboard computer can be called a control assembly, a control module, or a control unit. Electronic part names are set by the Society of Automotive Engineers OBD (onboard diagnostics) guidelines (covered later). According to OBD II guidelines, all new model computers that control things other than engine functions are called *powertrain control modules (PCMs).* This applies to many computers, as most cars with automatic transmissions now have electronic controls too. A typical system has one central computer and several modules. The modules deal with powertrain, ABS brakes, comfort systems, and body control.

A computer has four functions. It
- Gathers input
- Makes decisions and processes information
- Stores information
- Takes action by way of an output command.

The **microprocessor** is the calculating and decision-making chip in the computer. It does not actually think, but follows instructions programmed

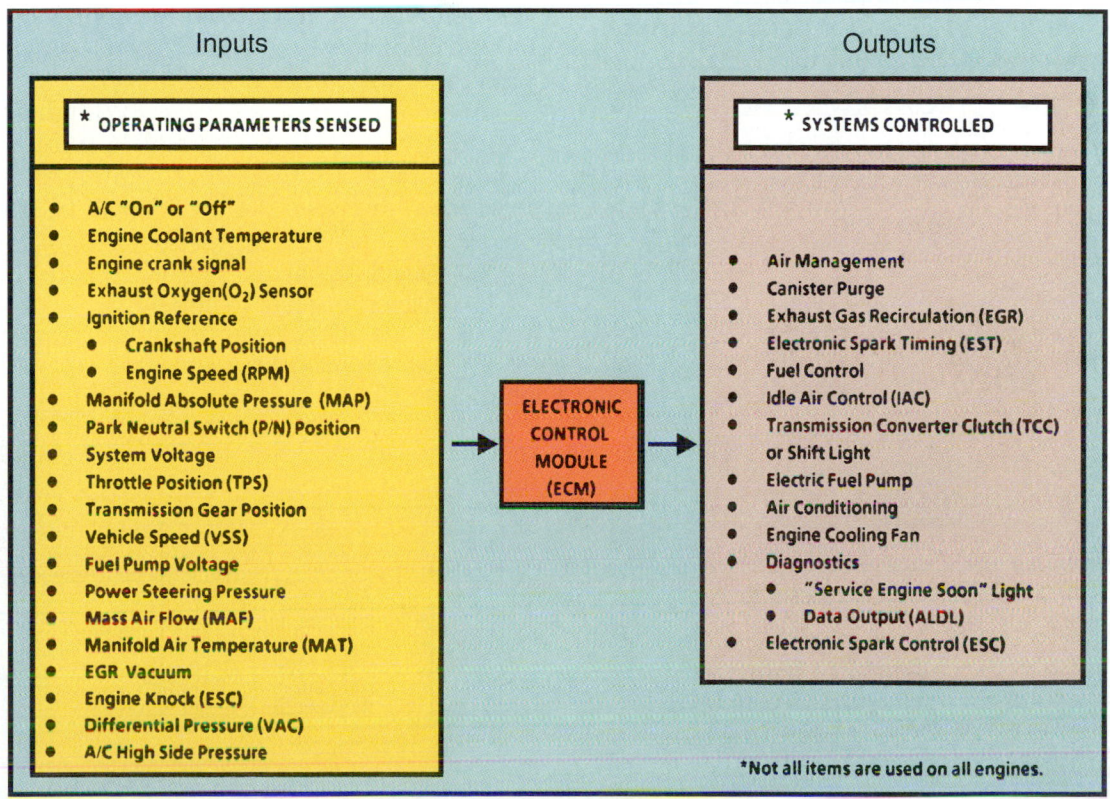

Figure 75.17 The many sensors send signals to the computer, which uses actuators to control engine and other vehicle functions. *[Courtesy of General Motors Corporation, Service Technology Group]*

into its memory. There are thousands of miniature diodes and transistors in the microprocessor. Voltage surges in the system are a problem so diodes are installed throughout the system.

■ COMPUTER NOTE ■

Transistors act like electronic switches and are either on or off. The computer interprets combinations of zeros and ones to form numbers or words. This binary (zeros and ones) information is processed to determine the meanings of the signals. Then, the processed information is stored or delivered from the computer.

When the computer receives a voltage of at least a certain value, it converts it to a one. When there is no voltage, the computer interprets a zero. Each zero and one represents a **bit** *of information. Eight bits makes a* **byte***, which is sometimes called a word. Electronic information is exchanged in bytes.*

Computers do all of the same things mechanical systems did. But they can do them much faster and more accurately. Newer cars make decisions in 600 thousandths of a second.

Items that make up a computer system include *hardware* and *software*. Hardware consists of the visible mechanical parts. Software is the magic stuff inside. It is information stored as electronic signals that can be modified. This is why it is called "soft." Software includes the *programs* for the computer.

Computer Electrical Control

A computer carries a very light electrical load. It controls grounds to output devices rather than power because it cannot turn heavy amp loads off and on.

This is called *ground side switching*. Typical amp loads like turning on the radiator fan relay, vacuum solenoids, the alternator field, and four or six injectors are loads that are too big for an electronic circuit. Also, if a short circuit was to occur in a powered circuit, the computer would act like a fuse and would be blown. For *power switching,* a logic module tells a power module what to do.

Automotive computer systems since mid-1983 use 5 volts to operate. This voltage is low enough to prevent damage to the circuits in the chip, yet it is high enough to provide consistent transmission of information. Reference voltage must be less than minimum battery voltage or signals will be inaccurate. Because battery voltage never dips this low, the computer will operate consistently.

■ INFORMATION PROCESSING

The signals from sensors must be conditioned or processed. The computer's logic circuits either turn these signals into output commands or store them for a short or long time in the computer's memory (Figure 75.18). The memory contains values programmed into the computer software that the computer checks incoming information against.

Sensor information can be digital (on/off), but is often *analog* when the sensed information varies gradually. Remember the earlier discussion on binary information. Automotive processors used since the eighties are digital. An analog signal is a sign wave and a digital processor must see digital input, which means it is something like yes or no, or on/off.

An analog temperature sensor might have an output voltage range from 0 V to 5 V. If the sensor's temperature range is from 0° to 250°F, then a voltage somewhere between 0 V to 5 V will be interpreted by

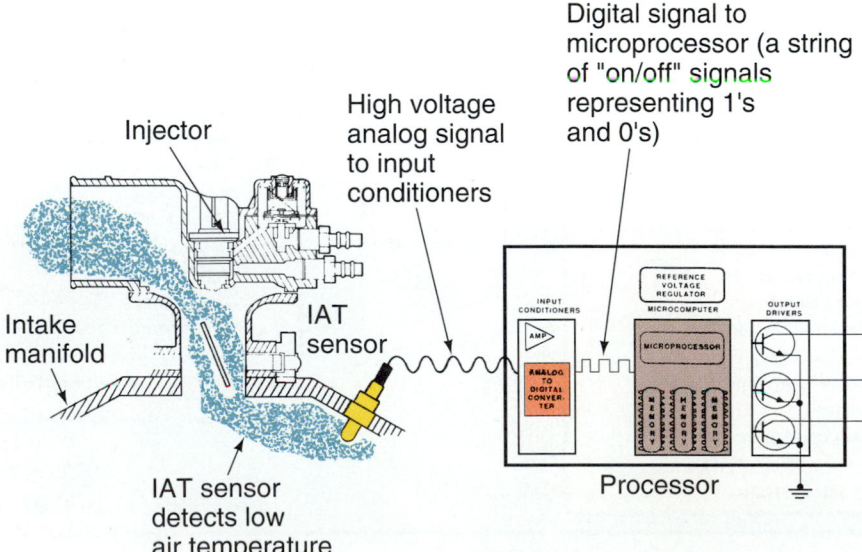

Figure 75.18 The signals from sensors must be conditioned or processed. *[Courtesy of Ford Motor Company]*

the computer as a specific temperature. Each 1 V change will be equal to a 50°F change in temperature.

■ COMPUTER MEMORY

The computer's memories contain programs that it refers to as it processes information. The microprocessor can read information from the memories or it can write information into them. Memory files are located in many places. Each piece of information has an address. When stored information is needed, the micro-processor looks it up at its address in memory.

Random Access Memory

Information that will be stored temporarily is sent from the microprocessor to **random access memory (RAM)**. RAM is like a notepad that you can read from and write to. Input from the sensors changes frequently, so it is stored in RAM (Figure 75.19). *Random access* means that the information can be retrieved in any order.

Volatile and Non-volatile RAM. When RAM is *volatile*, it is erased each time the ignition is turned off. RAM can also be *non-volatile*. This means that the information is not erased when the ignition switch is turned off. This can be compared to the station settings on an electronically tuned radio. When the ignition is off, the radio remembers the station settings, yet the radio turns off with the key. If the battery is disconnected, however, the radio will forget the station settings.

Read Only Memory

Read only memory (ROM) is permanently programmed information available to the microprocessor (Figure 75.20). This information is programmed into the chip during manufacturing, so disconnecting the battery does not change it. ROM contains *lookup tables,* which is program information on how the car is supposed to perform. One system might have 560 programmed instructions for different conditions. These *parameters* are what the engineers feel is the best adjustment for a particular operating condition. The computer compares sensor information to these parameters and makes adjustments as needed.

Programmable Read Only Memory (PROM)

ROM also contains the *calibration tables.* These are instructions for specific engines and drive trains. A car with a manual transaxle will have different program requirements than one with an automatic transaxle. Calibration table chips can be either PROMs or EEP-ROMs. A PROM is *programmable read only memory.* PROMs provide a specific vehicle with a program for its fuel and emission systems, including spark advance. PROMs contain instructions for the EGR valve, vapor canister purge, torque converter clutch, air conditioning, radiator electric fan, and others.

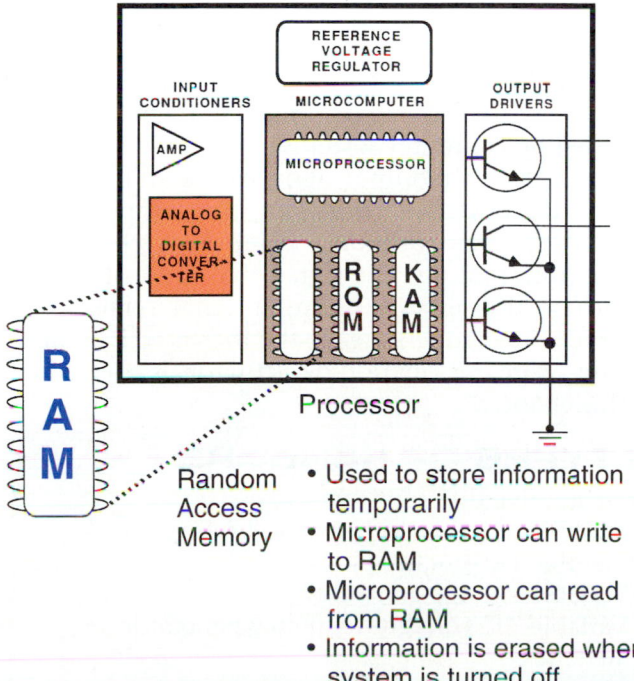

Figure 75.19 Random access memory. *(Courtesy of Ford Motor Company)*

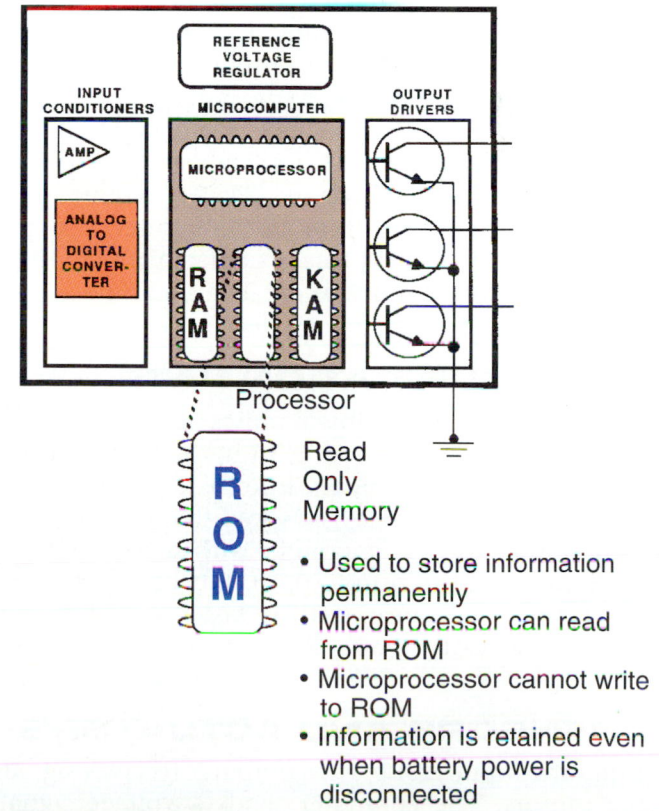

Figure 75.20 Read only memory. *(Courtesy of Ford Motor Company)*

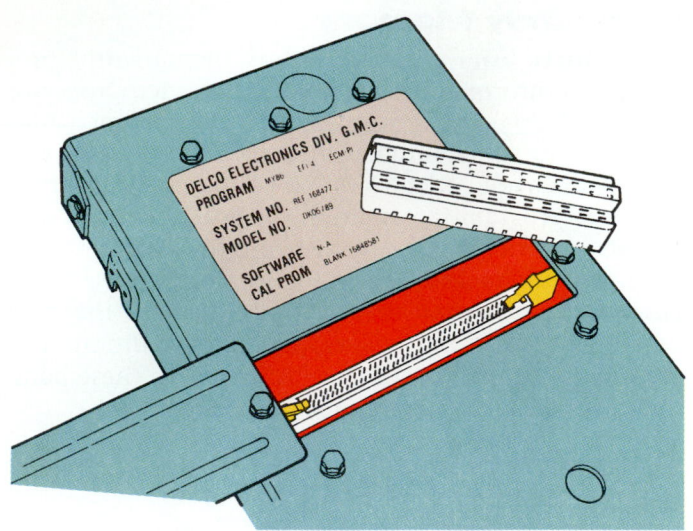

Figure 75.21 On some cars the PROM is replaceable. [Courtesy of General Motors Corporation, Service Technology Group]

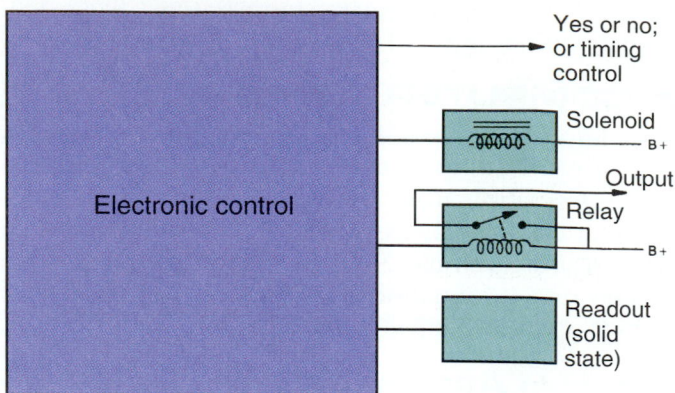

Figure 75.22 The computer processes information and controls the current flow to actuators. [Courtesy of Ford Motor Company]

There might be sixty specific parameters for that particular car.

In many GM cars, the computers are the same but the PROM is replaceable (Figure 75.21). Other manufacturers write the information into ROM and then replace the computer if there is a change or use EEPROM. An EEPROM is *electronically erasable read only memory*. When there is an update in the engineering of the computer program, these chips can be reprogrammed in the car, in the service bay, or by telephone modem from the manufacturer's facility. When the information in an EEPROM is erased and reprogrammed, this is called flashing the PROM.

An older chip, called an *EPROM,* required the PROM to be removed from the computer and put under an ultraviolet light for a specified period of time. Another kind of memory, keep alive memory, is covered later.

■ COMMUNICATION RATE

A quartz crystal uses timed pulses to maintain an orderly flow of information. During the time between each pulse, one bit of binary information is transmitted within the computer. The speed that this occurs is called **baud rate**. A computer with a baud rate of 56,000 (56K) can transmit 56,000 bits of information per second. Baud rate is getting faster as technology advances. An early 1980's GM computer had a baud rate of 160.

■ SENSORS AND ACTUATORS

During the input phase, the computer receives signals from sensors. The computer sends a voltage signal, called *reference voltage,* to many of the sensors. Most computers use 5 volts as a reference signal. Sensors monitor engine functions and modify the voltage signals that return to the computer.

An **actuator** is an electronic or magnetic relay that can perform a desired function. The computer processes information and controls the current flow to these devices (Figure 75.22). They act upon the commands when their electrical circuits are opened and closed. This results in adjustments that bring operating conditions back to within the programmed parameters.

Sensors and actuators are all transducers. A **transducer** converts energy from one form to another. Sensors convert energy (like temperature, light, or motion) to voltage signals. Actuators change electricity into a mechanical action (work). Transducers can also be mechanical, such as a vacuum advance unit. It reacts to changing air pressure on its diaphragm to move the advance plate in the distributor. Mechanical transducers like solenoids and diaphragms are often operated by computer systems.

Sometimes, an output signal can be an input signal for another device. For instance, the computer might turn on the air conditioner compressor clutch while the engine is idling. The signal that causes the clutch to turn on also serves as an input signal to raise engine idle speed to compensate for the increased load on the engine. More information on actuators is covered later in this chapter.

■ TYPES OF SENSORS

There are five different types of sensors.
- Variable resistors (voltage modifying)
- Variable DC frequency
- Variable voltage generators
- Variable AC voltage/frequency generators
- Switches

Specific operation of various types of sensors is covered in the chapters that deal with their systems. Some of the more popular ones are listed here.

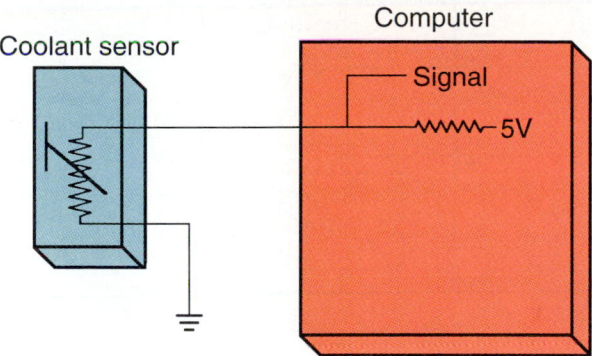

Figure 75.23 Varying current flow through a thermistor to ground in a coolant temperature sensor circuit. *(Courtesy of SPX Corporation, Aftermarket Tool and Equipment Group)*

■ THERMISTORS

A **thermistor** is a variable resistor made from semiconductor material. Its resistance changes predictably as its temperature changes. It is used for measuring air and water temperatures because even a small change in temperature will result in a change in its resistance. Its resistance decreases as temperature rises. Air temperature information is used by the computer to calculate fuel delivery. Thermistors are also used for coolant temperature sensors (Figure 75.23).

■ VOLTAGE DIVIDERS

A *voltage divider* circuit has three wires. Voltage dividers are variable resistors that produce a variable DC voltage signal. The signal varies according to the sensor's mechanical position. A potentiometer is a three-terminal variable resistor. A reference voltage is applied to one of the terminals. The other end of its circuit is ground. A movable center contact or wiper senses voltage between ground and its position on a wire-wound resistor. The ground connection completes the circuit to the wiper (Figure 75.24).

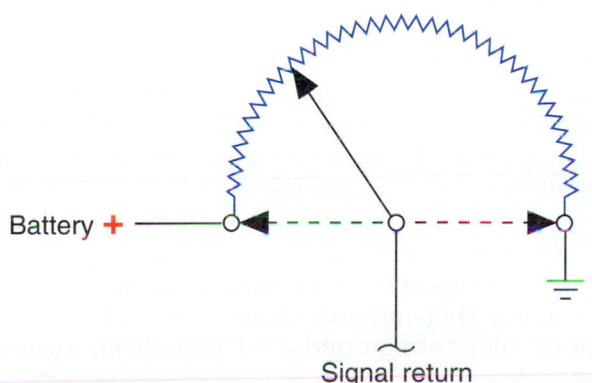

Figure 75.24 In a potentiometer, a movable center contact or wiper senses voltage between ground and its position on a wire wound resistor.

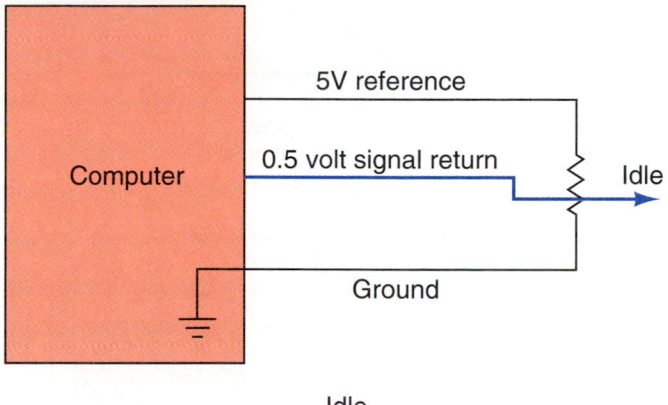

Idle

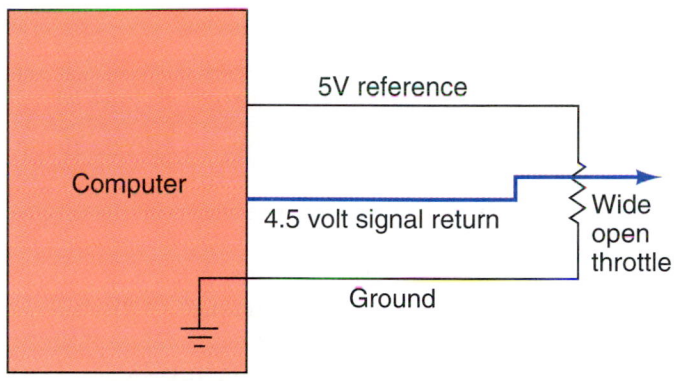

Wide open throttle

Figure 75.25 A throttle position sensor will read a high voltage (4.5 V) as wide open throttle. A lower voltage (0.5 V) means closed throttle.

The potentiometer measures linear or rotary motion in throttle position, air flow, and EGR valve position sensors. In a throttle position sensor, it is attached to one end of the throttle shaft. Moving the shaft moves the wiper to a different position on the wire-wound resistor coil. A high voltage (4.5 V) is read as WOT while a lower voltage (0.5 V) means closed throttle (Figure 75.25). Voltages in between give an indication of throttle position. When a potentiometer is used in an air flow meter, it reacts to pivoting of the sensor shaft also.

A rheostat differs from a potentiometer in that it carries current. It also has a movable arm, but has only two wires. There is no ground wire, as ground is supplied by the movable arm.

■ PIEZOELECTRIC AND PIEZORESISTIVE SENSORS

Piezoelectric crystals develop a voltage on their surfaces when pressure is applied to them. In late model vehicles, these devices are commonly used as switches for measuring pressure in engine oil, power

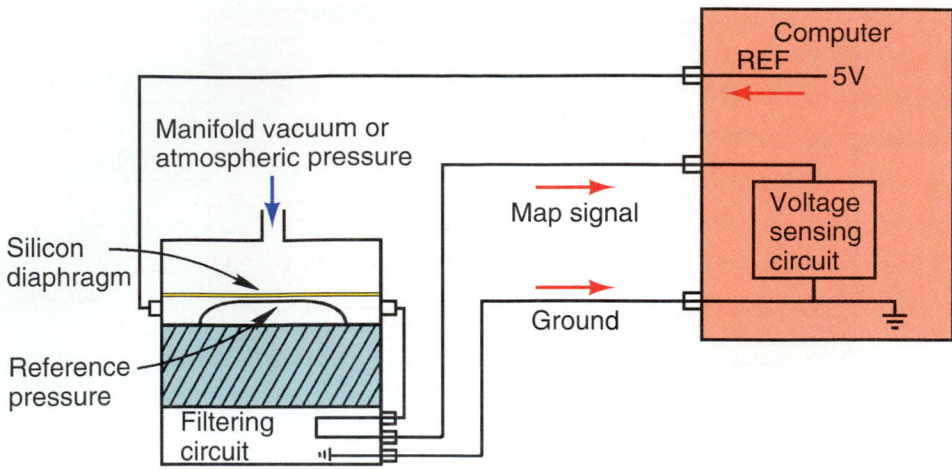

Figure 75.26 A piezoresistive sensor is made up of a silicon diaphragm sealed to a quartz plate.

steering, or air conditioning systems. In air conditioning and power steering systems, pressure switches give a signal to the computer so that it can control engine idle speed.

A similar crystal is commonly used in MAP sensors for measuring intake manifold pressure. It is called a *piezoresistive* sensor. It is made up of a silicon diaphragm sealed to a quartz plate (Figure 75.26). Doping the silicon creates four resistances around the edges of the diaphragm. There is a vacuum chamber between the quartz and the silicon diaphragm. When pressure is applied to the diaphragm it deflects. This causes a change in the resistance of the resistors. The reference voltage input is changed as it goes through the sensor, resulting in lower output voltage.

■ HEATED RESISTIVE SENSORS

A heated resistive **sensor** can be used to monitor the amount of air taken into the engine. An electric current is applied to a platinum wire or foil screen. The computer maintains the amount of current necessary for the wire to remain at a constant temperature. The current required to maintain this temperature is interpreted by the computer. From this information, it can calculate the amount of air flow into the engine. This is called a mass air flow sensor (MAF). Refer to Chapter 26 for more on the operation of MAP, BARO, and air density sensors.

■ VARIABLE DC FREQUENCY SENSORS

Frequency sensors are used for the same things as the previous sensors but they produce a digital signal. They have three wires like other MAP, BARO, and air density sensors. The signal return line produces a pulsed frequency measured in hertz (cycles per sec-

ond). Increased air flow causes the sensor to vary how fast it cycles off and on.

■ VOLTAGE GENERATORS

The sensors covered previously receive a reference voltage that they send a signal response to. Voltage generating sensors have no reference voltage, but create their own. There are both AC and DC voltage generators.

In a *variable AC voltage generator,* a magnetic pickup generates an AC analog signal (consisting of a positive and negative sine wave). *Pulse generators* are signal generators like those found in antilock brake wheel sensors, magnetic distributor triggers, or crankshaft or camshaft position sensors.

An oxygen sensor is a *variable DC voltage generator.* It is a galvanic battery that can generate a voltage of from 0.1 volt to 0.9 volt (100 to 900 millivolts) in response to the amount of oxygen in the vehicle's exhaust. A rich mixture (less oxygen) results in a higher voltage (450–900 millivolts). More oxygen results in a lower voltage. The operation of the O_2 sensor is covered in Chapter 26.

A knock sensor has a piezoelectric crystal sensing element that senses vibration and creates a voltage signal of 300 to 500 mV (Figure 75.27). When the voltage signal is generated, the computer retards ignition timing until the engine knock goes away.

■ SWITCHES

The computer receives some sensor signals from simple switches. There are three types of switches: switch-to-power, switch-to-ground, and Hall-effect. Switches are used for transmission gear position switches. A brake on or off switch gives a signal to the torque converter clutch and cruise control. A power steering switch senses when the wheel is turned against a lock

On/off switches, Hall-effect switches, and some air flow sensors generate digital signals, so they do not require the computer to convert their signal from analog.

■ TYPES OF ACTUATORS

Actuators are the devices that act upon processed signals received from the computer. They can be either solenoids, DC motors, relays, switches, or control modules. Actuators are always powered. When they are not operating it is because they do not have a complete circuit to ground. P-N-P transistors in the computer, called *output drivers,* supply the actuators with a ground to switch them on (ground side switching). Figure 75.29 shows a schematic of a computer with output drivers. When the output driver does not provide a ground to the actuator, the actuator does not operate.

One single module can have a group of four transistors. This is called a *quad driver.* This space-saving module can control up to four actuators.

■ SOLENOID ACTUATORS

A solenoid is a magnetic switch (see Chapter 32). Solenoids are used for several things.
- The air management system uses solenoids for controlling the flow of air from the smog pump.
- A solenoid opens to allow fumes to be purged from the vapor canister when the engine is warm and above idle speed.
- EGR vacuum control has two solenoids; one used for vacuum actuation and the other for bleeding off vacuum.
- Fuel injectors are solenoids whose pulse width (see Chapter 26) is controlled by the computer.
- An idle speed solenoid controls engine idle.
- Electronic transmissions have shift solenoids and a converter lockup solenoid.

■ RELAY ACTUATORS

Relays trigger the operation of high current loads such as the fuel pump, cooling fan, air conditioning compressor clutch, O_2 sensor heater, computer, alternator, or trunk latch. Most relays are open until they are energized. This is called a *normally open relay.* The computer controls the ground side of the relay (which has low current). When it is grounded, current can pass through the relay.

■ MOTOR ACTUATORS

An idle speed control motor is an example of a motor actuator, as is an electric fuel pump. Engine idle is controlled by the computer. Sensors detect anything that will put a load on the engine at idle and the computer makes corrections by powering a motor. Some other systems shut off the air conditioning or alternator at idle rather than trying to compensate for the load of these devices.

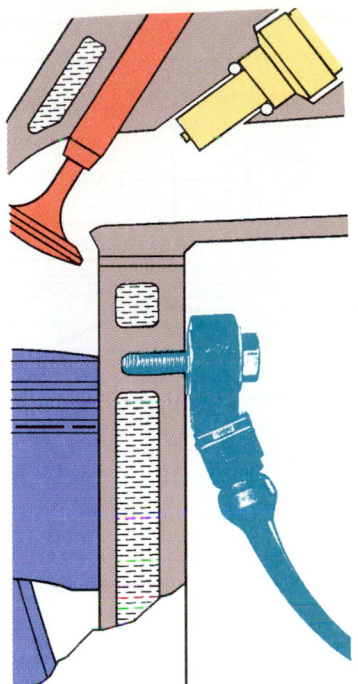

Figure 75.27 A knock sensor has a piezoelectric sensing element that senses vibration and creates a voltage signal of 300 to 500 mV. *(Reprinted with permission from Robert Bosch Corporation)*

to raise engine idle. Figure 75.28 shows examples of several types of simple switches.

Mechanical pressure switches can sense hydraulic or air pressure. Temperature switches have a bimetal element that opens and closes in response to heat.

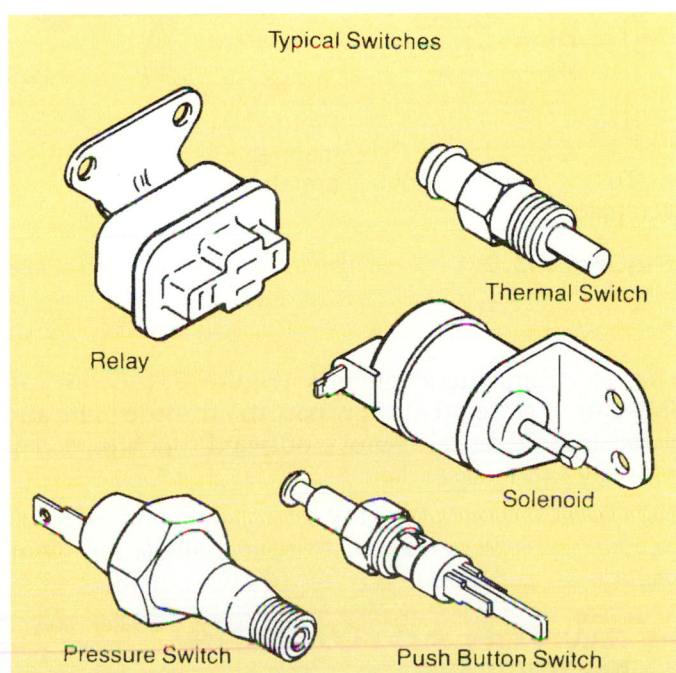

Typical Switches

Relay

Thermal Switch

Solenoid

Pressure Switch

Push Button Switch

Figure 75.28 Several types of switches. *(Courtesy of Cooper Automotive/NAPA Belden)*

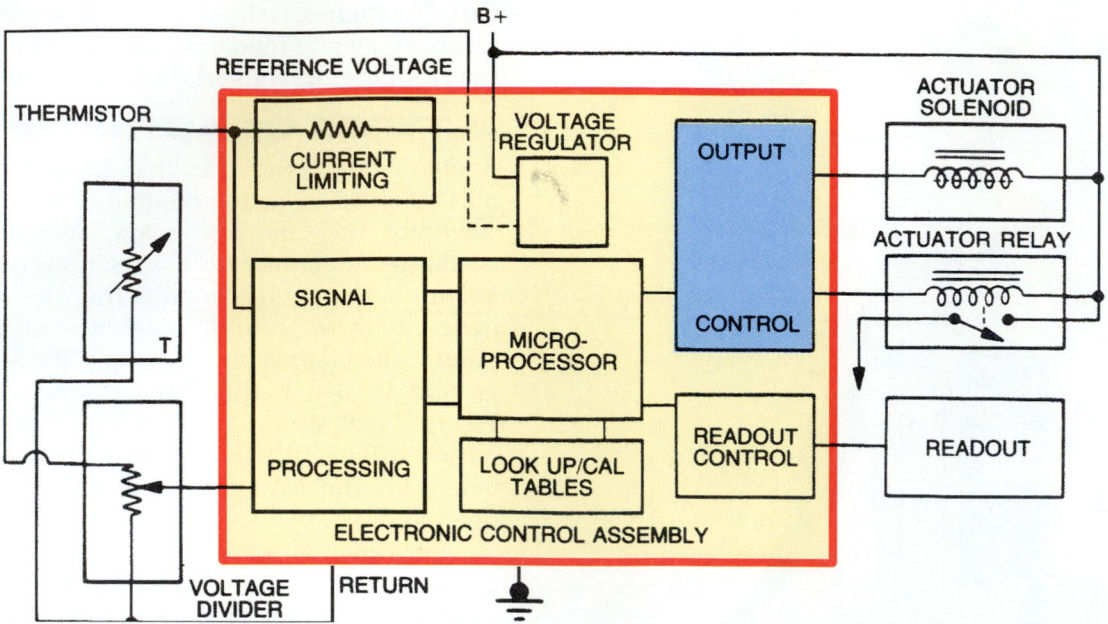

Figure 75.29 Output drivers supply the actuators with a ground to switch them on. This is called ground side switching. *(Courtesy of Ford Motor Company)*

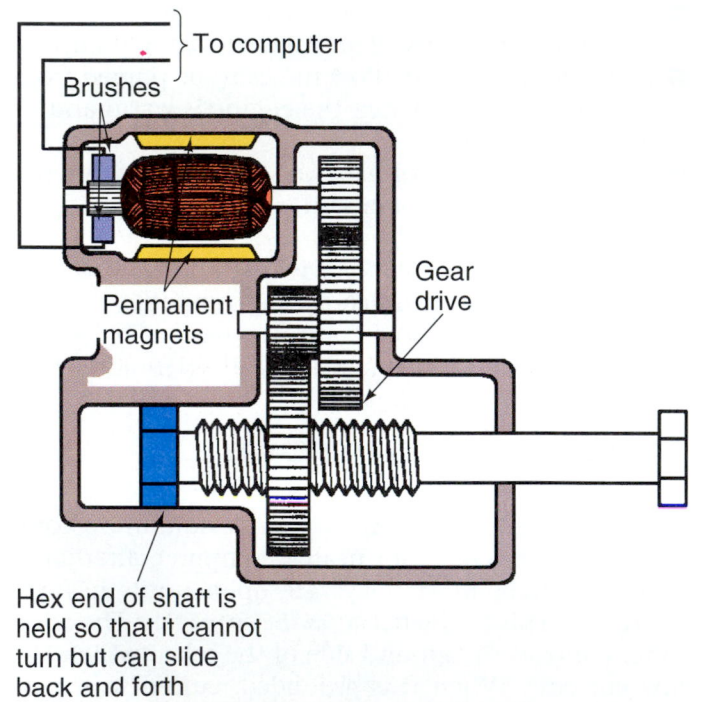

Figure 75.30 A permanent magnet DC reversible motor used to move the throttle shaft to control idle speed.

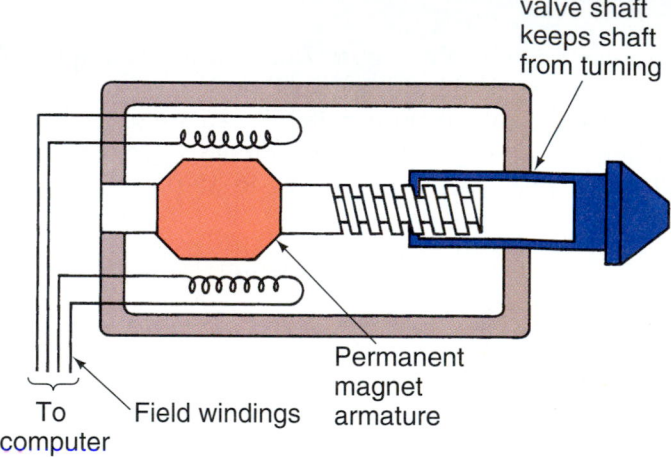

Figure 75.31 A stepper motor used to control idle speed.

There are two types of idle motors. One is used on throttle body injection and *feedback* (computer controlled) carburetors. It opens the throttle plate to control idle speed and is called an *idle speed control (ISC) motor* (Figure 75.30). The device used on fuel injection is an *idle air control (IAC) motor*. It is a *stepper motor* with

two electromagnetic circuits (Figure 75.31). It can move in to close off air flow past the throttle plate and lower idle. It can also move outward, opening a passageway to increase idle.

NOTE: *Most computer control devices are controlled by the computer through ground side switching. An idle air control motor is an exception to this rule.*

■ SWITCH ACTUATORS/ MODULES

The ignition module is one example of an actuator switch. It turns the primary ignition system on and off

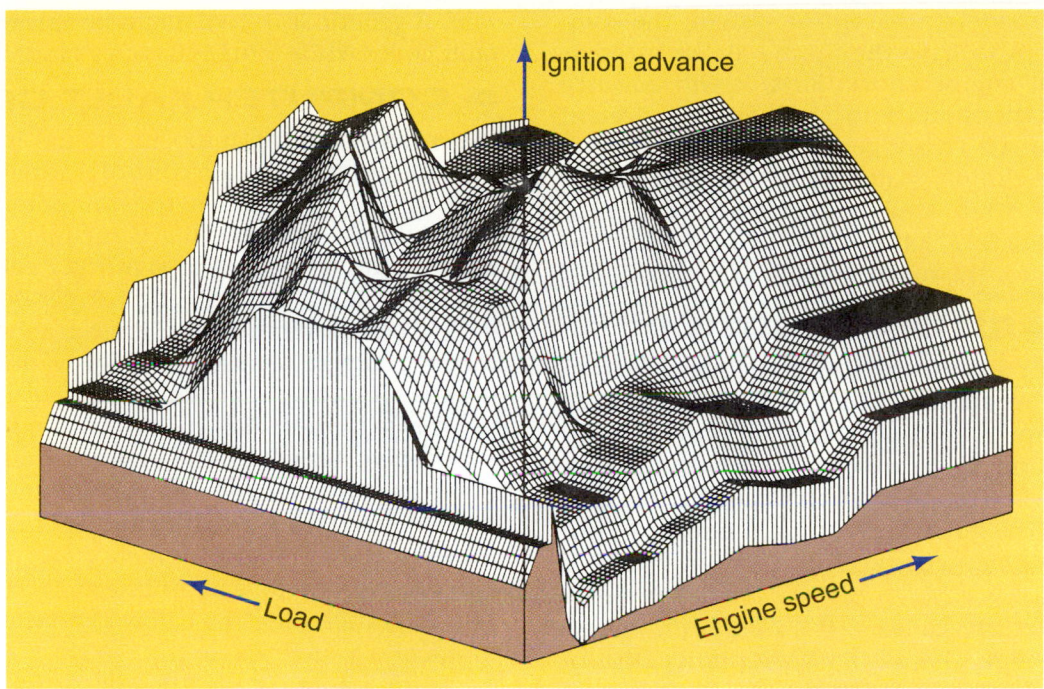

Figure 75.32 An ignition timing map. *(Reprinted with permission from Robert Bosch Corporation)*

in response to computer sensor inputs. On computerized systems, spark timing by the computer takes the place of vacuum and mechanical spark advance.

Two sensors that are part of the fuel system (see Chapter 26) are the MAP and BARO sensors. In dual function electronic ignition systems, these take the place of the vacuum advance. A chip in the ignition module analyzes information from these sensors, along with the coolant temperature sensor signal and tachometer signal (for the mechanical advance). A chip provides the correct total spark advance signal to the coil (Figure 75.32). Some computerized systems also use a throttle position sensor, air temperature sensor, and knock sensor.

Other modules control the operation of the air conditioning compressor. The cruise control is also engaged and disengaged by a module.

■ ADAPTIVE STRATEGY

Adaptive strategy uses **keep alive memory (KAM)** (Figure 75.33). Keep alive memory means the computer maintains power to random access memory when the ignition switch is off. This allows it to keep information as long as the battery is not disconnected. KAM allows the computer to adapt to different hardware from car to car. The computer will compensate for these differences and changes resulting from wear and aging. Variations in the options the vehicle is equipped with are also compensated for. Examples of types of adaptive strategies are idle air control, adaptive fuel trim, and electronic transmission shift schedules.

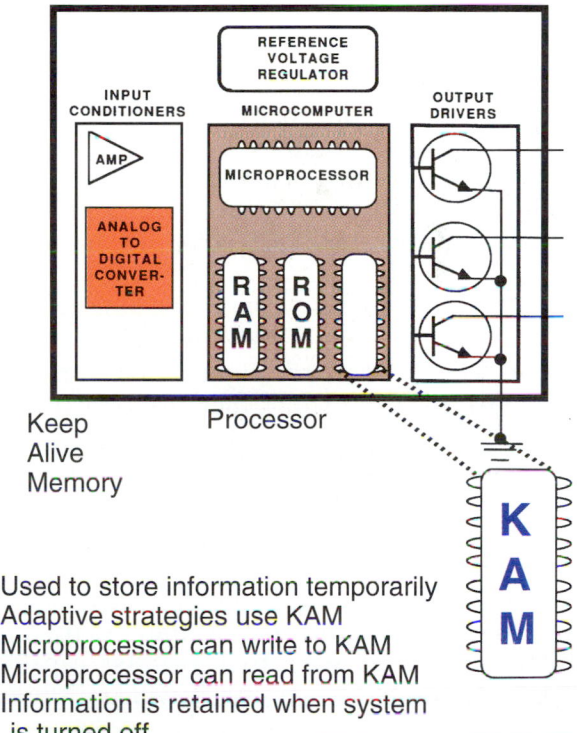

- Used to store information temporarily
- Adaptive strategies use KAM
- Microprocessor can write to KAM
- Microprocessor can read from KAM
- Information is retained when system is turned off

Figure 75.33 Keep alive memory. *(Courtesy of Ford Motor Company)*

Adaptive fuel trim allows the fuel system to operate at the correct air/fuel ratio, no matter what the differences in fuel systems are. This has a large effect on improving emission control.

With a late model knock sensor system, the computer will attempt to have the timing as advanced as possible without engine knock. First, it will advance the timing until the engine knocks. Then, it will retard the timing until the knock goes away. Once again it will advance the timing, searching for the best timing setting for that rpm, load, temperature, and fuel. When it finds the best setting, it stores the information in memory for the future.

ON BOARD DIAGNOSTICS

OBD II (onboard diagnostics) guidelines began in 1994 on some cars and are on all new cars built since 1997. These standards came about to provide standardization of terms and connections between makes. A universal data link connection is available for reading fault codes. The guidelines call for a universal scan tool that will be able to be used on different makes of vehicles.

Trouble codes are called *DTCs (diagnostic trouble codes)* or fault codes. There is an SAE-approved list of generic terms for electronic control system component names. The U.S. government has accepted this list as one that will be used in all automotive service publications in the United States since January 1995. An explanation of trouble codes is given in Chapter 76.

The scan tool will identify the vehicle automatically from a code received. The scan tool can also clear trouble codes and record snapshots during a test drive when a driveability symptom is felt. These provide a frame by frame recording of a period of several seconds before and after a button on the scan tool is pressed. Using this process can help identify intermittent faults.

The *malfunction indicator lamp (MIL)* located on the dash display must light if the emissions exceed 1½ times the federal standard (Figure 75.34). Emission-related codes are all stored in memory on OBD II systems.

The data link connector is located under the dash on the driver's side on all vehicles. They are all D-shaped 16 pin connectors that can be connected only one way. Seven of the sixteen pins are assigned according to OBD II standards. The manufacturers use information from the remaining pins as they see fit. A manufacturer's dedicated scan tool will be capable of locating more problems because it will use the other pins. A generic scan tool might be able to diagnose fifty fault codes while a manufacturer's tool might do 100.

COMPUTER FAULT CODES

When an electronic problem occurs in a circuit that the computer senses or controls, a diagnostic trouble code (DTC) is stored in the computer's non-volatile RAM. Terms used in OBD II include warm-up cycle, trip, and drive cycle. A **warm-up cycle** occurs every time the engine cools off and temperature rises at least 40°F. Coolant temperature must reach at least 160°F. DTCs are usually erased after 40 warm-up cycles if it does recur during that time.

A *trip* is when the ignition switch has been off for a period of time and the engine is restarted. During driving, various emission control monitors on the vehicle operate. These include monitors for engine misfire, catalyst efficiency, fuel system, O_2 sensor, EGR, evaporative system, and air injection. Five of the monitors (not including the catalyst monitor) are required to operate to define one trip.

During a *drive cycle,* the engine must enter closed loop and all five of the trip monitors must operate. The catalyst monitor must also operate. To generate a DTC, the same fault must be detected during two drive cycles. The code will be stored in the computer's memory and will light the malfunction indicator lamp (MIL). If the monitors do not detect a fault for three consecutive drive cycles, the MIL will turn off.

Newer systems are very sophisticated and can store many trouble codes. A misfire will be stored immediately as a code. Vehicles manufactured since 1996 have two oxygen sensors and are capable of monitoring catalytic converter operation. The catalyst efficiency monitor will set a code if the same fault happens on three drive cycles. A pending DTC is a code that has not occurred enough times to light the MIL. Some scan tools can read pending DTCs.

The battery supplies the power to the computer for memory when the engine is off. This results in very little drain. Computer code memory draws less than 0.005 amp (5 mA).

COMPUTER SELF-DIAGNOSTICS

Computer systems have become very complicated. As they have progressed, their self-diagnostic capabilities have become much better. Computers can diagnose over 90% of the faults that occur in electronic systems. The malfunction indicator lamp comes on when the key is turned on and during engine cranking. It is supposed to go out shortly after the engine starts. If it is on when the car is running, there is probably a fault code stored in the computer.

Whenever the key is turned to the "on" position, the system does a *self-check.* Some faults detected during this test cause the light to come on. Others are not

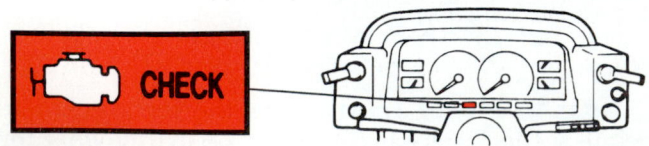

Figure 75.34 A malfunction indicator lamp on the dash display.

serious and are stored for later service. Hard faults are those codes that are present and are stored in memory at the time of the self test. Most of these codes will result in the MIL coming on and stay on. Others will only light the MIL while the problem is actually happening. An intermittent problem will set a code that will remain in memory. It might remain indefinitely or for up to fifty restart/warm-up cycles. This is called non-volatile RAM.

An **intermittent fault code** is one that only occurs occasionally for a short period of time and is not present in the system at the time of the fault test. Each system component has a number assigned to it as a fault code.

■ MULTIPLEXING

Multiplexing is a term used when several computers are linked together by one pair of wires twisted together to form what is called a *twisted pair* (Figure 75.35). This saves a great deal of wire. The computers share information. One sensor can provide information to several computers. A chip prevents digital information from being overlapped when it is transmitted through the twisted pair. Each signal starts with an identification code the chip uses to determine how high a priority the code has. When two computers are attempting to transmit at the same time, higher priority codes are transmitted before lower priority codes.

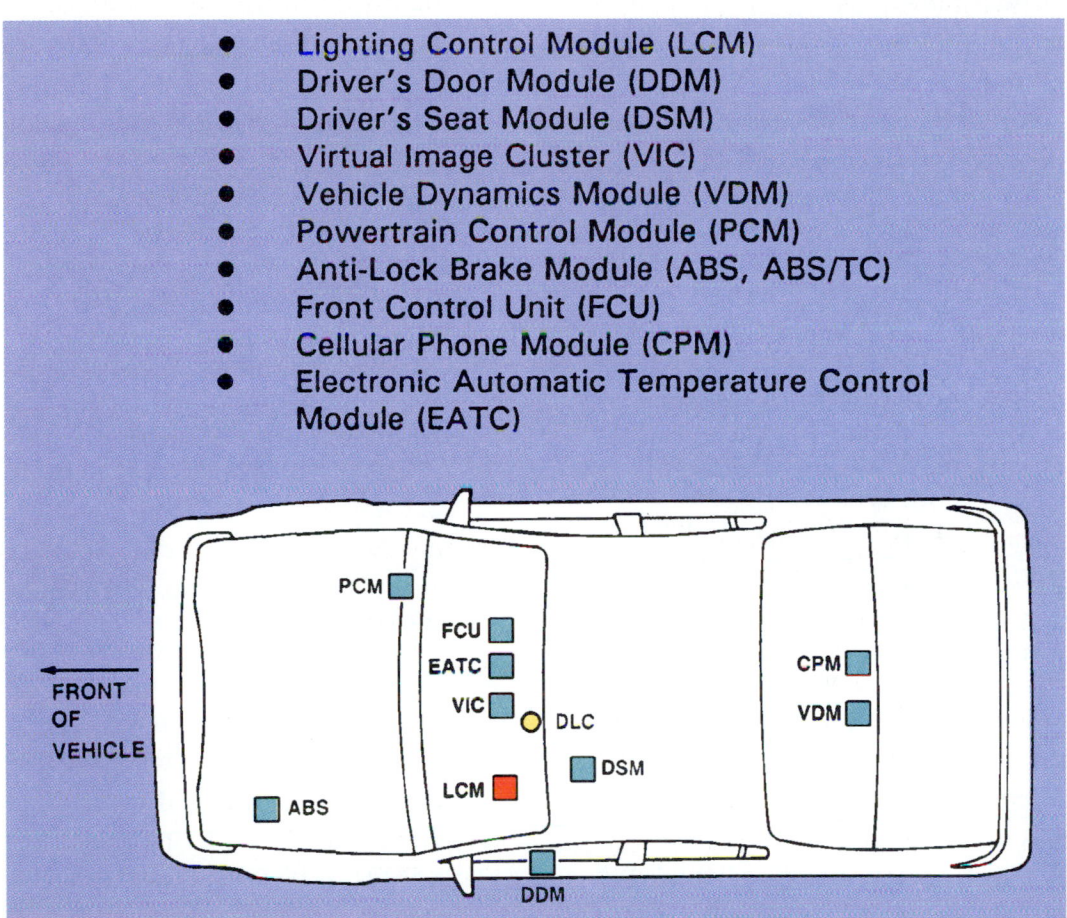

- Lighting Control Module (LCM)
- Driver's Door Module (DDM)
- Driver's Seat Module (DSM)
- Virtual Image Cluster (VIC)
- Vehicle Dynamics Module (VDM)
- Powertrain Control Module (PCM)
- Anti-Lock Brake Module (ABS, ABS/TC)
- Front Control Unit (FCU)
- Cellular Phone Module (CPM)
- Electronic Automatic Temperature Control Module (EATC)

Figure 75.35 This vehicle has ten computers. *(Courtesy of Ford Motor Company)*

■ REVIEW QUESTIONS

1. A _____ is a material that can be either a conductor or an insulator.

2. Conventional electrical theory as used in automobiles says that electricity flows from _____ to _____.

3. A small amount of current applied to a _____ causes it to relay larger amounts of electricity through it.

4. The three semiconductor crystal layers in a transistor are called the emitter, the base, and the _____.

5. A _____ diode is used to control a voltage regulator.

6. The three main parts to automotive computer systems are the computer, sensors, and _____.

7. If a temperature sensor's range is 0°F–250°F, each volt of difference will be interpreted by the computer as a _____° change in temperature.

8. The speed at which binary information is transmitted within the computer is called its _____ rate.

9. A _____ is a variable resistor made from semiconductor material whose resistance changes predictably as its temperature changes.

10. _____ is a term for the sharing of two wires for communication between several computers.

■ ASE STYLE REVIEW QUESTIONS

1. Technician A says the layer of a transistor that controls the switching function is called the base. Technician B says that according to electron theory, electricity flows from positive to negative. Who is right?
   - **a.** Technician A
   - **b.** Technician B
   - **c.** Both A and B
   - **d.** Neither A nor B

2. Technician A says that semiconductors can be either positively or negatively charged. Technician B says that some diodes conduct electricity from positive to negative and others conduct from negative to positive. Who is right?
   - **a.** Technician A
   - **b.** Technician B
   - **c.** Both A and B
   - **d.** Neither A nor B

3. Technician A says that voltage applied to a transistor causes the base layer to reverse its charge and become a conductor. Technician B says that a transistor that is forward biased allows current to flow. Who is right?
   - **a.** Technician A
   - **b.** Technician B
   - **c.** Both A and B
   - **d.** Neither A nor B

4. Technician A says that integrated circuits can carry large amounts of current. Technician B says that the programs for the computer are part of its hardware. Who is right?
   - **a.** Technician A
   - **b.** Technician B
   - **c.** Both A and B
   - **d.** Neither A nor B

5. Technician A says that computers usually control the power side of circuits. Technician B says that automotive computers use battery voltage for their reference voltage. Who is right?
   - **a.** Technician A
   - **b.** Technician B
   - **c.** Both A and B
   - **d.** Neither A nor B

6. Technician A says that temporarily stored information is located in RAM. Technician B says that permanently stored information that the computer refers to is called ROM. Who is right?
   - **a.** Technician A
   - **b.** Technician B
   - **c.** Both A and B
   - **d.** Neither A nor B

7. Technician A says that an EPROM can be reprogrammed in the car. Technician B says that actuators change electricity into mechanical work. Who is right?
   - **a.** Technician A
   - **b.** Technician B
   - **c.** Both A and B
   - **d.** Neither A nor B

8. Technician A says that piezoelectric crystals develop voltage in response to pressure applied to them. Technician B says that intermittent faults are those codes that are present and are stored in memory. Who is right?
   - **a.** Technician A
   - **b.** Technician B
   - **c.** Both A and B
   - **d.** Neither A nor B

9. Technician A says that when there is no voltage the computer sees a one. Technician B says that keep alive memory (KAM) means the computer maintains power to ROM when the ignition switch is off. Who is right?
   - **a.** Technician A
   - **b.** Technician B
   - **c.** Both A and B
   - **d.** Neither A nor B

10. Technician A says that a transistor has two P-N junctions. Technician B says that a diode has one P-N junction. Who is right?
   - **a.** Technician A
   - **b.** Technician B
   - **c.** Both A and B
   - **d.** Neither A nor B

Electronic and Computer System Service

■ INTRODUCTION

Computer systems have become very sophisticated. In order to repair complicated systems, you will need to consult a model-specific repair manual. They include step-by-step procedures for troubleshooting systems. The aim in this chapter is to cover items in a generic way so that you will have an understanding of the ways that the various general types of sensors and actuators are tested. You should read the fuel and emission systems chapters in this book before you study the information presented here.

■ INSPECTION SEQUENCE

Computers have self-diagnostic ability. The newer the computer, the more accurately it can diagnose problems. A simple problem like a loose or corroded wiring connection can cause problems that the computer tries to correct by compensating with other changes. This can make it seem like the computer is at fault.

When diagnosing all cars with computers, a logical diagnosis sequence must be followed before checking the computer. You cannot overlook the basics and just start replacing expensive electronic parts until a fix is made. Parts stores will usually not accept the return of electronic parts. Be sure to carefully question the customer about the symptoms and when they occur.

On older cars, mechanics could often make successful repairs by jumping to quick conclusions based on past experience. On newer cars, the computer responds to information it receives from the sensor inputs. The sensors are reacting to changes in engine operating conditions. Be careful to eliminate all of the mechanical conditions first before going on to a computer diagnosis. A large majority of cars will not actually have computer problems but can be fixed during the inspection and maintenance process.

Problems in a computer system are isolated using visual and diagnostic checks, a digital voltmeter, an ohmmeter, or a scan tool. Most sensors and actuators have specifications for resistance measurement. Voltage to and from devices can also be tested. A mechanical failure can also occur in come cases. Individual devices are not tested until other test procedures point in the direction of that area.

High Impedance Meters

A digital multimeter (DMM), also called a digital volt/ohmmeter (DVOM) (see Figure 29.41), is the instrument used to measure electricity in electronic circuits. A **high impedance voltmeter** must be used to perform tests on these systems. The integrated circuits in computer systems operate on very small amounts of current. An analog meter (one with a needle) (see Figure 29.38) that operates on magnetism can load down a computer circuit and actually change what is happening in the circuit. The meter seems like a short in the circuit, offering an easier path for electrical flow.

A high impedance meter has very high input resistance (usually about 10 million ohms). It has a digital display instead of a needle. The high input resistance prevents the meter from drawing current while it is connected to a circuit. This protects the circuit and keeps readings accurate.

Other advantages of DMMs are:

- They are often self-scaling (adjustments are not required).
- They do not have to be connected in the correct polarity (a plus or minus ID is displayed on the screen with the reading).

To check a meter to see if it has high impedance, turn the selector to the DC volts position. Measure the resistance through it with another ohmmeter. A meter of high enough impedence to use in automotive electronic circuits will read less than 1.5 ohms.

Visual Inspection

Electronic systems are quite dependable. The cause of a problem can often be determined during a careful visual inspection. If the engine runs but has poor driveability, perform a visual inspection under the hood for disconnected or broken vacuum lines or electrical connections. While the engine runs listen for vacuum leaks. Inspect vacuum hoses for horizontal cracks. High vacuum can cause a crack to suck together. Hard neoprene hoses can develop a restriction bubble. The result is poor throttle response due to trapped vacuum. Make sure vacuum hoses are routed correctly.

■ PERFORM DIAGNOSTIC TESTS

Analyze the cause of the problem rather than just fixing the problem's result. If there are no codes stored in memory, check for fuel or ignition problems with the scope and infrared analyzer. If a problem is in one of those areas, check any sensors that might contribute. Remember, a test drive takes time to perform. If you can solve the problem in a few short minutes in the service bay, you will be more productive.

Check Engine Condition

As a quick check of compression, listen during cranking for an even rhythm and then for a smooth idle. A vacuum gauge (see Figure 42.3) is easy to connect and can be used to qualify engine condition. It should produce between 17 and 21 inches of steady vacuum at idle and 2500 rpm.

An EGR valve is not open at idle during normal operation. A hand vacuum pump can be used to check that the EGR valve remains closed at idle. When the valve is opened by the vacuum pump, the engine should stumble and run roughly. This is due to the lean condition created by the open EGR valve.

Engines with electronic engine controls have catalytic converters. If a cylinder has a regular misfire, the converter will become very hot as it attempts to catalyze the unburned fuel that results. After a power balance test, run the engine at 1500 to 2000 rpm for 30 seconds to run air through the converter to give it a chance to get rid of extra fuel.

Ignition Advance Checks

Check the base timing setting. Follow manufacturer's instructions for disconnecting the computer for this test. Otherwise, even in open loop (when the car is cold) the computer will provide extra ignition advance. If base ignition timing is not within 3°, adjust it. Use an adjustable timing light (see Figure 39.15) to check total advance to confirm operation of the computer. Total ignition advance is usually 15° to 25° at idle. At 2500 rpm, expect to find 30° or more of ignition advance.

Idle Speed

On old cars that had ignition points, tuneup work included adjustment. That type of maintenance is mostly gone, especially with today's domestic cars. With computerized electronic fuel injection, idle speed is more important because it can cause changes in other settings. To set base idle speed, follow manufacturers' directions. When rpm is too low, the engine can stall. When idle rpm is too high, a surge can result. Most throttle position sensors are no longer adjustable except for the throttle stop that prevents the throttle plate from binding up when all the way closed.

Charging System Test

A charging system test is one of the core things to do before beginning a diagnostic procedure. A bad diode in an alternator can cause driveability problems and set other trouble codes.

If the battery is not fully charged, the computer system will still work, but related problems can affect the operation of the engine. For electronic fuel injection to work properly, new cars require a minimum of 11 volts. Older fuel injection systems would work on 8 to 9 volts. If voltage drops below about 11.6 V, the computer will raise the engine's idle to increase the alternator's charging rate.

NOTE: *When the charging system voltage borders on this amount, stepping on the brakes will turn on the brake lights, causing the computer to raise idle.*

When voltage drops below 10 volts, the system can go into limp-in mode. Normal computer operation will be lost, resulting in poor driveability.

■ ONBOARD DIAGNOSTICS

Computer systems today are very complicated. It would be very difficult and time consuming to diagnose problems if the computer did not have its own memory and self-diagnostic capability. Computers can detect incorrect electrical conditions and save diagnostic trouble codes to memory. On later model cars, there are codes for many malfunctions that the computer can diagnose.

Each time the key is turned on, the computer does a *self-check* of its circuits. The computer sends a test voltage signal to the sensors and actuators, checking for continuity and return signal voltage. Sometimes, a value shows up during the self-check that is outside of expectations. When this happens, the computer will either store a code to provide service information to a technician or it will turn on the malfunction indicator light (MIL) on newer cars that have them. The malfunction indicator light is also called a *check engine light*. MIL is the OBD II term.

Check to see if the MIL is lit when the engine is running. If it is, check for hard fault codes.

- A hard fault is one that is present at the time of the self-test.
- An intermittent fault is one that occurs occasionally, but does not remain at all times. It is the hardest to fix.

When the key is turned on, the MIL should come on for a few seconds and remain on during engine cranking. This provides a check of the bulb and circuit. The MIL should go out shortly after the engine starts. If it comes on when the engine runs, there should be a trouble code stored in the computer's memory.

NOTE: *The codes are only a starting point. They do not tell you what to replace. They simply inform you that something out of the ordinary has occurred within a certain circuit.*

To isolate a fault, use the correct test instrument and service information. A **diagnostic tree**, or flow chart, in the service manual provides a step-by-step diagnostic procedure to follow when troubleshooting hard fault problems (Figure 76.1).

A common technician error is to replace a component at the bottom of the flow chart. Always follow the tree in order. Going out of sequence can cause you to miss the problem. This is like following a road map. If you miss a turn at a street, you will be lost. If the last step on the tree indicates that the computer is the problem, be sure to thoroughly test power and computer grounds before replacing it.

Sensors are more often the cause of electronic control problems than actuators are. The logical test sequence would therefore be to check the sensors and wiring before testing actuators and wiring. The computer is tested last. Systems that do not have onboard diagnostics are diagnosed using a process of elimination. If there is a test procedure for a component, it is tested to eliminate it as a cause of a problem.

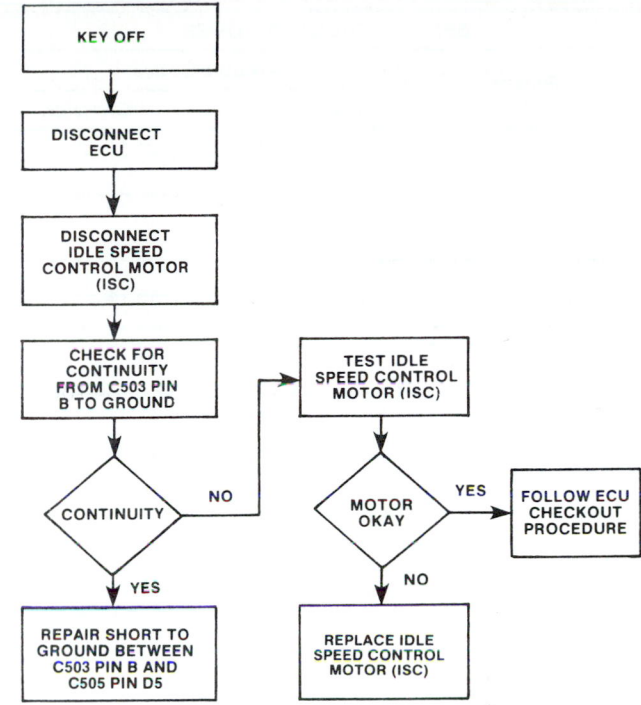

Figure 76.1 A diagnostic tree. *(Courtesy of Chrysler Corporation)*

■ READING TROUBLE CODES

There are different ways to read trouble codes, depending on the year and make of car. Modern OBD II systems have standardized connectors and procedures. Earlier systems had different procedures for retrieving codes.

Most systems have a diagnostic connector, called a **data link connector (DLC)** (Figure 76.2). A test instrument can be connected to it to read codes. The procedure for retrieving fault codes varies. The service manual gives the procedure for the particular make of vehicle and describes what each trouble code means.

Code numbers on OBD II cars are all the same for a particular sensor. Earlier cars had different codes. For instance, a coolant temperature sensor would be a code 14 or 15 on a GM car; a code 21, 51, or 61 on a Ford; and a code 17 or 22 on a Chrysler. Some manufacturers' trouble code listings are shown in Figure 76.3. When there is more than one trouble code in memory, the computer will give the lowest number first.

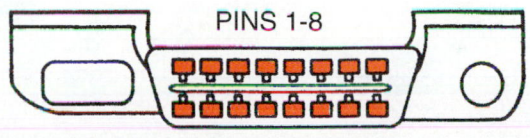

Figure 76.2 A standard OBDII data link connector (DLC). *(Courtesy of Ford Motor Company)*

SENSOR TROUBLE CODES			
Sensor	**Trouble Codes**		
	GM	**Ford**	**Chrysler**
Oxygen (EGO)	13,44, 45,55	43,91, 92,93	21,51,52
Throttle Position (TPS)	21,22	23,53, 63,73	24
Engine Vacuum (MAP)	31,33,34	22,72	13,14
Barometric Pressure (BARO)	32	—	37
Coolant (ECT)	14,15	21,51,61	17,22
Knock	42,43	25	17 (some only)
Vehicle Speed (VSS)	24	—	15
Air Temperature (MAT, VAT, ACT)	23,25	24,54,64	23
Airflow (VAF, MAF)	33,34, 44,45	26,56, 66,76	—
EGR Valve (EGR, EVP)	—	31,32,33, 34,83,84	31

Figure 76.3 A comparison of trouble codes from different manufacturers. *(Courtesy of Babcox Publications)*

Older systems call for using a test light, a voltmeter, or a more sophisticated scan tool to read codes. Some imported cars flash an LED located on the side of the computer. Others flash the check engine light on the dashboard. Some cars can display codes digitally on the dashboard display. The easiest way to read codes in all systems is to use a scan tool when possible (Figure 76.4).

OBD II Codes

OBD II codes have five characters.
- The first is a letter identifying the area of the vehicle the code relates to. For instance, B is for body, C is for chassis, and P is for powertrain.

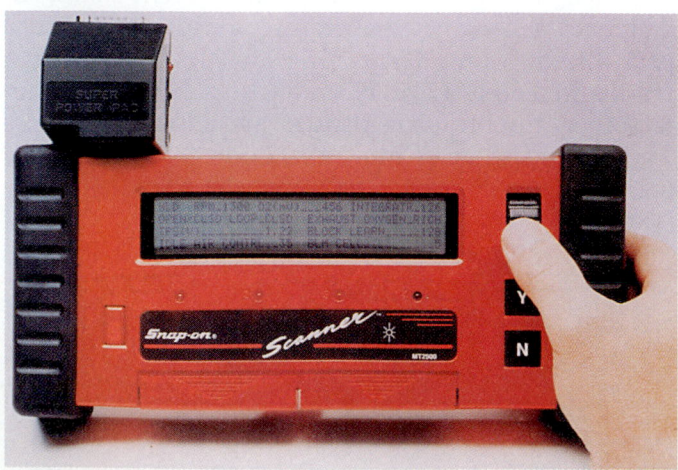

Figure 76.4 A handheld scan tool.

- The second character is a number from 0 to 3. Zero is an SAE code, 1 is a code assigned by the vehicle manufacturer, and 2 and 3 are reserved for future use.
- The third character is a number from 1 to 8 representing a system of the vehicle such as anti-lock brakes.
- The last two characters are numbers that represent the actual fault code listing the sensor or actuator circuit where the problem has occurred.

■ SCAN TOOLS

A scan tool is a portable computer that reads data from the car's onboard computer. It is connected to the vehicle's diagnostic connector and provides quick access to diagnostic information. Figure 76.5 shows typical scan tool input/output data.

Late model vehicles have *live serial data streams* that can be accessed through the data link connector. Some earlier computer systems had live data streams and others did not. If there is no data stream, the scan tool is simply an expensive way of reading trouble codes.

When interpreting information from the data stream, only some of the information available is useful. The most important use of the information is for comparing related sensor readings. For instance, readings on a fuel-injected engine related to idle rpm and an idle air control motor can be compared to see if they are within acceptable limits.

Software Cartridges

Service manuals have detailed step-by-step procedures for diagnosis. On newer scan tools, interchangeable

```
1993 FORD TRUCK               A/C

3.0L V6 - EECIV EFI           A/T

  ** SCROLL FOR DATA. OK TO DRIVE. **

   RPM_2544 O2S1(MV)_558 O2S2(MV)___133

TP=TPS(V) _____1.77    TP MODE _____P/T
TPCT _____0.9       ECT(V) _____0.48
ECT(dF) _____214     IAT=ACT(V) _____2.19
IAT-ACT(DF) ____104     IDLE AIR(%) ___98.8
MASS.AIR(V) ___2.93     DPFE( V) _____2.81
EVR(%) _____63.2     INJ PW1(MS) __11.60
INJ PW2(MS) __11.68     SFTRIM 1(%) ____2.5
SFTRIM 2(%) ____0.5     LFTRIM 1(%) ___-6.2
LFTRIM 2(%) ___-6.2     VPWR=BATT(V) __14.0
VREF(V) _____4.97     SPARK ADV(D) ____29
WAC=WOT A/C _____ON    FP=FUEL PUMP ____ON
CANP=PURGE _____ON    VEH SPEED (MPH) _30
PARK/NEU POS _-R-DL     BOO=BRAKE SW ___OFF
OPEN/CLSDLOOP _CLSD     ACCS=A/C _____OFF
```

Figure 76.5 Input and output data from a 1993 Ford Truck using a Snap-on Scan Tool and DCL cartridge.

program software cartridges contain service tips and bulletins, as well as specifications for the particular vehicle (Figure 76.6). The cartridges can be updated each year. This takes the place of manuals, saving the technician time that would have been spent looking up specifications. The most probable cause of a problem is given first.

Scan Tool Features

The scan tool is handheld and can be taken on a road test. While on the road, live serial data can be read, or when an intermittent problem occurs information can be saved for later interpretation. A scan tool can also be connected to a printer so that a hard copy of the data can be produced. This can be used to verify that a problem has been corrected. The data can also be saved to a personal computer for later reference.

Scan Tool Limitations

The scan tool is limited to diagnosing computer problems. Systems or parts of systems not controlled by the computer will not be diagnosed by the scan tool. When the scan tool says there is a problem, that just means to look in that area for a fault. A code could have been mistakenly set during an earlier repair or there could be a problem with the wiring in the circuit. The computer could also give a command to an actuator, but a problem with the actuator prevents it from carrying out the command. Symptoms must be followed up to diagnose these problems.

Bi-Directional Communications

The link between the scan tool and the computer can be uni-directional or bi-directional. **Uni-directional** means that the scan tool can read data, but cannot give commands to the computer. **Bi-directional** systems can receive commands from the scan tool. Anything that is computer controlled can be operated from the scan tool. For instance, the left window motor can be operated, one headlight can be turned on, or the horn can be honked. Engine actuators can also be operated to check their function. If they function correctly, the problem must be related to the signal on the input side.

■ BREAKOUT BOX

When a car does not have a data stream, a *breakout box* can sometimes be used to diagnose problems (Figure 76.7). The probes of a DMM are inserted into the pin holes in the breakout box to access various circuits' sensors and actuators through the pin connector to the computer. A breakout box gives a way of reading what the raw values within the system are. You cannot read processed data.

■ RETRIEVING TROUBLE CODES

Reading Chrysler Codes

Chrysler's procedure for retrieving codes is to cycle the ignition switch on three times within 5 seconds without starting the engine. The MIL will glow for a short time to test the bulb. Then, it begins to flash. The first set of flashes is counted as tens. After a short pause, the light flashes again. These flashes are counted as ones. The two sets of flashes are added together to get the trouble code. For instance, if the number of first flashes is two and the number of second flashes is four, this is a code 24, or throttle position sensor.

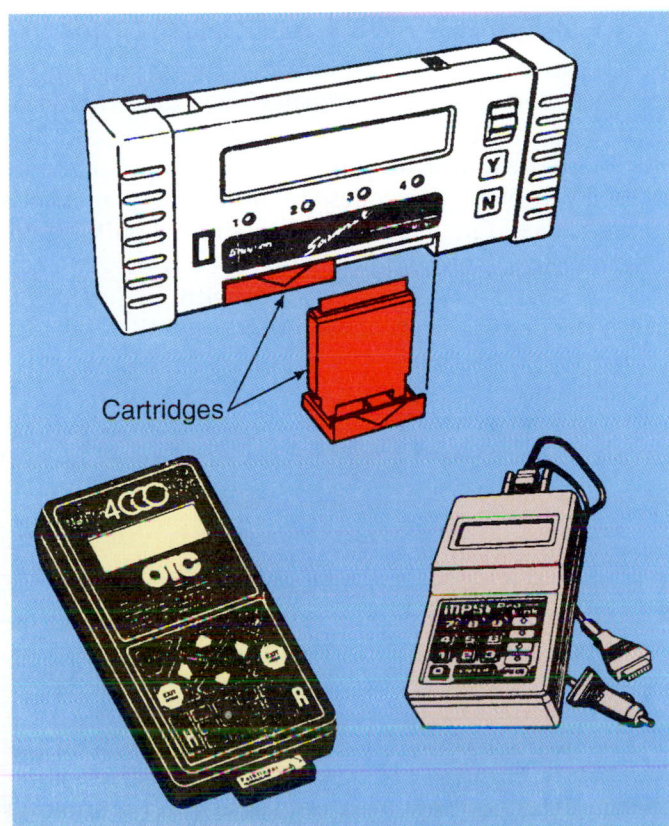

Figure 76.6 Scan tools have software cartridges.

Cartridges

Figure 76.7 A breakout box has pin holes for accessing various sensors and actuators. *(Courtesy of OTC-SPX Corp. Aftermarket Tool & Equipment Group)*

Reading GM Codes

On GM cars, stored codes are given in two digits. With the key on and engine off, shorting between the A and B terminals in the diagnostic connector results in flashes of light that are counted to determine the code. An indicator code will tell you that the system is ready to start flashing codes. Another indicator code lets you know that all of the codes have been shown. The computer will then repeat the showing of the codes. As an example, one flash followed by a pause and then two flashes would be a code 12 (Figure 76.8).

Reading Ford Codes

On Ford cars, codes can be read with an analog voltmeter or a special tester from a self-test connector in the engine compartment (Figure 76.9). You can use an analog voltmeter, counting the number of sweeps of the needle. During a sweep, the needle moves across the entire dial when on a 12-volt scale. A 2-second pause indicates the first number is finished. If the nee-

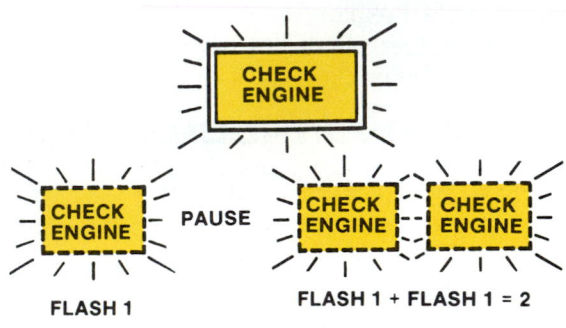

DIAGNOSTIC CODE DISPLAY

Figure 76.8   The check engine light is flashing a code 12. *(Courtesy of General Motors Corporation, Service Technology Group)*

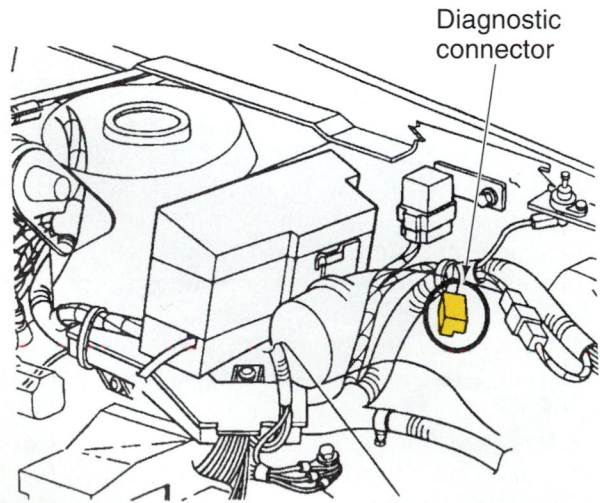

Figure 76.9   A data link connector in the engine compartment. *(Courtesy of Chrysler Corporation)*

dle moved two times before the pause, the first digit of the code is a two. After the 2-second pause, the needle moves again to count out the second digit of the code. This is followed by a 4-second pause. One more sweep of the needle would indicate a 1 for the second digit. This would be a code 21, a coolant temperature sensor. The code will repeat itself one more time before going on to the next code.

Figure 76.10 shows a code 21, followed by a code 11. The first code (21) is a hard code. A second code is a memory code (intermittent). Code 11 means no memory codes are stored at this time.

Reading OBD II Codes

On OBD II cars, a scan tool must be used to read codes. The scan tool can also be used on older models, but it simply reads codes. It cannot read serial data if the system did not have that available.

■ WORKING WITH CODES

When more than one code is given, fix the lower number code first. A lower code sets a higher code. The lower number code could have caused the higher number code(s) to set. Fix the problem first, then start again working with the lower number codes. Be sure to check power and grounds. After fixing the lower number code first, erase the codes and test drive the car to reset codes from any hard faults that are still present. Then, read the codes again.

■ ERASING TROUBLE CODES

A code can remain in memory, even though a problem has been corrected. After repairs have been made in response to codes stored in memory, clear the codes. A procedure that shuts off power to the computer can be followed to erase the codes. Manufacturers' methods for this vary. On some systems, simply pulling the computer fuse from the fuse panel for a short time can do it. A scan tool can erase codes without having to disconnect anything. Disconnecting the battery ground cable will erase the computer's memory, but will also erase the memory of the clock, radio, power seats, and other things. After erasing the codes, test drive the car to see if it still "sets codes."

Test Drive

Before taking a car on a test drive:
- Write down any codes that are present.
- On cars with over 50,000 miles, check the backside of the throttle plate for carbon or gum buildup. The buildup results from the cumulative effects of blow-by that can cause driveability problems.

  On adaptive strategy systems, disconnect the negative battery terminal for 10 minutes to clear computer volatile memory. This is so the system can reprogram itself. During a failure of part of the system, it might have relearned to compensate for a problem. Erasing

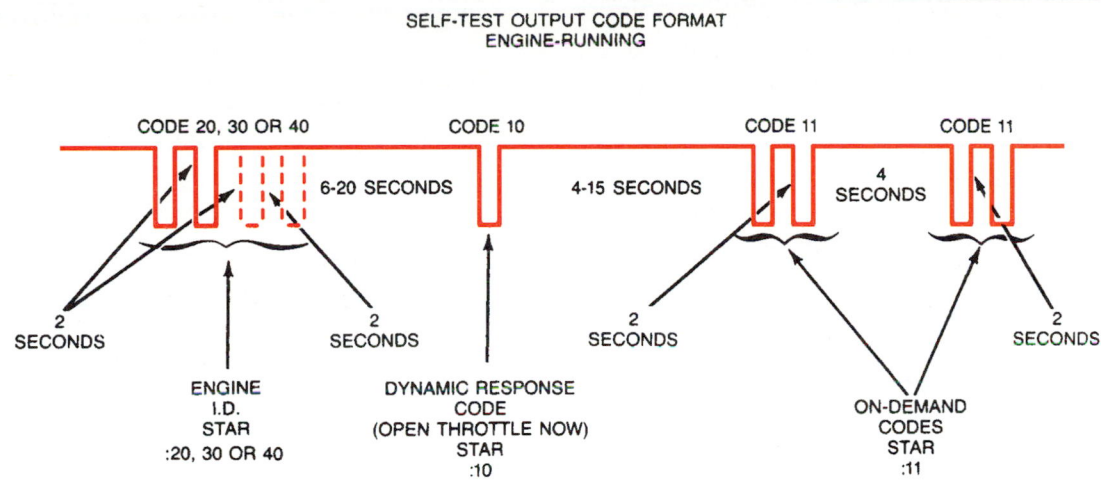

Figure 76.10 A Ford code 21, followed by a code 11. *(Courtesy of Ford Motor Company)*

the volatile memory is done before fixing computer problems. Be sure to make a note of radio stations and seat positions so you can reprogram them for the customer.

Test drive the car to see if any of the codes return. These are hard faults. During the test drive, try to duplicate the conditions that the customer highlighted in their complaint. If the problem cannot be duplicated, this is an intermittent problem.

■ SCAN TOOL SNAPSHOT

The scan tool has a feature like a flight recorder in an airplane. Information is frozen, even if the car has not set a code. This feature is handy for catching glitches and intermittent problems know as soft faults. A **snapshot**, also called an *event* or a *movie,* is like a roll of film. Each picture is a frame. Scan tools vary on the number of pictures that can be recorded. This depends on the computer and the individual scan tool.

A typical snapshot would be as follows. Imagine there are 61 frames on the roll of film. During a test drive, you feel an intermittent problem and press the record button or any number key on the scan tool. The tool freezes (memorizes) the information in the 61 frames that the machine reads, some before and some after you pressed the button.

There is also a setting that automatically records when any fault code occurs during the test drive. The tool can also be programmed to only record when a particular fault code sets. Some tools allow for multiple snapshots, while others will only record one.

Back at the service bay, you can read the information (Figure 76.11). Go back 10 frames from zero. Zero is when you pushed the button. A throttle position sensor (TPS) might show a value that is off 0.5 V or so during frame changes in the movie. A TPS is a potentiometer with a wiper and a coil. A glitch in the wiper coil could be the cause of the problem. If you do not

find a problem, go 10 frames ahead of zero. Most problems will be found in this 21-frame span. If a problem is not found, look for an *engine* problem that you might have missed.

Before a test drive with a scan tool attached, be sure to follow manufacturers instructions. A GM car that is driven in the wrong test mode can be damaged. Also, always check the **key-on, engine off (KOEO)** data before running the engine. You might find a 4.5-volt signal from a throttle position sensor, instead of 0.5 volt. This would happen if there was an open in the ground circuit to the TPS. It would have no way to reduce the reference voltage sent by the computer. The computer sees the throttle as being wide open and shuts off fuel while trying to "clear flood." The engine will not start.

Practicing with the scan tool and making comparisons with specifications will give you an understanding of the various acceptable readings to be expected.

■ CONFIRM CLOSED LOOP

The computer requires the correct inputs from sensors and the correct actions from actuators to be able to function properly. When the engine is cold, the system is in open loop operation (see Chapter 26). During this time, system operation is just like it was with older fuel injection and ignition timing computers. It watches all monitored signal, ignoring the O_2 sensor. Then it operates the fuel and ignition systems according to pre-programmed parameters.

When the engine has warmed up and goes into closed loop, the computer system receives a good deal of information from sensors and decides what action the actuators should take. The result is the best driveability, emission control, and fuel economy for the condition at the time.

Three things must occur for a computer system to go into closed loop. The O_2 sensor must be at operating

```
91VR3301 E-Z Event Plus
Chrysler 1991 3.3/3.8 MFI VIN
Event 1 Tag — Frame -30
Fault Codes: No

                   -15    -14    -13    -12    -11    -10    -9     -8     -7     -6     -5     -4     -3     -2     -1     0
Coolant Temp       120    120    120    120    120    120    120    120    120    120    120    120    120    120    120    120
RPM                0000   0000   0000   0000   0000   0000   0000   0000   0000   0000   0000   0000   0000   0000   0000   0000
Inj Pulse Width    14.7   14.7   14.7   14.7   14.7   14.7   14.7   14.7   14.7   14.7   14.7   14.7   14.7   14.7   14.7   14.7
MAP                29.4   29.4   29.4   29.4   29.4   29.4   29.4   29.4   29.4   29.4   29.4   29.4   29.4   29.4   29.4   29.4
O2 Volts           0.50   0.50   0.50   0.50   0.50   0.50   0.50   0.50   0.50   0.50   0.50   0.50   0.50   0.50   0.50   0.50
Throt Pos Sensor   0.66   0.66   0.66   0.66   0.66   0.66   0.66   0.66   0.66   0.66   0.66   0.66   0.66   0.66   0.66   0.66
Adaptive Fuel Fac  -03    -03    -03    -03    -03    -03    -03    -03    -03    -03    -03    -03    -03    -03    -03    -03
Baro Pressure      29.4   29.4   29.4   29.4   29.4   29.4   29.4   29.4   29.4   29.4   29.4   29.4   29.4   29.4   29.4   29.4
Baro Pressure Vol  4.76   4.76   4.76   4.76   4.76   4.76   4.76   4.76   4.76   4.76   4.76   4.76   4.76   4.76   4.76   4.76
Battery Temp       065    065    065    065    065    065    065    065    065    065    065    065    065    065    065    065
Battery Volts      12.2   12.2   12.2   12.2   12.2   12.2   12.2   12.2   12.2   12.2   12.2   12.2   12.2   12.2   12.2   12.2
Charge Temp        105    105    105    105    105    105    105    105    105    105    105    105    105    105    105    105
Charg Sys Targ Vol 13.9   13.9   13.9   13.9   13.9   13.9   13.9   13.9   13.9   13.9   13.9   13.9   13.9   13.9   13.9   13.9
Cruise Control Cut KEY    KEY    KEY    KEY    KEY    KEY    KEY    KEY    KEY    KEY    KEY    KEY    KEY    KEY    KEY    KEY
Cruise Control     OFF    OFF    OFF    OFF    OFF    OFF    OFF    OFF    OFF    OFF    OFF    OFF    OFF    OFF    OFF    OFF
Cruise Control Sta CRSW   CRSW   CRSW   CRSW   CRSW   CRSW   CRSW   CRSW   CRSW   CRSW   CRSW   CRSW   CRSW   CRSW   CRSW   CRSW
Cruise Set Speed   000    000    000    000    000    000    000    000    000    000    000    000    000    000    000    000
Dis Cam Status     NO     NO     NO     NO     NO     NO     NO     NO     NO     NO     NO     NO     NO     NO     NO     NO
Dis Crank Status   NO     NO     NO     NO     NO     NO     NO     NO     NO     NO     NO     NO     NO     NO     NO     NO
Electronic Adv/Re  016    016    016    016    016    016    016    016    016    016    016    016    016    016    016    016
Idle Air Control   041    041    041    041    041    041    041    041    041    041    041    041    041    041    041    041
Idle RPM x 10      078    078    078    078    078    078    078    078    078    078    078    078    078    078    078    078
Knock Active       NO     NO     NO     NO     NO     NO     NO     NO     NO     NO     NO     NO     NO     NO     NO     NO
Knock Volts        0.00   0.00   0.00   0.00   0.00   0.00   0.00   0.00   0.00   0.00   0.00   0.00   0.00   0.00   0.00   0.00
Map Volts          4.56   4.56   4.56   4.56   4.56   4.56   4.56   4.56   4.56   4.56   4.56   4.56   4.56   4.56   4.56   4.56
TPS-Minimum        0.84   0.84   0.84   0.84   0.84   0.84   0.84   0.84   0.84   0.84   0.84   0.84   0.84   0.84   0.84   0.84
Vacuum             00 I   00 I   00 I   00 I   00 I   00 I   00 I   00 I   00 I   00 I   00 I   00 I   00 I   00 I   00 I   00 I
Vehicle Speed Sen  000    000    000    000    000    000    000    000    000    000    000    000    000    000    000    000
```

Figure 76.11 A snapshot captured with an OTC scan tool.

temperature and start sending signals. Engine coolant must reach a specified minimum temperature and a specified period of time must have passed after the engine is first started. Coolant temperature and time period vary between manufacturers.

When a sensor whose input is needed for the engine to run properly fails to give the correct signal, the computer has a fail-safe or limp-in mode that allows the engine to continue to operate. It will not run at its best, but at least it will run. If the coolant temperature sensor or the oxygen sensor stops working, the system returns to open loop operation. In open loop, the fuel mixture will be slightly rich and will not vary. Emissions will be higher and fuel economy will be lower.

A faulty thermostat is a common reason for an engine not to operate in closed loop. A thermostat might be stuck so that it does not close all of the way. This can result in too slow a warmup when the car is not driven far enough on a normal trip. Because the engine does not warm up, the computer does not go into closed loop and give feedback to the fuel system. Some people mistakenly install a lower temperature thermostat in an attempt to solve an overheating problem. This causes the same result; no closed loop operation.

■ CHECK FOR CLOSED LOOP

If the system is working properly in closed loop the rest of the computer system is probably working well. Methods for confirming that the vehicle goes into closed loop vary with manufacturers and whether an engine has fuel injection or a carburetor. A digital multimeter (DMM), a dwell meter, or a scan tool are the three most common ways.

An artificially rich condition can be created by restricting the air intake. The computer should compensate by reducing the amount of fuel from the carburetor or fuel injectors. An artificially lean condition can be created by removing a vacuum hose. The computer should "drive" the fuel system rich.

Multimeter Test for Closed Loop

With a digital multimeter (DMM), hook a jumper wire in series with the connector on the oxygen sensor (Figure 76.12). Connect the positive lead to the jumper wire and the negative lead to a good ground. Set the meter to DC volts.

NOTE: *Do **not** use an ohmmeter to test an O_2 sensor. To measure resistance in a circuit, an ohmmeter puts voltage through the circuit. Putting voltage into an O_2 sensor can change the reading it responds to.*

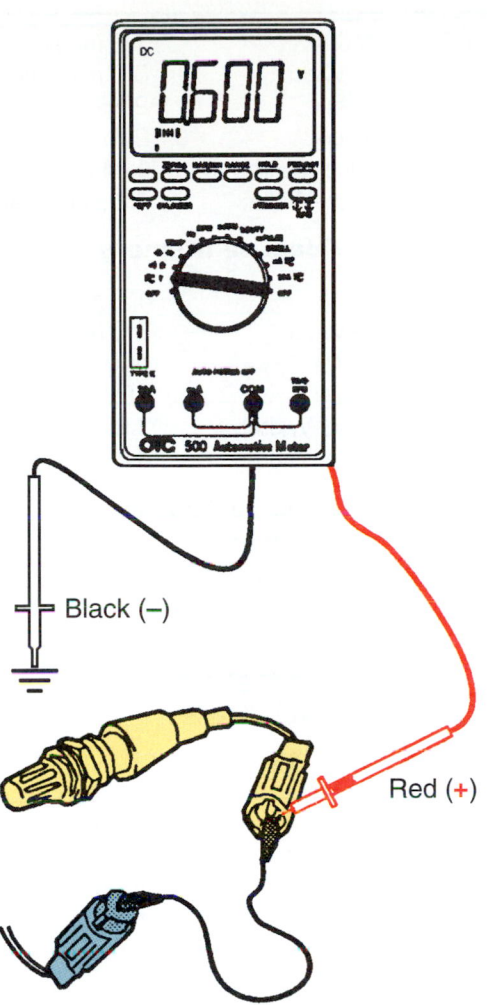

Figure 76.12 Monitoring O₂ sensor voltage. *(Courtesy of OTC-SPX Corp. Aftermarket Tool & Equipment Group)*

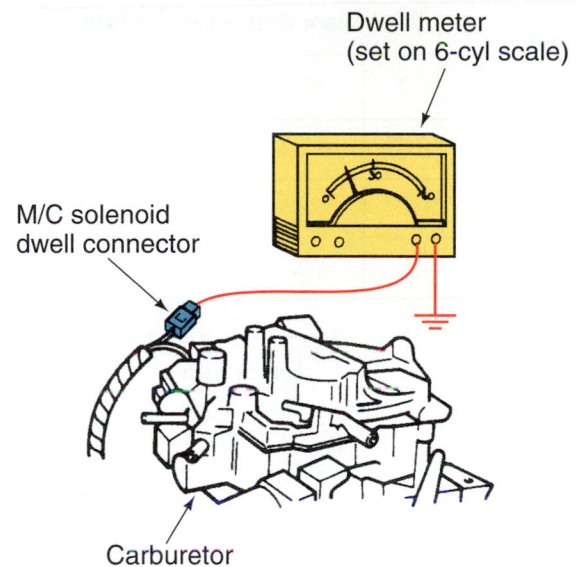

Figure 76.13 Using a dwellmeter to test for closed loop on a feedback carburetor. *(Courtesy of General Motors Corporation, Service Technology Group)*

Firmly apply the parking brake and run the engine at 2000 to 2500 rpm until the cooling fan comes on. On a fuel-injected car, when the vehicle goes into closed loop, the voltage will rapidly swing above and below 0.5 volt as it goes from rich to lean. Although you will not be able to see this with the DMM, a good system will switch between rich and lean several times a second. The speed at which this switching takes place is called cross counts. Late model computers will pick up an O_2 sensor that is of low quality or has become slower and set a code.

A slow O_2 sensor can cause a stumble during acceleration. After a period of slow operation, the oxygen sensor fails. When this happens, it usually goes rich or stays in open loop. The result is worse fuel economy.

Dwell Meter Test for Closed Loop

Carbureted cars can be checked with a dwell meter, voltmeter, or duty-cycle meter when the carburetor

has a two-wire mixture control solenoid. The meter is connected to the ground side of the solenoid. On a six-cylinder scale, the needle should float between 25° and 35° (Figure 76.13). If the dwell reading is above 30°, the computer is reducing fuel to compensate for a rich mixture. If it is below 30°, the computer is adding fuel. If the reading is above 54° or below 6°, the computer is not able to compensate for an excessively rich or lean condition. When there is a low reading, the computer sees lean and drives the system rich.

Scan Tool Test for Closed Loop

With a scan tool, data is displayed showing that feedback fuel control is occurring. Manufacturers use different names for fuel control. These include *mixture control/dwell*, *Integrator and Block Learn*, *Short and Long Term Fuel Trim*, and *Adaptive Memory and Additive Fuel Factor*. These systems allow the computer to change the air/fuel mixture average to either side of stoichiometric to compensate for small problems.

■ SCAN TOOL DIAGNOSIS OF O₂ FEEDBACK

Analyzing scan tool data from these systems is a good diagnostic tool. On a GM Block Integrator and Block Learn system, a reading of 128 on the scan tool is stoichiometric. Lower numbers mean the computer is attempting to correct for a rich condition. Higher numbers mean that the computer is correcting for a lean condition. The farther away the number is from 128, the worse the condition is. The system can only make a certain amount of correction. This is limited to a range of about 100 to 150. Block Learn is the long-term

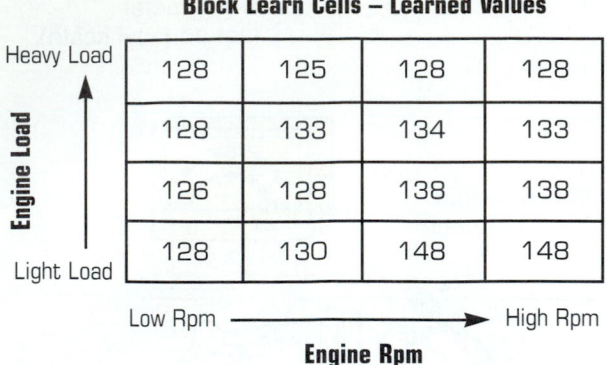

Block Learn Cells – Learned Values

Engine Load	Low Rpm → High Rpm			
Heavy Load	128	125	128	128
	128	133	134	133
	126	128	138	138
Light Load	128	130	148	148

Engine Rpm

Figure 76.14 Long term corrections are saved in the block learn cells.

memory part of the system (Figure 76.14). Block Integrator is short term only. It is part of volatile RAM so it is lost each time the key is turned off.

Ford's Short and Long Term Fuel Trim works the same way, but the reading is a percentage of correction. The equivalent to GM's 128 is 0%. A positive percentage indicates that the computer is adding fuel to compensate for a lean exhaust condition. A rich condition causes a negative percentage. If the short term trim cannot correct a problem, then the long term trim will try to bring the mixture back to 0%.

On Chrysler systems, the Adaptive Memory is figured into the injector pulse width reading as a positive percentage if it is lean and a negative percentage if it is rich (Figure 76.15). It changes with engine speed and load. The scan tool will show the long-term change as Additive Adaptive in microseconds. This results in a lengthening of the corrected pulse width reading in milliseconds (Figure 76.16). This system can make the largest change when the pulse width is low.

Some systems use dual oxygen sensors. The readings will be displayed on the scan tool and can be compared to see if there is a problem in one side of the system or the other. If both sides are the same, a problem is located in something common to them, like fuel pressure, battery voltage, grounds, or the computer. Chapter 27 covers diagnosis, service, and replacement of oxygen sensors.

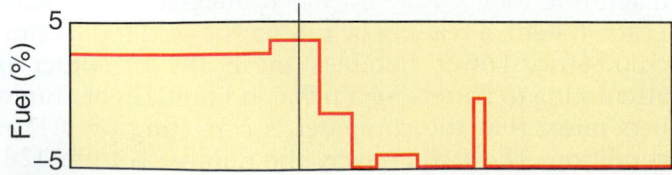

Figure 76.15 Adaptive memory is figured into the fuel reading. A positive percentage is lean. *(Courtesy of Chrysler Corporation)*

Idle Pulse–Width:	1.2 milliseconds
W.O.T. Pulse–Width:	6.0 milliseconds
Adaptive Memory:	20%
Adaptive Memory:	20%
Corrected Pulse–Width:	**1.44 milliseconds**
Corrected Pulse–Width:	**7.2 milliseconds**

Adaptive Memory

Idle Pulse–Width:	1.2 milliseconds
W.O.T. Pulse–Width:	6.0 milliseconds
Additive Adaptive:	400 microseconds
Additive Adaptive:	400 microseconds
Corrected Pulse–Width:	**1.60 milliseconds**
Corrected Pulse–Width:	**6.4 milliseconds**

Additive Adaptive

Figure 76.16 Additive adaptive has lengthened the corrected pulse width.

■ DIGITAL STORAGE OSCILLOSCOPE

An electronic lab scope, called a DSO (digital storage oscilloscope), is a popular tool for analyzing computer system problems. While digital voltmeters can sample information several times a second and scan tools can freeze information that the computer sees, a DSO can do both. A DSO reads in 40 billionths of a second (NS). It measures voltage and time like an ordinary oscilloscope.

The scale can be adjusted so that each division is equal to a certain voltage and time value. The lab scope can show AC, DC, RFI noise, and frequency all at once (Figure 76.17). When it can show two channels, it is called a *dual trace scope*. Lab scope patterns are shown during the sensor testing section of this chapter.

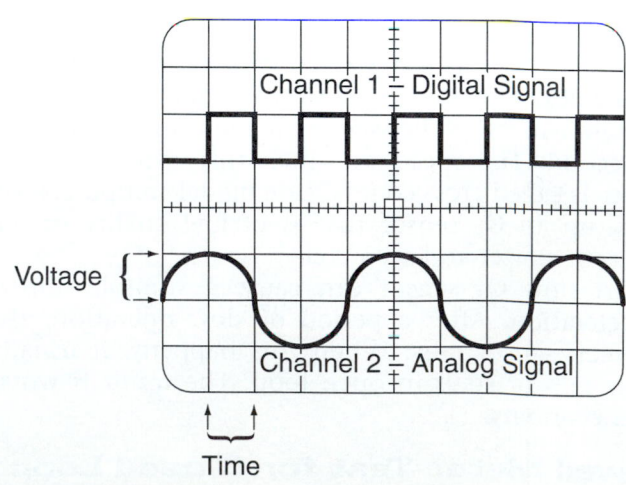

Figure 76.17 A dual trace scope showing analog and digital signals.

■ SENSOR AND ACTUATOR TESTING

Sensor Testing Strategy

No Start Conditions.   On a fuel-injected car, a no-start condition can be caused if a distributor reference or crankshaft sensor signal is lost. The computer cannot substitute for the missing information and the engine will not start. This is also a good place to look for a problem when a vehicle dies intermittently.

A defective or misadjusted throttle position sensor can send an excessive voltage to the computer that is interpreted as wide open throttle. The computer thinks that the engine is flooded because the throttle is being held to the floor during engine cranking, so it shuts off the fuel and the engine will not start.

NOTE: *Do not disconnect any electrical components with the key on unless instruction specifically says to do this.*

■ DIAGNOSING SENSOR PROBLEMS

Properly operating sensors are necessary for an engine to run smoothly during all operating conditions. The signals listed here are listed in the order of probability that they might cause a driveability problem. The higher on the list, the more probability of a problem (Figure 76.18).

When using a scan tool to diagnose sensors and actuators, use the following sequence to perform a quick check.
- First, check the input sensors.
- Then, perform a quick check of input switches. This includes checking in park, reverse, and neutral to see if the neutral safety switch is saying the engine is loaded in gear or unloaded in neutral. Work the power steering by turning the steering wheel to see if a load is sensed and corrected for by raising engine rpm.
- Next, check outputs. Check injector pulse width and rpm with the key on and engine off (KOEO). The idle air control needs a startup position so it should not be on "O." Check spark advance to see

---

- Engine rpm…
- Load Sensor…
- Throttle Position…
- Coolant Temperature…
- Barometic Pressure…
- Inlet Air Temperature…
- O_2…
- Switches…

Figure 76.18 Sensors listed in the order of probability of causing a driveability problem.

---

that it works. You do not need to check the accuracy of the advance curve. Just see that advance is working as the engine speed is raised. Check to see that the air switching valve is changing from upstream to downstream.

Passive Sensors

Passive sensors are ones that do not generate their own voltage. Examples of these are temperature sensors, position sensors, pressure sensors, and air flow sensors. These sensors respond to a reference voltage from the computer. Check to see that the reference voltage signal is the correct amount (usually 5 volts).

Active Sensors

Active sensors are those that generate their own signal. These include the O_2 sensor and magnetic sensors for engine speed, cam position, and vehicle speed and piezoelectric sensors like the knock sensor and solar sensors. Measure signal output from these sensors to test them.

The following are ways to diagnose sensor problems:
- Voltage generating sensors—digital meter
- Switching type sensors—ohmmeter and voltmeter
- Variable resistance sensors—ohmmeter and voltmeter, voltage drop

■ SENSOR TESTS

Engine Speed Sensor

A fuel-injected engine must have an rpm signal for the engine to run. If it stops, the engine stops. A carbureted engine would continue to run but set the check engine light. Either condition will be obvious.

Vehicle Speed Sensor

Vehicle speed sensors supply input for electronic speedometers and cruise control systems. They are also used to control the torque converter clutch, electronic transmission shift control, some emission controls, variable assist power steering, and electronically adjusted shock absorbers. There are two types of speed sensors. One is a photoelectric type and the other is a magnetic AC generator (see Figure 66.53).

A failed sensor or sensor circuit can cause premature or no converter clutch lockup, lack of change in steering assist, or inoperative or inaccurate cruise control and speedometer operation. The system is most easily tested by connecting a scan tool and running the drive wheels with the vehicle off the ground.

NOTE: *Front wheel drive cars should not have the front wheels run while they hang unsupported. Place jack stands under the lower control arms before running the drive train.*

A speed indication should be read when speed exceeds 3 mph. Codes are also set when the sensor does not work properly.

■ OXYGEN SENSOR

The oxygen sensor's job is to enrich the mixture enough so the reduction (NO_x) catalyst can work. It must also provide a lean enough mixture for HC and CO to oxidize. The mixture must go across stoichiometric from rich to lean in order for the catalytic converter to work efficiently. A scope or a scan tool is used to look at O_2 sensor operation to tell if it is going rich and lean in the correct amounts and for the right amount of time.

Because the oxygen sensor does not work during cold operation of the engine, a cold driveability problem (such as hard starting, poor idle, or a lack of full power) *cannot* be related to a defective O_2 sensor. Operation and service of the oxygen sensor is covered in Chapters 26 and 27.

The following are some items for consideration when testing an O_2 sensor:

■ On most models, an O_2 sensor can only compensate by changing the air/fuel mixture by about 5%. It cannot cause a no-start condition.

■ A voltage reading that is low could indicate a failing sensor, sending a "lean" signal, which will cause the computer to "drive" the fuel system rich.

■ A lean condition in the fuel system could also be the cause of a low voltage reading.

■ Smog pumps on cars with three-way catalytic converters have a switch that pumps air on either side of the converter (*upstream* or *downstream* of the CAT). If a smog pump is putting air into the system before the O_2 sensor (upstream) during closed loop, the sensor senses the extra air in the exhaust and registers a lean mixture. The computer responds by increasing the fuel injector pulse width.

CASE HISTORY

A Jeep blew an air pump and pieces of the fiber vanes blew into the air hose downstream of the pump. A small piece became stuck in the upstream valve on the diverter. Some air was always allowed to pump upstream even in closed loop. This resulted in the computer driving the fuel system rich whenever the engine was in closed loop.

■ The effect of a vacuum leak is much worse at idle on a four-cylinder engine than it would be with a six- or eight-cylinder. The O_2 sensor will stop working at idle and the system will lose its ability to correct the air/fuel ratio. It returns to operation when on the road.

■ When there is a bad spark plug, the misfire that results causes the engine to use less O_2. When there is more oxygen in the exhaust, there is a lower voltage from the O_2 sensor (Figure 76.19). Because it senses that the mixture is lean, the computer drives the system rich.

■ An engine running at the stoichiometric ratio (14.7:1) will produce a reading of about 1 to 1.5% O_2 in the exhaust.

Oxygen Sensor Condition

A "lazy" sensor is one that produces voltage slowly and then does not change back and forth between rich and lean signals fast enough. A properly operating sensor will move between 0.2 and 0.8 volt quickly. For an O_2 sensor to be considered to be fast enough on newer systems, it will require a specified cross count speed (Figure 76.20). Check manufacturers' specifications.

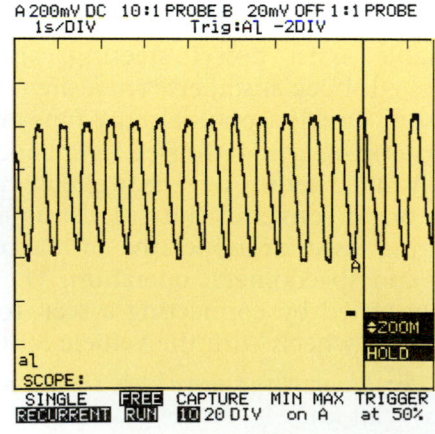

Normal

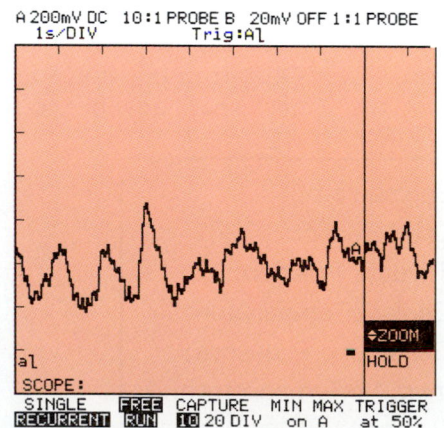

Open plug wire

Figure 76.19 A normal DSO oxygen sensor pattern and one when a plug wire is open. *(Reproduced with permission of Fluke Corporation)*

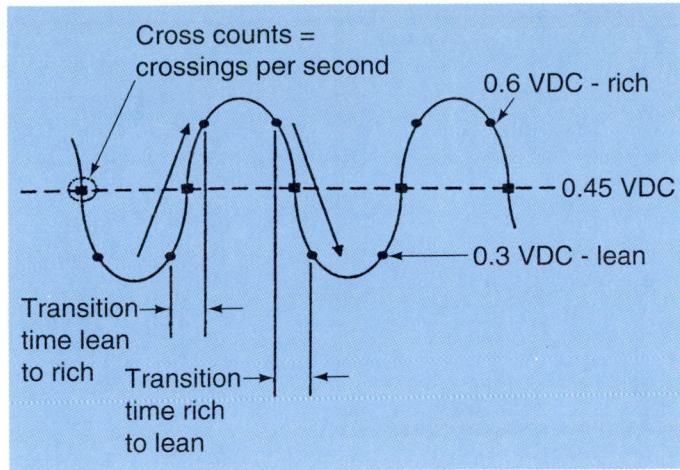

Figure 76.20 Oxygen sensor cross counts. *(Courtesy of OTC-SPX Corp. Aftermarket Tool & Equipment Group)*

Test an oxygen sensor's range by creating full rich and full lean conditions. When a large vacuum hose is pulled (from the power brake booster, for instance) the voltage should go to at least as low as 0.2 V. A propane enrichment test uses a propane bottle, hose, and metering device. Propane can be metered slowly through the end of the hose that was removed to create the vacuum leak. Add propane until the maximum voltage value is reached (rich mixture showing on the voltmeter). When the propane is pulled away from the hose, the voltmeter should respond immediately by dropping the voltage.

Replacement O₂ Sensors

Check the vents in the thimble of a replacement O₂ sensor. There should be the same number of holes and they should face clockwise or counterclockwise like the ones on the original sensor. Installing the wrong sensor can result in slower cross counts.

Titania dioxide O_2 sensors do not produce their own voltages, so they run from 0–5 volts. Some of them work backwards of normal, with a higher voltage being a leaner mixture. Check the manufacturer's service manual for the vehicle if a voltage higher than 0.9 V is encountered.

■ LOAD SENSORS

These sensors (MAP, vacuum, and mass air flow) tell the computer how much air is entering the engine. They affect ignition timing and air/fuel ratios. Mass air flow sensors are covered in Chapter 27.

■ MAP SENSOR

Basic fuel delivery to the engine is determined by the MAP sensor and rpm signal. The other sensors support these two. A MAP sensor takes the place of the power valve in a carburetor and vacuum advance diaphragm on a distributor. When engine load is high, the fuel injectors remain on longer. The sensor ignores the signal from the O_2 sensor under high load. You can cause the mixture to go very rich by disconnecting the vacuum supply to the MAP sensor. The engine will run extremely rich and possibly die. The computer also retards timing under high load.

When there is a problem in the MAP sensor, its circuit, or its vacuum connection, driveability symptoms can include detonation, power loss, stalling, a rough idle, and poor fuel economy. If there is a problem with the MAP sensor vacuum hose or an engine vacuum leak, the computer "sees" high load. It compensates by lengthening the fuel injection pulse width and retards the ignition timing.

A MAP sensor produces a voltage that drops as vacuum becomes higher (Figure 76.21). This means the voltage should drop when the engine is quickly accelerated and then returned to normal (Figure 76.22).

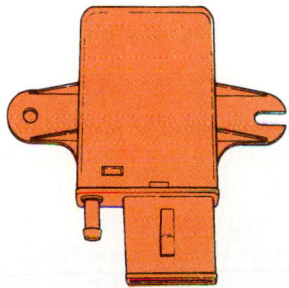

Manifold absolute pressure (MAP) sensor

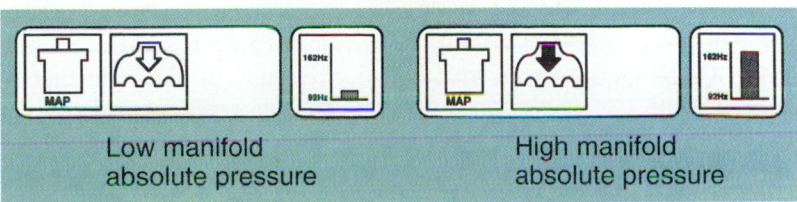

Low manifold absolute pressure | High manifold absolute pressure

Figure 76.21 MAP sensor voltages at low and high vacuum. *(Courtesy of Ford Motor Company)*

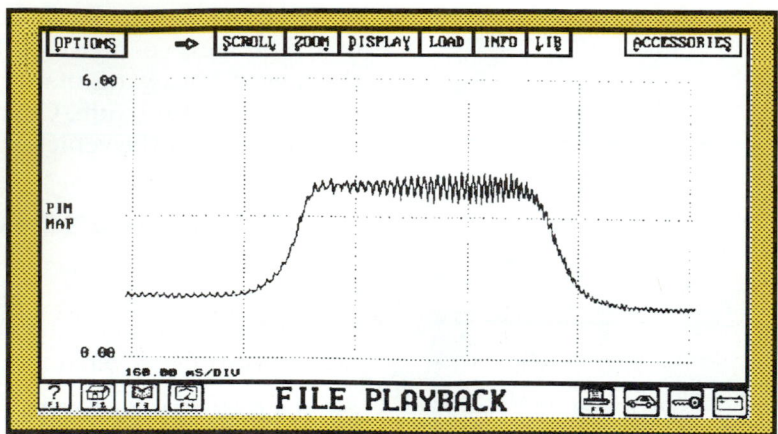

Figure 76.22 A normal DSO (digital storage oscilloscope) MAP sensor pattern. *(Courtesy of Edge Diagnostic Systems)*

Diagnostic codes will be set when readings are different from expectations. MAP sensor operation can be checked with a scan tool connected to the diagnostic connector. The sensor can also be tested with a digital voltmeter connected to it while applying vacuum to the sensor. With a 5-volt reference signal, a typical MAP sensor will read close to 5 volts with no vacuum signal. Voltage will drop in steps until it reaches about 1 volt with 20 inches of vacuum applied. There is also a digital MAP sensor (Ford) that produces a frequency signal (Hz) that changes based on vacuum conditions.

■ BARO SENSORS

Barometric pressure (BARO) sensors are used on some systems to monitor changes in the weather or altitude. Less air requires less fuel and less timing advance so the computer adjusts to compensate. Some systems combine the BARO sensor with the MAP sensor. When the key is turned on, altitude is read by the computer. If you drive from low altitude to high altitude the computer does not change until the key is cycled off and on again. When driving through a big change in altitude, it may be necessary to stop and shut off the engine. After restarting, the computer will compensate for the change in altitude.

There are different types of BARO sensors. Most work like a MAP sensor, producing a voltage signal that becomes less as altitude increases. Trouble codes can indicate a problem. Testing can be done with a scan tool. Some scan tools will give actual pressure readings.

A defective sensor or sensor circuit can cause poor high altitude performance or spark knock because the air/fuel ratio and timing will not be correct. If a sensor reading is in the correct range, apply vacuum to it and watch for a change in voltage output.

■ VACUUM SENSORS

Some systems have vacuum sensors that measure the difference between atmospheric pressure and intake manifold pressure (below the throttle plate). A MAP sensor compares the reading inside the engine to an absolute pressure calibration in the computer. Systems that use vacuum sensors must also use a barometric pressure sensor. Computer systems that do not have a MAP sensor use a throttle position sensor and an air flow sensor to determine the load on the engine.

If the engine has no BARO sensor, the MAP sensor is a BMAP (combination barometric and map) sensor. When the ignition key is on but the engine is off, the BMAP senses barometric pressure. In high altitude there is not as much air. This calls for less fuel and more spark advance. The BMAP gives information to the computer so it can calculate the correct spark timing and air/fuel mixture. A BMAP works as a MAP sensor when the engine is running. MAP and BARO sensors are not interchangeable.

■ THROTTLE POSITION SENSOR

The throttle position sensor (TPS) is a potentiometer (see Figure 75.24) mounted on the throttle shaft. It measures the angle of the throttle plate. When the rpm signal is under 600 rpm, the computer senses that the engine is cranking. The TPS also works as an accelerator pump. When the voltage signal changes quickly, the computer gives a signal for more fuel.

A defective or misadjusted TPS will cause a hesitation when accelerating, just like a bad accelerator pump on a car with a carburetor. When the engine is under a heavy load, the air conditioning compressor is shut off by the computer. A bad TPS can shut off an air conditioning compressor. Also, on computer controlled cars, the key should be turned on before depressing the throttle or the computer can receive a faulty TPS reading and the car might not start.

The TPS can be checked with either a voltmeter or an ohmmeter (Figure 76.23). To perform ohmmeter tests, the switch must be disconnected. That makes the voltmeter (or scope, which is actually a voltmeter, too)

Throttle position (TP) sensor

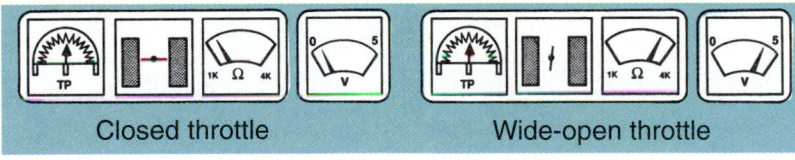

Closed throttle   Wide-open throttle

Figure 76.23 A throttle position sensor can be checked with an ohmmeter or a voltmeter. The elongated slots in this TPS allow for adjustment. *[Courtesy of Ford Motor Company]*

the tester of choice. Three tests are made for TPS operation. Reference voltage must be present at the switch with the key on. The base voltage is also compared to specifications. Finally, voltage should change gradually and evenly as the throttle is opened and closed. If the voltage does not rise, or there are skips (glitches) in the voltage measurement (Figure 76.24), the sensor is bad.

When checking voltage output of a TPS, typically there would be less than 0.1 V at idle. Typical voltage at half throttle would be 2.5 V. Remember, there is a 5-volt reference signal. That is why there is a 2.5 V signal when the throttle is half open. At WOT the signal would be 4.5 V.

Clear flood operation is to compensate in case the coolant temperature sensor is sending an incorrect signal. When the throttle is more than about 80% open, the computer sees wide open throttle. The TPS voltage will be more than 3.5 volts. The computer knows from the rpm signal that the engine is cranking and that a clear flood condition is needed. It stops the supply of fuel to the cylinders.

Some systems work the opposite way with high voltage at idle and low voltage at wide open throttle. Reference voltage can also be different. The principle of operation is the same, however.

Different kinds of problems can happen with a TPS. In cold weather a TPS gets cold and the feather (wiper arm) sometimes does not wipe. This results in high resistance in the switch. Hard starting, intermittent high idle, or a hesitation can happen when fuel vapors get into the switch. Remember, the switch is mounted on the throttle plate shaft. On carburetor and TBI systems, fuel can migrate into the sensor.

Replacing a Throttle Position Sensor

Some throttle position sensors are adjustable. Follow manufacturer's instructions. On older engines, correct adjustment is crucial to proper system operation. On newer engines, this is not as important because the computer uses whatever reading it takes at idle as base voltage (adaptive learn strategy).

■ COOLANT TEMPERATURE SENSOR

The *engine coolant temperature sensor (ECT)* affects how the engine operates in all conditions. The most common problem is when the computer system will not go into closed loop when the engine is warm. This causes poor fuel economy due to the rich mixture with no feedback control. Cold performance problems can also result from a bad ECT sensor. Symptoms can include poor idle, stalling, and hesitation or stumble on acceleration.

When the computer senses an ECT circuit problem on an OBD II car, it sets a code PO115. Problems are most often related to wiring or connectors, rather than failure of the sensor. When there is a failure in the coolant temperature sensor or its circuit, the result is

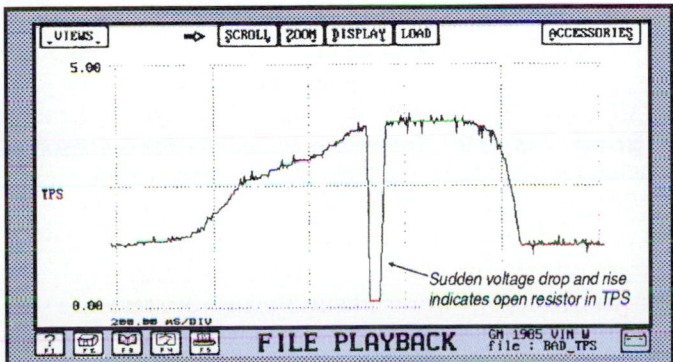

Figure 76.24 A glitch in a DSO pattern on a TPS test. *[Courtesy of Edge Diagnostic Systems]*

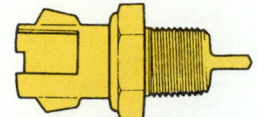

Engine coolant temperature
(ECT) sensor

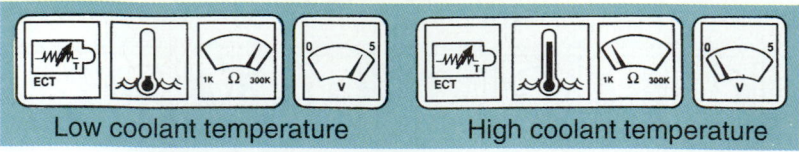

Low coolant temperature   High coolant temperature

Figure 76.25 The ECT sensor can be tested with an ohmmeter or a voltmeter. *(Courtesy of Ford Motor Company)*

like an automatic choke problem on a car with a carburetor. The computer sends out a small voltage signal (5 V) to the sensor and monitors the voltage drop in the circuit taking place based on the resistance of the sensor.

If the sensor circuit is open or the sensor has a bad ground, circuit continuity is lost. The voltage will not be able to be reduced. The computer will assume that the coolant temperature is −40°F and set fuel delivery and spark timing accordingly. Spark advance will be less than normal and the mixture will be very rich.

If circuit wiring becomes grounded, the 5 V reference signal will drop to zero. The computer assumes that the engine is very hot. A cold engine may not start. When the coolant temperature is too high, the air conditioning compressor is shut off by the computer.

Testing a Coolant Temperature Sensor

The ECT sensor can be tested with an ohmmeter or a voltmeter (Figure 76.25). Resistance readings change with temperature (see Figure 27.29). An ECT sensor is a thermistor and will have lower resistance as the temperature of the coolant rises. This is called a **negative coefficient thermistor**.

Check the resistance of the sensor and compare it to the actual temperature of the engine to see if it is working properly. The sensor can also be removed and tested on a hot plate in the same manner a thermostat is tested (Figure 76.26).

■ AIR TEMPERATURE SENSORS

Intake air temperature (IAT) sensors have been called a variety of names in the past (ACT, VAT, MCT, MAT, and ATS). Under OBD II guidelines, these parts are now called IATs. It is used to fine-tune the air/fuel mixture to compensate for the density of the air. The IAT sensor is mounted in the intake system with its sensor end exposed to the incoming airstream (Figure 76.27).

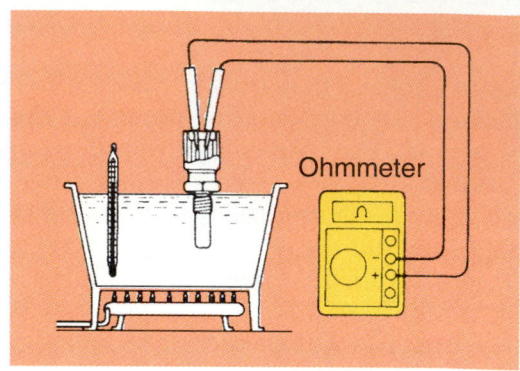

Ohmmeter

Figure 76.26 An engine coolant temperature sensor can be tested on a hot plate.

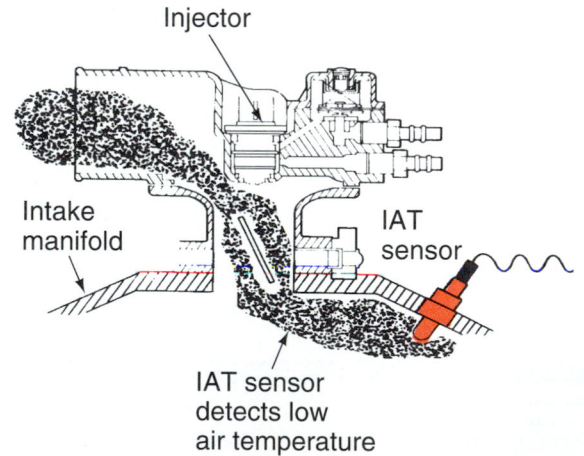

Injector

Intake
manifold

IAT
sensor

IAT sensor
detects low
air temperature

Figure 76.27 The intake air temperature sensor is mounted in the intake manifold with its sensor end exposed to the incoming air stream. *(Courtesy of Ford Motor Company)*

An IAT sensor works like a coolant temperature sensor. Its resistance lowers as temperature rises (Figure 76.28). It can be damaged by intake manifold popbacks and its operation can be hampered by contamination with oil and dirt. IAT sensor temperature should be 30°

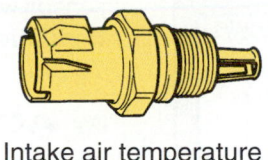

Intake air temperature
(IAT) sensor

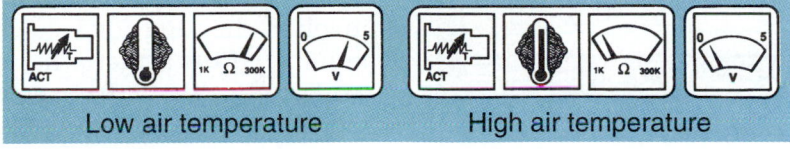

Low air temperature   High air temperature

Figure 76.28 An IAT sensor's resistance lowers as temperature rises. *(Courtesy of Ford Motor Company)*

to 40° cooler than CTS temperature. Too high a temperature of the air in the intake manifold could be due to EGR or hot air stove problems.

An IAT circuit problem on an OBD II car sets a code PO110. When testing with a scan tool, as the engine speed is raised, temperature should drop slightly as the sensor is cooled by the increased amount of air moving across it.

If the sensor is removed from the engine it can be tested with heated water and an ohmmeter the same way an engine coolant temperature sensor is tested (see Figure 76.26). Blow hot air on it with a hair dryer or heat gun. Do *not* use a propane torch. Watch for a drop in resistance as temperature increases.

■ AIR FLOW SENSOR SERVICE

Air flow sensor controlled systems react poorly to vacuum leaks. This would allow unmetered air into the engine, resulting in an excessively lean mixture. Once the engine is warm and is in closed loop, the O_2 sensor can compensate for small amounts of air leaking in. Because the O_2 sensor is in the exhaust, it comes after the leak. Even though it causes the fuel system to compensate for the leak, a hesitation on acceleration could result (especially when cold).

Dirt can also cause problems in a vane air flow sensor (VAF). A dirty air filter or housing left dirty when an air filter cartridge is changed can cause sticking of the shaft that the air door pivots on. Feel the operation of the door by pushing it open gently. If it is sticky or binding and spray carburetor cleaner does not free it up, it will have to be replaced.

An intake manifold popback can also cause the door to bend or break. A backfire valve is built into some sensor doors to prevent this. If the backfire valve begins to leak, it will cause a rich-running engine.

Trouble codes will set when there is a problem with a VAF sensor. Checking the operation of the electrical portion of the sensor is done with a voltmeter. Follow service manual instructions for the car you are

working on. Opening the air door should slowly increase the voltage until reference voltage is reached.

A mass air flow sensor (MAF) has no moving parts to fail like a vane air flow sensor. Trouble codes will set if the electronics of the sensor or its circuit gives an unexpected reading. Hot wire sensors produce a voltage that varies from about 0.5 volt at idle to reference voltage (5 V) at WOT. Glitches can be spotted using a DSO (Figure 76.29).

Hot film MAF sensors produce a variable frequency instead of a voltage. The frequency varies from about 30 Hz at idle to about 150 Hz at WOT. This can be read on a scan tool as grams per second. A bad MAF frequency sensor DSO pattern is shown in Figure 76.30. MAF sensor output can also be checked by monitoring frequency (Hz) on a digital multimeter.

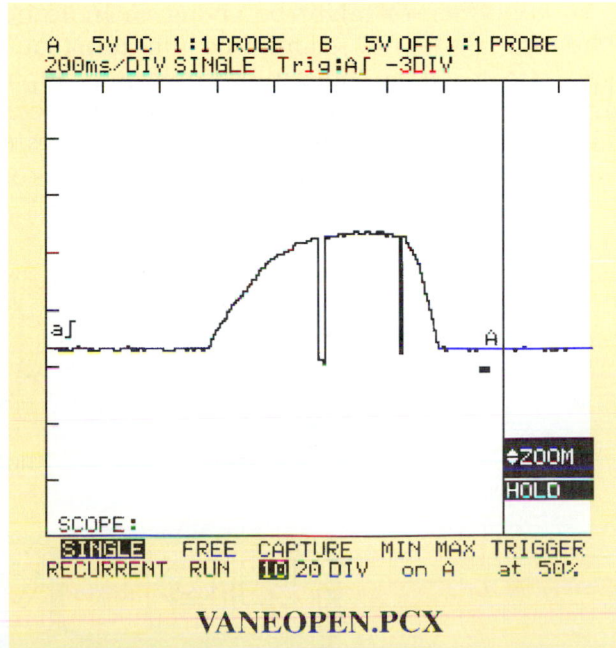

VANEOPEN.PCX

Figure 76.29 Glitches in a mass air flow sensor pattern. *(Reproduced with permission of Fluke Corporation)*

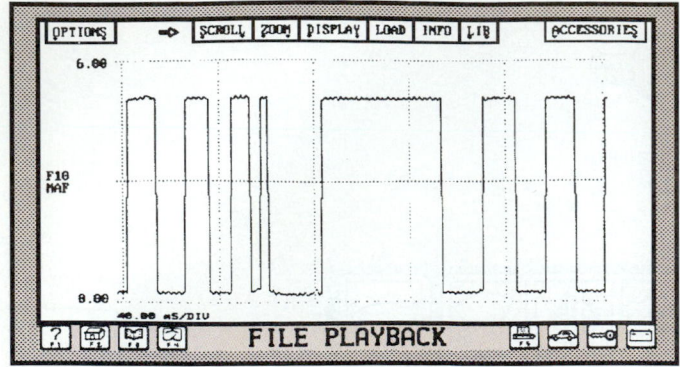

Figure 76.30 A good frequency-type MAF sensor waveform would show proportional increases in the digital signals. This one is bad. *(Courtesy of Edge Diagnostic Systems)*

■ KNOCK SENSOR SERVICE

A knock sensor only affects ignition timing. If a knock sensor fails, the computer will not retard the timing to prevent spark knock. A loose bracket or other vibration can also cause the timing to be retarded when it does not need to be. Sometimes, a main bearing knock or piston slap will cause the timing to retard too. Some inline five- and six-cylinder engines have two knock sensors.

The sensor can be tested by rapping on the engine near the sensor with a metal tool like a socket extension or a wrench. Use a timing light with the engine running on fast idle. A decrease in ignition advance should be observed. With a voltmeter, a typical knock sensor reading would be 300 to 500 millivolts, depending on how bad the knock is (Figure 76.31).

The knock sensor might be giving an indication but the timing does not retard. This indicates a faulty advance circuit. A scan tool can be used while rapping on the engine. Some systems will show the actual amount of spark retard and others will show a yes/no or on/off reading on the scan tool. A normal knock sensor waveform on a DSO is a group of closely spaced oscillations with a gradually decreasing voltage (Figure 76.32).

■ ACTUATOR SERVICE

Actuators include solenoids (including fuel injectors), stepper motors, and motors for electronic suspension hydraulic controls. Operation of these actuators is covered in their respective chapters. Testing an actuator involves checking for voltage at the actuator control terminal. If the actuator has voltage to it but does not work, be sure that it has a good ground path. Test individual actuators according to instructions in the repair manual.

■ REPAIR THE PROBLEM

After repairing a problem, road test the car again. If the problem caused a trouble code, be sure to erase the computer's memory before testing the car again. When a car has had previous problems, the computer learns to correct for them. The test drive will allow a late model computer with adaptive strategy to relearn its best adjustments.

NOTE: *Adaptive memory learns faster if the computer's memory is erased first. With old knowledge after a part is changed, it takes longer to adapt.*

After replacement of the computer or when a battery has been disconnected, poor driveability and performance can result until the computer relearns the best driveability settings. Hard starting, high idle, stumble, and stalling are possible symptoms.

■ COMPUTER WIRING SERVICE

Poor electrical connections are the most common cause of problems in computer systems. Electronics use very, very small amounts of electricity and are very sensitive to resistance in wiring. As little as 200 ohms resistance can cause a problem. Pin-type connectors should fit tight to a paper clip.

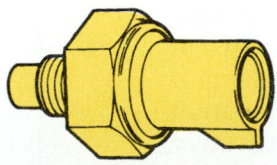

Knock sensor (KS)

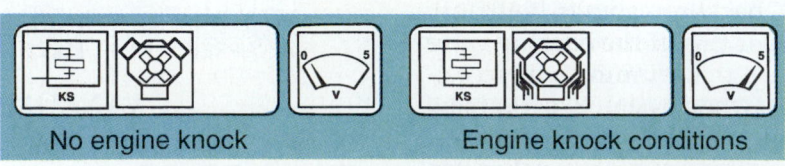

Figure 76.31 A knock sensor can be tested with a voltmeter. *(Courtesy of Ford Motor Company)*

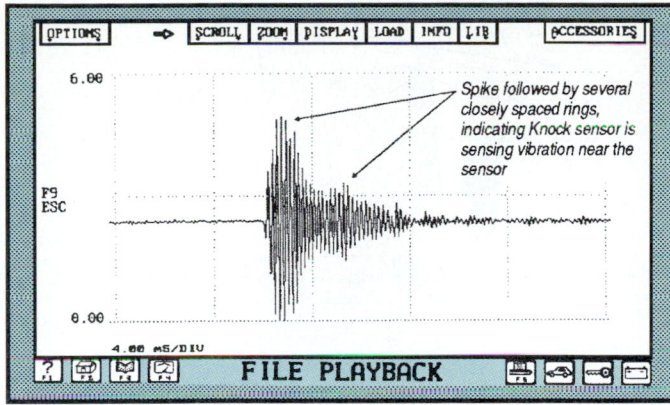

Figure 76.32 A normal knock sensor waveform. *(Courtesy of Edge Diagnostic Systems)*

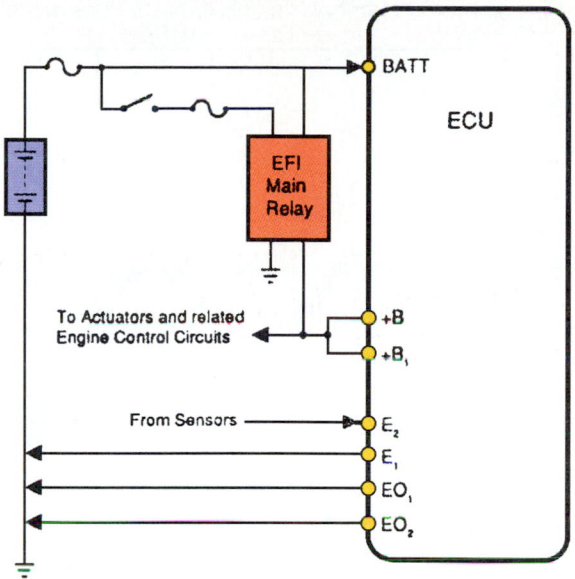

Figure 76.33 One computer lead always has power (BATT). The other two on this diagram are powered from the relay when the key is on.

Wiring problems include loose or corroded connections and grounded wires. An advantage of onboard diagnostics is that it helps keep the wiring connections from becoming corroded or damaged because wires are not removed. This maintains the integrity of the connectors.

Always use a wiring diagram when working on computer systems. First, check to see which parts are affected. When several devices are not working, look closer to the power source. When only one or two are not working, look in that part of the circuit for the problem. Be sure to ask the customer if the problem only occurs when hot or cold. Does it always happen or is it intermittent?

The computer must have good power and ground connections. Most computers have two or three power leads (Figure 76.33). One lead will always have power. This allows the computer to keep information in volatile memory. The second lead (and sometimes a third lead if used) supplies power when the key is on. Fuse links protect these wires. They are located near the battery, starter, or starter relay.

The number of grounds varies from two to six. They are either power or sensor grounds. Power grounds are used for actuators like motors or solenoids. Sensor grounds return to the computer and then to a ground source.

To test a ground circuit, measure voltage drop across the ground side of the circuit. Connect the meter to the ground side of the sensor or actuator. This is the lead that returns to the computer. Connect the negative side of the meter to the negative battery terminal. Voltage drop in a sensor's ground circuit should be less than 0.1 V. Power ground circuits should not exceed 0.3 V. Isolate the problem by working your way down the circuit (see Chapter 29). Ground side resistance that is too high will decrease voltage in the circuit. The computer will receive too low a return signal from the sensor.

NOTES:

■ *If you remove and repair a ground connection, be sure you reconnect it to the same place. Do not change the length of the wire.*

■ *If you clean the engine, do not use a steam cleaner. Temperatures of more than 300°F can ruin computers and electronics. High pressure cleaning results in corrosion.*

SHOP TIP A gentle way of cleaning the engine is to use car soap and a brush. Then hose off the engine with a garden hose. Some shops like to use laundry detergent sprayed on with a weed sprayer.

Examine wires to see that they are properly routed and are all original in regard to length and shape. *Radio interference (RFI)* is a problem with computer systems. If a computer feed wire is routed near the alternator or secondary ignition wires, signals that the computer receives will be jumbled. Figure 76.34 shows a DSO waveform of RFI. This can result in serious driveability problems. If you find a bare wire wrapped around the outside of wires running from the distributor to the computer, these are *shielding wires*. The shielding wire is connected to ground and absorbs RFI that might interfere with the rpm reference signal to the computer.

When a wiring problem is suspected, disconnect and clean the connector. Electronic connector cleaner can be used.

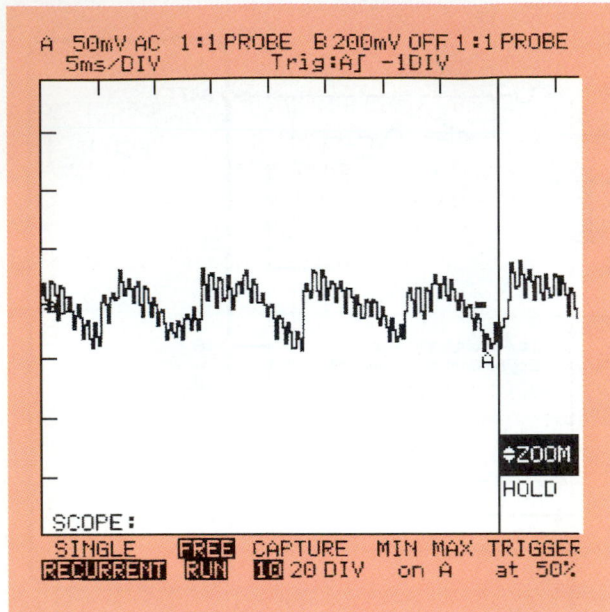

Figure 76.34 A DSO pattern showing AC superimposed over DC-causing radio interference. *(Reproduced with permission of Fluke Corporation)*

CAUTION
■ Do not use color TV cleaner. It contains acids that will cause further corrosion of the connection.
■ Do not use carburetor cleaner, brake cleaner, alcohol, or electromotive spray. The connector is held in plastic, which has a hydrocarbon base. A strong cleaner can ruin it.
■ Do not touch the metal parts of the connector. Acid from your hands can begin the corrosion process.

Clean any corrosion from the connection with a soft bronze bristle brush (available at electronic supply stores). After cleaning the connection, coat the wire ends with dielectric compound if it was used previously.

When connections are replaced, use solder rather than crimp connectors. In areas where the roads are salted in the winter, use shrink tubing over the soldered connection. Salt accelerates the corrosion process.

The quality of ground connections is very important too. One of the most prominent causes of ignition module failure is loss of ground. Check ground straps and wires that connect the engine and chassis. The eyelet of a ground wire where it bolts to the cylinder head must be tight and its bolt must not have paint or rust under its head.

■ COMPUTER SERVICE

The computer is rarely the cause of problems in the fuel system. When it is, the parts are not repaired. The

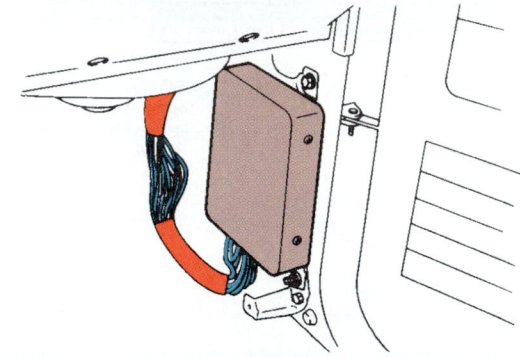

Figure 76.35 A computer located inside the right front kick panel. *(Courtesy of Mazda Motor Corporation)*

entire unit is replaced. Be sure to check the power and ground circuits before replacing a computer.

Remanufactured computers are widely available at a reduced cost for popular makes of cars. Some input and output devices can be adjusted or repaired. Processing and storage components are replaced. Some imported car computers are very costly because there are not enough cores available for exchange.

Some computers have a replaceable element called a PROM (see Figure 75.21). The old PROM is installed in the replacement computer. Other later model computers have EE Flash PROMS. These are programmed electronically, either in the dealership or by telephone modem from the manufacturer. See Chapter 75 for more about computer systems.

Computer Location

Electronic instruments can be damaged by excessive voltage, heat, moisture, or vibration, so computers are usually mounted in the driver's compartment. Sometimes, they are in the right-hand *kick panel* (in front of the passenger door) (Figure 76.35) or near the glove box. Other times they are located under the passenger seat. Newer computers are designed to dissipate heat more quickly. Some of these are installed in the engine compartment. To be safe, pull the computer fuse before unhooking its connectors.

■ STATIC ELECTRICITY

Static electricity from sliding around on the front seat is a concern when working around electronic components. People who work around sensitive components sometimes wear a ground strap attached to their wrist. You can eliminate the static electricity by touching ground before you touch a computer. Do not take the computer out of the container until you are already in the front seat. Backprobing the computer under the hood is the only time you need to wear a ground strap.

ELECTRICAL DAMAGE TO A CIRCUIT

Too much electrical current causes heat that can damage an electrical circuit. Voltage is limited to about 20 V. If a circuit never exceeds its voltage or current limits, it is very reliable. Damaged connections are usually the reason for failure of electronic components. Connector pins on chips can also be damaged or bent during installation.

Semiconductors are designed to handle a limited amount of current. A light bulb can draw 2.5 amps. Transistors in computers draw 200 milliamps (0.2 A). If too much current is applied in a reverse direction, it can force through and ruin a diode or transistor. Computers can tolerate a high current surge for about 5 seconds.

Voltage spikes are the biggest cause of electrical damage to an integrated circuit. If a sensor or wiring connection is unplugged while there is power to it, a spike can occur. Most automotive computer systems operate on 5 volts. A spike while disconnecting a powered circuit lead can result in a spike of 50 volts or so. This happens because electrons that were in motion before the circuit was broken back up at the connection. When they try to push their way across the gap, the spike is created. Manufacturers build safeguards for voltage spikes into their newer systems, but not enough to protect against disconnecting and reconnecting components. Voltage spikes can also result from an arc welder. Always disconnect the battery before doing any welding.

REVIEW QUESTIONS

1. A high _____ voltmeter is used to measure voltages in computer circuits.

2. If the MIL comes on when the engine is running, there is probably a trouble _____ stored in the computer's memory.

3. A _____ _____ is a portable computer that reads data from the car's onboard computer.

4. A _____ is a scan tool feature that is like a flight recorder.

5. A good oxygen sensor will cause the fuel system to switch between rich and lean about _____ times in ten seconds.

6. _____ sensors are ones that do not generate their own voltage.

7. Which tool would not require a throttle position sensor to be disconnected for testing, a voltmeter or an ohmmeter?

8. What is the name of the sensor that is tested by rapping on the engine with a tool?

9. Computer grounds are either power or _____ grounds.

10. Voltage _____ happen when electrical connections are pulled apart while there is a load.

ASE STYLE REVIEW QUESTIONS

1. Technician A says that an analog voltmeter is used to test computer systems. Technician B says that high input resistance prevents an analog meter from drawing current while it is connected to a circuit. Who is right?
   a. Technician A   b. Technician B
   c. Both A and B   d. Neither A nor B

2. Technician A says that when the base timing is too far advanced, the computer will attempt to raise the engine's idle to compensate. Technician B says that an intermittent fault is one that is present at the time of the self-test. Who is right?
   a. Technician A   b. Technician B
   c. Both A and B   d. Neither A nor B

3. Technician A says that sensors are more often the cause of electronic control problems than actuators are. Technician B says that when there is more than one trouble code in memory, the computer will give the lowest number first. Who is right?
   a. Technician A   b. Technician B
   c. Both A and B   d. Neither A nor B

4. Technician A says that a scan tool can be used to read serial data on older computer cars. Technician B says when there is more than one code, fix a higher number code first. Who is right?
   a. Technician A   b. Technician B
   c. Both A and B   d. Neither A nor B

5. Technician A says that a code can still remain in memory even though a problem has been corrected. Technician B says that a scan tool can erase trouble codes without having to disconnect anything electrical. Who is right?
   a. Technician A   b. Technician B
   c. Both A and B   d. Neither A nor B

6. Technician A says that a faulty thermostat can cause a lack of computer feedback fuel control. Technician B says to use an ohmmeter to test the oxygen sensor. Who is right?

   a. Technician A   **b.** Technician B

   c. Both A and B   **d.** Neither A nor B

7. Technician A says that a defective oxygen sensor can cause an engine not to start. Technician B says that a car that has lost its distributor reference or crankshaft sensor signal will not start. Who is right?

   a. Technician A   **b.** Technician B

   c. Both A and B   **d.** Neither A nor B

8. Technician A says that a lazy O_2 sensor will usually cause the system to run lean. Technician B says that a car that runs poorly when cold could have a bad oxygen sensor. Who is right?

   a. Technician A   **b.** Technician B

   c. Both A and B   **d.** Neither A nor B

9. Technician A says that a smog pump that blows air upstream during closed loop could cause a rich air/fuel mixture. Technician B says that a bad spark plug on a feedback car can result in a rich air/fuel mixture. Who is right?

   a. Technician A   **b.** Technician B

   c. Both A and B   **d.** Neither A nor B

10. Technician A says that disconnecting the vacuum supply to the MAP sensor will cause the fuel system to go rich. Technician B says that most computer systems use battery voltage for a reference signal. Who is right?

    a. Technician A   **b.** Technician B

    c. Both A and B   **d.** Neither A nor B

English-Metric Conversion Chart

CONVERSION FACTORS

Unit	To Unit	Multiply By	Unit	To Unit	Multiply By
LENGTH			**WEIGHT**		
Millimeters	Inches	.03937	Grams	Ounces	.03527
Inches	Millimeters	25.4	Ounces	Grams	28.34953
Meters	Feet	3.28084	Kilograms	Pounds	2.20462
Feet	Meters	.3048	Pounds	Kilograms	.45359
Kilometers	Miles	.62137	**WORK**		
Miles	Kilometers	1.60935	Centimeter		
AREA			Kilograms	Inch Pounds	.8676
Square			Inch Pounds	Centimeter Kilograms	1.15262
Centimeters	Square Inches	.155	Meter Kilograms	Foot Pounds	7.23301
Square Inches	Square Centimeters	6.45159	Foot Pounds	Newton Meters	1.3558
VOLUME			**PRESSURE**		
Cubic Centimeters	Cubic Inches	.06103	Kilograms/		
Cubic Inches	Cubic Centimeters	16.38703	Sq. Centimeter	Pounds/Sq. Inch	14.22334
Liters	Cubic Inches	61.025	Pounds/Sq. Inch	Kilograms/Sq.	
Cubic Inches	Liters	.01639		Centimeter	.07031
Liters	Quarts	1.05672	Bar	Pounds/Sq. Inch	14.504
Quarts	Liters	.94633	Pounds/Sq. Inch	Bar	.06895
Liters	Pints	2.11344	Atmosphere	Pounds/Sq. Inch	14.696
Pints	Liters	.47317	Pounds/Sq. Inch	Atmosphere	.06805
Liters	Ounces	33.81497	**TEMPERATURE**		
Ounces	Liters	.02957	Centigrade Degrees	Fahrenheit Degrees	$(C° \times 9/5) + 32$
			Fahrenheit Degrees	Centigrade Degrees	$(F° - 32) \times 5/9$

Inches	Decimals	MM	Inches	Decimals	MM
1/64	.016	.397	33/64	.516	13.097
1/32	.031	.794	17/32	.531	13.494
3/64	.047	1.191	35/64	.547	13.891
1/16	.063	1.588	9/16	.563	14.288
5/64	.078	1.984	37/64	.578	14.684
3/32	.094	2.381	19/32	.594	15.081
7/64	.109	2.778	39/64	.609	15.478
1/8	.125	3.175	5/8	.625	15.875
9/64	.141	3.572	41/64	.641	16.272
5/32	.156	3.969	21/32	.656	16.669
11/64	.172	4.366	43/64	.672	17.066
3/16	.188	4.763	11/16	.687	17.463
13/64	.203	5.159	45/64	.703	17.859
7/32	.219	5.556	23/32	.719	18.256
15/64	.234	5.953	47/64	.734	18.653
1/4	.250	6.350	3/4	.750	19.050
17/64	.266	6.747	49/64	.766	19.447
9/32	.281	7.144	25/32	.781	19.844
19/64	.297	7.541	51/64	.797	20.241
5/16	.313	7.938	13/16	.813	20.638
21/64	.328	8.334	53/64	.828	21.034
11/32	.344	8.731	27/32	.844	21.431
23/64	.359	9.128	55/64	.859	21.828
3/8	.375	9.525	7/8	.875	22.225
25/64	.391	9.922	57/64	.891	22.622
13/32	.406	10.319	29/32	.906	23.019
27/64	.422	10.716	59/64	.922	23.416
7/16	.438	11.113	15/16	.938	23.813
29/64	.453	11.509	61/64	.953	24.209
15/32	.469	11.906	31/32	.969	24.606
31/64	.484	12.303	63/64	.984	25.003
1/2	.500	12.700			

Appendix A.1 English-metric conversion chart. *(Courtesy of Mitchell International)*

Size Conversion Chart

Fractions	Decimal In.	Metric MM.	Fractions	Decimal In.	Metric MM.
1/64	.015625	.39688	33/64	.515625	13.09687
1/32	.03125	.79375	17/32	.53125	13.49375
3/64	.046875	1.19062	35/64	.546875	13.89062
1/16	.0625	1.58750	9/16	.5625	14.28750
5/64	.078125	1.98437	37/64	.578125	14.68437
3/32	.09375	2.38125	19/32	.59375	15.08125
7/64	.109375	2.77812	39/64	.609375	15.47812
1/8	.125	3.1750	5/8	.625	15.87500
9/64	.140625	3.57187	41/64	.640625	16.27187
5/32	.15625	3.96875	21/32	.65625	16.66875
11/64	.171875	4.36562	43/64	.671875	17.06562
3/16	.1875	4.76250	11/16	.6875	17.46250
13/64	.203125	5.15937	45/64	.703125	17.85937
7/32	.21875	5.55625	23/32	.71875	18.25625
15/64	.234375	5.95312	47/64	.734375	18.65312
1/4	.250	6.35000	3/4	.750	19.05000
17/64	.265625	6.74687	49/64	.765625	19.44687
9/32	.28125	7.14375	25/32	.78125	19.84375
19/64	.296875	7.54062	51/64	.796875	20.24062
5/16	.3125	7.93750	13/16	.8125	20.63750
21/64	.328125	8.33437	53/64	.828125	21.03437
11/32	.34375	8.73125	27/32	.84375	21.43125
23/64	.359375	9.12812	55/64	.859375	21.82812
3/8	.375	9.52500	7/8	.875	22.22500
25/64	.390625	9.92187	57/64	.890625	22.62187
13/32	.40625	10.31875	29/32	.90625	23.01875
27/64	.421875	10.71562	59/64	.921875	23.41562
7/16	.4375	11.11250	15/16	.9375	23.81250
29/64	.453125	11.50937	61/64	.953125	24.20937
15/32	.46875	11.90625	31/32	.96875	24.60625
31/64	.484375	12.30312	63/64	.984375	25.00312
1/2	.500	12.70000	1	1.00	25.40000

Appendix A.2 Size conversion chart. *(Courtesy of General Motors Corporation, Service Technology Group)*

General Torque Specifications Chart
(When SAE 10 oil is used as a lubricant)

Material	SAE 2 Mild Steel		SAE 5		SAE 8	Socket Head Cap Screws
Minimum Tensile P.S.I. Strength	74,000	60,000	120,000	105,000	150,000	160,000
Proof P.S.I. Load	55,000	33,000	85,000	74,000	120,000	136,000
Steel Grade Symbols						
Bolt Diameter Inches	Torque: pound foot					
¼	7	—	10	—	14	16
5/16	14	—	21	—	30	33
3/8	24	—	37	—	52	59
7/16	39	—	60	—	84	95
½	59	—	90	—	128	145
9/16	85	—	130	—	184	210
5/8	117	—	180	—	255	290
¾	205	—	320	—	450	510
7/8	—	200	515	—	730	825
1	—	300	775	—	1090	1235
1⅛	—	425	—	955	1545	1750
1¼	—	600	—	1345	2180	2000

**NOTE: Use only when manufacturer's specifications are not available. These values are for stiff metal-to-metal joints and are based on 90% of proof load. Do not use for gasketed joints or joints of soft materials.

Appendix A.3 General torque specifications chart (when SAE 10 oil is used as a lubricant). *(Courtesy of Snap-on Tools Company, Copyright Owner)*

General Torque Specifications Chart for I.S.O.* Metric Fasteners**
(When SAE 10 oil is used as a lubricant)

Minimum Tensile Strength	kg/mm²	40		50		60			80	100	120
	P.S.I.	56,900		71,100		85,340			113,800	142,200	170,700
Proof Load	kg/mm²	22.6	29.1	28.2	36.4	33.9	43.7	47.5	58.2	79.2	95.0
	P.S.I.	32,150	41,390	40,110	51,770	48,220	62,160	67,560	82,780	112,650	135,130
Property Class		4.6	4.8	5.6	5.8	6.6	6.8	6.9	8.8	10.9	12.9

Bolt Diameter metric	inch	Torque ☐ kilogram centimetre ☐ kilogram metre									
6mm	.236	49	63	61	79	74	95	103	126	172	206
8mm	.315	119	153	148	191	178	230	250	306	417	500
10mm	.394	235	303	294	379	353	455	495	606	8.2	10
12mm	.472	411	529	427	662	616	7.9	8.6	10.5	14	17
14mm	.551	654	8.4	8.2	10.5	10	12	13	17	23	27
16mm	.630	10	13	12	16	15	20	21	26	36	43
18mm	.709	14	18	17	23	21	27	30	36	49	59
22mm	.866	27	35	34	44	41	52	57	70	95	114

*I.S.O. = International Standardization Organization.
**NOTE: Use only when manufacturer's specifications are not available. These values are for stiff metal-to-metal joints and are based on 90% of proof load. Do not use for gasketed joints or joints of soft materials.

Appendix A.4 General torque specifications chart for I.S.O.* metric fasteners** (when SAE 10 oil is used as a lubricant). *(Courtesy of Snap-on Tools Company, Copyright Owner)*

Metric Conversion Chart

METRIC CONVERSION: lb. ft. to N•m

The chart below can be used to convert pound foot to newton metre. The left hand column lists pound foot in multiples of 10 and the numbers at the top of the columns list the second digit. Thus 36 pound foot is found by following the 30 pound foot line to the right to "6" and the conversion is 49 N•m

lb. ft.	0	1	2	3	4	5	6	7	8	9
	N•m	N•m	N•m	N•m	N•m	N•m	N•m	N•m	N•m	N•m
0	0	1.36	2.7	4.1	5.4	6.8	8.1	9.5	10.9	12.2
10	13.6	14.9	16.3	17.6	19.0	20.3	21.7	23.1	24.4	25.8
20*	27	28	30	31	33	34	35	37	38	39
30	41	42	43	45	46	47	49	50	52	53
40	54	56	57	58	60	61	62	64	65	66
50	68	69	71	72	73	75	76	77	79	80
60	81	83	84	85	87	88	90	91	92	94
70	95	96	98	99	100	102	103	104	106	107
80	109	110	111	113	114	115	117	118	119	121
90	122	123	125	126	127	129	130	132	133	134
100	136	137	138	140	141	142	144	145	146	148

* Above 20 lb. ft. the converted N•m readings are rounded to the nearest N•m.

METRIC CONVERSION: kg•cm to N•m

The chart below can be used to convert kilogram centimetre to newton metre. The left hand column lists kg•cm in multiples of 10 and the numbers at the top of the columns list the second digit. Thus 72 kg•cm is found by following the 70 kg•cm line to the right to "2" and the conversion is 7.1 N•m.

kg. cm.	0	1	2	3	4	5	6	7	8	9
	N•m	N•m	N•m	N•m	N•m	N•m	N•m	N•m	N•m	N•m
0	0	.098	.20	.29	.39	.49	.59	.69	.78	.88
10	.98	1.08	1.18	1.27	1.37	1.47	1.57	1.67	1.76	1.86
20	2.0	2.1	2.2	2.3	2.4	2.5	2.6	2.7	2.8	2.8
30	2.9	3.0	3.1	3.2	3.3	3.4	3.5	3.6	3.7	3.8
40	3.9	4.0	4.1	4.2	4.3	4.4	4.5	4.6	4.7	4.8
50	4.9	5.0	5.1	5.2	5.3	5.4	5.5	5.6	5.7	5.8
60	5.9	6.0	6.1	6.2	6.3	6.4	6.5	6.6	6.7	6.8
70	6.9	7.0	7.1	7.2	7.3	7.4	7.5	7.6	7.7	7.8
80	7.9	7.9	8.0	8.1	8.2	8.3	8.4	8.5	8.6	8.7
90	8.8	8.9	9.0	9.1	9.2	9.3	9.4	9.5	9.6	9.7
100	9.8	9.9	10.0	10.1	10.2	10.3	10.4	10.5	10.6	10.7

One oz. in. = 28.35 gms. in.
One lb. in. = 1.152 kg•cm
One lb. ft. = .138 kg•m

One kg•cm = .8679 lb. in.
One kg•cm = 7.233 lb. ft.
One N•cm = .0885 lb. in.
One N•m = .7375 lb. ft.

Courtesy of Snap-on Tools

Appendix A.5 Metric conversion chart. *(Courtesy of Snap-on Tools Company, Copyright Owner)*

Lubricants and Threaded Fasteners

Lubricants, when used on the threads of either "blind" or tapped holes, hex nuts, or on the cap screw itself may increase the effect of the applied torque. This is because the lubricant may reduce the thread friction and allow more of the applied torque to develop tension or clamping force, since less of the torque applied to the assembly goes toward overcoming friction.

The random use of thread lubricants frequently causes screw failures and fastener breakdowns. The reduced amount of friction in the fastener assembly and the torque applied may produce far greater tension than is recommended. In many cases the applied torque will actually result in screw tension that exceeds the strength of the fastener used, and the fastener may fail.

The uncontrolled use of a lubricant is a potential cause of assemblies being overloaded and, therefore, subject to failure unless torque corrections are made.

TORQUE REDUCTIONS REQUIRED FOR VARIOUS LUBRICANTS

TORQUE REDUCTION REQUIRED FROM 'DRY' TORQUE

LUBRICANT	SUPERTANIUM CAP SCREWS		ALL OTHER CAP SCREWS
	WITH NUT	BLIND HOLE	
Premier Thread Lubricant	0%	-20%	-45%
Premier ETP	0%	-20%	-40%
Never-Seize	0%	-20%	-45%
Premier Thread-Eze	0%	-20%	-40%
Moly-Cote (Molybdenum Disulphide)	0%	-20%	-45%
Heavy Oils	0%	-20%	-40%
Graphite	0%	-20%	-30%
White Lead	0%	-20%	-25%

Appendix B.1 Lubricants and threaded fasteners. *(Reproduced with permission of Premier Farnell Corporation)*

Size	Threads per inch			Outside Diameter Inches	Tap Drill Approx. 75% Full Thread	Decimal Equivalent of Tap Drill	Size	Threads per inch			Outside Diameter Inches	Tap Drill Approx. 75% Full Thread	Decimal Equivalent of Tap Drill
	NC	NF	NS					NC	NF	NS			
0	…	80	…	.0600	3/64	.0469	10	…	…	28	.1900	23	.1540
1	…	…	56	.0730	54	.0550	10	…	…	30	.1900	22	.1570
1	64	…	…	.0730	53	.0595	10	…	32	…	.1900	21	.1590
1	…	72	…	.0730	53	.0595	12	24	…	…	.2160	16	.1770
2	56	…	…	.0860	50	.0700	12	…	28	…	.2160	14	.1820
2	…	64	…	.0860	50	.0700	12	…	…	32	.2160	13	.1850
3	48	…	…	.0990	47	.0785	1/4	20	…	…	.2500	7	.2010
3	…	56	…	.0990	45	.0820	1/4	…	28	…	.2500	3	.2130
4	…	…	32	.1120	45	.0820	5/16	18	…	…	.3125	F	.2570
4	…	…	36	.1120	44	.0860	5/16	…	24	…	.3125	I	.2720
4	40	…	…	.1120	43	.0890	3/8	16	…	…	.3750	5/16	.3125
4	…	48	…	.1120	42	.0935	3/8	…	24	…	.3750	Q	.3320
5	…	…	36	.1250	40	.0980	7/16	14	…	…	.4375	U	.3680
5	40	…	…	.1250	38	.1015	7/16	…	20	…	.4375	25/64	.3906
5	…	44	…	.1250	37	.1040	1/2	13	…	…	.5000	27/64	.4219
6	32	…	…	.1380	36	.1065	1/2	…	20	…	.5000	29/64	.4531
6	…	…	36	.1380	34	.1110	9/16	12	…	…	.5625	31/64	.4844
6	…	40	…	.1380	33	.1130	9/16	…	18	…	.5625	33/64	.5156
8	…	…	30	.1640	30	.1285	5/8	11	…	…	.6250	17/32	.5312
8	32	…	…	.1640	29	.1360	5/8	…	18	…	.6250	37/64	.5781
8	…	36	…	.1640	29	.1360	3/4	10	…	…	.7500	21/32	.6562
8	…	…	40	.1640	28	.1405	3/4	…	16	…	.7500	11/16	.6875
10	24	…	…	.1900	25	.1495	7/8	9	…	…	.8750	49/64	.7656

Appendix C.1 A tap drill chart for inch sized drills. *(Courtesy of the L.S. Starrett Company)*

Bore/stroke engine displacement...

STROKE

BORE	3.000	.062	.125	.187	.250	.312	.375	.437	.500	.562	.625	.687	.750	.812	.875	.937	4.000	.062	.125	.187	.250	.312	.375	.437	.500
8.000	170	173	177	180	184	187	191	194	198	201	205	209	212	216	219	223	226	230	233	237	240	244	247	251	254
.062	177	180	184	188	192	195	199	203	206	210	214	217	221	225	228	232	236	239	243	247	250	254	258	261	265
.125	184	188	192	196	199	203	207	211	215	219	222	226	230	234	238	242	245	249	253	257	261	265	268	272	276
.187	192	296	299	203	207	211	215	219	223	227	231	235	239	243	247	251	255	259	263	267	271	275	279	283	287
.250	199	203	207	212	216	220	224	228	232	236	241	245	249	253	257	261	265	270	274	278	282	286	290	295	299
.312	207	211	215	220	224	228	233	237	241	246	250	254	259	263	267	271	276	280	284	289	293	297	302	306	310
.375	215	219	224	228	233	237	242	246	250	255	259	264	268	273	277	282	286	291	296	300	304	309	313	318	322
.437	223	227	232	237	241	246	251	255	260	264	269	274	278	283	288	292	297	302	306	311	316	320	325	329	334
.500	231	236	241	245	250	255	260	265	269	274	279	284	289	293	298	303	308	313	317	322	327	332	337	342	346
.562	239	244	249	254	259	264	269	274	279	284	289	294	299	304	309	314	319	324	329	334	339	344	349	354	359
.625	248	253	258	263	268	273	279	284	289	294	299	304	310	315	320	325	330	335	341	346	351	356	361	366	372
.687	256	262	267	272	278	283	288	294	299	304	310	315	320	326	331	336	342	347	352	358	363	368	374	379	385
.750	265	271	276	282	287	293	298	304	309	315	320	326	331	337	342	348	353	359	364	370	376	381	387	392	398
.812	274	280	285	291	297	303	308	314	320	325	331	337	343	348	354	360	365	371	377	382	388	394	400	405	411
.875	283	289	295	301	307	313	318	324	330	336	342	348	354	360	366	371	377	383	389	395	401	407	413	419	425
.937	292	298	304	310	317	323	329	335	341	347	353	359	365	371	378	384	390	396	402	408	414	420	426	432	438
4.000	301	308	314	320	327	333	339	346	352	358	364	371	377	383	390	396	402	408	415	421	427	434	440	446	452
.062	311	318	324	331	337	344	350	356	363	369	376	382	389	395	402	408	415	421	428	434	441	447	454	460	467
.125	321	328	334	341	348	354	361	368	374	381	388	394	401	408	414	421	428	434	441	448	454	461	468	475	481
.187	330	337	344	351	358	365	372	379	385	392	399	406	413	420	427	434	441	447	454	461	468	475	482	489	496
.250	341	348	355	362	369	376	383	390	397	404	411	419	426	433	440	447	454	461	468	475	482	489	497	504	511
.312	351	358	365	373	380	387	394	402	409	416	424	431	438	446	453	460	467	475	482	489	497	504	511	519	526
.375	361	368	376	383	391	398	406	414	421	429	436	444	451	459	466	474	481	489	496	504	511	519	526	534	541
.437	371	379	387	394	402	410	418	425	433	441	449	456	464	472	479	487	495	503	510	518	526	534	541	549	557
.500	382	390	398	406	414	421	429	437	445	453	461	469	477	485	493	501	509	517	525	533	541	549	557	565	573

The above chart is for 8 cylinder engines. For the dispacement of a 6 cylinder engine multiply the 8 cylinder figure by .75. For the dispacement of a 4 cylinder engine multiply the 8 cylinder figure by .5 or divide by 2

Appendix D.1 Bore/stroke engine displacement. *(Courtesy of Crane Cams, Inc.)*

AMERICAN WIRE GAUGE SIZES

Gauge Size	Conductor Diameter (Inch)	Cross Section Area (Circular Mils)
20	.032″	1,020
16	.051″	2,580
12	.081″	6,530
8	.128″	16,500
2	.258″	66,400
0	.325″	106,000
2/0	.365″	133,000

AWG Size	Metric Size
20	0.5
18	0.8
16	1.0
14	2.0
12	3.0
10	5.0
8	8.0
6	13.0
4	19.0

Appendix E.1 American wire gauge sizes.

Total Approx. Circuit Amperes	Total Circuit Watts	Total Candle Power	Wire Gage (For Length in Feet)											
12V	12V	12V	3'	5'	7'	10'	15'	20'	25'	30'	40'	50'	75'	100'
1.0	12	6	18	18	18	18	18	18	18	18	18	18	18	18
1.5		10	18	18	18	18	18	18	18	18	18	18	18	18
2	24	16	18	18	18	18	18	18	18	18	18	18	16	16
3		24	18	18	18	18	18	18	18	18	18	18	14	14
4	48	30	18	18	18	18	18	18	18	18	16	16	12	12
5		40	18	18	18	18	18	18	18	18	16	14	12	12
6	72	50	18	18	18	18	18	18	16	16	16	14	12	10
7		60	18	18	18	18	18	18	16	16	14	14	10	10
8	96	70	18	18	18	18	18	16	16	16	14	12	10	10
10	120	80	18	18	18	18	16	16	16	14	12	12	10	10
11		90	18	18	18	18	16	16	14	14	12	12	10	8
12	144	100	18	18	18	18	16	16	14	14	12	12	10	8
15		120	18	18	18	18	14	14	12	12	12	10	8	8
18	216	140	18	18	16	16	14	14	12	12	10	10	8	8
20	240	160	18	18	16	16	14	12	10	10	10	10	8	6
22	264	180	18	18	16	16	12	12	10	10	10	8	6	6
24	288	200	*18	18	16	16	12	12	10	10	10	8	6	6
30			18	16	16	14	10	10	10	10	10	6	4	4
40			18	16	14	12	10	10	8	8	6	6	4	2
50			16	14	12	12	10	10	8	8	6	6	2	2
100			12	12	10	10	6	6	4	4	4	2	1	1/0
150			10	10	8	8	4	4	2	2	2	1	2/0	2/0
200			10	8	8	6	4	4	2	2	1	1/0	4/0	4/0

HOW TO USE CHART

1
Measure Length of Wire in Circuit — Chart applies to ground return. Two-wire circuits will be total of both wire lengths. Be sure to include both vehicles on auto and trailer applications.

2
Find the total amperes, watts or candlepower and choose nearest value in proper column.

3
Move horizontally to proper footage column and find nearest wire gage.

Based on maximum of 10% voltage drop

*18 gage indicated above this line could be 20 gage electrically — 18 gage is recommended for mechanical strength.

Appendix F.1 *(Courtesy of Cooper Automotive/NAPA Belden)*

AUTO LIGHT REPLACEMENT GUIDE MINIATURE BULBS

	DESCRIPTION	GE LAMP NO.	REPLACES		DESCRIPTION	GE LAMP NO.	REPLACES
	Heavy Duty Turn Signal, Stop, Tail, Parking Light	1157	1034		Foreign Car Interior Dome Light	DE3022	–
	Heavy Duty Turn Signal, Parking Light	1157NA	1157A, 1034A, 1034NA		Foreign Car Interior Dome Light	DE3175	–
	Heavy Duty Turn Signal, Stop, Tail, Parking Light	2057	–		Dome, Courtesy Light	1004	–
	Heavy Duty Turn Signal, Parking Light	2057NA	–		Heavy Duty Instrument Indicator Light	1895	57
	Heavy Duty Back-Up Turn Signal Light High Mount Stop	1156	1073		License, Parking Light	97	67, 1155
	Back-Up Light	1141	–		Dome, Courtesy Light	89	–
					Dome, Courtesy Light	1003	93, 105
	Heavy Duty Indicator, Instrument, Side Marker, License Light	194	158		Indicator, Instrument Light	1445	53
		168	–				
	Dome, Courtesy Light	211-2	211-1		Dome, Map, Courtesy Light	90	–
	Dome, Map, High Mount Stop	561	–		Dome, Map, High Mount Stop	912	
						921	
	Turn Signal, Stop, Tail, Parking Light	*3057	2457, 2358		Turn Signal, Parking Light	3057NA	2457NA & 2358NA
		*3157	2458, 2359			3157NA	3458NA & 2359NA

*NOTE: The above lamp sockets allow interchangeability of the double filament lamp types (3057, 3157, 3057NA, 3157NA). To comply with the DOT photometric specifications however, it is necessary to replace lamps with the correct type. Double filament lamps (3057, 3157) will fit the single filament socket (3156, 2456, 2356) and can be substituted. They will provide the correct photometric output.

Appendix G.1 *(Courtesy of General Electric)*

TABLE I	LOAD CAPACITY INDEX CHART				
Load Index	Mx. Load Per Tire Kg.	Lbs.	Load Index	Mx. Load Per Tire Kg.	Lbs.
60	250	551	88	560	1234
61	257	566	89	580	1278
62	265	584	90	600	1323
63	272	599	91	615	1356
64	280	617	92	630	1389
65	290	639	93	650	1433
66	300	661	94	670	1477
67	307	676	95	690	1521
68	315	694	96	710	1565
69	325	716	97	730	1609
70	335	738	98	750	1653
71	345	760	99	775	1708
72	355	783	100	800	1764
73	365	804	101	825	1819
74	375	826	102	850	1874
75	387	853	103	875	1919
76	400	882	104	900	1984
77	412	908	105	925	2039
78	425	937	106	950	2094
79	437	963	107	975	2149
80	450	992	108	1000	2205
81	462	1018	109	1030	2271
82	475	1047	110	1060	2337
83	487	1074	111	1090	2403
84	500	1102	112	1120	2469
85	515	1135	113	1160	2535
86	530	1168	114	1180	2601
87	545	1201			

Appendix H.1 *(Courtesy of Pirelli Tire Corporation)*

PREALIGNMENT INSPECTION CHECKLIST

Owner _____ Phone _____ Date _____|_____|_____

LAST　　　　　　　　FIRST

Address _____ Ser. No. _____

Make _____ Model _____ Yr. _____ Lic. No. _____ Mileage _____

1. Road Test Disclosed

Vehicle Pulls		Yes	No	Right	Left
	Above 30 MPH				
	Below 30 MPH				
	Bump Steer				
	When Braking				

	Yes	No	Front	Rear
Steering Wheel Movement When Stopping From 2-3 MPH—Front Brakes				
Vehicle Steers Hard				
Steering Wheel Returns Normally				
Steering Wheel Position				
Vibration				

2. Tire Pressure   SPECS Front _____ Rear _____

RECORD PRESSURE FOUND
RF _____ LF _____ RR _____ LR _____

3. Chassis Height   SPECS Front _____ Rear _____

RECORD HEIGHT FOUND
RF _____ LF _____ RR _____ LR _____

	YES	NO
Springs Sagged		
Torsion Bars Adjusted		

4. Rubber Bushings   OK

	OK
Upper Control Arm	
Lower Control Arm	
Sway Bar/Stabilizer Link	
Strut Rod	
Rear Bushing	

5. Shock Absorbers/Struts

	Front	Rear

6. Steering Linkage

	Rear OK	Front OK
Tie-Rod Ends		
Idler Arm		
Center Link		
Sector Shaft		
Pitman Arm		
Gearbox/Rack Adjustment		
Gearbox/Rack Mounting		

7. Ball Joints   OK

	OK
Load Bearings	

SPECS		READINGS	
Right _____	Left _____	Right _____	Left _____

Follower	
Upper Strut Bearing Mount	
Rear	

8. Power Steering   OK

	OK
Belt Tension	
Fluid Level	
Leaks/Hose Fittings	
Spool Valve Centered	

9. Tires/Wheels   OK

	OK
Wheel Runout	
Condition	
Equal Tread Depth	
Wheel Bearing	

10. Brakes Operating Properly

11. Alignment

	Specs		Initial Readings		Adj. Readings	
	Right	Left	Right	Left	Right	Left
Camber						
Caster						
Toe						

Bump Steer	TOE CHANGE Right Wheel		TOE CHANGE Left Wheel	
	Amount	Direction	Amount	Direction
Chassis DOWN 3″				
Chassis UP 3″				

	Specs		Initial Readings		Adj. Readings	
	Right	Left	Right	Left	Right	Left
Toe-Out On Turns						
SAI						
Rear Camber						
Rear Total Toe						
Rear Indiv. Toe						

Wheel Balance	Front _____ Rear _____
Radial Tire Pull	Yes _____ No _____

FLAG READINGS

COMMENTS:

Presure vs. Vacuum			
PSIG	PSI (abs)	BAR	IN.Hg
2	16	1.14	–
1	15	1.07	–
0	14.2	1.00	0
–1	13	0.94	2
–2	12	0.87	4
–3	11	0.80	6
–4	10	0.72	9
–5	9	0.64	11
–6	8	0.57	13
–7	7	0.50	15
–8	6	0.44	18
–9	5	0.37	20
–10	4	0.30	22
–11	3	0.22	24
–12	2	0.14	26
–13	1	0.07	28
–14	0	0.00	30

Appendix J.1

°C																				
–30	-25	-20	-17.8	-15	-10	-5	0	5	10	15	20	25	30	35	40	50	60	70	80	90
100																				
–22	-13	-4	-0	-5	-14	-23	32	41	50	59	68	77	86	95	104	122	140	158	176	194

Appendix K.1

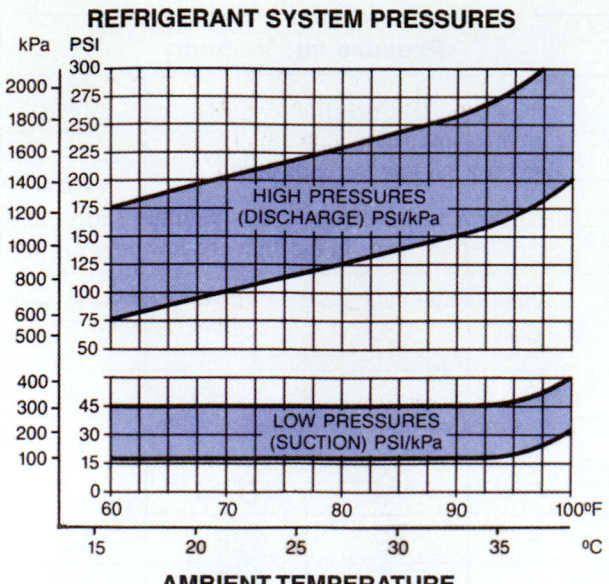

Ambient Temp °F	High Side PSIG, R-12	Low Side PSIG, R-12	High Side PSIG,R-134a	Low Side PSIG, R-134a
60	120–150	5–15	120–170	7–15
70	140–180	8–16	150–250	8–16
80	160–250	10–18	190–280	10–20
90	200–280	12–25	220–330	15–25
100	220–300	15–30	250–350	20–30
110	250–320	20–35	280–400	25–40

Appendix L.1 *(Top, Courtesy of Ford Motor Company)*

ENGINE REPAIR (TEST A1)	
A. GENERAL ENGINE DIAGNOSIS	Ch. 42 Ch. 43 Ch. 50
1. Verify driver's complaint and/or road test vehicle; determine necessary action.	Ch. 42 Ch. 43 Ch. 50 Road Test
2. Determine if no-crank, no-start, or hard starting condition is an ignition system, cranking system, fuel system or engine mechanical problem.	Ch. 42
3. Inspect engine assembly for fuel, oil, coolant and other leaks; determine necessary action.	Ch. 42 Ch. 43
4. Listen to engine noises; determine necessary action.	Ch. 43
5. Diagnose the cause of excessive oil consumption, coolant consumption, unusual engine exhaust color, odor and sound; determine necessary action.	Ch. 43
6. Perform engine vacuum tests;determine necessary action.	Ch. 42
7. Perform cylinder power balance tests; determine necessary action.	Ch. 42
8. Perform cylinder compression tests; determine necessary action.	Ch. 42
9. Perform cylinder leakage tests; determine necessary action.	Ch. 42
B. CYLINDER HEAD AND VALVE TRAIN DIAGNOSIS AND REPAIR	Ch. 15 Ch. 18 Ch. 44 Ch. 46
1. Remove cylinder heads, disassemble, clean, and prepare for inspection according to manufacturer's procedures.	Ch. 15 Ch. 18 Description Ch. 44 Ch. 46
2. Visually inspect cylinder heads for cracks, gasket surface areas for warpage, corrosion and leakage, and check passage condition.	Ch. 15 Ch. 18 Description Ch. 44 Ch. 46
3. Inspect and test valve springs for squareness, pressure and free height comparison; replace as necessary.	Ch. 15 Ch. 18 Description Ch. 46
4. Inspect valve spring retainers, rotators, locks and valve lock grooves.	Ch. 46
5. Replace valve stem seals.	Ch. 46
6. Inspect valve guides for wear; check valve guide height and stem-to-guide clearance; determine needed repairs.	Ch. 15 Ch. 18 Description Ch. 46
7. Inspect valves; resurface or replace according to manufacturer's procedures.	Ch. 46
8. Inspect and resurface valve seats according to manufacturer's procedures.	Ch. 15 Ch. 18 Description Ch. 46
9. Check valve face-to-seat contact and valve seat concentricity (runout).	Ch. 46

10. Check valve spring installed (assembled) height and valve stem height; service valve and spring assemblies as necessary.	Ch. 46
11. Inspect pushrods, rocker arms, rocker arm pivots and shafts for wear, bending, cracks, looseness and blocked oil passages; repair or replace as required.	Ch. 46
12. Inspect and replace hydraulic or mechanical lifters/lash adjusters.	Ch. 15 Ch. 18 Description Ch. 46
13. Adjust valves on engines with mechanical or hydraulic lifters.	Ch. 46
14. Inspect and replace camshaft drives (includes checking gear wear and backlash, sprocket and chain wear, overhead cam drive sprockets, drive belts, belt tension, tensioners and cam sensor components).	Ch. 15 Ch. 18 Description Ch. 46
15. Inspect and measure camshaft journals and lobes.	Ch. 15 Ch. 18 Description Ch. 46
16. Inspect and measure camshaft bore for wear damage, out-of-round and alignment; repair or replace according to manufacturer's specifications.	Ch. 15 Ch. 18 Description Ch. 44
17. Time camshaft(s) to crankshaft.	Ch. 46
18. Reassemble and install cylinder heads and gaskets; replace and tighten fasteners according to manufacturer's procedures.	Ch. 15 Ch. 18 Description Ch. 45 Ch. 46

C. ENGINE BLOCK DIAGNOSIS AND REPAIR

	Ch. 15 Ch. 19 Ch. 44 Ch. 45 Ch. 47 Ch. 48 Ch. 62 Ch. 63
1. Disassemble engine block and clean and prepare components for inspection.	Ch. 44 Engine Disassembly Ch. 47 Cleaning
2. Visually inspect engine block for cracks, corrosion, passage condition, core and gallery plug holes, and surface warpage; determine necessary action.	Ch. 15 Ch. 19 Description Ch. 47 Service
3. Inspect and repair damaged threads where allowed; install core and gallery plugs.	Ch. 47
4. Inspect and measure cylinder walls; remove cylinder wall ridges; hone and clean cylinder walls; determine need for further action.	Ch. 44 Ch. 47
5. Visually inspect crankshaft for surface cracks and journal damage; check oil passage condition; check crankshaft sensor reluctor ring (where applicable); determine necessary action.	Ch. 15 Ch. 19 Description Ch. 47 Service
6. Inspect and measure main bearing bores and cap alignment and fir.	Ch. 19 Description Ch. 47 Service
7. Install main bearings and crankshaft; check bearing clearances and endplay; replace/retorque bolts according to manufacturers' procedures.	Ch. 19 Description Ch. 47 Service
8. Inspect camshaft bearings for unusual wear; remove and and replace camshaft bearings; install camshaft timing, timing chain, and gears; check endplay.	Ch. 18 Description Ch. 46 Service

9. Inspect auxiliary (balance, intermediate, idler, counterbalance or silencer) shaft(s) and support bearings for damage and wear; determine necessary action.	Ch. 18 Description
10. Inspect, measure, service,repair, or replace pistons, piston pins and pin bushings; identify piston and bearing wear patterns that indicate connecting rod alignment problems; determine necessary action.	Ch. 15 Ch. 19 Description Ch. 47 Ch. 48 Service
11. Inspect connection rods for damage, alignment, bore condition, and pit fit; determine necessary action.	Ch. 19 Description Ch. 48 Service
12. Inspect, measure and install or replace piston rings; assemble piston and connecting rod; install piston/rod assembly; check bearing clearance and sideplay; replace/retorque fasteners according to manufacturers' procedures.	Ch. 15 Ch. 19 Description Ch. 48 Service
13. Inspect, reinstall, or replace crankshaft vibration damper (harmonic balancer).	Ch. 15 Ch. 19 Description Ch. 47 Service
14. Inspect crankshaft flange and flywheel mating surfaces; inspect, remove and replace crankshaft pilot bearing/bushing (if applicable); inspect flywheel/flexplate for cracks and wear (includes flywheel ring gear); measure flywheel runout; determine necessary action	Ch. 62 Description Ch. 63
15. Inspect and replace pans, covers, gaskets and seals.	Ch. 45
16. Assemble engine parts using formed-in-place (tube-applied) sealants or gaskets.	Ch. 45

D. LUBRICATION AND COOLING SYSTEMS DIAGNOSIS AND REPAIR

D. LUBRICATION AND COOLING SYSTEMS DIAGNOSIS AND REPAIR	Ch. 20 Ch. 21 Ch. 22 Ch. 23 Ch. 12 Ch. 47 Ch. 50
1. Perform oil pressure tests; determine necessary action.	Ch. 47
2. Disassemble, inspect, measure and repair oil pump (includes gears, rotors, housing and pickup assembly), pressure relief valves and pump drive; replace oil filter.	Ch. 12 Oil Filter Ch. 47 Inspection Ch. 50 Installation
3. Perform cooling system tests; determine necessary action.	Ch. 21
4. Inspect, replace and adjust drive belts, tensioners, and pulleys.	Ch. 21 Ch. 22
5. Inspect and replace engine cooling and heater system hoses.	Ch. 23
6. Inspect, test and replace thermostat, bypass and housing.	Ch. 20 Description Ch. 21 Service
7. Inspect coolant; drain, flush and refill cooling system with recommended coolant and bleed air as required.	Ch. 21
8. Inspect, test and replace water pump.	Ch. 20 Description Ch. 21 Service
9. Inspect, test and replace radiator, heater core, pressure cap, and coolant recovery system.	Ch. 20 Description Ch. 21 Service
10. Clean, inspect, test and replace fan (both electrical and mechanical), fan clutch, fan shroud, air dams and cooling-related temperature sensors.	Ch. 20 Description Ch. 21 Service
11. Inspect, test and repair or replace auxiliary oil coolers.	Not covered

E. FUEL, ELECTRICAL, IGNITION, AND EXHAUST SYSTEMS INSPECTION AND SERVICE

Ch. 14
Ch. 26
Ch. 27
Ch. 28
Ch. 30
Ch. 31
Ch. 32
Ch. 38
Ch. 39
Ch. 42

1. Inspect, clean or replace fuel and air induction system components, intake manifold, and gaskets.	Ch. 26 Description Ch. 27 Service
2. Inspect, service or replace air filters, filter housings and intake ductwork.	Ch. 28 Description
3. Inspect turbocharger/supercharger; determine necessary action.	Ch. 28
4. Test battery; charge as necessary.	Ch. 30 Description of Batteries and Parts Ch. 31 Service
5. Remove and replace starter.	Ch. 32 Description Ch. 33 Service
6. Inspect and replace positive crankcase ventilation (PCV) system components.	Ch. 40 Description Ch. 41 Service
7. Visually inspect and reinstall primary and secondary ignition system components; time distributor.	Ch. 38 Description Ch. 39 Service
8. Inspect, service and replace exhaust manifold.	Ch. 14 Inspection Ch. 28 Description and Service

AUTOMATIC TRANSMISSION/TRANSAXLE (TEST A2)

A. GENERAL TRANSMISSION/TRANSAXLE DIAGNOSIS

Ch. 50
Ch. 66
Ch. 67
Ch. 72

1. MECHANICAL/HYDRAULIC SYSTEMS

1. Listen to driver's concern and road test vehicle to verify mechanical/hydraulic system problems; determine necessary action.	Ch. 50 Road Test Ch. 67
2. Diagnose noise and vibration problems; determine necessary action.	Ch. 67 Noises Ch. 72 Vibration
3. Diagnose unusual fluid usage, type, level and condition problems; determine necessary action.	Ch. 66 Description Ch. 67 Description & Service
4. Perform pressure tests; determine necessary action.	Ch. 67
5. Perform stall tests; determine necessary action.	Ch. 67
6. Perform lock-up converter mechanical/hydraulic system tests; determine necessary action.	Ch. 66 Description
7. Diagnose mechanical and vacuum control systems; determine necessary action.	Ch. 66 Description Ch. 67 Service

2. ELECTRONIC SYSTEM

Ch. 66 Description

1. Listen to driver's concern and road test vehicle to verify electronic system problems; determine necessary action.	Ch. 67 & 76 Road Test

2. Perform pressure tests; determine necessary action.	Ch. 67
3. Perform lock-up converter electronic system tests; determine necessary action.	Not Covered
4. Diagnose electronic transmission control systems using appropriate test equipment; determine necessary action.	Ch. 76
5. Verify proper operation of starting and charging systems; check battery connections and vehicle grounds.	Ch. 31 Ch. 33 Ch. 35 Ch. 76

B. TRANSMISSION/TRANSAXLE MAINTENANCE AND ADJUSTMENT

B. TRANSMISSION/TRANSAXLE MAINTENANCE AND ADJUSTMENT	Ch. 62 Ch. 66 Ch. 67
1. Inspect, adjust and replace manual valve shift linkage and transmission range sensor/switch.	Ch. 66 Description Ch. 67 Service
2. Inspect, adjust and replace cables or linkages for throttle valve (TV), kickdown and accelerator pedal.	Ch. 62 Clutch Linkage Ch. 67 Kickdown Cable or Linkage
3. Adjust bands where applicable.	Ch. 66 Description Ch. 67 Service
4. Replace fluid and filter(s).	Ch. 67

C. IN-VEHICLE TRANSMISSION/TRANSAXLE REPAIR

C. IN-VEHICLE TRANSMISSION/TRANSAXLE REPAIR	Ch. 66 Ch. 67 Ch. 76
1. Inspect, adjust and replace vacuum modulator, valve, lines and hoses.	Ch. 66 Description Ch. 67 Service
2. Inspect, adjust, repair and replace governor cover, seals, sleeve/bore, valve, weight, springs, retainers and gear.	Ch. 66 Description Ch. 67 Service
3. Inspect and replace external seals and gaskets.	Ch. 66 Description Ch. 67 Service
4. Inspect and repair or replace extension housing; replace bushing.	Ch. 67
5. Check condition of engine cooling system; inspect, test, flush and replace transmission cooler, lines and fittings.	Ch. 66 Description Ch. 67 Service
6. Inspect and replace speedometer/speed sensor drive gear, driven gear and retainers.	Ch. 66 Description Ch. 67 Service
7. Inspect valve body mating surfaces, bores, valves, springs, sleeves, retainers, brackets, check balls, screens, spacers and gaskets; replace as necessary.	Ch. 66 Description Ch. 67 Service
8. Check/adjust valve body bolt torque.	Ch. 66 Description
9. Inspect servo bore, piston, seals, pin, spring and retainers; repair/replace as necessary.	Ch. 67 Service
10. Inspect accumulator bore, piston, seals, spring and retainers; repair or replace as necessary.	Ch. 66 Description
11. Inspect and replace parking pawl, shaft, spring and retainer.	Ch. 66 Description
12. Inspect, test, adjust or replace electrical/electronic components including computers, solenoids, sensors, relays, fuses, terminals, connector, switches and harnesses.	Ch. 66 Description Ch. 67 Service
13. Inspect, replace and align power train mounts.	Ch. 44 Engine Mounts

D. OFF-VEHICLE TRANSMISSION/TRANSAXLE REPAIR

D. OFF-VEHICLE TRANSMISSION/TRANSAXLE REPAIR	Ch. 47 Ch. 50 Ch. 66 Ch. 67 Ch. 69 Ch. 70 Ch. 71

1. REMOVAL, DISASSEMBLY AND ASSEMBLY	Ch. 67
1. Remove and replace transmission/transaxle; inspect engine core plugs, transmission dowel pins, and dowel pin holes.	Ch. 67
2. Disassemble, clean and inspect.	Ch. 67
3. Assemble after repair.	Ch. 67
4. Inspect converter flex (drive) plate, converter attaching bolts, converter pilot and converter pump drive surfaces.	Ch. 66 Description Ch. 67 Service
2. GEAR TRAIN, SHAFTS, BUSHINGS, OIL PUMP, AND CASE	Ch. 47 Ch. 50 Ch. 66 Ch. 67
1. Inspect, measure and replace oil pump housings and parts.	Ch. 47 Inspection Ch. 50 Installation (ch. 66-67 fluid pump)
2. Check end play and/or preload; determine needed service.	Ch. 67 Endplay Ch. 69 Preload
3. Inspect, measure and replace thrust washers and bearings.	Ch. 67
4. Inspect and replace shafts.	Ch. 66 Description
5. Inspect oil delivery seal rings, ring grooves and sealing surface areas, feed pipes, orifices, and ecapsulated check valves (balls).	Ch. 67
6. Inspect and replace bushings.	Ch. 67
7. Inspect and measure planetary gear assembly; replace parts as necessary.	Ch. 66 Description Ch. 69 Service
8. Inspect, repair and replace case(s) bores, passages, bushings, vents and mating surfaces and dowel pins.	Not Covered
9. Inspect, repair or replace transaxle drive chains, sprockets, gears, bearings and bushings.	Ch. 70 Description Ch. 71 Service
10. Inspect, measure, repair, adjust or replace transaxle final drive components.	Ch. 68 Differential Description Ch. 69 Service
3. FRICTION AND REACTION UNITS	Ch. 66 Ch. 67
1. Inspect clutch assembly; replace parts as necessary.	Ch. 66 Description Ch. 67 Service
2. Measure and adjust clutch pack clearance.	Ch. 66 Description Ch. 67 Service
3. Air test the operation of clutch and servo assemblies.	Ch. 67
4. Inspect one-way clutch assemblies; replace parts as necessary.	Ch. 66 Description
5. Inspect and replace bands and drums.	Ch. 66 Description Ch. 67 Service

MANUAL DRIVE TRAIN AND AXLES (TEST A3)

A. CLUTCH DIAGNOSIS AND REPAIR	Ch. 44 Ch. 47 Ch. 49 Ch. 62 Ch. 63
1. Diagnose clutch noise, binding, slippage, pulsation, chatter, pedal feel/effort, and release problems; determine needed repairs.	Ch. 63
2. Inspect, adjust, replace clutch pedal linkage, cables and automatic adjuster mechanisms, brackets, bushings, pivots and springs.	Ch. 62 Description Ch. 63 Service
3. Inspect, adjust, replace, and bleed hydraulic clutch slave and master cylinders, lines and hoses.	Ch. 62 Description Ch. 63 Service
4. Inspect, adjust and replace release (throw-out) bearing, lever and pivot.	Ch. 62 Description Ch. 63 Service
5. Inspect and replace clutch disc and pressure plate assembly.	Ch. 62 Description Ch. 63 Service
6. Inspect and replace pilot bearing.	Ch. 62 Description Ch. 63 Service
7. Inspect and measure flywheel and ring gear; repair or replace as necessary.	Ch. 62 Description Ch. 63 Service
8. Inspect engine block, clutch (bell) housing, and transmission case mating surfaces; determine needed repairs.	Ch. 63
9. Measure flywheel-to-block runout and crankshaft end play; determine needed repairs.	Ch. 47 Thrust Bearing Wear Ch. 49
10. Measure clutch (bell) housing bore-to-crankshaft runout and face squareness; determine needed repairs.	Ch. 63
11. Inspect, replace and align power train mounts.	Ch. 44
B. TRANSMISSION DIAGNOSIS AND REPAIR	Ch. 63 Ch. 64 Ch. 65
1. Diagnose transmission noise, hard shifting, jumping out of gear, and fluid leakage problems; determine needed repairs.	Ch. 63 Noise Ch. 65
2. Inspect, adjust and replace transmission external shifter assembly, shift linkages, brackets, bushings grommets, pivots and levers.	Ch. 64 Description Ch. 65 Service
3. Inspect and replace transmission gaskets, sealants, seals and fasteners; inspect sealing surfaces.	Ch. 64 Description Ch. 65 Service
4. Remove and replace transmission; inspect transmission mounts.	Ch. 65
5. Disassemble and clean transmission components; reassemble transmission.	Ch. 65
6. Inspect, repair and/or replace transmission shift cover and internal shift forks, bushings, levers, shafts, sleeves, detent mechanisms, interlocks and springs.	Ch. 64 Description Ch. 65 Service
7. Inspect and replace input (clutch) shaft, bearings, and retainers.	Ch. 64 Description Ch. 65 Service
8. Inspect and replace main shaft, gears, thrust washers, bearings and retainers/snap rings.	Ch. 64 Description Ch. 65 Service
9. Inspect and replace synchronizer hub, sleeve, keys (inserts), springs and blocking (synchronizing) rings; measure bocking ring clearance.	Ch. 64 Description Ch. 65 Service
10. Inspect and replace counter (cluster) gear, shaft, bearings, thrust washers and retainers.	Ch. 64 Description Ch. 65 Service

11. Inspect and replace reverse idler gear, shaft, bearings, thrust washers and retainers/snap rings.	Ch. 64 Description Ch. 65 Service
12. Measure and adjust shaft, gear and synchronizer end play.	Ch. 65
13. Measure and adjust bearing preload.	Not Covered
14. Inspect, repair and replace extension housing and transmission case mating surfaces, bores, bushings and vents.	Ch. 64 Description Ch. 65 Service
15. Inspect and replace speedometer drive gear, driven gear and retainers.	Ch. 64 Description Ch. 65 Inspection
16. Inspect, test and replace transmission sensors and switches.	Ch. 64 Description
17. Inspect lubrication devices; check fluid level, and refill with proper fluid.	Ch. 64 Description Ch. 65 Service

C. TRANSAXLE DIAGNOSIS AND REPAIR

	Ch. 64 Ch. 65 Ch. 68 Ch. 69 Ch. 70 Ch. 71
1. Diagnose transaxle noise, hard shifting, jumping out of gear and fluid leakage problems; determine needed repairs.	Ch. 69 Ch. 71
2. Inspect, adjust and replace transaxle external shift assemblies, linkages, brackets, bushings/grommets, cables, pivots and levers.	Ch. 70 Description (Ch. 64 Description and Ch. 65 Service for RWD Manual Transmission)
3. Inspect and replace transaxle gaskets, sealants, seals and fasteners; inspect sealing surfaces.	Ch. 64 Description Ch. 65 Service for RWD Manual Transmission
4. Remove and replace transaxle; inspect, replace, and align transaxle mounts.	Ch. 71
5. Disassemble and clean transaxle components; reassemble transaxle.	Ch. 71
6. Inspect, repair and/or replace transaxle shift cover, and internal shift forks, levers, bushings, shafts, sleeves, detent mechanisms, interlocks and springs.	Ch. 64 Description Ch. 65 Service for RWD Manual Transmission
7. Inspect and replace input shaft and bearings.	Ch. 70 Description Ch. 64 Description Ch. 65 Service for RWD Manual Transmission
8. Inspect and replace output shaft, gears, thrust washers, bearings and retainers/snap rings.	Ch. 70 Description Ch. 64 Description Ch. 65 Service for RWD Manual Transmission
9. Inspect and replace synchronizer hub, sleeve, keys (inserts), springs, and blocking (synchronizing) rings; measure blocking ring clearance.	Ch. 64 Description Ch. 65 Service for RWD Manual Transmission
10. Inspect and replace reverse idler gear, shaft, bearings, thrust washers, and retainers/snap rings.	Ch. 64 Description Ch. 65 Service for RWD Manual Transmission

11. Inspect, repair, and replace transaxle case mating surfaces, bores, bushings, and vents.	Ch. 64 Description Ch. 65 Service for RWD Manual Transmission
12. Inspect and replace speedometer drive gear, driven gear, and retainers.	Ch. 64 Description Ch. 65 Service for RWD Manual Transmission
13. Inspect, test, and replace transaxle sensors and switches.	Not Covered
14. Diagnose differential assembly noise and vibration problems; determine needed repairs.	Ch. 68 Description Ch. 69 Service
15. Remove and replace differential assembly.	Ch. 68 Description Ch. 69 Service
16. Inspect, measure, adjust and replace differential pinion gears (spiders), shaft, side gears, thrust washers, and case.	Ch. 68 Description Ch. 69 Service
17. Inspect and replace differential side bearings.	Ch. 68 Description Ch. 69 Service
18. Measure shaft end play/preload (shim/spacer selection procedure).	Ch. 68 Description Ch. 69 Service
19. Inspect lubrication devices; check fluid level and refill with proper fluid.	Ch. 68 Description Ch. 69 Service
D. DRIVE (HALF)SHAFT AND UNIVERSAL JOINT/CONSTANT VELOCITY (CV) JOINT DIAGNOSIS AND REPAIR (FRONT AND REAR WHEEL DRIVE)	Ch. 68 Ch. 69 Ch. 70 Ch. 71 Ch. 72
1. Diagnose shaft and universal/CV joint noise and vibration problems; determine needed repairs.	Ch. 69 Ch. 71 Ch. 72
2. Inspect, service and replace shafts, yokes, boots and universal/CV joints.	Ch. 68 Ch. 70 Description Ch. 69 Ch. 71 Ch. 72 Service
3. Inspect, service and replace shaft center support bearings.	Ch. 68 Description Ch. 69 Service
4. Check and correct propeller shaft balance.	Ch. 72
5. Measure shaft runout.	Ch. 72
6. Measure and adjust shaft angles.	Ch. 72
E. REAR WHEEL DRIVE AXLE DIAGNOSIS AND REPAIR	Ch. 55 Ch. 68 Ch. 69 Ch. 72
1. RING AND PINION GEARS	Ch. 68 Ch. 69 Ch. 72
1. Diagnose noise, vibration and fluid leakage problems; determine needed repairs.	Ch. 69 Ch. 72
2. Inspect and replace companion flange and pinion seal; measure companion flange runout.	Ch. 68 Description Ch. 69 Service

3. Measure ring gear runout; determine needed repairs.	Ch. 69
4. Inspect and replace ring and pinion gear set, collapsible spacers, sleeves (shims) and bearings.	Ch. 68 Description Ch. 69 Service
5. Measure and adjust drive pinion depth.	Ch. 69
6. Measure and adjust drive pinion bearing preload (collapsible spacer or shim type).	Ch. 69
7. Measure and adjust differential (side) bearing preload, and ring and pinion backlash (threaded cup or shim type).	Ch. 69
8. Perform ring and pinion tooth contact pattern checks; determine needed adjustments.	Ch. 68 Description
2. DIFFERENTIAL CASE ASSEMBLY	Ch. 68 Ch. 69 Ch. 72
1. Diagnose differential noise and vibration problems;determine needed adjustments.	Ch. 69 Ch. 72
2. Remove and replace differential assembly.	Ch. 69
3. Inspect, measure, adjust and replace differential pinion gears (spiders), shaft, side gears, thrust washers and case.	Ch. 68 Description Ch. 69 Service
4. Inspect and replace differential side bearings.	Ch. 68 Description Ch. 69 Service
5. Measure differential case runout; determine needed repairs.	Ch. 69
3. LIMITED SLIP DIFFERENTIAL	Ch. 68 Ch. 69
1. Diagnose limited slip differential noise, slippage and chatter problems; determine needed repairs.	Ch. 68 Description Ch. 69 Service
2. Inspect, flush and refill with correct lubricant.	Ch. 69
3. Inspect, adjust and replace clutch (cone/plate) pack.	Ch. 68 Description Ch. 69 Service
4. AXLE SHAFTS	Ch. 55 Ch. 68 Ch. 69
1. Diagnose rear axle shaft noise, vibration and fluid leakage problems; determine needed repairs.	Ch. 68 Description Ch. 69 Service
2. Inspect and replace rear axle shaft wheel studs.	Ch. 55
3. Remove, inspect,and replace rear axle shafts, seals, bearings and retainers.	Ch. 68 Description Ch. 69 Service
4. Measure rear axle flange runout and shaft end play; determine needed repairs.	Not Covered
F. FOUR-WHEEL DRIVE COMPONENT DIAGNOSIS AND REPAIR	Ch. 68 Ch. 69
1. Diagnose four-wheel drive assembly noise, vibration, shifting and steering problems; determine needed repairs.	Not Covered
2. Inspect, adjust and repair transfer case shifting mechanisms, bushings, mounts, levers and brackets.	Ch. 68 Description
3. Remove and replace transfer case.	Ch. 68 Description
4. Disassemble and clean transfer case and components; reassemble transfer case.	Not Covered
5. Inspect and service transfer case and internal components; check lube level.	Ch. 68 Description
6. Inspect, service and replace front-drive (propeller) shafts and universal/CV joints.	Ch. 69
7. Inspect, service and replace front-drive axle knuckles and driving shafts.	Not Covered

8. Inspect, service and replace front wheel bearings and locking hubs.	Ch. 68 Description
9. Check transfer case and front axle seals and remote vents.	Not Covered
10. Diagnose, test, adjust, and replace electrical/electronic components of 4WD systems.	Not Covered

SUSPENSION AND STEERING (TEST A4)

A. STEERING SYSEMS DIAGNOSIS AND REPAIR	Ch. 22 Ch. 23 Ch. 45 Ch. 58 Ch. 59 Ch. 76
1. STEERING COLUMNS AND MANUAL STEERING GEARS	Ch. 58 Ch. 59
1. Diagnose steering column noises, looseness and binding problems (including tilt mechanisms); determine needed repairs.	Ch. 58 Steering Column Description Ch. 59 Diagnosis
2. Diagnose manual steering gear (non-rack and pinion type)noises, binding, uneven turning effort, looseness, hard steering and lubricant leakage problems; determine needed repairs.	Ch. 58 Manual Steering Gear Description Ch. 59 Diagnosis
3. Diagnose rack and pinion steering gear noises, vibration, looseness and hard steering problems; determine needed repairs.	Ch. 58 Rack and Pinion Gear Description Ch. 59 Diagnosis
4. Inspect and replace steering shaft U-joint(s), flexible coupling(s), collapsible columns, steering wheels	Ch. 59
5. Remove and replace manual steering gear (non-rack and pinion type) (includes vehicles equipped with air bags and/or other steering wheel mounted controls and components).	Ch. 59
6. Adjust manual steering gear (non-rack and pinion type) worm bearing preload and sector lash.	Ch. 59
7. Remove and replace rack and pinion steering gear (includes vehicles equipped with air bags and/or other steering wheel mounted controls and components).	Ch. 59
8. Adjust rack and pinion steering gear.	Ch. 59
9. Inspect and replace rack and pinion steering gear inner tie rod ends (sockets) and bellows boots.	Ch. 59
10. Inspect and replace rack and pinion steering gear mounting bushings and brackets.	Ch. 59
2. POWER-ASSISTED STEERING UNITS	Ch. 22 Ch. 23 Ch. 54 Ch. 58 Ch. 59 Ch. 76
1. Diagnose power steering gear (non-rack and pinion) noises, binding, uneven turning effort, looseness, hard steering and fluid leakage problems; determine needed repairs.	Ch. 58 Power Steering Description Ch. 59 Diagnosis
2. Diagnose power rack and pinion steering gear noises, vibration, looseness, hard steering and fluid leakage problems; determine needed repairs.	Ch. 58 Power Steering Description Ch. 59 Diagnosis
3. Inspect power steering fluid level and condition; adjust level in accordance with vehicles manufacturer's recommendations.	Ch. 23 Ch. 59

4. Inspect, adjust tension and alignment, and replace power steering pump belt(s).	Ch. 22 General Information on Servicing Belts
5. Remove and replace power steering pump; inspect pump mounts.	Ch. 58 Power Steering Pump Description
6. Inspect and replace power steering pump seals and gaskets.	Ch. 59
7. Inspect and replace power steering pump pulley.	Ch. 54
8. Perform power steering system pressure and flow tests; determine needed repairs.	Not Covered
9. Inspect and replace power steering hoses, fittings and O-rings.	Ch. 59 Ch. 23
10. Remove and replace power steering gear (non-rack and pinion type) (includes vehicles equipped with air bags and/or other steering wheel mounted controls and components).	Ch. 59 Pittman Arm
11. Remove and replace power rack and pinion steering gear; inspect and replace mounting bushings and brackets (includes vehicles equipped with air bags and/or other steering wheel mounted controls and components).	Ch. 59
12. Adjust power steering gear (non-rack and pinion type) worm bearing preload and sector lash.	Ch. 59
13. Inspect and replace power steering gear (non-rack and pinion type) seals and gaskets.	Ch. 59
14. Adjust power rack and pinion steering gear.	Ch. 59
15. Inspect and replace power rack and pinion steering gear inner tie rod ends (sockets), seals, gaskets, O-rings and bellows boots.	Ch. 59
16. Diagnose, inspect, adjust, repair or replace components of electronically controlled steering systems.	Not Covered
17. Flush, fill and bleed power steering system.	Ch. 23 Ch. 59
18. Diagnose, inspect, repair or replace components of variable-assist steering.	Ch. 59
3. STEERING LINKAGE	Ch. 14 Ch. 58 Ch. 59
1. Inspect and adjust (where applicable) front and rear steering linkage geometry including parallelism and vehicle ride height.	Ch. 14 Inspection Ch. 58 Description Ch. 59 Service
2. Inspect and replace pitman arm.	Ch. 58 Description Ch. 59 Service
3. Inspect and replace relay rod (center link/intermediate rod).	Ch. 58 Description Ch. 59 Service
4. Inspect, adjust (where applicable) and replace idler arm and mountings.	Ch. 58 Description Ch. 59 Service
5. Inspect, replace and adjust tie rods, tie rod sleeves, clamps and tie rod ends (sockets).	Ch. 58 Description Ch. 59 Service
6. Inspect and replace steering linkage damper.	Not Covered
B. SUSPENSION SYSTEMS DIAGNOSIS AND REPAIR	Ch. 14 Ch. 56 Ch. 57 Ch. 59
1. FRONT SUSPENSIONS	Ch. 14 Ch. 56 Ch. 57

1. Diagnose front suspension system noises, body sway/roll, and ride height problems; determine needed repairs.	Ch. 14 Ch. 57
2. Inspect and replace upper and lower control arms, bushings, shafts and rebound bumpers.	Ch. 56 Description Ch. 57 Service
3. Inspect, adjust and replace strut rods/radius arm (compression/tension) and bushings.	Ch. 56 Description
4. Inspect and replace upper and lower ball joints (with or without wear indicators).	Ch. 56 Description Ch. 57 Service
5. Inspect and replace steering knuckle/spindle assemblies and steering arms.	Ch. 56 Description Ch. 57 Service
6. Inspect and replace front suspension system coil springs and spring insulators (silencers).	Ch. 56 Description Ch. 57 Service
7. Inspect and replace front suspension system leaf spring(s), leaf spring insulators (silencers), shackles, brackets, bushings and mounts.	Ch. 56 Description
8. Inspect, replace and adjust front suspension system torsion bars; inspect mounts.	Ch. 56 Description
9. Inspect and replace stabilizer bar (sway bar) bushings, brackets and links.	Ch. 56 Description Ch. 57 Service
10. Inspect and replace MacPherson strut cartridge or assembly.	Ch. 56 Description Ch. 57 Service
11. Inspect and replace MacPherson strut upper bearing and mount.	Ch. 56 Shock Absorber Mounts
2. REAR SUSPENSIONS	No separate section on rear suspension. Ch. 14 Ch. 56 Ch. 57 Ch. 59
1. Diagnose suspension system noises, body sway and ride height problems; determine needed repairs.	Ch. 14 Ch. 57
2. Inspect and replace rear suspension system coil springs and spring insulators (silencers).	Ch. 56 Description Ch. 57 Service
3. Inspect and replace rear suspension system transverse links (track bars), control arms, stabilizer bars (sway bars), bushings and mounts.	Ch. 56 Description Ch. 57 Service
4. Inspect and replace rear suspension system leaf spring(s), leaf spring insulators (silencers), shackles, brackets, bushings and mounts.	Ch. 56 Description Ch. 57 Service
5. Inspect and replace rear MacPherson strut cartridge or assembly, and upper mount assembly.	Ch. 56 Description Ch. 57 Service
6. Inspect non-independent rear axle assembly for bending, warpage and misalignment.	Not Covered
7. Inspect and replace rear ball joints and tie rod assemblies.	Ch. 57 Ball Joints Ch. 59 Replacing Tie Rods
3. MISCELLANEOUS SERVICE	Ch. 14 Ch. 53 Ch. 56 Ch. 57 Ch. 76
1. Inspect and replace shock absorbers.	Ch. 14 Inspection Ch. 56 Described Ch. 57 Service

2. Inspect and replace air shock absorbers, lines and fittings.	Ch. 56 Described Ch. 57 Service
3. Diagnose and service front and/or rear wheel bearings.	Ch. 53
4. Diagnose, inspect, adjust, repair or replace components of electronically controlled suspension system.	Ch. 56 Ch. 76 Electronic System Service

C. WHEEL ALIGNMENT DIAGNOSIS, ADJUSTMENT AND REPAIR

C. WHEEL ALIGNMENT DIAGNOSIS, ADJUSTMENT AND REPAIR	Ch. 60 Ch. 61
1. Diagnose vehicle wander, drift, pull, hard steering, bump steer, memory steer, torque steer and steering return problems; determine needed repairs.	Ch. 60 Ch. 61
2. Measure vehicle ride height; determine needed repairs.	Ch. 61
3. Check and adjust front and rear wheel camber on suspension systems with a camber adjustment.	Ch. 60 Camber Description Ch. 61
4. Check front and rear wheel camber on non-adjustable suspension systems; determine needed repairs.	Ch. 60 Camber Description Ch. 61
5. Check and adjust caster on suspension systems with caster adjustment.	Ch. 60 Caster Description Ch. 61
6. Check caster on non-adjustable suspension systems; determine needed repairs.	Ch. 60 Caster Description Ch. 61
7. Check and adjust front wheel toe.	Ch. 60 Toe Description Ch. 61
8. Center steering wheel.	Ch. 61
9. Check toe-out-on-turns (turning radius/angle); determine needed repairs.	Ch. 60 Description Ch. 61
10. Check SAI/KPI (steering axis inclination/king pin inclination); determine needed repairs.	Ch. 60 Description Ch. 61
11. Check included angle;determine needed repairs.	Ch. 60 Description Ch. 61
12. Check rear wheel toe; determine needed repairs or adjustments.	Ch. 60 Description Ch. 61
13. Check rear wheel thrust angle; determine needed repairs or adjustments.	Ch. 60 Description Ch. 61
14. Check for front wheel setback; determine needed repairs or adjustments.	Ch. 60 Description Ch. 61
15. Check front cradle (subframe) alignment; determine needed repairs or adjustments.	Ch. 60 Description Ch. 61

D. WHEEL AND TIRE DIAGNOSIS AND REPAIR

D. WHEEL AND TIRE DIAGNOSIS AND REPAIR	Ch. 14 Ch. 54 Tire and Wheel Description Ch. 55
1. Diagnose tire wear patterns; determine needed repairs.	Ch. 14 Ch. 55
2. Inspect tires; check and adjust air pressure.	Ch. 14 Ch. 55

3. Diagnose wheel/tire vibration, shimmy and noise problems; determine needed repairs.	Ch. 55
4. Rotate tires/wheels according to manufacturer's recommendations.	Ch. 55
5. Measure wheel, tire, axle flange and hub runout (radial and lateral); determine needed repairs.	Ch. 14 Ch. 55
6. Diagnose tire pull (lead) problems; determine corrective actions.	Ch. 55
7. Balance wheel and tire assembly (static and/or dynamic).	Ch. 55

BRAKES (TEST A5)

A. HYDRAULIC SYSTEM DIAGNOSIS AND REPAIR

Ch. 51
Ch. 52

1. MASTER CYLINDERS (NON-ABS)

Ch. 51
Ch. 52

1. Diagnose poor stopping or dragging caused by problems in the master cylinder; determine needed repairs.	Ch. 51 Description Ch. 52 Service
2. Diagnose poor stopping, dragging, high or low pedal, or hard pedal caused by problems in the step bore master cylinder and interval valves (e.g. volume control devices, quick take-up valve, fast-fill valve, pressure regulating valve; determine needed repairs.	Ch. 51 Description Ch. 52 Service
3. Measure and adjust pedal pushrod length.	Ch. 51 Description Ch. 52 Service
4. Check master cylinder for defects by depressing brake pedal; determine needed repairs.	Ch. 51 Description Ch. 52 Service
5. Diagnose master cylinder for secondary cup defects by inspecting for external fluid leakage.	Ch. 51 Description Ch. 52 Service
6. Remove master cylinder from vehicle.	Ch. 52 Service
7. Bench bleed (check for function and remove air) all non-ABS master cylinders.	Ch. 52 Service
8. Install master cylinder in vehicle; test operation of the hydraulic system.	Ch. 52 Service

2. FLUIDS, LINES AND HOSES

Ch. 14
Ch. 23
Ch. 24
Ch. 51

1. Diagnose poor stopping, pulling or dragging caused by problems in the brake fluid, lines, and hoses; determine needed repairs.	Ch. 51 Brake Fluid Description Ch. 52 Service
2. Inspect brake lines and fittings for leaks, dents, kinks, rust, cracks or wear; tighten loose fittings and supports.	Ch. 14 Inspection
3. Inspect flexible brake hoses for leaks, kinks, cracks, bulging or wear; tighten loose fittings and supports.	Ch. 14 Inspection Ch. 23 Description/ Service Ch. 51 Description Ch. 52 Service
4. Replace brake lines (double flare and ISO types), hoses, fittings and supports.	Ch. 24 Replace Tubing
5. Select, handle, store and install brake fluids (including silicone fluids).	Ch. 51 Description Ch. 52 Service

3. VALVES AND SWITCHES (NON-ABS)

Ch. 51
Ch. 52

1. Diagnose poor stopping, pulling or dragging caused by problems in the hydraulic system valve(s); determine needed repairs.	Ch. 51 Description Ch. 52 Service

2. Inspect, test and replace metering (hold-off), proportioning, pressure differential and combination valves.	Ch. 51 Description
3. Inspect, test, replace and adjust load or height sensing-type proportioning valve(s).	Ch. 51 Description
4. Inspect, test and replace brake warning light, switch and wiring.	Ch. 52
5. Re-set brake pressure differential valve (if necessary).	Not Covered
4. BLEEDING, FLUSHING AND LEAK TESTING (NON-ABS SYSTEMS)	Ch. 52
1. Bleed (manual, pressure. vacuum or surge) and/or flush hydraulic system.	Ch. 52
2. Pressure test brake hydraulic system.	Ch. 52

B. DRUM BRAKE DIAGNOSIS AND REPAIR

B. DRUM BRAKE DIAGNOSIS AND REPAIR	Ch. 51 Drum Brake Description Ch. 52
1. Diagnose poor stopping, pulling or dragging caused by drum brake hydraulic problems; determine needed repairs.	Ch. 52
2. Diagnose poor stopping, noise, pulling, grabbing, dragging or pedal pulsation caused by drum brake mechanical problems; determine needed repairs.	Ch. 52
3. Remove, clean, inspect and measure brake drums; follow manufacturer's recommendations in determining need to machine or replace.	Ch. 52
4. Machine drum according to manufacturer's procedures and specifications.	Ch. 52
5. Using proper safety procedures, remove, clean and inspect brake shoes/linings, springs, pins, adjusters/self-adjusters, levers, clips and other related brake hardware; determine needed repairs.	Ch. 52
6. Using proper safety procedures, clean and inspect brake backing (support) plates; determine needed repairs.	Ch. 52
7. Disassemble and clean wheel cylinder assembly; inspect parts for wear, rust, scoring and damage; hone cylinder (if necessary and recommended by mfr.); replace all cups, boots and any damaged or worn parts; reassemble.	Ch. 52
8. Lubricate brake shoe support pads on backing (support) plate, adjuster/self-adjuster mechanisms and other brake hardware.	Ch. 52
9. Install brake shoes and related hardware.	Ch. 52
10. Pre-adjust brake shoes and parking brake before installing brake drums or drum/hub assemblies and wheel bearings.	Ch. 52
11. Reinstall wheel, torque lug nuts and make final checks and adjustments.	Ch. 52

C. DISC BRAKE DIAGNOSIS AND REPAIR

C. DISC BRAKE DIAGNOSIS AND REPAIR	Ch. 51 Disc Brake Description Ch. 52
1. Diagnose poor stopping, pulling or dragging caused by disc brake hydraulic problems; determine needed repairs.	Ch. 52
2. Diagnose poor stopping, noise, pulling, grabbing, dragging or pedal pulsation caused by disc brake mechanical problems; determine needed repairs.	Ch. 52
3. Remove caliper assembly from mountings; clean and inspect for leaks and damage to caliper housing.	Ch. 52
4. Clean and inspect caliper mountings and slides for wear and damage.	Ch. 52
5. Remove, clean and inspect pads and retaining hardware; determine needed repairs, adjustments and replacements.	Ch. 52
6. Disassemble and clean caliper assembly; inspect parts for wear, rust scoring and damage; replace all seals, boots and any damaged or worn parts.	Ch. 52
7. Reassemble caliper	Ch. 52

8. Clean, inspect and measure rotor with a dial indicator and a micrometer; follow manufacturer's recommendations in determining need to machine or replace.	Ch. 52
9. Remove and replace rotor.	Ch. 52
10. Machine rotor according to manufacturer's procedures and specifications.	Ch. 52
11. Install, pads, calipers and related attaching hardware; bleed system.	Ch. 52
12. Adjust calipers with integrated parking brakes according to manufacturer's recommendations.	Ch. 52
13. Fill master cylinder to proper level with recommended fluid; inspect caliper for leaks.	Ch. 52
14. Reinstall wheel, torque lug nuts and make final checks and adjustments.	Ch. 52

D. POWER ASSIST UNITS DIAGNOSIS AND REPAIR

D. POWER ASSIST UNITS DIAGNOSIS AND REPAIR	Ch. 51
1. Test pedal free travel with and without engine running to check power booster operation.	Ch. 52
2. Check vacuum supply (manifold or auxiliary pump) to vacuum-type power booster with a vacuum gauge.	Ch. 52
3. Inspect the vacuum-type power booster unit for vacuum leaks; inspect the check valve for proper operation; repair, adjust or replace parts as necessary.	Ch. 51 Description Ch. 52
4. Inspect and test hydro-boost system and accumulator for leaks and proper operation; repair, adjust or replace parts as necessary.	Ch. 51 Description Ch. 52

E. MISCELLANEOUS (WHEEL BEARINGS, PARKING BRAKES, ELECTRICAL, ETC.) DIAGNOSIS AND REPAIR

E. MISCELLANEOUS (WHEEL BEARINGS, PARKING BRAKES, ELECTRICAL, ETC.) DIAGNOSIS AND REPAIR	Ch. 51 Ch. 52 Ch. 53
1. Diagnose wheel bearing noises, wheel shimmy and vibration problems; determine needed repairs.	Ch. 53
2. Remove, clean, inspect, repack wheel bearings or replace wheel bearings and races; replace seals; adjust wheel bearings according to manufacturer's specifications.	Ch. 53
3. Check parking brake system; inspect cables and parts for wear, rusting and corrosion; clean or replace parts as necessary; lubricate assembly.	Ch. 51 Description Ch. 52 Service
4. Adjust parking brake assembly; check operation.	Ch. 52
5. Test parking brake indicator light(s), switches, and wiring.	Ch. 52
6. Test, adjust, repair or replace brake stop light switch and wiring.	Ch. 51 Description Ch. 52 Service

F. ANTILOCK BRAKE SYSTEM DIAGNOSIS AND REPAIR

F. ANTILOCK BRAKE SYSTEM DIAGNOSIS AND REPAIR	Ch. 51 Ch. 52 Ch. 76
1. Follow accepted service and safety precautions during inspection, testing and servicing of ABS hydraulic, electrical and mechanical components.	Ch. 52
2. Diagnose poor stopping, wheel lock up, pedal feel, pulsation and noise problems caused by the ABS; determine needed repairs.	Ch. 51
3. Observe ABS warning light(s) at startup; determine if further diagnosis is needed.	Ch. 51 Description
4. Diagnose ABS electronic control(s) and components using self diagnosis and/or recommended test equipment; determine needed repairs.	Ch. 51 Description Ch. 52 Service Ch. 76 Electronic System Diagnosis
5. Depressurize integral (high pressure) components of the ABS following manufacturer's recommended safety procedures.	Ch. 51 Description Ch. 52 Service
6. Fill the ABS master cylinder with recommended fluid to proper level following manufacturer's procedures; inspect system for leaks.	Ch. 51 Description Ch. 52 Leaks are Covered under Bleeding

7. Bleed the ABS front and rear hydraulic circuits following manufacturer's procedures.	Ch. 51 Description Ch. 52 Service
8. Perform a fluid pressure (hydraulic boost) diagnosis on the integral (high pressure) ABS; determine needed repairs.	Not Covered
9. Remove and install ABS components following manufacturer's procedures and specifications.	Ch. 51 Component and System Description
10. Service, test and adjust ABS speed sensors following manufacturer's recommended procedures.	Ch. 51 Description Ch. 52 Service Ch. 76 Sensor servicing
11. Diagnose ABS braking problems caused by vehicle modifications (tire size, curb height, final drive ratio etc.)	Not Covered

ELECTRICAL/ELECTRONIC SYSTEMS (TEST A6)

A. GENERAL ELECTRICAL/ELECTRONIC SYSTEM DIAGNOSIS	Ch. 29 Ch. 31 Ch. 36 Ch. 37
1. Check electrical circuits with a test light; determine needed repairs.	Ch. 29
2. Check voltages and voltage drops in electrical/electronic circuits with a voltmeter; determine needed repairs.	Ch. 29
3. Check current flow in electrical/electronic circuits and components with an ammeter; determine needed repairs.	Ch. 29
4. Check continuity and resistances in electrical/electronic circuits and components with an ohmmeter; determine needed repairs.	Ch. 29
5. Check electrical/electronic circuits with jumper wires; determine needed repairs.	Ch. 29
6. Find shorts, grounds, opens and resistance problems in electrical/electronic circuits; determine needed repairs.	Ch. 29 Ch. 37
7. Measure and diagnose the cause(s) of abnormal key-off battery drain (parasitic draw); determine needed repairs.	Ch. 31
8. Inspect, test and replace fusible links, circuit breakers, fuses, and other current limiting devices.	Ch. 36 Description Ch. 37 Service
9. Read and interpret electrical schematic diagrams and symbols.	Not Covered
B. BATTERY DIAGNOSIS AND SERVICE	Ch. 13 Ch. 30 BATTERY FUNDAMENTALS Ch. 31
1. Perform battery state-of-charge test; determine needed service.	Ch. 31
2. Perform battery capacity (load, high-rate discharge) test; determine needed service.	Ch. 31
3. Maintain or restore electronic memory functions.	Ch. 31
4. Inspect, clean, fill or replace battery.	Ch. 13 Ch. 31
5. Perform slow/fast battery charge in accordance with manufacturer's recommendations.	Ch. 31
6. Inspect, clean and repair or replace battery cables, connectors, clamps and hold downs.	Ch. 31
7. Jump start a vehicle with jumper cables and a booster battery or auxiliary power supply.	Ch. 31

C. STARTING SYSTEM DIAGNOSIS AND REPAIR

C. STARTING SYSTEM DIAGNOSIS AND REPAIR	Ch. 32 STARTER FUNDAMENTALS Ch. 33
1. Perform starter current draw test; determine needed repairs.	Ch. 33
2. Perform starter circuit voltage drop tests; determine needed repairs.	Ch. 33
3. Inspect, test, and repair or replace switches, connectors and wires of starter control circuits.	Ch. 32 Description Ch. 33 Service
4. Inspect, test and replace starter relays and solenoids.	Ch. 32 Description Ch. 33 Service
5. Remove and replace starter.	Ch. 33
6. Differentiate between electrical and engine mechanical problems that cause a slow crank or no crank condition.	Ch. 33
D. CHARGING SYSTEM DIAGNOSIS AND REPAIR	Ch. 34 Ch. 35
1. Diagnose charging system problems that cause and undercharge, a no-charge or an overcharge condition.	Ch. 35
2. Inspect, adjust and replace generator (alternator) drive belts, pulleys and tensioners.	Ch. 34 Alternator Description Ch. 35
3. Perform charging system output test; determine needed repairs.	Ch. 35
4. Perform generator (alternator) output test; determine needed repairs.	Ch. 35
5. Inspect, test, repair or replace voltage regulator/regulating circuit; determine needed repairs.	Ch. 34 Description Ch. 35 Service
6. Perform charging circuit voltage drop tests; determine needed repairs.	Ch. 35
7. Inspect, repair or replace connectors and wires of charging circuits.	Ch. 35
8. Remove, inspect, and replace generator (alternator).	Ch. 34 Alternator Description Ch. 35
9. Disassemble, clean, inspect, test and replace generator (alternator) components.	Ch. 34 Alternator Description Ch. 35
E. LIGHTING SYSTEMS DIAGNOSIS AND REPAIR	Ch. 29 Ch. 36 Ch. 37
1. HEADLIGHTS, PARKING LIGHTS, TAILLIGHTS, DASHLIGHTS AND COURTESY LIGHTS	Ch. 29 Ch. 36 Ch. 37
1. Diagnose the cause of brighter than normal, intermittent, dim or no operation of headlight.	Ch. 29
2. Inspect, replace and aim headlights/bulbs.	Ch. 36 Description Ch. 37 Service
3. Inspect, test and repair or replace head-light and dimmer switches, relays, control units, sensors, sockets, connectors and wires of headlight circuits.	Ch. 26 Illustration Ch. 37 Dimmer Switches
4. Diagnose the cause of intermittent, slow or no operation of retractable headlight assembly.	Ch. 29
5. Inspect, test and repair or replace motors, switches, relays, connectors, wires and controllers of retractable headlight assembly circuits.	Not Covered

6. Diagnose the cause of brighter than normal intermittent, dim or no operation of parking lights and/or taillights.	Ch. 29
7. Inspect, test and repair or replace switches, relays, bulbs, sockets, connectors and wires of parking light and taillight circuits.	Ch. 36 Description Ch. 37 Service
8. Diagnose the cause of intermittent, dim, no lights or no brightness control of instrument lighting circuits.	Ch. 29
9. Inspect, test and repair or replace switches, relays, bulbs, sockets, connectors, wires, controllers, and printed circuit boards of instrument lighting circuits.	Ch. 36 Description Ch. 37 Service
10. Diagnose the cause of intermittent, dim or no operation of courtesy lights (dome, map, vanity).	Ch. 29
11. Inspect, test and repair or replace switches, relays, bulbs, sockets, connectors, and wires of courtesy light (dome, map, vanity) circuits.	Ch. 29
2. STOPLIGHTS, TURN SIGNALS, HAZARD LIGHTS AND BACK-UP LIGHTS	Ch. 29 Ch. 36 Ch. 37
1. Diagnose the cause intermittent, dim or no operation of stoplight (brake light).	Ch. 29
2. Inspect, test, adjust and repair or replace switch, bulbs, sockets, connectors and wires of stoplight circuits.	Ch. 36 Description Ch. 37 Service
3. Diagnose the cause of no turn signal and/or hazard lights or lights with no flash on one or both sides.	Ch. 36 Ch. 37
4. Inspect, test and repair or replace switches, flasher units, bulbs, sockets, connectors and wires of turn signal and hazard light circuits.	Ch. 36 Description Ch. 37 Service
5. Diagnose the cause of intermittent, dim or no operation of back-up light.	Ch. 29
6. Inspect, test and repair or replace switch, bulbs, sockets, connectors and wires of back-up light circuits.	Ch. 36 Description Ch. 37 Service
F. GAUGES, WARNING DEVICES, AND DRIVER INFORMATION SYSTEMS DIAGNOSIS AND REPAIR	Ch. 26 Ch. 29 Ch. 75 Ch. 76
1. Diagnose the cause of intermittent, high, low or no gauge readings.	Ch. 26 Fuel Gauge Ch. 29
2. Inspect, test and repair or replace gauges, gauge sending units, connectors, wires and printed circuit boards of gauge circuits.	Not Covered
3. Diagnose the cause(s) of intermittent, high, low or no readings on electronic digital instrument clusters.	Not Covered
4. Inspect, test, repair or replace sensors, sending units, connectors and wires of electronic instrument circuits.	Ch. 75 Description Ch. 76 Service
5. Diagnose the cause of constant, intermittent or no operation of warning lights, indicator lights and other driver information systems.	Ch. 76
6. Inspect, test and repair or replace bulbs, sockets, connectors, wires, electronic components, and controllers of warning light, indicator light and driver information system circuits.	Ch. 29
7. Diagnose the cause of constant, intermittent or no operation of audible warning devices.	Not Covered
8. Inspect, test and repair or replace switches, relays, timers, electronic components, controllers, printed circuits, connectors and wires of audible warning device circuits.	Not Covered
G. HORN AND WIPER/WASHER DIAGNOSIS AND REPAIR	Ch. 36
1. Diagnose the cause of constant, intermittent or no operation of horn(s).	Ch. 36 Horn Fundamentals

2. Inspect, test and repair or replace horn(s), horn relay, horn button (switch), connectors and wires of horn circuits.	Ch. 36 Illustration
3. Diagnose the cause of wiper problems including constant operation, intermittent operation, poor speed control, no parking or no operation of wiper.	Ch. 36 Wiper Description
4. Inspect, test and replace intermittent (pulsing) wiper controls.	Ch. 36 Wiper Description
5. Inspect, test and replace wiper motor, resistors, switches, relays, controllers, connections and wires of wiper circuits.	Ch. 36 Wiper Description
6. Diagnose the cause of constant, intermittent or no operation of windshield washer.	Not Covered
7. Inspect, test and repair or replace washer motor, pump assembly, relays, switches, connectors and wires of washer circuits.	Ch. 36 Wiper Description
H. ACCESSORIES DIAGNOSIS AND REPAIR	VERY BRIEFLY COVERED IN Ch. 13. GENERAL WIRING IS COVERED IN Ch. 36 & 37

HEATING AND AIR CONDITIONING (TEST A7)

A. A/C SYSTEM DIAGNOSIS AND REPAIR	Ch. 73 Ch. 74
1. Diagnose the cause of unusual operating noises of the A/C system; determine needed repairs.	Not Covered
2. Identify system type and conduct performance test on the A/C system; determine needed repairs.	Ch. 74
3. Diagnose A/C system problems indicated by refrigerant flow past the sight glass (for systems using a sight glass); determine needed repairs.	Ch. 74
4. Diagnose A/C system problems indicated by pressure gauge readings; determine needed repairs.	Ch. 73 Pressure Description Ch. 74 Diagnosis
5. Diagnose A/C system problems indicated by visual, smell and touch procedures; determine needed repairs.	Ch. 74
6. Leak test A/C system; determine needed repairs.	Ch. 74
7. Identify and recover A/C system refrigerant.	Ch. 73 Description Ch. 74 Service
8. Evacuate A/C system.	Ch. 74
9. Clean A/C system components and hoses.	Ch. 74 Flushing the system
10. Charge A/C system with refrigerant (liquid or vapor).	Ch. 74
11. Identify lubricant type; inspect level in A/C system.	Ch. 74
B. REFRIGERATION SYSTEM COMPONENT DIAGNOSIS AND REPAIR	Ch. 22 Ch. 73 Ch. 74
1. COMPRESSOR AND CLUTCH	Ch. 22 Ch. 73 Ch. 74
1. Diagnose A/C system problems that cause protection devices (pressure, thermal and PCM) to interrupt system operation; determine needed repairs.	Ch. 73 Description

2. Inpect, test and replace A/C system pressure and thermal protection devices.	Ch. 74
3. Inspect, adjust and replace A/C compressor drive belts and pulleys.	Ch. 22 Drive Belt Description Ch. 74 Inspection
4. Inspect, test, service and replace A/C compressor clutch components or assembly.	Ch. 73 Description Ch. 74 Inspection
5. Identify lubricant type; inspect and correct level in A/C compressor.	Ch. 74
6. Inspect, test, service or replace A/C compressor.	Ch. 73 Compressor Ch. 74
7. Inspect, repair or replace A/C compressor mounting.	Not Covered
2. EVAPORATOR, CONDENSER AND RELATED COMPONENTS	Ch. 23 Ch. 73 Ch. 74
1. Inspect, repair or replace A/C system mufflers, hoses, lines, filters, fittings and seals.	Ch. 23 Hoses Ch. 74
2. Inspect A/C condenser for proper air flow.	Ch. 73 Description Ch. 74 Service
3. Inspect, test and replace A/C system condenser and mountings.	Ch. 73 Description Ch. 74 Leak Detection
4. Inspect and replace reciever/drier or accumulator/drier.	Ch. 73 Description Ch. 74 Service
5. Inspect, test and replace expansion valve.	Ch. 73 Description Ch. 74 Service
6. Inspect and replace orifice tube.	Ch. 73 Description Ch. 74 Service
7. Inspect, test or replace evaporator.	Ch. 73 Description Ch. 74 Service
8. Inspect, clean and repair evaporator, housing and water drain.	Ch. 73 Ch. 74
9. Inspect, test and replace evaporator pressure/temperature control systems and devices.	Ch. 73 Description
10. Identify, inspect and replace A/C system service valves (gauge connections).	Ch. 74
11. Inspect and replace A/C system high pressure relief device.	Ch. 73 Description
C. HEATING AND ENGINE COOLING SYSTEMS DIAGNOSIS AND REPAIR	Ch. 20 Ch. 21 Ch. 22 Ch. 23 Ch. 73 Ch. 74
1. Diagnose the cause of temperature control problems in the heater/ventilation system; determine needed repairs.	Ch. 73 Description Ch. 74 Service
2. Diagnose window fogging problems; determine needed repairs.	Ch. 73 Description of why windows fog
3. Perform cooling system tests; determine needed repairs.	Ch. 21
4. Inspect and replace engine cooling and heater system hoses and belts.	Ch. 13 Inspection Ch. 22 Belts Ch. 23 Hoses Ch. 20

5. Inspect, test and replace radiator, pressure cap, coolant recovery system and water pump.	Ch. 13 Inspection Ch. 20 Description Ch. 21 Service
6. Inspect, test and replace thermostat, bypass and housing.	Ch. 20 Description Ch. 21 Service
7. Inspect, recover coolant; flush and refill system with proper coolant.	Ch. 13 Inspection Ch. 20 Description Ch. 21 Service
8. Inspect, test and replace fan (both electrical and mechanical), fan clutch, fan belts, fan shrouds and air dams.	Ch. 20 Description Ch. 21 Service
9. Inspect, test and replace heater coolant control valve (manual, vacuum and electrical types).	Ch. 73 service Ch. 74 Service
10. Inspect, flush and replace heater core.	Ch. 74

D. OPERATING SYSTEMS AND RELATED CONTROLS DIAGNOSIS AND REPAIR

D. OPERATING SYSTEMS AND RELATED CONTROLS DIAGNOSIS AND REPAIR	Ch. 23 Ch. 29 Ch. 73 Ch. 74 Ch. 75
1. ELECTRICAL	NO SPECIAL SECTION ON THIS TOPIC. Ch. 29 COVERS BASIC ELECTRICITY. ELECTRONICS ARE COVERED IN Ch. 75 AND Ch. 76.
1. Diagnose the cause of failures in the electrical control system of heating, ventilating and A/C systems; determine needed repairs.	Ch. 29
2. Inspect, test, repair and replace A/C-heater blower motors, resistors, switches, relay/modules, wiring and protection devices.	Not Covered
3. Inspect, test, repair and replace A/C compressor clutch, relay/modules, wiring, sensors, switches, diodes and protection devices.	Ch. 29 Ch. 75 Ch. 76 Sensors, Diodes, etc.
4. Inspect, test, repair, replace and adjust A/C-related engine control systems.	Not Covered
5. Inspect, test, repair, replace and adjust load sensitive A/C compressor cut-off systems.	Ch. 73 Description
6. Inspect, test, repair and replace engine cooling/condenser fan motors, relays/modules, switches, sensors, wiring and protection devices.	Ch. 29
7. Inspect, test, adjust, repair and replace electric actuator motors, relays/modules, switches, sensors, wiring and protection devices.	Not Covered
8. Inspect, test, service or replace heating, ventilating and A/C control panel assemblies.	Not Covered
2. VACUUM/MECHANICAL	Ch. 23 Ch. 73 Ch. 76
1. Diagnose the cause of failures in the vacuum and mechanical switches and controls of the heating, ventilating and A/C systems; determine needed repairs.	Ch. 73 Description
2. Inspect, test, service or replace heating, ventilating and A/C control panel assemblies.	Not Covered
3. Inspect, test, adjust and replace heating, ventilating and A/C control cables and linkages.	Not Covered
4. Inspect, test and replace heating, ventilating and A/C vacuum actuators (diaphragms/motors) and hoses.	Ch. 23 Hoses

5. Identify, inspect, test and replace heating, ventilating and A/C vacuum reservoir, check valve and restrictors.	Not Covered
6. Inspect, test, adjust, repair or replace heating, ventilating and A/C ducts, doors and outlets.	Not Covered
3. AUTOMATIC AND SEMI-AUTOMATIC HEATING, VENTILATING AND A/C SYSTEMS	NO SPECIFIC SECTION ON THESE TOPICS. SOME TOPICS ARE PARTIALLY COVERED IN Ch. 73, Ch. 74 AND Ch. 75
1. Diagnose temperature control system problems; determine needed repairs.	Ch. 73 Ch. 74
2. Diagnose blower system problems; determine needed repairs.	Not Covered
3. Diagnose air distribution problems; determine needed repairs.	Ch. 73 Description
4. Diagnose compressor clutch control system; determine needed repairs.	Ch. 74
5. Inspect, test, adjust or replace climate control and sunload sensors.	Ch. 73 Description
6. Inspect, test, adjust and replace temperature blend door/power servo.	Not Covered
7. Inspect, test and replace low engine coolant temperature blower control system.	Not Covered
8. Inspect, test and replace heater water valve and controls.	Ch. 73
9. Inspect, test and replace electric and vacuum motors, solenoids and switches.	Ch. 75 Descriptions of Solenoids and Switches
10. Inspect, test and replace ATC control panel.	Not Covered
11. Inspect, test, adjust or replace ATC microprocessor (climate control computer/programmer).	Ch. 75 General Discussion of Microprocessors.
12. Check and adjust calibration of ATC system.	Not Covered
E. REFRIGERANT RECOVERY, RECYCLING AND HANDLING	Ch. 73 Ch. 74
1. Maintain and verify correct operation of certified equipment.	Not Covered
2. Identify and recover A/C system refrigerant.	Ch. 73 Description Ch. 74 Service
3. Recycle refrigerant.	Ch. 73 Description Ch. 74 Service
4. Label and store refrigerant.	Not Covered
5. Test recycled refrigerant for non-condensable gases.	Not Covered

ENGINE PERFORMANCE (TEST A8)

A. GENERAL ENGINE DIAGNOSIS	Ch. 25 Ch. 27 Ch. 39 Ch. 41 Ch. 42 Ch. 43
1. Verify driver's complaint, perform visual inspection, and/or road test vehicle; determine needed action.	Ch. 42 Ch. 43 Ch. 50
2. Diagnose the cause of unusual engine noise and/or vibration problems; determine needed action.	Ch. 43 Noise

3. Diagnose the cause of unusual exhaust color, odor and sound; determine needed action.	Ch. 25 Ch. 43
4. Perform engine manifold vacuum or pressure tests; determine needed action.	Ch. 42
5. Perform cylinder power balance test; determine needed action.	Ch. 42
6. Perform cylinder compression test; determine needed action.	Ch. 42
7. Perform cylinder leakage test; determine needed action.	Ch. 42
8. Diagnose engine mechanical, electrical, electronic, fuel and ignition problems with an oscilloscope and/or engine analyzer; determine needed action.	Ch. 39
9. Prepare and inspect vehicle and analyzer for exhaust gas analysis; obtain exhaust gas readings.	Ch. 27 Ch. 41
B. IGNITION SYSTEM DIAGNOSIS AND REPAIR	Ch. 38 Ch. 39 Ch. 42
1. Diagnose no-starting, hard starting, engine misfire, poor driveability, spark knock, power loss, poor mileage and emissions problems on vehicles with distributor and distributorless ignition systems; determine needed repairs.	Ch. 42
2. Inspect, test, repair or replace ignition primary circuit wiring and components.	Ch. 38 Description of Primary and Secondary Wiring
3. Inspect, test and service distributor.	Ch. 38 Description Ch. 39 Service
4. Inspect, test, service, repair or replace ignition system secondary circuit wiring and components.	Ch. 38 Description of Primary and Secondary Wiring
5. Inspect, test and replace ignition coils.	Ch. 38 Description Ch. 39 Service
6. Check and adjust ignition system timing and timing advance/retard.	Ch. 38 Description Ch. 39 Service
7. Inspect, test and replace ignition system pick-up sensor or triggering devices.	Ch. 38 Description Ch. 39 Service
8. Inspect, test and replace ignition control module.	Ch. 38 Description Ch. 39 Service
C. FUEL, AIR INDUCTION AND EXHAUST SYSTEM DIAGNOSIS AND REPAIR	Ch. 13 Ch. 23 Ch. 24 Ch. 26 Ch. 27 Ch. 28 Ch. 40 Ch. 41 Ch. 45
1. Diagnose hot or cold no-starting, hard starting, poor driveability, incorrect idle speed, poor idle flooding, hesitation, surging, engine misfire, power loss, stalling, poor mileage, dieseling and emissions problems on vehicles with injection-type or carburetor-type fuel systems; determine needed action.	Ch. 27
2. Perform fuel system pressure and volume tests; determine needed action.	Ch. 27

3. Inspect fuel tank, tank filter and gas cap; inspect and replace fuel lines, fittings and hoses; check fuel for contaminants and quality.	Ch. 13 Inspection Ch. 23 Hoses Ch. 24 Tubing and Fittings Ch. 26 Description Ch. 27 Service
4. Inspect, test and replace mechanical and electrical fuel pumps and pump control systems; inspect, service and replace fuel filters.	Ch. 26 Description Ch. 27 Service
5. Inspect, test and repair or replace fuel pressure regulation system and components of injection-type fuel systems.	Ch. 26 Description Ch. 27 Service
6. Inspect, test, adjust and repair or replace cold enrichment system components.	Ch. 26 Description Ch. 27 Service
7. Inspect, test, adjust and repair or replace acceleration enrichment system components.	Ch. 26 Description Ch. 27 Service
8. Inspect, test and replace deceleration fuel reduction or shut-off system components.	Not Covered
9. Remove, clean and replace throttle body; adjust related linkages.	Ch. 26 Description Ch. 27 Service
10. Inspect, test, clean and replace fuel injectors.	Ch. 26 Description Ch. 27 Service
11. Inspect, service and repair or replace air filtration system components.	Ch. 13 Inspection Ch. 28
12. Inspect, clean or replace throttle body mounting plates, air induction system, intake manifold and gaskets.	Ch. 26 Description Throttle Plate Ch. 27 Service Throttle Plate Ch. 28 Intake Sys. Ch. 45 Gaskets
13. Check/adjust idle speed and fuel mixture.	Ch. 26 Description Ch. 27 Service
14. Remove, clean, inspect/test, and repair or replace vacuum and electrical components and connections of fuel systems.	Ch. 26 Description Ch. 27 Service
15. Inspect, service and replace exhaust manifold, exhaust pipes, mufflers, resonators, catalytic converters, tail pipes and heat shields.	Ch. 28 Exhaust System Ch. 40 Ch. 41 Catalytic Converter
16. Perform exhaust system back-pressure test; determine needed action.	Ch. 28 Description of Back-Pressure
17. Test the operation of turbocharger/super-charger systems; determine needed action.	Ch. 28
18. Remove, clean, inspect, and repair or replace turbocharger/supercharger system components.	Ch. 28
19. Identify the causes of turbocharger/supercharger failure; determine needed action.	Ch. 28

D. EMISSIONS CONTROL SYSTEM DIAGNOSIS AND REPAIR

	Ch. 40 Ch. 41

1. POSITIVE CRANKCASE VENTILATION

	Ch. 40 PCV FUNDAMENTALS Ch. 41 SERVICE
1. Diagnose emission or driveability problems caused by positive crankcase (PCV) system.	Ch. 40
2. Test the operation of PCV systems.	Ch. 40
3. Inspect, service and replace PCV filter/breather cap, valve, tubes, orifices and hoses.	Ch. 40

2. EXHAUST GAS RECIRCULATION

	Ch. 40 EGR FUNDAMENTALS
	Ch. 41 SERVICE
	Ch. 75
	Ch. 76

1. Diagnose emissions or driveability problems caused by the exhaust gas recirculation (EGR) system.	Ch. 41
2. Test the operation of EGR system.	Ch. 41
3. Inspect, test, service and replace valve, EGR tubing and exhaust passages of EGR systems.	Ch. 41
4. Inspect, test, service and replace vacuum/pressure controls, filters and hoses of EGR systems.	Ch. 41
5. Inspect, test, repair and replace electrical/electronic sensors, controls, solenoids and wiring of EGR systems.	Ch. 75 Ch. 76

3. EXHAUST GAS TREATMENT

	Ch. 40 FUNDAMENTALS
	Ch. 41 SERVICE
	Ch. 75
	Ch. 76

1. Diagnose emissions or driveability problems caused by the secondary air injection or catalytic converter system.	Ch. 41
2. Test the operation of secondary in injection systems.	Ch. 41
3. Inspect, test, service and replace mechanical components of secondary air injection systems.	Ch. 41
4. Inspect and replace electrical/electronically-operated components and circuits of secondary air injection systems.	Ch. 75 Ch. 76
5. Inspect, test and replace components of catalytic converter systems.	Ch. 41

4. EVAPORATIVE EMISSIONS CONTROLS

	Ch. 40 FUNDAMENTALS
	Ch. 41 SERVICE

1. Diagnose emissions or driveability problems caused by the evaporative emissions control system.	Ch. 41
2. Test the operation of evaporative emissions control systems.	Ch. 41
3. Inspect and replace components and hoses of evaporative emissions control systems.	Ch. 41

E. COMPUTERIZED ENGINE CONTROLS DIAGNOSIS AND REPAIR

	Ch. 4
	Ch. 75 COMPUTER & ELECTRONICS OVERVIEW
	Ch. 76

1. Diagnose the causes of emissions or driveability problems resulting from failure of computerized engine controls with no diagnostic trouble codes stored.	Ch. 76
2. Retrieve and record stored diagnostic trouble codes.	Ch. 76
3. Diagnose the causes of emissions or driveability problems resulting from failure of computerized engine controls with stored trouble codes.	Ch. 76
4. Inspect, test, adjust and replace computerized engine control system sensors, powertrain control module (PCM), actuators and circuits.	Ch. 76
5. Use and interpret digital multimeter (DMM) readings.	Ch. 76
6. Read and interpret technical literature information (service publications).	Ch. 4
7. Test, remove, inspect, clean, service and repair or replace power and ground distribution circuits and connections.	Ch. 76

8. Practice recommended precautions when handling static sensitive devices.	Ch. 76
9. Diagnose driveability and emission problems resulting from failures of interrelated systems (cruise control, security alarms, torque controls, suspension controls, traction controls, torque management, A/C and similar systems).	Not Covered
10. Diagnose the causes of emissions or driveability problems resulting from computerized spark timing controls; determine needed repairs.	Ch. 76

F. ENGINE RELATED SERVICES

Ch. 18
Ch. 20
Ch. 21
Ch. 46

1. Adjust valves on engines with mechanical or hydraulic lifters.	Ch. 18 Description Ch. 46 Service
2. Verify correct camshaft timing; determine needed action.	Ch. 18 Description Ch. 46 Service
3. Check engine operating temperature; determine needed action.	Ch. 21
4. Perform cooling system pressure tests; check coolant; inspect, test, radiator, thermostat, pressure cap, coolant recovery tank and hoses; determine needed repairs.	Ch. 20 Description Ch. 21 Service
5. Inspect, test and replace mechanical/electrical fans, fan clutch, fan shroud/ducting and fan control devices.	Ch. 20 Description Ch. 21 Service

G. ENGINES ELECTRICAL SYSTEMS DIAGNOSIS AND REPAIR

Ch. 31
Ch. 33
Ch. 35

1. BATTERY

Ch. 31

1. Inspect, service and replace battery, battery cables, clamps and hold-down devices.	Ch. 31
2. Measure and diagnose the cause(s) of abnormal key-off battery drain; determine needed repairs.	Ch. 31
3. Perform battery capacity (load, high-rate discharge) test; determine needed service.	Ch. 31
4. Perform battery charging procedures.	Ch. 31

2. STARTING SYSTEM

Ch. 33

1. Perform starter current draw test; determine needed action.	Ch. 33
2. Perform starter circuit voltage drop tests; determine needed action.	Ch. 33
3. Inspect, test, and repair or replace components and wires in the starter control circuit.	Ch. 33

3. CHARGING SYSTEM

Ch. 35

1. Diagnose charging system problems that cause and undercharge, overcharge or nocharge condition; determine needed action.	Ch. 35
2. Inspect, adjust and replace alternator (generator) drive belts, pulleys and fans.	Ch. 35
3. Inspect and repair or replace charging circuit connectors and wires.	Ch. 35

A

A-circuit—a charging system with the regulator in the ground circuit.

Circuito A—un sistema de cargar que tiene el regulador en el circuito a tierra.

abnormal combustion—also called spark knock, is the term for preignition and detonation, which can cause spark knock and engine damage.

combustión anormal—tambien llamado detonación de las bujías, es el término para el encendido anticipado y la detonación, que puede causar la detonación de las bujías y los daños al motor.

abrasion—rubbing.

abrasión—frotamiento.

ABS—antilock brake systems.

ABS—sistema de frenos antibloqueantes.

AC voltage generator—a magnetic pickup that generates an AC analog signal.

generador de voltaje AC—un captador magnético que genera una señal AC análoga.

accumulator (air conditioing)—a refrigerant reservoir and dryer that supplies refrigerant vapor to the compressor.

acumulador (aire acondicionado)—un suministro de refrigerante y un secador que proporciona el vapor de refrigerante al comprimidor.

accumulator (transmission)—a reservoir used in timing and cushioning gear shifts.

acumulador (transmisión)—un suministro usado en sincronizar y en suavizar los cambios de velocidades.

acid—solutions that are below 7 on the pH scale are acidic. A pH rating of 1 is a strong acid.

ácido—las soluciones que son menos del 7 en la escama de pH se consideran acídicos. Un rango Ph de 1 es un ácido fuerte.

Ackerman angle—also called turning radius or toe out on turns. The tires toe out during a turn because the steering arms are bent at an angle.

ángulo Ackerman—tambien se llama radio de viraje o divergencia en las vueltas. Los neumáticos se inclinan hacia afuera en una vuelta porque los brazos de dirección son torcidos en un ángulo.

active sensors—those that generate their own signals.

sensores activos—los que generan sus proprios señales.

active suspensions—an automatic suspension where each wheel has a hydraulic cylinder to keep the car body level during all driving conditions.

suspensiones activas—una suspensión en la cual cada rueda tiene un cilindro hidráulico para mantener nivel al coche en cualquier condición de manejo.

actuator—an electronic or magnetic relay that can perform a desired function.

actuador—un relé electrónico o magnético que puede cumplir una función deseada.

actuators—the devices that act upon processed signals received from the computer. They can be either solenoids, DC motors, relays, switches, or control modules.

impulsores—los dispositivos que actuan sobre los señales recibidos de la computadora. Pueden ser solenoides, los motores DC, los relés, los interumptores o los módulos de control.

adaptive fuel trim—it allows the fuel system to operate at the correct air/fuel ratio, no matter what the differences in fuel systems are.

compensador adaptivo de combustible—permite operar al sistema de combustible en la tasa correcta del aire/combustible, sin que importa las diferencias entre los sistemas de combustible.

adaptive suspension systems—suspension leveling systems that keep the vehicle at the same height when weight is added to parts of the car.

sistemas adaptivas de suspensión—sistemas aniveladores de suspensión que mantienen al vehículo en la misma altura cuando se añade el peso en las areas del coche.

AERA—Automotive Engine Rebuilders Association.

AERA—la Asociación de Reconstructores de Motores Automotrices.

aeration—also called cavitation, is when hydraulic fluid becomes mixed with air.

aeración—tambien llamado cavitación, es cuando el fluido hidráulico se mezcla con el aire.

aerobic chemical—a chemical that hardens only when exposed to air.

química aeróbica—una química que se endurece solamente al ser expuesta al aire.

aftercooler—also called an intercooler or charge air cooler, it is a sophisticated air cooler for blowers.

post refrigerador—tambien llamado un enfriador intermedio o un enfriador de aire cargado, es un enfriador de aire sofisticado para los ventiladores.

air conditioning—the process in which air inside the passenger compartment is cooled, dried, cleaned, and circulated.

aire acondicionado—el proceso por el cual el aire dentro del compartimento de pasajeros se enfría, se seca, se limpia y se circula.

air density systems—when an air flow sensor measures the volume of air entering the engine.

sistemas de densidad del aire—cuando el sensor del flujo del aire mide el volumen del aire entrando en el motor.

air filter capacity—the amount of dirt an air filter can hold before it becomes restricted.

capacidad del filtro de aire—la cantidad de partículas que puede contener un filtro de aire antes de obstruirse.

air injection system—a smog pump or aspirator valve that lowers HC and CO in the exhaust. It also provides air to the catalytic converter to help heat it up when it is cold.

sistema de inyección de aire—una bomba de esmog o una válvula aspiradora que disminuye la cantidad del HC y CO en el escape. Tambien proporciona el aire al convertidor catalítico para ayudar en calentarlo cuando hace frío.

air mix damper—also called a blend air door, it controls heating and air conditioning action.

compuerta de mezcla de aire—tambien llamado una compuerta combinadora, controla la acción de calentamiento y aire acondicionado.

air test—a test of the operation of various clutch packs and servos before disassembling a transmission fully.

prueba de aire—una prueba de la operación de varios equipos de embrague y los servos antes de completamente desarmar una transmisión.

alkaline—a strong alkaline cleaner has a pH rating above 10.

alcalino—una limpiadora alcalina fuerte tiene la clasificación Ph más alta de las 10.

all wheel drive (AWD)—when all four wheels are driven.

tracción a cuatro ruedas—cuando impulsan todas cuatro ruedas.

alloy—when two or more metals are combined to make one.

aleación—cuando se combinan dos o más metales para hacer uno.

alternating current (AC)—oscillating electrical current that surges from positive to negative and back again.

corriente alterna (AC)—un corriente eléctrico oscilante que surge del positivo al negativo y luego regresa.

alternator—an AC (Alternatic Current) generator that produces electrical current and forces it into the battery to recharge it.

alternador—un generador AC (de corriente alterna) que produce un corriente eléctrico y lo dirige a una batería para cargarla.

ambient air—surrounding air.

aire ambiente—el aire alrededor.

ambient temperature switch—keeps an air conditioning compressor from working when outside temperatures are cold.

interruptor de la temperatura del ambiente—previene de que funciona el comprimidor del aire acondicionado cuando las temperaturas externas son frías.

ampere—unit of electrical current flow, one coulomb per second of electron flow.

amperio—una unedad del flujo de un corriente eléctrico, un culombio por segundo de flujo de los electrones.

amplitude—the intensity of a vibration.

amplitúd—la intensdad de una vibración.

anaerobic chemical—a chemical that hardens only in the absence of air.

química anaeróbica—una química que se endurece solamente con la ausencia del aire.

analog meter—a meter with a dial whose needle is moved by a magnet.

medidor análogo—un medidor con un cuadrante cuyo aguja indicador se mueve por un imán.

anneal—to soften a metal, it is heated and allowed to cool slowly.

recocer—para ablandar un metal, se calienta y se permite enfriar lentamente.

anode—like the positive terminal on a battery.

ánodo—como el terminal positivo de una batería.

anodized aluminum—aluminum with a protective plating.

aluminio anodizado—el aluminio que tiene un blindaje protectivo.

ANSI—American National Standards Institute.

ANSI—Instituto Nacional de Normas Americanas.

anti-drainback valve—a valve in a horizontally mounted oil filter that prevents it from emptying when the engine is off.

válvula que evita el drenaje de regreso del aceite—una válvula en un filtro de aceite montado horizontalmente que previene que se vacia cuando esta apagado el motor.

anti-friction bearings—also called frictionless bearings, provide a rolling contact with either balls, rollers, or needles.

balero antifricción—tambien llamado balero sin fricción, provee un contacto rodante utilizando bolas, rodillos o agujas.

API—American Petroleum Institute.

API—Instituto Americano de Petroleo.

arc welding—welding using electricity.

soldadura con arco—la soldadura que usa la electricidad.

arcing—burning.

formación del arco—quemando.

ASME—American Society of Mechanical Engineers - 345 E. 47th Street, New York, NY 10017.

ASME—La Sociedad Americana de Ingenieros Mecánicos - 345 E. 47th Street, New York, NY 10017.

aspect ratio—a measurement of the height-to-width ratio of a tire.

relación de aspecto—una medida de la relación entre altura-a—anchura de un neumático.

aspirator valve air system—also called a pulse-air system uses the pulses created by the exhaust gases to blow fresh air into the exhaust.

sistema de aire de válvula aspirador—tambien llamado un sistema de aire de impulso usa los impulsos creado por los gases del escape para introducir el aire fresco dentro del escape.

ATF—automatic transmission fluid.

ATF—fluido para transmisión automática.

atomization—the name of the process by which fuel is suspended in air in a mist of tiny droplets like fog.

atomización—el nombre del proceso por el cual el combustible se suspenda en el aire en un vapor de gotitas pequeñas como una nieble.

atoms—the building blocks of life composed of protons, neutrons, and electrons.

átomos—los elementos esenciales de toda vida compuesta de los protones, los neutrones y los electrones.

ATRA—Automatic Transmission Rebuilders Association.

ATRA—Asociación de Reconstructores de Transmisiones Automáticas.

AWG—american wire gauge.

AWG—calibre americano de los alambres.

axial compressor—an air conditioning compressor that has pistons that move lengthwise in the compressor body.

comprimidor axial—un comprimidor de aire acondicionado que tiene pistones que muevan por la longitud del cuerpo del comprimidor.

axial movement—vertical movement.

movimiento axial—un movimiento vertical.

axle bearings—the term for bearings that are on live axles (those that drive wheels).

cojinete del eje—el término para los cojinetes que se encuentran en los ejes vivos (los que impulsan las ruedas).

B

B-circuit—a charging system with the regulator between the positive feed (insulated side) and the field coil (which is grounded inside of the alternator).

circuito B—un sistema de cargar que tiene el regulador entre la alimentación positiva (lado aislado) y la bobina inductora (que va a tierra dentro del alternador).

backfire—a term for combustion that occurs in the vehicle's exhaust (after it leaves the cylinder).

petardear—un término para la combustión que ocurre en el escape del vehículo (después de que sale del cilindro).

backlash—the clearance between meshing gear teeth.

juego libre—la holgura entre los dientes engrenados de un engrenaje.

backpressure transducer valve—part of an EGR valve that senses engine load and changes the amount of exhaust flow into the intake manifold.

válvula transductor de contrapresión—parte de una válvula EGR que percibe la carga del motor y cambia la cantidad del flujo del escape en el múltiple de admisión.

ball and trunnion—a type of constant velocity universal joint.

bola y muñón—un tipo de union universal de velocidad fija.

ball sockets—parts that connect steering linkage parts.

rótulas—las partes que conectan las partes de enlace de dirección.

banjo fitting—a fitting, resembling a banjo, that allows fluid to turn 90° in a short area.

montaje banjo—un montaje, parecido a un banjo, que permite que el fluido gira el 90° en una área muy corta.

banjo housing—a differential housing without the third member, called this because it resembles a two-sided banjo.

cárter de la caja del diferencial (banjo)—una caja de un diferencial sin tercer miembro, llamado "banjo" porque parece un banjo de dos lados.

barrel—a term for the number of throttle passages found in a carburetor.

cañón—un término para el número de los pasos del acelerador que se encuentra en el carburador.

barrier creams—creams that are rubbed on the skin prior to working in a greasy environment to aid in cleanup.

cremas decapantes—las cremas que se aplican al piel antes de trabajar en un ambiente con mucha grasa para ayudar en desengrasar.

base—a pH rating of above 7 is an alkaline cleaner, caustic or base.

base—una clasificación pH más del 7 es una limpiador alcalino, cáustico, o un base.

base circle—what the cam would be if there were no lobe.

círculo base—lo que sería la leva excéntrica si no tuviera lóbulo.

base timing—also called initial timing, it is the timing setting before the computer has a chance to make changes.

sincronización de referencia—tambien llamado el encendido inicial, es la colocación del encendido antes de que la computadora haya tenido la oportunidad de hacer los cambios.

battery terminal—the connection on the side or top of a battery that the clamp connects to.

terminal de la batería—la conexión en el lado o en la parte superior de una batería a la cual se conecta la abrazadera de la batería.

baud rate—the speed that timed pulses (bits) of information are transmitted within the computer. A computer with a baud rate of 28,800 (28.8K) can transmit 28,800 bits of information per second.

coeficiente baúd—la velocidad con que los impulsos (bits) cronometizados se transmiten con una computadora. Una computadora con una coeficiente baúd de 28,800 (28.8K) puede transmitir 28,800 bits de información por segundo.

BCA—Battery Council International.

BCA—Concilio Internacional de la Batería.

bearing cage—also called separator, is a stamped steel or plastic insert that keeps bearing balls or rollers properly spaced around the bearing assembly.

jaula del cojinete—tambien llamado un espaciador, es una inserción hecho del acero embutido o del plástico que mantiene un espacio adecuado entre las bolas o las rodillas del cojinete y la asamblea del cojinete.

bearing clearance—clearance between a bearing and bearing journal.

holgura del cojinete—la holgura entre un cojinete y un muñon del cojinete.

bearing crush—the bearing insert extends out of the bearing bore. When the two halves are tightened against each other, the bearing is held from turning in its bore.

compresión del cojinete—la inserción del cojinete se extiende más allá del agujero del cojinete. Cuando las dos mitades se aprietan la una contra la otra, el cojinete no puede girar en su agujero.

bearing retained axle—a pressed-fit axle with a bearing retainer ring.

eje sujetado por cojinete—un eje de ajuste a presión con un anillo de retén para el cojinete.

bearing spread—the spread of the bearing insert makes it possible for it to stay in place in its bore during engine assembly.

extensión del cojinete—la extensión de la inserción del cojinete hace posible quedarse en su posición en el agujero durante la asamblea del motor.

beat/boom vibration—one vibration interacting with another.

vibración latido/zumbido—una vibración que reciproca con otra.

belt—a cord structure made up of plies.

correa/banda—una estructura de cordón compuesto de telas.

bi-directional—when the computer can receive commands from the scan tool and anything that is computer controlled can be operated from the scan tool.

bidireccional—cuando la computadora puede recibir ordenes de un aparato explorador y cualquiera cosa que se controla por medio de la compututadora se puede operar desde el aparato explorador.

bias-ply—a tire structure, also called diagonal, or cross-ply, that has casing plies that cross each other at an angle of 35°-45°.

doble capa—un estructura de neumático, tambien llamada bandas diagonales, o telas al sesgo que tiene las capas que se cruzan en un ángulo de 35°-45°.

bimetal coil spring—a thermostatic coil consisting of two types of metal wound together. When the coil is heated it expands. When it cools it shrinks.

resorte helicoidal bimetal—un arrollamiento termostático que consiste de dos tipos de metal enredados. Cuando el arrollamiento se calienta, expande. Cuando se enfría, encoge.

bimetal engine—an engine made of iron and aluminum.

motor bimetal—un metal hecho de hierro y aluminio.

bimetal strip—two metal strips with different expansion rates.

banda bimetálica—dos bandas de metal con una coeficiente de expansión distintas.

binary information—transistors are electronic switches that are either on or off (zeros and ones). The computer interprets combinations of zeros and ones to form numbers or words, processing the information to determine the meanings of the signals.

información binaria—los transistores son los interruptores electrónicos que estan o prendidas o apagadas (ceros o unos). La computadora interprete la combinación de ceros y unos para formar los números o las palabras, procesando la información para determinar el significado de las señales.

bit—each zero and one represents a bit of information.

bit—cada cero y uno representa un bit de información.

black light test—a test for oil leakage that uses fluorescent dye and an ultraviolet light source.

prueba de luz negra—una prueba para fuga de aceite que usa un tinte fluoroscente y una fuenta de luz ultravioleta.

bleeding shocks—a process of removing air from shocks that have been stored on a shelf.

purgar los amortiguadores—el proceso de remover el aire de los amortiguadores que se ha almacenado en un estante.

blend air door—also called an air mix damper, it controls heating and air conditioning action.

puerta para mezclar el aire—tambien llamado una compuerta para mezclar aire, controla la acción de calentamiento y aire acondicionado.

blocktest—a test for an internal coolant leak in which a dye changes color when exposed to exhaust gas in the radiator.

prueba del bloqueo—una prueba para encontrar una fuga interior de fluido refrigerante en la cual un tinte cambia de color al exponerse al gas del escape en el radiador.

blocker ring synchronizer—the most popular style of synchronizer.

sincronizador tipo anillo—el estilo más popular de sincronizador.

blow-by gases—gases that leak past the piston rings into the crankcase during the power stroke. They contain acids as well as unburned and burned fuel.

fuga de gases a través de los anillos de los cilindros—la fuga de los gases desde los anillos de los pistones al cárter durante la carrera de fuerza. Contienen los ácidos tanto como el combustible no consumido y consumido.

blow-through—when a supercharger or turbocharger pressurizes the air cleaner above the carburetor.

inyección superior—cuando un sobrealimentador o un turbocompresor presuriza el limpiador de aire en la parte superior del carburador.

blower—a belt-driven pump, also called a supercharger.

ventilador—una bomba impulsada por banda, tambien llamado un sobrealimentador.

bob weights—are used when spinning the crankshaft to simulate the correct weight of the piston and related parts during machine shop testing.

pesos de contra balance—se usan al girar el cigüeñal para simular el peso correcto del pistón y sus partes relacionadas durante la prueba en el taller de máquinas.

bolt grade/property class—an SAE or metric classification system that rates the minimum tensile strength of a fastener.

grado de perno/ clase de propriedad—un sistema de clasificación SAE o métrica que clasifica la fuerza mínima de tensión de un sujetador.

bonded linings—linings that are held to their backing by a binder (glue).

forros aglomerados—los forros que se pegan al respaldo con un ligante (pegamento).

boost pressure—the amount of air density a turbo provides.

presión auxiliar—la cantidad de densidad del aire que provee un turbo.

bore—the diameter of the cylinder.

diámetro interior—el diámetro de un cilindro.

boundary lubrication—when an oil film becomes too thin or starts to break down under load.

límite de lubricación—cuando una película de aceite se espesa demasiado y comienza a deteriorarse bajo una carga.

brake band—an external brake planetary holding device.

banda de freno—un dispositivo planetario externo de sujeción.

brake bleeding—removing air and water from a brake hydraulic system.

purgando los frenos—removando el aire y el agua de un sistema hidráulica de frenos.

brake drag—when the brakes stay on after a stop.

arrastre de freno—cuando los frenos siguen funcionado después de parar.

brake fade—resulting from excessive brake heat, the drum expands and the pedal moves closer to the floor. The coefficient of friction of the lining also drops off.

amortiguamiento del freno—resulta de un calor excesivo del freno, el tambor expande y el pedal se acerca al piso. La coeficiente de fricción del forro tambien se disminuya.

brake grab—when brakes apply quickly and tend to stick on.

amarro del freno—cuando los frenos se aplican rapidamente y suelen quedarse aplicados.

brake horsepower (BHP)—the usable horsepower at the crankshaft.

caballos de fuerza del freno (BHP)—los caballos de fuerza disponibles en el cigüeñal.

brake mean effective pressure (BMEP)—a term relating to the pressure in the cylinder calculated from the horsepower reading on a dynamometer.

presión efectivo mediano del freno (BMEP)—un término relacionado a la presión en el cilindro que se calcula por la lectura de la potencia de fuerza en un dinamómetro.

brake pads—disc brake friction linings.

almohadilla de freno—los forros de fricción en un freno de disco.

brake pedal pulsation—when the brake pedal moves up and down during a stop due to out-of-round drum brakes or overheated disc brake rotors.

pulsación del pedal de frenos—cuando el pedal de freno brinca durante un paro debido a la eccentricidad de los frenos de tambor o a los rotores sobrecalentados de un freno de disco.

brake pedal reserve—pedal travel distance before brakes begin to apply.

reserva del pedal de frenos—la distancia que avanza el pedal antes de que comienzan a aplicar los frenos.

brake pull—when a vehicle pulls to one side during a stop.

jalón del freno—cuando un vehículo jala hacia un lado durante una parada.

brake shoes—the friction material backing on drum brake systems.

zapatos de freno—la material de forro de fricción en los sistemas de freno de tambor.

brake spoon—a tool used to adjust drum brake clearance.

cuchara de freno—una herramienta que sirve para ajustar la holgura del tambor del freno.

brake thermal efficiency—brake horsepower converted to BTUs, divided by the fuel's heat input in BTUs with the result multiplied by 100.

eficiencia termal del freno—los caballos de fuerza convertido a los BTUs, dividido por los BTUs de calor del combustible con el resultado multiplicado por 100.

breakaway torque—the amount of torque needed to make a side gear rotate the clutches in a limited-slip differential.

par máximo de salida—la cantidad del par requirido para hacer que un engrenaje lateral gira los embragues en un diferencial autoblocante.

breakout box—a tool in which the probes of a DVOM are inserted into its pin holes to access various sensors and actuators through the pin connector to the computer.

caja electróncia indicadora de códigos de problemas—una herramienta en la cual las sondas de un DVOM se insertan en los agujeros para pasador obteniendo aceso a varios sensores y impulsores por el conectador de púas en la computadora.

brinelling—when the bearing or race has indentations from shock loads.

efecto brinell—cuando el cojinete o la pista tiene muescas debido a los choque de la cargas.

British Imperial (U.S.) System—a measuring system that uses fractions and decimals, based on inches, feet, and yards.

Sistema Imperial Inglesa (U.S.)—un sistema de medida que usa las fracciones y los decimales, basado en las pulgadas, los pies y las yardas.

British thermal units (BTU)—One BTU is the amount of heat required to heat one pound of water by 1°F.

unedad térmica Inglesa (BTU)—Una BTU es la cantidad del calor requerido para aumentar el calor una libra de agua por 1°F.

brushes—parts, usually carbon, that provide electricity to a rotating part.

escobillas—las partes, tipicamente del carbón, que proveen la electricidad a la partes giratorias.

BTU—British Thermal Unit (BTU)—the amount of energy required to raise the temperature of one pound of water by 1°F.

BTU—Unedad Térmica Inglesa—la cantidad de la energía requerido para aumentar el calor una libra de agua por 1°F.

bulb trade number—the number that identifies a bulb for all manufacturers.

marca de fabricante de bombilla—el número que identifica una bombilla para todos los fabricantes.

bulkhead or firewall—the metal wall between the engine and passenger compartments.

tabique o mamparo de encendios—el muro de metal entre el motor y los compartimentos de los pasajeros.

bump steer—also called orbital steer, happens when a wheel with tie rods at unequal heights goes over a bump. The car momentarily steers the direction that the wheel turns as the toe on that one wheel changes.

dirección de topes—tambien llamado la dirección orbital, ocurre cuando una rueda cuyos tirantes de tracción son de niveles desiguales pasa sobre un tope. El coche se desvía momentariamente en la dirección en que gira la rueda mientras que cambia la divergencia de esa rueda.

butterfly valve—a valve that controls air flow.

válvula de mariposa—una válvula que controla el flujo del aire.

bypass or partial-flow oil filter—an oil filtering system that only filters about 10% of the oil at once, trapping very fine materials and allowing very little oil to flow through it.

filtro del aceite de desvío o de flujo limitado—un sistema para purificar el aceite que sólo filtra aproximadamente el 10% del aceite a la vez, atrapando a las partículas muy finas y permitiendo que pase muy poco aceite a través del filtro.

bypass valve—a valve in the oil filter that opens to let unfiltered oil flow to the bearings when the oil is cold and thick or the filtering material is plugged due to a lack of proper maintenance.

válvula de desvío—una válvula en el filtro de aceite que abre para permitir que el aceite no filtrado pase a los cojinetes cuando el aceite esta frío y espeso o cuando el material de filtro esta atascada por falta del mantenimiento adecuada.

byte—eight bits, sometimes called a word.

byte—ocho bits, a veces llamado una palabra.

C

C-lock or C-clip axle—an axle with a groove on the inside that a clip fits into to keep it from coming out.

eje de cerradura C o grapa C—un eje que tiene una muesca en el interior en la cual queda una grapa que previene que salga.

C.I. engine—a diesel compression ignition engine.

motor C.I.—un motor de diesel con encendido de compresión.

cables—large wires that allow more electrical current flow.

cables—los alambres grandes que permiten más flujo del corriente.

CAFE—Corporate Average Fuel Economy. Each manufacturer must meet a standard or pay a "gas-guzzler" penalty to the government.

CAFE—Normas de la Economía del Combustible indicadas por la Corporación. Cada fabricante debe adherir a una norma o pagar una multa de "consumidor de combustible" al gobierno.

cam ground piston—when a piston is cold its skirt is oval shaped. As it warms up it takes on a round shape.

pistón excéntrico—cuando esta frío su faldilla es de forma ovalada. Al calentarse toma una forma redonda.

camber—the inward or outward tilt of a tire at the top.

curvatura—la inclinación hacia adentro o hacia afuera en la parte superior de un neumático.

camber roll—when a cambered tire rolls in a circle, as if it were at the large end of a cone.

rodar en comba—cuando un neumático rueda en un círculo, como si estuviera en la extremidad de un cono.

candlepower—a rating for the intensity of a headlamp.

unidad de intensidad luminosa—una potencia indicada de la intensidad de un faro.

capacitor—also called a condenser, it stores electricity.

capacitor—tambien llamado un condensador, almacene la electricidad.

capscrew—an externally threaded fastener that is used without a nut in a threaded or blind hole application to hold components together.

casquete fileteado—un sujetador enroscado exteriormente que se usa sin tuerca en una aplicación de orificio enroscado o ciego para sujetar los componentes.

carbide—a harder metal than ordinary tool steel.

carburo—un metal más duro que el acero común de herramientas.

carbon blaster—a tool that uses crushed walnut shells blasted by compressed air to remove the carbon deposits.

chorro de carbón—una herramienta que usa las cascaras molidas de nueces chorreadas con el aire comprimido para despegar los depósitos del carbón.

carbon pile—alternating positive and negative layers of carbon that are compressed against each other when a large knob on the face of a volt amp tester is tightened to cause a rapid discharge of the battery.

pila de carbón—las capas alternas de carbón positivo y negativo que se unen bajo presión al apretar un gran botón en la cara de un comprobador de voltamperios para causar una descarga rápida de la batería.

carbon trail—a line of electrically conducting carbon that forms in a cracked distributor cap.

rastro de carbón—una linea de carbón que conduce la electricidad que forma en una tapa agrietada de un distribuidor.

carbonaceous deposits—also called cauliflower deposits, they are thin, hard carbon deposits that result from fuel.

depósitos carbonosos—tambien llamados depósitos de coliflor, son los depósitos de carbón delgados y duros que resultan del combustible.

carburetor—a mechanical device that mixes fuel and air in response to how much air is flowing through it.

carburador—un dispositivo mecánico que mezcla el combustible con el aire como una respuesta de cuánto aire lo atraviesa.

casewashers—they are hard on the surface only; the core of the washer is soft and will compress.

arandela cementada—son duras solamente en sus superficies; el núcleo de la arandela es blanda y comprimirá.

caster—the forward or rearward tilt of the spindle support arm.

ángulo del caster—la inclinación hacia afrente o hacia atrás del brazo de soporte del husillo.

catalyst—a substance that causes a chemical reaction to occur faster without going under any measurable change to itself.

catalizador—una sustancia que cause ocurrir más rapidamente una reacción química sin efectuar un cambio apreciable a si mismo.

catalytic converter—an emission device located in front of the muffler in the exhaust system. It looks very much like a heavy muffler and contains catalysts to clean up an engine's emissions before they leave the end of the exhaust pipe.

convertidor catalítico—un dispositivo de emisiones ubicado en la parte delantera del silenciador en el sistema del escape. Se parece mucho a un silenciador de trabajo pesado y contiene los catlizadores para limpiar las emisiones del motor antes de que salgan del tubo del escape.

cathode—like the negative terminal on a battery.

cátodo—parecido al terminal negativo de una batería.

caustic—a pH rating of above 7 is an alkaline cleaner, caustic or base.

cáustico—una clasificación pH de más de 7 es una limpiador alcalino, cáustico o un base.

cavitation—also called aeration, is when hydraulic fluid becomes mixed with air.

cavitación—tambien llamado aeración, es cuando el fluido hidráulico se mezcla con el aire.

CCA—cold cranking amps.

CCA—amperios de arranque en frío.

CCOT—cycling clutch orifice tube.

CCOT—tubo del orificio del embraque en ciclo.

CD-ROM—compact-disc-read-only-memory (CDs). CDs store a great deal of information that can be accessed by a personal computer with a CD player.

CD-ROM—disco-compacto-de-memoria-solo-lectura (CDs). Los CDs almacenen una gran cantidad de información que se puede acesar con una computadora personal con un tocadisco de discos compactos.

center of gravity—the point between the front and the rear wheels where the weight will be distributed evenly.

centro de gravedad—el punto entre las ruedas delanteras y traseras en done se distribuirá el peso igualmente.

centrifugal advance—also called mechanical advance, it senses the speed of the engine.

avance centrífuga—tambien llamado un avance mecánica, percibe la velocidad del motor.

cetane number—describes how easily a diesel fuel will ignite.

índice de ceteno—describe qué tal facilmente un combustible de deisel encenderá.

CFC—chlorofluorocarbon, CFCs are depleting the protective ozone layer through a chemical reaction.

CFC -cloroflorocarburo, los CFCs estan deteriorando la capa protectiva de ozono por medio de una reacción química.

chainfall—a chain hoist.

caída de cadenas—una montacargas con cadena.

charcoal canister—a small plastic or steel container filled with activated charcoal that can store gasoline vapors until the right time for them to be drawn into the engine and burned.

bote de carbón—un pequeño receptáculo de plástico o de acero llenado con el carbón activo que puede almacenar los vapores de la gasolina hasta el tiempo apropiado de que entran al motor para quemarse.

charge air cooler—also called an intercooler or an aftercooler, it is a sophisticated air cooler for blowers.

enfriador de aire cargado—tambien llamado un enfriador intermedio o postrefrigerador, es un enfriador del aire sofisticado para los ventiladores.

charging system output test—a test that loads the charging system and measures its current output.

prueba de salida del sistema de carga—una prueba que pone una carga en el sistema de cargar y mide su salida de corriente.

chassis—the group of parts that includes the frame, shocks and springs, steering parts, tires, brakes, and wheels. It supports the engine and the car body.

chasis—el conjunto de partes que incluyen el armazón, los amortiguadores y resortes, las partes de dirección, los neumáticos, y las ruedas. Soporta el motor y la carrocería del coche.

chassis lubricant—a grease of a consistency that allows it to be applied through a zert fitting with a grease gun.

lubricante del chasis—una grasa de una consistencia que permite que se aplica por un niple de grasa con una pistola engrasadora.

chattering clutch—when the pedal shakes as the clutch pedal is raised from the floor.

castañateo del embrague—cuando el pedal tiembla al subir el pedal del embrague del piso.

cherry picker—an engine hoist.

burro—un aparato para izar el motor.

chip—tiny sandwiched silicon wafers constructed by photographically reproducing circuit patterns onto them make up a chip. As many as 30,000 transistors are placed in an area that is only ¼" square.

ficha—las obleas diminutas de silicio intercaladas construidas con los patrones de circuitos reproducidos fotográficamente en ellos. Tantos como 30,000 transistores se colocan en una área que mide un cuadro de sólo un ¼ de una pulgada.

circuit—when a complete circle is provided for electrical flow.

circuito—cuando un círculo completo se provee para el flujo de la electricidad.

circuit breakers—electrical circuit protection devices that open in the event of an electrical overload to prevent damage from occurring elsewhere in an electrical circuit. They reset automatically or can be manually reset after they trip.

rompedor de circuito—los dispositivos de protección de los circuitos eléctricos que se abren en caso de una sobrecarga eléctrica para prevenir que ocurren daños en otras áreas del circuito eléctrico. Se vuelven a cerrar automáticamente o pueden ser cerrados a mano después de que se hayan desconectado.

circuit resistance tests—tests that check the resistance in the ground and insulated circuits by measuring voltage drop.

pruebas de resistencia del circuito—las pruebas que verifican la resistencia en la tierra y los circuitos aislados al medir la caída del voltaje.

clamping force—results from the tension or stretch on bolts. It also results from the springiness of a gasket between parts.

fuerza de apriete—resulta de la tensión o la estiración en los pernos. Tambien resulta de la elasticidad de un empaque entre las partes.

Class A fire—one that can be put out with water.

Encendio grado A—uno que se puede apagar con el agua.

Class B fire—one in which there are flammable liquids such as grease, oil, gasoline, or paint.

Encendio grado B—uno en el cual hay líquidos inflamables tal como la grasa, el aceite, la gasolina o la pintura.

Class C fire—an electrical fire.

Encendio grado C—un encendio eléctrico.

Class D fire—a fire involving a flammable metal such as magnesium or potassium.

Encendio grado D—un encendio que involucra un metal inflamable tal como el magnesio o el potasio.

clock spring—also called spiral cable, it connects the steering column to the air bag in the steering wheel.

resorte espiral—tambien llamado un cable espiral, conecta el tubo de dirección a la bolsa de aire en el volante de dirección.

close nipple—a small section of pipe that has male tapered threads on each end that join in the middle.

niple cerrado—una pequeña sección de tubo que tiene una rosca cónica macho en cada extremidad que se unen el en medio.

close ratio transmission—when there is a small difference between the ratios of a transmission's forward gears.

transmisión de relación aproximada—cuando hay una pequeña diferencia entre las relaciones de las velocidades de marcha en adelante de una transmisión.

closed loop—when a computer controlled engine reaches operating temperature and the computer is receiving feedback from the oxygen sensor and making adjustments to the air/fuel mixture.

bucle cerrado—cuando un motor controlado por computadora llega a la temperatura de operación y la computadora recibe la información del detector de oxígeno y hace los ajustes a la mezcla de combustible/aire.

closed ventilation system—a crankcase ventilation system that does not allow blowby to escape into the atmosphere.

sistema de ventilación cerrado—un sistema de ventilación del cárter que no permite que las fuga de gases a través de los anillos de los cilindros salgan a la atmósfera.

cloud point—when diesel fuel appears cloudy when the wax separates out in very cold weather.

punto de turbiedad—cuando el combustible de deisel tiene una aparencia nebulosa al separarse la cera en la clima muy fría.

clutch—a device that uncouples the powertrain from the engine.

embrague—un dispositivo que desenganche el tren de potencia del motor.

clutch cushion plate—a metal cushion in between the clutch facings that lets the facings compress.

placa amortiguadora del embrague—un amortiguador de metal entre los forros del embrague que deja que se comprimen los forros.

clutch facings—the friction material part of the clutch disc.

forros del embrague—la parte de materia de fricción del disco del embrague.

clutch fork—also called a throwout or release lever, it fits between the release bearing and the clutch cable or linkage.

horquilla del embraque—tambien llamado una palanca de desenganche o descarga, queda entre el cojinete de desenganche y el cable o la biela del embrague.

clutch free play—also called free travel, it is movement measured at the clutch pedal.

juego del embrague—tambien llamado carrera libre, es el movimiento medido en el pedal del embrague.

clutch hub—the inner part of a clutch disc.

cubo del embrague—la parte interior de un disco de embrague.

CNG—compressed natural gas.

CNG—gas natural comprimido.

CO—carbon monoxide emissions that result when gasoline is not completely burned.

CO—las emisiones del monóxido de carbono que resultan cuando la gasolina no se consume completamente.

CO_2—carbon dioxide.

CO_2—anhídrido carbónico.

coast—deceleration, power is on the concave side of the gear tooth.

desembrague (vuelo)—la deceleración, la potencia esta en el lado cóncavo del diente del engrenaje.

coast side—the concave side of a gear tooth.

lado del desembrague (vuelo)—el lado cóncavo del diente del engrenaje.

coefficient of friction—the ratio of the force holding two surfaces in contact to the force required to slide one over the other.

coeficiente de la fricción—la relación de la fuerza que sostiene el contacto entre dos superficies con la fuerza que se requiere en deslizar una sobre la otra.

coil saturation—when the magnetic field has finished its buildup inside of the coil.

saturación de la bobina—cuando se ha terminado la acumulación del campo magnético dentro de la bobina.

coil spring—a spring steel rod wound into a coil.

resorte helicoidal—una barra del acero para resortes que se ha enrollado en la forma de espiral.

cold solder connection—a defective electrical connection made when the solder is melted by the soldering iron, but the wire is too cold to bond to it.

conexión de soldadura fría—una conexión eléctrica defectuosa hecho cuando la soldadura se derrite con el hierro de soldadura, pero el alambre es demasiado frío para adherir con ella.

collet—a device that fits around the outside of a round part. When clamped around the outside, it holds that round part tightly.

collar de apriete—un dispositivo que queda alrededor de una parte redonda. Al apretarse al exterior, sujete esa parte fuertemente.

companion cylinders—two cylinders whose pistons come to TDC and BDC at the same time.

cilíndros acoplados—dos cilindros cuyos pistones llegan a PMS y PMI a la misma vez.

composite headlamp—a headlamp housing with a glass balloon that the halogen lamp fits inside of.

faro compuesto—una caja del faro que tiene un receptáculo de vidrio en el cual queda una lámpara halógena.

compound gauge—a gauge that reads in either pressure or vacuum.

indicador combinado—un medidor que lee bajo presión o en vacío.

compound planetary gear set—two planetary gear sets combined to provide more gear ratio possibilities.

conjunto compuesto de engranajes planetarios—dos conjuntos de engranajes planetarios combinados para proveer más posibilidades de relaciones entre velocidades.

compression fittings—fittings that compress the end of a nut or a sleeve to provide a seal.

guarniciones de compresión—los accesorios que comprimen en la extremidad de una tuerca o una manguilla para proveer un sello.

compression or jounce—when the wheel moves up as the spring compresses.

compresión o sacudo—cuando la rueda se mueva hacia arriba al comprimir el resorte.

compression pressure—the amount of pressure made by the piston moving up in the cylinder.

presión de compresión—la cantidad de presión creado por el pistón subiendo en el cilindro.

compression ratio—the difference between a cylinder's volumes with the piston at TDC and BDC.

relación de compresión—la diferencia entre los volumenes de un cilindro cuando el pistón esta en PMS y PMI.

compression test—a test of engine condition where a pressure gauge is inserted into a spark plug hole and registers a reading as the engine is cranked.

prueba de compresión—una prueba de la condición del motor en la cual un indicador de presión se introduce en un orificio de bujía y registra una lectura al arrancar el motor.

condensation—when a vapor changes to a liquid.

condensación—cuando el vapor se convierta en líquido.

condenser (air conditioning)—part of an air conditioning system, it is a radiator for refrigerant.

condensador (aire acondicionado)—parte del sistema de aire acondicionado, es el radiador para el fluido refrigerante.

condenser (electrical)—also called a capacitor, it stores electricity.

condensador (eléctrica)—tambien se llama un capacitor, almacene la electricidad.

conductors—materials with atoms that allow electricity to flow freely.

conductores—las materials cuyos átomos permiten fluir fácilmente la electricidad.

connecting rod resizing—a machine shop process that restores the big end of the connecting rod to original size and roundness.

recalibración de la biela—un proceso del taller de máquinas que restaura la extremidad grande de la biela a su tamaño y redondez original.

constant velocity universal joints—a universal joint whose output speed is constant like its input speed.

juntas universales de velocidad constante—una junta universal cuyo salida de velocidad es constante como su entrada de velocidad.

convection—when a body gives off heat, the surrounding air becomes warmer and moves upward.

convección—cuando un cuerpo despide el calor, el aire del ambiente se calienta y suba.

coolant—also called anti-freeze, is a combination of ethylene glycol and water used instead of pure water in the engine's cooling system.

fluido refrigerante—tambien llamado anticongelante, es una combinación del glicol etileno y el agua que se usa en vez del agua puro en el sistema de enfriamiento del motor.

coolant hydrometer—a tool that compares the weight of pure coolant to water to give an indication of coolant strength.

hidrómetro del fluido refrigerante—una herramienta que compara el peso del fluido refrigerante puro al del agua para dar una indicación de la concentración del fluido refrigerante.

copolymer—a combination of two polymers.

copolímeros—una combinación de dos polímeros.

core—a part that is returned for rebuilding.

núcleo—una parte que se entrega para reconstrucción.

countergear—one gear made up of a series of gears that mesh with the various gears on the mainshaft. It is often called a cluster gear.

engrane auxiliar—un engranaje compuesto de una serie de engranajes que se endientan con varios engrenajes en el árbol principal. Muchas veces se refiere como el conjunto de ruedas dentadas.

coupling speed—the point at which all of the converter parts and ATF all turn as a unit.

velocidad del acoplamiento—el punto en el cual todas las partes del convertidor y el ATF giran como si fueran una unedad.

CPU—central processing unit, a computer.

CPU—unedad de proceso central, una computadora.

cradle—also called the crossmember, it is the large steel part of the frame beneath the engine and between the front wheels.

cuna—tambien llamado una traviesa, es una parte grande del armazón debajo del motor y entre las dos ruedas delanteras.

crank angle sensor—a primary trigger that senses the position of the piston for the computer.

sensor del ángulo de arranque—un disparador primario que detecta la posición del pistón para la computadora.

crank polishing—a motor driven emery belt is used to polish the crankshaft, usually after grinding.

pulido del cigüeñal—una banda de esmeril impulsado por motor se usa en pulir el cigüeñal, normalmente después del esmerilado.

crankcase—the area surrounding the crankshaft.

cárter—el área alrededor del cigüeñal.

cranking vacuum test—a test to see how much vacuum an engine can produce when cranking.

prueba de arranque en vacío—una prueba para ver cuánto vacío puede producir un motor al arrancarse.

cross and yoke universal joint—also called Cardan, is the most popular universal joint design.

junta de horquilla y cruceta—tambien llamada Cardán, es el diseño más popular de junta universal.

cross counts—the speed at which the fuel signal from an O_2 sensor fluctuates from rich to lean.

salidas reversas—la velocidad con la que un señal de combustible de un sensor de oxígeno varia de rico a pobre.

cross fluid contamination—when oil and coolant are interchanged with each other.

contaminación entre los fluidos—cuando el aceite y el fluido refrigerante se intercambian.

cross groove plunge joint—also called a pancake joint, it has balls and grooves in the bearing races that would cross each other if they extended far enough.

junta profundizada de muesca en cruz—tambien llamado una junta achatada ,tiene las bolas y las muescas en las pistas de los cojinetes que se cruzarían si se extendieran lo bastante.

crossfire induction—when one spark plug firing induces a spark in the one next to it, causing it to fire before its time.

inducción de encendido cruzadi—cuando el disparo de una bujía induce una chispa en el que esta a lado, causando que dispare prematuramente.

crossflow head—a cylinder head that has the intake and exhaust ports on opposite sides of an inline engine.

cabeza de flujo transversal—una cabeza del cilindro que tiene las puertas de entrada y salida en lados opuestos de un motor en linea.

crossflow radiators—a radiator design where coolant flows from side to side.

radiadores de flujo transversal—un diseño de radiador en el cual el fluido refrigerante fluye de lado a lado.

crosshatch—the criss-cross finish left when the cylinder is honed.

marcas cruzadas—el acabado de razguños cruzados que resulta cuando se rectifica un cilindro.

crossmember—also called the cradle, it is the large steel part of the frame beneath the engine and between the front wheels. A crossmember is sometimes used to support a rear wheel drive transmission.

traviesa—tambien llamado una cuna, es la parte grande de acero de un armazón debajo del motor y entre la dos ruedas delanteras. Una traviesa a veces se usa para soportar una transmisión de tracción trasera.

crowned road—when the road is higher at the center than the outside.

carretera corcovada—cuando la carretera es más alta en el centro que en las orillas.

cruise—the car is maintaining its speed.

crucero—el coche mantiene su velocidad.

current—the amount of electrons flowing in a circuit.

corriente—la cantidad de los electrones que fluyen en un circuito.

current draw—the amount of current required to operate a load.

consumo de corriente—la cantidad del corriente que se requiere en operar una carga.

current limiting system—when the ignition module shuts back on current flow as soon as the coil primary winding is saturated.

sistema limitador de corriente—cuando un módulo de encendido disminuya el flujo de corriente tan pronto que se satura la bobina del encendido.

custom rebuild—when an engine is rebuilt as a unit, rather than assembled from a variety of parts.

reconstrucción especializada—cuando un motor se reconstruye como una unedad, en contraste de asemblarse de una variedad de partes.

CV joint—constant velocity universal joint used on both ends of front wheel drive axles, also used on some rear wheel drives.

junta CV—junta universal de velocidad constante usado en ambas extremidades de los ejes de tracción delantera, tambien usada en algunas de tracción trasera.

CVT—continuously variable transmission.

CVT—transmisión variable continua.

cylinder bank—a row of cylinders.

banco de los cilindros—una fila de cilindros.

cylinder glaze—the area where the rings ride in the cylinder develops a glazed appearance.

porcelana (barníz) del cilindro—el área en donde los anillos viajan en el cilindro toma una aparencia barnizado.

cylinder leakage test—a test where regulated compressed air is introduced into a cylinder through the spark plug hole.

prueba de fuga del cilindro—una prueba en que el aire comprimido regulado se introduce al cilindro por el orificio de la bujía.

cylinder power balance test—a test where the spark to a cylinder is shorted out, resulting in a drop in engine rpm.

prueba del balance de fuerza en el cilindro—una prueba en la cual se efectúa un corto en la chispa del cilindro, resultando en una caída del rpm en el motor.

D

DIS—distributorless ignition system or direct ignition system, it has multiple ignition coils that fire spark plugs based on a signal received from the computer.

DIS—sistema de encendido sin distribuidor o sistema del encendido directo, tiene múltiples bobinas del encendido que disparan las bujías según un señal recibido de la computadora.

dampened hubs—clutch hubs with torsional dampers that are either coil springs or rubber positioned between the disc plate and the clutch hub to absorb shock during engagement.

cubos amortiguados—los cubos del embrague que tienen amortiguadores de torsión que son de resortes helicoidales o de caucho entre la placa disco y el cubo del embrague para absorber los choques durante el enganchamiento.

data link connector (DLC)—a diagnostic connector that a test instrument can be connected to for read computer serial data.

conector de enlace de datos—un conector diagnóstico al cual se puede conectar un probador para leer los datos seriales de la computadora.

deep-cycle—when the battery is allowed to run almost completely dead, and then is recharged.

ciclo prolongado—cuando se permite que se descarga casi completamente una batería, y luego se recarga.

delta winding—an alternator stator winding that resembles the Greek letter delta (Δ).

devanado triangular (en delta)—un devanado del estátor del alternador que parece la letra griega delta (Δ).

dermatitis—irritated skin.

dermatitis—el piel irritado.

desiccant—a material that removes moisture from the system.

desicante—una materia que remueva la humedad del sistema.

detent—forced kickdown.

detención—un disminuyo forzado.

detonation—self-ignition due to excessive pressure in the cylinder.

detonación—el autoencendido debido a una presión excesiva en el cilindro.

detonation sensor—also called a knock sensor, it is a piezo—electric crystal that detects the frequency of the vibration caused by detonation and tells the computer to retard the spark timing until the vibration goes away.

sensor de detonación—tambien llamado un sensor de golpeteos, es un cristal piezo-elétrico que percibe la frecuencia de la vibración causada por la detonación y mande que la computadora retarda la chispa al encendido hasta que se desaparece la vibración.

Dexron III/Mercon—the latest and most popular ATF.

Dexron III/Mercon—el ATF más nuevo y más popular.

diagnostic tree—part of the service information that provides a step-by-step diagnostic procedure to follow when troubleshooting hard fault electrical problems.

esquema diagnóstica—parte de la información de servicio que provee un procedimiento paso-a-paso diagnóstico que uno puede seguir en diagnosticar los problemas de fallos eléctricos difíciles.

diagonal braking system—operates the brakes on opposite corners of the vehicle.

sistema de enfrenamiento diagonal—opera los frenos en las esquinas opuestas del vehículo.

diaphragm spring—also called a Belleville spring, replaces the release levers and coil springs in a diaphragm clutch.

resorte de diafragma—tambien llamado un resorte Belleville, reemplaza las palancas de desconexión y los resortes helicoidales en un embrague de diafragma.

dielectric—a nonconductive grease or coupling.

dielétrico—una grasa o acoplamiento aislante.

Diesel-cycle—an engine that ignites its air and fuel using compression.

ciclo diesel—un motor que encienda su aire y combustible utilisando la compresión.

dieseling—also called run-on, is when an engine continues to run even after the ignition key is turned off.

autoencendido—tambien llamado marcha continua, es cuando un motor continua a marchar aún después de que se haya apagado la llave del encendido.

differential—a device that allows the rear wheels to be able to rotate at different speeds as the vehicle goes around corners.

diferencial—un dispositivo que permite girar las ruedas traseras en velocidades diferentes cuando un vehículo da la vuelta en una esquina.

digital processed information—signals that are yes or no, or on/off.

información procesado digitalmente—los señales que son de si o no, o de prendido/apagado.

diode—a one-way electrical check valve made by placing P-type and N-type crystals back-to-back.

diodo—una válvula de detención eléctrica de una vía que se hace poniendo los cristales de tipo P y de tipo N espalda-a-espalda.

direct current (DC)—when electrons flow in only one direction.

corriente directo (DC)—cuando los electrones fluyen en una sóla dirección.

discard diameter—the maximum allowable size for a brake drum.

diámetro de rechazo—el tamaño máximo permitible de un tambor de freno.

discharge service valve—the high side service valve in an air conditioning system.

válvula de descarga de servicio—la válvula del lado alto de servicio en un sistema de aire acondicionado.

displacement—the volume that the piston displaces in the cylinder as it moves from TDC to BDC.

desplazamiento—el volumen que desplaza el pistón en el cilindro al moverse del PMS al PMI.

display pattern—also called parade pattern, it displays all of the cylinders next to each other, side-by-side so that the heights of the voltage spikes can be compared.

modelo indicativo—tambien llamado modelo desfile, demuestra todos los cilindros juntos, lado-a-lado para que se puede comparar las alturas de los impulsos de tensión.

diverter valve—a device that prevents a backfire during deceleration on carbureted cars.

válvula derivador—un dispositivo que previene un retorno de encendido durante la deceleración en los coches con carburadores.

DMM—digital multimeter.

DMM—multímetro digital.

dog teeth—little teeth around the circumference of the edge of the gear.

dientes pontiagudas—los pequeños dientes alrededor de la circunferencia en la orilla del engranaje.

DOHC—dual overhead cam.

DOHC—doble árbol leva sobre la cabeza.

dominant end—the end of a coil spring that aligns in the coil spring seat in the frame or lower control arm.

extremidad dominante—la extremidad de un resorte helicoidal que se alinea en el asiento del resorte helicoidal que se encuentra en el armazón o en el brazo de mando inferior.

doping—a process of adding an impurity to a crystal to make a semiconductor.

preparación del semiconductor—el proceso de añadir una impureza a un cristal para fabricar un semiconductor.

DOT—Department of Transportation.

DOT—Departamento de Transportación.

DOT dry specification—for new fluid.

DOT especificación seca—para un fluido nuevo.

DOT wet specification—for fluid that has absorbed 2% water.

DOT especificación en húmedo—para el fluido que ha absorbido el 2% del agua.

double flare—one way in which the ends of tubing are formed.

extremidad de bocinado doble—una manera en la cual se forman las extremidades de los tubos.

double offset plunge joint—has six balls evenly spaced by a cage. The outer housing is considerably deeper than a cross groove so it can plunge further. The races have straight grooves.

junta profundizada de desviación doble—tiene seis bolas separadas uniformemente por una pista. La caja exterior es de una profundidad considerable más de una muesca en cruz para que puede profundizar aún más. Las pistas tienen las muescas rectas.

double walled tubing—a type of tubing that has two layers of metal and can bend easier.

Tubo de doble muro—un tipo de tubo que tiene dos capas de metal y que puede doblarse más facilmente.

double-acting shock absorbers—control motion when moving both up and down.

amortiguadores de choque de doble acción—controlan ambos movimientos de hacia arriba y de hacia abajo.

downflow radiator—a radiator design where coolant flows from top to bottom.

radiador de flujo vertical—un diseño del radiador en el cual el fluido fluye desde arriba hacia abajo.

downshift—when the transmission shifts to a lower gear; second to first, for instance.

cambio decendente—cuando la transmisión cambia de velocidad a un engranaje más baja, del segundo al primero, por ejemplo.

dragging clutch—when a clutch does not release easily.

embrague arrastrante—cuando un embrague no quiere desconectarse facilmente.

draw-through—when a supercharger or turbocharger pressurizes the intake manifold after the carburetor and air cleaner.

aspirar—cuando un sobrealimentador o un turbocompresor presuriza el múltiple de admisión después del carburador y el filtro de aire.

drive—when under acceleration, power is on the convex side of the gear tooth.

impulso—al estar en aceleración, la potencia esta en el lado convexo del diente del embrague.

drive side—the convex side of a gear tooth.

lado del impulso—el lado convexo de un diente del engranaje.

driveline—a term that describes the parts that transfer power from the transmission to the rear wheels.

flecha motríz—un término que describe las partes que transferen la potencia de la transmisión a las ruedas traseras.

driveshaft—a hollow metal tube that has a universal joint at each end.

árbol de transmisión—un tubo metálico hueco que tiene una junta universal en ambas extremidades.

driveshaft phase—when trunnions of the front and rear U-joints are in the same plane (parallel).

fase del árbol de transmisión—cuando los muñones de las juntas en U delanteras y traseras estan en el mismo plano (son paralelos).

drop center—also called rim well, provides a means of removing and installing a tire from the wheel.

llanta de centro cóncavo—tambien llamado llanta cóncava, provee una manera de remover e instalar un neumático de la rueda.

dropping point—the temperature at which a grease turns into a liquid.

punto de caída—la temperatura en la cual una grasa se convierte en líquido.

dry park check—with the tires on the ground, an assistant turns the steering wheel back and forth a short distance. Looseness in the steering linkage or suspension is detected.

verificación de park en seco—con los neumáticos en contacto con el suelo, un asistente gira el volantede dirección de un lado al otro por una corta distancia. Se detecta así cualquier juego libre en las bielas de dirección o en la suspensión.

dry start—when the crankshaft rubs against the bearings before oil has been pumped to them.

arranque en seco—cuando el cigüeñal frota contra los cojinetes antes de que se les ha proporcionado el aceite.

DTC—diagnostic trouble code.

DTC—código diagnóstico de errores.

dual bed converter—a catalytic converter that has separate reduction and oxidation parts.

convertidor de doble alojamiento—un convertidor catalítco que tiene partes separadas para reducción y oxidación.

dual exhaust system—when the exhaust system has two pipes, two mufflers, and two catalytic converters.

sistema de doble escape—cuando el sistema de escape tiene dos tubos, dos silenciadores y dos convertidores catalíticos.

dual function electronic ignition system—spark timing by the computer takes the place of vacuum and mechanical spark advance.

sistema de arranque electrónico de doble función—la sincronización de la chispa reemplaza un avance de la chispa por medio del vacío o por avance mecánico.

dual pass machine—an air conditioning station that requires two procedures; first recover, then recycle.

máquina de dos pasos—una estación de aire acondicionado que requiere dos procedimientos; el primero de recobrar, luego de reciclar.

dual-plane manifold—when each barrel supplies half of the engine's cylinders.

múltiple de doble plano—cuando cada cañón provee para la mitad de los cilindros de un motor.

dummy shaft—a short shaft that is used to keep the bearings in place while removing a transmission shaft.

árbol falso—un eje corte que sirve para mantener en posición los cojinetes mientras que se remueva el árbol de la transmisión.

duration—the number of degrees of crankshaft travel while the valve is off its seat.

duración—el número de grados que avance un cigüeñal mientras que la válvula esta levantada de su asiento.

duty cycle—a measurement of the maximum practical capacity of a piece of equipment. If an air compressor's maximum capacity is when it runs for 7 out of every 10 minutes this is referred to as a 70% duty cycle.

ciclo de trabajo—una media de la capacidad máxima practica de una pieza de equipo. Si la capacidad máxima de un comprimidor de aire es cuando trabaja 7 de cada 10 minutos esto se llama un ciclo de trabajo de 70%.

DVOM—digital volt/ohmmeter.

DVOM—voltío/ohmiometro digital.

dwell—the length of time in degrees of distributor rotation that primary current is flowing in the coil primary winding.

ángulo de cierre—la cantidad del tiempo en grados de la rotación del distribuidor en que el corriente primario fluye en las bobinas primarias.

dye penetrant—a crack detection process that uses a dye and developer.

tinte penetrante—un proceso de detectar las grietas que usa un tinte y un revelador.

dynamic—moving.

dinámico—en movimiento.

dynamic and couple imbalance—types of imbalance that require adding or removing metal at two different places on the part. Computer balancers compute the combined amount of dynamic and force imbalance and tell where to remove metal to correct them.

desequilibrado de emparejado dinámico—los tipos de desequilibrado que requieren el añadir o remover el metal en dos lugares distinctos de la parte. Los equilibradores computerizados calculan la cantidad combinada del desequilibrio dinámico y forzado y indica en dónde hay que remover el metal para corregirlas.

dynamic seals—seals that work against moving parts.

sellos dinámicos—los sellos que funcionan contra las partes en movimiento.

dynamometer—a power absorption device for measuring engine power.

dinamómetro—un dispositivo de absorción de fuerza para medir la fuerza de un motor.

E

E.P.—extreme pressure lubricant.

E.P.—lubricante de presión extrema.

ECA—electronic control assembly, a computer.

ECA—asamblea de control electrónica, una computadora.

ECU—electronic control unit, a computer.

ECU—unedad electrónica de control, una computadora.

EDM—electrical discharge machining.

EDM—mecanizado por descarga eléctrica.

EEPROM—electronically erasable read only memory.

EEPROM—memoria solo lectura borrable electronicamente.

EFE—early fuel evaporation.

EFE—evaporación temprana del combustible.

EGR—the exhaust gas recirculation system that allows a small amount of exhaust gas to be routed into the incoming air/fuel mixture to reduce NO_X emissions.

EGR—el sistema de recirculación de los gases de escape que permite que una pequeña cantidad del gas del escape se desvía a la mezcla del combustible/aire para disminuir las emisiones NO_X.

elastic limit—this point is reached when a fastener will no longer return to its original shape when loosened.

límite elástico—este punto se alcanza cuando un sujetador no volverá a su forma original al ser aflojado.

electricity—the flow of electrons from one atom to another.

electricidad—el flujo de los electrones desde un átomo a otro.

electrolysis—using electricity to break down water into hydrogen and oxygen.

electrólisis—usando la electricidad para descomponer el agua al hidrógeno y el oxígeno.

electrolyte—a mixture of sulfuric acid and water used in automotive batteries.

electrólito—una mezcla del ácido sulfúrico y el agua que se usa en las baterías automotivos.

electromagnetic induction—when electricity is produced by moving a magnetic field over a conductor.

inducción electromagnética—cuando la electricidad se produce por medio de pasar un campo magnético sobre un conductor.

electromechanical voltage regulator—a regulator that operates using coils of wire and electrical contact points.

regulador de voltaje electromecánico—un regulador que opera utilizando las bobinas de alambre y los puntos de contacto eléctricos.

electron theory—electron flow is from negative to positive.

teoría de electrones—el flujo de los electrones es del negativo al positivo.

electronic computer advance—uses information available to the computer for the fuel and emission systems to control spark advance.

avance electrónico computerizado—usa la información disponible a la computadora de los sistemas de combustible e emisiones para controlar el avance de la chispa.

electronic ignition system—when a transistor triggers the buildup and collapse of the magnetic field in the coil primary.

sistema de arranque electrónico—cuando un transistor dispara la acumulación y el colapso del campo magnético en la bobina primaria.

element—a group of battery plates connected together in parallel.

elemento—un grupo de placas de batería conectado juntos en paralelo.

EMF—electromotive force, also called voltage, which pushes or pulls an electron out of its orbit.

EMF—fuerza electromotríz, tambien llamado voltaje, que empuja o jala un electrón fuera de su órbito.

end gap—the gap between the ends of the piston ring when it is installed in the cylinder.

holgura en la extremidad—la holgura entre las extremidades del anillo de pistón al instalarse en el cilindro.

end play—back and forth clearance.

juego en la extremidad—holgura lateral.

end thrust—side to side or front to rear force against a shaft.

empuje de la extremidad—la fuerza de un lado a otro o de frente a atrás contra un eje.

energy—the ability to do work.

energía—la habilidad a trabajar.

engine displacement—determined by multiplying the cylinder displacement by the number of cylinders.

desplazamiento del motor—se determine al multiplicar el desplazamiento por el número de cilindros.

engine kit—parts sold together as a group for rebuilding an engine.

piezas del motor sueltas—las partes vendidas juntos en grupo para reconstruir un motor.

engine knocks—noises that result from excessive clearance or abnormal combustion.

golpeteo del motor—los ruidos que resultan de una holgura excesiva o de la combustión anormal.

engine sling—a cable or chain used with an engine hoist.

eslinga del motor—un cable o una cadena que se usa con una grúa para el motor.

EP additives—extreme pressure additives added to lubricants to prevent welding between metal surfaces.

aditivo EP—los aditivos de presión extrema que se agregan a los lubricantes para prevenir la soldadura entre los superficies metálicos.

EPA—Environmental Protection Agency.

EPA—Agencia de Protección del Medio Ambiente.

EPR—evaporator pressure regulator valve.

EPR—válvula reguladora de la presión de evaporación.

EPROM—an older prom that had to be removed from the computer and put under an ultraviolet light for a specified period of time to reprogram it.

EPROM—un modelo prom más antiguo se tenía que remover de la computadora y ser puesta bajo un luz ultraviolet por un período específico para reprogramarla.

ester—polyalester oil for R-134A air conditioning systems.

ester—el aceite polialaster para los sistemas de aire acondicionado R-134A.

EVA—electronic vibration analyzer.

EVA—analizador electrónico de vibraciones.

evaporation—a method of heat transfer where moisture is vaporized as it absorbs heat.

evaporación—un metodo de tranferencia del calor en el cual la humedad se vaporiza al absorber el calor.

evaporator—part of an air conditioning system. A small, radiator-like device used to remove heat from the passenger compartment.

evaporador—parte de un sistema de aire acondicionado. Un dispositivo pequeño parecido a un radiador que sirve para remover el calor del compartimiento de pasajeros.

exhaust backpressure—resistance to an engine's air flow caused by restrictions in the exhaust system.

contrapresión del escape—la resistencia al flujo del aire del motor debido a las restricciones en el sistema del escape.

expansion tank—a small tank sometimes located inside of the main gas tank that allows for expansion of fuel in a full tank on a hot day.

tanque de expansión—un tanque pequeño que a veces se encuentra dentro del tanque de gas principal que permite

expandir el combustible dentro de un tanque lleno en un día caliente.

external combustion engine—an engine where the fuel is burned outside of the engine, a steam engine.

motor de combustión externa—un motor en el cual el combustible se consume fuera del motor, un motor de vapor.

F

face—also called top, the part of a gear tooth that is the area above the pitch line.

cara—tambien llamado parte superior, la parte de un diente de engranaje que esta en el área superior a la linea cero.

fan clutch—a temperature or torque sensitive clutch attached to a belt-driven cooling fan.

embrague ventilador—un embrague sensitivo a la temperatura o a la torsión que esta conectada a un ventilador impulsado por correa.

feedback systems—computer fuel systems that monitor the oxygen content in the exhaust.

sistema de retroalimentación—los sistemas de combustible computerizados que amonestran la cantidad del oxígeno en el escape.

ferrous—iron and steel.

férreo—hierro y acero.

ferrule—a sleeve used in a compression fitting.

férula—un manguito que se usa en un guarnición de compresión.

field coils—heavy copper ribbons wound around soft iron cores to form an electromagnet.

bobinas inductoras—las gruesas cintas del cobre enredadas alrededor de los núcleos de hierro blando para formar un imán.

filament—a wire in a light bulb that provides a resistance to electron flow. When it heats up, it causes light.

filamento—un alambre en un foco que provee una resistencia al flujo de electrones. Cuando se calienta, alumbra.

filings—the pieces of metal that come off of the workpiece during filing.

limaduras—los pedazos del metal que se desprenden de la pieza al limar.

filter sock—a filter inside of the gas tank.

filtro con colador—un filtro dentro del tanque de gas.

final drive ratio—the ratio between the transmission output shaft and the differential ring gear.

relación de mando final—la relación entre la salida de la transmisión y la corona.

firewall—the metal wall between the engine and passenger compartments.

mamparo de encendios—el muro de metal entre los compartimentos del motor y de los pasajeros.

firing line—the upward line that starts the scope pattern.

indicación del encendido—la linea ascendente con la cual comienza el patrón en el osciloscopio.

firing order—the order in which the spark plugs in a multicylinder engine fire.

orden del encendido—el orden en el cual disparan las bujías en un motor con múltiples cilindros.

first order vibration—anything that spins at driveshaft speed that vibrates only once every revolution.

vibración del primer orden—cualquier cosa girando con la velocidad del árbol de la transmisión que vibra solamente una vez por cada revolución.

fixed joint—the outboard CV joint that allows for a change in the angle of the axle in response to bumps and for steering.

junta fija—la junta CV fuera de borda que permite un cambio en el ángulo de un eje como respuesta a los choques y para la dirección.

flame front—when heat ignites molecules next to already burning molecules and a chain reaction takes place, which results in a flame expanding evenly across the cylinder.

frente de llama—cuando el calor encienda las moléculas juntas a las moléculas ya quemando y resulta una reacción en cadena, lo cual causa una llama que se expande uniformemente por el cilindro.

flank—also called root, the part of a gear tooth that is the area below the pitch line.

flanco—tambien llamado el raíz, la parte de un diente de un engranaje que esta en el área inferior a la linea cero.

flapper valve—also called a duckbill valve, a rubber valve that allows fluid to flow through its center in one direction only.

válvula de charnela—tambien llamado una válvula de aleta, una válvula de caucho que permite que el fluido fluye por su centro en solamente una dirección.

flare nuts—hollow fittings on fuel, brake, or hydraulic lines.

tuercas bocinadas—los guarniciones huecos en las lineas de combustible, de frenos o hidráulicas.

flash point—the temperature at which the vapors of a flammable liquid will ignite when brought into contact with an open flame.

temperatura de inflamabilidad—la temperatura en que los vapores de un líquido inflamable encenderán al ponerse en contacto con una llama expuesta.

flashing the PROM—when the information in an EEPROM is erased and reprogrammed.

limpiar el PROM—cuando la información en un EEPROM es borrada y reprogramada.

flat cam—when a cam lobe wears out.

leva aplanada—cuando se desgaste el lóbulo excéntrico de una leva.

flat-rate manual—also called a Parts and Time Guide, it lists the estimated cost of parts. A service writer or a business owner will use it to give a fairly accurate estimate to a customer.

manual de precio fijo—tambien llamado una Guía de Partes y Labor, cataloga los precios estimados de las partes. Un girador de servicios o un negociante lo usará para fijar un presupuesto bastante acertado para una clientela.

float—car speed is slowly dropping.

flotar—la velocidad del coche se disminuya poco a poco.

flooding—when an engine gets too much fuel to support combustion.

ahogar—cuando un motor recibe demasiado combustible para efectuar la combustión.

fluid coupling—a fluid clutch.

acoplador flúido—un embrague de fluido.

flux—rosin or acid used to clean metal so that solder will stick to it.

flujo eléctrico—la resina o el ácido que se usa para limpiar el metal para que la soldadura se adherirá.

follower ball joint—the nonweight carrying ball joint that keeps the steering knuckle in alignment.

junta de casquillo de rótula—la articulación esférica de no soportar peso que mantiene alineado el muñón de dirección.

foot-pound—a measurement of work where one pound is moved for a distance of one foot.

libra-pie—una medida del trabajo en la cual una libra se mueve la distancia de un pie.

foot-pounds/Newtons—torque readings are expressed in foot-pounds in the English system, Newtons in the metric system.

libras-pie/Newtons—las lecturas de torsión se describen en libras-pie en el sistema inglés, en Newtons en el sistema métrico.

footprint—the area of the tire tread that contacts the road.

huella—el área de la banda de rodamiento que toca la carretera.

force—any action that changes, or tends to change, the position of something.

fuerza—cualquier acción que cambia, o tiene tendencias a cambiar, la posición de algo.

force, static or kinetic imbalance—different names for the same thing. A type of imbalance that can be compared to balancing tires with a bubble (level) balancer.

fuerza, estático o desequilibrio cinético—los nombres distintos para la misma cosa. Un tipo de desequilibrio que se puede comparar al equilibrar los neumáticos con un compensador anivelador (de burbuja).

forward bias—causes a P-N junction to conduct current.

polarización directa—causa que un enlace P-N conduzca el corriente.

fouled spark plug—one with a buildup of carbon that shorts it out.

bujía ensuciado—uno que tiene una acumulación del carbón que causa un cortocircuito.

four corners scuffing—damage to the piston that usually occurs as a result of an external cause such as too lean an air/fuel mixture, which causes the top of the piston to run too hot.

rozamiento en cuatro puntos—el daño al pistón que suele ocurrir como resultado de una circunstancia externa, tal como una mezcla desmasiada pobre del combustible/aire, lo cual causa que la parte superior del pistón marcha con excesivo calor.

four-stroke cycle—also called the Otto-cycle, intake, compression, power, exhaust.

ciclo de cuatro carreras—tambien llamado el ciclo Otto, entrada, compresión, fuerza, escape.

four wheel alignment—a method of wheel alignment done on many newer cars that are equipped with rear wheel adjustment capability.

alineación a cuatro ruedas—un método de alineación en las ruedas que se efectúa en muchos de los coches de último modelo que se equipan con la capacidad de ajustar las ruedas traseras.

free air delivery—the best measurement of an air compressor's useful capacity measured in standard cubic feet per minute (SCFM).

entregada de aire libre—la mejor medida de la capacidad útil de un comprimidor de aire que se mide en pies cúbicos calibrados por minuto (SFCM).

freewheeling engine—an engine that will not experience piston-to-valve interference if the timing chain or belt skips or breaks.

motor de sobremarcha—un motor que no experiencia una interferencia de pistón-a-válvula si brinca o se rompa la cadena o correa de tiempo.

frequency—how many cycles take place in a period of time.

frecuencia—cuántos ciclos ocurren en un período del tiempo.

friction disc—the driven member of a clutch that is positioned between the driving parts.

disco de fricción—el miembro arrastrado de un embrague colocado entre dos partes impulsores.

frictional horsepower (fhp)—the power lost due to friction.

caballos de fuerza friccional (fhp)—la fuerza perdida debido a la fricción.

frictionless bearings—also called anti-friction bearings, provide a rolling contact with either balls, rollers, or needles.

cojinetes sin fricción—tambien llamado cojinetes antifricción, proveen un contacto rodante por medio de bolas, rodamientos, o las agujas.

full field test—a test that takes the regulator out of the circuit and causes the alternator to give full output.

prueba completa de campo—una prueba que remueva el regulador del circuito y causa que el alternador dé su salida completa.

full-floating axles—an axle design in which the bearings do not touch the axle but are located on the outside of the axle housing.

ejes flotantes—un diseño de eje en el cual los cojinetes no tocan al eje pero se encuentran al exterior de la caja del eje.

full-floating plain bearings—bearings that have oil clearance on both sides, they spin at about one-third shaft rpm.

cojinetes lisos flotantes—los cojinetes que tienen holgura para el aceite en ambos lados, giran una tercera parte de la velocidad rpm.

full-flow oil filter—a filtering system designed to flow all of the oil supplied by the pump to the oil filter on its way to the engine bearings.

filtro de aceite de pleno flujo—un sistema de filtración diseñado para dirigir todo el aceite suministrado por la bomba en su rumbo a los cojinetes del motor.

furnace brazing—a process used to make seamless tubing that fuses copper to steel. When the copper melts, the seam disappears.

cobresoldadura en horno—un proceso usado en la fabricación de los tubos sin cordón que funde el cobre en el hierro. Cuando se derrite el cobre, el cordón se desaparece.

fuse—a circuit protection device designed to melt when the flow of current becomes too high for the wires or loads in the circuit.

fusible—un dispositivo para la protección de un circuito diseñado a fundir cuando el flujo de corriente exceda lo que puede tolerar los alambres o las cargas de un circuito.

fuse link—a circuit protection device that is a small length of wire, smaller in diameter than the wire it is connected to.

cartucho de fusible—un dispositivo para la protección de un circuito que consiste en un trozo pequeño del alambre, de un diámetro más pequeño del alambre al cual esta conectado.

FWD—front wheel drive.

FWD—tracción delantera.

G

galling—wear caused by metal to metal contact in the absence of adequate lubrication. Metal is transferred from one surface to another.

desgaste por fricción—el desgaste causado por el contacto entre un metal y otro en la ausencia de la lubricación adecuada. El metal se transfiere de una superficie a otra.

gas shock—a shock with the oil column pressurized to keep the bubbles from forming in the fluid.

amortiguador de gas—un amortiguado con una columna de aceite bajo presión que previene que forman las burbujas en el fluido.

gasohol—a mixture of gasoline and alcohol.

gasohol—una mezcla de la gasolina y el alcohol.

Gauss meter or gauge—a meter or gauge that detects magnetism.

medidor o calibrador Gauss—un medidor o calibrador que detecta el magnetismo.

gear pattern—a pattern taken off of the ring gear teeth from a coating of colored paste painted on the gear teeth.

patrón de engranajes—una impresión que se toma de los dientes del corona por medio de una pasta de color pintado en los dientes del engranaje.

gear radius—the distance from the center of a gear to its outside edge. It is where the torque is measured from.

radio de engranaje—la distancia del centro de un engranaje a su orilla exterior. Es en dónde se mide la torsión.

gear ratio—the difference in speed between two gears calculated by dividing the number of teeth on the driven gear by number of teeth on the driving gear.

relación de engranajes—la diferencia en velocidad entre dos engranajes que se calcula al dividir el número de los dientes en el

engranaje arrastrado por el número de los dientes en el engranaje propulsor.

generator—an electrical device that produces alternating current that is rectified to DC current by the brushes and commutator.

generador—un dispositivo eléctrico que produce el corriente alterna que se rectifica en un corriente DC por medio de las escobillas y el conmutador.

geometric centerline—a line drawn between the center of the front axle and the center of the rear axle.

linea central geométrico—una linea dibujada entre el centro del eje delantero y el centro del eje trasero.

glitch—a momentary interruption of an electrical signal.

glitch—una interrupción momentánea de una señal eléctrica.

governor pressure—the kind of transmission pressure that results from increases in vehicle speed.

presión del regulador—el tipo de presión de transmisión que resulta de aumentar la velocidad del vehículo.

grabbing clutch—when the friction disc does not slip normally, but grabs all at once.

embrague agarrador—cuando el disco de fricción no se desliza normalmente, sino que se agarra de una vez.

granny gear—when a transmission has a very low first gear, it is sometimes called a granny gear.

engranaje abuelita—cuando una transmisión tiene la primera velocidad muy lenta, a veces se llama un engranaje abuelita.

gravity bleeding—bleeding a hydraulic clutch by opening the slave cylinder bleed screw and letting fluid flow out.

purgar con gravedad—purgar un embrague hidráulico abriendo el tornillo de purgar del cilindro secundario y dejando que salga el fluido.

grease—a combination of oil and a thickening agent.

grasa—una combinación del aceite y un agente espesativo.

greasesweep—an absorbent material, such as rice hull ash or kitty litter used for cleaning spills.

limpiagrasa—una material asorbente, tal como la ceniza de casco de arroz o la arena usado para limpiar los derrames.

gross horsepower (ghp)—engine power available with only the water pump and alternator using power.

caballo de fuerza bruto (ghp)—la fuerza del motor disponible cuando sólo usan potencia la bomba de agua y el alternador.

ground side switching—when the ground sides of circuits to output devices are controlled, rather than the power sides.

conmutación a tierra—cuando los lados a tierra de los circuitos a los dispositivos de salida son controlados, en vez de los lados de potencia.

ground spark plug electrode—the electrode on the end of the metal case, it can be bent toward or away from the center electrode to make the correct spark plug gap.

electrodo de bujía a tierra—el electrodo en la extremidad de una caja metálica, se puede torcer hacia o en dirección contraria del electrodo central para hacer una abertura de chispa correcta.

grounded circuit—when current goes directly to ground.

circuito a tierra—cuando el corriente va directamente a tierra.

guide pins—capscrews with the heads cut off. They have a hacksaw slot for a screwdriver.

clavijas de guía—los tornillos de casquete cuyos casquetes se han quitado cortandolos. Tienen una muesca cortada con una sierra en donde se puede meter un destornillador.

H

half shaft—also called axle shaft, it is a drive axle on a front wheel drive vehicle.

semi-eje—tambien llamado una árbol motor, es un eje propulsor en un vehículo de tracción delantera.

Hall switch—a stationary sensor and rotating trigger wheel that uses the Hall-effect to control the ignition primary.

interruptor Hall—un sensor estático y rueda disparadora rotativa que utiliza el efecto Hall para controlar el encendido primario.

Hall-effect—current is passed through a thin semiconductor material while a magnetic field passes through it. This produces a small voltage (about 0.4 V) in the semiconductor.

efecto Hall—el corriente pasa por una material delgada semiconductor mientras que lo atraviesa un campo magnético. Esto produce un pequeño voltaje (aproximadamente 0.4 V) en el semiconductor.

halogen headlamp—a brighter headlamp used on newer cars.

faro halógeno—un faro más luminoso que se usan en los coches de último modelo.

hand tools—a term for such tools as sockets, wrenches, and screwdrivers.

herramientas manuales—un término para las herramientas tales como las llaves de caja, las llaves de tuerca, y los destornilladores.

hard faults—those codes that are present and are stored in memory at the time of the self test.

fallos fijos—esos códigos que estan presentes y se almacenan en la memoria en el tiempo de autoprueba.

hard parts—metal parts such as gears, valves, pump bodies, clutch drums, and other things that would not normally be replaced in an automatic transmission rebuild.

partes duras—las partes metálicas tal como los engranajes, las válvulas, los cárteres de las bombas, los tambores de los embragues y otras cosas que normalmente no se reemplazan durante una reconstrucción de la transmisión.

hard spots—where the metallurgical composition of cast iron has been changed to steel by heat.

zonas duras—en dónde la composición metalúrgica del hierro colado se ha convertido en el acero por medio del calor.

hardware—the mechanical parts of an electronic system.

hardware—las partes mecánicas de un sistema electrónico.

harsh shift—the opposite of a mushy shift, it is when the transmission makes a gear change that is too fast.

cambio brusco—la opuesta de un cambio blando, es cuando la transmisión cambia de velocidades demasiado rapidamente.

Hazard Communication rules—In the United States, the Environmental Protection Agency (EPA) regulations that outline disposal requirements for cleaning chemicals, used oil, heavy metals, anti-freeze/coolant, asbestos, and other hazardous materials.

reglas de la Comunicación de Riesgos—En los Estados Unidos, las regulaciones de la Agencia de Protección del Medio Ambiente (EPA) que detallan los requerimientos de la disposición de las químicas de limpieza, el aceite usado, los metales pesados, los fluidos de anti-congelante o refrigerante, el amianto, u otros materiales peligrosas.

HC—hydrocarbon.

HC—hidrocarburo.

headers—aftermarket manifolds made of tube steel.

cabezales de tubo—los múltiples no originales fabricado del acero tubular.

heat exchanger—another name for a cooler.

cambiador de calor—otro nombre para un enfriador.

heat range—an indication of how fast heat can travel away from the spark plug center electrode to the cooling system's water jackets in the cylinder head.

gama de calor—una indicación de qué tal rápido puede viajar el calor del electrodo central de bujía a las camisas de agua del sistema enfriador de la cabeza del cilindro.

heat sink—a part that helps to dissipate heat caused by electrical flow.

fuente fría—una parte que ayuda en dispersar el calor causado por un flujo eléctrico.

heat transfer—the movement of heat that occurs whenever there is a difference in temperatures between two objects.

transferencia del calor—el movimiento del calor que ocurre cuando hay una diferencia de temperatura entre dos objetos.

heat-shrink tubing—tubing that shrinks to seal an electrical joint when heated.

tubo que se contrae con el calor—el tubo que se contrae para sellar una junta eléctrica al calentarse.

heater core—a small heat exchanger in the passenger compartment that engine coolant is circulated through.

núcleo del calentador—un pequeño intercambiador de calor en el compartimento de pasajeros por el cual se circula el fluido refrigerante del motor.

heel—the outer end of the gear tooth.

talón—la extremidad exterior de un diente de engranaje.

helical gears—gears that are machined at an angle that gives them a continuous flow of power across the gear teeth and makes them quieter in operation.

engranaje helicoidales—los engranajes que se maquinan en un ángulo lo cual les proporciona un flujo de fuerza atravéz de los dientes de engranaje y los hace más silenciosos durante la operación.

hemispherical combustion chamber—a nonturbulent combustion chamber design shaped like half a globe.

cámara de combustión hemiesférica—un diseño de cámara de combustión sin turbulencia que tiene una forma media esférica.

HEPA vacuum—a vacuum cleaner with a special filter that is used for asbestos.

apiradora HEPA—una aspirador con un filtro especial que se usa para el amianto.

Hertz—cycles per second.

Hertz—ciclos por segundo.

high cordline belt—a higher quality belt with the tensile cord above center.

correa de alta tensión—una correa de alta calidad que tiene la cuerda de tensión superior al centro.

high impedance meter—a voltmeter with a very high input resistance (usually about 10 million ohms). The high input resistance prevents the meter from drawing current while it is connected to a circuit.

medidor de alta impedancia—un voltímetro con una resistencia de entrada muy alta (normalmente aproximadamente 10 millónes de ohms). La resistencia muy alta de entrada previene que el calibrador jala el corriente mientras que esté conectado a un circuito.

high point—the point where the teeth of a conventional steering gear come closer together when the car is travelling straight ahead.

punto alto—el punto en el cual los dientes un engranaje de dirección convencional se acercan mientras que el coche viaja en una linea recta.

high side—the side of an air conditioning system that comes after the compressor.

lado alto—el lado del sistema de aire acondicionado después de que pase por el comprimidor.

high temperature pyrometer—a temperature measuring device that is touched against a surface to obtain a reading.

pirómetro de alta temperatura—un dipositivo de medir la temperatura que se toca contra una superficie para obtener una lectura.

hobbing—a rough machine process that leaves heavy machine marks on the surface of the gear tooth.

fresar—un proceso desbastador por máquina que deja las huellas de maquinar profundas en la superficie del diente de engranaje.

horseless carriage—an early vehicle built on the principle of the horse and wagon.

carruaje sin caballo—un vehículo primitivo fabricado sobre los fundamentales del caballo y carro.

horsepower—the measurement of an engine's ability to perform work.

potencia de caballo—la medida de la habilidad de un motor en efectuar el trabajo.

hot soak—when the engine is shut off on hot days and gasoline boils out of the float bowl into the engine.

evaporación por calor—cuando se apaga un motor en tiempo cálido y la gasolina escapa hirviendo de la taza flotante al motor.

Hotchkiss drive—an open driveshaft.

propulsión tipo Hotchkiss—un árbol motor descubierto.

howl—a noise like the wind across the top of a soda bottle. Its range is from 120~300 Hz.

aullido—un ruido parecido al viento pasando por la boca de una botella. Su gama es del 120~300 Hz.

hub-centric—means that the center of the wheel has a machined counterbore that pilots on a machined area of the hub.

cubo-céntrico—quiere decir que el centro de la rueda tiene un agujero abocardado que sirve de guía en un área maquinada del cubo.

humidity—the moisture content of the air. When humidity is 50% the air is holding half the amount of moisture that it is capable of holding at a given temperature.

humedad—el contenido de humedad en el aire. Cuando la humedad es del 50% el aire contiene la mitad de la cantidad del humedad que puede contener en una temperatura dada.

hunting gear set—when a pinion gear tooth will move around until it contacts all of the ring gear teeth.

conjunto de engranaje con diente suplementario—cuando un diente del engranaje piñón dará la vuelta hasta que se ha puesto en contacto con todo los dientes del corona.

hybrid vehicle—a vehicle that uses more than one type of energy for its power.

vehículo híbrido—un vehículo que usa más de un tipo de energía para su propulsión.

hydraulics—when liquid under pressure is used to transfer motion or apply force.

hidráulica—cuando un líquido bajo presión se usa para transferir el movimiento o para aplicar la fuerza.

hydrolocked engine—when coolant or fuel in a cylinder prevents an engine crankshaft from turning over.

motor bloqueado hidráulicamente—cuando el líquido refrigerante o el combustible en un cilindro previene que capota el cigüeñal del motor.

hydrometer—a device that compares the weight of electrolyte to the weight of pure water.

hidrómetro—un dispositivo que compara el peso del electrólito con el peso del agua pura.

hydroplaning—when water forms a wedge under a tire that can actually float the car.

efecto hidroavión—cuando el agua forma una cuña debajo de un neumático que puede en efecto hacer flotar a un coche.

hydrostatic lock—when a liquid is trapped in a blind hole when the fastener contacts the oil, it cannot compress it so the fastener cannot be properly tightened.

cerradura hidrestática—cuando un líquido se atrapa en un orificio ciego al ponerse en contacto con el aceite el sujetador, no

puede comprimir conque el sujetador no se puede apretar correctamente.

hygroscopic—when a material can absorb water.

hygroscópico—cuando una materia puede absorber el agua.

hypoid gears—a gear design with teeth that are spiraled and curved used when a ring gear and a pinion gear intersect below the centerline of the ring gear.

engranaje hipoide—un diseño de engranaje cuyos dientes son en espiral y curvadas usado cuando la corona y el engranaje de piñón cruzan debajo del linea central de la corona.

hysteresis—a term used by chemical engineers to describe a rubber's energy absorption characteristics.

histéresis—un término usado por los ingenieros químicos para describir las caracteristicas de absorción de la energía del caucho.

I

I-head—valve placement in modern engines is in the cylinder head, above the piston.

cabeza I—una colocación de las válvulas en los motores más modernos es en la cabeza del cilindro, superior al pistón.

I.D.—inside diameter.

I.D.—diámetro interior.

IAC motor—idle air control motor. It is a stepper motor with two electromagnetic circuits that open and close an air passage to control idle speed.

motor IAC—motor de control del aire de marcha en vacío. Es un motor paso a paso con dos circuitos electromagnéticos que abren y cierran el paso del aire para controlar la velocidad de la marcha en vacío.

IC—integrated circuit—a complete miniaturized electric circuit.

IC—circuito integrado—un circuito eléctrico en miniatura completo.

idler gear—a gear used between two other gears, the purpose of which is to change direction of rotation.

engranaje de piñón loco—un engranaje que se usa entre dos otros engranajes, su propósito es cambiar la dirección de rotación.

ignition system—creates and distributes a timed spark to the cylinder.

sistema del encendido—crea y distribuya una chispa sincronizada al cilindro.

ignition timing—the point at which ignition occurs in a cylinder.

tiempo del encendido—el punto en el cual ocurre el encendido en el cilindro.

impedance—a meter's internal resistance.

impedancia—la resistencia interna de un medidor.

inboard side—the inside end of an axle shaft. The opposite side is called the outboard side.

lado abordo—la extremidad interior de una flecha del eje. El lado opuesto se llama el lada fuera de bordo.

inches of mercury (in.Hg)—the measurement increment for vacuum or low pressure.

pulgadas de mercurio (in.Hg)—el incremento de dimensión de la presión en vacío o baja presión.

included angle—the combination of SAI and camber.

ángulo incluído—la combinación de SAI y combadura.

independent suspension—a type of suspension system that when it goes over a bump, only that wheel will deflect.

suspensión independiente—un tipo de sistema de suspención que al pasar por un tope, sólo se desvía esa rueda.

indicated horsepower (ihp)—the amount of pressure made in the combustion chambers.

potencia de caballo indicado (ihp)—la cantidad de pressión creado en las cámaras de combustión.

induction hardened valve seats—integral valve seats that are harder than the surrounding area of a cylinder head.

asientos de válvula endurecidos por inducción—los asientos de válvulas íntegros que son más duros que el área al su alrededor en la cabeza del cilindro.

inductive pick-up—the pickup for an electrical meter that wraps around the wire being sensed and measures current flowing through it using principles of magnetism.

captador inductivo—el captador de un medidor eléctrico que se enreda alrededor del alambre que se tiene que detectar y mide el corriente que lo atraviesa usando los fundamentales del magnetismo.

inert gas—a gas that does not burn and takes up space in the combustion chamber.

gas inerta—un gas que no se quema y ocupe el espacio en la cámara de combustión.

inert gas shield—an inert gas shield (argon, helium, or CO_2) is applied over the arc area during welding to prevent oxidation of the metal.

blindaje de gas inerta—un blindaje de gas inerta (argón, helio o el CO_2) se aplica al arco durante la soldadura para prevenir la oxidación en el metal.

inertia—the tendency of a body to keep its state of rest or motion.

inercia—la tendencia de un cuerpo de mantener su estado de descanso o movimiento.

inertia starter drive—also called Bendix drives, they operate like a heavy nut on a screw.

acoplamiento del arrancador de inercia—tambien llamado los acomplamientos Bendix, operan como una tuerca muy gruesa en un tornillo.

infrared thermometer—a thermometer that takes a temperature reading when it is aimed toward a surface.

termómetro infrarrojo—un termómetro que toma una lectura de temperatura cuando se apunta a una superficie.

initial timing—also called base timing, it is the timing setting before the computer has a chance to make changes.

tiempo inicial—tambien llamado tiempo base, es la regulación del tiempo antes de que la computador haya tenido una oportunidad de efectuar los cambios.

insulators—the opposite of conductors; they have no, or few, free electrons.

aisladores—lo opuesto de los conductores; no tienen ningunos, o muy pocos, electrones libres.

intake manifold vacuum—the pressure inside of the intake manifold when the engine is running.

vacío del múltiple de entrada—la presión en el interior de la entrada del múltiple mientras que esté en marcha el motor.

integral seats and guides—parts that are one with the head.

asientos y guías íntegros—las partes que son parte de la cabeza.

intercooler—also called a charge air cooler or an aftercooler, it is a sophisticated air cooler for blowers.

enfriador intermedio—tambien llamado enfriador de aire cargado o un postrefrigerador, es un enfriador sofisticado para los ventiladores.

interference angle—a difference between the valve and seat face angles that results in increased pressure to aid valve seating.

ángulo de interferencia—la diferencia entre los ángulos de la válvula y de la cara del asiento que resulta en una presión aumentada para ayudar en asentar las válvulas.

interference engine—also called non-freewheeling, an engine that will experience piston-to-valve interference if the timing chain or belt skips or breaks.

motor de interferencia—tambien llamado de no rodamiento libre, un motor que experiencerá la interferencia entre pistón-y-válvula si se brinca o quiebra la cadena o correa de tiempo.

interference fit—when two parts have a pressed fit.

ajuste de interferencia—cuando dos partes tienen un ajuste embutido.

intermittent fault code—one that only occurs occasionally for a short period of time and is not present in the system at the time of the fault test.

código de fallos intermitentes—uno que sólo ocurre a veces por muy corto tiempo y no se presenta en el sistema durante el tiempo de la prueba de fallos.

internal balancing—balancing that is done by drilling holes on the crankshaft counterweights.

equilibración interna—la equilibración que se efectúa taladrando los orificios en los contrapesos del cigüeñal.

inversion layer—when warm air becomes trapped within 1000 feet of the ground. The trapped air produces smog.

capa de inversión—cuando el aire tibio se atrapa a una altura de menos de 1000 pies de la tierra. El aire atrapado produce el esmog.

inverted flare nut—the common type of SAE flare used on automobiles.

tuerca bocinada invertida—un tipo de abocinamiento SAE usado en los automóviles.

ISC motor—idle speed control motor. It opens the throttle plate to control idle speed.

motor ISC—motor de control de velocidad en marcha en vacío. Abre la placa del acelerador para controlar la velocidad de marcha en vacío.

ISO (International Standards Organization) flare—one style of forming the ends of tubing, also called a bubble flare.

ISO (Organización Internacional de Normas) abocinamiento—un estilo de formar las extremidades de los tubos, tambien llamado un abocinamiento de burbúja.

isolated field—an alternator with two field leads that is energized and grounded externally.

campo aislado—un alternador con dos lineas del campo que se energetiza y va a tierra al exterior.

J

jackscrew—a screw for adjusting belt tension.

gato—una tuerca para ajustar la tensión de la correa.

jamb nut—a locknut.

tuerca de seguridad—una contra tuerca.

joule—an equivalent value that compares heat energy (Btu) to mechanical energy (ft.-lb.). 1 Btu = 778 ft.-lb.

julio—un valor equivalente que compara la energía calorífica (Btu) a la energía mecánica (ft.-lb.). 1 Btu = 778 ft.-lb.

K

keep alive memory—KAM, means the computer maintains power to random access memory (RAM) when the ignition switch is off. This allows it to keep information as long as the battery is not disconnected.

memoria de retención—KAM, quiere decir que la computadora mantiene la potencia al memoria de acceso aleatoria (RAM) cuando esta apagado el interruptor del encendido. Este lo permite mantener la información mientras que no se disconecte la batería.

kinetic energy—energy of motion.

energía cinética—la energía del movimiento.

Kirchoff's law—the total voltage drop in an electrical circuit will always be equal to the available voltage at the source.

ley de Kirchoff—la caída total del voltaje en un circuíto eléctrico siempre será igual al voltage disponible en la fuente.

knock sensor—also called a detonation sensor, it is a piezo—electric crystal that detects the frequency of the vibration caused by detonation and tells the computer to retard the spark timing until the vibration goes away.

sensor de golpeteo—tambien llamado un sensor de detonación, es un cristal eléctrico que detecta la frequencia de la vibración causado por la detonación y manda que la computadora retrasa la chispa del tiempo hasta que desaparezca la vibración.

knock-off type CV joint—an internal circlip CV joint retaining design.

junta CV de fácil desenganche—un diseño de retención de junta CV con grapa circular.

knurling—a metal displacement process that results in the enlargement of the surface of a metal.

moletear—un proceso de desplazamiento de metal que resulta en el ensanchamiento de la superficie de un metal.

KOEO—key-on, engine off.

KOEO—llave prendida, motor apagado.

L

L-head—also called flat head, this older engine design has the valves in the cylinder block, next to the cylinder.

cabeza L—tambien llamado una cabeza plana, este diseño más antiguo tiene las válvulas en el bloque del cilindro, junto al cilindro.

labor intensive—a process that requires a human's labor.

labor de mano intensivo—un proceso que require el labor de un ser humano.

laminated—more than one layer.

laminado—de más de una capa.

latent heat—the extra heat required before matter can change its state.

calor latente—el calor extra que se requiere antes de que la materia puede cambiar su estado.

latent heat of condensation—the heat released during condensation. When steam condenses back to water, it releases 970 Btus of heat per pound.

condensación latente del calor—el calor soltado durante la condensación. Cuando el vapor se condensa y se convierta al agua, suelta 970 Btus del calor por libra.

latent heat of vaporization—the heat required to heat water at 212°F to turn it into steam. It will require an additional 970 Btus of heat to make it boil.

calor latente de la vaporización—el calor requerido a calentar el agua de 212°F para convertirlo al vapor. Requerirá unos 970 Btus adicionales del calor para hacerlo hervir.

lateral runout—wobble of a part in a side-to-side direction.

corrimiento lateral—el bamboleo de una parte en una dirección de lado-a-lado.

launch shudder—a common second order vibration complaint in raised trucks with a high driveshaft angle.

temblor inicial—una queja común de vibración de segunda orden en los camiones elevados con un ángulo muy alto del árbol motor.

LCD—liquid crystal display.

LCD—presentación de cristal líquido.

lead oxidation—when lead is exposed to air it forms a black oxidized coating.

oxidación de plomo—cuando el plomo se expone al aire forma una capa negra de oxidación.

LED—light emitting diodes found in digital displays on the dashboard and in test instruments.

LED—los diodo emisores de luz que se encuentran en las presentaciones digitales en el tablero de instrumentos y en los intrumentos para efectuar pruebas.

lift—the height to which the lobe raises the lifter.

altura de izado—la altura a la cual el lóbulo levanta el aparato de izar.

light off—when the catalytic converter becomes hot enough and begins to oxidize pollutants.

luces apagados—cuando el convertidor catalítico se calienta lo suficiente y comienza a oxidar los contaminantes.

limited slip differential—it locks up the spider gears when one wheel starts to lose traction.

diferencial autoblocante—enclava los satélites del diferencial cuando una rueda comienza a perder la tracción.

line honing or line boring—methods of refinishing the main bearing bores to assure correct alignment.

barrenado en linea o rectificado en linea—los metodos de acabado de los diámetro interiores de los cojinetes principales.

liquid vapor separator—a part of the fuel tank or the expansion tank that keeps liquid fuel from being drawn into the charcoal canister.

separador de vapor líquido—una parte del tanque de combustible o el tanque de expansión que previene que el combustible en líquido se aspira al bote de carbono.

live axles—axles that turn with the wheels.

ejes vivos—los ejes que se giran con las ruedas.

load carrier ball joint—the ball joint in the control arm with the spring attached to it.

articulación esférica de carga—la articulación esférica en el brazo de mando que tiene un resorte conectado.

load index—the maximum load at the designated speed rating.

índice de carga—la carga máxima en la tasa de velocidad designada.

loaded caliper—a rebuilt caliper complete with new friction pads, hardware, and shims.

calibre cargado—un calibre reconstruído completo con las almohadillas, la herramienta, y las chapas nuevas.

lock to lock—when the steering wheel is turned all the way from one direction to the other.

cierre a cierre—cuando el volante de dirección se da la vuelta completamente de una dirección a la otra.

lock-up torque converter—a converter with a friction disc that locks the impeller and turbine together.

convertidor de par enclavado—un convertidor con un disco de fricción que enclava al impelador con la turbina.

locking hubs—allow a 4 wheel drive vehicle to be used in two wheel drive without the axles and differential being attached to the front wheels.

cubos enclavadores—permite que un vehículo de tracción en cuatro ruedas utilisa la tracción de dos ruedas sin que se conectan los ejes y el diferencial con las ruedas delanteras.

LOF—lube, oil, and filter.

LOF—la lubricación, el aceite, y el filtro.

long block—a complete engine assembly with heads.

bloque largo—una asamblea completa del motor con las cabezas.

long nipple—a nipple with a section of plain pipe separating the threads.

niple larga—un niple con una sección de tubo simple que separa las roscas.

longitudinal braking system—operates the front and rear brakes separately.

sistema de enfrenamiento longitudinal—opera los frenos delanteros y traseros por separado.

lookup tables—program information on how the car is supposed to perform.

tablas de verificación—información de la programa de cómo debe operar el coche.

low-maintenance battery—a battery that does not normally require water to be added.

batería de bajo mantenimiento—una batería que no suele requerir que se añade el agua.

low pressure cutout switch—shuts off the compressor clutch when air conditioning pressure drops too low, usually because the system is low on refrigerant.

interrumptor de corto-circuito en baja presión—apaga al embrague del comprimidor cuando la presión del aire acondicionado cae demasiado, normalmente debido a que le falta refrigerante al sistema.

low side—the side of an air conditioning system that comes after the flow control device before the evaporator.

lado bajo—el lado de un sistema de aire acondicionado después del dispositivo de control de flujo y antes del evaporador.

lower end—the parts that make up a short block.

extremidad baja—las partes que constituyen un bloque corto.

lower gear—when a small gear drives a larger gear.

velocidad baja—cuando un engranaje pequeño propulsa un engranaje más grande.

LPG—liquified petroleum gas.

LPG—gas de petroleo liquado.

lugging—lugging occurs when the load on the engine is greater than the rpm needed to develop enough horsepower to pull the load.

arrastre—el arraste ocurre cuando la carga en el motor supera los rpm que se requieren para desarrollar la potencia de caballo para tirar de la carga.

M

M+S, MS, M&S, M/S—any combination of the letters M and S on the tire sidewall means that the tire meets snow tire definitions set by the Rubber Manufacturers Association.

M+S, MS, M&S, M/S—cualquier combinación de las letras M y S en el refuerzo lateral de un neumático quiere decir que el neumático cumple con los requerimientos definidos por la Asociación de Fabricantes de Caucho.

MacPherson strut—a suspension design that incorporates the shock absorber into the front suspension.

poste MacPherson—un diseño de suspención que incorpora el amortiguador de choque en la suspensión delantera.

MAF—mass air flow sensor.

MAF—sensor del flujo del aire en masa.

maintenance-free battery—a battery with no provision for adding water.

batería de no mantenimiento—una batería que no tiene provisiones para añadir el agua.

malfunction indicator light (MIL)—an OBD II term for a light on the dash display, also called a check engine light, that tells when a hard code has been detected.

luz indicador de defecto (MIL)—un término OBD II para un luz en el indicador del tablero de instrumentos, tambien llamado el luz revisa el motor, que informa de que se ha descubierto un código fijo.

manifest—an EPA form for tracking hazardous wastes.

manifestación—una forma del EPA para rastrear los desechos tóxicos.

manifold heat control valve—a device also called a heat riser, it is a butterfly valve that fits between the exhaust manifold and exhaust pipe and routes exhaust gas under the floor of the intake manifold when the engine is cold to improve vaporization of the cold fuel.

válvula de control del calor del múltiple—un dispositivo que tambien se llama una columna de calor, es una válvula de mari-posa que queda entre el múltiple del escape y el tubo del escape que desvía el gas debajo del piso del múltiple de admisión cuando el motor está frio para mejorar la vaporización del combustible frío.

manual transmission—a transmission that is manually shifted and is used with a clutch. It is also called a stick shift or a standard transmission.

transmisión manual—una transmisión que cambia de velocidades manualmente que se usa con un embrague. Tambien se llama una palanca de cambiar velocidades o una transmisión estandard.

match mount—when runout of one part and runout of the other are considered during assembly to cancel each other out.

montaje adaptativo—cuando el corrimiento de una parte y el corrimiento de otra parte se consideran durante la asamblea para cancelarse mutuamente.

mechanical efficiency—describes all of the ways friction is lost in an engine. Brake horsepower (Bhp) divided by the indicated horsepower.

rendimiento mecánico—describe todas las maneras en que se puede perder la fricción en un motor. La potencia de caballo del enfrenamiento (Bhp) se divide por la potencia indicada en caballo.

mesh—when two gears run against each other.

endentado—cuando dos engranajes funcionan en contacto.

metallic linings—linings made of metal that are used in very heavy-duty and racing conditions.

forros metálicos—los forros hechos del metal que se usan bajo condiciones de trabajo pesado y en coche de carreras.

metering valve—shuts off pressure to the front brakes until about 125 psi builds up in the rear brakes. This keeps front pads from doing too much of the light braking and helps prevent dangerous skids that could result on slick surfaces if the front brakes were to apply before the rears.

válvula de medida—apaga la presión en los frenos delanteros hasta que acumula aproximadamente 125 psi en los frenos traseros. Esto previene que las almohadillas delanteras efectuan demasiado del enfrenamiento ligero y ayuda en prevenir los deslizamientos peligrosos que podrían resultar en las superficies resbaladizas si se aplicaran los frenos delanteros antes de los traseros.

Metric System—the international system (S.I.) of measurement based on the meter, which is 39.37" long, slightly longer than a yard.

Sistema Métrica—el sistema internacional (S.I.) de medida basado en el metro, que mide 39.37 pulgadas de longitud, un poquito más largo que una yarda.

microfiche—a small plastic film card that is magnified by a microfiche reader.

microfiche—una pequeña tarjeta de película de plástico que se magnifica con un lector de microfiche.

microinch—a microinch is one millionth of an inch.

micropulgada—una micropulgada es la millonésima parte de una pulgada.

microprocessor—the calculating and decision making chip in the computer.

mircroprocesador—la ficha que calcula y de forma las decisiones en una computadora.

MIG welding—metal inert gas welding.

soldadura MIG—soldadura de metal con blindaje de gas inerta.

MIL—malfunction indicator lamp located on the dash display.

MIL—luz indicador de defectos situado en el indicador del tablero de instrumentos.

min/max—a feature of some voltmeters that compares the maximum and minimum readings during a dynamic test.

min/max—una característica de algunos voltímetros que compara las lecturas máximas y mínimas durante una prueba dinámica.

mini-fuse—also called blade fuse it has a fuse element cast into a clear plastic outer body.

fusible miniatura—tambien llamado un fusible de hoja tiene el elemento de fusible incorporado en un cuerpo exterior de plástico transparente.

moan—also called drone, it is a noise that is compared to the hum of a bumble bee. Its range is from 60~120 Hz.

quejido—tambien llamado zumbido, es un ruido que se compara al zumbido de una abeja. Su gama es del 60~120 Hz.

molecular sieve—a material that will not crumble that has been used for desiccants since the early eighties.

criba molecular—una materia que no se desmenuza que ha sido usado como desecante desde los principios de la decada ochenta.

momentum—a body going in a straight line will keep going the same direction at the same speed if no other forces act on it.

impulso mecánico—un cuerpo viajando en una línea recta continuará viajando en la misma direccíon en la misma velocidad si ninguna otra fuerza actúa en su contra.

monolithic converter—a converter with honeycomb structure.

convertidor monolítico—un convertidor que tiene una estructura de nido de abeja.

Montreal Protocol—an agreement that set limits on the production of ozone depleting chemicals, with a total phase out by the year 2000.

Protocol de Montreal—un acuerdo que definó los límites en la producción de las químicas que deterioran la capa de ozono, con su prohibición completa por el año 2000.

morning sickness—a term applying to power steering or automatic transmissions that describes a lack of hydraulic pressure buildup when seals are cold and hard. A cold automatic transmission will hesitate from a few seconds to a minute, leaving the vehicle standing still as if in neutral while the engine races.

achaques—un término que aplica a la dirección hidráulica o a las transmisiones automáticas que describe la carencia de acumulación de presión hidráulica cuando los sellos son duros y fríos. Una transmisión automatica fría vacilará de unos segundos a un minuto, dejando parado al vehículo como si estuviera en neutral mientras que el motor acelera.

MSDS—material safety data sheet, it provides details of the composition of a chemical, lists possible health and safety problems, and gives precautions for its safe use.

MSDS—hoja de datos de seguridad de las materias, provee los detalles de la composición de las químicas, detalla los problemas de salud y seguridad y avisa de precauciones en el uso prudente.

multiple viscosity—also called multi-vis, a lubricant with a rating for cold and hot viscosity.

viscosidad múltiple—tambien llamado multi-vis, un lubricante con una clasificación de viscosidad fría y caliente.

multiplexing—a term used when several computers are linked together by one pair of wires allowing one sensor to provide information to several computers.

unir con multiplex—un término que se usa cuando varias computadores se eslabonan juntos con un par de alambres permitiendo que un sensor provee la información a varias computadoras.

mushroomed valve tip—a description of the shape of the tip of a valve stem that results from the repeated impacts due to excessive clearance.

boquilla de la válvula aplanada—una descripción de la forma de la boquilla del vástago de la válvula que resulta de choques repetidos debido a una holgura excesiva.

mushy shift—the engine increases in speed briefly during a shift until the clutch pack or servo piston finally seals and pressure builds up.

cambio de velocidades blanda—la velocidad del motor aumenta brevemente durante un cambio de velocidades hasta que el equipo de embragues o el servo del pistón sella finalmente y aumenta la presión.

N

N-type—a negative type crystal doped with an atom like phosphorous with extra electrons.

tipo N—un cristal de tipo negativo al cual se ha agregado un átomo como el fósforo que tiene electrones extra.

NA—a natural amber light bulb.

NA—un foco de luz de ámbar natural.

natural frequency—also called resonant frequency, it is the frequency at which a body vibrates. A tuning fork is an example.

frecuencia natural—tambien llamado una frecuencia resonante, es la frecuencia en la cual vibra un cuerpo. Un diapasón es un ejemplo.

NCFR—no cause for removal.

NCFR—no hay causa para remover.

needle bearing—a very small roller bearing used to control thrust or radial loads.

cojinete de agujas—un cojinete de rodamientos muy pequeño que se usa para controlar las cargas de empuje o radiales.

negative camber—when a tire is tilted in at the top.

comba negativa—cuando un neumático esta inclinada en la parte superior.

negative caster—the forward tilt of the steering axis.

ángulo de caster negativo—la inclinación hacia adelante del eje de dirección.

negative coefficient thermistor—will have lower resistance as the temperature of the coolant rises.

termistor de coeficiente negativo—tendrá una resistencia más baja al subir la temperatura del fluido refrigerante.

negative offset—when the wheel is offset in such a way as to increase the track width of the tires.

escentrado negativo—cuando la rueda esta descentrado en una manera que aumenta la anchura de la banda de los neumáticos.

neoprene—a kind of oil-resistant artificial rubber.

neopreno—un tipo de caucho artificial resistente al aceite.

net horsepower—the maximum power available from the engine when all the accessories are turned on.

potencia neta de caballo—la cantidad máxima de la fuerza disponible del motor cuando se han prendido todos los accesorios.

NLGI—National Lubricating Grease Institute.

NLGI—Instituto Nacional de Grasa Lubricante.

non-hunting gear set—when one tooth on the pinion gear will mesh with the same tooth on the ring gear again and again, rather than "hunting" for all of the rest of the teeth.

conjunto de engranaje sin diente suplementario—cuando un diente del engranaje del piñón se endenta con el mismo diente en la corona repetidamente, en vez de "buscar" todos los otros dientes.

non-retorque head gaskets—non-retorque gaskets compress less than conventional gaskets so the head does not require retorquing.

juntas de cabeza de tipo no retorsión—las juntas de no retorsión se comprimen menos que las juntas convencionales conque no se tiene que retorcer la cabeza.

non-volatile RAM—when the information in memory is not erased when the ignition switch is turned off.

RAM no volátil—cuando la información en la memoria no se borra al apagar el interruptor del encendido.

normally aspirated—an engine that is not supercharged.

aspirado normalmente—un motor que no es supercargado.

NO_x—oxides of nitrogen are produced when combustion temperatures are too high (above 2500°F), causing nitrogen to react with oxygen.

NO_x—los óxidos del nitrógeno se producen cuando las temperaturas de la combustión son demasiada altas (más del 2500°F), causando que el nitrógeno reacciona con el oxígeno.

O

O.D.—outside diameter.

O.D.—diámetro exterior.

O_2—oxygen.

O_2—oxígeno.

OBD—onboard diagnostics; regulations that require the computer to monitor the engine's oxygen sensor, EGR valve, and charcoal canister purge solenoid to see that all of these systems continue operating properly.

OBD—diagnósticos a bordo; las regulaciones que requieren que la computadora amonesta el sensor de oxígeno, la válvula EGR, y el solenoide de purga del bote de carbono del motor para verificar que todos estos sistemas continuan a operar correctamente.

OBD II—guidelines that began in 1994 on some cars and are on all gasoline-powered new cars built since 1996. These standards came about to provide standardization of terms and connections between makes.

OBD II—policía que comenzó en el 1994 en algunos coche y en todos los coches nuevos de combustión de gasolina desde el 1996. Estas normas se crearon para proveer una normalización de términos y conexiones entre los fabricantes.

octane—a measurement of a fuel's ability to resist explosion during combustion.

octano—una medida de la habilidad de un combustible en resistir la explosión durante la combustión.

OHC—overhead cam valve train design, also called cam-in-head.

OHC—diseño de tren de válvulas con leva en cabeza, tambien llamado leva-en-cabeza.

ohm—the unit of electrical resistance measurement. One ohm is the resistance that will allow one ampere to flow when pushed by one volt.

ohmio—la unedad de medida de la resistencia eléctrica. Un ohmio es la resistencia que permitirá un ampere a fluir al ser impulsado por un voltio.

Ohm's law—the law governing the relationship between volts, ohms, and amps.

ley de Ohm—el ley que gobierna la relación entre los voltios, los ohmios y los amperes.

oil analysis—a process for determining contaminants in engine oil.

analisis de aceite—un proceso para determinar los contaminantes presente en el aceite del motor.

oil—based carbon deposits—the traditional gummy, black carbon deposits like those found on valves. They result when oil and heat come together.

aceite—depósitos basado en carbón—los depósitos pegajosos tradicionales de carbón negro parecidos a los que se encuentran en las válvulas. Resulten cuando el aceite y el calor se juntan.

oil cooler—also called a heat exchanger, it is usually in the bottom or side of the radiator that has automatic transmission fluid running through it.

enfriador de aceite—tambien llamado un intercambiador de calor, suele estar en la parte inferior o en un lado del radiador por el cual fluye el fluido de transmisión automática.

oil pressure sending unit—the part that is threaded into the cylinder block oil passageway. It sends an electrical signal to the gauge or light on the dashboard.

unidad emisor de presión del aceite—una parte enroscada en el paso de aceite del bloque del cilindro. Manda una señal al indicador o al luz en el tablero de instrumentos.

oil wash—when excessive fuel washes oil from the cylinder walls.

lavado del aceite—cuando el combustible excesivo enjuaga el aceite de los muros del cilíndro.

on-board diagnostics—computers are capable of detecting incorrect electrical conditions and saving diagnostic trouble codes to memory.

diagnósticos a bordo—las computadoras son capaces de percibir las condiciones erróneas eléctricas y almacenar los códigos diagnósticos en la memoria.

one-way clutch—an overrunning clutch or the roller or sprag design that locks in one direction and freewheels in the other.

embrague de una vía—un diseños de embrague de sobremarcha o la rodilla o puntal que se enclava de una dirección y rueda libremente en la otra.

open circuit—when there is a break in the path of electrical flow in a circuit.

circuito abierto—cuando hay un corto en la vía del flujo eléctrico en un circuito.

open loop—when a computer controlled engine is cold and the computer is not receiving feedback from the oxygen sensor.

bucle abierto—cuando un motor controlado por computadora esta frío y la computador no esta recibiendo datos del sensor de oxígeno.

orifice—a restriction in a passage to slow down the flow of fluid.

orificio—una restricción en un paso para retardar el flujo de un fluido.

Otto-cycle—the four-stroke gasoline spark ignition engine design.

ciclo Otto—el diseño de motor de gasolina de cuatro carreras encendido por chispa.

overcenter spring—a spring that pulls one way during the first half of travel and the other way during the second half of travel.

resorte sobrecentro—un resorte que jala en una dirección durante la primera parte de la carrera y de la otra dirección en la segunda parte de la carrera.

overdrive—the opposite of gear reduction, it is when the output shaft turns faster than the input shaft.

sobremarcha—lo opuesto de reducción de velocidad, es cuando el eje de salida gira más rapidamente que el eje de entrada.

overhead cam engine—an engine with the cam above the cylinder head.

motor de levas sobre la cabeza—un motor que tiene la leva superior a la cabeza del cilindro.

overload spring—an extra spring leaf that does not work until the other leaves have deflected enough under load to allow them to come into contact with the overload spring.

resorte de sobrecarga—un resorte adicional de lámina que no trabaja hasta que las otras láminas hayan desviado lo bastante de la carga para permitir que se ponga en contacto con el resorte de sobrecarga.

overrunning clutch—a one-way clutch.

embrague de sobremarcha—un embrague de una vía.

oversquare—when an engine has a cylinder bore that is larger than its stroke.

sobrecuadrada—cuando un motor tiene un diámetro interior del cilindro que es más grande que su carrera.

oversteer—when a car turns too far in response to steering wheel movement.

sobregirar—cuando un coche da la vuelta demasiado en respuesta al movimiento del volante de dirección.

owner's manual—a booklet that comes with a new car.

manual del dueño—un librito que viene con un coche nuevo.

oxidation catalyst—a two-way catalyst that controls HC and CO.

catalizador de oxidación—un catalizador de dos vías que controla los HC y el CO.

oxides of nitrogen (NO_x)—the entire family of nitrogen oxides that form when oxygen and nitrogen bond under high heat and pressure in the engine. The "x" in NO_x means a variable number.

óxidos de nitrógeno (NO_x)—todo clase de óxidos de nitrógeno que forman cuando el oxígeno y el nitrógeno se unen bajo el alto calor y presión en el motor. El "x" en NO_x quiere decir un número variable.

oxidizing—burning of the fuel.

oxidando—combustión del combustible.

oxyacetylene welding—gas welding.

soldadura oxiacetileno—soldadura con gas.

P

P-type—a crystal like aluminum or boron that is positively charged and can carry electrical current.

tipo P—un cristal como el aluminio o el boro que tiene una carga positiva y puede transportar el corriente eléctrico.

PAG—polyalkylene glycol refrigerant oil for R-134A.

PAG—el aceite refrigerante polialquílico glicol para el R-134A.

parallel circuit—a circuit with different branches that current can flow through, starting from a common point.

circuito paralelo—un circuito con derivaciones diferentes por los cuales puede fluir el corriente, comenzando de un punto común.

parallelogram steering—linkage that makes the shape of a parallelogram during a turn.

dirección en paralelogramo—la biela que hace la forma de un paralelegramo al efectuar un viraje.

parasitic load—an electrical load when the key is off.

carga parásita—una carga eléctrica cuando la llave esta en posición apagada.

park pawl—a lever that locks the output shaft of the transmission when the shift lever is placed in park.

trinquete de park (estacionamiento)—una palanca que enclava el eje de la salida de la transmisión cuando la palanca selector esta en la posición de park.

part cores—used parts that are returned to the rebuilder for use in another rebuild.

almas—las partes usadas se regresan al reconstruidor para ser utilizados en otra reconstrucción.

partial non-hunting gear set—when one pinion tooth contacts two or three different ring gear teeth during one revolution and two or three different ones the next revolution.

conjunto de engranaje con diente suplementario parcial—cuando un diente del engranaje del piñón se endenta con dos o tres dientes de la corona durante una revolución y con dos o tres distinctos en la proxima revolución.

particulates—airborne microscopic particles or dust or soot that contain lead and carbon.

partículas—las partículas microscópicos transmitidas por el aire o el polvo o el hollín que contiene el plomo y el carbón.

Pascal's law—the hydraulic law that states that pressure in an enclosed system is equal in all parts of that system.

ley de Pascal—el ley de las hidráulicas que dice que la presión en un sistema cerrado es igual en cada parte de ese sistema.

passive sensors—those that do not generate their own voltage.

sensores pasivos—aquellos que no generan su propio voltaje.

patch—a piece of rubber vulcanized to the inner liner of a tire to repair a leak.

parche—un pedazo del caucho vulcanizado al forro interior de un neumático para reparar una fuga.

PCV—positive crankcase ventilation. It prevents the emission of blow-by gases from the crankcase.

PCV—ventilación positivo del cárter del cigüeñal. Previene la emisión de los gases de fuga del cárter del cigüeñal.

pH scale—a scale that gives the strength or alkalinity of a solution. The scale ranges from 1 through 14. Pure water is rated at 7.

escama pH—una escama que da la fuerza o lo alcalino de una solución. La escama varia del 1 al 14. El agua pura tiene una clasificación de 7.

photochemical smog—the name given to visible air pollution.

esmog fotoquímica—el nombre dado a la contaminación visible del aire.

piezoelectric crystals—crystals that develop a voltage on their surfaces when pressure is applied to them.

cristales piezoeléctricos—los cristales que desarrollan un voltaje en sus superficies cuando se les aplica la presión.

pinion bearing preload—tension between the pinion bearings to keep them at the correct clearance to their bearing races.

precarga del cojinete de piñón—la tensión entre los cojinetes del piñón para mantenerlas en la holgura correcta con respeto a las pistas del cojinete.

pinning or stitching—a crack repair process that uses threaded tapered plugs.

chavetear o ribetear—un proceso de reparación de grietas que usa los espárragos cónicos.

pipe dies—tools used to form the threads on the outside of pipe.

macho para tubo—las herramientas que sirven para cortar las roscas al exterior del tubo.

piston rings—parts that seal the piston to the cylinder.

anillos del pistón—las partes que sellan el pistón al cilindro.

piston slap—when a piston skirt fits the cylinder too loosely, the resulting noise is called piston slap.

palmada del pistón—cuando la falda del pistón queda muy floja en en cilindro, el ruido que resulta se llama una palmada del pistón.

piston stroke—is the distance that the piston moves from TDC to BDC.

carrera del pistón—es la distancia que mueva el pistón del PMS al PMI.

pitch diameter—the point where the teeth of the two gears meet. This is the effective diameter of the gear.

diámetro del paso—el punto en que se juntan los dientes de dos engranajes. Este es el diámetro efectivo del engranaje.

pitch line—a line that runs through the center of the tooth from heel to toe.

línea cero—una línea que corre por el centro del diente del talón al dedo (la punta).

placard—a tire information sticker.

cartel—una etiqueta de información del neumático.

plain bearings—a bushing type of bearing that uses no rolling parts and provides sliding contact between two mating surfaces.

cojinetes sencillos—un cojinete de tipo buje que no usa las partes rodantes y provee un contacto deslizante entre dos superficies apareadas.

plastigauge—a thin strip of plastic that deforms when crushed, it is used to measure bearing oil clearance.

plasticímetro—una lámina delgada del plástico que se deforma al ser aplastada, se usa para medir la holgura del aceite del cojinete.

plenum—the air space in the intake manifold below the carburetor.

pleno—el espacio del aire en el múltiple de admisión debajo del carburedor.

plunge joint—the inboard CV joint that changes in length like a slip yoke does in a rear wheel drive car.

junta profundizada—la junta CV abordo que cambia en su longitud parecido a como lo hace una horquilla deslizante en un coche de tracción trasera.

POA—pilot operated absolute valve.

POA—válvula absoluta operada por piloto.

points—called contact points or breaker points, they are an electromechanical trigger for the ignition primary circuit.

puntos—llamados los puntos de contacto o contactos platinados, son un disparador electromecánico para el circuito primario del encendido.

pole shoes—the soft iron cores of an electromagnet.

pieza polar—los núcleos de hierro blando en un electroimán.

polymer—a substance, called a plastic, consisting of giant molecules, formed from smaller molecules of the same kind.

polímero—una sustancia, llamado un plástico, consistiendo de las moléculas gigantescas, formadas de las moléculas más pequeñas del mismo tipo.

popback—a term for combustion occurring in the fuel induction system (before it enters the cylinder).

detonación—un término por la combustión ocurriendo en el sistema de inducción del combustible (antes de que entra al cilindro).

poppet style valve—the style of valve used by four stroke cycle internal combustion engines.

válvula estilo elevación—el estilo de la válvula usado en los motores de combustión interna de cuatro carreras.

port injection—a fuel injection system that uses a fuel rail with individual fuel injectors at each intake port.

inyección por lumbrera—un sistema de inyección del combustible que usa un carríl de combustible con los inyectores individuales en cada lumbrera de admisión.

ported vacuum—when the vacuum port is above the throttle plate.

vacío con lumbrera—cuando la lumbrera del vacío queda arriba de la placa del acelerador.

positive camber—when a tire is tilted out at the top.

comba positiva—cuando un neumático esta inclinada hacia afuera en la parte superior.

positive caster—the rearward tilt of the steering axis at the top.

ángulo del caster positivo—la inclinación hacia atrás del eje de dirección.

positive offset—when the wheel is offset in such a way as to decrease the track width of the tires, it is often found on FWD cars.

descentrado positivo—cuando la rueda esta descentrado en una manera que disminuya la anchura de la banda de los neumáticos, se suele encontrar en los coches FWD (tracción delantera).

positive spark plug electrode—the end of the conductor.

electrodo positivo de la bujía—en la extremidad del conductor.

positive stop—an automatic adjustment feature used with hydraulic valve lifters.

paro positivo—una característica de ajuste automático usado con las levantaválvulas hidráulicas.

potentiometer—a variable resistor that measures linear or rotary motion.

potenciómetro—un resistor variable que mide el movimiento linear o giratorio.

power—how fast work is done or how fast motion is produced against a resistance.

fuerza—el rapidez con el cual se efectúa el trabajo o el rapidez con el cual se produce un movimiento contra una resistencia.

preignition—also called ping, occurs when the air/fuel mixture ignites before the regular spark occurs.

autoencendido—tambien llamado "ping", ocurre cuando la mezcla de aire/combustible encienda antes de que ocurre la chispa normal.

preload—the clamping load on a fastener.

precargar—la carga de apriete en un sujetador.

preload adjustment—an adjustment where a bearing has an additional load placed against it after it is snug (zero-lash).

ajuste de la carga previa—un ajuste en el cual se impone una carga adicional en el cojinete después de que esté apretada (juego de cero).

pressed fit—when two parts with an interference fit are forced together.

ajuste con presión (embutido)—cuando dos partes con un ajuste de interferencia se unen bajo fuerza.

pressure bleeder—a canister that uses air pressure against a diaphragm to pressurize brake fluid.

purga de presión—un bote que usa la presión del aire contra un diafragma para presurizar el fludio de freno.

pressure cycling switch—a device that turns a magnetic air conditioning clutch on and off.

interruptor del ciclo de presión—un dispositivo que encienda y apaga un embrague magnético del aire acondicionador.

pressure plate assembly—also called the clutch cover assembly, it is bolted to (and rotates with) the flywheel. It compresses the clutch disc against the flywheel.

asamblea de placa de presión—tambien llamado la asamblea de la tapadera del embrague, es empernada a (y gira con) el volante. Oprime el disco del embrague contra el volante.

pressure priming—a method of priming the lubrication system with a pressure tank.

cebado de presión—un método de cebar el sistema de lubricación con un tanque de presión.

pressure relief valve—a valve in a hydraulic system that opens to allow excess pressure to bleed off to the fluid intake side of the pump.

válvula de descompresión—una válvula en un sistema hidráulico que abre para permitir que la presión excesiva se purga hacia el lado de admisión de fluidos de la bomba.

pressure test—a test to tell if the transmission is experiencing internal leakage or requires adjustment to linkages or the vacuum modulator.

prueba de presión—una prueba para verificar si la transmisión tiene una fuga inteior o si requiere un ajuste en las bielas o en el modulador del vacío.

primary trigger—the electronic or electromechanical device that controls the switching of current flow in the primary winding.

disparador primario—el dispositivo electrónico o electromecánico que controla el cambio del flujo del corriente en el devanado primario.

primary vibration—when the piston slows down as it approaches TDC, its force pulls the engine up. When it approaches BDC, it pulls the engine down.

vibración primario—cuando el pistón retarda al acercarse al PMS, su fuerza hace subir al motor. Cuando se acerca al PMI, hace jale al motor hacia abajo.

primary winding—the large winding in the ignition coil made up of 150 to 250 turns of wires.

devanado primario—el devanado grande la bobina del encendido que consiste de 150 a 250 vueltas de los alambres.

primary wiring—low voltage wiring.

alambre primario—el alambre de bajo voltaje.

production engine—an engine produced at the factory.

motor fabricado—un motor producido en una fábrica.

profile—a tire's height.

perfíl—la altura de un neumático.

PROM—programmable read only memory.

PROM—memoria programable de solo lectura.

prony brake—a simple friction brake dynamometer.

freno prony—un dinamómetro para los frenos de fricción simple.

propane enrichment test—a test where propane or acetylene gas is put into the oil filler opening to see if the engine idle changes in response to an intake manifold gasket leak.

prueba de enriquecimiento de propano—una prueba en la cual se introduce el propano o el gas acetileno en la apertura de agregar aceite para ver si se cambia la marcha en vacío como resultado de una fuga en el empaque del múltiple de admisión.

psi—pounds per square inch.

psi—libras por pulgada cuadrada.

pull—when the steering wheel is held straight ahead and the car goes to the side.

desviación—cuando el volante de dirección se mantiene recta y el coche va a un lado.

pulse width—the length of time that an injector remains open.

impulso en anchura—la cantidad del tiempo en que queda abierto un inyector.

pulse width modulation—turning the alternator rapidly on and off to regulate voltage.

modulación al impulso en anchura—prender y apagar rápidamente al alternador para regular el voltaje.

pumpkin—a removable differential third member.

calabaza—un tercer miembro móvil del diferencial.

purge—when a vapor canister empties itself of vapor storage when the engine is running.

purga—cuando un bote de vapor se vacía si mismo del vapor almacenado mientras que marcha el motor.

push fit—when a part slides into place by hand.

ajuste a mano—cuando una parte se puede instalar a mano.

pushrod engine—this valve train design uses pushrods and has the cam located in the cylinder block.

motor de varillas empujadoras—este diseño de tren de válvulas usa las varillas empujadoras y tiene la leva situada en el bloque del cilindro.

pyrotechnics—a term that describes fireworks.

pirotecnia—un término que describe los fuegos artificiales.

Q

quad driver—when one single space saving module has a group of four transistors and can control up to four actuators.

impulsor de cuatro—cuando un módulo eficiente en espacio tiene un grupo de cuatro transistores que pueden controlar hasta cuatros actuadores.

quill—the front transmission bearing retainer.

árbol hueco—el retenedor del cojinete delantero de la transmisión.

R

R&R—(remove and replace) job; the estimated time it should take a professional technician to perform a specified job.

R&R—un trabajo (remover y reemplazar); el tiempo estimado que debe tomar un técnico profesional en efectuar un trabajo especificado.

R-12 refrigerant—also called freon, it is a CFC (chlorofluorocarbon).

fluido refrigerante R-12—tambien llamado el freón, es un CFC (clorofluorocarbono).

R-134A refrigerant—a refrigerant that is used in newer vehicles. It is hydrofluorocarbon.

fluido refrigernate R-134A—un fluido refrigerante que se usa en los vehículos de ultima moda. Es un hidroflorocarbono.

race—a bearing cup.

pista—un anillo exterior de un cojinete.

rack and pinion steering—the end of the steering shaft has a pinion gear that meshes with a rack gear.

dirección de piñón y cremallera—la extremidad de un eje de dirección tiene un engranaje de piñón que endenta con un engranaje de cremallera.

radial compressor—an air conditioning compressor that resembles a radial aircraft engine. It has multiple cylinders with double ended pistons and one eccentric crankshaft throw.

comprimidor radial—un comprimidor del aire acondicionado que parece un motor de aeroplano. Tiene múltiple cilindros con pistones de doble cara y un radio del cigüeñal excéntrico.

radial load—a load that is in an up and down direction.

carga radial—una carga que mueve en la dirección de arriba a abajo.

radial movement—horizontal movement.

movimiento radial—un movimiento horizontal.

radial ply—a tire structure that has casing plies that run across the tire from bead seat to bead seat in the "radial" direction of the wheel.

carcasa radial—una estructura de neumático que tiene cordón de las telas atravesando el neumático desde el asiento de la ceja al asiento de la ceja en una dirección "radial" de la rueda.

radial runout—wobble of a part in an up and down direction.

corrimiento radial—un bamboleo de una parte en una dirección de arriba a abajo.

radiation—a method of heat transfer where heat bounces off of a surface. Dark car colors absorb and radiate heat better than light colors.

radiación—un metodo de transferencia del calor en el cual el calor bota de una superficie. Los coches de colores oscuros absorben y radian el calor mejor que los colores claros.

radius plates—an alignment gauge that fits under the front wheels and measures in degrees how far a wheel is turned to the right or left.

placas radiales—un medidor de alineación que queda debajo de las ruedas delanteras y mide en grados cuánto gira una rueda a la derecha o a la izquierda.

RAM—random access memory; like a notepad that you can read from and write to.

RAM—memoria de acceso aleatorio; como un cuadernito en que puede leer y escribir.

raster pattern—also called stacked pattern, it displays all of the cylinders vertically, one above the next.

patrón de cuadro—tambien llamado un patrón amontonado, presenta todos los cilindros verticalmente, el uno sobre el otro.

Ravigneaux gear train—a compound planetary gear design with long and short pinions.

tren de embragues Ravigneaux—un diseño de conjunto de embrague planetarios con los piñones largos y cortos.

reach—the length of the threaded area of a spark plug.

alcance—la longitud del área enroscado de una bujía.

rebound—when the wheel moves back down.

rebote—cuando la rueda descende de nuevo.

rebound bumpers—rubber bumpers bolted to the lower control arm or chassis that come into use when the suspension reaches the full limit of its travel.

parachoques del rebote—los parachoques de caucho empernados en el brazo de control inferior o en el chasis que funcionan cuando la suspensión llega al límite total de su extensión.

receiver dryer—a refrigerant reservoir and dryer that supplies liquid refrigerant to the expansion valve.

colector desecador—un recipiente de fluido refrigerante que proporciona el fluido refrigerante a la válvula de expansión.

reciprocating gasoline engine—an engine with a piston that goes up and down and a crankshaft that rotates.

motor de gasolina reciprocante—un motor con un pistón que suba y baja y un cigüeñal que gira.

recirculating ball steering gear—a popular low friction steering gear box. A sector gear meshes with a ball nut that rides on ball bearings on the worm shaft to provide a smooth steering feel.

embrague de dirección de bola recirculatoria—un cárter de engranajes de dirección de baja fricción popular. Un engranaje sectorial endenta con una tuerca esférica montado sobre los rodamientos de bolas en un eje fileteado para proveer una sensación de dirección uniforme.

reduction converter—a catalytic converter that catalyzes NO_x.

convertidor reductor—un convertidor catalítico que cataliza el NO_x.

regulator maximum voltage test—a test that checks to see that the voltage regulator is energizing the rotor field and is not allowing the alternator to overcharge.

prueba del voltaje máximo del regulador—una prueba para verificar que el regulador de voltaje proporciona la energía al campo del rotor y no permite que sobrecarga el alternador.

Reid vapor pressure—a standard by which volatility is measured.

presión de vapor Reid—una norma por la cual se mide la volatilidad.

relay—a magnetically controlled switch used when a large load must be controlled by a small wire.

relé—un interruptor controlado por magnetísmo que se usa cuando se tiene que controlar una carga grande con un alambre pequeño.

release bearing—also called a throw out bearing, it contacts the rotating clutch to release its disc.

cojinete de empuje del embrague—tambien llamado el cojinete de desembrague, se pone en contacto con el embrague giratorio para desengranar el disco.

release levers—also called fingers, a part of the coil spring clutch that when compressed pulls the pressure plate away from the flywheel.

palancas de desembrague—tambien llamado patillas, una parte del embrague de resorte helicoidal que al ser comprimido retira la placa de presión del volante.

reserve capacity—a measurement of the battery's ability to provide current when there is no electricity from the charging system.

capacidad de reserva—una medida de la habilidad de la batería en proveer el corriente cuando no viene electricidad del sistema de cargar.

resistance—when there is an obstruction to electrical flow.

resistencia—cuando hay una obstrucción al flujo elétrico.

resistors—electrical devices used to make heat or to control the intensity of a load.

resistorcs—los dispositivos eléctricos que se usan para producir el calor o para controlar la intensidad de una carga.

resonator—a second muffler in line with the other muffler.

resonador—un silenciador secundario en linea con el otro silenciador.

reverse bias—not allowing current to flow.

polarización reversa—no dejando fluir el corriente.

reverse bleeding—moving fluid through a hydraulic system from the slave cylinder to the master cylinder.

purga invertida—mudando el fluido a través del sistema hidráulico del cilindro secundario al cilindro maestro.

rheostat—varies current flow through the circuit as a movable wiper arm runs along a resistor.

reóstato—modifica el flujo del corriente atravesando un circuito cuando el brazo de contacto móvil desliza por un resistor.

riding the clutch—resting a foot on the clutch when driving.

desgaste continuo del engrane—descanzando el pie en el pedal del embrague mientras que uno maneja.

rigid axle—a non-independent axle, also called a straight axle or I-beam, found on most rear ends and some heavy truck front ends.

eje rígido—un eje no independiente, tambien llamado un eje recto o un hierro en T, encontrado en la mayoría de las asambleas traseras y en algunos asambleas delanteras de camiones de trabajo pesado.

ring ridge—the ledge that forms at the top of the cylinder as the cylinder wears.

nervadura en el pistón—la arruga formado en la parte superior de un cilindro al desgastarse el cilindro.

riveted linings—linings that are held to their backing by rivets.

forros con remaches—los forros que se adhieren al respaldo por medio de los remaches.

RMA—Rubber Manufacturers Association.

RMA—Asociación de Fabricantes de Caucho.

road horsepower—measures horsepower available at the car's drive wheels.

potencia de caballo en carretera—mide la potencia en caballos disponible en las ruedas de tracción del coche.

rocking couple—another form of vibration that occurs when there are only two cylinders that fire 180° apart instead of 360° apart.

oscilación con dos—otra forma de vibración que ocurre cuando sólo hay dos cilíndros que encienden en la posición de 180° en vez de en la posición de 360°.

rod bolt protectors—pieces of soft plastic or rubber used to protect the crankshaft from nicks during engine disassembly and assembly.

protectores del vástago—las piezas flexibles de plástico o caucho que se usan para evitar los daños al cigüeñal cuando se desmonta o se arma un motor.

rodding out—when a radiator becomes plugged, this is done by a radiator shop to clean its cooling passageways.

escariar—cuando un radiador esta atascado, un taller de radiadores efectúa esto para limpiar los pasos de enfriar.

roll over valve—a valve in the vent line from the fuel tank that blocks the line if the car is in an accident and rolls over preventing liquid fuel from being able to leak out.

válvula de seguridad—una válvula en la linea de ventilación del tanque de combustible que cierra la linea en caso de un accidente en el cual el coche se voltea previniendo una fuga del combustible en forma líquido.

rolled edge clamps—hose clamps used for fuel injection and other high pressure uses. They are designed so that they will not cut into a hose.

grapa con bordes redondeados—las grapas para tubos que se usan en la inyección del combustible y en otros aplicaciones de alta presión. Se han diseñado para no cortar un tubo.

ROM—read only memory is permanent programmed information available to the microprocessor programmed into the chip during manufacturing.

ROM—memoria de solo lectura es la información programada permanente disponible al microprocesador programado en la ficha durante la fabricación.

roots-type supercharger—also called a lobe-type supercharger, it is the most popular positive displacement supercharger.

sobrealimentador tipo roots—tambien llamado un sobrealimentador con lóbulo, es el tipo más popular de sobrealimentador de desplazamiento positivo.

rotary vane pump—a pump with blades used for power steering pump, air conditioning, or a smog pump.

bomba giratorio de aletas—una bomba con palas que se usa en la bomba de dirección hidráulica, el aire acondicionado, o una bomba de esmog.

RTV—room temperature vulcanizing silicone rubber.

RTV—el caucho de silicona que vulcaniza en temperatura ambiente.

rubberized cork—a combination of cork and artificial rubber that is a superior material to regular cork because it does not shrink.

corcho cauchotado—una combinación del corcho y el caucho sintético que es una materia superior al corcho natural porque no se encoje.

runners—the passages in an intake manifold.

conductos—los pasos en un múltiple de admisión.

runout—the amount that a part wobbles up-and-down or side-to-side when rotated.

corrimiento—la cantidad que oscila una parte de arriba-a-abajo o de un-lado-a-otro al ser girado.

RWD—rear wheel drive.

RWD—tracción trasera.

Rzeppa CV joints—a popular ball and socket type joint used in most outboard fixed CV joints.

juntas CV Rzeppa—una junta de articulación esférica popular que se usa en la mayoría de las juntas CV fuera de bordo.

S

S.I. engine—an engine that ignites its air and fuel with a spark.

motor S.I.—un motor que encienda su aire y combustible con una chispa.

SAE—Society of Automotive Engineers.

SAE—Asociación de Ingenieros Automotivos.

Salisbury axle—also called integral, its third member is not removable as a unit.

eje Salisbury—tambien llamado un integral, su tercer miembro no se puede remover como una unedad.

saturated vapor—a liquid that is in contact with its vapor within an enclosed space.

vapor saturado—un líquido que esta en contacto con su vapor en un espacio encerrado.

scale—this forms as minerals in the local water supply settle out of the water during heating and become deposited on heated parts.

incrustación—esto forma cuando los minerales del suministro del agua local sedimentan al ser calentados y se depositan en las partes calentadas.

scan tool—a portable computer connected to the vehicle's diagnostic connector that reads data from the car's on-board computer.

aparato de exploración—una computadora portátil conectada al conector diagnóstico que lee los datos de la computador abordo del coche.

SCFM—standard cubic feet per minute.

SCFM—estandar de pies cúbicos por minuto.

schematic—a wiring diagram.

esquemático—una diagrama del cableado.

scroll compressor—an compressor with a fixed and a moveable scroll.

comprimidor voluta—una comprimidora que tiene un espiral fijo y uno móvil.

scrub radius—the distance at the road surface between the centerline of true vertical at the center of the tire tread and the steering axis pivot center line.

radio viraje en rueda—la distancia en la superficie de la carretera entre la linea central vertical verdadera del centro de la banda de rodamiento y la línea central de pivote del eje de dirección.

scuff—due to incorrect toe, when the tires move sideways for a certain distance out of every mile travelled.

desgaste—debido a la divergencia incorrecta, cuando los neumáticos muevan lateralmente una cierta distancia por cada milla de viaje.

scuffing—the damage caused when parts momentarily weld to one another. It occurs due to excessive heat.

rozamiento—el daño causado cuando las partes se sueldan momentariamente la una a la otra. Ocurre debido al calor excesivo.

second order vibration—a vibration from a universal joint.

vibración del segundo orden—una vibración de una junta universal.

secondary winding—the smaller winding within the center of the coil primary winding made up of several thousand turns of very small, hairlike insulated wire.

devanado secundario—el arrollamiento más pequeño dentro del centro del devanado de la bobina primaria compuesta de varias miles de vueltas de alambre aislada pequeña y fina .

secondary wiring—high voltage ignition wiring.

cableado secundario—el cableado de encendido de alto voltaje.

seized engine—when a crankshaft will not turn.

motor agarrotado—cuando un cigüeñal no dará la vuelta.

Selden patent—a patent covering all gasoline powered self-propelled road vehicles.

patente Selden—un patente que se aplica a todos los vehículos de autopropulsión de combustión de gasolina.

self-energization—when the leading shoe on a drum brake is forced into the brake drum.

autoenergetización—cuando la zapata de guía de un freno de tambor se empuja contra el tambor del freno.

SEMA—Specialty Equipment Manufacturers Association.

SEMA—Asociación de Fabricante de Aparatos Especiales.

semi-floating axle—an axle with a bearing that rides on the axle.

eje semi-flotante—un eje con un cojinete que va montado sobre el eje.

semi-metallic linings—organic linings with sponge iron and steel fibers mixed into them to add strength and temperature resistance.

forros semi-metálicos—los forros orgánicos que tienen incorporado el hierro esponjoso y las fibras del acero para proporcionarles más fuerza y resistencia a la temperatura.

semiconductor—a solid state material that acts as both an insulator or a conductor.

semiconductor—una material de estado sólido que se comporta como un aislador o tambien un conductor.

sending unit—a device that sends a signal to a gauge regarding oil pressure or coolant temperature.

unidad emisor—un dispositivo que envía una señal a un indicador referente a la presión del aceite o la temperatura del fluido refrigerante.

sensible heat—the heat that goes into matter as its temperature increases.

calor sensible—el calor que entra en la materia en cuanto aumenta su temperatura.

sensor safe RTV—low-volatile silicone that does not give off vapors that can damage an oxygen sensor.

RTV para sensores—un silicio resistente a la volatilidad que no emite los vapores que pueden dañar un sensor de oxígeno.

sensors—devices that relay to the computer such information on throttle position, air and coolant temperature, air flow, manifold pressure, and barometric pressure.

sensores—los dispositivos que relevan a la computadora la información referente a la posición del acelerador, la temperatura del aire y del fluido refrigerante, el flujo del aire, la presión del múltiple y la presión barométrica.

series circuit—an electrical circuit where current flowing equally though all parts of a circuit flows first to one load and then on to the next one.

circuito en serie—un circuito eléctrico en el cual el corriente fluyendo igualmente por todas partes del circuito fluye primero a una carga y luego a la próxima.

serpentine belt—a belt path that is called serpentine because of the snake-like path that it follows.

correa serpentín—la trayectoria por la cual pasa una correa que se llama serpentín debido a que la senda que sigue parece la de una culebra.

servo action—when the leading shoe on a drum brake is forced into the brake drum.

acción servo—cuando la zapata de guía de un freno de tambor se empuja contra el tambor del freno.

set-back—the amount that one front wheel is behind the one on the other side of the car.

retroceso—la cantidad de que una rueda delantera queda atrás de la del otro lado del coche.

severe service—includes trailer towing and stop-and-go city driving in hot weather.

servicio severo—incluye usando remolque y viajando en tránsito pesado en clima cálida.

SFE—Society of Fuse Engineers.

SFE—Sociedad de Ingenieros de Fusibles.

shift quadrant—the readout on the gear selector that selects what gear the transmission is in.

sector de cambios—el indicador en el selector de velocidades que selecciona en cuál velocidad va la transmisión.

shock absorber control force—determines the softness or harshness of the ride.

fuerza del control del amortiguador—determina la uniformidad o la brusquedad del viaje.

shock absorbers—devices that dampen spring oscillations by converting the energy from spring movement into heat energy.

amortiguadores—los dispositivos que amortiguan las oscilaciones de los resortes al convertir la energía del movimiento de los resortes a la energía calorífica.

shock ratio—the difference between the amount of control on compression and extension.

relación de choques—la diferencia entre la cantidad del control en la compresión y la extensión.

short block—an engine without heads.

bloque corto—un motor sin las cabezas.

short circuit—when the electrical path has been shortened by a wire contacting another.

corto circuito—cuando una trayectoria eléctrica se ha cortado debido a una alambre tocando a otra.

short/long arm (SLA)—a suspension design that uses two control arms of unequal length.

brazo corto/largo (SLA)—un diseño de suspención que usa dos brazos de control de longitud disparejos.

siamese runners—when one runner feeds two neighboring cylinders.

conductos siameses—cuando un conducto alimenta dos cilindros que son juntos.

signal flasher—a device activated by the heat of the electricity travelling through it to cause the turn signal bulbs to flash.

pulsador indicador—un dispositivo activado por el calor de la electricidad que lo atraviesa para causar que destellan los focos indicatores de vueltas.

Simpson gear train—a compound planetary gear design that shares the same sun gear between the gear sets.

tren de engranajes Simpson—un diseño de engranajes planetarios combinados que comparten un engrane solar entre los conjuntos de engranajes.

sine wave voltage—an oscilloscope pattern showing positive and negative voltage.

voltaje de onda senoidal—un patrón del osiloscopio que muestra el voltaje positivo y negativo.

single exhaust system—when there is only one pipe to the rear of the car.

sistema simple del escape—cuando hay sólo un tubo en la parte trasera del coche.

single pass machine—a machine for air conditioning that removes the refrigerant and recycles it during the 15-minute minimum vacuum cycle.

máquina de un solo paso—una máquina para el aire acondicionado que remueva el refrigerante y lo recicla durante el ciclo de vacío mínimo de 15 minutos.

single-plane manifold—when each of a manifold's barrels serve all of the engine's cylinders.

múltiple monoplano—cuando cada uno de los cañones del múltiple sirve para todos los cilindros del motor.

size equivalents—when different tire sizes have the same diameter and load capacity.

equivalentes de tamaño—cuando los tamaños distinctos de los neumáticos tienen el mismo diámetro y capacidad de carga.

skin effect—the name for the thin area of fuel that does not burn when the surface area of the combustion chamber is cold so fuel burning is quenched.

efecto de película—el nombre de la superficia delgada del combustible que no quema cuando la superficie de la cámara de combustión esta fría lo cual apaga al combustible que esta quemando.

slave cylinder—also called the reaction or actuator cylinder, it is the output piston in a hydraulic clutch that is attached to the release lever at the clutch.

cilindro secundario—tambien llamado un cilindro de reaccíon o actuador, es el pistón de salida de un embrague hidráulico contectado a la palanca de desenganche en el embrague.

slip angle—the tendency during a turn for a tire to continue to go in the direction it was going, even though the rim has turned in response to steering wheel movement.

ángulo de desviación—la tendencia durante un viraje a que un neumático continua a viajar en la dirección en la cual iba, aunque la llanta se ha girado respondiendo al movimiento al volante de dirección.

sludge—a mixture of moisture, oil, and contaminants that occurs when engine temperatures are too low.

cieno—una mezcla de la humedad, el aceite y los contaminantes que ocurre cuando las temperaturas del motor son demasiadas bajas.

smog pump—an air injection system air pump.

bomba del esmog—una bomba de sistema de aire de aire inyección.

snapshots—during a test drive when a driveability symptom is felt, these record frame-by-frame a period of several seconds before and after the button on the scan tool is pressed. This helps diagnose intermittent faults.

instantáneos—durante un viaje de prueba cuando se siente una síntoma, estos registran cuadro-a-cuadro un período de varios segundos antes y después de que se haya oprimido el botón en el aparato explorador. Esto ayuda en el diagnósis de los fallos intermitentes.

soft parts—these are part of a rebuild kit, including all rubber seals and gaskets, bands, clutch friction discs, modulator, thrust washers, and bushings.

partes flexibles de repuesto—estas son parte de un juego de piezas de reconstrucción, incluye todos los sellos e empaques de caucho, las correas, los discos de fricción del embrague, el modulador, las arandelas de empuje y los bujes.

software—information stored as electronic signals that can be modified.

software—la información almacenado en forma de señales electrónicas que se pueden modificar.

SOHC—single overhead cam.

SOHC—árbol leva única sobre la cabeza.

solenoid—a combination magnetic switch and mechanical device.

solenoide—una combinación de interruptor magnético y dispositivo mecánico.

sound—vibration in the air.

sonido—una vibración en el aire.

spalling—when pieces break off of the bearing metal.

escamación—cuando se despegan pedazos del metal del cojinete.

spark line—a horizontal line that begins at the voltage level where electrons start to flow across the spark plug gap.

linea de chispa—una linea horizonal que comienza en el nivel de voltaje en dónde comienzan a fluir los electrones atravéz del entrehierro de la bujía.

spark plug deposit test—a test in which spark plugs are examined to see if an engine is using oil.

prueba de depósitos en la bujía—una prueba en la cual se examinan las bujías para verificar si el motor esta usando el aceite.

specific gravity—the weight (strength) of electrolyte.

gravedad específica—el peso (la fuerza) de un electrólito.

speed density systems—when the computer uses a MAP (manifold absolute pressure) sensor and engine rpm (tach) signal to calculate the amount of air entering the engine.

sistemas de densidad en velocidad—cuando una computadora usa un sensor MAP (presión absoluta del múltiple) y el señal del rpm del motor (tachímetro) para calcular la cantidad del aire entrando al motor.

speed rating—a tire rating that means that a properly inflated tire has been designed to operate at up to a certain mph for short periods of time.

capacidad de la velocidad—una clasificación del neumático quiere decir que un neumático inflado correctamente ha sido diseñado a operar en ciertas mph por cortos períodos de tiempo.

spider gears—the side gears and differential pinions.

satélites del diferencial—los engranajes laterales y piñones del diferencial.

spin test—when an engine is spun with a spin tester to check such things as oil pressure and compression.

prueba rotativa—cuando se gira un motor con una probador giratoria para verificar las cosas como la presión del aceite y la compresión.

spiral cable—also called clock spring, it connects the steering column to the air bag in the steering wheel.

cable espiral—tambien llamado un resorte espiral, conecta la columna de dirección a la bolsa de aire en el volante de dirección.

spongy brake pedal—a soft pedal feel due to air in the brake hydraulic system.

pedal de freno esponjoso—una sensación blanda en el pedal debido al aire en el sistema hidráulico de frenos.

spool valve—a valve that has lands, valleys, and faces.

válvula de carrete—una válvula que tiene partes planas, las acanaladuras y las caras.

spread—the difference between alignment measurements from side to side.

aplastamiento—la diferencia entre las medidas de alineación de un lado a otro lado.

sprung weight—the weight supported by the car springs.

peso del resorte—el peso soportado por los resortes de un coche.

spur gears—a simple gear design with straight cut teeth.

engranajes rectos—un diseño simple de engranaje que tiene dientes rectos.

squirrel cage fan—a fan used with a blower motor that resembles a squirrel cage.

ventilador caja de ardilla—un ventilador con un motor soplador que parece una caja de ardilla.

stall test—with brakes applied, the engine is accelerated until speed no longer increases. Note engine rpm at this point.

prueba de paro—con los frenos puestos, se acelera el motor hasta que no aumenta en velocidad. Nota el rpm en este punto.

static—at rest.

estático—en descanso.

static pressure—the pressure reading in the system when it is not operating.

presión estático—la lectura de la presión en el sistema mientras que no esta operando.

static timing—timing the distributor with the engine stopped.

regulación estática—regulando el distribuidor con el motor parado.

steering axis inclination (SAI)—the amount that the spindle support arm leans in at the top. SAI is also known as ball joint inclination (BJI) or king pin inclination (KPI).

inclinación del eje director (SAI)—la cantidad en que inclina hacia adentro el husillo del brazo de apoyo en la parte superior. El SAI tambien se conoce como la inclinación de la articulación esférica (BJI) o la inclinación de la clavija maestra (KPI).

steering damper—a sideways shock absorber for road shocks.

amortiguador de la dirección—un amortiguador que absorba los choques de la carretera lateralmente.

steering linkage—parts that connect the steering gear to the wheels.

biela de dirección—las partes que conectan el engranaje de dirección a las ruedas.

steering ratio—the number of teeth on the driving gear compared to the number of teeth on the driven gear.

relación de dirección—el número de dientes en el engranaje impulsor comparado al número de dientes en el engranaje arrastrado.

steering taper—a steering part connection.

ahusamiento de la dirección—una parte de conexión en la dirección.

stock—original equipment.

fondo—aparatos originales.

stoichiometric—the ideal air/fuel ratio, 14.7:1 by weight, in which all of the oxygen is consumed in the burning of the fuel.

estoquiométrico—la relación ideal del aire/combustible, 14.7:1 por peso, en la cual se consume todo el oxígeno al quemarse el combustible.

stratified charge—the spark is ignited in a richer air/fuel mixture. The flame from this mixture burns a leaner mixture in its path.

carga estratificada—la chispa se enciende en una mezcla más rica de aire/combustible. La llama de esta mezcla enciende la mezcla más pobre que esta en su ruta.

street elbow—a 90° connection with a male thread on one end and a female thread on the other.

codo roscado macho y hembra—un conector de 90°con una rosca macho en una extremidad y una rosca hembra en la otra.

stress raiser—when a metal surface is damaged it raises stress, weakening the part.

deformación de tensión—cuando la superficie del metal se daña crea la tensión, debilitando la parte.

stress test—when ignition parts like a module or pickup coil are cooled or heated while watching the scope pattern.

prueba de esfuerzo—cuando las partes del encendido, como un módulo o un devanado detector se enfrían o se calientan mientras que se observa el patrón en el osciloscopio.

stroke—the movement of the piston from the top of its travel to the bottom of its travel.

carrera—el movimiento del pistón desde el punto superior de su viaje al punto inferior de su viaje.

strut cartridge—a replacement shock absorber for MacPherson struts.

cartucho del poste—un amortiguador de reemplazo de los postes MacPherson.

stub shaft—the live axle end at the outside of the outboard CV joint on a front wheel drive axle shaft.

mangueta del eje—la extremidad viva del eje en la parte exterior de la junta CV fuera de bordo en un árbol motor de tracción delantera.

stud-centric—wheels that center using the wheel studs.

centrado en el husillo—las ruedas cuyos centros son los husillos de las ruedas.

STV—suction throttling valve.

STV—válvula de aceleración por aspiración.

sub-frame—a group of parts in the front of a front wheel drive vehicle that includes the engine, transaxle, and steering/suspension system.

bastidor—un grupo de partes en la parte delantera de un vehículo de tracción delantera que incluye al motor, el transeje, y el sistema de dirección/suspensión.

subcooling—a liquid with a temperature considerably beneath its boiling point.

subenfriar—un líquido con una temperatura bastante más baja que su punto de ebullición.

suction service valve—the low side service valve on the inlet to the air conditioning compressor (low side).

válvula de aspiración de servicio—la válvula de servicio del lado bajo en la admisión al comprimidor del aire acondicionador (lado bajo).

sulfation—when the battery is allowed to remain in a discharged state, the lead sulfate in the battery's plates becomes hard and resistant to recharging.

sulfatación—cuando se permite que la batería se queda en un estado de descarga, el sulfato de plomo en las placas de la batería se endurece y resiste la carga.

supercharger—a belt-driven pump, also called a blower.

sobrealimentador—una bomba conducida por correa, tambien llamado un ventilador.

superheat—temperatures above the liquid's boiling point.

supercalentar—las temperaturas más altas que el punto de ebullición.

superimposed pattern—a scope pattern used to compare all of the cylinders while their patterns are displayed one on top of the other.

patrón superimpuesta—un patrón de osciloscopio usado para comparar todos los cilindros mientras que sus patrones su muestran uno sobre el otro.

supplemental inflatable restraints (SIR)—air bags.

resguardos suplementales inflables (SIR)—las bolsas de aire.

surface charge—the charge on the surface of a battery plate.

carga superficial—la carga en la superficie de una placa de batería.

surface texture—measurements of the roughness or smoothness of a surface are rated on the ANSI (American National Standard Institute) scale.

textura de la superficie—las medidas de lo áspero o lo liso de un a superficie se clasifican en la escala ANSI (Instituto Americano de Normas Nacionales).

surface-to-volume ratio—a term for the amount of surface area exposed in the combustion chamber.

relación de superficie-a-volumen—un término para la cantidad del área de la superficie expuesta en la cámara de combustión.

surge—a driveability condition that results from a lean air/fuel mixture.

sacudida—una condición del manejo que resulta de una mezcla de aire/ combustible pobre.

suspension—a group of parts that supports the vehicle, cushioning the ride while holding the tire and wheel correctly positioned in relation to the road. Suspension system parts include the springs, shock absorbers, control arms, ball joints, steering knuckle and spindle, or axle.

suspensión—un grupo de partes que soportan al vehículo, suavizando el viaje mientra que mantiene correctamente posicionado la rueda y el neumático en su relación a la carretera. Las partes del sistema de suspensión incluyen los resortes, los amortiguadores, los brazos de mando, las articulaciones esféricas, el articulación y muñón de dirección, o el eje.

swaged—a means of permanently locking a part to another part by deforming part of it or the material that surrounds it.

estampada—una manera de enclavar permanentemente una parte a otra deformándo una porcion de ella o a la materia que la rodea.

swaged lug studs—a method of holding a lug stud to a wheel hub or axle.

agarradera estampada—un método de fijar un pasador saliente en el cubo o el eje.

swept volume—another means of determining a cylinder's displacement.

cilindrada—otra manera de determinar el desplazamiento de un cilindro.

swing axles—axles with a universal joint at one or both ends.

ejes articulados—los ejes que tienen una junta universal en una o ambas extremidades.

switch—a device used to interrupt a circuit, providing a means of turning it on and off.

interruptor—un dispositivo usado para interrumpir un circuito, proveyendo una manera de encenderlo y apagarlo.

synchronizer—also called a synchro, it keeps two meshing gears from clashing during a shift.

sincronizador—tambien llamado un sincro, previene que chocan dos engranajes endentados durante un cambio de velocidades.

T

TAC—thermostatic air cleaner.

TAC—limpiador de aire termostático.

tap drill size—a tap drill provides the correct hole size prior to tapping a thread.

tamaño del barreno—un barreno provee el tamaño correcto del agujero antes de que se cortan las roscas.

taper—a steering linkage connection.

transición ahusada—una conexión de la biela de dirección.

taper cylinder wear—wear occurring in the top inch of a cylinder.

desgaste del cilindro cónico—el desgaste que ocurre in la primera pulgada de la parte superior de un cilindro.

TBA—tires, batteries and accessories.

TBA—los neumáticos, las baterías y los accesorios.

temper—to toughen a metal, it is heated to a specific temperature and then quenched.

templar—para endurecer un metal, se calienta a una temperatura específica y luego se enfría rapidamente.

tempilsticks—a stick that melts at a predetermined temperature, giving an indication of a metal's temperature during heating.

varilla de templar—una varilla que se derrite en una temperatura predeterminada, proporcionando una indicación de la temperatura de un metal al calentarlo.

tensile cords—cords in the belt that provide strength.

cordón de tensión—los cordones en la correa que proveen la fuerza.

tensile strength—the maximum stress a material can withstand without breaking.

fuerza de tensión—el esfuerzo máximo que puede aguantar una materia sin romper.

thermal efficiency—the ratio of how effectively an engine converts a fuel's heat energy into usable work.

rendimiento termal—la relación de qué tal eficaz es un motor en convertir la energía calorífica del combustible al trabajo útil.

thermistor—a variable resistor made from semiconductor material. Its resistance changes predictably as its temperature changes. Although voltage varies in response to changes in resistance, current flow remains constant.

termistor—un resistor variable hecho de la material de semiconductor. Su resistencia cambia según cambia su temperatura. Aunque varía su voltaje en respuesta a los cambios en resistencia, el flujo del corriente queda constante.

thermoplastic seizure—when coolant has migrated into the engine oil, it can seize the engine.

agarrotamiento termoplástico—cuando el fluido refrigerante ha migrado al aceite del motor, puede agarrotar al motor.

thermostat bypass—a passage way that allows coolant to circulate when the thermostat is closed.

termostato de derivación—un paso que permite que el fluido refrigerante circula cuando esta cerrado el termostato.

thermostatic expansion valve (TXV)—a device that controls the amount of refrigerant allowed to flow to the evaporator.

válvula de expansión del termostato (TXV)—un dispositivo que controla la cantidad del fluido refrigerante que fluye al evaporador.

thickening agent—a constituent of grease at a concentration of from 5 to 25%, it is usually a soap such as aluminum, lithium, sodium, or calcium.

agente espesativo—un constituyente de la grasa en una concentración del 5 a 25%, suele ser un jabón tal como el aluminio, el litio, el sodio o el calcio.

thinwall—a valve guide repair process that uses a bronze bushing.

reparación aislada—un proceso de reparar las guías de las válvulas usando un buje de bronce.

third member—the differential assembly is called this. The engine is the first member and the transmission is the second member.

tercer miembro—así se llama la asamblea del diferencial. El motor es el primer miembro, y la transmisión es el segundo miembro.

three-phase electrical output—when the three windings of the stator produce their currents in phases.

salida eléctrica trifásica—cuando los tres devanados del estator producen sus corrientes en fases.

three-way converter—converts HC and CO, and also changes harmful NO_x into harmless nitrogen and oxygen.

convertidor tridireccional—convierta el HC y el CO, y tambien cambia el NO_x nocivo al nitrógeno y oxígeno innocuo.

three-way oxidation converter—used with computer cars, it has a tube in the center to provide air from the smog pump to the reduction part of the catalytic converter.

convertidor de la oxidación tridireccional—usado en los coches de computadora, tiene un tubo en el centro para proveer el aire de la bomba del esmog a la parte de reducción del convertidor catalítico.

throttle pressure—the kind of pressure that results when engine vacuum changes that gives an indication of engine load.

presión del acelerador—el tipo de presión que resulta cuando cambia el vacío del motor que da una indicación de la carga en el motor.

throttle-body injection—also called (TBI) or central fuel injection (CFI), it uses an intake manifold similar to a carbureted system.

inyección al cuerpo del accelerador—tambien llamado el (TBI) o inyección central del combustible, usa un múltiple de admisión parecido a un sistema carburado.

through-hardened washers—these are hardened throughout, not just on the surface.

arandelas de temple profundo—estos son totalmente templados, no sólo en la superficie.

throw—the amount that the rod journal moves out from the main bearing center line or one-half of the total stroke.

recorrido—la cantidad que mueva el muñón de la biela de la linea central del cojinete principal o la mitad de la carrera total.

thrust angle—the angle formed by the thrustline and the geometric centerline.

ángulo de empuje—el ángulo formado por el eje de empuje y la línea central geométrica.

thrust bearing—a bearing mounted at 90° to the load.

cojinete de empuje de bolas—un cojinete montado en 90° a la carga.

thrust load—a load in a front to rear or side to side direction.

carga de empuje—una carga en una dirección delantera a lado o de lado a lado.

thrustline—when the rear wheels are held in a fixed position, this defines the wheels' true straight ahead position.

eje de empuje—cuando las ruedas traseras se sujeten en una posición fija, esto define la posición verdadera de las ruedas directamente hacia afrente.

timing light—a strobe light that is triggered by the voltage going through the number one spark plug cable.

luz de encendido—una lámpara estroboscópica disparada por el voltaje atravesando el cable de la bujía número uno.

tinning a wire—when soldering, a wire is preheated and has solder applied to it.

preparación al estañado—al soldar, se precalienta un alambre y se le aplica la soldadura.

tire plug—a piece of rubber vulcanized into a hole in a tire.

tapón de neumático—un pedazo de caucho vulcanizado en un hoyo de un neumático.

toe alignment—a comparison of the distances between the fronts and the rears of a pair of tires.

alineación convergencia o divergencia—una comparación de las distancias entre las partes delanteras y las traseras de un par de neumáticos.

toe gear—the inner end of a gear tooth.

interior del engranaje—la extremidad interior de un diente de un engranaje.

toe change—when toe changes in response to changes in ride height.

cambio de divergencia—cuando la divergencia cambia como respuesta a los cambios de la altura del viaje.

toe-in—when the tires are closer together at the front.

convergencia—cuando los neumáticos son más cercas en la parte delantera.

toe-out—when the tires are further apart at the front.

divergencia—cuando los neumáticos quedan más lejos en la parte delantera.

toe-out-on-turns—also called ackerman angle or turning radius. The tires toe out during a turn because the steering arms are bent at an angle.

divergencia-en-viraje—tambien llamado ángulo ackerman o el radio de viraje. Los neumáticos divergen durante una viraje porque los brazos de dirección son doblados en un ángulo.

tolerance—the range of measurement that is allowable for a part.

tolerancia—el límite de medida que se permite en una parte.

ton—12,000 Btu/hr.

tonelada—12,000 Btu/hr.

tons rating—how much heat needs to be added in 24 hours to turn 1 ton of ice at 32° into water.

potencia indicada de tonelada—cuánto calor se requiere añadir en 24 horas para convertir una tonelada del hielo en 32° al agua.

torque—the tendency of force to rotate a body on which it acts.

torsión—la tendencia de la fuerza en girar un cuerpo en el cual obra.

torque converter—a fluid coupling that multiplies torque.

convertidor del par—un coplamiento fluido que multiplica la torsión.

torque steer—when there are different length shafts, the long one wraps up and there is a slight lag before it puts its torque to the wheel, causing the car to pull to the passenger side.

torsión de dirección—cuando hay ejes de distinctas longitudes, el eje largo se tuerza y hay una pequeña pausa antes de que aplica su torsión a la rueda, causando que el coche jale al lado del pasajero.

torque-to-yield—when head bolts are purposely torqued to within 2% of their yield point.

torsión-al-límite—cuando los pernos de la cabeza se tuerzan a propósito dentro del 2% de su límite de proporcionalidad.

torque tube—an enclosed driveshaft.

tubo de torsión—un eje encerrado.

torque turn—when a bolt is tightened to a specified torque and then turned an additional 35 to 180°.

viraje de torsión—cuando un perno se aprieta a un par específico y luego se le da una vuelta adicional de 35 a 80°.

torsion bar—a straight rod that twists when working as a spring.

barra de torsión—una varilla recta que se retuerza cuando funciona como un resorte.

torsional vibration—vibrations occurring in the crankshaft when force on the pistons is imparted to the crankshaft of a V-type engine.

vibración torsional—las vibraciones ocurriendo en el cigüeñal cuando la fuerza en los pistones se transfere al cigüeñal de un motor tipo V.

track—the side-to-side distance between an axle's tires.

pista—la distancia de un lado-a-lado entre los neumáticos de un eje.

tracking—a term that refers to the relationship between the average direction that the rear tires point, when compared to the front tires.

seguimiento—un término que refiere a la relación entre la dirección normal hacia cual apuntan los neumáticos traseros, al ser comparado con los neumáticos delanteros.

traction—how well a tire grips the road.

tracción—la habilidad con la cual un neumático agarra la carretera.

transaxle—a unit that combines a transmission and differential.

flecha de transmisión—una unedad que combina una transmisión y un diferencial.

transducer—a device that converts energy from one form to another. It is a magnet that is spring loaded on both sides. It moves back and forth in a coil of wire to send a signal to a meter in response to vibration.

transductor—un dispositivo que convierta la energía de una forma a otra. Es un imán cargado por resorte en ambos lados. Oscila en una bobina de alambre para mandar una señal a un aparato de medida como respuesta a la vibración.

transistor—an electronic relay.

transistor—un relé electrónico.

transmission heat exchanger—also called an oil cooler, it is usually in the bottom or side of the radiator that has automatic transmission fluid running through it.

intercambiador de calor de la transmisión—tambien llamado un enfriador del aceite, suele estar en la parte inferior o en un lado del radiador por el cual atraviesa el fluido de transmisión.

tread—the section of the tire that rides on the road.

banda de rodamiento—la sección de un neumático que viaja en la carretera.

tread wear indicators—also called wear bars, these are molded into the tire's tread and indicate when $\frac{1}{16}$" of tread is remaining.

indicadores del desgaste de la banda—tambien llamado barras del desgaste, estos estan moldeados dentro de la banda de rodamiento de un neumático e indican cuando le queda un $\frac{1}{16}$" de uso.

TSB—technical service bulletin.

TSB—boletín de servicio técnico.

tulip—the yoke or housing of a tripod CV plunge joint. It has three grooves that the rollers can roll in and out as the axle length changes.

tulipán—la brida o el cárter de una junta trípoda tipo CV. Tiene tres muescas en las cuales pueden rodar los rodillos adentro y afuera al alargar el eje.

turbo lag—the time required to bring the turbo up to a speed where it can function effectively.

retraso del turbo—el tiempo requerido en accionar el turbo a una velocidad en la cual puede funcionar eficazmente.

turbocharger—a small radial fan pump driven by the energy of the exhaust flow.

turbocompresor—un pequeño ventilador radial impulsado por la energía del flujo del escape.

turbocharger cartridge—a new or rebuilt replacement unit.

bote del turbocompresor—una unedad de reemplazo nueva o reconstruida.

turnbuckle—a part between two rods. Turning it one way shortens a rod assembly. Turning it the other way lengthens it.

tensor de tornillo—una parte entre dos varillas. Dándole vuelta en una dirección acorta la asamblea de varilla. Volteandolo en la otra dirección lo alarga.

turning radius—also called ackerman angle or toe-out-on-turns. The tires toe-out during a turn because the steering arms are bent at an angle.

radio de viraje—tambien llamado el ángula ackerman o divergencia-en-vueltas. Los neumáticos divergen en las vueltas porque los brazos de dirección son torcidos en un ángulo.

TVRS—television/radio suppression resistor spark plug cables made of braided aramid fiber impregnated with graphite and latex.

TVRS—los cables de bujía con resistores de supresión de la televisión/radio hechos de fibra aramid impregnada con el grafito y el látex.

TVS—thermal vacuum switch, also called a ported vacuum switch (PVS), it is found in a cooling passage on precomputer cars. It controls the flow of vacuum according to engine temperature.

TVS—interruptor de vacío termal, tambien llamado un interruptor de vacío con lumbrera, se encuentra en los pasos de enfriamiento en los coches precomputerizados. Controla el flujo del vacío conforme a la temperatura del motor.

two-stroke engine—an engine that completes its entire operation cycle in one turn of the crankshaft.

motor de dos carreras—un motor que completa su ciclo de operación completo con una vuelta del cigüeñal.

type I headlamp—has high beam only.

faro tipo I—sólo tiene la luz larga.

type II headlamp—has both low and high beams.

faro tipo II—tiene ambas luces cortas y largas.

U

underhood label—a label found on cars since 1972 that describes the size of the engine, ignition timing specifications, idle speed, valve lash clearance adjustment, and the emission devices that are included on the engine.

etiqueta bajo el capót—una etiqueta encontrado en los coches desde el 1972 que describe el tamaño del motor, las especificaciones del encendido, la velocidad de marcha en vacío, los ajustes de holgura del juego de las válvulas, y los dispositivos de emisión que se incluyen en el motor.

undersquare—when the bore is less than the stroke.

subcuadrado—cuando el diámetro interior es menos que la carrera.

understeer—when a car does not seem to respond to movement of the steering wheel during a hard turn.

subdirección—cuando un coche no parece responder al movimiento del volante de dirección durante un viraje abrupto.

uni-directional—when the scan tool can read data, but cannot give commands to the computer.

unidireccional—cuando un aparato explorador puede leer los datos, pero no puede enviar los ordenes a la computadora.

unibody design—a chassis design that has a floorpan and a small subframe section in the front and rear.

diseño monocasco—un diseño del chasis que tiene un piso y una pequeña sección del armazón en la parte delantera y trasera.

union—a connection that joins two pieces of tubing together often used on vacuum or air pressure lines.

unión—una conexión que une dos piezas de tubo que suele usarse en las líneas de vacío o de presión de aire.

unloaded caliper—a bare rebuilt caliper with no pads or hardware.

calibre descargado—un calibre reconstruido desprovisto de las almohadillas o la herramienta.

unsprung weight—the weight of parts of the suspension that are not supported by springs. It affects vehicle handling.

peso desprovisto de muelles—el peso de las partes de suspensión que no se soportan con muelles. Afecta el manejo del vehículo.

upper end—parts of the upper end of the engine include the cylinder head(s) and valve train.

extremidad superior—las partes de la extremidad superior del motor incluye la(s) cabeza(s) y el tren de válvulas.

upshift—when the transmission shifts from a low gear to a higher gear; second to third, for instance.

cambiar a una velocidad más alta—cuando la transmisión cambia desde una velocidad baja a una más alta; de la segunda a la tercera, por ejemplo.

UTQG—uniform tire quality grade system that rates tread wear, traction, and temperature resistance.

UTQG—sistema uniforme de calificar la calidad de los neumáticos que valúa el gasto de la banda, la tracción y la resistencia a la temperatura.

V

V-ribbed belt—a belt that has multiple ribs on one side and is flat on the other side.

banda con refuerzas en V—una banda que tiene varias nervaduras en un lado y esta plana del otro lado.

vacuum advance—part of the distributor that senses engine load and changes the timing to compensate.

avance de vacío—una parte del distribuidor que detecta la carga del motor y cambia el encendido para compensar.

vacuum motor—a vacuum-operated device that opens and closes doors or valves.

motor de vacío—un dispositivo operado por vacío que abre y cierre las puertas o las válvulas.

valve body—the hydraulic control assembly of the transmission.

cuerpo de la válvula—la asamblea del control hidráulico de la transmisión.

valve clearance (lash)—usually adjustable clearance in the valve train.

holgura (juego) de las válvulas—la holgura que normalmente se puede ajustar en el tren de válvulas.

valve overlap—the number of degrees of crankshaft rotation that occurs during the time that both the intake and exhaust valves are open at the same time.

solape de las válvulas—el número de grados de rotación del cigüeñal que ocurre durante el tiempo en que ambos válvulas de admisión y escape estan abiertas al mismo tiempo.

valve port—a passage that allows the flow of intake or exhaust into or out of the cylinder.

puerto de válvula—un paso que permite el flujo de admisión o del escape entrar o salir del cilindro.

valve spring installed height—a measurement that increases with machining of the valve and valve seat. Shims can be used to restore the correct height and spring tension.

altura instalada del resorte de la válvula—una medida que aumenta cuando la válvula y el asiento de la válvula se han mecanizado. Las chapas se pueden usar para restaurar la altura y la tensión del resorte correcta.

valve stem height—a measurement that reflects how much a valve and valve seat have been ground.

altura del vástago de la válvula—una medida que refleja cuánto se han rectificado la válvula y el asiento de la válvula.

valve train—the parts that open and close the valves.

tren de válvula—las partes que abren y cierran las válvulas.

vapor lock—a condition that occurs to a carbureted car when fuel boils in the fuel line or filter.

cierre de vapor—una condición que ocurre en un coche con carburador cuando hierve el combustible en la línea del combustible o en el filtro.

vaporization—when atomized fuel turns into a gas.

vaporización—cuando un combustible atomizado se convierta en gas.

variable DC voltage generator—a zirconia oxygen sensor is a galvanic battery that can generate a voltage of 0.1 volt to 0.9 volt (100 to 900 millivolts) in response to the amount of oxygen in the vehicle's exhaust.

generador de voltaje DC variable—un sensor de oxígeno zirconía es una batería galvánico que puede generar un voltaje de 0.1 voltaje a 0.9 voltaje (100 a 900 milivoltios) como respuesta a la cantidad del oxígeno en el escape del vehículo.

variable displacement turbo—a turbocharger that changes its nozzle opening in response to changes in engine load.

turbo de desplazamiento variable—un turbocargador que cambia la apertura de la boquilla como respuesta a los cambios en la carga en el motor.

variable dwell—in an electronic ignition system when at low engine rpm, dwell is shorter and at higher rpm it is longer.

parada variable de movimiento—en un sistema de encendido electrónico cuando en bajas rpm del motor, la parada es más corta y en rpm más altas es más larga.

variable rate spring—a tapered rod spring that gives a smoother ride when going over smaller bumps and still allows for heavier carrying capacity.

muelle de capacidad variable—un muelle de varilla cónico que proporciona un viaje más uniforme al pasar por los topes más chiquitos y que todavía permite una capacidad de cargas pesadas.

variable resistors—called rheostats or potentiometers, used to control speed and intensity of electrical loads.

resistores variables—llamados reostatos o potenciómetros, usados en controlar la velocidad y la intensidad de las cargas eléctricas.

varnish—forms in a transmission when the fluid has become oxidized from heat.

vidriado—forma en una transmisión cuando el fluido se ha oxidado debido al calor.

velocity—a combination of amplitude and frequency.

velocidad—una combinación del amplitud y frequencia.

venturi—a smaller area in the carburetor that restricts air flow resulting in lower pressure to draw fuel from the float bowl. Air also speeds up due to the restriction.

venturi—una área pequeña en el carburador que restringe el flujo del aire resultando en que la presión más baja aspira el combustible de la taza del flotador. El aire tambien se acelera debido a la restricción.

vibration—a part that is in motion in waves, or cycles.

vibración—una parte en movimiento en forma de ondas, o ciclos.

viscosity—a measurement of how easily a liquid can flow.

viscosidad—una medida de qué tal facilmente puede fluir un líquido.

viscosity index—an oil's ability to resist change in viscosity as it heats up.

índice de la viscosidad—la habilidad del aceite de resistir un cambio en viscocidad al calentarse.

viscous coupling—a sealed clutch unit that contains silicone fluid.

acoplador viscoso—una unedad hermética del embrague que contiene el silicio líquido.

volatile RAM—it is erased each time the ignition is turned off.

RAM volátil—se borra cada vez que se apaga el interruptor del encendido.

volatility—a measure of how easy a fuel evaporates.

volatilidad—una medida de qué tal facilmente se evapora un combustible.

voltage dividers—variable resistors that produce a variable DC voltage signal.

divisor de tensión—los resistores variables que producen una señal de voltaje DC variable.

voltage drop—a loss of voltage caused by current flow through a resistance.

caída de voltaje—la pérdida del voltaje causada por el flujo del corriente a través de una resistencia.

voltage drop testing—a test using a voltmeter to check for resistance in a loaded circuit.

prueba de caída de voltaje—una prueba usando un voltímetro para verificar la resistencia en un circuito cargado.

voltage regulator—a device that controls the strength of the electromagnetic field in the rotor of an alternator.

regulador del voltaje—un dispositivo que controla la fuerza del campo electromagnético en el rotor de un alternador.

voltage spike—a high voltage that occurs while disconnecting a powered circuit lead that happens because electrons that were in motion before the circuit was broken, back up at the connection. When they try to push their way across the gap, the spike is created.

impulsos de tensión—un voltaje alto que ocurre al desconectar un hilo de circuito bajo tensión que pasa porque los electrones estaban en movimiento antes de que se abriera el circuito, en la conexión. Cuando tratan a traversar la abertura, se crea el impulso.

volts—the measurement of electrical pressure.

voltios—la medida de presión eléctrica.

volumetric efficiency (V.E.)—the measurement comparing the volume of air flow actually entering the engine with the maximum that theoretically could enter it.

rendimiento volumétrico (V.E.)—la medida que compara el volumen del flujo del aire actualmente entrando al motor con lo máximo que podría entrar teoréticamente.

vortex—a spiral pattern of fluid flow from impeller to turbine.

torbellino—un patrón espiral del flujo del fluido del impulsor a la turbina.

vulcanized—heating rubber to make it stable.

vulcanizado—calentando al caucho para estabilizarlo.

vulcanizing—a process that cures rubber by heating it.

vulcanizando—un proceso que cura el caucho calentandolo.

W

Wankel engine—a rotary engine design.

motor Wankel—un diseño de motor rotativo.

warm-up cycle—occurs every time the engine cools off and temperature rises at least 40°F.

ciclo de recalentamiento—ocurre cada vez que un motor se enfría y la temperatura llega al menos a los 40°F.

waste gate—a relief valve for excessive turbocharger boost.

compuerta de residuos—una válvula de relieve para soplante excesivo del turbocompresor.

waste spark method—when a distributorless ignition system has both ends of each coil's secondary winding attached to one of a pair of companion cylinders spark plugs and both cylinders' spark plugs fire every revolution of the crankshaft.

metodo de encendido perdido—cuando un sistema de encendido sin distribuidor tiene ambas extremidades del devanado secundaria de cada bobina conectada a un par de bujías en cilindros acoplados y la bujía de ambos cilindros disparan cada revolución del cigüeñal.

watts—the horsepower equivalent in the metric system. One horsepower equals 0.746 kw.

vatio—la potencia de caballo equivalente en el sistema métrico. Una potencia de caballo igual a 0.746 kw.

wear indicator ball joint—a ball joint with a shoulder that indicates when it is worn out.

articulación esférica indicador del desgaste—una articulación esférica con un collar que indica cuando esta gastado.

wedge combustion chamber—a turbulent combustion chamber design shaped like a wedge.

cámara de combustión en forma de cuña—un diseño de cámara de combustión turbulente en forma de cuña.

wet compression test—a compression test where oil is squirted into the cylinder to test for worn rings.

prueba de compresión en húmedo—una prueba de compresión en la cual se chorrea el aceite dentro del cilindro para comprobar el gasto en los anillos.

wheel base—the distance between the front and rear tires.

batalla (empate)—la distancia entre los neumáticos delanteros y traseros.

wheel bearings—the term for non-drive front and rear wheel bearings.

cojinetes de las ruedas—el término para los cojinetes no impulsores de las ruedas delanteras y traseras.

wheel cylinder kit—a parts kit for rebuilding wheel cylinders.

piezas de reconstrucción del cilindro de la rueda—el conjunto de las partes sueltas para reconstruir los cilindros de las ruedas.

whine—a sound like a turbine or mosquitos. The range is from 300~500 Hz. It is usually related to gear noise.

gemido—un ruido como una turbina o los mosquitos. La gama es del 300~500 Hz. Suele ser relacionado al ruido de los engranajes.

wide ratio transmission—when there is a large difference between the ratios of a transmission's forward gears.

transmisión de relación extensiva—cuando hay una gran diferencia entre las relaciones de las velocidades de marcha en adelante de una transmisión.

work—when an object is moved against a resistance or opposing force.

trabajo—cuando un objeto se mueva contra una resistencia o una fuerza opuesta.

work harden—when a material is flexed or pounded on and it becomes harder.

endurecido por acritúd—cuando una materia es doblado o molido y se endurece.

worm gear clamp—one of the most popular hose clamps for all applications except fuel injection.

abrazadera graduable—uno de las abrazaderas de tubo flexible más populares para todas aplicaciones menos los de inyección del combustible.

WOT—wide open throttle.

WOT—Acelerador completamente abierto.

wye (Y) winding—an alternator stator winding in which three leads are branched out in a "Y" pattern.

devanado estrella (Y)—un devanado del estator del alternador en el cual tres hilos se extienden para formar una "Y".

Z

ZEV—zero emission vehicle.

ZEV—vehículo de cero emisión.

zener diode—a semiconductor used to control transistors and electronic voltage regulators that only conducts electricity when a certain voltage is reached.

diodo zener—un semiconductor usado para controlar los transistoras y los reguladores de voltaje electrónicos que sólo conduce la electricidad llegando a un cierto voltaje.

zerk fitting—a grease fitting threaded to fit into holes in the part to be lubricated.

niple de grasa—una válvula de lubricación enroscada que se ajusta en un agujero de una parte para que se puede lubricar.

zero-lash—when there is no clearance and no interference between parts of the valve train.

juego de cero—cuando no hay holgura ni interferencia entre las partes del tren de válvulas.